# ANIMALS

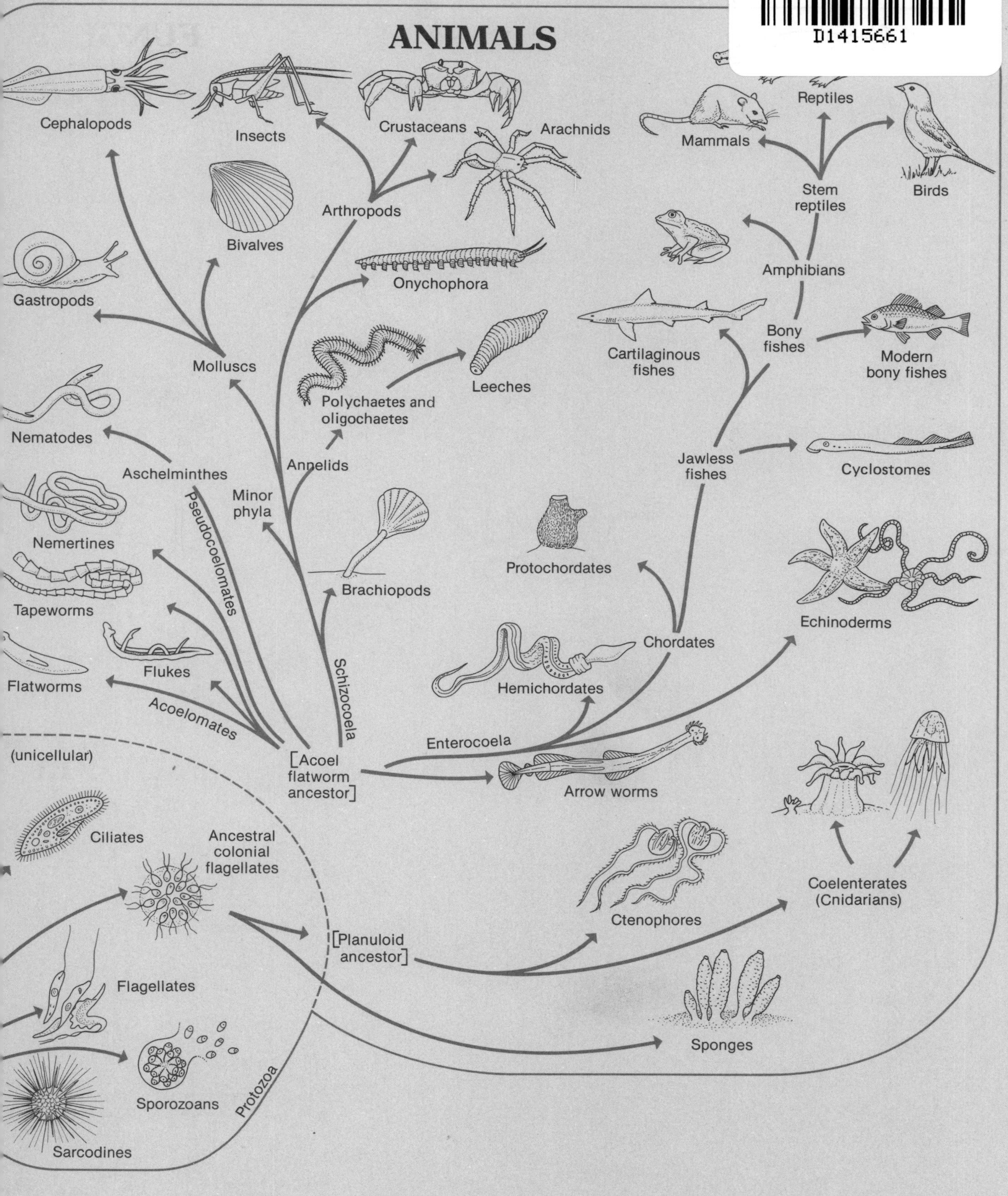
Cephalopods
Insects
Crustaceans
Arachnids
Reptiles
Mammals
Birds
Stem reptiles
Arthropods
Bivalves
Gastropods
Onychophora
Amphibians
Bony fishes
Cartilaginous fishes
Modern bony fishes
Molluscs
Leeches
Polychaetes and oligochaetes
Nematodes
Jawless fishes
Cyclostomes
Aschelminthes
Annelids
Minor phyla
Pseudocoelomates
Nemertines
Protochordates
Echinoderms
Tapeworms
Brachiopods
Chordates
Schizocoela
Flukes
Flatworms
Hemichordates
Acoelomates
Enterocoela
(unicellular)
[Acoel flatworm ancestor]
Arrow worms
Ciliates
Ancestral colonial flagellates
Ctenophores
Coelenterates (Cnidarians)
[Planuloid ancestor]
Flagellates
Sponges
Sporozoans
Protozoa
Sarcodines

# BIOLOGY

# BIOLOGY

**JAMES M. BARRETT**
**PETER ABRAMOFF**
**A. KRISHNA KUMARAN**
**WILLIAM F. MILLINGTON**
*Marquette University*

**PRENTICE-HALL**
*ENGLEWOOD CLIFFS, N.J.*

**Library of Congress Cataloging-in-Publication Data**
Main entry under title:
Biology.

Includes bibliographies and index.
1. Biology. I. Barrett, James M.
QH308.2.B56444 1985 574 85-12443
ISBN 0-13-076597-X



Printed in the United States of America

10 9 8 7 6 5 4 3 2 1

Prentice-Hall International (UK) Limited, *London*
Prentice-Hall of Australia Pty. Limited, *Sydney*
Prentice-Hall Canada Inc., *Toronto*
Prentice-Hall Hispanoamericana, S.A., *Mexico*
Prentice-Hall of India Private Limited, *New Delhi*
Prentice-Hall of Japan, Inc., *Tokyo*
Prentice-Hall of Southeast Asia Pte. Ltd., *Singapore*
Editora Prentice-Hall do Brasil, Ltda., *Rio de Janeiro*
Whitehall Books Limited, *Wellington, New Zealand*

**Credits**
Art Director: Florence Dara Silverman
Development Editor: Albert McGrigor
Production Editor: Jeanne Hoeting
Book Designer & Page Layout: Berta Lewis
Cover Designer: Berta Lewis
Line Art: Danmark and Michaels, Inc./Carole Russell Hilmer
Barbara L. Schmidt/Valerie K. Pawlak
Photo Researcher: Anita Duncan/Tobi Zausner
Manufacturing Buyer: Ray Keating
Cover Photos: *Top right,* sea otter feasting on urchin. (Scott Ransom) *Bottom right,* false-color TEM (electron micrograph) of DNA of a bacterial plasmid in which two genes (blue) have been mapped. Normal DNA is orange. DNA/RNA hybrid in red. (P. A. McTurk and D. Parker, Photo Researchers, Inc.) *Bottom left,* Queen Anne's lace with stinkbugs. (Florence Dara Silverman) *Top left,* diatoms, one-celled eukaryotic organisms. (Biophoto Associates/Photo Researchers, Inc.)

(Credits begin on p. 1128, which constitutes a continuation of the copyright page.)

ISBN 0-13-076597-X 01

To Mary, Terri, Jyoti, and Kay

# Contents in Brief

## UNIT IV The Biology of Organisms: Integration and Behavior



## UNIT V Reproduction, Genetics, Development, and Evolution



## UNIT VI Life's Diversity: The Products of Evolution

# UNIT VII Biology of the Environment

# Contents

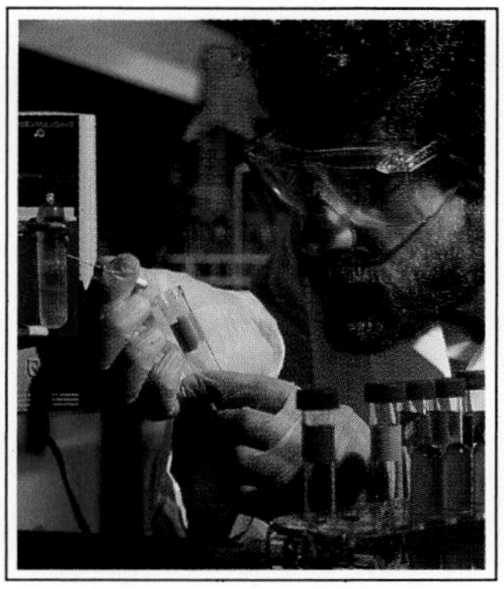

## UNIT I Biology as a Science



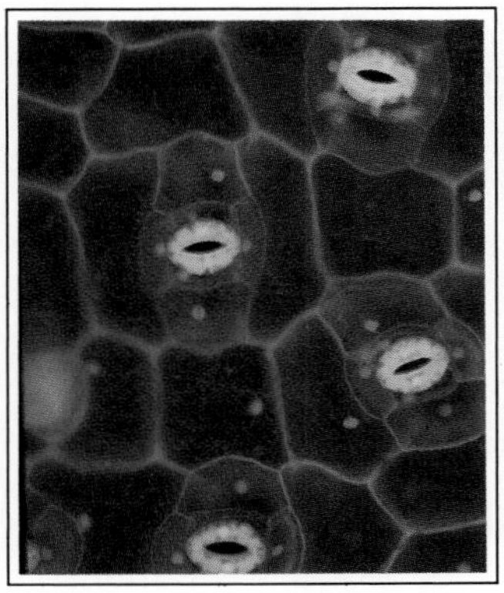

## UNIT II The Biology of Cells

# UNIT III Homeostasis

# UNIT IV The Biology of Organisms: Integration and Behavior

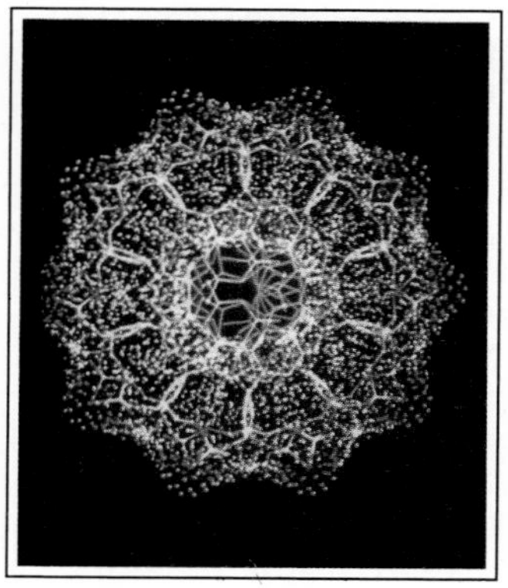

# UNIT V Reproduction, Genetics, Development, and Evolution

# UNIT VI Life's Diversity: The Products of Evolution

# UNIT VII Biology of the Environment

# Preface

This book is designed to convey the conceptual basis of modern biology without unduly burdening the novice with what could seem to be a bewildering multitude of scientific facts. By assuming little prior knowledge of science and by virtue of its level and style, the book is readily comprehended by the college freshman. Although suitable for courses for nonscience majors, it also provides an excellent introduction to the subject for the future professional biologist or health scientist, with its emphasis on the experimental basis of biological science. Prospective biology majors are given a solid basis in all major areas of specialization. The following are among the book's important features.

- An effort has been made to explain the experimental basis for current understanding of biological phenomena, thereby contributing to an understanding of how scientific knowledge is enlarged. Principles are presented in historical perspective, allowing a development of concepts from simple observations to rigorous experimental verification of theories. This is aimed at developing in the student the inductive and deductive skills required for scientific analysis.
- The flow of energy—at the cellular, organismal, and community levels—is followed throughout the book.
- The theme of organic evolution, the central unifying theory of all biology, is brought in wherever it is significant, rather than being compartmentalized into one or two chapters near the end of the book. For example, feedback loops and their particular setpoints are noted as being determined by natural selection, the same being true for behavioral mechanisms.
- Up-to-date coverage is provided in all areas, including introductory chemistry, molecular and cell biology, photosynthesis, intermediary metabolism, respiration, nutrition, plant and animal transport, excretion, ionic and osmoregulation, endocrinology, and Mendelian genetics.
- Expanded coverage is given to areas of growing research interest, such as neurobiology, immunobiology, molecular genetics, animal behavior, human evolution, and ecology. One full chapter is devoted entirely to human impact on world ecosystems, a topic that has usually been ignored or is only briefly covered in most introductory texts. Our chapter presents the available facts, allowing students to make their own judgments and draw their own conclusions.

- Organismal diversity, which for years had been slighted and is now gaining renewed interest, is accorded five chapters. As in chapters on physiology, these are comparative in their outlook and have been written so as to impress the student with the fundamental unity of all life as well as with the degree of biological diversity.
- Plant biology is fully covered in all its major aspects. In addition to several chapters on botanical topics, botanical concepts and principles are integrated into other chapters wherever appropriate.
- The book has been organized to permit its use in courses with various topic sequences. The first two sections give an overview of the science of biology and introduce the student to the chemistry and physics of life processes and the world of the cell. Any sequence of the five remaining sections may then be adopted according to the wishes of the instructor. Alternatively, Unit II, on chemistry and cell biology, can be deferred until the second half of the course and organismal biology emphasized in the first half.
- Because of the style, level, and organization, the teacher is able to assign many sections of the text for self-instruction by the student.
- Almost every chapter has panels that add to reader interest by presenting interest-generating material, such as classical experiments, modern problems, and controversies, that is not essential to an understanding of basic principles.
- Over three dozen reviewers, generalists as well as specialists, have ensured accuracy and currency of the principles presented in the text. A number of them teach introductory courses themselves and were able to make helpful suggestions regarding level and style. The book was classroom tested at two stages of its development.
- Appendices include a chronology of highlights in the history of biology.
- All chapters are generously illustrated with diagrams, tables, graphs, drawings, and photographs, which are usually self-explanatory and are well integrated with the text. The cell biology section is especially well illustrated. It makes excellent use of color and of original artwork and photographs.
- Extensive chapter summaries provide an excellent overview for student and teacher.
- The glossary, prepared by the authors, is probably the best in any introductory text. The definitions are of dictionary quality, rather than of the traditional, simplified, glossary type.
- The recommended readings at the end of each chapter are at a level understandable by a beginning student; they were chosen to instruct as well as stimulate.
- The high quality index includes over 6,000 entries.
- End papers include a detailed phylogenetic tree, a geological time chart, and conversion tables for metric and English units of measurement.

It is our intent to make the study of biology an intellectual experience. Besides presenting evidences that have led to our knowledge and understanding, we have communicated the flavor of some of the many controversies and unanswered questions in biology.

The *raison d'être* of introductory biology as part of a college curriculum is, in our view, that it imparts a knowledge of some of the more important principles of biology, described and discussed in a manner that achieves an understanding of the processes of scientific inquiry and logic. We view both this knowledge and the understanding of scientific methodology as an integral part of the education, not only of a scientist or health professional, but of any person who will shoulder his or her responsibilities as a citizen.

The Authors
Milwaukee, Wisconsin

# Acknowledgments

The preparation of this book involved the skills and willing cooperation of a large number of talented people. For the original suggestion and for seeing the project through its early phases, thanks go to Prentice-Hall's Paul Feyen. To copy editor Pamela Lloyd goes the credit for much of the clarity and logical flow of thought, to which final touches were added by Margot Otway. Others who have contributed to the effort include Product Development Director David Esner and his successor Raymond Mullaney and editors Harry Gaines, Jack Bruggeman, Logan Campbell, Robert Sickles, and Doug Humphrey and their competent assistants Jo Marie Jacobs and Jeanne Lyzell. The watchful eye of Edward H. Stanford, now College Book Division President, was important to the project's success at several junctures. A special acknowledgment is due Albert McGrigor, Product Development Editor, for the valuable contributions he made during the writing and for the role he played in overseeing the entire project. The skill, dedication, enthusiasm, and high standards of the Book Project Division in the persons of Jeanne Hoeting and Florence Dara Silverman ensured the quality of the final product, aided by Fay Ahuja's efficiency and accuracy in checking many stages of proof and art. Photo research was professionally executed by Anita Duncan, the enthusiastic Tobi Zausner, and by Lori Morris and Page Poore. Credits for photographs and sketches begin on page 1128. The extensive original artwork is the product of two studios—Danmark and Michaels, Inc. and Carole Russell Hilmer—and of artists Barbara L. Schmidt and Valerie K. Pawlak.

Probably the most important contributions to the book's quality by others were those of the many reviewers with their many fine suggestions for improvements in the text at each state of its preparation. Three who twice reviewed the entire manuscript as it approached final form were Judith Goodenough, John Harley, and Kenneth Thomulka. For an unusual contribution of knowledge and wisdom we especially thank Dr. Goodenough. A complete list of reviewers appears below.

Kenneth B. Armitage, University of Kansas
Robert R. Becker, Oregon State University
John P. Bihn, LaGuardia Community College
Mildred I. Brammer, Ithaca College
David Brauer, University of Kansas
Alan P. Brockway, University of Colorado-Denver
Herbert J. Caswell, Jr., Eastern Michigan University
Steven G. Clarke, University of California, Los Angeles
Morris G. Cline, Ohio State University
Steve S. Easter, Jr., University of Michigan
D. Craig Edwards, University of Massachusetts-Amherst

Edward N. Frang, University of New Hampshire
Douglas G. Fratianne, Ohio State University
Elizabeth A. Godrick, Boston University
Judith Goodenough, University of Massachusetts-Amherst
Steven N. Handel, University of South Carolina
John P. Harley, Eastern Kentucky University
Daniel T. Haworth, Marquette University
Stephen C. Hedman, University of Minnesota-Duluth
Herbert B. Herscowitz, Georgetown University
Norman Kerr, University of Minnesota-Minneapolis
Edward J. Kormondy, California State University-Los Angeles
Allen MacNeill, Cornell University
Gary L. Meeker, California State University-Sacramento
H. Gordon Morris, University of Tennessee-Martin
Norman C. Negus, University of Utah
Lowell, P. Orr, Kent State University
Elliot A. Stein, Marquette University
Eliot Stellar, University of Pennsylvania
Daryl C. Sweeney, University of Illinois-Urbana-Champaign
Thomas M. Terry, University of Connecticut
Kenneth Thomulka, Philadelphia College of Pharmacy and Science
William J. Weishar, Marquette University
Thomas Wentworth, North Carolina State University
Charles A. Wilkie, Marquette University
Clarence C. Wolfe, Northern Virginia Community College
John L. Zimmerman, Kansas State University

Most of the typing of the manuscript was done by the capable hands of Barbara DeNoyer, with substantial additional work being done by others, notably Marilyn Mutzenbauer and Rose Radetski. We are deeply grateful for their fine work.

The Authors
Milwaukee, Wisconsin

# BIOLOGY

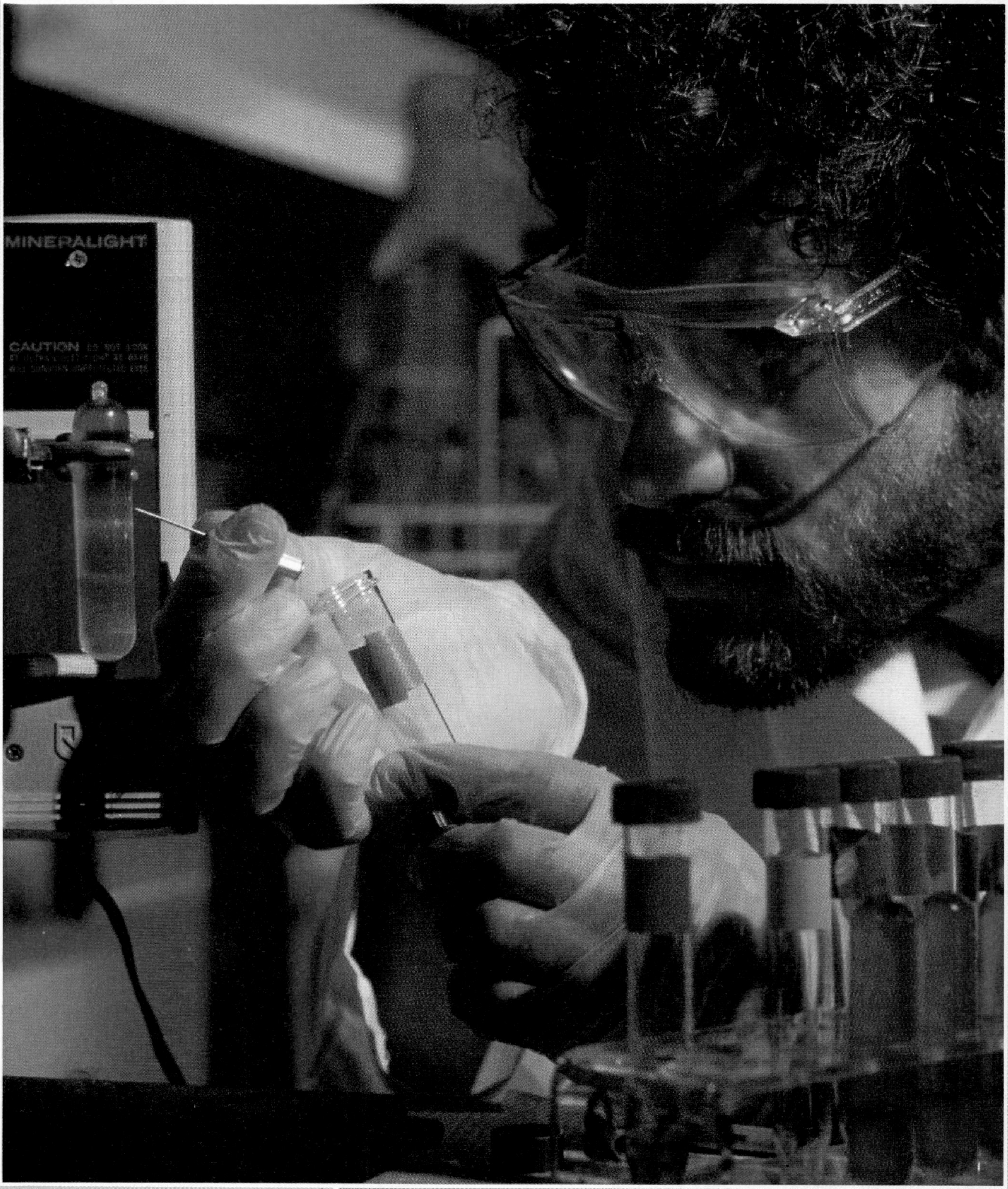
MINERALIGHT
CAUTION DO NOT LOOK

# UNIT I

# Biology as a Science

CHAPTER

THE SCIENCE OF BIOLOGY [1]

Biological science is the study of organisms—that is, living things. A science is a systematized body of knowledge based on observation and experimentation. In its basic methods and logical approaches to asking questions and solving problems, biology does not differ from other natural sciences, such as astronomy, geology, physics, and chemistry. Biological research is undertaken to investigate the structure and function of organisms and their parts as well as their physical and chemical nature. A physicochemical approach is as applicable to genetics (the science of heredity), evolution, reproduction, and the development of organisms as it is to physiology (the study of function). Yet biology is not merely the physics and chemistry of organisms. It concerns all phenomena exhibited by the highly organized systems of matter and energy called organisms.

Why study science, and, in particular, why study biology? Answers to these questions vary according to our individual interests, goals, and perceptions of the world around us. One excellent reason is to understand the world we live in. Scientific knowledge of our physical surroundings is useful to all of us. Another reason might be to prepare for a career in science. Many people enjoy science so much, however, that they study it mostly for the personal satisfaction that mastery of a demanding and interesting subject provides. But there are good reasons why all educated people should have a grasp of fundamental scientific principles and methods. Scientific study, particularly when it includes developing the analytical skills involved in laboratory research, is essential to a well-rounded education. By its continuing demand for exactness and objectivity, laboratory research is one of the best disciplinarians when it comes to developing habits of patience and reserve in forming judgments until all pertinent evidence is considered. Although the study of science is not unique in its ability to provide the benefits derived from scholarly endeavor, it is an important and exciting field of inquiry, as beneficial as any other in this regard.

But why study biology in particular? The study of biology, like that of any other natural science, provides all the benefits noted above, but it does more. It provides insights into the living world of which we humans are an integral part. We thus come to understand ourselves better through such a course of study, and being knowledgeable about a subject of growing social importance worldwide, we are able to become better citizens as well.

# 1

# The Science of Biology

OUTLINE



## THE UNIQUE CHARACTERISTICS OF LIVING THINGS

Simply defined, science is the systematic understanding or knowledge of the physical world of matter and energy. The science of biology seeks to know and understand the characteristics of matter and energy as they are manifested in living things, or **organisms.** Although the elemental matter and energy found in organisms do not differ from those in nonliving things, organisms are unique in the atomic arrangement of their matter and in the fact that they process energy. Not only do organisms acquire and use energy for maintenance, growth, and reproduction, but they regularly undergo heritable changes, the most suitable of which tend to persist; that is, they evolve. These two processes—energy utilization and evolution—are the major themes of this book.

### ORGANISMS AS PROCESSORS OF ENERGY

Energy is present on earth in many forms, some more useful to organisms than others. For example, in the process of photosynthesis, plants absorb the radiant energy of sunlight and convert it to chemically

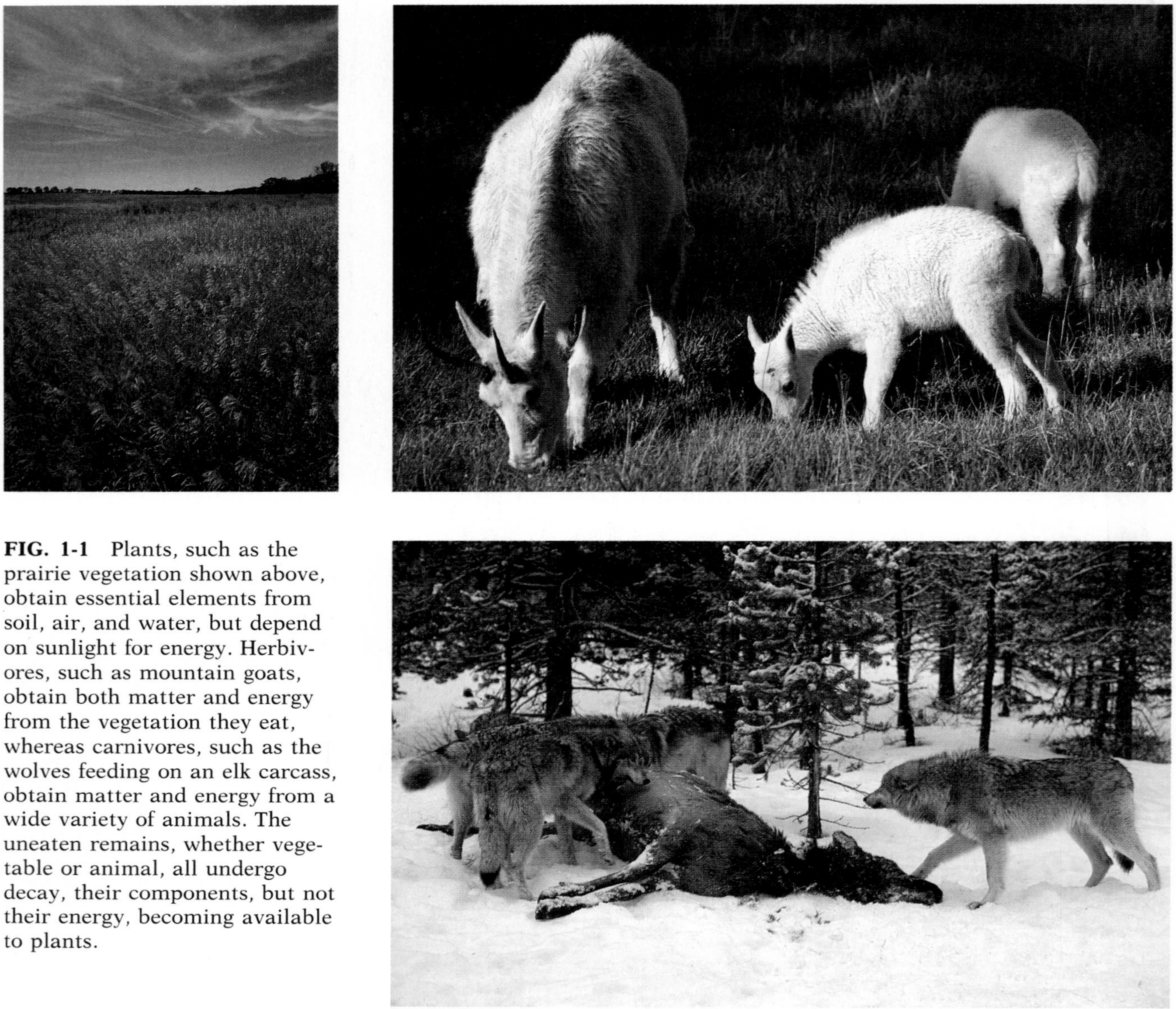

**FIG. 1-1** Plants, such as the prairie vegetation shown above, obtain essential elements from soil, air, and water, but depend on sunlight for energy. Herbivores, such as mountain goats, obtain both matter and energy from the vegetation they eat, whereas carnivores, such as the wolves feeding on an elk carcass, obtain matter and energy from a wide variety of animals. The uneaten remains, whether vegetable or animal, all undergo decay, their components, but not their energy, becoming available to plants.

bound energy. Animals eat plants or other animals and convert the chemically bound energy thus obtained to other forms (Fig. 1-1). Although some forms of energy, such as the energy of a fire or of molten lava, are not directly usable by organisms, one kind of organism or another is capable of acquiring and processing every form of energy compatible with life. This great variety of energy exploiters is the product of a very long evolution that began about 4 billion* years ago.

## ORGANISMS AS PRODUCTS OF EVOLUTION

From several lines of evidence, we know that the earth, like the solar system of which it is a part, is about 4.7 billion years old. The conditions present when the earth formed could not have supported life. But

* In this book, the word *billion* is used in the American sense, to mean a thousand million.

slowly, over many millions of years, those conditions changed. Seas formed and became not only capable of sustaining life, but are believed to have given rise to it. At first, those ancient seas would have contained only simple chemical substances. Concurrently, the earth's early atmosphere probably included water ($H_2O$) vapor, molecular hydrogen ($H_2$), and a number of other compounds, perhaps ammonia and probably the simple gas methane ($CH_4$). Molecular oxygen ($O_2$) was probably absent. Energy was available, mostly as sunlight, but also as lightning, volcanic action, and various types of radiations (Fig. 1-2a). This energy was probably sufficient to promote a variety of chemical reactions among the simple molecules then found on earth. Tide pools in which seawater became heated and concentrated by evaporation on hot days may have been especially favorable sites for such reactions (Fig. 1-2b).

***From Simple Substances to Building Blocks.*** Among the larger molecules produced in these early reactions were the **amino acids:** the constituent units of **proteins,** the largest and most complex of all molecules in organisms (see Panel 1-1). Other products of these reactions would have included the units that make up **nucleic acids,** particularly deoxyribonucleic acid or **DNA.** The DNA molecules would eventually form genes, the inherited molecules that constitute the link between one generation of organisms and the next. The molecular units within genes (DNA molecules) are so arranged as to form coded specifications for synthesizing proteins. Genes have been compared to the directions used to build an automobile or the recipe for a cake. Genes also serve as templates, or patterns, for making copies of themselves.

***The Formation of Macromolecules.*** From the simple molecules in the primeval seas, various larger and more complex molecules—macromolecules—would have formed, including many basic components of cells that are found in modern organisms. Although not every molecule formed would have been useful to living things, many would have been, including certain proteins called **enzymes.** These are molecules that facilitate various chemical reactions that would otherwise seldom, if ever,

(a) (b)

**FIG. 1-2** (a) Volcanic action continues today in many regions of the world, as this view of an eruption on the Kamchatka Peninsula shows. Volcanism, along with sunlight, cosmic radiation, and lightning, would have provided ample energy to drive the reactions that led to the origin of life. (b) A tide pool. On hot, dry, windy days, seawater left in tide pools high above the low-tide mark gradually evaporates, changing its water-to-salt ratio and becoming concentrated in the process. In ancient seas, this would have promoted a variety of chemical reactions. Lightning would have further accelerated the process.

take place at moderate temperatures. Besides macromolecules, other products of these randomly occurring reactions would have included such small, energy-rich phosphate molecules as **ATP.** Found in all organisms, ATP facilitates the transfer of energy among molecular systems. By now, nucleic acids, such as DNA, would also have begun to appear.

***Natural Selection and the Origin of Life.*** Most experts agree that the formation of enzymes and nucleic acids may have taken about 1 billion years. With the development of such macromolecules, the stage was set for the next important event: the development of highly structured molecular aggregates—tiny, membrane-enclosed systems of macromole-

PANEL 1-1

## Chemical Evolution in a Reaction Flask

Although in the early part of this century our ideas about the origin of life were almost entirely speculative, we are now much more certain of how life arose. In 1953, Stanley L. Miller, a graduate student at the University of Chicago, performed a landmark experiment. Using simple molecules of the type thought to be present in the ancient atmosphere—water, molecular hydrogen, ammonia, and methane—Miller subjected them to weather conditions simulated to match those that probably prevailed on earth during its first 1 or 2 billion years. Harold Urey, the American chemist in whose laboratory Miller performed his experiments, calculated that the ancient seas may have contained as much as a 10-percent concentration of these molecules long before the beginning of life.

In Miller's experiment the materials were repeatedly vaporized and condensed in a closed system. Intermittent electrical sparks were used to simulate lightning. After only a week of this treatment, Miller was excited to find that several kinds of amino acids had formed in the solution in the flask! He and others repeated the experiment many times with similar results.

Variations on this experiment, some of them simulating sunlight by the use of ultraviolet light, have produced additional kinds of amino acids, the energy-rich phosphate compound called adenosine triphosphate or ATP, and some of the molecular units of which nucleic acids are composed. In the first 2 billion years of the earth's existence, under conditions similar to those in Miller's reaction flask, a large number of these molecular building blocks and even some proteins could have formed.

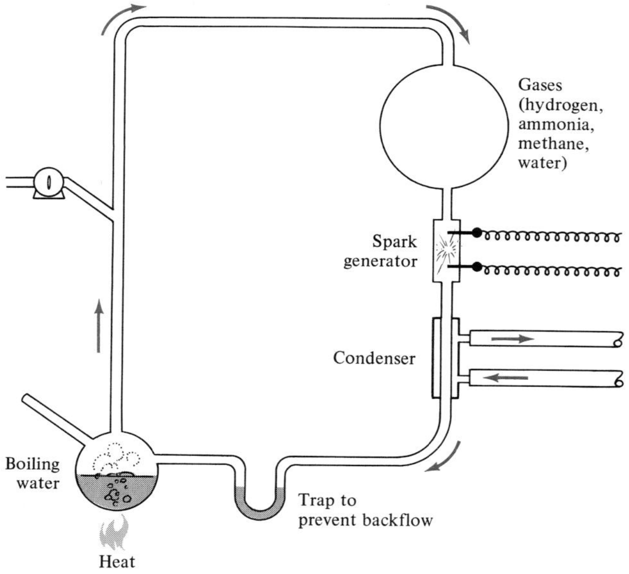


Sketch of the apparatus used by Stanley Miller in his pioneering experiments on the generation of amino acids under conditions such as those that probably prevailed on earth before life arose.

Recent evidence indicates that the earth's atmosphere might not have been as rich in ammonia when life arose as Miller believed. It now seems likely that the early atmosphere contained principally carbon dioxide, carbon monoxide, water vapor, nitrogen, and smaller amounts of ammonia, hydrogen, methane, and hydrogen sulfide. This mixture would have given rise to life on earth as early as 3.5 billion years ago.

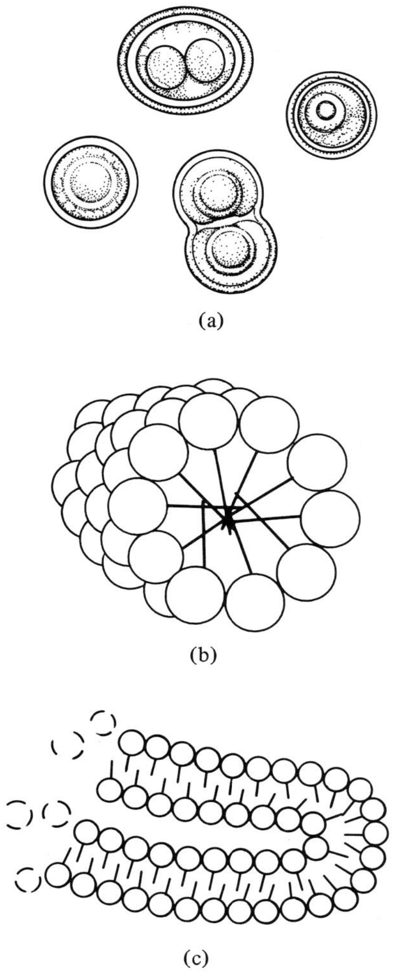

**FIG. 1-3** Sketches of proteinoids and micelles. (a) Proteinoids form when amino acid mixtures are heated and then cooled by contact with water. (b) Micelles form when amphoteric substances, such as phosphoglycerides, are mixed with water. (c) They may aggregate to form a bimolecular layer, as in cell membranes.

cules that increased in size by merging with some of the smaller aggregates with which they collided. Such aggregates could have arisen in more than one way. One possibility involves clay surfaces, which bear electric charges. Oppositely charged molecules become adsorbed to the clay; that is, they adhere to it by electrostatic attraction. By concentrating molecules in this way, such clay surfaces can act as catalysts, promoting reactions between adjacent molecules.

Another possible mechanism for the formation of molecular aggregates was demonstrated in 1960 by the American biochemist Sidney W. Fox in experiments in which he heated mixtures of amino acids to between 75 and 150 degrees centigrade (°C) and then cooled them. Such treatment resulted in the production of **polypeptides,** or chains of amino acids, the first step in protein formation. He called the resulting spherules **proteinoids** (Fig. 1-3a). Within some of these spherules, such molecules as enzymes were trapped.

A third type of macromolecular aggregate is the **micelle,** which results when substances whose molecules are soluble in water at only one end are mixed with water (Fig. 1-3b, c). A common example is the spherical micelle of a phosphoglyceride, the type of molecule of which biological membranes are primarily composed. As a micelle of phosphoglycerides forms, other kinds of molecules, such as water and amino acids, can be trapped within it.

When they reach a certain size, micelles fragment into similar, smaller units. At first this would have occurred very haphazardly, but over hundreds of millions of years, some units would have evolved an orderliness, probably influenced by the presence of enzymes and nucleic acids. Certain self-contained systems could have maintained their complexity and stability by utilizing various forms of energy in their environment. Those systems that were most efficient in the acquisition and processing of energy and materials would tend to survive in greater numbers than other, less efficient systems. Here, at a simple level, would have been a form of **natural selection,** the blind "selection" by the environment of more stable or efficient products over those less stable or efficient.

At some stage in this natural selection of chemical systems, at least one structured, self-reproducing macromolecular system would have evolved that could, by most definitions, be called living. The origin of life was thus not a single, dramatic event, but rather a stage in a long, continuous process of chemical evolution—a process that continues as part of the evolution of organisms themselves. Each step in this long process would have depended on a supply of energy.

***The Origin of Organismal Diversity and the Role of Natural Selection.*** Once life arose, those organisms that possessed the combinations of genes that allowed them to be the most efficient in capturing energy and materials, in maintaining their existence (for example, by escaping predators), and in passing these qualities on to their offspring were the ones to survive as less efficient types failed and became extinct. This is what is meant by natural selection in the usual sense: "nature selects" the ones to survive, the ones that at a given time and in a particular environment exhibit the highest degree of fitness for the existing conditions. Over thousands and millions of years, as other combinations of genes arose and were selected, even greater fitness was exhibited by some individuals. (Panel 1-2 examines one aspect of fitness, the relationship of structure and function.) However, whenever environments changed, even these types were replaced by others more suited to new environmental conditions.

PANEL 1-2

# The Relationship of Structure and Function

## FUNCTION DEPENDS ON STRUCTURE

One basic difference between organisms and nonliving things is that organisms are composed of very complex molecules in special arrangements: they are much more highly organized. Moreover, the precise structure of various complex organic molecules determines the special characteristics of each organism. In any study of a biological system, it soon becomes evident that the structure of any part of an organism, whether it be as small as a protein molecule or as large as an elephant's trunk, is directly related to the function of that part. Conversely, each part of an organism has a function dictated by its structure. For example, a leaf, with its large surface area and its pores, is highly suited to intercept rays of light and to absorb carbon dioxide from the atmosphere.

An African elephant spraying dust on itself (Amboseli, Kenya). Anyone who has watched an elephant use its trunk to pick up a peanut and eat it, drink water, spray dust on itself, or lift a large log, will have been impressed with the relationship of the structure of the trunk to the trunk's functions.

## WHEN THE WHOLE IS GREATER

The various characteristics and functions exhibited by any structure are usually attributes of the entire, intact structure but not of its isolated parts. The whole thus exhibits qualities not displayed by the parts alone. This is as true of molecules as it is of organisms. For example, a water molecule has properties not exhibited singly by either of its constituents, hydrogen and oxygen. In the same way, various proteins—which are composed almost entirely of carbon, hydrogen, oxygen, and nitrogen—exhibit properties not found in any of their constituent elements when they are studied alone.

## DIFFERENT ARRANGEMENTS, DIFFERENT PROPERTIES

The relationship of the parts to the whole is also illustrated by the fact that different arrangements of the same atoms can produce molecules with different properties. For example, both starch and the cellulose of plant cell walls are composed of molecules of the simple sugar glucose. The differences between starch and cellulose depend, not on their component units, but on the location of the chemical bonds between their glucose units. Just as the same kind of bricks can be used to construct different kinds of buildings, the same chemical components can be arranged to produce a variety of structures.

## EMERGENT PROPERTIES

New properties are sometimes said to "emerge" from atoms and larger units when they are combined with other types of atoms and units in various ways. However, this does not mean that some mysterious force is involved. Perhaps it is better to say that among the properties of hydrogen and oxygen, for example, are the properties of water, but that these can be manifested only when hydrogen and oxygen are chemically combined in a certain proportion. Similarly, a soccer player's abilities "emerge" only when the player is put into the company of other players and supplied with a soccer ball, a field, and a game.

## OTHER UNIQUE CHARACTERISTICS OF LIFE

Besides being able to process energy and to evolve, organisms exhibit several other unique properties. The most important are responsiveness to environmental stimuli; participation in the life cycle of growth, development, aging, and death; the ability to produce offspring like themselves; and the ability to maintain themselves.

***Responsiveness.*** All organisms respond to stimuli in their environments. For example, fish react to changes in light, temperature, and the oxygen content of water, and many plants develop flowers in response to changes in day length.

***Growth, Development, Aging, and Death.*** Every organism has a life cycle. This typically includes a period of growth and development marked by increasing size and complexity, followed by periods of maturity, aging, and death. As a multicellular organism develops, many new, specialized cell types gradually arise. The new types are different in appearance and in function from the original cells and from the other cell types in the organism.

***Reproduction, Heredity, and Variation.*** Organisms produce organisms like themselves. The offspring not only look like their parents, but they also inherit their parents' potential (Fig. 1-4). But despite the fact that parents give rise to offspring like themselves, heritable differences do occur between generations, sometimes very unusual ones (Fig. 1-5). Nearly all of these differences are based on **mutations,** occasional variations in **genes,** the DNA molecules that constitute the inherited, coded specifications for the structures and capabilities of an organism. Mutations are random events. As such, very few would be expected to be beneficial to the mutant organism, especially if it is already well adapted to its environment. Most mutations are harmful, many are neu-

**FIG. 1-4** Mature pine trees produce small pine trees like themselves. Adult moose produce young moose that resemble their parents not only in appearance but also in behavior.

**FIG. 1-5** An albino lowland gorilla (Barcelona Zoo). Although not always this obvious, variation from one generation to the next is the rule.

tral, and only a few out of many thousands are beneficial. Various combinations of mutant genes provide variety in every type of organism, the raw material on which natural selection acts in the ongoing process of evolution.

***Self-Maintenance.*** An organism must maintain its highly organized state. It must repair some types of damage, remove wastes, synthesize certain substances, acquire materials, and so on. As with all other vital functions, self-maintenance involves work, that is, it uses energy.

# TAXONOMY (SYSTEMATICS)

It was probably apparent even to the first humans that most kinds, or species, of organisms could easily be classified or placed into larger groups based on similarities. For example, the meaning of the lay term *cat family* is clear to anyone who has seen examples of lions, tigers, cougars, and house cats. It is as though organisms are related to each other—closely related to those they most closely resemble and distantly related to those they resemble only slightly. The genetic basis for these resemblances, however, was gradually realized only in the last two centuries.

## EARLY CLASSIFICATIONS

The earliest known classification scheme is that of the Greek philosopher Aristotle (382–322 B.C.). He placed organisms on a ladderlike scale of being, with nonliving things at the bottom, plants next, then animals, and finally humans at the top. Aristotle advocated the close comparison of organisms with one another. He was the first to classify whales and dolphins, the cetaceans, as mammals on the basis of their ability to breathe air and to bear and suckle live young. Some 2,000 years intervened before cetaceans were again classed in their rightful place with other mammals, instead of being included with the fishes.

## THE CONTRIBUTIONS OF LINNAEUS

The great Swedish taxonomist Carolus Linnaeus (1707–1778) rescued the cetaceans from the company of the fishes (Fig. 1-6). Linnaeus is considered the father of **taxonomy,** the formal scientific naming and classifying of organisms. Most of Linnaeus' life was devoted to the description, naming, and classifying of the world's organisms. He not only listed between the covers of a single volume the names and brief descriptions of most of the then known organisms, but he developed several major taxonomic principles as well.

First, he adopted and popularized a **binomial system of nomenclature,** which designated each kind of organism by both a **genus** name and a **species** name. Previously it had been common practice to designate an organism by its genus name followed by a short description, a genus being a group of two or more similar kinds of organisms. Linnaeus substituted a Latin adjective for the description, giving each species of organism a double name, or **binomial.** For example, *Musca* is the genus of many kinds of flies, and *Musca domestica* is the species name of the common housefly. Linnaeus thus directed attention to the very concept

**FIG. 1-6** Carolus Linnaeus (1707–1778), the father of taxonomy. Besides popularizing the binomial system of nomenclature, he published the names and descriptions of all species of organisms known at his time.

of species, to the idea of a species as a group of organisms whose members fit the same description. Perhaps Linnaeus' most valuable contribution was his emphasis on what then was regarded as a **natural system** of classification: one that took into account as many traits of an organism as possible. For example, based on a detailed comparison of various features, he correctly classified whales with four-footed mammals, or quadrupeds.

Linnaeus' popularization of natural systems of classification directed biologists' attention to just those traits most reasonably interpreted as indicating a common ancestry. The arrangement of species within Linnaeus' system thus eventually became an argument for the theory of evolution. He had classified organisms into the very kinds of groups that would have resulted if the members of each group had descended from common ancestors, such as all cats, tigers, etc. from one ancestral group of felines; all dogs, wolves, etc. from one ancestral group of canines; and so on.

***The Fixity of Species Versus Evolution.*** When Linnaeus began his work, like most other naturalists of his day, he believed that species were fixed, meaning that they had varied little, if at all, since the day of their creation. In fact, he asserted in the early editions of his great work *Systema Naturae* that God had created a single breeding pair of each species and that all living organisms were their virtually unchanged descendants. This view fit the widely accepted literal interpretation of the book of Genesis in Judaeo-Christian sacred scripture. It should also be remembered that at that time, few Europeans believed the earth to be more than about 10,000 years old.* The thoughts that any biologist might have had about evolutionary processes could not have been compatible with a conviction that the world was only a few thousand years old. Although evolutionary views had surfaced repeatedly since the ancient Greeks, there was little reason to believe them, and so the scriptural view of species as fixed entities prevailed.

As Linnaeus continued his work and observed more plant and animal forms, particularly intergrading forms, he changed his views about the fixity of species. Eventually, he came to believe that perhaps God had created only one species in each genus and that the rest had arisen by variation, or as we would say, had evolved. In the last edition of *Systema Naturae,* he deleted his assertion that no new species ever arise. A century later, again as a result of close study of a great variety of organisms, another man was also to abandon the view that species were fixed in favor of a conviction that new species had arisen and were still arising through evolution. That man was the English naturalist Charles Darwin (Fig. 1-7a).

## THE IMPACT OF THE NATURAL SELECTION THEORY ON TAXONOMY

By the mid-nineteenth century, data that potentially constituted overwhelming evidence for an evolutionary point of view had been gathering for over a hundred years. But there was no reasonable, unifying theory to give meaning to this mass of information. The principal data came from three fields of investigation:

* A point of discussion within 8th-century Hindu philosophy was over whether the earth was more than 2.2 billion years old.

1. The cataloging and classification of species from all over the world in itself had begun to make the kinships among organisms evident to more and more biologists.
2. A rapidly growing number of fossils was being discovered, and some of the prevailing explanations for them, such as their having resulted from the biblical deluge, were being reexamined. Moreover, it was becoming evident that the lowest strata of rock, presumably the oldest, contained simpler organisms than the higher strata.
3. Geological studies led to the conclusion that the age of the earth, and of the fossilized remains of organisms, was to be reckoned in millions, not thousands, of years (see inside back cover).

***Darwin's Natural Selection Theory.*** To the available knowledge, Charles Darwin added his own observations on geology and the diversity of life, based not only on a five-year voyage around the world as naturalist aboard H.M.S. *Beagle* (Fig. 1-7b) but on many other observations derived from his studies of animal and plant breeding. He was also

(a)

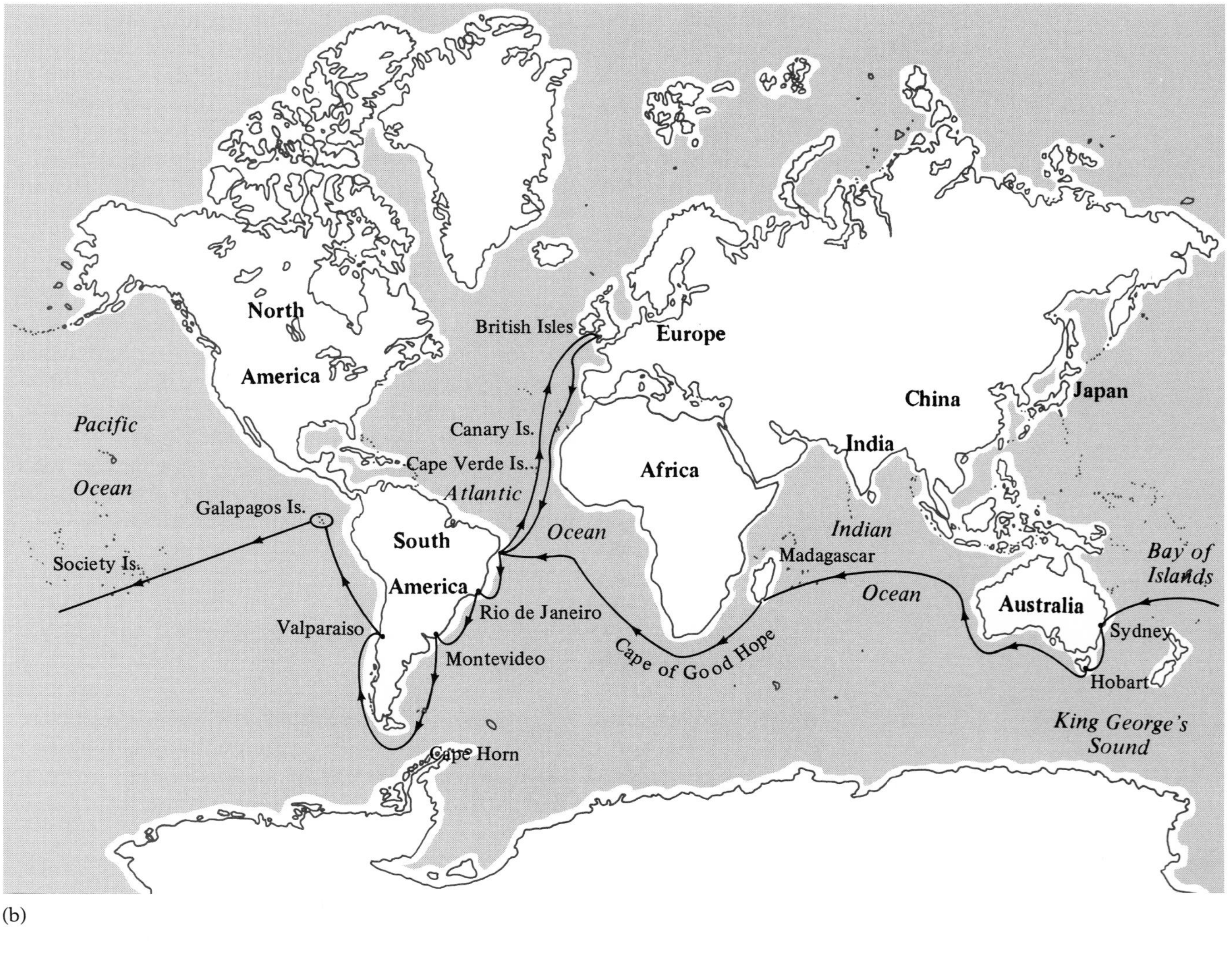


(b)

**FIG. 1-7** Charles Darwin and his voyage around the world. (a) The young Darwin four years after his return from the historic voyage. (b) Route of the voyage of H.M.S. Beagle (1831–1836).

struck by the implications for all species of the principle, enunciated by the economist Thomas Malthus, that human populations have the capacity to increase faster than their food supply. Darwin's careful observations and logical thought culminated in his theory that natural selection was the final determinant in the process of evolution. A summary of his findings was published in 1859 in *On the Origin of Species by Means of Natural Selection or the Preservation of Favored Races in the Struggle for Life.* The theory's basic components, outlined in modern form, are:

1. Every species can produce more offspring than can be supported by the environment. The survivors are those that have successfully competed for the available materials, energy, space, shelter, and mates.
2. Heritable differences arise in all species. Darwin did not know the source of heritable changes; he knew merely that they occurred.
3. Environments change over time. In each population, some individuals may possess heritable traits that enable them to survive better than others under such new conditions as changes in climate or the appearance of new predators, competitors, or foods. Changes in environment thus promote evolutionary change. In modern terms, a **selection pressure** is created. If a species that is well adapted to one environment does not include enough individuals with a genetic makeup suited to the new conditions, it may become extinct.
4. On occasion, members of a species cross barriers and migrate into new regions with different environments. On other occasions, new barriers arise, as when mountains are uplifted, rivers are created, or continents drift apart. Individuals that become isolated from their parent populations in one of these ways usually do not exhibit all the variations found in the parent population. Moreover, in time, new mutant genes will arise in both the new and the original population. Rarely would any of these chance mutations be the same in both populations. Because the environments on the two sides of a barrier nearly always differ in various ways, they will also exert different selection pressures on the two populations. Depending on conditions, this selection could happen rapidly or slowly. In any case, isolation of the two populations over time, produces two groups that include among their many differences an inability to interbreed. In time, all these factors will contribute to **speciation,** the development of one or more species from a single parent species. Over millions of years, this process leads to **adaptive radiation,** in which there evolves not only new genera (the plural of genus) but larger categories as well (Fig. 1-8), each adapted to a different environment.

***How Darwin Changed a Point of View.*** Neither Darwin nor any of his contemporaries understood the mechanisms of inheritance and genetic change.* Although Darwin was ignorant of the underlying processes, he grasped the revolutionary concept that a slow accumulation of small inherited variations that suited individual organisms to a particular environment would lead to evolutionary change.

Before Darwin, naturalists, including Linnaeus, tended to see spe-

* For an account of the rise of evolutionary thought and the role played in the development of the natural selection theory by Alfred R. Wallace, a younger contemporary of Darwin, see Chapter 27.

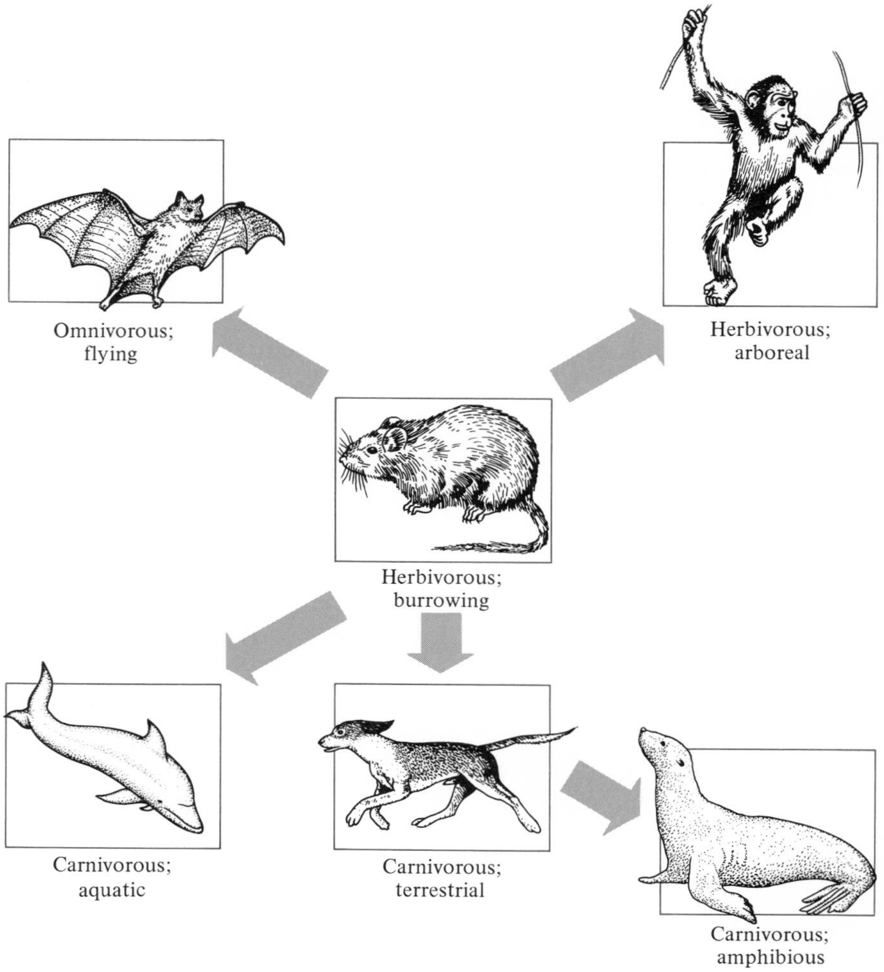


**FIG. 1-8** Adaptive radiation within the class Mammalia related to the evolution of types of locomotion and adaptations to different environmental situations.

cies as being typological: they viewed all individuals within a species as essentially alike, modeled after a perfect "type." Individual variations were interpreted as imperfections. Darwin, however, saw that variation was the rule. He regarded species "types" as averages or generalizations of the features of various populations, and he contended that the variation inherent in a population provided the basis for the finer and finer adaptation of species to particular environments. By the cogency of his evidence and arguments, Darwin led the scientific world to see that all organisms and nearly all their traits—their structures, physiology, development, and patterns of behavior—were probably the products of an evolution shaped by natural selection.

## MODERN TAXONOMY

Most of today's taxonomists, or **systematists,** as they are also called, use the term **natural** in a new sense—to refer to systems of classification that reflect the kinships of organisms.* An artificial system is thus one

* There is a difference of opinion on this. While using all available criteria—appearance, biochemistry, behavior, and the like—the **pheneticist** school of thought contends that it is usually impossible to determine kinships with any accuracy and we should not try to do so. Systematists who classify according to how long ago various groups have branched from each other are referred to as **cladists.** No matter which school is followed, the groups into which organisms get placed are usually the same.

that ignores such kinship relationships. Taxonomists draw upon comparative studies of anatomy, physiology, embryology, immunology, behavior, and biochemistry. They also look at details of life cycles and consider the results of breeding experiments to classify organisms. The term **species** is usually defined as a group of organisms that can interbreed with each other with full fertility, but not with any other group. Because breeding experiments are not always feasible (they never are with asexually reproducing species), systematists usually judge whether organisms *might* have been able to interbreed, by looking for similarities in all of their heritable traits. By using all inherited characteristics as a basis for classification, kinships are necessarily taken into account in constructing systems and in assigning organisms to places within them.

***The Phylogenetic Tree.*** When a taxonomist classifies an organism today, it is not just catalogued; it is put in its place on a kind of family tree called a **phylogenetic tree.** In this book, we follow the widely adopted taxonomic scheme proposed in 1969 by the American systematist Robert Whittaker and depicted inside the front cover. Whittaker divided the living world into five kingdoms, based primarily on such criteria as the presence of a nucleus in an organism's cells, whether an organism is unicellular or multicellular, and an organism's mode of nutrition:

- Monera: the prokaryotes, one-celled organisms without true nuclei (bacteria, including the cyanobacteria, formerly called blue-green algae)
- Protista: one-celled organisms with true nuclei, including some plantlike forms (protophyta) and some animallike forms (protozoans)
- Fungi: molds, mushrooms, and the like (nonphotosynthetic, multicellular, plantlike forms with cell walls)
- Plantae: plants (photosynthetic, multicellular organisms with cell walls)
- Animalia: animals (nonphotosynthetic, multicellular, generally motile organisms that lack cell walls and eat their nutrients)

Like all phylogenetic trees, the one inside the front cover employs three conventions:

1. The smallest groups are shown as the twigs of the tree's branches. Groups on twigs of the same branch are closely related.
2. The points at which two twigs or two branches connect suggests the point in the evolution of life on earth at which the particular groups represented by the twigs or branches arose from a common ancestor.
3. The lowest parts of the tree along the main branches and the trunk generally represent the most ancient times and the most ancient groups in the history of life on earth.

***Problems of Classification.*** Because evolution is a continuing process in which one group gives rise to two or more groups, some groups represent transitional stages and are difficult to classify. As a result, several classification schemes have emerged. For example, some authors propose only two kingdoms. In their scheme, bacteria, fungi, and all protophyta are included in the plant kingdom and all protozoans are placed with animals. Other schemes have three kingdoms, and still others, four kingdoms.

Due to the nature of evolution, disagreements and practical problems arise regardless of which scheme is used. An example of such a problem is provided by the algae, which are relatively simple, photosynthetic organisms of various descriptions. Almost all are aquatic. Many are one-celled and therefore protists, but many are multicellular and therefore plants. They range in size from microscopic specks to seaweeds over 30 meters long. In the Whittaker five-kingdom classification, algae thus show up in two of the five kingdoms: in Protista as protophyta and in Plantae as multicellular algae.* Although they are members of the plant kingdom, some groups of multicellular algae are obviously more closely related to certain one-celled protistan forms than to any other group in their own kingdom. Abolishing the kingdom Protista and classifying all photosynthetic organisms as plants would seem to solve this problem. It would, but then we are left with another problem, one that defining the kingdom Protista was intended to solve: some one-celled algae are identical to certain protozoans (one-celled animals in a two-kingdom classification), except that they possess chloroplasts and can photosynthesize. In fact, subjecting these green, one-celled algae to a simple treatment will, without harming them, rid them of their chloroplasts, whereupon they become completely indistinguishable from those protozoans. Such problems as these have long been accepted by most taxonomists as part of their job. For, if organisms are evolving, then such intergrading of groups is inevitable. Taxonomic systems can thus be regarded as progress reports on evolution.

***The Major Taxonomic Divisions of the Five Kingdoms.*** A brief examination of the major subdivisions of the five kingdoms will be helpful at this point. Later chapters refer to some of the subgroups listed in Table 1-1, making it a useful reference for placing subgroups in their major

* The so-called blue-green algae of the kingdom Monera are more commonly regarded as cyanobacteria.

**TABLE 1-1 Outline Classification of Some Major Subgroups of the Five Kingdoms***

| *Kingdom* | *Major subgroups* |
|---|---|
| Monera | Bacteria |
| | Cyanobacteria (formerly blue-green algae) |
| Protista | Green algae |
| | Yellow-green algae |
| | Yellow-brown algae |
| | Diatoms |
| | Dinoflagellates |
| | Euglenoid (photosynthetic) flagellates |
| | Protozoan flagellates |
| | Ciliates |
| | Sarcodine protozoans |
| | Sporozoans |

* Only major taxonomic divisions or phyla and in some cases, subphyla and classes, are listed here. The groups that are listed are ones most often referred to in the text. In all cases the anglicized, rather than the Latin name of the group, is used, followed by one or more examples whenever they are also known by common names.

**TABLE 1-1** ***(continued)***

| *Kingdom* | *Major subgroups* |
|---|---|
| Plantae | Green algae |
| | Stoneworts |
| | Brown algae |
| | Red algae |
| | Mosses |
| | Vascular plants |
| | Ferns, horsetails, club mosses |
| | Seed plants |
| | Seed ferns (extinct) |
| | Cycads |
| | Ginkgo (maidenhair tree) |
| | Conifers (pine, fir) |
| | Flowering plants |
| | Monocots (grasses, corn, lily, onion) |
| | Dicots (most common flowering plants) |
| Fungi | Slime molds |
| | Molds |
| | Sac fungi |
| | Club fungi |
| Animalia | Poriferans (sponges) |
| | Cnidarians (coelenterates) |
| | Ctenophorans (comb jellies) |
| | Platyhelminthes (flatworms) |
| | Nemerteans (ribbon worms) |
| | Nematodes (round worms) |
| | Annelids (earthworms) |
| | Brachiopods |
| | Molluscs (clams, snails, squids) |
| | Onychophorans |
| | Arthropods |
| | Arachnids (spiders) |
| | Crustaceans (lobsters, crabs, shrimps) |
| | Myriapods (centipedes, millipedes) |
| | Insects |
| | Arrow worms |
| | Echinoderms |
| | Hemichordates |
| | Chordates |
| | Protochordates |
| | Vertebrates |
| | Jawless fishes |
| | Cartilaginous fishes |
| | Bony fishes |
| | Amphibians |
| | Reptiles |
| | Birds |
| | Mammals |

**FIG. 1-9** The white oak, *Quercus alba*.

taxonomic category. The phylogenetic tree inside the front cover provides a more complete picture of the relationships of the various groups to each other.

Table 1-1 includes most of the major subdivisions of the five kingdoms. In order to accommodate the many known species, each kingdom is much more finely subdivided than is shown in the table. Two typical classifications of organisms are given below, the first for the white oak, *Quercus alba* (Fig. 1-9), and the second for the green frog, *Rana clamitans* (Fig. 1-10):

- Kingdom: Plantae (already defined)
- Division: Tracheophyta (plants with vascular tissues)
- Subdivision: Spermopsida (vascular plants that form seeds)
- Class: Angiospermae (flowering plants)
- Subclass: Dicotyledoneae (with two seed leaves)
- Order: Fagales (beeches, oaks, birches, alders, chestnuts, and several others)
- Family: Fagaceae (beeches, oaks, and a few others)
- Genus: *Quercus* (oaks)
- Species: *Quercus alba* (the white oak)

- Kingdom: Animalia (already defined)
- Phylum: Chordata (animals with notochords, etc.)
- Subphylum: Vertebrata (chordates with backbones)
- Class: Amphibia (salamanders, frogs, toads)
- Order: Anura (frogs and toads)
- Family: Ranidae (frogs)
- Genus: *Rana* (certain kinds of frogs)
- Species: *Rana clamitans* (the green frog)

Further subdivisions are also employed. For example, orders may be divided into suborders, each of which includes a group of similar genera. Note that each lower group belongs to all those groups above it;

**FIG. 1-10** The green frog, *Rana clamitans*.

for example, all frogs are both vertebrates and amphibians, and all oaks are both angiosperms and dicotyledons. A difference between plant and animal classifications is that the plant kingdom is divided into divisions, not phyla.

# THE NATURE OF SCIENTIFIC INQUIRY

The preceding descriptions of the nature and origin of organisms and their relationships to each other refer to specific ways in which scientists arrived at some of their conclusions. The way in which scientific understanding is advanced is actually a universally accepted, rather formal process. Viewed in broad perspective, biology shares its logical framework and its basic approach to investigation with other natural sciences, such as chemistry and physics.

## PRESUPPOSITIONS

In approaching any scientific study, the scientist has made some presuppositions, notably (1) that matter and energy will always and everywhere react in the same way under the same conditions, (2) that the scientist's senses and instruments are capable of conveying true impressions of objects and events, and (3) that human beings are capable of understanding the universe. While these presuppositions are not consciously alluded to on a regular basis, they underlie all scientific investigation. The investigation itself, the actual development of new scientific knowledge, involves making a series of observations of certain objects or correlated events followed by a judgment or inference regarding the nature of those objects or events.

## LOGICAL PROCESSES

Ideally inferences made by scientists are generalizations, such as, "All organisms contain carbon, hydrogen, oxygen, and nitrogen." This particular judgment, like most, is based on many observations. Such a statement is not merely the sum of many observations but is an inference based on them. This is commonly referred to as **inductive reasoning.** Indeed, almost all, if not all, generalizations made in science are **inferences:** from a certain number of facts, it is judged that what is true of them is also true in all similar cases. No one actually knows that such an inference, or induction, is absolutely true; but experience has taught us that such judgments are usually reliable. The scientist's inference is almost always tentative, however, to be changed as new information warrants it.

Lower animals are capable of primitive generalizations regarding shapes, sounds, tastes, and other factors, but humans go well beyond this level to an intellectual understanding of the nature of objects and events. The observations that scientists make are not merely observations; they are made and considered in an intellectual way, against a background of theory and understanding. Solar eclipses, for example, can be detected in some way by most animals and even by plants. But when an astronomer observes an eclipse, the data chosen to be recorded is received into a rich theoretical context. For example, it may be used to modify existing theory if the new data and the theory prove to be incongruous. Scientific observation is thus an intellectual activity.

The logical processes involved in developing new scientific knowledge include: (1) close and intelligent observation of natural events or phenomena; (2) constructing tentative explanations, or **hypotheses,** for them; (3) validating the hypotheses by testing predictions made from them; and (4) after adequate testing, coming to a conclusion as to the validity of the hypotheses.

Applied to scientific matters, such processes are often referred to as the scientific method. But as scholars of history, sociology, economics, and other subjects of intellectual inquiry know, there is nothing uniquely scientific about this approach. It is true that natural science, more than any other field, relies heavily on experimentation as a way of testing its hypotheses. Statistical analysis of data is another common scientific practice. But neither experimentation nor statistical analysis is uniquely scientific. Nor, for that matter, is all scientific knowledge acquired by this method. The basic difference between science and other studies is its subject matter. Its logical processes are common to all areas of scholarship.

## THE SCIENTIFIC INVESTIGATION

What goes on in a scientific program in which significant discoveries are being made? In its simplest and broadest terms, most experimental work seeks to test the truth of hypotheses that have been proposed for phenomena. Of course, before any hypothesis is devised, it is preceded by many observations and much thought (Fig. 1-11).

***Prerequisites for Productivity.*** But how does a scientist arrive at a hypothesis that purports to explain some puzzling phenomenon in the first place? There is no one way, but generally an hypothesis is the combined result of three factors: (1) a broad and deep knowledge of a particular area of study; (2) an accumulation of careful observations, including information on variable factors, or **variables,** regarding the particular phenomenon; and (3) a fertile imagination that suggests various possible explanations for the phenomenon and the kinds of experiments that would test the validity of those explanations.

This list of prerequisites probably sounds more formidable than it really is. Following their formal education, investigators stay abreast of current developments in their fields by reading articles in scientific periodicals that report such work. For investigators with a specialized education who are working within their specialty, drafting hypotheses is usually not a problem. What is difficult is designing and conducting investigations that adequately test the validity of a hypothesis.

***Need for Caution.*** Students will find it instructive to examine the way in which experimental results are stated and conclusions drawn from them in a scientific "paper," or article. Each word or phrase has usually been chosen with care. The language may even strike one as overcautious, but it rarely is. For example, most results are usually qualified as being obtained "under the conditions of this experiment" or "with this species and variety of organism." Moreover, the conditions under which the experiments were done and the materials used are explained in detail. Such detailed reporting enables other scientists to judge the

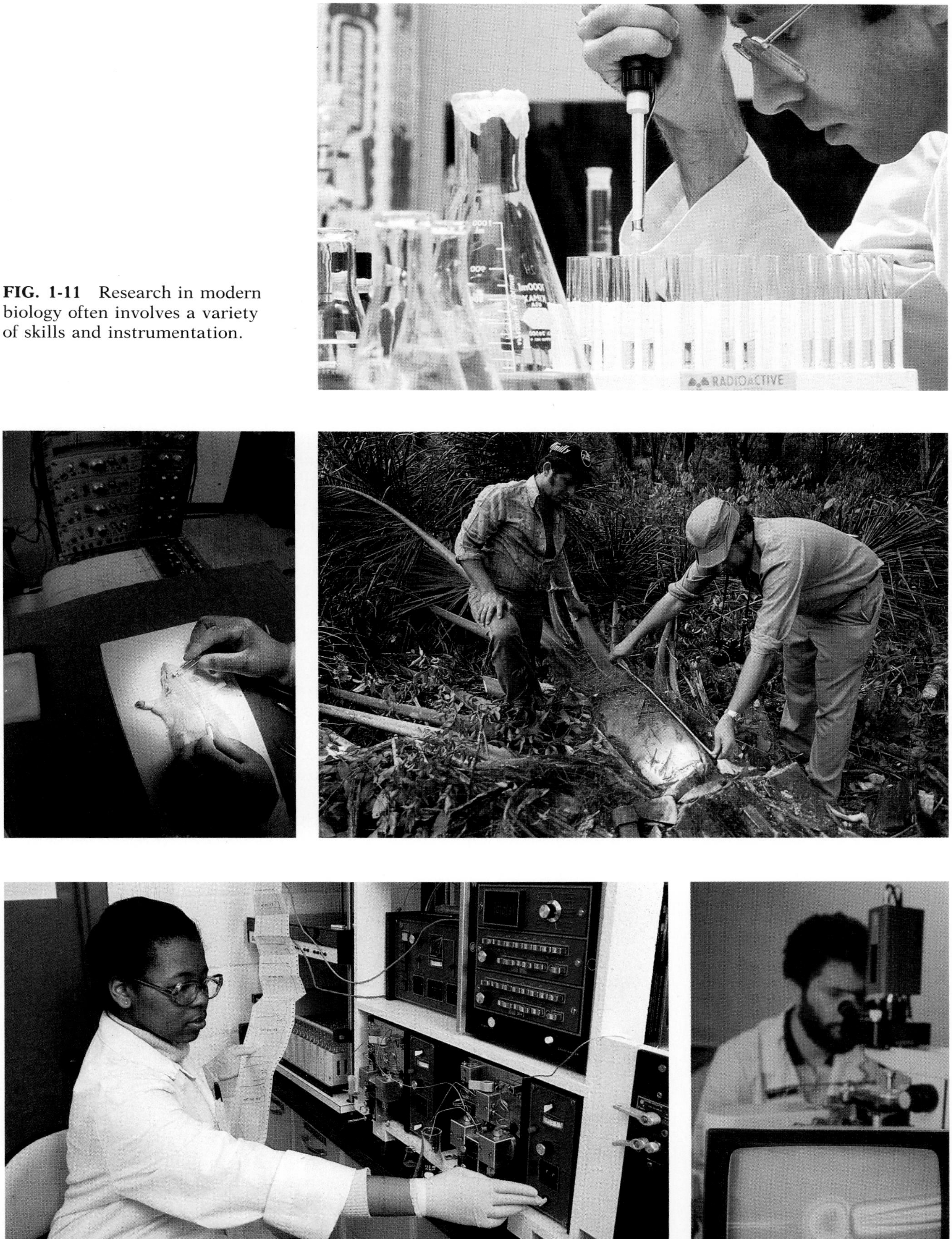


**FIG. 1-11** Research in modern biology often involves a variety of skills and instrumentation.

validity of the conclusions, to repeat the work to verify the results, or to conduct their own variations of the experiments.

In drafting conclusions, it is seldom said that one factor or event "caused" another. Strictly speaking, events do not have a single cause but, rather, are achieved by the combined effects of many factors or variables. More important, it is often impossible to be certain where the causes lie. Merely because one event always follows another, for example, does not mean that the first event caused the second. Certain North American flowers, for instance, bloom soon after the arrival of robins from the south; they have never been seen to bloom in the robins' absence. Without a knowledge of the factors that influence both events, someone might have been tempted to conclude that the arrival of robins caused the flowers to bloom.

Because we are so easily misled by appearances and by our prejudices, it is usually best to avoid assigning causes in most cases and to be content with stating that the presence of various factors is correlated with certain occurrences. Scientific reports are replete with such phrases as "the evidence suggests that . . . " or "the results are compatible with the hypothesis that . . . " and other such cautious statements. This is not the language of investigators being modest about the importance of their findings. Rather, it is usually the only safe way to express scientific findings, especially because of the possibility of hidden or unknown variables. To advise caution in assigning causes is not to suggest, however, that there are no causes for the events we study or that we lack interest in finding them. On the contrary, discovering the causes of phenomena is science's ultimate goal. (For a discussion of some philosophical aspects of scientific inquiry, see Panel 1-3.)

***A Mathematical Parallel and a Difference.*** There is an obvious parallel between scientific investigations and mathematical deductions in that when all but one variable are kept constant, it can be deduced that any difference or new result is due to the one changed variable. The logical process is exactly the same in both disciplines; but in science, especially in biological investigations, it is rarely possible to control or even identify all the variables in a situation. However, we rarely need to identify all the variables as long as we employ what are called experimental controls or, simply, controls.

***Designing and Conducting Experiments.*** The rules followed in designing and conducting experiments are few but important. An **experiment** consists of an alteration of the conditions of a state or process followed by a study of the results of the alteration. As simple as this sounds, it is usually impossible to make only one alteration in any situation. Someone interested in determining the effects of an injected drug on the function of the rat heart, for example, would need to inject the rat with the drug while studying the functioning of its heart. The number of variables that might affect the functioning of the rat's heart, in this case, are numerous: (1) being kept in a cage, (2) being fed artificial food, (3) having the cage cleaned, (4) having people occasionally walk by the cage, (5) being injected, and, not least important, (6) being subjected to the technique used to measure the heart's activity. Even the fluid in which the injected drug is dissolved might have an effect on the heart. If, as a result of such an experiment, the rat's heart function does change, how can the investigator be sure which of the many variables caused the change? The answer is found in the use of experimental controls.

EXPERIMENTAL CONTROLS. Nearly all biological experiments are conducted on at least two groups of organisms. One is the **experimental group,** or experimentals, which in our example are rats that receive the drug being studied. The experiment is also conducted on one or more **control groups,** which, as far as possible, are fed, frightened, cleaned, handled, injected, measured, and otherwise treated exactly like the experimentals with a single difference: the controlled variable. In our example, the control groups do not receive the drug being studied. Instead, they are injected with equal amounts of substances known to have no particular effect on heart function, for example, salt or sugar solution or merely the solvent in which the drug under study was dissolved (Table 1-2). In experiments in which the function of an organ is studied by removing it surgically, a sham operation is performed; the controls are anesthetized and cut open, incisions are made in the organ, as is done with the experimentals, but everything is then sewn back in place with nothing being removed. A comparable control in our example is piercing the rats with a hypodermic needle without injecting anything.

With adequate controls, the investigator at least avoids the error of attributing results to the wrong factor. Any changes due to unknown variables would most likely occur both in the controls and in the experimentals. If both groups exhibit a change, the results are inconclusive; the investigator is thereby alerted to the presence of an unknown variable and the need to design a new experiment.

Largely because of a lack of space, textbook accounts of research findings almost never mention the controls employed. This is not the case with the publication of original work in scientific journals, however. All such papers indicate the controls used. Results of experimental work presented at scientific meetings also include an account of the controls so that the audience of scientists can judge the value of the work. In reading textbook accounts of research findings, it is helpful, for someone wishing to develop a scientific habit of mind, to imagine what types of controls were used if they are not specifically mentioned.

SOURCES OF BIAS IN DATA GATHERING. Several potential sources of error in scientific investigation lie in the data-gathering process itself. Whatever is measured always represents a sample of a much larger population. When an experiment is performed on white rats or bean seedlings,

**Table 1-2 An Experiment to Study the Effects of a Chemical on Rats**

| *Group* | *Type* | *Number of rats* | *Injection* |
|---|---|---|---|
| 1 | Experimental | 24 | 2.0 ml 1-percent solution of the weed killer 2,4-D in water |
| 2 | Control | 6 | 2.0 ml 1-percent glucose solution |
| 3 | Control | 6 | 2.0 ml distilled water |
| 4 | Control | 6 | Pierced with a hypodermic needle but nothing injected |

NOTE: The heart function of all rats is monitored for 24 hours before and for 1 week after treatment.

PANEL 1-3

## Some Philosophical Aspects of Biology: Vitalism, Mechanism, and Teleology

### VITALISM VERSUS MECHANISM

Biology can be viewed as a study of the physicochemical properties of organisms. Ultimately, biological questions are answered in chemical and physical terms as questions dealing with matter and energy. But largely because we humans are living organisms whose lives seem to be so much more than chemistry, another view of biological processes, **vitalism,** has often been taken. In its most extreme form, vitalism states that all growth, development, and functioning of an organism is due to the controlling influence of the organism's "vital spirit." Organisms are said to exist, function, and possess their important qualities and capabilities by virtue of their vital spirits.

But spirit, by definition, lacks physical qualities, and so the question of vitalism has always been outside the purview of scientific inquiry. A spirit cannot be tested or measured and is not predictable under controlled conditions. Whether spirits even exist is not a question for scientific inquiry because no scientific means can verify their presence. Moreover, invoking the concept of vital spirit to answer scientific questions would discourage the advancement of scientific understanding.

Biology subscribes instead to a mechanistic view, which contends that scientific studies of the properties of living systems are to be made as studies of matter and energy, just as in chemistry and physics. **Mechanism** does not deny that there are other ways of studying plants, animals, and humans, nor does it contend that chemical and physical studies are the only ways of understanding them. For that matter, an individual scientist may well have vitalistic feelings about what ultimately controls such things as metabolism, the course of development, or evolution. But as a scientist, an investigator can only study the properties of life in terms of matter and energy.

### TELEOLOGY

Closely related to vitalism is **teleology,** a philosophical school of thought that sees purpose or design in nature. One form of this doctrine states

for example, it is conducted on only some white rats or some bean seedlings, not on all organisms of that kind. But from the results, inferences may often be drawn about all such rats or all such seedlings. Thus, it is important that the organisms used and the data collected from them be as representative as possible of such organisms. For example, if the animal's sex could be a factor in an experiment on rats, equal numbers of male and female rats should be used both in the control and in the experimental groups. The rats should also be representative in age, weight, and other characteristics. Some of the potential sources of bias in data gathering are fairly obvious and are easily avoided. Detecting and guarding against others sometimes taxes the imagination and ingenuity of the investigator.

***Statistical Analysis of Experimental Results.*** There are two kinds of statistics: (1) **descriptive statistics,** involving the quantitative aspects of various situations, such as daily temperatures, unemployment levels, and other factors, and (2) **inferential statistics,** involving the analysis and interpretation of numerical data. Inferential statistics have particular importance for the biologist. Many, but by no means all, biological studies require statistical analysis to determine the significance of the data obtained, that is, to decide exactly what can be inferred from them. Without this tool, progress in modern biological research would be severely hampered. A lack of statistical analysis can lead to one of two that God not only created the universe but has a specific purpose for everything in it. The purpose of the sun, for example, is to provide energy and light; the purposes of insects are to pollinate flowers, provide food for birds, and so on. Moreover, according to this view, each part of an organism has been developed for a purpose: horses developed ears to hear sounds, tails to swish flies, and so on. Teleology implies a desired result, or a purpose, for each organism and each of its parts—it implies a plan, an intended goal, toward which nature is working.

A biologist, however, regards the development of organisms in the context of evolutionary theory. Evolutionary change results from a genetic, or hereditary, change that is adaptive for a particular kind of organism at a given time. In time, such changes accumulate. For example, since hooves on a horse must have increased its ability to escape or otherwise defend itself against predators, early horses with hooflike feet must have survived and reproduced more successfully than horses without hooves. A horse is the result of many small evolutionary changes that occurred over a very long time. On the other hand, millions upon millions of organisms that were not well suited to their environments and whose parts did not work harmoniously became extinct only because they were not able to survive the competition they encountered from other organisms for food, mates, shelter, and other resources.

Some have contended that to explain the origin and evolution of life in mechanistic terms and thereby to reject vitalistic and teleological reasoning is to deny the existence of God. But in fact mechanism says nothing on that subject at all. What it does deny is that a god, a spirit, or someone's purpose must be invoked at every turn to explain the evolution or the functioning of each organ, enzyme system, or behavioral mechanism.

Confusion sometimes arises in the minds of nonscientists regarding teleology in science because of some biologists' habit of speaking of a structure, substance, or behavior as having a purpose. Some scientists call this "reasoning teleologically." However, to say that something has a purpose is not teleology in any real sense. Such statements simply mean that a particular trait, like most others, would probably have been retained in the course of evolution only if it had served a useful function. Indeed, the safest assumption biologists can make when faced with a new substance, structure, or behavior, is that, at least at one time, it probably served a useful function. Because competition has been so keen over the eons of time, organisms probably possess very few characteristics that did not in some way promote the survival of their ancestors. Although some inherited structures and traits probably arise as by-products of other structures or traits that are selected for, in time they too may acquire adaptive value in some circumstances. Nonadaptive traits are probably the exception, not the rule.

kinds of errors: (1) concluding more than is warranted by the results or (2) concluding less than can reasonably be inferred from them.

Statistical Analysis of Chance Events. A common use of statistical analysis is in the assessment of probabilities in situations involving chance or random happenings. For example, in flipping a coin equally weighted on both sides, the chance of its coming down heads is one-half. But only when a coin is flipped many times does it usually come down heads about half the time. Flipping it only two or three times may even result in heads every time. In any case, the 50:50 ratio is unusual with a small number of trials. Statistical analysis not only provides a way to treat such random events mathematically but it allows generalizations to be made on samples of minimal size. In the case of a coin toss, by taking into account the number of flips, the probability of the coin's not landing heads up half the time can be calculated mathematically. With two flips, for example, the chances would be estimated by multiplying its probability of occurring: $\frac{1}{2} \times \frac{1}{2} = \frac{1}{4}$. We would thus expect to get two heads in a row only once in every four series of two flips. That is, the probability of getting heads twice in a row is $\frac{1}{4}$, or 25 percent. What about three flips? Since $\frac{1}{2} \times \frac{1}{2} \times \frac{1}{2} = \frac{1}{8}$, we would expect, on the average, three heads in a row only once in every eight series of three flips. The chance of deviating from the 50:50 ratio thus decreases as the number of flips per series increases. The number of flips per series—or the number of organisms being used in a biological experiment—is thus an important factor in determining the probability of obtaining various kinds of results.

Suppose a coin was flipped 100 times and came down heads 65 times and tails 35 times; most people would suspect that the coin was unevenly weighted. Using a simple statistical method, it can be shown that the probability of an unweighted coin's landing heads up 65 times out of 100 is only 19 out of 10,000, or 0.0019.

Suppose the coin tossed was part of an experiment in which coins had been treated in some way before being tossed. If 100 flips of the treated coins resulted in 65 heads and 35 tails and 100 flips of untreated, control coins resulted in, say 54 heads and 46 tails, we would be able to conclude that the probability was 9,981 out of 10,000 that the 65:35 results were due to the treatment. In practice, such cumbersome figures are avoided. In this case, the probability would be stated as 0.99. A probability of 1.0 is the highest.

Biologists' Special Reason for Using Statistical Analysis. Unlike coins, organisms exhibit many natural variations from one to the next and, as Darwin noted, even in members of the same species. A good example of this is shown in Figure 1-12, a graph of body temperatures of a group of laboratory mice of the same strain maintained at a room temperature of 19 to 20°C. The purpose of the graph is to show how body temperature taken rectally varies with the level of activity of the animals, as measured by the rate at which they consume oxygen. Oxygen consumption was measured as milliliters of oxygen per gram of body weight per hour. A correlation between activity and temperature is evident: it can be seen from the graph that generally the rate of oxygen usage and body temperature are directly proportional to each other. But note the great individual variations: one mouse had a lower rate of oxygen consumption at 37°C than some others had at 33°C. Only statistical analysis of many individual determinations can give an investigator confidence that the conclusions drawn in such experiments are valid.

Given the variety among and within groups of organisms, it would

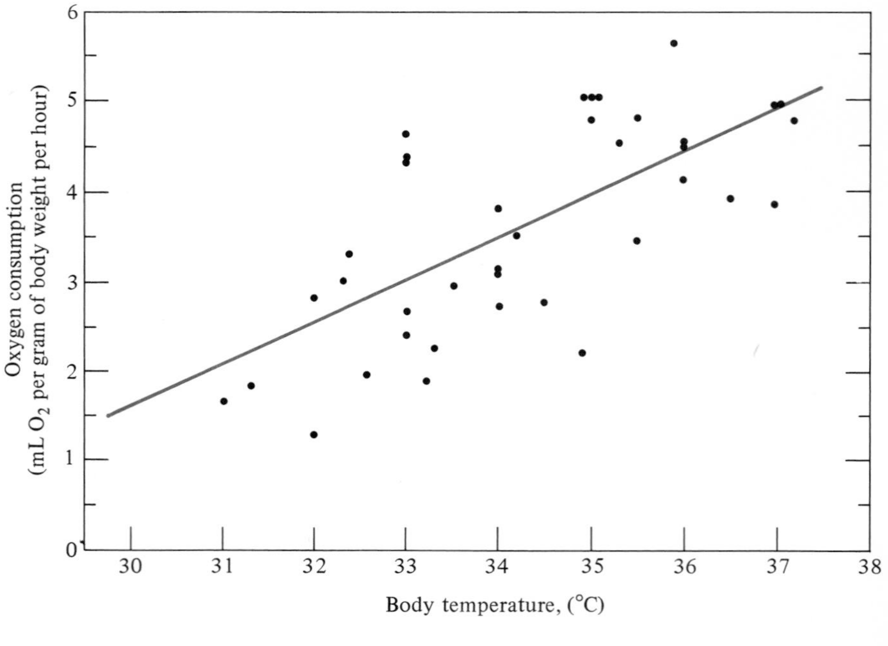


**FIG. 1-12** Variability within a group of organisms of the same age and species is illustrated here by the relationship between activity, as measured by oxygen consumption, and body temperature in white laboratory mice kept at 19 to 20°C.

be foolhardy to draw conclusions based on one experimental organism and one control. Suppose that the one individual given a drug died. That individual might have been about to die anyway. To be reasonably sure of the validity of results, all procedures should be performed several times or on several organisms, exactly how many to be determined by various considerations.

The larger the number of determinations made, the more reliable the conclusions that can be drawn. However, the investigator should set an upper limit on the number of determinations that is consonant with economy and common sense. No benefit is achieved by using hundreds of animals when two or three dozen prove the point. The principle of adequate numbers applies to controls as well as to experimentals, but a smaller number of controls may serve the purpose, especially if none of the controls exhibits any of the effects seen in the experimentals.

DO STATISTICS EVER REALLY PROVE ANYTHING? A proper statistical analysis can *reject* a hypothesis if it is found that the data from the two groups—experimentals and controls—are not significantly different. Therefore, use of a statistical method can prove a hypothesis wrong. However, no method can ever prove that a hypothesis is true or correct; by not rejecting a hypothesis, a statistical method merely shows that it *could* be the explanation of or is compatible with the data obtained.

It is easy to be overly impressed by statistical analysis. However, even the best statistical method cannot validate an experiment with poor design, inadequate controls, too small a sample size, or a biased gathering of data anymore than using the correct silverware and china can transform garbage into a good dinner. Fortunately, nearly all cases of faulty design that escape a scientist's notice are detected in the review process conducted by all high-quality scientific journals when an article is sent to them for publication.

***The Nature of Scientific Progress.*** The first hypothesis a scientist postulates to explain something seldom turns out to be correct. In fact, it is common in scientific laboratories to test a series of explanations by experiments only to find that none is correct, at least in its original form. After weeks or months of work in which the experimental results neither prove nor disprove a particular hypothesis, the hypothesis may be recast in a new form in light of the results obtained. There are times when an entire experimental program must be abandoned because the only experiments that can be brought to bear on the question have yielded inconclusive results, which neither support nor disprove the hypothesis. In the laboratory of the experienced and resourceful scientist, however, the experiment that fails may sometimes lead to a more important discovery than the one sought originally. Sometimes a new explanation evolves that is quite different from the first but that is compatible with all results. Further experiments aimed at verifying the new hypothesis may be conducted. Experimental results thus generate hypotheses as frequently as hypotheses lead to experimentation.

Published scientific reports usually do not mention the false starts that occurred before the hypothesis reached its final form. This is not a matter of dishonesty; rather, since such information is of little importance, costly space is not allotted in journals to descriptions of dead ends and incorrect hypotheses. Thus, usually only hypotheses supported by experimental results find their way into print.

Scientific progress seldom follows a direct route. Like mountain climbing, it typically involves false starts and retracing of steps as failures and unforeseen obstacles appear along the way. The simple path of observation → hypothesis → experiment → results → conclusions → established theory that is suggested by formal publications or textbook accounts of discoveries is almost never the way things actually happen. Almost all progress is achieved by tortuous routes and a slow accumulation of small pieces of large puzzles. The English biologist Joseph Priestley said in 1772: "Very lame and imperfect theories are sufficient to suggest useful experiments which serve to correct those theories and give birth to others more perfect. These then occasion farther experiments, which bring us still nearer to the truth, and in this method of approximation, we must be content to proceed, and we ought to think ourselves happy, if, in this slow method, we make any real progress."

## THE PRACTICAL VALUE OF BASIC RESEARCH

Much of the biological research in colleges and universities is basic: it is directed toward knowing and understanding life in all its manifestations rather than toward solving particular problems of human disease, hunger, weaponry, and the like. Indeed, much basic research may have no obvious medical, economic, or military value. Even from a strictly utilitarian point of view, however, nearly all important advances in such areas as medicine and agriculture have been made by pursuing both types of research—basic as well as applied. One reason for this is that we never know what to look for (Fig. 1-13). But perhaps even more important, an understanding of basic principles must almost always precede any useful application of them. For example, most of our important knowledge about disease organisms is based, not on studies of organisms that cause disease in humans, but on biological studies of organisms that are not harmful to humans. Similarly, our most basic understanding of human inheritance is the result of studying the genetics of garden peas, bacteria, molds, and the tiny red-eyed flies, *Drosophila,* that gather about ripe fruit.

**FIG. 1-13** Sir Alexander Flemming, discoverer of penicillin. His discovery of the first antibiotic to be widely used in medicine was a byproduct of his research, not its goal.

The ultimate control of disease can probably come only through a basic understanding of life. Much can be said, of course, for direct approaches to the discovery of ways to treat disease—for programs, for example, in which new drugs are tested for their effect in alleviating symptoms or curing diseases, without necessarily leading to basic understanding of the diseases or of their cure. Indeed, this direct approach has provided modern medicine with many drugs that, despite the fact that no one understands them, are very useful in treating diseases. But such blind, direct approaches are not nearly enough. We must understand diseases if we are to prevent, cure, or control most of them. The best practical approach to understanding disease is through first understanding normal organisms, an understanding that cannot be gained by studying humans alone.

It is fortunate for biologists that the life processes of different species are so similar; this means that mechanisms discovered in one species are often found in most others. Use of lower forms of life as subjects of research is probably the best practical route to an understanding of the biology of humans and other higher organisms. Because life has such variety, investigators can select the species of organism whose qualities make it the best choice of research material for particular experiments. Besides these considerations, all experiments must be based on a solid foundation of descriptive biology, such as anatomy and classification. Even in doing experiments on a particular plant or animal, it is essential to determine that all specimens used belong to the same identified species.

Basic biological research is important in another way: many of the statements appearing in textbooks of medicine are based on a knowledge of lower organisms. A section on embryonic development, for example, might appear in the book as much because one researcher collected, described, and classified African frogs as because another was curious about how eggs—whether of insects, birds, frogs, or humans—developed.

### BEYOND PRACTICALITY

A strong case can certainly be made for the support and pursuit of basic research because of its ultimate practical value. But advances in scientific knowledge have a worth beyond their economic, agricultural, medical, or military use. An understanding of our universe is good for the human spirit; not only do new discoveries enrich the lives of the investigators who make the findings, but they broaden the horizons and deepen the appreciation of the world about them of those who read accounts of these discoveries. In this sense, the readers become discoverers too and are enriched by their new-found knowledge, as was the poet John Keats upon reading George Chapman's translation of Homer: "Then felt I like some watcher of the skies,/When a new planet swims into his ken." Government, industry, and private foundations around the world have long supported scholarship in literature and the fine arts. Basic scientific research and the education of students for careers in scientific research deserve public support for the same reasons, as eloquently stated by the late American microbiologist René Dubos: "The acquisition, organization, and assimilation of scientific knowledge constitutes a major intellectual and artistic enterprise, together with philosophy, music, writing, painting, or acting. There is an aspect of science that transcends pedestrian utility yet has immense social relevance because it enhances humanness."

## Summary

In all the physical world, organisms are unique. They are highly organized units capable of acquiring and processing one or more forms of energy. All available evidence indicates that life began in the sea billions of years ago, the evolution of organisms having been preceded by a long chemical evolution in which available energy was used to drive chemical reactions that eventually led to the origin of life. Today's organisms use energy to maintain themselves in highly organized states. Their processing of energy also gives rise to other unique traits, including growth, development, and responsiveness to stimuli. Every species is also capable of reproducing, with heritable variations being manifested in the offspring of each generation. This pool of variability is the raw material upon which natural selection acts. Ultimately, it is an organism's environment that determines which of the many forms present will be the ones to survive; only the fit are "selected." As environments change or as organisms migrate to new situations, different adaptations are selected. As a result, new species evolve. In time, new genera, new orders, classes, and divisions or phyla arise through this selection process. Natural selection is believed to be the principal, although not the only, shaper of evolution.

Despite a lack of other evidence for evolution, the classifying of organisms led more than one biologist, including Carolus Linnaeus, to believe that at least some groups of organisms are related to each other through a common ancestor. However, the scientific world did not accept evolution as an explanation for the patterns of biological diversity until it became evident that the earth was many millions of years old and until the unifying theory of natural selection was developed. Taxonomy, like the rest of biology, was revolutionized by the emergence of the concept of evolution. The fact that taxonomic systems are but progress reports on evolution explains why there are so many differences of opinion in matters of classification. Robert Whittaker's five-kingdom tree is the scheme followed in this book.

Science is not a unique type of scholarly activity. Its logical processes and its use of hypothesis and evidence are employed in most fields of intellectual inquiry. Science relies heavily on experimentation, but other fields use experimentation too. Science differs from other studies only in regard to its subject matter.

Most scientific investigations proceed by conducting experiments or otherwise gathering evidence that will test the validity of tentative explanations called hypotheses, proposed to explain some phenomenon. To rule out the possibility that uncontrolled variables contribute to the results, controls are an essential component of most experiments, especially in biology. Statistical analysis of the results is often necessary to determine whether the results are due to the experimental treatment or could have been expected on the basis of chance alone.

Because of constraints of time and space, scientific articles seldom describe all the experiments that preceded the published accounts. Many projects undergo several changes in direction before the hypothesis being tested and the experiments used to test it achieve the form described in print. The progress of research is thus slow and painstaking.

Although there are other ways of looking at organisms, scientific study is, by definition, limited to looking at life only in terms of matter and energy. This mechanistic view is opposed to the philosophical doctrine of vitalism, according to which the life and actions of organisms can be explained only by influence of their "vital spirits." Teleology, the assignment of purposefulness to organisms and their organs or behavior, is also invalid as a scientific approach. Purpose implies the intention of an intelligent being. The "teleological reasoning" occasionally used by some scientists—for example, the identification of the purpose of a bird's wing as for flying—is with rare exception not true teleology. It is merely a shorthand way of saying that the structure probably had been useful, or adaptive, for an ancestor and was preserved by natural selection.

For many reasons, basic research—research not aimed at solving some practical problem in medicine, engineering, and so on—has been very important in the progress made in applied fields. Even applied research is conducted against a background of knowledge and understanding supplied by basic research. Beyond the importance of scientific research for its many practical uses from medicine to making war, there is another reason for pursuing and supporting it: it is a major intellectual and artistic enterprise as worthy as any human endeavor.

## Recommended Readings

AYALA, F. J. and T. DOBZHANSKY (eds). 1975. *Studies in the Philosophy of Biology.* University of California Press, Berkeley. An excellent collection of essays that describe modern biological research methods.

BRONOWSKI, J. 1964. *Insight.* Harper & Row, New York. A classic introduction to science, particularly to biology, by a master scientist, poet, and teacher.

BRONOWSKI, J. 1973. *The Ascent of Man.* Little, Brown, Boston. A beautifully illustrated history of the sciences and their interaction with human culture. Based on the popular public television series of the same title.

DARWIN, C. 1872. *The Origin of Species by Means of Natural Selection or the Preservation of Favored Races in the Struggle for Life,* 6th ed. Random House, New York. The major evidence for evolution known in 1872 as explained by the best-known proponent of the concept.

DUBOS, R. 1970. *Reason Awake: Science for Man.* Columbia University Press, New York. A wide-ranging, insightful essay on the importance of science for our life, past, present, and future, by an eminent bacteriologist and Pulitzer Prize–winning author.

HARRINGTON, J. W. 1981. *Discovering Science* (paperback). Boston: Houghton Mifflin. Explains the common approach to gaining knowledge used in all sciences and emphasizes the need to learn by understanding rather than by rote memory.

LEDERMAN, L. M. 1984. "The Value of Fundamental Science." *Scientific American,* 251(5):40–47. Education and research in basic science deserve support for their own sakes and for their great practical value.

MAYR, E. 1978. "Evolution." *Scientific American,* 239(3): 46–20. This well-illustrated, interesting piece is the lead article in an entire issue devoted to evolution.

MAYR, E. and W. B. PROVINE, (eds). 1980. *The Evolutionary Synthesis: Perspectives in the Unification of Biology.* Harvard University Press, Cambridge, Mass. A fascinating and informative history of the emergence of the theory of evolution, although not an introduction to the subject itself.

RACLE, F. 1979. *Introduction to Evolution,* Prentice-Hall, Englewood Cliffs, N.J. An introduction to the evidence and arguments for and the mechanisms of evolution in animals.

REMINGTON, R. D., and M. A. SCHORK. 1985. *Statistics with Applications to the Biological and Health Sciences,* 2d ed. Prentice-Hall, Englewood Cliffs, N.J. A brief but thorough introduction to biostatistics.

STEBBINS, G. L. 1977. *Processes of Organic Evolution* (paperback), 3d ed. Prentice-Hall, Englewood Cliffs, N.J. An excellent introduction to the evolution of plants and animals by a lifelong student and teacher of the subject.

VAN NORMAN, R. W. 1971. *Experimental Biology,* 2d ed. Prentice-Hall, Englewood Cliffs, N.J. Explains in interesting style how biological literature is used, how experiments are designed and conducted, and how scientific papers are published.

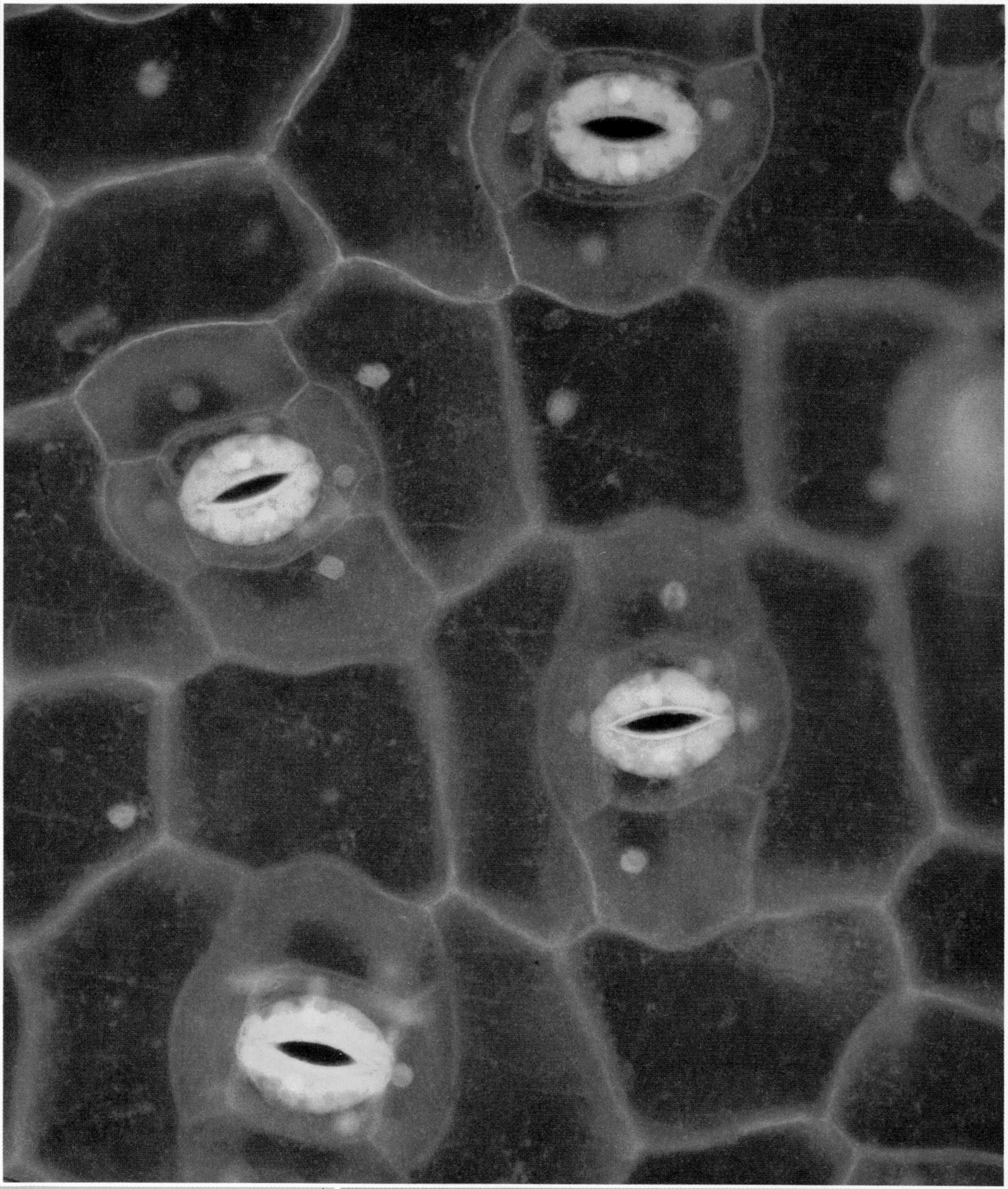

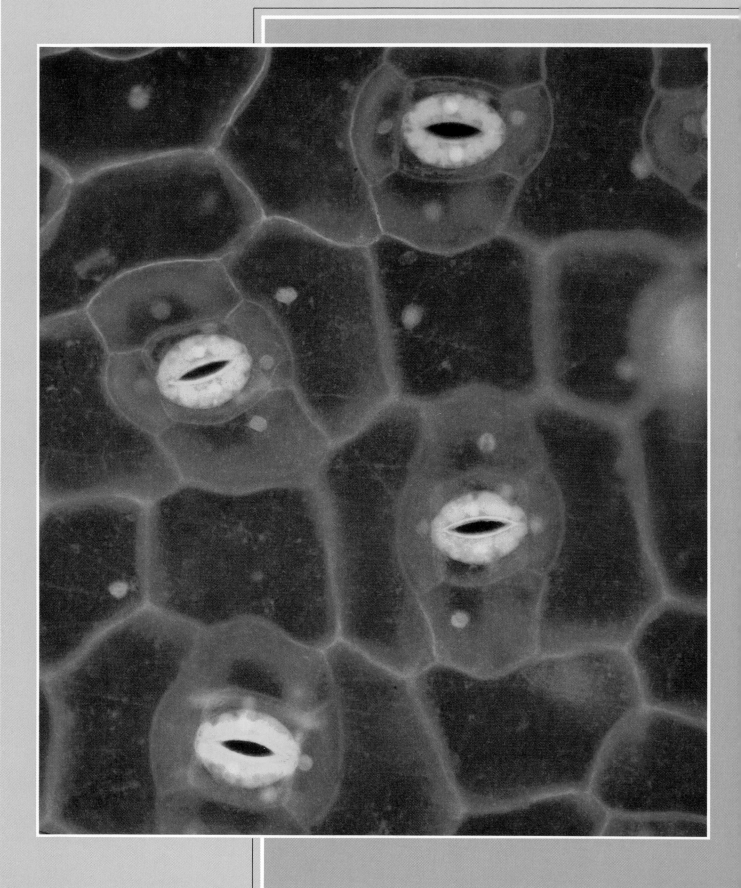

# UNIT II

# The Biology of Cells



One of the great scientific generalizations is that of the cell theory, or cell doctrine, which states that all organisms are composed of microscopic units called cells. Modern biologists view the cell as the unit of life's structure, function, origin, and development. Cells can be regarded as the simplest units of life and, indeed, can survive and multiply for an indefinite period of time apart from the organisms from which they were taken. The same cannot be said of any cell part.

Exceptions to the broadly stated cell theory—according to which all organisms are composed of cells and all cells have a nucleus—have long been recognized. For example, bacteria, while possessing DNA, lack organized nuclei. Should they thus be considered noncellular, or are they composed of one cell? Both alternatives contradict the cell theory. Moreover, some tissues are essentially cells with more than one nucleus. Despite these exceptions, the cell theory remains one of the fundamental unifying biological principles.

The six chapters in this unit explore various principles of cellular biology. The kinds of molecules that make up cells, the kinds of reactions in which these molecules engage, and the mechanisms by which materials enter and leave cells are described in three of the chapters. One chapter is devoted to descriptions of the structure and function of the microscopic parts of cells and the processes of cell and nuclear division by which cells reproduce. The major pathways by which chemically bound energy is transferred and released in organisms are described in another chapter. Finally, the means by which photosynthetic cells capture the energy of sunlight and convert it to chemically bound forms is examined. All of these processes form the very basis of life and therefore, literally, are vital functions.

2

# Basic Chemistry

OUTLINE



Matter and energy are fundamental to all living systems. Before the explosion of the first atomic bomb, it was generally believed that matter and energy were separate and distinct entities. However, the detonation of the atomic bomb at Alamogordo in the New Mexico desert in 1945 demonstrated Albert Einstein's contention that under certain conditions a small amount of matter can be converted into a large amount of energy (Fig. 2-1). In living systems, matter is not converted into energy but is used to transfer energy from energy-releasing to energy-using processes. The integration of this release and use of energy by living matter is what we call *life*. Death occurs when this flow of energy ceases.

## THE NATURE OF MATTER

Chemistry deals with the composition, structure, and properties of matter (anything that occupies space) and the changes it undergoes. All forms of matter may be classified as either mixtures or pure substances. Examples of each are given in Table 2-1.

### MIXTURES

A mixture is composed of two or more substances that can be separated from one another by physical means, that is, without undergoing chem-

**FIG. 2-1** Mushroom cloud from a nuclear explosion over Eniwetok Atoll in the Pacific, in July, 1956.

ical change. Mixtures are either homogeneous, with uniform composition throughout, or heterogeneous, with variable composition. Some homogeneous mixtures—saltwater, for example—may appear to be one substance because their components are not individually recognizable. Nevertheless, the components of saltwater can be separated by purely physical means; the salt, which has been dissolved in the water, can be recovered simply by evaporating the water. Homogeneous liquid mixtures such as saltwater are called *solutions*. Solder, an alloy of tin and lead, is an example of a homogeneous solid mixture, or solid solution.

## PURE SUBSTANCES

A pure substance is a homogeneous form of matter that can be broken down into two or more substances only by a chemical reaction, if at all. Pure substances are of two types. If a substance cannot be separated into two or more substances by chemical means, the substance is an element. The substance is a compound if it can be broken down into two or more chemical elements.

## ELEMENTS

In the past 150 years of research, 106 elements have been discovered. Many of the elements exist on earth only in small amounts. As indicated in Table 2-2, more than 99 percent of the earth's crust, atmosphere, and oceans is made up of just 10 elements, with less than 1 percent being made up of the other 80 naturally occurring elements.* Moon rocks are composed of the same elements as those found on earth, and as far as we know, so is all matter in the universe (Fig. 2-2).

Over 90 percent of the human body consists of just three elements—carbon, hydrogen, and oxygen. Other major components are nitrogen, phosphorus, and sulfur, which are found in lesser amounts (Table 2-3). Several other elements, including calcium, sodium, iron, potassium, and magnesium, are also necessary for life but are present in still smaller amounts. Finally, copper, manganese, and chromium are among the trace elements, found in minimal but essential quantities within cells.

* The remaining 16 elements can be produced only as a result of nuclear reactions and do not occur naturally on earth.

**TABLE 2-1 Classification of Matter**

| *Examples of pure substances* | | *Examples of mixtures* | |
|---|---|---|---|
| *Elements* | *Compounds* | *Homogeneous* | *Heterogeneous* |
| Gold | Salt | Salt water | Paper |
| Chlorine | Sugar | Air | Rocks |
| Lead | Alcohol | Brass | Smog |

**TABLE 2-2 Abundance of Elements in the Earth's Crust, Oceans, and Atmosphere**

| *Element* | *Chemical symbol* | *Percentage (%)* | | *Element* | *Chemical symbol* | *Percentage (%)* | |
|---|---|---|---|---|---|---|---|
| Oxygen* | O | 49.5 | 99.2 | Chlorine | Cl | 0.29 | 0.8 |
| Silicon | Si | 25.7 | | Phosphorus | P | 0.12 | |
| Aluminum | Al | 7.5 | | Manganese | Mn | 0.09 | |
| Iron | Fe | 4.7 | | Carbon* | C | 0.08 | |
| Calcium | Ca | 3.4 | | Sulfur | S | 0.06 | |
| Sodium | Na | 2.6 | | Barium | Ba | 0.04 | |
| Potassium | K | 2.4 | | Chromium | Cr | 0.03 | |
| Magnesium | Mg | 1.9 | | Nitrogen* | N | 0.03 | |
| Hydrogen* | H | 0.9 | | Fluorine | F | 0.03 | |
| Titanium | Ti | 0.6 | | Zirconium | Zr | 0.02 | |
| | | | | All others | | <0.1 | |

* The four most abundant elements in organisms.

## ATOMIC THEORY

The smallest unit of an element that can combine with another element is the atom. Atoms contain three primary components: protons and neutrons, which are located in the nucleus, and electrons, which move rapidly about in the rest of the space occupied by the atom. As indicated in Table 2-4, protons have a positive electric charge, neutrons are elec-

**FIG. 2-2** Rock on the surface of the moon as photographed by Apollo astronauts. Moon rocks are composed of elements familiar on earth, such as iron and calcium, and appear to be volcanic in origin.

**TABLE 2-3 Most Abundant Elements in the Human Body**

| *Major Elements* | *Symbol* | *Approximate percentage of the human body* |
|---|---|---|
| Oxygen | O | 65.0 |
| Carbon | C | 18.5 |
| Hydrogen | H | 9.5 |
| Nitrogen | N | 3.2 |
| Calcium | Ca | 1.5 |
| Phosphorus | P | 1.0 |
| Potassium | K | 0.4 |
| Sulfur | S | 0.3 |
| Chlorine | Cl | 0.2 |
| Sodium | Na | 0.2 |
| Magnesium | Mg | 0.1 |
| *Trace Elements* | | |
| Cobalt | Co | Less than 0.1 |
| Copper | Cu | |
| Fluorine | F | |
| Iodine | I | |
| Iron | Fe | |
| Manganese | Mn | |
| Zinc | Zn | |

trically neutral, and electrons have a negative charge. The electric charges of a proton and an electron are equal in magnitude. Because oppositely charged particles attract and neutralize each other, atoms normally have no net electric charge and have equal numbers of electrons and protons. Protons and neutrons have approximately equal mass, while electrons have only about 1/1,840 as much. Thus, the nucleus contains most of the atomic mass, and the electrons occupy most of the volume. A small matchbox filled with material as dense as an atomic nucleus would weigh about 2.5 billion tons!

***Atomic Number.*** An element can be defined as a substance with a given atomic number, equal to the number of protons in the nuclei of its atoms. For example, hydrogen, with only one proton, has an atomic

**TABLE 2-4 Components of the Atom**

| *Particle* | *Relative mass* | *Electric charge* |
|---|---|---|
| Electron | 1/1,840 | −1 (negative) |
| Proton | 1 | +1 (positive) |
| Neutron | 1 | 0 (neutral) |

number of 1. Carbon has six protons and an atomic number of 6. Neptunium, the largest of the natural elements, has 93 protons and an atomic number of 93. In the periodic table (Fig. 2-3, p. 40), the elements are arranged according to atomic number.

***Isotopes.*** Almost all elements occur in more than one form, each with the same number of protons but a different number of neutrons in its atomic nucleus. Hydrogen, for example, exists in three forms, with nuclei containing 0, 1, and 2 neutrons. Forms of an element distinguished by different numbers of neutrons—and consequently different masses—are called **isotopes** (from the Greek *isos,* "equal" and *topos,* "place"). With hydrogen, the addition of one neutron to the nucleus, which consists of one proton, doubles the atomic mass to produce the isotope deuterium. The addition of two neutrons triples the atomic mass to produce the isotope tritium. Each hydrogen isotope has an atomic number of 1, but the **mass number**—the total number of protons and neutrons—can be 1, 2, or 3.

The composition of an atomic nucleus can be indicated by writing the atomic number as a subscript and the mass number as a superscript at the left of the symbol of the element. Because all isotopes of a given element have the same atomic number, the subscript is redundant and is usually omitted. Thus, $^{235}U$ indicates a uranium atom with 92 protons and 143 neutrons, and $^{238}U$ represents a uranium atom with 92 protons and 146 neutrons. Deuterium, with one proton and one neutron, is written $^{2}H$; tritium, with one proton and two neutrons, is $^{3}H$. Normally, even the superscripts are omitted for all but an element's less common isotopes. Isotopes can also be represented with the element's name and mass number, as in uranium-235.

## RADIOACTIVITY

Some atomic nuclei, for example, those with equal numbers of protons and neutrons, are always stable and exhibit no tendency to disintegrate. However, in some cases—usually in nuclei with more neutrons than protons and in all elements with 84 or more protons—the atomic nucleus is unstable. The nucleus tends to disintegrate spontaneously, resulting in radiation of energy. Isotopes that spontaneously undergo this process of decay are called **radioactive.** For example, the nucleus of the tritium atom ($^{3}H$), with two neutrons and one proton, is radioactive; it spontaneously decays to a more stable form and in the process emits beta particles, which are high-speed electrons. Beta emissions result when one neutron disintegrates into a proton and an electron. By conversion of a neutron to a proton and an electron, the tritium atom is transformed into an element with an atomic number of 2 and a mass of 3. It thus has become an isotope of helium: $^{3}He$. (The common form of helium, $^{4}He$, has a nucleus with two protons and two neutrons.) This decay can be expressed by the following equation:

$$\underset{\text{tritium}}{^{3}H} \longrightarrow \underset{\text{helium}}{^{3}He^{+1}} + e^{-} \text{ (emitted electron, beta particle)}$$

The equation is balanced: the one proton, two neutrons, and one electron of $^{3}H$ become the two protons and one neutron of the helium isotope and the emitted electron (Fig. 2-4, p. 41).

**FIG. 2-3** The periodic table of elements with a list of the elements with their symbols and atomic numbers.

Key (Li cell): 3 ............. Atomic number; Li ............. Chemical symbol; 6.939 ............. Atomic weight (mass)

Active metals: 1A, 2A. Transition metals: 3B–2B. Nonmetals: 3A–8A. Inert gases: 7A, 8A.

| 1A | 2A | 3B | 4B | 5B | 6B | 7B | 8B | 8B | 8B | 1B | 2B | 3A | 4A | 5A | 6A | 7A | 8A |
|---|---|---|---|---|---|---|---|---|---|---|---|---|---|---|---|---|---|
| | | | | | | | | | | | | | | | | **1**<br>H<br>1.00797 | **2**<br>He<br>4.0026 |
| **3**<br>Li<br>6.939 | **4**<br>Be<br>9.0122 | | | | | | | | | | | **5**<br>B<br>10.811 | **6**<br>C<br>12.01115 | **7**<br>N<br>14.0067 | **8**<br>O<br>15.9994 | **9**<br>F<br>18.9984 | **10**<br>Ne<br>20.183 |
| **11**<br>Na<br>22.9898 | **12**<br>Mg<br>24.312 | | | | | | | | | | | **13**<br>Al<br>26.9815 | **14**<br>Si<br>28.086 | **15**<br>P<br>30.9738 | **16**<br>S<br>32.064 | **17**<br>Cl<br>35.453 | **18**<br>Ar<br>39.948 |
| **19**<br>K<br>39.098 | **20**<br>Ca<br>40.08 | **21**<br>Sc<br>44.956 | **22**<br>Ti<br>47.90 | **23**<br>V<br>50.942 | **24**<br>Cr<br>51.996 | **25**<br>Mn<br>54.9380 | **26**<br>Fe<br>55.847 | **27**<br>Co<br>58.9332 | **28**<br>Ni<br>58.71 | **29**<br>Cu<br>63.54 | **30**<br>Zn<br>65.37 | **31**<br>Ga<br>69.72 | **32**<br>Ge<br>72.59 | **33**<br>As<br>74.9216 | **34**<br>Se<br>78.96 | **35**<br>Br<br>79.909 | **36**<br>Kr<br>83.80 |
| **37**<br>Rb<br>85.47 | **38**<br>Sr<br>87.62 | **39**<br>Y<br>88.905 | **40**<br>Zr<br>91.22 | **41**<br>Nb<br>92.906 | **42**<br>Mo<br>95.94 | **43**<br>Tc<br>(99) | **44**<br>Ru<br>101.07 | **45**<br>Rh<br>102.905 | **46**<br>Pd<br>106.4 | **47**<br>Ag<br>107.870 | **48**<br>Cd<br>112.41 | **49**<br>In<br>114.82 | **50**<br>Sn<br>118.69 | **51**<br>Sb<br>121.75 | **52**<br>Te<br>127.60 | **53**<br>I<br>126.9044 | **54**<br>Xe<br>131.30 |
| **55**<br>Cs<br>132.905 | **56**<br>Ba<br>137.33 | **57**<br>La*<br>138.91 | **72**<br>Hf<br>178.49 | **73**<br>Ta<br>:80.948 | **74**<br>W<br>183.85 | **75**<br>Re<br>186.2 | **76**<br>Os<br>190.2 | **77**<br>Ir<br>192.2 | **78**<br>Pt<br>195.09 | **79**<br>Au<br>196.967 | **80**<br>Hg<br>200.59 | **81**<br>Tl<br>204.37 | **82**<br>Pb<br>207.19 | **83**<br>Bi<br>208.980 | **84**<br>Po<br>(210) | **85**<br>At<br>(210) | **86**<br>Rn<br>(222) |
| **87**<br>Fr<br>(223) | **88**<br>Ra<br>(226) | **89**<br>Ac†<br>(227) | **104**<br>Rf<br>(257) | **105**<br>Ha<br>(260) | **106**<br>Unh<br>(263) | | | | | | | | | | | | |

| | | | | | | | | | | | | | | |
|---|---|---|---|---|---|---|---|---|---|---|---|---|---|---|
| *Lanthanide series (Rare earth metals) | **58**<br>Ce<br>140.12 | **59**<br>Pr<br>140.907 | **60**<br>Nd<br>144.24 | **61**<br>Pm<br>(147) | **62**<br>Sm<br>150.35 | **63**<br>Eu<br>151.96 | **64**<br>Gd<br>157.25 | **65**<br>Tb<br>158.924 | **66**<br>Dy<br>162.50 | **67**<br>Ho<br>164.930 | **68**<br>Er<br>167.26 | **69**<br>Tm<br>168.934 | **70**<br>Yb<br>173.04 | **71**<br>Lu<br>174.97 |
| †Actinide series (Uranium metals) | **90**<br>Th<br>232.038 | **91**<br>Pa<br>(231) | **92**<br>U<br>238.03 | **93**<br>Np<br>(237) | **94**<br>Pu<br>(242) | **95**<br>Am<br>(243) | **96**<br>Cm<br>(247) | **97**<br>Bk<br>(247) | **98**<br>Cf<br>(249) | **99**<br>Es<br>(254) | **100**<br>Fm<br>(253) | **101**<br>Md<br>(256) | **102**<br>No<br>(253) | **103**<br>Lr<br>(257) |

NOTE: For elements all of whose isotopes are radioactive, parentheses enclose the atomic weight of the isotope with the longest half-life.

| *Element* | *Symbol* | *Atomic number* | *Element* | *Symbol* | *Atomic number* | *Element* | *Symbol* | *Atomic number* |
|---|---|---|---|---|---|---|---|---|
| Actinium | Ac | 89 | Hafnium | Hf | 72 | Praseodymium | Pr | 59 |
| Aluminum | Al | 13 | Hahnium | Ha | 105 | Promethium | Pm | 61 |
| Americium | Am | 95 | Helium | He | 2 | Protactinium | Pa | 91 |
| Antimony | Sb | 51 | Holmium | Ho | 67 | Radium | Ra | 88 |
| Argon | Ar | 18 | Hydrogen | H | 1 | Radon | Rn | 86 |
| Arsenic | As | 33 | Indium | In | 49 | Rhenium | Re | 75 |
| Astatine | At | 85 | Iodine | I | 53 | Rhodium | Rh | 45 |
| Barium | Ba | 56 | Iridium | Ir | 77 | Rubidium | Rb | 37 |
| Berkelium | Bk | 97 | Iron | Fe | 26 | Ruthenium | Ru | 44 |
| Beryllium | Be | 4 | Krypton | Kr | 36 | Rutherfordium | Rf | 104 |
| Bismuth | Bi | 83 | Lanthanum | La | 57 | Samarium | Sm | 62 |
| Boron | B | 5 | Lawrencium | Lr | 103 | Scandium | Sc | 21 |
| Bromine | Br | 35 | Lead | Pb | 82 | Selenium | Se | 34 |
| Cadmium | Cd | 48 | Lithium | Li | 3 | Silicon | Si | 14 |
| Calcium | Ca | 20 | Lutetium | Lu | 71 | Silver | Ag | 47 |
| Californium | Cf | 98 | Magnesium | Mg | 12 | Sodium | Na | 11 |
| Carbon | C | 6 | Manganese | Mn | 25 | Strontium | Sr | 38 |
| Cerium | Ce | 58 | Mendelevium | Md | 101 | Sulfur | S | 16 |
| Cesium | Cs | 55 | Mercury | Hg | 80 | Tantalum | Ta | 73 |
| Chlorine | Cl | 17 | Molybdenum | Mo | 42 | Technetium | Tc | 43 |
| Chromium | Cr | 24 | Neodymium | Nd | 60 | Tellurium | Te | 52 |
| Cobalt | Co | 27 | Neon | Ne | 10 | Terbium | Tb | 65 |
| Copper | Cu | 29 | Neptunium | Np | 93 | Thallium | Tl | 81 |
| Curium | Cm | 96 | Nickel | Ni | 28 | Thorium | Th | 90 |
| Dysprosium | Dy | 66 | Niobium | Nb | 41 | Thulium | Tm | 69 |
| Einsteinium | Es | 99 | Nitrogen | N | 7 | Tin | Sn | 50 |
| Erbium | Er | 68 | Nobelium | No | 102 | Titanium | Ti | 22 |
| Europium | Eu | 63 | Osmium | Os | 76 | Tungsten | W | 74 |
| Fermium | Fm | 100 | Oxygen | O | 8 | Unnilhexium | Unh | 106 |
| Fluorine | F | 9 | Palladium | Pd | 46 | Uranium | U | 92 |
| Francium | Fr | 87 | Phosphorus | P | 15 | Vanadium | V | 23 |
| Gadolinium | Gd | 64 | Platinum | Pt | 78 | Xenon | Xe | 54 |
| Gallium | Ga | 31 | Plutonium | Pu | 94 | Ytterbium | Yb | 70 |
| Germanium | Ge | 32 | Polonium | Po | 84 | Yttrium | Y | 39 |
| Gold | Au | 79 | Potassium | K | 19 | Zinc | Zn | 30 |
| | | | | | | Zirconium | Zr | 40 |

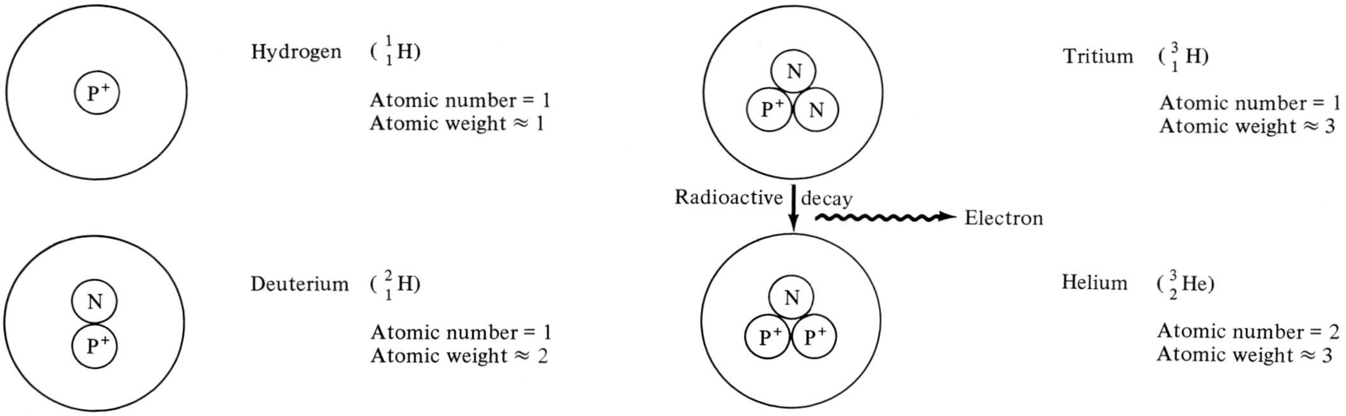


**FIG. 2-4** The isotopes of hydrogen, all of which have one proton and therefore have the same atomic number. Their atomic weights differ because of the number of neutrons present in the nucleus. Tritium, a radioactive isotope of hydrogen, undergoes radioactive decay to helium, with the release of an electron.

***The Biological Significance of Radioactive Emissions.*** All types of radioactive decay involve emission of one or more of the three major kinds of penetrating rays: alpha ($\alpha$), beta ($\beta$), and gamma ($\gamma$) rays. As indicated in Table 2-5, each type of radiation is characterized by a specific electric charge and a given penetrating ability. Alpha rays are streams of helium nuclei (two protons and two neutrons); each particle has a +2 charge. Alpha particles travel at about 1/20 the speed of light. Their penetrating power is slight; they can be stopped by a sheet of paper. Beta rays are streams of high-speed electrons traveling as fast as 99 percent of the speed of light and with 100 times the penetrating ability of alpha rays. Gamma rays are electrically neutral, high-energy electromagnetic radiations similar to X rays, but with somewhat shorter wavelengths. They travel at the speed of light ($3 \times 10^{10}$ cm/sec or 186,000 miles/sec) and have 1,000 times the penetrating ability of alpha rays. Gamma radiation nearly always accompanies other radioactive emissions. It represents much of the energy lost when the remaining components of a disintegrating atomic nucleus reorganize into more stable arrangements: atomic particles are lost, and the remaining atomic particles reorganize into different and stable atoms.

With the development of nuclear reactors and cyclotrons after World War II, many radioactive isotopes were produced artificially from some of the lighter, biologically important elements such as carbon, phosphorus, sulfur, and nitrogen. Some radioactive isotopes are useful in biological and medical research and in medicine. Because radioactive iso-

**TABLE 2-5 Properties of Alpha ($\alpha$), Beta ($\beta$), and Gamma ($\gamma$) Rays**

| | *Type of radiation* | | |
|---|---|---|---|
| *Property* | $\alpha$ | $\beta$ | $\gamma$ |
| Charge | +2 | −1 | 0 |
| Mass | $6.6 \times 10^{-24}$ g | $9.11 \times 10^{-28}$ g | 0 |
| Relative penetrating power | 1 | 100 | 1,000 |
| Identity | Helium nuclei ($^4_2He$) | Electrons | High-energy radiation |

topes emit radiation as they decay, even trace amounts of the isotopes can be detected. For example, beta rays and other forms of radiation can be detected by their ability to "expose" a sheet of photographic film much as ordinary light does. Radioactive emissions can also be detected and measured by a special device, the Geiger counter. Because of their detectability, trace amounts of radioactive atoms can be incorporated in molecules and then introduced into an organism, where their participation in biochemical reactions can be traced. For instance, the use of radioactively "labeled"* water ($H_2O$) and carbon dioxide ($CO_2$) has enhanced our understanding of the utilization of these molecules during photosynthesis. Radioactive tracers are also used to determine the site of action of drugs and hormones, the metabolic pathways of nutrients, and the course of many other biochemical reactions. Table 2-6 lists the radioactive isotopes, or **radioisotopes,** most commonly used as tracers in biology. Many common biological elements are readily available as labeled isotopes. These isotopes can be incorporated into molecules and used as tracers in biological reactions or in cells and tissues. The radiation emitted as the elements decay can be measured by a variety of techniques. One of these, called *autoradiography,* involves placing a specimen containing labeled radioactive material in close contact with photographic film. When the film is developed, it shows a pattern formed by radioactivity in the tissues that have incorporated the label.

All atoms of a particular radioactive isotope have the same likelihood of decaying, whatever their age may be. The time required for a radioactive isotope to decay completely cannot be specified, but the time required for decay of half the parent atoms can be determined. This time is the **half-life,** represented by the symbol $t_{1/2}$. Each radioisotope has its own characteristic half-life. For example, the half-life of strontium-90 ($^{90}Sr$) is about 28 years. If we started with 10 grams of $^{90}Sr$, 5 grams of that isotope would remain after 28 years. The other half

* Molecules in which one of the atoms has been substituted by a radioactive isotope of the same atom.

**TABLE 2-6 Commonly Used Isotopic Tracers**

| *Isotope* | *Half-life ($t_{1/2}$)* | *Type of radiation emitted* |
|---|---|---|
| $^{3}_{1}H$ (tritium) | 12.3 yr | $\beta$ |
| $^{14}_{6}C$ (carbon-14) | 5,730 yr | $\beta$ |
| $^{24}_{11}Na$ (sodium-24) | 15 h | $\beta, \gamma$ |
| $^{32}_{15}P$ (phosphorus-32) | 14.3 d | $\beta$ |
| $^{35}_{16}S$ (sulfur-35) | 87 d | $\beta$ |
| $^{36}_{17}Cl$ (chlorine-36) | $3 \times 10^5$ yr | $\beta$ |
| $^{40}_{19}K$ (potassium-40) | $1.3 \times 10^9$ yr | $\beta, \gamma$ |
| $^{45}_{20}Ca$ (calcium-45) | 165 d | $\beta$ |
| $^{59}_{26}Fe$ (iron-59) | 45 d | $\beta, \gamma$ |
| $^{60}_{27}Co$ (cobalt-60) | 5.2 yr | $\beta, \gamma$ |
| $^{131}_{53}I$ (iodine-131) | 8 d | $\beta, \gamma$ |

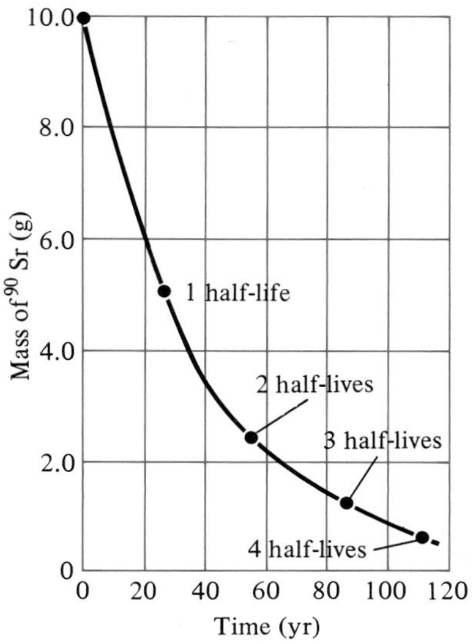


**FIG. 2-5** Radioactive decay of strontium-90 ($^{90}Sr$). Strontium has a half-life of 28 years.

would have been converted to yttrium-90 ($^{90}Y$). After another 28-year period, half of the remaining 5 grams of strontium-90 would decay (Fig. 2-5). The decay rate of a radioactive element is so predictable that it can be used as a molecular clock to determine the age of a substance; this dating procedure is described in Panel 27-3 of Chapter 27.

Radioactivity is biologically significant not only because it is useful but also because it is harmful to living systems. Several types of radiation increase the frequency of genetic mutations, almost all of which are harmful. All tissues, but especially growing tissues, can be damaged or killed by radioactive emissions. While radioactive emissions pose a threat to health and life, they have medical applications, such as the treatment of cancers with radioactive cobalt emissions (Fig. 2-6).

## QUANTUM MECHANICS AND ELECTRON ORBITALS

In 1913 the Danish physicist Niels Bohr postulated that the hydrogen atom consisted of a central proton around which one electron moved in a circular orbit. In larger atoms, each electron occupied its own orbit. The atom was, in this view, much like a tiny solar system, with the nucleus representing the sun and the electrons the orbiting planets.

According to this theory, each electron orbit was at a specified distance from the nucleus, and each had a specific energy level. Bohr's theory was thus consistent with the quantum theory of German physicist Max Planck, which held that electrons possess only certain discrete amounts, or quanta, of energy. Planck hypothesized seven discrete atomic energy levels, each of which could hold a certain maximum number of electrons.

Bohr's theory was very important because it introduced the idea of energy levels for electrons in atoms. However, the theory was deficient in part because it did not adequately account for repulsions between

**FIG. 2-6** Radiation from a radioactive cobalt source being used to treat a patient with cancer.

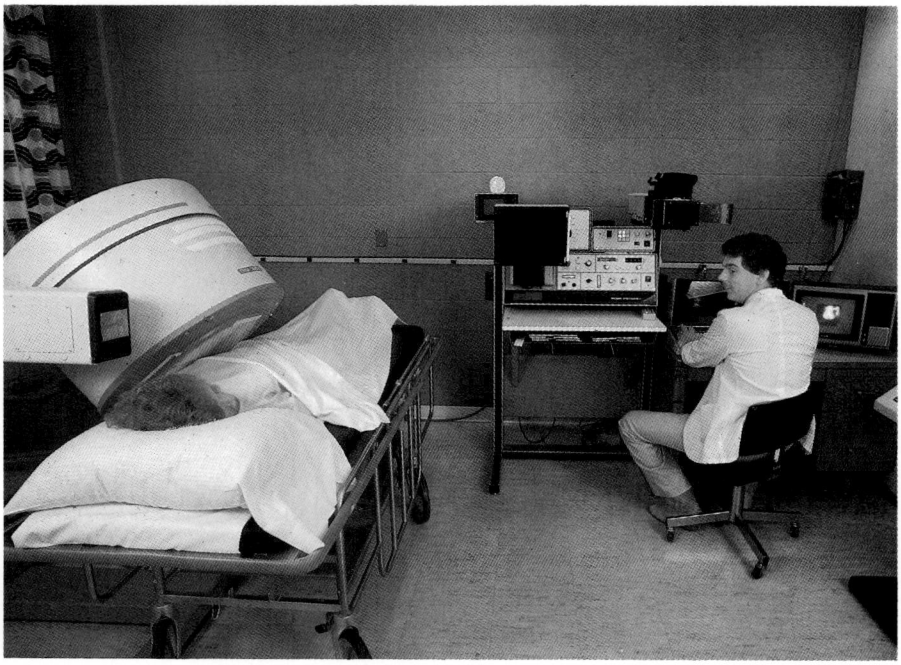

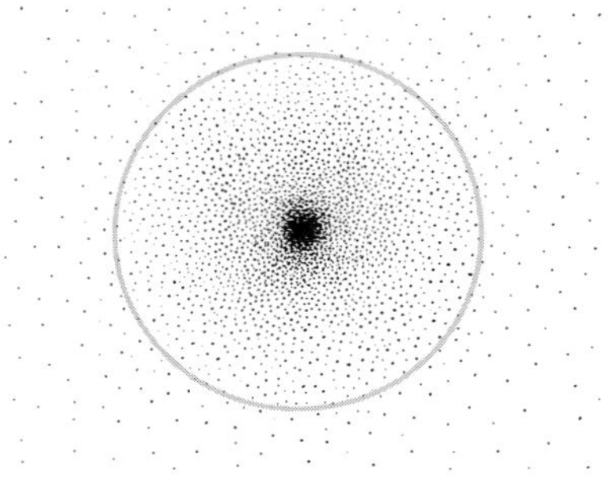

**FIG. 2-7** Electron density of the 1*s* orbital in a hydrogen atom. The circle drawn in the sketch represents the area in which the electron can be expected to occur 90 percent of the time.

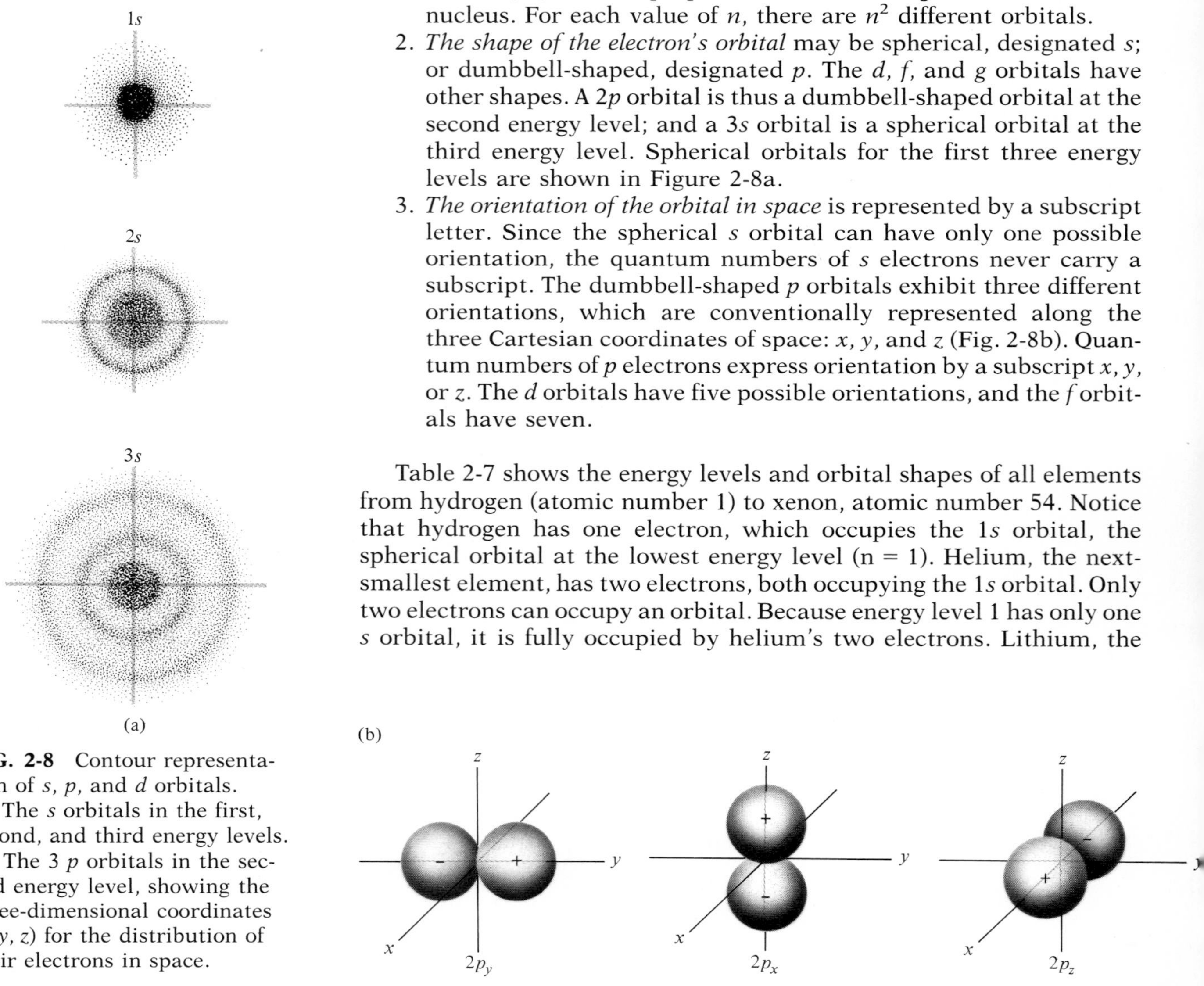


**FIG. 2-8** Contour representation of *s*, *p*, and *d* orbitals. (a) The *s* orbitals in the first, second, and third energy levels. (b) The 3 *p* orbitals in the second energy level, showing the three-dimensional coordinates (*x*, *y*, *z*) for the distribution of their electrons in space.

electrons. We know that charged particles of the same sign repel one another. Thus, in any atom containing two or more electrons, the electrons must "keep their distance" from one another while at the same time being attracted to the nucleus.

In time, Bohr's theory was replaced by a new model consistent with French physicist Louis de Broglie's proposal that the movements of an electron in an atom are wavelike and that the presence of a given electron in a certain region of space at a given instant is a matter of statistical probability. The circle in the sketch of a hydrogen atom in Figure 2-7 represents the spherical volume of space in which the atom's one electron will be found 90 percent of the time. Such a volume is called an **electronic orbital** or, simply, an **orbital.** The general mathematical theory describing the interactions of matter and radiation in terms of observable quantities is known as **quantum mechanics.**

***Quantum numbers.*** Every electron in an atom has a quantum number, which identifies three of the electron's characteristics:

1. *The electron's energy level* is defined by its principal quantum number, $n$, which is proportional to its average distance from the nucleus. For each value of $n$, there are $n^2$ different orbitals.
2. *The shape of the electron's orbital* may be spherical, designated *s*; or dumbbell-shaped, designated *p*. The *d*, *f*, and *g* orbitals have other shapes. A 2*p* orbital is thus a dumbbell-shaped orbital at the second energy level; and a 3*s* orbital is a spherical orbital at the third energy level. Spherical orbitals for the first three energy levels are shown in Figure 2-8a.
3. *The orientation of the orbital in space* is represented by a subscript letter. Since the spherical *s* orbital can have only one possible orientation, the quantum numbers of *s* electrons never carry a subscript. The dumbbell-shaped *p* orbitals exhibit three different orientations, which are conventionally represented along the three Cartesian coordinates of space: *x*, *y*, and *z* (Fig. 2-8b). Quantum numbers of *p* electrons express orientation by a subscript *x*, *y*, or *z*. The *d* orbitals have five possible orientations, and the *f* orbitals have seven.

Table 2-7 shows the energy levels and orbital shapes of all elements from hydrogen (atomic number 1) to xenon, atomic number 54. Notice that hydrogen has one electron, which occupies the 1*s* orbital, the spherical orbital at the lowest energy level (n = 1). Helium, the next-smallest element, has two electrons, both occupying the 1*s* orbital. Only two electrons can occupy an orbital. Because energy level 1 has only one *s* orbital, it is fully occupied by helium's two electrons. Lithium, the

**TABLE 2-7 Electronic Configurations of Elements with Atomic Numbers 1 (Hydrogen) Through 54 (Xenon)**

| | | *Energy level, or principal quantum number (n)* | | | | | | | | | | | | | |
|---|---|---|---|---|---|---|---|---|---|---|---|---|---|---|---|
| | | 1 | 2 | | 3 | | | 4 | | | | 5 | | | |
| | | Orbital shape | | | | | | | | | | | | | |
| Atomic number | Element | *s* | *s* | *p* | *s* | *p* | *d* | *s* | *p* | *d* | *f* | *s* | *p* | *d* | *f* |
| 1 | H | 1 | | | | | | | | | | | | | |
| 2 | He | 2 | | | | | | | | | | | | | |
| 3 | Li | 2 | 1 | | | | | | | | | | | | |
| 4 | Be | 2 | 2 | | | | | | | | | | | | |
| 5 | B | 2 | 2 | 1 | | | | | | | | | | | |
| 6 | C | 2 | 2 | 2 | | | | | | | | | | | |
| 7 | N | 2 | 2 | 3 | | | | | | | | | | | |
| 8 | O | 2 | 2 | 4 | | | | | | | | | | | |
| 9 | F | 2 | 2 | 5 | | | | | | | | | | | |
| 10 | Ne | 2 | 2 | 6 | | | | | | | | | | | |
| 11 | Na | 2 | 2 | 6 | 1 | | | | | | | | | | |
| 12 | Mg | 2 | 2 | 6 | 2 | | | | | | | | | | |
| 13 | Al | 2 | 2 | 6 | 2 | 1 | | | | | | | | | |
| 14 | Si | 2 | 2 | 6 | 2 | 2 | | | | | | | | | |
| 15 | P | 2 | 2 | 6 | 2 | 3 | | | | | | | | | |
| 16 | S | 2 | 2 | 6 | 2 | 4 | | | | | | | | | |
| 17 | Cl | 2 | 2 | 6 | 2 | 5 | | | | | | | | | |
| 18 | Ar | 2 | 2 | 6 | 2 | 6 | | | | | | | | | |
| 19 | K | 2 | 2 | 6 | 2 | 6 | | 1 | | | | | | | |
| 20 | Ca | 2 | 2 | 6 | 2 | 6 | | 2 | | | | | | | |
| 21 | Sc | 2 | 2 | 6 | 2 | 6 | 1 | 2 | | | | | | | |
| 22 | Ti | 2 | 2 | 6 | 2 | 6 | 2 | 2 | | | | | | | |
| 23 | V | 2 | 2 | 6 | 2 | 6 | 3 | 2 | | | | | | | |
| 24 | Cr | 2 | 2 | 6 | 2 | 6 | 5 | 1 | | | | | | | |
| 25 | Mn | 2 | 2 | 6 | 2 | 6 | 5 | 2 | | | | | | | |
| 26 | Fe | 2 | 2 | 6 | 2 | 6 | 6 | 2 | | | | | | | |
| 27 | Co | 2 | 2 | 6 | 2 | 6 | 7 | 2 | | | | | | | |
| 28 | Ni | 2 | 2 | 6 | 2 | 6 | 8 | 2 | | | | | | | |
| 29 | Cu | 2 | 2 | 6 | 2 | 6 | 10 | 1 | | | | | | | |
| 30 | Zn | 2 | 2 | 6 | 2 | 6 | 10 | 2 | | | | | | | |
| 31 | Ga | 2 | 2 | 6 | 2 | 6 | 10 | 2 | 1 | | | | | | |
| 32 | Ge | 2 | 2 | 6 | 2 | 6 | 10 | 2 | 2 | | | | | | |
| 33 | As | 2 | 2 | 6 | 2 | 6 | 10 | 2 | 3 | | | | | | |
| 34 | Se | 2 | 2 | 6 | 2 | 6 | 10 | 2 | 4 | | | | | | |
| 35 | Br | 2 | 2 | 6 | 2 | 6 | 10 | 2 | 5 | | | | | | |
| 36 | Kr | 2 | 2 | 6 | 2 | 6 | 10 | 2 | 6 | | | | | | |
| 37 | Rb | 2 | 2 | 6 | 2 | 6 | 10 | 2 | 6 | | | 1 | | | |
| 38 | Sr | 2 | 2 | 6 | 2 | 6 | 10 | 2 | 6 | | | 2 | | | |
| 39 | Y | 2 | 2 | 6 | 2 | 6 | 10 | 2 | 6 | 1 | | 2 | | | |
| 40 | Zr | 2 | 2 | 6 | 2 | 6 | 10 | 2 | 6 | 2 | | 2 | | | |
| 41 | Nb | 2 | 2 | 6 | 2 | 6 | 10 | 2 | 6 | 4 | | 1 | | | |
| 42 | Mo | 2 | 2 | 6 | 2 | 6 | 10 | 2 | 6 | 5 | | 1 | | | |
| 43 | Tc | 2 | 2 | 6 | 2 | 6 | 10 | 2 | 6 | 6 | | 1? | | | |
| 44 | Ru | 2 | 2 | 6 | 2 | 6 | 10 | 2 | 6 | 7 | | 1 | | | |
| 45 | Rh | 2 | 2 | 6 | 2 | 6 | 10 | 2 | 6 | 8 | | 1 | | | |
| 46 | Pd | 2 | 2 | 6 | 2 | 6 | 10 | 2 | 6 | 10 | | | | | |
| 47 | Ag | 2 | 2 | 6 | 2 | 6 | 10 | 2 | 6 | 10 | | 1 | | | |
| 48 | Cd | 2 | 2 | 6 | 2 | 6 | 10 | 2 | 6 | 10 | | 2 | | | |
| 49 | In | 2 | 2 | 6 | 2 | 6 | 10 | 2 | 6 | 10 | | 2 | 1 | | |
| 50 | Sn | 2 | 2 | 6 | 2 | 6 | 10 | 2 | 6 | 10 | | 2 | 2 | | |
| 51 | Sb | 2 | 2 | 6 | 2 | 6 | 10 | 2 | 6 | 10 | | 2 | 3 | | |
| 52 | Te | 2 | 2 | 6 | 2 | 6 | 10 | 2 | 6 | 10 | | 2 | 4 | | |
| 53 | I | 2 | 2 | 6 | 2 | 6 | 10 | 2 | 6 | 10 | | 2 | 5 | | |
| 54 | Xe | 2 | 2 | 6 | 2 | 6 | 10 | 2 | 6 | 10 | | 2 | 6 | | |

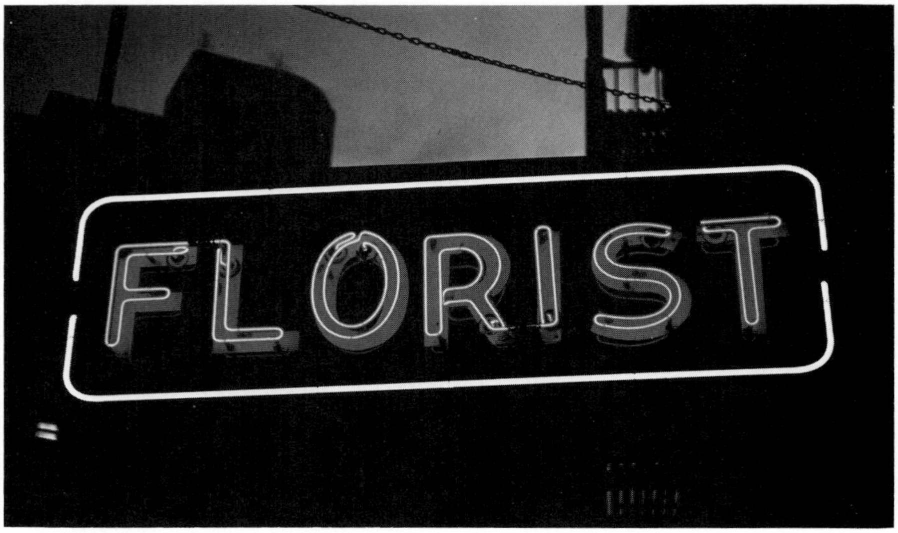


**FIG. 2-9** The glitter of a neon sign.

next element, has three electrons, two in the 1*s* orbital and one in a 2*s* orbital, an *s* orbital at the second energy level.

***Chemical Reactivity.*** One way of explaining chemical reactivity is in terms of the degree to which an atom attracts electrons of other atoms. An atom's attraction for such electrons is determined by the number of electrons in its outer energy level, or outer shell. Atoms with completely filled outer shells do not attract electrons and are relatively unreactive or inert. Atoms with incomplete outer shells tend to gain or lose electrons to achieve the electron configurations of the six inert, or noble, gases: helium, neon, argon, krypton, xenon, and radon, with 2, 8, 8, 8, 8, and 8 outer shell electrons, respectively.

An atom with only one electron in its outermost level tends to lose that electron readily. In losing the electron, the atom acquires an overall electric charge of +1, enabling it to be attracted to and react with atoms of opposite charge. Conversely, atoms with a single vacancy in their outer energy levels tend to capture an electron, becoming negatively charged. Except at the 1*s* level, atoms with two electrons in their outer energy levels are inclined to lose two electrons; atoms with six outer shell electrons tend to gain two. Atoms with four outer shell electrons tend neither to lose nor to gain electrons. Under most conditions—and particularly in living systems—atoms lose or gain electrons only if acceptor or donor atoms react with them.

**FIG. 2-10** When a neon atom is struck by an electric discharge, an electron absorbs the energy and leaps from its normal orbit around the nucleus to a new, but temporary, orbit. Upon returning to its normal orbit, this electron gives off the absorbed energy as the colored light we see in neon signs.

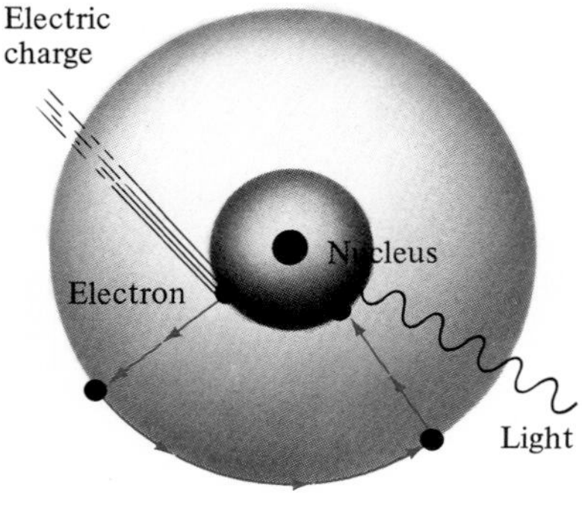


***Energy States.*** Under ordinary conditions, in the so-called unexcited state, an atom's electrons remain at the lowest energy level possible. When an atom absorbs energy, one or more of its electrons move from a lower to a higher energy level orbital. Such electrons are in an excited state.

An atom must absorb a specific amount of energy for one of its electrons to jump to the next higher energy level. The same amount of energy is released when the electron drops back to its former level, or **ground state.** The energy is released in precise amounts and at particular wavelengths in the form of visible light, radio waves, X rays, or other types of radiant energy. This fact is illustrated in the operation of a neon sign, a device consisting of neon gas contained in a closed tube and supplied with energy in the form of an electric discharge (Fig. 2-9).

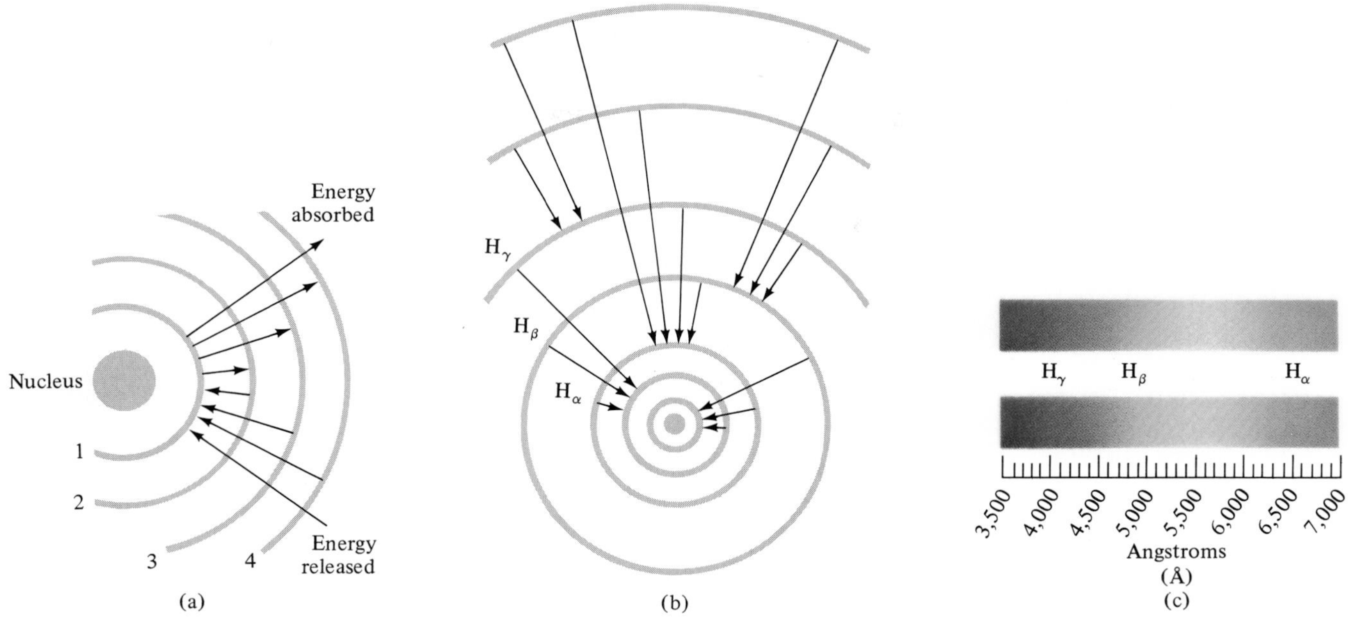


**FIG. 2-11** Absorption and emission of energy by atoms. (a) Possible energy transitions for electrons in an atom with four energy levels. (b) The hydrogen atom as seen by Bohr. Each group of arrows represents possible energy transitions for the single electron of hydrogen. Each group corresponds to (c) emissions of specific spectral lines.

When various electrons of the neon atoms absorb some of this energy, they jump to various higher energy levels within the neon atoms (Fig. 2-10). Almost immediately, they fall back to their more stable ground states, giving off the absorbed energy as light at the predominantly red wavelengths characteristic of neon. When excited atoms emit visible light in this way, they are said to **fluoresce.** The sign can be made to fluoresce in different colors by the addition of other elements to the tube; mercury, for example, adds blue, and sodium, yellow.

***Emission Spectra.*** The absorption and emission of energy by the electrons in an atom is represented in Figure 2-11. The wavelength of the radiation emitted is determined by the nature of the element, the number of excited electrons, and the number of energy quanta absorbed by its atoms. The entire range of possible wavelengths is indicated by the electromagnetic spectrum (see Fig. 2-12).

As Figure 2-13 (p. 48) illustrates, the patterns of emissions of excited atoms are as characteristic of each element as are fingerprints of people. This can be explained by the fact that the electrons about the nucleus of a particular element are at energy levels characteristic for the element. Closely related elements tend to have their electrons at similar energy levels. Hence, the more two elements are alike, the more similar their emission spectra will be. Such emissions can be used to identify excited elements, even those present in a distant star.

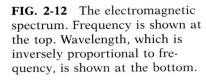

**FIG. 2-12** The electromagnetic spectrum. Frequency is shown at the top. Wavelength, which is inversely proportional to frequency, is shown at the bottom.

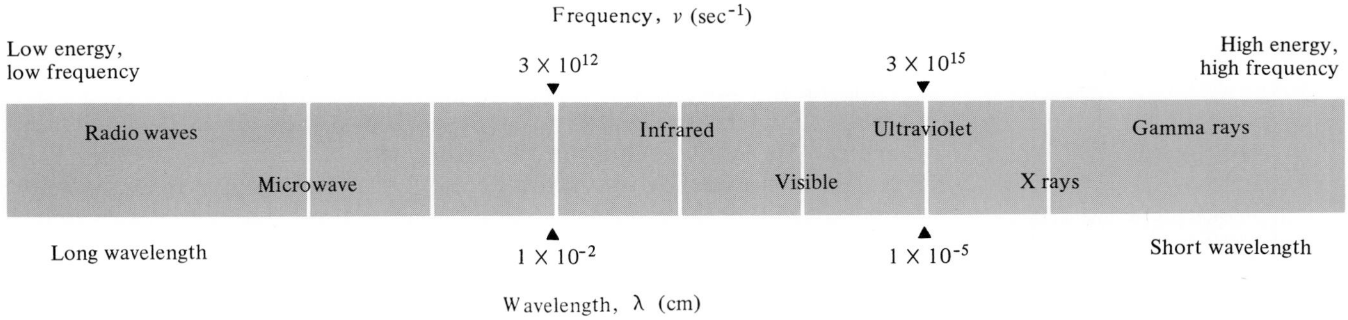

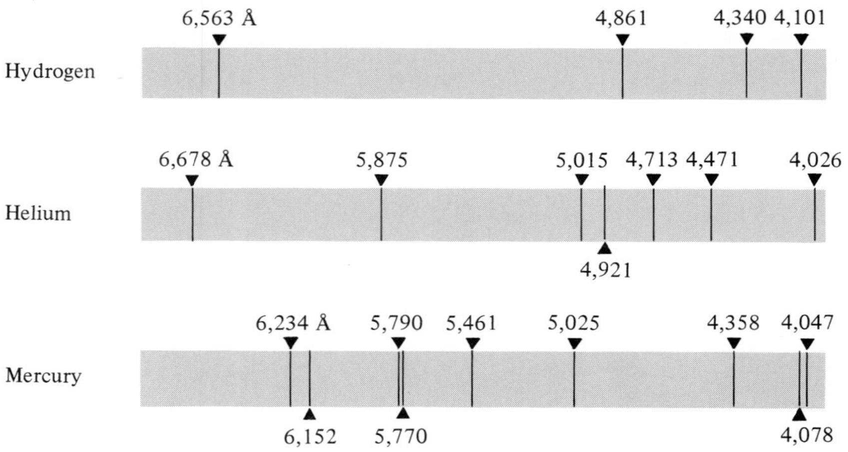


**FIG. 2-13** The emission spectra of the elements hydrogen, helium, and mercury. The wavelengths at which these emissions occur is shown in angstrom (Å) units.

# CHEMICAL BONDING

Except for the noble gases, atoms tend to join to each other by attractive forces known as **chemical bonds.** In some cases, a relatively stable entity known as a molecule is formed by the constituent atoms. In other cases, a less stable association called an **ionic compound** is established between atoms of different elements. Molecular, ionic, and other kinds of chemical bonding occur in the chemical reactions essential for life.

## IONIC BONDS

One of the most common substances on earth is the compound sodium chloride (NaCl), table salt. An examination of sodium chloride, either in aqueous (water) solution or in the dry crystalline state, shows it to be composed of equal numbers of sodium ions ($Na^+$) and chloride ions ($Cl^-$). Ions (from the Greek *ion,* "going") are atoms or molecules that bear electrical charges as a result of having gained or lost one or more electrons. An atom that loses one or more electrons becomes positively charged. Such an atom moves toward the negative electrode, or cathode, in an electrical field (the space between positively and negatively charged electrodes through which an electric current passes) and is known as a **cation.** An atom that gains electrons becomes negatively charged, moves toward the anode in an electrical field, and is known as an **anion** (Fig. 2-14). Being oppositely charged, anions and cations are also attracted to and tend to bind to each other. The binding of anions with cations constitutes an ionic bond. Because they are at a lower level of potential energy,* the atoms in ionic compounds such as sodium chloride are in a more stable arrangement than unbonded sodium and chlorine atoms.

**FIG. 2-14** Movement of anions and cations to the cathode and anode, respectively, in an electric field.

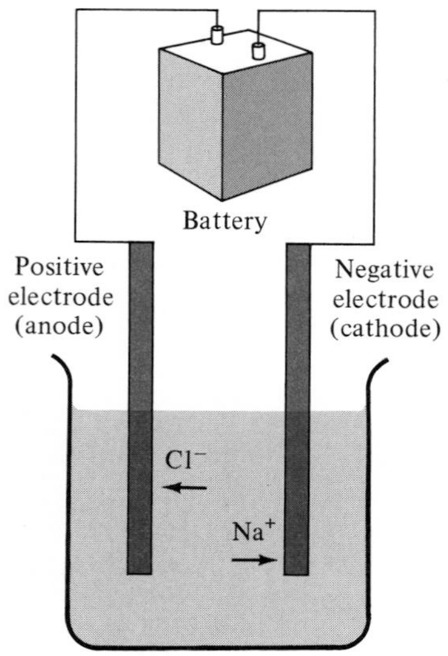


***How Ionic Bonds Are Formed.*** The chlorine atom has seven electrons in its outer energy level (Fig. 2-15), with one vacancy in its outermost orbital. Its nucleus thus has a strong attraction for any weakly held electron of a nearby atom and attracts it as strongly as it attracts the seven electrons already in its second energy level. The sodium atom has

* Potential energy is the energy stored in a body by virtue of its composition or its position in space.

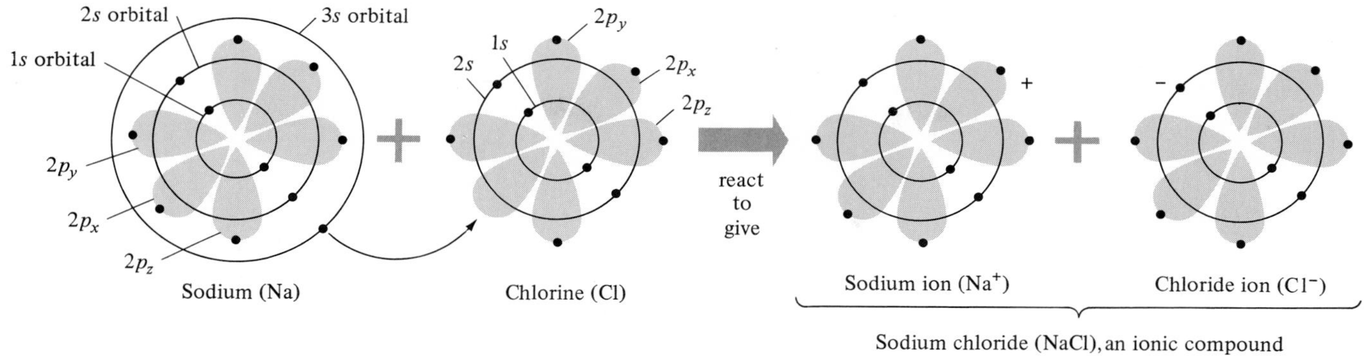


**FIG. 2-15** Formation of the ionic compound sodium chloride, NaCl, by donation of an electron from the sodium atom to the chloride atom.

one electron in its outer energy level. When sodium and chlorine atoms are mixed, the chlorine atoms attract the lone 3*s* electrons in the outer shells of the sodium atoms. By losing one electron, the sodium atoms become sodium ions. By gaining an electron, the chlorine atoms become chloride ions. The sodium and chloride ions are held to each other by opposite charges, forming the ionic compound sodium chloride.

In solution, no pair of sodium and chloride ions remains associated continuously. Individual sodium and chloride ions often have different partners from one moment to the next. In the crystalline state, however, as represented in Figure 2-16, the ions settle into a stable arrangement. Each sodium ion in a crystal is surrounded by several chloride ions and each chloride ion is surrounded by several sodium ions. Like all atoms, they are less chemically reactive with full orbitals in their outermost energy levels. They have achieved stable inert gas electron configurations.

## COVALENT BONDS

More common than ionic bonding is covalent bonding: the sharing of electrons by nuclei of two atoms. Like ionic bonds, covalent bonds are formed by the attraction of one atomic nucleus for the electrons of the other atom, but molecules, which usually result from covalent bonds, are more stable than the units that result from ionic bonds. If a molecule contains atoms of two or more elements, the substance is a molecule of a compound. If the joined atoms are of the same element, the substance is the molecular form of that element: molecular oxygen, molecular nitrogen, and so on.

**FIG. 2-16** Model of a sodium chloride crystal. The larger chloride ions surround and are surrounded by the smaller sodium ions.

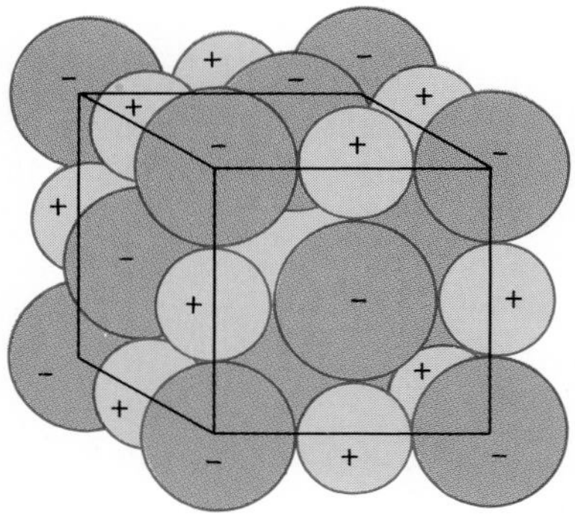

***How Covalent Bonds Are Formed.*** We can gain some understanding of covalent bonding by visualizing the orbitals of two or more atoms and imagining what happens when the atoms closely approach each other. Consider what happens, for example, when two hydrogen atoms unite with an oxygen atom in forming the familiar compound water ($H_2O$). Oxygen, with atomic number 8, has eight electrons, two electrons each in its 1*s*, 2*s*, and 2*p* orbitals and one in each of the other 2*p* orbitals (see Table 2-7 and Fig. 2-17a). Because each orbital can accommodate two electrons, oxygen has two vacancies in its outermost energy level. The hydrogen atom's only electron is in a 1*s* orbital, at the first energy level (Fig. 2-17b), thus hydrogen has one vacancy.

When two hydrogen atoms are close to an oxygen atom, the two half-filled 2*p* orbitals of the oxygen atom and the spherical *s* orbital of

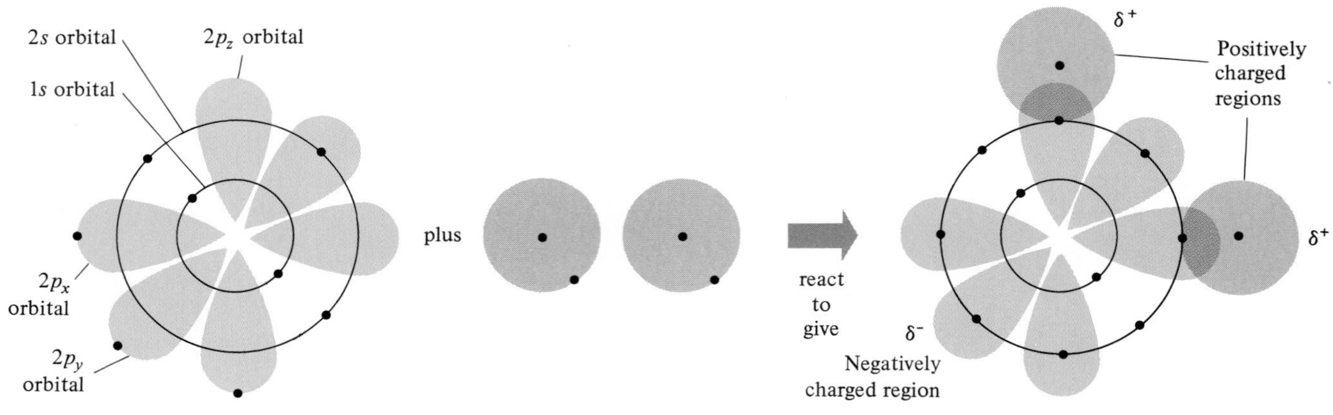


(a) One oxygen atom (b) Two hydrogen atoms (c) One water molecule

**FIG. 2-17** Electron orbital configurations of (a) oxygen and (b) hydrogen. (c) Formation of covalent bonds in water, resulting in a polar compound. The designations $\delta^+$ and $\delta^-$ represent differences in the electric charge of the two poles of the molecule.

each hydrogen atom overlap, as shown in Figure 2-17c. The outer orbitals of all three atoms can now be considered filled since they are now sharing electrons. In the newly formed molecule, two pairs of electrons are attracted by and actually reside for a substantial fraction of the time with all three nuclei. This sharing of electrons and their attraction to all the nuclei constitute the two covalent bonds that are formed. The nuclei cannot approach more closely because of their mutual repulsion of each other. The electrons and nuclei maintain those distances from each other at which all attractive and repulsive forces are at equilibrium, thereby forming a stable bond.

**FIG. 2-18** Orientation of polar molecules in an electric field.

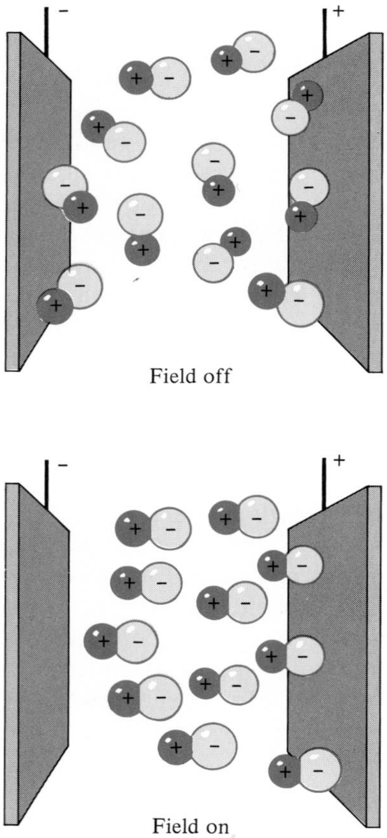


***Polar Covalent Bonds.*** In a water molecule, the oxygen nucleus attracts the shared electrons much more strongly than do the small hydrogen nuclei, because there are several protons in the oxygen nucleus and only one in each hydrogen. The result is an unequal distribution of charge within the molecule. The end with the oxygen atom is somewhat negative because it has more electrons moving about it at any given moment than there are protons in its nucleus. Conversely, the end of the molecule with the hydrogen atoms is slightly positive because of the general absence of electrons from that region. These small differences in charge are denoted in Figure 2-17 by the small Greek letter delta: $\delta^+$ and $\delta^-$. The pairs of oppositely charged ends in such molecules are called **electric dipoles.** In an electric field, the negative end of the molecule will be oriented toward the positive pole and the positive end of the molecule toward the negative pole (Fig. 2-18). Because there is a difference between the two ends, or poles, of these molecules, they are known as **polar molecules.** The bonds responsible for producing polarity are termed **polar covalent bonds.** The polarity of the water molecule is crucial to the biochemical reactions that underlie many biological processes.

In molecules such as $H_2$, $O_2$, and $F_2$, which are formed of pairs of identical atoms, the shared pair of electrons is visualized as spending as much time with one nucleus as with the other. The covalent bonds in these molecules are nonpolar.

## VAN DER WAALS FORCES

The weak electrostatic forces that sometime hold nonpolar molecules to each other are van der Waals forces, which result from the movement of

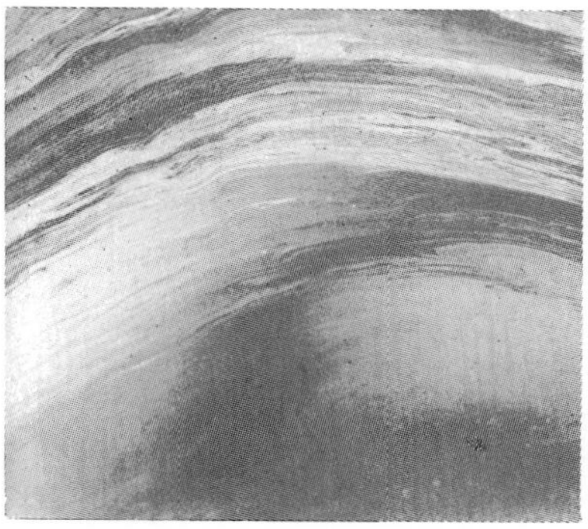

FIG. 2-19 A mixture of water and oil showing that they don't mix.

of electrons. At any given instant, the charges on a molecule may not be uniformly distributed, inducing the formation of equal and opposite electric charges in different parts of the molecule. Molecules of the same kind are often held together by van der Waals forces.

## HYDROPHOBIC INTERACTIONS

Another kind of bonding occurs among molecules that have little or no affinity for water. These molecules tend to group together when in an aqueous environment, forming droplets. For example, when oil is added to water, the two do not mix (Fig. 2-19). This kind of interaction is seen in biological membranes, which are composed largely of lipids. The oil molecules remain together not so much because they are attracted to each other as because they are forced together by surrounding water molecules.

## HYDROGEN BONDS

Chemical bonding can also occur between molecules containing hydrogen covalently bonded to oxygen, nitrogen, or fluorine. Examples of such molecules include water ($H_2O$), ammonia ($NH_3$), and hydrogen fluoride (HF). In a water molecule, as noted above, the hydrogen nuclei are largely deprived of their electrons and have a slight positive charge (see Fig. 2-20). As a result, a hydrogen nucleus of a water molecule attracts and is attracted by the negative portions of other polar molecules, such as the oxygen atom end of other water molecules. When two oppositely charged regions of polar molecules approach each other closely enough, a weak but definite bonding occurs between them. When many such bonds occur between two molecules, considerable stability can be achieved. Hydrogen bonding plays important roles in the structures and chemical behavior of many biologically important molecules, including water, proteins, and nucleic acids

# WATER

Of all the substances indispensable to life, water is by far the most important. Water is the principal constituent of living systems and makes up at least 70 percent of the weight of most forms of life. It serves as a medium for the cell's metabolic activities and for the transport of nutrients and other vital substances throughout the body. Three-quarters of the earth's surface is covered by water.

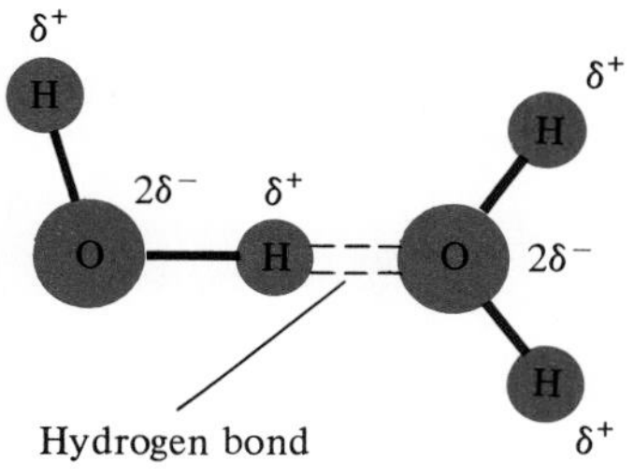


FIG. 2-20 Hydrogen bond between two polar water molecules. Hydrogen bonds have only about 1/20 the strength of a covalent bond. The designations $\delta^+$ and $\delta^-$ are the positive and negative charges, respectively, on these molecules.

## HYDROGEN BONDS AND THE PHYSICAL PROPERTIES OF WATER

Although it is a relatively small and simple molecule, water has a number of unique physical and chemical properties related to its ability to form hydrogen bonds (Fig. 2-20). While the hydrogen bonds between water molecules are much weaker than the covalent bonds within them, such hydrogen bonds are nevertheless quite strong. In effect, a container of water is a weak polymer, or single large molecule. Recall that the water molecule is a dipole: one part of the molecule is positively charged, the other negatively (see Fig. 2-20). For this reason, there is constant "changing of partners" between hydrogen atoms of adjacent water molecules; however, the rate of change is proportional to temperature. At one moment, a hydrogen atom of one water molecule is attracted to the oxygen atom of a nearby water molecule. A moment later, the same hydrogen atom may form a hydrogen bond with the oxygen atom of a different water molecule. Despite this changing of partners, the water molecules attract each other strongly.

***Water as a Solvent.*** Water is such an efficient solvent that, for example, 1 gallon of water can dissolve 70 pounds of the fertilizer ammonium nitrate. Because of this efficiency, perfectly pure water is very rare and may not even exist in nature. Even rain, as it condenses and descends through the atmosphere, dissolves such materials as dust and atmospheric gases. This fact is evidenced by the devastation caused by acid rain, which contains sulfuric and nitric acids derived from the gaseous oxides of sulfur and nitrogen. About half of all the chemical elements are soluble in water (Fig. 2-21). Water comes closer than any other substance to being a universal solvent.

***Molecular Dipoles.*** The dipolar nature of water molecules, illustrated in Figure 2-20, accounts for its capacity as a universal solvent. In many substances, such as sodium chloride, or table salt, the atoms are held

**FIG. 2-21** Every trickle, puddle, stream, river, lake, and sea on earth is an aqueous solution in which are dissolved a large variety of substances.

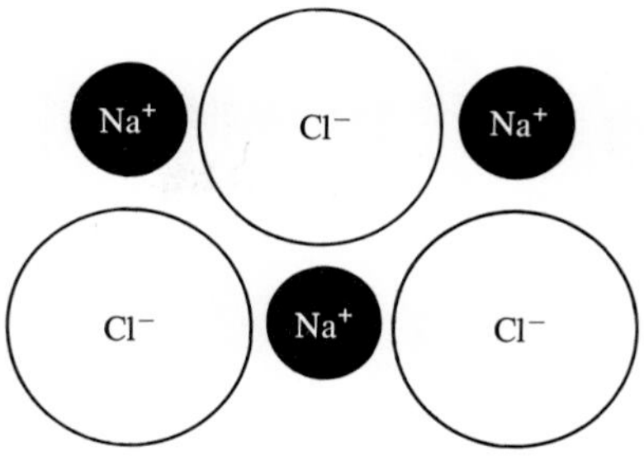


(a) Sodium chloride crystal

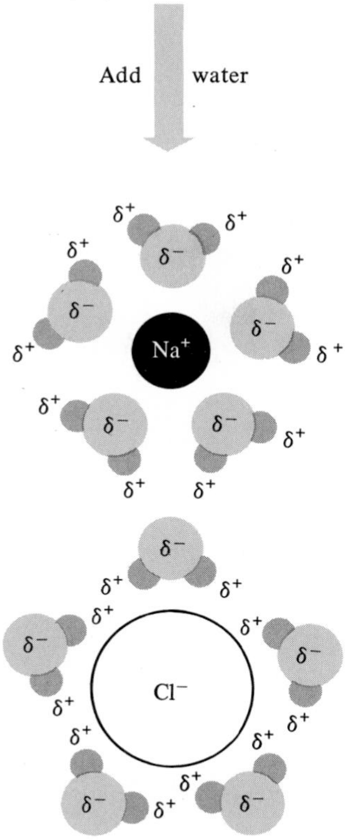


(b) Sodium and chloride ions surrounded by water molecules

**FIG. 2-22** Sodium chloride, or table salt, dissolving in water. (a) The opposite electrical charges of the $Na^+$ and $Cl^+$ ions are attracted to one another to form the stable sodium chloride crystal. (b) When water is added, the partial negative charges ($\delta^-$) on its oxygen molecule are attracted to the positively charged $Na^+$ ions. At the same time, the partial positive charges ($\delta^+$) of the hydrogen atoms of water are attracted to the $Cl^-$ ions. As a result, the water molecules are able to surround the $Na^+$ and $Cl^-$ ions and push them away from each other. They are now dissolved.

together by opposite electric charges (see Fig. 2-15). But when table salt is added to water, the sodium ions ($Na^+$) are attracted to the weak negative ($\delta^-$) charges of the oxygen atoms of water molecules. The opposite happens with the chloride ions ($Cl^-$): they are attracted to the $\delta^+$ charges of the hydrogen atoms of water. As the water molecules exert their electric tugs on the table salt molecules, they cancel some of the electrical attraction between the $Na^+$ and $Cl^-$ ions. As a result, the ions move apart and separate (Fig. 2-22). In this state, they are totally surrounded by water and are said to be dissolved.

In addition to its ability to dissolve ionic compounds, water is also able to dissolve polar compounds. These compounds have electric dipoles similar to those of water molecules. Water molecules are able to form hydrogen bonds with polar molecules much as they do with each other. Thus, for example, ammonia molecules readily dissolve in water (Fig. 2-23).

**FIG. 2-23** Formation of hydrogen bonds between ammonia and water molecules enables ammonia to dissolve in water.

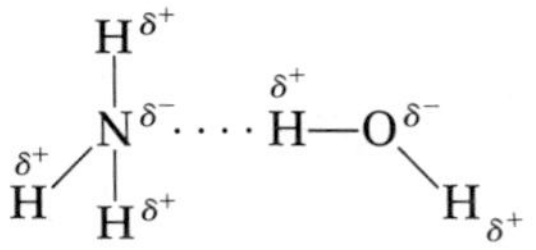


## COHESIVENESS

Tension between water molecules due to hydrogen bonds makes the surface of water behave as if it were an elastic membrane. This phenomenon is termed **surface tension** and is due to the **cohesiveness** (attraction between molecules of the same substance) of water molecules. This is evidenced by the ability of insects such as the water strider to walk on water (Fig. 2-24). Water droplets form on a leaf (Fig. 2-25) or a dripping faucet because the surface tension of the water molecules tends to pull the drops into shapes with the smallest possible surface area.

**FIG. 2-24** A water strider walking on the water surface.

***Climbing Molecules.*** Because of their cohesiveness and **adhesiveness** (attraction between molecules of different substances), water molecules are able to rise against gravity in small glass tubes by a process called **capillary action** (Fig. 2-26). Like cohesiveness, adhesiveness is due to hydrogen bonds. It enables water to bind to other substances, such as glass, clay, or soil. Thus, the surface of the colored water shown in Figure 2-26 is like a chain of water molecules held together by hydrogen bonds. As the water molecules at the edge of the capillary tube are attracted to the glass molecules above them, they pull up the rest of the molecules with them, and the water column rises to a new level. This process continues until the downward pull of gravity on the column of water is too strong for capillary action to overcome. Capillary action enables soil water to move up through soil, dissolving nutrients as it goes. This capillary action brings water and nutrients to plant root hairs, where they can be absorbed. Seed imbibition (from the Latin, *imbibere,* "to drink") also results from capillary action. The seed coat becomes soaked, and water penetrates to the embryo within the germinating seed. The seed coat ultimately splits, and the growing roots reach into the soil (Fig. 2-27).

**FIG. 2-25** Spherical water droplets on a leaf formed by the cohesiveness of the water molecules at the surface of the droplet. The surface tension created tends to form a sphere, a shape having the smallest surface area per unit volume.

## HEAT ABSORPTION

Raising the temperature of a liquid is a matter of increasing the average kinetic energy of its molecules, that is, their rate of motion. In water, this molecular motion is resisted by the hydrogen bonding between water molecules. Therefore, much of the heat energy absorbed by water is used to break the hydrogen bonds before the molecular motion of water molecules can be increased and the temperature of the water raised.

Water has a large **heat capacity,** or specific heat. In other words, it takes a lot of heat to raise the temperature of a given amount of water. This fact is obvious to anyone who has burned a hand on an iron pot handle while the water in the pot was only warm. Because the heat capacity of the iron handle is much less than that of water, the handle

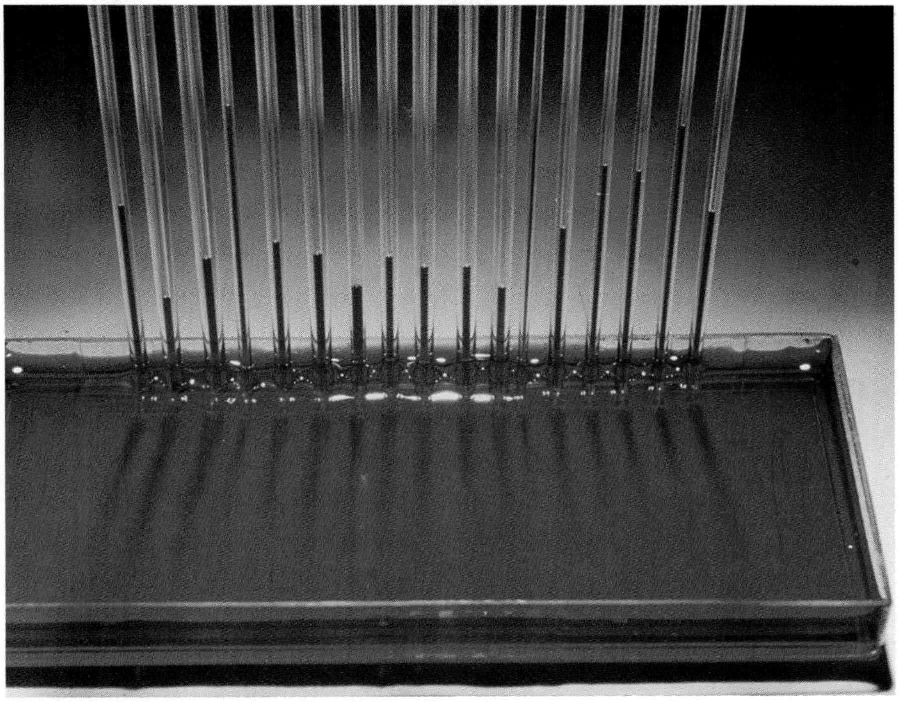

**FIG. 2-26** The capillary action of water as demonstrated by the rising of colored water in glass tubes of varying diameters.

**FIG. 2-27** A germinating seed showing its seed coat split as a result of a massive uptake of water by imbibition.

absorbs heat almost 10 times faster than water does. In fact, it takes more heat to raise the temperature of water than that of any other common substance except liquid ammonia.

The high specific heat of water is important to biological systems for several reasons. First, metabolic activity generates large amounts of heat, which is readily absorbed by the water in the system without greatly increasing its temperature. Thus, water helps regulate and stabilize body temperature. This is especially important for enzymes—the molecules that catalyze most of the chemical reactions in living systems—which function effectively only within narrow temperature ranges. Second, water regulates environmental temperature by preventing large fluctuations in it. We have all observed how slowly lakes and streams warm in the spring and cool in the fall.

Not only is water able to absorb large amounts of heat, but it can readily transfer this heat; that is, water has high **thermal conductivity.** Since water is distributed throughout an organism's tissues, it can readily transfer heat from one part of the body to another, thus maintaining a relatively constant temperature throughout. This is especially important to aquatic organisms, because most of them are unable to regulate their body temperatures and depend on the relatively constant temperatures of their home environment.

## HEAT OF VAPORIZATION

It takes a great deal of heat to raise the temperature of water from the boiling point to the point at which it becomes a gas by the process called **vaporization.** Water has the highest heat of vaporization of any liquid. At its boiling point, 100°C at a pressure of 1 atmosphere, it takes 540 calories of heat to evaporate 1 gram (g) of water. A calorie (c) is the amount of energy in the form of heat required to raise the temperature

**FIG. 2-28** Effective evaporative cooling: sweat releases heat absorbed by the body.

of 1 g of water by 1°C, from 14.5°C to 15.5°C. The Calorie (C) used in metabolic studies is the kilocalorie (1,000 c), the amount of heat required to raise the temperature of 1,000 g of water from 14.5°C to 15.5°C.

Water's high heat of vaporization, like its high specific heat, also serves to moderate environmental temperature extremes. The high rate of evaporation in tropical areas effectively removes large amounts of heat from those regions. Sweating and panting are two ways in which some animals can dissipate excess body heat through evaporation (Fig. 2-28). The principle of evaporative cooling is also used to cool homes and greenhouses.

## FREEZING

Water becomes less dense as it freezes to become ice, that is, as it changes from a liquid to a solid state. (Density is the mass* of a substance in a given volume.) In this respect, water is unlike most substances, which are denser in their solid states. Like all liquids, water shrinks in volume—and thus increases in density—as the temperature decreases. This is due to the fact that a decrease in the kinetic energy of the water molecules reduces their rate of movement and thus brings the molecules closer together. However, at 4°C (39°F), where the maximum density of water is reached, water ceases to contract. As the temperature decreases below 4°C, the hydrogen bonds cause the water molecules to move apart and form crystal lattices, and the molecules become less closely packed. Ice is therefore less dense than liquid water and, as a result, it floats (Fig. 2-29). Those who have experienced a sudden winter freeze are familiar with the results of the expansion of water in the form of broken water pipes and raised concrete slabs.

* It is often incorrectly assumed that mass and weight are the same. Mass is a measure of the amount of material in an object, whereas the weight of an object depends on both its mass and the effect of the attractive forces of gravity. In outer space, where gravitational forces are very weak, an astronaut may be weightless, but still has the same mass as on earth.

**FIG. 2-29** Floating icebergs. Water decreases in density as it freezes, and, therefore, ice floats.

The fact that ice floats is essential to the survival of life. If ice were denser than water, it would sink to the bottom of lakes and rivers and force water to the surface, where it would in turn freeze and sink. In all but tropical regions, all bodies of water would ultimately become frozen solid, with perhaps only thin layers of water forming over the ice during the warmest seasons. Under such conditions, most of our planet's water supply would be unavailable to plants and animals. Furthermore, the ability of large bodies of water to control large and rapid fluctuations in temperature would be lost, and most life as we know it could not have evolved on earth.

Because water expands and forms crystals when it freezes, freezing usually destroys cells. However, some organisms produce a kind of biological antifreeze that prevents crystal formation and enables them to survive in Arctic and Antarctic environments.

## ENERGETICS AND CHEMICAL REACTIONS

A supply of energy is needed to support any form of life. A simple definition of *energy is the capacity to do work.* The energy available to living cells is derived from chemical reactions. In order to understand the principles by which the chemical reactions essential to life proceed, we must examine the energy relationships in chemical processes.

We have already seen that atoms are able to react chemically with each other to form molecules and compounds. This chemical reactivity is at the core of all of life's processes. The energy by which atoms are held to one another in an ionic or a covalent bond—or the energy that would have to be applied to break them apart—is referred to as **bonding energy.**

When enough energy is supplied to break one or more of the chemical bonds of a substance, a chemical reaction occurs. The reaction consists of the separation of the atoms of the compound and, in most cases, their uniting to form a new compound. It is accompanied in many cases by a release of energy. Some of this energy is termed **free energy,** in the sense that it is available to do work. Free energy is comparable to the kinetic energy of a falling boulder that has been dislodged from its position at the top of a cliff. Once the boulder is dislodged, it falls by itself. Once enough energy is supplied to some compounds to break chemical bonds, the chemical reaction proceeds to completion with no additional input of energy.

A boulder at the bottom of a cliff has no potential energy and will not move to the top of the cliff unless a suitable form of energy is applied to it. Similarly, any chemical reaction that requires an input of energy to proceed will not proceed—even after enough energy is provided to overcome the bonding energy—unless a suitable form of energy is continually applied to push the reaction up its energy "hill."

To recapitulate: Free energy is important because it makes work in biological systems possible.

### EXERGONIC AND ENDERGONIC REACTIONS

Chemical reactions are either spontaneous or nonspontaneous. Once the bonding energies are overcome, spontaneous reactions proceed, even in isolation, without the need for additional energy. Nonspontaneous reactions, on the other hand, require a continual supply of energy from an outside source and could never occur in isolation. A reaction that requires continual energy input to proceed is **endergonic.** A reaction that releases energy is **exergonic.***

Exergonic reactions are comparable to the fall of the dislodged boulder to the bottom of the cliff. In such reactions, energy is spontaneously released once the process has begun. The energy supplied to an exergonic reaction to initiate it is comparable to the push needed to dislodge the boulder from its position at the top of the cliff. At normal temperatures and with oxygen available, even such a combustible substance as gasoline does not burn spontaneously. Some input of energy is required to set it afire. The fact that at normal temperatures many exergonic reactions require an input of energy to get started is explained by the collision theory of chemical reactions.

## THE COLLISION THEORY AND ACTIVATION ENERGY

Substances that react chemically to form products usually need some help to do so, at least to start the reaction. Mixing dry crystals, even those of some very reactive materials, usually produces no reaction. Even if we dissolve such crystals in water, a reaction may not occur. In solution, individual molecules are free to move about and do so at a rate directly proportional to their temperature. As they move, they collide with each other with a force dependent on their velocity. At cool temperatures their movement is comparatively sluggish and the force of their collisions relatively slight. But if the temperature is raised by placing the reaction flask on a burner for a time, the reaction might begin.

Because the outer portions of all molecules are composed of clouds of electrons, molecules tend to repel each other. Only when the force of their collisions is great enough to overcome the repulsion of their electron clouds will two reactive substances react. The amount of energy needed to overcome such repulsive forces sufficiently for a reaction to begin is the reaction's **activation energy.**† In an exergonic reaction, once the reaction has begun, sufficient heat is liberated by the reaction to keep the process going without the need for further input of energy from outside. If the reaction is endergonic, however, not only must activation energy be supplied, but a supply of energy must be continued for as long as the reaction is to occur. The amount of the bonding energy of the reactants and the amount of activation energy needed for a reaction to begin are independent of whether the reaction is exergonic or endergonic.

***Reversible Reactions and Chemical Equilibria.*** Almost all chemical reactions occurring in living systems are reversible; that is, it is possible for them to proceed both to the left and to the right as written. Reversibility is often indicated by two half-arrows, one pointing each way:

$$A + B \rightleftharpoons AB + \text{energy}$$

Reversible reactions are exergonic or endergonic, depending on the direction in which the reaction proceeds.

* The terms endothermic and exothermic are sometimes used, but their frame of reference is limited to energy in the form of heat. Because the energy involved in biochemical reactions includes other forms as well, the broader terms exergonic and endergonic are preferred.

† In reactions involving the breaking of chemical bonds, the activation energy also includes the amount of energy sufficient to overcome the bonding energy.

Reversible reactions rarely proceed to completion in either direction. Before all the reactants are converted to products, some of the products (by convention shown on the right side of the equation) begin to take part in the reverse reaction, and the supply of reactants (on the left) is replenished. Eventually, reversible reactions reach a state of **chemical equilibrium,** the point at which the reaction rate is the same in both directions.

For example, in an exergonic reaction that began with equal parts of reactants and products in a solution, the reaction would proceed, in conventional notation, from left to right until somewhat more than half the material was present in the form of products. The point at which equilibrium is reached differs from one reaction to another. In one exergonic reaction, for example, equilibrium concentrations might be 85 percent products and 15 percent reactants:

$$\underset{15\%}{A + B} \rightleftharpoons \underset{85\%}{AB}$$

If, instead of mixing equal amounts of reactants, we dissolved a sample composed of 95 percent *AB* and 5 percent *A* and *B* in water, a reaction would occur in the reverse (right to left) direction until the 15 percent *A* and *B* and 85 percent *AB* equilibrium concentrations were reached.

A particular reversible endergonic reaction might be represented as:

$$\underset{95\%}{P + Q} \rightleftharpoons \underset{5\%}{PQ}$$

Such a reaction might normally produce so little product as to be useless for most purposes. Even if the reaction were exergonic and its equilibrium point were slightly to the right—for example, at concentrations of 49 percent and 51 percent—it might proceed so slowly as to be equally useless. Many reactions of great importance to life are either endergonic or only slightly exergonic. However, the amount of products formed by such reactions can be increased in two ways. Both methods require the expenditure of energy.

***Coupled Reactions.*** The amount of products of an endergonic reaction can be increased in two ways, both of which exemplify the chemical law of mass action*:

1. *By increasing the amount of reactants present.* Increasing the amount of reactants increases the absolute amount of products accordingly. The percentages of reactants and products, of course, remain the same, but if the amount of reactants is doubled, so is the amount of products. Increasing the amount of products by adding to the reactants is referred to as "pushing" the reaction in the forward direction.
2. *By removing the products of the reaction as rapidly as they are produced.* In the example with equilibrium concentrations of 95 percent reactants and 5 percent products, the 5 percent concentration would never be achieved if the products were removed as rapidly as they formed. The reaction would, in such a case, continue to flow in the "forward" (left to right) direction. Increasing the amount of products by removing products is known as "pulling" the reaction in the forward direction.

* According to the law of mass action, as every chemical reaction reaches equilibrium at a given temperature there exists a definite ratio, specific for the reaction, of reactants and products.

To obtain products from endergonic reactions, living systems have evolved enzymes that catalyze both "pushing" and "pulling" mechanisms, resulting in what are termed **coupled reactions.** Enzymes promote coupled reactions by providing active sites at which coupled reactions occur side-by-side. Reactions with equilibrium points far to the left are coupled to exergonic reactions. With the pushing mechanism, an exergonic reaction produces a high concentration of a product that is a key reactant in the endergonic reaction. The continuing buildup of reactants drives the endergonic reaction to the right. With the pulling mechanism, a key product of the endergonic reaction is a reactant in the exergonic reaction to which it is coupled. The small amount of product is thus used up by the exergonic reaction as soon as it is produced. The endergonic reaction never reaches equilibrium and continues in the forward direction.

## IMPORTANT BIOCHEMICAL REACTIONS

The same thermodynamic principles underlie all chemical reactions, including those of greatest importance in biology. The most significant biochemical reactions are acid-base neutralizations, oxidation-reduction reactions, and dehydration-hydrolysis reactions.

***Acid-base Neutralization Reactions.*** Ionic bonds are formed between two elements when one yields electrons and the other accepts them. Ionic bonding underlies the chemical reactions that occur between the substances known as acids and bases.

Acids can be defined as substances that undergo ionization in aqueous solution, with a release of excess hydrogen ions. Acids that tend to ionize completely in dilute solutions are referred to as strong acids; weak acids ionize only slightly. For example, when hydrogen chloride (HCl), a strong acid, is dissolved in water, it ionizes by dissociating into hydrogen ions ($H^+$) and chloride ions ($Cl^-$):

$$\underset{\text{hydrochloric acid}}{HCl} \longrightarrow \underset{\text{hydrogen ions}}{H^+} + \underset{\text{chloride ions}}{Cl^-}$$

The tendency to dissociate into ions when dissolved in water is characteristic of any compound or part of a compound that is held together by ionic bonds.

A hydrogen ion, of course, is a hydrogen atom without its only electron; that is, it is a proton and thus has a positive charge. In aqueous HCl, the hydrogen electron is captured by the chlorine atom, giving the chlorine atom one extra negative charge, thus converting it to a chloride ion.

Bases can be defined as substances that ionize in water with the release of excess hydroxide, or hydroxyl, ions ($OH^-$) in aqueous solution. If a base, or alkali, is dissolved in water, it also undergoes ionization, as does an acid. But instead of hydrogen ions, it releases hydroxide ions. For example, the strong base sodium hydroxide, also known as caustic soda or lye, dissociates as follows:

$$\underset{\text{sodium hydroxide}}{NaOH} \longrightarrow \underset{\text{sodium ions}}{Na^+} + \underset{\text{hydroxide ions}}{OH^-}$$

The hydroxide ions are negatively charged, in this case by having captured electrons from sodium, leaving the sodium atoms positively charged.

When an acid and base are mixed, they do not merely ionize. In addition, the $H^+$ and $OH^-$ ions neutralize each other to the extent that they are present in equivalent amounts. The several specific properties

## PANEL 2-1

# pH and Buffers

By convention, the acidity, neutrality, or alkalinity of a solution is usually given in terms of its pH value, which is a way of stating its hydrogen ion concentration. The pH scale is based on the slight degree to which pure water dissociates into hydrogen ions and hydroxide ions: it has a concentration of $1 \times 10^{-7}$ hydrogen ions per liter and, of course, an equal number of hydroxide ions ($OH^-$), the other ion produced when water dissociates. Adding a moderate amount of acid to distilled water raises the hydrogen ion concentration slightly, for example, to $1 \times 10^{-6.5}$ or greater. Addition of a base to distilled water, on the other hand, lowers the hydrogen ion concentration and provides an excess of hydroxide ions. A solution with a hydrogen ion concentration of $1 \times 10^{-7.5}$ or less is slightly alkaline.

To avoid working with such cumbersome figures, the term *pH* was defined as the negative logarithm of the hydrogen ion concentration. (Using the negative of the negative value converts it to a positive value.) Thus, neutral solutions, such as distilled water, have a pH of 7; acidic solutions have pH values of less than 7.0; and alkaline solutions have pH values of greater than 7.0. The pH scale ranges from 0.0 to 14.0. In biological systems, most pH values are close to neutrality (pH 7). The pH within most cells is slightly acidic, about 6.8, and the pH of the body fluids, including the blood, of higher animals is about 7.4. Exceptional values are found in urine and gastric (stomach) juice. Depending on diet and other factors, urine may be as acidic as pH 4.0; with its high content of hydrochloric acid, human gastric juice has a pH of about 1.0.

The pH of cytoplasm and of body fluids is crucial to life. In the slightly acidic pH of the cytoplasm, weak organic acids and their salts tend to ionize slightly and are thus able to participate in reactions that could not occur at a pH above 7. The stability of the pH levels of 6.8 for cytoplasm and 7.4 for blood is maintained by **buffers,** which are substances that combine with $H^+$ ions when the hydrogen ion concentration increases or that release $H^+$ ions to combine with $OH^-$ ions when the hydrogen ion concentration decreases. Without buffering action, pH stability would be affected by the many metabolic reactions in which $H^+$ ions are released or bound, as well as by the ingestion of acidic or alkaline foods.

The pH values of some common substances. The two right-hand columns show the concentration of the $H^+$ and $OH^-$ ions at the indicated pH values.

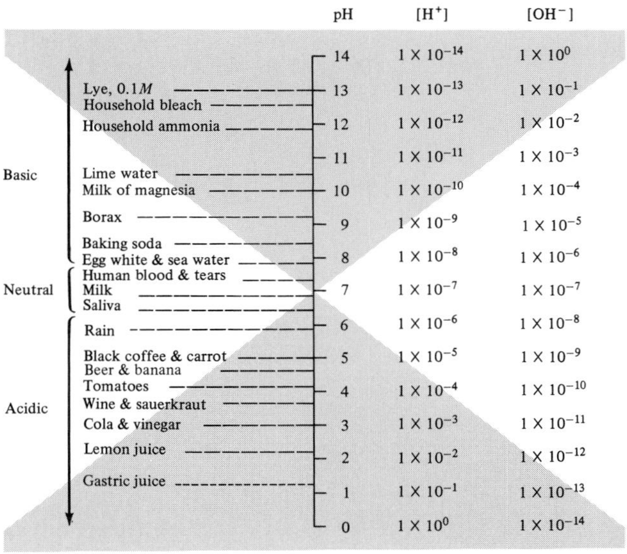

of acids and bases are thereby lost. One property of acids is that they react with reactive metals, releasing hydrogen gas:

| | | | | | | |
|---|---|---|---|---|---|---|
| $2HCl$ | + | $Mg$ | $\longrightarrow$ | $MgCl_2$ | + | $H_2$ |
| hydrochloric acid | | magnesium | | magnesium chloride (a salt) | | hydrogen |

When an acid is mixed with a base, a similar reaction occurs, except that the acid's hydrogen ions and the base's hydroxide ions (which have a strong affinity for each other) unite to form water:

| | | | | | | |
|---|---|---|---|---|---|---|
| $HCl$ | + | $NaOH$ | $\longrightarrow$ | $NaCl$ | + | $H_2O$ |
| hydrochloric acid | | sodium hydroxide | | sodium chloride | | water |

***Oxidation-Reduction Reactions.*** Another type of chemical reaction of prime importance to living systems is that in which a molecule or atom loses one or more electrons. Because the first known examples of this type of reaction involved substances that gained oxygen while losing electrons, the reaction came to be known as **oxidation.** Its reverse, the gain of electrons, is called **reduction.** In time, it became known that oxygen did not have to be present for the essential part of the reaction to occur. By then, however, the term oxidation was in wide use and is retained despite its inappropriateness.

A substance can lose, or donate, electrons only if another substance is at hand to accept them. Therefore, all oxidations are accompanied by corresponding reductions and vice-versa. When both aspects of such reactions are of concern, they are referred to as *oxidation-reduction,* or *redox,* reactions. With reference to the electron donor, the reaction is an oxidation. With reference to the electron acceptor, the reaction is a reduction.

Most oxidations are accompanied by the release of energy, but different compounds release different amounts of energy depending on the size of a compound's potential energy. For example, consider the situation of a number of boulders at various levels on the side of a hill. The potential energy of each boulder is proportional to both its mass and its distance from the bottom of the hill. Similarly, the amount of energy that can be derived from a compound depends on how much energy the compound possessed originally and how much oxidation it has already undergone.

The amount of energy that can be derived by oxidation of a substance can be expressed in terms of the tendency of that substance to donate electrons to a given standard electron acceptor—in boulder terms, an amount proportional to its mass and its elevation from the bottom of the hill. Electrons always tend to flow "downhill" from substances that yield them readily to those that accept them from the "stronger donors." The tendency of a substance to donate electrons is referred to as its *redox potential* or, less frequently, its *electron pressure.* In living cells, electrons are often rapidly passed along a series of donors and acceptors, with the energy released at each oxidation given off as heat or used to drive various energy-requiring reactions that are coupled to the oxidation.

The great importance of oxidation-reduction reactions for living systems lies in the fact that oxidations are generally energy-yielding reactions. Most reductions require an input of energy to proceed. Although

# PANEL 2-2

# Some Useful Chemical Conventions

A kind of chemical shorthand is used to simplify the written expression of complex chemical formulas.

## STRUCTURAL FORMULAS

The empirical formula of a compound—for example, $C_2H_4O_2$ for acetic acid—does not convey enough information for most purposes. A structural formula is then used. For example, the structural formula of acetic acid is

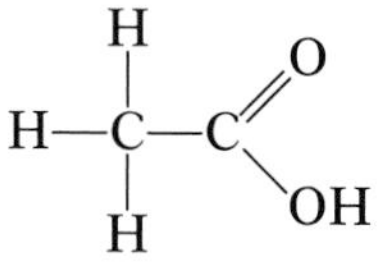

or, in abbreviated form, $CH_3COOH$.

## COVALENT BOND REPRESENTATIONS

Covalent bonding between two atoms in a compound can be represented in three ways:

1. A pair of dots between the two atoms can be used to represent the shared pair of electrons:

| | |
|---|---|
| H:O:H | O::C::O |
| water | carbon dioxide |

2. A more common convention is to represent each shared pair of electrons as a dash. The sharing of a second or third pair of electrons is represented by a double or triple dash:

H—O—H

water

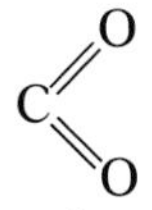

carbon dioxide

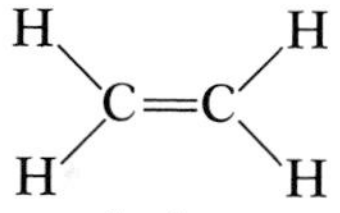

ethylene

H—C≡C—H

acetylene

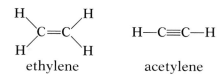

3. When the covalent nature of certain bonds is well known, some dashes may be omitted, as in ethylene: $H_2C{=}CH_2$.

## COMMON CHEMICAL GROUPS

Chemical groups are two or more atoms that tend to remain together during reactions, for example, the phosphate group (—$PO_4$), the methyl group (—$CH_3$), and so on. Chemical groups are not compounds but portions of compounds. Chemical groups common in biological reactions include the following:

—OH

alcohol group

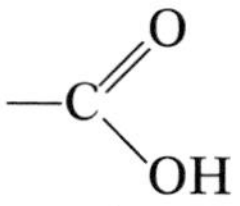

carboxylic group

## THE SYMBOL *R*

The symbol *R* is used in so-called general formulas to denote any atom or group or in some cases, a particular kind of group. Thus,

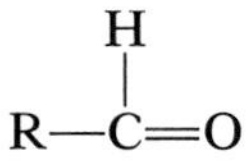

is an aldehyde, and

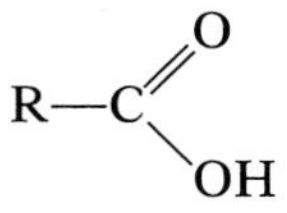

is a carboxylic acid. In these cases, *R* can represent a hydrogen atom or any kind of side chain.

all oxidations involve loss of electrons, there are three ways in which a substance can undergo oxidation:

1. By addition of oxygen, as in combustion:

$$\underset{\text{ethanol}}{2CH_3{-}CH_2OH} + \underset{\text{oxygen}}{O_2} \longrightarrow \underset{\text{acetic acid}}{2CH_3{-}COOH} + \underset{\text{water}}{2H_2O}$$

This is what happens when wine "goes sour" and turns to vinegar.

2. By removal of hydrogen, as in the oxidation of a hypothetical substance, $R$*—OH:

$$\underset{\text{reduced substance}}{R{-}\overset{\displaystyle H \atop |}{O}H} \longrightarrow \underset{\text{oxidized substance}}{R{=}O} + \underset{\text{hydrogen ions}}{2H^+} + \underset{\text{electrons}}{2e^-}$$

3. By direct removal of electrons, as with any substance capable of yielding electrons:

$$\underset{\text{ferrous (iron) ion (reduced form)}}{Fe^{2+}} \longrightarrow \underset{\text{ferric ion (oxidized form)}}{Fe^{3+}} + \underset{\text{electron}}{e^-}$$

***Dehydration-Hydrolysis Reactions.*** A third type of chemical reaction of great importance to life is a condensation reaction called **dehydration synthesis.** (You will see in Chapters 7 and 9 how these reactions are essential to life.) Like oxidation-reduction, it is a reversible reaction; the reverse reaction is known as **hydrolysis.** In dehydration synthesis, a large molecule is synthesized from two or more smaller components. These smaller units become joined by covalent bonding during dehydration, the removal of one or more molecules of water. An example of dehydration synthesis is the formation of an ester from an organic (carbon-containing) acid and an alcohol:

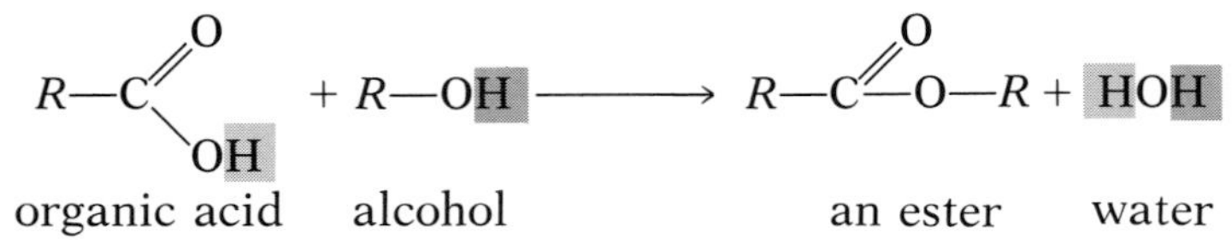


One hydrogen (shown in color) of the water molecule came from the acid and the other (shown in grey) from the alcohol. A compound formed from an organic acid and an alcohol is known as an **ester.** Certain esters are responsible for the pleasant odor of many fruits.

Some esters spontaneously undergo hydrolysis in the presence of water:

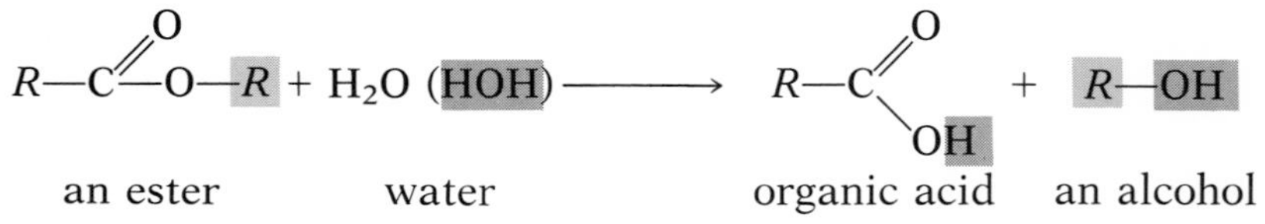


The reaction of hydrolysis in such cases can be thought of as a splitting (Greek *lysis*) of a compound *by* water (Greek *hydro*). Like oxidation, hydrolysis is generally an energy-yielding reaction. The amount of energy released varies with the compound.

* The symbol $R$ and certain other conventions commonly used in chemistry are discussed in Panel 2-2.

## Summary

Matter is anything that occupies space, and the amount of matter in a given object is its mass. All forms of matter may be classified as either mixtures or pure substances. Mixtures are characterized by variable composition and by the fact that they can be separated by physical means. A homogeneous mixture is uniform but can be separated without chemical change into two or more substances. A heterogeneous mixture is not uniform: identical samples cannot be obtained from each and every location of the sample.

A pure substance is a homogeneous form of matter whose composition and intrinsic properties are uniform throughout and that cannot be separated into two or more forms of matter without chemical changes. If two or more substances cannot be obtained from one pure substance, then the substance is classified as an element. Elements are the building blocks of all other pure substances.

Modern atomic theory has shown that the atoms of elements are composed of three primary building blocks, called protons, neutrons, and electrons. Protons are positively charged, neutrons are uncharged, and electrons are negatively charged. The protons and neutrons are concentrated in a very small volume within the atom known as the atomic nucleus. The atomic number of an element is the number of protons in the nucleus. Atoms of the same atomic number (number of protons) but different mass number (number of protons plus neutrons) are called *isotopes.* Some isotopes are unstable and break down spontaneously, giving off three types of emissions: alpha particles, beta particles, and gamma rays. Knowledge of the rate of radioactive decay of certain isotopes can be used as a clock to determine the age of rocks and of fossilized organisms.

Electrons surrounding atomic nuclei move within volumes of space called electron *orbitals.* Orbitals represent distinct energy states in which electrons exist within an atom. Orbitals are arranged in groups within energy levels. Movement of electrons from one energy level to another results in the addition or loss of discrete units of energy called *quanta.*

Two or more atoms may be joined to form a molecule, with chemical properties that differ from those of the original atoms. When two atoms come close enough for their electron orbitals to overlap, an energy exchange may occur, forming a chemical bond in which the nucleus of each atom attracts and holds electrons of the other atom. This rearrangement of electrons may occur in one of two ways: one atom can give up one or more of its electrons to the other, resulting in the formation of a polar, or ionic bond; or each atom can share one or more of its electrons with the other to form a nonpolar, or covalent bond.

Both polar and nonpolar molecules can be held together by such weak, noncovalent bonds as van der Waals forces and hydrogen bonds.

Chemical reactions can be either spontaneous (exergonic), requiring no input of energy, or nonspontaneous (endergonic), requiring a continuous outside source of energy. Most reactions at room temperatures require some input of energy (activation energy) to get them started. Chemical reactions in living systems are facilitated by enzymes, each of which functions most efficiently within its optimal temperature and pH range.

Almost all chemical reactions occurring in living substances are reversible. Moreover, they rarely go to completion in either direction. As in any chemical reaction, the amount of product formed can be increased by increasing the amount of reactants present or by removing products of the reaction as they are produced. Important endergonic reactions are able to proceed when coupled with exergonic reactions that perform one of these functions.

Among the types of chemical reactions of greatest importance to biological systems are acid-base neutralizations, oxidation-reduction reactions, and dehydration syntheses. In neutralizations, a mixture of acids and bases becomes neutralized when both compounds are present in equivalent amounts. Oxidation-reduction reactions involve the loss of electrons by one substance to another substance, which accepts them. In dehydration synthesis, a large molecule is synthesized from smaller components during the removal of water.

## Recommended Readings

BAKER, J. W., and G. E. ALLEN. 1981. *Matter, Energy, and Life,* 4th ed. Addison-Wesley, Reading, Mass. This paperback contains highly readable descriptions of the structure of matter and the formation of molecules.

BROWN, T. L., and H. E. LEMAY, JR. 1985. *Chemistry: The Central Science,* 3d ed. Prentice-Hall, Englewood Cliffs, N.J. An excellent treatment of general chemistry at the college level. This book has an especially good discussion of the development of modern atomic theory.

CURTIS, T. SEARS, and CONRAD L. STANITSKI. 1983. *Chemistry for Health-Related Sciences,* 2d ed. Prentice-Hall, Englewood Cliffs, N.J. An excellent introduction to biological chemistry.

EICHER, D. L. 1976. *Geologic Time,* 2d ed. Prentice-Hall, Englewood Cliffs, N.J. This small paperback has a good description of the various radioactive methods used for dating biologic matter and rocks.

# 3

# Carbon and the Molecules of Life

OUTLINE



Cellular structure and function depend on the organization of four major classes of large biochemical substances: carbohydrates, lipids, proteins, and nucleic acids composed of carbon (C), hydrogen (H), nitrogen (N), oxygen (O), sulfur (S), and phosphorus (P). The discussion of atomic structure, chemical bonding, and chemical reactions in Chapter 2 provides a basis for understanding why living systems are composed primarily of these six elements and why carbon plays a unique role in the organization of living things. These elements are not the most abundant in nature, yet, they are evolutionarily important and play important roles in living systems.

## ATOMIC SIZE AND BOND STRENGTH IN BIOLOGICAL MOLECULES

Carbon, hydrogen, oxygen, and nitrogen, which are among the smallest and most abundant elements in living systems, possess chemical characteristics that enable them to form highly stable molecules. For example, they generally form stronger chemical bonds than do larger atoms. The small number of electrons in small atoms provide only minimal repulsive forces to interfere with the attraction of the nuclei for electrons of other atoms and thus with the formation of bonds. In molecules made up of small atoms, the bonding orbitals overlap to a greater degree than in most molecules, resulting in a stronger bond. Larger atoms, such as phosphorus and sulfur, have more electrons at energy levels below their outer, or **valence,** shells. These electrons create repulsive forces that limit the effectiveness of the nucleus in attracting other atoms. As a result, bonds formed by larger atoms tend to be weaker and more easily broken than those formed by smaller atoms. As you read this book, you will come to realize the importance of molecular shape, or configuration, and bond strength in determining the behavior of molecules.

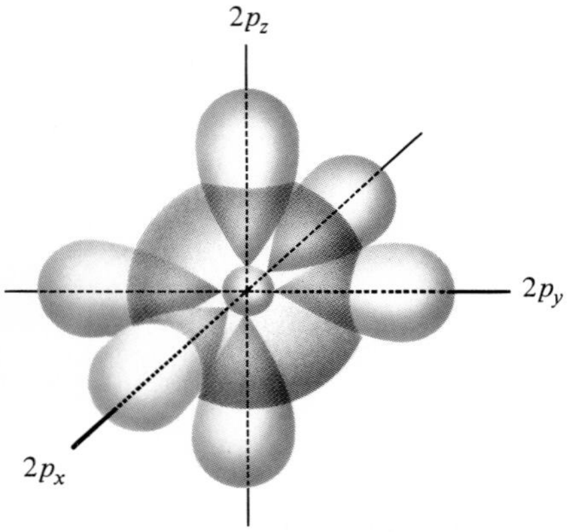


**FIG. 3-1** Orbital configuration of a carbon atom.

## CARBON: THE "BACKBONE" OF BIOLOGICAL MOLECULES

In addition to its small size, the carbon atom possesses other characteristics that make it capable of serving as the backbone of the most important molecules in living systems. The carbon atom has a unique arrangement of electrons that enhances its ability to form strong bonds and enables it to bond with a variety of atoms. Carbon, atomic number 6, has two electrons in its first energy level and four in its second (Fig. 3-1); two of the latter are in the 2*s* orbital, and each of the others occupies one of the three available 2*p* orbitals. Based on this orbital configuration, we might expect carbon to form up to two covalent bonds with other atoms, since there are two unpaired electrons in two of the *p* orbitals. However, experimental evidence shows that when carbon undergoes a chemical reaction it normally forms up to four covalent bonds. It is believed that when a carbon atom forms four bonds—for example, with four hydrogen atoms—there is a shift in the 2*p* and 2*s* electron orbitals of the carbon atoms, resulting in the formation of four equivalent orbitals. Apparently, one of the 2*s* electrons is "promoted" to the vacant 2*p* orbital. Carbon is thus effectively provided with an outer shell of four electrons, with four vacancies and four electrons to share with other atoms. As Figure 3-2 illustrates, each of the hybrid orbitals extends toward what would be the corners of a tetrahedron. A tetrahedral configuration provides the strongest, most stable bonds. Because the four hybrid orbitals are derived from one 2*s* and three 2*p* orbitals, they are termed ***$sp^3$* hybrid orbitals.** The formation of $sp^3$ hybrid orbitals enables the carbon atom to form four equivalent bonds in a molecule, as when hydrogen interacts with carbon to form methane (Fig. 3-3). Because the hybrid orbitals extend farther from the carbon nucleus than do the 2*s* and 2*p* orbitals, they overlap more with the orbitals of the atoms with which they are bonding. This overlap results in strong bonds and in molecules of great stability. It is thus not surprising that carbon compounds are relatively inert.

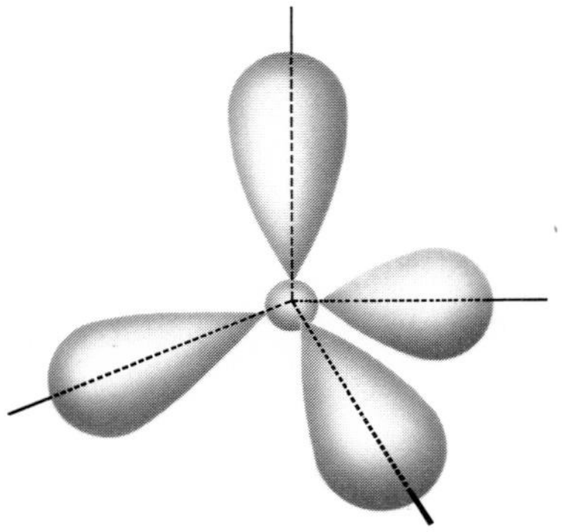

**FIG. 3-2** Hybrid $sp_3$ orbitals in carbon, formed from one *s* and three *p* orbitals.

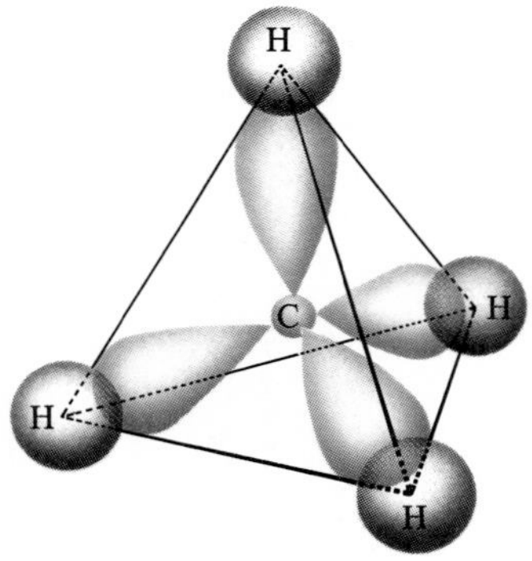


**FIG. 3-3** Geometric configuration of a methane molecule, formed by the sharing of electrons of four hydrogen 1*s* orbitals with the four $sp_3$ orbitals of a carbon atom.

***Carbon-to-Carbon Bonds.*** Because the carbon atom has a strong tendency to form covalent bonds, it can combine with a wide variety of atoms, including other carbon atoms. Straight, branching, or circular carbon atom chains of varying length can be formed. Carbon atoms can also react in various ways: they can share two or three electron pairs to form double (C=C) or triple (C≡C) bonds. Carbon, nitrogen, oxygen, phosphorus, and sulfur—five of the six most common elements in living systems—are the atoms that most commonly participate in the formation of such multiple bonds. Since multiple bonds are stronger than single bonds between the same atoms, molecules that possess them are very stable, a characteristic essential for some constituents of living systems.

### METABOLIC ROLES OF PHOSPHORUS AND SULFUR

Just as carbon, hydrogen, oxygen, and nitrogen possess properties that particularly suit them to form the stable molecules that play structural roles in living systems, phosphorus and sulfur possess unique properties that enable them to play important roles in the metabolism of cells. Because of their larger size and the correspondingly larger number of electrons below their valence shells, phosphorus (atomic number 15) and sulfur (atomic number 16) form somewhat weaker bonds than do carbon, hydrogen, nitrogen, and oxygen. In addition, phosphorus- and sulfur-containing molecules are unstable because electrons in the valence shells of both elements—three in phosphorus and four in sulfur—can rearrange so that additional unfilled third orbitals become available for covalent bonding. The resulting formation of additional bonds contributes to the instability of phosphorus- and sulfur-containing compounds because these bonds readily break down and release energy. By this mechanism, the cells in living systems are supplied with energy, without reactions that might destroy other important molecules. Thus, the weakness and instability of their bonds permit phosphorus- and sulfur-containing compounds to play a major role in the transfer of energy in living systems.

## CARBOHYDRATES

Carbohydrates serve a variety of functions in living systems. They are an important source of energy and function as structural elements within the cell. The name **carbohydrate** ("hydrate of carbon") is derived from the hydrogen-to-oxygen ratio in these compounds, which is, generally, 2:1, the same as that in water. Carbohydrates do not actually contain hydrated carbon; they merely happen to have the same ratio of hydrogen to oxygen atoms as water. Nevertheless, since the ratio of carbon to hydrogen to oxygen atoms is so constant, the general formula $(CH_2O)_n$ is used to describe carbohydrates; $n$ is the number of $CH_2O$ units.

Carbohydrates are divided into three major groups—monosaccharides, disaccharides, and polysaccharides—according to the number of sugar units in each molecule.

### MONOSACCHARIDES: THE ENERGY SUGARS

**Monosaccharides** (single sugars) cannot be broken down into smaller carbohydrate units by hydrolysis, the splitting of a larger molecule into smaller ones by the chemical addition of water. Monosaccharides are classified by the number of carbon atoms they contain. Thus, the 3-, 4-,

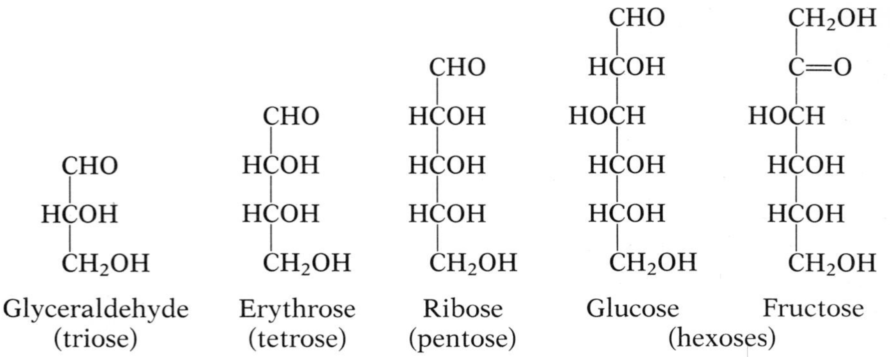


**FIG. 3-4** Monosaccharides with various numbers of carbon atoms.

5-, and 6-carbon monosaccharides whose structural formulas are shown in Figure 3-4 are called **trioses, tetroses, pentoses,** and **hexoses,** respectively. The most important monosaccharides from a biological standpoint are two hexoses, glucose and fructose, and two pentoses, ribose and deoxyribose.

Glucose and fructose are of central importance in the processes of energy transfer within cells and among living systems. Glucose is produced as a result of photosynthesis, discussed in Chapter 7. Plants and other photosynthetic organisms convert glucose into fructose and larger molecules, such as starch and cellulose. Animals, fungi, and other organisms, which are ultimately dependent on photosynthetic organisms for their survival, hydrolyze larger carbohydrate molecules into monosaccharides.

***The Structure of Glucose and Fructose.*** The chemical bonding in glucose ($C_6H_{12}O_6$), the most abundant monosaccharide, can be represented in two dimensions as shown in Figure 3-5. As noted in Panel 2-2, the

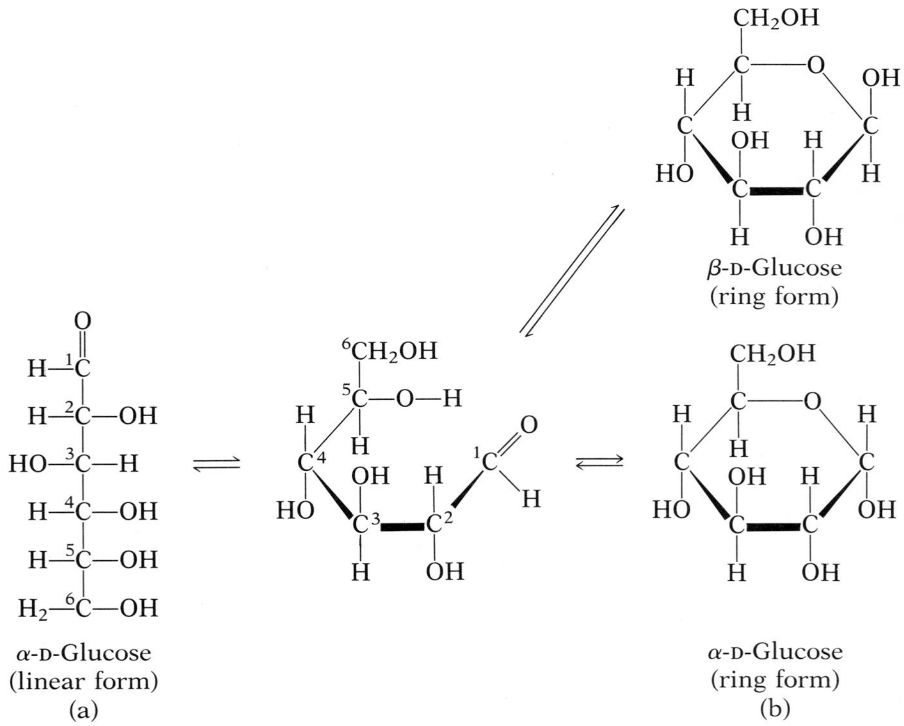


**FIG. 3-5** Linear and ring forms of glucose.

**FIG. 3-6** Linear and ring forms of fructose.

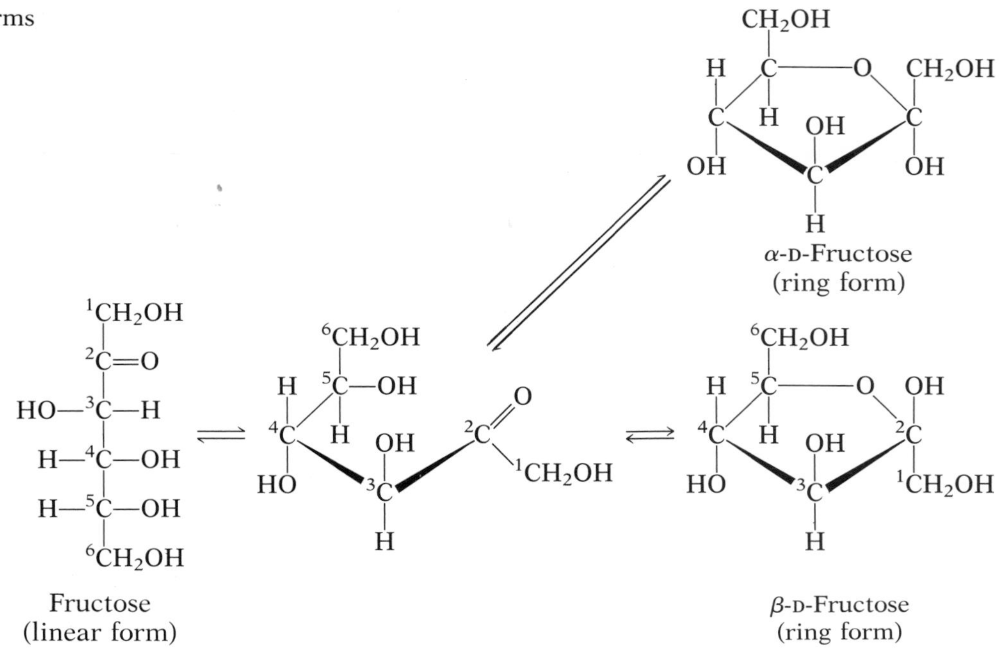


connecting lines between atomic symbols each represents a covalent bond, that is, a pair of shared electrons. The linear, open chain forms shown in Figure 3-5a exist only in dilute solutions. In solution, glucose tends to take on the two ring forms shown in figure 3-5b because of the tendency of the number 1 carbon atom to link up with the oxygen atom on the number 5 carbon atom. The ring formulas shown in Figure 3-5b are three-dimensional representations, with the heavier bonds at the bottom of the structures representing the part of the molecule nearest the reader. The hydrogen atoms and —OH groups connected to the carbon atoms are to be regarded as extending up and down in the plane of the paper, with the ring itself at right angles to them.

Fructose, another important hexose in living systems, is similar to glucose (Fig. 3-6). While glucose has a six-membered ring, fructose has a five-membered ring structure, formed by four of its carbon atoms and an oxygen atom. Fructose has the same number of carbon, hydrogen, and oxygen atoms as glucose, but because of their different structures, glucose and fructose have different chemical and physical properties; for example, they react differently, and fructose tastes sweeter than glucose. Compounds with the same chemical composition but different structural arrangements are called **isomers** of each other.

## DISACCHARIDES: THE TRANSPORT SUGARS

**Disaccharides** ("double sugars") are formed by the condensation of two monosaccharides. Three of the most common disaccharides are sucrose, maltose, and lactose. Sucrose (cane sugar), shown in Figure 3-7, is made up of a glucose unit linked to a fructose unit. Maltose (malt sugar) consists of two linked glucose units, while lactose (milk sugar) consists of linked units of glucose and galactose, another hexose (Fig. 3-8). All three disaccharides are formed from two monosaccharides by dehydration synthesis (see Chapter 2); each disaccharide can therefore be broken down into its component monosaccharides by hydrolysis. Although all

**FIG. 3-7** Dehydration synthesis of a sucrose molecule by the condensation of glucose and fructose.

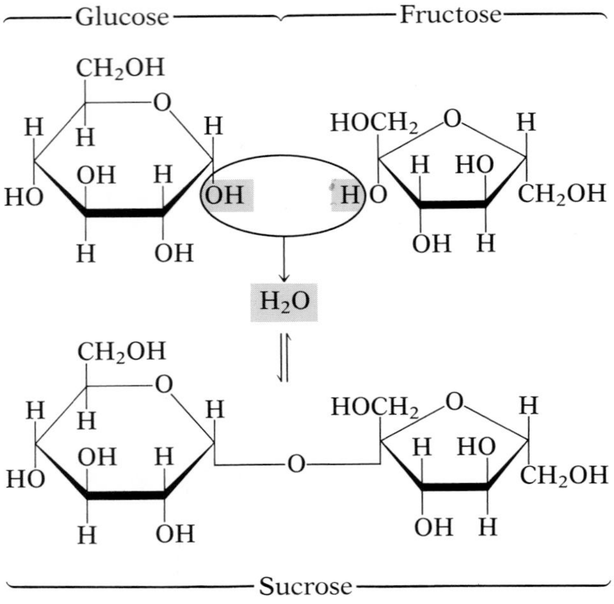


three of these molecules are sugars, their sweetness varies. Sucrose is about six times as sweet as lactose, about three times as sweet as maltose, slightly sweeter than glucose, and only half as sweet as fructose.

Although the monosaccharide glucose is the most commonly transported carbohydrate in vertebrates, most other organisms transport carbohydrates in the form of disaccharides. For example, plants produce sucrose in their photosynthetic cells, located primarily in the leaves, and transport it to other parts of the plant to be used as an energy source or converted to starch or cellulose. Lactose, found only in milk, is the form of sugar provided by a mammalian mother to its offspring (Fig. 3-9). Carbohydrate is transported through the blood of most insects as trehalose, another disaccharide.

## POLYSACCHARIDES: THE STORAGE AND STRUCTURAL SUGARS

Most carbohydrates stored in living systems are polysaccharides of high molecular weight. **Polysaccharides** are long chains, or polymers, of monosaccharides, with glucose the most common primary sugar unit. The most important polysaccharides in living systems are starch, glyco-

**FIG. 3-8** Structures of lactose and maltose.

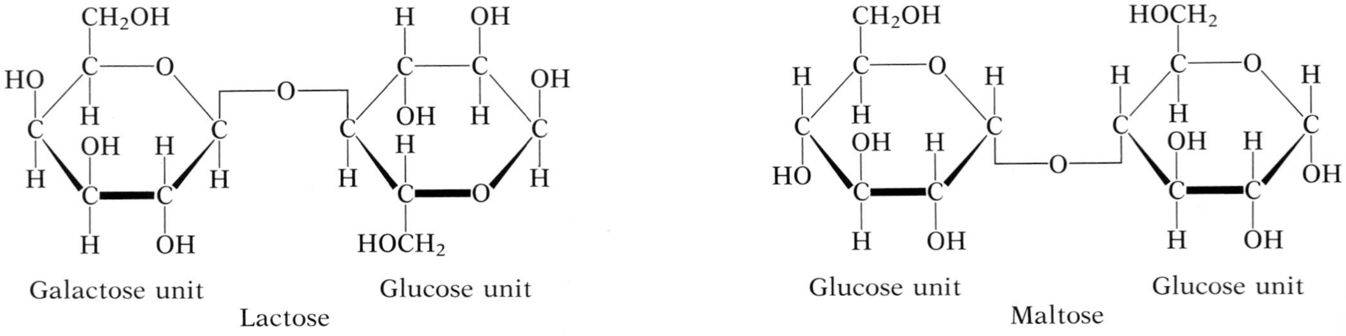

**FIG. 3-9** Cow feeding her calf milk, which contains lactose, a primary energy source for the developing calf.

gen, and cellulose. Although all three are composed of glucose units, these polysaccharides exhibit different properties because their glucose units are linked in different ways.

***The Starch of Seeds and Tubers.*** Plants store food primarily in the form of carbohydrates called **starches.** Cereal grains, nuts, and tubers (thickened, underground stems such as the potato), have high starch content (Fig. 3-10). The stored food in seeds, such as grains and nuts, provides the nutrients needed for seed germination; the starch in tubers gives the plant an early start on growth in the spring. Plant storage products also serve as major energy sources for many kinds of animals, including humans. The digestive tracts of animals contain enzymes that hydrolyze starches to glucose, which then enters the blood or other body fluids.

**FIG. 3-10** The potato, a primary source of nutrient in the American diet. Rootlike, and often mislabeled as such, the potato is actually a tuber, part of the underground stem of the potato plant.

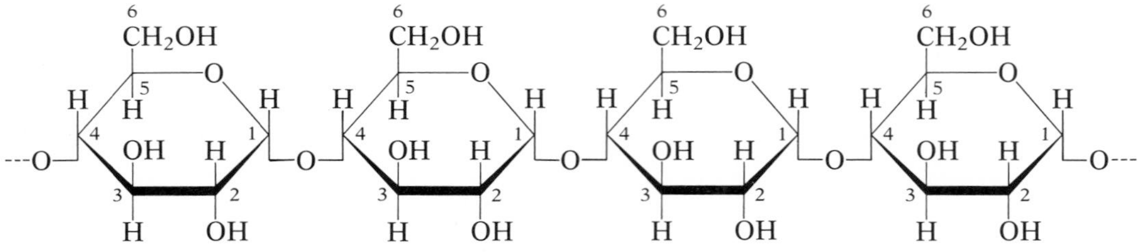

(a) Amylose

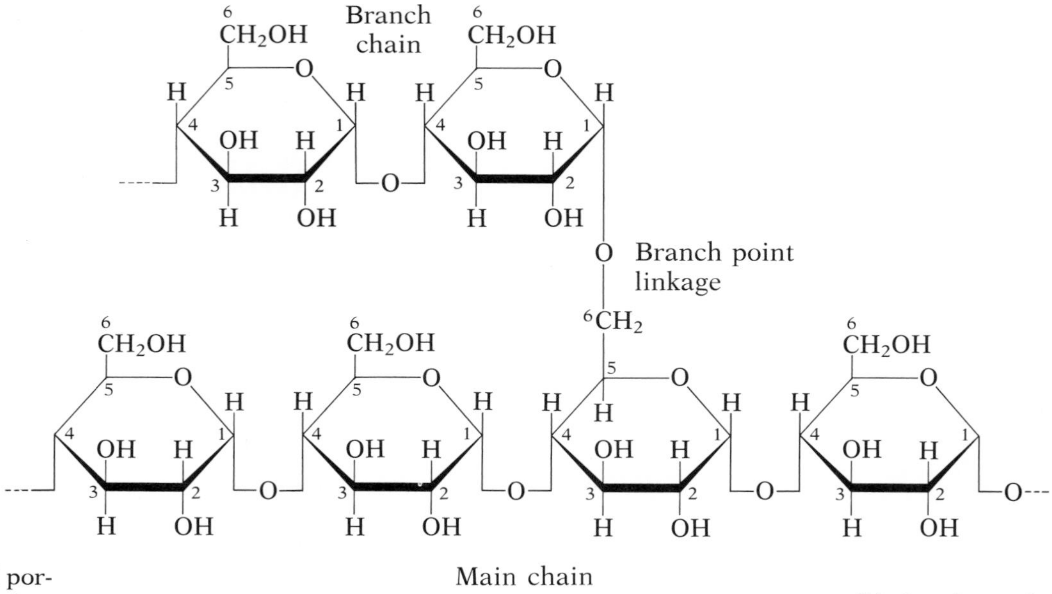


(b) Amylopectin

**FIG. 3-11** Structures of portions of the starch molecules (a) amylose and (b) amylopectin.

Starch usually exists as a mixture of two major types of glucose polymers. One of the two types, shown in Figure 3-11a, is an unbranched polymer of glucose units called **amylose;** the other (Fig. 3-11b) is a branched polymer called **amylopectin.** In amylose, the first carbon atom of each glucose ring is joined to the fourth carbon atom of the next glucose unit to form unbranched chains with molecular weights as great as 1 million. Amylopectin contains the same basic structure as amylose except that it has many branches of 24 to 30 glucose units formed by the bonding of the first and sixth carbon atoms of glucose molecules at the branch points. In both amylose and amylopectin, the individual chains of glucose units are not linear but, as shown in Figure 3-12, assume a helical configuration, with about six glucose units per complete turn of the helix. These helical chains do not form stable hydrogen bonds with neighboring starch chains but do form such bonds with other surrounding molecules. Consequently, starch is a nonfibrous polysaccharide that forms hydrated micelles in water.

***Glycogen.*** Sometimes called *animal starch,* glycogen is the main storage carbohydrate of animal cells. Its structure is similar to but more branched than that of starch. Glycogen molecules are usually very large and may be composed of as many as half a million glucose units. Glycogen is especially abundant in the liver and muscle cells of vertebrates

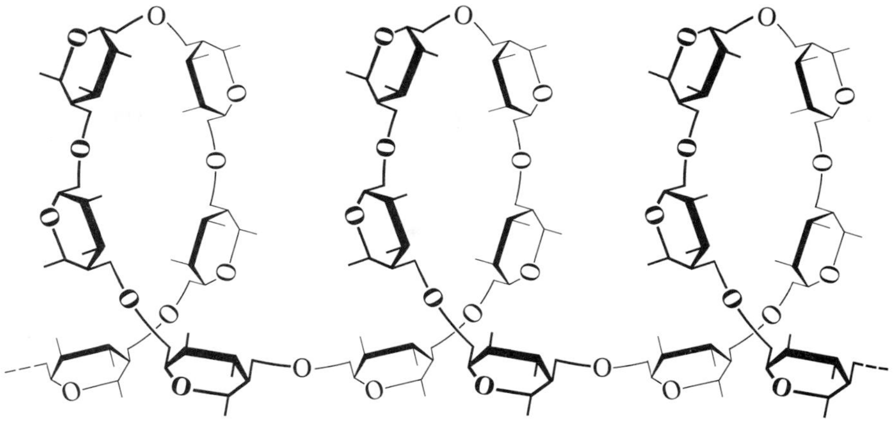

**FIG. 3-12** The helical nature of amylose and amylopectin chains.

(Fig. 3-13). In the liver, it serves as a stored source of glucose available to the bloodstream. As much as 10 percent of the wet weight of a mammalian liver may be glycogen. In mammalian muscle cells, glycogen serves as a readily available source of energy.

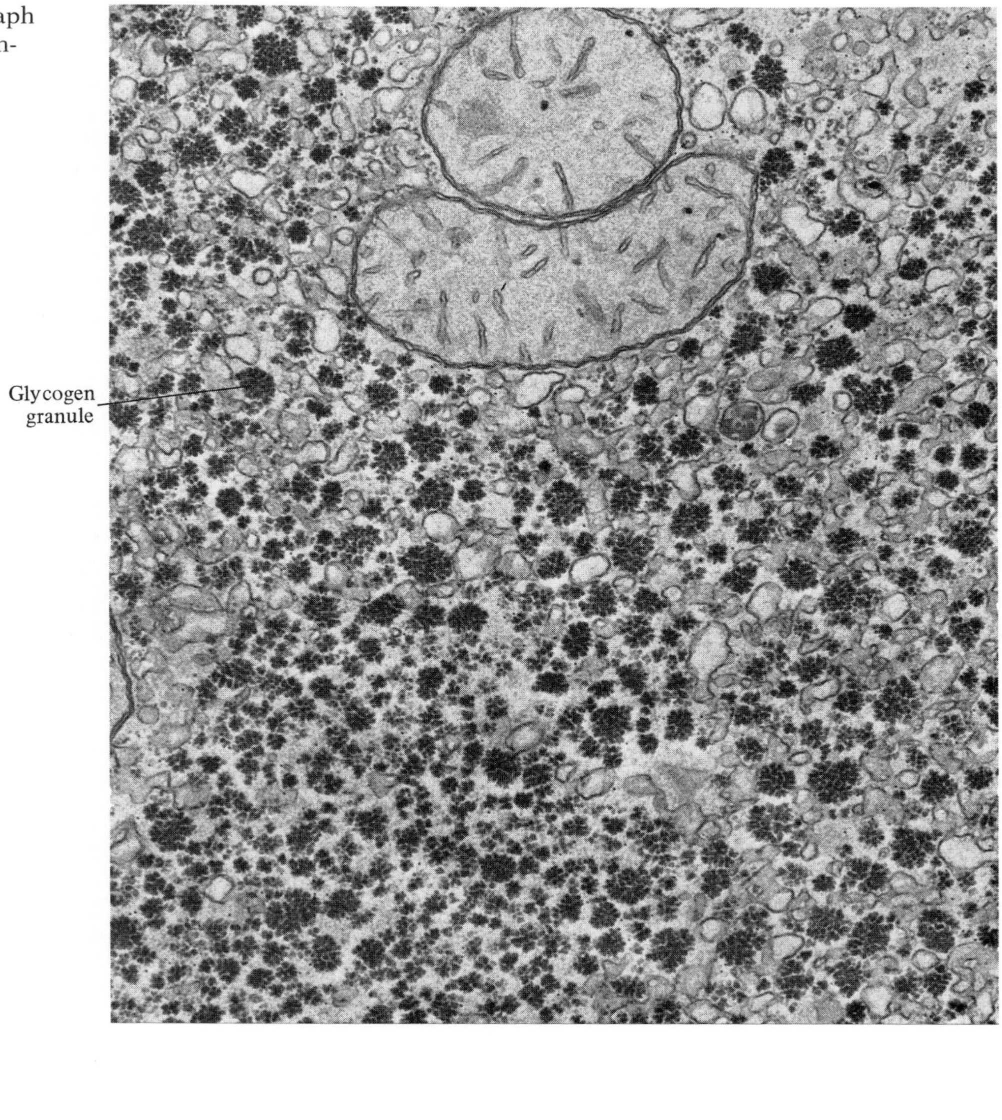


**FIG. 3-13** Electron micrograph of glycogen granules in a hamster liver cell.

***Cellulose.*** The polysaccharide cellulose is a major structural component of plant cell walls (Fig. 3-14). Wood is about 50 percent cellulose, dry leaves about 10 percent. Cotton fiber, which is about 98 percent cellulose, is the best industrial source of the pure material. Cellulose molecules are largely unbranched polymers of glucose units with molecular weights averaging more than 500,000 daltons. (The dalton is a unit of mass equal to the mass of one hydrogen atom or $1.6 \times 10^{-24}$ gram.) These molecules readily form hydrogen bonds between adjacent chains and are found in closely packed fibers of 100 to 200 parallel chains held together by a very large number of hydrogen bonds. Although each bond is weak, their combined strength is great. Such bonded chains are highly resistant to dissolving in water and in organic solvents. A comparison of Figures 3-15 and 3-9 shows that the bonds between the glucose units in cellulose are different from those in starch and glycogen; in cellulose the orientation of adjacent glucose units is reversed, while in starch all the glucose units are facing the same way. The linkage is more stable in cellulose.

THE INDIGESTIBILITY OF CELLULOSE The enzymes of animal digestive systems that hydrolyze starches are not effective on cellulose. As a result, this polysaccharide tends to pass through the digestive systems of most animals unchanged. Many bacteria and some protozoa, however, con-

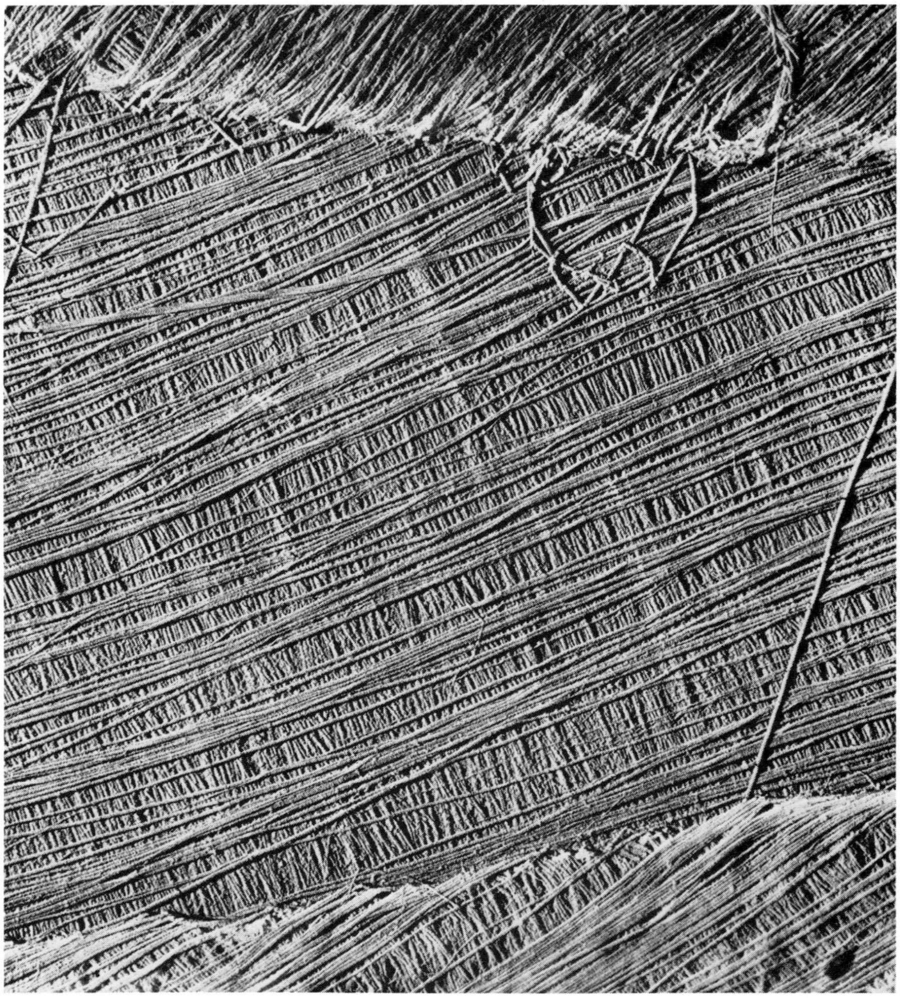

**FIG. 3-14** Cellulose fibrils in the cell wall of a green alga.

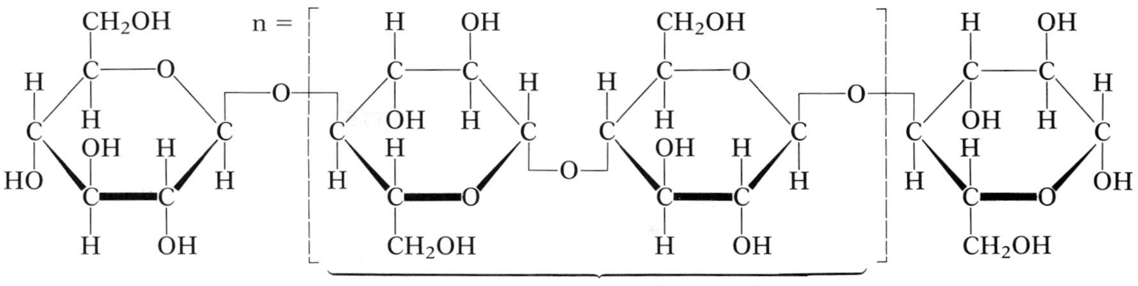

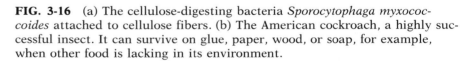

FIG. 3-15 Structure of a portion of a cellulose molecule in which n = repeating pair of glucose units.

tain enzymes that are able to hydrolyze cellulose to glucose. The presence of such bacteria and protozoans in the digestive tracts of grazing animals, such as cattle, horses, sheep, and deer, enables them to digest grass and other leaves, using the cellulose as an energy source (Fig. 3-16a). Termites and cockroaches rely on protozoans that live in their gut for the digestion of the wood they eat (Fig. 3-16b). There is evidence that these protozoans themselves may contain cellulose-digesting bacteria.

***Chitin: The Original Suit of Armor.*** Chitin is a structural polysaccharide that forms the tough exoskeletons of arthropods (Fig. 3-17) and the cell walls of molds. It is a linear, unbranched molecule consisting of a derivative of glucose, acetylglucosamine, joined by the same types of linkages found in cellulose. Chitin is difficult to hydrolyze, and few animals possess the enzymes to digest it.

## LIPIDS

Lipids comprise a diverse group of biochemical substances that are soluble in organic solvents such as acetone, benzene, ether, and chloroform. They are only slightly soluble or are completely insoluble in

**FIG. 3-16** (a) The cellulose-digesting bacteria *Sporocytophaga myxococcoides* attached to cellulose fibers. (b) The American cockroach, a highly successful insect. It can survive on glue, paper, wood, or soap, for example, when other food is lacking in its environment.

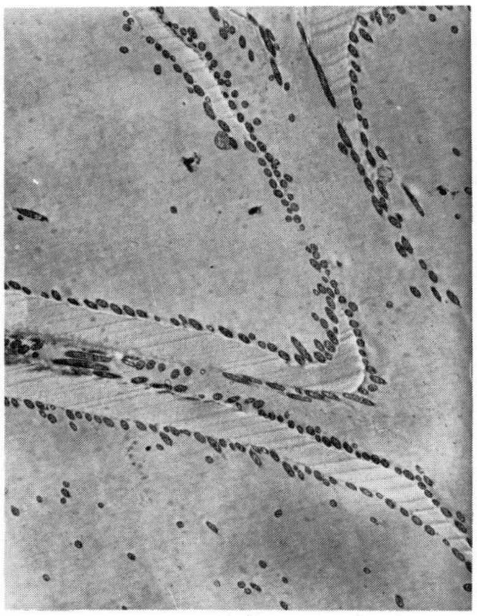

(a)

(b)

**FIG. 3-17** The tough exoskeleton of this spiny lobster is made of chitin.

water, a fact related to their nonpolar nature. Like carbohydrates, lipids contain carbon, hydrogen, and oxygen. However, the ratio of hydrogen to oxygen in lipids is greater than the 2:1 ratio characteristic of carbohydrates. Three categories include most, but not all, of the biologically important lipids: **fats** and **oils** serve as energy storage compounds, **phosphoglycerides** are important components of cellular membranes, and **steroids** are components of membranes or serve as vitamins and hormones.

## FATS AND OILS: THE STORAGE LIPIDS

Lipids are used for energy storage in both animals (Fig. 3-18) and plants. Fats are defined as lipids that are generally solid at room temperature and are produced by animals; lard and butter are examples. Oils, such as corn oil and olive oil, are produced by plants and are generally liquid at room temperature.

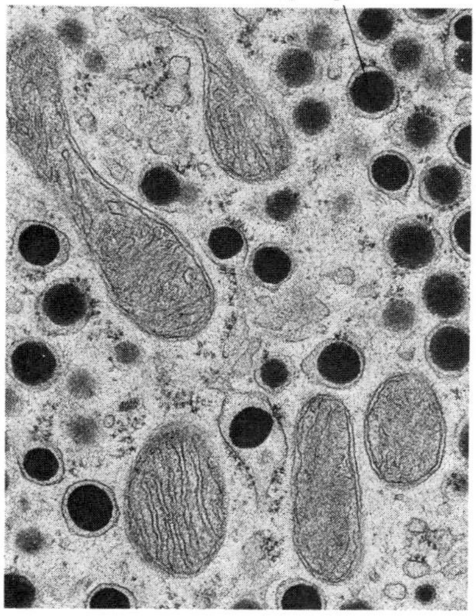


**FIG. 3-18** Lipid globules in the rat intestinal epithelium cells during fat absorption.

***Chemical Composition of Glycerol and Fatty Acids.*** Fats and oils belong to the chemical group called **esters,** compounds formed by a reaction between an alcohol and an acid. The complete hydrolysis of a fat or an oil yields **glycerol,** an alcohol with three —OH groups, and one to three long chain molecules known as **fatty acids** (Fig. 3-19). Because glycerol is a trihydric alcohol, it has three sites available for the formation, by dehydration synthesis, of ester linkages with fatty acids. Depending on whether one, two, or three fatty acids are thus joined to glycerol, the lipid is termed a **mono-**, **di-**, or **triacylglycerol** (formerly *mono-, di-,* or *triglyceride*). The three fatty acids in a triacylglycerol are seldom of the same length. In mammals, the major site of accumulation of triacylglycerols are the fat cells (Fig. 3-20).

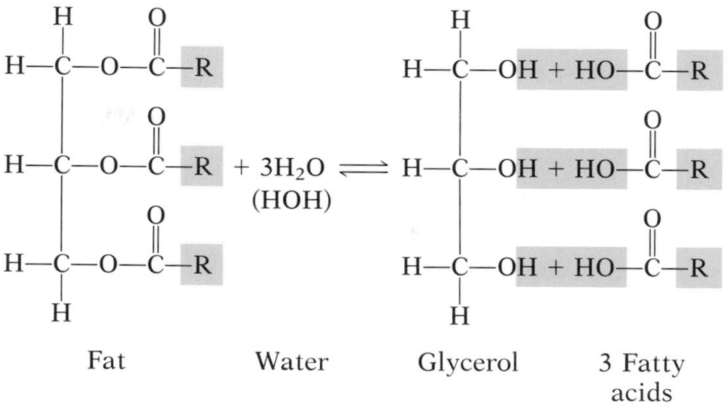


**FIG. 3-19** Hydrolysis of a fat molecule to yield glycerol and fatty acids. Side chains are represented by *R*.

***Fatty Acid Types.*** Fatty acids consist of long, zigzag chains of carbon atoms, usually with two hydrogens per carbon and always with a **carboxyl,** or **carboxylic acid, group** ($-\mathrm{C}(=\mathrm{O})-\mathrm{OH}$) at one end of the chain. Figure 3-21 shows five ways of representing a fatty acid, in this case butyric acid, the main component of butterfat.

If the maximum number of hydrogen atoms is incorporated in a fatty acid chain—that is, if there are no double bonds between any of the carbon atoms—the fatty acid and the fat containing it are said to be **saturated.** Fatty acids without the maximum number of hydrogen atoms and with at least some carbon atoms joined by double bonds are **unsaturated** (Fig. 3-22). If a fatty acid has more than one set of double bonds, it is **polyunsaturated.**

**FIG. 3-20** Scanning electron micrograph of fatty tissue showing fat-storing cells, called adipocytes, held together by connective tissue fibers. These adipocytes are primary storage centers for triacylglycerols. Ad: adipocytes; Fi: fibers.

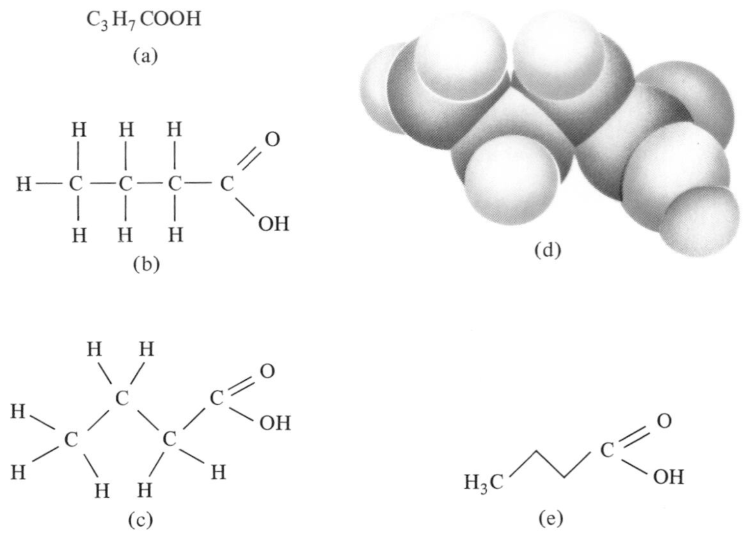


**FIG. 3-21** Five ways of representing butyric acid, a fatty acid. (a) The empirical formula. The carboxyl group (COOH) is shown, however, to indicate that it is an acid. The empirical formula is a handy way of representing large fatty acids, some of which have 18 carbon atoms. (b) A type of structural formula commonly used to represent short-chain acids or to show some detail in the chain, such as a double bond. (c) A more realistic structural formula that shows the bond angles characteristic of all fatty-acid chains. (d) A space-filling model. The carbon atoms are black, the hydrogen atoms the light color, and the oxygen atoms the darker color. (e) An abbreviated form of (c), each angle representing a carbon atom and, in most cases, two hydrogen atoms. The type of formula used in a given instance is dictated by the type of information being communicated.

Some of the more common saturated and unsaturated fatty acids found in fats and oils are listed in Table 3-1. Animal fats, such as beef fat, lard, and butter, contain mainly saturated fatty acids, while the oils obtained from such plant products as corn, peanuts, and soybeans contain a substantial amount of unsaturated fatty acids. Much of the vegetable oil sold as food, however, has been hydrogenated, that is, the unsaturated positions of the fatty acid molecules have been artificially filled with hydrogen; this "hardens" the oil, turning it into a solid at room temperature. Many of the margarines, peanut butters, and solid vegetable cooking products sold in the United States are made with hydrogenated oil. For a discussion about the benefits of unsaturated versus saturated fats and oils in the human diet see Panel 9-4.

**FIG. 3-22** (a) Two representations of stearic acid, a common saturated fatty acid. (b) Two representations of oleic acid, a common unsaturated fatty acid. Note the double bond, which creates a bend in the chain.

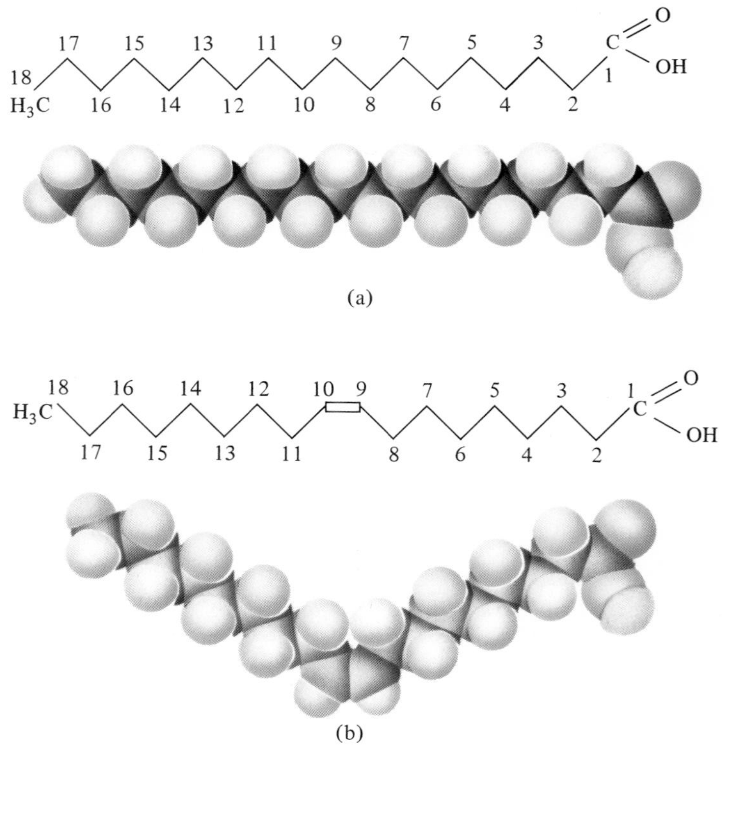

**TABLE 3-1 Common Fatty Acids Found in Fats and Oils**

| *Common name* | *Source* |
|---|---|
| Saturated fatty acids | |
| Butyric acid | Butter |
| Cerotic acid | Beeswax, wool fat |
| Lauric acid | Coconuts |
| Palmitic acid | Animal and vegetable fats |
| Unsaturated fatty acids | |
| Arachidonic acid | Synthesized from linoleic acid, not found in plants |
| Linoleic acid | Vegetable oils (e.g., corn, cottonseed, linseed) |
| Oleic acid | Olive oil |

***Prostaglandins: A Class of Modified Unsaturated Fatty Acids.*** The mammalian body produces a remarkable family of unsaturated compounds called **prostaglandins,** so named because they were discovered in the seminal fluid produced by the prostate gland of the male mammal. All are derived from a 20-carbon unsaturated fatty acid, such as arachidonic acid, shown in Figure 3-23a. Recent years have seen a growing interest in these substances, which occur in minute amounts in all mammalian tissues thus far examined. Prostaglandins act much like hormones in that only very small amounts are necessary to trigger metabolic activities. But unlike hormones, which are produced at one site and affect tissues elsewhere in the body, nearly all prostaglandins are synthesized and act intracellularly. They are important in blood clotting, respiration, kidney function, reproduction, and other vital functions.

***Fat as an Energy Source.*** Fats are the most concentrated source of biologically usable energy in animals. Oils constitute the major storage product in some plant organs (Fig. 3-24) and products, especially seeds. The complete oxidation of lipids releases twice as many calories of energy per gram as do carbohydrates and proteins because of their greater ratio of hydrogen to carbon and oxygen. Fats also produce more water when they are oxidized. This is a bonus to animals that have evolved the

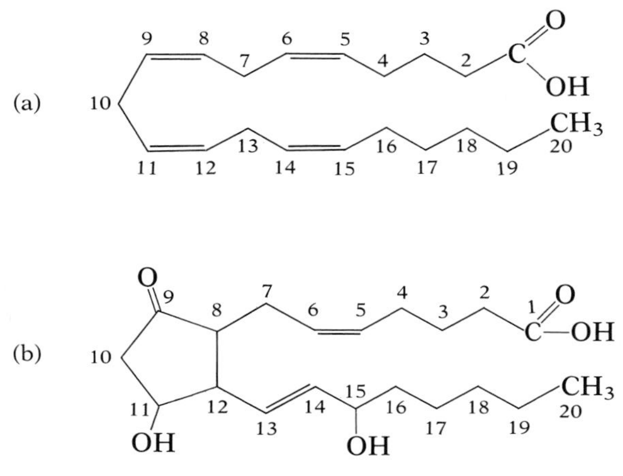


**FIG. 3-23** (a) Arachidonic acid, a polyunsaturated fatty acid with four double bonds. It is synthesized in animal tissues, using oleic acid as a precursor. (b) The prostaglandin $PGE_2$, obtained from arachidonic acid. It retains two of the four double bonds of arachidonic acid.

**FIG. 3-24** Peanuts, an excellent source of unsaturated fats in the form of peanut oil.

ability to store large quantities of fat prior to hibernation or migration. Animals store as fat the same amount of energy that plants store as starch in only half the mass. In mobile animals, as in automobiles, less weight means more kilometers per gram of fuel.

## PHOSPHOGLYCERIDES: THE STRUCTURAL LIPIDS

Phosphoglycerides are similar to triacylglycerols in form, except that a phosphoric acid is substituted for one of the three fatty acids and an additional group is added to the phosphate group. Phosphatidyl choline, a common phosphoglyceride found in most animal membranes, is shown in Figure 3-25. One of the two fatty acids has double bonds, while the other has none.

***The Polar Region of a Phosphoglyceride.*** The additional group attached to the phosphate group of the phosphoglyceride molecule provides a region of strong polarity. In phosphatidyl choline, the added group is choline, $(CH_3)_3—N^+—CH_2—CH_2—OH$, which has a positively charged region. Phosphatidyl choline has a negatively charged region in the phosphate group. Consequently, a molecule such as phosphatidyl choline attracts other charged or polar molecules. Although water molecules have no affinity for the fatty acid and alcohol chains of the molecule, they cluster around the charged part of a phosphoglyceride. Since the long-chain portions of these phosphoglycerides are uncharged and repel water, they are **hydrophobic** (from the Greek *hydro,* "water," and *phobia,* "fear"). The charged phosphate-containing ends that attract water are **hydrophilic** (from the Greek *philos,* "loving").

The polar nature of phosphoglyceride molecules is shown in Figure 3-26a, with circles representing the hydrophilic, **polar** end and a pair of zigzag lines the hydrophobic, **nonpolar** end. As might be expected, large phosphoglyceride molecules are not soluble in water. However, if they are vigorously shaken with water, they form a suspension of droplets called **micelles.** As illustrated in Figure 3-26b, the hydrophilic ends of the lipid molecules face outward to interact with the water, while the fatty acid chains face inward and tend to bind to each other. Since lipids are less dense than water, they tend to float on the surface. If a small amount is poured on water, it forms a monolayer, a layer that is one molecule thick. The hydrophilic ends of the molecule orient toward the water, as shown in Figure 3-26c, and the fatty acid chains toward the air. The polar nature of phosphoglyceride molecules is important in the formation of the various membranes of all living cells.

## STEROIDS: FATLIKE SUBSTANCES

Steroids are a group of lipid compounds consisting of four interconnected carbon rings, as exemplified by the structures of cholesterol, estradiol, and testosterone. As is obvious from comparing structural formulas, steroids have no structural relationship to fats and phosphoglycerides and are classed as lipids only because they are insoluble in water and are soluble in organic solvents.

One of the most abundant of the steroids is cholesterol, shown in

FIG. 3-25 Phosphatidyl choline, a common phosphoglyceride.

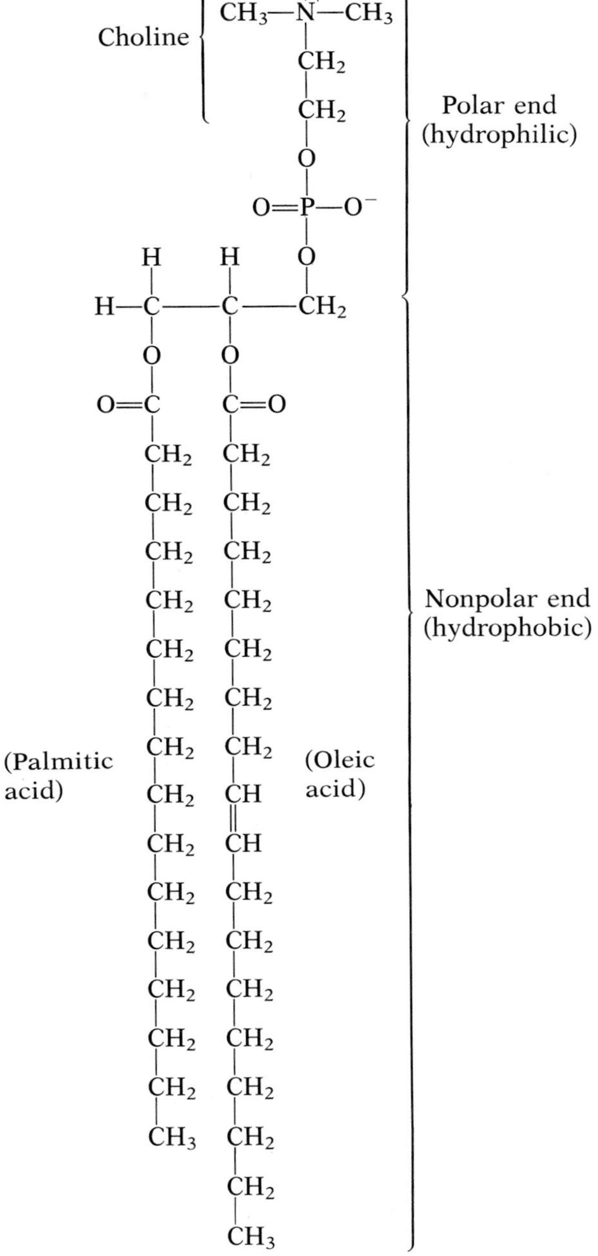


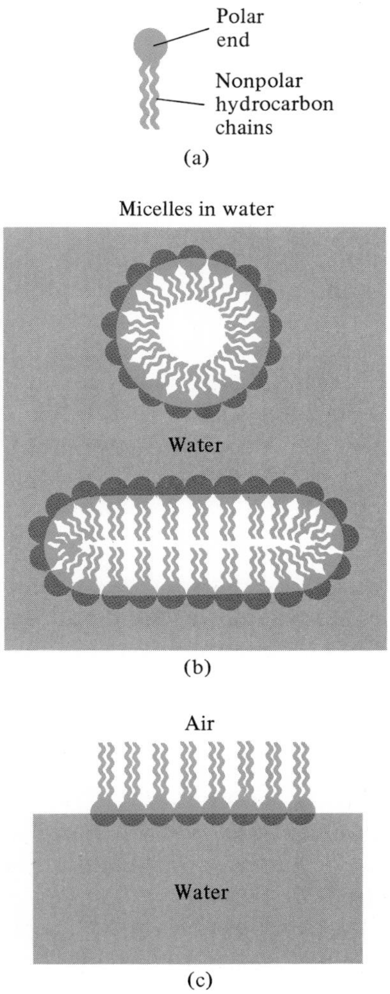


**FIG. 3-26** Behavior of phosphoglycerides in water. (a) Conventional representation of a phosphoglyceride molecule. (b) Micelles, or droplets, of phosphoglyceride formed in water. Note that the polar ends orient toward the water phase, the nonpolar chains toward the lipid phase. (c) The formation of a monolayer of phosphoglycerides at a water-air interface.

Figure 3-27. Cholesterol is found in the cellular membranes of all animals. For example, it is a major component of the membranes that wrap around and insulate many nerve fibers. It is also the chemical precursor of several important animal hormones; that is, it is used as a starting material for their synthesis, much as unsaturated fatty acids are the material from which prostaglandins are produced.

Cholesterol is normally present in considerable amounts in human beings; a 60-kilogram (132-pound) person may contain as much as a quarter of a kilogram (about a half pound). Because it is insoluble in water, cholesterol may settle out of solution when present in high con-

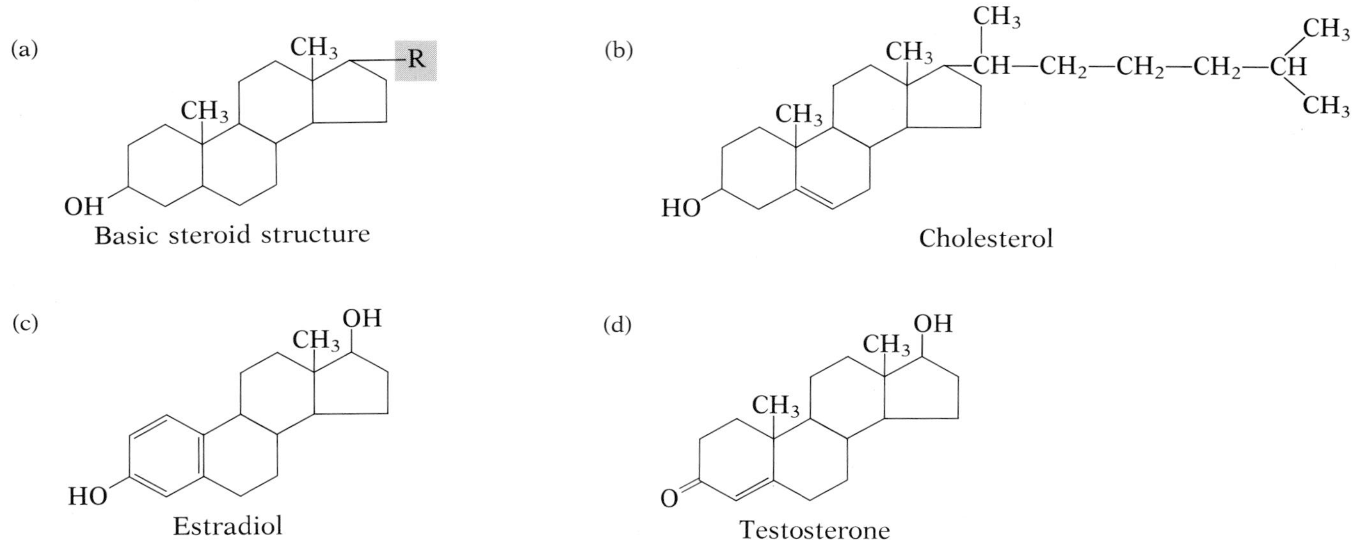


**FIG. 3-27** (a) Basic structure of the steroids. (b) Cholesterol. (c) Estradiol, a mammalian female sex hormone. (d) Testosterone, a mammalian male sex hormone.

centrations. It sometimes precipitates from bile in the gallbladder to form pebblelike, crystalline gallstones (Fig. 3-28). It may also accumulate in the walls of arteries, contributing to the narrowing of the arterial lumen, or cavity, in the disease atherosclerosis (from the Greek *athere,* "fatty," and *sklerosis,* "hardening") (see Panel 9-4). This condition contributes to high blood pressure, strokes, and heart attacks.

## PROTEINS

Among the most important and abundant of all of the substances that make up a living organism are the group of macromolecules called proteins (from the Greek *proteios,* "primary"). These compounds serve a number of functions essential to life:

1. *Structural support.* Proteins are a major structural component of bone, cartilage, feathers, fur, and hair and a component of all biological membranes and chromosomes.
2. *Enzymatic catalysis.* All enzymes are proteins. They play a unique role in determining the pattern of all chemical transformations and thus of all growth and function.
3. *Transport and storage.* Specific proteins serve as carriers of many important small molecules and ions, some in trace amounts. For example, iron atoms are carried in mammalian blood by the protein called transferrin and are stored in the liver as a complex with ferritin, another protein. Binding to proteins is a quick way of transporting and storing molecules; it also increases metabolic efficiency by ensuring the delivery of substances only to the appropriate receptors of specific cells. Another instance involves the proteins that transport cholesterol. These are known as high- and low-density lipoproteins as discussed in Panel 9-4.
4. *Movement.* Muscle fibers are composed primarily of protein filaments. Muscle contraction is accomplished by the sliding motion of two major kinds of protein filaments: actin and myosin; their structure and function are discussed in Chapter 18.

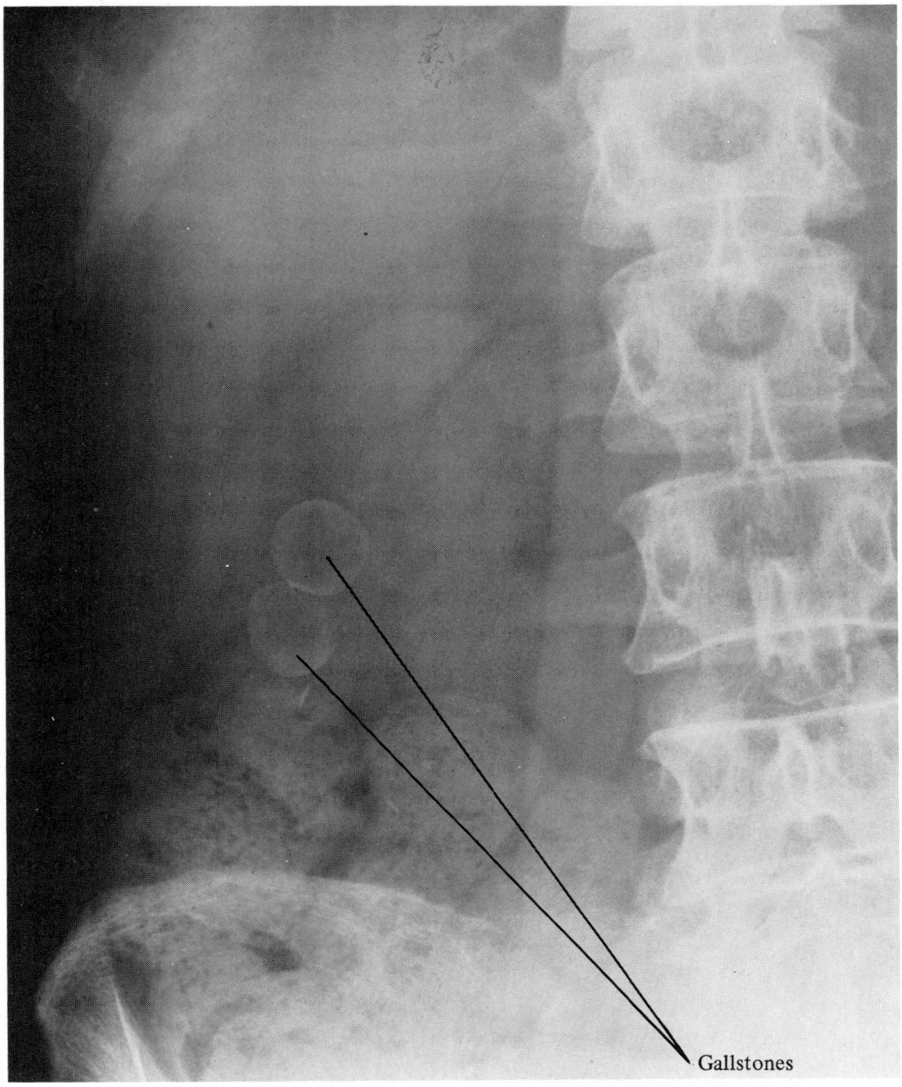


**FIG. 3-28** X-ray of a human gallbladder with two crystalline gallstones.

5. *Immune protection.* Immunoglobulins are a group of highly specific protein molecules that chemically "recognize" and combine with such potentially dangerous foreign invaders of the body as bacteria and viruses. The structural and functional properties of immunoglobulin molecules are described in Chapter 12.
6. *Hormonal regulation of physiological activities.* The physical activities of living systems are regulated and coordinated by hormones. Several hormones are proteins; others are transported by proteins. Hormones are discussed in Chapters 14, 21, and 22.
7. *Sensory perception and integration of the nervous system.* The responses of nerve cells to specific stimuli are often mediated by specific receptor proteins, such as the light-sensitive, pigmented proteins in the retinal cells of the vertebrate eye.

The functional diversity of proteins is derived from their practically unlimited potential for structural variation. This enormous variety is due to the fact that proteins are composed of a group of about 20 relatively simple but very versatile molecules—the $\alpha$-amino acids.

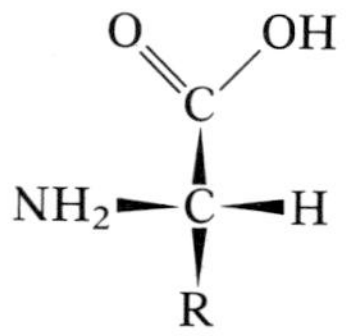


**FIG. 3-29** General formula for an amino acid molecule. "R" represents the variable groups that are attached to this basic molecule to make up the 20 common amino acids.

## AMINO ACIDS

**Amino acids** are organic acids that possess both an amino group ($—NH_2$) and a carboxyl group ($—C(=O)—OH$). The general formula for amino acids is shown in Figure 3-29. Amino groups are basic, and carboxyl groups are acidic. The nitrogen atom of the amino group can pick up an unionized water molecule and then release the $OH^-$ ion, thus acting as a base. In the structural formula for an amino acid, the carbon atom attached to the carboxyl group is called the **alpha** ($\alpha$) **carbon.** An amino acid in which the amino group is attached to the alpha carbon atom is called an $\alpha$-**amino acid.** The 20 $\alpha$-amino acids form the building blocks of proteins. Each of these amino acids is identified with a different side chain attached to its alpha carbon.

The size, shape, charge, hydrogen-bonding capacity, and chemical reactivity of the $\alpha$-amino acids are largely dependent on their side chains. Polarity and charge on the amino acids are summarized in Table 3-2. Side chains vary in complexity (Fig. 3-30) from a single hydrogen atom, as in glycine, to branching chains, as in leucine, to ring structures, as in tryptophan. The side chain may have one or two additional basic groups, as in arginine, and an additional carboxylic group, as in aspartic acid; or it may contain a sulfur atom, as in cysteine. As noted in the table, some amino acids are polar but uncharged, such as glutamine and tyrosine, whereas others are polar and, because of an additional amino or carboxyl group, are charged, such as aspartic acid and lysine. A basic protein has a predominance of basic amino acids, those with extra basic groups, while an acidic protein has a predominance of acidic amino acids, those with extra carboxyl groups.

Even in the case of a small protein, such as one with only 100 amino acids, the number of possible amino acid combinations is $20^{100}$—far greater than the number of atoms in the entire universe! The amount of variety possible can be visualized by thinking of proteins as constituting a language with an alphabet of 20 letters. In this language, a single word (protein molecule) may have thousands of letters (amino acids) in any combination. For example, the same amino acid can occur several times in a row in a protein. Complex organisms have several thousand different kinds of proteins ranging in size from small molecules with only 50 amino acids to very large molecules with several hundred. By virtue of the arrangement of its component amino acids, each of an organism's different proteins has structural properties that are specifically adapted to carry out a particular function.

**TABLE 3-2 Some Characteristics of the Amino Acids**

| *Nonpolar* | *Uncharged polar* | *Charged polar* | |
|---|---|---|---|
| Glycine | Serine | Aspartic acid | Acidic (−charge) |
| Alanine | Threonine | Glutamic acid | |
| Valine | Asparagine | Histidine | Basic (+ charge) |
| Leucine | Glutamine | Lysine | |
| Isoleucine | Tyrosine | Arginine | |
| Proline | Tryptophan | | |
| Phenylalanine | Cysteine | | |
| Methionine | | | |

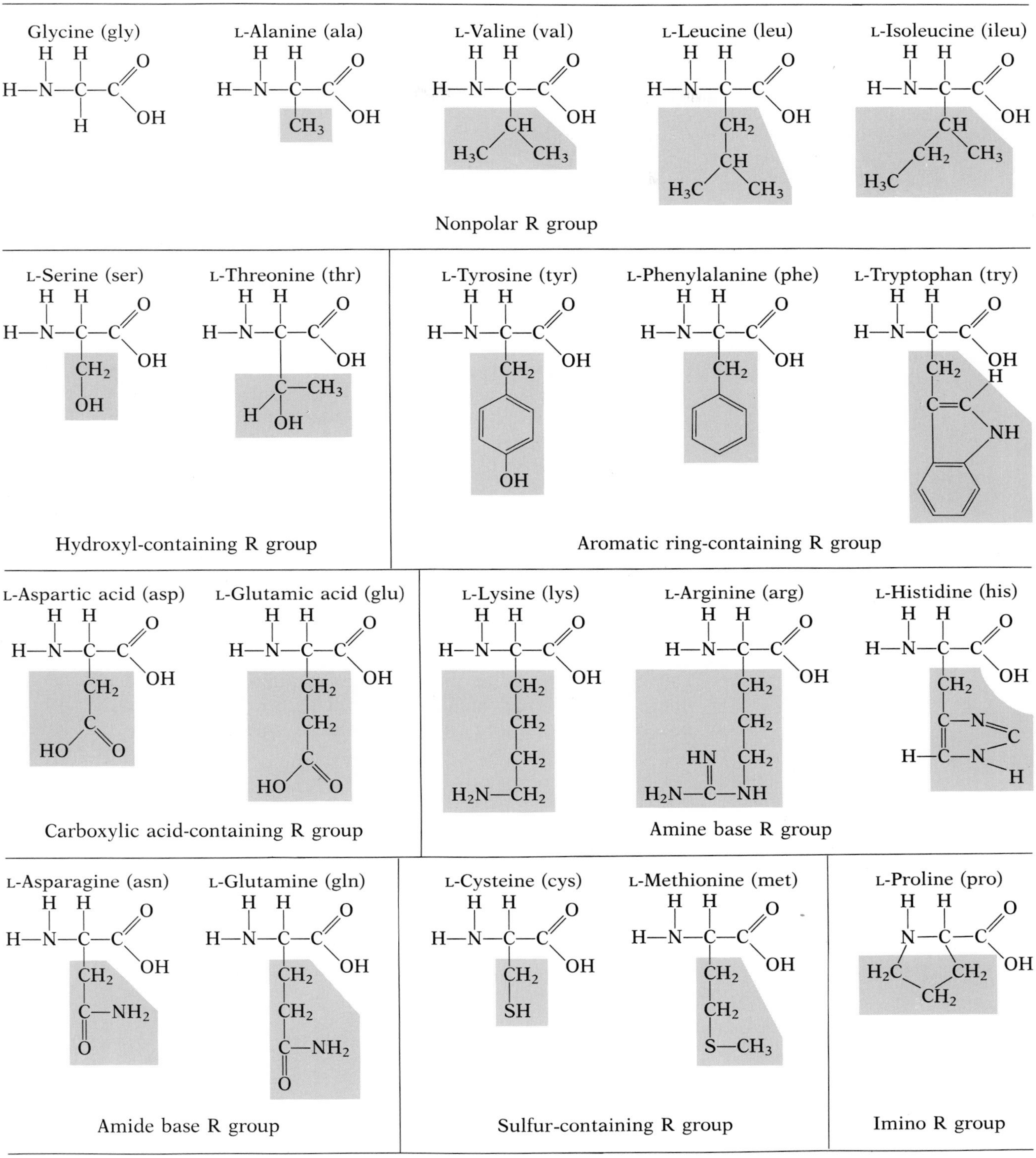


**FIG. 3-30** The structural formulas of 20 common amino acids showing the various kinds of side chains (R).

***The Peptide Bond.*** The amino acids of a protein are joined by a covalent bond between the carboxyl group of one amino acid and the amino group of a neighboring one, as shown in Figure 3-31. The formation of the bond is accompanied by the removal of a molecule of water. Since

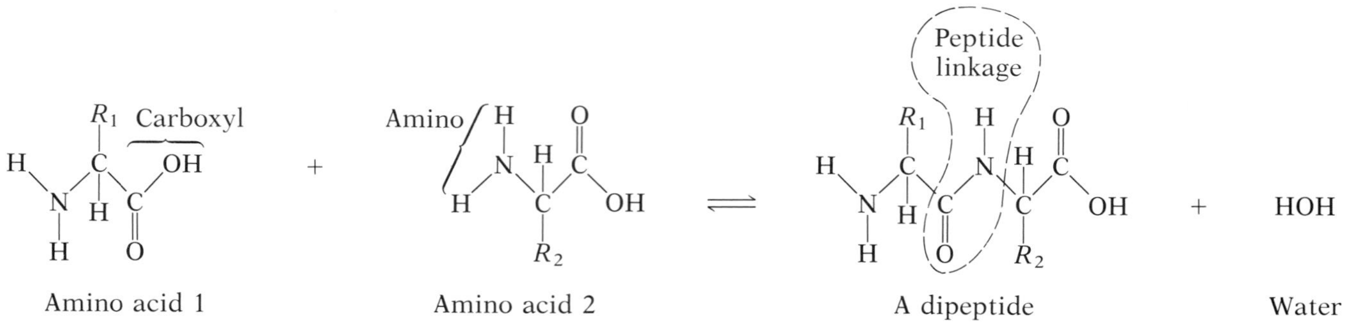


**FIG. 3-31** Formation of a peptide bond between two amino acids. A molecule of water is removed during the process.

this reaction is a dehydration synthesis, the bond formed can be broken by hydrolysis. The bond that links an amino group of one molecule to a carboxyl group of the next is called a **peptide bond.** If only two amino acids are linked by a peptide bond, the resulting molecule is called a dipeptide; if three, a tripeptide, and so on. If a large number are thus joined, the molecule is called a **polypeptide.** Peptide chains of any length have a definite direction or polarity because of the difference in the orientation of their component amino acids; moreover, there is always a free amino group—one not joined to another amino acid—at one end and a free carboxyl group at the other end (Fig. 3-32). The amino acid with the free amino group is known as the N-terminal group, and the amino acid with the free carboxyl group is called the C-terminal group. The free amino group end is taken as the beginning of a polypeptide chain; it is at this end that the cell's synthetic machinery begins the assembly of a polypeptide by adding amino acids. Accordingly, it is conventional to write the sequence of amino acids in a polypeptide chain (in cases in which it has been determined) beginning with the N-terminal group and ending with the C-terminal group.

## THE THREE-DIMENSIONAL STRUCTURE OF PROTEINS

The polypeptide molecules that make up proteins are folded into characteristic three-dimensional configurations determined by the chemical properties of the constituent amino acids. The structure of a protein may have four levels of organization: primary, secondary, tertiary, and quaternary.

**FIG. 3-32** A short polypeptide chain composed solely of glycine residues.

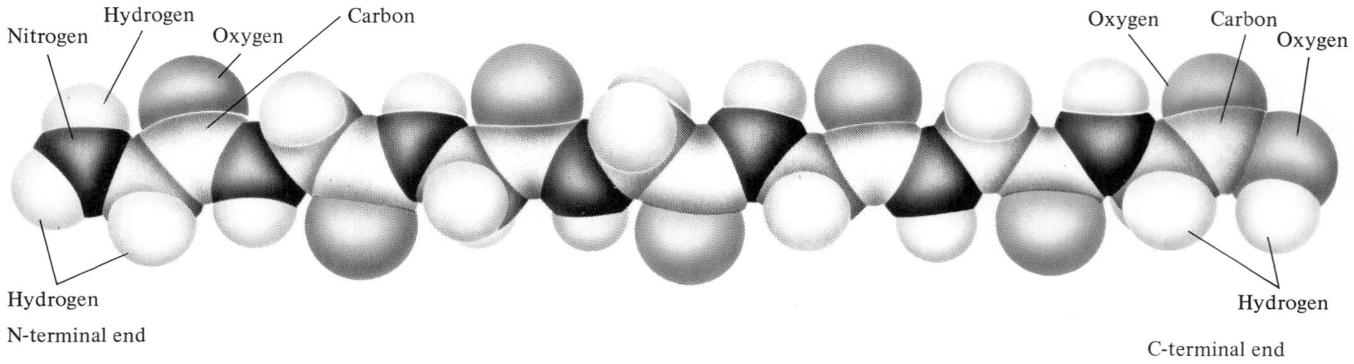

***Primary Structure.*** The primary structure of a protein is the basic order, or sequence, of amino acids in its polypeptide chains. Changes in primary structure can drastically alter the shape and therefore the biological activity of a protein. The shape of an enzyme, for example, is crucial to its ability to catalyze the reaction for which it is specific. The relationship between primary structure and biological activity is dramatically demonstrated by hemoglobin, the oxygen-carrying protein in vertebrate red blood cells and in the blood of many invertebrates (see Panel 3-1).

While some amino acid substitutions alter the biological properties of protein molecules profoundly, others have no detectable effect on the biological activity of a molecule. A given protein molecule has only certain regions where any change in the amino acid sequence is apt to affect its configuration and chemical properties.

PANEL 3-1

## Sickle-cell Anemia

Normal human hemoglobin contains four polypeptide chains: two identical alpha chains and two identical beta chains. The hemoglobin of people affected with sickle-cell anemia, an inherited disorder, has a single substitution of a valine residue for the sixth amino acid (glutamic acid) in one of the beta chains. All other amino acid residues and their sequences in the alpha and beta chains are identical to that in normal hemoglobin. This one amino acid substitution greatly affects the structure and biological function of the hemoglobin molecule, causing a decrease in the solubility of hemoglobin when oxygen is in low concentration. Such hemoglobin molecules change the shape of the red blood cell from that of a normal biconcave disc to crescent, or sickle, shape. These distorted cells block capillaries and interfere with normal oxygen flow to tissues. And, because of their shape, they are very fragile and are readily destroyed, reducing the number of oxygen-carrying cells in the circulation and resulting in the anemia associated with this disease. In its severe form, sickle-cell anemia can be fatal at an early age.

Scanning electron micrograph of human red blood cells. The longitudinal arrays of abnormal hemoglobin have the sickle shape characteristic of sickle cell anemia.

Sickle-cell anemia offers an interesting example of biological adaptation. The sickle-cell trait is common among Africans living within areas where malaria is prevalent. Malaria is a disease caused by a protozoan (*Plasmodium*) that invades the host's red blood cells, and anemia is one of the symptoms. It would be logical to assume that malaria would be very high among these people suffering from sickle-cell anemia. However, this is not the case. People who have inherited the sickle-cell trait appear to have a resistance to malaria: the condition actually protects them from malaria. Thus, the sickle-cell trait has been selected for in the course of evolution. People carrying the trait in malarial regions tend to survive in greater numbers than do those lacking it (see also Chapters 12, 23, and 24).

***Secondary Structure.*** As a polypeptide chain is synthesized, it is subjected to certain weak, noncovalent bonding forces that determine its precise three-dimensional conformation as it lengthens. The hydrogen bond is the most common of these weak bonding forces. Each amino acid residue contains a carbonyl group ($>C=O$) and a secondary amino (or imino) group ($>N-H$), derived from its carboxyl ($-C(=O)-OH$) and amino ($-NH_2$) groups, respectively. The oxygen and hydrogen atoms of these groups are available for hydrogen bonding (see Chapter 2). When such hydrogen and oxygen atoms are close enough, a hydrogen bond forms between them: $>C=O\cdots H-N<$ (the dotted line represents the hydrogen bond). As amino acids are added to the growing polypeptide chain, each amino acid residue twists slightly, forming a coiled, or helical, main chain, with any side chains projecting outward, away from the core of the molecule.

Not every protein is composed of helical chains. Silk fibroin polypeptides, for example, form sheets of protein with hydrogen bonds between adjacent polypeptide chains rather than within them.

***Tertiary Structure.*** In many protein molecules, especially those that perform special functions in cells, the polypeptide chains fold back and forth upon themselves to form compact, somewhat globular shapes. These three-dimensional foldings of a chain constitute the tertiary structure of a protein and define its specific three-dimensional configuration. The roughly spherical shape assumed by many proteins is determined by the attraction or actual bonding between amino acids that lie next to each other in the folded polypeptide chain. These attractions and bonds contribute to the stability of the structure. A protein's tertiary structure is also stabilized by noncovalent ionic bonds, hydrophobic interactions, hydrogen bonds between the atoms in adjacent side chains, and covalent bonds, such as those formed between two adjacent cysteine residues. Cysteine contains a sulfhydryl group (—SH), which can lose its hydrogen atom and form a covalent disulfide bond (—S—S—) with another cysteine molecule, as shown in Figure 3-33. These bonds are important because of their unusual strength; they tend to lock the tertiary structure of a protein molecule into a fairly fixed and stable configuration. The net effect of all of these forces is the formation of a unique three-dimensional configuration for each kind of protein.

When a particular protein is heated above a certain temperature (each protein has its own specific temperature tolerance) or when it is subjected to unusually acidic or basic conditions, its polypeptide chains begin to unfold, altering its secondary and tertiary structures. When this structural change occurs, the protein loses its unique biological capability and is said to be **denatured.** Proteins with their primary and secondary structures intact are in their **native state.** The polypeptide backbone and disulfide bonds are usually not broken during denaturation, but hydrogen bonds and other weak forces holding the protein in its native state are always disrupted. The protein may return to its original tertiary structure and regain its biological activity upon restoration of normal conditions. However, severe denaturation permanently

PANEL 3-2

# The Helical Structure of Proteins

X-ray diffraction, used in studies of crystals, is also a method of studying molecular structure. X rays transmitted through a crystalline substance are diffracted (diverted from their course) as they emerge from the spaces between the atoms of a crystal. The X-ray beam is thus scattered, creating a diffraction pattern on a photographic plate. The diffraction pattern differs among molecules. The angles at which the beams emerge can be determined from this pattern, and the atomic arrangement can be determined from these angles.

In 1951, the American scientists Linus Pauling and Robert B. Corey were able to show by X-ray diffraction and other studies that polypeptides can form a coil, or helix, of 3.6 amino acid residues per turn. This coiling arrangement allows a hydrogen bond to form between every three residues: first and fourth, second and fifth, third and sixth, fourth and seventh, and so on. These 1-4 overlapping hydrogen bonds provide considerable stability to the polypeptide chain, more than would be provided by any other combination, such as 1-3, 1-5, or 1-6. The 1-4 bonded polypeptide helix is called the **alpha helix.** It forms an important part of the structure of many polypeptides, whose side chains extend outward from the helix to form a kind of chemical spiral staircase, but with the "steps" anchored to only one continuous structure.

The alpha helical structure of a protein molecule.

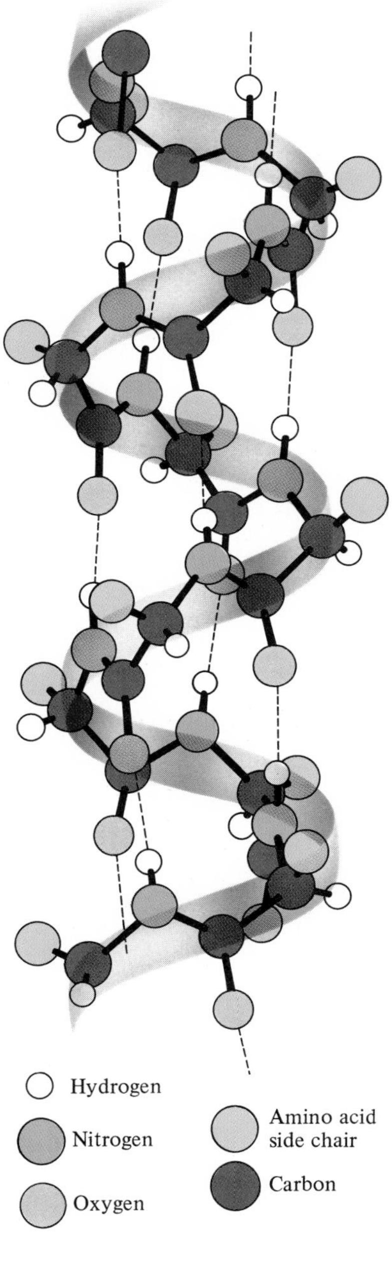


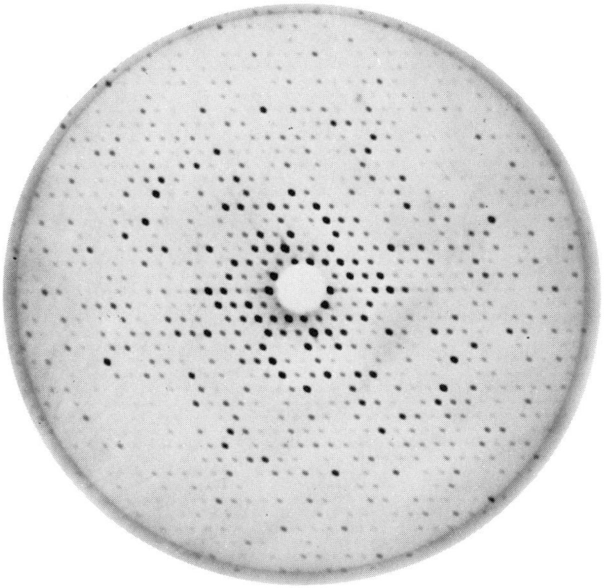

X-ray diffraction pattern of a single crystal of hemoglobin. The distances of the spots from the center of the photograph together with their size can be used to construct the molecular structure shown in Fig. 3-34.

FIG. **3-33** Formation of a disulfide bond (—S—S).

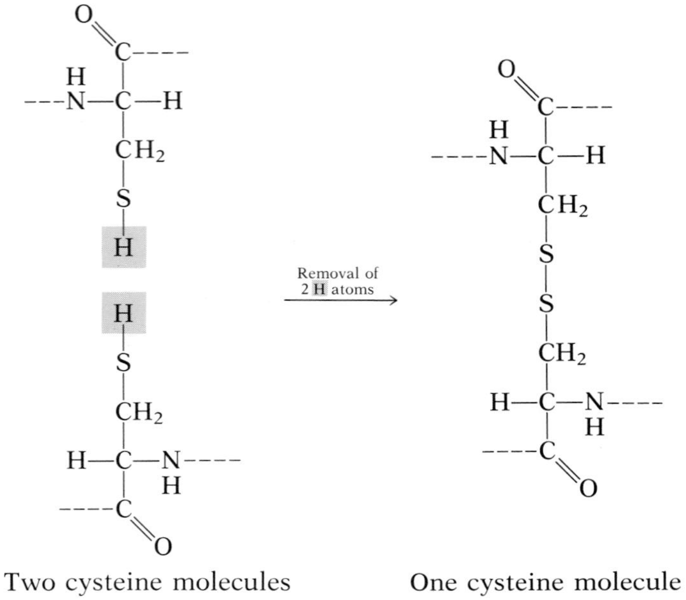


alters the structure and biological activity of a protein. It is an old joke among biochemists that it is impossible to unfry the white of an egg, which consists largely of the protein ovalbumin, which is denatured upon heating.

***Quaternary Structure.*** Most proteins larger than 50,000 daltons have quaternary structures, formed by two or more polypeptide chains. These chains may be linked by covalent disulfide bonds, noncovalent ionic bonds, hydrogen bonds, or hydrophobic forces. Each polypeptide chain or subunit has its own tertiary structure. The subunits may be alike or quite different. For example, each of the two types of chain in human hemoglobin has its own characteristic primary, secondary, and tertiary structures (see Figure 3-34). The four subunits of a quaternary structure are oriented in a specific configuration relative to one another.

**FIG. 3-34** The four chain (quaternary) structure of human hemoglobin.

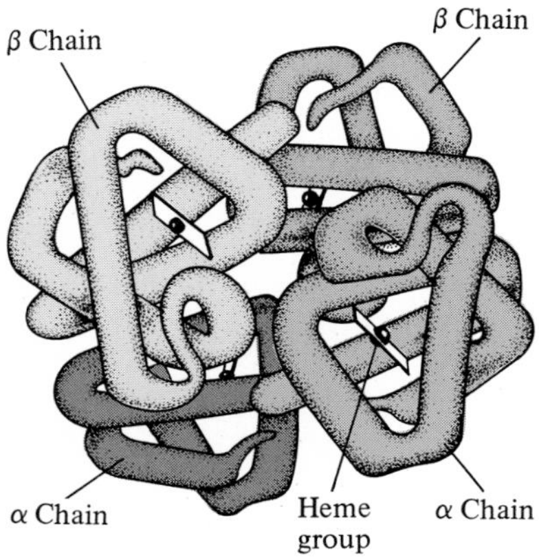


## ENZYMES

Enzymes are specialized protein molecules whose function is to facilitate chemical reactions within the body. A chemical reaction such as the oxidation of glucose releases a great deal of energy. This energy is captured in biological systems by a complex stepwise series of coupled reactions carried out at the relatively low temperatures that prevail within the body of most organisms (see Chapter 6). A substance that affects the rate of a chemical reaction is called a catalyst, one of its properties being that it remains unchanged by the chemical reaction in which it participates.

In order for any chemical reaction to take place, the reactant molecules must first be brought together in a way that will make them interact. At normal room temperature or even at the normal body temperature of 37°C, few molecules would come together with enough frequency or force to react with each other. Take, for example, a piece of paper and

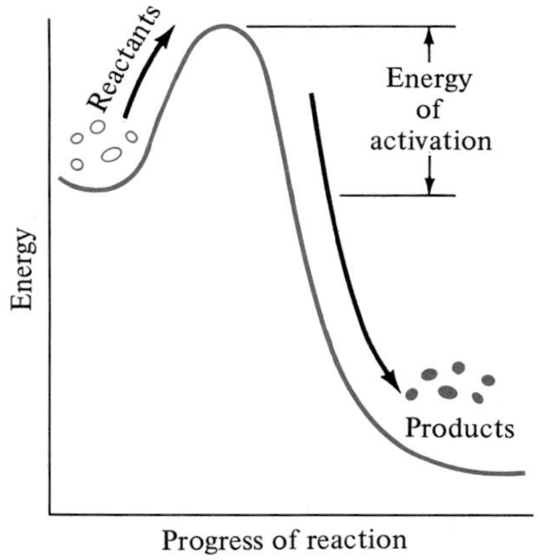


**FIG. 3-35** Activation energy for chemical reactions.

a supply of oxygen. Ordinarily, nothing will happen if the two are brought together. However, if we provide some energy in the form of a flame, a combustion will take place; the paper will burn. It is obvious that what enabled the paper to burn was the heat from the flame. The same occurs with the oxidation of glucose and other chemical reactions. They all need some form of heat in order to get started. The amount of energy required to bring the interacting molecules, or **reactants,** to an energy state conducive to a chemical reaction is known as the **activation energy** of the reaction (Fig. 3-35). This may be thought of as an "energy barrier" over which molecules must be raised before they can interact, form their products, and release energy.

Heat increases both the frequency and the force of the collisions among reacting molecules, thus promoting the reaction. But the amount of heat tolerated by a system outside of the body, as represented by a test tube, is much higher than that tolerated by the body itself. Biological systems cannot operate at the high temperatures of a flame, nor are their molecules usually found in such concentrations as to enable them to react with each other spontaneously. Therefore, the problem faced by a biological system is twofold: (a) molecules must be brought together; (b) energy must be made available to start a reaction.

## ENZYME-SUBSTRATE COMPLEXES

By combining with reactant molecules, called the **substrate** or **substrate molecules**, enzymes bring the reactant molecules into close contact with each other without the need for forceful collisions. The substrate molecules can then react with each other, the products of the reaction immediately being freed from the enzyme-substrate complex. In effect, the enzyme has lowered the activation energy required to bring the reactants together in order for them to undergo a reaction (Fig. 3-36). An enzyme-catalyzed reaction may be written as,

| A + B | + | E | | ES | | C + D | + | E |
|---|---|---|---|---|---|---|---|---|
| Substrate molecules | | Enzyme | ⟶ | Enzyme-substrate complex | ⟶ | Products | | Enzyme |

The enzyme-substrate complex dissociates rapidly; in fact, as soon as products C and D are formed. Once they are released from the enzyme, the enzyme can immediately catalyze another reaction, since, as we

**FIG. 3-36** Lowering the activation energy for chemical reactions by enzymes.

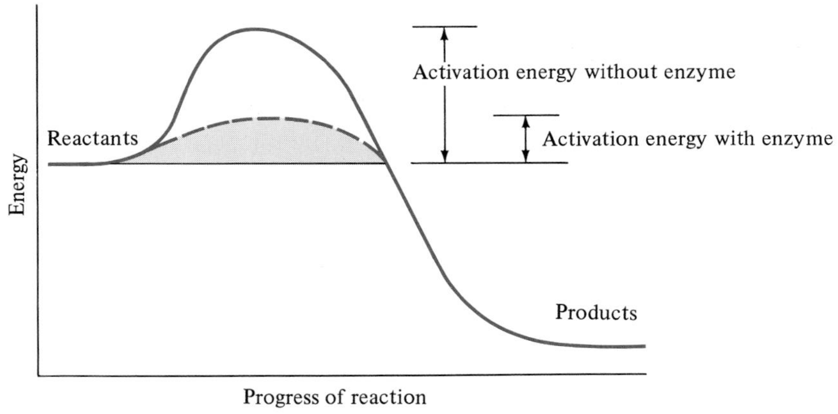

know, it remains unchanged. Therefore, a small amount of enzyme can repeatedly catalyze the reactions of many substrate molecules. Depending on the enzyme, the number of substrate molecules catalyzed by a single enzyme molecule per second (called the **turnover number**) varies from 100 to more than 3,000,000 molecules.

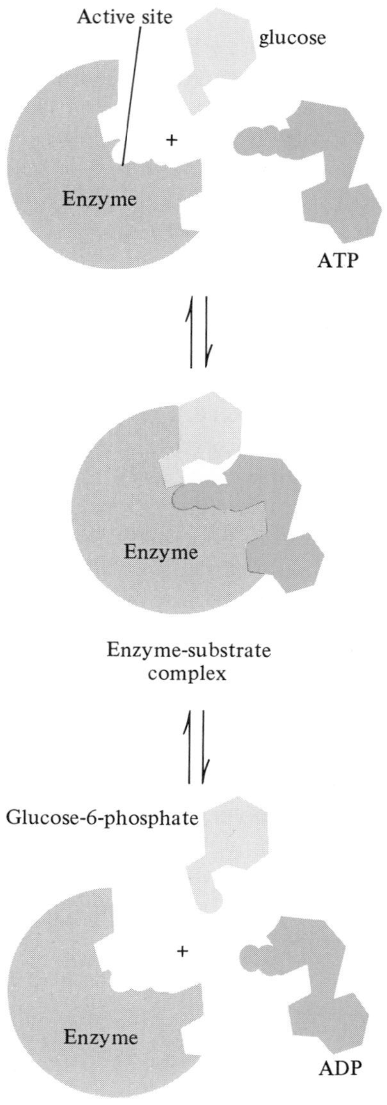


**FIG. 3-37** The lock-and-key model for the active site of an enzyme. The glucose and ATP molecules form an enzyme-substrate complex that enables the enzyme to catalyze the removal of one of the phosphate groups of ATP to form glucose-6-phosphate and ADP.

## SPECIFICITY OF ENZYME ACTION

Enzymes are specific in their activity, which means that they will usually catalyze only a single type of reaction. Since they are proteins, all enzymes consist of specific amino acid sequences, and these sequences determine an enzyme's three-dimensional shape. Since each kind of enzyme has a different amino acid sequence, every enzyme has a characteristic shape, or configuration. Enzymes have small regions on their surfaces called **active sites,** and these sites are complementary in shape to the substrate molecules with which they interact. In other words, the shape of the active site matches the shape of a specific part of the substrate molecule. And the charges on each of these regions are different and opposite from each other. Furthermore, these regions precisely match each other in lock-and-key fashion (Fig. 3-37). The specificity of enzymatic activity has long been explained by this lock-and-key model. However, more recently, the notion of a rigid lock-and-key fit between enzyme and substrate has proven inadequate.

Several lines of evidence now indicate that most, but perhaps not all, enzymes are flexible molecules that assume slightly different shapes under different physiologic conditions. This property helps explain some types of enzyme activation and inhibition and has led to the **induced fit hypothesis.**

***The Induced Fit Hypothesis.*** A substrate molecule, upon binding to the active site on an enzyme, apparently induces a change in the shape of the enzyme molecule. This explains why, in certain cases, two different substrate molecules can bind to an active site on the same enzyme but only in a certain order. The change in shape of the active site induced by the first substrate molecule is apparently what makes the enzyme receptive to the second substrate molecule.

Once a substrate molecule has made contact with the active site, and has initiated a change in the enzyme's shape (Fig. 3-38), the enzyme initiates the actual reaction of the substrate molecule. Once the reaction has occurred, the attraction between the substrate and enzyme ceases, thereby freeing the enzyme from the product and enabling it to catalyze a second reaction of the same type. The changed shape in the enzyme induced by the substrate occupying the enzyme's active site—called the transitional state—is thought not to correspond exactly to the normal shape of the substrate in either the reactant or the product form. In a reversible reaction, the transitional state is unstable, and a strain is placed on both the substrate and the enzyme. In this case, the strain is relieved by the reaction's going to completion in either direction, the direction being determined by the relative concentration of reactants and products.

***Enzyme Activation.*** Despite the presence of the appropriate substrate, many enzymes do not function unless an activator, called a **cofactor,** is

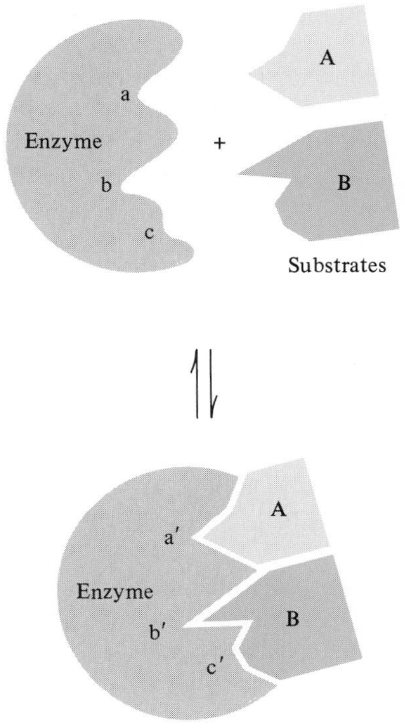


**FIG. 3-38** The induced-fit model for the interaction of substrates with the active site of an enzyme.

also present. In one type of activation, a small activator molecule enters the active site, inducing a conformational change in the enzyme. This change enables the substrate to enter and be acted upon. The cofactor molecule does not actually participate in the reaction, however. In another type of activation, the cofactor binds to a site on the enzyme other than the active site. This produces a conformational change in the enzyme molecule resulting in the active site's being made receptive to the substrate molecule. Enzymes whose action is modulated (either activated or inhibited) by binding of a molecule to a site other than the active site are termed **allosteric enzymes** (from the Greek meaning "other space").* The site to which the modulator binds is called the allosteric site.

A cofactor may be a metal ion or an organic molecule. The organic molecule is called a **coenzyme.** Some enzymes require both types of cofactors in order to function as biological catalysts. Among the metal ions that can act as cofactors are zinc ($Zn^{2+}$), iron ($Fe^{2+}$ or $Fe^{3+}$), copper ($Cu^{2+}$), potassium ($K^{+}$), and sodium ($Na^{+}$). In some cases, more than one metal ion is required for a particular enzyme-catalyzed reaction. Coenzymes are present in trace amounts in all cells, for, like enzymes, they function over and over again. Plants can usually synthesize all of the required coenzymes. But this is not the case with higher animals, which must obtain some coenzymes, or their precursors, from their diet. Analysis of coenzymes has shown that some are vitamins, some are synthesized from vitamins, and still others contain vitamins as structural components. Vitamins, by definition, are substances that animals require in trace amounts but cannot synthesize within their cells and must obtain from their diets.

***Cooperativity.*** Some enzymes can bind two, three, or more molecules of substrate at a time. In a phenomenon called **cooperativity,** the first substrate molecule to be bound induces a conformational change in the enzyme, making it more receptive to a second, third, or fourth substrate molecule than it was to the first. The enzyme rapidly binds the additional substrate molecules and the reaction reaches top speed. Whether the increased receptivity of an enzyme to a substrate is the result of cooperativity or of its binding a molecule of a non-substrate activator to an allosteric site, the effect is to make it highly sensitive to small changes in concentration of either its substrate or the activator.

***Inhibition of Enzymes.*** It is often just as important to "turn off" an enzyme as it is to activate it. The induced fit hypothesis explains various types of inhibition as readily as it explains activation.

1. *Negative cooperativity.* In this case, the binding of the first substrate molecule to the active site produces a conformational change that makes the enzyme less receptive to a second or third substrate molecule. This type of inhibition allows a reaction to proceed normally at a moderate speed, and to increase its rate further only if an unusually high accumulation of substrate occurs.
2. *Inhibition by a product.* A common type of inhibition involves one of the products of the reaction functioning as an allosteric inhibi-

* Besides enzymes, certain other regulatory proteins are modulated in this way. A classic example is hemoglobin, described in Chapter 11.

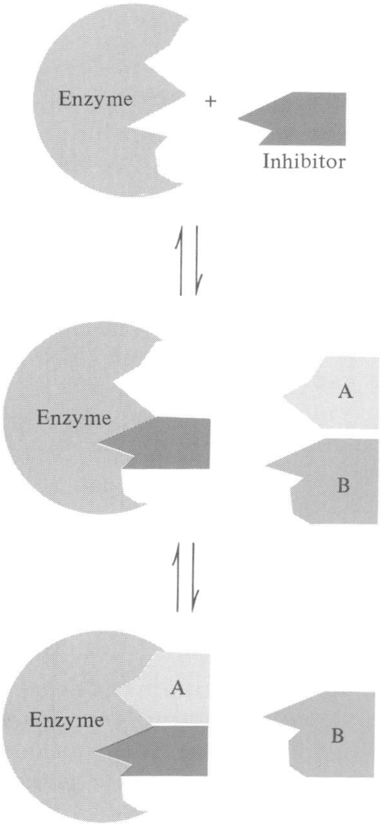


**FIG. 3-39** Action of a competitive inhibitor in blocking the interaction of an enzyme with its normal substrate.

tor. Another type involves the end product of a series of reactions being an inhibitor of the first enzyme in the series. It might be wondered how one enzyme molecule can be activated or inhibited by such a variety of molecules. Actually, only a very small portion of an enzyme's surface possesses active sites. The rest is available for binding activators and inhibitors.

3. *Competitive inhibition.* Another type of inhibition occurs when a substrate that is not the normal substrate but whose molecules resemble the substrate in shape and charge is able to occupy part or all of the active site of an enzyme (Fig. 3-39). Although in such instances the active site would be occupied, no reaction would take place. This substance is known as a competitive inhibitor. Because competitive inhibitors drift in and out of active sites, the degree of competitive inhibition is always proportional to the ratio of inhibitor concentration to normal substrate concentration. Competitive inhibition can always be reduced by increasing the concentrations of normal substrate and thus diluting the relative concentration of the inhibitor.

   Competitive inhibition is the basis for the antibacterial action of certain drugs. Among these are the sulfa compounds. These compounds compete with para-aminobenzoic acid (PABA), a component of the B vitamin complex, for the active site of a bacterial enzyme. This enzyme catalyzes the synthesis of a bacterial metabolite from PABA and other precursors that is essential for bacterial growth. But the sulfa compounds prevent the metabolites from forming, and the bacteria are killed. Since humans do not utilize PABA in the same manner as bacteria do, such sulfa compounds are relatively harmless to humans but deadly to bacteria. Competitive inhibitors which act in this manner are referred to as **antimetabolites.**

4. *Other types of enzyme inhibition.* Some inhibitors act upon the enzyme-substrate complex after the substrate has entered the active site. Others inhibit enzyme action by binding a metal ion that is essential for enzyme function. There are a number of inhibitors, properly called poisons, that irreversibly combine with the enzyme or the enzyme-substrate complex, rendering the enzyme incapable of further action. Their effect is directly proportional to their own concentration (not to the ratio of their concentration to the substrate as in competitive inhibition). If the enzyme being inhibited is important to life itself, such inhibitors are deadly poisons; examples are potassium cyanide, which binds to cytochromes (see Chapter 6), and carbon monoxide, which binds to hemoglobin.

## NUCLEIC ACIDS

Nucleic acids differ markedly from carbohydrates, lipids, and proteins both in structure and in the role they play in living systems. In general, we might say that proteins constitute much of the machinery of an organism, phosphoglycerides are the walls of the building housing the machinery, lipids and carbohydrates are the fuel for running it, and nucleic acids provide the blueprints that direct its construction and operation. Nucleic acids are the carriers of genetic information, the inherited molecules that direct the synthesis of all substances required for an organism's growth and vital processes.

FIG. 3-40 Friedrich Miescher, the discoverer of DNA.

## THE CHEMISTRY OF NUCLEIC ACIDS

In 1868, Johann Friedrich Miescher (Fig. 3-40), a German physician, discovered the existence of nucleic acids while serving in the Franco-Prussian war. He had become interested in the chemical nature of the pus found in surgical dressings. Pus contains a high concentration of white blood cells, which have very large nuclei or more than one nucleus. Miescher isolated the nuclei from a preparation of these cells and made extracts of their chemical contents. He found that they contained a substance that could not be classified as a protein, a carbohydrate, or a lipid. The nuclear extract contained a very high proportion of phosphorus and appeared to represent a new class of macromolecules, which Miescher called nuclein. When the extract was purified, it was found to consist primarily of what is now known as **nucleic acid.** But experimental proof that such acids form the molecular basis of heredity and evolution did not come about until the 1950s.

## THE STRUCTURE OF NUCLEIC ACIDS

Nucleic acids are composed of three molecular units: a pentose (five-carbon sugar), phosphate ($PO_4$) groups, and various nitrogen-containing ring compounds referred to as nitrogenous bases. The nucleic acid with deoxyribose (ribose with one oxygen removed) as its pentose sugar is known as **deoxyribonucleic acid** or, simply, **DNA.** DNA is the genetic material of most organisms and is usually confined to the nucleus. Another nucleic acid occurs in cells but is found mostly in the cytoplasm. It differs in several ways from DNA, one of them being that the pentose it contains is ribose; it has accordingly been named **ribonucleic acid,** or **RNA.**

***Nucleotides.*** Like proteins and polysaccharides, nucleic acids are composed of long chains of complex subunits. Nucleic acid subunits are called **mononucleotides.** Both DNA and RNA are **polynucleotides,** long-chain polymers of mononucleotides. As Figure 3-41 illustrates, each mononucleotide consists of a nitrogenous base bonded to a pentose sugar that is also bonded to a phosphate group. A given polynucleotide contains one of five different nitrogenous bases belonging to one of two groups known as purines and pyrimidines. All pentoses and all phosphate groups of a polynucleotide are identical; any difference from one mononucleotide to the next involves only the nitrogenous bases. The purines—adenine and guanine—are double-ring bases, while the pyrimidines—cytosine, thymine, and uracil—are single-ring structures (Fig. 3-42). Purine molecules are thus larger than pyrimidine molecules, a matter of some importance, as we shall see later. Since mononucleotides differ only in the nitrogenous base they contain, they are named according to the type of base they contain. Thus, the adenine-containing nucleotide shown in Figure 3-43 is adenylate, or adenylic acid. Because the portion composed of adenine and ribose is named adenosine, this same nucleotide is often referred to as *adenosine monophosphate,* or *AMP.* A nucleotide with guanine as its base is called guanylate, guanylic acid, or *guanosine monophosphate (GMP),* and so on.

FIG. 3-41 Basic structure of a nucleotide.

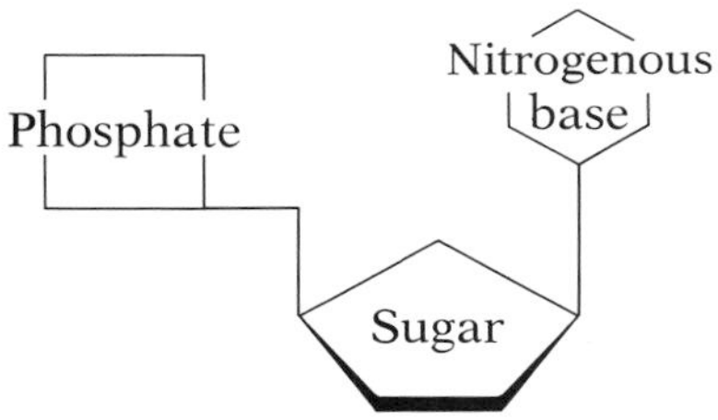


***Other Roles for Mononucleotides.*** Besides being components of nucleic acids, some mononucleotides have other important functions in cells. Some have one or two additional phosphate groups, forming mol-

**FIG. 3-42** Structural formulas of the purines and pyrimidines.

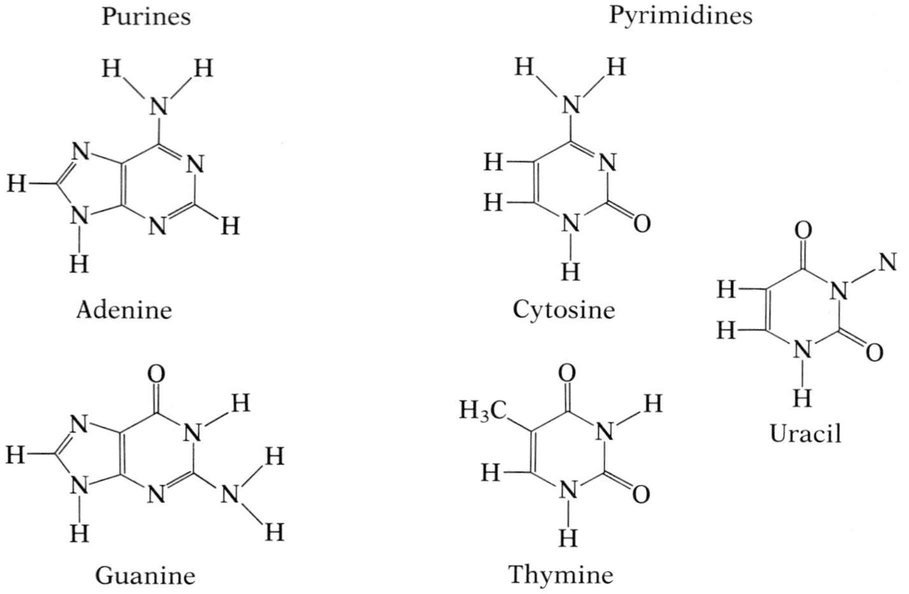


ecules important in cellular energy storage and transfer. Two such energy-rich compounds are adenosine diphosphate (ADP) and adenosine triphosphate (ATP) (Fig. 3-44a). Another energy-rich molecule is cyclic AMP, which is produced by the removal of pyrophosphate from ATP by the enzyme adenylate cyclase; this results in the closure of a six-member ring as shown in Figure 3-44b. Some nucleotides are also constituents of important coenzymes.

## THE STRUCTURE OF DNA: THE DOUBLE HELIX

The three-dimensional structure of DNA was established during the 1950s following an intensive series of studies on the chemical and physical properties of the molecule. In 1950, the American biochemist Erwin Chargaff extracted DNA from a variety of organisms and found that,

**FIG. 3-43** Structural formula of adenosine monophosphate (AMP).

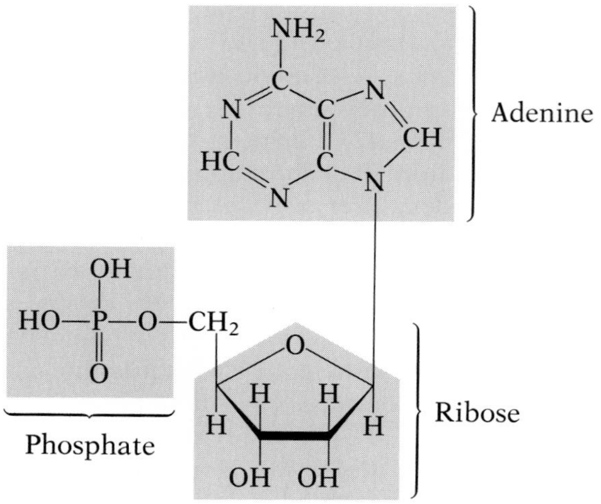

**FIG. 3-44** (a) Structural formula of adenosine triphosphate (ATP). (b) Formation of cyclic adenylate (cyclic AMP) by the removal of pyrophosphate from adenylate.

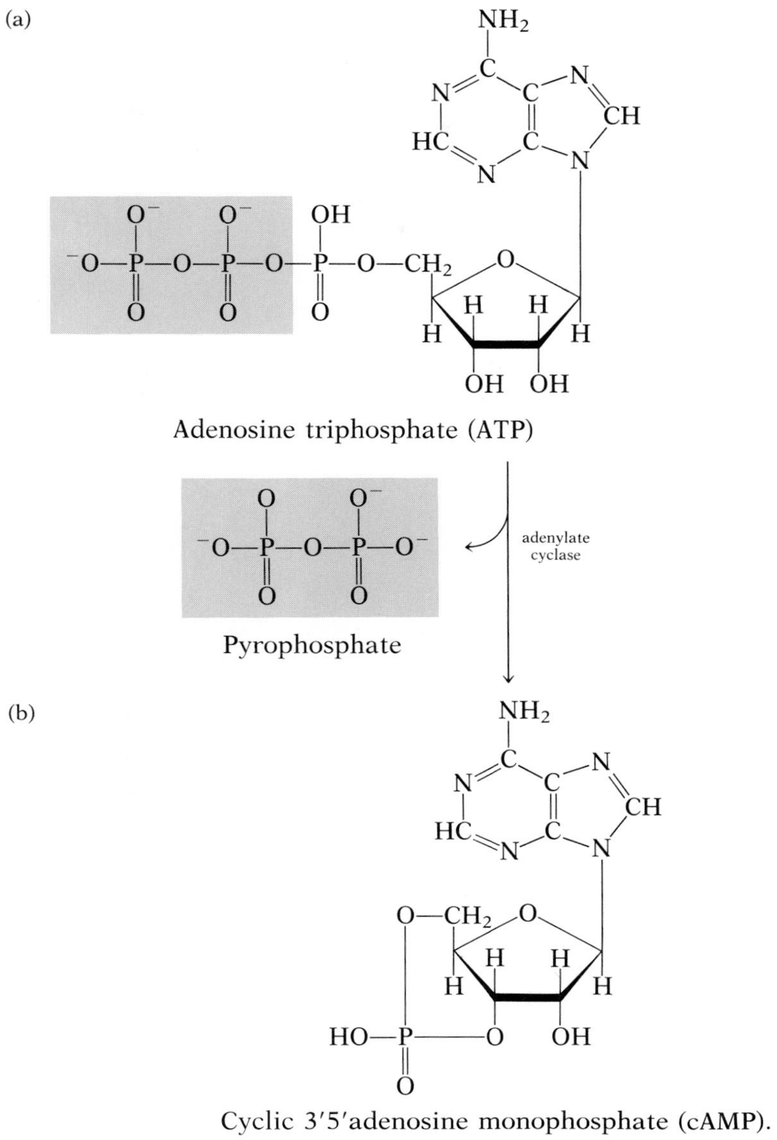


regardless of the source of the DNA, a ratio of approximately 1:1 existed between the number of adenine (A) molecules and the number of thymine (T) molecules, and between the number of guanine (G) molecules and the number of cytosine (C) molecules. He also found that usually the number of A—T pairs does not equal the number of G—C pairs. This would imply that the members of each pair of bases, A—T and G—C, are related to each other in some way within every DNA molecule. The regular occurrence of this ratio in the DNA of all organisms became known as *Chargaff's rules.*

Chargaff's rules, an awareness of the helical nature of the primary structure of proteins as developed by Linus Pauling and Robert Corey in 1951, and some X-ray diffraction data for DNA prepared by Rosalind Franklin and Maurice Wilkins formed the basis of a molecular model of the structure of DNA proposed in 1953. The proposal of a young American biologist, James D. Watson, and the British physicist Francis Crick

**FIG. 3-45** James D. Watson (left) and Francis Crick (right) clarified the molecular structure of DNA, one of biology's great riddles.

**FIG. 3-46** Watson-Crick double helix model of the DNA molecule. S = sugar; P = phosphate; A, T, G, and C = nitrogenous bases.

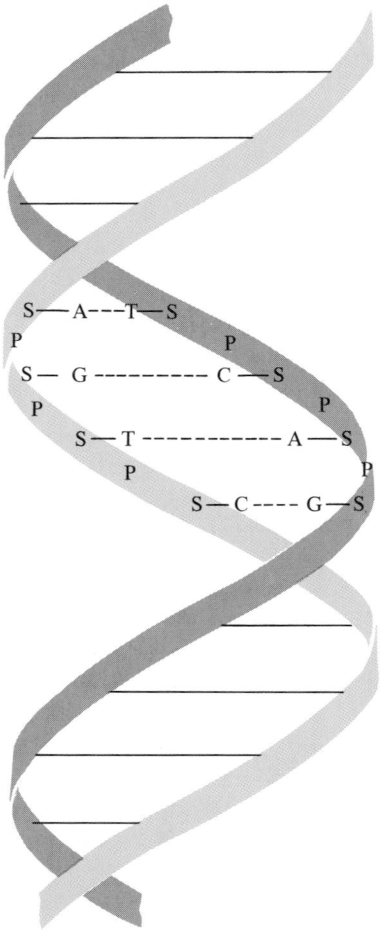


(Fig. 3-45), both working at the Cavendish Laboratories in Cambridge, England, suggested that DNA exists as a double helix, in which two nucleotide chains are twisted together much like a pair of polypeptide chains forming a long filament. In their well-known double helix model, illustrated in Figure 3-46, the nitrogenous bases opposite each other in the paired chains form hydrogen bonds with each other; because of their complementary sizes and matching bonding capabilities, adenine always pairs only with thymine (A with T) and guanine only with cytosine (G with C). As shown in Figure 3-47, the two strands of the double helix are actually held together by hydrogen bonds between these base pairs. The A—T bonding involves two hydrogen bonds; G—C bonding involves three. Although each bond is weak, the total number is great, as is their collective strength. The shape of the double helix that results is much like a twisted rope ladder or a spiral staircase. The two parallel supports of the ladder are the chains formed by the linked phosphate and deoxyribose residues. Each rung of the ladder (or step of the staircase) is composed of a base pair, either A—T or G—C.

The proposed double helix model explained the 1:1 ratio of A—T and G—C discovered by Chargaff. Because of the larger sizes of the purines (A and G), the possibility of bonding between them (A—G) was ruled out: there was not enough room within the DNA molecule to accommodate the pairing of two large bases. Pairings between two pyrimidines (T with C) would not work either: the two small bases would be too far from each other for effective hydrogen bonding. Only A—T and G—C pairings would allow the members of any nucleotide pair to form hydrogen bonds between them. The idea that the two chains are held together by hydrogen bonds also explained the ease with which the two strands of a DNA molecule can be separated by gentle heating.

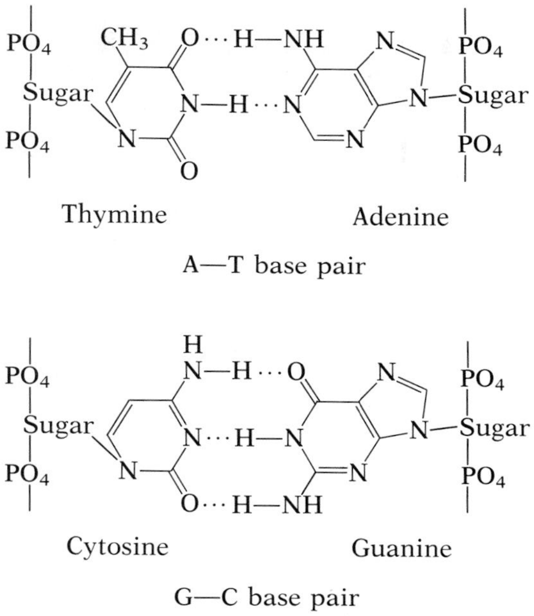


**FIG. 3-47** The hydrogen bonds (dotted line) which join the purine and pyrimidine base pairs in a DNA molecule.

Their model of DNA structure also suggested to Watson and Crick the mechanism for the replication of DNA molecules, an event that must occur at some time in the life of each cell if the cell is to hand on to subsequent generations copies of all the DNA molecules it possesses. They proposed that prior to replication, the hydrogen bonds between matching bases in the double helix are broken and each "rung" separates into two portions. As this occurs, the two chains unwind from each other and separate, as illustrated in Figure 3-48. By attracting available mononucleotides to the exposed bases—the separated pieces, or "rungs"—each chain can serve as a template, or pattern, for the formation of a new companion. Because this happens to both of the separated chains, two new pairs of DNA chains, two double helixes, are formed from a single old one. Because of the precise nature of base pairing, each old chain has a new companion chain identical to the one that had separated from it at the beginning of the process. In 1962, Watson and Crick shared a Nobel prize for their insightful model of DNA.

**FIG. 3-48** Mechanism of DNA replication.

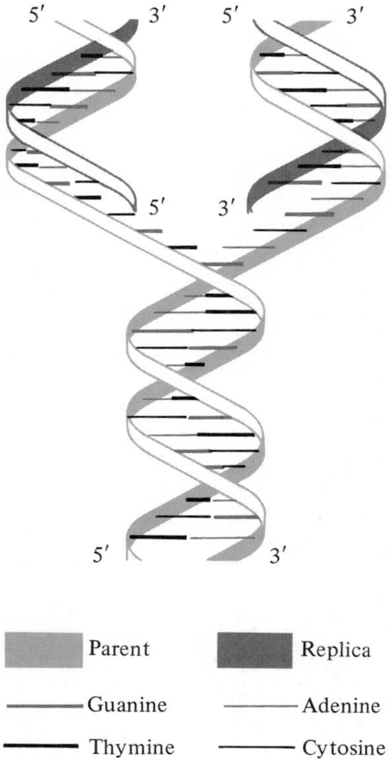


## RNA

Like DNA, RNA is a long, unbranched polynucleotide, but the nucleotides of RNA differ somewhat in composition from those of DNA. An RNA nucleotide is made up of the pentose ribose, a phosphate group, and a nitrogenous base, either adenine, guanine, cytosine, or uracil (U). Thus, the chemical composition of RNA nucleotides differs from that of DNA nucleotides in two respects: RNA contains ribose rather than deoxyribose and uracil rather than thymine. The only difference between uracil and thymine is that uracil has no methyl group.

Structurally, RNA molecules are single-stranded. Since they do not have the double helix structure of DNA, RNA molecules lack the complementary base ratios found in DNA; Chargaff's rules do not apply. In most RNA molecules, the proportion of adenine differs from that of uracil, and the proportion of guanine from that of cytosine.

The three kinds of RNA that exist in cells are different in size and in function. Transfer RNA, ribosomal RNA, and messenger RNA cooperate in the synthesis of proteins in the cell.

***Transfer RNA.*** Transfer RNA, or tRNA, was initially called soluble RNA because, as the smallest of the RNA molecules, it is much more soluble in water than the other two kinds of RNA. Each tRNA molecule contains only 70 to 80 nucleotides. The several kinds of transfer RNA are able to join temporarily to individual amino acid molecules within cells, delivering the amino acids to the **ribosomes,** the sites of protein synthesis. Since any one kind of tRNA molecule will bind to only one type of amino acid, it was deduced that there are at least 20 molecular versions of tRNA, one for each type of amino acid.

***Ribosomal RNA.*** Ribosomal RNA (rRNA) is the most abundant kind. It is a major component of ribosomes, the cellular structures on which proteins are synthesized (see Chapter 4). Protein synthesis occurs only on the surface of ribosomes.

***Messenger RNA.*** Messenger RNA is synthesized in the nucleus and transported to the cytoplasm, where it participates in protein synthesis with the ribosomes. Since protein synthesis is a cytoplasmic process, a message is needed to carry the coded information contained in the DNA molecules of the chromosomes from the nucleus to the cytoplasm (Fig. 3-49). Each messenger RNA (mRNA) molecule carries coded information for the synthesis of a specific protein. The code is in the form of a sequence of nucleotides in a particular mRNA molecule. Through chemical reactions taking place on the ribosomes, the nucleotide sequence becomes transformed into an amino acid sequence, as it is "translated" from the "language" of nucleic acids to the "language" of protein molecules. The precise sequence of bases in a given mRNA molecule is de-

**FIG. 3-49** Part of a chromosome of the bacterium *Escherichia coli* showing newly synthesized RNA. The ribosomes are attached to strands of messenger RNA.

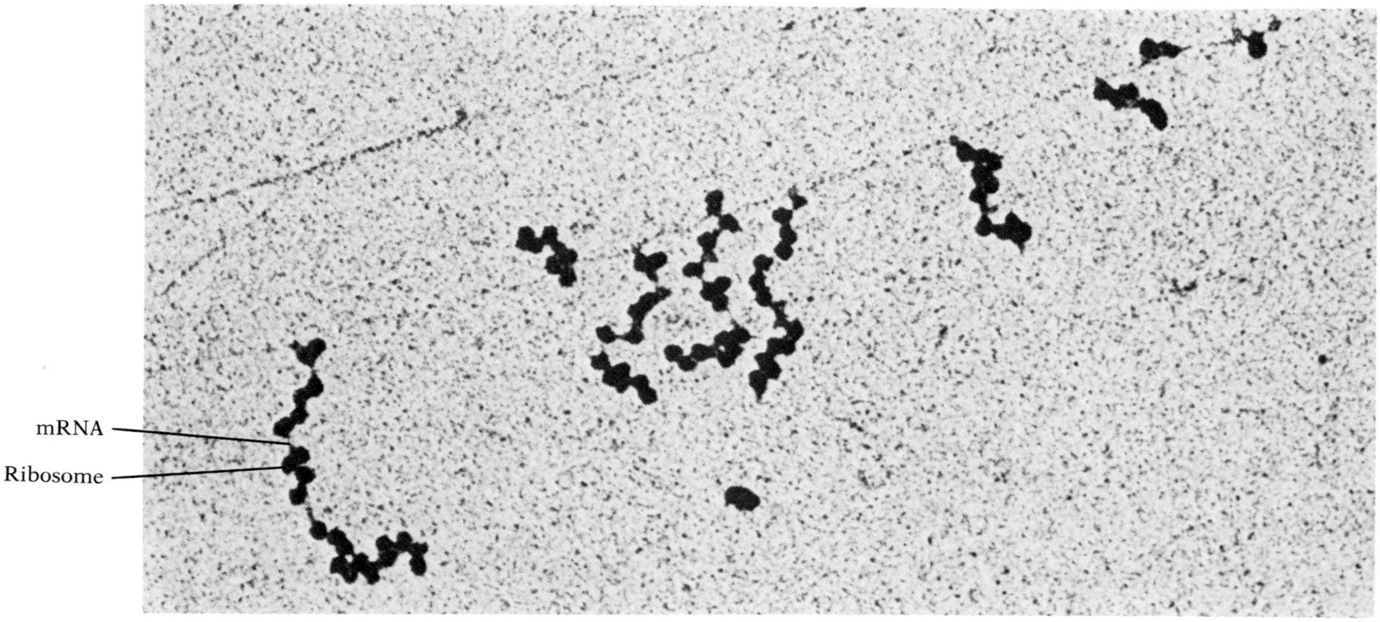

rived from the nucleotide sequence of the particular DNA molecule after which it was patterned when the mRNA was synthesized in the nucleus. The fact that DNA is an inherited molecule explains how each generation of a given species produces the same proteins—and the same species. The process by which mRNA and specific proteins are synthesized is described in greater detail in Chapter 24.

## Summary

The molecules of living systems determine the fundamental characteristics of those systems. An understanding of atomic and molecular structure and of the interactions of atoms and molecules is thus basic to an understanding of life. The main components of living systems—carbon, hydrogen, oxygen, and nitrogen—are elements of relatively small atomic size. Small atoms form short bonds and generally form relatively stable compounds. Some larger atoms, such as phosphorus and sulfur, which form less stable bonds, play major roles in biological energy exchanges. The major classes of compounds that comprise living things are carbohydrates, lipids, proteins, and nucleic acids.

Energy captured in photosynthesis is trapped in chemical form as carbohydrate. This class of compounds serves as the main source of energy, both for producers and for consumers. Carbohydrates are also the primary constituents of wood, and they form an essential part of nucleic acids and some proteins.

Lipids fall into three large classes: (1) the mono-, di-, and triacylglycerols, which comprise an important storage form of energy for most organisms; (2) the phosphoglycerides, of particular importance as the primary constituents of biological membranes; and (3) the steroids which comprise a large family of biologically active compounds that includes many important vertebrate hormones.

Because of their large size and the many different kinds of amino acids of which they are composed, proteins exhibit great variety. Various proteins serve such functions as structural support, enzymatic catalysis, muscle contraction, and immune protection, among others.

Because of their capability for self-replication in the presence of the necessary enzymes, nucleic acids are truly inherited molecules. Most important, the sequence in which their units, the nucleotides, are assembled specifies the sequence in which amino acids will be assembled during protein synthesis. An organism's DNA contains coded specifications for all the proteins the organism will produce, including the enzymes that determine what reactions can occur and thus how the organism will develop and behave.

## Recommended Readings

Alberts, B., D. Bray, J. Lewis, M. Raff, K. Roberts, and J. D. Watson. 1983. *Molecular Biology of the Cell.* Garland, N.Y. A concise and up-to-date description of the basic structural and functional characteristics of prokaryotes and eukaryotes. Contains a well-illustrated chapter on the chemical components of cells and protein structure.

Calvin, M., and W. A. Pryor. 1973. *Organic Chemistry of*

*Life*. Freeman, San Francisco. A collection of *Scientific American* articles that includes several papers relevant to topics covered in this chapter.

DICKERSON, R., and I. GEIS. 1985. *The Structure and Action of Proteins,* 2d ed. W. A. Benjamin, Menlo Park, Calif. A very clear description of the structure and function of proteins.

FRIEFELDER, D. 1983. *Molecular Biology*. Van Nostrand Reinhold, N.Y. This multipurpose text on molecular biology contains excellent chapters on the basics of biochemistry, macromolecules, nucleic acids, and proteins.

KENDREW, J. C. 1961. "Three-dimensional study of a protein." *Scientific American,* December 1961. An excellent, well-illustrated discussion of the methods by which the three-dimensional structure of the protein myoglobin was determined.

LEHNINGER, A. L. 1982. *Principles of Biochemistry*. Worth, N.Y. This new text by Lehninger is intended primarily for students taking their first course in biochemistry. It has excellent chapters on the structure and function of biologically important molecules.

SUTTE, J. W. 1977. *Introduction to Biochemistry*, 2d ed. Holt, Rinehart and Winston, New York. An introductory biochemistry text that contains excellent chapters on the chemistry of carbohydrates, lipids, proteins, and nucleic acids.

WATSON, J. D. 1968. *The Double Helix*. Atheneum, New York. A fascinating first-person account of the events leading to the discovery of the structure of DNA.

# 4

# The Cell: Unit of Life

OUTLINE



A human being is made up of a hundred trillion or more cells. We have about a million cells in every square inch of our skin, about 30 billion cells in our brain, and about 20 trillion red blood cells in our blood, 30,000 of which would fit into this letter O. On the other hand, bacteria and many other microscopic organisms consist of only a single cell.

Because it is impossible for the unaided eye to detect the cells of a tissue, the study of cells depended on the development of the microscope. Indeed, the development of our present understanding of the structure and function of the cell closely parallels the evolution of the microscope.

## THE CELL THEORY

The **cell theory**—the concept that the cell is the fundamental unit of life—has been accepted for about a century and a half. However, although this concept dates from the nineteenth century, its roots can be traced to the pioneering work of scientists working in previous centuries.

## THE EARLY MICROSCOPISTS

The word *microscope* means "seer of little things." The earliest microscopes were simple glass globes filled with water; Pliny the Elder left a record of their use in the first century A.D. The use of carved gemstones as magnifying glasses is thought to have been in use at least 500 years earlier. However, the discovery of cells awaited the development of the modern microscope in 1590 by the Dutch optician Zaccharias Janssen.

One of the first to use a microscope to describe the structure of living materials was the Englishman Robert Hooke (1635–1703). Looking at thin slices of cork through a microscope that he built, Hooke saw little boxes arranged somewhat in the manner of a honeycomb. His sketch of them is shown in Figure 4-1. Hooke coined the term **cell** to describe each tiny individual structure because it reminded him of the stark living quarters of a monk.

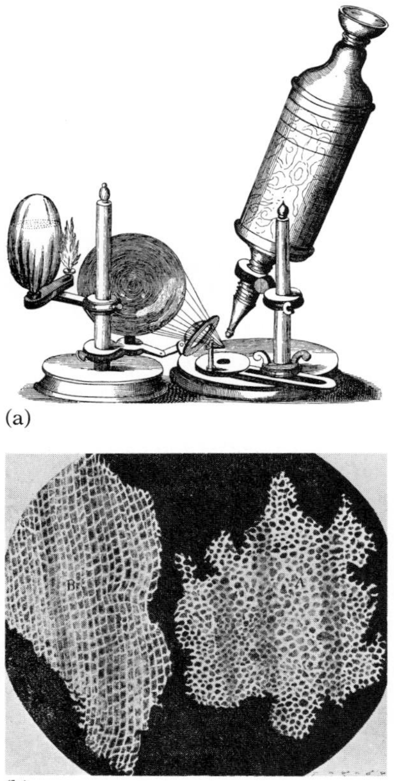

(a) (b)

**FIG. 4-1** (a) Robert Hooke's microscope and (b) his sketch of a thin slice of cork.

***Leeuwenhoek.*** One of Hooke's contemporaries, the Dutch merchant Antoni van Leeuwenhoek, used the time during slack hours in his dry goods store to build simple microscopes. He would mount a lens between two flat pieces of metal and add a point for holding the specimen; this point could be adjusted to bring the specimen into focus by means of a pivot and two motions controlled by screws (Fig. 4-2). Altogether, van Leeuwenhoek built 247 of these microscopes, grinding each lens himself. He was the first to describe living cells: protozoans collected from a rain barrel, spermatozoa from several animal species, red blood cells moving through the blood vessels of a fish tail, the connecting vessels (capillaries) between arteries and veins, and the bacteria from his own mouth. During the late 1600s, he communicated many of his findings in meticulous notes, complete with sketches and magnifications of each object, to the Royal Society of London, where a permanent record of his correspondence is kept.

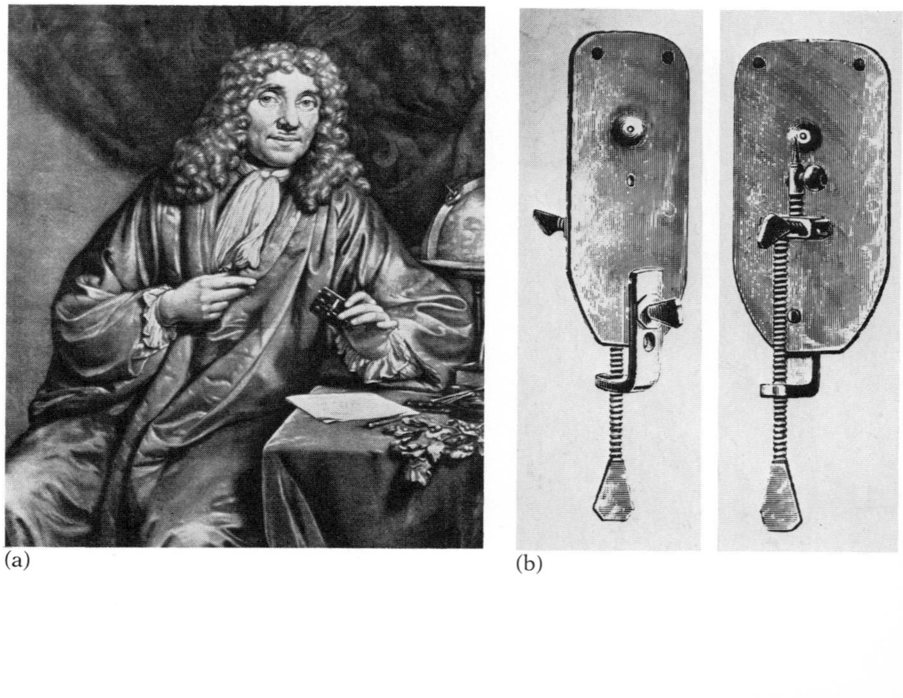

(a) (b)

**FIG. 4-2** (a) Antoni van Leeuwenhoek. (b) Microscope used by Leeuwenhoek to study the structure of cells and tissues.

***Dutrochet and Brown.*** French botanist Henri Joachim Dutrochet observed in 1824 that "the cell is the fundamental element of organization," and "growth occurs by the enlarging of old cells and formation of new ones." Moreover, he noted, "plants and animals are composed entirely of cells and of organs derived from cells." In 1831 British botanist Robert Brown added to this picture by describing the cell nucleus he had observed in a study of the cells of orchids. However, despite these various published observations, a wide appreciation of the fact that cells are the fundamental units of structure and function of living things was slow in coming. That understanding did not fully develop until a controversy arose over a related matter: the origin of new cells.

***Schleiden and Schwann.*** In late 1838, the German botanist Matthias Schleiden published an account of the process by which new plant cells arise. Schleiden contended that new cells arise by the formation of blisters within old cells and did not believe that old cells divide into new ones. However, his paper also strongly emphasized that cellular tissue is the rule in plants and that cells are the fundamental units of development; in these two points he was correct. Early in 1839 a German zoologist, Theodor Schwann, published a paper affirming that all of Schleiden's observations for plants were also true for animals. In applying the cell theory to all organisms, Schwann spoke of the cell as an organism and said that "entire animals and plants are aggregates of these organisms arranged according to definite laws." In promoting their erroneous theory of how new cells arise, Schleiden and Schwann attracted much attention and, although their theory of cell reproduction was soon disproven, they became rather well known for having "jointly announced" the cell theory.

***Virchow.*** Another important advance in the development of the cell theory came 20 years later when Rudolph Virchow, a renowned German pathologist, published his observation that "where a cell exists, there must have been a preexisting cell, just as the animal arises only from an animal and the plant only from a plant." He summarized this doctrine in an often quoted Latin phrase, "*Omnis cellula e cellula,*" meaning that all cells arise from other cells. The cell had now been identified as the simplest living unit, the unit that forms the basis of all structure, function, growth, and reproduction.

## TECHNIQUES IN THE STUDY OF CELLULAR ORGANIZATION

### THE MICROSCOPE

Although the function of a microscope is to magnify an object so that it becomes visible to the naked eye, its usefulness lies not only in its ability to magnify. Another characteristic of a magnifying device is its ability to separate clearly the individual parts of an image, a quality called **resolving power.** The human eye is capable of distinguishing two points as distinct from each other only if they are separated by at least 0.1 millimeter (mm). Two objects closer than this appear as one. An effective microscope must not only enlarge the image to dimensions that can be seen by the human eye, but it must also be able to resolve fine detail.

***The Limits of Light.*** Figure 4-3 shows the main features of the light, or optical, microscope. Although the light microscope was the basic instrument used for the study of cell structure for about three hundred years, its resolving power is not sufficient to reveal the fine details of cellular structure. No matter what the magnification achieved, light waves are too long to permit the clear resolution of two objects closer together than about 0.2 micrometers ($\mu$m), or 200 nanometers (nm) (see Panel 4-1). The smallest bacteria can barely be seen in the best light microscope (Fig. 4-4). Most internal cellular structures, or **organelles,** cannot be viewed by light microscopy.

***The Cell as Viewed in the Light Microscope.*** The limited resolving power of light microscopes resulted in a simplistic view of the cell. It was long thought, for example, that all cells were largely composed of a "fundamental substance," **protoplasm,** that was much the same from one cell to another. Differences between cell types were believed to be largely a matter of different "inclusions," or organelles. As knowledge of cellular chemistry and enzymes developed, the cell next was thought of as a "sac of enzymes" with little in the way of structure beyond the limits of light microscopy. This simple view of cell structure is represented by the well-known illustration of a generalized cell shown in Figure 4-5a; it first appeared in 1925 in *The Cell in Development and Heredity,* a now classic book by the American biologist Edmund B. Wilson. His sketch shows cell structure as seen under the light microscope after being treated with commonly used stains: clear **cytoplasm,** or "ground substance," containing various inclusions and a **nucleus.**

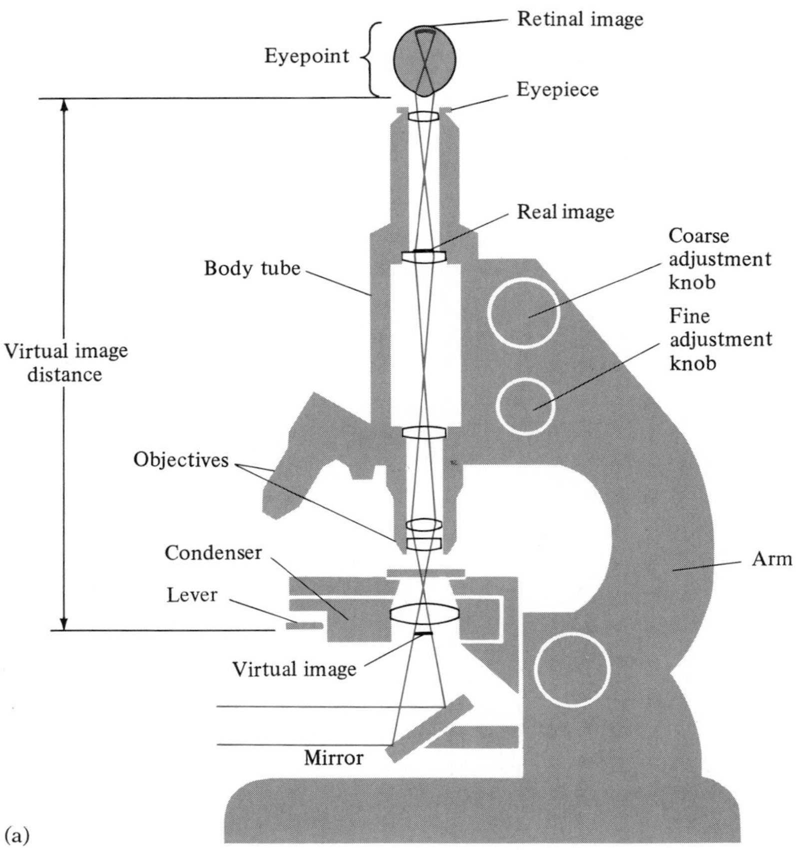


(a)

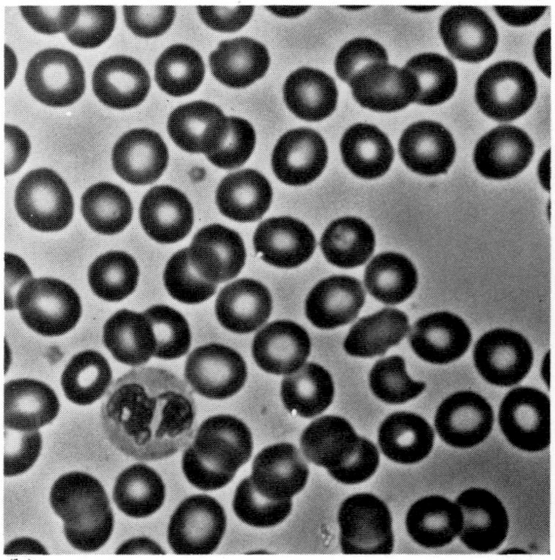

(b)

**FIG. 4-3** (a) Basic features of the optical, or light, microscope. (b) Photograph of human red blood cells as seen with a light microscope.

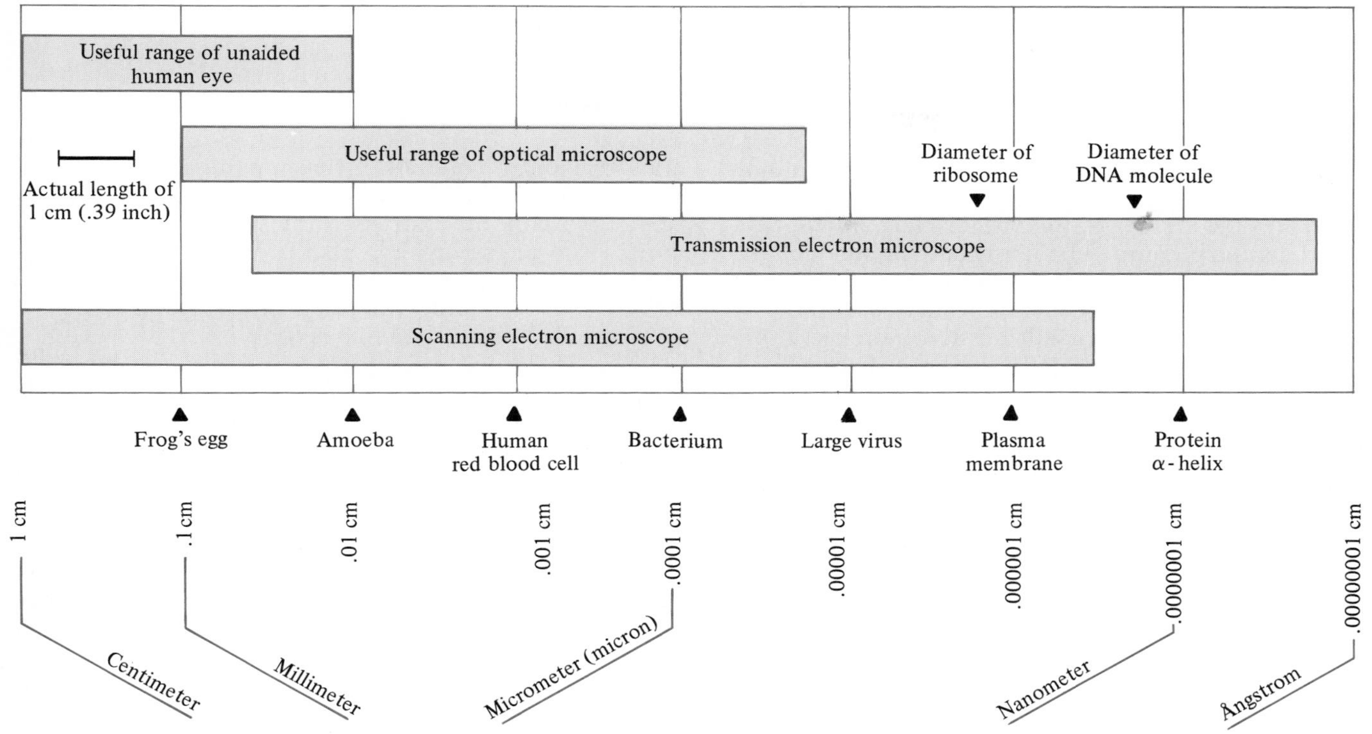


**FIG. 4-4** Scale showing the resolving power of the optical, transmission electron, and scanning electron microscopes. In eight steps, the scale moves from the centimeter to the submicroscopic Ångstrom, a hundred-millionth of a centimeter. Each unit of measurement is one-tenth the size of the preceding one. The size of a fermi is a ten-trillionth of a centimeter. See the inside back cover for a table of International System units.

## THE ELECTRON MICROSCOPE

The transmission electron microscope (TEM) uses a beam of electrons rather than light rays and focuses the beam by using electromagnets instead of glass lenses (Fig. 4-6a). The resolving and magnifying power of the TEM is as much as 400 times that of the light microscope.

The kind of cellular detail revealed by the TEM is usually referred to as a cell's fine structure, or ultrastructure. Because the human eye cannot detect electron beams, the image is focused on a fluorescent screen or a photographic plate. In practice, the screen is used to select the

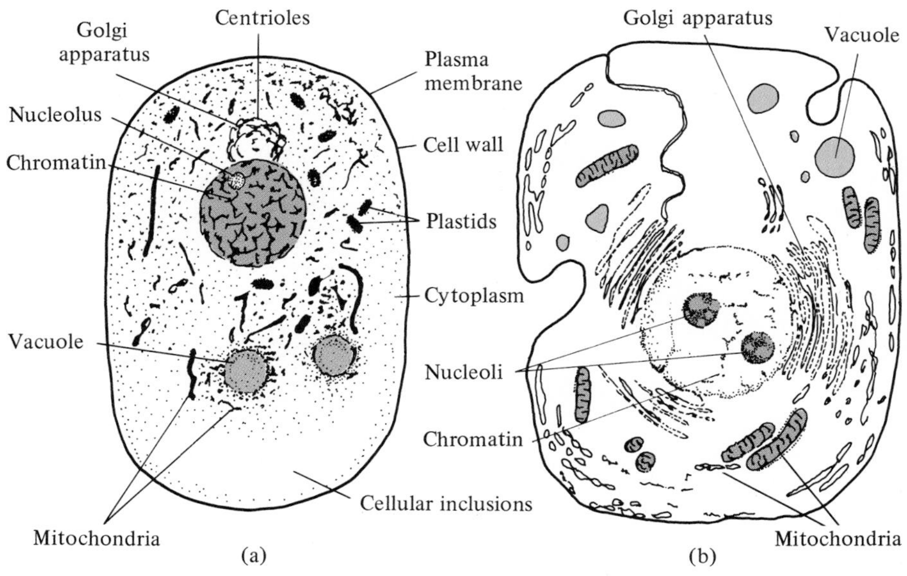


**FIG. 4-5** (a) Diagram of a generalized cell, based on the original sketch drawn by Edmund B. Wilson in 1925. It illustrates what can be seen in stained cells by light microscopy and represents the conception of cellular structure before development of the electron microscope. (b) Jean Brachet's contemporary cell is based on modern electron microscopy.

## PANEL 4-1

# Units of Measurement in Biology

The units of measurement most often used in biology are the micrometer (pronounced with the accent on the first syllable rather than on the second, as in the name of the measuring device) and the angstrom. The micrometer is symbolized $\mu$m and is identical to the older term micron ($\mu$), which is still widely used. A micrometer is a millionth of a meter ($1 \times 10^{-6}$ meter); the angstrom unit (Å or A) is $1 \times 10^{-10}$ meter. Although the angstrom unit is not part of the International System (SI) of Units adopted by international agreement in 1960, its use is being retained for the present in cell biology because of the great convenience of its small size: the most powerful microscopes cannot measure anything smaller than an angstrom, and any of the dimensions of the smallest subcellular structure can be expressed in whole angstrom units. The angstrom has also proven useful as a unit of molecular measurement: amino acids are about 10 Å long, and the adenosine triphosphate (ATP) molecule is about 20 Å long.

The SI unit nearest in size to the angstrom is the nanometer (nm), formerly called the millimicron (m$\mu$). The nanometer is one order of magnitude larger than the angstrom: 1 Å = 0.1 nm. The nanometer is $1 \times 10^{-9}$ meter in length. SI units are expected to replace all units eventually. See the inside back cover for a table of metric units.

appropriate sections and the plate to make a photograph, called an **electron micrograph.**

In a transmission electron microscope, a beam of electrons passes through a specimen that has been treated to make certain parts of it electron-dense. In order to improve contrast, the specimen is treated with reagents that make some parts of it absorb electron beams more than others (Figure 4-6b). This procedure requires that the material be sliced or sectioned very thinly or be very thin to begin with. Another technique is shadow casting, in which a very thin preparation is placed in an evacuated chamber and a beam of electron-dense material is sprayed on it from one side. This procedure highlights the parts of the specimen that are the thickest, that is, that protrude the most; these allow some surface details to be revealed when the specimen is placed in the TEM (Fig. 4-7).

***Updating the Generalized Cell.*** The development of the TEM, of thin sectioning, and of various methods of electron staining led to a view of the cell far different from the one represented in Wilson's sketch. Many cells proved to have a highly structured cytoplasm with complex organelles, several of which had never been seen in light microscopy. One generalized representation of this updated view of the cell is shown in Figure 4-5b. This sketch, prepared by the Belgian embryologist Jean Brachet in 1961, is now even more well known than Wilson's classical drawing.

***Interpreting Two-Dimensional Electron Micrographs.*** One of the disadvantages of light and transmission electron microscopy is that light must pass through a thin section of the specimen. The resulting image is flat and bears little resemblance to the three-dimensional appearance of the live specimen. With the light microscope, the observer can partly compensate for this shortcoming by "focusing up or down" and examining the specimen at various levels. The rather thick preparations used in light microscopy make this possible. But with the TEM it is different. After the thin section or shadowed thin object is focused on a fluores-

**FIG. 4-6** (a) The transmission electron microscope (TEM). (b) A human pancreas cell as seen in a TEM. (c) The scanning electron microscope. (d) Human red blood cells as seen in a SEM.

(a)

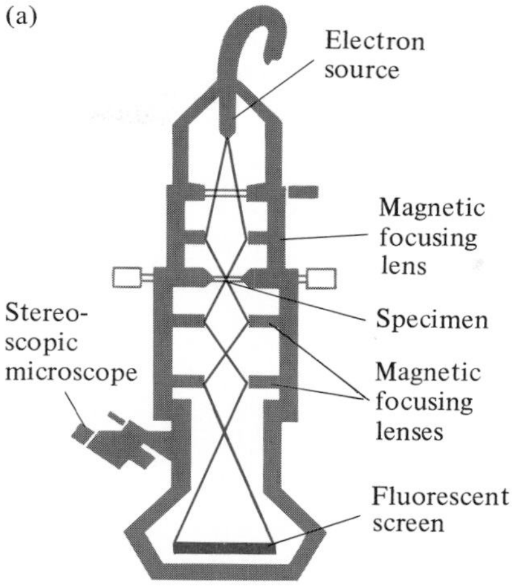


(a) In the transmission electron microscope (TEM), an electron gun provides a beam of electrons that are focused by a series of magnetic condenser lenses; these lenses are equivalent to the glass magnifying lenses in a light microscope. The beam of electrons passes through the specimen being examined and is focused onto a fluorescent screen. The greatly magnified image can be examined by a stereoscopic microscope, and any particular area can be photographed to obtain an electron micrograph as shown in Figure (b).

(c)

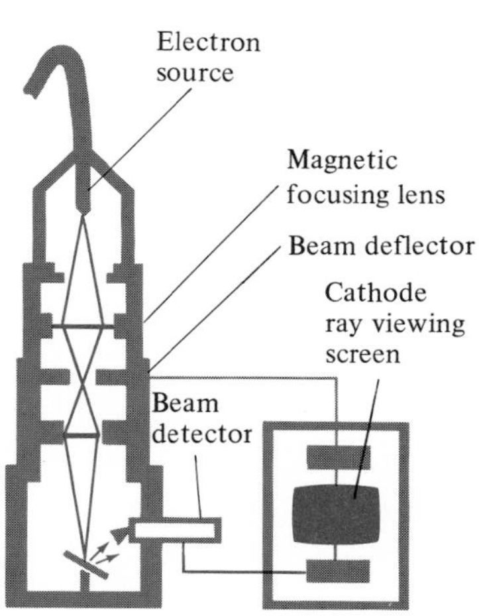


(c) In the scanning electron microscope (SEM), the magnetic lenses focus the beam of electrons from the electron gun into an exceedingly fine beam which is rapidly swept across the specimen surface. The specimen's molecules are excited to high energy levels by the beam's electrons; as a result, the specimen releases secondary electrons. These electrons are picked up by a detector and amplified and are then projected onto a cathode ray viewing screen, where the image can be viewed and photographed as shown in Figure (d).

(b)

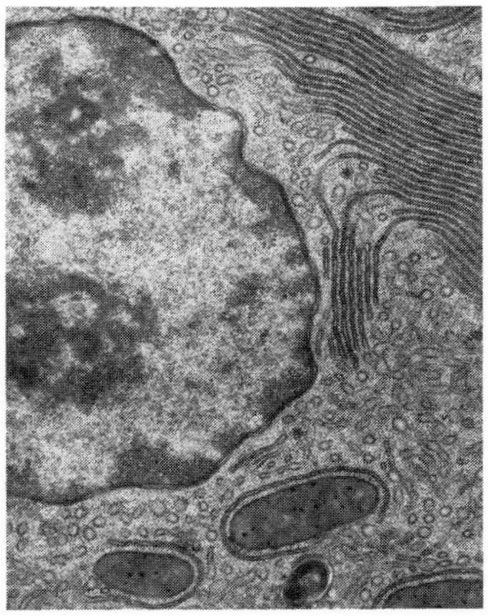

(d)

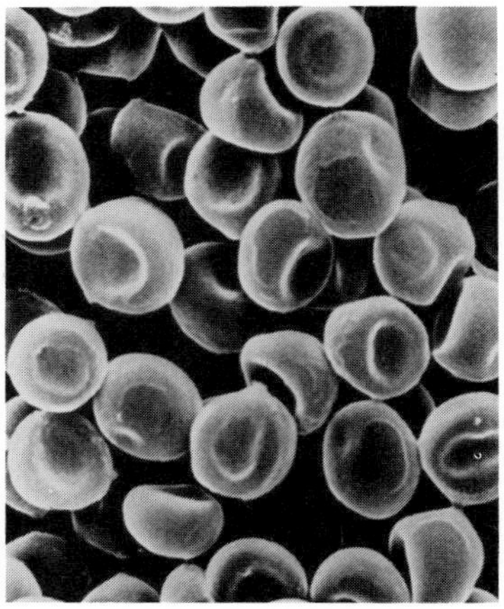

**FIG. 4-7** Viruses attached to a colon bacillus. "Shadowing" with uranium oxide makes the viruses stand out in relief.

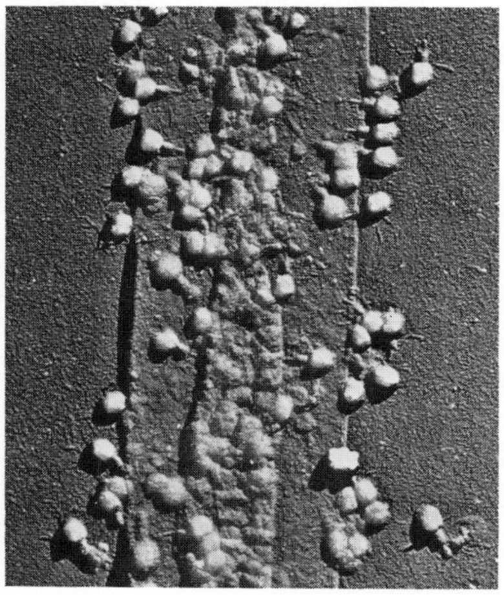

cent screen, a photographic plate is inserted in the instrument and an electron micrograph is taken. Because the result is two-dimensional, the novice microscopist often has difficulty in interpreting electron micrographs. This is also true for photomicrographs taken through the light microscope.

Another disadvantage of electron microscopy is that only dead cells and tissues can be viewed. With the light microscope, it is possible to view living material and see details within living cells. In electron microscopy, however, the specimen must be in an evacuated chamber so that electron beams can be employed as the transmitting medium; no moisture—essential for life—is permitted in the preparation.

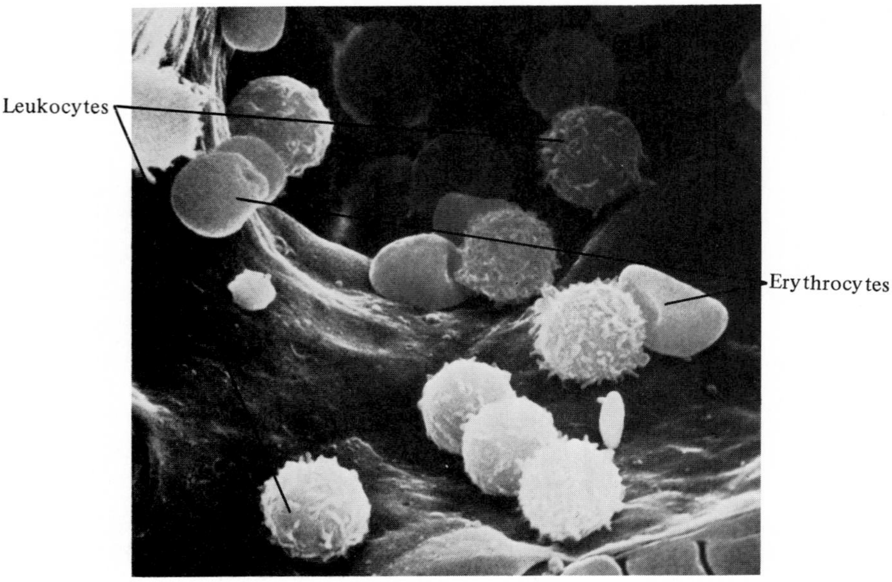


**FIG. 4-8** Scanning electron micrograph showing a number of biconcave erythrocytes and rounded leukocytes in an arteriole.

***The Scanning Electron Microscope.*** With the development of the scanning electron microscope (SEM), biologists could, for the first time, view the surfaces of cells and cell parts in three dimensions in the kind of detail previously limited to thin section viewing (Fig. 4-8). The SEM projects a beam of electrons that penetrates the specimen slightly, rather than passes through it. This causes the specimen to release a shower of its own electrons, which are collected by a signal detector and transmitted to a cathode tube viewing screen (Fig. 4-6c). As with the TEM, permanent electron micrographs can also be made (Fig. 4-6d). The magnification and resolution provided by the SEM is less than that of the TEM; but because of its three-dimensional image, its use complements that of the TEM.

## THE ISOLATION OF CELLULAR COMPONENTS

While the examination of cells with microscopes provides a visual image of the cells, the isolation of cellular components is another powerful tool of the cell biologist. With this technique one can study the biochemical properties of these components. Cellular components can be removed, isolated from other components, and studied by a wide variety of methods. However, all the methods involve the same basic steps.

### A NEED FOR MANY CELLS OF THE SAME KIND

The first step in a study of the components of a particular kind of cell is to obtain a large quantity of cells of the type to be analyzed, preferably all of the same age and condition. For example, cellular preparations made from liver, muscle, leaves, and other tissues are relatively homogeneous and easy to obtain in abundance. In addition, cultures of unicellular organisms such as algae, bacteria, or protozoa can be raised in large numbers without contamination.

## FRAGMENTATION

The cells of a tissue or cell culture must be fragmented before the cellular components can be separated from each other. Large tissues are put in a food blender, called a homogenizer, and reduced to a souplike homogenate (Fig. 4-9). Ultrasound, high frequency vibrations, and repeated freezing and thawing are other techniques used to break up cells and tissues. Tissues and cells must be kept chilled in a container of ice to keep cellular components from being digested by the enzymes of the cells.

## SEPARATION

Once a tissue or culture is reduced to a homogenate, or *brei* (German for "broth"), the next step is to separate the cell components from each other. The more soluble molecules can be removed by treating the material with a succession of different buffered salt solutions and solvents. The solutions can then be analyzed for their contents.

## CENTRIFUGATION AND SEPARATION OF THE COMPONENTS ON THE BASIS OF CELL DENSITY

To assist gravity in separating the denser particles from the homogenate, the material is placed in a tube and centrifuged at high speed. The denser particles settle to the bottom of the centrifuge tube, where they

**FIG. 4-9** The separation of cellular components by differential cell fractionation.

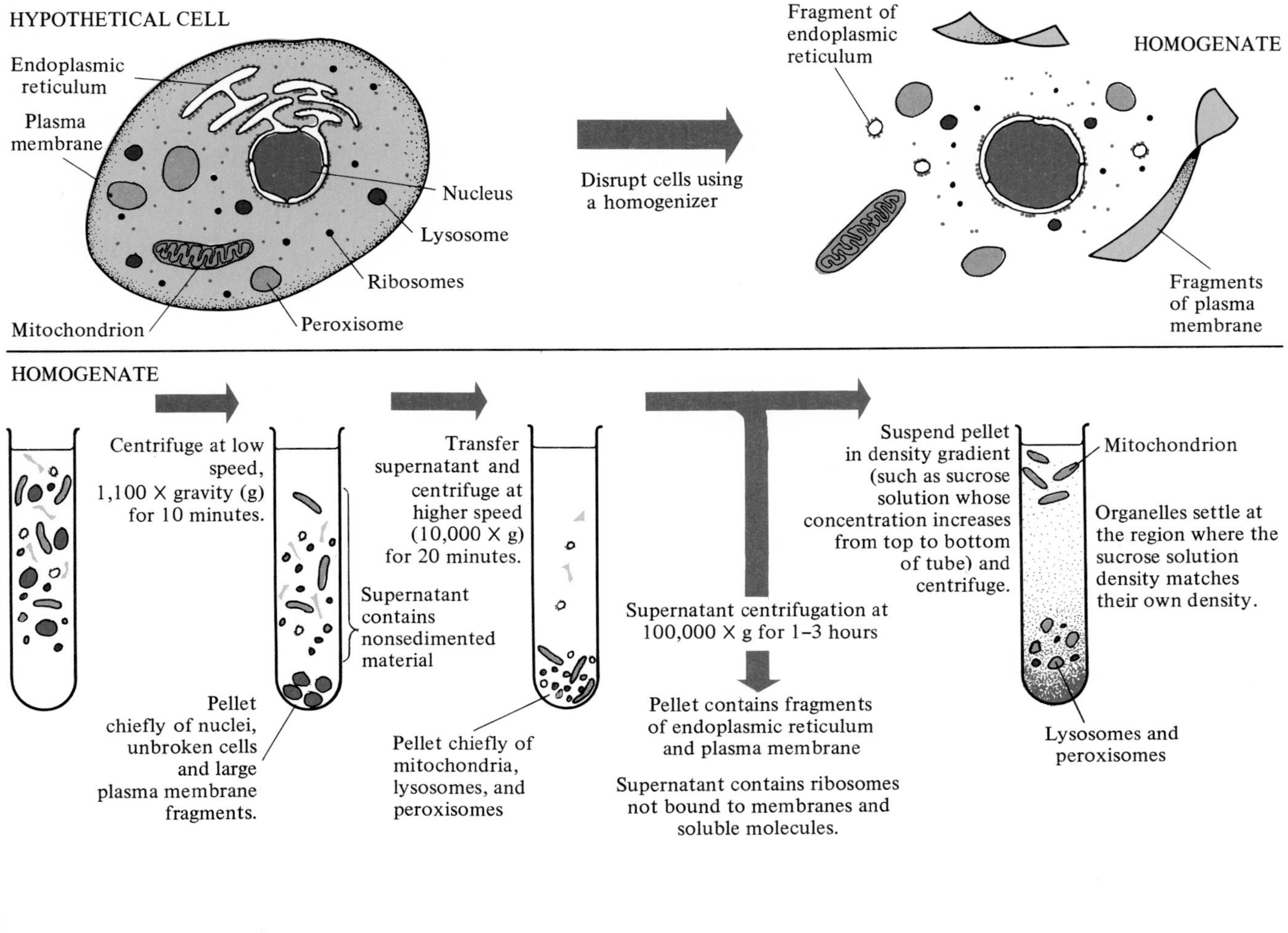
HYPOTHETICAL CELL
Endoplasmic reticulum
Plasma membrane
Nucleus
Lysosome
Ribosomes
Mitochondrion
Peroxisome
Disrupt cells using a homogenizer
Fragment of endoplasmic reticulum
HOMOGENATE
Fragments of plasma membrane
HOMOGENATE
Centrifuge at low speed, 1,100 X gravity (g) for 10 minutes.
Pellet chiefly of nuclei, unbroken cells and large plasma membrane fragments.
Supernatant contains nonsedimented material
Transfer supernatant and centrifuge at higher speed (10,000 X g) for 20 minutes.
Pellet chiefly of mitochondria, lysosomes, and peroxisomes
Supernatant centrifugation at 100,000 X g for 1–3 hours
Pellet contains fragments of endoplasmic reticulum and plasma membrane
Supernatant contains ribosomes not bound to membranes and soluble molecules.
Suspend pellet in density gradient (such as sucrose solution whose concentration increases from top to bottom of tube) and centrifuge.
Mitochondrion
Organelles settle at the region where the sucrose solution density matches their own density.
Lysosomes and peroxisomes

are compacted as a pellet. The supernatant fluid above the pellet containing the lighter particles and dissolved material is decanted (poured off), and the supernatant and pellet are analyzed separately by a variety of methods.

Fractionation of homogenates by centrifugation usually proceeds in two stages:

1. *Differential centrifugation,* in which the organelles settle out according to *size* and *density*. At low speeds, the pellet contains mostly nuclei and pieces of plasma membranes. At somewhat higher speeds, most of the mitochondria and lysosomes settle out, and, at the highest speeds, the finest particles are removed.
2. In *density gradient centrifugation* each fraction obtained as a pellet is resuspended in a sucrose density gradient and further fractionated, each component appearing as a band in the portion of the gradient corresponding to its own density. Lysosomes, mitochondria, and peroxisomes (microbodies) are separated from each other in this way.

The cellular pellet can be resuspended and analyzed by density gradient centrifugation, a technique that takes advantage of the fact that many cell components have different specific gravities. In a suspension, some organelles float and some sink more rapidly than others, just as when whole milk is centrifuged, it separates into a cream fraction on top and "skim milk" below. In each case, the material in the pellet is resuspended in a density gradient (Fig. 4-9). That is, a solution whose concentration—and therefore density—increases gradually from the top to the bottom of the tube is used as the medium for centrifuging a second time. The various cellular components form layers, the heaviest on the bottom, the lightest on top, and the rest in layers between them. A given layer can be removed and studied biochemically. One layer, or fraction, will be found to contain the nuclei, another the mitochondria, and so on. It is noteworthy that an important part of many cells, the endoplasmic reticulum (ER), was fragmented, isolated, and studied biochemically in this way for years before it was finally observed under the electron microscope. The fragments were too small to be seen by microscopy then available and were referred to as the "microsome fraction" until their true nature as fragments of the endoplasmic reticulum became known.

### THE ULTRACENTRIFUGE

The principle of density gradient centrifugation has been applied to the separation of molecules of different size. For this task, an ultracentrifuge is employed, a highly sophisticated machine with relatively frictionless rotors.* Ultracentrifuges can generate a centrifugal force over 400,000 times that of gravity.

## TWO BASIC CELL TYPES

The greatly improved resolution provided by the electron microscope allowed the classification of all living things into two distinct groups on the basis of cellular organization. The **prokaryotes**—a group made up of

* A frictionless rotor consists of a metal container in which tubes held at a fixed angle and suspended by magnetic force rather than on steel bearings, as in a conventional rotor, rotate in a vacuum to eliminate air friction.

the bacteria and the cyanobacteria, formerly called the blue-green algae—possess the simplest of all cellular organization. Their lack of an organized nucleus suggested the name prokaryote (from the Greek *pro,* "primitive," and *karyon,* "nucleus"). All other organisms are said to be **eukaryotes** (from the Greek *eu,* "true"); their cells have well-developed nuclei. Eukaryotes differ from prokaryotes in other important ways as well. In fact, the distinction between prokaryotic and eukaryotic organisms is so sharp that it is believed to represent one of the major evolutionary divergences in the history of life on earth.

## PROKARYOTIC CELLS

The prokaryotic cell is characteristic of the kingdom Monera, a group of organisms that includes the bacteria and the cyanobacteria. Although a knowledge of the differences between prokaryotes and eukaryotes is a modern development based largely on electron microscopy, the term Monera was first used by the German Ernst Haeckel in 1868. He applied it to what he regarded as the most primitive, that is, the least highly evolved of all organisms; those without nuclei.

Prokaryotic cells are usually surrounded by somewhat rigid cell walls composed mostly of polysaccharides and polypeptides (Fig. 4-10). Directly beneath the cell wall lies the very thin plasma membrane, which typically exhibits involutions, called **mesosomes,** that extend into the cell's cytoplasm. Prokaryotic cell organization is principally a system of such mesosomes. Respiration and photosynthesis have been found to be associated with special regions of these mesosomes.

Prokaryotic cells contain **ribosomes** (cell organelles involved in protein synthesis), which are small, rounded, membraneless granules in the cytoplasm, but they lack other organelles. Even the plastids (cell organelles in plant cells which are often pigmented) that are characteristic of all other photosynthetic cells are absent from the photosynthetic cyanobacteria and bacteria. Rather than being contained in a nucleus, the genetic information of prokaryotes is located on a single chromosome consisting of circular, double-stranded DNA. The region of the cell containing the DNA strand is called the **nucleoid** (from the Greek *oid,* "similar to").

**FIG. 4-10** (a) A clump of bacterial cells *(Escherichia coli)* as seen in a scanning electron micrograph. (b) A bacterium, *Bacillus subtilis,* showing the prominent cell wall surrounding the cell, mesosomes, and nucleoid. The mesosomes are infoldings of the plasma membrane.

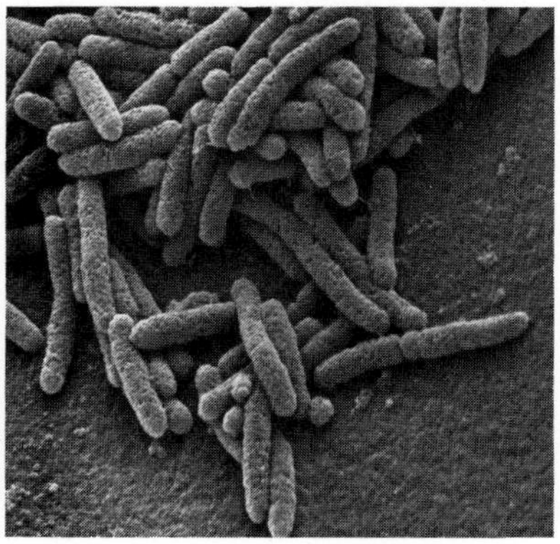

(a)

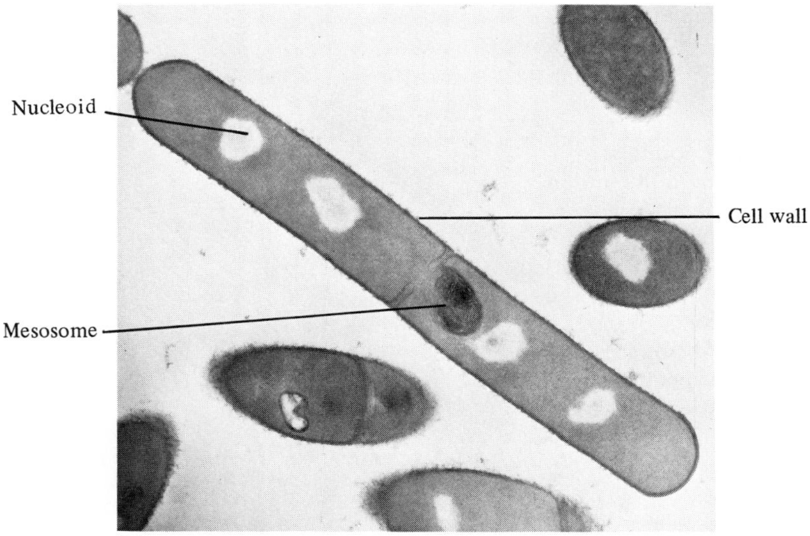


(b)

The absence of membrane-bound organelles is not an indication that prokaryotes are very simple. Biochemically, prokaryotes are very complex. The various prokaryotic species are able to carry out a wide range of biochemical activities essential to survival, growth, and reproduction in widely different environments; some prokaryotes thrive in situations that provide only the simplest materials as nutrients.

## THE "TYPICAL" EUKARYOTIC CELL

Members of the other four kingdoms—protists, fungi, plants, and animals—have eukaryotic cells. Although the various cells of all these organisms exhibit great structural variability, some broad generalizations can be made about eukaryotic cells.

In the "typical" eukaryotic cell, the nuclear contents are separated from the cytoplasm by a **nuclear envelope.** Although sometimes referred to as the nuclear membrane, this structure has usually been called an envelope since it was found to be composed of two membranes; both are clearly visible in Figure 4-11. The genetic information of the cell, which is encoded in large molecules of DNA, is contained in the nucleus and packaged in structures called **chromosomes,** which are usually visible only during nuclear division. A small dense body, the nucleolus, is located in the nucleus of eukaryotic cells and is the site of ribosomal RNA production.

The cytoplasm is made up of all of the material outside the nucleus. As shown in Figure 4-11, most of the organelles are located in the cytoplasm. The **mitochondria** are organelles that convert the chemical energy of carbohydrates and fats into ATP, the energy form most readily available for cellular work (see Chapter 2). The complex membrane systems of the **endoplasmic reticulum (ER)** synthesize proteins and lipids. Another type of organelle, not shown in the figure, is the **Golgi complex,** which processes proteins and glycoproteins and packages them in small vacuoles, called vesicles, prior to their secretion by the cell. The **chlo-**

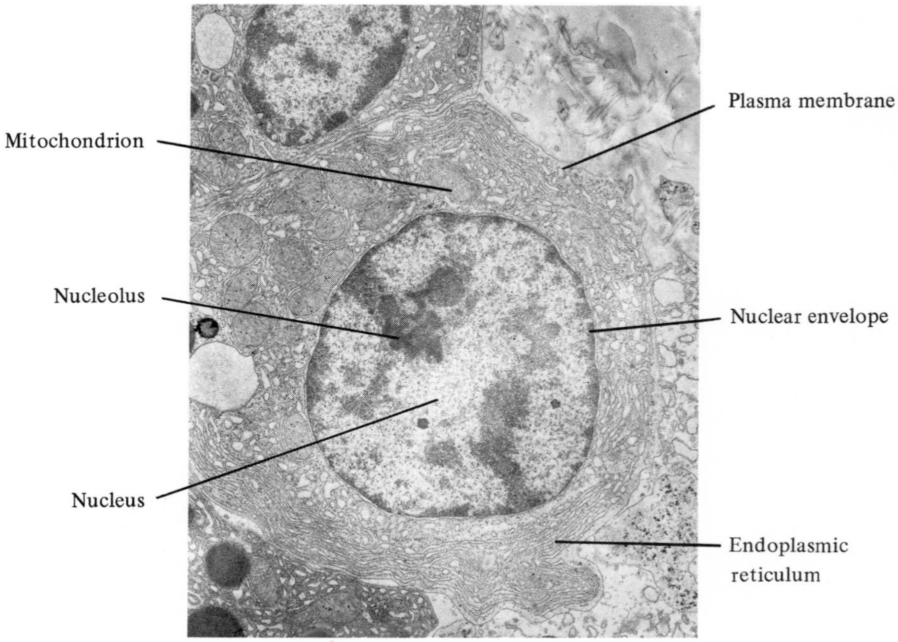


**FIG. 4-11** A pancreas cell from a frog. Among the structures visible are the nucleus, with its diffuse chromatin and prominent nucleolus. The nucleus is bounded by the nuclear envelope, a double-membrane structure. Prominent cytoplasmic structures include the endoplasmic reticulum, studded with ribosomes, as is the outer membrane of the nuclear envelope. The prominent round structures are mitochondria, most of which are sectioned transversely. A plasma membrane separates the cell from a neighboring cell, which contains a prominent mitochondrion, sectioned longitudinally.

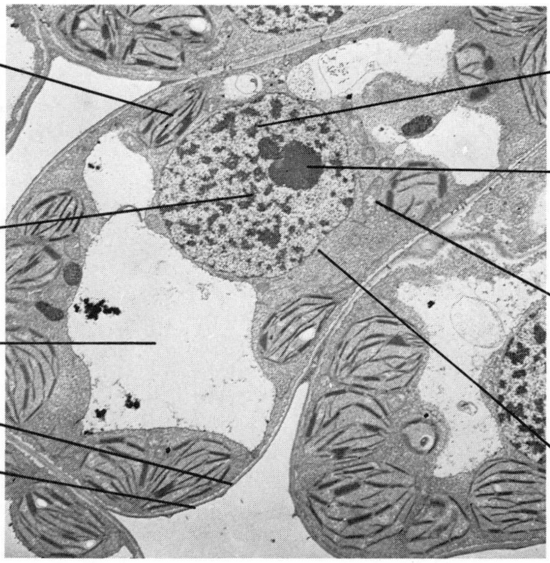


**FIG. 4-12** Structure of a higher plant cell.

**roplasts,** organelles found only in photosynthetic cells, trap the energy of sunlight and convert it to chemical energy stored as carbohydrates (Fig. 4-12). **Vacuoles, lysosomes,** and **microbodies** are other types of membrane-bound organelles found in eukaryotic cells.

**Ribosomes** are organelles without membranes and are found in prokaryotes as well as eukaryotes. Ribosomes are important in the synthesis of proteins. Some are found singly or in clusters called polysomes in the cytoplasm. Others are attached to the ER, giving it the rough, studded appearance seen in Figure 4-11. **Microfilaments** and **microtubules** also occur in the eukaryotic cytoplasm; they are involved in the maintenance of cell shape and in cell movements.

All of these organelles are discussed in greater detail later in this chapter.

# THE CELL SURFACE

## THE PLASMA MEMBRANE

It was known for many years that all cells are bounded by a membrane too thin to be seen by light microscopy; this is the plasma membrane. Its existence was inferred in part from the behavior of cells in different solutions: cells shrink in certain solutions and swell in others. Moreover, when the surface of a cell is torn with a fine needle, the cell contents pour out unless the rupture is so small that it closes almost immediately.

Another line of evidence pointing to the membrane's existence, as well as some of its properties, is the differential rate at which various chemical substances enter cells. Methanol, or methyl alcohol ($CH_3OH$), for example, enters at about twice the rate of ethanol ($CH_3CH_2OH$). If the penetration rate of methanol into an algal cell is rated 1.00, the rate for ethanol is 0.56, while it is 0.0076 for glycerol and 0.00031 for glucose. The selectivity of the plasma membrane thus favors substances that dissolve readily in lipids. This observation suggested that the plasma membrane had a major lipid component and led to some early experiments aimed at determining the chemical structure and molecular composition of the membrane.

***Gorter, Grendel, and Ghosts.*** A landmark finding regarding the membrane's chemical composition was made by the Dutch scientists E. Gorter and F. Grendel in 1925 in studies of the plasma membranes of human red blood cells. When such red blood cells, which lack nuclei, are placed in distilled water, they rapidly absorb it. As the cells swell, their plasma membranes rupture. The cytoplasm and its hemoglobin pour out in a process called **hemolysis.** The hemolyzed cells are repeatedly washed and centrifuged to provide a preparation composed almost entirely of plasma membranes. The membranes are called "ghosts" because they are pale and hard to see microscopically compared to intact cells filled with hemoglobin. These ghosts have provided a favorite material for the investigation of the characteristics of the plasma membrane. Gorter and Grendel found that the membrane was largely phospholipid and that it was most likely in the form of a bimolecular layer. They concluded that the plasma membrane was composed, at least in part, of a bimolecular layer, or bilayer, of lipid molecules. By calculating the average size of the kind of lipid molecules involved, they estimated the thickness of the bilayer to be about 4 nm.

**FIG. 4-13** Danielli-Davson models of the structure of the plasma membrane: (a) the 1935 model; (b) the 1954 model.

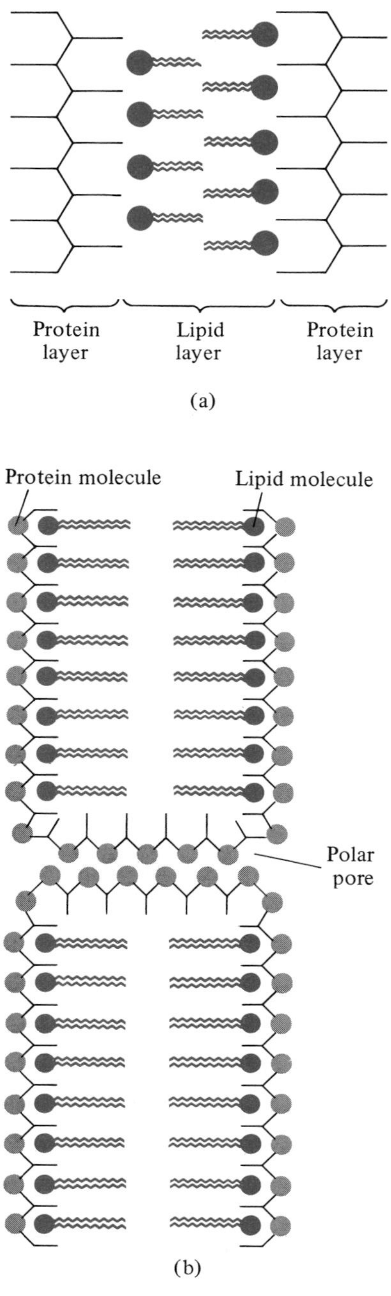


***The Danielli-Davson Model.*** Similar investigations of the plasma membrane continued in many laboratories, including those of the British biologists James F. Danielli and Hugh Davson. They studied not only red cells but eggs and other cell types. Their studies led to the conclusion that the plasma membrane contained a substantial protein fraction in addition to its lipid bilayer base.

Danielli and Davson made a series of determinations of the surface tension of artificially constructed lipid bilayers, which they compared to the surface tension of intact plasma membranes. They found that such lipid bilayers had five to 30 times the tensile strength of plasma membranes. But when they added protein to a lipid film, its surface tension was reduced to the same value found in plasma membranes. Danielli and Davson thus concluded that the plasma membrane is a lipid bilayer, probably with protein on both the inner and outer surfaces. The lipid fraction would be hydrophobic; the protein fraction, hydrophilic. They summarized their theory in the model of the plasma membrane shown in Figure 4-13a.

Danielli and Davson proposed their model in 1935. This model was widely accepted immediately after it was proposed, but from the beginning it was acknowledged that the membranes of cells could not be uniformly as simple as the model suggests, since the model did not explain how some substances—for example, materials insoluble in lipids—could pass through the membrane. In 1954 Danielli and Davson modified their model, as shown in Figure 4-13b, adding pores to provide for the passage of polar compounds through the membrane.

***Robertson's Unit Membrane Concept.*** As staining methods and thin sectioning for electron microscopy improved over the next few years, the Danielli-Davson model gained credibility. The plasma membrane and various other membranes within the cell were revealed in electron micrographs to be tripartite. With permanganate staining of the plasma membrane, for example, a light, nonstaining layer 3.5 nm thick is re-

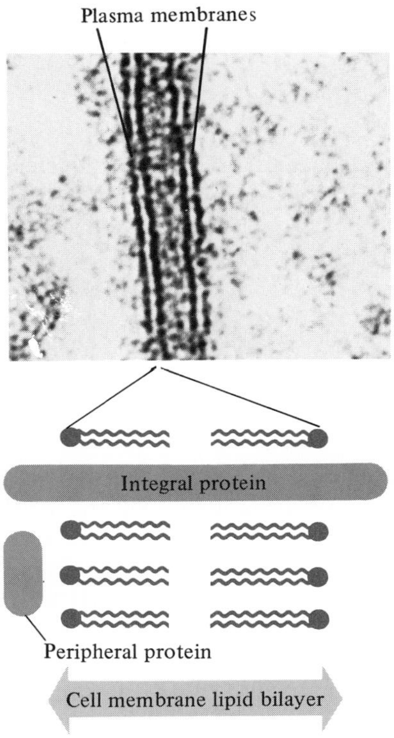


**FIG. 4-14** Electron micrograph of the plasma membrane of two adjacent cells, each of which shows the three-part structure that led to Robertson's concept of the unit membrane.

vealed between two electron-dense layers that are 2 nm thick (Fig. 4-14). It was tempting to assume that the dark layers were the protein and the light layer the lipid bilayer in the Danielli-Davson model.

In 1959, the Harvard biologist J. David Robertson extended the Danielli-Davson concept to include all biological membranes of cells. He suggested that a "unit membrane" formed the basis for all biological membranes; the principal difference between the membrane structure he proposed and that of the Danielli-Davson model is that, in Robertson's model, the inner and outer surfaces have a different protein composition.

***Singer-Nicolson Fluid Mosaic Model.*** By the mid-1960s, both the Danielli-Davson model and the unit membrane concept were considered inadequate because they could not account for all of the permeability characteristics and physical properties of membranes. Some membranes were discovered, for example, to be 80 percent lipid and 20 percent protein, some 25 percent lipid and 75 percent protein, and many 50 percent each. Thicknesses were found to vary from 5 to 9 nm or more. Moreover, high-resolution electron micrographs revealed the membranes of some organelles to be largely composed of tiny, globular subunits, an observation that did not fit the unit membrane model. Such findings suggested that the membrane was not a sandwich of a lipid bilayer between two protein layers but, rather, consisted of a single layer of globular units, much like a layer of marbles, with each unit composed of a protein granule in a lipid bilayer. In 1972, S. Jonathan Singer and Garth Nicolson proposed a model to account for all of the properties of biological membranes and for the globular subunits observed in membrane structures.

In the Singer-Nicolson fluid mosaic model, the primary framework of all membranes is provided by a lipid bilayer, as in the older model. However, in the fluid mosaic model, the membranes differ in the number and type of proteins embedded in their bilayers. Because the lipid layer is in a fluid state, these proteins can move within the membrane (Fig. 4-15). A variety of proteins occur in various membranes either singly or in pairs. Integral proteins may be located at or near the membrane surface, extending only a short way into the lipid bilayer, or they may bridge both sides of the membrane. Peripheral proteins are loosely

**FIG. 4-15** One version of the Singer-Nicolson fluid-mosaic model of the structure of plasma membranes. Integral proteins are embedded in the lipid layer, while peripheral proteins are merely associated with the membrane surface. The carbohydrates attached to the proteins and lipids form the membrane glycoproteins and glycolipids; they are always found on the outer surface of the membrane.

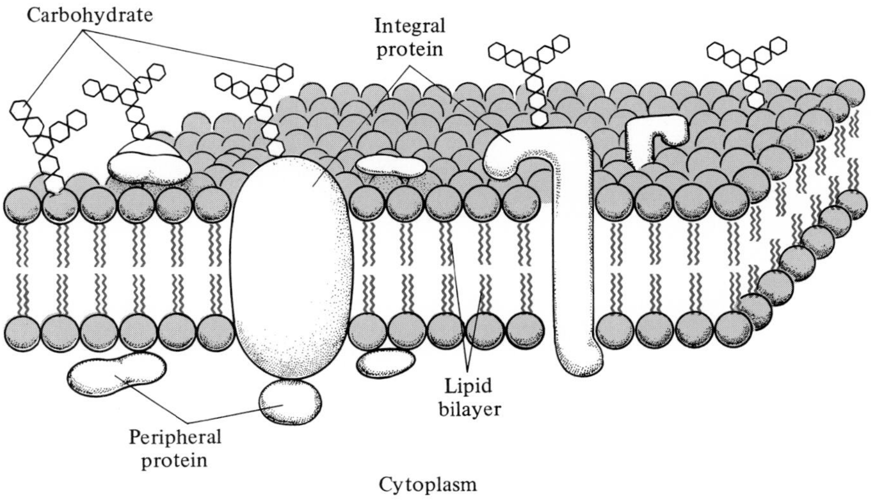

attached to proteins on the membrane surface. No model explains all known membrane properties, but of those thus far proposed, the fluid mosaic model seems to come the closest.

## PLANT CELL WALLS

A cell wall is a distinguishing characteristic of all plant cells (it is also found, by the way, in the cells of prokaryotes, protophyta, protists, and fungi. The cell wall confers upon plants a kind of structural rigidity that distinguishes plants from animals. This rigidity is appropriate to the sedentary mode of life that is characteristic of most plants.

Most cells in hypotonic solutions take up water by a process called **osmosis.** As a result, they tend to swell, and would burst if they could not expel excess water from their cells via water expulsion vesicles called contractile vacuoles. In cells with walls, however, the cell wall limits the net uptake of water. When the pressure exerted by the wall equals or exceeds the tendency to take up water osmotically, water molecules leave the cell in equivalent numbers to those that enter. The wall thus restricts enlargement of the cell, and an internal pressure, called **turgor pressure,** results. Turgor pressure causes plant cells to be swollen and turgid, and this in turn makes the plant body rigid. When turgor pressure is reduced by loss of water, the cells become flaccid, and the plant wilts.

The wall between two plant cells consists of three layers: a cementing layer called the middle lamella and on either side of it a primary wall layer (Fig. 4-16a, b). Strands of cytoplasm, called **plasmodesmata,** extend through the wall and connect the two cells (Fig. 4-17). Primary walls are formed while cells are still growing, at which time they are flexible and can stretch. In certain cell types, a secondary wall may be deposited internal to the primary wall (Fig. 4-16b), usually after growth of the cell has stopped.

The properties of the cell wall are related to its organization. A skeleton of cellulose microfibrils provides structural support and fine pores permit the flow of solutions; the wall is permeable to ions and most small molecules, such as plant hormones. Cellulose microfibrils are cylinders made up of about 40 pairs of cellulose chains (Fig. 4-18). (But fungal cell walls have a framework of chitin.) In turn, the cellulose

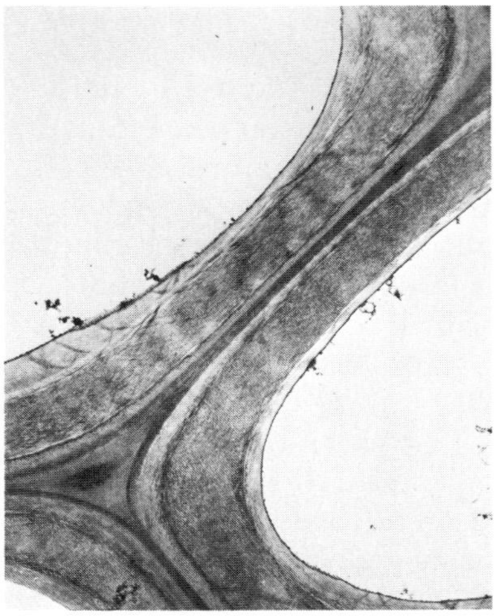
(a)

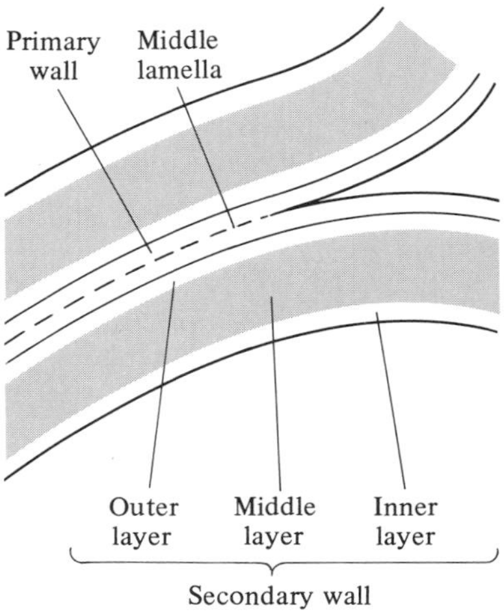

(b)

**FIG. 4-16** (a) Electron micrograph of two adjacent cell walls of vessel elements, the cells that form the tubes that conduct water in plants. The cells shown are from the wood of a black locust tree; the protoplasts are no longer present. (b) Diagram of plant cell walls. The cells are joined by the middle lamella.

**FIG. 4-17** Numerous plasmodesmata running through a primary cell wall between two cells of the alga *Chara.*

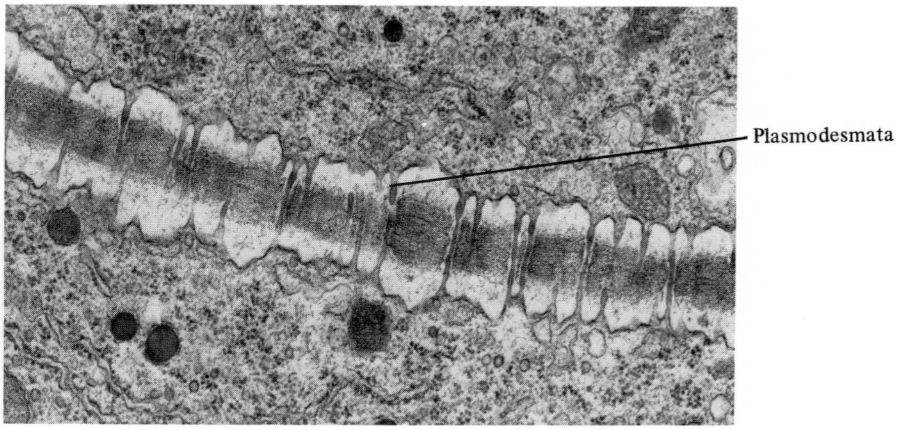

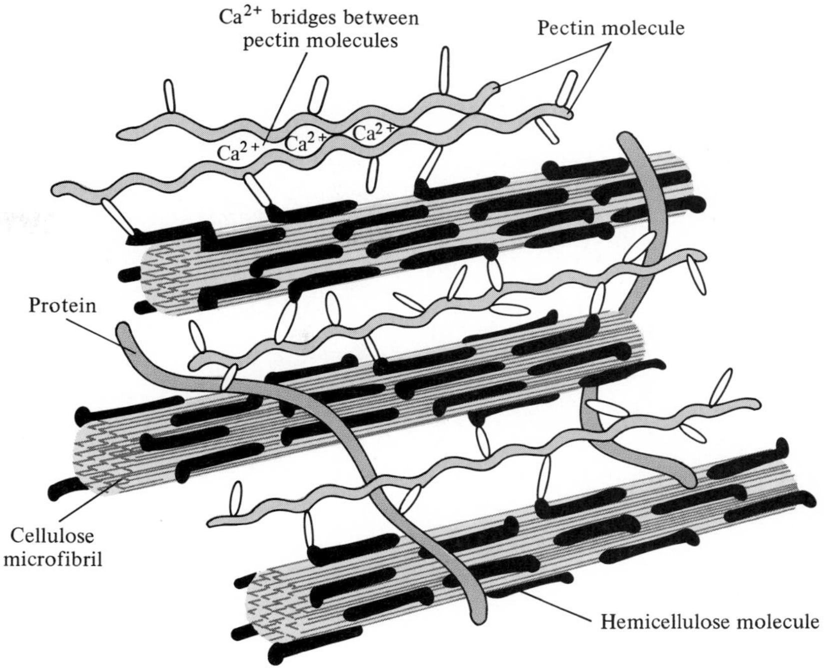


**FIG. 4-18** A molecular model of the cell wall showing cellulose microfibrils composed of paired chains of cellulose. The microfibrils are held together by hydrogen bonds between hemicellulose molecules, which in turn are cross-linked to pectin. Pectin molecules are joined by calcium ion bridges. Protein molecules weave through the matrix linked to the other matrix molecules.

chains consist of several hundred glucose molecules. Two other components are hemicellulose, a polymer of 5-carbon and 6-carbon sugars, and pectin, a polymer of a 6-carbon sugar acid and a 5-carbon sugar. Except for the protein molecules interspersed among the cellulose microfibrils, all cell wall components are thus made up of carbohydrates. The microfibrils of cellulose are dispersed in a matrix of hemicellulose, pectin, and protein, sometimes randomly arranged in a "brush heap" or sometimes in parallel arrays. The cellulose microfibrils are held in position by hemicellulose molecules, which, in turn, are bonded to the pectin molecules.

The cellulose molecule does not stretch and has the tensile strength of a steel cable of comparable dimensions. Therefore, cell walls can only expand by a loosening of the hydrogen bonds between the microfibrils. The direction in which plant cells may be elongated is thus determined by the orientation of the microfibrils; in a growing cell, microfibrils tend to be perpendicular to the direction of cell elongation.

## CELL ORGANELLES

### THE NUCLEUS

The nucleus is the largest and most prominent organelle in many cells. Under a light microscope, it is easiest to see when stained with a basic dye. The nucleus is enclosed by the double-membraned nuclear envelope (Fig. 4-19a). The outer membrane is continuous with a system of membranes within the cytoplasm, and the outer surface is often studded with ribosomes. In most places, the two membranes are separated by the perinuclear space, but in some places they are continuous with each other, forming nuclear pores through which the nucleus and cytoplasm communicate (Fig. 4-19b). Although the presence of pores may suggest an easy passageway for molecules of all sizes, high-resolution

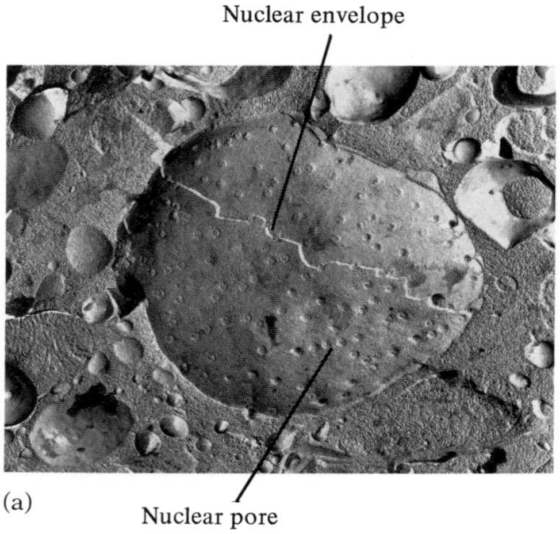


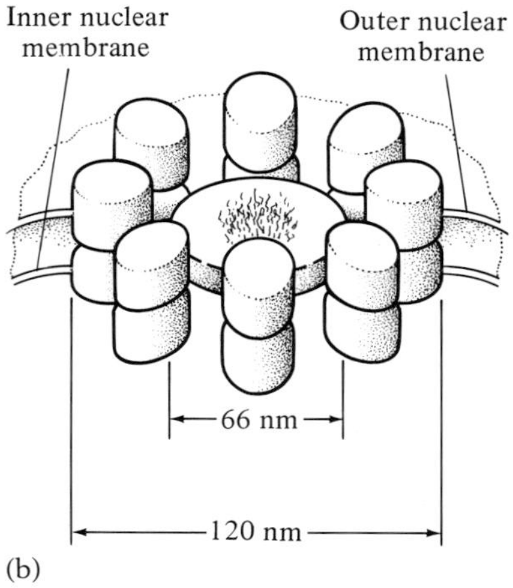


**FIG. 4-19** (a) Preparation of a nucleus in an onion root-tip cell. In this face view of the nuclear envelope, its pores are very evident. The pores extend into the interior of the nucleus. (b) Diagram showing the character and dimensions of a pore in the nuclear envelope.

electron microscopy shows that the pores are often guarded by eight proteinaceous granules. This pore complex probably regulates the passage of small ions and macromolecules through the pores.

***Nucleoli.*** A stained, nondividing nucleus shows one or more small, dark bodies considerably denser than the surrounding nucleoplasm. These are nucleoli; the singular is nucleolus, Greek for "small nucleus" (Fig. 4-20a). Nucleoli are often attached to and in fact are formed by specific regions of particular chromosomes. The regions are called **nucleolar organizers** (Fig. 4-20b). The nucleolus is the site of synthesis of ribosomal RNA and ribosomes, which are transported to the cytoplasm. In nearly all cells, the nucleoli disintegrate during nuclear division, and reform later in each daughter nucleus.

***Chromosomes and Chromatin.*** Most of the nucleus is occupied by the threadlike chromosomes, which are not readily distinguishable except during nuclear division. The material of the chromosomes stains read-

**FIG. 4-20** (a) Insect gland cell showing two prominent nucleoli. (b) Preparation of a cell in a corn plant. The cell was squashed before being treated to separate the organelles and permit observation of some details. The threadlike structures are pairs of closely associated chromosomes. The prominent, dense sphere is a nucleolus attached to its nucleolar organizer.

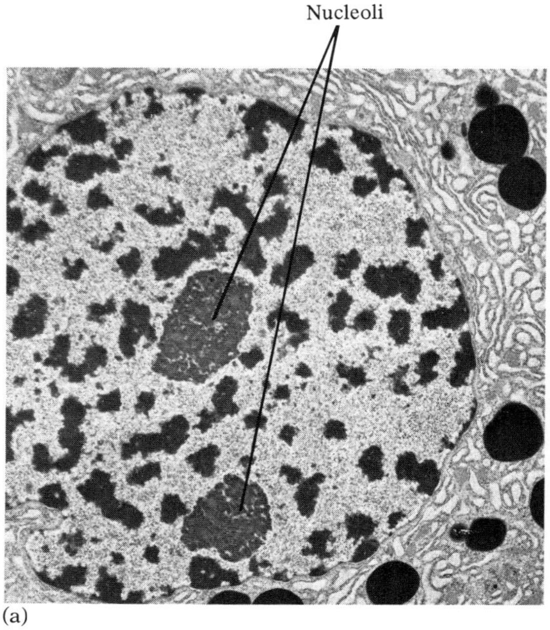


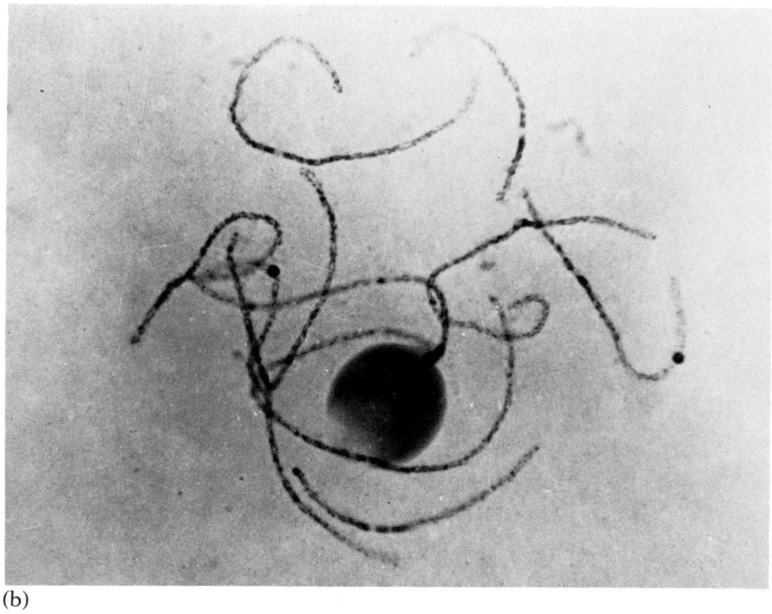

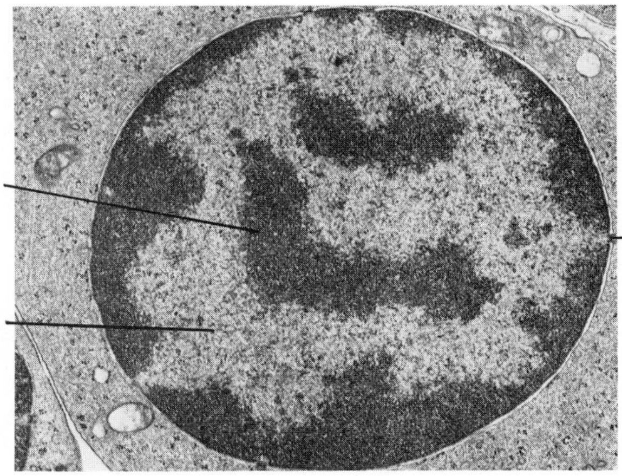


**FIG. 4-21** The nucleus of an interphase cell from guinea pig bone marrow. The visible chromatin is aggregated into extremely dense masses of heterochromatin. Also visible are a few areas of less densely coiled chromatin, euchromatin, mostly near the nuclear pores.

ily with various basic dyes and is therefore called **chromatin** (from the Greek *chromos,* "color"). Chromatin consists of double-stranded DNA to which large amounts of protein and small amounts of RNA are bound. Chromatin tends to coil spontaneously to form strands from 10 to 30 nm in diameter. The chromatin fibers are often attached to the inside of the nuclear envelope. In an interphase cell part of each chromosome's chromatin is tightly coiled and remains visible (Fig. 4-21). The rest is loosely coiled and distributed throughout the nucleus. These less dense portions, called **euchromatin,** are involved in the production of RNA. The tightly coiled, electron-dense material, called **heterochromatin,** is not active in RNA synthesis.

## THE ENDOPLASMIC RETICULUM AND THE RIBOSOMES

To continue growing and dividing, cells must fabricate additional cell membranes and duplicate their organelles and other cellular structures. Those cells that are not growing or dividing need to replace nonfunctioning organelles, to synthesize the many enzymes needed for normal functioning and to manufacture such products as hormones, immunoglobulins, mucus, saliva, and milk. In most eukaryotic cells, the site of many of these chemical syntheses is the endoplasmic reticulum. The ER consists of a system of interconnecting flattened membranous sacs. These sacs usually take the form of large, flattened vesicles called **cisternae,** shown in Figure 4-22, but the ER may also occur as inflated vesicles or tubules. Like the plasma membrane, ER membranes usually have a "unit membrane" appearance, although they are thinner than the plasma membrane (5–6 nm compared to 9–10 nm). In some cells, the ER is continuous with the plasma membrane and the nuclear envelope, with all membranes of the cell forming a continuous system.

***Rough ER and Smooth ER.*** In a region of a cell that is actively synthesizing protein, the outer surfaces of the ER are covered with ribosomes. Membranes of this type of ER have a rough appearance and are thus called granular or rough ER (Fig. 4-22a). Because ribosomes function as the sites of protein synthesis, an extensive rough ER is characteristic of such animal cells as plasma cells, which synthesize immunoglobulins; pancreatic cells, which produce hormones and digestive enzymes; and

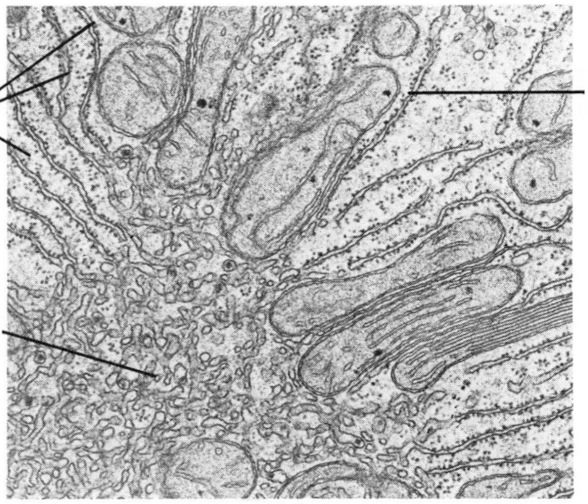


(a)

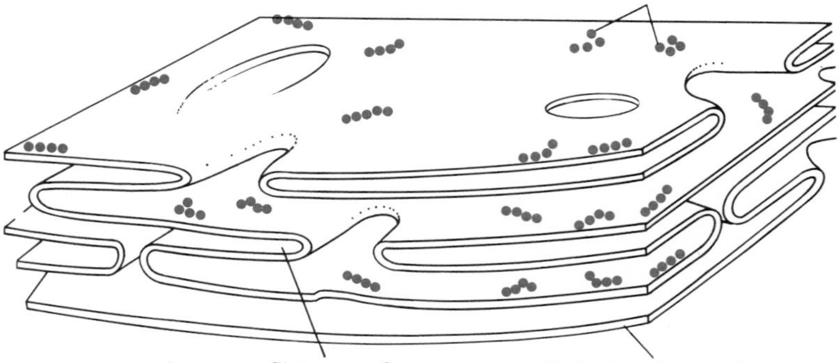


(b)

**FIG. 4-22** (a) Electron micrograph of a hamster liver cell showing granular or rough and agranular or smooth endoplasmic reticulum. The tubular elements of the smooth endoplasmic reticulum are continuous at many sites with cisternae of the rough endoplasmic reticulum (see arrows). (b) The structural organization of the endoplasmic reticulum.

fibroblasts, which produce collagen. The proteins synthesized by the ribosomes on the outer surface of the ER pass directly through the membrane and into the cisternal space of the ER, which is continuous within the cytoplasm. The newly synthesized protein is enclosed in vesicles, which make their way through the ER cisternal spaces to other regions of the cell, particularly the Golgi complex. The newly synthesized proteins may be used in the cell or secreted to the outside of the cell.

Endoplasmic reticulum that lacks ribosomes is called agranular or smooth ER (Fig. 4-22a). Smooth ER is especially prevalent in cells that synthesize and secrete steroid hormones, such as the cortical cells of the adrenal gland and certain cells of the testes and ovaries. Smooth ER also appears to contain the enzymes necessary for the synthesis of fats that are to be stored.

***Ribosomes.*** In addition to synthesizing protein on the ER, ribosomes occur free in the cytoplasm, where they often form clusters of two to eight, called **polyribosomes** or **polysomes** (Fig. 4-23). Such clusters are held together by a single molecule of messenger RNA, and the attached ribosomes are engaged in translating the base sequences of mRNA molecules into the amino acid sequences of polypeptides. Ribosomes also

**FIG. 4-23** Electron micrograph of polyribosomes from a culture of rabbit reticulocytes. Individual globular units in the clusters are ribosomes, which are believed to be held together by strands of messenger RNA.

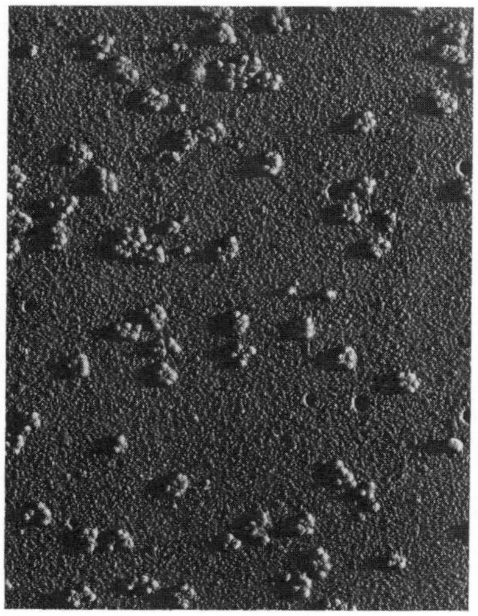

occur in chloroplasts of photosynthetic cells and in mitochondria, discussed below.

## THE GOLGI COMPLEX

Like the endoplasmic reticulum, the Golgi complex is a system of membranes found in almost all eukaryotic cells. The Golgi complex is made up of units called **Golgi bodies (dictyosomes).** The structures are named after their Italian discoverer, Camillo Golgi.

***Golgi Fine Structure.*** The Golgi complex usually takes the form of several layers of flattened cisternae (Fig. 4-24a). From the edges of these sacs, spherical vesicles form by a "pinching-off" process (Fig. 4-24b). In some algae, the Golgi bodies are located close to the nuclear envelope. The number of Golgi bodies in a cell varies from one to 20, depending on the cell type. In some algal cells, however, these are thousands of dictyosomes. In animal cells the greatest number is found in cells involved in secretion.

The main function of the Golgi complex is packaging of the newly synthesized proteins and carbohydrates (or polysaccharides) in mem-

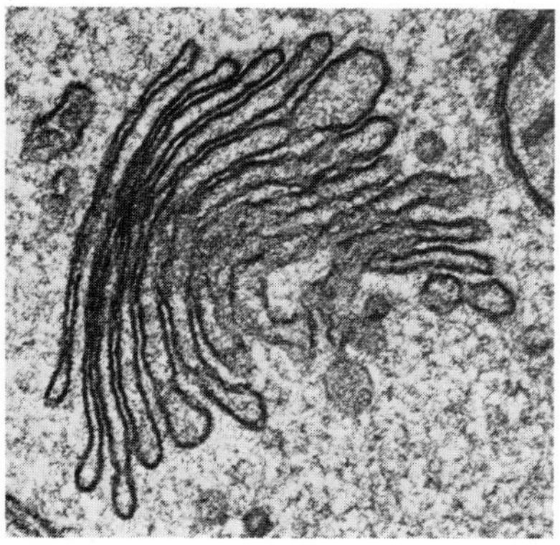

(a)

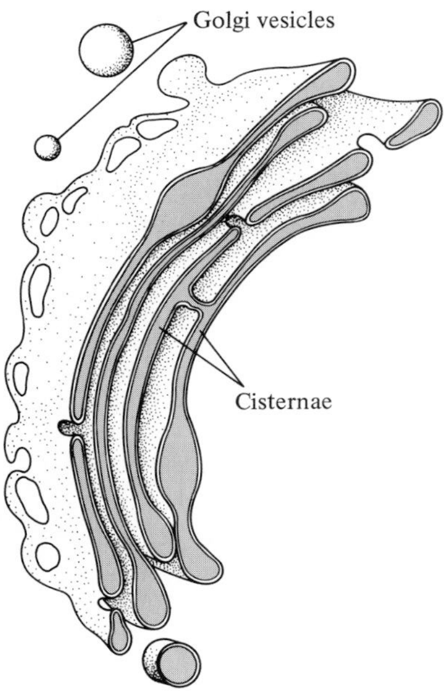


(b)

**FIG. 4-24** (a) Electron micrograph of a Golgi body (dictyosome) from the protist *Chlamydomonas.* (b) Diagram showing the structural organization of a Golgi body.

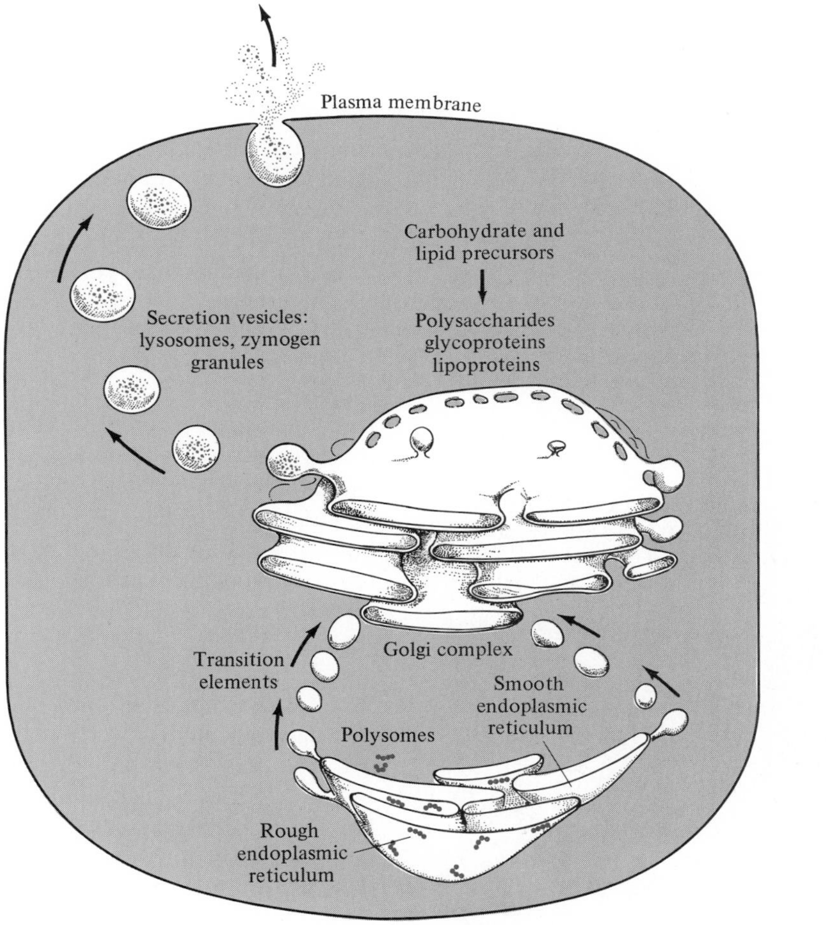


**FIG. 4-25** Summary of the role of the endoplasmic reticulum and the Golgi complex in the synthesis of membrane-bound products.

brane-bound vesicles for transport to their final destination inside and outside the cell. In animal cells, proteins synthesized on the ribosomes of the rough ER accumulate in vesicles; then they migrate to a Golgi body, where they are complexed with carbohydrates to form glycoproteins, and may be further modified. The glycoproteins are concentrated in the vesicles at the edges of the Golgi bodies. As the vesicles pinch off, or detach, they move to other parts of the cytoplasm or to the plasma membrane, where their contents are discharged to the outside of the cell. Figure 4-25 summarizes this process.

The formation of plant cell walls also depends on the activity of the Golgi complex. During plant cell division, vesicles pinch off from the Golgi complex and move to the area where the new cell walls will form. There, the vesicles fuse with each other, their membranes contributing to the formation of a pair of new plasma membranes and the middle lamella. The new cell walls then form between the plasma membranes of the daughter cells.

## LYSOSOMES

Two of the processes carried out within animal cells are the digestion of macromolecules that enter the cell and the breakdown of old, nonfunctioning organelles. Both are accomplished by **lysosomes,** which can be thought of as the disposal and recycling units of the cell. Each lysosome is a membrane-bound vesicle containing various hydrolytic enzymes. Different types of lysosomes contain different enzymes. Lysosomal enzymes are produced on the rough ER, and vesicles containing them migrate to the Golgi complex. After fusing with Golgi membranes, the lysosomes are pinched off from the edges of the Golgi cisternae.

Normally, lysosomal enzymes are isolated from the rest of the cell by a membrane. In a severely stressed or dying cell, however, these membranes disintegrate, releasing the destructive lysosomal enzymes into the cytoplasm, where they digest all organelles indiscriminately, hastening the cell's death by this self-digestion, or autolysis.

About 50 diseases involving malfunction or absence of lysosomal enzymes have now been described. Each of these diseases is characterized by a specific enzyme deficiency and shows a characteristic pattern of accumulation of polysaccharides or lipids in tissues of nerve cells, muscle, liver, or spleen. The first of these diseases to be described was Pompe's disease, an inherited disease characterized by the absence in lysosomes of maltase, an enzyme involved in glycogen metabolism. As a result, glycogen accumulates in muscles in very high amounts, causing paralysis and death by the age of two.

**FIG. 4-26** Scanning electron micrograph of a human macrophage in the process of phagocytizing red blood cells.

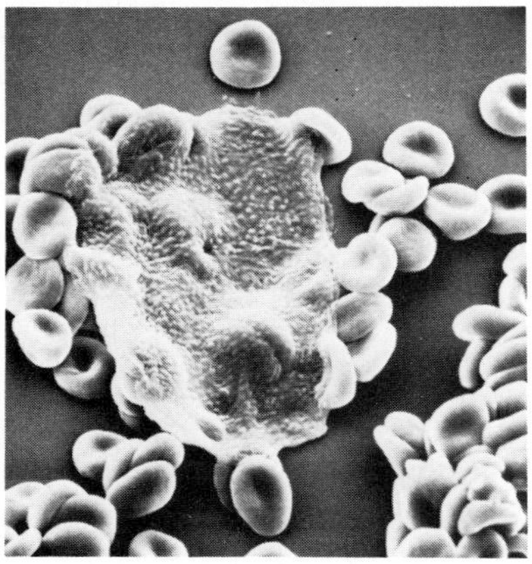

***The Lysosomes of Phagocytic Cells.*** The major function of lysosomes in phagocytic cells is the digestion of various foreign materials taken into the cell. Examples of this lysosomal function include the digestion of engulfed food organisms or particles by protozoans such as amoeba, the digestion of bacteria by white blood cells, and the digestion of bacteria, parasites, and worn-out blood cells by cells of the vertebrate liver (Fig. 4-26). The ability of certain animal cells to destroy bacteria and other disease-causing organisms is a major line of defense against infection. In the process of **endocytosis,** such cells engulf materials by forming an invagination, or inward depression, of the plasma membrane. Then the invagination pinches off as a vesicle, which enters the cyto-

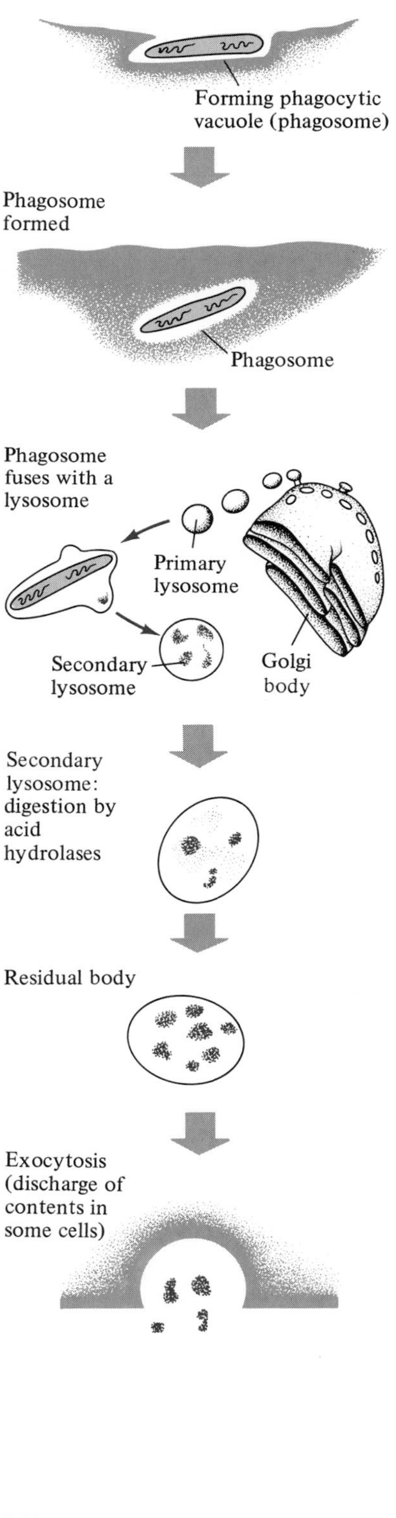


**FIG. 4-27** The role of lysosomes in intracellular digestion. There are several stages in the life of lysosomes, each representing a stage in the digestion process. The storage granule is the initial stage; it is a product of the endoplasmic reticulum and Golgi complex. When the cell ingests substances by phagocytosis, a food vacuole, or phagosome, is formed. Several phagosomes may fuse to form one large vacuole. A storage granule fuses with the phagosome to form a digestive vacuole. In the digestion of cell organelles, the organelle or debris to be digested enters the lysosome. Digestion products diffuse through the lysosomal membrane into the cytoplasm. Indigestible material is ejected by the cell.

plasm (Fig. 4-27). The process of taking solid materials into cells in this manner is called **phagocytosis** (Greek for "cell eating"); the vesicles formed are known as **phagosomes.** Taking in water containing only dissolved matter in this way is called **pinocytosis** (Greek for "cell drinking"), and the vesicles are referred to as **pinocytotic vesicles.**

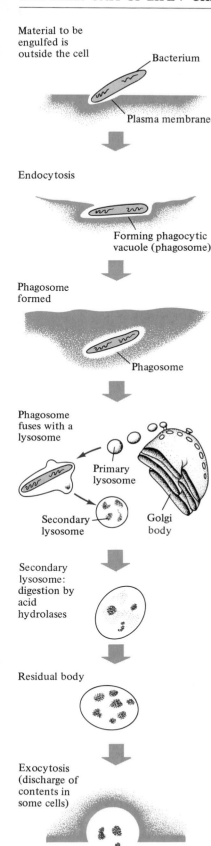

Following phagocytosis, several phagosomes fuse with lysosomes to form a secondary lysosome, which functions as a digestion vacuole. The lysosomal enzymes released into this vacuole digest the bacteria or other materials. Small molecules resulting from this digestive process then diffuse out of the vacuole through its membrane and are used by the cell to synthesize new molecules. Any indigestible material is eliminated by the fusion of the vacuolar membrane with the plasma membrane in such a way that the undigested material is deposited outside the cell.

***The Role of Lysosomes in Recycling Cell Components.*** In addition to breaking down macromolecules that enter the cell, lysosomes digest worn-out or damaged organelles and macromolecules of their own cells. Such digestion is normally done on a very selective basis. For example, damaged mitochondria and those not required at the time in respiration, are engulfed and digested. The small molecules, such as amino acids, resulting from digestion of cell components are recycled: they diffuse out of the lysosome and are used in the production of new organelles, enzymes, and other cellular components.

During prolonged starvation, lysosomes digest some organelles to provide an emergency energy source for the most essential life-sustaining activities. In some cases, cells programmed to die as a normal phase of development digest themselves completely in this way. Examples of such programmed cell death are the resorption of the frog tadpole's tail as it metamorphoses into an adult, and the web between the fingers and toes in human and chick embryos during early development (Fig. 4-28). Such selective cellular death plays a part in the development of many animals.

**FIG. 4-28** Distribution of cellular death (shaded areas) during the development of the leg bud in the chick embryo.

## PEROXISOMES

The **peroxisome,** or **microbody,** is one of the most recently defined cell organelles. They are membrane-bound sacs of enzymes found in the cytoplasm of many eukaryotic cells; unlike lysosomes, they are produced by the smooth ER system (Fig. 4-29).

**Peroxisomes** contain one or more enzymes that digest amino acids, lactic acid, and other substances into smaller products. They are active in yeast, protozoans, kidney cells, and mammalian liver cells. In certain cells of certain kinds of plants they are associated with chloroplasts with which they interact to recirculate photosynthetic products. Peroxisomes consume as much as 20 percent of the oxygen in liver cells. They are thought to use this oxygen largely for the conversion of hydrogen peroxide—a highly toxic product of metabolism—to oxygen and water. The enzyme involved in this conversion is catalase. **Glyoxisomes** are peroxisomes that contain enzymes that convert fat into carbohydrates. They are found in microorganisms and in plant cells, for example, in the seeds of oil-storing plants such as the castor bean.

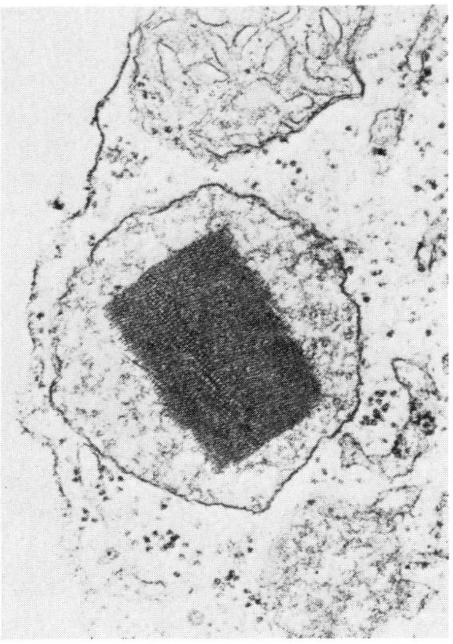

**FIG. 4-29** Electron micrograph of a hamster liver cell with a microbody containing a crystalloid.

## MITOCHONDRIA

The immediate source of energy for most exergonic reactions in cells is ATP. This energy-rich compound can be generated by anaerobic oxidation and by photosynthesis. However, in aerobic organisms most ATP is formed using energy released in the complete oxidation of fats and carbohydrates to carbon dioxide and water. These oxidations occur within the **mitochondria** (singular, mitochondrion), often referred to as the powerhouses of the cell (Fig. 4-30a).

When viewed in the light microscope, most mitochondria appear as simple threadlike or granulelike structures, hence their name (from the Greek *mitos,* "thread," and *chondrium,* "granule"). As Figure 4-30b illustrates, mitochondria are bounded by a double membrane consisting of an outer membrane and an inner membrane that is folded to form inward projections called **cristae.** The cristae are bathed in a gellike matrix that fills the inner spaces of the mitochondrion. The mitochondria of cells whose energy requirements are low—for example, fat storage cells—usually have relatively few cristae, whereas those of cells

**FIG. 4-30** (a) Electron micrograph of a typical mitochondrion from mouse epididymis cells. (b) Diagram showing the fine structure of a mitochondrion.

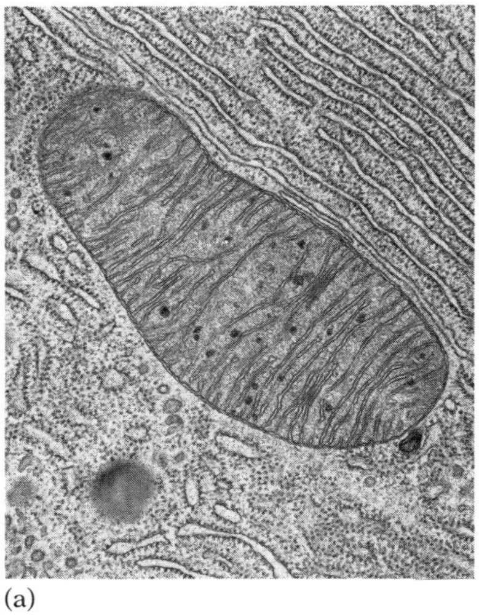

(a)

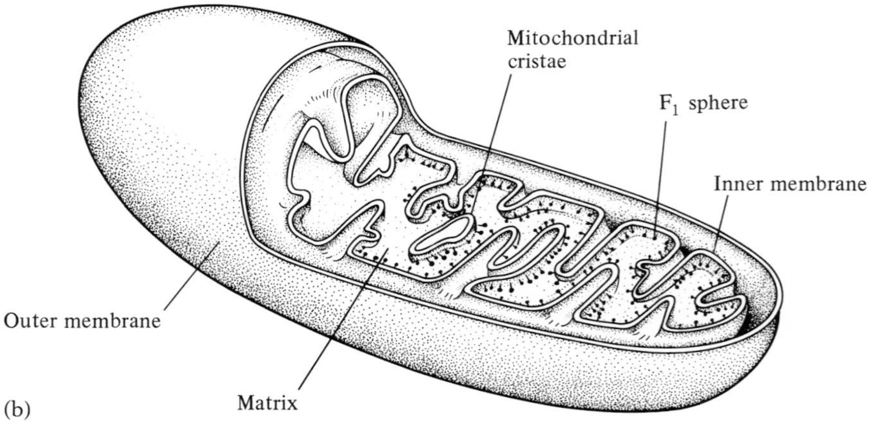


(b)

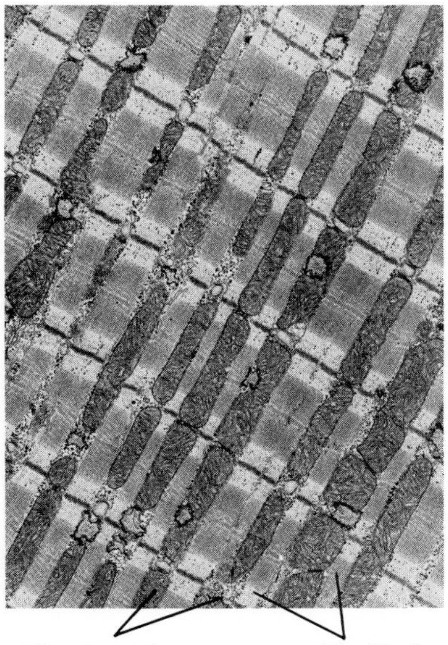


**FIG. 4-31** Heart muscle showing a concentration of mitochondria between the myofibrils.

with high energy requirements—such as cells in the flight muscles of birds and insects—are packed with cristae. Most mitochondria range in size from 3 to 10 micrometers (m$\mu$) in length and range in shape from spheres to simple or branching rods. Mammalian liver cells have as many as 1,000 mitochondria each, and the egg cell of the frog has several million.

Mitochondria are usually found in regions of the cell where energy expenditure is greatest. For example, in sperm cells the mitochondria are concentrated at the base of the flagellum, or tail, of the sperm, where they provide the energy necessary for locomotion. In skeletal muscle, the mitochondria are in close contact with the fibrils responsible for muscle contraction (Fig. 4-31).

***Were Mitochondria Organisms?*** The mitochondrion is a self-contained chemical plant that synthesizes some of its own enzymes; some of these are dissolved in the matrix and some are arranged in the membranes of the cristae. The more extensive its cristae, the more such enzymes a mitochondrion contains. In fact, mitochondria contain miniature versions of much of a cell's essential biochemical machinery. A mitochondrion contains very small ribosomes, RNA, and a small quantity of circular DNA. Its RNA and ribosomes are capable of protein synthesis; its DNA contains the specifications for the synthesis of some of the mitochondrial enzymes. Possessing its own DNA allows a mitochondrion to grow and reproduce by simple division—mitochondria are not produced by another part of the cell, as are most organelles. These observations, coupled with the fact that mitochondrial DNA resembles that of prokaryotic cells, have lent credence to the suggestion that mitochondria might once have been independent prokaryotic organisms that have survived in an intimate, symbiotic relationship with eukaryotic cells.

## PLASTIDS

**Plastids,** found in eukaryotic photosynthetic cells, have some of the characteristics of mitochondria. Chloroplasts, for example, are bounded by a folded double-layered membrane, they contain DNA, and they reproduce independently of the cell nucleus.

A **leucoplast** is a colorless plastid; leucoplasts that contain starch are called **amyloplasts.** Amyloplasts are particularly common in storage roots and stems of such plants as turnips and potatoes and in seeds. **Chromoplasts** are plastids that contain pigments, such as the carotenoids, which give sunflowers and daisies their characteristic yellow and orange colors. Especially important are the **chloroplasts,** which are green because they contain the photosynthetic pigment chlorophyll and the biochemical machinery necessary for photosynthesis.

Most chloroplasts are oval bodies about 4 to 10 $\mu$m long. Higher plants have about 20 to 40 per cell, while some algae have only one or a few large chloroplasts per cell. A typical chloroplast consists of parallel, flattened sacs called **thylakoids** or thylakoid discs (see Fig. 4-32). The membranes of the thylakoids bear the photosynthetic pigments. In certain regions of the plastid, the thylakoids are arranged in stacks called **grana,** (singular, granum) (see Fig. 4-32b and c). Thylakoids are surrounded by a matrix called the **stroma.** The structure of the chloroplast and its relationship to photosynthetic activity is fully discussed in Chapter 7.

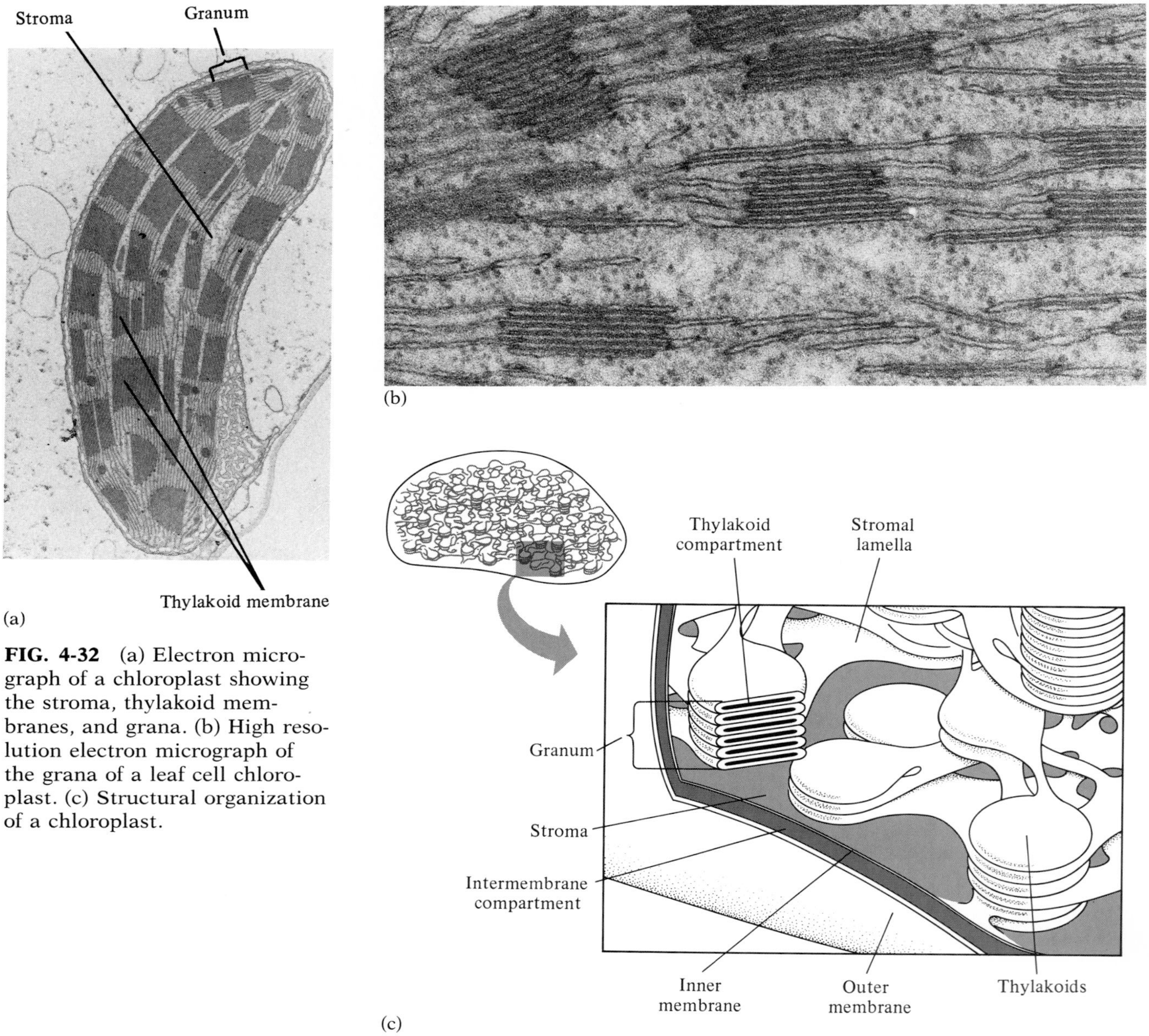


**FIG. 4-32** (a) Electron micrograph of a chloroplast showing the stroma, thylakoid membranes, and grana. (b) High resolution electron micrograph of the grana of a leaf cell chloroplast. (c) Structural organization of a chloroplast.

## MICROFILAMENTS AND MICROTUBULES

Eukaryotic cells have two kinds of long, narrow filamentous structures: microfilaments and microtubules. **Microfilaments** are threadlike fibrils involved in cellular and cytoplasmic movements. Some, such as the microfilaments of muscle cells, participate in contraction.

**Microtubules** are long, cylindrical structures. The microtubule wall is composed of 13 protofilaments (Fig. 4-33). Each microtubule is constructed of a polymer of $\alpha$ and $\beta$ tubulins, which are globular proteins. Microtubules are associated with a variety of cellular processes, one of the most familiar being the part played by spindle fibers in the movement of chromosomes during mitosis. They are also involved in maintaining the shape and rigidity of the cell, in cytoplasmic movements such as the extensions and contractions occurring in phagocytosis, and

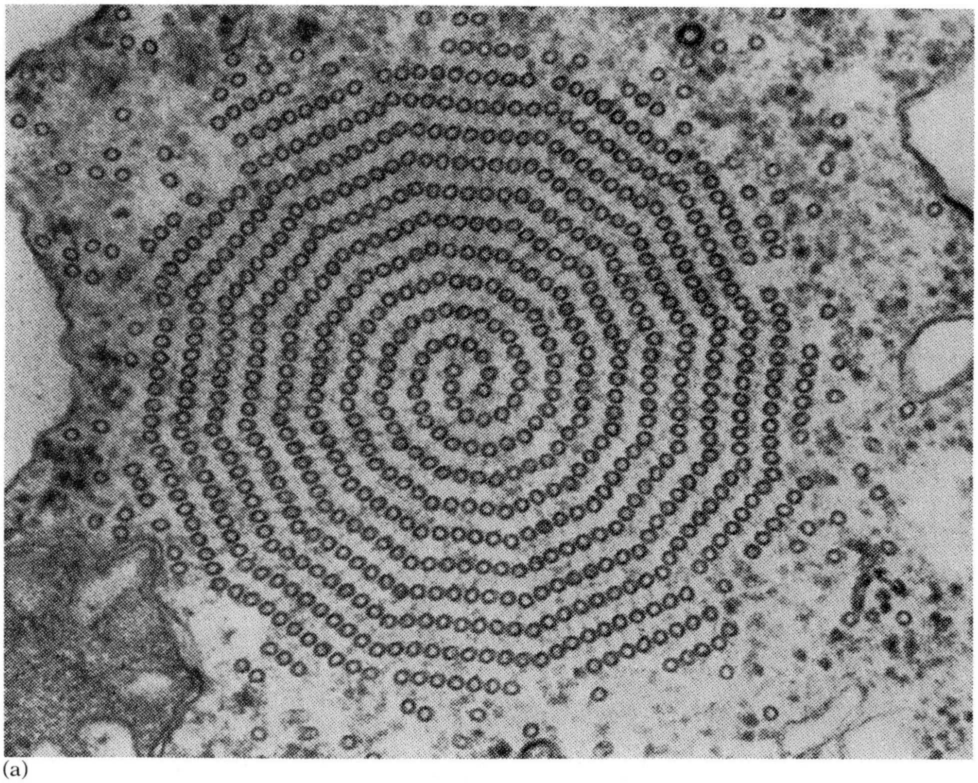

(a)

FIG. 4-33 (a) Electron-micrograph showing spiral arrangement of microtubules in the protozoan *Echinosphaerium nucleofilum.* (b) Model of microtubule structure showing protofilaments.

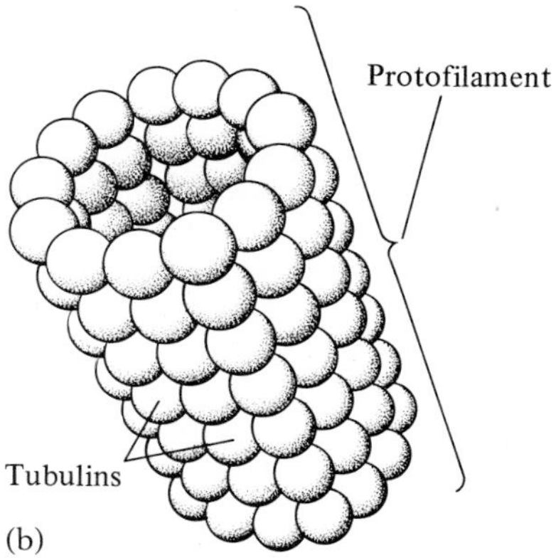


(b)

the formation of secondary wall thickenings of certain plant cells. Microtubules also appear to effect secretion, such as the release of insulin from pancreatic cells. They are the major components of cilia, flagella, ciliary basal bodies, and centrioles.

## CILIA AND FLAGELLA

Extending from the surface of many types of eukaryotic cells are fine, hairlike processes, or projections, known as **cilia** and **flagella.** Cilia differ from flagella only in that they are shorter and usually more numerous. Their internal structure is precisely the same in all eukaryotic cells—from protists, such as the ciliate *Paramecium,* to the ciliated cells of the human oviduct (Fig. 4-34).

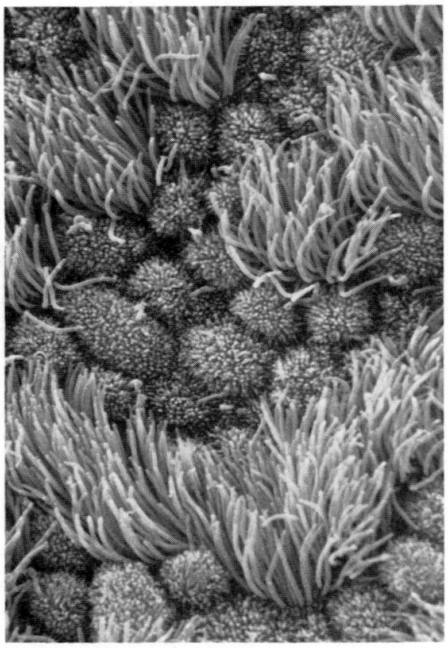

**FIG. 4-34** Scanning electron micrograph of cilia in the human oviduct.

A number of unicellular organisms move about with the aid of numerous cilia or one or more flagella (Fig. 4-35a). Some also employ cilia to stir their immediate environment and create currents that will funnel food into the gullet. Many small, multicellular aquatic animals also move by ciliary activity. The surfaces of many animal organs are ciliated. For example, the lining of the Fallopian tubes, which extend from the ovaries to the uterus, is covered with cilia, which help transport the ovum (egg) to the uterus after it has been released from the ovary. The cilia lining the surfaces of the respiratory tract beat upward, removing dust, pollen, spores, soot, and other inhaled debris. Such particles are moved up to the back of the throat, where they are swallowed. Tobacco smoke slows the action of the cilia or damages them permanently.

Cilia and flagella consist of two central microtubules surrounded by nine pairs of microtubules (Fig. 4-35a, b). This nine-plus-two arrangement is characteristic of all eukaryotic cells. The microtubules are embedded in a matrix enclosed by an extension of the cell membrane. At the base of every cilium or flagellum is a **basal body** with its own characteristic arrangement of microtubules (Figs. 4-35c and 4-36). Basal bodies are involved in the formation of cilia and flagella; they form a template for the nine plus two arrangement of the ciliary microtubules.

## CENTRIOLES

Two or more **centrioles** are located near the nucleus of most animal cells and in the cells of algae and other organisms with motile gametes, reproductive cells which fuse and develop into a new organism. As shown in Figure 4-37a, centrioles occur in pairs, each member of the pair consisting of a short cylinder arranged at right angles to the other. Each cylinder has nine sets of microtubules and each set has three members (Fig. 4-37b, c). The same structure is characteristic of basal bodies of cilia and flagella. Centrioles and basal bodies are essentially the same in structure but differ in location.

**FIG. 4-35** (a) Electron micrograph of the flagella from a marine green alga. (b) Electron micrograph of a cross section of several flagella showing the arrangement of the microtubules. The internal structure of the flagellum consists of a central pair of fibrils surrounded by nine other pairs of fibrils. (c) Cross section of a group of basal bodies of the protozoan *Stentor*.

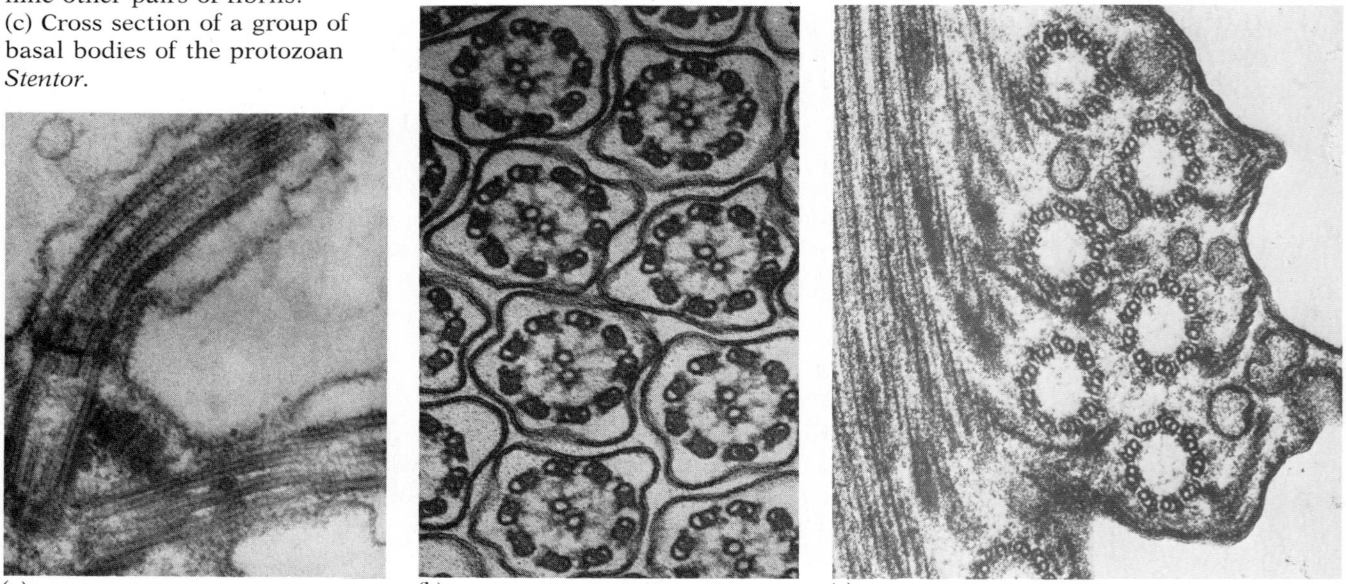

(a) (b) (c)

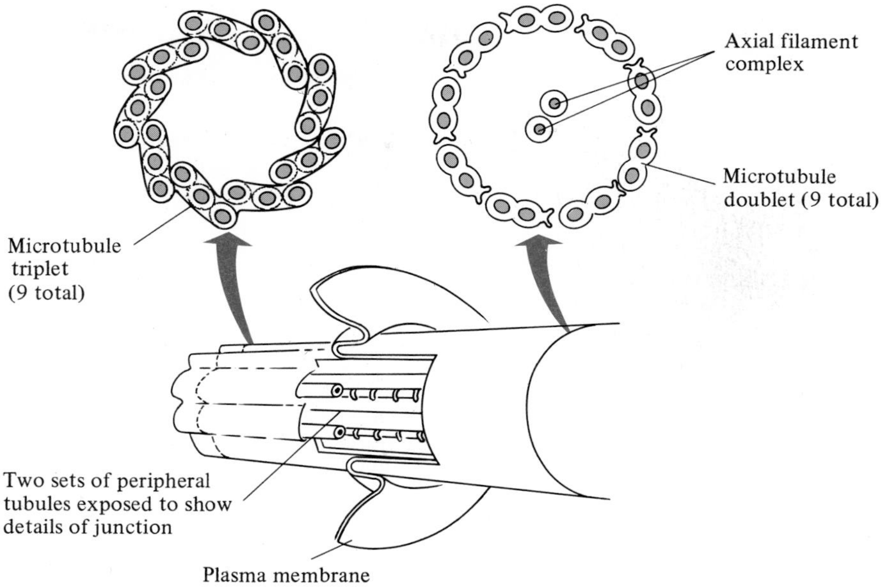


**FIG. 4-36** Diagram illustrating the three-dimensional structure of a flagellum basal body.

During the reproductive process of mitosis, the centrioles, located near the nucleus of some cells, divide; each then becomes associated with the system of microtubules called the **spindle.** This close association of centrioles with the mitotic spindle led to the obvious suggestion that centrioles are essential for the formation of the spindle. However, it was known that this could not be universally true because many cells, such as those of all higher plants, lack centrioles yet undergo normal mitosis. It was postulated that animal cells with centrioles need them to form their spindles. However, if centrioles are removed from various animal cells, the cells continue to form spindles during mitosis. Other than to act as basal bodies and to give rise to cilia or flagella, centrioles are not known to have any other function. There is thus so far no evidence that centrioles play a role in spindle formation or mitosis. A more plausible suggestion, first made by Edmund B. Wilson early in this century, is that the division of centrioles and their association with the two ends of the spindle may have evolved as a way of ensuring that both daughter cells receive a centriole, ultimately providing a centriole to those cells, such as sperm cells, that are to become ciliated or flagellated.

## VACUOLES

Most plant cells and some animal cells have **vacuoles,** cytoplasmic sacs filled with various materials. A plant cell vacuole contains salts, organic molecules, and waste products of cellular metabolism. In a mature plant cell, the single large vacuole is surrounded by a membrane, the **tonoplast.** Young plant cells may have many small vacuoles, which fuse as the cell matures to form a large central vacuole, which may constitute as much as 90 percent of the cell's volume. The cytoplasm, nucleus,

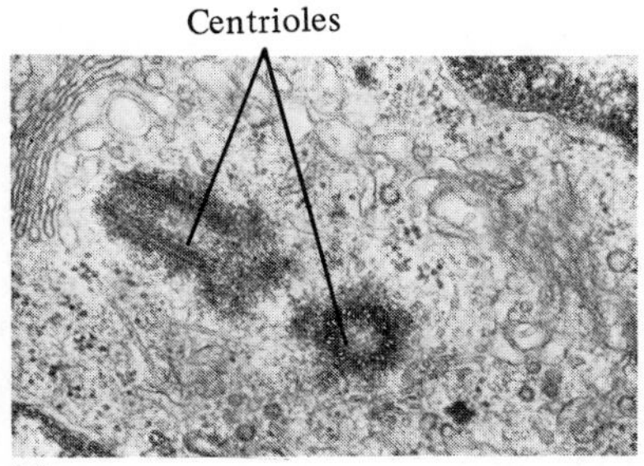


(a)

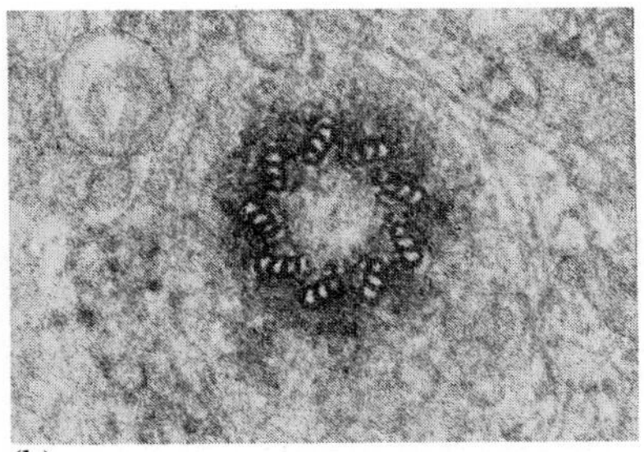

(b)

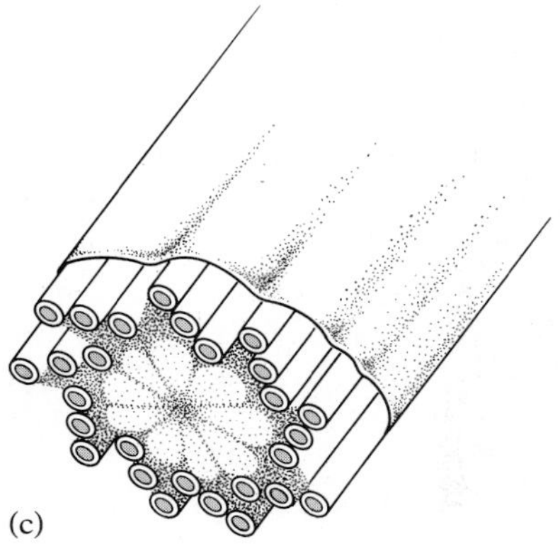

(c)

**FIG. 4-37** (a) A pair of centrioles from a guinea pig bone marrow cell. Note that each member of a pair is typically at right angles to the other. (b) Cross section of a centriole showing the nine sets of microtubules, each set having three tubules. (c) Diagram of the cross sectional structure of a centriole.

and organelles become pushed against the plasma membrane, as shown in Figure 4-38.

The plant cell vacuole has at least two major functions. It can serve as a storage tank for dissolved sugars, amino acids, various water-soluble pigments, and other organic molecules. It also allows cells to become turgid, thereby providing structural support. The tonoplast has the property of **differential permeability,** allowing some molecules to pass through the membrane readily, some with difficulty, and others not at all. In dilute solutions water readily enters plant cells. The vacuole swells, exerting pressure on the surrounding cytoplasm and thus on the cell wall. A young cell, in which the walls are still elastic, elongates as a result. In all plant cells, it is this turgor pressure that gives the cell rigidity and causes the plant body to be turgid.

The cellular structures described in this chapter are summarized in Figure 4-39.

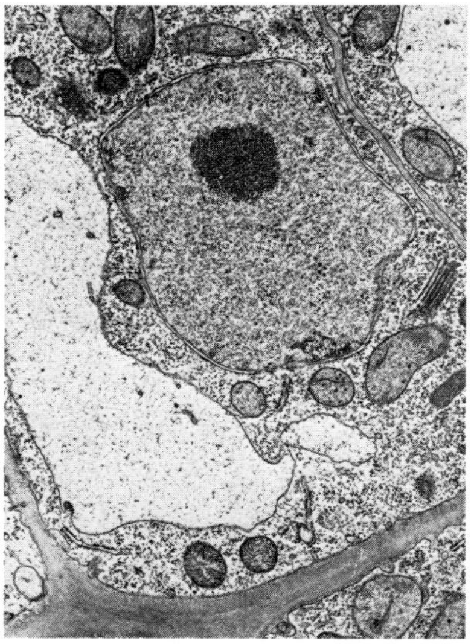

**FIG. 4-38** Plant root tip cell showing a large vacuole.

## CELL AND NUCLEAR DIVISION

As indicated earlier in this chapter, one of the greatest advances in the cell theory was the recognition by Virchow that cells arise only from preexisting cells. Indeed, the cells of a sexually reproducing multicellular organism originate by a series of cell divisions that start with but a single cell: the fertilized egg, or **zygote.** The union of a sperm with an

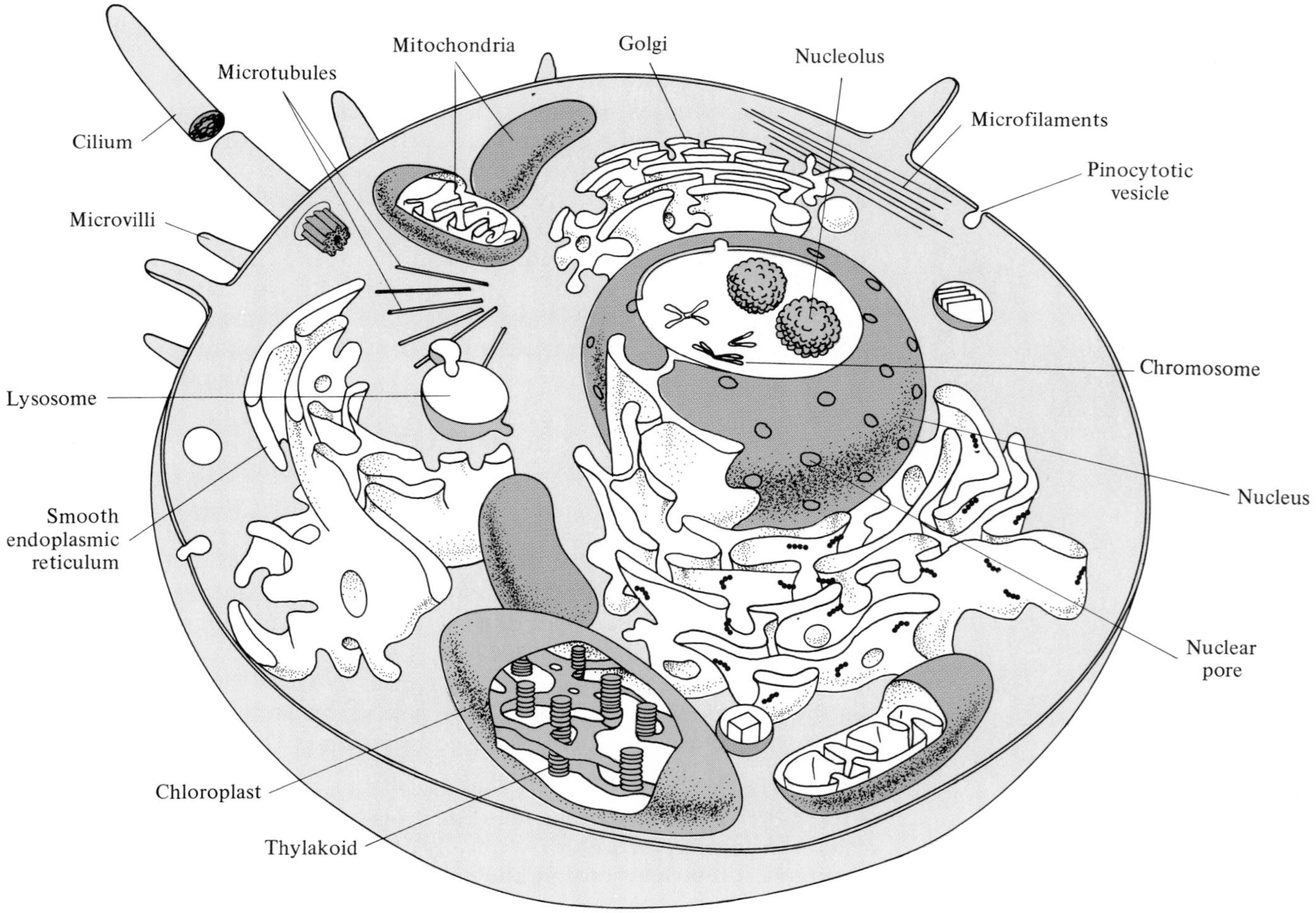


**FIG. 4-39** A generalized view of the cell, showing organelles in ultrastructural detail.

egg initiates this series of divisions that eventually produce the mature multicellular organism. These cells gradually become specialized, a process known as **differentiation,** until an organism recognizable as a member of its species is formed. Its growth and differentiation continue in a highly coordinated way until maturity is reached. Although growth and differentiation may then cease, not all cell division stops. In some tissues, cells continue to divide, and many worn-out and damaged cells are replaced throughout the life of the organism. Cell and nuclear division are the processes by which all new cells are produced.

**FIG. 4-40** (a) Septum formation in the bacterium *Bacillus licheniformis.* (b) Completion of cell division in this bacterium.

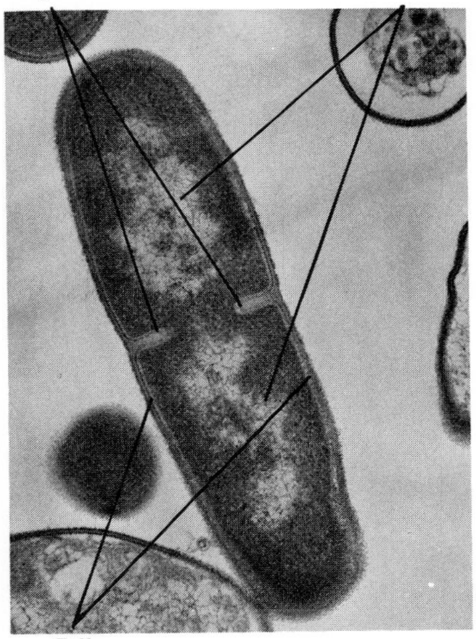


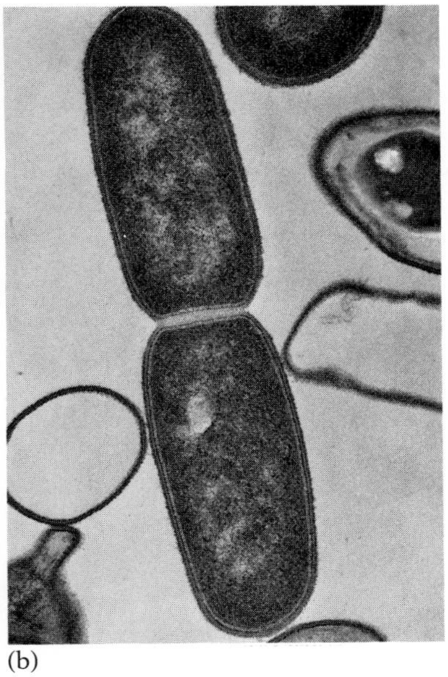


## CELL DIVISION IN PROKARYOTES

The division of prokaryotic cells involves two overlapping stages: (1) replication and separation of the cell's DNA molecules so that each daughter cell receives one complete copy of all the hereditary material of the original parent cell and (2) division of the cytoplasm into two approximately equal parts through the process of **cytokinesis** (from the Greek *cytos,* "cell," and *kinesis,* "movement").

***DNA Replication and Attachment.*** The hereditary material of prokaryotes usually consists of a single, very long molecule of DNA that is circular, with no free ends. In prokaryotes, replication of DNA and cytokinesis are not as closely linked as they usually are in eukaryotic cells. During the growth of a prokaryotic organism, DNA replication usually occurs well before cytokinesis. Thus, the DNA has replicated even before the onset of cell division. Then, before division commences, the two daughter DNA molecules become attached to different regions of the plasma membrane. The attachment of the two DNA molecules to different regions of the membrane is apparently important in ensuring distribution of one of them to each of the daughter cells.

***Cell Division.*** Cell division occurs when new plasma membranes and a new cell wall, or septum, become synthesized at the middle of the cell. The initiation of cytokinesis first becomes evident with an inward growth of the plasma membrane between the two DNA molecules and the secretion next to it of new cell wall material. The plasma membrane and cell wall continue to grow inward, and a complete wall eventually forms. As shown in Figure 4-40, this separates the daughter DNA molecules from each other while dividing the cytoplasm into roughly two equal parts. The wall then splits to form two new cells. In some prokaryotes, wall cleavage is incomplete. As a result, long chains of joined cells form, as in certain species of filamentous cyanobacteria.

## CELL DIVISION IN EUKARYOTES

As with prokaryotes, nuclear and cell division in eukaryotes occur in two fairly distinct stages; these stages may even occur separately but in most cells occur in coordination with one another. The division of the eukaryotic nucleus is termed **mitosis** or, less commonly, karyokinesis. Mitosis results in two daughter nuclei, each with the same number and kinds of chromosomes as the parent nucleus. As in prokaryotes, cell or cytoplasmic division is termed cytokinesis. Mitosis is a complex process that has evolved as a way of evenly distributing the hereditary material, that is, the cell's DNA, to daughter nuclei. Because mitosis occurs in a similar way in all eukaryotes, it is thought to have evolved very early in the history of eukaryotic organisms.

***Chromosomes.*** In eukaryotic cells, the genetic material is not contained in a single DNA molecule, as it is in prokaryotes. Instead, the hereditary material is divided among a number of separate large molecules. Strands of these molecules of DNA, with associated proteins called histones, constitute the chromosomes.

***Interphase Chromosomes.*** During interphase, when the nucleus is not dividing, only the tightly condensed heterochromatin regions of the chromosomes are usually visible. These heterochromatic regions appear as chromatin granules distributed throughout the nucleoplasm. However, at the beginning of mitosis, the uncoiled, threadlike regions of the chromosomes, known as euchromatin, become more and more tightly coiled, condensing into stainable, visible structures. At this time, the chromosomes are discernible as a group.

***The Diploid Phase in Sexually Reproducing Organisms.*** At some stage of the life cycle of every sexually reproducing eukaryotic individual, each of its body cells contains two sets of chromosomes, the number and appearance of which are characteristic of its species. One of these sets was inherited from the individual's mother, the other from its father. The two chromosomes of a pair are called **homologous** chromosomes or are said to be **homologs** of one another. Cells that have two sets of chromosomes are said to be **diploid** (from the Greek *di,* "two," and *ploid,* "form"). Mitosis results in each daughter cell nucleus receiving both sets of chromosomes.

## THE STAGES OF MITOSIS

In describing mitosis, it is convenient to divide the process into four defined stages: **prophase, metaphase, anaphase,** and **telophase.** The period between successive mitoses is called **interphase.** Mitosis does not actually occur in discrete stages but is a continuous process with each stage merging with the next.

***Interphase.*** Between divisions, nuclei are in interphase. Because cell division usually accompanies mitosis, it is common to refer to the entire cell as being in interphase, prophase, metaphase, anaphase, or telophase. However, it is the nucleus, not the cell, that undergoes mitosis.

Most DNA, RNA, and protein synthesis required for cell division occurs during interphase. Thus, interphase is a period of high metabolic activity. DNA replication occurs during interphase, but only during a discrete interval, called the **S** (for synthesis) **phase** of the cell cycle (Fig. 4-41).

***Determining the Time of DNA Replication.*** In the 1920s, Robert Joachim Feulgen, a German biochemist, developed a highly specific test for the presence of chromatin; actually, it turned out to be a test for DNA. The test is semiquantitative, since the magenta (red) coloration produced in the presence of DNA is twice as intense in nuclei with a diploid set of chromosomes than in comparable nuclei with only one set. By precisely comparing the density of Feulgen-stained, diploid, somatic (body) cell nuclei with stained nuclei of the same organism's gametes (eggs and sperm), it was found that body cell nuclei stained

**FIG. 4-41** Stages of the cell cycle.

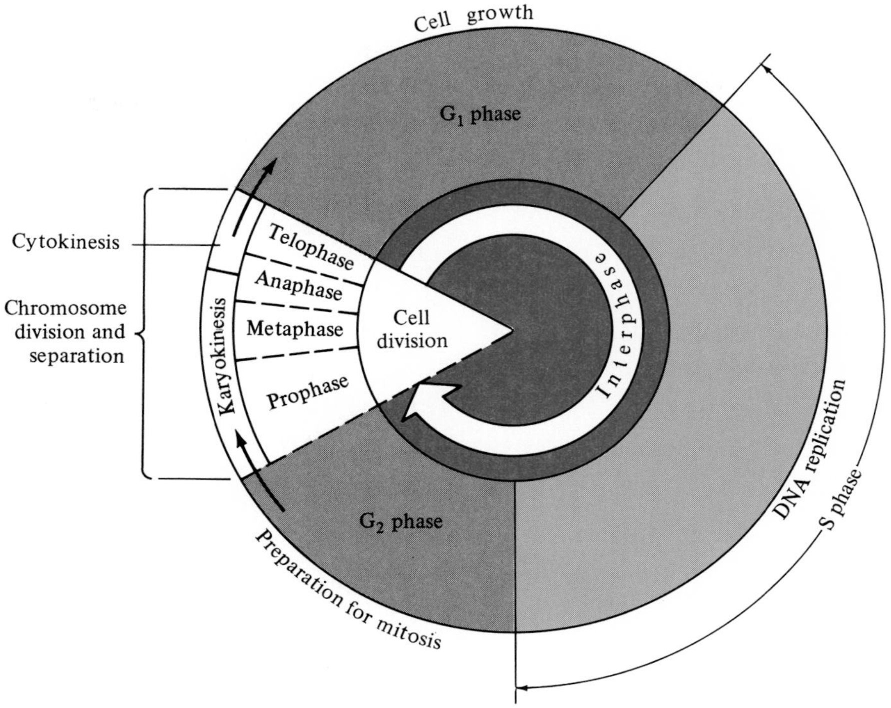


twice as densely as gamete nuclei. Gamete nuclei have long been known to be **haploid** (from the Greek *haploides*, "single"), that is, to contain but one set of chromosomes. By determining the Feulgen-staining intensity of somatic cell nuclei during the entire cell cycle, it was found that the doubling in staining intensity—representing the actual replication of DNA molecules—did not occur in prophase but much earlier, during interphase; the time of DNA synthesis later became known as the S phase.

The two periods before and after the S phase are called G (for gap) phases. The first of these, the **$G_1$ phase,** is the period between the end of telophase of one mitotic division and the beginning of the S phase of the next division (Fig. 4-41). The period from the end of the S phase to the beginning of prophase of the next mitosis is the **$G_2$ phase.** During the $G_1$ phase, growth of the cytoplasmic material, including replication of the various organelles, takes place. During the $G_2$ phase, the structures involved directly with mitosis, such as the spindle fibers, are assembled. The combination of $G_1$, S, $G_2$, and M (for mitosis) phases comprise the cell, or mitotic, cycle.

The durations of individual phases or of the complete cell cycle vary greatly from one cell type and species to another. For example, a certain strain of animal cells grown in artificial medium, a procedure called tissue culture, has a cell cycle of about 24 hours. Its $G_1$ phase lasts about 10 hours, its S phase about 8 hours, its $G_2$ phase about 5 hours, and its M phase about 30 minutes. In the growing root tips of some plants, the cell cycle is often about 14 hours, with 2.5 hours for $G_1$, 6 hours for S, 3.5 hours for $G_2$, and 2 hours for mitosis. In cells in which division has ceased, as in most of the body cells of adult animals, the $G_1$ phase is prolonged.

***Prophase.*** During prophase, shown in Figure 4-42a and b, the nuclear chromatin, which had duplicated during the S phase of the preceding interphase, begins to condense by coiling and folding into discernible chromosomes. In the light microscope, these can be seen as separate, threadlike strands within the nucleus. With appropriate staining, it is sometimes possible to see that each chromosome consists of two strands, the chromatids. The two chromatids of each chromosome are firmly attached to each other at a specific point of indentation called the primary constriction, or **centromere.** The centromere may be located anywhere along the chromosome, even at its end.

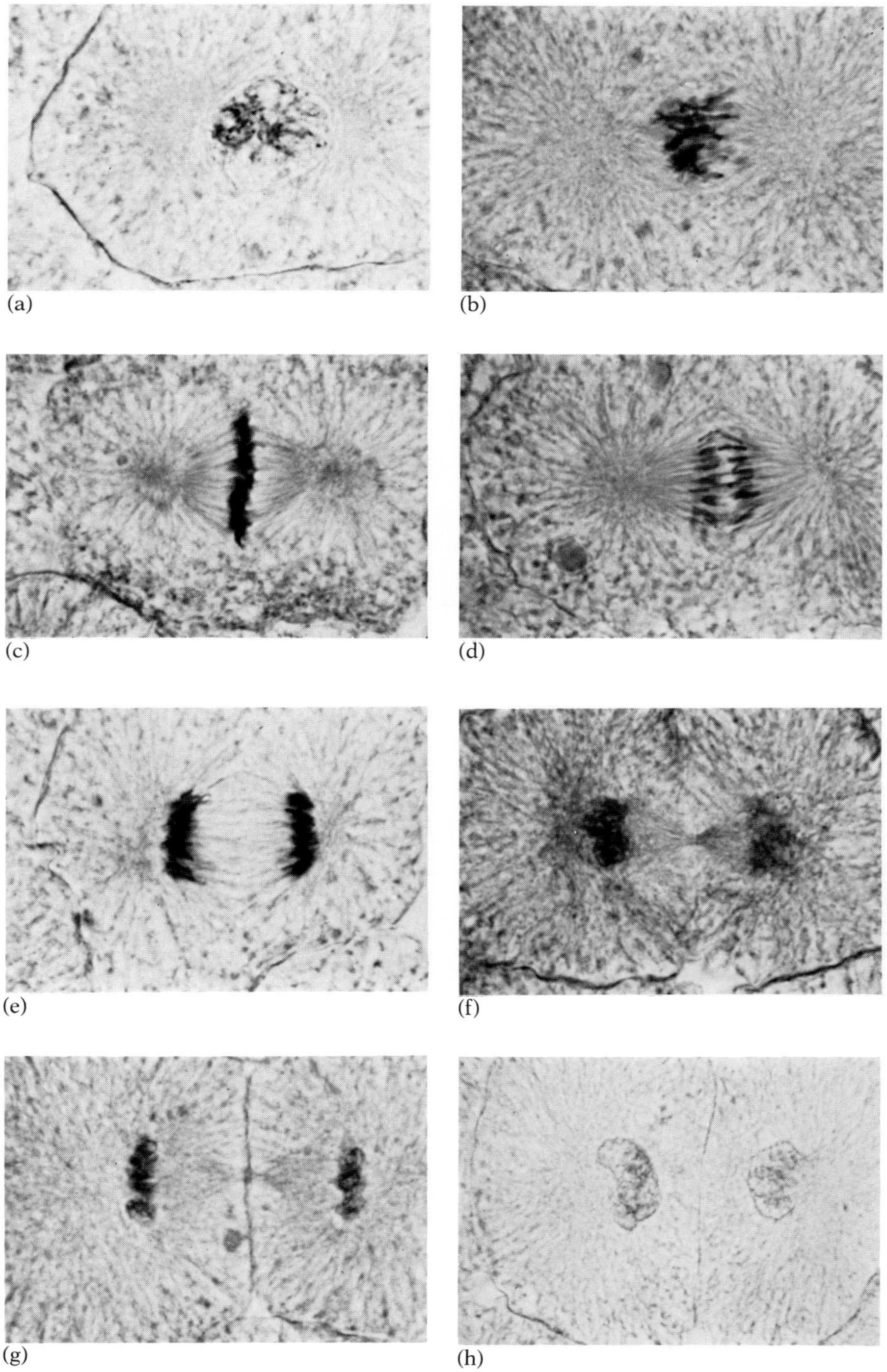

**FIG. 4-42** Mitosis in the developing whitefish embryo: (a) early prophase, (b) prophase, (c) metaphase, (d) early anaphase, (e) late anaphase, (f) telophase, (g) late telophase, (h) daughter cells.

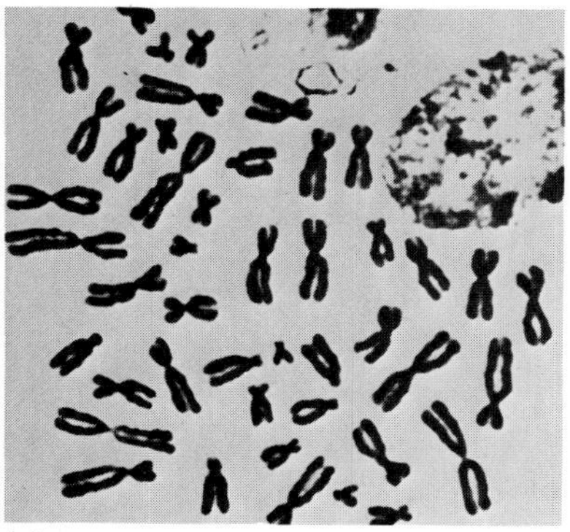

(a)

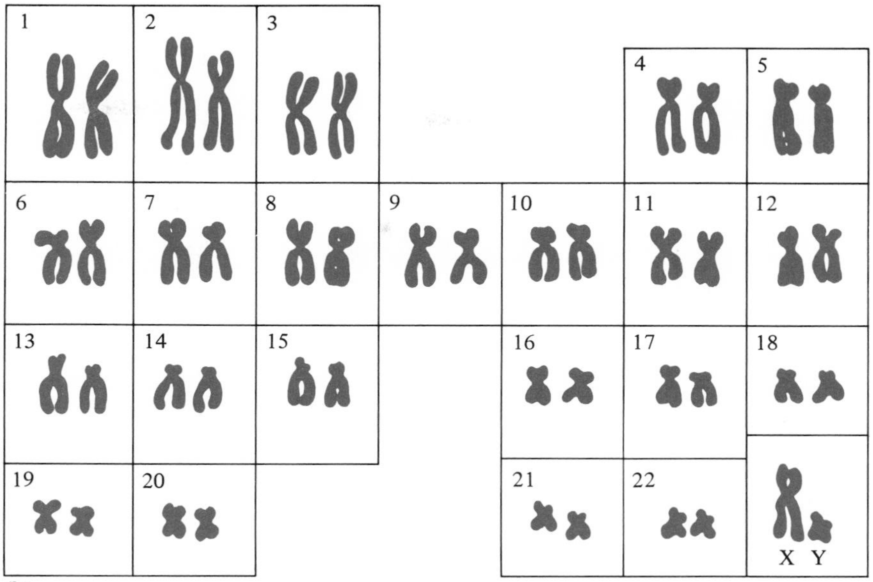


(b)

**FIG. 4-43** (a) Normal human female chromosomes from a cell grown in tissue culture. Mitosis was stopped in metaphase with colchicine, and the cells squashed and stained. The two chromatids held together by the centromere, or kinetochore, can be seen. (b) The metaphase chromosomes of the human male.

As condensation continues, the chromosomes become shorter and thicker and their nucleoli become smaller. By late prophase, the nucleoli have disappeared, and the chromosomes may begin to exhibit characteristic secondary constrictions. The differences in general shape and length of chromosomes, the position of their centromeres, and the disposition of their secondary constrictions distinguish the individual chromosomes of a species from each other (Fig. 4-43a). The identification and numbering of the chromosomes of a species according to size is known as that species' **karyotype.** The karyotype for a human male is shown in Figure 4-43b.

During prophase, the system of microtubules called the mitotic spindle assembles in the cytoplasm next to the nucleus (Fig. 4-44). Near the end of $G_2$, in cells that have centrioles, this paired structure becomes evident as two pairs of centrioles. During prophase, the two pairs of centrioles begin to migrate, as indicated in Figure 4-45a, b, and c, to

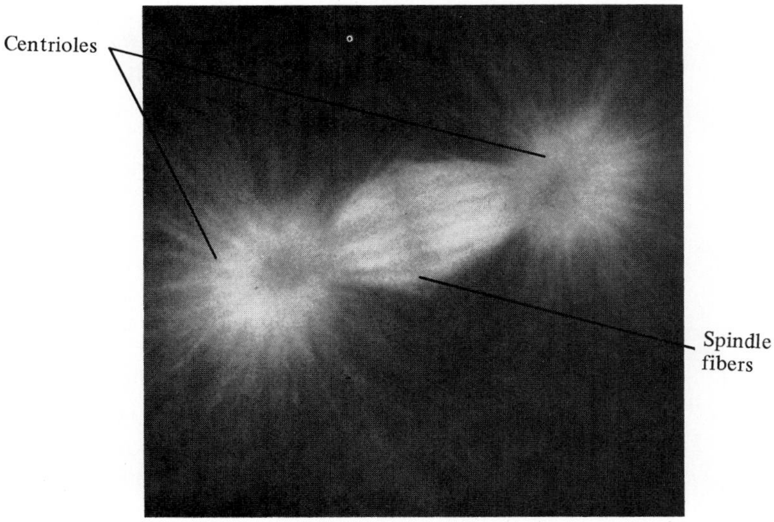


**FIG. 4-44** An isolated mitotic apparatus. Polarized light and special optics were used to allow the viewing of these details in live material.

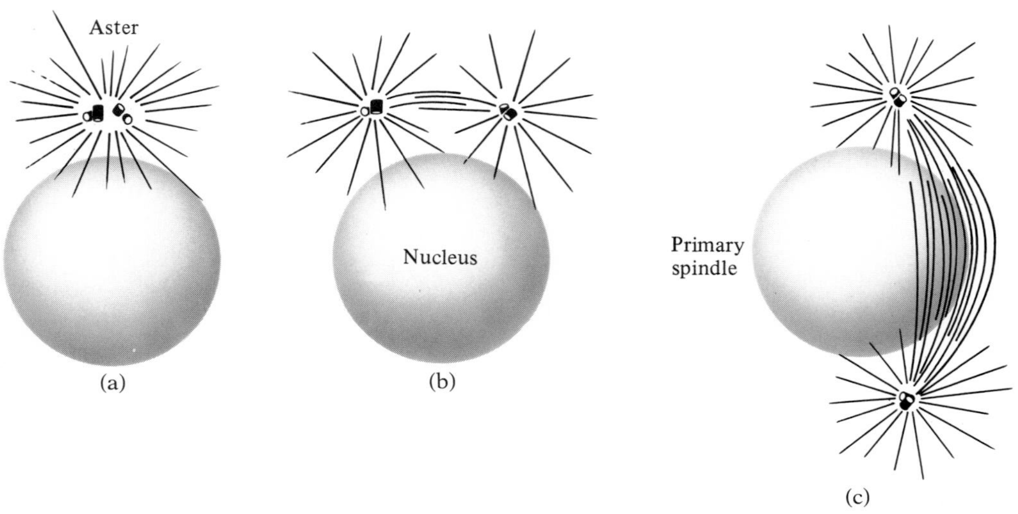


**FIG. 4-45** Movements of the spindle and centrioles during prophase.

opposite sides of the nucleus. As the centrioles move, the system of microtubules shown in the figure radiates in all directions from each pair of centrioles to form a starlike structure, the **aster.** Longer microtubules extend between the two pairs of centrioles, forming the spindle. The centrioles complete their migration when they near the opposite ends of the cell. Most plant cells lack centrioles but otherwise form a spindle in a similar fashion.

In most cells, as the centrioles migrate to the poles, the nuclear envelope begins to disintegrate. By the time the envelope has disappeared, each member of a chromatid pair has become attached at its centromere to two fibers, each of which reaches one of the poles, as illustrated in Figure 4-46.

**FIG. 4-46** Spindle fiber microtubules attached to a chromosome at the centromere.

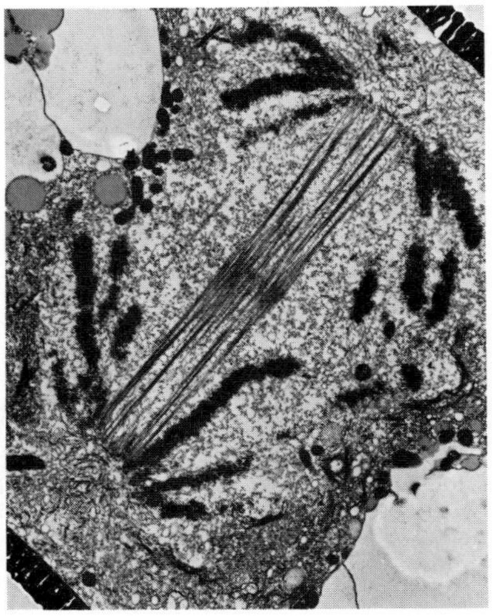

***Metaphase.*** In most types of cells, metaphase is a relatively brief stage. It is characterized by the arrangement of the chromosomes in a plane that lies in the middle of the spindle and at right angles to the fibers (Fig. 4-42c, d, e, p. 138). This region is called the **metaphase plate region** in dividing plant cells. At this stage, the chromosomes reach their maximum contraction. Therefore, the chromosomes at this stage of mitosis are normally used for karyotype analysis.

By the time the chromosomes have assembled at the metaphase plate, one or more spindle fibers, each composed of a bundle of four to eight or more microtubules, have attached to the centromere of each sister chromatid. These are known as the chromosome spindle fibers. This microtubule attachment occurs in such a way that one chromatid of each pair is connected to one pole of the mitotic spindle and its sister chromatid is attached to the other pole (Fig. 4-47). Other spindle fibers, called continuous fibers, stretch from one pole toward the opposite pole beyond the center of the spindle without attaching to any chromosomes. At one time it was believed that the continuous fibers extended from one pole to the other.

***Anaphase.*** During the brief stage known as anaphase, the sister chromatids of each metaphase chromosome separate, one going to one pole and the other to the opposite pole (Fig. 4-42d–g). Anaphase begins when the first chromatids separate and begin their movement toward opposite poles. Once the two chromatids of a pair separate, they are daugh-

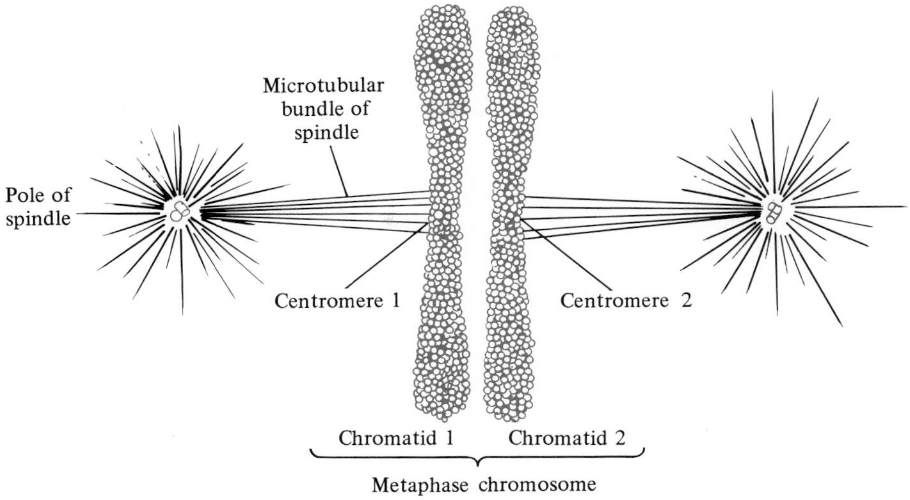


**FIG. 4-47** Illustration of the attachment of spindle microtubules to metaphase chromosomes during mitosis.

ter chromosomes. When all members of the two groups of chromosomes have reached the opposite ends of the spindle, anaphase has ended and telophase begins.

Evidence indicates that the microtubules are responsible for the movement of the chromatids to the poles. Experimental evidence for this view includes the fact that treatment of cells with the drug colchicine—which specifically prevents the formation of spindle fibers—prevents anaphase movement of chromosomes. Evidence now indicates that the chromosome spindle fibers depolymerize at their polar ends, thus pulling the chromosomes to the poles.

***Telophase.*** Telophase begins when all of the newly formed daughter chromosomes have reached the poles of the mitotic spindle. The chromatin then gradually becomes diffuse again as each chromosome uncoils, appearing longer and thinner. Since it is now less stainable, it appears less distinct in the light microscope. Meanwhile, the spindle disintegrates, the nucleolus or nucleoli reappear, and a nuclear envelope reforms around the uncoiling chromosomes. The nuclei slowly resume their interphase appearance and telophase ends.

## CYTOKINESIS

Although the stages of mitosis are essentially the same in plant and animal cells, the events of cytokinesis are different.

In most animal cells and in most protozoans, cytokinesis is coordinated with mitosis and usually begins about mid-anaphase. However, mitosis without accompanying cytokinesis is not uncommon. For example, the development of the insect zygote proceeds by repeated mitotic divisions without cell divisions until several thousand nuclei have been formed in the egg. Only then do cytoplasmic membranes appear between the nuclei to segregate each nucleus into a separate cell; many new cells thus result in a very short time.

***Cytokinesis in Animal Cells.*** Usually, cytokinesis is coordinated with mitosis, and, soon after, the telophase stage of cell division begins with the formation of a cleavage furrow (Fig. 4-42f, g). At first, this furrow is only a ring around the cell surface at the equator. The furrow then gradually deepens. Earlier, beginning in mid- to late anaphase, an accumulation of dense material occurs at the cell's equator, forming a platelike midbody; this dense area is visible under the light microscope by late anaphase. As the furrow continues to deepen, it meets the midbody and compresses it, after which the midbody disappears.

The process of cleavage furrow formation is believed to be controlled by a ring of contractile microfilaments that concentrate under the plasma membrane in the region where the furrow is to form. Furrowing and cytokinesis are blocked if dividing cells are treated with the drug cytochalasin B, an antibiotic that acts by disrupting microfilaments. If the treated cell is then freed of the drug, both processes resume. The microfilaments are composed of the protein actin, one of the proteins involved in muscle contraction. At telophase, the microfilament ring appears to contract, pinching the cell into two daughter cells.

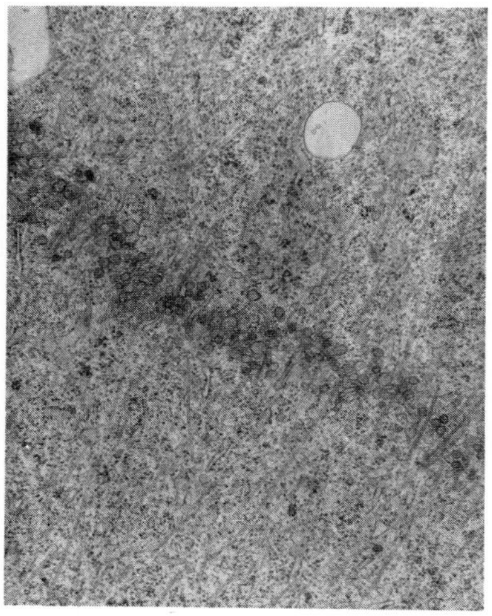

**FIG. 4-48** Cell plate formation during late telophase in an immature lily ovary cell. Note vesicles lining up to form the cell plate. Portions of the spindle fibers can be seen trapped in the forming cell plate.

***Cytokinesis in Plant Cells.*** As in most animal cells, cytokinesis in plants and most protistan algae and fungi usually accompanies mitosis. Mitosis without cytokinesis, however, is common in some protistan algae and certain fungi. This produces single-celled, multinucleate stages called **coenocytes.** A similar situation occurs regularly during certain phases of reproduction in seed plants.

In plants, furrow formation does not occur in cytokinesis. Instead, in late anaphase the cytoplasm becomes divided by the formation of a cell plate at the equator of the cell (Fig. 4-48). An important contribution to the formation of the plate is made by Golgi vesicles, which assemble among the long spindle fibers to form a platelike array after migrating to the region previously occupied by the metaphase plate. The fusion of these vesicles forms the cell plate, which eventually consists of two membranes derived from the vesicles and separated by the middle lamella derived from the vesicle contents. The plate then expands toward the edges of the cell by the addition of vesicles. Finally, the rim of the plate fuses with the lateral plasma membranes of the parent cell. At this point, each daughter cell is surrounded by a continuous plasma membrane and is separated from its sister cell by the pectin material that makes up the cell plate. The primary wall is deposited on the cell plate material by the plasma membrane of each daughter cell. The matrix material is derived from Golgi vesicles; the cellulose microfibrils are assembled by the plasma membrane.

## Summary

The first simple cells to arise in evolution were probably prokaryotes, cells lacking organelles and organized nuclei. Their modern representatives are the bacteria and cyanobacteria (blue-green algae), which, in general structure, have not progressed much beyond the primitive condition. Although the cyanobacteria have extensive thylakoids and both they and the bacteria have ribosomes, these cells lack other organelles and a discrete nucleus. Instead, all biological functions are conducted within the single compartment. Metabolically, however, modern prokaryotes are diverse and highly specialized, being adapted to a great number of environments, including some that provide only simple carbon and nitrogen sources and inorganic ions as a source of nutrients. Largely because of their cellular organization, prokaryotes are severely limited in size and in structural and functional diversity. Their biological activities consist primarily of metabolism, growth, and reproduction.

A much more successful line of evolution has been that of the eukaryotes. Eukaryotic cells, as their name suggests, have "true" nuclei; that is, their DNA, organized into discrete chromosomes, is enclosed during most or all of the cell cycle within a membranous nuclear envelope. Membranes abound in eukaryotic cells. They form the endoplasmic reticulum, which synthesizes proteins and various other molecules, and the Golgi complex, which modifies, stores, and packages the cell's secretory products. The specialized membrane systems of the mitochondria and chloroplasts contain enzyme complexes that catalyze the key biological processes of respiration and photosynthesis.

The plasma membrane surrounding eukaryotic cells is composed of a phosphoglyceride bilayer in which float various protein complexes. These proteins move freely within the lipid bilayer in response to changing conditions in the cellular environment, suggesting that the plasma membrane is like a fluid mosaic.

The largest and most prominent feature of all eukaryotic cells is the membrane-bound nucleus. Inside the nucleus are one or more nucleoli, which are organized around specific chromosomal regions. Nucleoli are the sites of synthesis of ribosomal RNA and the assembly of ribosomes. During interphase most of the nucleus is filled with chromatin, which condenses into chromosomes during mitosis.

Most eukaryotic cells have a system of interconnecting membranes, the endoplasmic reticulum. Regions of a cell that are actively synthesizing protein have ribosomes attached to the ER. Regions of a cell that are primarily involved in the synthesis and secretion of hormones or production of storage fats have a smooth ER.

Among the organelles in eukaryotic cells, one of the most prominent is the Golgi apparatus, a series of flattened sacs whose main function appears to be the packaging and transport out of the cell of certain cell products, especially glycoproteins. Less conspicuous organelles in animal cells are the lysosomes, which digest food materials taken into the cell as well as old, nonfunctioning cell organelles. Peroxisomes, or microbodies, are membrane-bound sacs which contain one or more enzymes that can break down amino acids, lactic acid, and other substances; one type, the glyoxosomes, contain enzymes that can convert fats into carbohydrates.

The immediate source of most of the energy needed by animal cells to carry out their metabolic activities is the ATP generated by specialized organelles called mitochondria. ATP in plant cells is also produced by another organelle, the chloroplast, which is able to capture energy from the rays of the sun to drive the process of photosynthesis.

Eukaryotic cells have two kinds of very long, narrow filamentous structures: microfilaments and microtubules. Microfilaments are involved in ameboid movement, muscular contraction, and cytoplasmic streaming. Microtubules are associated with a variety of cellular structures and processes; they are components of spindle fibers, centrioles, basal bodies, cilia, and flagella and are essential to the formation of secondary wall thickenings in certain plant cells and (in cooperation with microfilaments) to the maintenance of cell shape and rigidity.

All the cells of a multicellular organism are derived by a successive series of cell divisions. The division of prokaryotic cells involves two overlapping processes: replication and distribution of the cell's DNA, with each daughter cell receiving a complete set of the hereditary material found in the parent cell; and division of the cytoplasm by the process of cytokinesis.

Division in eukaryotes is similar in principle to that of prokaryotes, except that the stages are more complex. In most eukaryotes, cellular division is also divided into two processes, which normally occur together: mitosis, or nuclear division, and cytokinesis, or cell division. The mitotic process is conventionally divided into four stages: prophase, metaphase, anaphase, and telophase. The interphase is the period between successive mitoses.

## Recommended Readings

Albersheim, P. 1975. "The walls of growing plant cells." *Scientific American,* 232:80–95. A comprehensive, well-written description of the biochemistry of plant cell wall growth.

Alberts, B., D. Bray, J. Lewis, M. Raff, K. Roberts, and J. D. Watson. 1983. *Molecular Biology of the Cell.* Garland, New York. Concise up-to-date description of the basic structural and functional characteristics of prokaryotes and eukaryotes. The illustrations are particularly well done.

Capaldi, R. A. 1975. "A dynamic model of cell membranes." *Scientific American,* 230:26–33. An excellent description of the fluid mosaic model of the structure of cell membranes.

DeWitt, W. 1977. *Biology of the Cell: An Evolutionary Approach.* Saunders, Philadelphia. A fine introductory textbook on cell biology. Highly readable and well organized.

Fawcett, D. W. 1981. *An Atlas of Fine Structure,* 2nd ed. Saunders, Philadelphia. This classic text contains a superb collection of electron micrographs illustrating all aspects of cell ultrastructure.

Holtzman, E., and A. B. Novihoff. 1984. *Cells and Organelles.* 3rd ed. Saunders, Philadelphia. This text integrates the structural, molecular, and bi-chemical knowledge of cells and organelles into a comprehensive introduction to modern cell biology.

Jensen, W. A., and F. B. Salisbury. 1984. *Botany,* 2nd ed., Wadsworth, Belmont, CA. An outstanding introductory botany text that has a particularly good treatment of the morphology of cells and their organelles.

Kessell, R. G., and R. H. Kardon. 1979. *Tissues and Organs.* Freeman, New York. An atlas of transmission and scanning electron micrographs of mammalian tissues and organs. Detailed descriptions accompany this outstanding collection of micrographs.

Swanson, C. P., and P. L. Webster. 1985. *The Cell,* 5th ed. Prentice-Hall, Englewood Cliffs, N.J. An outstanding elementary paperback on cell structure and function.

Wolfe, S. L. 1981. *Biology of the Cell,* 2d ed. Wadsworth. Belmont, Calif. A very well-written and -illustrated introductory text.

5

# Transport of Materials Across Cell Membranes

OUTLINE



All cells are bounded by a delicate plasma membrane. Moreover, most organelles in the cytoplasm of eukaryotic cells are membrane bound. Besides holding in the contents of the cell or organelle, these membranes play other vital roles in the life of the cell. They control internal concentrations of substances essential to cell life and regulate the entrance and exit of water, wastes, nutrients, and other important substances. Because not all substances pass through cell membranes with equal ease, membranes are described as **differentially** or **selectively permeable;** that is, they allow the passage of some substances, while blocking the passage of others. Some membranes are **semipermeable:** they are permeable to certain **solvents**—the fluid portion of a solution —but impermeable to **solutes**—substances that dissolve in solvents. Although semipermeable membranes can be constructed in the laboratory, none are known to exist in living systems.

## PASSIVE TRANSPORT

**Diffusion** is a process of passive transport by which molecules and other small particles move spontaneously from a region in which they are in high concentration to one in which their concentration is lower. This

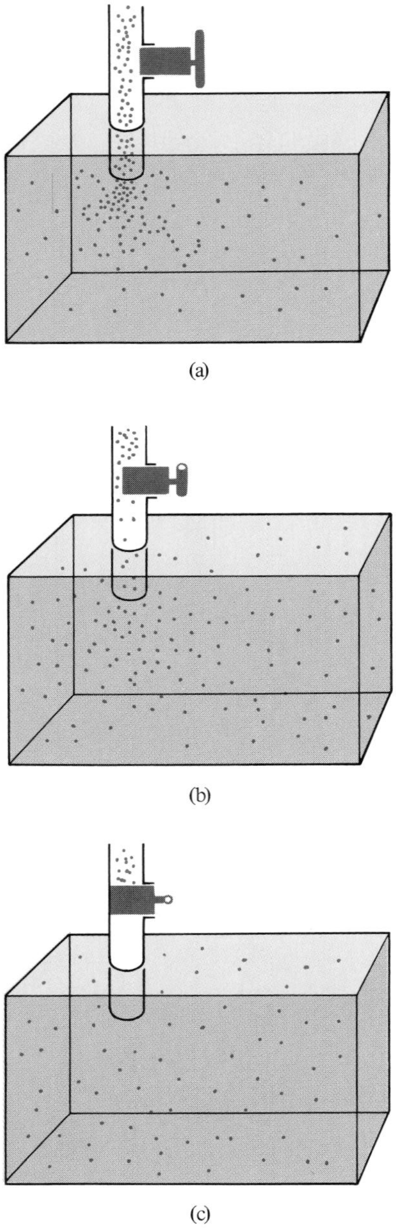

FIG. 5-1 Diffusion of gas molecules released at one end of a closed container. (a) The greater the initial difference between concentrations, the more rapid the initial rate of diffusion. (b) The rate of diffusion slowly decreases as the concentrations become equalized. (c) Diffusion ceases when the concentrations in the two areas have become equalized.

process is familiar to anyone who has opened a container of a substance such as ammonia. The ammonia molecules gradually become equally distributed throughout the room (Fig. 5-1). Another example of diffusion can be observed when a drop of dye solution is placed in a jar of water. The solute, dissolved dye in this case, diffuses through the water and eventually becomes equally dispersed throughout the jar.

## THE MECHANISM OF DIFFUSION

To understand the mechanism of diffusion it is necessary to understand the motion of molecules. Theoretically, at a temperature of absolute zero (0 Kelvin), which is equivalent to −273° Celsius (C), molecules have no motion at all. At any temperature above absolute zero, molecules move or vibrate at a rate proportional to the temperature and inversely proportional to the substance's molecular weight, the smallest molecules moving the fastest. For example, at 90°C, hydrogen molecules ($H_2$) move at an average of 45 kilometers (km) per minute (28 miles per minute), sulfur dioxide molecules ($SO_2$) at an average of 19 km/min, and the molecules of all the gases in air at an average of 32 km/min. Heat thus provides molecules with the energy of motion, or kinetic energy.

***Kinetic Energy Provides the Basis for Diffusion.*** Molecules in motion tend to move in random directions but in straight lines. Typically, whenever such molecules collide with one another, they rebound and move off in new directions. If ammonia vapor is released at one end of a chamber from which the air has been evacuated, it can be detected at the other end of the chamber almost immediately. In a vacuum, few collisions occur among the molecules at the edge of the mass of released vapor; consequently, the molecules are unimpeded and quickly reach the other end of the chamber. However, vacuums are uncommon. If ammonia molecules are released at one end of a classroom, it takes quite some time before they can be detected at the other end of the room. This delay occurs because the released molecules collide not only with each other but also with the molecules of nitrogen, oxygen, carbon dioxide, and with molecules of water vapor and any other substances in the room air. A given ammonia molecule can travel only a short distance in any direction before colliding with other molecules. This explains why diffusion is usually a slow process.

The diffusion of any substance is normally independent of the concentration of any other substance in the area. A high concentration of ammonia at one end of a room, for example, is not balanced by a high concentration of some other substance at the other end. Both substances will diffuse and in time become equally distributed throughout the room.

The greater the initial difference between the concentrations of a substance in any two areas, the more rapid will be the initial rate of its diffusion (Fig. 5-1a). As the concentrations of the substance in the two areas become equalized, the rate of diffusion slowly decreases (Fig. 5-1b). Finally, when the concentrations in the two areas have become equal, diffusion ceases (Fig. 5-1c). The individual molecules of the substance continue to move and to collide at the same rate as before, but the number of molecules that move out of an area now equals the number that move in. At this point, the molecules are said to be in a state of **equilibrium.** The equilibrium they have achieved is a dynamic equilibrium, not a static equilibrium. A chemical system in dynamic equilibrium is similar to a college that maintains a constant level of enrollment over a period of years, with students entering and leaving in equal numbers each semester.

***Solutes Diffuse Very Slowly.*** The behavior of dissolved particles in solution parallels that of the gas molecules described above. But solute particles diffuse much more slowly than do vapor molecules. One reason, of course, is their frequent collisions with the more abundant molecules of the solvent in which they are dissolved. The movement of solute molecules is also usually impeded by attractive forces that normally exist between them and the molecules of solvent.

In biological systems, diffusion is effective for the transport of materials over a distance of only about half a millimeter. Despite this limitation, it is of major importance in moving materials within cells, between cells, and in exchanges between cells and their immediate environment.

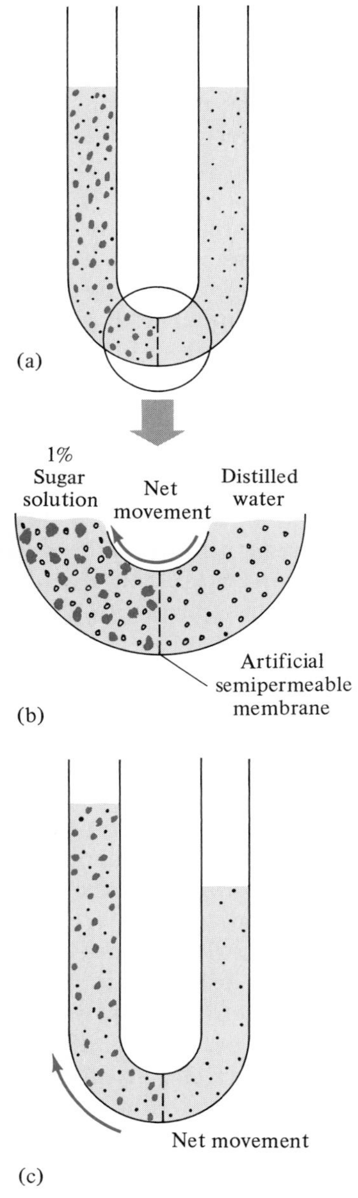


**FIG. 5-2** Osmosis across an artificial semipermeable membrane, resulting in an increase in volume of the solution on one side of the membrane and a decrease in volume of distilled water on the other side.

## OSMOSIS

The principles that govern the diffusion of molecules in an open system also operate when a membrane permeable to both solute and solvent is interposed between solutions with different concentrations of a solute. However, if the membrane is impermeable to the solute but permeable to the solvent, the result is different, as can be demonstrated by an apparatus such as that shown in Figure 5-2. In this device, a glass U tube is divided at its base by an artificial membrane that is permeable to water but is impermeable to some solutes, such as sugar. Under these conditions, the water molecules are able to move through the membrane in either direction (Fig. 5-2b). However, the water molecules on the right, where there are no solute molecules, are less restricted in their movement than are those on the left, whose movement is impeded by the dissolved sugar molecules. The result is a net movement of water molecules into the sugar solution. Such movement is termed **osmosis.**

In this example, the process of osmosis is retarded for at least three basic reasons:*

1. Because of the mutual attraction that exists between water molecules and the solute molecules dissolved in them, solute molecules in water, or aqueous, solutions develop a close association with the nearest water molecules; they are said to be hydrated, that is, combined with water. Thus, the presence of the solute molecules restricts the movement—and hence the diffusion—of the water molecules.
2. Solute molecules take up space that water molecules might otherwise occupy. In the sugar solution, for example, some of the water molecules that might have passed through the membrane do not because solute molecules are in their way.
3. In many solutions, especially concentrated ones, a given volume of solution (solvent plus solute) contains fewer water molecules than does an equal volume of pure solvent. In such cases the solution can be thought of as having a lower concentration of water than does pure water alone.

* More water usually crosses membranes during osmosis than can be accounted for by these factors. Apparently other forces are also involved.

For these reasons, although water molecules move in both directions across a semipermeable membrane such as the one shown in Figure 5-2b, at any given time more molecules will be moving from the side of the U tube containing water into the side containing 1-percent sugar solution than in the reverse direction. In other words, a net movement of water occurs from the distilled water into the 1-percent sugar solution.

***Osmosis Does Not Require Pure Solvent on One Side of the Membrane.*** If enough sugar molecules are added to the distilled water side of the tube shown in Figure 5-2 to make a 2-percent solution, the osmotic flow would be reversed because the greater restriction of water molecule movement would occur on the side of the membrane containing the higher concentration of sugar molecules. The more highly concentrated, 2-percent solution of sugar, with its lower concentration of water, would then gain water at the expense of the 1-percent sugar solution, with its higher concentration of water. Osmosis can therefore be defined as the movement of water or another solvent by diffusion from a region of low or no solute concentration, across a membrane, to a region of higher solute concentration. Note that osmosis applies only to the passage of water and other solvents through the membrane. If a solute to which the membrane is permeable, for example, sodium chloride, is introduced into one arm of the U tube, it diffuses through the membrane until it becomes equally distributed on both sides. In this case, its passage through the membrane is an example of diffusion, not of osmosis.

***Osmotic Pressure Is Proportional to the Concentration of Dissolved Particles.*** Consider again the example depicted in Figure 5-2, in which a semipermeable membrane separates a 1-percent sugar solution from water. As the sugar solution gains water osmotically, the column of water on the left side will rise (Fig. 5-2c). Eventually, the weight of this rising column will become great enough to force water molecules back through the membrane at the same rate at which they entered the sugar solution by osmosis. However, if a watertight piston is placed in the left side of the tube before osmosis begins and just enough pressure is applied to the piston to prevent any upward movement of water, the force generated by the difference between a 1-percent sugar solution and pure water can be measured by an appropriate device. This force is referred to as **osmotic pressure** and is usually measured in atmospheres, the units used for air or water pressure. A solution's osmotic pressure is best defined as the tendency of a solvent to enter it by osmosis through a perfectly semipermeable membrane from pure solvent, which contains no dissolved solute particles. A solution's osmotic pressure is directly proportional to the concentration of all solute particles it contains. Thus, describing a solution as 1-percent sugar says little about its osmotic pressure. For example, if the sugar in question is glucose, a 1-percent solution (1 gram of glucose in 100 grams of solution) would generate about twice the osmotic pressure as a 1-percent solution of sucrose. The reason is, of course, that sucrose molecules are about twice the molecular weight of glucose molecules, and therefore 1 gram of sucrose contains only about half the number of molecules as 1 gram of glucose. This suggests that a convenient index to a solution's osmotic pressure would be its **molarity**, the concentration of molecules it contains.* However, this will not suffice because some substances, called electrolytes, undergo dissociation into ions when dissolved in water. A dilute solution of sodium chloride (NaCl), for example, dissociates almost completely into sodium ions ($Na^+$) and chloride ions ($Cl^-$). Thus, a 0.1-molar solution of NaCl would be expected to generate about twice the osmotic pressure of a 0.1-molar solution of glucose. A 0.1-molar solution of calcium chloride ($CaCl_2$), which contains one calcium ion ($Ca^{2+}$) and two chloride ions ($Cl^-$) for each of the $CaCl_2$ molecules, would generate about three times the osmotic pressure of an equimolar solution of a nonelectrolyte such as glucose.

***Osmolarity Defines Particle Concentration.*** An **osmole** is $6.02 \times 10^{23}$ molecules of a solute or mixture of solutes of any kind. An osmole of solute dissolved in 1,000 ml (1 liter) of solution constitutes a **1 osmolar solution.** The important consideration in determining the osmolar concentration of a solution is the effective concentration of all of its particles. A 0.1-osmolar† solution could be a mixture of salts, ions, sugars, amino acids, and protein molecules. Large constituents, such as red blood cells and very large molecules, may not contribute to a solution's effective osmolarity; such components are osmotically inert or inactive. The **osmolarity** of a solution is always measured in an ideal situation, for example, with the solution separated from pure water by a perfectly semipermeable membrane.

Osmolarity is one of the factors that determines the osmotic activity of various fluids of biological interest, such as soil water, lake water, seawater, cytoplasm, and blood. For example, the osmolarities of root cell cytoplasm and of soil water can be used to predict whether osmosis will occur between roots and soil water, and in which direction. Two solutions are **isosmolar** if their effective particle concentrations are equal. If solution A has a higher concentration of osmotically active particles than solution B, A is **hyperosmolar** to B, and B is **hypoosmolar** to A.

***Tonicity Indicates Whether Osmosis Will Occur in Specific Cases.*** Osmolarity of an unknown solution can be measured by using a semipermeable membrane; however, such membranes are not known to exist in biological systems. Instead, biological membranes are often differentially permeable: they permit the free passage of some solutes, allow others to pass in small amounts, and are completely impermeable to yet others. The mere presence of some solutes can change a membrane's permeability to various other solutes and even to water. Thus, the fact that two solutions separated by a membrane are isosmolar does not necessarily mean that no osmosis will occur. The osmotic effectiveness of a given solution in a specific situation is indicated by whether a given cell or other membrane-bounded structure swells, shrinks, or does nei-

* A molar solution of a substance is a mole, or gram molecular weight (the molecular weight of the substance expressed in grams), dissolved in a liter of solution. A molar solution of any substance contains the same concentration of molecules: $6.02 \times 10^{23}$ molecules per liter of solution. This figure is known as Avogadro's number, after the Italian chemist Amedeo Avogadro, whose hypothesis—that equal volumes of gases at the same temperature and pressure contain an equal number of molecules—is a cornerstone of modern chemistry.

† The terms *osmolar* and *osmolarity* refer to osmoles per liter of solution. The terms *osmolal* and *osmolality* refer to osmoles per kilogram of water. In dealing with biological fluids such as sap and blood, which are mixtures of many different solutes, molalities are more convenient to use than are molarities.

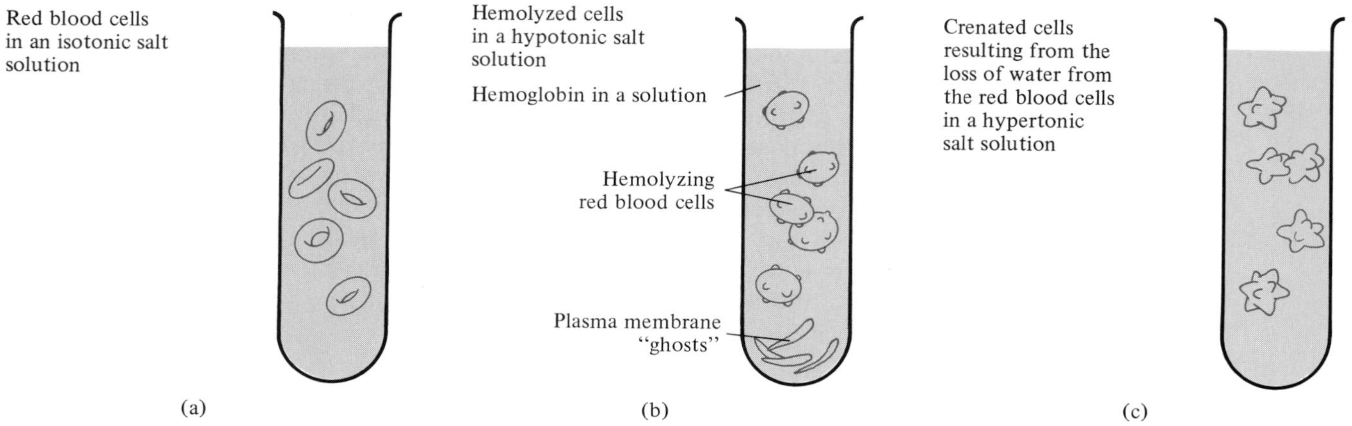


**FIG. 5-3** Hemolysis and crenation in red blood cells. (a) Normal red blood cells in isotonic solution. (b) Hemolyzed red blood cells in hypotonic solution. (c) Crenated red blood cells in hypertonic solution.

ther when placed in the solution. For example, if human red blood cells are placed in a 0.16-molar solution of NaCl, as shown in Figure 5-3a, they neither swell nor shrink. The 0.16-molar salt solution is therefore said to be **isotonic** to the cells' cytoplasm. If distilled water is added to the salt solution (Fig. 5-3b), diluting it, water enters the cells by osmosis and they swell. Lacking the support of a rigid cell wall such as that of plant cells, the plasma membrane becomes stretched to the breaking point and the cell contents—consisting mostly of the protein hemoglobin—flow out. In red blood cells, this process is called **hemolysis** because the cells burst, or lyse (Fig. 5-3b). In this case, the dilute salt solution is **hypotonic** to the cell contents, and the cell contents are **hypertonic** to the dilute salt solution. If salt is added to the 0.16-molar solution, the cells shrink; the solution is thus hypertonic to the cell contents, and the cell contents are hypotonic to the more concentrated salt solution. In hypertonic solutions, red blood cells become crenated, or wrinkled, in a characteristic way (Fig. 5-3c). Some cells have evolved mechanisms that enable them to live in environments characterized by high salt concentrations. Some plants, such as *Spartina* grass, are specialized to grow in or near a salt water environment (Fig. 5-4).

**FIG. 5-4** *Spartina* grass growing at the seashore. This grass is specialized to grow in or near a salt water environment.

***Osmotic Pressure Is Important to Living Systems.*** The cytoplasm of nearly all freshwater and terrestrial plants and algal protists is maintained at a higher osmolar concentration than the solute concentration of the lake, stream, or soil water in which they live. Therefore, their cells tend to absorb water from their environment. Because these cells are surrounded by fairly rigid cell walls, excess swelling is prevented once the cell has become turgid. In hypertonic solutions, however, the plant cell protoplast—all the material bounded by the plasma membrane—shrinks away from the cell wall as a result of water loss, as depicted in Figure 5-5. This process is called **plasmolysis.** If plasmolysis is prolonged, the plant will wilt and die. This is why plants die along roadways to which salt is applied in winter and why a heavy application of fertilizer to a lawn kills the grass.

***Sea Urchin Eggs as Osmometers.*** The membrane that encloses a sea urchin egg shows little permeability to most solutes but is very permeable to water. When sea urchin eggs are placed in various dilutions of seawater, they absorb water osmotically and adjust their volume in each dilution until their cytoplasm is isotonic to that dilution (Table 5-1). When they are returned to full-strength seawater, their volume returns to normal.

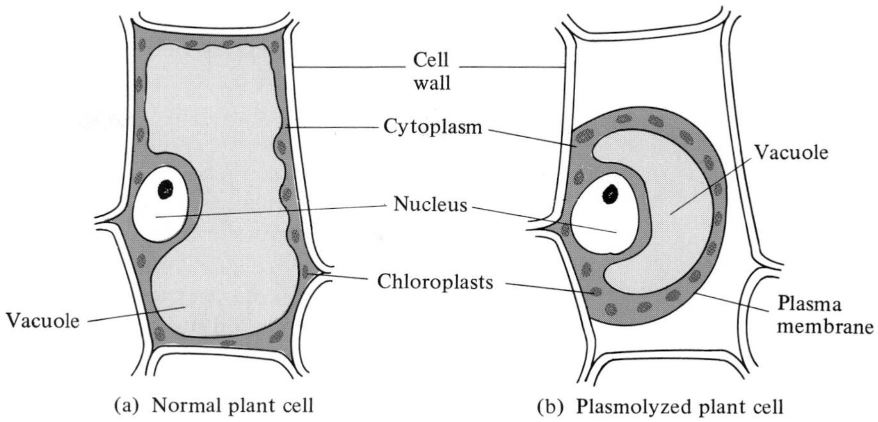


**FIG. 5-5** Plasmolysis in a plant cell in hypertonic solution. (a) Normal plant cell. (b) Plasmolyzed plant cell. The cell plasma membrane has shrunk away from the cell walls.

The magnitude of the response of a sea urchin egg to different dilutions of seawater is not exactly what would have been predicted from the egg's volume, considering that the cell contents are isotonic to 100-percent seawater and that the egg's outer membrane shows little permeability to solutes. However, when a correction is made for the osmotically inactive 12.5 percent of the cell contents, composed of insoluble proteins and lipids, the cells behave as nearly perfect osmometers—devices for measuring osmotic pressure.

## MEMBRANE PERMEABILITY

Using a variety of methods, it is possible to measure the rate of penetration of various substances into cells and to arrive at some conclusions about the permeability of plasma membranes and the mechanisms by which materials are transported across them. Of the many factors that determine whether a particular solute can cross a membrane, three appear to be of major importance.

First, the rate of penetration of a substance is primarily related to its solubility in lipids. This factor is known as the **solubility coefficient** of a penetrating substance. The more soluble a substance is in lipid or in lipid solvents, the more readily the substance will penetrate the plasma membrane. In a series of compounds with decreasing solubility in lipids, such as methyl alcohol, glycerol ethyl ether, glycerol methyl ether,

**TABLE 5-1 Changes in the Volume of Sea Urchin Eggs in Various Dilutions of Seawater**

| *Strength of seawater (%)* | *Volume of eggs (% of normal)* |
|---|---|
| 100 | 100.00 |
| 90 | 109.19 |
| 80 | 121.17 |
| 70 | 137.77 |
| 60 | 161.24 |
| 50 | 188.68 |

Data from B. Lucke and M. McCutcheon, *Physiological Reviews,* 12:68, 1932.

glycerol, and erythritol, the relative rates of penetration through plasma membranes decreases in the same order. This was one of the observations that led to the concept of the lipoidal nature of the plasma membrane (see Chapter 4).

A second factor affecting membrane permeability to solutes is the molecular size of the solute. Small molecules usually pass through the plasma membrane more readily than do large molecules of the same type. Among a series of compounds with increasing molecular size, such as urea, glycerol, arabinose, glucose, and sucrose, the rate of transport through a plasma membrane decreases in the same order. However, the lipid solubility of a molecule appears to be more important than its size in determining its rate of penetration. Thus, a large molecule with a high lipid solubility will probably penetrate the plasma membrane more readily than a smaller molecule with a low lipid solubility.

Third, plasma membranes are more permeable to uncharged particles than they are to those that are charged. Electrolytes thus generally penetrate plasma membranes more slowly than do nonelectrolytes of the same size. Furthermore, strong electrolytes—those that tend to ionize completely in solution—generally pass through membranes more slowly than do weak electrolytes; moreover, the greater the charge of an ion, the slower its rate of penetration. Of the two types of ions, anions (negatively charged particles) penetrate membranes far more readily than do cations (positively charged particles). Reduced permeability to charged particles appears to be due both to the fact that ions in aqueous solution are intimately surrounded by clouds of water molecules called **hydration shells,** which increase their effective size, and to the fact that ions are generally less soluble in lipid than are uncharged particles.

These three factors affecting membrane permeability have a great bearing on the physiological functioning of the cells. Many substances essential for the survival of cells are not soluble in lipids. Among them are such molecules as sugars and amino acids, and several important elements, such as K, Na, Cl, Ca, Mg, P, all of which are available to cells only in an ionized state (as anions or cations). Plasma membranes have specialized mechanisms that facilitate the movement of these vital ions into and out of cells.

In the 1930s it was assumed that the plasma membrane had specialized channels or pores that worked as gates; this would be a means by which charged molecules that are insoluble in lipids could cross the membrane. Later, in the 1950s, biologists J. F. Danielli and H. Davson refined this concept (Fig. 5-6). We now know that such pores exist, but their structure is different from that proposed earlier. (For a discussion of membranes, see Chapter 4.)

**FIG. 5-6** The 1954 Danielli-Davson model of the pores in the plasma membrane: a sandwich of two layers of lipids between two layers of protein.

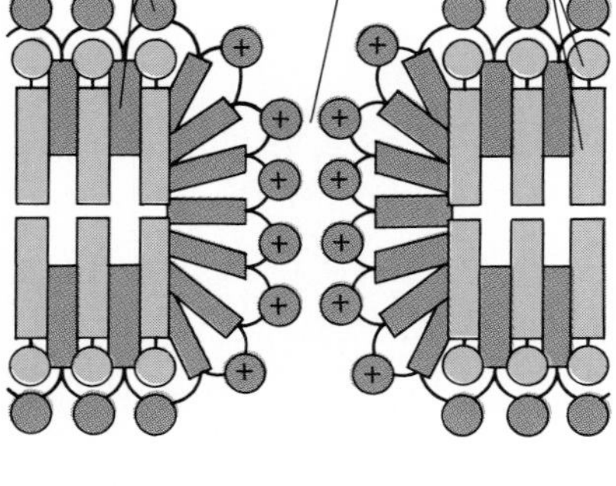


## FACILITATED DIFFUSION

Certain amino acids, sugars, and other compounds can move across plasma membranes in response to a concentration gradient. However, the rate of movement is proportional to the difference in concentration of the substances on the two sides of the membrane only up to a point. After that point, further increases in the concentration difference do not increase the rate of transport across the membrane. This observation suggests that the transport of such compounds is not a case of simple diffusion but is facilitated by some membrane mechanism that becomes saturated at high solute concentrations. In other words, diffusion appears to be facilitated by proteins in a mechanism analogous to that of

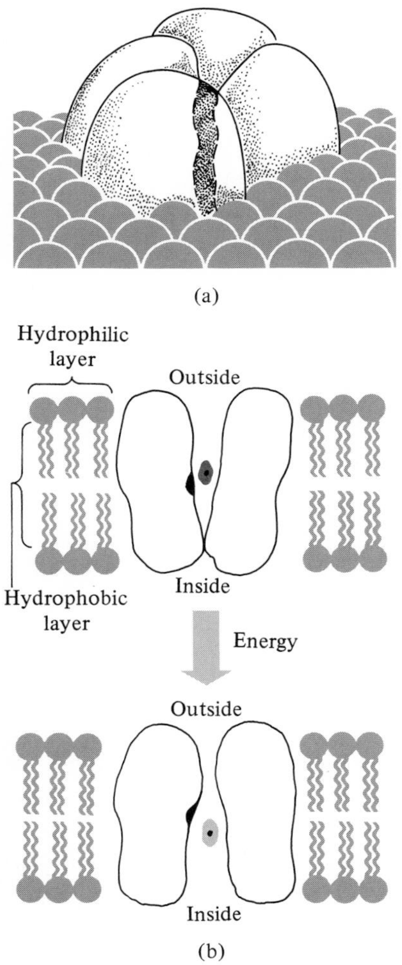


**FIG. 5-7** (a) A model showing detail of a protein complex of the type involved in facilitated diffusion. The complex contains a channel through which small molecules can pass. (b) Facilitated diffusion involves a conformational change in the protein that allows the diffusing molecules to enter on one side and be released on the other. The energy for the passage of the molecule is provided by the concentration difference on the two sides of the membrane. But in active transport, an exergonic reaction usually involving the hydrolysis of ATP is coupled to the process, allowing the substance to be transported against a concentration gradient.

enzyme reactions (see Chapter 3). It is believed that the diffusing molecules combine to form complexes with these specific carrier protein molecules within the membrane. We now know that membrane-bound carrier-protein complexes form channels through which the diffusing molecules pass (Fig. 5-7a). For this reason, they are sometimes referred to as **permeases.** First, the shape of the protein changes in response to the diffusing molecule: the protein parts, so to speak, allowing the molecule to cross the membrane to the other side (Fig. 5-7b). This shape, or conformational, change also lowers the affinity of the protein for the diffusing molecule once it has reached the other side, allowing the molecule to be released. Once the diffusing molecule is released, the protein resumes its original shape and can bind to another molecule. Thus, **facilitated diffusion,** as this process is called, enables molecules to cross an otherwise impermeable or poorly permeable membrane. Except for the limitations imposed by the temporary union with a membrane protein, facilitated diffusion proceeds according to the principles of ordinary diffusion: transport takes place only from a region of high to a region of low concentration and requires no expenditure of energy. Because diffusion is a passive process, it can proceed in either direction, depending on which side of the membrane has the higher concentration of the diffusing substance.

## ACTIVE TRANSPORT

The chemical composition of a cell differs from that of its surroundings in many ways. As noted in Chapter 3, the amount of potassium inside a cell is higher than that in the surrounding environment. Similarly, the concentration of sugars and amino acids inside a cell is usually higher than that in the medium in which the cell grows, at least in the protists and free-living prokaryotes. On the other hand, the amount of sodium in the cell is lower than that in its surroundings. Thus, it is clear that cells are capable of moving molecules and ions against concentration gradients that normally exist in their environments. Since such transport requires energy, it is called **active transport.** Active transport may also take place in the direction favored by a concentration gradient. In such cases, the gradient serves to facilitate the active transport process.

Active transport can be distinguished from simple or facilitated diffusion by its dependence on energy. Thus, it is relatively simple to determine whether the transport of a particular substance across a membrane is truly active: if the process continues in the absence of an energy source, such as adenosine triphosphate (ATP), it cannot be active transport. (See Chapter 6 for a discussion of ATP as an energy source.)

### THE SODIUM-POTASSIUM PUMP

Among the most intensively studied examples of active transport is the mechanism by which sodium ions ($Na^+$) and potassium ions ($K^+$) are transported across the plasma membranes of animal cells. Plant and animal cells tend to maintain low internal concentrations of $Na^+$ and high internal concentrations of $K^+$ in spite of the fact that the surrounding fluids are high in $Na^+$ and low in $K^+$ (Table 5-2). The maintenance of differential ionic concentrations is vital to cells, and about one-third of

**TABLE 5-2 Concentrations of $Na^+$ and $K^+$ (millimoles/1) for Various Cell Types**

| | *Intracellular fluid* | | *Fluid surrounding cells* | |
|---|---|---|---|---|
| *Cell* | *$Na^+$* | *$K^+$* | *$Na^+$* | *$K^+$* |
| Human erythrocyte | 11 | 91 | 138 | 4.2 |
| Rat erythrocyte | 12 | 100 | 151 | 5.9 |
| Frog muscle | 16 | 127 | 106 | 2.6 |
| Rat muscle | 8 | 160 | 147 | 7.3 |
| Dog erythrocyte | 16 | 99 | 158 | 4.1 |
| Cow erythrocyte | 70 | 25 | 142 | 4.8 |
| *Valonia* (marine alga) | 35 | 576 | 460 | 10 |

Data calculated from H. B. Steinbach, in M. Florkin and H. S. Mason, eds., *Comparative Biochemistry*, vol. 4, part B, pp. 677–720, Academic Press, New York, 1963.

the energy expended by most cells is used to maintain $Na^+$ and $K^+$ gradients. Enzyme activity, protein synthesis, the conduction of nerve impulses, and the contraction of muscle all require different concentrations of these ions.

The active transport system that maintains high levels of $K^+$ and low levels of $Na^+$ within the cell is called the sodium-potassium pump ($Na^+/K^+$ pump). Although understanding of this pump is based largely on studies of the mechanism in red blood cells, this knowledge is applicable to all $Na^+/K^+$ pumps. An important step in elucidating this mechanism was the discovery of the enzyme sodium-potassium adenosine triphosphatase ($Na^+/K^+$ ATPase) in the red blood cell membrane. This enzyme is capable of hydrolyzing ATP to adenosine diphosphate (ADP) and inorganic phosphate ($P_i$) only in the presence of sodium, potassium, and magnesium ions by the following reaction:

$$\text{ATP} + \text{H}_2 \xrightarrow[\text{Na}^+/\text{K}^+\ \text{ATPase}]{\text{Na}^+,\ \text{K}^+,\ \text{Mg}^{2+}} \text{ADP} + \text{P}_i + \text{energy}$$

The demonstration of the presence of this enzyme in the plasma membrane led to the idea that the hydrolysis of ATP is probably coupled to the active transport of $K^+$ into and $Na^+$ out of the cell. This role for the enzyme is strongly suggested by the fact that $Na^+/K^+$ ATPase activity and the functioning of the $Na^+/K^+$ transport system in intact cells are specifically inhibited by a plant extract called ouabain. For example, ouabain interferes with normal heart function because it inhibits active transport of cations.

The functioning of the $Na^+/K^+$ carrier enzyme has the following characteristics:

1. The pump functions only if both $Na^+$ and $K^+$ are present, and the same pump transports both ions.
2. The pump works only in one direction for a given ion type: $K^+$ must be present on the outside and $Na^+$ on the inside of the membrane so that $K^+$ is pumped into the cell and $Na^+$ is pumped out.
3. Ouabain blocks cation transport only if applied to the outside of the membrane.

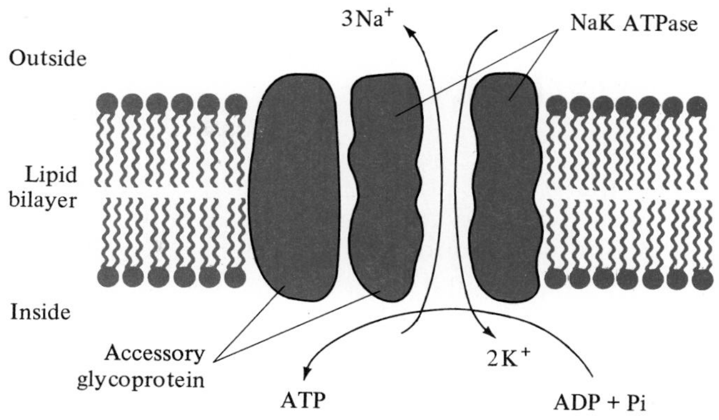


**FIG. 5-8** Diagram illustrating the sodium-potassium pump in cell membranes. The pump creates a gradient of ions across the cell membrane and is an example of active transport. ATP is required to release $Na^+$ outside the cell and to transport $K^+$ to the inside of the cell.

4. The ratio of transport of the two ions is 3 $Na^+$ to 2 $K^+$ for each ATP molecule hydrolyzed by the enzyme (Fig. 5-8). This causes an electrical imbalance between the two sides of the membrane, the inside being negative to the outside; the electrical imbalance is maintained by the membrane's impermeability to some ions, particularly to $Cl^-$ and $Na^+$.
5. The membrane-bound protein involved in this pump is a dimer, a molecule formed by the union of two molecules of a simpler compound; this dimer consists of two separate polypeptide chains joined by a covalent bond.
6. Although the mechanism involved in this transport process is not fully understood, it apparently involves the transfer of an energy-rich phosphate group from ATP to the membrane-bound protein in response to the presence of the $Na^+/K^+$ ATPase enzyme. This reaction, called **phosphorylation**, seems to induce conformational changes in the enzyme, resulting in a high affinity of the protein at the inner surface of the membrane for $Na^+$. Simultaneously, the affinity of the protein at the outer surface of the membrane for $K^+$ also increases. Once the $Na^+$ has moved to the outer surface of the membrane and the $K^+$ to the inner surface, a second conformational change in the protein results in a much lowered affinity for these ions as well as to the phosphate. Then $Na^+$ is released to the outside of the cell and $K^+$ enters the cell and the phosphate is released as $P_i$. The pump is now ready for another transport cycle.

## OTHER PUMPS

Potassium and sodium ions are not the only ions transported across plasma membranes by an active transport mechanism involving a carrier molecule. Calcium ions ($Ca^{2+}$), for example, are transported across the plasma membranes of skeletal muscle cells by a calcium pump that uses a calcium-activated ATPase carrier. The $Ca^{2+}$ ATPase enzyme regulates the amount of free calcium near the muscle fibers and thus regulates muscle contraction.

Several other membrane carrier mechanisms regulate the transport of other ions as well as molecules such as sugars and amino acids across plasma membranes. One example is the active absorption of glucose by the cells that line the lumen of the vertebrate small intestine (Panel 5-1). Some of these mechanisms are known to involve a membrane-bound ATPase pump.

PANEL 5-1

## Cotransport

While the energy released by the hydrolysis of ATP is used directly in the active transport of $Na^+$ and $K^+$ in the ion pump, other systems use energy indirectly to transport a substance against a concentration gradient. For example, glucose can be actively transported across the plasma membrane of the epithelial cells lining the intestinal wall against a concentration gradient. However, the energy expended is used, not for glucose transport directly, but to pump $Na^+$ out of the cells, where it is maintained in high concentrations. Special proteins in the membranes of these cells facilitate the diffusion of $Na^+$ from the lumen of the intestine into the cells. The same proteins act as carriers of glucose; they transport $Na^+$ only if they bind and transport glucose at the same time. Because the $Na^+$ concentration in the lumen is higher than that in the cytoplasm of the cells, $Na^+$ ions diffuse into the cells along the gradient, carrying glucose with them, despite glucose being in higher concentration inside the cell. This type of transport of one substance along with another one across a plasma membrane is called **cotransport.** The energy utilized in this process is used to pump $Na^+$ ions out of the cell to maintain the $Na^+$ gradient between the intestinal lumen and the cytoplasm of the epithelial cells.

Because these transport proteins move $Na^+$ and glucose in the same direction, they are called **symports.** Proteins that transport one molecule into a cell while transporting another molecule out, as in the $Na^+/K^+$ pump, are called **antiports.** Proteins specialized to move only one kind of substance across a cell membrane are called **uniports.** Some cells, such as those lining the urinary tubules of the kidney, have many transport proteins of all three types associated with their plasma membranes.

## BULK TRANSPORT

In addition to transporting small molecules and ions by regular and facilitated diffusion and active transport, the plasma membrane can transport large molecules and liquids into and out of cells. The general term for the process by which such bulk materials are transported into the cell is **endocytosis.** When the substance taken into a cell is a solution, such as a solution of nutrient substances, the process is called **pinocytosis** (Greek for "cell drinking"). When the substances taken in are particulate, such as bacteria, the process is termed **phagocytosis** (Greek for "cell eating"). The transport of bulk materials out of the cell is called **exocytosis.** The secretion of cellular products such as hormones and the removal of cellular wastes involves exocytosis. All these processes require metabolic energy and, like active transport, can be inhibited by substances that block the generation of such energy (see Chapter 4).

### ENDOCYTOSIS

Pinocytosis occurs in all animal cells that regularly take in large molecules. It can be induced in many cells merely by adding protein to their culture medium. Pinocytosis can be demonstrated in many cells by a tracer technique using fluorescent-dye-labeled protein as an inducing agent. Upon being irradiated with ultraviolet light in a special microscope, the labeled protein gives off a colored light, or fluoresces. Following, or tracing, the movement of the fluorescent molecules pinpoints the precise location of the pinocytotic activity. Pinocytosis can also be demonstrated by another tracer technique using the iron-containing protein

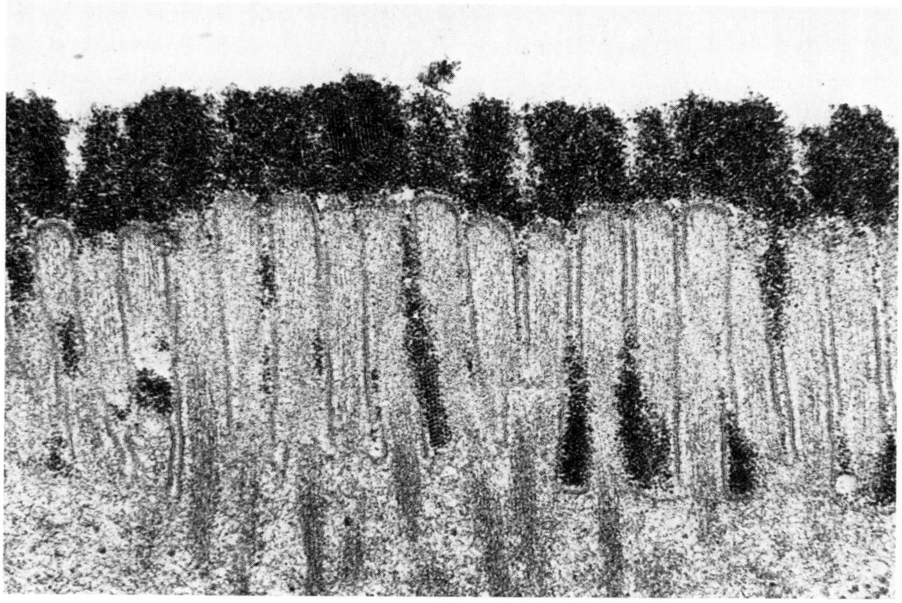

**FIG. 5-9** Pinocytosis can be demonstrated with ferritin. In this electron micrograph ferritin molecules appear attached to the fine filaments of the plasma membrane of the cells of the intestinal mucosa.

ferritin, which shows up in electron micrographs at those locations on the plasma membrane at which it is being taken in (Fig. 5-9). Because of their large number of iron atoms, ferritin molecules scatter electrons much more effectively than do most other cell components. They are therefore more readily visualized in electron micrographs.

***The Four Stages of Pinocytosis.*** Through the use of tracer techniques, it has been shown that pinocytosis occurs in the four steps illustrated in Figure 5-10: (1) binding of the protein or other inducing molecules to the plasma membrane, (2) invagination of the membrane to form a channel, (3) formation of vesicles and their movement into the interior of the cell, and (4) utilization of the materials that have been brought into the cell.

The fact that the binding stage is apparently unaffected by temperature and metabolic poisons suggests that the first step in pinocytosis is purely passive. The uptake of inducing proteins and other large mole-

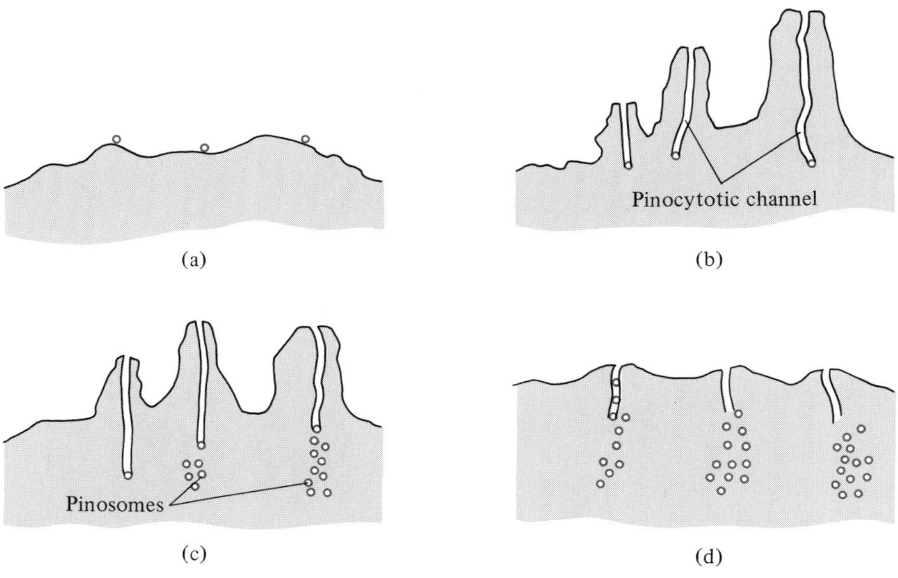


**FIG. 5-10** Diagram of the four stages of pinocytosis as viewed in an ameba. (a) Binding of molecule to plasma membrane. (b) Invagination of membrane to form channels. (c) Formation of pinosomes. (d) Disappearance of channels.

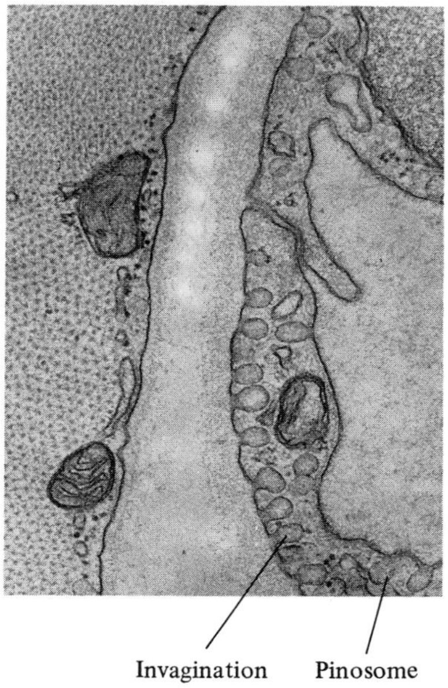


**FIG. 5-11** Electron micrograph showing pinocytosis in the capillary endothelium of mammalian cardiac muscle.

cules does appear to be affected by both the size of the molecules and their electric charge. For molecules with the same positive charge, size is an important factor; larger molecules are more readily taken in by pinocytosis. Otherwise, the size of the charge may be the determining factor, molecules with a higher positive charge being more readily absorbed into cells. The invagination stage is often accompanied by a bulging , or projection, of the cytoplasm around the invagination (Fig. 5-11). Little is known about how the channels are formed except that treatment with respiratory poisons interferes with their formation, which indicates that at least this step requires a metabolic energy supply.

Pinocytotic vesicles, or **pinosomes,** are formed at the cell surface or are pinched off from the ends of the channels. They then migrate into the cytoplasm. Once inside the cell, pinosomes fuse with lysosomes, forming secondary lysosomes. The enzymes of the lysosomes hydrolyze ingested material into small molecules, which diffuse out through the lysosomal membrane into the cytoplasm. Later, the vesicle, along with any residual undigested material, may be extruded from the cell by the process of exocytosis.

***The Many Functions of Phagocytosis.*** The uptake, or phagocytosis, of large, solid particles by cells has long been known to be a widely occurring phenomenon. It forms the basis for the nutrition of many protists, particularly the Protozoa. It is readily seen by light microscopy and has been reported in a large variety of animal cells. In mammals, phagocytosis by certain white blood cells and other phagocytes (Greek for "eating cells") is an important means of defense against the invasion of the body by such foreign substances as bacteria, parasitic organisms, and dust particles. Besides white blood cells, mammalian phagocytes include several types of cells of the liver, spleen, lymph nodes, connective tissue, brain, bone marrow, lungs, and other tissues.

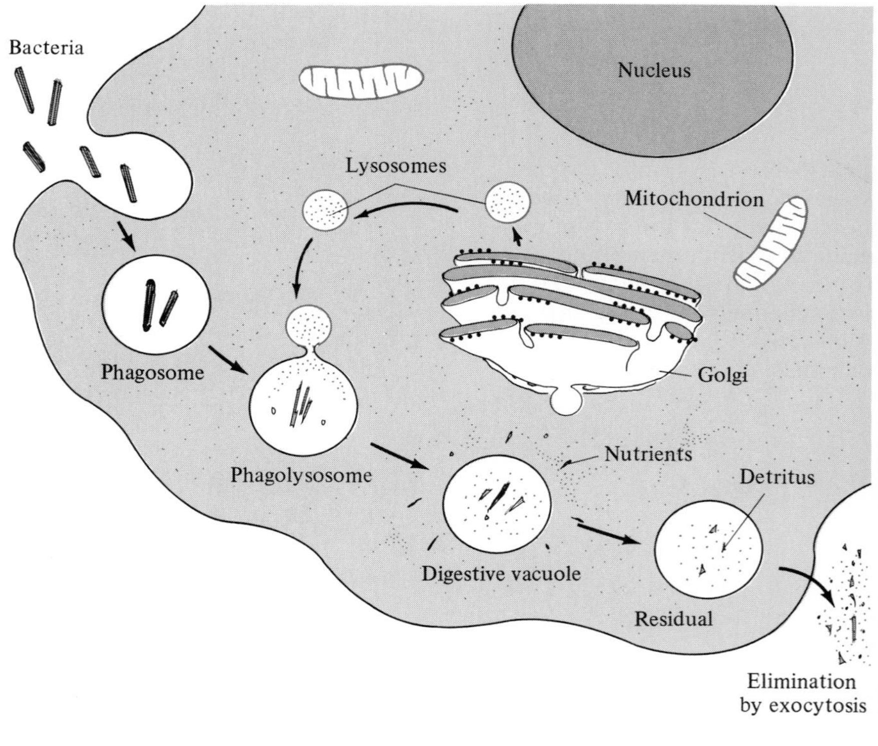


**FIG. 5-12** Schematic representation of phagocytosis, showing ingestion, intracellular digestion, and exocytosis.

**FIG. 5-13** Electron micrograph showing exocytosis of mucus into the lumen of the human colon from one of the secretory granules in the apical cytoplasm of a mucus-secreting cell.

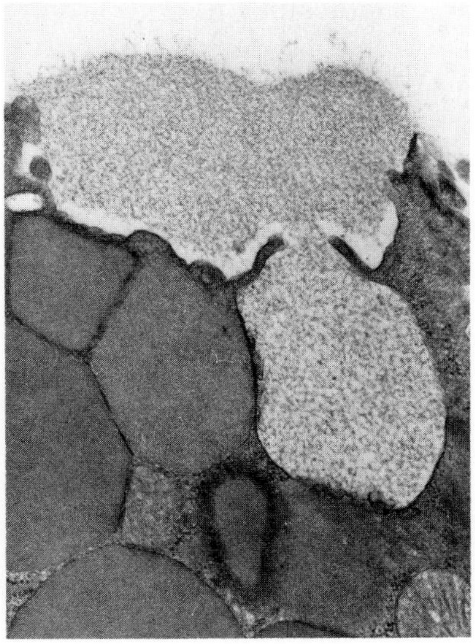

The process of phagocytosis is sketched in Figure 5-12. The basic steps in phagocytosis are (1) attachment of the foreign particle to the cell surface; (2) engulfing of the particle by **pseudopodia** (Greek for "false feet") extended from the cytoplasm; (3) pinching off of the vesicle, called a **phagosome;** (4) fusion of lysosomes with the phagosomes to form **phagolysosomes;** (5) digestion of the particles by the lysosomal enzymes; and (6) extrusion of the undigested debris from the cell by exocytosis.

## EXOCYTOSIS

The process of exocytosis, by which cells transport substances in bulk from the cytoplasm out of the cell, is highly developed in the secretory gland cells of animals (Fig. 5-13). The secretory cell products are synthesized in the rough endoplasmic reticulum and accumulate in vesicles, which are pinched off into the cytoplasm. These vesicles move to a Golgi complex, where the secretory products are packaged into secretory vesicles. In salivary, pancreatic, and adrenal glands, the secretory vesicles move to the plasma membrane, and the vesicle membrane fuses with the plasma membrane, releasing the vesicle contents outside the cell.

An interesting variation of this method of secretion is seen in the oil glands of mammalian skin. Secretion occurs when the cell disintegrates, releasing oil onto the surface of the skin. Such a secretory process in which the entire secreting cell, along with its accumulated secretion, forms the secreted matter is called **holocrine secretion.** A similar process takes place in milk-producing glands, but here only the part of the cell in which the secretions have accumulated is pinched off by a process called **apocrine secretion.**

## Summary

By a number of mechanisms, a wide range of molecules are transported across the plasma membrane, which acts as a selective barrier against the free movement of materials into and out of cells.

Passive transport is accomplished by the physical process of diffusion, by which molecules move from a region in which they are highly concentrated to one in which their concentration is lower. Since diffusion depends on the kinetic energy of molecules, the concentration of the molecules and the temperature have a significant effect on the rate at which molecules diffuse. The higher the temperature and the steeper the concentration gradient of the molecules, the greater the rate of diffusion.

The same principles that govern the diffusion of molecules in an open system operate when a concentration gradient between two solutions is established across a membrane that is impermeable to solutes but permeable to solvents. Diffusion of solvents through a membrane in such a case is called osmosis. The plasma membranes of cells are selectively permeable; water passes through them relatively easily, and most solutes pass through at lesser rates or not at all. Because plasma membranes are permeable to water, animals have evolved a variety of mechanisms that regulate water balance by osmosis. The osmotic behavior of plant cells is comparable to that of animal cells except that plant cells possess rigid cell walls. Because of their cell walls, plant cells develop turgor pressure, which gives the plant body rigidity.

The plasma membrane is selectively permeable to various solutes, depending on the chemical characteristics of the solute molecules. The nature of the plasma membrane itself and the concentration of substances on both sides of the membrane are factors affecting the transport of materials across cell membranes. Because of the lipid nature of plasma membranes, they are much more permeable to nonelectrolytes than to electrolytes and to small, uncharged molecules than to larger, charged molecules.

Glucose, certain amino acids, and some other compounds are transported across plasma membranes in response to a concentration gradient, but their movement is facilitated by binding to carrier protein molecules, which enable them to move across the membrane. Like other passive mechanisms, such facilitated diffusion proceeds according to the principles of diffusion and does not require an expenditure of metabolic energy.

Active transport is the movement of molecules across the plasma membrane that can occur against a concentration gradient and that involves an expenditure of energy. For example, sodium and potassium ions are transported across animal plasma membranes by a transport system called the $Na^+/K^+$ pump. This mechanism involves the existence of a carrier enzyme, $Na^+/K^+$ ATPase, that combines with sodium ions on one side of the membrane and combines with potassium ions on the other side. The potassium ions are then transferred to and released on the inside while the sodium ions are released on the outside.

In addition to its role in passive diffusion and active transport, the plasma membrane participates in the bulk transport of large molecules and other substances into and out of cells through the processes of endocytosis and exocytosis, respectively. When the substance taken into a cell is a solution that may contain large molecules or other nutrients, the process is called pinocytosis. When the substances taken in are particulate materials, such as bacteria, the process is termed phagocytosis.

## Recommended Readings


Alberts, B., D. Bray, J. Lewis, M. Raff, K. Roberts, and J. D. Watson. 1983. *Molecular Biology of the Cell.* Garland, N.Y. Concise and up-to-date description of the structure of plasma membranes and how they function in the movement of materials in and out of cells. The illustrations are particularly well done and serve to explain the various mechanisms described.

De Witt, W. 1977. *Biology of the Cell: An Evolutionary Approach.* Saunders, Philadelphia. A basic cell biology textbook designed for students with a minimal background in biology and chemistry.

Hoffman, J. F. 1974. "Ionic transport across the plasma membrane." *Hospital Practice,* 9:119–127. An excellent discussion of the experimental evidence for one of the theories regarding the function of the sodium potassium pump.

Holter, H. 1961. "How things get into cells." *Scientific American,* 205:166–180. A well-illustrated discussion of osmosis, active transport, and pinocytosis.

Karp, G. 1984. *Cell Biology,* 2nd ed. McGraw-Hill, New York. Contains a concise, up-to-date history of our knowledge of the structure and function of the plasma membrane and other cell membranes. The experiments by which the knowledge was acquired are described in an interesting fashion.

Swanson, C. P., and P. L. Webster. 1985. *The Cell,* 5th ed. Prentice-Hall, Englewood Cliffs, N.J. This well-illustrated text contains a chapter that describes in clear language the exchanges of materials between cells and their surroundings.

6

# Energy: Life's Driving Force

## OUTLINE



All organisms need energy to fuel the many kinds of tasks they perform. Energy is defined as the capacity to do work. There are many kinds of energy, including mechanical, chemical, thermal, radiant, electrical, and atomic. All energy can be classified as either **potential energy,** which is stored, or **kinetic energy,** the energy of motion. A boulder at the edge of a cliff or the water in a reservoir (Fig. 6-1) represent potential energy because they both can do work. If the boulder falls, its potential energy becomes kinetic energy. If the boulder is attached by a line and pulley to a smaller boulder at the bottom of the cliff and given a push, the falling boulder can perform work by lifting the smaller boulder to the top of the cliff.

The falling of the larger boulder and the raising of the smaller one constitutes a transfer of energy from one object to another. But that is not all that happens. The stretching of the line, its rubbing on the pulley, the turning of the pulley on its bearings, and the friction of the boulder moving through the air all generate some heat. This frictional heating represents a conversion of mechanical energy to another form: thermal energy. A device that accomplishes the conversion of one kind of energy to another is called a transducer. Energy conversion is also

**FIG. 6-1** A water wheel being driven by a stream of water. This energy can be used to grind corn or generate electricity.

accomplished by organisms. For example, the radiant energy of the sun is converted to chemically bound energy by plants. Animals obtain this energy when they eat plants. They then convert some of this energy to other forms of chemically bound energy, some to the mechanical energy of muscle contraction, some to the thermal energy of body heat, and, in animals like the firefly, some back to visible radiant energy.

## ENERGY FLOW

In using energy to perform its various tasks, an organism is comparable to a small boulder being raised by the falling of a larger one. All work done by organisms, including merely maintaining their organized state, requires a continual supply of energy. Moreover, the source of the energy used for any task must always be greater than the energy used in performing the task. Even in a frictionless system, a boulder falling from a cliff cannot raise another of equal weight to the top of the cliff. Unless the falling boulder is heavier than the other boulder, no transfer of energy will occur.

**Thermodynamics** is the branch of physics that concerns transfers and conversions of energy. It owes its name to the fact that its laws were first formulated for exchanges of heat within and between systems. However, its principles have since been applied to exchanges of all forms of energy.

## THE FIRST LAW OF THERMODYNAMICS

According to the First Law of Thermodynamics, the energy lost by bodies within an isolated system is equal to the energy gained by other bodies in the system or to its equivalent in work.* One result of this—or another way of stating the law—is that *energy is neither created nor destroyed during either its transfer or its conversion from one form to another.* This means that the total amount of energy within an isolated system remains constant. The law applies to the entire universe as well. The one exception to the first law is Einstein's statement that, under special conditions, mass and energy of atomic nuclei are interconvertible, that is, one can be converted to the other. Einstein expressed the relationship of mass and energy as $E = mc^2$, where $E$ is energy, $m$ is mass, and $c$ is the velocity of light. Modified accordingly, the first law states that the total mass-energy of the universe is constant.

The case of the falling boulder is a good illustration of the first law. Although energy is transferred from one object to another (from the large boulder to the small one) and to some extent is even converted from one form (mechanical) to another (thermal), no energy is created or destroyed by transfer or conversion. The total amount of energy in this system remains constant.

The first law sounds simple; it even seems rather obvious. However, it is of profound significance for life. For example, energy cannot be transferred from a lower level, or potential, to a higher one. Energy can be built up within an organism only at the expense of some source with higher potential energy outside the organism. This principle also applies to chemical reactions that occur within organisms. If a biochemical reaction needs energy to proceed, it will take place only if a source of energy greater than that used in the reaction is at hand to drive the reaction. Moreover, the reaction will proceed at a rate proportional to the difference in the two energy levels. Similarly, if two bodies of water at slightly different levels are connected, water flows slowly from one to the other.

## THE SECOND LAW OF THERMODYNAMICS

Energy exchanges involving heat are governed by the Second Law of Thermodynamics. Heat can be defined as the random motion of atoms or molecules that occurs in a substance at any temperature above absolute zero (−273°C). One way of stating the second law is as follows: *energy within an isolated system tends to be converted to heat.* It is impossible to convert heat—that is, the random motion of a system—to ordered motion without adding energy from another source. Heat can be used to do work, but the energy transfer is always accompanied by a loss of order or organization. Moreover, the transformation of heat to mechanical work is, by its very nature, always less than 100-percent efficient. Since it is theoretically possible to have a frictionless mechan-

* Work can be defined as the movement of an object against some force; for example, separating a positively charged ion from a negatively charged one.

ical system, such as lines and pulleys operating in a perfect vacuum, it is theoretically possible to transfer mechanical energy from one object to another with 100-percent efficiency. But in transfers of energy involving heat, 100-percent efficiency is not even theoretically possible.

***Entropy: The Concept of Unavailable Energy.*** Many energy exchanges involving heat result in the loss of most of the transferred energy as heat. The "lost" heat at first warms the immediate environment, that is, increases the rate of random motion of nearby atoms and molecules. The heat then continues to radiate, ultimately dissipating in outer space. Although this energy is not lost in the sense of being destroyed, for practical purposes it might as well be. For none of this dissipated heat could ever be drawn upon to drive any process that required energy.

A swimming pool full of warm water may contain far more energy than a cup of boiling water. However, unless work is done to extract this energy, it is useless for the purpose of heating even a spoonful of water to any temperature above that of the pool. **Entropy,** symbolized S and often referred to as randomness, is the energy within a given system that is no longer available to do work. Thus, within any isolated system, as energy exchanges occur, entropy increases. Eventually, in the absence of external input of energy, all further exchanges of energy within the system will cease.

***Do Organisms "Violate" the Second Law of Thermodynamics?*** It has sometimes been suggested that organisms represent a "violation" of the second law. It has even been proposed that the decrease in entropy within growing, developing organisms is due to some mysterious life force. But organisms, like any other physical system, increase their level of organization, or order, only at the expense of some outside, higher source of energy. Ultimately, the sun is this source of energy.

## FOOD CHAINS

An organism's utilization of the sun's radiant energy begins with the "capture" of that energy by photosynthetic organisms, which convert light energy to chemical energy in the form of carbohydrates and then use this energy to carry out all their activities, including the synthesis of complex organic molecules. These organisms are termed **producers.** The **primary consumers** are the plant-eating animals, or herbivores, who obtain these complex organic molecules by eating producers or their products (Fig. 6-2). Primary consumers are in turn preyed upon by other animals, the **secondary consumers,** and so on in what is termed a **food chain.** As each of these organisms dies, its constituents are broken down by various **decomposers,** such as bacteria and fungi. This producer-consumer-decomposer sequence in a food chain represents a flow of both energy and matter. However, although elemental matter—carbon, hydrogen, oxygen, and so on—is usually recycled, energy is not.

The flow of energy through organisms has three stages: (1) the photosynthetic conversion of light energy into chemically bound energy; (2) the conversion of the molecules containing this chemically bound energy to forms useful for cellular work; and (3) the utilization of the released energy for such tasks as nerve impulse transmission, muscle contraction, secretion, growth, and reproduction. The first of these three stages, photosynthesis, will be described in Chapter 7. The third

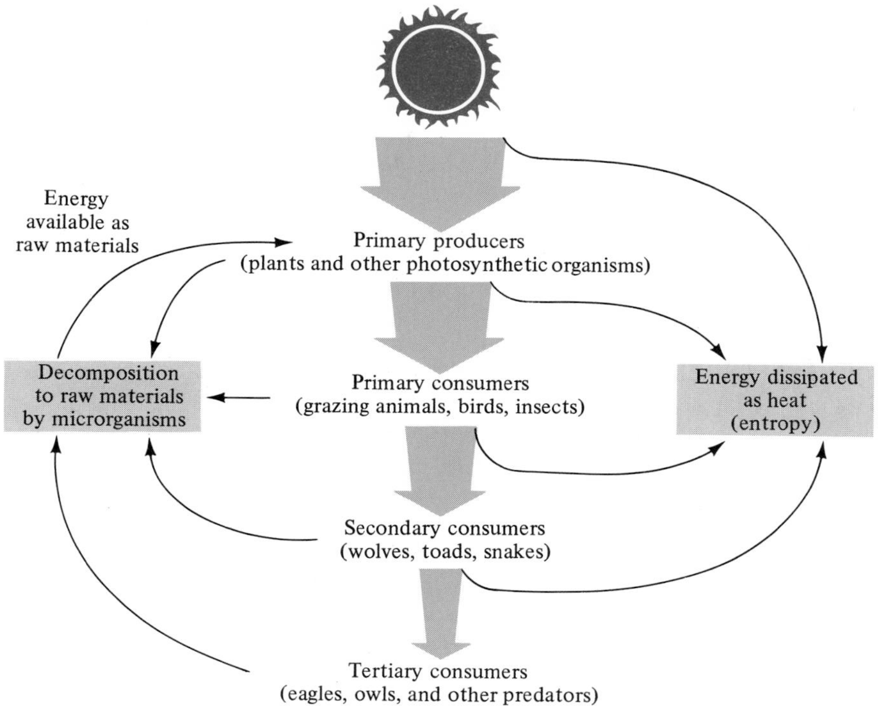


**FIG. 6-2** The flow of energy and materials through living systems. The vegetation (producers) receives energy from the sun, $CO_2$ from the air, and minerals and water from the soil. The herbivores (primary consumers) consume vegetation for its content of energy and materials. Eventually organisms die or are eaten. What is not eaten by other consumers is decayed to inorganic form by decomposers such as bacteria and fungi. In time, the materials, even of such hard parts as skeletons, are recycled when returned to the soil, lakes, and so on. Because energy cannot be recycled, all communities of organisms need a continuing outside supply of it to support their vital activities.

stage, the ultimate use of the energy by the organism, is discussed in later chapters. The balance of this chapter is devoted to the second stage: cellular metabolism—how the organism breaks down energy-containing molecules.

## AN OVERVIEW OF METABOLISM

**Metabolism,** the sum total of all chemical reactions occurring in an organism's cells and tissues, consists of the synthetic processes of **anabolism** and the degradative reactions of **catabolism.** Among the catabolic reactions are the oxidations of fats and carbohydrates, which yield most of the energy that drives anabolism and other energy-requiring processes.

The amount of energy contained in fats and carbohydrates would be far too great to be used by any cellular mechanism if it were released all at once. Accordingly, each molecule of the major foodstuffs—fats, carbohydrates, and proteins—is oxidized in a stepwise series of reactions, many of which release amounts of energy small enough to be useful in anabolic reactions. Much of this energy is conserved by the coupling of these energy-releasing, or exergonic, oxidations of foodstuffs to energy-requiring, or endergonic, reactions in which certain small, high-energy molecules are synthesized (Fig. 6-3a). Among the most important of these small, energy-rich molecules is adenosine triphosphate (ATP), shown in Figure 6-3b. It is produced by a reaction involving adenosine diphosphate (ADP) and inorganic phosphate ($P_i$). ATP is valuable as a source of energy because (1) it is a relatively small, mobile molecule that readily diffuses about the cell and can thus reach any area in which energy is needed for cellular work; and (2) it possesses energy-rich bonds that release an unusually large amount of energy when broken in an exergonic reaction (Panel 6-1).

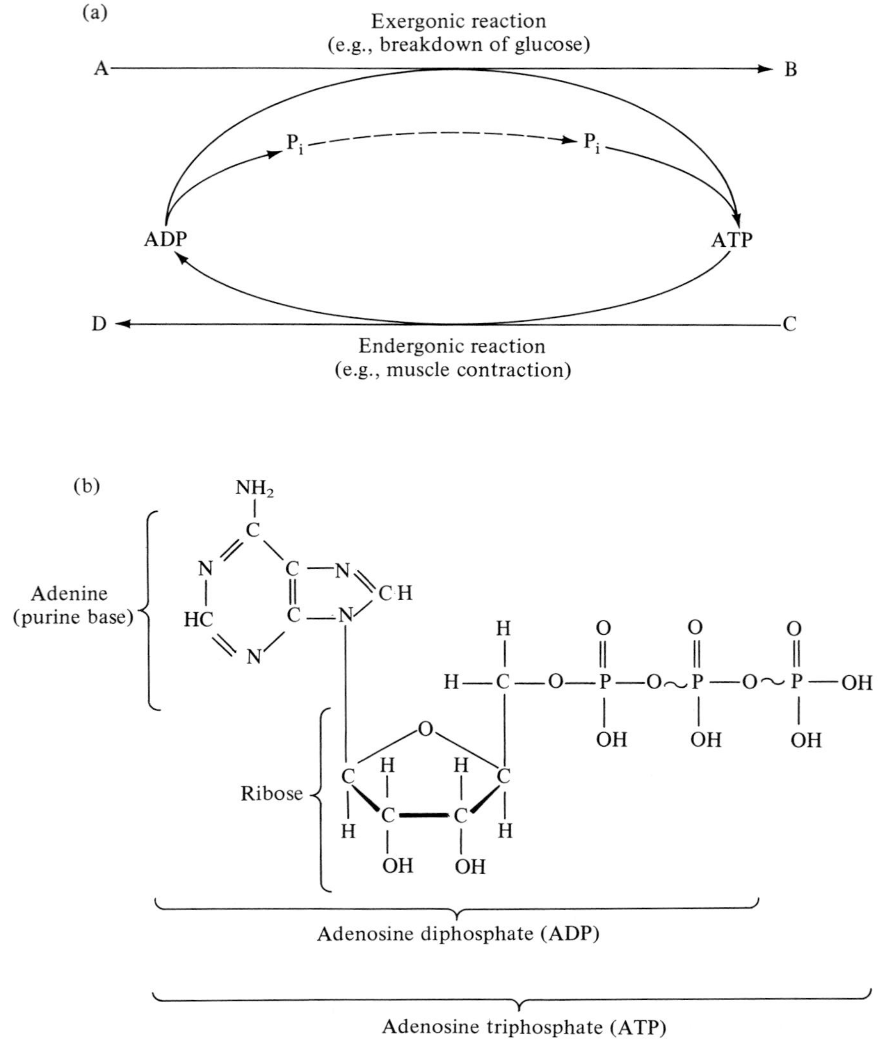


**FIG. 6-3** (a) A coupled reaction. The energy released in the exergonic reaction (A $\longrightarrow$ B) is stored in ATP molecules to be used subsequently to drive the endergonic reaction (C $\longrightarrow$ D). (b) Structure of adenosine triphosphate (ATP).

## THE SOURCES OF ENERGY

Cells require a continuous supply of energy. To obtain energy, the body must burn fuel. Just as a furnace generates heat by burning wood, coal, gas, or oil, the body obtains energy by oxidizing various kinds of foods. Some of these foods contain more energy than others.

High on the list of energy foods are the carbohydrates: sugars and starches. Most dietary carbohydrates are converted to fats and stored by the body. The body also consumes fatty foods directly. Fats, or lipids, can release more energy per unit weight than carbohydrates can because they contain less oxygen than do carbohydrates and are therefore capable of undergoing more oxidation. Compared to carbohydrates, fats thus contain more **calories,** or units of heat. Proteins are lowest on the list of energy-producing foods and, under normal circumstances, are not an important source of energy; their primary function is not energy storage.

In order to be used, the complex molecules in foods must be broken down, or digested, into simpler molecules that can be absorbed by the cells. For example, starches are broken down into glucose, proteins into amino acids, and fats into glycerol and fatty acids. Animals then metabolize the food molecules through chemical reactions in which the energy stored in the chemical bonds of the foods is used for various energy-requiring processes. In such energy transfers, electrons are passed from one molecule or atom to another through oxidation-reduction reactions (see Chapter 2). The molecule that loses an electron is oxidized, and the process of electron loss is **oxidation.** The acceptance of an electron by an atom or molecule is a **reduction** reaction. Oxidation and reduction take place simultaneously so that the electron given up by the oxidized atom or molecule can be immediately accepted by another atom or molecule, which is thereby reduced.

When oxygen is not freely available, many organisms can obtain energy from oxidations without oxygen, or **anaerobic oxidations.** In fact, some organisms depend entirely on anaerobic oxidations for all their energy requirements. As might be expected, such organisms normally live in oxygen-poor environments; intestinal parasites and the bacterium that causes botulism (Fig. 6-4) are examples. In anaerobic oxidation, the final electron acceptor, or reducing agent, may be any of a variety of molecules. Oxidation with oxygen, or **aerobic oxidation,** is a better source of energy than any anaerobic process because the complete oxidation of an organic substance can be accomplished only in the presence of oxygen, the final products being carbon dioxide and water.

## PANEL 6-1

# Energy-Rich Phosphate Bonds

Although biochemists speak of energy-rich bonds, often designated by a wavy line (~), the strength of a particular bond in an organic compound may have little to do with the amount of energy released when it is broken. It is useful to regard the energy released by a chemical reaction as a function of the equilibrium point of that reaction. Strongly exergonic reactions, that is, those for which the point of equilibrium is far to the right, release far more energy than do those with equilibrium points only slightly to the right. For example, a reaction that reaches equilibrium when 12 percent of the reacting material is present as reactants and 88 percent as products releases far more energy than a reaction that reaches equilibrium when the material is 45 percent reactants and 55 percent products. Endergonic reactions (those with equilibrium points to the left) will, if the starting point is equal amounts of products and reactants, release no energy at all, no matter how much energy is contained in the bond being broken. Such reactions require an input of energy if they are to move in the forward (left to right) direction.

The amount of energy released when ATP molecules split into ADP and $P_i$ is about 7.3 kcal (7,300 calories) per mole of ATP, which is sufficient to drive most of the endergonic reactions of the cell. ATP thus serves as a convenient currency in the cell's energy economy. However, for an energy-requiring process to use the energy of ATP, the process must be coupled to the reaction ATP $\rightarrow$ ADP + $P_i$ + energy. Organisms have evolved not only ATP, but also enzymes capable of coupling such reactions. The enzyme that couples the reactions in the sodium-potassium pump, described in Chapter 5, is an example.

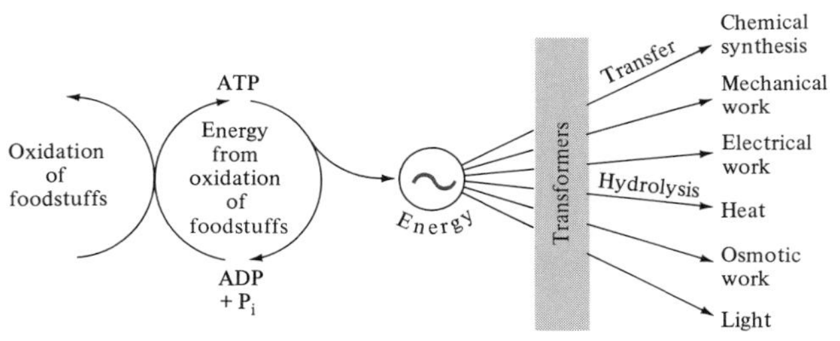

The ATP "energy dynamo."

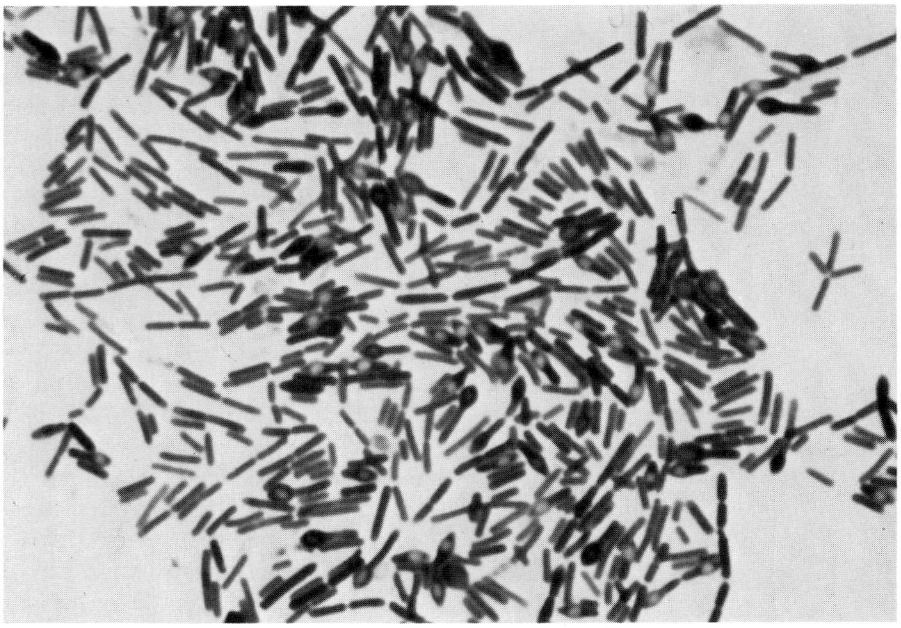

**FIG. 6-4** *Clostridium botulinum,* the bacterium that causes botulism. These bacteria depend entirely on anaerobic oxidations for all their energy requirements.

# CATABOLISM OF GLUCOSE

The catabolism of carbohydrates is one of the major sources of energy used in the generation of ATP. Carbohydrates in various forms, such as glucose, are the primary dietary source of energy for the great majority of nonphotosynthetic organisms. And the synthesis of carbohydrates that takes place in photosynthetic organisms is the ultimate source of organic compounds for nearly all organisms. Furthermore, the major pathways of glucose degradation connect with other important catabolic and anabolic pathways at points involving common intermediates. Carbohydrate metabolism is so important to living organisms that its pathway can be considered the core of cellular metabolism.

## GLYCOLYSIS: THE INITIAL DEGRADATION OF GLUCOSE

**Glycolysis** (from the Greek *glykys,* "sugar," and *lysis,* "dissolution") is the sequence of enzyme-catalyzed reactions involved in the breakdown of glucose to pyruvate* with a concomitant production of ATP. The initial reactions are the same for most organisms. Glycolysis can proceed to completion either aerobically or anaerobically, but the final products will differ. Once glucose has been broken down into pyruvate, two possible anaerobic pathways can be followed, each with a different final product. One pathway is used by vertebrate muscle cells when functioning with a shortage of oxygen. The other pathway, alcoholic fermentation, is used by yeast cells growing under anaerobic conditions (Fig. 6-5).

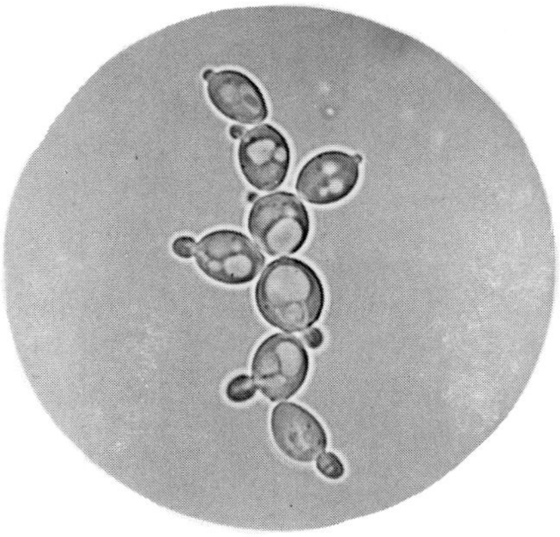

**FIG. 6-5** Yeast cells. The study of the biochemistry of yeasts has contributed enormously to our understanding of the energy pathways of the eukaryotic cell.

* By convention, organic acids in the slightly alkaline pH of the cell are referred to in their ionized form. In this case, pyruvic acid is referred to as the pyruvate ion, or, more conveniently, pyruvate. In the same way, acetic acid becomes acetate, lactic acid becomes lactate, and so on.

## ALCOHOLIC FERMENTATION

Alcoholic fermentation is a conversion of glucose to ethanol (ethyl alcohol) and carbon dioxide. In the absence of oxygen, two ethanol molecules and two $CO_2$ molecules are produced for every glucose molecule fermented in a yeast cell.

$$\underset{\text{glucose}}{C_6H_{12}O_6} \longrightarrow 2CO_2 + \underset{\text{ethanol}}{2CH_3\text{—}CH_2OH}$$

The chemical nature of this process was first described by the French chemist Louis Pasteur, who defined fermentation as "life without oxygen." Pasteur came to believe that the fermentation of carbohydrates could occur only in the presence of living yeast cells. But in 1897, the German chemists Hans and Eduard Büchner discovered that a yeast extract fermented the sugar they had added as a preservative even though no live yeast cells were present. The Büchners called the fermentative factor present in yeast juice an enzyme (Greek for "in yeast").

## COENZYMES

Work by the British biochemists Arthur Harden and William Young in 1905 established that small, heat-stable molecules, which they called coenzymes, were also required for fermentation. **Coenzymes** are molecules that are essential to the effective functioning of enzyme-catalyzed reactions, in which they often act as a donor or acceptor of substances necessary for the reaction.

The first coenzyme to be identified was nicotinamide adenine dinucleotide, or **NAD,** which was isolated in 1933. Although this coenzyme contains two phosphate groups, it is not a phosphate donor, as is ATP. Rather, it is a hydrogen acceptor. Its structure is shown in Figure 6-6. Like many coenzymes, it is a derivative of one of the water-soluble vitamins, in this case, nicotinic acid, sold commercially as niacin. The following year, the German biochemist Otto Warburg isolated a similar

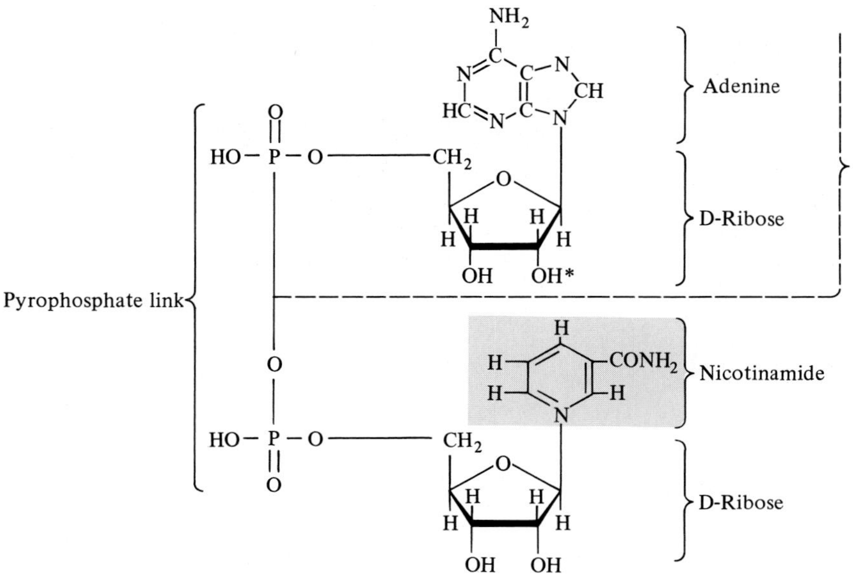


**FIG. 6-6** Structure of NAD. This hydroxyl (OH*) group in NAD is replaced with a phosphate group in NADP. The shaded area is the group involved in electron transfer.

coenzyme, nicotinamide adenine dinucleotide phosphate or **NADP.** NADP differs from NAD by the presence of an additional phosphate group on one of the ribose components, as indicated in Figure 6-6.

NAD can be reduced by accepting electrons from various substrates. As will be seen in the next chapter, NADP plays a similar role in photosynthesis. The reduction of NAD takes place in conjunction with one of the dehydrogenase enzymes. Reduced NAD can pass electrons on to other acceptors. Like NADP, NAD thus exists in either a reduced or an oxidized form. These two states are commonly represented in the following manner:

$$NAD_{red} \longrightarrow NAD_{oxid} + 2 \text{ electrons}$$

In some notations, reduced NAD is represented as $NADH_2$ or, more correctly, $NADH + H^+$, and the oxidized form as $NAD^+$.

## STAGES IN THE ANAEROBIC OXIDATION OF GLUCOSE

There are nine steps in the breakdown of one molecule of glucose to two molecules of pyruvate. These steps are grouped into two major stages (Fig. 6-7). In the initial reaction of the first stage, glucose is activated, or prepared, for its oxidation by being phosphorylated by the addition of one phosphate group from ATP to form glucose 6-phosphate (step 1). After a rearrangement of the atoms of this molecule to form fructose 6-phosphate (step 2), and at the expense of a second ATP molecule, another energy-rich phosphate (~P) group is added to this molecule to form fructose 1,6-diphosphate (step 3). Each fructose 1,6-diphosphate molecule splits to form two molecules of glyceraldehyde 3-phosphate (step 4). In these first four steps, no energy has been extracted from the breakdown of glucose. On the contrary, two molecules of ATP have been used. The first stage in the anaerobic oxidation of glucose is therefore endergonic.

The second major stage involves a series of steps that harvest some of the energy contained in glyceraldehyde 3-phosphate; this stage is therefore exergonic. In the initial step of the second stage, each glyceraldehyde 3-phosphate molecule acquires a second phosphate group, obtained from inorganic phosphate ($P_i$) (step 5). The enzyme involved in this reaction is glyceraldehyde 3-phosphate dehydrogenase, appropriately named because it removes a pair of hydrogen atoms from each glyceraldehyde 3-phosphate molecule.* This requires the presence of the coenzyme NAD, which is reduced from the oxidized state, $NAD_{oxid}$, to the reduced state, $NAD_{red}$. The addition of a second phosphate group to glyceraldehyde 3-phosphate and the simultaneous removal of two of its hydrogen atoms produces an oxidized molecule, diphosphoglycerate, but still capable of releasing a substantial amount of energy. Some of this energy is conserved when the molecule donates an energy-rich phosphate group to ADP, ATP being formed as a result (step 6). This process is known as **substrate level phosphorylation.** Following two re-

* This constitutes an excellent example of the coupling of two reactions by one enzyme, as described in Chapter 2. One reaction (glyceraldehyde 3-phosphate + $P_i \rightarrow$) is strongly endergonic. However, it is "pushed" by the other reaction (glyceraldehyde 3-phosphate − 2H →), which is even more strongly exergonic than the first is endergonic. The overall reaction is thus slightly exergonic and proceeds in the forward direction. Further facilitating this reaction is the next reaction in the series, in which ATP is generated (step 6). It is strongly exergonic and therefore "pulls" the first one along.

**FIG. 6-7** The endergonic and exergonic stages of the anaerobic glycolysis of glucose to pyruvate.

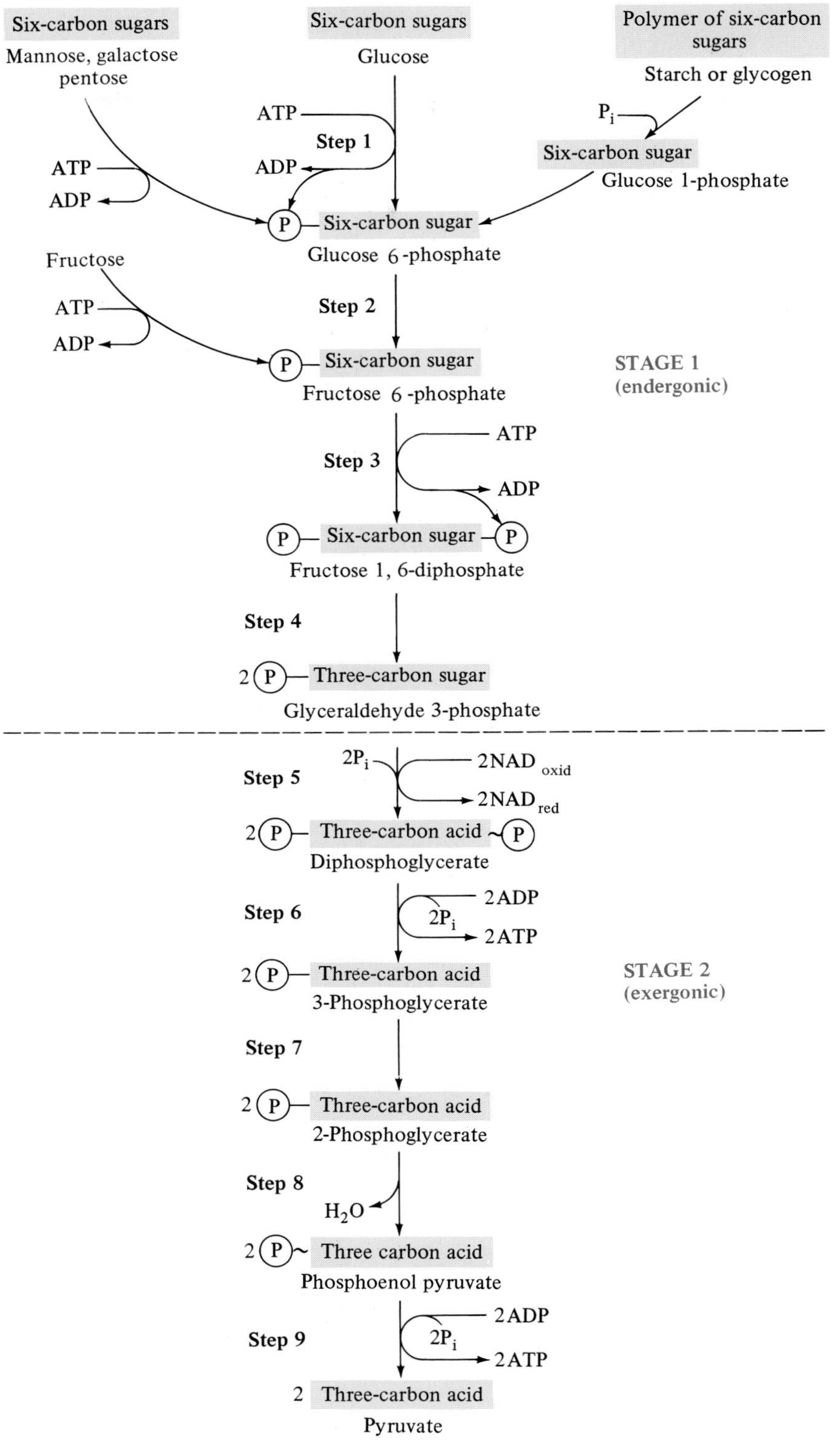


arrangements of the atoms in the diphosphoglycerate molecule (steps 7 and 8), a similar dephosphorylation occurs at step 9, with another ATP being generated. The second stage thus couples an oxidation—the removal of hydrogens—with the subsequent generation of ATP from ADP. The product resulting from these two dephosphorylations is pyruvate.

An alternative source of glucose 6-phosphate is the breakdown of starch or glycogen. This process, which is also shown in Figure 6-7, occurs in two steps. Initially, starch or glycogen is enzymatically broken down and phosphorylated to molecules of glucose 1-phosphate. However, unlike the direct phosphorylation of glucose utilizing ATP, the phosphorylation of starch and glycogen requires the input of energy since it utilizes inorganic phosphate. In the second step, glucose 1-phosphate is converted to glucose 6-phosphate by another specific enzyme.

Other hexoses (six-carbon sugars), such as fructose, galactose, and mannose, enter the glycolytic pathway by being converted to one of the intermediates in the first stage of this pathway. Such conversions usually involve several steps and utilize ATP as a source of energy.

## FATE OF PYRUVATE UNDER ANAEROBIC CONDITIONS

Organisms that live under anaerobic conditions, such as yeasts and certain species of bacteria, cannot use oxygen to extract the large amount of energy still available in the pyruvate molecules derived from the breakdown of glucose. These organisms convert pyruvate to lactate or ethyl alcohol.

In the last step of the simplest kind of fermentation, the hydrogen in the $NAD_{red}$ that was generated in step 5 of glycolysis is added to pyruvate to form lactate (Fig. 6-8). This reaction regenerates $NAD_{oxid}$, which can participate in the glycolysis of additional sugar molecules. If $NAD_{oxid}$ were not regenerated by this reaction, the glycolytic pathway could not continue to operate in those species of bacteria that rely upon anaerobic glycolysis as a major source of their energy because $NAD_{oxid}$ would soon be in short supply.

In cells of higher organisms, aerobic metabolism of glucose operates when oxygen is abundant, and anaerobic glycolysis operates when oxygen is in short supply. This is particularly true of skeletal muscle,

**FIG. 6-8** The fate of pyruvate under aerobic and anaerobic conditions.

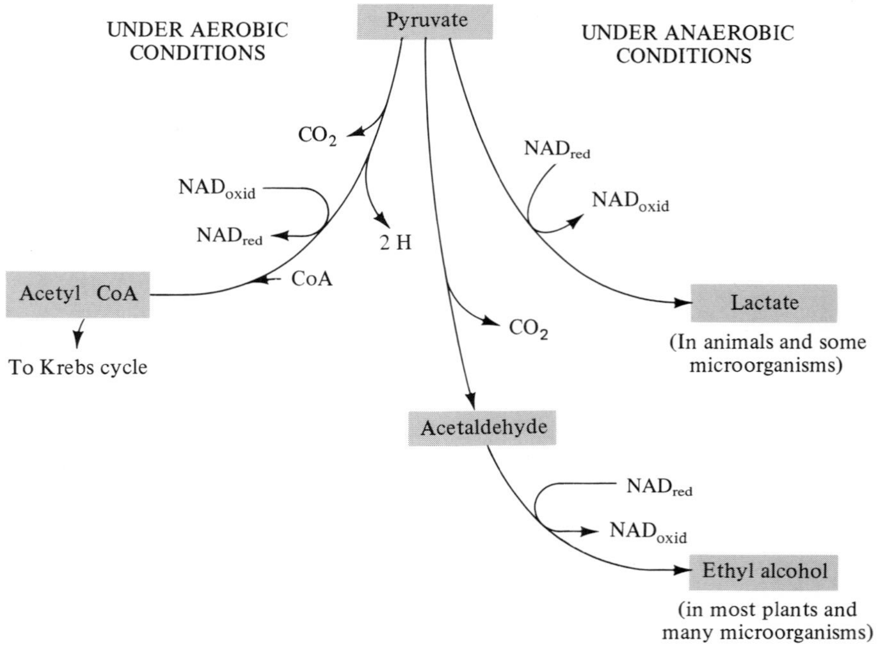

FIG. 6-9 Marathon runners collapsing from exhaustion after crossing the finish line.

which, under extreme exertion, can continue to function when the supply of oxygen reaching the muscles cannot keep up with the demand. Such muscles can go into what is known as "oxygen debt" by converting much of the glycogen that is normally stored in muscles into lactate. It is the excess accumulation of lactate that contributes to the sensation of muscle fatigue and soreness. Following muscle exertion, when oxygen again becomes available, the oxygen debt is repaid when the accumulated lactate is converted back to pyruvate, which is then completely oxidized to $CO_2$ and $H_2O$.

The energy requirements of a sprinter are not the same as those of a marathon runner. Since sprinters must give their all within a short span of time, they need a fuel that will burn fast enough and provide them with bursts of energy as needed. Their requirements are immediate. Marathon runners, on the other hand, need staying power—enough energy to keep them going in their long-distance running and to keep them away from the so-called wall, the state of exhaustion caused by lactate build-up that hits athletes engaged in endurance competitions (Fig. 6-9).

## FERMENTATION IN YEAST CELLS

In yeasts, pyruvate is first broken down to acetaldehyde and carbon dioxide (Fig. 6-8). The second step involves the conversion of acetaldehyde to ethyl alcohol by $NAD_{red}$. By this process of alcoholic fermentation, carried out in the absence of oxygen, yeasts can convert the juices of grapes and other fruit into wine by converting their glucose into ethyl alcohol (Fig. 6-10). Since yeasts, like all organisms, have a limited tolerance for alcohol, the yeast cells die and fermentation ceases when the ethyl alcohol concentration reaches about 12 percent.

FIG. 6-10 Winemaker sampling wine in a wine cellar.

The fact that the anaerobic metabolism of sugars takes place in many primitive one-celled organisms has led to the suggestion that the earliest organisms on earth depended solely on this process for their energy needs. When life began on this planet, our atmosphere may have had large amounts of free hydrogen and probably very little oxygen. This lack of oxygen necessitated the utilization of glycolysis and fermentation as an energy source for many millions of years. It was not until large numbers of photosynthetic organisms had produced enough oxygen in the atmosphere that the present highly efficient aerobic metabolism of glucose could have evolved.

## WHY IS ANAEROBIC OXIDATION IMPORTANT?

The chief importance of anaerobic pathways lies in their ability to provide energy in the absence of oxygen. They enable anaerobic organisms, or **anaerobes,** to exploit environments totally devoid of oxygen. For aerobic organisms, or **aerobes,** such pathways provide key intermediates (pyruvate and lactate) as well as an alternative pathway when $O_2$ is in short supply. The major reactions of these anaerobic pathways can be summarized as follows.

The reduction of $NAD_{oxid}$ to $NAD_{red}$ accompanying the loss of hydrogen atoms by glyceraldehyde 3-phosphate is followed by the coenzyme's reoxidation when alcohol or lactate is formed. This transfer of hydrogen atoms enables the process to continue. It would otherwise stop as soon as the small amount of $NAD_{oxid}$ present in a cell had been reduced.

Glucose phosphorylation requires an investment of two energy-rich phosphate groups (~Ps). This investment is made by the dephosphorylation of two molecules of ATP to ADP for each molecule of hexose diphosphate formed. Hexose diphosphate splits into two glyceraldehyde 3-phosphate molecules, and each receives an inorganic phosphate at the same time that glyceraldehyde 3-phosphate is oxidized by the loss of two hydrogen atoms. The resulting product undergoes two dephosphorylations in which two ATP molecules are formed from the available ADP. Thus, for an original investment of two ~P groups as ATP, a return of four ~P groups is realized, also in the form of ATPs, yielding a "profit" of two ~P groups.

## AEROBIC OXIDATION AS CELLULAR RESPIRATION

Like molecules of gasoline, molecules of fats and carbohydrates contain much chemically bound energy. This energy becomes fully available only when each molecule is completely oxidized to $CO_2$ and water. Achieving that end is the task of the cell's respiratory pathways.

Two pathways are involved. One is the **Krebs cycle,** in which two-carbon fragments derived from the pyruvate produced by glycolysis or, more often, from fats, are degraded in a series of steps that split off $CO_2$ and hydrogen atoms. The electrons of the hydrogen atoms, which are at a high energy level, are then accepted by a second pathway, the respiratory chain, or **cytochrome system.** This system uses that energy to synthesize ATP. At the end of the chain, the hydrogen atoms combine with oxygen, and water is formed as a result. Unless oxygen is present, no electrons can be transported by the respiratory chain or even received by it from the Krebs cycle. In the absence of oxygen, both pathways cease operating, which is why oxygen is essential to those organisms that depend on cellular respiration for energy.

Unlike the enzymes of the glycolytic pathway, the enzymes and coenzymes involved in cellular respiration are contained within the mitochondria. The Krebs cycle enzymes and the reactions they catalyze are located in the matrix of the mitochondrion, and the respiratory chain enzymes are components of the inner mitochondrial membrane.

Knowing that oxygen and hydrogen ultimately combine to form water in aerobic respiration, early biochemists tried to get oxygen to react directly with glyceraldehyde 3-phosphate, pyruvate, and other intermediates of glycolysis. All attempts failed. It eventually became clear that there are many biochemical steps between the removal of hydrogen from pyruvate and the addition of hydrogen to oxygen to form water. The first of those steps is the formation of a complex with coenzyme A (CoA) when the pyruvate derived from glycolysis loses two hydrogen atoms, thereby reducing $NAD_{oxid}$ and undergoes decarboxylation (loss of $CO_2$) within the mitochondrion. You can see this if you look back at Figure 6-8. With the loss of two hydrogen atoms and one $CO_2$ molecule, the pyruvate becomes acetyl, a two-carbon compound related to acetic acid. This compound forms a complex with coenzyme A; the complex is referred to as **acetyl CoA.** This two-carbon complex then enters the oxidative pathway known as the Krebs cycle.

## THE KREBS CYCLE

The Krebs cycle was named after the German born, British scientist, Nobel prize winner Sir Hans Krebs, who worked out the last of the cyclic series of reactions involved in the pathway and showed in 1937 that it was a cycle. In the Krebs cycle, the two-carbon acetyl group of acetyl CoA undergoes a series of oxidative degradations. The cycle can be regarded as occurring in the following six major steps (Fig. 6-11):

1. Acetyl CoA joins with oxaloacetate, a four-carbon compound present in all eukaryotic cells, producing citrate (citric acid), a six-carbon compound; the coenzyme is split off during the formation of citrate. Citrate contains three carboxylic groups. The Krebs cycle is thus also known as the citric acid or tricarboxylic acid (TCA) cycle.

2. In a series of four reactions, citrate undergoes a type of molecular rearrangement known as isomerization and a partial oxidation, in which it loses a pair of hydrogen atoms, which are used to reduce $NAD_{oxid}$ to $NAD_{red}$; this reaction is catalyzed by a dehydrogenase enzyme. A molecule of $CO_2$ is split off in this stage as well. The final product is a five-carbon compound, $\alpha$(alpha)-ketoglutarate.

3. Coenzyme A participates in a reaction in which a second $CO_2$ and a second pair of hydrogen atoms are split off, with another NAD being reduced to $NAD_{red}$. A four-carbon compound, succinyl CoA, is formed.

4. The product of step 3 is converted to succinate, another four-carbon compound, in a strongly exergonic reaction that is usually coupled to the production of guanosine triphosphate (GTP) from guanosine diphosphate (GDP) and $P_i$. (In some organisms, ATP is produced directly from ADP and $P_i$, without the intervention of GDP and GTP.) GTP then yields a phosphate group to ADP, thereby forming ATP. Because no oxidation is immediately coupled to this formation of ATP, it is called a substrate-level phosphorylation rather than oxidative phosphorylation.

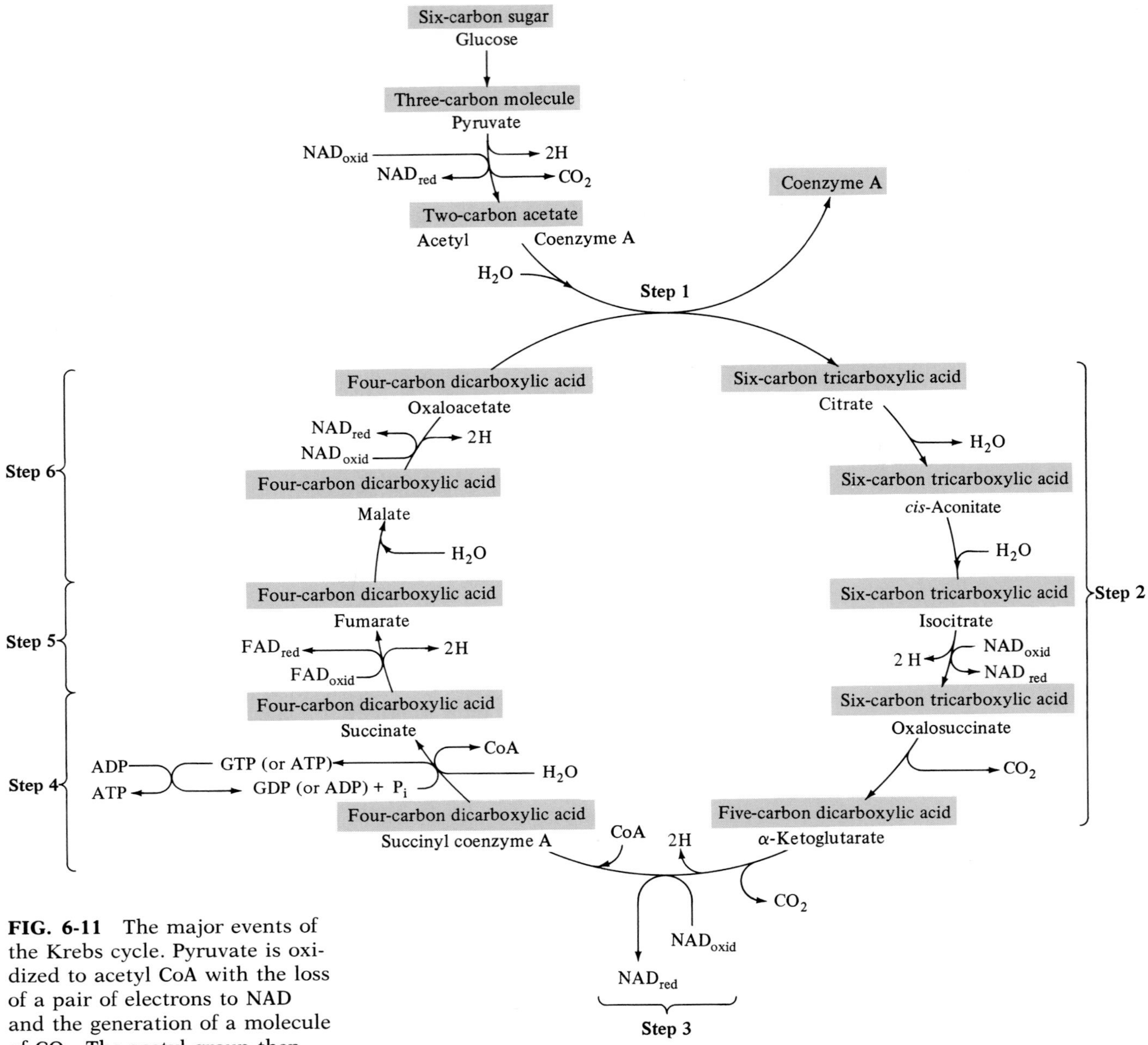


**FIG. 6-11** The major events of the Krebs cycle. Pyruvate is oxidized to acetyl CoA with the loss of a pair of electrons to NAD and the generation of a molecule of $CO_2$. The acetyl group then enters the cycle by joining to oxaloacetate to form citrate, which, in a series of reactions loses two $CO_2$ molecules and four pairs of electrons. The electrons are passed on to the respiratory chain. The free energy made available by the passage of the electrons down the chain is conserved by the formation of ATP molecules.

5. The four-carbon compound, succinate, that is produced at this point in the cycle has a relatively low energy level. It lacks sufficient energy to reduce $NAD_{oxid}$ and it instead reduces another electron acceptor, flavine adenine dinucleotide, or **FAD,** to form fumarate, another four-carbon dicarboxylic acid. FAD is reduced to $FADH_2$, and succinate is converted to fumarate.

6. Fumarate, after a molecular rearrangement to form malate, is oxidized to oxaloacetate with the release of two more hydrogen atoms; these are captured by $NAD_{oxid}$, which is reduced to $NAD_{red}$. The oxaloacetate is now available to interact with another acetyl CoA complex.

Thus, it takes two turns of the TCA cycle to oxidize the equivalent of two molecules of pyruvate produced from each glucose molecule. In summary, the oxidation of one molecule of pyruvate—one turn of the Krebs cycle—involves the transfer of five pairs of electrons to NAD or

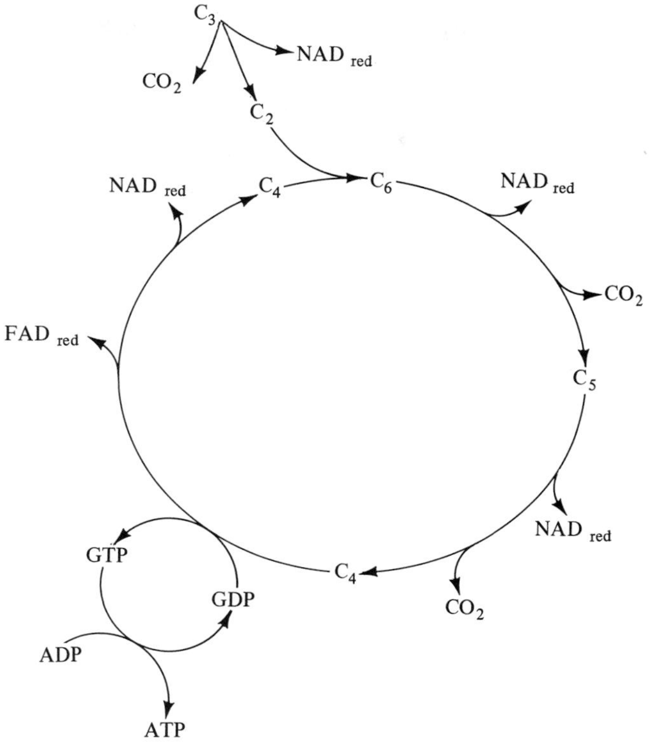


**FIG. 6-12** An overview of the Krebs cycle.

FAD, the formation of three $CO_2$ molecules, and the generation of one ATP molecule (Fig. 6-12).

## THE ROLE OF THE MITOCHONDRION

As we have seen, the machinery for aerobic respiration in eukaryotes is contained, not in the cytoplasm, but within the mitochondrion. Mitochondria, which we studied in Chapter 4, are bounded by two membranes. The outer membrane is of uniform structure and appearance, and the inner membrane is folded at intervals. These folds, or cristae, extend inward as irregular shelves in most animal mitochondria (Fig. 6-13). In plants, they are more tubular in structure. Cristae are most numerous in mitochondria found in actively respiring cells (Fig. 6-14).

Because of the arrangement of membranes and cristae, there are two principal compartments within the mitochondrion. Both are shown in

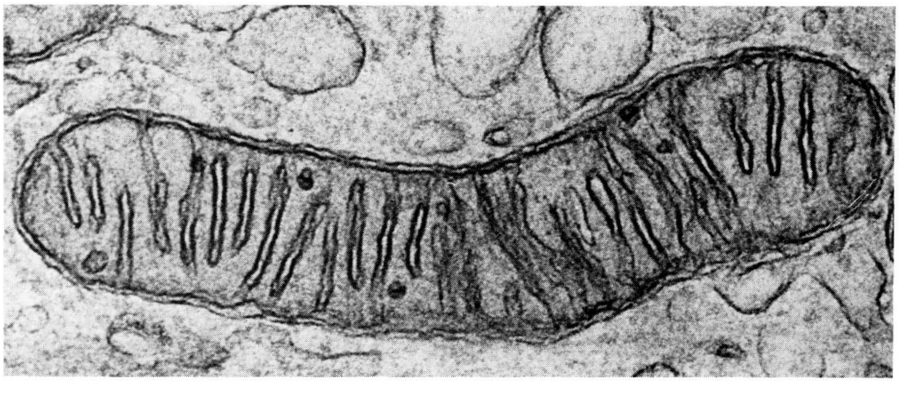

**FIG. 6-13** Electron micrograph of a thin section of a mitochondrion from the epididymis of a mouse.

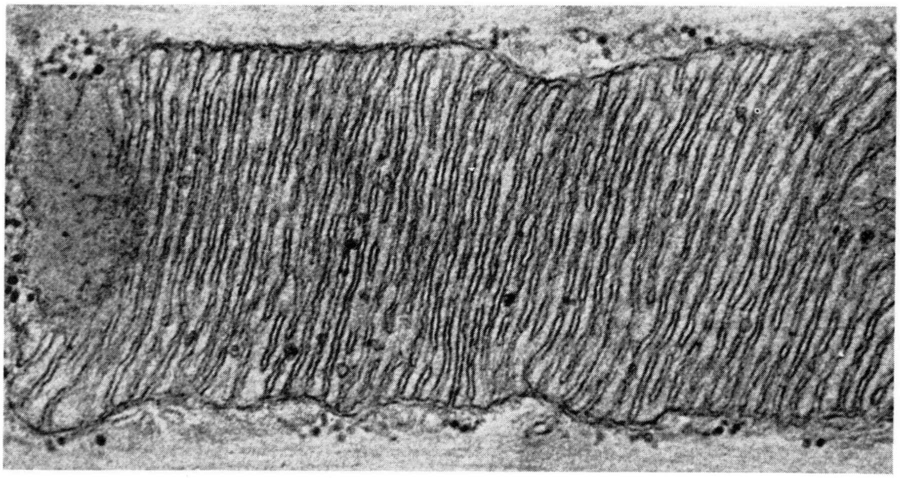

**FIG. 6-14** Mitochondria from bat muscle showing the large numbers of cristae associated with an actively respiring cell.

Figure 6-13: the centrally located matrix and the intermembrane space, between the outer and inner mitochondrial membranes.

***The Structure and Function of the Membranes.*** The outer membrane is about 50 percent lipid and is permeable to molecules of 10,000 daltons or less. The inner membrane is 70 to 80 percent protein and is highly specialized, containing at least 60 different proteins. It is highly impermeable to all but small, uncharged molecules. Ions and other substances normally enter only by way of highly specialized active transport mechanisms mediated by the membrane proteins. High-resolution electron micrographs of inner membranes reveal a series of spheres on stalks resembling lollipops attached to the membrane's inner surface (Fig. 6-15). With one exception, all the Krebs cycle enzymes are contained within the mitochondrial matrix—the area enclosed by the inner membrane into which the spheres project. The exception among the enzymes is succinic dehydrogenase, which is part of the inner membrane.

Sonication (disruption by ultrasound—sound with a frequency higher than 20,000 cycles/second) and differential centrifugation of mitochondria have produced fractions that include pieces of the inner membrane. As with most membrane fragments, the broken ends tend to fuse, producing roughly spherical enclosures known as submitochondrial vesicles. Unlike the normal orientation of the inner membrane, however, the vesicles form with their lollipops pointing outward, as illustrated in Figure 6-16a. When washed free of all matrix, the vesicles alone do not catalyze TCA cycle reactions. However, when submitochondrial vesicles are provided with $NAD_{red}$ and molecular oxygen, electrons are accepted from the $NAD_{red}$ and passed to oxygen, and water is formed. If ADP and $P_i$ are present, ATP is synthesized as well. The inner mitochondrial membrane is the site of the respiratory electron transport chain and the coupled mechanism that phosphorylates ADP to produce ATP.

**FIG. 6-15** Electron micrograph of lollipop spheres (arrows) in the mitochondria of beef heart muscle.

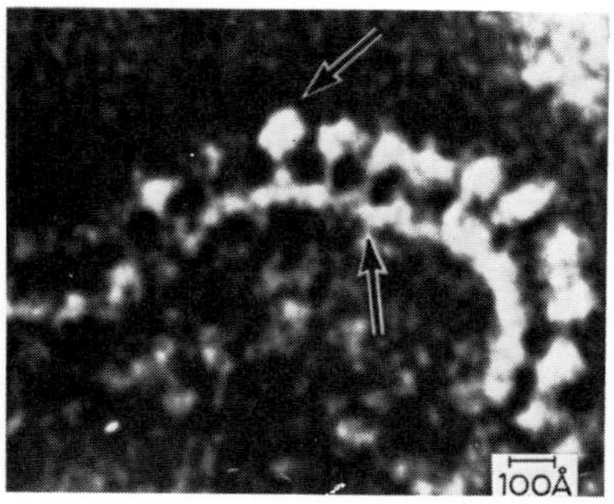


***"Lollipops" Contain an ATP Synthetase.*** What is the significance of the lollipop spheres? It was believed that if a way to detach them from the membrane could be found, their precise function might be determined. Treatment of submitochondrial vesicles with urea separated the lollipops from the vesicles (Fig. 6-16b). The two components of the inner membrane could thus be studied separately.

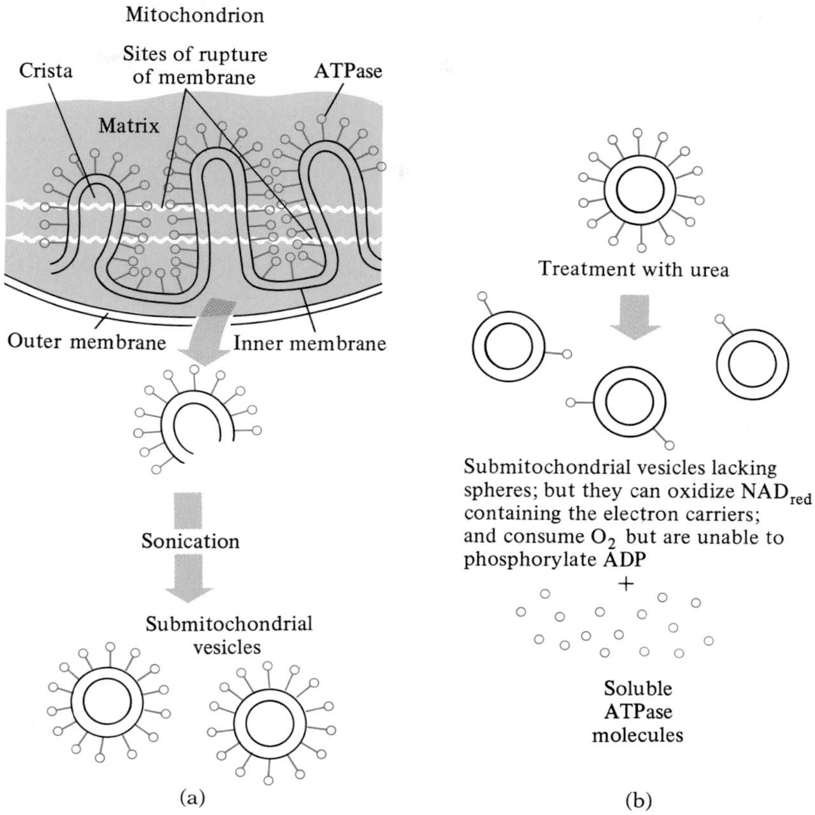


**FIG. 6-16** (a) Sketch representing the relationship among lollipop spheres, the inner membrane, and the matrix space of a mitochondrion and sketches of submitochondrial vesicles as obtained by sonication.
(b) Sketch of the results of treating lollipop spheres with urea. Stalks dissolve, leaving preparations that contain spheres only or, alternatively, vesicles nearly devoid of spheres.

In the presence of molecular oxygen, ADP, and $P_i$, vesicles without spheres readily oxidized $NAD_{red}$, water being formed in the process. But no phosphorylation of ADP to ATP occurred. The spheres showed ATPase activity, that is, they readily hydrolyzed ATP to ADP and $P_i$.

When the isolated spheres were added to the vesicles free of spheres, they became reattached to the membrane. The reconstituted submitochondrial vesicles, when presented with $NAD_{red}$, ADP, and $P_i$, in the presence of molecular oxygen, could once more carry out oxidative phosphorylation, with $NAD_{red}$ being oxidized and ATP and water being produced.

It seems clear from these results that the inner mitochondrial membrane contains the respiratory chain and that the lollipop spheres contain the enzymes capable of forming ATP from ADP and $P_i$ (or of splitting ATP to ADP and $P_i$ if an oxidation is not coupled to the mechanism so that the reaction will proceed in the other direction).

The complex machinery of the inner membrane contains a remarkable array of electron-transporting molecules. Of the known components, only some have been mentioned. Yet by no means are all the roles of all of the membrane's components fully understood. The most generally accepted theory of how oxidative phosphorylation is accomplished is briefly described in Panel 6-2.

## THE RESPIRATORY CHAIN AND OXIDATIVE PHOSPHORYLATION

Energy release and its coupling to the synthesis of ATP is initiated whenever the coenzymes NAD and FAD, which have been reduced by receiving electrons, undergo oxidation. This process occurs in the respiratory chain, located in the inner membrane of the mitochondria. The series of enzymes and coenzymes forming the respiratory assembly is a highly specialized "bucket brigade" of electron-transferring agents. Many of the enzymes of the respiratory chain are also cytochromes, all of which contain a form of the iron-bearing pigment heme as a prosthetic group (a nonpolypeptide unit attached to a protein and essential for its biological activities). Their name derives from their red color in concentrated solution, similar to that of the hemoglobin in red blood

PANEL 6-2

# The Chemiosmotic Theory

One of the unsolved problems concerning the inner mitochondrial membrane is the exact mechanism by which the energy released by the passage of electrons down the respiratory chain is coupled to the phosphorylation of ADP. According to the model proposed by the British Nobel prize winner, biochemist Peter Mitchell, electron transfer works to create a concentration gradient of $H^+$ (protons) and therefore of electrical charge between the inside and outside of the inner mitochondrial membrane. Once established, this gradient would normally be maintained by the inner membrane's high impermeability to ions. However, at certain places in the inner membrane, there appear to be channels that permit protons to reenter the matrix space. At these places the lollipop spheres are located, constituting a mechanism that couples the phosphorylation of ADP with the passage of the protons down the electrochemical concentration gradient, from outside the membrane into the matrix.

Evidence for the theory is as follows:

1. Active mitochondria increase the hydrogen ion concentration in the medium around them. Moreover, when engaged in electron transport, submitochondrial vesicles, which are inside out, accumulate protons ($H^+$). These two lines of evidence indicate that the inner membrane normally ejects protons to its outside.
2. Some components of the electron transport chain can carry only electrons but not the hydrogen ions.
3. Sophisticated tests indicate that some components of each respiratory assembly are located either on one side or the other of the membrane. Other components may extend through the membrane or even move from one side of it to the other. This results in the asymmetry necessary to generate a gradient of proton concentration.
4. Proton gradients experimentally set up in submitochondrial vesicles can be shown to drive ATP synthesis in the vesicles in the absence of $NAD_{red}$.
5. ATP synthesis can be inhibited by compounds that allow protons to cross the inner membrane. These compounds, proton ionophore reagents, are known as uncouplers of oxidative phosphorylation.

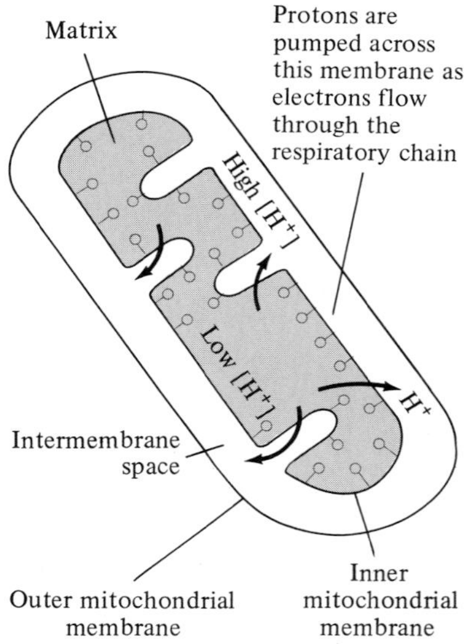


Diagram illustrating the generation of a proton gradient across the inner mitochondrial membrane by electron transfer through the respiratory chain.

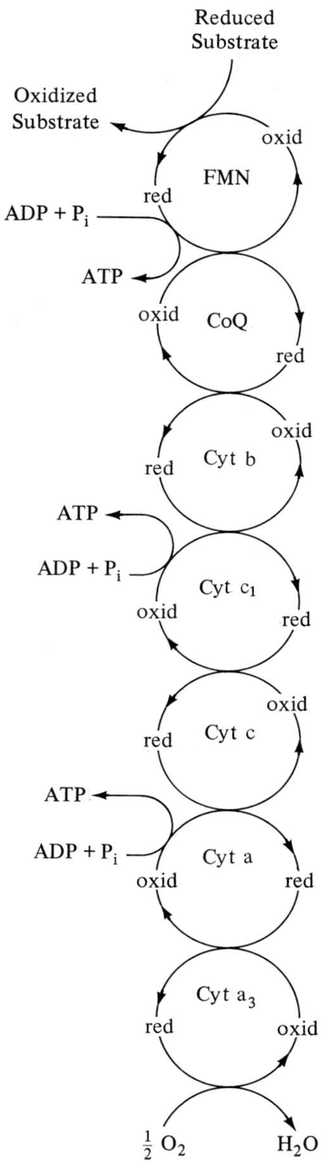


**FIG. 6-17** The electron transport respiratory chain, which occurs within the inner membrane of the mitochondrion. Note that phosphorylation of ADP, producing ATP, occurs at three oxidoreductive levels. For each pair of electrons passed to the flavoprotein by $NAD_{red}$, three ATP molecules are generated. Although entire hydrogen atoms are accepted by the flavoprotein and, at the end of the chain, by oxygen, most carriers in the chain transfer only electrons. The rest of the hydrogen atom—a proton, or hydrogen ion ($H^+$)—moves slowly through the matrix, catching up at the end of the chain.

cells. Each cytochrome molecule alternately acts as an oxidizing agent, then as a reducing agent. Each readily accepts electrons from the next stronger reducing agent in the system, readily passing them on to the next stronger oxidizing agent.

Some of the known components of the respiratory chain are shown in Figure 6-17. Cytochromes are designated by lower-case, subscript letters and numbers, indicating the order of their discovery. The component to receive electrons from $NAD_{red}$ is a flavoprotein, a dehydrogenase enzyme containing a tightly bound riboflavin (vitamin $B_2$) derivative as a prosthetic group. The flavoprotein of the respiratory chain contains flavin mononucleotide (FMN) as its prosthetic group. Besides flavoprotein, there are other members of the respiratory chain that are not cytochromes. Some are not even proteins. One known example of a nonprotein member is coenzyme Q (CoQ), also known as ubiquinone.

Although some members of the respiratory chain can be reduced by receiving hydrogen atoms, most are reduced only by receiving electrons. A moment after being reduced, the component is reoxidized as it passes on the electron or hydrogen atom to the next member in the "bucket brigade." In some of these oxidations, energy is given off as heat. But much of the energy released by the oxidations is indirectly coupled to the endergonic reaction ADP + $P_i \rightarrow$ ATP. This generation of ATP by oxidations is termed **oxidative phosphorylation.** For every pair of hydrogen atoms accepted in this process by NAD and passed on to oxygen via the respiratory chain, three ATP molecules are generated. The electrons derived from the oxidation of succinate—which are carried by FAD, not NAD—are an exception. FAD is the prosthetic group of succinic dehydrogenase that is reduced by succinate oxidation, is unable to feed electrons to the flavoprotein of the respiratory chain. Instead, it reduces coenzyme Q directly. This "short-circuiting" of the first step in the chain means that only two ATP molecules are generated by the passage of a pair of hydrogen atoms from $FAD_{red}$ to oxygen.

## THE COST/BENEFIT RATIO IN AEROBIC OXIDATION

For every pyruvate molecule that undergoes aerobic oxidation, 15 ATP molecules are generated. Considering that two pyruvates are produced from each glucose and that a net profit of two ATPs had already been realized from the glycolytic pathway, this means that at least 32 ATPs are derived from the oxidation of one glucose molecule. However, more than 32 ATP molecules are generated.

The pairs of hydrogen atoms that NAD accepts from each of the two

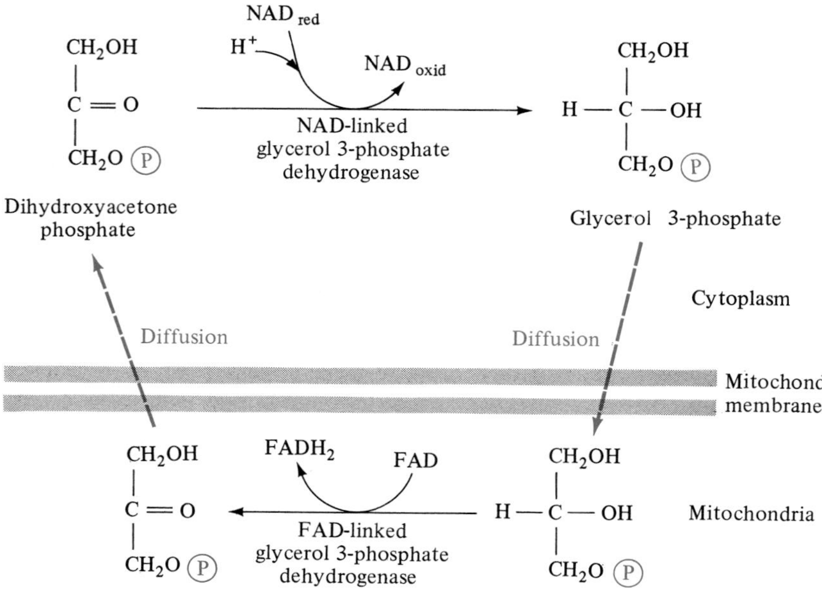


**FIG. 6-18** The electrons accepted by NAD during glycolysis can be transferred to the respiratory chain under aerobic conditions. This can only be done indirectly, however, because the mitochondrial membrane is impermeable to NAD. The indirect route generates two, rather than three, ATP molecules, because the electrons are passed to coenzyme Q rather than to the chain's flavoprotein.

glyceraldehyde 3-phosphate molecules formed from each glucose molecule during glycolysis are also capable of being passed down the respiratory chain. If each pair of hydrogen atoms passed from $NAD_{red}$ down the chain generates three ATPs, it seems that this would add another six ATPs, for a total of 38 ATPs formed by the oxidation of each glucose molecule. However, there is a difficulty in obtaining all 38.

The TCA cycle enzymes and the respiratory chain are located inside the mitochondria, while the $NAD_{red}$ formed by oxidation of glyceraldehyde 3-phosphate is in the cytoplasm. Mitochondrial inner membranes are impermeable to NAD. However, they are permeable to small molecules such as pyruvate and glycerol phosphate. $NAD_{red}$ can be oxidized to $NAD_{oxid}$ in a reaction that transfers two hydrogen atoms from $NAD_{red}$ to dihydroxyacetone phosphate, reducing it to glycerol phosphate. This glycerol phosphate then enters the mitochondrion, where it is converted to dihydroxyacetone phosphate once more,* donating the two electrons to the FAD of the flavoprotein enzyme (Fig. 6-18). $FAD_{oxid}$ is thereby reduced to $FAD_{red}$ and passes the pair of hydrogen atoms to coenzyme Q of the respiratory chain. As before, this "short-circuit" results in the generation of just two ATP molecules per pair of hydrogen atoms.

Dihydroxyacetone then diffuses out into the cytoplasm, where it can be reduced again (see Fig. 6-18). The cost of having to penetrate the mitochondrial membrane in this roundabout way is thus one ATP molecule for each pair of hydrogen atoms originally obtained by NAD from the glycolytic pathway. The total profit for aerobic oxidation of one glucose molecule is therefore only 36, reduced from what might have been 38 ATPs if there were no mitochondrial membrane barrier.† But compared to a two-ATP "profit" from the anaerobic oxidation of a glucose molecule, 36 ATPs is a handsome return. These energy relationships are summarized in Table 6-1.

* This is one of several ways in which hydrogens can cross the inner mitochondrial membrane.

† Aerobic prokaryotes are an exception. Since their respiratory chain is not contained in mitochondria, they realize 38 ATPs per glucose molecule as a result.

**TABLE 6-1 ATP Tally Sheet for Eukaryotic Cells**

| *Anaerobic oxidations (fermentation, glycolysis)* | *Aerobic oxidations (TCA cycle + respiratory chain)* |
|---|---|
| Invested: | Invested: |
| 2 ATPs/glucose | 2 ATPs/glucose |
| Gross return: | Gross return: |
| 4 ATPs/glucose | 38 ATPs/glucose* |
| Net return ("profit"): | Net return ("profit"): |
| 2 ATPs/glucose | 36 ATPs/glucose |

* Itemization of gross return under aerobic conditions:
4 ATPs, glycolytic pathway
30 ATPs, 2 pyruvates oxidized to $CO_2$ and $H_2O$
4 ATPs, 2 $FAD_{red}$ donating electrons to respiratory chain at Q

## FAT METABOLISM

In addition to carbohydrates, lipids also serve as an energy source, ultimately being oxidized in the TCA cycle. Indeed, except in periods of strenuous activity, most of the energy required by human beings is obtained from the oxidative phosphorylation of stored fat. Dietary fats, or triacylglycerols, are digested by hydrolysis to glycerol and fatty acids. This can occur within cells and in the digestive tracts of animals. The glycerol that is split off in the hydrolysis is phosphorylated, consuming an ATP molecule; dehydrogenated, with the reduction of NAD; and converted to glyceraldehyde 3-phosphate. Once this molecule enters the glycolytic pathway, it is either further oxidized, as previously described, or is converted to glucose. Fat can thus be a source of carbohydrate, just as carbohydrate can be a source of fat.

Fatty acids, whether from the diet or from stored fat, follow a different route. Fatty acids are able to penetrate the mitochondrial membrane only after combining with CoA and carnitine to form acyl carnitine. In the mitochondrion, the fatty acyl carnitine compounds undergo a process called $\beta$ (beta)-oxidation, which repeatedly removes two-carbon fragments from the fatty acid chain. ATP, coenzyme A, FAD, and NAD are involved in this process, which yields half as many acetyl CoA molecules as there are carbon atoms in the particular fatty acid. Acetyl CoA can immediately enter the TCA cycle and be oxidized to $CO_2$ and water. However, when more acetyl CoA is derived from pyruvate oxidation than is needed as an energy source for the body's activities, acetyl CoA is converted to fatty acids, joined to glycerol, and stored as fat. This explains why a substantial excess of dietary carbohydrates can lead to obesity.

## PROTEIN METABOLISM

Considering that there are 20 different kinds of amino acids in proteins, in addition to various side groups, the metabolism of proteins is indeed complex. Various portions of about half of the 20 amino acids can enter

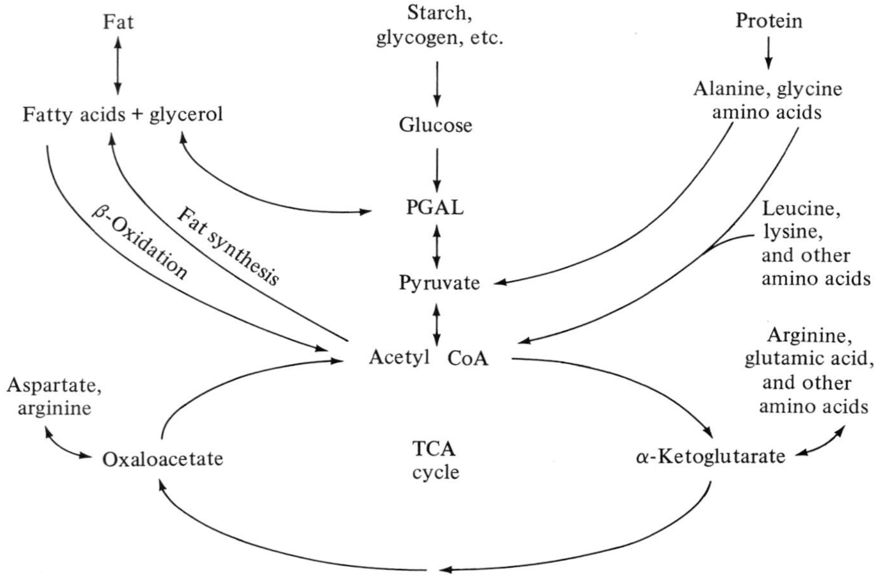


**FIG. 6-19** Convergence of fat, carbohydrate, and protein metabolism at acetyl CoA "crossroads."

the oxidative pathways of the cell when in dietary excess. In such cases, the amino acid first loses its amino (—$NH_2$) group in a process called **deamination.** The amino group is converted to urea in terrestrial vertebrates and excreted.

In undergoing further degradation, fragments of some of the amino acids are converted to acetyl CoA. At this crossroads, they can either be oxidized in the Krebs cycle or converted to fat. Portions of other kinds of amino acids are converted to pyruvate, which may then be converted to glucose or to acetyl CoA. Besides entry into the Krebs cycle as acetyl CoA, portions of some amino acids can be converted directly to TCA cycle intermediates, such as aspartate to oxaloacetate.

A sketch showing the major interrelationships in the metabolism of carbohydrates, fats, and amino acids appears in Figure 6-19.

## Summary

Energy exchanges in living as well as physical systems are governed by the laws of thermodynamics. The first law states that energy can be changed from one form to another, but it cannot be created or destroyed. The second law states that all physical and chemical processes tend to proceed toward equilibrium, that is, they proceed from a state of orderliness to one of disorder or randomness. Living systems are able to maintain or even increase their highly ordered states by taking up energy from the environment and utilizing a portion of it for carrying on those processes that are characteristic of life. A remarkable feature of living systems, therefore, is that they function quite differently from the remainder of the universe in that they are order-creating both in their structure and in their management of energy. This is not an exception to the second law, however, since organisms increase their orderliness only at the expense of outside sources of energy.

The cells of an organism are able to capture and utilize the stored energy in foodstuffs by the process of cellular respiration, in which the oxidation of glucose takes place in a complex sequence of reactions that release small amounts of energy. Much of this released energy is captured by coupling the steps in glucose or fatty acid breakdown with the simultaneous formation of high-energy molecules, such at ATP, which can store energy for later use as needed.

The oxidation of carbohydrates and lipids is the major source of energy used in the generation of ATP molecules. Glucose oxidation begins with the degradation of the six-carbon glucose molecule by the process of glycolysis. This process can proceed aerobically or anaerobically. Through a series of enzyme-catalyzed reactions, glucose is broken down to two three-carbon molecules of pyruvate, with the formation of two new molecules of ATP and two reduced molecules of NAD. In the absence of oxygen, yeasts and certain species of bacteria convert pyruvate to lactate or ethyl alcohol. In the cells of higher organisms, pyruvate is also converted to lactic acid when oxygen is in short supply, as in muscle cells under extreme exertion.

Under aerobic conditions, the final breakdown of glucose molecules or of acetyl CoA molecules derived from fats or proteins involves two stages—the Krebs or tricarboxylic acid cycle and the electron transport system—in the overall process of cellular respiration. For glucose, this process begins with the breakdown of each three-carbon pyruvate molecule to a two-carbon acetyl group, which enters the Krebs cycle. For each pyruvate molecule, three carbon dioxide molecules, one ATP molecule, four $NAD_{red}$ molecules, and one $FAD_{red}$ molecule are generated.

Hydrogen atoms accepted by NAD and FAD in the TCA cycle are passed down the electron transport system, a series of electron carriers and enzymes embedded in the inner membranes, or cristae, of mitochondria. Energy in the form of the pairs of electrons in the hydrogens bound to reduced NAD and FAD are passed along these carriers and generate a proton gradient across the inner membrane, which in turn causes the synthesis of ATP from ADP and phosphate ($P_i$). The net result of the breakdown of one molecule of glucose is the formation of six $CO_2$ and six $H_2O$ molecules and the generation of stored energy in the form of 36 molecules of ATP.

Lipids, which serve as the body's chief energy source, are also oxidized in the Krebs cycle. Fats derived from the diet or stored fats are first hydrolyzed to glycerol and fatty acids. Fatty acids enter the mitochondria, where they are oxidized to two carbon compounds, which enter the Krebs cycle as acetyl CoA, where they are oxidized. The metabolism of proteins is more complex because they are composed of 20 amino acids and various prosthetic groups. Some amino acids, following deamination, are transformed into carboxylic acids; the carboxylic acids participate in the Krebs cycle.

## Recommended Readings

ALBERTS, B., D. BRAY, J. LEWIS, M. RAFF, K. ROBERTS, AND J. D. WATSON. 1983. *Molecular Biology of the Cell.* Garland, New York. This is a well-written and illustrated text that thoroughly covers the subject of molecular biology from the structure and function of macromolecules to cell morphology.

BALDWIN, E. 1967. *Dynamic Aspects of Biochemistry,* 5th ed. Cambridge University Press, New York. This classic prize-winning book is the most interesting of all introductions to biochemistry. It provides a wealth of experimental evidence for our understanding of basic metabolic pathways and is written in an engaging style.

De Witt, W. 1977. *Biology of the Cell: An Evolutionary Approach.* Saunders, Philadelphia. This text has excellent descriptions of chemical thermodynamics and enzyme chemistry.

Dickerson, R. E. 1980. "Cytochrome C and the evolution of energy metabolism." *Scientific American,* 242(3):136–153. In tracing the evolution of the metabolism of modern organisms, including human beings, to bacteria, the author also reviews basic cellular respiration.

Freifelder, D. 1983. *Molecular Biology. A Comprehensive Introduction to Prokaryotes and Eukaryotes.* This text emphasizes basic molecular processes (such as protein, DNA, and RNA synthesis) and genetic phenomena in both prokaryotic and eukaryotic cells.

Lehninger, A. L. 1982. *Principles of Biochemistry.* Worth, New York. One of the best of the introductory biochemistry texts. It includes comprehensive descriptions of energy flow in biological systems and stresses an understanding based on experimental evidence.

7

# Energy Capture: Photosynthesis

## OUTLINE



In Chapter 6, we saw how organisms are able to derive energy for life processes from carbohydrates and other foods. Now we look into the question of how carbohydrate molecules are synthesized by organisms and how energy is stored in the process.

The sun's radiant energy is captured and stored by organisms that possess specialized pigments: plants, algae, and a few species of pigmented bacteria. These organisms have the ability to manufacture carbohydrates in the presence of light through the process of photosynthesis. Only a fraction of the light energy reaching earth is captured; however, when captured and stored through photosynthesis, this fraction forms the basis for nearly all vital activity on our planet. While much photosynthesis takes place in the leaves of higher plants, a good deal occurs in algae, particularly in the algae of the oceans.*

The organic matter produced as a result of photosynthesis provides the molecular building blocks and the energy source for some bacteria, all fungi and animals (Fig. 7-1). It is quite possible for humans to live on a diet consisting entirely of plant products. Even the energy we derive from eating meat originates with the sun. We also owe our fossil fuels and firewood to photosynthesis. But photosynthesis plays a further essential role: it provides the oxygen we breathe and uses the carbon dioxide we produce.

* A few bacteria are photosynthetic, but except for the cyanobacteria (blue-green algae) they do not release oxygen in the process, and their photosynthesis pigments differ from those of algae and plants.

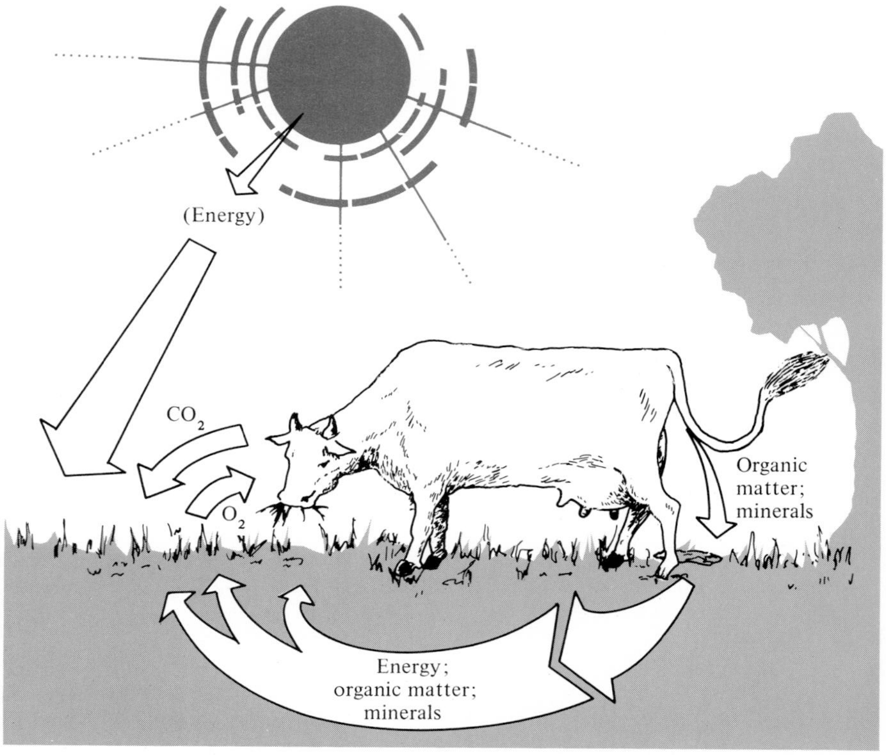


**FIG. 7-1** The role of photosynthesis in the utilization of energy and nutrients in organisms. Matter is recycled whereas energy is not. The $CO_2$ released in respiration is utilized in photosynthesis; the oxygen produced in photosynthesis is consumed in respiration.

## UNDERSTANDING PHOTOSYNTHESIS

In the process of photosynthesis, **chlorophyll** (from the Greek *chloros*, "green"), the green pigment of plants, absorbs the radiant energy of light, which increases the energy level of some of its electrons. This increases the oxidoreductive, or oxidation-reduction ("redox"), potential of chlorophyll; the energy is ultimately stored in the form of carbohydrates synthesized from $CO_2$ and $H_2O$. The process in which a sugar molecule or other carbohydrate ($CH_2O$) is formed from carbon dioxide and water, with the release of oxygen, is represented by the summary reaction:

$$\underset{\text{carbon dioxide}}{CO_2} + \underset{\text{water}}{H_2O} + \text{light energy} \xrightarrow{\text{chlorophyll}} \underset{\text{carbohydrate}}{(CH_2O)} + \underset{\text{oxygen}}{O_2}$$

Our understanding of photosynthesis has resulted from the efforts of many investigators. Until about 300 years ago, it was believed that plants obtained food from the soil. This idea persists today in the erroneous reference to fertilizers as "plant food." A classic experiment by the Belgian philosopher-physician Jean-Baptiste van Helmont demonstrated in 1640 that plants obtain very little substance from the soil. He found that a potted willow tree had gained over 1,300 times as much weight as was lost from the soil in the pot. Van Helmont concluded erroneously that the additional weight had come from the water he had added to the soil.

Investigators did not realize that air was involved in the process of photosynthesis until after oxygen and carbon dioxide were discovered. In 1727, Stephen Hales, an English minister, found that plants grown under glass jars caused a decrease in the volume of air in the jar. This observation was not understood until after the discovery of oxygen in 1771 by Joseph Priestley, the English clergyman and chemist. He observed that the air inside a jar in which a candle had burned for a time before going out would not support the life of a mouse. But when he "put a sprig of mint into a quantity of air, in which a wax candle had burned out . . . another candle burned perfectly well in it." "Nor was [such air] at all inconvenient to a mouse."

Plant life, Priestley said, "reversed the effects of breathing." He believed that animals *added* something to the air and that plants purified the air by *removing* this material from it. It was widely believed that burning and breathing added a substance called phlogiston to the air and that only "dephlogisticated" air could support a flame or the life of an animal. We know now that combustion and breathing produce carbon dioxide and use oxygen and that plants produce oxygen and remove carbon dioxide during photosynthesis.

That the "purification" of the air by a plant occurs only after the sun has risen was observed in 1779 by the Dutch physician Jan Ingenhousz. Ingenhousz also observed that, except for the "purification" of air by their green parts, plants had the same "contaminating" effect on air as did animals, that is, they produced carbon dioxide. He concluded that in light plants removed carbon dioxide from the air, split off the carbon, and returned the oxygen to the air. This belief regarding the source of the oxygen produced in photosynthesis seemed entirely logical: one gas was being slightly modified to produce another. It would be many years before it was proven that water, not carbon dioxide, was the source of the oxygen.

By the early 1800s, it was clear that photosynthesis was a chemical process that required light, carbon dioxide, and water and produced oxygen and organic matter. How photosynthesis occurred became the quest of scientists in several countries in the following years.

## LOCATION OF PHOTOSYNTHESIS IN THE CELL

Ingenhousz had noted that photosynthesis was associated only with the green parts of plants. The green color is due to the presence of the pigment chlorophyll. Microscopic examination has shown that chlorophyll is not dissolved in the cell cytoplasm but is contained within the structures known as **chloroplasts.** These are membrane-bound organelles containing chlorophyll-bearing membranes called **thylakoids.** Chloroplasts vary in form from football-shaped bodies found in plants to assorted ribbons, discs, and other shapes found in various species of algae. Ingenious experiments by the German microbiologist Theodor Engelmann, published in 1882, demonstrated that the chloroplasts are the sites of photosynthesis.

Engelmann chose a multicellular, filamentous alga with chloroplasts large enough to be plainly seen under the microscope and placed several filaments in water on a microscope slide. To detect whether photosynthesis was occurring, he used the motile bacterium *Pseudomonas,* which responds to the presence of $O_2$ by moving toward it. Engelmann added *Pseudomonas* to the slide and sealed off the edges of the preparation so that no oxygen could enter from the atmosphere.

Engelmann kept the preparation out of the light until its oxygen was depleted and the bacteria had stopped moving and were randomly dispersed. Then he placed the slide under the microscope and turned on the light. As the algae began to carry on photosynthesis and produce oxygen, the bacteria nearest the chloroplasts began to move. Only as oxygen diffused outward from the chloroplasts did other bacteria begin to move. After a time, the swimming bacteria gathered about the algal cells, as close as they could get to the chloroplasts. This was the earliest demonstration that the chloroplast was the site of photosynthesis.

## LIGHT AND PHOTOSYNTHESIS

How a plant absorbs radiant energy from the sun and transforms it into sugars has been one of the most challenging problems in biology. It is also, of course, the key to life on this planet. For a very long time, this key was a mystery, and, even today, we don't have quite all the answers. But we know that in a complex series of reactions light energy is converted to chemical energy and stored in the chemical bonds of sugars. In this process oxygen is released.

***Oxygen from Water.*** It had long been assumed that the oxygen released in photosynthesis came from carbon dioxide. In the 1930s, C. B. van Niel, a graduate student at Stanford University, made the bold suggestion that the oxygen produced in photosynthesis was derived from water, not carbon dioxide. His reason was based on a study of the photosynthetic "sulfur bacteria." These unusual microbes thrive in sunlit stagnant pools in which hydrogen sulfide ($H_2S$) is produced by decay of organic matter. Unlike most bacteria, they carry on photosynthesis. But instead of producing oxygen as a by-product, they release sulfur in a reaction that can be summarized as follows:

$$\underset{\substack{\text{hydrogen}\\\text{sulfide}}}{2H_2S} + CO_2 \xrightarrow{\text{light}} 2S + H_2O + \underset{\text{carbohydrate}}{(CH_2O)}$$

Van Niel pointed out the parallel between this type of photosynthesis and that which occurs in algae and plants. He suggested that it was most logical to assume that one type of photosynthesis had evolved from the other and was therefore only a slightly different version of it. If that were true, then the oxygen produced in photosynthesis came from water, not from carbon dioxide. For all photosynthesis, he proposed a general reaction in which hydrogen is provided by a hydrogen donor, designated *A*, in the following reaction:

$$\underset{\text{H donor}}{2H_2A} + \underset{\text{H acceptor}}{CO_2} \xrightarrow{\text{light}} \underset{\substack{\text{oxidized}\\\text{H donor}}}{2A} + \underset{\substack{\text{reduced}\\\text{H acceptor}}}{(CH_2O)} + H_2O$$

In sulfur bacteria, the hydrogen donor is $H_2S$; in common photosynthesis, van Niel suggested, it is $H_2O$, as in the following reaction:

$$2H_2O + CO_2 \xrightarrow{\text{light}} O_2 + (CH_2O) + H_2O$$

***Energy Transfer.*** An important question in the study of photosynthesis is how light energy is captured and incorporated into carbohydrate. One likely possibility for the initial interaction of light and matter in photosynthesis is that when chlorophyll absorbs light energy, the energy level of its electrons is raised and the excited electrons are given to one or more electron acceptors. In chemical terms, the chlorophyll is oxidized by, or loses electrons to, an electron acceptor, or oxidizing agent, which is reduced by accepting the electrons.

In the late 1930s, Nobel prize winner Robin Hill of Cambridge University, showed that suspensions of chloroplasts prepared by crushing leaves and filtering out the chloroplasts carried on photosynthesis in vitro. In the initial experiment, Hill found that the normal acceptors apparently had been lost in preparing the chloroplast suspension. It was necessary to add a suitable electron acceptor, such as a ferric ($Fe^{3+}$) salt, to the medium to obtain oxygen evolution in the light. The ferric atoms were reduced by accepting electrons and were thus changed to the ferrous ($Fe^{2+}$) state; oxygen was produced at the same time in what came to be known as the Hill reaction:

$$4Fe^{3+} + 2H_2O \xrightarrow[\text{light}]{\text{chloroplast}} 4Fe^{2+} + 4H + O_2$$

Hill's work was important for several reasons:

1. He used cell-free preparations of chloroplasts for the first time, which suggested that the chloroplast may contain the entire photosynthetic mechanism.
2. His work focused attention on electron transfer as an essential part of photosynthesis.
3. By providing an unlimited supply of electron acceptors in the form of ferric salts, he could sustain production of oxygen by the chloroplasts as long as light was present. It was notable that no carbohydrate was formed; there was no detectable assimilation of $CO_2$. The release of oxygen in photosynthesis was, it seemed, independent of the formation of carbohydrate.

When the heavy isotope of oxygen, $^{18}O$, became available, as a result of atomic research in the late 1930s, it was finally possible to demonstrate the source of the $O_2$ released in photosynthesis. All doubt that the oxygen came from water was removed by the experiments of Samuel Ruben and his coworkers at the University of California at Berkeley in 1941.

Ruben's work involved a preparation of water in which nearly all the oxygen was the heavy isotope $^{18}O$. Algae were suspended in the $H_2{}^{18}O$ and illuminated. Then carbon dioxide in which all the oxygen was the common form $^{16}O$ was introduced into the flask. After several minutes of photosynthesis, the oxygen that had evolved was collected and analyzed in a mass spectrometer, a device used to determine the proportions of isotopes in a mixture. The oxygen produced in the experiment proved to have the same proportion of $^{18}O$ as did the oxygen in the water that was used. The oxygen of carbon dioxide remained in the carbohydrate produced.

$$2H_2{}^{18}O + CO_2 \xrightarrow{\text{light}} {}^{18}O_2 + (CH_2O) + H_2O$$

A reciprocal experiment was then conducted in which the water used contained only $^{16}O$ and the $CO_2$ used contained mostly $^{18}O$. When the oxygen formed was analyzed, only a trace was found to be $^{18}O$; the $^{18}O$ was in the carbohydrate produced.

## LIGHT-DEPENDENT REACTIONS AND LIGHT-INDEPENDENT REACTIONS

Early experiments on the rate of photosynthesis had indicated that the process involved two sets of reactions. In the 1950s, Robert Emerson at the University of Illinois investigated the rate of photosynthesis in algal cells grown in intermittent as opposed to continuous light. He found that cells grown in very briefly flashing light could photosynthesize as fast and thus produce as much oxygen as cells grown in continuous light of the same intensity. When the cells were grown at low temperatures, the difference was even more marked: the intermittently illuminated cells used light 400 times more efficiently than the continuously illuminated cells. How are these results explained? Evidently photosynthesis involves two sets of reactions: a set of light-dependent, or photochemical, reactions ("light reactions") and a set of reactions that do not use light and can occur in darkness ("dark reactions"). The dark reactions are much slower than the light reactions, and limit the rate of photosynthesis as a whole; thus a cell can process in a few microseconds enough light to keep its dark reactions busy for much longer. The fact that lowering the temperature slowed the dark reactions dramatically but did not affect the light reactions suggests that only the dark reactions are enzymatic (because enzymatic but not photochemical reactions are strongly temperature dependent).

Different teams of investigators turned their attention to one or the other of the two sets of reactions. A boon to such biochemical studies was the ready availability of radioisotopes at the close of World War II. Now specific reactants, such as $CO_2$, could be radioactively labeled and more easily followed through various reactions. Radioisotopes were to be of crucial importance in determining the major pathway of carbon, from $CO_2$ to glucose, in the dark reactions.

## THE LIGHT REACTIONS

After Hill's work was published, scientists devoted much effort to discovering the normal electron acceptors in the chloroplast, the pathways along which the electrons were passed, and the end products of the light reactions. Biochemists were aided in this research by their knowledge of the electron acceptors and donors in cellular respiration (Chapter 6).

***The Role of NADP.*** In cellular oxidations, the coenzyme nicotinamide adenine dinucleotide (NAD) was known to be reduced to $NAD_{red}$ by accepting electrons and a hydrogen atom. In respiration, as we saw in Chapter 6, $NAD_{red}$ passes these electrons and hydrogen atoms to oxygen through a system of electron-transferring substances called the respiratory chain, with water formed as an end product. Adenosine triphosphate (ATP) is generated as a result. The small but energy-rich ATP molecule serves as a convenient energy source for a large number of cellular functions.

In studies of respiration, the coenzyme nicotinamide adenine dinucleotide phosphate (NADP) had also been found to serve as an electron acceptor and donor in certain important reactions. NADP had also been found in chloroplasts. The two coenzymes NAD and NADP act in a similar manner: they are converted from their oxidized forms ($NAD_{oxid}$ and

$NADP_{oxid}$) to their reduced forms ($NAD_{red}$ and $NADP_{red}$) by accepting an electron and one hydrogen from suitable donors; the second hydrogen atom from the donor forms a hydrogen ion, or proton ($H^+$).

In 1951, it was discovered that NADP was reduced during photosynthesis. Moreover, it was shown that $NADP_{oxid}$ readily substituted for the ferric salts as an electron acceptor in the Hill experiment; equivalent amounts of oxygen were produced for each mole of $NADP_{oxid}$ reduced to $NADP_{red}$. The Hill reaction could therefore be written:

$$2H_2O + 2NADP_{oxid} \xrightarrow[\text{light}]{\text{chlorophyll}} 2NADP_{red} + 2H^+ + O_2$$

It appeared that in chloroplasts, oxidations known to occur in all cells were made to "run backwards." Using the energy of sunlight, electrons and hydrogen atoms were removed from water and passed uphill, thermodynamically speaking, to reduce $NADP_{oxid}$ to $NADP_{red}$, thus producing a rich and mobile store of energy.

***The Formation of ATP.*** A second form of stored energy resulting from the light reactions was found to be none other than ATP. Using a suspension of chloroplasts from spinach leaves, Daniel Arnon and others at the University of California at Berkeley demonstrated in 1954 that:

$$ADP + P_i \xrightarrow[\text{light}]{\text{chloroplasts}} ATP$$

Arnon and other investigators were discovering the source of what Arnon later called the chloroplast's "assimilatory power" and has come to be called "reducing power": ATP and $NADP_{red}$.

In 1958, again using spinach chloroplasts, Arnon provided an elegant demonstration of the independence of the light and dark reactions. Chloroplasts isolated from spinach leaves were illuminated in the absence of $CO_2$, then broken up and separated into two fractions, a membrane fraction consisting of the grana and a remaining fraction, the stroma. The membranes were removed by differential centrifugation and discarded, leaving the stroma and anything soluble that might have entered the stroma from the grana during the period of illumination. Then, in the dark, $CO_2$ was added to the solution containing the stroma materials. Apparently using the energy that had been captured and stored earlier when the intact chloroplasts were illuminated, the stroma solution converted the $CO_2$ into sugar. The substances in the stroma that provided the energy for the dark reactions were found to be ATP and $NADP_{red}$. The role of the light reactions was thus established as the production of ATP and $NADP_{red}$. In the dark reactions, ATP and $NADP_{red}$ provide a source of energy, reducing power, for producing carbohydrate from $CO_2$.

***Plant Pigments and Light Absorption.*** One of the earliest findings in the investigation of photosynthesis was that not all wavelengths of light are equally effective in promoting the process. A plot of the visible part of the spectrum showing the relative effectiveness of the various wavelengths of light in promoting photosynthesis is called an **action spectrum** of photosynthesis.

Light energy represents only a small portion of the electromagnetic spectrum of radiation received on earth, the portion our eyes perceive (see Fig. 2-12). The energy at any given region of the spectrum is inversely related to the wavelength. Thus a unit, or photon, of blue light has more energy than a photon of light in the red part of the spectrum.

Light energy must be absorbed to be utilized in photosynthesis; light absorption is the role of chlorophyll and some other plant pigments.

***Plant Pigments.*** Chlorophyll is green in color because it reflects light of that wavelength, absorbing light in the blue and red regions of the spectrum. Chlorophyll occurs in three forms, chlorophylls *a*, *b*, and *c*. (Some bacteria have another form, bacteriochlorophyll.) Chlorophyll *a* is the most common: it occurs in all cyanobacteria, algae, and plants. Chlorophyll *b* is found in green algae and all other plants. Chlorophyll *c* occurs only in certain algae. The various chlorophylls all have the same basic molecular structure: a flat ring (porphyrin ring) of four nitrogen-containing pyrroles with a magnesium atom at the center and a long carbon side chain called phytol. The three forms differ from each other only in the structure of another small side chain, two of which are shown in Fig. 7-2a.

In addition to chlorophyll, all photosynthetic organisms have pigments of other colors that absorb sunlight in different parts of the spectrum and thereby aid in photosynthesis. Carotenoids (Fig. 7-2b, c) are yellow to red in color and consist of molecules that form long chains. They include the carotenes, present in all photosynthetic cells, and the closely similar carotenols (also called xanthophylls) that contain an additional oxygen atom and occur in assorted forms in various algae and plants.

Phycobilins (Fig. 7-2d), still another type of photosynthetic pigment, are similar to bile pigments in structure and are found only in cyanobacteria and red algae. They are blue to red in color, and thus absorb in other parts of the spectrum.

A separate class of pigments, the anthocyanins, occurs in higher plants. They are responsible for many flower colors and for the reddish stems and leaves of some species, but they are not chloroplast pigments and do not participate in photosynthesis. (Pigments are also discussed in Chapter 28.)

Each of the pigments found in photosynthetic organisms absorbs energy in a specific part of the spectrum, termed its absorption spectrum (Fig. 7-3 and Panel 7-1). For chlorophyll, the regions of maximum absorption are in a band in the blue region and a band in the red. The action spectrum for chlorophyll in photosynthesis matches its absorption spectrum. This indicates that chlorophyll is the photosynthetic pigment (Fig. 7-3).

The action spectrum of photosynthesis was demonstrated in experiments of Theodor Engelmann similar to those described earlier. Engelmann had a special appliance built for his microscope that enabled him to shine light upon a specimen after the beam had been dispersed by a prism into a spectrum. Specimens placed under the microscope thus were illuminated from one end of the microscopic field to the other by red, orange, yellow, green, blue, or violet light. Under the microscope he placed filaments of algae and added *Pseudomonas* bacteria, which become motile in the presence of oxygen (Fig. 7-4). After a period of darkness, during which any oxygen in the water was depleted and the bacteria stopped moving, the microscope light was turned on. The bacteria adjacent to the alga began to move wherever oxygen was being evolved. Most gathered in the regions exposed to red and blue wavelengths.

The absorption spectra of chlorophyll molecules that have been ex-

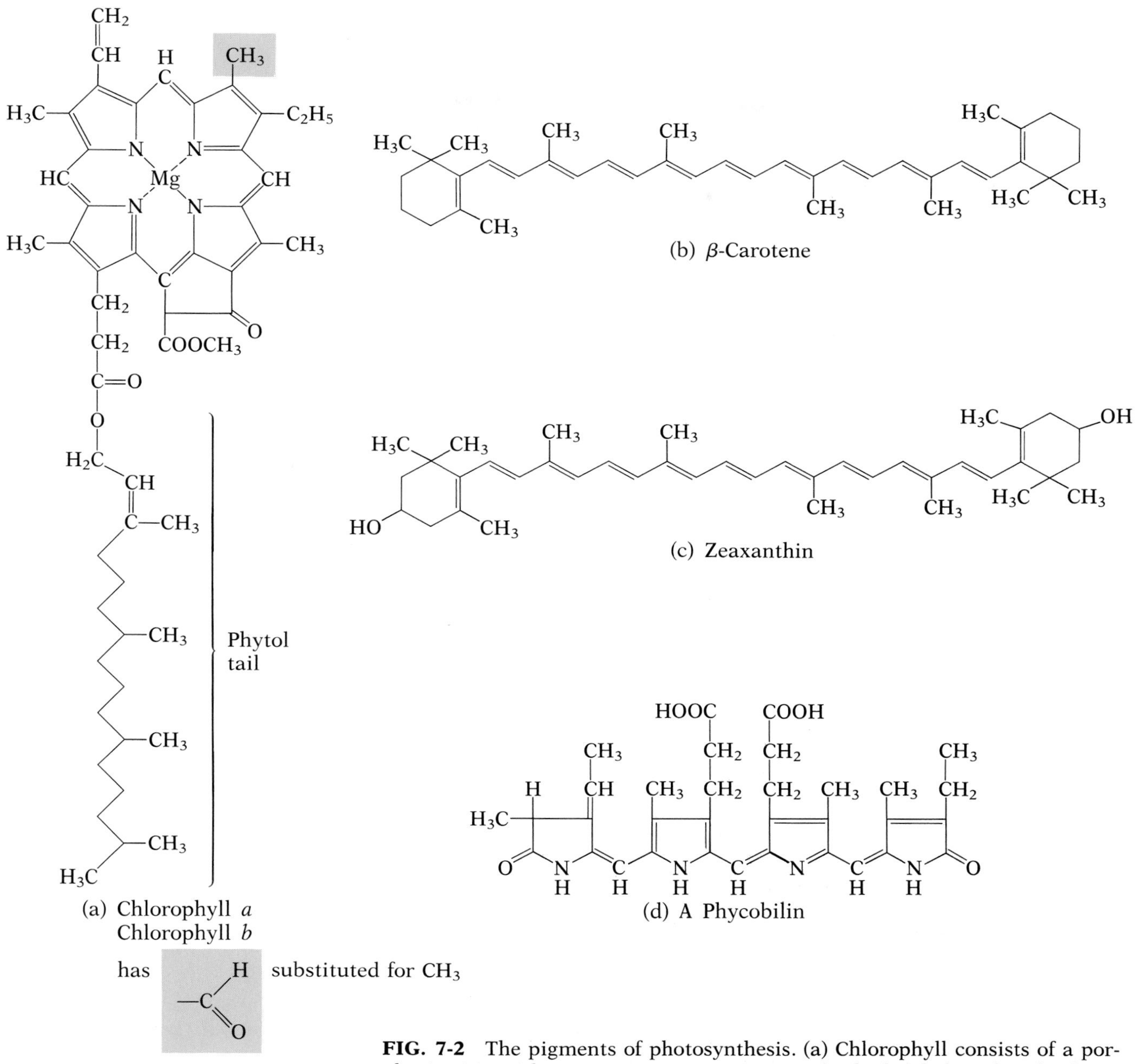


**FIG. 7-2** The pigments of photosynthesis. (a) Chlorophyll consists of a porphyrin ring with magnesium at its center and a hydrophobic phytol side chain. Chlorophyll *a* and *b* differ in a side chain on the porphyrin ring, as indicated. (b) A number of carotenes are found among plants, all of which are isoprenoids, with the basic structure of beta carotene. (c) Xanthophylls are carotenoids with —OH groups, as shown for zeaxanthin, and occur in great variety. (d) The phycobilins, with the basic structure shown, are found in the cyanobacteria and red algae as accessory pigments.

tracted from leaves or algal cells and put in solution do not correspond exactly with the action spectra for photosynthesis of intact cells. Specifically, there is photosynthetic activity in portions of the spectrum where chlorophyll *a* and *b* do not absorb. This activity is due to accessory photosynthetic pigments, such as the phycobilins of cyanobacteria and red algae and the carotenoids (Fig. 7-3). These pigments absorb energy of wavelengths not absorbed by chlorophyll; then they transfer

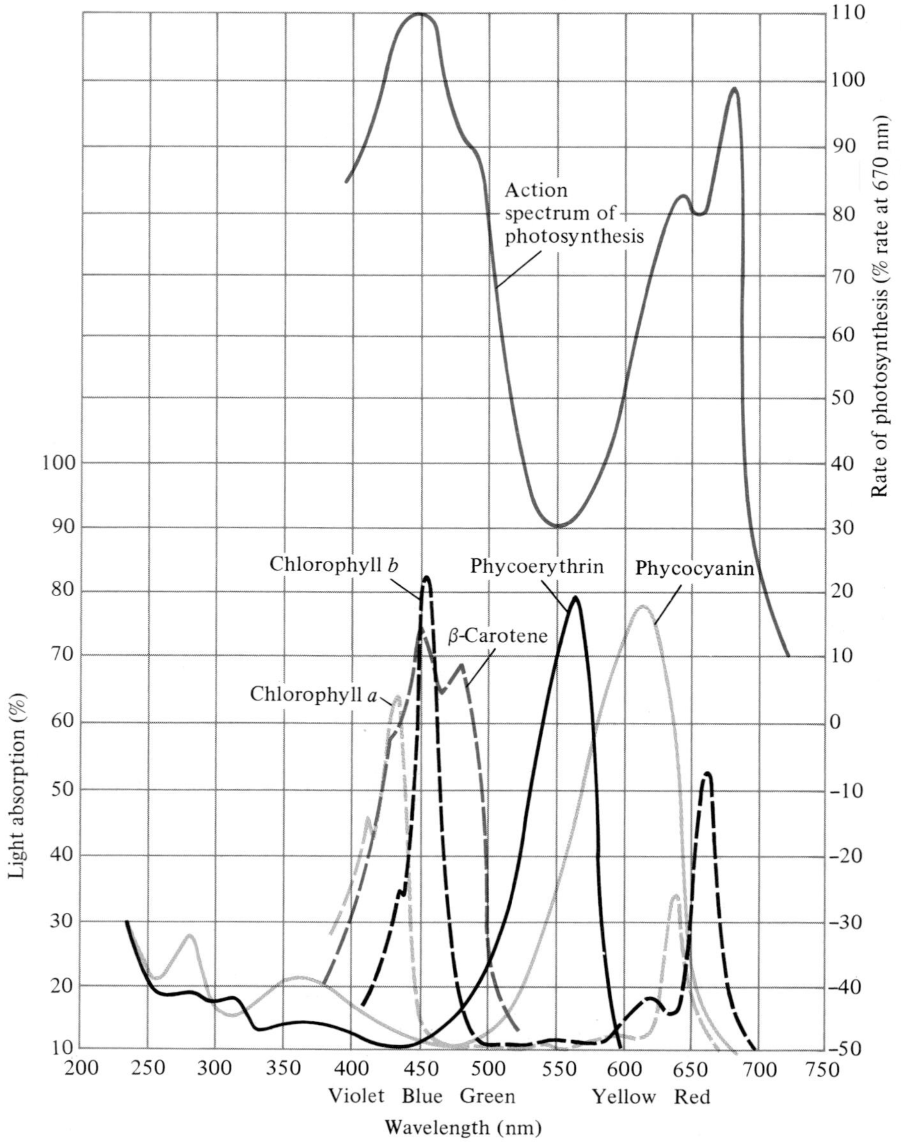


**FIG. 7-3** Absorption spectra of solutions of the three principal classes of pigments of higher plants and those of cyanobacteria, contrasted with the action spectrum for intact leaves. Above 550 nm, the chlorophylls alone in leaves seem to be involved in light absorption. Phycocyanins (solid line) and phycoerythrin (dotted line), phycobilin pigments of the cyanobacteria, can utilize additional regions of the spectrum.

the energy to chlorophyll, extending the regions of the spectrum utilized in photosynthesis.

***Energy Capture.*** An assortment of pigments with different absorption spectra is found in the cell in bacteria and in the thylakoid membranes of the chloroplasts of plants and algae. When combined with different proteins in the thylakoids, even the same pigments have different absorption spectra. As we have noted, the presence of various pigments in an organism enables it to absorb light of a greater variety of wavelengths than if it had only one or two pigments. Submerged algae with accessory pigments may thus absorb wavelengths not filtered out by water, which gives them an advantage over algae lacking such pigments. The advantages of having several pigments are illustrated in Figure 7-3, in which the absorption spectra of accessory pigments and chlorophyll *a* and *b* are compared with the action spectrum of a leaf.

Robert Emerson's investigations led to an understanding of energy conversion in photosynthesis. He noted that not all the chlorophyll

## PANEL 7-1

# Absorption Spectra and Photosynthesis

All atoms, and therefore all molecules, are capable of absorbing and emitting energy. When a substance absorbs electromagnetic radiation in the ultraviolet and visible regions of the spectrum, its electrons go from lower energy orbitals to vacant higher energy orbitals. The electrons, the atom, and the molecule of which they are a part are said to have gone from their electronic ground state to an excited state.

According to Einstein's particle theory of light, light is made up of particles of energy called photons. The amount, or quanta (sing. quantum), of energy in a photon is a function of the wavelength of the radiation, not of its intensity: the shorter the wavelength (or greater the frequency), the greater the energy of the photon. For example, a photon in the violet region of the spectrum carries nearly twice the energy of one in the far red portion. The greater the intensity of the radiation, the greater the number of photons in the beam and the more electrons and atoms it can excite. Absorption of wavelengths of 200 nm or less, the shortest of the ultraviolet rays, produces ionization; that is, the electrons of atoms absorbing energy from these wavelengths are driven completely out of their atoms or molecules. Radiation of such short wavelength is called *ionizing radiation.* Absorption of such high-energy radiation—X rays, for example—is destructive to living systems because it breaks the chemical bonds of molecules important to life, including DNA.

Most of the radiation absorbed by living systems on our planet is in the ultraviolet and visible region of the spectrum. The long wavelength, low-energy, infrared rays of the sun are mostly absorbed by the water and carbon dioxide of the atmosphere. They are also readily dissipated as heat when they enter organisms, which are almost all water. Most of the dangerous, high-energy radiation of short wavelength arriving from the sun and other stars is absorbed by atmospheric oxygen and by a form of oxygen called ozone in the upper atmosphere.

Whenever an electron absorbs energy and moves to a higher energy orbital, it is in an unstable condition. Either it drops back to its ground state immediately, in the process emitting the energy it had absorbed as heat or light, or another electron in the higher energy level drops down to take its place, emitting energy. The wavelength of the emitted radiant energy corresponds exactly to the energy difference between the ground state orbital and the higher energy orbital to which the electron had moved. Such emissions are therefore uniquely characteristic of each kind of atom or molecule. The emission spectrum is as unique to each substance as are fingerprints to a human.

Just as atoms and molecules differ in their emission spectra, they differ in the amount of energy they are capable of absorbing, that is, in their absorption spectra. Each electron of a particular atom is capable of moving to various higher energy orbitals, but only if it absorbs the precise amount of energy needed to move to one of the new positions. A given electron must therefore absorb not only the minimum amount of energy needed to boost it to a particular higher energy orbital but *exactly* that amount. Any other wavelength is ineffective. Other electrons in the same atom respond to other wavelengths that meet their particular requirements. As a result, a given type of atom will absorb only photons of certain wavelengths, containing specific quanta, or amounts, of energy. Therefore, each element and each kind of molecule has a unique absorption spectrum.

molecules in a chloroplast are involved in diverting light energy into a photosynthetic pathway. Actually, there are about 300 times more chlorophyll molecules present in chloroplasts than are required in photosynthesis. Pigment molecules exist in large units, and only one of every 300 is capable of transferring excited electrons to the primary electron acceptor, the initial step in energy conversion. This special chlorophyll molecule, which acts as the intermediary between the pigments and the acceptor, is known as the **reaction center** (Fig. 7-5).

All the pigment molecules in each photosynthetic unit, including the other chlorophyll *a* molecules and the accessory pigments, the carotenoids, phycobilins, and other chlorophylls are called antenna pigments because they feed energy into the reaction center, much as a radio an-

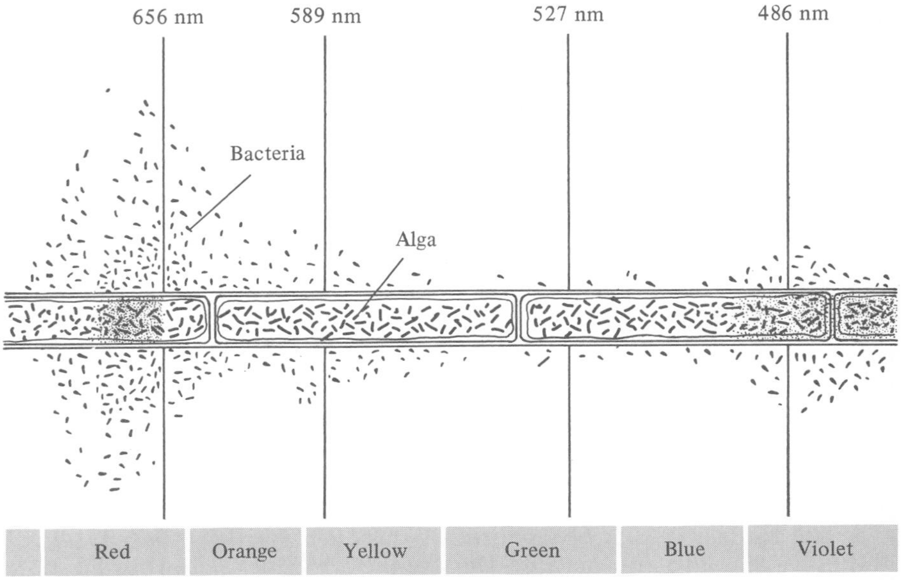


**FIG. 7-4** Engelmann's action spectrum for photosynthesis in the alga *Cladophora*. The motile, aerobic bacteria *Pseudomonas* are shown to have gathered in greatest numbers in those regions of the spectrum where oxygen is being produced in the greatest amounts, that is, where photosynthesis is proceeding most rapidly.

tenna absorbs radio waves and feeds them into the receiver.* Two kinds of reaction centers are found in the thylakoid membranes of the chloroplast. Because the chlorophyll pigments of the reaction centers have different absorption maxima, one at 700 nm and the other at 680 nm, they are referred to as P700 and P680.

The various pigments funnel the energy they absorb from light or from other pigments to the reaction center. A pigment can accept energy from another pigment only if the energy being received is of the same or a shorter (more energetic) wavelength than the absorption maximum of the receiving pigment. The reaction center pigment has an absorption maximum at the longest wavelength of any pigment in the unit. Thus it can absorb energy from every one of them but will not lose energy to any of them. As a result, all energy absorbed by any of the pigments in a photosynthetic unit is inevitably funneled down to the reaction center, no matter which pigment had originally absorbed the energy as light.

* While the carotenoids contribute energy to chlorophyll in the algae, in plants they appear to play largely a protective role in leaf chloroplasts, preventing chlorophyll destruction in bright light.

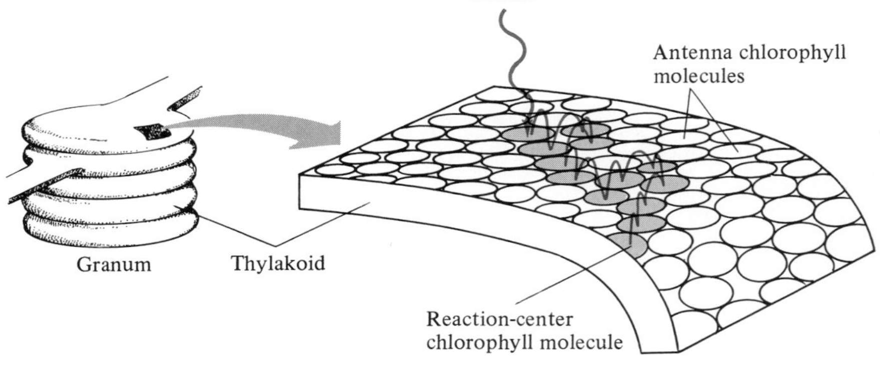


**FIG. 7-5** Pigment systems contain an assortment of pigments. Out of every 300 pigment molecules, only one, the reaction center, transfers energy in the form of an excited electron to the primary electron acceptor. The other pigments absorb light of various wavelengths and transfer the energy to the reaction center.

## PANEL 7-2

# Evidence for Two Photosystems

The work of Robert Emerson that confirmed the nature of the "light reactions" also revealed that these reactions involve the participation of two separate photosystems. Emerson compared the absorption spectrum of algal cells with the action spectrum obtained by successively irradiating the cells with light of different wavelengths within the absorption spectrum of chlorophyll. The cells photosynthesized at a moderate rate for tested wavelengths up to 680 nm. However, when cells were exposed only to red light of wavelength greater than 680 nm, there was an abrupt drop in photosynthesis (even though chlorophyll absorbs light up to 700 nm). When the red light over 680 nm was supplemented by a second beam of less than 680 nm, for example 650 nm, the wavelength efficiency increased dramatically. In fact, the rate of photosynthesis was much greater in the presence of the combined light than the sum of the rates when the wavelengths were used separately. This showed that the cells could use light of greater than 680 nm in wavelength if light of a shorter wavelength was also supplied.

Emerson then illuminated algae, first with light above 680 nm, then with light of 650 nm. For about a second after the switch, there was a burst of oxygen production similar to that occurring when lights of both wavelengths were turned on at the same time. Emerson then performed the experiment with cells cooled to the temperature of liquid nitrogen (−195°C), conditions that arrest chemical reactions but not photochemical events. The result was dramatic: by stopping chemical events from occurring, the overall process was stopped. This demonstrated the existence of two separate pigment systems in the chloroplast, each absorbing light at different wavelengths, one system supplying products to the other by one or more chemical steps.

***Energy Conversion.*** Two sets of photochemical reactions, or **photosystems,** are associated with the reaction centers (Panel 7-2). Photosystem I (so designated because it was discovered first) contains P700, has a higher proportion of chlorophyll *a* than of chlorophyll *b*, and is sensitive to longer wavelengths of light. Photosystem II contains P680, has a greater proportion of chlorophyll *b* than of chlorophyll *a*, and is sensitive to shorter wavelengths.

As we have noted, the "light reactions" produce the $NADP_{red}$ and ATP that drive the "dark reactions." The reduction of $NADP_{oxid}$ to $NADP_{red}$ is accomplished by using the light energy absorbed by chlorophyll to drive the loosely held electrons of NADPs from a high (positive) oxidoreductive potential to a low (negative) one. This is called quantum conversion. Light energy is thereby converted to chemical energy and reducing power is generated, which is used in forming carbohydrate. If it were possible, it would of course be economical for the cell to generate $NADP_{red}$ directly from light. Instead, a series of reactions is required to transfer the energy of the excited chlorophyll to NADP. The energy transfer is accomplished in the two photosystems situated in the thylakoid membranes, the pigment-bearing membranes of the chloroplasts. When a photon elevates an electron of a chlorophyll molecule to a high energy state, the electron is transferred through a series of other chlorophyll molecules to the reaction center, the special chlorophyll molecule of the "antenna" (Fig. 7-5). The electron is then transferred to an acceptor molecule and passed through a series of steps to $NADP_{oxid}$ in two series of reactions taking place in photosystems I and II. This series of events is sketched in Figure 7-6 and occurs as follows.

First, light is absorbed by photosystem II (PS II). As a result of the absorption of photons, electrons are driven out of the chlorophyll molecule, each leaving a vacancy, or "electron hole." The excited electrons

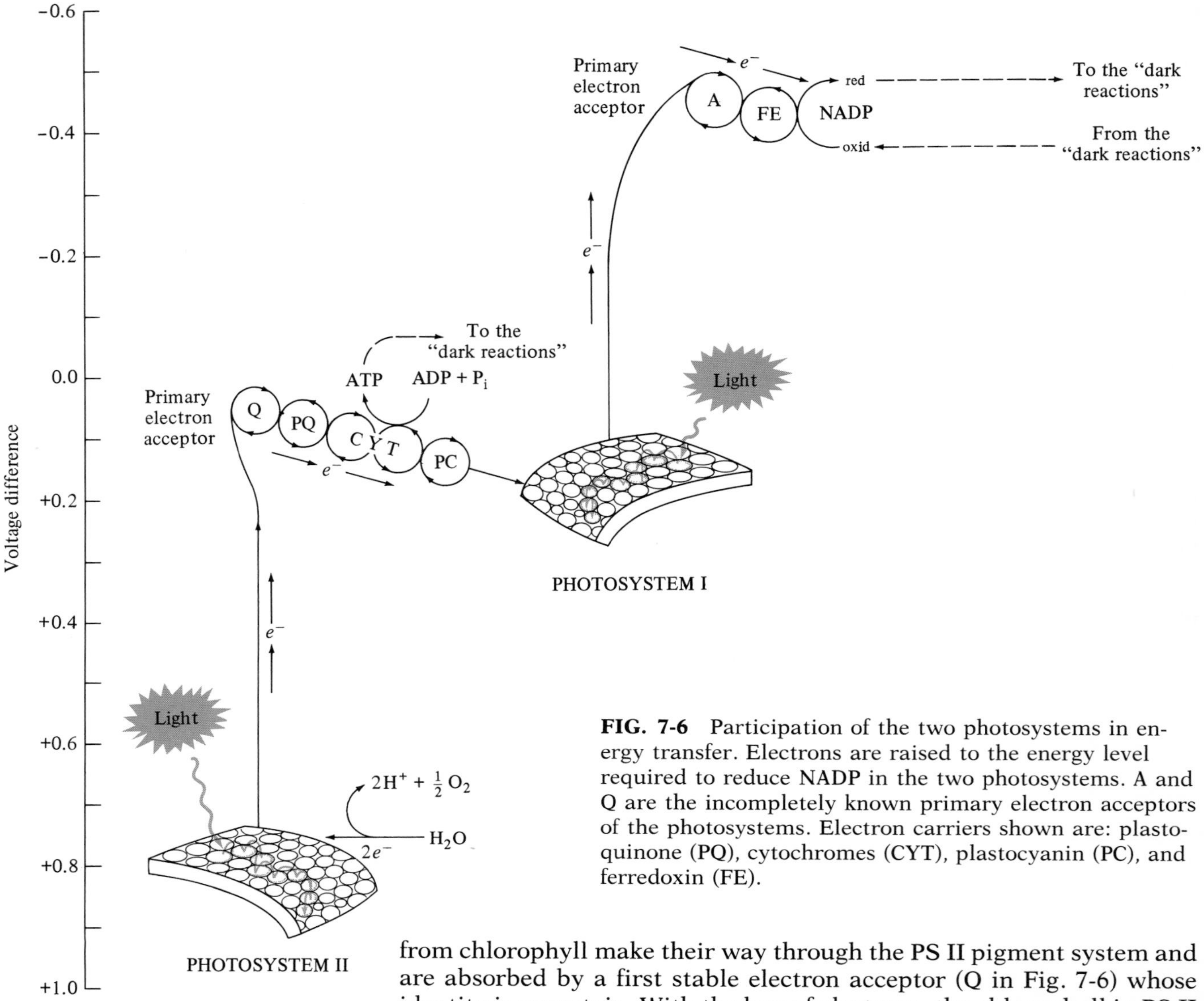


**FIG. 7-6** Participation of the two photosystems in energy transfer. Electrons are raised to the energy level required to reduce NADP in the two photosystems. A and Q are the incompletely known primary electron acceptors of the photosystems. Electron carriers shown are: plastoquinone (PQ), cytochromes (CYT), plastocyanin (PC), and ferredoxin (FE).

from chlorophyll make their way through the PS II pigment system and are absorbed by a first stable electron acceptor (Q in Fig. 7-6) whose identity is uncertain. With the loss of electrons, the chlorophyll in PS II becomes an oxidizing agent powerful enough to remove electrons from water, releasing $O_2$. These electrons fill the "electron holes" in chlorophyll created by the loss of electrons to the acceptor (Q). Acceptor Q is converted to a weak reducing agent (donor of electrons), but it is still strong enough to donate electrons to a special series of electron acceptors and carriers, including quinones, cytochromes, and a copper protein called plastocyanin. The molecule transfers electrons to PS I.

Meanwhile, a photon of light has also been absorbed by PS I, causing it to lose an electron to another stable electron acceptor, one or more iron-sulfur proteins (A in Fig. 7-6). When PS I loses an electron, it becomes a weak oxidizing agent; this allows it to accept an electron from the plastocyanin. By accepting an electron, its primary electron acceptor becomes a strong reducing agent. It is thereby able to donate electrons to NADP, which has a more positive oxidoreductive potential than does the reduced acceptor. This donation is not accomplished directly, but via a pathway that includes an iron-containing molecule, ferredoxin, and a flavoprotein.

As mentioned above, the electron hole left in PS I by the reduction of NADP is filled by electrons from PS II. Thus, the electrons split off from

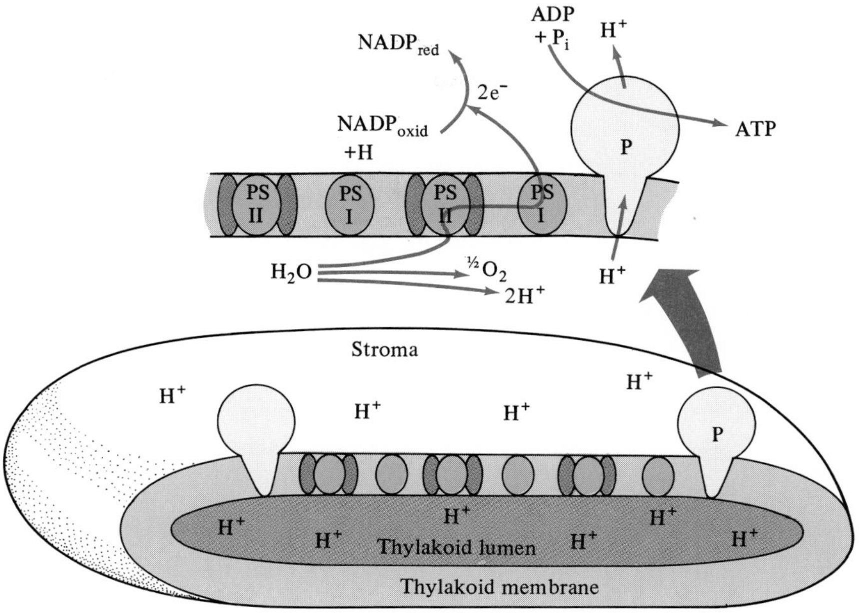


**FIG. 7-7** During the light reactions of photosynthesis, the thylakoid sacs of the chloroplast grana separate charges and thus create a proton gradient across the thylakoid membranes with the resulting generation of potential energy. At the outer surface of the membrane, NADP$_{oxid}$ is reduced to NADP$_{red}$, with electrons transported through the electron transport chains to the surface of the thylakoid membrane. With dissociation of water into protons ($H^+$), oxygen and electrons in the thylakoid, protons are pumped across the membrane and accumulate in the sac, generating the proton gradient. Large protein molecules (P) in the thylakoid membrane make possible the outflow of protons and the synthesis of ATP from ADP and $P_i$ (photophosphorylation).

the hydrogen atoms of water are transferred to NADP in two stages: they are removed from water and raised in energy by PS II, and then given a second boost of energy by PS I so that they can reduce NADP. As we will see below, these electrons will be used to reduce $CO_2$ in the process of turning it into sugar.

***Photophosphorylation.*** The electron transport pathway between acceptor Q of PS II and PS I is of special interest. The passage of electrons along the electron transport chain, that is, the successive reduction and oxidation of the electron-transferring molecules, is coupled to the generation of ATP by the phosphorylation of adenosine diphosphate (ADP). In contrast to oxidative phosphorylation, this process depends on light; hence it is termed **photophosphorylation.** Light energy is converted to chemical energy in the formation of ATP. In the process, electrons, driven by light energy along the electron transport chains in the chloroplast membranes, cause $H^+$ to be transported across the membrane to the interior of the thylakoid (Fig. 7-7). The resultant pH gradient across the membrane provides the energy for converting ADP to ATP, just as in

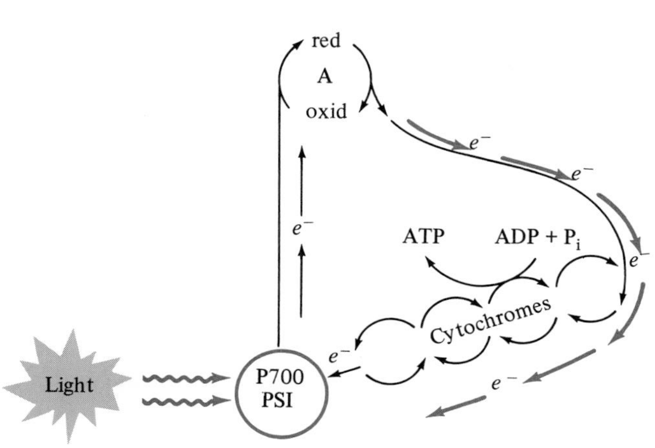


**FIG. 7-8** Cyclic photophosphorylation. By following a short circuit, electrons excited by light can generate ATP even when no NADP$_{oxid}$ is available to accept them. The cytochrome pathway is used as in ordinary photophosphorylation. The cycle is valuable because ATP has many other uses besides driving the dark reactions.

the chemiosmotic process that occurs in mitochondrial phosphorylation (Chapter 6). As the hydrogen ions diffuse outward as a result of the gradient, they move along certain protein channels in the membrane and activate the formation of ATP through the enzymatic activity of ATP synthetase.

***Cyclic Photophosphorylation.*** Electrons passed from PS I to acceptor A and then to ferredoxin reach a branch point; the main route leads to $NADP_{oxid}$, as we have seen. But, as is shown in Figure 7-8, a side route forms a short circuit that leads back to the cytochrome system. Via this pathway, ATP is generated as described above, but without the reduction of NADP. Thus, the chloroplast can supplement the production of ATP associated with photosystem II as long as light is absorbed by photosystem I and as long as ADP and $P_i$ are available. This cyclic generation of ATP by light energy is called **cyclic photophosphorylation.**

## THE PATHWAY OF CARBON: LIGHT-INDEPENDENT REACTIONS

The phases of photosynthesis related to the processing or "fixing" of $CO_2$ into carbohydrates do not require light. These dark reactions were elucidated primarily by the Nobel-prize-winning work of Melvin Calvin and his coworkers at the University of California at Berkeley in the 1950s. Their objective was to identify the first-formed products of photosynthesis and to trace the pathway of carbohydrate synthesis. The approach was to introduce $^{14}C$, the radioactive isotope of carbon, into cells by bubbling $^{14}C$-labeled $CO_2$ ($^{14}CO_2$) through cultures of the one-celled alga *Chlorella*. Extracts of the cells were then analyzed by paper chromatography at intervals during the experiment (Fig. 7-9). Paper chromatography involves separating the components of a mixture on a sheet of paper and staining to reveal their identity, as shown in Fig. 7-9. Each spot on the paper was identified by removing, or eluting, it from the paper and analyzing it. A duplicate chromatogram was placed in contact with X-ray film (in the dark). Any of the spots that were radioactive because they had incorporated some of the $^{14}C$-labeled $CO_2$ would expose the film at a corresponding place, producing an autoradiograph.

Calvin found that after 30 seconds of photosynthesis almost every organic compound on the chromatogram was radioactive (Fig. 7-10a). Besides amino acids, these included a variety of carbohydrates, most of which were familiar as intermediates in glycolysis, the anaerobic degradation of glucose. But one sugar found was not a glycolytic intermediate: the pentose (five-carbon sugar) ribulose 1,5-bisphosphate (RuBP).

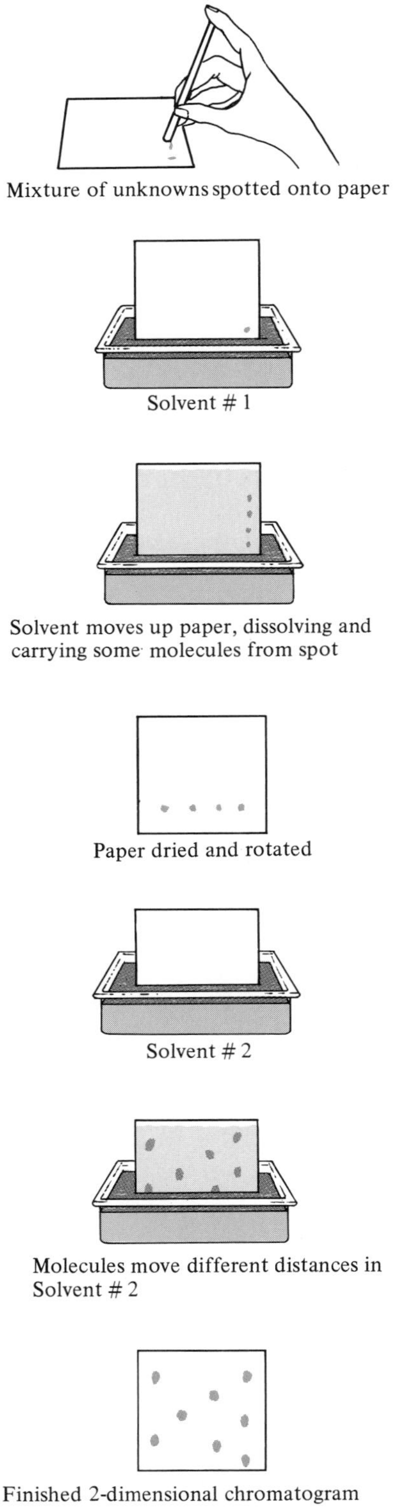

Mixture of unknowns spotted onto paper
Solvent # 1
Solvent moves up paper, dissolving and carrying some molecules from spot
Paper dried and rotated
Solvent # 2
Molecules move different distances in Solvent # 2
Finished 2-dimensional chromatogram


**FIG. 7-9** The identification of the first products of photosynthesis. In the presence of $CO_2$ labeled with the radioisotope $^{14}C$, algal cells were allowed to carry on photosynthesis for brief periods. Aqueous extracts of the cells were applied to a corner of a filter paper, the edge of which was then immersed in a solvent mixture. Components of the extract are carried upward and spread out according to their size and differential solubility. The paper is then removed, dried, and rotated 90°. After being rotated, it is put in another solvent mixture to further separate the components of the extract. When a photographic film is applied to the dried paper in the dark for several hours, or days, and then developed, products containing $^{14}C$ are localized on the resulting autoradiograph.

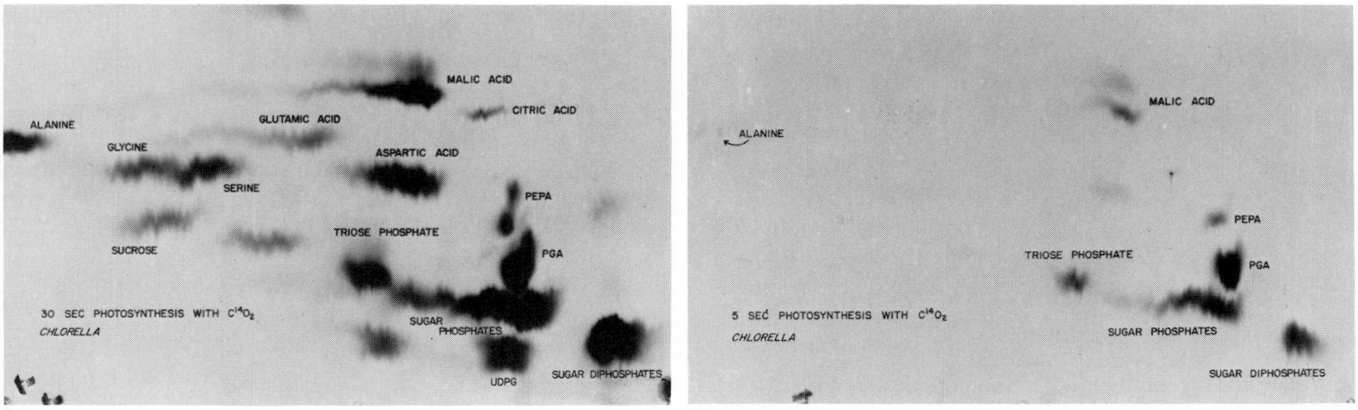

(a) (b)

**FIG. 7-10** (a) An autoradiograph of a chromatogram similar to that in Fig. 7-9, except that the exposure time to $^{14}CO_2$ was 30 seconds. In addition to sugars, several amino acids, a Krebs cycle intermediate, and a nucleotide now contain labeled carbon. The nucleotide UDP-glucose is an intermediate in the formation of cellulose.
(b) Autoradiograph of a chromatogram prepared from organic compounds extracted from algae *(Chlorella)* after they were exposed to $^{14}CO_2$ in the light for 5 seconds. Radioactivity is found in several molecules, including PGA, sugar phosphates, and amino acids.

Shortening the time during which photosynthesis was allowed to proceed in the presence of $^{14}CO_2$ to less than 5 seconds showed that the labeled carbon appeared only in the glycolytic intermediate 3-phosphoglyceric acid (PGA) (Fig. 7-10b). PGA therefore appeared to be the first product of photosynthesis to be formed from $CO_2$. When the experiment was allowed to run for a longer time, the label appeared in other glycolytic intermediates: first phosphoglyceraldehyde (PGAL) and then hexose phosphates, such as glucose phosphate.

In analyzing the slightly radioactive PGA produced in the first 3 to 5 seconds of photosynthesis, it was found that only the molecule's carboxylic group (—COOH) contained the radioactive carbon atom, suggesting a reaction such as this:

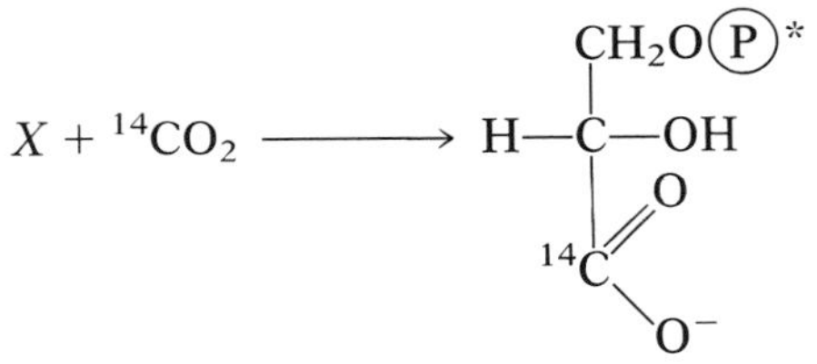


This observation suggested that $^{14}CO_2$ had joined with some phosphorylated two-carbon compound to form PGA as the first step of the "dark reaction" sequence. The investigators searched in vain for such a two-carbon molecule in the algal cells. Finally they came to the conclusion that the $CO_2$ was probably being added to a larger (five-carbon) molecule that was rapidly split into two (three-carbon) molecules of PGA. This view was confirmed by analyses of the products of successively longer light exposure of the cells to $^{14}CO_2$. In succeeding seconds, as the radioactivity became distributed to more compounds, the number of $^{14}C$ atoms found in PGA increased until all three carbons of PGA produced were radioactive. Significantly, the five-carbon compound ribulose bisphosphate also had first one and then two $^{14}C$ atoms per molecule and so on until all five of its carbons were radioactive.

It is the five-carbon RuBP that accepts the $^{14}CO_2$, forming an unstable six-carbon compound. This molecule is immediately hydrolyzed into two PGA molecules, one of which contains the newly incorporated $CO_2$ molecule, as shown in the following equation:

* The symbol Ⓟ is used to designate a phosphate group, $—PO_4$.

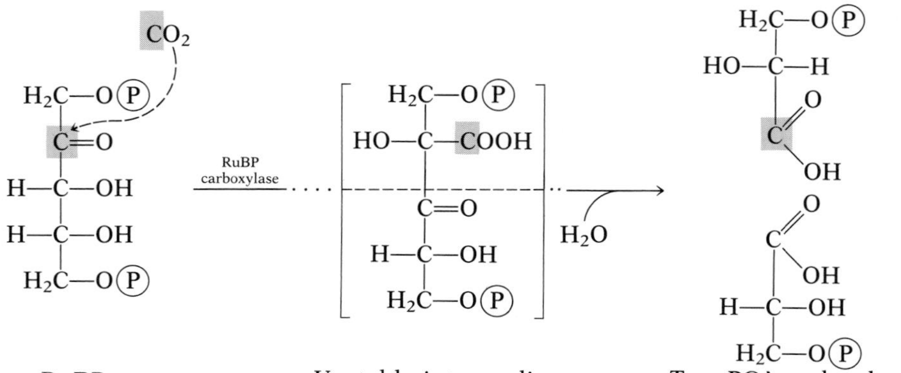


RuBP    Unstable intermediate    Two PGA molecules

The enzyme that catalyzes the addition of $CO_2$ to RuBP is ribulose biphosphate carboxylase, one of the most important enzymes in plants and one of the most significant to life. It is also the most abundant enzyme in nature.

These results also showed that RuBP is continually regenerated. Although some PGA is converted to glucose, five out of six PGA molecules are used to regenerate RuBP molecules. This became clear in a series of experiments in the 1950s in which Calvin, Benson, Bassham, and others elucidated the rest of the cycle. The Calvin cycle, also known as the Calvin-Benson cycle or $C_3$ pathway, is summarized in Figure 7-11. Two major pathways are open to PGAL. A hexose (six-carbon sugar) is produced from two (three-carbon) PGAL molecules in one pathway. In the other pathway, the cyclic part of the process, six five-carbon RuBP molecules are formed from every 10 molecules of PGAL. Note that for every six $CO_2$ molecules absorbed, 12 PGA molecules are produced, and from every 12 PGA molecules, one hexose molecule and six RuBP molecules are synthesized. The hexose formed, hexose biphosphate, may be converted to glucose phosphate, several molecules of which may in turn be polymerized to starch. Triose phosphate is exported from the chloroplast and in the cytoplasm is converted to sucrose, the form in which most sugar is transported within the plant, and to cell wall polysaccharides. Three steps in the Calvin cycle are of particular importance

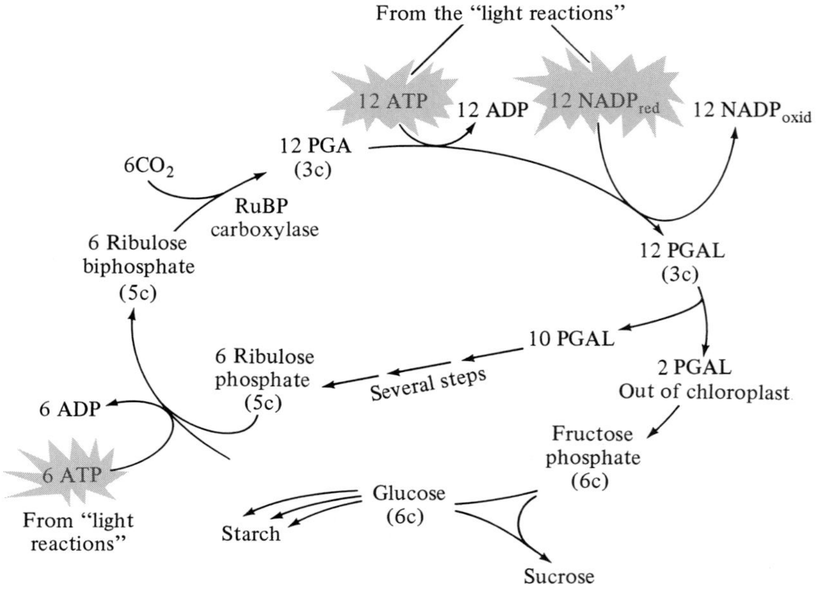


**FIG. 7-11** Simplified diagram of the Calvin cycle showing where ATP and $NADP_{red}$ furnish energy to the cycle.

because they require an input of energy to proceed. The energy is provided by the splitting of ATP and the oxidation of $NADP_{red}$, the energy-rich substances generated in the light reactions.

## CHLOROPLAST STRUCTURE AND FUNCTION

Except in photosynthetic bacteria, where the thylakoids are dispersed in the cells, photosynthesis takes place within chloroplasts, which vary in form from the spiraled ribbons of some algae to the football-shaped plastids of the flowering plants. Electron micrographs of chloroplasts show that they are bounded by an envelope, or double membrane, and consist of a system of chlorophyll-bearing membranes, or lamellae, distributed in the **stroma** (Fig. 7-12). At intervals in the chloroplast, the lamellae are tightly appressed into stacks of discs called **grana** (Fig. 7-12a, b). Each disc is a flattened, membrane-enclosed sac called a **thylakoid.** The grana are interconnected by thylakoids in the stroma, called intergrana lamellae.

The grana can be separated from the stroma by fractionating chloroplasts, allowing their separate functions to be studied, as noted earlier. The grana carry out the "light reactions." The stroma fraction incorporates $CO_2$ into carbohydrate if provided with energy in the form of ATP and $NADPH_{red}$.

The structure of the chloroplast is important in maintaining the sequence of reactions in photosynthesis. Electrons in the electron trans-

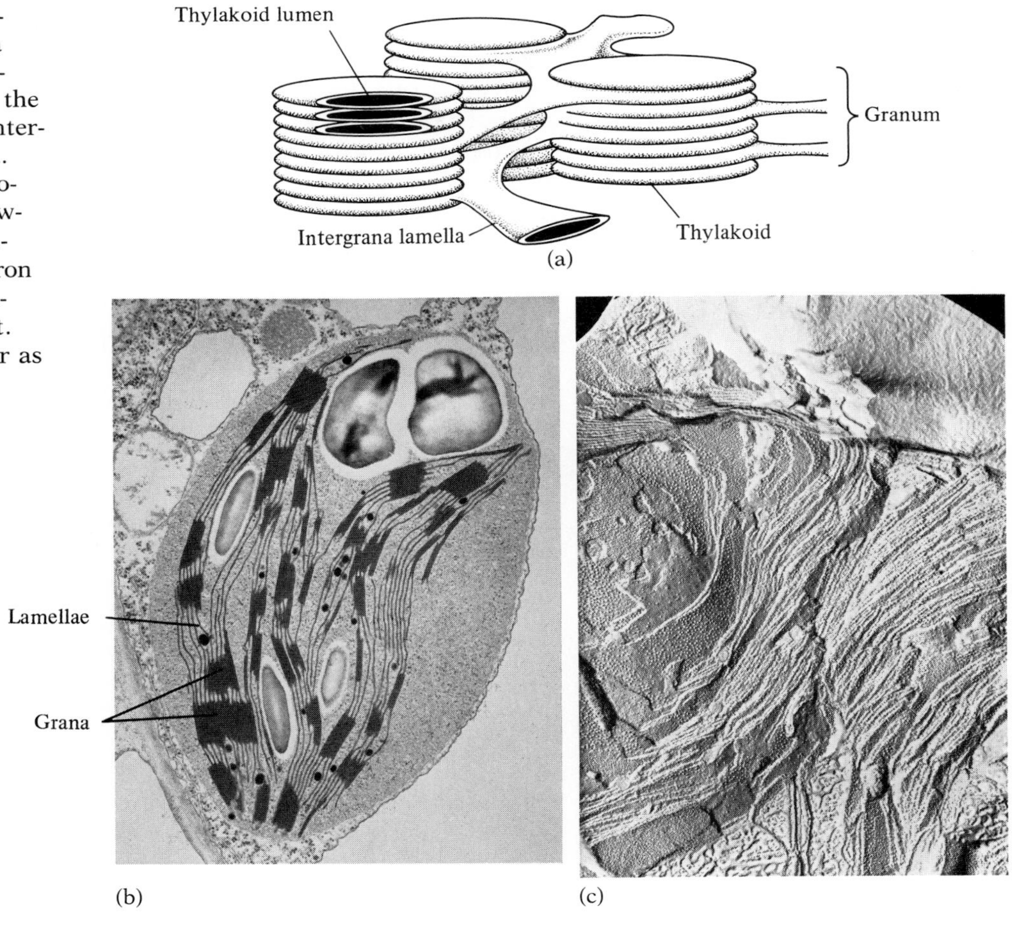


**FIG. 7-12** Chloroplast structure. (a) Diagram of part of a chloroplast, showing arrangement of stacks of thylakoids, the grana, and interconnecting intergrana lamellae in the stroma. (b) A chloroplast from a mesophyll cell of a dicot leaf, showing numerous grana interconnected by lamellae. (c) Electron micrograph showing a freeze-fractured spinach chloroplast. Thylakoid membranes appear as a series of lines.

port pathway are thought to flow across the thylakoid membrane from within the thylakoid to the stroma, where $NADP_{oxid}$ is reduced to $NADP_{red}$ as discussed earlier. Hydrogen ions diffusing outward in the chemiosmotic process across the thylakoid membranes generate ATP.

One of the techniques used in the study of chloroplasts is the freeze-fracture technique. It involves rapidly freezing cells at the temperature of liquid nitrogen, then fracturing them under vacuum in a special apparatus. Often the fracture splits a membrane, revealing its internal faces. Replicas of these surfaces are made by depositing a film of platinum-carbon over the fractured cells in the apparatus. The replicas are then viewed by electron microscopy. Electron micrographs of chloroplast thylakoid membranes prepared by the freeze-fracture technique show particles in the membranes (Fig. 7-12c). Although the function of the various particles has not been established, they are being studied for possible correlation with the light-capturing and energy-transferring processes. It has been suggested that PS I may be associated with one size of particle and PS II with another.

# THE LEAF AS A PHOTOSYNTHETIC ORGAN

## STRUCTURE OF THE LEAF

Being flat, the leaves of most plants present a maximum surface for the reception of sunlight. Within the leaf, the photosynthetic tissues are arranged in such a way that absorption of sunlight is maximized. Leaves contain a number of specialized cells. Epidermal cells are commonly free of chloroplasts, except for the paired **guard cells** that border each of the **stomata,** the many openings into the leaf, which are often concentrated on the underside in plants with horizontal leaves. Because of a waxy surface layer, the **cuticle,** very little water vapor loss, or **transpiration,** occurs through the epidermis. Light, however, passes relatively unimpeded through the epidermis to the photosynthetic cells of the **mesophyll,** the tissue between the upper and lower epidermis. These cells contain chloroplasts.

Toward the upper surface of a typical leaf are one to three layers of columnar cells, the **palisade parenchyma** layer (Fig. 7-13). The small chloroplasts in these cells usually circulate about the cell in the streaming cytoplasm, taking turns, so to speak, at maximal light exposure. Beneath the palisade cells is the **spongy parenchyma** layer, whose cells often have fewer chloroplasts. The extensive intercellular spaces of this layer are continuous with smaller spaces between the palisade cells and with the outer atmosphere by way of the stomata. $CO_2$, which is consumed in photosynthesis, enters through the stomata and, following a concentration gradient, diffuses into the intercellular spaces among the mesophyll cells. $CO_2$ enters the mesophyll cells after first dissolving in the thin film of water covering their surfaces. Oxygen escapes through this same film, diffusing out into the intercellular spaces and then into the surrounding air by way of the stomata. It is inevitable that while thus facilitating the exchange of carbon dioxide and oxygen, the stomata also allow the escape of water vapor. This stomatal transpiration is controlled by the guard cells, which regulate the opening and closing of the stomata. Stomata are found on photosynthetic stems and fruits as

**FIG. 7-13** (a) Sketch of a portion of a dicot leaf. (b) Scanning electron micrograph of a transverse section of a dicot leaf. Trichomes and glands cover the leaf surface.

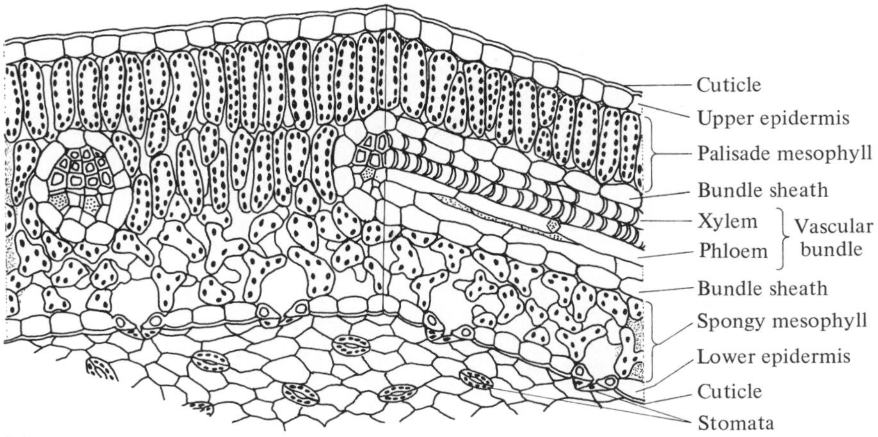


(a)

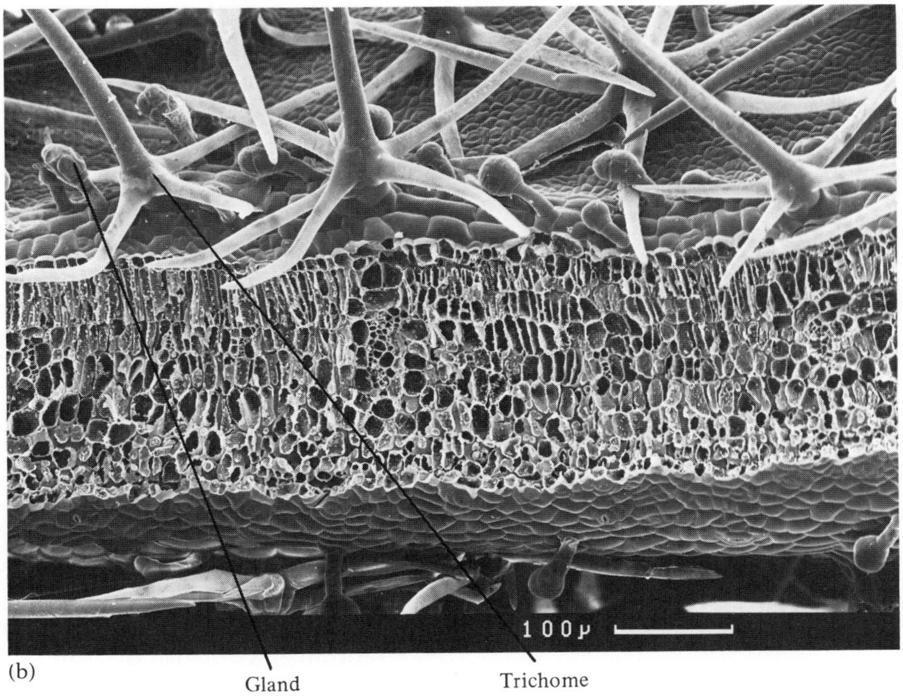


(b)

well as on leaves. The guard cells of dicots are paired, bean-shaped cells (Fig. 7-14a, c); those of grasses and some other monocots are dumbbell-shaped (Fig. 7-14b, c; see also Chapter 30).

***Guard Cell Function.*** While permitting the exchange of $CO_2$ and $O_2$ through the stomata during photosynthesis, guard cells conserve water by closing the stomata in the dark when $CO_2$ is not being used, and under conditions of water stress. (The less complex mosses and liverworts, plants that live in moist environments, have permanently open pores). Stomata close under water stress when the guard cells lose water to adjacent cells of the epidermis and become flaccid. Water uptake causes the guard cells to become turgid and swell, and the stomatal pores open (Fig. 7-14c). Differential thickening of the guard cell walls and the orientation of cellulose microfibrils, the structural framework in the walls, determine how the cells of dicots and monocots swell in

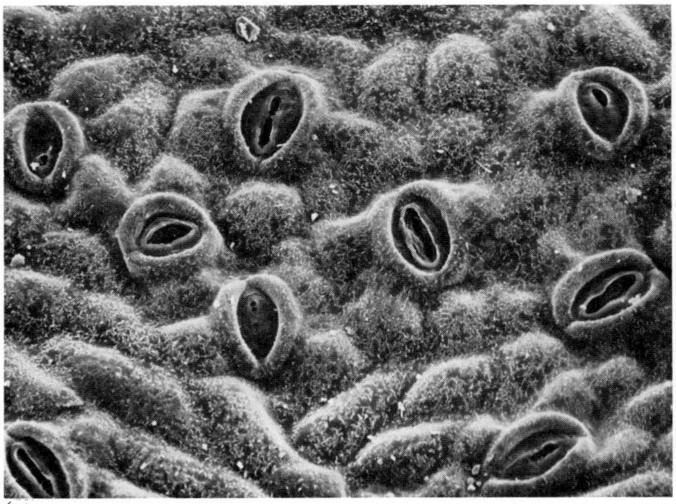
(a)

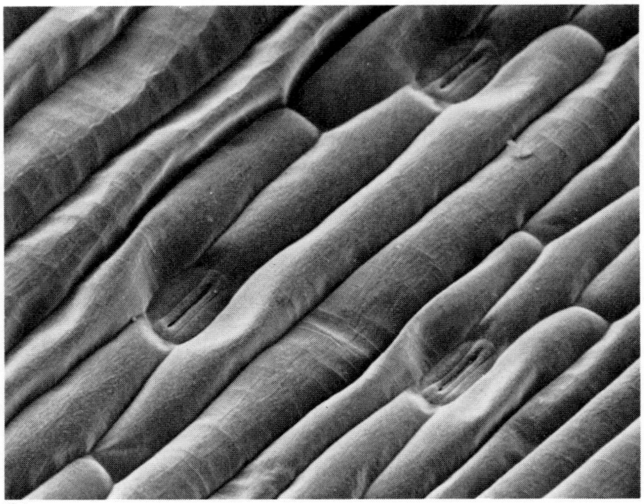
(b)

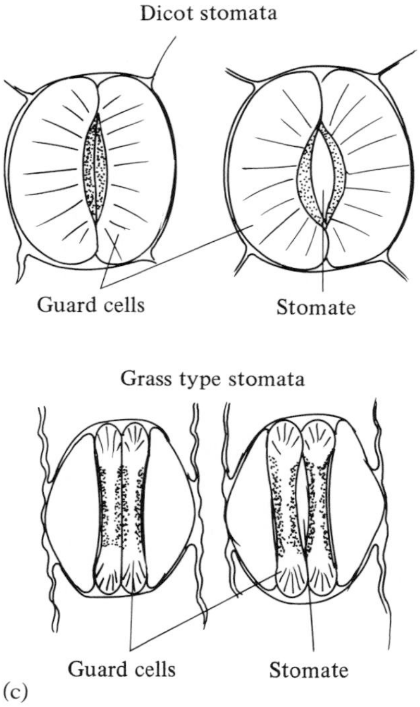

(c)

**FIG. 7-14** Scanning electron micrographs of stomata showing guard cells. (a) Bean-shaped guard cells of a dicot leaf. (b) Dumbbell-shaped guard cells of barley. (c) The opening and closing of the stomata in response to water movement. Water entering guard cells causes them to swell and pull apart at the thicker inner walls.

response to water uptake. In both cases, swelling of the guard cell pair causes them to pull apart where their adjacent, thickened walls are in contact.

Movement of water in and out of the guard cells is an osmotic phenomenon. There is a change in solute concentration in response to changes in light or $CO_2$ levels in the leaf. It was once thought that the interconversion of starch and glucose caused the osmotic water movement. (Starch is converted to glucose in light). Now the osmotic differences are explained in terms of inflow (influx) and outflow (efflux) of potassium ions ($K^+$) derived from adjacent cells. An increase of $K^+$ in the guard cells causes osmotic entry of water and consequent swelling of the cell. Exit of $K^+$ leads to loss of water. The mechanism that controls $K^+$ flux in the guard cells thus regulates the opening and closing of stomata. One model suggests that the chloroplasts of the guard cells produce ATP, which provides the energy to pump $K^+$ into the guard cells. But how the process is regulated is still under investigation.

Stomatal opening and closing is triggered both by light and by $CO_2$ concentration. It has been observed that blowing $CO_2$ across a leaf with open stomata, for example, immediately causes the stomata to close. The effect of light is indicated by the fact that stomata are ordinarily closed in the dark, conserving water at night. Interaction of light and $CO_2$ concentration is an effective regulating system. In the dark, $CO_2$ concentration in the intercellular spaces increases as a result of cellular respiration in the absence of photosynthesis. Illumination initiates photosynthesis in the mesophyll, and $CO_2$ is consumed. The resulting decline in $CO_2$ triggers the mechanism producing turgor in the guard cells, and they open the stomata, permitting $CO_2$ to diffuse into the leaf. The triggering effect of light on stomatal opening is little understood but involves blue light and a pigment other than chlorophyll.

Whenever a plant is subjected to the stress of dessication, as at noon on a warm, clear, dry day, the $CO_2$-sensing system is overriden. Loss of water from the guard cells causes a loss of turgor and consequent closing of the stomata, thereby conserving water in the plant. The plant hormone abscisic acid (see Chapter 26) has also been shown to induce stomatal closing and seems to be involved in the response of guard cells to water stress.

***Supply and Export Routes for Dissolved Materials.*** Like a factory, the leaf requires a supply of raw materials and a system for exporting products. Gases are exchanged through the stomata; water and dissolved minerals are supplied to the leaf by the xylem vessels. The carbohydrates produced are transported to other parts of the plant by the phloem sieve tubes. The xylem and phloem comprise the vascular bundles (veins), which branch throughout the middle of the leaf, between the palisade and spongy layers. The vascular bundles of dicots extend outward from the central midrib and major veins, subdividing into finer and finer veins until they terminate amid islands of mesophyll cells (Fig. 7-15). Because of this fine network, most mesophyll cells lie close to a vein. In monocots the vascular bundles run parallel to the midrib, but they, too, are interconnected by fine veins. Vascular bundles are enclosed by a cylinder of parenchyma cells, the **bundle sheath,** through which materials enter and leave the vascular tissues (Fig. 7-15).

## THE PHOTOSYNTHETIC EFFICIENCY OF LEAVES

The efficiency of plants, algae, and a few bacteria in converting the solar radiation the earth receives to organic matter is between 1 and 2 percent. One reason for this inefficiency is that only 20 percent of the solar energy received by the earth is in wavelengths that can be utilized in photosynthesis. Furthermore, the maximal efficiency of energy conversion (energy input/energy content of product) of algal cells is only about 28 percent. Photosynthetic efficiency of plants is further reduced by several other factors: (1) photorespiration, supplemental respiration which releases fixed $CO_2$, (2) the position of the leaves in relation to the angle of incidence of sunlight, (3) the internal structure of the leaf, such as its thickness (deeper cells receive less light), (4) the amount of sunlight available to shaded leaves (many ground plants in a forest never receive direct sunlight, (5) the concentration of $CO_2$ in the area, (6) temperature, (7) wind velocity, (8) humidity, and (9) soil moisture.

Plants vary in their light intensity requirements. Most crop plants are "sun plants" and require full sunlight. One group, called the $C_4$ plants, to be discussed, are 50 percent more efficient than other plants at high light intensity. Others are shade-tolerant "shade plants," such as the herbaceous plants of the forest floor. Because of their shade toler-

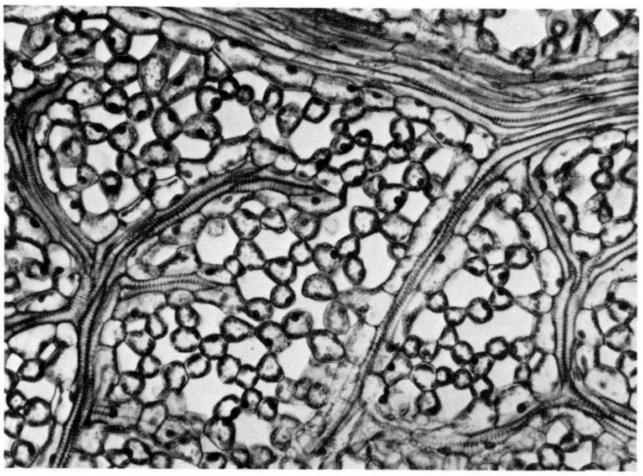

**FIG. 7-15** The extensive branching of the vascular bundles, terminating in fine vein endings, is evident in a leaf section. A vein ends in a group of mesophyll cells, none very distant from the vein. Note bundle sheath cells enclosing xylem and phloem.

ance, many species of the understory in the tropical rain forest have been bred as houseplants.

Photosynthetic efficiency of leaves involves more than efficiency in light capture; utilization of $CO_2$ is also important. And associated with the efficient processing of $CO_2$ is the problem of water conservation. In the evolution of flowering plants, or angiosperms, various adaptations have emerged that enhance photosynthetic efficiency while limiting water loss.

***Photorespiration.*** From an agricultural standpoint, photorespiration is an important factor reducing the photosynthetic efficiency of plants. All living plant cells respire, producing $CO_2$, but in some plants the rate of $CO_2$ release is higher in the light. Such light-stimulated $CO_2$ release, however, wastes energy. Over 50 percent of the $CO_2$ ordinarily fixed in photosynthesis may be lost in photorespiration without the formation of ATP.

Photorespiration is induced by high levels of oxygen, which inhibit plant growth. When plants are placed in an environment with an oxygen content of only 1 to 2 percent that of normal air, they show considerably more growth than do controls. One reason is that oxygen competes with $CO_2$ for sites on the enzyme RuBP carboxylase, reducing the amount of $CO_2$ that can be fixed. In addition, in photorespiration, RuBP is split into PGA and a two-carbon molecule, glycolic acid, which is not available to the $C_3$ or Calvin cycle. Glycolic acid diffuses out of the chloroplast and is picked up in special microbodies called peroxisomes (also called glyoxysomes) that are associated with chloroplasts. In the peroxisomes, two molecules of glycolic acid are converted to PGA and $CO_2$, an oxidation not coupled to ATP synthesis. The energy is lost to the plant, and the $CO_2$ is released into the environment. The $C_4$ plants do not conduct photorespiration, one of the reasons they are more efficient in photosynthesis than are $C_3$ plants in certain environments.

***The $C_4$ Pathway.*** A number of angiosperms have evolved a pathway of carbon fixation that has adapted them to the hot, bright, dry conditions of the grasslands of the subtropics. This is known as the **$C_4$, or Hatch-Slack, pathway.** These $C_4$ plants are able to concentrate $CO_2$ in certain photosynthetic cells, thus continuing photosynthesis when the stomata are nearly closed and $CO_2$ uptake and water loss are restricted. Included among the $C_4$ plants are several important crop plants that grow rapidly in bright light: corn, sugarcane, sorghum, and millet.

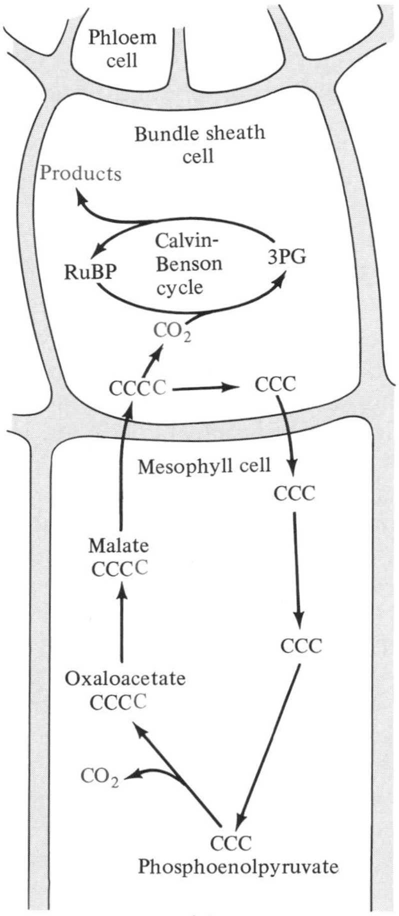


(a)

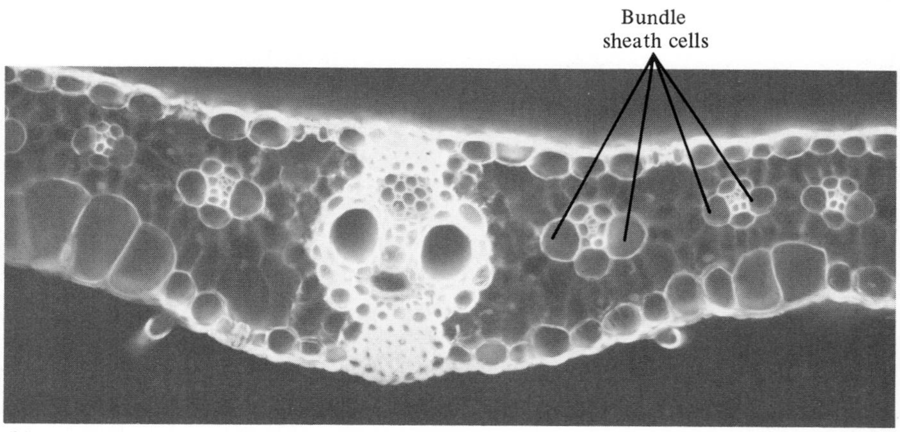


(b)

**FIG. 7-16** (a) $CO_2$ enters a mesophyll cell and is incorporated into oxaloacetate by its addition to pyruvate by phosphoenolpyruvate (PEP) carboxylase. Oxaloacetate is then converted to malate and transferred into a bundle sheath cell, where $CO_2$ is released. The $CO_2$ enters the Calvin cycle, and the pyruvate generated returns to the mesophyll cell, where it is converted to PEP. (b) Transverse section of the leaf of corn *(Zea mays)*, a $C_4$ plant, showing large bundle sheath cells surrounding the numerous leaf veins.

When $^{14}CO_2$ is supplied to $C_4$ plants, it shows up initially in four-carbon (hence $C_4$) organic acids before it appears in the three-carbon PGA of the Calvin cycle. $C_4$ plants absorb carbon dioxide more efficiently than $C_3$ plants and are more conservative in their use of water. The efficiency in $CO_2$ fixation is possible because the mesophyll cells contain the enzyme phosphoenolpyruvate (PEP) carboxylase. PEP carboxylase has a strong affinity for $CO_2$, which binds to the three-carbon compound phosphoenolpyruvate, forming the four-carbon organic acid oxaloacetic acid. Oxaloacetic acid is converted to two other four-carbon acids, malic acid and aspartic acid, which diffuse out of the mesophyll cells (Fig. 7-16). In $C_4$ plants, the bundle sheath cells enclosing the leaf veins take up the $C_4$ acids from the adjacent mesophyll cells. $CO_2$ is released from the $C_4$ acids and is then processed in the usual way through the $C_3$ Calvin cycle. The three-carbon acid residue is returned to the mesophyll cells, where it takes up more $CO_2$. The $CO_2$ concentration in the bundle sheath cells may be 100 times that in the mesophyll cells. It is this high $CO_2$ concentration that suppresses $O_2$ binding to RuBP carboxylase and thereby prevents photorespiration in $C_4$ plants. Although the stomata of $C_4$ plants are nearly closed under hot, dry, bright conditions, greatly limiting water loss, $CO_2$ continues to enter. This is due to the strong affinity of PEP carboxylase for $CO_2$. The gas is incorporated into oxaloacetate as rapidly as it enters the leaf. The resulting steep concentration gradient of $CO_2$ between the outside and the inside of the leaf ensures its continuing diffusion through the narrowed stomatal openings.

Although $C_4$ plants thrive under bright light and dry conditions, they are less energy efficient than $C_3$ plants. Fixation of a molecule of $CO_2$ in $C_4$ photosynthesis requires expenditure of five molecules of ATP, whereas the $C_3$ cycle uses three. Although $C_4$ plants have the advantage by being able to maintain high concentrations of $CO_2$ in their bundle sheath cells when light energy is abundant but water is not, $C_3$ plants have the advantage in cooler, wetter environments.

**FIG. 7-17** Crassulas are familiar window sill and garden plants having fleshy leaves and often beautiful flowers.

***Water Conservation in CAM Plants.*** Succulent plants of the family Crassulaceae, usually found in dry areas, have evolved a mechanism that conserves water by storing $CO_2$ at night (Fig. 7-17). Not only do the succulent plants store water in their tissues and conserve it with thick cuticles, but many also have the stomata closed during the day and open at night. Although this feature reduces water loss during the extreme drying conditions of the desert day, it limits the supply of $CO_2$. The crassulacean acid metabolism (CAM) plants, as they are called, circumvent this limitation by taking up $CO_2$ at night and storing it in organic acids, particularly malic acid. During the day, the $CO_2$ is released and incorporated into carbohydrate as the malic acid breaks down.

## Summary

Through the process of photosynthesis, plants, algae, and a few bacteria capture light energy and convert it to chemical energy in the form of carbohydrate. As a result of photosynthesis, plants provide energy for all other organisms which must directly or indirectly consume plants. In the process of photosynthesis, $CO_2$ is consumed and $O_2$ released, making aerobic life possible.

An understanding of the process of photosynthesis became possible only after years of study by a great number of investigators. Photosynthesis involves two groups of processes, one requiring light, the other not. In the light-dependent, or "light reactions," photons of energy are absorbed by photosynthetic pigments, including chlorophylls, carotenoids, and phycobilins, and the energy is transferred to special chlorophyll *a* molecules known as reaction centers. The excited chlorophyll molecules transfer electrons to acceptor molecules, which pass them along a sequence of electron-carrier molecules in a series of oxidation-reduction steps to $NADP_{oxid}$, ATP also being formed. The $NADP_{red}$ and ATP produced are used to reduce $CO_2$ to carbohydrate in a series of light-independent, or "dark reactions."

In the energy capture and transfer process, electrons are removed from water, releasing $O_2$ and $H^+$. Two photosystems, each with energy-gathering pigments, a reaction center, and electron acceptors and carriers, are required to move the electrons to the energy level needed to reduce $NADP_{oxid}$. Photosystem II feeds electrons into photosystem I, and along the pathway, ATP is formed in photophosphorylation. The two photosystems are situated in the thylakoid membranes of chloroplasts in precise arrangement. $NADP_{red}$ and ATP are used to incorporate $CO_2$ into carbohydrate in the Calvin cycle reactions. $CO_2$ is added to a five-carbon molecule (RuBP) by the enzyme RuBP carboxylase; two three-carbon molecules (PGA) are formed and are then reduced to a triose sugar by energy provided by $NADP_{red}$ and ATP. Two triose molecules are combined to make a hexose, which can be converted to other sugars. Five-sixths of the trioses formed, however, regenerate RuBP.

The "light" and "dark" reactions are carried out in the chloroplasts, in which the photosynthetic membranes form flattened vesicles called thylakoids arranged in interconnected stacks called grana. The "light reactions" occur in the thylakoid membranes, the "dark reactions" in the surrounding stroma.

The typical leaf is a photosynthetic organ adapted to the interception of light and the uptake of $CO_2$ while conserving water. $CO_2$ enters through the stomata, which are usually open during periods of photosynthesis and closed at other times and under stress of dessication. Opening and closing of the stomata is controlled by the guard cells. Swelling of the guard cells, which results in stomatal opening, occurs as a result of the osmotic influx of water molecules in response to an uptake of potassium ions by the cells. $K^+$ flux is regulated by $CO_2$ concentration, light and abscisic acid. Water is supplied to the cells of the leaf through the xylem vessels of the vascular bundles. Carbohydrate, made in the mesophyll cells, is transported in the phloem to other parts of the plant.

The photosynthetic efficiency of plants varies with a number of internal and external factors. Photorespiration is a wasteful process in which previously fixed carbon is lost by oxidation of RuBP to PGA and $CO_2$, without the formation of ATP. CAM plants have developed a system that allows them to photosynthesize while conserving water in their arid environment. They store $CO_2$ at night in four-carbon organic acids, and then use it in photosynthesis during the day with closed stomata. Another group of subtropical plants, the $C_4$ plants, have evolved a mechanism that enables them not merely to conserve water, but also to photosynthesize rapidly at high light intensities and high temperature—conditions under which $C_3$ plants do poorly because they cannot get enough $CO_2$ and have a high rate of photorespiration. In $C_4$ plants, $CO_2$ is stored in four-carbon organic acids in mesophyll cells and transferred to bundle sheath cells, where the $CO_2$ is released to enter the Calvin cycle. $C_4$ plants owe their success in hot, dry conditions to efficient incorporation of $CO_2$, permitting it to enter leaves through nearly closed stomata. $C_4$ plants do not conduct photorespiration.

## Recommended Readings

ARNON, D. I.: The light reactions of photosynthesis. *National Academy of Sciences.* 68:2883–2892, 1971. A historical account of this aspect of photosynthesis by a major investigator.

BASSHAM, J. A. 1962. The path of carbon in photosynthesis. In *Bio-Organic Chemistry: Readings From Scientific American.* 1968. W. H. Freeman and Company, San Francisco. This reprint of a June 1962 article by one of Calvin's coworkers is an excellent introduction both to techniques used in photosynthetic research and the basic principles of the process itself.

BJORKMAN, O., and J. BERRY. High efficiency photosynthesis. *Scientific American,* 229 (6): p. 80, October 1973. An interestingly written and well illustrated article on $C_4$ plants and their role in nature and agriculture.

CALVIN, M. 1962. The path of carbon in photosynthesis. *Science,* 135: pp. 879–889. A fascinating account of the unraveling of the Calvin-Benson cycle.

CLAYTON, R. D. 1980. *Photosynthesis: Physical Mechanisms and Chemical Patterns.* Cambridge University Press, New York. The basics of photosynthesis and their discovery in a concise treatment.

GABRIEL, M., and S. FOGEL. 1955. *Great Experiments in Biology.* Prentice-Hall, Inc., Englewood Cliffs, N.J. Annotated reprints of excerpts of classic experiments in biology, including those of Engelmann, de Saussure and others.

GOLDSWORTHY, A. 1976. Photorespiration. *Oxford Biology Readers,* No. 80. Carolina Biological Supply, Burlington, N.C. A 16 page summary of photorespiration and $C_4$ photosynthesis.

GOVINDJEE, and R. GOVINDJEE. The absorption of light in photosynthesis. *Scientific American,* 231: pp. 68–82, December, 1974. A detailed coverage of the light reactions and the status of this aspect of photosynthetic research by eminent investigators.

HATCH, M. and N. BOARDMAN. 1981. *Photosynthesis in the Biochemistry of Plants,* Vol. 8., ed. by P. K. Stumpf and E. E. Conn. A recent comprehensive treatise with chapters on various topics by authorities in the field.

HEATH, O. V. S. 1981. Stomata. *Carolina Biology Reader.* Structure and function of stomata in 32 pages.

MILLER, K. R. The photosynthetic membrane. *Scientific American,* 241: pp. 102–113, October, 1979. Recent information on thylakoid membranes and the light reactions.

WHITTINGHAM, C. P. Photosynthesis. *Carolina Biology Reader.* 1977 (2nd ed.). 16 pages. A short summary of photosynthesis.

# UNIT III Homeostasis

CHAPTERS



Wherever on earth the conditions and materials essential for life are present, some form of life exists. Organisms are found in conditions as extreme as steaming hot springs, the frozen Arctic, and the hottest, driest deserts. The evolutionary adaptations that have made it possible for organisms to succeed under such conditions have long fascinated biologists and nonscientists alike. Yet, as dramatic as these cases are, all organisms occasionally contend with such heavy stresses as nutrient shortages, infectious diseases, osmotic stress, and limited oxygen supplies. Internal stresses, such as the production of toxic waste products of metabolism, are also endured by all. The success of a species often depends on the ability of its members, in the face of such stresses, to so regulate their internal environment as to provide stable, optimal conditions for life's various needs.

The chapters in this unit concern the principal mechanisms by which organisms maintain a degree of internal stability despite being subjected to a variety of stresses.

## CLAUDE BERNARD AND THE CONCEPT OF THE INTERNAL ENVIRONMENT

The cells of protists and other simple organisms are in rather intimate contact with their environment, usually the water of rivers, lakes, or the sea. Such organisms are affected by fluctuations in temperature, light, and the chemical composition of the water in which they live. Only those organisms that have evolved an ability to produce and regulate their own seawaterlike internal fluids, thereby gaining a measure of independence from their environment, have

**Walter B. Cannon, the American physiologist whose researches led him to formulate the concept of homeostasis.**

been able to evolve beyond the life of a protist, alga, sponge, or jellyfish. The maintenance of comparatively stable internal conditions by mammals was first termed **homeostasis** by the American physiologist Walter Cannon early in this century. This concept has since been extended to include any internal stability maintained by any organism. Even protists have regulatory enzymes and control their cytoplasmic pH and concentrations of other ions. In all cases, the equilibrium maintained in homeostasis is dynamic, not static, involving a continual exchange of molecules between the organism and its environment.

Plants also have homeostatic mechanisms. For example, the opening of stomata—the microscopic pores on the undersides of leaves that admit carbon dioxide from the atmosphere—occurs whenever the concentration of carbon dioxide in leaf tissue drops below a certain level. Another mechanism closes the stomata whenever excessive water loss occurs. Other homeostatic controls are important to plants because their growth depends on the provision of essential nutrients to the growing tissues in certain proportions and within certain limits of concentration. The solutes in xylem sap, for example, are maintained at fairly constant concentrations, despite variations in the concentration of solutes in soil water.

Homeostasis is thus not an exclusive characteristic of the mammal or of the vertebrate or even of animals. But it is within the birds and mammals that it has evolved to its highest degree. This distinction between higher and lower animals was first emphasized 50 years before Cannon by the French physiologist Claude Bernard (1813–1878). He noted that in "higher animals," by which he usually meant vertebrates, most cells are relatively isolated from the outside environment and are bathed by the fluids of the body. Even the living cells of our skin are isolated from the environment by a tough outer layer of dead cells. About 98 percent of their oxygen is obtained from the blood, which also supplies their nutrients and removes their wastes.

Only after animals had evolved the ability to produce an internal fluid medium similar in many respects to seawater were they able to colonize fresh

Claude Bernard at age 53.

water and land and thereby utilize the many sources of energy and materials available in those environments. In time, such animals also improved the composition of their "internal seawater," further suiting it to various needs of life and permitting the adoption of more new habits of life. Eventually, their development of a complex internal medium also provided an exquisite instrument of communication and coordination among the different parts of the body.

Bernard's concept of an internal environment formed a foundation upon which much of modern **physiology,** the study of the functioning of organisms and their parts, was to be built. Cannon was the first to show how each regulatory mechanism depends on the existence of appropriate sensory receptors that detect the external or internal changes that necessitate adjustments. He observed that a sufficiently rapid means of communication (often the nervous system) that triggered the proper corrective action is always a part of the mechanism. The number of internal conditions that are controlled seems endless, including, for example, the concentrations of virtually every substance found in the body fluids.

## FEEDBACK LOOPS AND SETPOINTS

The sensitive receptors, structures excited by specific stimuli, that monitor each condition under homeostatic control are linked in some way to mechanisms capable of correcting changes in particular states. A linked circuit of receptors and corrective mechanisms of this sort is known as a **feedback loop.** Usually the action that results when a receptor is stimulated also ultimately inhibits the receptor, in which case the mechanism is called a **negative feedback loop.**

A useful analogy can be drawn between a biological negative feedback loop and a thermostatically controlled furnace. As the thermostat cools below a certain temperature, determined by where its dial is set, it completes an electric circuit and turns on the furnace. The heat from the furnace then warms the building—and the thermostat. When the tempera-

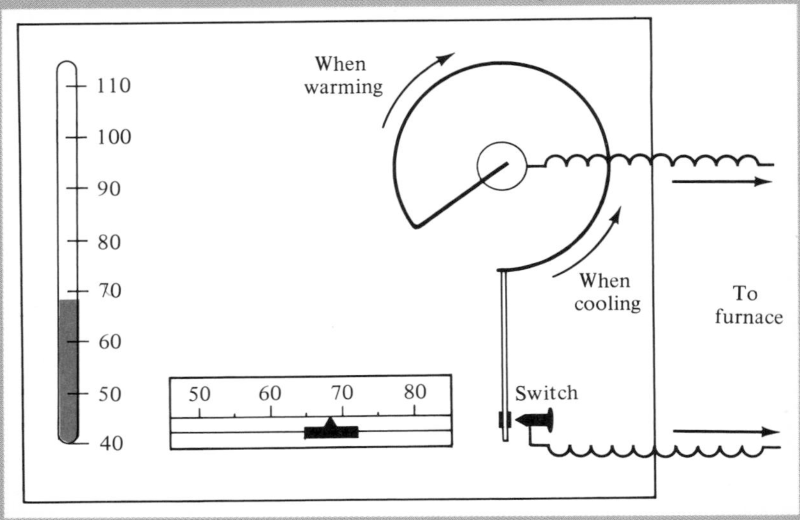


**A furnace thermostat is an example of a negative feedback mechanism. When the thermostat cools to below 68°F (the setpoint, at which the dial has been set) the bimetallic ring expands, closing the switch and turning on the furnace. As the furnace warms the house above 68°F, the thermostat opens the switch, turning off the furnace.**

ture of the thermostat rises above the set temperature, the circuit is broken and the furnace is turned off. In organisms, as with furnaces, there may be several controls for any one feedback loop.

Biological negative feedback loops have **setpoints.** These are qualities of some component of a loop, often of the receptor, that determine what particular concentration, pH level, temperature, or other value is to be regulated. When the value deviates somewhat from the setpoint, a stimulating or inhibiting effect automatically occurs. (In the furnace thermostat analogy, the setpoint is the point on the dial at which the thermostat has been set.) Like other biological traits, the setpoints of feedback loops are subject to natural selection. Those that a species has inherited are those that made it possible for their ancestors to survive.

# 8 Gas Exchange

OUTLINE



To take advantage of the rich source of energy represented by the organic matter on our planet, aerobic organisms must solve two practical problems. First, they must break down, or digest, the organic matter so that it can enter the cells that are to metabolize it (the digestive process is described in Chapter 9). Second, they must provide the cells with both an adequate supply of the molecular oxygen required for respiration and a way of eliminating the carbon dioxide produced by respiration. As explained in Chapter 6, oxygen is consumed and carbon dioxide and water are produced during aerobic respiration. The more active an aerobic organism is, the more oxygen it needs and the more carbon dioxide it produces and must eliminate. This process of gas exchange with the environment, also called **external respiration,** is the subject of this chapter.

All higher plants and animals and many species of monerans, protists, and fungi exchange oxygen and carbon dioxide with their environment. The same general principles of gas exchange apply to all. The physical routes by which oxygen reaches the tissues of an aerobic organism are usually the same routes by which carbon dioxide is eliminated. Of the two aspects of gaseous exchange, oxygen delivery is usually more crucial than carbon dioxide elimination to the survival or success of an organism in a particular environment. Temporary shortages of oxygen are likely to be fatal, whereas temporary accumulations of carbon dioxide can usually be tolerated.

**FIG. 8-1** In this scene from Western Samoa, in the south central Pacific, the persons on the horse, the horse, and the surrounding vegetation all need a supply of oxygen to maintain respiration in their tissues. Carbon dioxide is formed as a by-product of respiration. Although photosynthesizing plant tissues use $CO_2$ and produce an excess of oxygen, all plant tissues otherwise use oxygen and produce $CO_2$, their rate of respiration never being as great as that of a large, active animal.

# THE DELIVERY OF OXYGEN

Oxygen reaches the cells of a multicellular aerobic organism in four steps:

1. Oxygen comes into contact with the body of the organism, particularly with its respiratory organs or tissues.
2. Oxygen enters the body.
3. Oxygen is transported from the point of entry into the body to the rest of the body.
4. Oxygen enters the cells.

Inefficiency in any of these steps constitutes a limiting factor in the ecological distribution of organisms; that is, it tends to determine the environments in which a species can succeed.

The practical problems of accomplishing the four steps of oxygen delivery arise from three physical factors: pressure, solubility, and the rates of diffusion of oxygen in air and in water.

## PRESSURE

The total number of molecules of oxygen and the other gases of air in a given volume at a particular temperature can be expressed in terms of the pressure they exert. Gas pressure is caused by the collisions of gas molecules against the surfaces with which a gas is in contact. Such collisions determine how much air dissolves in a body of water or in the body fluids with which air comes in contact. The higher the pressure, the more air dissolves.

The atmosphere at sea level is normally under pressure due to gravity equivalent to 760 millimeters of mercury (mm Hg), or 14.7 pounds per square inch as measured by a barometer. That is, the pressure is sufficient to raise a column of mercury 760 mm (29.9 inches). About 21 percent (160 mm Hg) of this pressure is exerted by oxygen and most of the rest by nitrogen (see Panel 8-1). In a gaseous mixture such as air, each gas exerts pressure independently of the others. The amount of pressure exerted by each gas is referred to as the **partial pressure** of that

**FIG. 8-2** (a) Barometric pressures and partial pressures of oxygen at various altitudes. (b) The Peruvian Andes. The tourists are undoubtedly experiencing light-headedness and possibly headaches and nausea, whereas the natives and their flock of alpacas are adapted to the rarefied atmosphere and experience little or none.

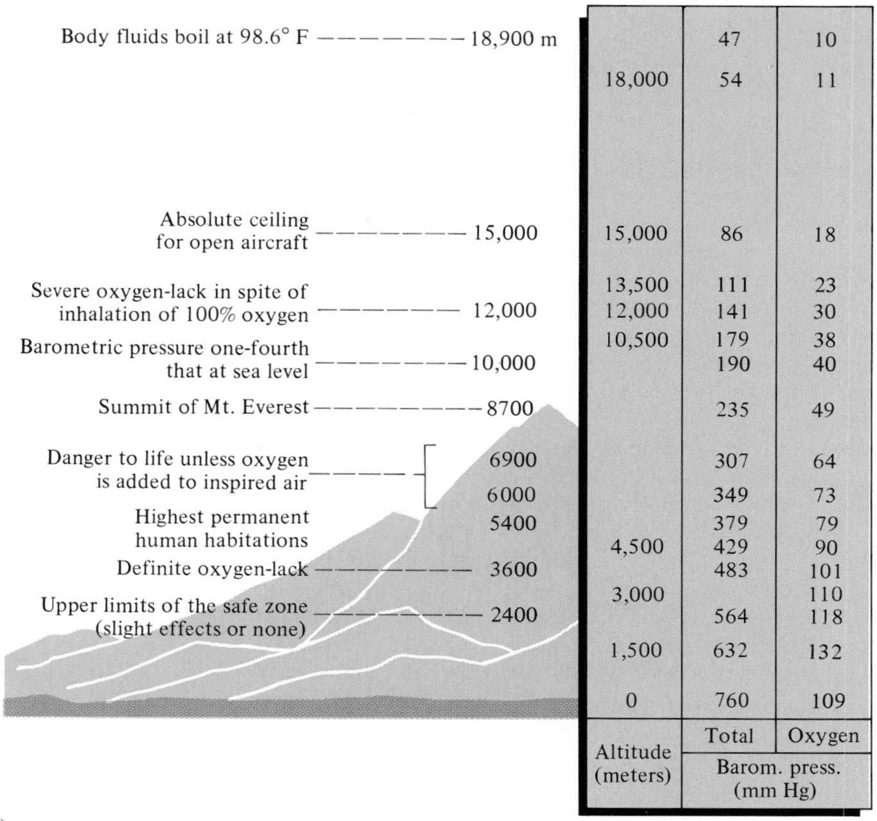


(a)

(b)

gas. As we climb a tall mountain or ascend in an airplane, oxygen continues to make up 21 percent of the atmosphere, but at a lower and lower partial pressure (Fig. 8-2a). For example, in the Peruvian village of Morococha, at 4.5 kilometers (2.8 miles) above sea level, the partial pressure of oxygen drops to 85 mm Hg, little more than half that at sea level. The stress caused by such marginal oxygen pressure can cause chronic "mountain sickness"—with symptoms of lightheadedness, headaches, weakness, nausea, and mental dullness—even among many of the highly adapted people living there (Fig. 8-2b).

## PANEL 8-1

# Decompression Sickness: "The Bends"

Except for the conversion of molecular nitrogen ($N_2$) to ammonia by nitrogen-fixing bacteria, atmospheric nitrogen is inert, biologically speaking, that is, it does not react chemically with the body. Although it makes up over 79 percent of the atmosphere, we breathe it in and out day after day without metabolizing it. It is found in simple solution in all the cells and fluids of the human body, amounting to about 1.5 liters in an average-sized adult.

However, nitrogen, despite its chemically inert nature, poses a threat to human life under certain circumstances. For example, the blood and tissues of underwater divers and others who work under high pressure contain much more air—80 percent of it nitrogen—than they did at sea level atmospheric pressure. Rapid decompression, which occurs when a diver rises quickly to the surface, releases the nitrogen from solution in the form of gas bubbles, as when the cap is removed from a bottle of carbonated beverage. The result is called **decompression sickness,** caisson disease, or "the bends", marked by pain and paralysis as the gas bubbles block the flow of blood to various organs.* An aviator who ascends rapidly above 7,600 meters (25,000 feet) in a nonpressurized airplane experiences the same result. The effect can be avoided by a slower ascent, or slower **decompression.** Divers experiencing the bends are usually rushed to pressure chambers, resubjected to high pressure, and then brought more slowly back to sea level conditions.

* The term caisson disease derives from the fact that people working in caissons, the pressurized, watertight chambers that are used in the construction of underwater foundations for bridges, often suffered from the disease after returning to normal pressures. As they doubled over in pain, they were said to have "the bends." The problem is now avoided by having such workers undergo slow decompression when returning to normal atmospheric pressure.

## SOLUBILITY OF OXYGEN IN WATER AS A LIMITING FACTOR

Like most gases, oxygen is poorly soluble in water. In contrast to the atmosphere, with an oxygen concentration of 210 milliliters (ml) per liter of air, the surface waters of a lake at sea level at 20°C contain only 5 to 10 ml of oxygen per liter. Because gases diffuse slowly in water, deeper waters often contain even less oxygen. Any other substances dissolved in water further decreases its ability to dissolve oxygen. This property affects not only the seawater in which many organisms live but also such oxygen-carrying body fluids as blood. The solubility of gases in water also declines with a rise in temperature. For example, although a liter of distilled water holds 6.5 ml of oxygen at 20°C, it holds only 0.27 ml at 32°C. The effect of temperature on the solubility of gases in water can be observed by comparing cold and warm carbonated beverages. The latter lose their carbonation ($CO_2$) readily.

The amount of oxygen dissolved in water is often barely sufficient to meet the requirements of some of the more active aquatic animals. Trout, for example, survive only in cool, well-aerated waters. Any factor that lowers the oxygen content of a body of water, such as thermal pollution or heavy bacterial growth, may result in massive deaths of fishes, other aquatic animals, and plants.

## OXYGEN DIFFUSION

The process of diffusion is inefficient for moving substances farther than half a millimeter. Nonetheless, it is the only process by which oxygen can move across respiratory membranes and into the bodies of organisms. Moreover, after oxygen is transported within the body to the tissues using it, diffusion is the only process by which oxygen enters the respiring cells. Because diffusion of oxygen in water is so slow, the evolution of large or active animals depended on their development of methods for bringing oxygen into contact with particular membranes.

## GAS EXCHANGE IN PLANTS

All aerobic organisms carry on the same reactions of cellular respiration. A difference between the energy metabolism of plants and animals is that plants draw primarily on carbohydrates as an energy source whereas animals depend mostly on fats. In plants, respiratory gas exchange occurs along with the uptake and utilization in photosynthesis of the atmospheric $CO_2$ that enters plants by diffusion and the $CO_2$ they produce as a result of cellular respiration. (These processes are described in Chapters 6 and 7). Compared to most animals and many protists, however, plants lead relatively inactive lives. As a result, they do not demand as much oxygen as do vertebrates, for example. This limited demand for oxygen is satisfied by the transport of air through an extensive but simple network of intercellular air spaces in loosely arranged plant tissues. Because oxygen diffuses 3 million times faster in air than it does in water, oxygen readily reaches all the cells of the plant tissue without the need for a complex circulatory system. In potato tissue, for example, which is 1 percent air space, oxygen diffuses 430 times faster than through water. In addition to the development of intercellular air spaces, another plant adaptation to respiratory needs was the development of fine pores, or **lenticels,** in the bark of stems and in the surfaces of many fruits; the lenticels admit oxygen to the deeper tissues.

**FIG. 8-3** The "knees" of the bald cypress allow these trees to survive in swamps and waterlogged soil.

### THE SPECIAL CASE OF PLANT ROOTS

In regard to respiration, roots occupy the most vulnerable position of any plant organ. Although soil with good structure normally contains many air-filled spaces, in poorly drained soil the spaces may remain filled with water for days after a heavy rain. Root tissues may die under such conditions because water contains little dissolved oxygen. Overwatering house plants and trees can kill them for the same reason. The vulnerability of roots to lack of oxygen can also be seen after extensive flooding or where subsoil excavated from a construction site has been spread thickly under nearby trees. Such conditions create a barrier between underlying root systems and the atmospheric oxygen that normally diffuses down to them. In warm weather, when respiration rates rise and less oxygen is dissolved in soil water, the trees may die as a result.

Some species of land plants have adapted to living in wet soil or, in some cases, in water. For example, the roots of rice plants have evolved large air-conducting spaces through which oxygen diffuses to the submerged tissues. In water lilies, the pads and stems conduct oxygen to the rest of the plant. The roots of bald cypress trees growing in swamps develop conical "knees" that give the submerged roots access to atmospheric oxygen (Fig. 8-3). The mangrove and some other species of tropical plants that grow in water or wet soil have **pneumatophores** (Greek for "air-carriers"), which are specialized, spongy structures that extend up into the air from the buried horizontal roots and permit gas ex-

FIG. 8-4 Pneumatophores, or "air-carriers," of the red mangrove.

change between root tissue and the atmosphere (Fig. 8-4). However, except for the few adaptations of this sort and the important exchanges of $CO_2$, $H_2O$ vapor, and $O_2$ through their stomata (see Chapter 7), plants generally lack special structures for gaseous exchange.

# EVOLUTIONARY PATTERNS OF GAS EXCHANGE IN ANIMALS

The transition from simple aquatic respiratory mechanisms to the most advanced, air-breathing respiratory systems in animals is not a single evolutionary sequence, with the lower aquatic forms at one end and air-breathing birds and mammals at the other. Rather, many of the respiratory mechanisms found in animals evolved independently. In some cases, evolutionary convergence occurred, with some organisms evolving a mechanism independently acquired by an unrelated group.

## DELIVERY OF OXYGEN TO THE RESPIRATORY ORGANS OF AQUATIC ANIMALS

Because diffusion in water is effective only over short distances, it is essential that oxygen-bearing water regularly be brought into immediate contact with aquatic organisms. This poses few problems in surface waters. Waves and currents may stir the surface waters of rivers, lakes, and oceans sufficiently to maintain an adequate renewal of oxygen in the upper water layer.

Many aquatic animals, however, also need to engage in **ventilation:** some regular action that maintains a continual supply of oxygen by replacing the water immediately surrounding them or that bathes their gills or other respiratory organs. In many species, ventilatory movements may serve only a respiratory function. In others, ventilation is a result of such activities as feeding or locomotion. Sponges and many corals possess cilia that create currents of water that bring food and oxygen-laden water not only into the vicinity of the animal but into, through, and out of its body again, carrying away $CO_2$ and other wastes (Fig. 8-5). Swimming worms are ventilated when the water in contact with them is replaced merely by their swimming through it. Tube-dwelling worms undulate in a swimming motion, thereby creating a respiratory current.

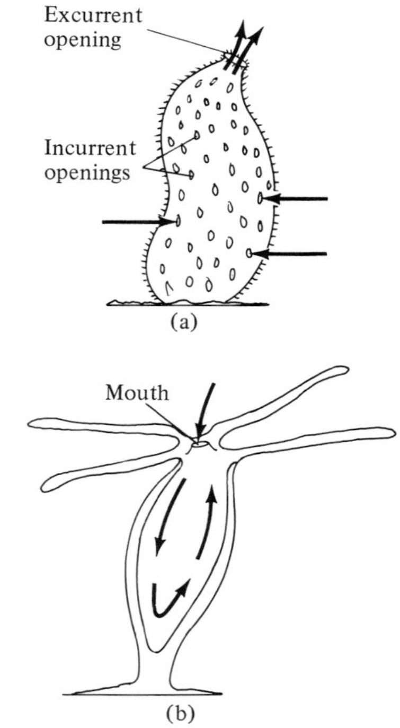


FIG. 8-5 Body plans of (a) a sponge and (b) a coelenterate, or cnidarian, showing circulation of seawater through the body. The currents, produced by cilia, serve several functions, including nutrition and respiration.

Animals with specialized respiratory structures often perform corresponding specialized ventilatory movements. For example, one way that fishes ventilate their gills is by swimming with their mouths open. When not swimming, most teleosts (bony fishes) ventilate their gill chambers by first closing the gill coverings, or **opercula,** while opening the mouth, and then expanding the mouth and gill cavity, an action that draws in water through the mouth. They then close the mouth, open the opercula, and, by a contraction of the mouth cavity, expel the water across the gills. Some fishes, such as the mackerel, have lost this ability; unless they keep swimming or position themselves in a current of water, they die of oxygen deprivation. The same is true of some cartilaginous fishes, including most sharks.

***Aquatic Respiratory Organs.*** Oxygen usually enters the bodies of higher animals through specialized respiratory membranes that are readily permeable to oxygen. In many small animals, however, the gen-

**FIG. 8-6** The delicate tentacles of a feather-duster worm, a marine annelid. The tentacles serve both a respiratory and a feeding function; they are covered with cilia that produce a water current when the tentacles are expanded.

eral body surface serves as the only respiratory membrane. For a small, thin, or inactive animal, this can be adequate. But in larger or more active animals, the total body surface cannot absorb oxygen as rapidly as needed. The critical value is the ratio of body surface to body volume. In some animals, an increase in the surface area through which oxygen can enter—and thus an increase in the ratio—is provided by structures that also serve other than respiratory functions, such as the feeding tentacles of the feather-duster worm, a type of tube-dwelling marine annelid (Fig. 8-6). Swimming marine annelids have **parapodia** (Greek for "side feet") that aid their locomotion and provide the needed additional respiratory surface. Other examples are found in such crustaceans as the brine shrimp, *Artemia,* in which elaborate outgrowths of the locomotory appendages function as respiratory structures (Fig. 8-7).

**FIG. 8-7** *Artemia salina,* the brine shrimp. The lateral appendages function in both locomotion and respiration. This shrimp is one of the few organisms that can live in the highly saline environment of the Great Salt Lake, Utah.

**FIG. 8-8** The cerata, or modified gills, of the nudibranch, or sea slug, *Flabellina,* a marine gastropod mollusc.

A **gill** is any organ of an aquatic animal that has as its primary function the exchange of oxygen and $CO_2$ between the animal and the water. In some animals with gills, the amount of additional surface provided is rather small, as in the gills on the skin of the starfish or the simple gills of the slow-moving nudibranchs (Greek for "naked gill") (Fig. 8-8), the colorful marine cousins of the garden slug. But aquatic animals that are large, active, or both need extensive gill surfaces to provide an adequate supply of oxygen. Among invertebrates, examples of elaborate gills are seen in clams and oysters (Fig. 8-9) and in such large crustaceans as crabs, shrimps and lobsters. The gills of fishes, however, which possess

**FIG. 8-9** The respiratory system of clams and oysters. (a) Diagram of a clam with the right shell, mantle, and gills removed to show one of the gill lamellae on the left side. Arrows indicate overall direction of ciliary current, which serves both a nutritional and respiratory function. The gill lamella is sketched as if sectioned, to show its water channels. (b) Diagram of a portion of a gill lamella showing the ostia through which the water enters the gill, driven by a ciliary current.

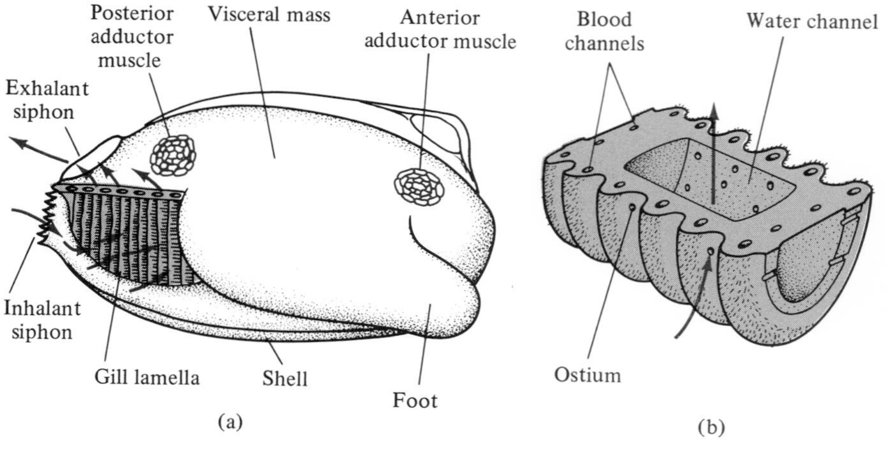

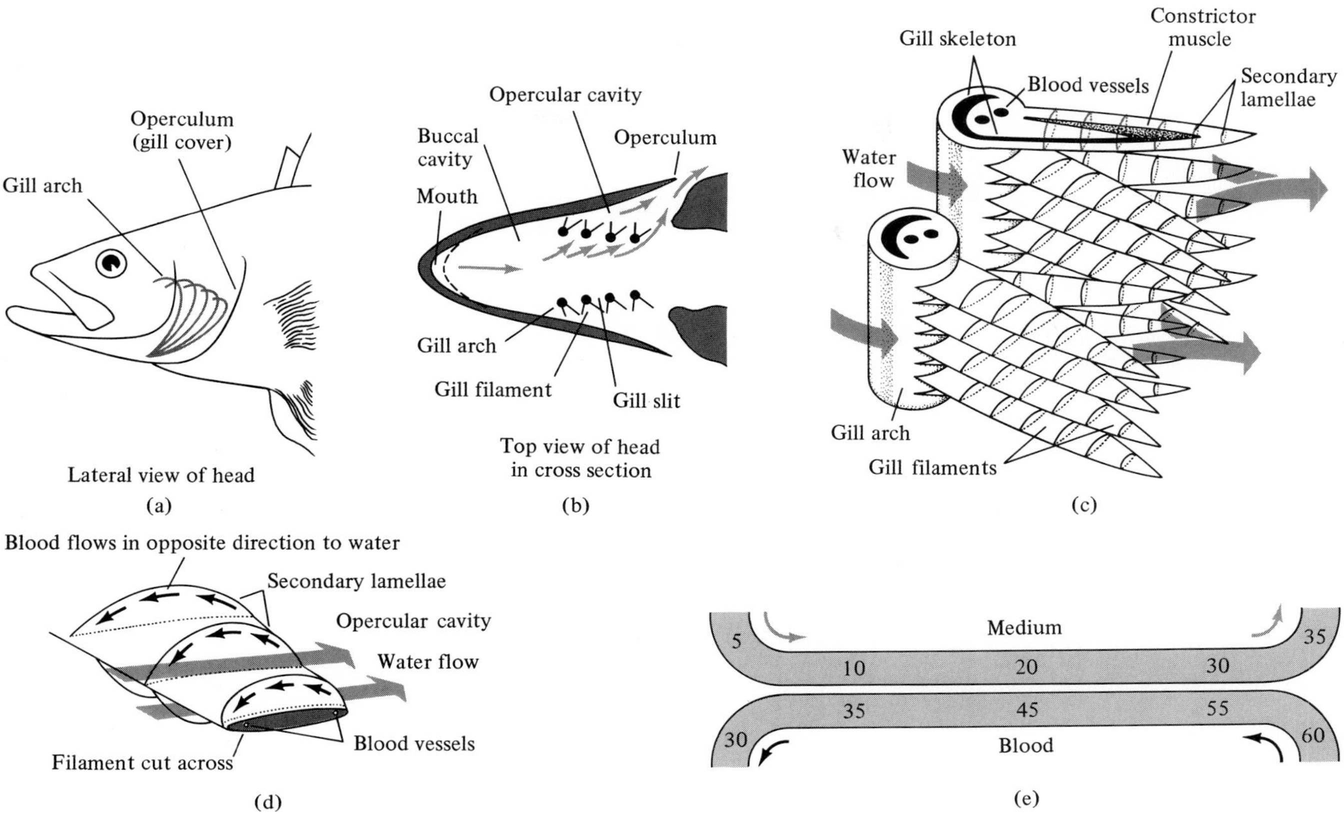


**FIG. 8-10** The gills of a fish. (a) Position of the gill arches beneath the gill cover, or operculum. (b) Top view shows direction of water flow; it enters through mouth and leaves by gills. (c) Two gill arches shown in detail. Filaments of adjacent rows touch at their tips. Water flow is indicated by arrows. (d) Diagram of part of one gill filament, showing how the water current crossing the gill membrane flows in the opposite direction to that of the blood, providing the efficiency of a countercurrent exchange. (e) The principle of the countercurrent exchanger is at work in organs in which an efficient exchange of heat, oxygen, and the like is accomplished.

a **countercurrent exchange mechanism,** (Panel 8-2) are the most highly evolved of all (Fig. 8-10).

## AIR BREATHING: AN EVOLUTIONARY BREAKTHROUGH

Air-breathing animals are descendants of aquatic ancestors that adapted to breathing air. The nature of this evolutionary transition is suggested by some fishes and other aquatic animals now in existence that can survive for a time in air, some for hours, by breathing through modified gills or other surfaces that are moist and well vascularized (supplied with blood vessels).

Some of the more primitive vertebrates alive today are especially interesting because they suggest some of the stages that our own remote ancestors passed through as fish gave rise to the ancient amphibians that evolved into reptiles, birds, and mammals. Before we consider the vertebrates, however, it will be instructive to examine air-breathing members of three invertebrate groups: molluscs, arachnids, and insects.

PANEL 8-2

# Countercurrent Exchange

The principle of countercurrent exchange applies to several independently evolved physiological mechanisms in animals. These include the production of gas by fishes in swim bladders (referred to later in this chapter), the exchange of respiratory gases in the lungs of birds (also described in this chapter), the control of water loss by the vertebrate kidney (Chapter 13), and the control of heat loss in the extremities of birds and various mammals. The exchange of respiratory gases by the gills of fishes is another example.

The countercurrent exchange principle, which applies to heat exchange as well as to the diffusion of gases, is a simple one. The rate at which a substance diffuses from a region of high concentration to one of low concentration is directly proportional to the difference in the two concentrations. In the fish gill, the blood flows through the gill blood vessels in the opposite direction to the water flowing through the gills (see Fig. 8-10, in text). From the standpoint of oxygen uptake, this pairs the blood entering the gills from the body, which is low in oxygen, with water that has already passed over the gills but still retains a substantial amount of oxygen. This allows more oxygen to be absorbed by the blood than if the water exiting the gill had been paired with blood that had already absorbed a lot of oxygen. At the opposite end, water that has just entered the gill cavity, and is, therefore, rich in oxygen, is paired with blood that has already passed through the gills and, having absorbed oxygen, is about to return to the body. Again, because of the concentration differential, oxygen enters the blood. Experimentally, water has been made to flow through fish gills in the same direction as blood flow; this reduced the uptake of oxygen to about one-eighth its normal amount.

## AIR BREATHING IN INVERTEBRATES

Prominent anatomical features of all molluscs—chitons, clams, snails, slugs, squids, and so on—are the **mantle,** a thick sheet of glandular tissue, and the **mantle cavity,** the space enclosed by it. In most species of molluscs the mantle serves primarily to secrete the shell and protect the gills. In **cephalopods,** such as squids and octopods, however, it is used in locomotion, and in **pulmonate gastropods,** such as land snails and garden slugs (Fig. 8-11), the gills are gone and the mantle cavity has been

**FIG. 8-11** A garden slug. Note the prominent opening into the lung cavity.

converted into a lung that is suited for breathing air. When it serves as a lung it has but one external opening, which opens and closes from time to time. Its lining is folded which maximizes its surface area. Like the respiratory surfaces of all terrestrial animals, the lining is thin, well-vascularized, moist, and in contact with a constant supply of oxygen. This type of mantle cavity is termed a **diffusion lung** because it depends largely on diffusion rather than ventilation for renewing its air supply. However, some ventilation does occur when the size of the cavity changes as the animal extends or contracts its body during locomotion. Although pulmonate mantle cavities usually function as lungs and are suited only to air breathing, transitional forms occur; in tropical apple snails, for example, the mantle cavities are divided into two sections, one containing a gill and the other functioning as a lung.

In the course of their long evolution, some pulmonate gastropods have returned to an aquatic life. Some of these have retained their lungs as water lungs, "breathing"water with them; but these pulmonates also rise to the surface from time to time to inspire air. In another aquatic pulmonate group, the lung has been reduced to a nonfunctional structure. These animals have met their respiratory needs by evolving a new kind of gill—a heavily pleated and folded outward extension of the mantle that in no way resembles the original gill found in other molluscs.

***Book Lungs.*** Another interesting modification of an essentially aquatic organ for use as an air-breathing lung is found in Arachnida, a group of invertebrate animals that includes spiders, ticks, mites, and scorpions. The evolutionary origin of such an organ is suggested by the respiratory organ of the horseshoe crab, *Limulus,* a close cousin of the arachnids. This primitive animal possesses remarkable aquatic respiratory organs, the **book gills** (Fig. 8-12). These gills consist of five pairs of broad, flat, abdominal appendages fused at their centers and covered by a sixth pair that forms a protective operculum. The animal waves the

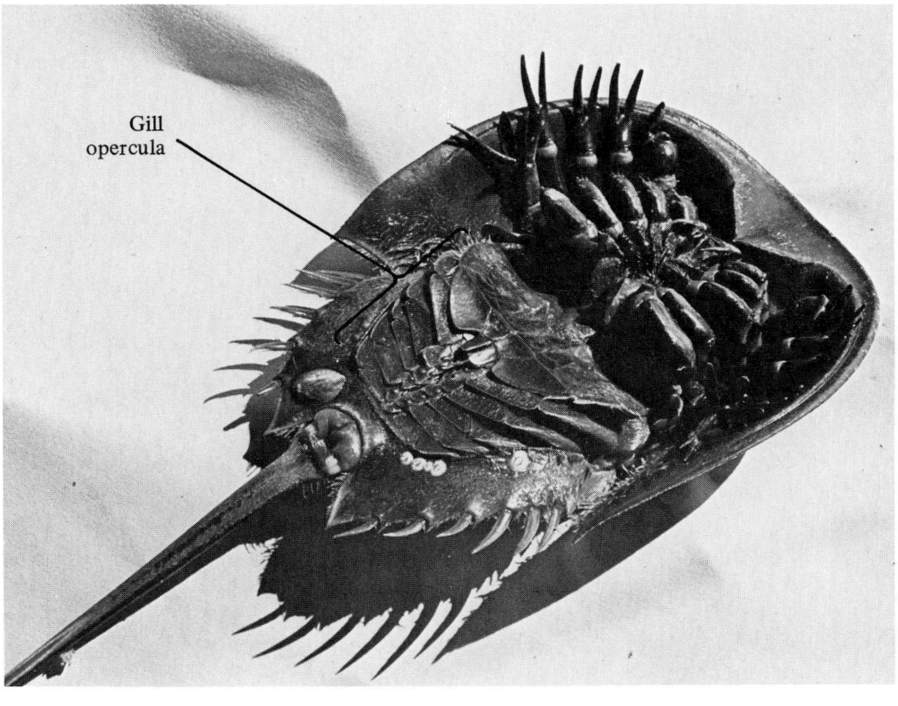

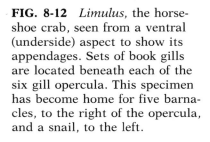

**FIG. 8-12** *Limulus,* the horseshoe crab, seen from a ventral (underside) aspect to show its appendages. Sets of book gills are located beneath each of the six gill opercula. This specimen has become home for five barnacles, to the right of the opercula, and a snail, to the left.

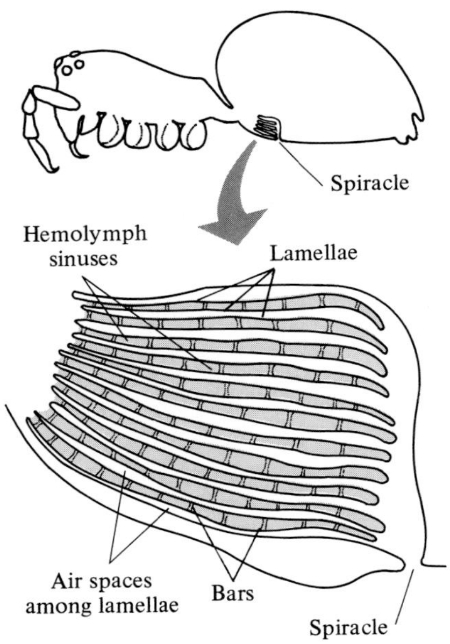


**FIG. 8-13** Diagram of a section through the book lung of an arachnid.

appendages back and forth, causing water to flow over their outer surfaces and a body fluid called **hemolymph** to flow through them. The underside of each appendage is deeply folded into membranous, leaflike pleats resembling the pages of a book and suggesting the name of the organ.

The **book lungs** of arachnids, air-breathing relatives of *Limulus*, appear to be modified versions of book gills (Fig. 8-13). The major differences between these two organs seem to be related to the necessity of preventing the membranous "leaves," or lamellae, of book lungs from collapsing on each other in air and of protecting them from damage and desiccation. The leaves are prevented from collapsing by being held apart by small bars and are protected by enclosure in lung cavities. The lungs open to the exterior through **spiracles,** small slits in the ventral abdominal wall. Book lungs are ventilated by special muscles that alternately dilate and contract a small chamber adjacent to each lung.

Book lungs are the only respiratory organs of scorpions, tarantulas, and trap-door spiders. Other spiders, however, supplement the respiration of their book lungs with a tracheal respiratory system like that of insects.

***The Insect Tracheal System.*** The tissue cells of most terrestrial animals cannot absorb oxygen unless it is dissolved in an aqueous body fluid such as blood or hemolymph. Despite the presence of respiratory pigments, which allow blood to carry large amounts of oxygen, the low solubility of oxygen in water severely reduces the efficiency with which any fluid can supply oxygen to cells. If oxygen could be brought directly to each cell without having to be dissolved in and transported for some distance by a fluid medium, this inefficiency could be avoided. In the tracheae of insects and a few other arthropods, such a mechanism has evolved. Insect **tracheae** are a system of highly branched air tubules that bring the atmosphere into close contact with all cells of the body.

THE STRUCTURE OF TRACHEAL SYSTEMS. Tracheae open to the exterior of the body by a series of spiracles, usually one pair per body segment (Fig. 8-14a). Tracheal walls are kept from collapsing by rings or spirals of a stiff material, the **cuticle.** The tracheae of large insects may widen into elastic air sacs (Fig. 8-14b). When the insect moves, the air sacs are compressed and expanded, achieving some ventilation. The respiratory exchange with tissue occurs in the fine, thin-walled tracheal branches called **tracheoles.** Less than 1 mm thick, tracheoles surround and may even penetrate individual cells.

SPIRACULAR VALVES. Tracheoles are permeable to liquids as well as gases, and their fine terminal branches are filled with tissue fluid where they come in contact with cells (Fig. 8-14c). Oxygen in the tracheoles diffuses through this tissue fluid into the cells. However, the fluid is subject to evaporation and can leave the tracheae in the form of water vapor by the same route through which the oxygen enters. Most spiracles have valves that restrict the loss of this vital moisture by opening only in response to $CO_2$ accumulation. Such control is essential. Not only is moisture needed for the final transfer of oxygen to tissues, but water loss is a life-and-death matter to insects, particularly in arid climates.

Tracheal respiration is a highly efficient process in small insects. But because tracheal function depends on diffusion and diffusion is efficient

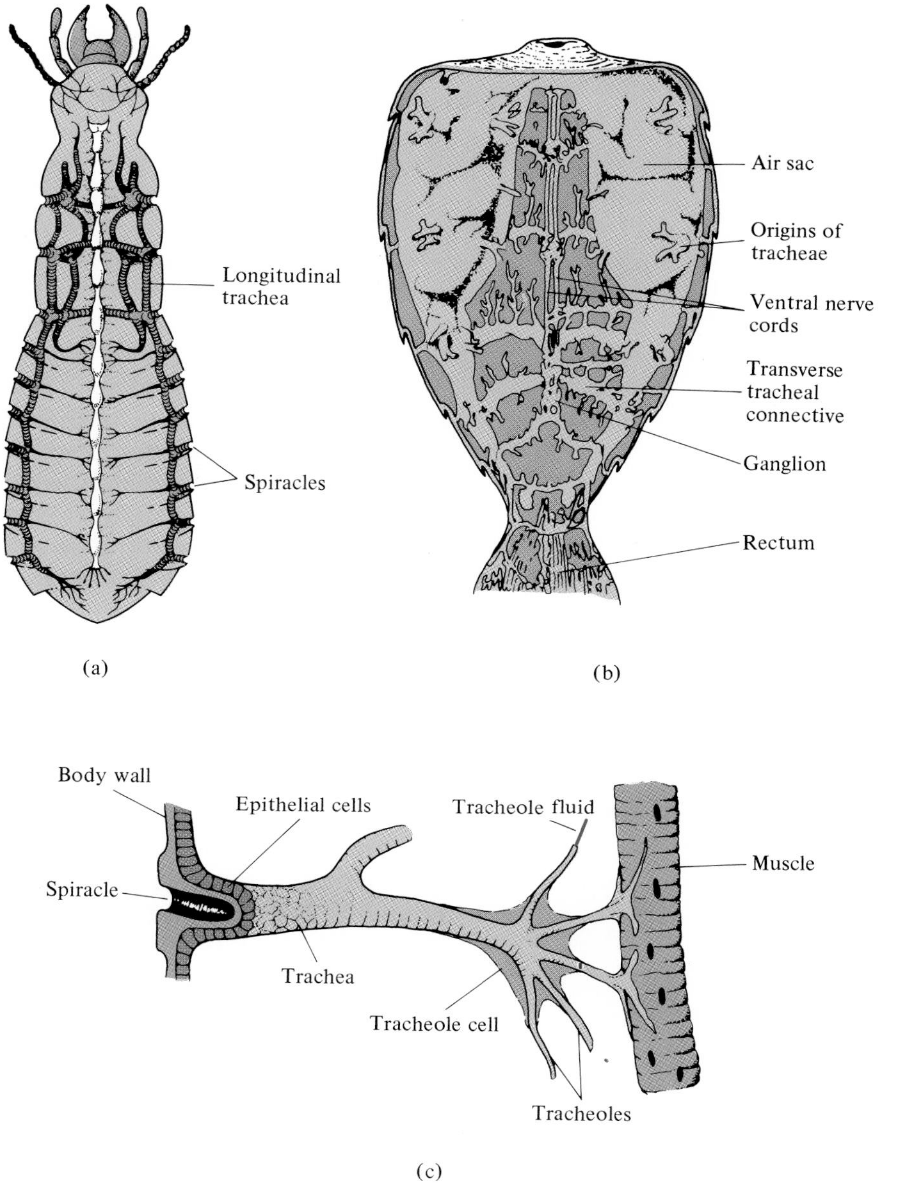


**FIG. 8-14** Insect tracheae. (a) Diagram of the tracheal system of an insect. (b) Dorsal view of the dissected abdomen of a worker honeybee, showing the air sacs and associated tracheae. The gut has been folded back (downward) to show the tracheal supply to the rectum. (c) Diagram showing the relationship of spiracles to tracheae and tracheoles.

for only short distances, it has long been assumed that this factor has been a restriction on the evolution of insect body size.

## WHEN FISHES BREATHE AIR

Fishes that breathe air fascinate biologists who seek to trace the course of evolution from an aquatic to a terrestrial existence. Anyone accustomed to thinking of a fish out of water as being in serious trouble will be surprised to learn that some fishes have functional accessory lunglike structures in their gill, or branchial, chambers (Fig. 8-15). Such adaptations not only supplement gill function by providing for air breathing but enable some species to move about on land for extended periods. Other such fishes never leave the water but have become so dependent on breathing that they suffocate if prevented from coming to the surface and inhaling air.

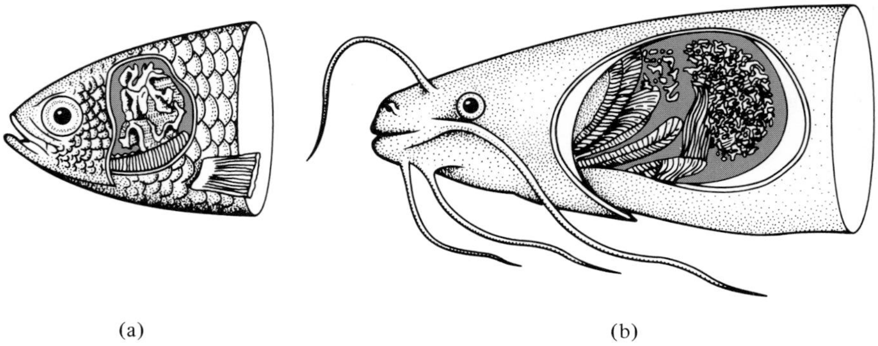


**FIG. 8-15** Special respiratory organs in (a) climbing perch, *Anabas,* and (b) African catfish, *Clarias.*

***Lungfishes.*** While some fishes have gill chambers that function in aerial respiration, **lungfishes** breathe air through one or two lungs (Fig. 8-16). One type of lungfish, the most primitive of all bony fishes, has changed very little throughout vertebrate evolution. All three genera of lungfishes alive today are tropical animals that live in ponds and backwaters of rivers, where the oxygen content is too low for fishes that depend on gills for respiration. It is believed that such environmental conditions prevailed during the mid-Paleozoic era, 400 million years ago, when bony fishes first evolved.

***Which Came First, the Swim Bladder or the Lung?*** Much of the interest in lungfishes is derived from interest in how the lungs of four-limbed vertebrates, or **tetrapods,** evolved. Most bony fishes have a **swim bladder,** the major function of which is to regulate overall body density. The fish regulates its buoyancy by secreting gas (mostly oxygen) into the bladder or absorbing gas from it, thereby rising or sinking with a minimum of effort. In most fishes, the bladder is connected to the mouth by way of the digestive tube. In this, the swim bladder is similar to the lungs (Fig. 8-17). Hence, for many years, it was assumed that the lungs of tetrapods had evolved from the swim bladder of fishes.

This assumption was wrong. Although swim bladders and lungs may be homologous—that is, they develop in a similar way and have a common evolutionary origin—lungs proved to be the older of the two. Lungs occur in the most primitive members of that group of bony fishes with swim bladders. The modern view is that swim bladders may even have evolved from the lungs of ancient fishes.

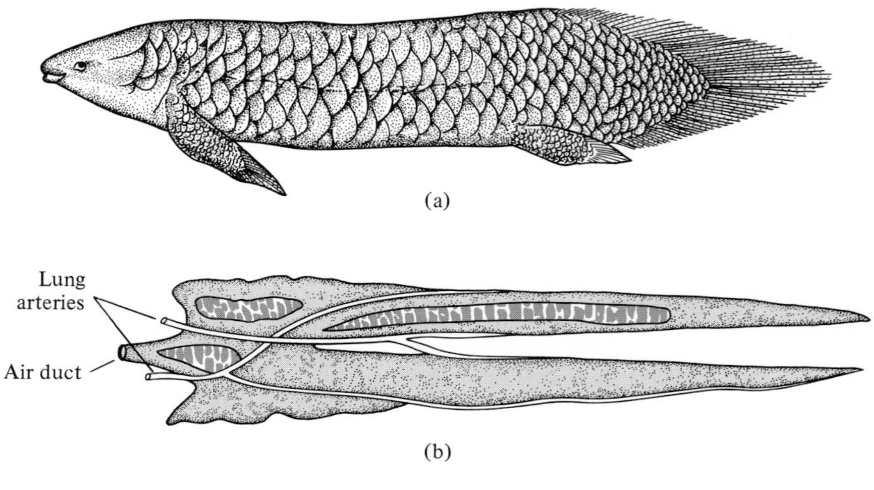


**FIG. 8-16** (a) *Epiceratodus,* an Australian lungfish. (b) Lungs of the African lungfish *Protopterus.*

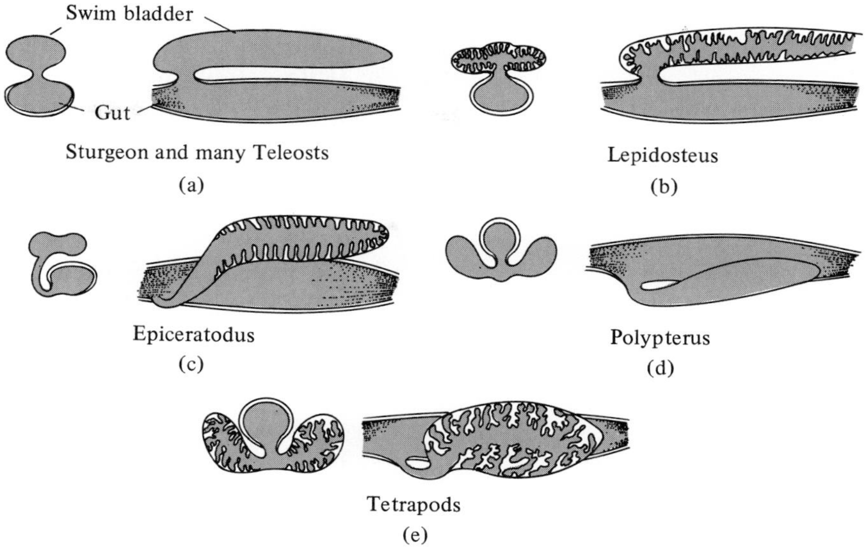


**FIG. 8-17** Diagrams of cross and longitudinal sections of swim bladders or lungs of vertebrates. (a) Swim bladder of the type found in most modern fishes. (b) Modified swim bladder of the gar pike, *Lepidosteus,* capable of some gaseous exchange. (c) Lung of Australian lungfish, *Epiceratodus.* (d) Lung of the primitive African fish *Polypterus.* (e) The tetrapod lung. Note that lungs branch off the ventral side of the gut, whereas swim bladders branch off the dorsal surface.

## THE INEFFICIENCY OF AMPHIBIAN LUNGS

Amphibians begin larval life as true aquatic animals, as dependent on gills as are fishes. In most, at metamorphosis, the gills are resorbed and lungs develop. Yet, because their inner respiratory surfaces have only a few minor folds, the lungs of amphibians are not very efficient (Fig. 8-18). Some adult amphibians even lack lungs.

The amphibian's principal respiratory organ is the skin. The adult frog has a richly vascularized skin that must always be kept moist, providing a much larger respiratory surface than its simple, saclike lungs. The need to keep its skin moist, however, greatly limits a frog's freedom, restricting it to moist environments. Its lungs provide oxygen only when the frog is active during the warm summer months. Even in summer, frogs can remain below water for extended periods, especially when they are inactive and the water is cool. Then, in late fall, the frog takes its last breath of the year and burrows into the mud of the lake bottom, depending entirely on its skin for respiration until spring.

## THE LUNGS OF TERRESTRIAL VERTEBRATES

The typical lung of the terrestrial vertebrate comprises one or more internal blind pouches into which air is either drawn or forced. The respiratory epithelium of lungs is thin, well vascularized, and typically divided into a large number of small units, which greatly increase the surface area for gaseous exchange between the lung air and the blood. The blind pouch construction, however, limits the efficiency with which oxygen and $CO_2$ are exchanged with the atmosphere because only a portion of the lung air is ever replaced with any one breath. Birds are exceptional in that they have very efficient lungs.

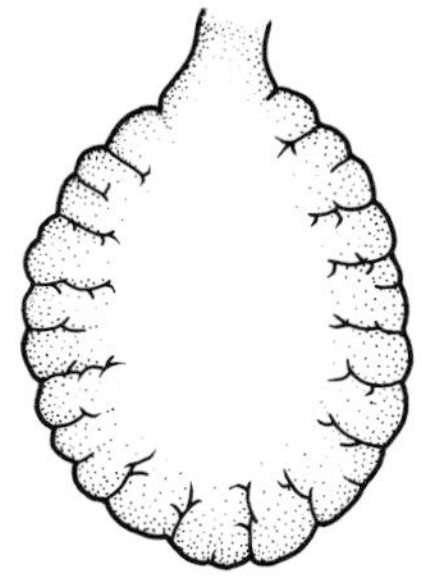

**FIG. 8-18** Lung of the frog in cross section. Note the limited surface area available for gas exchange within the lung.

## THE RESPIRATORY SYSTEM OF BIRDS

Birds use enormous amounts of oxygen. Not only do they maintain a high metabolic rate even while at rest, as evidenced by their high body temperature, but the exertion involved in active flight, although not as great as in running, requires a high respiratory rate. Moreover, the body size of birds is generally small. The consequent high ratio of surface area to body volume in a homeotherm results in a rapid loss of body heat and requires an increased oxygen consumption to maintain body temperature. Even though the oxygen requirements of birds are high, the sustained migratory flights of waterfowl dwarf endurance records by human athletes. Such flights are greatly facilitated by the structure of a birds' respiratory system.

The structure of a bird's "windpipe," or **trachea,** is much like that of the human trachea. But the rest of its respiratory system is strikingly different from that of the mammal. The bird's trachea divides into a right and left **primary bronchus.** Each primary bronchus branches into anterior and posterior **secondary bronchi** before terminating in a large abdominal air sac (Fig. 8-19a, b).

The secondary bronchi on each side connect with seven more air sacs (three paired and one unpaired) as well as with each other, the latter by means of the **parabronchi,** which contain numerous fine **air capillaries** (Fig. 8-19c). The air sacs are poorly vascularized and serve mainly to store air temporarily, but the air capillaries are surrounded by blood capillaries and make up most of the lung tissue.

The diagrammatic sketch in Figure 8-19b indicates how the avian lung functions. The anterior sacs are inflated during inspiration by air that has been diverted through the air capillaries. The posterior sacs are also inflated during inspiration but by air that reaches them directly via the primary bronchi. During expiration, the air in the anterior sacs bypasses the parabronchi and air capillaries as it exits, whereas air leaving the posterior air sacs is exhaled through them. Thus, oxygen-laden air passes through the parabronchi and air capillaries, always in the same direction, during both inspiration and expiration, rather than undergoing dilution, stagnation, or interruption of flow, as in mamma-

**FIG. 8-19** The avian lung as seen in a duck. (a) Main respiratory passages. (b) Air streams in bronchi, air sacs, and lung during breathing. Arrows show directions of air flow; lung volume changes are small relative to those of the air sacs; lung volume diminishes during inspiration; the lungs expand during expiration. (c) Scanning electron micrograph of a section through the lung of a 14-day old chicken, showing the tubular parabronchi and the small air capillaries branching from them.

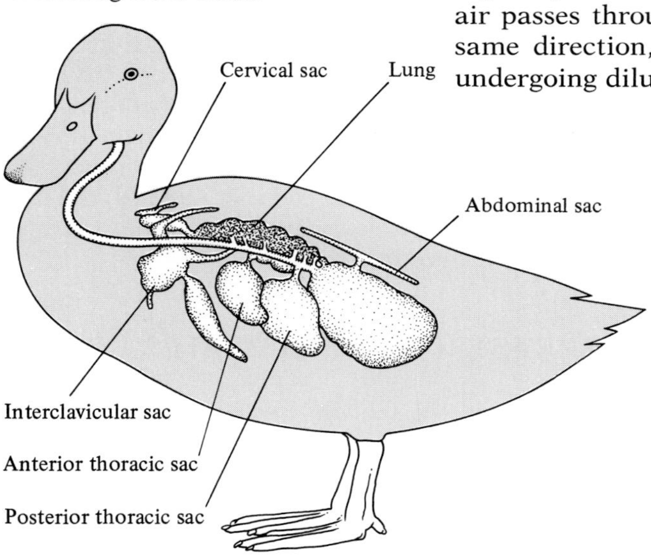


(a)

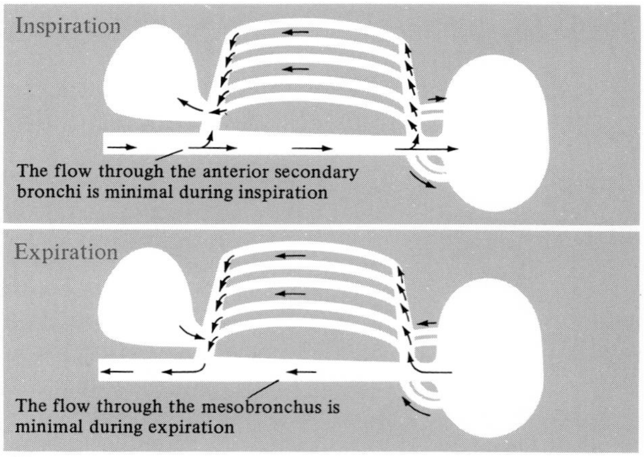


(b)

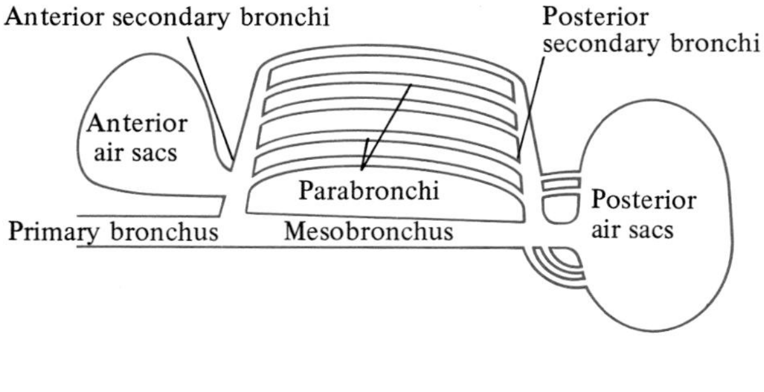


(c)

lian lungs. The direction of blood flow in the vessels serving the air capillaries is opposite the direction of the air flow, thus providing the high exchange efficiency of a countercurrent exchange mechanism (see Panel 8-2).

Besides contributing to respiration, the air sacs help cool the body, a matter of special importance during sustained flight. Because of air spaces between their feathers, birds' bodies are well insulated against heat loss. Since birds lack sweat glands, another avenue of heat loss is required during long periods of flight, and it is provided by the air sacs. The air sacs also contribute to the buoyancy of the body, greatly reducing the amount of work required for flying. Both buoyancy and cooling are enhanced by the extension of the air sacs into hollow air cavities in the long bones. Because of these air cavities, a bird with a blocked trachea but a broken wing could continue breathing—through its broken wing bone!

## THE INTERNAL TRANSPORT OF OXYGEN TO TISSUES

Oxygen enters the body of an animal through its respiratory surfaces and immediately dissolves in a body fluid, such as blood. The amount of oxygen absorbed by the fluid depends on the same factors that affect its solubility in any other aqueous medium: the fluid's temperature, salinity, and the partial pressure of oxygen. The first two factors determine the absolute amount of oxygen that can be absorbed, whereas partial pressure determines the rate of absorption. Unless a sufficiently steep gradient is maintained between the partial pressure of oxygen in the environment and that in the body fluids, too little oxygen may enter the blood in a given period (Fig. 8-20). The same principle applies at the

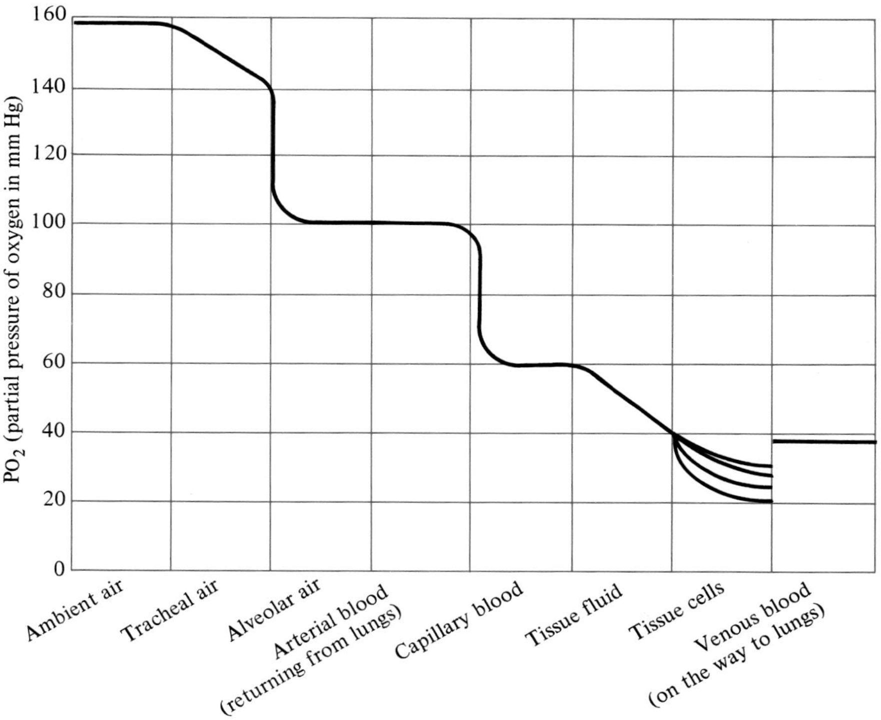


**FIG. 8-20** Oxygen tension gradients between the surrounding, or ambient, atmosphere and metabolizing tissue cells and for the various body fluids between the two. Values for the cells vary with respiratory rate and type of cell. Most data represent typical average values for humans.

tissues: if the difference between the **oxygen tension*** of the body fluid and that of the tissue cells is small, the rate of oxygen diffusion into the cells will be too low to supply actively respiring tissue.

The role of body fluids in respiration is to absorb oxygen from the environment in high enough concentration so that, when transported to the metabolizing tissues, it will diffuse rapidly into them.

When we consider all the factors that adversely affect the transfer of oxygen from the atmosphere to the cells of an organism, it is a wonder that enough reaches them. Indeed, most animals could not survive without the help of vascular systems or their equivalent and respiratory pigments.

***Vascular and Other Fluid Systems.*** Most animals have body fluids that function in the transport of nutrients and wastes. Insects, for example, have **hemolymph.** In the simplest animals, these fluids usually do not circulate within a system of vessels, or **vascular system.** Rather, they move back and forth in response to muscular contraction of the body. In a small animal, this movement is sufficient for adequate distribution of the fluid.

Larger animals require that materials be distributed by a blood vascular system, a necessity for active animals, even those of modest size. Typically, such vascular systems have a pumping device, such as a heart, that moves the blood along its roughly circular movement through the body. This process, called **circulation,** transports oxygen from where it is taken into the body to the tissues that require it.

***Respiratory Pigments.*** Because oxygen is poorly soluble in blood and other body fluids, little is ever carried in solution. In order to transport oxygen, active animals thus depend on substances in their body fluids to bind it temporarily in a loose chemical union. In humans, for example, only about 3 percent of the oxygen that is carried from the lungs to metabolizing tissues is dissolved in the **plasma,** the fluid part of the blood. All the rest is bound to **hemoglobin,** the respiratory pigment that gives blood its red color.

**Respiratory pigments** are common in animals. They occur in body cavity fluids, in blood, or in both. They may occur in cells, as in human red blood cells, or be dissolved in the plasma. They even occur in other tissues; for example, **myoglobin,** which gives beef its red color, is found in muscle.

Not all respiratory pigments are red. **Hemocyanin,** for example, found in snails, clams, and lobsters, is bluish. Nor are all blood pigments respiratory; some have no known function.

Of the several types of respiratory pigments, hemoglobin is the most common. This is not surprising, since the prosthetic group of this conjugated protein†—the iron-containing pigment **heme,** a porphyrin—is a component of one of the cytochromes found in all aerobic organisms (Fig. 8-21). The ability to synthesize heme was probably one of the early evolutionary developments that made aerobic respiration possible. The protein, a globin, to which heme is joined differs somewhat from one animal group to another, although its overall shape and much of its structure are nearly the same in many cases. The heme part of the molecule is always the same.

Because of the special chemical properties of hemoglobin and other

* *Tension* in this sense denotes the partial pressure of a gas dissolved in a liquid.

† A **conjugated protein** consists of a protein molecule joined to a nonprotein molecule, called the prosthetic group.

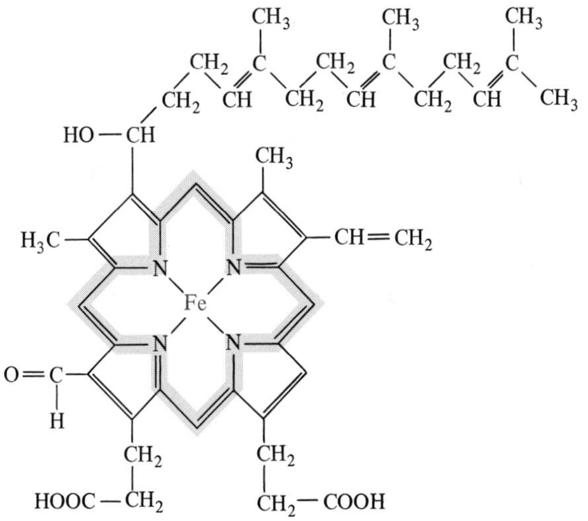


**FIG. 8-21** The structure of the heme molecule. The iron atom is in the center of the porphyrin ring.

respiratory pigments, blood not only can absorb a large amount of oxygen at the lungs and gills but also can release much of the oxygen at points where it is utilized. This release creates a steep concentration gradient and ensures the efficient entry of oxygen into the tissues that need it.

## HUMAN RESPIRATION

The mode of external respiration in humans is typical of mammals. It also embodies several physiological principles that apply to all air-breathing animals. These include the mechanics of breathing, oxygen transport by hemoglobin, carbon dioxide transport, and the control of breathing by the central nervous system.

### TRANSPORT INTO THE BLOOD

Air normally enters the human body by way of the **nasal cavities** (Fig. 8-22). The upper surfaces of these cavities contain specialized olfactory nerve fiber endings, the receptors of the sense of smell. Besides warming and moistening the air, the nasal passages filter out particles from the air by means of cilia and coarse hairs just inside the nostrils. Opening into the nasal passages are four pairs of **paranasal sinuses** that drain their secretion into the nose. They are notorious for their ability to become inflamed or infected.

The nasal cavities and the mouth lead into a common passageway, the throat, or **pharynx.** The pharynx functions as a traffic circle of sorts, normally routing air and food to the trachea and esophagus, respectively. At the sides of the pharynx, the **eustachian tubes** connect with the ears, enabling pressure in the middle ear to adjust to changes in atmospheric pressure.

At the entrance to the trachea is the **larynx,** the voice organ, supported by a flexible skeleton of nine cartilages and lined with a mucous membrane (Fig. 8-23). The mucous membrane consists of two shelflike folds that divide the cavity in two. The slit between these folds is the **glottis.** One of the laryngeal cartilages, the **epiglottis,** covers the glottis during swallowing and opens it during breathing. The **vocal folds**—"vocal cords," in popular terms—are shelflike ligaments embedded in

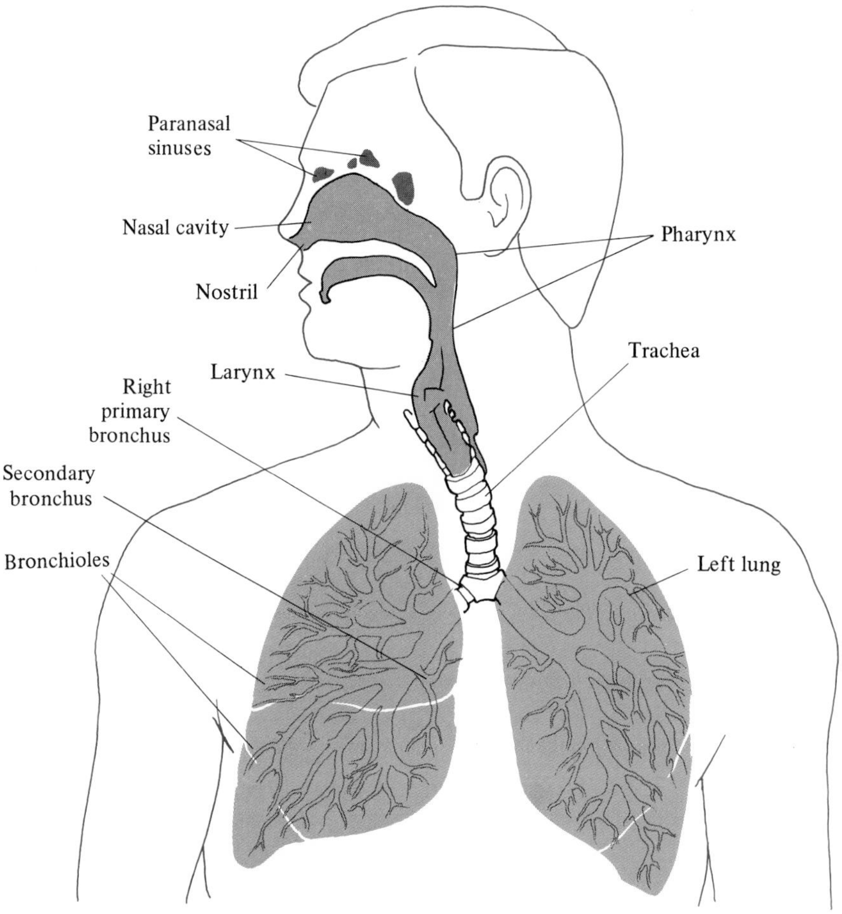


**FIG. 8-22** Organs of the human respiratory system.

**FIG. 8-23** The larynx, dorsal aspect.

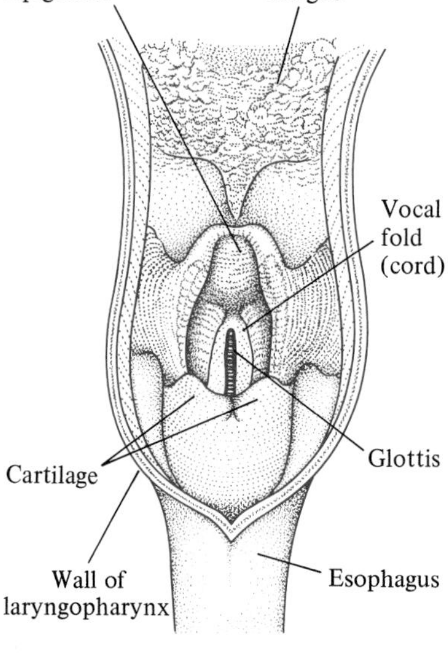


the edges of the glottis. The vocal folds are involved in production of sound and are controlled by a complex set of special muscles.

The glottis opens into the **trachea,** a ciliated, membrane-lined muscular tube about an inch in diameter. The trachea is kept from collapsing by C-shaped tracheal cartilages. The trachea branches into the left and the right **bronchi,** each of which branches again and again into smaller bronchi as it leads to a lung. The bronchi further branch into **bronchioles,** which in turn subdivide repeatedly. Each bronchiole ends in several extremely thin-walled **alveoli,** small groups of which constitute the **alveolar sacs.** It is the spasmodic contractions of the smooth muscles in the walls of the finest bronchioles that constitute the distressing attacks of bronchial asthma.

***The Alveoli.*** The exchange of oxygen and carbon dioxide takes place across the membranes of the alveoli (Fig. 8-24). Alveoli are rich in capillaries, and their cavities are separated from the bloodstream by walls that are often less than 0.5 $\mu$m thick (Fig. 8-25).

In contrast to the alveoli, the walls of the nasal cavities, the trachea, bronchi, and bronchioles exchange only a negligible amount of gas. They are said to enclose an anatomical "dead space," so-called because air occupying that space contributes nothing to respiration. However, this region warms and filters the air that passes through it and saturates it with moisture. The entire respiratory tract, except for the alveoli, is lined with cilia that maintain an upward current of fluid. When inhaled foreign particles strike the sides of the tract, they are trapped in

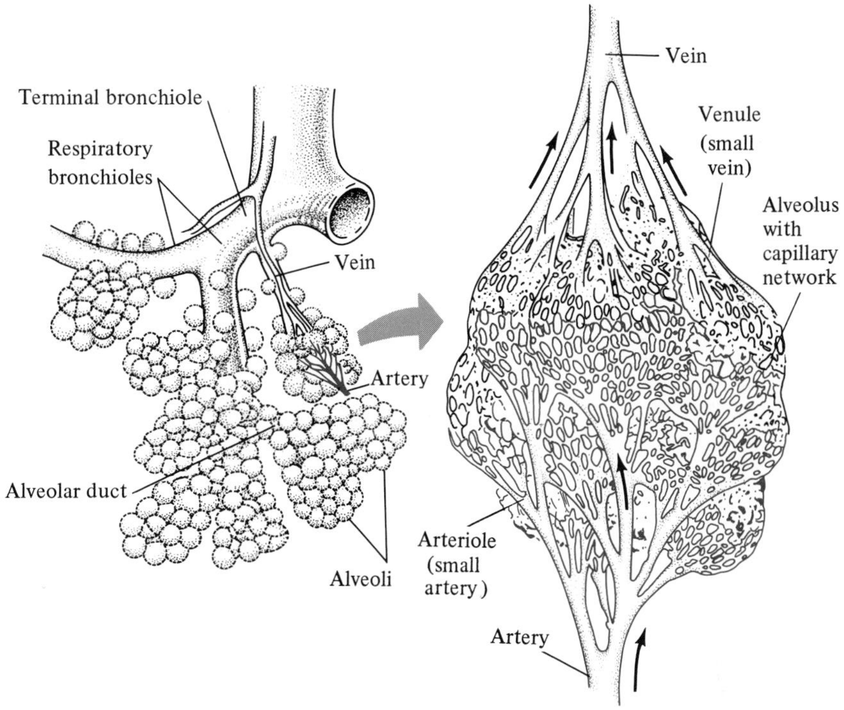


**FIG. 8-24** View of several alveoli of the mammalian lung showing their blood supply.

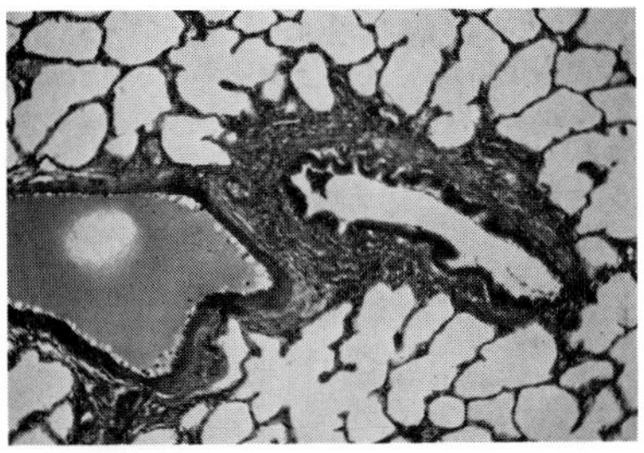

**FIG. 8-25** A photomicrograph of a section through alveoli of a human lung. Also shown are a bronchiole, on the left, and a blood vessel, on the right.

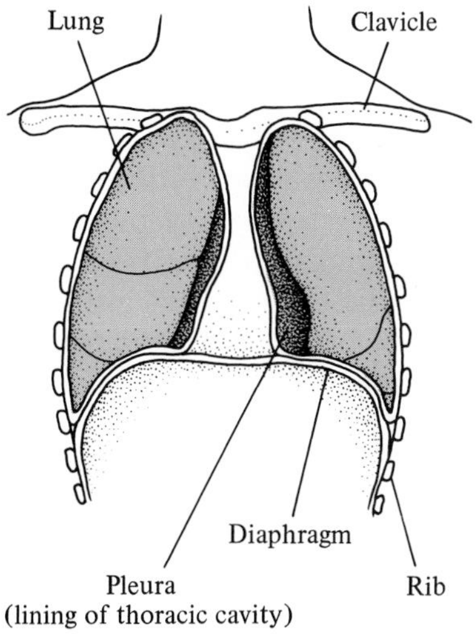


**FIG. 8-26** Anterior view of the opened thorax showing the position and extent of the diaphragm.

the film of moisture. The ciliary current returns them to the throat, where they are coughed up or swallowed (see Panel 8-3).

## THE MECHANICS OF BREATHING

Breathing has two phases, inspiration, the intake of air, and expiration, air expulsion.

***Inspiration.*** The only opening into the chest, or thoracic cavity is by way of the trachea, which does not open into the cavity itself, but into the lungs. For air to enter the lungs, pressure within the thorax must be reduced while the glottis is kept open to the atmosphere. Pressure can be reduced in two ways: (1) by raising the ribs and thus increasing the depth of the thorax, and (2) by lowering the floor of the thorax, which is composed of a broad sheet of muscle, the **diaphragm** (Fig. 8-26). After expiration, the rib cage is somewhat collapsed, but upon stimulation by nerve impulses, the rib, or **intercostal, muscles** contract. This raises the ribs and deepens the chest. The diaphragm, which at rest is curved upward like an inverted soup bowl, flattens upon contraction, substantially increasing the volume of the thorax. These two actions increase the chest volume and, if the glottis is open, air is drawn in, with a consequent expansion of the lungs (Fig. 8-27).

***Expiration.*** Expiration usually occurs passively when the intercostal muscles and the diaphragm relax. Once the intercostal muscles relax, the ribs slump again, their collapse helping to force air out the trachea. The elastic contraction of the lungs creates a lowered pressure within the thorax between the lungs and the walls of the pleural cavities. This pulls the relaxed diaphragm upward and the chest wall inward. The strength of lung elasticity is dramatically demonstrated when one side of the chest wall is punctured and air is thus allowed to enter the pleu-

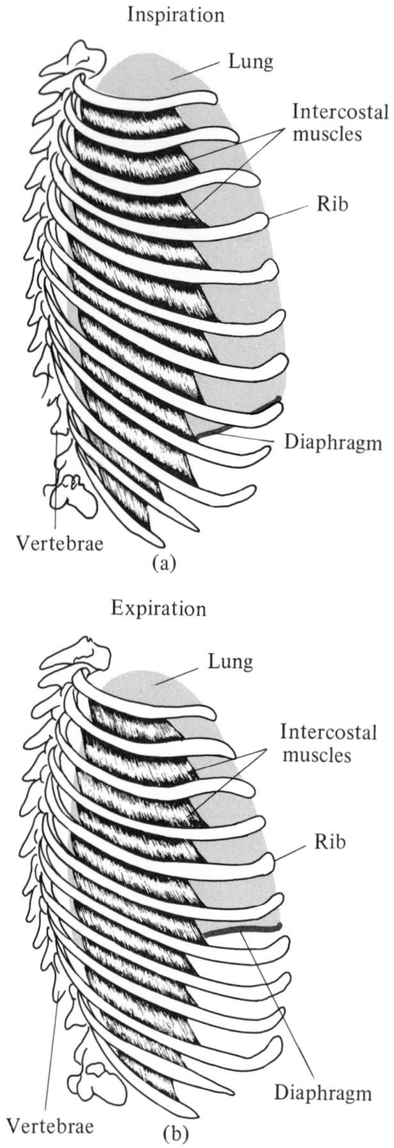


**FIG. 8-27** (a) Mechanics of expansion of the thoracic cavity by inspiration by the raising of the ribs and especially by the lowering of the diaphragm. Enlargement of the cavity lowers the thoracic pressure and air is inhaled. (b) Expiration occurs passively when the muscles relax, allowing gravity to pull down on the ribs and the elasticity of the lung tissue to contract it.

ral cavity. Much like a toy balloon, but one filled with spongy tissue, the lung promptly collapses,* a condition known as pneumothorax. Although pneumothorax may result from an injury, this condition is sometimes therapeutically induced to rest a diseased lung so that it can heal without expanding and collapsing.

***Other Muscular Action.*** The intercostal muscles and the diaphragm are usually the only muscles involved in normal respiration. It is possible to exhale additional air by contracting several other muscles, but no matter how forceful the expiration, some air always remains in the lungs. Thus, inhaled "fresh" air is always diluted by the residual air, which amounts to about a liter in healthy adult lungs. During normal breathing, the amount of new air introduced into the lungs at each inhalation is less than 10 percent of that present in the lungs. The alveolar lung is thus somewhat inefficient when compared to that of birds.

## THE TRANSPORT OF OXYGEN BY THE BLOOD

As oxygen leaves the alveoli to enter the capillaries, it dissolves in the fluids of the cells through which it passes. Its solubility in human plasma or any body fluid is quite low, largely because of the high temperature and solute concentrations of these fluids. Indeed, blood would be a poor fluid for carrying oxygen if it were not for its numerous red blood cells, or **erythrocytes,** which comprise over 40 percent of the blood's volume. These flexible red discs contain mostly hemoglobin, the pigmented protein with an affinity for oxygen. Red cells are so abundant in blood—5 million/mm$^3$—as to make it viscous. This fact, coupled with their short life span, averaging 120 days, means that they are produced by the bone marrow at the incredible rate of 2.5 million per second. They are also removed from the circulation at the same rate by phagocytic cells of the liver and spleen; these cells digest aging erythrocytes and recycle most of their component materials, with the exception of the heme portion of the hemoglobin molecule. In the liver, heme's iron atom is removed and retained, but the rest of the molecule undergoes several oxidations to form the bile pigments. These pigments, which give bile its color, are eliminated with the feces.

***Adaptations to Environmental Oxygen Shortages.*** The bone marrow produces erythrocytes in response to the body's need for oxygen. If oxygen tension falls because erythrocytes are destroyed or removed in greater than usual numbers—for example by hemorrhage or a donation of blood—they are produced in greater than usual numbers until the balance is restored. However, the mechanism responds to lack of oxygen rather than to erythrocyte numbers.† Consequently, people who move to high altitudes slowly adapt to the lower air pressure and less oxygen in part by acquiring and maintaining a higher concentration of erythrocytes in their bloodstream. For example, people who live at 5.3 kilometers (3.3 miles) above sea level have an erythrocyte count of 7.3 million/mm$^3$, compared to 5 million/mm$^3$ for those living at sea level. Dogs born and raised at high altitudes also have a much higher myoglo-

* In emphysema (see Panel 8-3), the lung remains inflated when the chest cavity is opened.

† The mechanism involves the kidney, which in the absence of normal amounts of oxygen, releases a hormone, erythropoetin, that stimulates the marrow to produce erythrocytes.

bin* concentration than do similar dogs raised at sea level. Such adaptations permit those who live at high altitudes to do a day's work without distress while the visitor feels exhausted, faint, and even ill upon slight exertion. Among the many adaptations of aquatic diving animals for prolonged breath holding are an unusually high myoglobin and erythrocyte concentration and an unusually high blood volume.

***The Efficiency of Hemoglobin as a Respiratory Pigment.*** Hemoglobin is exquisitely suited to the task of carrying oxygen. Whole blood—consisting of plasma and blood cells—carries 70 times more oxygen than could an equivalent volume of plasma alone. Hemoglobin thus represents a great economy for animals, both in the amount of blood they require to sustain life and in the work their hearts must perform.

Such a capability to bind oxygen, however, would be useless unless the oxygenated hemoglobin, called **oxyhemoglobin,** could readily release the oxygen when it reached the tissues. But it releases it most efficiently. As blood leaves the lungs, its hemoglobin is usually about 95 to 97 percent saturated with oxygen. From 24 to 45 percent of this oxygen is unloaded at the respiring tissues as blood passes through them, depending on the amount required by the tissues at any given moment. The efficiency of this highly evolved protein is truly remarkable.

* Myoglobin is a smaller, modified form of hemoglobin found in muscle tissue.

PANEL 8-3

## Self-Induced Emphysema

The habitual inhalation of tobacco smoke, even in rather moderate amounts, anesthetizes and finally may destroy many of the cilia of the respiratory tract. This renders the entire tract highly vulnerable to damaging materials in inhaled air, including tobacco smoke itself. But the tissues most severely affected by tobacco smoke and similar air pollutants are the delicate alveoli. Overexposure to such pollutants causes the thin alveolar walls to break down progressively, causing adjacent alveoli to fuse into larger and larger chambers. This process of erosion greatly reduces the surface area available for gaseous exchange. Moreover, the remaining alveolar walls become thickened and are less permeable to gases. The tissue gradually loses much of its elasticity, reducing the normal expulsion of air from the lungs. In addition, capillaries and arterioles become clogged, obstructing blood flow, further reducing lung efficiency. This general condition, characterized largely by the loss of lung elasticity, is called **emphysema.** Although nonsmokers can develop emphysema, especially in large cities where air pollutants abound, it is 13 times more prevalent in smokers. Respiratory diseases of this sort put a heavy burden on the heart because the heart must work harder to compensate for diminished oxygen absorption, thereby contributing greatly to heart disease. The incidence of lung cancer is also higher in smokers.

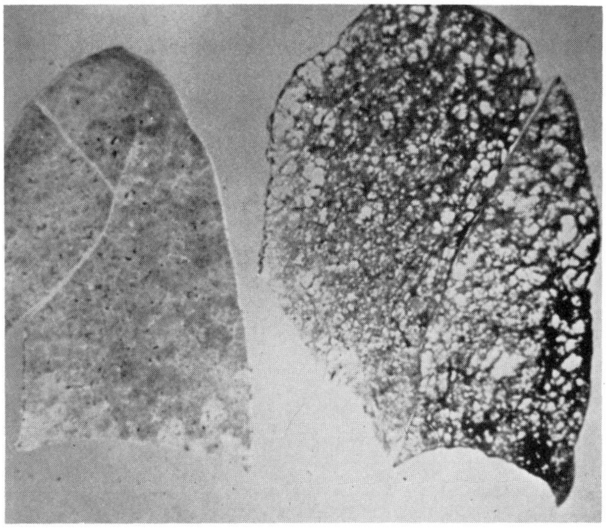

Lung disease. Normal lung tissue is shown on the left and lung tissue with emphysema is on the right.

Hemoglobin is composed of four polypeptide chains, two of one type of structure (the alpha chains) and two of another (the beta chains) (Fig. 8-28). Each of the four chains in each hemoglobin molecule binds one heme group. In each case, the iron atom of the heme group is covalently bonded to the protein on one side by one of two adjacent histidine residues (histidine is an amino acid). The other histidine is on the opposite face of the heme molecule, where it helps bind an oxygen molecule to the iron atom. Because there are four chains, each with a heme group, a total of four oxygen molecules can be carried by each hemoglobin molecule.

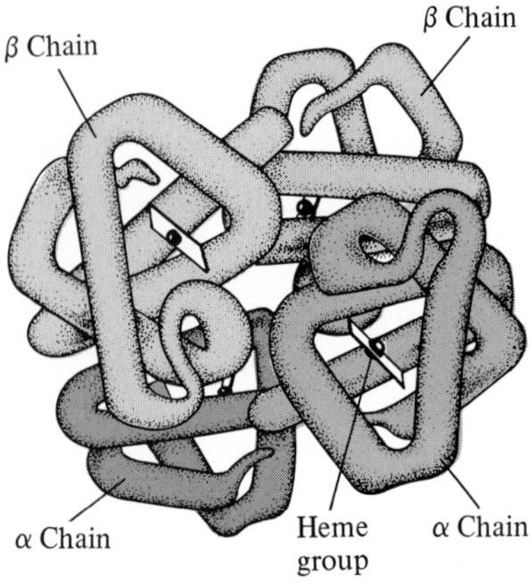

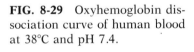

**FIG. 8-28** The three-dimensional structure of the hemoglobin molecule. The prosthetic group containing the heme portion is surrounded by the α and β chains.

The precise configuration of these chains specifically suits hemoglobin to be an oxygen carrier.* For, as each oxygen molecule is bound to a hemoglobin molecule, it induces in the protein a conformational change in its shape that increases the affinity of the remaining vacant sites for oxygen. The effect is an example of **cooperativity,** a type of allosteric regulation described for some regulatory enzymes in Chapter 3.

Cooperativity has two aspects. As hemoglobin molecules begin to bind oxygen at the lungs, where oxygen tension is high, their affinity for it increases sharply, and they bind much more as a result. (A similar allosteric effect, but with an unfortunate difference, is involved when carbon monoxide is inhaled; see Panel 8-4). As oxyhemoglobin loses oxygen at the tissues, where oxygen levels are low, its affinity for oxygen decreases and much more is released. When these events are plotted graphically, the result is the S-shaped, or sigmoid, oxygen dissociation curve shown in Figure 8-29, rather than the straight line relationship that would exist without the allosteric effect. An examination of the curve shows that hemoglobin is half saturated at an oxygen tension of 26 mm Hg. This is the steepest part of the curve, the region in which the allosteric effect renders hemoglobin most efficient as a respiratory pigment. During normal activity, however, hemoglobin is not called upon

* How finely adapted the hemoglobin molecule is to this role is illustrated by the disease sickle-cell anemia, which is due to a mutation in which one amino acid has been substituted for another (valine for glutamic acid) at the sixth position from one end of the beta chains. The resulting effect on the shape of the hemoglobin molecule and its ability to bind oxygen is severe. The red blood cells may collapse and death may result.

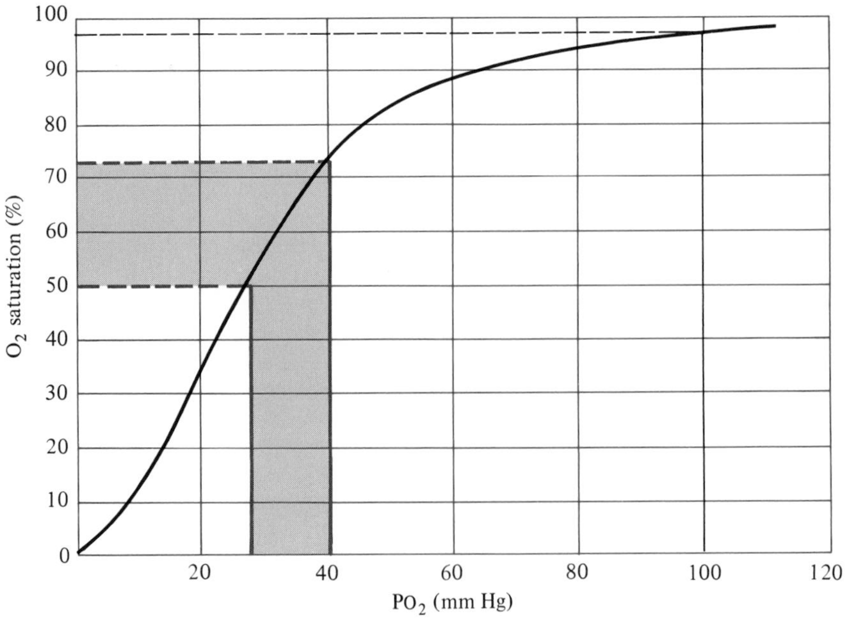


**FIG. 8-29** Oxyhemoglobin dissociation curve of human blood at 38°C and pH 7.4.

PANEL 8-4

# The Lethal Qualities of Carbon Monoxide

Carbon monoxide (CO) is not just another air pollutant or even just another poison. Two of its chemical attributes combine to make it a very deadly poison. First, hemoglobin has more than 200 times the affinity for CO as it does for $O_2$. Carbon monoxide thus competes effectively with $O_2$ for hemoglobin, and once it binds to hemoglobin as carboxyhemoglobin, it remains bound indefinitely, preempting sites that oxygen might have used. Moreover, when CO combines with one or two of the four heme groups in a hemoglobin molecule, an allosteric effect greatly increases the affinity of the remaining heme groups for $O_2$, with the result that any oxygen that is bound to the same hemoglobin molecule is not readily released at the tissues. Because of these two effects, a concentration as low as 0.1 percent CO in inspired air inactivates 50 percent of the body's hemoglobin in a few hours—this occurs even sooner at higher concentrations of CO. Fifty percent inactivation of an individual's hemoglobin is usually fatal.

to do its best; the vertical lines at 40 and 99 mm Hg represent the oxygen tensions of a resting individual's venous and arterial blood, respectively. The pigment's peak performance is reserved for times of strenuous muscular activity or other periods of oxygen stress.

***Other Factors Affecting Oxygen Delivery.*** The efficiency of oxygen delivery to tissues also depends on heart rate and rate of ventilation. Heart, or cardiac, rate may triple during vigorous activity. The blood supply to active muscles improves as the activity continues—one reason athletes "warm up" before an event. Warming up diverts blood to the active muscles and away from some other parts of the body, such as the digestive tract. The rate and efficiency of ventilation also increase during activity, more effectively renewing the lung air, although oxygen cannot be absorbed by the blood at the same peak efficiency (up to 98-percent saturation) that is possible at rest, when it travels more slowly through the lungs. Nevertheless, the total amount of oxygen that can be transferred from lung air to the blood is capable of increasing from 200 to 250 ml/minute at rest to as much as 5,500 ml/minute during extremely vigorous exercise.

## THE TRANSPORT OF CARBON DIOXIDE

The problems involved in removal of $CO_2$ from respiring cells are in some ways the reverse of those involved in the supply of oxygen. Diffusion again plays an essential role, with the steepness of the concentration gradients determining the rate, and therefore the efficiency, of $CO_2$ transfer at each step along the way. A graph very similar to that shown for oxygen transport could be constructed for $CO_2$ transport.

***Blood's Efficiency as a Vehicle for Transporting $CO_2$.*** Carbon dioxide is an unusual gas in that its effective solubility in water is quite high. Moreover, besides dissolving readily in plasma, $CO_2$ combines directly with hemoglobin. The pigment thus is a carrier for $CO_2$ as well as $O_2$. However, carbon dioxide does not compete with oxygen for a combining site on heme; rather, it binds to free amino groups (mostly of the amino acid histidine) on the protein (globin) part of the molecule. The complex thus formed is called carbaminohemoglobin. In the resting body, as much as 27 percent of the $CO_2$ delivered to the lungs is transported as carbaminohemoglobin, a smaller amount (about 9 percent) being carried in simple solution. By far the greatest amount (60 to 70 percent) of $CO_2$ carried by blood, however, is combined with water to form carbonic acid ($H_2CO_3$) or its ionized form, the bicarbonate ion ($HCO_3^-$).

The ability of the blood to carry $CO_2$ as $H_2CO_3$ is greatly enhanced by the presence of the enzyme carbonic anhydrase, found in erythrocytes:

$$CO_2 + H_2O \xrightarrow{\text{carbonic anhydrase}} H_2CO_3$$

Carbonic acid ionizes to form bicarbonate and hydrogen ions:

$$H_2CO_3 \longrightarrow HCO_3^- + H^+$$

Although the equilibrium point of this reaction is well to the left, the tendency of carbonic acid to dissociate, coupled with the considerable ability of the hemoglobin molecules to absorb hydrogen ions ($H^+$), moves the equilibrium point somewhat to the right. For, as carbonic acid ionizes, the hydrogen ions it releases are removed from solution by hemoglobin. It is probably the basic side groups of hemoglobin's amino acids, such as histidine, that absorb, or **buffer,** the hydrogen ions. Thus, despite the volume of $CO_2$ and carbonic acid that the plasma carries, its pH changes little (see also Panel 8-5).

The binding of both $CO_2$ and $H^+$ to hemoglobin in the blood serving the tissues in which $CO_2$ is produced, contributes to another important quality of hemoglobin. Oxygen is more readily released by hemoglobin after it has absorbed $CO_2$ and $H^+$. The result of this so-called **Bohr effect,** another allosteric phenomenon, causes oxyhemoglobin to release up to 15 percent more oxygen when passing through an actively respiring tissue (Fig. 8-30). At the lungs, the reverse occurs; the binding of oxygen by hemoglobin reduces the pigment's affinity for $CO_2$ and $H^+$, thus increasing the amount of $CO_2$ it releases into the lung air. Thus, it can be said that oxygen drives out $CO_2$ from hemoglobin in the lungs and $CO_2$ drives out oxygen at the tissues.

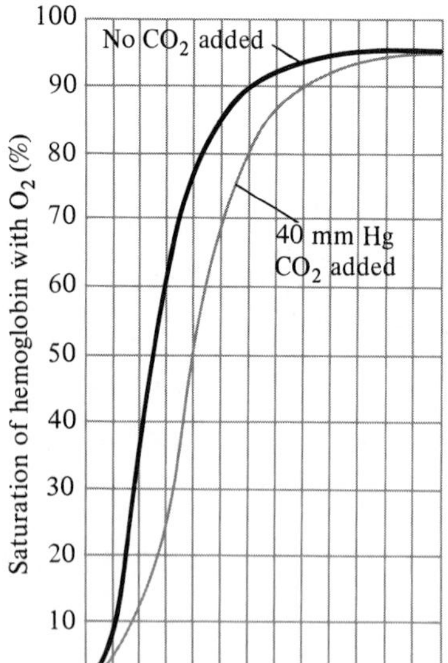


**FIG. 8-30** The effect of the presence of $CO_2$ on the oxygen dissociation curve of hemoglobin, known as the Bohr effect.

***Carbon Dioxide at the Lungs.*** As blood enters the lung capillaries, the processes that took place in the tissue capillaries are reversed:

1. Carbon dioxide in solution rapidly diffuses from the blood into the lung air, where the $CO_2$ concentration is much lower.
2. As rapidly as dissolved $CO_2$ leaves the blood, it is replaced by $CO_2$ that had been combined with hemoglobin as carbaminohemoglobin and $CO_2$ that had been combined with water to form carbonic acid. The decomposition of $H_2CO_3$ into $CO_2$ and $H_2O$ is speeded by carbonic anhydrase.
3. As rapidly as plasma $H_2CO_3$ is depleted, more of the ionized form ($HCO_3^-$ and $H^+$) reassociates to replace it. The reverse Bohr effect promotes this reaction when oxygen binds to hemoglobin, causing additional release of $H^+$.
4. As bicarbonate ions within the erythrocyte combine with $H^+$ to form carbonic acid, bicarbonate in the plasma diffuses in to replace it; $Cl^-$ diffuses out into the plasma in a reversal of the chloride shift (see Panel 8-5).

The net result of these events is a rather rapid release of $CO_2$ as the blood passes through the lungs. As a result, the $CO_2$ in the blood leaving the lungs is usually in equilibrium with the $CO_2$ in the lung air.

## PANEL 8-5

# How the Chloride Shift Maintains Ionic and Electrical Balance

It was noted earlier (Panel 2-1), that the pH of cytoplasm and body fluids, which is crucial for the functioning of enzymes or other molecules with a content of organic acids, is controlled by various buffers. An example of this is seen in the absorption of excess hydrogen ions by hemoglobin.

As $CO_2$ diffuses into erythrocytes, carbonic anhydrase quickly converts it into $H_2CO_3$, which dissociates into $HCO_3^-$ and $H^+$, with the $H^+$ rapidly being neutralized by hemoglobin. As $HCO_3^-$ accumulates within erythrocytes, it diffuses into the plasma and is carried to the lungs. The erythrocyte plasma membrane is quite permeable to $HCO_3^-$, as it is to anions in general. But plasma membranes are not generally permeable to cations. If this were not true, any bicarbonate ions that diffused out would be accompanied by such cations as the potassium ion ($K^+$), which are relatively abundant in cells. Because of the membrane's low permeability to $K^+$, an electrical imbalance results between the inside and the outside of the erythrocyte. This electrical difference causes a flow of chloride ions ($Cl^-$), the most abundant anion in plasma, into the cells. The chloride ions replace many of the bicarbonate ions that left, restoring a degree of electrical balance between the inside and outside of the cell membrane. This flow of chloride ions is called the **chloride shift,** and is also known as the Hamburger shift, after H. J. Hamburger, the German physiologist who elucidated it.

## THE CONTROL OF BREATHING

The rhythmic inspiration and expiration that characterizes normal, quiet breathing is regulated by the coordination of (1) the inherent rhythm of the respiratory center in the brain; (2) the direct and indirect influence of components of the body fluids, or humoral factors, namely, $CO_2$, $H^+$, and $O_2$ carried in the blood; and (3) inhibition of the respiratory center by nerve impulses generated in special receptors in the lung tissue that are sensitive to being stretched. This regular inhibition that occurs at each expansion of the lungs is essential: without it the respiratory muscles would remain contracted indefinitely.

***The Role of the Respiratory Center.*** Certain lower regions in the brain control breathing automatically. Breathing can also be controlled consciously by higher regions of the brain. When coordinated with such actions as sneezing, talking, coughing, laughing, singing, swimming, or playing a wind instrument, consciously controlled breathing can be a complicated activity. The automatic control of breathing is much simpler.

**FIG. 8-31** Cross section of the human brain showing the location of the pons and the medulla, sites of the nervous control of automatic breathing.

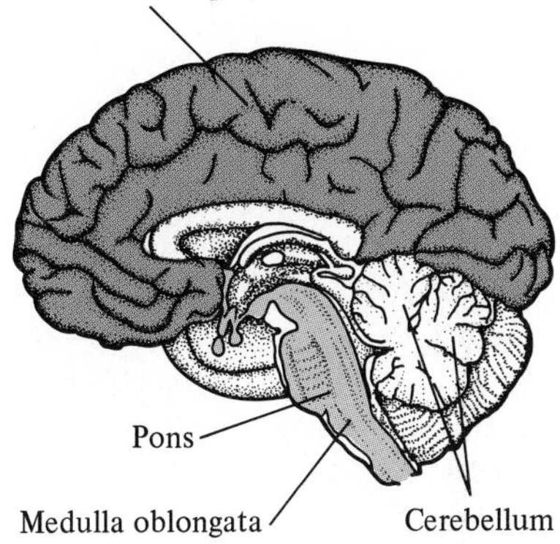


Normal quiet breathing depends primarily on a spontaneous series of nerve impulses arising in two adjacent groups of nerve cells, or **neurons,** in the **pons** and the **medulla oblongata,** the hindmost portions of the brain. These two groups of cells constitute the **respiratory center** (Fig. 8-31). One group of cells controls inspiration, the other expiration. Interaction between them forms the basis of the rhythm of respiration; they usually alternate in inhibiting each other. But their normal functioning depends on stimuli from outside the center.

***The Role of Humoral Factors.*** The principal stimulation of the respiratory center is a rise in arterial $CO_2$ concentration. The precise nature of this effect is not fully understood. In simplest terms it can be said, however, that normal breathing is a response to $CO_2$ accumulation in the arterial blood. Recent studies suggest that the respiratory center is affected by arterial $CO_2$ concentration as follows: $CO_2$ molecules dissolved in the blood diffuse through the special membranes that separate the blood from the brain tissue fluid known as **cerebrospinal fluid,** or CSF. Once in the CSF, some of the $CO_2$ combines with water to form carbonic acid, which dissociates into bicarbonate and hydrogen ions. Because of the absence of hemoglobin, CSF has a much lower buffering capacity than blood, so a given amount of $CO_2$ produces a higher $H^+$ concentration (lower pH) in CSF than in blood. This is believed to stimulate chemoreceptors located on the surface of the medulla. These cells in turn stimulate the nearby respiratory center.

In the final analysis, it is the $H^+$, not the $CO_2$, that stimulates the brain. But if that is the case, why does an increase of $H^+$ in the arterial blood affect the central chemoreceptors less than a rise in $CO_2$ does? The answer lies in a simple fact: membranes that separate the blood from the CSF are relatively impermeable to both anions and cations, including $HCO_3^-$ and $H^+$. Nevertheless, a rise in arterial $H^+$ does stimulate respiration, for two reasons: (1) some of the peripheral chemoreceptors in the blood vessels are sensitive to pH, and (2) the brain has some direct sensitivity to arterial $H^+$, although not nearly as much as to arterial $CO_2$. Thus, in normal breathing the respiratory center responds primarily to increases in arterial $CO_2$ concentration and to a lesser degree to increases in arterial $H^+$ concentration.

***The Role of the Peripheral Chemoreceptors.*** Why should the principal stimulation of breathing be provided by $CO_2$ accumulation rather than by lack of oxygen? After all, the greater danger to the body lies in lack of oxygen, not in $CO_2$ accumulation. In the absence of oxygen, death of brain tissues begins almost immediately. The answer is that sensitivity to a $CO_2$ increase anticipates, and thus avoids, the risk of oxygen shortage. But in fact, mammals have a separate mechanism that monitors each condition. Besides detecting and responding to some degree to high $CO_2$ and $H^+$ concentration, certain groups of cells, the **peripheral chemoreceptors,** located in the circulatory system, guard against the danger of lack of oxygen. They are stimulated and trigger appropriate responses whenever oxygen tension falls to a level of 50 to 60 mm Hg, corresponding to a saturation of arterial hemoglobin of 80 percent.

FIG. 8-32 Location of the carotid and aortic bodies.

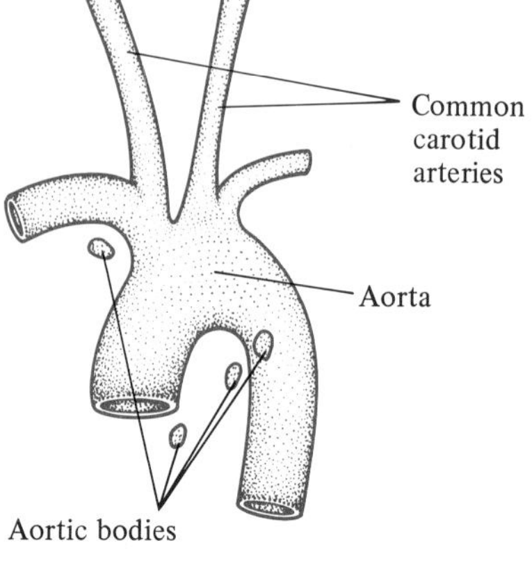


The peripheral chemoreceptors thus play a role in two control systems, one, like the respiratory center, giving an "early warning" due to $CO_2$ and $H^+$ accumulation and the other signaling a "red alert" if oxygen tension falls too low. Usually the accumulation of $CO_2$ stimulates ventilation sufficiently to prevent the stimulation of the peripheral chemoreceptors by lack of oxygen. But in some situations—on a high mountain, for example—oxygen tension can be insufficient even when $CO_2$ concentration is low. In such cases, the oxygen shortage, or **hypoxia,** triggers the peripheral chemoreceptors, sending excitatory nerve impulses to the respiratory center via the vagus nerve.

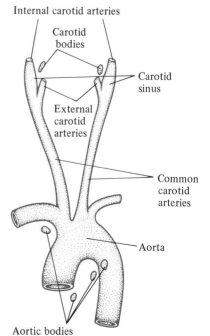

Two of the oxygen-sensitive chemoreceptors,* the **carotid bodies,** are located at the base of the internal carotid arteries, which supply the brain. Others, the **aortic bodies,** are located at the base of the right common carotid artery and in the aortic arch (Fig. 8-32). The carotid and aortic bodies respond to lowered $O_2$ tension in two ways: (1) by stimulating the respiratory center and (2) by causing constriction of

* These should not be confused with the baroreceptors, pressure-sensitive receptors found in the artery walls at the same locations, where they monitor another vital factor, blood pressure, ensuring that it does not drop too low. Baroreceptors are discussed in Chapter 11.

blood vessels, or vasoconstriction, especially of the arterioles, thus raising blood pressure and thereby increasing the efficiency of the blood's delivery of oxygen. The low oxygen tension at which these reflexes are activated, however, suggests that this mechanism does not usually contribute to the normal respiratory rhythm in healthy individuals at sea level conditions.

As noted above, the carotid and aortic bodies are also sensitive to $CO_2$, but the part this plays in normal respiratory control is not clear. It has been suggested that this sensitivity constitutes a standby system that responds to high levels of $CO_2$ and $H^+$ only when the central mechanism in the brain fails to respond adequately.

## Summary

Oxygen delivery and carbon dioxide elimination are universal requirements of aerobic organisms. The problems of supplying oxygen are associated with bringing it into contact with the body and transporting it into the body, to the tissues, and into the cells. The major limiting factors involved are the partial pressure of the oxygen, its solubility in water, and its rate of diffusion in water.

Although plants produce oxygen during photosynthesis, all tissues not carrying on photosynthesis need a regular, though small, supply of oxygen. Extensive intercellular air spaces provide adequate routes for oxygen supply in plants. Roots are especially vulnerable to lack of oxygen, however; land plants adapted to aquatic life have evolved specialized structures that provide oxygen to their roots.

Animals above a certain size or level of activity engage in ventilation, an action that regularly replaces the air or water at their respiratory surfaces with oxygen-carrying air or water. Because of oxygen's low solubility in water, the need for efficient ventilation is especially important for aquatic animals. Ventilation is often a result of other activities, such as feeding and locomotion. Specialized respiratory structures are necessary for large or active animals in which the amount of gaseous exchange possible through the body surface is inadequate. Such structures include many types of gills and lungs, modified mantle cavities, and, in insects and some spiders, the unique tracheae, which bring atmospheric air directly to the cells they serve.

Numerous forms that suggest what the transitional stages in evolution between aquatic respiration and air breathing might have been are represented on earth today. For example, amphibians have gills as larvae and lungs as adults. Three genera of lungfishes and certain other fishes have gills but are also able to breathe air. Some fishes even venture onto land for extended periods.

In contrast to the lungs of amphibians, those of terrestrial vertebrates are highly efficient, with large surface areas through which exchange of $O_2$ and $CO_2$ can occur between the lungs and the adjacent bloodstream. The blind-pouch construction of most vertebrate lungs, however, is inefficient compared with the unique, complex respiratory system of birds, in which oxygen-laden air passes through air capillaries in the same direction during inspiration and expiration. The direction of blood flow is opposite that of air flow, creating a highly efficient countercurrent exchange mechanism in the avian respiratory system.

The efficiency of oxygen transport within the bodies of animals is enhanced in two ways. One is by vascular and other fluid systems, many of which circulate the body fluids. The second contribution to efficiency is made by highly evolved respiratory pigments that enable body fluids to carry much more oxygen (and $CO_2$ in some cases) than would otherwise be possible and to unload it at the tissues as well. Of the several types of respiratory pigments, hemoglobin is the most common.

The human respiratory system is typical of mammals. Both the mouth and nose communicate with the trachea, which branches into several bronchi and then into many bronchioles, which end in blind, delicate pouches, the alveoli, within which gas exchange occurs with the blood. Inspiration is accomplished by expansion of the thorax upon contraction of the diaphragm and intercostal muscles. Expiration occurs when these muscles relax and the thorax partially collapses, mostly in response to the elastic recoil of the lung tissue.

Most oxygen transported by the human blood is bound to the hemoglobin contained in erythrocytes. Because it contains hemoglobin, whole blood can carry 70 times more oxygen than plasma can. Up to 45 percent of the oxygen absorbed at the lungs is unloaded at the tissues. This level of efficiency is due to an allosteric cooperativity in which, as each molecule of $O_2$ is bound to hemoglobin (which binds up to four), the hemoglobin molecule's affinity for oxygen increases, the reverse being true as it unloads its oxygen.

Carbon dioxide transport in human blood is accomplished by several mechanisms: (1) 7 percent is held in simple solution; (2) 50 to 70 percent is combined with $H_2O$ to form $H_2CO_3$, most of which dissociates to form bicarbonate and hydrogen ions, the latter being absorbed by various buffers, including hemoglobin; and (3) up to 27 percent is bound to hemoglobin itself. The binding of $CO_2$ and $H^+$ to hemoglobin produces the allosteric Bohr effect, enabling hemoglobin to unload 15 percent more of its oxygen at the tissues than if $CO_2$ were not present.

Breathing is automatically controlled from the respiratory center in the medulla oblongata and pons, with one group of cells controlling inspiration and another expiration. The principal stimulus to increased rate and depth of breathing is through a rise in the $CO_2$ concentration of the blood. Carbon dioxide molecules diffusing from the blood into the cerebrospinal fluid combine with water to form carbonic acid. The $H^+$ ions formed by its dissociation stimulate receptors of the medulla. Receptors in blood vessels also respond to a rise in $CO_2$ and $H^+$ concentration, besides detecting and responding to lowered oxygen concentration.

## Recommended Readings

COMROE, J. H., JR. 1966. "The Lung." *Scientific American*, 214(2):56–68. An interesting, semipopular account of human respiratory structure and function.

HOAR, W. S. 1983. *General and Comparative Physiology*, 3d ed. Prentice-Hall, Englewood Cliffs, N.J. Contains brief, up-to-date accounts of respiration in a variety of animals.

HUGHES, G. M. 1979. *The Vertebrate Lung*, rev. ed. Carolina Biology Readers Series. Carolina Biological Supply Co., Burlington, NC. A brief introduction to the anatomy and physiology of the mammalian lung.

KROGH, A. 1941. *The Comparative Physiology of Respiratory Mechanisms*. University of Pennsylvania Press, Philadelphia. This classic work remains one of the best introductions to the physiology of respiration.

MOUNTCASTLE, V. B., ed. 1980. *Medical Physiology*, vols. 1 and 2, 14th ed. Mosby, St. Louis. This excellent, upper-level work contains several chapters on human respiration written by experts in their respective fields.

NICHOLLS, P. 1982. *The Biology of Oxygen*. J. J. Head, ed. Carolina Biology Reader Series. Carolina Biological Supply Co., Burlington, N.C. An introduction to respiration.

SCHMIDT-NIELSEN, K. 1981. Countercurrent systems in animals. *Scientific American*, 244(5):118–128. Countercurrent systems, from camels to whale flukes.

# 9 Nutrition and Digestion

**OUTLINE**



Picture a prairie at dawn in late spring. As the sunlight strikes the leaves of the grasses and other plants, their rate of photosynthesis increases and their stomata open wide, admitting carbon dioxide and emitting water vapor. The loss of water is more than compensated for by the absorption of water from the soil by the roots of the plants. Important **nutrients**—substances required for growth and maintenance—are also absorbed from the soil, in the form of mineral ions. Rain and sunlight have been plentiful, and the meadow has experienced a lush growth since the arrival of spring.

As the sun warms the meadow, a young field mouse slowly emerges from its burrow and begins quietly to feed on the tenderest of the green leaves. Besides water, the leaves contain many kinds of nutrient materials needed by the mouse: simple salts such as sodium and calcium chlorides and complex organic matter such as proteins, carbohydrates, and lipids. The mouse, as it has each morning in the last few weeks, moves across the meadow, munching as it goes. Then, from the sky, swoops a

**FIG. 9-1** Three nutritionally distinct ways of life: autotroph (the brush and grasses in the background), herbivore (the mouse), and carnivore (sparrow hawk).

hawk. The mouse races for its burrow in a dodging, zigzag run that taxes the skill of the hawk to the utmost. But the mouse has wandered a few feet too far from safety. Just as it reaches its burrow, it is seized and quickly killed. The hawk's breakfast, like the mouse's, contains water, inorganic salts, and a great variety of organic substances (Fig. 9-1).

## THE EVOLUTION OF NUTRITION

Besides providing materials necessary for growth and maintenance, nutrients meet another need of animals: the need for energy. Whereas the energy requirements of plants are supplied by sunlight, in animals these needs are met by the energy contained in the food they eat.

The nutritional requirements of an organism are inversely related to its ability to synthesize the molecules essential to life: the fewer such **biosynthetic abilities** an organism has, the more kinds of nutrients it must obtain from its environment. Plants have the fewest such nutritional requirements. Besides their source of energy, light, they need only simple materials: $CO_2$, $H_2O$, inorganic nitrogen compounds, and various inorganic salts. Because nearly all plants synthesize all their own complex molecules from simple inorganic substances, they have long been called **autotrophs** (from the Greek *auto,* "self," and *trophe,* "nourishing"). Animals are described as **heterotrophs** (from the Greek *heteros,* "another" or "different") because they cannot synthesize many of their own organic molecules and must obtain them by eating other organisms or their products. Moreover, all the energy of animals is derived from organic matter that they eat. Animals, such as rabbits, that subsist entirely on plant matter are called **herbivores.** Those, such as hawks, that eat only meat are **carnivores.** Those that eat both are referred to as **omnivores.**

Much of the evolution of organisms has been marked by losses in biosynthetic abilities. Once an organism routinely obtains essential, complex organic molecules in its diet, it can afford to lose the ability to synthesize those materials. Moreover, loss of this ability confers a selective advantage on the animal because the animal stops expending energy and resources to synthesize substances that are always included in its diet. Thus, as the diet of organisms became more varied, they tended to lose their abilities to synthesize such widely available molecules as some of the amino acids, for example. This tendency led to the evolution of a variety of nutritional types, many of which do not fit the simple autotrophic-heterotrophic classification. Examples are the many algae that require one or more organic growth factors, such as vitamin $B_1$ (thiamine), to live. Such algae were formerly classified as autotrophs—yet autotrophs by definition need only inorganic matter to live. New systems of nutritional classification have been devised to deal with this fact (see Panel 9-1).

### NUTRIENT AVAILABILITY SELECTS FOR NUTRITIONAL CAPABILITY

Merely because all the chemical elements that an organism needs are present in its environment does not guarantee its survival. These materials must be in a form that can be utilized by that organism. For example, the nitrogen in air exists as molecular nitrogen, $N_2$, a form that can

## PANEL 9-1

# Types of Nutrition: A Modern View

One of the simpler ways of accommodating the several types of nutrition now known to exist is to classify all organisms according to their principal source of energy. Thus, those that use light energy are termed **phototrophs,** and those that use chemically bound energy are **chemotrophs.** Phototrophs can be subdivided according to whether they need *only* inorganic matter and $CO_2$. If they can grow on such simple materials, which is the case with most plants, they are termed **photolithotrophs** (from the Greek *lithos,* "stone"). If organic growth factors are required, as in the case of many algae, they are called **photoauxotrophs** (from the Greek *auxo,* "increase").

The chemotrophs are subdivided according to whether their energy source is an *inorganic* substance, as is the case with sulfur bacteria; such organisms are classified as **chemolithotrophs.** If, as with animals, organic matter is their primary source of energy, the organisms are classified as **chemoorganotrophs.**

be used by only a few species of bacteria and cyanobacteria. Large amounts of calcium, another universal requirement of organisms, are present in many rocks, but few organisms can utilize it in this form. Petroleum is another rich source of energy that very few organisms can utilize. Wood is an excellent source of nutrient matter, but only for the few species of plants and animals that have evolved the enzymes to digest it. Although termites eat wood, it is actually digested by certain protozoans found in their gut. Each kind of environment thus harbors those organisms with the physiological, behavioral, and biochemical capabilities that allow them to utilize the nutrients available in that environment.

## PLANT NUTRITION

Algal protists and plants occupy a very special place in nutrition. They take water, the soluble nutrients of the soil (or ocean, lake, or river, in the case of aquatic forms), and the $CO_2$ of the air and convert them into the complex nutrients upon which nearly all other forms of life depend. They synthesize all of their complex molecules from simple compounds. Except for their $CO_2$ requirement, the nutrition of plants and other photosynthetic organisms is essentially mineral nutrition: it depends upon inorganic compounds obtained from rocks and soil. It was not until well into the nineteenth century that plants were raised in the absence of soil by placing their roots in water to which had been added various combinations and proportions of salts. This method of plant culture, now called **hydroponics** (Fig. 9-2), showed that for best growth, plants need certain proportions of calcium nitrate, $Ca(NO_3)_2$; potassium dihydrogen phosphate, $KH_2PO_4$; and magnesium sulfate, $MgSO_4$. From such experiments it was found that the principal mineral elements needed by plants are nitrogen (N), phosphorus (P), potassium (K), calcium (Ca), magnesium (Mg), and sulfur (S). These major nutritional elements are called **macronutrients** because they are required in relatively large amounts compared to other nutrients. Of these, N, P, and K are used in greatest amounts by plants and therefore are most readily depleted from farm or garden soil. They must often be added to the soil to make it

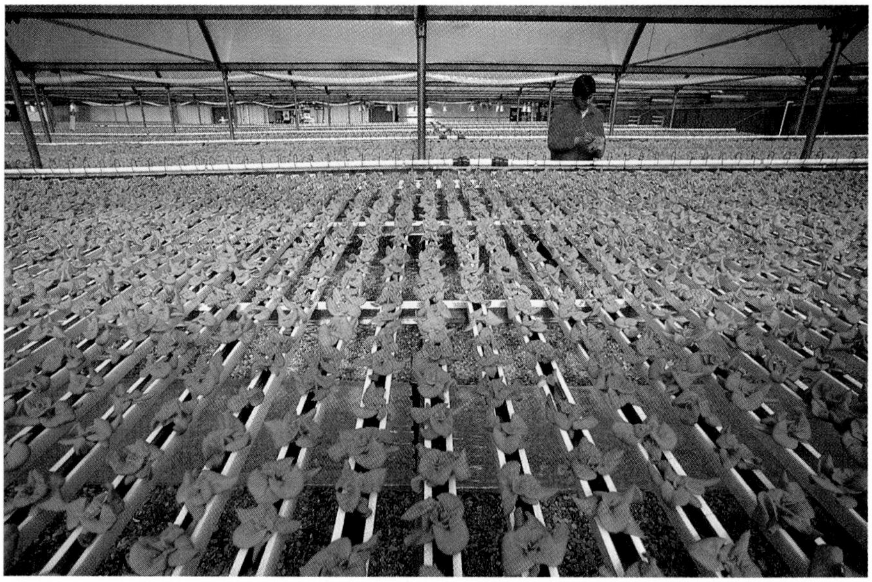

**FIG. 9-2** Growing lettuce by hydroponics. It has been known since the middle of the nineteenth century that plants can be grown in the absence of soil as long as all essential nutrients are added to the water in which the roots are immersed.

highly productive. Accordingly, N, P, and K are called fertilizer elements. A fertilizer rated 20-5-3, for example, contains 20 percent N, 5 percent P, and 3 percent K.

## MICRONUTRIENTS

Determining an organism's requirements for minerals that are needed only in minute amounts has been a difficult problem to resolve because such minerals always occur as trace contaminants in the purest chemical reagents. These plant **micronutrients,** or minor nutritional elements, include iron (Fe), manganese (Mn), boron (B), copper (Cu), zinc (Zn), chlorine (Cl), and molybdenum (Mo). Some plants also require trace amounts of selenium (Se), silicon (Si), aluminum (Al), cobalt (Co), gallium (Ga), or vanadium (V).

Despite the large number of mineral nutrients, 95 percent of plant tissue is composed of three other elements, carbon, hydrogen, and oxygen, which are derived from water or atmospheric carbon dioxide, the basic ingredients for photosynthesis. Because water and carbon dioxide are usually abundant, the limiting nutritional factors in plant growth are almost always the soil nutrients. Thus, the study of plant nutrition deals almost exclusively with the supply and use of the mineral nutrients dissolved in the water of the soil, rivers, lakes, and oceans in which plants live. A list of elements found essential for most higher plants is given in Table 9-1.

**FIG. 9-3** Surface cells of a root, showing root hairs.

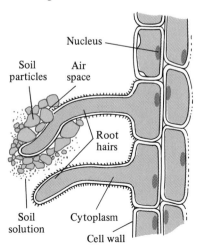

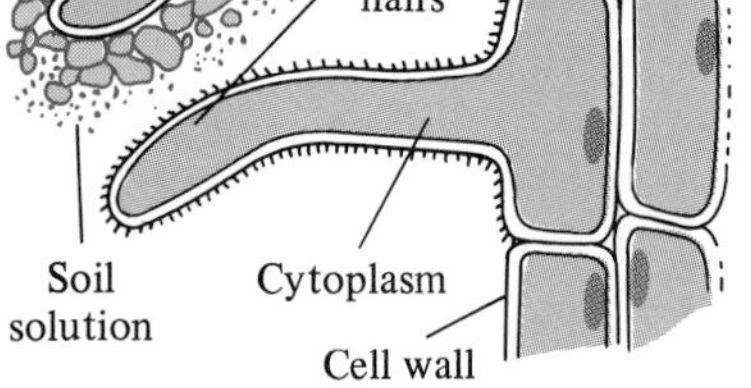


## HOW MINERALS ENTER PLANTS

Water and minerals are absorbed from the soil through the walls of the epidermal cells of a plant's roots, particularly through the root hairs, microscopically fine extensions of the epidermal cells (Fig. 9-3). Most water also enters by way of the minute pores in the cell walls enclosing the epidermal cells. The roots of many plants have symbiotic fungi called **mycorrhizae** growing on and in their root tissue; these facilitate the entrance of water and nutrients. For some plant species, mycorrhi-

**TABLE 9-1 Elements Essential for Most Higher Plants**

| *Element* | *Chemical symbol* | *Percent of dry weight of plant tissue** |
|---|---|---|
| Oxygen | O | 44.4 |
| Carbon | C | 43.6 |
| Hydrogen | H | 6.2 |
| Nitrogen | N | 1.5 |
| Potassium | K | 0.9 |
| Boron | B | 0.2 or less |
| Calcium | Ca | |
| Chlorine | Cl | |
| Iron | Fe | |
| Magnesium | Mg | |
| Manganese | Mn | |
| Molybdenum | Mo | |
| Phosphorus | P | |
| Sulfur | S | |

* Dry weight, about 10 percent of fresh weight, values are for a mature corn plant, *Zea mays.*

zae are essential for survival. Plants normally absorb water from the soil by osmosis and under the tension (pull) generated by transpiration, the loss of water vapor through the stomata of leaves (Chapters 7 and 10). In absorbing water, the plant also passively absorbs some dissolved ions. Mineral uptake is thereby slightly increased because of water uptake. Contrary to an earlier belief that minerals dissolved in soil water were merely absorbed along with osmotically imbibed water, plants absorb most minerals from soil water by diffusion and active transport.

## NUTRITIONAL PROBLEMS IN FARMING AND GARDENING

In farming and gardening, shortages of micronutrients are rarely a problem. They are contained in many fertilizers, in organic matter such as manure and compost, and in clay, a component of most soils. However, as continued heavy cropping is practiced with the use of artificial fertilizers and less manure and compost, micronutrient depletion is becoming a problem on more and more farms.

***Macronutrients.*** Plant nutrient shortages are nearly always a question of a lack of nitrogen, phosphorus, or potassium, the so-called fertilizer elements. Nitrogen is the macronutrient most often in short supply in farm and garden soils, while phosphorus is the one usually lacking in lakes and rivers. A nitrogen-poor soil yields stunted plants with yellowed leaves. Adding nitrogen stimulates vegetative growth, producing deep green, succulent leaves, and plump grains. The other three macronutrients—calcium, magnesium, and sulfur—if not already abundant in farm soil, are usually added with other substances, such as lime, phosphate, and manure.

***Does Organic Matter Have a Role in Plant Nutrition?*** Besides inorganic matter, average farm or garden soils contain 3 to 5 percent organic matter. If intensively cropped for years without the addition of organic matter, however, a soil may contain less than 1 percent organic matter. A virgin prairie topsoil, on the other hand, may contain up to 12 percent organic matter. Exceptional cases are the organic soils, called "muck" soils, of ancient bogs, or overgrown swamps, that contain up to 80 percent or more organic matter (Fig. 9-4).

At least 5 percent organic matter is needed to maintain optimum **soil structure** for root and air penetration and for moisture retention. Soil structure refers to the way soil particles cling to each other to form aggregates. Soil structure is influenced by **soil texture** (its proportion of sand, silt, and clay particles), by wet-dry and freeze-thaw cycles, by types of roots and soil organisms, by pH, by tilling, and, of course, by the soil content of organic matter. Besides benefiting soil structure, organic matter decays under the action of bacteria, protists, and fungi, slowly releasing inorganic nutrients that are thus provided to plants as they are needed. Organic matter also binds important ions, keeping them from being leached out of the soil before plant roots can absorb them.

***Sources of Organic Matter.*** Organic matter arises naturally from the many organisms that live and die on or in the soil. Animals return much organic matter to the soil as feces and urine. Large plants improve soil structure by loosening it with their roots, by bringing materials up from the subsoil, and, because roots continually shed surface cells, by contributing their own tissues to the organic layer of the soil. A plant's roots release substances into the soil that condition the soil for the plant itself as well as for soil organisms. Various protists, fungi, bacteria, and small animals (such as insects, sowbugs, centipedes, spiders, mites, slugs, snails, earthworms, nematodes, and rotifers) subsist on other organisms or their residues, returning them to the soil in various stages of decay. Organisms or parts of organisms that have decayed to such an extent that they can no longer be recognized are called **humus.**

**FIG. 9-4** Organic, or "muck," soil. This type of soil, containing as much as 80 percent organic matter, is preferred for a few types of crops, such as the celery being planted here in a South Bar, Florida, farm.

***Humus.*** As far as plants are concerned, soil particles are mostly insoluble and inert, nutritionally speaking. But fine humus and, to a lesser extent, clay particles have negative electric charges on their surfaces. At the right soil pH, these particles bind, or adsorb, cations such as $H^+$, $Ca^{2+}$, $Mg^{2+}$, $K^+$, and $Na^+$. Because $H^+$ ions compete for binding sites with other cations, very acid soils—pH 5.5 or less—are poor in other cations. Although humus is insoluble, its slow decay eventually results in the release of inorganic matter that becomes available to roots. Almost without exception, plants absorb only inorganic ions (see Panel 9-2). Only a very small amount of the simpler organic molecules, such as amino acids and sugars, are ever absorbed from the soil.

## HETEROTROPHIC HIGHER PLANTS

Although most plants are photolithotrophic—autotrophic in the strict sense—and require no intake of organic matter, a few genera of flowering plants augment their nutrient supply with organic matter. Others are true parasites and depend entirely on other plants for nutrients.

### PANEL 9-2

## Organic Gardening

In response to various abuses of agricultural soil by some segments of the crop-raising industry, a movement supporting what is called organic gardening and farming has arisen in recent years. Its adherents have generally tried to break away from practices they see as harmful to the soil or crops, such as the use of inorganic fertilizers. Most insist on the incorporation of organic matter into the soil as the basis for good gardening and farming. In the extreme version of the concept, all insecticides and herbicides are avoided and nothing but organic matter and water are added to the soil. However, only the most avid "organic gardeners" exclude all limestone or dolomite. Farm manure may be supplemented with phosphate, although perhaps only with untreated ground rock phosphate. Some organic farmers and gardeners use insecticides, but sparingly, and only from a small list of **biodegradable** plant derivatives, that is, those that readily decompose under the action of bacteria, fungi, and other soil organisms. Most organic gardeners also avoid treating harvested foods with preservatives. The phrase "organic living" has been used to describe this extension of the philosophy.

Because organic farming on a small scale has proven workable, many claims have been advanced on its behalf. The most extravagant of these assertions, however, are quite untenable. Two examples are the notions that plants absorb organic matter from the soil and that inorganic chemicals are generally harmful to plants and to anyone who eats crops grown in soil containing inorganic chemicals. Neither is true. Organic matter, such as compost or manure, must be converted to *inorganic* form before it can be absorbed by plants. Without adequate amounts of inorganic nutrients, plants cannot survive. Plants grown on rich prairie soil containing enough nutrients to last for years of gardening without adding anything may have misled people into believing that plants can be grown on organic matter alone. Organic matter is, of course, beneficial to plants, but *not* because it is absorbed as such. Rather, its nutrients can be utilized by the plant only after the matter is metabolized (decayed) by soil organisms, which then release inorganic salts into the soil. Organic matter also acts to bind ionized nutrient salts, which are thus made available to the plant as needed instead of being washed away, as can happen when some commercial fertilizers are used. Moreover, organic matter is essential to good soil structure, contributing importantly to water retention and to the soil porosity needed for aeration and root penetration.

(a)

(b)

**FIG. 9-5** Venus's flytrap, an insectivorous plant. (a) A fly has landed. If it touches two or more of the 6 hairs inside the trap or the same one twice, the trap closes quickly. (b) A trap is closing on a fly, which may not die for several hours. Digestion and absorption require about 10 days; after that time, the trap opens and releases the remains.

***Insectivorous Plants.*** Some species of plants absorb organic matter obtained from the capture and digestion of insects (Fig. 9-5). These **insectivorous plants** usually produce a nectar that attracts insects, a sticky substance that entraps them, and digestive enzymes that reduce the prey to a broth of organic molecules. Nitrogenous matter, particularly amino acids, appear to be the most important of the substances absorbed by the plant. Most insectivorous plants are adapted to living in acid bogs and soils poor in available nitrogen. The evolution of their "insect-eating" ability has helped provide this vital substance.

***Parasitic Flowering Plants.*** Several genera of flowering plants obtain their nutrients by parasitizing another plant, termed the host. Mistletoe is an example. The parasite penetrates the tissues of its host, often a tree, with specialized organs called suckers that absorb organic nutrients such as sugar. Some parasitic flowering plants are entirely devoid of chlorophyll.

**FIG. 9-6** All fungi depend for most of their nutrients on the absorption of decaying organic matter.

***Saprotrophs.*** Two kingdoms that were included with plants under older taxonomic systems are the fungi and bacteria, both of which are heterotrophic with the exception of a few species of bacteria. These organisms depend on the absorption of partially decayed material from their immediate environment for their survival (Fig. 9-6). Nutrition that involves the absorption of organic matter is termed **saprotrophy** (from the Greek *sapros,* "rotten"). Fungi and some bacteria secrete enzymes into their immediate environment. These enzymes decompose large organic molecules to smaller ones, thus permitting their absorption. Some fungi supplement their diet by actually entrapping and digesting soil organisms. In this manner, some fungi, like insectivorous plants, are able to enhance their uptake of nitrogen. When the weather is warm, fertile soil and ponds normally teem with saprotrophic microorganisms—100,000,000 bacteria per gram of soil is common—that digest, absorb, and degrade almost any kind of naturally occurring organic matter.

# THE METABOLIC FATES OF NUTRIENTS IN HETEROTROPHS

Biologists have long had a special interest in discovering the major roles of specific nutrients needed by organisms. Although it has been known for a long time that certain nutrients are essential to human health—for example, Lancaster showed in 1601 that scurvy could be prevented in British sailors by eating fresh fruits or vegetables—most of our knowledge of dietary needs has been acquired in this century. The science of nutrition has established a long list of mineral and organic dietary requirements for mammals in particular and heterotrophs in general. But it is one thing to state what is needed for normal growth and health and another to say why. We will now consider some of the principal functions of important nutrients.

## METABOLIC NEED FOR MINERALS

Minerals play a major role in nutrition. For example, phosphorus is a component of ATP and nucleic acids. Magnesium forms an essential complex with ATP in most enzymatic reactions in which ATP is a phosphate donor. Iron is a part of heme and is thus found both in the cytochromes and in hemoglobin. Calcium is another mineral that is very important to the body. Examples of body functions that require calcium are nerve impulse transmission, muscle contraction, and various active transport mechanisms.

## DIETARY NEEDS FOR ORGANIC MOLECULES

With a few notable exceptions, heterotrophs require such organic substances as protein, starch, and fat in their diets, not for these substances in themselves, but for their component molecules. Only when the macromolecules of food are broken down into their components can they be used as "building blocks" or as sources of energy.

***Protein Requirements.*** Proteins are needed for their amino acids, which organisms use to build their own body proteins. Of the 20 or so kinds of naturally occurring amino acids—those found in proteins—about half cannot be synthesized by animals and must be obtained from foods in their diet. They are therefore called the essential amino acids. The precise list differs slightly from one major group of heterotrophs to another. Remarkably, some ciliate protozoans require 10 of the same amino acids that are needed by rats, chickens, and probably humans. Some are needed only to support growth.

When some amino acids are said to be essential, it does not mean that the others are not needed to build proteins. The other amino acids are also needed, but they can be synthesized from other materials, such as other amino acids, and are not essential in the diet as such.

***Carbohydrates.*** The chief dietary source of energy for most heterotrophs is carbohydrate. However, there are no specific essential dietary carbohydrates. The dietary need that is normally satisfied by carbohydrate can be met by various polysaccharides, any of a variety of sugars, or even noncarbohydrates, such as amino acids. Carbohydrates also serve as a major carbon source. The body needs carbon atoms usually in the form of methyl ($CH_3$—) or acetyl ($CH_3$—$CH_2O$—) groups, for incorporation into important organic compounds.

***Unsaturated Lipids.*** Many animals have an absolute dietary requirement for lipids, sometimes for specific types. **Unsaturated fatty acids,** those with double bonds in the chain (such as linoleic acid, linolenic acid, and arachidonic acid) are required in the diets of several animals, from protozoans to mammals. Their most obvious known functions are as precursors of the prostaglandins (Chapter 3). We have seen that the phospholipids are the primary constituents of biological membranes (Chapter 5). Because they can be synthesized from other materials by all organisms, they are not a specific dietary requirement.

***Steroids.*** A large family of **steroids,** multiple-ring compounds with some lipid characteristics, plays important roles, particularly as hormones, in vertebrate physiology. Other known functions include acting as precursors of insect hormones and as constituents of the plasma membrane and other cellular membranes. Important among them as a precursor of other steroids is cholesterol. However, steroids can be synthesized by vertebrates from acetate or fatty acids and other lipids and therefore are not required in the diet. But, a steroid requirement has been demonstrated for many invertebrates and several protozoans.

***Vitamins.*** The most widely known specific human dietary requirements are the vitamins, organic compounds needed in small amounts for normal growth and metabolism. Some are needed only in small amounts because they function as cofactors in metabolic reactions. Like enzymes, they are not used up in reactions. They are divided into two groups: the fat-soluble vitamins, A, D, E, and K, and the water-soluble vitamins, which include vitamin C and the vitamin B complex. Most water-soluble vitamins are prosthetic groups of enzymes or components of coenzymes essential for important metabolic reactions. For example, nicotinic acid (niacin) is a component of the coenzyme NAD, and riboflavin (vitamin $B_2$) is a component of FAD, the prosthetic group of succinic dehydrogenase (Chapter 6). Pantothenic acid is a component of coenzyme A. When vitamins were first discovered they were referred to as accessory factors. When the first few to be chemically identified proved to be amines, they were called vital amines or vitamines. The name is retained in the shortened form vitamin, although it is now known that many vitamins are not amines. Table 9-2 lists some of the vitamins needed in the human diet, their known functions, and diseases resulting from dietary deficiencies.

Christiaan Eijkman, a Dutch physician who had been sent to Java in 1890 on a government commission to investigate an epidemic of beriberi, was the first to develop the idea of a need for a "complete" diet. Beriberi, a degenerative disease of the nervous system, had long been thought to be induced by eating old, spoiled, or contaminated rice. In 1882 a Japanese naval physician, Takaki, had observed that sailors whose rice diet was supplemented with barley, fresh vegetables, meat, and fish, did not get the disease. This prompted Eijkman to study the effects of various diets on human health. In a series of experiments on chickens, he found that when fed only polished rice the birds developed

**TABLE 9-2 Key Vitamins for Humans**

| | *Letter designation and name* | *Major sources* | *Function* | *Clinical deficiency symptoms* |
|---|---|---|---|---|
| *Fat-Soluble Vitamins* | A, carotene | Egg yolk, green or yellow vegetables and fruits | Forms visual pigments; maintains normal epithelial structure | Night blindness, skin lesions |
| | $D_2$, calciferol | Fish oils, liver | Increases Ca absorption from intestine; important in bone and tooth formation | Rickets (defective bone formation) |
| | E, tocopherol | Green leafy vegetables | Maintains resistance of red blood cells to hemolysis | Increased red blood cell fragility |
| | K, naphthoquinone | Synthesis by intestinal bacteria; liver | Enables prothrombin synthesis by liver | Failure of blood coagulation |
| *Water-Soluble Vitamins (= "B complex" + Vitamin C)* | $B_1$, thiamine | Brain, liver, kidney, heart; whole grains | Forms enzyme involved in decarboxylation (Krebs cycle) | Stoppage of carbohydrate metabolism at pyruvate; beriberi, neuritis, heart failure, mental disturbance |
| | $B_2$, riboflavin | Milk, eggs, liver, whole cereal grains | Flavoproteins in oxidative phosphorylation (electron transport) | Photophobia, fissuring of skin |
| | Niacin | Whole grains | Coenzyme in electron transport (NAD, NADP) | Pellagra, skin lesions, digestive disturbances, dementia |
| | $B_{12}$, cyanocobalamin | Liver, kidney, brain; synthesis by intestinal bacteria | Nucleoprotein synthesis (RNA), prevents pernicious anemia | Pernicious anemia, malformed erythrocytes |
| | Folic acid (folacin or pteroyl glutamate) | Meats | Nucleoprotein synthesis, formation of erythrocytes | Failure of erythrocytes to mature, anemia |
| | $B_6$, pyridoxine | Whole grains | Coenzyme for amino acid metabolism and fatty acid metabolism | Dermatitis, nervous disorders |
| | Pantothenic acid | ? | Forms part of a coenzyme A (CoA) | Neuromotor disorders, cardiovascular disorders, gastrointestinal distress |
| | Biotin | Egg white; synthesis by intestinal bacteria | Concerned with protein synthesis, $CO_2$ fixation, etc. | Scaly skin, muscle pains, weakness |
| | Inositol | ? | Aids in fat metabolism, prevents fatty liver | Fatty liver |
| | C, ascorbic acid | Citrus fruits | Vital to collagen connective tissue | Scurvy (failure to form connective tissue fibers) |

After J. E. Crouch and J. R. McClintic. *Human Anatomy and Physiology,* p. 450, Wiley, New York, 1971.

symptoms similar to those observed in beriberi victims. Controls (chickens fed on half-polished rice) remained healthy. Eijkman made extracts of rice polishings and isolated a fraction that prevented beriberi. Eventually, the antiberiberi factor was characterized chemically and named thiamine (vitamin $B_1$).

An organism requires a vitamin or any other complex organic molecule in its diet because it cannot synthesize that compound. It is not surprising that animal species differ in their vitamin requirements just as they do in their needs for essential amino acids. However, as with amino acids, many vitamin requirements are almost universal. For example, some algal protists require three or four of the same vitamins that are needed by humans. But some vitamins needed by humans are not required by lower animals and vice versa.

## DIGESTION IN ANIMALS

In several ways, digestion is quite similar in all animals, largely because of certain chemical similarities in foods. For example, diets of most heterotrophs include proteins, lipids, and carbohydrates. This means that each group must possess digestive enzymes that can break apart, or hydrolyze, the specific kinds of chemical bonds found in these foodstuffs. Hydrolysis is the reverse of the dehydration synthesis reactions in which the food macromolecules were formed from simpler units such as amino acids and monosaccharides. Organisms as widely diverse as protozoans, sponges, worms, insects, fishes, and humans can hydrolyze many of the same biochemical bonds and have many of the same types of digestive enzymes.

In contrast to these similarities in digestive capabilities, animals exhibit much variation in digestive organ structure and in physiological methods of acquiring, eating, and mechanically processing food. For example, a large stomach or a crop (a chamber that stores food before digestion) is needed in an animal that takes large, infrequent meals. Herbivores need larger and longer digestive tracts than do carnivores to compensate for the long time it takes to digest plant material. Other examples of variation are seen in mouth parts, which are adaptations to eating certain kinds of food (Fig. 9-7).

**FIG. 9-7** The roseate spoonbill of the Florida Everglades. Its highly specialized bill allows it to probe the mud of mangrove swamps for small aquatic animals.

### THE DIGESTIVE SYSTEM AS A WRECKING COMPANY

The ultimate role of the digestive system is the chemical breakdown of food macromolecules. Digestion yields molecular components that (1) are small enough to be soluble in water and to be absorbed from the digestive tract and (2) are in the form most useful to the tissues as a source of energy and raw materials and thus for growth and repair. All chemical phases of digestion are achieved by hydrolysis of the bonds that join the component molecules of foodstuffs. The digestive system is therefore not unlike a wrecking company that deals in recycled materials suitable for construction, repair, and fuel. What is "wrecked" are the macromolecules of the diet. The "bricks" and "boards" are the amino acids, sugars, and other chemical components of the diet. The "customers," of course, are the body's cells and tissues.

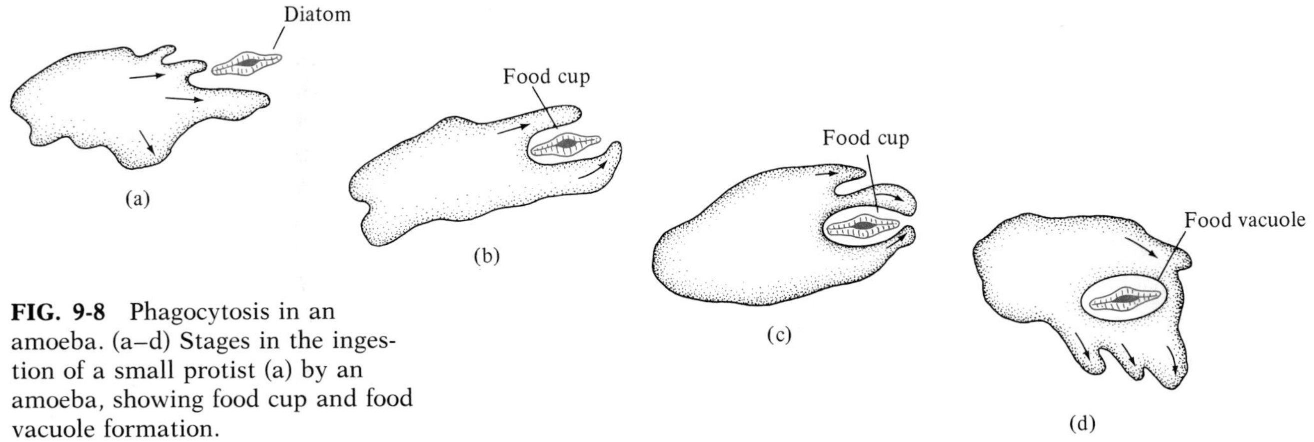


**FIG. 9-8** Phagocytosis in an amoeba. (a–d) Stages in the ingestion of a small protist (a) by an amoeba, showing food cup and food vacuole formation.

## PROTOZOA

Although protozoans (Greek for "primitive animal") are protists, not animals, they are closely allied to animals and were included with them in older classifications. Their digestion is usually intracellular, within food vacuoles, the food usually being ingested by phagocytosis (Fig. 9-8). Lysosomes secrete digestive enzymes into the vacuoles. As in most animals, digestion in the vacuole begins with an acid phase, followed by an alkaline phase. The acid, which may reach a high concentration, kills many bacteria and other live food, dissolves calcium salts, and provides an optimum pH for the functioning of some enzymes. In many protozoans, the final phase of digestion is completed close to neutrality. As the food is reduced to molecular size, it is absorbed into the cytoplasm. Finally, indigestible materials are expelled by exocytosis as the vacuolar membrane fuses with the plasma membrane in a process essentially the reverse of food vacuole formation.

## LOWER INVERTEBRATES

These organisms have no specialized digestive organs. In many, digestion takes place within individual cells.

***Sponges and Coelenterates.*** Sponges have no digestive cavity. Instead, specialized cells engulf food particles, enclosing them in food vacuoles, much as in protozoans. But before digestion is complete, the material is passed along to other cells called **amoebocytes,** which complete the process and distribute the products of digestion to other cells.

Coelenterates, or cnidarians, possess a **gastrovascular cavity,** which serves for both digestion and circulation. It has but one opening, the mouth, into which all food enters and out of which all residue is expelled. The vascular inefficiency of this cavity is unimportant because, as in sponges, most digestion is intracellular. However, before the intracellular phase, the gastrodermis, the cell layer lining the cavity, secretes digestive enzymes into the gastrovascular cavity where gross degradation occurs. The digestion of the resulting small particles is then completed intracellularly in food vacuoles. In an evolutionary sense, coelenterates are very little advanced over sponges in their mode of digestion (Fig. 9-9).

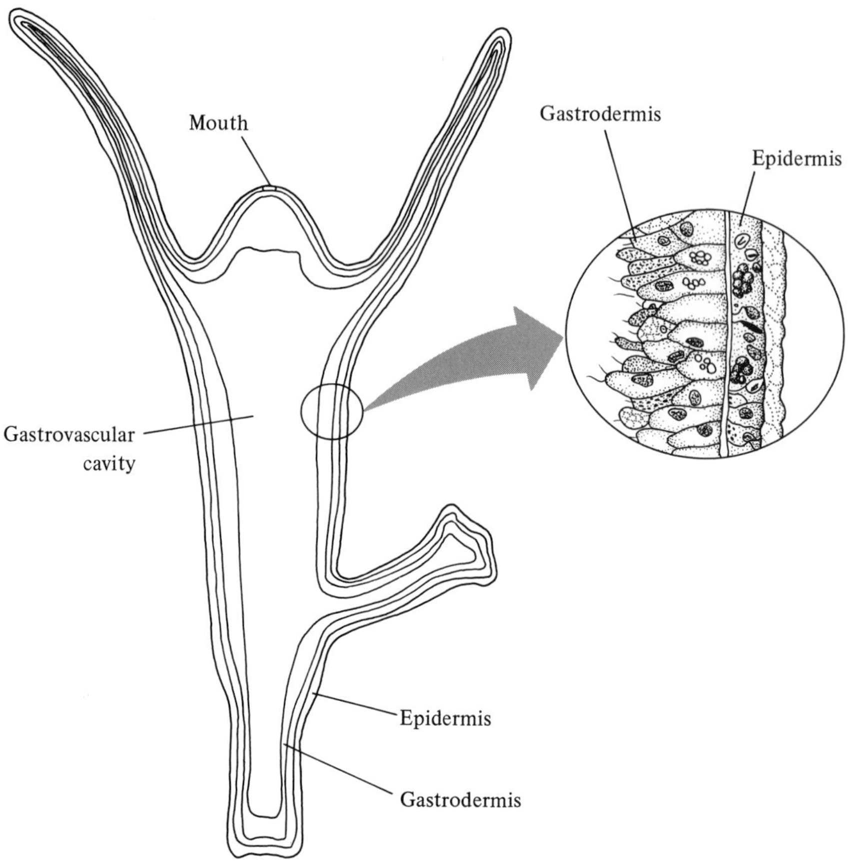


**FIG. 9-9** The gastrovascular cavity of a cnidarian (coelenterate), *Hydra,* shown in longitudinal section.

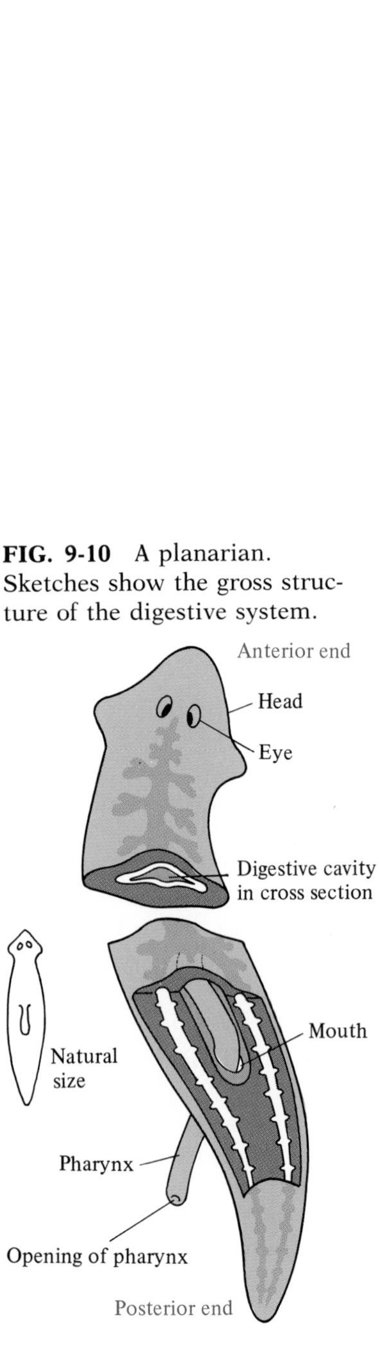


**FIG. 9-10** A planarian. Sketches show the gross structure of the digestive system.

***Flatworms.*** Platyhelminthes, or flatworms, exhibit an interesting array of approaches to nutrition. Among free-living species, or nonparasites, some have a mouth but no digestive cavity, the food particles being digested by the cells of a loosely organized tissue near the mouth. Other free-living forms have a simple digestive cavity, essentially a coelenteratelike gastrovascular cavity, which usually has multiple branches (Fig. 9-10). Among parasitic flatworms, the flukes have a gut with only two branches (Fig. 9-11a). Its simplicity is probably related to the fact that these parasites have an unvarying diet, the nature of which depends on which organ of their host they parasitize. Finally, tapeworms have no digestive organs at all, depending entirely on products of digestion furnished by their host—which is why adult tapeworms never live anywhere but in the digestive tracts of other animals (Fig. 9-11b). The outer surfaces of tapeworms, through which they absorb the products of digestion, possess fine projections called microvilli similar to those lining the absorptive surface of the mammalian small intestine.

## DIGESTION IN HIGHER ANIMALS

Beginning with the nemerteans and nematodes, the digestive tracts of all higher animals are one-way traffic routes, an important evolutionary breakthrough that greatly increased nutritional efficiency. In primitive systems, the mouth is used to discharge undigested matter. But with the evolution of a mouth and an anus, one specialized for receiving food and the other for discharging residue, new possibilities for effi-

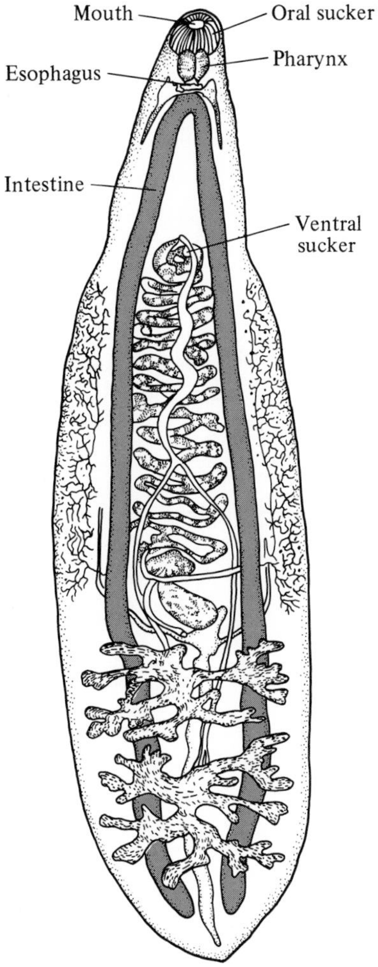


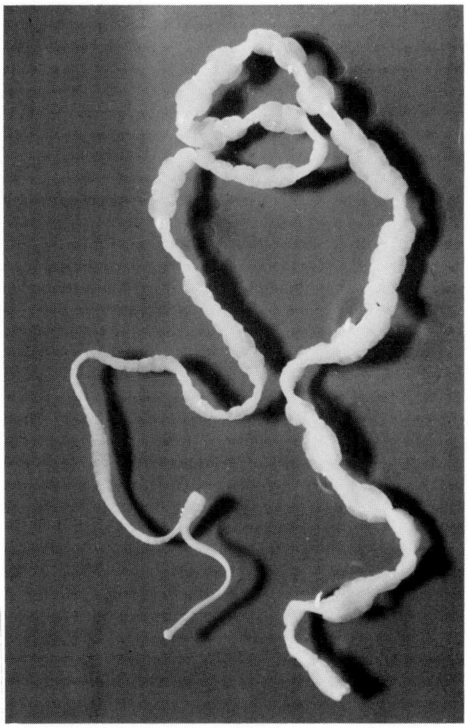


**FIG. 9-11** (a) An adult trematode (the Chinese liver fluke). The simple intestine can be seen dividing into two tubelike branches just below the mouth at the worm's anterior end. (b) *Taenia pisiformis,* a tapeworm of the dog.

ciency emerged. Various regions of the digestive tract became specialized in a kind of reverse assembly line that performed food-processing tasks in a far more orderly and efficient way than was otherwise possible (see Fig. 9-12 and Table 9-3).

## THE SAGA OF A SANDWICH

Merely serving a sandwich to a hungry person sets in operation a number of preparatory actions within the digestive system, the most obvious of which is that the mouth "waters," or produces an unusual amount of saliva. Taking a bite of the sandwich immediately involves other senses and prompts some decisions by the diner regarding the food. The mouth is well suited to monitor what enters the body because of its rich supply of touch, temperature, and chemical receptors that permit decisions over whether the sandwich is fit for consumption. Is it too hot? Is it toasted? Does it taste good? The sense of smell also plays an important role in the choice of food.

Once a bite of the sandwich is accepted into the mouth, it may be moistened if it is too dry or chewed into small pieces if it is too large to be easily swallowed. Chewing, or mastication, not only makes food small enough for swallowing but actually constitutes the first stage of

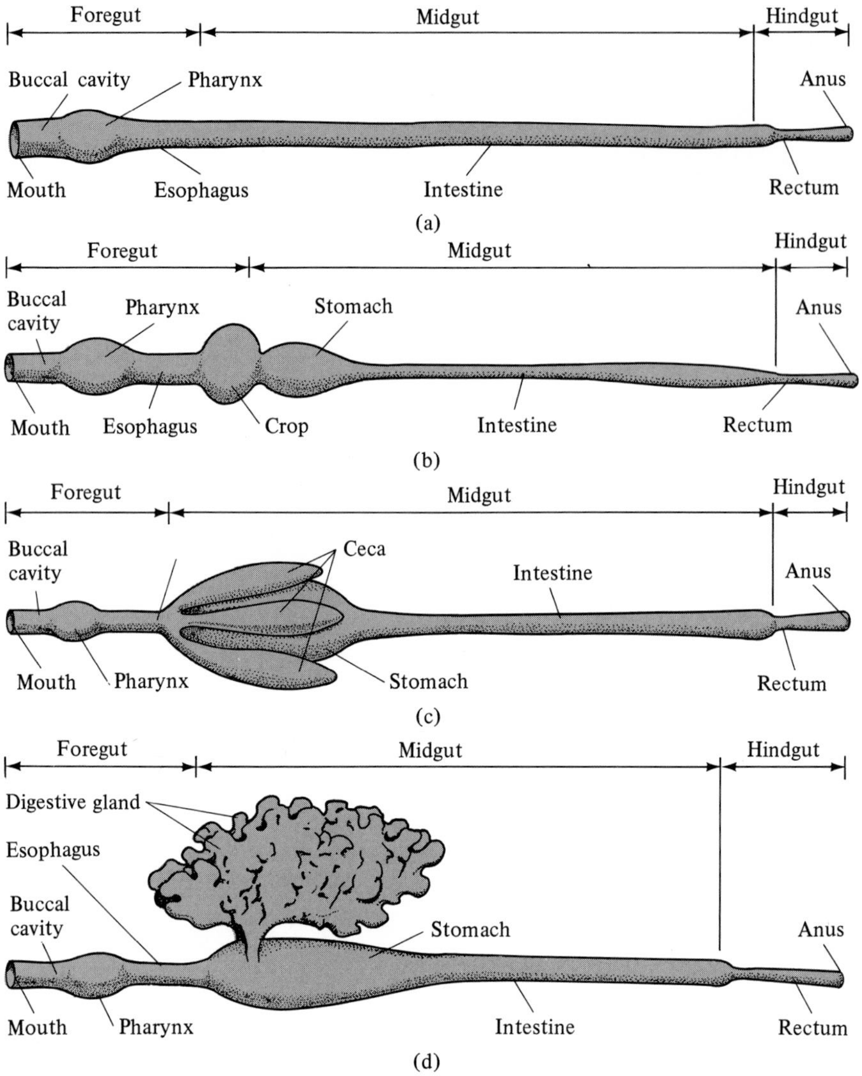


**FIG. 9-12** Sketches of one-way traffic digestive systems in representative invertebrate animals. (a) The simplest system, found in nematodes and nemerteans. (b) Slightly modified version of (a), similar to that in earthworms. (c) A further modification with a crop consisting of a number of blind ceca, as found in insects that eat vegetation. (d) The basic plan found in molluscs and arthropods other than insects.

**TABLE 9-3 Functions of Regions of One-Way Digestive Tracts**

| *Function* | *Region* |
|---|---|
| Mechanical degradation | Mouth parts, teeth, etc. |
| Pretreatment with salivary secretions such as poisons, enzymes, and anticoagulants | Mouth |
| Swallowing | Pharynx (sucking), esophagus (peristalsis), stomach (pumping) |
| Temporary storage of food | Crop |
| Additional mechanical degradation | Gizzard |
| Storage, gross digestion (acid phase), and preparation for later stages of digestion | Stomach, glands, ceca |
| Completion of digestion (alkaline phase) | Midgut and associated glands, such as liver and pancreas |
| Absorption of the products of digestion | Midgut epithelium |
| Reclamation of water | Hindgut |
| Formation and temporary storage of feces | Hindgut |
| Defecation | Anus |

NOTE: Digestion occurs within the cavity of the tract and not within the body itself.

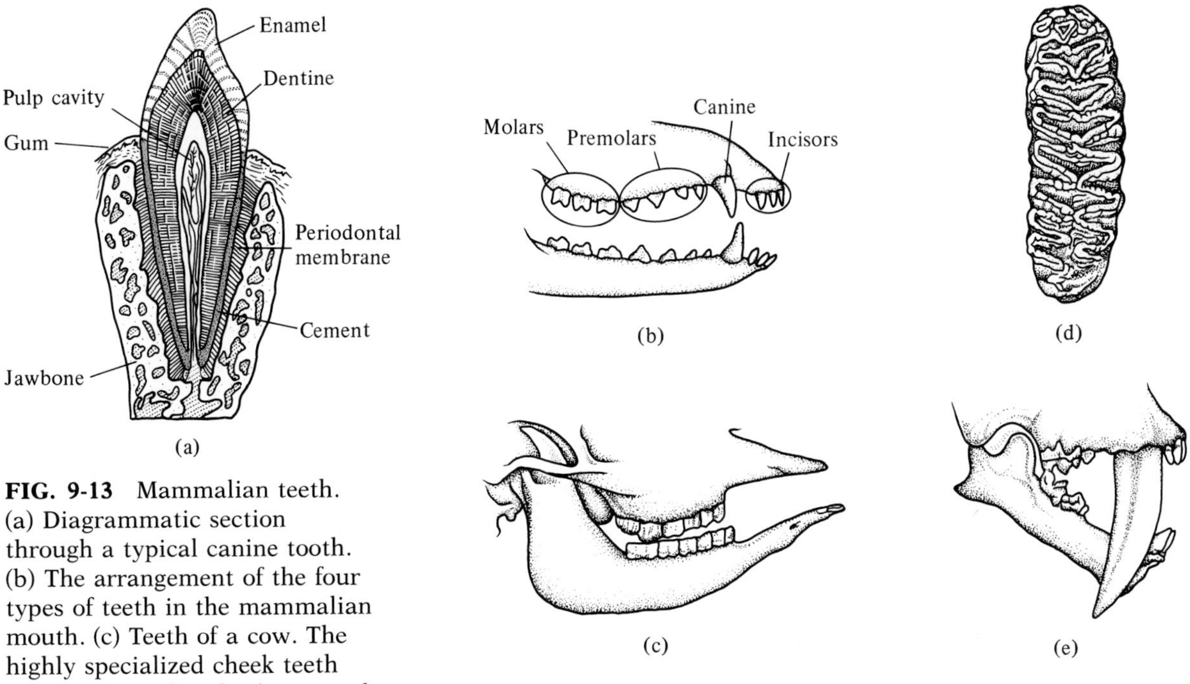


**FIG. 9-13** Mammalian teeth. (a) Diagrammatic section through a typical canine tooth. (b) The arrangement of the four types of teeth in the mammalian mouth. (c) Teeth of a cow. The highly specialized cheek teeth (premolars and molars) are used for grinding grass. (d) An elephant's upper molar, showing the grinding surfaces of dentin (dark areas) ringed with sharp-edged enamel. (e) The stabbing upper canine of the extinct saber-toothed tiger.

the digestive process, mechanically "opening the gates" to a large array of enzymes. Without chewing, digestion of some meals would take many hours longer.

Teeth accomplish the mechanical breakdown of foods and help mix food with saliva, preparing it for the chemical stages of digestion to follow (Fig. 9-13a). In the human, two sets of teeth develop, the 20 deciduous, or "milk," teeth of childhood and the 32 permanent teeth that replace them. The permanent teeth of mammals are of four types, each type adapted to a different function (Fig. 9-13b). Humans have eight incisors, used for biting; four canines, used for tearing, and eight premolars, or bicuspids, and 12 molars, or tricuspids, used for grinding. The last four tricuspids to erupt in humans are called the "wisdom teeth."

In teeth, as in other anatomical parts, form is related to function. For example, rodent incisors, which are specialized for gnawing, are continually worn away and never cease growing. Teeth of herbivores such as the elephant are broad and flat—highly suited to grinding plant tissue with thick cell walls (Fig. 9-13c, d). In several mammals, some teeth are elongated as tusks, highly efficient in offense and defense. Elephant tusks are incisors, whereas the tusks of boars and the now extinct saber-toothed tiger are canines (Fig. 9-13e).

Each tooth is composed of a **crown** and **root.** Most of the tooth is composed of bonelike **dentin,** but the crown is covered with **enamel,** the hardest material in the body (Fig. 9-13a). A good supply of dietary calcium is needed in early youth for optimum tooth development.

**FIG. 9-14** The human salivary glands.

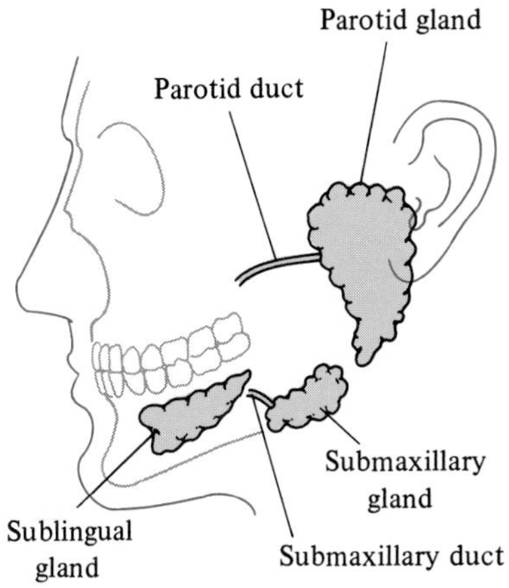


***Saliva.*** The mouth is served by three pairs of salivary glands. One pair is at the angle of the jaw, another pair just below the lower jaw at the base of the tongue, and a third pair below the tongue (Fig. 9-14). The pair at the angle of the jaw produce a watery secretion rich in enzymes. The others produce a mucus as well. Different foods elicit different types of salivary secretion.

The most important enzyme in saliva is **ptyalin,** or salivary amylase, which partially digests starches before they are swallowed. A more important function of saliva is to moisten and lubricate food materials. This not only makes them easier to swallow but dissolves the soluble ones and suspends others, thereby readying them for action by the digestive juices of the stomach and intestine (Fig. 9-15a).

***The Esophagus.*** Once the bite taken out of a sandwich has been moistened and chewed, it is moved by the tongue to the pharynx at the rear of the oral cavity. Here its presence induces a swallowing reflex. By peristalsis, the food—now referred to as a food bolus—is passed down the esophagus to the stomach.

**FIG. 9-15** (a) The human digestive tract. (b) Peristalsis. In a tubular organ such as the esophagus, stomach, or intestine, tonus (a state of partial contraction) is normally maintained. The peristaltic wave begins as a wave of relaxation of the wall muscles, followed immediately by a wave of contraction, thus moving the contents of the organ in the direction of the wave.

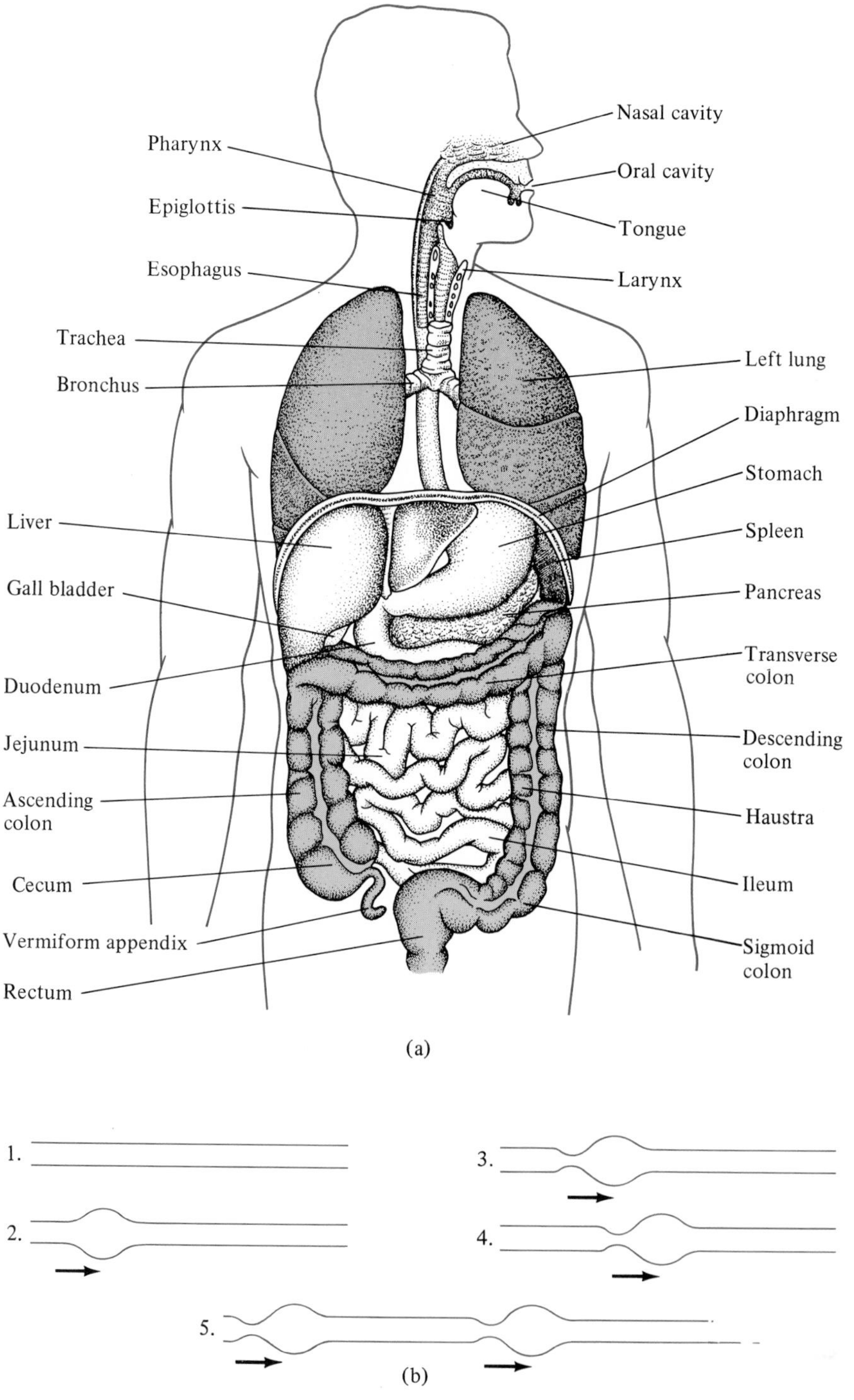

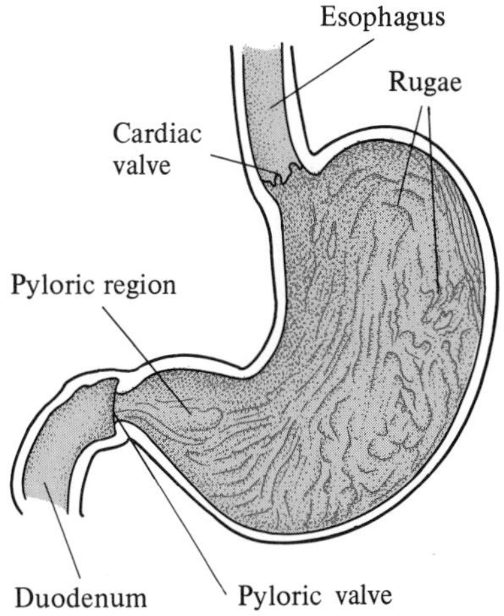


**FIG. 9-16** Drawing of a section through the human stomach to show the rugae and the cardiac and pyloric valves.

**Peristalsis** is characterized by a relaxation-contraction sequence of the muscles in an organ's wall that passes in waves down the length of the organ in a "milking" action that moves the contents along (Fig. 9-15b). Both the circular and longitudinal muscles are normally in a state of partial contraction called tonus (1). Relaxation of the circular musculature results in a widening of the lumen, the cavity within (2). This is followed immediately by a contraction of adjacent circular fibers, pushing the contents into the relaxed portion (3). The entire process is repeated down the length of the organ (4, 5). When properly stimulated, organs capable of peristalsis can usually conduct it in the reverse direction as well, as in vomiting.

***The Stomach.*** The food next enters the stomach, a curved, muscular, saclike organ, the internal surface of which is thrown into coarse folds called **rugae,** which allow the stomach to expand and thereby accommodate food (Fig. 9-16). The inner surface is marked by numerous, small gastric pits at the bases of which are millions of microscopic **gastric glands** (Fig. 9-17). The mammalian stomach serves to some degree as a storage organ, an important function, considering the time-consuming nature of digestion. Very few nutrients are absorbed here. It also functions to dissolve or suspend food particles. The warmth and churning action of the stomach serve to reduce to fine droplets, or emulsify, lipids and lipid-containing foods such as butter, cheese, oils, or bits of fat, all of which might be found in a meat sandwich. The most important digestion that occurs in the stomach is that which is accomplished through the action of **pepsin,** the enzyme that initiates the breakdown of the proteins found in food. The sole role of pepsin is to hydrolyze peptide bonds between certain amino acids. This converts large proteins to somewhat smaller molecules called peptones. One early insight into the nature of this process was provided by William Beaumont (Panel 9-3).

**FIG. 9-17** Photomicrograph of a stained section through the human stomach. The gastric glands appear as long tubes at the base of the gastric pits.

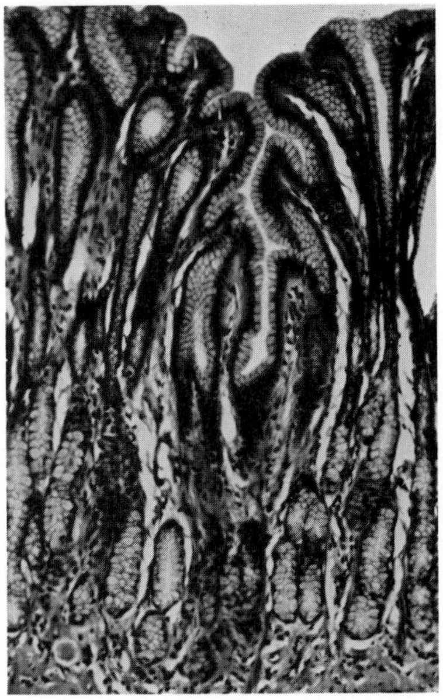

The secretory epithelium lining the stomach includes three cell types: **chief cells,** which secrete **pepsinogen,** an inactive precursor of the enzyme pepsin; **parietal cells,** which secrete hydrochloric acid (HCl); and **clear columnar cells,** which secrete **mucus,** a ropy proteinaceous material that in normal amounts protects the gastric epithelium from the ulcers or open sores that can result from its being digested by its own secretions.

Even in the absence of food, the stomach of the average adult contains at least a liter of gastric juice. Because of its HCl content, the pH of gastric juice is extremely low: 0.9 to 2.0. This creates a formidable barrier to most bacteria and other live cells, quickly killing them as they enter. The acid has other roles as well. It hydrolyzes some foods. It coagulates proteins, thereby preparing them chemically for hydrolysis by various enzymes. It also provides the optimum pH for the action of the powerful protease, pepsin. The inactive form of pepsin, pepsinogen, is activated by HCl.

Just as the sight and smell of food promotes a flow of saliva, the taste of the first bite of a sandwich prompts increased flow of gastric juice into the stomach. The arrival in the stomach of a food bolus resulting from that first bite also promotes an increase in gastric secretion. This phase of gastric secretion has been shown to be due in part to a direct, triggering effect of dissolved food molecules on the gastric epithelium. A similar secretory response is elicited in the small intestine by the arrival of food. Gastric secretion involves a combination of nervous and hormonal controls.

## PANEL 9-3

# The Remarkable Story of Beaumont and St. Martin

Although several important contributions to our understanding of gastric digestion foreshadowed the experiments of the United States Army surgeon William Beaumont (1785–1853), his studies are the most celebrated of them all. By rare circumstance, he was able to study a young patient with an accidentally produced **gastric fistula,** an opening directly into the stomach from outside the body. With such convenient access to the stomach, Beaumont was able to study the course of gastric digestion of various foods. By testing a variety of foods placed in samples of gastric juice that had been removed through the fistula, he confirmed that digestion is essentially a chemical process.

The accident, which occurred in June of 1822 on Mackinac Island in the Michigan Territory, involved the wounding of an 18-year-old guide, Alexis St. Martin. A musket discharge had severely wounded the young man's abdomen, lungs, and stomach. St. Martin, under Beaumont's care, survived the injury, his convalescence taking nearly two years. Although St. Martin fully recovered his health, the wounded stomach had healed to the body wall so as to form a fistula. The young man refused to undergo an operation to correct the defect.

Beginning in May 1825, nearly three years after the accident, Beaumont, with St. Martin's permission, began a series of experiments on gastric digestion.

In August 1825, Beaumont was transferred to Burlington, Vermont, and then to Plattsburg, New York, each time taking St. Martin with him. At this point, the patient absconded to Canada "without obtaining my consent," as Beaumont put it. For the next four years, he related, "I did not remit, however, my efforts to obtain information about his place of residence and condition." St. Martin continued in excellent health, married,

The result of all this gastric activity is that the food is reduced to a thick broth called **chyme.** At this stage, the **pylorus**—the narrow, muscular, funnel-shaped posterior end of the stomach—squirts the chyme through the pyloric valve into the small intestine. The pyloric valve normally prevents material that has entered the intestine from reentering the stomach. The cardiac valve at the entrance to the stomach performs a similar function for food that enters from the esophagus.

***Tissues of the Small Intestine.*** The small intestine is divided into three regions: the **duodenum,** the **jejunum,** and the **ileum** (Fig. 9-15). The duodenum acquired its name (from the Latin *duodecim,* "twelve") because it is about twelve finger widths in length (the fingers of an average anatomist, presumably). It is the most highly specialized region of the gastrointestinal tract. Among the secretions that empty into it are those of the liver, the pancreas, and its own epithelial glandular cells. The cells lining the lumen are capable of absorbing many kinds of materials, some passively and some by active transport.

Fine, fingerlike **villi,** (sing. villus), 0.5 to 1.5 mm long, cover the lining of the small intestine, thereby increasing by about eight times the surface area available for secretion and absorption. Even more important, as electron microscopy has revealed, the cell surfaces of the epithelial cells forming the villi have "brush borders" of **microvilli** (Fig. 9-18). This brings the total increase in surface area of the intestinal lining to about 150-fold. This tissue is richly supplied with blood and lymph vessels (Fig. 9-19).

At the base of the intestinal villi are the pits or crypts into which empty the thick mucus secretion of Brunner's glands, providing a coat-

Doctor William Beaumont

and fathered two children. Then Beaumont learned of his whereabouts and sent an agent "to find and engage him for my service if practicable." "After considerable difficulty and at great expense," Beaumont succeeded in rehiring St. Martin and transported him, his wife, and his children some 2,000 miles to Beaumont's new station at Prairie du Chien (Fort Crawford) in the Wisconsin Territory on the Mississippi. Beaumont was pleased to note that "the aperture was open and his health good." Then, except for a leave of 18 months—with permission—St. Martin continued in Beaumont's employ from 1829 to 1833, when the surgeon published a summary of his findings in a monograph entitled *Experiments and Observations on the Gastric Juice and the Physiology of Digestion.*

Beaumont not only demonstrated the chemical nature of digestion but also established the beneficial effects of chewing and cooking food, especially fibrous material. He studied hunger and factors that allayed it. He studied the muscular action of the stomach during digestion. He showed that the presence of food in the stomach elicits copious secretion of gastric juice, samples of which he had analyzed, establishing that it contained HCl and "other active chemical principles." The monograph so impressed the scientific world that it at once became the basis for our understanding of gastric digestion.

ing to the cell surfaces that is regarded as critical in the protection against duodenal ulcer. The epithelial lining of the intestine is continually renewed, being completely replaced every three to six days. It is estimated that 250 g (8.8 ounces) of tissue are shed into the intestine of the average adult each day.

**FIG. 9-18** Electron micrograph of the small intestine, showing the "brush border" of microvilli on the surface of two adjacent villi.

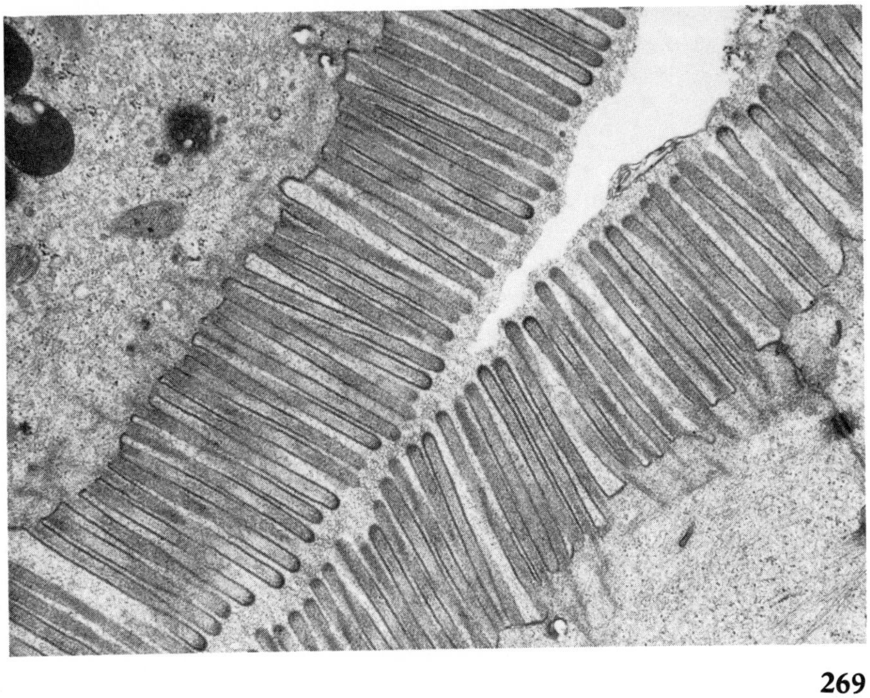

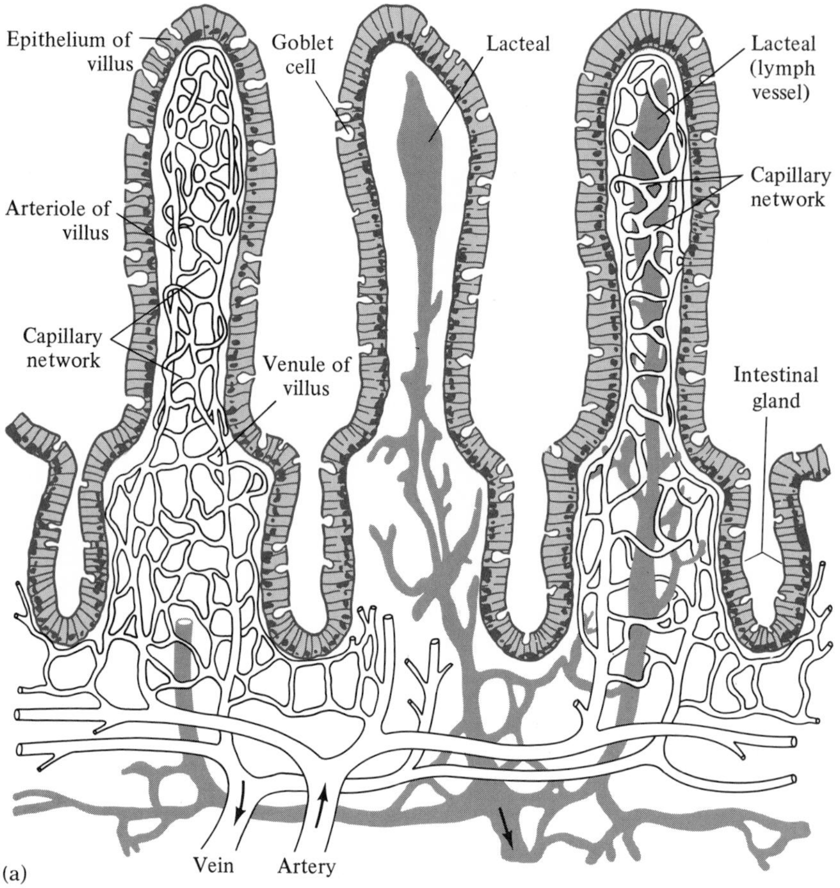

**FIG. 9-19** (a) Diagrammatic sketch of intestinal villi showing capillaries and lymphatic vessels. The villus on the left shows only blood vessels, the center one only a lymphatic vessel, the one on the right both. (b) Several intestinal villi of the mouse, with the tips removed. Below the epithelial layer can be seen the core containing nerves, blood vessels, and lymphatics.

***Small Intestine Secretions.*** As soon as the chyme enters the small intestine, it mixes with the secretions found there, and the pace of digestion quickens. The secretions of the small intestine are of three types.

1. *Bile* is produced by the liver, stored in the gallbladder, and released when fatty foods enter the duodenum. It contains **bile salts,** derivatives of the steroid cholesterol that act as emulsifying agents, lowering the surface tension of fats and reducing them to microscopic droplets. This not only prepares fats for digestion by enzymes called **lipases** but prevents fats from coating proteins and other foods, thereby shielding them from the action of their own specific enzymes. Later, bile salts combine chemically with the products of fat digestion, thus facilitating their absorption.
2. *Pancreatic juice* is alkaline (pH 7.2–8.2) and neutralizes the chyme as it leaves the acid stomach, abruptly halting the action of pepsin, which can only function in an acid pH. The alkaline pH is also optimal for action of the enzymes in the duodenum, many of which are also pancreatic secretions. The pancreatic enzymes include trypsin, chymotrypsin, carboxypeptidase, deoxyribonuclease, ribonuclease, pancreatic amylase (amylopsin), and pancreatic lipase.
3. *Other digestive enzymes* originate in the epithelium of the intestinal wall, including peptidases, a lipase, amylase, maltase, sucrase, lactase, and one called enterokinase. Careful studies have indicated that only enterokinase, amylase, and possibly maltase are secreted into the lumen. The others are found there only be-

cause they occur in the disintegrating mucosal epithelium that is shed into the lumen.

A list of digestive enzymes, their sources, actions, and products is given in Table 9-4.

***Digestion and Absorption in the Small Intestine.*** Most digestion and nutrient absorption takes place in the small intestine. The nature of these processes is typified by the digestion and absorption of the three major classes of foodstuffs: proteins, lipids, and carbohydrates.

PROTEINS. Any digestion of the proteins in a sandwich that had begun in the stomach continues in the duodenum under the action of trypsin, chymotrypsin, and carboxypeptidases from the pancreas and at least two peptidases from the intestinal wall. The combined action of these enzymes reduces proteins to smaller and smaller polypeptides, then to tri- and dipeptides and finally to individual amino acids, as indicated in Table 9-4.

The amino acids released by hydrolysis of proteins are rapidly absorbed by the cells lining the intestinal wall. This absorptive capability, found throughout the small intestine, is due to several active transport mechanisms, each specific for a certain class of amino acids.

**TABLE 9-4 Some Digestive Enzymes**

| *Enzyme* | *Source* | *Action (hydrolysis of)* | *Product* |
|---|---|---|---|
| PROTEASES | | | |
| Pepsin | Chief cells of gastric mucosa | Peptide bonds at the $NH_2$-group of aromatic and acidic amino acids | Proteoses, peptones |
| Trypsin | Pancreas | Peptide bonds at the COOH-group of the amino acids arginine and lysine | Proteoses, peptones, some free amino acids |
| Chymotrypsin | Pancreas | Peptide bonds at the COOH-group of the amino acids tryptophan, phenylalanine, tyrosine | Proteoses, peptones, some free amino acids |
| Carboxypeptidase | Pancreas | COOH-terminal peptide bonds on polypeptides | Free amino acids |
| LIPASES | | | |
| Pancreatic lipase | Pancreas | Triacylglycerols | Diglycerides, monoglycerides, free fatty acids, glycerol |
| Intestinal lipase | Intestinal epithelium | Tri-, di-, and monoglycerides | Free fatty acids, glycerol |
| CARBOHYDRASES | | | |
| Ptyalin | Saliva | Starch | Dextrins, some maltose |
| Amylopsin | Pancreas | Starch | Dextrins, maltose |
| Intestinal amylase | Intestinal epithelium | Starch | Dextrins, maltose |
| Maltase | Intestinal epithelium | Maltose | Glucose |
| Sucrase | Intestinal epithelium | Sucrose | Glucose, fructose |
| Lactase | Intestinal epithelium | Lactose | Glucose, galactose |

LIPIDS. Triacylglycerols (triglycerides), such as are found in mayonnaise, olive oil, butter, meat, or cheese—at least one of which is likely to be found in most sandwiches—are first emulsified by the bile salts secreted by the liver. Lipases then partially digest the triacylglycerols to fatty acids and monoglycerides, glycerol molecules with only one fatty acid. These two products and the bile salts form a complex that becomes highly emulsified. The fatty acids and monoglycerides then enter the cells, but the bile salts remain in the intestinal lumen. Lipid absorption occurs principally in the duodenum and upper jejunum. The mechanism of absorption is apparently passive, a matter of diffusion rather than of active transport.

Within the epithelial cells, the newly absorbed products of lipid digestion are immediately resynthesized into triacylglycerols and phosphoglycerides. This part of the process requires energy and expends ATP. The newly synthesized lipid is coated with albumin, a low molecular weight protein, and then secreted, as very fine droplets called chylomicrons, into the lymphatic vessels, giving the lymph a milky appearance after a meal.* Hence, the small lymphatic vessels are called lacteals (from Latin *lacteus,* "milk"). Short-chain fatty acids, however, make their way from the intestinal lumen, through the epithelial cells, through the lymph, and into the bloodstream without being resynthesized into lipid. A brief account of the forms in which lipids are transported in the blood is found in Panel 9-4.

CARBOHYDRATES. Any sandwich contains an abundance of starch, the main polysaccharide in bread. If the sandwich contains meat, then another polysaccharide, glycogen, may also be present. If the bite of the sandwich is followed by a swallow of milk, then lactose, a disaccharide, is included. If the drink is sweetened, another disaccharide, sucrose, may be ingested. Appropriate enzymes digest these carbohydrates to monosaccharides such as glucose, fructose, and galactose. Salivary amylase (ptyalin), pancreatic amylase (amylopsin), and an amylase secreted into the lumen by the intestinal epithelium break down starch, at first into smaller polysaccharides called dextrins and then into the disaccharide maltose. Because ptyalin is inactivated by gastric HCl, most starch is digested by amylopsin and intestinal amylase. The maltose produced is reduced to glucose by a maltase secreted by the intestinal mucosa. Sucrose (table sugar) and lactose (milk sugar) are hydrolyzed by sucrase and lactase, respectively, which are also secreted by the intestinal mucosa. The products are glucose and fructose from sucrose and glucose and galactose from lactose.

It has long been known that glucose is actively absorbed from the lumen of the small intestine. Although the process has been studied intensively, it was only recently elucidated. It is complex and investigators do not yet fully agree on its details. Glucose dissolved in the intestinal juice enters the "brush border" of the mucosal epithelium, where it binds to a symport carrier protein (see Panel 5-1) that also binds $Na^+$. The glucose and $Na^+$ are released into the cytoplasm whereupon the $Na^+$ is pumped out again. The process, which involves a carrier in the epithelial cell plasma membrane, is affected by differences in concentration of sodium and potassium ions between the inside and outside of the cells. This active transport mechanism is capable of absorbing all

* As it passes through the blood vessels serving the tissues, some of the fluid portion of the blood is squeezed out into the intercellular spaces, where it is called tissue fluid. Through a network of vessels known as the lymphatics, this fluid, now called lymph, slowly returns to the bloodstream.

glucose present in the intestinal lumen. As the absorbed glucose accumulates in the epithelial cells, it passes by diffusion into the bloodstream. Although galactose appears to be absorbed by the same mechanism as glucose, fructose is exceptional in being absorbed by diffusion alone. Despite this ability to absorb monosaccharides, all three hexoses usually cross the plasma membranes in disaccharide form and are only then digested to monosaccharides within the cells of the intestinal epithelium.

***What Happens in the Colon.*** The colon is separated from the ileum by the ileocecal valve. The human **cecum** is a large, blind pouch below the entrance of the ileum. The **vermiform appendix,** which appears to be a vestige of a more extensive cecum, joins the cecum proper. The large intestine, or **colon,** extends upward from the cecum, crosses the abdomen from right to left, and then courses downward as the descending colon, sigmoid (S-shaped) colon, and rectum, opening to the exterior at the anus. The anus is normally kept closed by external and internal sphincter valves.

The colon has three principal functions: (1) promotion of bacterial decay of undigested residues; (2) reclamation of water (about three-fourths is resorbed); and (3) temporary storage of feces prior to defecation. Contrary to a common belief, rather than being composed principally of undigested food, human feces are almost entirely a product of the digestive system.

Feces consist principally of water, bacteria, and epithelial cells from the intestinal mucosa. Also included are inorganic salts and whatever undigested food has not been metabolized by the bacteria. Mucus is also secreted into the lumen and acts as a lubricant for the feces. The composition of feces is almost entirely unaffected by diet, and, indeed, normal feces continue to be produced during early stages of fasting. In humans, eating a large amount of vegetables or cereal grains, however, results in the appearance in the feces of undigested fiber, small amounts of which are normally digested by bacteria. Although the role of bacteria is not nearly as important in humans as in herbivores, these microorganisms do contribute to our nutrition, most notably by producing vitamin K, a blood-clotting factor.

As the feces pass through the colon, most of the water is reabsorbed. If allowed to remain too long in the colon, too much water is reabsorbed and constipation results. When irritants such as raw vegetables are included in the diet, they stimulate peristalsis, preventing constipation. Overirritation, however, results in such rapid movement and insufficient absorption of water that diarrhea results.

By weight, normal feces are usually 40 percent water but can be as much as 75 percent. Of the dry weight, about 5 percent is nitrogen, up to 20 percent salts, up to 18 percent lipids, and up to 50 percent bacteria. The color of feces is due to bile pigments derived from the heme of hemoglobin.

## CONTROL OF GASTROINTESTINAL ACTIVITY AND SECRETION

Like most internal organs, the organs involved in digestion are innervated by the autonomic nervous system, a part of our nervous system that controls our automatic functions (see Chapter 15). Reflex peristal-

## PANEL 9-4

# Diet and the Healthy Heart

As is implicit from the preceding account of dietary needs, animals require specific nutrients in adequate amounts if they are to maintain good health. But in diets, as in other matters, there can be too much of a good thing. Diet has received a lot of attention as a factor in **coronary artery disease (CAD).** Coronary artery disease results from **atherosclerosis,** the deposition of lipids and cholesterol in the walls of the coronary arteries, the vessels that supply blood to the heart. As the walls of a vessel become thick with plaques of lipids and cholesterol, its lumen gradually narrows. As a result, the vessel's capacity for blood is reduced; consequently, the tissues that it serves receive less blood. Atherosclerosis begins to develop in infancy in populations of North America, North and Central Europe, Australia, and New Zealand. It progresses throughout life, usually without symptoms. In many individuals, however, some of the vessels may close completely or their roughened linings may be the site of a life-threatening blood clot, known as a coronary thrombosis, the most common kind of heart attack.

### THE LIPOPROTEINS

Lipids combine with certain blood proteins to form **lipoproteins;** nearly all lipid is carried in the bloodstream as lipoprotein. Lipoproteins play an important role in atherosclerosis and are classified according to their densities and functions:

a. **Chylomicrons.** These are aggregates of triacylglycerols and cholesterol derived from digestion. Of the lowest density, they contain less than 2% protein.
b. **Very low density lipoprotein (VLDL).** Synthesized in the liver, these lipids are transported to fat storage (adipose) tissue.
c. **Low density lipoprotein (LDL).** They constitute the primary supply of cholesterol to all tissues.
d. **High density lipoprotein (HDL).** Rich in phospholipids and cholesterol, these lipids are transported from outlying tissues to the liver.

There appears to be a correlation between the incidence of coronary artery disease and the concentrations of some of these lipoproteins in the blood. For example:

a. A strong correlation has been shown between coronary artery disease and high blood levels of total cholesterol or LDL.
b. A strong inverse correlation has been shown between the level of HDL and coronary artery disease; that is, the higher one's HDL level, the

tic responses to the presence of food are an example. Hormones* supplement the action of the nervous system in controlling digestive activity (see Chapter 14). Of the two, hormones are more important in this regard. The control of pancreatic secretion is one of the best examples.

When a substantial amount of hydrochloric acid enters the duodenum from the stomach, the duodenal mucosa is stimulated to secrete the hormone secretin into the bloodstream. When the blood-borne hormone reaches the pancreas, it stimulates the release of pancreatic secretions into the pancreatic duct. The bicarbonates in the pancreatic juice neutralize the HCl from the stomach. The mucosa detects the concentration of acid to be neutralized and responds accordingly.

## SPECIAL CASES: RUMINANT AND NONRUMINANT HERBIVORES

Perhaps the most unusual modifications of the stomach are seen in **ruminant mammals,** animals that "chew their cud" (Fig. 9-20). This method of nutrition has evolved in animals that need to eat large

* A hormone is a substance that is produced by a ductless gland or a limited area of tissue and controls or coordinates the function of another tissue to which it is transported or diffuses.

less chance of developing coronary artery disease.

## DIETARY FACTORS IN CORONARY ARTERY DISEASE

Some of the factors contributing to coronary artery disease are discussed in Chapter 11. But a few points about diet are appropriate here. Since our understanding of this matter is still incomplete, generalizations about the effect of diet on coronary artery disease should be made with caution, because what is true for a population is not necessarily true for every individual in that population; and diet is but one of several factors.

However, the following generalizations can be made:

a. Total blood cholesterol level (TC) can usually be reduced by a dietary reduction in fat and an increase in the ratio of polyunsaturated lipids (P) to saturated lipids (S). If an average Western diet of 40% fat calories and a P:S ratio of 0.4 is replaced by one that includes 30% fat calories and a P:S ratio of 1.0, an individual with a TC of 225 mg/dl* can expect a decrease of 20 mg/dl in TC and an increase in HDL level. Variations occur in individuals: some barely respond to a change of diet, whereas others increase their HDL fraction by as much as 28%.
b. For individuals on a moderate dietary cholesterol intake of 100 to 700 mg/day, a 5 mg/dl increase in TC will occur for each increase of 100 mg of cholesterol intake. For someone with a high intake of cholesterol, dietary increases have little effect on TC.
c. Saturated fatty acids in the diet raise TC levels twice as much as the same weight of polyunsaturated fatty acids lower them.

* mg/dl = milligrams per deciliter (a tenth of a liter)

## SOME RECOMMENDATIONS

Although too much fat in the diet is to be avoided, fat-free diets are not recommended. Fat is required in the diet if only to permit absorption of fat-soluble vitamins. Moreover, unsaturated fatty acids obtained from vegetable oils are essential as precursors of the prostaglandins. Given those comments, the following is recommended:

a. Limit intake of saturated fatty acids from such sources as pork, beef, coconut oil, and chocolate and from dairy products like whole milk, most cheeses, ice cream, and butter.
b. Where feasible, substitute unsaturated fatty acids for saturated. Use unsaturated vegetable oils such as safflower oil and corn oil in frying, baking, and in soft margarine spreads. Avoid lard or hardened (hydrogenated) vegetable oils.
c. Limit total lipid intake and total caloric intake.
d. Limit the number of egg yolks, which are rich in cholesterol, to three a week, including any included in bakery products.

amounts of food relatively quickly, chewing being done later at a more comfortable or safer location away from the feeding grounds. But more important, by providing an opportunity for hordes of microorganisms to digest the thick cellulose cell walls of grass and other vegetation, ruminants can utilize the vast amounts of forage available to herbivores. Animals generally lack the ability to produce a **cellulase,** an enzyme needed to digest cellulose. The intestinal bacteria and protozoans that have such enzymes have made the herbivorous lifestyle possible. In ruminants, the upper portion of the stomach expands to form a large

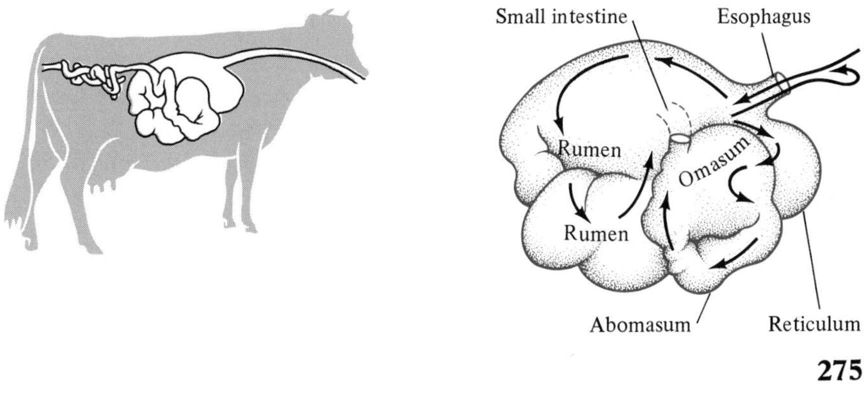


**FIG. 9-20** Sketch of a ruminant's (cow) stomach.

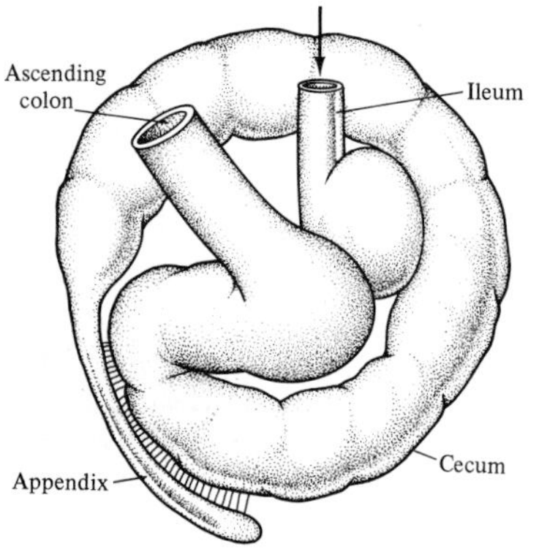


**FIG. 9-21** The extensive cecum of a nonruminant herbivore (rabbit).

pouch, the **rumen,** and a smaller, honeycomblike **reticulum.** The lower portion of the stomach consists of a small antechamber, the **omasum,** with the "true" stomach, or **abomasum,** behind it. Food first enters the rumen, where it encounters a remarkable community of specialized anaerobic bacteria and protozoans. Aided by copious secretions of fluid, body heat, and the churning and kneading caused by the contractions of the muscular wall of the rumen, the food undergoes partial digestion by the microorganisms and is reduced to a pulpy mass. Later the food moves into the reticulum, from which mouthfuls are regurgitated as "cud" and are thoroughly chewed for the first time. When reswallowed, the food usually reenters the rumen. Another cud is then passed up to the mouth to be chewed. This process continues, and the material becomes more completely digested and more liquid in consistency. Only then does it begin to flow out of the reticulum and into the lower parts of the stomach, first the omasum and then the glandular region, the abomasum. It is here that the food first encounters the animal's digestive enzymes, and digestion continues in the usual fashion.

Another remarkable benefit of the ruminant way of life is that nonprotein sources of nitrogen can be utilized as a raw material for the synthesis of protein through the agency of the microorganisms of the rumen. The best example of such a nitrogen source is urea, usually a nitrogenous waste product of mammals. In cattle, urea diffuses from the blood into the rumen where bacteria use it in protein synthesis. Thus recycled, urea becomes a valuable source of "dietary" amino acids. The rumen bacteria, moreover, synthesize all vitamins except A and D, making their hosts nearly self-sufficient in regard to dietary vitamin requirements.

Whereas the food of ruminants is attacked by bacteria before gastric digestion, in the typical nonruminant herbivore, bacterial action on cellulose occurs after gastric digestion. Rabbits, horses, and rats accomplish this feat by maintaining a population of microorganisms in their unusually large cecum, the blind pouch that extends from the colon (Fig. 9-21). Adding further to nutritional efficiency, a few nonruminant herbivores, such as mice and rabbits, can eat their own feces and process the remaining materials in them, thus having the best of both types of herbivore worlds.

## Summary

Nutrients provide organisms with the raw materials needed for growth and maintenance. Except for plants, which get their energy from sunlight, organisms also depend on nutrients as a source of energy.

Plants and other photosynthetic organisms convert carbon dioxide, water, and various minerals into the complex materials upon which heterotrophs depend for their nutrition. The principal mineral elements needed by plants are nitrogen, phosphorus, potassium, calcium, magnesium, and sulfur. Of these, nitrogen, phosphorus, and potassium are used in greatest amounts and often become depleted from farm soil. Some 95 percent of plant tissue, however, is composed of carbon, hydrogen, and oxygen, derived from air and water. Water and minerals are absorbed by plant roots, the former by osmosis and the latter mostly by active transport and diffusion. Carbon dioxide enters leaves from the air.

The importance of organic matter to farm and garden soil is its contribution to good soil structure and its slow decay to inorganic forms that can be absorbed by plants. Humus also tends to bind cations of H, Ca, Mg, K, and Na, preventing their being leached out of the soil before plants can absorb them. Exceptions to the rule that plants do not absorb organic matter are the insectivorous plants, which absorb the amino acids released by digestion of trapped insects.

Many saprotrophs, such as fungi, bacteria, and many protists, absorb the small organic molecules resulting from the decay they promote by secreting enzymes into their immediate environment. As a group, saprotrophs digest almost any organic compound arising in nature.

The many kinds of nutrients needed for normal growth, for maintenance, and for life itself have a wide variety of specific metabolic roles. Besides various mineral requirements, essential nutrients of heterotrophs include many specific organic compounds. Heterotrophs require such large molecules as protein, fats, and carbohydrates for certain of their component molecules as well as for a general carbon source. For normal growth and maintenance, all heterotrophs require about ten of the 20 or so amino acids in their diets, although they do not all require the same ten. Certain unsaturated fatty acids are also usually essential, and some organisms require a steroid source as well. An organism requires in its diet those essential substances it is unable to synthesize. Organic dietary components required in trace amounts are called vitamins. All heterotrophs and most algae have vitamin requirements, the exact list differing from one phylum to the next.

Digestion is essentially a matter of hydrolyzing macromolecules to smaller units that can be absorbed by the organism. Their smaller size means that these molecules are also more readily synthesized into the organism's own unique macromolecules or used as a source of energy.

Protozoans engulf particulate food into food vacuoles in which digestion proceeds through an acid and then a neutral phase, the products being absorbed into the cytoplasm. Such intracellular digestion also occurs in sponges and coelenterates, but it is supplemented by an initial extracellular digestion in the latter. Most free-living flatworms have coelenteratelike digestive cavities that may be highly branched. Of the parasitic flatworms, most trematodes, or flukes, have simple, two-branched digestive cavities, whereas tapeworms have none. Neither coelenterates nor flatworms have an anus.

Beginning with nemerteans and nematodes, nearly all animals have one-way digestive tracts with mouth and anus, allowing for the evolution of highly specialized digestive tract regions and organs of great efficiency. Examples of such specializations include teeth, for mechanical degradation; salivary glands, for moistening and dissolving food and for starch digestion; crop, for temporary storage; stomach, for storage and dissolving some foods and for gross digestion of proteins; midgut, for completion of digestion and for absorption of products of digestion; and hindgut, for reabsorption of water and for forming and storing feces.

Human digestive tract secretions and some of their roles include (1) ptyalin, in saliva, for partial digestion of starch; (2) pepsin, from the stomach, for initial digestion of proteins; (3) trypsin and other proteases, from the pancreas, for completion of protein digestion to amino acids; (4) bile salts, from the liver, for emulsification of lipids; (5) lipases, from the pancreas and intestinal epithelium, for hydrolysis of lipids; and (6) various carbohydrases, from the pancreas and intestinal epithelium, for completion of carbohydrate digestion to monosaccharides.

The principal roles of the colon, or large intestine, are promotion of bacterial decay of food residues, reabsorption of water, and formation and storage of feces. Feces are composed principally of water, bacteria, and epithelial cells.

Gastrointestinal activity and secretion are under both nervous and hormonal control. Peristalsis, for example, occurs as a nervous reflex in response to the presence of food in the esophagus, stomach, or intestine. An example of hormonal control is the release of the hormone secretin into the bloodstream by the duodenal mucosa in response to HCl entering the small intestine from the stomach. Secretin reaching the pancreas stimulates it to release pancreatic juice into the duodenum, one effect being the neutralization of the HCl.

With little chewing, ruminant mammals swallow large amounts of grass or other vegetation, which then enters the rumen. There it is partially digested to a pulpy mass by specialized bacteria and protozoans. It then moves to the reticulum, from which it is regurgitated as a "cud" and thoroughly chewed. When swallowed, it reenters the rumen, and another cud is passed up to the mouth. When the material has reached a liquid consistency, it enters the rest of the stomach, the omasum and the abomasum, where final digestion by the animal's own digestive enzymes occur. In nonruminant herbivores, the large cecum, the blind pouch of the colon, serves a somewhat similar function. Here, bacteria and protozoans digest plant material after the animal's own digestive enzymes have acted on the food. Mice and rabbits may then reprocess this material by eating their own feces.

## Recommended Readings

BALDWIN, R. L. 1984. Digestion and metabolism of ruminants. *BioScience,* 34(4): 244–249. A brief, up-to-date discussion of the efficiency of digestion and metabolism of ruminants and the role of these animals in producing food for humans.

BEAUMONT, W. 1959. *Experiments and Observations on the Gastric Juice and the Physiology of Digestion* (1833), reprint edition. Dover Publications, New York. 150 years after it was written, Beaumont's monograph still makes interesting reading.

BRILL, W. J. 1977. Biological nitrogen fixation. *Scientific American,* 236(3):68–81. An interesting description of how certain bacteria and blue-greens are able to convert atmospheric nitrogen into ammonia.

CALLOWAY, D. H. and K. O. CARPENTER. 1981. *Nutrition and Health.* Saunders College Publishing, Phila. A thought-provoking and up to date account of digestion, absorption, and nutrition and their relation to human health.

DAVENPORT, H. W. 1978. *A Digest of Digestion,* 2nd ed. Year Book of Medicine. Year Book Medical Publishers, Inc. Chicago, IL

GARDINER, M. S. 1972. *The Biology of Invertebrates,* McGraw-Hill Series in Organismic Biology. McGraw-Hill, New York. Contains a summary of most of the major types of food acquisition and digestion by invertebrate animals.

HOAR, W. S. 1983. *General and Comparative Physiology.* 3rd ed. Prentice-Hall, Englewood Cliffs, N.J. An upper level textbook with an excellent chapter on the comparative biochemistry and physiology of nutrition and digestion.

MOOG, F. 1981. The lining of the small intestine. *Scientific American,* 245(5):154–176. A description of the cells that line the small intestine and their many roles in digesting and absorbing nutrients.

RUCKEBUSCH, Y. and P. THIVEND. 1980. *Digestive Physiology and Metabolism in Ruminants.* AVI Publishing Company. Westport, CT. Up to date insights into the world of ruminant digestion and nutrition.

10

# Transport in Vascular Plants

OUTLINE



Only after plants had evolved vascular tissues were they able to achieve a large size. A large surface exposed to light and the carbon dioxide of the air provides a distinct photosynthetic advantage. But such an exposed surface loses water and requires an extensive root system to tap the available water and minerals in the soil. They must be transported to the aerial parts, and carbohydrate sustenance must be delivered to the living cells of the roots. Evolution of vascular tissues made this possible. The vascular tissues are the xylem and phloem, which extend throughout the roots and stems as a rod or cylinder of vascular bundles called a **stele** and into the leaves of vascular plants (Fig. 10-1). **Xylem** transports water and minerals from the soil to all parts of the plant. **Phloem** transports the **photosynthate**—carbohydrates produced in the leaves as a result of photosynthesis—to sites where it is stored or utilized by the plant in cellular respiration.

## XYLEM

The xylem consists of water-conducting cells, the **tracheids** and **vessel elements,** with scattered living parenchyma cells and supporting fibers. Tracheids and vessel elements are elongated cells reinforced with secondary walls (Fig. 10-2). They are dead and devoid of cytoplasm at maturity and therefore hollow. Their walls are perforated with pits through which water may pass laterally from cell to cell. Vessel ele-

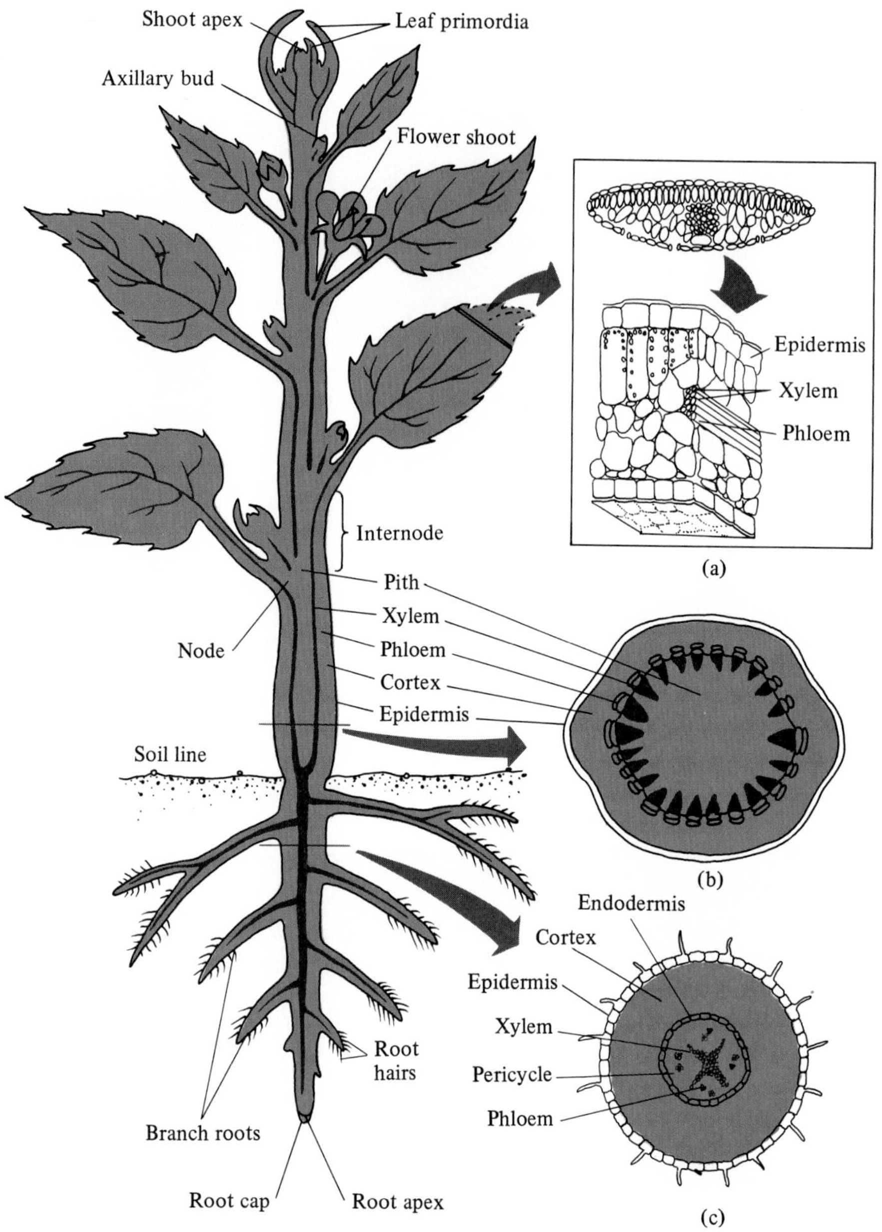


**FIG. 10-1** The vascular plant body, showing the paths of water and dissolved minerals and photosynthate. Water containing mineral ions enters the root through the root hairs, crosses the root cortex to the xylem vessels and tracheids, and is pulled upward into the leaves. Photosynthate produced in the leaves enters the sieve tubes of the phloem and is transported to "sinks" in the roots and fruits. Diagrams of cross sections of (a) leaf, (b) stem, and (c) root, showing the location of xylem and phloem.

ments have evolved from tracheids. They are found, with few exceptions, only in flowering plants. They are shorter and wider than tracheids and are joined end to end in a series forming a vessel; the end walls between the elements are largely dissolved, leaving a perforation, or pore. Vessels are thus more effective water conduits than a series of tracheids since water passing from one tracheid to another must pass through the pits in their lateral walls.

## PHLOEM

The phloem tissue consists of food-conducting cells, parenchyma cells, and fibers, thick-walled cells that lend support to stems and leaves. The food-conducting cells are **sieve tube elements** in angiosperms and **sieve**

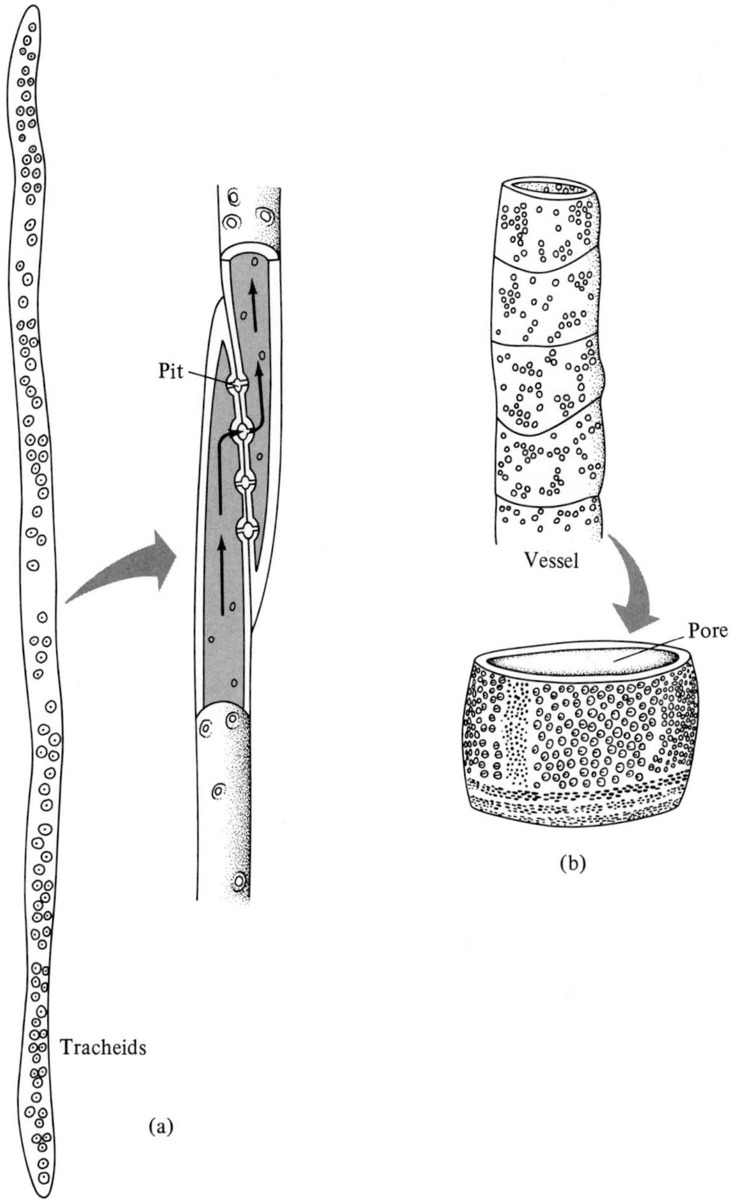


**FIG. 10-2** Water-conducting cells of the xylem, showing tracheids (a), which overlap at ends where water passes from cell to cell through pits, and (b) a vessel consisting of vessel elements aligned end to end with the intervening walls digested away, leaving a pore.

**cells** in other vascular plants (Fig. 10-3). Adjacent sieve cells and successive sieve tube elements are interconnected by sieve areas in their walls through which connecting strands of cytoplasm—large plasmodesmata—extend. Sieve tube elements are arranged end to end, forming **sieve tubes.** At the ends of the sieve tube elements, where the walls overlap, one or more **sieve areas** are clustered in a **sieve plate,** providing a greater area and more efficient translocation.

Functional sieve cells and sieve tube elements are living, although they have lost their nuclei, ribosomes, many organelles, and their internal and vacuolar membranes. The cytoplasm of adjacent cells is continuous through the connecting strands, forming a continuous living pathway for transport through either a sieve tube or chain of sieve cells. The cytoplasm is rich in P protein, a kind of fibrous protein that collects at the sieve plate when a sieve tube is injured or cut. The fibrils may act as a plug at the site of the wound and prevent the sieve tube contents from oozing out. Under certain circumstances, the sieve tubes may become

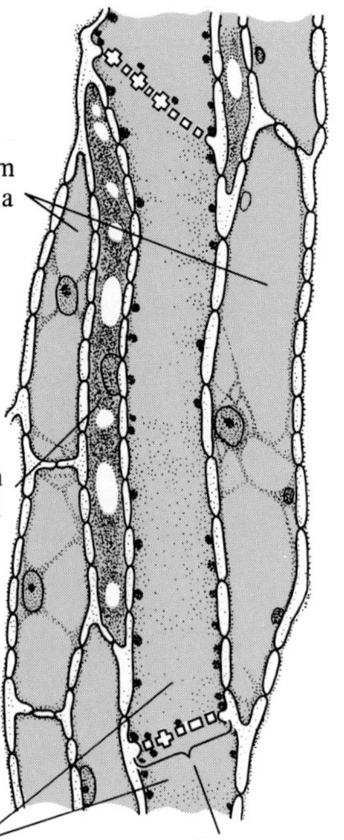


(a)

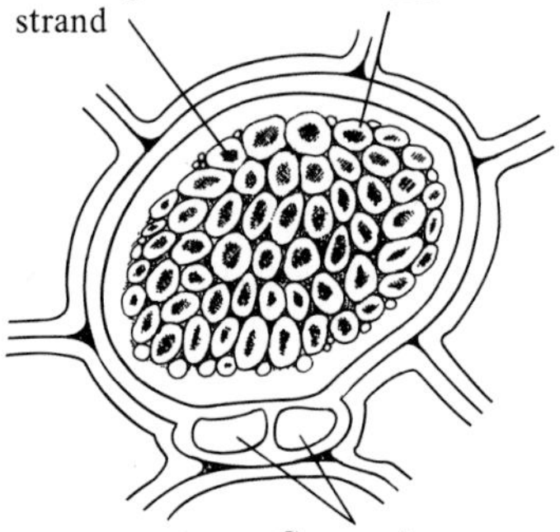


(b)

**FIG. 10-3** (a) Sketch of a sieve tube element showing sieve plates in end walls, with adjacent companion cell and phloem parenchyma cells. (b) Sieve plate in surface view. Clear areas consist of callose. Dark areas consist of cytoplasmic connecting strands.

blocked at the sieve plate with deposits of callose, a glucose polysaccharide. In perennial plants, for example, the sieve plates become blocked at the end of the growing season. New sieve tubes form in the spring and assume transport functions. The companion cell associated with each sieve tube element is thought to contribute to the transport process.

## VASCULAR TISSUE ORGANIZATION

The vascular tissues occur in characteristic arrangements in the roots, stems, and leaves. In the dicot root, the vascular tissues are located in a solid cylinder internal to the cortex at the center of the root (Fig. 10-4a), although a pith of living parenchyma cells is found at the center of monocot roots (Fig. 10-4b). Strands of phloem alternate with the radiating arms of xylem of the cylinder. The cylinder of vascular tissues is surrounded by a layer of living cells, the **pericycle.** Because of the alternate arrangement of xylem and phloem, it is possible for water to enter the xylem from the cortex without crossing the phloem. In the stem, the xylem and phloem are found together in vascular bundles. These are arranged in a cylinder in some groups of vascular plants (Fig. 10-4c), whereas in others they are scattered throughout the stem (Fig. 10-4d). In most woody species, the xylem and phloem are in concentric cylinders, with the phloem outward. The vascular bundles extend into the leaves from the cylinder of bundles in the stem, branching into finer and finer veins in the blade of the leaf (Chapter 7).

## ROOTS AND THE UPTAKE OF WATER AND MINERALS

As a general rule, a plant's root system is about as extensive as its shoot system. Water and minerals enter the plant from the soil through the surface of the fine branches of its roots. Several centimeters of the tip of each root are covered with numerous root hairs (Fig. 10-5). Each root hair is a tubular projection of an epidermal cell. Root hairs provide the plant with an extensive surface area and maximum access to the water supply. For example, a single rye plant was found to have a root system that occupied 0.08 $m^3$ of soil. The plant had an estimated 13.8 million roots covered with 14 billion root hairs, providing a combined surface area of 632 $m^2$ (6800 $feet^2$), one-seventh the area of a football field. As the root tip grows into the soil, new root hairs emerge behind it, and thus new reservoirs of water are continually being tapped. Meanwhile, older root hairs usually die.

Mineral ions essential to plant growth are taken up from the soil solution. Ions removed from soil water are replaced in the soil solution from negatively charged soil particles which bind cations (positively charged ions, such as $Ca^{2+}$, $K^+$, $Mg^{2+}$). Soils that are high in organic matter are more fertile because they have more negatively charged binding sites and can store and release more cations. Soils deficient in organic matter lose more minerals by leaching than fertile soils do and require frequent fertilizing. This is one reason why organic matter im-

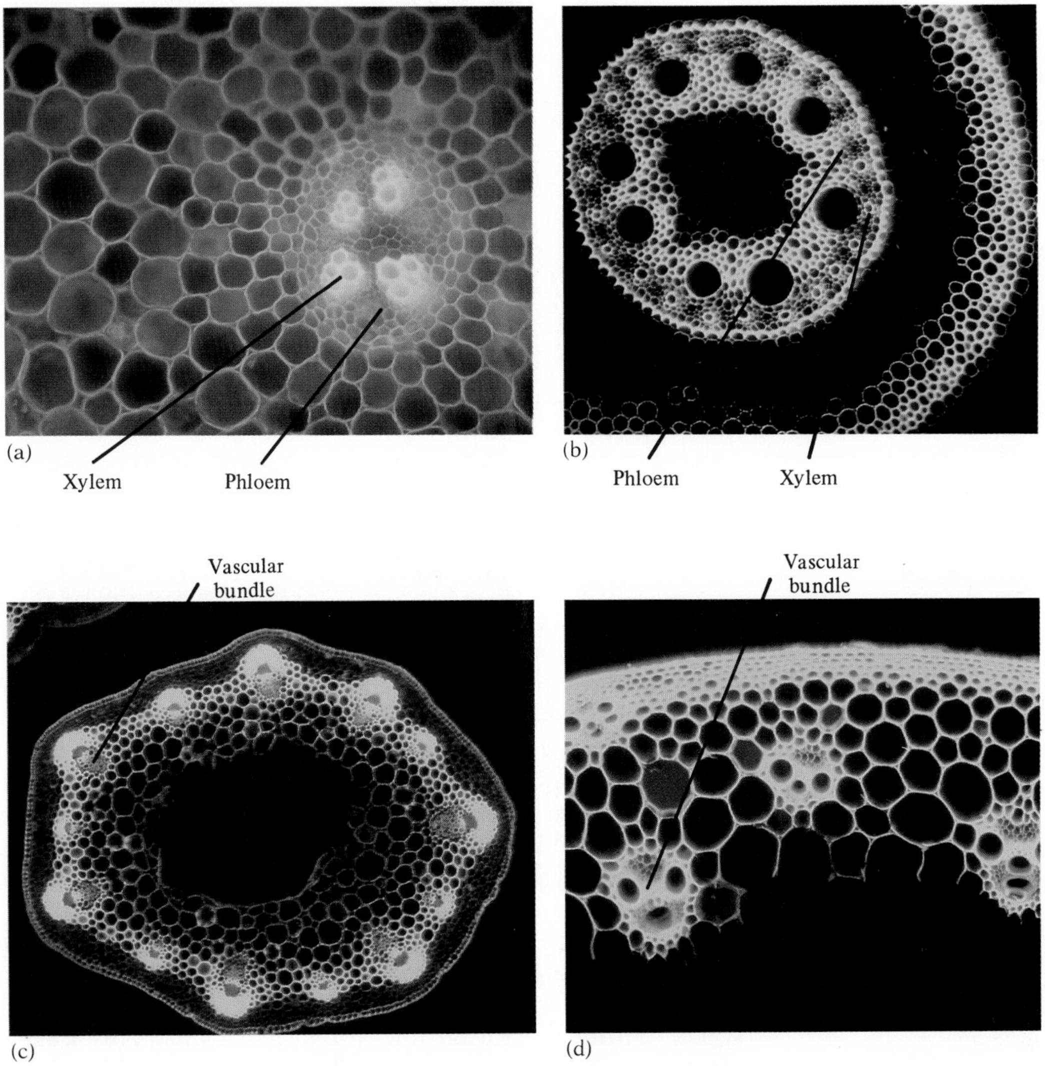


**FIG. 10-4** The root and stem of a dicot (a and c) and a monocot (b and d), showing the characteristic distribution of vascular tissues. The buttercup root (a) has a solid vascular rod; the corn root (b) is a cylinder with pith at the center. The alfalfa stem (c) has the vascular bundles in a cylinder; corn stem (d) has the vascular bundles scattered throughout the stem.

**FIG. 10-5** Root hairs are in direct contact with the soil particles and soil solution from which they take up minerals and water.

proves soil (Chapter 9) and why tropical rain forest soils, deficient in organic matter because of rapid cycling of litter from fallen vegetation, make poor agricultural land (Chapter 36).

Water and dissolved minerals enter roots principally through the freely permeable cell walls of the root hairs. Root hairs have no cuticle and therefore can absorb water readily. The water may traverse plasma membranes to enter the protoplasts (the body of protoplasm of the cell exclusive of the wall) of the epidermal cells by osmosis or move from cell to cell through the cell walls. Ions in the water cross the epidermal cell plasma membranes and enter cells by active transport. Current evidence indicates that protein carriers in epidermal cell plasma membranes selectively transport specific ions into or out of the cell.

From the epidermis, water and minerals pass through the cortex, a tissue of rounded living cells with extensive intercellular spaces, to the vascular cylinder (Fig. 10-6). Root cells through which water passes are

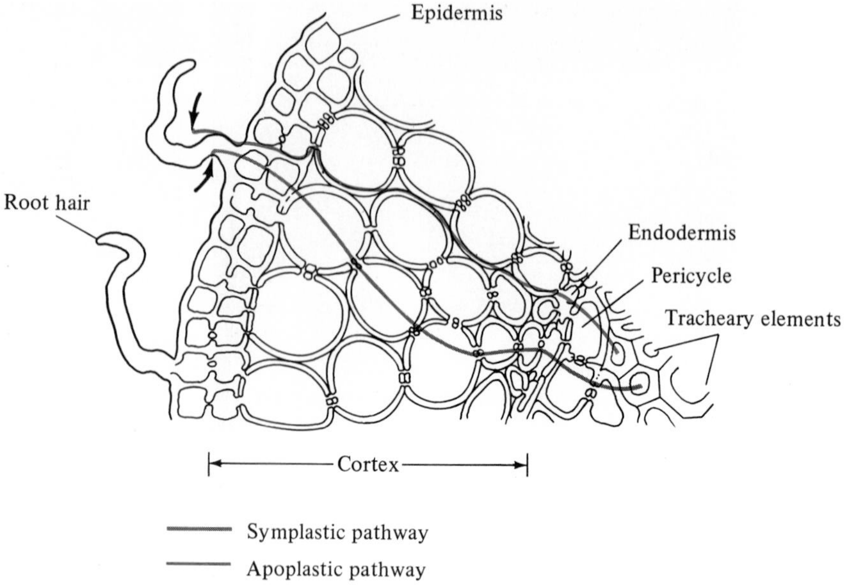


**FIG. 10-6** The alternative pathways for movement of water and inorganic ions from the soil and into the xylem (tracheary elements) of the root, via the root hairs, epidermis, cortex, and endodermis. In its peripheral tissues, two routes are open to water entering a root: the apoplastic pathway (shown in color) between the cells, a route made possible by the porosity of the cell walls; and the symplastic pathway (in black) through the cell protoplasts. When water reaches the endodermis, however, the Casparian bands surrounding these cells force all water to follow the symplastic pathway through the cells.

interconnected by plasmodesmata, the fine cytoplasmic strands that extend through small pores in the cell walls and connect adjacent cells. Water and minerals passing through the cortex via the protoplasts and plasmodesmata follow the **symplastic pathway,** which must be accessed through plasma membranes. Alternatively, they may move through the cell walls along the **apoplastic pathway.** Most water transport across the root is thought to be apoplastic, while ion transport is largely symplastic.

At the **endodermis,** the inner layer of the cortex, water and minerals can pass only through the protoplasts of the endodermal cells because these cells are encircled by a strip of **suberin,** a waterproof fatty substance within the radial and transverse walls. This **Casparian band** blocks water flow through the walls between the cells, restricting water movement to the protoplast (Fig. 10-7). Water and minerals emerging from the cytoplasm of the endodermal cells pass through the pericycle cells and into the xylem vessels at the ends of the xylem arms (Fig. 10-6). Endodermal cells may acquire a continuous layer of suberin and a secondary wall containing lignin, thickening the walls and making them impervious to water. This occurs first at the phloem centers and progresses laterally toward the xylem arms, where the cells, however, may remain permeable as "passage cells."

Once in the xylem of the root, the water and dissolved minerals are

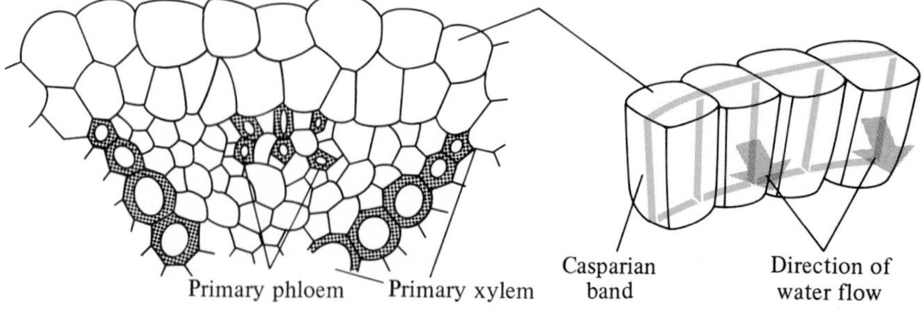


**FIG. 10-7** Movement of water through the protoplasts of endodermal cells. The Casparian band in the wall is composed of suberin, making it impermeable to water.

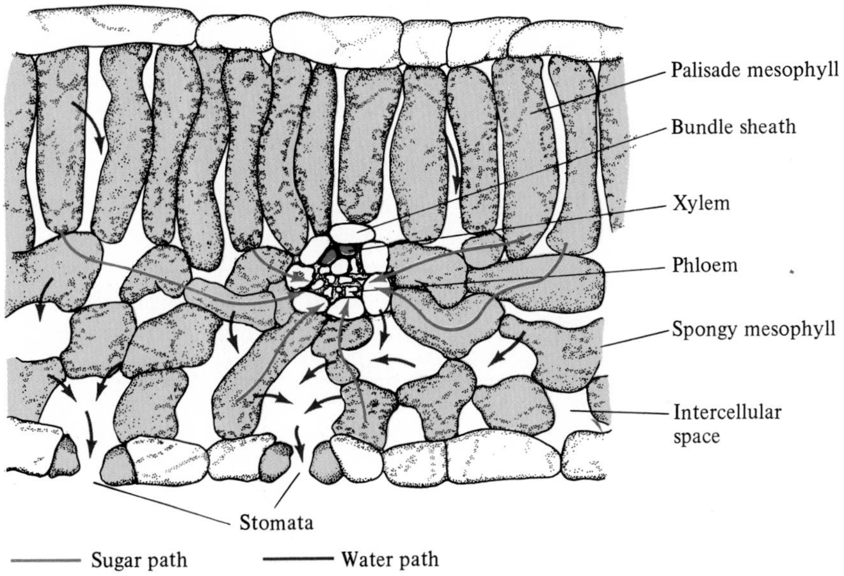


**FIG. 10-8** Diagram of a leaf showing routes of water molecules from xylem through the bundle sheath to mesophyll cells, from which water leaves in the form of vapor that diffuses out of the leaf via the stomata. Also shown are routes of sugar molecules from mesophyll cells to the phloem.

transported upward in the tracheids and vessels to the stem and its branches and thence into the leaves (Fig. 10-8). Cells along the pathway absorb water from the xylem as required.

# MECHANISMS OF WATER TRANSPORT IN XYLEM

For many years, one of the major questions in plant physiology concerned the mechanism by which water reaches the tops of trees, some of which may be over 100 m tall. The **transpiration-cohesion theory** (or cohesion-tension theory), proposed in the 1890s and supported by experimental studies since then, provides the most widely accepted explanation.

## TRANSPIRATION-COHESION AND TENSION

According to the transpiration-cohesion theory, transpiration, or the loss of water in the form of vapor, exerts a pulling force that moves water upward through the plant. Central to this theory is the fact that water molecules within the xylem vessels are held together by **cohesion** (attraction between like molecules as a result of hydrogen bonding), thereby imparting a great tensile strength, or ability to resist rupture, to the fine columns of water in the vessels and tracheids. **Adhesion** (attraction between unlike molecules) of water molecules to the cell walls of the xylem vessels also contributes to the tendency of water to rise in the xylem.

On a dry day, water diffuses out of the intercellular spaces of the leaf through the stomata. Closing of the stomata by the guard cells, the paired cells bordering the stomata (Chapter 7), reduces this loss of water. It is also reduced when the diffusion gradient is small, as when the relative humidity is high. As water leaves the intercellular spaces, evaporation from cell surfaces within the leaf increases, creating a water deficit in these cells. They in turn absorb water by osmosis from their neighbors, in a chain extending back to the bundle sheath cells of the veins. The bundle sheath cells take up water from the xylem by osmosis, exerting a pull on the columns of water in the vessels and tracheids. It is this tension, generated by transpiration and communicated from cell to cell by osmosis, that pulls water up the stem and into the leaves.

The hydrogen bonding of water molecules to each other transmits tension down the column of water in the xylem into the roots. The pull of transpiration is the primary factor in water absorption into the root. Water also enters root cells in response to solutes in the root cells and moves inward along an osmotic gradient.*

The fact that trees in the daytime often have a measurably smaller trunk diameter than they do at night reveals the tension in the xylem during transpiration and provides evidence to support this theory.

While transpiration is thus a major force in moving water in the plant, transpired water is lost by the plant, although cells along the path have had access to the water in the "transpiration stream." Transpirational loss of water is inevitable, since the stomata must be open for entry of carbon dioxide into the intercellular spaces of the leaf during photosynthesis. How the guard cells limit transpiration and water loss to a minimum is discussed in Chapter 7.

A problem to be considered in relation to the transpiration-cohesion theory is what happens when gas bubbles (embolisms) form within the vessels, as when the water in trees freezes in the winter. Part of the solution is that in the spring new vessels are formed whose water columns are intact. But it is also thought that the pits in the side walls protect against such embolisms. The pits are too small to allow the passage of the vapor bubbles, although water can move from cell to cell through the pits unimpeded. Ultimately the gas is redissolved in the water, and the water columns are no longer diverted.

## ROOT PRESSURE

When transpiration is limited, as in high humidity or when the stomata are closed, water moves by root pressure in some plants. This force is evident in **guttation,** the exudation of water droplets from the leaves of plants such as grass on cool mornings (Fig. 10-9). Root pressure is also evident when the top of a small plant is cut away and sap "bleeds" from the exposed vessels.

Root pressure develops when solutes or ions accumulate in the xylem vessels in the roots in the absence of transpiration. An osmotic

**FIG. 10-9** Leaves showing guttation as a result of root pressure.

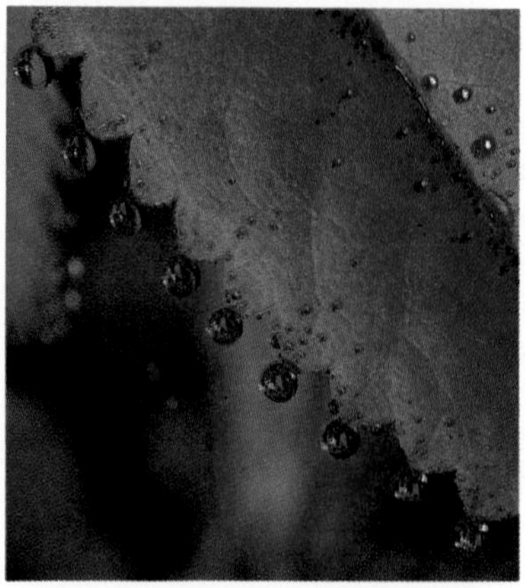

* Plant physiologists discuss water relations in terms of **water potential** ($\psi$), an expression of the free energy of water in a system. In this concept, water moves from a high to a low water potential. The water potential of pure water is set arbitrarily at 0. In a solution or under tension, the water potential is lower, a negative value. Plant water transport is readily explained in terms of water potential. Water enters a root hair cell from the soil because cell solutes lower the water potential ($\psi$). In the xylem vessels, water is under tension and has a still lower $\psi$, and water is pulled from the endodermis and across the root. Up in the leaf, the bundle sheath cells contain solutes and have a low $\psi$; they pull water from the xylem. Photosynthetic cells in the leaf, with their high sugar content, have a still lower $\psi$ and draw water from adjacent cells in a chain extending to the bundle sheath. As water evaporates from the leaf cells and diffuses out of the leaf, the lowered $\psi$ of these cells pulls water from adjacent cells, generating the transpiration pull. Outside the leaf, if the relative humidity is low, $\psi$ is low and water diffuses out of the intercellular spaces.

gradient is thus generated across the root, with water entering the root hairs and diffusing through the cells of the cortex to the xylem. The active secretion of sucrose into the xylem vessels may also generate root pressure, as in sugar maple trees in the spring.

PANEL 10-1

# Experimental Support for the Transpiration-Cohesion Theory

The basic principle underlying the transpiration-cohesion theory is readily demonstrated with an apparatus consisting of a narrow water-filled tube capped with a porous clay sphere and with its lower end immersed in a jar of mercury that is open to the atmosphere. When the clay vessel at the top of the water column is subjected to drying, water molecules evaporate from its surface. Cohesive forces of the water molecules at the surface thereupon pull neighboring molecules into position, replacing the evaporated water molecules. The resulting tension is transmitted down the column of water. This pull upon the column lowers the pressure at the bottom of the tube. As long as the water and mercury columns remain intact, the apparatus gives an indication of the combined tension upon the water column, the cohesive forces between water molecules, and, incidentally, the cohesive forces between the mercury atoms.

This experiment demonstrates the principle that a combination of evaporation and cohesion can pull a column of water up a tube. However, the tension that can be demonstrated with this device is limited; it is far less than the force needed to pull a column of water to the top of a tall tree. That the magnitude of the cohesive force of water was adequate to raise water the necessary height was demonstrated by a simple experiment showing that the diameter of the tube is a very important factor. Spinning S-shaped, water-filled capillary tubes in a centrifuge reveals the force needed to overcome the cohesive attraction between water molecules. Although water in tubes of wide diameter is pulled apart at low tensions, in the finest tubes great force is required to break the water columns. These results indicate that the effective cohesiveness of a column of water in such fine tubules as chains of xylem tracheids is more than enough to allow water to reach the top of the tallest giant redwood tree.

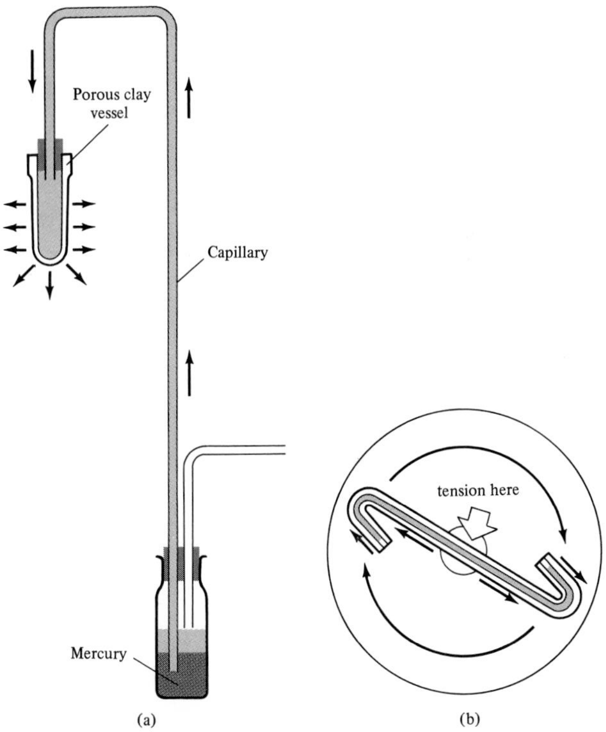


(a) Apparatus used to demonstrate transpirational pull. A capillary system (comparable to xylem vessels) connected to a clay vessel is filled with water. As water evaporates through the pores in the vessel, mercury is pulled up the capillary by a combination of cohesive and adhesive forces, providing a measure of the amount of force generated. (b) S-shaped tube, water filled, is spun in a centrifuge to show cohesion of water in a fine tube. Water tends to pull apart at the middle of the column by centrifugal force. The finer the tube, the greater the tensile strength of the column.

Root pressure is a negligible factor in moving water since the force is inadequate to elevate water to the tops of even small trees. It does not even occur in many plants, including the conifers, and in others it develops only under certain conditions, as when transpiration is limited.

## TRANSPORT IN PHLOEM

Sucrose produced in the photosynthesizing mesophyll cells of the leaf diffuses from cell to cell or via plasmodesmata and thence into the bundle sheath cells that enclose the veins of the leaf. From the bundle sheath it may diffuse into the parenchyma cells of the phloem, which secrete the sucrose into the sieve tubes. Secretion of sucrose into the sieve tubes, called **phloem loading,** is an active process requiring energy. In some plants, the cells involved in phloem loading, called transfer cells, have special infoldings of the plasma membrane that increase the membrane surface area and appear to aid in secretion. Once in the phloem sieve tubes, the sucrose is translocated to other parts of the plant.

An ingenious method for studying the translocation of materials in the phloem involves the use of aphids, insects that feed on plant juices by selectively tapping phloem vessels with their mouth parts, or stylets (Fig. 10-10). These miniature syringes, produced by snipping off the body of a feeding aphid from the mouth parts, permit the collecting of sieve tube contents. Such experiments have revealed information about the rate of flow in the sieve tube and about the sugars present, most of which is sucrose.

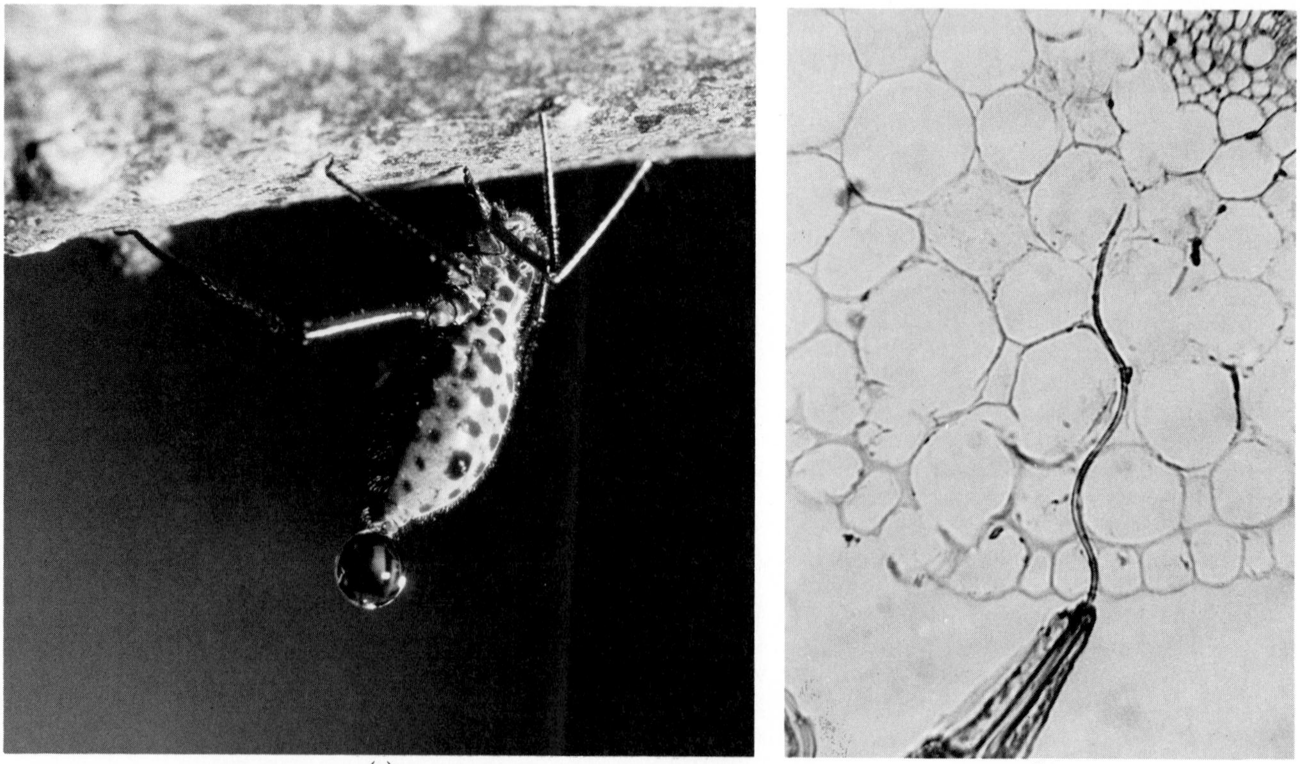

(a) (b)

**FIG. 10-10** The use of aphids in the study of phloem exudate composition and sucrose transport. (a) The aphid injects its stylet (mouth parts) into the stem. A drop of excess fluid (honeydew) is eliminated from its posterior end. (b) The tip of the stylet is seen penetrating a stem.

## THE PRESSURE-FLOW THEORY

The rate of sucrose flow in the phloem cytoplasm is remarkably rapid, in many cases exceeding 100 cm/hr. Because this rate is faster than can be accounted for by diffusion, other explanations have been sought.

A widely accepted theory is the pressure-flow hypothesis first proposed in 1927 by Ernst Münch, a German botanist. He postulated the occurrence of a mass flow of solution through sieve tubes in response to pressure differences in different regions of the phloem. According to the theory, a concentration gradient of sucrose develops between regions where sucrose is being produced, called "sources," and the sites of sucrose utilization or storage, called "sinks". One source is the leaves, where sugar is produced. Another is the storage cells of roots, where stored starch is converted to sugar. One type of sink is represented by cells using sugar for growth. Another exists in regions where sugar is stored, such as fruits, seeds, and roots.

The pressure-flow hypothesis is illustrated diagrammatically in Figure 10-11a. Theoretically, when sucrose is secreted into the sieve tubes at one of the sources, water enters the sieve tubes osmotically from nearby xylem vessels, generating pressure. At sinks elsewhere in the plant, sucrose is absorbed from the phloem, and its concentration decreases. The absorbed sugar may be used or converted to a storage product. As the sugar concentration drops at the sink, water diffuses out of the sieve tube into nearby cells. The essence of the pressure-flow theory is that the resulting pressure gradient within the phloem causes the flow of solution from sucrose source to sink. The application of Münch's model is shown in Figure 10-11b. According to this theory, no energy is expended directly in the sieve tubes. However, energy in the form of ATP

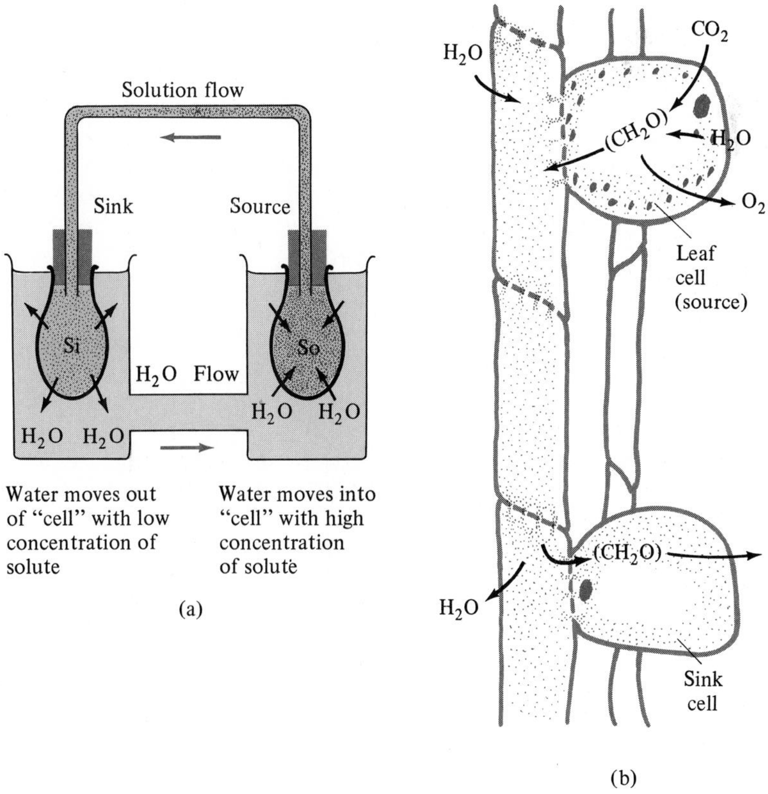


**FIG. 10-11** The pressure-flow theory of translocation. (a) A model suggesting how translocation occurs. Where sucrose is being produced at a source (on the right), such as leaves, osmotic pressure will be high and water will enter the phloem from the xylem by osmosis. Where sucrose is being used or stored (the "sink" at the left), osmotic pressure will fall and water will leave to reenter the xylem. The resulting difference in hydraulic pressure produces a flow of sucrose-laden fluid from the source to the sink. (b) The process depicted as it occurs in the plant. Leaves are a principal source of sucrose ($CH_2O$), and roots and fruits are sinks.

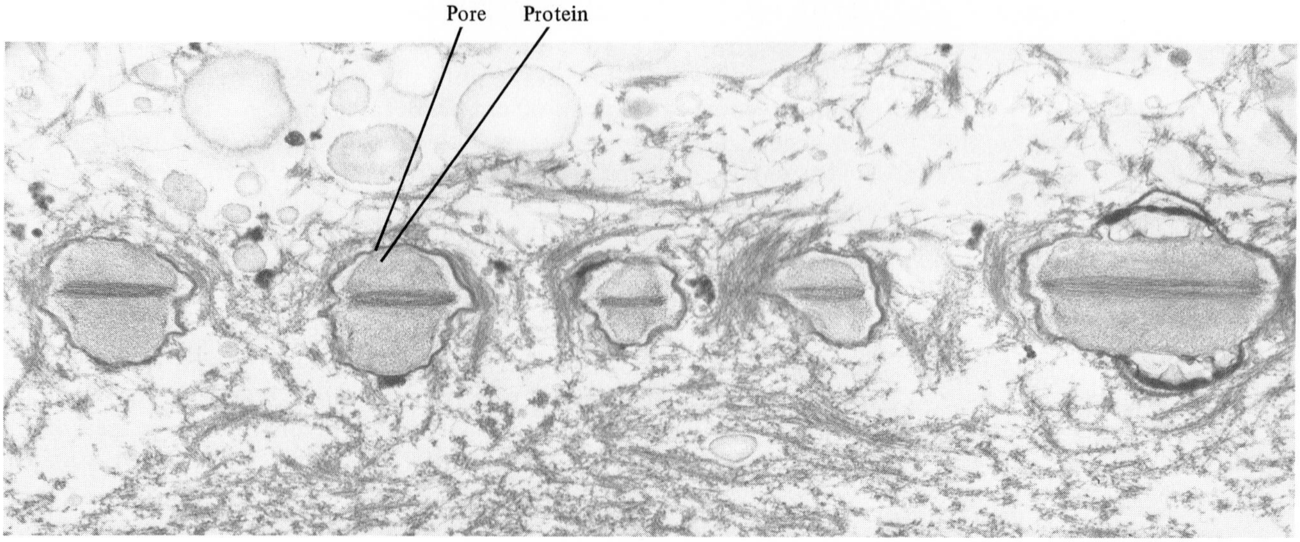


**FIG. 10-12** Sieve plate from phloem sieve tube element of the tobacco plant. The sieve plate pores are plugged with P-protein.

is utilized in the parenchyma cells that secrete sucrose into the sieve tubes. Thus the overall process is active; that is, it requires an input of energy.

A criticism of the pressure-flow theory is that it cannot explain phloem transport unless the pores in the sieve plates at the ends of sieve tubes are open. Yet nearly all chemically preserved preparations examined by electron microscopy show the sieve plate pores to be blocked by protein fibrils or wall material (Fig. 10-12). It is thought that any wounding of the phloem, as in cutting for tissue preparation, alters the sieve tubes. However, tissues prepared by the freeze-fracture method—in which cells are rapidly frozen and split open and a thin replica is made of the fracture faces—appear in the electron microscope to be loosely filled with protein, with no dense plugging. Only if the sieve plate pores are normally open is pressure-flow possible. If the pores are filled, then another explanation is necessary. The current consensus is that the pores are open while the sieve tubes are alive and functional. Blockage may be a protective, leak-preventing mechanism triggered when the phloem transport system is injured; it may be analogous to the clotting mechanism that arrests bleeding when an animal is wounded.

## CIRCULATION IN PLANTS

Whereas most water moves upward in the xylem, water also moves downward in the phloem in the aqueous sucrose solution. This water enters the phloem by osmosis from the xylem in the fine veins of the leaves. Some water thus may circulate within a plant by moving up through the xylem, from xylem to phloem in the leaf, and down through the phloem; in the root or other storage depots, it may reenter the xylem from the phloem. Radioisotope studies have confirmed movements of inorganic ions between xylem and phloem. Another variation from the general pathway is the transport of sucrose in the xylem, as in the spring, when xylem sap rises from the roots where sugar has been stored over winter. The spring rise of xylem sap is induced largely by the secretion of sugar into the xylem by the root cells. The hypertonic solution thus produced greatly enhances the absorption of water from the soil, thereby supplying nutrients and water to the rapidly expanding buds. Tapping sugar maple trees early in the spring by cutting through the bark to the xylem releases the sweet, flavorful solution, which, when concentrated by evaporation of the water, produces maple syrup and maple sugar.

## Summary

Transport in plants occurs in the xylem and phloem tissues. The xylem transports water and minerals taken up by the roots to the shoots; the phloem transports the sugar produced in photosynthesis throughout the plant. These tissues are found in characteristic arrangements in the plant organs. The water-conducting cells of the xylem are the hollow, dead tracheids and vessel elements. In the phloem, sieve cells and sieve tube elements translocate sugars in the living cytoplasm; the cells are interconnected by strands of cytoplasm called connecting strands, which extend through the sieve plates.

Water and minerals are taken up from the soil in the root hairs and cross the root cortex to the xylem. Water may pass through the cell walls, the apoplastic pathway, or through the cells, the symplastic pathway. Mineral ions tend to move through the cells. At the endodermis, water is forced to pass through the protoplasts because of Casparian bands in the cell walls. Water is pulled up and into the leaves by the tension generated at cell surfaces in the leaves when water evaporates into the intercellular spaces and is lost from the plant in transpiration. Cohesion of water molecules in the vessels and tracheids transmits the tension into the roots and may pull water across the root and even into the plant from the soil, although water molecules also enter epidermal cells by osmosis. Mineral ions enter epidermal cells by active transport.

Sucrose produced in leaf mesophyll cells diffuses from cell to cell and then to the bundle sheath cells surrounding the veins. In the process called phloem loading, sucrose is secreted into the sieve tubes. Sucrose is then translocated to roots or fruits, where it is used or stored. Mass flow of the sucrose solution in the sieve tubes is thought to be caused by a pressure gradient. Pressure is generated where water is taken up osmotically in response to the increased solutes, and it is reduced where sucrose is used or stored and water lost.

## Recommended Readings

*Annual Review of Plant Physiology.* This periodical is an important source of articles on current aspects of topics in plant physiology by authorities in their fields.

BORNMAN, C. H., and C. E. J. BATHA. 1973. "The role of aphids in phloem research," *Endeavor,* 32:129. An interesting account of translocation and the history and results of the use of aphids in phloem research.

DEVLIN, R. M., and F. H. WITHAM. 1983. *Plant Physiology,* 4th ed. Willard Grant Press, Boston. A widely used textbook of plant physiology with clear coverage of transport in plants.

JENSEN, W. A., and F. B. SALISBURY. 1984. *Botany,* 2nd ed. Wadsworth, Belmont, Calif. A revised edition of a popular textbook with clear, current treatment of cells, structure, and physiology. Well diagrammed and illustrated.

RAVEN, P. H., R. F. EVERT, and H. CURTIS. 1981. *Biology of Plants,* 3d ed. Worth, New York. An interesting, well-illustrated textbook of botany.

RAY, P. M., T. A. STEEVES, and S. A. FULTZ. 1983. *Botany.* Saunders, Philadelphia. An authoritative textbook of botany by recognized investigators.

SAIGO, R. H., and B. W. SAIGO. 1983. *Botany: Principles and Applications.* Prentice-Hall, Englewood Cliffs, N.J. Incorporates scientific, technological, and agricultural aspects of botany and aesthetic and recreational uses of plants in a highly readable and well-illustrated introduction to botany at the college level.

SALISBURY, F. B., and I. Ross. 1985. *Plant Physiology*, 3rd ed. Wadsworth, Belmont, Calif. A revised edition of a standard plant physiology text with a clear description of plant transport.

WOODING, F. B. P. 1978. *Phloem*. Carolina Biology Readers. Carolina Biol. Supply. A brief account of phloem structure and function illustrated with effective diagrams, photographs, and electron micrographs.

ZIMMERMAN, M. H. "How sap moves in trees." 1963. *Scientific American*, 208(3):132–140. An interesting consideration of the mechanisms of internal transport in plants and their study.

ZIMMERMAN, M. H., and J. A. MILBURN, eds. 1975. "Phloem transport." *Transport in Plants:* I. Vol. 1. Encyclopedia of Plant Physiology. Springer-Verlag. N.Y. A series of review articles by investigators, providing current information on the phloem and translocation.

# 11 Transport in Animals

OUTLINE



A city requires trucks and a system of streets to supply its various regions with materials and to remove the wastes generated there. Similarly, the body of an animal requires a system that distributes nutrients and oxygen and removes wastes. The blood or a similar fluid in an animal's body is analogous to the trucks in a city, and the vessels that contain the body fluid are analogous to the streets.

## INTERNAL TRANSPORT SYSTEMS

In many simple aquatic animals, such as sponges and cnidarians, the external medium—usually seawater—is circulated through the body by ciliary or muscular action. In these animals, as in protistan cells, every cell is bathed directly by the external medium. In the more highly evolved roundworms, or nematodes, the body cavity is filled with a fluid that differs somewhat in composition from that of the external medium and may even contain specialized cells (Fig. 11-1a). This simple

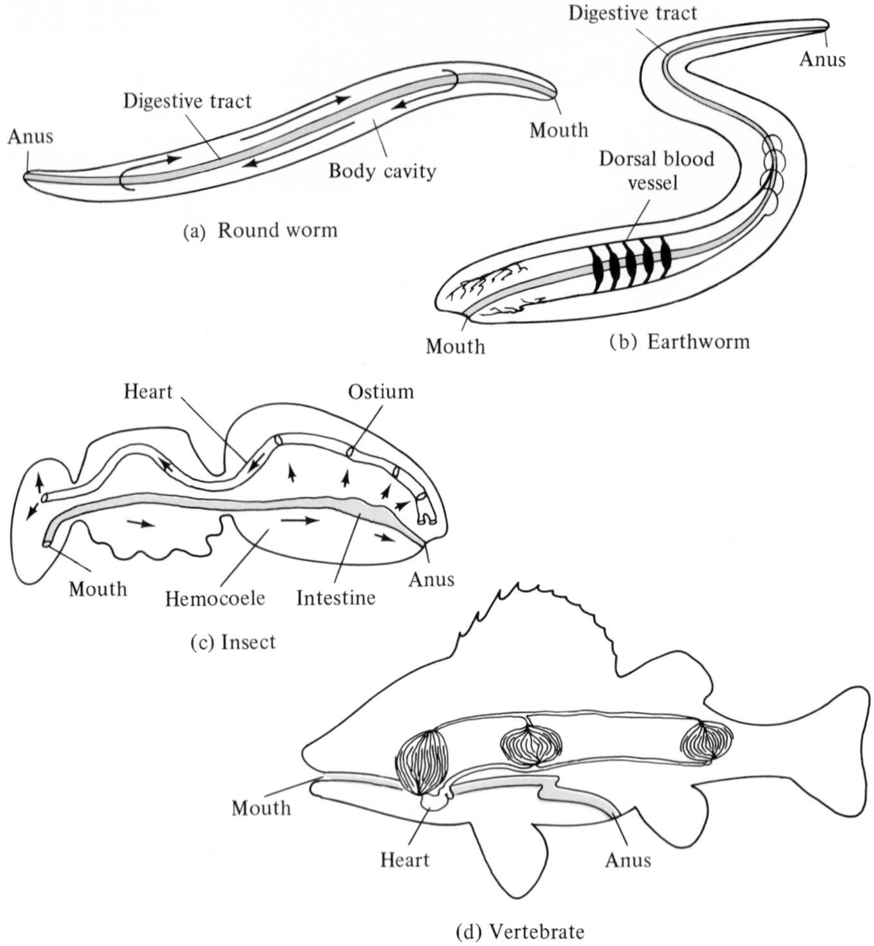


**FIG. 11-1** Diagrams of internal transport systems. (a) A roundworm, with a fluid-filled body cavity. (b) An earthworm, with both a fluid-filled body cavity and a closed circulatory system. (c) An arthropod, with a partially open system of sinuses in lieu of capillaries. (d) A vertebrate, with a body cavity, and closed blood system.

arrangement, although inefficient as an internal transport system, is adequate for a small animal.

In larger or more active animals, the role of the body cavity in the distribution of materials is supplemented by a system of vessels, the **blood vascular system.** If the movement of the blood through a vessel or chamber is always in the same direction and describes a generally circular route through the body, the network is called a **circulatory system.** As the blood flows through the body, it continually absorbs materials from some tissues and releases them to others. The vessels that carry blood away from the **heart,** if the organism has such a hollow, muscular pumping organ, are termed **arteries;** those returning blood to the heart are **veins.** The thin-walled vessels that ultimately connect arteries to veins are **capillaries.**

The earthworm has an almost completely **closed circulatory system,** in which most of its tissues and organs are supplied by capillaries (Fig. 11-1b). Many molluscs and arthropods have a largely **open circulatory system,** in which most organs are surrounded by **blood sinuses,** greatly enlarged blood vessels that replace most of the body cavity. In such animals, the organs are bathed directly by a kind of body fluid called **hemolymph,** rather than being supplied by a bed of capillaries (Fig. 11-1c). The vertebrate circulatory system (Fig. 11-1d) is highly developed and consists of a completely closed blood vascular system, a well-developed heart, and a supplemental network of vessels known as the **lymphatic system** (page 315).

## PUMPING MECHANISMS

The metabolic needs of highly developed animals require an efficient means of supplying tissues with oxygen and nutrients as well as a means of waste removal. Efficiency is promoted by pumping mechanisms, which may be of two types: muscular body activity and hearts.

### MUSCULAR ACTIVITY

Where blood vessels pass through muscle, a contraction of the muscle surrounding the vessels causes a displacement of the fluid within them. Rhythmic locomotory movements of such animals as worms thereby contribute indirectly but significantly to their circulation. Locomotion also mixes and distributes body cavity fluids. Most vascular systems contain numerous **one-way valves** that prevent back flow of fluids. Thus, any pressure produced on vessels with such valves forces the fluid to circulate through the system, always in the same direction.

### HEARTS

As efficient as muscular body activity can be in promoting fluid transport in active animals, additional pumping action is usually required, especially when the animals are not moving. In some cases, this is provided by peristaltic action of one or more of the vessels themselves. In earthworms, for example, peristaltic movements of the **dorsal blood vessel** pump blood in a forward direction. As shown in Figure 11-2, most of this blood then enters the large ventral blood vessel by way of five pairs of contractile vessels, referred to as lateral hearts, that connect the two major vessels. The blood then flows posteriorly in the ventral vessel and, after passing through capillaries, returns to the dorsal vessel. The lateral hearts, however, are not as important to the movement of the blood as is the dorsal blood vessel.

In animals, the pumping organs called hearts vary greatly in complexity of structure and function. For example, in one group of primitive chordates, the sea squirts, the heart is a simple tubular structure (Fig. 11-3). By peristaltic contractions similar to the milking action that moves food materials through the digestive tract, the sea squirt's heart pumps blood through its vessels in a clockwise direction for a series of about 10 contractions. After a slight pause, the heart reverses direction

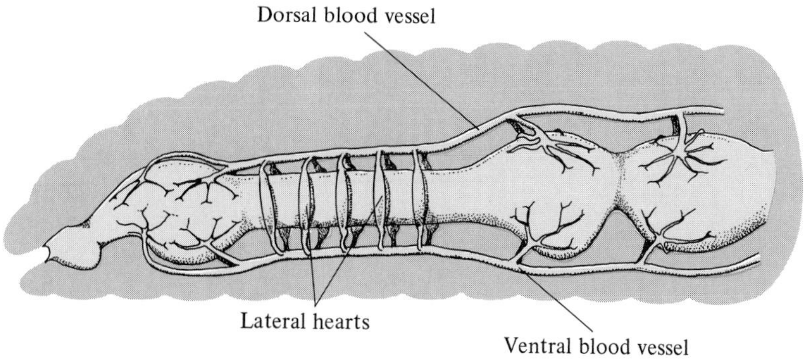


**FIG. 11-2** The anterior end of an earthworm, showing major blood vessels.

**FIG. 11-3** (a) Ascidians (sea squirts). (b) Diagram of circulation in the sea squirt. The direction of circulation is reversed from time to time.

(a)

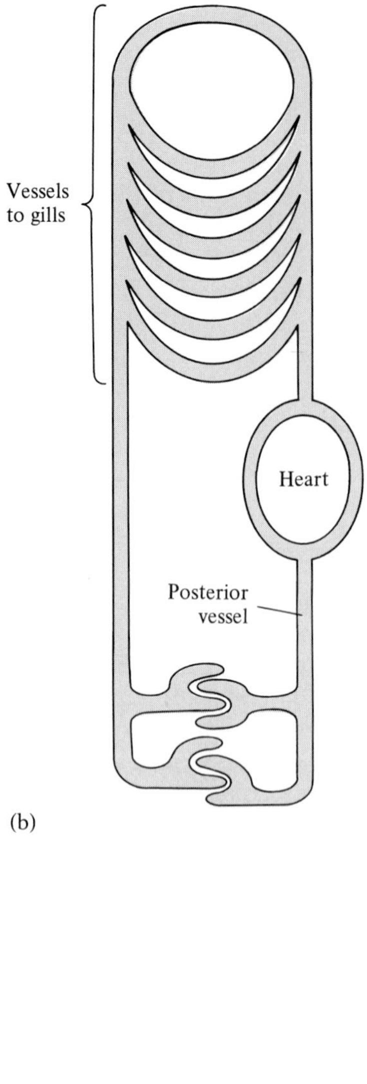


(b)

and pumps the blood counterclockwise for about the same number of contractions.

For small, inactive animals such as sea squirts, which spend their adult lives attached to stones in shallows at the edge of the sea, simple hearts that reverse direction every few contractions are adequate. But for large, active animals such as vertebrates, a more efficient pump is needed to circulate blood throughout the body. For example, the typical fish heart is essentially a single, S-shaped tube composed of four consecutive chambers through which blood is pumped in a posterior-to-anterior direction (Fig. 11-4). As in all vertebrate hearts, the chambers are equipped with one-way valves that prevent back flow. The valves function much as do doors that open in only one direction. Back pressure closes them. The chambers of the fish heart through which the blood passes are the sinus venosus, atrium, ventricle, and conus arteriosus. The **ventricle** is the most muscular of these chambers and provides the major force that sends the blood through the body. The **atrium** fills the ventricle to capacity just before each ventricular contraction, thus allowing the ventricle to function at top efficiency. The **sinus venosus** has a similar function in filling the atrium with blood returning from the body's tissues. The **conus arteriosus,** by the elasticity of its muscular wall, aids in regulating back pressure of the blood as it is forced into the ventral aorta leading to the gills. During the embryonic development of fishes, the heart arises as a straight tube (Fig. 11-4a) that later folds upon itself to form the S-shaped adult organ, in which the atrium has been displaced to a position dorsal and anterior to the ventricle (Fig. 11-4b).

With the evolution of air breathing in higher vertebrates, oxygenated blood was routed to the heart before being pumped to the tissues of the body. The heart became divided into two compartments, thereby separating the unoxygenated blood arriving from the body tissues and the oxygenated blood coming from the lungs. Complete separation of the two types of blood resulted in increased efficiency of the system.

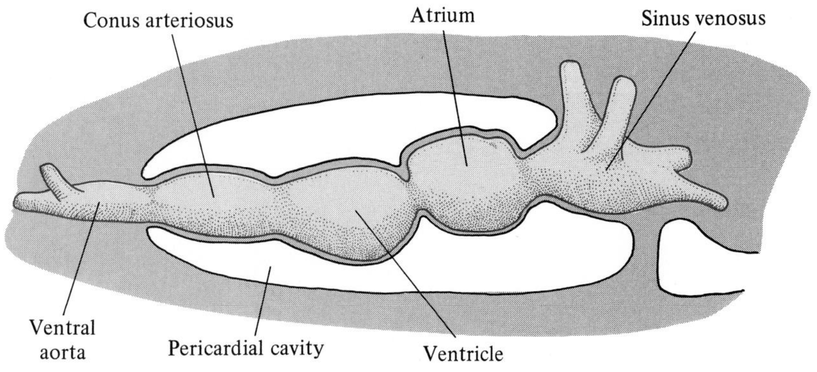


(a)

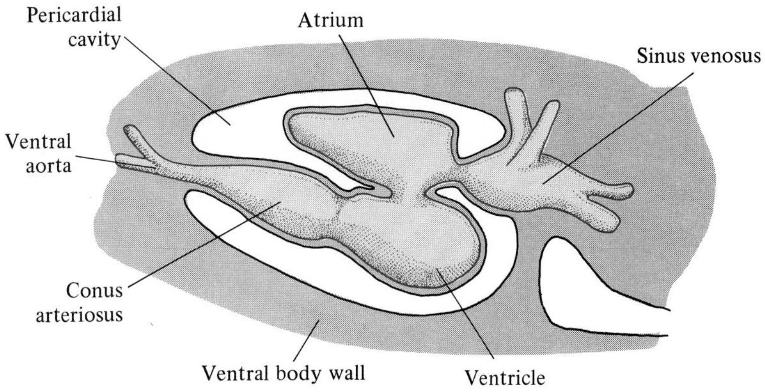


(b)

**FIG. 11-4** Probable evolution of the vertebrate heart. (a) Hypothetical ancestral condition, in which the four primitive chambers are sequentially arranged; this is also an embryonic stage of the modern vertebrate heart. (b) Shark heart, in which the atrium has pushed forward above the ventricle.

This led to an interesting evolutionary series of what have been called **double hearts,** culminating in the hearts of birds and mammals.

***Evolution of Double Hearts.*** The separation between oxygenated and unoxygenated blood is imperfect in the primitive air-breathing vertebrates: amphibians and reptiles. In lungfishes and amphibians, blood from the lungs is at first kept separate by bypassing the sinus venosus, which receives only the unoxygenated blood entering it from the largest of the body's veins, the two **anterior venae cavae** and the one **posterior vena cava.** The atrium is divided into two chambers: the right atrium receives blood from the sinus venosus, and the left atrium receives blood from the lungs (Fig. 11-5). In lungfishes this atrial separation is partial, whereas in amphibians it is complete.

Although the ventricle is also partially divided in lungfishes, it is a single compartment in modern amphibians. This lack of separation between the two sides of the ventricle in amphibians is related to the fact that their skin is a more important respiratory organ than their lungs in absorbing oxygen. In amphibians there is thus no lessening of efficiency if blood returning from the body—including the skin—is mixed with blood from its lungs.

The single ventricle of amphibians is not as inefficient as it first was assumed to be by biologists. An analysis of the dynamics of blood flow entering from the atria showed that mixing of blood from the two sources in the single ventricle is minimal. A **spiral valve** in the conus

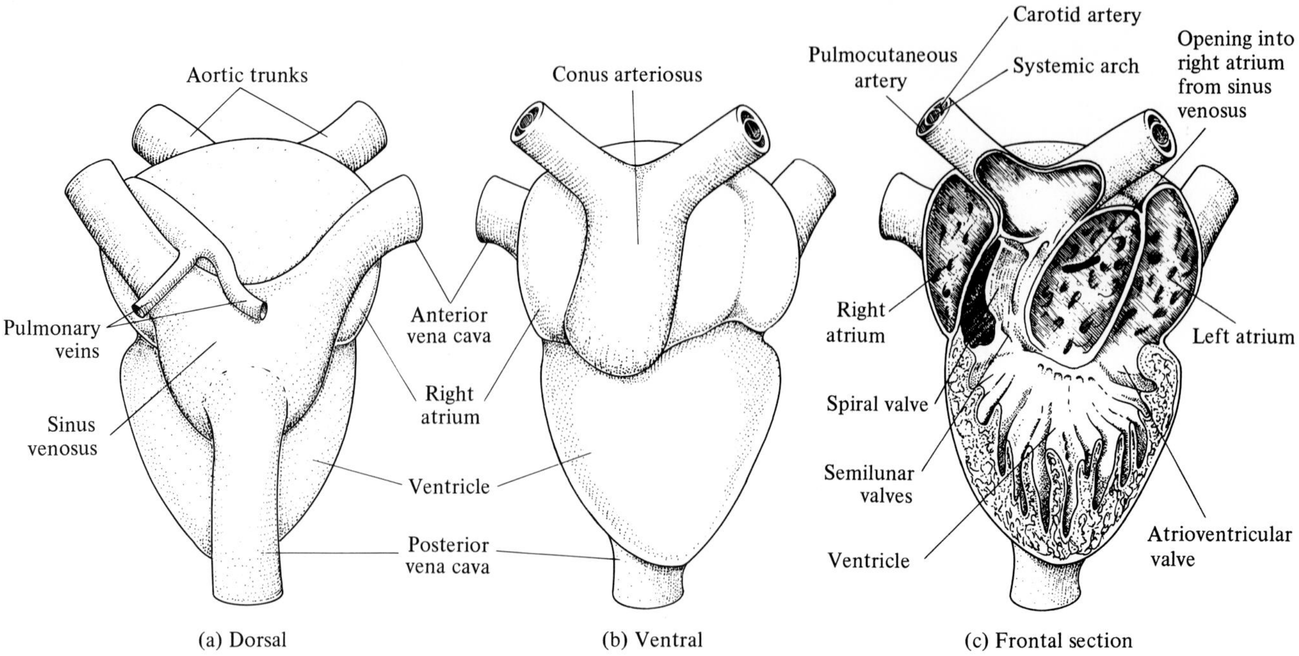


**FIG. 11-5** The frog heart. Note the single ventricle.

arteriosus directs the blood exiting the ventricle into two channels (Fig. 11-5c). When the heart contracts, the blood that enters the conus first is diverted to the **pulmonary arteries,** those that lead to the lungs. Most of this blood has come from the body tissues via the right atrium. As pressure builds during contraction of the ventricle, the spiral valve flops over. The rest of the blood entering the conus during this phase of contraction—primarily blood returning from the lungs via the left atrium—is directed to the vessels leading to the rest of the body.

In reptiles, whose lungs are the only important respiratory organ, the ventricle is almost completely divided into two chambers. In birds and mammals, the separation is complete: all blood from the body is sent to the right atrium and to the lungs via the right ventricle, and all blood from the lungs returns to the left atrium and is sent to the body via the left ventricle. A study of the embryonic development of birds and mammals has shown that they evolved this system independently of each other, that is, they did not inherit it from a common ancestor. As in fishes, the hearts of birds and mammals arise during embryonic development as a tube. These hearts seem to pass through, or recapitulate, their evolutionary stages during embryonic development. First, four sequentially arranged chambers are formed. These then fold in upon themselves to form a heart like that of a fish (Fig. 11-6a). Finally, the heart divides into right and left halves, the conus and sinus being incorporated into the tissues of the left ventricle and right atrium, respectively (Fig. 11-6b, c).

## THE MAMMALIAN HEART

The heart can be thought of as a highly specialized set of muscles within which are several cavities, or chambers. Mammals have a four-chambered heart composed of a **right atrium,** a **right ventricle,** a **left atrium,**

**FIG. 11-6** Stages in embryonic development of the human heart (ventral aspect). (a) Three millimeter embryo. (b) Five millimeter embryo. (c) Nine millimeter embryo. Note changing position of the atrium.

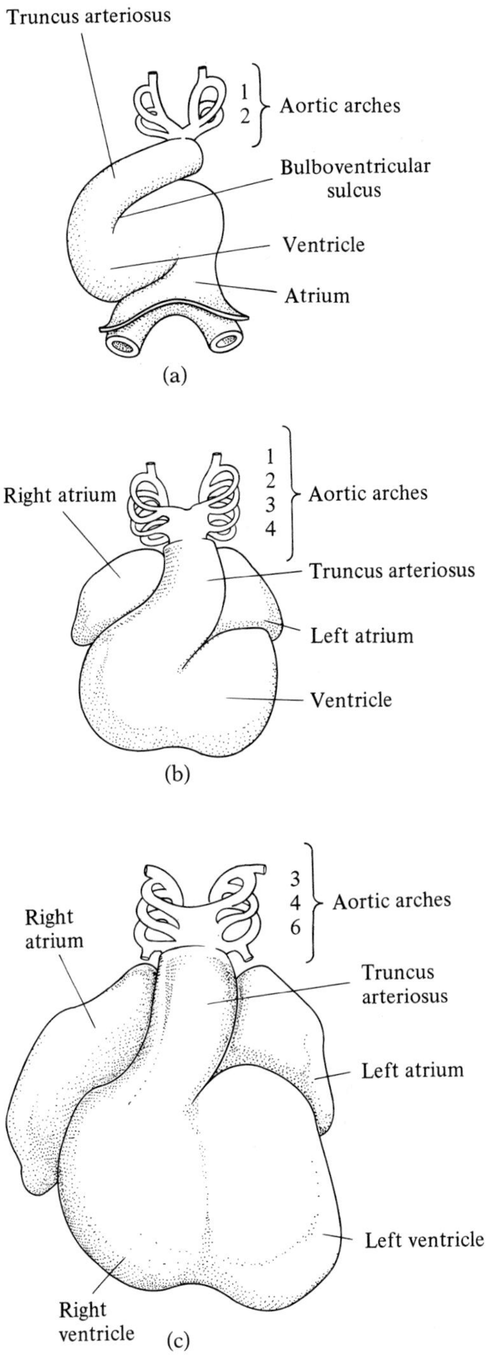


and a **left ventricle** (Fig. 11-7). This is said to provide a double circulation. The atria collect blood and fill the ventricles. Then the more muscular ventricles pump the blood out of the heart. The left ventricle sends blood via the **aorta,** the largest of the arteries, to all parts of the body except the lungs. The right ventricle sends blood via the **pulmonary arteries** to the lungs. Separating each atrium and the ventricle with which it connects is an **atrioventricular** (AV) **valve.** The valve between the left atrium and ventricle has two cusps, or flaps, of tissue and is called the **bicuspid** or **mitral valve.** Its three-flap counterpart on the right side is the **tricuspid valve.** The two AV valves are strengthened by the presence of cords of tissue, the **chordae tendinae,** attached to their

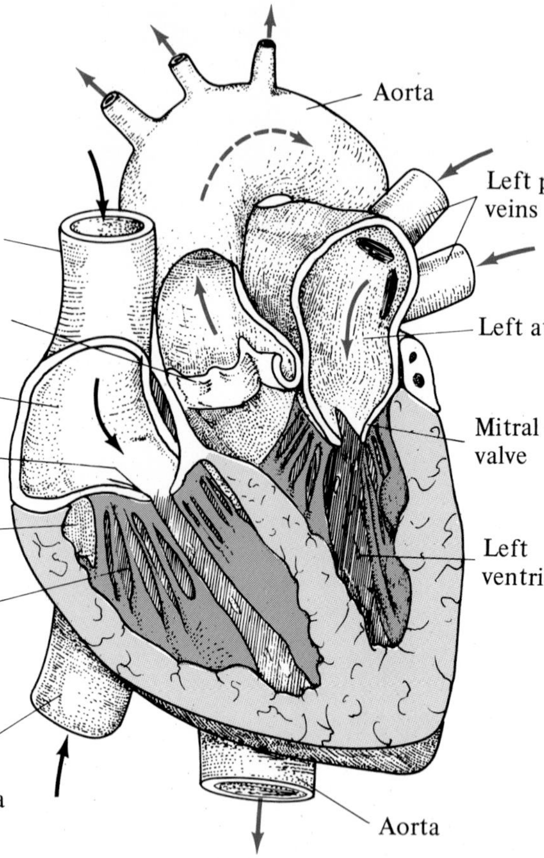


**FIG. 11-7** The four-chambered heart of a mammal. The pulmonary arteries are not shown.

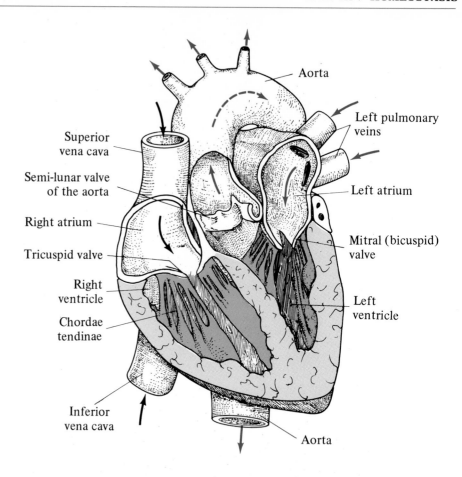

edges and anchored in the papillary muscles of the ventricle wall. These prevent the large flaps of tissue from opening into the atria when the ventricles contract. Other valves guard the exits from the ventricles. These are the **semilunar valves,** so-called because of their half-moon shape. The one located at the entrance to the aorta is the **aortic valve;** the one in the pulmonary artery is the **pulmonic valve.** All these valves keep the blood going in one direction and so prevent back flow in the system. The semilunar valves prevent blood that has left the ventricles from flowing back into the ventricles when they relax.

## MAJOR BLOOD VESSELS

As the pulmonary artery emerges from the right ventricle it immediately branches into the left and right pulmonary arteries. Blood that has returned from the body's tissues to the right atrium and ventricle and that is low in oxygen is thereby directed to the lungs. Oxygenated blood returns from the lungs via the right and left **pulmonary veins,** which empty into the left atrium. From there, blood enters the left ventricle, which, as we have seen, it leaves by way of the aorta. The aorta gives off the small right and left **coronary arteries,** which provide the heart muscle with a rich supply of blood under high pressure. The aorta ascends, then arches in a left, dorsal direction, opening into three major arteries as it proceeds: the brachiocephalic artery, which immediately branches into the right subclavian artery, to the right arm, and the right common carotid artery, to the right side of the head; the left common carotid artery; and the left subclavian artery (Fig. 11-8). The aorta continues in a posterior course just to the left of the vertebral column, in the dorsal portion of the chest cavity. As it continues posteriorly, it opens into other large arteries to major body organs, such as the renal arteries to the kidneys, and finally branches into the two common iliac arteries in the legs.

Blood returns to the heart from the lower part of the body by way of a large vein, the **posterior vena cava,** in humans called the **inferior vena cava,** which empties into the right atrium. The **anterior vena cava,** or

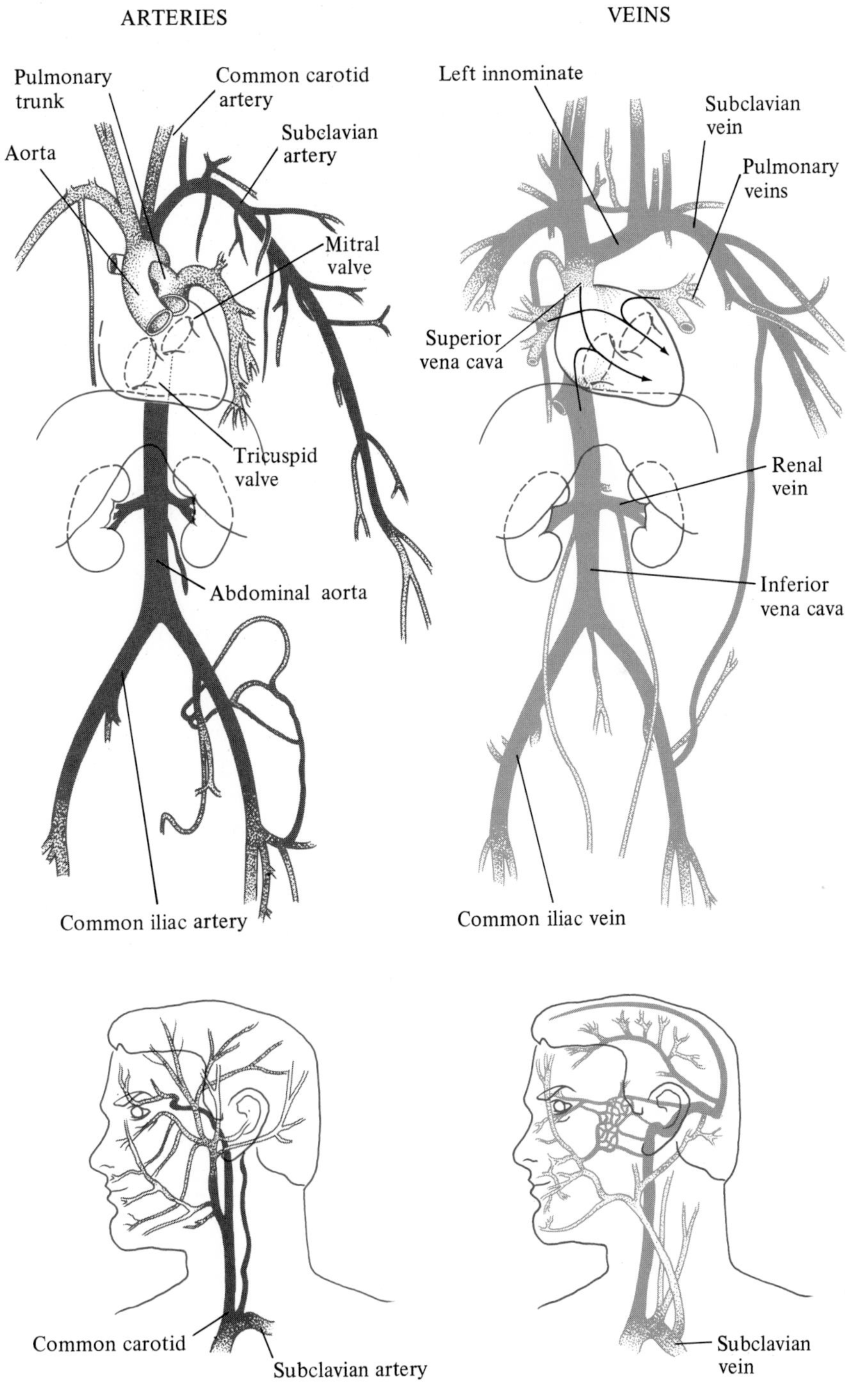


**FIG. 11-8** The arteries and veins of the human body. Major vessels noted in the text are labeled.

**superior vena cava,** drains the head and shoulder region, receiving tributaries that roughly correspond to the branches of the aortic arch. The anterior vena cava also empties into the right atrium, as does a smaller vein, the coronary sinus, which drains the heart tissue.

## THE CARDIAC CYCLE

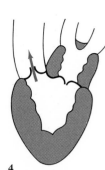

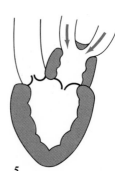

Heart, or cardiac, action has two phases: contraction, or **systole,** and relaxation, or **diastole** (Fig. 11-9). Close observation of a heart undergoing contraction reveals that atrial systole occurs first, ventricular systole following a moment later. The two atria contract at the same time, filling the ventricles as they do so. The ventricles then contract simultaneously, expelling the blood from the heart. Atrial systole is followed by diastole, during which both atria slowly and passively expand as they fill with blood. Ventricular diastole is also passive, being facilitated by the blood that is driven into the ventricles by atrial systole.

## CARDIAC OUTPUT

Throughout life, the human heart contracts an average of 72 times per minute, some 100,000 times a day, or over 2.5 billion times in a 70-year lifetime, resting only briefly between contractions. The amount of blood pumped in 70 years is some 155 million liters or 150 tons. During any one day, however, the output of the heart varies greatly. When a human being is asleep or at rest, about 5.5 liters of blood are pumped by the left ventricle into the aorta each minute. Up to 30 liters per minute course through the left ventricle during vigorous exertion. The amount of blood pumped out of a ventricle in 1 minute is called the **cardiac output** (CO), which is the product of the volume ejected at each contraction, or the **stroke volume** (SV), and the frequency, or the **heart rate** (HR). Thus, $CO = SV \times HR$. The stroke volume depends in part on how efficiently the ventricle is filled by the atrium. The heart rate depends on body temperature (it increases with a rise in temperature), the presence in the blood of certain substances (for example, it is increased by **epinephrine,** also known as adrenaline, and decreased by **acetylcholine**), and, most important, on inhibition or acceleration effected by the nerves that serve the heart (Fig. 11-10). Nervous control of cardiac out-

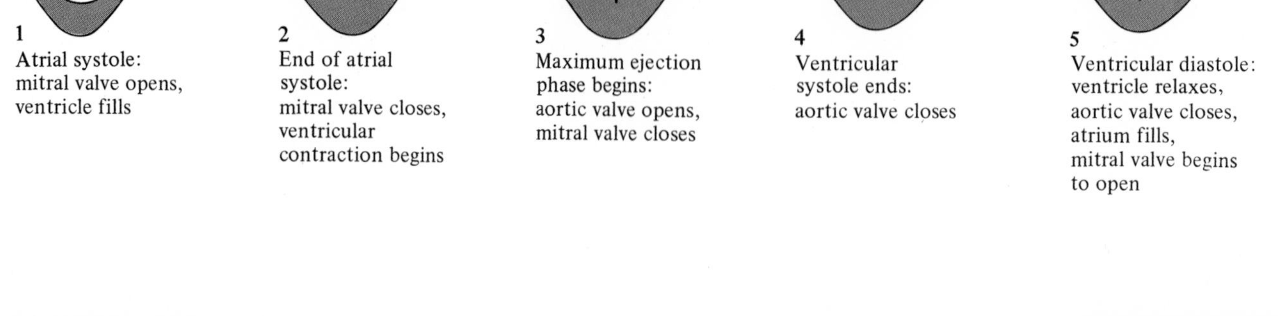

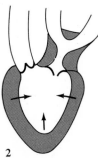


**FIG. 11-9** Left side of the mammalian heart in systole (1–4) and diastole (5). Arrows indicate directions of blood flow through the valves open at the time.

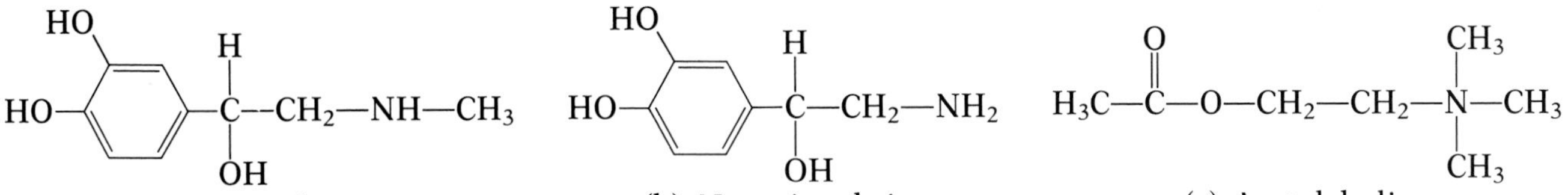

(a) Epinephrine (b) Norepinephrine (c) Acetylcholine

**FIG. 11-10** Epinephrine (a), sold under the trade name Adrenaline, is one of a class of molecules known as catecholamines that also includes norepinephrine (b), or noradrenaline, a slightly modified form of epinephrine. Acetylcholine (c) is produced at the endings of certain types of nerve cells. When released into an organ's tissues, it has an opposite, or antagonistic, effect to that of epinephrine or norepinephrine.

put is mediated by two specialized centers located in the brain: the cardiac acceleratory and inhibitory centers.

### NERVOUS CONTROL OF STROKE RATE

Fibers from the inhibitory center extend from the brain as part of the tenth cranial nerve, called the **vagus nerve.** Fibers from the acceleratory center exit the brain as part of spinal nerves in the neck, called the cervical sympathetic nerves. The basic function of these nerves is easily demonstrated. If the vagus nerve cells serving the heart are cut, heart rate increases. If the sympathetic nerves to the heart are cut, the rate decreases. If the vagus nerve is stimulated, the heart slows. Stimulating the sympathetic nerve to the heart increases its rate. How nerves can achieve such different effects is explained in part by the fact that the vagus fibers release acetylcholine into the heart tissue, whereas the sympathetic fibers release **norepinephrine,** a slightly modified derivative of epinephrine. Acetylcholine slows heart action, whereas norepinephrine speeds and strengthens it. It appears that under resting conditions the two sets of nerves exert a continuing antagonistic control of heart rate, that is, their actions have opposite effects. Such a system is capable of rapid, finely controlled responses.

### HEART SOUNDS

Two principal heart sounds can be detected easily by placing an ear to the chest. They sound much like "lubb, dup," the second sound being slightly higher pitched and briefer than the first. The first heart sound, "lubb," is due primarily to the sharp closing of the AV valves and the vibrations of the chordae tendinae as the atria complete their systole and the ventricles begin their contraction. The "dup" sound is principally composed of sounds made by the closing of the semilunar valves at the end of ventricular systole.

## PACEMAKERS

If all nerves connected to a mammalian heart are cut, the heart still continues to beat at a regular rate. The heart tissue primarily responsible for this internal rhythmicity is a small group of cells in the wall of the right atrium near the entrance of the superior vena cava. If this tissue, called the **sinoatrial node** (SA node), is surgically removed, along with a small amount of surrounding muscle, and placed in a dish of isotonic salt solution, it continues to contract about 75 to 80 times per minute. The rest of the heart will also continue to beat but at a distinctly slower rate. A similar result occurs if the SA node is merely prevented from communicating with the rest of the intact heart by the tying of a thread tightly around the heart below the node. This creates a "heart block," in which the SA node and the tissue above the thread continue to contract at the usual rate but the ventricles, below the thread, contract at a slower rate.

The contractions of the ventricles originate in another node of tissue, the **atrioventricular node** (AV node), in the dorsal wall of the right atrium. Each of the two pacemakers thus tends to set a different rate of contraction, with the SA node acting as the **primary pacemaker.** Only in "heart block," when the SA node fails to stimulate the AV node, however, does the AV node set the rhythm for the rest of the heart.

The SA node communicates with the rest of the heart by means of fibers of nodal tissue that extend from the node into the muscle of the atrial wall. These fibers, although derived from muscle, do not contract. Instead, they are specialized in the conduction of impulses, or waves of electrical disturbance. Impulses originating in the SA node rapidly spread throughout the atrial muscle along these fibers, triggering the contraction of both atria at about the same time.

When an SA impulse passing along the nodal fibers reaches the AV node, there is a brief delay before the AV node responds by generating an impulse of its own. This new impulse travels down a thick AV bundle of nodal tissue, the **bundle of His** (named for a nineteenth-century German anatomist Wilhelm His). The bundle divides into a multitude of **Purkinje* fibers,** which supply the walls of both ventricles (Fig. 11-11). Impulses thus pass to all parts of both ventricles in less than 0.07 second from the time they leave the AV node (Fig. 11-12). The result is a smooth, nearly simultaneous contraction of both ventricles.

* Named for the Czech physiologist Johannes E. Purkinje (1787–1869), who first described them.

**FIG. 11-11** The pacemaker system of the heart.

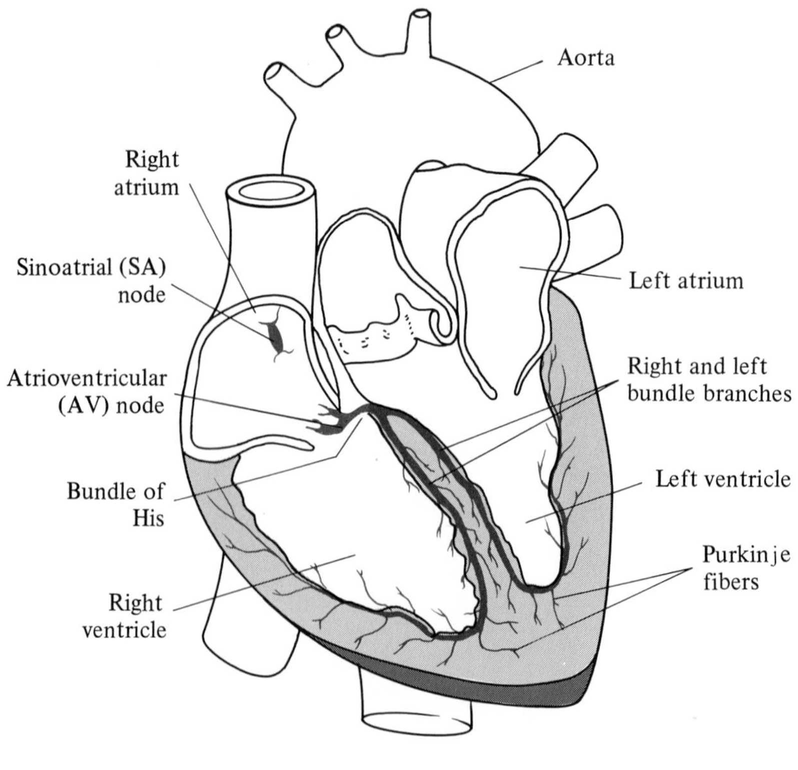

**FIG. 11-12** The rate of transmission of the pacemaking impulse, in seconds. Arrows point to path of diffusion.

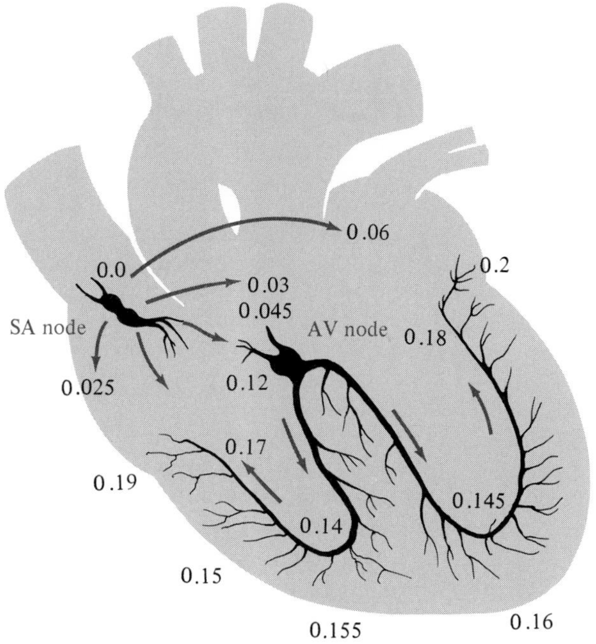


# THE ELECTROCARDIOGRAM

By attaching electrodes to suitable locations on the surface of the body, it is possible to obtain, through electrical amplification, a recording of the most prominent electrical events associated with the heart cycle. The resulting record is called an **electrocardiogram** (ECG) (or EKG, based on the German spelling), which consists of a series of tracings that reflect the activity of the heart (Fig. 11-13). Although there is a limit to what can be learned about the state of the heart from ECGs, much information regarding various heart defects can be acquired from them. For example, poor circulation to a region of the heart as a result of an embolism, that is, the blockage of a vessel by a blood clot or other obstruction, results in an abnormal ECG pattern. Depending on its sever-

**FIG. 11-13** By convention, each of the waves in an electrocardiogram is designated by a successive capital letter of the alphabet, from P through U, thereby permitting convenient references to various parts of the tracing. Each component of the ECG reflects a part of the cardiac cycle. For example, the P wave occurs at the beginning of atrial systole. Ventricular systole begins with the Q wave and lasts until the end of the T wave. (a) A normal heart electrocardiogram. (b) Components of an ECG.

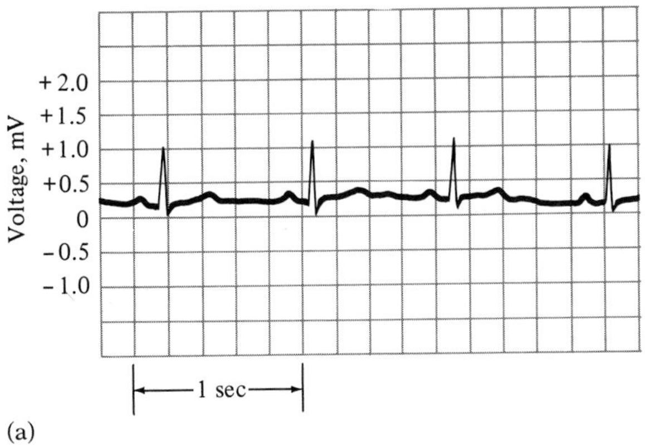


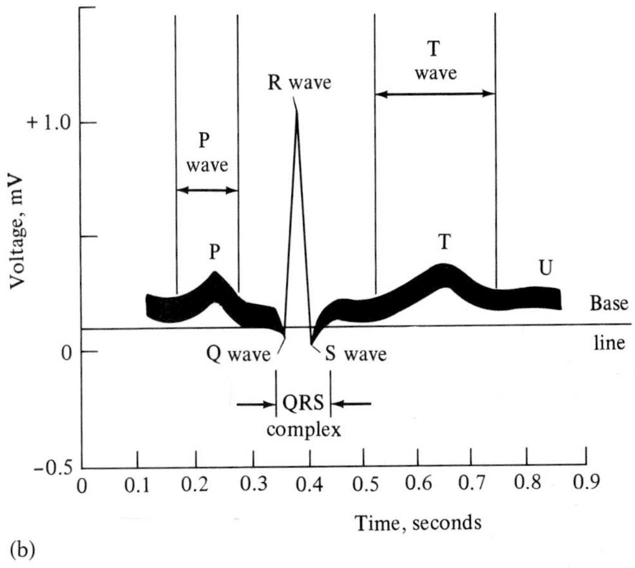

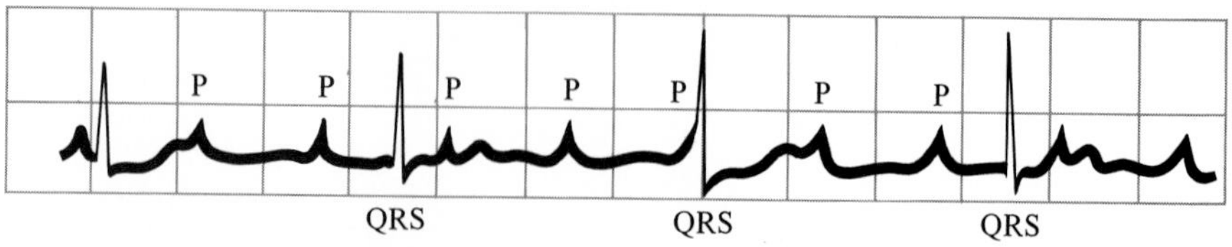


Complete heart block (atrial rate, 107; ventricular rate 43)

**FIG. 11-14** An electrocardiogram showing complete heart block, in which the AV node fails to respond to SA node stimulation and maintains its own independent, somewhat slower rhythm. Note the dissociation of the P waves from the QRS complex, compared with the normal pattern in Figure 11-13. Note also that there are often two P waves, reflecting atrial contractions, for every one QRS complex, reflecting ventricular contractions.

ity, a myocardial infarction or heart attack, in which some of the heart muscle dies when suddenly deprived of its blood supply, may also be reflected in ECG tracings and even followed through its course by this means. Permanent damage caused by a heart attack may also result in a changed ECG tracing.

In some individuals an impulse generated by the SA node occasionally fails to stimulate the AV node. In such cases, the AV node generates an impulse anyway, but at its own, much slower rate. This serious condition produces a heart block and is very evident in an ECG (Fig. 11-14). If the condition persists, an artificial pacemaker can be installed under the skin of the chest and connected to a probe inserted into the ventricle. The instrument monitors SA node impulses and, if the AV node fails to generate an impulse on time, delivers a slight shock to the ventricles, causing them to contract.

## BLOOD PRESSURE

One of the most important characteristics under precise homeostatic control is **blood pressure,** the force exerted by the blood against the vessel walls. At a given moment, about 5 percent of the blood in the body is found in the capillaries, only a small fraction of which are filled with blood. Yet the capillaries have more than enough capacity to accommodate all the blood in the body at one time. Although all cells of the nervous system continuously receive a rich blood supply, most tissues do not. While resting after a meal, for example, a person's circulation to the leg muscles is much reduced, whereas the digestive tract is well supplied. During a foot race, the reverse is true. However, no matter what tissue is being supplied, blood pressure must be sufficient to force the blood into the capillaries.

Blood pressure is controlled principally by constriction and dilatation, or expansion, of the body's many small, unnamed arteries, called **arterioles,** and to some extent by the pumping action of the heart. Adequate blood pressure is essential for adequate blood flow, but the body lacks receptors to detect flow rate. Instead, because blood flow depends on blood pressure, **baroreceptors,** which detect hydraulic pressure, indirectly perform this function. The arterioles constrict or dilate and the heart increases or decreases its output in response to changes in pressure detected by these baroreceptors.

# HEART DISEASES

Of the many types of heart disease, a large number may result in heart failure, which is characterized by weakened heart muscle. Some heart disorders are characterized by murmurlike sounds heard through the listening instrument called a stethoscope, usually caused by turbulence as the blood passes from one chamber to another or exits the heart. Although murmurs may occur in individuals with no detectable abnormalities, they sometimes result from serious heart defects, particularly from damaged valves. Rheumatic fever may so damage a heart valve that it no longer closes all the way, allowing some of the blood that has exited the left atrium, for example, to reenter it, producing the gurgling "murmur." The *Streptococcus* bacterium that causes rheumatic fever may also produce a narrowing of the opening guarded by the mitral valve, a condition known as **mitral stenosis;** less commonly, it affects the opening guarded by the aortic valve, resulting in **aortic stenosis.**

## CORONARY ARTERY DISEASE

By far the most common heart disease involves the coronary arteries. Like the brain, the heart muscle, called the **myocardium,** must be continually supplied with oxygen and nutrients by the blood. This is accomplished by a rich capillary network that branches from the coronary arteries and is drained by the coronary veins (Fig. 11-15). The coronary arteries are unusually susceptible to what is called **coronary artery disease** (CAD) that results from **atherosclerosis,** a condition characterized by a thickening and loss of elasticity of artery walls, the thickened regions being known as atherosclerotic plaques. The lumens of the affected arteries gradually narrow (Fig. 11-16). Eventually, such an artery may close entirely. Whether or not it does, the affected region is

**FIG. 11-15** The major coronary vessels.

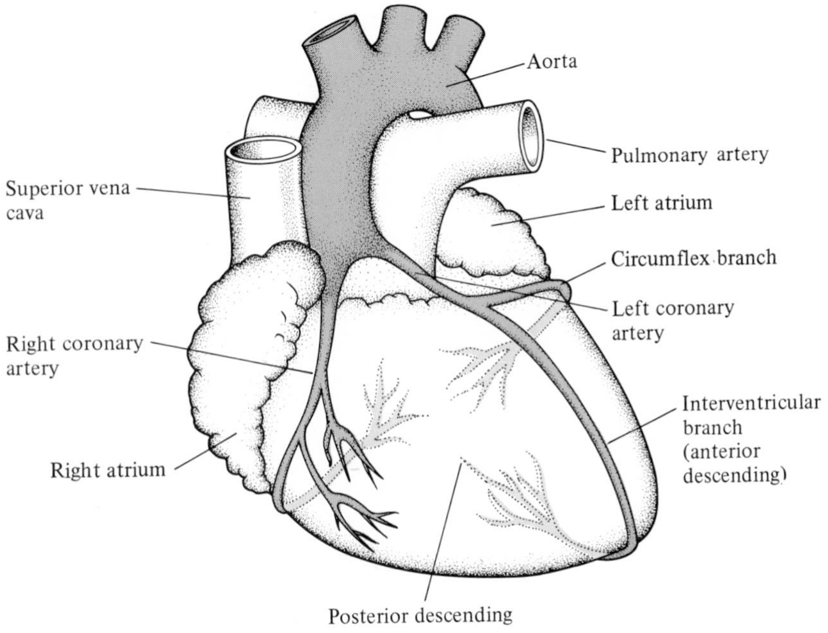

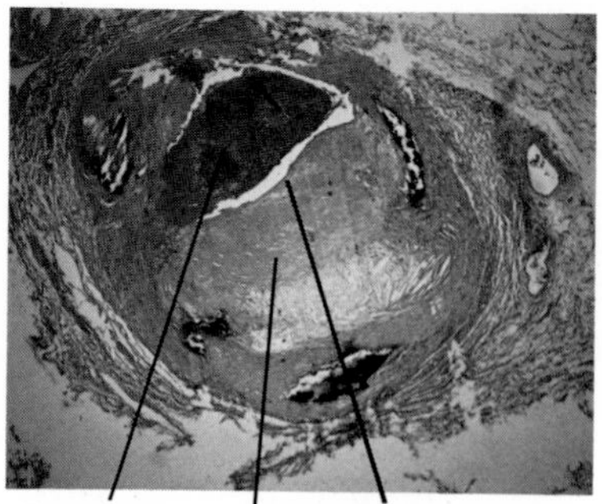


**FIG. 11-16** Section through a coronary artery showing coronary artery disease and a thrombosis. Visible in the photograph are an atherosclerotic plaque below the intima, the inner lining of blood vessels, and a thrombus (blood clot) that is almost completely blocking the artery.

inadequately supplied with blood, a condition known as **ischemia.** The condition can sometimes be remedied by surgery (Panel 11-1). See also Panel 9-3.

Under stress or physical exertion, chest pains, or **angina pectoris,** may result in affected individuals and there is an increased chance that a blood clot or thrombus, may form. If so, the affected myocardium dies, a condition called a **myocardial infarction,** the most common type of "heart attack." By the 1970s, coronary artery disease (CAD) had become the leading cause of human death in Europe and North America.

A classic study of about 5,000 residents of Framingham, Massachusetts, conducted from 1948 to 1960, demonstrated that a person's

PANEL 11-1

# Heart Surgery

Among the most striking medical advances in the second half of the twentieth century have been various surgical procedures performed on the heart and its associated blood vessels.

## VALVE SURGERY

The functioning of valves crippled with scar tissue adhesions due to rheumatic fever can be improved through surgery. In one procedure, a slit is made in the heart, and a finger is inserted and pressed against a valve that no longer opens fully. In this way, the cusps can sometimes be broken free of each other, thereby restoring normal function. When this procedure does not work, the valve may be repaired or replaced by a plastic valve. In all such operations, the heart action is stopped by stimulation of the cardioinhibitory nerves, and the body is maintained on a "heart-lung machine." Because the rate of any chemical reaction falls with a drop in temperature, the body is also cooled during such procedures to lessen the need for oxygen and to minimize cellular damage.

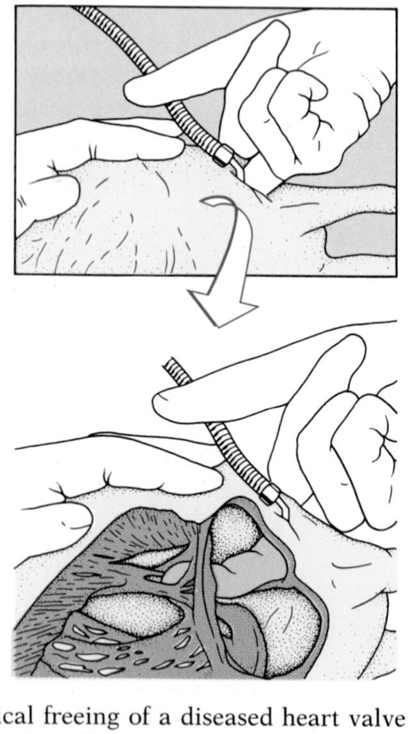

The surgical freeing of a diseased heart valve from its scar tissue adhesions.

## CARDIAC ANGIOGRAPHY

Although damaged valves are relatively easy to diagnose, detailed analysis of heart sounds or expert studies of ECGs provide only vague indications of the exact nature of many cardiac abnormalities. This is particularly true of CAD. In 1958, however, an American surgeon, F. Mason Sones developed a new technique, **cardiac angiography,** which not only provided an excellent means of studying heart function and diagnosing chances of dying of CAD were substantially increased by any one of the following factors: (1) cigarette smoking, a two- to sixfold increase; (2) lack of regular, vigorous exercise, a twofold increase; (3) being overweight, a two- to threefold increase; (4) a diet rich in saturated fat or cholesterol, such as red meat, eggs, whole milk, cheese, and butter; (5) being under tension and stress; and (6) having diabetes, a two- to fourfold increase.

The fact that the life-styles of many people in Europe and North America are characterized by several of these factors suggests why CAD is the leading cause of death in these regions. The effect of two or more of these factors is multiplicative, not additive. Thus, the combined effects of cigarette smoking, lack of exercise, and being overweight increases one's risk of developing CAD some eight- to thirty-sixfold, not merely six- to ninefold. Heredity also plays an important role and explains why some people are more readily or more severely affected by smoking, lack of exercise, and the like, than are other individuals.

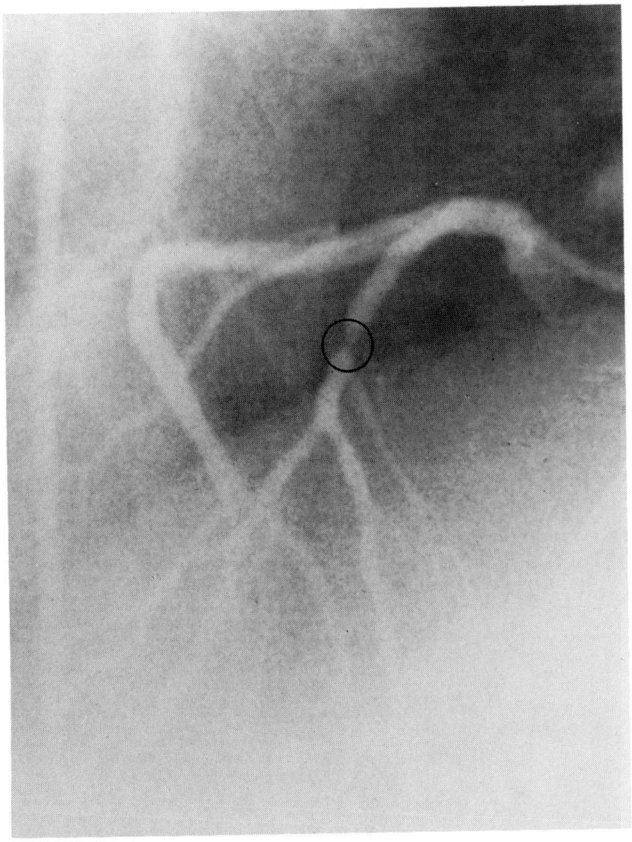

Coronary artery disease seen in an angiogram. The circled portion of the artery shows 95 percent narrowing.

various cardiac diseases but also made possible surgical procedures to remedy them. In cardiac angiography, a plastic tube called a catheter is inserted into an artery of the right arm or leg and passed up into the aorta and thence into the left side of the heart. By releasing a radiopaque substance, or "dye," from the catheter into the heart's chambers or associated vessels, the flow of blood through these structures can be visualized by a technique called fluoroscopy* or by X-ray photography, and abnormalities can be studied in detail. Similarly, the right side of the heart can be studied by catheterizing a vein.

## CORONARY VASCULAR SURGERY

Surgical procedures to remedy blocked coronary arteries include **myocardial revascularization** in which bypasses are placed around blocked regions of coronary arteries. The bypasses usually consist of portions of a large leg vein, which can be entirely removed from the leg without harm to the patient. Studies made a year after such operations show that as many as 85 percent of the patients experience a reduction in severity and frequency of anginal attacks. The degree to which myocardial infarctions are reduced as a result of such surgery, however, is not yet clear. Over 200,000 patients a year undergo bypass surgery in the United States.

In another procedure, **endarterectomy,** narrowed arteries are stripped of atherosclerotic plaques. Although this technique immediately improves coronary circulation, its effects on longevity are unknown.

* A fluoroscope is similar to an X-ray machine except that instead of taking "still" photographs of the X-rayed organ, the organ is viewed directly and continuously on a fluorescent screen similar to a television picture tube. A picture can be taken from time to time for a record that can be studied later by a surgical team to determine whether surgery is needed.

# BLOOD

The importance of blood to life would be difficult to overestimate. Indeed, it would be hard to imagine a general health checkup that did not include at least one blood test, because much can be learned about the status of the body from an analysis of blood. Although not usually thought of as such, blood is actually a tissue. It is a medium of transport for the necessities of life—oxygen and nutrients—and for carbon dioxide and other wastes. It is also a medium of communication and regulation. The average adult human has 5 to 6 liters of blood, which constitutes about 8 percent of the total body weight. Blood consists of two fractions: **plasma**—the fluid portion—and three classes of cells or other formed elements—**erythrocytes** (red blood cells), **leukocytes** (white blood cells), and **platelets.** Plasma comprises about 56 percent and the formed elements about 44 percent of the total volume.

## PLASMA

A solution of inorganic and organic substances, plasma is 91 to 92 percent water. Nearly all the organic matter consists of various **plasma proteins.** Other components include enzymes, hormones, glucose, amino acids, lipids, and nitrogenous wastes in the form of urea and uric acid (see Table 11-1).

## ERYTHROCYTES

The principal formed elements in the blood are the red blood cells, or erythrocytes. Their main function is in transport of respiratory gases, oxygen and carbon dioxide, which are bound to hemoglobin. Human erythrocytes are biconcave, that is, concave on both sides, disk-shaped cells without a nucleus. They are about 8.5 $\mu$m in diameter and are produced in the marrow of certain bones (Fig. 11-17). As rapidly as new erythrocytes are produced by the bone marrow, old ones are destroyed and their constituents recycled by phagocytes in the liver, spleen, and bone marrow itself. There are about 5 million red blood cells per cubic millimeter of blood. A concentration much below this or an insufficient amount of hemoglobin in the cells is termed **anemia.**

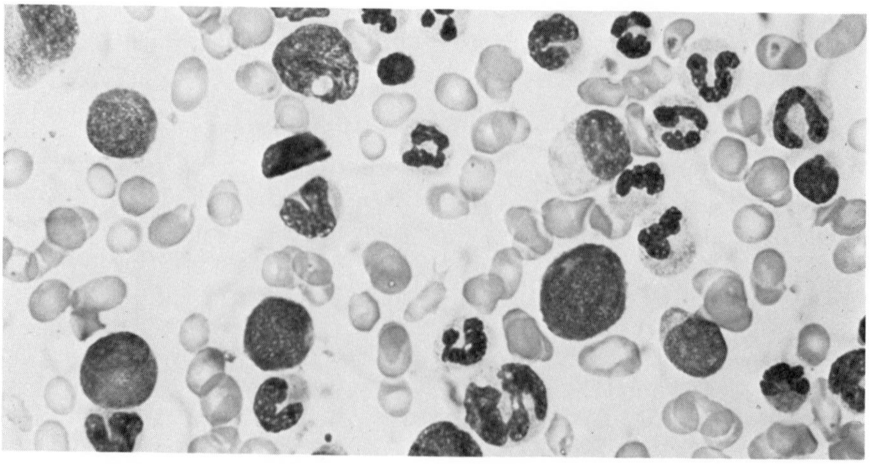

**FIG. 11-17** Stained preparation of normal human blood cells obtained from bone marrow. Besides the numerous erythrocytes, several types of leukocytes (the nucleated cells) can be seen.

**TABLE 11-1 The Major Components of Plasma**

| *Component* | *Percent of plasma** | *Examples of functions* |
|---|---|---|
| Water | 91–92 | Solvent, heat absorption |
| Electrolytes ($Na^+$, $K^+$, $Ca^{2+}$, $Mg^{2+}$, $Cl^-$, $HCO_3^-$) | 500 mg | Act as buffers, maintain osmotic pressure |
| Gases | | |
| Oxygen: $P_{O_2}$ | 80–90 mm Hg/liter of whole blood | |
| Carbon dioxide: $P_{CO_2}$ | 35–45 mm Hg/liter of whole blood | |
| Plasma proteins | 7–9 | |
| Albumins | | Contribute significantly to osmotic pressure, miscellaneous other functions |
| Globulins | | |
| Alpha | | Miscellaneous functions |
| Beta | | Miscellaneous functions |
| Gamma | | Antibodies to inactivate foreign substances, bacteria, etc. |
| Fibrinogen | | Forms the basis of blood clots |
| Other organic substances | | |
| Enzymes | | Catalyze various reactions |
| Hormones | | Control body functions |
| Nitrogenous wastes: urea, uric acid, etc. | | Waste products to be excreted |
| Glucose | 1.2 mg | Energy source |
| Lipids | | |
| Triacylglycerols | 10–190 mg | Energy source |
| Cholesterol | 115–340 mg | Steroid hormone precursor |
| Amino acids | 4.9 mg | Protein constituent |
| Vitamins | trace | Enzyme cofactors |

* Except where otherwise noted, percentages are for plasma. The term mg% indicates the number of milligrams per 100 ml of plasma.

## LEUKOCYTES

White blood cells, or leukocytes, are motile, nucleated, and ameboid, that is, of changing shape (Fig. 11-17). Like red blood cells, most originate in bone marrow. One type, the **lymphocyte,** is primarily located in lymphoid tissues, such as the tonsils, and is one of the cells involved in immunity, or resistance to infections (see Chapter 12). The others engage in phagocytosis of bacteria and various types of cellular debris. Leukocytes are usually found in the spleen, liver, bone marrow, and lymph nodes rather than in the circulating blood. They usually number from 5,000 to 9,000 per cubic millimeter of blood, although this figure rises sharply during bacterial and viral infections. The life spans of various types of leukocytes vary from a few hours to 200 days.

Large numbers of phagocytic leukocytes are attracted to substances released into the blood in areas of inflammation, tissue damage, or infection, where they perform their phagocytic functions. The whitish **pus** that results at the site of a wound is actually the remains of dead and dying leukocytes, which at this stage are called pus cells.

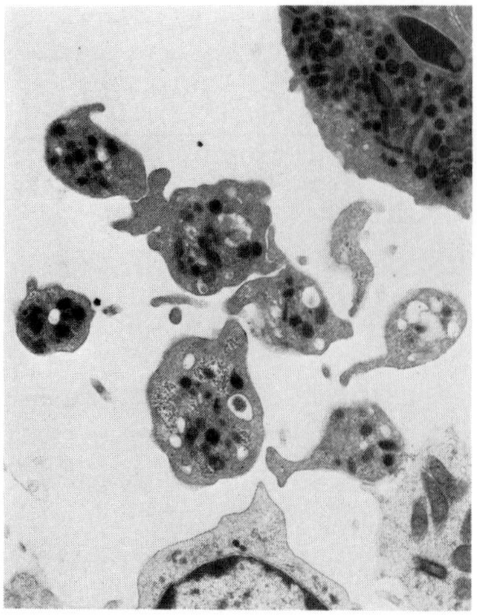

FIG. 11-18 Blood platelets.

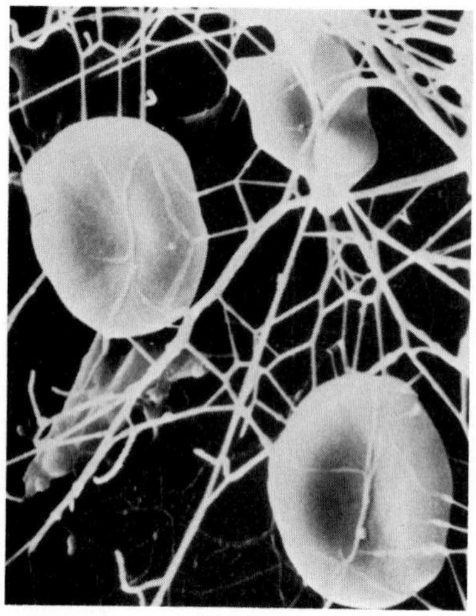

FIG. 11-19 Red blood cells trapped in a mesh of fibrin strands.

In leukemia, a cancer of the white blood cells, the normal control of leukocyte production fails, and the bone marrow releases many times the usual numbers into the bloodstream.

### PLATELETS

The third type of formed element in the blood is the platelet. Platelets are actually cytoplasmic fragments of cells produced in the bone marrow; they are involved in blood clotting. They are tiny, only 2 $\mu$m in diameter, and number from 150,000 to 200,000 per cubic millimeter of normal blood (Fig. 11-18). Besides contributing chemical factors to the clotting process, platelets reduce blood loss by aggregating at the site of a wound. Exposure to the air at the wound site induces them to aggregate, thereby forming a thick mat even before the clot itself forms. In less than two seconds after a blood vessel is damaged, large numbers of platelets accumulate and begin to close the wound.

### BLOOD CLOTTING

**Blood clotting,** or **coagulation,** is a complex biochemical process that involves the conversion of **fibrinogen,** a soluble plasma protein, into **fibrin,** an insoluble polymer. Fibrin precipitates to form a meshwork of needlelike filaments in which blood cells become trapped, forming the clot (Fig. 11-19). To prevent blood clotting during surgery or to dissolve a life-threatening clot, various **anticoagulants** are used to neutralize or remove one or another of the factors required for coagulation. In the inherited condition known as hemophilia, there is an insufficiency of one of the blood clotting factors, resulting in a dangerously long clotting time.

## CIRCULATION

The circulation is composed of the **pulmonary circulation,** serving the lungs, and the **systemic circulation,** serving the rest of the body. Each circulation is composed of arteries, which transport blood under high pressure to the tissues; arterioles, conveying blood to the capillary beds where materials are exchanged between the blood and tissues; and venules, which collect blood from the capillaries and fuse to form progressively larger veins returning blood to the heart.

### ARTERIES

There are three sizes of arteries: the large arteries are the aorta and pulmonary arteries and their immediate branches; the medium size arteries are the named arteries of the arms, legs, internal organs, and so on; and the small, unnamed arteries are the arterioles. The walls of large arteries contain numerous elastic connective tissue fibers, which

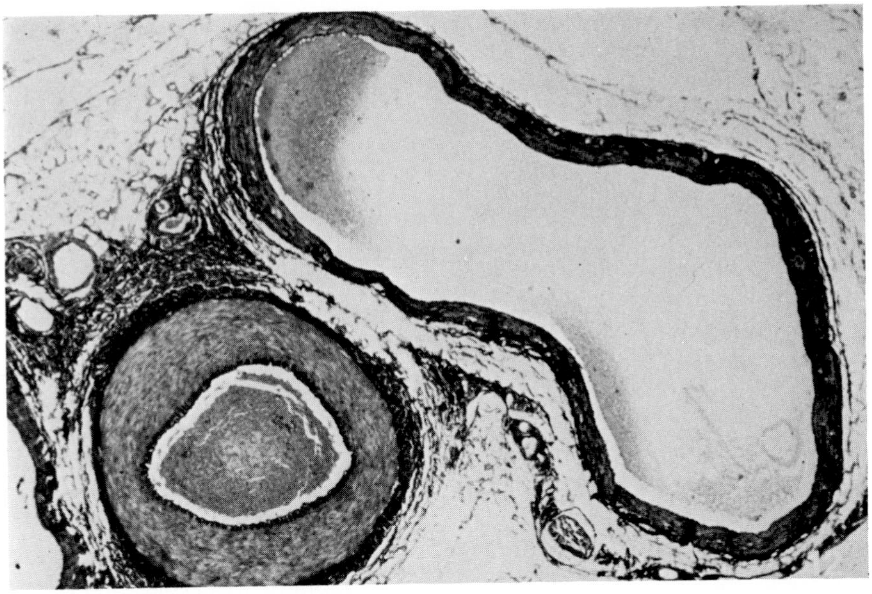

**FIG. 11-20** Section through an artery and adjacent vein (artery at left, vein at right). Notice the small vessels that make up the vasa vasorum.

allow these vessels to absorb much of the force of the pulse by expanding with each systole (Fig. 11-20). They then recoil between beats, forcing the blood into the small arteries. Although medium-size arteries have fewer elastic fibers than do large arteries, they are capable of active contraction, due to as many as 60 layers of smooth muscle fibers, all under nervous control.

Arterioles are the most numerous of the three types of arteries. They are equipped with five to 10 layers of muscle cells and some elastic fibers. Because of their large numbers, arterioles play very important roles in regulating the body's blood pressure as well as the flow of blood to specific organs under various conditions.

## VEINS

Although veins are generally larger in diameter than arteries, their walls are much thinner because they have fewer muscle cells and elastic fibers (Fig. 11-20). Like arteries, veins are of three size classes, the smallest being known as venules. Many of the small and medium-sized veins have one-way valves that permit blood to flow only toward the heart. Such valves are especially abundant in the legs, where they are important aids in the return of blood against gravity. Movement of the blood toward the heart depends partly on the muscles through which the veins pass. As the surrounding muscles contract, they exert pressure on the veins, forcing the blood through them by a milking action in the only direction in which the valves permit it to go—toward the heart (Fig. 11-21a). (Conversely, if a person stands still for long periods of time blood will congest in the feet and lower parts of the legs.) Blood pressure in most veins is about one-tenth that in arteries (Fig. 11-21b). Largely because of the ease with which vein walls expand, about 75 percent of the body's blood is contained in the veins, compared to about 20 percent in the arteries.

The walls of arteries and veins greater than 1 mm in diameter are themselves well supplied with small arterioles and venules, the **vasa**

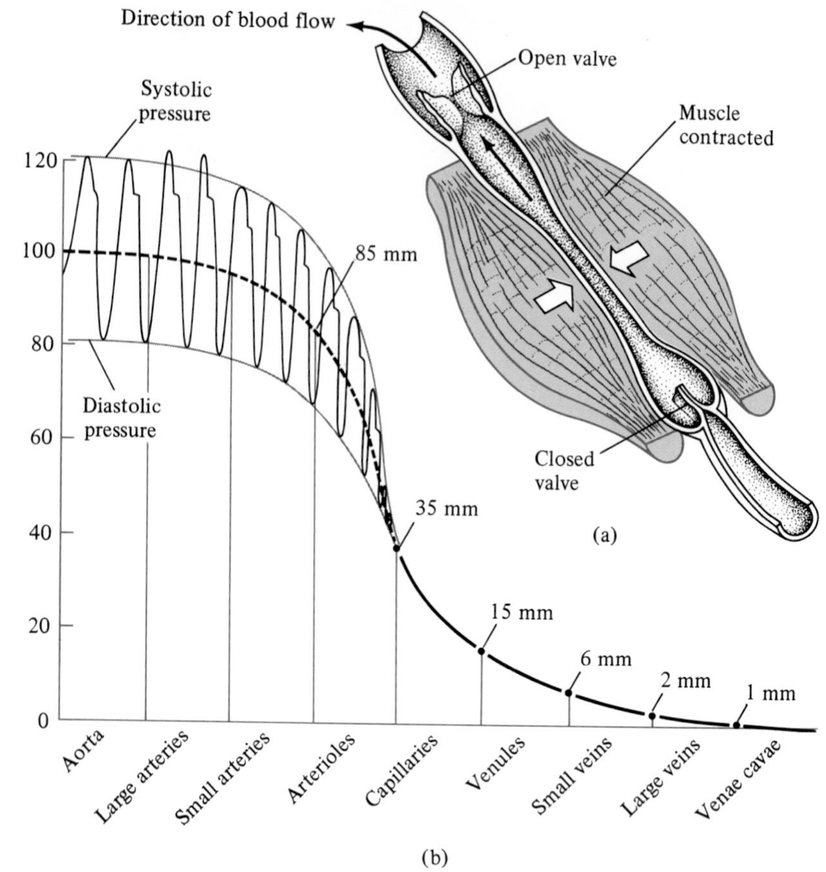


**FIG. 11-21** (a) How vein valves and surrounding muscle cooperate to move blood upward in leg veins. (b) Blood pressure in the various blood vessels of the body. The dotted line represents the mean (average) systolic pressure in the arteries.

**vasorum** (Fig. 11-18). Large blood vessels are also richly supplied with nerves.

## CAPILLARIES

The wall of a blood capillary is very thin, being composed of but a single layer of flat epithelial cells (Fig. 11-22). Several types of specialized capillaries serve particular organs, such as the kidney and endocrine glands. By one estimate, the body of an adult human has nearly 100,000 km of capillaries, enough to stretch more than twice around the world at the equator. The diameter of a capillary is so small that only a single red blood cell can pass through at a time. The larger white cells are temporarily deformed when passing through capillaries. Because of the thinness of their walls, the large amount of surface area they provide, and the slow rate at which blood flows through them, capillaries accomplish a very efficient exchange of fluids and dissolved materials between the bloodstream and the tissues they serve. If we compare the arteries and veins to the express routes in a large city, the capillaries are comparable to the city's congested side streets, where materials are picked up and delivered.

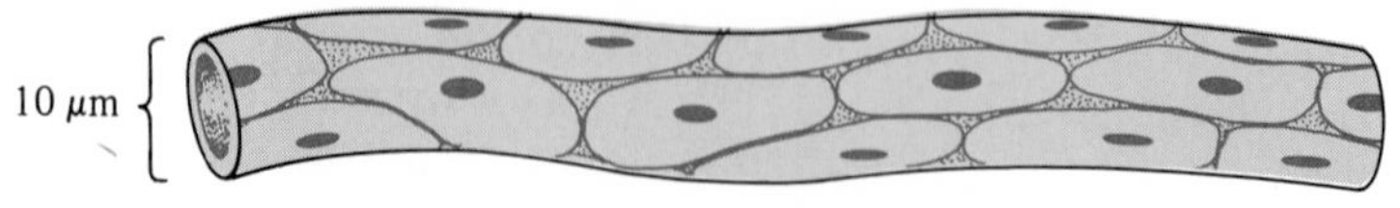


**FIG. 11-22** Sketch of a portion of a capillary from a frog. A characteristic of all capillaries is that they consist of a single layer of flat cells.

## TISSUE FLUID

Tissues supplied by capillaries do not come in direct contact with the blood itself. Rather, they are bathed in a fluid derived from the blood, the **interstitial fluid,** or **tissue fluid,** that always forms as blood flows through a bed of capillaries. Hydraulic pressure (blood pressure) force-filters some of the fluid of blood through the capillary walls into the interstitial spaces (Fig. 11-23). Although containing all the small molecules found in blood—glucose, amino acids, urea, $O_2$, and so on—interstitial fluid differs from blood in two ways: it lacks red blood cells and has little of the protein found in blood. Except for its low protein content, interstitial fluid is thus similar to plasma. Red cells are unable to leave blood vessels except in bleeding when the latter are ruptured. White cells, on the other hand, are motile and squeeze between the cells of the capillary wall by amoeboid movement, thereby entering the intercellular spaces.

Because the passage of protein through capillary walls is generally restricted—with only the smaller albumin molecules able to penetrate—the concentration of protein in interstitial fluid is only about half that in blood. This difference is significant. Because the plasma has a higher concentration of protein than does the tissue fluid on the other side of the capillary wall, fluid tends to flow osmotically from the tissue fluid through the capillary walls back into the bloodstream (Fig. 11-23). However, blood pressure is greater than this osmotic pressure, with the net result that more tissue fluid is formed than can be recovered by osmosis. Normally, the excess fluid slowly flows into the connecting lymphatic vessels, where it becomes known as **lymph.** If, for any reason, such as an abnormally low concentration of proteins in the blood, a large excess of tissue fluid accumulates, the tissues become swollen with fluid, a condition known as **edema.**

## THE LYMPHATIC SYSTEM

The lymphatic system is a network of vessels nearly as extensive as the blood vascular system. Like blood vessels, lymph vessels are of several sizes, the smallest being known as **lymph capillaries.** Although similar to blood capillaries, they are of somewhat larger diameter. Lymph capillaries drain tissue fluid from all tissues of the body except the brain, spinal cord, and eyes. Unlike walls of blood capillaries, lymph capillary walls are highly permeable to large molecules such as various proteins that enter the lymph from the tissues through which it passes.

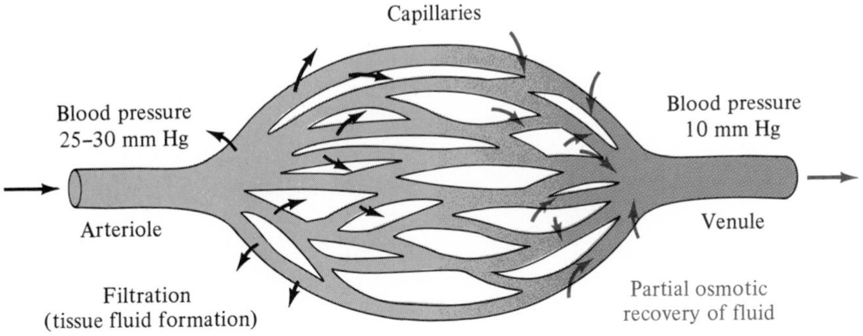


**FIG. 11-23** The formation of tissue fluid. Fluid is squeezed from the capillaries where the blood enters the capillary bed, at which point blood pressure is relatively high (25–30 mm Hg). As pressure drops along the length of the capillary bed to 10 or 15 mm Hg, much of the fluid is able to return to the blood by osmosis.

Lymph capillaries join to form large veinlike **collecting vessels** in which are found **lymph nodes,** glandular complexes of fine channels and sinuses lined with phagocytes through which the lymph filters. Because the human lymphatic system lacks a pumping mechanism, the flow of lymph is much slower than that of blood. This flow depends in part on the slight hydraulic pressure generated in the tissue spaces and, more significantly, on the numerous one-way valves in the collecting vessels. As with the flow of blood in veins, when muscle masses surrounding lymphatic vessels contract, the lymph is squeezed through the vessels in the only direction the valves permit—toward the lymph ducts. Two large **lymph ducts,** the largest of the lymphatic vessels, receive lymph from all the collecting vessels and empty it into two large veins in the chest area. In this way, all the interstitial fluid formed in the body's tissues is returned to the blood from which it was derived, along with the various materials collected by the lymph from organs or tissues along the way. (The lymphatic system and its role in immunity is discussed in the next chapter, Chapter 12.)

## Summary

Although lower organisms may not require a special system of internal transport, large or active animals do. In some animals the body cavity alone performs this function; in higher forms the body cavity is aided in this role by the more efficient blood vascular system. These systems may be largely open, with many large cavities, or sinuses, as in arthropods; largely closed, consisting mostly of vessels, as in the earthworm; or completely closed, as in vertebrates. Vertebrates also have a lymphatic system, which returns tissue fluid to the bloodstream.

Circulation of blood or a similar fluid is achieved in two ways: by muscular activity and by pumping organs such as hearts. Most circulatory systems have one-way valves that prevent back flow. Hearts vary from simple, tubular structures to the four-chambered "double hearts" of birds and mammals. The evolution of double hearts is represented in a series beginning with the folded-tube, single heart of fishes, through the three-chambered heart of amphibians and nearly four-chambered heart of reptiles, culminating in the four-chambered hearts of birds and mammals. During embryonic development, the mammalian heart seems to recapitulate this evolutionary history.

In the mammalian heart, the left atrium receives blood from the lungs by way of the pulmonary veins, the right atrium from the venae cavae, which drain all other parts of the body. Each atrium empties into its respective ventricle, the right ventricle sending blood to the lungs via the pulmonary arteries, and the left to all other tissues via the aorta and its branches. Back flow from the ventricles into the atria is prevented by the atrioventricular (AV) valves, the left one being the mitral, or bicuspid, valve, the right being the tricuspid. The semilunar aortic and pulmonic valves prevent back flow into the ventricles.

The cardiac cycle can be regarded as beginning with atrial contraction, or systole. The resulting expulsion of blood from the atria fills the ventricles, which passively expand during diastole. Diastole is immediately followed by ventricular systole, which sends blood out of the heart. Cardiac output (CO) is the product of the stroke volume (SV) and the heart rate (HR): $CO = SV \times HR$. Heart rate varies with activity, temperature, and effects of chemicals, including hormones, and is largely controlled by nerve impulses originating in the cardiac accelerator and inhibitory centers of the brain. These centers increase or slow the basic rate set by the heart's primary pacemaker, the sinoatrial (SA) node. Impulses from the SA node stimulate the atria to contract and also trigger the atrioventricular (AV) node, whose impulses stimulate contraction of the ventricles. When the AV node occasionally fails to respond to impulses from the SA node, the stimulation may be supplemented by an artificial pacemaker. The impulses generated by the heart's own pacemakers can be amplified and recorded as an electrocardiogram (ECG) by attaching recording electrodes to certain parts of the body.

Blood pressure is important for maintaining blood flow through the body's organs. Although heart action contributes to blood pressure, it is maintained principally by constriction of the arterioles.

Various heart diseases result in heart failure, characterized by weakened heart muscle. Heart valve defects resulting from rheumatic fever may include stenosis, which is a narrowing of a valve opening; the failure of one or more of the heart's valves to close completely; or both. Of all types of heart disease, coronary artery disease (CAD), resulting from atherosclerosis of coronary artery walls, is the most common. It is marked by thickening and loss of elasticity of the artery walls and narrowing of the artery lumen. This reduces blood flow to the affected area and increases the chance of a "heart attack." Surgical methods of repairing heart defects include valve surgery and replacement; reconstruction of coronary circulation through bypass surgery; and endoarterectomy, the removal of atherosclerotic plaques from artery walls.

Blood is a tissue that serves as a vehicle for transport of nutrients, respiratory gases, and wastes and for communication and regulation of many body functions. The fluid portion of blood, plasma, is a rich solution of inorganic and organic solutes. The formed elements, erythrocytes, leukocytes, and platelets, constitute 44 percent of the blood's volume. Erythrocytes and leukocytes are generated in the bone marrow. Human erythrocytes, which number about 5 million per cubic millimeter, lack nuclei and live about 120 days. Their primary function is the transport of respiratory gases. The primary role of the motile leukocytes, which have nuclei, is fighting infections. Platelets contribute to blood clotting by forming clumps at the site of a wound and by releasing chemical factors essential for clotting. The complex process of blood clotting involves the conversion of the soluble plasma protein fibrinogen to the insoluble form, fibrin.

The blood vascular system has five major types of vessels: arteries, large- and medium-sized vessels that conduct blood away from the heart; arterioles, small, unnamed arteries; capillaries, thin-walled vessels through which virtually all exchange occurs between blood and tissues; venules, small, unnamed vessels that conduct blood toward the heart; and veins, medium-to-large vessels that conduct blood toward the heart. A milking effect is produced on the veins, most of which have many one-way valves, by contractions of the surrounding body musculature. This effect is of chief importance in returning blood to the heart.

Tissue fluid is formed by the filtration of blood through capillary walls as a result of blood pressure. Tissue fluid contains all the constituents of blood except red blood cells and most blood proteins. Immediately upon forming, much tissue fluid is returned to the blood through the capillary walls by osmosis resulting from the difference in the protein concentration of the blood and the tissue fluid. The remainder, known as lymph, slowly returns to the bloodstream via the lymphatic system, an extensive supplementary system of vessels that drain the body's tissues and empty the lymph into large veins in the chest.

## Recommended Readings

ADOLPH, E. F. 1967. "The heart's pacemaker." *Scientific American,* 216(3):32–38. An interesting account of how a group of specialized cells regulates the basic rhythm of the heart, by one of America's most respected physiologists, authors, and teachers.

DOOLITTLE, R. F. 1981. Fibrinogen and fibrin. *Scientific American,* 245(6):126–135. An interesting and up to date account of the process by which blood clots form.

MAYERSON, H. S. 1963. "The lymphatic system." *Scientific American,* 208(6):80–90. The essential role of the body's second vascular system is described by a leading American heart physiologist.

NETTER, F. H., preparer, and F. F. YONKMAN, editor. 1969. *Heart.* Vol. V. CIBA Collection of Medical Illustrations. CIBA Pharmaceutical Company, Summit, N.J. Excellent illustrations and clear explanations of the heart's functioning in health and in disease, with descriptions of some modern surgical procedures.

SCHER, A. M. 1961. "The electrocardiogram." *Scientific American,* 205(5):132–139. A clear explanation of how, with the electrocardiogram, the heart's function in health and disease can be studied without invading the body itself.

STALLONES, R. A. 1980. The rise and fall of ischemic heart disease. *Scientific American,* 243(5):53–59. An analysis of the decline in United States heart attack fatalities.

VANDER, A. J., J. H. SHERMAN, and D. S. LUCIANO. 1980. *Human Physiology,* 3d ed. McGraw-Hill, New York. Among the best introductory textbooks of human physiology. Clear, well illustrated.

WIGGERS, C J. 1957. "The heart." *Scientific American,* 196(5):74–84. A clear explanation of heart function by an award-winning researcher and teacher of cardiac physiology.

ZUCKER, M. B. 1980. "The functioning of blood platelets" *Scientific American,* 242(6):86–103. An account of the several functions performed by platelets.

ZWEIFACH, B. 1959. "The microcirculation of the blood." *Scientific American,* 200(1):54–60. Insight into the functioning of the body's capillaries by a long-time student of and expert on the subject.

12

# The Body's Defense Mechanisms

## OUTLINE



During its life, every organism, great or small, comes in contact with a variety of microorganisms and parasites. Most of these are beneficial or harmless, but some are capable of causing disease, even death. Every major aspect of an organism's existence is affected, for better or for worse, by one or more species of bacteria, fungi, protozoans, or viruses that are capable of multiplying upon or within it. This chapter concerns the means by which the body defends itself against potentially harmful microorganisms and parasites. The multicellular organisms in existence today are those that, over millions of years, have evolved several kinds of defense mechanisms, collectively called immunity, that allow them to resist or at least survive such attacks.

# WHAT IS IMMUNITY?

The word *immune* was originally used to refer to people who were exempt from military service or the payment of taxes (from the Latin *immunis*, "free of taxes or burden"). The term was later applied to individuals who did not suffer further attacks of smallpox or plague once they had the disease and survived. In the broader sense, **immunity** refers to the resistance of a host organism to invasion by foreign organisms or to the effects of their products. There are two main types of immunity: nonspecific immunity, also called natural resistance, and specific immunity (Fig. 12-1).

# NONSPECIFIC IMMUNITY

Nonspecific immunity refers to the capacity of normal organisms to remain relatively unharmed by, or immune to, a variety of harmful agents present in the environment. This resistance to infections by microorganisms (bacteria, viruses, fungi, and protozoa) is due to anatomic, biochemical, and physiologic features of a species that represent primary and secondary lines of defense.

## HUMAN SKIN

Because it covers the whole body, the skin is probably the body's most effective defense system against infection. The skin is an organ, the body's largest and one of its most important. Like all organ systems, it consists of various kinds of tissues. It is a complex organ consisting of several layers. Figure 12-2 shows that human skin is composed of two main layers, the **dermis** (from the Greek, *derma*, "skin") and the overlying **epidermis.**

***The Dermis.*** The dermis is a connective tissue layer well supplied with blood vessels, nerves, and sensory receptors. Glands and hair follicles are located deep in the dermis, although both have dermal and epidermal components. Hair follicles are supplied with smooth muscle fibers

**FIG. 12-1** Types of immunity.

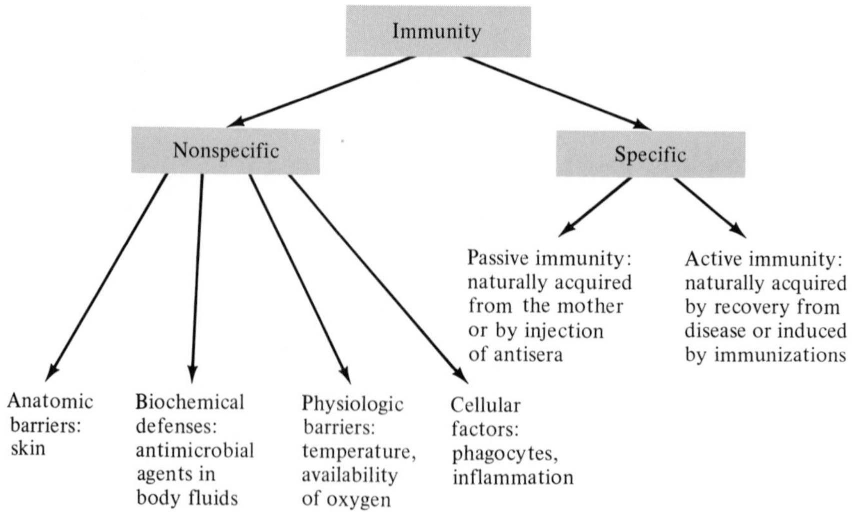

**FIG. 12-2** Cross section of human skin.

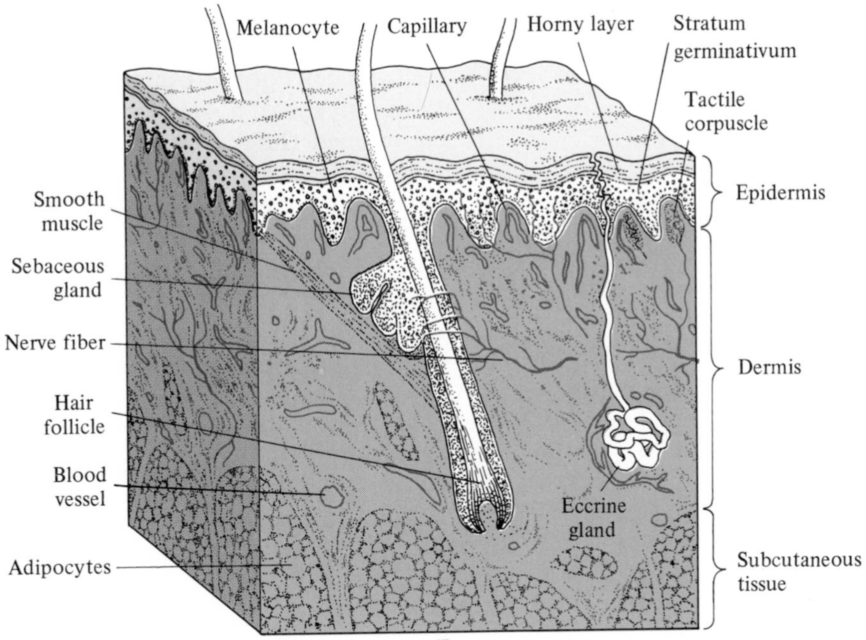


capable of making hair "stand on end" and producing "goose flesh." Underlying the dermis is a subcutaneous layer (from the Latin *sub,* "below," and *cutis,* "skin") of loose connective tissue containing many adipocytes (fat cells).

***The Epidermis.*** The epidermis is composed of several layers that intergrade with each other (Fig. 12-3). All of them are the products of the innermost layer, the **stratum germinativum.** The cuboidal cells of lowermost portion have central nuclei, and their organelles exhibit no specializations. The layer just above them, the **stratum spinosum,** is also composed of cuboidal cells. The cytoplasm of these cells contains bundles of fibers that converge on specific areas of the plasma membrane to form desmosomes. These intercellular connections are believed to be

**FIG. 12-3** The epidermal layers of skin from the sole of the human foot.

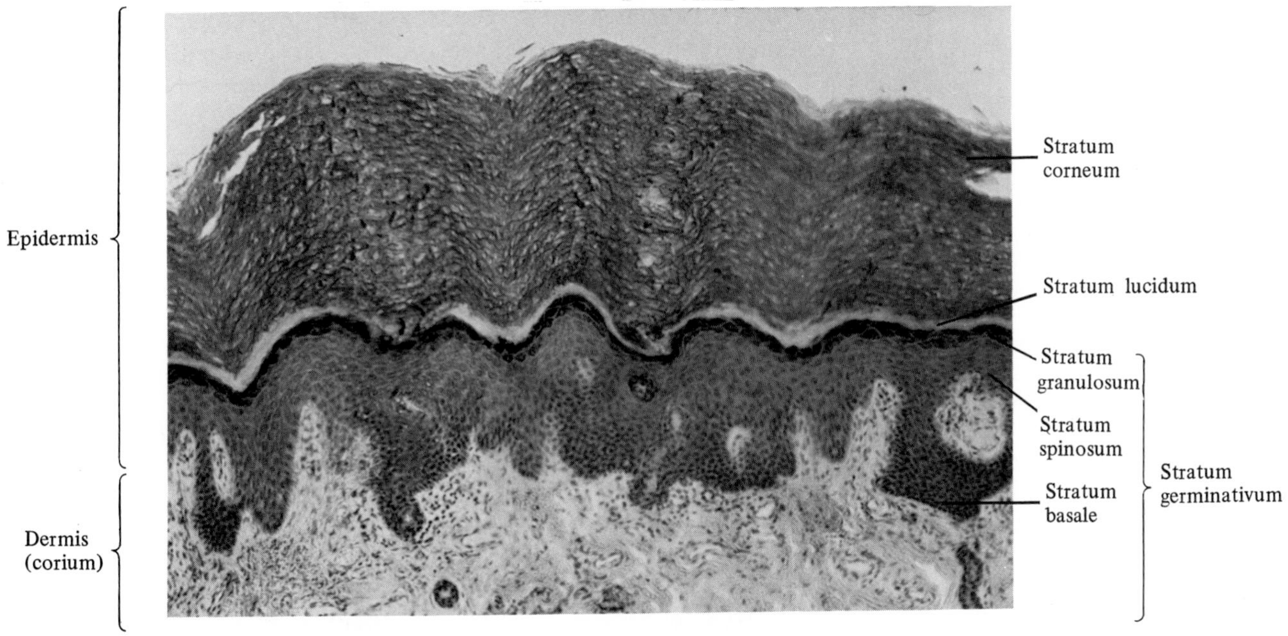

responsible for the cohesion among cells in the epidermis and to protect against loss of epidermal cells by abrasion.

The cells of the third layer, the **stratum granulosum,** are flattened and polygonal, also with centrally placed nuclei. They contain a well-developed Golgi complex and endoplasmic reticulum and produce a special sulfur-containing protein, **keratin.** This tough, water-repellent material is the primary constituent not only of the outer covering of the skin but of hair, nails, claws, and other secretory products of the epidermis. The stratum granulosum also synthesizes other complex molecules. But, unlike keratin, they are stored in membrane-bound granules.

The two outer layers of the skin, the **stratum lucidum** and the **stratum corneum,** are composed of flat, dead cells. They are continually worn away and replaced from below. This is what happens in the case of dandruff, a normal phenomenon. Dandruff is nothing but the dead cells that flake off. These outer layers are rich in keratin, mucopolysaccharides, and phospholipids which render them resistant to abrasion and relatively impermeable to water. Thus, they form a protective layer over the vulnerable living cells beneath them.

## THE SKIN AS A PHYSICAL BARRIER

The scaly skin of lizards and snakes, the feathery skin of birds, and the hairy skin of mammals have a common evolutionary origin and serve many of the same functions. The skin has the following remarkable qualities and important functions:

1. It protects against invasion by microorganisms, parasites, fungi, and other disease organisms. Its tough second major layer, the dermis, also protects against cuts and blows. The dermis of animal skins can be tanned and used as leather.
2. When cut or otherwise injured, the skin regenerates.
3. The skin helps regulate body temperature. A layer of fat cells is located in the loose connective tissue just below the skin (Fig. 12-2). This layer is especially prominent in some animals where it acts as an insulator. In animals with sweat glands, the body is cooled by the evaporation of sweat from the skin's surface. The blood vessels in the skin dilate in warm weather, bringing blood to the body surface and allowing body heat to be radiated into the environment. They constrict in cold weather, conserving body heat. The skin is thus an important homeostatic organ for the control of body temperature.
4. The skin blood vessels also contribute significantly to the control of blood pressure.
5. In mammals, epidermal cells produce cells that are modified to form hair, spines, nails, claws, hooves, horns, and scales. In birds, such specializations include scales, feathers, beak, and claws.
6. Some regions of the skin have special functions. For example, the friction ridges that form fingerprints and palm prints permit tight gripping of objects, especially valuable to an animal that swings by its hands from branches. Calluses form in response to irritation and thereby protect the hands and feet against excessive wear and injury.
7. Sebaceous glands at the base of hairs produce a fatty substance

called sebum which keeps the skin and hair supple and prevents damage caused by drying (Fig. 12-2).

8. The skin can be stretched. For example, it is stretched during pregnancy but comes back to its original form after delivery.
9. Human skin is pigmented. Pigment-bearing cells called **melanocytes** are located deep in the epithelium (Fig. 12-2). The melanocytes expand and inject granules of the pigment melanin into epidermal cells in response to exposure to sunlight, shielding the deeper layers against ultraviolet rays. Pigmentation varies among human beings according to the amount of pigment synthesized by their melanocytes, not the number of melanocytes they possess. Pigmentation in human populations is directly related to the annual amount and intensity of sunlight and is the result of evolution. Dark skin is an advantageous trait in tropical regions. An albino, an individual unable to synthesize any pigment, born in the tropics is strongly selected against; only extreme care to avoid the sun can save an albino's life in such an environment.
10. The skin is richly supplied with sensory nerve endings, receptors that obtain information about the outside world and convey it to the nervous system.
11. Human skin has two types of sweat glands. The most numerous are the exocrine glands (those which secrete their products into ducts) (Fig. 12-2), which secrete water containing some dissolved material. The less numerous apocrine glands (those which secrete their products directly into tissue fluid or blood) exude a milky, odorous secretion, usually in response to stress or sexual excitement, and are found only in the armpits, navel, anogenital area, nipples, and ears. They are far more numerous in women than in men. Mammary glands, from which mammals derive their name, are believed to be modified sweat glands.

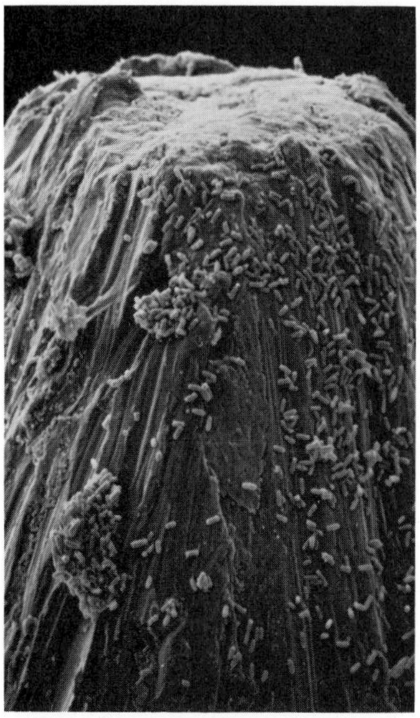

**FIG. 12-4** Just as bacteria cover the point of this household pin, so they cover the surface of the skin.

The primary defense system of the skin is enhanced by various kinds of microorganisms that live on the skin (Fig. 12-4). These microorganisms provide what is known as **bacterial antagonism,** the inhibition of growth and proliferation of certain other microorganisms that might be harmful. Bacterial antagonism is based on chemical by-products of cellular metabolism, including several kinds of acids. Sweat, another kind of substance found on the skin of mammals, also has microbicidal properties (i.e., the ability to destroy very minute organisms, or microbes).

Moist mucous membranes found in various parts of the body, especially those lining the respiratory tract, act to trap invading microbes. Mucus also contains microbicidal substances. As a result of ciliary action of the linings of the respiratory tract which brings the trapped material up into the mouth and the sucking action of the mouth, the mucus is normally carried back into the throat, where it is swallowed.

The gastrointestinal tract provides an internal barrier against microbes, supplemented by microbicidal substances in saliva and other digestive secretions. Furthermore, the high acidity of gastric juice kills most organisms that enter the stomach.

## BIOCHEMICAL DEFENSES

If an organism is able to penetrate the skin, it encounters a series of secondary defense mechanisms, including a variety of biochemical de-

# PANEL 12-1
# Interferons

A tissue or culture experimentally infected with one kind of virus is often able to resist infection by a second type of virus. This phenomenon is known as **viral interference,** and its action is associated with a group of low-molecular-weight proteins called **interferons.** These proteins are produced by the body within hours following an infection. In the presence of a virus, interferon binds to the surface of noninfected cells and stimulates them to produce an antiviral protein that blocks viral reproduction.

Interferon is currently one of the most effective weapons known to combat viral infections and may also be effective in controlling the growth of certain cancers (see p. 349). It is for these reasons that considerable effort is being made to isolate and purify interferon. Because it is produced in such minute quantities by mammalian cells, the cost of interferon preparations is so high as to make it virtually unavailable for the routine treatment of viral infections and the control of tumor growth. However, it is now possible to induce certain bacteria, such as *Escherichia coli* (*E. coli*), to synthesize interferon. This is accomplished by transferring the genes responsible for the synthesis of interferon from mammalian cells to *E. coli*. These bacteria can then be grown in quantities sufficient to provide yields of interferon large enough to be commercially available for medical use.

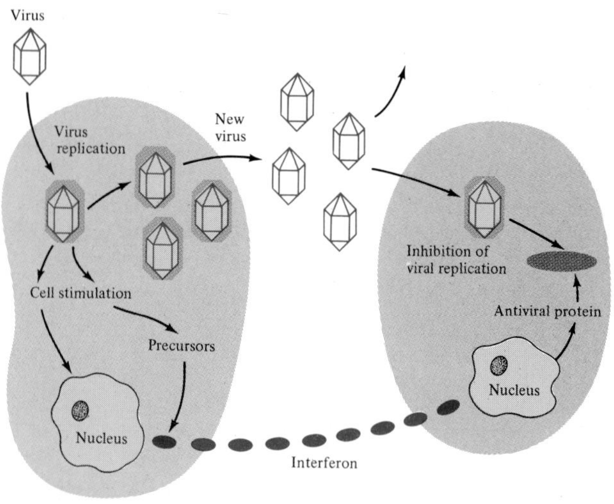


Infection of cells by viruses induces the production of interferon, which activates the synthesis of an antiviral protein that prevents the multiplication of viruses in host cells.

**FIG. 12-5** Lizards, which normally have body temperatures lower than mammals, are resistant to many microorganisms to which humans, with their higher body temperature, are susceptible.

fenses. The body contains a number of substances that kill microbes. They comprise a second line of defense. One of these substances is **lysozyme,** an enzyme that is found in tissues and is highly effective in breaking down a particular kind of chemical in the cell walls of many bacteria. Another defensive substance is **beta lysin,** which is active against aerobic bacilli that form reproductive bodies called spores. Beta lysin is usually present in low levels in human blood, but its concentration greatly increases during the acute phase of an infectious disease. A third defense molecule, which has received considerable attention in recent years, is **interferon,** described in Panel 12-1.

## PHYSIOLOGIC BARRIERS

In many animals their normal body temperature is effective in combating infection by inhibiting the growth of organisms that would otherwise flourish in their tissues. In some cases, however, temperature changes favor infection. For example, the chicken, which is normally resistant to the disease anthrax, has a normal body temperature of 40°C (105°F). However, it becomes susceptible to the bacterium that causes anthrax if its body temperature is lowered to 37°C (98.6°F), the body temperature of mammals. Lizards, which have a lower body temperature than mammals and are usually resistant to anthrax, become susceptible if their body temperature is raised to that of mammals (Fig. 12-5).

In some cases, the absence of certain essential nutrient molecules prevents growth and reproduction of infectious organisms. For example, most bacteria are unable to grow in the absence of iron or calcium ions. The availability of oxygen is another important factor influencing bacterial infection because, if the amount of oxygen is low, aerobic microorganisms will not multiply. Conversely, for anaerobic microorganisms, such as *Clostridium tetani,* the agent of the disease tetanus, lowered oxygen levels are favorable to multiplication. For such microorganisms, high oxygen levels are usually inhibitory.

## PHAGOCYTOSIS AND THE INFLAMMATORY PROCESS

Another secondary line of defense against microorganisms is a group of cells, called **phagocytes,** that engulf and digest foreign substances, including microbes. Phagocytic cells are primarily of two kinds: the circulating phagocytic cells of the blood, which are the polymorphonuclear leukocytes (PMNs) (Fig. 12-6a) and monocytes; and the fixed tissue phagocytes, such as the Kupffer cells of the liver, the microglial cells of the brain, alveolar macrophages of the lung (Fig. 12-6b), and the macrophages of the lymphatic tissue, such as the spleen and lymph nodes. The PMNs represent a major internal defense in the destruction of microbes. During the early stages of tissue injury, chemical factors, or **chemotaxins** are released from the injured tissues, attracting circulating PMNs to the site of injury. As a result, large numbers of PMNs are produced by the bone marrow and released into the circulation during systemic infection or significant tissue damage.

Inflammation at the site of a wound is another secondary defense system. The term *inflammation* is suggested by the reddening of the affected areas. Histamine, a molecule secreted by certain white blood cells that causes contraction of smooth muscles and dilation of blood vessels, is released, blood vessels dilate, and phagocytic cells and plasma move into the affected tissues. At the same time, fibrin causes the blood to clot, closing the wound to further microbial penetration with a scab. Microbicidal factors in the plasma also inhibit microbial growth and reproduction and thereby mediate the repair of damaged tissues.

**FIG. 12-6** (a) Electron micrograph of a polymorphonuclear leukocyte (neutrophil) phagocytizing staphylococcal bacteria. (b) Alveolar macrophage extending a pseudopod in order to phagocytize a bacterium.

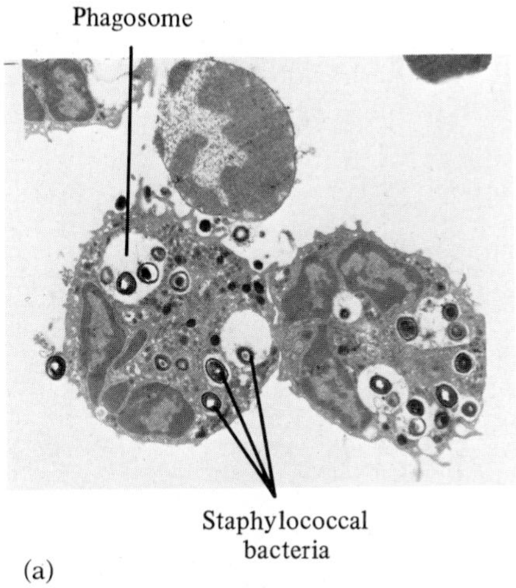


(a)

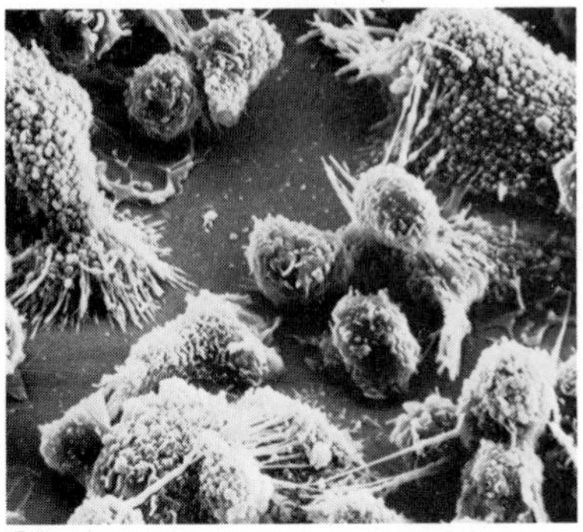

(b)

## SPECIFIC IMMUNITY

Although immunobiology is a relatively recent scientific discipline, the concept of immunity as a means of resistance to infection is an ancient one. Since the time of the Greeks it has been recognized that those who recover from plague, smallpox, yellow fever, and various other infectious diseases rarely contract the same disease again.

The first scientific attempts at artificial immunization were made in the late eighteenth century by the English physician Edward Jenner (Fig. 12-7). Jenner investigated the basis for the widespread belief of peasants in the rural areas in England that anyone who had had vaccinia, or cowpox (from the Latin *vacca,* "cow"), a disease that affected

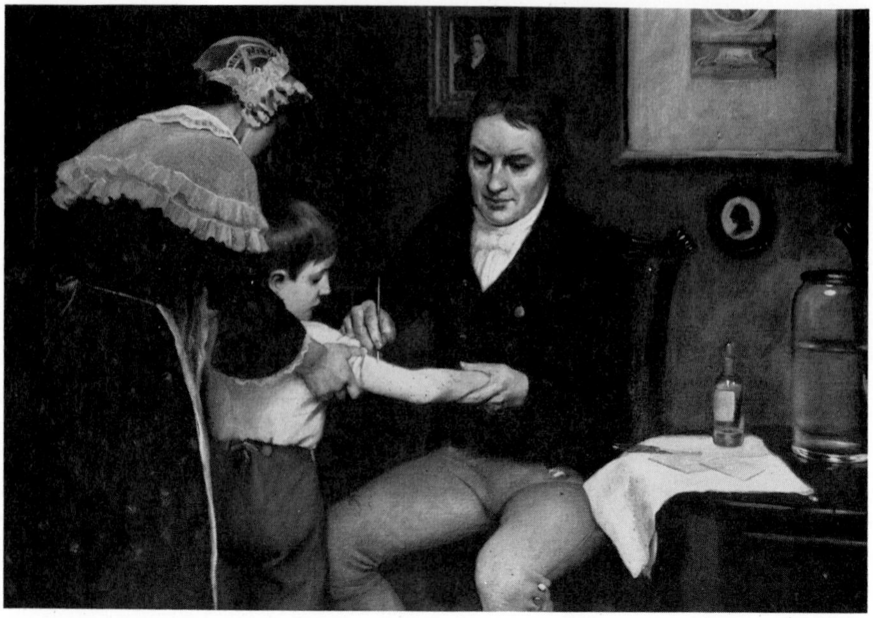

**FIG. 12-7** Edward Jenner (1749–1823), the English physician who developed a method of artificial immunization against smallpox.

both dairy cattle and humans, never contracted smallpox. Smallpox was not only often fatal—10 to 40 percent of those who contracted it died, and children were especially susceptible—but those that recovered usually had disfiguring pockmarks. Yet most British milkmaids, who were readily infected with cowpox, had clear skins because cowpox was a relatively mild infection that left no scars.

After some 20 years of close observation, including several deliberate attempts to give smallpox to people who had contracted cowpox, Jenner began to immunize people by deliberately infecting them with cowpox. His first subject was a healthy, eight-year-old boy known never to have had either of these two related diseases. As Jenner had expected, immunization with the cowpox virus caused only mild symptoms in the boy. When Jenner subsequently inoculated him with smallpox virus, the boy showed no symptoms of the disease.

Jenner subsequently inoculated patients in large numbers with cowpox pus, as did other physicians in England and on the European continent. By 1800, the practice, known as **vaccination,** had begun in America, and by 1805, Napoleon commanded all French soldiers to be vaccinated.

Further work on immunization was carried out by Louis Pasteur (1822–1895), the French physician who established the scientific basis for the germ theory of disease (the fact that microorganisms are responsible for disease) and developed techniques for the maintenance and growth of bacteria in test tubes. Pasteur discovered that neglected, old cultures of chicken cholera bacilli, which had not been placed in a fresh culture medium on a regular basis, produced only a mild attack of this disease in chickens inoculated with it. He then discovered that fresh cultures of the bacteria failed to produce the disease in any chickens that had been previously inoculated with such old cultures. The organisms in the old cultures had somehow become less pathogenic, or **attenuated.** They had lost their ability to cause damage to cells and tissues, a change that Pasteur later found he could regularly produce in cultures of other kinds of aerobic bacteria by growing them for long periods of time under anaerobic conditions. To honor Jenner, Pasteur gave the name **vaccine** to any preparations of a weakened pathogen, or

**FIG. 12-8** Louis Pasteur watches as his assistant inoculates a boy for "hydrophobia" (rabies).

infective microbes, that was used as was Jenner's "vaccine virus," to immunize against infectious disease.*

Pasteur used vaccination to protect animals against anthrax and people against rabies (Fig. 12-8). Following Pasteur's discovery, other investigations showed that not only weakened, living microorganisms but also microorganisms killed by treatment with formalin, merthiolate, phenol, or heat could induce immunity.

## ACQUIRED IMMUNITY

The immunity acquired following the penetration of the body by foreign substances depends on the body's ability to recognize the substances as foreign and to produce an immune response to them. Such substances are called **antigens.** This ability to distinguish self from nonself is one of the hallmarks of immunity. Such immunity may develop as a result of contracting a disease and recovering from it. This will usually result in a long-lasting immunity to another attack of the same disease, that is, the immunity is specific to the particular agent that engendered the immune response. Specificity is another hallmark of immunity.

Acquired immunity can also be artificially induced in individuals as a means of developing protection against the possibilities of future infections by specific microorganisms. This is accomplished by the injection of attenuated or killed microorganisms, or some of their metabolic products that are known to cause disease. These "shots," as they are commonly called, have been effective in developing resistance to microorganisms that cause cholera, diphtheria, measles, mumps, whooping cough, rabies, smallpox, tetanus, typhoid, yellow fever, and poliomyelitis (Fig. 12-9). Because of a massive and concerted worldwide campaign of immunization against smallpox, the World Health Organization declared this once dreaded disease eradicated from the world in 1980.

* The word *virus* (from the Latin *virus,* "poison"), as used in Pasteur's time, meant any agent capable of causing a disease. The current definition of a virus is a submicroscopic, noncellular particle composed of a nucleic acid core and a protein coat, which is able to reproduce only within living cells.

**FIG. 12-9** Child receiving a DPT shot, which is designed to provide immunity against diphtheria (D), whooping cough (or pertussis, P), and tetanus (T).

***Boosting the Immune Response.*** You may recall from your own experience that when you received a polio vaccine you had to have more than one shot. The second, or booster, shot is designed to increase immunity to a higher level than that which can be obtained with only one shot, called a primary immunization. This occurs because the immune system can recognize when it has previously encountered a given foreign substance and responds to a second immunization by producing a much larger immune response in a shorter period of time (Fig. 12-10). This secondary response is referred to as the **anamnestic response** (from the Latin *anamnesis,* "to recall") and is the third major characteristic of immunity: immunologic memory. Immunity developed by this procedure is long-lasting and may even last for the life of the individual. This accounts for the fact that many of the shots we received—such as inoculations against diphtheria, pertussis (whooping cough), tetanus (DPT), and polio—were given to us only when we were young and have not been repeated.

## PASSIVE IMMUNITY

Passive immunity is achieved by injecting individuals with **antibodies,** the specific protein molecules involved in immunity, that have been produced by another individual in response to a foreign substance. The protection afforded by this type of immunity usually lasts only a short time, since the antibodies are rapidly broken down and soon disappear from the circulation. Some diseases are treated by injection of a serum (the clear yellow fluid obtained from blood that has clotted) containing antibodies, called an **antiserum,** from animals that have been immunized or from humans that have experienced a disease and recovered. Passive immunization is used against such diseases as poliomyelitis and infectious hepatitis for exposed persons who have not been actively immunized to the microorganisms that cause these diseases. Tetanus antitoxin (antibody against toxic molecules, called toxins, produced by the bacterium *Clostridium tetani*) is used for people who risk infection by tetanus bacilli introduced into a dirty wound and who have not recently been actively immunized to tetanus toxoid (the pathogenic toxin produced by tetanus bacilli and treated so that it can no longer cause pathogenesis but can still induce an immune response).

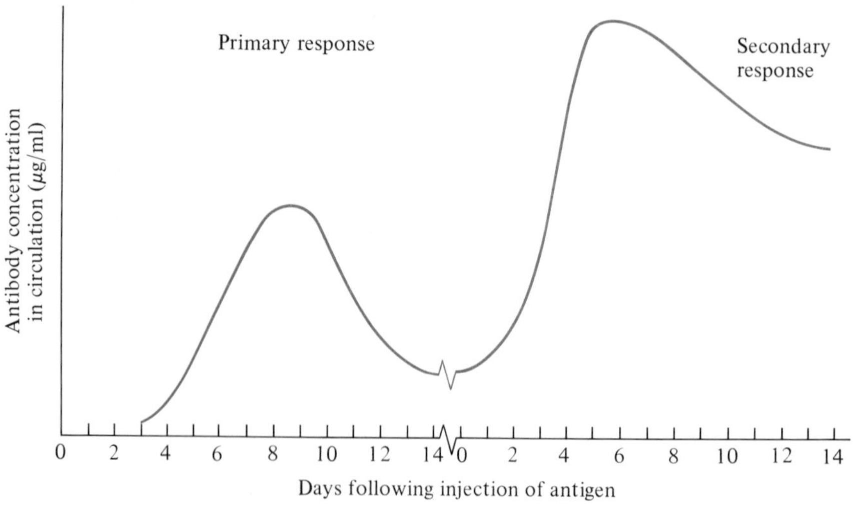

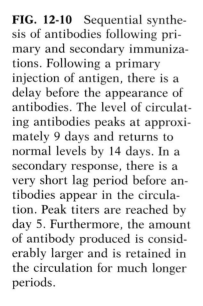

**FIG. 12-10** Sequential synthesis of antibodies following primary and secondary immunizations. Following a primary injection of antigen, there is a delay before the appearance of antibodies. The level of circulating antibodies peaks at approximately 9 days and returns to normal levels by 14 days. In a secondary response, there is a very short lag period before antibodies appear in the circulation. Peak titers are reached by day 5. Furthermore, the amount of antibody produced is considerably larger and is retained in the circulation for much longer periods.

Passive immunity is also naturally acquired by the transfer of antibodies from mother to child across the placenta during pregnancy. The antibodies are also absorbed through the child's intestine from the first milk of the mother, called **colostrum.** This passively acquired immunity protects newborns in the first few months of life, when their own immune system is not yet fully developed.

## THE CELLULAR BASIS OF IMMUNITY

The cells involved in immune responses are localized in the organs and tissues known as the lymphoid system. This system is comprised of encapsulated organs, such as the spleen and lymph nodes, and of diffuse lymphoid tissue, such as the Peyer's patches embedded in the wall of the small intestine. During development, stem cells from the bone marrow migrate into the primary, or central, lymphoid organs, where they differentiate, and mature, into immunocompetent lymphocytes (cells capable of carrying out an immune response); see Fig. 12-12.

### PRIMARY LYMPHOID ORGANS GIVE RISE TO TWO POPULATIONS OF LYMPHOCYTES

As we saw in Chapter 11, lymphocytes are a kind of white blood cell (Fig. 12-11). Without them, there would be no immune system. A milestone in the development of modern immunology was the discovery that what appeared to be a homogeneous population of lymphocytes is in fact made up of two functionally different populations: **T cells** and **B cells.** One population of lymphocytes develops into a cell type that can respond to foreign substances when they pass through the **thymus,** a

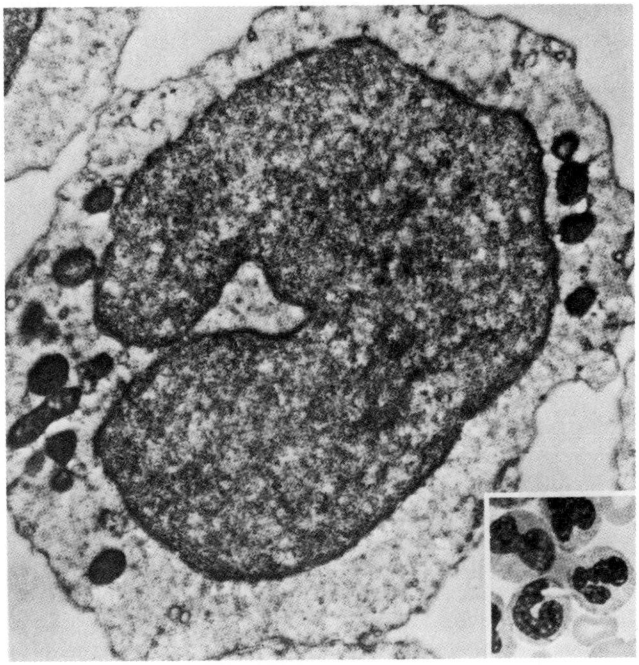

**FIG. 12-11** An electron micrograph of a human lymphocyte. The inset shows a group of lymphocytes photographed in a light microscope.

primary lymphoid organ located in the thoracic region of human beings (Fig. 12-12). These thymus-derived lymphocytes are the T cells. Another population of lymphocytes develops in the bone marrow or its equivalent, such as the **bursa of Fabricius,** a primary lymphoid organ associated with the cloacal (a common passage for fecal, urinary, and reproductive discharges) region of the gut of the chicken and other birds (Fig. 12-12b). These bursa-derived lymphocytes are the B cells. The two types of lymphocytes are responsible for a dual kind of immunity. Because B cells produce antibodies that are released in the circulation of body fluids, or humors, the immunity they generate is called the **humoral immune response.** The main role of T cells is the destruction of microorganisms and such tissues as those from a tumor or a foreign tissue or organ graft. Since the immunity involving T cells is associated with the cells themselves, this type of immunity is called **cell-mediated immunity.** Although T cells do not produce antibodies, a population of T cells known as helper T cells cooperates with the B cells in the synthesis of antibodies.

What makes lymphocytes develop into two different kinds of cells? Experiments with laboratory animals have confirmed that the thymus is necessary for the maturation of one population of lymphocytes into T cells. For example, newborn mice whose thymus glands have been removed are unable as adults to reject skin grafts. When these mice are injected with thymus cells from normal adults of the same strain, all the immune functions return. Studies have shown that lymphocytes found in the thymus gland originate in the bone marrow as **stem cells,** some of which migrate to the thymus. Normally, these cells then move to the spleen, lymph nodes, adenoids, tonsils, Peyer's patches, and other lymphoid tissues distributed throughout the body. (Figure 12-13 shows the human lymphatic system.) During their stay in the thymus or their passage through it, stem cells are converted to immunologically active, or immunocompetent, lymphocytes under the influence of a variety of factors in the thymus. The T cells normally found in the spleen and lymph nodes depend on the thymus gland for the development of their particular immune capabilities.

Similarly, B cells undergo a period of differentiation in order to be-

**FIG. 12-12** (a) The human thymus gland is responsible for the differentiation of bone marrow stem cells to immunocompetent T lymphocytes. These cells then localize in such lymphoid organs as the spleen and lymph nodes. (b) The lymphoid organs of the chicken. The bursa of Fabricius controls the differentiation of bone marrow stem cells to immunocompetent B lymphocytes. They also then localize in the spleen and other lymphoid tissues.

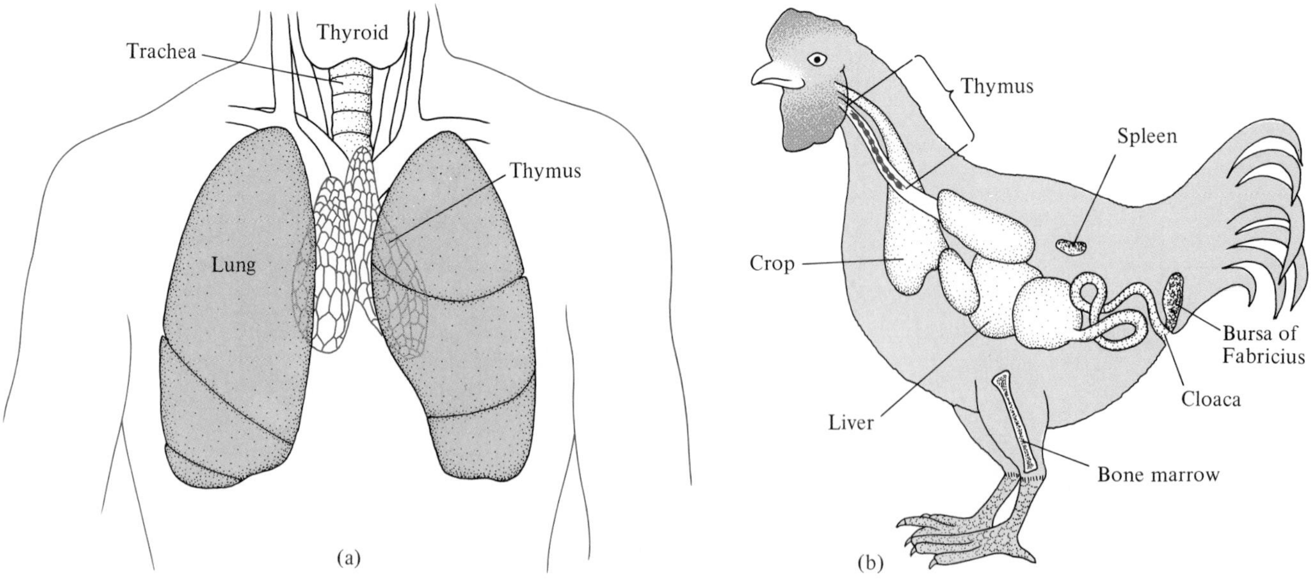

**FIG. 12-13** The human lymphatic system and its relationship to the major organs and tissues of the immune system.

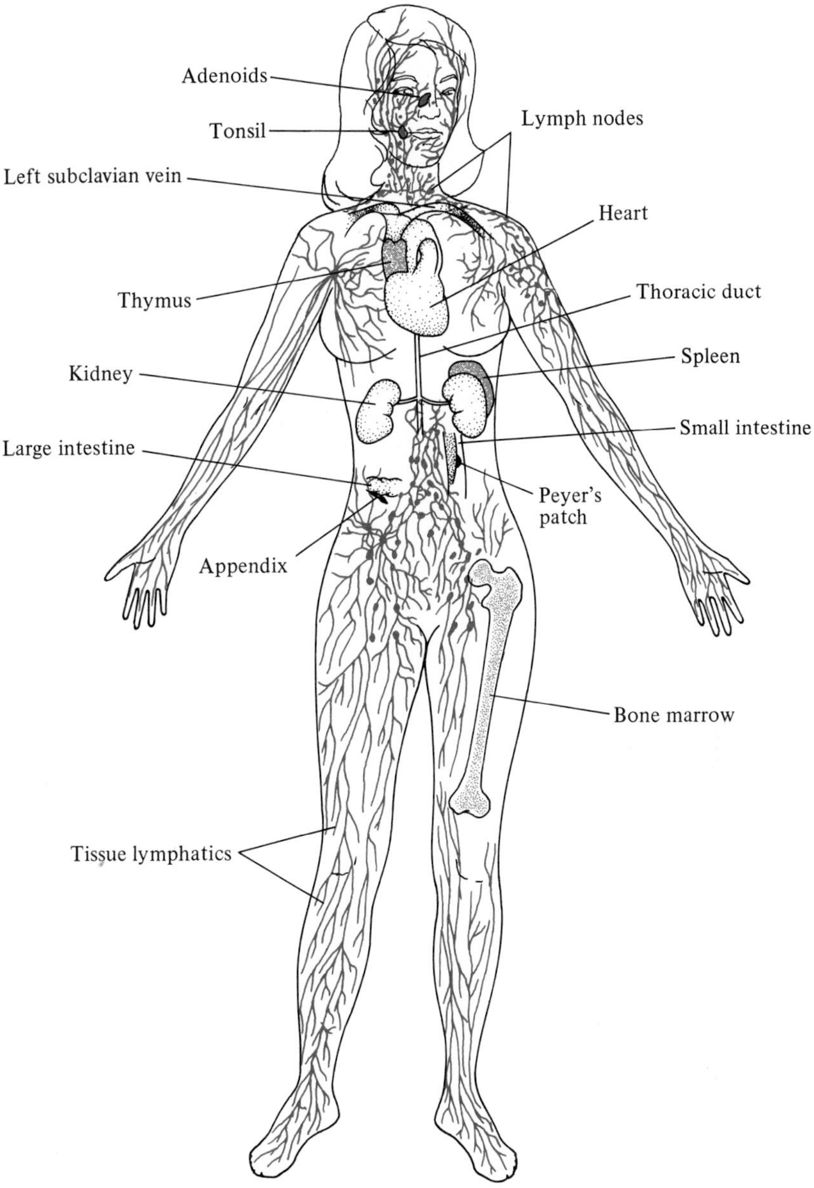


come immunocompetent. The tissue in which this maturation occurs was discovered by the American scientist Bruce Glick in 1954 while he was doing research on the bursa of Fabricius. He surgically removed this gland (bursectomy) from a number of chicks, which he later injected with bacteria. Most of the bursectomized chickens died, and those that survived had impaired B-cell function. However, if the thymus was intact, these animals had normal cell-mediated immune T-cell function, for example, graft rejection. Later research showed that B cells populate specific areas of the spleen and lymph nodes other than those occupied by T cells, and also originate as bone marrow stem cells.

Experiments in which immunologically deficient, or immunodeficient, animals are injected with either thymus cells or bone marrow cells show that both kinds of cells are necessary for these animals to develop immunocompetent B and T cells. Only those animals receiving injections of both kinds of cells have both humoral and cell-mediated

**FIG. 12-14** Schematic representation of the development of the humoral and cell-mediated immune responses. Immunocompetent T and B lymphocytes localized in the spleen originate in the bone marrow as stem cells and pass through the thymus and bursa of Fabricius, respectively. In man, the homologs of the bursa appear to be the bone marrow and liver.

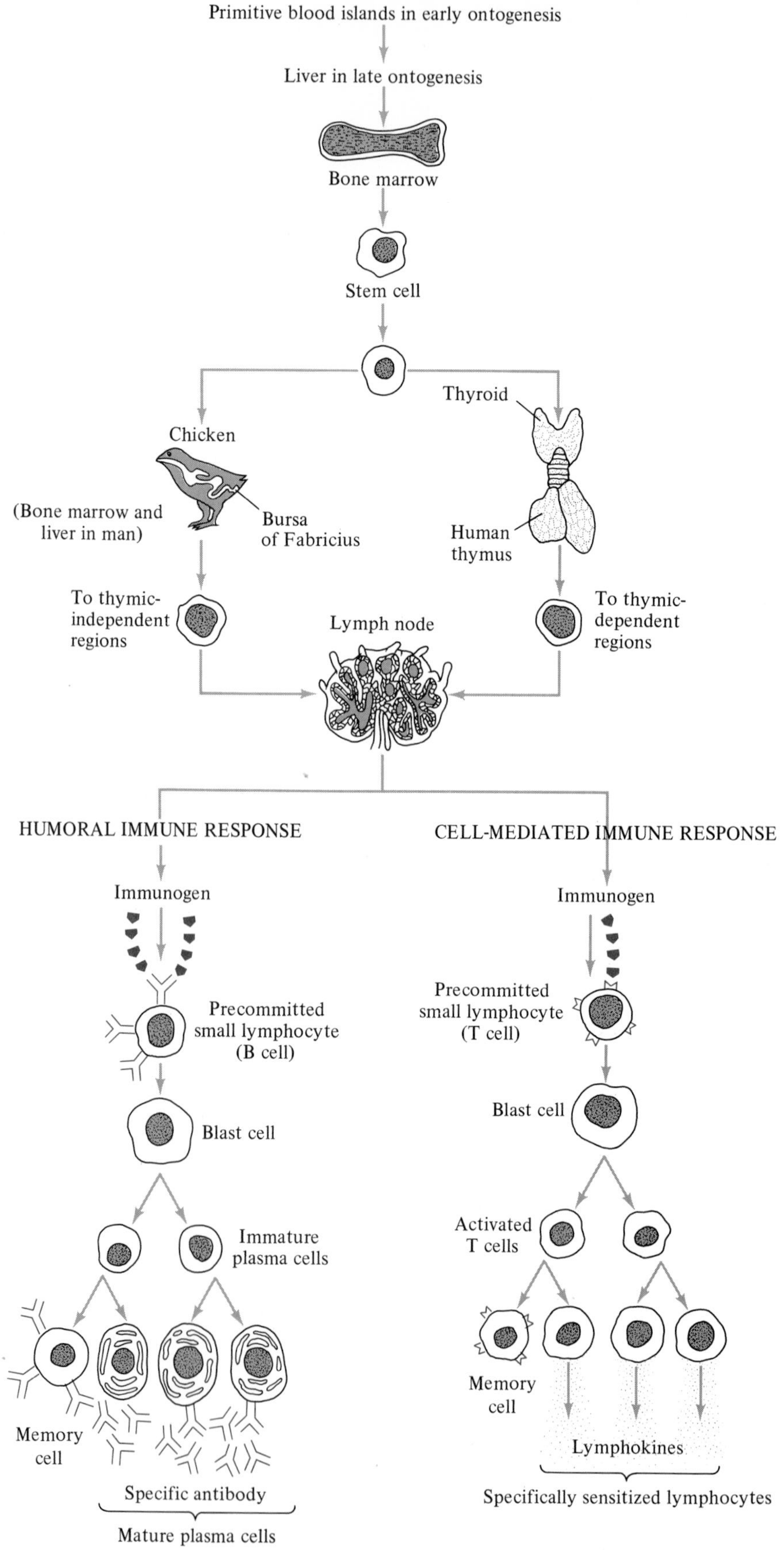

immunity. This proves that there is a division of labor between T and B cells and that both kinds are necessary for antibody formation while only T cells are necessary for graft rejection and immune responses to fungi, viruses, protozoa, and certain pathogenic bacteria (Fig. 12-14).

## SECONDARY LYMPHOID ORGANS

After becoming immunocompetent, B and T lymphocytes migrate to the spleen, lymph nodes, tonsils, adenoids, and appendix as well as the diffuse lymphoid tissues distributed throughout the body.

Although lymph nodes are found throughout the body, most of them are localized in particular areas such as the neck, armpits, and the groin region.

A typical lymph node is a bean-shaped organ covered by a capsule of connective tissue (Fig. 12-15). Just under this capsule is the subcapsular sinus, which is lined with macrophages, large phagocytic cells. Lymphocytes and foreign substances pass into the sinus by way of the afferent lymphatics, and any particulate matter contained within them is removed by the macrophages. Beneath the subcapsular sinus lies the **cortex,** which is densely packed with lymphocytes organized into spherical **follicles.** It is in these follicles that the B lymphocytes localize following their maturation in the primary lymphoid organs. The lymph nodes in your neck can readily be located when you have an infection because of the swelling caused by the cellular proliferation of the B lymphocytes in the follicles.

Between the cortex and the inner core of the lymph node, called the **medulla,** is the **paracortex.** Here the immunocompetent T lymphocytes localize following their maturation in the thymus. The paracortex is lined throughout with a network of fibrous tissue called the **reticulum.** This network also contains macrophages, which trap and phagocytize foreign materials and present them to the B and T cells for antibody formation.

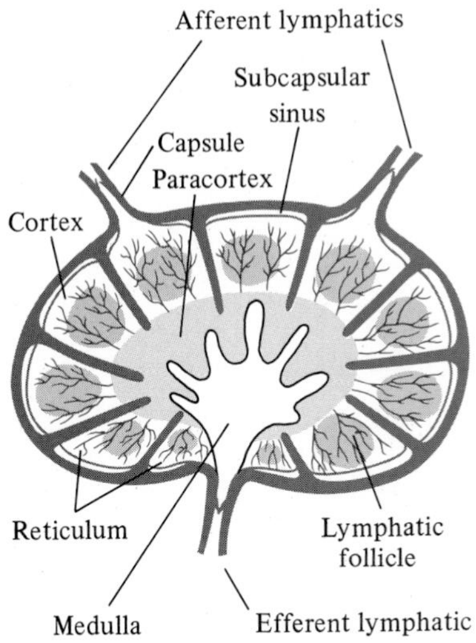


**FIG. 12-15** The structure of a lymph node. Beneath the connective capsule is the subcapsular sinus, which is lined with phagocytic macrophages. As lymphatic fluids pass into the sinus, the macrophages remove foreign particles by phagocytosis. Below the sinus is the cortex, which has large numbers of B lymphocytes within the follicles. The T lymphocytes, which have migrated into the lymph node from the thymus, are primarily concentrated in the paracortex.

## MACROPHAGES

A third kind of cell associated with the immune response is the **macrophage,** also known as an accessory cell (see Fig. 12-6b). Macrophages are phagocytic cells found in certain body tissues—lymph nodes, spleen, liver, lungs, and connective tissue—and in the circulation. They not only destroy invading bacteria by ingesting them but also play a supporting role as accessory cells in humoral and cell-mediated responses.

# THE HUMORAL IMMUNE RESPONSE

Basic to the humoral immune response is the formation of **antibodies,** protein molecules generated in response to the presence of foreign substances within the body. Foreign substances that induce an immune response and interact with antibodies are called **antigens.** In the humoral response, antibody combines with small chemical groupings, called **antigenic determinants** on the antigen to form an antigen-antibody complex. This reaction is characterized by specificity. For exam-

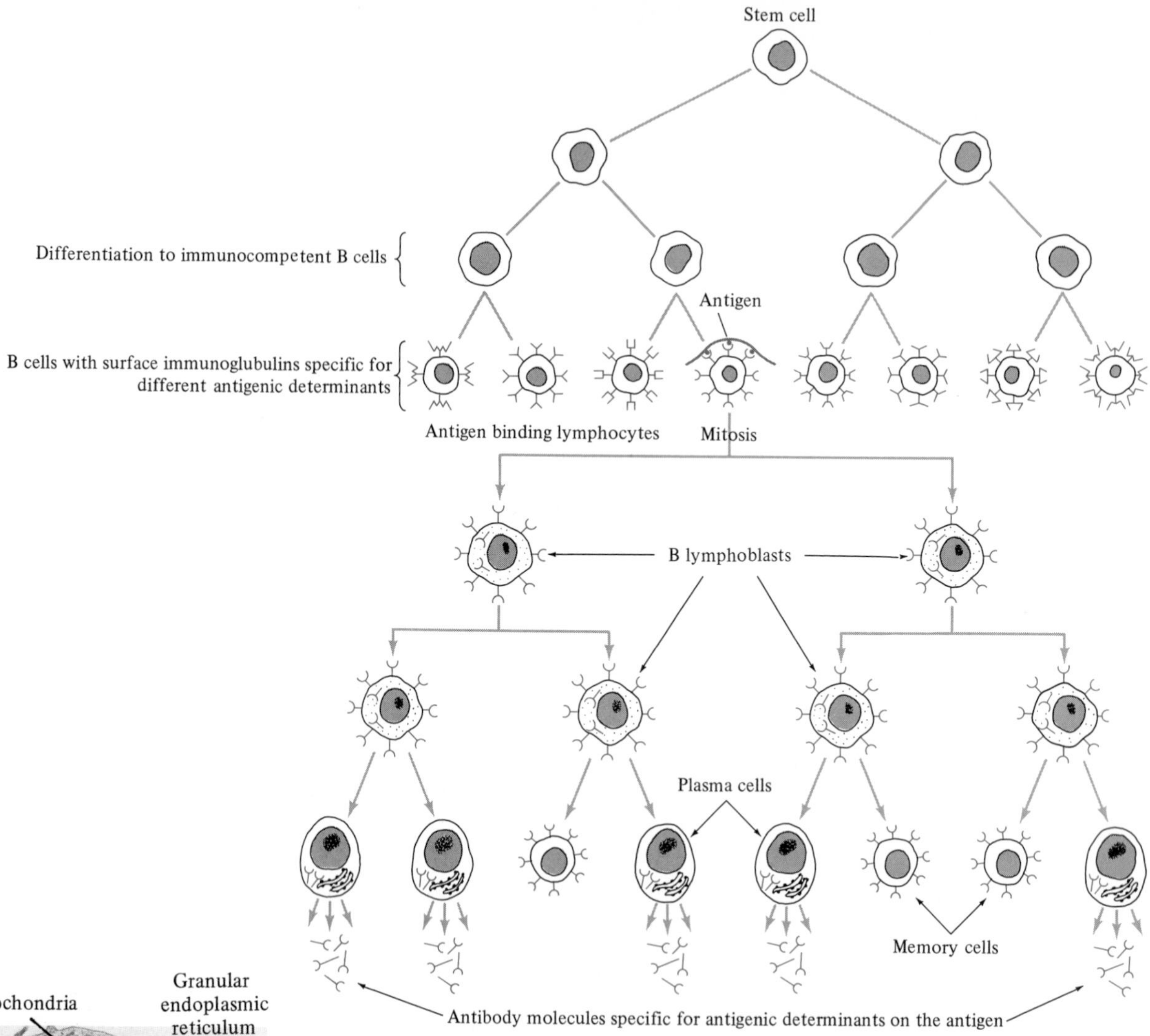


**FIG. 12-16** Schematic representation of the clonal selection theory. Undifferentiated stem cells originating in the bone marrow are multipotent and have the ability to generate approximately one million B lymphocytes, each with antibody molecules on their surface (called a receptor) directed against a different antigenic determinant. Following the interaction (selection) of an antigenic determinant with the appropriate antibody on the surface of a B cell, the B cell undergoes a number of divisions to form a large number of identical cells, called a clone. This clone of cells then differentiates into plasma cells, which secrete the same antibodies they carried on their surface. Not all of these cells become plasma cells. Some become memory cells, which are able to give a secondary (anamnestic) response when they encounter the same antigen a second time.

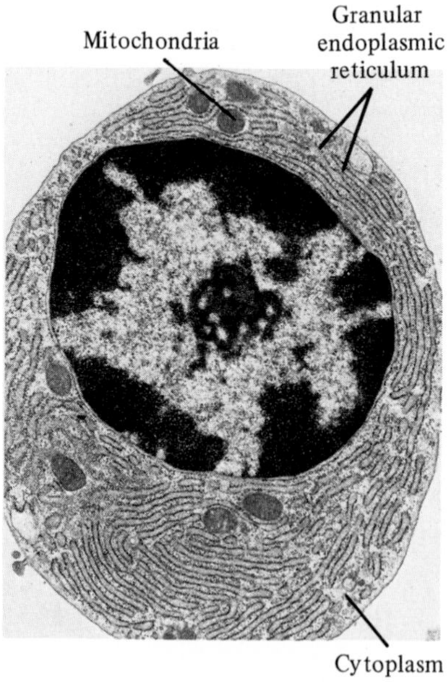


**FIG. 12-17** Electron micrograph of a guinea pig plasma cell. The abundant cytoplasm is packed with a granular endoplasmic reticulum and ribosomes, the sites of antibody synthesis.

ple, if a rabbit is injected with red blood cells from a horse, the rabbit's immune system will produce anti-horse-red-blood-cell antibodies. These specific antibodies will react only against horse red blood cells and not against any other kind of cell that might be injected.

## THE CLONAL SELECTION THEORY

The most satisfying theory yet to explain the ability of certain cells to produce an apparently limitless number of antibodies was proposed by

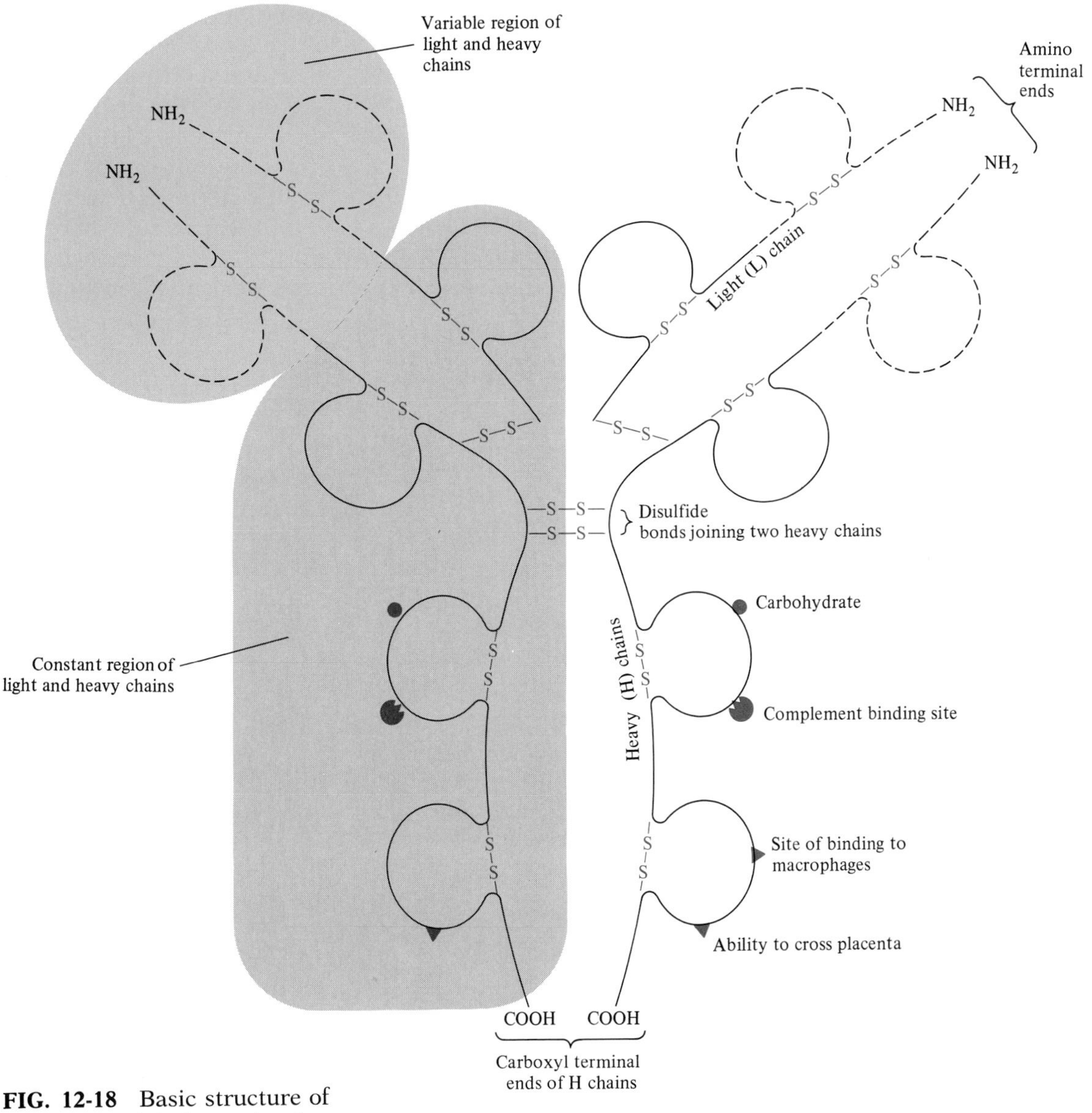


**FIG. 12-18** Basic structure of an immunoglobulin molecule. The basic four-chain structure consists of two heavy (H) and two light (L) chains joined by disulfide bonds (—S—S—). Each heavy and light chain has variable regions, which differ from one antibody molecule to another. The variable region gives each antibody molecule its specificity to a given antigen, and it is in this area where the antigen combining site is located. The constant regions are similar from one antibody molecule to another even though the antibodies are specific to different antigens. The biologic properties of the immunoglobulin molecule are associated with the constant region of the heavy chain.

the Australian physician MacFarlane Burnet in 1957 (Fig. 12-16). According to this theory, a small number of B lymphocytes is genetically programmed to respond to a particular antigen. Once introduced into the organism, the antigen makes contact with specific receptor sites found on the plasma membrane of B lymphocytes. This triggers the formation of antibodies against the antigen. After contact, these cells proliferate into a population of identical cells (clone) with the same specificity. This population in turn differentiates into a special kind of cell called a **plasma cell** (Fig. 12-17). The plasma cells synthesize and secrete large amounts of the antibody to that particular antigen into the circulation.

## THE ANTIBODY MOLECULE

Each antibody molecule is made up of two pairs of polypeptide chains joined by disulfide (sulfur-to-sulfur) bonds. One pair, the **light** (L) **chains,** is small, while the other pair, the **heavy** (H) **chains,** is much larger (Fig. 12-18). A unique feature of the heavy chains is that one segment of each chain contains essentially the same amino acid sequence as is found in a whole range of antibodies with different specificities. This end of the chain is called the **constant** (C) **region.** The arrangement of amino acids at the other end of each H chain is different in each antibody molecule of a different specificity; it is therefore called the **variable** (V) **region.** This variability confers upon the organism the capability of dealing with many different kinds of antigens. This extraordinary variability in amino acid sequence is unique to the antibody molecule; no other molecule in the body is quite like it. The constant region of an antibody molecule is responsible for the molecule's biologic activity, such as the ability to cross the placenta, to enhance phagocytosis by macrophages, and to activate a series of molecules, collectively called **complement,** that can cause **lysis,** or rupture, of cells.

***Immunoglobulins.*** Most antibodies belong to a class of serum proteins called **gamma globulins.** However, because some antibodies belong to other classes of globulins, it is common practice to describe all proteins with antibody function as **immunoglobulins (Ig).**

Based on minor amino acid differences in the constant regions of their heavy chains, all human immunoglobulins are placed in one of five classes: IgG, IgM, IgA, IgE, and IgD. Minor differences in structure among immunoglobulins of different classes are reflected in differences in biological properties. Moreover, the five types are found in different locations in the body (Table 12-1).

IgG is the most abundant immunoglobulin and is found in blood and other body fluids. It provides a major line of defense against microorganisms. IgG is the only immunoglobulin that can cross the placenta and is therefore important in the defense of newborns against infection.

IgM, the largest of the immunoglobulin molecules, is a complex of five immunoglobulin molecules resembling IgG. It is found mainly in blood, and it is the first to appear following initial contact with an antigen. Both IgM and IgG can activate complement.

IgA is found mostly in the secretions of exocrine glands, for example, in milk, saliva, and tears. It acts as a barrier against invasion by microbes through the epithelial lining of the respiratory, genital, and digestive tracts.

IgE is also found in the secretions of exocrine glands. It is the immunoglobulin associated with allergies, such as hives, eczema, hay fever, and asthma, as will be described later in this chapter.

IgD molecules are found on the surface of B lymphocytes, where they may have an important regulatory role in the synthesis of immunoglobulins by these cells.

**TABLE 12-1 Classes of Human Immunoglobulins**

| *Immunoglobulin (Ig) class* | *Molecular weight (daltons)* | *Amount in serum (mg/ml)* | *Primary location* | *Ability to bind complement by classical pathway* |
|---|---|---|---|---|
| IgG | 150,000 | 12.0 | Most abundant Ig in internal body fluids, including serum | + |
| IgM | 900,000 | 1.2 | Serum | ++ |
| IgA | 160,000 | 3.5 | Major Ig in mucus secretions, also found in serum | – |
| IgE | 200,000 | 0.03 | Bound to mast cells in lung and digestive tract | – |
| IgD | 185,000 | 0.00001 | Surface of B lymphocytes | – |

With the realization of the specificity of immunoglobulin molecules, a concerted effort was made to isolate and purify antibody molecules that could act as "magic bullets" to seek out and destroy an invading microorganism without harming the patient. As described in Panel 12-2, monoclonal antibodies serve as such magic bullets.

## ANTIGEN-ANTIBODY COMPLEX

Interaction between antigen and antibody is limited to certain areas of contact on the surface of each molecule. On the antigen molecule, these areas, called antigenic determinants, consist of specific amino acid sequences. When contact with an antibody is made, these determinants occupy the binding sites on the antibody molecule in a lock-and-key

### PANEL 12-2

## Monoclonal Antibodies: The Magic Bullets

Monoclonal antibodies are produced by identical cellular descendents—**a clone**—of a single cell that is genetically programmed to produce antibodies of a particular specificity. The production of monoclonal antibodies begins with the injection of a mouse with a particular antigen. A major advantage of this procedure is that the antigen used need not be purified, as is the case in traditional procedures for producing a specific antiserum. In a few days, when the B lymphocytes begin proliferating in response to the antigen, the spleen, a primary source of these lymphocytes, is removed from the mouse. Among the many lymphocytes isolated from this mouse spleen are a few B cells able to produce antibodies specifically for the antigen. These spleen cells are mixed with myeloma cells, a special type of lymphocyte tumor cell, a kind of cell that multiplies unchecked. Like most cancer cells, these myeloma tumor cells can proliferate indefinitely when grown in culture. Next, the lymphocytes and myeloma cells are fused by adding polyethylene glycol, which acts as a chemical glue. The result is a **hybridoma,** a cell that combines the antibody-producing capability of a B lymphocyte with the myeloma cell's capacity to divide and reproduce virtually forever—a combination that assures a nearly endless source of antibodies. Among the numerous hybridomas formed will be a few cells producing antibody of the appropriate specificity.

The desired cells are found and separated in a three-step process: selection, screening, and cloning. The first step is used to select only those cells that have fused to form a hybridoma; all unfused myeloma and spleen cells are eliminated. Once properly fused hybridoma cells have been selected, they are screened by one of various immunological procedures to identify those hybridomas that are producing antibody molecules of the desired specificity. At this point, the hybridoma cells are usually growing as small colonies, each in a separate well of a culture dish with as many as 96 wells. The final step is to ensure that all cells in the culture are producing the same antibody molecules: monoclonal antibodies with the desired specificity. This is done by diluting the cells from each well so that only one cell from each well can be isolated and deposited in a tissue culture vessel. In culture, this single cell divides, producing clones of identical hybridoma cells. This hybridoma cell line can be used indefinitely to produce its particular antibody. These cells can be stored in liquid nitrogen for further use whenever these antibodies are needed.

Monoclonal antibodies are like "magic bullets" because they go directly to a target cell or tissue when injected in an individual. Since they are produced against specific antigens, they attach only to those cells or tissues that bear these antigens. By tagging them with radioactive tracers monoclonal antibodies can be used to determine the exact location of a cancer or where it has spread to other areas of the body, a process called metastasis. If attached to anti-cancer drugs or toxic molecules, the antibodies can also be used to kill tumor cells without damaging healthy cells.

*(continued)*

How monoclonal antibodies are made. Production of monoclonal antibodies begins with the injection of antigen into a mouse. Antibody-producing spleen lymphocyte cells are removed from the mouse after a few days and mixed with mouse myeloma cells, a type of cancer cell that can readily be grown in tissue culture. The lymphocytes are then fused with the cancer cells to produce a hybridoma, a hybrid cell that proliferates at the high rate of its parent cancer cell and produces the antibody characteristic of its lymphocyte parent. The hybridomas are cultured and then grown in large numbers to form clones. The clones are used to produce large quantities of monoclonal antibody of known specificity.

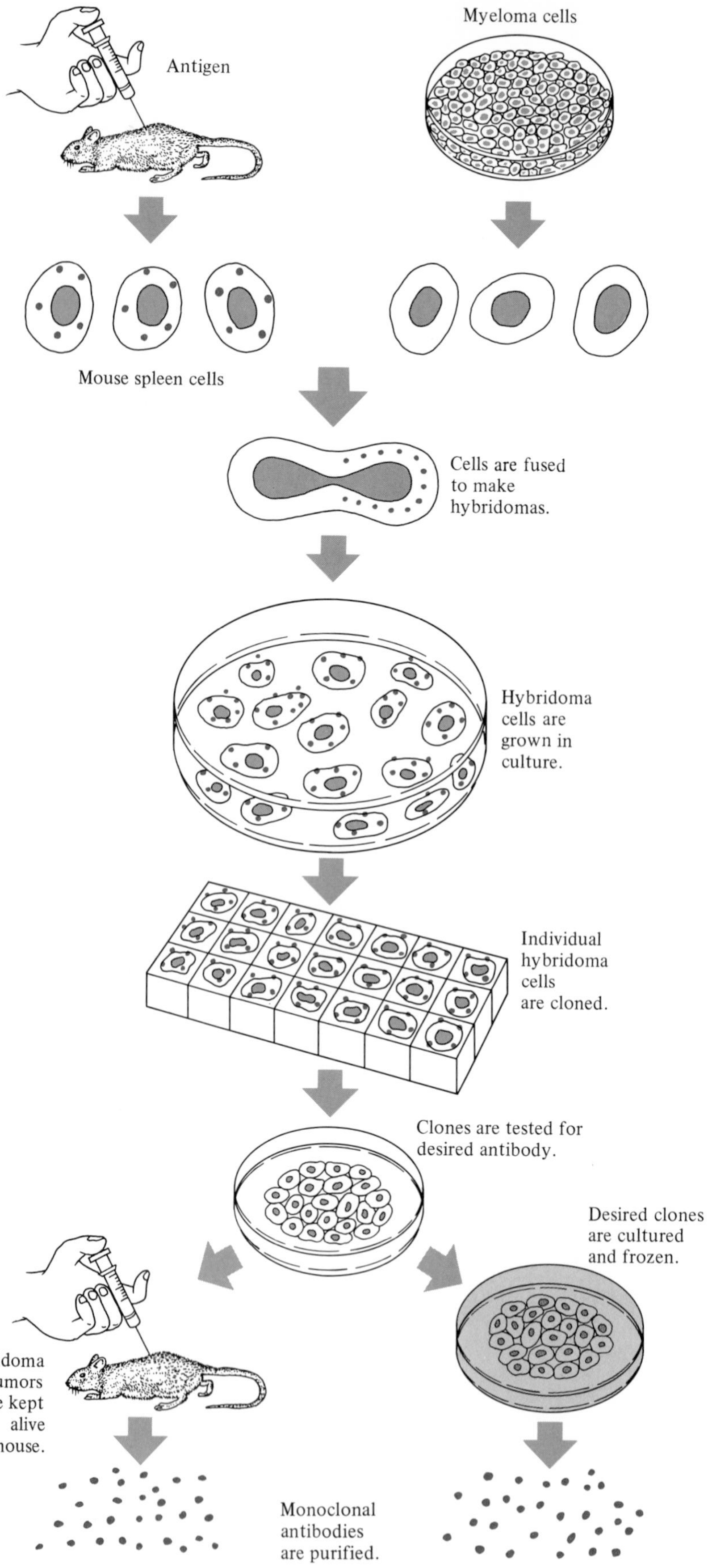
Antigen
Myeloma cells
Mouse spleen cells
Cells are fused to make hybridomas.
Hybridoma cells are grown in culture.
Individual hybridoma cells are cloned.
Clones are tested for desired antibody.
Desired clones are cultured and frozen.
Hybridoma tumors are kept alive in mouse.
Monoclonal antibodies are purified.

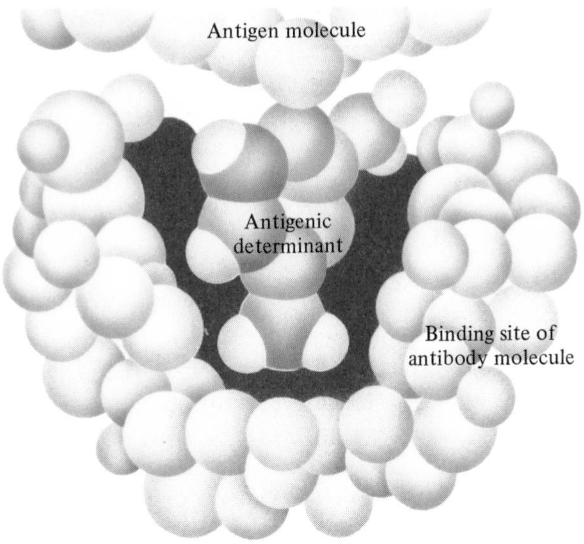


**FIG. 12-19** The binding of an antibody molecule in a lock-and-key fashion to the antigenic determinant on an antigenic molecule.

fashion (Fig. 12-19). Binding is determined by the particular amino acid sequence and the three-dimensional shape of the protein molecule.

In some cases, the combination of an antibody molecule with antigen causes clumping, or agglutination, of cells (Fig. 12-20), and the clumps are engulfed and digested by macrophages. In other cases, this

**FIG. 12-20** (a) Agglutination of human red blood cells. (b) Light micrograph showing the agglutination of human red blood cells. The red blood cells agglutinate as a result of binding to antigenic determinants on the surface of antibody molecules.

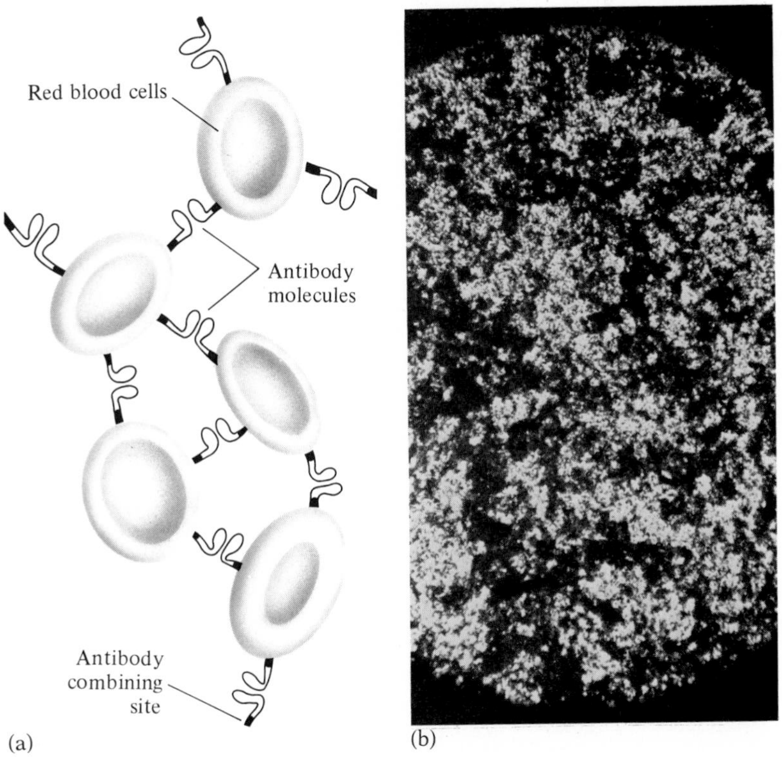

reaction leads to the activation of complement. All such reactions against antigens are the result of the cooperative efforts of B cells and T cells; the T cells act as helper cells for the B cells, which synthesize and secrete the antibodies. The cooperation of both is necessary for the formation of antibodies.

### COMPLEMENT

We have already noted that complement interacts with the constant regions of certain classes of immunoglobulins. **Complement** is a nonspecific defense mechanism that involves at least 20 different protein molecules found in serum. Once complement is activated by the formation of antigen-antibody complexes attached to the surface of cells, breakdown products of some of the complement components are released (Fig. 12-21). This stepwise activation of the complement components is called the **classical complement pathway.** Complement can also be activated by a variety of factors such as bacterial products that do not involve antibody, via what is called the **alternative pathway of complement activation.** When activated, some complement components (C3a, C5b) enhance the phagocytosis by macrophages of the cells to which they are attached. Other complement components (C8 and C9), when activated, attack and make holes in cell membranes, causing the lysis of these cells (Fig. 12-22). Such ruptured cells can then be attacked and devoured by macrophages.

## THE CELL-MEDIATED IMMUNE RESPONSE

We have already seen that T cells help B cells rid the body of foreign molecules by synthesizing antibodies. But some immune reactions do not involve antibodies at all. These reactions are carried on by T cells and are known as cell-mediated immune responses. They also involve cooperation between macrophages and T cells, as in humoral immunity. However, cell-mediated immunity, unlike that resulting from antibodies, cannot be transferred from one individual to another with serum. It can usually be transferred only with sensitized T cells.

The main role of T cells is the recognition and destruction of certain kinds of microorganisms (viruses, fungi, protozoa, and certain pathogenic bacteria) and tissues. These lymphocytes are also involved in the rejection of tissue (blood) and organ (kidney, liver, heart) grafts, in some skin reactions involving contact with irritating substances (such as poison ivy), and in certain reactions to tumors and cancer cells. The surfaces of such cells and tissues contain antigenic markers that T cells recognize as foreign, just as antigenic determinants on molecules trigger the formation of antibodies in the humoral immune response. When the T cells involved in cell-mediated immunity interact with the antigen for which they have a specific cell-surface receptor, they release several kinds of molecules known as **lymphokines,** which have a number of biological effects on other cells (Fig. 12-23). One of these lymphokines, **transfer factor,** can induce normal lymphocytes to destroy other cells, thereby meriting the name cytotoxic, or "killer," lymphocyte. Transfer factor has provided a means of treating people who are unable to produce antibodies to certain bacteria, viruses, or fungi because of some hereditary defect in their cell-mediated immunity. For example,

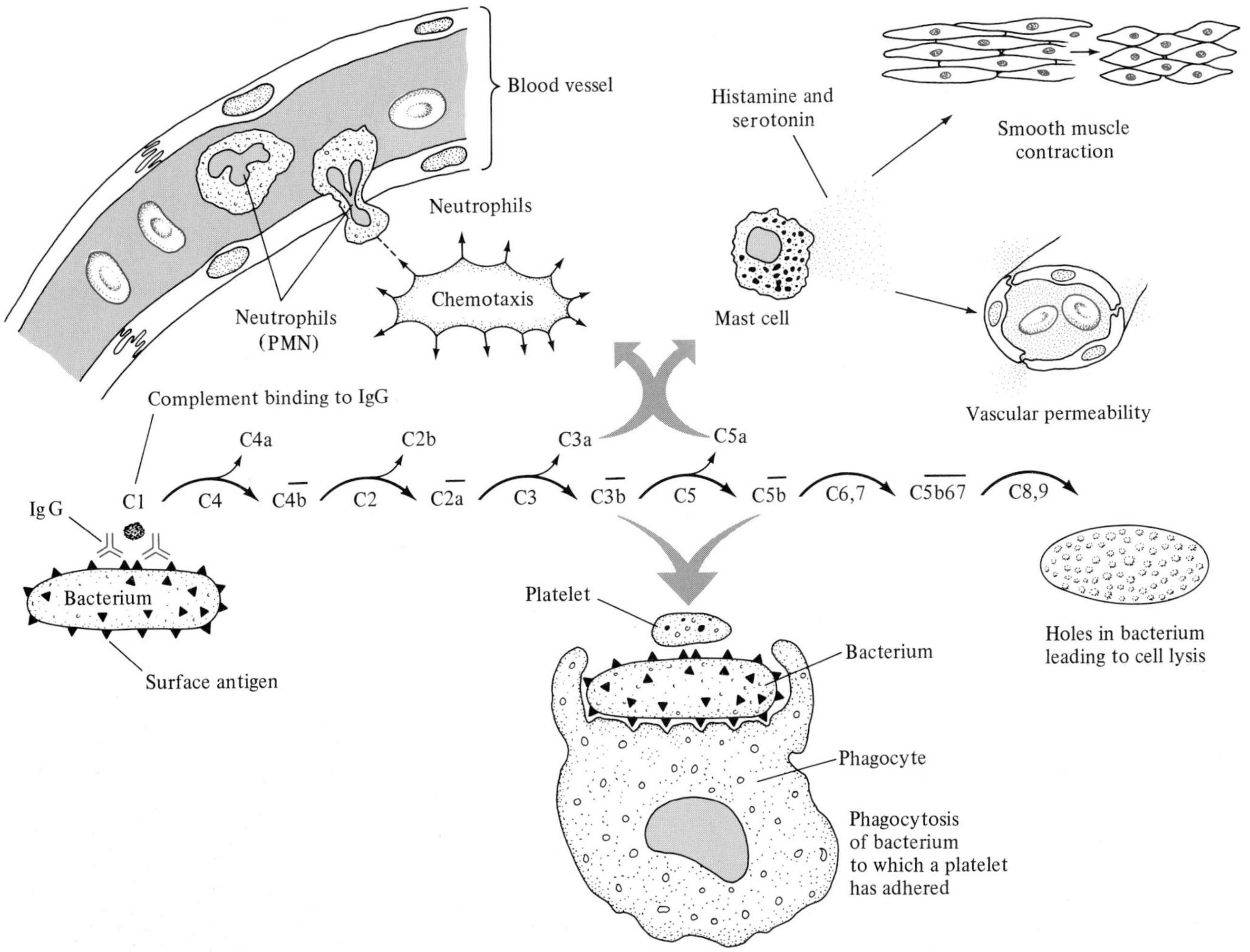


**FIG. 12-21** Activation of the classical complement pathway following the binding of IgG-specific antibodies to the antigenic determinants on the surface of a bacterial cell. Once complement component C1 binds to two adjacent IgG molecules, it becomes enzymatic and splits C4, the next component in the series. As subsequent complement components become involved in the series, biologically active components are split off. Some of these (C3a and C5a) interact with mast cells (a kind of white blood cell containing granules of histamine and serotonin). The mast cells then degranulate, releasing histamine and serotonin, substances that cause smooth muscle contraction and increased vascular permeability. These complement components also attract, through chemotaxis, polymorphonuclear leukocytes, another kind of white blood cell. Polymorphonuclear leukocytes are phagocytic and therefore attack and engulf bacteria. Still other complement breakdown products (C3b and C5b) cause macrophages to adhere to bacteria, which they then phagocytize. Finally, the last two complement components (C8 and C9) cause bacteria to lyse by puncturing bacterial cell membranes.

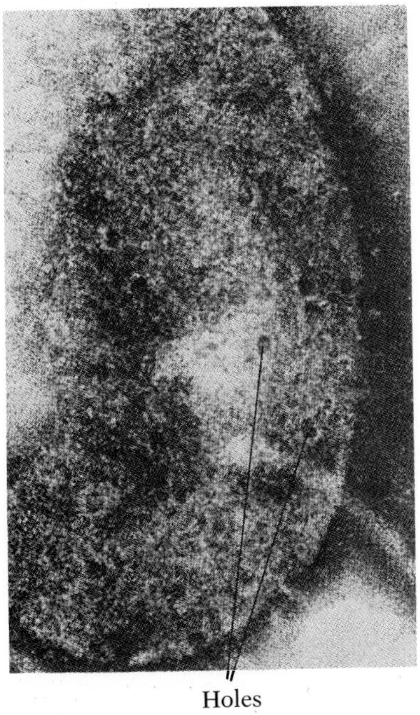


**FIG. 12-22** Electron micrograph showing holes produced by complement in the plasma membrane of human red blood cells.

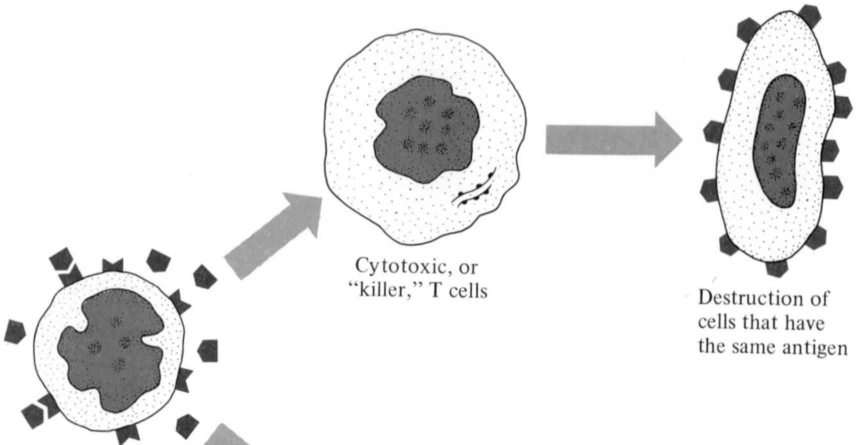

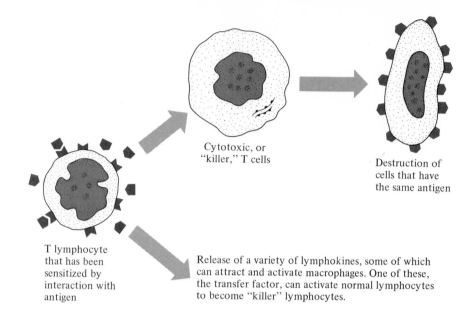


**FIG. 12-23** Cell-mediated immunity involves the differentiation of some T lymphocytes into cytotoxic, or "killer," lymphocytes. Killer lymphocytes release a number of lymphokines having a variety of biologic functions.

candidiasis, the disease found in individuals who are unable to mount an immune response to the pathogenic fungus *Candida albicans,* has been successfully treated with transfer factor.

## TRANSPLANTS AND IMMUNITY

Most of us are aware that the success of kidney, liver, and heart transplants is associated with controlling the rejection of the organs by the immune response of the graft recipient. The most successful of all transplants is a blood transfusion, which involves the transfer of blood from one individual into another (Fig. 12-24).

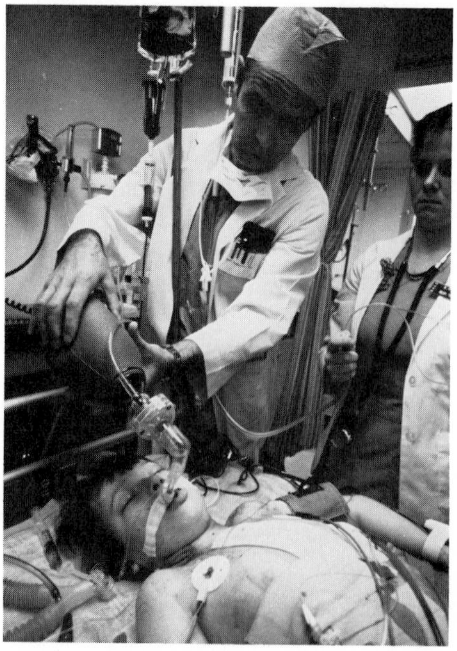

**FIG. 12-24** Patient receiving a blood transfusion. Massive replacement of a person's blood may be required during surgery. Blood must be carefully typed so that it matches that of the recipient.

## THE RELATIONSHIP OF ABO BLOOD GROUPS TO BLOOD TRANSFUSION

The success of blood transfusions is based on the fact that the surfaces of our red blood cells contain special protein molecules called glycoproteins. Because these molecules are antigenic, they can stimulate the production of antibodies against erythrocytes. Such antibodies can combine with the red blood cells, causing them to clump, or agglutinate.

In 1900, the Austrian-born pathologist Karl Landsteiner discovered that there are two major kinds of antigens on the surface of red blood cells. These two antigens were designated A and B. If you have type A blood, the surfaces of your red blood cells contain the A glycoprotein; if you have type B blood, your red blood cells contain the B glycoprotein; if you have type AB, your red blood cells have both antigens; and if you have type O, your blood cells have neither A nor B antigens. Furthermore, human serum contains antibodies against those antigens not present on its red cells. These antibodies are called **isohemagglutinins** because they agglutinate the red blood cells from other individuals of the same species. Thus, type A blood contains isohemagglutinins against type B red cells. Similarly, type B blood contains antibodies against type A; type O contains antibodies against types A and B; and type AB contains neither antibody. These antigens define what is commonly called the ABO blood group system.

Often during surgery or following a severe accident, patients require transfusions of blood to replace the lost blood. A transfusion of the wrong type of blood can cause massive hemagglutination, with severe and even fatal consequences in some cases. The donor's blood—usually 0.5 l per transfusion—is diluted when transfused to the recipient's circulatory system, which can accommodate an average of 4.7 to 5.7 l of blood. Because of this substantial dilution, the isohemagglutinins in the donor's blood cause little agglutination of the recipient's red blood cells. Thus, only the antibodies in the recipient's blood and the type of red cell of the donor are important in transfusions. In practice, the blood types of the donor and recipient are tested, or cross matched, to be sure that their blood is compatible for a transfusion. Thus, a person with type A blood can be safely transfused with either type A or type O blood, and a person with type B blood can receive type B or type O. The red blood cells of type O individuals have neither A nor B antigens and therefore will normally not be agglutinated by the blood of any recipient. People with type AB blood have no isohemagglutinins to type A or B red cells and therefore can be transfused with any blood type (see Table 12-2).

## IMPORTANCE OF RH BLOOD GROUPS TO PREGNANCY

Besides the A and B antigens, a number of other antigens have been identified on the surface of red blood cells. One of the most important of these is the **Rh factor** (named after rhesus monkeys, in whom the antigens were first discovered), which is crucial in pregnancies. The red cells of individuals who are Rh positive contain what are termed D antigens, while the red cells of Rh-negative individuals lack this antigen. A serious problem may occur when an Rh-negative woman and an

**TABLE 12-2 The Human ABO Blood Group**

| *Patient's blood group* | *Antigens on red cells* | *Isohemagglutinins in the blood* | *Blood type that can be received in a transfusion* |
|---|---|---|---|
| A | A | Anti-B | A or O |
| B | B | Anti-A | B or O |
| AB | A and B | Neither Anti-A nor Anti-B | A, B, AB, and O |
| O | None | Anti-A and Anti-B | O |

Rh-positive father have an Rh-positive child. The child produces red cells with D antigen, which gains access to the mother's circulation when the placenta ruptures during childbirth or in some cases during an abortion. Responding to the foreign D antigens, the mother produces anti-D antibodies of the IgG class of immunoglobulins, which are able to cross the placenta. During any subsequent pregnancy, the Rh-negative mother may produce enough anti-D antibodies to agglutinate the red cells of the Rh-positive child in the next pregnancy, producing a disease called **hemolytic disease of the newborn** (Fig. 12-25). This disease is characterized by anemia (reduction in circulating red blood cells) and jaundice (yellowing of the skin from the breakdown products of the red cells).

The most common treatment for the prevention of this problem is to inject the Rh-negative mother with anti-D antibody within 72 hours

**FIG. 12-25** Hemolytic disease of the newborn resulting from the transport of antibodies produced by the mother across the placenta. Such antibodies attach to the red blood cells of the fetus and activate complement, which results in the lysis of these cells.

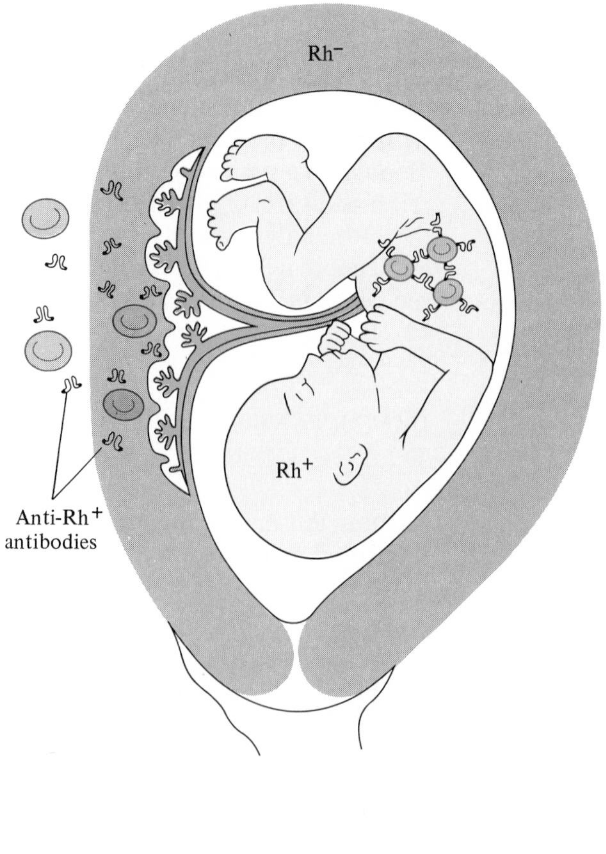

after delivery of the child. These passively administered antibodies prevent the mother from developing antibodies against the D antigen of the Rh-positive red cells that have entered her circulation.

## ORGAN TRANSPLANTS AND IMMUNITY

During World War II, Peter Medawar, an English anatomist, worked as a medical assistant grafting skin to severely burned soldiers. He became intrigued by the fact that strips of skin taken from one individual, the donor, and grafted to a burn patient, the recipient, initially appeared to "take" but would turn red within a few days and then blacken and drop off. In contrast, when skin was grafted from one part of a patient's body to another or from one identical twin to another, the graft healed perfectly. It was "accepted." On the basis of these and further studies, Medawar eventually was able to establish that rejection of grafts was due to an immune reaction.

The fate of a transplant depends less on surgical skill in grafting than on whether the graft is rejected by the recipient's immune system (Fig. 12-26). In recent years, a number of methods have been used to reduce the risk of rejection. Grafts are usually maintained by subjecting the patient to a mixture of immunosuppressive drugs, such as azathioprine (Imuran) and cyclosporin A, which inhibit the immune response elicited by the graft. The drug treatment is usually supplemented with injections of an antiserum against human lymphocytes that is produced by injecting lymphocytes into a horse. Such an antilymphocyte serum (ALS) attacks and destroys the T cells that might be involved in rejection of the graft. The problem with using such immunosuppressive procedures is that they inactivate the total immune defense, necessitating special precautions to avoid serious risks of infection by even the most common bacteria.

The need for immunosuppressive drugs in organ transplants can be overcome to some extent by matching the tissues of the donor and recipient, somewhat as is done in matching blood types in transfusions. The tissues and organs of any individual can be distinguished from those of other individuals by unique cell-surface protein markers known as **histocompatibility antigens,** the human leukocyte antigen, or HLA. When the components of this HLA antigen system have been analyzed in

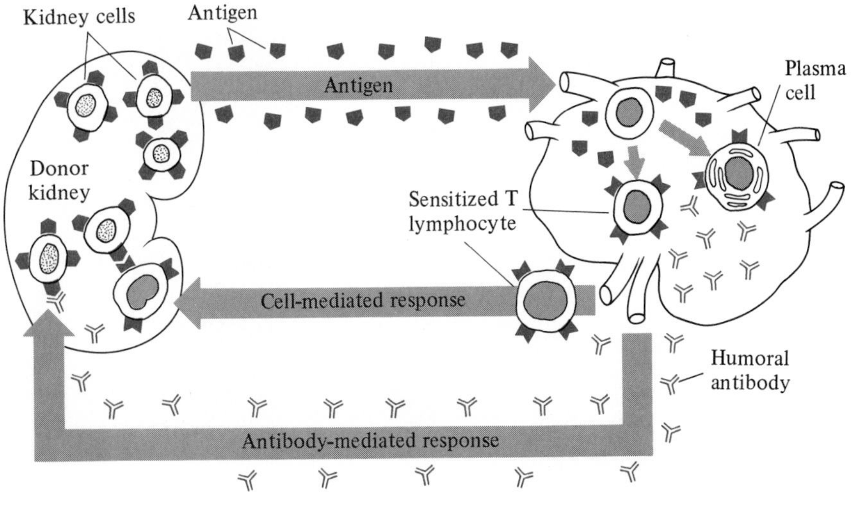


**FIG. 12-26** Mechanisms of kidney graft rejection between individuals who are not genetically identical. Donor kidney antigens react with the recipient's lymph nodes and induce sensitized T lymphocytes and humoral antibodies. These are then carried back to the kidney, where they cause cell- and antibody-mediated destruction of the kidney.

greater detail, it may be possible to transplant organs and tissues much more successfully by properly matching the HLA antigens of the donor and recipient.

## ABNORMALITIES OF THE IMMUNE RESPONSE

At times, the immune system can fail or cause immune reactions detrimental to the host. Such abnormalities may have genetic causes; for example, a person can be born with some kind of immune deficiency. Or the abnormalities may have environmental causes, such as stresses that overload the organism and depress immune functions.

### IMMUNODEFICIENCY

When the immune response is inadequate, the individual is said to be immunodeficient. Immunodeficiency results in a susceptibility to repeated infection with disease organisms. For example, people with immunodeficiencies are unusually susceptible to microbial infections and, in some cases, to cancer (Fig. 12-27). An immunodeficiency may be due to impaired function of B or T cells or both, as well as of other cells involved in immunity.

Experiments with thymectomized and bursectomized animals show that they have impaired T and B cell function, respectively. When a normal level of B and T cells is restored, the immune functions are restored. Some human immunodeficiencies are treated with thymus and bone marrow transplants, which restore their immune capabilities.

One of the most serious immunodeficiencies is **acquired immune deficiency syndrome** (AIDS), described in Panel 12-3.

**FIG. 12-27** This boy was born without an immune system and suffered from severe combined immunodeficiency. Therefore, he was defenseless against infection. The special suit kept the boy isolated from his environment.

PANEL 12-3

# AIDS

One of the most baffling human diseases known to the medical profession is the recently identified acquired immune deficiency syndrome, a disease characterized by a deadly breakdown of the body's immune system. Victims of this disorder are unable to fight infection caused by bacteria and viruses that rarely cause disease in healthy individuals. This mysterious and fatal syndrome came into prominence in 1981, when it was discovered among a number of homosexual males. However, since that time AIDS has appeared in other high-risk groups, including hemophiliacs, intravenous drug users, as well as in healthy individuals not in any of these high-risk categories.

AIDS patients are characterized by a high incidence of cancer, the most common being Kaposi's sarcoma (a malignant tumor associated with the skin, most often on the toes and feet), pneumonia, and a variety of protozoan (e.g., *Pneumocystis carinii*), fungal (*Candida albicans*), and viral (cytomegalovirus [CMV] and Epstein-Barr virus [EBV]) infections. The major cause of AIDS appears to be dysfunction of the patient's cell-mediated immune system, with an imbalance in the ratio of T helper cells to T-suppressor cells (a population of T cells capable of suppressing both humoral and cell-mediated immune responses), two T lymphocytes involved in the regulation of normal immune responses.

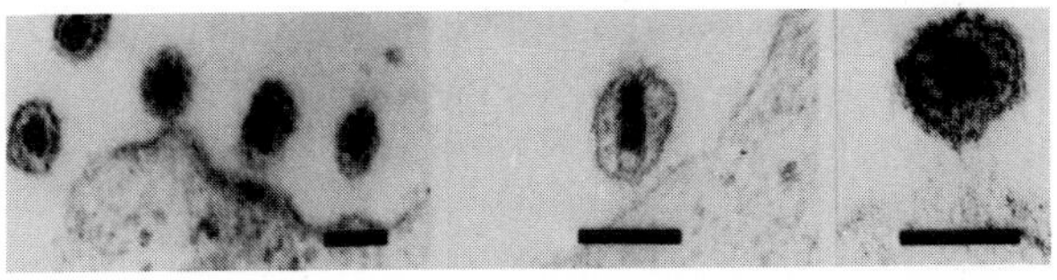

Isolated forms of the HTLV-III virus, a retrovirus, obtained from sera of patients with AIDS. Scale bar, 100 nm.

A virus called the human T-cell lymphotropic virus type III (HTLV-III) has now been identified as the causative agent of this disease. This virus is transmitted in the same manner as hepatitis B, that is, by serum, saliva, and blood or blood products. Other means of contagion remains to be demonstrated.

## HYPERSENSITIVITY

Normally, during the course of our lives, we become sensitized to a number of substances, and these sensitivities are stored as part of the body's immunologic memory. But, whereas immunity confers increased resistance to disease, hypersensitivity makes us more susceptible to potentially dangerous substances.

The numerous reactions we call **allergies** represent one type of hypersensitivity. Common allergies are hives, eczema, hay fever, and asthma. The sensitizing particle is called an **allergen.** Antibody combines with allergen, triggering certain cells in the body to release various substances that have pathogenic effects on the body. When the IgE that is bound to the surface of **mast cells** binds to a molecule from pollen or a particle from dog dander, for example, it activates these cells, a type found in connective tissue throughout the body, or basophils, a type of white blood cell (Fig. 12-28). These cells then release histamine and some related molecules into the blood stream. Histamine causes the contraction of smooth muscle, such as those that make up the bronchioles, and increases the secretion of mucus by nasal and bronchial glands. It also increases the permeability of small blood vessels, producing inflammatory reactions, with burning and itching sensations in the affected parts. The use of antihistamines during the hay fever season is designed to control the effects of histamine.

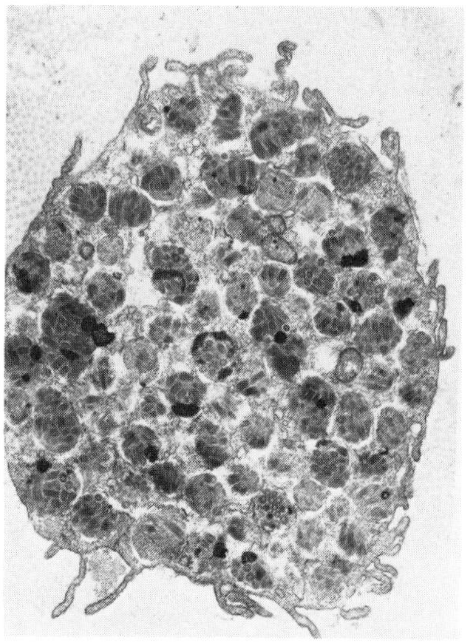

**FIG. 12-28** A human mast cell from the lung showing a large number of granular inclusions. These granules release a number of biologically active molecules following the binding of antigen (allergen) to IgE molecules attached to their surfaces. One of these molecules is histamine, which causes some of the symptoms associated with allergies.

## ANAPHYLAXIS

Hypersensitivity can induce a reaction so violent as to be fatal. In such cases, the whole organism is affected to such a degree that it requires emergency treatment. **Anaphylaxis** is a violent allergic reaction that some individuals experience to certain antigens, for example, after being stung by a bee or a wasp or after a penicillin injection. The skin becomes red and itchy, the eyes tear, the chest feels heavy, the throat becomes constricted, breathing becomes difficult because of congested lungs, and there is a sharp drop in blood pressure. These reactions are brought about by a dilation of blood vessels, increased capillary permeability, and contraction of smooth muscles. Unless something is done quickly to stop or reverse these reactions, such as the injection of adrenaline, the patient may die of asphyxiation caused by the marked constriction of the bronchioles.

Anaphylaxis is an immediate reaction, taking place in a matter of minutes. In sensitized individuals—those who are allergic to a particular allergen—IgE antibodies specific for that allergen are attached to the surface of the tissue mast cells and circulating basophils. When such individuals are exposed to these allergens, the allergens become bound to the IgE molecules and transmit a signal to the mast cells and basophils that causes them to release histamine and other molecules that cause the various signs of anaphylaxis. Because of the serious effects of such reactions, people with known hypersensitivities should wear Medic Alert badges, which identify any substances to which they have become sensitized (Fig. 12-29).

## AUTOIMMUNITY

Sometimes the immune system attacks the body's own cells and tissues as if they were foreign. For some reason, the body fails to recognize itself and reacts as if it were foreign. The body begins to manufacture antibodies to certain of its own antigens, creating a sort of biological civil war, often with dire consequences. This is known as an **autoimmune response.** Several theories have been offered as possible explanations. One possibility is that a normal body antigen may change as a result of an alteration in the tissue that produces it. One cause of such a change may be a somatic mutation, a change in the genetic material of a tissue's cells that is reflected in an alteration in the structure of one or more of the antigens synthesized by that cell. Another possibility is that injury or infection may release into the bloodstream antigenic materials that previously had been isolated from the immune system and therefore had never become recognized as "self." Bacterial infections sometimes precipitate an autoimmune syndrome. One called Bright's disease is briefly discussed in Chapter 13. Some autoimmune diseases can be treated, but their treatment is difficult.

**FIG. 12-29** Medic Alert badge worn by people with known hypersensitivities to specific foreign substances.

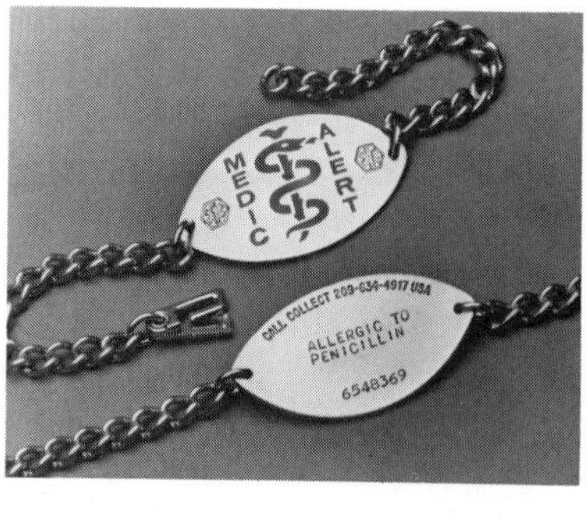


## CANCER

One of the functions of the immune system is the recognition and elimination of abnormal body cells that arise as the result of mutations, such as cancer cells. A **cancer cell** is one whose unusual growth pattern leads to the formation of a mass of similar cells, or a **tumor.** These cells appear to have the potential for unlimited growth, eventually forming large tumors within the body. Cancer cells may arise spontaneously or may be induced to develop by certain kinds of viruses (see Panel 24-4, p. 666), mutation-causing (mutagenic) chemicals (carcinogens), or radiation. Under normal circumstances, the immune system would recognize these mutant cells as foreign, or not-self, and would eliminate them, a process known as **immune surveillance.** But at times this process may fail, for reasons not well understood, leading to the unchecked growth of these cells and to the development of certain types of cancer.

Clinical evidence gathered from the use of immunosuppressive procedures to prevent the rejection of transplanted tissues and organs has provided important clues to the relationship between immunity and cancer. For example, undetected cancers in some patients undergoing organ transplants may begin to grow after the patients receive immunosuppressive drugs. Once immunosuppressive therapy is discontinued, the cancers begin to get smaller, or regress.

Investigators have attempted to harness the body's own immune system to fight cancer. By artificially stimulating the immune system, it can have a better chance of destroying tumors. The great advantage of such immunotherapy in treating cancer is that, when it works, it destroys every last cancer cell, something that none of the other modes of treatment usually can do. Unlike chemotherapy (destruction of cancer cells by chemicals) and radiotherapy (destruction by irradiation), immunotherapy does not destroy normal cells in the body. The only targets are the cancer cells. However, when the tumor mass is quite large, it is often necessary to reduce it by surgery, chemotherapy, or radiotherapy before immunotherapy can be used effectively.

Just as important as is the control of established cancers with immunotherapy are the attempts to develop immunological methods for their diagnosis and prevention. In the area of diagnosis, research is being carried out to identify and characterize tumor antigens specific for cancer cells. This has proven to be a difficult task since there are so many kinds of tumors. Nevertheless, researchers have been able to identify some specific tumor antigens. If such antigens can be isolated and purified, it may even be possible to immunize (vaccinate) humans against some types of cancer.

## Summary

Immunity in vertebrates includes all the physiologic mechanisms that give an animal's body the ability to recognize foreign substances and to neutralize or degrade them without injuring its own tissues. The immune system not only functions to defend against invasion by foreign microorganisms but removes damaged or dying cells and recognizes and eliminates mutant cells that constantly arise within the body.

In birds and mammals, the immune response depends on a specialized population of white blood cells, the lymphocytes, which are contained largely within a system of lymphoid organs, including the spleen, lymph nodes, Peyer's patches, bone marrow, and thymus. Lymphocytes are of two types, depending on where they originate during development and on their subsequent functional characteristics. One type, the T cells, originates in the bone marrow as stem cells and, after passing through the thymus, becomes localized in specialized areas of the various lymphoid organs and tissues. The B cells also originate as bone marrow stem cells. B cells depend on the bursa of Fabricius in the chicken and the bone marrow and liver in man for their maturation to immunocompetent cells. B cells also localize in the major lymphoid organs and tissues. The B cells are primarily responsible for the production of antibody molecules, which circulate through the blood and lymph, whereas the T cells are associated with cell-mediated immunity.

Antigens are foreign substances that, when introduced into an organism, will induce an immune response consisting of the production of circulating antibodies (humoral immunity) or cell-mediated immunity. Antigens react with the antibodies or the sensitized lymphocytes produced in response to their presence.

Antibodies, or immunoglobulins, are large, complex protein molecules. In humans, there are five classes of immunoglobulins, each with specific structural and functional characteristics.

A variety of human diseases are associated with various forms of deficiency of the immune system. Some of these immunodeficiency diseases are associated with abnormal development of B cells, T cells, or both. A variety of procedures, including bone marrow and thymus transplants, have been used to correct such immunodeficiencies, with varying degrees of success.

Hypersensitive immune systems are responsible for reactions that are harmful to the individual. Among the best-known hypersensitivity reactions are the allergic reactions, which occur when certain kinds of antigens, such as pollen, react with IgE molecules bound to histamine-producing mast cells in sensitized individuals.

Some of the most baffling and destructive human diseases are caused when the immune system attacks the body's own cells and tissues (autoimmunity). The rejection of transplanted organs and tissues is also due to an immune response to cell-surface antigens that are recognized by the immune system as foreign to the body. In order to avoid graft rejection, various drugs and antisera are used to suppress the immune response of the graft recipient. Immunosuppression has many harmful side effects, however, and attempts are being made to match more closely the antigens of the donor and recipient.

The immune system is also important in the defense against some cancers. Many types of cancer are believed to arise through a failure of the normal immune surveillance that usually eliminates mutant cells as they develop. Some procedures have been useful in enhancing this immune surveillance system.

## Recommended Readings

BURNET, F. M., ed. 1976. *Immunology.* Freeman, San Francisco. A collection of well-written and beautifully illustrated *Scientific American* articles dealing with most of the topics covered in this chapter.

BURNET, F. M. 1976. *Immunology, Aging and Cancer.* Freeman, San Francisco. Written by a Nobel laureate, this paperback is a stimulating discussion of the immune response and its relationship to aging and cancer.

COLLIER, R. J., and D. A. KOPLAN. 1984. Immunotoxins. *Scientific American,* 251(1):56–64. A toxic agent linked to a monoclonal antibody and binding to a tumor antigen forms a "magic bullet" against cancer cells.

EDELSON, R. L., and J. M. FINK. 1985. The immunologic function of the skin. *Scientific American,* 252(6):46–53. An account of how specialized epidermal cells present foreign antigens to skin lymphocytes.

KIMBALL, J. W. 1983. *Introduction to Immunology.* Macmillan, New York. A comprehensive and well illustrated treatment of the basic principles of immunobiology.

ROITT, I. M., J. BROSTOFF and D. K. MALE. 1985. *Immunology.* This book has excellent diagrams, is eminently readable, and is suitable as an introduction to the important principles of immunology as they relate to health and disease.

# 13 Excretion, Osmoregulation, and Ionic Regulation

OUTLINE



Excretion, the elimination of useless or toxic metabolic wastes, is essential for all organisms. Two of the metabolic wastes produced in abundance by heterotrophs are carbon dioxide, formed in respiration, and ammonia ($NH_3$), resulting from protein and nucleic acid metabolism. These same metabolites are also produced in photosynthetic organisms, in which they are recycled and almost never occur in excess. In plants, carbon dioxide and ammonia are often limiting factors in photosynthesis or growth, rather than being waste products. Only in animals is excretion of $CO_2$ (discussed in Chapter 8) and $NH_3$ critical. This chapter deals with excretion and **osmoregulation,** the control of the body's water content, in animals.

Because metabolic wastes typically are excreted in aqueous solution, the process often leads to the loss of much water and valuable solutes, particularly inorganic ions. Excretion, osmoregulation, and ionic regulation thus tend to be intimately interrelated in many animals, often being conducted by the same organs. Indeed, an organ such as the kidney may be primarily an excretory organ in one animal and an organ of osmoregulation in another. Moreover, depending on environmental conditions, such an organ may even switch primary roles from time to time in the same animal.

Animals need to engage in osmoregulation and ionic regulation not only because water and salts are lost in the course of excretion of wastes, but also because the water in an animal's environment is of a different ionic composition than its body fluids. The need for osmoregulation is most important in such aquatic animals as freshwater fish, which are continually flooded with osmotically absorbed water, or in such animals as marine fish and desert animals, which need to replenish the water lost by osmosis or evaporation. Osmoregulation is not a problem for marine invertebrates, however, because their body fluids are isotonic to seawater—that is, they neither gain nor lose water by osmosis.

All animals need to monitor and adjust their content of inorganic ions. This is true even of marine invertebrates, whose body fluids, although isotonic to seawater, differ from it in ionic composition. One important aspect of ionic regulation is the maintenance of the body's acid-base balance by keeping the pH of body fluids within the narrow range compatible with life, for example, between 7.0 and 7.8 for mammals. Precise acid-base regulation is needed largely because the rate of activity of most enzymes is sharply altered by small changes in pH. Enzyme activity, of course, is the basis of nearly all biological activity.

Because the processes of excretion, osmoregulation, and ionic regulation are so closely interrelated, they can be viewed as three aspects of a single major homeostatic process.

## WASTE PRODUCTS OF NITROGEN METABOLISM

The two main metabolic sources of nitrogenous wastes are proteins and nucleic acids. Proteins usually contribute over 90 percent and nucleic acids about 5 percent of the total amount of excreted nitrogen. Excretion is mostly a matter of getting rid of the ammonia that forms when amino groups ($—NH_2$) are split from amino acids derived from metabolized protein or from excess dietary protein. Ammonia is very toxic and is kept in very low concentration, with only traces occurring in the blood of most mammals (0.001 to 0.003 mg/100 ml). Higher concentrations can be lethal. For example, a rabbit will die if the $NH_3$ level of its blood reaches 5 mg/100 ml, or 0.005 percent.

Most higher large animals convert ammonia to a less toxic form soon after it is produced, thus allowing it to be retained until it can be excreted. An exception is the teleost fish, which excretes nearly all its nitrogen as ammonia, mostly via its gills but to some extent in its copious urine. The excretion of most of the nitrogen as ammonia is called **ammonotelic excretion.** Most aquatic invertebrates, such as sponges and flatworms, also belong to the category of animals with ammonotelic excretion. Because so many of the cells of such animals come in contact with their watery external environment, the small ammonia molecules readily diffuse into the external medium and never accumulate in the body to toxic concentrations.

Rather than being surrounded by an overabundance of water, most mammals need to drink water or eat food containing it. Thus, they need to conserve water. Mammals produce only a moderate amount of urine, and most of the ammonia formed from amino acid metabolism is excreted as urea, which is less toxic than ammonia (Fig. 13-1). The excre-

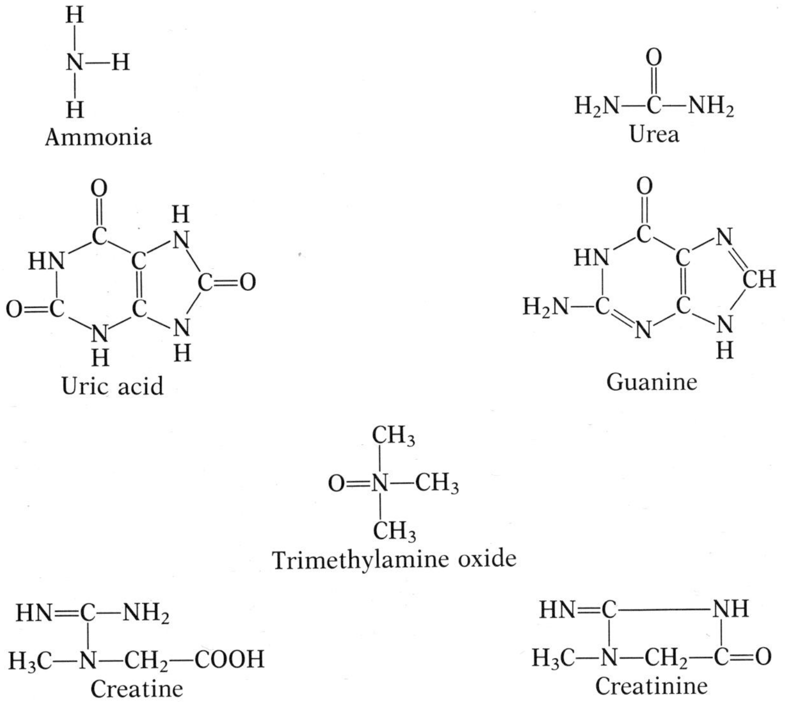


**FIG. 13-1** Some commonly excreted nitrogenous compounds.

tion of most of the nitrogen as urea is known as **ureotelic excretion.**

In a third category are animals that, because they inhabit dry environments or for other reasons of water economy, secrete a minimal amount of urine that contains little or no water. To this group belong the land snails, insects, snakes, lizards, and birds. These animals usually convert ammonia to uric acid (Fig. 13-1) or other highly insoluble substances. Because the substances are insoluble, only small amounts are retained in solution, thus greatly limiting their toxicity.* Most such **uricotelic** animals excrete their waste protein nitrogen as a solid or semisolid urine of uric acid crystals.

## CHEMISTRY OF NITROGEN EXCRETION

Ammonia, urea, and uric acid are the most common but not the only excreted end products of nitrogen metabolism. For example, spiders and scorpions excrete guanine, and some marine teleosts (bony fish) excrete trimethylamine oxide (TMO) (Fig. 13-1). Creatinine and creatine are also excreted in small quantities by a variety of animals (Fig. 13-1). Some animals even wastefully excrete some of their excess amino acids as such.

The formulas for some common nitrogenous excretory products are given in Figure 13-1. Because urea, uric acid, and TMO are less toxic than ammonia, it might be wondered why all animals do not excrete most of their protein nitrogen in these less poisonous forms. The answer is a matter of biological economics. The synthesis of these materials from ammonia involves expenditure of energy. For example, conversion of ammonia to urea requires three ATP molecules. Any organism using energy for an unnecessary function would tend to be selected against

* For any substance, toxic or not, to achieve a biological effect, it usually must be in solution.

during evolution. Therefore, an organism usually excretes its nitrogen in the form requiring the least expenditure of energy, given the environment in which it lives.

This rule does not apply as firmly to the 5 percent of an animal's excreted nitrogen that comes from the metabolism of the purine (adenine and guanine) fraction of its nucleic acids.* Purines can be broken down to ammonia only if the animal has the necessary enzymes, which are present in only a few aquatic invertebrates (Fig. 13-2). Most other species excrete purine nitrogen as uric acid or as one or more of the intermediate products in this chain. As indicated in Figure 13-3, some animals excrete both amino acid nitrogen and purine nitrogen in the same form, but mammals are an important exception. Urea is produced from $NH_3$ by a series of enzymatically catalyzed steps that occur in the liver of vertebrates that excrete urea; the enzymes involved are generally absent from the liver of animals that excrete uric acid.

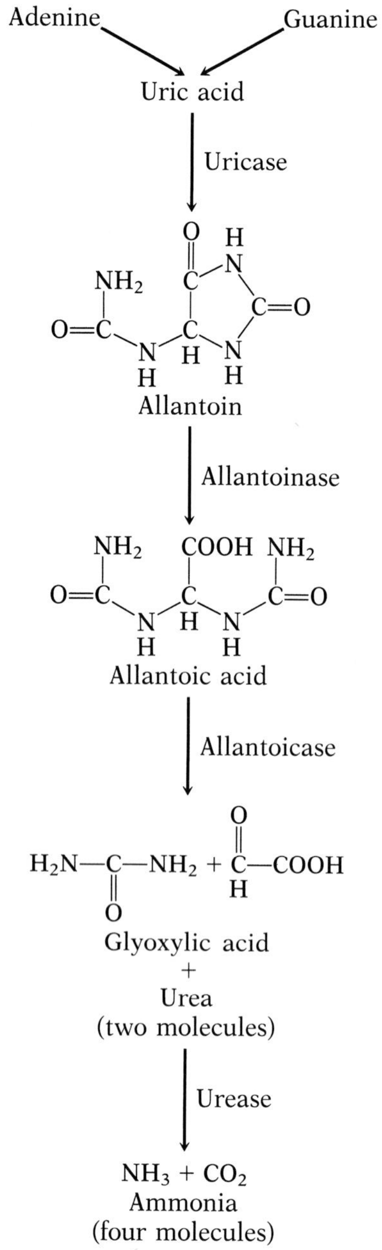


**FIG. 13-2** Intermediates in the metabolism of uric acid to ammonia, and the enzymes involved in this conversion.

# EVOLUTION OF THE KIDNEY

One of the most interesting evolutionary histories is that of the kidney. The excretory organs of primitive worms are one of two general types of tubules: the coelomoduct and the nephridium (from the Greek *nephros*, "kidney," and *idion*, "small"). The **coelomoduct** arises as an outward pocket, or evagination, of the body cavity, while the **nephridium** arises as an inward pocket, or invagination, of the body wall. The primitive nephridium has no opening into the body cavity and ends blindly in many simple structures called flame cells, which may float freely in the body cavity or become embedded in the tissues. Such organs are called **protonephridia** (Fig. 13-4). The rhythmic beating of the cilia in the lumen of the flame cells and solenocytes creates a gentle current that draws fluid from surrounding tissues or from the coelom, in which the flame cells float. The function of protonephridia is primarily osmoregulatory, as evidenced by the fact that, while they are a prominent feature of many modern freshwater worms, they are absent or much less abundant in marine species of the same genus.†

**Metanephridia** are a more advanced type of nephridium; they open into the coelomic cavity. A good example of a metanephridium is found in the common earthworm. As Figure 13-5 illustrates, this excretory organ has a ciliated funnel-shaped inner opening, the **nephrostome,** through which coelomic fluid passes down a long, richly vascularized tubule, various regions of which differ markedly in their cellular composition, that is, histology, and function. The useful substances in this fluid are reabsorbed by the tubular wall. Toxic and useless substances are allowed to remain and are thus excreted. The tubule finally expands into a widened bladder, which empties to the exterior by a tiny **nephridiopore.**

* Nucleic acids also contain another class of nitrogenous bases, the pyrimidines (thymine, cytosine, uracil). However, pyrimidines are completely degraded to $CO_2$, $H_2O$, and $NH_3$, the ammonia being excreted or detoxified in the same manner as that derived from amino acid metabolism.

† An analogous situation exists in freshwater Protozoa and sponges; these organisms possess water expulsion vesicles called **contractile vacuoles,** which function primarily in osmoregulation. Marine and parasitic species of Protozoa either lack these organelles altogether or employ them only to expel water that has been ingested with food. Marine sponges also lack them.

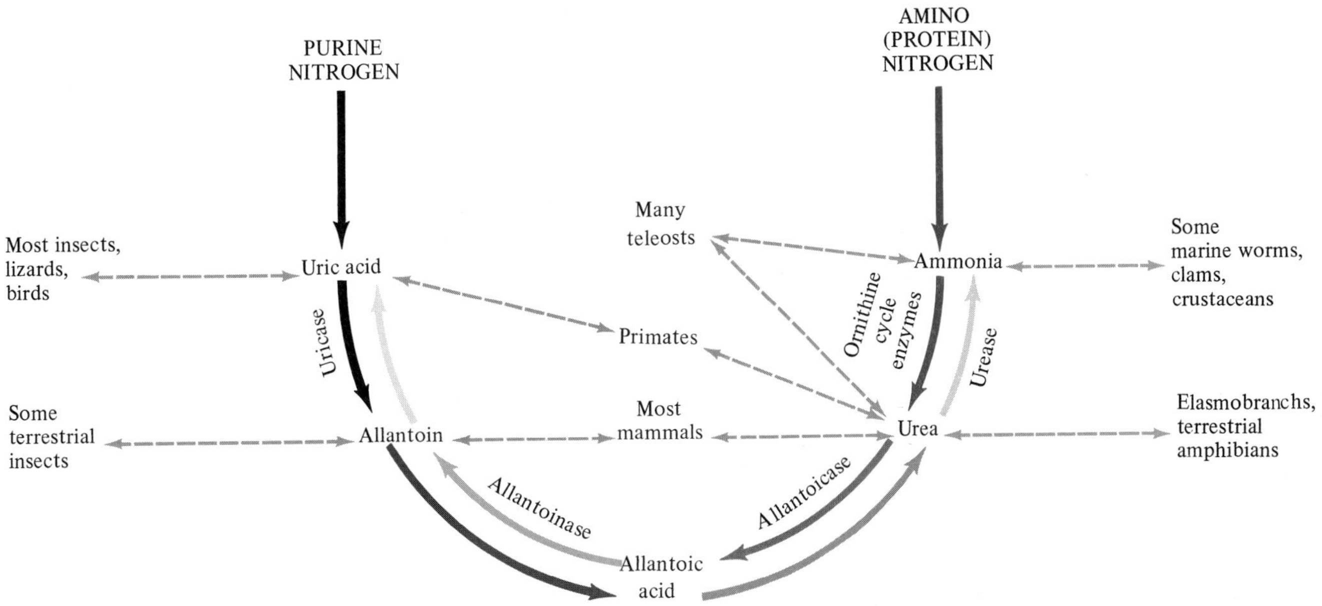


**FIG. 13-3** The end products of protein and purine nitrogen metabolism in various groups of animals. Note both protein and purine nitrogen tend to be excreted in the same end product.

Most invertebrates have excretory organs derived from the ancient nephridia or coelomoducts. The so-called nephridia of molluscs and noninsectan arthropods have evolved from coelomoducts. Examples of the arthropod "nephridium," the so-called antennal, or green, gland of the crayfish, and of molluscan "nephridia" are shown in Figure 13-6.

The coelomoducts and nephridia of many modern invertebrates are the products of various evolutionary modifications. The major modification is the fusion of parts of each of the ducts. In some, coelomoducts have fused with protonephridial tubules. In others, coelomoducts have fused with modified metanephridia. Thus, these organs evolved as hybrids of the two primitive structures, coelomoducts and nephridia.

Besides the organs derived from nephridia or coelomoducts, several types of excretory organs have independently evolved in invertebrates. Among the most studied and interesting are the **Malpighian tubules** of insects and a few other arthropods. All have closed inner ends, as in the typical example shown in Figure 13-7. The tubules, which number up to 250 in some species, are suspended in the hemolymph, and their open ends empty into the hind gut or rectum.

**FIG. 13-4** The protonephridial type of excretory organs in a planarian, a free-living flatworm.

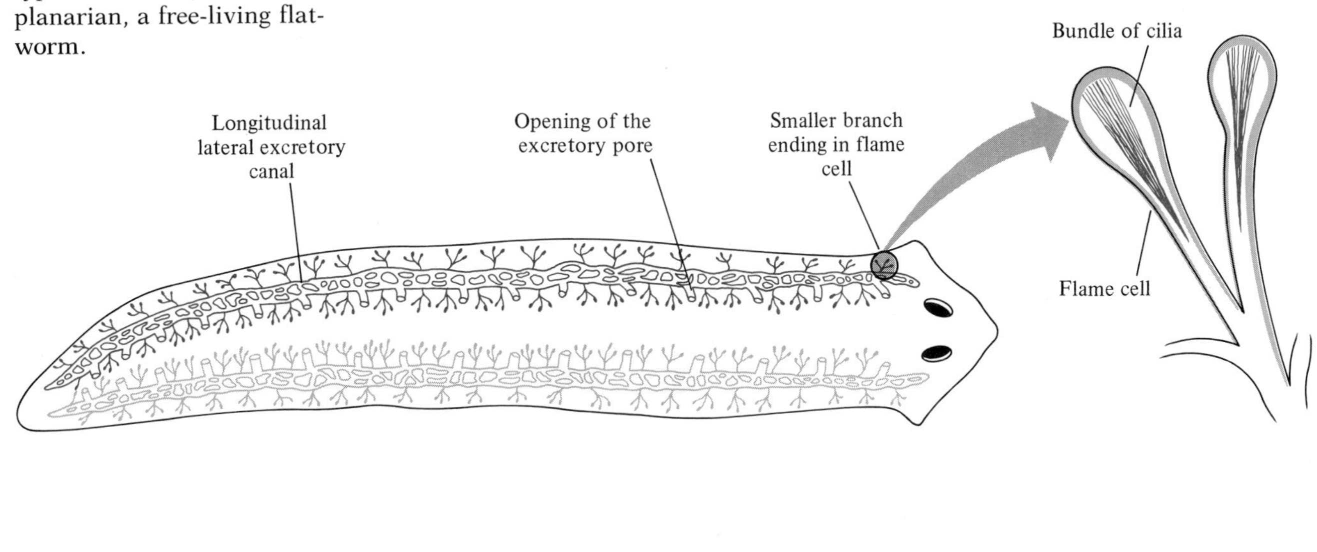

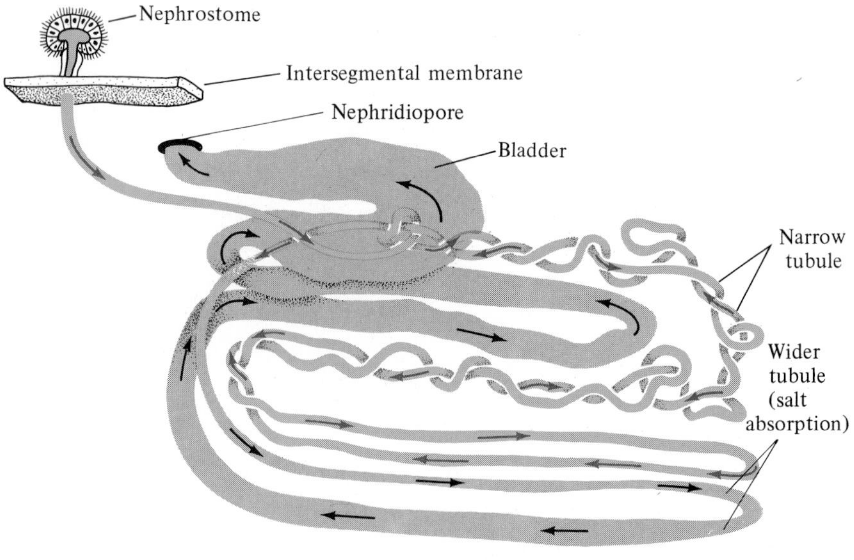


**FIG. 13-5** The metanephridium of the common earthworm.

## ORIGIN OF THE VERTEBRATE KIDNEY

The first vertebrates were jawless fishes of a type now long extinct. These hypothetical ancestral forms, referred to as protovertebrates, probably possessed a jawless mouth, pharyngeal gills, and a tiny brain and nerve cord; and they were distinctly segmented in several respects. They had a cartilaginous rod that served as a backbone. Each segment of the primitive vertebrate body probably had pairs of nerves, reproductive organs, and coelomoducts. These primitive ancestors of modern vertebrates probably lived in brackish waters, a mixture of salt water and fresh water.

***Elimination of Water in Protovertebrates.*** According to this hypothetical model, the coelomoducts of the protovertebrates would have provided a convenient route for elimination of excess fluid that, in the brackish water of estuaries, entered osmotically through the membranes of the gills and mouth (Fig. 13-8a). In time, evolutionary modifications of the coelomoduct's tubular epithelium permitted the reabsorption of valuable salts from the tubule fluid, thus improving the efficiency of the first protovertebrates, that is, fish having a notochord,

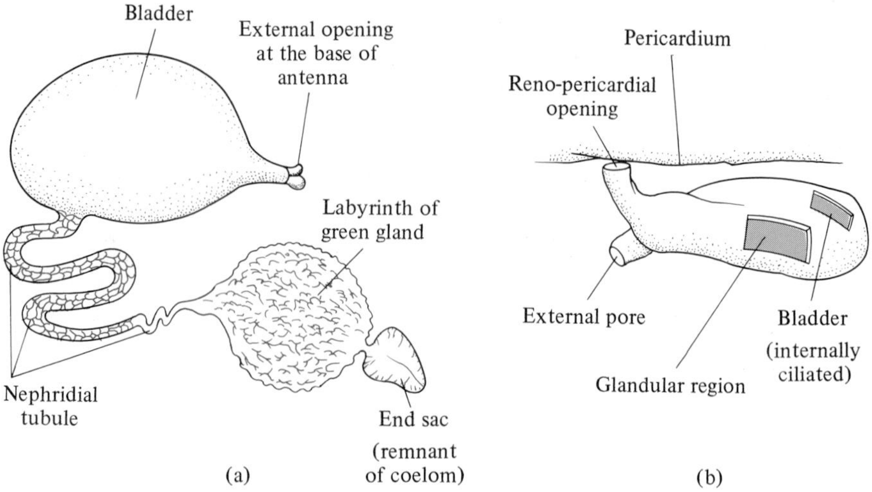


**FIG. 13-6** Excretory organs of a crustacean and of a mollusc. (a) The green or antennal gland of a crustacean. (b) The nephridium of a mollusc. These organs, like the vertebrate nephron, are considered to have evolved from the coelomoducts.

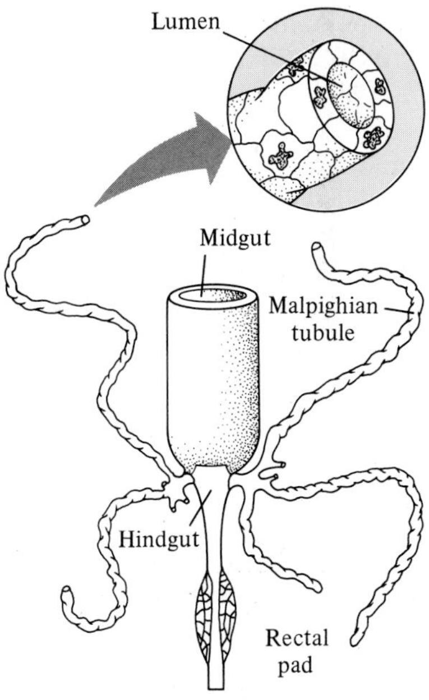


**FIG. 13-7** The Malpighian tubules of a grasshopper.

or primitive spinal column, in coping with the dilutions of seawater found in estuaries. Some modifications of the tubular epithelium might even have allowed the secretion of waste materials into the tubule.

***The Role of the Circulatory System in Osmoregulation.*** Because protovertebrate excretion depends on simple diffusion, there is a limit to the rate at which a system such as this can eliminate osmotically absorbed water. A way to improve the efficiency of osmoregulation was needed, and evolutionary modification of excretory organs allowed vertebrates not only to leave the estuaries and exploit the resources of freshwater habitats but eventually to colonize the land as well. The modification involved the positioning of a network of capillaries close to the mouth of each coelomoduct, the coelomostome (Fig. 13-8b). This change made a higher rate of elimination possible. Blood pressure within the capillaries produces a filtrate of the blood. Through a tuft of capillaries next to the entrance to the coelomoduct, the pumping action of the heart produced a filtrate of the blood that was free of blood cells and most plasma proteins and that was delivered directly to the tubule (Fig. 13-8c). Similar capillary tufts develop in the embryonic or kidney tubules of all modern vertebrates. The coelomostome is secondarily lost

**FIG. 13-8** Hypothetical stages in the evolution of the vertebrate nephron. (a) The coelomoduct of the hypothetical protovertebrates. (b) The coelomoduct with a capillary network associated with the coelomostome. (c) The coelomoduct with a persistent coelomostome and the addition of a glomerulus in a Bowman's capsule. (d) The modern nephron with no coelomostome but with a Malpighian corpuscle.

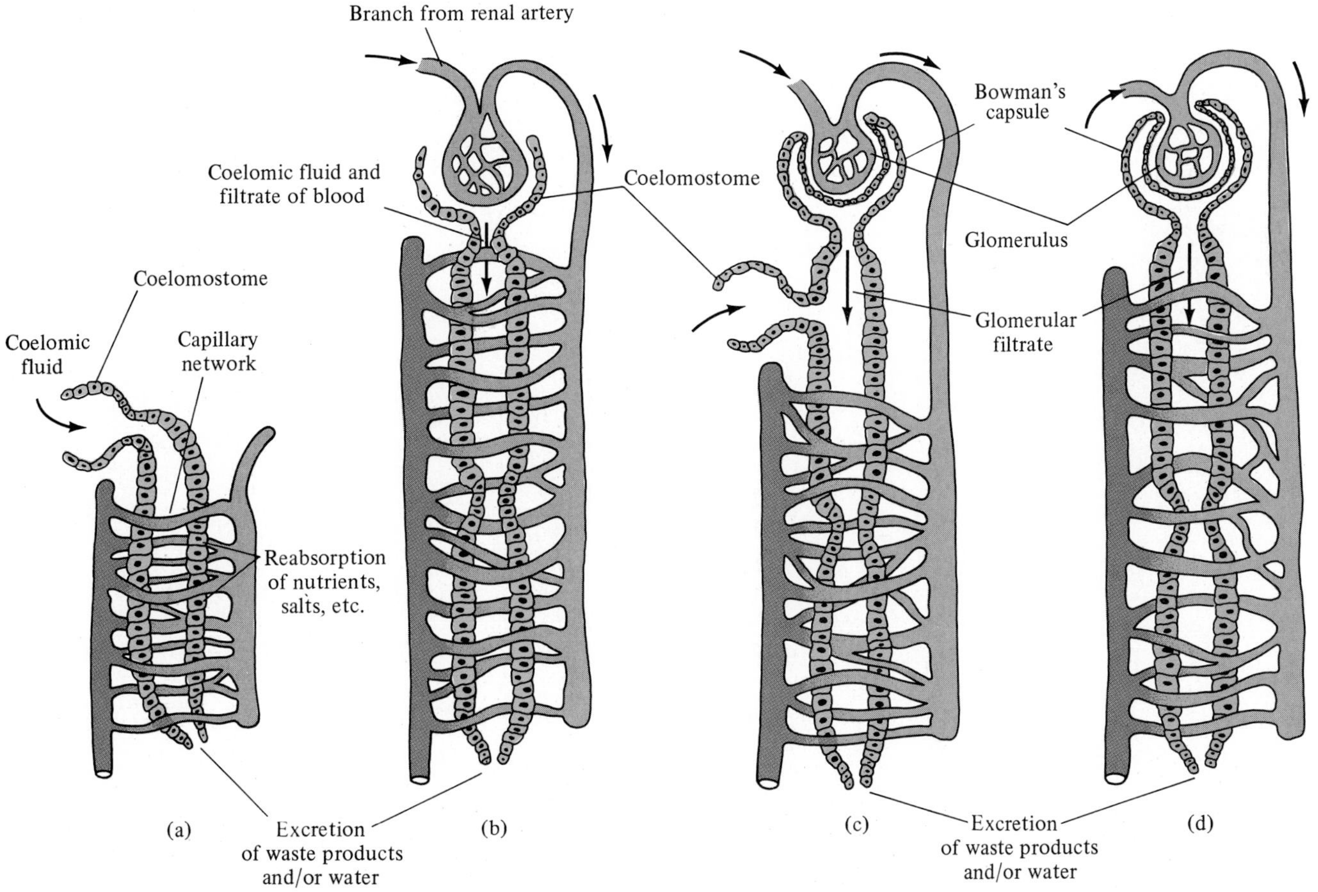

and the capillary tuft, termed the **glomerulus** (from the Greek for "little ball"), is inserted into a blind end of the tubule, further improving the vertebrate kidney's efficiency (Fig. 13-8d). The enlarged end of the nephric tubule that contains the glomerulus is called the **Bowman's capsule.** The kidney of higher vertebrates is a compact cluster of a large number of tubules, each with its own capillary tuft. These kidney tubules communicate with the exterior by a common duct, the **ureter.**

## STRUCTURE OF THE NEPHRON

The evolution of the functional unit of the vertebrate kidney, the **nephron,** is directly related to the demands placed upon various vertebrates by the environments in which they lived. The nephron's basic components are the glomerulus, located in the Bowman's capsule, and a **renal tubule** (Fig. 13-9). The glomerulus is a specialized network of capillaries that arise from a branch of the renal artery called the afferent renal arteriole. The capillaries rejoin to form the efferent renal arteriole, which supplies blood to the renal tubule. Blood filtered through the thin walls of the glomerular capillaries provides a cell-free ultrafiltrate that lacks most of the plasma proteins. Surrounding the glomerulus is the enlarged end of the renal tubule, termed the Bowman's capsule, which receives the filtrate. The renal tubule is a highly specialized structure that not only reclaims most of the water and solutes contained in the filtrate but also secretes additional materials into the tubule fluid.

Structural differences in the glomerulus and tubule among various species of vertebrates are generally related to whether the animals need to eliminate large or moderate amounts of water or to greatly conserve it. In all vertebrates below the mammals, the tubule usually has only

**FIG. 13-9** Sketch of a typical mammalian nephron.

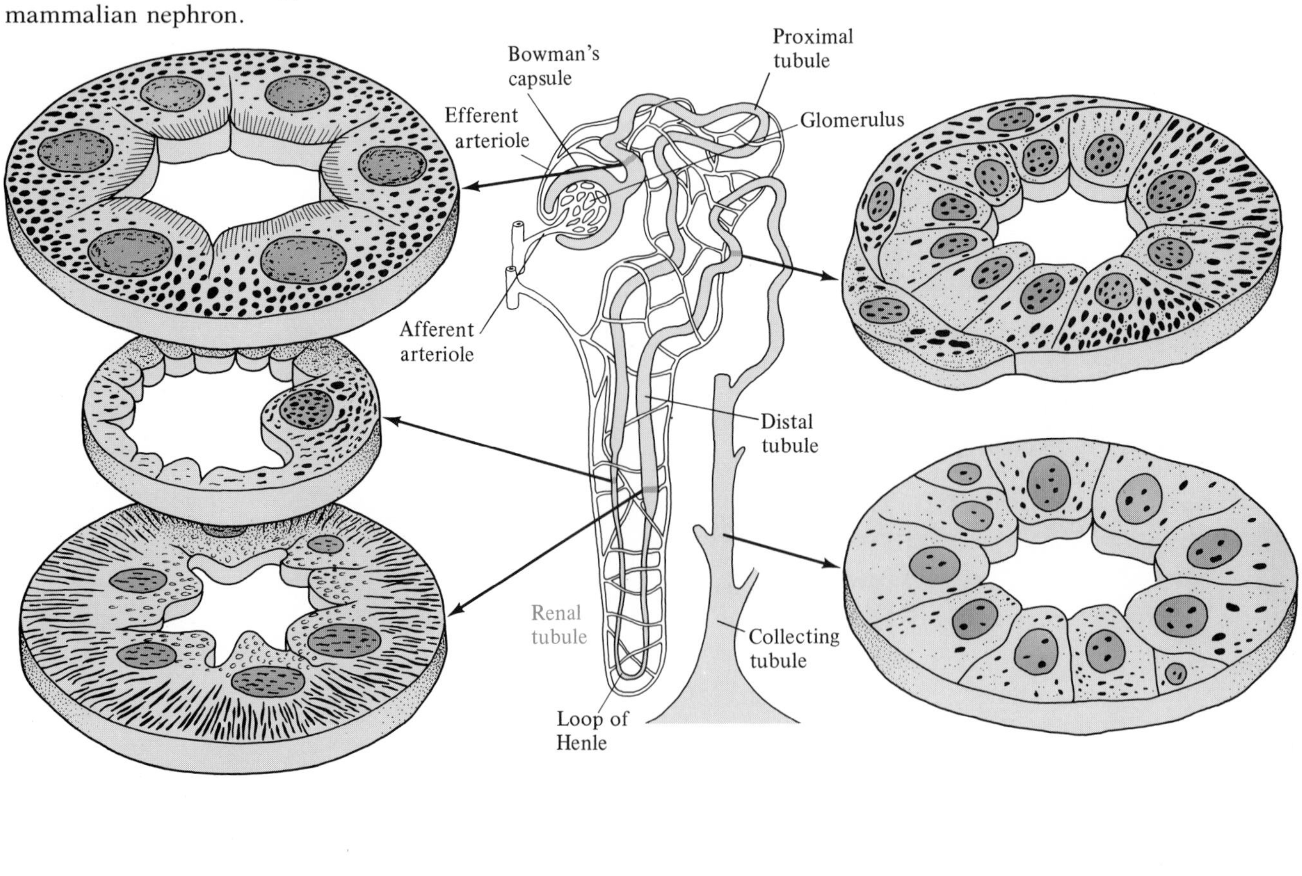

two prominent, distinct segments, connected by a short, narrow, intermediate segment. The two prominent segments are the **proximal tubule,** next to the glomerulus, and the **distal tubule,** which empties, along with tubules of other nephrons, into a **collecting duct.** Both of these principal tubule segments, as well as the collecting duct, contribute to urine formation. As seen in figure 13-9, the urinary tubule is composed of different kinds of cells.

In mammals, the proximal tubule reabsorbs all the glucose and amino acids and up to 85 percent of the $Na^+$, $Cl^-$, $HCO_3^-$, and water from the glomerular filtrate. Most of the remaining 15 percent of the NaCl, water, and other molecules are reabsorbed by the distal tubule and the collecting duct.

In some birds and all mammals, the intermediate segment between the proximal and distal tubules is especially long and is termed the **loop of Henle** (Fig. 13-9). The primary function of this loop is to facilitate reabsorption of water, leading to an unusually high salt concentration of the urine. Consequently, the total volume of urine output is low, thus conserving body water. The mechanism involved in increasing urine salt concentration is discussed later in the chapter. The loop of Henle is surrounded by capillaries that arise from the same blood vessel that gives rise to the glomerular capillaries. The length of the loop is directly related to the animal's need to conserve water and the degree to which it concentrates its urine. A comparison of nephron types and the habitats of the animals that possess them is shown in Figure 13-10.

**FIG. 13-10** The nephron structure in diverse chordates, and the relationship between its structure and the habitat of the organism. Acorn worm; pipe fish; perch; rattlesnake; kangaroo rat.

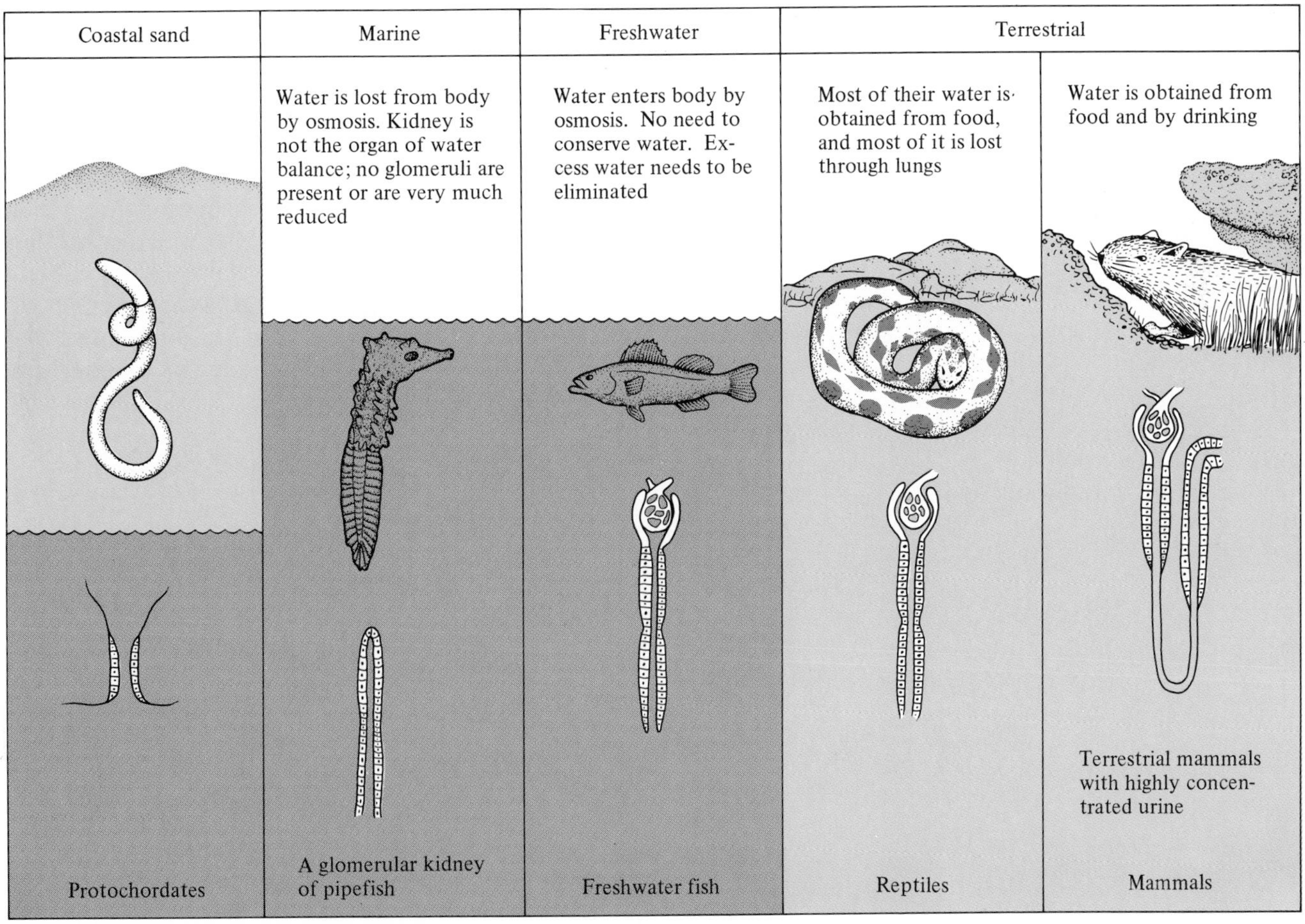

Coastal sand
Marine
Freshwater
Terrestrial
Water is lost from body by osmosis. Kidney is not the organ of water balance; no glomeruli are present or are very much reduced
Water enters body by osmosis. No need to conserve water. Excess water needs to be eliminated
Most of their water is obtained from food, and most of it is lost through lungs
Water is obtained from food and by drinking
Terrestrial mammals with highly concentrated urine
Protochordates
A glomerular kidney of pipefish
Freshwater fish
Reptiles
Mammals

## OSMOREGULATION AND IONIC REGULATION IN TELEOSTS AND OTHER ANIMALS

The vertebrate kidney arose in early freshwater fishes as an osmoregulatory organ. Its function as an organ of excretion was only incidental to its role in the elimination of excess water. A far more important excretory organ in the fishes is the gill. Most of the ammonia produced from nitrogen metabolism escapes through the gills by diffusion into the water that passes across them.

Although the osmoregulatory efficiency of their kidneys might have allowed the ancestors of freshwater fishes to make their way up estuaries into the waters of streams and lakes, a new problem, loss of salts, arose. This problem needed to be solved immediately if these animals were to survive in fresh water. The reabsorption of salts by the tubule of the nephron in freshwater fishes is quite efficient. However, the production of a large volume of urine (for example, 14 ml/kg/hour in the goldfish), though dilute, inevitably leads to an excessive loss of valuable ions, especially sodium and chloride ions, which are in very low concentrations in freshwater lakes and rivers. This problem was solved with the evolution of the ability of the gills to absorb sodium chloride by active transport against the steep concentration gradient existing between a fish's blood and its freshwater environment (Fig. 13-11a).

Long after they became established in fresh water, various species of teleosts entered the sea. There they encountered the opposite of what they had experienced in fresh water, because sea water is hypertonic to the body fluids of teleosts. Now, instead of being flooded with osmotically absorbed water entering through its gills, the marine teleost lost water by osmosis through its gills. Thus, instead of being overabundant, water was in short supply. Those teleosts that successfully entered the marine environment solved this problem by a combination of adaptations involving the gills, the kidneys, and the animals' behavior (Fig. 13-11b). To compensate for lost water, for example, marine teleosts drink water. (Freshwater fishes never drink water except for that which is taken in with food.) Drinking seawater, however, creates another problem, namely, an excess of salts. How teleosts solved the problem might have been anticipated: the same cells of the gills that in freshwater teleosts actively absorb salt from lakes and streams adapted in marine teleosts to excrete salt. All other excess ions—such as magnesium,

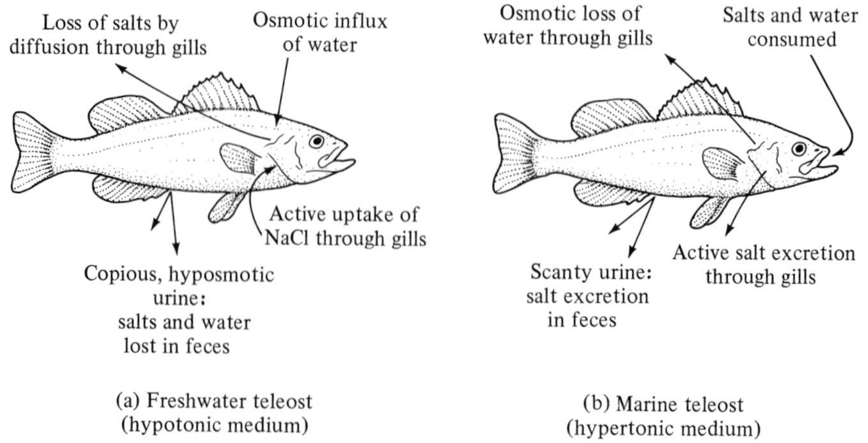


**FIG. 13-11** Ionic and osmoregulation in (a) freshwater and (b) marine teleosts.

calcium, and sulfate—as well as urea and other substances, are excreted by tubular secretion in the nephron.

## SALT SECRETION, SALT ABSORPTION, AND SALT BALANCE

All living cells are able to maintain ionic differences between their cytoplasm and the surrounding medium. It is thus not surprising that fish gills are specialized for the active transport of ions. Nor are fishes unique in having salt-absorbing cells. Most freshwater animals have organs that are able to actively absorb salts from their environment. Examples include the gills of crabs and crayfishes and the skin of frogs.

Marine and desert animals, on the other hand, have cells specialized to excrete salt. For example, elasmobranchs, the modern cartilaginous fishes, have a special gland that excretes excess salt. Other sea animals, such as fish-eating birds and marine turtles, which either eat food with a high salt content or drink seawater, excrete the excess salt through special salt glands located near the eyes. Sea turtles produce unusually salty tears. In the gull, the salt gland, which is composed of several lobules, empties into the nasal cavity (Fig. 13-12a). Soon after a gull takes a drink of seawater, over 80 percent of the water's salt content is excreted by the salt glands; only about 10 percent is excreted by the kidneys. Some desert reptiles also have salt glands. Excretion of salt against a concentration gradient is an energy-requiring active transport process. The cells in the lobules of salt glands involved in salt excretion are rich in mitochondria, particularly at the end of the cell at which salt excretion occurs (see Fig. 13-12d).

**FIG. 13-12** The salt gland of marine birds. (a) Location of the gland in the head. (b) Cross section of salt gland. (c) Enlarged section of lobe. (d) A salt gland cell showing membrane infolding and associated mitochondria.

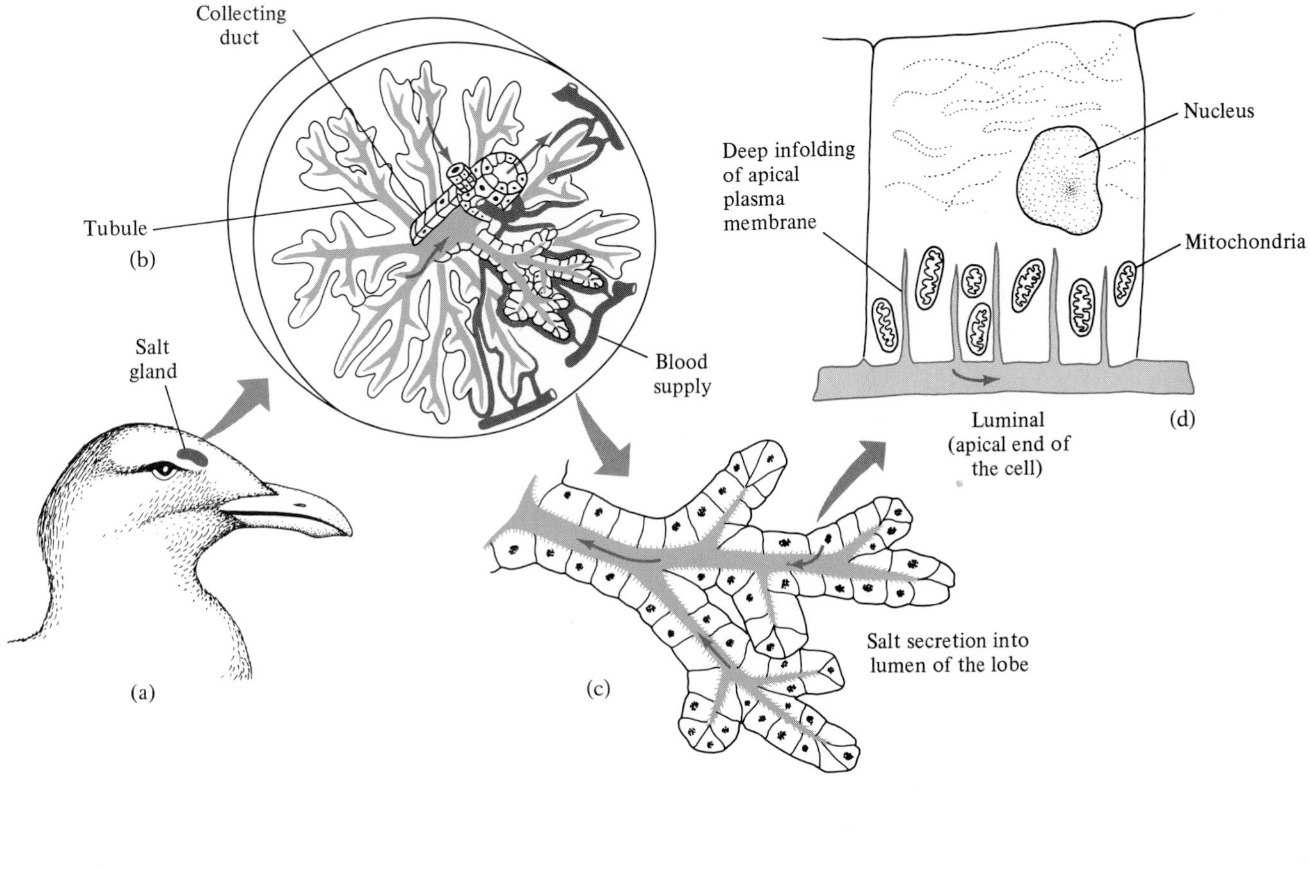

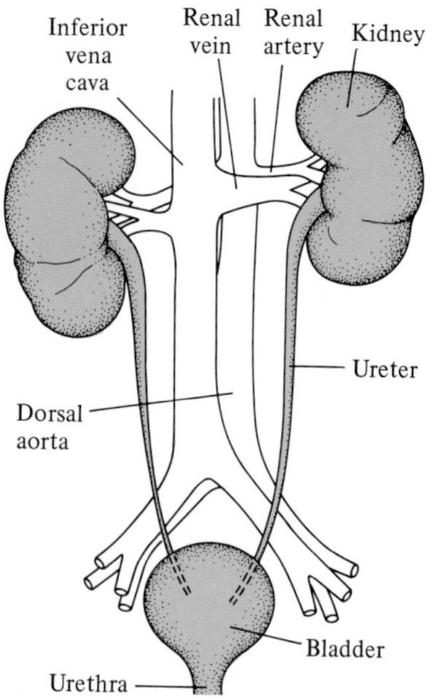


**FIG. 13-13** The human urinary system.

# STRUCTURE AND FUNCTION OF THE MAMMALIAN KIDNEY

In mammals, the kidneys are bean-shaped, paired organs located against the dorsal body wall next to the membranous peritoneum, which lines the abdominal cavity (Fig. 13-13). The right kidney is slightly posterior to the left one in most mammals, including humans. Urine produced by each kidney is released from the tips of the 10 or more **renal papillae,** which project into the cup-shaped depressions in the wall of a large collecting chamber, the **renal pelvis,** on the inner face of the kidney (Fig. 13-14). From the renal pelvis, the urine empties into a tubular ureter on the concave, medial side (toward the midline of the body) of the kidney (Fig. 13-14). The two ureters empty into the **urinary bladder** in which the urine is stored until voided by way of a single duct, the **urethra.**

Each kidney receives a rich blood supply from a large renal artery, branches of which encircle the renal medulla. The blood is drained by a renal vein (Fig. 13-14). The arterial branches further subdivide at the border between the outer portion, the **cortex,** and the inner portion, the **medulla,** of the kidney. These branches are called the arcuate arteries.

## HISTOLOGY

The arterial branches, termed **afferent arterioles,** supply blood to the functional units of the kidney, the nephrons. The human kidney has about 1 million nephrons. Each nephron consists of a **Malpighian corpuscle** made up of a glomerulus and an associated Bowman's capsule, the U-shaped uriniferous, or nephric tubule, and a portion of the collecting duct. Malpighian corpuscles are located only in the cortex, or outer layer, of the kidney. The two types of nephrons, cortical and juxtamedullary, can be distinguished by their position and the length of their slender Henle's loop. Cortical nephrons, which in humans outnumber

**FIG. 13-14** A longitudinal section through a mammalian kidney showing the cortex, medulla, calyx, ureter, and artery.

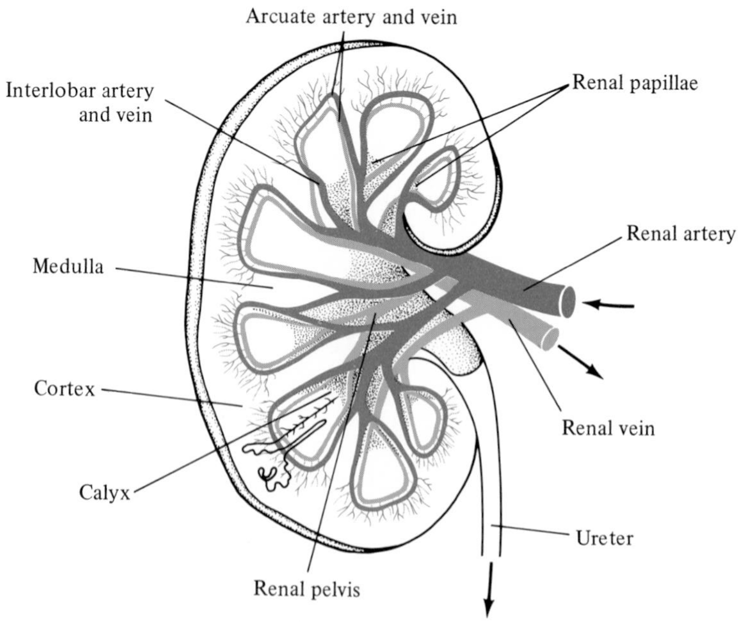

juxtamedullary nephrons seven to one, have short Henle's loops that barely dip into the medulla, or inner layer of the kidney. Juxtamedullary nephrons have long Henle's loops that extend deep into the medulla. In mammals that produce a very concentrated urine, the juxtamedullary nephron is more common; it is the only kind found in the kidney of the desert rat.

Each afferent arteriole supplies blood to one glomerulus, the network of capillaries that actually produces the ultrafiltrate within each nephron. The capillaries of each glomerulus empty into an **efferent arteriole,** the diameter of which is smaller than that of the afferent arteriole. Because of this difference in the size of the arterioles, blood pressure in the glomerulus is high and contributes to its production of filtrate. The efferent arterioles form a capillary network, known as **vasa recti,** around the renal tubule. The arcuate veins empty the capillary bed that supplies the renal tubules. Thus, the blood that loses about a fifth of its fluid and solutes as a filtrate in a glomerulus reabsorbs most of this fluid and solutes while releasing additional materials into the nephric tubule fluid. The capillary networks arising from an efferent arteriole not only supply several adjacent nephrons but also send special branches along the renal tubules deep into the papillae (Fig. 13-15).

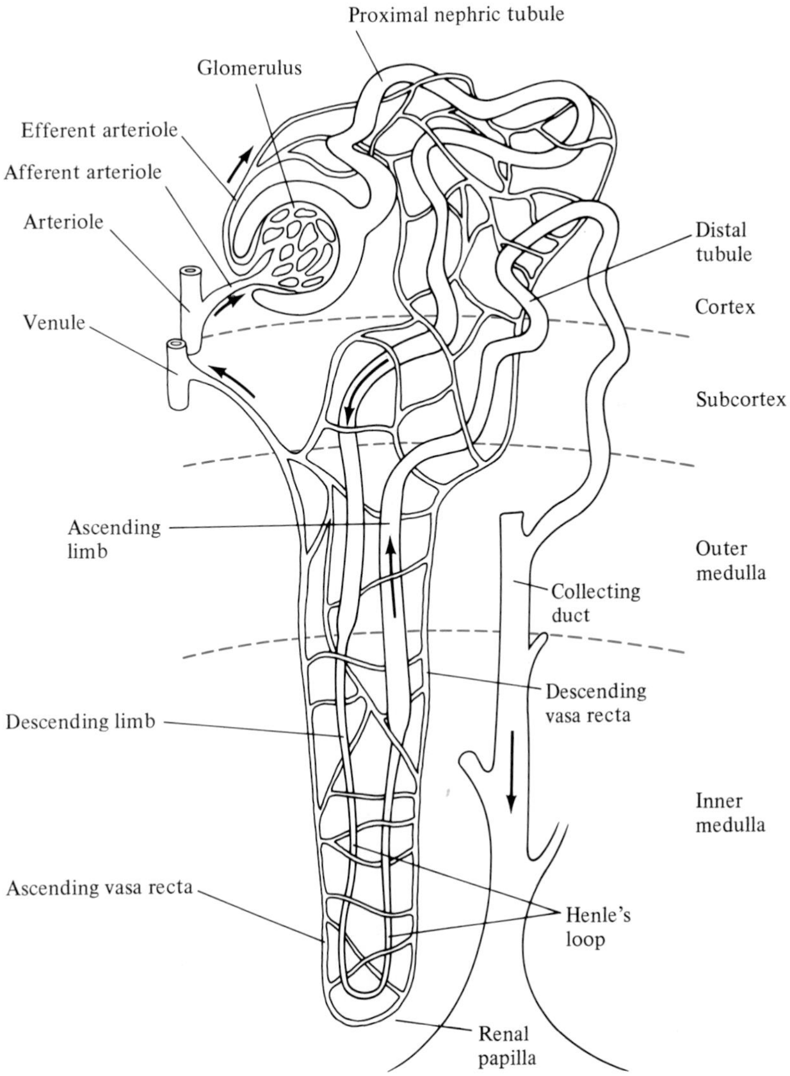


**FIG. 13-15** Sketch of the blood supply to a nephron in the mammalian kidney.

## FINE STRUCTURE OF THE GLOMERULUS AND BOWMAN'S CAPSULE

Glomerular capillaries are so numerous that they provide about twice the surface area for filtration as do capillaries in other tissues, such as muscle (Fig. 13-16a, b). Glomerular capillary walls are also more permeable than other capillaries, allowing freer passage of molecules of diameters up to 7.5 to 10 nm. Electron micrographs of a Bowman's capsule show that the highly branched cells lining it, called **podocytes,** are closely connected to the thin endothelial cells of the capillary wall (Fig. 13-17). Glomerular capillary walls are very thin (Fig. 13-18) and are highly permeable to water and to small solute molecules. An additional barrier that prevents loss of proteins and other large molecules is the interdigitating network present on the podocytes (Fig. 13-19). Because the glomerular capillaries present such a large surface area for filtration, the rate of glomerular filtration far exceeds that across the walls of most capillaries. Normally, about one-fifth of the plasma passing through the capillaries of a glomerulus is filtered into its Bowman's capsule. Some idea of the amount of filtrate the kidneys produce is gained by the realization that one-fifth of the cardiac output of blood enters the renal arteries each time the blood makes a circuit through the body.

The glomerular filtrate contains all that is of value in plasma, as well as excess salts, urea, uric acid, and other waste products that need to be excreted. The selective reabsorption of nearly all this filtrate is the task of the renal tubule and the collecting duct.

Evidence that the glomerulus produces a filtrate of the blood was first obtained in 1924 by the experiments of Joseph Wearn at Harvard and Alfred Richards at the University of Pennsylvania. Using a procedure known as the micropuncture technique, they analyzed the chemical composition of samples of filtrate taken through extremely fine quartz tubing (about 10 $\mu$m in diameter) inserted into the Bowman's capsule of a single nephron (see Fig. 13-20a). Their work was done with amphibians, but similar experiments have since been conducted with

**FIG. 13-16** Photomicrograph of glomeruli in a human kidney. The blood vessels entering and leaving the glomeruli can be clearly seen. (a) Low magnification. (b) High magnification.

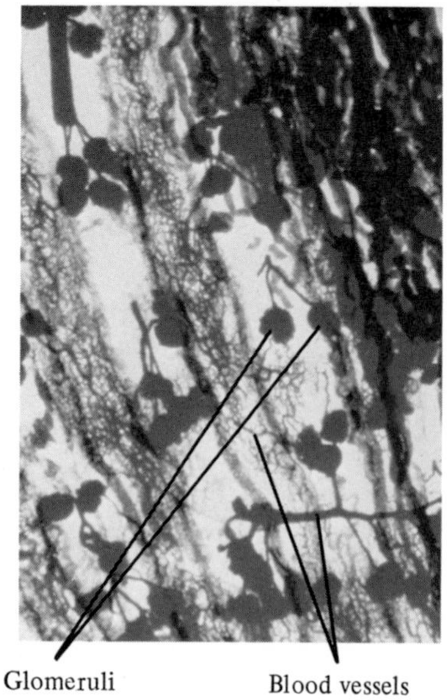

(a)

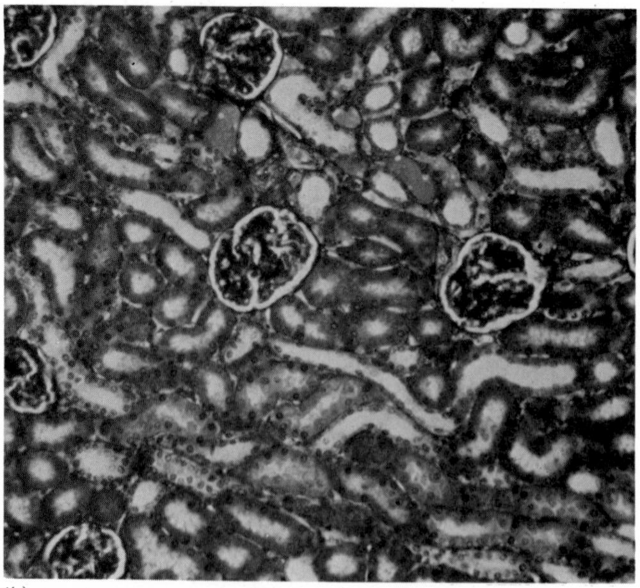
(b)

**FIG. 13-17** Electron micrograph of the Malpighian corpuscle showing the podocytes of the Bowman's capsule.

rats, dogs, monkeys, and other species. These classic studies established the fundamental role of the glomerulus in producing a virtually protein-free filtrate from the blood. Such experiments also established that the ultrafiltrate, although almost protein-free, is rich in glucose and sodium chloride. The fact that the urine in the urinary bladder lacks glucose

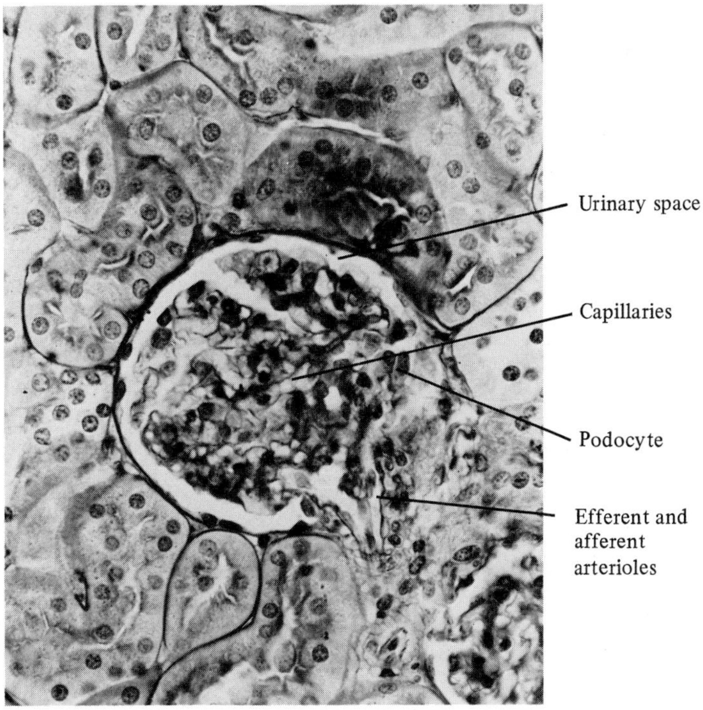


**FIG. 13-18** Electron micrograph of a cross section of a Bowman's capsule showing the double wall and the urinary space.

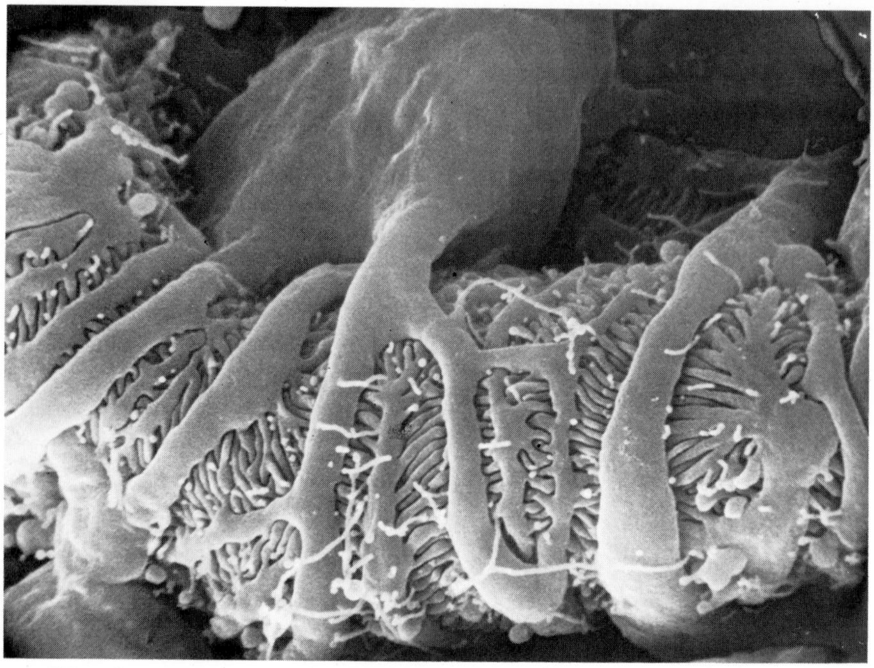

**FIG. 13-19** Electron micrograph of the podocytes. The interdigitating terminal branches of the podocytes through which blood is filtered are visible.

and has a much reduced concentration of sodium chloride suggests that the nephron reabsorbs these filtered solutes. Because sodium chloride and glucose are absorbed at different rates, Wearn and Richards also suggested that there may be more than one mechanism of tubular reabsorption of these solutes. Since 1924, there has been a continuing refinement of the micropuncture technique as well as the development of other methods, all of which have led to a more detailed knowledge of the functions of the various regions of the nephron.

## TUBULAR REABSORPTION

In a normal human adult, the glomerular filtrate amounts to about 180 l (45 gallons) a day. Over 99 percent of this is reabsorbed, and about 1,500 ml a day of urine are produced. The proximal nephric tubule does the largest share of the work of reabsorption, with the loop of Henle, the distal nephric tubule, and the collecting duct doing the rest.

Analysis of the fluids obtained by micropuncture shows that the fluid in the Bowman's capsule and the fluid remaining in the proximal tubule after as much as 85 percent of the volume has been reabsorbed are both isosmotic to blood plasma; that is, the total concentration of solutes in these fluids is the same as in the blood plasma. To the early investigators, this meant one of two things: either water was being actively absorbed by the tubule, with the solutes following passively, or one or more of the solutes were being actively absorbed, with the water following passively by osmosis because of the imbalance in tonicity created by the solute absorption. In a simple but elegant experiment on the urinary tubules in the mud puppy, *Necturus*, a proximal tubule was filled with castor oil by micropuncture of the glomerulus (Fig. 13-20a, b). A test solution was then introduced into the middle of the oil column, which prevented it from mixing with filtrate and other tubular fluids (Fig. 13-20b). After various test solutions had moved along the

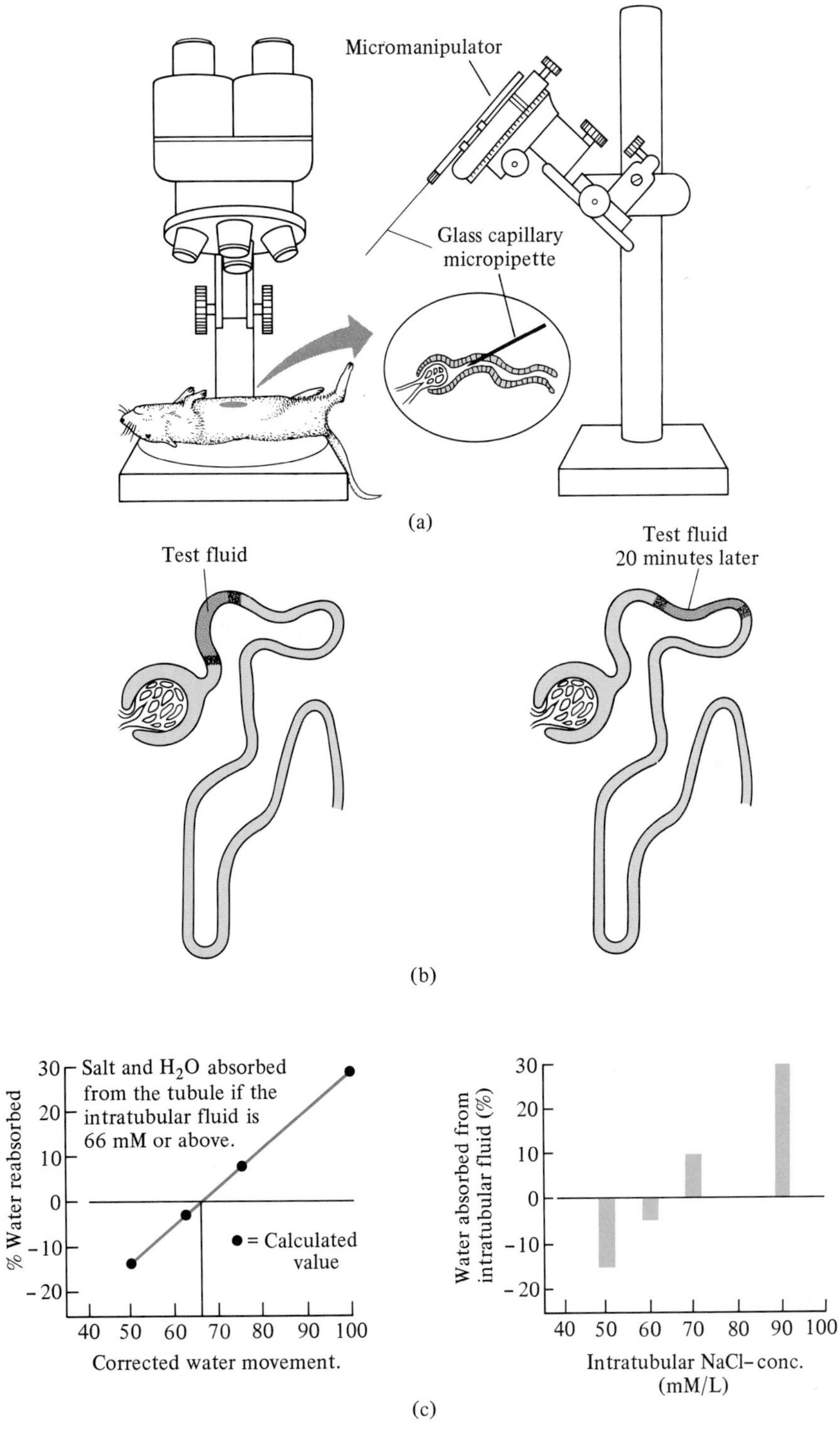


**FIG. 13-20** Micropuncture technique used to determine the absorptive properties of the urinary tubule. (a) Micropuncture apparatus. The inset shows the position of the needle in relation to a tubule. (b) Diagrammatic presentation of the stages in the experimental procedure: path of the test solutions. (c) Graphs showing the relationship between the salt concentration in the tubular fluid and reabsorption of water. Water is reabsorbed only if the salt concentration of the tubular fluids is 66 mM or higher. Water follows the active uptake of salts.

nephric tubule, they were withdrawn and analyzed for changes in solute concentration (Fig. 13-20b).

Each test solution consisted of known concentrations of NaCl and radioactive inulin, a fructose polysaccharide that is not absorbed by cells. Enough mannitol, a glucose derivative that also is not absorbed by cells, was added to each test solution to make it isosmolar to plasma. Thus, any change in the inulin concentration of a test solution in the tubule could be due only to changes in its water content. Furthermore, changes in water content could occur only by reabsorption or secretion of water into the tubule. Results of these experiments showed that if the initial concentration of NaCl in the test solution is 66 mMoles/liter (mM/l), both water and NaCl are reabsorbed from the tubule. When the concentration of NaCl is above 66 mM/l, the salt is absorbed by active transport against a concentration gradient (plasma has a NaCl concentration of 100 mM/l in *Necturus*) and water follows passively. However, if the NaCl concentration is less than 66 mM/l, the gradient between plasma and nephric tubule NaCl concentration is too steep to be overcome by active transport, and NaCl actually enters the tubule. Moreover, whenever the NaCl concentration in the tubular fluid is less than 66 mM/l, both NaCl and water enter the tubule lumen, as evidenced by the dilution of inulin, indicating that water is free to move passively in either direction along with any movement of salt (see Fig. 13-20c).

Further studies showed that, of the two ions $Na^+$ and $Cl^-$, only the sodium ion is actively absorbed from the proximal tubule. When $Na^+$ is reabsorbed by the proximal tubule cells, an electrical imbalance is created between the cytoplasm of tubule cells and the tubule fluid, as $Cl^-$ and $HCO_3^-$ are momentarily left behind in the tubule fluid. Because the tubule wall is permeable to both anions, they immediately follow the reabsorbed $Na^+$ by electrostatic attraction. Water passively follows the sodium, chloride, and bicarbonate ions by osmosis. The magnitude of this recycling of salts in the human kidney is amazing. For example, up to a kilogram (over 2 pounds) of sodium chloride is recycled per day. As water is reabsorbed osmotically, urea is also passively reabsorbed (up to 85 percent of it) in the proximal tubule, the cells of which are freely permeable to this small molecule.

Besides sodium, other substances that are actively absorbed by the proximal tubule cells include potassium (93 percent of the amount contained in the filtrate), calcium, phosphate, magnesium (90 to 99 percent), uric acid (98 percent),* glucose (100 percent), amino acids (100 percent), and trace amounts of protein (100 percent). Interestingly, of 15 sugars tested, only the naturally occurring hexoses, such as glucose, are reabsorbed by the tubule, suggesting the presence of a selective transport mechanism. In fact, three separate active transport mechanisms have been identified in amino acid reabsorption, one for each class of amino acid: dibasic, diacidic, and neutral.

## TUBULAR SECRETION

Among the materials added to the glomerular filtrate by tubular secretion after it has entered the renal tubule are several materials foreign to the body, such as the drug penicillin. But some normal metabolites are also excreted, including $NH_3$ and $H^+$, by a process discussed later in this chapter as an aspect of acid-base balance. Although most tubular secretion occurs in the proximal tubule, some occurs in the distal tubule and collecting duct. As we have noted, two-thirds of the uric acid normally appearing in the urine is secreted by the distal tubule. The collecting duct also secretes urea, uric acid, and other excretory products into urine.

* The remaining 2 percent that is not reabsorbed by the proximal tubule is one-third of the entire amount of uric acid excreted in mammalian urine. The other two-thirds is added by tubular secretion in the distal tubule.

## THE CONCENTRATED URINE OF BIRDS AND MAMMALS

Since 1909, investigators have known that birds and mammals are the only animals that can produce urine that is hypertonic to their blood and that they are the only animals that have nephrons with loops of Henle. As more species were examined, it became clear that the longer the loops, the more concentrated is the urine. The full significance of this relationship eluded physiologists until 1942, when the Swiss scientist Werner Kuhn and his associates proposed that the loop was part of a countercurrent system. Their complicated model of how the nephron functions was ignored until 1960; by then, the evidence for their explanation had become so overwhelming that it was at last generally accepted.

**FIG. 13-21** Illustration of the principle of countercurrent exchange of heat. Efficiency of heat exchange in (a) a parallel-flow heat exchanger and (b) a countercurrent heat exchanger. Difference in heat loss from the blood vessels of the limbs of a mammal in an ambient temperature of 5°C in the hypothetical absence of (c) and in the presence of a countercurrent exchange (d).

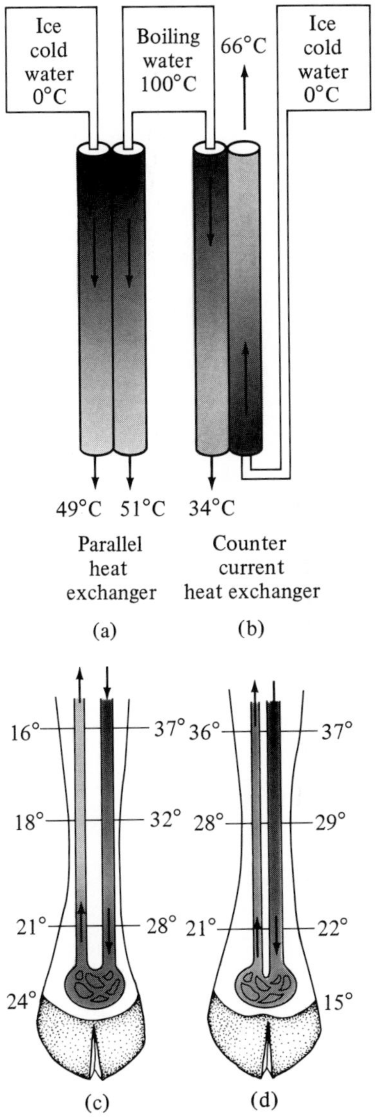


***The Principle of Countercurrent Exchange.*** Industrial heat exchangers, such as those in nuclear power plants, have long made use of the principle of countercurrent exchange. If two similar pipes, one carrying water at, say, 100°C and the other carrying water at 0°C, are placed next to each other, heat will pass from one to the other through the walls of the pipes. The hot water will get colder and the cold water will get hotter. The efficiency of such an exchange is greatly enhanced, however, if the water in the two pipes is flowing in opposite directions.

Compare, for example, the difference in the two situations illustrated in Fig. 13-21a. Where the water in the two pipes flows in the same direction, the temperature is brought close to equilibrium (51°C/49°C) at a given rate of flow and pipe length. But at the same rate of flow and within the same distance, the cold water will be warmed to 66°C and the hot water chilled to 34°C if the water in the two tubes is made to flow in opposite directions (Fig. 13-21b). This occurs because the differences in temperature—the temperature gradients—are much greater in the latter case, enabling heat to transfer more rapidly and efficiently.

This same principle of countercurrent exchange of heat operates in the limbs of birds and mammals adapted to living in cold climates. In caribou (reindeer), for example, the temperature of the lower leg can drop to 8°C while their body temperature is kept at 38°C. If blood going to the feet of such an animal were allowed to lose most of its heat to the environment, the animal would soon perish. Instead, most of the heat of the arterial blood going to the feet is transferred to the blood returning from the feet through veins that run parallel to the arteries adjacent to them. Moreover, the major arteries and veins are deep within the tissues of the legs, further minimizing heat loss (Fig. 13-21c, d). By the time it reaches the feet, arterial blood is thus only a little above freezing and loses very little heat to the environment. Another countercurrent system in fish gills operates to good advantage in the exchange of oxygen and carbon dioxide between the fish's blood and the water flowing in the opposite direction to the blood flow across its gills (see Chapter 8).

***The Loop of Henle as a Countercurrent Multiplier.*** Henle's loop is not a simple countercurrent exchanger. It is specialized as a countercurrent multiplier. Its function depends on the ability of the ascending limb of the loop to secrete NaCl actively while remaining relatively impermeable to water.* The descending limb, however, is very permeable to water and only somewhat less permeable to solutes. As the filtrate passes down the descending loop, water leaves by osmosis, entering the hypertonic interstitium, the space surrounding the descending and ascending loops. As a consequence, the tubule fluid becomes highly concentrated. This concentration of fluid is possible because the salt concentration of interstitial fluids exhibits a gradient along the length of the descending loop, with the highest concentration at the medullary end of the loop (Fig. 13-22). As the fluid starts moving up the ascending loop, the reverse happens. Salt is actively secreted into the interstitium from the tubular fluid. The secretion of salt from the fluid in the ascending limb is achieved gradually, with some solute being transported at each level of the ascending limb into the interstitium against a slight concentration gradient (Fig. 13-22).

The transport of relatively small amounts of salt along the length of the ascending limb gradually decreases the solute concentration within

* Contrary to what happens in the proximal tubule, in the ascending limb of Henle's loop, $Cl^-$ ions are actively reabsorbed, and $Na^+$ ions follow by electrostatic attraction.

PANEL 13-1

## Management of Renal Insufficiency

As we have read in this chapter, the kidneys are organs whose efficiency as a selective filter we can only marvel at and whose role in health is incalculable. As regulators of concentrations of metabolic by-products in the body fluids, they have no equal: they control the concentration of salts and ions in body fluids within precise limits and excrete metabolic waste products. Hence kidney malfunction, known by the general term renal insufficiency, is cause for concern. Diseased or damaged kidneys cannot perform their ion-salt balance function properly and cannot adequately eliminate metabolic waste products from the blood, causing a number of major alterations in ionic and salt balance in blood. These alterations may result in altered blood acid-base balance. For example, the blood's hydrogen ion concentration may increase beyond its normal range as a result of a decrease in renal tubular secretion of $H^+$ ions. This leads to a drop in blood pH. The blood is said to exhibit **acidosis.** When uncontrolled, acidosis may lead to coma and then death. In other instances, excessive loss of sodium because of distal tubular dysfunction can lead to dehydration. This can be corrected by increasing the intake of sodium chloride (table salt). However, excessive accumulation of sodium caused by insufficient glomerular filtration leads to edema, the accumulation of fluids in body tissue. This condition can often be alleviated by a restricted intake of sodium. Water and salt balance must be carefully maintained and monitored in all cases involving alterations of acid-base metabolism.

Renal infection or obstruction may lead to other kinds of symptoms. Among these is **uremia,** the retention of nitrogeneous waste products, such as urea, in the blood. Restricted protein intake can often reduce blood urea concentration to a tolerable level.

### DIALYSIS

When kidney function falls to 5 or 3 percent of the normal level, toxic wastes begin to accumulate in the blood. In such cases, an artificial kidney, to remove toxic wastes by dialysis of body fluids, is required to make up for the loss of kidney function. Dialysis is a process in which the excess solutes in the body fluids are removed by passing the fluids through a dialyzing membrane that allows the solutes to diffuse along the concentration grathe tubule. The final concentration achieved is proportional to the length of the limb. Thus, a countercurrent exchange occurs, but because the ascending limb secretes salt and is impermeable to water, its fluid becomes more and more dilute. By the time the fluid enters the distal convoluted tubule, it is actually hyposmotic to plasma and glomerular filtrate.

In humans and most mammals, the concentration of interstitial fluid at the point where the descending loop connects to the ascending loop reaches about 1,200 mOsm/kg*; in desert rats, it is even higher. This high salt concentration of the interstitial fluid of the medulla makes the production of hypertonic urine possible. For, as the urine makes its second and final trip through the medulla, this time within the collecting duct (Fig. 13-22), it loses water to the interstitium by osmosis, and its solute concentration reaches the high levels found in the interstitium of the papillae.

* Osmolality is an expression of concentration of a solute in grams per 1,000 grams of water. One milliosmolal (mOsm) is one thousandth of one gram-molecular weight of a substance dissolved in one thousand grams of water.

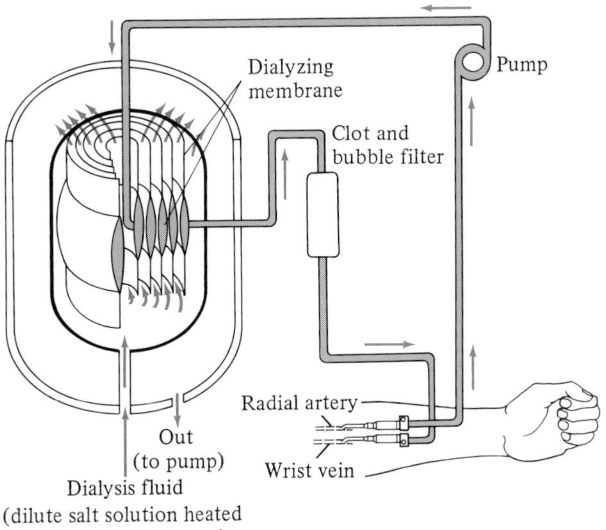


Blood flow through dialysis machine.

dient. In **peritoneal dialysis,** fluids are introduced into the peritoneal, or abdominal, cavity by means of a tube. In this case, the peritoneal membrane itself serves as a dialyzing membrane. Following the concentration gradient, toxic wastes and excess salts diffuse into the peritoneal fluid. After about 20 minutes, the fluid is siphoned out. This process is not very efficient but can be repeated as often as needed. A more efficient method is **hemodialysis,** which requires a dialysis machine. A dialysis machine consists of a cellophane tube immersed in a solution or bath. The tube acts as an artificial dialyzing membrane that lets the blood's metabolic waste products, excess salts, and ions through while retaining its proteins. The purified blood is then returned to the patient. Hemodialysis involves pumping blood directly from an artery into the cellophane tube. Once the blood is purified, it is returned to a vein. In individuals with chronic renal insufficiency, permanent tubes are inserted into an artery and a vein. These can be connected to a hemodialysis machine as the need arises.

## TRANSPLANTATION

If kidney damage is irreparable, a kidney transplant may be considered. The first successful kidney transplant, between identical twins, was performed in 1955. As noted in the preceding chapter, tissues of each individual bear very specific cell surface antigens. Identical twins, of course, bear the same antigens. But this is not the case with most people. Unless donor and recipient antigens are very closely matched, the transplant will be rejected by the recipient's immune response. Unfortunately, not all persons with permanent kidney damage can be transplant recipients. Only those who are free of systemic diseases, such as diabetes, atherosclerosis, and arteriosclerosis, are good prospects for kidney transplantation.

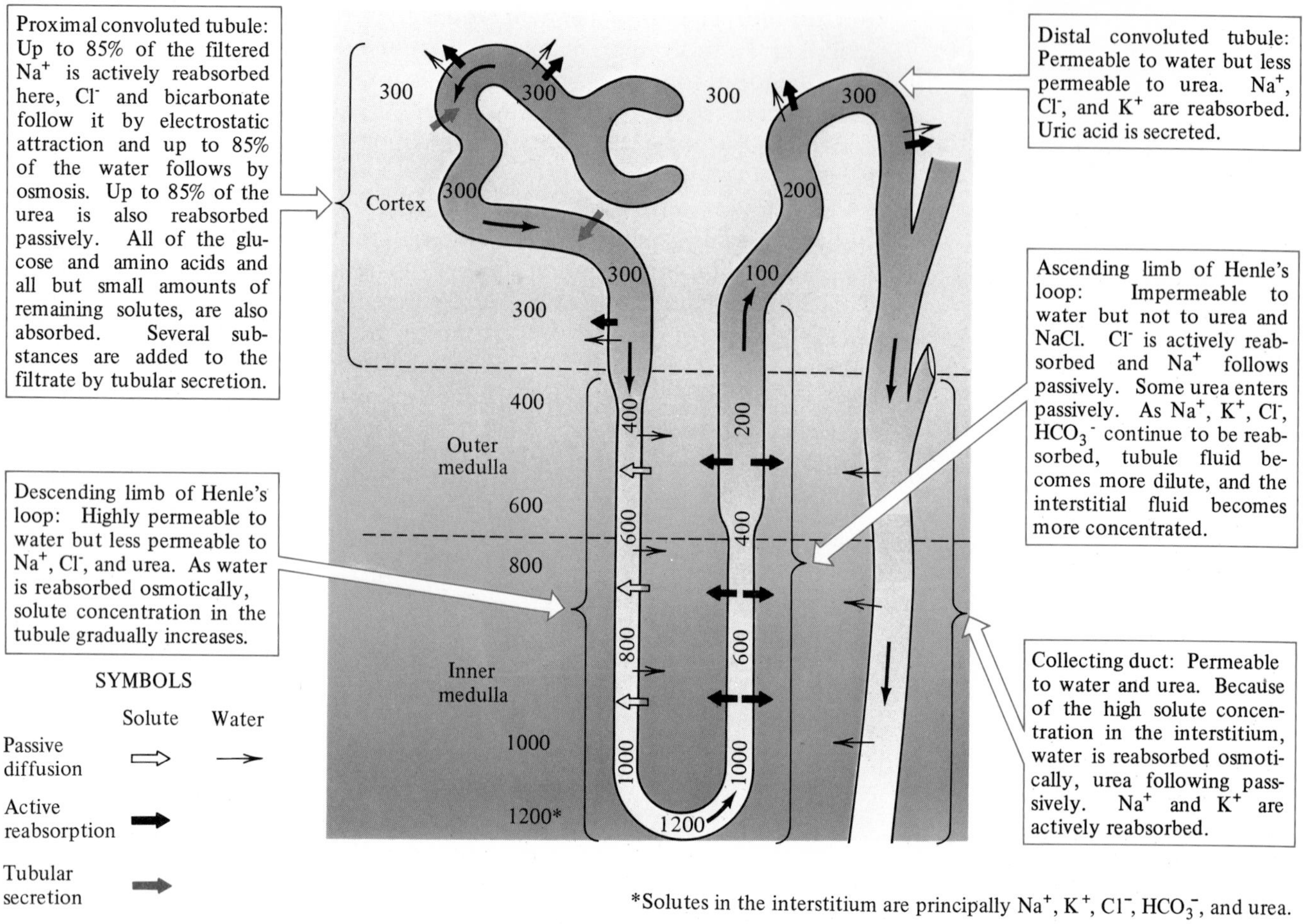
Proximal convoluted tubule: Up to 85% of the filtered Na+ is actively reabsorbed here, Cl- and bicarbonate follow it by electrostatic attraction and up to 85% of the water follows by osmosis. Up to 85% of the urea is also reabsorbed passively. All of the glucose and amino acids and all but small amounts of remaining solutes, are also absorbed. Several substances are added to the filtrate by tubular secretion.
Distal convoluted tubule: Permeable to water but less permeable to urea. Na+, Cl-, and K+ are reabsorbed. Uric acid is secreted.
Ascending limb of Henle's loop: Impermeable to water but not to urea and NaCl. Cl- is actively reabsorbed and Na+ follows passively. Some urea enters passively. As Na+, K+, Cl-, HCO3- continue to be reabsorbed, tubule fluid becomes more dilute, and the interstitial fluid becomes more concentrated.
Descending limb of Henle's loop: Highly permeable to water but less permeable to Na+, Cl-, and urea. As water is reabsorbed osmotically, solute concentration in the tubule gradually increases.
Collecting duct: Permeable to water and urea. Because of the high solute concentration in the interstitium, water is reabsorbed osmotically, urea following passively. Na+ and K+ are actively reabsorbed.
Cortex
Outer medulla
Inner medulla
300
400
600
800
1000
1200*
200
100
SYMBOLS
Solute
Water
Passive diffusion
Active reabsorption
Tubular secretion
*Solutes in the interstitium are principally Na+, K+, Cl-, HCO3-, and urea.

**FIG. 13-22** Diagrammatic summary of reabsorption, tubular secretion, and urine concentration processes in the mammalian nephron. Solute ($Na^+$, $K^+$, $C^-$, $HCO_3^-$, and urea) concentrations are in milliMoles.

As noted, the solute concentration in the medullary interstitium reaches 1,200 mOsm/kg. The osmolality of blood, however, is only 300 mOsm/kg. If the medulla were supplied by blood vessels similar to those in most tissues, the blood would lose water osmotically to the interstitium and gain solutes from it by diffusion, thus greatly diluting the interstitial fluid. Something must be done to prevent this situation, which is comparable to the need to guard against heat loss in the limbs of arctic animals; and, indeed, the problem is solved in a similar way. The arrangement of the vasa recta, the pairs of blood vessels that lie beside each other and penetrate deep into the medulla, helps solve this problem (Figs. 13-9 and 13-15). The two parallel blood vessels serve as countercurrent exchangers. Water is lost from the blood and solutes are absorbed as the blood vessels pass into the medulla, but water is reacquired and solutes are lost again as they return. The blood returning to the outer medulla is thus only a little richer in solutes than when it entered.

## ACID-BASE BALANCE REGULATION

Like the kidney of most vertebrates, the mammalian kidney is an important organ in regulating the acid-base balance of body fluids. The kidney shares this role with the respiratory system and other buffering systems of the body. Whereas the respiratory system maintains a constant carbonic acid ($H_2CO_3$) level in the blood, the kidney stabilizes the level of the bicarbonate ($HCO_3^-$) ion. This, in turn, stabilizes the pH of the plasma and cell contents at a level optimal for most enzymes to function.

The kidney accomplishes its part of the task by excreting any excess bicarbonate or, if bicarbonate has been depleted by an intake or production of acid, by excreting ammonium and phosphate ions which absorb $H^+$ ions, the other ion derived from the ionization of carbonic acid. $H^+$ ion is also secreted by renal tubule cells into the tubular fluid, and the resulting electrical imbalance is restored by absorption of $Na^+$ from the tubular fluid.

The fate of $H^+$ once it is secreted into the tubule depends on the concentration of particular buffers in the tubular fluid at the time. When buffer concentration is low, $NH_3$ is secreted into the tubule, where it combines with $H^+$ to form $NH_4^+$, in which form it is excreted in the urine.

## SOME KIDNEY DISEASES

The kidney is subject to many diseases, which are classified on the basis of the type of tissue affected. If the glomerulus is affected, the disease is termed glomerulonephritis (*nephritis*, inflammation of the kidney), and if the urinary tubule is affected, it is called tubulonephritis. If the interstitial tissue is affected, the disease is called interstitial nephritis. Inflammation or infection of the urinary tract and medullary portions of kidney associated with it is designated pyelonephritis. Damage to kidney tissues is usually caused by bacterial infections, but some diseases result from damage to kidney cells by antibodies produced in response to a bacterial infection in another part of the body. Bright's disease can be caused in this manner as an aftermath of streptococcal infections of the throat.

Kidney stones, or renal calculi, are crystalline precipitates that sometimes form when the urine becomes too concentrated. They usually contain uric acid, calcium oxalate, and calcium phosphate and generally occur in the pelvis of the kidney. If small, the stones pass through the ureters, causing intense pain. Recent studies indicate that these stones can be broken up by ultrasound thereby eliminating the need for surgery.

Toxemia of pregnancy occurs in some women during the last five months of pregnancy. The syndrome is characterized by hypertension (high blood pressure), albuminuria (protein in the urine), and edema (swelling of body tissues by an excessive accumulation of tissue fluid). In this disease, the glomeruli of the kidney are enlarged and the basement membranes swollen, resulting in the loss of plasma proteins into the urine. The cause is unknown, but an antigen-antibody reaction is suspected as a factor. Typically, the symptoms disappear soon after delivery, although occasionally some evidence of the disease, such as hypertension, persists.

## Summary

Excretion is the elimination of toxic and useless by-products of metabolism, ammonia being one of the most abundant. Ammonia, which is highly toxic, is excreted as such in primitive aquatic animals. Metabolic pathways that convert ammonia into various less toxic nitrogen compounds, such as urea and uric acid, have evolved in most higher animals. These metabolic steps require energy. The end product of nitrogen metabolism is dictated not only by the evolutionary history of a species but also by the demands of the environment in which it lives.

Invertebrate kidneys are one of two or a combination of two general types: (1) coelomoducts, which appear to have evolved as evaginations from the body cavity, and (2) nephridia, which arose as invaginations of the body wall. Protonephridia, which are primarily osmoregulatory in function, are examples of the second type. Protonephridia, common in freshwater flatworms, end internally in blind pouches that bear flame cells. Metanephridia, the excretory organs of the common earthworm and its relatives, open internally through the nephrostome. Some modifications of coelomoducts and nephridia are seen in the so-called kidneys of molluscs and crustaceans. These are essentially combinations of parts of a coelomoduct and a nephridium. In insects and a few other terrestrial arthropods, Malpighian tubules serve as excretory organs.

The vertebrate kidney appears to have evolved in ancient freshwater teleosts from the ancient coelomoducts of the hypothetical protovertebrates as an osmoregulatory organ. These animals probably arose in estuaries and made their way into fresh water as the efficiency of their kidneys in removing excess water and the ability of their gills to absorb NaCl improved. Later, some species of teleosts left fresh water and entered the sea. Their glomeruli, no longer an asset but now a liability, became reduced in size and function or were eliminated entirely, the renal tubule being retained for its ability to excrete by tubular secretion.

Arterial blood is filtered as it passes through the glomerulus. Composition of the glomerular filtrate is more or less identical to that of blood plasma without protein. As the filtrate passes through the proxima tubule, most of the water and all nutrients, such as glucose, are reabsorbed. In addition, water enters the interstitial space surrounding the descending limb. But as the fluid moves up the ascending limb of Henle's loop, salt is actively secreted into the interstitium making the fluid hyposmotic. In the final movement of the renal fluid through the collecting duct more water is lost into the interstitium by osmosis, the urine thereby achieving its final concentration.

When terrestrial vertebrates evolved from amphibians, the vertebrate kidney, despite its original primary function as an organ of osmoregulation, was remodeled to become the terrestrial vertebrate's only important organ of excretion. Reptiles in arid environments conserved water by evolving uricotely, allowing them to excrete their protein nitrogen wastes in crystalline form. Birds, however, in addition to sharing uricotely with the reptiles from which they evolved, developed a loop of Henle, as did mammals. These two groups thus became the only animals able to excrete urine that was hypertonic to their own blood, a valuable asset when water needs to be conserved.

Like all organs of the human body, kidneys are subject to diseases. The diseases are manifestations of malfunction of glomeruli, urinary tubules or ureters. The malfunction of glomeruli is designated glomerulonephritis, and that of tubules, tubulonephritis. The major causes of such malfunctions are bacterial or other infections, although destruction of kidney tissues by antibodies produced by the body against some other infection, such as strep throat, is not uncommon. Crystallization of salts along the path leading from a urinary tubule to the urethra is another cause for dysfunction, disease, or both.

## Recommended Readings

BAUMANN, J. W., and F. P. CHINARD. 1975. *Renal Function.* Mosby, St. Louis. A comprehensive and clearly written account of the physiology of excretion, written in fairly simple language.

MARSH, D. J. 1983. *Renal Physiology.* Raven Press, New York. A simple but comprehensive account of kidney function described in 176 pages.

MOFFAT, D. B. rev. ed. J. J. HEAD. 1978. *The Control of Water Balance by the Kidney.* Carolina Biological. A well-illustrated, 16-page description in clear language of the functioning of the mammalian kidney. Some historical perspective is given.

SCHMIDT-NIELSEN, K., and B. SCHMIDT-NIELSEN. 1953. "The desert rat." *Scientific American,* 189(1): 73–78. The role of the kidney in the survival of a remarkable little animal that lives on dry food and no water.

SMITH, H. W. 1953. *From Fish to Philosopher.* Little, Brown and Co., Boston. A delightful account of the evolutionary aspects of excretion. A classic monograph.

VANDER, A. J. 1980. *Renal Physiology.* 2nd ed. McGraw-Hill, New York. An excellent account of normal and pathological state of kidney function.

# UNIT IV

# The Biology of Organisms: Integration and Behavior

CHAPTERS



An organism's ability to respond to certain changes in internal or external conditions gives it an advantage over organisms unable to make such responses. Ultimately, this ability can make the difference in its success as a species. The processes by which an organism responds specifically to a given change comprise what is termed integration. All organisms detect and react to some environmental and internal changes on a continual basis. The term *integration*, however, is applied almost exclusively to certain processes in animals. Thus, the chapters in this unit deal only with animal biology. Comparable responses of plants include the control of stomatal opening and closing (Chapter 7) and seed germination and flowering (Chapter 26).

Integration is accomplished principally by the nervous system, often in coordination with the endocrine, or hormone, system. Only a few integrative mechanisms of animals are strictly hormonal.

Recent years have seen an explosion of knowledge in the areas of nervous integration and animal behavior as well as a growing interest in these subjects, not only on the part of most biologists, but also on the part of the educated public in general. Both subjects promise to command an even larger share of our attention in the coming years.

14

# Hormonal Coordination

OUTLINE



In multicellular organisms, the activities of individual cells must be coordinated for the good of the organism. The degree of coordination is minimal in colonial protists, clusters of several one-celled individuals, and in the loosely organized sponges. Yet some sort of chemical communication system operates even in these organisms. In organisms with well-organized tissues and organ systems, the coordinating systems are necessarily more complex. Vertebrates, especially mammals and birds, with their highly evolved homeostatic systems, exhibit the most complex chemical coordinating systems of all. Consider, for example, a frightened antelope that attempts to flee a source of danger. The muscular contraction and relaxation cycles essential for its flight are coordinated by the nervous system. But other physiological changes associated with flight from danger, such as increased blood glucose concentration and the rate of energy metabolism and respiration, are regulated by hormones.

Animal **hormones** are often called chemical messengers. They are produced by certain tissues or organs in minute quantities and evoke specific responses in target tissues and organs elsewhere in the body. Growth and reproduction, metabolic rate, ion concentration in the blood, blood pressure, blood sugar concentration, production of milk and its ejection from the mammary glands, and water balance are examples of physiological activities controlled by animal hormones. In plants, hormones may be produced by any active cell and may affect the cell that produces them. Plant hormones act principally in regulation of growth and development. This chapter describes the hormonal control of bodily functions in vertebrates and invertebrates. The role of hormones in plant growth and development is covered in Chapter 26.

## ENDOCRINE GLANDS

By definition, **endocrine glands** lack ducts through which their secretory products can be discharged. Endocrine glands release their secretions directly into the circulatory system, which distributes them throughout the body. On the other hand, **exocrine glands,** such as salivary and sweat glands and the liver, release their secretions upon a surface, usually into a duct.

Like the endocrine glands, certain nerve cells, the **neurosecretory cells,** secrete substances that play hormonal roles. Neurosecretion as a means of chemical coordination seems to have appeared very early in evolution and is functional in all animals from hydra to humans. It is apparently the more primitive form of hormonal control. Discrete endocrine glands are found only in animals with circulatory systems: vertebrates and such higher invertebrates as insects, crustaceans, and cephalopods, a class of molluscs including squids, cuttlefishes, and octopuses. The study of endocrine glands and neurosecretory cells, their secretions and their effects is termed **endocrinology.**

## ROLES OF HORMONES

The physiological effects of animal hormones and neurosecretions fall into four categories: (1) the control of body growth; (2) the regulation of reproduction, including the development of secondary sexual characteristics such as sexual differences in plumage, size, and behavior; (3) the maintenance of homeostasis; and (4) the functions of integration and coordination, which are shared with the nervous system. In the cnidarians, or coelenterates, such as corals and jellyfish; the platyhelminths, or flatworms; and the annelids, such as earthworms and leeches, hormones regulate only growth and reproduction, whereas in higher invertebrates and vertebrates, homeostasis is also under hormonal control.

## ENDOCRINE-REGULATED ACTIVITIES IN INVERTEBRATE ANIMALS

In invertebrates, the endocrine integration of growth, development, and metabolism has been well established. Diffuse neurosecretory cells are the sole source of hormones in cnidarians and annelid worms. Our knowledge of these hormonal systems is very limited. More advanced invertebrates, such as the arthropods and cephalopod molluscs, possess discrete endocrine organs. In contrast to what we know of lower forms,

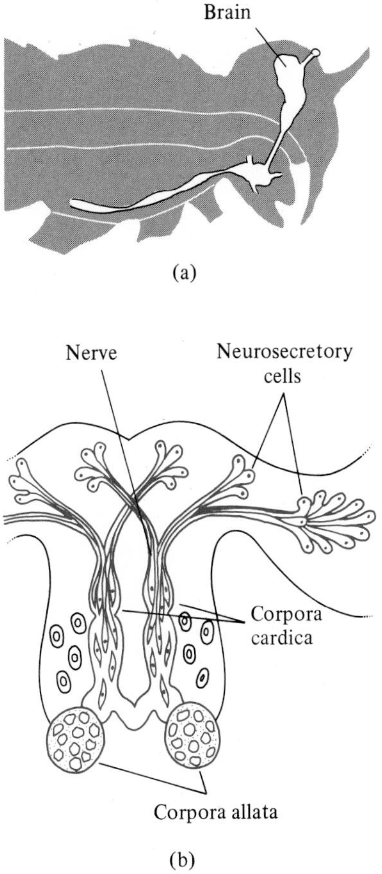


**FIG. 14-1** Arrangement of neuroendocrine glands in insects. (a) Location of the insect brain. (b) Details showing location of neurosecretory cells of the insect brain and associated neuroendocrine glands.

our knowledge of the endocrine controls of arthropods, particularly of crustaceans and insects, is extensive.

The four major sources of hormones in insects are the neurosecretory cells of the brain, various ganglia (clusters of nerve cell bodies), and two other structures composed of epithelial tissue, the prothoracic glands and the corpora allata (corpus allatum, singular) (Fig. 14-1). The neurosecretory cells of the brain produce many kinds of peptide (short amino acid chain) hormones, which pass down axons to the **corpora cardiaca,** where they are released into the circulation. These hormones, collectively called **brain hormones,** control many physiological processes. One such hormone is the **prothoracicotropic hormone** (PTTH), which regulates the activities of the prothoracic glands. When stimulated by PTTH, the prothoracic glands secrete a steroid hormone called **ecdysteroid,** or ecdysone, which regulates molting activity. Another neurosecretory product, the **allatotropic hormone,** regulates the **corpora allata.** When stimulated by this hormone, the corpora allata secrete **juvenile hormone** (JH), a lipid. Among other roles, JH controls differentiation of the adult organs in all insects and reproduction in some groups of insects. The metamorphosis of insect larvae, such as maggots and caterpillars, into pupae and then into adults, such as flies and butterflies, occurs only when JH is absent from the circulation.

Besides these developmental hormones, various peptide hormones regulate such aspects of metabolism as water balance, mobilization of lipid reserves, control of blood sugar concentration, and amino acid metabolism.

## VERTEBRATE ENDOCRINOLOGY

Although our main point of reference in the following description is human endocrinology, there is great similarity in the organization and function of the endocrine glands of all vertebrates. The molecular structure of hormones, even of protein hormones such as insulin, a regulator of carbohydrate and fat metabolism, is also strikingly similar among mammals. This has permitted the use of insulin derived from other mammals in the treatment of human diabetes. The locations of the principal human endocrine glands are shown in Figure 14-2, and their products and functions are listed in Table 14-1. Some of the functions of the more important of these glands are described in this section.

### THE PITUITARY GLAND

The pituitary is a relatively small gland, weighing 0.5 g in humans. Attached to the **hypothalamus,** at the bottom of the brain, by a slender stalk, it lies above the roof of the mouth. This gland has two major parts, the anterior lobe and posterior lobe (Fig. 14-3). The latter arises during development as an outgrowth of the floor of the brain, whereas the former arises as an ingrowth of the roof of the mouth. A third part, the intermediate lobe, is found in amphibians, reptiles, and most mammals. The posterior lobe secretes two hormones: **vasopressin,** also known as antidiuretic hormone (ADH), and **oxytocin.** The anterior lobe secretes at least seven hormones: (1) **growth hormone,** also called soma-

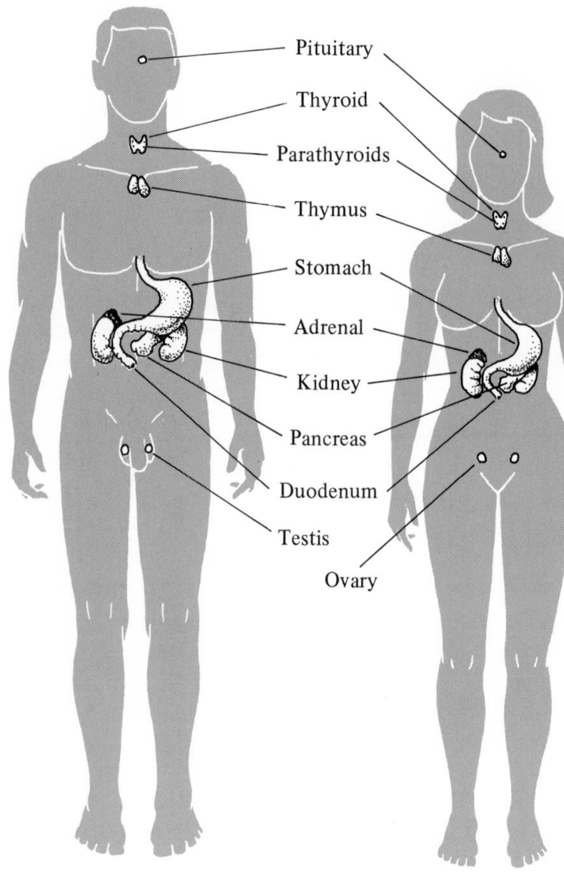


**FIG. 14-2** Location of the principal human endocrine glands.

totropin (STH); (2) **adrenocorticotropin** (ACTH); (3) **thyrotropin,** also called thyroid-stimulating hormone (TSH); (4) **intermedin,** also called melanocyte-stimulating hormone (MSH)*; (5) **luteinizing hormone** (LH), in males also called interstitial cell-stimulating hormone (ICSH); (6) **follicle-stimulating hormone** (FSH); and (7) **prolactin,** also known as lactogenic hormone or luteotropin. Because of their stimulatory effect on the gonads—the ovaries and testes—the last three are collectively termed the gonadotropins. All of the pituitary hormones are peptides.

The role of the pituitary and of any other endocrine gland can be experimentally demonstrated by ablation (surgical removal) and replacement therapy studies. For example, when the pituitary gland of an immature rhesus monkey is removed, an operation called hypophysectomy, the animal remains small in size and sexually immature and never loses its juvenile coat of hair (Fig. 14-4). From observations such as these, we know that the pituitary is essential for the control of normal growth and maturation. Similar results have been obtained with rats, dogs, fishes, and a variety of other vertebrates. Replacement therapy–the injection of active pituitary extracts into hypophysectomized animals, restores normal function. Such studies demonstrate that the pituitary gland in vertebrates is the source of the growth hormone. Besides growth, the pituitary has been shown to control sexual

* MSH is apparently produced by the anterior lobe in birds. It is produced by the pituitary's intermediate lobe (hence the term intermedin) in amphibians, reptiles, and most mammals. Although the intermediate lobe is present in the human fetus, both the lobe and MSH are absent in adult humans.

**TABLE 14-1 Some Human Hormones**

| *Source* | *Hormones* | *Target cells and principal actions* |
|---|---|---|
| Anterior lobe of pituitary | Somatotropin (STH, or growth hormone) | Growth of bone and muscle; promotes protein synthesis; affects lipid and carbohydrate metabolism |
| | Adrenocorticotropin (ACTH) | Stimulates secretion of adrenocortical steroids |
| | Thyrotropin (TSH) | Stimulates thyroid gland to synthesize and release thyroid hormones |
| | Gonadotropins: | |
| | Luteinizing or interstitial cell-stimulating hormone (LH or ICSH) | In ovary: formation of corpora lutea; secretion of progesterone; probably acts in conjunction with FSH |
| | | In testis: stimulates the interstitial cells of Leydig, thus promoting the secretion of androgen |
| | Follicle-stimulating hormone (FSH) | In ovary: growth of follicles; functions with LH to cause estrogen secretion and ovulation |
| | | In testis: possible action on seminiferous tubules to promote spermatogenesis |
| | Prolactin (lactogenic hormone or luteotropin) | Initiation of milk secretion by mammary glands; acts on crop sacs of some birds; stimulates maternal behavior in birds |
| Intermediate or posterior lobe of pituitary | Melanocyte-stimulating hormone (intermedin, MSH) | Expansion of amphibian melanophores; contraction of iridophores and xanthophores; melanin synthesis; darkening of the skin |
| Posterior lobe of pituitary | Vasopressin (ADH or anti-diuretic hormone) | Elevates blood pressure by acting on arterioles; promotes reabsorption of water by kidney tubules |
| | Oxytocin | Affects postpartum mammary gland, causing ejection of milk; promotes contraction of uterus; possible action in parturition and in sperm transport in female tract |
| Hypothalamus | Thyrotropin releasing hormone (TRH) | Stimulates release of TSH by anterior pituitary |
| | Adrenocorticotropin releasing hormone (CRH) | Stimulates release of ACTH by anterior pituitary |
| | Gonadotropin releasing hormone (GnRH) | Stimulates gonadotropin release by anterior pituitary |
| | Prolactin release inhibiting hormone (PIF) | Inhibits prolactin release by anterior pituitary |
| | Somatostatin | Inhibits release of STH by anterior pituitary |
| Thyroid gland | Thyroxine | Growth; amphibian metamorphosis; molting; metabolic rate in birds and mammals |
| Pancreas, islet cells | Insulin (from $\beta$ cells) | Promotes glycogen synthesis and glucose utilization |
| | Glucagon (from $\alpha$ cells) | Promotes glycogenolysis and raises blood glucose concentration |
| Adrenal cortex | Glucocorticoids (e.g., cortisol, corticosterone) | Promote synthesis of carbohydrate; protein breakdown; anti-inflammatory and antiallergic actions |
| | Mineralocorticoids (e.g., aldosterone) | Sodium retention and potassium loss through kidneys |
| Adrenal medulla | Epinephrine | Mobilization of glycogen; increased blood flow through skeletal muscle; increased oxygen consumption; heart rate increase |
| | Norepinephrine | Adrenergic neurotransmitter; elevation of blood pressure; constricts arterioles and venules |
| Testes | Androgens (e.g., testosterone) | Male sexual characteristics |
| Ovaries | Estrogens (e.g., estradiol) | Female sexual characteristics |
| Corpus luteum | Progesterone | Maintains pregnancy |
| Stomach wall cells | Gastrin | Promotes secretion of HCl by stomach wall |
| Duodenum wall cells | Secretin | Promotes bicarbonate secretion and digestive juices by the pancreas |

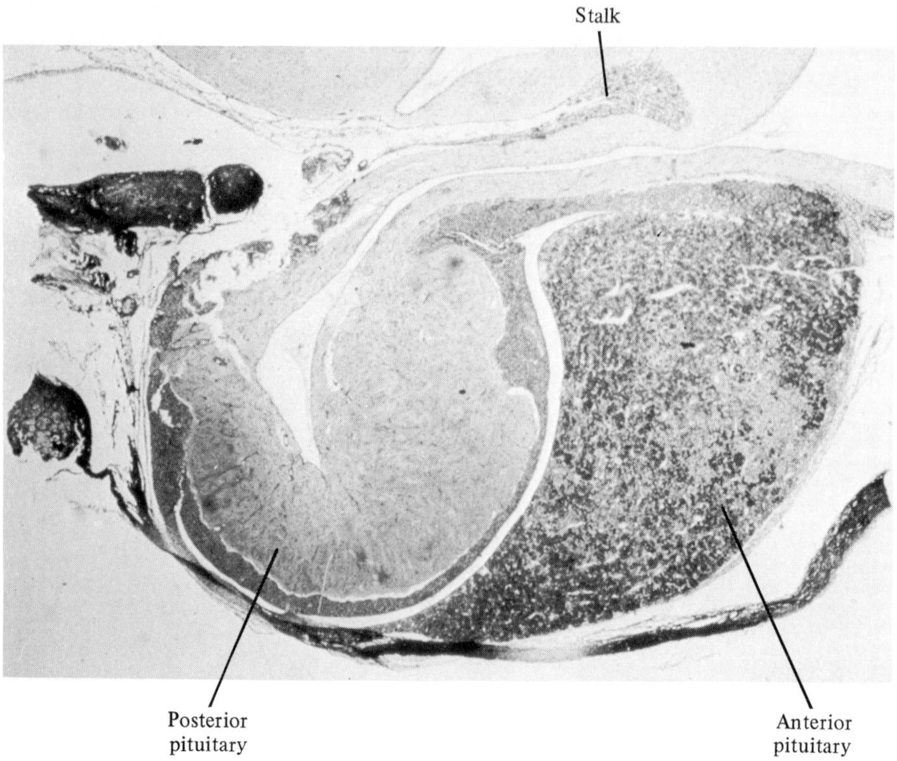


**FIG. 14-3** Longitudinal section through the pituitary gland of a human adult.

maturation. The nine pituitary hormones may be divided into two major groups, depending on the glands or tissues on which they act. One group, the tropic hormones, includes ACTH, TSH, and the gonadotropins FSH and LH, all products of the anterior pituitary. These act on other endocrine glands, the adrenal glands, the thyroid, and the gonads, respectively, to coordinate the functions of these glands. Hormones of the other group, including growth hormone, intermedin, prolactin, oxy-

**FIG. 14-4** Effects of hypophysectomy on growth and maturation of the rhesus monkey. Although the two monkeys weighed the same (about 2.7 kg) at the time of the operation, 2 years later the operated animal still weighed 2.7 kg, while the unoperated control animal weighed 8.4 kg.

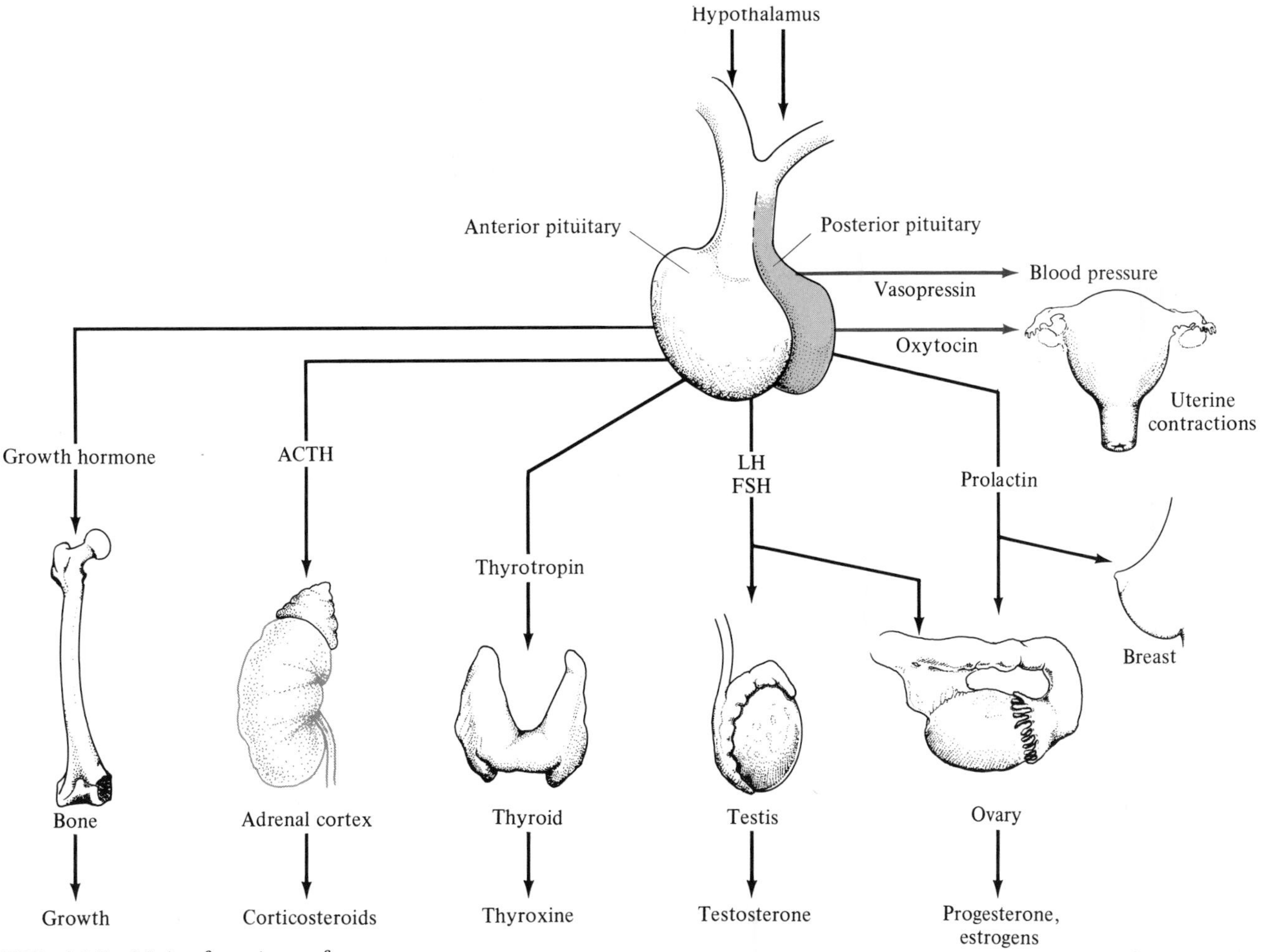


**FIG. 14-5** Major functions of the pituitary hormones and their target organs and tissues.

tocin, and vasopressin, act directly on nonendocrine target tissues (see Fig. 14-5 and Table 14-1).

## FUNCTIONS OF THE HYPOTHALAMUS

The posterior lobe of the pituitary, which arises from a part of the embryonic brain called the **hypophysis,** maintains close connection to and is controlled by the hypothalamus, a part of the brain rich in neurosecretory neurons. One hypothalamic role is to produce precursors of oxytocin and vasopressin, which are activated within the posterior lobe. The hypothalamus also produces several short-chain polypeptides known as the **hypophysiotropic factors.** These include several releasing hormones, or factors, that stimulate release of various anterior lobe hormones; other factors inhibit release of certain anterior lobe hormones. Included are **thyrotropin releasing hormone** (TRH), **gonadotropin releasing hormone** (GnRH), **(adreno)corticotropic releasing hormone** (CRH), a somatotropin inhibiting hormone, **somatostatin,** and a **prolactin release inhibiting factor** (PIF).

Besides being affected by the hypothalamic secretions, release of each of the anterior pituitary hormones that stimulate other endocrine glands, such as the thyroid gland, is inhibited by the target gland secretions whenever they reach a certain concentration in the bloodstream, a

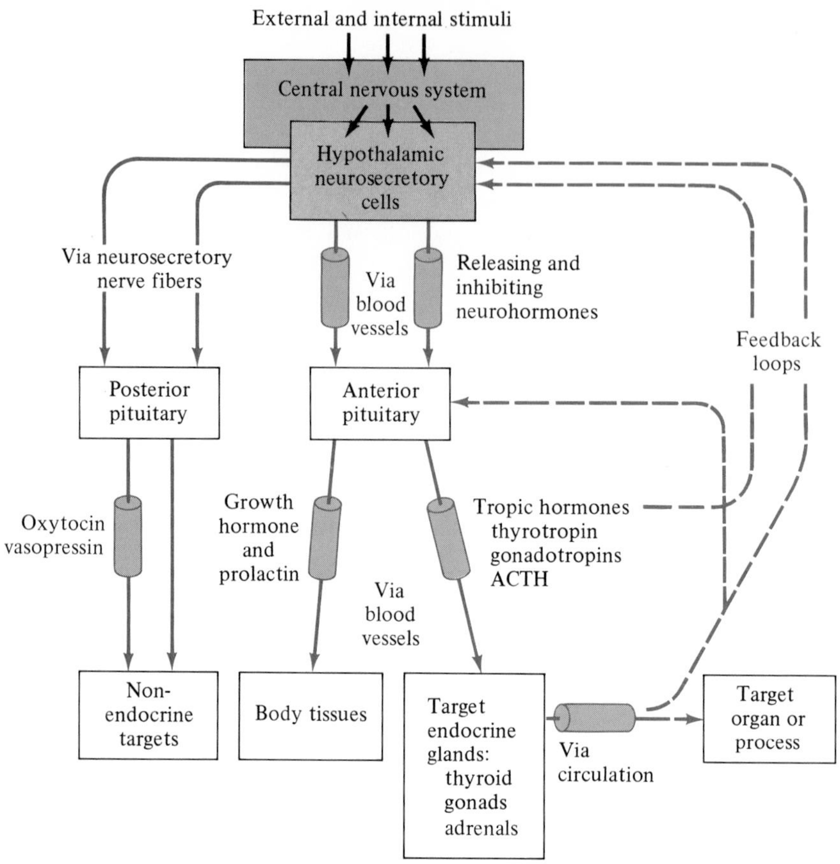


**FIG. 14-6** Summary of some of the hormonal interactions mediated by the hypothalamus and the pituitary gland in mammals.

classic example of negative feedback. These various relationships are summarized in Figure 14-6.

**FIG. 14-7** Hypothyroid goiter. The hypothyroid goiter of Marie de Medici (1573–1642, Queen of Henry IV of France) is evident in this portrait of her by Rubens. Until it was found to be due to hypothyroidism, a plump goiter was regarded as a mark of beauty in women. Regular inclusion of a small amount of iodine in the diet will prevent this disease.

## THE THYROID GLAND

The human thyroid gland is a paired structure straddling the trachea just below the larynx (Fig. 14-2). **Thyroxine,** the hormone secreted by the gland, is an iodinated derivative (that is, one to which iodine has been added) of the amino acid tyrosine. It is released into the circulation bound to thyroglobulin, a protein.

Thyroxine has multiple effects on the metabolism of adult vertebrates and affects development in embryonic and juvenile stages as well. Thyroidectomy—the surgical removal of the thyroid—or thyroid insufficiency—a lack of adequate amounts of the hormone—reduce the body's basal metabolic rate* while increasing the concentration of extracellular sodium and water and of cholesterol in the blood. These symptoms can be alleviated by administration of thyroxine.

The secretory activity of the thyroid gland varies, depending on the season of the year, diet, and reproductive state of the animal. Humans and other nonhibernating "warm-blooded" animals, or homeotherms, produce more thyroxine in the colder months of the year. On the other hand, emotional stress and experiences such as hemorrhage, trauma,

* The basal metabolic rate of an individual is the overall rate at which the chemical reactions needed to maintain life proceed. It is usually defined as the kilocalories of energy expended by the resting, fasting individual per square meter of body surface per hour. It can be determined by measuring the rate of $O_2$ consumption or $CO_2$ production.

and exposure to toxic substances decrease thyroid secretion. The iodine content of the diet also influences thyroxine production. If dietary intake of iodine is insufficient, the thyroid responds by increasing in size, resulting in a hypothyroid goiter, common in some regions of the world before iodized salt came into general use (Fig. 14-7).*

It has been experimentally established that thyrotropin, a glycoprotein hormone secreted by the anterior pituitary, controls thyroid activity. Following hypophysectomy, only traces of thyroxine appear in the blood, and the thyroid gland exhibits evidence of reduced activity. This situation can be alleviated either by injection of pituitary extracts or by injection of purified thyrotropin. Normal thyroid function depends on secretion of thyrotropin by the pituitary. But what controls the production and release of thyrotropin? The neurosecretory cells of the hypothalamus secrete thyrotropin-releasing hormone (TRH), which prompts the synthesis and release of thyrotropin by the anterior lobe of the pituitary. When a mammal is exposed to cold or certain other stresses, the sensory input to the brain results in nervous stimulation of the hypothalamus, which responds by producing more TRH.

Control of thyroxine output is achieved by a negative feedback loop comparable to the mechanism of a thermostat that turns off a furnace once the temperature reaches a set level. When the thyroxine level in the circulation reaches optimal concentration, the cells in the pituitary that produce thyrotropin and the neurosecretory cells in the hypothalamus that produce TRH are inhibited. Thyrotropin release is correspondingly reduced. Thyroxine is kept at normal levels by this feedback control.

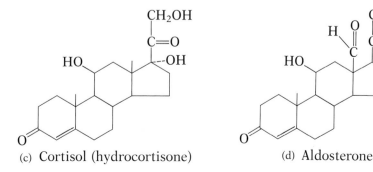

**FIG. 14-8** (a) Stained section through the adrenal gland, showing parts of the cortex and of the medulla. (b) Epinephrine, principal hormone of the adrenal medulla. Norepinephrine differs by the absence of the methyl group ($CH_3$). (c) Cortisol (hydrocortisone), one of the glucocorticoids of the adrenal cortex. (d) Aldosterone, a mineral corticoid.

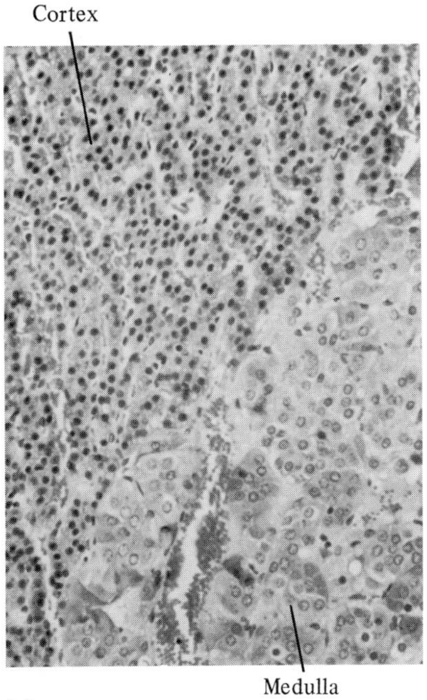

(a)

## THE ADRENAL GLAND

The adrenal glands in higher mammals consist of two distinct portions: an outer cortex and an inner medulla (Fig. 14-8a). These two parts have separate developmental origins, different modes of control, and distinctly separate functions. The cortex secretes several steroid hormones, collectively called the **corticosteroids.** Cortical hormone production is under the control of adrenocorticotropin from the anterior pituitary.

***The Adrenal Medulla.*** The medulla of the adrenal gland is composed of modified nerve cells. The principal product of these cells, amounting to 80 percent of the total, is epinephrine, commercially known as adren-

* People on a salt-free diet should not depend entirely on food raised in an inland region with iodine-poor soil, but should frequently include some iodine-containing food in their diet, such as seafood or vegetables raised in coastal states.

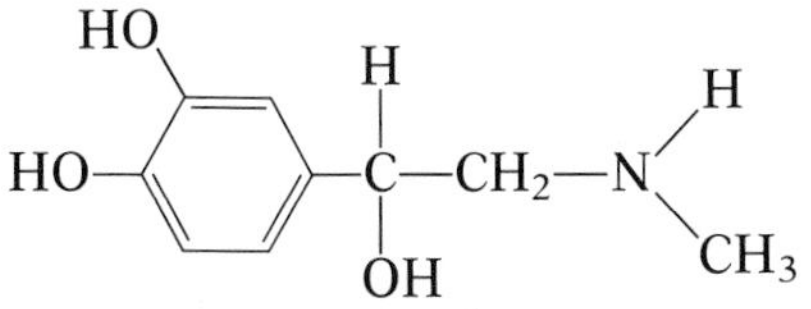

(b) Epinephrine

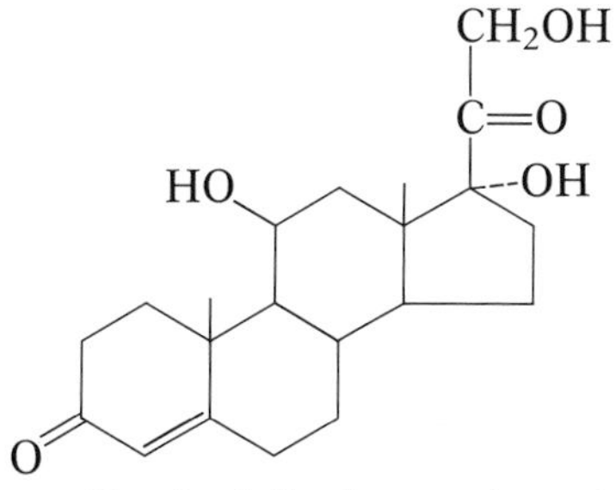

(c) Cortisol (hydrocortisone)

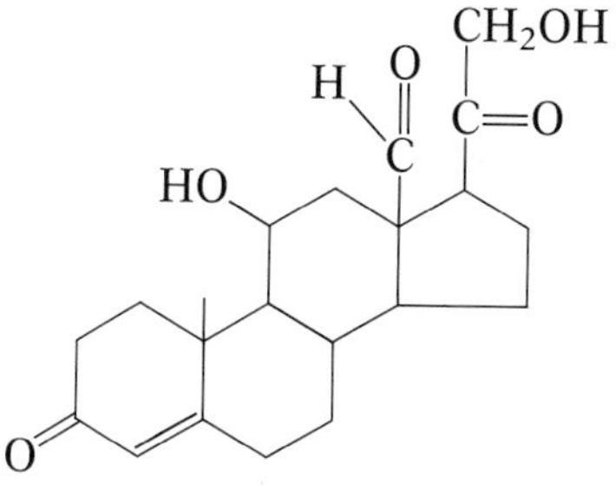

(d) Aldosterone

aline (Fig. 14-8b), and the remainder is a similar substance, norepinephrine, or noradrenaline. The two are collectively termed **catecholamines** and have similar effects.

When released into the bloodstream, epinephrine creates conditions throughout the body that prepare the animal for "fight or flight," that is, to meet a stressful situation. A sudden release of epinephrine—in response to fright or anger, for example—raises blood pressure, increases the blood supply to the liver and muscles, increases blood sugar concentration, stimulates respiration, dilates air passages, and increases cardiac output. The high blood pressure of people who are regularly subjected to stress is often due to increased output of catecholamines.

Because the adrenal medulla arises embryonically from nervous tissue, it is not surprising that it can react almost as rapidly as a nervous reflex in an emergency, instantaneously increasing its output of epinephrine. Unlike most hormone actions, the effects of adrenaline are fully achieved within a few seconds. Epinephrine, moreover, is short-lived, being inactivated by the liver within 5 minutes. In contrast, the responses of the body's tissues to increased or decreased thyroxine levels may extend over several weeks.

***The Adrenal Cortex.*** The cortex of the adrenal gland secretes two groups of corticosteroids: the glucocorticoids, such as cortisol (hydrocortisone) (Fig. 14-8c), and the mineralocorticoids, such as aldosterone (Fig. 14-8d). The **glucocorticoids** regulate sugar and protein metabolism. Because they also reduce the number of lymphocytes in the body, they are sometimes used as immunosuppressive agents to inhibit graft rejection in organ transplant patients (Chapter 12). The **mineralocorticoids** regulate water and electrolyte balance of body fluids.

The hormones of the adrenal cortex are essential for life. The symptoms of corticoid deficiency include extreme muscular weakness, hypoglycemia (low blood sugar concentration), reduced blood pressure and body temperature, dehydration, a higher than normal concentration of blood cells, and kidney failure.

Corticosteroid production by the adrenal cortex is under the control of adrenocorticotropic hormone from the anterior pituitary. The synthesis and release of ACTH is controlled by a cortical-releasing hormone (CRH) of the hypothalamus, which in turn is controlled by other parts of the brain and is affected by levels of various hormones and other substances in the bloodstream.

## REPRODUCTIVE HORMONES

Besides possessing different gonads, males and females of most species differ in various other ways, referred to as their secondary sexual characteristics. They may differ in pigmentation, skeletal size and shape, patterns of body hair distribution, pitch of voice, and, especially where reproductive activities are concerned, in behavior. Species in which adult males and females differ substantially in body structure, or morphology, are said to exhibit **sexual dimorphism** (from the Greek *morphe,* "form").

Development of all such secondary sexual characteristics is regulated by the sex hormones testosterone and estradiol. **Testosterone,** the male sex hormone, which is produced by the testes, promotes develop-

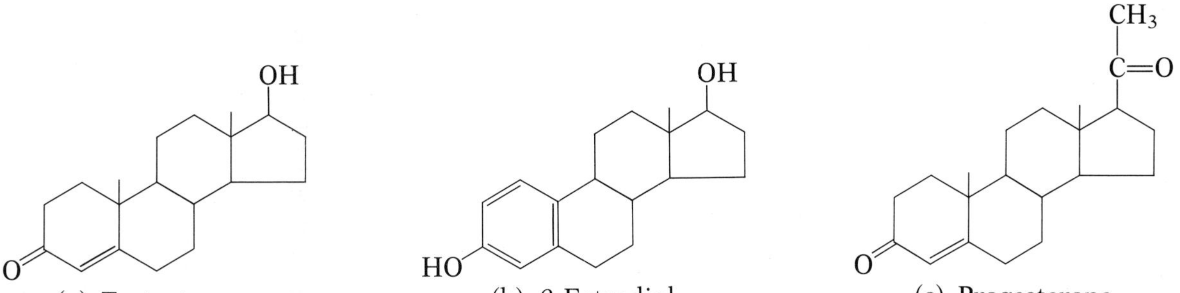


(a) Testosterone (b) β-Estradiol (c) Progesterone

**FIG. 14-9** The major sex hormones of vertebrates. (a) Testosterone, produced by the testes, is responsible for development of all secondary sexual characteristics in males. (b) Estradiol, an estrogen, is produced by the ovaries and is required for development of female secondary sexual characteristics. It also plays a role in the menstrual cycle and in pregnancy. (c) Progesterone, produced by the corpus luteum, a specialized tissue in the ovary, is also required for maintenance of pregnancy. Both estrogen and progesterone are also produced by the placenta.

ment of male secondary sexual characteristics (Fig. 14-9a), whereas **estradiol,** an estrogen produced by the ovaries, is necessary for development of female structures, appearance, and behavior (Fig. 14-9b). Like thyroxine secretion, secretion of the sex hormones is regulated by hormones from the anterior pituitary, in this case, FSH and LH, and their corresponding releasing factors produced by the hypothalamus. Reproductive cycles are influenced by input from the brain and may be synchronized with certain times of the year, month, or day.

Another steroid sex hormone is **progesterone,** produced by a special tissue within the ovary called the **corpus luteum** (Fig. 14-9c). Progesterone is important in maintaining a pregnancy. The **placenta,** the membranous complex on the wall of the uterus, or womb, to which the embryo is attached by the umbilical cord and through which it receives nourishment, also functions as an endocrine gland. It produces the gonadotropins estrogen and progestin throughout the pregnancy.

Details of the hormonal control of menstruation and pregnancy in humans are discussed in Chapter 22.

## INSULIN AND GLUCAGON

The pancreas is a gland that contains exocrine cells, in this case ones that elaborate digestive secretions, as well as patches of endocrine tissue called **islets of Langerhans** (Fig. 14-10). The islets are composed of two types of cells, $\beta$ and $\alpha$ cells. They secrete the hormones insulin and glucagon, respectively.

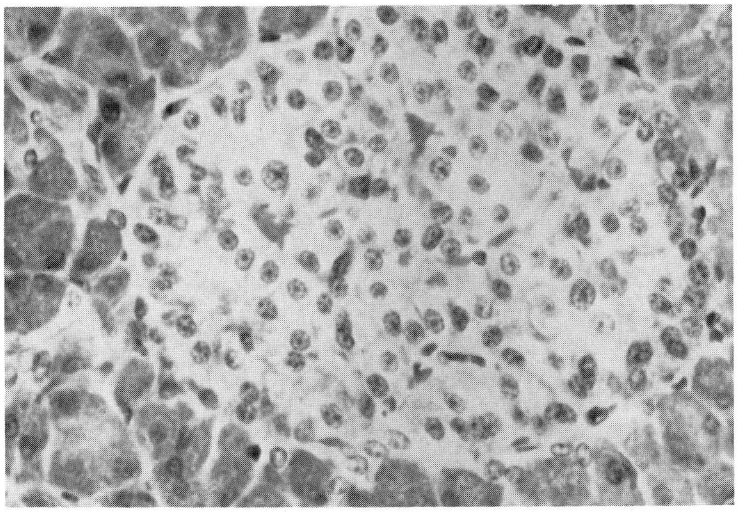

**FIG. 14-10** An islet of Langerhans, one of the groups of pancreatic cells that produce insulin and glucagon. The more darkly stained surrounding tissue produces the pancreatic digestive enzymes.

**Insulin** directly or indirectly affects many kinds of biochemical processes. Its major functions are to promote synthesis of glycogen by liver and muscle and to facilitate utilization of glucose by all cells. In addition, insulin increases movement of amino acids and glucose into cells and promotes protein and lipid synthesis. In insulin-deficient individuals, the amino acid and glucose levels in the blood remain unusually high. If the blood glucose level rises above the ability of the kidneys to reabsorb it from the glomerular filtrate, the excess is lost via the urine. This diseased state, called diabetes, is characterized by the passing of abnormally high amounts of urine, a constant feeling of thirst, a high intake of fluids, a constant feeling of hunger, and a strong tendency to eat excessively.

Because of protein and lipid breakdown, the acutely diabetic individual gradually becomes weak, and the condition is aggravated by a lowering of glycogen stores in the liver and muscle. Ketone bodies, consisting of partially oxidized fat, increase in the blood as a consequence of an abnormally high dependence on fat as a source of energy, making the blood dangerously acidic. As a consequence of the abnormal amount of metabolic work involved in reabsorbing excess sugar and amino acids, the kidneys cannot maintain the blood's normal ionic balance. Eventually, if these acute conditions persist, diabetic coma occurs, the individual goes into shock, and death ensues.

Glucose utilization is controlled in another way by **glucagon,** the other pancreatic hormone. As an antagonist of insulin, glucagon promotes the breakdown of glycogen, or glycogenolysis, in the liver and increases the glucose concentration of blood. Insulin and glucagon thus exert opposing influences to maintain the blood glucose level within a narrow range.

We have noted that epinephrine (adrenaline) and the glucocorticoids also influence blood sugar levels. The interplay of insulin, glucagon, and these hormones in the homeostasis of blood sugar is diagrammatically represented in Figure 14-11.

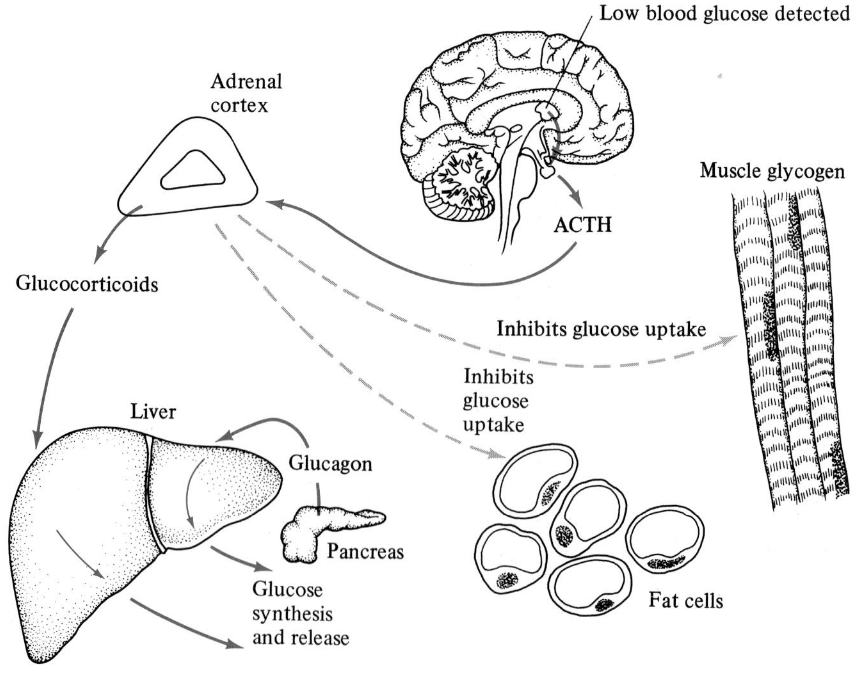


**FIG. 14-11** Endocrine control of glucose metabolism in response to a fall in blood sugar (glucose) concentration. ACTH stimulates release of glucocorticoids, which restore normal glucose concentration by inhibiting uptake of glucose from the blood by muscle and fat cells while promoting gluconeogenesis (synthesis of glucose) by the liver.

### GASTROINTESTINAL HORMONES

Hormones, rather than the nervous system, are the chief regulators of the activities of the digestive system. For example, **gastrin,** secreted by certain cells of the stomach wall, stimulates other stomach wall cells to secrete hydrochloric acid. **Secretin,** produced by cells lining the duodenum in response to the presence of HCl, stimulates the pancreas to release exactly the amount of bicarbonate that will neutralize the amount of HCl present.

## MODES OF HORMONAL ACTION

From the foregoing discussion, it is clear that hormones affect only specific target tissues. For example, TSH stimulates only the thyroid gland, ACTH only the adrenal cortex, and FSH and LH only the gonadal tissues. However, certain hormones may affect more than one target tissue. For example, estrogen, the principal female sex hormone, causes the growth of both the uterine lining, or endometrium, and the mammary glands; vasopressin acts on both the kidney and the smooth muscles of the blood vessels; and oxytocin stimulates contractions of the smooth muscles of both the uterus and the mammary gland ducts. Some hormones influence the activity of an even greater variety of tissues. For example, insulin influences glucose metabolism in adipose tissue, liver, skeletal muscle, and the glucagon-producing $\alpha$ cells of the pancreas.

Studies have shown that target cell specificity is due to the presence in target cells of various receptor proteins that bind only to specific hormones. Receptors for the female sex hormone, estrogen, for example, are present in the cells of uterine tissues, mammary glands, the pituitary, and the hypothalamus. All of these tissues respond, each in its own way, to this same hormone. Cells that lack receptors for a particular hormone are unaffected by that hormone.

### THE LOCATION OF HORMONE RECEPTORS

The receptor proteins are located either within the target cells or on their surface, depending on the kind of hormone involved. For example, steroid hormones, such as estrogen, progesterone, testosterone, and the corticosteroids, readily diffuse through the plasma membranes of target cells, then enter the nucleus. Within the nucleus, they bind with a protein receptor molecule to form a receptor-hormone complex. This complex activates specific genes.

Because the plasma membrane is impermeable to protein hormones, such as insulin, and glucagon and to the catecholamines, these hormones do not enter their target cells. Instead, the hormone molecules bind to specific receptor molecules on the outer surface of the plasma membrane of the target tissue cell. When the hormone binds to it, the molecular configuration of the protein receptor changes, producing appropriate responses by the cell. For example, following the binding of insulin to plasma membrane receptors, the plasma membranes of the target cells become more permeable to glucose. Insulin thus stimulates the entry of glucose into cells without entering the cells itself.

## THE MODE OF ACTION OF STEROID HORMONES

As described in Chapter 4, the synthesis of enzymes and other proteins occurs within cells on the ribosomes found both on the rough endoplasmic reticulum and free in the cytoplasm. Protein synthesis is initiated when the ribonucleic acid molecules known as messenger RNA (mRNA) join to ribosomes. All mRNA molecules are synthesized in the nucleus. They are actually copies of certain portions of the deoxyribonucleic acid (DNA) molecules, or genes, that constitute the genetic material of the chromosomes. This copying process is termed transcription. When they become bound to ribosomes, mRNA molecules serve as patterns, or templates, for the synthesis of polypeptide chains. The sequence of nucleotide units in the mRNA strand (copied from a gene) specifies the sequence in which various of the 20 kinds of amino acids will occur in the newly synthesized polypeptide strand. This nucleotide-sequence-to-amino-acid-sequence conversion process is known as translation. (These genetic mechanisms are discussed in Chapter 24.)

Steroid hormones, such as the sex hormones and corticosteroids, affect transcription within the nuclei of the target tissue cells. Being highly soluble in lipids, steroids make their way readily through the plasma membranes of cells (Fig. 14-12) and enter the nucleus. In the nucleoplasm of target tissue cells, they encounter specific protein receptor molecules with which they combine to form hormone-receptor complexes. The hormone-receptor complex binds to specific sites on the chromatin of the chromosomes, triggering the synthesis of specific mRNA molecules. These mRNA molecules are transported out to the

**FIG. 14-12** The probable sequence of events by which steroid hormones affect target cells. The blood-borne steroid enters the cell by passing through the lipid bilayer of the plasma membrane, then enters the nucleus where it binds with a receptor protein molecule. The resulting complex stimulates transcription of a specific region of DNA, generating a strand of mRNA that, upon diffusing into the cytoplasm, directs the synthesis, or translation, of polypeptides by the cell's ribosomes.

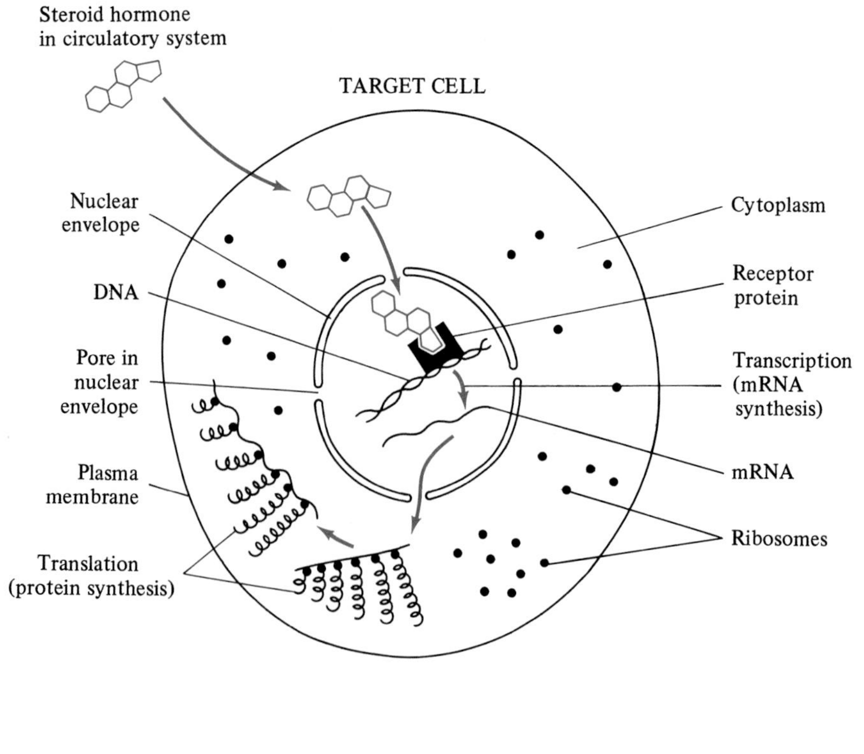

cytoplasm, where, on the ribosomes, they provide the specifications for the synthesis of specific enzymes or other proteins.

## THE MODE OF ACTION OF A NONSTEROID HORMONE, EPINEPHRINE

Hormones such as epinephrine which do not enter their target cells, bind to a receptor protein in the cell's plasma membrane (Fig. 14-13). The formation of this hormone-receptor complex activates **adenylate cyclase,** an enzyme associated with the receptor protein. The effect is known to be mediated by a third molecule called the G protein, which becomes active when it binds to guanosine triphosphate (GTP). Once activated, adenylate cyclase catalyzes the synthesis of molecules of **cyclic adenosine monophosphate (cAMP)** from ATP (Fig. 3-44b). The cAMP thereupon activates the protein kinases found in that kind of target cell. Kinases are enzymes that modulate (activate or inhibit) the activities of various proteins by phosphorylating (adding phosphate to) them. In some or all cases, cAMP may act as an allosteric effector (see Chapter 6) of the protein kinase in achieving the activation or inhibition.

Hormonal effects can be greatly amplified when cAMP acts as a second messenger. One hormone molecule entering the receptor site results in the production of many molecules of cAMP. The protein kinases that are activated provide further amplification. In some cells, still further amplification occurs when substances known as prostaglandins are also activated and stimulate the production of more cAMP. Such a reaction of cellular mechanisms is somewhat like the mobilization of an entire military unit in response to the receipt of a message at battalion headquarters.

## PROSTAGLANDINS

Prostaglandins are a unique group of 20-carbon fatty acids, each with a five-carbon ring. They are derived from unsaturated fatty acids, which have one or more double bonds between their carbon atoms. The pros-

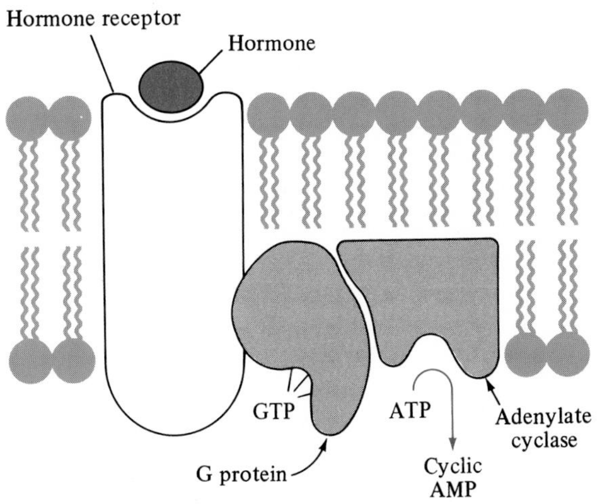


**FIG. 14-13** The mode of action of a hormone, such as epinephrine or glucagon, that does not enter the cell but binds to a receptor on the surface of the plasma membrane. Activation of the enzyme adenylate cyclase is achieved through the interaction of the receptor molecule and an intermediate molecule, G protein (so-called because it binds another nucleotide, guanosine triphosphate, GTP). Adenylate cyclase catalyzes the synthesis of cAMP from ATP.

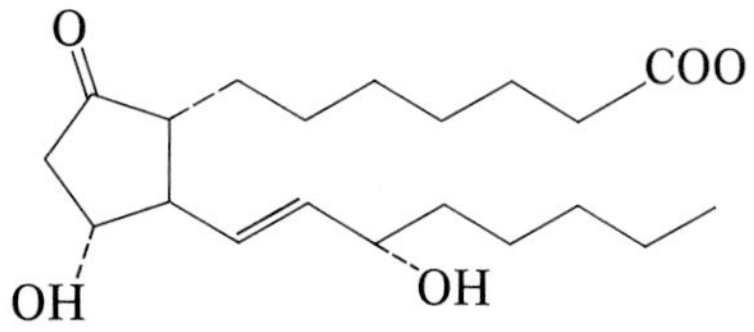


**FIG. 14-14** The prostaglandin $PGE_1$, an inhibitor of lipolysis in adipocytes. The subscript denotes its single set of double bonds.

taglandins are divided into four classes, designated by the letters E, F, A, and B; the names are abbreviated $PGE_1$, $PGE_2$, $PGF_1$, and so on, the subscripts denoting the number of double bonds in the molecule.

Prostaglandins were first discovered in human seminal fluid and, although produced primarily by the seminal vesicles, were so named because they were believed to be produced by the prostate gland, which produces most of the fluid. But they have since been found in nearly every tissue of the body. Although some function as hormones and can be detected in the blood, nearly all function only within the same cells in which they are produced, where they modulate hormonal effects.

The prostaglandin illustrated in Figure 14-14 is $PGE_1$. One of its effects is to inhibit lipolysis, or fat breakdown, in adipose (fat) tissue. $PGE_1$ thus antagonizes the action of such hormones as epinephrine and glucagon in this type of cell.

Prostaglandins enhance the inflammatory reaction of many tissues. The effect of aspirin in reducing inflammation is apparently achieved by inhibiting the synthesis of certain prostaglandins. In 1982 a Nobel prize was awarded for work in elucidating the action of prostaglandins.*

## Summary

The physiological functioning of organs and tissues must be closely coordinated for the survival and well-being of the organism. In animals, this is achieved by the nervous and hormonal systems. The hormonal system usually consists of several endocrine tissues and organs and their hormones. Animal hormones regulate four categories of biological activity: (1) growth, (2) reproduction and development, (3) homeostasis, and (4) integration and coordination. Plant hormones regulate only growth and development.

A hormone is a chemical substance that is produced and released in relatively small amounts in one part of the body and regulates tissues of another part of the body. Two types of animal tissues—the neurosecretory cells (modified neurons) and the epithelial endocrine glands (ductless glands) or tissues—produce hormonal agents. Whereas neurosecretory cells are found in all animals, epithelial endocrine glands are present only in vertebrates and those highly evolved invertebrates with circulatory systems. Hormones are chemically varied and include modified amino acids, polypeptides, proteins, glycoproteins, steroids, and catecholamines.

Among the invertebrates, the hormones of insects and crustaceans have been studied most extensively. An example of an important arthropod hormone is ecdysteroid or ecdysone, a steroid that controls both growth and molting. In insects, ecdysone is produced by the prothoracic glands under the influence of a brain hormone. The juvenile hormone controls development of adult structures and reproduction in most insects.

The pituitary gland of vertebrates coordinates several biological activities. Some of its hormones regulate the secretory activities of three other endocrine tissues: the thyroid, the adrenal cortex, and the gonads.

* The 1982 prize was shared by Sune K. Berström and Begnt I. Samuelsson of Sweden and John R. Vane of England.

Other pituitary hormones coordinate such functions as growth, blood pressure, and urine excretion, acting directly on nonendocrine target tissues. Production of hormones by the pituitary is subject to negative feedback control from the products of the target tissues or glands.

Thyroxine, the hormone produced by the thyroid gland, has multiple effects on metabolism, including increasing its rate, and is needed for normal development. Thyroxine output varies with the seasons, diet, and reproductive state and is reduced following trauma. An enlarged thyroid gland, or goiter, may result from an insufficiency of dietary iodine, a constituent of thyroxine. Thyroxine secretion is controlled by thyrotropin, or thyroid-stimulating hormone (TSH), from the anterior pituitary; TSH output is controlled by thyrotropin-releasing hormone (TRH) from the hypothalamus. The inhibition of TSH and TRH output by higher than normal concentrations of thyroxine in the bloodstream is an example of a negative feedback loop.

Fine control of reaction to stressful conditions is achieved by responses of the nervous system and by an interplay of corticosteroids from the adrenal cortex, epinephrine and norepinephrine from the adrenal medulla, and insulin and glucagon from the pancreas.

The two groups of corticosteroids, the glucocorticoids and the mineralocorticoids, from the adrenal cortex are produced in response to adrenocorticotropin (ACTH) from the anterior pituitary and cortical-releasing hormone (CRH) from the hypothalamus. The glucocorticoids regulate sugar and protein metabolism, while the mineralocorticoids regulate water and electrolyte balance of body fluids.

Reproduction in most animals is a seasonal activity and involves the coordination of the function of various organs and appropriate behavior. All of these are controlled by steroid gonadal hormones produced by the ovaries and testes. Production of sex steroids is regulated by gonadotropins from the pituitary, which in turn are controlled by hypothalamic neurosecretions.

A striking feature of all hormones is their target specificity. Although hormones circulate throughout the body, only specific target tissue cells respond to them, each in its own way. This specificity is attributed to the presence in the target tissue cells of hormone-specific protein receptor molecules. Steroid hormones bind to receptor molecules after entering the nucleus, where they activate specific genes. Other hormones react with receptor molecules on the cell surface. In some cases, this reaction promotes the synthesis of cAMP by an enzyme within the plasma membrane. The cAMP in turn activates protein kinases, which activate or inhibit various proteins by phosphorylating them.

Prostaglandins are modified 20-carbon, unsaturated fatty acids found in nearly every type of tissue cell; they modulate hormonal effects. They have a large variety of roles, either enhancing or inhibiting various cellular functions.

## Recommended Readings

CREWS, D. 1979. "The hormonal control of behavior in a lizard." *Scientific American,* 241(2):180–187. The anole lizard is the subject of this study of how the brain and the sex glands interact in the control of sexual behavior.

GUILLEMIN, R., and R. BURGUS. 1972. "The hormones of the hypothalamus." *Scientific American,* 227(5):24–33. An excellent account of research on the chemistry of the hypothalamic releasing factors by a Nobel laureate and a coworker.

HADLEY, M. E. 1984. *Endocrinology.* Prentice-Hall, Englewood Cliffs, N.J. An easily read, interesting, and up-to-date introduction to the study of vertebrate hormones.

LEVINE, S. 1971. "Stress and behavior." *Scientific American,* 224(1):70–86. Studies on the interaction of the pituitary and adrenal cortex hormones suggest that stress may play a major role in behavior.

O'MALLEY, B. W., and W. T. SCHRADER. 1976. "The receptors of steroid hormones." *Scientific American,* A description of how the receptor-hormone complex activates certain genes in target cells.

PASTAN, I. 1972. "Cyclic AMP," *Scientific American,* 227(2):97–105. Descriptions of some of the many functions performed by this small molecule.

TURNER, C. D., and J. T. BAGNARA. 1976. *General Endocrinology,* 6th ed. Saunders, Philadelphia. A well-written, interesting introduction to the subject.

WELLER, H. and R. L. WILEY. 1985. *Basic Human Physiology,* 2nd ed. Prindle, Weber & Schmidt, Boston. Homeostasis is the theme of this successful text.

# 15

# The Organization of Nervous Systems

OUTLINE



In Chapter 14, we saw how hormones regulate integration and coordination of activities in animals. Nearly all hormonal control is achieved through slow, prolonged responses that can extend over many days. Rapid reactions, especially those that last only briefly, are controlled by the nervous system. The organization of nervous systems in representative animal groups is discussed in this chapter.

The basic structure of the nervous system is a network of nerve cells called **neurons** (Fig. 15-1). Each neuron is composed of a **soma** (plural, somata), or nerve cell body, which contains the cell's nucleus, and an **axon,** a fiber of variable length—up to 2 meters in some animals—that extends from the soma. Besides the axon, one or more additional fibers, termed **dendrites,** may also project from the soma. In general, dendrites conduct impulses toward the soma, whereas the axon conducts them away from the soma. Shown in Figure 15-1 are three types of neurons: (a) a motor neuron, one that conducts impulses to muscles or glands; (b) a sensory neuron, one that conducts impulses from a sense receptor to the central nervous system (CNS), that is, the brain and spinal cord; and (c) an example of a type of neuron found in the CNS.

According to the neuron theory, developed early in this century, each neuron is a discrete unit separated from adjacent neurons by minute gaps called **synapses** (Fig. 15-2). Synapses are too small to be seen by the

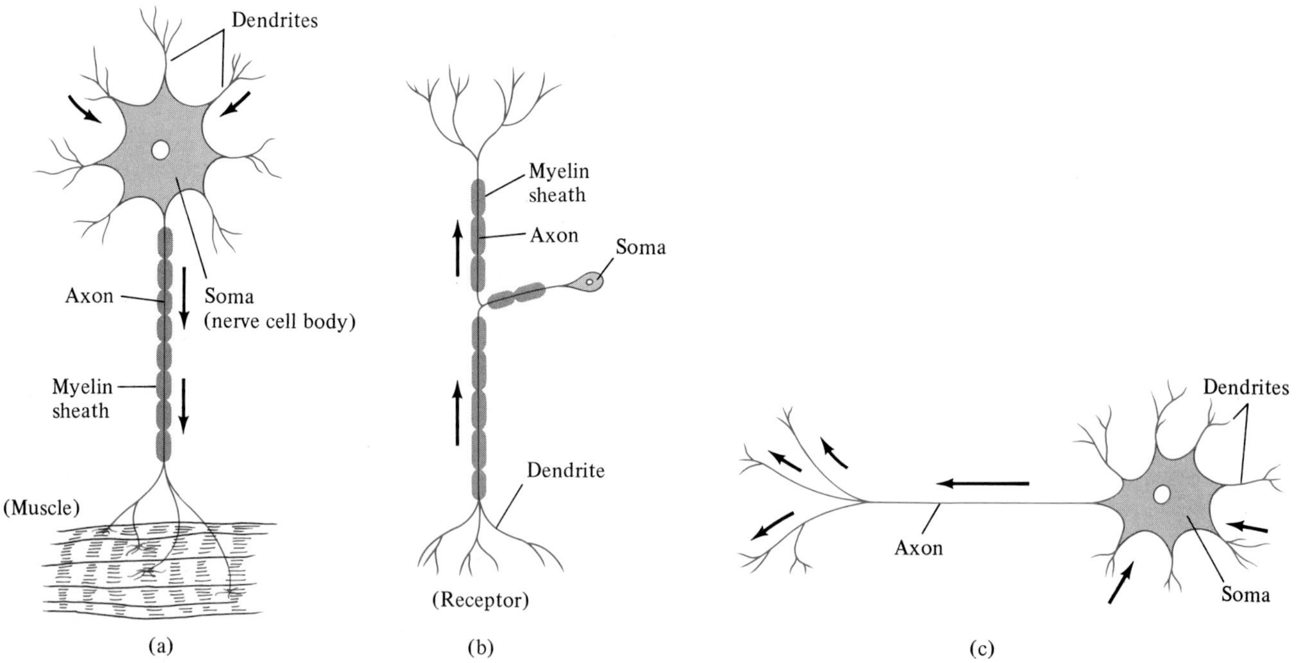


**FIG. 15-1** Three major types of neurons: (a) a motor, or efferent, neuron; (b) a sensory, or afferent, neuron, the dendritic tips of which may be specialized as receptors; and (c) one of the types of neurons found in the central nervous system (CNS), that is, the brain and spinal cord. Arrows indicate direction of nerve impulses.

light microscope. Yet, because of circumstantial evidence, the neuron theory came to be widely but not universally accepted for years before synapses were actually seen by electron microscopy. One of the most convincing pieces of early, circumstantial evidence that neurons were not fused with each other was the phenomenon of **synaptic delay,** a lapse of 0.5 to 1.0 milliseconds (msec) in the passage of an impulse from one neuron to another. Electron microscopy and thin sectioning has revealed that almost all connecting neurons are indeed separated by minute spaces called **synaptic clefts.** With certain exceptions, nerve impulses cross vertebrate synapses in one direction only, from axon to dendrite or soma. Concentrations of somata and synapses are called **ganglia** (singular, ganglion).

As Figure 15-2 illustrates, there are three types of synapses between neurons: (1) **axodendritic,** in which the synaptic knobs at the tips of an

**FIG. 15-2** Diagram of three different types of synapses: axosomatic, axodendritic, and axoaxonic.

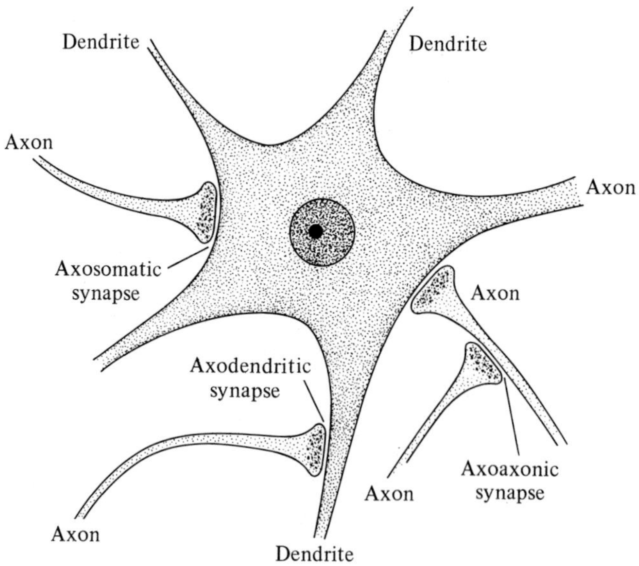

axon are adjacent to the surface of a dendrite; (2) **axosomatic,** in which the synapse is located between an axon and the soma of the next cell, called the **postsynaptic cell;** and (3) **axoaxonic,** in which the axon of one cell communicates directly with the axon of an adjacent cell. The first two types are the most common in vertebrates. This variety of types constitutes three of the many variables involved in nervous integration.

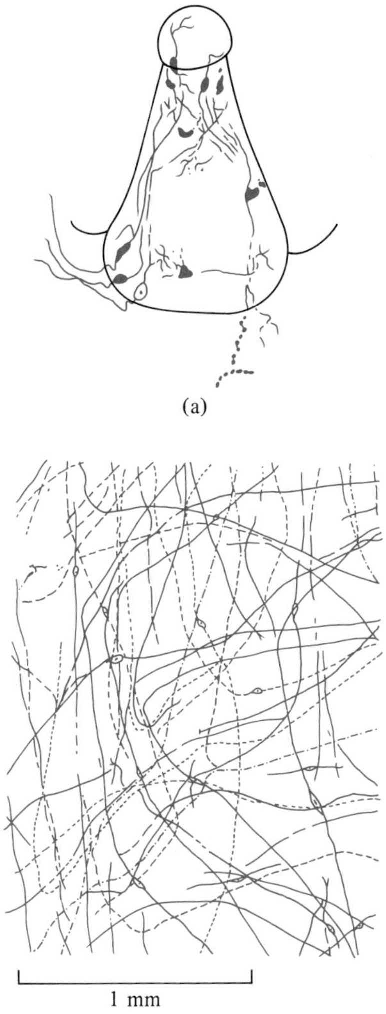


**FIG. 15-3** Cnidarian nervous systems. (a) The simple type of ganglion found at the edge of the bell of a jellyfish. (b) Portion of a sea anemone nerve net.

## INVERTEBRATE NERVOUS SYSTEMS

The lowest animals with nervous systems are the Cnidaria, or coelenterates. Cnidarians have no central region of nervous control, a reflection of their radial symmetry: they have a circular body arranged around a central axis. Some jellyfish possess marginal ganglia along the edge of their bells (Fig. 15-3a). Much of the coordination of a jellyfish's movements is controlled by the many synapses within these ganglia. Although the marginal ganglia of jellyfish contain various kinds of neurons, they are not nearly as highly organized as are the ganglia of higher invertebrates. Some jellyfish have rings of neurons, nervelike groups of parallel fibers encircling the margin of the bell that function much as ganglia do.

Many cnidarian neurons are organized into systems of one or more separate **nerve nets** (Fig. 15-3b). Each system responds to a different type of stimulus, and each causes the contraction of different muscle cells. Although in some nets the neurons are fused to each other, in most they are connected by two-way synapses, which allow impulses to travel in either direction.

### THE NERVOUS SYSTEMS OF FLATWORMS

The most primitive animals with **central nervous systems**—consisting of a brain and a longitudinal nerve cord—are the flatworms, which belong to the phylum Platyhelminthes. Many of the nonparasitic forms have a nervous system of the ladder type, so named because of the appearance of the longitudinal series of cross-connecting nerves between two main nerve cords (Fig. 15-4). The most primitive flatworms have several pairs of longitudinal cords in the epidermis that form the basis of a cnidarianlike nerve net. Flatworms are bilaterally symmetrical, with the anterior end directed forward in locomotion and possessing a concentration of sense receptors. These worms have a minute brain composed of an anterior pair of fused ganglia. *Notoplana* has one of the most complex of all flatworm brains; it is shown in Figure 15-5. The central core contains nerve fibers and groups of parallel fibers called nerve tracts. The outer cortex contains ganglia.

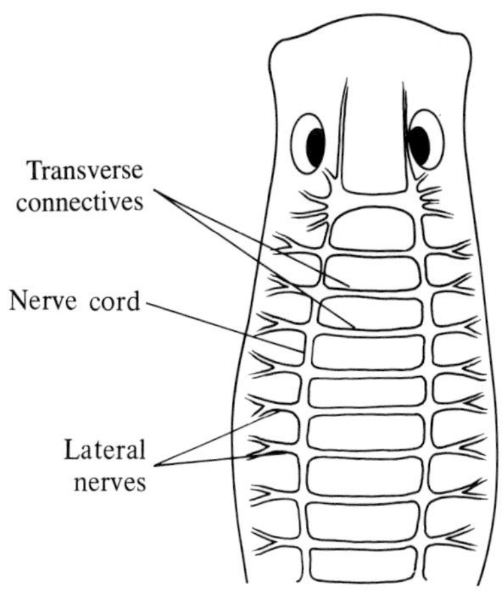


**FIG. 15-4** The ladder type of nervous system of a free-living flatworm.

### THE CENTRAL NERVOUS SYSTEMS OF SOME ANNELIDS AND ARTHROPODS

The annelids—earthworms and their relatives—and arthropods—crabs, insects, and related groups—exhibit remarkable similarity of general body plan, evidencing the closeness of their kinship. Basically, their nervous systems seem to have evolved from the ladder type found in flatworms. In the most primitive annelids, as in flatworms, the ven-

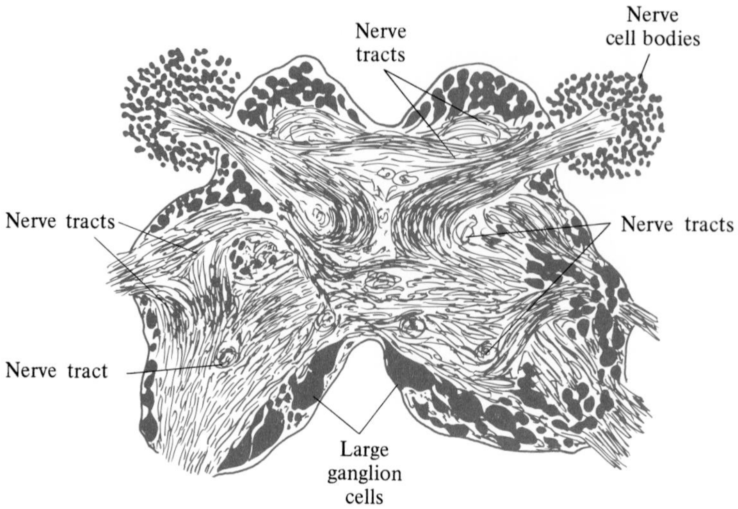


**FIG. 15-5** Section through one of the most complex brains of any flatworm *(Notoplana)*. Note the presence of nerve fibers and tracts in the central region, or core, and ganglion cells in the outer layer, or cortex.

tral nerve cord is a double cord near the body surface (Fig. 15-6a). In most annelids and arthropods, however, the pair of cords, while ventral, is found deeper in the body (Fig. 15-6b). In more highly evolved forms, such as the earthworm and many arthropods, the pair of nerve cords is partially or completely fused into a single cord (Fig. 15-6c). Typically, the annelid nerve cord has a thickened, ganglionlike region in each segment. In many annelids, however, the entire nerve cord is much like an elongated ganglion, with a central region rich in synapses and an outer cortex of nerve cell bodies.

In the course of annelid and arthropod evolution, highly specialized antennae and mouth parts arose, apparently from the locomotory appendages of various anterior body segments. Fusion, specialization, and enlargement of ganglia and nerve cord regions serving these structures were also part of this evolutionary process. As functional abilities evolved, particularly those associated with sense organs and mouth parts, there evolved a corresponding increase in brain complexity.

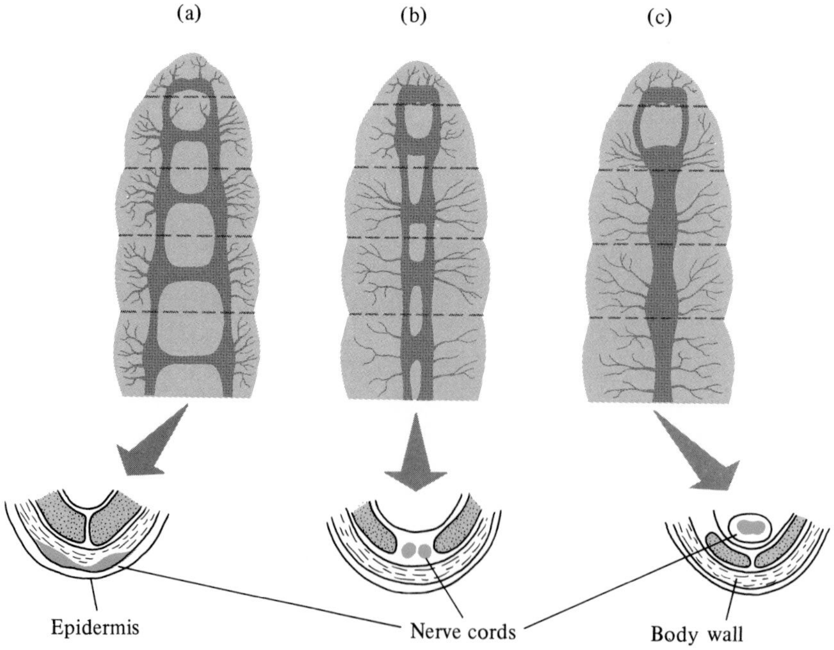


**FIG. 15-6** Evolutionary advancements in annelid nervous systems. (a) Primitive, ladder-type system reminiscent of that of flatworms. The double cord lies just below the epidermis. (b) A somewhat advanced type. The nerve cords are nearly fused and are located in the muscle layer of the body wall. (c) The most advanced type. The double nerve cord is fused into a single cord and is located in the coelom, the body cavity.

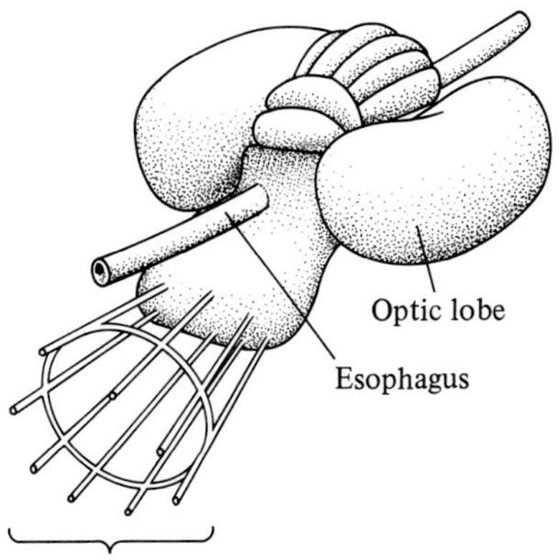


**FIG. 15-7** Brain of the octopus. Each of the various regions of the brain controls different functions.

**Cephalization** (from the Greek *kephalos*, "head") is the term given to the evolutionary process by which sense organs and nervous elements become concentrated in a distinct body part at the anterior end of animals, leading to the formation of a head or a headlike structure. The process is apparent in both annelids and arthropods, but especially in the arthropods.

## THE COMPLEX BRAIN OF THE OCTOPUS

Members of the invertebrate phylum Mollusca exhibit a great range of complexity of nervous systems. The most primitive molluscs have nervous systems as simple as those of flatworms. At the other end of this spectrum are the cephalopods, such as squids and the octopus. The octopus brain is as complex—that is, contains as many neurons and synapses—as those of many fishes (Fig. 15-7). The octopus, accordingly, exhibits an extensive repertoire of learning abilities. The various levels of complexity in the nervous systems of molluscs do not constitute a graded evolutionary series. Rather, their differences are correlated with the many adaptations to particular modes of existence that various members of this successful phylum of over 88,000 living species has achieved. Some idea of the complexity reached by some noncephalopod molluscan brains is indicated by the example of a snail nervous system shown in Figure 15-8.

# ORGANIZATION OF VERTEBRATE NERVOUS SYSTEMS

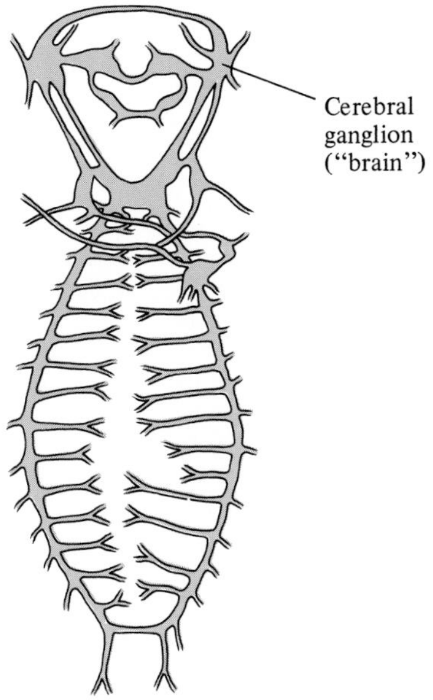


**FIG. 15-8** Nervous system of the snail *(Patella)*. The various ganglia are named according to the body part they regulate: labial (lips), buccal (mouth), pedal (foot), pleural (lung), and so on.

The entire vertebrate nervous system consists of two major sections: the central nervous system (CNS) and the peripheral nervous system (PNS). Their major subdivisions are as follows:

- Central nervous system
  - Brain
  - Spinal cord
- Peripheral nervous system
  - Cranial nerves
  - Spinal nerves
  - Autonomic nervous system
    - Sympathetic division
    - Parasympathetic division

As members of the phylum Chordata, all vertebrates possess a dorsal, tubular, nerve cord, called the **spinal cord.** This is in contrast to the ventral, solid cord typical of invertebrates. The tubular nature of the vertebrate cord is manifested by its **central canal,** which extends into the brain as a system of cavities called **ventricles.** Both the canal and the ventricles are filled with **cerebrospinal fluid,** an ultrafiltrate of plasma formed by filtration of blood from the **choroid plexus,** a delicate network of blood vessels in the ventricles of the brain. Both the brain and the spinal cord are enclosed within three membranes, the **meninges:** the outer dura mater, the middle arachnoid, and the innermost pia mater.

The spinal cord is continuous with the base of the brain and, like the brain, is composed of nervous tissue: neurons and their supporting cells, the **neuroglia,** or glial cells. The spinal cord is encased within a structure, the spinal column, or backbone, formed by numerous vertebrae. In humans, there are 33 vertebrae: 7 cervicals (neck), 12 thoracics (chest), 5 lumbars, (lower back), 5 sacrals (pelvis), and the coccyx (tail bone), formed by 4 fused coccygeals. A pair of spinal nerves leave the column at the level of each vertebra. The spinal cord in adult humans is about 44 cm (18 inches) long, with an average diameter of about 1 cm. It tapers to a point at the level between the first and second lumbar vertebrae of the spine and is anchored to the base of the spine by a thin nonnervous thread, the **filum terminale.** The lower part of the spinal column contains the last of the spinal nerves—the lumbar, sacral, and coccygeal nerves—collectively called the **cauda equina** (Latin for "horse's tail") because of the resemblance to the tail of a horse.

A cross section of the vertebrate spinal cord reveals distinct white and gray regions (Fig. 15-9). The gray matter is centrally located in the form of a crude letter H. **Gray matter** is composed of neuronal cell bodies, neuroglia, dendrites, and nerve fibers. In general, these fibers are intrinsic, that is, they do not leave the region of their soma. Because these fibers are unmyelinated—that is, they lack sheaths of **myelin,** a white, fatty substance—the tissue is gray in color. **White matter** occupies the remaining regions and is made up of ascending and descending fiber tracts, the fibers of which are covered with myelin. Myelin acts as an insulator that, by preventing leakage of the electrical current that is associated with nerve impulses, greatly accelerates the rate at which the impulses are conducted. The ascending, or **afferent,** tracts of the

**FIG. 15-9** The basic components of the nervous system. The sensory neurons are afferent neurons, and the motor neurons are efferent neurons.

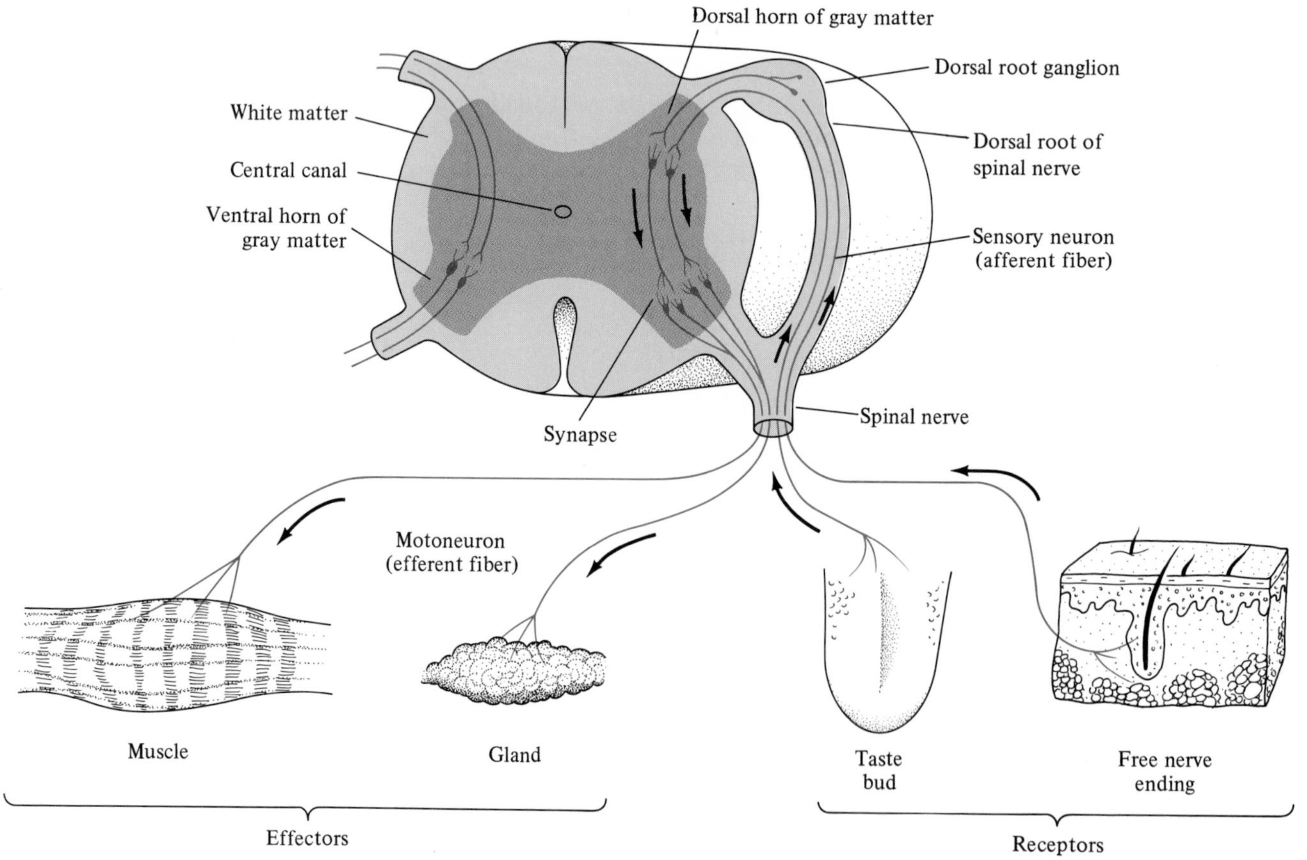

white matter are sensory and carry information toward the brain. Descending, or **efferent,** tracts are motor neurons, carrying impulses away from the brain to effectors—muscles and glands—in various parts of the body.

The CNS contains the cell bodies of most of the motor neurons and interneurons (Fig. 15-9). **Motor neurons** innervate and activate muscles and glands. **Interneurons** link sensory neurons with motor neurons. **Sensory neurons** convey information to the CNS regarding various environmental or internal stimuli.

The peripheral nervous system is composed of the axons of motor neurons and sensory neurons and their cell bodies. It also includes **plexuses,** which are loose tangles of nerve cells and fibers, and various ganglia. The term **nerve** applies to any group of parallel nerve fibers found in the peripheral nervous system. Within the CNS, such groups are referred to as **nerve tracts.**

## REFLEX ARCS

The simplest functional unit of nervous integration, the decision-making process by which stimuli are evaluated and acted on by the nervous system, is the simple **reflex arc.** Its minimal components are a receptor (usually the specialized ending of a sensory neuron), the sensory neuron itself, and a motor neuron with which it synapses. Motor neurons innervate effectors, usually muscles. Two types of simple reflex arcs are shown in Figure 15-9. If an arc has only two neurons, it is termed **monosynaptic.** If, as in most arcs, a third neuron, called an interneuron, is located between the sensory and motor neuron, it is **disynaptic.**

As shown in the sketch of a section through the spinal cord in Figure 15-9, the cell bodies of sensory neurons are located outside the CNS in a thickened region of a nerve root, the **dorsal root ganglion.** Motor neuron cell bodies are located within the spinal cord.

Simple reflex arcs are of two major types, depending upon the location and the nature of the receptors and effectors involved. When the arc is located in outlying areas, such as the limbs, skin, and body wall, the unit is known as a **somatic arc.** If the arc involves internal organs, it is a **visceral arc.** Although the two types of reflex arcs have much in common, the differences between them warrant separate descriptions.

**FIG. 15-10** Functioning of a simple somatic reflex arc, simplified. Actually, many more synapses and arcs are involved in the withdrawal reflex.

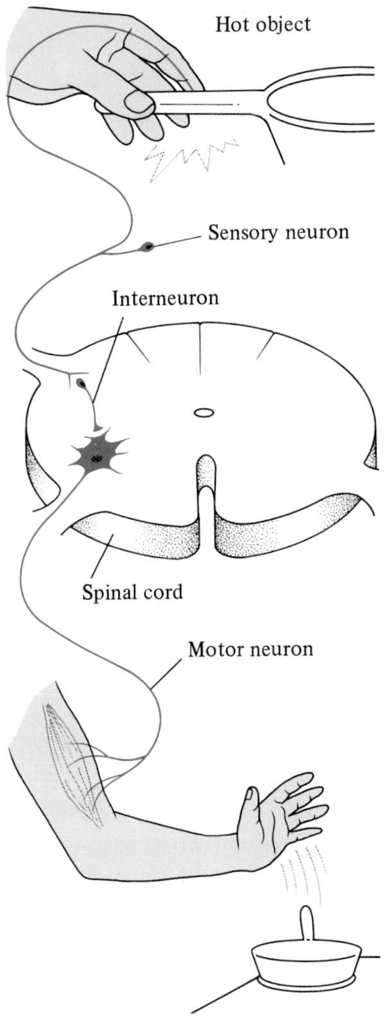


***Simple Somatic Reflex Arcs.*** Simple somatic reflex arcs function in emergencies, such as when the hand touches a very hot object (Fig. 15-10) or a bare foot steps on a sharp object. In such instances, the hand or foot is immediately withdrawn, whereas the hand or foot would be pulled back after some delay if the objects were merely uncomfortable to the touch. A delayed determination to withdraw is a decision made in the brain, and simple arcs are not involved. But the quick response mediated by a simple reflex is accomplished by a region of the nervous system at no higher a level of integration than the spinal cord.

If another neuron and synapse is added to the monosynaptic arc illustrated in Figure 15-11a, additional pathways will be involved (Fig. 15-11b). If enough receptors and neurons are involved, the entire body may draw back from a painful stimulus. No stimulus ever activates *only* a single simplex reflex arc, that is, a single pathway composed of but two or three neurons. Nor, if it did, would stimulation of a monosyn-

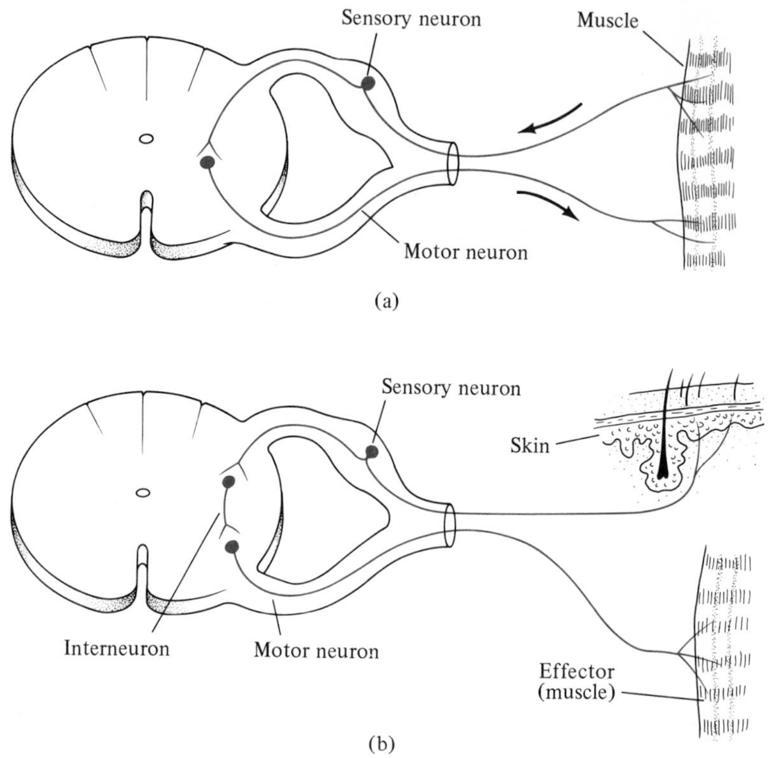


**FIG. 15-11** Simple somatic reflex arcs. (a) Monosynaptic arc. (b) Disynaptic arc.

aptic or disynaptic arc be sufficient to evoke a withdrawal reflex. All sensory neurons have several branches and communicate with many other neurons. This is evidenced by the fact that after a reflex withdrawal from a painful stimulus, the individual becomes conscious of the experience.

Visceral reflex arcs, which involve the internal organs, are somewhat more complex than somatic arcs. They are best understood as units of the autonomic nervous system and are described later in this chapter.

## THE PERIPHERAL NERVOUS SYSTEM

Extending from the brain and spinal cord are a number of nerves. These are the cranial and spinal nerves, respectively, each of which contains thousands of fibers.

***Cranial Nerves.*** Eleven pairs of nerves of various sizes leave the ventral surface of the human brain (Fig. 15-12). A twelfth pair exits from the brain's dorsal surface. Although most of these are mixed nerves, containing both sensory and motor neurons, some are purely sensory or purely motor. As a group, the cranial nerves receive information from and control the actions of various organs and regions of the head: the eyes, ears, nose, tongue, and face. The tenth (Xth) nerve, called the vagus nerve, innervates the larynx, or voice box. But unlike most cranial nerves, it also sends fibers to other organs, including the heart, lungs, and gastrointestinal tract.

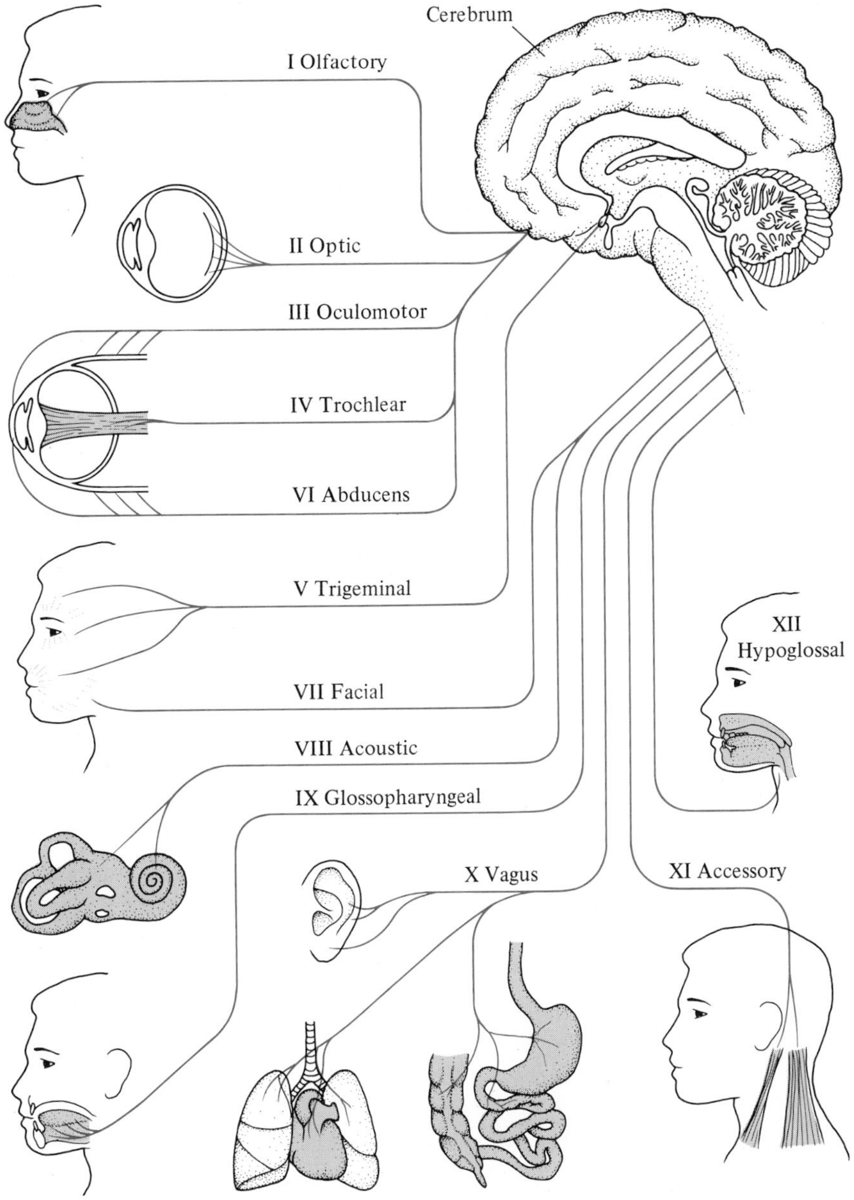


**FIG. 15-12** The cranial nerves and the organs they supply.

***Spinal Nerves.*** The vertebrate spinal cord gives off many pairs of nerves, the number differing from one major taxonomic group to another (Fig. 15-13). In the human, there are 31 pairs. Each nerve passes through an opening between the two vertebrae surrounding that part of the cord (Fig. 15-14). It then branches to serve various parts of the body in that general region. Each spinal nerve also connects by means of two **communicating rami,** or branches, with the sympathetic trunk, the chain of autonomic ganglia paralleling each side of the cord. Each spinal nerve originates from a dorsal root and a ventral root, the two roots fusing to form the nerve. If such a nerve is severed, both sensation and motor control of the area it serves are lost.

Although all spinal nerves are mixed nerves, carrying both afferent (sensory) and efferent (motor) fibers, this is not true of their roots. Dorsal roots are composed solely of afferent fibers, that is, fibers carrying impulses into the CNS. Ventral roots contain efferent fibers and a few afferent fibers. As noted earlier each dorsal root is marked by the pres-

**FIG. 15-13** Spinal nerves. Dorsal view of the spinal cord, spinal nerves, and spinal ganglia and plexuses.

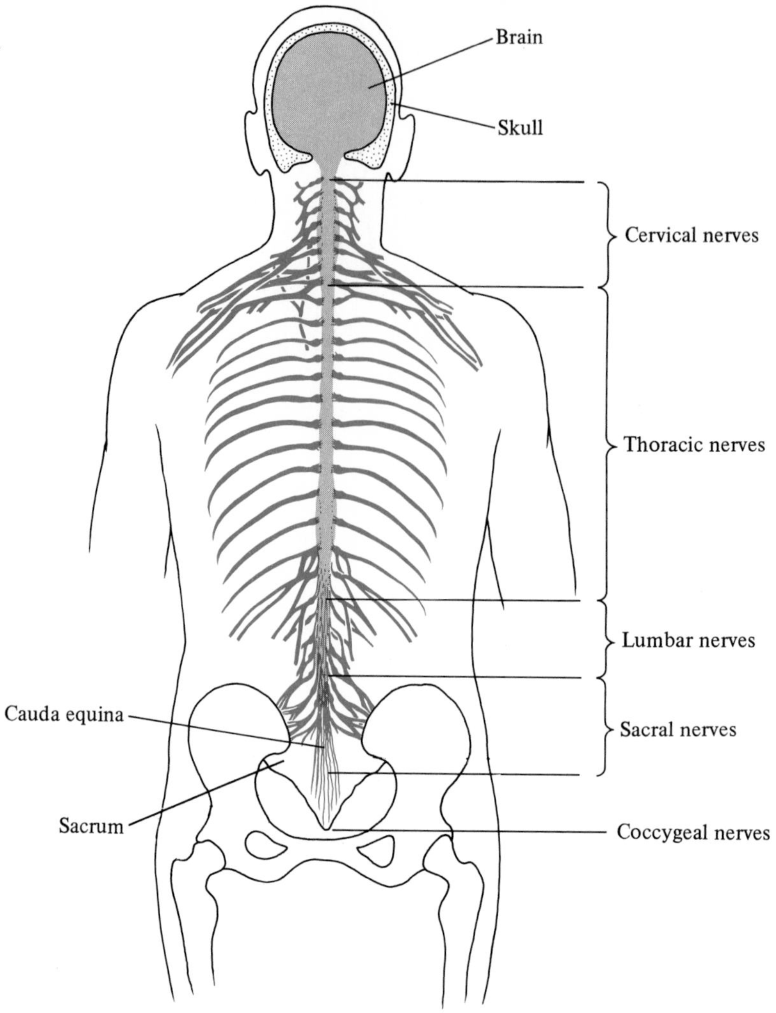


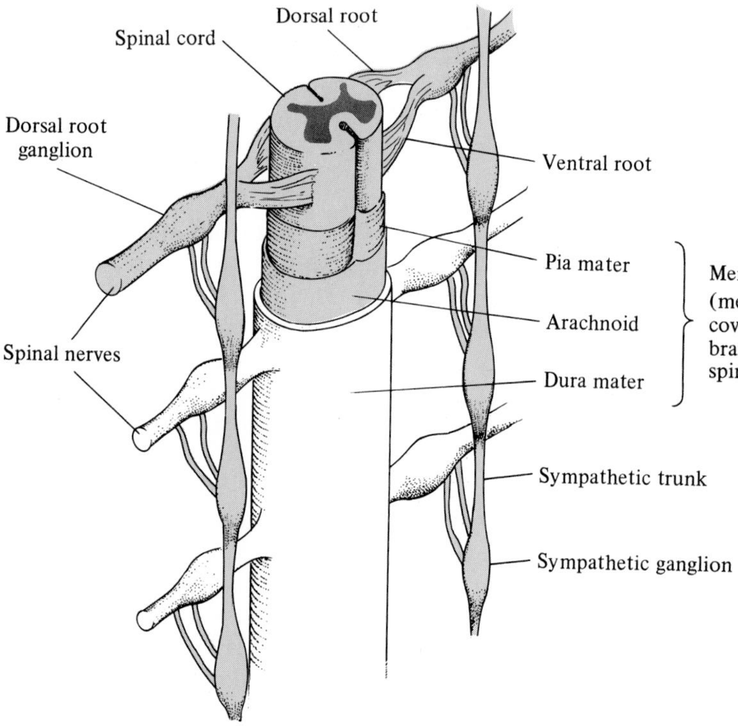


**FIG. 15-14** Spinal cord and its relationship to the meninges (covering membranes), spinal nerves, sympathetic trunk (chain), and ganglia.

ence of a dorsal root ganglion, containing the nerve cell bodies of the sensory neurons in the root.

## THE AUTONOMIC NERVOUS SYSTEM

The autonomic nervous system is the group of nerve cells that regulates the body's automatic activities, including the control of heart rate, blood pressure, respiration, digestion, sweating, and so on. As the term *autonomic* suggests, these functions are not generally under conscious control. We usually do not decide to raise or lower our blood pressure; nor are we usually even conscious of its happening. Many bodily functions, however, are under a combination of autonomic and somatic control.

The autonomic nervous system is not really an independent system, that is, it is not fully autonomous. Indeed, its centers of control reside in the CNS and in ganglia that are directly or indirectly linked to the rest of the CNS. Actions controlled by the many visceral reflexes harmonize with and support actions of the rest of the nervous system. For example, when the starter's gun signals the beginning of a foot race, in each runner a vast array of visceral receptors, reflex arcs, centers of integration, and effectors respond appropriately to the sudden demands placed on the body. These same autonomic components, whether we are asleep or awake, continually monitor such vital factors as blood pressure level, heart rate, stroke volume, and respiratory rate, and adjust them to the prevailing situation.

**TABLE 15-1 Some Functions of the Autonomic Nervous System**

| *Tissue* | *Location* | *Effect of parasympathetic stimulation* | *Effect of sympathetic stimulation* |
|---|---|---|---|
| Smooth muscle | Iris | Contraction of circular fibers, constricting pupil | Contraction of radial fibers, dilating pupil |
| | Stomach wall | Contraction and increased motility | Inhibition of contraction |
| | Intestinal wall | Increased tone and motility | Inhibition of motility |
| | Anal sphincter | Inhibition of contraction | Contraction |
| | Bladder wall | Contraction | Inhibition of contraction |
| | Bladder sphincter | Inhibition of contraction | Contraction |
| | Bronchioles | Constriction | Dilatation |
| | Hair follicle | | Contraction |
| Gland | Mouth (salivary) | Secretion of copious thin saliva | Secretion of scanty thick saliva |
| | Stomach (gastric) | Secretion | Inhibition of secretion |
| | Liver | Inhibition of glycogen breakdown | Increased glycogen breakdown |
| | Adrenal medulla | | Secretion |
| | Skin (sweat) | | Secretion |
| Blood vessel | Heart (arteries) | Constriction | Dilatation |
| | Skin | | Constriction |
| | Heart | Deceleration of rate | Acceleration of rate |

Based on W. S. Beck, *Human Design: Molecular, Cellular and Systematic Physiology;* p. 467, Harcourt, New York, 1971.

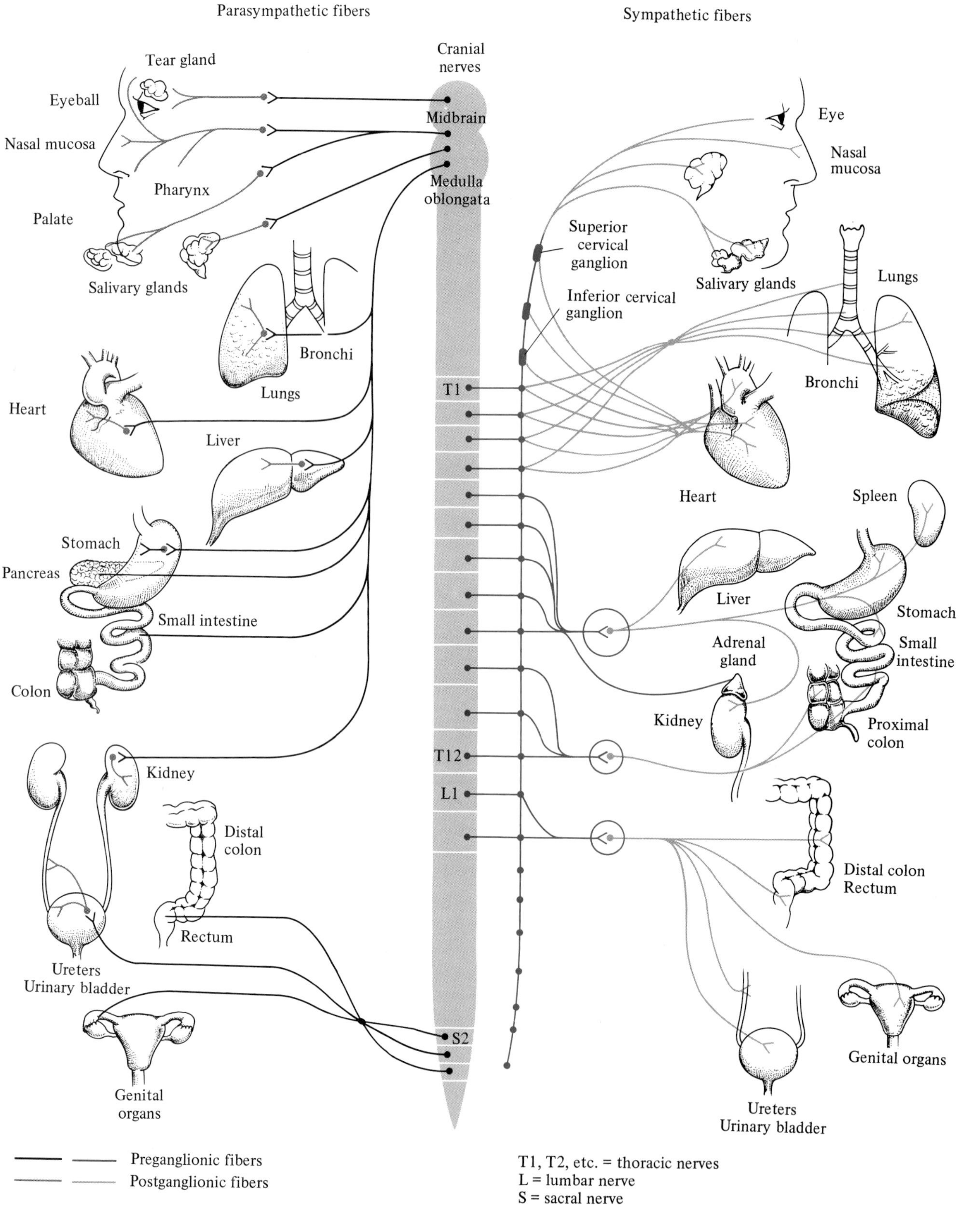


**FIG. 15-15** The autonomic nervous system. Note that nearly all viscera (internal organs) receive both a sympathetic and a parasympathetic nerve supply; that the ganglia and the postganglionic fibers of the parasympathetic division are all located entirely within the innervated organs; and that most sympathetic ganglia are located in the sympathetic trunks.

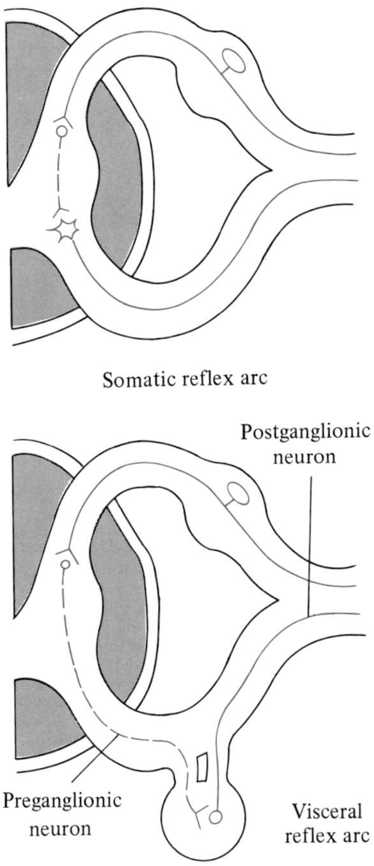


**FIG. 15-15 (*continued*)** The visceral reflex arc compared with the somatic reflex arc. The preganglionic neuron (dashed line) makes synaptic contact with a postganglionic neuron outside the CNS, either in a synaptic ganglion, as in this case, or in a ganglion within the innervated organ.

***The Divisions of the Autonomic Nervous System.*** The autonomic system is divided into the sympathetic and parasympathetic divisions. Most organs are innervated by fibers from both divisions. The effects upon an organ of activating its sympathetic nerve supply are generally the opposite of the effects produced by parasympathetic stimulation. The two divisions are thus usually antagonistic in their effects. The physical basis for the difference is twofold. First, in general, different chemicals, called **neurotransmitters,** are released into the innervated organ by each of the two types of fibers. Second, the receptor molecules on the innervated effector also differ for the two types. In the case of sympathetic fibers, the transmitter released onto the effector is generally norepinephrine (noradrenaline). Parasympathetic fibers release acetylcholine. Based on this difference, sympathetic fibers are termed **adrenergic** and parasympathetic fibers **cholinergic.**

Table 15-1 (page 407) lists the contrasting effects of the two divisions of the autonomic system. Parasympathetic nerves are said to conserve and restore bodily functions. For example, stimulation of parasympathetic nerves to the liver causes it to store glycogen and stop breaking down glycogen into glucose. Other effects of parasympathetic nerve stimulation include slowing the heart, constricting the bronchioles, and promoting digestion by stimulating contractions of the stomach. In contrast to these effects, stimulation of sympathetic nerves has the same general effect as the release of epinephrine into the bloodstream by the adrenal gland. This is not surprising because norepinephrine, the neurotransmitter released by sympathetic fibers, differs only slightly in structure from the hormone epinephrine and has the same effect as epinephrine on most organs. The fact that epinephrine prepares the body for emergencies—for "fight or flight," as noted in Chapter 14—indicates the nature of most actions of the sympathetic division. Thus, in contrast to the actions of the parasympathetic division, sympathetic stimulation results in such emergency preparations as promoting glycogen breakdown to glucose, dilatation of the bronchioles and the pupils of the eyes, quieting the activity of the stomach and intestine, and increasing the heart rate and stroke volume. All of these prepare the body for strenuous physical activity.

The origin and distribution of sympathetic and parasympathetic fibers is shown in Figure 15-15a. Sympathetic fibers originate from cell bodies in the gray matter of the thoracic (chest) and lumbar (lower back) regions of the spinal cord. Parasympathetic fibers originate from (1) parts of the brain and (2) from the gray matter of the sacral region of the spinal cord (the region supplying the hips and legs). Parasympathetic fibers originating in the brain exit via cranial nerves III, VII, IX, and X. All other fibers of both divisions exit via the spinal nerves.

***The Visceral Reflex Arc.*** There are some distinct differences between a somatic reflex arc and a visceral reflex arc. Most important, the visceral arc contains a chain of two fibers between the synapse in the spinal cord and the point of innervation of the particular organ. Only one fiber is found in the comparable part of a somatic arc.

We have already mentioned that in somatic arcs, an afferent (sensory) fiber, with its cell body in the dorsal root ganglion, makes synaptic contact—either directly or through an interneuron—with a motor neuron in the gray matter of the spinal cord. The motor neuron, in turn, leaves the cord via the ventral root and innervates an effector directly, not through the mediation of another neuron. In the visceral arc, the afferent fiber is essentially the same as a somatic afferent. It too has its cell body in the dorsal root ganglion and makes synaptic contact with another neuron in the gray matter. This other neuron also leaves the cord via the ventral root. But in visceral arcs, this neuron is not a motor neuron, that is, it does not innervate an effector. Instead, as Figure 15-15b shows, it communicates synaptically, always by means of the neurotransmitter acetylcholine, with another neuron, the cell body of which is located outside the CNS, in a ganglion. These two neurons unique to the visceral arc, and both of which are part of the autonomic nervous system, are termed the **preganglionic** and **postganglionic** neurons. The postganglionic neuron innervates the organ being supplied.

## EVOLUTION OF THE VERTEBRATE BRAIN

The vertebrate brain probably arose by modification of the primitive neural tube. Early in chordate evolution, the tube must have become divided into its three major regions—the forebrain, the midbrain, and the hindbrain—as it still does during embryonic development in all vertebrates (Fig. 15-16). In primitive vertebrates, each of these areas corresponds to one of the three main senses: chemoreception (smell and taste), sight, and hearing/balance. In the course of further evolution, the forebrain, midbrain, and hindbrain gave rise to additional regions.

### MAJOR DIVISIONS OF THE VERTEBRATE BRAIN

The major parts of the vertebrate brain are as follows:

- Forebrain
  - Telencephalon
    - Cerebrum
    - Basal nuclei
    - Olfactory bulbs
  - Diencephalon
    - Epithalamus
    - Thalamus
    - Hypothalamus
- Midbrain
  - Tectum (includes the optic lobes)
- Hindbrain
  - Pons
  - Cerebellum
  - Medulla oblongata

### FUNCTIONS OF SOME BRAIN SUBREGIONS

Many of the brain's functions can be traced to specific regions, as demonstrated by the fact that experimentally induced injury or removal of portions of the brain abolishes specific functions. Moreover, when portions of the brain are stimulated, corresponding organs, muscles, or glands respond. Stimulating the sense organs also evokes electrical potentials in specific brain areas.

***The Brainstem.*** The oldest part of the brain, evolutionarily speaking, is the **brainstem,** which includes the hindbrain and midbrain, virtually

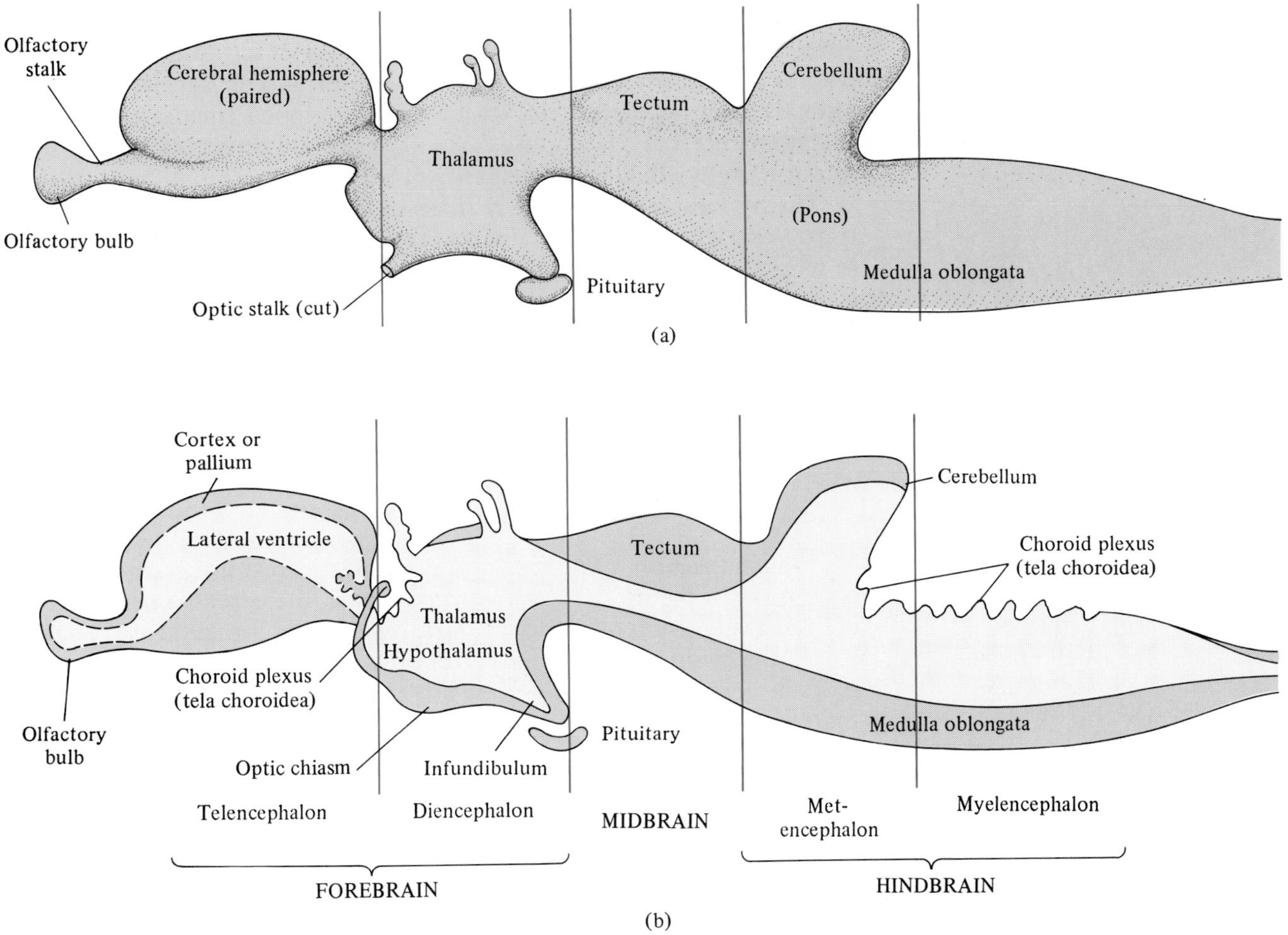


**FIG. 15-16** The principal divisions of the vertebrate brain as seen in (a) surface view and (b) longitudinal section.

all parts of the brain except the cerebellum and cerebrum (Fig. 15-17). The lowest parts of the brainstem consist of the **pons** and **medulla oblongata,** which, besides containing major sensory and motor pathways, are the sites of clusters of neurons called **centers.** The centers regulate circulatory and respiratory functions. The ventricles, or cavities, of the hindbrain and midbrain are surrounded by layers of neurons, the central gray matter cells. Around these cells is a large mass of **brain nuclei,** clusters of nerve cell bodies in the brain. They are similar to ganglia in that they are rich in synaptic contacts and integrative capabilities. Within the hindbrain and midbrain is a structure known as the **reticular formation,** a core of tissue within the brainstem extending from the medulla oblongata to the hypothalamus. Its role appears to be the promotion of arousal and alertness. The entire brainstem contains many nuclei, long regarded merely as relay centers, but now known to serve, as do all synaptic fields, as important integrative, or decision-making, centers. The **cerebellum** evolved from the brainstem but is not considered to be a part of it. It controls fine coordination, but not initiation, of muscular movements. It is largest and most highly developed in animals capable of the most intricate and graceful movements.

***The Diencephalon.*** The **thalamus** is very rich in "relay" nuclei that pass signals from external sense receptors through thalamic synapses to the cerebrum (Fig. 15-17). The tracts of nerve fibers to the cerebrum

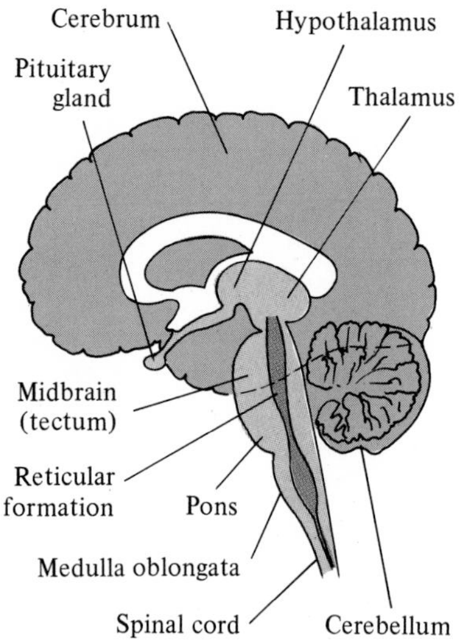


**FIG. 15-17** The brainstem (lightly shaded), showing location of the reticular formation (heavily shaded) within the brainstem. The cerebrum and cerebellum, not part of the brainstem, are also shown.

from the thalamus are referred to as the **cortical projections** of the thalamus. The dorsal thalamus, which communicates with the **cerebral cortex,** the cerebrum's outer layer of gray matter, is well developed only in animals, such as mammals, with a well-developed cortex. The ventral thalamus receives descending cortical fibers and thus aids in control of **motor activity,** that is, movements of skeletal muscles.

The **hypothalamus** is a tiny mass of tissue forming the floor of the third ventricle, just below the thalamus. Its importance is disproportionate to its size. As an autonomic center, it regulates such functions as sweating, shivering, vasodilation and vasoconstriction (the expansion and narrowing of the lumen of blood vessels), thirst, hunger, appetite, and satiety. Its cells monitor temperature, blood osmotic pressure, and other internal conditions. It produces posterior pituitary gland hormones, secretes factors controlling activity of the anterior pituitary, and is involved in behavioral drives, such as hunger, and in emotional behavior.

***The Telencephalon.*** The region of perhaps greatest interest to us is the cerebrum. In the most primitive vertebrates, the cerebral hemispheres are relatively small **olfactory** centers, responsible for the sense of smell. But in the evolution of higher vertebrates, as the sense of smell became more important, these lobes grew larger and, as a consequence, the gray matter expanded. (In most mammals, but not in humans, the sense of smell is highly developed.) In mammals, this region also came to underlie the center for the most complex of integrative activities, the cerebral cortex.

In the higher vertebrates, the gray matter is located at the surface of the hemispheres, where it forms the **pallium** (Latin for "cloak"), or cerebral cortex. Gray matter consists of unmyelinated axonal and dendritic endings and nerve cell bodies. Gray matter regions are thus regions of much integration, or decision making. The proliferation of this thin layer of cerebral gray matter in mammals far surpassed the development of the underlying white fibrous matter with which it was associated.

The problem of accommodating an increasingly extensive covering layer of gray matter while maintaining its association with the underlying regions was solved by the folding of the cortical layer upon itself. The folds are called **gyri** (singular, gyrus), and the furrows, or fissures, between them are **sulci** (singular, sulcus). Although the surface of the hemispheres is smooth in small and primitive mammals (Fig. 15-18a–d), in the most advanced species it is highly convoluted, with many sulci (Fig. 17-18e). In higher mammals, the cerebrum covers and overlaps other brain structures, achieving a size greater than all the other parts of the brain together.

***Structural Relationships.*** Besides the great longitudinal fissure that marks the division between the two cerebral hemispheres, there are two other sulci of significance. One, the **central sulcus,** or fissure of Rolando, separates the **frontal lobe** from the parietal lobe (Fig. 15-19). Another, the **lateral sulcus,** also called the fissure of Sylvius, separates the temporal lobe from the rest of the cerebrum above it. A **parietooccipital sulcus,** not visible in Figure 15-19, separates the parietal and **occipital lobes.** Special functions are localized within certain regions of each lobe, as described in the next section.

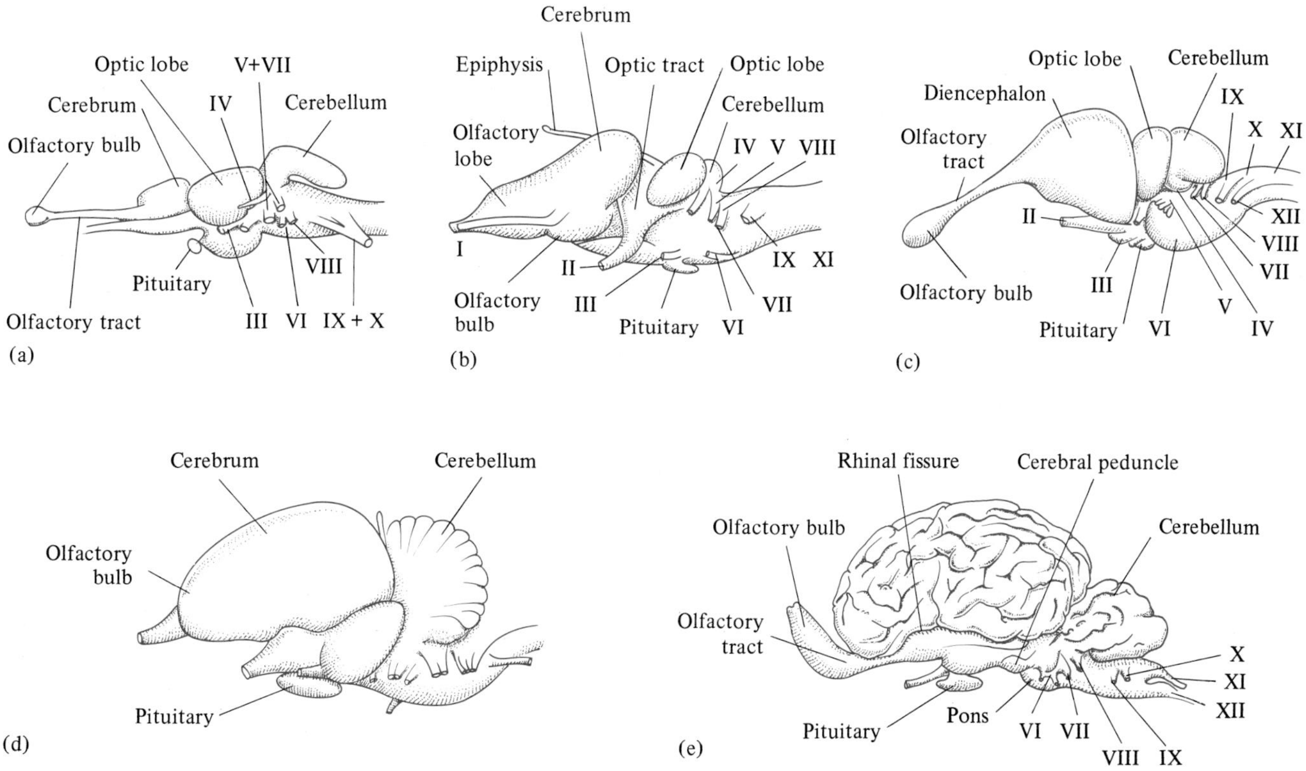


**FIG. 15-18** Lateral views of the brains of (a) codfish, (b) frog, (c) alligator, (d) goose, (e) horse.

## SOME SPECIFIC FUNCTIONS OF CEREBRAL CORTICAL REGIONS OF HIGHER MAMMALS

A description of most of the known functions of various regions of the mammalian cerebral cortex is beyond the scope of this book. A few examples will suffice to illustrate certain principles.

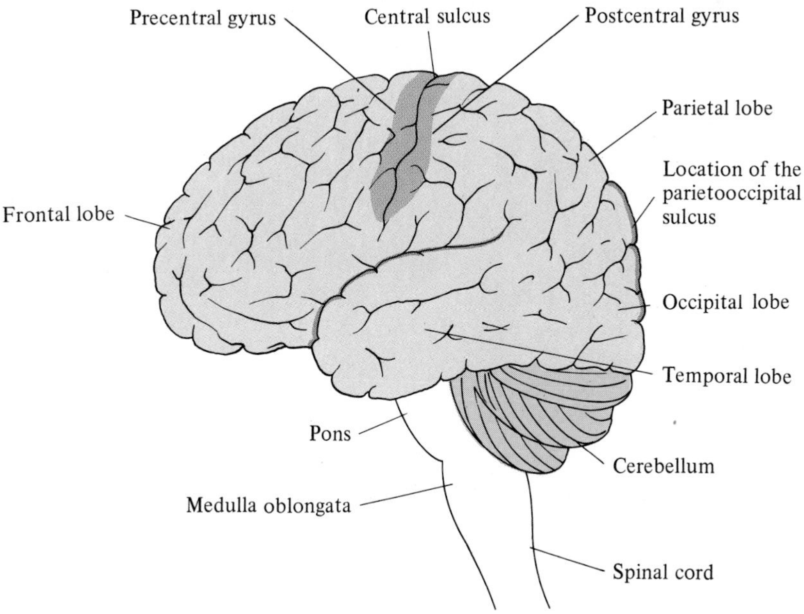


**FIG. 15-19** Gross anatomy of the human brain, surface view.

***Mapping of the Human Cortex.*** Several discrete, functional divisions can be identified in the cortex, using the sulci as reference points. Sensory areas of the cortex have been identified by recording electrical activity from the surface of the exposed brain when various of the body's sense receptors are stimulated. Motor areas are plotted by electrically stimulating the cortical area itself and observing the responses of muscles, glands, and other organs.

***Sensory Areas Become Active When Receptors Are Stimulated.*** Sensory areas are present in all but the frontal lobes (Fig. 15-20). The primary visual projection area, that is, the region to which information received from the optic nerves is relayed, is in the **occipital lobe,** while an auditory area, responsible for the hearing sense, is found in the upper **temporal lobe.** Visceral sensations, such as taste or a stomach ache, are localized in nerve tissue at the bottom of the lateral sulcus. Somatic, or body, sensations, such as those of pain, temperature, and touch, are primarily referred to an area of each parietal lobe just behind the central sulcus. The distribution of neurons in this region, known as the **primary somatosensory cortex,** reflects the density of sensory receptors in the various parts of the body. The more highly innervated the body part, the greater the amount of brain tissue devoted to it in this region. In Figure 15-21a, a sensory homunculus (Latin for "little man") illustrates both the areas and relative amounts of the cortex devoted to each body part. Note that, except for the facial components, the arrangement of brain areas is upside down compared to the arrangement of body parts. This same arrangement of sensory areas in the tissue behind the central sulcus exists, in mirror image, in each cerebral hemisphere. Because of crossover arrangements of fibers, a stimulus to a receptor on one side of the body is first registered in the opposite hemisphere of the brain.

***Motor Areas Send Commands to Muscles and Other Effectors.*** The area just anterior to the central sulcus in both hemispheres is the **primary motor cortex.** Thus, while the primary somatosensory cortex is at

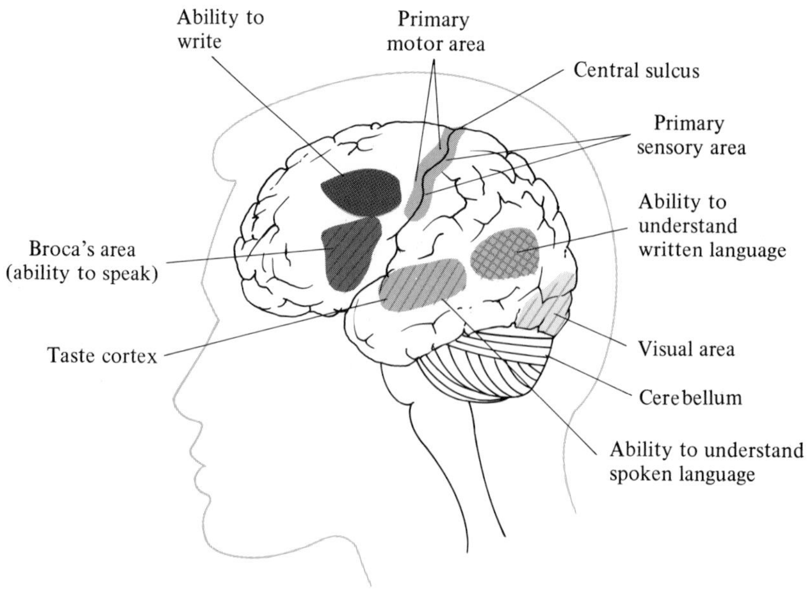

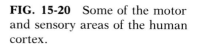

**FIG. 15-20** Some of the motor and sensory areas of the human cortex.

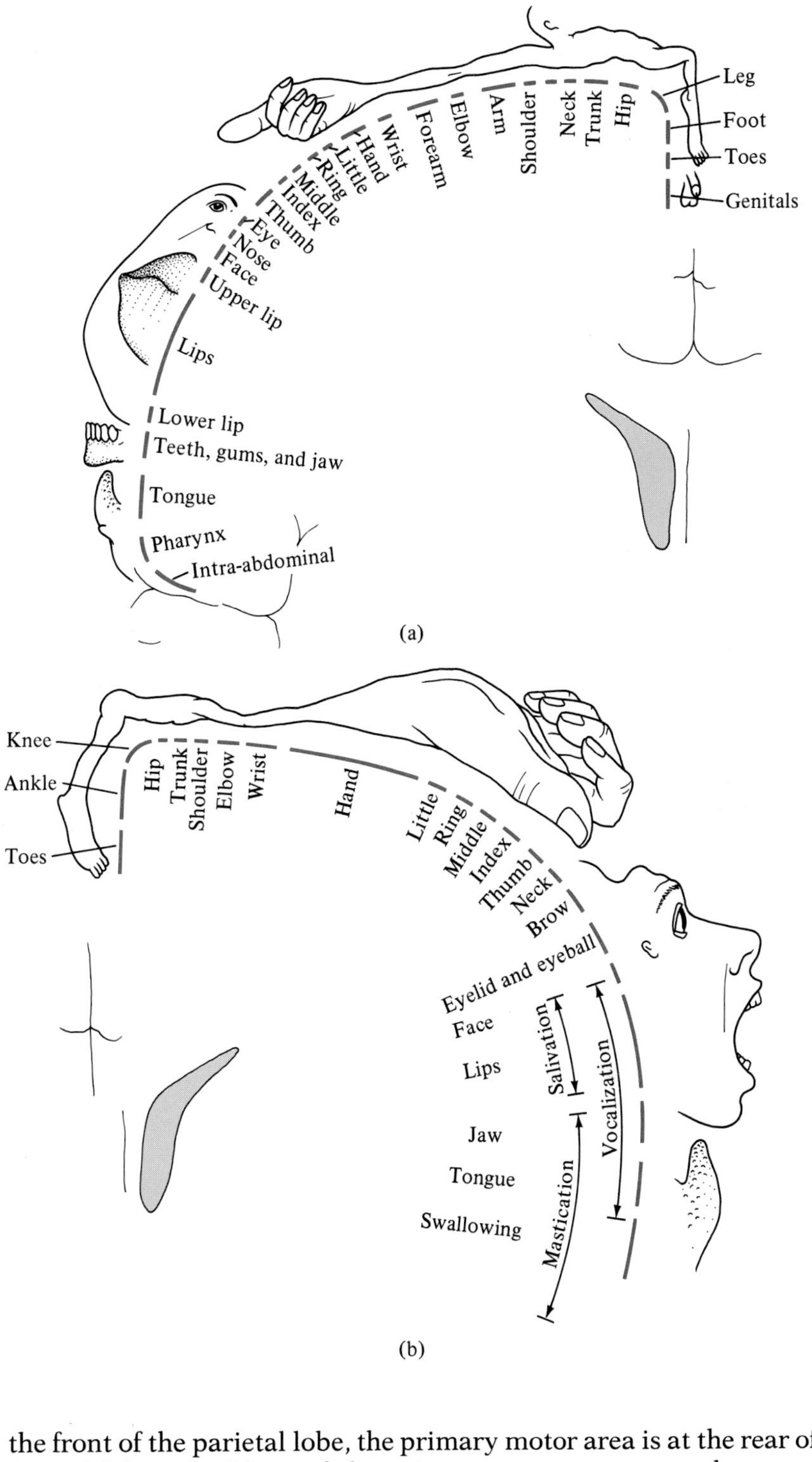


**FIG. 15-21** (a) The sensory homunculus and (b) the motor homunculus of the human cerebral cortex. The brain is represented as a frontal section, that is, divided into front and rear halves along the central sulcus. The sensory homunculus shows the anterior face of the right parietal lobe, the cortical area just behind the central sulcus; this is the primary somatosensory cortex. The motor homunculus shows the posterior face of the left frontal lobe, the cortical area just in front of the central sulcus; this is the primary motor area. The amount of cortex devoted to each part of the body is indicated. Notice the large emphasis on hand parts in the motor area.

the front of the parietal lobe, the primary motor area is at the rear of the frontal lobe. The fibers of the primary motor cortex send commands that ultimately control various skeletal muscles. The arrangement of each group of neurons along the cortex from top to bottom of the area corresponds to the anatomical location of the muscles they serve. In addition, those body parts with the largest number of motor fibers are represented by the largest amount of cortical tissue. In Figure 15-21b, a motor homunculus, comparable to the sensory homunculus, illustrates the relationship of cortical tissue to body parts. As in the sensory homunculus, except for parts of the face, the arrangement in the cortex is upside down compared to the actual arrangement of body parts. Again, because of crossing over of fibers, the control of the left side of the body resides in the right hemisphere and vice versa. Thus, a "stroke" (a blood clot or hemorrhage affecting the brain) on one side of the brain results in paralysis of the opposite side of the body.

Besides the primary motor areas, other motor areas, most of them in adjacent regions, affect various activities. For example, in humans, articulate speech capability resides in **Broca's area,** located near the bottom of the frontal lobe, but usually only in the left hemisphere. Unlike most mammals, humans have a strong tendency to localize some functions only on one side of the brain. Most people are right-handed, which reflects another kind of dominance. Besides speech and handedness, most people also have a dominant ear, eye, and leg.

***The Association Cortex Cannot Be Mapped Precisely.*** Besides the primary sensory areas, some other cortical regions exhibit electrical potentials when various receptors are stimulated. They are collectively termed the **association cortex,** or homotypical cortex, and include portions of the frontal, anterior temporal, and posterior parietal lobes. One way in which such areas differ from the primary sensory and motor areas is that they cannot be mapped precisely. Typically, they respond to stimuli applied to any of several places, such as to any of the limbs. Moreover, the same area might also respond to the administration of more than one type of stimulus, such as touch, sound, or light. Neighboring areas might also respond to the same stimuli. Another characteristic of these regions is that, after a first response to a particular stimulus, repeated administration of the stimulus may fail to produce further activity. These areas apparently receive their input by a less direct route than do the primary areas, allowing for some integration of the information before it reaches the association cortex. In the case of repeated stimuli, some other part of the nervous system apparently determines that the association cortex is not to be informed about the repetition. Presumably, the association cortex is too important to be informed about every little stimulus.

***Stimulation of the Temporal Lobes May Trigger Recall of Old Experiences.*** Among the most interesting areas of the cortex are the temporal lobes. When some regions of these lobes are stimulated electrically in a conscious subject, memories are sometimes evoked, complete with the emotions experienced when the event first occurred. For example, events of childhood or accidents, as well as the joys or fears associated with them, are "reexperienced." Repeating the stimulation may produce "reruns" of the same experiences and emotional reactions.

## BRAIN AND BEHAVIOR

One of the most interesting and energetically researched areas of modern neurobiological research concerns the relationships between the activities of various areas of the brain and various aspects of animal behavior. Several major brain regions are known to be involved in specific behaviors. Furthermore, the left and right halves of the vertebrate brain, although representing considerable redundancy, actually have somewhat different functions.

## THREE MAJOR REGIONS OF CONTROL

Some of the functions performed by the reticular formation, limbic system, and cerebrum illustrate two basic principles of brain function: (a) no function is localized entirely in one brain region and (b) the function of one region can be assumed by another after injury to the region normally responsible for the function.

***The Reticular Formation.*** Much of the central core of the vertebrate brainstem, the reticular formation, apparently contributes to alertness; if this region is damaged, the person sleeps much of the time (Fig. 15-17). The reticular formation is found in all vertebrates from fishes to humans.

***The Limbic System.*** The **limbic system** is a group of structures found in the cortex or just below it in higher vertebrates from reptiles to birds and mammals (Fig. 15-22). Some of its components are named for things that early investigators were reminded of by their shapes: amygdalae (Greek for "almonds"); cingulate (Latin for "girdle" or "belt"); gyrus, mammilary bodies (Latin for "nipples"); and hippocampuses (Greek for "sea horses"). Also included in the limbic system is the hypothalamus.

Emotions, such as fear or rage, which play a greater role in the lives of higher vertebrates than in those of lower vertebrates, are generated in the limbic system. Stimulation of the amygdalae, for example, can generate terror in a normally fearless animal. The removal of amygdalae from fierce wildcats tames them. In humans, various other emotions can be generated merely by stimulation of certain limbic regions. Interestingly, just before an epileptic seizure, the limbic system often becomes swept by a wave of stimulation. At that time, the person may experience a series of such sensations as intense hunger and thirst, rage, grief, fear, and profound joy. The seizure follows, with the person possibly losing consciousness or having convulsions. Hunger and thirst are known to be controlled by the hypothalamus. Other regions of the limbic system control taste, other oral functions, and sexual functions. The hippocampus may be essential to memory. Without it, the individual cannot remember anything that happened the previous day or even the previous few minutes.

**FIG. 15-22** The principal structures of the limbic system (shaded areas).

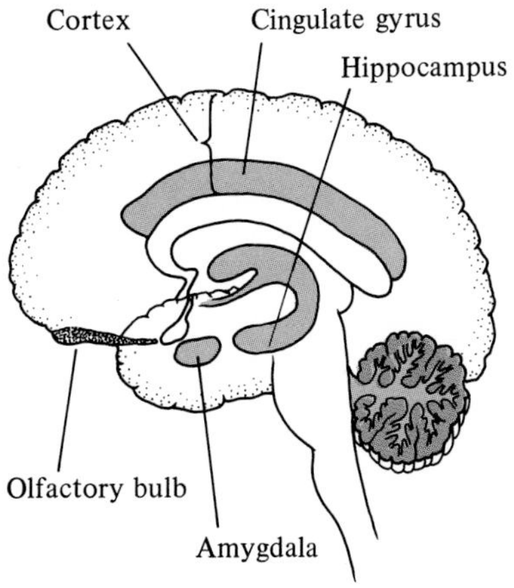


***The Cerebral Hemispheres.*** The most complex of the three major controlling regions of the higher mammal brain are the cerebral hemispheres, known collectively as the cerebrum. The cerebral hemispheres contain various sensory and motor areas, such as Broca's area, which controls articulation of speech, and the association cortex, much of the function of which is unknown. The parietal lobes, more than any other, seem to be involved in human intelligence or reasoning ability.

***The Separation of Functions.*** The foregoing are simplified descriptions of localization of functions in various brain regions. There is, however, no perfect separation of function among the three main brain regions in mammals. During the course of evolution, the limbic system and the neocortex could not have been merely added on to the reticular formation. Each must have evolved by expansion and elaboration of existing brain components. The actual behavior of an animal therefore represents a meld of influences from all three complexes.

***The Extent of Brain Redundancy.*** Many brain regions exhibit remarkable capacities to perform what are normally the functions of other regions. This is best seen when a part of the brain is destroyed. In time, other regions may assume the functions of the missing part. However, there are distinct limits to this ability. If damage is very extensive or occurs in elderly individuals, permanent disabilities may result.

One basis for the brain's redundancy is that, like so many other parts of the body, it is a double organ. However, studies of people with injuries to one side of the brain indicate that, in the adult human, the two sides of the cerebral cortex normally differ to some extent in their functions.

## THE TWO BRAINS OF *HOMO SAPIENS*

Although the two cerebral hemispheres represent a great amount of redundancy, it is now well-established that the two sides of the human brain differ considerably regarding several of their functions. Studies of persons with certain types of brain damage have fully documented this conclusion.

***Early Evidence of "Two Brains" in Humans.*** The first evidence that the two human cerebral hemispheres constitute a "left" and "right" brain and normally perform somewhat different tasks came from studies of persons with damage to one hemisphere. Damage to the left temporal lobe, for example, impairs the ability to read, write, speak, or do calculations. Comparable damage to the right temporal lobe usually does not have this effect. When a person's right parietal lobe was damaged, there was a lessened ability to recognize faces, to use maps and other patterns, and to accomplish three-dimensional vision. Damage to the left parietal lobe had little such effect. A person with injury to the posterior parietal lobe on his or her dominant side (in most people, the left brain) often exhibited such unusual behavior as combing the hair only on the right side of the head, making up or shaving only the right side of the face, and otherwise ignoring the left side of the body. In response to a stimulus such as a light pinprick to the left side of the body, such a person might shriek or weep in an exaggerated way, whereas the same stimulus given to the right side might be accepted without objection.

Damage to the right side of the brain also makes it difficult to engage in abstract reasoning and to perceive such relationships as are expressed by poetry. Studies of brain-damaged persons also indicate that humans remember melodies with their right hemisphere but read music with the left. Damage to the right brain impairs the ability to recall or recognize tunes, while reading of musical notations is unaffected. The right hemisphere apparently is important for composing good poetry, but is not involved in rhyming.

Other studies of brain-damaged persons, usually accident victims or wounded military personnel, indicated that in humans the right hemisphere normally does not retain information represented by words; instead, it usually seems to transmit language information immediately to the left hemisphere, where the words are processed. Also, whereas the left hemisphere processes information sequentially, the right hemisphere seems able to access several inputs at once. Remarkably, the left hemisphere is dominant in 90 percent of all people. Left-handedness,

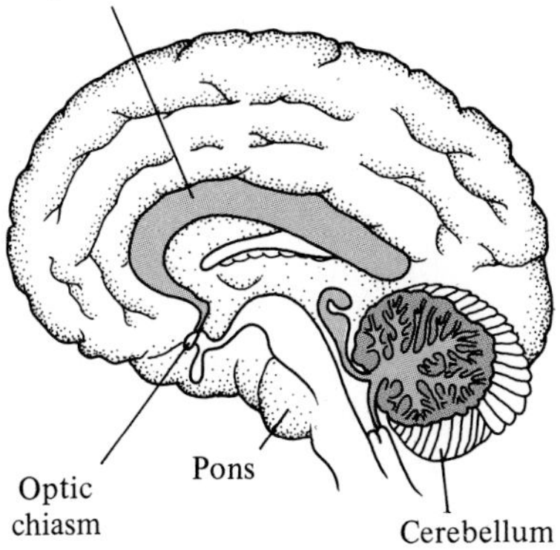


**FIG. 15-23** The corpus callosum and optic chiasm, two of the most important commissures severed in various split-brain patients.

however, present in 20 percent of all people, apparently has no necessary relationship to which hemisphere is dominant.

As valuable as were these "natural experiments" (accidental injuries to various parts of the brain) for mapping the different functions of the two cerebral hemispheres, much more remarkable findings came from the study of "split-brain" individuals, those on whom certain operations were performed.

***The Role of Commissures in Unoperated Individuals.*** Much of the information gained by the right hand and some other receptors on the right side of the body is normally immediately conveyed only to the left cerebral hemisphere. The converse is also true: much of the information obtained by the body's left side is immediately conveyed only to the right hemisphere. However, information received by either side is normally shared a moment later with the other hemisphere. This sharing interaction is made possible by a number of **commissures,** large neural connections between the two cerebral hemispheres (Fig. 15-23). The most important of these is the great commissure, or **corpus callosum.** In humans, this massive connection contains some 200 million fibers. Another important commissure is the **optic chiasm** (from the Latin *chiasma,* "crosspiece"). This crossover of optic nerves creates an exception to the general rule: it directs about half of the information received from each eye to the opposite side of the brain and the remainder to its own side. As a result, in the normal individual, anything seen in the right half of the visual field by either eye is first registered in the left side of the brain and vice versa (see Fig. 15-24). The information is normally then transferred via the corpus callosum to the other side.

**FIG. 15-24** Sketch of split-brain testing arrangement showing how a different image can be seen by each eye and registered in that side of the brain only. Note that normally the right visual field is registered in the left hemisphere only and the left visual field in the right hemisphere only.

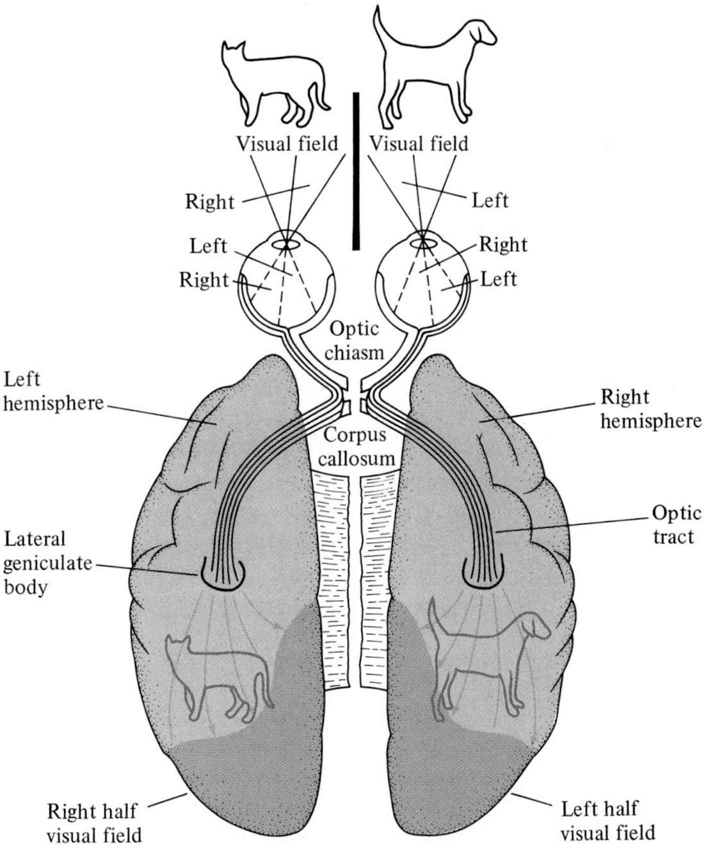

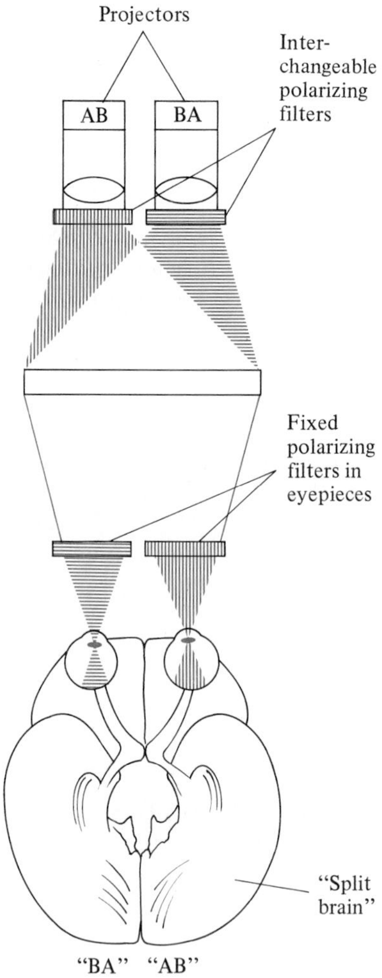


**FIG. 15-25** Sketch of a device by which each eye and hemisphere can be conditioned to choose one of two choices as correct. For testing, the filters are reversed so that both eyes see the images in the same order.

***Split-Brain Individuals.*** In recent years, it has been possible to study split-brain individuals. This has come about in the following way. In some epileptic patients, grand mal seizures, the most severe type, occur incessantly and frequently, as often as twice an hour, day and night. Because it is evident that the "neuroelectric storm" of the epileptic seizure spreads from one hemisphere to the other, it was decided as a last resort in one acute case to surgically sever the patient's major commissures: the corpus callosum and the optic chiasm. The expectation was that by confining the seizures to one hemisphere, their severity would be greatly reduced. The result was a happy one: unexpectedly, the seizures stopped entirely. The operation was subsequently performed in other acute cases with similar results. Some of these patients later agreed to be studied; the findings have added greatly to our understanding of the roles not only of these commissures but of the cerebral hemispheres themselves. In 1981, the American physiologist Roger Sperry received a Nobel prize for this work.

***Results of the Operation on Humans.*** When split-brain human patients were tested in certain ways, they displayed an unusual behavior. The tests utilized an apparatus that prevented the patient's left eye from seeing what the right eye saw and vice versa. With this device, a certain word or object could be shown in the right or left visual field of the patient's right eye while a different word or object was shown in the right or left visual field of the left eye. As illustrated in Figure 15-24, such tests revealed that objects seen in the left visual field register only in the right cerebrum and objects seen in the right visual field register only in the left cerebrum. Remarkably, when a split-brain person was shown an object, such as a pencil, only in the left visual field—that is, when only the right side of the brain received the information—the person could describe what a pencil is used for but could not name it. Retaining a memory of the names of objects seems to be a function reserved to the left hemisphere. If allowed to touch the pencil with the right hand, for example, the person was immediately able to name it.

When subjects with split brains were given a glimpse of a pair of figures simultaneously, one in each visual field, and asked to pick out a certain one, they would point to the correct one with their left hand while simultaneously either pointing to the correct one with their right hand or naming it. Normally, of course, both hands would not have participated in the response; normally the left hemisphere not only controls speech but dominates the right hemisphere as well. When asked about the responses they had made, the subjects said that they had no recollection of pointing with their left hand. It was as though two different individuals had been given the same command and both had followed it.

In a further series of experiments, split-brain individuals were trained so that one hemisphere regarded one of two choices as correct while the other hemisphere was trained to choose the other. The training was given simultaneously to both hemispheres by using the optical device sketched in Figure 15-25, so that the images seen by one eye were seen in reverse order by the other. After the training had established a consistent response, the optical device was rotated so that both eyes—and hemispheres—perceived the images in the same order. In earlier experiments with split-brain monkeys, no difficulties had resulted: after a brief hesitation, apparently one hemisphere or the other always assumed command and a consistent series of choices of the same object or action was made. Occasionally, however, the monkey's dominant hemisphere appeared to relinquish control, allowing the other to assume command for a time. Similar tests of human split-brain subjects usually had the same results. But from time to time a kind of conflict arose in the humans between the two hemispheres. In some cases, the right hand would even reach across and grab the left hand to force it to make what the right hand and left hemisphere regarded as the right choice. However, the task of arranging blocks in certain patterns was done better by the left hand (right hemisphere). At times, when the right hand had failed to execute a pattern correctly, the left hand would twitch or start or interfere with the right.

***What Is the Significance of Having "Two Brains"?*** It is tempting to speculate on the significance of the results of studies on split-brain humans. However, whether the two sides of some peoples' personalities—or at least the two modes by which we seem to arrive at conclusions—have a physical basis in differences between the right and left cortex is not a burning scientific question among neurobiologists. For a time, however, there was actually a movement among some enthusiastic educators to encourage United States schools to give more attention to the creative, poetic (right hemisphere) side of human nature, rather than to continue to put such strong emphasis on good performance by the verbal, mathematical side (left hemisphere). Our many questions about the precise differences between the right and left hemispheres will probably be answered in time. It may well be significant that so many functional differences have been found to exist between the right and left halves of that part of the body by which we differ so strikingly from the other animals: our cerebral cortex.

## Summary

Rapid, brief reactions almost invariably are mediated by the nervous system, the unit of which is the neuron or nerve cell. Each neuron is composed of a soma, or nerve cell body, containing the nucleus, and an axon, a fiber of variable length. One or more dendrites may also extend from the soma. With certain exceptions, nerve impulses pass along neurons only from dendrites to soma to axon and across synapses to other neurons, in that order. Synapses may be (1) axodendritic, (2) axosomatic, or (3) axoaxonic. The first two types are most common in vertebrates.

Cnidarians, the lowest animals to have a nervous system, lack a region of central control. However, the anterior end of bilaterally symmetrical animals (those above Cnidaria) typically exhibit a concentration of sense receptors and of nervous elements, the latter organized as a brain. The brain and spinal cord comprise the central nervous system. The brains of the most highly evolved invertebrates are remarkably complex. That of the octopus, a cephalopod, is as complex as that of many fishes.

All chordates possess a dorsally located nerve cord of tubular construction. Its fluid-filled central canal is continuous with a system of ventricles in the brain. Connected to the CNS is the peripheral nervous system, comprising the cranial nerves, serving principally the head and some internal organs, and the spinal nerves, serving the body musculature, the skin, bones, and glands. Also connected to the CNS is the autonomic nervous system, which monitors and regulates the functioning of the body's internal organs.

The smallest functional unit of integration is the reflex arc. The simplest reflex arcs are the monosynaptic and disynaptic arcs, which contain only two and three neurons, respectively: a sensory (afferent) neuron, the tip of which is usually specialized as a receptor; a motor (efferent) neuron, which forms a special synapse with an effector; and, in disynaptic arcs, an interneuron, located entirely within the spinal cord. Synapses between neurons constitute a major physical basis for integration.

The autonomic nervous system consists of the sympathetic and parasympathetic divisions, both of which serve most of the body's viscera, or internal organs. The actions of the two divisions on any one organ are nearly always antagonistic. Postganglionic axons of sympathetic neurons release norepinephrine onto the organs they innervate, whereas corresponding axons of parasympathetic fibers release acetylcholine.

The vertebrate brain has a long evolutionary history, stages of which are represented by different regions of the mammalian brain. The oldest region is the brainstem, divided into the midbrain, pons, and medulla. In today's primitive vertebrates, three main divisions can be discerned, the forebrain, midbrain, and hindbrain, each corresponding to one of the three main sense capabilities: smell and taste, which constitute a single category; sight; and hearing/balance. The first of the three regions gave rise to the cerebrum and diencephalon, the second to the tectum, and the third to the cerebellum and other lower centers.

Sensory areas are found in all regions of the cerebral cortex except the frontal lobes. When various receptors are stimulated, corresponding nervous activity occurs in the sensory cortex to which the receptors are directly connected, making precise mapping of these areas possible. Stimulating various regions of the frontal lobes results in contraction of the muscles or secretion by the glands controlled by those regions. Mappings of the "motor cortex" are based on such experiments.

Some cortical regions, such as the parietal lobes, cannot be mapped precisely. These areas of the "association cortex" apparently receive input by indirect routes, with a certain amount of filtering and integration occurring before any signal reaches them. The function of the left and right cerebral hemispheres appears to differ somewhat, not only in humans but in all vertebrates; in most people, for example, abilities to recognize patterns and to become oriented in space reside in the right hemisphere.

## Recommended Readings

The following readings are for Chapters 15, 16, 17, and 18. There are a large number of good introductory textbooks on neurobiology, most of which contain descriptions of reception. A growing number of introductory texts on physiology or anatomy and physiology also have good descriptions of nervous systems, receptors, and effectors. Several textbooks are listed here, together with some recent articles on special topics written at an undergraduate level.

### Books

BECK, W. S. 1971. *Human Design: Molecular, Cellular and Systematical Physiology.* Harcourt, New York. One of the better introductory books on human anatomy and physiology. Clear descriptions of important technical aspects and excellent illustrations.

BULLOCK, T. H. 1977. *Introduction to Nervous Systems.* Freeman, San Francisco. Probably the best introductory text in general neurobiology. Extraordinarily clear descriptions and abundant illustrations of very high quality.

HOAR, W. S. 1983. *General and Comparative Physiology,* 3d ed. Prentice-Hall, Englewood Cliffs, N.J. An excellent text for the advanced undergraduate. Contains clear, concise, and well-illustrated descriptions of the major topics in animal physiology.

HUXLEY, A. F. 1980. *Reflections on Muscle.* Princeton University Press, Princeton, N.J. Insights provided by a pioneer and one of the world's most famous muscle biochemists.

KANDEL, E. R. 1976. *Cellular Basis of Behavior: An Introduction to Behavioral Neurobiology.* Freeman, San Francisco. One of the best introductions to neurobiology. The book demonstrates the value of studying invertebrate nervous systems as the best way to understand many of the problems of the neurobiology of higher animals, including that of humans.

SHEPHERD, G. M. 1983. *Neurobiology.* Oxford University Press, New York. A well-illustrated, well-written, introductory textbook that presents neurobiology in a broad context—evolutionary, cellular, behavioral, and historical.

SQUIRE, J. 1981. *The Structural Basis of Muscle Contraction.* Plenum. New York. An authoritative summary of current understanding of the process of muscle contraction.

STEBBINS, W. C. 1983. *The Acoustic Sense of Animals.* Harvard University Press, Cambridge, MA. A clear, concise, nontechnical summary of the anatomy and physiology of hearing.

STEVENS, L. A. 1971. *Explorers of the Brain.* Knopf, New York. This work contains fascinating accounts of the history of investigation into the brain's secrets, including descriptions of methods used, all described in layperson's language.

### Articles
### Nervous Systems

AXELROD, J. 1974. "Neurotransmitters." *Scientific American,* 230(6):58–71. A discussion of types of transmitters and how they function.

DUMONT, Y. and M. ISRAEL. 1985. The release of acetylcholine. *Scientific American* 252(4):58–66. Elegantly conducted experiments involving electric fishes show how this major neurotransmitter conveys nerve impulses across synapses.

IVERSEN, L. L. 1979. "Chemistry of the brain." *Scientific American,* 241(3):134–149. An interesting summary of what is known of the brain's neurotransmitters and descriptions of some of the methods by which they are being investigated. This entire issue of *Scientific American* is devoted to articles written by experts on the structure and function of the brain.

KIMURA, D. 1973. "The asymmetry of the human brain." *Scientific American,* 228(3):70–81. Although it has been known for a long time that the left cerebral hemisphere is dominant in most people, it appears that the right hemisphere has some specialized functions of its own.

LLINÁS, R. R. 1975. "The cortex of the cerebellum." *Scientific American,* 232(1):56–71. The cerebellum is remarkably complex. It contains about $10^{11}$ neurons, one type of which receives synaptic input from 100,000 other cells. This maze of structural complexity is nevertheless understood to a remarkable degree.

LLINÁS, R. R. 1982. Calcium in synaptic transmission. *Scientific American* 247(4):56–65. The role of calcium ion currents in synaptic transmission as studied in the squid.

MORELL, P. and W. T. NORTON. 1980. Myelin. *Scientific American* 242(5):88–118. How myelin nerves acquire their sheaths and how the sheaths speed conduction of impulses.

### Reception

KNUDSEN, E. I. 1981. Hearing in the barn owl. *Scientific American* 245(6):113–125. A fascinating description of how barn owls can determine the orientation and direction of movement of a mouse in total darkness.

PARKER, D. E. 1980. The vestibular apparatus. *Scientific American* 243(5):118–135. How our inner ear functions in balance and orientation.

PETTIGREW, J. D. 1972. "The neurophysiology of binocular vision." *Scientific American,* 227(2):84–93. How the optic chiasma and certain neurons in the visual cortex of the brain provide us with binocular, three-dimensional vision.

SCHNEIDER, D. 1974. "The sex attractant receptor of moths." *Scientific American,* 231(1):28–35. Insights into how one molecule of sex attractant can trigger an action potential in the receptor cells of the male moth.

WERBLIN, F. S. 1973. "The control of sensitivity in the retina." *Scientific American,* 228(1):70–79. How the pigments in the eye's retina accommodate a wide range of light intensity while preserving the eye's sensitivity to detail.

### Response

DUSTIN, P. 1980. Microtubules. *Scientific American* 243(2):66–76. A well illustrated and interesting description of roles of microtubules in mitosis, cell movements, and in maintaining cell shape.

MARGARIA, R. 1972. "The sources of muscular energy." *Scientific American,* 226(3):84–91. An interesting description of the energy pathway that supports muscular work, from food to ATP. Practical suggestions for training for athletic contests are included.

MERTON, P. A. 1972. "How we control the contraction of our muscles." *Scientific American,* 226(5):30–37. The automatic feedback provided by the intrafusal fiber system in our striated muscles is explained.

MURRAY, J. M., and A. WEBER. 1974. "The cooperative action of muscle proteins." *Scientific American,* 230(2):58–71. How troponin, tropomyosin, actin, and myosin interact in the presence of calcium ions to produce contraction of a muscle.

SATIR, P. 1974. "How cilia move." *Scientific American,* 231(4):44–63. The evidence for the beating of cilia as a result of the active sliding of microtubules with an energy assist from ATP.

# 16 Conduction and Transmission of Nerve Impulses

OUTLINE



The twentieth century has brought an explosion of knowledge in neurobiology, due in large part to the perfection of microchemical methods—the analysis of minute quantities of material—electronic technology, such as recording voltages from within small cells, and electron microscopy. As in other areas of cell biology, refined techniques have been used to analyze minute details of cellular composition and function. In neurobiology, the cell most often studied is the neuron (Fig. 16-1).

## THE ELECTRICAL BASIS OF NERVE IMPULSE CONDUCTION

Just as it would be impossible to understand how a telegraph or telephone works without some knowledge of electric currents and magnetism, an understanding of resting potential is needed for any appreciation of how neurons carry information. The **resting potential** is the difference in electrical charge between the inside and the outside of a neuron's plasma membrane. In effect, the neuron is a tiny, weak battery, that is, a device that produces electrical energy by chemical action. Comparing the inside of the neuron to the outside, the potential difference, or voltage, maintained by most neurons is about −70 to −90 millivolts (mV) (Fig. 16-2). Most cells have the ability to maintain a

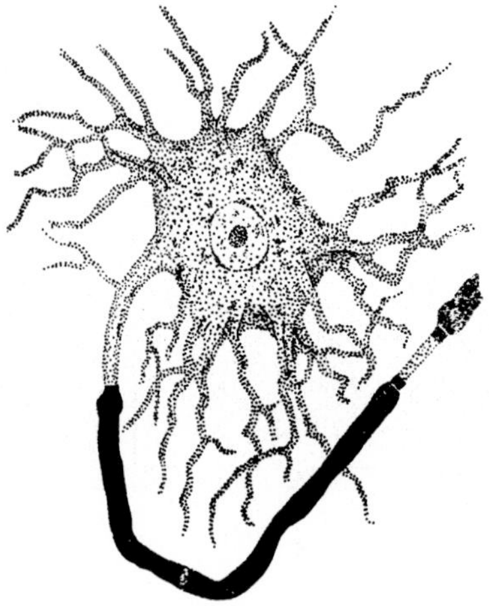

FIG. 16-1 Portion of a motor neuron of the electric ray *Torpedo*, showing its soma, bases of numerous dendrites, and its myelinated axon.

FIG. 16-2 Resting potential. Because of the imbalance in distribution of various ions between sides of the neuron's plasma membrane, the inside is about −70 to −90 millivolts compared to the outside. This condition in the resting cell, that is, one not conducting a nerve impulse, can be demonstrated with two electrical conductor tips, called electrodes, one of which is placed on the cell surface and the other in the cytoplasm and which are linked through a sensitive voltmeter.

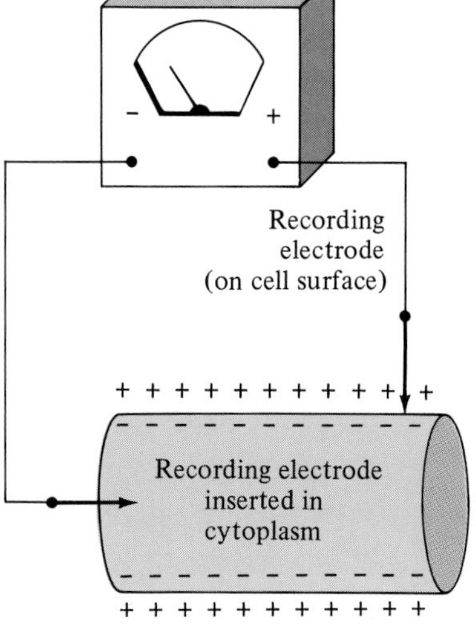


potential difference between the two sides of their plasma membrane. However, **excitable cells,** such as neurons and muscle fibers, can use this potential difference to convey information from one end of the cell to the other and even, in some cases, to communicate information from one cell to the next.

## ION DISTRIBUTION

To understand the special situation that exists in a neuron, consider these principles that apply to all cells:

1. The total number of all kinds of diffusible particles, such as molecules or inorganic ions, on one side of a plasma membrane tends to be balanced by an equal number of such particles on the other side of that membrane. However, because of the differing permeability of plasma membranes to various solutes (Chapter 5), few, if any, particles of the same kind are ever evenly distributed on the two sides of the same membrane.
2. The positively charged particles (cations) on either side of the plasma membrane also tend to be balanced by an equal number of negatively charged particles (anions) on the same side. However, whenever one type of ion, such as charged proteins, cannot diffuse through a membrane, the smaller ions that are free to cross it will become unequally distributed on the two sides of the membrane.

***Potassium, Sodium, and Other Important Ions.*** In neurons, as in cells generally, the cytoplasm contains a number of dissolved amino acids and proteins, most of which are anions. These large, negatively charged molecules cannot diffuse out of the cell because they cannot penetrate the plasma membrane. However, because the plasma membrane of the neuron is quite permeable to potassium ions ($K^+$), these small ions are free to enter the cell, where, to some extent, they neutralize the cell's organic anions. In addition, the plasma membrane has a mechanism called a pump that actively transports $K^+$ into the cell despite an opposing concentration gradient (see Chapter 5). As a result, potassium ions are about 25 times more concentrated inside the cell than in the tissue fluid surrounding it. The distribution of ions on the two sides of the plasma membrane of a giant axon* of the squid is given in Table 16-1.

Because the plasma membrane of a neuron that is not conducting a nerve impulse—the so-called resting neuron—is more permeable to $K^+$ than to any other ion, the neutralization of the electrical charge of the proteins and other nondiffusible anions inside the cell is almost entirely achieved by $K^+$. If $K^+$ were the only cation to enter, the resting potential would be about −90 mV, a value known as the equilibrium potential for $K^+$. However, because other ions, principally $Na^+$, also enter to some extent, the resting potential achieved in most cells is somewhat less than −90 mV. (Two other ions present in or about the neuron in significant amounts are $Cl^-$ and $Ca^{2+}$. Both are more concentrated outside the cell than inside.)

* The ventral nerve cord in several groups of invertebrates, notably the annelids and cephalopod molluscs, such as the squid, contain unusually thick (up to 1 mm in diameter) axons, called giant axons for that reason. Because of their thickness and the consequent ease with which they can be penetrated by recording electrodes, they were the first nerve fibers to be investigated in detail and remain a favorite object of study.

**TABLE 16-1 Ionic Concentrations on Both Sides of the Plasma Membrane of a Squid Giant Axon**

| | *Outside (mM/liter $H_2O$)* | *Inside (mM/liter $H_2O$)* | *Ratio (o/i)* |
|---|---|---|---|
| $K^+$ | 15 | 400 | 1/25 |
| $Na^+$ | 450 | 50 | 10/1 |
| $Cl^-$ | 545 | 50–100 | 7/1 |
| $Ca^+$ | 10 | 0.4 | 25/1 |
| Organic anions | 0 | 355 | |

The distribution of $Na^+$ on the two sides of the membrane is strikingly different from that of $K^+$. $Na^+$ is about 10 times more concentrated outside the cell than within it, resulting in large part from the action of the same ion pump that transports $K^+$ into the cell. Another reason for this sodium distribution is that, in the resting state, the neuron's plasma membrane maintains an almost complete impermeability to $Na^+$. In other words, the sodium "gates" in the sodium channels within the membrane are almost completely closed.

Some inward leakage of $Na^+$ and some outward loss of $K^+$ down their respective concentration gradients occurs continually, as has been demonstrated by radioactive tracer studies. It could be said that the neuron "battery" continually leaks some of its charge. This leakage is fully compensated for by the sodium-potassium pump, however.

***The Sodium-Potassium Pump.*** As long as the plasma membrane of a neuron is intact and has an energy supply in the form of ATP—both being absolute requirements for active transport—the neuron continually discharges as much $Na^+$ as enters the neuron. That the process is truly an active one, requiring energy, can be shown experimentally. For example, when a giant axon of a squid is perfused with dinitrophenol (DNP), a poison that blocks the formation of ATP, the active transport of $Na^+$ out of the neuron stops. When the axon is washed free of DNP, normal active transport of $Na^+$ resumes.

When the mechanism that actively transports $Na^+$ was first discovered, it was called the sodium pump. Eventually, it was found that $K^+$ is also actively pumped into the cell by the same transport mechanism. Accordingly, the mechanism was named the sodium-potassium pump or the ion pump.

The pump works only if both $Na^+$ and $K^+$ are present. And it works only in one direction for each ion: $Na^+$ out and $K^+$ in. The ratio of exchange is three $Na^+$ ions out for every two $K^+$ ions in, for every ATP molecule utilized. The $Na^+$-$K^+$ pump thus contributes somewhat to the maintenance of the resting potential (Fig. 16-3).

In summary, the distribution of ions in the resting neuron is as follows:

1. High concentrations of $Na^+$, $Ca^{2+}$, and $Cl^-$ exist outside the cell.
2. High concentrations of $K^+$ and organic anions exist inside the cell.
3. The plasma membrane is highly impermeable to $Na^+$ and $Cl^-$, less so to $Ca^{2+}$, somewhat permeable to $K^+$, and altogether impermeable to organic ions.

**FIG. 16-3** The sodium-potassium pump. This active transport mechanism in the plasma membrane by which sodium is secreted and potassium absorbed is partly responsible for maintenance of the resting potential of neurons.

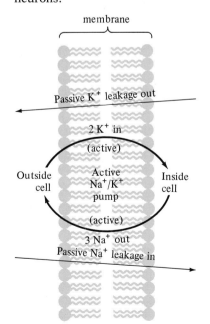

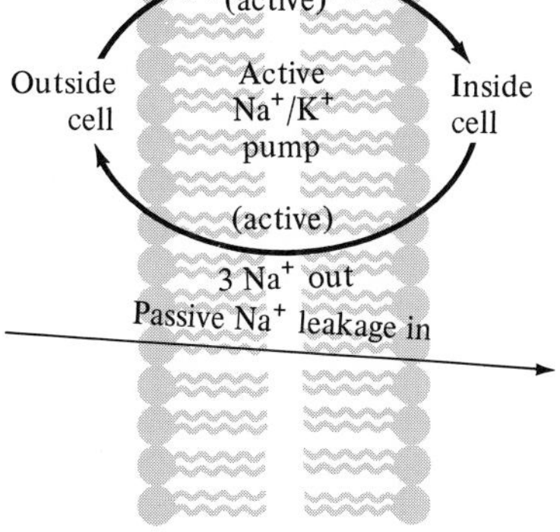

4. The $Na^+$-$K^+$ pump actively maintains the $Na^+$ and $K^+$ concentrations at a steady "resting" level.
5. Following its concentration gradient, $K^+$ tends to leak out of the cell. However, $K^+$ ions are held within the cell and to the outside of the membrane by attraction to the organic anions that cannot follow them through the membrane. The result is that the outside of the membrane is electrically positive, and the inside negative.

These conditions provide the physical basis for the resting potential. The electrical signals by which parts of the nervous system communicate with each other can be regarded as disturbances produced in the resting potential. Such signals are of two types: electrotonic potentials and action potentials, or nerve impulses.

## ELECTROTONIC POTENTIALS

If the electric charge inside a neuron is slightly lowered by introducing cations into the cell, a partial **depolarization** of the membrane occurs; that is, the inside becomes less negative in relation to the outside (Fig. 16-4a, b). The degree of depolarization depends on how many cations enter the neuron. If an electrode carrying a direct electric current is used to induce depolarization, the amount of depolarization is proportional to the strength, or voltage, of the current. If the administered current flows in the opposite direction, however, the membrane becomes **hyperpolarized;** that is, the inside becomes more, not less, negative and is less likely to generate a nerve impulse.

However, no matter whether the stimulus is depolarizing or hyperpolarizing, the effect on the neuron is to produce a disturbance known as an **electrotonic potential,** also referred to as a local, graded response. Such a disturbance weakens with both time and distance as it spreads from the point where the stimulus was given, along the inside and outside of the membrane, before fading away entirely after a short distance. Because of this decrease in intensity with time and distance, electrotonic potentials are useful only for local signaling.

## THE NEURON AS CAPACITOR

Because the neuron's plasma membrane is thin and relatively impermeable to various ions, the neuron has the properties of an electrical **capacitor,** or condenser, a device that accumulates and stores electric charges. In the neuron's role as capacitor, the plasma membrane serves as a thin layer of insulation separating the two oppositely charged elements. The positive charges on the outside of the membrane accumulate there because they are attracted to and held to the outer surface of the membrane by the negative charges just inside the membrane. The negative charges are provided by the organic anions that cannot diffuse out of the cell, whereas the positive charges are provided mostly by $Na^+$ and $K^+$ ions that are held to the outer surface of the membrane by electrostatic attraction to the organic anions on its inner surface. If the negative charge inside the membrane is reduced, by a depolarizing current, for example, the positively charged ions on the outside of the cell are no longer held to the membrane's surface. These cations promptly disperse into the surrounding tissue fluid. This flowing away of the membrane's outer charge is the first effect of a reduction of the negative charge inside a neuron (Fig. 16-4c). Together, the reduction of the

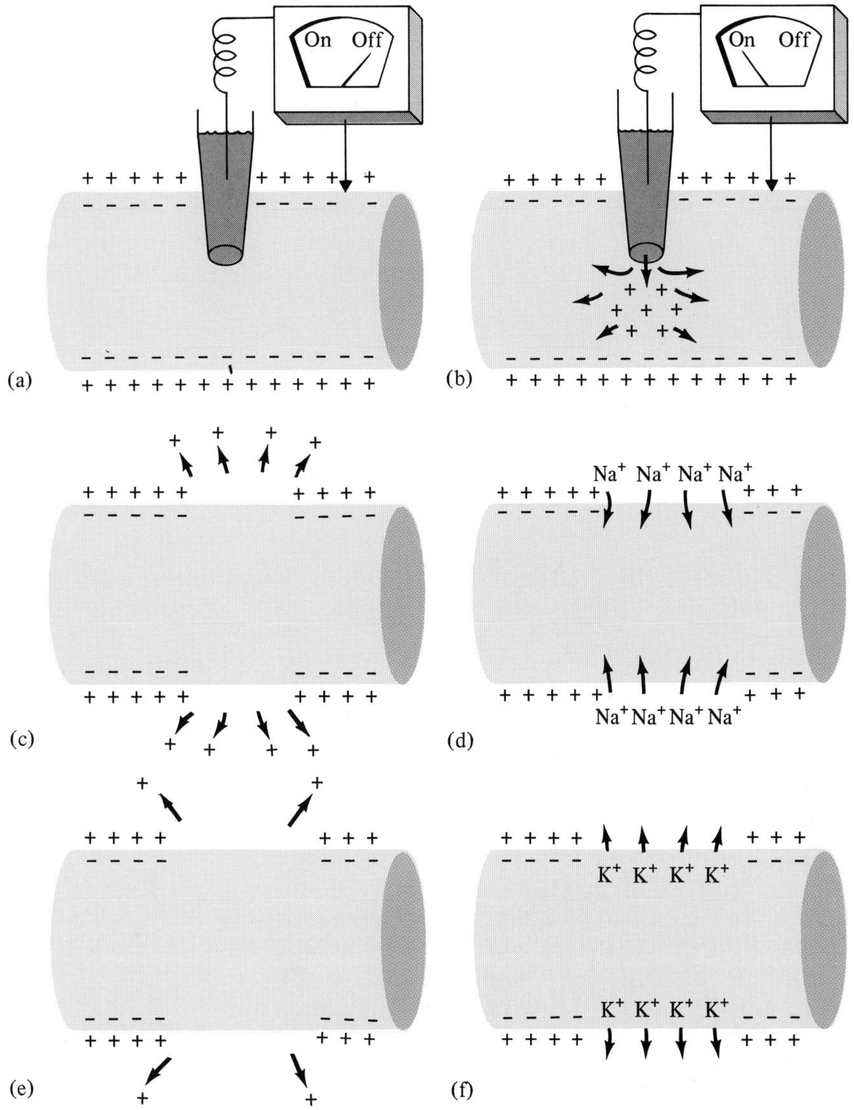


**FIG. 16-4** Generation of an electrotonic potential by a depolarizing electric current. (a) Portion of an axon showing the distribution of charges of the resting potential. A stimulating electrode has been inserted into the axoplasm. (b) Injection of a depolarizing electric current, producing a local neutralization of inner negative charges. (c) Flow of capacitative current occurs when positive charges are no longer held to the outer surface of the cell membrane because of the neutralization of inner negative charges. (d) Partial depolarization results in the membrane's becoming somewhat more permeable to $Na^+$. Entry of $Na^+$ further depolarizes the region and also contributes to the partial depolarization of adjacent regions. (e) A partially depolarized neuron. What happens next depends on whether the depolarization has exceeded the threshold. If it has not, the cell will return to the resting potential condition. (f) The electrotonic potential fades away if stimulus is not of threshold intensity. Repolarization occurs when the outflow of $K^+$ overcomes the effect of $Na^+$ entry.

inner charge and the resulting loss of outer charge comprise the depolarization.

***Sodium Ionic Current.*** Any depolarization of the resting neuron's plasma membrane has a special effect on the membrane: it immediately increases its permeability to $Na^+$. Depolarization causes the gates in the sodium channels to open. Because sodium is in higher concentration outside the cell than within it, the increased permeability to sodium immediately allows a current of sodium ions to flow into the cell in the depolarized region (Fig. 16-4d). The amount of initial depolarization and the strength of the resulting sodium current are directly proportional to the strength of the depolarizing stimulus. A weak stimulus is followed by the opening of only a few sodium gates and the entry of only a small amount of $Na^+$. But even the entry of a weak cationic current further neutralizes the negative charge inside the cell, that is, it further depolarizes the membrane, including the adjacent regions as well (Fig. 16-4e). This additional depolarization increases the membrane's permeability to $Na^+$ and, for a time, produces a **positive feedback effect,** in which the entry of $Na^+$ as a result of depolarization thus itself becomes a depolarizing stimulus. However, there is a check upon this process. After the membrane has undergone depolarization for a time in a given region, the sodium gates in that region close. They remain closed until that part of the membrane has once again become polarized by returning to the resting potential.

***The Potassium Channels and Repolarization of the Membrane.*** When depolarization begins and the sodium gates begin to open, the gates in the potassium channels also start to open, but much more slowly. Ac-

PANEL 16-1

# The Cathode Ray Oscilloscope

## ELECTRONIC PRINCIPLES OF CRO OPERATION

The principles upon which the cathode ray oscilloscope (CRO) is based are relatively simple. The main component is the **cathode ray tube,** an evacuated glass tube much like the picture tube of a television receiver. At one end of the tube is a cathode, or negatively charged electrode, (designated C in the figure), which, when heated by an electric current, emits a stream of electrons. At the opposite end of the tube is a glass screen coated with a chemical that can conduct a current and that fluoresces (glows) wherever it is struck by a sufficiently intense beam of electrons. By being connected to the same circuit as the cathode, the fluorescent screen acts as an anode and thus attracts the electrons emitted by the cathode.

The electrons leaving the cathode form a beam, referred to as the cathode ray, which is narrowed, or focused, by passing through a cylindrical anode ($A_1$) and accelerated to high speed by passing through a second anode ($A_2$). The electron beam next passes between the first of two pairs of deflection plates, which can be electrically charged. The first of the two pairs of plates (horizontal, or X, plates) are connected to a device that causes the beam to sweep across the screen from left to right when activated, producing a glowing straight line. As the beam reaches the end of the sweep and the charges on the horizontal plates are abruptly reversed, the beam returns instantaneously to the left side and begins another left-to-right sweep. If the speed of the sweep is fast enough, the straight line will remain visible, being renewed over and over again.

The second set of two plates (vertical, or Y, plates) is attached to a pair of recording electrodes that can be placed at appropriate points on a nerve or nerve fiber. Any electrical changes occurring in the fiber are thus communicated to the Y plates. Because the potentials generated by nerves are weak, it is necessary to amplify them electronically so that they are strong enough to deflect the cathode ray as it sweeps across the CRO screen.

## HOW THE CRO IS USED TO RECORD POTENTIALS

When an electrical disturbance (potential) passes down a nerve to which recording electrodes are attached as described above, the vertical plates become charged, causing a vertical deflection—up or down, depending on the potential—in the sweeping cathode ray.

By stimulating a nerve fiber repeatedly, a series of such potentials is traced on the screen. If the sweep of the beam from left to right is synchronized with the electrical stimulations to the nerve, each potential tracing on the screen becomes superimposed on the previous one, creating the appearance of a stationary wave. This allows the pattern to be studied in great detail. For example, when a suitable grid is placed over the face of the screen, the vertical deflections can be calibrated in terms of millivolts, and displacements along the Y axis can be calibrated in milliseconds. This allows for detailed study of various types of electrical events associated with the functioning of muscles and other organs as well as of nerves.

## CRO RECORDING OF RESTING, DEPOLARIZING, AND HYPERPOLARIZING POTENTIALS

When a CRO is turned on and the beam set to sweep, a bright horizontal line appears on the cordingly, $K^+$ begins to flow out of the cell at the point of stimulation (Figure 16-4f). The potassium current attains its peak outflow about half a millisecond after the sodium gates have closed. (Weak electrical currents of as brief duration as a millisecond or less can be amplified and displayed as tracings by the device known as a **cathode ray oscilloscope,** described in Panel 16-1.) The closing of the $Na^+$ gates, the opening of the $K^+$ gates, and the resulting outflow of $K^+$ repolarize the membrane in the affected region, restoring it to the resting potential condition.

screen. When an axon is pierced by one of two recording electrodes and the other electrode is placed on the surface of the fiber a little distance from the first, the bright line on the screen reflects the voltage difference between the inside and outside of the neuron's plasma membrane, shown in (b). The CRO thus records the resting potential of the axon. In the example shown in (b), the inside is 80 mV electronegative to the outside.

If the nerve fiber is stimulated electrically by a depolarizing stimulus (that is, one that lowers the resting potential) of less intensity than is needed to generate a nerve impulse, an electrotonic potential is generated, but no nerve impulse results from it. If the direction of the stimulating current is reversed (see text Figure 16-5) the result is another electrotonic potential, but one that is hyperpolarizing rather than depolarizing. As Figure 16-5 shows, increasing the stimulus intensity produces greater hyperpolarization and a more persistent electrotonic potential each time. If the stimulus is depolarizing, a similar series results until the **threshold level** of intensity is reached. At that point, an action potential is generated.

The cathode ray oscilloscope and CRO records. (a) The CRO. (b) The effect of inserting one recording electrode into a squid giant axon. The two electrodes thus record the resting potential, the difference in potential across the plasma membrane.

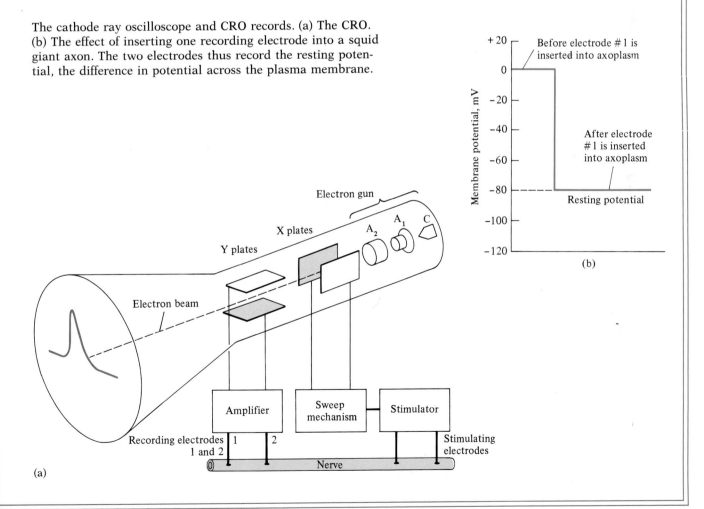

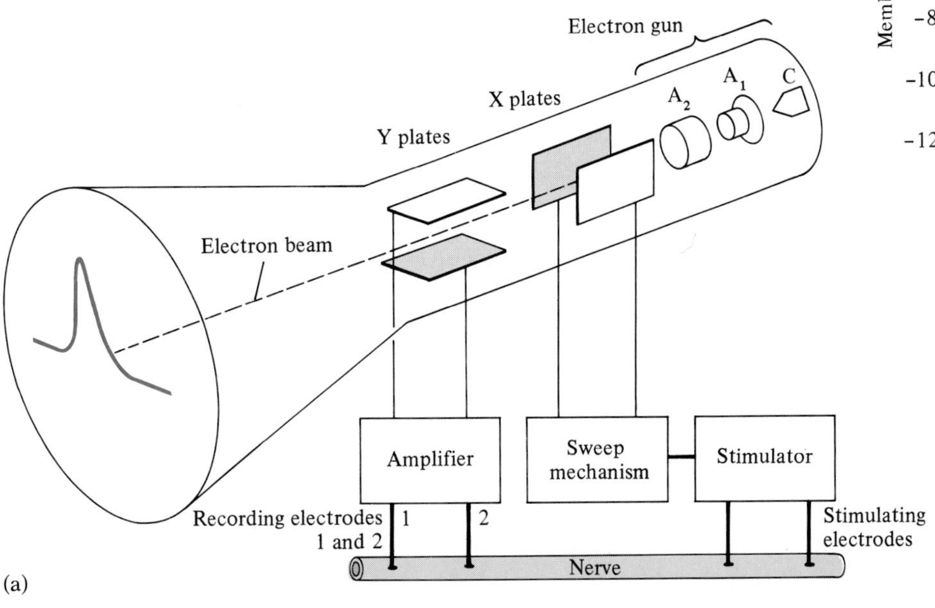


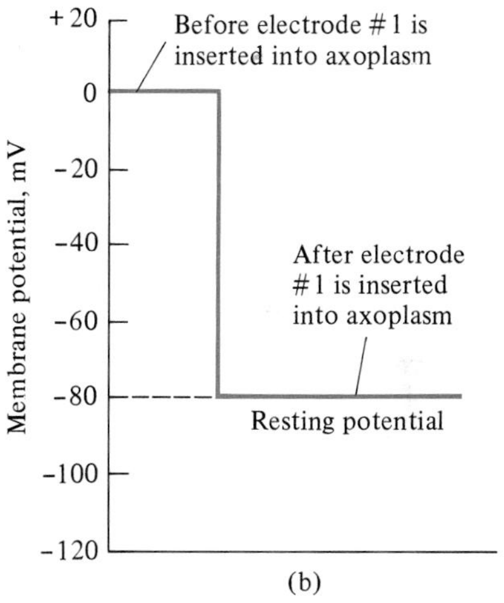

If the stimulus given to a neuron is weak and does not persist very long, only a few sodium channels are affected and few sodium ions enter the cell. The effect on the neighboring regions of the neuron is therefore minimal, and the electrotonic potential generated is limited to the immediate region and soon dies out. However, if a strong stimulus is given, opening many sodium gates, a considerable amount of $Na^+$ will enter, causing depolarization not only at the point of stimulation but of the immediately adjacent regions as well. The slowly reacting potassium gates are not able to repolarize the membrane before the strong sodium current has begun to flow through the cytoplasm of the neuron, completely depolarizing adjacent regions of the membrane. The process renews itself again and again, spreading rapidly down the fiber. An **action potential,** or **nerve impulse,** has been generated.

## ACTION POTENTIALS

In sharp contrast to the strength of the electrotonic potential, the strength of the action potential does not decrease with distance or time. Once triggered, action potentials pass along the entire length of the longest axon. Action potentials are thus well suited to long-distance communication.

The CRO record of an action potential is a spike-shaped curve. The upslope of the action potential curve displayed by a cathode ray oscilloscope results from increased membrane permeability to $Na^+$; the downslope results from increased permeability to $K^+$ with sodium channel inactivation (Fig. 16-5). In the initial phase of an action potential, the sodium gates open wide and $Na^+$ is free to enter the cell. As a result, the process does not end at complete depolarization, or zero membrane potential. Instead, $Na^+$ continues to enter and the electrical change continues beyond, or overshoots, the zero point. The inside becomes about 30 to 50 mV positive in relation to the outside, a value near the point at which the sodium concentrations inside and outside the membrane are at equilibrium. The sodium current has now run its course and the countercurrent of $K^+$ exerts its full effect by repolarizing the membrane.

As the $K^+$ gates open wide, the $Na^+$ gates close and cannot open again for 1 or more milliseconds, no matter how strong a depolarizing stimulus is given. As a result, this region of the neuron is temporarily inexcitable to further stimulation; it has entered what is termed the **absolute refractory period.** The rapid exit of $K^+$ that then occurs quickly returns the membrane to near the resting potential level. The entire action potential takes only a millisecond or so. The shape of the record of an action potential on the CRO screen suggested the term **spike potential** or simply spike.

***The Threshold-Level Stimulus.*** As we have seen, an electrotonic potential or other stimulus strong enough to depolarize a resting potential to about −40 mV causes an extensive opening of sodium gates followed by a strong inrush of $Na^+$ (Fig. 16-6). Instead of the disturbance fading and the membrane returning to its resting potential, a complete depolarization occurs and spreads to neighboring regions of the neuron. The minimum stimulus strength needed for an action potential to occur is said to be of **threshold-level intensity.** As with squeezing the trigger of a loaded gun, once the process reaches a certain point, it continues to completion even if the triggering stimulus is discontinued.

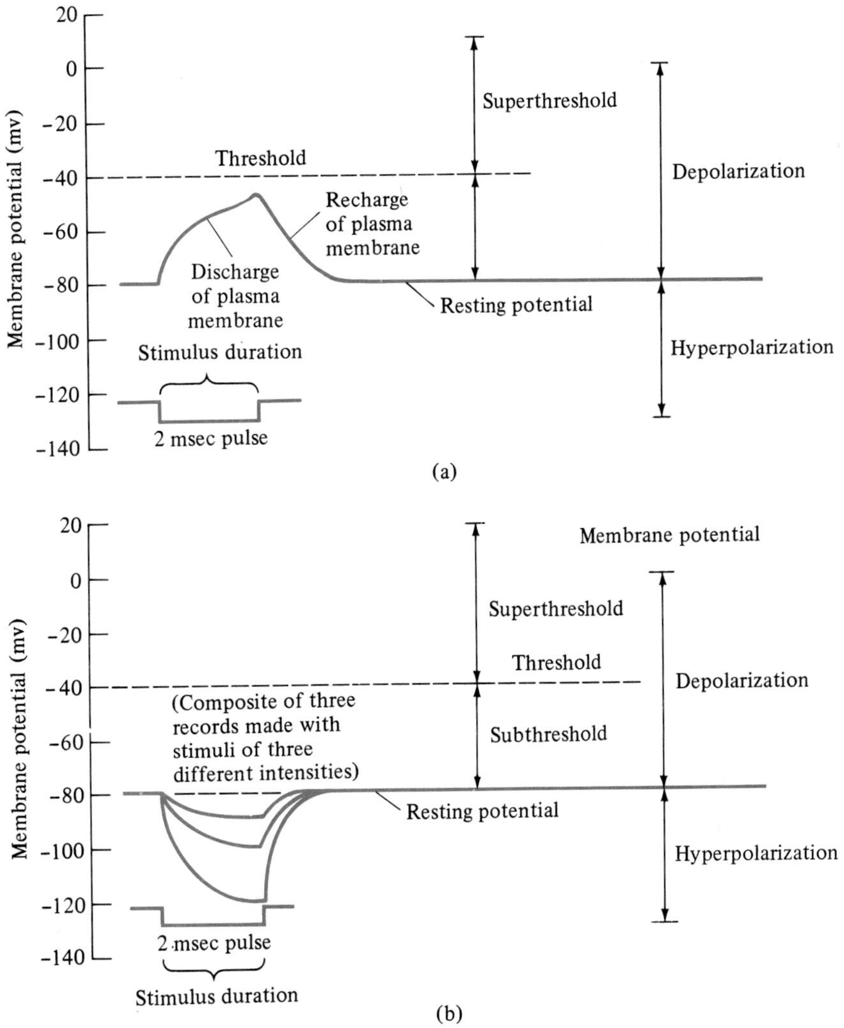


**FIG. 16-5** Cathode ray oscilloscope records of electrotonic potentials. (a) Effect of a subthreshold depolarizing stimulus in producing an electrotonic potential. (b) Effects of hyperpolarizing stimuli on the resting potential. Records of three stimuli of different intensities are superimposed for comparison.

As noted earlier, one way to cause a depolarization of threshold-level intensity is to introduce cations into the cytoplasm of a neuron. An action potential can also be generated in any part of a neuron by the arrival of a potential from another part of the same neuron. Because the

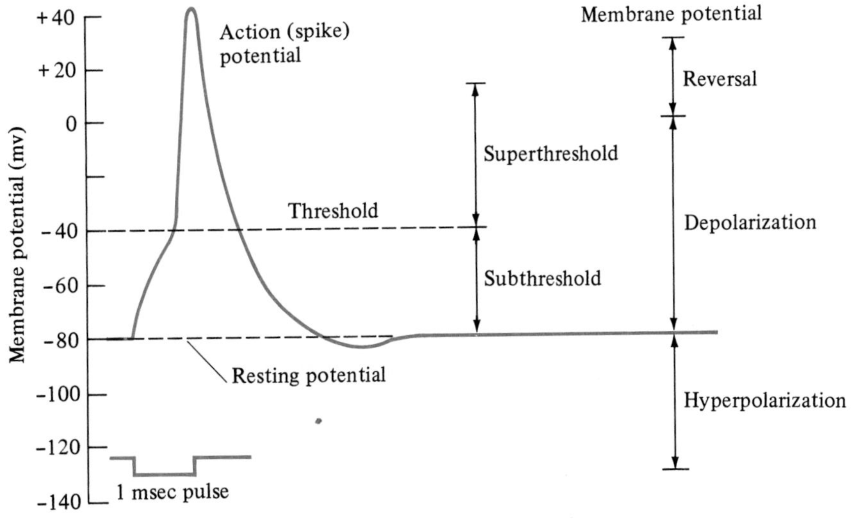


**FIG. 16-6** An action (spike) potential record produced by a depolarizing stimulus of greater than threshold intensity.

action potential depolarizes the neuron, it produces a strong electrotonic potential in the region of the membrane immediately ahead of it, one great enough to generate another action potential. It is this process that occurs over and over again down the entire axon. These events are summarized in Figure 16-7.

The generation and passage of an action potential is not altogether unlike the way a fuse, such as one attached to a stick of dynamite, burns after being ignited. Once ignited, it continues to burn down to its end. The action potential is a classic example of an "all-or-none" phenomenon, one that, like the burning of a fuse or the firing of a gun, either occurs entirely or not at all.

Of the various parts of a neuron, only the axon normally initiates action potentials. They arise at the region called the **axon hillock,** where the axon joins the soma (Fig. 16-8). Regions of the neuron such as the dendrites and cell bodies normally initiate only electrotonic potentials.

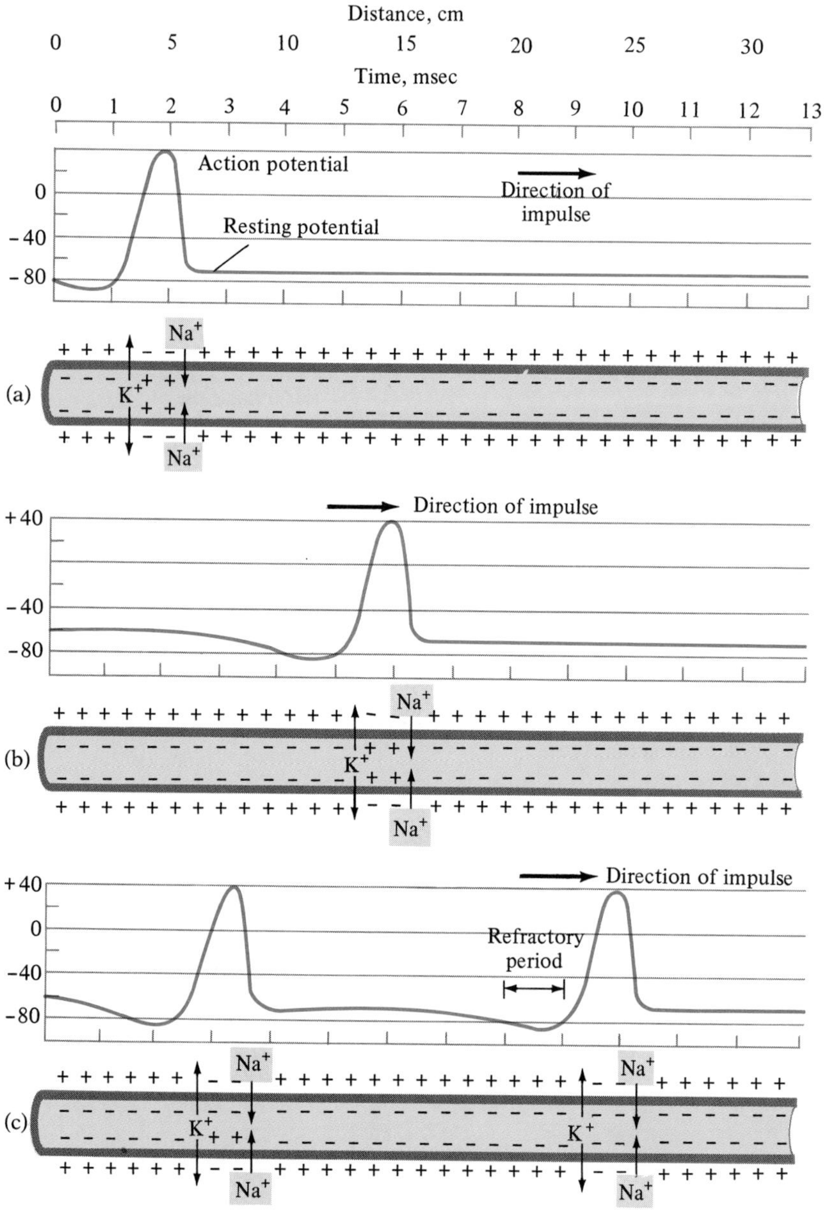


**FIG. 16-7** Major events of an action potential. (a, b) Depolarization of threshold level opens $Na^+$ gates and allows an inrush of $Na^+$. The polarity of the membrane is reversed. Inactivation closes the $Na^+$ gates, leaving the axon refractory to further stimulation for a time. $K^+$, now no longer being held inside the axoplasm, diffuses outward to repolarize the membrane. (c) A brief refractory period follows each impulse, during which time the axon is unresponsive to further stimulation. (Note that the shape of the spike appears to be reversed.) If the impulse were traveling in the opposite direction, the spike would be as it is usually represented, such as in Fig. 16-6.

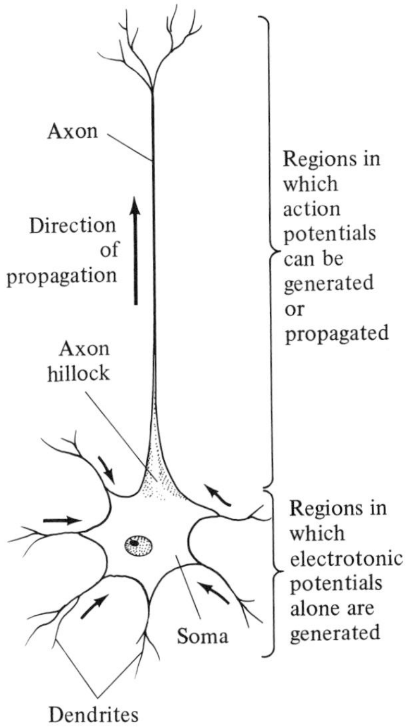


**FIG. 16-8** Action potentials normally arise only at the axon hillock and propagate down the axon.

Although these areas can also generate spikes, they seldom do because of their high thresholds. Thus, when a depolarizing stimulus is received from another neuron by a dendrite or soma, only an electrotonic potential is generated. It spreads passively and with decreasing strength across the cell surface. The stronger the stimulus, the longer the resulting electrotonic potential lasts, and the farther it spreads before it dies out. Only if its remaining strength is sufficient to depolarize the membrane of the axon hillock when it arrives there will a spike potential be generated in the neuron. Action potentials arising in an axon hillock, however, are propagated in both directions—down the axon and back to the soma and dendrites where they die out uneventfully.

## AXON DIAMETER AND CONDUCTION VELOCITY

Several invertebrate phyla have evolved the same general method for increasing the conduction velocity of action potentials in key axons: by increasing their diameter. The giant axons of earthworms, crustaceans, insects, and squids are good examples. Probably the best studied of these is the giant axon of the common squid. As in all neurons, a flow of ionic current in the axoplasm (cytoplasm of the axon) precedes the generation of each action potential. Just as a large-diameter pipe permits a more rapid flow of water than does a pipe of small diameter, ions flow more freely and thus more rapidly in a giant fiber, increasing the velocity of nerve impulse propagation.

## MYELINATION AND SALTATORY CONDUCTION

Vertebrates have evolved a different approach to increasing conduction velocity of nerve impulses. Surrounding many neurons are layers of a whitish, fatty substance called myelin, which acts much like the insulation around an electrical wire. Each **myelin** layer is actually layer upon layer, up to 100 layers, of plasma membranes of the adjacent neuroglial cell (Fig. 16-9a). Such cells associated with peripheral nerves are known as **Schwann cells.** Because myelin is an effective insulator, it prevents leakage of ionic currents through the plasma membrane. Moreover, $Na^+$ and $K^+$ are virtually absent in the myelinated regions of the axon. As a result, no spike potentials can be generated in any region of an axon covered by a myelin sheath. The myelin sheath, however, is interrupted at regular intervals by bare regions called **nodes of Ranvier.** At each node, the axon plasma membrane is exposed on all sides to the surrounding tissue fluid. At the nodes, the axon's plasma membrane has a very high concentration of $Na^+$ and $K^+$ channels.

Whenever an action potential is generated at the unmyelinated axon hillock, it is propagated along the neuron until it reaches the myelinated region of the axon. There the ionic currents of $Na^+$ and $K^+$ generated by the spike are unable to flow through the myelin-covered membrane. Ionic currents can flow through the axoplasm, however, as well as through the tissue fluid surrounding the axon. When an action potential arises at the axon hillock, the potential difference (voltage difference) between the hillock and the first node is great enough to trigger the opening of the $Na^+$ gates at the node, producing an action potential there. This immediately generates an ionic current, which travels be-

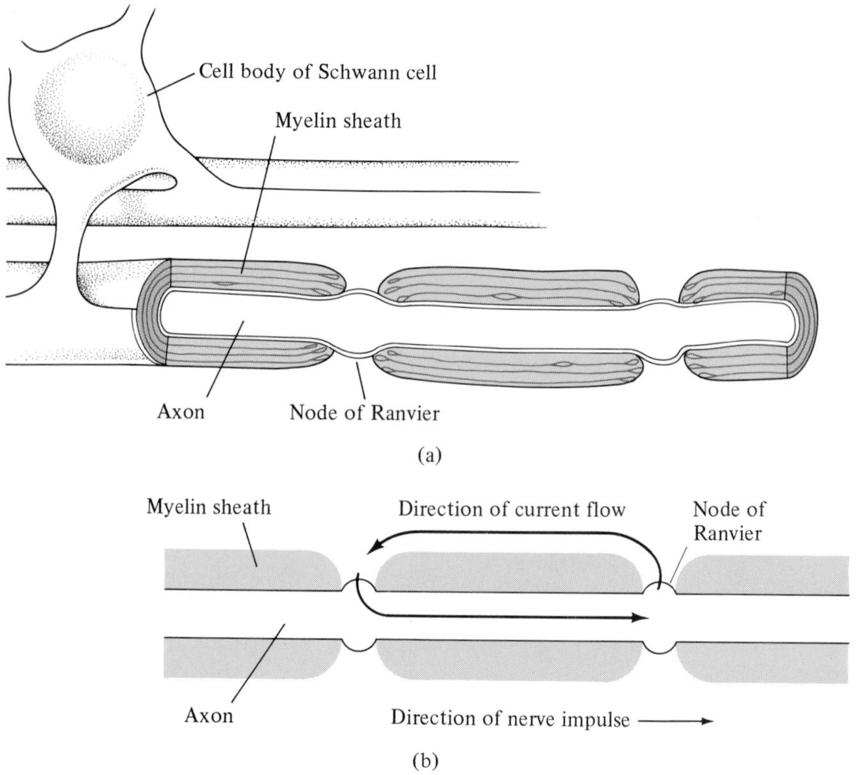


**FIG. 16-9** Saltatory conduction. (a) Relationship of a neuroglial cell to the myelin sheath in a myelinated fiber in a cat's spinal cord. (b) Conduction of the nerve impulse between nodes in a myelinated nerve is accomplished by a flow of current. Current flow is represented by heavy arrows. Action potentials are generated at the bare nodes, which in turn generate another current, which produces another action potential at the next node.

tween the first and the second node (Fig. 16-9b). Another spike is thereupon produced at the second node and so on at each node down the entire myelinated axon. Because the action potential thus appears to leap from one node to the next, this type of propagation has been called **saltatory conduction** (from the Latin, *saltare,* "to leap").

The nodes actually act as booster stations at which new ionic currents are produced as each new spike is generated by the previous ionic current. The passage of an impulse down a myelinated fiber is thus essentially the passage of an ionic current that is renewed at each node by spike potentials. Because currents travel so swiftly, the overall passage of the impulse down a myelinated fiber is very rapid, reaching 200 m per second (450 mph) in some neurons. Because current strength decreases with time and distance, it needs to be renewed at each node if it is to travel down an entire axon.

For an unmyelinated axon to achieve the same velocity of conduction as a myelinated one, it would need to be some 100 times thicker in diameter. Myelination thus permits "miniaturization" of nerves. In addition, because $Na^+$ influx and $K^+$ outflow occur only at nodes, much less energy is needed by the ion pumps of myelinated fibers to maintain their resting potentials.

## COMMUNICATION BY MEANS OF NERVE IMPULSES

It might be wondered what, if anything, can be communicated by nerve impulses. Recall that the principal means of communication between sense receptors and the brain is by action potentials, all of which, in any one nerve fiber, have the same magnitude and travel at the same velocity regardless of the stimulus. They are all-or-none phenomena. However, there is a way that stimulus intensity can be communicated, using the action potential as the communication medium. To see how this is accomplished, consider what happens after the generation of each spike potential.

We have seen that for a very brief time after an action potential has been generated in an axon, it is impossible to generate a second impulse. This interval is the absolute refractory period (Fig. 16-7c). If a neuron is continually stimulated by a threshold-level stimulus, it will generate and conduct impulses at a certain constant frequency. The time between the generation of one impulse and the next is the total refractory period for that neuron.

At the end of a neuron's absolute refractory period it is possible to generate another nerve impulse only if the stimulus intensity is above the normal threshold level. Viewed in another way, following the absolute refractory period, the threshold level for a neuron is higher than normal; this time period is known as the **relative refractory period**. Thus, if a stimulus that is more intense than the threshold level is continuously applied, the total refractory period is shorter than it is with continuous application of a stimulus that is barely at threshold level. Impulses are thus generated more frequently when the stimulus is appreciably above threshold intensity. If the stimulus intensity is increased even more, the refractory period will shorten further, generating impulses even more frequently (Fig. 16-10). But there is a limit to how much a neuron's refractory period can be shortened by increasing the stimulus intensity. Once that limit is reached, no matter how much the stimulus intensity is increased, no further increases in the frequency of impulses result.

This capability of responding differently to stimuli of different intensities means that within a certain range of intensities, neurons translate stimuli into impulses whose frequencies correspond to the intensity of the stimulus. Therefore, the rate at which impulses are received by the brain is directly related to the intensity of the stimulus. Another index of the intensity of a stimulus is the total number of neurons stimulated.

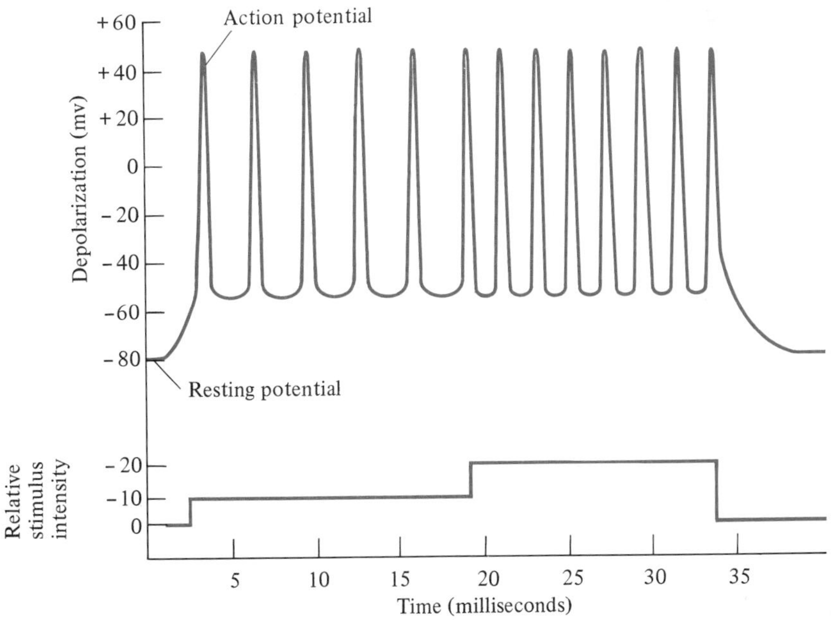


**FIG. 16-10** A CRO record illustrating that stimulus intensity is coded as, or translated into equivalents of, nerve impulse frequency. Note that even a doubling of stimulus intensity has no effect on the height of the action potential in reference to the resting potential, but it does markedly increase the frequency with which action potentials are generated; that is, it shortens the refractory period.

A more intense stimulus will recruit additional receptors, and therefore neurons, to signal information to the central nervous system.

## SYNAPTIC INTEGRATION

**Nerve impulse conduction** occurs along the length of a neuron, while **nerve impulse transmission** occurs from one neuron to another. Transmission occurs at synapses, defined as regions of communication between neurons (see Chapter 15). In referring to impulse transmission from one neuron to another, the first neuron along which the impulse passes is called the **presynaptic cell,** the second, the **postsynaptic cell.** It is the function of the postsynaptic cell to receive the input of dozens, hundreds, or even thousands of fibers, to integrate this information, and then to determine whether any signals will be sent to other units in response to receiving this information.

### BASIC MECHANISMS OF SYNAPTIC TRANSMISSION

Synapses are not unlike executive centers in a large organization. By a combination of chemical and electrical signals, incoming data are evaluated in computerlike fashion by the postsynaptic cell, which immediately determines what information is to be sent on to other parts of the body.

***Transmission by Neurotransmitters.*** With nearly all synapses, action potentials arriving at the axon tips, called **synaptic knobs** (Fig. 16-11a), of a presynaptic cell result in the release of precise amounts of a **neurotransmitter,** a special chemical substance contained in minute synaptic vesicles (Fig. 16-11b). The released neurotransmitter enters the synaptic clefts, the spaces between the axon tips and the postsynaptic

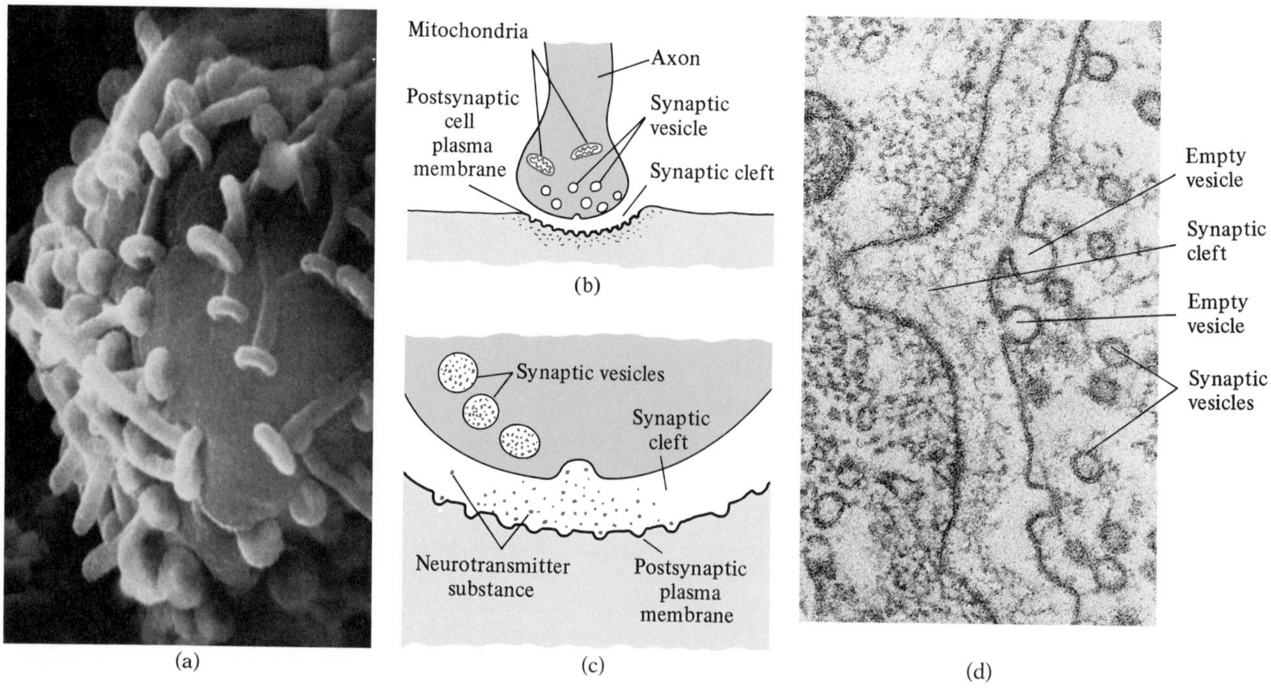


**FIG. 16-11** (a) Synaptic knobs of axons terminating on somas in an abdominal ganglion of a marine snail, as seen by scanning electron microscopy. (b, c) Release of neurotransmitter into the synaptic cleft by exocytosis from synaptic vesicles. (d) Electronmicrograph showing release of neurotransmitter at synapse.

cell's plasma membrane. The neurotransmitter acts on receptors on the plasma membrane of the postsynaptic cell to generate electrotonic potentials. The entire process, including the diffusion of the transmitter across the cleft, usually takes about a half a millisecond but may be slightly longer. It was this "synaptic delay" that first led physiologists to suspect that a chemical transmitter was involved and that a minute space separated the two cells.

In one common type of synapse, the arrival of action potentials at synaptic knobs produces a depolarizing **excitatory postsynaptic potential** (EPSP). If the EPSP or the sum of multiple EPSPs spreads as far as the axon hillock of the postsynaptic cell and results in a threshold depolarization of the hillock, one or more action potentials are generated in that cell (Fig. 16-12a). The best known of the neurotransmitters involved in generating EPSPs is acetylcholine (Ach).

Another type of postsynaptic potential hyperpolarizes the postsynaptic cell, making it more negative and therefore less likely to generate an action potential. This type is accordingly termed an **inhibitory postsynaptic potential** (IPSP) (Fig. 16-12b).

At most synapses at which EPSPs are generated, a small amount of neurotransmitter continually leaks into the synaptic cleft. Although it is not enough to induce an action potential in the postsynaptic cell, the small postsynaptic potential thus created makes it easier for a spike to be generated when action potentials do arrive at the synapse.

Calcium ions apparently play a role in synaptic transmission. The arrival of the action potential is thought to open the $Ca^{2+}$ gates in the plasma membrane of the presynaptic cell. Because $Ca^{2+}$ is over 100

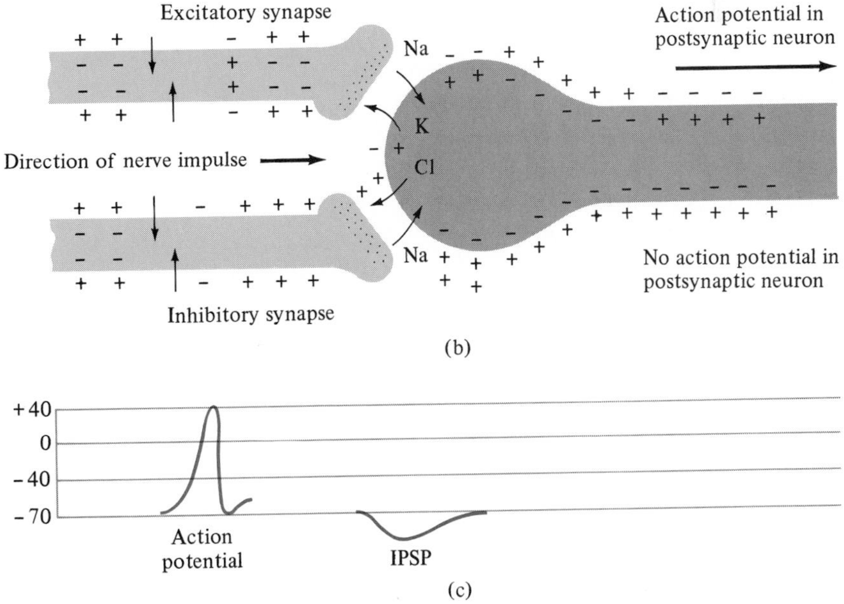


**FIG. 16-12** Comparison of synaptic excitation and inhibition. An excitatory postsynaptic potential (EPSP) is generated at the synapse of one presynaptic cell, and an inhibitory postsynaptic potential (IPSP) is generated at the synapse of the lower of the two presynaptic cells (b). CRO records are shown at (a) and (c). (As in Figure 16-7, the shapes of the potentials are reversed compared to the usual depiction for impulses traveling from left to right.)

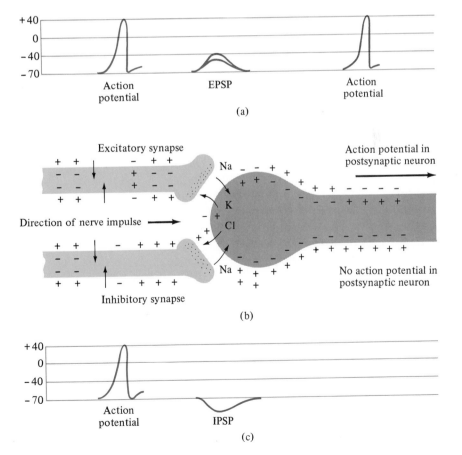

times as concentrated outside the cell as inside, a rapid influx of $Ca^{2+}$ into the presynaptic cell occurs. $Ca^{2+}$ plays an essential part in causing the membranes of the synaptic vesicles to fuse with the plasma membrane of the presynaptic cell, thereby releasing neurotransmitter into the synaptic cleft.

In a typical case, the arrival of one or more action potentials at the tips of axon branches causes the release of Ach into the synaptic cleft. The plasma membrane of the postsynaptic cell immediately opposite the points of release carries on its surface special Ach receptors, molecules with an enzymelike specificity for Ach molecules. When Ach unites with these receptors, it triggers the opening of both $Na^+$ and $K^+$ gates. However, the $Na^+$ gates open about 30 percent more widely than do the $K^+$ gates. Consequently, more $Na^+$ enters the cell than $K^+$ leaves it. The result is a rapid depolarization of the postsynaptic cell membrane. Although this might exceed the threshold for inducing an action potential in the axon of the postsynaptic cell, the threshold level for generating action potentials in dendrites and nerve cell bodies is much higher. Only electrotonic potentials are therefore induced. Unlike an action potential, an electrotonic potential decreases in intensity as it passes along the dendrites and nerve cell body. Only if it is still of threshold intensity by the time it reaches the axon hillock will it generate an action potential in the axon of the postsynaptic cell.

***Destruction of Acetylcholine.*** It might be wondered what would prevent acetylcholine, once released into a synaptic cleft, from continuing to stimulate the postsynaptic cell. Actually, nothing, if it were not for the presence of an enzyme: acetylcholinesterase. In the presence of this enzyme, acetylcholine is hydrolyzed to choline and acetic acid. Once this is accomplished, the material is no longer capable of stimulating the receptors in the plasma membrane of the postsynaptic cell. The effect of the enzyme is not so rapid, however, as to prevent the neurotransmitter from acting. The enzyme merely ensures that acetylcholine does not persist and continue to stimulate the postsynaptic cell. Similarly, $Ca^{2+}$ that has caused the release of Ach from a presynaptic cell promptly becomes sequestered by calcium binding proteins, mitochondria, or both. Consequently, for each action potential that reaches a synaptic knob, only a brief burst of Ach release occurs.

***Postsynaptic Inhibition.*** As noted, release of a transmitter substance by the tips of axons does not always produce EPSPs. Sometimes the result is an IPSP, in which case the postsynaptic cell is temporarily rendered less capable of firing, that is, of generating an action potential.

Because of the minute size of most interneuronal synapses, they are very difficult to study, and some of the most intriguing questions about inhibitory synapses remain unanswered. This is understandable when we consider the problems involved in determining variations in the nature of receptor molecules from one synapse to another among the hundreds present on a single postsynaptic cell. Despite this, evidence is accumulating that transmitters other than acetylcholine are released at many synapses. Some of these are inhibitory. One example is **gamma-aminobutyric acid** (GABA). The effect of GABA on a postsynaptic cell is to open the $Cl^-$ channels; the $Na^+$ and $K^+$ channels remain closed. Because $Cl^-$ ions are in higher concentration outside the cell than within, a rapid influx of $Cl^-$ ions occurs. This hyperpolarizes the membrane, thereby producing an IPSP. In the peripheral nervous system, however, the inhibitory quality of a particular synapse is often due to a difference in the nature of the receptor in the plasma membrane of the postsynaptic cell. The same neurotransmitter could thus generate either an EPSP or an IPSP, depending on the kind of receptor molecule affected. These differences in response depend on which channels are opened by the particular receptor-transmitter complex. When $Na^+$ channels are opened, an EPSP results; when anion channels are opened, the result is an IPSP.

At any given moment, the potential generated across the entire surface of a postsynaptic cell is a composite of the effects—both excitatory and inhibitory—achieved at the several synapses on its surface.

***Kinds of Chemical Synapses.*** Electron microscopy has shown that there is an enormous variety of synaptic types, providing a wealth of opportunities for integration. Most vertebrate synapses are chemical synapses, characterized by the presence of synaptic vesicles and mitochondria in the cytoplasm of the synaptic knobs of presynaptic cells. These organelles are absent from the synaptic region of the postsynaptic cell. Although any one chemical synapse usually possesses but one kind of vesicle, there are at least four kinds of vesicles and over 30 known or suspected transmitters in the CNS (see Panel 16-2).

***Electrical Synapses.*** Some unusual synapses lack vesicles, and the space between pre- and postsynaptic cells is only 20 to 50 Å wide, rather than the usual 200 Å or greater. Such cells are connected by intercellular channels that provide a low resistance to the passage of ions. Such a synapse is termed a **gap junction,** or nexus (Latin for "a fastening"). Because an action potential can pass without delay and without a chemical mediator between such cells, they are also known as electrical synapses. In effect, they unite the two cells into one long cell that can usually conduct impulses in either direction. Chemical synapses, on the other hand, are always polarized, that is, they transmit in one direction only, from the tips of axons, which contain synaptic vesicles, to dendrites or nerve cell bodies.

It is important to distinguish gap junctions, or electrical synapses, from so-called tight junctions. The latter are found between a variety of cells, including neurons. But they are not synapses. They do not have intercellular channels. Instead, they form high-resistance seals between cells and prevent electrical conduction between cells. They insulate rather than communicate.

## THE INTEGRATIVE CAPABILITIES OF THE INTERNEURONAL SYNAPSE

By their wide diversity, interneuronal synapses provide great opportunity for integration. For example, in some parts of the brain, hundreds or even thousands of synaptic knobs may form synapses on the dendrites and soma of a single neuron. While many of these may be the branches of a single axon, many represent input from multiple neurons. Moreover, while many presynaptic neurons are excitatory, many, if not most, are inhibitory. The postsynaptic cell integrates the input from these various sources and generates action potentials accordingly. A few of the many types of postsynaptic integration are described in the following sections.

***Summation of EPSPs.*** A single action potential arriving at an excitatory synapse almost never releases enough neurotransmitter to generate an action potential in the postsynaptic cell. Instead, a large number of EPSPs usually must be produced within a short time—usually at many synapses impinging on the same postsynaptic neuron—for the resulting depolarization to be sufficient to produce a spike potential at its axon hillock. The properties of the axon hillock determine whether action potentials are generated in the postsynaptic cell. The process by

PANEL 16-2

# Recent Findings About CNS Neurotransmission

Some of the most interesting findings of recent years concern the following categories: neurotransmitters of the CNS, normal transmitter function in the CNS, neurological disease, and pharmacological effects of certain drugs on CNS neurotransmission.

## NEUROTRANSMITTERS OF THE CNS

New techniques have recently raised to over 30 the number of substances known or suspected of being neurotransmitters within the brain. Several families of compounds are involved:

1. amines, such as dopamine, norepinephrine, serotonin, acetylcholine, and histamine;
2. amino acids, such as gamma-aminobutyric acid, glutamic acid, glycine, and taurine; and
3. neuropeptides, chains of two to 29 amino acid residues.

All three groups include some substances that perform other functions elsewhere in the body. The isolation provided by the blood brain barrier, the impermeability to many substances that characterizes the membranes separating the brain tissue from the blood stream, prevents cross-signaling in such cases.

## NORMAL TRANSMITTER FUNCTION

1. Many of the norepinephrine-containing cells of the brain are concentrated in regions that communicate with areas concerned with arousal, reward, dreaming, and mood regulation.
2. Serotonin-containing fibers communicate with the hypothalamus and thalamus and are thought to be involved in temperature regulation, sensory perception, and falling asleep.
3. As many as a third of all brain synapses are estimated to employ GABA as an inhibitory transmitter. Glutamic acid, which differs from GABA by a single chemical group, is an excitatory transmitter.

## NEUROLOGICAL DISEASE

1. Death of cells in the substantia nigra, an area in the brainstem rich in dopamine-containing fibers, is associated with the muscular rigidity and tremors of Parkinson's disease.
2. Analysis of brain tissue of autopsied schizophrenics shows their brains to have abnormally high concentrations of both dopamine and dopamine receptors, especially in those brain regions involved in emotional behavior. Drugs that have been found useful in treating schizophrenia, such as chlorpromazine (Thorazine), bind to dopamine receptors, preventing dopamine from exerting its effect.
3. The inherited disease Huntington's chorea is characterized by uncontrolled movements caused by degeneration of a region of the brain, the corpus striatum, in middle age. The brain damage includes a loss of inhibitory neurons that contain GABA.

which EPSPs are added to each other to generate action potentials is termed **summation;** it has both temporal and spatial dimensions.

TEMPORAL SUMMATION. If the time interval between impulses arriving at an excitatory synapse is sufficiently brief, the EPSP produced by the first impulse will not have died away completely before a second is generated and added to the residue of the first. As a result, a greater EPSP is produced than from a single impulse. If a third impulse arrives before the summed potentials induced by the first two have decayed, the third will be added to them, resulting in an even greater EPSP. Such addition of successive potentials to the remnants of previous ones is termed temporal summation.

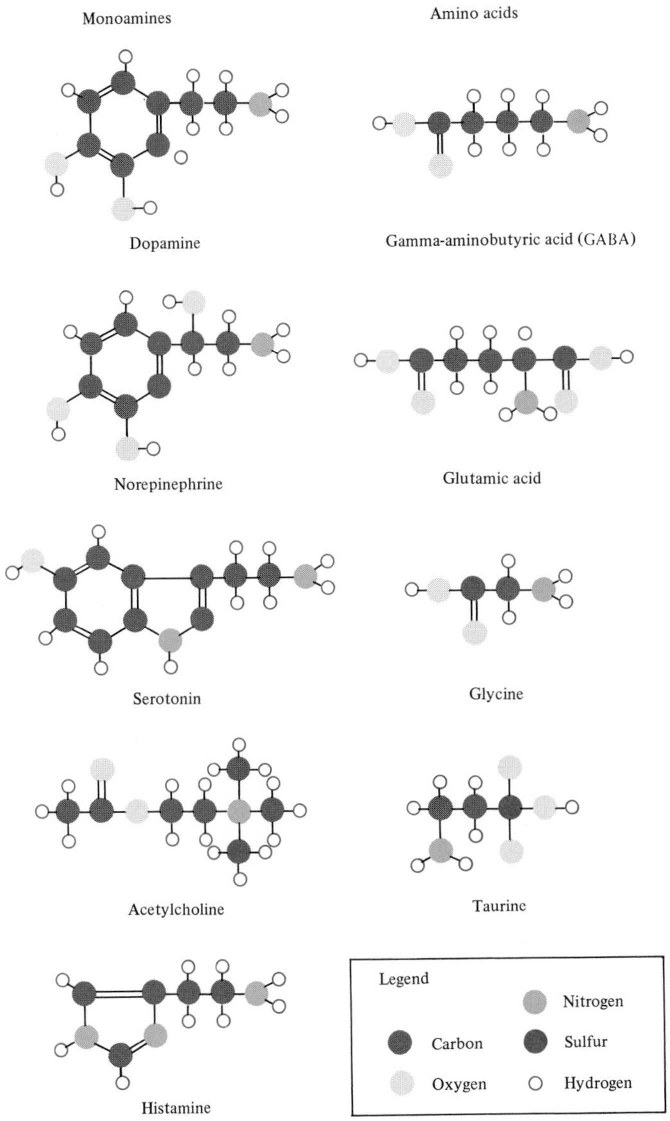


Transmitter chemicals are mostly small molecules that possess a positively charged nitrogen atom. Most are characteristically excitatory or inhibitory, but ultimately, their effect depends on the receptor molecule at the postsynaptic cell; different parts of the brain respond differently to the same transmitter. Histamine and taurine are regarded as putative transmitters, but conclusive experimental proof is lacking.

## ENKEPHALINS AND ENDORPHINS: SPECIAL TYPES OF NEUROPEPTIDES

The neuropeptides include some well-known substances, such as gastrin and vasopressin, and two newly discovered subclasses, the enkephalins and endorphins. Both of these are very similar to the opiate* morphine. It is probably significant that regions of the CNS involved with perception and integration of pain and with emotional experience have been found to strongly bind opiates. Enkephalins bind to these same receptors. Other experimental results suggest that enkephalins and endorphins, the latter found in the pituitary gland and the brain, can be released by acupuncture and by hypnosis. Drugs that block the binding of morphine to its receptors also reduce the effectiveness of acupuncture and hypnosis in the relief of pain.

* Opiates are derivatives of opium, a drug obtained from the opium poppy. Besides morphine, they include heroin and codeine.

SPATIAL SUMMATION. In temporal summation, EPSPs produced at close intervals of time by the same synapse are added to each other. In spatial summation, EPSPs produced at nearly the same time by neighboring synapses are added to each other. The effect is similar: a stronger EPSP is produced in either case. The principles of temporal and spatial summation apply to the production of IPSPs as well.

***Facilitation and Antifacilitation.*** Facilitation, another type of integrative activity occurring at synapses, is easily confused with temporal summation. As in temporal summation, facilitation depends on the arrival of two or more impulses at the synaptic surface or surfaces of the same postsynaptic cell within a certain space of time. But the result in this case is that the second EPSP generated is greater than the first, the third is greater than the second, and so on. In each case, moreover, the previous EPSP has died away completely before the next develops; so this is not an instance of adding to the residue of the previous potential, that is, it is not a case of temporal summation. Somehow, in facilitation, the first impulse to arrive at a synapse increases the sensitivity of the mechanism by which neurotransmitter is mobilized and released; the arrival of the second further enhances this sensitivity, and the effect recurs at the arrival of each successive impulse to a point at which no further enhancement results. The length of time that facilitation lasts may vary greatly from one type of synapse to another. In some, the effect persists only for milliseconds, in others for several minutes, and in some for a day or longer.

In sharp contrast to what happens in facilitation, in some postsynaptic cell junctions, the arrival of several impulses at short intervals results in decreased sensitivity to further impulses. This phenomenon is appropriately termed antifacilitation.

Both facilitation and antifacilitation can be useful integrative activities under certain circumstances. Facilitation probably plays a part in perfecting an animal's responses to specific stimuli. As such, it would probably be one of the benefits provided by a "warm-up practice" just before an athletic contest. One elegant refinement of facilitative integration is seen in certain types of synapses in which the junction can be temporarily changed from one that does not exhibit facilitation to one that does. In such cases, the postsynaptic cell becomes more sensitive to the arrival of impulses from a certain type of presynaptic cell if impulses have recently arrived at neighboring synapses formed by presynaptic cells of another type. Understandably, the phenomenon is difficult to distinguish from—and sometimes has been confused with—spatial summation. That it is not simply spatial summation is shown by the fact that the phenomenon does not occur when the two different types of presynaptic cells are stimulated in reverse order. Antifacilitation has a conservative function. It is one of the reasons we can become insensitive after a time to many recurrent or continuous stimuli—sounds, odors, even pain.

A quite different basis for decrease in sensitivity exists in **sensory adaptation,** a phenomenon seen in sensory receptors that become insensitive and fail to generate impulses when stimulated continuously or repeatedly at short intervals. Antifacilitation occurs at synapses, whereas sensory adaptation occurs at receptors. Such receptors, together with synaptic junctions capable of antifacilitation, spare animals the need to process much useless information.

Most of the mechanisms by which the various types of facilitation and antifacilitation are achieved are not yet understood.

## Summary

As with most cells, nerve and muscle cells maintain an electrical imbalance or polarization between the inside and outside of their plasma membranes. This is partly the result of the membrane's impermeability to proteins, most of which are negatively charged. Another contributing factor to this condition is an active transport mechanism, the sodium-potassium pump. These factors, together with a control of the permeability of the plasma membrane to sodium and potassium ions, results in the maintenance of sodium at 10 times the concentration outside the membrane as within and the maintenance of potassium at 25 times the concentration within the cell as without. The overall imbalance produced by these three factors creates a 70 to 90 mV negative charge within the cell, compared to the charge outside. This voltage, or potential difference, is termed resting potential.

Any stimulus causing a reduction of the resting potential of a neuron increases the plasma membrane's permeability to sodium. $Na^+$ influx causes the membrane in that region to depolarize further, becoming even more permeable to $Na^+$. Within a millisecond, however, the $Na^+$ gates have closed completely. An outflow of $K^+$, which began at the same time as the $Na^+$ influx, peaks at this point, restoring the resting potential. If the initial stimulus was strong enough to open many $Na^+$ gates, the resulting complete depolarization triggers this same process in adjacent regions of the neuron. The disturbance is then propagated as an action potential along the entire length of the neuron.

Stimuli too weak to generate action potentials produce electrotonic potentials only: local, graded (partial) depolarizations that decrease in intensity in both time and space. Electrotonic potentials can also be hyperpolarizing, in which case they inhibit or lessen the likelihood of the neuron's firing.

The rate at which nerve impulses are conducted by a neuron is directly proportional to the diameter of the fiber. The rate is also greatly increased in myelinated fibers, in which action potentials are generated anew at each node of Ranvier in the process known as saltatory conduction.

Action potentials themselves have limited ability to communicate information, although stimulus intensity is indicated by the frequency with which action potentials are generated. Stimulus intensity is also communicated by the number of receptors and connecting neurons recruited, or affected, by the stimulus.

Most integration—the determination of what actions to take in response to various events, stimuli, and so on or whether to act at all—occurs at the synapses between neurons. Most are chemical synapses, in which a neurotransmitter, the best known of which is acetylcholine, is released from synaptic vesicles at the tips of axons. Although some transmitters tend to be excitatory and others inhibitory, the effect achieved also depends on the nature of the receptor molecules in the plasma membrane of the postsynaptic cell. Only a few transmitters are well known, although at least 30 different substances are believed to function as transmitters within the CNS.

The variety of postsynaptic transmitter receptor molecules and of transmitters themselves provides a great wealth of integrative capabilities at the myriad synapses within the nervous system. Other bases for integration are provided by temporal and spatial summation of impulses arriving sequentially or from different neurons and by facilitation and antifacilitation of neurons to incoming nerve impulses.

## Recommended Readings

A list of recommended readings appears at the end of Chapter 15.

# 17 Sensory Reception

OUTLINE



Many remarkable receptive capabilities exist in the animal world. For example, bees can distinguish the faint odor of one type of flower among hundreds of other species in bloom at the same time. Dogs have a thousandfold greater sensitivity in discriminating smell than do humans. A hawk can see a mouse move in a meadow from well over a kilometer away. By detecting the echoes of its own cries with a type of sonar, a bat can avoid crisscrossed wires strung across a pitch-dark room. An owl can hear a mouse move and unerringly glide down upon it in pitch darkness from many meters away. Such abilities to smell, see, and hear are examples of sensory reception, the subject of this chapter.

## GENERAL PRINCIPLES OF RECEPTION

How does an animal receive information from its surroundings? How is it able to distinguish information of different types? In what form is information conveyed to the parts of the nervous system where decisions about it are made? The answers to such questions are crucial to understanding the nervous integration of information received through the various senses.

### RECEPTORS

Animals possess a great variety of cells called sensory receptors, which are specialized to receive stimuli. One classification of receptors is based on their location and general function. For example, **exterocep-**

**FIG. 17-1** A Vermont barred owl *(Strix varia)* catching a deermouse *(Peromyscus)*. An owl can hear a mouse or rat move at distances of several meters and silently glide down upon it in the dark.

**tors,** located on or near the surface of the body, perceive odors, tastes, colors, vibrations, textures, and other external stimuli. **Interoceptors,** within the body, are sensitive to $O_2$ and $CO_2$ concentrations in the blood, to pH and to osmotic and hydraulic pressure of blood and tissue fluid. **Proprioceptors,** located in muscles, tendons, and joints, provide information about the movements and positions of the body and appendages. Sense receptors are also classified according to the kinds of stimuli to which they are sensitive: mechano-, chemo-, and photoreceptors.

The information received by the receptors is conveyed by sensory nerve fibers leading from each receptor to a specific group of cells in the brain via the spinal cord or brain stem. Thus, whenever a receptor in a particular part of the body receives a stimulus, the information represented by the stimulus is conveyed in the form of nerve impulses to a particular region of the brain. This arrangement is somewhat like having all of a house's doorbells, alarm bells, and telephone bells, each with a recognizably different tone, located in the same room, each wired to and responsive only to its own button or sensor. Such an arrangement enables an animal with specialized receptors not only to distinguish different types of stimuli, but also—because of the location of the receptors and their "wiring"—to learn to distinguish the location of stimuli received almost anywhere on the body.

As discussed in Chapter 16, the intensity of a stimulus can be communicated along neuronal pathways to the brain by the frequency with which nerve impulses are generated. Strong stimuli generate more frequent action potentials than do weak stimuli. And in most cases, the more intense the stimulation, the greater the number of receptors stimulated and neuronal pathways involved.

Besides being able to distinguish stimuli of various intensities, many receptors alter their response—their pattern of impulse generation—when stimulated continuously. Such **sensory adaptation,** as it is called, involves two kinds of receptors. **Phasic receptors** respond initially to a stimulus by generating a short burst of action potentials. With continued stimulation, a phasic receptor responds further only if the stimulus intensity is increased or if a period of rest intervenes. **Tonic receptors,** on the other hand, generate action potentials throughout periods of continual, steady, intense stimulation. However, slight drops in frequency occur over a prolonged period of continuous stimulation.

Many receptors generate nerve impulses only in response to changes in the intensity of specific stimuli. Some receptors in the mammalian eye, for example, generate impulses in their sensory neurons only in response to an increase in light intensity. Other receptors respond only to decreases in stimulus intensity, and still others to any change in intensity, whether it is an increase or a decrease.

The failure of a particular receptor to generate an impulse means that no stimulus of adequate intensity for which the receptor is specific is detectable in the receptor's vicinity. "No news" is a kind of news, just as the fact that a fire alarm in working order and not ringing conveys the news that there is no fire.

***Receptors as Transducers.*** A transducer is any device that converts one form of energy to another. Some receptors transduce the energy of a given stimulus to the energy that generates nerve impulses. However, the amount of energy in many stimuli is small, as, for example, that in a faint beam of light detected by the eye, which could not be the sole source of energy for generating nerve impulses in optic nerves. Generation of an action potential in a sensory neuron requires that the region of the receptor cell membrane that is specialized to receive stimuli (1) be sensitive to the particular stimulus and (2) be able to produce an electrotonic potential known as a **generator potential,** which, if strong enough, evokes an action potential in its axon.

The relative unimportance of the amount of energy in the stimulus is also apparent in **chemoreception**—the ability to sense and distinguish various chemicals—in which the stimulus itself is probably never a significant source of energy. Chemoreception requires only that a molecule of the substance to be detected fit into a receptor molecule on the surface of the receptor cell, much as a substrate molecule fits into the active site on the surface of an enzyme. Presumably, it is the uniting of the two molecules that constitutes the stimulus in chemoreception. If enough receptor molecules are thus stimulated, the ultimate result is the generation of an action potential. But all the energy involved in chemoreception is furnished by the receptor, not the stimulus.

## PRODUCTION OF AN ACTION POTENTIAL BY A SENSORY NEURON

When a stimulus impinges on a receptor cell membrane such as that of one of the pressure receptors in the skin called Pacinian corpuscles, an action potential does not result immediately (Fig. 17-2a). Instead, a depolarizing generator potential is produced, the strength of which is proportionate to the strength of the stimulus that produced it. It spreads along the neuron to the axon region, becoming weaker as it spreads.

Except for its dendritic portion, the typical vertebrate sensory neuron is myelinated. When a generator potential spreads as far as the axon proper, an ionic current flows between the unmyelinated (dendritic) portion and the first node of Ranvier in the myelinated region (Fig. 17-2b). Weak generator potentials, however, die away before they generate an action potential, and no nerve impulses are generated. But if the current is of threshold strength, it will generate an action potential at

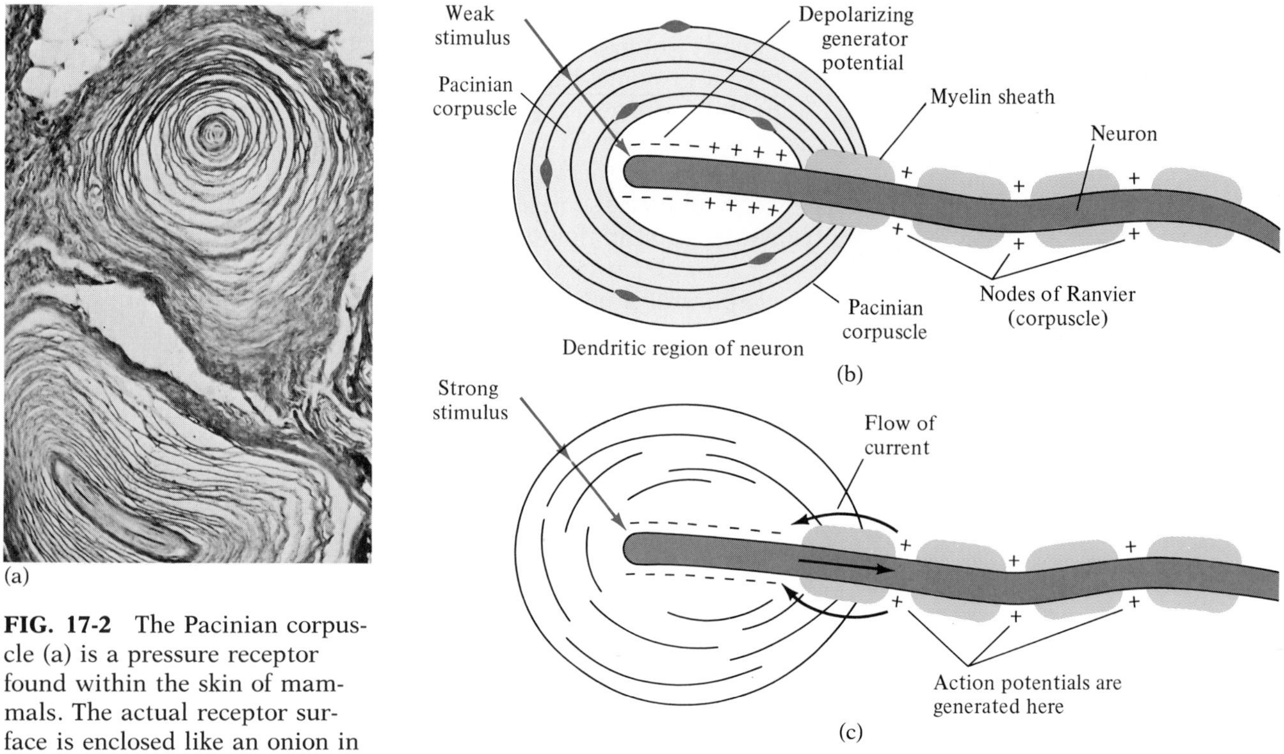


**FIG. 17-2** The Pacinian corpuscle (a) is a pressure receptor found within the skin of mammals. The actual receptor surface is enclosed like an onion in multiple layers of connective tissue. Shown here are two corpuscles from human skin, one in cross section, above, and the other in longitudinal section. A neuron can be seen at the very center of each. (b) A very weak stimulus causes only local depolarization of the receptor surface. (c) A stimulus of threshhold strength triggers an action potential at the first node of Ranvier. Action potentials are then generated by saltatory conduction at each subsequent node.

the first node and then be rapidly conducted down the axon by saltatory conduction.

## MECHANORECEPTION

If an axon is gently compressed or stretched or if its plasma membrane is otherwise distorted, it is likely to generate an electrotonic potential and then an action potential. Apparently, membrane deformation causes ion gates to open in the axon's plasma membrane. Similar deformations of dendrites and nerve cell bodies tend to produce electrotonic potentials that may also lead to action potentials. Such production of electrotonic or action potentials in response to mechanical stimulation probably explains our "seeing stars" when our optic nerves are mechanically stimulated by a severe blow to the head.

Although most neurons can probably respond to mechanical distortion by generating an electrotonic potential, the receptor membranes of **mechanoreceptor cells** are highly specialized for such reception. In the case of some of these cells in the skin, merely touching the surface of the skin with a fine hair generates action potentials.

### KINDS OF MECHANORECEPTORS

Mammals have at least a dozen types of mechanoreceptors in their skin alone. For example, hairs on the skin are associated with receptors that are quite sensitive to the slightest movement but stop generating impulses when the hair stops moving. Other kinds of receptors respond to light touch but not to stretching, while others respond to both kinds of stimuli.

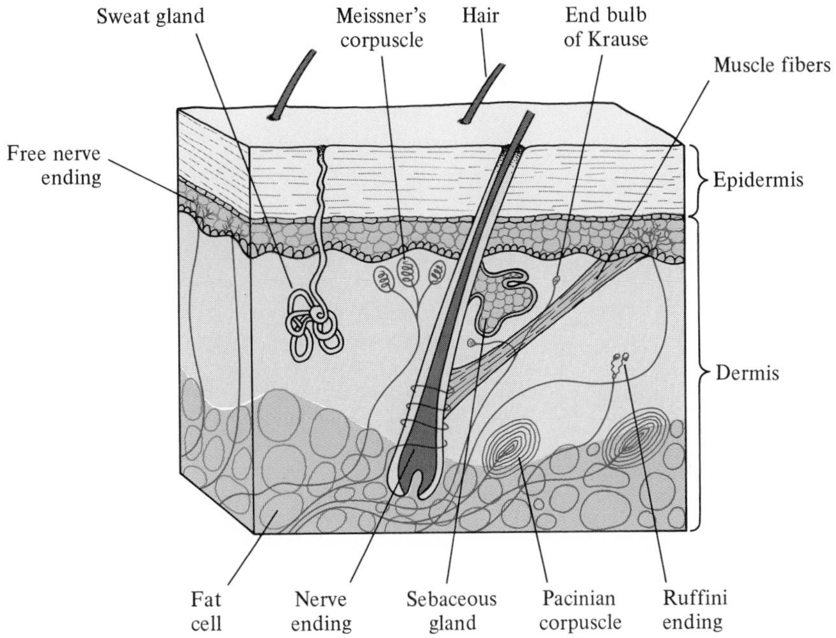


**FIG. 17-3** Schematic representation of some of the receptors found in mammalian skin. Not all are found in any one area. Nerve fibers are represented as heavy lines if myelinated, as fine lines if not. Meissner's corpuscles are touch receptors found in the most sensitive areas of the body. Ruffini endings are pressure receptors that are slowly adapting, that is, they continue to generate impulses for as long as they are stimulated. The function of the end bulbs of Krause is unknown.

Some of the specialized nerve endings and receptors found in human skin are illustrated in Figure 17-3. The function of only a few of these has been clearly established. It is known that in the Pacinian corpuscle, for instance, the unmyelinated tip of the neuron is surrounded by an onionlike capsule of cell layers that are not themselves receptive but serve to "filter" the stimulus, that is, they limit the amount and kind of stimulus that is allowed to reach the sensitive receptor surface. The Pacinian corpuscle is extremely sensitive to pressure and adapts very rapidly. It responds to a change in compression of only half a $\mu$m and can produce a single spike potential for each such stimulus. This receptor thus responds to the slightest changes in pressure on the skin.

The internal organs, or viscera, also have mechanoreceptors. The baroceptors of the aorta and carotid arteries that monitor blood pressure are examples (see Chapter 11). Another example is the discomfort occasioned by distension of the stomach or intestine by food or gas, which is due to stimulation of stretch receptors in these organs.

Attached to the muscles, tendons, joints, ligaments, and other connective tissues are a host of proprioceptors that are sensitive to stretch. They signal the CNS regarding the movements and positions of the limbs and the body. Other proprioceptors generate impulses only during movement. We are normally conscious of the latter but not of the former. While lying in bed, for example, if a limb in a comfortable position is kept motionless for a time, we soon become unaware of its exact position. But if we move it slightly, we become conscious of its location once more.

***Neuromuscular Spindles.*** The most thoroughly studied of all proprioceptors are the **neuromuscular spindles,** present in nearly all muscles. A spindle is made up of several specialized muscle fibers and their innervation, enclosed in a connective tissue sheath (Fig. 17-4a). These spindles generate action potentials whenever the muscle in which they are embedded is stretched. Each unit contains two functional types of receptors, called **intrafusal muscle fibers.** One type senses the length to which the muscle is stretched, and how long it remains stretched, the other the rate at which the stretching occurs. When a muscle is

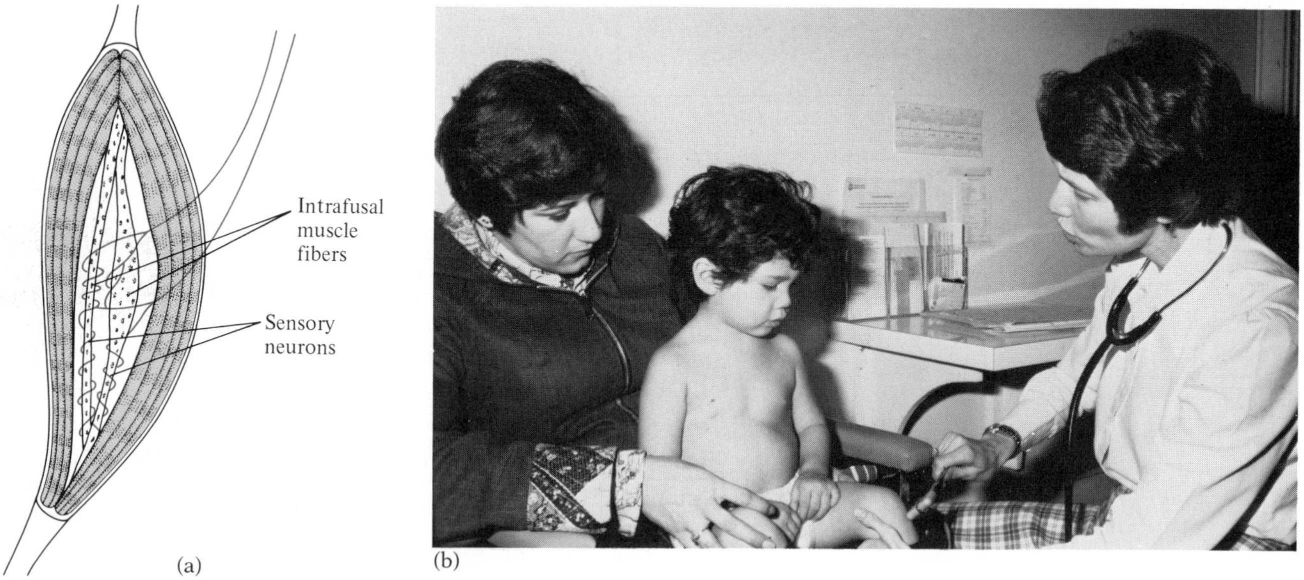


FIG. 17-4 (a) Highly schematized representation of a neuromuscular spindle in a limb muscle, with the spindle disproportionately enlarged and with only two intrafusal muscle fibers and their sensory neurons. (b) A tap on the patellar tendon of the knee elicits a reflex contraction of its muscle in normal individuals.

stretched, these three types of information are translated into action potentials sent to the CNS. Muscle contraction relieves the tension on the stretched spindle, and it stops firing. With rapid or severe stretching of spindles, a reflex is triggered that stimulates motor neurons innervating the stretched muscle, causing it to contract. The process is easily demonstrated by the sudden stretching of a muscle that can result from sharply tapping it or its tendon. The tapping activates an associated reflex arc in the lower spinal cord, resulting in a twitch of the muscle. This is the basis for the knee-jerk reflex. Tapping the tendon that crosses a flexed knee produces a sudden stretch of the thigh muscle to which the tendon is attached (Fig. 17-4b). When its spindles are thereby suddenly stretched, the reflex is activated and the muscle contracts.

The chief importance of the neuromuscular spindles lies in the continual feedback of information they provide to the brain about the movements of the various parts of the body. The feedback from the digits, limbs, and trunk makes possible such exquisitely controlled movements as those involved in playing the violin. The brainstem, cerebrum, and cerebellum receive this information and integrate it, thereby guiding the precisely coordinated movements.

FIG. 17-5 A single neuromast from a fish.

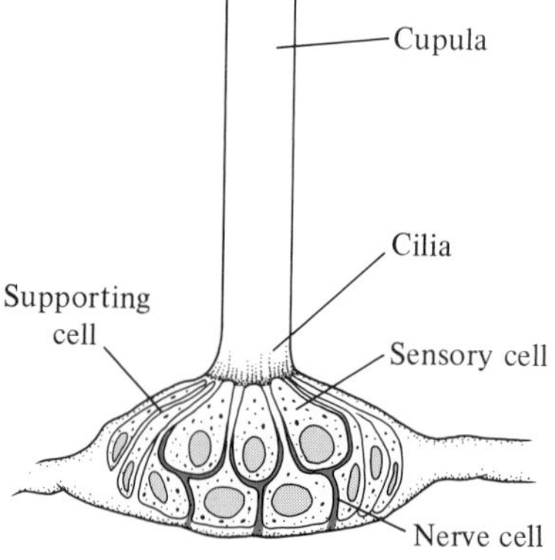


***Neuromasts.*** A common type of mechanoreceptor in aquatic vertebrates is the **neuromast,** which consists of sensory epithelial cells surrounding a clump of secondary, nonneuronal sensory receptor cells bearing fine hairs (Fig. 17-5). The hairs of the receptor cells project up into a **cupula** (Latin for "cup," plural, cupulae), a pendulous mass of gelatinous material secreted by the supporting cells of the neuromast. Any movement imparted to a cupula, such as by a vibration in the water, stimulates the sensitive hairs at its base, producing generator potentials in the associated neurons. In some fishes, each cupula is associated with several hundred hair cells.

***Lateral Line Organs.*** In fishes, a small canal, the **lateral line,** runs the length of the animal on both sides of its body (Fig. 17-6a). Although the canal system is exposed in primitive vertebrates such as sharks, in most modern fishes it is enclosed by an overlying layer of skin. The canals of the lateral line system are heavily lined with neuromasts and are well innervated. It has been shown that a continual, spontaneous nervous

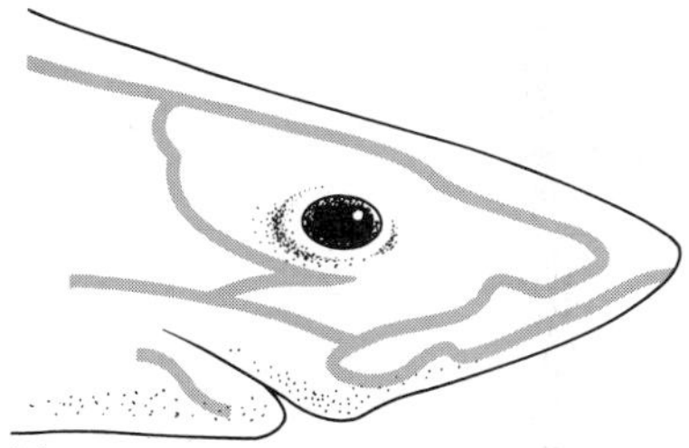

(a)

**FIG. 17-6** (a) A portion of the lateral line system of a shark. (b) A school of cottonwicks *(Haemulon melanorum)* over Molasses Reef, Key Largo, Florida.

(b)

activity originates in the lateral line neuromasts. Mechanical stimulation of the lateral line neuromasts produces variations in the basic pattern of this activity. Presumably, such differences are perceived and acted upon by the CNS.

The lateral line system is no doubt sensitive to a variety of disturbances, especially the pressure of currents of water against the body and the presence or action of other animals and obstructions in the surrounding water. By informing fishes of the presence of other fishes, this system is believed to help maintain the spatial configuration characteristic of a school of fishes (Fig. 17-6b).

Some lateral line neuromasts are also sensitive to temperature changes, and some respond to certain chemicals. Because similar information is also detectable by other receptors, it has generally been difficult to assign roles to the lateral line system with any degree of certainty. What primary roles they actually play in nature is unknown at this time.

**FIG. 17-7** Statocysts of (a) a clam and (b) a shrimp.

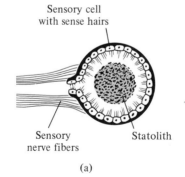

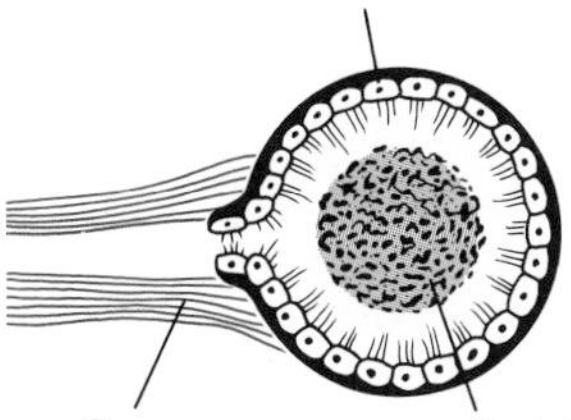


(a)

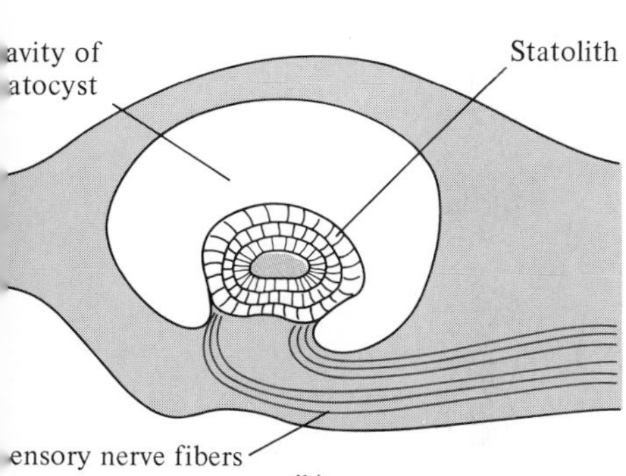


(b)

## ORGANS OF EQUILIBRIUM

One of the simplest of organs of equilibrium is the **statocyst,** found in many invertebrates (Fig. 17-7). It consists of a cavity lined by epithelial cells bearing fine sensory hairs and containing statoliths. **Statoliths** are loose calcium salt crystals or grains of sand that are free to move about the cavity. Movements of the statoliths against the hairs stimulate the hairs whenever the animal changes position. This allows the animal to perceive even slight changes in orientation and to make any necessary adjustments.

Crayfish lose their statoliths, which are composed of sand grains, at each molt but acquire more within a few hours, usually when the animal buries its head in sand. Some crustacea actually place the sand grains into their statocysts. If a crayfish molts in an aquarium free of sand grains but containing iron filings, it replaces its lost sand grains with filings. If a strong magnet is positioned above the animal, the animal responds to the upward pull on the iron filings—which press upon the hairs on the top of the statocyst—by promptly turning over on its back.

Sensory cells of statocysts are neuromasts without cupulae. Similar units are also found in the vertebrate ear, where they serve two distinctly different functions, equilibrium and hearing, both of which are based on mechanoreception.

## STRUCTURE AND FUNCTION OF THE MAMMALIAN EAR

The mammalian ear is composed of three sections: the external ear, the middle ear, and the inner ear. The external ear and middle ear are involved only in hearing, while the inner ear serves functions related to both hearing and equilibrium.

***The External Ear.*** The outer, flexible, visible portion of the ear is the **auricle,** or **pinna** (Fig. 17-8). In most mammals, it serves to funnel sound waves and to amplify them about tenfold. Mammals capable of hearing high-frequency sounds have some muscular control over the shape and the direction in which the pinnae are turned, thus increasing their efficiency. At the center of the pinna is the **external auditory opening,** which leads to the **external auditory canal.** Stretched across the inner end of the canal is the thin, vibratile tympanic membrane, the **tympanum,** commonly called the **eardrum.** The eardrum is the outermost part of the middle ear.

***The Middle Ear.*** Sheltered within a cavity of the temporal bone of the skull, the middle ear is detectable as a bulge just behind the pinna. The middle ear is closed at both ends, externally by the tympanum and internally by the **oval window,** a membrane-covered opening to the inner ear. The **Eustachian tube** connects the space of the middle ear to the pharynx and thus to the outside of the body. Although this tube is normally closed, it can be opened by yawning or swallowing, thereby equalizing pressure on both sides of the eardrum, an important condition for sound conduction.

The principal structures of the mammalian middle ear are three small, movable bones known as the **ear ossicles** (Latin for "little bones"). Because of their shapes, the three have been named the **malleus** (Latin for "mallet"), **incus** (Latin for "anvil"), and **stapes** (Latin for "stirrup") (Fig. 17-8). Each is separate from, but touches—and can

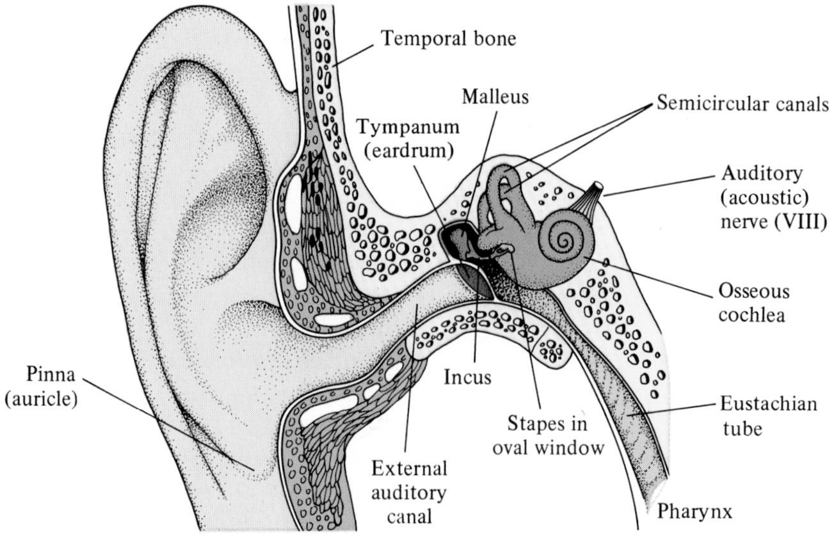


**FIG. 17-8** Principal structures of the human ear.

move—the next one, in a side-by-side arrangement. The malleus is positioned in close apposition to the tympanum, while the stapes fits snugly against the membrane covering the oval window, at the other end of the middle ear.

The ear ossicles are arranged in a way that transmits any vibrations received by the tympanum to the oval window, which marks the outer boundary of the inner ear. Because the tympanum has many times the surface area of the oval window, the result of such transmission is a significant amplification of sound. The airborne sound waves received at the tympanum are large-amplitude, low-force vibrations. They are transformed by the ossicles into low-amplitude, high-force vibrations, which are carried by the fluid of the inner ear and detected by the actual organ of hearing in the inner ear. Loss of hearing is usually due to loss of mobility of the ossicles or damage to the tympanum.

***The Inner Ear.*** The bony framework of the inner ear is called the **osseous labyrinth,** consisting of the **vestibule,** the three **semicircular canals,** and the **osseous cochlea.** Within the osseous labyrinth is the **membranous labyrinth** consisting of (1) the **sacculus** and **utriculus,** two saclike pouches enclosed in the vestibule; (2) the three **semicircular ducts,** which arise from the utriculus and are contained within the bony semicircular canals; and (3) the **cochlea,** arising from the sacculus and contained within the osseous cochlea (Fig. 17-9).* The space between the membranous labyrinth and the surrounding osseous labyrinth is filled with a fluid called **perilymph.** The fluid in the spaces within the membranous labyrinth is called **endolymph.** Both fluids are somewhat like tissue fluid in composition.

* The term cochlea is applied to either the membranous, coiled structure or to its bony encasement. Here the term refers only to the membranous portion.

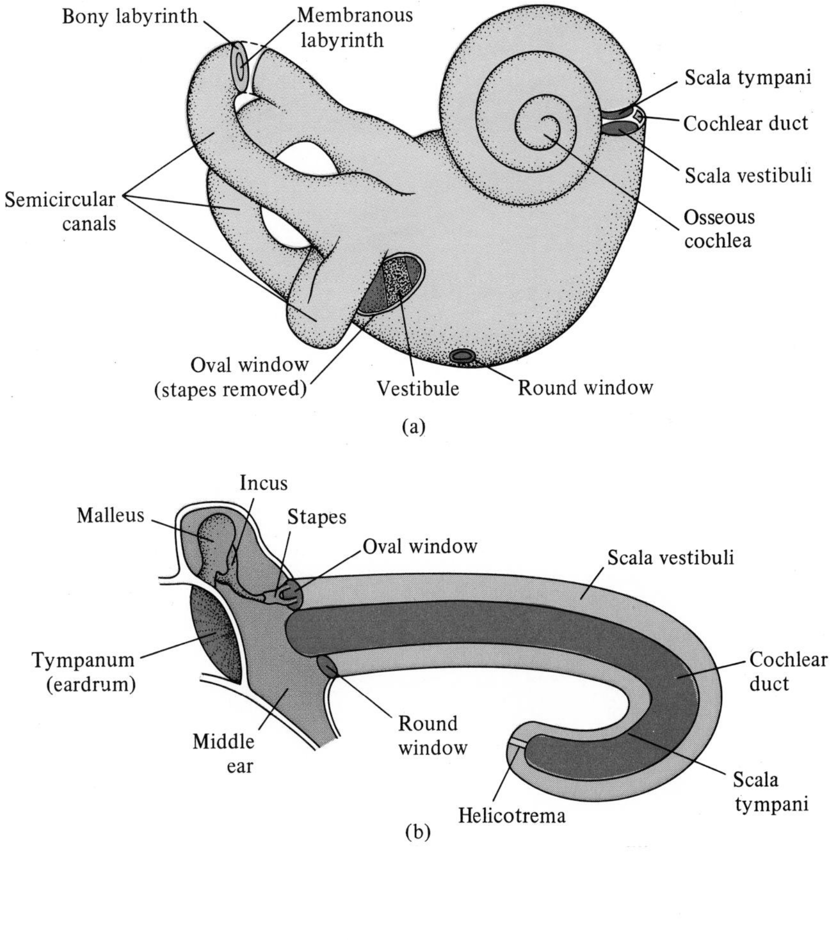


**FIG. 17-9** (a) Osseous and membraneous labyrinths of the mammalian inner ear. Arrows indicate direction of sound waves. For explanation see text. (b) Schematized drawing of the mammalian inner ear with the cochlea shown as if uncoiled.

## THE SENSE OF HEARING

The sense of hearing is localized within the **cochlea,** a snaillike, coiled tube that is divided lengthwise into three compartments (Fig. 17-9a). One of these, the **scala** (Latin for "ladder") **vestibuli,** runs from the oval window to the tip of the cochlea, where it communicates by a tiny pore, the **helicotrema,** with a similar chamber, the **scala tympani** (Fig. 17-9b) which extends from the helicotrema to the round window. Both windows are covered by membranes. Between the two chambers is a third compartment, the **cochlear duct,** also called the **scala media,** which is continuous with the sacculus. On the basilar membrane, which forms the floor of the cochlear duct, is the actual organ of hearing, the **organ of Corti.** It is made up of four or five long rows of hair cells (Fig. 17-10a). Hairlike cilia project up from the cells through a thin **reticular membrane** and make contact with the overlying **tectorial membrane** (Fig. 17-10b).

***How We Hear.*** When sound waves strike the tympanum, the vibrations are transmitted to and amplified by the middle ear ossicles. Vibrations of the stapes are communicated to the membrane of the oval window and thence to the endolymph of the scala vestibuli. The vibrations of the endolymph produce undulations in the basilar membrane. The variations in pressure in the fluid of the scala vestibuli are transmitted through the basilar membrane to the endolymph of the scala tympani, below it. Each time the membrane of the oval window bulges inward as a result of a motion of the stapes, the round window membrane bulges outward (Fig. 17-9b). The flexibility of the round window membrane allows it to move in and out in opposite phase with all vibrations of the membrane of the oval window. As a result, the basilar membrane, within the cochlea, also oscillates with each vibration. The amplitude of this oscillation differs with sounds of different pitch, the membrane acting as a resonator, that is, an intensifier of the vibrations. The stimulation of the sensitive hair cells of the organ of Corti is due to shear

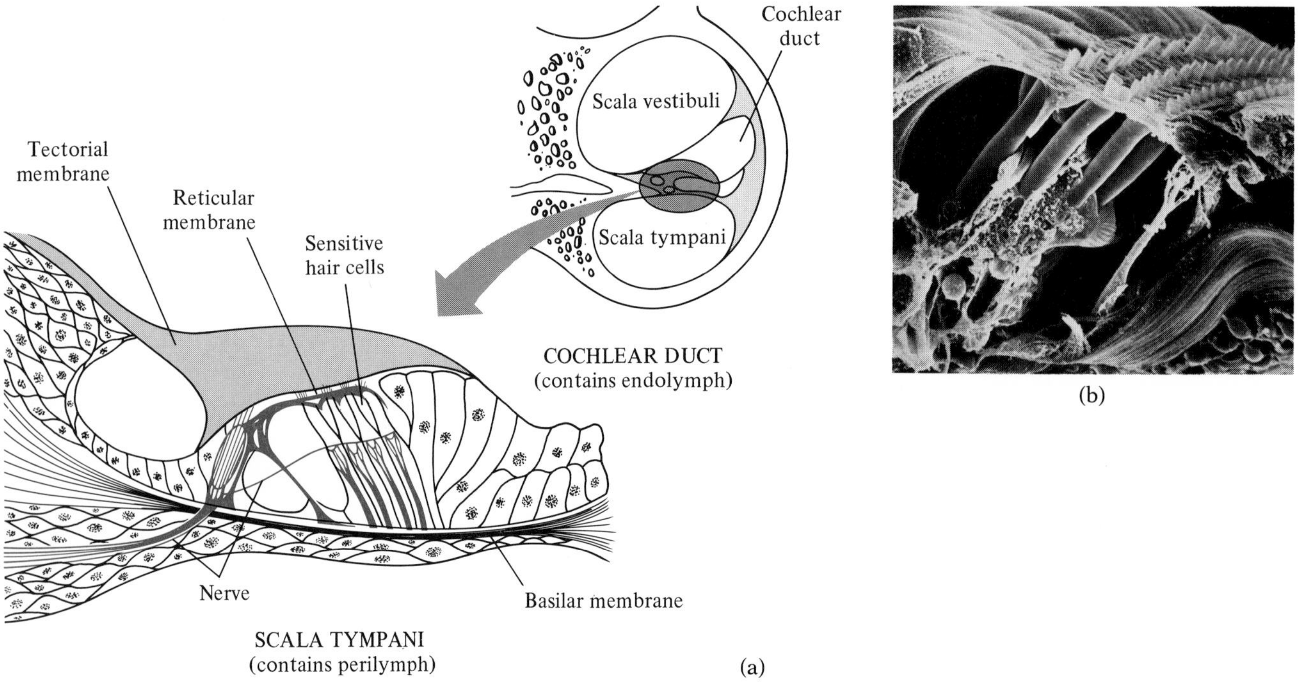


**FIG. 17-10** (a) Cross section through a portion of the cochlea and the organ of Corti, showing their relationship to each other. (b) Scanning EM of a section through the organ of Corti, showing rows of hair cells with their cilia projecting through the reticular membrane. A bundle of nerve fibers is seen at the lower right.

forces exerted upon the hairs by the undulations of the three membranes involved (Fig. 17-10). The shearing effect, in which the membranes slide in parallel but opposite directions, is produced at each undulation.

According to one theory, analysis of the patterns of nerve impulses generated in the organ of Corti suggests how information regarding the frequency, or pitch, and intensity of sounds is sent to the CNS. As with other receptors, neurons from different parts of the organ of Corti are each connected to a different region or group of cells in the hearing, or auditory, center of the brain. For a given sound, the regions of the membrane that oscillate with the greatest amplitude are those that provide the CNS with an indication of the pitch of that sound. In this way, particular regions of the brain are stimulated only when sounds of particular pitches are received. Moreover, evidence indicates that the frequency with which nerve impulses are generated and sent to the brain by the cells of the most strongly stimulated region of the membrane is also indicative of the pitch that is evoking the stimulus.

## THE SENSE OF EQUILIBRIUM

In all but the most primitive vertebrates, the sense of equilibrium is mediated by the utriculus, the sacculus, and the three semicircular canals. Each of these structures contains patches of neuromast cells. The patches in the utriculus and sacculus are known as **maculae** (Latin for "spots"). Over the combined tips of all the sensitive hairs of each macula is a massive cupula, made heavy by deposits of calcium carbonate crystals, forming an **otolith** (Greek for "ear stone"). Tilting the head in one direction or speeding up or slowing down the gait moves these cupulae, stimulating the sensitive hairs. Such body movements are thereby signaled to the CNS.

Each of the three semicircular canals lies in a plane at right angles to the others, one in each of the three planes of space (Figs. 17-8 and 17-9). The patches of neuromast cells in their ducts are termed **cristae** (Latin for "tufts" or "crests"). Each crista has a tall gelatinous cupula, which is capable of moving in either of two directions, like a swinging door (Fig. 17-11). Whenever the fluid in any one of the three canals is displaced, it forces the cupula to bend. Signals are generated by the sensitive hairs at the cupula's base. It is apparent from the structure and orientation of the three canals that they signal the turning movements of the head and body.

**FIG. 17-11** A crista of the mammalian ear.

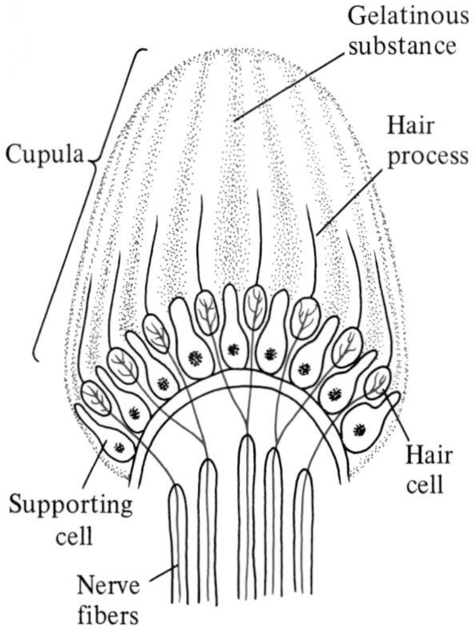


# CHEMORECEPTION

Like other organisms, all animals sense and distinguish various chemicals in their surroundings. For example, the body surfaces of the earthworm, as well as the lining of its mouth and pharynx, are well supplied with chemosensitive cells. All the appendages of shrimps, crabs, and other aquatic crustaceans bear many delicate hairs that are sensitive to chemicals dissolved in the water. Insects detect airborne molecules by means of patches of sensitive neurons; there are nearly 3,000 of these organs on each antenna of bees.

Sensitivity to airborne substances is classified as **olfaction,** the sense of smell, whereas the detection of chemicals in solution may be regarded as **gustation,** the sense of taste. Both depend on the matching of detected substances with specific receptor molecules in sensory cells.

## THE SENSE OF TASTE

On the surface of the mammalian tongue are numerous round, peglike **papillae,** visible to the naked eye. All these papillae contain **taste buds.** Although all taste buds are alike in structure, they are specialized to distinguish only certain tastes (Fig. 17-12a). Receptors sensitive to sweet substances are found at the front of the tongue, salt receptors just behind them at the tongue's edge, and sour and bitter receptors at the rear, with each area overlapping the adjacent ones.

Each taste bud contains receptor cells bearing microvilli at their surface (Fig. 17-12b). Each cell synapses with an afferent (sensory) neuron. Apparently, when a molecule of a soluble substance reacts with a receptor molecule specific for it in the surface of a taste bud cell, there is a lowering of the cell's plasma membrane permeability. This triggers a generator potential in the sensory cell, which induces a train of impulses in the afferent neuron with which it is in synaptic contact.

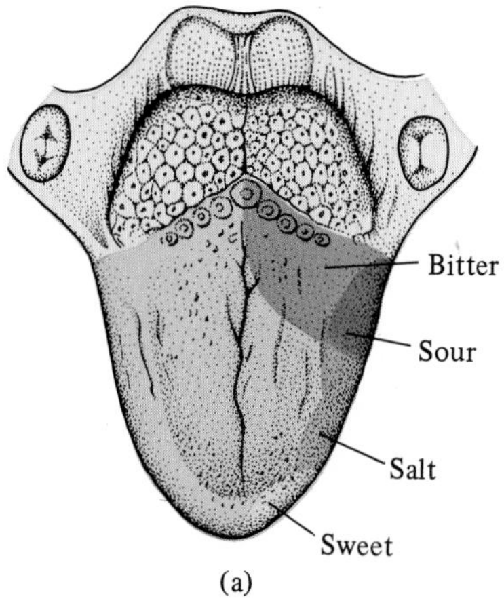


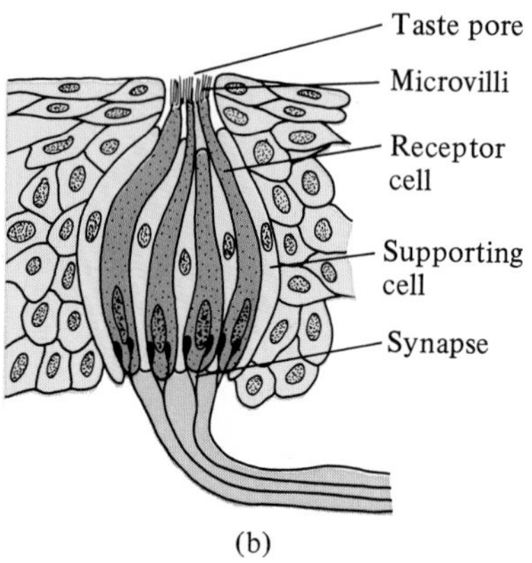


**FIG. 17-12** Taste receptors of the human tongue. (a) Upper surface, showing areas of maximum sensitivity to sweet, salt, sour, and bitter taste. (b) Longitudinal section of a taste bud.

## THE OLFACTORY SENSE

The physiology of olfaction is the least understood of the senses, partly because smell is a relatively unimportant sense in humans, compared to other mammals.

For a mammal to detect a substance as an odor, the substance must be volatile and soluble—in other words, it must be airborne and must have the ability to be dissolved in both water and lipids. A small patch of olfactory epithelium is located high in the roof of each nasal cavity. Because of the shape of nasal passages, odor-bearing air normally bypasses this receptor area, where scent molecules can be trapped by moisture on the surface of the cells. This is why airborne odors are most efficiently sampled by sniffing, an action that brings the air into contact with these receptors. As with taste, it is presumed that chemical substances that react with various receptor molecules on the sensitive cells somehow alter the membrane's permeability and produce nerve impulses (Fig. 17-13). Exactly how this is accomplished is not known.

## PHOTORECEPTION

Most major animal groups have one or more types of specialized photoreceptive cells. Such cells are often found in clusters, forming eyes (Fig. 17-14). Some relatively simple eyes consist of a cuplike arrangement of light-sensitive cells oriented to receive light from different directions. The compound eyes of arthropods are image forming and have an extraordinary ability to detect slight movements (Fig. 17-15). The most highly developed invertebrate eye is found in cephalopod molluscs such as squids, octopods, and cuttlefish. Like the vertebrate eye, the cuttlefish eye is covered by an eyelid, possesses a complex retina and contractile iris, and has a lens under muscular control.

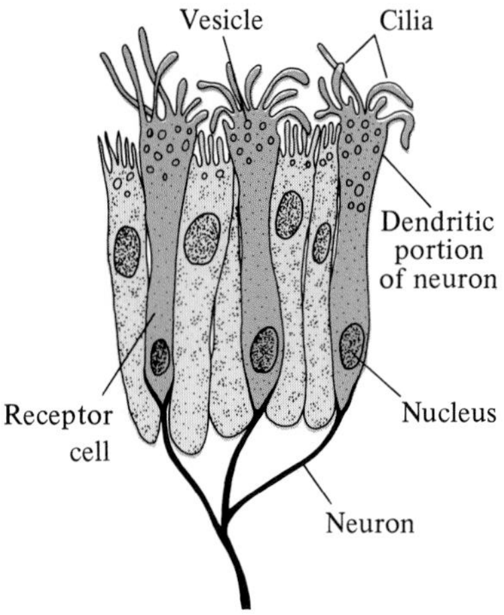


**FIG. 17-13** A sketch of a section of the human olfactory epithelium as seen by electron microscopy.

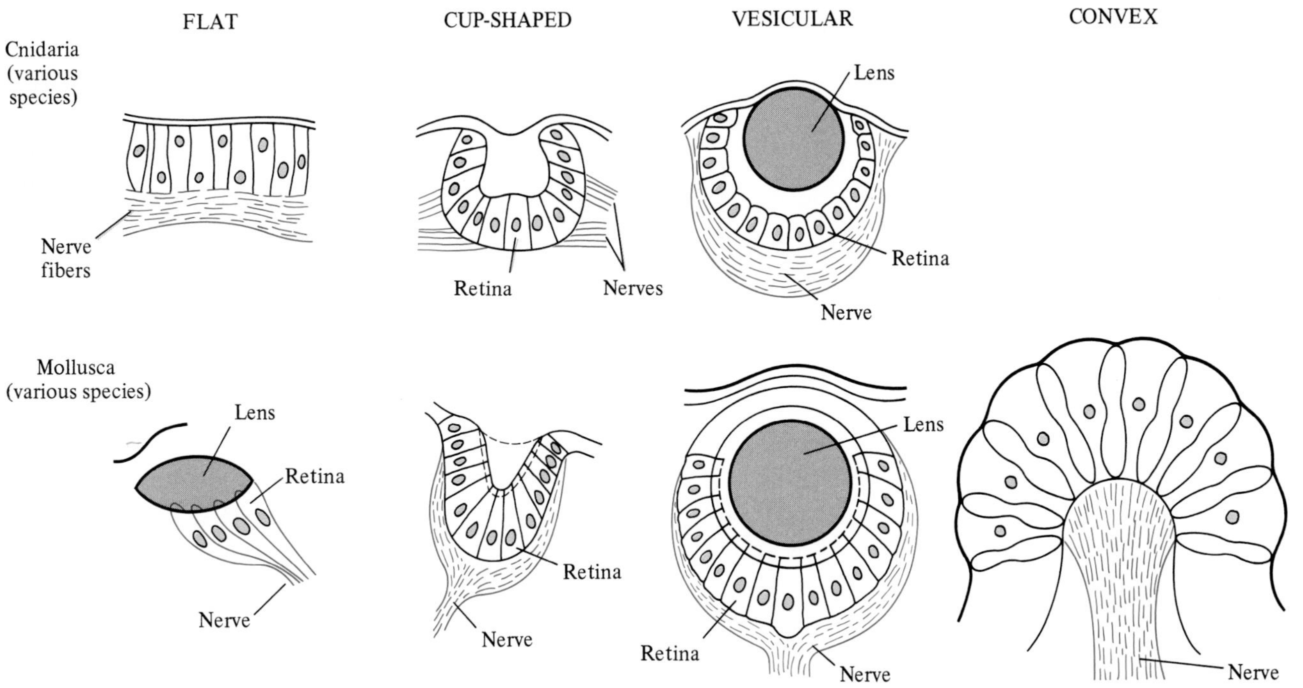


**FIG. 17-14** Four eye types found in almost all major animal groups.

## THE VERTEBRATE EYE

The vertebrate eyeball is bounded anteriorly by the convex, transparent **cornea,** which is continuous posteriorly with a tough, white coat, the **sclera,** making up the outer layer of most of the eye (Fig. 17-16). Behind the cornea is the **iris,** the colored portion of the eye. The opening in the iris is the **pupil.** The pupil can be widened or narrowed in response to changes in light intensity. At the edges of the iris is the **ciliary body.** Suspended from the ciliary body by means of the **suspensory ligaments** behind the iris is the crystalline **lens.** The cavity behind the lens is occu-

**FIG. 17-15** Sketch of the arthropod eye in longitudinal section. Each unit is termed an ommatidium.

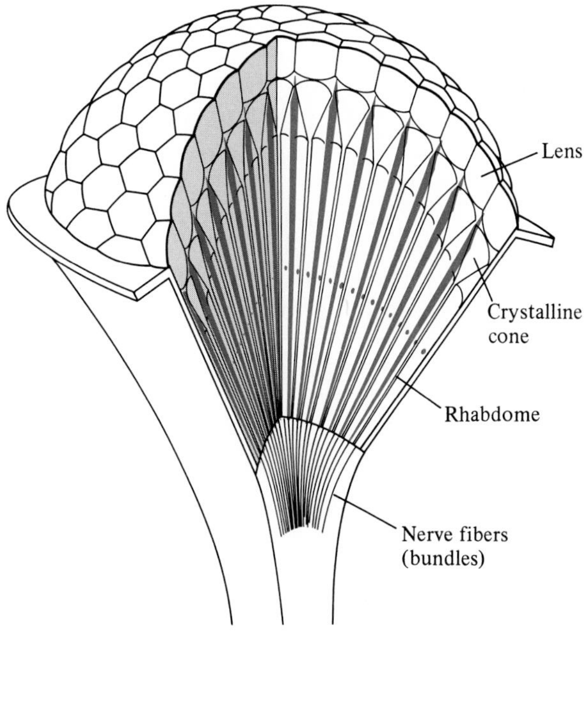

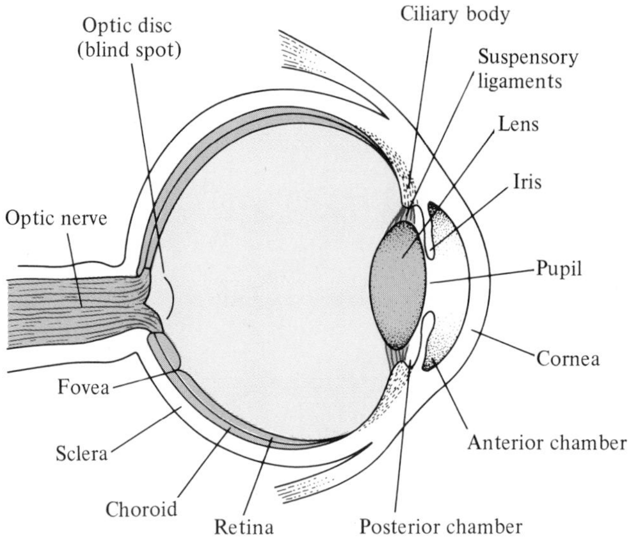


**FIG. 17-16** Sketch of the human eye in section.

pied by the **vitreous body,** a gelatinous material, the fluid component of which is termed **vitreous humor.** Inside the sclera, which covers most of the eye, is the **choroid layer,** composed almost entirely of blood vessels. A third and innermost layer of the eye is the **retina,** which contains the receptors of vision.

***Accommodation.*** Light entering the eye normally is focused on the retina, which is analogous to the film of a camera. Focusing, which produces on the retina a distinct, upside-down, reversed image of the object being viewed, is the result of the optical qualities of three structures: the cornea; its underlying fluid, the **aqueous humor;** and the crystalline lens, which lies just behind the iris and is attached by the suspensory ligament to the ciliary body. Because the corneal surface is convex and of a much higher index of refraction ($n$) than air ($n = 1.37$ compared to 1.0 for air), it and the aqueous humor ($n = 1.33$) act as a lens. (The index of refraction of an optical medium is a statement of the degree of bending of light rays that occurs as light obliquely enters the medium from air.) Parallel rays of light coming from a distance and entering the eye normally are focused on the retina. Nearby objects present a problem, however, because the optical qualities of the cornea tend to bring them into focus behind the retina, resulting in a blurred image.

Although the curvature of the cornea cannot change, the same is not true of the **lens.** A system of ligaments called zonal fibers and ciliary muscles associated with the lens controls its curvature and thus its focal length,* allowing objects at various distances to be brought into focus at will, a process called **accommodation.** It is accomplished by contraction of the ciliary muscles, which reduces the tension that the zonal fibers otherwise exert on the lens, enabling the elastic lens to increase its curvature and thereby shorten its focal length. In later adult life, the lens slowly loses its elasticity, making accommodation to near objects more difficult. As a result, corrective lenses may be required for close work or reading.

* The focal length of a lens is the distance from the lens to the point at which parallel rays of light converge after passing through the lens. It determines the distance from the lens at which the image projected by the lens will be in focus.

***Myopia and Hypermetropia.*** To some people, distant objects appear blurred because they become focused, not on the retina, but in front of it. This condition, called myopia, or nearsightedness, is usually due to abnormally elongate eyeballs. Myopia may require correction by glasses or contact lenses that diverge light rays coming from a distance. The opposite condition, hypermetropia, or farsightedness, is due to abnormally short eyeballs. Although in mild cases normal accommodation by the crystalline lens can compensate for hypermetropia, reading glasses or lenses are usually required or become necessary at an early age.

***The Retina.*** The retina exhibits a complex arrangement of receptors, glial cells, and neurons. The various layers of neurons achieve an astonishing amount of integration, comparable to that performed by the brains of many lower animals.

Like many other receptors, the receptors of the retina, called **rods** and **cones** because of their shapes, are actually sensitive regions of neurons (Fig. 17-17). Impulses generated in the rods and cones are relayed through a series of neurons whose axons converge at the **optic disk,** just off center of the retina, and leave the eye as the **optic nerve.**

Paradoxically, through one of evolution's anomalies, the rods and cones are at the back of the retina and face away from the light source (Fig. 17-18). To reach them, light must travel through all the retinal layers. These layers are highly transparent and transmit light about as readily as the aqueous and vitreous humor do.

***Distribution of Rods and Cones.*** At the periphery of the retina, only rods are found; on the rear wall, mainly cones. Both are present between these two regions. Cone vision predominates in bright light and is

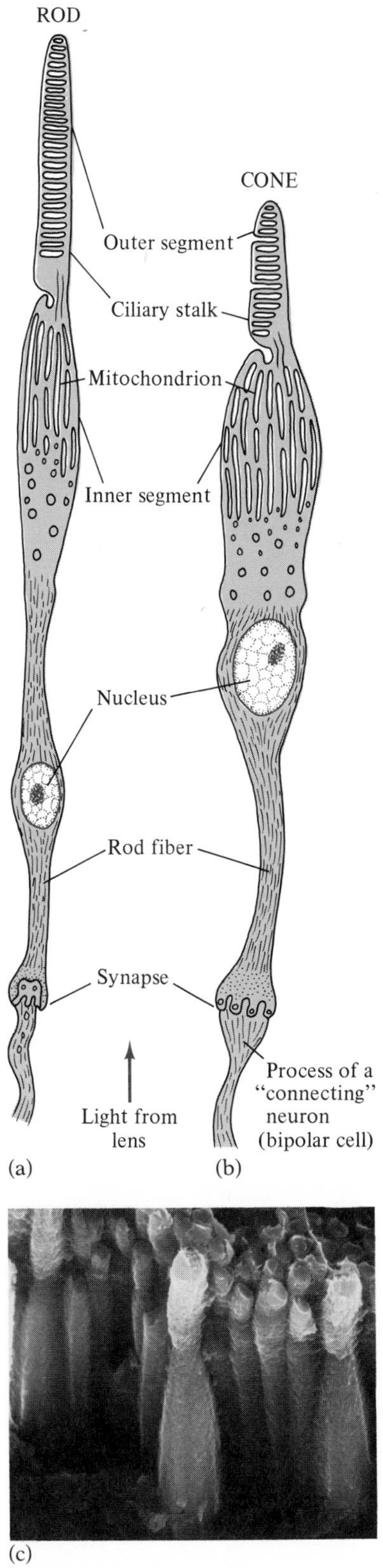


**FIG. 17-17** Sketches of the fine structure of (a) a rod and (b) a cone. (c) Scanning electron micrograph of rods and cones of the human retina.

**FIG. 17-18** Schematic representation of the principal cell types and their synaptic connections with each other in the retina of the primate eye.

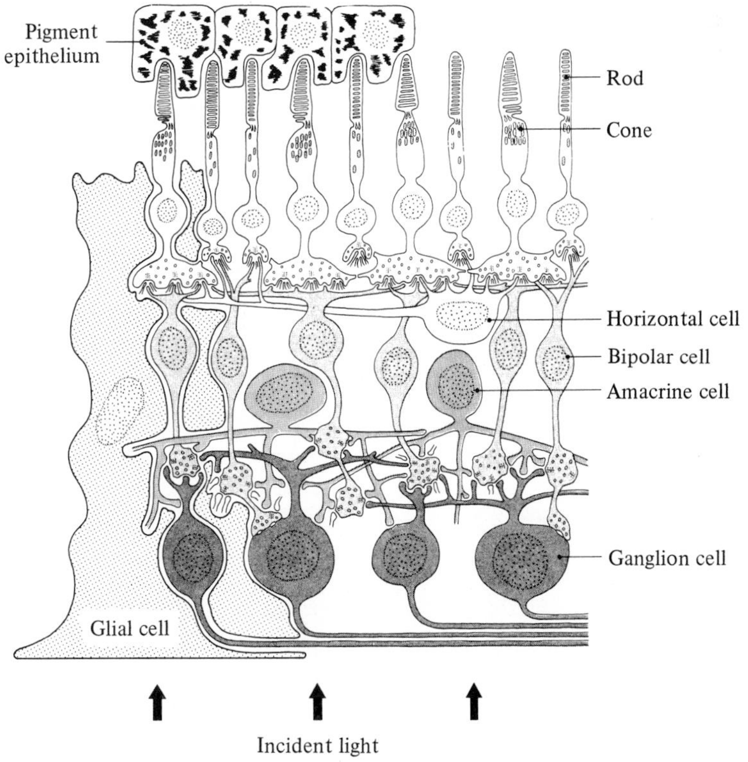

**FIG. 17-19** Birds of prey have unusually acute vision, provided by concentrations of up to a million cones per $mm^2$ of retinal surface (as compared to about 150,000 per $mm^2$ in human eyes). Shown here with its prey and young is a peregrine falcon, a predator on other bird species.

useful for distinguishing colors. Rods, which contain a different photosensitive pigment than do cones, are nonfunctional in bright light but can detect faint light in the dark. Rods are more sensitive than cones, but because of their large numbers, cones are much better able to distinguish details. A few diurnal animals—those active chiefly in the daytime—have only cones; a number of strictly nocturnal species have only rods.

***The Area Centralis and the Fovea.*** The center of the retina is the **area centralis,** the region of highest visual acuity, where fine details are best resolved, that is, distinguished from each other. It has the densest concentration of cones—some 150,000/$mm^2$ in humans and as many as 1,000,000/$mm^2$ in some birds of prey (Fig. 17-19). In humans and a number of other animals, a depression called the **fovea** exists at this spot. Here the retina is thinned and lacks blood vessels. Because the nerve fibers diverge at the fovea, light reaches the cones without having to first pass through the other cell layers and blood vessels.

It is generally believed that because of its high concentration of cones and the local absence of blood vessels and nerve fibers, the fovea is extremely sensitive to light. Another suggestion is that the foveal concavity in the retinal tissue produces an outward refraction (bending) of the light rays that strike the area. This might allow the clear retinal tissue to act as a lens (it has a different index of refraction than does the vitreous humor), thereby broadening or magnifying the image, involving more cones than would otherwise be stimulated, thus increasing the area's sensitivity. Birds of prey have more extensive foveas than other animals.

***The Blind Spot.*** At the spot where the optic nerve and blood vessels exit the retina, there are no rods or cones, and a blind spot exists. The existence of the blind spot can be demonstrated by drawing a small cross on a plain sheet of paper and a small dot 3 cm to the right of it. The left eye is covered and the right eye looks directly at the cross as the paper is slowly brought closer to the eye. At about 10 to 12 cm away from the eye, as the dot becomes focused on the blind spot, it disappears. If the exercise is done with the right eye covered and the left eye looking at the dot, the cross disappears. Because the blind spots of the eyes are both off center and on opposite sides of center, vision is not affected as long as both eyes are functional.

## Summary

An animal's ability to detect important changes occurring both within and without its body provides the informational basis for nervous integration. The great variety of sense receptors found in animals is essential to this process. Classified as to location and general function, they include exteroceptors, interoceptors, and proprioceptors. Categorized as to type of stimulus which they detect, they include mechanoreceptors, chemoreceptors, and photoreceptors.

Animals have evolved a wide spectrum of sensitivities to stimuli of importance from their external and internal environments. Although all receptors communicate by means of similar signals—action potentials—the messages conveyed by different types of receptors are received by different parts of the brain; thus, in addition to being able to distinguish stimuli of different intensities, animals can distinguish different types of stimuli as well as stimuli of the same type received in different parts of the body. Some receptors differ in responsiveness to steady and intermittent stimulation, whereas others respond only to changes in stimulus intensity.

There are many types of mechanoreceptors, all of which generate nerve impulses in response to deformations of the plasma membranes of the receptor cells. Proprioceptors are stretch receptors that signal movements and positions of the limbs and body. In the neuromast of aquatic vertebrates, the receptor is a clump of secondary (nonneuronal) sensory cells bearing fine hairs. The hairs may be made more sensitive to vibrations either by being enclosed in a pendulous, gelatinous cupula or by projecting through several membranes subject to shearing forces during vibrations, as in the mammalian ear. The incompletely understood lateral line organs of fishes are systems of neuromasts. Modified neuromasts also form the basis for vertebrate organs of equilibrium. In the sensitive elements of invertebrate statocysts, the receptor is a tuft of modified cilia.

The external ear and middle ear in mammals are concerned only with hearing, the inner ear with equilibrium as well. The external ear funnels and amplifies sound waves. The middle ear, which includes the tympanum (eardrum), the ear ossicles (malleus, incus, and stapes), and the oval window, opens to the pharynx by way of the Eustachian tube. The large tympanum, the small oval window, and the arrangement of the ossicles all serve to greatly amplify sound waves. Among the components of the fluid-filled inner ear are the three semicircular canals and the cochlea, which has three compartments. On the basilar membrane, which forms the floor of the central compartment (the cochlear duct), is the organ of Corti, composed of four or five long rows of sensitive hair cells. Cilia of these cells project through the overlying reticular membrane and contact the tectorial membrane above it. When the membrane of the oval window vibrates in response to sound waves, it causes undulations of the basilar membrane. These translate into shear forces that act on the cilia of the hair cells, thereby generating action potentials in associated neurons. Sounds of different frequencies produce the greatest oscillations in different regions of the basilar membrane.

The two major modes of external chemoreception, taste and smell, appear to depend on the interaction of molecules of the substance being detected and molecules of the receptor surfaces. Taste usually requires that the substance tasted is in solution, whereas, in mammals, olfaction requires that the substance be volatile enough to be airborne and soluble enough to dissolve in the thin film of mucus covering the olfactory receptor.

Photoreception is a capability of almost all living systems. The abilities to detect different intensities and wavelengths of light, to determine the direction from which light is coming, and, in many cases, to perceive images, have evolved in various animal groups. Two groups, the cephalopod molluscs and the vertebrates, have independently evolved image-forming eyes of high complexity and discriminatory ability.

Among the principal parts of the vertebrate eye is the anterior, transparent cornea, which is continuous with the outer layer, the sclera, at the back of the eye. The iris, whose opening is the pupil, can be widened or narrowed in response to changes in light intensity. At the iris's edges is the ciliary body, a system of fibers and muscles associated with the lens, which lies just behind the iris. Contraction of the ciliary muscles reduces the normal tension exerted by the fibers on the lens, thereby allowing it to round up and thus shorten its focal length. This permits visual accommodation to objects at various distances. Myopic, or nearsighted, people usually have abnormally elongated eyeballs, whereas hypermetropic, or farsighted, people have unusually short eyeballs. At the rear of the eye is the retina and behind it, the highly vascular choroid layer. Behind the choroid layer is the sclera. The multilayered retinal complex of receptors, neurons, and glial cells has integrative capabilities comparable to that of some invertebrate brains. The receptors are of two types: rods, which predominate at the edges of the retina and are especially effective in dim light; and cones, which predominate toward the retina's center, are especially effective in bright and colored light, and, because of their numbers, contribute most to visual acuity (detailed vision).

## Recommended Readings

A list of recommended readings appears at the end of Chapter 15.

# 18

# Muscle and Other Effectors

OUTLINE



The body's physiological actions and responses to stimuli are actually performed by its muscles, glands, and other **effectors.** Movement being the major such action of animals, this chapter examines the means by which muscular contraction is accomplished. Brief consideration is also given to two other actions: ciliary movement and the production of currents by the electric organs of certain fishes. Although the principal subject of this chapter is the morphology and function of vertebrate skeletal muscle, the basic principles presented are generally applicable to other types of vertebrate muscle and to the wide variety of muscle cell types found in invertebrates.

## MUSCLE STRUCTURE AND ARRANGEMENT

Vertebrates have three major types of muscle: cardiac muscle, of the heart, smooth muscle, of the viscera, and striated muscle, of the skeletal system.

## CARDIAC MUSCLE

Cardiac muscle cells are unusual in that they are all laterally fused to, or anastomosed with, each other (Fig. 18-1a). Such fusion is thought to provide more efficient conduction of the action potentials generated by the heart's pacemaker than if there were no fusion of cells. Also unique to the heart muscle, or **myocardium,** are numerous **intercalated discs** of tough connective tissue that join the muscle cells to each other at their ends. These discs give great tensile strength to the entire myocardium.

## SMOOTH MUSCLE

The cells of smooth muscle, also known as involuntary or nonstriated muscle, are spindle-shaped, being tapered at both ends (Fig. 18-1b). Each fiber, as a muscle cell is also called, has a single, centrally located nucleus and faint longitudinal striations called **myofibrils,** the contractile units of the muscle cell. Smooth muscle is the major muscle type in the internal organs of the vertebrate body; it makes up the muscular component of artery walls, gland-ducts, and digestive tract and is controlled by the autonomic nervous system. Smooth muscle constitutes but a small fraction of the vertebrate body's musculature. In many invertebrates, such as the common earthworm, smooth muscle is the only muscle tissue in their bodies.

## STRIATED MUSCLE

By far the greatest mass of the vertebrate body is composed of striated muscle (Fig. 18-1c). Striated muscle is so named because its fibers are prominently striped with alternate dark and light bands. These bands are readily visible under the light microscope, even without staining. Like smooth muscle, striated muscle is made up of many individual muscle fibers. Striated muscle is under conscious control and therefore is also known as skeletal or voluntary muscle.

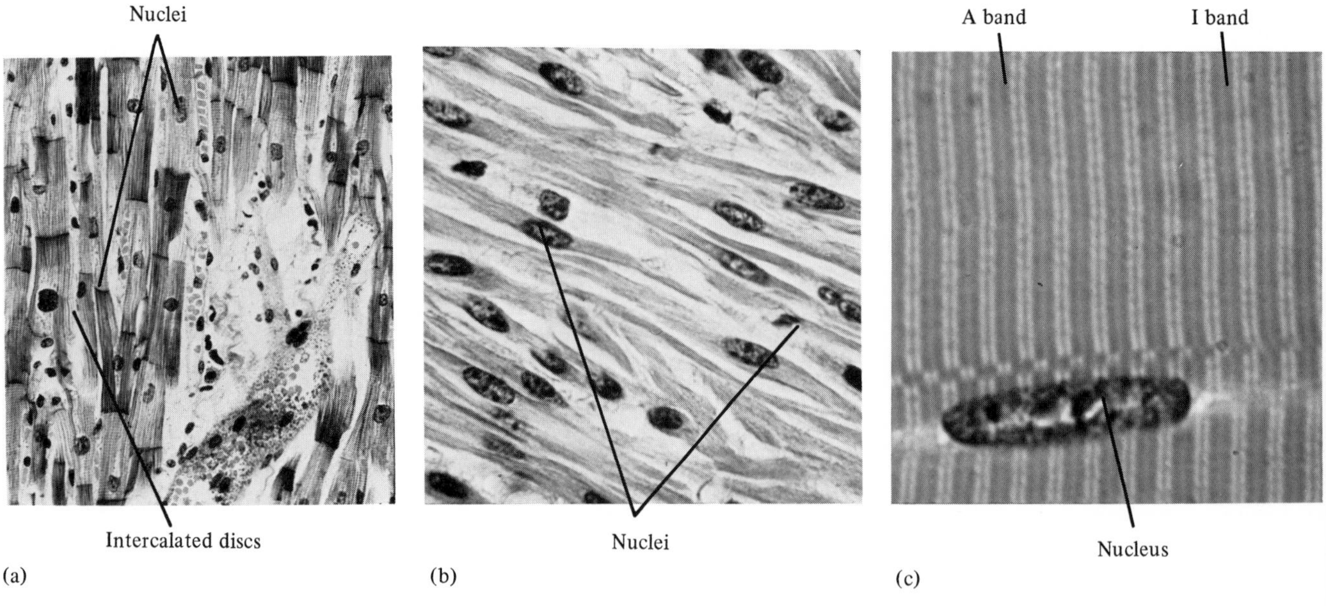


**FIG. 18-1** Types of vertebrate muscle tissue. (a) Cardiac muscle. Note dense intercalated disks of connective tissue that bind cells together lengthwise. Several anastomoses of adjacent cells can be seen. (b) Smooth muscle cells. (c) Portions of two striated muscle fibers. The ovoid object is a nucleus.

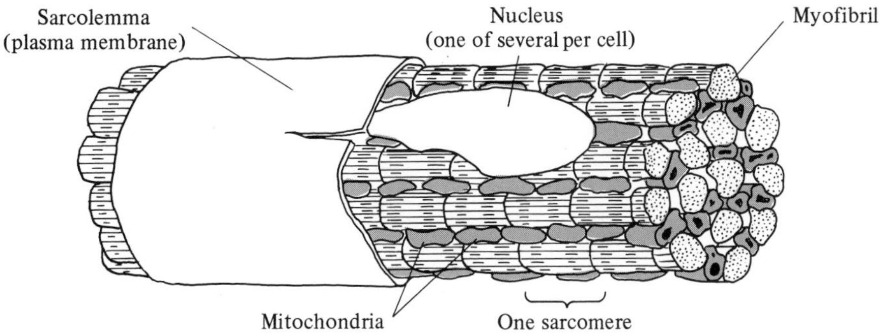


**FIG. 18-2** Sketch of a portion of a single striated muscle cell, or fiber, showing details visible in electron microscopy.

Striated muscle fibers are multinucleate, with the nuclei lying along the cell periphery, just inside the plasma membrane, which is known as the **sarcolemma** (Fig. 18-2). Mitochondria in striated muscle cells are large and numerous, as is appropriate for cells that expend great amounts of energy. Each muscle fiber contains many myofibrils. Striated muscle cells are found throughout the animal kingdom, from sponges to insects to mammals. Even some protozoans have striated contractile fibrils. Striated muscles are most common, however, in arthropods and vertebrates.

Groups of striated muscle fibers are organized into muscle bundles, or **fasciculi.** A muscle is composed of several to many fasciculi (Fig. 18-3). The bundles, as well as the entire muscle are enclosed in a sheath of tough connective tissue called **perimysium** or **fascia.** Individual muscle fibers are also ensheathed in connective tissue. In many skeletal muscles, the connective tissue sheath at one end is continuous with a **tendon,** a glistening white cord of tough connective tissue that attaches the muscle to a movable part of the skeleton, as in the lower leg (Fig. 18-4). The end of the muscle attached by tendon to the more movable skeletal part is termed its **insertion.** The other end of such a muscle, called its **origin,** is attached by short connective tissue **ligaments** to a less movable part of the skeleton. For example, the insertions of the lower leg muscles, which move the foot, are on the bones of the foot; their origins are on the upper and lower leg.

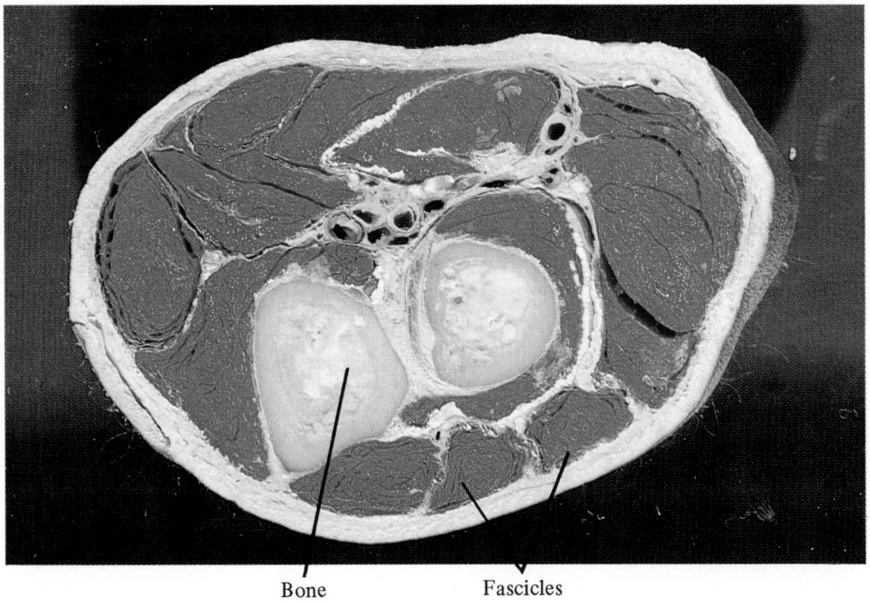


**FIG. 18-3** A cross section of human forearm muscles, showing the investment of the muscle fascicles with connective tissue.

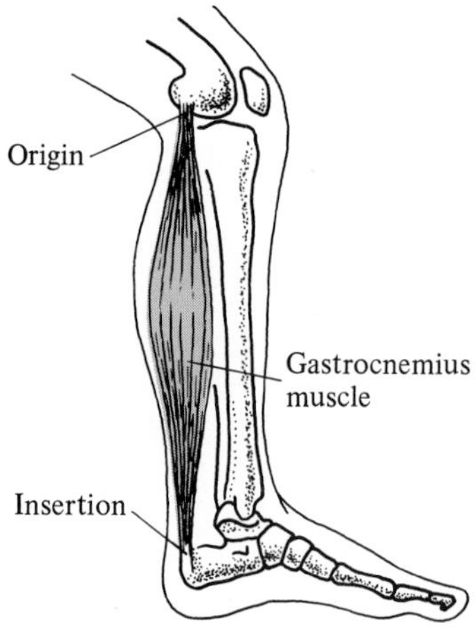


**FIG. 18-4** Sketch of the lower leg showing origin and insertion of the gastrocnemius (calf) muscle.

## THE ARRANGEMENT OF MUSCLES WITHIN THE BODY

Most, but not all, muscles are arranged in antagonistic sets, in which the action of one or more muscles is opposed by that of others. For example, the muscles that lower the jaw are antagonized in their action by those that raise the jaw. The only movement possible for a muscle is **contraction,** that is, shortening. For a structure to be moved in any direction, it must be pulled in that direction by one or more muscles. Thus, all moving body parts have at least two sets of muscles. One of the clearest examples of the antagonism of two striated muscles is provided by the actions of the **biceps femoris** and the **triceps femoris** of the upper arm. The biceps flexes the forearm, whereas the triceps extends it. The circular and longitudinal smooth muscles of the intestine and other tubular organs are also arranged in antagonistic sets (Fig. 18-5). Contraction of the longitudinal fibers of the intestine shortens and thickens it at the same time, stretching the circular muscles. When the circular muscles contract, the intestinal diameter is reduced, forcing it to lengthen—thereby stretching the longitudinal muscles.

**FIG. 18-5** Longitudinal section, from anterior to posterior, of the human small intestine showing layers of circular and longitudinal muscles.

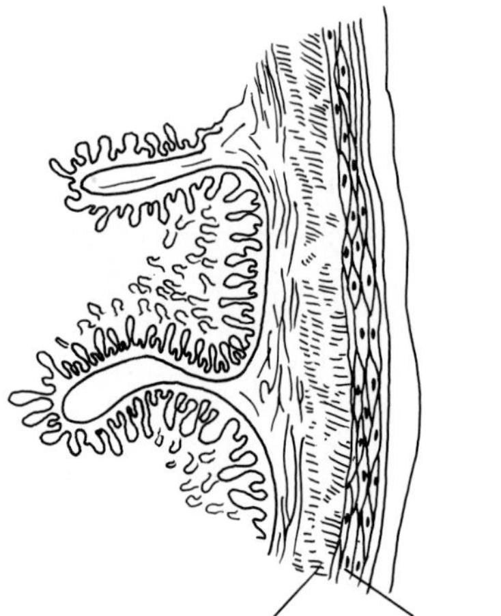


# PHYSIOLOGY OF STRIATED MUSCLE

Muscle function has long been investigated at two levels, one being the classical physiology and mechanics of entire muscles, the other being the biochemistry and microscopic structure of the muscle cell. Classical muscle physiology studies the bases for muscular fatigue, physical factors in contraction, and the varying responses of muscles to stimuli of different strengths and frequencies. Although the principles discussed in this and the following section on microscopy and biochemistry of muscle contraction refer to striated muscle, most of them apply to cardiac and smooth muscle as well.

## THE MOTOR UNIT

When a particular muscle, such as the calf muscle of the leg, contracts, seldom do all of its fibers respond at the same time. Instead, the fibers

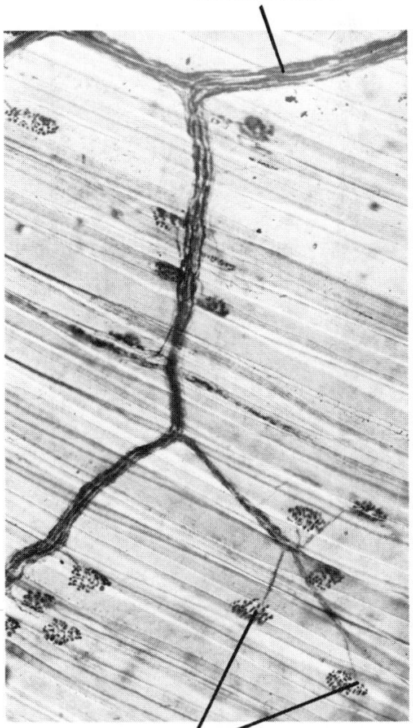


**FIG. 18-6** Photomicrograph of human muscle motor end plates, the motor neuron connections to striated muscle fibers.

work in groups, and, as fatigue occurs in one group, another group takes over; the various groups usually take turns in the execution of a given task. In Chapter 15 it was explained that each motor neuron, with its cell body in the spinal cord, sends out an axon to a skeletal muscle. At the muscle, the axon gives off many branches, each branch innervating a separate muscle fiber through its **motor end plate** (Fig. 18-6).

The muscle fibers served by the branches of a single motor neuron are all stimulated whenever that neuron generates a train of impulses. When an action potential reaches a branch point of an axon—always at a node of Ranvier—impulses normally go down both of the branches. As a result, all the muscle fibers served by the branches of one axon are stimulated and contract together. A motor neuron, its branches, and all the muscle fibers they serve form a **motor unit.** For example, a motor unit of a leg muscle may contain as many as 150 muscle fibers. The muscles that control such delicate actions as the movements of the eye, on the other hand, have an average ratio of only three muscle fibers to each motor neuron serving them, providing a much finer degree of control.

## FATIGUE

When a muscle performs a task for a long time, such as lifting and holding a heavy weight, a large number of motor units is involved. As time goes on, motor units continually take turns, gaining a period of rest when not contracting. Eventually, depending on the physical condition of the individual, the muscle is unable to do any further work; at this point fatigue occurs. One factor in fatigue is the adequacy of the muscle's blood supply, of importance because blood supplies the oxygen and glucose needed for energy, besides removing the waste products of muscle metabolism. One benefit of a warm-up before an athletic contest is an improvement in blood supply to the muscles. Engaging in a program of regular exercise strengthens the muscles and improves their circulation on a more lasting basis. As a result, fatigue does not occur as readily in people who exercise regularly.

The physiological basis of fatigue at the level of the motor unit is of two types. In one type, continual direct stimulation of a nerve attached to a particular muscle results in the depletion of the acetylcholine-containing synaptic vesicles at its motor end plates. The muscle then fails to respond until more acetylcholine is synthesized. That the muscle itself is not fatigued and is still capable of contracting can be shown by stimulating it directly. It contracts, showing that in such a case, fatigue has occurred at the synapse. Another cause of fatigue occurs in an excised muscle deprived of a blood supply. Continued stimulation will deplete the glycogen it contains and will result in an ultimate failure to respond to any stimulation, direct or not. Although both these types of fatigue are readily demonstrated in the laboratory, they are not thought to occur under normal circumstances.

## ELASTIC COMPONENTS AND THE SMOOTHNESS OF MUSCULAR ACTION

Although isolated muscle fibers contract in a rapid, jerky fashion when stimulated, intact muscles in the body contract very smoothly. One major reason for this is the **elasticity** of muscles, that is, their ability to be stretched. When a given muscle contracts, neighboring muscles that are not contracting, especially ones that are antagonistic to the contracting muscle, resist and slow its contraction. Another important contribution to muscle elasticity is made by the connective tissues associated with a muscle. Their natural elasticity also slows and thereby imparts smoothness to muscular contraction. The combined elasticity of muscles and associated connective tissues causes the tension, or stretching force, of contraction to be shared by these other tissues. Such internal friction in other tissues slows not only the rate of contraction but of relaxation as well. Although additional factors are involved, the sharing of tension by other tissues is a major reason for the fact that muscles remain contracted for a time after the stimulus that caused their contraction is discontinued. Even an isolated muscle fiber undergoing a single contraction lasting but a few milliseconds may take a full second—1,000 times as long—to relax again. Without the elasticity of muscle and connective tissue, our movements would be quite jerky. This tissue elasticity also keeps contracting muscles from literally tearing themselves apart.

## TYPES OF MUSCULAR CONTRACTION

The contraction of muscles can be classified in various ways. Descriptions of some types of muscular contraction follow.

1. In **isometric contraction,** the muscle is not permitted to shorten. Examples include the muscular activity involved in maintaining posture or that involved in attempting to pick up an object that is bolted to the floor. In such cases, the muscle actually does shorten somewhat but is not perceived to do so because of the elasticity of the attached tendons and other noncontractile connective tissue.
2. In **isotonic contraction,** the degree of tension within the muscle remains the same while the muscle contracts. Theoretically, a steady pull of the arm muscles during a slow swimming stroke could constitute an isotonic contraction. In reality, during ordinary movements of the body's muscles, the load or tension on a muscle continually changes. True isotonic contractions therefore rarely occur normally. Because such contractions can be induced experimentally, they have been a favorite subject of study by physiologists.
3. In **tonic contraction,** a muscle engages in sustained, partial contraction. This is actually a type of isometric contraction maintained over a long period of time. The term is applied to the normal but slight amount of contraction or tension exhibited by all healthy muscles, also referred to as muscle tone. An individual exhibits less muscle tone when deeply anesthetized than when awake.
4. In a **twitch,** a skeletal muscle responds to a single volley of nerve impulses by briefly contracting all of its fibers. A twitch can be produced isometrically or isotonically (Fig. 18-7a). The time and total duration of a twitch varies in different animals and in different types of muscles in the same animal. For example, the gastrocnemius (calf) muscle of a domestic cat contracts in 0.039 second and relaxes in 0.040 second, whereas the adjacent soleus muscle

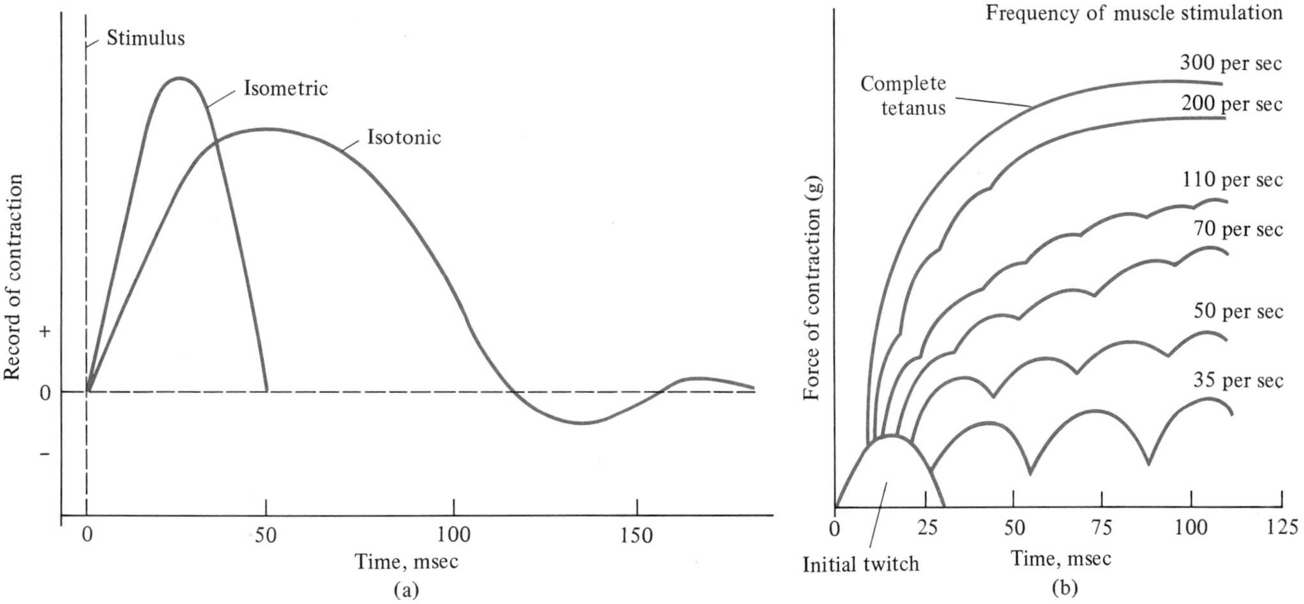


**FIG. 18-7** A comparison of single and repeated muscle twitches under different conditions. (a) Isometric and isotonic twitches. In isometric contraction, in which the muscle is not permitted to shorten, the contractile elements shorten somewhat, but only by stretching the elastic components. When a maximal stimulus is applied, full tension develops quickly and, because of the "snap" of the elastic components, quickly returns to normal. In isotonic contraction, in which constant tension is maintained on the muscle, there is a delay in the contraction due to the drag of the load on the muscle. A momentum then develops that prolongs the contraction and also causes an unusual lengthening upon relaxation. With a single, brief, maximal stimulus, the period of contraction (20 to 40 msec) is insufficient for the muscle to develop peak tension. (b) Summation and tetanus. If, under isotonic conditions, a second maximal stimulus is applied to the muscle before relaxation is complete, summation (treppe) occurs. If the muscle is stimulated 100 or more times per second, complete fusion or tetanus occurs. Tetanus is a continued state of maximum contraction.

contracts in 0.077 second and relaxes in 0.079 second. Like some other types of contraction, a single twitch rarely occurs in a normally functioning animal.

5. In **tetanus,** the muscle produces sustained, maximal contraction. This type of contraction is discussed more fully in the following section.

## THE EFFECTS OF REPEATED MAXIMAL STIMULUS

A maximal stimulus is one that achieves the fullest possible contraction; even if the strength of the maximal stimulus is doubled, no greater shortening of the muscle occurs. In a twitch, the fibers of a muscle contract following a single maximal stimulus. However, if a second maximal stimulus is applied before the first contraction cycle is completed, the second contraction is added to the residue of the first and a greater shortening of the muscle results (Fig. 18-7b). The two contractions are said to have summed, and the process is referred to as **summation.** The additional shortening results from the recoil of the noncontracting but elastic connective tissue components of the muscle. Even as the muscle fibers begin to relax, these elastic elements contribute to the twitch by "pulling back," much as a stretched rubber band resists being stretched. When a second stimulus is applied while these elastic components are still recoiling, the second contraction is added to this shortening.

If a third maximal stimulus is applied before the end of the second cycle, it is added to the first two, and an even greater shortening of the muscle results. Because the effect, as recorded by a measuring device, appears as a tracing resembling a series of steps on a staircase, it was referred to by early muscle physiologists as treppe (German for "stairs"), or the staircase effect (Fig. 18-7b).

What happens if a long series of maximal stimuli is applied to a muscle at short intervals? At first, the muscle continues to shorten with each new stimulus. However, the muscle shortens less and less with each new stimulus until it remains contracted at a constant length, even

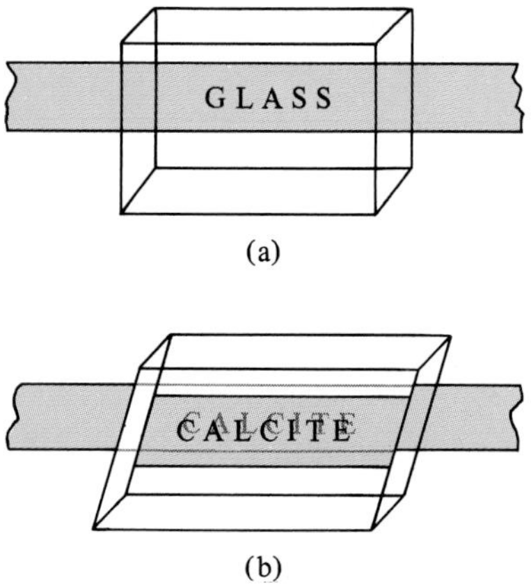


**FIG. 18-8** Birefringence (double refraction). (a) Single refraction by a glass plate. A single image is produced. (b) Birefringence by a calcite crystal produces two images.

if stimuli are continued at the same rate. The muscle is then in complete tetanus, or fusion.* For most muscles, the tension generated by a tetanic contraction is about four times that of a twitch. In normal functioning, most muscular contractions reach incomplete tetanus.

# THE STRUCTURE AND BIOCHEMISTRY OF STRIATED MUSCLE CONTRACTION

Muscle fiber structure and function have been intensively investigated by light and electron microscopy and at the molecular level. Our present understanding of muscular contraction reflects the combined results of observations made at all three of these levels of investigation.

## MICROSCOPIC APPEARANCE OF STRIATED MUSCLE

Birefringent, or doubly refractive, materials split a light beam into two beams as they refract, or bend, it (Fig. 18-8). In contrast to this reaction, a beam of light entering an isotropic substance such as glass, water, or air at an oblique angle from another medium is refracted to a new direction but still appears as a single beam. When examined by light microscopy in polarized light—light vibrating in only one plane—the dark cross striations of skeletal muscle fibers are found to be anisotropic, or birefringent, while the lighter bands are isotropic. This observation led to the designation of the dark striations as **A bands** and the lighter striations as **I bands.** Close examination under the light microscope or by electron microscopy reveals that the banded appearance of each fiber is due to the A and I bandings of the many fine, longitudinal myofibrils of which the fiber is principally composed (Fig. 18-9). Because the myofibrils are closely aligned with each other, the entire fiber appears striped. Each **sarcomere,** or unit of a myofibril, extends from the dark **Z line** in the middle of one I band to the Z line in the middle of the next I band. The lighter band often visible under higher magnification in the middle of the A band is the **H zone.** The dark line at its center is the **M line.**

Viewed by electron microscopy, the endoplasmic reticulum of a striated muscle cell, called its **sarcoplasmic reticulum** (SR), is seen to be an extensive system of smooth tubules, closely investing groups of myofibrils (Figs. 18-2 and 18-10). Cross connections called **terminal cisternae** overlie most of the I bands along the myofibrils.

As can be seen in Figures 18-2 and 18-10, a slender transverse **T tubule** runs between each pair of terminal cisternae at each Z line. The membranes of the T tubules are continuous with the sarcolemma, and their interior is open to the tissue fluid surrounding the fiber. These deep, tubular incursions of the plasma membrane into the muscle cell provide efficient communication between the plasma membrane and the underlying sarcomeres. The exact location of terminal cisternae and the number of T tubules per sarcomere varies in different species of vertebrates.

* The term *tetanus* has another application. The toxin produced by a certain bacterium can induce massive tetanic contractions of the body's skeletal muscles. In extreme cases, the contractions are so strong that bones are actually broken. Because of its effect on the jaw muscles, it is also called lockjaw. In another disease, induced by a low blood level of $Ca^{2+}$ ions, the skeletal muscles exhibit tetany, or spasms of contraction.

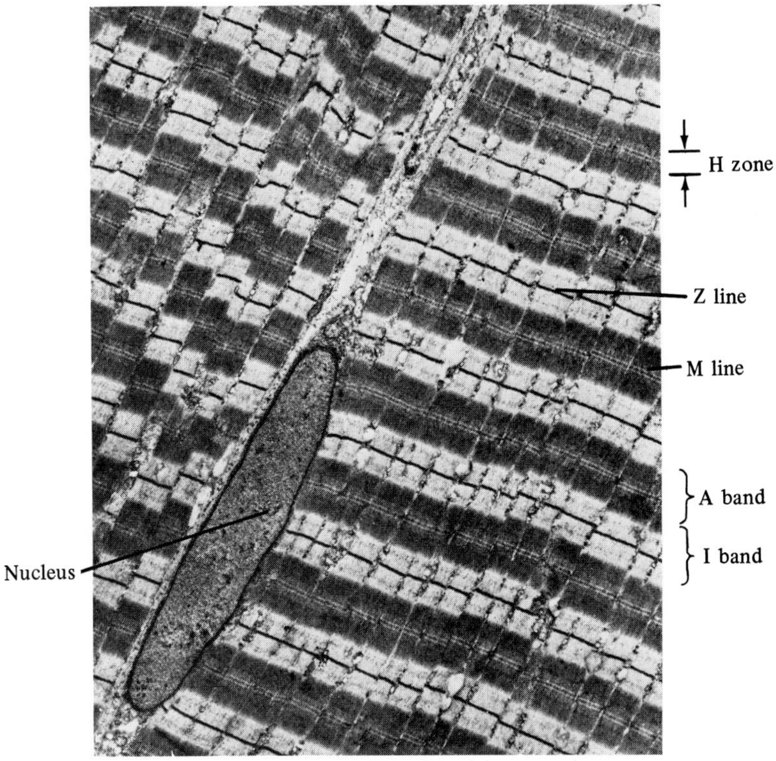


**FIG. 18-9** Electron micrograph of a nucleus and myofibrils of two adjacent human striated muscle fibers. The light band in the middle of the dark A band is the H zone. The line in the middle of the H zone is the M line. The heavy dark line in the middle of the light I band is the Z line, which delimits where one sarcomere ends and the next begins.

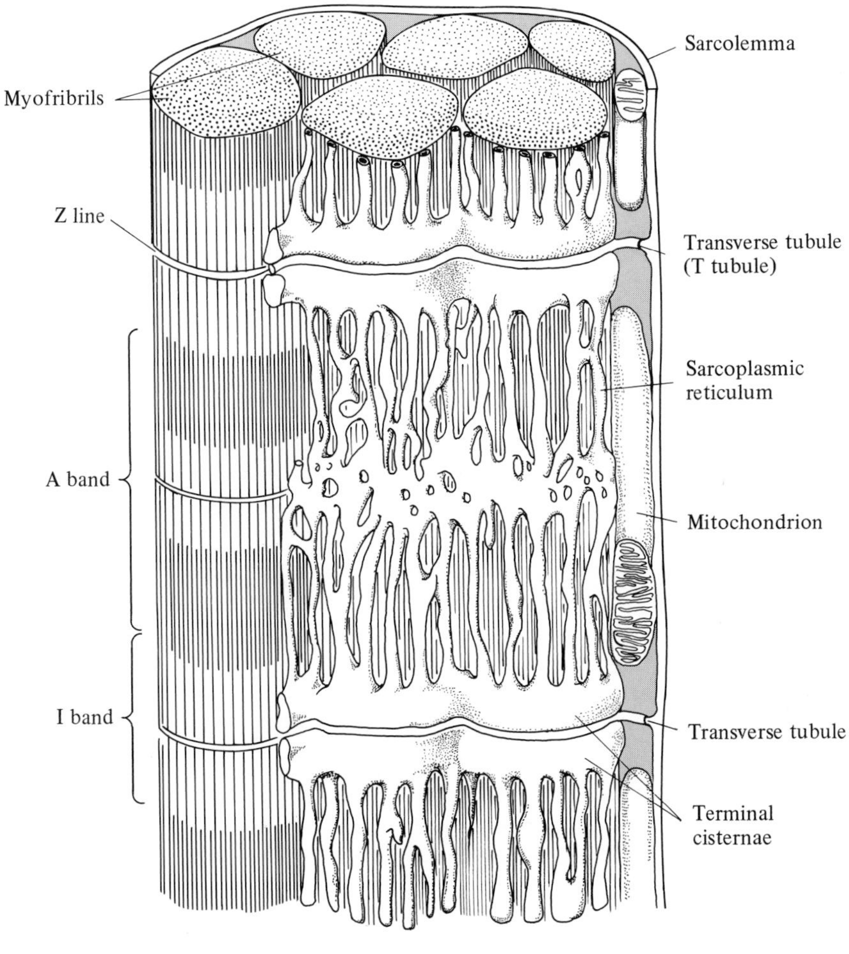


**FIG. 18-10** Relationship of the sarcoplasmic reticulum to the myofibrils of striated muscle. At the ends of each sarcomere, the SR forms terminal cisternae in this species. Each of the T tubules, inward extensions of the sarcolemma, are flanked by terminal cisternae.

## EARLY BIOCHEMICAL STUDIES

In the early 1930s, it was found that a moderately large protein molecule, **myosin** (molecular weight 500,000), could be extracted from muscle. Myosin accounts for about half of a muscle's protein content. Another fraction, containing the enzymes of glycolysis, was first removed by washing minced muscle in distilled water. The myosin fraction was then removed by prolonged washing in cold 0.6-molar potassium chloride solution. In time, it was found that myosin has enzymatic activity: it can hydrolyze ATP to ADP and inorganic phosphate, that is, it functions as an ATPase.

Some years later, another, smaller protein (molecular weight 42,000) was extracted and given the name **actin.** In the presence of ATP, the small, globular molecules of actin, or G actin, could be polymerized to form a long fibrous protein, F actin. Under the right conditions, myosin could also be made to form a long filament in the presence of actin. This substance was called **actomyosin.**

Actomyosin fibers—prepared by squirting the substance through a fine pipet—contracted when placed in a solution of $Ca^{2+}$ and ATP. This landmark achievement of the 1940s excited the scientific world. For the first time, the principal biochemical components of the contractile mechanism were clearly identified; they had been separately removed from the muscle cell, reassembled, and then made to perform their essential role of contraction.

In the years since the identification of actin and myosin as the principal molecules of vertebrate muscle, contractile elements composed of actin and myosin molecules have been discovered not only in all animals but in many protists and some plants as well. Actin and myosin appear to have evolved very early in the history of life on earth.

## CHANGES IN SARCOMERE STRUCTURE DURING STRETCHING AND CONTRACTION

Several important observations were made possible when comparisons were made of muscles prepared for examination by light and electron microscopy while (1) being stretched, (2) undergoing contraction, and (3) relaxed. Both light and electron microscopy of these preparations showed that A bands remained the same length under all three of these conditions. I bands, on the other hand, lengthened when the fiber was stretched and shortened during its contraction. Similar findings had been reported since the late 1800s but were not generally accepted until the 1950s.

***The Fine Structure and Chemistry of Myofibrils: Two Types of Filaments.*** Myofibrils that were sectioned transversely and viewed by electron microscopy were found to be composed of filaments of two thicknesses (Fig. 18-11). Extracting myosin from the preparations resulted in the disappearance of the thicker of the two filaments; removal of actin caused the disappearance of most of the thin filament structure.

In many cross sections of myofibrils, each thick filament (myosin) was surrounded by several fine filaments (actin). Exceptions were cross sections that passed through the center of the A band or through most regions of the I band. Sections through the center region of the A band usually revealed only the thick myosin filaments. Sections through the I band contained only the thin filaments of actin.

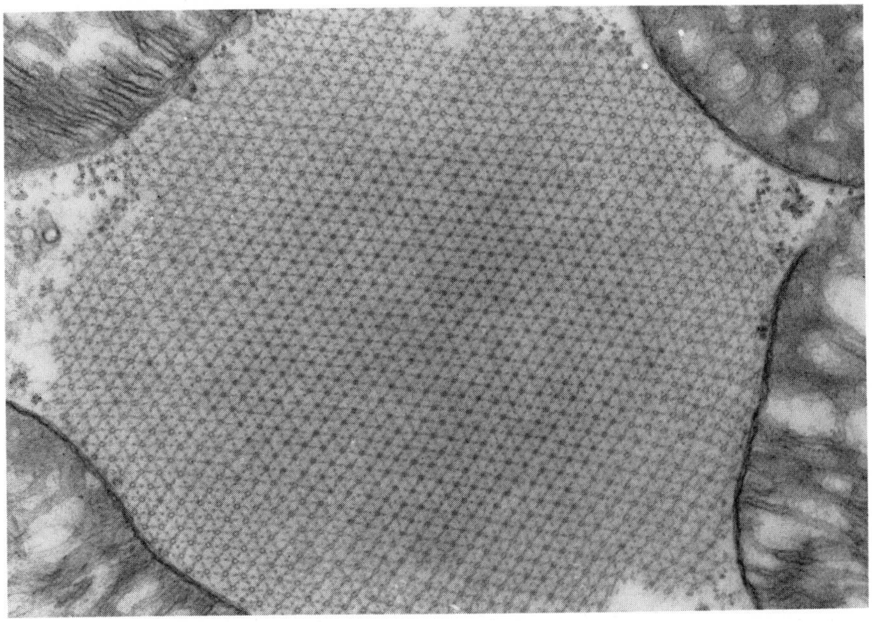

**FIG. 18-11** Cross section of striated muscle from a rat, uncontracted, or relaxed. The section was made through a dark portion of the A band and shows each thick filament to be surrounded by several thin filaments.

***The Constant Length of the A Band.*** Whether relaxed or contracted, the A band remains the same length (Fig. 18-12). When a myofibril in a relaxed (resting) condition is viewed by electron microscopy, the myosin and actin filaments are arranged as shown in Figure 18-12a. In the darkest regions of the A band, filaments of myosin and actin overlap. The light (H zone) area within the A band in rest-length myofibrils lacks actin filaments. The dark (M) line at the center of the H zone is due to cross connections between myosin filaments (Fig. 18-12a, b).

***The Changing Length of the I Band.*** Only actin filaments are found in the I band regions. If a fibril contracts, the I band shortens—as does the H zone—because of the further insertion of the actin filaments among the myosin filaments (Fig. 18-12b). Note that the Z lines, denoting the ends of the sarcomere, are much farther apart at rest than when the myofibril is contracted.

In the fully contracted state—usually achieved only if the myofibril has been freed of its connective tissue—the H band completely disap-

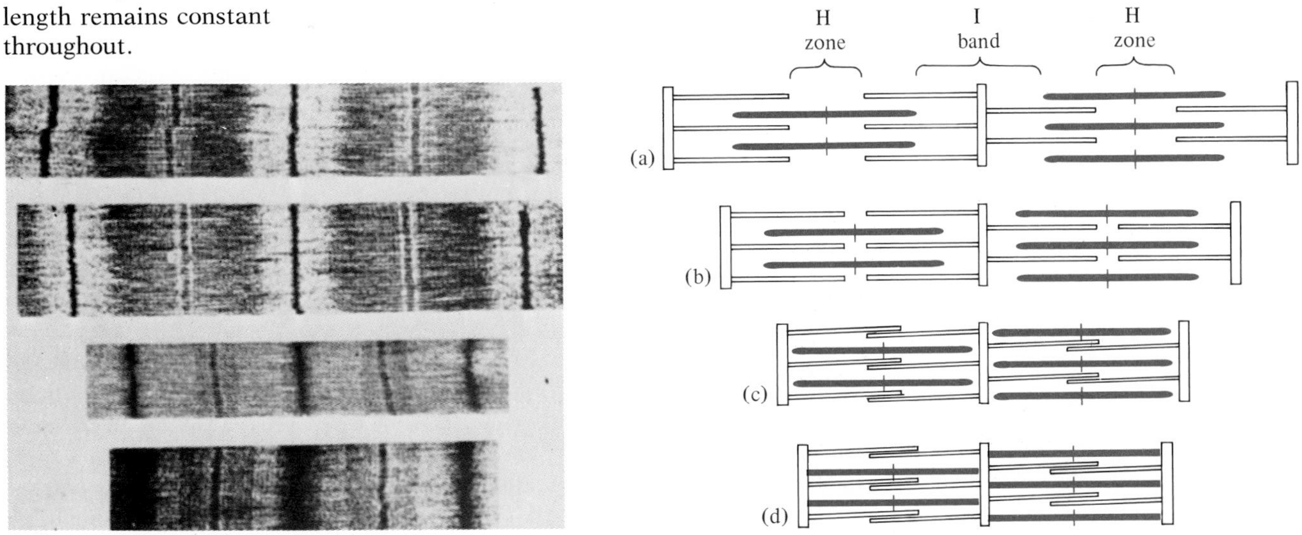


**FIG. 18-12** Changes in band length during contraction of striated muscle. (a, b) The H zone and I band shorten. (c, d) A new dense zone develops within the A band as thin filaments overlap each other, but the A band length remains constant throughout.

pears (Fig. 18-12c). In full contraction, the actin filaments meet and may even overlap at the center of the sarcomere (Fig. 18-12d). The I band almost completely disappears as well, and the Z lines approach the edge of the A band.

## THE SLIDING FILAMENT HYPOTHESIS

From these observations of the changes in sarcomere structure during stretching and contraction, largely the work of British biologists Andrew F. and Hugh E. Huxley, it became clear that the contraction of muscles is not due to a shortening of any of its component molecules—a view once widely held. Instead, myosin and actin filaments shorten the myofibril by actively sliding past each other. The myosin filaments can be regarded as retaining their position while the actin filaments move in reference to them. These discoveries changed existing theories of contractility and provided the molecular basis for what is known as the sliding filament theory of muscle contraction.

How can chemical energy be transduced into the mechanical energy needed to move the filaments past each other? What forces could cause such movement? What connections, if any, do the parallel thick and thin filaments have with each other? Some of these questions will be considered next.

***Cross Bridges Between Myosin and Actin.*** High-resolution electron micrographs of myofibrils show the existence of fine cross bridges between the myosin and actin filaments (Fig. 18-13). The bridges project at right angles from the surface of the myosin filaments at regular intervals except in a sizable central region in the H zone from which they are absent. They do not project in only one plane but form a helical (coiled)

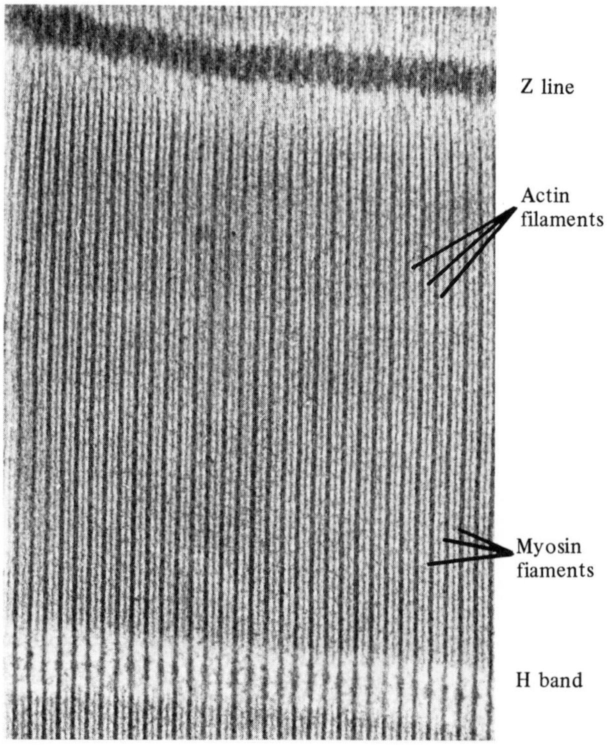


**FIG. 18-13** High voltage electron micrograph of a portion of a myofibril of a contracting flight muscle of a fly *(Drosophila)*. Cross bridges at regular intervals between actin and myosin filaments are barely discernible. Note the unusually long A bands and short I bands as compared to vertebrate muscles.

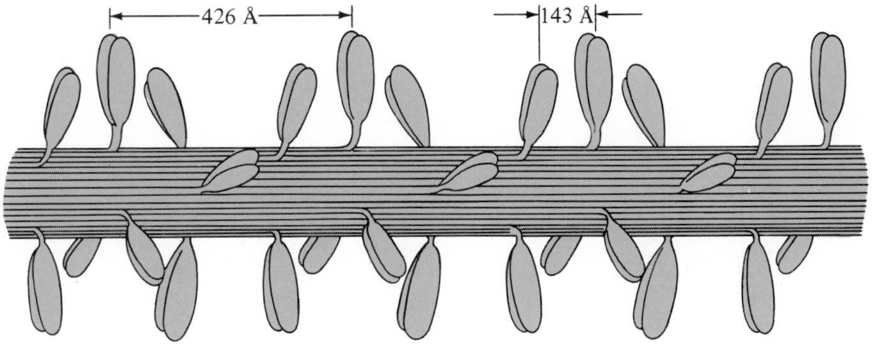


**FIG. 18-14** Diagram of cross bridges on a thick (myosin) filament of a striated muscle myofibril.

arrangement, with six cross bridges for each full turn of the helix (Fig. 18-14). Each of the six thin (actin) filaments surrounding any one thick (myosin) filament is touched by that thick filament as well as by the thick filaments on its other sides. Each of the six cross bridges in one full turn of the helix is bonded to the nearest of the six actin filaments surrounding it.

***The Proteins of Muscle.*** The myosin molecule as seen in an electron micrograph is rodlike, with a globular head at one end (Fig. 18-15). Each complex, made up of a globular head and a small part of the shaft, termed **heavy meromyosin,** has ATPase activity. Most of the rodlike part of the myosin molecule is called **light meromyosin.** Units of several myosin molecules become assembled in solution with the rodlike portions at the center and the globular heads projecting at both ends of the unit in the helical arrangement referred to above. This suggests that filaments of myosin in the intact muscle are formed in a similar way.

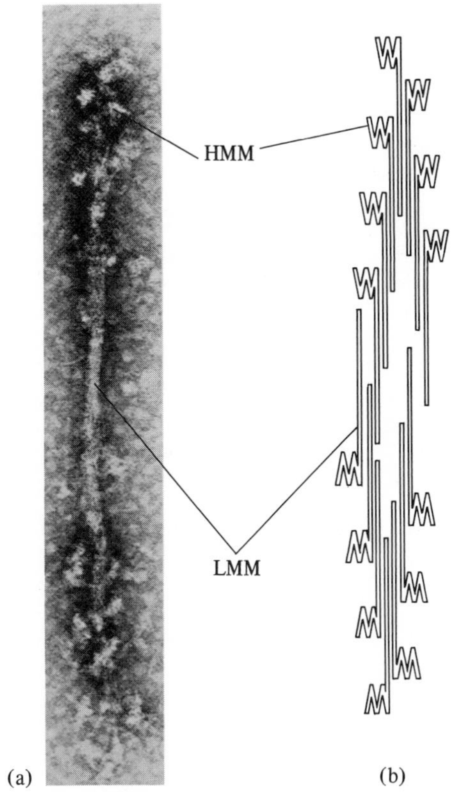


**FIG. 18-15** The structure of myosin molecules. (a) Electron micrograph of an aggregation of myosin molecules showing cross bridges at both ends of the aggregation. (b) Diagrammatic representation of possible molecular structure of the aggregate. The heavy meromyosin (HMM), which forms the globular head, is represented by a zigzag line; the light meromyosin (LMM) portion, which forms the rod, is shown as a straight line. When muscle contracts, molecules slide past each other.

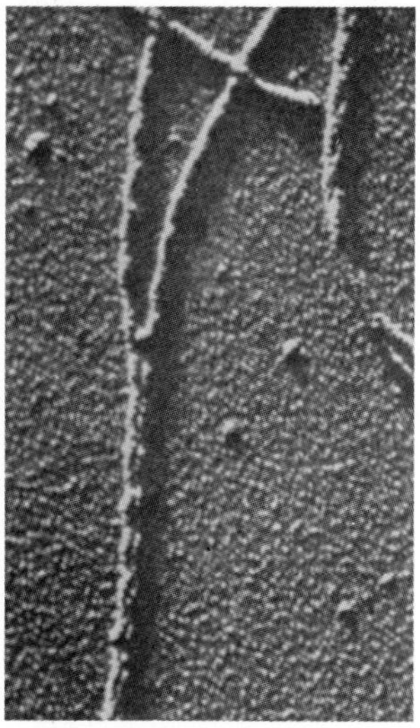

FIG. 18-16 Electron micrograph showing the helical form of actin molecules.

Actin filaments are made up of a double helical chain of globular actin molecules, resembling two strings of pearls twisted about each other (Fig. 18-16). The relationships described thus far are summarized in Figure 18-17.

Biochemical analysis of actin filaments has revealed the presence of two additional proteins: tropomyosin and troponin. Both are present in small amounts in the intact muscle. **Tropomyosin** is a long, fibrous protein, one molecule of which lies in each of the two grooves formed by the two twisted filaments of actin. The presence of tropomyosin apparently gives rigidity to the actin molecules. The **troponin** molecules are complexes of three polypeptides incorporated into the actin-tropomyosin filament at regular intervals (Fig. 18-18). In the absence of $Ca^{2+}$, actin-myosin cross-bridge formation is blocked by tropomyosin. But when $Ca^{2+}$ concentration is high, $Ca^{2+}$ binds to troponin, causing tropomyosin to move aside and expose binding sites at which the actin-myosin cross bridges can form.

***Summary of the Sliding Filament Theory.*** From the known facts about muscle proteins, the sliding filament theory of muscle contrac-

FIG. 18-17 Diagrammatic representation of the structure of striated muscle from the gross to the molecular level.

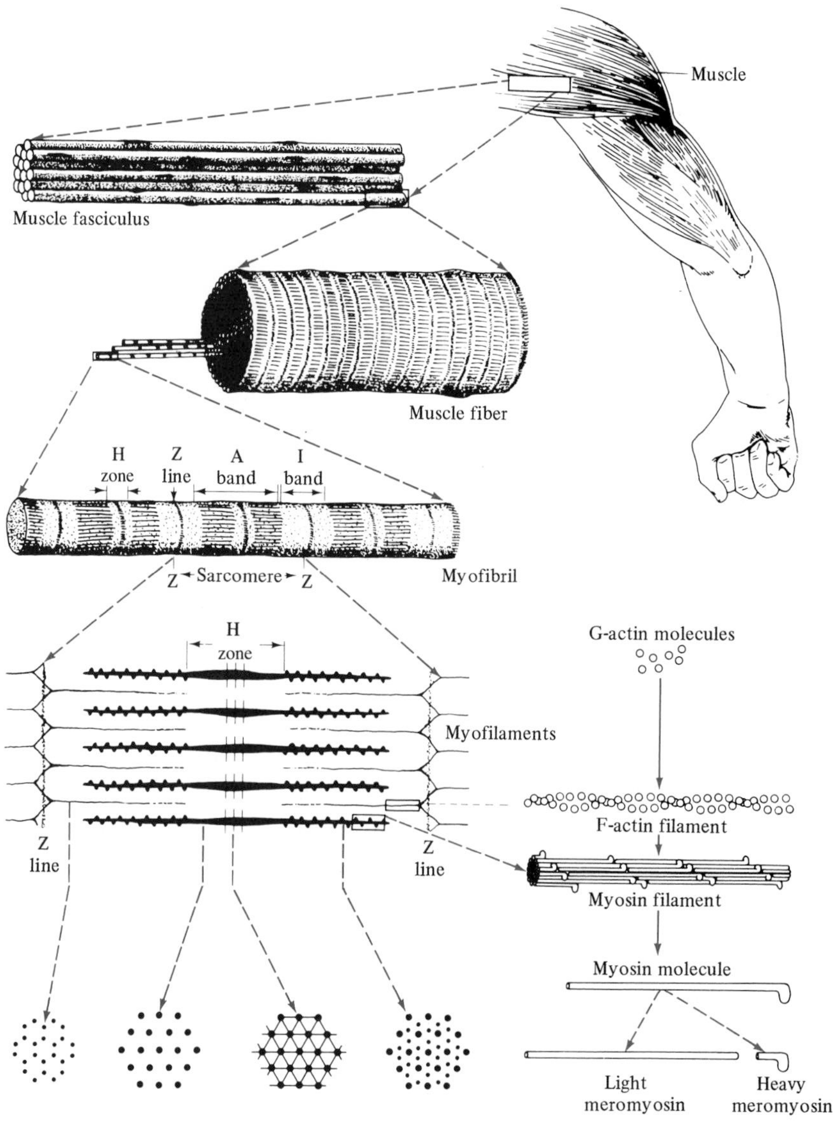

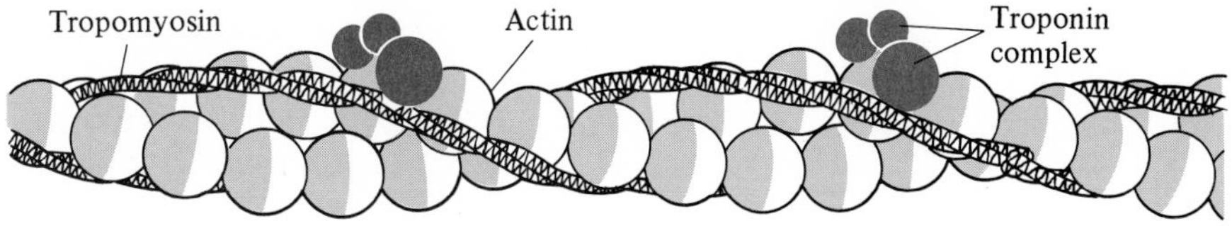


**FIG. 18-18** Diagram of the molecular structure of a thin (actin) filament, showing the possible arrangement of tropomyosin and the complexes of troponin molecules along the actin molecule.

tion has been developed. According to this theory, muscle contraction is believed to occur in the following way:

1. A series of nerve impulses arrives at a motor end plate, releasing acetylcholine into its specialized synaptic cleft (Fig. 18-19a).
2. A local depolarizing electrotonic potential, the end plate potential, is generated, followed, in every case, by an action potential that sweeps across the entire surface of the muscle fiber, including its invaginations, the T tubules (Fig. 18-19b).
3. The depolarization of the T tubule membrane affects the adjacent terminal cisterna of the sarcoplasmic reticulum by increasing its permeability to $Ca^{2+}$. Because $Ca^{2+}$ is 500 times more concentrated within the SR than in the sarcoplasm, the increase in the membrane's permeability to $Ca^{2+}$ results in a rapid diffusion of $Ca^{2+}$ into the sarcomere (Fig. 18-19c).
4. The flood of $Ca^{2+}$ binds calcium to troponin, which displaces tropomyosin, permitting the globular heads of the myosin molecules to extend outward to form temporary bridges with adjacent actin filaments (Fig. 18-20).

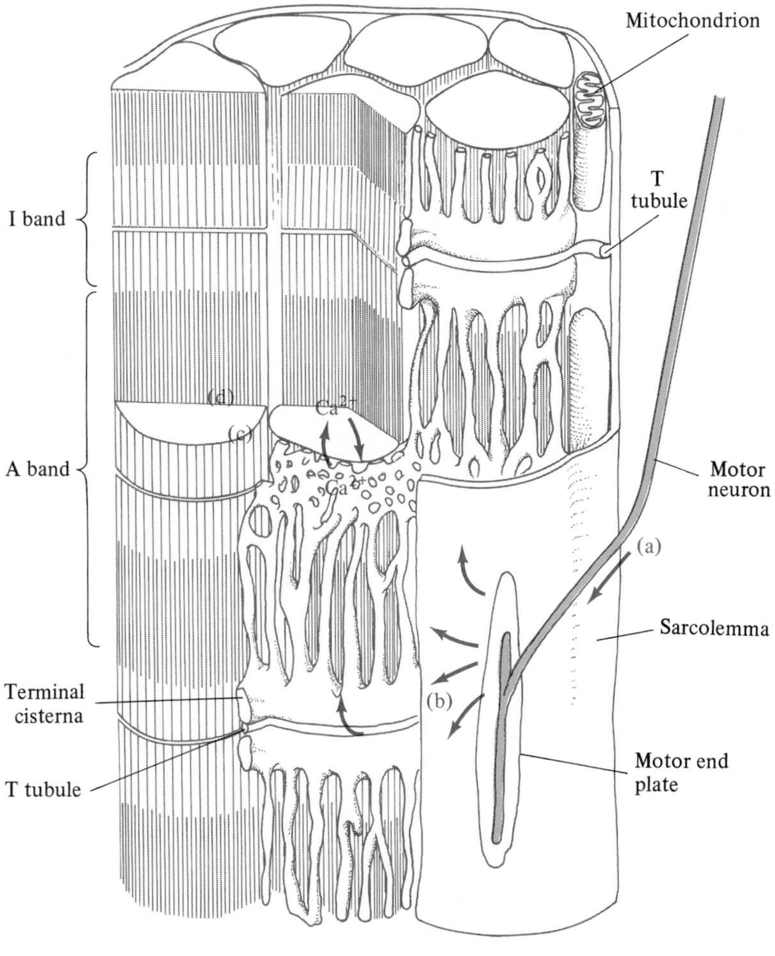


**FIG. 18-19** The sequence of events resulting in muscle contraction. Arrows indicate the course of events. (a) A nerve impulse arrives at the terminal branch of a motor neuron. (b) An action potential is generated in the plasma membrane of the muscle fiber and sweeps into its T tubules. (c) As a result, the permeability to $Ca^{2+}$ of the adjacent terminal cisternae of the sarcoplasmic reticulum is increased; a flood of $Ca^{2+}$ is released from the SR into the sarcoplasm. The $Ca^{2+}$ binds to troponin, displacing tropomyosin, thereby permitting actin to join temporarily with the myosin heads: ADP and Pi attached to the heads are released, creating stress on the heads that is relieved by a slight bending of the heads. This process is repeated in the bundle of myosin molecules 50 to 60 times per second. (d) A few milliseconds after its release, $Ca^{2+}$ begins to be reabsorbed by the SR.

5. When the globular head of a myosin molecule comes in contact with an actin filament, ADP and inorganic phosphate molecules associated with the myosin head are released. This changes the shape of the myosin molecule, causing the myosin head to bend backwards. This in turn pulls the actin molecule to which it is attached toward the center of the A band.
6. As all the cross bridges exert this pull at essentially the same time throughout the myofibril—and, indeed, throughout the entire muscle fiber—the actin filaments are forced to slide past the myosin filaments. The result is that each sarcomere contracts very slightly, no more than the length of a cross bridge (Fig. 18-20). This interaction between actin and myosin is immediately repeated, however, dozens of times per second.
7. In a healthy, rested muscle, ATP is regenerated as rapidly as it is split. As soon as an ATP molecule binds to the myosin head, the cross bridge formed by the head detaches from the actin and immediately snaps back to its original shape. The myofibril, however, at least for a moment, remains slightly contracted. If $Ca^{2+}$ is still present, as it normally is, the process is immediately repeated, occurring at an estimated rate of 50 to 60 times per second. The result is that the actin filaments continue to slide past the myosin filaments and each sarcomere continues to shorten. This shortens the myofibril, thereby producing contraction of the muscle.

## RELAXATION OF MUSCLE

A few milliseconds after a muscle has received a stimulus adequate to generate a contraction, the flood of $Ca^{2+}$ that is released from the SR

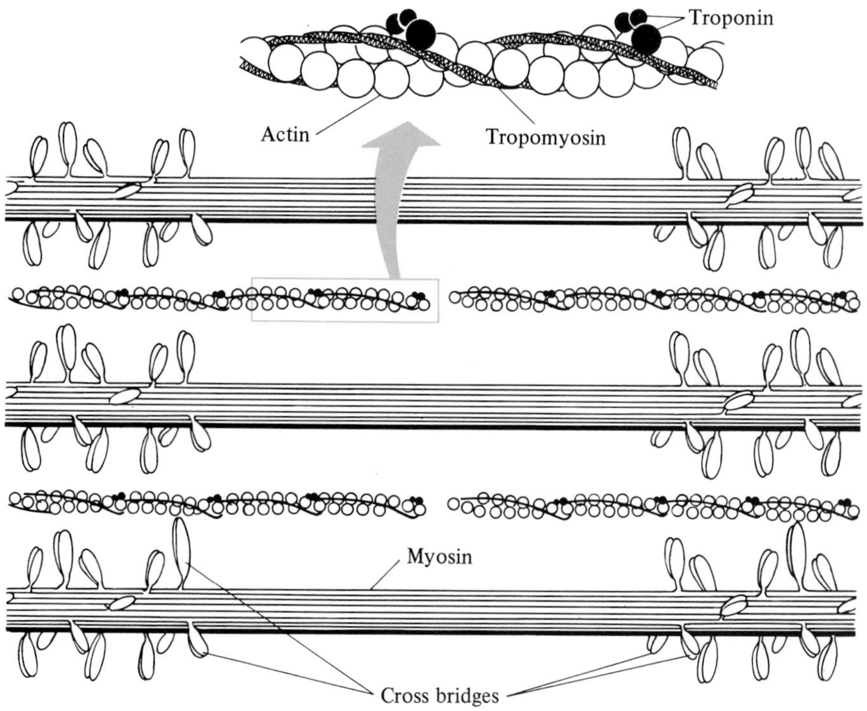


**FIG. 18-20** Probable relationship of actin and myosin filaments to each other during contraction. Note the two connecting cross bridges in the lowest figure.

begins to be reabsorbed if no further stimulation is applied. As a result of this withdrawal of $Ca^{2+}$ from the sarcoplasm, troponin no longer prevents tropomyosin from shielding the binding sites on actin. The actin and myosin filaments stop interacting, and the muscle slowly relaxes. The mechanism by which $Ca^{2+}$ is reabsorbed involves an ATP-using active process similar to the $Na^+/K^+$ pump. Thus, both contraction and relaxation require energy.

## MOVEMENT OF CILIA AND FLAGELLA

Ciliary movement has probably been a source of fascination and curiosity ever since Antoni van Leeuwenhoek watched protozoans swimming about under his microscopes in the late 1600s. A brief description of what is known of the basis of this type of movement is appropriate here.

### MICROTUBULES

Chapter 4 explained that cilia and flagella exhibit a characteristic internal arrangement in which two single **microtubules** are surrounded by nine double microtubules, or doublets (Fig. 18-21). Microtubules are composed of two similar globular proteins called **$\alpha$ and $\beta$ tubulins.** These proteins have the general property of polymerizing into fine, threadlike structures called **protofilaments.** Thirteen protofilaments form a complete microtubule. The doublets, however, contain one complete microtubule, called subfiber A, and one incomplete tubule, subfiber B, containing 10 protofilaments instead of the usual 13. At intervals along the edges of subfiber A—except at the very tip and base of the cilium—are projections composed of the protein **dynein,** which exhibits ATPase activity. According to several lines of evidence, these projections, called dynein arms, are involved in the generation of ciliary movement. The actual beating of the cilia, however, is more directly accomplished by the microtubules.

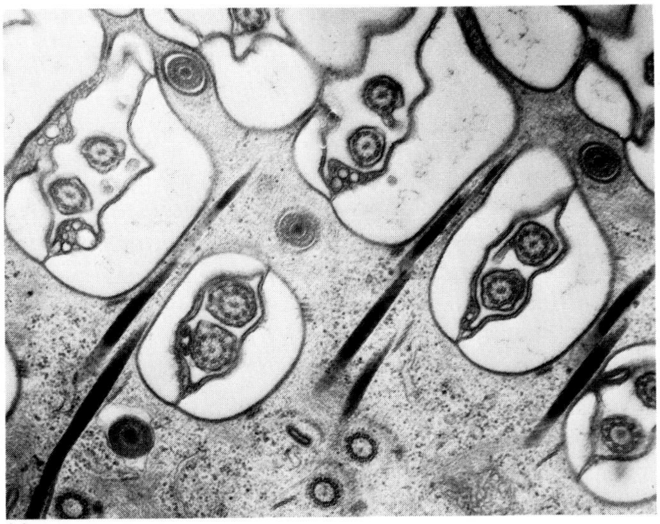
(a)

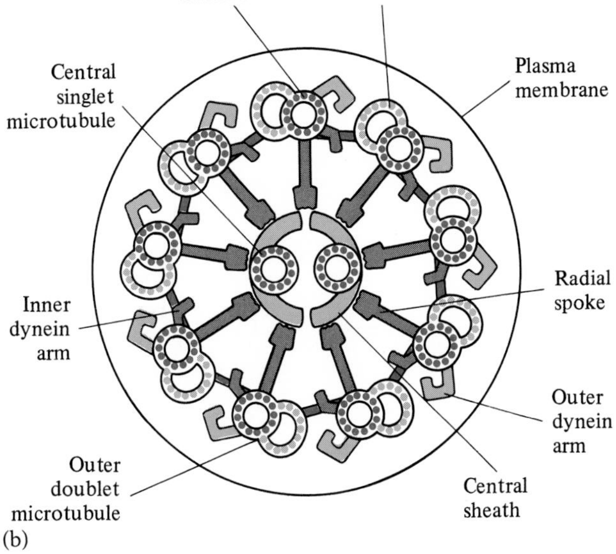

(b)

**FIG. 18-21** (a) The paired cilia of a protozoan *(Paramecium)* seen in cross section. Each pair emerges from a depression in the pellicle. At the lower center can be seen a pair of basal bodies which, like centrioles, contain only 9 triplet sets of microtubules whereas the cilia contain 9 doublets plus a central pair of microtubules. (b) Sketch showing the anatomy of a microtubule.

PANEL 18-1

# Microfilaments and Microtubules Involved in Nonmuscular Movements

## MICROFILAMENTS

As noted earlier, actin and myosin, often in a combined form, occur in nonmuscular cells of many eukaryotic organisms. These cells may contain as much as 10 percent of actin, myosin being found in smaller quantities. A few examples of such cells will be of interest here.

a. **Microfilaments** of about 70 Å in diameter occur in nearly all eukaryotic cells. Antibodies prepared against actin or tropomyosin extracts and labeled with a fluorescent dye are used to detect the presence of such microfilaments. When a tissue culture of embryonic cells (fibroblasts) of human skin is treated in this way, long filaments become dramatically evident, as seen in the first of the accompanying micrographs. Antibodies prepared against myosin have a similar effect on nonmoving cells. Actively moving cells are seen to have a diffuse distribution of myosin granules, however.
b. Microfilaments constitute the so-called **cytoskeleton,** which maintains the shape of many cells, sometimes in configurations that closely resemble geodesic domes.
c. The microvilli of the brush border of the intestinal lining contain a network of microfilaments; the microfilaments are attached to the inside of the plasma membrane all along the length of each microvillus, as seen in the second of the accompanying micrographs. Thick filaments of myosin are also present in this case. This system is involved in contractions of microvilli associated with the absorption of nutrients (see Chapter 9).

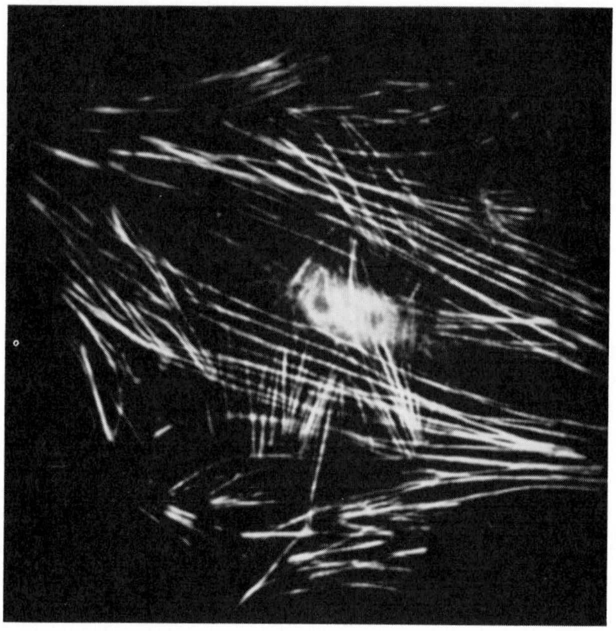

Distribution of microfilaments as revealed by preparing antibodies against actin extracts, labeling them with a fluorescent dye, and then reacting them with cells in tissue cultures of embryonic cells (fibroblasts) of human skin. Tropomyosin is apparently associated with the actin. When the process is repeated by preparing dye-labeled antibodies against myosin, a similar picture results in the case of nonmoving cells. In a moving cell, however, myosin staining is diffuse. These filaments are not permanent, but rapidly reorganize during cell movements.

## CILIARY MOVEMENT

Although an earlier theory had suggested that movement of cilia (Fig. 18-22) was due to contraction of their microtubules, electron micrographs have failed to show any evidence of tubular shortening in cilia "frozen" during movement by treating them with a killing reagent that instantaneously coagulated their proteins. On the contrary, micrographs support another theory, which proposes that the microtubule doublets actively slide past each other within the cilium. Examination of a series of cross sections of bent cilia shows that, indeed, the tubules on the concave side of a bent cilium project farther into the tip than do the others (Fig. 18-23). If they had contracted, they would be shorter and probably would have thickened and at least would not extend farther

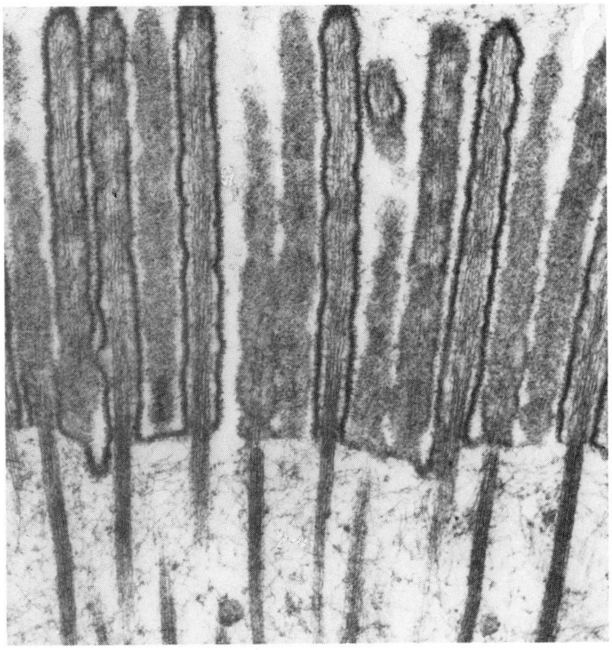

Filaments of the brush border of the intestinal lining of the chicken. Higher magnification shows the filaments to be attached to the inside of each villus by cross bridges.

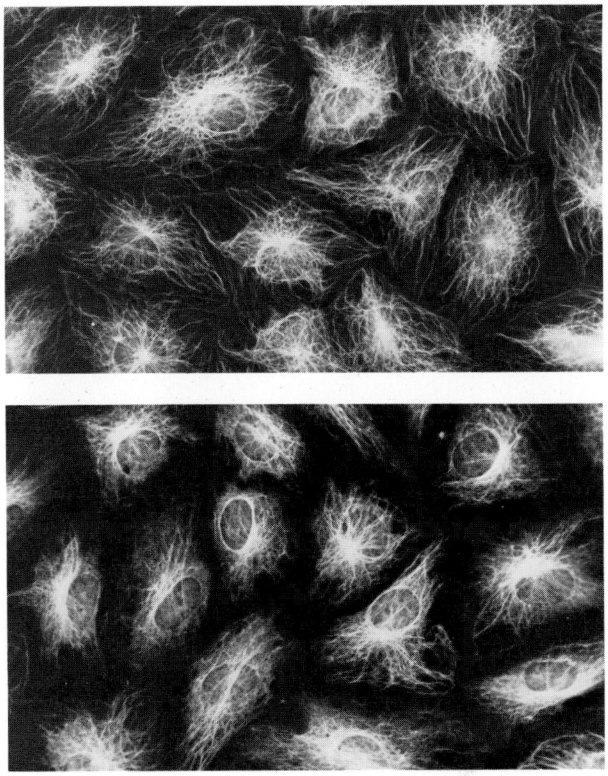

Microtubules and microfilaments of hamster cells in tissue culture. In the upper picture the cells have been reacted with fluorescent dye-labeled antibodies prepared against tubulins, the proteins of which microtubules are composed. In the lower picture, cells from the same strain were reacted with antibodies prepared against the protein of microfilaments. Do you see any differences in the two?

## MICROTUBULES

Microtubules, such as those that provide the basis for ciliary movement, also play contractile roles in nearly all eukaryotic cells. As described in the section on ciliary movement, microtubules are composed of proteins called $\alpha$ and $\beta$ tubulins. The outside diameter of microtubules is about 240 Å, as compared to about 70 Å for microfilaments. In addition to their functions in ciliary movement, microtubules constitute the spindle fibers seen in mitosis of most eukaryotic cells. They also compose the contractile, axial filaments found in the pseudopodia of certain protozoans. Motile cells, such as the fibroblasts in tissue cultures of embryonic cells of birds and mammals, have microtubules distributed throughout them as seen in the third of the accompanying illustrations.

than the others. Furthermore, cilia that have been briefly treated with protein-digesting enzymes, so as to disrupt some of the connections between the microtubular elements, bend in the presence of ATP. In similarly treated preparations of isolated cilia from which the plasma membranes have also been removed, the microtubules slide past each other when ATP is added. It thus appears that the microtubules of cilia are capable of active sliding movements in the presence of ATP.

In some preparations, spokelike projections can be seen to extend from the peripheral doublets to the central pair of microtubules. What look like spokes apparently act as ratchets during ciliary movement, that is, devices that prevent back-slipping. The precise mechanism by which the sliding movements of microtubules are converted into the bending movements of cilia is not fully understood.

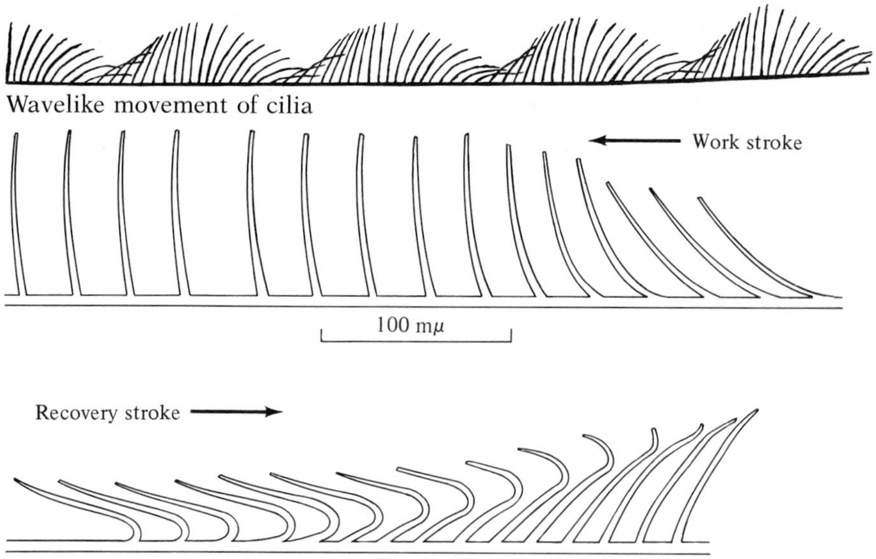


**FIG. 18-22** Sketches based on high-speed photographs of ciliary movement. The cilium is kept stiff during the work stroke (from right to left in the center sketch). During the recovery stroke (from left to right in the bottom sketch), the cilium is flexed, offering less resistance to the water.

## PRODUCTION AND DISCHARGE OF ELECTRICITY

One of the most fascinating of all abilities of animals is the ability to generate an electric charge. Several quite unrelated groups of fishes have developed this capability. In many cases, the currents produced are weak and are useful only for electrolocation, that is, determining the size and location of other objects by detecting the disturbances they produce in an electric field. But in several marine and freshwater fish, the current is powerful and stunning. The ability to generate such elec-

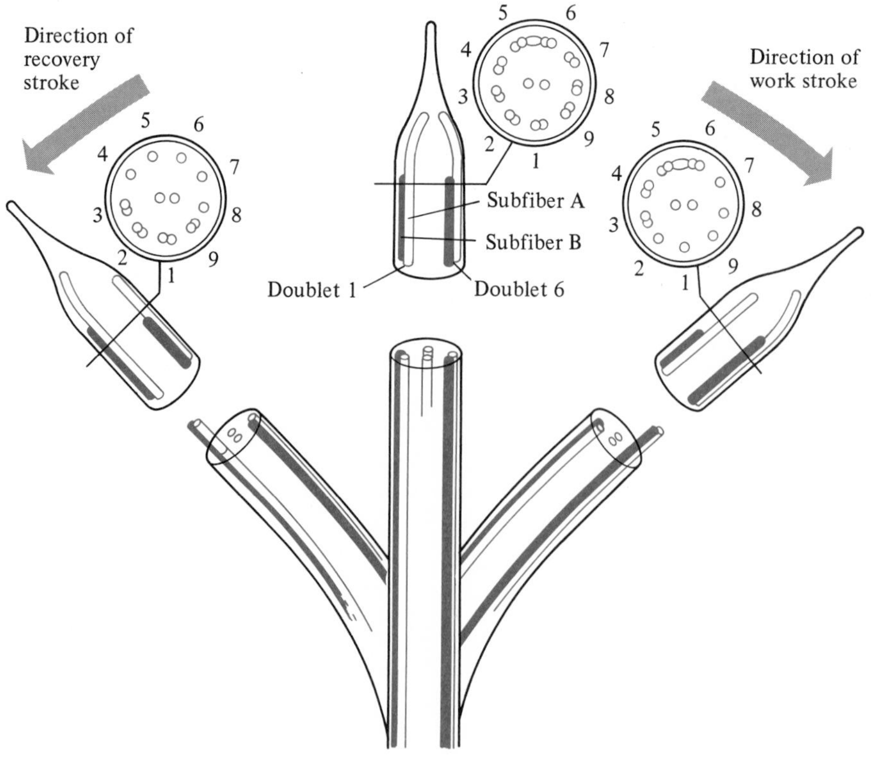


**FIG. 18-23** Evidence that the beating of cilia is accomplished by the active sliding of microtubules. Sketches of sections through the tips of cilia are shown for the recovery stroke (left), straight up (center), and work stroke (right). Compare doublets 4, 5, and 6 in each of the three positions. Only single microtubules appear in the tip on the left (recovery stroke) side, complete doublets in the other two. Also compare doublets 9, 1, and 2 with each other in each of the three positions.

trical discharges has evolved independently in at least six different groups of fishes, a striking example of evolutionary convergence.

## THE ELECTRIC EEL

A type of effector that is quite unlike muscle in its function is the **electric organ,** a structure that has been independently evolved by several genera of fishes. The organ is of interest here because it is actually composed of modified muscle.

***Structure of the Electric Organ.*** In the South American electric "eel" *Electrophorus* (which is not really an eel at all, but an elongate fish with a superficial resemblance to an eel) (Fig. 18-24a), the electric organ constitutes 40 percent of the body volume and is served by some 200 nerves (Fig. 18-24b). The organ consists of units called **electrocytes,** modified muscle cells arranged in longitudinal columns parallel to the spine and running from head to tail (Fig. 18-25). There are up to 10,000 cells per column, and the fish has about 60 of these columns on each side of its center line (Fig. 18-26). Electrocytes are plate-shaped cells that have two surfaces. One surface, the electric layer, is richly supplied with axon branches; the opposite surface, called the papillary layer, is well supplied with blood vessels.

***Mode of Action of the Electric Organ.*** The electrocytes of an organ are all stacked facing the same way: in one common arrangement, each cell's electric layer is on the side toward the tail, whereas the papillary layer faces the head. When not being stimulated, the electrocytes maintain a resting potential of about 84 mV (the inside of the cell is negatively charged) for both their upper and lower surfaces. Upon stimulation, only the electric layer responds. It generates a spike potential that reverses the −84 mV to a +67 mV. In other words, it not only depolarizes but becomes oppositely charged. The papillary layer does not depolarize at this time, but maintains its resting potential of −84 mV. This results in a potential difference of about 150 mV between the two sides

(a)

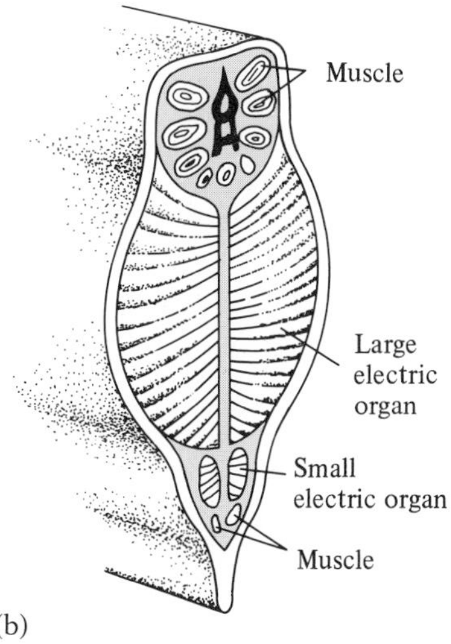

(b)

**FIG. 18-24** (a) *Electrophorus,* the electric eel of South America. (b) Section through the tail of the South American electric eel, showing the electric organ. Typical muscle is found dorsally and ventrally, but about half of the tail musculature is modified to form the electric organ.

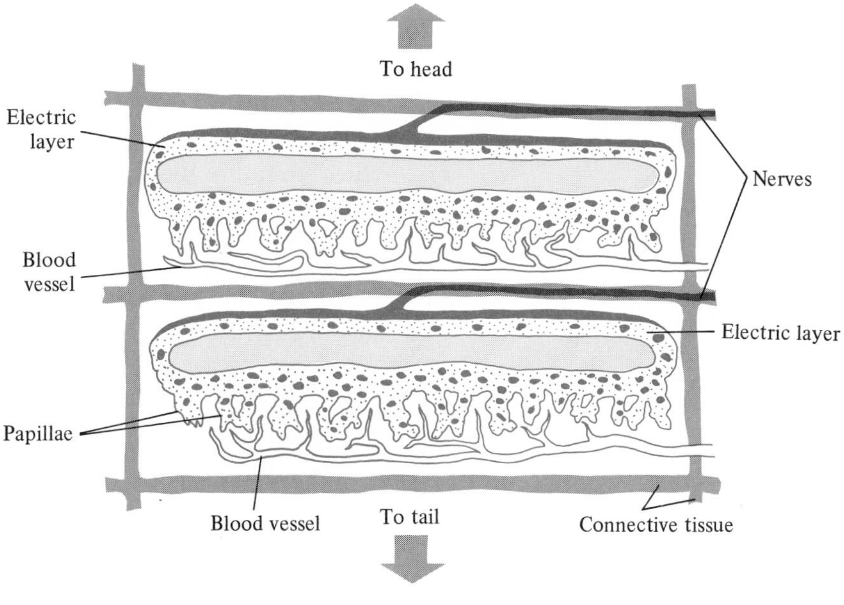


**FIG. 18-25** Structure of two neighboring electroplates of an electric organ.

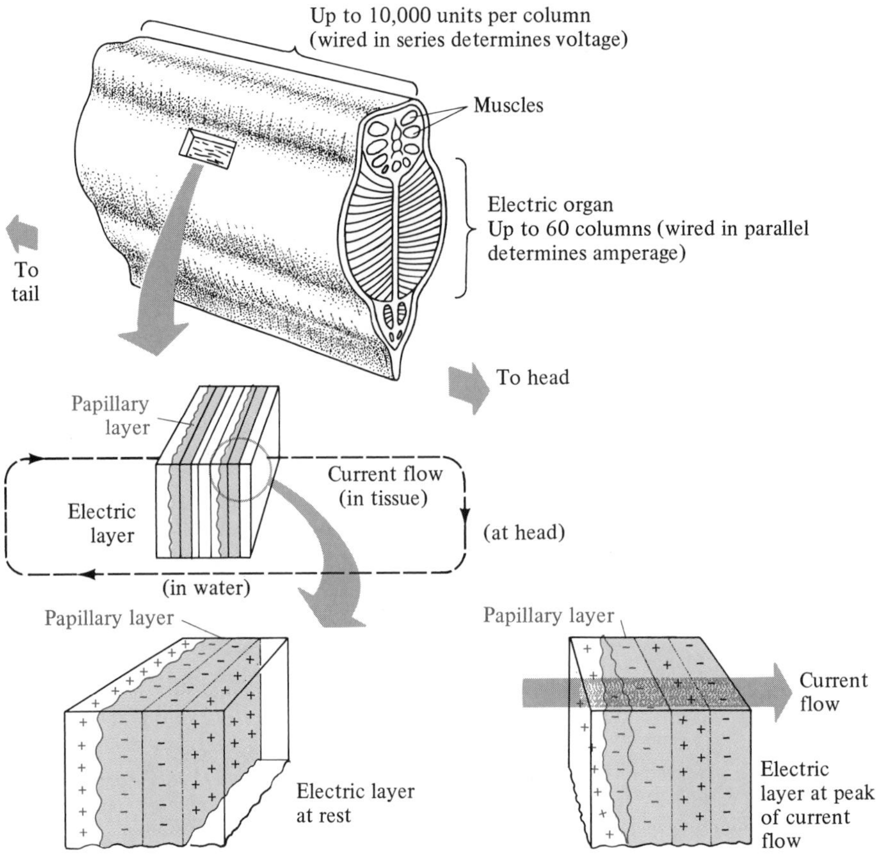


**FIG. 18-26** The discharge circuit of an electric eel.

of the same cell. As a consequence, a current flows from one side of the electrocyte to the other.

Immediately above and below each cell responding as described above is another cell exhibiting a current flow in the same direction. The electrical resistance of the plasma membranes of electrocytes is extraordinarily low, only 0.2 ohms/cm$^2$ in the papillary layer side.* This provides a conductance (the readiness with which an electric current flows through a conductor) up to 1,000 times greater than that of the ordinary plasma membrane of a muscle cell. Because of the way in which they are arranged, all the cells in a column are "wired in series" with each other. Like a long row of tiny batteries wired in series, the individual voltages they produce are added together. The result is a considerable flow of current from tail to head within the organ and from head to tail in the surrounding water—or through any animal that comes close enough to serve as a conductor of the discharge.

***Synchronization of Discharge.*** One of the questions that occurred to physiologists studying electric organs was, how does an 8-foot long fish manage to activate all its electrocytes at precisely the same moment? (If they fired out of synchrony with each other, the system would fail.) The problem is solved by a combination of methods, including the following: First, the neurons serving the electrocytes nearest to the brain are longer than they need to be to reach those cells. This delays the arrival

* Resistance is the property by which a conductor opposes a flow of electric current through it. The unit of resistance is the ohm, defined as the potential difference in volts divided by the current, in amperes. The resistance of a given conductor would be determined by the voltage needed to drive a given amount of current through it.

of impulses at these cells. Second, the more distant of the electrocytes are supplied by axons of unusually large diameter, which shortens the time required for action potentials to reach the more distant cells.

### VOLTS AND AMPERES

The freshwater electric eel, *Electrophorus*, has up to 10,000 cells per column, making it possible to generate up to 500 volts, nearly five times the power of the average household current, which is 110 volts. Most contacts with *Electrophorus* do not result in a shock of this magnitude, but far less is usually enough to stun their prey or predators. Marine electric fish, however, do not need this much power; because the electrical conductivity of seawater is much higher than that of fresh water, weaker voltages suffice.

The arrangement of electrocytes in series accounts for the high voltage produced. The amount of current, measured in amperes, that can flow through a shocked animal depends on the number of "parallel wired" columns in the electric organ. Because of the relatively low number in the electric "eel"—60 columns of 10,000 electrocytes on each side of the spine—only about 1 ampere can be generated. By contrast, while the electric ray, a marine relative of the shark, has only about 1,000 electrocytes per column, it possesses 2,000 columns. Thus, although it generates only about 20 to 30 volts, it produces a very high amperage, generating a power of up to 1,000 watts (volts × amperes). Most lamps ("light bulbs") used in homes draw only 60 to 100 watts of power.

## Summary

Vertebrate muscle is of three types: cardiac, or heart muscle; smooth muscle, also known as visceral or involuntary muscle; and striated muscle, also known as skeletal or voluntary muscle. Most muscles are arranged in antagonistic sets, in which the action of one set of muscle fibers is opposed by that of another. Muscle fibers are organized into motor units of three to 150 fibers, served by the branches of one axon. Motor units usually alternate in effecting muscular contraction, temporarily resting when fatigue occurs at motor end plates.

Muscle contractions are of several types, depending on the conditions under which they are performed. Isometric contractions are those in which the muscle is not allowed to shorten, as in the tonic contractions normally maintained by healthy muscles. In isotonic contractions, it is the degree of tension that is constant. Tetanus is a sustained, maximal contraction of a muscle. Most muscular contractions are sustained but incompletely tetanic; brief twitches, or single, full contractions, are rare in normal, healthy muscle. Tetanic contractions sum. Because each new contraction is added to the residue of the previous one, a muscle undergoing tetanic contraction shortens much more than if only one brief volley of nerve impulses were delivered to it. If stimulation is applied at a sufficiently high frequency, complete tetanus, or fusion, occurs.

Under the light microscope, skeletal muscle exhibits cross striations of dark (A) and light (I) bands, so named because they are anisotropic (birefringent) and isotropic, respectively. The banding pattern is due to the arrangement of banded myofibrils within each fiber. Each myofibril is composed of units called sarcomeres, each of which extends from the dark Z line in the middle of one I band to the Z line of the next I band.

Electron microscopy reveals that the endoplasmic reticulum (sarcoplasmic reticulum) of striated muscle fibers is an extensive system of tubules, which is in intimate contact with the sarcomeres of each myofibril. Myofibrils are composed of two major types of filaments: thin filaments of the protein actin and thick filaments of the protein myosin. Comparison of relaxed, stretched, and contracted myofibrils shows that during contraction the thin filaments slide past the thick ones. Fine cross bridges extending from the myosin filaments to the actin filaments form the basis for the movement. Two other proteins, troponin and tropomyosin, are involved in this sliding process. At rest, tropomyosin covers the sites on actin to which the cross bridges bind.

Muscle cells are induced to contract by the release of acetylcholine at their motor end plates. This depolarizes the sarcolemma (plasma membrane), generating an action potential that sweeps across the muscle fiber in all directions. The action potential apparently increases the sarcoplasmic reticulum's permeability to $Ca^{2+}$, which promptly causes $Ca^{2+}$ to enter the sarcomere. The entering $Ca^{2+}$ binds to troponin, with the result that tropomyosin no longer covers the binding sites that permit actin and adjacent myosin molecules to form cross bridges. At the same time, ADP and inorganic phosphate are released, inducing a change in the shape of the bridges. This forces the actin filaments to slide along the myosin filaments, thus shortening the fibril. The process is repeated throughout the fiber 50 to 60 times a second, shortening the entire muscle fiber. As $Ca^{2+}$ is actively reabsorbed from the sarcoplasm, the filaments cease interacting and the muscle fiber slowly returns to its relaxed length.

Some animals and some animal tissues respond to stimuli, not with muscular contraction, but with ciliary movement. Present evidence suggests that ciliary movement is based on the active sliding of microtubules past each other within the cilium.

In addition to the weak currents associated with the generation and propagation of the nerve impulse, certain fishes can produce a powerful external electric current capable of stunning prey and predators. The electric organ is usually a modification of some component of the neuromuscular apparatus: motor neurons, motor end plates, or muscle fibers. In the electric "eel," the organ is formed from the modified ventral half of the tail musculature; this fish can generate a shock of about 1 ampere of current delivered at up to 500 volts. The shock produced by the marine electric ray, while of low voltage, is a high-amperage current of up to 1,000 watts of power.

## Recommended Readings

A list of recommended readings appears at the end of Chapter 15.

# 19 Principles of Animal Behavior

OUTLINE



The actions of animals have probably always interested humans. For humans to kill or capture animals for food and to escape from predators required some knowledge of animal behavior (Fig. 19-1). Moreover, the qualities we find in domesticated animals were probably selected not only with a keen awareness of behavioral traits but with some knowledge of their heritability.

Ancient knowledge of animal behavior, however, was often derived from anecdotes passed down through generations. These were not always accurate and had a strong element of anthropomorphism, the attribution to animals of the emotions and motives of humans. Ultimately, it was out of this kind of lore, under the strong influence of Charles Darwin's *The Origin of Species* (1859),* that the science of animal behavior, called **ethology**, arose.

* The full title of Darwin's best-known book is *The Origin of Species or the Preservation of Favoured Races in the Struggle for Life.*

**FIG. 19-1** Early humans used animals as a source of food, clothing, and other materials. A knowledge of animal behavior was important in capturing animals and escaping predators.

## THE SUBJECT MATTER OF ETHOLOGY

Ethology (from the Greek *ethos,* "habit" or "convention") is the study of the various causes of specific animal behaviors such as those exhibited in feeding, reproduction, aggression, and migration. Individual behavior patterns are usually described from a stimulus-and-response point of view by noting what specific stimuli trigger specific reactions. Ethologists also try to learn what motivates a particular response, its underlying mechanisms, its survival value to the animal exhibiting it, and its evolutionary history and significance. Traditionally, ethologists have studied **stereotyped behavior,** commonly called instinctive behavior, which varies little from individual to individual and from generation to generation within a given species. In recent years, however, some ethologists have also become interested in learned behavior.

Formerly, the study of **learned behavior,** the modified behavior that develops as a result of experience, was left entirely to psychologists. At one time, the two schools of thought were sharply divided on which type of behavior, stereotyped or learned, was of most—or of any—importance in the day-to-day activities of animals. At times, strong criticisms were exchanged. Now, however, psychologists and ethologists sometimes find themselves studying the same phenomena, although usually from different points of view.

The branch of psychology of most interest to the ethologist is the comparative study of learned animal behavior. Such studies examine particular phenomena as they are manifested in a wide range of forms, from the more primitive to the more highly evolved species. Their focus is on the evolutionary process. This and the following chapter will note some of the contributions of comparative psychology to our understanding of behavior and their possible significance for human behavior.

## SIMPLE BEHAVIORAL RESPONSES AND PATTERNS

Behavior is a complex mixture of patterns of activity. For example, a mother cheetah chasing an antelope is exhibiting a variety of reflex actions while at the same time responding both to a hunger drive and an ancient urge to train her young in the catching of prey. But rather than begin our analysis of behavior by examining all of the components of a highly complex behavior at once, we will start with its simple aspects and proceed to more complex behavior. Each example cited will illustrate an important behavioral principle.

## SIMPLE REFLEXES

The physical basis of reflex action is the reflex arc described in Chapter 15. A simple reflex is a prime example of what is called a **closed program,** a behavioral response that cannot be modified by learning. It is usually unlearned, almost invariably elicited by its adequate stimulus, and unvarying in character. The simple reflex, such as blinking when something comes close to our eyes, is a classic example of a particular response to a particular stimulus.

It is interesting to note that despite the fact that most reflexes are automatic, it is possible to inhibit some of them by learned behavior. We can, for example, deliberately keep from blinking as an object approaches our eye. Or, with much effort, it is possible to hold our hand in a flame until it is severely burned. With practice, we can also control our blood pressure and heart rate to a degree.

## TAXES

An automatic movement made by an animal toward or away from a stimulus is a **taxis** (taxes, plural). A taxis is said to be positive if the movement is toward the stimulus, negative if away from it. Taxes are closed programs and are most readily demonstrated in lower animals. Examples of taxes include the migration of an earthworm to the surface after a heavy rain has saturated the soil and the flight of a moth toward a light (Fig. 19-2). A response that is pure taxis is very rare in vertebrates.

## KINESES

A response to an environmental condition by changing the rate at which an activity is performed is known as a **kinesis.** Examples include the response in sow bugs (wood lice) of traveling faster in the light than in the dark and the response in chitons (primitive marine molluscs) of moving faster when exposed to air than when submerged in water. Kineses, like taxes, are usually closed programs.

**FIG. 19-2** A moth flying toward a light is a classic example of a taxis.

## ORIENTATIONS

Many animals orient, or make a simple adjustment of their position, in reference to an environmental stimulus. For example, locusts in the morning light turn sideways and thus intercept the maximum amount of sunshine. In the middle of a hot day, however, locusts turn to face the sun, thereby intercepting the fewest rays.

## THE SIMPLE COMPONENTS OF MODERATELY COMPLEX BEHAVIOR

When analyzed, some moderately complex behavior patterns have been found to consist entirely of a combination of quite simple components, such as reflexes and taxes, each readily described as a specific response elicited by a specific stimulus. A classic example is the behavior of a chiton in response to exposure to air and sunlight. Chitons are often found in groups attached to the undersides of stones that are exposed at low tide. When such a stone is overturned and thereby exposed to air and sunlight, the animals begin to creep across its surface until all are again clustered on its underside (Fig. 19-3a). It might seem that their gathering together in this way denotes communication, social structure, or even that the animals were seeking the safety of the underside of the rock because they knew they were in danger. However, experimental laboratory analysis has indicated that chitons respond merely by means of two simple kineses and a taxis.

Chitons exposed to light in a flat, level dish move in random directions until they encounter a shaded area of the dish, whereupon they stop. If placed in the light again, they resume random movements until they encounter shade once more (see Fig. 19-3b). Chitons on a flat surface move slowly when submerged and more rapidly when exposed to air. While exposed to air, they do not cross a dry surface if they encounter one; they continue active movements until submerged again. Chitons exposed to air on the side of a glass plate or a rock, however, always move toward gravity (Fig. 19-3c). Once submerged, they do not respond to gravity.

By combining these few, relatively simple automatic responses in a larger pattern of behavior, the primitive chiton can cope very well with several life-threatening situations.

## DRIVES

We know that behavioral ability is not all that is required for the behavior to be exhibited. An animal must somehow feel an urge to perform the action. For example, placing a dish of food before a cat, dog, or human does not always elicit feeding behavior. In referring to an animal's varying responses to the same stimuli, ethologists speak of the animal as possessing a drive—in the example just cited, a hunger drive. The intensity of drives ranges from low, or weak, to high, or strong. For example, a bird experiencing a strong nest-building drive will go through the motions of nest building even if the materials for a nest are not available. The fact that an animal has a drive does not mean that the animal wants to do something. We have no way of knowing how it feels, for example, to be a bird at nest-building time.

***Appetitive and Consummatory Components of Drives.*** A useful way in which many ethologists analyze behavior that contains both variable and stereotyped elements is to classify the components as either appetitive or consummatory. The appetitive stage usually consists of highly variable restless or searching movements, such as for food, a mate, or nesting materials. The consummatory stage is a stereotyped, often species-specific, response to finding the object of the search. A particular behavior, such as finding and eating food or building a nest, often consists of a series of alternating appetitive and consummatory activities until the final task, such as finishing the meal or completing the nest, is accomplished.

***Drives as "Black Boxes."*** When a complex of poorly known mechanisms and factors operates in predictable ways, investigators find it useful to treat the complex as a "black box": a set of mechanisms ac-

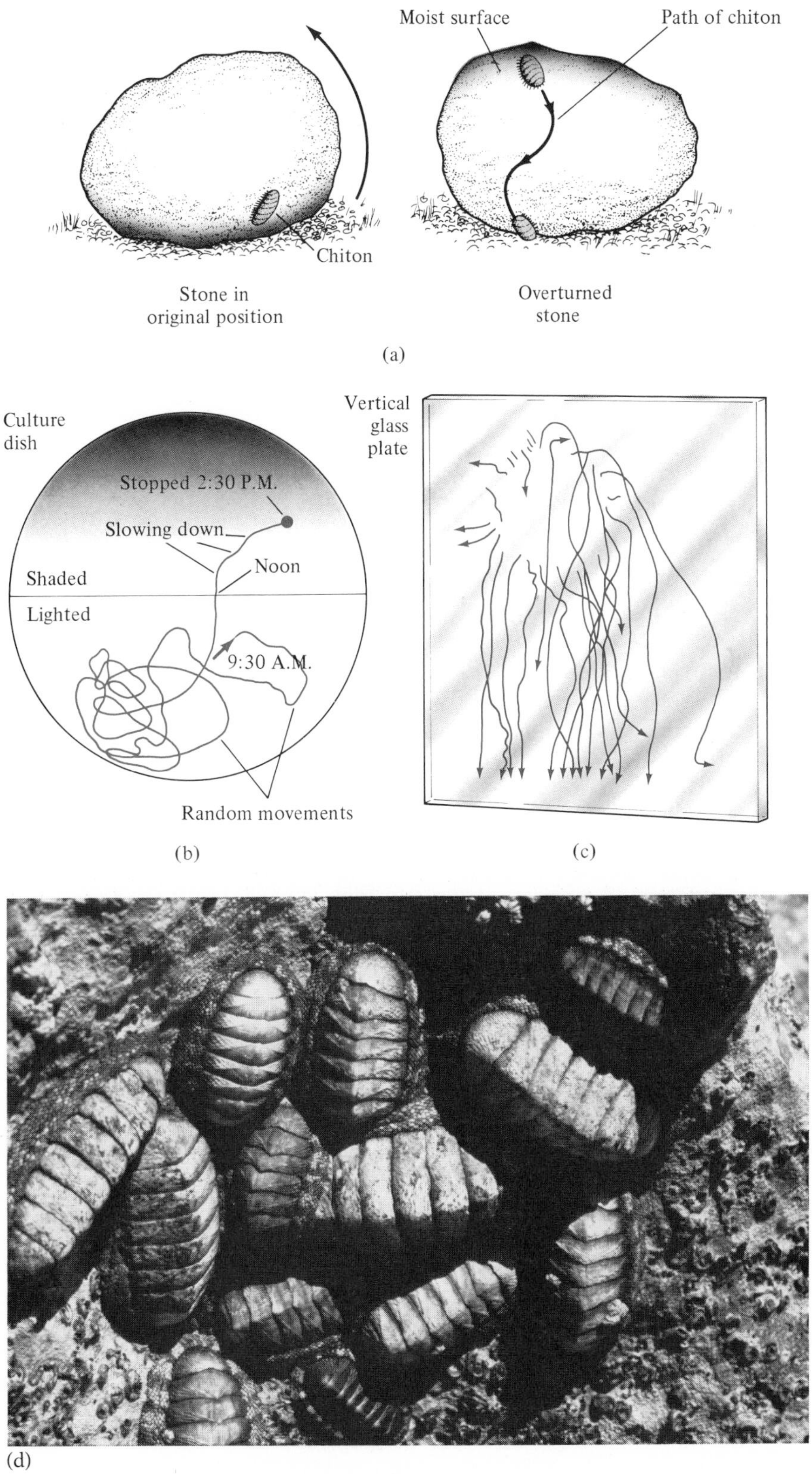


**FIG. 19-3** Analysis of behavior in chitons. The reaction of chitons to exposure to light and air when the stone to which they are attached is overturned is composed of simple reflexes, taxes, and kineses in response to light, air, moisture, and gravity. (a) When a chiton is exposed to light and air by overturning a stone to which it is attached, it moves downward until it is again submerged and in the shade. (b) Chitons on a submerged, horizontal surface, such as a Petri culture dish, remain active and exhibit random movements when exposed to light but slow to a stop when they enter shade. (c) When exposed to air on a vertical, smooth surface, such as a glass plate, chitons move toward gravity until they are submerged. Arrows indicate the orientation of the animal's heads after two and one-half hours. (d) Chitons.

knowledged as functioning in a situation but the full nature of which is unknown. The various drives of animals have been looked at in this way to good advantage. Just as someone can acquire much information and some grasp of principles concerning the operation of automobiles without knowing anything about internal combustion engines or transmissions, so have scientists learned much about behavior while knowing little of the mechanisms underlying drives.

***Different Drives.*** For a time, it was widely believed that within each animal there might be but one drive mechanism or source of "nervous energy" that could be activated by various kinds of stimuli. The "energy" from the one drive mechanism, it was thought, could be directed toward the accomplishment of any of the "goals" of the various drives. In time, however, it became clear that drives are specific. For example, a hungry animal directs almost all its effort and attention to finishing its meal—unless a predator threatens it, whereupon it stops eating and looks to its safety. It is true that fear, sexual arousal, and other strong emotions share some qualities, such as increased cardiac output, elevated blood pressure, and other physiological changes. Nevertheless, each drive has distinctive features. Largely because of some of the older, false connotations of the word *drive,* the term *motivation* is preferred by many ethologists.

***The Hunger Drive.*** As our knowledge of the central nervous system has improved, we have gained insights into mechanisms involved in the hunger drive. The hypothalamus (Chapter 15) contains a satiety (satisfaction) center. If a rat's satiety center is destroyed, the rat overeats, if food is readily available, until it is obese. However, rats without satiety centers will not work hard, endure shocks, or tolerate bitter taste to obtain food. Such rats are neither hungry nor satiated. Near the satiety center in the hypothalamus is the feeding center. If it is destroyed, the animal declines food even when starving. If the satiety center is also destroyed, the animal continues to decline food. This suggests that the satiety center normally functions by inhibiting the feeding center.

The hypothalamic feeding center is subdivided into two regions, one providing a basic feeding drive, or appetite, and the other a hunger drive.* The hunger region is apparently inhibited by the satiety center. The hunger center can be stimulated electrically or by administering epinephrine, whereupon the animal will endure pain to eat food.

Neither the nature of the stimuli affecting the hypothalamus nor the manner in which the hypothalamus creates the symptoms and sensations of hunger is known. In regard to feeding behavior, however, the hypothalamic centers apparently affect lower brain centers that control reflexes concerned with food ingestion, causing the centers to become either more or less responsive to stimuli.

***Reproductive and Migratory Drives.*** Reproductive behavior is typically prompted by a complex of drives, such as the sex drive, or inclination to mate (Fig. 19-4); nest building; brooding of eggs; and care of the young. Some of these have been studied in great detail. Usually, however, it is the behavioral pattern that has been studied, not the physical basis for the drive. The sex drive, for example, is more complex and more difficult to study than the hunger drive and is not well understood. It is known that hormones play a more prominent role in sexual behavior than in the hunger drive.

Migratory drives are also influenced by hormones. But even where something is known of the hormonal involvement, we are a long way

* Hunger and appetite are not the same. An animal may have an appetite for some foods even after its hunger has been satisfied.

**FIG. 19-4** A bull Hooper's sea lion and his harem, Auckland Island, New Zealand. During the mating season, the reproductive drive of the male is strong; he goes without eating until he has mated with every member of his harem, usually a matter of weeks.

from understanding the entire complex of inherited traits that prompts an animal to migrate, often for great distances, to precise locations and then to "lose its drive" when it reaches its destination.

The topic of drives, or motivation, and the elucidation of their components is one of the most interesting and important fields of study in animal behavior, yet it has also been one of the least explored.

## MAYR'S CLOSED AND OPEN PROGRAMS

In 1974, the American geneticist-evolutionist Ernst Mayr proposed an interesting and useful way for ethologists to look at behavioral patterns. He called behavioral programs that cannot be modified by learning genetically closed. Experience plays no role in them. Any changes that occur would involve changes in the genes. Simple reflexes, taxes, and kineses are types of closed programs. He called behavior that can be modified **open programs.** It is common, however, for behavior to have both stereotyped and learned (closed and open) components. Various learned "insertions" in an otherwise closed program are changes made in the nervous system, not the genes. Of course, both types of program have some genetic basis. In closed programs, however, a particular behavior invariably results from specific stimuli. In open programs, what is inherited is a capacity to learn appropriate responses to various stimuli.

## GENETIC BASES OF BEHAVIOR

The evidence that much behavior has a genetic basis is abundant and convincing. This is most apparent in the case of stereotyped behaviors, ones performed in the same way by all members of the same species, in many cases with near perfect execution at the first attempt. But learned behavior also has a genetic component in that the ability to learn various skills is inherited. As with any inherited traits, these are subject to natural selection. Several lines of evidence suggest that over many generations natural selection may replace some adaptive behaviors that

FIG. 19-5 A female weaver bird putting the finishing touches on her nest. Her complex actions are similar to those of all weaver birds, regardless of whether she has ever seen a nest built.

are learned anew at each generation with innate versions of the same behavior. Whether this happens or not depends on whether the particular conditions select for an innate or a learned behavior as the most adaptive. "Phylogenetic learning" is a term often used to describe an innate behavior that has substituted for a learned one.

***Stereotyped Behavior.*** The existence of highly stereotyped behavior—behavior that almost never varies within a species from individual to individual or from generation to generation—is a strong argument for the genetic basis of that behavior. For example, blackbird nests are always built the same way. The birds begin by laying a foundation of large twigs; next they form the sides with small twigs; then they make a cuplike lining of mud. Finally, they line the nest with fine grass and hair. Many of the motions involved in building the nest are highly stereotyped. An even more dramatic example of stereotyped behavior is the complex nest building of the weaver bird. The male builds the outer shell, then attracts a female to it. She finishes it (Fig. 19-5); her intricate actions, reminiscent of those of an expert tailor, are similar to those of all females of her species. Another example is provided by the egg-laying digger wasp, which always digs two or three burrows after mating. She then searches for caterpillars of a certain species of moth, paralyzes them with a sting, and places one in each burrow. She next lays an egg in the body of each caterpillar. After providing each burrow with an additional stock of paralyzed caterpillars, the female seals off each entrance. Upon hatching, the wasp larvae feed on the paralyzed caterpillars. The female offspring, at maturity, repeat exactly the same behavior without ever having witnessed it. Many other species of wasps exhibit similar behaviors, involving other types of nests and other species of insect larvae (Fig. 19-6).

***Inheritance of Learning Abilities.*** Much behavior is learned, especially in higher animals, but the ability to learn is also inherited. Some species of animals can learn to solve certain types of problems, but those species lacking the native capacity to do so never learn to solve these problems at all. Entire races of honeybees are consistently much better at learning landmarks near food sources than other races of the same species. Determining the role of heredity in such cases is not an easy matter. Although it is clear that the basic learning ability is inherited, even the ability to learn can be improved by experience.

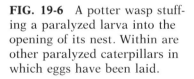

FIG. 19-6 A potter wasp stuffing a paralyzed larva into the opening of its nest. Within are other paralyzed caterpillars in which eggs have been laid.

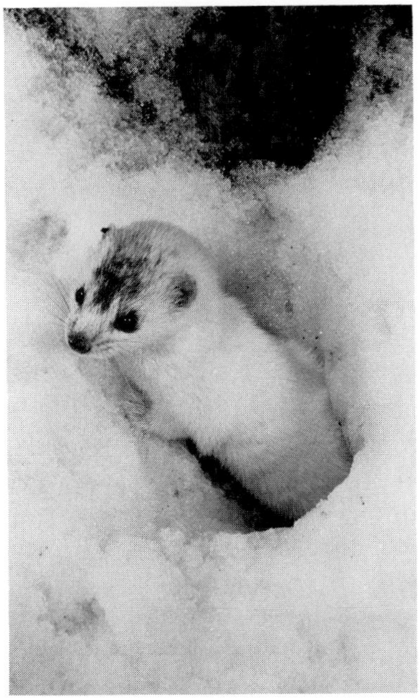

**FIG. 19-7** The least weasel escapes the worst of winter storms by burrowing. Its coat changes from brown to white in winter, enabling it to blend in with a snowy background and thus avoid detection by birds of prey.

***Behavior and Natural Selection.*** Behavioral abilities, like other inherited traits of animals, are subject to natural selection. Like other inherited traits, the stereotyped behavior of animals may vary to some extent from one individual to another within a species. This behavioral variability is of adaptive value because it enables a population to adapt to such things as changes in climate, invasions by competing or predatory species, or the advent of a new food species. For example, a long-term increase in the length or severity of winters would favor animals that burrow for the winter (Fig. 19-7) or that migrate.*

***The Physical Basis of Behavior: What Exactly Is Inherited?*** We know very few of the inherited components underlying animals' behavior patterns. But among them are possession of the sensory receptors by which information is obtained, the nervous and endocrine systems that accomplish integration of information, and the effectors—such as glands, muscles, and other organs—that execute the actions that comprise the behavior. A group of genes controlling egg-laying behavior in the sea hare *Aplysia,* a large marine snail, has recently been isolated from that species. The genes produce peptide neurotransmitters that trigger the snail's egg-laying behavior. Our present knowledge of integrative mechanisms, however, is inadequate to explain even such seemingly simple behavior as feeding. A directive by the nervous system to acquire and eat food is not simple. Moreover, as the muscles execute their tasks, their actions are continually monitored so that the motions are accomplished accurately. Furthermore, despite failures and obstacles, the end of such a task as the construction of a complicated nest is usually pursued over several days until finally achieved. This means that various "set points" within the nervous and endocrine systems must be "satisfied" with the results of the effectors' performances as each stage of the activity is completed. For example, a half-completed bird's nest acts as a stimulus that elicits from the bird the next stage of nest building. Only when the nest is completed will the bird cease its activity. If the nest is subsequently damaged, the activity is likely to resume until the nest is completely repaired. The physical basis for this type of behavior is largely unknown.

## FIXED ACTION PATTERNS

Some of the earliest ethological studies showed that many animals respond to certain stimuli with specific, stereotyped patterns of behavior. For example, frogs respond with feeding behavior to any small, dark object moving near them. This makes frogs good at catching insects. Much ethological research has been devoted to studying such innate, unvarying components of behavior, known as **fixed action patterns.** A fixed action pattern is not a simple reflex response but a series of actions.

Fixed action patterns appear to be the outward manifestations of genetically closed behavioral programs. They are usually alike in all members of the same species, at least in those of the same sex. The lives

* Of course, such climatic changes also create selection pressures favoring nonbehavioral adaptations that would enable an animal that neither burrowed nor migrated to withstand more severe weather. Growing a heavier coat of fur or feathers is such an adaptation.

(a)

(b)

**FIG. 19-8** Sex recognition in flicker woodpeckers. The male flicker (a) has a "mustache," and the female (b) does not. If a mustache is painted on his mate, the male will drive her away as if she were another male until the mustache is removed.

of most flies are almost totally governed by such stereotyped responses to food, water, and mates. In vertebrates, however, fixed action patterns are incorporated into learned behavior. They comprise the stereotyped parts of programs that are otherwise open.

## SIGN STIMULI

One key quality of fixed action patterns is that they are usually triggered by simple stimuli, called sign stimuli or releasers, to which the animal is quite sensitive. The releaser acts as a cue; it is normally always associated with the object of the response. The existence of only one cue for each object is appropriate and economical because there is no need to inform the CNS of a long list of characteristics of the object when one will suffice. For example, scallops, a kind of clam, will rapidly move away from the vicinity of a starfish, their predator. The same response is elicited if water taken from an aquarium containing starfish is poured over the clams. The escape behavior is triggered by a chemical associated with starfish. Its presence normally would signal the presence of a starfish.

Another example of how a single, simple cue serves to trigger an action pattern is seen in flickers, a type of woodpecker, in which the male differs from the female by the possession of a small "mustache," as shown in Figure 19-8. If a mustache is painted on a female flicker, her mate will violently attack her as he would an intruding male. If her mustache is removed, he will resume courting her.

## STEREOTYPED BEHAVIOR IN HUMANS

Although nearly all human behavior can be regarded as learned, that is, based on genetically open programs, our drives and motivations to perform certain types of behavior under certain conditions are evidently inherited. Like many other animals, we tend to respond in predictable ways to certain stimuli. Whether any such responses are innate, however, can be demonstrated only in very young children. It appears that most of whatever innate behavior we have can be overridden by learned behavior.

***Young Flirts.*** Some of the suggestive evidence of stereotyped behavior in humans has been obtained from comparative studies of different cultures. For example, in studies done in the 1960s, Irenaus Eibl-Eibesfeldt of the Max Planck Institute for Behavioral Physiology in Germany has shown that the manner of flirting with the eyes is quite similar in very young girls in such widely diverse cultures as those of Samoa, New Guinea, France, Japan, and various tribes of Africa and South America.

The basic behavioral pattern involves a smile with eye and eyebrow movement. The head is then turned aside, the gaze is lowered and the eyelids drop. The face may then be covered with the hand, and the girl may smile or laugh in embarrassment. She continues to sneak looks at the other person out of the corners of her eyes. The basic components of this behavior seem to be universal, although many variations are superimposed upon them (Fig. 19-9).

**FIG. 19-9** Flower girl smiles at ring boy in an example of stereotyped behavior in humans.

***Fear of Falling.*** Among the possible types of phylogenetic learning in humans are certain fears. Our fear of falling is shared with all other mammals and with birds. A fear of falling is demonstrable in young infants, although whether it is innate to any extent has not been demonstrated beyond question (Fig. 19-10). It may be significant that the fear of falling is coupled with the child's strong tendency to grasp hair. This combination would have been selected for in infants during the early history of human evolution when, millions of years ago, our ancestors swung through the trees high above the ground.

## DUCK COURTSHIP

Thus far we have considered several relatively simple responses to stimuli. While a casual observation of such complex behaviors as nest building or courtship and mating would suggest that they are not based on stereotyped responses, this is not the case. In fact, in both invertebrates and vertebrates, courtship typically is composed of a series of stereotyped actions that are part of a larger pattern that includes learned components as well.

**FIG. 19-10** Avoidance reaction of a "visual cliff." (a) Like all young mammals, a crawling infant will avoid the brink of a "cliff," even after finding that the cliff is a solid surface, in this case a sheet of plate glass. (b) The infant refuses to crawl to its mother over the visual cliff but does crawl to her where the visual evidence indicates a safe route. (c) A young kitten exhibits similar behavior.

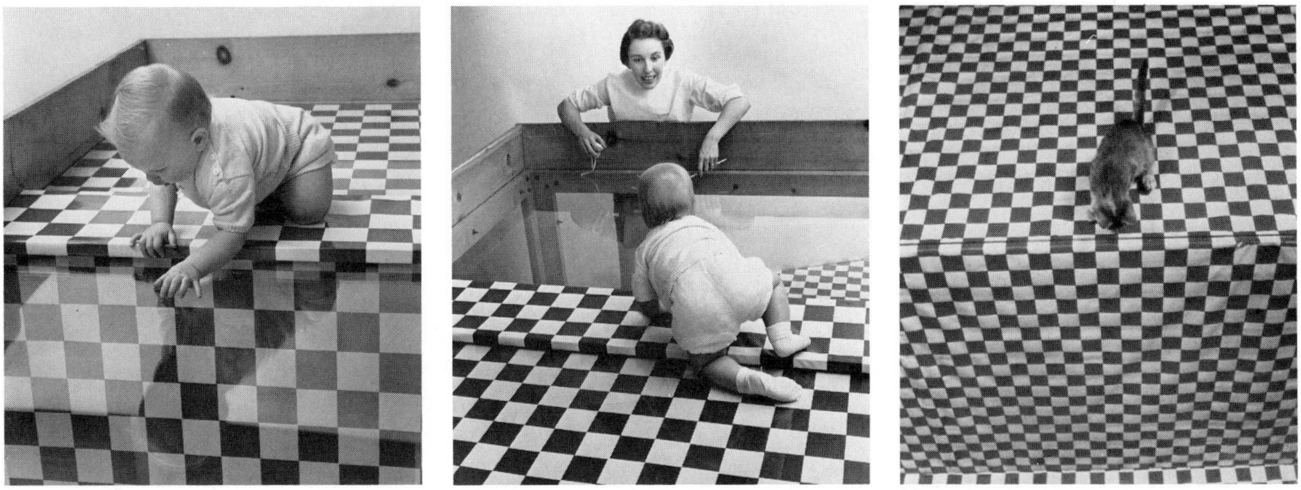

(a) (b) (c)

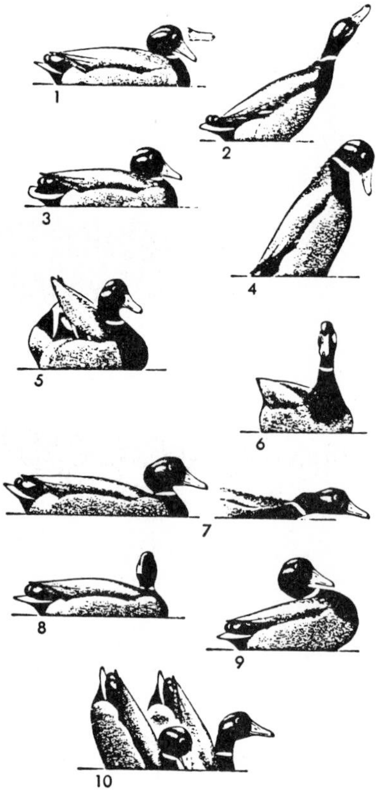


**FIG. 19-11** Various behavioral components inherited by surface-feeding ducks, as exhibited by the mallard: (1) initial bill shake, (2) head flick, (3) tail shake, (4) gruntwhistle, (5) head up, tail up, (6) turn toward female, (7) nod-swimming, (8) turning the back of the head, (9) bridling, and (10) down, up. The species-specific courtship sequence of these poses or actions is described in the text for three types of ducks.

One of the best examples of such behavior has been found in ducks. During courtship, various species of surface-feeding ducks exhibit combinations of some 20 separate, simple actions that are also performed in other contexts. The combinations, however, are unique to courtship.

Each display of one of these actions by the male acts as a releaser and elicits a certain fixed action pattern of response in the female. Her response then acts as a releaser for him and prompts his next movement, and so on. Only if the entire sequence characteristic of courtship in their species is completed will the two individuals finally mate with each other.*

Of the 20 or so stereotyped actions exhibited by ducks during courtship, three duck species possess the 10 shown in Figure 19-11. In these three species, moreover, some courtship actions are always linked to each other. For example, if both 4 and 3 are displayed, they are always linked in the 4-3 sequence. Movements 5 and 6 are also always displayed one after the other. However, each of the three species displays certain of the movements in a different combination. In the mallard, the courtship display sequence is 3-2-3, 1-4-3, and 5-6-7-8. The European teal, however, excites its mate with a 3-2-3-10 and 4-3-2-5-6-8 sequence, and the gadwall exhibits a series of 4, 3-2-3, and 5-6-10-6.

Although courtship patterns are constant within a given species of ducks, when experimental laboratory crossbreedings (ones not known to occur in nature) are performed among teals, mallards, and gadwalls, they generally produce individuals that display new behavioral combinations during courtship. In subsequent generations derived from the hybrids, other new behavioral combinations appear, sometimes combining traits of both parents, sometimes suppressing traits of one parent, and often resulting in combinations found in neither species, all in accordance with the basic laws of inheritance. Some of the novel combinations of traits of the hybrids of these three species, however, were found in nature in other species of ducks.

## RHYTHMIC AND CYCLIC BEHAVIOR

When various internal conditions of almost any organism are carefully monitored over an extended period of time, they are found to exhibit regular oscillations. These cycles are endogenous, or internally generated, and are not merely responses to external conditions. The endogenous nature of the rhythms can be shown by isolating the organism from all contact with cyclic changes in its environment by placing it, for example, deep in a mine. The rhythms continue.

### RHYTHMIC BEHAVIOR AS ADAPTATION TO EARTH'S CYCLES

Much animal behavior is rhythmic. Moreover, much of all life is coordinated with such cycles of nature as the day-night cycle of 24 hours (Table 19-1). Other cycles include the lunar cycles of 24.8 hours from one moonrise to the next and 29.5 days from one full moon to the next, the tidal rhythm of 12.4 hours, and, of course, the annual rhythm of the seasons (Fig. 19-12). Such cycles have radically influenced life on earth

* The adaptive value of courtship behavior is discussed in Chapter 20.

**TABLE 19-1 Some Biological Activities Subject to Rhythmic Oscillations**

| *Organism* | *Activity* |
|---|---|
| Plants | Seed germination |
| | Flowering |
| | Dormancy |
| Insects | Mating |
| | Egg laying |
| | Feeding (daily cycles) |
| | Emergence from pupa |
| | Migratory behavior |
| Birds or mammals | |
| Annual cycles | Feeding |
| | Nesting |
| | Breeding |
| | Hibernation |
| | Pelt thickening and thinning |
| | Color changes and molting |
| | Migratory behavior |
| Daily and lunar cycles | Body temperature |
| | Heart rate |
| | Breeding |
| | Sleep and wakefulness |
| | Ovulation and menstruation |

for billions of years. Organisms have evolved their own rhythms that coordinate their activities with environmental rhythms. Most of these are rigid and automatic responses to various signs that precede or accompany particular environmental changes. Many plants and animals, including humans, have daily cycles of increased metabolic activity and dormancy. The leaves of bean seedlings rise each morning, then droop at night. Fiddler crabs darken in color during the day and become pale at night. The blooming of flowers and the migration of many animals are cyclic activities that may begin or end near or even on the same day of each year.

In such cases, it is not sufficient for the organism merely to be sensitive to variations in light intensity, temperature, or the abundance of food. For migratory birds to respond to such an erratic occurrence as one chilly, overcast day would not be highly adaptive. One reliable index of seasonal change to which most of these organisms are sensitive is the average day length, termed the **photoperiod,** a value that changes slowly and regularly throughout the year. (Actually, only some of these organisms are sensitive to the length of the light period; most measure the dark period.) By responding to changes in photoperiod, some flowers bloom only in the spring, others only in the fall. To perceive changes in average day or night length, however, requires a standard of comparison: a biological clock. **Biological clocks** are internal mechanisms that provide a means of measuring time. They probably exist in all organisms except prokaryotes. Such clocks do not make organisms aware of

(a)

(b)

(c)

(d)

(e)

(f)

**FIG. 19-12** Some biological rhythms. Much of life is rhythmic in its manifestations. The rhythms are usually adaptations to natural daily, lunar, tidal, or annual cycles. (a) Wildflowers in bloom (Alaskan irises, *Iris setosa,* in the Barren Islands, Alaska). (b) Migrating wildebeests in Africa. (c) A herd of grazing bison in Yellowstone National Park. Many animals feed only at certain times. (d) Two bull elk contending for mastery of a harem of does before mating. Like migratory behavior, reproductive behavior is noticeably cyclic in most species. (e) A Maine hardwood forest in autumn. One of nature's more impressive displays of color is seen for a few weeks each year before the leaves begin to fall. (f) Like this arctic fox, many animals undergo seasonal changes in coat color. As with the arctic fox, such changes usually serve to camouflage the animal.

time; they merely regulate its activities automatically. The biological clock is essential for efficient control of any cyclic behavior.

## CLOSED CYCLIC BEHAVIORAL PROGRAMS AND INTERNAL CLOCKS

A biological clock is a necessity for any organism that is to adapt its activities to one or more of the rhythms of our planet. To be of use, the clock must be autonomous. For example, it cannot vary its timekeeping property in response to changes in the metabolic rate of the body. In addition, it must be linked with the various mechanisms that are to be regulated, such as wakefulness, migratory behavior, or emergence at low tide. It must also be capable of receiving input from at least one of the rhythms of the planet, such as the day-night cycles or tides. This input allows the clock to be set, or **entrained.**

Entrainment is analogous to setting a watch to the correct time. A biological clock does not have to be wound, but it must occasionally be entrained to the correct time or it will give its signals at the wrong times. Many biological clocks become precisely entrained by the 24-hour cycles of day and night. The phenomenon of jet lag, or its equivalent, results when an individual entrained by one schedule is suddenly subjected to a new day/night cycle (see Panel 19-1). If an animal is put out of touch with the day-night cycle by being placed in a darkened room or a room with continuous low-intensity illumination, its rhythmic activities continue but are no longer on an exact, 24-hour schedule. Instead, the cycle becomes slightly shorter or longer than 24 hours. The clock is said to have become free running. Cockroaches kept in continuous dim light, for example, might exhibit a 23.8-hour cycle of activity and quiescence instead of their usual 24-hour cycle (Fig. 19-13). Bean plants kept in total darkness continue to raise their leaves on a daily basis, but not at exactly the same time.

**FIG. 19-13** As anyone who has ever turned on the lights in the middle of the night in a cockroach-infested room might suspect, these insects are nocturnal. Studies have shown that they regulate their feeding and other activities to the 24-hour, day-night cycle.

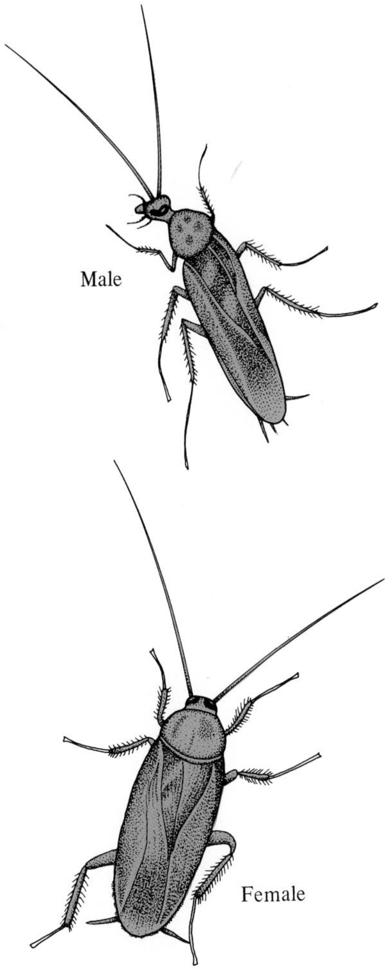


***Circadian Rhythms and Zeitgebers.*** Because the daily rhythms controlled by a free-running internal clock are only approximately a day in length, they are termed circadian rhythms (from the Latin *circa,* "about," and *dies,* "day"). In the case of roaches or other animals kept in continuous light, as time goes on, their clocks and their periods of activity become more and more out of phase with the days and nights. But even brief exposure to the regular day-night sequence reentrains the clocks. The environmental stimulus that entrains a clock is called its **zeitgeber** (from the German, *Zeit,* "time," and *Geber,* "giver"). An animal has more than one biological clock, each of which controls some physiological function, such as body temperature or the release of a hormone. Some of the clocks are master clocks and tend to dominate the others. Some zeitgebers are especially important in entrainment of the dominant clocks. Besides the light-dark cycles, cycles of rest and activity are especially effective in this regard, as are cycles of fasting and eating.

If roaches are subjected to artificial illumination cycles of, say, 22 hours (such as 12 hours of light and 10 of dark), their clocks become set to 22-hour days. The same kind of adjustment is also made if they are subjected to days of 26 hours. If a much longer artificial day is provided, it is ignored by the clock, which reverts to its internal free-running rhythm.

## PANEL 19-1

# Easing Jet Lag Problems

Probably all of us have experienced it: being overcome with drowsiness when we had to be alert and being wide awake when we needed to get to sleep. It is not merely a matter of being short on sleep in the one case and being "slept out" in the other. Sleepiness can occur after having ten hours of sleep the night before and alertness can follow a night of only a few hours of sleep.

What causes our clocks to get out of synchrony? Staying up late on weekends, then sleeping until noon and eating 1 P.M. brunches is one way to do it. An abrupt return to 6 A.M. rising and 6:30 A.M. breakfasts will probably be followed by several sleepy days, wide-awake evenings, and restless nights until our clocks are reentrained to the regular schedule. We can avoid such problems by taking into account some of the zeitgebers that entrain our various biological clocks. The zeitgebers include, besides the light and dark cycles, the time we take our rest, our periods of activity, and the foods and beverages we consume. The regulated physiological cycles include highs and lows of body temperature, heart rate, blood pressure, hormone release, sleepiness, alertness, and reaction time, among others. Although, when free-running, many of these cycles differ regarding their high and low periods, normally all are dominated by a few master clocks that play the principal role of adapting us to the cycles of our environment.

Despite the fact that biological clocks can be reentrained to new cycles, it appears that individuals differ in regard to the time of day or night when they tend to be most alert and energetic. Many find it difficult, if not impossible, to become entrained to night shift work, while others, the so-called night persons, readily take to night work and may even prefer it.

Almost everyone has trouble adjusting to changes in shifts. For many years some cities had their fire fighters follow a schedule in which 24 hours of duty alternated with 24 hours of time off for about a week, followed by three full days off. The biological clocks of fire fighters in companies with frequent night runs probably never became adjusted to a normal 24-hour day.

Being made to work during a period of sleepiness is not just a matter of discomfort. People are more prone to accidents when their biological clocks are out of synchrony with their work schedule. It may be significant that the accident at Three Mile Island nuclear power facility occurred at 4 A.M. with a crew that had just been rotated to the night shift. Many industries are now taking these facts into account in setting work schedules.

The best known example of maladjustment of

***Where Is the Master Clock Located?*** Efforts have been made to locate the master clocks of animals in the nervous system. In one study, destruction of a group of the neurosecretory cells in the roach brain in the region called the pars intercerebralis (PIC) abolished the insect's clock. Yet, further work showed that the same effect was achieved by merely cutting the connections between the optic lobes of the brain and the rest of the brain. Yet, if only the nerves from the eyes to the optic lobes were cut, the rhythm persisted but became free running, just as if the animal had been placed in continuous light or dark. The rhythm therefore originates, not in the PIC, but in the optic lobes. What generates the rhythm is as yet unknown.

The many circadian rhythms studied in vertebrates are similar to those of invertebrates. Light and sound are common vertebrate zeitgebers. Studies of sparrows that exhibit rhythmic activity have revealed that the pineal body (a small, cone-shaped body of nervous tissue attached to the roof of the third ventricle of the brain) and a group of nearby cells are the sources of the rhythm. Sparrows that have their pineal gland removed and are then kept in darkness continue their activity, but no longer in a rhythmic pattern. However, if exposed to day-night cycles, the activity rhythm resumes.

biological clocks to human activity is the phenomenon of jet lag. Flying west, to a time zone where the day begins and ends later than one's regular schedule, is the easiest to adjust to. Flying east, such as from Chicago to Paris, is more traumatic. Upon arriving in Paris at 7 A.M. (1 A.M. Chicago time), the temptation is to go to bed for a good rest, especially if you had had a typical work day in Chicago before departure. If you get to bed by 9 A.M. in Paris and have 8 hours of sleep you will arise at 3 P.M. and perhaps breakfast around 4 P.M. On the other hand, a business executive who needs to stay awake on the first day in Paris will be miserable and inattentive as he or she struggles through the day, still entrained to Chicago time. But there is a way to ease the effects of jet lag.

Knowing that one's food intake, activities, and the day and night cycles, all act to entrain our biological clocks, Charles F. Ehret of the Argonne National Laboratory near Chicago has prepared a special program for people expecting to experience jet lag. Ehret recommends that if you are planning such a trip you should:

a. Determine the time you expect to eat breakfast at your destination, for example, 8 A.M. Paris time.

b. Beginning four days before that Paris breakfast follow a schedule of alternate days of feasting and fasting, as follows:

On feasting days (days 1 and 3), eat a substantial breakfast and lunch consisting of such foods as eggs, high protein cereals, steak, hamburgers, and green beans. Eat a high carbohydrate dinner of such foods as spaghetti or other pasta, potatoes, starchy vegetables, and pastries, but no meat. No coffee or tea should be consumed except between 3 and 5 P.M. Decaffeinated beverages are permissible at all times.

On fasting days (days 2 and 4), eat light meals with few calories, consisting of such foods as skimpy salads, light soups or broths, unbuttered toast, fruit, and juices. No caffeinated beverages should be consumed except between 3 and 5 P.M.

On day 4, if caffeinated beverages are taken, they should be drunk in the morning before departure if travelling west, or between 6 and 11 P.M. Chicago time if going east.

Day 4, the day of the flight, is a fasting day. Drink no alcohol on the plane. On a long flight, try to sleep until the destination breakfast time at which time the fast is broken with a high-protein breakfast, again without caffeinated beverages. Do not sleep after arriving. Stay active during the day and follow the mealtime schedule of the destination, taking a large lunch and dinner.

This program may not completely eliminate the effects of jet lag but it is known to substantially ease the transition.

This same approach has been used to good advantage by shift workers in preparing for a change of shift. Someone expecting to change shifts beginning on a Monday can use the weekend to reentrain his or her clocks to the new schedule. If the switch is from the day shift to the evening shift or night shift it is comparable to flying west to the Orient. Switching from evenings to days is comparable to flying east.

In the absence of their internal clock, the sparrows merely adjusted their activity to the zeitgeber cycle. Remarkably, neither the pineal body nor the eyes were necessary for them to make this adjustment. Birds with both eyes and the pineal body removed were somehow able to sense the day-night cycles and follow their rhythms. It seems that several parts of the brain can sense light directly, through translucent, thin regions of the skull. This nonoptic sensitivity to light is not peculiar to sparrows. It has been found in other birds and in insects, crayfish, fish, frogs, salamanders, and lizards. Implants of luminescent pellets in quail brains have shown certain regions to be especially responsive to light.

***What Is the Nature of the Biological Clock?*** The prevailing opinion about the nature of the biological clock is that it is probably a biochemical mechanism. It is thought not to be an external signal from any regular fluctuation of the earth's magnetic field or from any variation in cosmic radiation. But what kind of biochemical mechanism does not slow down in cold weather or speed up on warm days—especially in a plant or in a poikilothermic, or cold-blooded, animal? Whereas the rate of biochemical activities usually doubles with each rise of 10°C, internal clocks of animals are only slightly affected by temperature change. Indeed, in some cases, they have been known to decrease slightly at high temperatures and increase slightly at low temperatures. This suggested to J. Woodland Hastings of Harvard University in the 1960s that the clock may be composed of what is termed a biochemical loop pathway. Various products of the pathway's reactions trigger the activities of the cell or individual. Like all other biochemical systems, the loop pathway would respond to temperature changes. But in the Hastings model, another reaction—also dependent on temperature—produces a specific inhibitor of a key step in the clock pathway. Whenever the temperature increases, more inhibitor is produced, thus putting a continual brake on the clock pathway, preventing it from ever increasing its rate. Conceivably, inhibitor might be produced in such large amounts at the highest temperatures as to slow the clock somewhat or in such small amounts at the lowest temperatures as to increase the clock's rate slightly.

## BIRD MIGRATIONS

Among the many known types of rhythmic or cyclic animal behavior, the annual migration of many species of birds has long been of interest to biologists (Fig. 19-14). Besides requiring a biological clock, bird migration usually involves special physiological and behavioral adaptations for long flights and the ability to use navigational references.

***Preparation for Migration.*** Shortly before an annual migration, birds become restless and begin to increase their food intake, resulting in a rapid deposition of body fat. Some birds double their weight within a

**FIG. 19-14** Flocks of up to a million birds of the same species can be seen in some regions at migration time. This flock, estimated at 100,000 birds, was photographed near Stockton, California in the winter of 1957. Similar flocks appear annually.

few days. The accumulated fat is consumed as fuel during the migration. The efficiency in the use of this energy store is remarkable: the Pacific golden plover, for example, cruises at over 100 km per hour. A 150-g bird has been known to cover 3,800 km nonstop in 37 hours, utilizing 18 g of fat in the process. No less remarkable is the migration of a hummingbird of the southern United States. In the fall, this 3½-g (⅛-ounce) bird makes a nonstop crossing of the Gulf of Mexico in 10 hours, a speed of 80 km per hour. In the spring, it stores enough fat to make the return trip.

***The Evolution of Navigational Ability.*** Among animals that evolved migratory behavior as a simple response to adverse seasonal conditions, some individuals would have been better able to recognize and learn landmarks and migratory routes than others. The earliest migrants to learn a route could then lead the way for others, including their own offspring; these offspring may then have learned, not only to migrate with their parents, but what route to take. Later, they would be able to migrate on their own and to lead the way for others. If this migratory behavior increased their chances for survival, such individuals would survive in larger numbers than others. Animals unskilled at such learning would have been selected against, of course. The ability to navigate probably first arose in any species as an open behavioral program. Only later in some species would innate stereotyped migrational behavior have evolved, replacing the learned behavior.

In some species of migratory birds, the ability to learn a route would have been succeeded, in time, by some kind of an inherited sense of the route and its landmarks. As a result of appropriate mutations, modifications within the animals' brains, perhaps creating *déjà-vu*–like impressions similar to memories, would come to be a part of their inheritance.

***Astronavigation.*** Some birds navigate by the stars and need to see only one or a few constellations to become oriented in the right direction. For example, when warblers are brought into a planetarium building at migration time, they orient in the direction in which they normally would migrate at that time of year. However, if the planetarium "sky" is rotated 180°, the birds all swing around and face what appears to them to be the correct direction.

In other planetarium experiments, performed in 1969, Stephen T. Emlen of Cornell University has tested the abilities of indigo buntings (Fig. 19-15) to orient if all southern stars plus the Big Dipper, Polaris, Cassiopeia, the Milky Way, and various other sectors of the sky are blotted out as they might be if clouded over. His results showed that the birds apparently do not detect only one or even a few stars or constellations, but rather are sensitive to the entire sky. In all cases, they oriented as long as they could see a good portion of the sky. But portions with very prominent star clusters proved to be more important than others. And, a point of evolutionary significance, some individual variation occurred: one bird in Emlen's study, for example, relied on southern stars more than the others did.

Members of many species of migratory birds have a sense of direction upon hatching and need not be taught the route by older birds. For example, upon reaching maturity and without guidance from experienced birds, young Manx shearwaters fly from the British Isles and Scandinavia to Brazil, across 9,600 km (5,760 miles) of open sea. The

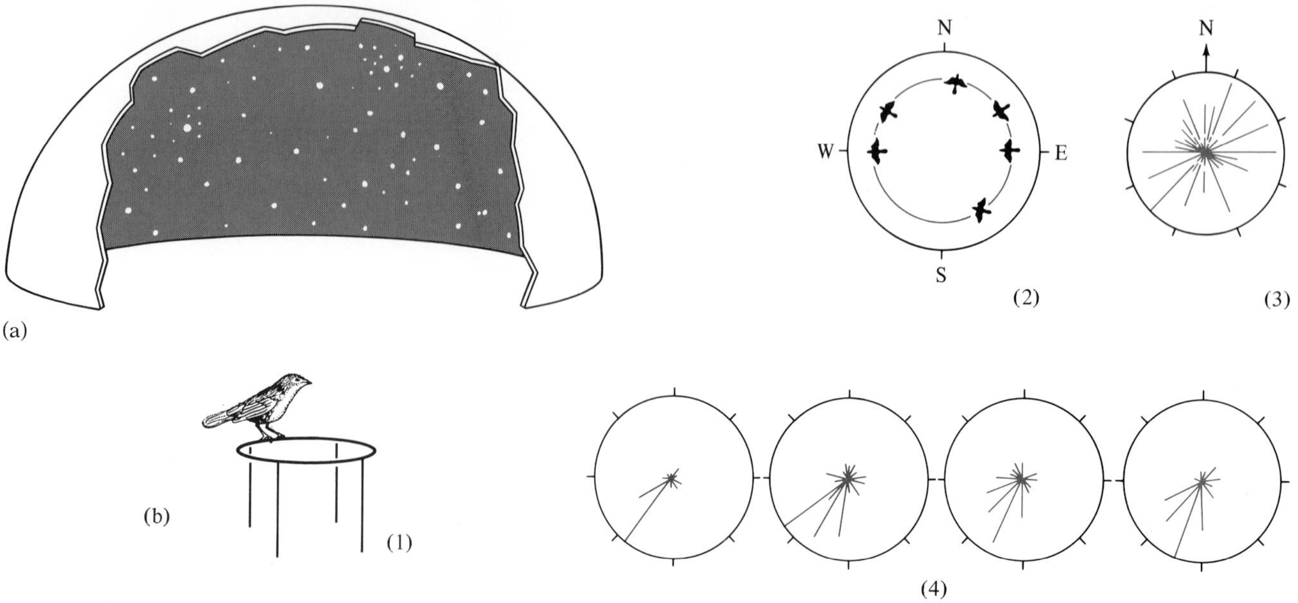

**FIG. 19-15** Method of studying orientation of birds during migratory restlessness. (a) A planetarium with cages placed to allow the birds to see the night sky. (b1, 2) Each bird stands on a circular perch from which it can observe a portion of the night sky. (3) When the sky is overcast, the fluttering movements of the birds are not oriented in any one direction. (4) When a clear part of the sky is visible, the birds in one experiment consistently fluttered their wings while facing southwest, the direction their population migrated at that time of the year. (c) An indigo bunting, the species of bird studied by Stephen Emlen.

"phylogenetic memory" by which such birds inherit an ability to navigate to certain locations by the stars, sun, and other guides is analogous to the inherited sense that many animals have of sources of danger to their species, such as the ability to recognize the animals that usually prey upon them.

***Migrants' Special Need for a Biological Clock.*** An obvious problem associated with using the sun as a navigational reference point is that it constantly changes its relative position in reference to the eastern and western horizons. But by marking the passage of time, the biological clock somehow automatically produces within the bird's nervous system a compensation for each change in the relative position of the sun. In an experiment in which starlings were allowed to orient to a lamp as their sun in a windowless room, it was noted that if the lamp was kept stationary, the birds changed their migratory position by 15° each hour, the rate at which the sun would be moving. This constituted one of the best demonstrations of the existence and function of the clock. Bees are known to have a similar ability, being able to adjust to the fact that the apparent motion of the sun is much more rapid at midday than near sunrise or sunset. If bees remain inside their hive for an hour or two, as often happens during summer showers, they nevertheless orient correctly when they resume their foraging. Experiments show that the direction in which they then fly to reach the same food source, and the compensation made for the time that had passed since they had last seen the sun, is based on how fast the sun seemed to be moving when they last saw it.

***Navigation by the Earth's Magnetic Field.*** When first suggested, the idea that birds could detect the small variations that exist around the world in the earth's magnetic field was met with incredulity. But when European robins in a windowless room were found in 1961 by H. G. Fromme of Frankfurt, Germany, to exhibit a migratory restlessness in which they hop about within their cages but with correct orientation in the direction of their fall flight, he decided to investigate that suggestion. In Frankfurt, the normal strength of the earth's magnetic field is 0.41 gauss (a unit of magnetic force). By shielding the bird's cage, Fromme was able to reduce the strength of this magnetic field. When the field was reduced below 0.14 gauss, the birds became disoriented while continuing to exhibit their migratory restlessness. Then, by using magnetic coils, the direction of the magnetic field around the birds' cage was artificially altered. When strengths similar to those in the earth's field and at the same inclination from the horizontal as the earth's field were employed, the birds' orientation could be manipulated just as had the orientation of the warblers and buntings in the planetarium. The finding that birds could detect and use the earth's magnetic field as a navigational aid explained how robins and a number of other species are able to continue their migratory flights in overcast weather, which "grounds" many other birds.

A possible explanation of how some birds detect magnetic lines of force and their direction lies in the discovery that the bodies of some birds contain magnetite. This is the iron ore that, in the form of lodestones, was used as a compass by sailors in ancient times. Magnetite has been found in the heads of pigeons and the abdomens of bees and in several other animal species.

Compass needles are usually suspended on bearings that allow rotation only in a horizontal plane. However, if a magnetized needle is suspended by a thread and allowed to rotate in a vertical plane, it comes to rest at an angle to the earth's surface called its angle of inclination, or dip. The closer the needle is to either pole, the greater its angle of inclination (Fig. 19-16). At the north magnetic pole, the north pole indicator (north-seeking pole) of the needle points straight down; at the south pole, it points straight up; and at the equator, it is horizontal to the earth's surface. The needle's angle of inclination thus indicates its position anywhere on the earth's surface, as far as north or south latitude are concerned.

In 1975, some species of motile bacteria that grow on bottom sediments in the sea were found to concentrate magnetite within them (Fig. 19-16b). It was theorized that this enables them to detect the vertical component of the earth's magnetic field. It was suggested that the bacteria use the vertical component of the earth's magnetic field as a guide in moving downward into the sediments, since the pull of gravity is practically undetectable to an organism as small as a bacterium. These particular strains of bacteria had been found in the northern hemisphere, where the downward inclination of the earth's magnetic field is in a northerly direction. In an artificial magnetic field, such bacteria travel northward. From this observation, it was predicted that any such bacteria in the southern hemisphere would tend to travel southward, for that would be downward in that hemisphere. In 1980, magnetic

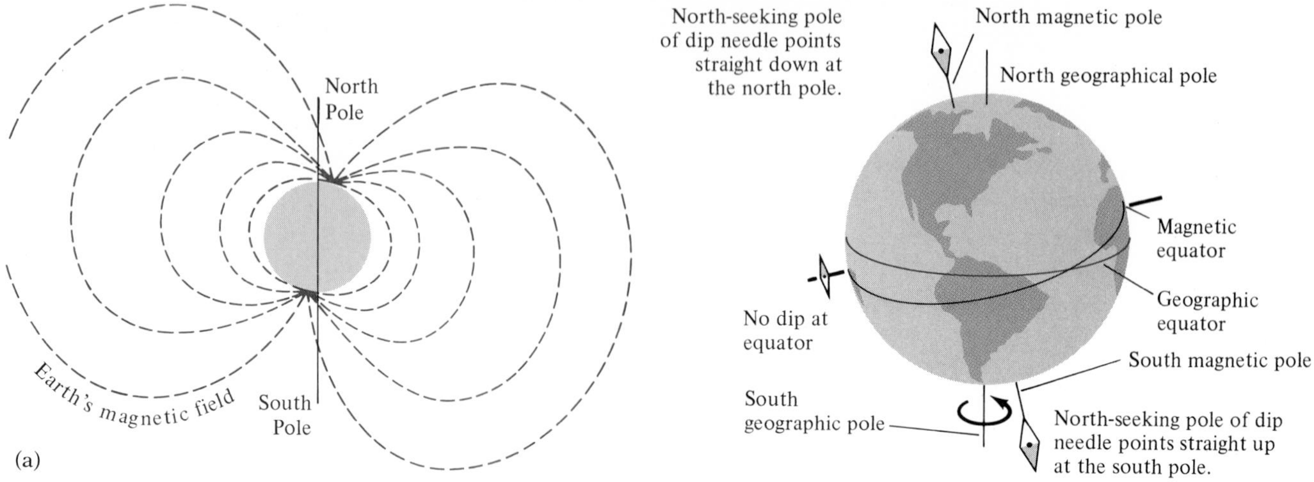


(a)

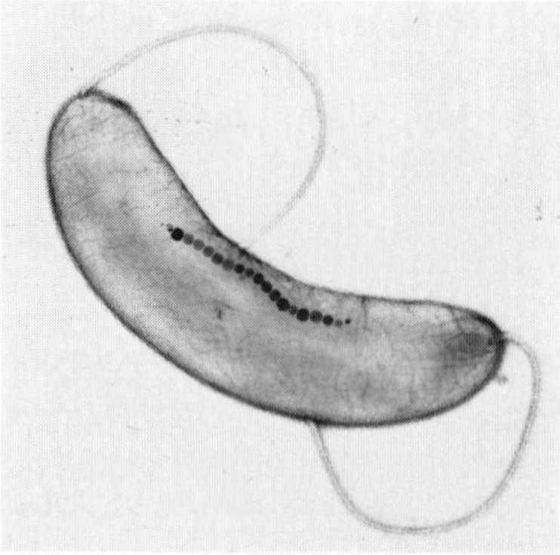

(b)

**FIG. 19-16** (a) Earth's magnetic field. Note the variation in the angle of inclination at various distances from the equator. (b) Electron micrograph of a freshwater magnetic bacterium. Note the row of magnetite granules.

bacteria were found in sediments near Australia. As expected, they tend to travel southward, and thus downward, in that locale.

## ETHOLOGICAL ANALYSES OF LEARNED BEHAVIOR

A practical definition of **learning** is the modification of a response as a result of experience or observation. To be beneficial, learning should have survival value. If an important environmental relationship is unchanging, the appropriate response is always the same and is best provided by unvarying behavior through an unlearned, genetically closed, innate behavioral program. Much migratory behavior and such reproductive behavior as the major components of courtship, nest building, and parental care are activities usually best performed by unlearned behavior. But when there are changes in an organism's relationship with its environment, such as the changes that often occur in the type, relative abundance, location, or escape reactions of prey, learning ability becomes crucial to a predator's survival and tends to be selected for. The same principle applies to any important relationship that is subject to change. Types of learned behavior include habituation, imprinting, and latent learning.

### HABITUATION

**FIG. 19-17** An alert hare. If the noise it hears continues and is made by the grass or leaves rustling in the wind, it soon comes to ignore it. This phenomenon is an example of habituation.

An animal often takes alarm when it receives a certain stimulus, such as a strange sound. But if the stimulus is given continuously or at short intervals, it comes to be temporarily ignored by the animal. The animal is said to have become **habituated** to it. This is not to say that the organism does not detect the stimulus.* Consider, for example, a hare that is startled by the rustling of leaves when the first breeze of the morning springs up (Fig. 19-17). As the breeze continues, the rustling is ignored. It could be said that the animal has learned that rustling leaves do not, at least at this time, pose a threat to it. Most ethologists regard habituation as the lowest level of learning. The regaining of responsiveness to a

* Habituation should be distinguished from sensory adaptation, in which a receptor becomes insensitive to a stimulus after having been continuously stimulated by it for a time.

stimulus to which an animal has become habituated is termed **dishabituation.** It usually requires that the stimulus be interrupted for a time.

## IMPRINTING

During a sensitive period early in its life, a young animal may become influenced in a special way, called **imprinting,** by another individual, usually its mother. This form of learning results in a rigid behavior pattern in response to specific stimuli. In one type of imprinting, the imprinted individual tends to follow and to develop an attachment for the other. The phenomenon was given its name by the German, Oskar Heinroth in 1911. The full development of the concept is credited to the Austrian, Konrad Lorenz in 1937. Examples of imprinting have been reported in many kinds of animals, from insects to primates. However, it has not been shown to constitute a substantial amount of learning in any species.

According to Lorenz, imprinting is a unique form of learning because it (1) takes place only during a brief sensitive period early in life; (2) has great stability, often lasting for the rest of the animal's life; and (3) influences the animal's choice of a sexual partner when it matures, through the animal's imprinted image of its mother. Examples of imprinting have been most fully described for birds (Fig. 19-18).

In birds, the sensitive, or critical, period during which imprinting occurs is quite brief, sometimes little more than a week. Birds have been experimentally imprinted by as little as 30 minutes exposure to the sight of a surrogate mother. Once a young bird has become imprinted on an individual or other object, the bird will thereafter follow it around whenever it is encountered. Although normally the object of the imprinting is the bird's own mother, it can be almost anything—a member of another species, including a human, or even an inanimate object, such as a balloon or a box. Studies of ducks have shown that because ducklings remain in the nest during their sensitive period, most of them become imprinted upon each other. Consequently, although only one or two ducklings may have been imprinted on the mother, when these follow her, the rest are sure to tag along. Wood ducks undergo voice imprinting also, learning the sound of their mother's quack from inside the nest.

As birds mature, they become independent of their mother. But when they reach sexual maturity, they often court and attempt to mate with an individual resembling their mother. This has led to some interesting experimental results. In one study, cockerels were imprinted by being shown cardboard boxes during their sensitive period. As mature roosters, they courted and even mounted cardboard boxes at mating time.

Despite the fact that some of the claims made about imprinting have been controversial, the concept has weathered criticism well. Although

**FIG. 19-18** Imprinting. (a) Konrad Lorenz demonstrating imprinting (on himself) of greylag goslings. (b) A bantam game cock on whom 11 guinea fowls had been imprinted during the first week after hatching.

(a)

(b)

it was long believed that imprinting did not occur in most species of animals, recent findings suggest that much behavior long regarded as innate actually has some imprinted components. Studies of monkeys show certain imprintinglike effects of early experience some of which may even prove to have important implications for human behavior (see Chapter 20).

### LATENT LEARNING

One type of behavior that obviously does not fit the conceptual framework of stimulus-response theory is what is called, for lack of a better term, **latent learning.** When introduced into a novel environment, many animals carefully explore their surroundings without any apparent benefit except the knowledge of the new area. Many animals seem ill at ease until they have explored their surroundings. Consider the example of a rat placed in an unfamiliar maze. The animal appears to learn something from exploring because, to obtain food, the same rat later learns to negotiate the maze more rapidly than rats not permitted to explore the maze beforehand. Latent learning is so named because it is not used immediately. It is easy to see how such learning ability would have adaptive value to an animal in the wild, enabling it to avail itself of the best escape routes when attacked or to capture its prey more efficiently.

## PSYCHOLOGICAL ANALYSES OF LEARNED BEHAVIOR

One area of study that ethologists had long left to comparative psychologists was that of conditioned learning. In recent years, however, ethologists have turned their attention to this interesting field of research.

### CLASSICAL CONDITIONING

**Classical,** or **Pavlovian, conditioning,** named after the Russian physiologist Ivan Pavlov (Fig. 19-19), who first drew attention to it, results from a training period in which two different types of stimuli are given in

**FIG. 19-19** The Russian physiologist Ivan Pavlov (1849–1936) and his research staff at a demonstration of classical conditioning.

quick succession to an animal. The first stimulus is called the conditioned stimulus and the second, the unconditioned stimulus. To demonstrate the effect, the **unconditioned stimulus** should be one that elicits a definite, measurable response, referred to as the **unconditioned response.** The **conditioned stimulus** should be one that does not normally elicit any observable response. One classic example of Pavlovian conditioning employed, as the unconditioned stimulus, the spraying of powdered meat into a dog's mouth. The dog's normal, unconditioned response to this stimulus is salivation. As the conditioned stimulus, the clicking of a metronome, which ordinarily does not cause a dog to salivate, preceded the giving of the meat powder. After a series of trials in which the clicking of the metronome preceded the application of meat powder, merely clicking the metronome resulted in salivation. The animal had thus learned to associate the clicking with the subsequent provision of food and salivated much as it would have upon smelling the appetizing food.

## OPERANT CONDITIONING

The American psychologist B. F. Skinner (Fig. 19-20a) conducted extensive studies of a type of learning called **operant conditioning** (1957). In an animal's natural state, operant conditioning is much more common and important than classical conditioning. It is based on positive and negative reinforcement.

***Positive and Negative Reinforcement.*** Operant conditioning results when the performance of a particular action is followed by a rewarding consequence. The phenomenon is most conveniently studied in a Skinner box (Fig. 19-20b). If the frequency with which a behavior is performed increases when it is followed by a certain consequence, that

**FIG. 19-20** (a) B. F. Skinner. (b) A rat in a Skinner box, a soundproof observation chamber equipped with keys, or levers, wired to electrical devices that release a food pellet when the key is pressed.

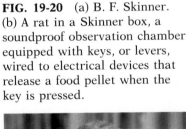

(a)

(b)

consequence is said to be a **positive reinforcer.** If the consequence relieves or abolishes an unpleasant situation, the consequence is a **negative reinforcer.** In either case, the animal tends to repeat the action. In many instances, both positive and negative reinforcement are achieved by a single action. When a thirsty animal drinks a pleasant-tasting drink, it rewards its taste buds while relieving its thirst; the former result is positive reinforcement, the second negative.* Extremely persistent conditioning is usually the result of such dual reinforcement. Operant conditioning in which a hungry animal is rewarded with small amounts of food forms the basis of trained animal acts involving dogs, seals, dolphins, and whales.

***Extinction of Conditioning.*** After pigeons are conditioned to obtain corn by pecking at a lever, the apparatus can be set so that no further reinforcement (no corn) is given for pecking at the lever. Nevertheless, the conditioned birds continue to peck at the levers at a rapid rate. From time to time, the unrewarded birds turn away from the machines. Once conditioning has been achieved, it may undergo slow extinction, but only if the conditioned individual continues to perform the operant behavior from time to time without ever receiving reinforcement.

***Intermittent Reinforcement.*** The persistence of operant behavior is greatly enhanced if not every response is rewarded. One experimental schedule of intermittent reinforcement employs regular time intervals, such as once per minute. Pigeons peck keys at a high rate if rewarded once a minute, more slowly if rewarded every five minutes. After undergoing conditioning in such a program, an animal usually slows its response rate immediately after reinforcement and increases the rate just before the time for the next reward. However, if reinforcement is provided at random intervals of time, the rate of pecking becomes steady. In one series, pigeons continued pecking at a rate of two to three pecks per minute for 15 successive hours. Moreover, when rewards were given on an average of every 50 pecks but with great irregularity, the animals pecked rapidly at a steady rate. Some pecked as fast as 18,000 times per hour for many hours.

It might be wondered what adaptive value such behavior has. Actually, the irregular, or variable, ratio of trial to reinforcement is exactly the situation found in nature when animals search for food. Food is seldom found on a regular time schedule but rather on a random basis from minute to minute, hour to hour, and day to day. The remarkable response of animals to a variable ratio of reinforcement is exactly suited to situations encountered in nature.

## DOES OPERANT CONDITIONING SHAPE HUMAN BEHAVIOR?

The question of what role operant conditioning plays in human affairs is controversial. But in many types of human activity, it has been useful and often economically profitable to proceed as if people are indeed conditioned by a variety of reinforcements. Much advertising, particularly on television, seems based on this philosophy. Appeals to buy a

* Negative reinforcement is not the same as punishment. It has been suggested that negative reinforcement *generates* a particular behavior (the thirsty animal is inclined to drink), whereas punishment *discourages* a certain behavior. The original inclination to engage in the punished activity persists after punishment is given, although the individual may modify its behavior to avoid punishment.

particular product are often made by suggesting, usually without expressly saying so, that use of the product will result in the user's becoming more virile or feminine, healthier, more respected, more sexually attractive, or more economically powerful. Parents often induce their children to study or to perform chores by regular rewards of money and approval, and parents themselves may be encouraged to work at jobs, often ones distasteful to them, for similar reasons. A concert pianist may practice 8 or more hours a day, day after day, to achieve a reward that may be years away. Many people are known to lead a long life of self-denial for a reward expected only after death. While operant conditioning may not be quite as important as B. F. Skinner suggests (he seems reluctant to concede any weight to genetic influence on behavior), it does seem to be a significant factor in human affairs.

## MALADAPTIVE OPERANT CONDITIONING IN HUMANS

While serving us in many lifesaving ways, operant conditioning also appears to promote much human misery. Prominent among the behavioral patterns in which operant conditioning seems to play an important part are compulsive gambling, habitual overeating, and various types of drug dependence, such as alcoholism. These are difficult problems for many millions of people. Their economic cost is staggering, and their cost in human suffering is incalculable. Although in a sense we have only begun to understand these behaviors, enough information is available to formulate a hypothesis explaining such behavior as largely due to operant conditioning.* While the validity of this hypothesis is argued by many, the greatest success in treating these behavioral problems has been by methods that, in effect, regard such behavior as conditioned.

In the case of each of these problem behaviors, a reinforcement, usually some kind of pleasure, has at one time followed its performance: the gambler won, the drug user had a temporary feeling of euphoric escape from serious problems, the eater experienced pleasure and the relief of hunger, and so on. Once an individual has been reinforced, or rewarded, one or more times after performing a particular behavior, he or she is thereby subconsciously conditioned to expect that reward for future performance of the behavior. The individual is said to have developed a psychologic dependence on the behavior. The strength of the dependence, or conditioning, depends on many factors. Some people have become compulsive gamblers after one big win; this is unlikely to occur when a millionaire wins a hundred dollars. Many people have become psychologically dependent on heroin after their first use of the drug. One person in 15 who begins to drink becomes an alcoholic. Although in most alcoholics, dependence develops only after 10 or more years of drinking, 10 percent of all alcoholics become alcoholics in their first year of drinking.

Among the conditions contributing to psychologic dependency are such factors as age, health, and stress arising over debts, job, marriage, or prospects for the future. The more serious these concerns are to an individual, the more he or she is likely to value the temporary satisfaction or escape provided by a drug, eating, or the bright promise of "win-

* This is not to say that operant conditioning is the only factor in such behavior. There is considerable evidence, for example, that people differ genetically in their susceptibility to overeating, alcoholism, and narcotic dependence.

ning big." Thus, the psychological state of the individual at the time of the experiences is one of the important factors in the conditioning effect they experience.

## MEMORY

While itself not a part of behavior, memory is essential to learning and thus to learned behavior. When an animal learns a task, such as how to run a particular maze, and demonstrates its learning upon subsequent testing, several events occur, as follows:

1. Certain information is received and associations are made, usually between one set of choices and positive reinforcement or between another set of choices and aversive (unpleasant) consequences or at least a lack of positive reinforcement.
2. The learning is stored in the central nervous system in a relatively permanent form, referred to as the memory trace, or **engram.**
3. The learning is recalled or drawn upon as a basis for action. (Recall is not automatic; some evidence suggests that a human, for example, permanently stores some information about most experiences of an entire life but is normally able to gain access to only a small part of the record.)
4. A decision is made to choose the positively reinforced behavior or avoid the alternative situation, and a directive is given to motor areas of the CNS to act on that decision with appropriate movements.

### TO LEARN IS NOT NECESSARILY TO REMEMBER

Much evidence indicates that learning and the establishment of a permanent engram of it are distinctly different processes, one following the other and separated from it by a considerable period of time in many cases. In other words, we can learn something and promptly forget it entirely. Most of our minute-by-minute experiences and impressions do not seem to be permanently recorded in our memory. Evolving such a means of handling information makes good sense: an animal making its way through a field or woods does not need to remember every branch, stick, leaf, and boulder for the rest of its life, but only long enough to negotiate its way around any that block its path.

***Protein Synthesis and the Establishment of a Long-Lasting Engram.*** Recently, it has been proposed that because chemical transmission at synapses depends on the synthesis of enzymes, experimental blockage of synthesis of brain proteins, and therefore of enzymes, would curtail memory to the extent that memory depends on synaptic transmission—or at least on protein synthesis. To test the theory, puromycin, a protein synthesis inhibitor, has been employed in a number of experiments.

In one experiment mice were trained to run a maze; before training, puromycin was injected into their brains at doses capable of reducing protein synthesis. One set of control animals, before being trained, was injected with salt solution or drugs known not to affect protein synthesis. A second control group was injected with puromycin a few days after the end of training.

The experimental animals (those receiving puromycin before training) ran the maze without difficulty. They were tested just after the training period, and their immediate recall was unaffected. But, as later testing showed, 15 minutes after training had ended, these mice had begun to forget; in 3 hours, they had forgotten almost all they had learned. Both control groups, however, retained the memory of the maze many days later.

This experiment demonstrated that blockage of protein synthesis had no effect on learning but profoundly affected the ability to consolidate learning in a long-term engram. This and similar experiments tend to confirm the view that memory is a two-stage process and that protein synthesis—perhaps the synthesis of enzymes and other structural proteins of nerve cells—is required for the second stage. The second stage may continue for several days and could depend on protein synthesis in more than one way. There is probably a lesson in such observations for students and all others whose business it is to remember things for more than a few hours. The evidence that memory is a two-stage process suggests that "cramming" the hour before an exam, while it may produce a short-term memory of what was read, does not ensure a long-lasting memory of it.

***Precisely Where Do Learning and Memory Reside?*** Most neurobiologists today believe that memory occurs through a modification of existing synapses. This belief is part of what is known as the synaptic theory. There is little support for one alternative idea, which suggests that learning promotes development of new nerve fibers. It is true, however, as experiments have proved, that neurons will not develop in a young individual or be maintained in an adult unless they are allowed to function normally by interacting with the environment. For example, when the eyes of newborn kittens are sewn shut, their optic nerves deteriorate. There is no proof, however, that use of neurons promotes their proliferation, beyond the normal early development characteristic of the animal's species. Except for such points as these, little is known about the biological basis of memory.

## Summary

Ethology is the study of the mechanisms and evolution of behavior. Of particular interest to ethologists are the stereotyped, unlearned components of behavior involved in such behaviors as defense, courtship, mating, nest building, brooding of eggs, and care of the young. The study of learned behavior, once the sole domain of psychologists, now occupies many ethologists as well.

The simplest behavioral response is the simple reflex, which is usually stereotyped and nearly always elicited by a specific stimulus. Taxes are automatic movements toward or away from a stimulus. A kinesis is a change in the rate of an activity in reference to a stimulus, whereas an orientation is a change in a position. Some moderately complex behavior of such lower animals as chitons has been shown to be composed entirely of combinations of simple responses such as taxes and kineses.

Drives, or motivations, are urges that animals have to accomplish certain goals, such as satisfaction of hunger, mating, nest building, and care of the young. Although little is known about most drives, it has been found that the hypothalamus contains a satiety center and an adjacent feeding center, the latter divided into two regions, one providing a feeding drive, or appetite, and the other a hunger center. Stimulation of the hunger center drives the animal to endure pain to get at food. If the satiety center is destroyed, the animal overeats but will not endure discomfort or pain to get at food. If the entire feeding center is destroyed, the animal declines food while starving. Animals whose satiety center and feeding center have both been destroyed also decline food, suggesting that the satiety center normally inhibits the hunger region of the feeding center.

Behavioral patterns have been classified by Ernst Mayr as genetically closed, if they cannot be modified by learning, and open, if they can be modified by learning. Much behavior is open, being a mixture of innate and learned components. Ethologists have sometimes classified components of a mixed behavior as appetitive, when consisting of searching movements, and consummatory, when composed of stereotyped, species-specific responses to finding the object of the search.

Both unlearned and learned behavior have a genetic basis. In unlearned behavior, species-specific, stereotyped responses are usually elicited by specific stimuli, called sign stimuli or releasers, whether or not the animal has ever before seen the actions performed. The ability to learn certain behaviors is also inherited and differs among animals, even among members of the same species. Like all inherited traits, both unlearned behavior and the ability to learn are shaped by natural selection. Among the inherited components essential for any behavior are an animal's sensory receptors, its nervous and endocrine systems, and its various effectors, such as glands, muscles, and other organs.

A fixed action pattern is a species-specific series of stereotyped actions in response to a specific releaser. Escape reactions and attack responses are examples. Humans appear to exhibit few fixed action patterns, all of which are usually overridden by learned behavior.

Complex behaviors such as nest building and courtship are typically composed of a series of stereotyped responses to individual stimuli and are part of a larger pattern of behavior that includes learned components. For example, during courtship, three species of ducks exhibit combinations of 20 separate, simple actions that are also performed individually in other contexts. The combinations are unique to courtship, however, and are species-specific. New combinations arise in hybrids produced by interbreeding.

Nearly all organisms exhibit regular oscillations of internal conditions, many of which are reflected in outward behavior. Most such rhythms are keyed to such cycles of nature as day-night, tidal, lunar, or seasonal cycles, thereby enabling adaptations to be made to the changing conditions these cycles produce. Biological clocks, which probably exist in all organisms except prokaryotes, are essential as an automatic standard of comparison for an organism engaged in a rhythmic or cyclic behavior. Their nature and location are not known.

Bird migration is a response to seasonal changes that requires a biological clock and involves one or more of a wide variety of navigational abilities, such as the ability to navigate by sun or stars, by following landmarks such as seacoasts and major rivers, and by the earth's magnetic field.

Learned behavior is the modification of response as a result of experience and is adaptive whenever an important relationship with the environment, such as the availability of food supplies, is changing. The simplest type of learning is habituation, learning to ignore repeated stimuli. Another type of learning is imprinting, in which an animal, during a brief, sensitive period early in its life, becomes influenced in a special long-lasting way. The object of the imprinting, usually the animal's mother, may influence the imprinted individual for the rest of its life, including in its choice of a mate. Latent learning is learning that lacks an obvious immediate reward. An example of latent learning is the knowledge gained by animals that explore an area when placed in new surroundings.

The psychological concept of conditioning is often used in the analysis of learned behavior. Classical, or Pavlovian, conditioning results from a training period in which a conditioned stimulus, which normally elicits no observable response, is immediately followed by an unconditioned stimulus, which normally does produce a response. After training, the conditioned stimulus alone produces the response normally elicited only by the unconditioned stimulus. In operant conditioning, the training consists of following the performance of a particular action with some kind of reward, classified as either a positive reinforcer (a pleasant sensation) or a negative reinforcer (removal of an aversive condition). Operant conditioning can undergo a slow extinction only if no reinforcement is ever given again. Intermittent reinforcement, especially if provided at random intervals, can result in frequent repetition of the behavior over a long time. It has sometimes been of practical value to regard much human behavior as conditioned. Much of modern advertising is based on this concept, as are certain approaches to curbing abusive eating, gambling, use of drugs, and other such behaviors.

Memory is of two sorts, short-term and long-term. Apparently, unless a process of consolidation occurs, most of our experiences leave no lasting traces, or engrams, in our nervous systems. Protein synthesis is involved in establishing a long-lasting engram. The blocking of protein synthesis in mice, while it has no effect on their learning, seems to result in a nearly complete loss of learning in three hours. Although there is evidence that memory occurs by the modification of existing neuronal synapses, not by a proliferation of new nerve fibers, little of its biological basis is known.

# Recommended Readings

(For Chapters 19 and 20.)

ALCOCK, J. 1983. *Animal Behavior: An Evolutionary Approach,* 3rd rev. ed. Sinauer Associates, Sunderland, Mass. One of the most interesting introductions to animal behavior; attention is given to human behavior, and evolution is emphasized.

BLAKEMORE, R. P., and R. B. FRANKEL. 1981. "Magnetic navigation in bacteria." *Scientific American* 245(6):58–65. An interesting account of how bacteria use magnetite to orient their burrowing into the mud.

CLUTTON-BROCK, T. H., and P. H. HARVEY, eds. 1978. *Readings in Sociobiology.* Freeman, San Francisco. An excellent assemblage of original essays by experts with various views on sociobiological questions.

DYER, F. C. and J. L. GOULD, 1983. "Honey bee navigation." *American Scientist* 71(6):587–597. A description of what is currently known about how bees detect the position of the sun, compensate for its apparent movement while in the hive, and communicate this in the dark to other bees.

FRISCH, K. VON. 1967. *The Dance Language and Orientation of Bees.* Harvard University Press, Cambridge, Mass. One of several fascinating books by a pioneer researcher of bee communication.

GAUTHREAUX, S. A., JR., ed. 1981. *Animal Migration, Orientation, and Navigation.* Academic Press, New York. A broad assessment of the ecology and evolution of migration and such underlying mechanisms as biological clocks and sensory systems. Each of the five chapters is written by an expert in the particular topic.

GHIGLIERI, M. P. 1985. "The social ecology of chimpanzees." *Scientific American* 252(6):102–113. The uniquely flexible social order of chimpanzees allows them to adapt to fluctuations in the abundance of food.

JOLLY, A. 1985. *The Evolution of Primate Behavior,* 2nd ed. The Macmillan Series in Physical Anthropology. Macmillan, New York. Very interesting accounts of primate behavior. Available in paperback.

MORSE, D. H. 1982. *Behavioral Mechanisms in Ecology.* Harvard University Press, Cambridge, Mass. An updated summary of the subject in paperback.

RUSE, M. 1979. *Sociobiology: Sense or Nonsense?* D. Reidel, Boston. A philosopher provides a reasoned assessment of sociobiology and some of the criticisms of it.

SAUNDERS, D. C. 1978. *An Introduction to Biological Rhythms.* Halstead Press (Wiley), New York. A brief, interesting account of adaptations to various environmental cycles by corresponding physiological rhythms.

SAUNDERS, D. S. 1982. *Insect Clocks,* 2d ed. Pergamon, New York. Excellent!

SCHELLER, R. H., and R. AXEL. 1984. "How genes control an innate behavior." *Scientific American* 250(3):54–62. A family of genes code for egg-laying behavior in a marine snail.

SHETTLEWORTH, S. J. 1983. "Memory in food-hoarding birds." *Scientific American* 248(3):102–110. One species of birds remembers where it has put thousands of seeds for as long as several months.

SKINNER, B. F. 1976. *About Behaviorism.* Random House, New York. Paperback edition. Skinner expounds upon his controversial thesis concerning the effect of operant conditioning on human behavior.

TINBERGEN, N. 1984. *Curious Naturalists.* University of Massachusetts Press, Amherst. Paperback edition. A delightful classic essay by a Nobel laureate ethologist.

WALLACE, R. A. 1979. *The Ecology and Evolution of Animal Behavior,* 2d ed. Goodyear Publishing, Santa Monica, Calif. One of the best introductory books on behavior. Brief, well-illustrated, and very interesting.

WILSON, E. O. 1975. *Sociobiology.* Harvard University Press, Cambridge, Mass. The book that forms the basis of a controversy that may continue indefinitely.

WILSON, E. O. 1978. *On Human Nature.* Harvard University Press, Cambridge, Mass. Wilson extends sociobiology to humans in an exciting essay.

# 20

# Social Behavior

OUTLINE



Animal **sociality,** or living in groups, exemplifies several interesting principles of evolution and behavior that warrant special consideration. As will be noted from the following accounts of behavior in social animals, sociality has no unique behavioral components. Social animals, from insects to mammals, exhibit fixed action patterns as well as learned behavior. They have drives and engage in reproductive behavior. Almost all social animals can learn at least a little from experience; some learn a great deal. Perhaps the one thing that might be said to be unusual about the behavioral components of social life is the degree to which social animals have evolved methods of communication. Selection pressures for adequate communicative behavior are quite strong for animals that live in groups.

## THE IMPORTANCE OF BEING SOCIAL

### THE COMMON DEFENSE

Many species of animals that form groups assume defensive formations when attacked by predators. Their combined force is usually enough to discourage predators from attacking their preferred prey: undefended

**FIG. 20-1** A musk ox herd drawn up in a defensive ring against a wolf attack. This sketch was made by D. B. MacMillan of International Wildlife Protection, from a photograph.

young or ailing adults. A classic example of such defense is provided by the arctic musk ox, which normally travels in herds. When attacked by wolves, the adult males form a ring around the females, calves, and any injured males; they face outward in a defensive stance toward the enemy (Fig. 20-1).

Many birds and fish that form flocks and schools, respectively, exhibit special avoidance reactions when attacked. These actions confuse predators by preventing them from devoting all their attention to one individual. Starlings, for example, mass tightly together when attacked, making it difficult for a falcon or other bird of prey to attack without being injured by colliding with one of them (Fig. 20-2). The tendency of most predators to attack only stray animals exerts a strong selection pressure against any inherited tendency of a prey animal to stray from the group, while exerting a strong selection pressure in favor of the tendency to form a group, at least when under attack.

## DIVISION OF LABOR

Adaptive advantages also accrue to a society in which the tasks of individual members are specialized. Where competition between species is keen, such a society, by being more efficient in finding food, for instance, is more likely to survive than one in which each member must perform all essential tasks.

The most highly evolved social insects not only have a division of labor among the members of a colony, but each colony has several morphologically distinct types of individuals. Some termite species, which represent an extreme example, include up to five or more types of individuals (such as workers, soldiers, and reproductives), each with distinct morphological and behavioral characteristics. In insect societies, the various jobs performed by individual workers are unlearned and are under fairly rigid genetic control. In some species, an individual may progress from one set of duties to another at various stages in its life.

**FIG. 20-2** A flock of starlings (a) before the appearance of a falcon and (b) afterward. The tight formation makes it dangerous for the falcon to attack, because it might be injured in diving into the flock.

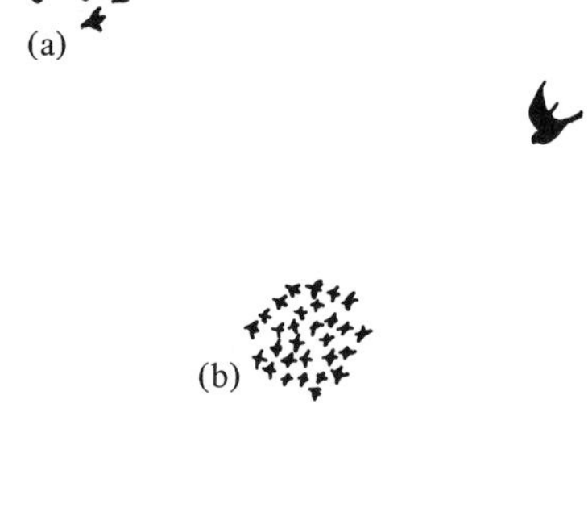


## REPRODUCTION AS A GROUP ACTIVITY

There are reproductive advantages to societal grouping. An obvious one is that most individuals are virtually certain to encounter a mate of the same species. In many social animals, moreover, reproductive behavior is triggered in an individual by its presence in a group of a certain size or larger. Conversely, in several species, when the population size drops below a certain figure, the group stops reproducing. But this is not a hard and fast rule; in some species, there is evidence that when the population drops below a certain minimum size, the breeding season is extended, thus providing greater compensation for predation and other adverse factors. Moreover, it also appears, at least in some species, that when a population reaches a certain maximum size, breeding is curtailed.

Reproductive behavior can be regarded as a group activity in which the group is the family, in some cases made up of only the two parents. Indeed, many ethologists believe that most social behavior evolved from family behavior. In this sense, the school, flock, or herd is a kind of extended family. Some social groups, such as prides of lions, are extended families in the literal sense.

## COMMUNICATION

Because some form of communication is essential for social behavior, it is not surprising that several means of communication have evolved in social animals. Visual, auditory, and chemical communication are three major types; they are based on sensory perception by the eyes, ears, and chemoreceptors, such as those of the nose, respectively.

### VISUAL SIGNALS

Instances in which visual signals are the primary means of communication have been observed in a wide range of species. For example, North American deer communicate alarm by displaying their white tails (Fig. 20-3). Showing the tail appears to aid in a coordinated escape of family groups. Rabbits and many birds also display otherwise hidden markings when fleeing. Their display appears to act as a releaser, triggering others to follow.

**Aggressive displays** are usually made by males to other males of the same species. Such displays occur when one male encroaches on a feeding ground or other defended territory of another. The display is often

**FIG. 20-3** The white tail of the white-tailed deer is normally not visible. When the deer is alarmed, the tail is displayed, apparently as a signal to others in the group. The deer on the left has just begun to raise its tail.

**FIG. 20-4** Threat display by a male mandrill baboon *(Papio sphinx)* of West Africa.

largely visual and involves an attack motion in which the organs of attack, such as teeth, are exhibited, as in the case of the baboon shown in Figure 20-4. Sometimes the same gesture is used at breeding time as part of a courtship ritual.

## SOUND SIGNALS

Probably without important exceptions, animal sounds all have special functions. Contrary to some of our fondest beliefs, pleasant animal sounds, particularly the more delightful bird songs, are not expressions of joy and gladness. In every case that has been analyzed, they are specific vocal signals. For example, a male bird singing from a prominent song post near his nest is proclaiming to other male birds of his species that the area within earshot is his territory.

Birds emit a rich variety of sounds, as do mammals. Like visual signals, vocal signals vary in meaning from alarm and threat to care and courtship. Auditory signals play several roles in reproduction. A few examples of communication by sound are described below.

***The Evolution of Stridulation in Insects.*** **Stridulation,** the production of sound by rubbing appendages together, is typical of many arthropods. The stiff exoskeleton of insects, well suited to the production of audible vibrations, made the evolution of stridulation a most likely occurrence in this group. Studies have indicated that stridulation probably arose independently several times in arthropods. The behavior of some species suggests that stridulation first occurred in species in which visual courtship displays by males were sometimes accompanied by sound production. It is thought that eventually, through a process called evolutionary substitution, the sounds were produced in place of the visual display. For animals that are camouflaged as a defensive adaptation against predation, auditory signals are especially useful in locating a mate.

***Tone-Deaf Insects.*** The sounds emitted by many insects have been recorded and analyzed with sophisticated techniques. It was found that these sounds do not vary in frequency (pitch), nor do insect sound receptors detect frequency variations. In other words, insects are literally tone deaf. Insects do vary the rate at which a sound pulse is repeated. Chirps (consisting of a few pulses) in one species of cricket, for example, will differ from those of another species by the length of the interval between the pulses that comprise the chirp. Trills (long series of pulses) vary in two ways: in the length of intervals between pulses and in the total number of pulses in each trill. Both types of variation can be detected and responded to. Surprisingly, experiments have shown that katydids learn to lengthen or shorten their chirps in response to hearing longer or shorter chirps around them. Unlike katydids, crickets never seem to learn anything new when it comes to chirping.

***Frog and Toad Talk.*** Many frog and toad calls, especially those of "cricket frogs," sound to human ears like insect calls. However, analyses of the sounds have shown that, unlike insects, frogs and toads emit and detect sounds of varying frequencies. It is thought that they distinguish species partly on this basis. In addition, like crickets, they vary

**FIG. 20-5** Male toads, such as this *Bufo americanus*, call from shallow water.

and detect variations in sound amplitude, or intensity. Some can be as loud as 100 decibels—the intensity of a riveting hammer! Because of the nature of their vocal cords, the larger males emit the deepest sounds. Frog and toad croaks function mainly to attract females to the males (Fig. 20-5). When a male frog in breeding condition approaches any other frog of its species, it mounts it. If the mounted animal is a male or a female that has already shed its eggs, it promptly emits a vigorous croak, whereupon the other promptly dismounts. If it is a female heavy with eggs, it remains silent. In frogs, silence means assent.

***Sounds of Bird Land.*** Birds rely heavily on auditory communication. Each species emits and responds to a variety of sounds. A herring gull on its nest is alerted by the mew of its returning mate; it can distinguish and respond to the call from among all the similar mews it hears from hundreds of other nearby gulls.

The males of many species of birds occupy a territory at mating time and, by means of song, proclaim it as theirs (Fig. 20-6); such singing is especially evident at dawn and dusk. Mature birds normally have developed a song that is very characteristic of their species. Studies have shown that in some species much of the song is learned from the song of parents and other birds of the same species in the vicinity. When birds of such species are raised in isolation or with birds of other species, they sing songs that are quite atypical of their species.

**FIG. 20-6** A male mocking bird proclaiming his territory by singing from a prominent song perch.

***Mammalian Vocal Signals.*** Most mammals communicate with other members of their species to some extent by sound; in almost all cases, however, their signals are less complex than those of birds. This is especially true in temperate zones, where mammals rely most heavily on olfactory communication. But tropical mammals, especially higher primates, carry on a considerable amount of vocal communication in addition to their visual displays. Both bats and rodents can communicate within species by ultrasound, frequencies too high for the human ear to detect. Porpoises and whales also apparently carry on extensive auditory communication, often at distances of several miles, by means of a great variety of sounds that travel well through water.

## CHEMICAL COMMUNICATION

**FIG. 20-7** An Atlantic salmon leaping a waterfall of the Matamek River as it returns from the sea to spawn.

Only recently have scientists begun to gather much precise data on the important medium of chemical communication. Although neither of the following examples involves communication between animals, they dramatically illustrate how exquisitely refined an animal's chemical sense can be.

After a life spent in the sea, salmon find their way back, by a chemical sense, not just to the same river, but to the particular small tributary in which they began life years before (Fig. 20-7). They not only discriminate well among combinations of different chemicals, but, at least in this regard, have a good memory besides.

The second example is proverbial. Bloodhounds not only can discriminate among the body odors of all humans, but can follow for many miles the faint trail of body odor left hours earlier on dry surfaces by someone wearing shoes.

PANEL 20-1

# Communication in Bees

In no invertebrate is communication as highly developed as in the honeybee. The various signals that bees give each other involve touch, hearing, taste, and odor, all of which are useful in the darkness of a hive. The functions served by bee communication include defense of the colony, scouting for a new colony site when swarming, confrontations between queens, mating, differentiation into worker castes, and foraging for food. The last of these is the subject of this brief summary.

In the pioneer work of the Austrian zoologist Karl von Frisch,* much of it done with simple materials during World War II, dishes of scented sugar water were set out at various distances and in various directions from an observation hive. When bees visited any dish, they were marked with distinctive paint spots that, in code, gave each a number. Bees returning from the various feeding stations could thus be distinguished from each other after they had entered the observation hive.

## THE ROUND DANCE MEANS "LOOK NEARBY"

In a typical experiment, bees would not visit any of the dishes for quite some time. But after the first one or two forager bees had stopped at a dish, it was soon visited by a number of bees. If the dish was no more than about 50 m (55 yards) from the hive and contained a rich sugar solution, the foragers on their return to the hive performed a vigorous "round dance" on the surface of the honeycomb. In this dance, the bee turns in a tight circle almost a full 360°, then traces a similar circle in the opposite direction. The dance is performed repeatedly. (All dances are normally performed on the vertical face of the comb. The observation hives were usually kept in dim light inside huts from which the bees could not see to the outside.)

Von Frisch noted that whenever a forager returned and did a vigorous round dance, other bees gathered around it and became excited. These bees, the recruits, then left the hive and began to search in the vicinity of the hive. They appeared not only at the station visited by the first foragers, but at all of the dishes placed within 50 m or so of the hive. Apparently, then, a vigorous round dance is a signal that triggers searching behavior in other bees, prompting them to look for food near the hive.

* For his work with bees, von Frisch shared a 1973 Nobel Prize with two other ethologists: Konrad Lorenz of Austria and Nikolaas Tinbergen of the Netherlands.

Karl von Frisch.

## TAIL WAGGING DENOTES GREATER DISTANCE

In another experiment, von Frisch allowed a number of bees to begin feeding at a station close to the hive. He then began to move the dish farther and farther from the hive. At first, the round dances continued as before. Then, when the dish had been moved to between 50 and 100 m from the hive, two curious things happened. First, the bees returning to the hive changed to another type of dance. Now they made a half circle to the left, following it with a short, straight run across the surface of the comb during which they waggled from side to side; this was then followed by a half circle to the right and another body-waggling run in the same direction as the first. Von Frisch dubbed this the tail-wagging dance.

The second change in the bee's behavior when the dish had been moved to greater distances was just as interesting. Now, instead of searching in all directions, the recruits concentrated their search in the direction of the food.

## IN WHAT DIRECTION?

Apparently, tail-wagging dances contained additional information, namely, the direction in

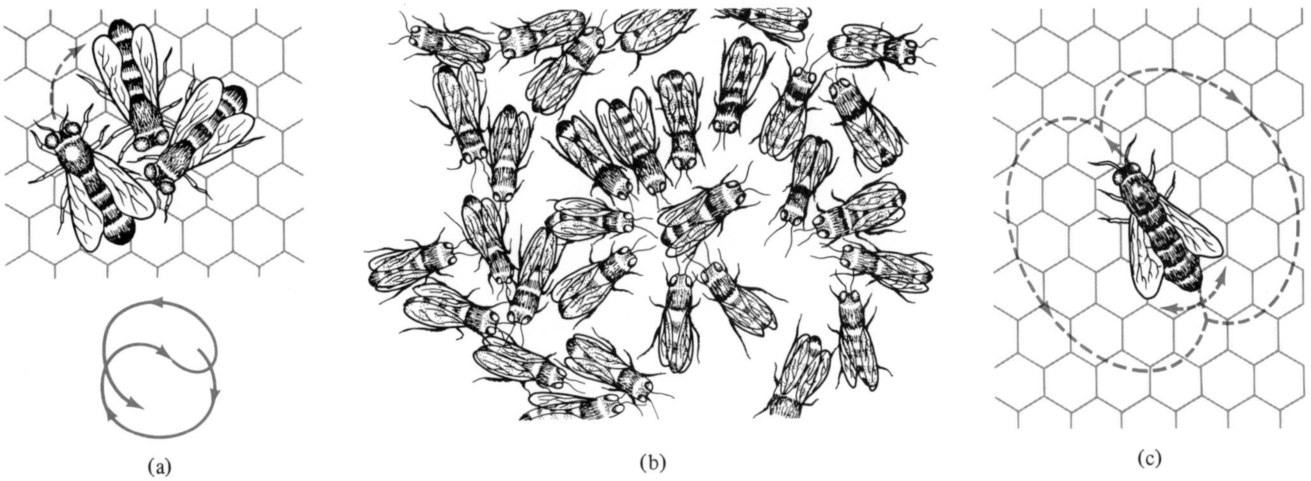


(a) The round dance of the worker honeybee. (b) The bee performing the dance is the one at the center of the picture. Most of the other bees have gathered around it. Worker honeybees gathering around a returned forager can be seen to receive regurgitated nectar from her. (c) The tail-wagging dance of the worker honeybee.

which to look. But how could a bee, in the dark, and on a vertical surface, communicate a horizontal direction? Von Frisch's first clue to this puzzle came when he noticed that two groups of marked bees returning from stations in opposite directions from the hive performed the straight runs of their dance (the part in which they waggle their bodies) in opposite directions. Somehow, he said, the bees must be communicating the direction to go by the direction of the tail-wagging part of their dance.

But a bee would need even more information than this to make use of the cue. It would need some reference point for the signal (the tail-wagging in the dark on the vertical surface of the honeycomb) to be translated into which direction to fly in order to reach the food. What reference point could be used inside the dark hive? And what sort of compass could be used outside the hive?

The key clue to the compass problem was uncovered when von Frisch noticed one day that bees returning from a certain feeding station—which had not been moved—gradually changed the direction of the tail-wagging part of their dance during the course of the day. This suggested to him that the reference point they were using outside the hive to determine direction was one that was slowly changing its position during the day. That reference point, of course, would be the sun! (As explained in Chapter 19, animals using the sun as a compass must have an internal biological clock as well.)

Two problems remained if von Frisch was to break the code used by the bees to communicate directions. How could a direction in reference to

*(Continued)*

Correspondence of the direction of the tail-wagging dance on the vertical surface of the honeycomb and the location of the food source in reference to the sun's position. In bee language, if the straight (tail-wagging) part of the dance is away from gravity, it prompts the recruited bees to fly toward the sun.

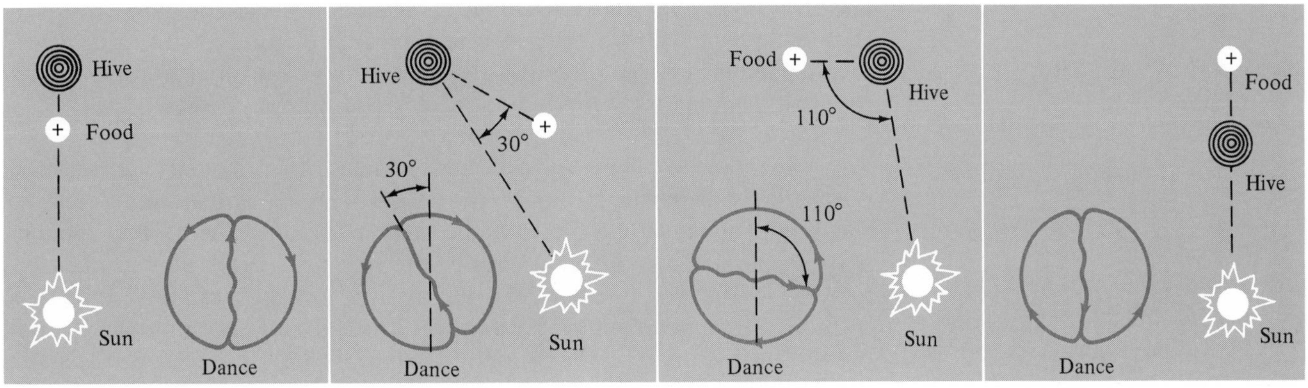

## PANEL 20-1 *(continued)*

the sun be indicated inside a dark hive? And how could any horizontal direction be indicated on a vertical honeycomb? The answer to both questions lay in a simple fact: the bees make use of gravity, something that can be sensed in the dark on a vertical surface. In bee language a straight, tail-wagging run away from gravity conveys the stimulus to search for food by flying toward the sun. A straight, tail-wagging run toward gravity signals "fly away from the sun." The system of indicating direction proved to work for any other point on the compass as well. For example, if the food is 75° left of the sun, the tail-wagging runs will be 75° to the left of the next vertical.

There was one direction for which the system could not be used, and that was up. Von Frisch tested this possibility by placing an observation hive at the foot of a tall radio tower. Then, after bees had begun to visit a dish of food near the hive, he gradually moved it up the ladder to the top of the tower. Bees that had already been recruited continued to find the food as it was moved to the top of the tower. But when they returned to the hive at the foot of the tower, the dances they did were hopelessly confused. It seems that bees have never evolved a way to tell each other about good sources of food to be found on plants several hundred meters tall.

### HOW FAR AWAY?

Bees routinely communicate other information about food sources to each other. One of the most important of these is the distance of the food source from the hive. In von Frisch's early experiments, he noted that bees returning from distant food sources did a slower tail-wagging dance than did foragers who had visited a nearby source. For example, if the source was only 100 m (328 feet) away, the bees danced at the rate of 40 complete cycles per minute, whereas those returning from 1,000 m (3,280 feet) away completed only 18 cycles per minute.

In recent years, it has also been found that during the straight part of the run, a dancing bee emits a buzzing sound, a vibration of about 200 to 250 Hertz, the frequency to which the receptors

***Pheromones.*** Chemicals released by animals and which influence the behavior of others of their species are known as **pheromones** (from the Greek *pherein*, "to carry"). Many of the functions served by various pheromones are similar to those mediated through visual and auditory signals. Some moths, for example, employ pheromones as sexual attractants. When female silkworm moths are sexually receptive, they release a pheromone that enables males to find them as far as 60 m (about 200 feet) away. The males fly upwind in a zigzag pattern, recrossing the stream of scented air again and again, much as an airplane pilot follows a radio beam for an instrument landing.

Knowledge of the role of pheromones as sex attractants has been used to some advantage in the control of such pests as the gypsy moth in the northeastern United States and the Mediterranean fruit fly in California. Traps are baited with the species-specific pheromone, luring unwitting males to their deaths in large numbers.

***Sex Attractants of Vertebrates.*** Many vertebrates emit sex attractants, particularly at mating time. In the musk deer, the substance known as musk is produced by a gland located near the anus. Musk has a penetrating, persistent odor that even some humans, with their modest olfactory sense, can detect in the air at dilutions greater than one to a million. It has been suggested that an evolutionary explanation for the presence of human axillary (armpit) hair is that it readily traps the musklike secretions of the apocrine glands of the skin. These modified on bee antennae are most sensitive. The duration of the run and the number of such buzzes given (buzzing is done only during the straight part of the run, not during turns) is even more precisely correlated with the food's distance from the hive. The farther away the food, the longer the runs and the more buzzes per run. The recruits, who gather closely around the returned forager, probably can detect the buzzing even if they might be prevented from actually touching the forager. Apparently, the recruits can determine the direction of the straight run from the sound of the buzzing and are thus stimulated to fly in the right direction. In any case, this method of communicating in the dark is not perfect; a recent series of experiments indicates that a recruit must follow a forager's dance for at least six complete cycles before it consistently flies in the right direction and for the right distance.

Forager bees were also shown by von Frisch to communicate both the scent and the taste of the nectar to the recruits. Unless the journey is very long, the scent clings to the forager's body; other bees then detect it with their antennae. The taste of the food, however, is communicated by the simple expedient of regurgitating some of it and feeding it to the other bees during the dance. This activity only became obvious from studying slow-motion films of the bees' dances.

Von Frisch also showed that a bee feeding at an excellent food source emits a pheromone from a gland at the rear of its abdomen. Passing bees are attracted by the odor and drop down to join in the meal.

It is worth emphasizing that there is no evidence, nor any reason to believe, that what bees or other social insects "learn" from each other is learning in the same sense that dogs learn tricks or that humans acquire knowledge. In other words, bee behavior is largely based on genetically closed programs. Young bees do not need to learn that a run away from gravity means "fly toward the sun." Nor do recruits need to "keep in mind" any of the rules on how to use information from others to find food. Rather, recruits respond automatically and correctly to each cue, just as a young chipmunk instinctively scurries for cover at the first sight of a hovering hawk. Nor does a returning forager bee learn to perform the right dances. The performance of the appropriate dance is automatically triggered by what the bee found and where it found it. The bee does what it does without giving it what we would call a thought.

sweat glands are concentrated in the axillary and genital regions and are 75 percent more plentiful in women than in men.

***Some Pheromones Are Alarming.*** When injured, many species of animals release substances that frighten away members of their own species from the area. Minnows will avoid an area for many hours where another minnow has been injured. Many snails, earthworms, and sea urchins also release alarm substances when injured. Social insects, without themselves having been injured, typically release alarm pheromones when their colony is invaded; this triggers defensive behavior in other members. When frightened, deer emit a special scent that evokes alarm in nearby deer.

***Labeling with Scents.*** In many animals, special scents are used to designate trails, food sources, home territories, and the like to other members of their own species. Many social insects—ants and some bees for example—release scents that guide other colony members to good food sources. Mammals are notorious for marking their territories with glandular scents and with urine or feces. Deer, for example, identify other members of the herd and occasionally stake out territories with scents produced by glands on the forehead and hind legs. Urine is also used. The sniffing of trees and urinating against them by dogs (Fig. 20-8) is a functional equivalent to some of the listening and singing done by male birds. In marking territories, birds sing and dogs urinate.

FIG. 20-8 A dog sniffing the base of a tree.

## AGGRESSION

Aggression is among the most conspicuous of all animal activities and one that has long fascinated students of behavior. It is found throughout the animal kingdom but is best known among vertebrates, where it has been the most thoroughly studied. Although the examples cited here are all drawn from vertebrate behavior, similar examples could be described for many invertebrates.

### TYPES OF AGGRESSION

**Aggressive action,** as the term is used here, includes attacks, threats, and other overt signs of hostility between two or more animals. There are several types of aggression. One is the attack of a predator upon its prey. But this is not typical aggression in the ethological sense. The prey usually does not evoke hostility in the predator any more than the sight of our dinner does in us. In many cases, predators have been kept together with their traditional prey without displaying signs of aggression—at least as long as the predators were well fed. Two other kinds of aggressive behavior are of much more interest to ethologists: that of prey against its predator and that between members of the same species.

***When the Hunter Becomes the Hunted.*** When flight is not possible, many animals will counterattack a predator. "Fighting like a cornered rat" is a proverbial example. One type of counterattack that epitomizes aggressive behavior is the ferocious response of a parent, especially a mother, to any threats to its young (Fig. 20-9a).

Another interesting aggressive response to predators is the phenomenon of mobbing, which is the attack upon a traditional predator by a group of prey animals whether or not the predator is threatening any of them at the time. Zebras have been known to "gang up" on a leopard

(a)

(b)

FIG. 20-9 (a) Hartebeest mother defending her calf against an attack by cheetahs. (b) A mounted crow being attacked by a redwing blackbird. The crow often raids other bird's nests for eggs and young.

when a herd of them has caught one away from its usual cover. Jays or crows on nest-robbing forays are usually set upon by the birds whose nests are being robbed. The alarm cries of the robbery victims attract others of the same prey species; together the group may be able to drive the robbers from the neighborhood. Jays and crows that are merely passing by may also be mobbed by birds of other species (Fig. 20-9b). Jays and crows, for their part, have been seen to mob a cat on the ground whether or not the cat was stalking a bird at the time. But these instances of interspecific aggression, despite the interesting evolutionary questions they pose, are somewhat unusual; they are not the types of aggression with which ethology is principally concerned.

***Intraspecific Aggression.*** By far the most common application of the term *aggression* in ethology is to intraspecific aggression. This form of aggression is typified by the battle of two males of the same species for control of a feeding or nesting site or of a mate. It has been suggested that this type of aggression may to some degree underlie some of the fights within human families, between neighbors, in barrooms, and in wars.

Most fights between members of the same species have a distinctive quality: they are usually not fights to the death, although they can be. Usually, at some point, one animal "wins" when the other gives in. In many species, the fight is more ritualistic than real and may consist entirely of threat displays, bellowing, or other harmless demonstrations of strength and vigor. Submission, or acknowledgment of the other's dominant position, is signaled to the victor by leaving the area or by a certain species-specific, ritualized behavior.

In groups of social animals, such as flocks of birds or packs of wolves, in which the animals live close to each other within a small area, final submission is usually signaled by a ritualized display by the defeated animal, often one in which the most vulnerable part of its body is presented to the victor. A defeated wolf, for example, will expose its throat to the victorious wolf, lower its tail, and give other signs of submission. Subsequent encounters between the victor and the vanquished seldom result in fighting. Instead, the dominant wolf assumes a threatening posture, whereupon the other usually adopts a submissive pose, as exemplified in Figure 20-10. Once superiority has been established, it is thus usually merely acknowledged by both sides thereafter, thereby avoiding the expenditure of energy and the danger to both individuals that repeated fighting would involve.

**FIG. 20-10** Postures assumed by dominant and subordinate female tundra wolves *(Canis lupus tundarum)* during mating season. Note the raised tail and ears of the dominant wolf on the right.

Because of the potential of intraspecific aggression for self-destructiveness, a question that comes to mind when we consider it is, What function does it serve? Actually, as is briefly described in the following sections, it has several adaptive functions within the two principal contexts in which aggression is exhibited: dominance hierarchies and territoriality.

## DOMINANCE HIERARCHIES

Whenever several individuals of the same species live in a group, certain rules of behavior and lines of authority are needed if optimal efficiency is to be achieved. Put another way, in social groups, a selection pressure exists in favor of modifying aggressive behavior and promoting cooperation, usually in the form of a **dominance hierarchy.**

***Dominance in a Wolf Pack.*** An obvious advantage of cooperative action can be seen when wolves hunt larger animals that might readily escape a single individual. A successful hunt often demands not only cooperation, but that one individual be in charge. If each male were to hunt alone, laying claim to an area and keeping all others out, the animals' food intake would be greatly reduced. Because aggressive tendencies are very strong in wolves, the submission of a somewhat less aggressive animal to the pack leader becomes important to the cooperative stalking of the prey and the success of the hunt.

In a wolf pack, when the established leader appears, the others gather around him as shown in Figure 20-11, wagging their tails, whining, licking his face, and gently biting his muzzle, all signs of subservience. These displays act to quiet the leader's aggressive tendencies. He responds with a show of ferocity, however, when other males challenge his position of dominance.

Important as is the submission of the vanquished to prevent his being severely injured or killed, it is just as important that the desire of the victor to attack and kill be suppressed whenever the vanquished individual displays the signs of submission. These complementary behaviors appear to have evolved together.

The behavior of wolves sheds some light on the behavior of many domestic dogs. To the dog, its master is the dominant member of its pack. Once the relationship is established, the dog is anxious to do the master's bidding. In some breeds of dogs, the relationship with its first master persists for the rest of the dog's life, even if ownership passes to a new master. Members of such breeds are known as one-person dogs.

***Pecking Orders and Other Dominance Hierarchies.*** Most societies, whether of birds, wolves, or primates, are based on dominance hierarchies. Each animal in a group is usually dominated by others above it in the hierarchy while dominating those below it. The only exceptions are the individual at the very top, called the alpha individual, and the one at the very bottom, the omega individual.

**FIG. 20-11** A tundra wolf pack leader (center) surrounded by his subordinates.

(a)

(b)

**FIG. 20-12** Dominance hierarchy in a flock of three hens. (a) The alpha hen drives away beta by pecking at its head. Omega leaves the scene. (b) When alpha leaves, omega approaches but is driven away by a peck from beta.

An unusual but often cited dominance hierarchy is the pecking order normally established in flocks of chickens. Chickens within a flock usually display aggression by motions to peck other birds, especially on their heads. Submission is signaled by lowering the head. The behavior is unusual in that each bird has a position in the flock: the alpha bird has the privilege of pecking all others, whereas the omega individual may be pecked by any of the others (Fig. 20-12). Moreover, if a new bird is added to the flock, the birds spend much time in threats and confrontations until the newcomer's place in the pecking order is established.

***Other Benefits of Dominance Hierarchies.*** Besides eliminating unnecessary fighting within the group, a dominance hierarchy tends to unify the group and increase its efficiency. The dominant members are usually the older and more experienced individuals, whose leadership is valuable to group unity and success. In monkey colonies, signals of alarm may even be ignored unless given by a higher-ranking member. All this contributes to the efficiency of food acquisition, competition with other groups of the same species, and defense against predation or other dangers.

## TERRITORIALITY

Natural selection favors individuals that are most fit, by reasons of their physiology and behavior, to survive and reproduce. Only then can they pass their traits on to the next generation. Fitness includes an individual's ability to obtain an adequate supply of food, not only for itself, but also for its offspring until they are able to get their own. In many vertebrate species, this requirement has been met in part by behavior known as territoriality (Fig. 20-13a). In American robins, a fairly typical example, each family maintains control, enforced by the male parent, over the immediate area around the nest. Depending on the abundance of local food resources, the territory may include up to a fourth of a hectare (nearly half an acre). Other robins are attacked and driven away whenever they enter the area.

Territoriality has a number of benefits. First, it ensures an effective distribution of the food supply. Because no more than a single family of

(a)

(b)

**FIG. 20-13** (a) Gannets in a territorial dispute, Cape St. Marys, Newfoundland. (b) A male American robin feeding his nestlings.

robins lives on the resources of a given territory, there is ordinarily enough food to support them through the summer. Second, territoriality tends to achieve an optimal distribution of the breeding population over all areas suitable for robins to occupy; this means that the animals that win territories have the best possible chances of successful breeding. Territoriality may also result in population control. A pair of American robins (Fig. 20-13b) can hatch 12 offspring in 1 year. This means that an area that supported two robins and their 12 offspring one year might need to support 98 birds the next year alone if all 14 returned to the same nesting site. All other robin families at all other nesting sites in North America have the same potential for reproduction. During the year, of course, weather, predators, various diseases, and perhaps starvation claim many of the birds, but by no means all. In some years, over half of them survive. But were the robin population to double every year, it would soon deplete the resources on which it depends. However, in any given year, robins that do not acquire a territory in which they can establish a family do not breed that year. A large number of robins is thus placed on a reproductive "waiting list." In times of severe weather or drought, birds may die by the thousands. Yet the next year's population may be of nearly average size after the breeding season because the robins' reproductive potential is so great. Part of that potential is the "reserve" of birds that are capable of breeding but normally do not because of exclusion by the territorial behavior of others.

Despite the fact that territoriality may benefit a species as a species, the benefit is an indirect result of individual selection. Like other adaptive qualities, territoriality is selected for when territorial individuals leave more offspring than do nonterritorial individuals. The fact that other members of the same species also benefit is a side effect of the behavior.

***Laying Claim to a Territory.*** Territoriality is most evident to human observation in an animal such as the robin just before the bird's breeding time. When male robins arrive at their nesting areas in the spring, each spends much of its time laying claim to an area of up to about a quarter hectare. In the first few spring days, there are many aggressive encounters between the males, who chase each other with threat displays and loud cries. As in many species of animals, these fierce encounters do not result in the wounding of either participant. Between skirmishes, a victorious male sings from a prominent post that usually commands a view of much of the territory. Thus he announces that this is his territory and that he intends to defend it against any other male robin that would dare to intrude.

It has been shown that other robins can judge from the quality of the song whether a particular male is capable of being readily dispossessed of his territory or whether he is probably too strong or too belligerent to be challenged. If another male robin enters the area, he is immediately attacked. But sparrows and most other bird species, even in the same tree, are ignored. Usually, only those other bird species that compete for the same resources are attacked.

***Small Territories.*** Many animals obtain their food outside the immediate vicinity of their nesting sites. Fish-eating birds, such as gulls and penguins, are examples. Such animals would not benefit by defending a large territory. The area they defend may be less than a meter in diameter or only as large as the nest itself. This is little more than **individual**

**FIG. 20-14** Spacing of nests in sea birds (Cape gannets, South Africa). When food is not obtained in the immediate area, nests can be very close together.

**space:** the area immediately around an animal that it aggressively defends against the intrusion of others. The nests of gulls, cormorants, and other fish-eating birds are sometimes so tightly clustered together that it is difficult for a human to walk among them (Fig. 20-14).

Although territorial size is reduced in animals with closely arranged nesting sites, their nests, mates, eggs, and young are almost always defended against intruders; in such cases, specialized aggressive behavior is awakened at nesting time.

***Some Special Types of Territoriality.*** Pet dogs are at their fiercest when defending their home (their territory) against the intrusion of strangers or another dog. Away from its home, a dog's behavior is noticeably different. Both dogs and cats, as well as many other mammalian species, exhibit territoriality when they mark a tree or area with urine, feces, or other scents. Cats have been seen to avoid a trail where another cat had preceded them by only a few minutes. However, if the scent is old enough, they ignore it. Lorenz has compared this behavior to the functioning of the block signals on a railway, which prevent rear-end collisions between trains. Such behavior amounts to temporal rather than spatial territoriality. Most mammals, however, stake out relatively permanent territories, which they remark with pheromones on a regular basis.

### OTHER BENEFITS OF AGGRESSIVE BEHAVIOR

Aggressive behavior has benefits other than those described thus far. Because only the most aggressive and successfully competitive males have offspring, the health and vigor of the next generation is thereby promoted. Moreover, the strength and fighting skills of the male contribute to the family when predators threaten them. In social animals, the strongest, most aggressive males usually lead counterattacks on predators.

## REPRODUCTIVE BEHAVIOR

Many of the examples of behavior described thus far have involved some aspect of sexual reproduction: securing of a nesting area and a territory, nest building, courtship, mating, and care of eggs and young. Reproduction, whether sexual or asexual, is an absolute requirement for the preservation of an individual's genes in the form of offspring, that is, for its evolutionary fitness and success. It is thus not surprising that much of the life of most animals involves reproductive behavior.

The basic requirements for sexual reproduction are much the same in every species: the production of gametes (eggs and sperm), the fertilization of eggs, and the survival of new individuals to reproductive maturity. In the evolution of a given species, there is continual selection pressure for optimal reproductive performance in each of these areas.

Among the prominent components of reproductive behavior are genetically closed programs that consistently deliver appropriate actions or responses. Learned behavior plays an important role only when learned responses are the most adaptive. In the examples of reproductive behavior that follow, both types of programs are noted.

## THE BASIC REQUIREMENT: UNION OF GAMETES

In the simplest types of sexual reproduction, exhibited by many sedentary marine invertebrates, gametes are merely released in large numbers into the water. In such situations, the gametes of many individuals often mature simultaneously, increasing the chances that **syngamy,** the union of gametes, will occur. Simultaneous release of gametes is often triggered by a signal perceived by all individuals. For example, in some species of worms, molluscs, and starfish, the release of gametes is accompanied by release of a pheromone that triggers all members of that species in the area to follow suit. As a consequence, many eggs are fertilized.

Many motile aquatic species exhibit swarming at mating time and may appear at the surface of the sea or at its edge in great numbers. The biological clock of each member of the species prompts it to arrive at the breeding place at a precise time of day and year. For example, Samoan palolo worms, which are marine annelids, swarm at dawn during the last quarter of the moon in October and November. Grunions, a type of fish, spawn, that is deposit their eggs and sperm, at night, twice a month, between March and September on the beaches of southern California. Their spawning runs coincide with the highest night tides and immediately follow the full and new moons (Fig. 20-15).

In terrestrial species, individuals of both sexes must not only be in breeding condition simultaneously, but one or both may need to be sexually excited for gametes to be released at about the same time. In species with strong aggressive behavior, one individual may also need to quiet the other's natural tendency to attack. In some spiders, it is necessary for the male to spend hours distracting, and even in tying down the female with silk strands, before mating is safely possible. The solution to such problems is provided by courtship.

**Courtship** is a series of precopulatory stimulations aimed at a member of the opposite sex of the same species. Both individuals participate in the process, which may involve the release of scents; the display of body shape, organs, and markings; and a great variety of other actions (Fig. 20-16). Each of these components need not be unique to courtship, but usually the combination of components is. Mating cats actually

**FIG. 20-15** Grunion spawning on a California beach. Spawning coincides with the highest night tides between March and September and always immediately follows the new and full moons of each month.

(a) (b) (c) (d)

**FIG. 20-16** Courtship displays take many forms. (a) A peacock displaying his plumage to a peahen. (b) A male emperor bird of paradise *(Paradisaea guilielmi)* displaying (New Guinea). (c) Courtship display of the throat fan, or dewlap, by a male green Anolis lizard *(Anolis carolinensis)*. (d) Mutual courtship display of the Layson albatross *(Diomedea immotabilis)*.

fight each other before copulating, even to the point of inflicting some bodily injury. But these fights are seldom as "serious" as those between males and soon give way to typical mating behavior. The fight increases the female cat's sexual excitement and appears to be necessary for ovulation to occur.

In species in which parental cooperation is important for the young's survival, there is often a tendency to form strong attachments between mates during courtship. Such pair bonding has been selected for in many species of birds and mammals and seems to be represented in human societies (Fig. 20-17).

Nest building is often a part of courtship, especially in fish and birds. In some cases, a female is not at all attracted to a male until he has established a territory and built a nest. In some species, the male and female build the nest together, which contributes to pair bonding; in such cases, the two usually share brooding and parental care later on. In a few species, the male watches the female build the nest by herself, after which they copulate and he leaves, never to return. In birds that make their nests in hollowed-out trees, the male usually arrives first, establishes his territory, selects and perhaps improves or prepares the hole in the tree, and then advertises his availability to passing females. Since this cannot be done from deep inside the nest, a special effort is

**FIG. 20-17** Pair bonding in Hamadryas baboons. The male is grooming the female as part of a courtship ritual.

required. For example, the male redstart either displays his brightly colored head plumage from the opening or turns around and displays his red tail feathers to passersby (Fig. 20-18a, b).

***Brooding Birds.*** The incubating, or brooding, of eggs is almost always necessary in birds. Because birds are warm-blooded, or homeothermic, the embryo will die if the egg is allowed to cool for long. Brooding the eggs not only ensures a rapid development of the embryo by utilizing parental body heat but also provides whatever protection the parent can give. In many cases, only the female broods the eggs, leaving for short periods to obtain food or having it brought to her by the male. In some species, the parents share the brooding, with the female ordinarily doing most of it. In phalaropes, a type of wading bird, however, the male does all the brooding. In these birds, the female signals the male that she is about to lay an egg by an unusual sequence of motions; then she leads the male to the nest and lays the egg. At this point, the male assumes responsibility for brooding.

In birds, the basic behaviors of brooding eggs and caring for young are stereotyped reactions; each component of these behaviors is triggered only by the proper cues, or releasers. In herring gulls, for example, the laying of the first egg triggers an abrupt change in the pattern of activity of the breeding pair. Instead of going off to feed together as before, one of the parents always remains behind. As a second and then a third egg is laid, even more time is spent in the nest. Herring gull eggs are laid 2 to 3 days apart in clutches of three. If the eggs are removed experimentally or by a predator, another clutch of three is begun a week later. Otherwise, no more are laid, and the three are incubated until hatching in about 30 days. During brooding, the parent occasionally turns the eggs with its beak. This not only warms them evenly but prevents adhesions of embryonic structures to the inside of the egg shell.

***The Stickleback.*** Although many fish produce eggs by the hundreds of thousands, and only a few eggs reach maturity, some fish, such as the stickleback, produce few eggs and give them much personal attention. In sticklebacks, nest building is done solely by the male. From algal strands that he glues together with kidney secretions, the male stickleback constructs a rather elaborate nest of the sort shown in Figure 20-19. By displaying the proper courtship behavior, he then leads a compatible female to the nest and induces her to enter it and spawn; when

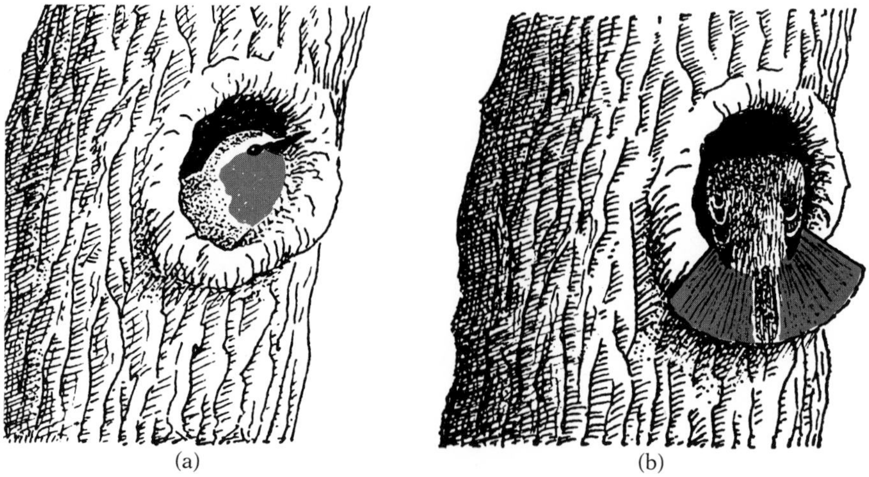


**FIG. 20-18** Male redstart advertising his nest hole to passing females (a) by showing his red head and (b) displaying his red tail.

FIG. 20-19 Courtship and mating in the three-spined stickleback. (a) Male courts a gravid female (one with eggs). (b) He leads her to the nest he built. (c) He induces her to enter and spawn. She then leaves the nest, after which he enters and fertilizes the eggs. He remains in the vicinity, repelling intruders and fanning the water (d) until the eggs hatch.

this is done, she leaves (Fig. 20-19a–c). The male then enters the nest and fertilizes the eggs, after which he chases away the female if she is still in the vicinity; he then repairs any damage the two have done to the nest, carefully concealing the eggs.

In the next few days, the male may court and mate with three or more females in the same way, adding to the clutch of eggs in the nest each time. Once they have deposited their eggs, the females never return; all care of the eggs is left to the male. He remains at the nest, repelling all intruders (Fig. 20-19d) and fanning the eggs, thereby providing them with oxygen, until they hatch.

## DIMENSIONS AND EXPRESSIONS OF PARENTAL CARE

As we have noted, in some species natural selection favors the ability to care for offspring until they can care for themselves. Unless some offspring are given a chance to reach reproductive age, the survival of the parent would have no evolutionary significance. Because, in such species, caring for its offspring promotes survival of the parent's genes, parental care is truly a selfish activity.

As we have seen, there are two approaches to maximizing the number of offspring that survive and reproduce. One is to produce a large number of young to whom little or no care is given but at least some of whom are, by chance, likely to reach maturity. The second alternative is to protect and nourish a small number of offspring, most or all of whom mature. Just as selection pressures exist for self-preserving behavior, so do they exist—in those species with few offspring—for parental care of those offspring. Species in which parents produced few offspring but did not care for them would become extinct if insufficient offspring reached reproductive age. This principle of parental care is amplified later in this chapter, in the discussion of sociobiology.

Most parental care is probably composed of both instinctive and learned behavior. Over the years, the genetically closed components have held the greatest fascination for ethologists. They include some classic examples of fixed action patterns and their releasers, some of which are described below. The examples chosen present a spectrum of various degrees and types of parental care.

(a)

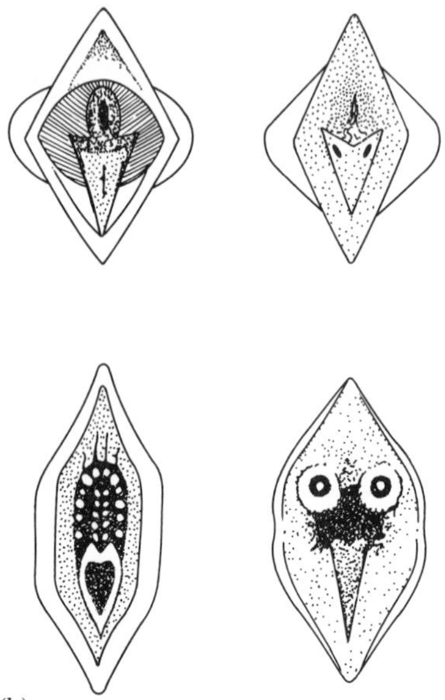

(b)

**FIG. 20-20** (a) A mother European redstart feeding insects to its young. (b) Throat markings of various species of nestlings, as seen by the parents when the nestlings gape and beg for food.

***Some Fish Care, but Not Many.*** Although in most fish species no care is provided to eggs or offspring, some fish do protect and aerate their eggs. A very few species do more. The male stickleback stays around the nest and protects its young for the first few days after hatching. In a few fish species, the parents protect their young in their own mouths, into which the newly hatched fish, or fry, scurry at a sign of danger. This behavior by the parent requires that its natural tendency to eat fish the size of the fry be inhibited by the presence of its offspring in its mouth. At the other extreme, many species not only neglect their young but regard them as food. As keepers of tropical fish know, unless most newly emerged young have a place to hide, they will be eaten by their parents.

***All Birds Care at Least a Little.*** Even in precocial birds, those whose young are independent enough to secure their own food, the hen at least provides protection in time of danger, as well as warmth on chilly days and nights. In flocks, this parent-offspring relationship has sometimes been found to merge into a dominance hierarchy as the chicks mature.

Birds with altricial, or helpless, young exhibit classic genetically closed programs of parental care. In such species, the parent must provide food and safety to the young for a considerable period; the parent often teaches the young various skills as well. In feeding its young, the parent's behavior depends on stimulation by such things as open mouths, cries, and various other signs of juvenile status (Fig. 20-20a). The signs often include special markings in the nestling's mouth or throat, such as those shown in Figure 20-20b. These can be seen by the mother from the nest's edge as she returns with food. The parent is usually cued only by the markings of its own species.

***Mammals Care the Most.*** The closest and most complex relationships between parents and young are seen in mammals. Females of most mammalian species, including herbivores, eat the placentas, or afterbirth, of their young at birth, then repeatedly lick their offspring, cleaning their nose, eyes, mouth, and ears. Feces and urine of the newborn are also usually eaten. If newborn sheep or goats are removed from the mother before she has cleaned them, she will reject and refuse to nurse them when they are returned a few hours later. This suggests that she normally undergoes a kind of chemical imprinting during the first hours after their birth. In herds of hooved mammals, the association that begins in the hours after birth often lasts throughout life, with offspring attached to mothers and mothers to grandmothers in a true matriarchy in which the oldest females maintain social control. But in most mammals, the association between mother and offspring is broken by the end of a year or two at most. This dissociation may involve the mother's driving the young away as her drive to care for them lessens. In male-dominated social groups larger than a family, the dominant males usually drive the young males away from the group when they reach sexual maturity. This sometimes leads to the formation of bachelor groups whose members remain near but do not mingle with the main colony until they can assert their own dominance, thereby gaining a place within the group and the privilege of mating.

***Parental Care in Primates.*** By far the greatest amount of care giving by parents and care soliciting by offspring occurs among the higher primates, with humans leading all others. In the human, the need for prolonged care giving by parents is related to the long period of immaturity. When one primate is compared to another, as in Figure 20-21, we note that the higher the species, the greater the length of gestation, infancy, the juvenile phase of development, and the life span. Primates as a group have an unusually long gestation period. In humans, longevity is greatest, and the juvenile stage is correspondingly prolonged, being more than twice that of the apes. It has been suggested that the prolonged juvenile stage may also be related in part to the evolutionary

**FIG. 20-21** Comparison of age spans and developmental periods in primates.

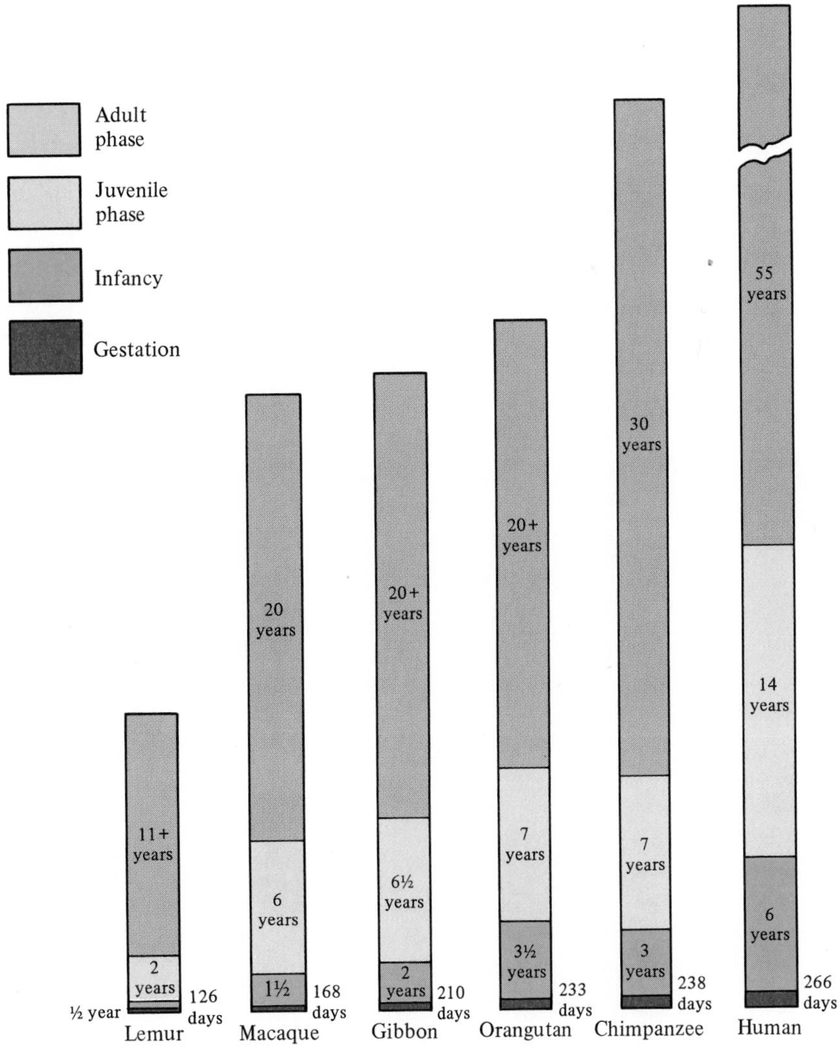

direction taken by humans, involving great enlargement of the brain and head in proportion to the rest of the body, in comparison to other primates. A large part of the growth of the brain occurs early in life, but the amount that can occur before birth is severely limited by the size of the mother's pelvis. This means much human brain development must occur after birth, in contrast to the development of many lower mammals, in which the newborn are precocial, that is, well developed and somewhat independent of the parents at birth. Much of human infancy corresponds to developmental stages passed in the uterus by lower mammals.

No primate young are precocial. Although a few primates place their newborn young in nests, most begin to carry them about from birth. The babies usually must hang on to their mother's fur and have well-developed, grasping hands that accomplish this (Fig. 20-22). Gorilla and chimpanzee mothers support their young with one hand during the first two months of the uncoordinated infants' lives. Hanuman langur monkeys groom their babies frequently and rescue them when they cry for assistance. Among the many observations of Jane Goodall, who studied chimpanzees in natural situations at the Gombe Stream Research Center in Tanzania, Africa was that chimpanzee mothers were not receptive to males for up to 5 years after giving birth. During that interval, the mother would spend much of her time caring for her offspring. But variations in parental care are great, not only from one primate species to another, but also within species. This variety is an indication of the genetic openness of parental care in higher primates.

THE ADAPTIVE VALUE OF CUTENESS. Studies of primate behavior show that there is something very attractive about an infant to young adult females called "aunts" when they are not the infant's mother. The infant's appearance, near helplessness, and care-soliciting behavior elicit a strong desire in the aunt to care for it (Fig. 20-22b). Occasionally, at least in captivity, a baby monkey has been literally pulled apart by females attempting to mother it. The infant's large head, protruding forehead, large eyes, rounded face, clumsiness, and short, thick limbs may combine to elicit what Lorenz called the cute response. Pictures of babies, whether of humans or other primates, also normally elicit feel-

**FIG. 20-22** (a) A mother Hamadryas baboon with a baby riding on her. (b) A baboon "aunt" attempting to mother another female's infant.

(a) (b)

**FIG. 20-23** A baby rhesus monkey shows its preference for a soft, furry mother by clinging to the terry cloth surrogate mother while feeding from the wood and wire mother.

ings of parental affection in humans. Probably some of the cute characteristics of most primate infants (for example, plump cheeks) have been selected for as specific releasers of parental care-giving behavior. The success of those species in which prolonged care of the young is necessary would have been promoted by the evolution of care-giving responses by adults to cuteness.

THE IMPORTANCE OF BODY CONTACT. Among the most significant findings in primate behavioral studies have been those of the University of Wisconsin biologist Harry F. Harlow. In his first experiments, in the 1950s, which were intended to determine those characteristics of their mothers to which infant rhesus monkeys responded, an infant rhesus monkey was taken from its mother a few hours after birth. It was then confined to a cage with two surrogate mothers, one constructed of bare wire screening and fitted with a wooden head, the other a similar structure but covered with terry cloth toweling. In one series of experiments, the wire screen mother had a nursing bottle of milk protruding from its "breast." In a second series, the terry cloth mother furnished the milk. All infants spent far more time clinging to the terry cloth mother than they did the bare wire mother, even when the milk was furnished by the wire mother. As Figure 20-23 illustrates, when the milk was available only from the wire mother, they would, if possible, maintain contact with the cloth mother while nursing from the wire mother. The cage floor was kept warm, but the monkeys preferred the cold cloth mother; as they grew older, they spent more time clinging to it.

The monkeys apparently looked to the soft cloth dummy as their mother, even though it was cold and even when the wire mother provided their food. This tended to discount an old theory that the affection of an infant for its mother develops primarily from her feeding it. Apparently, even warmth is less important to the infant than the soft feeling of the mother's body.

THE PRICE OF PARENTAL NEGLECT. At this point in Harlow's experiments, it seemed that an ideal mother had been provided for the infant monkeys—one that was available 24 hours a day and never scolded or punished. When frightened by anything, the monkeys rushed to the cloth mother, apparently for comfort and confidence. Only after a time would they venture to leave it and examine the object of their fear. As they matured, however, something proved to be lacking in monkeys raised in the presence of surrogate mothers. As the weeks progressed, the young monkey would huddle in a corner of the cage or room, especially if its surrogate mother was absent. Such young monkeys often sat and rocked in the manner of autistic (withdrawn) children, especially when confronted with a novel situation. Moreover, when placed with other monkeys of their age after being deprived of a real mother for an extended period, their behavior was strikingly abnormal.

Adult motherless monkeys were unable to mate. The males were inept at mounting any female, and the females rejected all males. When it was arranged that such females became pregnant despite their resistance to mating, they not only rejected babies born to them but usually treated them brutally. However, these motherless mothers improved with experience; they treated their second and third babies better.

In some experiments, motherless infant rhesus monkeys of the same age were placed together for varying lengths of time. Again, their behavior was abnormal; they spent much time clinging to each other in unusual arrangements such as shown in Figure 20-24. If such peer contact

**FIG. 20-24** "Together-together" monkeys. Lack of mothering results in an unusual togetherness among juveniles; this behavior compensates to some extent for lack of a mother.

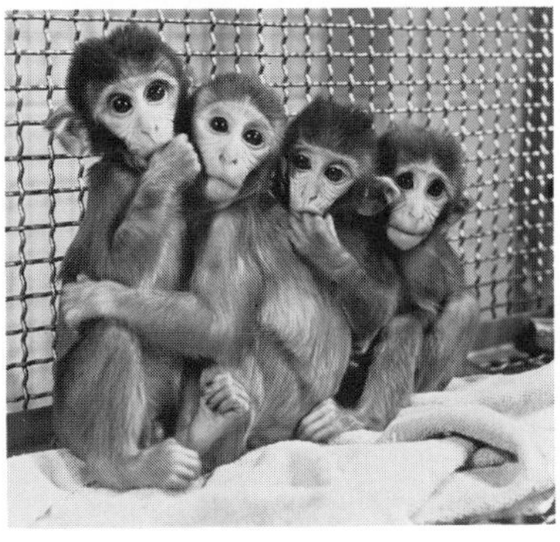

was permitted for even as little as 15 minutes a day, especially during the first 6 months of the monkeys' lives, there was a decided reduction in the severity of the abnormalities exhibited at maturity.

Other abnormalities that occur in monkeys and chimpanzees kept in isolation in infancy and youth include:

1. unusual fear of even the mildest novel stimuli
2. poor learning ability accompanied by poor attention
3. poor social communication (signals are misunderstood, and wrong signals are given)
4. extreme fear in social situations when young and inordinate aggressiveness when older (huge males are attacked and infants are beaten)
5. brutal treatment of offspring by mothers who were brutally treated by their motherless mothers

The possible significance of these findings for an understanding of human behavior is fairly obvious. Early experiences in human youth have, of course, long been thought to be important factors in maturation of a balanced personality. How important is difficult to determine and probably varies from one individual to another.

## SOCIOBIOLOGY

In recent years, the Harvard biologist Edward O. Wilson has given new impetus to attempts to unify the study of all social behavior, including that of humans. As Wilson defines it (1975), this hybrid field, called sociobiology, deals primarily with the population structure, castes, and communication of animal societies and the physiology underlying their social adaptations. Because it also concerns whatever can be learned about the biological basis of social behavior in humans, the field has become controversial. At present, the human aspect is being researched mostly through studies of early cultures and of existing primitive societies. Wilson has expressed the hope that in the next 20 to 30 years a unified theory of sociobiology could be formulated, based on application of genetic and evolutionary principles to human social behavior. His hope seems ambitious. With the limited possibilities that exist for adequately controlled experimentation on human behavior, the prospects of acquiring the necessary data for solid scientific progress in this field seem dim. Wilson himself noted that "whether the social sciences can be truly biologicized in this fashion remains to be seen." Nevertheless, rapid progress has been made in many other areas of sociobiology.

### THE EVOLUTION OF ALTRUISM

***Kin Selection.*** We have noted that when a parent provides for its young, it is really providing for an extension of itself through its genes. In evolutionary fitness, the individual's genes are "cared for" both in itself and its offspring. Parental care is thus an extension of one's **individual fitness.** Natural selection tends to preserve any behavior that safeguards the continuation of one's genes, whether they be present in offspring, siblings, cousins, or other close relatives. This extension of fitness is termed **inclusive fitness.** For such selection to operate, there must normally be a mechanism that enables the individual to recognize its relatives as such* or some other mechanism to ensure that care is given mostly to an individual's kin. The mechanism need not be perfect; it is sufficient that it usually functions in this way. The evolutionary process of selection for behavior that benefits close kin is termed kin selection.

***Is There Such a Thing as Altruism?*** Altruism is usually defined as unselfish behavior, actions aimed at furthering the welfare of others rather than the individual's own interests. An altruist is also regarded as forfeiting something of value to its own fitness, such as its own safety, for another's benefit. It appears that, even if we exclude parental care and kin selection from the definition, there may be cases of altruism in nature. One frequently cited example is the fact that in some bird species, the first one in a flock to detect a predator gives a warning cry; this allows the other birds to avoid the danger. However, the crier presumably is endangered by attracting attention to itself. Another example is that some species of shrimp and small fish risk being eaten alive when they enter the mouths and gill chambers of large fish to cleanse them of parasites. Both behaviors appear to have the elements of altruism: the actions entail cost or risk for the individual performing them, and they benefit another. Both of these classic cases bear more detailed examination.

***Is There Such a Thing as Group Selection?*** There are two ways to regard birds that warn their flocks at risk to themselves. For years, many biologists assumed that true altruistic behavior was being generally selected for; a species' chance for survival and success, it was thought, would certainly be increased if, under certain conditions, some of its members acted altruistically. This concept, according to which the altruistic behavior of an individual is aimed at furthering the welfare of a large group or even the animal's entire species, is known as group selection. Its most prominent protagonist has been a Scottish biologist V. C. Wynne-Edwards. According to group selection theory, every member of a group would normally exhibit a particular behavior any time it was "its turn"; in the bird example, this would be whenever any one of the birds was the first in the flock to see a predator.

The problem with this theory is that even if such behavior did become established, eventually genetic changes could result in cheating. A cheater in this case would be an individual that accepted the benefits of altruism without itself engaging in altruistic behavior. A cheater bird, for example, would seek cover when warned but would never give a warning itself. Once cheaters arise in a group, they would tend to be selected over altruists for long-term survival simply because they do not incur any of the costs, such as risks to life, that altruists do. Eventually the group would consist only of cheaters, and the altruistic behavior would disappear. In some cases, it is possible that the altruistic behavior might have been essential to the group's survival. If so, the entire group would then become extinct in the region it had occupied.

Before this objection was raised, the original explanation for group selection had much logical appeal. However, when mathematical models of group selection were constructed in accordance with what is known of population genetics, tests of the model showed that group

* This is not to suggest that because an animal recognizes its young and thereupon cares for them, the animal understands that it is promoting the survival of its genes.

selection would occur very infrequently if at all. It thus cannot be the explanation for any but rare instances of altruistic behavior. But a problem remains: what of birds that give warning cries?

One body of opinion suggests that birds that warn their companions may not be altruistic after all. In work reported during the 1970s, the American sociobiologist Robert Trivers pointed out that, by giving an alarm, a bird may be gaining as much in increased safety as it risks. When a bird gives a warning call, its own risk of being killed by the predator may be reduced, for several reasons. First, if the predator is allowed to catch and eat even one of the flock, not only will it be kept alive another day, but it will be more likely to eat others of the same species the next day, perhaps including the alarm giver. This is so because the predator will have been conditioned to search for the same kind of prey again; it is said to have developed a search image for that prey. Furthermore, the predator may have improved its predatory skills for catching that species. It will also be more apt to frequent the same area, and it is likely to learn useful information about the area, such as the escape routes of its prey. Therefore, warnings, especially those announcing the arrival of inexperienced predators, tend to prevent predators from preying on the caller's species and area of residence and thus on the caller itself. This explanation assumes that the alarm call is generally effective and that all of the flock escape the predator. When we add to all this the fact that the source of warning calls of birds are often difficult to locate, the risk seems rather small compared to the benefits likely to accrue to the caller. Similar analyses of several other cases of apparent altruism have produced similar results and conclusions. At least these cases, which could be exceptional, appear to be explainable in terms of individual selection or kin selection, not group selection, and thus are not altruistic in the true sense of the word.

***Reciprocal Altruism: Another Approach.*** Some of the most dramatic examples of altruistic behavior are the cleaning symbioses.* In these relationships, one animal, the cleaner, performs the beneficial task of ridding another animal, often one not at all closely related, of its dangerous parasites. The parasites serve as food for the cleaner; for some cleaners, the parasites are their only food. Such a relationship in which benefits are exchanged between nonrelated animals is termed reciprocal altruism.

Among the best-studied cases of cleaning symbiosis are those of large fish whose gill parasites are regularly removed by smaller fish or by shrimp. Figure 20-25 depicts a large grouper being cleaned by a small wrasse. Over 45 species of fish and six species of shrimp are known to be cleaners. A vast number of fish species are recipients of their cleaning. Small cleaners often enter the mouth of the host fish and clean it of ectoparasites, usually small crustaceans. If cleaning is prevented, the hosts soon develop ulcerated sores and frayed fins.

Stomach analyses of host species show that although cleaners are eaten occasionally, most are not swallowed by the host; several cleaner species have never been found in their beneficiary's stomachs. Yet the small cleaners are of the same size as many of the fish that serve as food for the host. Moreover, if a cleaner is in a host's mouth when the host

* A symbiosis is a relationship between members of different species in which one lives in intimate association with (usually on or in) the other, called the host. In the symbiotic relationship known as commensalism, the two merely share the same food; if the relationship is beneficial to both, it is called mutualism, and if the host is harmed, parasitism.

**FIG. 20-25** A wrasse cleaning a grouper. The grouper also allows the small fish, of a size it normally eats, to enter its mouth and gill chambers to remove ectoparasites without harming it.

must suddenly depart, it delays its departure to warn the cleaner. It does this by partially closing and opening its mouth, then shaking its body. This allows the cleaners to exit and take cover. Hosts have even been seen to drive away fish that threaten their cleaners. The behavior of both host and cleaner is largely innate, being based on a closed program. It is thus the product of a long period of natural selection.

## THE CONTROVERSIAL SOCIOBIOLOGY OF *HOMO SAPIENS*

From the discussion of kin selection and reciprocal altruism, it can be seen that some types of altruism can be selected for under certain conditions; but this occurs only if the altruist or its close kin are sufficiently compensated for the altruistic behavior. Some have argued that this is not truly altruism and that another term should be applied to such phenomena as cleaning symbioses. Indeed, many suggest that there is no such thing as altruism, even in humans; even acts that are seemingly most altruistic are viewed as cases of kin selection or as being performed for some kind of reward or inducement.

***Human Nature as a Product of Evolution.*** Few subjects have been studied as intensively as human nature. It has been tirelessly analyzed from many approaches, notably those of psychology, psychiatry, psychoanalysis, anthropology, sociology, and philosophy. Thus, it is to be expected that attempts to extend ethological or other biological principles to human behavior will be greeted by some as rude intrusions. Indeed, where biological science abuts or overlaps any of these other fields, controversies and angry contentions almost always arise.

One central issue seems to be whether there are inherited factors in human behavior. If there are, then they have been subject to selection; they are products of evolution that exist in us probably because they have adaptive value or at least were adaptive for our ancestors. If this is true, the question then is not whether anything about human behavior is inherited, but what.

In the rich fabric of human behavioral patterns, stereotyped responses to stimuli are rare exceptions; most of our actions are learned, and many are probably conditioned. The particular form taken by our behavior, whether during feeding, reproduction, or defense, is molded by our culture and our experiences. But our ability to behave in various ways and our drives to eat, to reproduce, and to care for our young are deeply ingrained in our nature and have apparently been inherited; that is, they are products of natural selection.

The ultimate focus of sociobiology is on the underlying biological basis for behavior. For example, sociobiology would inquire into the inherited features of the human endocrine and nervous systems that are responsible for altruistic behavior, for cheating, and for self-control of cheating. What is it that constitutes our capacity for the emotions of joy, rage, grief, and fear? What physical and chemical characteristics underlie our ability to talk rationally, to perform mathematical computations, and to reason intuitively? It will be a long while before answers to such questions are forthcoming. At the moment, we have only begun to guess at a few of the major features of the behaviors we have inherited and where in our brain they reside. Some of the current speculation centers on the evolution of human altruism and emotions.

***Are Humans Ever Altruistic?*** There is no doubt that in all cultures, humans often help each other; frequently, they do so with no possibility or hope of being repaid by the recipient of the help. The benefactor may even remain anonymous. Such behaviors seem to be classic examples of altruism. Of course, in light of the principles of kin selection, there would have been a strong selection for behavior that keeps members of one's own family, and thereby one's own genes, in existence. But what of all the cases of human altruism in which kin selection could not possibly be the explanation? The answer may lie in what Edward Wilson calls the two types of altruism: hard-core and soft-core. In hard-core, or one-way, altruism—for example, in parental care—equal return to the giver of the benefit normally does not occur. Hard-core altruism is likely to have evolved, suggests Wilson, through kin selection or the operation of natural selection on entire family or other closely related tribal units (Fig. 20-26). Hard-core altruism would be most intense with close kin and diminish with more distant relatives. Hard-core altruism is thus a form of kin selection.

**FIG. 20-26** It is common among many cultures to mark the body with distinctive markings. Often these can serve to identify the individual at a distance as a member of one's own group. In most cultures, behavior toward "one's own" is usually empathic. Shown here are the tribal markings of a man of the Mossi tribe of Bourkina-Fasso (Upper Volta), West Africa.

By contrast, soft-core altruism, Wilson suggests, is practiced only if some compensation is forthcoming immediately or in the future. It is one of the ways of benefiting from being in a group. The individual may be conscious of the expected reward, but not necessarily. There is always a consciousness of a desire to perform the altruistic act, however. Like other instincts aimed at self-preservation, soft-core altruism would thus be a product of individual selection and can be viewed as a form of reciprocal altruism.

***The Sociobiological Controversies.*** One of the strongest criticisms of sociobiological research is that it has resurrected **biological determinism,** a doctrine that attempts to show that the present states of human societies are the result of biological forces and the biological nature of the human species. There is some truth to this charge. We humans have descended from other life forms by natural selection, and, as with other organisms, much of our basic nature is an expression of the interaction of our genes with the environment. And it is precisely the main point of sociobiology that our emotions and many of our tendencies, including our inclinations toward altruism, have been selected for in the course of our evolution. But some see a serious consequence in this view.

Perhaps the most serious criticisms seem to arise from a fear that if sociobiological inquiry is pursued, it may be proved, or appear to be proved, that hereditary factors do play important roles in human behavior. Such discoveries, critics fear, could be used to justify existing political and economic injustices or be put to other evil use by demagogues.

Wilson denies these basic accusations and insists that to inherit a tendency to act in a socially unacceptable way does not require that we give in to it. Yet it is true that demagogues have misused good information in the past. But, Wilson contends, we should discourage the perversion of ideas, not ideas themselves. It is difficult to see why some truths are ones we ought not to know. Such a belief despairs of our capacity to treat each other with dignity despite any inherited propensity we may have for other behavior.

The evolution of *Homo sapiens* has also been the subject of heated controversy, ever since it was first raised. In examining human behavioral evolution, sociobiologists have stirred this controversy anew. In exploring the possible implications of sociobiological principles for humans, Wilson and others have been criticized by some for religious reasons. But to explain the physical basis for the human conscience, for instance, would not be to explain it away. If, through effort, talent, and skill, one comes to understand how the liver functions, it does not mean that there is no liver or that it is unimportant. Nor would a scientific understanding of the biological basis of human nature abolish religion. Wilson believes that the predisposition to religious belief is a most powerful force in the human mind and probably an ineradicable part of human nature.

Human understanding of the development of organisms advanced in the fourth century B.C. when Aristotle described the early development of the squid. In the fifteenth century, our understanding of physiology became less vitalistic* with William Harvey's description of circulation. Another veil of mystery began to lift in the mid-1800s when Gregor Mendel described inheritance as transmission of discrete unit factors. This century has already witnessed giant strides in the understanding of the evolution of the universe and the solar system and of the origin and evolution of life and *Homo sapiens*.

And for one who believes in God as creator of the universe—as the one responsible for the existence of matter and energy—a deeper understanding of that universe, including the biological basis of our behavior, could reveal nothing to deny the existence of God. An honest pursuit of truth by those who believe in God, while it may modify their personal approach to religion, should only provide a deeper knowledge of their God.

## Summary

Among the adaptive advantages of sociality, or living in groups, is safety; the combined force of many individuals discourages predation. When various tasks are divided among the members of a group, the resulting efficiency also benefits all group members. In addition, reproduction is enhanced by societal grouping, if only because the chance of meeting a receptive mate of the same species is increased.

* Vitalism was a belief that the functioning of the various organs was controlled by the body's vital spirit (see Chapter 1).

Communication, which is highly evolved in many social animals, includes visual, auditory, and chemical signals. Visual signals often consist of displays of teeth, claws, markings, or coloration and may be used in competition for food, breeding areas, or mates, or in courtship. Vocal signals serve many functions. They can have the same uses as visual displays and may replace them in the course of a species' evolution. Insects, amphibians, birds, and mammals all employ them. Chemical communication by pheromones between members of the same species is widespread. Pheromones are used as sex attractants, to signal alarm, and to mark out territories. Some social insects, such as ants and termites, conduct much of their lives in response to the pheromones of other members of their colonies. Bees communicate with each other by a variety of means, including various types of dance, scents, and the taste of regurgitated food.

Although aggressive displays and actions are sometimes performed in interactions with members of other species, in ethology the term *aggression* is usually applied to such behavior between members of the same species. Most such aggressive intraspecific encounters, including vigorous fights, are not usually fights to the death. They may even be entirely ritualistic. Usually, long before serious damage is done, one animal submits and allows the other to have the disputed mate, food, or territory. The vanquished individual often signals defeat by a ritualized display that quells the aggressive tendencies in the victor.

In close groups, such as wolf packs and flocks of chickens, a dominance hierarchy is often established in which one member is dominant over all the others; levels of dominance may exist within the group. The arrangement tends to unify the group, prevent unnecessary fighting, and greatly increase efficiency in food acquisition and defense.

Aggression is an integral part of territoriality, a behavior in which one animal, usually a male, maintains control over a limited area in which mating, nest building, and the raising of young occur. An early arrival in the mating season lays claim to the territory and attempts to defend it from later arrivals by threats, cries, and other hostile acts, including fighting. A stronger, more aggressive male may succeed in driving out the first occupant and becoming the one in charge. The victor thereafter occasionally defends the area from other intruders. The size of territories varies among different species.

Reproduction, an absolute requirement for the survival, or selection, of an individual's genes in the form of offspring, involves a variety of behaviors that further this end. Their many manifestations function to ensure (1) production of gametes, (2) fertilization, by fusing with other ripe gametes of the same species, and (3) successful development of the embryo and the young to reproductive maturity. The two extremes in animals' approaches to these requirements are (1) to produce many eggs and offspring, even millions per individual, and give them little or no care and (2) to produce one or a few offspring at a time and give them much care. A great variety of intermediate approaches also exist. The most complex of all relationships between parents and young are seen in mammals, especially the higher primates. Depriving monkeys of normal parental and sibling relationships during youth leads to abnormal behavior in youth and as adults.

Sociobiology deals primarily with the population structure and communication of animal societies and the underlying physiology of their social adaptations. One of its major proponents, Edward O. Wilson, hopes that eventually behavior can be analyzed in terms of genetic and evolutionary principles. Among the principles of sociobiology is kin selection, in which individuals provide for the survival of their genes even though the genes are present in other individuals, such as their offspring, cousins, and other relatives. Theoretically, all apparent cases of altruism can be viewed as providing as much care, defense, or other benefits for the "altruist" as for the recipient of its actions. Group selection, in which the actions of individuals are seen as adaptations for group, rather than individual, survival, is shown by mathematical models to have only short-term significance, if any. Cases of so-called reciprocal altruism are clearly instances of mutualism, in which both individuals receive important benefits.

Humans are products of an evolution stretching over millions of years. It is to be expected that inherited components of human behavior have usually been selected for. Most or all stereotyped human behavior is overridden by learned behavior. But the drives that are common to all of us, such as those for self-preservation and care of our young, have inherited components, as do our capacities for joy, fear, grief, and so on. Human altruism has been classified by Wilson as hard core and soft core. Hard-core altruism, typified by care for members of one's family, is a type of kin selection. Soft-care altruism, according to Wilson, is usually performed only if a reward of some sort is ultimately expected; it is a kind of reciprocal altruism.

## Recommended Readings

A list of recommended readings appears at the end of Chapter 19.

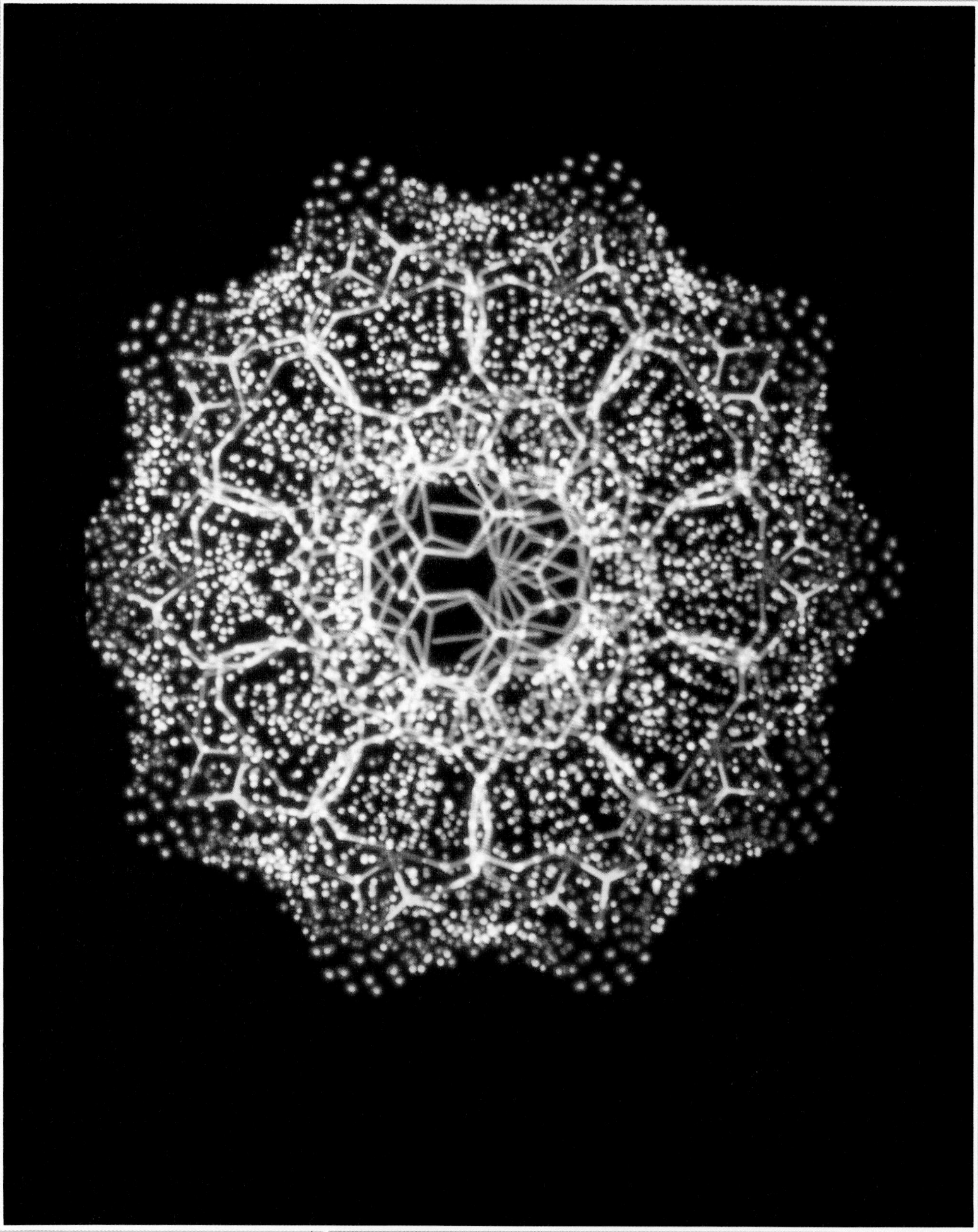

# UNIT V

# Reproduction, Genetics, Development, and Evolution

CHAPTERS



Life, the property that enables organisms to acquire and make use of energy, to grow and perceive and respond to their surroundings, is a continuum. But organisms, the bearers of this property, have a finite life span. The continuity of life is made possible by another of its fundamental characteristics: its ability to reproduce, or generate more of its own kind. Two basic means of reproduction, asexual and sexual, and their variations, have come to exist, each adapted to the organizational level of the organism and to the environment in which it lives.

Reproduction, whether asexual or sexual, depends on the physical transfer of parent material from one generation to the next. In asexual reproduction it is the transfer of as little as a single cell to as many as thousands of them, whereas in sexual reproduction it is the transfer of a single cell, the zygote. The new individual results from information packed into a single cell; the cell carries the blueprint from which to fashion a new individual of a particular species.

Understanding the physical basis of this transfer of information between generations is one of the towering achievements of biology. Genes, in the form of deoxyribonucleic acid, or DNA, are present in the cells of all organisms, and in all of the cells of a particular one. Encoded in genes are all the biological properties of an organism. It was Gregor Mendel, an Austrian monk, who first showed that each characteristic of an organism, or trait, can be inherited independently of all the others. It is now known that these traits are carried as information encoded and stored in DNA.

The details of the mechanism underlying the transmission of genetic information are now largely understood. The gradual unfolding of the developmental potential of the zygote in the elaboration and organization of an individual depends on the specifications provided by genes. The presence of variations among the individuals of a species, whether in the color of hair, skin, or eyes, or in height, weight, or shape, or any other feature, is due to variations in the genetic material. These variations are due to errors that arise when the genetic material is duplicated in the process of generating a new individual; they account for the diversity seen within species and are the raw material upon which natural selection operates.

The chapters in this unit explore some of the means of reproduction available to organisms, the basic principles of genetics and development, and the theories and mechanisms underlying the origin of biological diversity.

# 21

# General Principles of Reproduction

OUTLINE



The ability to reproduce is uniquely biological. Because the life span of every individual organism is finite, reproduction is a prerequisite for the continuity of any species. No matter how well it maintains homeostasis, every organism ages and ultimately meets a natural death. In the course of evolution, the only forms of life to avoid extinction have been those that could generate new individuals like themselves, that is, those that could reproduce.

Every kind of reproduction involves the detachment of a viable portion of the reproducing organism. The detached fragment, which can be as simple as a single cell, always contains at least one set of the organism's genes, (DNA) the units of inheritance.

The first organisms to evolve probably reproduced by pinching in two, much as do the simplest organisms today. This is a form of **asexual reproduction,** reproduction without the union of gametes, or sex cells. In the first 2 billion years or more of organic evolution, forms of asexual reproduction were probably the only means by which organisms increased in numbers. While asexual reproduction is effective in increasing the numbers of a species, those species reproducing asexually evolve very slowly because all offspring of any one individual tend to be alike, providing less diversity for evolutionary selection.

**Sexual reproduction** always involves the union of gametes, usually contributed by two different individuals. In populations of sexually reproducing individuals, variation continually arises through numerous new combinations of parental traits. The great diversity thus produced forms most of the raw material upon which natural selection operates. From this diversity of types, only those well suited to their environment survive. By producing diversity, sexual reproduction greatly increases the chances for some forms of a species to survive, while simultaneously accelerating the process of evolution.

## ASEXUAL REPRODUCTION

Asexual reproduction may be as simple a process as cell division, resulting in two separate individual cells, as in the prokaryotes and simpler eukaryotes, or it may involve the production of spores, special reproductive cells each capable of producing a new organism, common in the algae and fungi. At a more complex level, as in higher plants and some simple animals, asexual reproduction involves the formation of a complete multicellular individual, which becomes detached from its "parent." In each case, the new individuals are identical in appearance and genetic composition and are members of a clone.

### ASEXUAL REPRODUCTION IN MONERA, PROTISTA, FUNGI, AND ALGAE

Prokaryotes—bacteria and cyanobacteria—can multiply rapidly by **binary fission**—the division of one cell into two. In this process, the cell pinches in two by a furrowing inward of the cell membrane. The two cells may stay together and form filaments, which may elongate by further division; they undergo fragmentation into shorter filaments, asexually increasing the population. Protists, one-celled eukaryotic organisms, may also exhibit binary fission, with the production of two daughter individuals (Fig. 21-1a). Binary fission is common in protozoans; for some, it is their only means of reproduction. Simple cell division, involving mitosis (see Chapter 4) and cytokinesis, is a common means of asexual reproduction in protistan algae. In some cases, the nucleus divides repeatedly before cleavage forms several daughter cells. The daughter cells, when released, may remain together or assemble as a new colony (Fig. 21-1b), sometimes embedded in a gelatinous matrix. Filamentous algae, in the manner of some cyanobacteria, asexually increase the population of filaments by fragmentation.

Fungi utilize asexual reproduction very effectively. The cells of yeasts, for example, commonly reproduce asexually; mitosis is followed by a cell division in which a small cell is pinched off from the larger parent cell in a process called **budding** (Fig. 21-1c). Most other fungi propagate asexually by forming great numbers of **spores** at the tips of special **hyphae,** the elongate cells that comprise the body of a fungus (Fig. 21-1d). In terrestrial species, these spores may be carried great distances in the air. They are part of the "mold spore count" noted in summertime reports of air quality. Because each spore has the potential to start a new hypha, sporulation is an effective means of reproducing in great numbers. Aquatic molds also produce spores asexually; such spores are motile, flagellated cells called **zoospores,** which are able to germinate into new hyphae. Many algae also reproduce asexually from zoospores (Fig. 21-2).

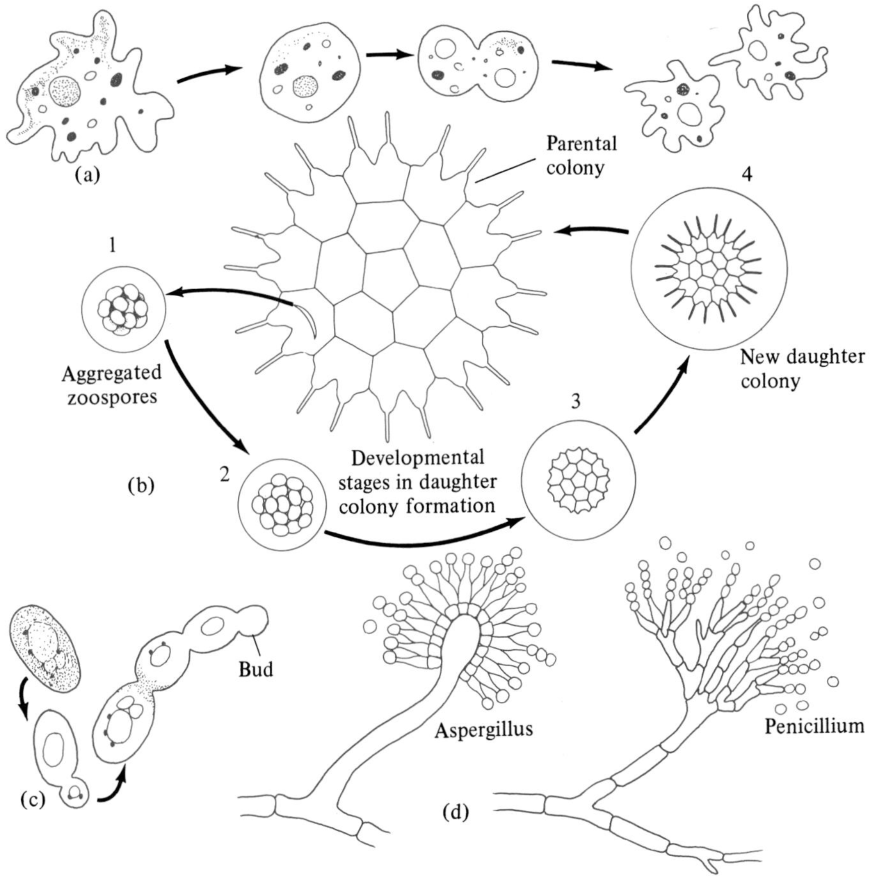


**FIG. 21-1** Asexual reproduction in lower organisms. (a) Division of a protozoan, *Amoeba proteus*, by binary fission. (b) Colony formation in the green alga *Pediastrum* by the aggregation of zoospores. (c) Brewer's yeast undergoing budding. (d) Formation of asexual spores in the common molds *Aspergillus* and *Penicillium*.

## ASEXUAL PROPAGATION OF PLANTS

Many plants reproduce both asexually and sexually. Some **bryophytes,** or nonvascular land plants, produce numerous **gemmae,** miniature plants, in special gemma cups on the surface of the plant (Fig. 21-3). When rain washes the gemmae to the soil, they develop into replicas of the parent. In some species of flowering plants, plantlets are produced asexually along the edges of the leaves (Fig. 21-4a). The plantlets eventually drop off and develop into new plants. Some vascular plants, such as the strawberry, generate new plants from the tips of **runners,** thin horizontal shoots that turn upward and form roots where the tip contacts the soil.

Horizontal shoots growing under or above the ground and known as **rhizomes** provide a similar method of asexual reproduction in some vascular plants. New plants can arise from branches along the rhizome when they become separated from the parent. Similar branches account for much of the reproductive success of grasses and many of the common weeds that compete with them in lawns. In some perennial plant species, **corms** and **bulbs**—short, vertical, underground shoots—develop from axillary buds, the buds that arise where the leaf joins the stem. Corms, such as those of crocuses, are shoots with thin, scaly leaves and thick stems in which food is stored. In bulbs, such as those of lilies, food is stored in thick, scaly leaves. A corm or bulb separated from the parent plant can grow into a complete new individual. Such asexual

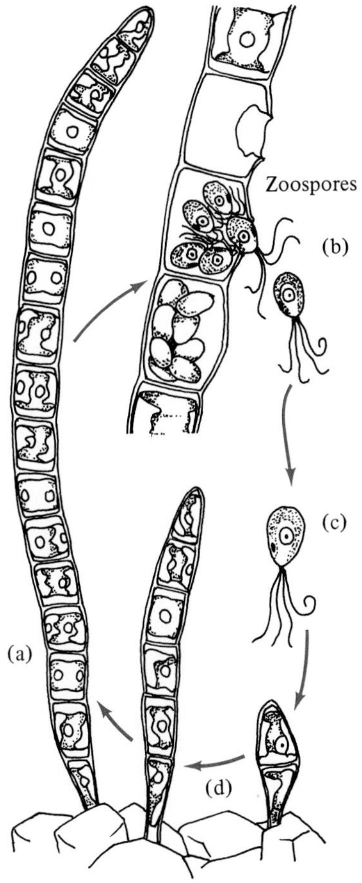


**FIG. 21-2** Asexual reproduction in a filamentous green alga, *Ulothrix*.

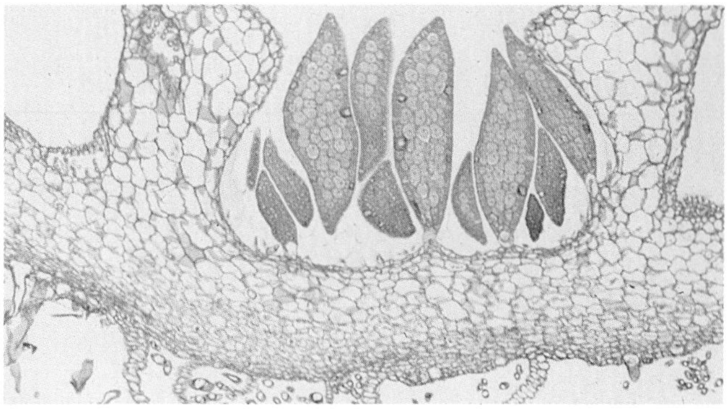

(a)

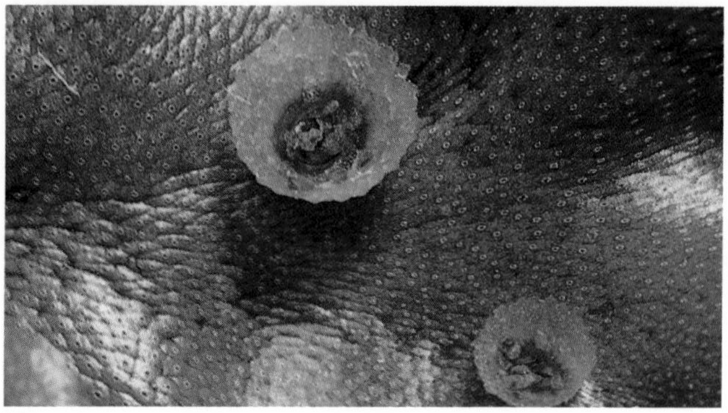

(b)

**FIG. 21-3** Gemma cups of a liverwort, *Marchantia.* (a) Cross section showing a gemma cup with several gemmae. (b) A surface view of the thallus showing gemma cups containing gemmae.

propagules are utilized by gardeners to reproduce favorable horticultural varieties.

The ability to regenerate lost parts is a widespread capability of higher plants. Gardeners have long taken advantage of this capacity by performing **vegetative propagation** of their desirable specimens, a form of asexual reproduction. This is usually done by placing cuttings, which are pieces of shoots or leaves, in a rooting medium such as moist sand. Roots soon form, and new shoots develop from buds on the cutting. All plants derived from cuttings of one plant are part of a clone; they are genetically identical. For generations, horticulturists have used cuttings to produce large numbers of special plant varieties. Today, great numbers of genetically identical plants are generated by **in vitro** culture in flasks containing nutrients (Fig. 21-4b).

The practice of **grafting** is another application of asexual propagation. In a typical case, a stem or bud, called the **scion,** is cut from a prized fruit tree and grafted to the shoot, called the **stock,** of a hardy variety. The scion fuses to the stock and develops into the trunk and branches of a tree on the root system of the stock. The fruits produced have the characteristics of the scion.

## ASEXUAL REPRODUCTION IN LOWER ANIMALS

Asexual reproduction is common in lower invertebrate animals such as sponges, coelenterates, flatworms, and many annelids and even in some

**FIG. 21-4** (a) Asexual reproduction by plantlet formation at the edge of leaves in *Kalanchoe.* (b) Carrot callus *in vitro* (will give rise to plants).

(a)

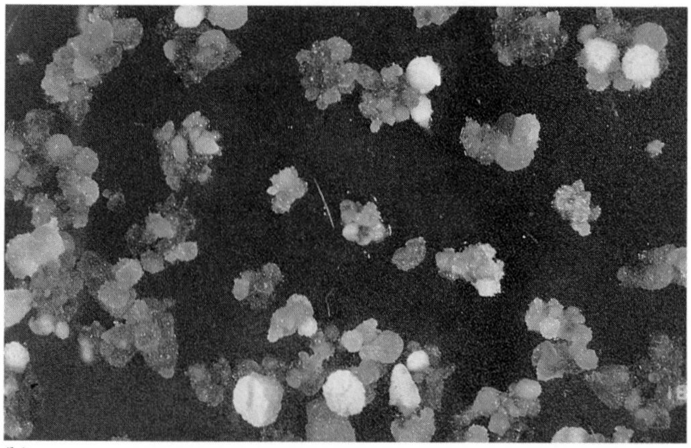

(b)

primitive chordates. The ability to reproduce asexually is often correlated with a marked capacity for regeneration. Although most of these animals also reproduce sexually, in the most highly evolved species, sexual reproduction normally supplants asexual reproduction completely. All vertebrates and most arthropods (insects, spiders, crabs, and so on) normally reproduce only by the sexual method.

In free-living flatworms, such as planarians, asexual reproduction often takes the form of **fission,** in which a transverse constriction occurs in the middle of the body, dividing it into two. The front portion then regenerates the missing posterior parts, while the rear regenerates the missing anterior parts. Some flatworms and annelids reproduce asexually by forming numerous constrictions along the length of the body; a chain of daughter individuals results (Fig. 21-5a). This type of reproduction is termed **multiple fission.**

Another method of asexual reproduction in lower animals is the process of budding (distinct from the budding of yeast cells). In the coelenterate *Hydra* and in many species of sponges, certain cells divide rapidly and develop on the body surface to form a bud (Fig. 21-5b, c). The bud's cells proliferate to form a cylindrical structure, which develops into a new individual, usually breaking away from the parent. Buds also arise in the primitive chordates. In sea squirts, for example, several buds arise from a specialized tubular organ, the **stolon.**

The predominance of asexual reproduction in lower animals can be explained on the basis of the environment in which they live, which is almost always aquatic. The marine environment is very stable and fluctuates very little. In other habitats, asexual reproduction is only seasonal. The season during which asexual reproduction occurs coincides with the period when the environment is hospitable. Under such conditions, it is advantageous for the organisms to produce asexually a large number of progeny with identical characteristics. A large number of organisms, well adapted to the specific environment, can be produced even if only one individual is present. Furthermore, the products of budding, fission, and other asexual methods of reproduction are less vulnerable to environmental stresses than the embryos and larvae produced by sexual reproduction.

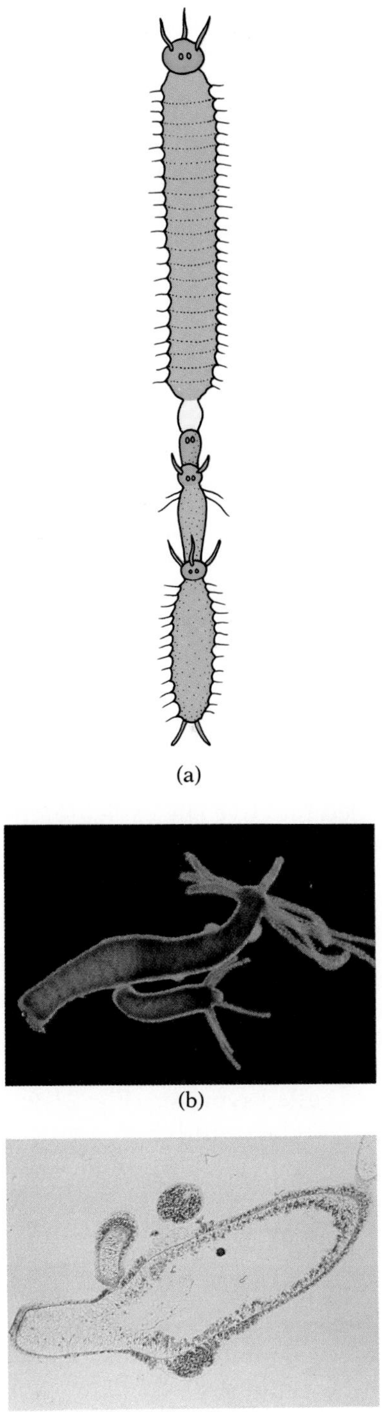

(a)

(b)

(c)

**FIG. 21-5** Asexual reproduction in invertebrates. (a) Budding in the annelid *Autolytus.* (b) Budding in *Hydra,* whole mount. Note also the testis above the bud. (c) Budding in *Hydra,* longitudinal section.

## SEXUAL REPRODUCTION

The main reason that sexual reproduction regularly results in diversity is that each new individual usually represents a combination of traits derived from two parents. This is accomplished by the union of two gametes in **syngamy,** or fertilization, one gamete being contributed by each parent. Each gamete is **haploid** (from the Greek *haploeides,* "single"), that is, it carries only one of the two sets of chromosomes characteristically found in the cells of that species. The number of chromosomes in a haploid set varies in different species. But this haploid number of chromosomes is constant for a species. The **zygote,** the cell that results from syngamy, is **diploid;** it contains a double set of chromosomes, one set having been contributed by each parent. The zygote, in due course, usually divides and forms an embryo which eventually develops into a diploid individual.

Because the cells of the new individual are diploid, it is essential that at some time in the individual's life cycle the diploid number of chromosomes be reduced to the haploid number, or the chromosome number would double with each generation. This reduction occurs in the process of **meiosis** (Greek for "diminution") (see Fig. 21-6). Sexual reproduction thus requires both fertilization and meiosis. Fertilization brings together the genetic traits of the two parents; meiosis, directly or indirectly, apportions the genes to the gametes. An understanding of meiosis is thus essential to comprehending not only sexual reproduction, but heredity and evolution as well.

## MEIOSIS

As we have noted, meiosis is a cellular activity essential to sexual reproduction. Meiosis shares several features with mitosis: formation of a microtubular spindle, alignment of chromosomes at the equatorial region of the spindle, and separation of chromosomes and their movement to the poles of the spindle. However, meiosis differs from mitosis in several important ways. Meiosis consists of two successive divisions, meiosis I and meiosis II. As with mitosis, meiosis I begins after replication of DNA; but between the two meiotic divisions, there is no further DNA synthesis. **Prophase I,** the prophase of the first meiotic division, is much prolonged compared to the prophase of the mitotic division; in human oocytes, it lasts several years. Based on the appearance of the chromosomes, meiotic prophase I is divided into five stages: **leptotene, zygotene, pachytene, diplotene,** and **diakinesis** (Fig. 21-6).

### PROPHASE I

Chromosomes coil, or condense, at the beginning of prophase. At the leptotene stage, they appear as single threads, although they have already duplicated (Fig. 21-6a). At the zygotene stage, the homologous chromosomes begin to pair up in a process called **synapsis** (Fig. 21-6b). The homologous chromosomes, which contain the same linear sequence of genes and are each inherited from a different parent, align themselves along their length in zipperlike fashion. They are held together by a complex tripartite structure, the **synaptonemal complex** (Fig. 21-6c). There is no equivalent structure during mitosis, and the whole process of synapsis of chromosomes is unique to meiosis.

In the next stage the chromosomes of each pair condense further and thicken, the basis for the name *pachytene* (Greek: thick thread). Distributed along the length of the synaptonemal complex at the pachytene stage are a variable number of granules, the **recombination nodules** (Fig. 21-6c). These granules seem to play an important role in exchanging parts of chromatids between homologous chromosomes in the process called **crossing over** (Fig. 21-6d). Chromatids that cross over are linked by **chiasma** (pl. chiasmata; Greek for "cross piece"). As we shall see in Chapter 23, crossing over is of great importance in heredity and evolution. At the next stage, diplotene, each of the homologous chromosomes separates into sister chromatids and, in addition, parts of the chromosome decondense to form lateral loops. Consequently, each chromosomal pair appear to be composed of four chromatids; hence, the chromosomes are designated **tetrads**. At this stage, parts of a chromatid from one chromosome may be exchanged for a homologous part

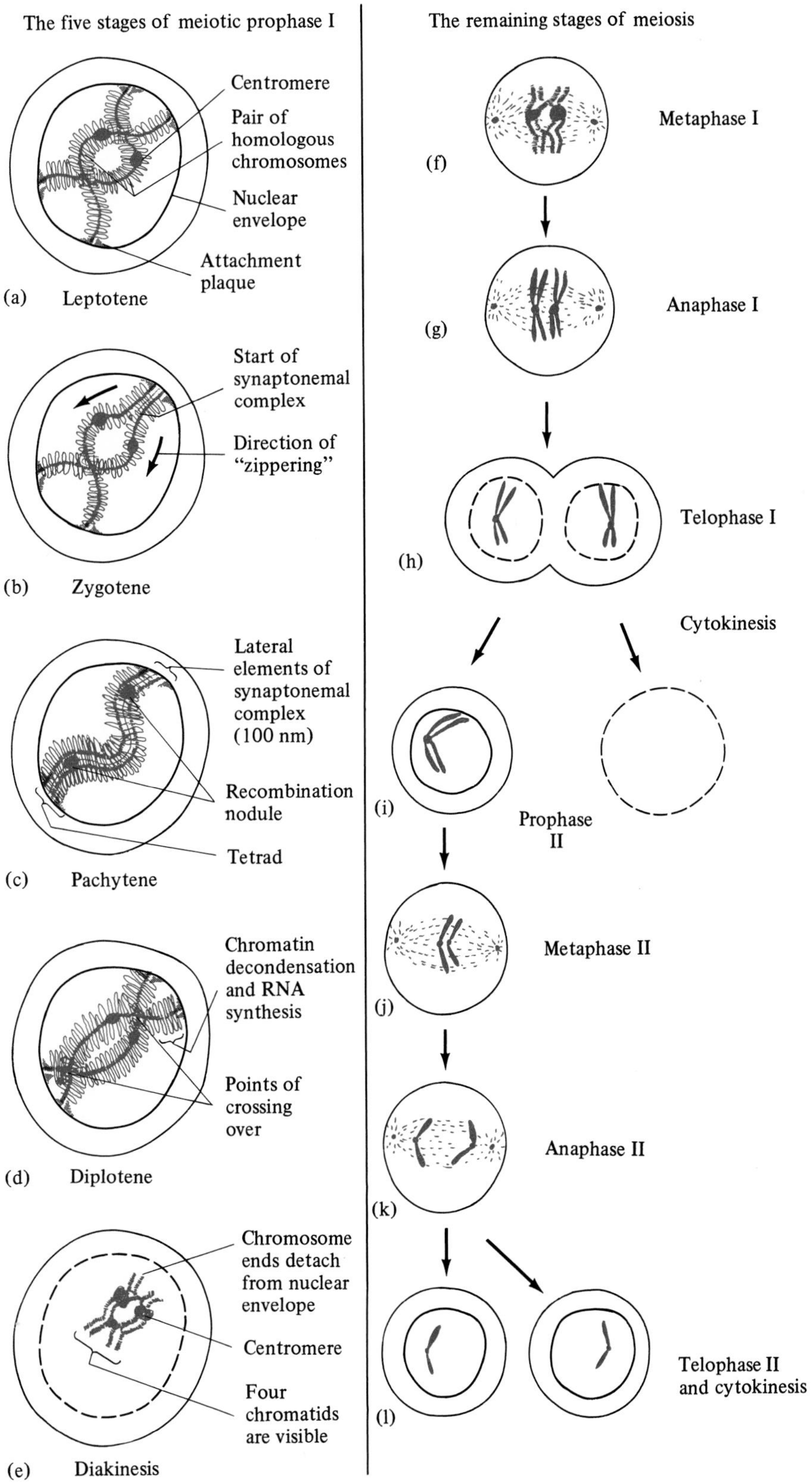


**FIG. 21-6** Stages in meiosis. Changes in chromosome appearance and distribution of chromatids. (a) Leptotene stage. Condensation of chromosomes and attachment at their ends to the nuclear membrane. (b) Zygotene stage. Homologous chromosomes begin to synapse. (c) Pachytene stage. Completion of synapsis and further condensation. (d) Diplotene stage. Chromosomes are desynapsed but are held at crossing over (chiasma) points. (e) Diakinesis. Condensation of the chromosome and tetrad nature (four chromatids) of the chromosome pair is distinctive. (f) Metaphase I. Alignment of tetrads on the equatorial plane of the meiotic spindle. (g) Anaphase I. Separation of the homologous chromosomes into dyads and movement toward the poles. (h) Telophase I. Movement of the chromosomes to the poles followed by cytokinesis. (i, j, k, l). Prophase II, Metaphase II, Anaphase II, and Telophase II of the second meiotic division resulting in haploid cells.

from its partner, thus giving the appearance of the crossing over of parts of the chromatids. During diakinesis, the sister chromatids disengage from each other, but the homologous chromosomes remain together (Fig. 21-6e).

## COMPLETION OF MEIOSIS I

Although prophase I may extend for weeks, months, or even years, the remaining events of meiosis occur rapidly. They may not take more than a few minutes. These events occur in three stages: **metaphase I, anaphase I,** and **telophase I** (Fig. 21-6f, g, h) after which, **cytokinesis,** or cell division, ordinarily takes place.

At the end of prophase I, the homologous paired chromosomes move toward the equatorial plane of the meiotic spindle. At metaphase I, the chromosomes line up along the equatorial plane of the meiotic spindle. Each homologous pair now consists of four chromatids, aligned parallel to each other.

During anaphase I, the homologous chromosomes separate and move to the poles. Unlike the corresponding stage in mitosis, the sister chromatids remain attached to one another. Thus, reduction in the chromosome number truly occurs at this stage. Each daughter cell receives one member of each homologous pair of chromosomes, each chromosome being made up of two chromatids. Each chromosome is called a **dyad.**

Telophase I marks the end of the first meiotic division. At this stage, a nuclear membrane may form around the chromosomes, which now consist of half the number of those found in the original nucleus. Cytokinesis takes place before cells enter to meiosis II. As in mitotic division, the plane of cell division always occurs perpendicular to the long axis of the spindle, exactly halfway from either pole.

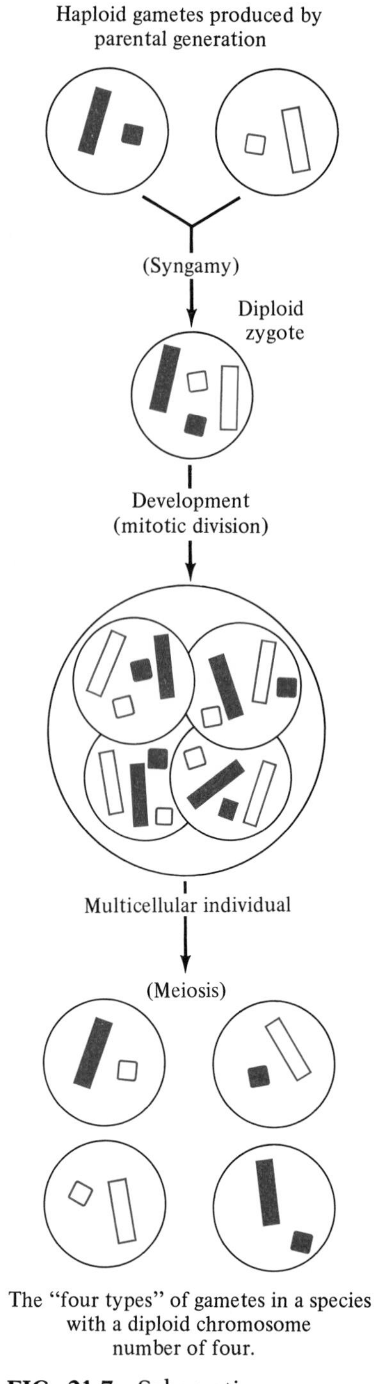


**FIG. 21-7** Schematic representation of the independent assortment of parental chromosomes into gametes during meiosis.

## MEIOSIS II

After the first meiotic division, the two daughter cells enter the second meiotic division without delay (Fig. 21-6i, j, k, l). At the second division, the two sister chromatids separate, and each of the daughter cells receives a haploid set of chromosomes.

An important consequence of meiosis results from the manner of distribution of the partners of the homologous pairs of chromosomes to the cells formed by meiosis I. It is a matter of chance as to which member of a homologous pair goes to a pole. Thus, a daughter cell may receive some of the paternal set of chromosomes and some of the maternal set. In addition, because of crossing over, some of the chromosomes are composite mosaics made up of parts of maternal and paternal chromosomes. In an organism with a haploid set of two chromosomes, four types of gametes may be formed. Two of these gametes are identical in their chromosome composition to the gametes that fused to form the individual. The other two types of gametes bear new combinations of the chromosomes (Fig. 21-7). The significance of this independent assortment of chromosomes and crossing over will become clear as we learn more about heredity and evolution.

Some algal protists are diploid only as zygotes; meiosis occurs at the first cell division (Fig. 21-8). In such species, the resulting haploid cells may then reproduce asexually (by mitosis) for an indefinite period. Eventually, gametes are formed, however, and syngamy occurs once again. In plants, the diploid zygote divides mitotically and develops into a diploid spore-producing individual called a **sporophyte.** The sporophyte, in a process called **sporogenesis,** produces haploid spores; sporogenesis thus involves meiosis. The haploid spores then divide by mitosis and develop into a second stage in the life cycle, the **gametophyte,** or gamete-producing generation, all cells of which are haploid.

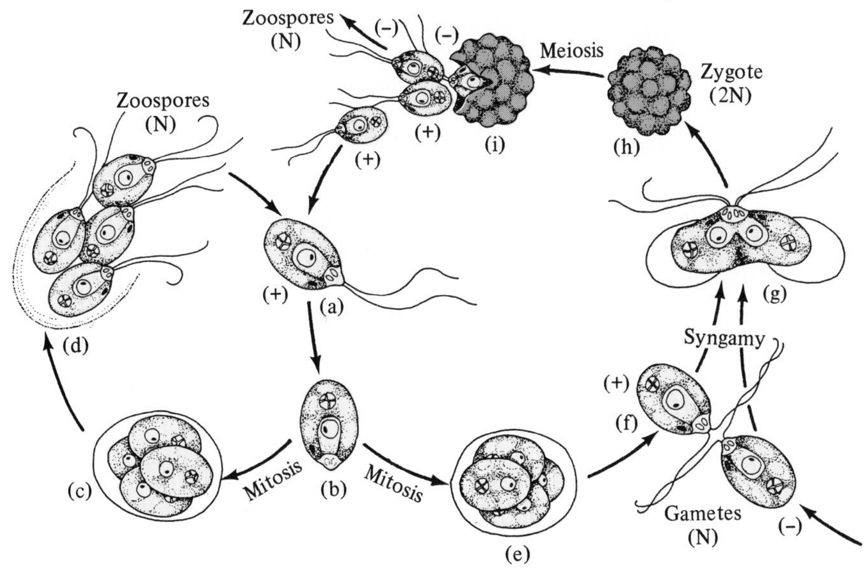


**FIG. 21-8** Asexual and sexual reproduction in the protistan *Chlamydomonas.* (a–d) Asexual reproduction by cell division. (e–i) Sexual reproduction. Pairs of opposite mating type (+ and −) fuse (g), forming a zygote (h). Under favorable conditions, the zygote undergoes meiosis (i), to form four haploid zoospores.

In some plants, the gametophyte is the most prominent stage in the life cycle; in seed plants, it is microscopic in size and consists of only a few cells.

## THE ORIGIN OF SEXUAL REPRODUCTION IN EUKARYOTES

Because protistan (one-celled) green algae are believed to be similar to early eukaryotes, as well as to the forerunners of plants, it is instructive to study these algae for clues to how sexual reproduction in eukaryotic organisms may have originated.

***Primitive Gametogenesis and Sexual Reproduction.*** Protistan green algae exhibit a wide variety of both sexual and asexual reproduction. As evolution progressed, some of these successful protistan mechanisms of reproduction may have been retained in the emerging lines of plants and animals. However, another possibility is that sexual reproduction arose independently on more than one occasion in the course of evolution.

A possible clue to the evolution of gametes is found in the reproduction of the protistan green alga *Chlamydomonas,* which exhibits both sexual and asexual reproduction during its life cycle (Fig. 21-8). Under optimal environmental conditions, this algal cell reproduces only asexually. In this mode, the algal cells undergo one to three mitoses without cytokinesis, then divide into two to eight new cells, each smaller than but otherwise identical to the parent cell (Fig. 21-8b, c). These new cells are motile forms, called zoospores, which grow into new adult individuals (Fig. 21-8d, a).

Under adverse environmental conditions, as when a pond begins to dry up, *Chlamydomonas* reproduces in a different manner. Rather than the usual one to three mitotic divisions, the cells undergo additional mitoses, dividing into a correspondingly greater number of still smaller cells. Although the resultant cells are identical in appearance to the zoospores, they differ significantly from them: they unite in pairs, eventually fusing. They are thus gametes. At first, the gametes formed from many individuals gather together into large clumps. Pairs of fusing cells then begin to separate from the clumps. As syngamy progresses (Fig. 21-8f, g), the nuclei unite and a diploid zygote ($2n$) with a resistant cell wall is formed (Fig. 21-8h). The zygote remains dormant until favorable conditions return. When they do, meiosis is initiated, producing four haploid ($n$) cells and thus completing the life cycle (Fig. 21-8i, j). These haploid cells continue to reproduce asexually until adverse conditions occur again.

This simple type of sexual reproduction, found in many algae, aquatic fungi, and protozoans, suggests that it would have required relatively minor evolutionary changes to enable asexual forms of a protist to function as gametes and thus give rise to sexual reproduction. It would have required the development of a chemical attractant to bring the gametes together, the ability of cell membranes and of nuclei to fuse, and modifications of mitosis to produce some simple form of meiosis.

***Cross-fertilization and Diversity.*** If the gametes uniting in fertilization were genetically identical, as they would tend to be if both gametes came from the same parent, very little diversity would result in the population. Only union of gametes from parents of different genetic types could be expected to produce many new combinations of genes in the progeny. Any mechanism that prevented self-fertilization would tend to ensure variety in offspring and thus promote the evolutionary success of a species. Many such mechanisms have evolved. For example, in some species of *Chlamydomonas,* gametes from the same strain are self-incompatible; zygote formation occurs only between gametes from two compatible strains. In certain sponges, male and female gametes of any one sponge do not mature at the same time. Production of different gametes at different stages of the life cycle is a common mechanism of preventing self-fertilization among hermaphroditic, or **monoecious,** invertebrate animals, that is, those in which both sexes are contained within the same individual. Thus, cross-fertilization is the rule, and self-fertilization is rare.

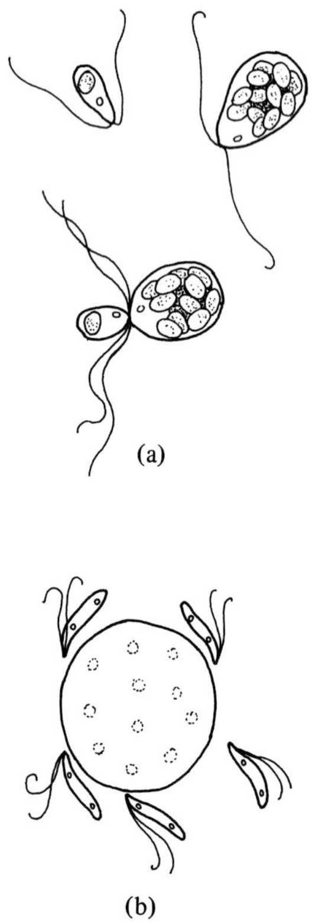

FIG. 21-9 Specialization of gametes. (a) Anisogametes of the green alga *Halicystis.* (b) Sperm and ovum, the oogamic gametes of the green alga *Volvox.*

***Specialization of Gametes.*** Another advance in the evolution of sexual reproduction was the specialization of gametes in which one gamete, the male, became small and motile, and the other, the female, became larger and nonmotile. This enabled the male gamete to specialize in reaching the female and the female gamete to specialize in storing food to be used in the early stages of growth of the resulting zygote. This trend can be seen even among such protists as *Chlamydomonas.* In some species of *Chlamydomonas,* the gametes are identical in size and appearance (Fig. 21-8e), a condition known as **isogamy** (from the Greek *iso,* "same"). In other species, one gamete is larger than the other, although the two are morphologically alike; this is termed **anisogamy** (Fig. 21-9a). In the extreme degree of gamete differentiation seen in all higher organisms, the gametes are morphologically distinct enough to be designated **sperm** (male) and **ovum** (female), a condition known as **oogamy** (Fig. 21-9b). In higher plants and in most animals, oogamy is the rule. However, oogamy occurs even in some species of *Chlamydomonas.* This single genus of protists thus represents the complete range of major gamete types found in most organisms.

## LIFE HISTORY VARIATIONS IN SEXUALLY REPRODUCING SPECIES

The life history, sometimes called the life cycle, of every sexually reproducing organism begins with the formation of a zygote as a result of syngamy. The zygote may develop into an embryo as in most animals and many plants and then into a sexually reproducing adult. The life history of every sexually reproducing group of organisms has a characteristic pattern. For example, the organism that produces gametes could be either haploid or diploid. If the organism is diploid, it produces gametes after meiotic divisions. In still others, the life history of an organism includes successive diploid and haploid phases, both of which may be capable of independent life and may even propagate asexually. But the time of meiosis varies with the organism.

***Three Basic Life Histories.*** Life histories of sexually reproducing organisms can be classified as three basic types:

1. The individual occupying the greater part of the life cycle is haploid, and the zygote is the only diploid phase (Fig. 21-10a). The zygote undergoes **zygotic meiosis** to produce four haploid cells. Ultimately, gametes are produced during the process of **gametogenesis;** these unite to form a zygote, as described in *Chlamydomonas* above. The first sexually reproducing organisms to evolve were probably of this type.
2. The only individual in the life cycle is diploid. Meiosis occurs in these organisms just before gamete formation, called **gametic meiosis** (Fig. 21-10b). This type of life cycle is characteristic of all animals, some aquatic fungi, and certain algae, including *Fucus,* a common brown multicellular alga.
3. There is an alternation of diploid and haploid individuals in the life cycle (Fig. 21-10c). In such species, the diploid individual (the sporophyte) produces haploid spores by **sporic meiosis;** these develop into multicellular haploid individuals. The haploid individuals, or gametophytes, produce gametes, which unite to form diploid zygotes. The zygotes develop into diploid sporophytes. Because completion of the life cycle requires alternation of a haploid and diploid generation, this type of life cycle is called **alternation of generations.**

## SEXUAL REPRODUCTION IN PLANTS

The life cycles of all plant groups involve alternation of generations. It is found in algae and is thus presumed to have evolved before plants moved onto the land. In the past, it was popular to speculate on how

**FIG. 21-10** Life histories of sexually reproducing organisms. (a) Life history with zygotic meiosis as in *Chlamydomonas.* (b) Life history with gametic meiosis as in the brown alga *Fucus* and most animals. (c) Life history with sporic meiosis as in most plants.

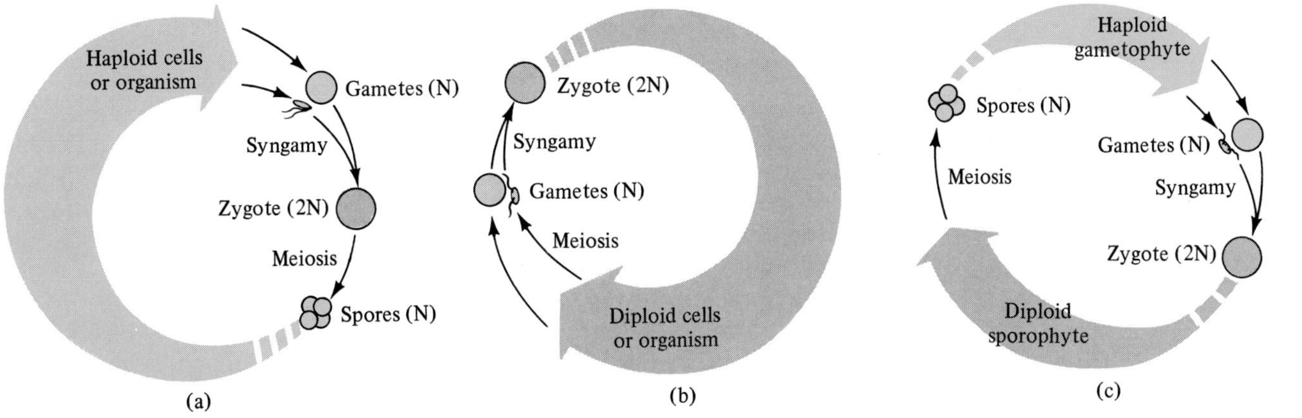

alternation of generations first became established. The answer is still not known.

Variations on the basic life cycle occur in different plant groups. In some algae, the haploid and diploid generations are identical in appearance; such algae are said to exhibit isomorphic alternation of generations (Fig. 21-11a). In others, the two generations are entirely different in morphology, exhibiting heteromorphic alternation of generations (Fig. 21-11b). In the two major lines of land plants, nonvascular and vascular plants, two distinct patterns were established (see Chapter 30). In nonvascular plants, the gametophyte generation predominates; for example, in the mosses, the gametophyte is the prominent, long-lived, leafy generation. In all vascular plants, the sporophyte generation with its roots, stems, and leaves predominates.

The opportunity for genetic diversity is greater in the diploid generation, because pairs of sometimes contrasting **alleles** determine the physical and physiological features of the organism. While one allele is dominant, the other can become masked, and its effect may not be expressed until subsequent generations. Evolution of the sporophyte of vascular plants led to an assortment of plant forms, new characteristics, and distinct plant divisions. The vascular plant divisions can be separated into two groups: those with seeds, the seed plants; and those without, the nonseed plants. Evolution of the seed as a reproductive device was an adaptation to existence on land; the seed plants became the most successful plants. Evolution of seeds made fertilization possible without an aqueous medium. In nonseed plants, an aquatic environment was required for syngamy: it was the only means by which swimming male gametes could reach the egg. The two groups of vascular plants and their reproductive differences are explored in Chapter 30.

## SEXUAL REPRODUCTION IN ANIMALS

Sexual reproduction is the predominant mode of reproduction in animals. As we have noted, in the more advanced groups of animals, such as molluscs, arthropods, echinoderms (sea urchins and starfishes), and vertebrates, sexual reproduction is the sole means by which new individuals are produced. Lower forms—sponges, coelenterates, flatworms, annelids, and the like—exhibit both asexual and sexual modes of reproduction. In these organisms, asexual reproduction is predominant during much of the year. Yet they engage in periodic sexual reproduction, often on a seasonal basis or linked to environmental changes such as depletion of nutrients. Seasonal sexual reproduction is responsible for most of the diversity in their phenotypes, that is, in their morphology, physiology, and behavior.

As we have seen, sexual reproduction involves the production of sex cells, or gametes, called ova and spermatozoa (sperm). The gametes are produced in many animals in specialized organs called **gonads.** The female gonad is the **ovary,** and the male, the **testis** (testes, plural). Unlike the situation in most flowering plants, the sexes in most animal species are separate: each individual bears either ovaries or testes. Such species are said to be **dioecious** (from the Greek *die,* "two," and *oikos,* "house"). In dioecious species, two individuals with unique genetic and physiological makeup together produce offspring. Thus, the chances for maintaining physical and physiological diversity and for recombining the

**FIG. 21-11** Isomorphic (a) and heteromorphic (b) alternation of generations in the algae.

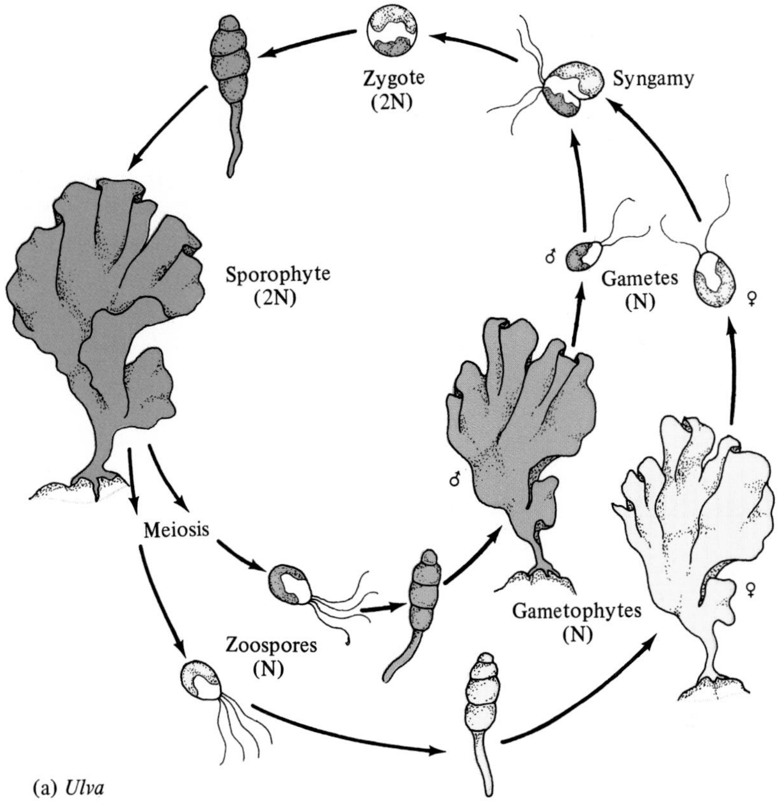


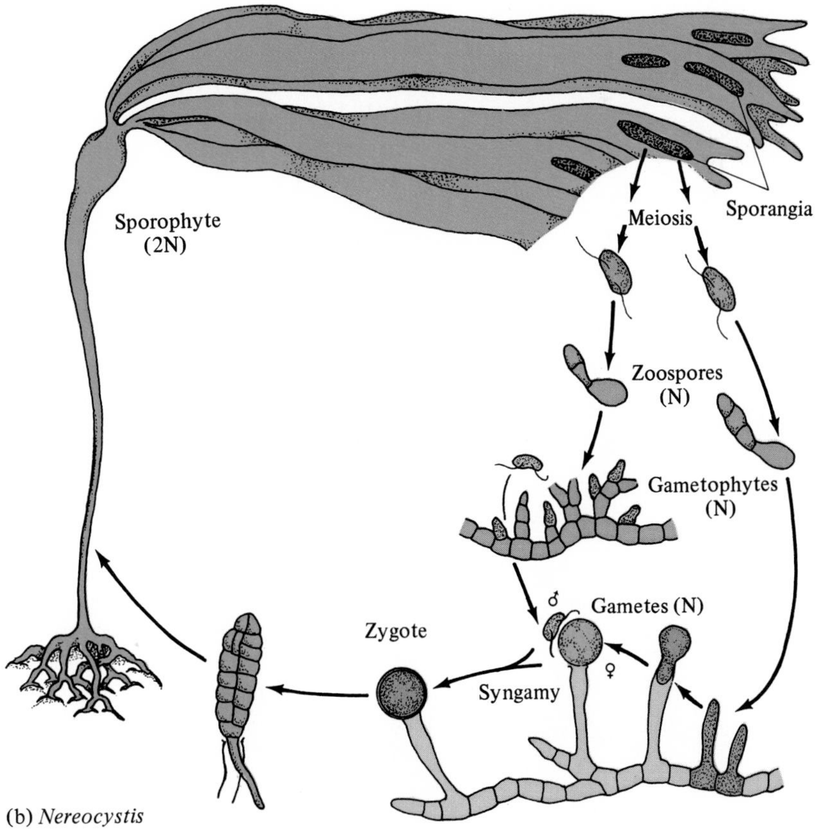

favorable traits of two separate individuals are increased by this arrangement.

Most annelids, arthropods, and molluscs; all echinoderms; and almost all vertebrates are dioecious. Most lower animals, including most sponges, coelenterates, and flatworms, are monoecious, that is, each individual bears both types of gonads and is capable of producing ova and spermatozoa. Yet, as we have seen, self-fertilization is rare even in these animals. Usually, the ova and sperm of any one individual mature and are released at different times, or, in many cases, the genital openings are so arranged that self-fertilization is physically impossible. This arrangement ensures that two separate individuals participate in sexual reproduction.

***Reproductive Organs.*** There are two types of reproductive organs in animals: the **primary sex organs,** or gonads, which produce gametes; and the **accessory sex organs,** which aid in bringing the two gametes together. Although all sexually reproducing animals bear gonads, accessory reproductive organs are usually present only in the more advanced groups of animals. Because early morphologists did not know the functional significance of some accessory reproductive organs, their nomenclature is quite diverse.

PRIMARY SEX ORGANS. The gonads in most animals are permanent, discrete organs set aside very early during embryonic development. However, the lower animals, such as sponges and coelenterates, lack definitive permanent primary or secondary sex organs. In sponges, groups of primordial germ cells, called **spermatogonia** and **oogonia,** arise from other types of cells. This may occur in any part of the sponge body. After repeated mitoses, the spermatogonia undergo meiosis to produce numerous spermatozoa, which are released into the surrounding water. Ova arise similarly from oogonia. The oogonia undergo a substantial increase in size at the expense of nearby cells, then mature by meiosis into haploid ova. This form of gametogenesis also occurs in some flatworms and annelids. Fertilization in sponges is achieved when a sperm from another individual of the same species happens to be brought by water currents to the ovum, resulting in their fusion.

In most coelenterates, the tissues that give rise to ova or spermatozoa are located in specific parts of the body. Although simple tissues, they are often referred to as gonads. In most coelenterates, there are no accessory reproductive organs; both ova and spermatozoa are merely released into the water, where syngamy occurs (Fig. 21-12). Certain species of coelenterates alternate an asexual stage with a sexual one. For example, *Obelia* is a colonial form made up of many hydralike polyps that reproduce asexually by budding. However, some of these polyps are able to produce jellyfishlike medusae that develop gonads. The gonads produce gametes following meiosis, and the zygote develops through embryonic and larval stages into an asexually reproducing polyp (Fig. 21-12).

Monoecious coelenterates produce only spermatozoa for a period of time, then undergo a "sex change," whereupon they produce only ova. Thus, although hermaphroditic, these coelenterates function so as to ensure that fusing gametes are derived from different individuals.

Most animals above the sponges and coelenterates possess rather well-defined, more or less permanent gonads, either ovaries or testes, or, in monoecious individuals, both. Depending on the species, the num-

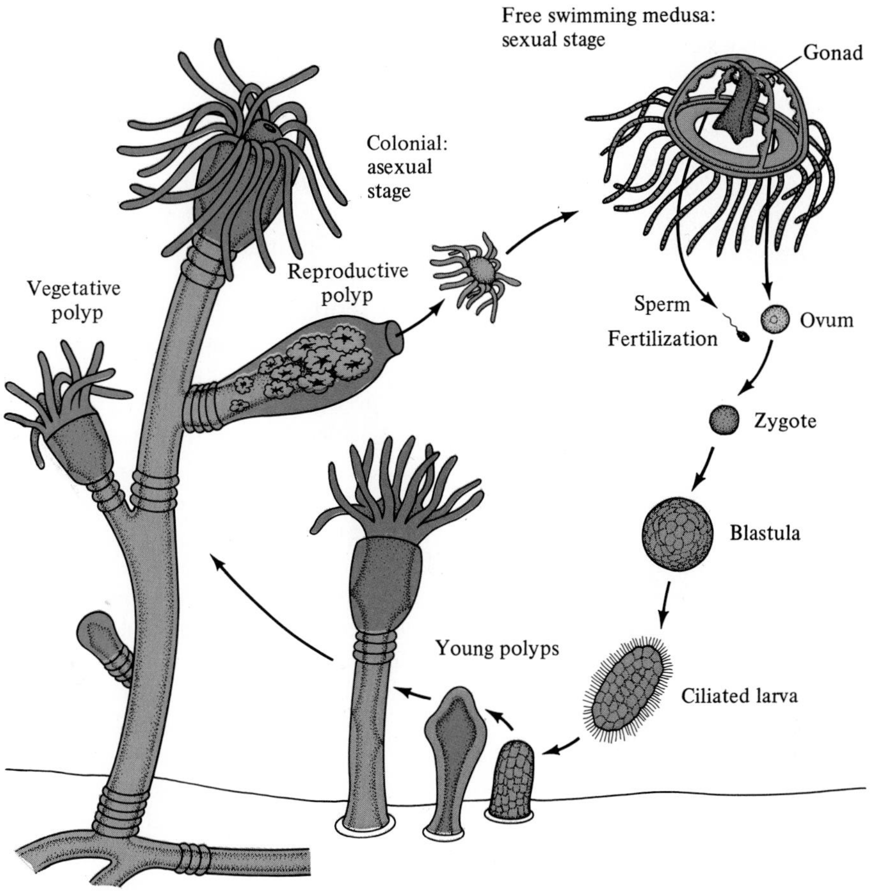


**FIG. 21-12** Life cycle of a hydrozoan coelenterate, *Obelia*.

ber of gonads in an individual varies from one to dozens, although within a species the number is constant. In any one class of animals, the gonads in different species are of similar morphology.

There are two major types of cells in the gonads of most, if not all, higher animals. One type, the **germ line cells,** produces gametes (Fig. 21-13). These are the cells that perpetuate the species. These cells are almost always derived from primordial germ cells (cells that give rise to spermatogonia and oogonia) set aside very early in embryonic development. In some insects, for example, the separation of germ line cells from body, or somatic (other than germ cells) cells, occurs by the third or fourth division of the zygote (the equivalent of 8- or 16-cell stage of development). In chickens and frogs, the germ line cells also become segregated from the somatic cells at an early stage. Although they may multiply by mitosis indefinitely, both oogonia and spermatogonia eventually undergo meiosis to form mature ova and spermatozoa.

The second category of gonadal cells is less easy to define. These are the somatic cells that surround the oogonia or spermatogonia and never give rise to gametes. In many species, these somatic gonadal cells have been shown to contribute to the development or nutrition of the gonial, or germ line, cells. In vertebrates, where somatic gonadal cells fill the spaces between blocks of developing gametes, they are termed **interstitial cells.** In the testes these interstitial cells produce the male sex hormone, **testosterone,** and in the ovary they produce the female sex hormone, **estradiol.** Both hormones affect reproductive development and function.

Somatic cells: die
without leaving descendants.
Gonadal cells: give
rise to gametes.

**FIG. 21-13** Diagrammatic sketch to illustrate that the somatic and germ cells are segregated early in development.

ACCESSORY REPRODUCTIVE ORGANS. Since sexual reproduction usually involves the fusion of two specialized cells produced in two different individuals, various organs that aid in this process have evolved in the different groups of animals. These organs, collectively known as the accessory reproductive organs, include the ducts that convey gametes from the gonads to the exterior, the organs for the delivery of gametes to the sites of their fusion, the organs that store gametes or fertilized eggs, and the glands that provide protective coverings or nutrients to the gametes. No group of animals possesses all the different types of accessory reproductive organs. But most animals above the organizational complexity of sponges and coelenterates bear one or more types of these organs. Typical examples of some accessory reproductive organs in a few groups of animals are shown in Figure 21-14.

Almost all animals above the coelenterate organization bear gonadal ducts, such as **vasa deferentia** from the testes and **oviducts** from the ovaries, which convey the gametes from the gonad to the exterior. In some animals there may be a duct, the **vas efferens,** that leads sperm from the testis to the vas deferens. In many species, the vas deferens has a pouch, the **seminal vesicle,** in which the sperm are stored before being released. In the females of many species there is a pouchlike outgrowth, often associated with the oviducts, the **seminal receptacle,** in which sperm are stored for a time following copulation. Many insects and annelids are examples of animals that bear these types of organs. In most terrestrial animals and several aquatic forms, specialized copulatory organs, usually the **penis** in the male, and the **vagina** in the female, have evolved. These ensure the placement of the male and female gametes in close proximity to one another, usually in a compartment within the body of the female. This arrangement greatly increases the chances that fertilization will occur. In many species, the vagina is represented by a terminal fusion of the two oviducts. In mammals, the vagina usually forms as a result of invagination of the body wall and is connected to the uterus by the cervix. Similarly, the two vasa deferentia in the male fuse and end in a common duct, the urethra, that passes through an elongate organ of intromission, the penis. In all mammals, the reproductive and urinary systems are in close association and thus form the **urogenital system.** In mammals, the penis is a specialized organ that develops in association with the urogenital organs. Associated with the penis are a number of accessory glands that aid in copulation. The **prostate** secretes an alkaline fluid that aids sperm motility, and a pair of **bulbourethral (Cowper's) glands** secretes a mucus that lubricates the urethra.

In some fish and various aquatic invertebrates, the copulatory organ is a modified leg (in lobsters, crayfish), fin (in sharks), or tentacle (in cuttlefish and octopus, in which the tentacle itself is a modified foot). Birds and amphibians lack penises and vaginas. In these animals, the urogenital ducts open into the rectum to form a common duct, the cloaca. In birds, the male transfers the sperm to the female's cloaca by apposing his cloacal opening to hers, thus allowing internal fertilization. But the rich diversity of anatomical and physiological specializa-

**FIG. 21-14** Comparative morphology of sex organs in animals. (a) Liver fluke *(Opisthorchis)* with both testes and ovary in the same individual. (b) The common nematode *Rhabditis.* (c) A generalized sketch of female and male reproductive organs of insects. (d) Reproductive organs of male and female pigeons. (e) Reproductive organs of the male rat. (f) Reproductive organs of the female rat.

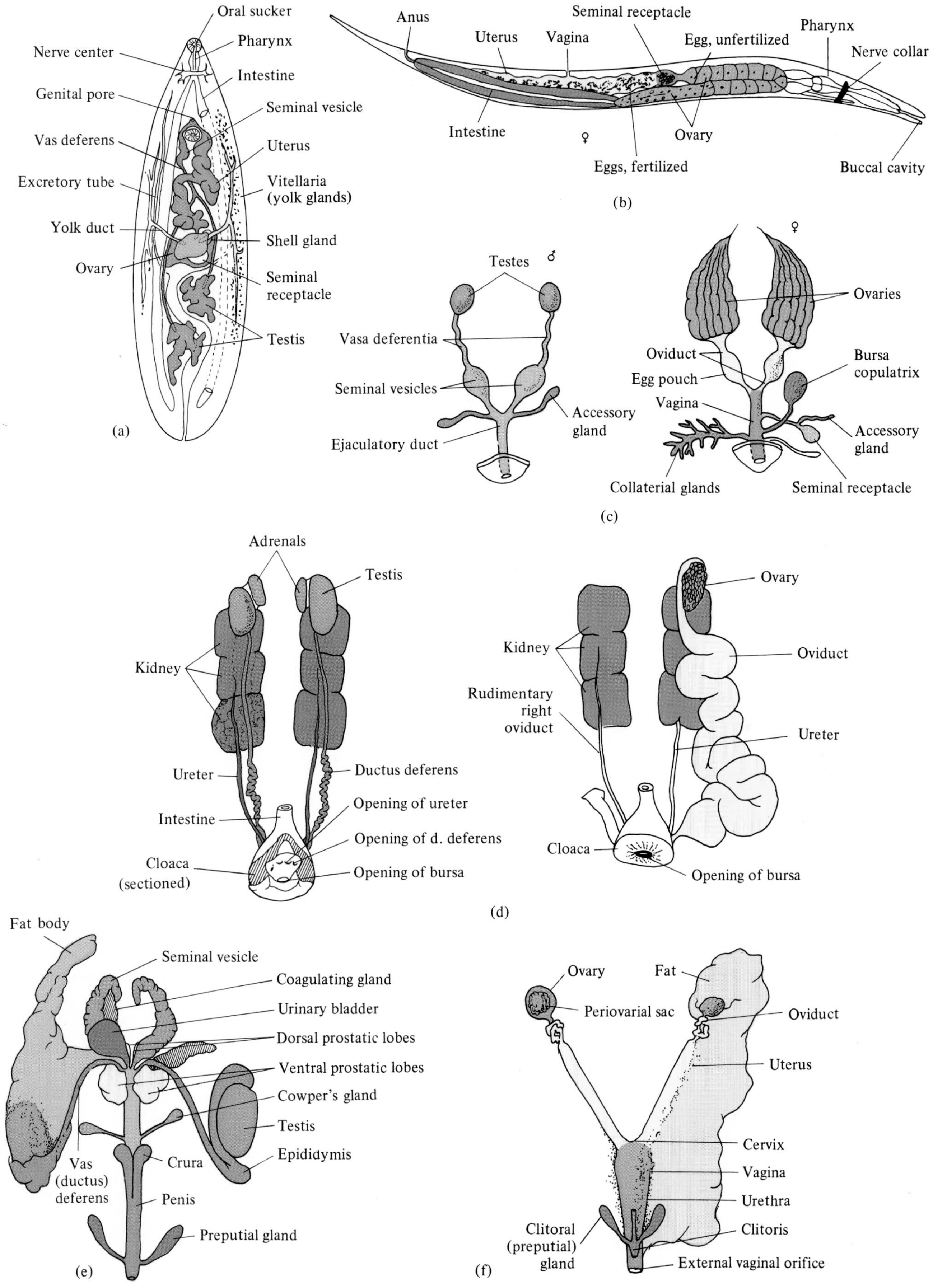

Oral sucker
Pharynx
Nerve center
Intestine
Genital pore
Seminal vesicle
Vas deferens
Uterus
Excretory tube
Vitellaria (yolk glands)
Yolk duct
Shell gland
Ovary
Seminal receptacle
Testis
(a)
Anus
Seminal receptacle
Pharynx
Uterus
Vagina
Egg, unfertilized
Nerve collar
Intestine
♀
Ovary
Eggs, fertilized
Buccal cavity
(b)
Testes ♂
Vasa deferentia
Seminal vesicles
Accessory gland
Ejaculatory duct
♀
Ovaries
Oviduct
Egg pouch
Bursa copulatrix
Vagina
Accessory gland
Collaterial glands
Seminal receptacle
(c)
Adrenals
Testis
Kidney
Ureter
Ductus deferens
Intestine
Opening of ureter
Opening of d. deferens
Cloaca (sectioned)
Opening of bursa
Ovary
Kidney
Oviduct
Rudimentary right oviduct
Ureter
Cloaca
Opening of bursa
(d)
Fat body
Seminal vesicle
Coagulating gland
Urinary bladder
Dorsal prostatic lobes
Ventral prostatic lobes
Cowper's gland
Testis
Epididymis
Crura
Vas (ductus) deferens
Penis
Preputial gland
(e)
Ovary
Fat
Periovarial sac
Oviduct
Uterus
Cervix
Vagina
Urethra
Clitoris
Clitoral (preputial) gland
External vaginal orifice
(f)

tions in the copulatory organs in different groups of animals clearly suggests that internal fertilization as a method of ensuring syngamy has evolved independently in many different groups of animals.

Another category of capabilities that enhance the survival of a species are those that provide nutriment and protection to the developing embryo. In coelenterates, flatworms, and annelids, the oocyte produces yolk and stores it. This yolk serves as nutriment for the developing embryo after fertilization. In higher animals, yolk is produced in other tissues—for example, in the liver in vertebrates and in the fat bodies in insects—but it is sequestered in the oocyte. Because these organs have many other metabolic functions unrelated to reproduction, they are not classified as accessory reproductive organs. However, in birds, the specialized glands in the oviducts that secrete the white of the egg serve exclusively in reproduction. The **shell gland,** which provides the porous, calcareous shells enclosing the eggs of reptiles and birds, and certain glands of cockroaches and related insects that secrete a protective covering over groups of eggs are truly accessory organs of reproduction.

The best protection for a developing embryo is the mother's body. Since the defenseless embryo is the most vulnerable of all prey, it is not surprising that in many animal species organs that shelter the embryo have evolved. Even in many of the lower animals, several adaptations have evolved that provide a sheltered space for the developing embryo in the mother's body. The most effective of such organs is the **uterus** of mammals, a modified part of the oviduct. The human uterus is the most highly advanced of this type of organ; it is described in Chapter 22. The uterus in mammals and in some insects (for example, the tsetse fly) provides space and nutrition to the developing embryo.

***The Disposition of Fertilized Eggs.*** The disposition of fertilized eggs is accomplished in diverse ways. In many species, the eggs are merely released into the environment, often (as in frogs) even before being fertilized. This is usually the case when an individual produces numerous eggs, sometimes thousands or millions. But in many species, even those that produce many eggs, the female selects a favorable site for their deposition or even attaches them to an object in the water. This assures good aeration and prevents their drifting to less favorable areas. Many species deposit their eggs in capsules or masses of jelly that afford them some protection. Frog eggs, for example, are enclosed in a mass of jelly that swells and congeals soon after fertilization. The eggs of several species of worms are also embedded in jelly. Many insects attach their eggs to leaves or twigs; some bury them. Many lay them on or close to a food supply suitable for their larvae after hatching—which is why garbage cans may be infested with larvae (maggots) in hot weather and why apples may have worms. Beyond the choice of a suitable environment for the deposition of their eggs, many animals provide them no further care. Some animals, however—including certain fishes, some amphibians and reptiles, and all birds—guard their eggs until they hatch and may even care for their young after hatching. Other animals, including lobsters, crayfish, crabs, and some amphibians, attach the eggs to their own bodies, carry them around until hatching, and in some cases provide parental care to the hatchlings.

An animal that sheds, deposits, or lays its eggs is said to be **oviparous.** Most invertebrates, fishes, amphibians, and reptiles, and all birds, and a small subclass of mammals that includes the duck-billed platypus are oviparous. Oviparity is a primitive condition. However, in many species of invertebrate animals, such as freshwater clams and various crustaceans, the eggs are retained within the mother's body until they hatch. Species in which the developing eggs are retained within some part of the reproductive tract are said to be **ovoviviparous.** The part of the body may be a special brood pouch, as in sea horses and pipefish, or a body cavity such as the mantle cavity, as in oysters. If, in addition, nutrients are derived from the tissues of the mother during the embryo's development, the animal is said to be **viviparous.** Placental mammals are classic examples of viviparous animals. However, viviparity also occurs among invertebrates, including some scorpions, one genus of annelid, and a few insects. An interesting borderline case of ovoviviparity is seen in the tsetse fly, the transmitter of African sleeping sickness. The female produces but a single egg, which is fertilized internally; the egg then migrates to the uterus, where it hatches into a larva. There the larva remains, nourished by accessory glands of the mother and passing through all but the last of its larval stages. The last larval stage emerges from the mother and metamorphoses into a pupa (a stage between larva and adult in some groups of insects) immediately.

***Finding a Mature Mate of the Right Kind.*** Finding a mature mate of the same species is a problem common to sexually reproducing animals, whether they are dioecious or cross-fertilizing, monoecious individuals. Many different types of solutions to this problem have evolved; the type of solution is dependent to some degree on the mode of fertilization. If the individuals release gametes to the exterior, where fertilization depends on a chance meeting of gametes, the solutions are different than those evolved in animals in which internal fertilization is the rule.

Aquatic invertebrates with external fertilization, such as corals, worms, oysters, and others, release their gametes into water during a very limited breeding season. One extreme example of such synchronization is seen in the palolo worm. Hundreds of thousands rise to the surface of the sea to release their gametes during 2- to 4-hour periods on 2 or 3 specific, successive days of the year. By synchronizing the release of gametes, the chance that ova and sperm of the same species will meet is greatly increased. Some species, such as sea urchins, release their gametes together with a special chemical signal, a pheromone, that triggers nearby members of the same species to release their gametes as well. Other animals form permanent or temporary social groups, such as schools of fish, that ensure the proximity of the two sexes.

Terrestrial animals and animals with internal fertilization have other problems. They must physically locate a mate and achieve a transfer of sperm from male to female. The likelihood of such transfer may be enhanced by the formation of permanent or temporary social groups, such as pairs, flocks, and herds. When no permanent or temporary social bond is maintained, other means have evolved. For example, some female insects release sex pheromones that attract mates from great distances. When tethered in a field the female moth of some species attracts all males of her species present in a hectare (2.5 acres) or more. One extreme example of ensuring the presence of a male in close proximity to a female is seen in a primitive worm, *Bonnellia*. In this genus, the female is much larger than the male, which is usually found as a parasite on the body of the female. When a larva of *Bonnellia* settles in the vicinity of a female, she releases a pheromone that causes the larva to develop into a male. By thus determining what sex its neighbor will be, the female ensures the presence of a male.

***Control of Reproductive Behavior in Vertebrates.*** Since most animals are seasonal reproducers, it is essential that both males and females be in reproductive condition at the same time of the year. Animals also need to provide for their own safety when mating and for that of their eggs and their young following hatching or birth. The problems are especially complicated when hatching or birth occurs a considerable period of time after mating. In many migratory birds, the place and time of year at which breeding and brooding occur coincides with the presence of the fewest predators and an abundance of food. In many species, however, the parents do not eat during this time. In mammals with relatively long gestation periods between mating and birth, provision must be made for the safety of the mating pair and of the mother and offspring at the time of birth. The length of gestation period in many animals appears to be correlated with this requirement.

In some mammals, including bats, bears, and badgers, implantation of the embryo in the mother's uterus—and thus the beginning of its development—is delayed for a time after fertilization, thus allowing birth to occur at the optimal time for raising the young. In many other mammals, the spermatozoa are stored within the female, and the time of fertilization is delayed.

Members of all species in which reproductive behavior occurs only at particular times of the year have an ability to detect the time of year by detecting changes in the average day length. This is true in plants as well, in which the phenomenon of **photoperiodism** was discovered. The photoperiod is coupled to the animal's sense of elapsed time, derived from its biological clock. As noted in Chapters 19 and 20, these behaviors are innate: the animal does not know that it would be wise for it to mate at a particular time. Mating at a specific time is not a conscious choice but an automatic response that has evolved in reference to a certain group of environmental cues.

Experiments have shown that photoperiodic effects are of major importance to the breeding of almost all vertebrates. For example, when certain species of mammals are placed in either continual darkness or continual light, their reproductive cycles cease. When spring-breeding species are placed under artificial spring lighting conditions (increasing day length) in the fall, they exhibit breeding behavior. Fall-breeding species can be induced to breed in the spring by being subjected to artificially shortened day lengths. When these mammals are transported from the northern to the southern hemisphere, their breeding seasons become reversed.

More sophisticated experimentation has established that the full development and functioning of vertebrate gonads require the presence of two glycoprotein hormones from the pituitary gland: follicle-stimulating hormone (FSH) and luteinizing hormone (LH). These pituitary gonadotropins act on gonads to produce steroid sex hormones: testosterone in males and estradiol in females. These sex steroids are the immediate cause of gonadal growth, gametogenesis, and breeding behavior. Both castration—the removal of the gonads—and hypophysectomy—the removal of the pituitary gland—abolish breeding behavior. For the pituitary to release its gonadotropins, releasing factors produced by the hypothalamus must also be present. In many seasonally breeding vertebrates, the hypothalamus must receive nerve impulses from other parts of the brain in order to produce these factors. Experiments have shown that whatever the environmental cue, the same neural and biochemical path—leading from the sensory receptors to the hypothalamus and then to the pituitary gland and to the gonads—is involved in controlling reproductive behavior.

## Summary

Organisms reproduce either sexually, by union of the specialized sex cells called gametes, or by one or more asexual means. Asexual reproduction involves the production by a single individual, the parent, of one or many offspring, each of which is usually genetically identical to the parent. The principal adaptive value of asexual reproduction is that it provides a means whereby a single individual not only can reproduce in the complete absence of any other member of its species, but can, within a short time, produce large numbers of individuals of a type already proven to be adapted to a particular environment. Although asexual reproduction can rapidly increase a population, it provides very little of the diversity essential to evolution. Because sexual reproduction involves a union of gametes usually derived from two individuals, each new individual is genetically different from either of its parents. Because the zygote is diploid, that is, contains two sets of chromosomes, the process of meiosis must occur somewhere during the life cycle of every sexually reproducing organism if haploid gametes, containing one set of chromosomes, are to be produced. While ensuring that each gamete has but one set of chromosomes, meiosis also constitutes a mechanism for creating diversity. As a result of the segregation of the homologous chromosomes in meiosis, the cells produced usually differ greatly in genetic composition. Further contributing to the genetic diversity of gametes is the process called crossing over, in which portions of homologous chromosomes are exchanged with each other during the initial stages of meiosis. The cells produced by meiosis are gametes if meiosis is associated with gametogenesis, as in animals, or spores if meiosis is associated with sporogenesis, as in plants. The plant spores germinate to become gametophytes and thus lead ultimately to the production of gametes. Eventually, the chromosomes of the gametes are brought together in syngamy, and individuals differing genetically from their parents are generated.

Sexual reproduction has great adaptive value by providing new genetic diversity to a species at each generation. Thus, no matter what environmental changes may occur or what new environments may be entered, the chances that some members of the species will survive are optimized. The advantages of diversity for evolutionary change have favored devices that promote cross-fertilization.

Although many species reproduce only by sexual means and several only by asexual methods, many major groups of organisms exhibit both forms of reproduction during their life cycles, thus deriving the special benefits that both types of reproduction provide.

All available evidence indicates that asexual reproduction evolved before sexual reproduction. The simplest form of sexual reproduction involves isogametes, gametes identical in size and form. Anisogamy, in which the gametes differ in size, occurs in algae and fungi. Oogamy, in which the gametes are differentiated into sperm and egg, is the most common pattern of sexual reproduction. Three basic types of life histories are found among organisms, and these are all represented among the green algae, thought to be among the most primitive of the eukaryotes. These life histories differ in the time of meiosis in the cycle. In one type of cycle, common in animals, the individuals are diploid. In another, common in some algae, the individuals are haploid. In the third type, two types of individuals alternate in the life cycle. In this alternation of generations, one generation (the gametophyte generation) produces gametes, and the other (the sporophyte generation) produces spores; the latter process is accompanied by meiosis. Alternation of generations is the common life cycle pattern of plants.

The primary sex organs of animals consist of the gonads: the testes in the male and the ovaries in the female. In all higher animals, each type of gonad is composed of two types of cells. The primordial germ cells, which are set apart from somatic cells very early during development, give rise to gametes. The somatic gonadal cells aid in the development or the nourishment of the gametes or both. These cells may also secrete sex hormones. In addition to gonads, most animals bear accessory reproductive organs of diverse types. The gonadal ducts—the vas deferens leading from the testis and the oviduct leading from the ovary—are almost universal above the level of coelenterates. Accessory reproductive organs that serve to store sperm include the seminal vesicles of males and the seminal receptacles of females. Organs involved in the transfer and receipt of sperm are the penis and the vagina. Other accessory organs include glands that secrete nutrients or protective coverings for the eggs.

Oviparous animals release their ova into the environment either before or after fertilization. In ovoviviparous animals, the egg is retained in a part of the body cavity or a specialized pouch, where it completes its development in the protective custody of its parent. In the viviparous, modern mammals, the embryo and its mother develop an organic connection through which the developing embryo is nourished.

Because sexual reproduction involves the collaboration of two individuals, several different means of ensuring such communion have evolved. For example, members of both sexes may be synchronized to reproduce at certain times of the year. Or chemical attractants may bring males and females together. Chemicals may also play a part in synchronizing the release of gametes. Regulation of seasonal reproduction in birds and mammals has been studied in detail. Photoperiod, temperature, and availability of food all control the time of functional activation of the gonads. The hypothalamic neurosecretory releasing hormones initiate these events. Triggered by specific hypothalamic releasing hormones, the pituitary glands secrete two gonadotropic hormones: follicle-stimulating hormone and leutinizing hormone. These hormones in turn activate the gonadal somatic cells to secrete the steroid hormones, testosterone in the male and estradiol in the female. These steroid hormones activate the gonads and initiate the growth of the accessory reproductive organs.

## Recommended Readings

Bold, H. C., C. J. Alexopoulos and T. Delevoryas. 1980. *Morphology of Plants and Fungi,* 4th ed. Harper & Row, New York. A survey of reproduction and life cycles in the various plant groups, including an account of plant morphology.

Graham, L. E. 1985. "The origin of the life cycles of land plants." *Amer. Sci.* 73; 178–186. A cogent analysis of the theoretical basis of the diversity in the life histories of land plants suitable even for nonspecialists.

John, B., and K. R. Lewis. 1984. *The Meiotic Mechanism,* 2nd ed. Carolina Biology Reader. A well illustrated, clear account of meiosis, a very important process involved in sexual reproduction.

Sadler, R. M. 1973. *The Reproduction of Vertebrates.* Academic, New York. This book presents a survey of the patterns of reproduction and mechanisms of reproductive control in the major groups of vertebrates.

Shields, W. M. 1982. *Philopatry, Inbreeding and the Evolution of Sex.* State University of New York Press, Albany. This short paperback book discusses the theories of the evolution of sex. The retention of inbreeding as a reproductive method under some conditions is explained.

22

# Human Reproduction

OUTLINE



Most mammals exhibit a seasonal reproductive phase. But humans, like the other higher primates, copulate not only at the time of ovulation but at other times as well. Over the course of the 4 to 5 million years in which humans have diverged from other modern anthropoids, they have evolved a year-round pattern of sexual activity. Although the development of this trait was accompanied by many changes in the social behavior of humans, the social aspects of human sexual evolution are beyond the purview of this text. This chapter describes the morphology and physiology of the reproductive organs of the human male and female, and discusses methods of fertility control in humans.

## REPRODUCTIVE ORGANS

The human reproductive system is organized according to the same general plan as that of all mammals. The primary and secondary reproductive organs of the human male and female are almost identical in organization and arrangement to those of the other higher primates.

### PRIMARY AND SECONDARY SEX ORGANS: MALE

The primary reproductive organs of the human male are the two testes. The testes are retained outside the abdominal cavity in a sac of skin called the **scrotum** (Fig. 22-1). This location is 2°C cooler than normal

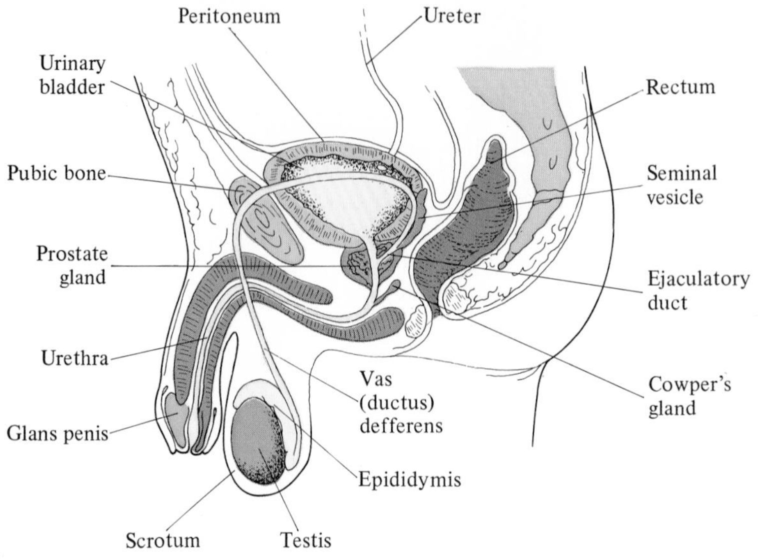


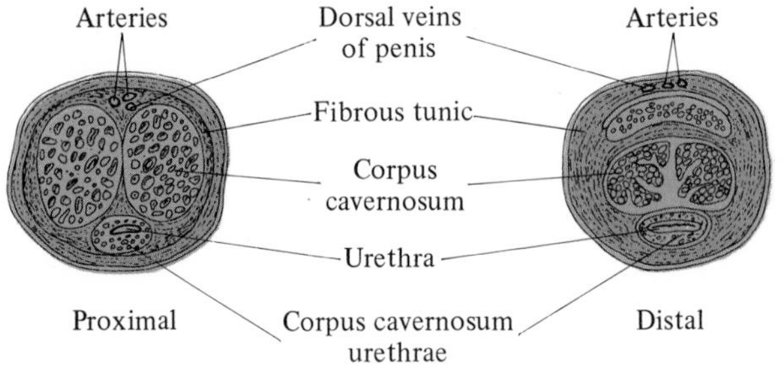


**FIG. 22-1** (a) The human male genitalia, shown in sagittal section. (b) Cross section of the penis showing the arrangement of blood vessels and erectile tissue.

body temperature—a necessary condition for the development and functioning of the testes. The testes normally descend from the body cavity, through a passageway called the **inguinal canal,** and into the scrotum just before birth; individuals with undescended testes are sterile. Each testis is ovoid and about 5 cm long. Attached to its upper and inside border is a crescent-shaped structure, the **epididymis** (from the

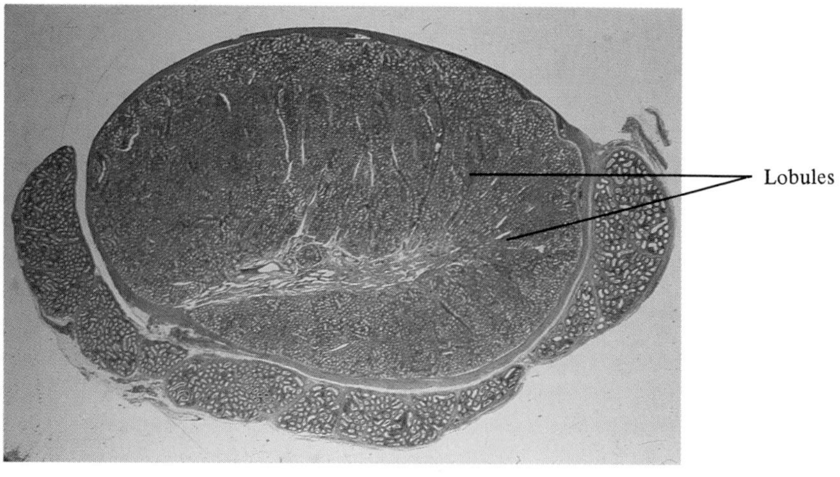


**FIG. 22-2** The human testis in section.

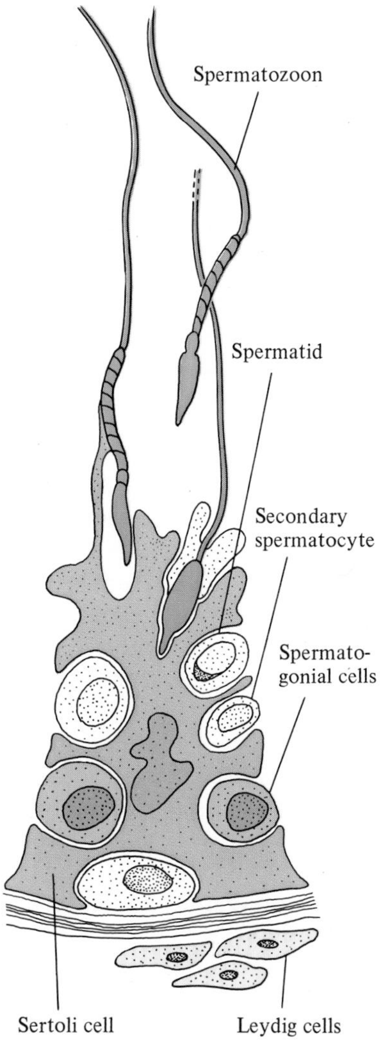

(a)

**FIG. 22-3** (a) Structure of seminiferous tubule. (b) Transverse section of a seminiferous tubule showing spermatogenesis. (c) Transverse section through a group of seminiferous tubules.

Greek *epi,* "above," and *didymis,* "twin"). The combination of the testis and epididymis is known as a **testicle.** Within each testis are somatic and reproductive cells arranged in about 250 lobules, each containing one to three convoluted **seminiferous tubules** (Fig. 22-2). The walls of the seminiferous tubules, or spermatic cords, are lined with the somatic **Sertoli** cells, which nurture the **spermatogonia,** or reproductive cells (Fig. 22-3a, b). The sperm develop from spermatogonial cells (Fig. 22-3). Among the seminiferous tubules are **Leydig,** or **interstitial, cells,** which secrete testosterone, the male sex hormone. These tubules lead into a system of ducts that join the long and tightly coiled epididymis, in which the sperm are stored. If uncoiled, the human epididymis would extend to about 6 m (20 feet); in some mammals, swine, for example, the epididymis is four to five times longer than in humans.

The epididymis is continuous with the vas deferens, which enters the body cavity by way of the inguinal canal, ending as the **ejaculatory duct** (Fig. 22-1). The ejaculatory ducts from the two testes empty into the urethra near the base of the urinary bladder. Each vas deferens receives viscous secretions from a **seminal vesicle.** Other accessory organs are the walnut-sized **prostate gland,** which provides a less viscous fluid containing nutrients for the sperm and the smaller **Cowper's gland,** located below the prostate gland, which secretes a fluid that serves as a lubricant during copulation. Although all three glands contribute to the seminal fluid in which the sperm are carried, most of the seminal fluid is secreted by the prostate gland.

The penis is external in humans, as in other primates. It consists of three elongate masses of spongy, erectile tissue, the paired **corpora cavernosa** and the unpaired **corpus cavernosa urethra,** and is covered with loose skin. The central spongy tissue is enclosed in a fibrous tunic. The tip of the penis, the **glans penis,** is formed by a mushroom-shaped expansion of the spongy tissue. In the natural condition—that is, in an uncircumcised individual—the glans is covered by a fold of skin, the foreskin, or **prepuce.** In circumcision, the prepuce is surgically removed. The glans is richly supplied with sensory receptors.

Sexual excitement, which can be induced by touch, friction, or a state of mind, produces a vasoconstriction in the blood vessels in the penis. There is an increase in the arterial blood supply to the spongy tissue while venous drainage is restricted. As a result, the spongy tissues enlarge and harden, leading to an erection of the penis.

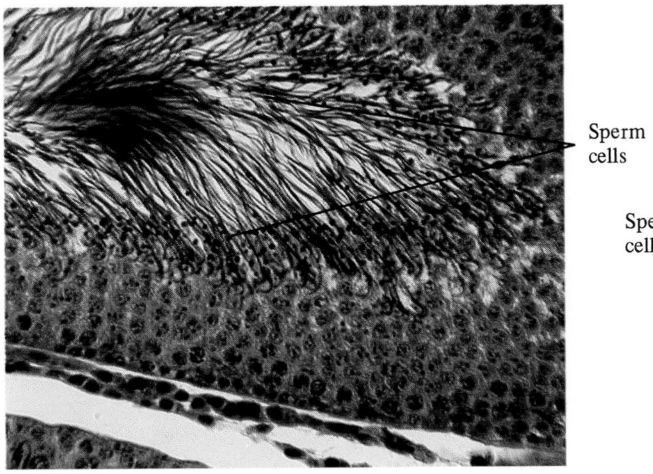

(b)

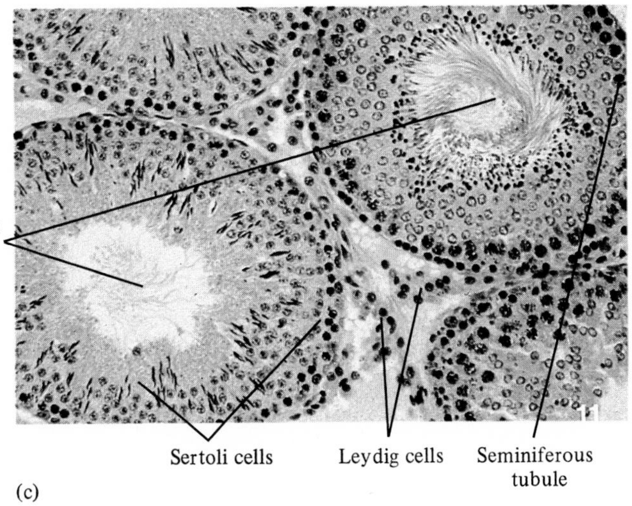

(c)

## PRIMARY AND SECONDARY SEX ORGANS: FEMALE

The human ovaries are paired, small, elongate (3 cm long) structures located in the pelvic region of the abdominal cavity (Fig. 22-4). Each ovary leads into an oviduct, termed the **fallopian tube.** The ciliated, flared-out opening of each tube partially surrounds its ovary in a way that facilitates the entrance of ova into the tube at the time of ovulation, when the ovum is released from the ovary. Both tubes lead to the unpaired, thick-walled and muscular, pear-shaped uterus, commonly called the womb, within which the embryo develops.

The richly vascularized epithelial layer that lines the uterine cavity is termed the **endometrium.** The ciliated cells lining the endometrium produce a continual downward current. The fluid, which originates in the body cavity, enters the fallopian tube at its open end and flows out of the body through the uterus. The tapered, lower end of the uterus, the **cervix** (Latin for "neck"), protrudes into the vagina. The vagina is the female copulatory organ; it is also known as the birth canal, through which the fetus emerges at the time of birth. Once the species-specific features are distinguishable, the developing mammalian embryo is known as a fetus. In young women, the vaginal orifice may be partially occluded by a delicate fold of mucous membrane known as the **hymen.** The hymen can be easily torn by insertion of the fingers, athletic activity, sexual intercourse, and so on.

The vagina opens into a shallow **vestibule.** The vaginal lips are termed the **labia minora** (labium minus, singular). Enclosing the vestibule are the **labia majora** (labium majus, singular). Immediately in front of the vagina is the urethra, which also opens into the vestibule. In front of the urethral orifice and also within the vestibule is the **clitoris,** the tip of which is homologous to the male glans penis. The external female genitalia are collectively termed the **vulva.**

Although as much as two-thirds of the male erectile tissue is external, the clitoris is the only external erectile tissue of the female. Its glans, like the glans penis, is richly supplied with sense receptors. The remaining female tissues that are homologous to the male erectile tissues lie within the labia minora and surround the outer third of the vagina. In the female, sexual excitement causes an initial erection of the glans and shaft of the clitoris, the release of a lubricating secretion into the vagina, a lengthening and "elevation" of the uterus (tilting toward the spine), enlargement of the labia minora, and separation and elevation of the labia majora.

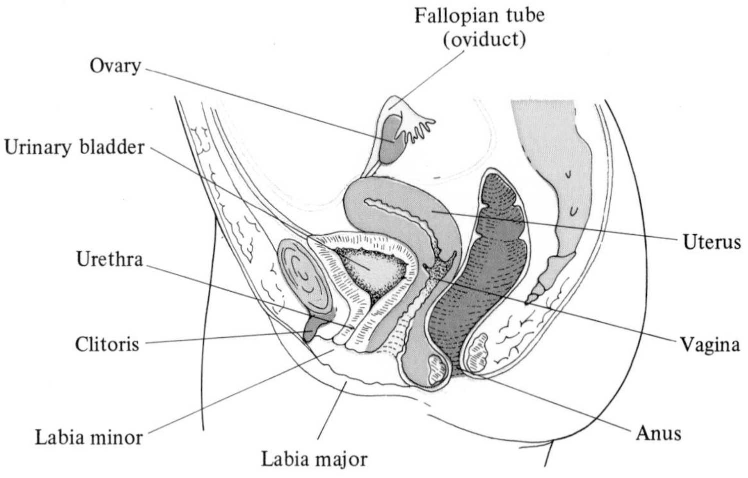


**FIG. 22-4** The human female reproductive organs shown in sagittal section.

### COPULATION, EJACULATION, AND ORGASM

Copulation is accomplished by the insertion of the erect penis into the vagina. Stimulation of the penis by repeated thrusting within the vagina normally results in orgasm of the male, an intense sensation accompanied by the ejaculation of seminal fluid into the vagina. Ejaculation consists of a series of rapid muscular spasms of the urethra, prostate gland, seminal vesicles, and ejaculatory ducts; this ejects the seminal fluid in a series of spurts. The muscular spasms associated with ejaculation are under the control of autonomic, that is, the sympathetic and parasympathetic, nervous system. Although the amount is variable, the total ejaculate averages about 4 ml in volume and usually contains an average of 120,000,000 sperm per ml. The female usually releases one ovum approximately each month and about 400 ova in her lifetime.

Women undergo orgasm similar in general respects to that of men. It is characterized by a series of rhythmic contractions of the vagina and uterus, occurring at the same frequency as the ejaculatory contractions that accompany male orgasm. Some women may never achieve orgasm during intercourse. The reasons are varied, including cultural attitudes toward sex, a particular state of mind, and the complex pattern of preliminary and accompanying stimulations by the partner. Some women normally experience multiple orgasms, two or more during the same period of arousal.

### CONCEPTION

The changes in the shape of the vagina and in the tilt of the uterus, coupled with the muscular spasms that accompany female orgasm, probably increase the chances of fertilization of the ovum by the sperm, or conception. However, neither female orgasm nor sexual arousal are necessary for conception, since the seminal fluid induces peristaltic contractions of both the uterus and fallopian tubes, and sperm are chemically attracted to the ovum. If ovulation has recently occurred and a viable ovum is present in one of the fallopian tubes, a sperm may succeed in fertilizing it. Fertilization of the ovum and early development of the zygote normally occur far up in the fallopian tube; the embryo then slowly moves down the tube to the uterus, where it becomes implanted in the uterine lining within a week after fertilization. Implantation of the embryo in the uterine wall signals the rest of the body that conception has occurred. The uterine wall allows implantation of the embryo only under certain physiological conditions, during a limited part of the cyclic period of uterine tissue growth and regression called the menstrual cycle.

## THE MENSTRUAL CYCLE

The uterine tissues in women and other female primates display a unique periodicity known as the **menstrual cycle.** Commencing at about the age of 13 years in the majority of women, the endometrial lining of the uterus undergoes a cycle about every 28 days, characterized by a period of growth and vascularization followed by a partial disintegration and a loss of blood, body fluid, and cellular debris. The flow of blood, fluid, and sloughed-off cells is termed **menstruation,** and the entire cycle is known as the menstrual cycle.* A generalized depiction of the human menstrual cycle is given in Figure 22-5. Following menstruation, the endometrial cells lining the uterus rapidly divide and become active in protein and nucleic acid synthesis. As a consequence, the tissue thickens and becomes supplied with an extensive network of blood vessels. This period of endometrial thickening usually continues for about 2 weeks. Meanwhile, several developing ova in each ovary have also begun to grow (Fig. 22-6). Only one, however, will usually have matured into an ovum 2 weeks later. Each developing ovum in the ovary is surrounded by numerous follicle cells, which grow along with it (Fig. 22-7). When one of the ova reaches full maturity, at about day 14 of the menstrual cycle,† the expanding follicle, now known as a Graafian follicle, rises to the surface of the ovary and ruptures; the ovum is thereby released, a process known as **ovulation.** The other ova become atretic; that is, they regress and disintegrate.

Even as it is being drawn into the fallopian tube by the current generated by the cilia of cells lining the reproductive tract (Fig. 22-8a), the released ovum continues the process of meiosis that began years previously, during the prenatal development of the ovary. If the ovum encounters sperm in the fallopian tube within the next 48 hours, fertiliza-

* A comparable cycle involving a periodic growth and loss of endometrial cells occurs in mammals other than primates, but no blood is lost. The shedding of cells is termed estrus and the cycle is called the estrus (from the Greek *oistros;* "mad desire") cycle.

† Menstrual cycles are said to begin with the beginning of menstrual flow, although menstruation is a result of ovulation and the subsequent failure of implantation of an embryo.

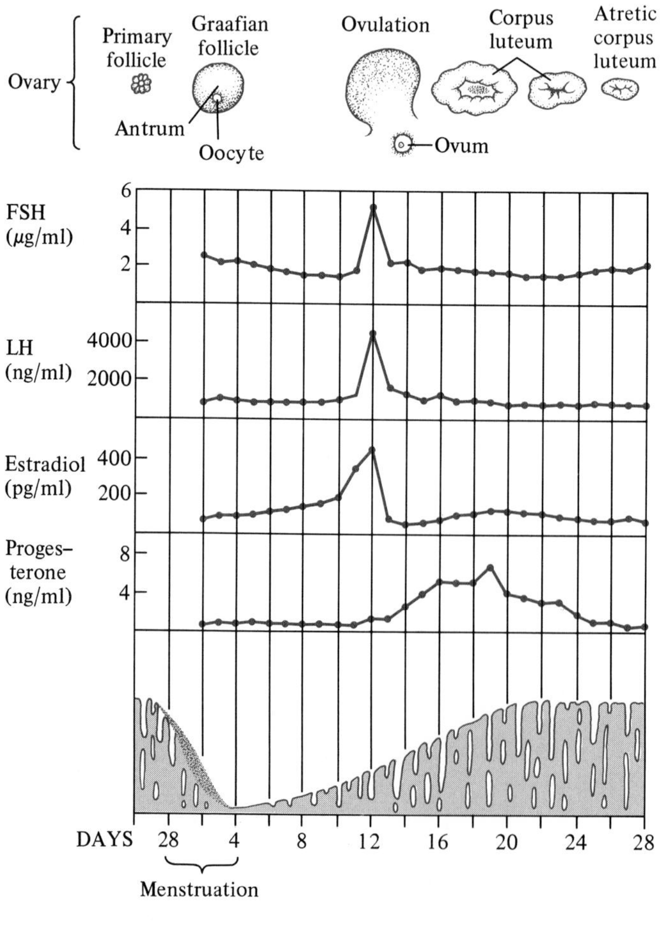


**FIG. 22-5** Diagrammatic representation of the human menstrual cycle, showing the relative concentrations of FSH, LH, estradiol, and progesterone and the corresponding states of the ovary and the endometrium.

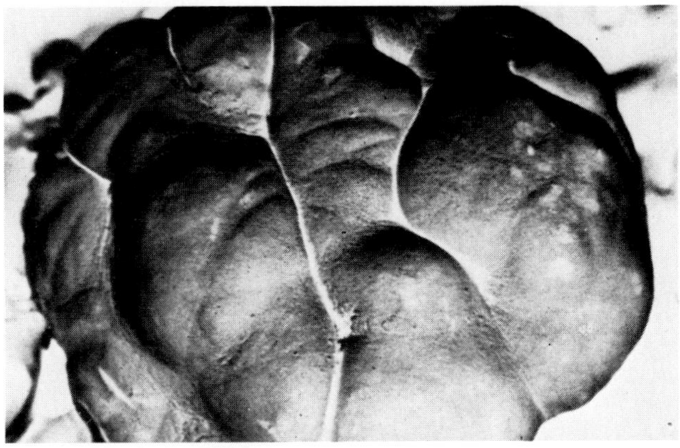

(a)

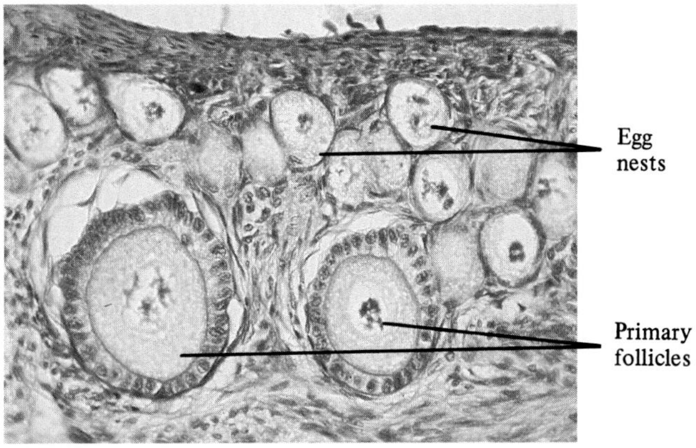


(b)

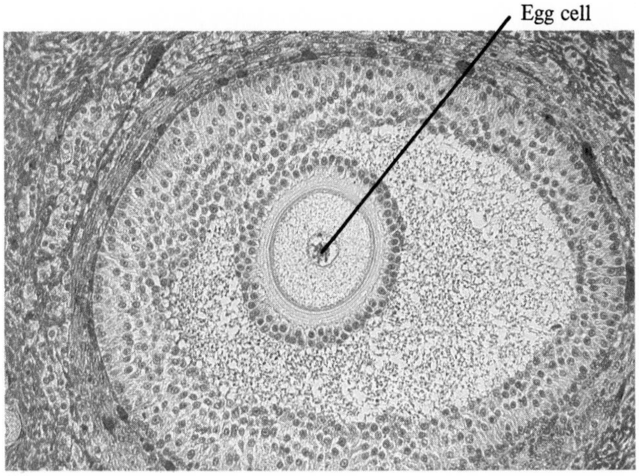


(c)

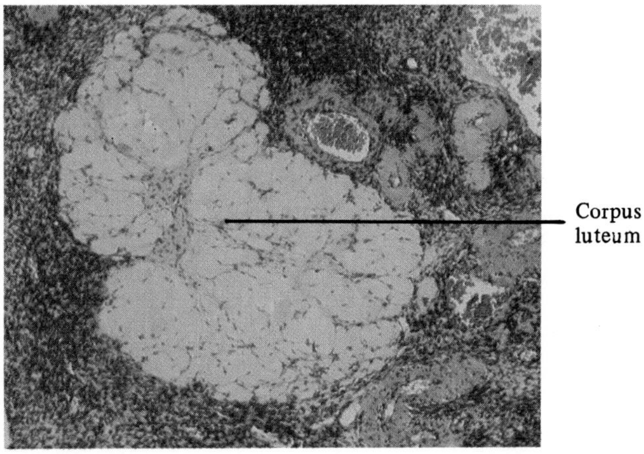


(d)

**FIG. 22-6** (a) Scanning electron micrograph of human ovary. (b) Cross section of human ovary showing two primary follicles and several egg nests. (c) Human Graafian follicle in section showing egg cell. (d) Graffian follicle after ovulation and degeneration of corpus luteum. (e) Sketch showing stages in ovulation.

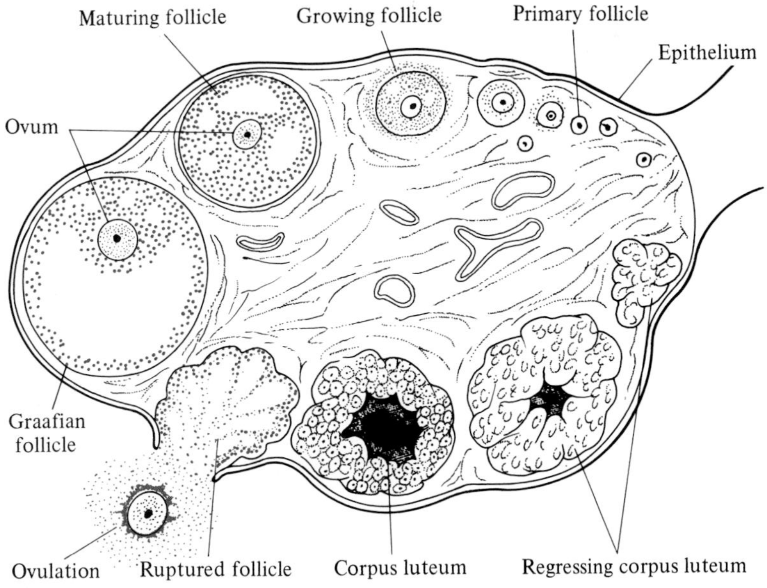


(e)

tion occurs. Meiosis in the ovum is completed only after its fusion with a sperm. Ova older than 36 to 48 hours (postovulation) at the time of fertilization tend to be nonviable, probably because of defects in the meiotic and mitotic divisions that follow the fusion with a sperm. Sperm remain viable for about 2 to 3 days in the female reproductive tract. Thus, whether sperm are introduced into the vagina as early as 2

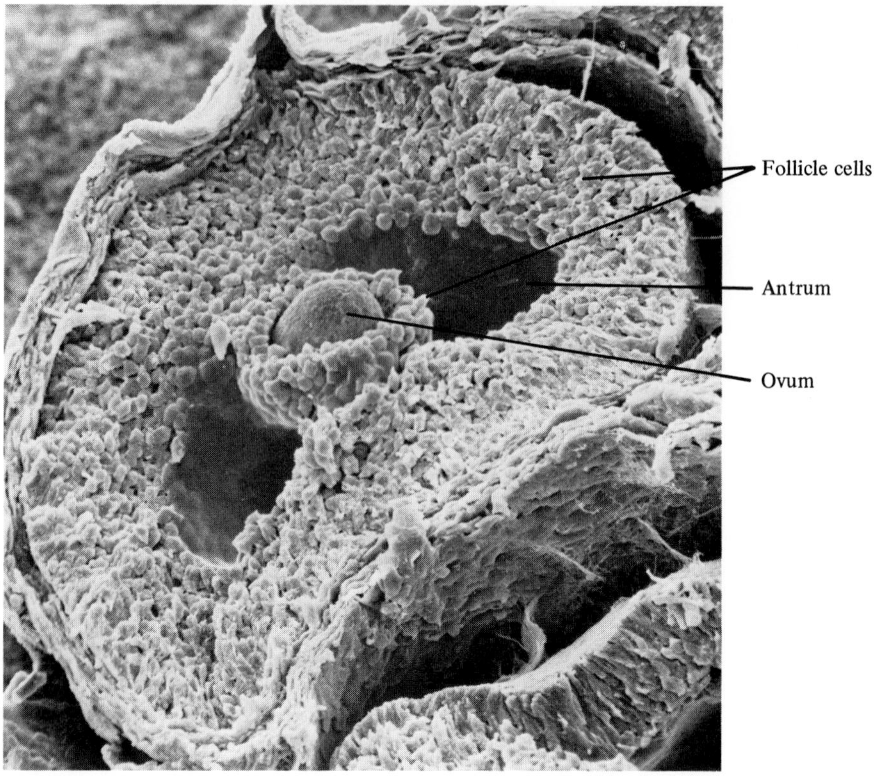


**FIG. 22-7** Scanning electron micrograph of ovarian follicle.

to 3 days before ovulation, or as late as two days after it, conception may result. A woman's fertile period is thus about 5 to 6 days in length during each menstrual cycle. After fertilization, it takes several days for the embryo to travel down the fallopian tube and into the uterus. Implantation of an embryo in the endometrium (Fig. 22-8b) normally occurs about the sixth day after fertilization, or on days 19 to 22 of that menstrual cycle.

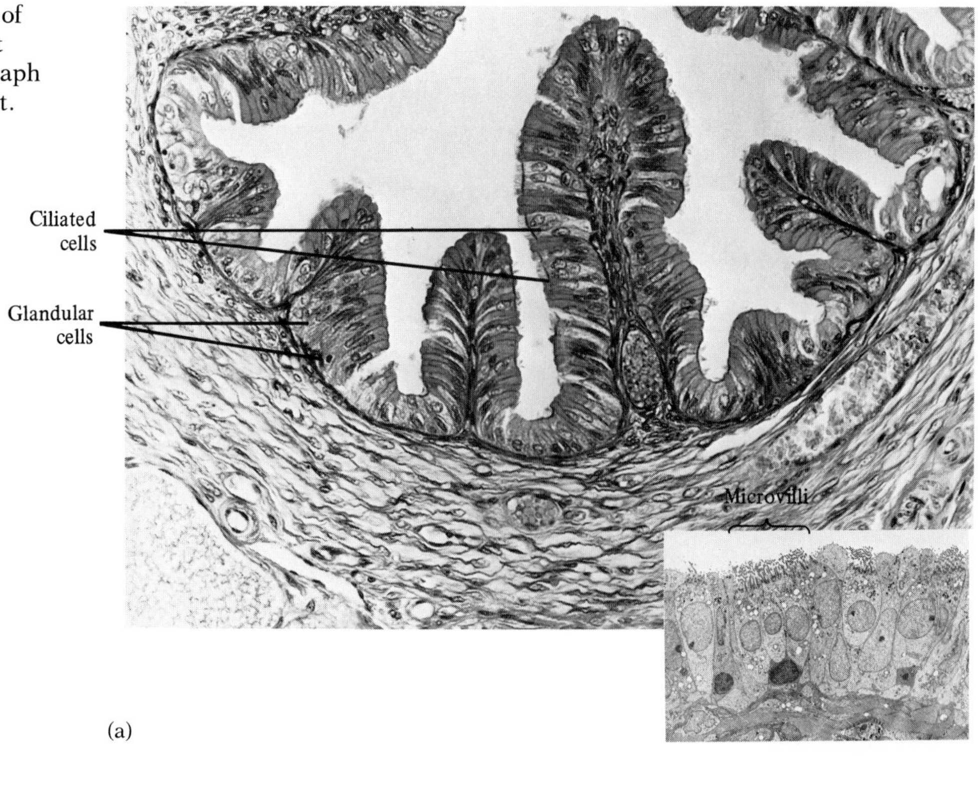


**FIG. 22-8** (a) Cross section of human oviduct with an inset showing an electron micrograph of the cells lining the oviduct.

**FIG. 22-8 (*continued*)** (b) Endometrium section of human uterus.

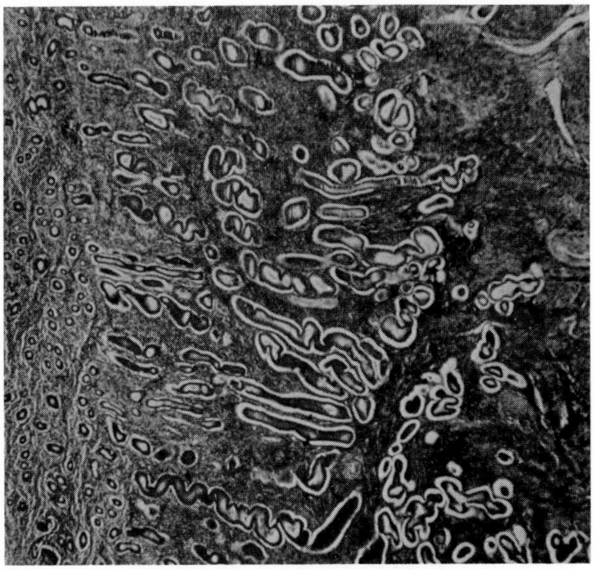

(b)

If implantation of an embryo has not occurred by about day 21, the outermost layer of the endometrium begins to disintegrate. In the next 4 to 6 days, the cellular debris, lost blood, and secreted fluid slowly flow out of the uterus into the vagina. As the flow of menstrual fluid ends, a new cycle of endometrial growth and vascularization commences.

Women commonly experience a variety of pains, discomforts, or emotional disturbances before or during menstruation. In some women, the premenstrual period may be characterized by lethargy, irritability, or deep sadness. In many cases, the pains are severe enough to interfere with daily routine. A fringe benefit of the first pregnancy in some women is that they never again experience menstrual pain. These effects are believed to be caused by an increased production of one type of prostaglandin. These chemicals cause the smooth muscles to contract and expel the endometrial debris. This group of hormones has a wide variety of other effects on blood flow and cell metabolism.

The length of the menstrual cycle varies among individuals; in most women, however, it is 28 days from the beginning of one menstruation to the beginning of the next. Although cycles shorter than 16 days or longer than 40 days occur, 90 percent of all women have cycles of from 22 to 38 days. The cycles of many women may also vary in length from one menstrual period to the next. In such irregular cycles, it is usually the time from the end of one menstruation to the next ovulation that varies. The time from ovulation to menstruation tends to be regular in almost all women. In 65 percent of all women, this interval from ovulation to menstruation is from 12 to 16 days. In almost all of the other 35 percent, it ranges from 3 to 19 days. Because of such irregularities, it may be difficult or even impossible to predict when ovulation will occur in some individuals.

## HORMONAL CONTROL OF THE MENSTRUAL CYCLE

Because of the abiding interest in understanding the functions of the human body and, in recent years, the possible use of this knowledge in the control of human fertility, the physiological basis for the regularity of the menstrual cycle in the human female has been studied intensely.

These studies, which initially used animal models but later involved human volunteers, elucidated the intricate details of the hormonal control of menstrual cycles in human females. The pituitary gland, the hypothalamus, and the hormones secreted by the ovarian tissues are involved in the regulation of the menstrual cycle. Figure 22-5 summarizes the relationship of the four regulatory hormones to the various stages of the menstrual cycle.

## SOURCES AND ACTIONS OF FSH, LH, ESTROGEN, AND PROGESTERONE

The follicle-stimulating hormone (FSH), one of two gonadotropins (gonad-stimulating hormones) produced in abundance by the anterior lobe of the pituitary gland at the beginning of each cycle, stimulates the development in each ovary of several follicles and the oocytes they contain. Within each of these follicles, the oocyte enlarges and the follicle cells begin to produce the female sex hormone, estrogen. This steroid hormone promotes follicular development and stimulates the growth of the endometrium. The concentration of estrogen in the blood increases and reaches a peak on about day 12 of a 28-day cycle. The prolonged action of increased estrogen concentration has two effects. First, the production of FSH increases precipitously by enhancement of the production of receptors in the pituitary for the FSH-releasing factor, a neurosecretory hormone produced by the hypothalamus. Second, high estrogen levels cause the production of a second pituitary gonadotropin, luteinizing hormone (LH), by promoting the production of an LH-releasing factor by the hypothalamus. Thus, both FSH and LH suddenly increase to a peak concentration on or about day 12, after which the concentrations of both decline. As the FSH concentration falls, there is a corresponding decline in estrogen synthesis. Ovulation occurs on about day 14 or 15, probably induced in part by the sudden increase in LH concentration (Fig. 22-5). After ovulation, the collapsed, vacated follicle fills with blood vessels and is known as the **corpus hemorrhagicum.** Now, under the influence of LH, cells begin to proliferate within the collapsed follicle. The cells accumulate a yellow lipid material and begin secreting progesterone, another steroid hormone. Because of its yellow color, the mass of cells, which is now functioning as a gland, is termed the **corpus luteum** (Latin for "yellow body").

Together with estrogen, progesterone promotes further proliferation of cells in the endometrium and stimulates its synthesis of lipid and glycogen. Progesterone also quiets the **myometrium,** the muscular layer of the uterus. In addition, estrogen and progesterone promote the growth and development of the mammary glands in preparation for **lactation,** or milk production.

By day 20, progesterone concentration in the blood has reached its peak, and the endometrium is fully capable of nourishing an embryo should one become implanted in it. The high progesterone concentrations in the blood inhibit the continued production of LH by the pituitary.* If implantation has not occurred by day 21, progesterone, by inhibition of LH production, causes a decline in progesterone output by the corpus luteum. The corpus luteum now begins to decrease in size and

* In each case involving the stimulation or inhibition of the pituitary, the effect is mediated by releasing factors of the hypothalamus and/or modulation of receptors for the releasing factors in the pituitary.

soon becomes nonfunctional. Estrogen concentration, which had reached a second peak on day 20 or 21, also now declines. As the concentration of the two steroid hormones drops, degenerative changes begin to occur in the endometrium. The special blood vessels undergo spasms and then rupture, releasing blood into the uterine cavity, or lumen. The cells of the endometrium lining the uterine lumen die and are sloughed off. The debris that accumulates in the uterus stimulates contractions of the uterine smooth muscles, thus accelerating the menstrual flow and sometimes producing painful cramps. About 40 ml of blood and the same amount of secreted fluid are usually lost during the 4 to 6 days of menstrual flow. Even as the menstrual flow continues, however, the decline in the level of the two steroids triggers a release of FSH-releasing factor by the hypothalamus. As a result, the anterior pituitary begins to increase its secretion of FSH once more, initiating another cycle.

## THE ENDOCRINOLOGY OF PREGNANCY

If implantation of an embryo does occur (day 19 to 21), the pituitary is signaled to continue its output of LH, the corpus luteum remains in existence, estrogen and progesterone output continue, and the degenerative changes in the endometrium are prevented. Complex changes now occur rapidly in the uterus as it accommodates the growing embryo. The **chorion,** the most prominent of the extraembryonic membranes that surround the embryo proper, eventually contributes most of the tissue of the **placenta,** the structure that forms the connection between the embryo and the uterus. Soon after implantation, the chorion acquires the ability to produce hormones. One of these, **human chorionic gonadotropin** (hCG), has an LH-like effect on the corpus luteum.* The mammary glands are also stimulated to further development in preparation for lactation at the time of birth. Two other chorionic hormones are similar to estrogen and progesterone in their structure and effects on the uterine tissues. Because of the hCG from the chorion, by the end of the second month of pregnancy, the amount of estrogen found in the blood is 50 times the normal concentration, and progesterone is 20 times that of normal. The LH-like effect of chorionic gonadotropin is responsible for maintaining the corpus luteum and for promoting an increase of both progesterone and estrogen by the ovary. As a further result, endometrial growth continues, supplying the embryo with a supportive medium in which to develop.

## MENOPAUSE: THE CESSATION OF MENSTRUAL CYCLES

In most women, menstrual cycles continue until sometime between the ages of 45 and 50, when, over the course of a year or so, estrogen production declines and ovulation and menstruation become irregular and then cease altogether. This period is termed the **climacteric,** and the

* By the end of 2 weeks following implantation, so much chorionic gonadotropin is present in a pregnant woman's blood, and therefore in her urine as well, that it was long used as the basis for pregnancy tests. When pregnancy urine was injected into a test animal (usually a hamster, rabbit, mouse, rat, frog, or toad), it induced precocious sexual development, formation of corpora lutea, or similar responses. These tests have now been superseded by various immunological tests for the presence of hCG that are highly accurate and take much less time to reveal the result.

cessation of menstruation is called **menopause.** Although menopause usually occurs at about the age of 45, it is highly variable, and its onset does not appear to be correlated with the age of **menarche,** onset of the first menstruation.

The climacteric and menopause result from a growing failure of the ovarian interstitial cells to produce sufficient estrogen. At this age, the primary follicles (those capable of producing ova) have been depleted, and estrogen production is accordingly reduced. In the absence of previously normal levels of estrogen, the endometrium no longer undergoes its typical cyclical growth and deterioration. Also associated with the decreased output of estrogen and progesterone is an unusually high output of gonadotropins by the pituitary, which is no longer inhibited as much as it had been before. Women commonly experience a variety of recurrent symptoms at this time. The most common are hot flashes, characterized by sudden, brief rushes of blood to the skin of the upper body and accompanied by sweating. Menopause also causes irritability, depression, insomnia, and headaches. Although the precise physiological basis of these menopausal symptoms is uncertain, they are known to be caused by changes in hormonal balance. Many myths have surrounded menopause, including the erroneous belief that sexual enjoyment is no longer possible from that time on.

## HORMONAL CONTROL OF TESTICULAR FUNCTION

Male sexual activity is under the control of androgens, which are steroid hormones that stimulate the growth of male tissues. Among the hormones that regulate male sexual activity are found the pituitary gonadotropins, or hormones that affect gonadal growth. A schematic view of the endocrine control of testicular function is shown in Figure 22-9. A number of parallels can be drawn between the sexual physiology of

**FIG. 22-9** Schematic diagram of the endocrine control of human testicular function.

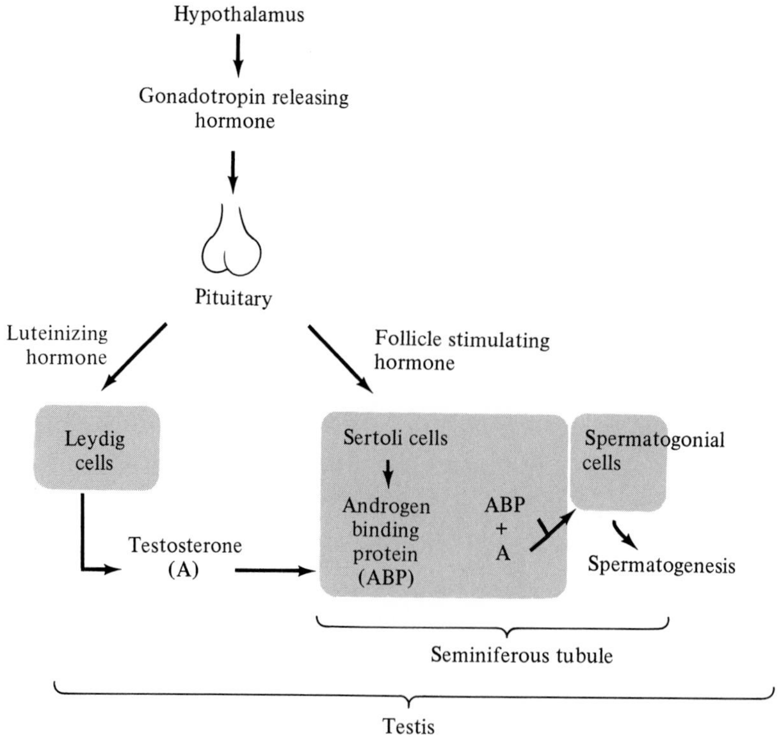

males and that of females. Among these is the fact that many of the same hormones are involved. The testes have two major functions: production of testosterone, the male sex hormone, and of spermatozoa.

***Pituitary Gonadotropins and Testes Growth.*** Two pituitary gonadotropins together with corresponding releasing factors from the hypothalamus team up with testosterone to regulate testicular function. The pituitary hormones are LH and FSH, the same hormones that regulate ovarian function.

Surgical removal of the mammalian pituitary gland, called a hypophysectomy, leads to testicular degeneration (see Chapter 14). But testicular function can be restored by the administration of pituitary extracts or purified gonadotropins. These observations suggest that gonadotropins are necessary for normal testicular function. Orchidectomy, or removal of the testes, promotes gonadotropin secretion, whereas injection of testosterone into the normal animal decreases gonadotropin production. These results indicate a feedback control mechanism by the testes on the pituitary.

Gonadotropin concentrations in blood do not exhibit any recognizable cyclic variations, but vary over a wide range within relatively short periods of time. In the adult human male, the release of LH occurs in pulses at approximately 90-minute intervals. But such periods occur irregularly from day to day, and the precise nature of the regulatory mechanism is not known. In mice, LH is released by the pituitary when the male encounters a female. After successive encounters by a male mouse with the same female, LH releasing factors are not produced and LH is not released. On the other hand, encounters with a different female will reinitiate LH secretion.

***Role of Testosterone.*** Leydig cells release testosterone into the circulation upon stimulation by LH; testosterone is then taken up by target cells throughout the body. It also diffuses into the seminiferous tubules, where it enters the Sertoli cells. Sertoli cells then convert testosterone into dihydrotestosterone (DHT), which is the active form of the hormone. FSH is also required in the hormonal control of spermatogenesis. It probably induces spermatogonia to divide and spermatids to become spermatozoa (spermiogenesis). But much remains to be learned about the specific roles of LH and FSH in male sexual physiology.

***Androgens and the Development of Male Sexual Characteristics.*** In the prepubertal human male, androgens promote the development of the physical characteristics of the male. Androgens also affect the vocal cords, resulting in a deepening of the voice; and the skin, stimulating hair growth on the face, chest, and pubic area during adolescence, and hair loss later on in life. These hormones also stimulate bone growth, and after sexual maturity cause the closure of the epiphysial centers, the regions at which long bones grow in length. Androgens also stimulate the growth of the scrotum, penis, vas deferens, and such male accessory reproductive glands as the prostate gland and seminal vesicles. They also affect the central nervous system, in a manner not fully understood, to increase libido (sexual desire) and, moreover stimulate the seminiferous tubules of the testis to begin spermatogenesis.

***Male Climacteric.*** Generally speaking, testosterone production declines with age. In certain cases, reduced testosterone concentration results in symptoms that are known as male climacteric and parallel the female menopausal state. The symptoms include a decreased libido and sperm production and a decline in nitrogen balance and muscle function. The causes are varied and include a reduced production of gonadotropins and a reduced sensitivity of Leydig cells to pituitary gonadotropins, both of which decrease the concentration of testosterone in the circulation.

# THE ARTIFICIAL CONTROL OF HUMAN FERTILITY

Darwin observed that all organisms have a reproductive capacity far in excess of that needed for the survival of their species and far greater than could be supported by the earth's resources. Humans are no exception. Most healthy, young couples could, for example, manage to have 15 children during their reproductive lifetime. At this rate, one city block with 30 such families, could, if all descendants survived and reproduced, add over 10 million people to their city's population in only 200 years. However, most families limit their size in some way. Because of the growing realization that the world's human population is threatening to exceed the earth's capacity to support it, there is currently an unprecedented interest in developing effective means of fertility control.*

## SOME COMMON METHODS OF FERTILITY CONTROL

The fertility control methods described here are limited to those involving some application of our knowledge of reproductive physiology. They are classified according to their use-effectiveness (as distinguished from theoretical effectiveness), as reported by those who have used them.

***Tubal Sterilization and Vasectomy.*** Tubal sterilization involves sectioning (cutting) and ligation (tying off) of the fallopian tubes (Fig. 22-10a). Ova are thus prevented from reaching the uterus, and sperm from reaching the ova. Although it is possible to restore tubal function, the success rate is very low. Vasectomy is similar to tubal ligation except that it is performed on the vasa deferentia of the male, thus preventing sperm from entering the seminal fluid (Fig. 22-10b). Restoration is even more difficult than in tubal ligation. These are the most reliable methods of fertility control and cause the fewest ill side effects.

***Oral Contraceptives.*** There are a great variety of oral contraceptives (OCs) on the market, and new ones are constantly being developed (Fig. 22-11). The type most commonly used, and also the most effective, contains a balanced proportion of synthetic estrogen and progesterone. If a sufficiently high concentration of these steroids is maintained in the bloodstream, the pituitary is inhibited from releasing FSH and LH in sufficient quantities to cause ovulation. In the usual program, "the pill" is taken daily, beginning with day 5 of the menstrual cycle and continuing for 20 days. On day 25, due to cessation of the supply of exogenous progesterone, as when no embryo implantation occurs, the endome-

* The current worldwide population explosion is due, not to increase in family size, but to recent medical progress in the control of infectious diseases. This has permitted most children to survive to reproductive age and has lengthened the life span of individuals.

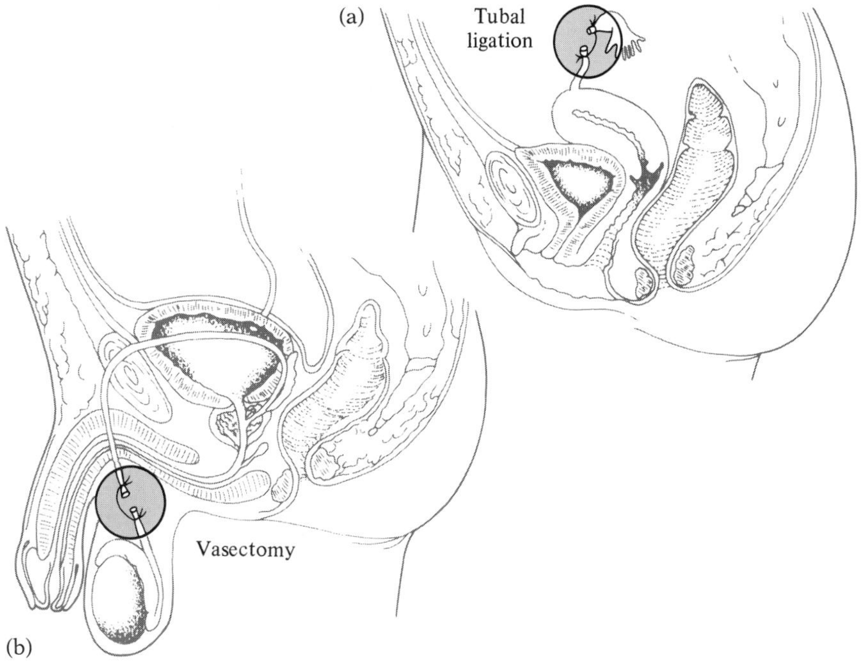


**FIG. 22-10** Methods of preventing fertility. (a) Tubal ligation. (b) Vasectomy.

trium begins to disintegrate. Menstruation usually follows on the twenty-eighth day. Five days later, the program is begun once more. The pill method is reliable and very effective, with less than 1 percent failure rate.

OC use has been correlated with an increased incidence (three to four times normal in some studies) of some types of circulatory disease. Cigarette smoking greatly enhances these side effects, and the risk of heart attack is as much as 23 times greater in women who use OCs and smoke than in women who do neither. However, in one study, women who had used OCs for 4 years showed only half the incidence of breast cancer and rheumatoid arthritis as did nonusers. In the general popula-

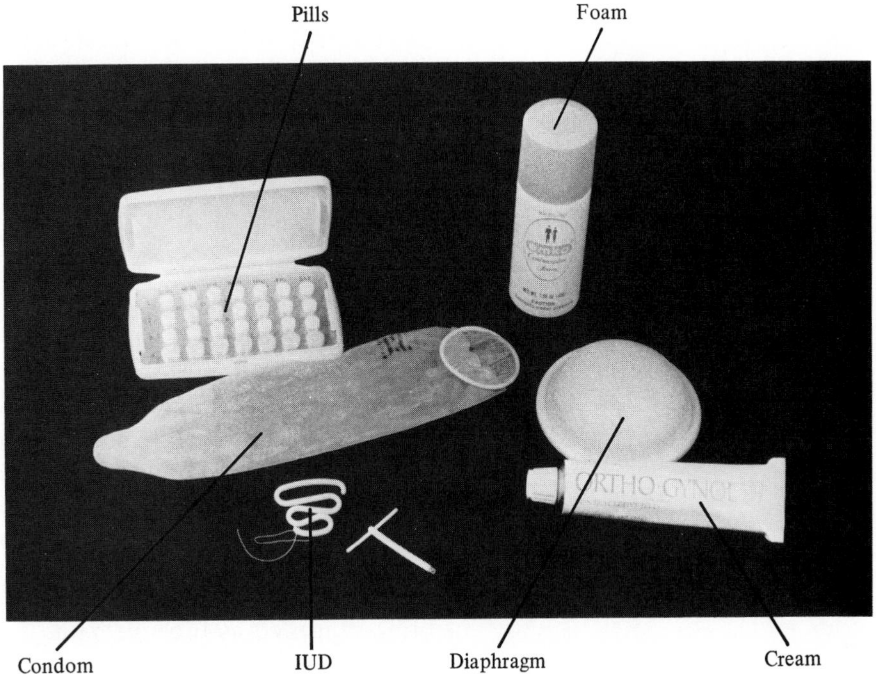


**FIG. 22-11** Methods of reducing fertility include oral contraceptives (pills), spermicidal creams, jellies, and foams, condoms, intrauterine devices, and the diaphragm.

tion, the overall increase in health risks associated with the use of oral contraceptives is less than the increased risk to health posed by pregnancy. However the pill is not a recommended method of birth control for women over 30 and for women with a family history of circulatory disease.

Although at least in theory male fertility could be controlled by a "male pill" that disrupts the hormonal balance, this has not materialized as yet. The physiological approaches attempted include blocking testosterone synthesis or its uptake by target cells. But lowering the blood concentration of testosterone results in a decreased libido, atrophy of accessory glands, and loss of muscle. However, a novel male pill containing gossypiol, a phenolic compound from cotton plants, holds some promise for the future. Gossypiol lowers sperm cell lactate dehydrogenase and renders the sperm less motile.

***Intrauterine Devices.*** It is thought that intrauterine devices (IUDs) were used by the ancients. The Greek physician Hippocrates (circa 460–377 B.C.) used a lead tube to insert devices into the uterus, although to what purpose is not clear. Tradition holds that camel drivers inserted pebbles into their beasts' uteri to prevent unwanted pregnancies. The modern devices, most often made of plastic, are inserted into the uterus by a physician (Fig. 22-11) at the end of a woman's menstrual period. They appear to cause a permanent inflammatory response that somehow creates an unfavorable condition for the implantation or survival of the embryo. The most commonly held view as to the mechanism of the IUD's action is that the embryo undergoing implantation is engulfed by the numerous leukocytes associated with the inflammatory response caused by the IUD. These leukocytes, which include macrophages, also phagocytose sperm in large numbers, thus reducing the likelihood of conception. The fluid contents of the uterus become acidic and are unsupportive of the life of the embryo.

According to a worldwide estimate, over 15 million IUDs were in use in 1978; in 10 countries, mostly in the Third World, they are the principal approach to fertility control. Difficulties involved in their use include an annual 2- to 4-percent pregnancy rate, an expulsion rate of 3 to 15 percent, and most troublesome, the need to remove from 5 to 20 percent because of excessive menstrual bleeding, pain, or both. Their principal advantage is that they can usually be inserted once and need no further attention, at least through a few menstrual cycles. This is regarded as an especially valuable consideration in populations with marginal medical care. However, because of the high risk to health this method of fertility control is no longer recommended.

***Diaphragm Plus Spermicide.*** The diaphragm is a modern version of another ancient method of contraception: a physical barrier inserted into the vagina to prevent the entry of sperm into the uterus. It consists of a soft rubber cup with a rubber-covered metal spring reinforcing the rim (Fig. 22-11). It is inserted in the vagina by the user so as to cover the cervix and is removed several hours after intercourse. Because even the best-fitting diaphragms or cervical caps cannot block the path of all sperm, a seal of spermicidal cream or jelly must also be applied. The diaphragm does not carry any of the health risks of IUDs or OCs. It does require a certain amount of skill and motivation to use correctly and is often regarded as inconvenient.

***The Condom.*** The condom has been employed with mixed success for centuries. Almost all modern condoms are made of latex, or natural rubber (Fig. 22-11). The thin rubber sheath is placed over the penis before insertion into the vagina. At ejaculation, the seminal fluid is collected in a pouch at the end of the sheath and does not enter the vagina. Except for a chance tear in the sheath, the condom serves as an effective means of preventing pregnancy. Although young couples in their first year of marriage have reported only 80- to 90-percent success with diaphragms or condoms, older couples have reported 97.4-percent success with the same techniques.

***Calendar Rhythm.*** The knowledge that a fertilizable ovum is present in the female reproductive tract for only a brief period (1 to 3 days) in each menstrual cycle led to the development of the rhythm method of birth control. In 65 percent of all women who have a 28-day menstrual cycle, ovulation occurs on about the fourteenth day after the commencement of the previous menstrual flow. Because sperm are viable only up to about 2 to 3 days in the female reproductive tract, it is usually possible for a woman with, for example, a regular 28-day cycle to avoid pregnancy by not having sexual intercourse during days 10 to 17 of the cycle, the so called unsafe period. The days just after menstruation, up to day 9, and the days just before the next menstruation, beginning with day 18, are termed the safe period.

For women with variable cycles, the number of safe days is proportionately reduced by subtracting 18 days from the shortest cycle to obtain the last safe day following menstruation; subtracting 11 days from the longest cycle gives the first safe day after the unsafe period. Careful records must be kept for at least a year before determining the degree of variation in menstrual cycles. Women with irregular cycles may find that they have no safe days.

Another method can also be used to determine the time of ovulation. This variation of the rhythm method, which is more reliable and useful than calendar rhythm, is based on changes in the character of the mucus secretions of the uterine cervix during the course of the cycle. In 70 percent of all women, cervical mucus usually changes from a sticky secretion to a more fluid one at about the time of ovulation; in women in whom this happens on a regular, predictable basis, it can be used to judge the time of ovulation. Another means of determining ovulation consists of taking the daily body temperature and keeping a record of it. Body temperature rises a few degrees just after ovulation.

***Breast-Feeding.*** Breast-feeding constitutes a natural means of suppressing ovulation by inhibiting release of the pituitary gonadotropins. To be effective, however, it requires that the baby be a "strong nurser" and that no supplemental feeding of cereal or bottled milk be given. Although it has no doubt contributed to the 2-year spacing of children seen in many families in which babies are breast-fed for 2 years before being weaned, it is a highly unreliable means of birth control.

## METHODS OF INCREASING FERTILITY

A knowledge of reproductive physiology can also be used to enhance the reproductive capability in infertile couples. In fact, the development of oral contraceptives grew out of early attempts to regularize the menstrual cycles of women who, because of irregularity, had been unable to conceive children. The approach is widely used today with great success. Through appropriate doses of estrogen and progesterone, the menstrual cycle and the time of ovulation are made more regular. Then the unsafe period is deliberately chosen as a time for sexual intercourse.

PANEL 22-1

## In Vitro Fertilization: Embryo Transfer

Perhaps the most dramatic success in promoting fertility involves what the popular press has called test tube babies. An estimated 500,000 women in the United States are sterile because their fallopian tubes are blocked or incompletely formed. Such blockage can result from an infection or a developmental aberration. In these individuals, the normal route for the transport of the ovum from the ovary to the uterus is blocked, and therefore the sperm cannot reach the ovum.

In 1978, reproductive physiologists and gynecologists in England and India made the first public announcement of in vitro fertilization methods of improving fertility in such women. The in vitro fertilization procedure is direct and simple. It involves the surgical removal of the ovum from the ovary; maintenance of the ovum in a suitable external medium, where it is exposed to the sperm; incubation of the zygote in a controlled environment until it develops to an appropriate stage (hence the name test tube baby); and the return of the embryo to the uterus of the natural mother for completion of development.

Women with blocked fallopian tubes may exhibit normal menstrual cycles, and ova may be released on schedule. However, to achieve external fertilization, the ovum must be collected before it is released into the body cavity. Therefore, ovulation is induced by the injection of gonadotropins to ensure the precise day of ovulation. An ovum is then surgically removed just before ovulation and fertilized outside the body in a balanced salt solution containing required growth factors. After the embryo has developed in an incubator to the stage equivalent to a normal 6 day embryo, it is introduced into the woman's uterus through the cervix, where it implants. Pregnancy then proceeds to full term. While the embryo is developing outside the woman's body, her uterus is brought into a receptive state for implantation by careful manipulation of her progesterone and estrogen levels. In this manner, even two embryos can be implanted in a uterus at the same time. In mid-1981, a woman gave birth to fraternal twins and, in 1984, another woman gave birth to quadruplets, conceived in this manner.

By 1983, in vitro fertilization methods had been used in 14 institutions throughout the world, and over 2,000 women had been screened as potential candidates for in vitro fertilization. Of those, 1,110 received one or more embryo transfers, and 184 became pregnant. Since about 16.5 percent of the women receiving embryos became pregnant, this method is now regarded as an effective means of solving some infertility problems. But it is expensive. Nevertheless in 1985, more than 20 in vitro fertilization centers were in operation in the U.S.A., and another 10 to 20 were in planning stages to meet the demand in the United States.

A number of variants of the in vitro fertilization method are potentially feasible in overcoming other types of sterility. For example, male infertility resulting from a decreased number of sperm (oligospermia) or low sperm motility can be overcome by collecting an ovum and fertilizing it in vitro with the sperm. Low motility or low numbers of sperm do not affect the fertilizing capacity of sperm in vitro. Similarly, a woman incapable of producing ova, as a result of the removal of tumorous ovaries or due to certain infections, may be a recipient of an embryo obtained by the fertilization of an ovum from another woman, thus "adopting" the embryo.

Any discovery of this magnitude is subject to abuse. For example, for eugenic purposes—that is, to control reproduction in order to improve the hereditary characteristics of a race—a government might dictate which genetic traits of embryos may be favored for propagation. Like involuntary sterilization, legislating embryo transfer procedures would impose restrictions on the exercise of a fundamental biological function of the human body, namely, reproduction.

Another approach to counteracting infertility has been the use of fertility drugs; these are actually pituitary gonadotropins derived from menopausal urine or synthesized in the laboratory. An overdose of such drugs, however, can have the undesirable effect of causing maturation and release of several ova in the same menstrual cycle. Multiple births, up to septuplets, have sometimes resulted because of treatment with fertility drugs. There is not space enough in the human uterus, however, for more than four or five embryos to reach a viable birth size. The chances of survival of even five embryos in a uterus are very low.

## Summary

The primary reproductive organs in the human male, as in all mammals, consist of a pair of testes. Each testis is made up of a large number of spermatic cords in which the spermatogonial and the somatic, hormone-producing cells are arranged. Spermatocytes arise by mitotic divisions of spermatogonial cells, and sperm are generated by meiotic division of spermatocytes. The sperm mature while anchored in the somatic cells lining the spermatic cords and are stored in the epididymis. The latter is continuous with the vas deferens. The accessory male reproductive organs are two paired seminal vesicles, and the unpaired Cowper's and prostate glands; the prostate gland is the major contributor to seminal fluid. The vasa deferentia join the urethra, which enters the penis, carries both urine and seminal fluid. The penis is composed of the urethra that passes through a spongy erectile tissue, the paired spongy corpora cavernosa and the glans.

The primary female reproductive organs are the pair of ovaries, which, in the course of the reproductive life of a woman, produce about 400 mature ova. The paired fallopian tubes, with their widened internal openings, collect the ovum released into the body cavity approximately every 28 days. The fallopian tubes join to form the muscular, pear-shaped uterus, which opens to the exterior by way of the vagina. Lining the luminal surfaces of the oviduct and the uterus is a ciliated epithelium. A current of fluid is constantly carried from the abdominal cavity to the exterior by its ciliary motion. This flow of fluid carries the ovum to the oviduct and thence to the uterus, where, if fertilized and implanted, it grows into a fetus.

A reproductively mature woman exhibits a menstrual cycle that lasts approximately 28 days. During this cycle, the endometrial lining of the uterus exhibits proliferation, growth, and vascularization, followed by disintegration, sloughing, and discharge to the exterior. The menstrual cycle is correlated with the release of a mature ovum, approximately halfway through the cycle, and preparation of the endometrium for reception, implantation, and nourishment of an embryo should the ovum be fertilized. The onset of the first menstruation is known as the menarche.

The menstrual cycle is regulated by the interaction of two pituitary hormones—the follicle-stimulating hormone (FSH) and the luteinizing hormone (LH)—and two steroid hormones—estrogen and progesterone—produced by the ovarian tissues. The cycle starts with the production and release of FSH, which stimulates an egg follicle to grow and produce estrogen. Estrogen stimulates proliferation and vascularization of the endometrium, triggers the production and release of LH, and inhibits the production of FSH. The LH causes rupture of the egg follicle, release of the oocyte, and production of progesterone by the cells left behind in the egg follicle. After ovulation, cells within the collapsed follicle begin to proliferate, forming the corpus luteum.

Should an embryo implant in the endometrium, the cells covering the embryo, called the chorion, produce a hormone that acts like LH. This hormone is called human chorionic gonadotropin. It stimulates production of progesterone by the corpus luteum. The amounts of progesterone and estrogen in the bloodstream remain high, and the endometrium persists. In the absence of these hormones, however, the endometrium disintegrates and is ejected to the exterior as part of the menstrual discharge. When blood levels of estrogen and progesterone are low enough, the pituitary gland produces and releases FSH again, starting a new menstrual cycle. The menstrual cycle continues until between the ages of 45 and 50. At this time, estrogen production declines, and ovulation and menstruation cease. This period of the human female life-cycle is known as the climacteric; the cessation of menstruation is known as the menopause. Males may also go through a climacteric, as a result of reduced production of testosterone.

Testicular function is also regulated by FSH and LH from the pituitary and testosterone from the testes. FSH and LH are believed to stimulate Sertoli cells and Leydig cells to produce androgens. Androgens induce spermatogenesis and development of secondary sexual characteristics such as voice change and hair growth. Unlike in the female, FSH and LH in males are released at approximately 90 minute intervals.

A knowledge of the physiology of reproduction has helped in devising methods of preventing pregnancy as well as methods of increasing fertility. Surgical methods physically prevent release of the gametes to their normal anatomical theaters of action. For example, vasectomy prevents the release of sperm into the seminal fluid, and tubal ligation prevents the ovum from reaching the uterus. Oral contraceptives, hormonal or pill treatments, are used to inhibit ovulation. The intrauterine device seems to prevent the implantation of the embryo. Temporary physical methods of preventing the union of the gametes involve the use of the condom and the diaphragm. The rhythm method is based on avoiding intercourse during potentially fertile days. Methods used to promote fertility, including the procedure resulting in the birth of so-called test tube babies, further demonstrate the practical benefits of a thorough knowledge of the biology of reproduction.

## Recommended Readings

EDWARDS, R. G., and J. M. PURDY, eds. 1982. *Human Conception In Vitro*. Academic, New York. An account of the test tube method of pregnancy is given in detail.

HART, G. 1977. *Human Sexual Behavior*. Oxford Biology Readers, Burlington, N.C. An excellent summary of various aspects of human sexual behavior.

HUFF, R. W., and C. J. PAUERSTEIN. 1979. *Human Reproduction: Physiology and Pathophysiology*. Wiley & Sons, New York. An account of the normal and diseased state of reproductive physiology.

JONES, R. G. 1984. *Human Reproduction and Sexual Behavior*. Prentice-Hall, Englewood Cliffs. This comprehensive book on human reproduction is an advanced text useful for the student seriously interested in this subject.

SEGAL, S. J. 1974. "The physiology of human reproduction," *Scientific American*, 231:53–62. A summary of the human reproductive process and of some contraceptive methods.

# 23

# Mendelian Genetics

OUTLINE



The ability to reproduce is unique to living systems. Moreover, all organisms arise only from preexisting living individual organisms. Thus, every individual organism must be viewed as the current member of an uninterrupted line of successive generations leading back to the origin of its species and ultimately to the dawn of life itself. This chapter discusses the basic principles of **genetics**, the science of **heredity**—the transmission of characteristics from one generation to the next—and **variation**—the changes that frequently occur in heritable characteristics.

It has been observed from ancient times that the offspring of all organisms resemble their parents. The essence of heredity is this recurrent appearance in successive generations of the characteristics of a species, defined as one or more populations of an organism that interbreed but do not in nature exchange genes with other organisms. Besides expressing the general features of its species, however, each organism exhibits individual traits. Many of these can also be traced from one generation to the next, since they too are governed by the laws of heredity.

**FIG. 23-1** Gregor Mendel (1822–1884), the Austrian monk known as the father of genetics.

# MENDELISM

Most progress in the discovery of the laws of nature is accomplished through the efforts of many workers, and the same is true of genetics. However, because an extremely significant contribution was made in 1866 by a single worker, Gregor Mendel, that man is often called the father of genetics, and the basic laws of inheritance are called Mendelian genetics (Fig. 23-1). One factor that may have contributed to Mendel's discoveries was his experience with plants during his youth on a farm. He also carried on plant breeding experiments that led to the development of prize-winning new varieties of fruits and vegetables. As a young, largely self-taught scientist, he joined a teaching order of Augustinian monks near Brünn, Austria (now Brno, Czechoslovakia), partly because it was the only way for him to pursue a scholarly career. He taught science in local secondary schools for a time before being sent to the University of Vienna for formal study in botany, physics, and applied mathematics. Upon returning to Brünn, he established 34 pure-breeding lines of garden peas on the monastery grounds in order to experiment with them (Fig. 23-2). (Pure-breeding lines always produce similar offspring.) Mendel obtained most of the varieties through seed catalogs.

## CONCEPTS OF INHERITANCE IN THE MID-1800S

When Mendel began the experiments that would make him famous, almost nothing was known of the physical basis of inheritance. Concepts such as haploidy, diploidy, and meiosis were unknown at the time, and it had not even been established that nuclei carried genetic information. Understanding the physical basis of inheritance was to await the further development of the light microscope and staining techniques near the close of the century. However, many biologists, among them Charles Darwin, had been **hybridizing,** that is, mating different strains of plants and animals in attempts to gain insight into the

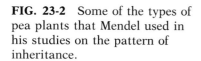

**FIG. 23-2** Some of the types of pea plants that Mendel used in his studies on the pattern of inheritance.

mechanisms of inheritance and, more important, into the basis for the formation of new species.

It was known at this time that in sexually reproducing organisms, both parents might contribute traits to their offspring. Moreover, in crosses between dissimilar parents, it has been noted that some of the traits of one parent might be expressed to the exclusion of the alternative traits of the other. Some workers had also reported that hybrids, when interbred, were inclined to revert to the parental forms.

## MENDEL'S INNOVATIONS

Mendel's innovative approach to studying inheritance included the following considerations:

1. He deliberately searched for a plant species (a) the varieties of which differed from each other in constant, distinct, and mutually exclusive ways and lacked intermediate varietal types; (b) whose flowers normally underwent self-pollination and were naturally protected from chance pollination from another plant; and (c) that would continue to breed with full fertility for many generations after being hybridized. He finally settled on the common garden pea *(Pisum sativum)* as possessing the necessary qualifications.
2. When he crossed different strains of pea plants, he followed the inheritance of only one set of contrasting traits at a time, ignoring at first the inheritance of all other traits.
3. He counted and recorded the thousands of progeny from his experiments and then analyzed his results statistically. (This was a rare approach to biology at the time. Mendel's interdisciplinary training in botany, physics, and mathematics probably contributed to this statistical approach.)

To perform an experiment, Mendel chose two stocks with contrasting traits. He then opened the immature flowers of the strain he chose to be the seed-bearing parent and removed all its anthers, the expanded tips of stamens containing pollen, before they produced any pollen. Then he dusted pollen from the other strain, the pollen-bearing parent, on the stigma of the seed-bearing parent and covered the flower to prevent further pollination. To cross two members of the same genetic line, he repeated this process with two similar individuals; alternatively, he permitted **selfing** to occur by allowing the anthers of a flower to mature and pollinate the carpels (where the ovules are found) of the same flower, thus, in effect, crossing two identical individuals.

## MENDEL'S LAWS

The discovery of the pattern of inheritance in pea plants by Mendel led to the formulation of the basic principles of inheritance. These principles have been confirmed in later studies on a variety of organisms and have been recognized as the basis for heredity. In recognition of Mendel's contribution to the field of genetics, a termed coined by William

Bateson in 1906 to describe the scientific study of heredity, these laws are called Mendel's laws or Mendel's laws of heredity. Mendel observed three consistent inheritance patterns, which were later designated as the laws of dominance, segregation, and independent assortment. These laws have been enriched by later knowledge, but they remain valid. We will describe in this chapter first these laws as formulated by Mendel and then recent discoveries that further extended our understanding of heredity.

***The Concept of Dominance and Recessiveness.*** Mendel reported on crosses between individuals differing in one or more of seven sharply contrasting traits, including round or wrinkled ripe seeds, green or yellow seed endosperm, inflated or wrinkled ripe seed pods, and several other characteristics. When he crossed two of his pure-breeding lines, he found that in the resulting hybrids, one of the two contrasting traits was always expressed to the exclusion of the other. He termed the unexpressed trait **recessive** and the expressed trait **dominant.**

For example, Mendel crossed pure-breeding plants with round seeds with pure-breeding plants with wrinkled seeds. All the offspring had round seeds. Similar crosses were made between plants with yellow endosperm and those with green endosperm and between plants with smooth seed pods and those with wrinkled seed pods. All the offspring of these crosses had yellow seeds and smooth seed pods, respectively. Reciprocal crosses showed that the same traits were always dominant, no matter whether contributed by the pollen-bearing or the seed-bearing parent.

***The Segregable Particles of Inheritance.*** Mendel found that the complete dominance of one trait over the other was always expressed only when two pure-breeding lines of plants were crossed. He designated such parental stocks the parental (P) generation and their immediate offspring the first filial ($F_1$) generation.

When members of the $F_1$ generation were crossed with each other, however, a different result was obtained. For example, pure lines of round-seeded and wrinkled-seeded varieties (the P generation) were crossed with each other to yield an $F_1$ generation of round-seeded individuals. When these were selfed or crossed with each other, about 75 percent of the offspring (termed the $F_2$ generation) had round seeds and about 25 percent had wrinkled seeds (Fig. 23-3). Similar results were obtained with crosses of $F_1$ individuals resulting from crosses between the other pure-line varieties. The $F_2$ individuals expressing the recessive trait of one of their grandparents (the P generation) were indistinguishable in appearance from the recessive grandparents.

This last result, consistently obtained for all seven pairs of contrasting traits, was an historical event of great magnitude. Previously, inheritance had generally been considered a mixing or blending of parental traits, much like the mixing of paints of two colors, an idea that still persists in many popular beliefs. With many traits, this does seem to be the case. For example, crosses of tall and short plants or animals often produce intermediate-sized offspring. Similarly, children born to Negro and Caucasian parents are usually intermediate in pigmentation to their parents. But at least in the cases cited by Mendel, the recessive traits were able to reappear in unchanged form after a generation of being masked by the presence of a dominant contrasting trait. From these observations, Mendel made the following conclusions:

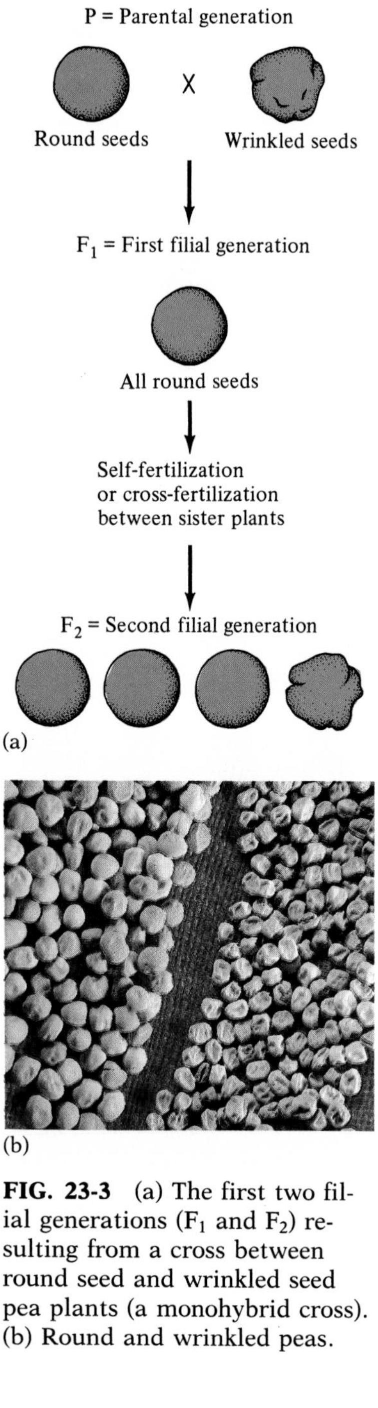


**FIG. 23-3** (a) The first two filial generations ($F_1$ and $F_2$) resulting from a cross between round seed and wrinkled seed pea plants (a monohybrid cross). (b) Round and wrinkled peas.

1. For each trait inherited, both parents contribute hereditary factors to their offspring, whether or not the factors are expressed by the offspring. Mendel called these unit factors; we now call them **genes.**
2. The appearance of an individual is not necessarily indicative of its genetic constitution; it can bear recessive traits that are masked by dominant ones.
3. Although its presence may be masked for one or more generations, a recessive factor can separate, or segregate, from the dominant factor and again be as fully expressed as it had been originally. This concept is known as Mendel's Law (or Principle) of Segregation.

The implications of the Law of Segregation were both novel and revolutionary. Mendel's first law implied that inheritance was based on the transmission of discrete particles, or unit factors. Moreover, each individual possessed at least two such factors for each inherited trait, one derived from each parent. Mendel thus had shown the nature of inheritance to be both particulate and duplicate.

Further support for the duplicate nature of inheritance came from Mendel's experiments with members of the $F_2$ generation. The 25 percent of the $F_2$ members displaying the recessive trait always bred true whenever selfed or crossed with another individual displaying the recessive trait. But the three-fourths of the $F_2$ generation displaying the dominant trait turned out to be of two kinds: one-third of them (that is, one-fourth of all $F_2$ progeny) also bred true when selfed. However, the remaining two-thirds (one-half of all $F_2$ progeny) when selfed yielded a generation in which three-fourths were of the dominant variety and one-fourth exhibited the recessive trait. In other words, half of the $F_2$ generation members were similar in genetic makeup to the $F_1$ generation.

The significance of these results did not escape Mendel. He proposed not only that each offspring received one of the two factors from each parent, but also that it was a matter of chance which of the two factors a hybrid parent (one possessing factors for both of a pair of contrasting traits) would transmit to any one of its offspring. Not only could the factors segregate from each other from one generation to the next, but they did so independently of each other, entering a particular gamete entirely by chance. Mendel deduced this from a statistical analysis of the number of offspring bearing the different traits. He found that the hybrids had transmitted to their offspring the two contrasting factors they possessed in approximately equal numbers.

***Mendel's Law of Independent Assortment.*** Having elucidated the laws of inheritance pertaining to plants differing in only one character, Mendel then investigated whether these laws applied "when several diverse characters are united in the hybrid." To do this, he performed a series of **dihybrid crosses,** which are crosses between individuals differing in two pairs of traits. In perhaps his most often cited dihybrid cross, Mendel began by crossing two pure lines, one with round, yellow seeds and the other with wrinkled, green seeds. As we have noted, round and yellow are dominant over wrinkled and green, respectively.

When Mendel crossed a pea plant with round, yellow seeds with a plant with wrinkled, green seeds, all of the seeds produced by the $F_1$ generation plants were round and yellow. Fifteen plants were then

Parental genotype: Aa
Parental phenotype: Round seeds.
Types of gametes produced by each parent: A and a.
Possible combinations of genotypes among the offspring: AA, Aa, and aa.

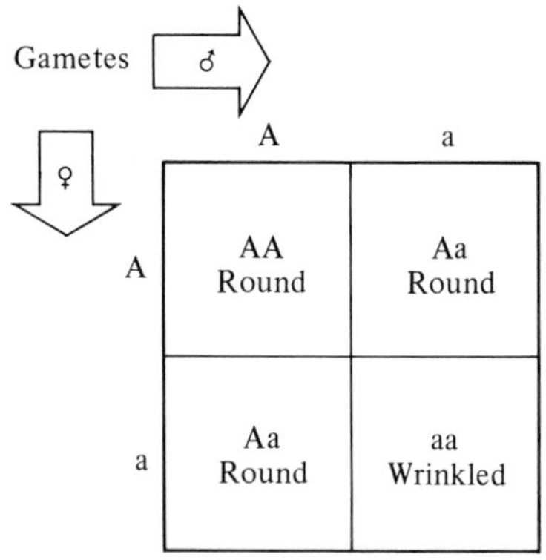


Genotypic ratio: 1AA, 2Aa, 1aa
Phenotypic ratio: 3 round, 1 wrinkled.

**FIG. 23-4** Punnett square representation of the pattern of inheritance in a monohybrid cross.

raised from these seeds and were either crossed with each other or were self-pollinated. A total of 556 seeds were produced, of four sorts, in the following distribution:

- 315 round, yellow
- 101 wrinkled, yellow
- 108 round, green
- 32 wrinkled, green

The ratio of the four combinations was approximately 9:3:3:1. Mendel noted that this dihybrid ratio could be represented as a square of the monohybrid ratio $(3:1)^2$. When each of the two pairs of contrasting traits was considered separately, it was seen that three-fourths of the seeds exhibited the dominant trait and one-fourth its recessive counterpart (423 round to 133 wrinkled and 416 yellow to 140 green). When $F_2$ plants with dominant traits were selfed, they were found to be of two kinds: those that bred true and those that produced a 3:1 ratio of dominant to recessive traits among their offspring. Again, there were about twice as many of the latter as of the former. Using the letters *A* for round, *a* for wrinkled, *B* for yellow, and *b* for green, a Punnett square, named after the American geneticist Reginald C. Punnett, who used this convenient method to represent the genes of parents and offspring in particular crosses, can be constructed for this experiment by placing the different types of gametes produced by each parent along the two axes, as shown in Figure 23-4.

It can be seen from Figure 23-5 that the $F_2$ generation includes nine genotypic classes: *AABB, AABb, AAbb, AaBB, AaBb, Aabb, aaBB, aaBb,* and *aabb*. Phenotypically, these individuals constitute a ratio of nine round, yellow to three round, green to three wrinkled, yellow to one wrinkled, green. In any prediction of the expected genotypic or phenotypic ratios, it must be remembered that these are statements of probabilities. As with flipping coins, if the number of cases (or progeny, in this instance) is small, the ratios seldom come close to expectations; with hundreds or thousands of offspring, however, the results are very close to the expected ratios.

These dihybrid crosses confirmed the Principle of Segregation. In addition, a new law had emerged: the Law of Independent Assortment. Although the genes for round and yellow had originated with one of the grandparents and those for wrinkled and green with the other, each pair of factors was inherited independently of the other pair and formed new combinations (called recombinants) according to mathematical laws of chance, or probability.

Mendel not only continued his crosses through several generations but repeated them with each of the seven pairs of contrasting traits he had chosen to study. He found that no matter which two or more of the seven pairs were involved, all pairs of factors were inherited independently of one another. This tendency of contrasting pairs of factors to segregate independently of pairs of factors for other traits became known as Mendel's Law of Independent Assortment.

Mendel also performed trihybrid crosses, with similar results, producing a correspondingly larger number of genotypes and phenotypes in the $F_2$ generation. You may find it instructive to work out a trihybrid cross between two individuals, in which both have a genotype of *AaBbCc*, where *C* stands for smooth seed pods and *c* for wrinkled pods.

(Because each individual produces eight kinds of gametes, the Punnett square (see Fig. 23-5) should have 64 boxes, and the $F_2$ generation should exhibit a total of eight phenotypic combinations in the ratio 27:9:9:9:3:3:3:1).

## THE FATE OF MENDEL'S PAPER

Mendel's findings, which were published in 1866 (in an 1865 issue of a Brünn science journal), were ignored for 34 years. Then, within a 90-day period in 1900, three botanists, Hugo De Vries of Holland, Carl Correns of Germany, and Erich von Tschermak-Seysenegg of Austria, each independently published experimental results similar to those reported by Mendel. These studies revealed the same laws of inheritance described by Mendel years earlier. Before publishing their observations, these researchers had searched the literature for reports of similar work and were led to Mendel's pioneering experiments, which they duly acknowledged in their papers.

**FIG. 23-5** Punnett square representation of a dihybrid cross: pea plants with yellow, round seeds were crossed with plants with green, wrinkled seeds.

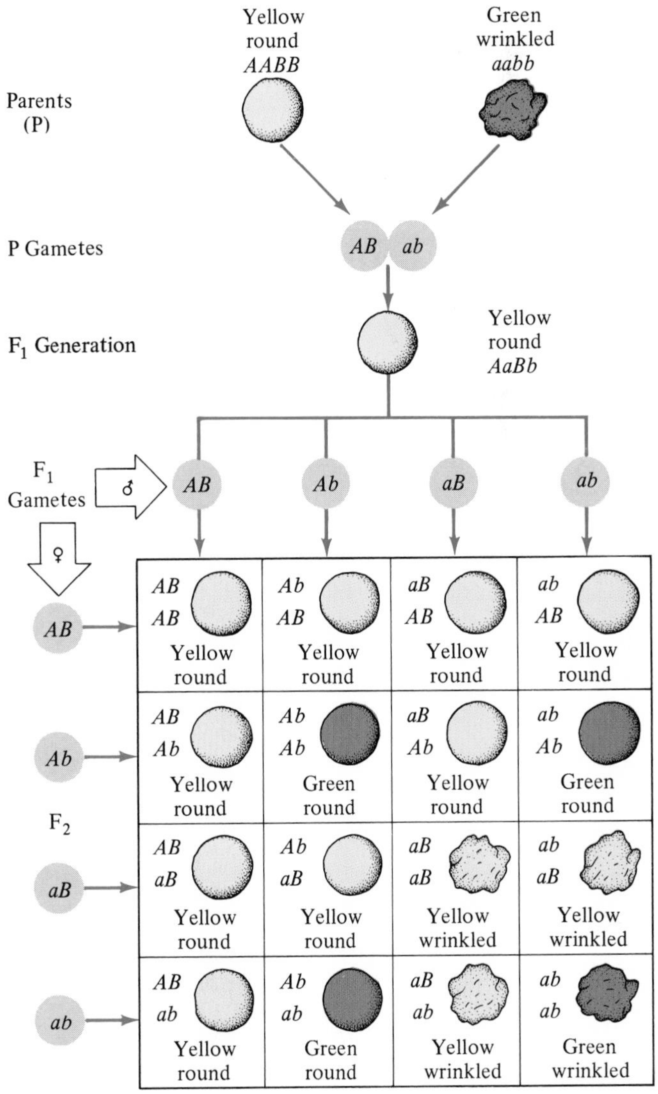

## THE CONCEPTS OF GENOTYPE AND PHENOTYPE

Some of the symbols and specific terminology that Mendel used to describe patterns of inheritance of dominant and recessive traits are used today, sometimes with minor changes. For example, he represented dominant traits with capital letters and recessive traits with corresponding lower-case letters. Thus, for seed shape, the dominant gene for round seeds might be represented by *A*, and the recessive alternative, wrinkled seeds, by *a*. Endosperm color might be designated *B* for the dominant trait, yellow and *b* for the recessive trait, green.

In modern terminology, alternative varieties of a gene are called **alleles** of each other. Thus, the gene for seed color is represented by two alleles, one for yellow seeds and the other for green.

Eventually, to distinguish between an individual's physical appearance and its genetic constitution, the terms **phenotype,** for the observable properties of an organism, and **genotype,** for the genetic constitution of an organism, were proposed. A statement of an individual's genotype can be made in terms of previously defined symbols, such as *AaBb*, or in words, such as, in this case, **heterozygous** (with dissimilar alleles). An individual with the genotype, AABB, will be described as **homozygous** (with identical alleles). A statement about the phenotype for both the AaBb and AABB individuals would note merely that they had round, green seeds. Phenotype is easily determined by simple inspection; except for individuals exhibiting a recessive trait, genotype can usually be determined only by breeding experiments. If an individual exhibits a recessive trait, the individual is homozygous for that trait. An individual exhibiting a dominant trait could be either homozygous or heterozygous for that trait. An individual that is heterozygous for a trait is a heterozygote; one that is homozygous is a homozygote.

## THE PATTERN OF SEGREGATION

In the cross between individuals of different seed shape described earlier, the three generations can be summarized as follows:

| | | | | |
|---|---|---|---|---|
| P | *AA* homozygous (round) | × | *aa* homozygous (wrinkled) | |
| | | ↓ | | |
| $F_1$ | *Aa* heterozygous (round) | × | *Aa* heterozygous (round) | |
| | | ↓ | | |
| $F_2$ | *AA* homozygous (round) | *Aa* heterozygous (round) | *Aa* heterozygous (round) | *aa* homozygous (wrinkled) |

$F_1$ individuals produce two kinds of gametes, usually in approximately equal numbers, 50 percent with the dominant allele *(A)* and 50 percent with the recessive allele *(a)*. During fertilization, each type of gamete produced by an $F_1$ parent has an equal chance of joining with each type produced by the other parent, resulting in an $F_2$ generation with a probable composition of 25 percent *AA*, 50 percent *Aa*, and 25 percent *aa*. The distribution in this cross is shown in the Punnett square in Figure 23-4 and that of a dihybrid cross in Figure 23-5.

***A Test Cross to Determine Genotype.*** From the foregoing discussion, it is obvious that an organism with a dominant phenotype could be either a heterozygote or a homozygous, pure-breeding line. The genotype of an individual that produces both male and female gametes is indicated by the ratio of dominant to recessive offspring when the gametes are self-fertilized. Where the sexes are separate however, particularly if the number of offspring produced by an organism at each breeding cycle is low, say four or five, the recessive phenotype may not be expressed even if both parents are heterozygous for the recessive gene. By chance, only one in four offspring of such a cross should exhibit the recessive phenotype. To increase the chances of finding a recessive phenotype and to allow testing of organisms in which sexes are separate, a backcross, also known as a test cross, is normally used. In a test cross, the suspected heterozygous individual is mated to an individual with the recessive phenotype. Half of the offspring from such a cross would be expected to exhibit the recessive phenotype if the tested organism is heterozygous. If the test organism is homozygous for the dominant trait, all the offspring of the test cross would be expected to be of the dominant phenotype. A test cross thus greatly increases the chances of revealing the genotype. You may find it instructive to draw a Punnett square distribution of the gametes in a test cross in which the tested individual is heterozygous for the trait. The Punnett square will show how 50 percent of the offspring are expected to bear the recessive phenotype.

## THE CHROMOSOME THEORY OF HEREDITY

By 1900, improvements in microscopy had given rise to the new science of cytogenetics through a merger of cytology, the microscopic study of cells, and genetics. It soon became clear that Mendel's "unit factors" were not only carried in the nuclei of gametes but were part of and, indeed, aligned along the length of the chromosomes. Several lines of evidence also indicated that each somatic cell of an animal and each cell of the sporophyte stage of a plant had two sets of chromosomes, one derived from each parent. Meiosis was soon shown to reduce the diploid set of chromosomes to the haploid number (see Chapter 21). Furthermore, the distribution of chromatids among the four daughter cells produced by meiosis resembled the distribution of Mendel's inheritable factors, forming a physical basis for segregation and independent assortment (Fig. 23-6).

The parallel between the segregation of chromosomes in meiosis and the segregation of Mendel's "unit factors" in his hybridization experiments can be illustrated by depicting the random assortment and recombination of paternal and maternal chromosomes in a Punnett square (Fig. 23-7).

### LINKAGE AND INDEPENDENT ASSORTMENT

If an individual is heterozygous for two pairs of traits, with a genotype such as *AaBb*, the two pairs of genes might be carried on different chromosomes, for example *A* and *a* on the long chromosomes and *B* and *b* on the short chromosomes shown in Figure 23-6. Alternatively, both types

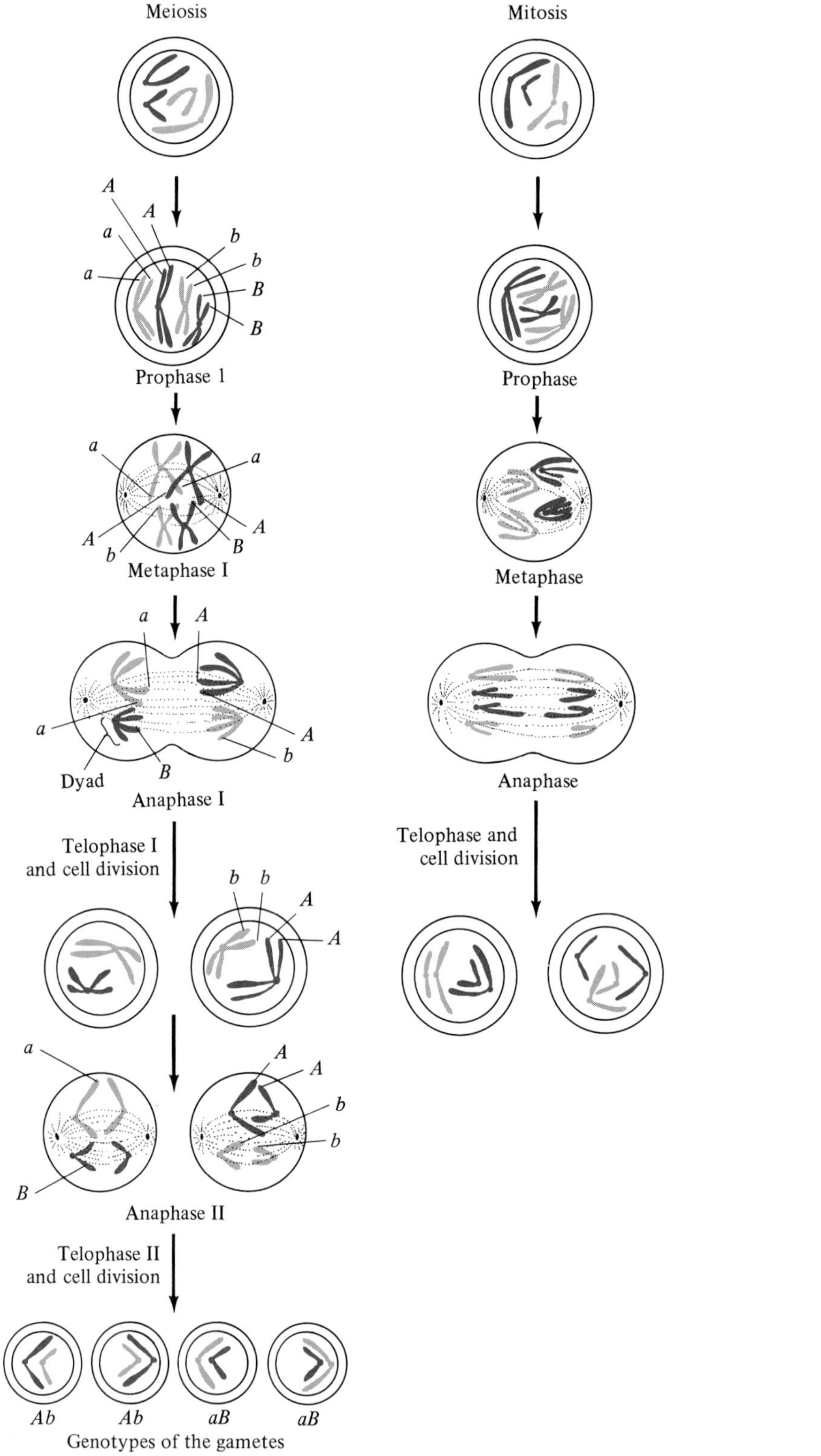


**FIG. 23-6** Pattern of distribution of chromosomes during meiosis compared to that during mitosis. Note independent assortment of chromosomes in meiosis. Alternatively, the final meiotic products in this example could have been *AB* and *ab*. Thus, with a diploid (2n) number of only four, four genetically different gametes would be produced. Only two are shown here.

of genes might be carried on the same chromosomes, for example, both on the long ones (see Fig. 23-7a) or both on the short ones. In such cases, particular combinations of genes would tend to be inherited together. The gametes produced by the individual in the example shown in Fig. 23-7a would tend to contain either *A* and *B* or *a* and *b*, with few, if any, containing recombinants, that is, either *A* and *b* or *a* and *B*. In other words, independent assortment does not occur for genes carried on the

**FIG. 23-7** Pattern of distribution of paternal and maternal chromosomes during meiosis in an organism with a diploid number of four. (a) Four types of chromosomal combinations are produced in gametes. (b) Chromosomal combinations possible in offspring are presented in a Punnett square.

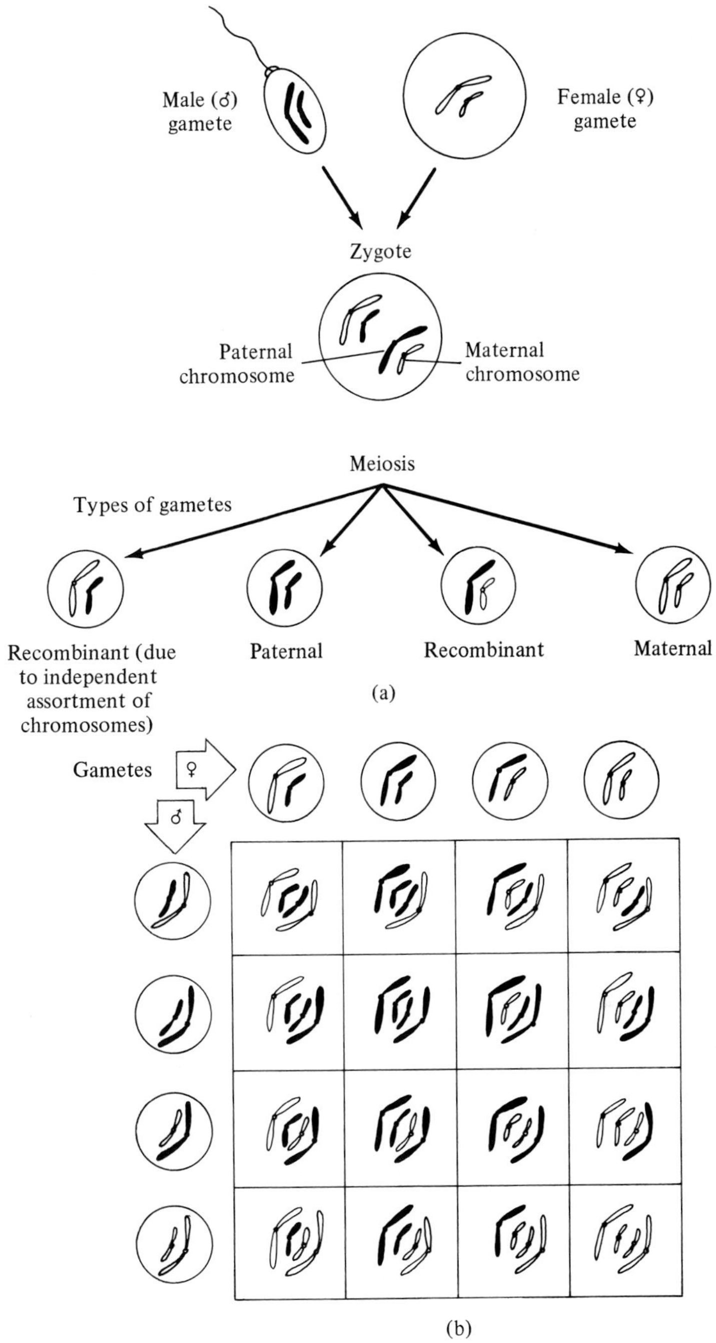


same chromosome. Thomas Hunt Morgan of Columbia University proposed in 1910 that such genes are linked because of their location on the same chromosome.

As the Mendelian laws were applied to more and more organisms, it became clear that each species had a characteristic number of linkage groups, that is, groups of genes that tended to be inherited together. It was also found that the number of linkage groups in a species was equal to the haploid number of chromosomes it possessed. For example, maize (*Zea mays*, or American corn) has 10 linkage groups and a diploid number of 20; and the common fruit fly *(Drosophila)* has four linkage groups and a diploid number of eight. The pea has seven pairs of chromosomes (a diploid number of 14); fortuitously, the genes for each of the

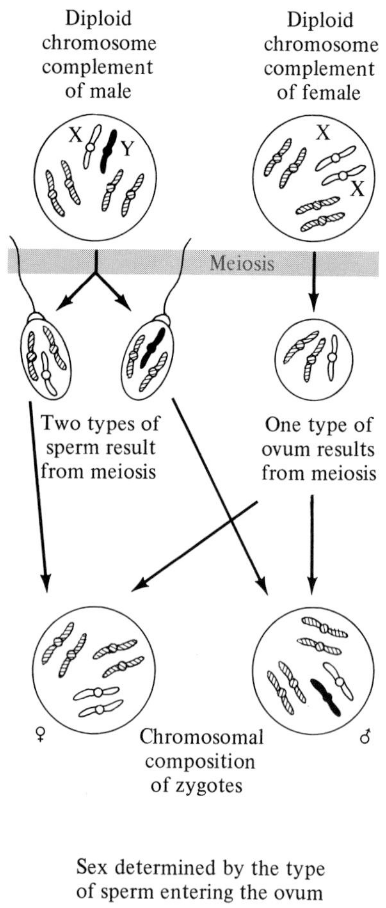


**FIG. 23-8** Chromosomal basis of sex determination. The species represented is one with a diploid chromosome number ($2n$) of six.

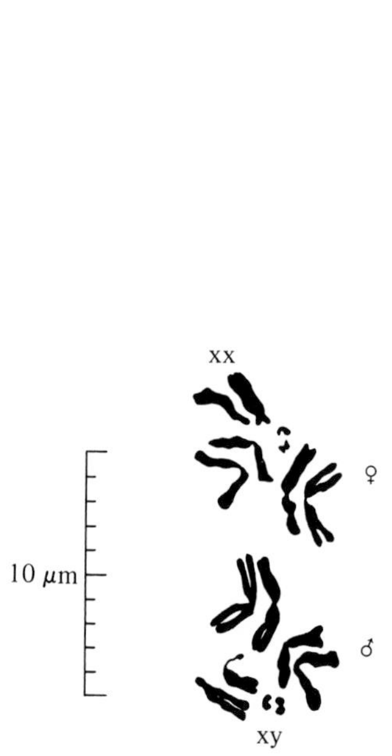


**FIG. 23-9** Karyotype of male and female *Drosophila.*

seven pairs of contrasting traits reported by Mendel apparently occurred on different chromosomes. Otherwise the different traits would not have assorted independently, and Mendel would have missed this important concept. For another possible explanation, see Panel 23-3.

## SEX DETERMINATION AND SEX LINKAGE

The concepts that each pair of chromosomes in a diploid cell is different from the others and carries different traits, that genes are borne along the length of the chromosomes, and that of the two sets of chromosomes, one is paternal (inherited from the individual's father) and the other maternal (inherited from the mother) came to be known as the chromosome theory of heredity. The first evidence for the theory came from studies that showed that the sex of an animal is determined by possession of certain specific chromosomes. Most of an animal's chromosomes are termed **autosomes.** Besides these, each individual in most animal species has two sex chromosomes. The female in many species possesses two X chromosomes, while the male has an X and a smaller Y chromosome (Fig. 23-8). An individual's sex thus depends on whether the egg is fertilized by a sperm with an X chromosome or one with a Y chromosome. Ova contain only an X chromosome because females bear two X chromosomes. But a sperm has an equal chance of containing an X or Y chromosome. Offspring, therefore, should be about equally divided as to sex, and in almost all species this is the case.

Figure 23-9 shows the entire chromosome complement, karyotype, for *Drosophila,* the common fruit fly. The female has two X chromosomes, and the male has an X and a Y chromosome. Although this is true of most animals, there are many exceptions. For example, in birds and butterflies, females have two dissimilar sex chromosomes, while males have a pair of identical sex chromosomes. In these species, the sex chromosomes in the females are designated ZW and those in the male ZZ. The presence of sex chromosomes is by no means universal. Some fishes, for example, have no physically identifiable sex chromosomes, but they do have genetic factors that determine sex. All of these observations confirm the fact that, with rare exception, the distribution of sex chromosomes or sex-factor–bearing chromosomes determines the sex of an individual. It was this finding that firmly established the chromosome theory of heredity.

Because males and females differ in their sex chromosome composition, cells of males and females of mammals can be distinguished by their cytological features. These cytological differences between the cells of males and females have implications not only in criminal investigations but also in studying regulation of chromosome function. Panel 23-1 describes the chromosomal basis of the visible difference between the cells of male and female mammals.

***Sex-Linked Genes.*** Genes for traits not related to sex are also carried on the sex chromosomes, particularly on the X chromosome. This discovery was made by Thomas Hunt Morgan at Columbia University in the first decade of this century. He had begun his genetic studies by looking for an organism that was convenient to raise in large numbers and that had a short generation time. He chose *Drosophila,* the common fruit fly. One female can lay up to 1,000 eggs in her lifetime, and it takes only two weeks for an egg to develop into a sexually mature individual.

## PANEL 23-1

# Recognizing Female Cells by the Presence of Barr Bodies

In 1960, Murray L. Barr, a Canadian physician, observed a dark-staining clump of chromatin in some of the nuclei of the cells of female mammals. These bodies were absent from the nuclei of cells from males. The number of Barr bodies, as these chromatin bodies came to be called, was always one less than the number of X chromosomes in organisms with more than two X chromosomes.

The origin of Barr bodies was explained by Mary Lyon's studies in England of female mice bearing a certain gene for hair color on the X chromosome. Unlike males with a mutant allele or females homozygous for it, heterozygous females were mottled in appearance even though the mutant allele was recessive to its wild-type allele which is defined as the most frequently observed and thus arbitrarily regarded as the normal allele. On close inspection, the fur proved to be a mosaic of patches of the two colors grey and white specified by the individual's genotype. Lyon hypothesized in 1961, that the Barr body was an inactivated X chromosome. Which of the two X chromosomes of a female was inactivated, she suggested, was a matter of chance. However, once an X chromosome in a cell became inactivated, the same X chromosome remains inactive in all cells descended from that cell. Apparently, the inactivation occurs early in development, when the embryo is composed of a few hundred cells. The result is a **mosaic individual,** one with a mixture of tissues in which one X or the other of the chromosomes is inactivated. In some cells of the tissue, the X chromosome with the mutant allele was inactivated, and the grey color was expressed. In other cells of the same type of tissue, the X chromosome bearing the wild type allele was inactivated, and the mutant white color was expressed.

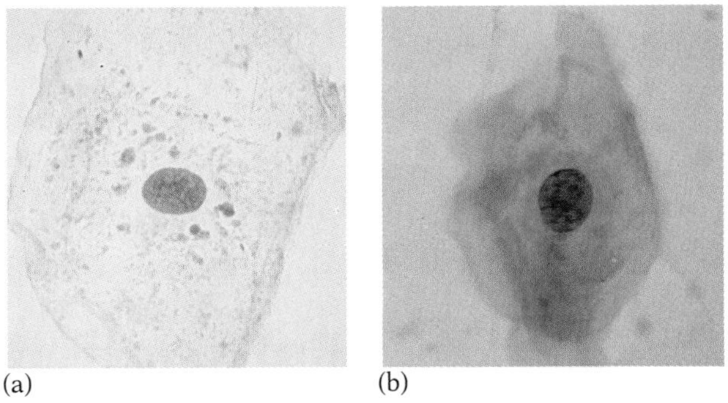


Barr bodies in nuclei of human epithelial cells. Cells from females (a) contain a dark staining body, but male cells (b) do not.

What function is served by inactivation of one of the sex chromosomes of a female? The answer probably lies in the fact that the X chromosome bears many genes that control synthesis of important enzymes. A female with two X chromosomes possesses twice as many copies of such genes as a male, with a single X. The female would accordingly be expected to produce twice as much of those enzymes and other proteins whose production is controlled by the genes borne on the X chromosome. Inactivation of one of the X chromosomes prevents this imbalance from occurring. Whatever its function, the presence of a Barr body can be exploited to identify the sex of a fetus. The presence of a dark body in the nuclei of cells floating in the amniotic fluid, collected through a process called amniocentesis, indicates a female embryo.

Although at first all the flies in his collection had similar phenotypes, Morgan kept looking for any with heritable variations. The first new variety he found suddenly appeared one morning in a culture of one of his pure-breeding stocks. It was hard to overlook since it was a white-eyed fly in a jar of flies with the typical red eyes of the wild type. The most frequently observed allele is regarded as the wild type allele, and the individual bearing this allele is regarded as the wild type individual. Arbitrarily, such an allele is also considered as the normal allele whereas the infrequent allele is the mutant or abnormal allele. Wild-type alleles are almost invariably dominant over their mutant alleles. One of the important conclusions to which Morgan was led by his discovery was that genes can spontaneously change and the changed form can then be inherited. Such a change, or mutation, was important raw material upon which natural selection could act.

The white-eyed fly that showed up in Morgan's culture jar was a male. When it was crossed with a wild-type female, all of the offspring had red eyes, like their mother. White eye color was thus recessive to red eye color. But when the $F_1$ flies were crossed with each other, the result was a surprise. Although the $F_2$ generation exhibited a 3:1 red-eyed to white-eyed ratio, as expected, all the females were red-eyed and half the males were white-eyed. Further studies showed the reason: the trait for white eye color was carried on the X chromosome, a condition referred to as **sex linked.** Hemophilia and color blindness are examples of sex-linked traits in humans and are discussed in Panel 23-2. Only when a female was homozygous for white eyes (by inheriting the gene from both her mother and her father) did she have white eyes. A male, however, would be white-eyed merely by inheriting one gene for the trait from his mother, because males have only one X chromosome. Of course, a male receives, not an X chromosome, but a Y chromosome from his father. The male is said to be a **hemizygote,** or **hemizygous** for the sex-linked traits. The X chromosome is homologous to only a small portion of the Y chromosome; that is, most genes carried on the X are not carried on the Y. The genes carried on the Y chromosome control **sex-limited traits:** those found only in males.

The discovery that the X chromosome carried genes for eye color in *Drosophila* was the first demonstration that a specific set of alleles was associated with a particular chromosome. At the same time, it showed that the sex chromosomes also carried genes unrelated to sex determination. Between 1910 and 1915, Morgan and his associates isolated 85 mutations in *Drosophila.* Some of these were sex-linked, but most occurred on the three autosomal chromosomes.

## CROSSING OVER AND THE INDEPENDENT ASSORTMENT OF GENES ON A CHROMOSOME

Linkage is not absolute. Two genes that usually are inherited together sometimes show up in separate offspring. For example, in 1906, in an experiment with sweet peas, Bateson and Punnett crossed plants differing in two sets of alleles known to be in the same linkage group, that is, on the same chromosome.

When Bateson and Punnett crossed sweet peas bearing purple flowers and long pollen grains with ones having red flowers and round pollen grains, all $F_1$ members had purple flowers and long pollen grains. But when $F_1$ members were selfed, the $F_2$ ratio was about 3 purple, long to 1 red, round to 0.15 purple, round to 0.15 red, long. Similar experiments with other combinations of alleles and other organisms produced similar results. However, the frequency of such exceptions to absolute linkage varied, depending on which two sets of alleles were being compared. But for any two sets, the percentage of exceptional recombinants was constant. Thus, in *Drosophila,* the heterozygous individuals with alleles for white eyes and yellow bodies (both on the X chromosome) always produced about 1 percent new combinations when they were interbred while the individuals with white eyes and miniature wings (also on the X chromosome) always exhibited about 34 percent recombinants.

PANEL 23-2

# Color Blindness and Hemophilia in Humans: Two Sex-Linked Traits

Some sex-linked mutant alleles cause atypical or diseased states such as color blindness and hemophilia more frequently in human males than in females. Color blindness is far more common in men than in women; because males have only one X chromosome, if the chromosome bears the mutant gene it will be expressed. As is typical of sex-linked inheritance, color blindness tends to skip a generation, from grandfather to grandsons, as did recessive genes in the Mendelian crosses. A woman will be color blind only if both her X chromosomes carry the mutant alleles for color blindness. Such a situation arises only if she had a color-blind father and a mother who carried the color-blind gene on at least one of her chromosomes. Therefore, all sons of a color blind woman will express the trait, and all her daughters will be carriers of it.

Hemophilia is inherited in the same manner. The disease is characterized by the absence of a clotting factor, which makes the individual susceptible to hemorrhage, or prolonged bleeding. The trait is apparently lethal early in development in women homozygous for it.

A well-known example of the sex-linked inheritance of hemophilia was described in 1948, in the genealogy of the royal families of Europe by John Burdon Sanderson Haldane, a British geneticist. Queen Victoria of England was a carrier of the hemophilia gene, apparently a mutation in a gamete that gave rise to the queen. Being heterozygous, she was not affected. But she transmitted the gene through her daughter Alice to her grandson Prince Frederick William of Prussia and granddaughter Princess Alexandra of Russia. Through Princess Alexandra, some members of the Romanovs, the Russian Royal family were also affected.

Inheritance pattern of color blindness, a sex-linked human trait.

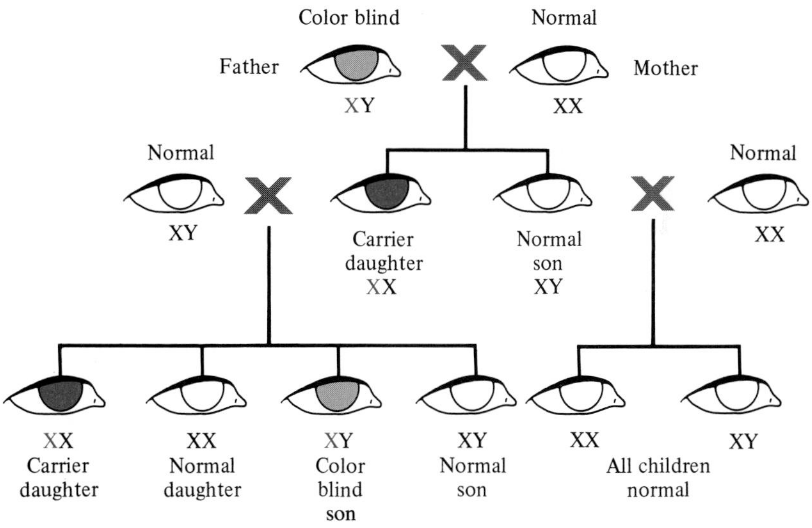


Such studies demonstrated two important and related facts:

1. Because two genes are on the same chromosome does not prevent them from being inherited independently to some degree.
2. The degree to which linked genes are inherited independently (the percentage of recombinants) is constant for any two sets of alleles in a linkage group but differs depending on which two genes are being compared.

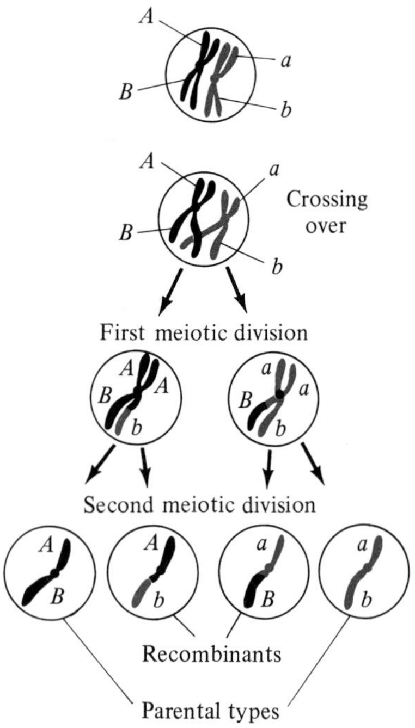


**FIG. 23-10** Diagrammatic presentation of the events in crossing over between homologous chromosomes.

**FIG. 23-11** The polytenic chromosomes of the salivary glands of *Drosophila melanogaster.*

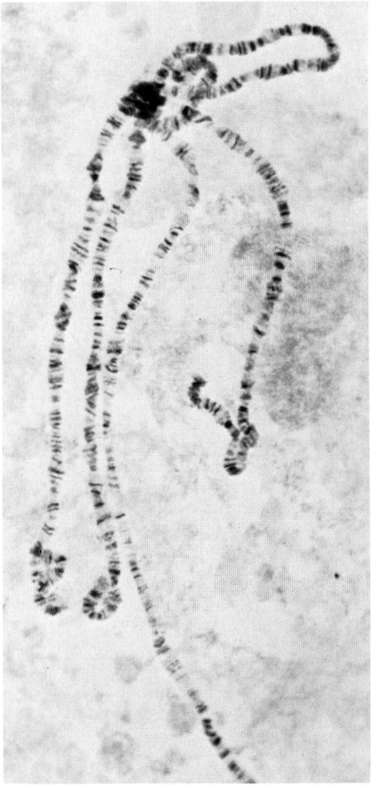

In 1910, Morgan proposed that during synapsis in prophase I of meiosis, portions of the two homologous chromosomes were exchanged with each other (Fig. 23-10). This exchange of genes during prophase I was termed **crossing over.** Here, then, was the solution to the puzzle of what function was served by two successive nuclear divisions during meiosis. The two divisions provided a way in which portions of homologous chromosomes could be exchanged. The pairing of homologous chromosomes at the first meiotic division serves to permit exchange of parts of chromosomes between the homologs, that is, permits crossing over. The first division itself serves merely to separate homologs from each other and to segregate them into separate nuclei. The second division provides for the division of the chromosomes that had undergone replication before the first division.

Morgan now made a further suggestion: that the percentage of new combinations between two pairs of alleles was proportional to the distance between the sets of alleles on the chromosome. His insight was right on target. Alfred Sturtevant capitalized on the concept by assigning positions to genes along the length of each chromosome, based on the frequency of crossing over between them.* Another implication of this theory is that the genes are arranged on the chromosomes in linear order. Using a technique called chromosome mapping, he was able to determine not only how far apart various genes were from each other but also in what sequence they occurred (see Panel 23-3).

*Drosophila* proved to be a remarkably useful experimental organism for this purpose. Like many related insects—houseflies and midges, for example—some of its tissues have nuclei with giant polytene chromosomes in which the genetic material is present in many copies arranged in parallel. The best example is found in the nuclei of larval salivary gland cells (Fig. 23-11). Not only is each giant chromosome distinct from the others, but each bears an irregular, species-specific pattern of prominent cross bands, constrictions, and enlargements along its length; a total of about 5,000 bands are discernible in all four chromosomes. This made topographical mapping of genes possible, allowing gene loci (locus, singular) to be identified within specific chromosomal regions (Fig. 23-12). A **locus** is a fixed position on a chromosome occupied by a specific gene or one of its allelic forms.

## MUTATIONS AND GENETIC EXPRESSION

Although chromosomes and the genes they bear usually replicate with great accuracy, deviations occasionally arise. Hugo De Vries, one of the rediscoverers of Mendel's laws, noted in studies of the evening primrose, *Oenothera lamarckiana,* that the offspring of a cross sometimes exhibited a trait that was not present in either parent, despite the fact that the parents were from pure-breeding stocks. He termed such sudden heritable changes mutations. Mutations, later reported in a wide variety of organisms, may be of two major types: those due to changes in the molecular structure of genes, called **point mutations,** and those due to various types of **chromosomal aberrations.**

Point mutations, the most common type of mutation, arise as chemical changes in the DNA molecules that constitute the genes. During the

* At an early stage in such a study, it must be determined (1) on *which* chromosome the genes of a particular linkage group are located and (2) the precise location of at least one of them, usually determined by studies involving deletions.

## PANEL 23-3

# Constructing Chromosome Linkage Maps

The distances between genes are reflected by the frequency of crossing over between them, as shown by the percentage of recombinants in the $F_2$ generation. The distances are stated as crossover units. Thus, if new combinations consistently occur in 15 percent of the $F_2$ generation individuals between genes *A* and *B*, they are 15 crossover units apart. If, in breeding experiments involving genes *B* and *M*, recombinants occur 10 percent of the time, the two genes are 10 crossover units apart. With only this information, it is impossible to tell whether *M* is 5 or 25 crossover units from *A*. The question can be answered with a second two-factor cross, involving *A* and *M*. If such a cross results in 5 percent recombinations, the relative positions are approximately as seen in the accompanying figure.

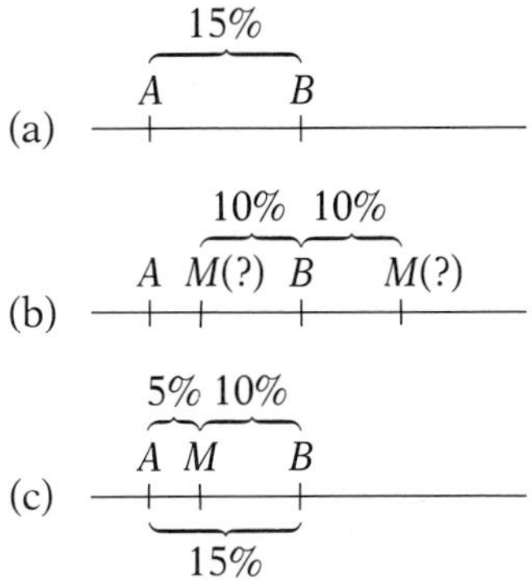

The principle underlying construction of chromosome linkage maps.

A further determination of relative position in such a case can be made using a three-factor cross. In this case, stocks with genotypes of *AAMMBB* and *aammbb* are crossed to produce an $F_1$ population of *AaMmBb*. When $F_1$ individuals are crossed with each other, the $F_2$ generation is composed mostly of individuals with genotypes of *AaMmBb*, *AAMMBB*, and *aammbb*. About 10 percent of the $F_2$ generation are recombinants, such as *Aammbb* or *AaMmBB*. In addition, there might be a few with genotypes such as *AammBb* or *aaMmbb*. These would represent double crossovers.

Crossovers of 50 percent result in independent assortment. An interesting aspect of linkage is presented by genes that are about 50 crossover units apart. The results of two-factor crosses involving such genes are actually similar to those of crosses involving unlinked genes: they assort independently. The fact that such genes are linked is evident only in crosses involving other, more closely linked genes. There is a strong probability that the seven pairs of alleles with which Mendel worked were carried on only five or six of the seven pairs of chromosomes in the garden pea. The fact that all seven assorted independently could be explained by the fact that the linked genes were located some distance from each other on the chromosome. The strains of peas with which Mendel worked are probably no longer in existence, and the question cannot be resolved with certitude.

process of replication of a DNA molecule, a base is sometimes omitted or a "wrong" base substituted (see Chapter 3 for the structure of DNA). If the resulting change produces a detectable phenotypic effect, the gene is said to have mutated.

Morgan's discovery of a white-eyed mutant was the first of many discoveries of mutant forms. Once he and his coworkers acquired a detailed knowledge of *Drosophila* anatomy, they were able to observe many instances in which wild-type genes, such as those for bristle pattern or for body color, had mutated to recessive alleles. Besides those that occurred in their own stocks, they were able to collect a large number of other mutants from nature. (Nature in this case was often a fruit market or a garbage can.) It is generally believed that the existing alleles of wild-type genes were derived from the wild-type genes by random mutations. With rare exceptions, a mutant allele is recessive to the wild-type allele.

**FIG. 23-12** Linkage map of three of the four chromosomes of *Drosophila melanogaster* as known in 1935. About 10,000 loci are now known.

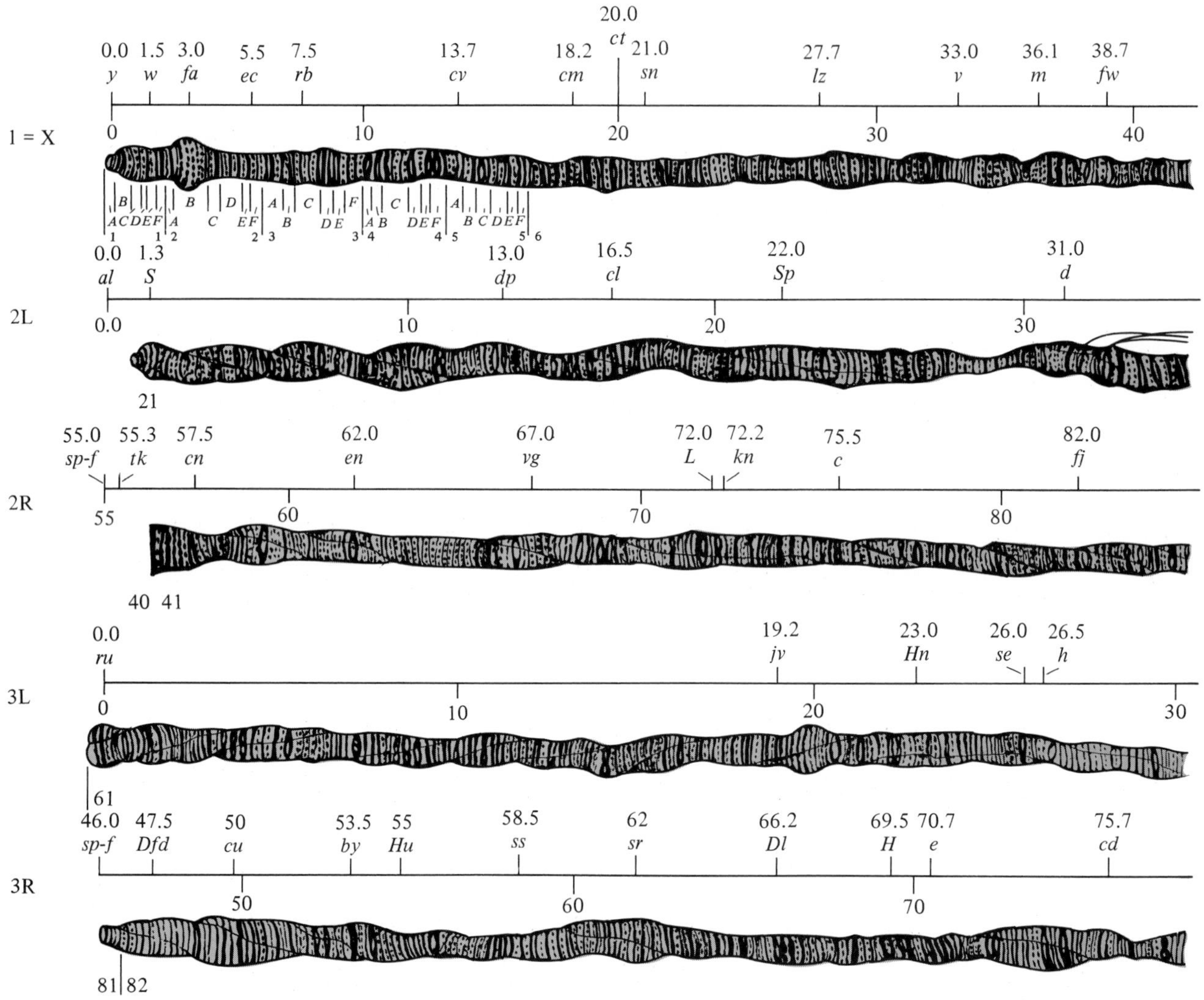

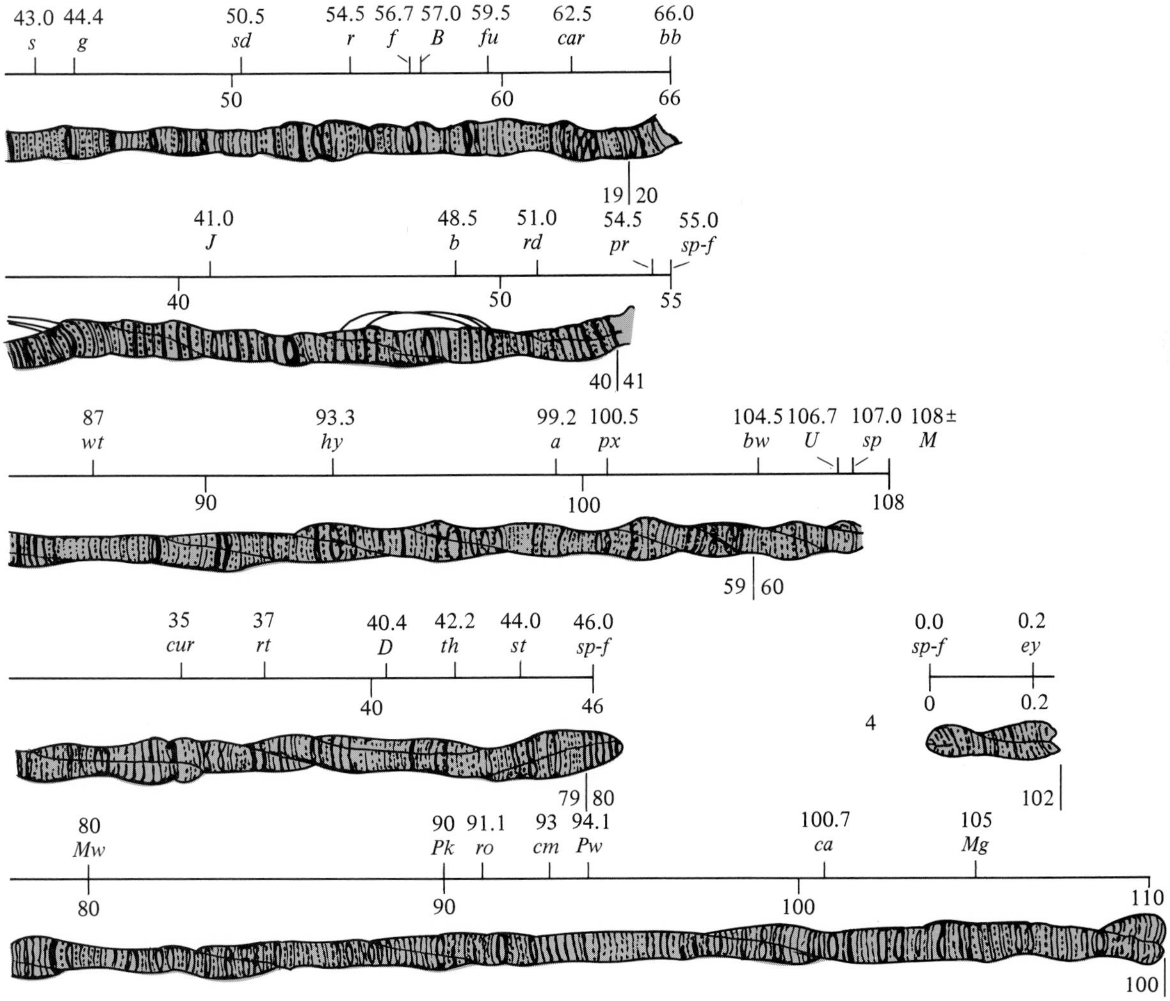
43.0 44.4 50.5 54.5 56.7 57.0 59.5 62.5 66.0
s g sd r f B fu car bb
50 60 66
19 20
41.0 48.5 51.0 54.5 55.0
J b rd pr sp-f
40 50 55
40 41
87 93.3 99.2 100.5 104.5 106.7 107.0 108±
wt hy a px bw U sp M
90 100 108
59 60
35 37 40.4 42.2 44.0 46.0
cur rt D th st sp-f
40 46
0.0 0.2
sp-f ey
0 0.2
4
79 80
102
80 90 91.1 93 94.1 100.7 105
Mw Pk ro cm Pw ca Mg
80 90 100 110
100

Close observation has indicated that the genes on the X chromosome of *Drosophila* mutate at a frequency of one in a thousand, that is, for every 1,000 ova or sperm produced, one of them will, on the average, contain a newly mutated gene on its X chromosome. Other chromosomes have comparable gene mutation rates. However, genes do not all mutate with the same frequency. Some genes mutate with relatively high frequency and have done so repeatedly for many years in otherwise pure-breeding cultures. Other mutants have been found to arise less frequently. Many genes appear so stable that their mutant alleles have never been found. Probably all mutations of these genes are lethal. A wild-type characteristic can be added to the list of known genes and its locus can be determined on a chromosome map only when it is also found in an allelic form.

It is possible to produce new mutations and to increase the rate of occurrence of mutations artificially. In 1927, Herman J. Muller, a former student of Morgan, was able to show that irradiation with X rays could increase mutation rates up to 150-fold. The rate of mutation was found to be dosage-dependent within a certain range. That is, if the organism was irradiated with a higher dose of X rays, a greater number of mutants appeared among the offspring. With one mutation that he had induced, Muller found that irradiation of flies possessing the mutant gene could also induce a reversion to the wild type. His results provided an early clue as to the molecular nature of the gene. (It was not yet known that DNA was the genetic material.)

Since Muller's pioneering work, for which he received the 1946 Nobel prize in physiology, many others agents have been shown to be mutation producing, or mutagenic, including ultraviolet light and a variety of chemicals. It is now known that the ability of some chemicals and certain viruses to cause cancer resides in their ability to induce certain types of mutations in body cells.

Mutations, whether of genes or resulting from chromosomal aberrations, are mostly harmful to the organism expressing them. This is not surprising, when we consider that most organisms in existence are highly adapted to their environments. As with a machine that has undergone a century of improvements to suit it to its task, it would be an unusual change that further improved an organism's performance. Unlike the changes in design that an engineer might make in a machine, mutations are random events; they occur haphazardly, bearing no necessary relationship to what might be useful in further adapting their possessors to the present or any other environment. Hence it is not surprising that a very large proportion of mutations are lethal. Considering the numbers of individuals involved and the time scale in which evolutionary changes operate, the small percentage of beneficial mutations that occur are adequate to explain the diversity in the phenotypes of organisms and among the individuals of a species. Because the majority of mutations are recessive, organisms may harbor the mutant genes in the population for several generations. Some of these apparently deleterious mutations may even serve a useful function in the heterozygous condition. Humans, for example, are estimated to harbor an average of four lethal genes each. Because these genes are recessive, it is difficult to detect their presence. Some lethal genes in humans, when homozygous, cause spontaneous abortion at early stages of embryonic development; while others produce offspring with serious defects, which are usually fatal in early postnatal life.

## SEMIDOMINANCE

The traits with which Mendel demonstrated that genes are discrete, particulate entities apparently were chosen deliberately. In those traits, the heterozygotes could not be distinguished phenotypically from the homozygous dominant individuals. Yet, many other cases were known at that time in which the hybrids formed between two contrasting pure-line individuals exhibited a phenotype that was intermediate between that of the parents. Contrasting traits that hybridize to produce intermediate phenotypes are called **semidominant.*** Some cases of semidominance are especially interesting in light of the Mendelian laws. In snapdragons, for example, a cross between red- and white-flowered individuals produces offspring all of which have pink flowers. However, as illustrated in Figure 23-13, a cross between two pink-flowered individuals produces an $F_2$ generation of which one-fourth has red flowers, one-half has pink flowers, and one-fourth has white flowers. Many similar cases are known. In such cases, the genotype is always reflected in the phenotype, and the heterozygotes can be distinguished from the homozygotes by simple inspection. The laws of segregation and independent assortment of genes nevertheless apply.

* Semidominance is also known as incomplete dominance, partial dominance, or blending inheritance.

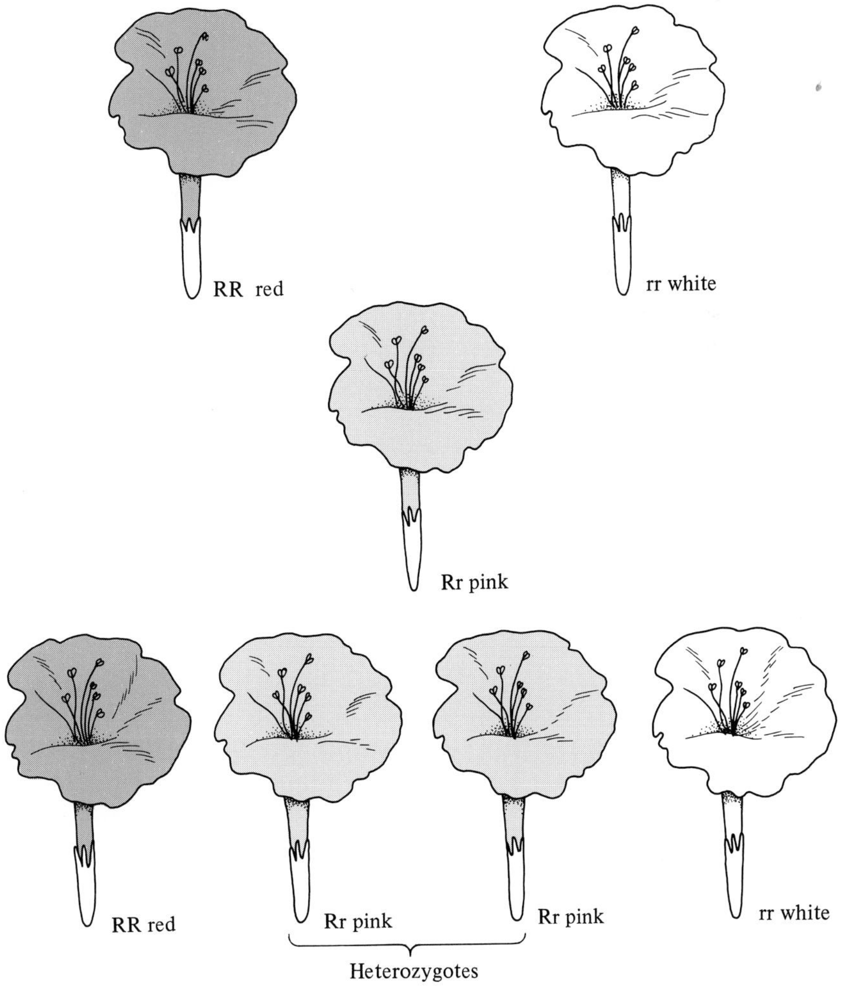


**FIG. 23-13** Pattern of inheritance of semidominant genes. Note that the pattern of inheritance of genotypes is similar to the pattern of inheritance of normal dominant genes. Only the phenotype of the hybrid is intermediate between the recessive and dominant traits.

## CODOMINANCE

When both alleles of a pair are fully expressed in a heterozygote, they are termed **codominants.** Human blood groups (ABO series) are examples of codominant alleles. In individuals with the blood type AB, both the $I^A$ gene and its allele, $I^B$, are expressed. Each of the phenotypes is characterized by a particular sugar unit present on the glycoproteins that are exposed on the surface of the erythrocytes.

Individuals with blood type A could be genetically homozygous for the allele $I^A$, or they could be heterozygous, with a genotype of $I^Ai$. Similarly, type B individuals are either homozygous for $I^B$ allele or have a genotype of $I^Bi$. The recessive allele, *i*, is inactive, and individuals homozygous for the *i* allele have type O blood.

## MULTIPLE ALLELES

As in the case of the A, B, O blood types, a given gene can exist in more than two allelic forms and such a group is called **multiple alleles,** heteroalleles, or allelic mutations. Thus, a particular locus on a chromosome might contain a wild-type gene for a certain trait, or one of the many allelic forms of that gene. However, normally only two alleles are ever present in one individual, because the same locus is represented only twice in an individual, once on each of the two homologous chromosomes that carry it.

It is common to designate a locus by a symbol for a well-known allele and to represent other alleles in the same series with the same symbol and a superscript. The wild-type locus is always represented by a superscript plus sign or a simple plus sign. Thus, the white-eye allele in *Drosophila* is designated *w*. Its alleles are designated $w^+$ (wild type), $w^a$ (apricot color), $w^{ch}$ (cherry), and so on. An individual homozygous for white eyes is represented as *w/w*, one with red eyes (wild type) but heterozygous with white is $w^+/w$ or $+/w$, and so on. Other eye color mutants that are allelic to white in *Drosophila* include ivory ($w^i$), pearl ($w^p$), tinged ($w^t$), honey ($w^h$), blood ($w^{bl}$). It is important to note that all eye colors other than red (the wild type) are not necessarily due to alleles at a single white locus. But it is easy to find out which ones are, by what is called a complementation test. If, for example, apricot (a/a) is mated to cherry (ch/ch), all the offspring will be heterozygotes with a genotype of a/ch, but only if the genes are alleles of each other, that is, are at the same locus. If they were not alleles, but at different loci, all the offspring would have the red, wild-type eyes, with a genotype of +/a plus +/ch. The principle underlying the complementation test is illustrated in Figure 23-14.

## POLYGENES

Mendel's task in discerning that inheritance is based on the transmission of pairs of particles was especially difficult because most of the quantitatively variable traits of organisms are not controlled by single pairs of alleles or even one series of multiple alleles but, rather, by many nonallelic genes, most of which exert a small but cumulative semidominant effect. For his demonstrations, Mendel chose exceptional traits, those, as it turned out, that were controlled by single pairs of factors.

Polygenes, or multiple factors, are a set of nonallelic genes that singly exert a slight effect but collectively control a quantitatively variable

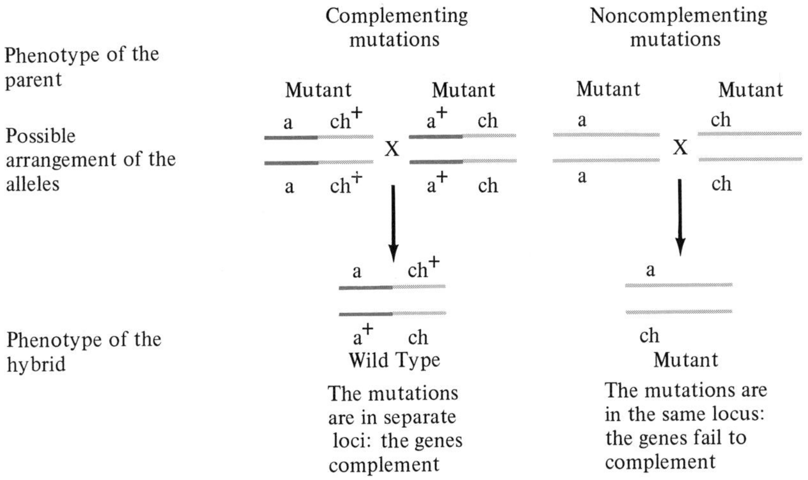


**FIG. 23-14** The genetic complementation test determines whether certain phenotypic mutations are allelic.

character. Because each gene has only a small effect, these traits do not exhibit a clear-cut segregation into typical Mendelian ratios. Furthermore, the expression of polygenes, like that of other genes, is quite sensitive to environmental influences, such as nutrition, temperature, and so on. To study polygene effects, environmental factors must be very carefully controlled. Most of the genetics of domesticated plants and animals involves polygenic inheritance and expression. Many traits in humans, such as skin color and body size, are also regulated by polygenes. Special statistical methods have been devised to analyze this complex type of inheritance.

## EPISTASIS

Any gene that masks the expression of another, nonallelic gene is said to be **epistatic** to it, and the masked gene is said to be **hypostatic** to the masking one. For example, in fruit flies, a recessive mutation called *apterous,* in the homozygous condition, suppresses the formation of wings. Mutations that affect wing morphology, such as curly wing, are hypostatic to apterous. No wing is formed in individuals homozygous for apterous, and hence the curly gene cannot be expressed even when it is in the homozygous condition.

## PLEIOTROPY

The production of more than one obvious effect by a single gene is termed pleiotropy. In higher organisms, it can usually be assumed that most genes exert a "domino series" of effects. A particular gene directs the synthesis of an enzyme, which in turn affects a reaction in which the product becomes a reactant in yet another reaction, and so on. In most cases, however, the many intermediate effects are at the biochemical level and are not readily recognized or no one has attempted to look for them. In some cases, however, several of the effects of one gene are dramatic and obvious. The cystic fibrosis syndrome in humans, for example, is due to a single autosomal recessive gene that causes the production of an unusual form of glycoprotein in mucous secretions. Mucus containing this protein is highly viscous and, as a result, interferes with a large number of normal functions, thereby producing a long list of readily apparent abnormalities, all of them due, indirectly or directly, to the modified glycoprotein in the mucus. There are several such examples of pleiotropic effects among human inherited syndromes. For example, the Lesch-Nyhan syndrome, characterized by spastic, destructive, self-mutilating children with low intelligence, is caused by a defective enzyme that is involved in purine metabolism.

## ENVIRONMENTAL INFLUENCES ON GENETIC EXPRESSION

All phenotypes are actually products of the interaction of genes with their environments. A gene is inherited as a factor that opens up a range of potential capabilities. For example, the inheritance of a gene for red flowers in primroses does not guarantee red flowers; if the plants are grown at temperatures above 34°C (86°F), white flowers are produced. Some species of sponge have very different shapes and colors depending on whether they develop in shallow water with heavy wave action or in deeper, quieter water. In the twentieth century, improvements in human infant and childhood nutrition in Europe and the United States have significantly increased the average height of those populations over nineteenth-century averages. Famine-stricken countries produce a generation of people with stunted growth.

Principles elucidated by the study of less complex organisms, under carefully controlled conditions, can generally be assumed to apply to humans. Among the principles learned from experiments with plants is that environmental factors affecting polygenic expression are so important that they often outweigh hereditary differences. Under ideal environmental conditions, plants with inferior genotypes often develop phenotypes equal to or superior to individuals with better genotypes that had been raised under poorer conditions. Thus, although differences in potential probably exist among human individuals, a proper environment in which to develop might be of more practical importance to the welfare of a given individual than the majority of such genetic differences. Because studies under rigidly controlled conditions are rarely possible with humans, this topic of "nature versus nurture" as the most important factor in variability among humans will continue to be controversial.

It should be well understood that environmental effects on gene expression are not inherited; genes are. Genes denied an adequate environment in which to reach full expression in one generation can, in an adequate environment, achieve their full potential in a later generation.

## CHROMOSOMAL ABERRATIONS

Because chromosomes can break and reunite, many unusual configurations, or chromosomal aberrations, occur. In some cases, a segment of a chromosome is lost, producing a **deletion,** or deficiency. If two chromosomes break at approximately, but not exactly, the same place and the broken ends of each then unite with the broken ends of their homolog, a deletion results in one chromosome and a corresponding **duplication** occurs in the other. In another type of aberration, a chromosomal fragment unites with a nonhomologous chromosome; this is termed a **translocation.** Another common aberration is an **inversion,** in which a detached segment becomes reunited with its chromosome in reverse orientation.

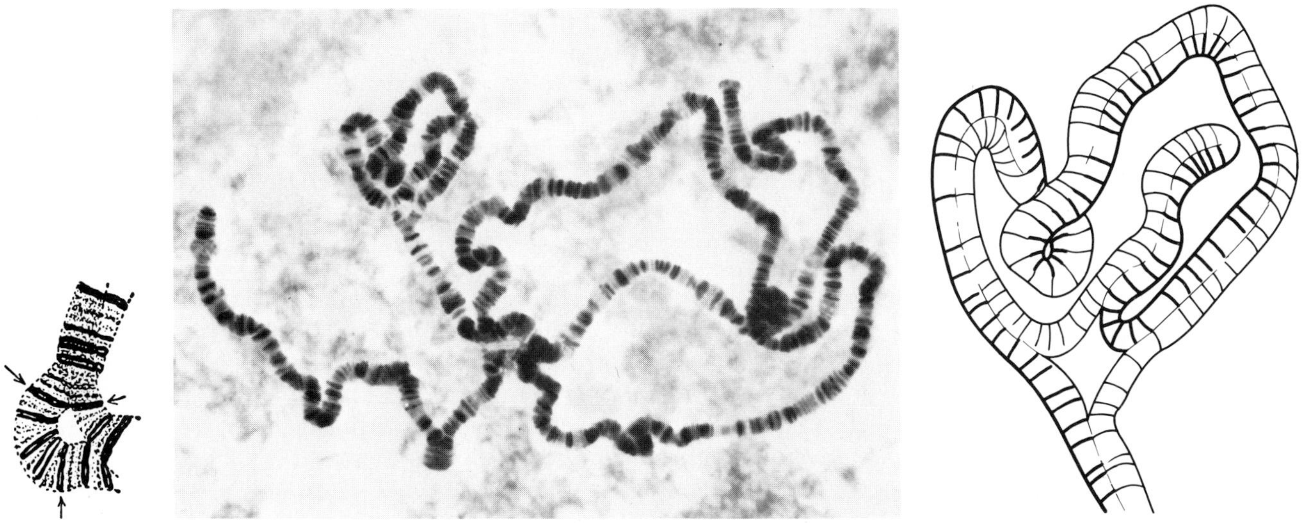

(a) (b) (c)

**FIG. 23-15** Chromosomal aberrations as seen in polytene chromosomes. (a) Deletion. The arrows indicate the missing segment in one of the chromosomes. (b) Inversion. The inverted region of the chromosome is at top, just left of center. (c) A diagram of the inverted region of part (b).

Chromosomal aberrations are most easily detected by examining chromosomes undergoing synapsis, or pairing of homologous chromosomes, as in prophase I of meiosis. But meiotic chromosomes are very slender threads and hence difficult to examine under a light microscope. However, the somatic pairing of chromosomes in the giant polytene chromosomes of *Drosophila* tissues provides valuable material for the study of chromosomal aberrations.

Deletions are denoted by the presence of unpaired loops in a chromosome in which all parts of the two homologs are paired, except where a piece is missing from one of them (Fig. 23-15a). A duplication presents a similar appearance except that the unpaired loop is identical in appearance to an adjacent segment. With proper staining, inversions can be distinguished from deletions and duplications because the loop contains two strands instead of one (Fig. 23-15b). The pattern of pairing of the homologous chromosomes is also affected in inversions. The pattern of chromosome pairing in an inversion is shown in a diagrammatic sketch (Fig. 23-15c). Chromosome pairing in a case with translocation is shown in Figure 23-16.

When particular inversions have become established in a population, they sometimes act as isolating mechanisms. That is, they effectively prevent exchange of genes between sister populations by producing infertility between them and thus promoting speciation, with each of the populations going its own evolutionary way. Reciprocal translocations (exchange of chromosome segments between nonhomologous chromosomes without any loss or gain of portions) act in a similar way in some species. But, translocations that consist of additions of chromosome fragments lead to abnormal development and will be considered below.

***Euploidy.*** Variations in the number of entire sets of chromosomes are termed euploidy (from the Greek *eu,* "true," or "even," and *ploid,* "unit"). Examples include **haploidy** (one set, or $1n$ number of chromosomes), **triploidy** (three sets, or $3n$), **tetraploidy** ($4n$), and so on. The general term **polyploidy** denotes an unspecified number of extra sets of chromosomes. When such variations first occur in nature or are induced experimentally, they may be lethal or result in progeny with low viability or various phenotypic anomalies. On the other hand, many poly-

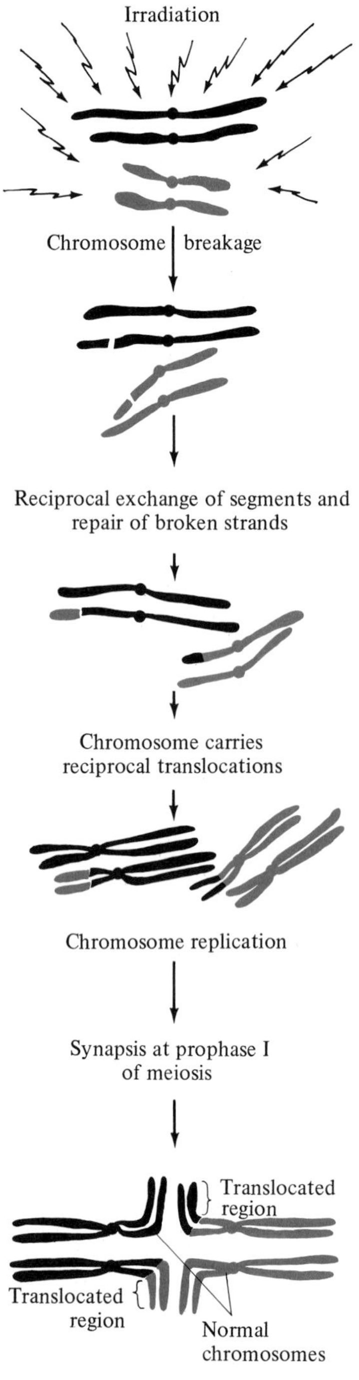


**FIG. 23-16** Chromosomal aberration, a part of one chromosome is translocated to another chromosome. The pattern of pairing in such chromosomes during meiosis is shown.

ploid varieties of crop plants are more vigorous and produce larger and better fruits than the parent stocks. The chromosome numbers of many domesticated plant species strongly suggest that they had their evolutionary origin in polyploidy. Bread wheat *(Triticum aestivum)* has a $2n$ number of 42 ($n = 21$). For emmer wheat *(T. dicoccum),* however, $n = 14$, and for einkorn wheat *(T. monococcum)* $n = 7$. Bread wheat thus appears to be a hexaploid species. Polyploidy, although important in plant evolution, does not seem to have played a major role in animal evolution.

***Aneuploidy*** (from the Greek *aneu,* "uneven"). The possession of one or more fewer or extra chromosomes, termed aneuploidy, is responsible for a large variety of phenotypic effects in humans, depending on which chromosomes are involved. Aneuploidy usually arises from nondisjunction of one of the chromosomes during meiosis (Fig. 23-17a). In a classic type of nondisjunction, one of the pairs of homologous chromosomes undergoing synapsis fails to separate during anaphase I; one daughter nucleus receives no copies of that chromosome, and the other receives two. At fertilization, the resulting zygote thus has either only one copy of the chromosome (monosomy) or three (trisomy) (Fig. 23-17b).

## GENETIC CAUSES OF BIRTH DEFECTS

Several diseases are known to be caused by a single gene. Some are sex-linked, that is, the genes are borne on the X chromosome. As a consequence, these diseases are rare in females, almost all known cases being found in males. A mother carrying a sex-linked gene should expect half of her sons to exhibit the trait and half of her daughters to be carriers of it. Color blindness and hemophilia, classic examples of sex-linked genetic defects, have been described earlier in this chapter. Here we shall describe some other common genetic disorders caused by single gene mutations.

***Phenylketonuria.*** Some individuals lack the ability to produce the liver enzyme phenylalanine hydroxylase, which metabolizes the amino acid phenylalanine. This condition is known as phenylketonuria (PKU). Normally, phenylalanine is converted to tyrosine, an important metabolite. People with PKU are deprived of tyrosine, and their phenylalanine is converted to other phenyl compounds, such as phenyl pyruvic acid, which accumulate up to 100 times their normal concentration in blood and tissue fluid. These toxic by-products produce irreversible brain damage and mental retardation. If the condition is diagnosed early enough by detecting phenyl pyruvic acid in the urine, an infant can be placed on a special diet that contains no more phenylalanine than is needed for protein synthesis; brain damage can thus be avoided.

***Tay-Sachs Disease.*** Common among Jews of Russian origin, Tay-Sachs disease involves an accumulation of fat in nervous tissues, resulting in early death, usually between 3 and 6 years of age. The disease is caused by a recessive mutation that results in the deficiency of an enzyme, N-acetylhexosaminadase, that cleaves hexosamine (an amino sugar) on a brain ganglioside, a lipid containing sugars.

***Sickle-Cell Anemia.*** More common among blacks than in other races, sickle-cell anemia results from a single change in the amino acid sequence of the beta chain of the hemoglobin molecule. A result of this

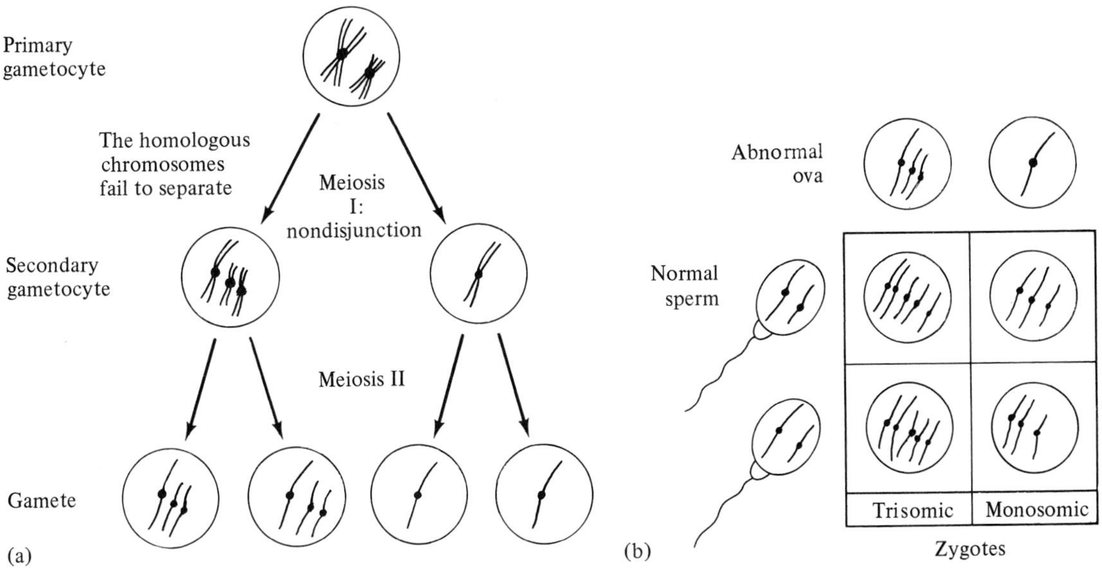


**FIG. 23-17** (a) Principle of chromosomal nondisjunction and (b) the formation of monosomy and trisomy.

change in the primary structure of the polypeptide is an abnormal configuration of the red blood cells when they are exposed to low oxygen tensions such as those that exist in veins. Instead of being circular biconcave discs, the red blood cells become sickle shaped, suggesting the name of the disease. Individuals who are homozygous for this trait experience a collapse of their red blood cells, which then clog the body's small blood capillaries. Homozygotes seldom reach maturity. Heterozygotes exhibit a mild impairment of circulation, which permits their survival. But they too suffer under conditions of exertion when demand for oxygen by the tissues is high. At first, it seemed surprising that such a deleterious gene had survived over several hundred generations. A clue to why the mutant sickle-cell allele has been preserved was provided by the discovery of the advantage conferred by the heterozygous condition in a particular environment. In malarious zones, sickle-cell heterozygotes have a higher survival rate from malarial infections than do individuals homozygous for the "normal" gene. In such areas, the heterozygous condition is thus adaptive and is selected for by the environment.

***Trisomy in Humans.*** Aneuploidy in humans has come under intensive study since Down's syndrome, formerly called mongolism, was shown to be due to trisomy-21, the possession of three copies of the twenty-first chromosome. Such individuals are mentally retarded and have a high susceptibility to respiratory diseases (Fig. 23-18a). They usually have a grossly defective heart, among other symptoms. The incidence of this syndrome is directly related to the age of the mother; the number of live births exhibiting the syndrome increases sharply after the age of 35 and reaches 2.0 percent for women in their mid-forties. The cause apparently lies in the increasing percentage of nondisjunctions occurring in meiosis II of oogenesis. In humans, oogenesis of all oocytes is halted in prophase I while the woman is still a fetus. The process in a given ovum is not completed until after ovulation, 12 to 45 years later. The longer the wait, the more nondisjunctions occur at anaphase II.

Nondisjunction of chromosome 21 results either in a nonviable ovum with 22 chromosomes or a viable one with 24. If an ovum with 24 chromosomes is fertilized, the resulting zygote thus has 47 chromosomes, three of them being number 21 (Fig. 23-18b). As a result of the

(a)

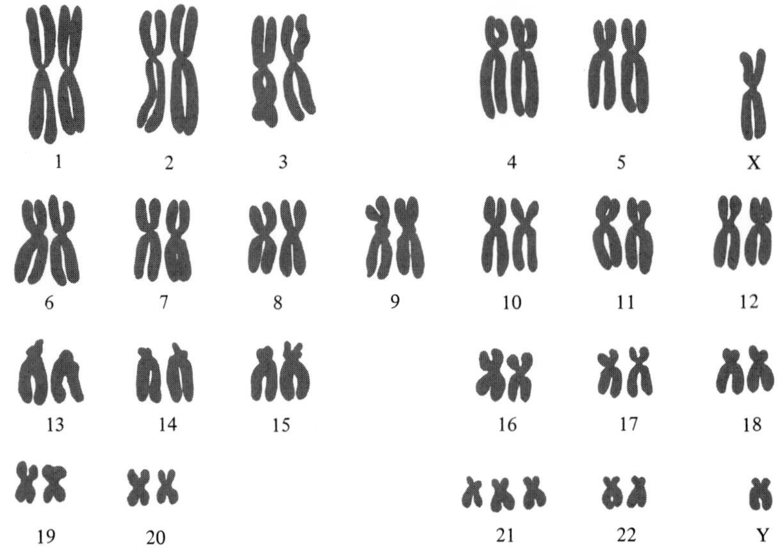


(b)

**FIG. 23-18** (a) A child with Down's syndrome. (b) Karyotype of a male with Down's syndrome. Metaphase chromosomes.

imbalance this creates, the rates of growth and differentiation of various embryonic components are altered, and the normally harmonious stages of development become disorganized. The final result is one or more abnormalities in every tissue and organ. The present average life expectancy is 16.2 years.

Other cases of viable trisomy known in humans include trisomy-13, -18, and -22. Each produces its own special syndrome and almost all individuals die within 6 months of birth.

***Aneuploidy Involving the Sex Chromosomes.*** Two major types of aneuploid syndrome involving the sex chromosomes (number 23) occur in humans, one type involving each sex. In Turner's syndrome, caused by monosomy of the X chromosome, the affected individual is a short, sterile female with several distinctive features but is usually not mentally deficient. In Klinefelter's syndrome, seen in males, the individual has, in addition to his normal complement of an X and Y chromosome, an extra X chromosome. All but a small percentage of such individuals exhibit normal behavior, although an unusual number are above average height.

## GENETIC COUNSELING

Biochemical tests have been developed for identification of many genetically inherited diseases. Prospective parents can seek genetic counseling and be tested for the possibility that they may be carriers of deleterious alleles in the heterozygous condition. If both parents are found to have a particular defective gene, they may, for example, decide to adopt children rather than to risk having their own. The chances of a child's being homozygous for the trait if both parents are carriers is, of course, one in four. By use of **amniocentesis**—a sampling and analysis of cells in the amniotic fluid—several genetic diseases, such as Tay-Sachs disease and hemophilia, as well as chromosomal aberrations such as trisomy 21, can be diagnosed even before birth, and the severe consequences on the parents and the unborn can be anticipated or, if the parents choose, prevented by termination of pregnancy.

Although research in this area is progressing, no cure for inherited diseases of humans has been developed. Like treating diabetes with insulin, present treatments at best only alleviate the symptoms. One direction in research is toward genetic engineering. It is anticipated that normal genes or clones—that is, descendents of a single cell—of normal cells might someday be introduced into deficient individuals to accomplish basic changes in their DNA legacy. The subject is fraught with legal and moral difficulties, but it may be many years before these need be addressed. A description of the procedures of genetic engineering is given in Chapter 24, on the molecular basis of heredity.

## Summary

Although the groundwork had been laid 35 years earlier with Mendel's experiments, it was not until 1900 that the field of genetics became a science, with the discovery of Mendel's paper on hybridization in peas and the simultaneous confirmation of his work by three investigators. Mendel's principal contribution was the establishment of the fact that inheritance involves the transmission of discrete, segregable particles—his "unit factors," now called genes—and that in sexually reproducing organisms, each individual receives at least one pair of these factors for each inherited trait, one factor from each parent. He also showed that some traits are dominant to others but that recessive traits could retain their identity, even when not expressed. These observations put an end to the concept of an irreversible mixing of traits as an explanation for the fact that some offspring exhibit shape, size, color, and so on that are intermediate to those characteristics of their parents.

After the rediscovery of Mendel's laws of inheritance, simple methods of determining the pattern of segregation of genes were devised. However, there was still no knowledge of the physicochemical basis of inheritance. But soon, thanks to the efforts of several cytologists, chromosomes were identified as the bearers of heredity. Particularly the correlation between the distribution of the sex chromosomes, designated X and Y in some species and Z and W in others, and the determination of sex convinced many scientists that chromosomes were indeed the bearers of Mendel's factors. The females of many species have two X chromosomes in each cell but only one is functional. The other, inactive X chromosome forms the Barr body. The distribution of chromosomes during meiosis, in which each chromosome pairs with its homolog and exchanges parts with it before separation and segregation into gametes, further suggested that the chromosomes were the bearers of genes, each gene occupying a specific locus on the chromosome. Thus, the concept that each individual receives one set of chromosomes from each parent was established. Each of a pair of genes that occupy the same chromosomal locus constitutes an allele. One allele of a series is usually dominant to the others, which are recessive. Consequently, the observable traits of an organism, the phenotype, may be different from its genotype, or genetic composition. The presence of a recessive allele in an organism with a dominant phenotype can often be determined by a test cross in which the organism is crossed to an individual with the recessive phenotype.

Although all paired sets of characters in peas reported by Mendel followed the rules of independent assortment, this is not the case with many sets of characters. Genes on the same chromosome form a single linkage group and do not always assort independently. Usually linkage is not absolute and some recombination of traits borne on the same chromosome does occur; the rate at which they recombine is proportional to the distance between the two genes. Such recombination is explained by crossing over, the exchange of chromatid segments that occurs during the first meiotic division. From the frequency of gene recombinations, linkage maps are constructed.

The science of genetics aided enormously in our understanding of evolution. A study of the common fruit fly, *Drosophila melanogaster*, led to our understanding of mutations, or sudden, heritable changes in the phenotype within a pure line. The first white-eyed fly paved the way to understanding how new phenotypes arise and thus possibly provide the raw material for natural selection and evolution. Other studies showed many other subtle features in the laws of heredity. For example, although Mendel's studies dealt only with allelic pairs, more than two alleles may be found at a given gene locus. Some alleles cannot be classified as dominant or recessive; some are codominant and are expressed together while others are semidominant and produce an intermediate phenotype. The expression of a gene may also be affected by a nonallelic gene. A gene is said to be hypostatic if its expression is dependent on the nonexpression of another gene. The second gene which exerts this influence is said to be epistatic.

Some quantitatively variable traits, such as height, weight, and skin color in humans, are controlled, not by one pair of alleles, but by several nonallelic genes. The pattern of inheritance of such polygenes, or multiple factors, is different from that of traits regulated by a single gene. Polygenic inheritance is usually expressed as a blending or mixing of traits. A gene may affect many phenotypic traits, and in such cases the gene is said to have pleiotropic effects. Some phenotypic traits, however, can be traced to environmental rather than purely genetic factors. In almost all organisms, food, humidity, and temperature can have an effect on the phenotype. It is particularly important to keep this in mind when one considers the apparent pattern of inheritance of certain traits for which no specific genes can be identified.

Knowledge of genetics helps in our understanding of some inborn errors of metabolism and other genetic diseases in humans. Hemophilia, color blindness, cystic fibrosis, phenylketonuria, sickle-cell anemia, and Tay-Sachs disease are examples of defects caused by simple gene mutations. Birth defects such as Down's, Klinefelter's and Turner's syndromes are caused by aberrations in chromosome numbers. Down's syndrome is caused by the presence of three copies of chromosome 21 (trisomy-21). Turner's syndrome, in females, is the result of the presence of a single X chromosome instead of the usual two X chromosomes, whereas Klinefelter's syndrome, in males, is the result of the presence of two X chromosomes and a Y chromosome instead of the usual one X and one Y chromosome.

## Recommended Readings

CARLSON, E. A. 1966. *The Gene: A Critical History.* Saunders, Philadelphia. A good account of the origins of the concepts of the gene theory and the chemistry of genes.

GOODENOUGH, U. 1984. *Genetics.* Saunders, Philadelphia, 3d ed. A clear presentation of the principles of Mendelian genetics and the chemical basis of heredity. It is a suitable text for students interested in an indepth study of heredity.

NOVITSKI, E. and S. BLIXT. 1978. *Mendel, Linkage and Synteny.* Bioscience, 28:34–35. An interesting discussion of linkage maps and Mendel's selection of traits.

PETERS, J. A. 1959. *Classic Papers in Genetics.* Prentice-Hall, Englewood Cliffs, N.J. A collection of original papers in various fields of genetics.

SUZUKI, D. T., A. J. GRIFFTHS, and R. C. LEWONTIN. 1981. *An Introduction to Genetic Analysis.* Freeman, San Francisco, 2d ed. This medium-level college text explains the fundamentals of genetics well and can be used as a general reference.

24

# The Molecular Basis of Heredity

OUTLINE



One of the striking achievements of modern biology is an understanding of life processes in terms of chemical reactions. The chemical basis of heredity is the most recent addition to this knowledge. Even at the time of Mendel's discovery of the laws of heredity, biologically oriented chemists were inventorying the chemicals in living cells. The techniques then available were adequate to identify some of the more abundant small molecules, such as sugars, but not to distinguish between varieties of the more complex substances, especially those in low concentrations. Biochemists of the late nineteenth century, however, were aware that the larger, more complex macromolecules, such as proteins, are important for life processes.

One of the first ideas proposed in an attempt to understand the chemical basis of heredity was that genes somehow regulate the synthesis of enzymes. In 1909, a British physician-biochemist, Archibald Garrod, published a book called *Inborn Errors of Metabolism,* in which he described several metabolic diseases, such as alkaptonuria, a disease affecting tyrosine metabolism, that were inherited according to Mendelian laws.

One line of research in the early part of this century pursued differences in the enzyme composition of eye color mutants of the fruit fly and color mutants of flowering plants. It was found that the mutant types lacked an enzyme necessary to make a certain pigment. Such studies confirmed the belief that there is a definite relationship between genes and enzymes. Since enzymes, like all proteins, are made up of chains of amino acids and each type derives its special properties from its own unique sequence of amino acids, it seemed likely that genes somehow determine the primary structure of enzymes.

## THE NATURE OF HEREDITARY MATERIAL

For most of the first half of this century, biologists believed that genes probably were proteins. Not only were enzymes proteins, but the proteins appeared to be the only class of molecules sufficiently varied and complex to transmit the inherited information needed to specify the many characteristics of organisms. Lipids and carbohydrates, for example, were regarded as far too simple and repetitious in structure to contain enough information. It was known, moreover, that chromosomes, the bearers of genes, contained substantial amounts of protein. True, they also contained deoxyribonucleic acid (DNA). However, the nucleic acids were known to be composed of a repetitious backbone of only four kinds of nucleotides. Although Tobjörn Caspersson, of Sweden, had shown in the 1930s that careful extractions of cells yielded DNA molecules of over 500,000 daltons in size, larger than many proteins, nucleic acids still seemed to lack sufficient variability to contain genetic information.

One piece of early evidence that did suggest that DNA might be the genetic material was that the most mutagenic region of the ultraviolet spectrum was 254 nm, precisely the region at which maximum absorption of energy by DNA occurs. However, it was argued that the energy absorbed by DNA from ultraviolet light was merely passed on to the protein to which it was bound. By thus being raised to a higher energy level, the gene mutated. At this stage, however, biologists began to think of the gene as a nucleoprotein, a substance made up of proteins as well as nucleic acids.

### PROTEIN STRUCTURE AND THE NATURE OF GENES

It had become quite clear by the mid-1940s that heredity required a methodical ordering of the primary structure of enzymes and other proteins. It was understood that this ordering must be a primary function of the genes. However, it was still a matter of speculation whether the DNA or the protein of the nucleoprotein carried the hereditary information. As the structure of proteins and the nature and action of enzymes became better understood, it was suggested that proteins could not contain all the information needed to determine the structure of the multitude of proteins in the body. It was argued that if the information for ordering the sequence of amino acids was specified by other proteins acting as enzymes, this would require a separate enzyme for each amino acid in the new protein. To synthesize each of these enzymes would require as many enzymes, and hence genes, as there were amino acids in each enzyme—a truly enormous number. Assuming that each protein is made up of only 100 amino acids this number would be $100^{100}$.

Other reasons for doubting that genes were proteins were derived from new knowledge of enzyme-substrate complementarity. The most logical way to assemble new proteins would be to form them against a template or mold that, by its shape, would specify the exact sequence of amino acids to be linked together. From what was known about the interactions that would be necessary between a protein template and the amino acids of any newly forming chain, a template of protein would be poorly suited to such a task. Too many "errors" in specification of primary structure of a protein would result, because structural differences between some amino acids (e.g. glycine and alanine or valine and leucine differ by a single methyl carbon [$CH_2$]) are too small to be recognized by a protein template.

## DNA AS THE GENETIC MATERIAL

Besides the logical considerations that excluded other macromolecules from possibly being genetic material, the early evidence for DNA being the stuff of which genes are made had all been indirect. For example, almost all DNA is located in the chromosomes, whereas RNA and protein are found in abundance in both the nucleus and the cytoplasm. One way of distinguishing the distribution of DNA within the cell is by using a cytochemical test called the Feulgen (pronounced foil-ghen) reaction, named after Robert Feulgen, the American biochemist who originated it in 1923. Cytochemical tests help identify the chemical components of cellular organelles, and the Feulgen reaction is specific for DNA. When stained with the Feulgen reagent, only the nucleus shows a strongly positive response for DNA. Moreover, the intensity of staining of a nucleus following the Feulgen reaction is proportional to the amount of DNA present. Quantitative studies of the DNA content of cells showed that most somatic cells of diploid organisms contain a constant amount of DNA that is twice as much as is found in their gametic (haploid) cells, thereby exactly paralleling the presence of haploid and diploid number of chromosomes. Also, the proportion of purine to pyrimidine bases in DNA, unlike in RNA, is the same in all types of cells of an organism, and in all members of a species, suggesting that the type of DNA present in all cells of an organism might have identical base composition or even identical base sequence.

***The First Direct Evidence.*** The first direct evidence that DNA contains genetic information was provided by experiments of Oswald T. Avery, Colin M. McLeod, and Maclyn McCarty at New York's Rockefeller Institute in 1943. It had been known since 1928 when Frederick Griffith, a British researcher, showed that some species of bacteria, such as the pneumonia organism, *Diplococcus pneumoniae,* better known as *pneumococcus,* acquire new heritable traits upon exposure to dead cells of a different strain, a process known as bacterial **transformation.** One such trait is the bacterial cell's ability to produce an outer polysaccharide coat of a certain type. Encapsulated bacteria of a specific type with a smooth surface (strain S) are pathogenic: they cause pneumonia. Encapsulated bacteria with a rough surface (strain R) are nonvirulent, or nonpathogenic. The smooth surface of the capsule prevents it from being phagocytosed by leukocytes. When mice were injected with the nonpathogenic strain of bacteria along with dead cells of the pathogenic strain, the mice contracted the disease and died. Furthermore, the bacteria isolated from the dead mice were of type S, even though no viable type S cells were injected.

Avery and his colleagues set out to find the transforming principle. By means of a series of ingenious experiments, they were able to transform bacteria of one genetic strain (R) into another strain (S) by treating them with DNA extracted from the second strain (Fig. 24-1). The preparations that were treated with DNase, an enzyme that hydrolyzes DNA, were ineffective in transforming bacteria. On the other hand, RNase, which hydrolyzes RNA, had no effect on the ability of the extract to transform. RNA and protein extracts both failed to produce the same effect. Thus, in these strains of bacteria, DNA from one strain of organisms appeared to be incorporated into the genome of another. More important, this DNA was the "transforming principle" responsible for the hereditary change.

***Proof That Some Viral Genes Are DNA.*** Most biologists were unconvinced of the universal applicability of Avery's findings, suggesting that the researchers had studied a unique situation and that genes were still most likely proteins. In 1952, further proof that DNA is the hereditary material came from studies of Alfred D. Hershey and Martha Chase of the Carnegie Institute of Genetics, at Cold Spring Harbor, on certain viruses that infect bacteria: **bacteriophages,** or, simply, phages. Like other viruses, phages are obligate parasites of cells. Outside of a cell, they exist as particles composed of a DNA core and a protein coat and are visible only with electron microscopy. Once a virus enters the right kind of cell, it "comes to life," in the sense that it begins to reproduce, but it is completely dependent upon the host cell's nutrients, ribosomes, energy sources, and so on. Bacteriophage T2, the virus with which Hershey and Chase worked, is specific for the common colon bacterium, *Escherichia coli (E. coli)*. Upon entering a host cell, the virus reproduces repeatedly, producing numerous copies of itself, all at the expense of the host cell's materials and mechanisms (Fig. 24-2). The host cell is finally destroyed in the process, releasing the new generation of virus particles, which are now free to invade other cells.

In their experiments, Hershey and Chase set out to learn whether the protein, the DNA, or both of these components of an infecting virus actually enter the host cell. They prepared two stocks of T2 virus. One was allowed to reproduce in the presence of $^{32}P$, a radioisotope of phosphorus that is a constituent of DNA but not usually of proteins. The other stock was grown in the presence of $^{35}S$, a radioisotope of sulfur that is a constituent of proteins but not of DNA.

Viruses of each stock were then allowed to infect separate bacterial cultures, after which the bacteria were immediately washed and examined for cytoplasmic radioactivity. The results were dramatic: when $^{35}S$-labeled viruses were used to infect the cells, the bacteria did not

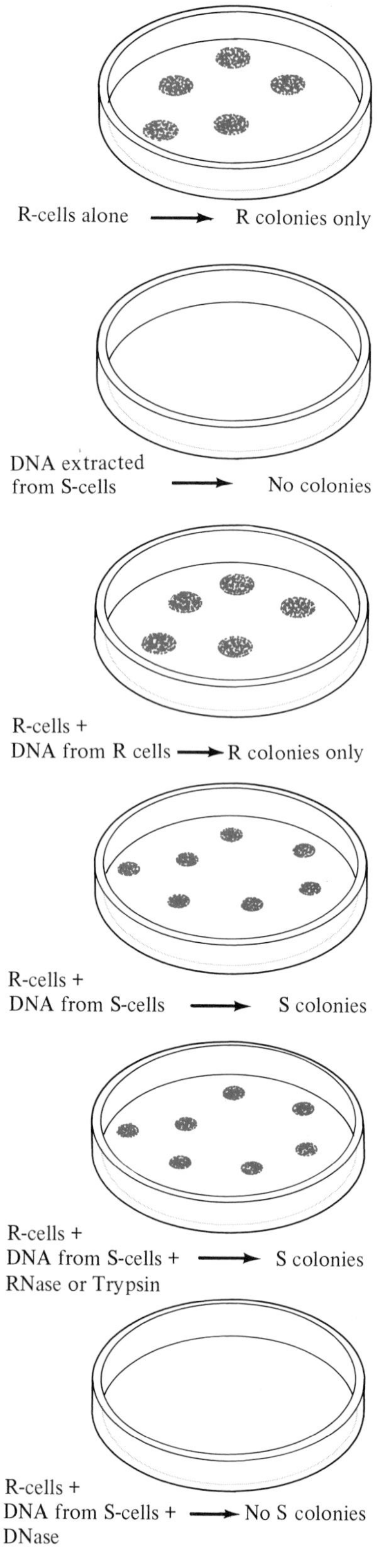


**FIG. 24-1** Experimental protocol to determine the role of DNA in the transformation of pneumococcus bacteria. R strain cells treated with DNA from S strain cells transformed to S strain cells. Treatment of the DNA preparation with RNase or trypsin had no effect on the ability of the preparation to transform R strain cells into S strain. DNAse treatment, however, abolished the ability of the extract to transform R cells. DNA from S cells alone or R cells alone does not produce S strain cells.

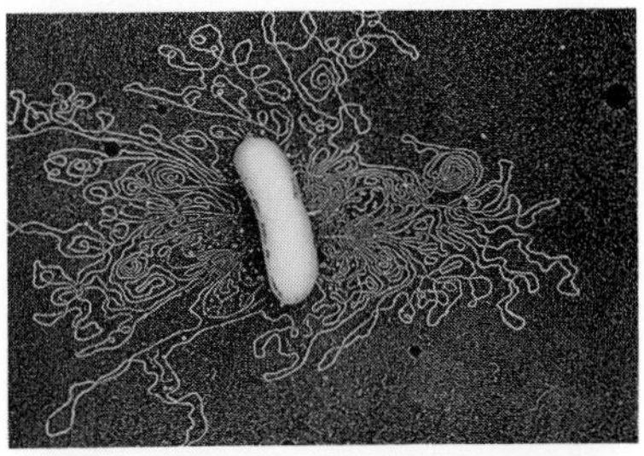

(a)

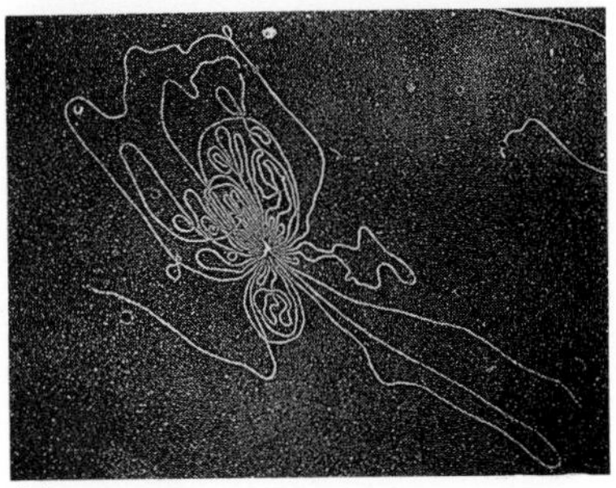

(b)

**FIG. 24-2** (a, b) Electron micrographs of *E. coli* and a virus to show the presence of DNA in these organisms. (c) The normal life cycle of $T_2$ bacteriophage that lives in the bacterium *E. coli.*

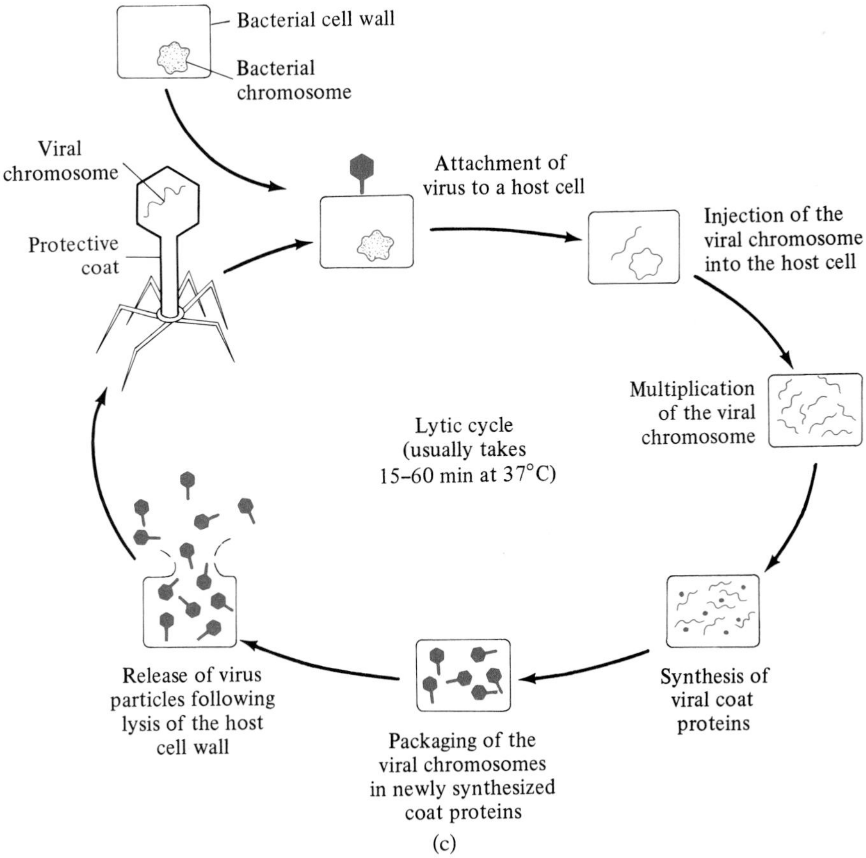


(c)

become radioactive; when $^{32}P$-labeled viruses were added, however, the result was a culture of radioactively labeled bacteria (Fig. 24-3).

From these results it was clear that the viral DNA enters the host cell and the protein coat is shed, remaining outside the cell. Here was convincing evidence that viral DNA contains genetic information, even when isolated from its protein coat. The viral DNA molecules entering a bacterial cell thus had instructed the protein-making machinery of their host to produce viral proteins. In the years that followed, both viruses and bacteria were to play other important roles as experimental subjects in unraveling the many puzzles of molecular genetics.

***Tobacco Mosaic Virus.*** Another important discovery was made by Heinz Fraenkel-Conrat and his associates at the University of California, Berkeley, in studies of the tobacco mosaic virus (TMV), which produces a mottling of the leaves in infected tobacco plants. TMV is one of several viral types that has RNA rather than DNA as its nucleic acid. These investigators treated strains of TMV so as to separate their protein and RNA components. Then, under certain conditions, they completely reconstituted the dismantled viruses by mixing the RNA and the protein. Infective virus particles were also formed by reconstitution of virus from the RNA of one strain and the protein of another. When these "new" strains were used to infect tobacco plants, the particles produced in the leaves were always phenotypically and genotypically identical to the parent strain from which the RNA, not the protein, had come. Once again, it was the nucleic acid that contained the genetic information, not the protein. But something else was learned as well: RNA can also carry genetic information.

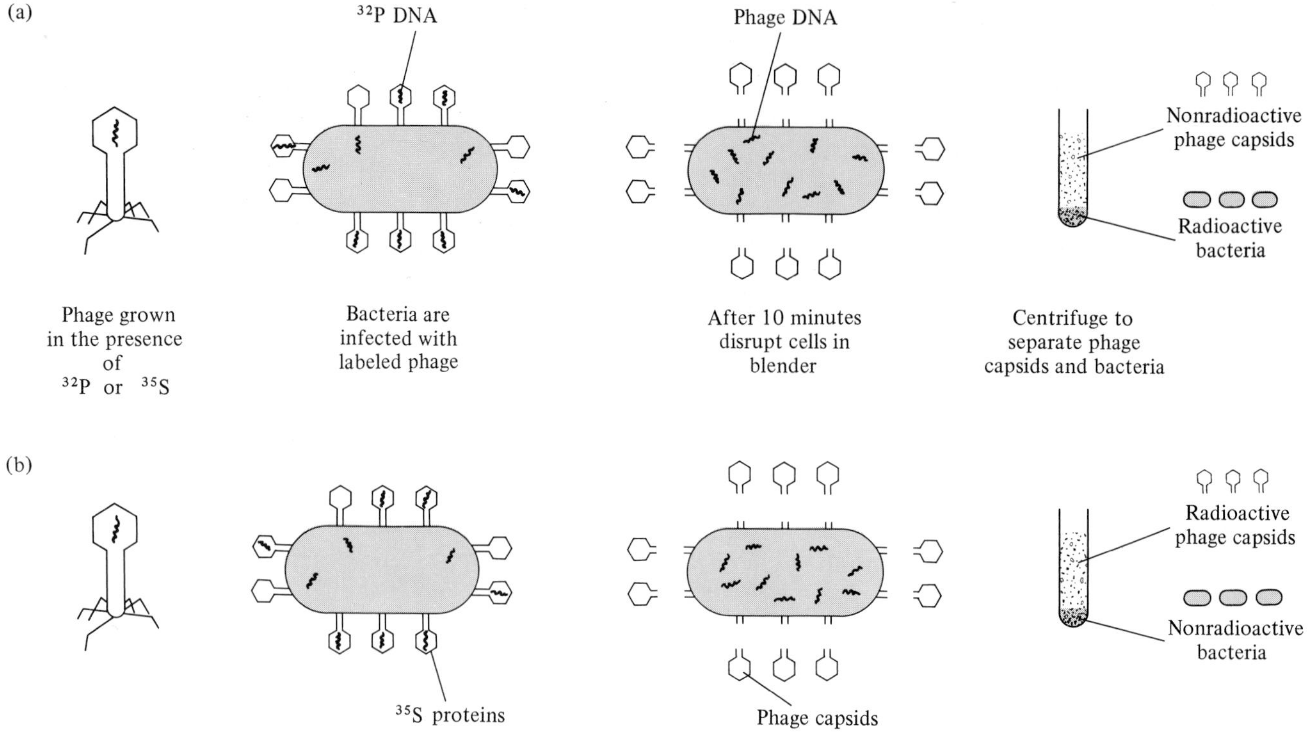


**FIG. 24-3** Design of the experiments that demonstrated the role of DNA in hereditary information. (a) Phage grown in the presence of $^{32}P$ was used to infect *E. coli.* The phage coats were separated from the bacteria 10 minutes after infection. Almost all radioactivity was found in the bacterial cytoplasm. (b) When phage labeled with $^{35}S$ was used in a similar experiment, almost no radioactivity was found in the bacterial cytoplasm.

## THE WATSON-CRICK MODEL: DOES DNA REPLICATE AS PREDICTED?

As noted in Chapter 3, in 1953, the American biochemist James Watson and the British biophysicist Francis Crick, then at the Cavendish laboratories at Cambridge University in the United Kingdom, proposed a molecular model for DNA that accounted for all known information about this nucleic acid. Not only did their theory explain the double helix structure of the hereditary molecules, but it also provided a method for their replication. It remained to find a way to test their theory, at least to see whether DNA replication took place by separation of the two strands followed by synthesis of a complementary strand by each of the two parent strands.

***The Semiconservative Nature of DNA Replication.*** Working with the bacterium *E. coli,* Matthew S. Meselson and Franklin W. Stahl, of the California Institute of Technology, made ingenious use of nonradioactive isotopes to follow the process of DNA replication. As reported in their 1958 paper, they first grew *E. coli* cells for many generations in a culture medium containing a heavy isotope of nitrogen, $^{15}N$. Purines and pyrimidines in which $^{15}N$ is substituted for the usual isotope, $^{14}N$, are heavier molecules than those containing the common isotope. It should be noted that the bases in nucleic acids are rich in nitrogen: each purine molecule contains five nitrogen atoms, and each pyrimidine has two or three. When a concentrated solution of a heavy salt, cesium chloride (CsCl), is centrifuged at very high speed (40,000 to 50,000 rpm) for 48 to 72 hours, a **density gradient** is produced in which the heaviest concentration of the salt in the solution, hence the highest density, is at the bottom of the tube and the lightest at the top. When DNA molecules are centrifuged in such a gradient the centrifugal force pulls the DNA molecule toward the bottom of the tube while the bouyancy of the molecule in relation to the density of the CsCl solution causes the molecule to be suspended wherever its density equals that of the solution. Meselson and Stahl found that a mixture of heavy, $^{15}$N-containing DNA and light DNA, containing only $^{14}$N nucleotides, stratified into two bands, the lower band being that of heavy DNA with the lighter DNA above it (Fig. 24-4a).

*E. coli* was allowed to grow in the presence of $^{15}$N-labeled nucleotides for many generations, then washed free of the heavy isotope and introduced into a medium containing unlabeled nucleotides. After these cells had undergone one cell division (and their DNA had replicated once), their DNA was extracted and centrifuged in a CsCl gradient. This DNA exhibited a density that was intermediate between that of DNA extracted from cells grown for several generations in $^{15}$N medium and of DNA from cells grown only in $^{14}$N medium. Such DNA molecules were therefore hybrid molecules, containing equal amounts of $^{14}$N- and $^{15}$N-containing bases.

In a further experiment, cells grown for a long time in $^{15}$N medium were washed and then allowed to grow in $^{14}$N medium long enough for *two* cell divisions (and two cycles of DNA replication) to occur. Their DNA was then extracted and centrifuged in a CsCl gradient as before. In this case, two bands were obtained, one of the hybrid variety and the other a lighter one with a density equal to the DNA containing only $^{14}$N.

These results, diagrammed in Figure 24-4b, clearly confirmed that the two strands of the double helix separated and that each strand then served as a template for the synthesis of a new complementary strand (see Figure 24-5a).

Two of the three possible ways in which replication might occur were unequivocally excluded by these results. One possibility was that DNA replicated conservatively, that is, that the original double helix, after opening up and serving as a template for the synthesis of two new strands, would rezipper, so to speak, retaining its integrity (Fig. 24-5b). If this had been the case, no hybrid DNA molecules would have been formed; only two bands, one each of the heavy and light DNA would have formed. Another excluded possibility is termed dispersive replication. It was conceivable that some segments of old strands intermingled with newly synthesized segments in daughter molecules (Fig. 24-5c). The actual method of replication that does occur, as demonstrated by the work of Meselson and Stahl, is termed semiconservative. Each new DNA molecule consists of one "conserved" (the original) strand from the parent molecule and one entirely new strand (Fig. 24-5a).

***Cytological Confirmation of Semiconservative Replication.*** Following the biochemical demonstration of the validity of the semiconservative replication of DNA there came another confirmation in 1963, this time from a cytological approach. The Australian John Cairns cultured *E. coli* in a medium containing tritiated thymidine, the nucleoside thy-

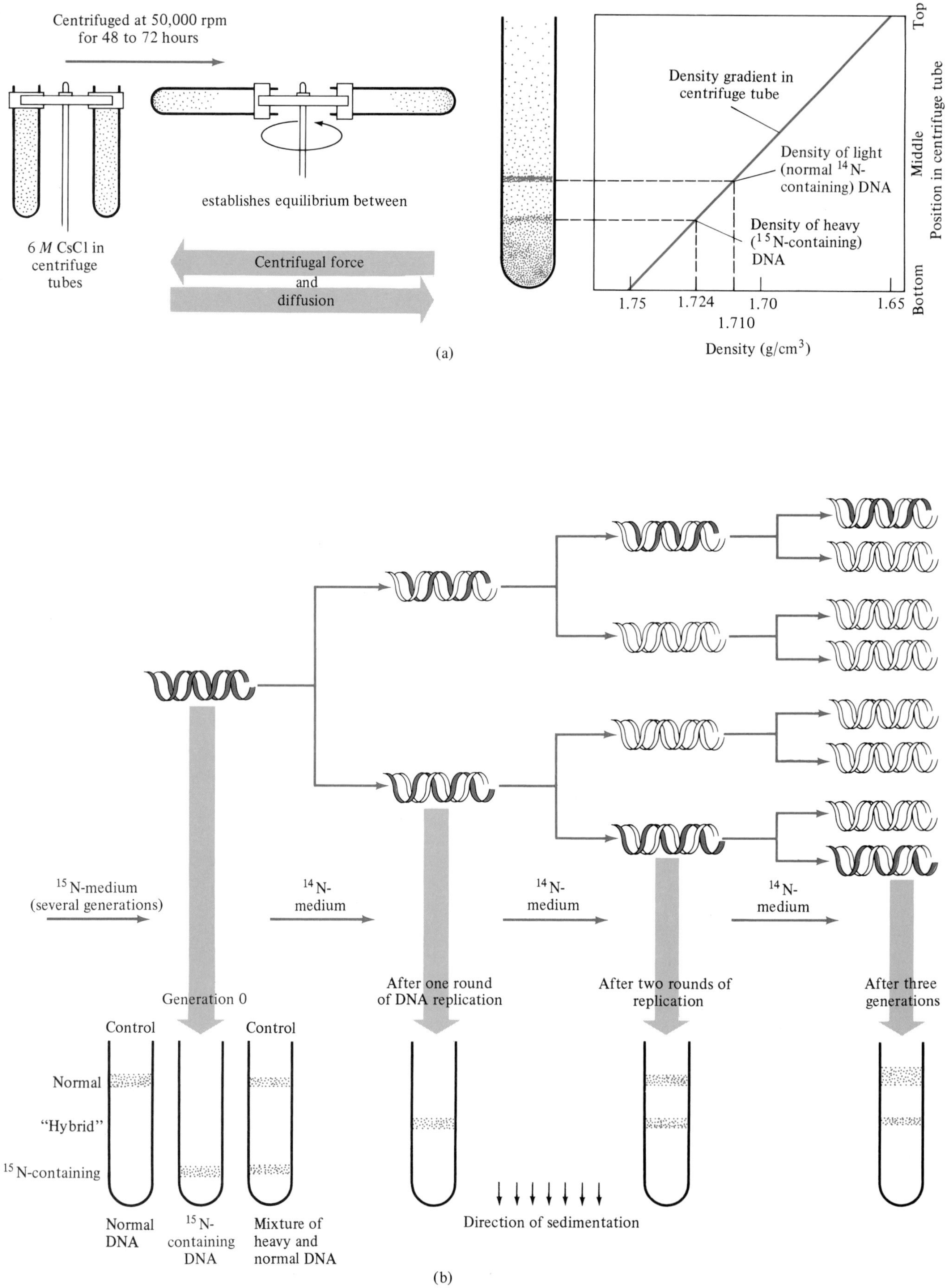
Centrifuged at 50,000 rpm for 48 to 72 hours
6 M CsCl in centrifuge tubes
establishes equilibrium between
Centrifugal force and diffusion
Density gradient in centrifuge tube
Density of light (normal $^{14}N$-containing) DNA
Density of heavy ($^{15}N$-containing) DNA
1.75
1.724
1.70
1.710
1.65
Density (g/cm$^3$)
Top
Middle
Bottom
Position in centrifuge tube
(a)
$^{15}N$-medium (several generations)
$^{14}N$-medium
$^{14}N$-medium
$^{14}N$-medium
Generation 0
After one round of DNA replication
After two rounds of replication
After three generations
Control
Control
Normal
"Hybrid"
$^{15}N$-containing
Normal DNA
$^{15}N$-containing DNA
Mixture of heavy and normal DNA
Direction of sedimentation
(b)

**FIG. 24-4 (opposite)** Density gradient (cesium chloride) centrifugation procedure to separate DNA molecules on the basis of their density. (a) Normal DNA has a density of 1.71 whereas that extracted from *E. coli* grown in the presence of $^{15}N$ labeled medium has a density of 1.724. The width of the band indicates the relative amount of DNA in each band. (b) Meselson and Stahl's confirmation of the semiconservative model of DNA replication. All DNA from *E. coli* grown for several generations in a $^{15}N$-containing medium, has a density of 1.724. After allowing these cells to divide once in the presence of normal medium their DNA has a density midway between normal and "heavy" DNA. After a second round of DNA replication, equal proportions of "light" and "hybrid" DNA were found. Only one-fourth of the third generation DNA has the density of the hybrid variety.

midine into which had been incorporated the radioisotope tritium ($^3H$).* He applied the principles of autoradiography to visualize the pattern of distribution of the labeled strands after various numbers of divisions in the absence of isotopically labeled nucleoside. In this technique, the microscope slide bearing the bacterial chromosome is covered with a photographic emulsion and kept in the dark; the labeled chromosome strands "expose" the emulsion grains wherever tritium atoms undergoing radioactive decay are located. If both strands of DNA in a chromosome are labeled, there are twice as many "exposed" grains as there would be if only one strand had been labeled. In Cairns' experiments, at the end of one generation, only one strand in each of the double helix DNA molecules contained the radioactive label (Fig. 24-5d). At the end of a second division of the cells, only half of the cells had one labeled DNA strand. (In bacteria, each cell has only one chromosome, and each chromosome has only a single DNA molecule.) After a third division, the label was present in only one-fourth of the cells, further confirming the semiconservative replication of DNA.

The double helix model of DNA predicts the potential for the self-replicating property of the molecule. The experiments of Meselson and Stahl described above prove that DNA replication is semiconservative. A variety of enzymes and other DNA binding proteins are necessary for DNA replication. Some, but not all, of these events in DNA replication have been elucidated. In the accompanying panel (24-1) some of the details of the DNA replication process are described.

* Thymidine is the nucleoside formed by the base thymine and the sugar deoxyribose. Nucleosides included in a culture medium become incorporated into the nucleic acids produced by the cultured organisms. Thymidine is chosen for the DNA label because it is a constituent of DNA and not of RNA. The radioactive label is thus incorporated only into the DNA being synthesized, not into the RNA.

**FIG. 24-5** The three theoretical models to explain DNA replication: (a) Semiconservative; (b) conservative; and (c) dispersive. (d) Electron micrograph confirming the semiconservative replication of DNA.

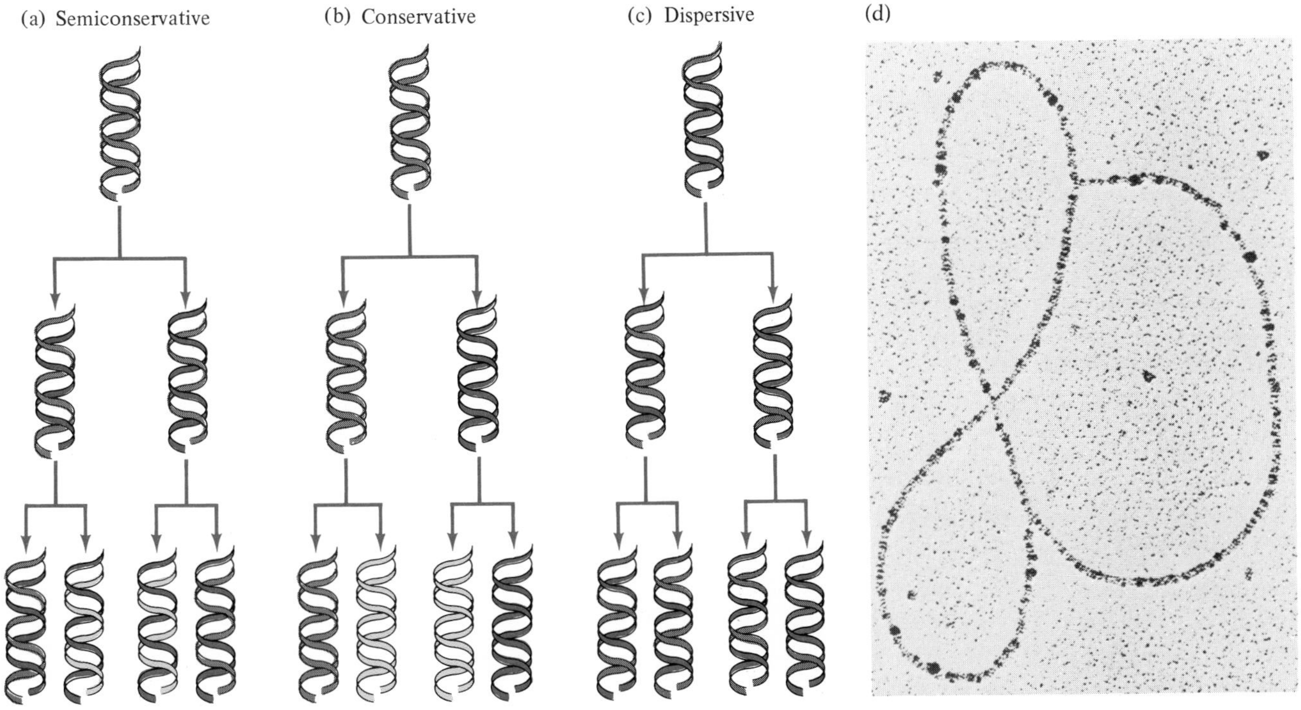

PANEL 24-1

# Machinery for DNA Synthesis

Among the most crucial stages of prebiotic evolution, which took place during the first billion years of the earth's existence and ultimately led to the origin of life itself, was the self-replication of a DNA molecule. The molecular mechanisms of DNA replication have attracted the attention of molecular biologists not only because of the fundamental importance of replication in understanding the origin and evolution of life on earth, but also because of its practical value in the control of cancer, a disease characterized by unrestricted DNA replication and cell division.

## DNA POLYMERASE

The first in vitro demonstration of DNA replication was accomplished by Arthur Kornberg and his colleagues at Washington University, St Louis, in 1956. Their demonstration involved fractionating cells into various organelles, macromolecules, and other components, then reconstituting the cell components in various combinations in vitro. They found that the following substances were required for DNA replication to proceed in vitro:

1. a mixture of the 5′ triphosphates* of the four deoxyribonucleosides: deoxyadenosine triphosphate (dATP) and dTTP, dGTP, and dCTP (the thymine, guanine, and cytosine equivalents)
2. the magnesium ion ($Mg^{2+}$)
3. a DNA molecule which is largely single stranded but containing some double-stranded regions, that serve as sites of addition of nucleotides (template)
4. a particular enzyme, which they named DNA polymerase

## DIRECTION OF DNA ASSEMBLY

As noted, DNA polymerase cannot synthesize new DNA molecules from their components unless a preexisting single-stranded DNA molecule is present to act as a template. The preexisting DNA molecule thus acts much like an essential coenzyme. It also specifies the structure of the new DNA molecule to be synthesized.

DNA strands have polarity. One end terminates in a phosphate group that is attached to the 5′ carbon of the pentose deoxyribose; the other end terminates in a hydroxyl (—OH) group that is attached to the 3′ carbon of the sugar. For convenience, the two ends are referred to as the 5′ and 3′ ("5-prime and 3-prime") ends, respectively. In the double helix DNA molecule, the two strands have antiparallel orientation, one being 3′ to 5′, the other 5′ to 3′. The enzyme DNA polymerase can catalyze the addition of nucleotides to the 3′-hydroxyl end of the preexisting primer DNA strand, but it cannot add nucleotides to the 5′ end. The *direction* of synthesis of a newly forming DNA strand is thus always from 5′ to 3′, using an existing 3′-to-5′ strand as a template.

But how can two oppositely oriented DNA strands, unzippered at only one end, both grow in the 5′-to-3′ direction? This paradox became more and more perplexing as biochemists searched for a DNA polymerase that would catalyze the synthesis of DNA chains in the 3′-to-5′ direction. No such enzyme has ever been found. The problem was finally resolved in 1969 by the use of tritiated thymidine. Very brief exposure to the labeled nucleoside during replication showed that DNA synthesis in one of the two daughter strands was discontinuous, that is, only very short segments (of about 1,000 nucleotides) were formed at a time. These **Okazaki fragments,** named after their discoverer, Reiji Okazaki of Japan, were each synthesized from the 5′ to the 3′ end after each short length of the double strand was unzippered. In the other strand, however, DNA synthesis was taking place in a continuous fashion. The two strands came to be known as the **leading strand** (with continuous 5′-to-3′ growth of the new chain) and the **lagging strand.** It was found that another enzyme, **DNA ligase,** subsequently joined the short segments to each other, resulting in the overall growth in the 3′-to-5′ direction of the lagging strand. The same enzymes also participate in repair of DNA.

## OTHER ENZYMES INVOLVED IN DNA REPLICATION

Machinery for DNA synthesis has been found to include other enzymes or proteins for such tasks as the separation of the two strands, the stabilization of single-stranded DNA, the cutting of single strands of DNA, and the rotation of DNA to relieve the tension caused by the separation of strands.

*(continued on page 638)*

* A nucleotide triphosphate is designated as 5′ or 3′, depending on the specific carbon atom in the sugar to which the phosphoric acid residues are attached. In DNA synthesis 5′ nucleotides are used.

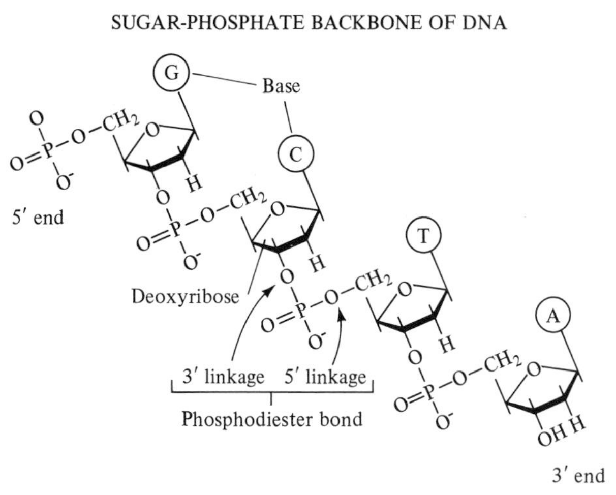


(a)

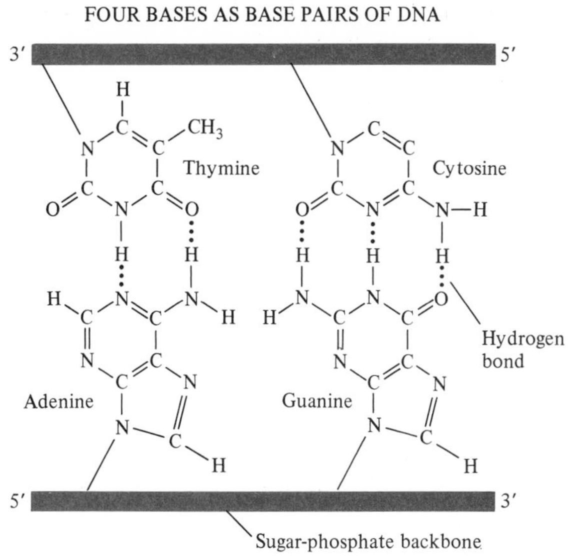


(b)

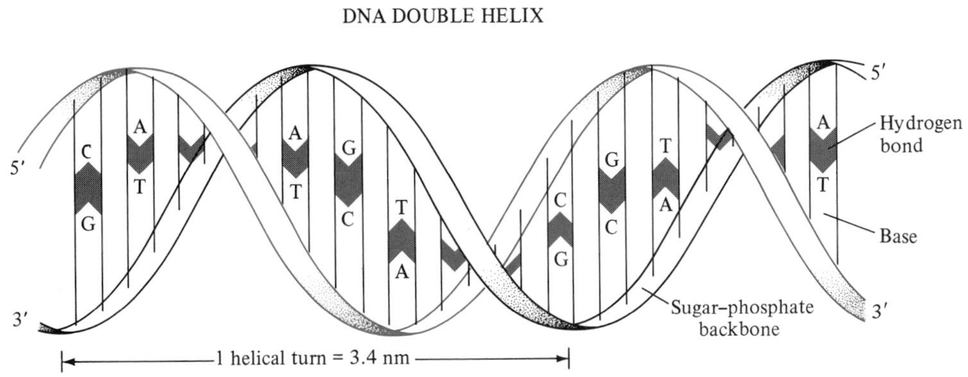


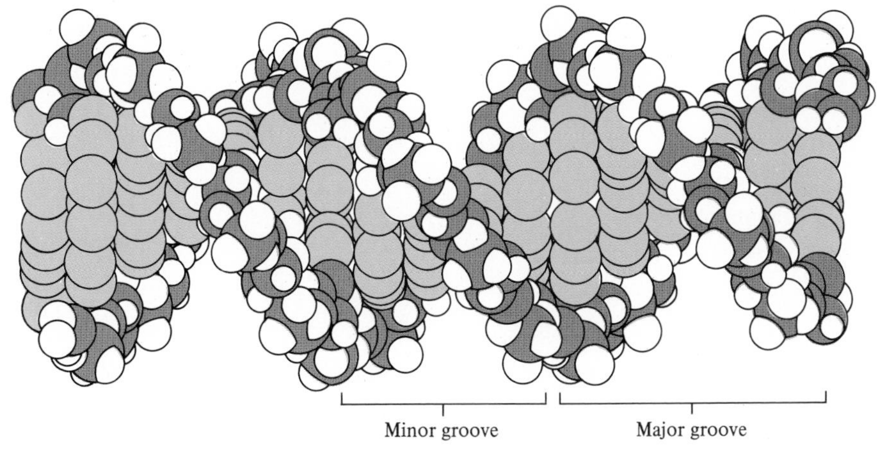


(c)

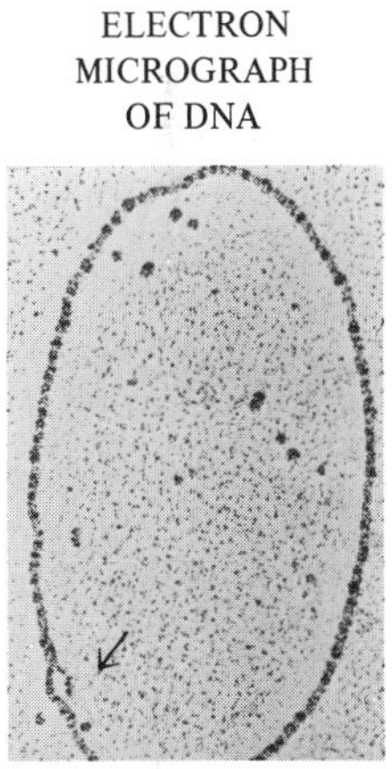


(d)

The structural organization of DNA. (a) Molecular structure of the sugar-phosphate backbone of a tetranucleotide. (b) Hydrogen bonding between the bases in a DNA molecule. (c) Space filling and ordinary three-dimensional models of a DNA double helix. (d) Electron micrograph of a DNA molecule of a plasmid with a replication site (arrow).

# PANEL 24-1 *(continued)*

These steps are catalyzed by single-strand binding proteins, gyrases, topoisomerases, or helicases, and nucleases. We have already seen how the enzyme DNA polymerase can add nucleotides only to an existing 3′ end of the polynucleotide. Since the lagging strand has no such starting point to add on nucleotides, DNA synthesis in this strand (synthesis of Okazaki fragments) requires another enzyme, **RNA primase,** which synthesizes a short RNA primer to which DNA polymerase adds deoxynucleotides. Finally, the RNA primer must be removed, the space must be filled with deoxynucleotides, and the neighboring Okazaki fragments must be bound, or ligated, to form a continuous DNA strand. The first two steps are probably carried out by another DNA polymerase, which has a repair function, and the last step by DNA ligase.

DNA molecules are very long. There is probably a single DNA molecule in each chromosome. Thus, human cells, each of which contains 23 pairs of chromosomes, actually have 46 long DNA molecules. If DNA replication were to begin at one end of the molecule and proceed to the other end, it would probably take a long time. Moreover, the tension produced by the unwinding of a double helix is rather high and would probably cause breaks within the DNA molecule. DNA replication, instead, initiates at many sites along the DNA molecule. Electron micrography showing the bubblelike replication sites demonstrates that this is the way it happens (see photo on p. 637).

DNA replication. Schematic model of DNA replication showing the role of DNA polymerase, DNA ligase, RNA primase, DNA helicase, and gyrase. Note the pattern of copying the leading and lagging strands.

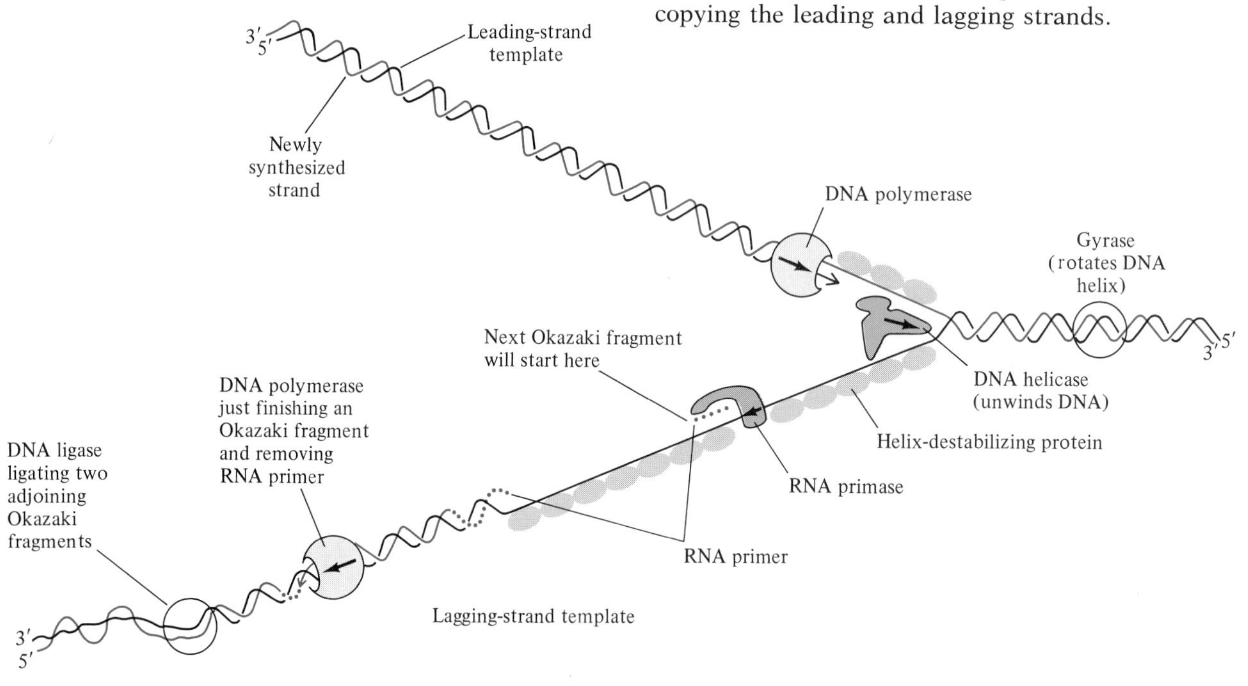


## THE GENE IN ACTION

While the structure of DNA molecules and their capability for self-replication were still being determined, investigators were attempting to learn the biochemical pathways by which genes achieve their various functions. These pathways have proved to be very complex, and many of their details have yet to be resolved.

## ONE GENE, ONE POLYPEPTIDE

Although Garrod had suggested in 1909 that individual steps in metabolic pathways were under the control of genes, it was not until 1941, with the work of George Beadle and Edward L. Tatum, then of Stanford University, that the full significance of Garrod's prophetic hypothesis was appreciated. Wild-type strains of the pink bread mold *Neurospora crassa* can usually grow on a very simple, or "minimal," culture medium containing only inorganic salts, a simple sugar as a carbon and energy source, and a single vitamin, biotin. The mold is conveniently grown on the surface of nutrient agar in test tubes. Beadle and Tatum reasoned that the mold must possess genes that specify all the enzymes needed to synthesize its many essential materials from the simple components of the minimal medium. They further reasoned that mutations in such genes would probably create "growth requirements" in the mutant, that is, mutants would be able to grow only in a medium that contained the essential metabolites they could no longer synthesize. With this as a rationale, they performed a series of experiments that won them a Nobel prize in 1958.

Beadle and Tatum irradiated spores of wild-type *Neurospora* with ultraviolet or X rays. The spores are haploid, as is the organism growing from a spore. Colonies resulting from germination of these spores were first grown on a "complete" medium containing amino acids, vitamins, purines, pyrimidines, and other nutrients. Each of these colonies was then tested to see whether it could also grow on a minimal medium or on a minimal medium supplemented with only one or another of the metabolites present in the complete medium (Fig. 24-6). Any strains that grew only in the presence of an added metabolite on the minimal medium were studied further. The strains that required a particular added nutrient were considered deficient in their ability to synthesize that specific substance. These mutants are called **auxotrophic** mutants or auxotrophs.

The metabolic pathways involving the nutrients required for growth by the mutant strains were analyzed. In several cases, it was shown that one mutation resulted in the loss of only one biosynthetic capacity. Since *Neurospora* is haploid, these mutations, which would be recessive in a diploid organism and would probably have been masked by the dominant wild-type allele, were always expressed in the cultures.

Soon after the publication of this work, similar cases were made in many other organisms for what came to be called the "one gene, one enzyme" concept. In time, many proteins, including many enzymes, were found to be composed of two or more different polypeptide chains. It was then shown that a single gene difference could produce a change in only one chain of such a protein, while a second mutation could account for a change in another of its polypeptide chains. As a result, the concept was modified to "one gene, one polypeptide chain."

The concept of a one-to-one correspondence between a particular gene and synthesis of a particular polypeptide received a clear boost in 1957 from the studies on sickle-cell anemia by Linus Pauling and Vernon Ingram of the California Institute of Technology. As noted in Chapter 23, sickle-cell anemia is caused by a recessive autosomal mutation. The mutation affects the oxygen-carrying capacity of hemoglobin. Each hemoglobin molecule contains one pair each of two different polypeptides, its $\alpha$ and $\beta$ globin chains. Amino acid analysis of the two polypeptides from the mutant-type hemoglobin clearly showed that only the $\beta$

**FIG. 24-6** Experimental design to study the relationship between genes and enzymes in the auxotrophic mutants of *Neurospora crassa*. Spores of the pink bread mold were irradiated to induce mutations. Individual haploid spores that were able to germinate and grow on a complete medium but not on a minimal medium were regarded as auxotrophs and thus deficient in their ability to synthesize one or more essential amino acids or vitamins. (a) Screening procedure to identify the major class of nutrients that the auxotroph cannot synthesize. (b) Experimental protocol to identify the loss of ability to synthesize a specific vitamin in the auxotrophs.

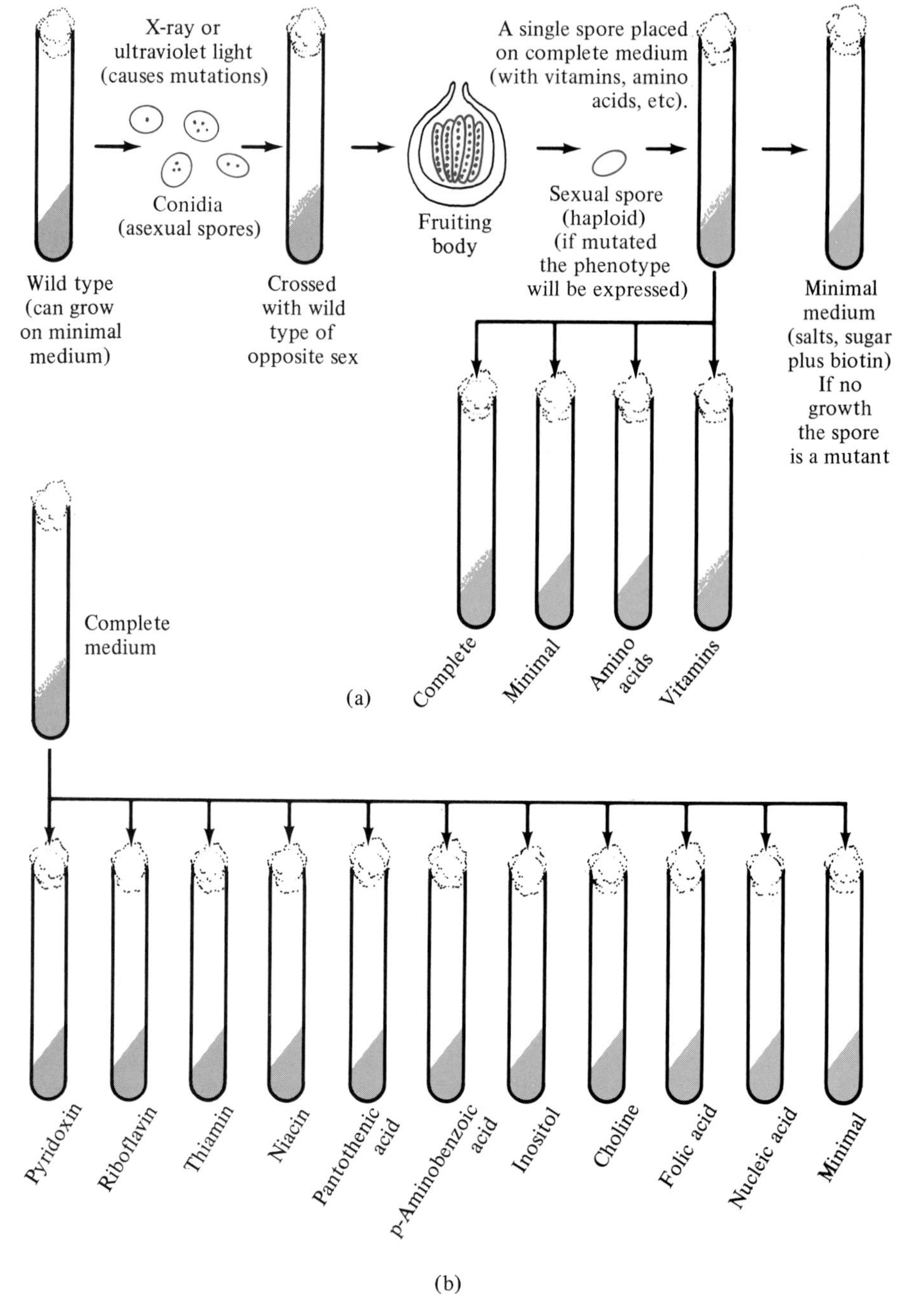


chain was affected by the mutation. The sixth amino acid from the amino terminal end of the polypeptide was changed in the mutants from a glutamic acid to valine. It was later shown that the mutation involved a change in a single nitrogenous base in the DNA molecule coding for this polypeptide.

## COLINEARITY OF THE NUCLEOTIDE AND AMINO ACID SEQUENCES

Another line of clear evidence for the relationship between the base sequence in DNA and the amino acid sequence in a polypeptide came from a determination of the relative positions of a mutation in a specific gene and of the position of the substituted amino acid in the polypeptide that the gene controls. In the early 1960s, two genes were studied in detail. One of them controls the synthesis of the enzyme tryptophan synthetase in *Escherichia coli.* Charles Yanofsky and his associates at Stanford University studied this enzyme and showed that the gene and its corresponding polypeptide chain are **colinear,** that is, the base sequence at a given distance from one end of the DNA molecule corresponds to the amino acid sequence in the polypeptide chain at a proportionate distance from the end of the polypeptide chain. This was, however, shown in an indirect manner because methods to determine the base sequence were not yet available. Although methods to determine the amino acid sequence were laborious and time consuming, the relative positions of an intragenic mutation and of the substituted amino acid in the polypeptide chain could be determined relatively easily by genetic analysis.

Yanofsky and his associates collected many mutants of the tryptophan synthetase gene in *E. coli.* From a recombination frequency, analogous to the crossing-over phenomenon discussed in the preceding chapter, they determined the relative position of the mutation in the gene. They also purified the mutant enzyme and prepared tryptic digests of it, that is, short polypeptide fragments produced by digestion of the protein by trypsin. From a simple chromatographic analysis of the peptides, a procedure called fingerprinting, they were able to determine the approximate region of the polypeptide chain in which the substitution occurred. If a mutation resulted in the substitution of one amino acid for another, only the peptide containing the substituent migrated to a different position in the chromatogram.

As it became increasingly evident that genes contained the information that specified the structure of polypeptides, attention became focused on the remaining questions:

1. How could DNA, which resided in the nucleus, specify the synthesis of proteins that were assembled in the cytoplasm?
2. How could the nucleotide sequence of DNA, involving only four bases, specify the structure of polypeptides that contained 20 different kinds of amino acids?

## TRANSCRIPTION AND mRNA

The question of how DNA, located in the nucleus, could control protein synthesis in the cytoplasm involved elucidation of the roles of the various types of RNA that had been found in both the nucleus and the cytoplasm. Like DNA, RNA is a polynucleotide (a series of joined mononucleotides), although RNA molecules are shorter than those of DNA. There are, however, three more important differences between DNA and RNA: (1) RNA has the sugar ribose in place of deoxyribose; (2) it has the base uracil in place of thymine; and (3) it is single-stranded, not a double helix.

Experimental studies showed that one of the two strands of a DNA molecule, when temporarily "unzippered" from its antiparallel strand, can act as a template on which a complementary strand of RNA is synthesized by a process known as **transcription** (Fig. 24-7a). Like DNA replication, transcription of an RNA molecule requires the presence of a special enzyme, RNA polymerase.

**FIG. 24-7** (a) Diagrammatic sketch of the transcription machinery. (b) Time course of distribution of newly synthesized RNA between the cytoplasm and nucleus.

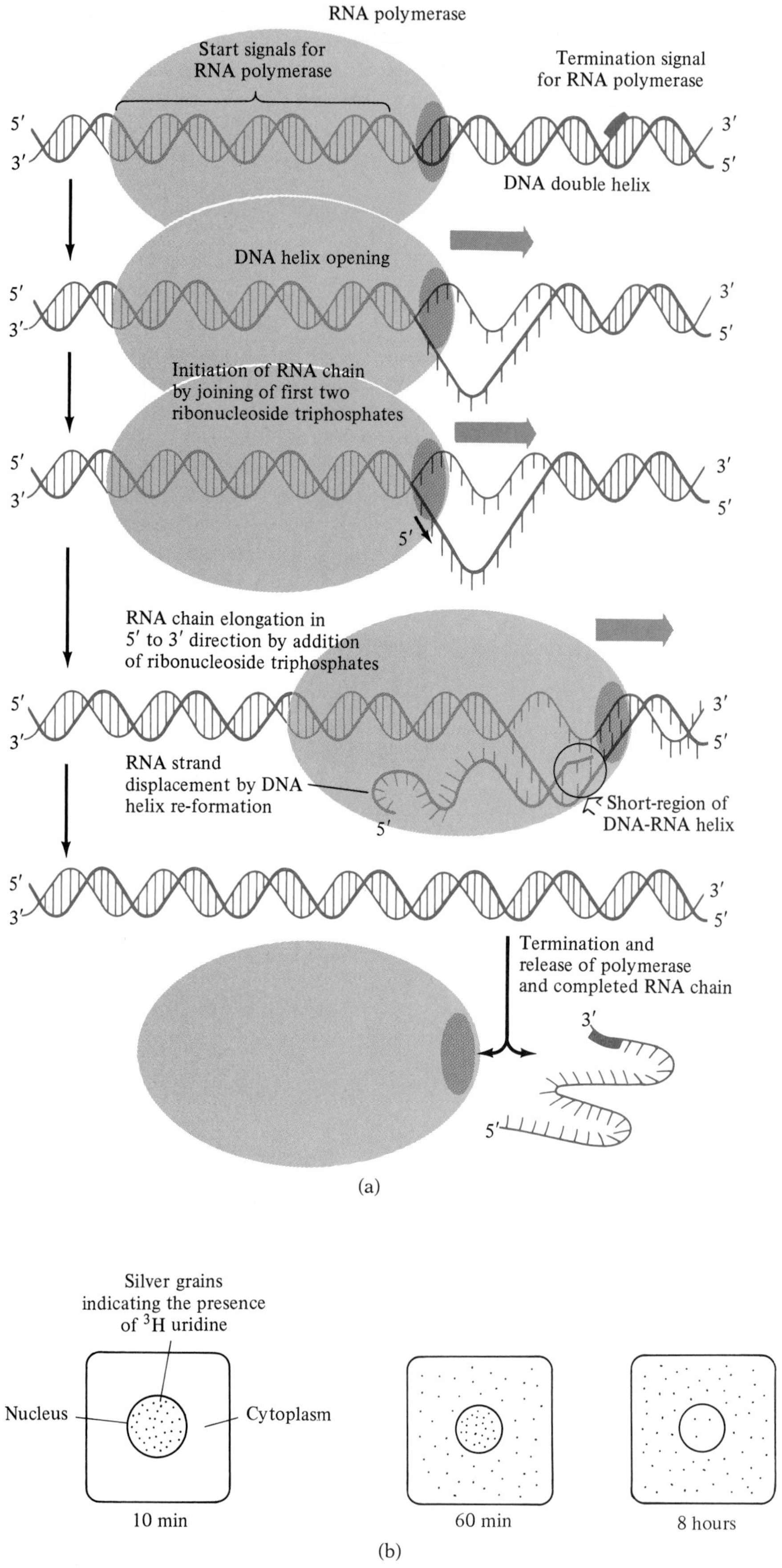

All organisms contain three classes of RNA: **ribosomal, messenger,** and **transfer.** All three types are single-stranded, although they may contain some double-stranded regions due to folding of the molecule. Ribosomal RNA (rRNA) derives its name from its location in the cell's ribosomes. Four subclasses of rRNAs—28S, 18S, 5.8S, and 5S—are present in all eukaryotes. S refers to sedimentation rate, and the notation is based on the approximate sedimentation rates during high-speed centrifugation of these molecules; it is an indication of their size. Although the sedimentation values of rRNAs vary among different groups of animals and plants, 18S and 28S rRNA are used as generic terms. In prokaryotes, all of which lack 5.8S rRNA, the major rRNAs have a sedimentation value of 16S and 23S, corresponding to the eukaryote 18S and 28S RNA, respectively. The rRNAs of eukaryotes are highly conserved, that is, the base sequence of rRNA from different groups of animals and plants is very similar, thereby suggesting a common ancestry for all eukaryotes and a common function for these molecules. The genes coding for 18S, 28S, and 5.8S rRNAs in eukaryotes are located in the nucleolus.

Molecules of messenger RNA (mRNA), unlike those of rRNA, are heterogeneous in size and in base composition. The base sequence in a given mRNA molecule precisely reflects the information encoded in the base sequence in the gene from which it was transcribed. Its size thus depends on the information content of that gene. If the protein coded for by a given gene is large, the mRNA transcribed from the gene is long as well.

Transfer RNA (tRNA) molecules are relatively small in size (3.8S) and have a cloverleaflike appearance because of the double-stranded sectors they contain (Fig. 24-8). These sectors are formed by the pairing

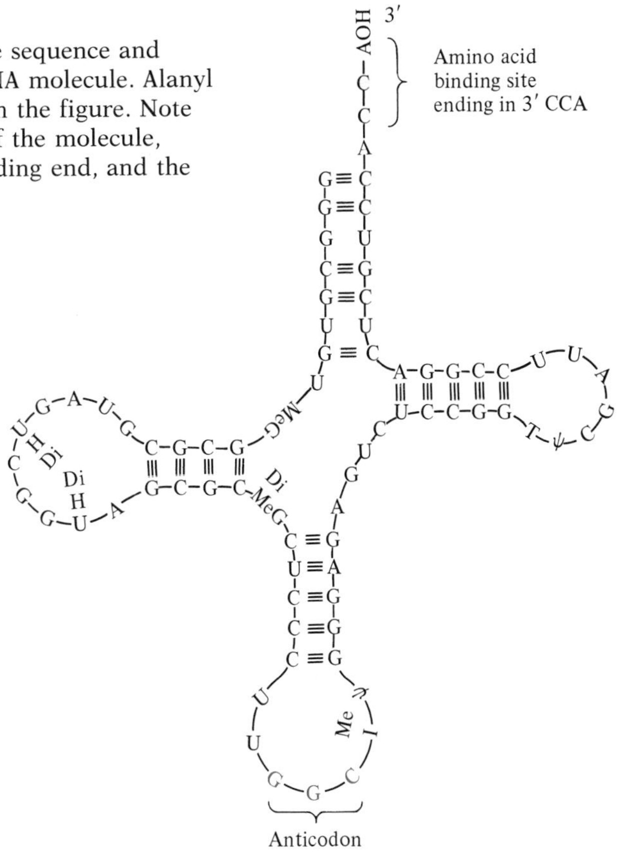


**FIG. 24-8** Nucleotide sequence and overall shape of a tRNA molecule. Alanyl tRNA is represented in the figure. Note the cloverleaf shape of the molecule, the 3′ amino acid binding end, and the anticodon.

of bases in closely adjacent sections of the molecules. There are 20 types of tRNA, each with a specific affinity for a different amino acid.

We shall now briefly describe the role of each of these RNAs in the synthesis of specific proteins. Newly transcribed molecules of mRNA diffuse into the cytoplasm, where, in association with ribosomes, they provide the detailed specifications for the synthesis of polypeptide chains. Molecules of tRNA "ferry" the appropriate amino acids to the ribosomes in the course of protein synthesis. The step in which the coded instructions contained in the base sequence of an mRNA molecule specify the synthesis of a polypeptide chain with a certain amino acid sequence is termed **translation.**

Evidence that the DNA of chromosomes is indeed transcribed into mRNA, which enters the cytoplasm where its message is translated by the ribosomes into specific polypeptides, was provided in part by experiments using radioisotope labeling and autoradiography. These are described below.

***Experimental Evidence for Transcription.*** When embryonic rabbit cells are incubated in a medium containing tritiated uridine* for 10 minutes, then examined by autoradiography, almost all labeled RNA is found in the nucleus (Fig. 24-7b). If incubated for 60 minutes in the same medium, the label appears in both the nucleus and the cytoplasm (Fig. 24-7b). When similar cells are incubated in a "chase" of nonradioactive medium for 8 hours following 1 hour of exposure to the tritiated uridine, then washed, killed, and examined, almost all of the radioactivity is located in the cytoplasm (Fig. 24-7b). Labeled RNA has moved from the nucleus to the cytoplasm.

When *E. coli* cells are infected with T2 phage, a large burst of RNA synthesis soon occurs in the bacterial cells. The newly synthesized RNA molecules, which, in the living cell, have a half-life of only a few minutes before being degraded, have nucleotide compositions and sequences reflective of the T2 phage DNA and unlike that of the host cell's DNA. The newly synthesized phage RNA is localized on the host cell's ribosomes and is engaged in synthesis of the phage proteins.

A simple and elegant proof that the base sequence in DNA is transcribed into the base sequence in mRNA comes from an application of **molecular hybridization** methods and electron microscopy (Figs. 24-9, 10).

* Uridine is the nucleoside containing uracil, the base found only in RNA and not in DNA. Thus, if incorporation of uridine occurs, labeling would be limited to RNA.

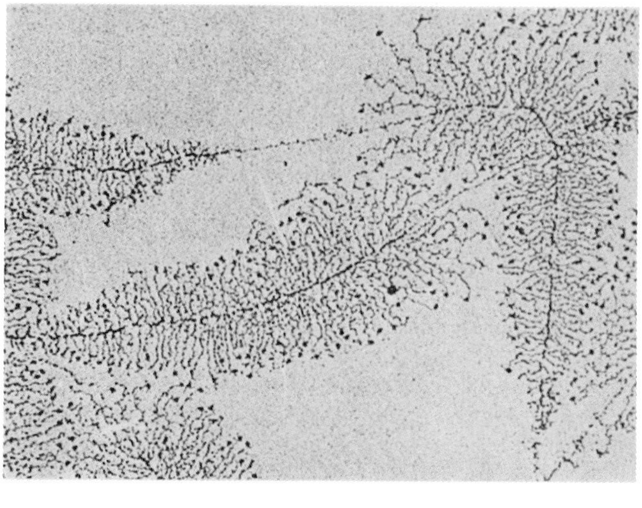

**FIG. 24-9** Electron micrographs of transcription. The long backbone is DNA and the lateral projections are the transcripts.

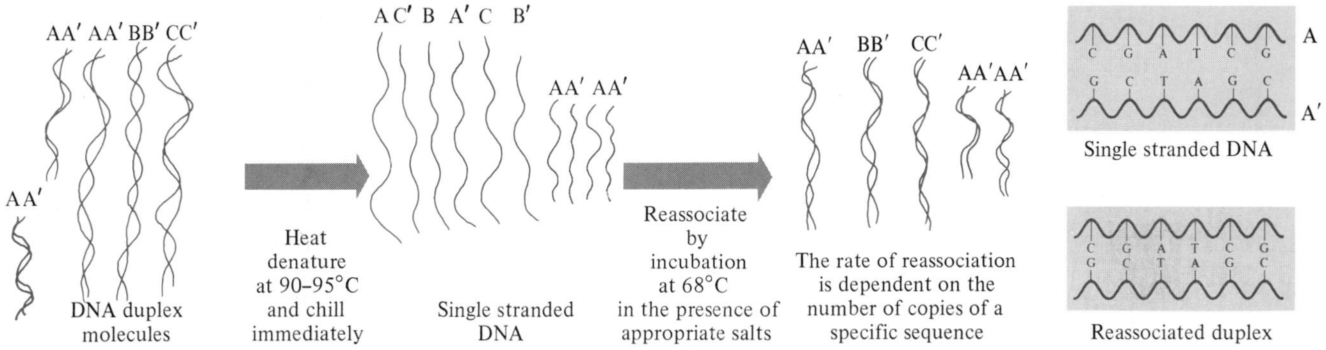


**FIG. 24-10** The principles of molecular hybridization.

The principles of molecular hybridization are simple. Complementary hydrogen bonding can form only between the bases adenine and thymine (or uracil) and between guanine and cytosine. This can happen whether the complementary bases are part of a DNA molecule or of an RNA molecule. The two strands of a DNA molecule can be separated experimentally by heating a DNA solution to its melting temperature, the temperature at which the hydrogen bonds break. Upon cooling, the complementary single-stranded DNA molecules reassociate to form normal double-stranded DNA. If, however, RNA complementary to one of the strands of DNA is present at the time that reannealing (reassociation) of the separated DNA strands would occur, hybrid duplex molecules containing one DNA strand and one RNA strand can form. The chances of formation of hybrid duplexes can be increased by increasing the amount of RNA present. Hybrid duplexes, which can be photographed by electron microscopy, possess triple-stranded regions in which one strand of the DNA molecule is separated from the DNA-RNA hybrid duplex. This hybrid molecule forms what is designated as an R loop. The photographic procedure, termed R loop mapping, has proved beyond all reasonable doubt two aspects of transcription: (1) the base sequence in RNA is complementary to the base sequence in DNA, and (2) only one of the two DNA strands is transcribed, as indicated by the fact that RNA is complementary to only one of the two DNA strands.

## TRANSLATION

Translation of the genetic code in an mRNA molecule into the primary sequence of a polypeptide is a complex process that occurs in the ribosomes and involves many kinds of macromolecules.

**FIG. 24-11** Diagrammatic sketch of a eukaryotic ribosome attached to an mRNA molecule.

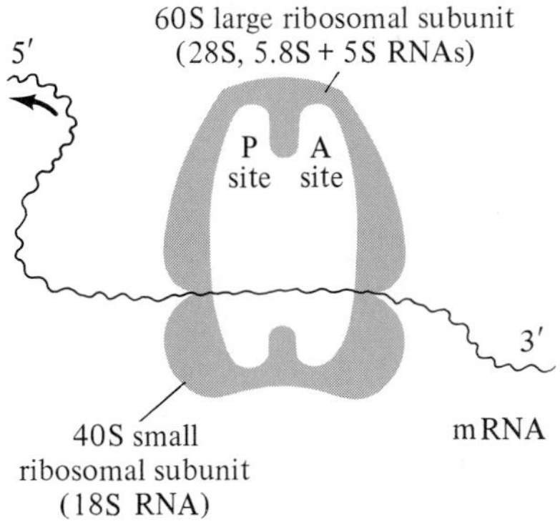


Each ribosome is composed of two subunits, one smaller than the other. For example, the two subunits of *E. coli* ribosomes are 30S and 50S (Fig. 24-11), and those of eukaryotes are 40S and 60S. Ribosomes are composed of about half protein and half RNA. The smaller subunit contains the 18S or 16S rRNA, and the larger subunit contains the other types of rRNA. In *E. coli,* the smaller ribosomal subunit has 19 proteins and the larger 30. The ribosome is nonspecific and can serve as the locale for synthesis of any protein, depending on the particular mRNA being supplied to it at the time.

Also needed for polypeptide synthesis are tRNA molecules (Fig. 24-8), each kind bound to its own particular kind of amino acid. When an amino acid is attached to its tRNA, the entire complex is called an **aminoacyl-tRNA.** For a tRNA to become bound to its amino acid requires the cooperation of an amino acid–activating enzyme, **aminoacyl**

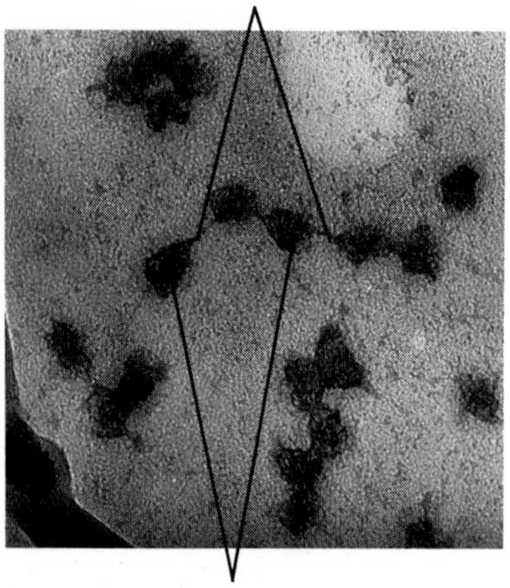


**FIG. 24-12** Electron micrograph of a polysome. Several ribosomes (spherical structures) simultaneously translate an mRNA molecule (threadlike structure).

**synthetase.** Each of these enzymes is also specific for one kind of tRNA molecule and for the one kind of amino acid to which the tRNA molecule becomes bound.

As Figure 24-8 indicates, the tRNA molecule has one triplet of bases capable of "recognizing" and binding to a complementary set of three bases in an mRNA molecule. A set of three bases in DNA or mRNA specifying an amino acid is called a **codon.** Such a triplet in a tRNA molecule is termed an **anticodon.** If the anticodon on a tRNA molecule were, for example, cytosine, guanine, and cytosine (CGC), it would be able to match with the complementary codon GCG on an mRNA molecule. On the arm of the tRNA molecule opposite its anticodon is the binding site of the amino acid specific for that particular type of tRNA. In our example, the tRNA with the anticodon CGC carries the amino acid alanine.

The large and small subunits of ribosomes are normally dissociated from each other, only associating to form complete ribosomes in the presence of mRNA. Normally, mRNAs are translated simultaneously by several ribosomes. Usually, ribosomes are spaced about 90 nucleotides apart along the RNA molecule, forming a "beaded necklace" called a **polysome** (Fig. 24-12). As the mRNA molecule is translated, a growing polypeptide chain issues from each ribosome; the chain on each successive ribosome toward the trailing end of the mRNA molecule is about 30 amino acids longer than the previous one (Fig. 24-13). In prokaryotes, an mRNA molecule may be undergoing translation at its 5′ end while its 3′ end is still being added to by transcription from the DNA molecule.

Each ribosome has two adjacent tRNA binding sites (Fig. 24-13), each with a different function. The **A site** (aminoacyl binding site) binds the tRNA carrying the *next* amino acid to be added to the growing polypeptide chain. The **P site** (peptidyl binding site) binds the tRNA to which the *last* amino acid that was added to the polypeptide chain is attached. As an mRNA molecule moves across the ribosome, it accepts the specific tRNA with a matching anticodon for the mRNA codon adjacent to the A site at that moment.

Initiation of synthesis of a polypeptide always begins with the amino acid methionine specified by the codon AUG. In prokaryotes, this molecule is actually formyl methionine. Each polypeptide chain in prokaryotes thus has a modified methionine as its first amino acid; in many cases it is later removed.

***Initiation of the Translation Process.*** The first event in translation is the association of the small subunit of a ribosome with the 5′ end of an mRNA molecule and with an initiator tRNA carrying a formyl methionine (Met) molecule, which binds to the first AUG codon at the 5′ end of the mRNA (Fig. 24-14a). For all this to occur requires the activity of three initiation factors and a guanosine triphosphate (GTP) molecule

**FIG. 24-13** A diagram of a polysome showing the arrangement of the A and P sites on the ribosome, mRNA, and the growing polypeptide chains.

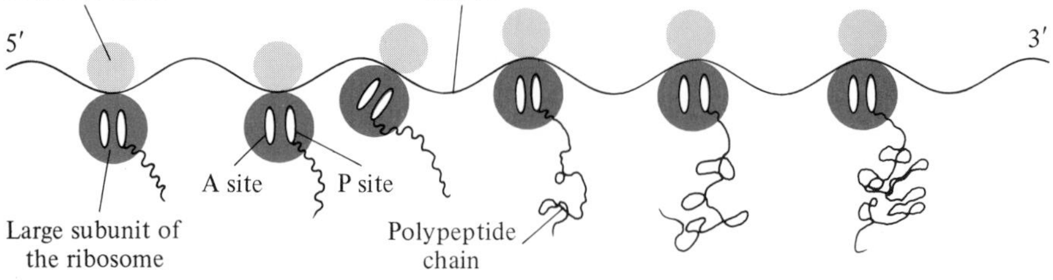

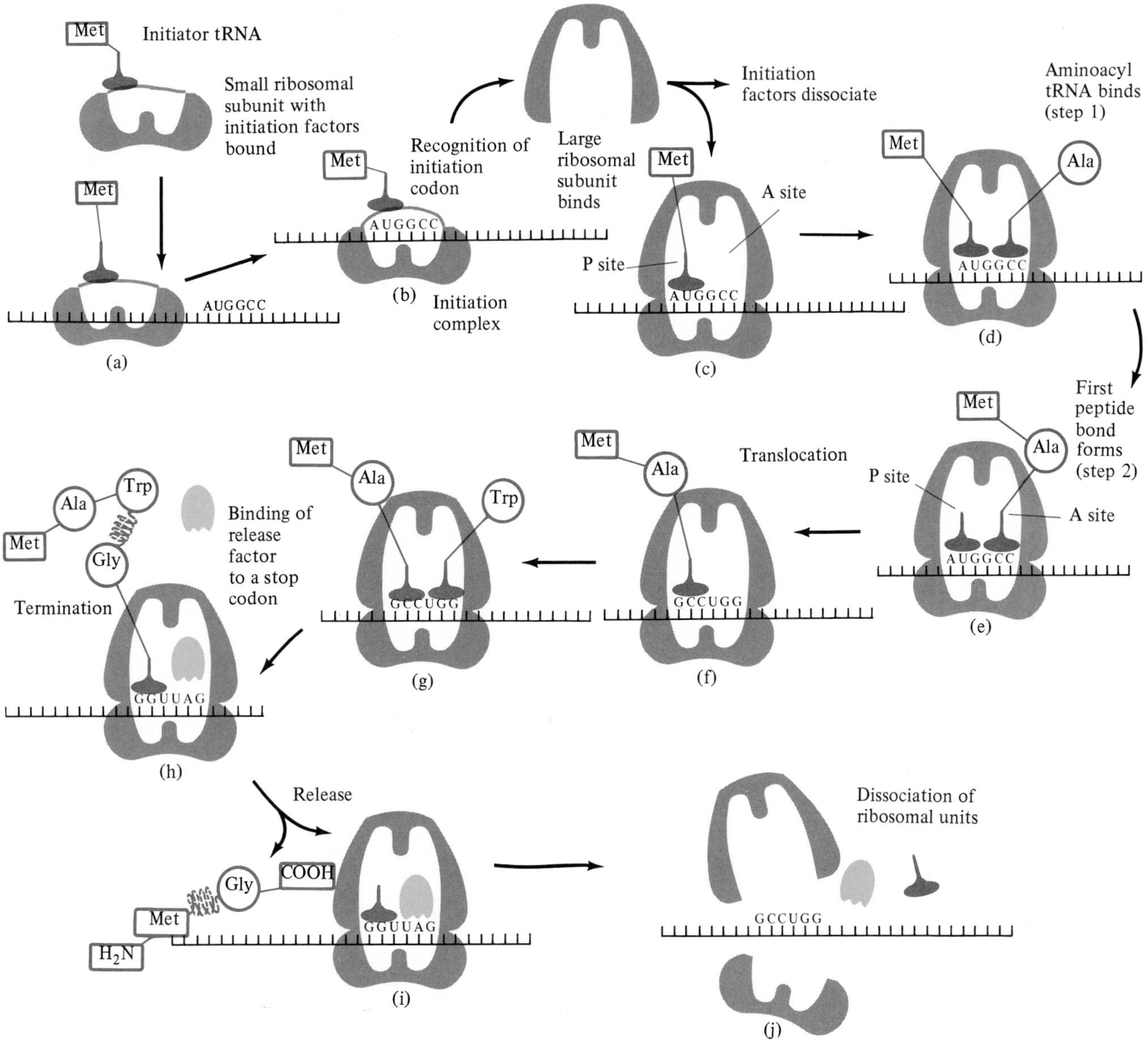


**FIG. 24-14** A diagrammatic presentation of the major steps in translation of an mRNA into the corresponding polypeptide. See the text for a description of the translation events.

(Fig. 24-14b). This complex then combines with a large ribosomal subunit. The initiator tRNA with its attached methionine becomes bound to the P binding site of the ribosome (Fig. 24-14c). The A binding site has now become aligned with the codon of the mRNA molecule next to AUG. The codons at the P and A sites indicated in Figure 24-14c are AUG and GCC, specifying tRNA molecules with the anticodons UAC and CGG. They bear methionine and alanine, respectively, as their amino acids.

The binding of each tRNA in the A site is an endergonic reaction, the energy being supplied each time by the hydrolysis of a molecule of GTP to GDP and inorganic phosphate. Additional protein factors are also needed. One is an enzyme that is a constituent of the large ribosomal subunit; this catalyzes the formation of a peptide bond between the two amino acids held by the tRNAs in the A and P sites (Fig. 24-14d). In the example illustrated, this produces the peptide bond between methionylalanine (or Met-Ala) dipeptide. At this same moment, the bond between methionine and its tRNA breaks, the dipeptide now attached only to the tRNA is still held in the A binding site (Fig. 24-14e).

The next events are a movement of the dipeptide and its attached tRNA from the A site to the P site and a movement of the ribosome across the mRNA strand a distance of three nucleotides (Fig. 24-14f). The mRNA codon previously in line with the A site is now opposite the P site, and, in the illustrated example, the next codon, UCC (for the amino acid serine) moves to align with the A site. For these movements, called translocation, another protein factor is needed, and energy for the work is again supplied by the hydrolysis of a molecule of GTP.

As soon as the serine codon of the mRNA is in place, a tRNA with an anticodon of AGG and bearing a molecule of serine binds to the A site, and the process of peptide bond formation is repeated (Fig. 24-14f–g). The entire sequence, called elongation, is repeated for each of the codons on the mRNA molecule until a termination codon comes opposite the A site. This leads to the entrance of a protein known as release factor RF-1 into the A binding site (Fig. 24-14h). When this occurs, the newly formed polypeptide is released from the tRNA in line with the P site (Fig. 24-14i). The two subunits of the ribosome thereupon dissociate and release mRNA and tRNAs from the ribosome (Fig. 24-14j). Two additional protein factors participate in the termination of polypeptide synthesis and dissociation of the ribosome. A polypeptide has been synthesized.

An interesting benefit of the studies that elucidated these steps is the rationale for antibiotic use in the control of bacterial infections. Antibiotics such as streptomycin, erythromycin, tetracycline, and chloromycin bind to bacterial ribosomes and prevent one of the steps in translation but do not affect translation in eukaryotic ribosomes. For example, in bacteria, streptomycin binds to the small ribosomal subunit and causes the ribosome to release or misread the message in the mRNA. Similarly, erythromycin binds to the large subunit and blocks the P site.

## THE GENETIC CODE

What needed to be explained was how a language with only four "letters"—nucleotides differing only in the bases, adenine (A), thymine (T) (in RNA the base thymine is replaced by uracil [U]), guanine (G), and cytosine (C)—could contain coded equivalents for "words" (polypeptides) in a language with 20 "letters" (amino acids). It was apparent that a one-to-one translation (base for amino acid) was out of the question. Moreover, a code word or codon, of two bases in length also fell short of requirements: with codons of two bases in length and only four kinds of bases to choose from, the number of combinations is only 16 ($4 \times 4$). George Gamow, the University of Colorado astrophysicist, suggested in 1955 that codons would prove to be a triplet of bases. This would give 64 combinations ($4 \times 4 \times 4$) to choose from, far more than would be needed.

***Deciphering the Genetic Code.*** The first major progress in deciphering the genetic code came almost simultaneously in 1961 in the laboratories of Marshall Nirenberg of the National Institutes of Health and Har Gobind Khorana of the University of Wisconsin, who shared a Nobel prize in 1968. In the early experiments, protein synthesis was achieved in a test tube containing a mixture including ribosomes, enzymes, aminoacyl-tRNAs, and certain soluble proteins found to be necessary for translation. The mixture contained no naturally occurring mRNA molecules. When an artificially synthesized RNA, polyuridylic acid (UUUUUUUUU and so on) was added to the mixture, polypeptides composed only of phenylalanine were synthesized.

In further experiments, synthetic RNA molecules containing random sequences of known nucleotides in equal amounts, such as 50 percent U and 50 percent G, were used. These particular RNA molecules would thus contain 12.5 percent of each of the following eight codons: UUU, UUG, UGU, GUU, UGG, GUG, GGU, and GGG. In this case, the resulting polypeptides were found to contain the amino acids leucine, cysteine, valine, tryptophan, and glycine, in addition to phenylalanine. To gain insight into which of these codons specified which of the amino acids, RNA molecules with random sequences of bases were again prepared, but with differing proportions of each base, for example, 75 percent G and 25 percent U, or vice-versa. As expected, this affected the frequency with which each of the amino acids was synthesized.

Further insight into the nature of the genetic code was gained by using synthetic messengers with specific sequences. Khorana used a synthetic polymer of the sequence UCUCUCUC, that is, one containing alternating codons of UCU and CUC. When this polymer was added to the cell-free translation system, a polypeptide containing only serine and leucine was produced. The repeating trinucleotide UCUUCUUCU produced three different polypeptides: polyserine (UCU), polyleucine (CUU), and polyphenylalanine (UUC).

It was clear from these studies that a messenger RNA molecule could bind to the ribosome at any site on the strand (that is, at any codon) and that once it was attached, it read in groups of three nucleotides without any punctuation. Every nucleotide formed a part of a code word. Furthermore, the codons were nonoverlapping, that is, each nucleotide was part of only one amino acid code word. These observations of Nirenberg and Khorana and their associates established clearly that the genetic code is a triplet, commaless (that is, all nucleotides are parts of codons), nonoverlapping code. Although the genetic code is commaless, the gene itself may be in bits and pieces, interrupted by DNA that does not code for a protein. For example, the gene that codes for the protein ovalbumin, which is found in egg white, is interrupted seven times. This unexpected, and fascinating, feature of genes is described in Panel 24-2. But the interruptions are not to be mistaken for commas in the genetic code. There are no commas in the coding region.

All these early studies were made with synthetic messengers and in the presence of relatively high concentrations of magnesium in a cell-free, protein-synthesizing system, leaving some doubt whether they truly represented what occurred in living systems. But subsequent studies on the base sequence of mutant genes and the altered amino acid sequence of the proteins that resulted from them confirmed all these conclusions. Studies on mutant hemoglobin described earlier were especially important in this regard. Similar studies on mutants of the bacterium *E. coli* and other organisms also proved the universality of the genetic code.

In later studies, molecules were prepared, each composed of only three known nucleotides in known sequences. When any of these were bound to ribosomes and then added to a mixture of tRNAs, it was found that each type of "mini-mRNA" bound only a specific kind of aminoacyl-tRNA to the ribosomes. Specific codons for termination of polypeptide synthesis were also identified. This technique allowed the testing of every one of the 64 possible codons. The complete genetic code that emerged is shown in Table 24-1.

PANEL 24-2

# Split Genes and RNA Processing

Discovery of colinearity between nucleotide and amino acid sequences and of the universal commaless, nonoverlapping genetic code at first gave the impression that each gene was an uninterrupted sequence of nucleotides in the DNA that codes for the polypeptide. It was also known that each structural gene (one coding for a particular polypeptide) may contain some regulatory nucleotide sequences at either of its ends. But there was a great surprise awaiting. In 1977, in the course of a study of the nucleotide sequence of the globin gene and of globin mRNA, it was observed that the gene contains two other stretches of nucleotide sequences not represented in the mRNA transcribed from it. The two stretches of noncoding sequences became known as the intervening sequences, or **introns.** They divide the coding region into units termed **exons.**

## IDENTIFICATION OF EXONS AND INTRONS

The procedure involved in the identification of introns was relatively simple. The base sequence in globin mRNA was used as a template to synthesize what is called **complementary DNA,** or **copy DNA** (cDNA), using an RNA-dependent DNA polymerase called **reverse transcriptase,** an enzyme that catalyzes synthesis of DNA on RNA templates. Then the cDNA was ligated to DNA from a plasmid an extrachromosomal circular DNA present in bacteria, designated pBR322, and was cloned in *E. coli* (the cloning procedure is described later in this chapter). The cloned cDNA represented all the nucleotides present in the mature globin mRNA. A bacterial cell containing this plasmid was allowed to multiply. After amplification, that is, the making of many copies, the cDNA-containing plasmid was isolated from the bacterial cells. The cDNA was separated from the plasmid DNA and hybridized to DNA from mouse cells (genomic DNA of mice). The globin cDNA reassociated only with that part of the genomic DNA that has a nucleotide sequence complementary to the cDNA. Hence, only the globin gene DNA formed a hybrid molecule with the globin cDNA.

Electron microscopic examination of the hybrid molecule revealed a surprising feature of the globin gene DNA. It contained some sequences that were not present in globin cDNA. These unmatched sequences formed single stranded loops, while the rest of the molecule formed a duplex consisting of one cDNA strand and one genomic DNA strand. The loops were the introns, or intervening sequences, that were not represented in the mature globin mRNA. Later studies have shown that introns are present in many eukaryotic genes. A description of the experiment is shown in the figure.

## RNA PROCESSING

Existence of fragmented genes would seem to create problems for the cell in synthesizing mRNA that is complementary to the exons but not the introns. Theoretically, two possible solutions exist. RNA polymerase may skip introns and selectively transcribe only the exons when forming the mRNA. Alternatively, a large precursor molecule of RNA, representing both the exons and introns, may be transcribed, after which regions complementary to the exons would be spliced together end-to-end and the intron complementary portions "edited out." Experiments showed that the initial RNA transcript in eukaryotes is indeed much larger than the final RNA product. This precursor RNA, called **heterogeneous nuclear RNA** (hnRNA), is found only in the nucleus, and only its processed products are transported to the cytoplasm.

The hnRNA is also usually much more complex than the RNA found in the cytoplasm because it represents a transcription product of the DNA adjacent to the gene as well. Direct proof of the precursor-product relationship between hnRNA and mRNA was adduced from a molecular hybridization analysis of these two types of molecules (see the figure). Nuclear RNA from reticulocytes, precursors of red blood cells, and globin cDNA were used in these studies.

The hnRNA molecule that has sequences complementary to globin cDNA is about twice the length of the globin mRNA molecule. A heteroduplex formed by annealing globin cDNA and globin hnRNA contains unmatched single-stranded regions in the hnRNA similar to the loops in the globin gene-globin cDNA duplex.

The steps involved in the processing of hnRNA into the final cytoplasmic RNA are collectively termed RNA processing. Many details of this process have been elucidated. Besides the globin gene, many other genes in eukaryotes have in-

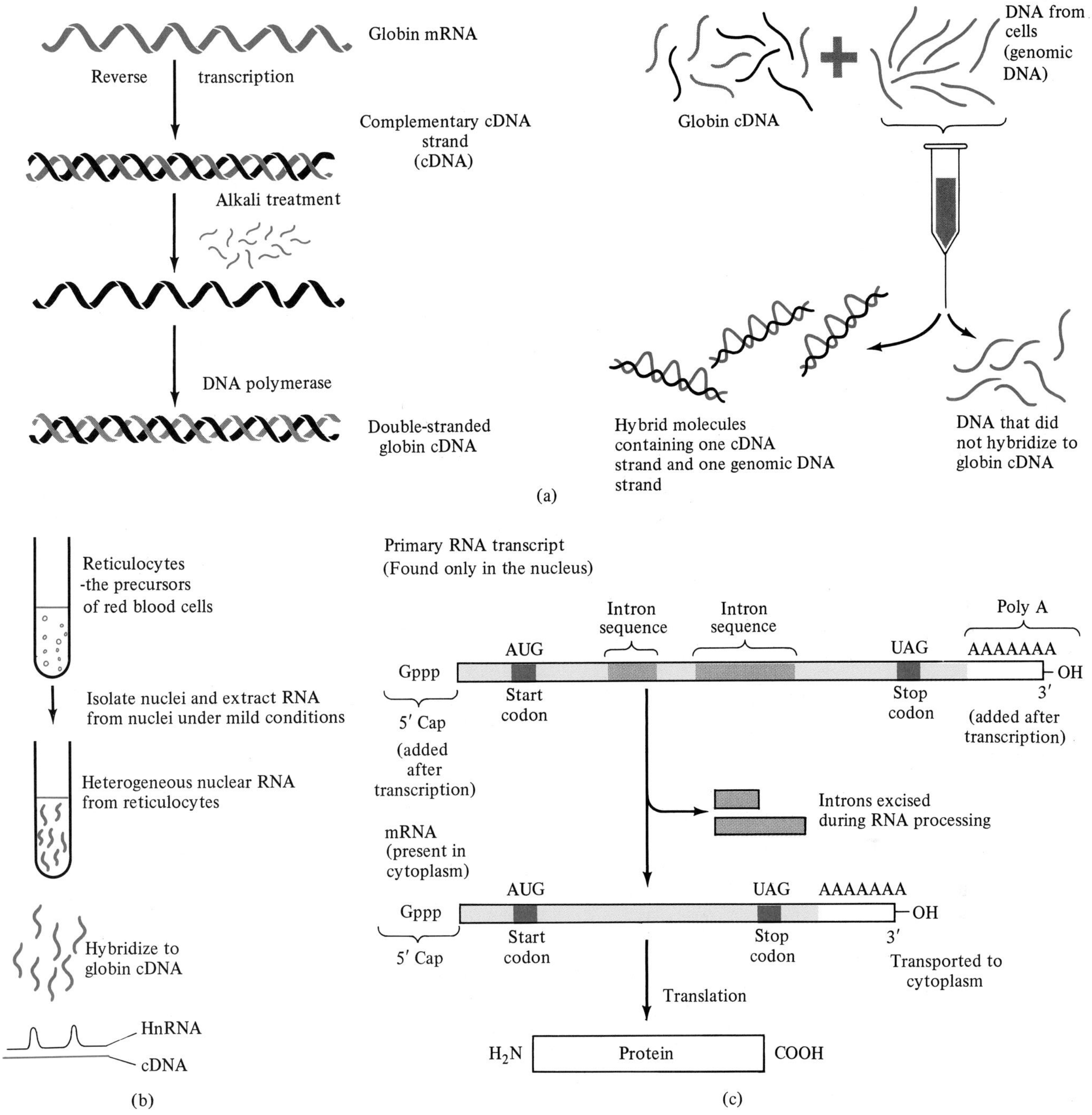


Experimental protocol used to identify introns in a gene RNA processing. (a) Steps in the preparation of globin cDNA and its hybridization to globin gene and the relationship between hnRNA and mRNA. (b) The experimental procedures to isolate and identify globin hnRNA. (c) Relationship between globin hnRNA, globin mRNA, and globin.

trons. All eukaryotes produce hnRNA, which is then processed. Genes in prokaryotes, on the other hand, do not contain introns. Their transcripts do not require processing. The significance of this difference between these two major groups of organisms is not clear.

**TABLE 24-1 Sequence of Nucleotides in the Genetic Code**

| *First position (5′ end)* | *Second position* | | | | | | | | *Third position (3′ end)* |
|---|---|---|---|---|---|---|---|---|---|
| | U | | C | | A | | G | | |
| U | UUU | Phe | UCU | Ser | UAU | Tyr | UGU | Cys | U |
| | UUC | Phe | UCC | Ser | UAC | Tyr | UGC | Cys | C |
| | UUA | Leu | UCA | Ser | UAA | Stop | UGA | Stop | A |
| | UUG | Leu | UCG | Ser | UAG | Stop | UGG | Trp | G |
| C | CUU | Leu | CCU | Pro | CAU | His | CGU | Arg | U |
| | CUC | Leu | CCC | Pro | CAC | His | CGC | Arg | C |
| | CUA | Leu | CCA | Pro | CAA | Gln | CGA | Arg | A |
| | CUG | Leu | CCG | Pro | CAG | Gln | CGG | Arg | G |
| A | AUU | Ile | ACU | Thr | AAU | Asn | AGU | Ser | U |
| | AUC | Ile | ACC | Thr | AAC | Asn | AGC | Ser | C |
| | AUA | Ile | ACA | Thr | AAA | Lys | AGA | Arg | A |
| | AUG | Met | ACG | Thr | AAG | Lys | AGG | Arg | G |
| G | GUU | Val | GCU | Ala | GAU | Asp | GGU | Gly | U |
| | GUC | Val | GCC | Ala | GAC | Asp | GGC | Gly | C |
| | GUA | Val | GCA | Ala | GAA | Glu | GGA | Gly | A |
| | GUG | Val | GCG | Ala | GAG | Glu | GGG | Gly | G |

Except for methionine and tryptophan, all the amino acids are specified by more than one codon. Three amino acids are each specified by six codons, one by three codons, and the other 14 by either two or four codons. Such codes are said to be degenerate, or redundant. Multiple codons were also found for initiation (AUG and GUC) and termination (UAA, UAG, UGA) of synthesis of a polypeptide. However, the pattern of the degeneracy appears to have been a product of natural selection for the following reason: most multiple codons for a particular amino acid differ only in their third base (the one on the 3′ side of the codon). Thus, many "accidental" base substitutions at the third position that might have resulted in a mutation do not because the amino acid specified is the same one. Because this arrangement provides genetic stability, it was probably selected for early in evolution. A second feature of the genetic code is that amino acids with closely similar properties are specified by codons differing in only one base. Thus, even if one of these amino acids is substituted for another in the group, the effect upon the polypeptide may be small or even undetectable.

## MUTATIONS AT THE MOLECULAR LEVEL

The elucidation of the genetic code permitted an understanding of the molecular basis of point mutations, those involving changes in a few base pairs. A mutation in a codon that results in a replacement of one amino acid by another is called a **missense mutation,** which is the most common type of point mutation. Missense codons produce a wide range of possible effects. For example, the mutant polypeptide that results may behave in any of the following ways:

1. It may function, but not as well as the wild-type polypeptide.
2. It may not function at all.
3. It may function only under special conditions.
4. It may function better than the wild-type (a rare occurrence).
5. It may perform a different or an additional function (an extremely rare occurrence).

One form of alternative 2 involves a mutation that produces one of the termination codons. When this occurs somewhere within the coding portion of a gene, the resulting mRNA will specify a shorter polypeptide; such shortened polypeptides are unlikely to be functional. Mutations that produce termination codons are called **nonsense mutations.**

***Frameshift Mutations: The Loss or Addition of One or More Nucleotides Within the DNA Molecule.*** Among the most drastic of mutations are **deletion mutations,** in which only one or two nucleotides are lost. This results in a frameshift mutation, altering the "reading frame" for the entire gene. From that point on, the mRNA molecule transcribed from the gene will have altered codons to the end of the molecule. For example, if one letter, such as G, is dropped from the sentence THE BIG PET CAT ATE THE FAT RAT and the rest of the letters are moved so that the "sentence" is still composed of three-letter "words," it would make no sense: THE BIP ETC ATA TET HEF ATR AT. The severity of the effect upon the polypeptide depends on how far from the end of the gene the deletion occurs. A similar effect occurs in an **insertion mutation,** when one or two nucleotides are inserted within a gene.

***Mutagens.*** Some chemical mutagens (mutagenic agents) such as bromodeoxyuridine (Brdu) are molecules that are analogs of (are similar in structure to) nucleoside bases and achieve their effect by becoming substituted for a nucleotide. When this happens, during the next round of replication, it may lead to a deletion or an insertion mutation. Still another class of chemical mutagens are alkylating agents, such as poison gas (mustard gas), that replace an active hydrogen atom in a compound with methyl or a similar alkyl radical. Such changes in a nucleotide alter the properties of the base, resulting in miscopying during replication and misreading during transcription—resulting in a mutation. Nitrous acid, formed from nitrites added to red meats and hot dogs to retain color of the meats, and nitrosamines formed during cooking such meats, convert amino groups of purines and pyrimidines into hydroxyl groups. Cytosine without an amino group on the sixth carbon resembles a thymine very closely. The result can be a point mutation.

In addition to chemical mutagens causing point mutations, there is another category of mutagens in the form of radiations, such as ultraviolet light and x-rays. The energy inherent in these radiations is absorbed by atoms on the nucleic acids, causing these atoms to lose electrons. As a result, the molecule becomes chemically active, which in turn leads, in some cases, to breaks the sugar-phosphate chain of nucleic acids. Such a mutagen is called a **clastogen.** Cells have the ability to repair such breaks. But errors often occur during repair synthesis of DNA, resulting in the loss or addition of a base.

In any case, these mutagens can cause changes in the gene's nucleotide sequence and produce point mutations. If the mutation occurs in a germline cell, for example, the mutation may be transmitted to the next generation via the gametes. However, if the mutation occurs in a somatic cell, designated as a **somatic mutation,** only the progeny of the cell in the individual is affected. The severity of the effects of a somatic mutation depends on the function of the gene in which it occurs. Many remain unnoticed. But mutations in genes that regulate cell growth or cell division may lead to the formation of cancers. We will describe the role of genes in carcinogenesis later in the chapter.

A sizable portion of the DNA in each cell has apparently no known function. Mutations in this DNA that does not code for any proteins (see Panel 24-3) cannot be recognized. Thus many mutations may be occurring as a result of radiations and environmental chemicals not all of which are recognized.

Lastly, in the late 1970s, a completely new class of mutagens has been identified. These are pieces of DNA that move from one location on a chromosome to another, in the process inactivating genes in the vicinity of their insertion. These mobile genetic elements are known as insertional elements. They are a type of DNA that does not seem to have a specific coding function and are described in Panel 24-3.

## THE REGULATION OF GENE EXPRESSION

Despite the fact that most of the cells of a multicellular organism contain the same sets of chromosomes, and thus the same sets of genes, the phenotype of different types of cells, tissues, and organs can be strik-

### PANEL 24-3

# DNA That Does Not Code for Proteins

The amount of DNA present in each cell is constant within a species. Furthermore, with a few exceptions, the amount of DNA per cell in a species is proportional to the complexity of the organism. Thus, the amount of DNA per cell in a bacterium is one-tenth that in the simplest eukaryotes, such as yeast. Algal cells have a little more DNA than yeast, the protozoans and sponges still more. Mammals have more DNA per cell than most fish, frogs, or reptiles. At first it seemed as though complexity of organization was proportional to the information content in the cell.

However, exceptions to this rule became apparent when more organisms were examined. Salamanders bear 25 times more DNA per cell than do humans. Yet salamanders are no more complex than humans. It gradually became clear that the sole function of each DNA molecule could not be to code for a different protein. If this were the case, the human genome (the total amount of genetic material in a cell), consisting of $3 \times 10^9$ base pairs, could code for 3 million different proteins, each with about 300 amino acids. This number appeared to be too large to have a meaningful function in an organism. In other words, this evidence alone indicated that not all DNA in a chromosome codes for different proteins. One possibility was that other functions may be carried out by some of the DNA. Alternatively, some genes may be represented by many copies.

In the late 1960s, the molecular hybridization methods described earlier in this chapter were applied to determine the copy frequency of genes. When the two strands of a DNA duplex in a solution were separated, or denatured, by warming the mixture, the rate at which the complementary strands reassociated upon cooling to form duplex molecules was proportional to the number of copies of the gene or particular DNA sequence present in the genome. If a particular DNA sequence were present in multiple copies, the chances of the two complementary strands finding each other was greater than if the sequence was represented only once per haploid genome. This assumption ingly different. Comparing bone, liver, skin, nerve, and blood cells of a vertebrate illustrates the point. Yet all of these cells and tissues orginate from a single cell, the zygote. Why are the cells different if all contain the same sets of directions?

Another aspect of this paradox is evident during development. A 16-cell stage of a vertebrate, for example, has all the genes to specify the synthesis of macromolecules characteristic of bone, liver, skin, neurons, and leukocytes. Yet these cell-specific products are not produced at the 16-cell stage. Instead, each of the enzymes and other proteins that function in the embryo are appropriate to the embryonic stages; many of these disappear as the individual matures. Even more striking are the differences in genic expression represented by a caterpillar and the butterfly into which it eventually metamorphoses. Clearly, some kind of control or regulation of gene expression is at work in such instances. Merely as a matter of efficiency, such a regulation makes sense. It would be wasteful for each cell to synthesize the products of all the genes present in it. It is economical and efficient for each cell to turn on only those genes necessary at particular times to express the phenotype of the specific cell.

Regulation of gene expression can occur in eukaryotes at several levels: transcription, mRNA processing (some mRNA molecules need to be modified before they are functional), mRNA stability, translation, and the activation and inhibition of enzymes. Most of the regulation of gene expression in bacteria, however, is achieved by control of mRNA transcription.

---

was based on the random chance meeting of two complementary strands as the rate-limiting step in reassociation of the two strands.

The rates of reassociation of DNA from different organisms showed characteristic differences. For example, in bacteria, the rate of DNA reassociation indicated the presence of only one copy of each different gene. In eukaryotes, on the other hand, three classes of DNA were distinguishable, based on the rate of reassociation. A rapidly reassociating class, which constituted about 10 to 15 percent of the total DNA, was designated as the highly repetitive fraction because, in this fraction, several thousand copies of each DNA sequence were represented. A moderately repetitive class, with an intermediate rate of reassociation, contained genes whose copy number ranged in the hundreds. About 20 to 25 percent of the genome belonged to this class. The bulk of DNA (60 to 70 percent) was slowly reassociating. This was the nonrepetitive, or unique sequence fraction and contained most of the genes that code for the various polypeptides the organism could synthesize. These sequences were usually represented only once per haploid genome.

It was found that not even all the DNA in the unique fraction coded for proteins, because it also included the DNA that regulates gene expression. The moderately repetitive class included those sequences coding for ribosomal RNAs, tRNAs, histones (the proteins associated with DNA in eukaryotic chromosomes) and a few other proteins that are needed in large quantities in all cells. The significance of the highly repetitive DNA is obscure. Some of this DNA is located physically in the centromere region of a chromosome. Most of the highly repetitive DNA appears to be nonfunctional in the genetic sense. In a normal cell, this DNA appears as the intensely staining clumps of chromatin called heterochromatin.

Recent studies have revealed that some of the repetitive sequences are mobile, that is, they may not occupy the same position in a chromosome in all individuals of a species. These **mobile genetic elements** affect gene expression. When they insert in or near some genes, the genes stop functioning, behaving as though they had mutated.

The functions of the various types of repetitive DNA are unknown. Some scientists have suggested that much of it is parasitic "selfish DNA" whose sole function is to maintain itself. So far, this is only a hypothesis. Until we find a more rational biological function for this DNA, this explanation is as good as any.

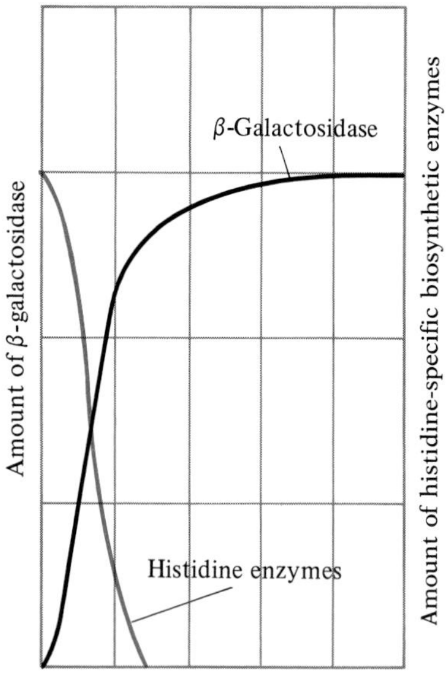


**FIG. 24-15** Concentrations of the inducible enzyme β galactosidase and repressible enzymes for synthesis of histidine in *E. coli*. The inducer is lactose and the end product is histidine.

***Induction and Repression of Gene Expression in Prokaryotes.*** *E. coli* can utilize any one of several carbohydrates as an energy and carbon source, for example, glucose, sucrose, galactose, lactose, and arabinose. Whichever of these sugars is present is metabolized, any one of them being capable of supporting growth. If lactose is present, it is hydrolyzed by the enzyme, β-galactosidase, and the resulting galactose and glucose are further metabolized to release energy. However, for the bacterium to live on lactose, three special enzymes are needed: β-galactosidase, which hydrolyzes lactose to yield glucose and galactose; galactoside permease, which facilitates absorption of lactose into the cell; and acetyl transferase, which transfers an acetyl CoA to lactose. Synthesizing these three enzymes involves expenditure of both energy and materials. In the absence of lactose, none of these enzymes are produced in any substantial amount. Yet, if *E. coli* cells being grown on a medium containing glucose as the only carbohydrate are abruptly transferred to one containing lactose but no glucose, they immediately begin synthesizing the three enzymes, one of which is β-galactosidase (Fig. 24-15). In this situation, lactose is termed an **inducer,** and the three enzymes are called **inducible enzymes.** It has been shown that most bacterial enzymes needed for catabolism of molecules are inducible enzymes. Induction occurs at the level of the gene; the inducer increases the rate of transcription of the mRNAs that code for these enzymes.

A second type of control is **repression,** in which the normal transcription of a particular gene is inhibited by a particular substance. If repression ceases, the gene is said to be **derepressed.** Repression and derepression are typical of many enzymes of anabolism. A classic example of such enzymes are those needed for the synthesis of the amino acid histidine (Fig. 24-15). When histidine is present in the immediate environment of a bacterium, the nine genes that code for the enzymes needed to synthesize it are repressed. If the supply of histidine is depleted, these genes become derepressed and the genes are transcribed into mRNAs. The corresponding enzymes are synthesized again, leading to the biosynthesis of histidine by the cell.

***The Operon.*** The operon is a detailed model of the organization of related groups of genes, devised to explain the molecular basis of induction and repression of gene expression in bacteria. It was first proposed in 1961 by the French scientists François Jacob, Jacques Monod, and Jean-Pierre Changeaux to explain the regulation of the ***lac* operon,** the complex of genes essential for lactose utilization by *E. coli*. An operon consists of the first three of the following elements, shown in Figure 24-16a:

1. *One or more structural genes:* the genes that actually code for the polypeptides regarded as the gene's products. All the structural genes are usually transcribed into a composite molecule called the polycistronic messenger, which contains the mRNAs of all the structural genes in the operon. (In the case of the *lac* operon, there are three structural genes, coding for β-galactosidase, galactoside permease, and acetyl transferase. The genes are labeled *z, y,* and *a,* respectively.)
2. *The operator:* the element immediately adjacent to the structural gene or genes, and the binding site of a regulatory molecule that determines the functional state of the structural gene or genes (this molecule is described in the following section).
3. *The promoter:* the region of DNA immediately adjacent to the op-

erator. RNA polymerase must bind to the promoter if transcription is to occur. The promoter and operator may overlap.

4. *The regulator gene:* also called the repressor gene in the *lac* operon, located somewhere near the operon on the DNA molecule (immediately adjacent to the promoter in the case of the *lac* operon). A promoter site where RNA polymerase binds to begin transcription, as may be expected of any gene that is transcribed, is also present next to the regulator gene. The regulator gene codes for a regulatory protein called the **repressor,** which, by binding to the operator, prevents transcription from occurring. The regulator gene is transcribed into mRNA, which is then translated to produce the repressor molecules.

***Positive and Negative Control of Gene Expression.*** The functioning of an operon depends on the presence or absence of specific effector molecules, such as lactose or histidine, in a bacterium's immediate environment (the culture medium in a laboratory situation). If the operon is inducible, the effector molecule, for example, lactose, is called an inducer. In the case of repressible operons, such as the *his* (histidine) operon in *E. coli* the effector molecule, histidine in this case, is called a corepressor. But in both cases, the operon is functional when the repressor is not bound to the operator. The genes are not transcribed when the repressor is bound to the operator. Such an operon is said to be under negative control. In inducible operons such as the *lac* operon, lactose binds to the repressor (Fig. 24-16b), preventing the repressor from binding to the operator or, if the repressor is already bound to the operator, causing it to detach. Transcription is then free to occur. In a repressible operon, the effector molecule—that is, the corepressor, such as histidine—binds to the repressor, which in this case is unable to bind to the operator by itself. But the two together can bind to the operator and thus repress transcription (Fig. 24-16c).

An operon is said to be under positive control if the genes in the operon are transcribed only when some regulatory molecule is bound to the operon. The *lac* operon, which is under the negative control of a repressor, surprisingly, is also under the positive control of another molecule. That this is so was discovered after it became apparent that under certain conditions the *lac* operon is transcribed in the presence of both glucose and lactose. It was observed that normally in the presence of glucose, the *lac* operon is not functional even if lactose is present in the medium. It was also observed that in the presence of glucose, the cyclic AMP (a regulatory nucleotide) content of *E. coli* is low. Most important, it was found that in the presence of cyclic AMP, the *lac* operon becomes functional even if glucose is abundant. This anomalous situation was explained by the presence of a cyclic AMP binding protein (CAP), (also known as the catabolite gene activator) which binds to the *lac* operon DNA in the presence of cyclic AMP. Only when this occurs, can the structural genes be derepressed and transcribed (Fig. 24-17).

In summary, control of bacterial operons is of two types: negative and positive. In negatively controlled operons, the regulatory molecule turns off transcription on binding to the operator region. In positively controlled operons, genes are transcribed only when the regulatory molecule is bound to the operator region. Furthermore, negatively controlled operons may be induced by an inducer or repressed by a corepressor; that is, an inducer binds to a regulatory molecule thus reducing its affinity to the operator. Similarly, a corepressor may enhance binding of the regulator to the operator region.

**FIG. 24-16** The operon model. (a) The essential components of all operons: operator, promoter, structural genes, and a regulatory gene. (b) The mode of activation of a negatively controlled operon by inducers: the *lac* operon.

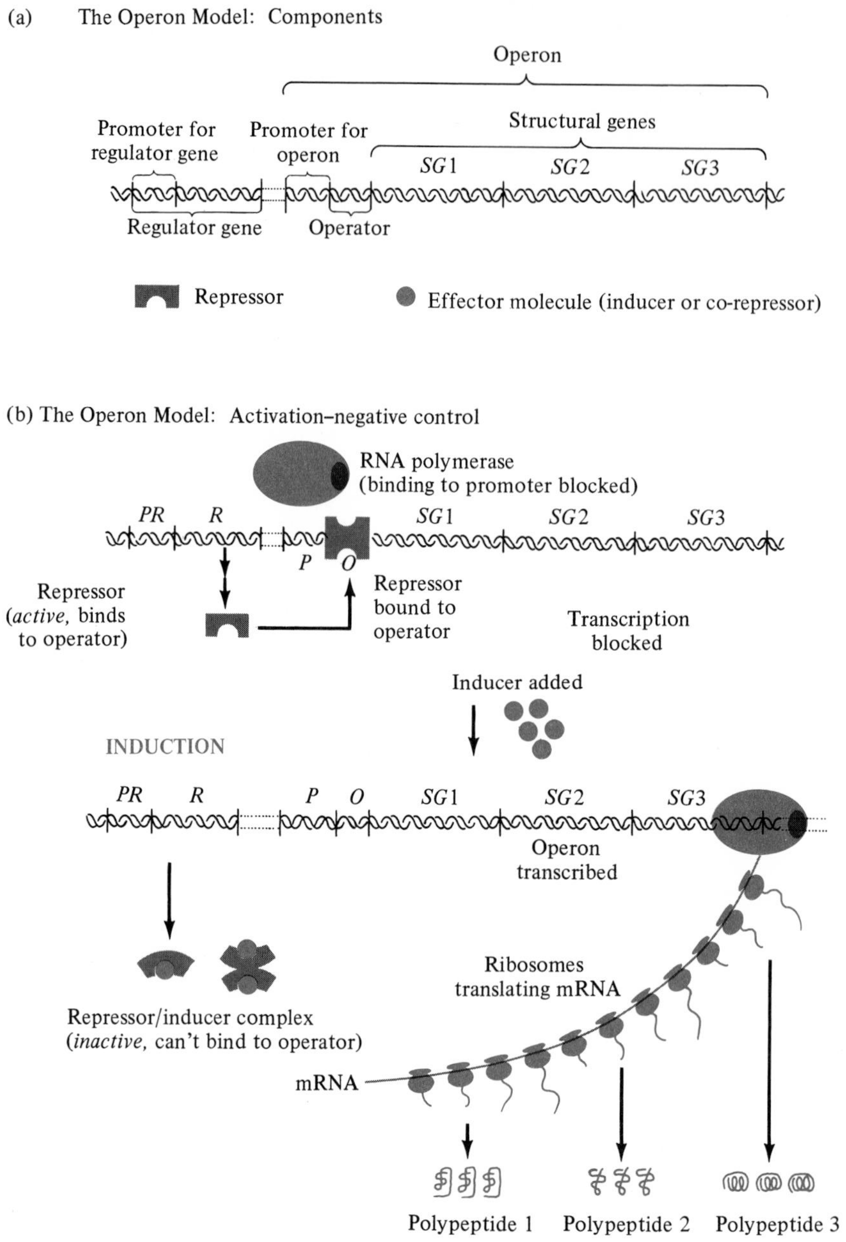


***Operons in Higher Organisms.*** Although there is evidence for the existence of operonlike units in fungi and other lower eukaryotes, operons appear to be either very rare or nonexistent in higher eukaryotes. All mRNAs of higher organisms thus far characterized have proved to code only for single, structural genes. Control of genic expression in higher organisms apparently is much more complex. One factor in control is that DNA in eukaryotic chromosomes is always associated with histones. These positively charged basic proteins help pack DNA into compact supercoils. Such supercoiled DNA is usually inactive, probably because it is inaccessible to enzymes. In addition to histones, there are a wide variety of nonhistone proteins in the eukaryotic chromosome. Some of these proteins may include specific activators of genes, which aid in triggering activation of sets of genes through mediation of hormones or other substances. (This topic is discussed briefly in the next chapter as one aspect of development.)

**FIG. 24-16 *(continued)*** (c) The mode of repression of a negatively controlled operon by corepressors: the histidine operon.

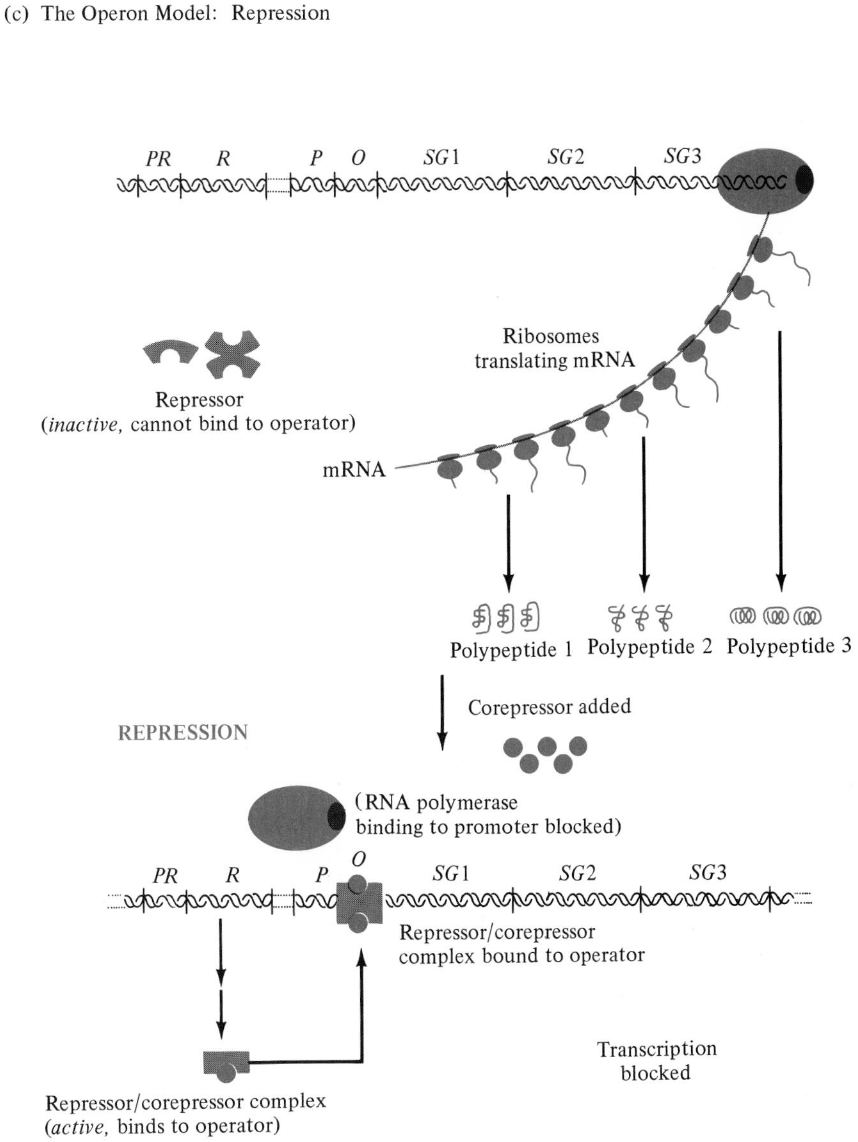


## BACTERIA, VIRUSES, AND MOLECULAR GENETICS

Bacteria have been found to be superior to all other experimental subjects as a means of advancing our understanding of molecular genetics. Compared to other organisms, they are easy and inexpensive to raise in large numbers, and most have a generation time of about 20 minutes. The species investigated most thoroughly has been *Escherichia coli.* A further boon to investigators studying bacterial genetics has been the existence of various phages (bacterial viruses) and certain naked DNA molecules called plasmids. Both can parasitize bacteria while also acting as couriers of genetic material (DNA) from one cell to another. Properly manipulated, they can be used in remarkable ways to study the structure and function of genes of almost any organism.

Mutants are easily obtained in *E. coli,* and thousands of them have been isolated. This has allowed a detailed analysis of many biochemical pathways. Although at one time it was thought that whatever might be learned of bacterial metabolic pathways might have little significance

**FIG. 24-17** Positive control of an operon. Cyclic AMP in concert with CAP protein binds to the CAP site of the operon and activates the lactose operon.

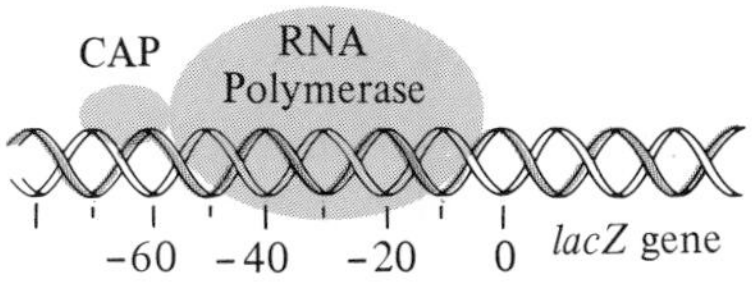

for the biochemistry of higher organisms, the opposite proved to be true. Most of the principles revealed by the study of microbial biochemistry and genetics have been found to apply to all organisms, including humans.

Besides allowing us to learn about the chemistry of such matters as resistance to antibiotics and the ability of a bacterium to metabolize various nutrients, mutant genes that require special conditions to be expressed have proved useful as genetic markers. In a mixed population, a microbe with a particular marker gene can be isolated from among many billions of others because of its special traits.

## GENETIC RECOMBINATION IN BACTERIA

Despite the relative ease with which the biochemical pathways of bacteria could be studied, bacterial genetics appeared to have a somewhat bleak future until it was discovered that two strains of bacteria sharing the same culture (or infecting the same host) are sometimes able to transmit DNA from one cell to another. This knowledge provided the basis for experimental studies comparable to breeding experiments with higher organisms and thus opened up a vast array of research opportunities.

Three major ways are now known in which genetic material can be transferred from one bacterium to another: (1) **transformation,** the absorption into a cell of DNA molecules that had originated in another cell; (2) **transduction,** in which a bacterial gene or genes is carried from one cell to another by an infecting bacteriophage; and (3) **conjugation,** in which DNA is transferred from one bacterium to another through a special tube that develops while the two cells are in contact with each other.

***Transformation.*** The process of transformation provided the basis for the pioneering work of Avery and coworkers, in which DNA proved to be the "transforming principle" that can change the phenotype of a bacterium that absorbs another's DNA. It is an active process (that is, an energy-requiring process) involving a special enzymatic machinery possessed only by some species of bacteria. It can be induced artifically in other bacterial species, however.

Following the cell's uptake of DNA, the newly acquired DNA is apparently incorporated into the host cell's circular chromosome, replacing a segment of the host cell's double helix. Because only short segments of DNA are incorporated, this has sometimes permitted a determination of which genes are close to each other on the bacterial chromosome, thus aiding in the preparation of chromosome maps. A more important mapping tool, however, is conjugation, described below.

***Transduction.*** In the process of transduction a bacteriophage (a bacterial virus with a DNA core and a protein coat), while residing in a bacterial cell, incorporates a segment of bacterial chromosome into its own DNA. Transduction occurs when this bacterial DNA is subsequently transferred to another host cell when the virus infects it. This situation occurs when the phage genome is incorporated into the host (bacterial) chromosome and maintained in what is known as the **lysogenic cycle** of the phage (see Fig. 24-18). Under various conditions of stress, the viral genome is excised from the host chromosome, and the phage prolifer-

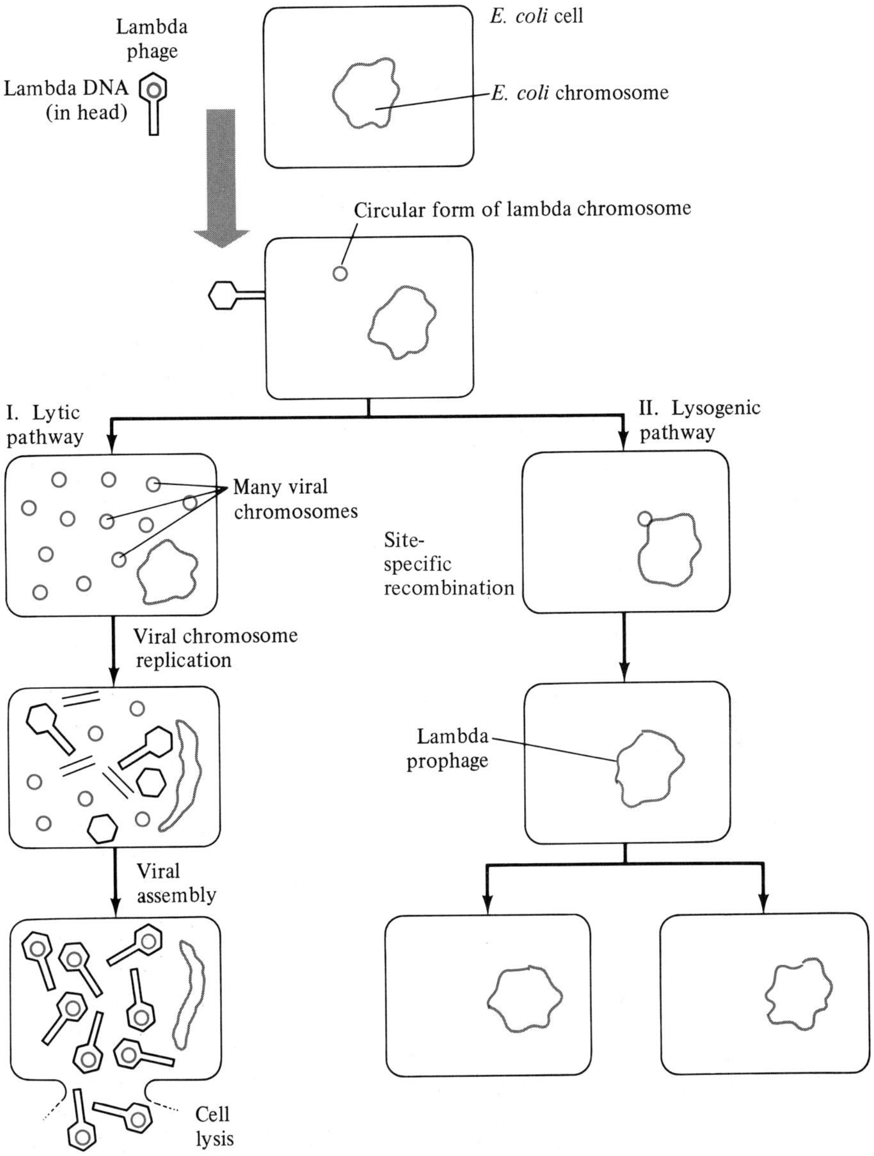


**FIG. 24-18** Diagrammatic representation of the lytic and lysogenic life cycles of a typical bacteriophage in a bacterium.

ates rapidly, eventually leading to lysis of the host cell, a condition called the **lytic phase** (Fig. 24-18). In the process of excision, some host genes may be carried by the phage and transferred to the next host.

***Conjugation.*** Bacteria are haploid and have only one chromosome, made up of a single, circular double helix. To be able to transfer DNA to another cell during conjugation, bacteria require the assistance of a particular type of plasmid. Like phages, plasmids are capable of independent replication within a host cell's cytoplasm. There are several types of plasmids, including some that confer resistance to antibiotics on the host cell. One type, the **F plasmid,** contains the F (for fertility) factor. This gene is necessary for bacterial conjugation to occur. A bacterial cell with F plasmids in its cytoplasm is called an **$F^+$ cell** (or male cell). A bacterial cell lacking the F plasmids in its cytoplasm is an **$F^-$**, or female, **cell** (Fig. 24-19a, b). Among the genes possessed by the F plasmid are some that direct the synthesis of **pili:** whiskerlike appendages on the outer surface of the $F^+$ cell. Possession of pili makes the $F^+$ cell capable of adhering to another, uninfected *E. coli* cell, the $F^-$ cell

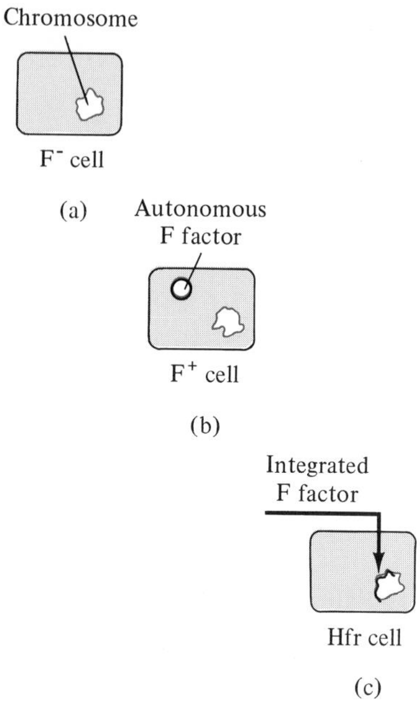


**FIG. 24-19** The three states of an *E. coli* cell in regard to the conjugation factors. (a) $F^-$ cell (no fertility factor). (b) $F^+$ cell (contains the F factor as a plasmid). (c) Hfr cell (the F factor is integrated into the chromosome).

(or female cell) (Fig. 24-19b). When this occurs, a tube is formed between the two cells, through which the plasmid DNA enters the $F^-$ cell, converting it into an $F^+$ cell.

In some cases, the entire F plasmid becomes integrated into the host cell chromosome (Fig. 24-19c). The host cell then becomes an **Hfr cell,** for high frequency of recombination, so-called because conjugation of such cells results frequently in genetic recombination. If such a cell engages in conjugation, its chromosome, or a segment of it, becomes transferred to the $F^-$ cell, starting with a part of the plasmid DNA. Only if the entire $F^+$ chromosome is transferred, however, does all of the plasmid DNA, including the entire F factor, also enter. A diagram of the method of transfer of part of an Hfr chromosome is shown in Figure 24-20.

It takes about 89 minutes for an entire bacterial chromosome to be transferred. If, for any reason, conjugation is interrupted before completion, which is nearly always the case, only a portion of the bacterial chromosome is transferred. Although first viewed by researchers as an annoying inconvenience because bacterial characteristics are altered in the process, this phenomenon came to be exploited to great advantage when it was discovered that in a given Hfr strain, a particular region of the male chromosome always entered the female cell first. This allowed conjugation, and thus transference of specific lengths of the chromosome, to be precisely controlled; it thus helped in mapping the bacterial chromosome in the following way.

In these bacterial chromosome mapping experiments, cells capable of conjugation were first mixed together. Then, at various times, from 1 to 89 minutes after the beginning of conjugation, they were violently agitated in a food blender. This was done to break apart the conjugating cells after various lengths of their chromosomes had been transferred; the maximum chromosomal length entering the female cells depended directly on the number of minutes that had elapsed since the beginning of conjugation. By mating strains of *E. coli* differing in one of a variety of alleles, it was possible to determine where on the chromosome a locus existed by the number of minutes it took for an allele at that locus to be transferred to an $F^-$ cell. In contrast to eukaryotic chromosome maps, which are marked in crossover units, on bacterial chromosome maps the distances are given in minutes (Fig. 24-21). The linkage map of *E. coli* is standardized to a total length of 100 minutes. Among many other interesting benefits derived from use of this technique were the determination of the intragenic mutation site in tryptophan synthetase and the determination of the role of the regulator gene in the *lac* operon.

## RECOMBINANT DNA TECHNOLOGY AND GENE CLONING

One of the most exciting of all the scientific developments to occur in this century has been the development of recombinant DNA technology and gene cloning. The techniques have been made possible by a knowledge of phage and plasmid biology and of bacterial genetics. The principles of gene cloning are relatively simple. The gene to be cloned is ligated, that is, spliced to a plasmid or bacteriophage DNA. Because plasmid DNA is circular, if a single cut is made in the circle, it becomes linearized, or opened up. When the DNA fragment to be cloned is then ligated to the ends of the linear plasmid DNA, a circular DNA molecule containing both the plasmid DNA and the DNA fragment to be cloned is

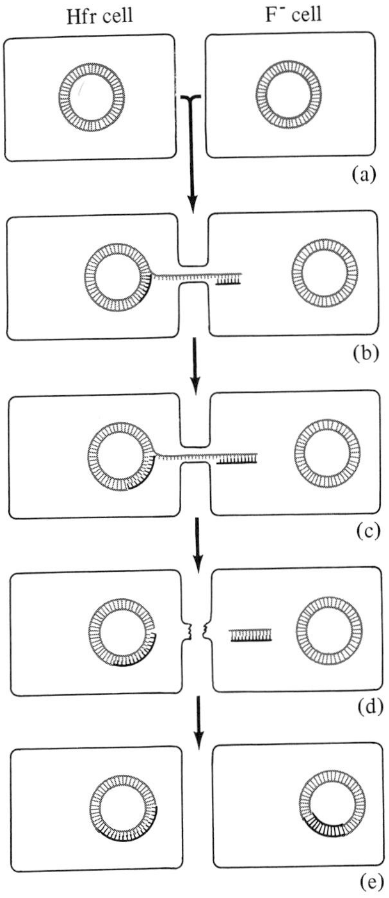


**FIG. 24-20** The steps in the transfer of a part of the chromosome from an Hfr cell to F⁻ cell during conjugation. After a certain amount of DNA is transferred from Hfr to F⁻ cell (c), the junction between the two cells is broken. The transferred DNA fragment recombines with the F⁻ cell chromosome (e).

formed. Similarly, phage DNA from which the genes nonessential for phage survival have been removed can be used. In either case, the recombinant molecules are used to infect or transform appropriate bacterial host cells. Only one out of every 10,000 bacterial cells or so takes up the plasmid and becomes transformed. However, the transformed cells can be isolated by a screening procedure using antibiotics. The plasmid contains a gene that confers antibiotic resistance. When grown on a medium containing the antibiotic, only the transformed bacteria bearing the plasmid survive, thus making it easy to isolate and grow them.

As the transformed bacteria containing the recombinant plasmid multiply, the plasmid also multiplies. Because a bacterial cell is likely to harbor a single recombinant plasmid, each cell in a bacterial colony produced from that cell will contain the same cloned DNA fragment. Similarly, because only one recombinant phage particle is likely to infect a single bacterial cell, all the phage progeny from that cell will contain the same cloned DNA fragment. The goal of cloning is to amplify (produce multiple copies of) a particular DNA sequence from an organism. Multiplication of a plasmid or phage that carries the particular DNA sequence enables the sequence to amplify.

Gene splicing has been aided considerably by the existence of a class of enzymes called **restriction endonucleases.** These enzymes have apparently evolved in bacteria as a protection against the entry of foreign (viral) DNA. Each restriction enzyme cleaves DNA molecules at specific sites, called **restriction sites,** which are characterized by specific nucleotide sequences. The nucleotide sequence cleaved by each type of restriction enzyme is different. (As an adaptation that avoids degrading their own DNA by their own endonucleases, bacteria modify one or more nucleotides in each restriction site sequence, usually by attaching methyl groups to cytosines that precede a guanine residue. Such meth-

**FIG. 24-21** The linkage map of *E. coli,* strain K-12. Only a few of about 1,000 known genes are shown in the outer circle. The numbers inside the circle refer to map position in minutes arbitrarily set at 100 minutes. The letters and numbers on the inner circle refer to the Hfr strains used in construction of the map.

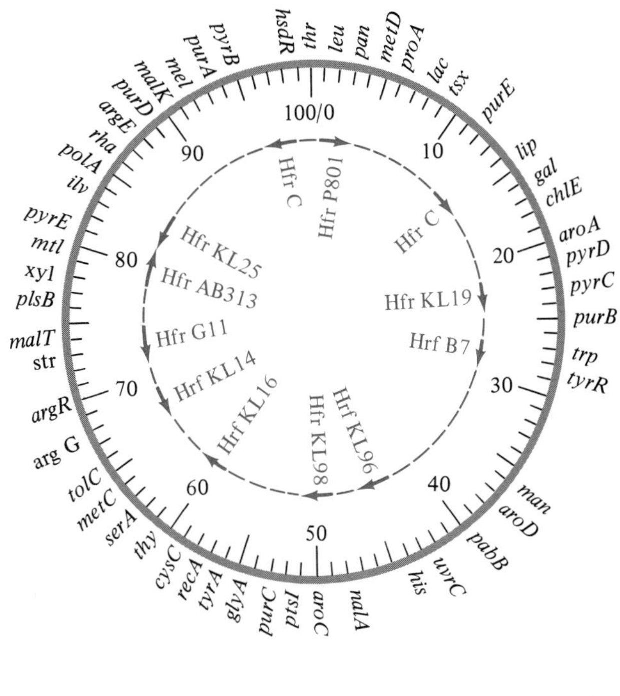

ylation in the new DNA strand occurs a short time after replication of the chromosome.) Many restriction endonucleases are specific for **palindromes,** which are base-pair sequences that read the same forward or backward, such as the word *madam*. In such cases, the ends of the cut DNA molecule will be as shown in Fig. 24-22.

In a gene cloning experiment, DNA molecules are cleaved at the spe-

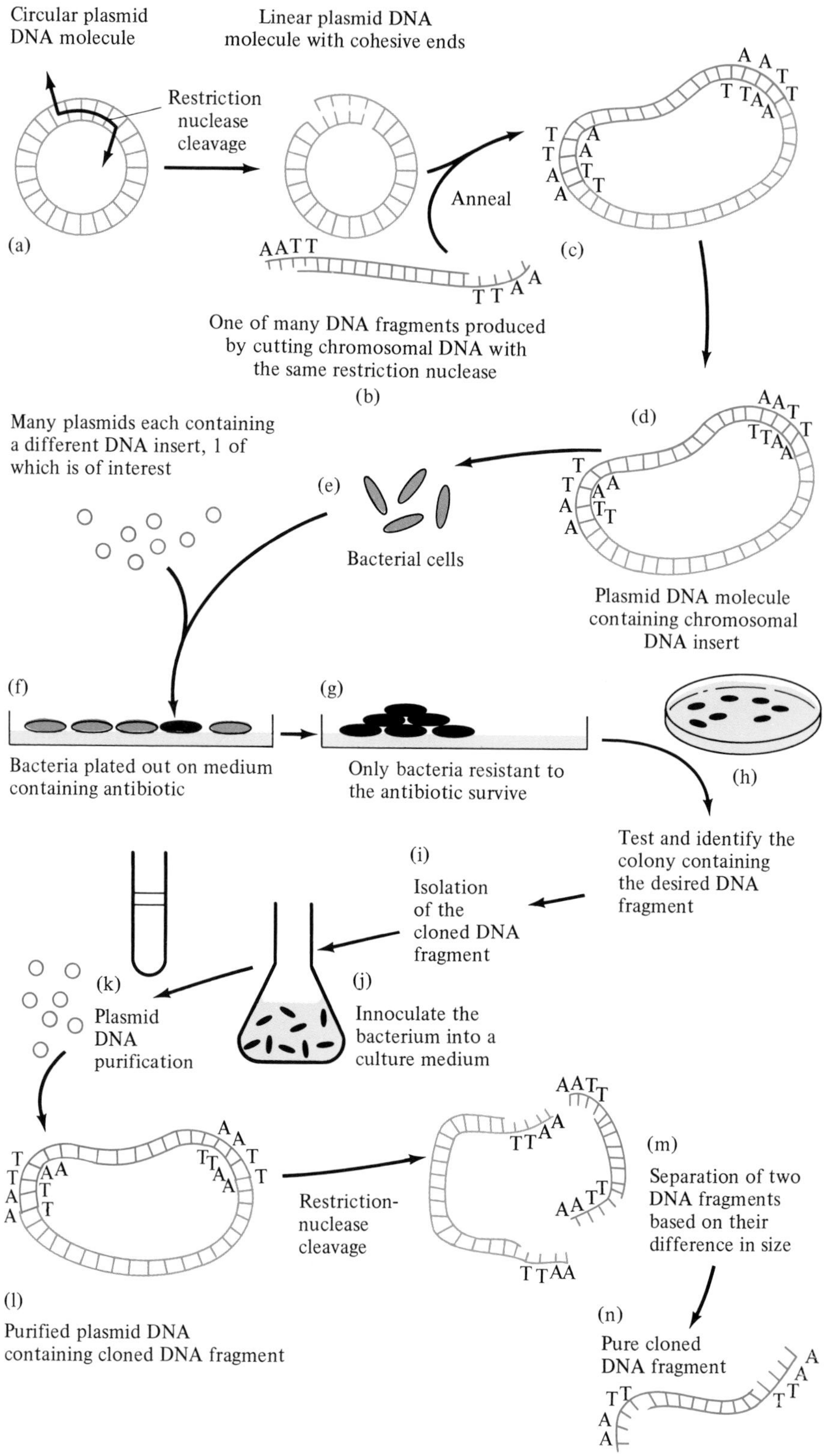


**FIG. 24-22** Diagrammatic presentation of the methods whereby a recombinant molecule is produced and specific genes are cloned and isolated.

cific restriction sites into small fragments, their number and size depending on the number of such restriction sites in the DNA molecule and the amount of enzyme present. When DNA fragments of phage or plasmid are mixed with fragments from higher organisms that have been cleaved by the same restriction enzyme of DNA, the staggered cut ends randomly attach to each other. Recombinant molecules containing phage or plasmid DNA, called vector DNA, and DNA from human chromosomes have been generated in this manner. As noted, it is a chance event as to which human DNA sequence becomes ligated with a vector DNA fragment. But the size of the fragment can be adjusted in such a way that each recombinant vector contains only one genomic DNA fragment. These recombinant molecules are used to infect bacteria if the vector is a phage or to transform bacteria if the vector is a plasmid. The entire bacterial culture containing all the different recombinant phages or plasmids is called the **genomic library.** (If cDNA prepared from mRNA is used in the preparation of recombinant plasmids, the preparation is called a **cDNA library.**) All the sequences of the genomic DNA are represented in these recombinant molecules.

From this library, the desired human gene, or that of any other organism whose DNA is cloned, can be isolated by special screening procedures. The principles of screening are based on the complementarity of the gene and its transcript (RNA). For example, to locate the human insulin gene, bacteria from the genomic library, that is, the bacteria carrying the recombinant plasmids, are grown individually, and screened with radioactive insulin mRNA or a radioactive cDNA prepared from insulin mRNA. Because only the insulin gene DNA will hybridize to the mRNA coding for insulin, the hybrid molecules can be identified by their radioactivity. Since each bacterium is likely to contain only one plasmid and hence only one type of recombinant molecule, all the progeny of such a cell also contain the same cloned DNA fragment.

Once a particular clone has been identified as carrying the desired gene, this technique allows a particular fragment to be multiplied without the presence of any other contaminating human DNA. That in itself provides an advantage for studying the various properties of genes. Among the immediate benefits is that bacteria containing recombinant DNA molecules can be manipulated to produce products useful in medicine, agriculture, and industry. Human genes for insulin, blood clotting factor, and growth hormone have been cloned to produce pure forms of these biological substances for medical purposes. Interferon, a substance that has antiviral and anticancer properties, is also being produced in this manner. One problem is that bacterial cells cannot modify the resulting proteins, for example, by adding sugars; such modifications are possible only in eukaryotic cells. In 1984, gonadotropin (a glycoprotein hormone) gene was cloned in a vector which was used to transform human cells in culture. From such transformed human cells functional gonadotropic hormones were produced.

Other benefits are also envisaged from recombinant technology. Genetically manipulated microorganisms are currently being used to clean oil spills, digest certain toxic substances in the environment, and to produce such commercially useful substances as vitamins and enzymes. Studies are also in progress to alter plants genetically to permit nonleguminous plants to serve as hosts for nitrogen-fixing bacteria and render plants resistant to insect pests. In the not-too-distant future, we may anticipate gene therapy for some inherited human abnormalities, such as sickle-cell anemia and Tay-Sachs disease.

PANEL 24-4

# Oncogenes

One anticipated benefit of the study of the molecular biology of the gene is an understanding of the etiology, that is, origin and causes, of some types of cancers. Peyton Rous, a pioneer virologist at the Rockefeller Institute for Medical Research, observed in 1915 that a chicken sarcoma, a cancer of connective tissues, is associated with infection by an RNA virus called the avian sarcoma virus (ASV) also called the Rous Sarcoma Virus (RSV). In the 1970s, the gene responsible for tumor transformation in ASV-infected chick embryo fibroblasts, connective tissue cells that secrete collagen, was identified as a viral gene. The gene *src* (also designated as *onc*), which codes for a phosphoprotein, a protein containing phosphorylated amino acids, which has protein kinase activity. Further studies revealed that all healthy chickens and other birds as well, bear a gene, *sarc,* which also codes for a similar protein kinase. The *src* gene in ASV and the native *sarc* gene have very similar nucleotide sequences. When the native *sarc* gene is activated to produce more than the normal amount of its product, tumors develop. The *sarc* gene is regarded as a protooncogene, or a cellular oncogene, and the *src* gene of ASV is believed to be a protooncogene translocated onto the viral genome. Either activation of the protooncogene or incorporation of an active *src*

***Possible Dangers of Recombinant DNA.*** When the techniques for gene splicing and cloning were first being developed, some concern arose. It was suggested that someone might accidentally create a strain of *E. coli* or similar common inhabitant of humans that would carry a gene for cancer or other deadly disease. After a period during which such research was briefly halted, an elaborate set of guidelines was drafted and promulgated by the largest funder of basic biological research, the United States National Institutes of Health. After careful study, it was generally agreed that so far there is no real cause for concern. In fact attempts to produce such strains of bacteria all failed, because highly adapted wild strains quickly won out in the natural selection contest. It has been suggested that perhaps such strains as were feared probably have already arisen through evolution several times and have all been unsuccessful.

Recombinant DNA technology thus promises to be a beneficial tool, unless, of course, misused. In the ultimate analysis, a detailed knowledge of the mode of action of genes is essential to understand life processes such as growth, development, and even behavior. All activities of cells seem to be regulated by genes, and the basis of all biological activity is cellular function. Thus normal as well as most abnormal cell functions, such as diseased states and cancers, are to be understood in terms of gene action. Available evidence suggests that cancer—the unrestrained division of cells—results from an abnormal regulation of a small group of genes. These genes are believed to control some as yet unidentified but vital syntheses essential for normal control of cell division. This group of genes, present in all organisms, is suggested to be the source of viral **oncogenes,** genes present in cancer-causing viruses, also called tumor viruses. When tumor viruses infect normal cells, oncogenes reenter the genome and cause cancer. A detailed account of oncogenes is presented in panel 24-4.

gene into chick fibroblast cell genome is thought to cause tumorous transformation of these cells.

The concept of protooncogenes has been extended to humans and other mammals. Several oncogenes have been identified in the genomes of viruses associated with certain tumors in humans, rats, and mice. The protooncogenes that have been identified from human, rat, and mouse genomes have similar nucleotide sequences. Two examples of human cancer illustrate how protooncogenes are activated, thereby resulting in cancer.

One protooncogene, the *myc* gene, is normally located on chromosome 8 in the human genome. But in people with the cancer of the lymphoid tissue of the immune system called lymphoma, the *myc* gene has been translocated to chromosome 14, 22, or 2 in close proximity to the immunoglobin genes located there. By this translocation, the *myc* gene becomes activated to a high level of expression, and is always accompanied by carcinogenesis of the lymphoid system associated with the immune system.

A point mutation in the protooncogene *Ha-ras* is responsible for human urinary bladder cancer. The gene has mutated in such a way that the twelfth amino acid residue in its polypeptide product is changed from a glycine to valine. Clear proof of the effect of this mutation has come from in vitro transformation studies of fibroblasts; when this mutated form of DNA was incorporated into normal cells, they became cancerous. This observation supports the view that somatic mutations caused by chemical and other mutagens are responsible for some cancers. Whether by translocation or point mutation, it is now clear that most cancers, if not all, are caused by abnormal gene function.

## Summary

The unprecedented progress made since 1953 in determining the chemical nature of the gene, in unraveling the code by which it communicates its inherited message to the rest of the cell, and in understanding the molecular mechanisms by which genic expression is regulated appears to have been but the opening act in a great play that has just begun. With the use of such powerful research tools as the ability to manipulate bacterial transformation, transduction, and conjugation, the rate of progress is expected to continue accelerating in the years ahead.

Early studies of human inborn errors of metabolism suggested that genes influence the production of enzymes. Careful scientific studies on the bread mold *Neurospora* and the fruit fly, *Drosophila,* confirmed that genes affect the structure and function of polypeptides. Thus arose the concept that each gene controls the synthesis of one particular polypeptide.

Several lines of evidence indicated that the chemical basis of heredity resides in deoxyribonucleic acid (DNA), a large molecule made up of two antiparallel polydeoxynucleotide strands. The amount of DNA present in each cell of a species is constant. Most of the cellular DNA is located in the nucleus where chromosomes, the bearers of heredity, are found. The first clear indication of the role of DNA in heredity came from the identification of the chemical responsible for viral infections and for the transformation of nonpathogenic bacteria into virulent strains. Elucidation of the three-dimensional structure of the double stranded helical DNA molecule and its potential to make an exact copy of itself by separation of the two strands and their synthesis of the complementary antiparallel strands expedited its acceptance as the genetic material.

Soon after DNA was identified as the chemical transmitter of heredity, molecular mechanisms of information transfer from DNA to protein molecules were studied. It was found that the nucleotide sequence of a DNA molecule corresponds to the amino acid sequence of the polypeptide for which it codes. The information content of DNA, coded into its nucleotide sequence, is transcribed to an intermediate molecule, ribonucleic acid (RNA). There are three classes of RNA: ribosomal, messenger, and transfer. Ribosomal RNA (rRNA) is a major constituent of the cytoplasmic organelles called ribosomes, which serve as the site of synthesis of polypeptides. Transfer RNA (tRNA) serves as a conveyor of specific amino acids to the site of assembly of the polypeptide chains. The genetic information that specifies the amino acid sequence of the polypeptides is transcribed as nucleotide sequences in messenger RNA (mRNA). The mRNA binds to ribosomes and is then sequentially translated into polypeptides.

The information content of the nucleotide sequences of an mRNA molecule is read in groups of three bases called codons. The genetic code is thus a triplet code that is commaless and nonoverlapping.

All cells in an organism bear the same genetic information. However, the phenotype of each cell type is different, which suggests that only certain groups of genes are selectively activated in each cell type. The molecular basis of selective gene expression in eukaryotic organisms is unknown. In prokaryotes, selective gene expression is mediated by repressors and activators of genes. Functionally related groups of genes are organized into units of function called operons. Each operon consists of structural genes, which code for a functionally related group of enzymes; an operator, which regulates transcription of the structural genes; and a promoter, the site where the transcriptional machinery, that is, RNA polymerase, binds. A fourth component of the operon is the regulator gene, whose product binds to the operator to regulate gene expression.

A remarkable aspect of what may yet be a boon to human welfare is recombinant DNA technology. This new knowledge has arisen out of research on the biology of submicroscopic parasites of bacteria, a subject that at first appeared to be of little practical value. Yet, among the prospective benefits of this new technology are the production of inexpensive, high-quality drugs, such as human insulin and growth hormone, the cleaning up of environmental oil spills, and the genetic improvement of cattle and crops.

## Recommended Readings


Alberts, B., D. Bray, J. Lewis, M. Raff, K. Roberts and J. D. Watson. 1983. *Molecular Biology of the Cell.* Garland, New York. A comprehensive and up-to-date survey of cellular functions with special emphasis on the molecular aspects. The chapters on nuclear function describe the chemical basis of heredity in detail. This is a very good reference book.

Miller, O. L., Jr. 1973. "The visualization of genes in action." *Scientific American,* 228(1):34. A visual documentation of transcription and translation by a pioneer who developed the methods of study.

Nirenberg, M. W. 1963. "The genetic code: II." *Scientific American,* 228(1):80. An eminently clear account of the discovery of the genetic code by a discoverer of the code word.

Watson, J. D. 1976. *Molecular Biology of the Gene,* 3d ed. W. A. Benjamin, Menlo Park, Calif. A clear and logical account of the molecular basis of heredity by one of the codiscoverers of the structure of DNA. The descriptions can be easily understood by any interested student of general biology.

25

# Animal Development

OUTLINE



Every individual of a sexually reproducing animal species begins life as a single cell: a fertilized egg, or zygote, or, rarely, as a **parthenogenetic egg,** which develops without the need for fusion with a sperm. Successive cleavages of the zygote produce a multicellular structure, the embryo, which grows into a mature individual that reproduces, ages, and ultimately dies. Development includes all of these changes as well as the molecular, cellular, and organismal events that underlie them.

## EARLY EMBRYONIC DEVELOPMENT

Although the developmental changes in the life of an individual comprise a continuum, it is convenient to subdivide early embryonic development into four distinct stages: (1) **syngamy,** or **fertilization,** the fusion of the male and female gametes; (2) **cleavage,** the rapid divisions of the zygote into cells called **blastomeres,** which typically become arranged into a fluid-filled sphere; (3) **gastrulation,** the rearrangement of the blastomeres to form a multilayered stage called the **gastrula;** and (4) **organogenesis,** the establishment of organ rudiments. Then, by a process of **differentiation,** the various types of cells, tissues, and organs characteristic of an adult gradually emerge from the blastomeres of the

**FIG. 25-1** Mammalian spermatozoa. (a) A diagrammatic sketch of the organelles in a sperm cell. (b) Low-power photograph of spermatozoa. (c) Scanning electron micrograph of human sperm head.

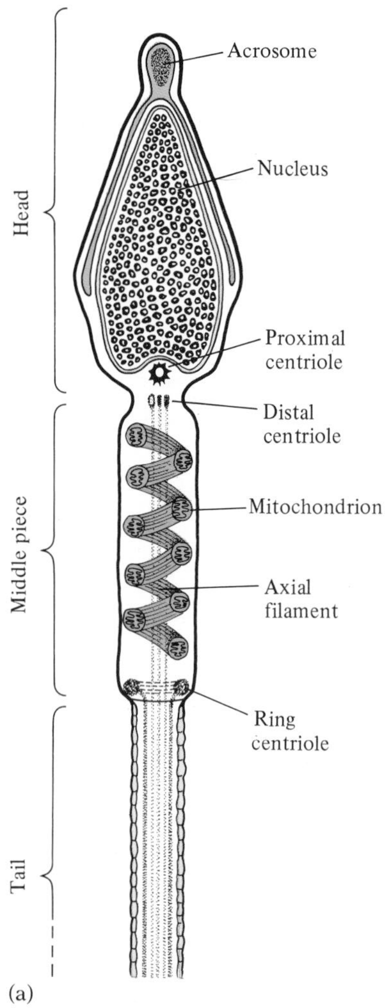


(a)

early embryo. During differentiation, many profound changes occur in the functional organization of the embryo's cells.

## GAMETES: HIGHLY SPECIALIZED CELLS

In sexually reproducing organisms, every individual usually has two parents, each of which has contributed a single specialized cell, the gamete. The gamete produced by the female is the egg, or ovum; the male gamete is the sperm, or spermatozoon.

Highly specialized cells, both spermatozoa and ova are prime illustrations of the intimate relationship that exists between structure and function in biological systems. Ova are usually large and immotile and contain a full complement of organelles. Most ova also contain nutritional reserves called **yolk.** In these and other ways, they are highly suited to be the founding cells of embryos. To embark on development, an ovum usually needs only to receive a second set of chromosomes and to be activated to begin cleavage.* Sperm has evolved in quite a different direction from that of the ovum. The sperm must be motile if it is to reach the ovum. Its essential components are its nucleus, which contains a haploid set of chromosomes, and its flagellum and associated organelles required to deliver its nucleus into the cytoplasm of the egg. Almost all spermatozoa are flagellated. They are streamlined and lightweight, containing nothing that is not essential to the efficient performance of their limited task. A closer look at both the ovum and the spermatozoon will emphasize these differences between the two gametes.

***Spermatozoa: Specialized to Deliver DNA.*** A sperm cell has three distinct parts: a spherical or ovate head, a cylindrical middle piece, and a flagellum (Fig. 25-1). Spermatozoa contain an enzyme-filled vesicle called the **acrosome,** a nucleus, numerous mitochondria, a pair of centrioles, and the flagellum. Like all cells, a spermatozoon is bounded by a

* Eggs of many species (for example, echinoderms, amphibians, and mammals) can be artificially induced to undergo parthenogenesis, or development without fertilization, by such nonspecific stimuli as the prick of a needle or brief contact with a hypertonic salt solution. In several species of animals (for example, bees and water fleas), parthenogenesis is a natural occurrence. All male bees, for instance, develop from unfertilized eggs.

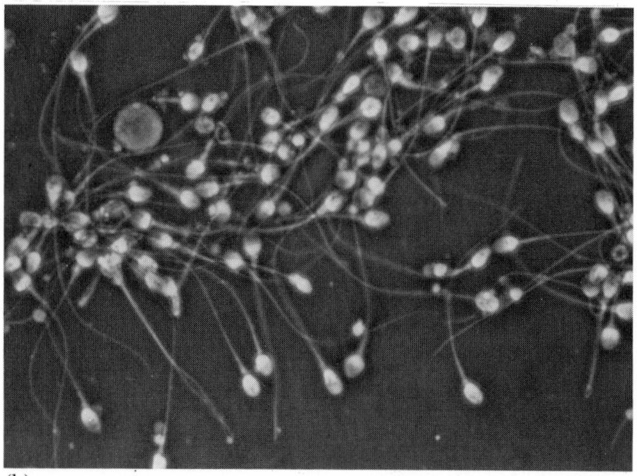

(b)

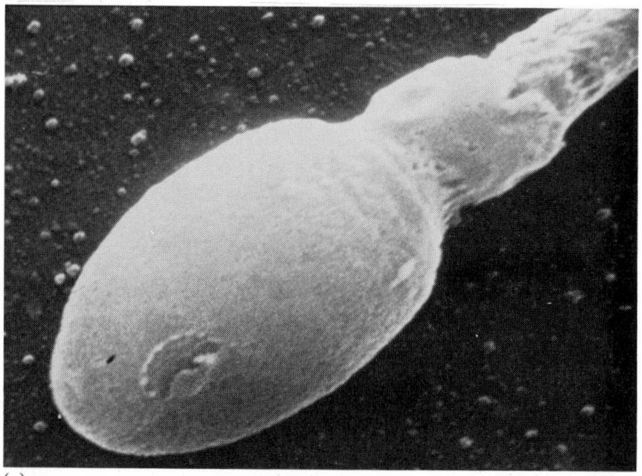

(c)

plasma membrane. Unlike most cells, however, sperm cells lack ribosomes, Golgi bodies, and cytoplasmic membrane systems.

Spermatozoa begin their development, called **spermatogenesis,** as diploid spermatogonial cells (Fig. 25-2a). Spermatogonia are located at the periphery of the seminiferous tubules of the testes, where they continually divide by mitosis (see Chapter 22). The spermatogonia that move toward the lumen of the tubule undergo enlargement, forming primary spermatocytes. These divide meiotically to form four haploid spermatids. The cells produced by the first meiotic division (meiosis I) are termed secondary spermatocytes. The chromosome number is reduced to the haploid number by this division, although the resulting cells contain twice the haploid amount of DNA. Division of the two

**FIG. 25-2** Gametogenesis in an animal with a diploid chromosome number of six. (a) Spermatogenesis. (b) Oogenesis.

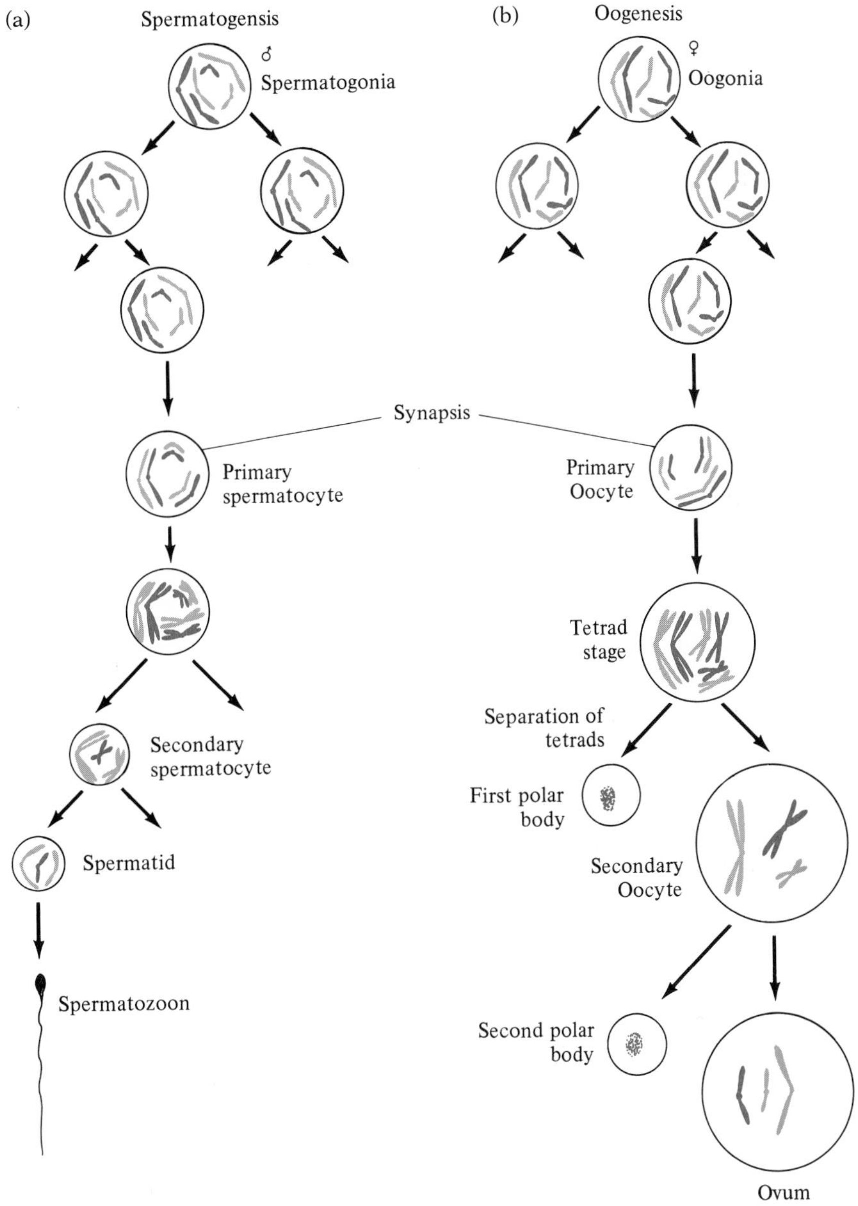

secondary spermatocytes (meiosis II) then occurs, yielding four haploid spermatids.

Except for their haploid chromosome complement, spermatids generally resemble spermatogonial cells and other cells of the body. However, during the next, and final, developmental phase of spermatogenesis, called **spermiogenesis,** the spermatid undergoes many cytological changes. A large part of the spermatid cytoplasm, containing endoplasmic reticulum (ER), ribosomes, and the Golgi complex, is left behind as the sperm develops. The mature sperm cell is less than one-hundredth the size of a spermatid and has a flagellum and an acrosome. The acrosomal enzymes of the sperm can digest a hole in the membranes covering the ovum. The plasma membrane of the sperm also possesses special molecules that allow it to fuse only with female gametes of its species.

***Egg Specializations.*** Ova are unique cells in that most contain (1) relatively large stores of ribosomes and messenger RNA (mRNA) molecules, (2) a large supply of lipid and protein nutritional reserves called yolk, and (3) a specialized plasma membrane envelope that can "recognize" and fuse with the plasma membrane of a sperm of the same species. The cellular changes that transform an oogonial cell (the diploid female reproductive cell, comparable to the spermatogonial cell) into a mature ovum comprise the process of **oogenesis** (Fig. 25-2b). The first step in this long series of developmental events is meiosis. An oogonial cell that enters meiosis is called the primary oocyte, corresponding to the primary spermatocyte in spermatogenesis. The two types of gametes develop in rather different ways, however, a fact related to the very different roles of their final products, the egg and sperm.

From the time of onset of oogenesis, oogonial cells undergo a lengthy period of growth and development. During this time, the primary oocytes accumulate the materials that will be required in some abundance during the early stages of embryonic development. In meeting the requirements for a rapid early development, it would be highly inefficient for an oogonial cell to accumulate its large stores of yolk and other materials only to divide meiotically into four equal-sized cells before fertilization. Early in evolution, ova solved this problem through the agency of **polar bodies,** or **polocytes.** As Figure 25-2b illustrates, at each of the two meiotic divisions, rather than dividing into daughter cells of equal size, as spermatocytes do, oocytes extrude tiny cells, the polar bodies, each containing a nucleus and an insignificant amount of cytoplasm. Polar bodies disintegrate, often within a few hours. The remaining cell, termed the **ootid,** or ovum, thus retains all but a very small amount of the materials that had accumulated in the primary oocyte.

Unlike the spermatid, the ootid does not need to undergo further changes before it is functional. In fact, the terms *ootid, ovum,* and *oocyte* are often used interchangeably, and for good reason. From one animal group to another, ovulation, and even fertilization, may occur at almost any stage of oogenesis. In some species, both meiotic divisions are completed before ovulation. In other species, meiosis is not completed until after ovulation but is complete before entry of the sperm. In still other species, including humans, meiosis II is initiated only after a sperm has entered the oocyte.

The growth phase of the primary oocyte is characterized by the following nuclear and cytoplasmic activities:

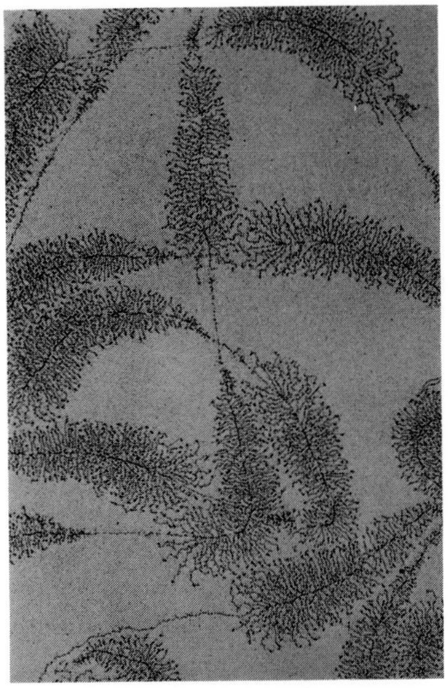

**FIG. 25-3** Nucleolar genes in amphibian oocytes. The central strands are chromosomes and the lateral "branches" are RNA molecules in different stages of transcription.

1. Amplification of the genes that code for ribosomal RNA (rRNA), by production of multiple copies of those genes
2. Synthesis of rRNA used for biogenesis of ribosomes and synthesis of mRNA that will be used in protein synthesis during early cleavage stages
3. Accumulation of yolk and various other nutrients

***Gene Amplification.*** Ribosomal genes are selectively replicated during the early stages of oogenesis, by a process called amplification. As a result of the ribosomal gene amplification, the DNA content of a primary oocyte is several times the diploid amount. In *Xenopus*, the clawed toad, the DNA content is 1,000 times greater, although in most animal species it is not nearly that great.

The elucidation of the phenomenon of gene amplification revealed the role of oocyte nucleoli, the densely staining organelles found in most nuclei. Whereas *Xenopus* cells other than oocytes have but two nucleoli, the mature oocyte has over 1,000 nucleoluslike bodies.* Electron microscopy showed each of them to consist of several connected, featherlike structures (Fig. 25-3). The shafts of the "feather" proved to be strands of DNA, that is, chromosomes. From the DNA strands extend a series of RNA molecules of differing lengths that are being transcribed from the chromosomes. Their graded lengths represent different stages in transcription.

The shortest RNA fibrils are those just beginning transcription; the longest have practically completed the process. Between each "feather" is a length of **spacer DNA,** which appears not to be transcribed. The dots at the base of each RNA strand are probably molecules of RNA polymerase. As each rRNA strand is formed, it becomes associated with protein molecules that enter the nucleus from the cytoplasm. Some of these proteins will become incorporated into the ribosomes along with the rRNA.

***Lampbrush Chromosomes.*** Another type of nuclear activity that occurs in the early stages of oogenesis is the transcription and accumulation of mRNA molecules. Because this activity occurs during late prophase I, when the chromosomes are compactly coiled, it results in a dramatic change in the morphology of those regions of the chromosomes undergoing transcription. Such chromosomal regions form numerous pairs of loops; each member of a pair belongs to one of the homologous chromatids that paired for meiotic division (Fig. 25-4). Because of their general appearance, chromosomes undergoing such transcription are called lampbrush chromosomes.†

The main axis of a lampbrush chromosome is a **chromonema** (chromonemata, plural), the chromatin thread composed of DNA and proteins that have remained compactly coiled. The lateral loops are the uncoiled and extended sections of the chromonemata, and, apparently, they change constantly. Lampbrush loops are regions of chromonema

* The discovery in 1958 of a chromosomal deletion in *Xenopus* led the way to our understanding of nucleoli and gene amplification. A wild-type *Xenopus* has two nucleoli. Heterozygotes with the deletion have only one. Individuals homozygous for the deletion have no nucleoli. Such embryos make no rRNA and die at an early stage of development.

† When these chromosomes were named, lampbrushes, which resemble test tube brushes or bottle brushes, were common household items, used to clean soot from the inside of the glass chimneys of kerosene lamps. They were even more common than the then-familiar thread-wound spindles of spinning wheels, after which the mitotic spindle was named.

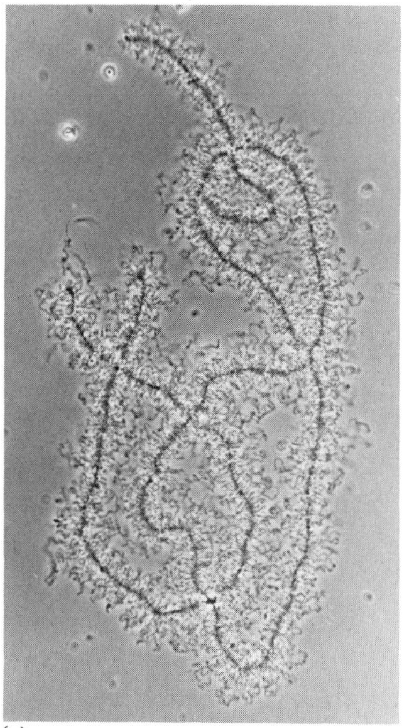

(a)

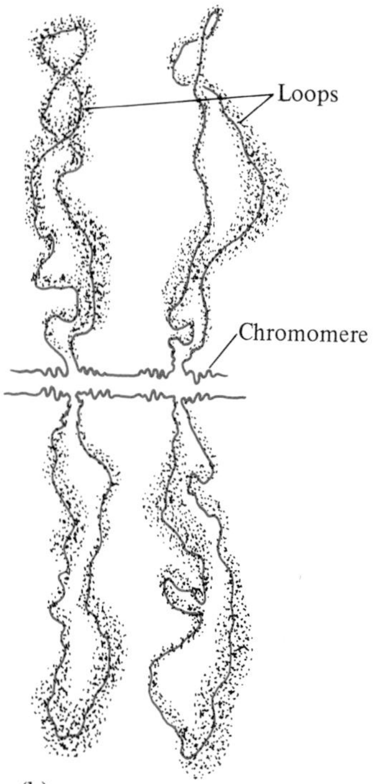


(b)

**FIG. 25-4** Lampbrush chromosomes from amphibian oocytes. (a) Photomicrograph showing hundreds of loops extending from the main axis of the chromosome. (b) A diagrammatic sketch of a few lampbrush loops.

engaged in RNA synthesis, as indicated by isotopic labeling. When prophase I oocytes are incubated with tritiated uridine, the lampbrush loops selectively incorporate the radioactive label, indicating that they are engaged in transcription of RNA. Pulse-chase experiments in which cells are briefly incubated in isotopically labeled precursors, followed by incubation in unlabeled precursors, have shown that much of the RNA accumulates in the oocyte during its enlargement. Some of these RNA molecules, which become translated into proteins during the early stages of embryonic development, are referred to as **maternal messages.**

***Yolk Accumulation.*** Following the period of nuclear activity, oocytes accumulate yolk, or **vitellin,** a phospholipoprotein nutrient reserve. Vitellin is a specialized protein containing unusual amounts of phosphorylated amino acids and lipids. This phase is called the **vitellogenic phase.** In higher animals, the bulk of the yolk is synthesized in precursor form outside the ovary—in the liver in vertebrates, for example. Developing primary oocytes take up these molecules from the bloodstream and process them into yolk.

Both the amount of yolk and its distribution within the oocyte are highly variable from one animal group to another. For example, the eggs of modern mammals lack yolk and are described as **alecithal;** the eggs of echinoderms and protochordates have a small amount of yolk and are termed **oligolecithal;** and the eggs of fishes, amphibians, reptiles, birds, and egg-laying mammals accumulate a great quantity of yolk at one end of the egg and are classed as **telolecithal.** The amount and distribution of yolk in an egg profoundly influences the pattern of cleavage it exhibits.

In addition to the modifications already noted, many other changes occur in oocytes during oogenesis. The number of mitochondria increases enormously. In the mature oocyte of a frog, for example, there are about 300 million mitochondria, while a typical body cell contains only a few hundred. Another change is the formation of **cortical granules.** These are initially formed in association with the Golgi complex at the center of the cell and then gradually move to the periphery, where they become arranged just beneath the plasma membrane.

Toward the end of oogenesis, the follicle cells surrounding the oocyte secrete a specialized protective covering, the **vitelline membrane,** which envelopes the oocyte. This membrane is just outside the plasma membrane of the oocyte and bears specific macromolecules important for sperm recognition.

## FERTILIZATION

The gametes of most animal species are short-lived. Human ova, for example, are viable for only about 24 hours after ovulation, and sperm for about 48 hours after ejaculation. However, when a spermatozoon fuses with an ovum of the same species, the resulting zygote acquires great developmental potential. A new individual has come into existence.* The zygote has all the genetic information and potential needed to interact with an appropriate environment and thus to develop into a mature organism with all the morphological, biochemical, and behav-

* To state, in the case of the human ovum, that a new individual has come into existence is not to define, either legally or philosophically, whether the zygote is a human being with all the legal and natural rights of a human being. These are legal and philosophical, not scientific, questions.

ioral attributes typical of its species. In the absence of fertilization, the ova of most species will perish; fertilization is thus the first and the most crucial of a series of cellular interactions necessary for normal development. Since fertilization is an important step in development of an individual organism, in the course of evolution several mechanisms that aid the successful completion of this event have been selected for in evolution. Some of these mechanisms are described in Panel 25-1.

PANEL 25-1

# Mechanisms That Promote Fertilization

Because fertilization is a crucial step in the initiation of a new individual, it is no surprise that animals have evolved mechanisms that enhance the likelihood of its occurrence. These specializations have occurred at the level of behavior and physiology of the organism and at the cellular level. Some of the behavioral and physiological aspects are discussed in Chapter 21. Here, we will consider the properties of animal gametes that enhance fertilization.

First and foremost is the release of numerous spermatozoa. Each ovum may be surrounded by hundreds of spermatozoa. However, only one spermatozoon fuses with each ovum.

In addition, the membranes covering the ova and spermatozoa of each species bear specific molecules that aid in mutual recognition. Some of the same molecules seem to be involved in the chemotactic attraction of spermatozoa to ova of their species. For example, sea urchin eggs release a chemical, **fertilizin,** that promotes the rapid movement of sperm of the same species toward the eggs. The same substance also causes the clumping and immobilization of spermatozoa when they are exposed to a concentrated solution. Normally, when sperm are some distance from eggs, they are stimulated by fertilizin to swim faster toward the eggs. Once the sperm have reached the surface of an egg, they cease further movement, apparently by encountering a higher concentration of fertilizin. Fertilizin has such great specificity that the sperm of one species rarely, if ever, responds in these ways to the presence of eggs of another species.

The sperm reacts in still another way to an egg of its species. As it nears the egg, its acrosome extrudes a filament, which attaches to the surface of the egg, releasing the acrosomal enzymes, e.g. hyaluronidase in mammalian sperm. The egg's coverings and plasma membrane disintegrate at the point of contact, allowing the sperm nucleus to enter the egg cytoplasm. Where the egg is contacted by the sperm, it bulges out to form the **fertilization cone.** The sperm head enters this cone and is engulfed by the egg cytoplasm. Just as there is a mechanism making possible the entrance of a sperm into the egg, there is one that prevents the egg from being fertilized twice. The electrical charge on the egg membrane changes after fertilization takes place; the membrane repels any additional sperm. In this manner, the correct diploid condition of the zygote is preserved.

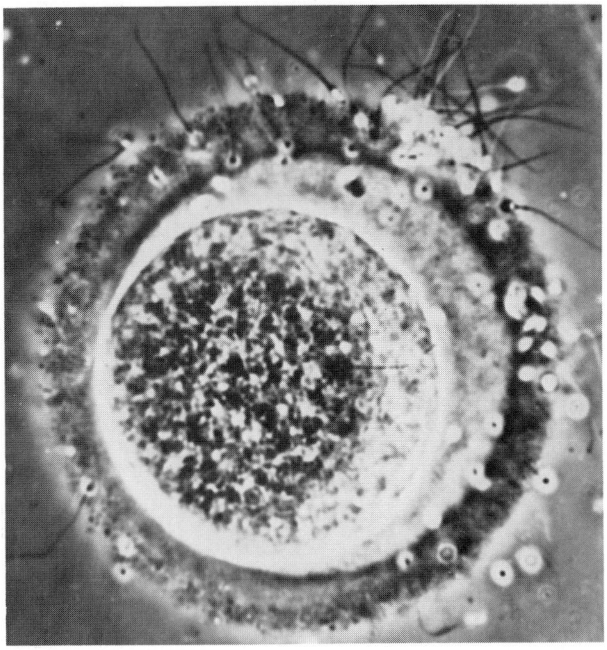

Scanning electron micrograph of an ovum with numerous sperm on its surface.

(continued)

# PANEL 25-1 *(continued)*

Changes in the oocyte and sperm membranes during fertilization. (a) Oocyte before entry of the sperm. The successive stages in the dissolution of the oocyte and sperm membranes are diagrammatically presented in illustrations b–g. (h) Oocyte after entry of the sperm nucleus.

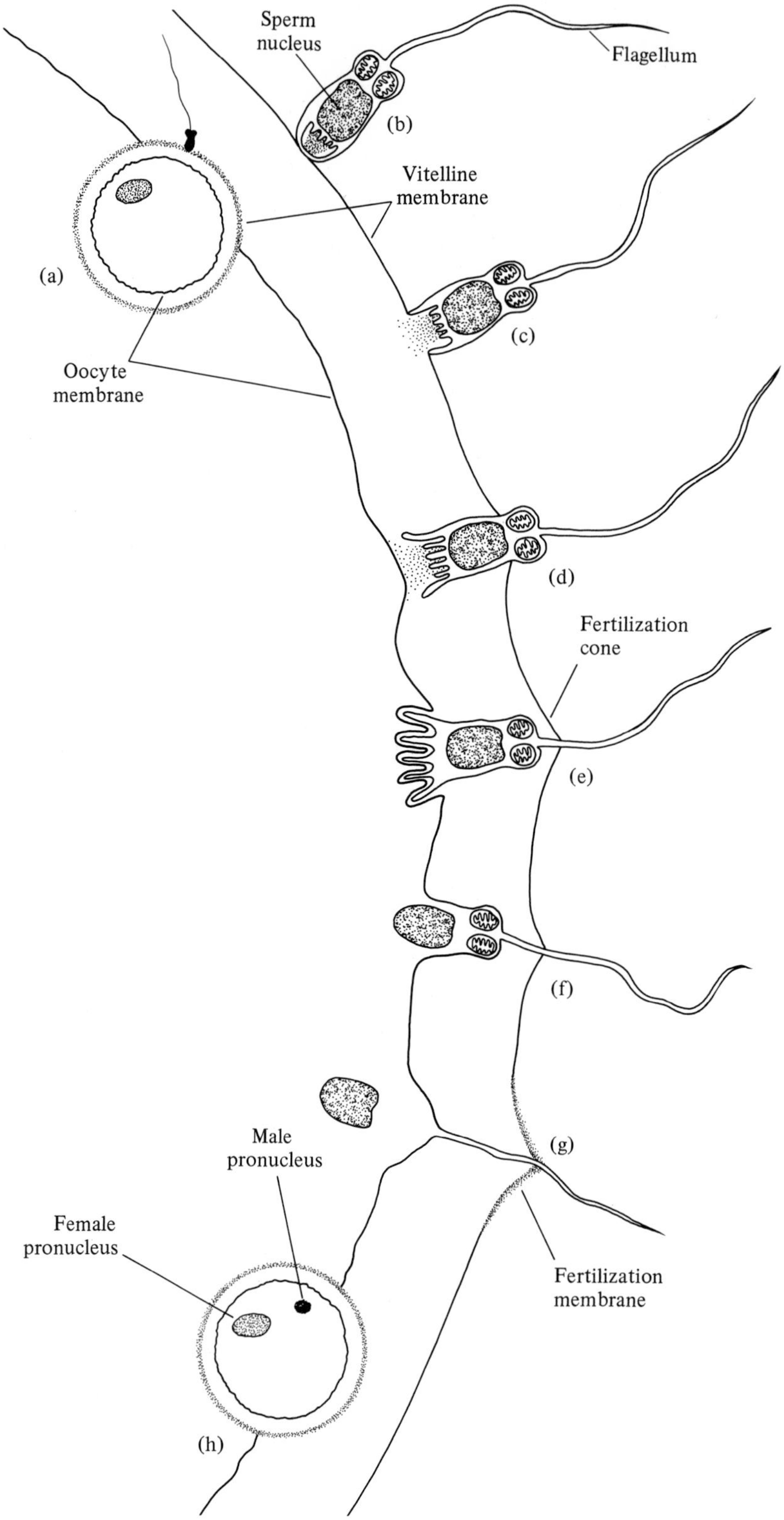

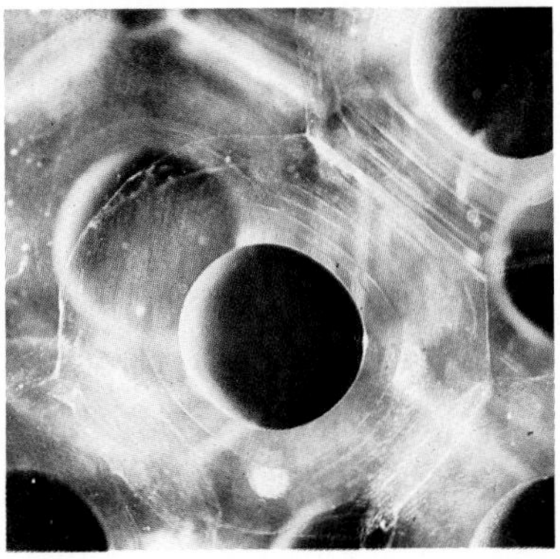

**FIG. 25-5** A frog zygote.

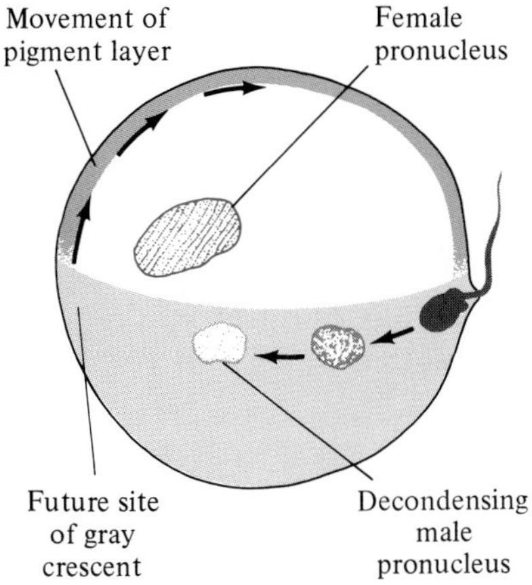


**FIG. 25-6** Diagram of a section through a fertilized egg showing the penetration path of the sperm and the movement of the pigment layer.

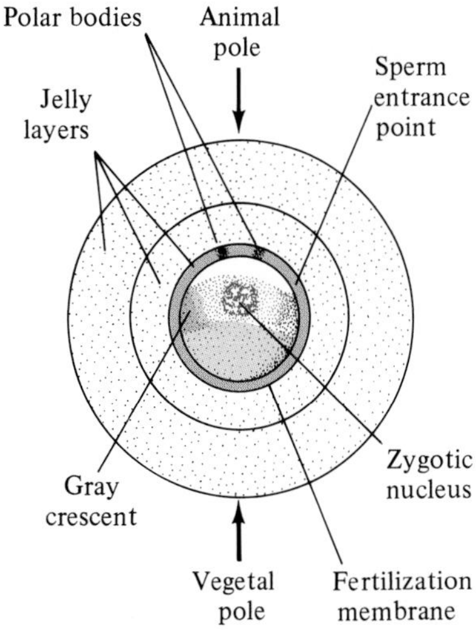


**FIG. 25-7** Diagram of the membranes covering a frog's egg.

***Processes Triggered by the Entry of a Sperm.*** Upon entry of a sperm, rapid, dramatic changes occur in the vitelline membrane, the plasma membrane, and the cortex of the egg. Starting from the spot where the sperm entered the egg and spreading over the entire egg surface, cortical granules first fuse with the plasma membrane, then extrude their contents into the space between the plasma membrane and the vitelline membrane. This causes the vitelline membrane to detach from the underlying plasma membrane, forming a **perivitelline space.** The entry of the cortical granule contents into this space induces a chemical change in the vitelline membrane, making it impervious to the entry of any additional sperm; it is thereafter designated the **fertilization membrane.**

Fertilization also sets in motion many biochemical activities in the egg. Just before fertilization, an oocyte exhibits low rates of oxygen consumption and protein and RNA synthesis. DNA synthesis is halted. Immediately after fertilization, however, the egg shows increased rates of respiration and protein and RNA synthesis. DNA replication also resumes.

***Fertilization of the Frog Egg.*** Although the details differ from one animal group to another, a description of fertilization in amphibians illustrates several general principles of the process.

In frogs, the egg nucleus is located at the **animal pole,** at the top of the darkly pigmented upper, or animal, hemisphere. The sperm enters the egg in the animal hemisphere, just above the equator (Fig. 25-5).* After entry, the sperm nucleus migrates toward the center of the egg, leaving a cone-shaped trail of pigment behind it (Fig. 25-6). Apparently as a result of this movement of the sperm nucleus, there occurs a streaming of pigmented surface cytoplasm from other equatorial areas toward the point of sperm penetration. This streaming produces an area of light pigmentation, called the **gray crescent,** diagonally opposite the entry point (Fig. 25-7). The gray crescent has developmental significance: the sperm entrance point will become the anterior end and the gray crescent area the posterior end of the future embryo. Thus, even before the gametic nuclei have fused, the fertilized frog's egg has changed from a radially symmetrical to a bilaterally symmetrical sphere.

Sperm entrance triggers the resumption of meiosis by the egg. The egg nucleus had previously risen to the surface near the animal pole. There it divides for the second time (meiosis II), producing the second polar body and what can now be termed the **female pronucleus.** The entire process of fertilization, from entrance of the sperm through meiosis II, takes about 20 minutes.

After moving inward from the point of entrance (the penetration path), the sperm nucleus, now termed the **male pronucleus,** moves directly toward the female pronucleus (the copulation path). Whereas the penetration path can be said to determine the future anterior-posterior axis of the embryo, the copulation path usually determines the plane of the first cleavage. About 50 percent of the time the two paths are in the same plane. In these cases, the first cleavage exactly bisects the gray crescent.

* Because yolk is dense, the vegetal hemisphere is the heavier of the two and is thought of as the lower half, while the lighter animal hemisphere is considered the upper half of the egg and the early embryo.

## CLEAVAGE TYPES

As noted, the amount and distribution of yolk within an egg profoundly affect its cleavage pattern. In alecithal and oligolecithal eggs, each of the early cleavages completely divides the cells into approximately equal-sized blastomeres. This type of cleavage is termed **holoblastic equal cleavage** and is typified by the primitive chordates such as *Amphioxus* and by echinoderms, such as the starfish and the sea cucumber (Fig. 25-8). The first cleavage (Fig. 25-8a) is meridional, that is, it passes through the animal and vegetal poles. The second cleavage (Fig. 25-8b) is also meridional and at right angles to the first, dividing the egg into four equal-sized blastomeres, as the cells of the early embryo are called. The third cleavage (Fig. 25-8c) is equatorial, dividing the four cells produced by the first two cleavages into eight: an upper and a lower quartet.

By the fourth cleavage division (the 16-cell stage) (Fig. 25-8d), the blastomeres have pulled away from each other at the center of the cell mass, thus forming a central cavity, the **blastocoel.** As the blastomeres continue to divide, the blastocoel becomes more prominent. The embryo is now known as a **blastula** (Fig. 25-8d–f). It consists of a single layer of cells surrounding a fluid-filled cavity.

***Cleavage in the Frog Egg.*** In a moderately telolecithal egg (one with a moderate amount of yolk concentrated at one end), such as the frog egg (Fig. 25-9), the cleavages are holoblastic but unequal. Due to the resistance provided by the yolk mass in such eggs, the blastomeres of the

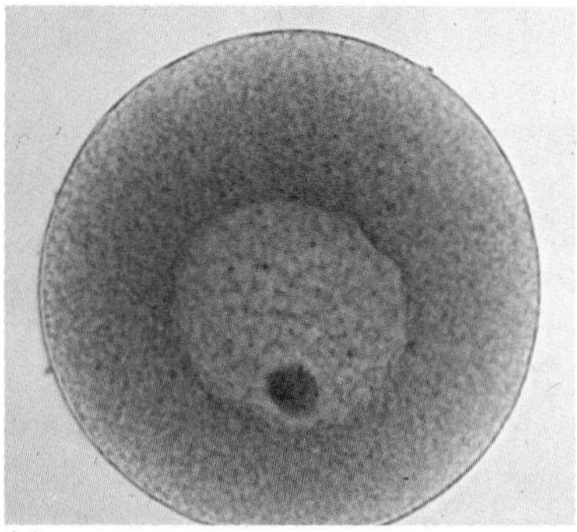
(a)

**FIG. 25-8** Early development of a starfish ovum showing holoblastic, equal cleavage. (a–f) Stages from a fertilized ovum through a 32-cell stage embryo. (g) Section through the blastula.

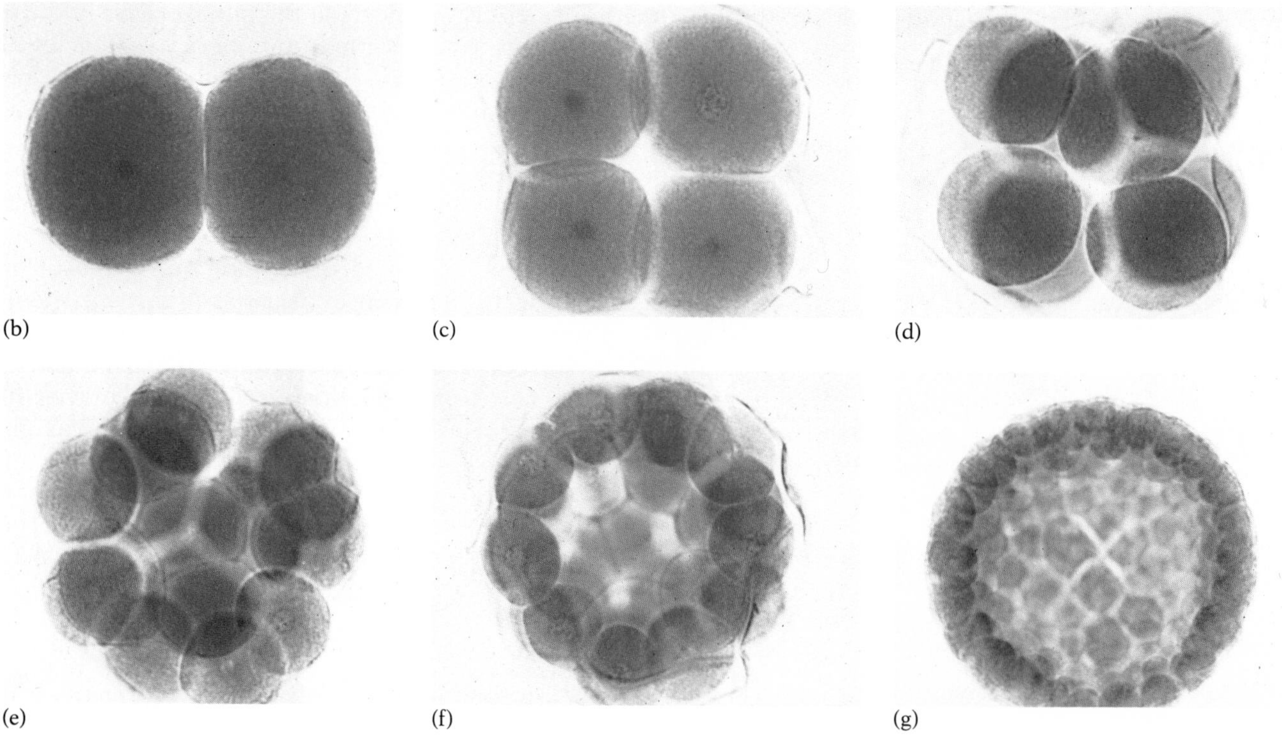

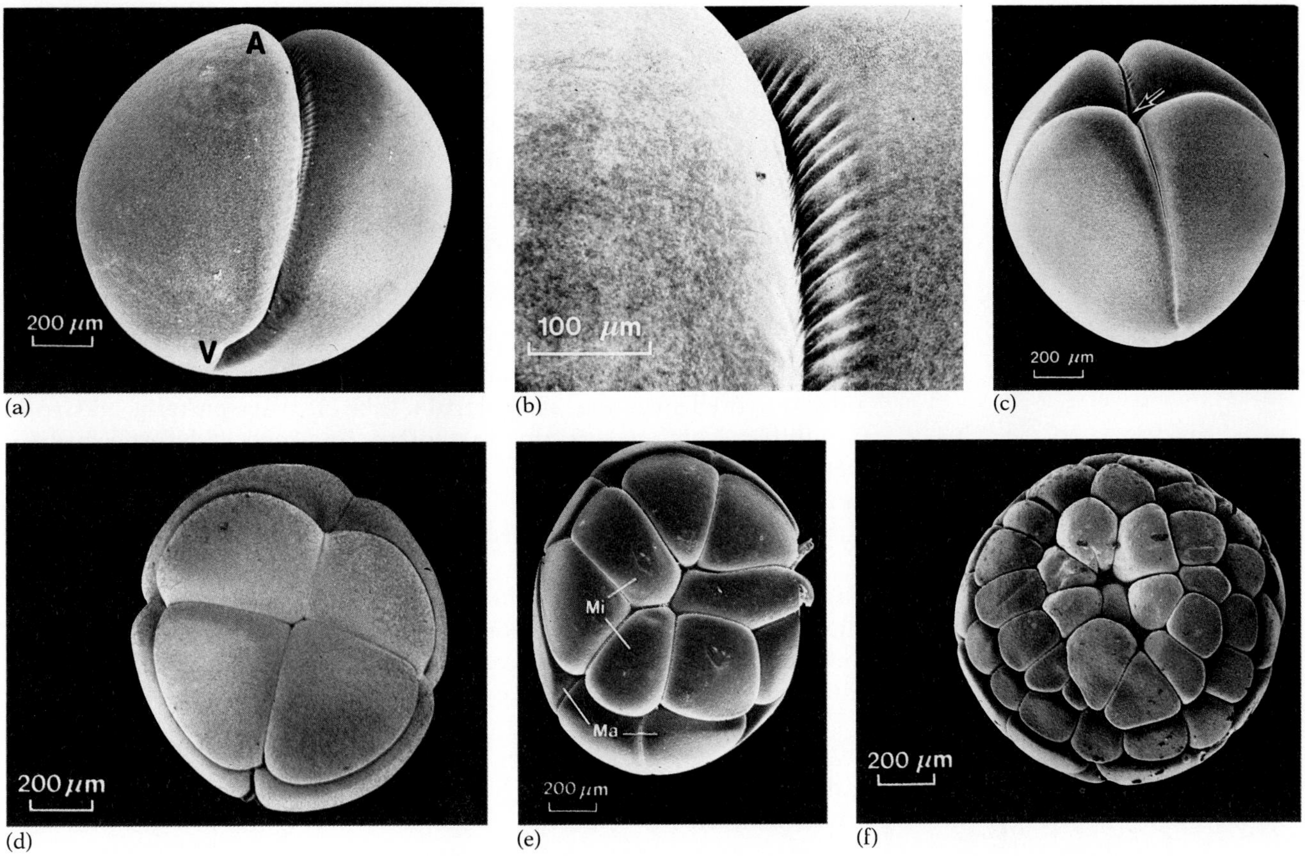


**FIG. 25-9** Early development of frog egg showing holoblastic, unequal cleavage: (a, b) 2-, (c) 4-, (d) 8-, (e) 16-, and (f) 56-cell stages. A: Animal pole; Ma: Macromeres; Mi: Micromeres; V: Vegetal pole. Arrow points to the cleavage furrow.

vegetal hemisphere cleave much more slowly than those in the animal hemisphere, resulting in cells of much larger size. In such eggs, the blastocoel is located entirely within the animal hemisphere (Fig. 25-10).

***Cleavage in Meroblastic Bird Eggs.*** A third type of cleavage, is exhibited by the extremely telolecithal eggs of birds and reptiles. Their very large amount of yolk makes holoblastic cleavage physically impossible.

**FIG. 25-10** Blastula of the frog. (a) Whole mount. (b) Cross section. Bl: Blastocoel; Ma: Macromeres; Mi: Micromeres

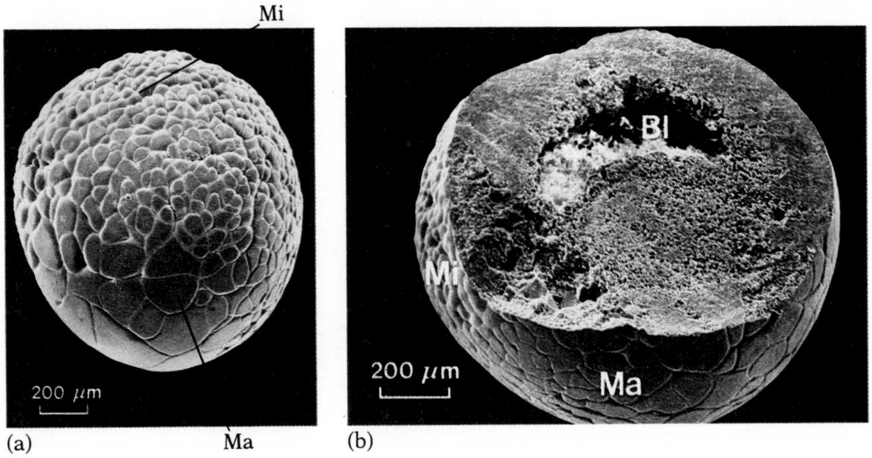

Such eggs exhibit a strikingly different type of cleavage termed **meroblastic cleavage.** Although each of the first cleavage furrows begins to cut into the yolk, none proceeds very far. The next cleavages occur at right angles to the first, forming a cap of cells, the **blastoderm,** atop the yolk. The cells at the blastoderm's center soon lift free of the yolk mass, giving rise to a blastocoel with a blastoderm roof and a yolk floor (Fig. 25-11).

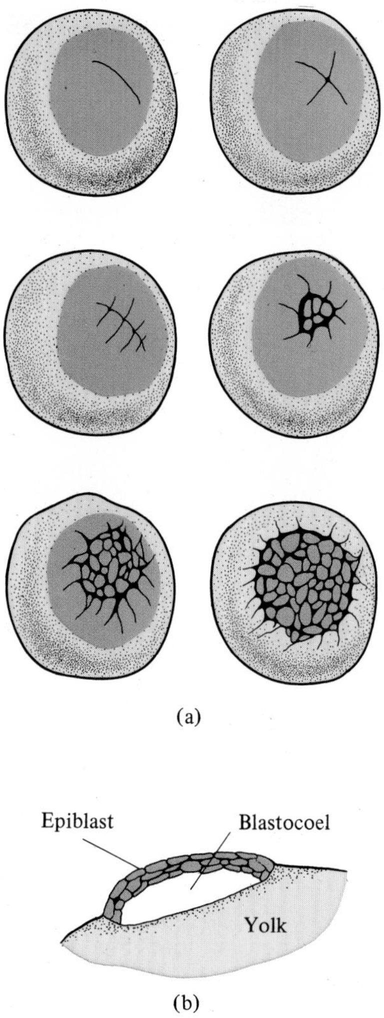


**FIG. 25-11** Cleavage in a chick egg showing meroblastic cleavage. (a) Successive cleavage furrows. (b) Section through blastula.

## GASTRULATION

Gastrulation can be defined as the process by which a blastula becomes a multilayered structure, the **gastrula.** Like cleavage patterns, gastrulation differs from one egg type to another, and most differences are apparently related to differences in the amount and distribution of yolk. Among the simplest patterns are those of embryos developing from oligolecithal eggs, as in the sea urchin (Fig. 25-12).

***Gastrulation in the Sea Urchin.*** The vegetal pole of the sea urchin blastula undergoes invagination, an inward movement of cells from the surface, forming first a depression, then a cavity, termed the **archenteron,** or primitive gut. The opening of the archenteron to the exterior is called the **blastopore.** The embryo now consists of two cellular germ layers: an outer **ectoderm,** an inner **endoderm,** and a few loose cells called **mesenchyme.** A while later, a third germ layer, the **mesoderm,** arises between the ectoderm and endoderm (Fig. 25-13). From each of the three germ layers, specific tissues and organs will develop.

**FIG. 25-12** Gastrulation in a sea urchin embryo. (a) Initiation of invagination. (b–d) Origin of the archenteron, blastopore, endoderm, and ectoderm.

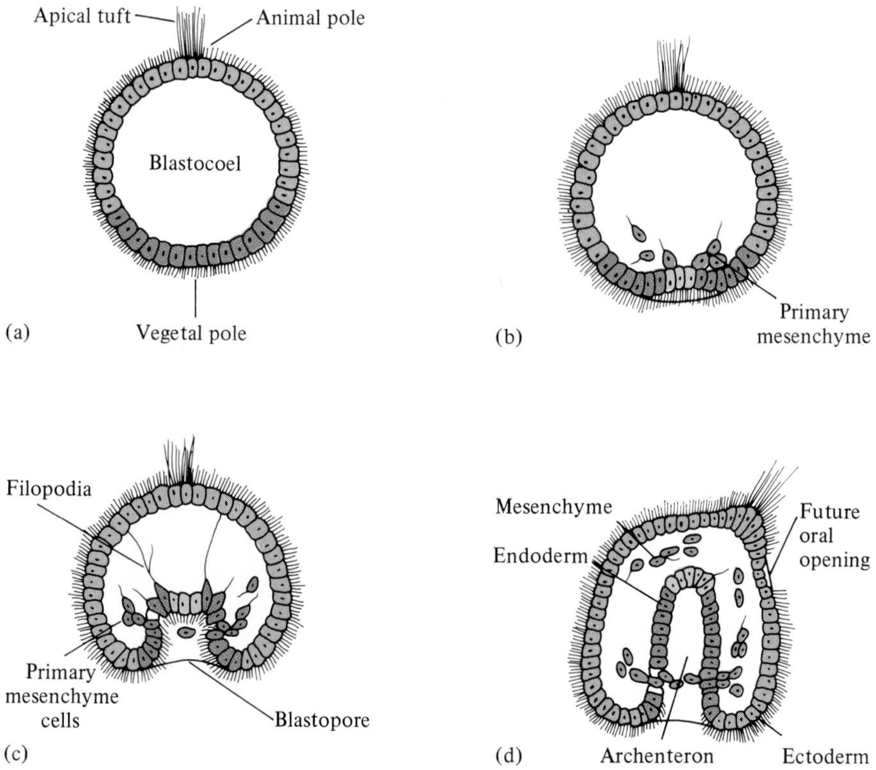

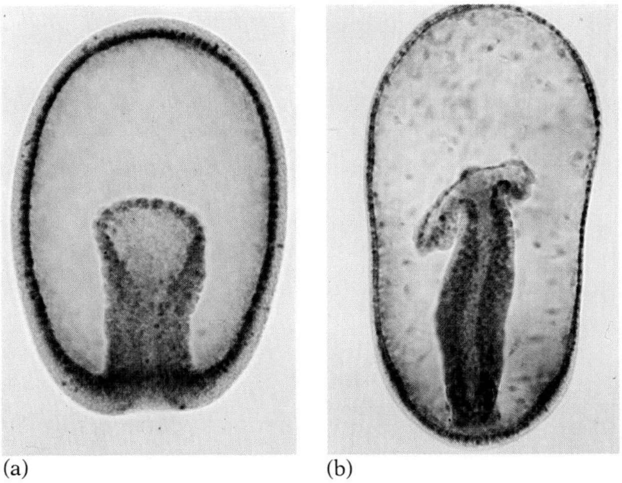


**FIG. 25-13** Enterocoelic mode of coelom formation in an echinoderm (starfish) embryo. (a) Gastrula in an early stage of gut formation. (b) A later stage with coelomic pouches forming from the archenteron.

***Gastrulation in the Frog.*** Although gastrulation in the frog has similarities to echinoderm gastrulation, it exhibits some important differences. It too begins with an invagination to form an archenteron and a blastopore. The large, yolk-laden cells of the vegetal hemisphere, however, render invagination at the vegetal pole physically impossible. Instead, invagination occurs a little below the equator, in the region of the gray crescent (Figs. 25-14 and 25-15). A portion of the layer of pigmented surface cells above the gray crescent now moves downward and then

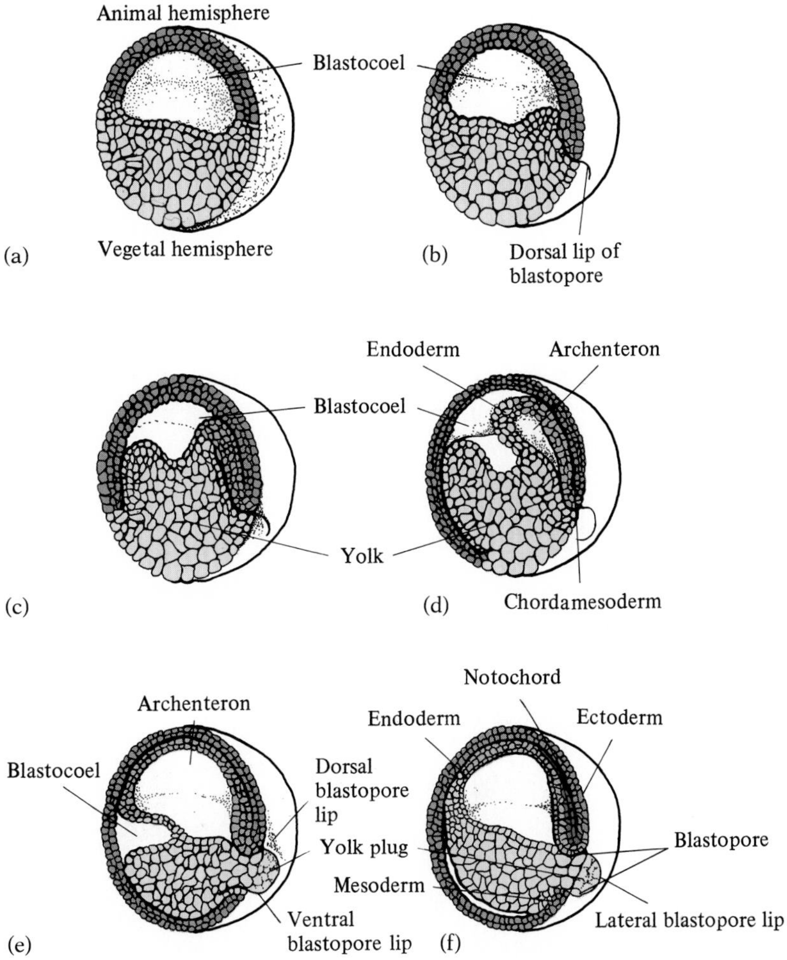


**FIG. 25-14** Diagrammatic gastrulation in the frog embryo. Sketches of (a) Blastula. (b) Early gastrula. (c–f) Later stages in gastrulation. The blastocoel gradually reduces as the gastrocoel enlarges.

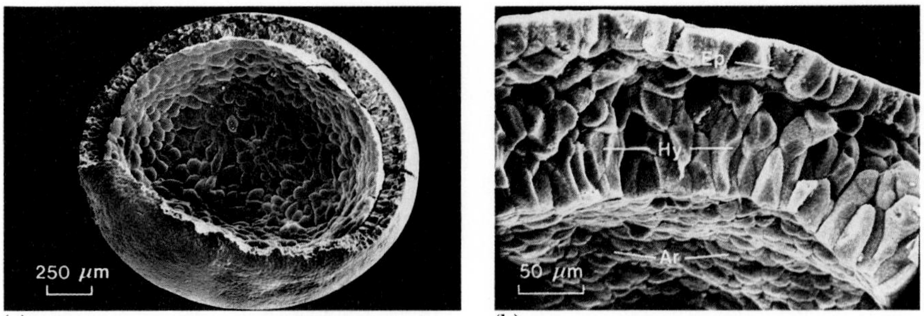


(a) (b)

**FIG. 25-15** Scanning electron micrograph of an amphibian gastrula. (a) The inner layer is exposed. (b) Cross section through hypoblast (Hy) and epiblast (Ep).

flows inward, disappearing from the outer surface into the blastopore, in a process called **involution.** As a broader expanse of cells becomes involved in this inward movement, the resulting blastopore acquires a semicircular shape (Fig. 25-16c). As surface cells continue their involution, the newly formed archenteron continues to enlarge; as it pushes upward toward the animal pole, it creates a bulge in the floor of the blastocoel. As the bulge enlarges, the blastocoel is slowly obliterated (Fig. 25-14e, f).

While continuing their involution and thus adding to the enlargement of the archenteron, the surface cells above the blastopore begin to flow downward more rapidly than they disappear into the blastopore. The result is a movement called **epiboly,** a flowing over the yolk-laden cells of a folded layer of pigmented cells (Fig. 25-16b, c).

As the archenteron continues its enlargement at the expense of the shrinking blastocoel, it pushes its way to both sides, where it courses ventrally, in a motion not unlike that of a pair of hands scooping up the yolk-laden cells (Fig. 25-14f–h). These lateral extensions of the archenteron, and thus of the lips of the blastopore as well, finally meet ventrally, completely encircling the yolk-laden cells at the vegetal pole. Externally, the combination of epiboly and involution that has been occurring on all sides of the blastopore (Figs. 25-14d–h and 25-17a) re-

**FIG. 25-16** Surface view of gastrulation in amphibian embryo. (a) Early stage of blastopore lip formation. (b) A wide circular blastopore. (c) Late gastrula with a small circular blastopore.

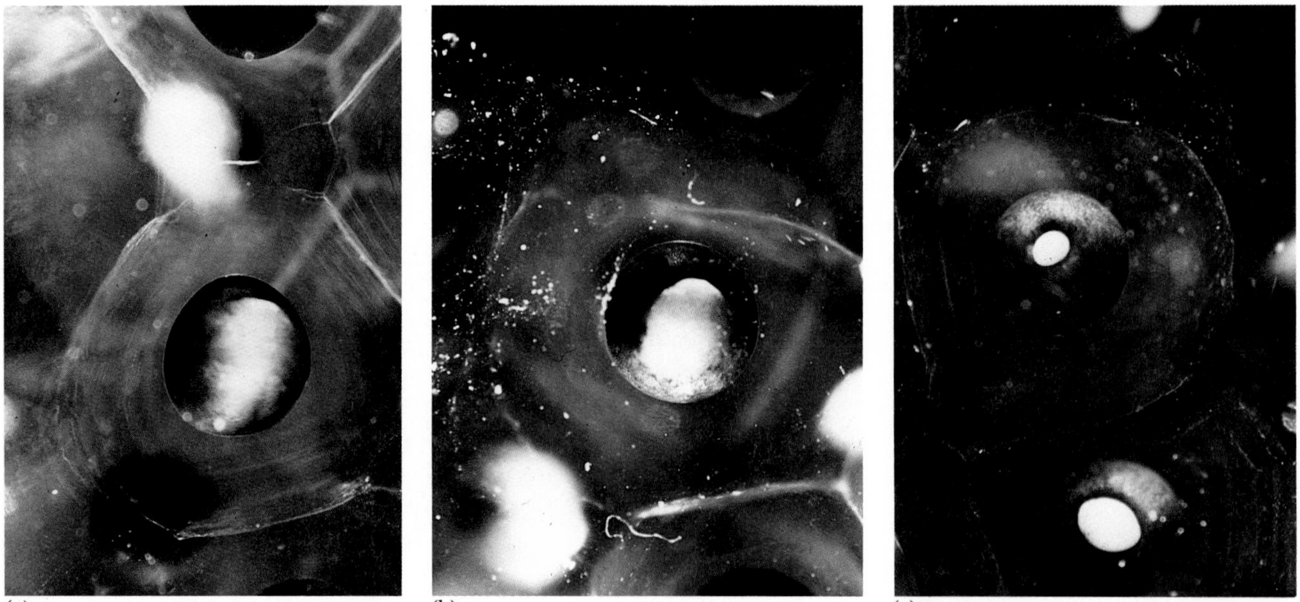

(a) (b) (c)

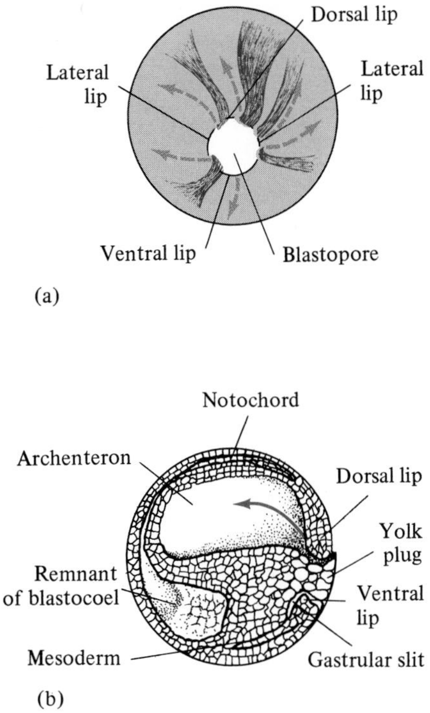


**FIG. 25-17** Gastrulation movements in amphibian embryos. (a) Diagrammatic presentation of the direction of movement of cells in gastrulation. (b) Sagittal section through a late gastrula.

**FIG. 25-18** Early gastrulation in chick embryo showing hypoblast separation. (a) Blastoderm before hypoblast cell accumulation. (b) Accumulation of hypoblast cells at the posterior end of the blastoderm. (c) Spreading of the hypoblast.

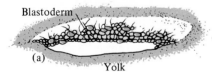

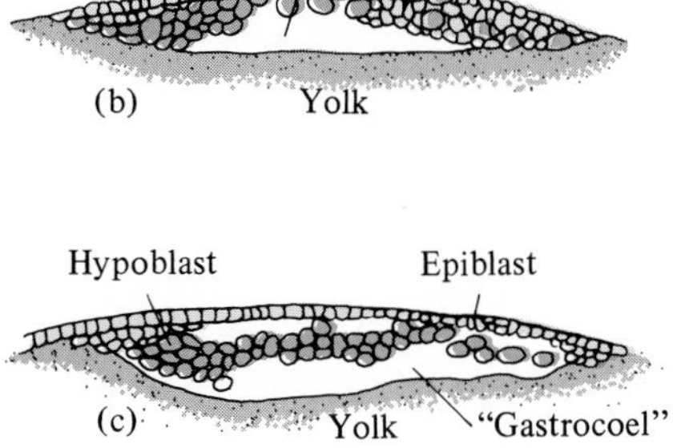


duces the amount of yolk-laden cells visible from the exterior to a very few; these form the so-called **yolk plug** (Fig. 25-17b). The blastopore at this stage presents a circular appearance; its edges are termed its dorsal, ventral, and lateral lips. Eventually, the blastopore closes over completely.

***The Germ Layers of the Frog.*** In a frog embryo, as soon as the archenteron begins to form, the layer of cells that comprises its floor can be termed the endoderm. Before long, different regions are distinguishable within the archenteron: a foregut, midgut, hindgut, and a liver diverticulum. As noted the outer layer of cells of the embryo is called the ectoderm. The mesoderm in a frog embryo arises from the mass of cells that involute into the blastocoel in the region of the dorsal lip of the blastopore (Figs. 25-14d and 25-17b). Because a part of the forming mesoderm will develop directly into the notochord, this tissue is known as the **chordamesoderm.** Chordamesoderm cells proliferate both laterally and ventrally, pushing their way in between the ectoderm and endoderm. The chordamesoderm tissue in the region immediately underlying the middorsal surface of the embryo condenses to form the **notochord,** a cartilagelike, supportive, rod-shaped structure found along the dorsal midline of all chordate embryos. In vertebrates, it is replaced by the vertebral column later in development (see Figs. 25-14f and 17b). As in all chordates, the ectodermal cells overlying the notochord develop into the neural plate and eventually into the central nervous system.

***The Unique Features of Gastrulation in Birds.*** Like the formation of the blastula, gastrulation in bird embryos differs from that in frogs and sea urchins in ways to accommodate the great mass of yolk. As discussed earlier, in the chick, the blastula is formed when the central part of the cap of cells, called the **blastoderm,** lifts free of the yolk mass at the center while remaining attached to it along the edges (Fig. 25-18a). At this stage, a central clear area, the **area pellucida,** and a peripheral area, the **area opaca,** are distinguishable. Gastrulation, which occurs in two phases, begins when cells at the posterior end of the blastoderm rapidly multiply and then migrate inward and anteriorly into the blastocoel to form a sheet of cells underlying the blastoderm (Fig. 25-18b, c). The blastoderm is now termed the **epiblast,** the new layer below it the **hypoblast.** The hypoblast gives rise to the endoderm; the epiblast contributes cells to the endoderm and mesoderm, and those that remain behind become the ectoderm.

During the second phase of chick gastrulation, cells on the surface of the epiblast migrate from the sides toward the median line, where they involute to produce an elongate depression, the **primitive groove.** The groove and the two primitive ridges are together called the **primitive streak** (Fig. 25-19a, b); the streak is homologous to the blastopore of the amphibian embryo. The streak lengthens anteriorly, at its full extent occupying the posterior two-thirds of the spreading epiblast. It ends anteriorally in a pit in the center of a raised portion, **Hensen's node** (Fig. 25-19a, c). Cells extending anteriorly from Hensen's node develop into the notochord. The overlying cells develop into the neural plate, as in all chordate embryos.

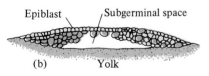

Like the dorsal lip of the blastopore in amphibian embryos, the primitive streak is not a static structure. Its two ridges are, in fact, produced by an accumulation of migrating epiblast cells. These cells involute, thus contributing to an additional internal layer, the meso-

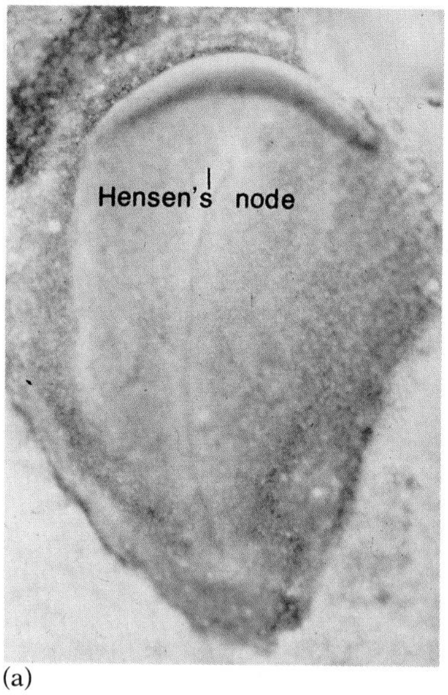

(a)

(b)
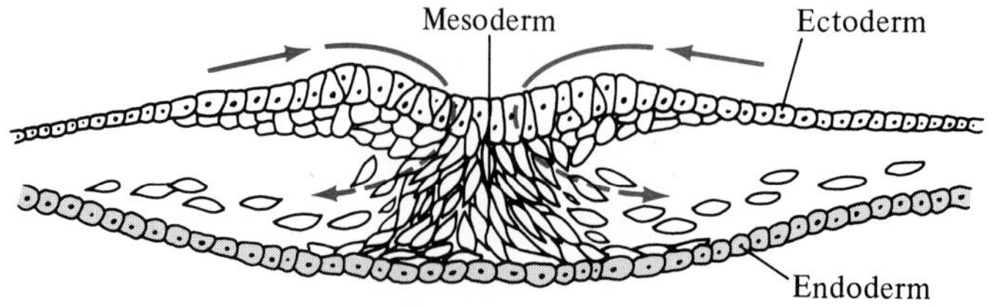


(c)
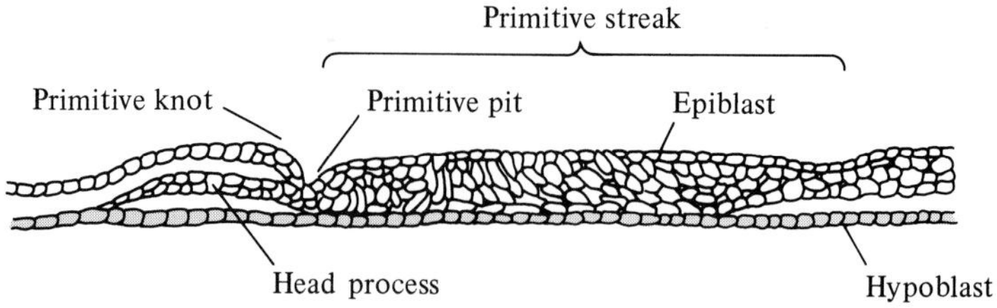


**FIG. 25-19** Late gastrulation in a chick embryo. (a) Surface view of an early embryo. (b) Cross section through the primitive streak. (c) Longitudinal section adjacent to the primitive streak. (d) Neural plate stage.

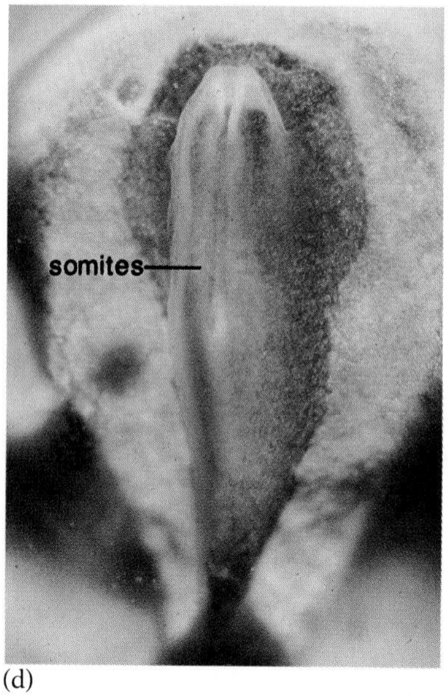

(d)

derm, which now spreads to both sides of the midline between the epiblast and the hypoblast (Fig. 25-19b–d). Like the blastopore, the primitive streak disappears after the completion of gastrulation.

Fishes and amphibians lay their eggs in water, and embryonic development takes place in the aquatic medium. Being immersed in water, the embryos weigh much less than in air. Consequently, invagination, overgrowth, evagination, and other processes can occur with relative ease. They do not have to overcome the problem of gravity. The eggs of land vertebrates, on the other hand, are laid on land. As noted earlier, like the chick egg they are protected by a calcareous though porous shell that protects the embryo from desiccation. But the problem of gravity and weight can be overcome only by a fluid medium. An "artificial pond" constructed from cells that do not participate in embryo formation provides such a fluid medium surrounding the embryo. The structures that form these investments outside the embryo are called the extraembryonic membranes and are described in Panel 25-2.

The development of the extraembryonic membranes of all amniotes is very similar. They arise from either the extraembryonic ectoderm or endoderm and the associated mesoderm. There are four extraembryonic structures: **yolk sac, amnion, chorion,** and **allantois.** The yolk sac arises from endoderm and, as the name implies, encloses the yolk if present. Even when yolk is absent, as in the eggs of modern mammals, a small yolk sac develops and serves transiently in nourishing the embryo by transferring nutriment from the uterine fluids. The amnion and the chorion are twin membranes that arise as a single fold of the ectoderm. The fluid-filled amniotic cavity is the "artificial pond" in which the embryo develops. The allantois is an outgrowth from the hind end of the gut and thus arises from the endodermal layer.

## ORGANOGENESIS

Although a detailed account of organ formation is beyond the scope of this presentation, a brief description of the major events leading to or-

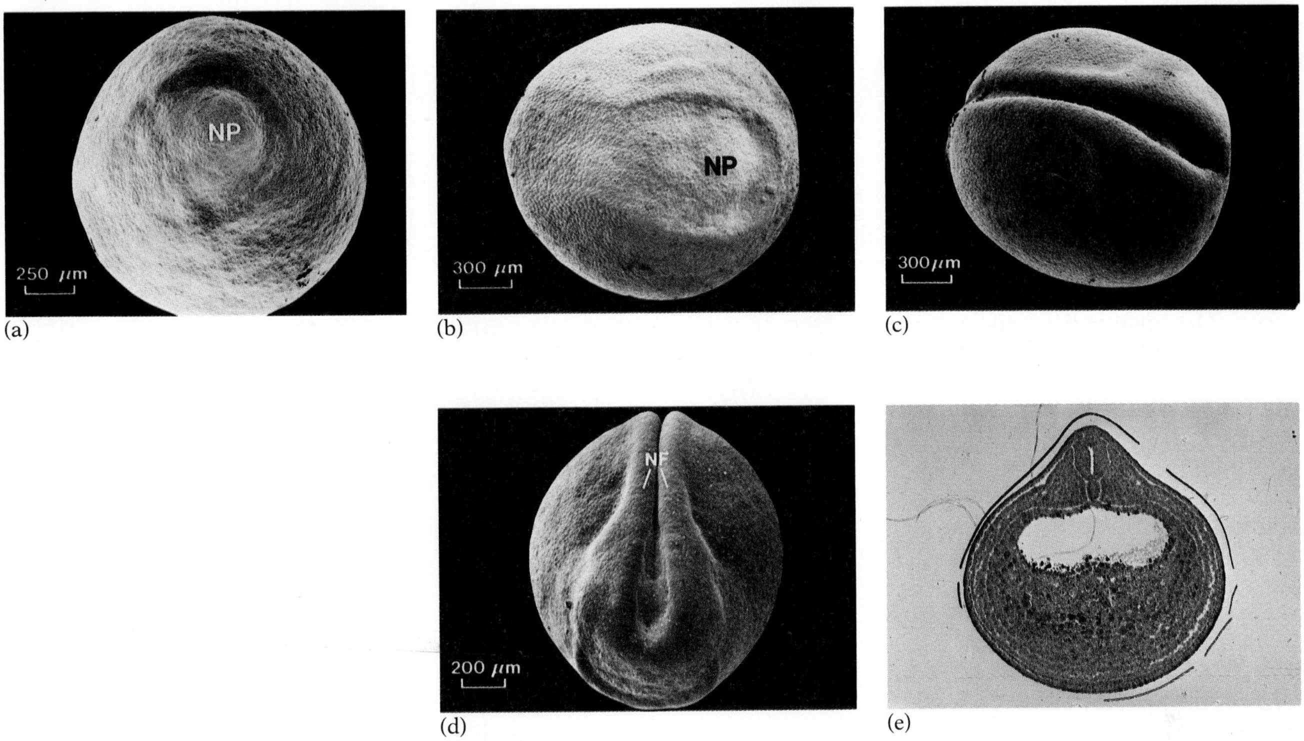


**FIG. 25-20** Neurulation in a frog embryo. (a, b) Early and late stages in neural plate (NP) formation. (c, d) Early and later stages of neural fold formation. (e) Cross section showing neural tube formation.

**FIG. 25-21** Neurulation in a chick embryo.

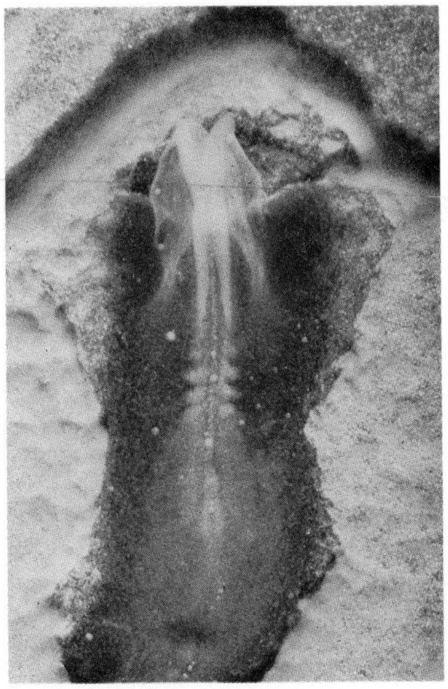

ganogenesis is appropriate. The shaping of organs occurs through a combination of morphogenetic movements, differential rates of growth of various tissues, and a process called **embryonic induction,** in which the proximity of a particular germ layer or tissue elicits specific developmental responses from adjacent tissues. A classic example of organogenesis is the formation of the vertebrate brain and spinal cord, exemplified by the developmental process that occurs in the frog embryo.

In the frog, after gastrulation is completed, there occurs a flattening of the ectoderm all along the dorsal surface, forming what is called the neural plate. Next, there develops an elongate neural groove along the plate's midline, while a pair of ridges, the neural folds, simultaneously arise along its edges (Fig. 25-20a–d). As the folds continue to rise, they arch toward each other, finally roofing over the groove to form the neural tube (Fig. 25-20e). From the neural tube arise the brain and spinal cord; the tube lumen persists as the cerebrospinal canal and the ventricles of the brain. Even as the neural tube is forming, the underlying chordamesoderm completes its differentiation into a notochord. To either side of the notochord, mesodermal cells aggregate into a series of segments, or blocks of cells, called **somites.** These eventually give rise to the segmentally arranged vertebrae and to the skeletal muscles of the body.

Changes very similar to these in the frog also occur in other vertebrate embryos. In the chick, for example, mesoderm anterior to the primitive streak gives rise to the notochord, while its overlying epiblast, now called the ectoderm, forms the neural plate, neural groove, and neural tube (Fig. 25-21) from which arise the brain and spinal cord.

PANEL 25-2

# The Extraembryonic Membranes of Amniotes

The three highest classes of vertebrates—reptiles, birds, and mammals—are collectively referred to as **amniotes,** and comprise the superclass Amniota. The term is derived from the amnion, one of several membranes that enclose the embryo in these animals. The eggs of lower vertebrates, or **anamniotes,** are laid in and develop in water. Reptiles, however, evolved a shelled egg, which could be laid on land, thus freeing them from the aquatic environment in this regard. As birds and mammals evolved from the ancient reptiles, they retained the same general pattern of embryonic development, even though some reptiles and almost all mammals came to bear their young alive. The extraembryonic membranes are cases in point. All these membranes are found in all amniotes, although some have come to serve different functions in live-bearing amniotes than they do in amniotes that lay eggs.

***The yolk sac.*** By the sixteenth hour of incubation, the gut, or **enteron,** of the chick embryo is represented by a flat, circular cavity with a roof of endoderm and an overlying mesoderm that arch over a floor of noncellular yolk. At its edges, the roof of the developing gut is continuous with the edges of the blastodisc, where a triple layer of ectoderm, mesoderm, and endoderm overlies the yolk to form a region called the area opaca (Fig. 25-19a). Here the endoderm is closely applied to the surface of the yolk. As growth continues, the area opaca continues to proliferate, spreading across a wider expanse of the yolk. Eventually this tissue forms the yolk sac, which, by the completion of embryonic development, will completely enclose the yolk mass.

By the twenty-fourth hour of incubation, an anterior pocket has developed in the archenteron, forming the foregut in the head region. A day

(a) A 36-hour chick embryo showing the amnion, optic vesicles, regions of the brain, and somites. (b) Early stages in the formation of the extraembryonic membranes in a chick embryo. (c) The fully formed extraembryonic membranes in a chick embryo. (d) The extraembryonic membranes of an egg laying mammal, the duck-billed platypus.

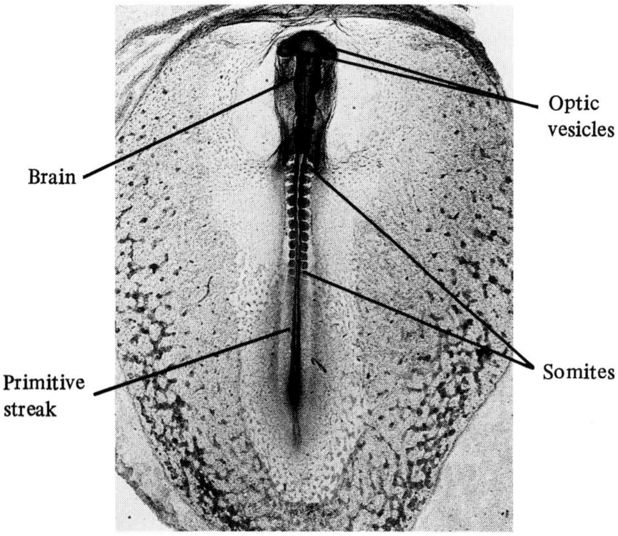

(a)

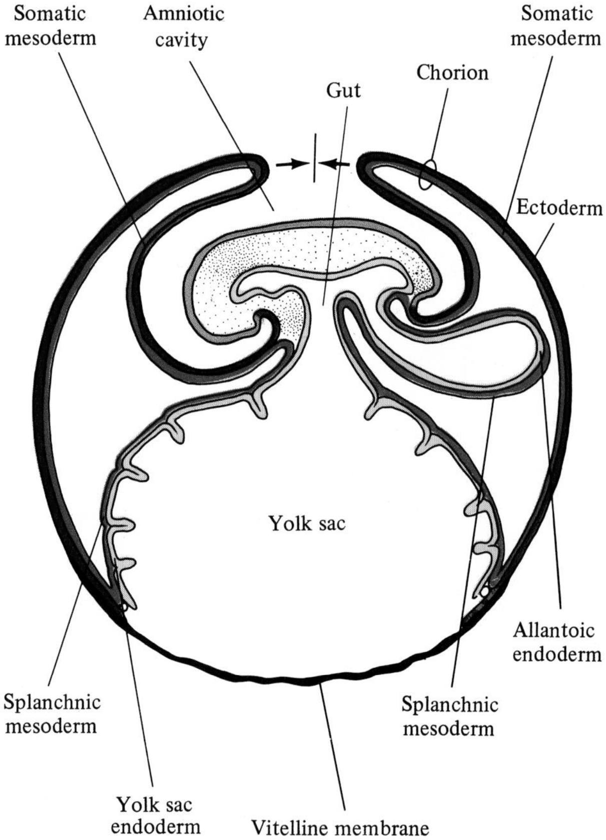

(b)

later, the hindgut has also formed. By the fourth day, the yolk sac has covered most of the yolk. Where it connects with the gut, it is referred to as the **yolk stalk.** Its enzymes digest the underlying yolk, and the released nutrients are transported to the embryo via numerous blood vessels. As the embryo grows, the yolk mass and the yolk sac diminish in size until 2 to 3 days before hatching (which occurs on the twenty-first day), when its remnants are drawn in and incorporated into the gut. A vestige of the yolk sac is detectable in the small intestine several days after hatching.

***The amnion and the chorion.*** During the second day of incubation, the well-defined embryonic head of the chick juts over and then sinks into the underlying yolk mass. Immediately in front of the head, a crescent-shaped fold of the blastoderm, termed the amniotic head fold, rises and grows back in hoodlike fashion over the head. During the third day, a similar amniotic tail fold appears behind the tail bud and begins to grow forward. Later, two lateral amniotic folds appear at the sides of the embryo. The membrane forming the amniotic folds is double-layered, each of the two layers itself composed of a layer of ectoderm and a layer of mesoderm. All four folds eventually meet above the embryo, completely enclosing it in two double-layered membranes.

Where the amniotic folds meet above the embryo, they fuse, whereupon the outer membrane, now termed the chorion, becomes free of and separate from the inner membrane, now termed the amnion. The space enclosed by the amnion, and within which the embryo lies, is the **amniotic cavity.** The space enclosed by the chorion is the **extraembryonic coelom,** or exocoelom. (A coelom is defined as a body cavity entirely lined by mesoderm or mesodermally derived tissue. The extra-

*(continued)*

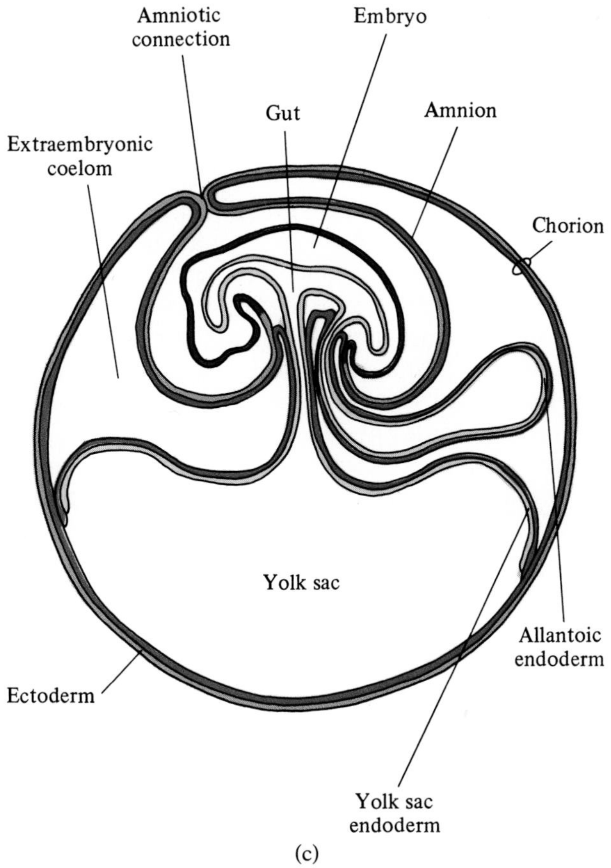


(c)

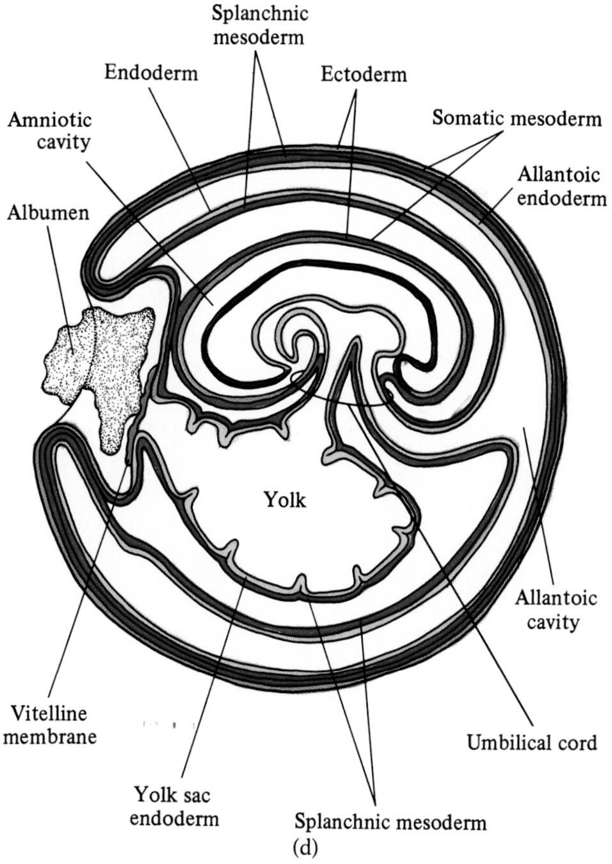


(d)

## PANEL 25-2 *(continued)*

embryonic coelom is such a cavity but is outside the body of the embryo.)

***The allantois.*** Even as the embryo is enclosed by the amniotic folds, a fourth extraembryonic membrane arises. At 72 hours of incubation, a saclike extension of the hindgut, the allantois, arises just behind the yolk stalk. From the fourth to sixth days, it expands into the exocoelom between the amnion and chorion, gradually replacing that cavity with its own, the **allantoic cavity.** Where it comes in contact with the chorion, it fuses with it, forming the **chorioallantoic membrane,** which underlies the eggshell.

The richly vascularized allantois serves several functions in the chick, including aiding in respiratory exchange through the porous eggshell. The allantoic cavity serves as a storage vessel for uric acid crystals excreted during the chick's embryonic development. The portion of the allantoic stalk that lies within the embryo gives rise to tissues of the urinary bladder.

The fate of yolk sac in the three classes of mammals provides an instructive evolutionary series. In monotremes, the eggs have a large yolk, and the yolk sac is well developed, as in birds and reptiles. In marsupials, the embryo receives its early nourishment from the mother's uterus. These embryos have a poorly-developed yolk sac that helps absorb nutrients from the uterus. The little yolk that is present in the egg does not get enclosed by the yolk sac. Eggs of placental mammals bear no yolk; yet a yolk sac forms as if by evolutionary recapitulation.

# DIFFERENTIATION

Identification of the factors that influence or control development of a more or less homogeneous egg into a complex individual composed of various strikingly different organs is a subject of inquiry that dates back to the time of Aristotle. Modern investigations of the matter have taken two major approaches: (1) analysis of cell and tissue interactions at the cell and tissue level and (2) biochemical studies. The first type of approach has dealt mainly with the developmental effects of certain adjacent cells and tissues upon each other. The experiments in the first part of this century were almost exclusively concerned with this approach. Because many of the questions analyzed in this way remain unanswered, such investigations continue today. Despite their lack of sophisticated biochemical methods, some of the early investigators also sought to explain developmental phenomena in chemical terms. Not until the advent of modern molecular genetics, however, did this aspect of developmental investigation, called **developmental genetics,** begin to come of age. Only as our understanding of gene action has deepened have we gained insights into the molecular events underlying development.

## PREFORMATION VERSUS EPIGENESIS

One of the oldest questions about development, one debated by the ancient Greeks, was whether an adult organism was already present in miniature form, that is, preformed, in the egg. The alternative view, **epigenesis,** held that the complexity seen in the adult arose during development, by a gradual adding on of detail.

The preformation theory proposed that the zygote, if not the unfertilized egg, was a mosaic, each region of which was capable of developing only into certain parts of the adult. An experiment that seemed to confirm this view was performed by Wilhelm Roux, a pioneer German experimental embryologist in the late 19th century. Roux destroyed one of the cells of a two-cell–stage frog embryo by pricking it with a hot needle. The remaining viable cell continued to develop—into a half embryo. Although the preformationists were encouraged by this, further experiments showed that if the two cells of the two-celled stage were separated from each other, each developed into a complete, although somewhat smaller than normal, frog embryo. Similar experiments on several other species of animals gave similar results. In sea urchins, for example, when the four-cell stage was separated into four cells, each cell developed into a complete individual. These results established that, in these species at least, there was no rigid preformation. Even a fourth of the egg was capable of developing into an entire embryo with no defects. The part of an embryo that a particular portion of an egg developed into thus seemed to depend on its relative position within the developing embryo (Fig. 25-22). Another type of experiment that reinforced the epigenetic view of development was one in which two young embryos were fused with each other. Depending on the orientation of the fused embryo, the fusion produced a single, but unusually large, embryo.

As these studies continued, however, not all embryos were found to follow this pattern. In some species, removal of specific blastomeres in the two- or four-cell stage resulted in corresponding defects in the embryo (Fig. 25-23). Such eggs came to be termed **mosaic eggs.** Those, such

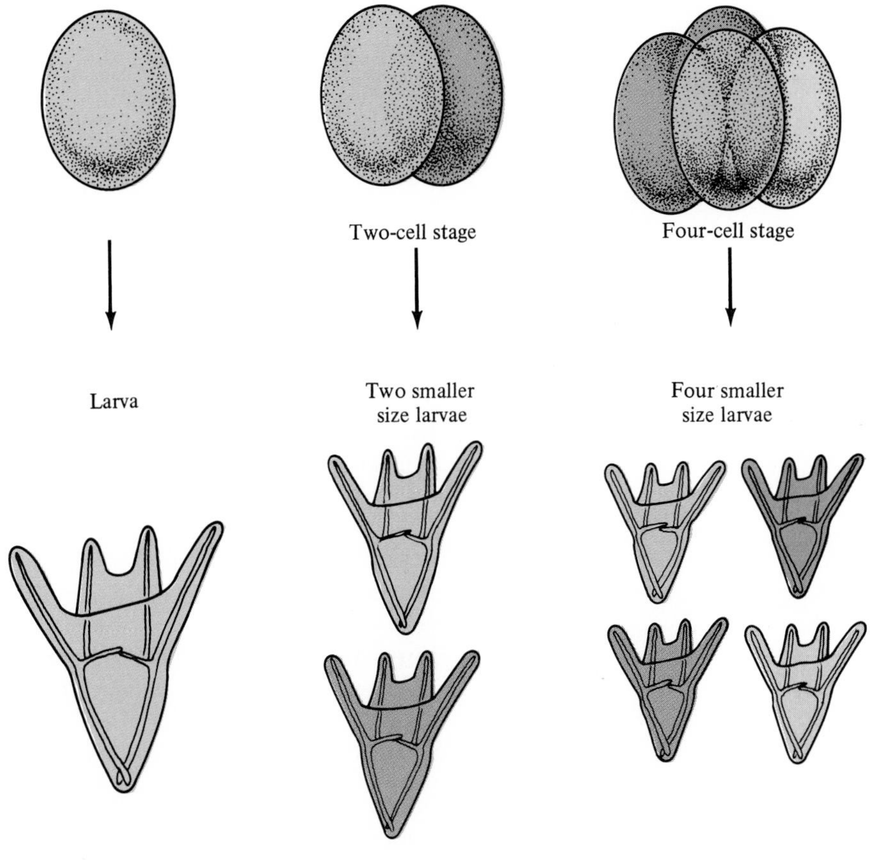


**FIG. 25-22** Experimental protocol used to identify a regulative egg with indeterminate cleavage.

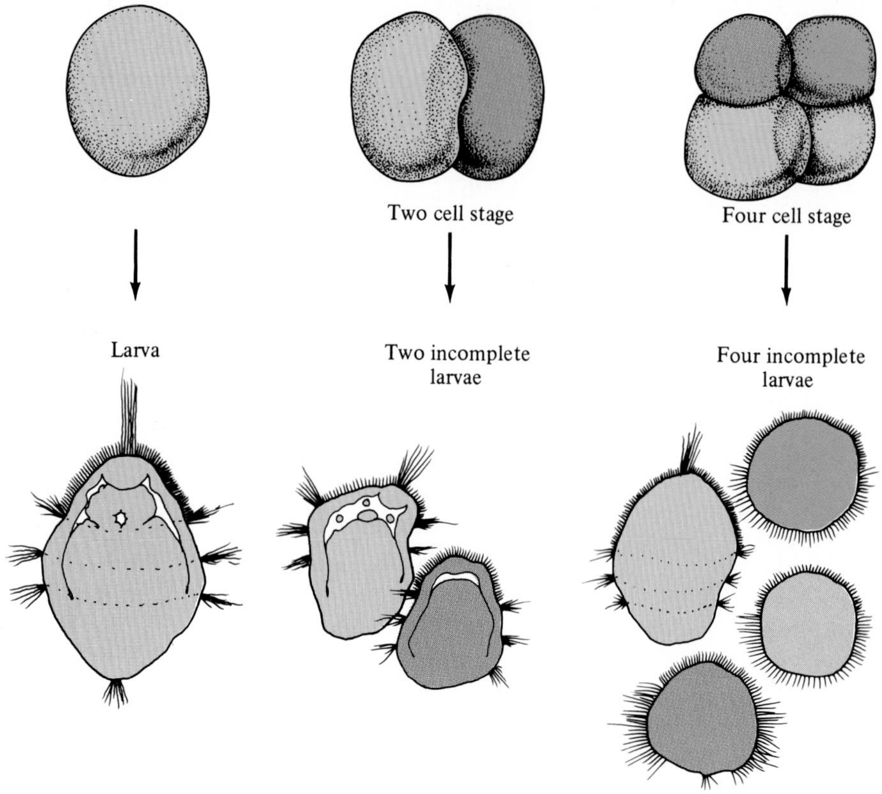


**FIG. 25-23** Experimental protocol used to identify a mosaic egg with a determinate pattern of development. The type of embryo formed after separation of blastomeres at the 2- or 4-cell stage is shown.

as sea urchin's and frog's eggs, that adjusted to early defects were called **regulative eggs.** Mosaic eggs were said to exhibit **determinate development,** whereas regulative eggs were said to have **indeterminate development.**

As experiments were extended to older embryos of regulative eggs, in every case a stage of development was reached at which removal of parts from the embryo did produce a lasting defect. In the eight-cell stage of sea urchins, for example, when the top quartet of cells was separated from the bottom quartet, both halves exhibited defects as development proceeded. Nor could a single blastomere isolated from an eight-cell stage of the sea urchin embryo develop into a complete one. This stage of determination is reached at different points in different species. In some coelenterates, it is reached only at the 64-cell stage. These results thus showed that there is no absolute difference in eggs in this regard. At some stage in the development of all eggs, various regions have become irrevocably committed as to what they can become.

It was from such studies that the concept of embryonic determination arose. A cell or tissue becomes determined to be a certain structure at a stage in development that precedes its actual differentiation into that structure. If, for example, ectoderm from the dorsal surface of a very early gastrula is removed to another location, it does not develop into a neural tube, although its cells may undergo some morphological changes. But if the experiment is performed somewhat later, after the neural plate has been formed, the transplanted piece of tissue proceeds to develop into a recognizable portion of the neural tube, just as it would have if left in place. Thus, although each tissue has a broad prospective potency in early stages of development, there comes a stage when it is determined, that is, irrevocably committed to but one of the several developmental pathways it previously was capable of following.

## INDUCTIVE INTERACTIONS IN ORGANOGENESIS

One of the most exciting discoveries of the early part of the century was that adjacent tissues interact with each other in ways that influence organ formation. The German biologist Hans Spemann, in a 20-year series of experiments, showed that the chordamesoderm underlying the ectoderm of the newt gastrula not only induces the ectoderm to develop into the nervous system, but sets in motion an entire series of developmental events that produce the larva of this amphibian. His most dramatic demonstration was an experiment done in collaboration with Hilde Mangold in 1924. Spemann was awarded the 1937 Nobel Prize in Physiology or Medicine for his work on inductive phenomena.

In the newt, a tailed amphibian, cells of the dorsal lip of the blastopore give rise to the chordamesoderm, which underlies the dorsal ectoderm that differentiates into neural plate and neural tube. In their classic experiment, Spemann and Mangold transplanted the dorsal lip of the blastopore of a heavily pigmented species of newt into the blastocoel of a blastula stage of a lightly pigmented species where, at gastrulation, it came to underlie the belly ectoderm of the host embryo (Fig. 25-24). There it differentiated into chordamesoderm, as it would have if left in place. Because of the differences in natural pigmentation, the tissues of the donor and host were distinguishable from each other throughout their development. In some of these experiments, the belly ectoderm of the host, under which lay the implanted chordamesoderm tissue, developed into a neural tube. In some cases, a nearly complete secondary embryo arose where the belly normally would have been. The

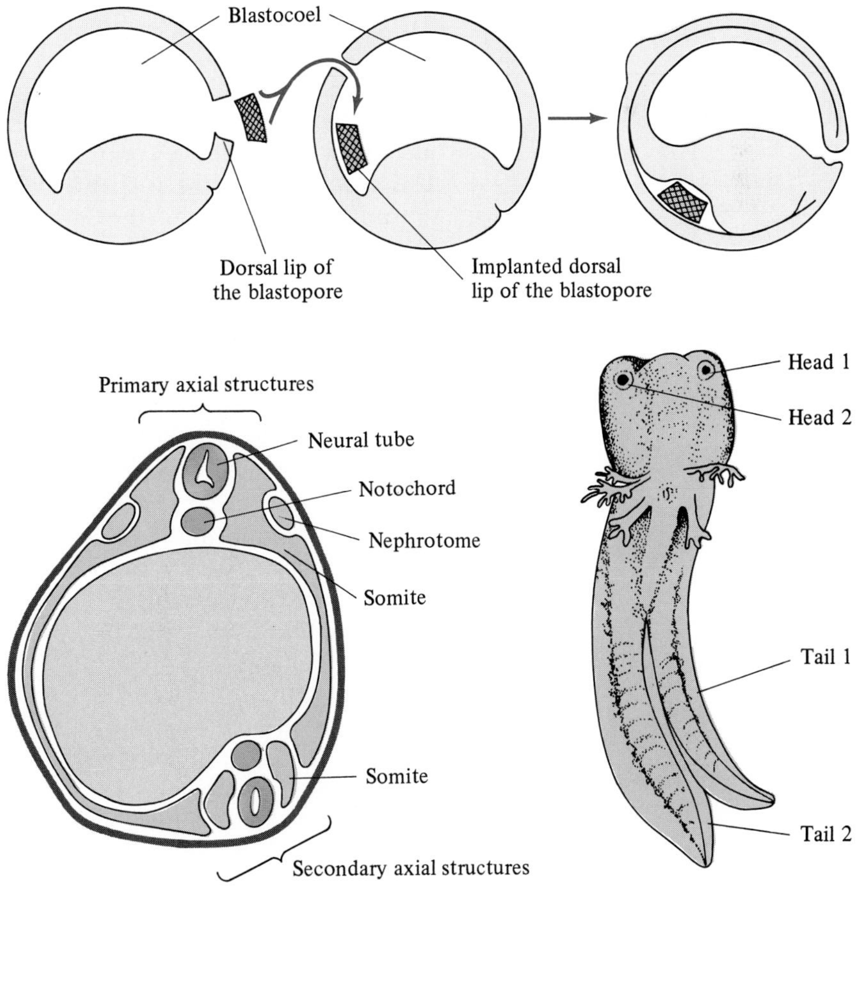


**FIG. 25-24** The inductive influence of the dorsal lip of the blastopore. Note the secondary set of axial structure in the section, and the second head in the tadpole.

secondary embryo is largely composed of host tissue that had been destined to become belly. Meanwhile, the host's own chordamesoderm and ectoderm underwent normal development into the primary embryo (Fig. 25-24).

The inductive interaction between chordamesoderm and ectoderm demonstrated by Spemann proved to be but one of many such inductions that occur during development. Many structures were shown to induce specific responses in previously undetermined tissues. Regions of the prospective brain, for example, induce the formation of eye parts in the ectoderm that overlie them.

In one series of experiments, it was also possible to show that the inductive stimulus given by one tissue layer to another need not be highly specific in order to elicit the appropriate response. The specific nature of the response is established and limited by the particular genetic capability of the responding tissue. Both of these principles were dramatically demonstrated in an elegant experiment performed by Hans Spemann and Oscar Schotté in 1932.

***Tissue Response to Inductive Commands.*** Frog tadpoles possess two suckers just behind their mouth. Newt larvae lack suckers but have two rodlike balancers in approximately the same location. When Spemann and Schotté transplanted undetermined presumptive belly ectoderm from a frog to the ventral surface of the head of a newt embryo, the graft developed suckers, not balancers, in the place where suckers usually arose in a frog larva. When, in a reciprocal experiment, undetermined belly ectoderm from a newt embryo was transplanted to the ventral surface of a frog embryo's head, it developed balancers, not suckers.

These experiments demonstrated once again that the ectoderm of an early embryo is undetermined and thus is capable of more than one alternative course of development. But they also demonstrated another principle: that induction is not necessarily a highly detailed message, that it does not itself need to carry much information. In the experiment just cited, it need have commanded no more than "make mouth parts." The specificity of the response, that is, the kind of mouth parts, was within the province of the induced tissue. Analogously, the inducing tissue is like a band conductor who gives a signal merely to play a waltz. When the signals were received, each tissue played the only waltz it knew.

***Properties of Inductive Interactions.*** The seemingly simple question, what are the properties of inductive interactions, has a variety of complex answers that differ from one organ to another and from one animal group to another. Some of the properties thus far discovered are described below.

In some amphibian embryos, parts of the brain can be induced to form from prospective epidermis by such nonspecific agents as pH changes, steroids, or the dye methylene blue, thereby suggesting that this inductive interaction is nonspecific and can be replaced by any traumatic change.

The mammalian pancreas develops from an evagination of the embryonic gut after it has been surrounded by mesenchyme (loose associations of mesodermal cells). An incipient pancreatic rudiment will not differentiate into pancreatic tissue when cultured outside the embryo if the mesenchyme surrounding it is removed from it before 30 hours have elapsed from its formation as a separate outgrowth. Despite the essential nature of this requirement it has no species specificity; any pancreatic mesenchyme will do. Even chick mesenchyme will satisfy this requirement. After the first 30 hours, differentiation into pancreas will occur in vitro in the absence of the surrounding mesenchyme, but only if the other tissues normally adjacent to it, the embryonic gut, remain attached to it. Finally, a stage of determination is reached at which the endoderm will differentiate into pancreas in the absence of both the mesenchyme and the normally adjacent tissues, yet only if combined with mesoderm from any source. It need not be from the region of the pancreas.

In the development of the limbs in the chick embryo, it has been shown that the inductive interactions are very specific and instructive. At one stage of development, the presumptive dermis (a mesodermal derivative) of a limb will command the overlying presumptive epidermis (an ectodermal derivative) to make leg or wing parts, as the case may be (Fig. 25-25a, b). The epidermis responds by doing so, but it makes the scales, feathers, and other structures appropriate to its particular location within that limb. Thus, when mesoderm of the tip of the leg bud is transplanted to the prospective thigh, it induces the formation of foot parts (scales, spur, and so on) and not thigh feathers (Fig. 25-25c).

***What Is the Nature of the Inducing Agent?*** Spemann had shown that in the induction of the eye, no induction occurred unless the two tissues concerned in the induction were in contact with each other. Furthermore, even if the two layers were separated only by a sheet of cellophane or a thin layer of connective tissue, induction failed to occur. However, when millipore filters, which permit the passage of macromolecules but not of cells, were used to separate two tissues, induction could often be achieved. The induction of the mouse pancreas, for example, was shown to involve the transfer of molecules small enough to pass through pores 200 nm in diameter. This type of observation led to the

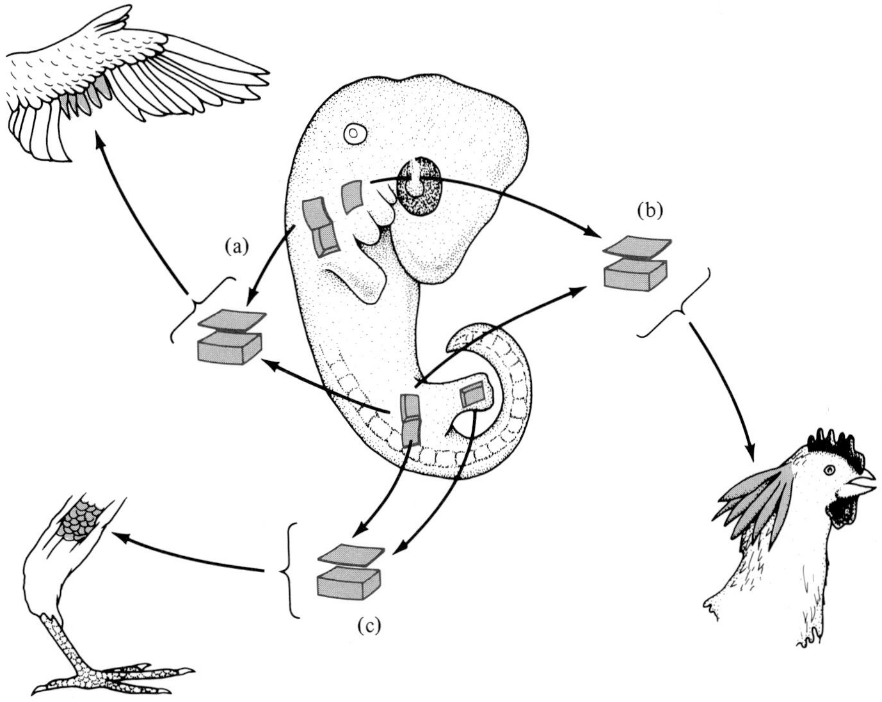


**FIG. 25-25** Inductive interactions between mesoderm and ectoderm. (a) Leg mesoderm transplanted to a wing bud. (b) Thigh mesoderm interacting with nonfeather-bearing ectoderm. (c) Foot mesoderm interacting with thigh ectoderm.

suggestion that the inducing agent is a chemical transmitted from the inducer to the induced tissues. But recent studies suggest that the two cell types maintain a physical contact with the help of fine pseudopodia that pass through these tiny pores. Thus the nature of the signal could be some sort of cell contact.

These few examples show that induction is not a single phenomenon, but differs widely among tissues and species in its characteristics. This area of research continues to involve many unanswered questions.

**FIG. 25-26** Totipotency of early cleavage nuclei in amphibian embryos. (a) Ligature of a zygote to prevent movement of the cleavage nucleus to one of the blastomeres. (b) After one of the blastomere nuclei had migrated into the noncleaving blastomere. (c and d) Developmental fate of the normal and "delayed" embryonic fragments.

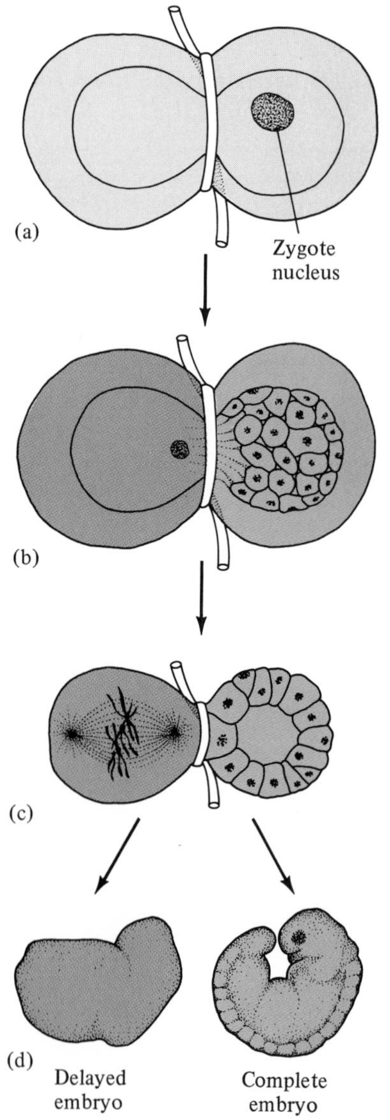


## THE MOLECULAR BASIS OF CELLULAR DIFFERENTIATION

Much of the research carried out in the second half of the twentieth century has been directed at an analysis of the role of genes and the molecular basis of cellular differentiation. Although it became evident that the genetic composition of differentiated cells is identical to that of the zygote, only a small fraction of the genes present in a differentiated cell are functional. These findings confirm the suggestion that cell differentiation is the consequence of selective gene activation.

***The Identical Genetic Composition of All Cells of an Organism.*** In the late 1800s, it was believed that cell differentiation may result from a parceling out of the genes of the early embryo among the nuclei of the different blastomeres. The theory then appeared logical. It was obvious that cells of the different tissues of an adult were different from each other, although all had come from the same cell.

One of the earliest attempts to show the equivalence of nuclei from older embryos was made by Spemann. He tied a newt zygote with a fine hair so that only a narrow isthmus connected the two halves. Consequently, one of the blastomeres received both of the nuclei at the first cleavage (Fig. 25-26). The isthmus connecting the two halves was so small that none of the nuclei of the early cleavage stages could pass through it. As a result, the nonnucleated blastomere did not cleave, whereas the other blastomere divided more or less normally (Fig. 25-26a, b). After this blastomere had completed four or more divisions, a nucleus was small enough to pass through the isthmus to the uncleaved half of the embryo. The latter then began to cleave (Fig. 25-26c). If the two halves of the embryo were then completely separated by tightening the knot, each half developed into a normal embryo (Fig. 25-26d). A single nucleus from as late as a 32-cell–stage or a 64-cell–stage embryo supported normal development. This experiment clearly demonstrated that, at least at this stage of development, all nuclei were still equivalent in their potential to direct development.

More recently, this conclusion was found to apply to still later developmental stages as well. In the early 1950s, Robert W. Briggs and Thomas J. King, while working at the Cancer Research Institute in Philadelphia, devised nuclear transplantation techniques to test this hypothesis. They removed the nucleus from a mature frog's egg and replaced it with a nucleus from a cell in a frog blastula. When activated, the egg containing the implanted nucleus often developed into a normal frog (Fig. 25-27a). The nucleus from a blastula cell is obviously qualitatively identical to a zygote nucleus in the frog.

In subsequent years, when nuclei from cells of gastrulae and tail-bud-stage embryos were implanted into enucleated frog eggs, they were

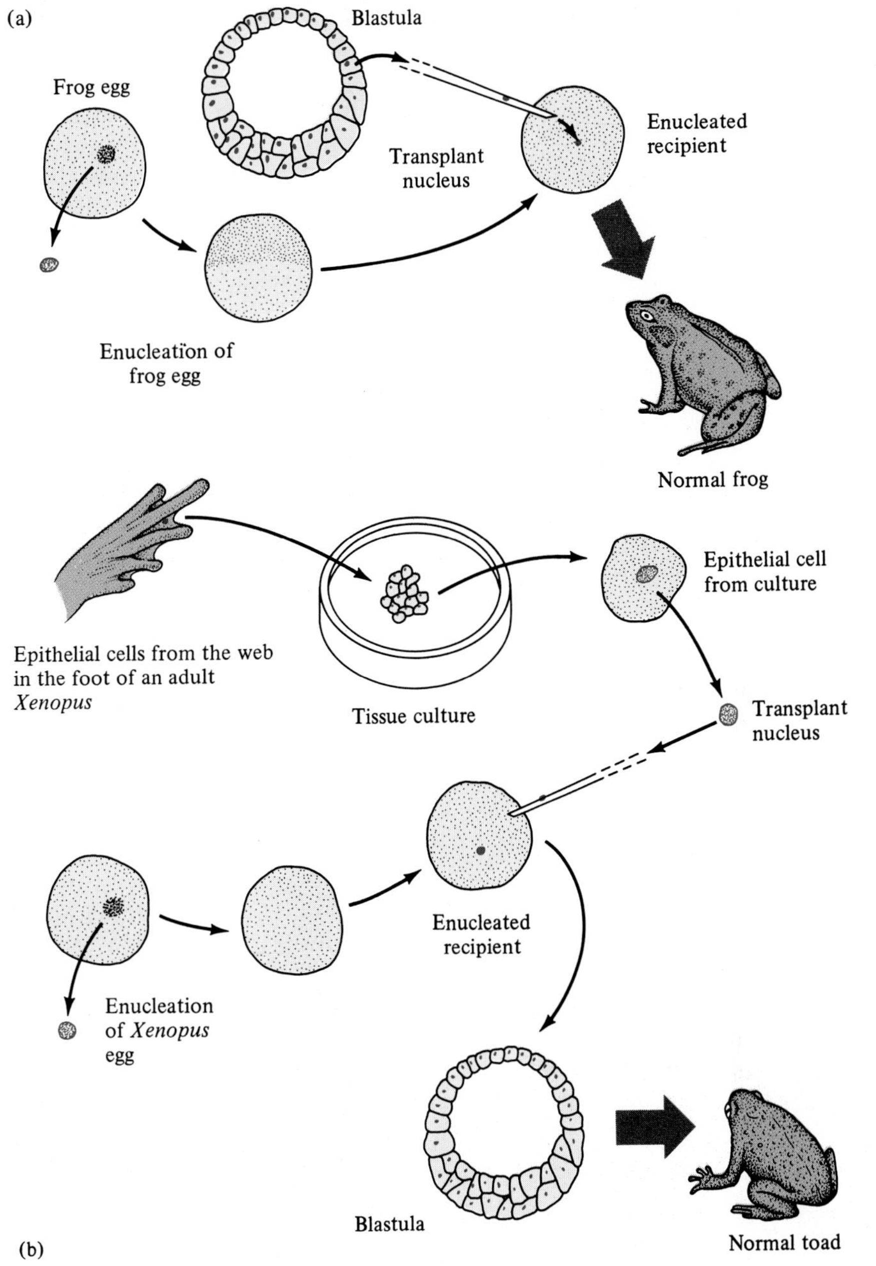


**FIG. 25-27** Totipotency of blastula and adult cell nuclei. (a) Protocol to test totipotency of blastula cell nuclei. Blastula cell nuclei were implanted into enucleated eggs. (b) Protocol to test the totipotency of epithelial cell nuclei from the web of the foot of an adult toad. Nuclei from cells that were cultured in vitro were implanted into enucleated eggs. Nuclei from the cells of the resulting blastulae were retransplanted into enucleated eggs.

also found to support normal development. Finally, in 1976, John Gurdon, a British biologist, showed that nuclei from epidermal cells derived from the web of an adult toad foot were able to support the normal development of enucleate eggs (Fig. 25-27b). In other words, nuclei from gastrula and tail-bud–stage embryonic cells and from adult epidermis are qualitatively identical to zygotic nuclei. Similar studies carried out on *Drosophila* embryos also showed that cleavage nuclei have the same developmental capacities as the zygote nucleus. These studies suggest that cell differentiation is not a result of qualitative changes in the genetic composition of nuclei.

A further line of evidence for this view has come from DNA extractions of adult tissues. Such studies have shown that the nucleotide sequence of DNA from different adult tissues is identical to that of an embryo.

***Differential Gene Activity in Differentiated Cells.*** If the nuclei of all the somatic cells of an organism are identical in genetic composition, the origin of differences in the structure, chemistry, and function of differentiated cells can be explained only by differential gene activity. In fact, different sets of genes are known to be functional in each type of differentiated cell. In muscle cells, for example, the genes coding for myosin, actin, creatine phosphatase, and others are active. However, these genes are inactive, or active at a very low level, in other tissues. Similarly, in the exocrine cells of the pancreas, the genes coding for amylase and lipase are functional. Those same genes are not active to the same degree in any other cell type in the organism, including the endocrine cells of the pancreas, which produce insulin and glucagon.

A gene is said to be active if mRNA molecules are being transcribed from it. By comparing the variety of RNA molecules present in one cell type with those in another, it is possible to estimate differences in the composition of active genes between the two cell types. For example, mRNA molecules extracted separately from myoblasts (cells that give rise to muscle cells) and erythroblasts (precursors of red blood cells) and translated into proteins in vitro yield myosin and globin, respectively. Myoblast nuclei thus must be synthesizing mRNA coding for myosin but not for globin, and vice versa. A more elaborate approach to this problem involves methods of molecular hybridization.

As noted in Chapter 24, if a solution of DNA is heated to about 90°C, the hydrogen bonds between the two strands of the double helix dissociate. The DNA molecule thereby becomes denatured, that is, the two polynucleotide strands separate. Upon cooling, the hydrogen bonds reform. As a result, the separated strands reassociate to form double-stranded molecules again. However, if RNA from the same species is introduced into the mixture before allowing the two strands to reassociate, RNA molecules with base sequences complementary to those of any regions of the DNA molecules will hybridize with the DNA to form RNA-DNA duplexes, double-stranded molecules composed of one strand of DNA and one of RNA.

Molecular hybridization methods are useful in two ways: (1) they can provide a measure of the variety of RNAs present in a particular cell type, and (2) they allow a determination of the percentage of the animal's DNA from which RNA has been transcribed in a particular cell type.

Experimental results strongly suggest that all but about 2 to 6 percent of the genes in any one cell type in an adult animal are inactive. The actual percentage apparently varies slightly from one cell type to another. The hybridization experiments in which RNA from two cell types is simultaneously reassociated with DNA indicate that at least some of the genes that are active in the different cell types are specific for each cell type.

Another line of research also supports the theory of selective gene activity. Giant polytene chromosomes, such as occur in some cells of some dipteran insects, or flies, are composed of up to 1,000 chromatin fibers arranged in parallel arrays. These chromosomes remain condensed in a permanent meiotic synapsis (see Chapter 23). When genes of a particular region of a polytene chromosome engage in transcription, the tightly coiled chromatin fiber in that region uncoils, giving a diffuse appearance to that part of the chromosome. Such regions are called **puffs.** From comparative studies of puffing patterns in cells of larval salivary glands, Malpighian tubules, gut tissue, and so on it has been

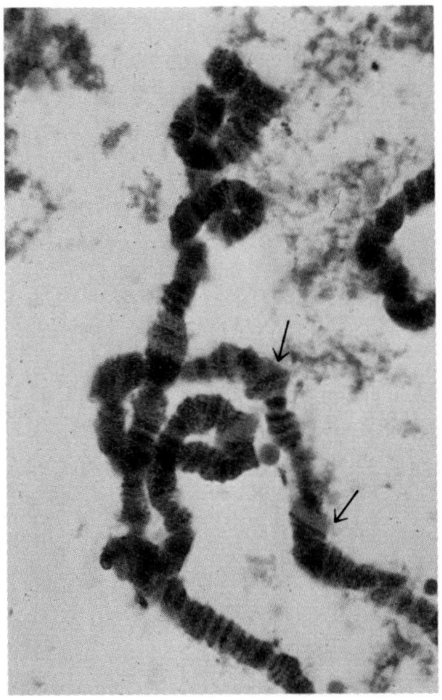

**FIG. 25-28** Polytene salivary gland chromosomes showing puffs indicated by the arrows.

possible to relate the puffing of specific chromosome regions with the specific developmental or physiological activities in which these various types of cells are engaged at the time. For example, changes in puffing patterns have been related to the secretion of ecdysone, a developmental insect hormone. Puffing patterns in polytene chromosomes from the fruitfly, *Drosophila*, are shown in Figure 25-28. The appearance of specific puffs is associated with the activation of certain genes by hormones.

In summary, molecular hybridization studies and studies of puffing patterns of polytene chromosomes in different tissues of flies support the theory of differential gene activity as an explanation for the origin of cellular differentiation.

***A Molecular Basis for Differential Gene Activity.*** From the accounts given above, it is evident that only a fraction of the genes in any given differentiated cell are functional. The question that naturally arises concerns the mechanisms by which selective gene activation is accomplished. How do genes get turned on and off? It is one thing to show that cell types differ in the sets of genes that are expressed but another to show how they do it. The problem has been approached by examining the various types of molecules found in the nucleus, where transcription occurs.

Besides RNA, five types of small basic proteins, the histones, are found in association with DNA. Also present are some 500 different nonhistone chromosomal proteins. All histones have a strong affinity for DNA and bind to it readily, forming nucleoprotein complexes. The fact that some types of histones are nearly identical in such widely diverse groups as plants and animals suggests their universal role of organizing DNA into compact chromosomes.

When histones are added to DNA, the DNA molecules undergo a very tight coiling. In addition, histones inhibit the in vitro transcription of RNA from DNA. DNA preparations from which all histones have been removed synthesize three to four times the amount of RNA as DNA in the native chromatin. Such evidence strongly suggests that histones act to repress gene expression, that is, to repress the transcription of RNA from DNA.

The nonhistone chromosomal proteins of the nucleus also seem to play a role in selective gene activation. In a series of experiments, chromatin from various tissues, such as chicken liver and oviduct, was fractionated into its major components: DNA, RNA, histones, and nonhistone proteins. In each case, it was possible to reconstitute the chromatin in vitro. Moreover, the reconstituted chromatin was found capable of transcribing mRNA with a base sequence characteristic of the RNA from the cells from which the extract was made. However, when chromatin was reconstituted in novel recombinations—for example, DNA and histone from liver and nonhistone proteins from oviduct—the base sequences of the RNA molecules transcribed from these preparations were always characteristic of the cell type from which the nonhistone protein component had been taken. In one case, for example, when the reconstituted chromatin included nonhistone proteins from oviduct nuclei, it promoted synthesis of mRNA coding for ovalbumin, or egg white, a substance that is normally synthesized by oviduct cells of the chicken.

Although many questions remain to be answered, it appears that nuclear histones have important roles in gene repression, or "turning off," while nonhistone proteins are involved in selective activation, or "turning on" specific genes. One of the important remaining questions concerns how these proteins control gene expression. More precisely, how is the configuration of chromatin modified to render specific genes available for transcription? An even more fundamental question—what determines that a particular cell type (for example, ectoderm cells) synthesizes one type of nonhistone protein, while another (for example, endoderm cells) produces a different type?—remains unanswered.

## POSTEMBRYONIC DEVELOPMENT

Embryonic development proceeds within the confines of the egg membranes and other protective coverings of the eggs until the nutrient reserves of eggs are exhausted. Once the nutrient reserve of the egg is depleted, the embryo hatches and seeks independent nourishment. The developmental state of the embryo at the time of hatching is highly variable and depends on the species and the amount of nutrient reserve. As may be expected in species with a limited yolk supply, the embryo is in very early stages of embryogenesis at the time of hatching. In sea urchins, for example, the embryo hatches as a mesenchyme blastula even before completion of gastrulation. But in frogs, which have a more abundant yolk supply in their eggs, the embryo has not only completed gastrulation and early organogenesis but is a fully formed tadpole larva.

Irrespective of the developmental stage at which they emerge, larvae must continue their development before reaching adulthood. This stage of development is called postembryonic. The larva differs from the embryo in the source of its nourishment. While an embryo obtains its nutritional requirements from the stored reserves of the egg, a larva obtains its nourishment from its environment. Because larvae lead an independent life and because the larvae usually differ from the adults of the species in their morphology, early systematists classified many larval types as independent species and named them accordingly. For example, the sea urchin larva is called a pluteus, that of a freshwater clam, a glochidium, and so on. These names now persist for those larval stages.

Developmental changes continue in the larva. At some point in its development, a larva undergoes a sudden and drastic change in its morphology; this change is called **metamorphosis.** Since these changes take place relatively slowly, studies on the regulation of metamorphosis yield valuable insights into the molecular basis of cell differentiation. In Panel 25-3, the role of hormones in insect and amphibian metamorphosis is described.

# HUMAN DEVELOPMENT

To many people, the most fascinating aspect of the subject of development is our own: human development. There is nothing truly unique about our development, however. Its major events are homologous to the development of embryos of other vertebrates in general and of amniotes and placental mammals in particular.

PANEL 25-3

# The Role of Hormones in Regulating Postembryonic Development

In amniote embryos, the rudiments of adult organs are established during embryonic stages. The organism that hatches from an egg resembles the adult in all features except size and sexual maturity. In many species, such as frogs and many invertebrates, the embryo emerging from an egg, called a larva, does not bear close resemblance to the adult of its species. In addition, the larva has a life-style distinctly different from that of the adult. Often the larval stage is adapted to life in an environment different from that of the adult. As a general rule, larvae also help disseminate a species. This is extremely helpful for organisms in which the adults are sessile (attached to a substratum) or sedentary (move only very slowly).

How is the development of organs and tissues typical of an adult regulated in such larvae? Two groups, amphibians and insects, have been studied extensively in this respect. In both groups the developmental change from a larval phenotype to adult morphology, called metamorphosis, as well as its associated differentiation are regulated by hormones.

During metamorphosis of frog tadpoles, some tissues, such as the tail and gills, disappear. Other tissues, such as limbs and lungs, develop. A frog tadpole will metamorphose prematurely into a froglet on being fed thyroxine, a hormone secreted by the thyroid gland. Tadpole metamorphosis is also affected by another hormone, prolactin, which can delay metamorphosis.

Insect metamorphosis has been studied in greater detail. Larvae of butterflies and moths, called caterpillars, bear no resemblance to the adults of the species. The caterpillar is merely a feeding machine that voraciously engorges itself. After hatching, a caterpillar increases some 5,000-fold in size in a matter of days. In the course of this growth, it periodically sheds, or molts, its old cuticle, replacing it each time with a larger, new cuticle. Then it suddenly stops feeding and spins a

*(continued)*

(a)

(b)

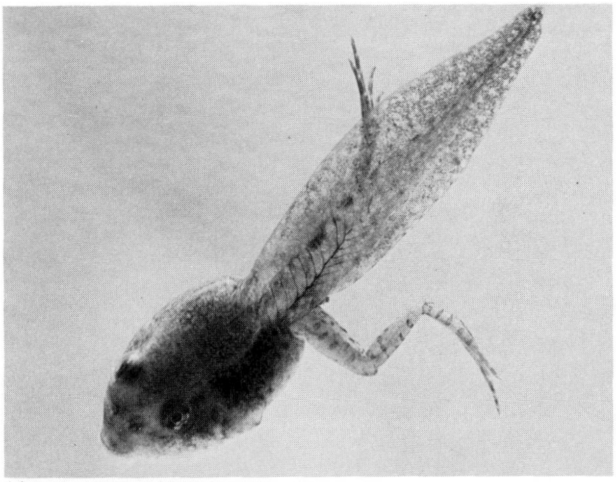
(c)

(d)

Stages in the metamorphosis of the tadpole larva. (a) Embryos about to hatch. (b) No legs have yet developed at this stage. (c) Hindlegs have now developed. (d) Fully-formed tadpole, with complete set of legs.

## PANEL 25-3 *(continued)*

cocoon, within which it metamorphoses into a pupa. The pupa then develops into an adult moth or butterfly. The adult moth in most species is primarily a reproductive machine. In many species, it mates soon after its wings have expanded and hardened, lays eggs, and then dies. Thus, the postembryonic life of these insects consists of a variable number of larval stages, followed by a pupal stage from which the adult emerges.

In insects, all these developmental changes are regulated by two major hormones: ecdysone and juvenile hormone, secreted by the prothoracic glands and corpora allata, respectively. When both hormones are present in a moth larva, it feeds, grows, and molts into a larger larva. At the end of larval life, the glands that secrete juvenile hormones are shut off. As a result, tissues are exposed to ecdysone alone. The larva then stops feeding, spins a cocoon, and makes a cuticle of a different type. At the succeeding molt, which occurs within the cocoon, the appearance of the organism has changed completely: it has become a pupa. After the formation of the pupa, ecdysone is again produced in the absence of juvenile hormone. As a result of the action of ecdysone on pupal tissues, these tissues develop into adult tissues and at the next molt, a moth emerges.

When juvenile hormone is applied to a full-grown larva before it begins pupal development, the developmental events associated with metamorphosis are blocked. Thus, juvenile hormone prevents metamorphosis, and the insect remains a juvenile. Ecdysone, on the other hand, promotes development toward the adult stage if it acts alone. These interactions between the two hormones were elucidated in the 1950s by Vincent Wigglesworth of Cambridge University and Carroll Williams of Harvard University. Williams has proposed that juvenile hormones could be used to control insect pests, without harm to other animals, because application of juvenile hormone affects only the insects. With no adult stage, no reproduction could occur, and the pest population would be reduced.

## THE FIRST TRIMESTER

Traditionally, human development has been divided into three periods, the first, second, and third trimesters, each of three months duration. Although development is a continuous process, it was found convenient for learning purposes to divide the developmental events into discrete units. These divisions are not based on any specific developmental landmarks but are merely convenient time points and have no other significance for development.

***Early Cleavage Stages.*** The human zygote is minute, 0.1 mm in diameter (Fig. 25-29a, b); it is barely visible to the unaided eye as a tiny mote when viewed in good light against a dark background. Fertilization usually occurs high in the Fallopian tube a day or two after ovulation. As in all vertebrates, the first cleavage is along the meridional axis, transecting the animal and vegetal poles (Fig. 25-29c). The second division is also meridional, at right angles to the first, with one blastomere usually dividing a bit earlier than the other (Fig. 25-29d). Subsequent development is typical of alecithal eggs (Fig. 25-29e, f). The divisions are rapid and seem to lack precise orientation. After several cleavages, the blastomeres are arranged in a solid ball of cells resembling the fruit of the mulberry tree, a stage long referred to as the **morula** (Latin for "mulberry") (Fig. 25-29f).

The cells of the morula continue dividing, and by the fifth day, they surround a cavity equivalent to a blastocoel (Fig. 25-29g). This stage of development, termed the **blastocyst** (Fig. 25-29g, h), consists of an outer wall, called the **trophoblast,** and an inner cluster of cells at one pole, the **inner cell mass.** Some of the cells of the inner cell mass will form the

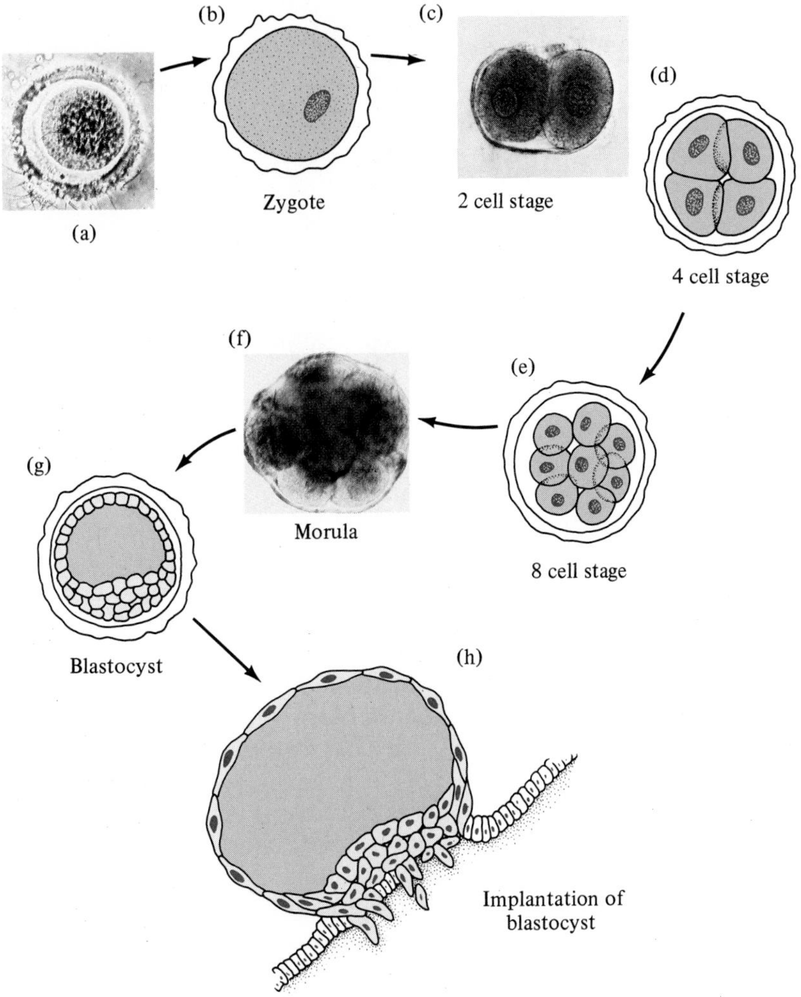


**FIG. 25-29** Early stages of human development. (a) Egg with sperm. (b) Zygote. (c) 2-cell stage. (d) 4-cell stage. (e) 8-cell stage. (f) Morula. (g) Blastocyst. (h) Implantation of blastocyst.

embryo. As in chick development, the embryo proper is only a small part of the embryonic structures. Structures external to the embryo proper, such as the trophoblast, are said to be extraembryonic.

***Implantation.*** By the sixth day of development, the blastocyst has reached the upper portion of the uterus. Its outer covering, the **zona pellucida,** which had covered the ovum even before ovulation, is now shed, allowing the trophoblast to come into contact with the endometrium (Fig. 25-29h). (See Chapter 22 for a description of the female reproductive organs.) The trophoblast now secretes a protease that digests an opening in the uterine tissue, allowing the embryo to implant beneath the endometrial surface, usually high on the posterior wall of the uterus.

As the blastocyst sinks into the endometrium, the trophoblast cells rapidly proliferate, forming a mass of cells without plasma membranes, called the **syntrophoblast.** This tissue continues to digest endometrial tissue while absorbing nutrients from the resulting fluid. Because maternal blood capillaries are also eroded by the trophoblast, the embryo becomes bathed in a mixture of tissue fluid and blood that fills the

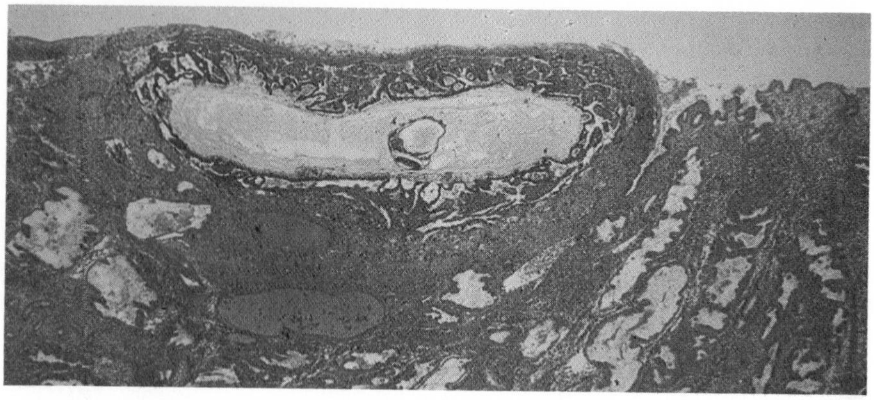

**FIG. 25-30** Human development continued. The embryonic and uterine tissues at 14 days.

spaces between the fingerlike villi; then the trophoblast begins to extend into the uterine tissue.

By days 7 to 8, cells of the inner cell mass have rearranged to form a central layer of columnar ectoderm cells; these soon become lined on the blastocoel side by loosely arranged endoderm cells. It is from this double layer of cells, called the **embryonic disc,** that the embryo develops. Above the embryonic disc, the amniotic cavity appears, enclosed by the amnion.

By days 10 to 12, implantation is complete, and the point of entry has healed over, leaving a slight bulge on the inner surface of the uterine wall. By this time, ectodermal cells of the trophoblast have developed into a continuous layer, the **cytotrophoblast,** which encircles the embryo.. Unlike the syntrophoblast, the cytotrophoblast is composed of discrete cells. At this stage, the endoderm cells below the inner cell mass have formed a vesicle called the yolk sac. The yolk sac has no known function in placental mammals. It is probably an evolutionary relic. Mesoderm cells have also appeared by now and begin to line the inside of the cytotrophoblast and the outside of the yolk sac.

***The Second and Third Months.*** By day 12, the yolk sac is fully formed, with a layer of extraembryonic mesoderm lining its outer surface and that of the amnion. Mesoderm also covers the inner surface of the cytotrophoblast. When the blastocoel becomes completely lined by mesoderm, it is called the extraembryonic coelom, or exocoelom. During the next 24 hours, an outpocketing arises at the posterior end of the embryo. This later develops into the allantois. As in the chick, the allantois arises from the presumptive hindgut region, near the yolk sac stalk. It grows out into the space between the amnion and the cytotrophoblast at the rear of the embryo. At first, the allantois is a vesicle, much like that in the chick, but then it changes into a rodlike structure, forming a substantial part of what is called the **embryonic stalk.** The blood vessels that arise in the embryonic stalk develop into the blood vessels of the umbilical cord, the lifeline between the embryo and the maternal tissues.

When the cytotrophoblast becomes lined internally by the mesoderm, it produces a double-layered extraembryonic membrane, the chorion. The chorion plays a crucial role in anchoring the embryo to the uterine wall and in the absorption of nutrients from the maternal tissues. The chorion is also the source of hormones important for the maintenance of pregnancy. The chorion develops numerous projections, the chorionic villi, which increase its absorptive surface area (Fig. 25-31a–

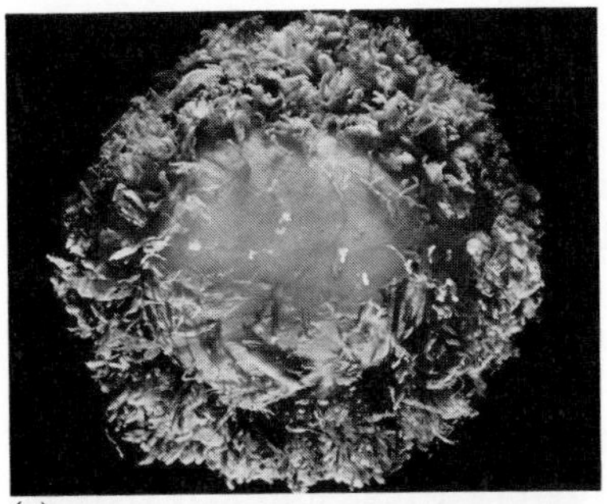
(a)
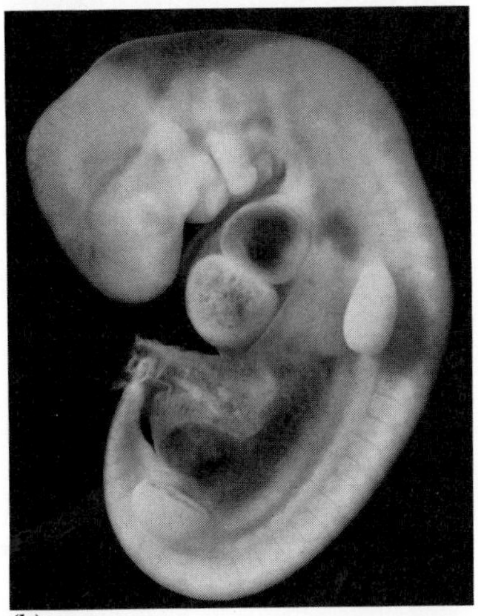
(b)
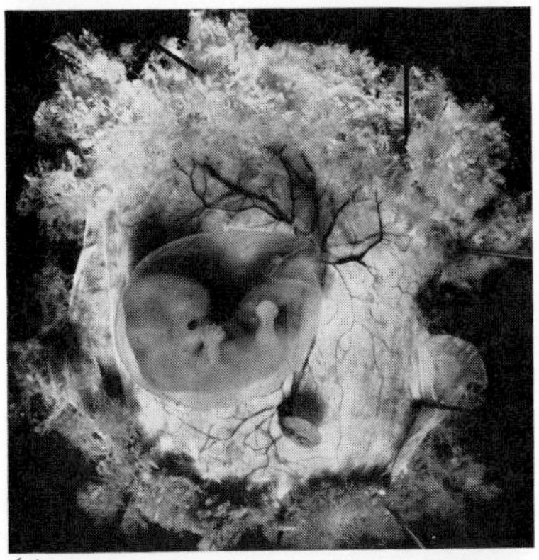
(c)
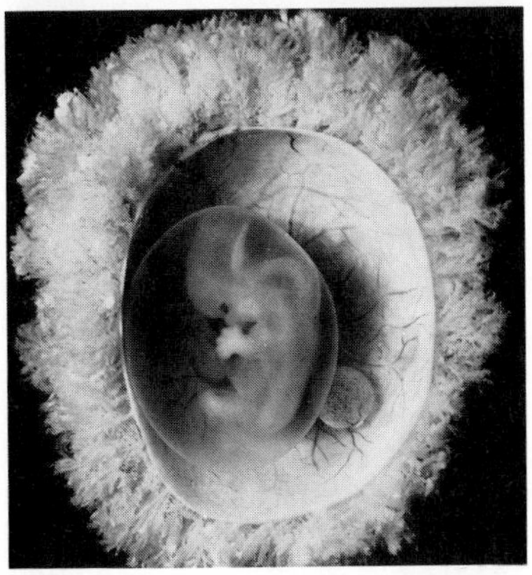
(d)

**FIG. 25-31** The first 45 days of human development showing (a) 28-, (b) 30-, (c) 39-, and (d) 45-day-old embryos along with the placenta.

c). In time, the villi become highly vascularized, their capillaries eventually forming connections with the blood vessels that arise in the allantois and, thus, with those of the embryo. Nutrients and oxygen are thereby supplied to the embryo from the maternal blood, in which the chorionic villi are bathed. The developing embryo's carbon dioxide and nitrogenous wastes diffuse from its bloodstream into the maternal circulation. No direct connection becomes established between the bloodstream of mother and embryo.

In a 2-week-old embryo, the embryonic disc consists of two layers of cells, of which the "upper" one, facing the amniotic cavity, is the ectoderm and the "lower" one, facing the cavity of the yolk sac, is the endoderm. In the next day or two, the primitive streak, similar to that of the chick embryo, arises. It is in the posterior, or caudal, half of the embryo and is composed of a groove flanked by ridges. As in the chick, an elevation at its anterior end, Hensen's node, marks the site of involution of notochordal tissue. The cells that involute through the primitive streak apparently develop into embryonic mesoderm, which gives rise to all the mesodermal structures of the embryo proper.

At this stage, the embryo is 1.5 mm long, an oval structure with the primitive streak denoting its long axis. Its three germ layers appear as simple, solid sheets and are distinguishable by their appearance from the extraembryonic germ layers. The ectoderm overlying the region anterior to Hensen's node now markedly thickens to form the neural plate, which almost immediately folds into a neural groove and then into the neural tube. The head end of the neural tube, which is wider than the tail end, gives rise to the brain. By the end of the third week of development, the neural tube is fully closed except at its anterior end.

Also by the end of the third week, the mesodermal cells that entered through the primitive streak have begun to develop into somites, very similar in appearance to those of the chick. By day 20, about three pairs of somites are distinguishable. Eye and ear rudiments, or primordia, are also discernible in the head region by this stage.

By the fourth week, the human embryo is about 4 mm long, bears about 30 pairs of somites, and has a distinct tail and signs of gill clefts.* By the end of the first month, the primordia of the eyes, nose, and ears have formed, and a tubular structure that will become the heart has differentiated and begins to pulsate irregularly. Lung primordia are now forming, and minute buds can be seen where arms and legs will arise. A tail bud is also evident.

By the end of the second month, the embryo is 3 cm (1.25 inches) long. Its gonads have differentiated, although its sex cannot yet be determined. The limbs are now evident, but the fingers and toes have not yet separated from each other and have webbing between them. The amniotic fluid in which the embryo floats is continually secreted and reabsorbed by the fetal circulation, being replaced an average of once every 3 hours. The embryo now bears distinctive human features and from this point on is referred to as a **fetus.**

All major organ systems are present by the end of the first trimester,

* In contrast to the development of fish embryos, however, the clefts do not break through to form gills. Moreover, the rudimentary gill supports between the clefts develop into such structures as ear ossicles, the bony elements in the middle ear, and tracheal cartilages, rather than gill supports, as they probably did in our fish ancestors.

the first 3 of the 9 calendar months of gestation. The 7.6 cm (3 inch)-long fetus now regularly inhales and swallows amniotic fluid and begins to add to it with its urine. The skeletal parts that had been forming from cartilage rapidly begin to undergo **ossification,** or replacement by bone. Facial hair is now evident. Although half the length of the fetal body is composed of head, from this point on, the rest of the body begins to gain in size on the head. At birth, however, the head will still constitute one-tenth of the body's length.

By the end of the fourth month, the fetus is 18 to 20 cm (7 to 8 inches) long and has begun spontaneous stretching of its legs and arms. Some movements may be detected by its mother at this stage. The heart beats regularly, although the fluid it pumps is only partly blood. The heartbeat can be detected with a stethoscope placed over its mother abdomen. Brain convolutions have begun to form, and the eyes are sensitive to light shined on the eyelids.

**FIG. 25-32** Fetal stages of human development showing (a) 8-, (b) 10-, (c) 13-, and (d) 32-week-old fetuses.

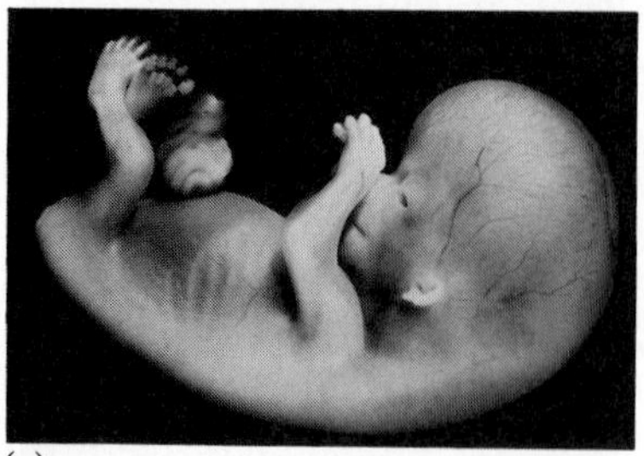

(a)

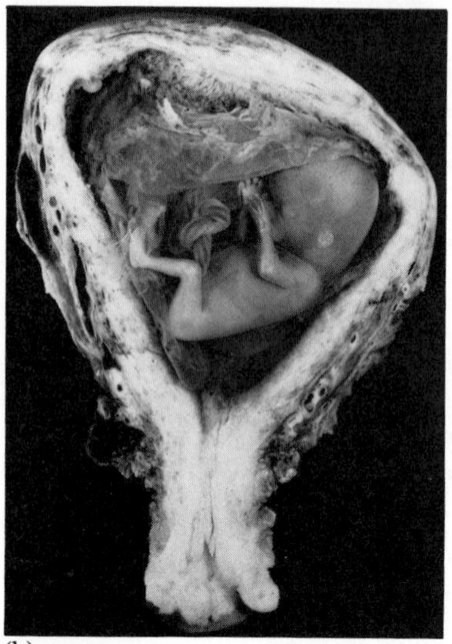

(b)

## THE SECOND AND THIRD TRIMESTERS

In the fifth month, the fetus reaches a length of 23 to 28 cm (9 to 11 inches) and a weight of 225 g (half a pound). A liter of amniotic fluid is now present and is recycled at a rate of over 20 liters a day. Fingers and toes are well formed and have begun to develop nails. The fetus may be quite active from now on, kicking, stretching, and turning, although it is quiet when it is sleeping. The gastrointestinal tract and the lungs are well formed, but neither is yet functional. If born now, the baby probably would not be able to survive.

By the close of the second trimester (the sixth month), the embryo is 30 to 35 cm (12 to 14 inches) long. Only a small percentage of fetuses born at this stage survive. Skeletal formation has been extensive during the sixth month. Sweat glands and hair follicles are now numerous in the skin. Diagrams of 3-, 5-, 8-, 9-, 10-, 11-, 12-, 16-, and 24-week-old human embryos in Figures 25-31 and 25-32 give a pictorial view of the progressive changes in the embryonic body.

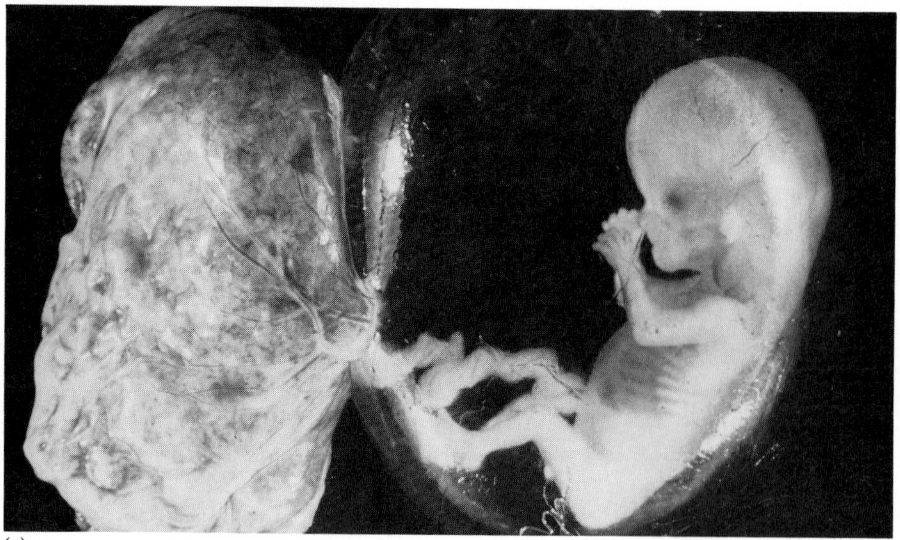

(c)

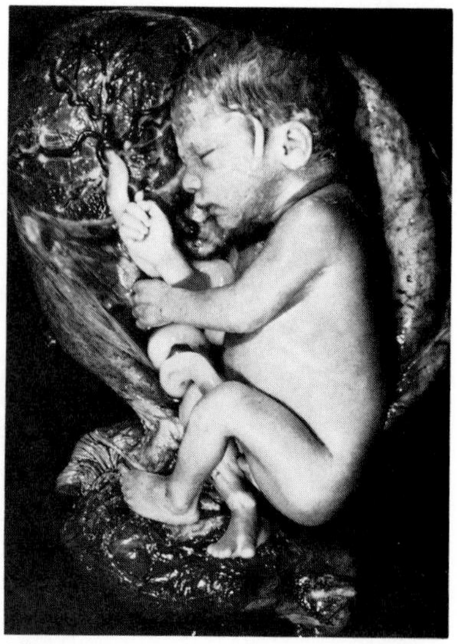

(d)

The last trimester places the heaviest burden on the pregnant mother as the fetus becomes fully formed (see Fig. 25-23). Besides the added weight, a growing pressure is exerted on her internal organs. Her blood volume is now 30 percent above normal. During this time, an increasingly larger intake of food is needed. Most of the dietary calcium and iron now go to form the skeleton and blood cells, respectively, of the growing embryo. Amino acid intake is important at this stage for proper development of the brain and other organs. The viability rate of babies born during the last trimester increases weekly. Although only 10 percent of those born in the seventh month survive, the figure rises to 70 percent during the eighth month and 95 percent during the ninth. The uterus now has stretched to 60 times its normal size, and the baby becomes less active as its growing size confines it within the limited space.

Sometime during the several weeks preceding birth, the mother's abdomen changes shape as the upper part of the uterus descends toward the pelvis and the baby's head enters the lower portion of the uterus. Because the fetus does no longer press hard against the diaphragm, this lightening, as it is sometimes termed colloquially, allows the mother to take deep breaths for the first time in several weeks, although her movements may be more difficult and the desire to urinate more frequent.

As seen in this chapter, each individual arises as a single cell, the zygote. But the mature individual is made up of trillions of cells. Moreover, all cells in the human body, or that of any organism, are not alike, but are specialized in structure and function. The ultimate goal of studies in developmental biology is a causal analysis of the origin of these different cell types. Attainment of this goal is far in the future, although we now have a complete list of the diverse cell types present in the human body, about 200 types in all, each unique in its structure, chemistry, and function. Students of histology, the branch of biology that studies the structure and function of tissues, observed that the different cell types are organized into four types of tissues: epithelia, connective tissue, nervous tissue, and muscle. Each of these tissues may be assembled from more than one cell type. A brief account of cell and tissue specializations is given in Panel 25-4.

## PARTURITION

Parturition, or birth, is induced by strong rhythmic contractions of the uterus, known as labor. In most cases, parturition occurs approximately 280 days after the last menstruation, or 256 to 276 days after fertilization. Labor is composed of three stages: **dilatation,** the **expulsion phase,** and the **placental phase.**

The onset of cervical dilatation, which is caused by the baby pushing out, also marks the beginning of uterine contractions. The duration of this stage is highly variable and normally lasts 2 to 16 hours. The contractions are marked at first by slight labor pains lasting 25 to 30 seconds and occurring at intervals of 15 to 20 minutes or less. They become stronger, more painful, and more frequent as labor progresses, eventually resulting in crowning, the emerging of the crown of the head through the cervical and vaginal openings.

The expulsion phase begins when the entire head has emerged, by which time the contractions last 50 to 90 seconds, with 1 to 2 minute

PANEL 25-4

# Cell and Tissue Specializations

The different cell types present in an organism are organized into four major types of tissues: epithelia, connective tissue, nervous tissue, and muscle. Each type of tissue has a characteristic arrangement of its component cells. For example, epithelial cells are arranged in discrete sheets to form epithelia, such as in the skin that covers our body, while connective tissue, such as that found as a filler in many tissues, is characterized by an extensive network of cells embedded in an intercellular matrix. However, there are several kinds of tissues and epithelia. The cells that constitute one kind of epithelium differ, in structure and function, from those of another type of epithelium. In the accompanying figures some examples of the tissues are shown.

***Epithelia.*** Epithelia are sheets of cells that line the outer and inner body surfaces. Epithelial cells are closely packed and are oriented so that the two ends of the cells differ in their biological properties. The outer, or apical, end is exposed to the surface, while the basal, or inner, end overlies a basement membrane. Epithelia resist tearing because their cells are held together by intercellular junctions of various types, such as tight junctions, desmosomes, and gap junctions. These help hold the cells tightly together. Tight junctions render the epithelium impermeable to small molecules due to the fact that the outer, lipid layer of the cell membrane is fused to that of adjacent cells. Consequently, material present in the cavity, or lumen, of an organ (e.g., urinary bladder) cannot seep back into the body. Gap junctions, which are characterized by the presence of protein-lined channels between neighboring cells, act as a means of chemical and electrical communication between epithelial cells in addition to binding the cells together.

The cells of epithelial tissue differ in shape and function thus giving the epithelia their characteristic feature. The cells may be taller than broad (*columnar*); cube-shaped (*cuboid*); or thin and flat (*squamous*). The cells may have different functions: ciliated, helping in movement of fluid, as in bronchii and trachea; glandular, secreting specific products, as in the liver, pancreas, thyroid, and so on; absorptive, provided with microvilli that absorb nutrients, as in the intestine; or sensory, with modified cilia, as in olfactory and retinal epithelia.

***Connective tissue.*** The term connective tissue is derived from the fact that this tissue binds or connects all cells of organs and tissues. Connective tissue is characterized by the presence of specialized cellular components in an amorphous matrix. The latter is usually a network of protein fibers secreted by specialized cells called fibroblasts. Connective tissue is classified according to its function and to the nature of the intercellular matrix.

Blood is also regarded, for lack of better classification, as a kind of loose connective tissue consisting of a fluid intercellular matrix with a variety of cells, called formed elements, floating in it: red cells, or erythrocytes, white cells, or leukocytes, and platelets. Adipose tissue is made up of fat cells, or adipocytes; these cells are among the largest in the human body and are specialized for the synthesis and storage of fat. The connective tissue surrounding organs is composed of fibroblasts and of collagen and elastin, two major types of extracellular fibrous proteins. This type of connective tissue is designated as loose or dense, depending upon the amount and arrangement of its fibrous proteins. Ligaments and tendons are dense connective tissue while the rest is classified as loose connective tissue. Cartilage and bone are composed of chondrocytes and osteocytes surrounded by an extensive extracellular matrix. In bone the extracellular matrix is calcified into a hard material, the bone. Osteocytes are irregular, star-shaped structures with long branching processes. Unlike chondrocytes, the cells that secrete cartilage are typically rounded cells with no branching processes.

***Nervous tissue.*** Nervous tissue is composed of

intervals between them. Usually, as soon as the head appears, the shoulders and the rest of the body quickly follow. The umbilical cord is then tied off and cut, producing an immediate build-up of carbon dioxide within the baby's blood and strongly stimulating its respiratory center. The first breath and a loud cry follow. If the first breath is delayed, stroking or otherwise stimulating the baby usually prompts it to begin neurons, or nerve cells, and supporting glial cells. Neurons are specialized for the transmission of electrical impulses. Each neuron typically consists of a cell body, or soma, dendrites, which receive the stimuli, and an axon through which the cell communicates with its neighbors. The organization of the nervous system was discussed in depth in Chapter 16.

***Muscle.*** Muscle tissue is specialized for contraction. There are three types of vertebrate muscle: cardiac, smooth, and striated; all have been described in detail in Chapter 18. By their contraction, muscles can produce mechanical force. Each striated muscle fiber is a multinucleate cell formed by the fusion of undifferentiated precursor muscle cells, or myoblasts. The smooth muscle fibers are unicellular.

***Developmental origins of tissues.*** In the description of early embryonic development, we described the formation of three germ layers, ectoderm, endoderm, and mesoderm. Ectoderm gives rise to the nervous system, sensory organs, and also to the epidermis, the outermost layer of the skin. Endoderm gives rise to the digestive tract, the associated glands, and to the lungs. Mesoderm gives rise to connective tissue, the dermis of the skin, muscle, skeleton, and vascular system as well as to the kidneys and gonads (but not the reproductive germ cells). Germ cells are set aside very early in development even before the three germ layers are separated.

However, bear in mind that classification of tissues into epithelia and so on has no specific relationship to the primary germ layers. An epithelium may originate from ectoderm (skin), endoderm (intestinal lining), or mesoderm (urinary tubule). Similarly, stomach may be composed of endodermal epithelium as well as muscle, and connective tissue that are derived from mesoderm, in addition to various kinds of neurons that are derived from ectodermal cells. Thus each organ system may be composed of cells from more than one primary germ layer.

Examples of the epithelial and connective tissue types in the human body and the cytoskeleton of a cell.

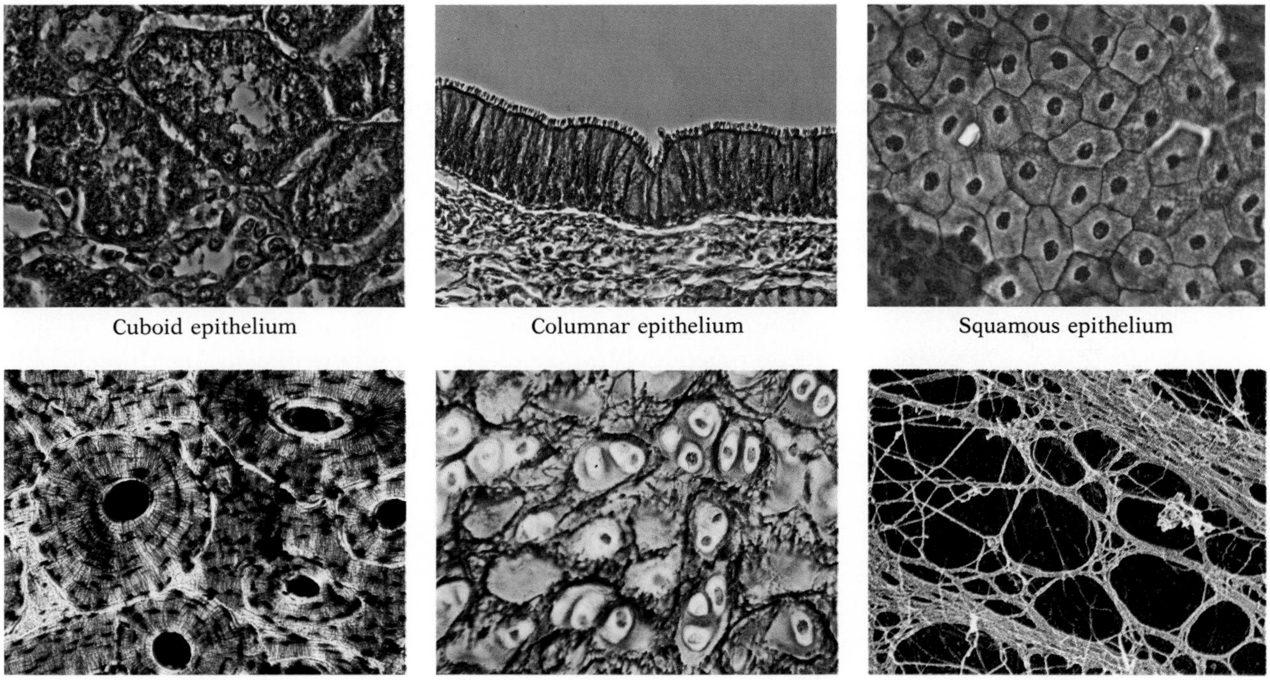

Cuboid epithelium — Columnar epithelium — Squamous epithelium

Bone — Cartilage — Cytoskeleton

breathing. The child is now cleaned and carefully examined while the mother is attended to.

The placental stage of labor now occurs. The uterus continues to undergo contractions, shrinking in size, expelling blood, fluid, and the placenta, known as the afterbirth.

A newborn baby has a traumatic experience as it leaves a warm, dark, quiet, watery, secure environment and emerges into the air of a strange, luminous, noisy world full of other individuals and a host of objects and stimuli. Yet it begins adapting at once, interacting with its mother and other persons and objects that it encounters. As it enjoys its first meal, the baby continues the development that began 9 months earlier.

## Summary

Animal development, which embraces the molecular, cellular, and organismal changes that characterize the various phases of an individual's life span, is a dramatic biological phenomenon. Each individual begins as a single cell, the zygote. The zygote is formed by the fusion of two very specialized cells, the ovum, or female gamete, and the spermatozoon, or male gamete. The gametes are themselves the end products of a long series of developmental changes. The cytological and biochemical changes involved in the formation of ova are collectively called oogenesis. These preparations include accumulation of yolk, the nutrient store, ribosomes, messenger RNA, mitochondria, and other material used in the early stages of development. In addition, specific molecules that help recognize sperm from the same species are synthesized and embedded in the protective coverings. Spermatogenesis, or the formation of sperm, involves the loss of cytoplasmic organelles, such as the endoplasmic reticulum and the Golgi apparatus, and the formation of a flagellated cell containing a haploid nucleus, centrioles, and a few mitochondria.

The zygote undergoes cleavage divisions and produces a number of more or less identical cells called blastomeres. The blastomeres usually arrange themselves in a layer at the periphery of a hollow sphere, as in sea urchins and frogs, to form a blastula. In chick embryos, early development is somewhat different. The zygote does not cleave completely. Due to this meroblastic cleavage, a number of cells, partitioned only at the sides and not completely separated from the underlying yolk, are formed. Eventually, these cells develop membranes that separate them from the yolk. The cell layer is called a blastoderm. In mammalian eggs, the cleavage pattern is somewhat different from both that of frogs and chickens. The blastomeres become arranged into a hollow sphere called a blastocyst. One group of cells, known as the inner cell mass, which develops into the embryo, is distinguishable from the rest of the embryo, called the trophoblast.

At the blastula stage of lower chordates and of echinoderms, the embryo does not exhibit visible indications of future organization. The blastoderm of the chick is also a rather uniform layer of cells. The blastula in sea urchins undergoes dramatic changes that convert the single-layered structure into a two-layered gastrula. This process, called gastrulation, begins as an invagination of cells on one side of the blastula; as they push inward, they slowly obliterate the blastocoel. The exact details of gastrulation differ in different groups of animals, although in all there is a conversion of a single-layered structure into a multilayered embryo. The major differences between gastrulation in sea urchins and in frog and chick embryos are due to the large amount of yolk present in the latter two. Yolk not only retards cleavage in the frog, but the yolk-laden cells of the vegetal hemisphere are too large to be invaginated during gastrulation. As a result, gastrulation in the frog embryo occurs about a third of the way up from the vegetal pole. Overgrowth, or epiboly, of the outer layer also plays a role in frog gastrulation. In chick embryos, where the yolk does not cleave at all, gastrulation is accomplished entirely by cell migrations. At the site of migration of cells from the outer layer into the space underlying it, the transient primitive streak develops. It is equivalent to the region of invagination in sea urchins and frog's eggs.

At the end of gastrulation, three layers of cells—the outer ectoderm, the middle mesoderm, and the inner endoderm—are distinguishable. Ectoderm gives rise to the epidermis of the skin, the nervous system, sensory receptors, and others. Endoderm gives rise to the gut lining and to other organs of the digestive system. Mesoderm gives rise to bone, muscles, blood vascular system, and the dermis of the skin.

During gastrulation in the frog, some of the invaginating tissue induces formation of a neural plate in the middorsal ectoderm. This process, called primary induction, begins organogenesis. The neural plate undergoes topological changes to form the neural tube, which by differential growth develops into the brain and spinal cord. Meanwhile, the mesoderm immediately beneath the neural plate forms the notochord and other derivatives, while the endoderm begins its development into the digestive tract.

The formation of all organ rudiments proceeds in two steps: determination and differentiation. Cells are said to be determined when the developmental potential of the cells is irrevocably committed to a particular path of development. Thus, for example, an eye rudiment is said to be determined if cells excised from the prospective eye region of an embryo and transplanted to a different part of the embryo develop into an eye. Optic cells are said to be differentiated when the cells in the eye rudiment actually exhibit the cytological and biochemical features characteristic of eye cells.

The embryos of amniotes (reptiles, birds and mammals) have three extraembryonic membranes that perform various functions for the developing embryo. The amnion and the chorion, which are folds of extraembryonic ectoderm and mesoderm, form an artificial pond in which the developing embryo is cradled. The third membrane, the allantois, is an outgrowth of the endoderm covered by mesoderm. It is a highly vascularized sac that stores the metabolic wastes in bird and reptilian embryos. In mammals, the allantois participates, along with some maternal tissues, in the formation of the umbilical cord and placenta. The fourth extraembryonic membrane, the yolk sac, is well developed in reptiles and birds but is very much reduced in placental mammals.

An old, fundamental question about development concerns the molecular mechanisms underlying cellular differentiation. From a series of experiments in which nuclei from progressively older embryos were transplanted into enucleated oocytes, no evidence for the occurrence of qualitative changes in the genetic formation of the differentiated cells could be found. Furthermore, studies using DNA hybridization methods have shown that nuclei from different cell types have identical DNA sequences. On the basis of these studies, it appears that differential gene expression is the underlying cause of cell differentiation.

Postembryonic development in higher animals is direct, that is, it does not involve an intermediate stage between the embryo and the juvenile. In mammals and birds, when the embryo is born or hatched it resembles a miniature adult. Its habitat and mode of life are also the same as the adult's. But in most invertebrates and in frogs and other amphibians, a larval stage intervenes. Besides not resembling the adult, a larva usually occupies a different ecological niche as well. Then, after a period of growth, it undergoes a sudden change, called metamorphosis. In insects such as moths and butterflies, metamorphosis is regulated by the interaction of two hormones, ecdysone and juvenile hormone. Metamorphosis is inhibited by juvenile hormone and promoted by ecdysone. In frog tadpoles, metamorphosis is promoted by thyroxine and inhibited by prolactin. Except for its end product, human development does not differ fundamentally from that of other mammals.

## Recommended Readings

BALINSKY, B. I. 1981. *An Introduction to Embryology,* 5th ed. Saunders, Philadelphia. A classical textbook of embryology with emphasis on vertebrate development. Clearly written and well organized.

BERRILL, N. J., and G. KARP. 1981. *Development,* 2d ed. McGraw-Hill, New York. General concepts of developmental biology are discussed at the undergraduate level.

CORLISS, C. E. 1976. *Patten's Human Embryology.* McGraw-Hill, New York. A well-organized, brief synopsis of human development in which both early development of the embryo and development of individual organ systems are carefully traced.

SLACK, J. M. 1983. *From Egg to Embryo: Determinative Events in Early Development.* Cambridge University Press, London. A brief discussion of the causality of developmental events in animal embryos.

TRINKAUS, J. P. 1984. *Cells into Organs,* 2d ed. Prentice-Hall, Englewood Cliffs, N.J. An excellent discussion of the forces that shape the organs during morphogenesis in animal embryos.

# 26

# Plant Development

## OUTLINE



Plant development varies in most respects from animal development, a reflection of the basic structural differences between the members of the two kingdoms. As photosynthesizing organisms, plants expose maximum surface to light, their source of energy, and to air, their source of carbon dioxide. As nonmotile organisms, they obtain water and minerals from a root system, which also anchors them in the soil. As plants grow and develop, they further extend their photosynthesizing leaves and stems into the light and the carbon dioxide-laden air and their roots into the soil, with its supplies of water and minerals (Fig. 26-1).

The presence of a cell wall in plant cells accounts for a major difference between the pattern of development of plants and that of animals. In cell wall formation, a **cell plate** is formed during cytokinesis, and upon the cell plate each of the daughter cells resulting from cytokinesis deposits a **primary wall,** as described in Chapter 4. The pectic material of the cell plate forms the **middle lamella** between the walls and cements them together. In some cell types an additional or **secondary wall** is deposited internally to the primary wall in various patterns (see Fig. 26-8). Cell walls limit the growth of plant cells and ultimately affect the form of the plant by determining how much, how fast, and in which direction the cells can elongate. Because plant cells tend to remain in place following cell division, the shaping of organs, that is, the roots, stems, and leaves, is also affected by the number of cell divisions and the plane in which they occur. In plants one does not find the inductive interactions seen in developing animal embryos as they undergo morphogenetic movements. Plant tissues do not move, and one tissue does not induce another to differentiate. Development of plants is instead

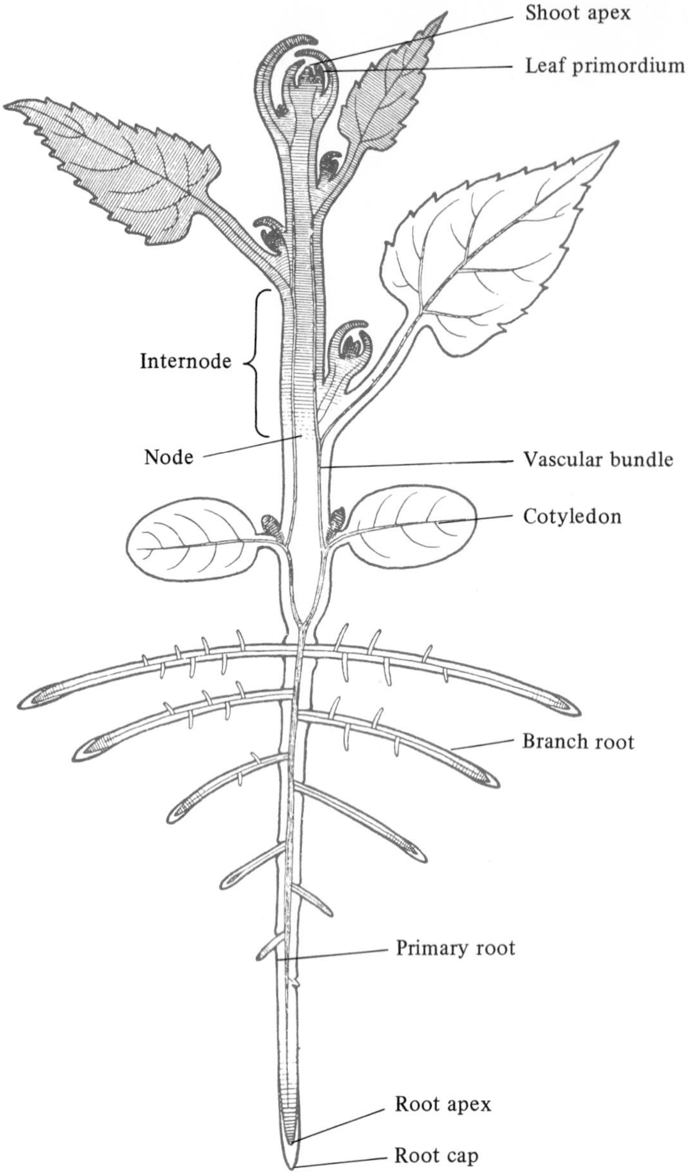


(a)

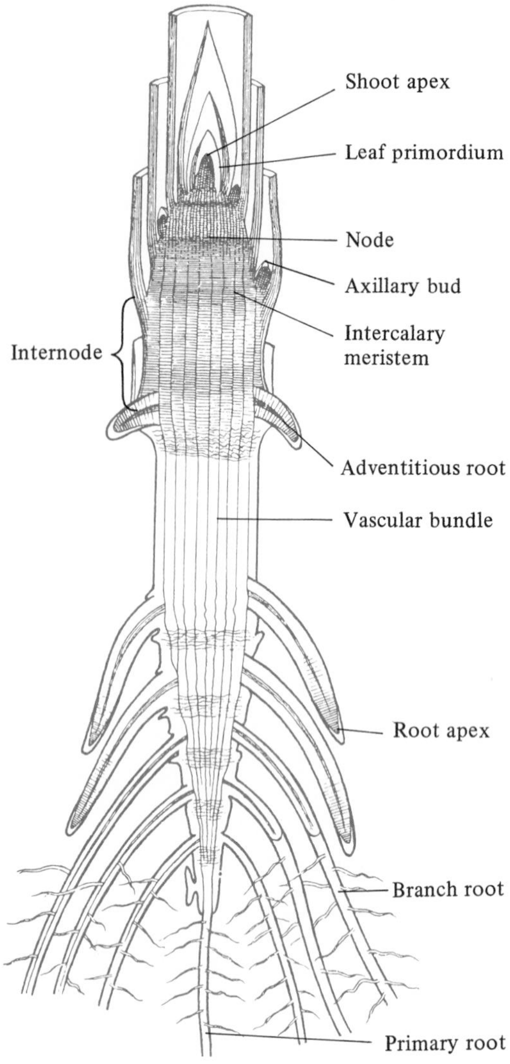


(b)

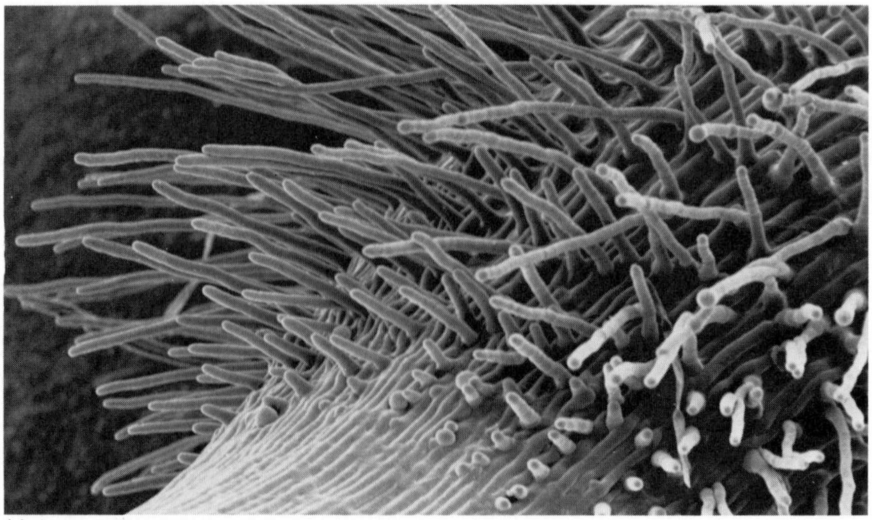

(c)

**FIG. 26-1** Development of the plant. (a) A typical dicot. The organs (roots, stems and leaves) originate in apical meristems in the tips of roots and shoots, the root apex and shoot apex. Branch roots grow from new root apices that arise in the parent root; new shoots or branches originate from axillary buds in the axils of leaves. (b) A monocot of the cereal type, as in maize. Leaf bases ensheath the stem and shoot tip; vascular bundles are scattered through the stem. (c) Development of root hairs on Cress *(Lepidium sativum)*.

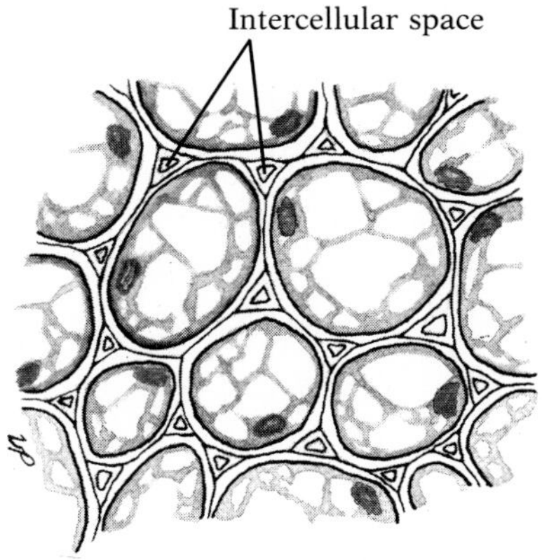


**FIG. 26-2** Parenchyma.

regulated by plant hormones, factors that affect cell elongation and cell division, as well as how the cells differentiate into tissues.

Cell division occurs primarily at the tip, or apex, of growing roots and shoots, in regions called **apical meristems** (Fig. 26-1). The dividing cells in apical meristems are organized in such a way that as new cells form at a root or shoot apex, the older cells remain behind in regular arrangements. These cells differentiate into the **primary tissues** of the plant. New parts are continually added by the apical meristems throughout the life of the plant in a continuing embryogeny, unlike in most animals, which acquire all their organs before birth, during embryogenesis. Among **perennial plants,** which live on from year to year, this continual development requires mechanisms that enable the plants to adapt to climates with distinctly different seasons. For example, perennial plants have evolved the ability to detect signs of forthcoming seasonal changes and trigger such appropriate responses in the meristems as making bud scales instead of leaves to form a winter bud.

As plants get taller, their stems also get thicker, thereby providing support and helping fulfill the need for increased transport of water and food. This is accomplished when **secondary tissues** are added to the primary tissues derived from the apical meristems. The source of secondary tissues is a **lateral meristem,** the vascular **cambium,** a layer of dividing cells within the vascular cylinder of a plant's roots and shoots.

**FIG. 26-3** Collenchyma in stem sections of elderberry (a) and squash (b). Note thickenings in the angles between cells.

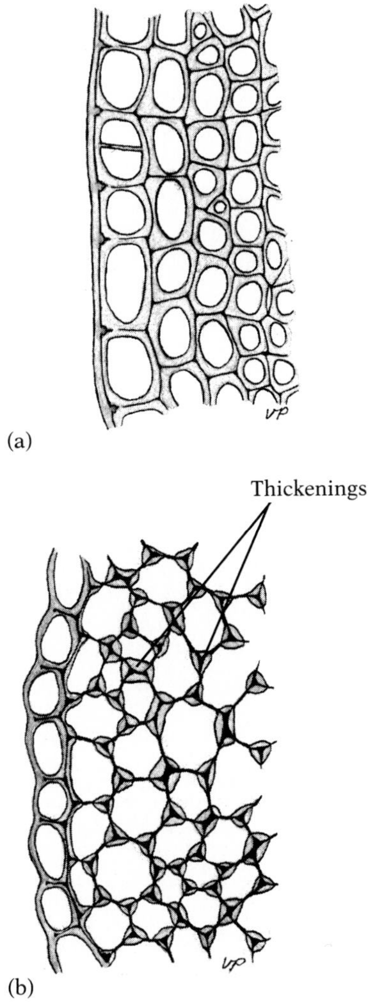


# PLANT CELLS AND TISSUES

Three major tissue systems or groups of tissues make up the plant body, the **dermal tissues,** the **vascular tissues,** and the **ground tissues.** Dermal tissues, like the skin in animals, form the surface of the plant and thus regulate uptake of materials and protect against water loss and invasion by pathogens. Vascular tissues provide support, but mainly transport water and dissolved minerals, photosynthetic products, and hormones. The ground tissues form the bulk of the roots, stems, and leaves, and function in food storage, support, and photosynthesis. Each tissue consists of differentiated cells specialized for particular functions, and within the organs of the plant, tissues are arranged in characteristic patterns in each tissue system.

## CELL TYPES

Three types of cells are common to most plant tissues, **parenchyma, collenchyma,** and **sclerenchyma:**

- *Parenchyma:* more or less spherical living cells with thin walls (except in wood, where they often have secondary walls) (Fig. 26-2). In some tissues they store food; in the mesophyll (from the Greek for "middle of the leaf") of leaves they contain chloroplasts.
- *Collenchyma:* Elongate living cells with their walls thickened irregularly in columns that provide support (Fig. 26-3a, b). They are found in growing stems and leaf petioles.
- *Sclerenchyma:* Dead cells with thickened secondary walls, hollow because the cytoplasmic contents have been lysed away to leave a lumen

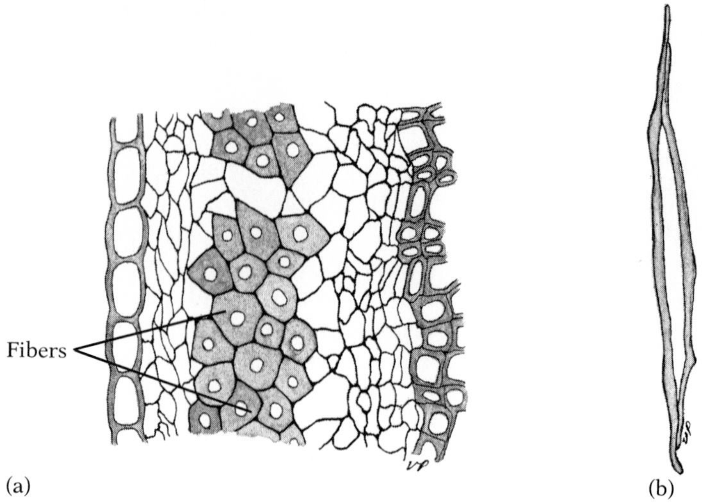


**FIG. 26-4** Sclerenchyma. (a) Fibers in a stem transection of flax. (b) A single fiber is shown in longitudinal view.

(space) within (Figs. 26-4 and 26-5). Lignin, a complex substance dispersed amidst the cellulose, makes the walls hard and durable. Channels, called simple pits, extend through the walls. Sclerenchyma includes:

- *fibers:* elongate, needle-shaped supporting cells found in stems (Fig. 26-4a, b) and leaves, particularly in vascular bundles.
- *sclereids:* "stone cells," ovoid or spherical cells clustered in stems or fruits, or single, branched cells in leaves and stems (Fig. 26-5a, b). They form the hard shells of nuts, the stones of such fruits as the cherry and peach, and seed coats.

Clusters of each of the above types are regarded as simple tissues, named for the cell type. Thus, the ground tissue system is often made up of parenchyma tissue filled with parenchyma cells.

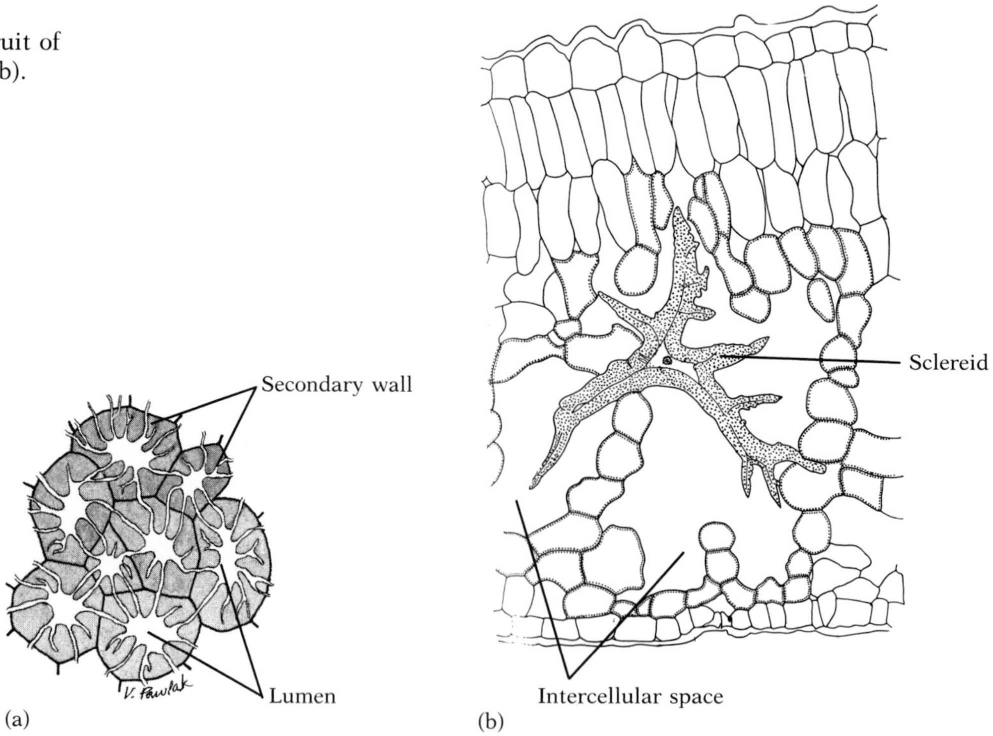


**FIG. 26-5** Sclerenchyma. Sclereid cluster from the fruit of the pear (a), and in a leaf (b).

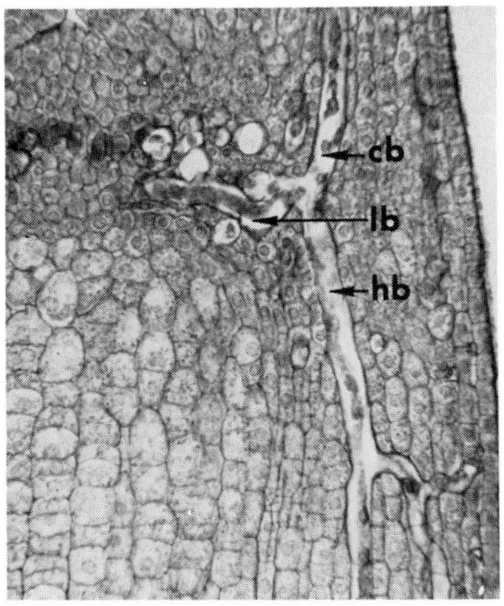


**FIG. 26-6** Portion of a laticifer in the embryo of *Euphorbia*. Arrows show branches of the cell.

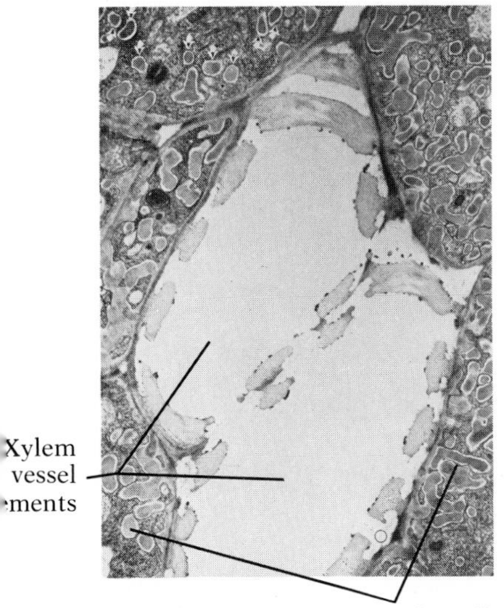


**FIG. 26-7** Transfer cells surrounding two xylem elements in a vascular bundle of a leaf. The numerous wall ingrowths in the transfer cells increase the surface of the plasma membrane.

In addition to the above types, two specialized kinds of living cells occurring in special tissues or in certain species should be noted for their uniqueness:

- *Laticifers* are highly branched cells that grow at their tips (Fig. 26-6) and laticifers often ramify throughout the plant. They are thus large cells that become multinucleate. They produce and contain latex, a milky juice. A tree such as *Ficus elastica* is rich in latex, the source of natural rubber; cutting into the bark causes the latex to bleed from the laticifers. In this manner it can be collected for processing.
- *Transfer cells* are parenchyma cells in which protuberances of the cell wall press the plasma membrane inward. The increased surface area that results aids in secretion, the major function of a transfer cell (Fig. 26-7).

## DERMAL TISSUES

Initially, the plant is covered by the epidermis, usually a single layer of cells, but multiple layered in certain roots and leaves. The root epidermis takes up water and minerals. In the shoot, the function of the epidermis is to conserve water; here, the epidermis is covered by a lipid layer, the **cuticle** (formed of a waxy substance, cutin), often interspersed with waxes, both of which retard water loss from the cells beneath. Single- or multicelled hairs, or trichomes, are scattered in the shoot epidermis of many species. It is thought that they reduce water loss, though some are glandular and secrete aromatic products, as do those of the tomato. Later in development, the epidermis may be supplanted by periderm, a tissue derived from a lateral meristem, the cork **cambium** (phellogen), which produces cork (phellem). Cork cell walls contain another waxy substance, suberin, which retards passage of water.

## VASCULAR TISSUES

Vascular tissues occur as a solid rod (protostele) found in many roots and some stems, as a cylinder of bundles (siphonostele) in many stems, as scattered vascular bundles in the stems of monocots (atactostele), and in leaves, where they form the veins. The vascular tissues include the **xylem,** the water conducting tissue, and the **phloem,** the food conducting tissue. Specialized conducting cells in addition to the basic cell types are found in the xylem and phloem.

- *Xylem:* The water conducting cells are the **tracheids** (Fig. 26-8a), elongate cells with tapered ends that overlap in the tissue, and **vessel elements** (Fig. 26-8b), shorter, wider cells joined end to end in a **vessel.** Both kinds of cells have secondary walls internal to the primary walls that reinforce the cells against collapse under the tension generated by transpiration pull (Chapter 10). In cells that are elongating as they differentiate, the secondary walls may be in the form of rings, or spirals, or horizontal rods like the rungs of a ladder (Fig. 26-8c). If the cells had completed their growth when they differentiated they have **pits,** channels that penetrate the wall. The pits of adjacent cells match up in **pit pairs,** permitting the lateral passage of water between cells. Lignin may harden the secondary walls as is the case in sclerenchyma

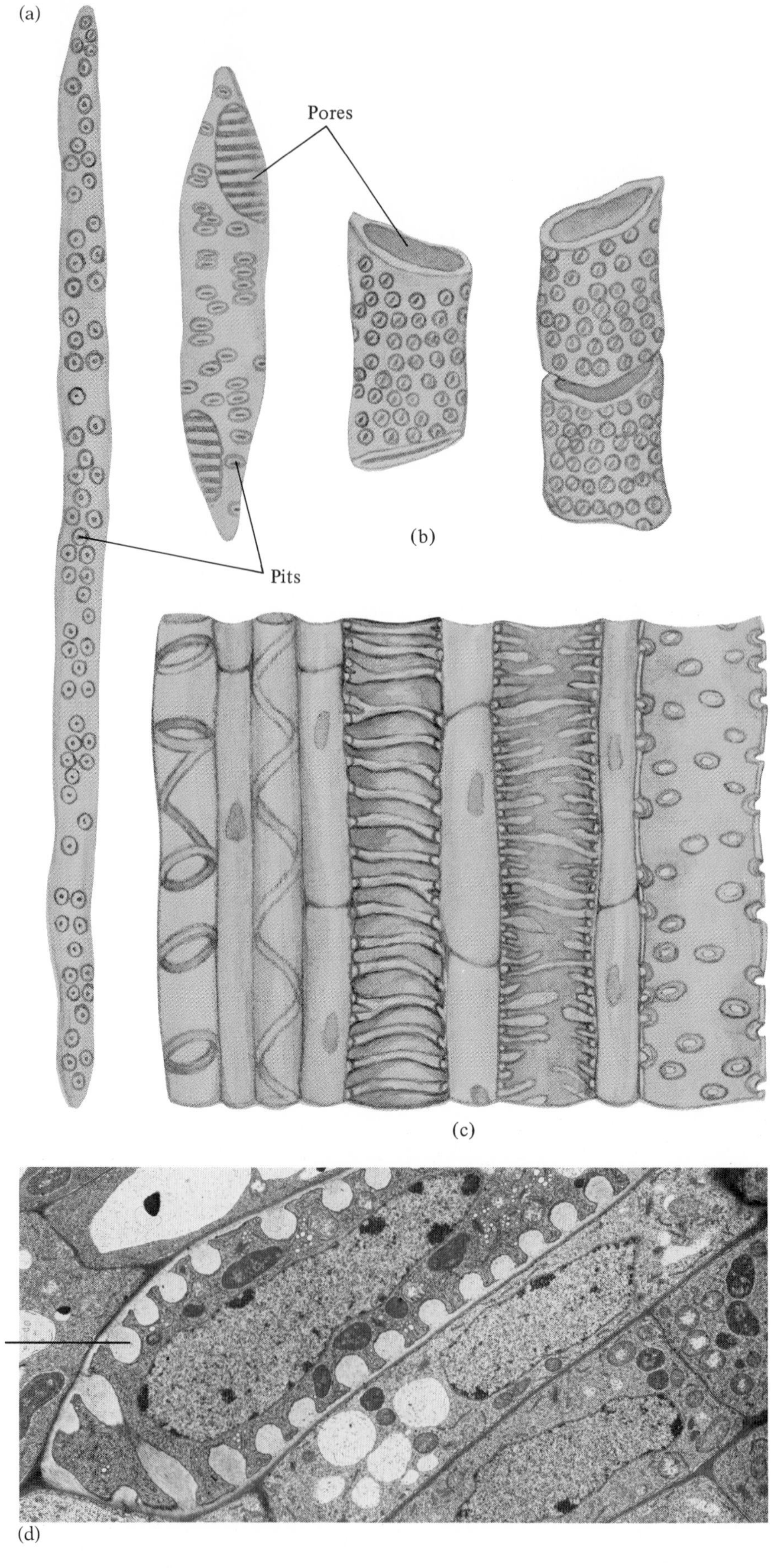


**FIG. 26-8** (a) Tracheids and (b) vessel elements of the xylem. (c) Vessels in the xylem showing differences in the wall pattern related to the timing of differentiation. (d) Electron micrograph of a differentiating vessel.

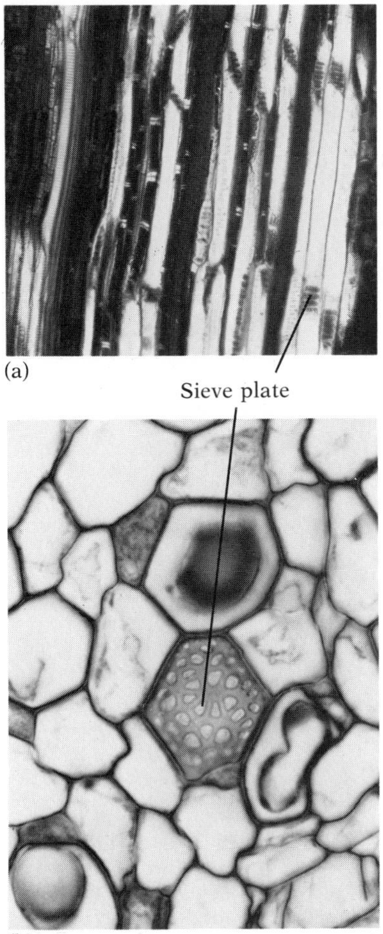


**FIG. 26-9** (a) Longitudinal section through phloem showing sieve tube elements, companion cells, and fibers. (b) Face view of sieve plate between two sieve tube elements.

cells. Where vessel elements overlap at the ends, the intervening wall dissolves, leaving a pore, sometimes interrupted by bars. When the cytoplasm disintegrates, the vessel becomes a hollow conduit composed of dead empty cells. Except for rare occurrences in a few other vascular plant groups, vessels are found only in angiosperms. In addition to tracheids and vessels, parenchyma cells and fibers occur in the xylem.

- *Phloem:* The food conducting cells are represented by **sieve cells** in the non-flowering plants, and by **sieve tube elements** in the angiosperms (Fig. 26-9). Sieve tube elements are aligned end to end to form **sieve tubes.** Where the elements overlap, the walls are modified into a **sieve plate,** consisting of one or more **sieve areas** through which connecting strands of cytoplasm extend. Sieve areas are also found on the lateral walls of adjoining sieve tube elements. Sieve tube elements lack a nucleus and a vacuolar membrane. The cytoplasm is continuous from cell to cell, providing a living transport pathway via the connecting strands. Special parenchyma cells, **companion cells,** are associated with each sieve tube element, and additional parenchyma cells are scattered in the phloem. Fibers are associated with the phloem, usually grouped in bands that lend support to the stem.

### GROUND TISSUES

The tissue filling the root and stem between the epidermis and the vascular tissues is the **cortex.** It consists predominantly of parenchyma, but at its periphery, beneath the epidermis, there is commonly a multicellular layer of collenchyma, or of fibers. At the interior, adjacent to the xylem in roots and horizontal stems is a special layer, the **endodermis.** Cells of the endodermis are encircled by a strip of suberin (Casparian band) embedded in the cell walls at the ends and sides of the cell. This strip, impervious to water, forces water passing through the endodermis to the xylem to go through the protoplasts and not the walls of the endodermal cells.

Roots and stems with siphonosteles have ground tissues at the center consisting predominantly of parenchyma and called **pith.** Occasionally, in what is called chambered pith, the pith contains bands of sclereids in regular patterns.

In the leaf the ground tissues consist of the photosynthetic parenchyma cells that make up the **mesophyll.** Columnar mesophyll cells are the palisade mesophyll; the loosely compacted cells make up the **spongy mesophyll.** Interspersed are the vascular bundles, which subdivide and branch into smaller and smaller veins.

## THE DEVELOPMENT OF A SEED PLANT

Although plant development, like animal development, begins with the fertilized egg, studies of plant development are commonly made with seedlings or the growing parts, such as the tips of roots and shoots. This is possible because of their continuing embryogeny: the study of plant growth and development does not require that one work with an embryo. Instead, a germinating seed of a flowering species, in which the embryo within has resumed growth following its formation and dormant period, or the apical meristems are good subjects in which to study plant development.

**FIG. 26-10** (a) Germination of the pea. Development of the seedling progresses from the apical meristems (arrows) in the seedling. (b) Germination of the bean.

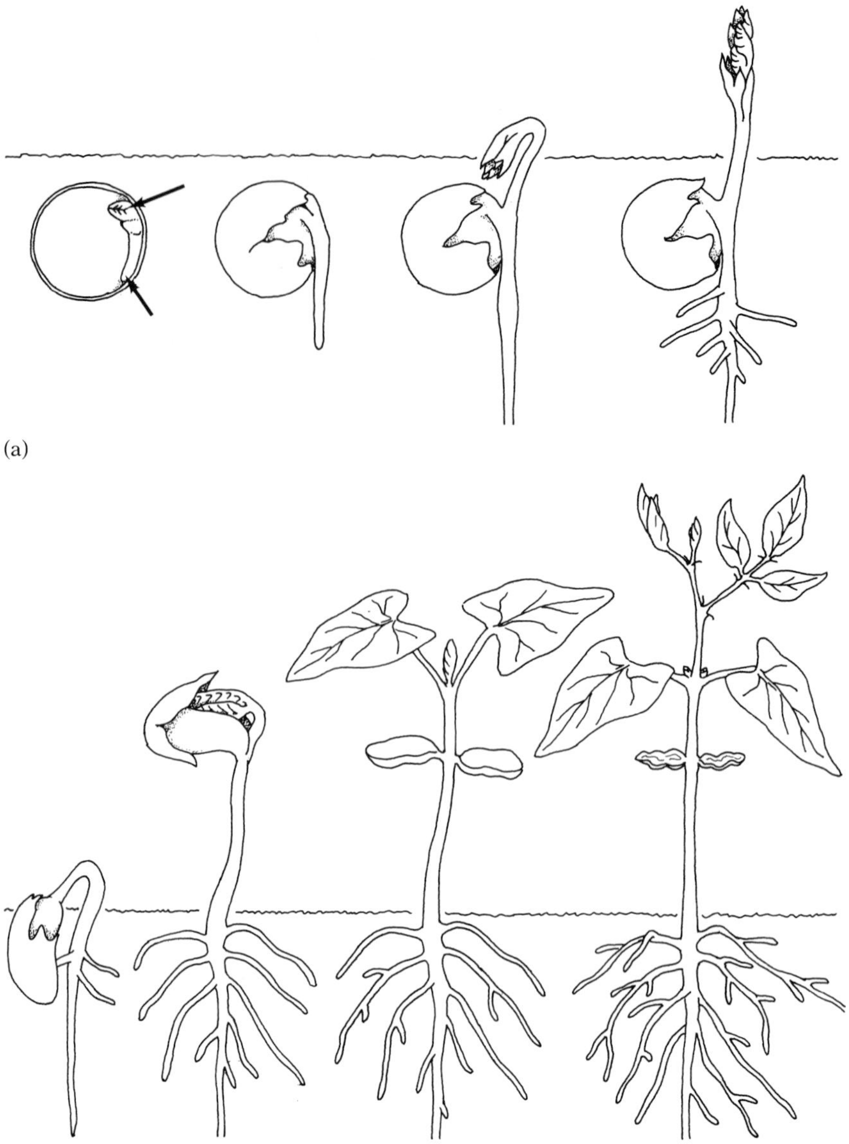

(a)

(b)

## GERMINATION AND SEEDLING DEVELOPMENT

A **seed** is a package containing a dormant embryo and its food supply, both wrapped in the **seed coat.** The food is either stored in the **endosperm tissue** adjacent to the embryo or it has been transferred to the **cotyledons,** or seed leaves, of the embryo and stored there. The embryo in most seeds already has a root apex and a shoot apex with a leaf or two (Fig. 26-10a). As their name suggests, **dicotyledons,** or dicots, have two cotyledons; **monocotyledons,** or monocots, have a single cotyledon.

Different types of seeds require different kinds of stimuli and conditions to end dormancy and begin growth, or germination. In some seeds, a chemical that inhibits growth must be degraded or removed before germination can begin. Some seeds cannot germinate until exposed to light. Many seeds must have their seed coats scuffed or partially digested, whereas others need only to be moistened, whereupon they begin absorbing water. Once a certain amount of water is absorbed, germination begins: the embryo starts to grow and the seed coat splits open. Hormones produced in the embryo activate or promote the synthesis of hydrolytic enzymes, which digest the stored starches, lipids, and proteins, releasing nutrients that are absorbed and utilized by the developing embryo.

The pattern of seed germination and emergence of the seedling shoot from the soil varies among different plant groups. In some species, such as the pea, the cotyledons remain in the soil when the seedling shoot emerges (Fig. 26-10a). In others, the cotyledons emerge as leaves attached to the base of the shoot, and either wither and drop off when their food has been absorbed, as in the bean (Fig. 26-10b), or are retained as functional leaves, as in the radish. In monocots, such as grasses, corn, and other cereal grains, the single cotyledon absorbs nutrients from the endosperm. These nutrients are transferred to the developing root and shoot system. Such cotyledons neither function as leaves nor have the appearance of leaves.

## THE ROOT APEX AND ROOT DEVELOPMENT

As primary growth begins, the first part of the embryo to emerge from the ruptured seed coat is the root (Fig. 26-10a). Primary growth of the root progresses from the root apex, where rows of cells fan out from the tapered tip (Fig. 26-11). Transverse cell divisions in the root tip add new cells, which elongate, pushing the root tip into the soil; divisions perpendicular to the transverse divisions add new rows of cells, thereby thickening the root. The root apex organizes the tissues of the root, which is evidenced by the fact that when a 0.5-mm root tip is excised and cultured in suitable nutrients, it continues to develop a complete root.

Specific meristematic regions in the root apex give rise to specific root tissues. Thus, cells of the **protoderm,** at the periphery of the apex, become the epidermis. Cells in the central apex, called the **procambium,** or **provascular tissues,** become the vascular tissues—xylem and

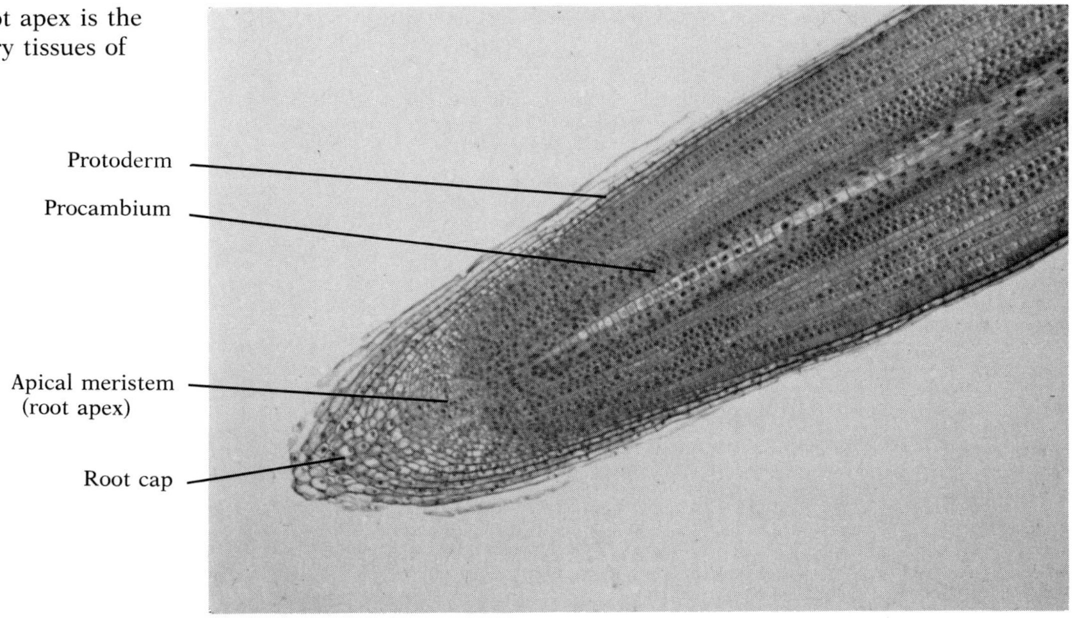


**FIG. 26-11** The root apex is the source of the primary tissues of the root.

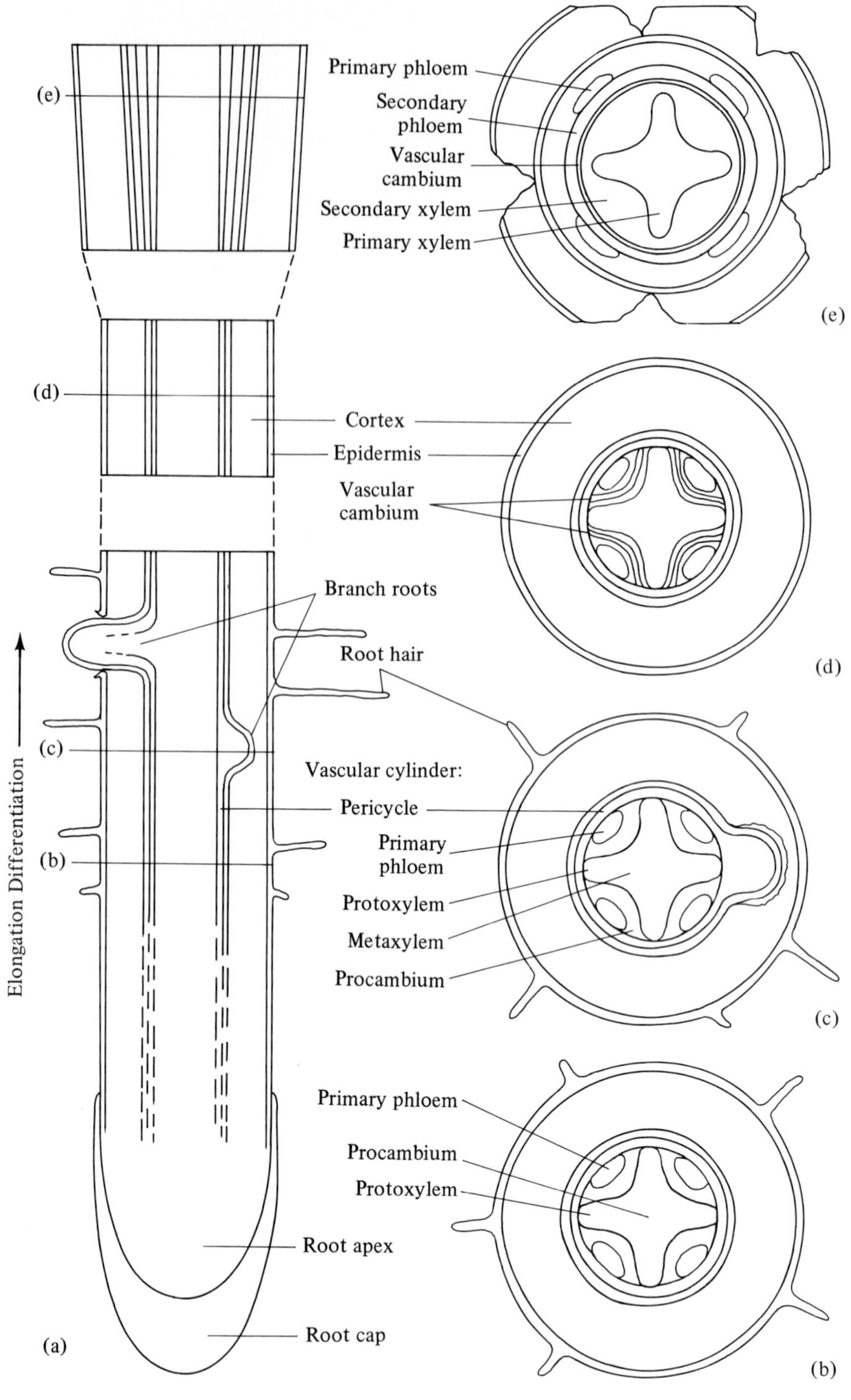


**FIG. 26-12** (a) Development of the root. (b) Differentiation of the primary tissues occurs after elongation of the cells produced at the root apex. (c) Branch roots originate in the pericycle. (d, e) The cambium originates from procambium situated between the primary xylem and phloem.

phloem—and may also produce the cambium. The cells of the root apex between the protoderm and procambium form the cortex. Mitotic divisions at the very tip of the apex produce a cellular sheath, the **root cap** (Fig. 26-11), whose cells secrete a polysaccharide slime. As the root tip pushes into the soil, the cap cells are continually eroded but just as rapidly replaced from the underlying meristem. The cap cushions and protects the advancing root tip as it penetrates the soil, and the slime acts as a lubricant.

## GROWTH OF THE ROOT

Behind the root apex is a region of cell elongation (Figs. 26-11 and 26-12). Since much of the increase in the size of plant organs is the result of cell elongation, the terms *growth* and *elongation* are often used interchangeably in botanical literature. It is understood, however, that the synthesis of protoplasm implied in the term *growth* does occur in these elongating cells. Cell elongation depends largely upon an intake of water. The resultant turgor pressure within the cell produces elongation if the cell wall is plastic, and once the cell is stretched, it remains extended.

The shape taken by a growing cell is controlled in part by the orientation of the cellulose microfibrils of the cell wall, which determines the direction in which the wall may stretch. There is some evidence that microtubules affect the orientation of the cellulose microfibrils as they are deposited by the plasma membrane in the wall. As the cell grows, it synthesizes more cell wall materials. Precursors of wall molecules other than cellulose are made within the Golgi complex. Vesicles released from the Golgi complex convey the wall precursors to the cell surface (see Chapter 4). Cellulose is produced at the plasma membrane.

If a developing seedling is placed horizontally, the growing shoot turns upward, while the growing root turns downward; both processes are examples of **geotropism,** or **gravitropism,** the growth response to gravity. In roots, the response occurs in the elongation zone, where the upper cells grow more rapidly than those on the bottom side. Recent studies indicate that the hormone **abscisic acid,** which acts as a growth inhibitor, is produced in cells of the root cap of a horizontally oriented root and moves backward on the lower side to the elongation zone, slowing the growth of the cells there. More recently, asymmetric distribution of $Ca^{2+}$ has been reported to act in root curvature, perhaps through an effect on the hormone auxin. The stimulus of gravity is sensed in the root cap cells, where starch plastids settle to the lower side of the horizontally placed root and in some unknown way cause the asymmetric distribution of abscisic acid or other factor.

## DIFFERENTIATION OF ROOT TISSUES

Following their elongation, cells begin to differentiate in patterns characteristic of roots and commensurate with their function of taking up water and minerals from the soil (Chapter 10). Ahead of the root apex, the root cap parenchyma cells are formed. Behind the elongation zone tissues differentiate.

***The Epidermis.*** Certain cells in the protoderm form the filamentous extensions called root hairs, which enhance water absorption by virtue of their extensive surfaces (Fig. 26-12a). Root hairs persist for only a short time, but new ones are continually added as the root grows.

***The Cortex.*** Between the epidermis and the vascular cylinder, the cells differentiate as the cortex, a tissue through which water must pass on its way to the vascular tissues, and later largely devoted to storage of food. The inner layer of the cortex, the endodermis, regulates the passage of water into the xylem.

***The Vascular Cylinder.*** At the center of the root are the procambial cells, which differentiate into either primary xylem or primary phloem,

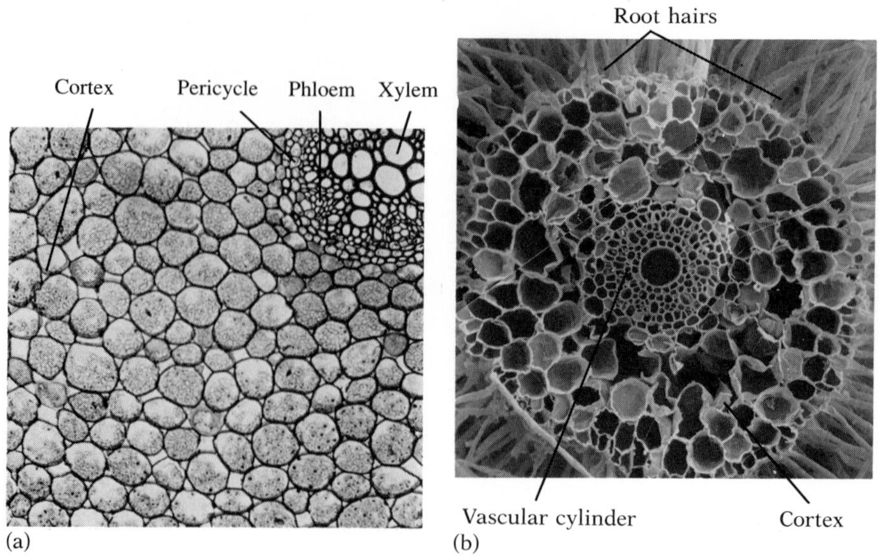


**FIG. 26-13** (a) Transection of a buttercup root in the differentiated region. (b) SEM of the root of barley, a monocot, showing root hairs, cortex, and vascular cylinder.

depending on their position. Columns of xylem and phloem alternate around the vascular cylinder, or **stele** (Figs. 26-12 and 26-13). This alternate arrangement permits water to enter the xylem from the endodermis without passing through phloem.

***The Primary Phloem.*** The first procambial cells to differentiate become sieve tube elements of the primary phloem. Located at points around the periphery of the vascular cylinder, just internal to its outer boundary, or **pericycle** (Fig. 26-13), these sieve tubes will transport nutrients to the root apex. Adjacent procambial cells will differentiate as sieve tube elements and companion cells to form the groups of primary phloem that alternate with the primary xylem in the root.

***The Primary Xylem.*** Procambial cells internal to the pericycle and located between the groups of phloem cells differentiate as primary xylem vessel elements (Fig. 26-13). These vessel elements differentiate while they are elongating; their secondary walls are pulled apart into rings or spirals. They are the first xylem cells to differentiate and are called protoxylem. Procambial cells toward the center of the root differentiate in sequence; the later ones have pitted walls and form the metaxylem. Differentiation of xylem in the root thus progresses toward the center of the root, or centripetally.

## THE VASCULAR CAMBIUM AND THE THICKENING OF ROOTS

As the primary vascular tissues of the root complete differentiation a few centimeters back from the root apex, cells located between the xylem and phloem begin dividing, initiating a lateral meristem called the **vascular cambium.** The dividing bands extend and meet to form a cylinder of cambial cells. Since the plane of division of these cells, called **initials,** is parallel to the root surface, the root thickens accordingly (Fig. 26-12d, e). Cells produced by these divisions differentiate according to their positions. Those toward the center, next to the primary xylem, become secondary xylem. Cells produced toward the outer side of the vascular cambium, nearest the phloem, become secondary phloem. At each cell division, the non-differentiating daughter cell is capable of dividing again, thus maintaining the cambium as a meristematic layer.

Cell division in the cambium and differentiation of the cells it produces are regulated by the plant hormones **auxin** and **gibberellin.** If a dormant woody stem is severed near the tip and an auxin is applied to the cut surface, the cambium begins dividing as auxin progresses down the stem. Applied auxin also promotes differentiation of xylem elements and affects the type of xylem elements formed. Applied gibberellin promotes phloem differentiation of the cambial products.

Pressure also affects differentiation of the cambial products. If in the spring a flap of bark is pried free at one end by cutting and lifting with a chisel, the cambium produces an amorphous mass of parenchyma cells called **callus** in the wound. But if the flap is pressed back in place with an impermeable membrane inserted between flap and wood to block diffusible substances, and the flap is bound tightly, the cambium continues to produce xylem and phloem.

***Parenchyma Cells and the Storage of Food.*** Not all cells of the secondary xylem and phloem are conducting cells. Within both tissues are parenchyma cells, which may store food. In edible roots, such as carrots, parenchyma cells predominate in the secondary xylem and phloem. In contrast, a woody tree root has a high proportion of tracheids and xylem vessels and proportionally more sieve tubes than parenchyma in the phloem.

***Branch Roots.*** The well-developed root system of a seedling has numerous branches that feed water and minerals into the main root. A section cut through a young root shows that it is the **pericycle,** the layer of parenchyma cells at the periphery of the vascular cylinder, that is the source of branch, or lateral, roots. Cell divisions occur in sites adjacent to the xylem and phloem, resulting in a bulge of cells (Fig. 26-14a). Within the bulge, a root apex becomes organized and lays down root tissues. As it grows, the branch root primordium forces its way outward through the parent tissues (Fig. 26-14b, c), its xylem and phloem arising in continuity with the vascular tissues of the parent root. As the branch root grows, provascular tissues added by its root apex continue to extend the vascular system within it.

In parasitic angiosperms, the lateral roots are modified as haustoria. Haustoria (sing. haustorium) are a kind of modified root that invades the tissues of the host plant and takes up nutrients from its vascular tissue. Recently, a factor, called an elicitor, has been isolated from the exudate of host tissues that induce haustoria to form when applied to

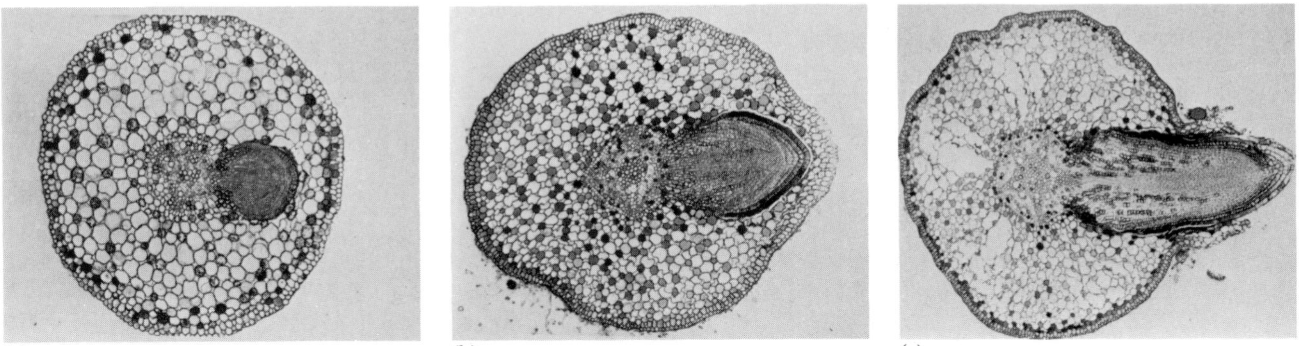

(a) (b) (c)

**FIG. 26-14** Stages in the origin of branch roots, as seen in sections of a willow tree root. (a) A developing root primordium. (b) Primordium penetrating the cortex. (c) Root primordium emerging from the root surface.

the root of its parasite. Such elicitors not only enable haustoria to form but determine the specificity between host and parasite.

## THE SHOOT APEX AND SHOOT DEVELOPMENT

While the root apex produces the primary tissues of the root, the shoot apex has the more complicated task of producing the shoot, which involves the formation of the stem primary tissues as well as the initiation of leaves. The shoot apex must thus regulate the timing of formation of the leaves and their spacing on the stem, called **phyllotaxy,** (or phyllotaxis) as well as regulating tissue patterns in the stem.

***Initiation of Leaves.*** Examination of the growing shoot of the embryo in a germinating seed or dissection of the tip of any growing shoot reveals successively younger and younger leaves arranged in order about the apical meristem, or shoot apex (Fig. 26-15). These new leaves, as well as the stem on which they occur, arise in the shoot apex. The earliest indication of leaf origin in the shoot apex is the appearance of localized cell divisions in the second or third outer cell layers of the shoot apex. Continued divisions form first a bulge and then an erect peg that may have lateral extensions at its base which curve around the apex like a collar. (The youngest leaf at the apex is designated P-1, the next older P-2, etc.) From its inception, the peg contains one or more provascular, or procambial, strands, which are continuous with vascular strands in the stem below them (see Fig. 26-15b). Paired outgrowths at the base of a leaf, the **stipules,** or a leaf sheath may develop from the leaf base in some species; then provascular strands may encircle the shoot and extend into the developing leaf and its cylindrical base. Upon further development, the provascular strands differentiate into the vascular bundles of the leaf.

Following formation of the peg, the embryonic leaf, or leaf primordium, assumes its final form in miniature, the shaping of which is related to the distribution of cell divisions. At first, divisions are concentrated in the tip as the peg grows in length. Meristematic activity then

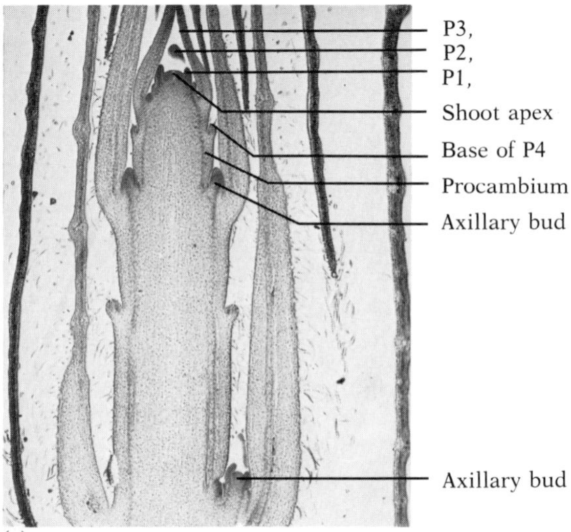


(a)

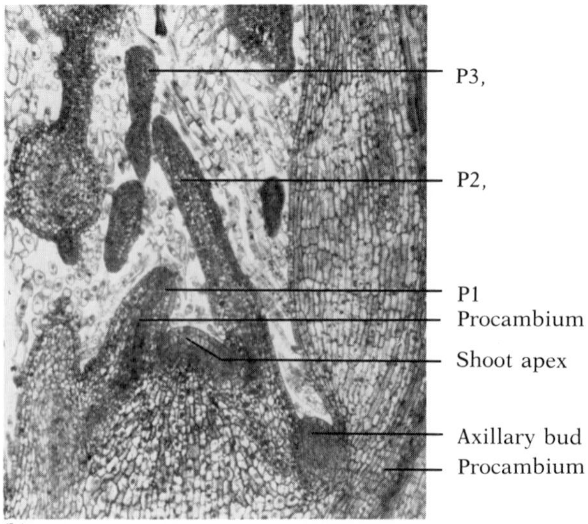


(b)

**FIG. 26-15** Longitudinal sections of shoot tips of (a) honeysuckle, a plant with opposite leaves (P1, P2, etc.), and (b) cocklebur, showing shoot apex and leaves in progressive stages of development. Each leaf has a procambial strand. An axillary bud is present at leaf 6. Cells above apex and between leaves are sections of epidermal hairs of leaves.

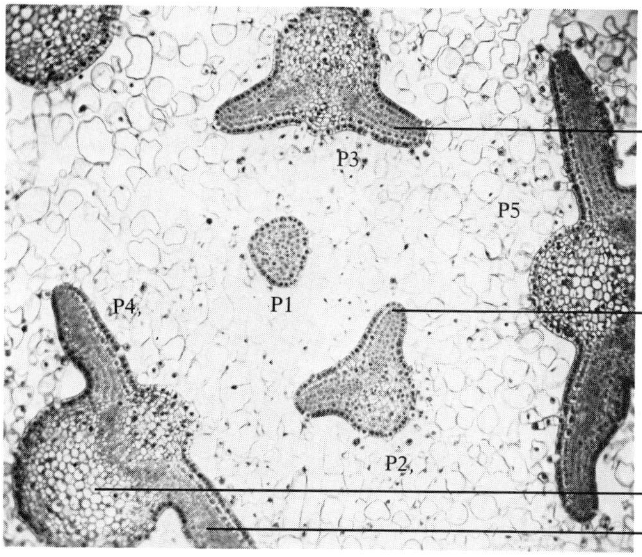


**FIG. 26-16** Successive leaf primordia in a cross section of the shoot tip of tobacco show the phyllotaxy as well as development of the leaf blade. The layers of the leaf blade are established by the marginal meristems and maintained by cell division in the plate meristems. Cell divisions along the major veins thicken the veins and midrib of the leaf blade.

shifts to the **marginal meristems,** which are ridges along each side of the primordium (Fig. 26-16). Within the marginal meristems, a fixed number of cell layers (usually five to seven) is produced, and through the addition of cells within the layers, the flattened **blade** broadens at both sides of the central **midrib.** The activity of the marginal meristems shapes the leaf blade. If additional cells are produced in alternating sites along the margins, lobes or teeth arise at the edges of the leaf. Regions in which few cells are produced form indentations, or sinuses. The midrib and major veins of the leaf are thickened by cell division. In plants with compound leaves, leaflet primordia arise along the leaf margin and develop just as the simple leaf does. More complex leaf forms, such as the cup-shaped leaf of the pitcher plant, result from localized regions of division in the peg.

***Tissue Differentiation in the Leaf Blade.*** The layers of the young leaf blade develop into the tissues of the mature leaf (see Chapter 7). Regular, columnar cells develop adjacent to the upper surface layer, or upper epidermis, and elongate to form the photosynthesizing **palisade cells.** Cells of the lower layers are drawn apart as the leaf expands and become the **spongy mesophyll,** the spaces between cells permitting exchange of gases during photosynthesis.

The surface layers, particularly the lower epidermis, develop stomata with guard cells that regulate their opening and closing. Except for the guard cells, epidermal cells lack chloroplasts and transmit light to the leaf interior. As noted earlier, the waxy cuticle, and in many species, trichomes, retard water loss.

***Leaf Arrangement.*** The site of leaf attachment to the stem is termed a **node,** and the stem region between two nodes an **internode** (see Fig. 26-1). Viewing a shoot tip from above reveals the precise phyllotaxy of the leaves (Fig. 26-16). Each new leaf is spaced a specific number of degrees around the apex from its next older neighbor, the precise arrangement differing among species. Variations include spiral, or alternate phyllotaxis, in which one leaf occurs at each node; opposite phyllotaxis, in which leaves are paired opposite each other at a node; and whorled phyllotaxis, in which three or more leaves occur at a node. The phyllotaxis of a given species is so regular that, given the species and the position of the youngest leaf at the shoot apex, the site of origin of the next leaf can be predicted before it arises.

***Leaf Abscission.*** Leaves drop off, or undergo abscission, in the autumn and under such other conditions as senescence (old age) and mineral deficiency. Separation occurs where the **petiole** or leaf stalk joins the stem and in some cases where the blade joins the petiole as well. It involves a zone of cells across the leaf base, including a separation layer, in which the cell walls separate between cells, and a protective layer, in which an outer covering called a periderm develops and seals off the exposed surface. Abscission is controlled by hormonal interaction involving auxin and other plant hormones called **cytokinins,** which counter abscission, and **abscisic acid** and the gas **ethylene,** which promote it.

***Growth and Differentiation of the Stem.*** As each new leaf arises, the cells below the shoot apex grow and differentiate into the characteristic tissue patterns of the stem (Fig. 26-17). The protoderm becomes the epidermis. The procambial strands associated with the leaf primordia become the vascular bundles, arranged in a cylinder in dicots and scattered through the stem in monocots. The ground tissues differentiate within and around the vascular cylinder. In dicotyledons, they form the pith at the center and the cortex between the vascular tissues and epidermis. In the stems of monocotyledons, the vascular bundles are dispersed in the parenchyma (Fig. 26-17c).

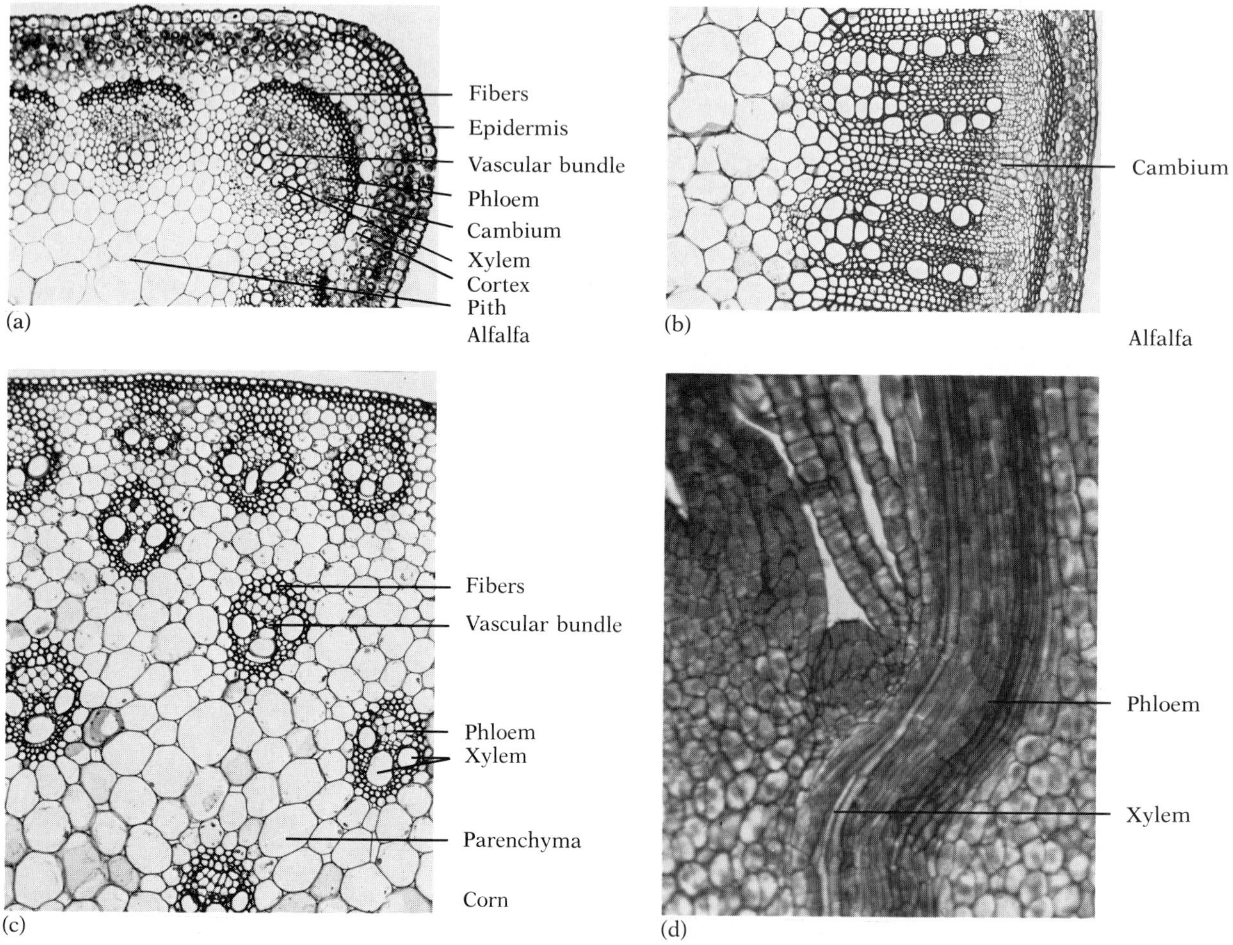


**FIG. 26-17** The stems of (a, b) a herbaceous dicot, alfalfa, and a monocot, (c) corn, in transverse sections. The vascular bundles form a cylinder in the dicot but are scattered throughout the stem in a monocot. In the older alfalfa stem (b), the cambium has extended between the vascular bundles to form a complete cylinder. (d) Differentiation of primary xylem and phloem in the provascular strand of a cocklebur.

Stems exhibit various patterns of growth. While most grow vertically, showing negative geotropism, the stems of many species of plants trail along the ground unresponsive to gravity or climb structures they encounter. In most species, stem elongation between the nodes leads to a spacing of leaves along the stem (see Fig. 26-1). In others, little or no internodal elongation occurs, and the leaves are tightly clustered, either in a rosette or in a head, as in cabbage and lettuce.

As the stem elongates, not only do cells enlarge, but cell division continues for a considerable distance back from the shoot apex. While many of the cell divisions are transverse, adding cells lengthwise, some are parallel to the axis, increasing the number of cell rows, thus widening the stem.

As the provascular strands differentiate into the xylem and phloem of the stem (Fig. 26-17d), cells of the protoderm differentiate into the various cells of the epidermis, including guard cells, trichomes, and glands, as in the leaf. And, like the leaf, the stem is covered by a layer of cutin, secreted by the epidermal cells.

In the cortex and pith, most of the cells differentiate as parenchyma cells, but other cell types may differentiate singly or in clusters. Clusters of ovoid sclereids bound open chambers in the pith, for example. In addition, elongate fibers form supporting columns around the vascular bundles or a rigid cylinder in the outer cortex (Fig. 26-17) providing mechanical support to the stem.

***Origin and Development of Branches.*** The branching of a plant's shoot system increases the number of shoots and thus the potential number of leaves, thereby exposing additional leaf surfaces to light and increasing the rate and amount of photosynthesis. One simple type is dichotomous branching, characteristic of lower vascular plants such as club mosses. In this type, the shoot apex divides to create a Y-shaped fork with shoot apices at the tips of each branch. In most vascular plants, however, branches originate from meristems in the **leaf axils**, the angles between the leaves and the stem (see Figs. 26-1 and 26-15). A mound of cells forms in the leaf axil, and within it a new shoot apex arises to form a bud (Fig. 26-18). Whether such an axillary bud primordium develops

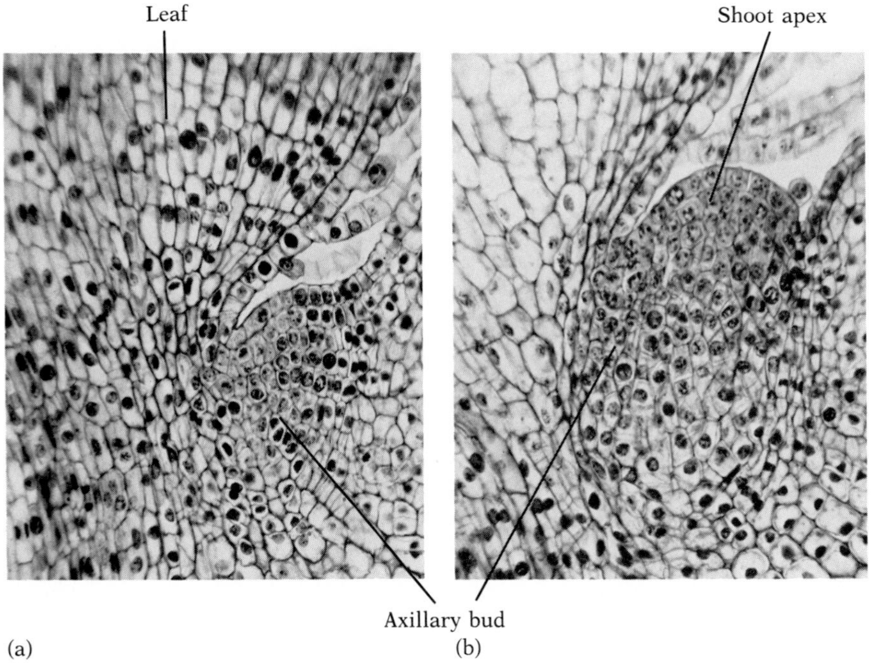


**FIG. 26-18** (a) Axillary buds originate near the shoot apex in the axil of a leaf as a small bulge. Cell divisions contribute cells to the bulge and it grows. (b) Cells of the bud become organized as a shoot apex, which soon begins the production of leaves.

into a shoot or is suppressed, it is controlled by the parent shoot tip, a phenomenon called **apical dominance.**

Apical dominance is partly mediated by the inhibitory effect of the hormone auxin on axillary buds. Auxin produced in the growing tips is transported down the stems to the leaf axils. Removing the growing tip of a shoot removes auxin's inhibiting effect and permits the buds nearest the decapitated apex to develop into branches. When the growing tips of a bush or hedge are pruned, the higher axillary buds are released from apical dominance and develop into new branches. Experiments discussed later, in which another hormone, a cytokinin, is added to the axillary buds, show that this hormone can overcome apical dominance despite the presence of auxin, and indicate that lateral bud growth is regulated through an interaction of auxin and cytokinin, with the possible influence of the hormone gibberellin as well.

## VASCULAR CAMBIUM AND THE FORMATION OF WOOD AND BARK

The thickening of stems, like the thickening of roots, is the result of the activity of a lateral meristem, the vascular cambium. As the source of the wood of trees, the cambium is of considerable importance.

The cambium of the stem originates from cells of the procambial strands situated between the differentiating xylem and phloem. This strip of cells across the middle of each vascular strand retains its meristematic potential (Fig. 26-17a, b). A cylinder of dividing cambial cells, formed from the strips in each bundle, thus remains between the primary xylem and phloem. If the bundles are spaced apart and there are gaps in the cylinder, the intervening parenchyma cells may resume division, making the cambial cylinder continuous (Fig. 26-17). In trees and some other woody plants, the vascular bundles are usually close together, forming a continuous cambial cylinder close to the shoot tip.

The inner derivatives of the cambial initials differentiate as secondary xylem, the wood of the stem (Fig. 26-19). The cells of the xylem include the water-conducting tracheids and vessel elements; the thick-walled, supporting fiber tracheids; and the food-storing wood parenchyma. In addition, abundant wood rays composed of horizontal parenchyma cells extend outward through the xylem. The nature and relative numbers of these diverse elements and the composition and thickness of their cell walls determine the properties of the wood. The patterns in which the cells of the wood differentiate vary among species. The wood anatomy of a species is constant enough to enable identification of the species by microscopic inspection of a cross section of a piece of wood (Fig. 26-20).

The influence of environment on plant development is evident in the wood of temperate zone trees. Seasonal changes influence the formation of secondary xylem from the cambium. Cells produced by the cambium in the spring, called spring or early wood, are larger than those produced later, called summer or late wood; this size difference reflects the seasonal changes in climate. The small late wood elements adjacent to the larger spring wood cells create the **annual ring,** or growth ring (Figs. 26-19 and 26-20a). Drought, excessive insect infestations, fires, and other environmental factors also affect cambial activity, creating false rings, and may be revealed by inspection of the annual rings of an old tree. Not only are tree rings thus testimony to climates and other conditions of the past, but they have been found useful in historical anthro-

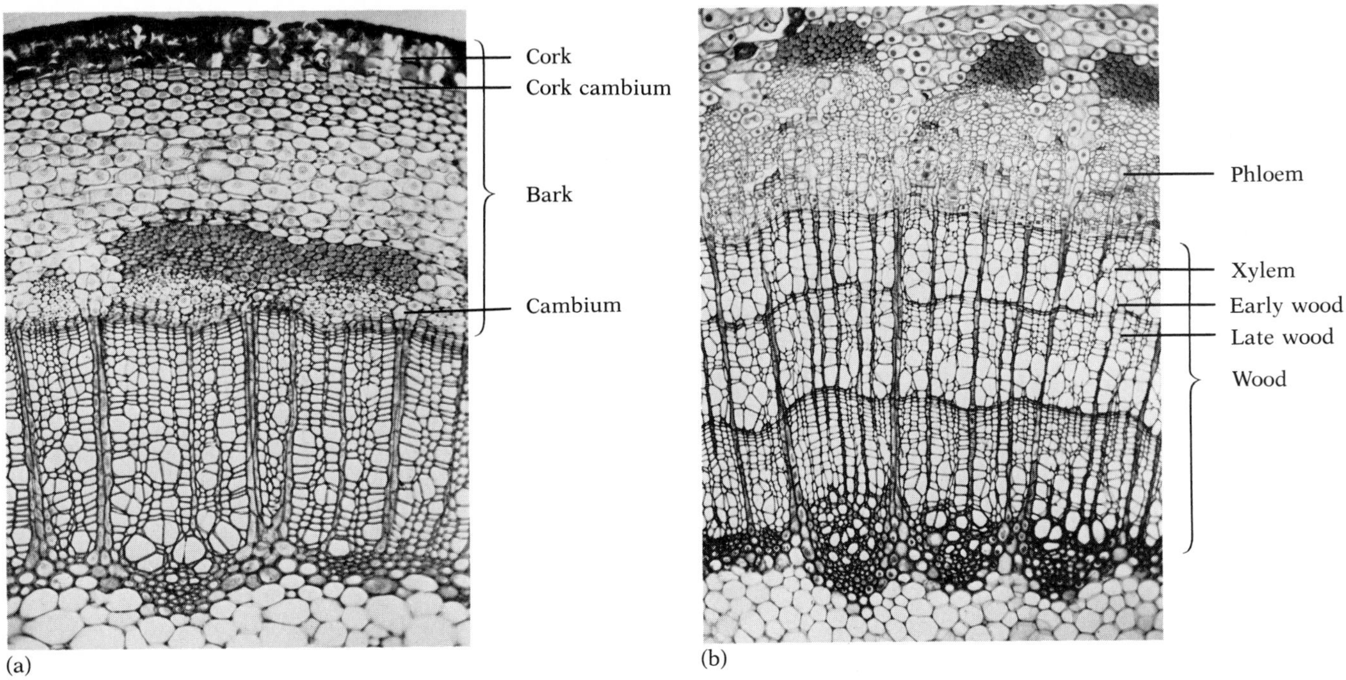


**FIG. 26-19** Sections through (a) 1- and (b) 3-year-old stems of *Magnolia*. Secondary xylem forms the wood of the tree. All tissues external to the wood make up the bark. Annual, or growth, rings are evident where early, or spring, wood is adjacent to the smaller cells of the late, or summer, wood. In the bark, a cork cambium develops beneath the epidermis and produces cork, which forms the outer covering of the stem.

pology. Tree rings in logs used to construct ancient shelters can be matched with rings of old living trees of the same region. This often allows a precise determination of the year in which the tree was felled and gives a clue to the climate and other conditions of the times.

The outer derivatives of the vascular cambium form the secondary phloem of the bark. The phloem consists of alternating bands of thick-walled fibers and food-conducting sieve tubes with their companion cells and parenchyma (Fig. 26-19). The youngest sieve tubes function in sugar translocation. Older phloem elements cease functioning and become compressed as a result of stem expansion.

**FIG. 26-20** (a) Transverse section of oak wood. (b) SEM of wood of hazel.

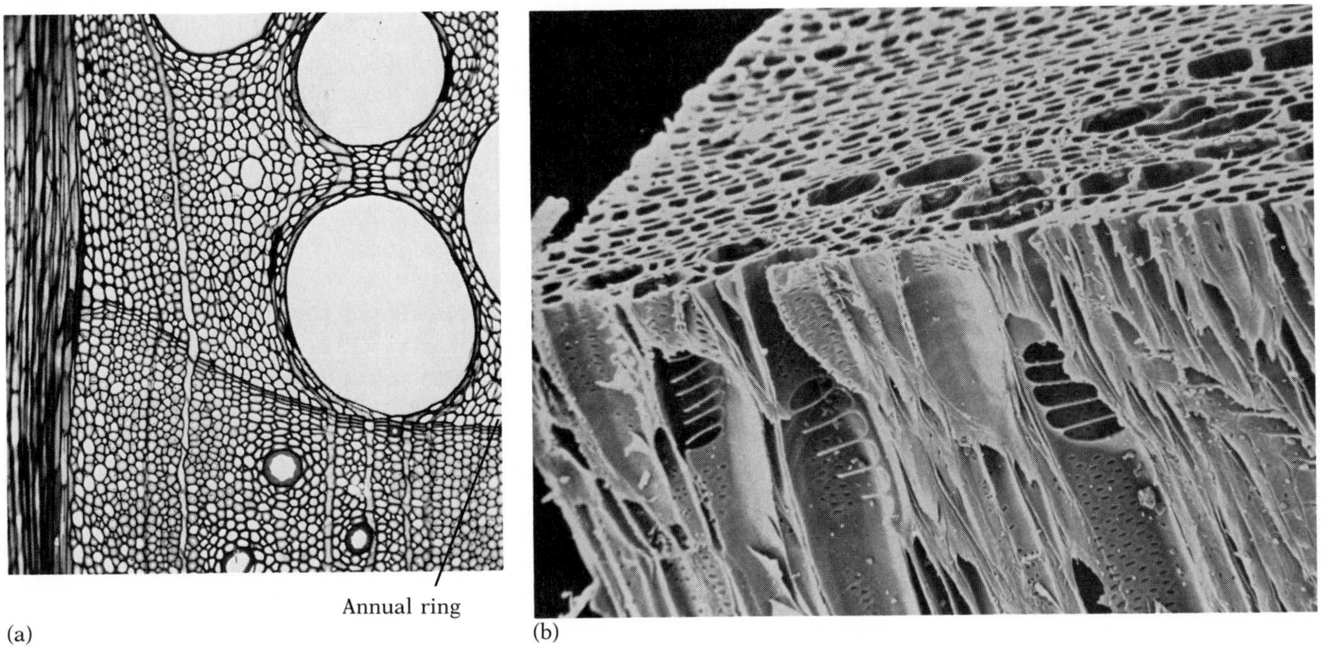

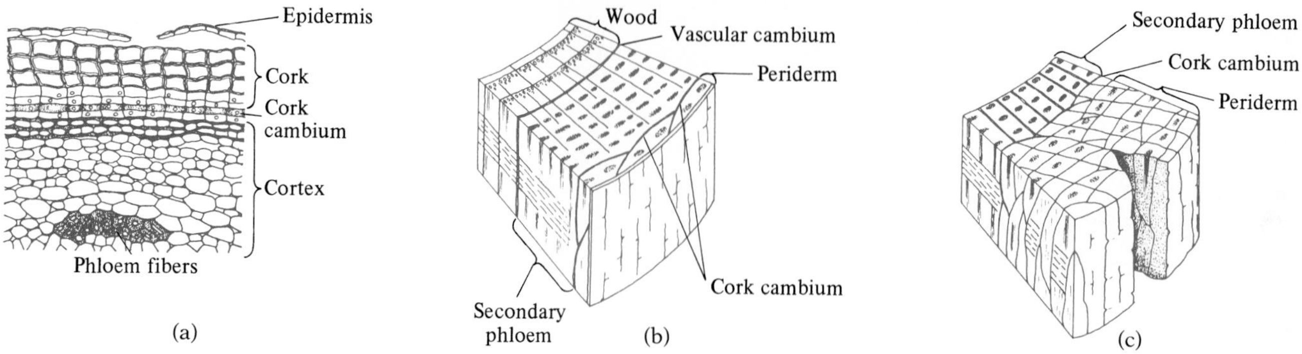


**FIG. 26-21** Periderm formation. (a, b) The cork cambium originates beneath the epidermis and produces layers of cork. (c) Subsequently, new cork cambiums arise in the phloem in arcs, which produce sheets of cork that flake off the surface of the tree.

## THE PERIDERM: A PROTECTIVE COVERING

As a tree continues to increase in girth, its outer tissues, no longer able to divide, split. Beneath the epidermis a protective layer, called the **periderm** has already started to form. The periderm includes layers of cork cells, the walls of which contain a waxy substance called **suberin,** which blocks passage of water (Fig. 26-21a, b). The periderm of an older tree thus assumes the role of the epidermis and cuticle in reducing water loss. The cork cells are produced from a lateral meristem, the **cork cambium.** As a tree increases in girth, more splits and fissures appear in its expanding surface as a result of the death of the outer cells, which are no longer able to obtain water. New arcs of cork cambium arise deeper and deeper in the bark as parenchyma cells begin to divide. These layers of cork cambium produce cork. Ultimately, cork cambium arises in the phloem of the bark (Fig. 26-21b, c). The arcs of cork cambium originate as sheets of a form characteristic for each particular species of tree. The patches of cork they produce conform to their shape. As a result, the pattern in the cork surface, or "bark," of the tree is specific enough in many cases to enable identification of the species.

# THE CONTROL OF FLOWERING

At some stage in the development of a flowering plant, a striking change takes place at its shoot apex. Instead of production of leaves, new developmental programs are set in motion. The sequence includes flower formation, meiosis and sporogenesis, gametogenesis, fertilization, and, finally, fruit and seed formation. Gene sets hitherto repressed now become activated. New plant structures form at the apex—sepals, petals, stamens, and carpels—and a variety of new molecules is synthesized. Some of the more obvious new substances are the pigments and aromatic substances of petals. When a fruit is formed still other new molecules and structures are elaborated, including additional pigments and aromatic and flavor molecules attractive to animals. All of these events require the expression of many heretofore quiescent genes.

The reorganization that takes place in the shoot apex at the time of flower initiation first becomes evident as a redistribution of cell divisions in which a mantle of dividing cell layers comes to cover a central growing core. Sepals and petals arise in whorls (Fig. 26-22) in the mantle layers, much as do the leaf primordia. This is followed by the differentiation of stamens and carpels.

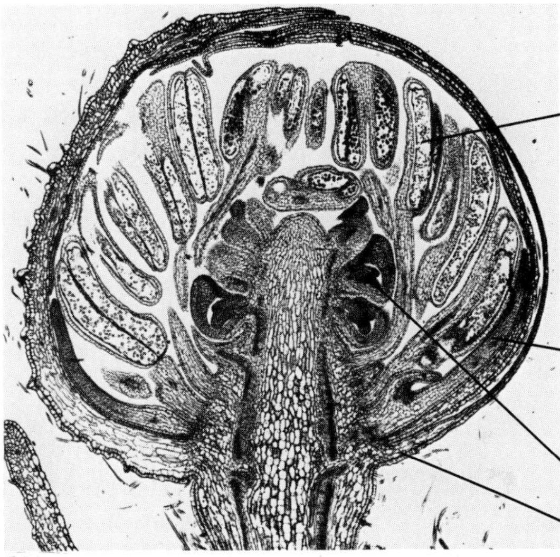


**FIG. 26-22** The initiation of a flower. (a) Scanning electron micrograph of early flower of *Genista* with rudiments of petals, sepals, stamens, and carpel. (b) A section through a young buttercup flower, showing cellular organization.

A large variety of species-specific patterns occurs among flowering plants. The number of parts in a whorl, for example, differs significantly among families of angiosperms. In some species, the parts of a petal whorl are fused, whereas in others the parts are separate. The components of successive whorls may also be fused. These constant structural differences in flowers are used in the classification of flowering plants.

## PHOTOPERIODISM AND THE INDUCTION OF FLOWERING

The factors that determine whether and when a shoot apex begins the initiation of a flower have long been of interest to plant physiologists and horticulturists. Although many plant species undergo flowering upon reaching a certain size and stage of maturity, in many others, specific environmental factors and stimuli are essential to bring about the physiological changes that evoke flowering. This process is called induction, or evocation, of flowering. One of the additional factors essential to the success of flowering and fruiting in some species is exposure to cold temperature. Some biennials (plants requiring 2 years to flower and form seed) and some cereals, such as winter wheat, require a cold period before flowering. Many plant species must be subjected to an appropriate day length, or light-dark cycle, before flowering is induced. First discovered in plants, this phenomenon, termed **photoperiodism,** was eventually shown to be a factor in several other developmental phenomena in plants as well as in a variety of rhythmically occurring activities in invertebrate and vertebrate animals, such as mating, metamorphosis, molting, and migration (see Chapters 19–21).

***The Discovery of Short-Day and Long-Day Plants.*** The discovery of the role of photoperiodism in flowering was made by W. W. Garner and H. A. Allard between 1915 and 1920 in studies of new varieties of tobacco and soybeans being developed at the United States Department of Agriculture Station at Beltsville, Maryland. To the researchers' surprise, successive plantings of certain varieties of soybeans made at intervals in the spring and early summer almost all flowered on the same date. With tobacco plants, a new, broad-leaved, and therefore commercially desirable, mutant, called Maryland Mammoth, failed to flower and set seed by late fall. But if brought into a greenhouse before being killed by the first frost, Maryland Mammoth plants would flower and set seed in the short days of winter, allowing the mutant strain to be propagated. The clue to the role of photoperiod came when Garner and Allard observed the new shoots that had developed in the greenhouse in the longer days of late spring from the stumps of tobacco plants that had been cut back after flowering earlier in the year. Rather than flowering, the new shoots remained vegetative throughout the summer. Because temperature was constant in the greenhouse, another factor was sought. The only constantly varying factor in the plants' environment was day length.

Not only did photoperiod appear to explain the responses of the tobacco plants, but it also accounted for the flowering response of the soybeans. From subsequent studies of several plant species raised in growth chambers under various combinations of light-dark cycles, Garner and Allard classified plants into three types according to their response to photoperiod.

1. Short-day plants, in which the mature plants required a day-night cycle of short days and long nights before they would flower (Fig. 26-23). This group included the Maryland Mammoth tobacco strain, ragweed, and some varieties of soybeans.
2. Long-day plants, in which the mature plants flowered only if subjected to long days and short nights. Spinach and lettuce were among the plants in this category.
3. Day-neutral plants, which flowered whenever they reached maturity, without reference to day length. This group included maize (corn) and cucumber.

Subsequent studies of the photoperiod requirements by others have shown the photoperiod response to be more complicated than first appeared. A few species have dual requirements for flowering. For example, some plants require first long days and then short days to flower; these were called long-short-day plants. Short-long-day plants have also been identified. Furthermore, while some short-day plants, such as cocklebur (*Xanthium*) and soybeans will flower only if exposed to short-days, other short-day species eventually flower even in a non-inductive photoperiod.

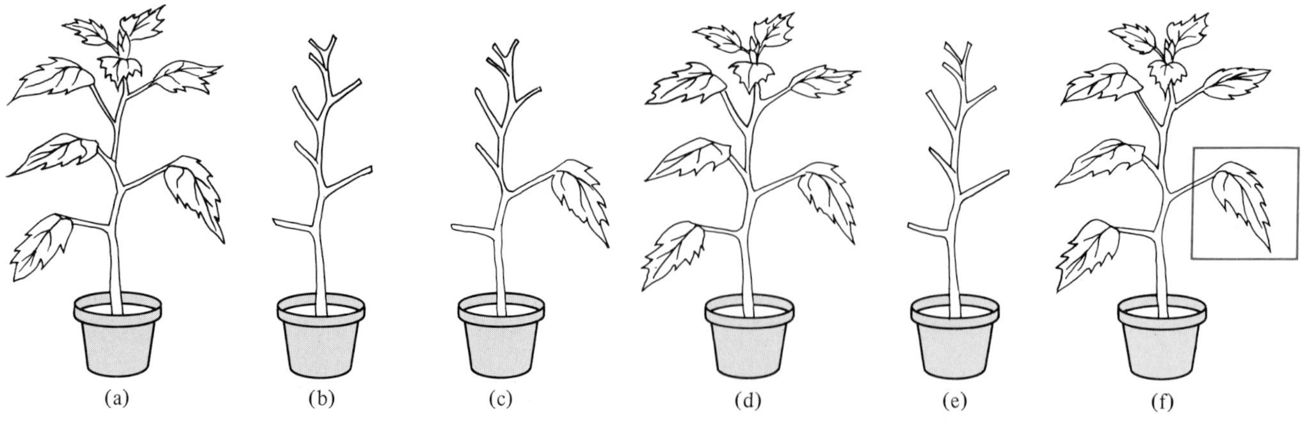


**FIG. 26-23** Response of the SDP cocklebur *(Xanthium)* to photoperiod. All plants were grown under a noninductive photoperiod (16 hours light, 8 hours dark) before the experiment. Plants a–c were given a single photoinductive period of 15 hours light and 9 hours dark before which all leaves were removed from b and all but one leaf from c. Plants a and c flowered. Plants d–f were all continued under the noninductive photoperiod; all leaves were removed from plant e. One leaf of plant f, however, was given a single inductive photoperiod (15 hours light, 9 hours dark) after which flowering occurred in the entire plant. (The presence of other leaves opposes this effect in many species. It is most dramatic if all other leaves have been removed.) No flowering occurred in plants d and e.

## MECHANISMS INVOLVED IN PHOTOPERIOD-REGULATED FLOWERING

The discovery that flowering could be controlled through photoperiodism provided an experimental tool for the study of the mechanism of flowering. Now the perception of the stimulus could be separated from the actual initiation of the flower at the shoot apex and events in the process could be analyzed.

***The Critical Day Length Concept.*** A short-day plant (SDP) is defined as one that flowers if the light period is shorter than what is called the critical day length for that variety or species. For Maryland Mammoth tobacco, for example, the critical day length is 12 hours; for the cocklebur (*Xanthium*), it is 15.5 hours. Long-day plants (LDP) are those that flower when the light period is longer than their critical day length. Studies in which the length of the light and dark periods were varied showed that it is actually the length of the dark period, not of the light period, that is critical in flowering. If the dark period is interrupted with a flash of light, a short-day plant fails to flower. Briefly interrupting a long dark period promotes flowering in a long-day plant.

***Phytochrome: The Light Sensor in the Photoperiod Response.*** When it was recognized that the flowering response could be blocked in SDPs by a light break during the dark period, experiments were undertaken to discover the participating pigment. Interrupting the dark period in the SDP cocklebur with light of various wavelengths revealed that red light (660 nm) was as effective in blocking flowering as white light. However, if the flash of red light was immediately followed by a flash of far-red light (730 nm), the red-light effect was neutralized and flowering occurred as if the dark period had not been interrupted. If the sequence was reversed and a flash of red light followed the flash of far-red light, flowering did not occur, an effect similar to that of using red light or white light alone to interrupt the dark period.

These responses implicated a pigment, now called **phytochrome,** later found to also be involved in light-sensitive seed germination and other plant phenomena. Phytochrome is a pigment that exists in two forms, $P_r$, the form that absorbs red light, and $P_{fr}$, the far-red–absorbing form. In the dark, the $P_{fr}$ pigment undergoes a slow conversion to the $P_r$ form. $P_r$ is converted to $P_{fr}$ in red or white light, so that $P_{fr}$ is the form that predominates during the day. During the dark period, $P_{fr}$ is either slowly converted to $P_r$ or is slowly destroyed. If the dark period is interrupted by red or white light, phytochrome is immediately converted to the $P_{fr}$ form. $P_{fr}$ appears to be the physiologically active form; the proportion of $P_{fr}$ to $P_r$ is critical in plant responses. These changes in the form of phytochrome are summarized below:

$$\text{Synthesis} \longrightarrow P_r \underset{\text{FR}}{\overset{\text{R}}{\rightleftharpoons}} P_{fr} \longrightarrow \text{Decay}$$
$$P_{fr} \xrightarrow{\text{Darkness}} P_r$$

Phytochrome thus enables the plant to detect light and dark. But photoperiod-sensitive plants are able to measure the *length* of the dark period. It might be expected that the conversion of $P_{fr}$ to $P_r$ during the dark period acts like an hourglass clock, measuring time by the accumulation or destruction of a product. However, this turned out not to be true—$P_{fr}$ reversion to $P_r$ is too fast for this mechanism to be the timer. Although phytochrome enables the plant to detect light and dark, *dura-*

*tion* of the dark period requires participation of the plant's internal biological clock (see Chapter 19). Precisely how phytochrome and the biological clock interact to control induction of the flowering stimulus remains to be discovered. Not only is the role of $P_{fr}$ in determining flowering in SDP and LDP unclear, but how phytochrome regulates formation of the flowering stimulus in leaf cells is not yet understood.

***What Is the Flowering Stimulus?*** The discovery of photoperiodism provided a tool for studying the mechanism of flowering, since in some plants flowering could now be induced at will. It was shown early that in an SDP, the leaves perceived the light stimulus, and it was transmitted to the shoot apex, where flower initiation takes place. In cocklebur, exposure of a single leaf is adequate to induce flowering (Fig. 26-23). This suggested the involvement of a hormone. The postulated flowering hormone has been called florigen. The hormone appears to be the same in long-day and short-day plants, as evidenced by the fact that in a graft of an LDP onto an SDP, the LDP partner is induced to flower by short days. However, no one has yet been able to isolate a single substance with florigenlike activity that is effective in all plants.

The plant growth substances, gibberellin and auxin, described later in this chapter, have been shown to affect flowering, in some cases promoting flowering, in others inhibiting it, but they do not act as florigens, inducing flowering, but merely affect the floral response. It now seems improbable that a single hormone is involved in flowering in all plants. Some workers have suggested that flowering is the result of an interaction of the various plant growth hormones. The hormonal control of flowering remains an unsolved problem.

## ROLES OF HORMONES IN PLANT DEVELOPMENT

In plants, hormones act primarily in directing growth and differentiation. They are less involved in the integration of physiological activities than are animal hormones. Unlike hormones in general, some hormonelike growth substances in plants achieve their effects in the region of the plant in which they are produced. All known plant hormones have more than one developmental effect. For example, a given hormone may regulate both growth and differentiation. Furthermore, most developmental responses in plants are regulated through the interaction of two or more hormones.

Five major groups of plant hormones or growth substances are generally recognized, although additional molecules with hormonelike activity have been reported and others may yet be discovered. These major groups are the auxins, gibberellins, cytokinins, abscisic acid, and ethylene.

### EVIDENCE OF HORMONAL CONTROL OF CELL GROWTH

The first indication of hormonal activity in plants came from a study of tropisms, curvatures involving growth toward or away from a stimulus such as light. Curvature toward a stimulus is a positive response or tropism; curvature away from a stimulus is negative. Such curvatures

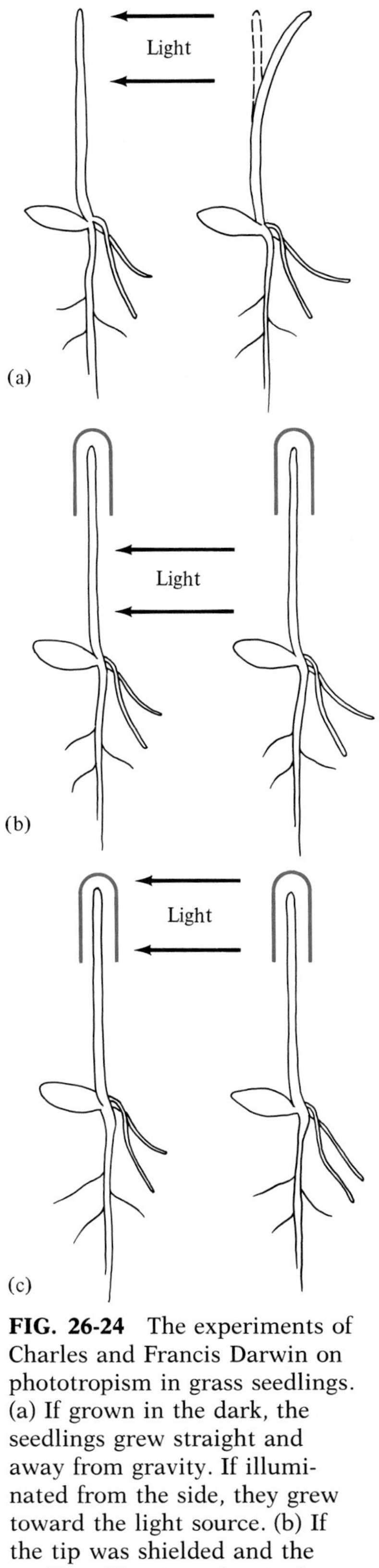


**FIG. 26-24** The experiments of Charles and Francis Darwin on phototropism in grass seedlings. (a) If grown in the dark, the seedlings grew straight and away from gravity. If illuminated from the side, they grew toward the light source. (b) If the tip was shielded and the seedling illuminated at its base, no bending occurred. (c) If the tip was shielded and the seedling illuminated from the side, as in (a), no bending occurred.

occur as a result of greater growth of cells on one side of an organ such as a stem.

***The Discovery of Auxin.*** The first investigations that led to the eventual discovery of plant hormones were conducted by Charles Darwin and his son, Francis, in 1880. Charles Darwin had been interested in a phenomenon commonly seen in houseplants in which the shoots grow toward light coming from a nearby window. This curvature toward light is called **phototropism.** When the Darwins germinated grass seedlings in the dark, the seedlings grew erect (away from gravity), but when illuminated from one side, they grew toward the light source (Fig. 26-24a). However, when only the base of the seedling was illuminated from the side, no bending occurred (Fig. 26-24b). Bending also failed to occur when the tip of the seedling was covered with a light-tight shield of metal foil (Fig. 26-24c). The Darwins concluded that the tip sensed the direction from which the light came and generated "an influence" that passed down the stem, resulting in the bending seen in the unshielded seedlings. Cells on the shaded side grew more than those on the illuminated side.

Subsequent experiments by investigators in several countries revealed that the "influence" in phototropism was a diffusible substance, a hormone. As was later shown, this influence could pass through a layer of gelatin placed between the excised tip of a coleoptile (the first leaf of a grass seedling) and the stump and still promote cell growth. Without a tip, the seedling ceased growing. If the coleoptile tip was removed and placed off center (asymmetrically) on the cut surface of the stump, the coleoptile curved away from the side on which the tip was placed as the result of accelerated growth in the tissues immediately beneath the tip. The uneven distribution of the influence suspected of originating in the tip had caused the same kind of curvature response seen in shoots illuminated from one side.

In 1928, the Dutch plant physiologist Frits W. Went established that a hormone was involved in phototropism. Tips of dark-grown *Avena* (oat) coleoptiles were removed and placed on an agar sheet for several hours (Fig. 26-25a). The tips were then discarded and the agar cut into blocks. When one of these blocks was placed on a cut stump, the seedling resumed growth (Fig. 26-25b). When the block was placed to one side, the coleoptile tip bent away from the side with the block as it grew (Fig. 26-25c). The degree of bending was found to be directly proportional to the concentration of the substance in the block (as determined, for example, by the length of time a severed tip had been in contact with the block).

This first plant hormone to be discovered was named **auxin** (from the Greek *auxein,* "to increase"). It was subsequently identified as indoleacetic acid (IAA) (Fig. 26-26). Eventually, several analogs (molecules with functional similarities) of auxin that also promote coleoptile elongation were synthesized. Some of these synthetic auxins, such as 2,4-D and 2,4,5T, are used as weed killers and defoliants because of their adverse effects on certain plants. The name *auxin* is now a general term, referring to IAA or any related substance with the same effects.*

The observation that seedling curvature was proportional to auxin concentration led to development of a bioassay, that is, an analytic

* Although the name *auxin* connotes an excitatory effect, some of the actions of auxins are inhibitory.

method utilizing a biological test system, for the determination of auxin concentration. For a time, this highly sensitive *Avena* curvature test became the standard assay for auxin; the hormone normally occurs in concentrations of micrograms per kilogram of tissue, originally too low to be detected by chemical tests. In this assay, extracts made and applied under standardized conditions are tested for the amount of curvature produced in *Avena* seedling stumps when the sample is placed to one side of the cut surface. Although this curvature test remains the most sensitive assay for auxin, other, more convenient, methods are now employed. One of these tests measures the straight growth, or elongation, produced in sections of dark-grown seedlings that are floated on the test solution. Low concentrations of auxin can also be measured by the sensitive chemical method of high-performance liquid chromatography (HPLC), in which mixtures may be separated by chromatography under pressure.

Although the Darwins had set out to explain phototropism, their studies led ultimately to the discovery of an important plant hormone. Auxin is the principal regulator of cell elongation. Thus, geotropism in shoots is also explained on the basis of an asymmetric distribution of auxin on the top and bottom of a horizontally oriented shoot, in this case in response to gravity.

There is still some question today as to whether all phototropism can be ascribed to an effect of light on the distribution of auxin coming from the tip of a shoot. Until recently, most of the studies of phototropism were done with the seedling coleoptiles of oats or maize. Recent studies with other seedling types indicate that many seedlings curve without their tips; in these cases, other explanations are needed.

Several other growth phenomena are also being reexamined today by plant physiologists concerned with how hormones act in plants. Information on hormone concentration in tissues in relation to response has suggested that the factor of sensitivity of the tissue to the hormone or the presence of hormone carriers may play a more important role than the actual concentration of auxin.

***How Auxin Promotes Cell Elongation.*** The ultimate effect of auxin upon an elongating cell is to increase the plasticity of the cell wall, thereby allowing elongation to occur under the force of turgor pressure. It has been proposed that auxin contributes to the plasticity of the cell wall by causing an extrusion of hydrogen ions into the wall. Acidification of the wall weakens cross linkages between molecules within the wall, thereby permitting the microfibrils to slide past one another. Auxin's role in $H^+$ extrusion is thought to involve activation of a proton pump in the plasma membrane. Auxin also promotes RNA and protein syntheses and in this way may affect the addition of new carbohydrates to the cell wall. The entire process is complex and still not fully understood.

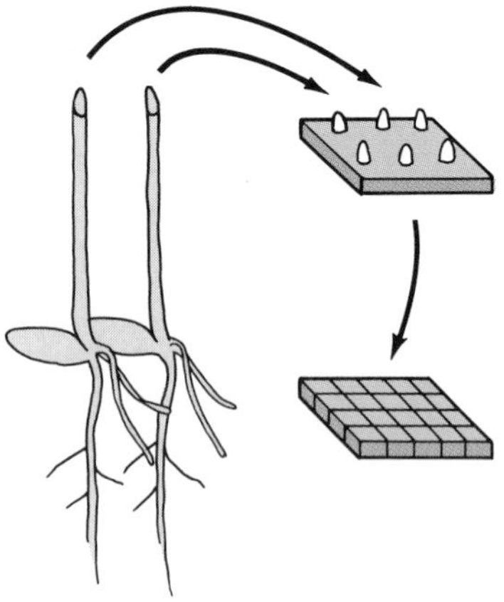

(a)

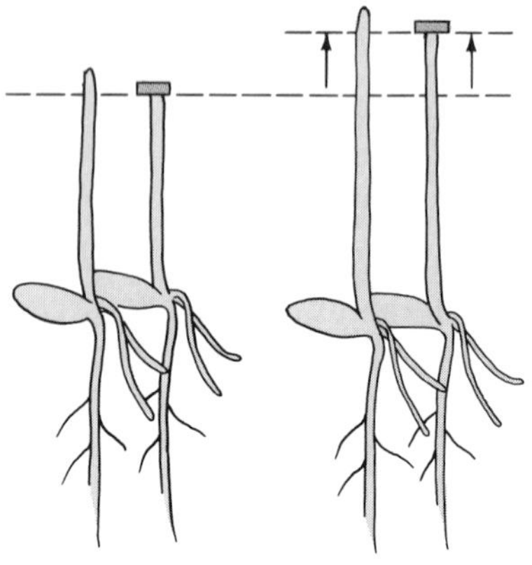

(b)

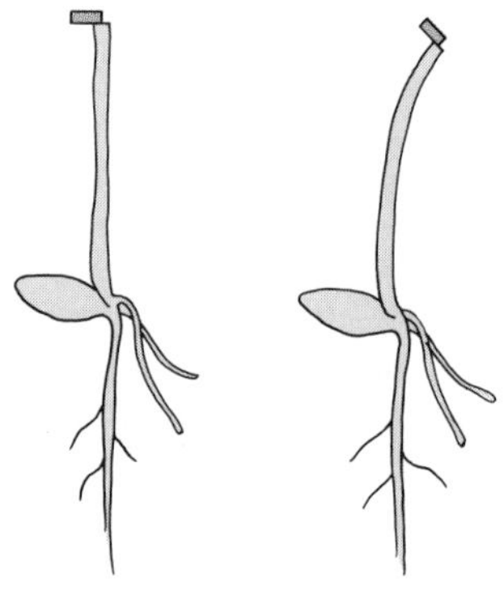

(c)

**FIG. 26-25** Demonstration of the action of the hormone auxin in the *Avena* (oat) coleoptile. (a) Coleoptile tips excised and placed on agar; auxin diffuses into agar blocks. (b) Auxin in agar block causes comparable growth of coleoptile to that in control. The curvature obtained when an agar block containing auxin was placed off center on the cut surface of the stump (c) formed the basis for the *Avena* coleoptile curvature test, a bioassay for auxin in small concentrations.

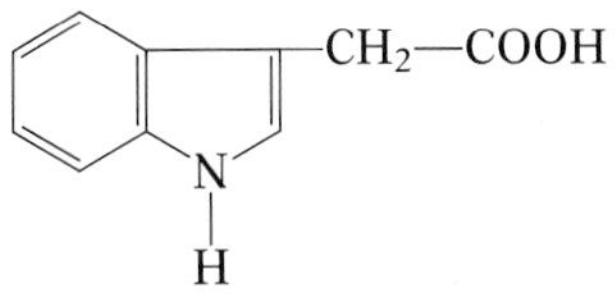


**FIG. 26-26** The chemical formula for auxin.

## OTHER ROLES FOR AUXINS

Aside from its role in growth, auxin participates in many other developmental processes. Most of the other effects of auxin on development result from interactions with other plant hormones, to be discussed in the next section of this chapter. The hormones cytokinins, gibberellins, abscisic acid, and ethylene all interact with auxin in the regulation of various developmental phenomena.

Other developmental phenomena in which auxin participates include cell division, cell growth, and the differentiation of cells and tissues. Among the best known are the following:

1. *Root initiation:* small amounts of auxin stimulate initiation of branch roots; larger amounts promote the formation of new (adventitious) roots in stem tissues. This stimulatory effect of auxin on cell division associated with root initiation is used in horticulture to promote rooting of cuttings.
2. *Abscission:* leaf drop is correlated with a decline in auxin production in leaves. Spraying leaves with auxin can often delay leaf drop. Ethylene, on the other hand, which interacts with auxin, promotes abscission. The auxin effect in this case may be an effect on the release of ethylene from the tissues. In contrast, cytokinins inhibit leaf fall. Cytokinin applied to a leaf prevents its senescence, or aging, and thus delays its abscission. Fruit drop follows a pattern similar to that of leaf drop and can be delayed by spraying young fruits with auxin, a valuable tool of the commercial grower. Abscission of cotton fruits in response to abscisic acid in the buds suggests that hormonal interaction may be involved.
3. *Fruit growth:* developing seeds have been shown to produce auxin, which stimulates the ovary wall to develop into a fleshy fruit. A striking demonstration of this is in the strawberry (Fig. 26-27), where auxin from the single seeded pistils causes the fleshy part of the fruit to develop. When unpollinated tomato or cucumber flowers are sprayed with auxin, the ovaries develop into fruits without fertilization, and thus without seeds. This also has commercial applications.
4. *Cell differentiation:* auxin not only stimulates the cambium to divide, but it also promotes differentiation of xylem vessels and tracheids. In experiments in which a vascular bundle of a *Coleus* stem was severed, new vessels differentiated from adjacent cells to bridge the gap. However, this occurred only if buds were present as an auxin source above the wound, or if auxin was applied in their place (Fig. 26-28). It has been demonstrated in decapitated tree stems that auxin applied to the cut surface affects whether tracheids and vessels produced in the secondary xylem are of the early or late wood type. This indicates a role of auxin in expressing the effects of the environment on annual ring formation in wood.

   Similar experiments with gibberellin and auxin have shown that the differentiation of cambial derivatives is affected by the relative amounts of the two hormones. Higher auxin concentrations promote differentiation of xylem elements; higher gibberellin favors phloem differentiation.

   In a subsequent section we shall see that relative concentrations of auxin and cytokinin influence whether roots or shoots develop in plant tissue cultures.

5. *Lateral bud inhibition:* apical dominance of the shoot apex over axillary buds has already been described. It was seen that growth of the buds is suppressed by auxin from the terminal bud (apical dominance), but this may be overcome by a cytokinin applied to the axillary buds (Fig. 26-29). Commercial advantage is taken of the inhibitory effect of auxin on bud growth by spraying potatoes with synthetic auxin, which allows them to be stored for long periods without sprouting.

**FIG. 26-27** Auxin from the single pistil of a strawberry induces development of fleshy tissue of the receptacle that forms the "berry." The classic experiment of French botanist Jean Nitsch showed that removal of pistils prevented berry formation (d). If one pistil were left in place, a small patch of berry tissue beneath it developed (a); if three pistils remained, three patches developed (b). If all pistils were left in place, the berry developed normally (c). If auxin was substituted for the pistils, a "seedless" strawberry developed (e). (The "seeds" on strawberries are actually one-seeded fruits, or achenes.)

**FIG. 26-28** Auxin regulates the differentiation of xylem. If the stem of a *Coleus* shoot is wounded so as to sever one or more vascular bundles, parenchyma cells adjacent to the wound differentiate as vessel elements, forming strands of xylem about the wound. If the shoot apex and young leaves, sources of auxin, are removed before wounding, no bridge of vessels is formed. But if the apex and leaves are replaced by auxin, the strands of vessels form.

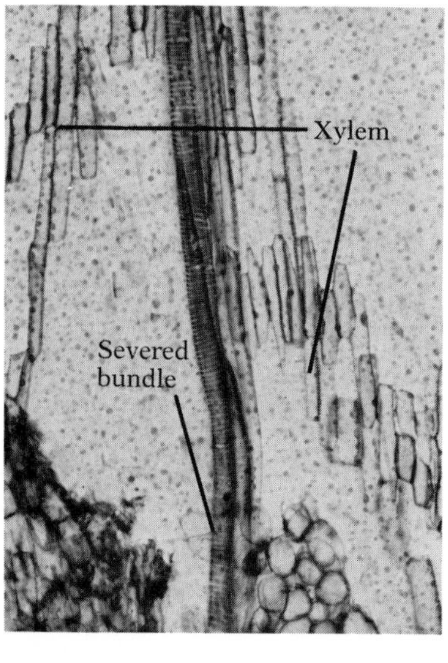


## OTHER HORMONES AND GROWTH SUBSTANCES

While a knowledge of auxin as a plant hormone was developed in the 1920s, it was several years before the other plant hormones were discovered.

***The Gibberellins.*** In the 1920s, in Japan, a fungus disease of rice plants that caused their rapid growth and collapse was found to be caused by a substance produced by the fungus *Gibberella.* The substance was named gibberellin. Japanese biochemists later isolated and identified a series of such substances. However, the limited scientific exchange between western plant physiologists and the Japanese investigators during World War II effectively prevented work on gibberellin by western botanists until the early 1950s. Then, in 1951, a growth-promoting substance identified as a gibberellin was found in extracts of bean seeds. Gibberellins were subsequently found to be normal constituents of

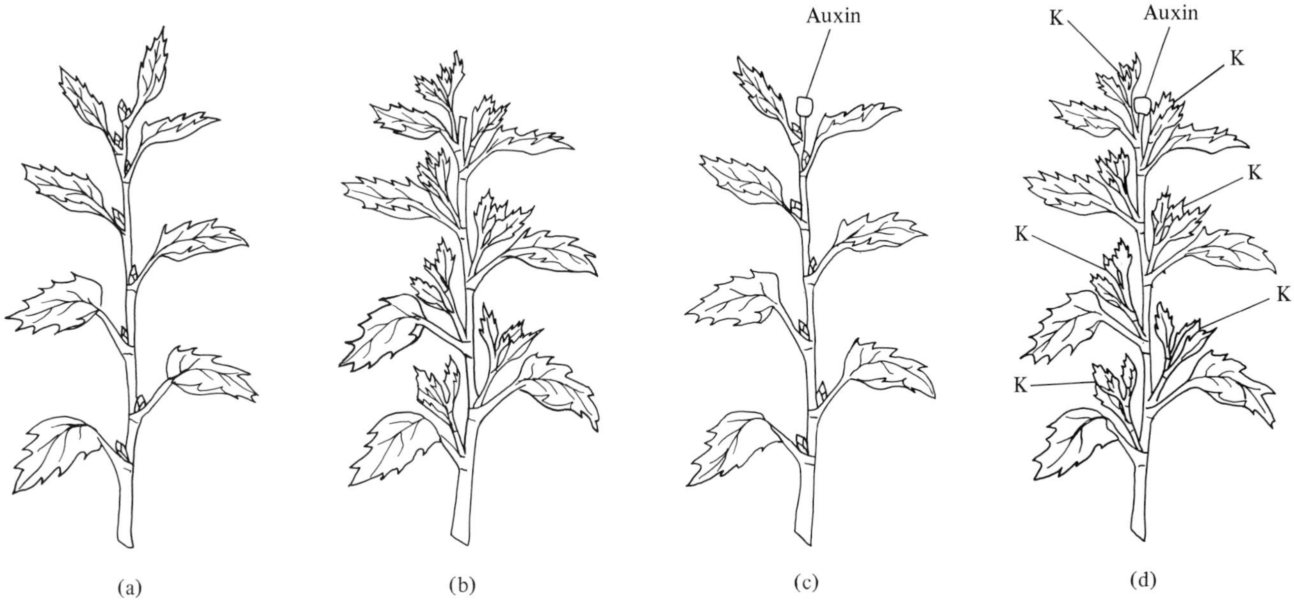


**FIG. 26-29** Auxin and cytokinin interact in controlling lateral bud development. (a) If the terminal bud is present, apical dominance is evident: lateral bud growth is suppressed in some species. (b) If the terminal bud is removed, lateral buds develop. (c) If auxin is applied to the decapitated stem, the buds fail to develop. (d) If cytokinin (K) is applied directly to the buds, they develop into branches.

many plant seeds. Over 50 different gibberellins, or gibberellic acids (GAs), have been isolated and identified (Figure 26-30).

GA enhances stem elongation by stimulating both cell division and cell elongation. For example, biennial plants, which grow their first year as rosette plants, with their leaves telescoped together because internodes fail to elongate, are stimulated by GA to elongate without undergoing the cold treatment usually required to produce this effect. A most dramatic effect of GA is seen when it is applied to genetic dwarf varieties, particularly of corn; normal, elongated growth results. This suggests that such mutations result in an inability to synthesize GA in normal amounts. Dwarf corn mutants are used as a bioassay tool for GA.

GA also promotes flowering in some species by producing the effects normally achieved by cold temperatures and specific photoperiods. It has also been shown to be responsible for juvenile characteristics in some plant species, such as leaf shape in English ivy. In its absence, adult characters develop. Fruit development without fertilization can also be induced by GA in some species that do not respond in this way to auxin.

Application of GA to dormant seeds and buds that normally require special conditions of temperature or light to germinate or to develop has caused germination and bud growth in the absence of these stimuli.

A most important discovery was the role of GA in the regulation of germination in cereal grains, such as barley. GA, which is released from the embryo when the grain is soaked in water, diffuses into the outer layer, or aleurone, of the endosperm, the starch-containing, food-storage tissue. In response to GA, the aleurone cells release hydrolytic enzymes (hydrolases) that diffuse into the endosperm, where they digest the stored food. In the case of the starch-digesting enzyme $\beta$-amylase, GA frees a bound, inactive form already present in the seed. In the case of $\alpha$-amylase, GA induces synthesis of the hydrolase. The products of digestion then diffuse to the embryo, enabling it to begin growth.

**FIG. 26-30** The chemical formula for gibberellic acid.

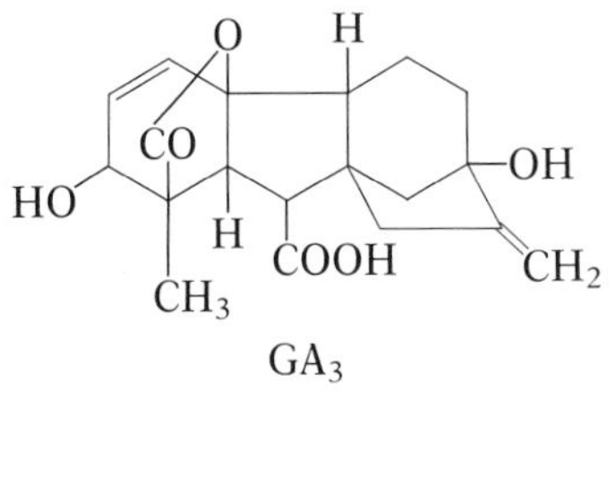


$GA_3$

***The Cytokinins.*** In studies of the growth of pith tissues isolated from tobacco plants and cultured in vitro, Folke Skoog and his colleagues at

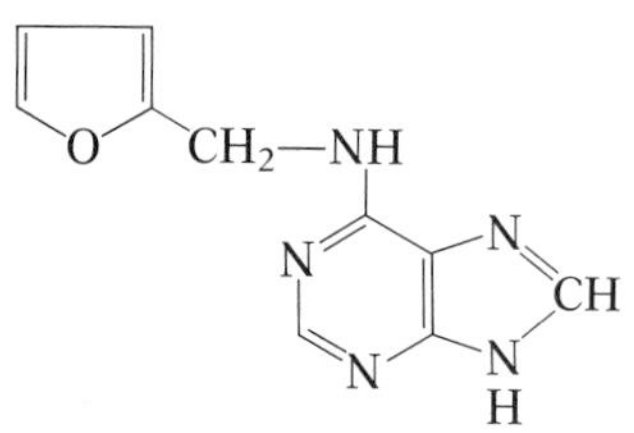

(a)

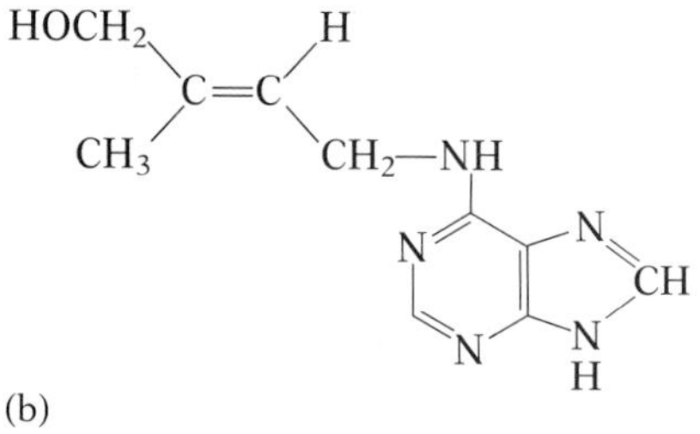

(b)

**FIG. 26-31** The chemical formulas for (a) kinetin and (b) zeatin.

the University of Wisconsin found, late in the 1940s, that the presence of vascular tissues stimulated cell division in the pith cells, producing a mass of undifferentiated cells called **callus.** In searching for the active factors that specifically stimulated cell division in these cultures, Skoog and Carlos Miller were led to try preparations of nucleic acids. They found that an old, deteriorating sample of DNA contained a strong promoter of cell division. Analysis of such degraded DNA showed the active principle to be an analog of the purine adenine; they named it **kinetin** (Fig. 26-31a). Subsequently, other substances with kinetinlike activity were isolated from plant tissues, thereby establishing a new class of plant hormones, the **cytokinins.** The first naturally occurring, and the most active, cytokinin to be isolated was **zeatin** (Fig. 26-31b), from corn (*Zea mays*). Cytokinins are now known to occur in dividing tissues of several species of plants.

In addition to stimulating cell division in callus, cytokinins are also known to promote bud development and seed germination in various species and to delay the aging of leaves. Moreover, in studies with cultures of tobacco pith, Skoog and his colleagues discovered an important interaction of cytokinins with auxin: the auxin/kinetin ratio determined whether a callus culture of tobacco pith formed roots or buds or merely grew as undifferentiated callus (Fig. 26-32). Relatively high levels of auxin promoted root formation; high levels of kinetin favored bud formation. Auxin and kinetin in balanced concentrations caused callus growth without differentiation.

***Abscisic Acid.*** For several years, it has been known that dormant buds of some plants contain substances that are inhibitory to growth. When

**FIG. 26-32** The formation of roots (above) and shoots (below) from callus in response to different concentrations of auxin and kinetin.

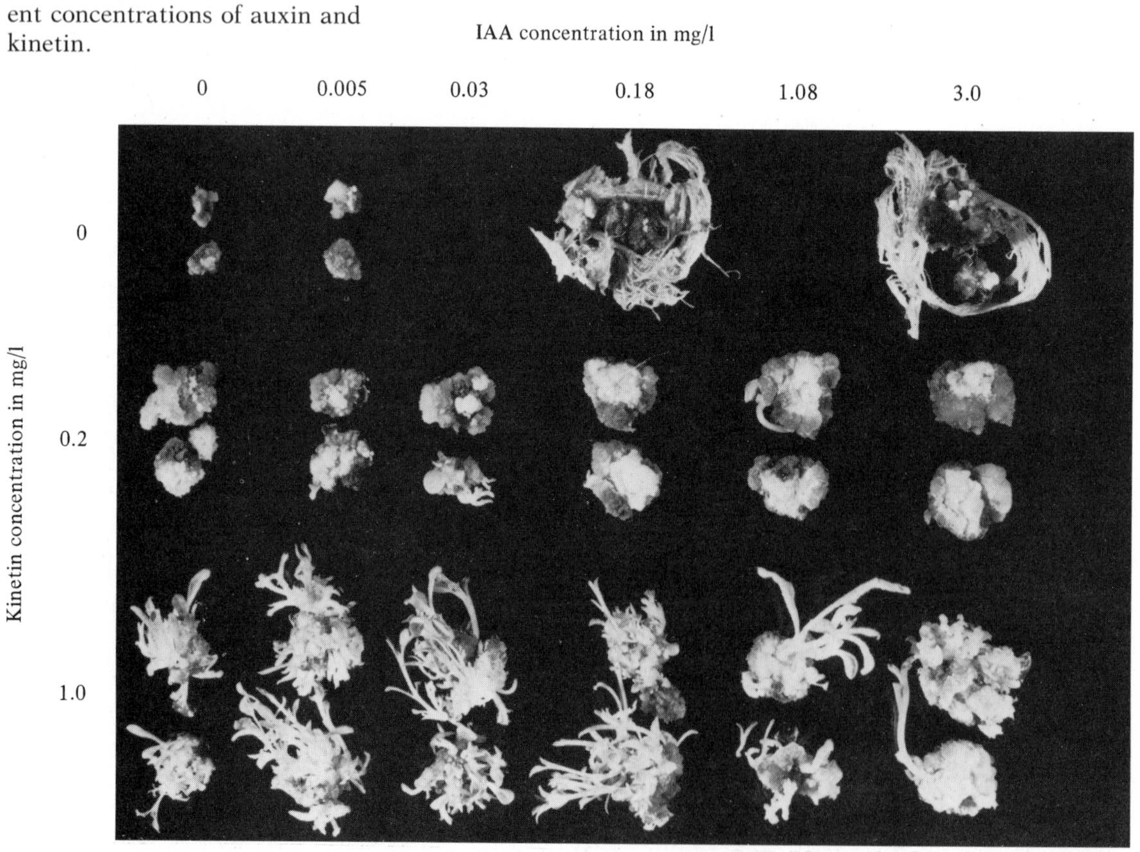

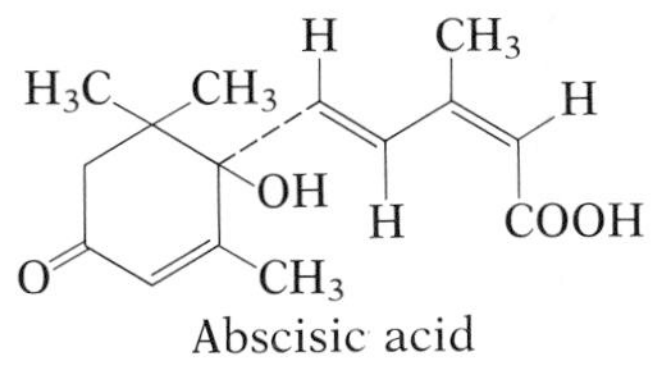

FIG. 26-33 The chemical formula for abscisic acid.

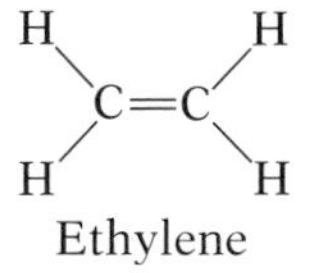

FIG. 26-34 The chemical formula for ethylene.

bud dormancy is broken, the amount of these inhibitors within the buds has declined. In 1965, Philip F. Wareing and his colleagues in Wales isolated from dormant maple buds a growth inhibitor that they characterized chemically and called dormin. At the same time, in the United States, Fredrick T. Addicott and his colleagues isolated a factor from cotton plants that caused abscission of cotton fruits. They called it abscisin. Then it was discovered that dormin and abscisin were identical; it was agreed to call the substance **abscisic acid** (ABA) (Fig. 26-33). The choice of names has turned out to be inappropriate, since the major effect of ABA in plants is in regulation of dormancy in buds and seeds, rather than in abscission. In addition it has also been shown to promote the closing of stomata in leaves during water stress.

***Ethylene.*** **Ethylene** is an exceptional growth substance in that it is a gas at normal temperatures (Fig. 26-34). In a strict sense, it is not a hormone; it achieves its effect at the site of its production. It is a simple hydrocarbon that has several effects, including acceleration of leaf abscission and ripening of fruit. Some of auxin's apparent effects on fruits and flowers appear to be achieved by promoting the production of ethylene, a kind of "second messenger."

## PLANTS IN VITRO

The pioneering studies of Philip R. White of the Rockefeller Institute in the early 1940s led to the formulation of a culture medium that could sustain the growth of isolated tomato roots in culture flasks. Modifications of White's medium have made it possible to grow a great variety of plant tissues and plant organs in vitro on a chemically defined nutrient medium consisting of minerals, sugar, certain vitamins, and, in some cases, plant hormones. Thus a culture medium devised by Toshio Murashige and Folke Skoog has served as a basis for the culture of various plant parts of many different species, from leaf cells to pollen grains.

When White aseptically explanted the tips of tomato roots and successfully subcultured them, he demonstrated that the root apex had the unlimited potential to generate the root. Since that time, it has also become possible to explant the shoot tip and have it develop an entire shoot. It is not uncommon to induce root formation on shoots in tissue culture, after which the plants may be transferred to pots and develop flowers, fruit, and seeds. When pieces of plant parts lacking apical meristems are explanted to an appropriate culture medium, cells begin to divide and form callus. As Skoog and Miller showed in their studies with tobacco callus cultures, roots and shoots may be induced by manipulating the concentrations of auxin and cytokinin. Clones of plants may thus be derived from such cultures (Fig. 26-35). Genetic variation often appears in the cells of callus as a result of increase in chromosome number, gene mutation, formation of aneuploids, etc. Plants derived from callus cultures may thus show phenotypic variation, and desirable qualities may be selected and propagated. Favorable mutants are selected to develop new horticultural lines, or crossed with established breeding lines. Increased sucrose yield in sugar cane and increased resistance to disease and environmental stress in sugar cane and tomatoes have already been obtained through use of these methods.

FIG. 26-35 Tobacco buds that have arisen from callus and developed into shoots in the flask.

While shoots and roots can be induced to form from callus, it is also

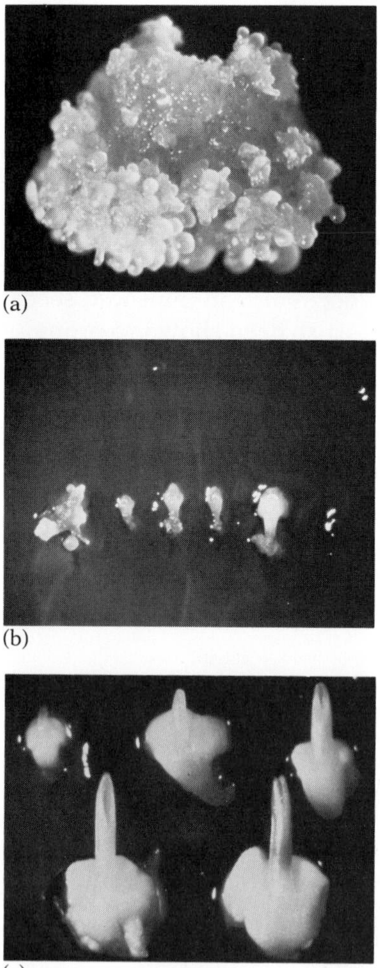

(a)

(b)

(c)

**FIG. 26-36** (a) Formation of corn embryoids in callus and (b, c) isolated somatic embryos.

possible to generate plantlets from embryolike structures called somatic embryos, or **embryoids** (Fig. 26-36) that appear in the callus. These can be transplanted to soil and develop directly into plants. This was shown by F. C. Steward at Cornell University in studies with carrot root explants cultured in nutrients with coconut milk, a natural source of growth-promoting substances.

Tissue culture, or organ culture, is widely used today to propagate desirable plant varieties or in attempts to create new lines of horticultural plants. A monumental achievement would be to transfer into crop plants the nitrogenase genes that enable bacteria in the soil and in root nodules to fix nitrogen (Chapter 28), thus reducing the requirements for nitrogen fertilizers. Introduction of new genes into plants will require transferring the desired genes into a plant cell, and proliferation of the cell into callus, and subsequent differentiation into roots and shoots, or into embryoids. Using techniques such as described in Chapter 24, genes may be transferred via plasmids, which can be introduced into plant cells by some naturally occurring agent, or vector, such as bacteria that cause plant disease. One of these diseases, the crown gall disease, characterized by tumorlike outgrowths on shoots, is caused by the bacterium *Agrobacterium tumefaciens*. *Agrobacterium* carries a plasmid that is transferred into host cells and induces the disease symptoms. Viruses that cause other diseases are also potential vectors for use in the transfer of genes.

It is also theoretically possible to produce new, desirable combinations of genes by fusion of two unrelated cells. The plant cell wall is a barrier to this process, but methods have been devised to circumvent this problem. Enzymes capable of digesting cellulose and other wall components, have been used to digest away the cell walls. Adding polyethylene glycol, or exposure to an electric field then permits the wall-less protoplasts to fuse into one cell containing the nuclei and genes of both parental cells (Fig. 26-37). Such protoplast fusion has been accomplished in several instances. For example, hybrids have been formed from fusion of a protoplast of the Russet Burbank potato and one of the Rutgers tomato. The great potential these techniques have for generating new plant phenotypes and the consequent implications for agriculture have led to an explosion of research in plant genetic engineering in universities and industrial enterprises around the world.

**FIG. 26-37** Photomicrographs of oat *(Avena sativa)* mesophyll protoplasts fusing. (a) Cells align. (b, c) Membranes open and cell contents begin to fuse. (d) New cell.

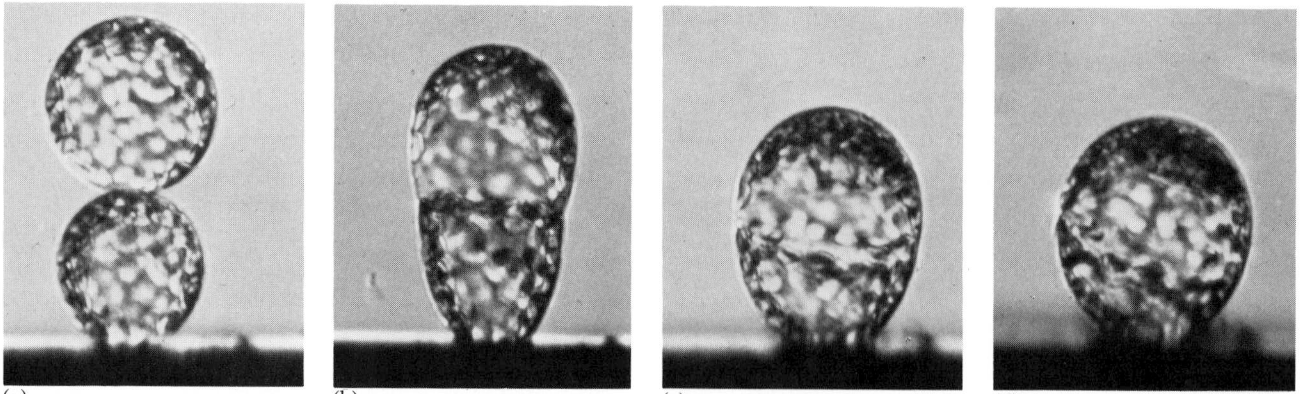

(a) (b) (c) (d)

## Summary

Plant development is a continuing process, with the plant making new roots as it extends down into the water and mineral supplies of the soil and new leaves and stems as it extends upward into the light and carbon dioxide supply of the atmosphere. The new organs originate in the apical meristems, perpetually embryonic regions at the tips of the root and shoot, from which the primary tissues originate. The plant responds to environmental changes, such as occur in the progression of the seasons, by developing new parts at its apical meristems. As plants grow taller, they expand in girth through the activity of a lateral meristem, the vascular cambium, which produces the secondary tissues.

Apical meristems are present at the tips of the embryo in the dormant seed, ready to begin development upon seed germination. The cotyledons of the embryo provide energy for the young seedling from stored food, from photosynthesis, or by transmitting food from the endosperm tissue of the seed.

The root apex generates the primary tissues of the root and the root cap. Cells immediately behind the growing apex elongate, forcing the root apex downward into the soil. Further back from the tip, cells differentiate into tissues in characteristic patterns, with the vascular cylinder at the center. Root hairs extend from epidermal cells. Branch roots originate internally in the pericycle of the vascular cylinder and grow out through the cortex.

The shoot apex produces the stem and on it the peg-shaped leaf primordia in a regular leaf arrangement. Marginal meristems along the edge of the leaf primordium produce the tissues of the leaf blade and shape the leaf. A procambial strand in the leaf primordium is continuous with the vascular cylinder of the stem and becomes the vascular bundle of the leaf. The cells left behind by the growing shoot apex differentiate into the primary tissues of the stem, including the xylem and phloem of the vascular cylinder, a collection of vascular bundles that supply the leaves. Branches develop from axillary buds that originate in the axils of the leaf primordia. Whether buds develop into branches is regulated by plant hormones.

The vascular cambium, between the primary xylem and phloem of the vascular cylinder, produces inwardly the secondary xylem or wood and outwardly the secondary phloem of the bark. Seasonal changes in hormones are reflected in the formation of early and late wood and the resultant annual rings of the stem in perennials. A cork cambium originates in the subepidermal cells of the cortex and produces the cork, or periderm.

By mechanisms not yet understood, a shoot apex may be converted from its role of producing leaves to one of forming flowers. Photoperiod and low temperatures are among factors affecting the initiation of flowering in various angiosperms; the process is mediated by a pigment, phytochrome. There is some evidence for the existence of a flowering hormone, called florigen, that is produced in the leaves and induces the shoot apex to change from producing leaves to making flower parts, but florigen has never been isolated.

Plant hormones each affect development in various ways. Auxin may promote or inhibit growth, regulate cell differentiation, and stimulate root formation and fruit development. The gibberellins stimulate shoot growth, affect flowering and juvenility, and play an important role in seed germination. Cytokinins stimulate cell division and bud development and delay leaf senescence. Abscisic acid causes abscission and shoot dormancy and affects stomatal closing. Ethylene, a gas, promotes leaf abscission and fruit ripening.

The in vitro culture of plant tissues and organs holds great promise for the cloning of plants engineered genetically for various purposes.

## Recommended Readings

BERNIER, G., J. KINET, and R. SACHS. 1981. *The Physiology of Flowering,* I and II. CRC Press, Boca Raton, Fla. A review of various aspects of flowering by several investigators.

CUTTER, E. 1982. *Plant Anatomy,* 2d ed. Addison-Wesley, Reading, Mass. An up-to-date textbook of plant development by a major contributor to the field. Treats plant structure from the standpoint of causative factors and experimental approaches.

HESLOP-HARRISON, J. 1979. "Darwin and the movement of plants: a retrospect." In Skoog, F., ed., *Plant Growth Substances.* Springer-Verlag, New York. An especially interesting examination by an outstanding British plant developmentalist of Darwin's research on tropisms.

JACOBS, W. P. 1979. *Plant Hormones and Plant Development.* Cambridge University Press, Cambridge. The role of hormones in plant development is examined by a major contributor to the study of plant development.

*Mosaic,* 13(3). 1982. The entire issue is devoted to genetic engineering in plants.

RAY, P. M., T. A. STEEVES, and S. A. FULTZ. 1983. *Botany.* Saunders, Philadelphia. An introductory botany text with an excellent treatment of development.

SALISBURY, F., and I. ROSS. 1985. *Plant Physiology.* Wadsworth, Belmont, Calif. A popular textbook with current treatment of growth and development, guest essays by eminent investigators, and a collection of special summaries of research peripheral to plant physiology.

STEEVES, T., and I. SUSSEX. 1972. *Patterns in Plant Development,* Prentice-Hall, Englewood Cliffs, N.J. Integration of experimental results and structure in an account of research over the past few decades in the development of vascular plants.

THIMANN, K. V. 1979. "The development of plant hormone research in the last 60 years." In Skoog, F., ed., *Plant Growth Substances,* Springer-Verlag, New York. A personal account of the early work on plant growth substances in which the author played a major part.

WAREING, P. F., and I. D. J. PHILLIPS. 1981. *Growth and Differentiation in Plants,* 3d ed. Pergamon, Elmsford, N.Y. A morphological and physiological approach to advances in the study of plant development, with emphasis on plant hormones and their roles.

# 27 Organic Evolution: Theories and Mechanisms

OUTLINE



Prehistoric humans must have wondered at much of what they saw of the living world about them: how the offspring of each kind of organism resembled their parents yet differed from them in some ways; how small creatures such as mice and maggots apparently rose spontaneously from old rags and decaying matter; and how, at times, one kind of organism apparently changed into another. Much like the changes that occurred in nonliving things, various kinds of change seemed to be a part of life. But how the diverse kinds of animals and plants had come to exist on earth was an enigma. To primitive humans, the simplest explanation for the diversity in the living world was the God who created all the different organisms. This view, one version of which is presented in the Old Testament, was one of the earliest explanations for animal and plant diversity. However, thoughtful consideration of the nature of change led the ancient Greek philosophers to suggest that change in morphology of organisms was not only possible but probable. These philosophers laid the philosophical foundation for a concept of organic evolution that was not to be developed by biologists for another 2,000 years.

## HISTORY OF THE CONCEPT OF ORGANIC EVOLUTION

Although the fourth-century B.C. Greek philosopher, Aristotle, classified organisms in what appeared to be natural groups, he said little about the origin of the various groups of organisms. In the western world, the biblical account of Genesis was literally interpreted and accepted. Each species was believed to have been created by God, with all present-day individuals tracing their ancestry to the individual created by God. All species were said to be fixed, or immutable. St. Augustine (354–430), bishop of Hippo, in North Africa, modified this belief, suggesting that God created the "seminal reasons," or elements, from which individual creatures arose. Thus, St. Augustine's view was that the earth had the power to generate new creatures. But another 1,500 years would pass before this view was to gain general acceptance.

Then, the English philosopher Francis Bacon (1561–1626) called for investigation of the causes of variations in organisms and noted such intermediate forms as flying fish and bats. The French mathematical philosopher René Descartes (1596–1650) strongly pressed the point that the entire universe, down to the nervous system, could be explained by physical principles. Gottfried Leibniz (1646–1716), a German mathematical philosopher, speculated on the relationship between fossil and living forms, suggesting that habitat had a role in causing changes in species. Compte de Buffon (1707–1788), a French naturalist, noted that deep (old) fossils were less like modern forms than shallow (recently fossilized) ones. He asserted two principles: one erroneous, that environmental changes produced changes in organisms; and the other valid, that humans and apes, like other groups of species that resembled each other, had common ancestors (without explanation, he later denied this). Buffon also suggested that a species was not immutable, and he spoke of a struggle for existence and the survival of the fittest, but not in the context of a coherent theory of evolution.

**FIG. 27-1** Jean-Baptiste de Lamarck, proponent of the theory of inheritance of acquired characters as a cause for evolutionary change in organisms.

Jean-Baptiste de Lamarck (1744–1829), of France, was the first naturalist to present a unified, well-developed theory of evolution in a forceful manner (Fig. 27-1). His basic contribution was the theory of the inheritance of acquired characteristics, which attempted to explain the observed changes in organisms and, more important, the origin of organisms with unique characteristics. Although Lamarck's theory proved wrong, it stimulated research on the origin of diversity among organisms. According to this theory, features acquired by an organism during its lifetime, especially effects achieved by an unusual amount of use or of disuse of an organ, were inherited by its offspring. Furthermore, the organisms that developed a need for a particular structure tended to develop the structure to such an extent that a unique organism with this special characteristic eventually appeared. This process was said to be very slow, requiring several hundreds of generations of use or disuse of an organ for a new species to evolve. According to Lamarckianism, giraffes arose from animals stretching their necks to reach leaves on trees, and moles lost their eyes because they had not been using them. Although Lamarckian concepts of mutability of species and origin of new species as adaptations to the environment were modern ideas, the mechanism that he proposed was eventually proven wrong. As a result, his whole concept of evolution was discredited.

While over the centuries philosophers and biologists continued to offer their views on the origin of organic diversity, the knowledge of organic diversity itself continued to grow. As the anatomy of fossils and living forms became more thoroughly known and were compared with each other, scholars found themselves forced into opposing camps represented by two schools of thought: organic evolution and special creation.

Among the most important protagonists of special creation was the French anatomist Georges Cuvier (1769–1832). He proclaimed that all similarities in anatomical structure merely reflected the general plans designed by the Creator. Fossils were explained as organisms that had died in a series of catastrophes. The fact that different types of fossils were found at different depths was explained in part by their having migrated to the region from elsewhere after the catastrophe. Other creationists asserted that the new forms were the results of successive acts of creation. Furthermore, according to the biblical account, the world was only a few thousand years old, and therefore there would not have been enough time for all of the varieties of organisms to have evolved. It was also assumed that each organism suited to its environment must have been created by God to occupy that particular environment.

A major obstacle to the acceptance of any theory of organic evolution was the widespread belief that the earth was only a few thousand years old, too little time for much evolutionary change to have occurred. That obstacle was removed largely through the efforts of two men, James Hutton and Charles Lyell.

Until the end of the eighteenth century, almost all geologists believed in some form of Cuvier's catastrophic theory to explain the presence of fossils. In 1795, however, a Scot, James Hutton, presented arguments that fossil-bearing sedimentary rock was formed by the same slow accumulation of sediments in lakes, at river mouths, and at seashores that could be observed in his day as well as today. If he was correct, the earth would be very much older than just a few thousand years. Although the publication of Hutton's theory was scorned and ridiculed by theologians, it was also read and thought about by geologists. Among them was the Scotland-born Charles Lyell (1797–1875) who had turned from the practice of law to the study of geology (Fig. 27-2). One of Lyell's findings was that the most recently formed sedimentary rocks contained among its fossils the highest proportion of living species of molluscs, a confirmation of Buffon's observations. In his multivolume masterwork, *Principles of Geology,* Lyell presented evidence that the earth's age had to be reckoned in millions, not thousands, of years. Although Lyell was a deeply religious person, he made a strong plea for the use of reason as the basis for making judgments about the earth's antiquity.

**FIG. 27-2** Sir Charles Lyell, known as the father of geology.

## THE THEORY OF ORGANIC EVOLUTION BY NATURAL SELECTION

In 1859, the British naturalist, Charles Darwin (1809–1882), presented a comprehensive theory of the origin of species by natural selection. According to this theory, all the diverse organisms seen today are the result of a long evolutionary history. All organisms are constantly changing, and the changes generally preserved in each species are suited to

PANEL 27-1

## Charles Darwin: Life of a Naturalist

Charles Darwin was born into a wealthy family in Shrewsbury, England, in 1809. His father, like his grandfather, was a physician; it was planned that Charles would continue the tradition, and at 16 he was sent to the University of Edinburgh, in Scotland, to study medicine. When Darwin's father discovered that he was neglecting medicine for lectures on zoology and geology, he was withdrawn from Edinburgh and sent to Cambridge University, in England, to study for the Anglican ministry. His main interest, however, continued to be natural history, and his closest friendships were with two biology professors: the botanist John Stevens Henslow and the zoologist Adam Sedgewick.

At Henslow's suggestion, the 22-year old Darwin joined an exploratory expedition aboard *H.M.S. Beagle* as assistant ship's naturalist and Captain Robert Fitz-Roy's mealtime companion. They departed England on December 27, 1831, not to return until October 2, 1836, nearly 5 years later, a voyage that became one of the most historic of all times.

When he left England, the young Darwin shared with almost everyone else the view that the earth was young and that there had been little, if any, evolution. Both views were to change profoundly. One contributing factor to the changes was a book he took with him: the recently published first volume of Lyell's *Principles of Geology*. The other was the combined impact of all that he observed on the voyage.

In its trip around the world, the *Beagle* stopped at several ports along both coasts of South America; at the Galapagos Islands, off the coast of Ecuador; and in New Zealand, Australia, South Africa, and South America again before returning to England. Darwin collected and made notes on a wide assortment of specimens: rocks, fossils, plants, insects, birds, mammals, and other organisms. He took long trips into the South American mountains, noting such things as the presence of marine fossils and signs of glaciation (the action of glaciers) high in the mountains. He witnessed the geological changes wrought by a severe earthquake that occurred while he was in Chile. One of his most impressive experiences was his 4-week visit in the Galapagos Islands, where each island had its own types of finches and land tortoises. The larger islands had different types of finches and tortoises in different parts of the islands. Many of the different kinds of finches occupied different types of environmental settings, with the result that some were seed eating, others insect eating, and so on. Darwin was deeply impressed with the fact that, despite the many unique species he found on oceanic islands, some of them hundreds of miles from the nearest mainland, all the fauna (animals) and flora (plants) of such islands always bore the distinct stamp of the nearest mainland—a "deep organic bond," as he put

The H.M.S. Beagle in the Straits of Magellan.

the environment in which it lives. An important corollary of this theory is that there is no need to hypothesize a supernatural power that created the diverse organisms on earth. Among the general properties of organisms is their capacity for heritable change. These variations provide the raw material for evolution. A brief account of Darwin's life is given in Panel 27-1 as a background to the development of the most important concept in biology: that of natural selection as the shaper of evolutionary change.

it. The bond between them was, of course, a common inheritance; they were kin to each other.

Upon his return, Darwin attacked his work with great energy. Within 6 months, he had sorted and identified his specimens. He also wrote a popular account of the voyage, a book on coral reef formation (1842), one on volcanic islands (1844), and one on the geology of South America (1846). In none of these books did he broach the topic of evolution. Privately, however, he was hard at work on a theory of evolution that would change the thinking of the world. By 1838, he had completed three of four notebooks on the subject, and in 1842 he wrote a 35-page sketch of his theory. He also began work on a major publication that was to present, not only his theory, but the major lines of evidence for evolution itself.

But Darwin published none of this material, dreading the storm of controversy he knew it would provoke. The essay was given to his wife, with instructions to publish it if he should die unexpectedly.

Late in 1857, after having worked on his theory for 20 years, Darwin received a letter from a young biologist, Alfred Russel Wallace. Enclosed was an 11-page essay outlining an idea that had occurred to Wallace: the theory of natural selection. Wallace asked Darwin, if he thought well of the essay, to forward it for publication by the Linnean Society, a society of naturalists. Darwin, of course, thought very well of his own brainchild. He sent the essay to Lyell and offered to burn his own manuscript for fear it might be thought that he had copied the idea from Wallace.

Knowing of Darwin's years of work, Lyell and others insisted that Darwin and Wallace both present their theory at a Linnean Society meeting and that Darwin publish his half-completed book as soon as possible. Darwin's and Wallace's brief papers, jointly presented in July 1858, drew little attention. Darwin's book *On the Origin of Species by Means of Natural Selection, Or the Preservation of Favoured Races in the Struggle for Life,* published in 1859, was another matter.

Darwin's *Origin* masterfully outlined the evidence for the earth's great antiquity and for evo-

Alfred Russel Wallace, coauthor of the theory of evolution by natural selection.

lution, all in clear language readily understood by any educated person. The theory of natural selection, while it did not account for the origin of variations (about which, said Darwin, "our ignorance . . . is profound"), explained the mechanism by which certain variations persisted. The effect of the publication of the *Origin* was truly revolutionary, extending not only to every branch of biological science, but to social and economic thinking as well. It has been called the most important contribution of the nineteenth century to the world's intellectual history.

Darwin later authored several other publications, some of them elaborations on the *Origin.* He lived quietly at his country estate, shunning publicity and leaving the defense of his controversial positions to others. He died in 1882 at the age of 73.

Darwin's theory of evolution by natural selection is said to consist of three facts and two deductions. The first fact is that there is great prolificity in nature. For example, a salmon lays 3 to 5 million eggs, and an oyster 60 million eggs. Even the reproductive capacity of an elephant, a comparatively slow breeder, is enormous. A single pair of elephants reproducing at normal rates will multiply to 19 million descendants in 750 years, if all the offspring survive. The protistan *Paramecium,* which reproduces at the rate of three divisions every day, would, if enough food were available and all the progeny survived, produce in 5 years a mass 10 times as large as that of the planet Earth. From many observations of this nature, Darwin concluded that each organism tends to produce more offspring than are needed to replace the number of parents.

The second fact is that, despite the tendency for geometric increase in the numbers of organisms within a given species, the numbers of individuals of that species remains more or less constant. For many species, there may be cyclical increases and decreases in numbers of individuals associated with the seasons, availability of food, or density of predator and prey populations, but in general the numbers of a given species remain constant.

From these two facts emerged a deduction that Darwin called the struggle for existence, which included not only the survival of the individual but also of the species. Thus, there is a struggle for existence between the millions of offspring produced by a fish, such as a salmon, and also between the salmon fingerlings, or fry, and other fish species that inhabit the same area.

Darwin's third fact of nature relates to the individual variation that occurs within a species. For example, the physical characteristics of no two humans are totally alike, except perhaps for those of identical twins. Indeed, there is an infinite variation among individuals within all animal populations. Although at first sight all cattle might look alike, a careful observer can recognize individual variations in shape, size, color, and temperament within a herd.

From these facts, Darwin made his second and more important deduction: survival of the fittest occurs by natural selection. From among the variant individuals in a population, those most suited to their environment survive in larger numbers to reproductive age and leave more offspring who also bear those traits. Consequently, a higher proportion of the population comes to bear the characteristics that adapt members of a species to the environment in which they live.

Two other postulates of Darwin were novel ideas. Darwin postulated that all similar animals evolved from a common ancestor. For example, all mammals evolved from some common ancestor. Second, he suggested that all living organisms might have evolved from a few or even one common ancestor that existed many millions of years ago.

## THE EVIDENCE OF EVOLUTION

Proof of evolutionary theory, in the rigorous sense of an experimental verification or demonstration, is impossible for several reasons. The most important reason is that evolution is a historical phenomenon. Most of evolution must be examined as a historian examines the evidence that the American Revolution, for example, actually occurred.

The entire process, in the case of evolution, took billions of years. Despite this handicap, several lines of scientific evidence strongly support evolutionary theory.

## EVIDENCE FROM THE SIMILARITY OF FUNDAMENTAL LIFE PROCESSES IN ALL ORGANISMS

One of the tenets of the theory of evolution is that all the diverse plant and animal varieties that we observe today have evolved from a common ancestral stock. As noted in Chapter 1, life is thought to have arisen on this planet by chance under a special set of physical and chemical conditions that existed in the early stages of the formation of the earth. From this "early protoplasm," all varieties of organisms evolved gradually by natural selection. If this theory is correct, many basic life processes—release of energy, synthesis of ATP, transfer of genetic information, and so on—should be similar in all organisms. This is in fact the case. For example, in cellular respiration, the cytochromes are involved in the transport of electrons and thus the release of energy. From the primitive protistan *Amoeba* to the gigantic baleen whales and from the primitive single-celled algae to the giant sequoias, we find the cytochromes to be involved in the transport of electrons (Fig. 27-3). The amino acid sequence of the cytochrome c molecule from yeasts to humans is remarkably similar, so much so that one can learn something about the genetic relationship of organisms by examining the amino acid sequences of their cytochrome c (Table 27-1). This sequence tends to change with time and shows that the gene for human cytochrome c differs from that of a dog by 13 nucleotides, from that of a bread mold by 63, but only by one from that of a monkey.

Another example that supports the principle of common ancestry of all organisms is the chemical basis of information transfer. In our discussion of the chemical basis of heredity (Chapter 24), DNA was noted as the carrier of heredity in all prokaryotes and eukaryotes, that is, in all organisms on earth. Furthermore, the information encoded in DNA is transcribed in the form of RNA molecules and then translated into the amino acid sequence of proteins. The genetic code that specifies the sequence of amino acids in a polypeptide chain is also identical in all organisms except for some variations in that of mitochondria and chloroplasts. Moreover, the enzymes involved in the transfer of this information are similar in all organisms. For example, RNA polymerase, an enzyme required for the transcription of RNA from DNA in the bacterium *E. coli,* is capable of transcribing RNA from the DNA of other species as well, including that of higher plants and animals. This similarity in the process of information transfer in all living beings is best explained by assuming their common ancestry.

In eukaryotes, DNA is always associated with basic proteins called histones. Together with DNA, the histones form the beaded arrays called nucleosomes; nucleosomes of all eukaryotes have the same organization. The amino acid sequence of some of the histones from all animals and plants is also identical or nearly so. The similarity in these molecules, which reflects a corresponding similarity in their genes, is best explained by assuming that all plants and animals have a common ancestry. All such reasoning invokes what is called the principle of homology.

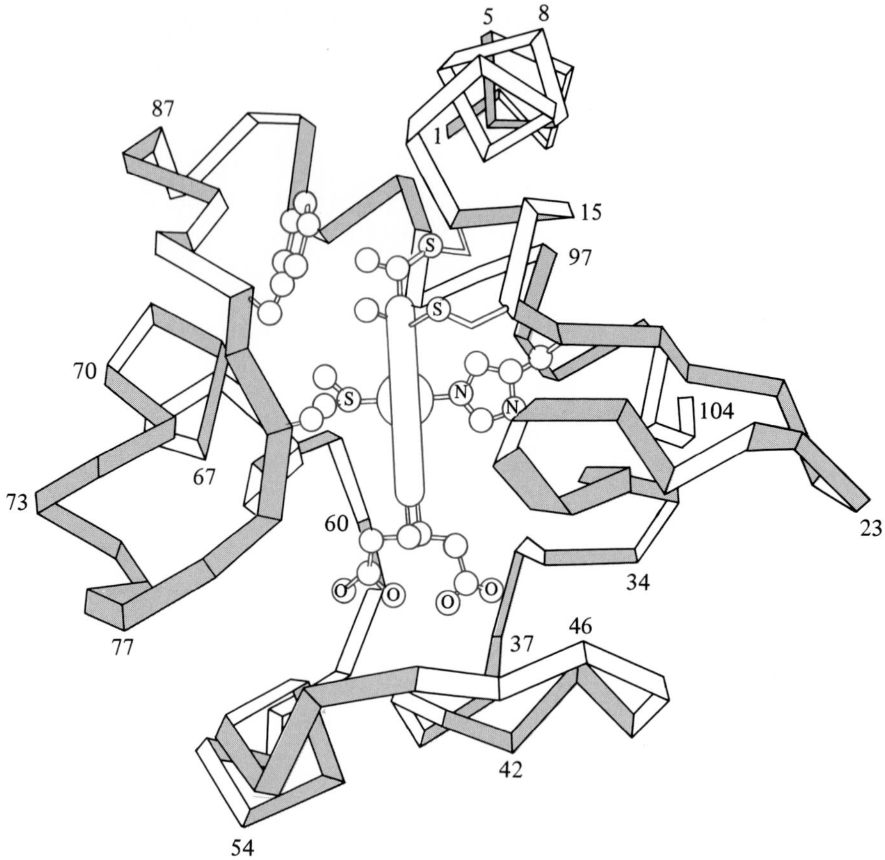


**FIG. 27-3** Three-dimensional structure of cytochrome c showing variable and invariable amino acid sequences. The numbers in the figure refer to the amino acid residues in the molecule. The position of the heme is also shown. The variable and the invariable regions are shown in different shades.

The **principle of homology** states, in its simplest form, that inherited similarities in organisms indicate kinship. In other words, organisms that resemble each other are related, and, as a corollary, the more closely two organisms are alike, the more closely they are related. We apply the principle frequently in our everyday lives, whenever we judge that two persons with similar appearance are members of the same race or the same family. If they look exactly alike, we would of course surmise that they are identical twins.

In biology, the principle of homology is applied when we judge that all trees are more closely related—that is, have a more recent common ancestor—than, say, a tree and a moss or a seaweed. At the same time, we judge trees, mosses, and seaweeds to be more closely related to each other than any of them are to grizzly bears or elephants. At another level, we judge the various members of the cat family—from lions and leopards to house cats—to be more closely related to each other than they are to such other carnivores as dogs and wolves. In making these judgments, it is possible to be mistaken at times, just as we might mistake two look-alikes for brothers because of a superficial resemblance. The history of biology lists many such mistakes (whales were classified

**TABLE 27-1 Minimum Numbers of Nucleotide Differences (Inferred from Amino Acid Sequences) among the Genes Coding for Cytochrome c of Twenty Organisms**

| | 1 | 2 | 3 | 4 | 5 | 6 | 7 | 8 | 9 | 10 | 11 | 12 | 13 | 14 | 15 | 16 | 17 | 18 | 19 |
|---|---|---|---|---|---|---|---|---|---|---|---|---|---|---|---|---|---|---|---|
| 1. Man | | | | | | | | | | | | | | | | | | | |
| 2. Monkey | 1 | | | | | | | | | | | | | | | | | | |
| 3. Dog | 13 | 12 | | | | | | | | | | | | | | | | | |
| 4. Horse | 17 | 16 | 10 | | | | | | | | | | | | | | | | |
| 5. Donkey | 16 | 15 | 8 | 1 | | | | | | | | | | | | | | | |
| 6. Pig | 13 | 12 | 4 | 5 | 4 | | | | | | | | | | | | | | |
| 7. Rabbit | 12 | 11 | 6 | 11 | 10 | 6 | | | | | | | | | | | | | |
| 8. Kangaroo | 12 | 13 | 7 | 11 | 12 | 7 | 7 | | | | | | | | | | | | |
| 9. Duck | 17 | 16 | 12 | 16 | 15 | 13 | 10 | 14 | | | | | | | | | | | |
| 10. Pigeon | 16 | 15 | 12 | 16 | 15 | 13 | 8 | 14 | 3 | | | | | | | | | | |
| 11. Chicken | 18 | 17 | 14 | 16 | 15 | 13 | 11 | 15 | 3 | 4 | | | | | | | | | |
| 12. Penguin | 18 | 17 | 14 | 17 | 16 | 14 | 11 | 13 | 3 | 4 | 2 | | | | | | | | |
| 13. Turtle | 19 | 18 | 13 | 16 | 15 | 13 | 11 | 14 | 7 | 8 | 8 | 8 | | | | | | | |
| 14. Rattlesnake | 20 | 21 | 30 | 32 | 31 | 30 | 25 | 30 | 24 | 24 | 28 | 28 | 30 | | | | | | |
| 15. Tuna | 31 | 32 | 29 | 27 | 26 | 25 | 26 | 27 | 26 | 27 | 26 | 27 | 27 | 38 | | | | | |
| 16. Screw worm fly | 33 | 32 | 24 | 24 | 25 | 26 | 23 | 26 | 25 | 26 | 26 | 28 | 30 | 40 | 34 | | | | |
| 17. Moth | 36 | 35 | 28 | 33 | 32 | 31 | 29 | 31 | 29 | 30 | 31 | 30 | 33 | 41 | 41 | 16 | | | |
| 18. *Neurospora* | 63 | 62 | 64 | 64 | 64 | 64 | 62 | 66 | 61 | 59 | 61 | 62 | 65 | 61 | 72 | 58 | 59 | | |
| 19. *Saccharomyces* | 56 | 57 | 61 | 60 | 59 | 59 | 59 | 58 | 62 | 62 | 62 | 61 | 64 | 61 | 66 | 63 | 60 | 57 | |
| 20. *Candida* | 66 | 65 | 66 | 68 | 67 | 67 | 67 | 68 | 66 | 66 | 66 | 65 | 67 | 69 | 69 | 65 | 61 | 61 | 41 |

as fish at one time), but closer study, as in the case of unrelated "brothers," usually reveals the error for what it is.

## EVIDENCE FROM TAXONOMY: THE PHYLOGENETIC TREE

In Chapter 1, we briefly described the major taxonomic groups of organisms. The same classification is represented in the form of a phylogenetic tree (See inside front cover). That classification is based on fundamental similarities among organisms. For example, although one group is warm-blooded and the other is not, birds and reptiles share many general anatomical and physiological attributes and are represented on the tree as having a common ancestor. Similarly, the arthropods, with their segmented bodies, are regarded as modified annelids; they differ from the annelids in that they have evolved rigid exoskeletons and jointed legs. Whenever organisms have been carefully studied, it has sooner or later become apparent that they naturally fall into groups with fundamentally similar traits. Moreover, each major group has always been readily divided into subgroups in which the members resemble each other more closely than they do any other organism. But how do such taxonomic relationships provide evidence for evolution?

The strength of the argument from taxonomy lies in the fact that any classification of organisms that is based on fundamental similarities invariably arranges the groups of organisms as if they were related to each other, that is, in a kind of family tree. No other arrangement makes much sense. The only logical explanation for such a natural classification of a group of organisms is the probable common ancestry of all organisms in the group. For example it is unlikely that the vertebral column evolved independently in each group of vertebrates or the chitinous exoskeleton in each group of arthropods. All vertebrates have a vertebral column because all vertebrates are descendants of a group of early animals that by chance evolved a vertebral column.

This argument may at first appear circular because organisms are classified on the basis of similarity in their structure, and the classification is used as evidence for evolutionary kinship. Yet this is a valid argument in favor of evolution. The organisms fall *naturally* into these taxonomic groups. Evolution from a common ancestor seems the most logical explanation for these similarities among the organisms in a taxonomic group.

## EMBRYONIC FEATURES AS EVIDENCE OF EVOLUTIONARY RELATIONSHIPS

The early development of an organism often reveals something of its evolutionary past. For example, vertebrates exhibit several progressive changes that can be organized into a graded series starting with the most primitive jawless fishes through bony fishes, amphibians, reptiles, birds, and mammals. The more primitive of these groups are judged to be ancestral to the more advanced ones. This view is held despite the obvious differences between fishes, which are aquatic and have gills and fins, and adult reptiles, birds, and mammals, which are air-breathing tetrapods, lack gills, and are usually highly adapted for terrestrial life. But when we compare the embryonic stages of fishes with those of the higher vertebrates, a fundamental relationship between the fishes and the higher vertebrates becomes obvious (Fig. 27-4). Embryos of all major groups of vertebrates, from fishes to mammals, possess gill pouches and gill furrows in addition to their common gross morphology. But only in fishes do these develop into functional gills. Why would the embryos of reptiles, birds, and mammals possess gill pouches and furrows even if they do not develop into gills? Ernst Haeckel, a nineteenth-century German biologist, suggested that the embryonic stages of advanced organisms briefly recapitulate some of the morphological characters of their ancestors. Thus, the reason that a bird embryo bears gill pouches and furrows is because a group of ancient fishes were ancestral to birds. As Haeckel put it, "ontogeny [development] recapitulates phylogeny [evolution]." This does not, however, imply that all embryonic features can be interpreted as recapitulatory. For example, embryos of reptiles, birds, and mammals develop the embryonic membranes amnion, chorion, and allantois. Rather than representing equivalent structures possessed by ancestors, these are newly evolved structures important for survival of embryos on land.

Another example of evidence from embryology of a common ancestry is seen in the larvae of annelids and molluscs. The adults of these phyla bear no resemblance to each other. Yet the larvae are surprisingly similar (Fig. 27-5). They resemble each other not only in general appearance but also, for example, in the position and development of mouth and anus and even in the orientation of the future dorsal and ventral sides and anterior and posterior ends. They are so much alike that it would be hard to believe that the two phyla did not have common ancestors and that it was pure coincidence that the two larvae are so similar. There are many other such examples: the adults and the nauplius larvae of *Sacculina*, a parasite on crabs, and of *Cyclops*, a common fresh-

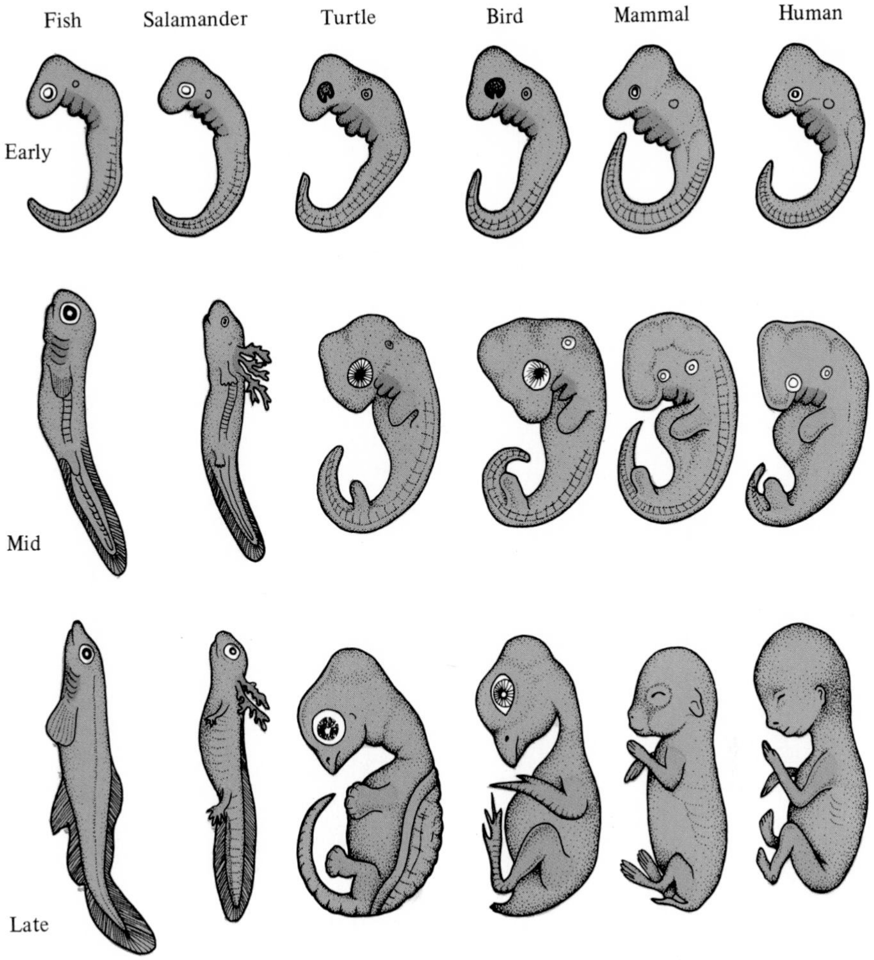


**FIG. 27-4** Embryonic stages of diverse vertebrates. Note similarity in the external features of early embryos of all vertebrates.

water crustacean are illustrated in Figure 27-6. *Sacculina,* which is nothing but a bag of tissue attached to a crab (Fig. 27-6a), bears no resemblance to its host. But both are classified under crustacea because the larva of *Sacculina* (Fig. 27-6b) is anatomically similar to the larva of other crustaceans (Fig. 27-6c) such as *Cyclops* (Fig. 27-6d).

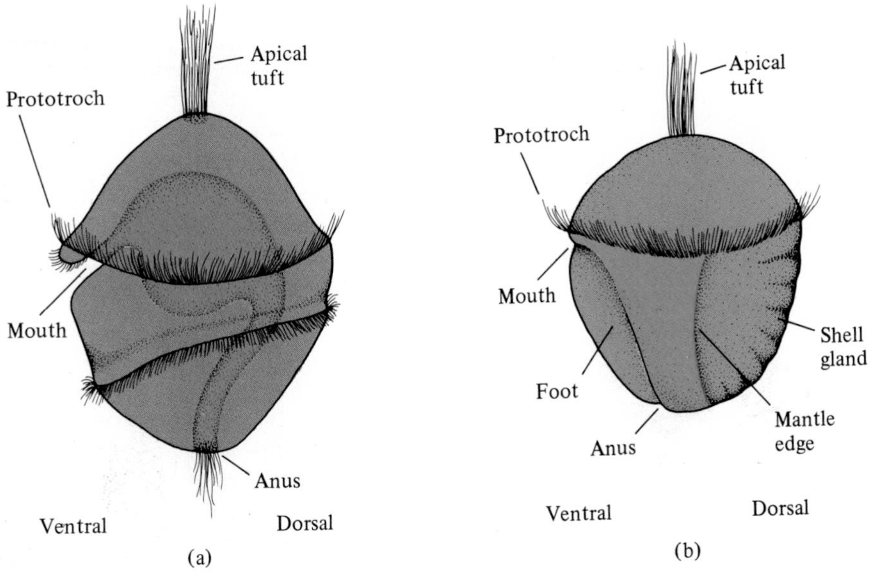


**FIG. 27-5** Trochophore larvae of an annelid (a) and a mollusc (b).

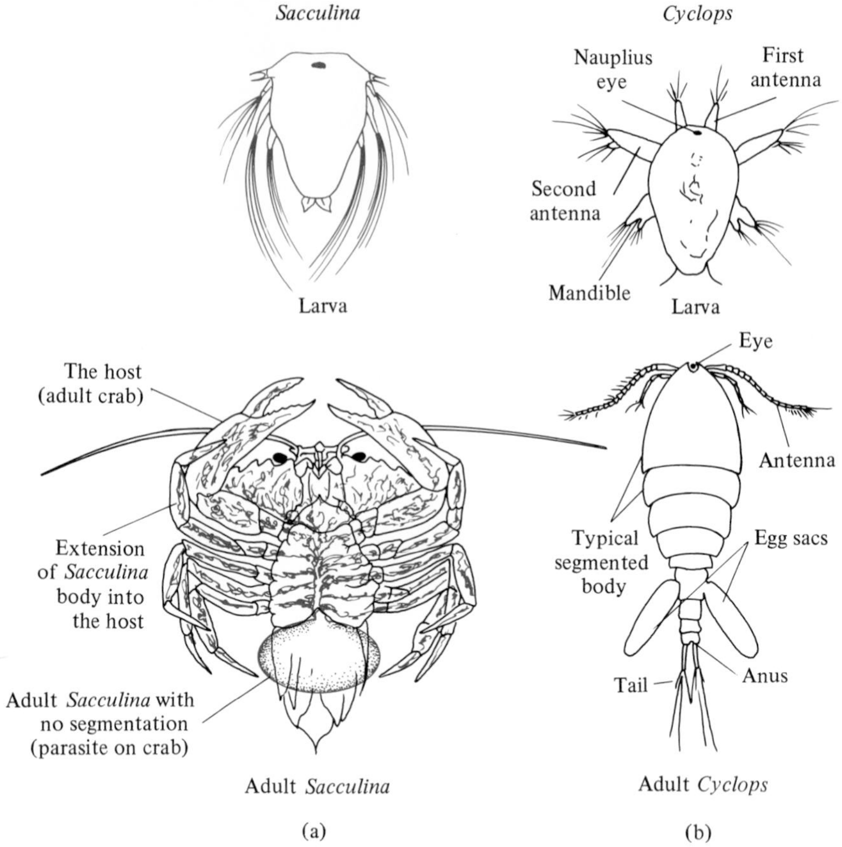


**FIG. 27-6** Larval and adult stages of (a) *Sacculina*, a parasitic, highly degenerate crustacean, and (b) *Cyclops*, a freeliving crustacean showing similarity in larval stages despite great differences in adult morphology.

## COMPARATIVE ANATOMY AS EVIDENCE OF EVOLUTION

The origin of specialized organs superbly adapted to an organism's mode of life has often puzzled those thinking about evolution. Examine the wing of a bird, the foreleg of a frog, the arm of a human being, the wing of a bat, and the pectoral fins of a whale (Fig. 27-7). They differ in gross morphology, and each is highly adapted for the mode of life the animal leads. But a study of the structure and development of these organs shows that they develop from the same rudiments, have the same relationship to the body, and exhibit a similarity in the pattern of bones and muscles despite their dissimilar functions. They are said to be homologous to each other. Like all tetrapod forelimbs, these structures all contain a single upper arm bone, the humerus, and a pair of lower arm bones, the radius and ulna. The carpal (wrist), the metacarpal (hand), and the bones of the digits (fingers) in these animals are also similar to one another, although the degree of development of the bones varies in each of the specialized structures. The forelimbs of several different vertebrates are illustrated in Figure 27-7 for comparison.

There are many examples of homologous structures that serve different functions. For example, from a developmental analysis of the mammalian incus and malleus (two of the three ossicles of the ear), it is clear that they are homologous to the quadrate and articular of reptiles (Fig. 27-8). These bones are small and help in articulation of the lower jaw to

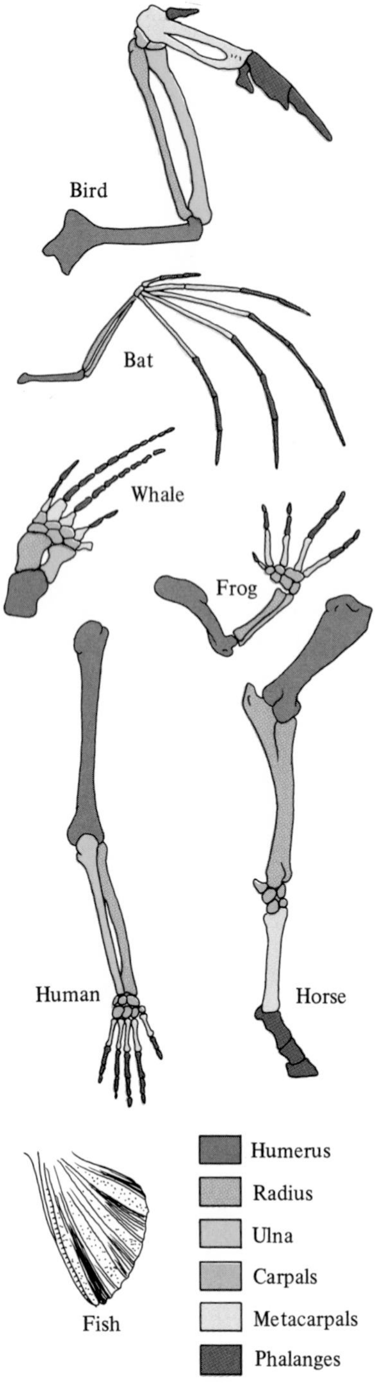


**FIG. 27-7** Forelimb skeleton of diverse vertebrates demonstrating uniformity of construction despite variation in function.

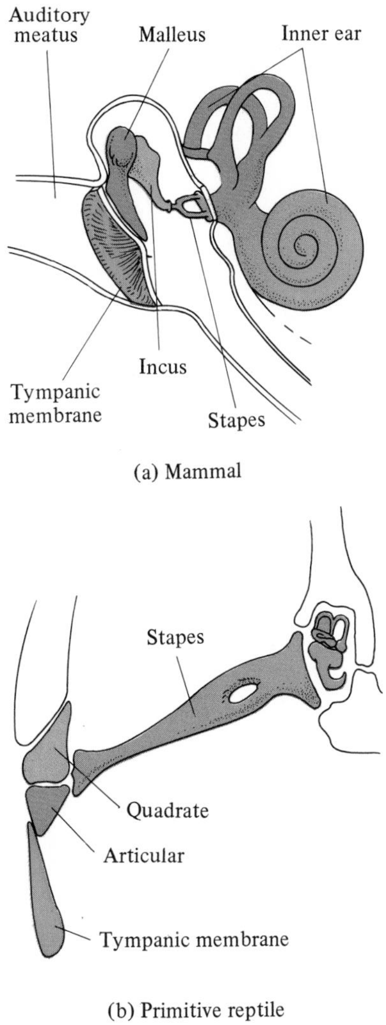

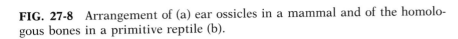

**FIG. 27-8** Arrangement of (a) ear ossicles in a mammal and of the homologous bones in a primitive reptile (b).

the skull in one extinct group of reptiles, the cyanodonts, but are longer in other reptiles, amphibians, and teleost fishes. In teleosts, they are involved in the articulation of the mandibles, the lower jawbones to the skull and are believed to be homologous to the skeletal elements of jaws in sharks. Skeletal elements of shark jaws are apparently evolutionary descendants of the skeletal elements of the gill arches in jawless fishes.

Such homologies have been explained by some as the design of creation. Others interpret them as the result of the common ancestry of such animals. The existence of homologous vestigial organs (degenerate or atrophied remnants of once well developed organs) supports the latter position: evolution by common descent. Take, for example, the presence of remnants of hind limbs in some whales, which have bony and muscular elements comparable to those of the hind limbs of tetrapods. They are not functional and are very much reduced in size. They do not exhibit functional similarity to the hind limbs of tetrapods. Homology of vestigial organs is not readily explained as the result of design in creation. Such homologous structures are more readily explained on the basis of common descent.

Evolution is the most logical explanation for the presence of vestigial organs in a series of animals that exhibit a gradation in the degree to which a certain organ has degenerated. For example, snakes normally lack paired limbs and limb girdles. Pythons are an exception. They bear vestigial hind limbs. Snakes are believed to have evolved from lizards. Among most lizards the limbs are fully functional. But in skinks the limbs are reduced. Movement is mainly by serpentine body movements. In *Amphisbaena*, a limbless lizard, the limbs are altogether absent. Yet these lizards have normal girdles, the bony elements to which the limbs are attached. The presence of limb girdles that serve no present function can best be explained by assuming their evolution from a group with functional limbs.

(a)

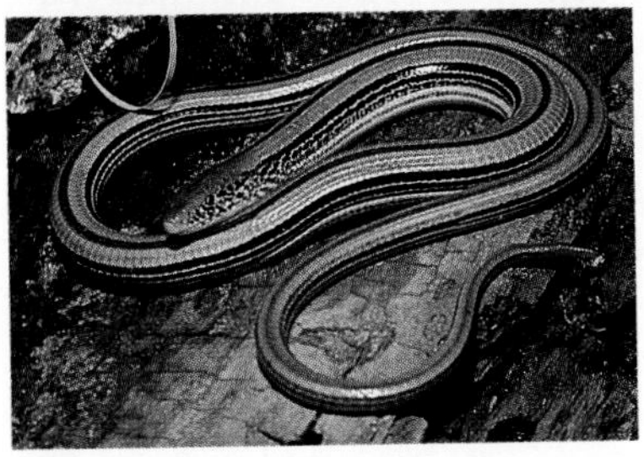
(b)

**FIG. 27-9** Convergent evolution in external features of (a) a snake and (b) a limbless lizard.

## CONVERGENT EVOLUTION

When unrelated organisms have adapted to the same specific habitat and mode of life, they often exhibit similar morphological features even though their underlying anatomy may be different. This phenomenon is known as **evolutionary convergence.** For example, the pectoral fins of fishes and whales look alike, but they are of very different construction. In fishes, the fins are supported by fin rays, whereas in whales, the fin is supported by the finger bones. Developmentally, fin rays are different from the bones of the fingers. Thus the fins in the two groups are only superficially similar. Similarly, the wings of birds and the wings of bats are similar and serve the same function but have a different underlying organization. Snakes, limbless lizards, blind worms, and apodan (legless) amphibians, such as cecilians, are adapted to a burrowing life, in which legs are a distinct disadvantage (Fig. 27-9). Although closely related to animals with functional limbs, these animals secondarily lost their limbs as an adaptation to a life in burrows. Similarly, despite their different ancestry, fishes and whales both have streamlined body shapes well adapted to swimming (Fig. 27-10). There are many such examples of convergent evolution, clearly suggesting that the morphology of organisms can and does change over the course of generations. The theory of evolution predicts that such changes would occur as adaptations to various environmental situations, and investigators must be careful not to interpret such superficial similarities between unrelated groups as evolutionary relationships.

(a) (b) (c)

**FIG. 27-10** Streamlined body of (a) shark, (b) perch, and (c) dolphin, adaptations for active aquatic life.

## GEOGRAPHIC DISTRIBUTION OF ORGANISMS AS EVIDENCE OF EVOLUTION

The present-day geographic distribution of organisms constitutes persuasive evidence for evolution. Consider, for example, the many unique species of animals and plants found on the various oceanic islands of the world. Even where soil and climate are alike, species differ. Darwin found 13 endemic (unique to the area) species of finches on the Galapagos Islands. If they had come from the South American mainland, where finches abound, why had all individuals of these 13 types migrated, leaving none behind? Why are marsupials and monotremes (egg-laying mammals) the only kinds of mammals found in Australia, except for humans and some imported placental mammals? It cannot be that placental mammals are not suited to Australia. The overwhelming population explosion that has occurred among imported rabbits proves that.

The only reasonable explanation for such geographic distribution of organisms in discontinuous regions is that a few individuals were the ancient ancestors of the present populations of those areas. In time, they evolved into the variety we see there today.

The geographical distribution of marsupials is particularly instructive. In the past, marsupials were distributed all over the world, as evidenced by the fossil record. However, with the exception of a few, such as the various species of opposum in the Americas, they now survive only in Australia and adjacent islands. The explanation for the marsupial survival and radiation, or diversification into many different types, on the Australian continent is Australia's separation from the land masses of the rest of the southern hemisphere before the appearance of modern mammals. In the absence of competition from placental mammals in Australia, marsupials there evolved into a large variety of species, successfully occupying diverse ecological niches. In other parts of the world, marsupials became extinct after the more efficient modern mammals evolved.

Geographical distribution of animals suggests that the world can be divided into five major realms: the holarctic (palearctic and nearctic), oriental, neotropical, Ethiopian, and Australian (Fig. 27-11). Each region has its own characteristic fauna, with intergradations evident wherever land masses have been connected from time to time.

Besides the occasional appearance and disappearance of land bridges between continents, the continents themselves have undergone major changes in their relative positions over many millions of years. From many lines of evidence, it appears that at one time there existed a single large land mass that subsequently broke up into the smaller masses represented by today's continents (see Panel 27-2). These then drifted away from each other. This theory of **continental drift** explains the close similarity between certain flightless birds found only in Aus-

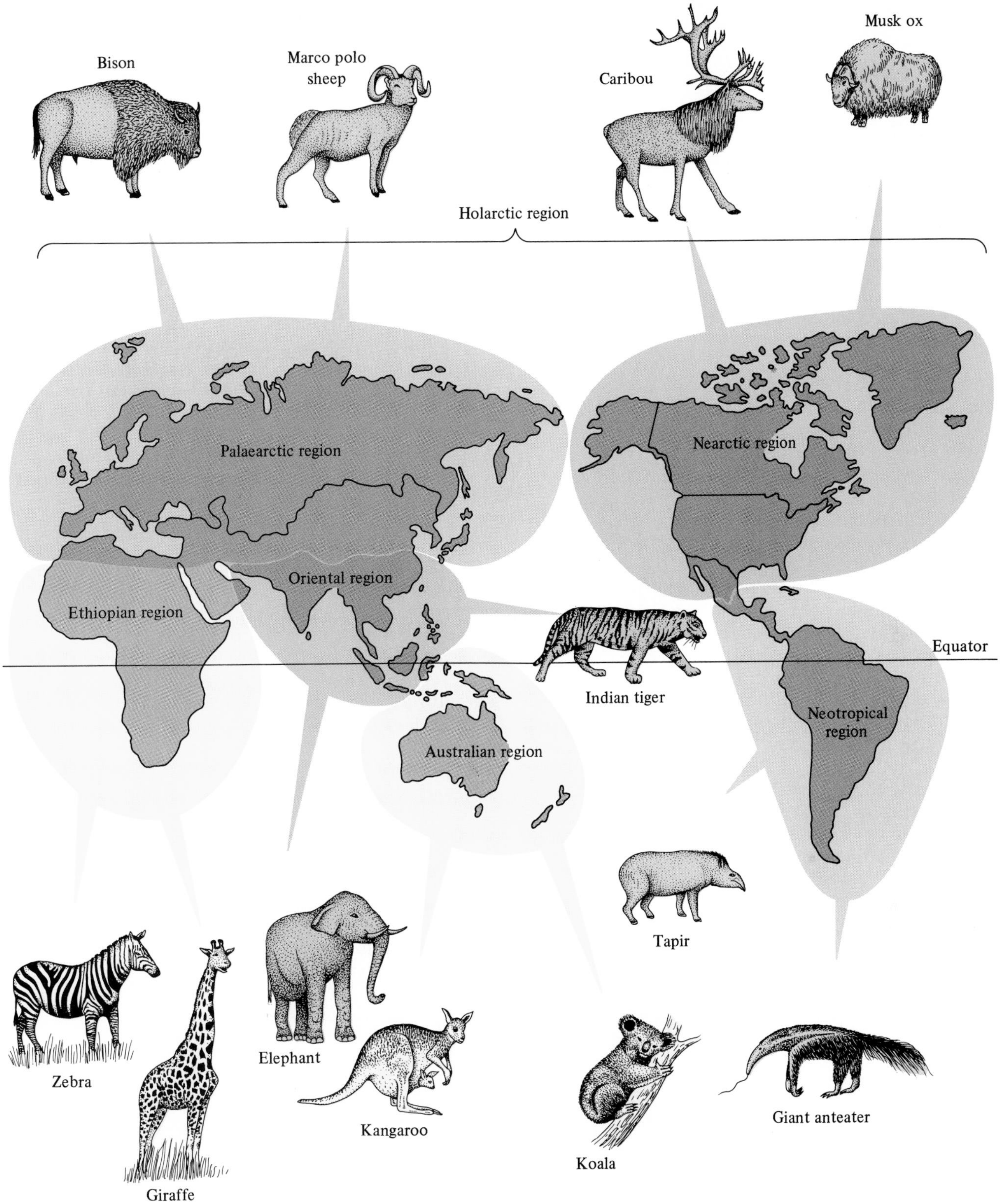


**FIG. 27-11** Biogeographical realms of the world, and a few selected representative mammals of each region.

tralia, South America, and Africa, as well as similarities in various invertebrates, freshwater fishes, and plants found only on those continents. As we have already seen, the distribution of both fossil and living marsupials is also explained by this theory.

PANEL 27-2

# Continental Drift and Plate Tectonics

The theory of continental drift hypothesizes that the continents have not always occupied their present-day positions throughout the history of the earth. As the name of the theory suggests, the continents and the underlying plates are believed to have drifted apart from a single original land mass called **Pangaea.**

Biologists were interested in the theory of continental drift well before it became accepted by geologists because it so logically explained the distribution of many plants and animals and their fossils around the world. This seemingly outlandish theory that the continents actually moved apart from one land mass is not new. Francis Bacon had noted in 1620, when the first world maps were published, that the shape of the continents was such that if they could be moved together like pieces of a jigsaw puzzle, the fit was quite remarkable. Alfred Wegener, a German geologist, is credited with the modern theory of continental drift.

Continental drift was discredited until after results of certain sonar and magnetometer (an instrument that measures magnetic forces) surveys of the land and ocean floor were published in the mid-1960s. Then, in 1965, **plate tectonics,** the study of the movements of the crustal plates that make up the earth's surface, was born. Since then, abundant evidence has been accumulated by geologists that validates the theory of continental drift.

Magnetometer surveys of the continents have revealed that the orientation of local magnetic fields in igneous rocks varies with the age of the rock. In igneous rock, iron is aligned at the time of crystallization of the magma (molten basalt), in accordance with the earth's magnetic field.* The survey findings indicated that either the earth's magnetic field had shifted dramatically since the magma had cooled, or the continents had been moving. It was concluded that the continents had indeed been drifting. From the changed magnetic field orientations in the continental rocks, the "wandering paths" of the continents could be reconstructed. Tracing these paths backward led to the conclusion that all continents had originated in a single land mass, named Pangaea.

Geological and biological evidence indicates that late in the Paleozoic era, Pangaea began to separate into two land masses, which were named Laurasia (later to become North America, Europe, and northern Asia) and Gondwanaland (later to become South America, Africa, India, Australia, New Zealand, Malaysia, and Antarctica), separated by the Tethys Sea. The Mediterranean is thought to be a remnant of the Tethys Sea.

As numerous sonar studies of the ocean floor have shown, a mountain ridge exists in the mid-Atlantic Ocean. Another ridge, the East Pacific Rise, lies off the southern coast of California. A similar ridge is located off the west coast of South America. As a physical map of the world shows, a system of such ridges exists on ocean floors around the earth. These ridges are the centers of volcanic action and innumerable earthquakes. When samples of the rock of the ocean floor were obtained by drilling, newly formed basalt was found along the ridges. The ages of the rock samples collected progressively farther from the mid-Atlantic Ridge, on both sides, were found to be progressively older. In other words, the sea floor is spreading away from the ridges. The rate of spreading in the Atlantic is 2 cm per year; in the Pacific, 9 cm per year.

Further studies of the mid-Atlantic ridge showed that large blocks of crust were sliding past one another. When the ocean floors and their shifting blocks were mapped, it became apparent that the earth's crust is composed of several large, shifting plates. North America, for example, is part of what is called the North American Plate, South America is part of the South American Plate, and so on.

As the plates move apart at the ridge, they collide at the advancing edges, and one plate pushes under, or subducts, the other at a subduction zone, or trench. As one plate subducts, the other is uplifted. The Himalayan Mountains are thought to have been formed when the Indian Plate subducted under the Asian Plate. Subduction zones are the sites of deep earthquakes and volcanic action. The crust of a subducting plate edge becomes molten as it penetrates beneath the mantle and is incorporated into the earth's magma, which circulates and extrudes at the ridges.

In addition to the fit of the continents, other geological evidence also supports the continental drift theory. Geological formations that are now on separate continents but were once part of the same continental plate would be aligned if the continents were placed together in a Pangaea configuration. Not only would igneous rocks of common age be aligned, but so would sedimentary

* Where magma reaches the surface of the earth, it is called lava. When lava solidifies, it is called igneous rock.

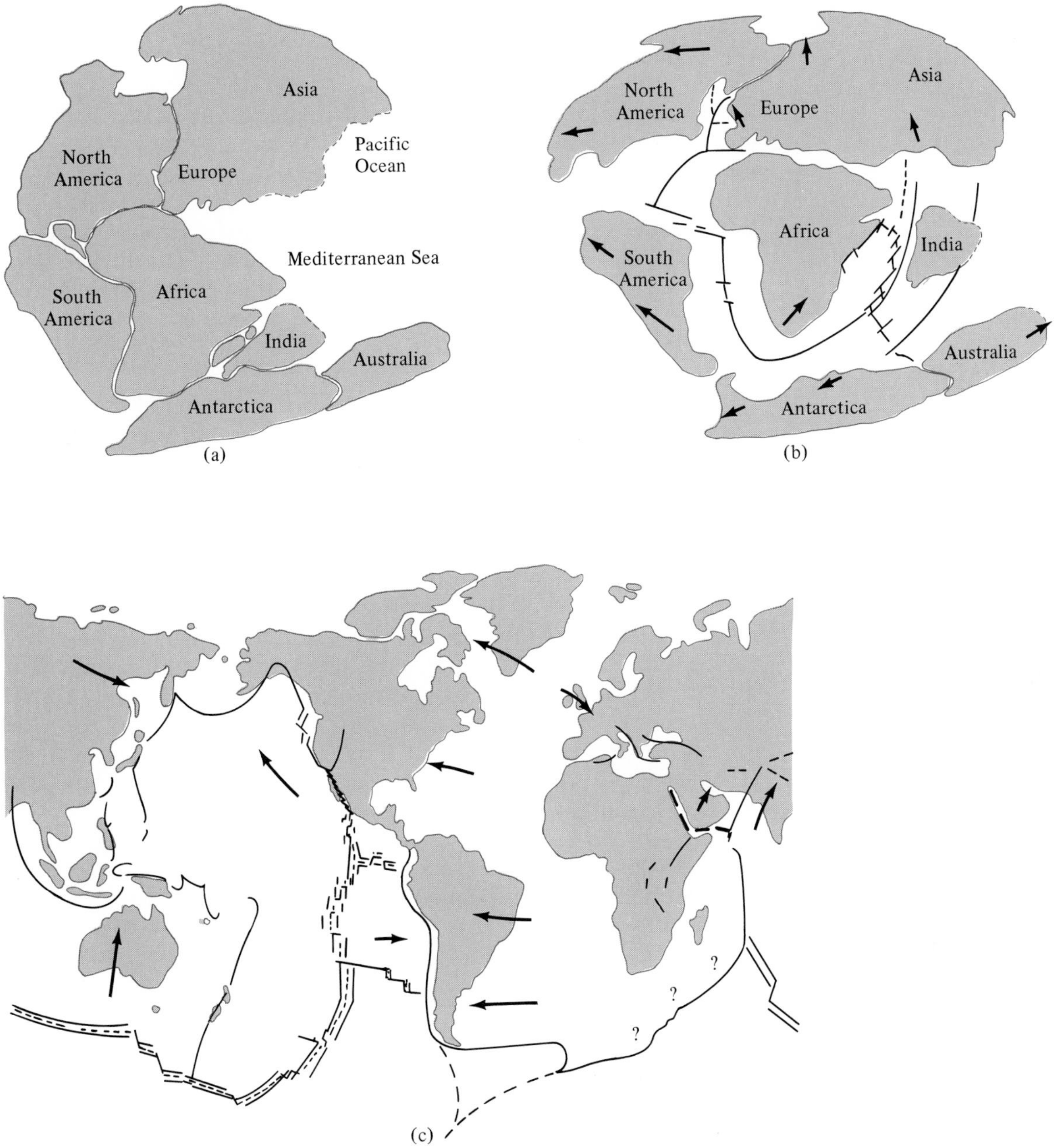


Distribution of the continents and crustal plates (a) 200 million years ago, (b) 60 million years ago, and (c) at the present time. The arrows indicate the direction of movement of the crustal plates.

strata; for example, the sedimentary strata of New England, England and Scandinavia would align, so also those of Australia, India, Antarctica, and South Africa. Moreover, the presence of the volcanic islands in the Pacific ocean, which form a chain of "sea mounts" of increasing age from the Hawaiian Islands (3.5 million years) to the Midway Islands (15 to 25 million years) to the Emperor Chain (70 million years), can be explained as the result of the drifting of the Pacific Plate over a "hot spot," or plume of magma.

For these and other compelling reasons, the continental drift theory, once a subject of jest and ridicule, is now a well-established scientific theory.

## THE FOSSIL RECORD AS EVIDENCE OF EVOLUTION

If primitive organisms appeared early in the history of life on earth and more advanced ones later, it is to be expected that the lowest strata of the earth's crust would contain only primitive organisms. This is in fact true. As noted in Chapter 1, the earth is about 4.6 billion years old. The first billion or so years were without life. During the next 2.5 billion years, organic molecules appeared and life arose. The earliest known fossils are of bacteria and cyanobacteria (blue-green algae) found in rocks 3 billion years old. Algal filaments, protists, jellyfishes, and sponges have also left fossil remains in younger Precambrian rocks (older than 600 million years).

Fossils of all major groups of invertebrates are found in rocks of the Cambrian era (600 to 500 million years ago). Only during the mid-Ordovician era (500 to 440 million years ago) do we find the first fish fossils, the earliest vertebrates. In progressively more recent rock strata, more advanced forms appear: amphibians late in the Devonian era (400 to 350 million years ago) and reptiles in the Permian era (270 to 225 million years ago). (See inside back cover for the geological time chart.)

A similar series exists for various stages in plant evolution. While fossil algae and fungi that resemble modern members of these divisions can be traced back to Precambrian times, vascular plants appeared only late in the Silurian (440 to 400 million years ago) period. Fossils of gymnosperms and ferns became abundant in the strata of a still later era, the Carboniferous (350 to 270 million years ago), when the earth's coal deposits were formed. The first angiosperms did not appear until the Cretaceous era, 135 to 70 million years ago. Again, in each case the more highly evolved forms appear in the more recent strata.

This progressive appearance of more advanced forms clearly suggests an ancestor-descendant relationship between the primitive and the advanced groups of organisms. These conclusions are dependable, however, only if the methods of determining the age of fossils are reliable. These methods are explained in Panel 27-3.

## EVOLUTION IN THE CONTEXT OF MODERN GENETICS

The chief tenet of the theory of organic evolution is that organisms change in their anatomical and physiological features over the course of time. These changes are very slow and gradual. Hence the changes are discernible generally only when we study changes in a group of organisms over millions of years. For this reason evolutionary change is not obvious when we apply the usual methods of observation. Although the concept of gradual change in organisms as a result of fixation of hereditary variations had been accepted by animal and crop plant breeders even before Darwin's time, the concept of evolutionary change had met many obstacles on its way to more or less universal acceptance.

The major difficulty in accepting Darwin's theory of evolution by natural selection, at the time, was a lack of knowledge of the causes of variation. As Darwin himself noted, the laws of inheritance were not well known. It was generally believed that the traits of parents blended to produce intermediate features in the offspring. As noted in Chapter

PANEL 27-3

## Methods of Determining the Age of Rocks

The validity of paleontological evidence is sometimes questioned by people unfamiliar with radiometric methods of determining the age of igneous rock strata, rocks formed by cooling of lava, which do not contain fossils. Once the age of igneous rocks has been determined by this method, the age of the overlying sedimentary rocks can be estimated. Then index fossils, which are unique to a particular sedimentary rock stratum, over or beneath an igneous rock stratum, are used to identify strata of equivalent age in different parts of the world and to determine their age.

The radiometric method is based on the fact that radioactive elements steadily decay to different forms over the course of time. The rate of decay varies from one radioactive element to another but is constant for each element; and it is independent of the physical and chemical properties (temperature, nature of the rocks, exposure to sun, and other factors) of the rocks in which the elements occur. The age of a rock is estimated by determining the ratio of the mass of the parent radioactive element—such as uranium, potassium, or rubidium—to that of the product of decay—lead, argon, or strontium, respectively—present in intimate association with it. We know the rates at which various elements undergo radioactive decay. For example, uranium takes 4.5 billion years for half of the atoms in a given sample to undergo radioactive decay to lead. This length of time is the radioactive half-life of uranium. By determining the ratio of the uranium to the lead in a sample, we can calculate the time that elapsed from the formation of that particular rock to the present. Using these procedures, the age of igneous rocks can be determined.

23, Mendel's laws of the particulate nature of inheritance were discovered in Darwin's lifetime, but they were totally ignored until the beginning of the twentieth century.

The chemical identification of the physical basis of heredity and the elucidation of the mechanisms underlying the transfer of information from DNA to protein have put the concepts of particulate heredity on a firm footing. Knowledge of modern genetics has had a profound influence in promoting the acceptance of evolution as the source of biological diversity. Modern genetics can explain not only the origin of this diversity but can also explain how this diversity is maintained.

Evolutionary change by natural selection, as proposed by Darwin, occurs within a population of organisms. Thus evolution is a communal event. But it is the individuals within the population that are subject to this change. Darwin correctly perceived this phenomenon and hypothesized that natural selection operates through the differential reproductive success of individual organisms. In this section evolution will be described in the context of modern genetics.

### SPECIES: A GENETIC PERSPECTIVE

Although a species is usually defined as one or more populations of organisms, members of which can potentially interbreed to produce fertile offspring, the biological problem of defining a species is old and as yet unresolved. From a genetic perspective, a species may be viewed as a population of organisms that "exchange" genes within the population. The individuals within a population may differ from one another in some of their physical or genetic features. These differences are the variations that Darwin talked about. Yet these individuals can interbreed and produce fertile offspring. That is, the genes of one individual can intermingle with those of any other individual in the population to produce viable and fertile offspring. The genes are said to be interchangeable between individuals. One-half of the genes of an individual combine with half of the genes of another individual in the production of an offspring.

As a corollary, a species may be viewed as the sum total of all the variant forms of genes in a population, known as the **gene pool.** Thus evolutionary change that results in the origin of a different species from an existing one is a change in the gene pool of a subpopulation of individuals such that individuals of this population cannot interbreed with members of the parental population. No gene flow occurs between the two populations once the "daughter" and "parental" populations have become established as two separate species. How can natural selection aid the origin of new species? For an understanding of this question we shall first describe the role of variations and of natural selection in the propagation of a favorable trait in a population.

## VARIATION: THE RAW MATERIAL OF EVOLUTIONARY CHANGE

As Darwin noted, there is great variation in structure, function, and behavior within each species. Although the evolutionary significance of many of these variations in heritable characteristics is unknown, some variations are adaptive in that they increase their possessor's chances of surviving long enough to reproduce under particular environmental conditions. For example, the appearance of a dark-colored variant of a normally light-pigmented insect in an industrial area blackened by soot may be of survival value to that population. In highly industrialized districts of England in which soft coal was used as fuel, the trees and other objects in the area gradually became darkened by soot. (Soot killed the lichens that normally lived on tree trunks, and consequently the dark color of the bark was exposed.) During the past 200 years, certain dark-colored mutants have appeared among a species of moth living in the industrial city of Manchester, England. Museum collections and records show that a dark variant of the peppered moth *(Biston betularia)* first appeared in Manchester about 1850 A.D. (Fig. 27-12). Collection records show that since that time, the dark mutant variety spread widely throughout the industrial belt of England.

(a) (b)

**FIG. 27-12** *Biston betularia,* the peppered moth. The normal, white with speckled black, and the mutant black forms are shown on (a) lichen-covered and (b) soot-blackened tree trunks.

Unlike the mottled ancestral form, the dark mutant blends with lichen-free, dark tree trunks and is not as easily spotted by its predators (Fig. 27-12b). It is therefore more likely to survive long enough to reproduce. Like all mutations, the dark mutation was a fortuitous occurrence, which under other circumstances might have been of no advantage to the moth; for that matter, against a light background, the mutation has been shown to be detrimental. Equal proportions of mottled and solid-colored moths were released in various environments and the survivors collected after a day or two. When the experiment was carried out in smoke-blackened regions of England, the percentage of mottled moths that survived was much lower than the percentage of dark-colored ones. Direct observations confirmed that birds were most successful in preying upon any moths that rested on sharply contrasting backgrounds, whether they were mottled moths on dark tree trunks or dark moths on lichen-covered, light-colored tree trunks.

The mottled and the solid-colored moths are variant types of this species. The two variant types can interbreed if they coexist in the same region. However, each variant has better chances of surviving in a different type of habitat. These observations clearly show that a variant phenotype that is favored by a specific type of habitat becomes a predominant or even an exclusive type in such an area. But in nature we often find a heterogeneous population coexisting within the same region. This heterogeneity may persist if the variation confers no survival advantage or disadvantage. In some instances, heterogeneity is maintained because natural selection favors the heterozygote. This is the reason for the persistence of the allele for sickle-cell anemia.

As noted in Chapter 23, sickle-cell disease is a genetic disorder. The hemoglobin *s* allele, when homozygous, is usually lethal. Yet in some areas of Africa, the Mediterranean, and Asia where malaria is prevalent this allele has survived many thousands of years. This is due to the selective advantage of heterozygotes. Individuals homozygous for the normal allele of the globin gene are susceptible to malaria. Individuals homozygous for hemoglobin *s* die of sickle-cell anemia. Being spared both malaria and sickle-cell anemia, heterozygous individuals survive to reproduce, thus maintaining the lethal allele.

(a)

(b)

**FIG. 27-13** The banded snail *(Cepaea nemoralis)* showing two color phases: (a) dark and (b) light.

Heterogeneity in a population can also be maintained because of the survival value of an allele in different seasons or in different ecological niches. The banded snail, *Cepaea nemoralis,* lives in woodlands, hedgerows, and green meadows in England. Three variations in the color pattern of their shells are common. Some are uniformly dark or light colored and others are banded with alternating dark and light color stripes, each variety camouflaged in a different kind of ecological niche (Fig. 27-13). The color pattern of their shells is genetically determined and is inherited in simple Mendelian fashion. Thrushes, their predators, find them by sight. A color pattern that is conspicuous against its background is selected against by nature. Thrushes seem to find these snails easily and eat them. Then how do all the three patterns persist? The answer lies in the fact that the snails live in woodlands, hedgerows, and green meadows, three diverse ecological niches. A shell of solid dark color is inconspicuous in woodlands. Light colored shells are camouflaged better in green meadows. In hedgerows, with shadows and rays of light, banded shells merge with the surroundings. Thus all three phenotypes coexist in nature.

Were the woodland population of snails not to interbreed with the hedgerow population for several generations and if all the offspring of the woodland population are limited to woodlands, the alleles for banded pattern could disappear from the population. But as long as the three populations interbreed, and their offspring are dispersed in all three habitats, the variant phenotypes will be maintained. Whenever two populations of a species do not interbreed for several generations, each population accumulates its own set of variant phenotypes and may evolve into a separate species.

## IF EVOLUTION IS A CONTINUUM, HOW DO SPECIES ARISE?

As noted earlier, the term *species* is applied to one or more populations of organisms, members of which can potentially interbreed to produce fertile offspring. Two populations closely resembling each other in physical features may sometimes be distinguished as separate species on the basis of their members' inability to interbreed or to produce fertile offspring. If the two are closely related, they are termed sibling species, which, by definition, must have evolved from a common ancestral population of a single species. This means that there was a transition phase during which one species was becoming two. In deciding whether two populations belong to the same species breeding experiments can seldom be conducted to determine whether they can produce fertile offspring. The great majority of species listed in taxonomic catalogs are, by necessity, assigned largely on the basis of their morphological features. This is always true for fossil species.

Although members of one species can usually be distinguished from those of most other species—for example, humans from monkeys or mice from rats—this is not always possible. For instance, many organisms are structurally similar but are known never to interbreed. Although we cannot tell them apart morphologically, they are, by definition, different species. In many other cases, however, the reverse is true: two populations with distinctly different morphology may interbreed and produce fertile offspring and, thus, by definition, belong to the same species. Such a species, which exists in more than one form, is termed **polymorphic.**

Many cases are known in which a species extends over many thousands of kilometers, and all adjacent populations are capable of interbreeding with full fertility. Yet those at the opposite ends of the range, if brought together, may not be able to interbreed. Here, our definition seems to fail. However, this is exactly the kind of situation we should expect to encounter if species do indeed arise by evolution.

The problem of defining a species is further complicated when we regard each species as a gene pool whose members potentially exchange genes with each other but not with members of any other species. If it were practical, all taxonomic classifications should be based on differences in DNA nucleotide sequences between populations. Studies on DNA nucleotide sequence of individual members of a population indicate great genetic heterogeneity within a species. However, the "genetic distance," that is, the differences in their nucleotide sequences, between what are judged by other criteria to be two sibling species is always less than that between two more distantly related species.

From the foregoing discussion, it is obvious that our classification of organisms into particular species is usually a matter of individual judgment based on available genetic, biochemical, morphological, and ecological information. Because most species exhibit a great range of variation in morphology, it is often difficult to decide whether to classify a particular group into one, two, or more species. However, if evolution has occurred—and is still occurring—such problems are exactly what would be expected. If new species do indeed arise from older species, we would expect to find transitional forms, and we do.

Thus, evolution of species by natural selection may be viewed as the fixation into separate species of variations that arise among individuals of a population. Three major mechanisms—mutation, isolation, and genetic drift—aid in the formation of new species. These are discussed in the following pages. But before considering the mechanisms of evolution, let us briefly consider the rate of evolution.

## GRADUALISM VERSUS PUNCTUATIONISM IN EVOLUTION

If evolution is a continuum, we would expect the rate of evolution of new species through the ages to be more or less uniform. Origin of species by natural selection involving differential reproductive success among individuals suggests that evolution would be gradual.

If we consider changes in amino acid sequence of ubiquitous proteins, such as cytochrome c, in different organisms, evolution appears to have occurred at a uniform rate. Cytochrome c, a protein component of the respiratory chain, is present in all organisms. The gene coding for this protein probably appeared very early in evolution and was favored by natural selection. In the course of time the gene for cytochrome c, like all other genes, must have mutated. Some of these mutations, but not all, altered the amino acid sequence of the protein. The amino acid sequence of cytochrome c from different organisms has been determined as well as the number of amino acid substitutions between the proteins of different species. On the average, cytochrome c from horses and other mammals differ in 5.1 amino acids. Since horses diverged from other mammals about 90 million years ago, it is estimated that on the average one amino acid substitution occurs every 17.6 million years ($90 \div 5.1$). The average amino acid difference between reptilian and mammalian cytochrome c is 14.8. From paleontological evidence, these two vertebrate groups diverged about 300 million years ago. Their average difference should be $300 \div 17.6$, or 17.0 amino acids. Similarly, the average amino acid difference between cytochrome c of higher plants and animals is 45. At the rate of one amino acid per 17.6 million years, plants and animals must have diverged at least 792 million years ago. This figure is close to the 800 million years estimated by some paleontologists. This type of calculation suggests that molecular evolution occurs at a more or less uniform rate if we consider it over long periods of time. Some scientists believe this so-called molecular clock is a real phenomenon. However other scientists suggest that this apparent correlation is fortuitious and of no evolutionary significance.

Studies of the fossil history of many organisms suggest that some new species have appeared abruptly in the past and then survived with little change until extinction millions of years later. For example, in the Turkana basin, in East Africa, 10 of the 13 species of fossil molluscs found there appeared within the relatively short geological period of 5,000 to 50,000 years and then remained unchanged for 3 to 5 million years. Thus, the evolutionary history of some species, at least, seems to be characterized by abrupt punctuations of rapid speciation in otherwise prolonged periods of unchanging morphology.

It has been charged that such **punctuated equilibrium,** as it has been termed, is contrary to the Darwinian principle of slow, gradual origin of species by selection. However, it must be emphasized that a given period of speciation may be characterized as rapid only in relative terms. In the molluscan example cited above, 5,000 to 50,000 years for the appearance of many species is a long period, comprising nearly as many generations. It is enough years and generations for differential reproductive success to select new species suitable to their environment. Thus, although the causes for rapid speciation are unknown, rapid speciation itself is not contradictory to the concept of evolution by natural selection.

## MUTATIONS: THE SOURCE OF VARIATIONS

As noted in Chapter 23, almost all physical, chemical, and behavioral traits of organisms have a genetic basis. In the fruit fly, *Drosophila,* the most widely studied animal in genetics, over 10,000 mutant varieties are known. Most inherited variations probably arise as gene mutations, that is, changes in the purine and pyrimidine base sequences of DNA molecules.

Not all variations in individuals are heritable. For example, an animal grown on a deficient diet may become a weakling. But it and its offspring still bear the genes for normal growth. The progressive increase in the physical stature of certain human populations in the twentieth century is attributed to their improved nutritional state, compared to that of their recent ancestors. Plants growing at higher altitudes are often smaller than those of the same variety grown in lower-lying plains. When grown in the plains, seeds from the same smaller plants grown at high altitudes reach the typical size of the plains specimens of that species. Variations in size and shape of individual organisms that result from environmental influences are called **acquired characters.** They are not heritable. Only variations in genes are heritable, and, as Darwin observed, only heritable variations are the raw material of evolutionary change. Although evolution appears to follow certain trends, all available evidence indicates that mutations themselves occur at random. As we have noted, only a few mutations prove to be adaptive, or advantageous to survival; the great majority are harmful, and some are neutral. It is thus quite unlikely that mutations direct evolutionary change along the lines that it actually follows. What directs the course of evolution is natural selection. Furthermore, natural selection does not perfect an organ or organism in the sense that it selects a variant with a forethought. Natural selection favors individuals with the traits favorable to that environment. If the variation is neutral, natural selection has no effect on it. Only if the variation is either beneficial or deleterious, is it favored or eliminated by natural selection through the survival of the individuals bearing the variant traits.

When a certain mutation or a combination of mutations increases the chances that an individual bearing the traits will leave a larger number of fertile offspring, the fraction of the population containing the variant allele or alleles tends to increase. By competition, the individuals lacking this trait are ultimately eliminated. In this way, a new subspecies may emerge, depending upon the magnitude of the changes and the degree to which the variants exchange genes with the main population.

## ISOLATION AIDS SPECIATION

What are the factors that help establish and maintain a mutant variety until it becomes a new species? Consider the Galapagos Islands, some 600 miles off the coast of Ecuador, where there are 13 species of finches (Fig. 27-14). Some of these finches are specialized to feed on cacti, others on various types of seeds, and still others on insects. Some of the insectivorous finches have even evolved behavioral features common to woodpeckers, including the ability to climb up and down tree trunks and to probe into the bark for insect larvae with the aid of a twig held in their beaks. None of these birds is similar in appearance and habit to the seed-eating finches of the South American mainland from which they most probably originated.

(a)

(b)

**FIG. 27-14** One of 13 different kinds of Galapagos finches (a). Although at first sight all the species may appear to resemble each other, each species is characterized by having a distinctive bill shape. The shape is correlated with the bird's diet. Three characteristic shapes are shown in (b).

It would appear that a few typical seed-eating finches, or perhaps even a single previously mated female, had reached the Galapagos Islands from the mainland at some time in the distant past. In the absence of competition from most types of other birds, this population evolved into diverse species that adapted to many of the environmental situations, termed **ecological niches,** available on these islands. Some of the mutations that made these adaptations possible had probably also arisen occasionally among mainland finches, but only in the absence of competition from other highly adapted birds would such new mutant types be able to succeed in evolving into new species.

In the absence of gene flow between any two isolated populations of a species, in the course of time, the populations usually diverge into separate species. Because mutations occur at random, two isolated populations of the same species tend to accumulate different mutations by chance alone. This tends to occur irrespective of whether the two environments are similar. Eventually the number of such mutations reaches a point at which the two populations differ so much that they are no longer able to interbreed. At this point, they have become separate species.

The availability of ecological niches that differ in climatic or biological features (including prey, predators, and competing organisms of other species) or both can also enhance speciation. Consider again the Galapagos finches. The first few generations of finches to populate the islands had very little competition, either from each other, or from almost any other kind of birds. They had plenty of seeds on which to feed, and in the absence of competition and predators, they greatly increased in number. Eventually, however, as their numbers grew, intraspecific struggle for survival occurred. Then, only those individuals that procured sufficient food and breeding grounds were able to reproduce. At this point, any mutations that allowed birds to feed on underutilized kinds of seeds or other food materials, such as insects, would increase the chances of survival of the individuals bearing the mutations. Such mutant individuals would then also increase in numbers until they, too, would begin to compete with each other. Further speciation would then involve the type of seeds or insects that these finches fed on, or whether they fed on the ground or in the treetops, or in arid regions instead of well-watered areas, and so on. This process of diversification could go on for millions of years until all the ecological niches suitable for occupation by small birds such as finches were filled. Finch evolution is probably continuing today on the relatively young Galapagos Islands.

In addition to the isolation provided by islands, mountain ranges and deserts also serve as barriers for organisms. The Himalayas, for example, isolated populations of organisms in the Indian subcontinent from those of the rest of Eurasia. From the Oligocene (about 50 million years ago) when the Himalayas began to form, the populations on either side of it have been isolated from each other. Similarly, the formation of the Sahara desert isolated the flora and fauna of sub-Saharan Africa from that of north Africa.

In the absence of gene flow, each population evolves in independent directions. Reproductive isolation of populations may also occur because of mutations that affect mating behavior or mating season. Under these conditions, new species may arise when populations become separated from the parental population while both are occupying the same region. This type of speciation, called **sympatric speciation,** differs from **allopatric speciation** in that in the latter type the two populations that evolved into sibling species are geographically separated.

## GENETIC DRIFT

Occasionally, as a result of chance events in which natural selection is not involved, changes can occur in gene frequency, leading to speciation. This process, termed **genetic drift,** can result when a small, nonrepresentative population, often one of less than 100 individuals, becomes isolated from its parent population. A particular allele of a gene or genes present in 75 percent of the parent population might, by chance, be present in all members of the smaller group. Alternatively, all members of the population may lack that allele but possess the other allele instead. Genetic drift, then, is a kind of sampling error due to the smallness of the sample; therefore, speciation resulting from genetic drift is not due to natural selection.

Genetic drift is believed to accelerate speciation by giving a newly isolated population a "head start" in an evolutionary direction away from that of its parent population. For example, if the first finches to arrive on the Galapagos Islands had, by chance, been quite different from the great majority of mainland finches, they would have been well on their way to forming at least one new species. Such cases in which the first members of a species to populate an isolated area are markedly different from most of their parent population and subsequently give rise to a new population of individuals like themselves is known as the **founder effect.** It is a dramatic example of genetic drift.

Although few have argued with the validity of the concept of genetic drift, that is, whether it could occur, exactly how important genetic drift has been in evolution is a matter of controversy.

## THE HARDY-WEINBERG LAW OF GENETIC VARIATION

A new species is said to have arisen, according to the views described in the preceding pages, when two populations of a species can no longer share the same gene pool. If natural selection is supposed to eliminate the alleles that are deleterious, how is heterogeneity in the genetic composition of individuals maintained in a population? Sir Godfrey Hardy, a British statistician, and Wilhelm Weinberg, a German physician, independently discovered a mathematical relationship between the genotypes and the frequency of distribution of alleles within randomly interbreeding populations. This relationship is called the Hardy-Weinberg law.

Suppose that two alleles, *A* and *a*, can occur at a particular chromosomal locus and that the fractions of the two alleles in a population are $p$ and $q$, respectively. It can then be said that the total of the two alleles is $p + q = 1$. Assuming that mating is random, the probability that an individual receives allele *A* from one of the parents is proportionate to the fraction of the population bearing this allele, which in this case is $p$. The probability of receiving this same allele *A* from both parents is $p \times p$, or $p^2$. Similarly, the probability that an individual receives the other allele, *a*, at that particular locus from both parents is $q \times q = q^2$. Thus, $p^2$ and $q^2$ represent the fraction of individuals homozygous for the alleles *A* and *a*, respectively. The fraction of individuals that receive allele *A* from one parent and the allele *a* from the other parent is $2pq$. Hence, in a randomly breeding population, the proportion of the genotypes *AA*, *aa*, and *Aa* are $p^2$, $q^2$, and $2pq$, respectively.

If the fraction of the population with allele *A* is 0.5, then the fraction bearing allele *a* will be 0.5. The probable genotypes of the succeeding generation are then 0.25 (*AA*), 0.25 (*aa*), and 0.50 (*Aa*). The frequency of the *A* (dominant) phenotype in this case is 0.75, and that of the *a* (recessive) phenotype is 0.25. This mathematical relationship allows us to predict that if both phenotypes have equal chances of survival, the gene frequencies will remain the same as that of the parental generation; such a condition is referred to as a **Hardy-Weinberg equilibrium.**

However, if a particular phenotype is subject to selection pressure, the fraction of the population bearing the less suitable allele will gradually decrease. Suppose that in our example, 20 percent of the recessive phenotype and 100 percent of the dominant phenotype survive. Under these conditions, the proportion of surviving phenotypes will be 94 to 6. Yet the frequency of distribution of the two alleles is 0.63 to 0.37. (The calculations are shown in Table 27-2.) In one generation, the deleterious recessive alleles decreased only 25 percent even though 80 percent of

**TABLE 27-2 Frequency Distribution of Allelic Genes in $F_1$ Generation Under Various Levels of Selection Pressure**

| | *Phenotype* | | *Genotype* | | | *Gene frequency* | |
|---|---|---|---|---|---|---|---|
| | Wild type | Recessive mutant | *AA* | *Aa* | *aa* | *A* | *a* |
| Parental generation | 0.75 | 0.25 | 0.25 | 0.5 | 0.25 | 0.5 | 0.5 |
| Both phenotypes survive equally well | 0.75 | 0.25 | 0.25 | 0.5 | 0.25 | 0.5 | 0.5 |
| 20% of the recessive phenotype survive | 0.94 | 0.06 | 0.32 | 0.62 | 0.06 | 0.63 | 0.37 |
| Recessive phenotype lethal | 1.00 | 0.0 | 0.34 | 0.66 | 0.0 | 0.67 | 0.33 |

the recessive phenotype died without leaving progeny. Even if the recessive phenotype is lethal, and no individuals of that phenotype survive to reproductive age, the frequency of distribution of the recessive allele would be 0.33 (see Table 27-2). In other words, even a lethal allele can remain in the population for a large number of generations.

The Hardy-Weinberg law can also explain the gradual change in the phenotype of a population following the appearance of an advantageous mutation. For example, in the right environment, a single mutation, such as the dark pigment pattern in the moths discussed earlier in this chapter, can confer a selective advantage on the phenotype. Under such conditions, the frequency with which the advantageous allele appears in successive generations slowly increases. The British biologist J. B. S. Haldane calculated that a dominant beneficial mutant allele that appeared by chance would increase gradually and become the common phenotype in about 300 generations. Therefore, if the generation time of a moth is 1 year, in 300 years a mutant phenotype that was once unique in the population could become the most common one. Yet the mutant allele would not be expected to completely replace the recessive allele for many more generations. Assuming the human generation time to be about 25 years, a rare, selectively advantageous mutant allele would probably become the common allele in 7,500 years.

Thus, deleterious alleles can persist almost indefinitely in natural populations, and the gene frequencies of advantageous alleles increase very slowly.

## MOLECULAR DRIVE

Natural selection explains evolutionary changes in regard to an organism's mode of adaptation to its environment; genetic drift explains fixation of mutations that are not necessarily adaptive; and the Hardy-Weinberg law explains the tendency in a population to maintain its genetic heterogeneity over many generations. But these theories do not explain the genetic mechanisms underlying speciation, the origin of new species. A species, as defined above, is a population whose members are capable of interbreeding, whereas speciation is diversification of one such population into two or more distinct populations incapable of interbreeding with each other to produce fertile offspring. Since the genes of one species cannot be exchanged with those of another species, two sibling species represent a genetic discontinuity in the gene pool of the original, or parental, population. In 1982, Gabriel Dover, a British geneticist, proposed a novel genetic mechanism, called **molecular drive,** as one explanation of the origin of evolutionary novelty and speciation. Molecular drive is a fixation of variants in a population by genetic mechanisms that cause rearrangement of genes in the genome.

It is now generally accepted that in each eukaryotic species and some primitive bacteria a substantial fraction of the genome is represented by multicopy families of genes. Some of these multicopy genes code for proteins (e.g., histones and immunoglobulins), some are transcribed into functional RNAs (e.g., ribosomal RNA), whereas others have no known function. The multicopy ribosomal genes exhibit a remarkable sequence homogeneity, usually possessing identical nucleotide sequences for all individuals of the species and even among individuals of closely related species. However, the spacer DNA between neighboring ribosomal genes on a chromosome, although homogeneous within a species, is substantially different among closely related species. How do all individuals of a newly diverged species come to bear the same mutant form of DNA?

The position of genes on eukaryotic chromosomes is in constant flux. The genes undergo continual, albeit slow, change in position by transposition and translocation. The number of copies of multicopy genes in the genome may also be affected by unequal exchange between chromatids during crossing over. Moreover, in a process called gene conversion, a gene is sometimes favored over its allele, that is, one allele may be replicated more frequently than the other, resulting in a non-Mendelian distribution of the alleles in offspring. As a result of gene conversion, a heterozygous individual (Aa) may produce gametes in the ratio of 3A:1a instead of in the usual ratio of 2A:2a. According to the theory of molecular drive, the genetic mechanisms of transposition, translocation, unequal crossing over, and gene conversion could affect fixation of variant genes in a population and thus affect speciation.

The biological consequence of molecular drive is to maximize homogeneity of the gene pool within a species while increasing interspecific discontinuity in DNA sequence homology. Because of such genetic discontinuity, individuals belonging to different species cannot produce fertile offspring. When two individuals differing, for example, in the spacer DNA present between the ribosomal genes mate, they may produce a hybrid individual. But because of marked differences in nucleotide sequence of paternal and maternal chromosomes, homologous chromosomes of the hybrids fail to pair at meiosis and consequently cannot produce viable gametes. Thus molecular drive, which makes all members of a breeding population homogeneous with reference to the variant DNA sequence, would aid in speciation. But the precise factors involved in bringing about the specific changes in the genome remain to be elucidated.

## NATURAL SELECTION AND THE COURSE OF EVOLUTION

The theory of evolution by natural selection predicts the evolution of new species as well as that of new genera, families, orders, classes, and phyla. In discussing the role of mutations, isolation, and genetic drift in evolution, we were mainly concerned with speciation, both sympatric and allopatric. In these cases, we explained the probable mechanisms of origin of diverse species belonging to a single genus. Evolutionary theories should also explain the origin of phyla. This type of evolution is usually called **macroevolution.**

In the study of macroevolution we focus on the origin of major groups, such as arthropods and molluscs from annelids, chordates from invertebrates, and vertebrates from protochordates. Even macroevolution, despite a different terminology, is evolution at the species level. But the difference is that the changes occurred over millions of years when environmental conditions, both physical and biological, were often much different from those of today.

One must always bear in mind that natural selection has no direction. Humans often regard themselves as the culmination of evolutionary progress. But there is no factual basis to believe that "progress" or progressive improvement is a goal of evolution. Appearance of a specific trait at any one time in the history of evolution is a chance event. Selection of the specific trait is merely an expression of its suitability for a specific time and place.

Let us examine the probable sequence of events and the possible role of natural selection in the course of evolution in the vertebrates. All evidence supports the evolution of terrestrial vertebrates from fishlike ancestors. What needs to be explained is the evolution of such mammalian characteristics as hair and mammary glands; and such primate traits as bipedal gait, binocular vision, and opposable thumb (the first digit of the human hand can be placed opposite the others in the hand); and such unique traits as speech, written communication, and intelligence.

It is generally believed that at least four major evolutionary phases would have had to occur in sequence for humans to have arisen from fishes. Specifically, certain characteristics should have evolved in order: (1) tetrapod limbs, lungs, dry skin, and associated changes that would enable an animal to live on land; (2) an embryo enclosed in fluid-filled embryonic membranes and an egg shell, enabling the organisms to breed out of water; (3) homeothermy (warm-bloodedness), combined with a placental mode of nourishing the embryo and extensive parental care; and (4) a bipedal gait, freeing the forelimbs from their role in locomotion and thereby enabling the eventual development of the human brain to its present state. How might these events have occurred as a result of random variation, most of which are known to be either neutral or harmful?

The Devonian epoch was characterized by a great variety of fishes, some of which lived in fresh water. Since ponds and streams are subject to drying up, any mutations that bestowed upon a fish the ability to breath air or to move about on land would increase its chances of survival. The fossil record, as well as our knowledge of present-day fishes, reveals that various groups of fishes exhibited one or more of such accessory air breathing organs. For example, many modern tropical freshwater fishes have accessory respiratory organs that they use in aerial respiration. Examples of such organs include modified gill chambers, esophagus, rectum, and lunglike evaginations from the esophagus. Lungs have thus been only one of several types of organs that evolved as adaptations to aerial respiration.

Evolution of lungs in the group of fishes that eventually also developed tetrapod limbs and thus gave rise to the first amphibians would, like other evolutionary developments, have also resulted from chance events. The first fossil amphibians appear in Devonian strata and are intermediate between certain fishes and modern amphibians in their skeletal features. During the next 50 million years, several types of amphibians evolved and began to utilize the resources of the land adjoining the water masses. Yet their terrestrial adaptation was incomplete. They had to return to water for breeding, as they must to this day. Early amphibian fossils are difficult to distinguish from those of the ancestral fishes. The evolutionary change may be of the same magnitude as that between two species. But one species could move on land. Thus the two species became reproductively isolated.

During the Carboniferous epoch, some 345 million years ago, the first reptiles arose. Although we have no fossil record of any intermediate egg types between those of reptiles and amphibians, there is abundant evidence of a gradual change from the amphibian skeletal system and skin to those of land-adapted reptiles, such as snakes and lizards. Several different evolutionary lines can be recognized from fossil records of the early amphibians, one of which culminated in the reptiles.

Having acquired a completely terrestrial mode of life and finding no rivals on land, reptiles rapidly diversified into many kinds, including the giant dinosaurs. The presence of land plants, mainly ferns and gymnosperms, provided a source of nutrients. But there were problems. Unlike the temperatures of water, land temperatures fluctuated widely. Reptiles, being poikilothermic (cold-blooded), had a low metabolic rate during cool nights and the colder seasons of the year. Moreover, vast areas of the globe where cold temperatures prevailed were uninhabitable by cold-blooded animals. The occurrence of mutations that allowed them to maintain a warm body temperature at all times would therefore have been of great advantage to these animals. It would not only make them capable of responding to emergencies at all times but would enable them to colonize areas as far north or south as the fringes of the polar ice caps and to survive periods of glaciation.

Homeothermy seems to have evolved independently in many groups of reptiles. Some dinosaurs and related reptiles are believed to have been homeothermic. Eventually, as the fossil record indicates, some reptiles evolved into birds and others into mammals.

One of the last major changes that culminated in the emergence of *Homo sapiens* was a bipedal gait. By the end of the Cretaceous period, the dinosaurs and other great reptiles, for some as yet unknown reason, became extinct, and the flowering plants had become the dominant group of plants. Because of their potential superiority over reptiles, probably by virtue of their homeothermy, mammals eventually emerged as the dominant land vertebrate group with few competitors. In the absence of competition, mammals evolved in a variety of directions, not only on land but even in the water and the air. The evolution of large trees provided a specialized environment for any animals that could adapt to a tree-living, or arboreal habitat. One group of mammals to evolve in this direction was the primates, the group that came to include monkeys, apes, and humans. Among the evolutionary changes that adapted the primates to arboreal life between 30 and 65 million years ago were modifications of the hands, feet, and eyes. Then, 15 to 20 million years ago, a change in climate appears to have occurred in Africa, resulting in the widespread disappearance of forests and the establishment of extensive grasslands, thereby creating a new environment to which anthropoids could adapt. Mutations that resulted in bipedal locomotion would have adaptive value for such animals. It was from such anthropoid, or humanlike, primates that human beings, *Homo sapiens,* arose only a few million years ago.

In the evolution of the vertebrates, as with all other groups, not one, but many different phenotypes appeared at each phase of evolutionary change. However, within a competing group, only phenotypes most suited for a particular habitat will survive and successfully propagate. There is no reason to believe that each of the events described above was predetermined to occur. Rather, what occurred was a natural selection, by environmental conditions, of the best-adapted organisms at each stage over some 400 million years. From the age of fishes to the age of humans, each selection was made from among the variety of mutant forms in existence at each stage.

## FOSSIL EVIDENCE OF THE EVOLUTION OF *HOMO SAPIENS*

Anatomically and physiologically, humans possess many typically mammalian traits, including homeothermy, hair, mammary glands, mammalian types of teeth, and the bearing of live young. *Homo sapiens* is classified in the order primates because of the presence of an opposa-

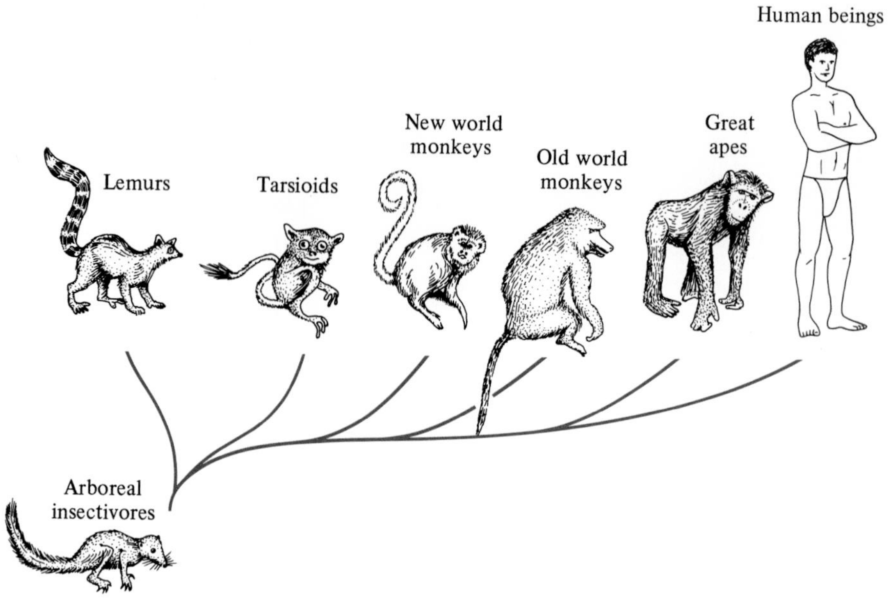


**FIG. 27-15** Evolutionary relationship between different primate groups.

ble thumb, nails instead of claws, and binocular vision. Besides humans, the primates include apes, new world (American) and old world monkeys, lemurs, and other close relatives such as tarsiers (Fig. 27-15). As in all other primates, the limbs in humans are not specialized, and the olfactory organs are poorly developed. Perhaps the most unique anatomical feature of the advanced primates is complexity of brain structure. Of all the primates, the anthropoid apes—chimpanzees, orangutans, gorillas, and gibbons—resemble humans most closely in anatomical and physiological features.

Of all the great apes, our closest relatives are the chimpanzees (Fig. 27-16). This closeness is evidenced in numerous ways. For example, the amino acid sequence of the hemoglobin molecules in humans and chimpanzees do not differ at all. Humans and gorillas differ in regard to only two of the 141 amino acids in hemoglobin. Differences between apes and humans in terms of the molecular structure of cytochrome c, insulin, and serum albumins are also minimal. All such close similarities in molecular structure of proteins reflect an evolutionary closeness. Despite all these similarities, however, it is obvious that apes are distinctly different from humans in their anatomy and behavior. One of the most conspicuous differences is the completely bipedal gait in humans, with only the hindlimbs being used for locomotion; apes usually walk on all fours, using the knuckles of the hands. Associated with the evolution of a bipedal gait are changes in the arrangement of the bones of the vertebral column and pelvic girdle. Many other minor anatomical differences between apes and humans are associated with this basic difference in gait.

Another striking difference in anatomy between humans and apes is the disparity in the size of the brain. The human brain is much larger, per kilogram of body weight, than that of the apes. Associated with this are two other unique features of humans: their communication abilities and mental activities. Many other animals, such as chimpanzees, porpoises, and insects, exhibit some form of language, but no other animal has complex language or a culture, the ability to train, develop and refine mind and manners.

(a)

(b)

**FIG. 27-16** Chimpanzee, the closest primate to humans, (a) holding a baby and (b) using a twig as a tool.

## THE EVOLUTIONARY HISTORY OF *HOMO SAPIENS*

Human evolutionary history may be arbitrarily divided into two periods. The first period involves the evolution of the prehominid primates, and the second the evolution of hominids, or members of the family Hominidae, that is, humans and their immediate predecessors.

***Early History of the Primates.*** The species *Homo sapiens* may be partially defined in its anatomical characters as follows: a cranial capacity of about 1,350 cc (82 cubic inches); a nearly vertical face, with jaws and teeth that are small compared to those of the apes; small, flat canine teeth that do not overlap to form shears; a pointed chin; a rounded occipital region (back of the skull) with only a limited area for the attachment of neck and jaw muscles; skeletal features associated with an erect posture; and bipedal locomotion. These human characteristics are limited to skeletal features because the study of fossils can yield abundant information only about skeletal elements. Information regarding human activity, such as tools and signs of a culture, is also regarded as essential to the classification of a particular fossil hominid as *Homo sapiens*. By using the skeletal features and the signs of a culture as identifying characteristics, paleoanthropologists, students of fossil humans, have been able to trace the evolution of *Homo sapiens*.

Mammals appeared toward the end of the Mesozoic era, about 180 million years ago. But the major radiation of the mammalian stock took place during the Tertiary period of the Cenozoic era. Near the beginning of the Tertiary period—65 to 70 million years ago—primates were represented by shrewlike animals that barely fit the definition of primates. Only because they have nails instead of claws and an arrangement of eyes that provides binocular vision do today's tree shrews fit the primate description. But most of the shrewlike primates of the past possessed chisel-shaped incisors, like that of rodents, and probably competed with rodents for food and shelter. Nearly all of these early rodentlike primates lost the struggle for survival and gradually became extinct. However, some evolved into arboreal forms, thereby probably saving the primate stock from total extinction. During the Paleocene and Eocene epochs (which lasted about 26 million years) the primate stock evolved into lemur- and tarsier-like forms (Fig. 27-15). No hint of the characters typical of monkeys, apes, and humans had yet appeared in these primates of 38 million years ago.

Evidence of the presence of primitive monkeys and primitive apes during the Oligocene epoch (38 to 26 million years ago) has come from fossils excavated from deposits in the Fayum region of Egypt. Fossils of two genera believed to be ancestral to modern monkeys were collected from these deposits. From the same epoch, other fossil genera, one believed to be an ancestral gibbon and another that may have been ancestral to the modern apes, were also collected.

In 1856, a large fossil ape was collected from Miocene (26 to 7 million years ago) deposits in France. This specimen, called the oak ape, *Dryopithecus*, has typical apelike features, large canine teeth, short legs, and long arms. However, these features in *Dryopithecus* are not as accentuated as in the modern apes. It is not agreed whether the population of primitive apes represented by *Dryopithecus* could be our direct ancestors or merely an offshoot of the ancestral stock from which hominids and pongids (members of the ape family) both evolved.

***The First Hominids.*** The exact age of the hominid family is uncertain. But a fragment of an upper jaw fossilized in the Siwalik hills of India some 6 to 10 million years ago has features of a primitive hominid. The significance of this fragment, assigned to the genus *Ramapithecus* at the time of its discovery in 1934, was not recognized until more fossils belonging to the same genus were later collected from western Asia, Ethiopia, and East Africa. Examination of these fossils showed that *Ramapithecus* had a short face, thickly enameled cheek (molar and premolar) teeth, and rounded lower jaw. However, *Ramapithecus* and its relative *Sivapithecus* possessed many pongid traits: they had large teeth and small brain cases and possibly walked on their knuckles. The dating of these specimens has also been controversial. The Indian specimen was attributed to the upper Miocene epoch (about 10 million years ago), whereas the East African and Kenyan specimens were placed in the early Pliocene epoch (about 7 million years ago). *Ramapithecus* is thus the earliest hominid candidate (Fig. 27-17). Another fossil, *Proconsul africanus,* dating from 17 to 22 million years ago, is also being considered as hominid. But it has many apelike or monkeylike features, including its elbow and shoulder joints, wrist, and vertebrae. Yet, other of its features suggest hominid relationship.

Even if we were to accept the suggestion that *Ramapithecus* represented an early phase of hominid evolution, many anatomical features of hominids were yet to evolve. *Australopithecus* (Fig. 27-17), a fossil find first announced by Raymond A. Dart in 1925, represents such an intermediate stage between the ramapithecines and the genus *Homo*.

**Australopithecus.** Dart, of the University of Witswatersrand, in South Africa, had received a shipment of fossils in 1924 from a quarry at Taung, in South Africa. These fossil-bearing rocks included a fossilized brain cast, or mold, and a skull embedded in rock. The skull proved to be that of a 6-year-old child with a full set of deciduous ("milk") teeth. The brain case could not have belonged to any ape because its capacity was three times that of a modern baboon and larger than that of an adult chimpanzee. The face of the skull exhibited a distinct forehead instead of the receding brow typical of apes. The upper jaw was short and retracted under the skull, another hominid feature. Although the child had deciduous teeth, its permanent teeth were about to emerge at the time of its death. The canine teeth were relatively small, as in humans, and, from its shape, the skull would have been positioned on the vertebral column in a manner suggestive of true bipedal gait and posture. Dart created a new genus and species, *Australopithecus africanus* (from the Latin *australis,* "south," and *pithecus,* "ape"), to include individuals answering this description.

*Australopithecus* attracted the attention of many other anthropologists, including Louis S. B. Leakey, who subsequently contributed substantially to our knowledge of human evolution. Leakey's studies in the Olduvai gorge, in what is now Tanzania, yielded many fossils that showed East Africa to be the cradle of human evolution. In 1937 and thereafter, many more fossils with the same general features as *Australopithecus* were discovered at various sites in Transvaal, South Africa. In 1959, similar australopithecine fossils were discovered from late Pleistocene (about 1.7 million years ago) beds at the Olduvai gorge. More recently, similar fossils were found in the Pliocene beds of the Omo and Afar regions of southern Ethiopia and in northern Kenya. In 1979, skeletal remains of several hominids were found in Ethiopia and named *Australopithecus afarensis.* These were radiometrically dated as 3 million years old and are the oldest undisputed hominids yet uncovered. More

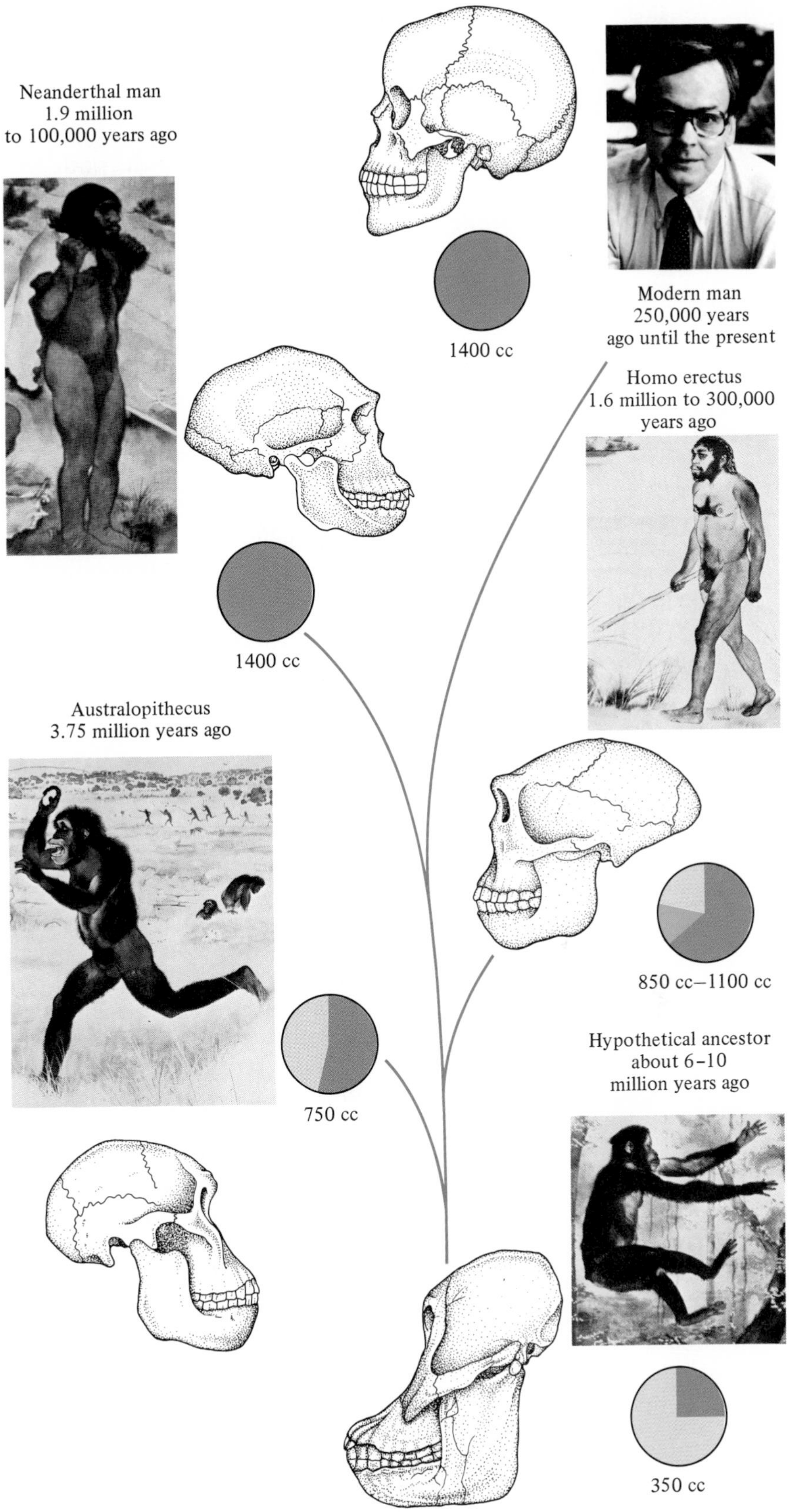


**FIG. 27-17** Evolutionary history of *Homo sapiens*. Reconstructions of the body and of the skull, geological age, and brain capacity of a hypothetical ancestor, *Australopithecus*, *Homo erectus*, *Homo neanderthalensis*, and *Homo sapiens*.

important, *Australopithecus afarensis* is believed to be ancestral to all later hominids of the genus *Homo* and to all other australopithecines.

The major characteristics of australopithecines are a cranial capacity equivalent to that of the modern large apes (450 cc) but probably exceeding that of apes in proportion to body size. In regard to their large jaws, large molar and premolar teeth, and large pelvis and limb bones (Fig. 27-17), *Australopithecus* resembles the apes while differing from modern *Homo sapiens*. However, australopithecines also exhibit many features not found in apes, including the small size of the bulge of the occipital region of the skull, the height of the cranium, the arrangement of the bones at the base of the skull, and the small size of the canine and incisor teeth. The features of the pelvis and thigh bones combined with those of the base of the skull show that *Australopithecus* was probably adapted for erect, bipedal posture and gait. *A. afarensis* was between 4 and 5 feet tall.

Many other Pleistocene fossil specimens with similar features have been collected from sites in eastern and southern Africa. Although paleontologists have created new genera to accommodate many of these finds, prevailing opinion holds that all belong to the genus *Australopithecus* and may be mere geographical variations of a single species of that genus.

Two varieties of australopithecine fossils, called **gracile** and **robust,** can be clearly distinguished. The gracile variety, called *Australopithecus africanus,* was a small, upright, bipedal hominid with relatively light skeletal elements. The robust variety, *Australopithecus robustus,* was less efficiently bipedal and was more adapted to a vegetarian diet. Some paleontologists, however, have suggested that the two varieties belong to two separate genera and have named the gracile fossils *Homo habilis,* suggesting that it was an immediate ancestor of modern humans.

The argument in favor of including such individuals in the Hominidae and assigning one of them to the species *H. habilis* is strengthened by the finding of tools in the Olduvai deposits of Tanzania along with the bones of australopithecines.

**Homo Erectus: *The Forerunner of Modern Humans.*** In 1891, Eugene Dubois, a physician with the Royal Dutch Army in Sumatra (now part of Indonesia) was seeking fossil remains of humanoid ancestors on the island of Java. Dubois had heard from early Dutch travelers and traders that fossil human bones were being found in the Indonesian islands, particularly on Java. He suspected that the cradle of human evolution might be in these islands. In September 1891, his search uncovered a chocolate-brown cranium of an apelike creature. Although the cranium was too low, flat, and small to be that of a modern human, the contours of the skull were such that it could not have belonged to a modern ape. Dubois continued his excavations and the next year found a thigh bone, which he recognized as typically human. After a 2-year study of these bones, Dubois announced that the humanlike leg bone and the apelike cranium belonged to the same type of individual and named it *Pithecanthropus* (from the Greek *pithekos,* ape, and *anthropos,* man, "ape-man"). Colloquially, it has been referred to as Java man. Dubois' announcement was roundly derided by antievolutionists but greeted by others as "the missing link."

In 1929, while exploring caves for human fossils 40 miles southwest of Peking, a team lead by Canadian physician Davidson Black found a fossilized prehuman skull cap at Choukoutien. The skull and numerous fossil teeth were designated to be from *Sinanthropus pekinensis* ("Chinese man of Peking"), colloquially called Peking man. The skull of Peking man exhibited many apelike features but housed a brain of about 1,000 cc, much more human than apelike (average brain size in the gorilla is 450 cc). The many cultural artifacts found with Peking man fossils firmly places it in the human family.

Subsequent study of the casts and bones of specimens of Peking man and Java man have shown that the two do not differ from each other more than do different races of modern mankind. The two specimens were later classified as species of the genus *Pithecanthropus: P. erectus* for Java man and *P. sinensis* for Peking man. The deposits in which they were found have been estimated to be 1.9 million years old in the case of Java man and between 500,000 and 200,000 years ago in the case of Peking man.

In a recent revision of nomenclature, all pithecanthropine individuals have been assigned to the species *Homo erectus* (Fig. 27-17), as intermediate between *Homo habilis* (the gracile australopithecines) and *Homo sapiens* (modern humans).

Numerous specimens assigned to *Homo erectus* have also been found in Europe and Africa. The so-called Heidelberg jaw, the jaw of an individual that lived between 700,000 and 100,000 years ago, found in a sand pit at Mauer, Germany, in 1907, is one example. Similar jaws were found in 1954 at Ternifine, Algeria, and at Casablanca, Morocco. The Casablanca specimen appears to be only 200,000 to 100,000 years old. Other such specimens have been found in the Olduvai gorge and in South Africa. Crude tools have been found in association with many of these *Homo erectus* fossils. The species is believed to have lived in Eurasia and Africa from about 1.9 million years ago to about 100,000 years ago.

***The Antiquity of* Homo Sapiens.** Although all skeletal features of *H. erectus* suggest that some advanced members of this species could be ancestors of modern humans, it is very difficult to determine with any degree of exactness when the species *Homo sapiens* actually arose. Among the more important finds that bear on the question are the fossil remains of humans found in 1965 near Budapest, Hungary. Radiometric evidence suggests that they lived some 350,000 years ago. Vertesszollos man, as it is called, appears to be an intermediate stage between *Homo erectus* and *Homo sapiens*.

Other fossils of modern humans dating back 200,000 years are the so-called Swanscombe man, discovered in gravel beds at Swanscombe, in Kent, England, and Steinheim man, found in 1933 near Stuttgart, West Germany. Many other more recent fossil human bones have been collected from the late Pleistocene beds of France, Italy, and Germany. The gene pool of modern humans apparently emerged from these populations of near-modern humans about 250,000 years ago.

***Neanderthal Man.*** In the Neander valley near Düsseldorf, Germany, the remains of many human bones were found beginning in 1856; these came to be known as Neanderthal man (Fig. 27-17). The Neanderthals were a short, thickset people with powerful muscles who appear to have been well adapted to cold conditions. They were cave dwellers who used fire and made flint tools, which they used in hunting. They were large-brained, with thick cranial bones and an elongate skull but with a low cranial vault and prominent brow ridges, as in *Homo erectus*. Their faces sloped instead of rising abruptly, and the cheek arches sloped backward, giving the appearance of high cheekbones. Judging from the ritualized way in which their dead were buried, the Neanderthals might have had concepts of religion in which the cave bear seems to have played a prominent part.

Neanderthal fossils have been found in fossil beds in France, Italy, Belgium, Czechoslovakia, Greece, Russia, North Africa, and western Asia (the Middle East). The Neanderthals were apparently a successful group that lived for many thousands of years in Eurasia and Africa. Their fossil record, however, suddenly ended with strata more recent than 40,000 years ago. In Europe, Neanderthals were replaced by Cro-Magnon man. Cro-Magnon fossils have been found in France, Italy, England, and many other places. The Cro-Magnon people had long skulls with upright foreheads and distinctive chins. They were bipedal, erect people with a modern skeletal form. They had a complex culture and have left us a legacy of beautiful artwork in the caves they occupied.

The causes for the sudden disappearance of Neanderthal people are unknown. Several theories have been offered to explain their extinction. Carleton S. Coon, of the University of Pennsylvania, has suggested that Neanderthals might have succumbed to the bitter cold of the last glacial age. Others have hypothesized that Neanderthals might have been killed by the more efficient newcomers, such as Cro-Magnon man, just as the Australian Bushmen and the American Indians were nearly all killed by European settlers. Other theories attribute their demise to diseases brought by the newcomers. One theory suggests that Neanderthal genes might have been absorbed into the gene pool of modern man by interbreeding. If the last explanation is true, then the Neanderthals are in the direct line of ancestry of modern humans.

Despite the great diversity in fossil records of human ancestors, evidence from human mitochondrial DNA sequence analysis suggests that hominids passed through a genetic bottleneck about 500,000 to a million years ago. In other words, all present members of the species *Homo sapiens* evolved from a geographically restricted, possibly small population. Fossil data is incomplete, but the available fossil evidence suggests a sub-Saharan African origin for the modern human. By about 30,000 years ago, humans had evolved cultural traits that definitely identify them as modern *Homo sapiens*. The more precise course of evolution of *Homo sapiens* from a million years ago to about 50,000 years ago can be reconstructed only when more fossil evidence becomes available.

## Summary

The concept of evolution by natural selection, put forth by Charles Darwin and Alfred Russel Wallace in 1858, revolutionized scientific concepts regarding the origin of diversity in organisms. In predarwinian times, most people believed that each species or genus had an independent origin and was probably created as such by supernatural force. According to Darwinian evolutionary theory, species are not immutable, and diversity in organisms arises by the slow accumulation of heritable variations, now called mutations. It is difficult to define a species precisely. A species is usually regarded as one or more populations whose individual members can potentially interbreed to produce fertile offspring. Every species exhibits great variety. However, only the individuals with traits that are favored by the specific environment in which they live will, in the long run, survive to reproduce.

Several lines of evidence support the evolutionary theory. Similarity in fundamental biological processes such as the chemical basis of heredity, mechanisms of RNA synthesis, information transfer and release of energy, and ATP formation shared by all organisms strongly suggest a common descent. Geological evidence for the successive appearance of more complex organisms in the course of our planet's history also clearly suggests an ancestor-descendant relationship between the earlier, simpler organisms and the more complex ones that followed. Geographic distribution of groups of organisms can be logically explained only by a limited degree of dispersal and organic evolution. Comparative anatomy and embryology also support the concept of common descent. The presence of vestigial organs and convergence in the anatomy of diverse organisms adapted to a specific mode of life also support the theory of evolution by natural selection.

The principal mechanisms of evolutionary change are fairly well understood. Mutation of genes, a largely random event, causes heritable variations in structure and function. Those variations that enhance the survival of an organism to reproductive age increase its chances of leaving offspring of the same type. The environment determines, or selects, which variations will persist. When the environment changes or an organism migrates to a new environment, the traits that determine survival may also change. Thus, the selection of suitable traits and the shaping of evolutionary progress is done by nature through the process known as natural selection.

Mutations provide the genetic basis of variations. Speciation is aided by geographic, climatic, or behavioral isolation of subpopulations of a species. Genetic drift, a chance isolation of a nonrepresentative subpopulation from its parent population, is also believed to contribute to speciation.

Human evolution appears, from all lines of evidence, to have followed a path similar to that of other species and to have been shaped by similar causal factors. Primate evolution possibly began about 65 million years ago with the origin of shrewlike animals. Not until the appearance of *Dryopithecus,* a primitive ape, 20 million years ago and *Ramapithecus,* at least 7 million years ago, did any primate exhibit traits typical of modern humans. The apparent descendants of *Ramapithecus,* distributed all over Eurasia and Africa, have been given the name *Australopithecus* and appear to be ancestors of *Homo sapiens.*

Australopithecines lived from about 4 million years ago to about 1.5 million years ago. Some might have used tools. All were bipedal. About 2.5 million years ago, populations of the primitive *Australopithecus afarensis* appear to have evolved into *Homo habilis,* which a million years later evolved into *Homo erectus. Homo sapiens,* modern humans, probably appeared about 250,000 years ago, possibly from a population of *Homo erectus.*

## Recommended Readings

DOBZHANSKY, T. J., F. J. AYALA, G. L. STEBBINS, and J. W. VALENTINE. 1977. *Evolution.* Freeman, San Francisco. An excellent account of various aspects of evolution.

GODFREY, L. R., ed. 1983. *Scientists Confront Creationism.* Norton, New York. Evidence from geology, anthropology, and astronomy is cohesively combined to explain evolution in this valuable guide.

JOHANSON, D. C., and T. D. WHITE. 1979. "A systematic assessment of early African hominids." *Science,* 203(1):321–330. A summary of all recent hominid fossils found in Tanzania and Ethiopia and their phylogenetic position in the evolution of *Homo sapiens.*

PILBEAM, D. 1984. "The descent of hominoids and hominids." *Scientific American,* 250(1):84–96. A clear account of the evolution of apes from Old World monkeys and of human beings from apes.

*Scientific American.* 1978. 239(3). This issue, exclusively devoted to evolution, is an excellent collection of well-written articles on various aspects of the subject by experts in the field.

WILSON, A. C. 1985. "The molecular basis of evolution." *Scientific American,* 253(4):164–173. A clear account of the new insights into evolution at the molecular and organismal levels.

UNIT VI

# Life's Diversity: The Products of Evolution

CHAPTERS



During Charles Darwin's renowned 5-year voyage aboard *H.M.S. Beagle,* he was deeply impressed by several truths: the great age of the earth, as described in Charles Lyell's newly published *Principles of Geology,* which Darwin took with him; the enormous and seemingly infinite variety of both living and fossil forms that he encountered around the world, especially in tropical regions, where the greatest variety occurs; the close resemblances, always accompanied by distinct differences, that he observed among mainland species and those on the nearest islands; and the same relationships—of resemblances with differences—that he saw between living forms and fossils in the most recent rock strata. The great diversity itself made a deep impression on Darwin. The resemblances bespoke "a deep organic bond," as he put it, between island species and those of the nearby mainland. He concluded that the bond was heredity: that these similar organisms were related to each other and had shared a common ancestor. Later, when Darwin noted that despite the ability of all organisms to outreproduce their supply of nutrients their populations remained about the same over the years, he concluded that those individuals from each generation best suited to life in a given environment were the ones "selected" by the environment to survive.

Being deeply impressed by life's great variety of adaptations to different environments was probably essential to the reasoning that culminated in Darwin's natural selection theory. Some 20 years later, the same kind of experience led to the same conclusions on the part of Alfred Russel Wallace. Earlier, even Linnaeus withdrew his insistence that all species were fixed as he encountered more variety accompanied by similarities. Perhaps if he had had the benefit of Darwin's knowledge of the age of the earth, he might have gone further in his reasoning that there had been some evolutionary change at the level of species.

What was true for Darwin and Wallace is probably true for modern students of biology: a knowledge of life's great variety and its fundamental similarities is essential for an appreciation of the extent of evolution. The brief descriptions of the major groups of organisms in the next five chapters may not alone provide such an appreciation, but they are a start.

# 28 Prokaryotes and Viruses

OUTLINE



Chapter 1 noted that a major evolutionary discontinuity separated all organisms into two groups: those with nuclei and cellular organelles, known as eukaryotes, and those without either, the prokaryotes. The prokaryotes include bacteria; some allied forms, such as the smaller mycoplasmas; and cyanobacteria, or blue-green bacteria, long classified as blue-green algae because, like the algae, they contain chlorophyll *a*. This chapter discusses the distribution and characteristics of the prokaryotes and their influence on humans and the environment. Viruses, self-replicating structures that grow only in living hosts, are also included in this chapter for convenience, even though they do not fit in the classification of living organisms.

## PROKARYOTES

The prokaryotes are the oldest group of organisms on earth and the most abundant. Fossil prokaryotes have been identified in Precambrian rocks 3 billion years old. Today, these single-celled organisms occupy all the environmental niches in which eukaryotes are found and some where no other organisms can survive. For example, some bacteria grow in the dark and near-freezing cold of ocean depths; others live in hot springs where the temperature approaches that of boiling water. Some cyanobacteria live in snow, and others can tolerate almost as much heat as thermophilic, or heat-loving, bacteria. They live at the edges of hot springs and form a trail of different species down the temperature gradient in the streams that flow from the springs, where the water temperature is well above that at which milk is pasteurized to limit bacterial growth.

## CHARACTERISTICS OF PROKARYOTIC CELLS

While the main distinction between prokaryotes and eukaryotes is the lack of nuclei and membrane-bound cellular organelles in prokaryotes, characteristics associated with these differences are significant in relating prokaryotes to the evolution of living organisms.

### THE PROKARYOTIC CHROMOSOME

The DNA of the prokaryotic cell is found in a central nuclear region, or **nucleoid,** that is not separated from the cytoplasm by a nuclear membrane (Figs. 28-1 and 28-2). Although referred to as a chromosome, the DNA lacks the associated RNA and proteins characteristic of the eukaryotic chromosome. The DNA of prokaryotes is a single, circular double-stranded molecule. Prokaryotes divide by simple fission: the DNA replicates, the strands separate, and the cytoplasm pinches in half. Lacking the spindle fibers that move the chromosomes in most eukaryotes, prokaryotes have another mechanism to separate their DNA. Apparently, the daughter DNA molecules are attached separately to the cell membrane and pull apart as the cell expands.

**FIG. 28-1** The bacterium *E. coli,* a typical prokaryote.

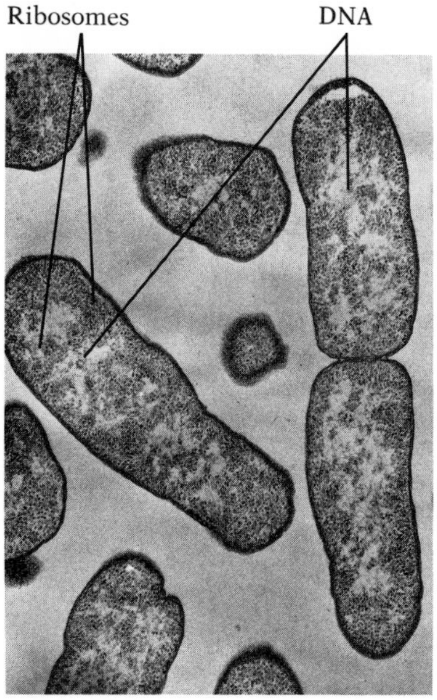


### CYTOPLASMIC COMPONENTS

Although the organelles characteristic of eukaryotes are missing from the prokaryote cytoplasm, ribosomes (smaller than those of eukaryotes) and membranes are present, as well as various cellular inclusions such as granules of stored materials (Figs. 28-1 and 28-2). The plasma membrane is similar to that of eukaryotic cells, except that it lacks steroids. Within the membrane is the electron transport system (respiratory chain) of the cell, which in eukaryotes is localized in the mitochondria. The similarity of bacteria to mitochondria has suggested the endosymbiont theory, which postulates that mitochondria originated by the invasion of a cell by a bacterium, which reproduced and continued to live as a symbiont in the host cell (see discussion of theory, Chapter 30). Some bacteria have photosynthetic pigments in the outer membrane, although in most photosynthetic prokaryotes the pigments are found in internal membranes of the cell. The photosynthetic membranes are not

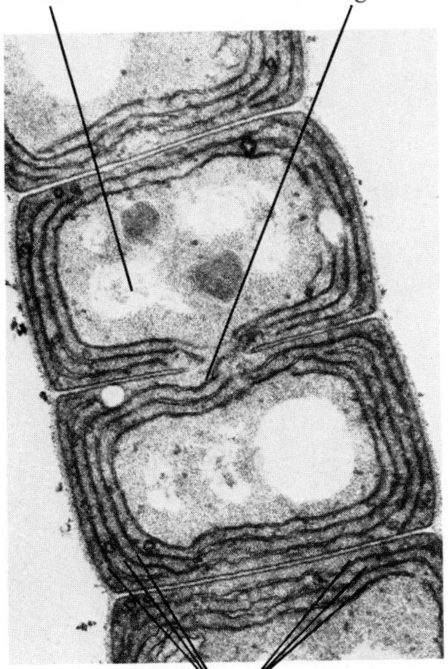


**FIG. 28-2** Electron micrograph of the cyanobacterium *Oscillaatoria.*

packaged in chloroplasts. In fact, the entire cell resembles a eukaryotic chloroplast, just as a nonphotosynthetic bacterial cell resembles a eukaryotic mitochondrion. In further support of the endosymbiont theory, some cyanobacteria live within other cells. Some cytoplasmic inclusions of bacteria are products of cellular metabolism, such as the storage polysaccharide glycogen and granules of hydroxybutyric acid, a storage form of lipid. The cytoplasm may also contain fats, granules of polyphosphate, and, in some bacteria, crystals of sulfur.

## THE ROLE OF GAS VACUOLES

Gas vacuoles form in some bacteria and cyanobacteria that occupy aquatic habitats and play a part in the fluctuations in population there (Fig. 28-3). Under certain conditions of light and temperature, the accumulation of gases in the vacuoles causes the cells to become buoyant and float to the surface. The buoyancy of the cells, coupled with their rapid multiplication leads to the periodic occurrence of algal "blooms" in nutrient-rich lakes. The Red Sea was so named because of the recurrent blooms of a red-pigmented species of cyanobacteria in that body of water. Unpleasant odors result when the cyanobacteria decompose. Ultimately, the remaining cyanobacteria settle to the bottom when the gas is dispersed and the vacuoles disappear. In the laboratory, cyanobacteria in jars sink when they are compressed, as when a stopper is forced into the jar, which collapses the gas vacuoles. The membranes that enclose gas vacuoles are different from other cell membranes in that they are composed entirely of protein.

## THE CELL WALL

**FIG. 28-3** Gas vacuoles in a cell of the cyanobacterium *Anabaena flos aqua.*

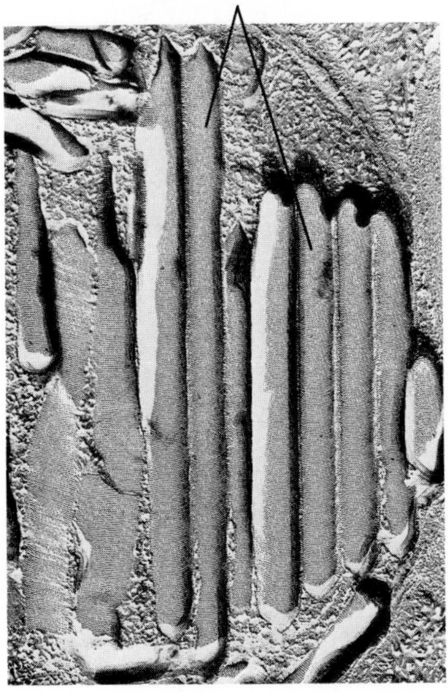


Prokaryote cell walls are biochemically different from those of eukaryotic plant cells. Instead of cellulose, the main component of eukaryotic plant cell walls, the prokaryotic cell wall is composed of a peptidoglycan, a polymer formed from building blocks derived from sugars and cross linked by amino acids. The antibiotic penicillin kills bacteria by preventing cross linking in the peptidoglycan layer, thus weakening the wall and leading to lysis of the cell. The role of the cell wall in maintaining cell shape becomes evident when bacterial cells are experimentally treated with the enzyme lysozyme. Lysozyme hydrolyzes the polysaccharides of the wall, releasing the protoplast, which thereupon assumes a spherical shape (provided it is not in a hypotonic solution, which would make it swell and burst).

In some bacteria and in the cyanobacteria, the cell wall is multilayered (Fig. 28-4). The outer layer is high in lipopolysaccharide. The capacity of this layer to retain the dye crystal violet is the basis of the Gram stain, a test used in bacterial identification since its development in 1884 by the Danish bacterologist Hans Christian Gram. Gram found that some species of bacteria retain the purple stain (Gram positive), while others lose it in alcohol and absorb a second stain, safranin or acid fuchsin, and become red or pink (Gram negative) (Fig. 28-4e).

Many prokaryotes secrete an outer layer of polysaccharide or polypeptide, which forms a slimy sheath around the cell. The sheath becomes visible under the microscope when the organisms are suspended

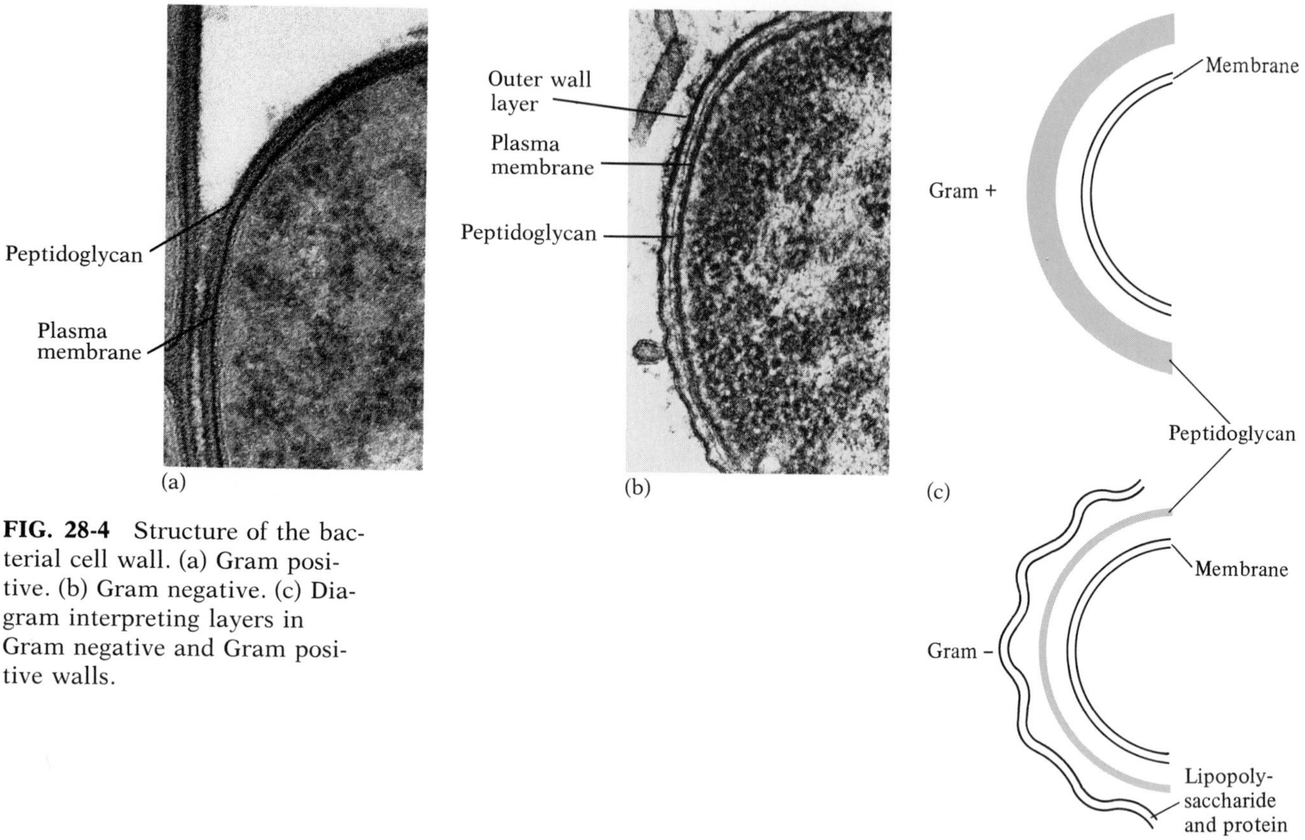


**FIG. 28-4** Structure of the bacterial cell wall. (a) Gram positive. (b) Gram negative. (c) Diagram interpreting layers in Gram negative and Gram positive walls.

in dilute India ink, which outlines the sheath (Fig. 28-5). In some bacteria, an outer layer of polysaccharide fibers, called the **glycocalyx,** is involved in infection, permitting adhesion of the bacteria to various substrates, such as the enamel of teeth. In still other bacteria, the outer wall layer is a resistant capsule that protects the bacteria under adverse conditions (Fig. 28-6). For example, encapsulated forms of the pathogen

**FIG. 28-5** Outer wall layer or sheath of cyanobacteria outlined by India ink.

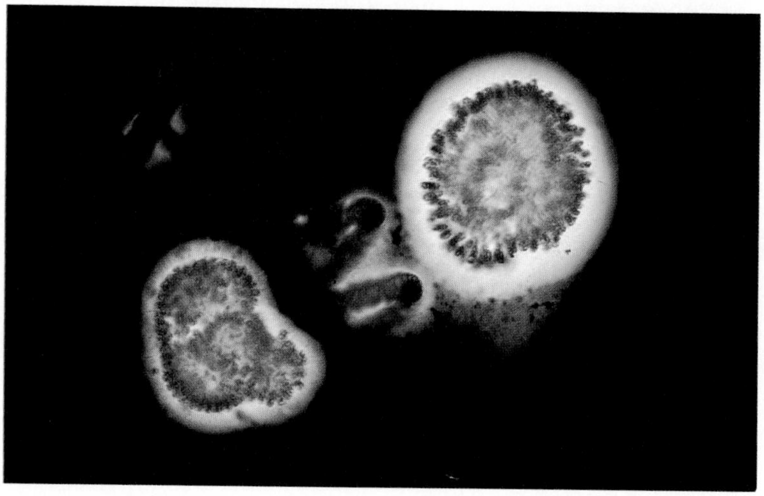

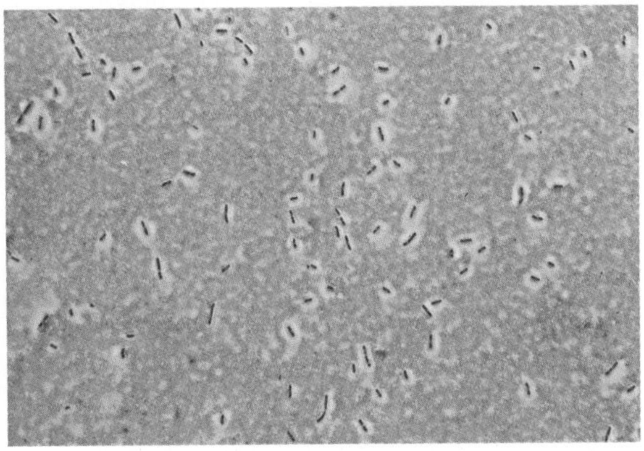

**FIG. 28-6** Encapsulated cells of *Diplococcus pneumoniae.*

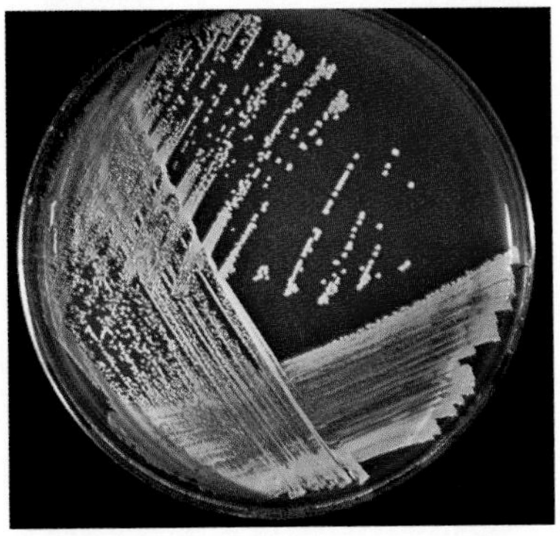

**FIG. 28-7** A laboratory culture of *Staphylococcus* growing in a blood agar medium.

that causes pneumonia in humans are virulent because they are resistant to phagocytosis by the host leukocytes. Strains that do not form capsules are phagocytized and eliminated by the host and therefore are usually not pathogenic.

## BACTERIA

Classifying the bacteria and cyanobacteria as prokaryotes emphasizes their basic similarity and distinguishes them from the eukaryotes. Although bacteria and cyanobacteria have long been recognized as being fundamentally similar, the difference in their nutrition was emphasized in early classification schemes. The bacteria are largely heterotrophic, which is associated with their role as pathogens and saprotrophs in nature; some are autotrophic.

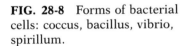

**FIG. 28-8** Forms of bacterial cells: coccus, bacillus, vibrio, spirillum.

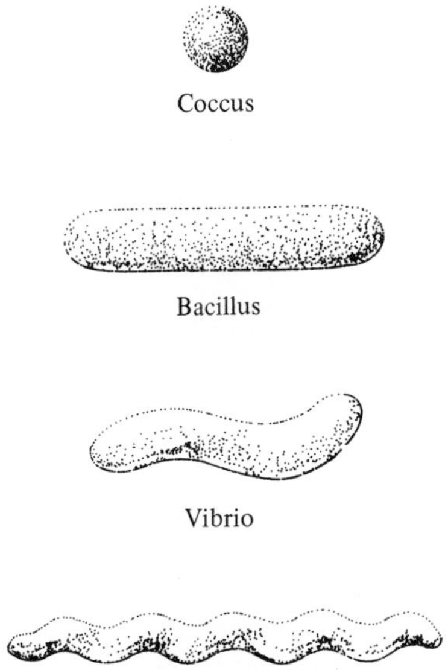


### THE FORMS OF BACTERIA

Because they are too small to be individually visible with the unaided eye, bacteria were discovered only after the microscope was invented. Without magnification, we can, of course, see masses of them growing on culture plates in a laboratory (Fig. 28-7) or on contaminated food, and we can even distinguish the color and texture of the plaques, or patches, they form. The simple microscope of Antoni van Leeuwenhoek first revealed that such masses, or colonies, were made up of minute rods and spheres, which he sketched in a paper published in 1684 (see Chapter 4).

As far as the shape of bacterial cells is concerned, the van Leeuwenhoek microscope revealed essentially all there is to know. Bacteria exist as spheres, or **cocci** (coccus, singular); rods, or **bacilli** (bacillus, singular); and comma- or corkscrew-shaped, motile cylinders, **vibrios,** or **spirilla** (spirillum, singular) (Fig. 28-8, 9). When they fail to separate after cell division, cocci may be held together in pairs, called **diplococci** (Fig. 28-6); chains, called **streptococci;** or bunches, called **staphylococci** (Fig. 28-10). The arrangement of the cells in filaments, plates, or clusters is determined by the plane of cell division.

In addition to the **eubacteria,** or true bacteria, a number of orga-

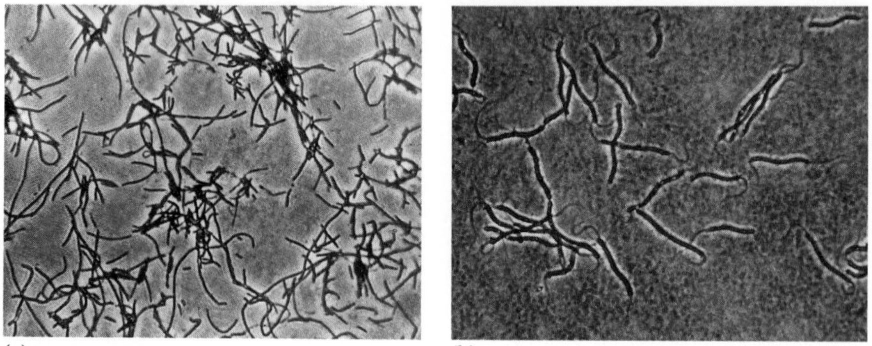

(a) (b)

**FIG. 28-9** (a) *Lactobacillus acidophilus* sours milk and produces lactic acid from carbohydrates. (b) *Spirillum volutans.*

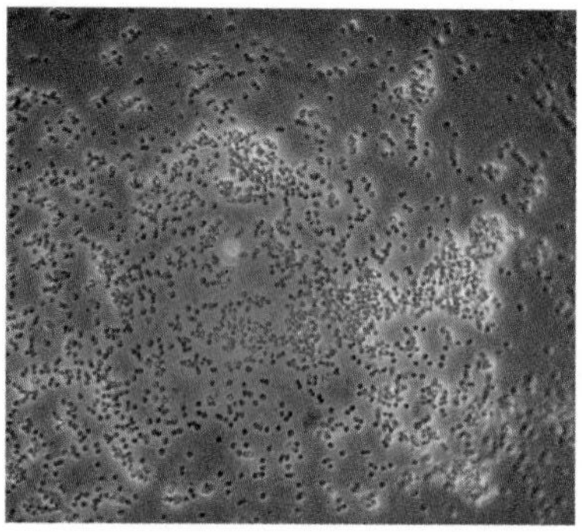

**FIG. 28-10** *Staphylococcus.*

nisms with bacterialike physiological characteristics but of different form are also classed as bacteria. These include the **actinomycetes,** filamentous forms found largely in soil; **mycoplasmas,** often filamentous but thinner than bacteria; and **rickettsiae,** small pathogenic organisms that survive only in host cells. These affiliates are described and illustrated in Panel 28-1.

## BACTERIAL MOTILITY

Many bacterial species are motile, moving by means of flagella distributed over the cell surface or located just at the ends of cells (Fig. 28-11a). Bacterial flagella are much thinner (10 to 20 nm) than the flagella or cilia of eukaryotes and differ radically from them in structure. They are assembled from units of the protein flagellin, which is distinct from the tubulins that make up the microtubules of eukaryotic flagella (Chapter 4). Bacterial flagella have hollow cores, unlike the eukaryotic flagella, with their nine-plus-two arrangement of microtubules. In addition to flagella, some bacteria have smaller, finer protuberances called pili (Fig. 28-11a). Special hollow pili are involved in sexual union, or conjugation, in *E. coli* (Fig. 28-11b).

Some bacteria move by means other than flagella, gliding over a surface without the aid of locomotor organelles. In spirochaetes, filaments attached to ends of the cell somehow cause motion.

**FIG. 28-11** Bacterial flagella and pili (a) and (b) sexual conjugation in *E. coli,* showing conjugation tubes.

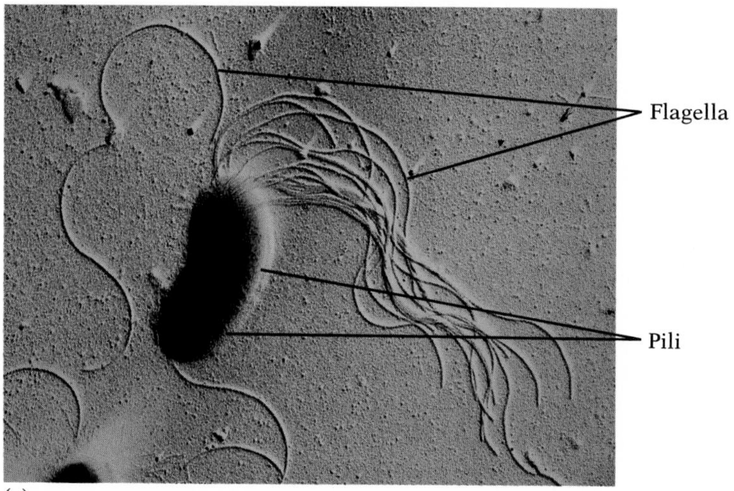


(a)

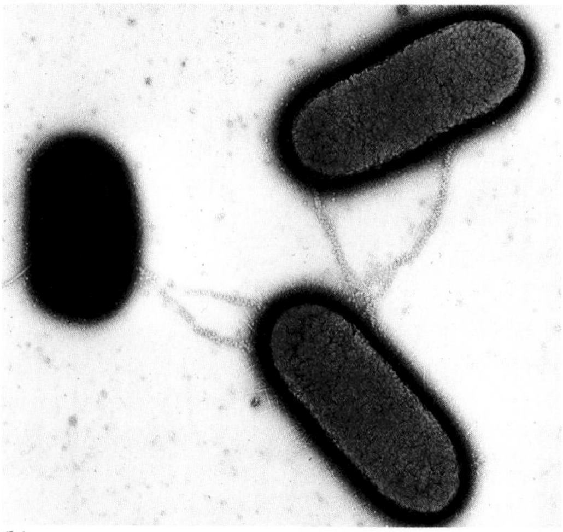

(b)

PANEL 28-1

# Some Bacterial Affiliates

Included with the bacteria in many classification schemes are some prokaryotic organisms similar to them but distinct in form or other features. These include some soil organisms of benefit to humans as well as some disease causing organisms.

## ACTINOMYCETES

The actinomycetes differ from true bacteria in that they form branching filaments. They resemble fungi in this regard, but their filaments are narrower. Like fungi, actinomycetes also produce spores. Most are found in the soil. Actinomycetes such as *Streptomyces* are the sources of many useful antibiotics. When breathed into the lungs, the spores of certain species of actinomycetes can cause a disease called farmer's lung.

## SPIROCHETES

Spirochetes are long, helical forms that move with a unique spiral rotation and flexing motion, induced by fine fibrils attached at the ends of the cells. The spirochete *Treponema pallidum* causes syphilis, and another species causes the disfiguring tropical disease yaws.

## MYCOPLASMAS

Mycoplasmas are the smallest organisms capable of growing autonomously. They pass through materials commonly used to filter bacteria and have been studied effectively only since the electron microscope became available. They are irregular in form because they lack a cell wall. Often filamentous, they sometimes resemble minute fungi, hence their name (from the Greek *myco,* "fungus"). Mycoplasmas have been found in animals, plants, soil, and organic matter such as sewage. Some that invade plant and animal hosts cause disease.

## RICKETTSIAE

Although rickettsiae (named for Howard Ricketts of the University of Chicago, who discovered them and died from typhus fever, a disease caused by rickettsiae) are like cocci or bacilli in form, they are known only as obligate parasites of animals; that is, they require living host cells for survival. Uniquely, they can multiply within the phagocytes of animal hosts. Rickettsiae cause several significant human diseases, including typhus, Rocky Mountain spotted fever, and Q fever. Rickettsiae are transmitted to humans by the bites of blood-sucking insects. For example, typhus can be transmitted to humans from rats by fleas, and from person to person by body lice. The rickettsiae that cause Rocky Mountain spotted fever are transmitted to humans by ticks, which acquire it from feeding on other hosts such as rodents.

Actinomycetes.

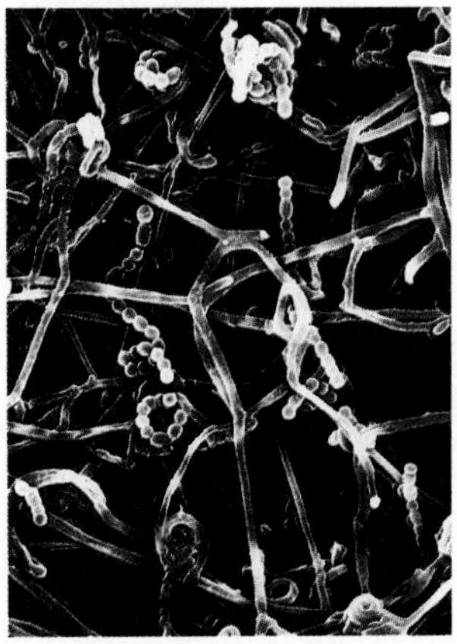

Spirochetes.

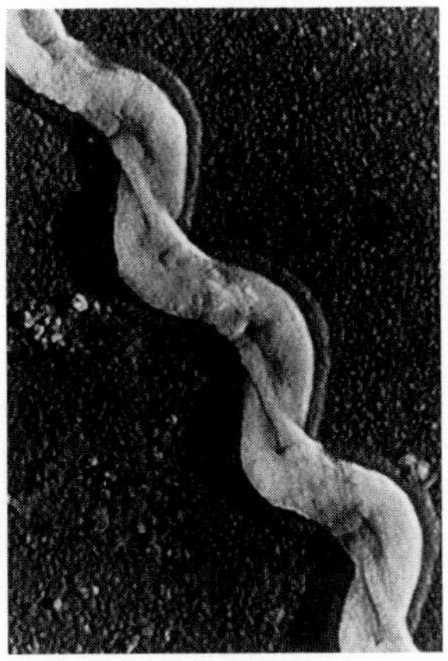

Mycoplasma.

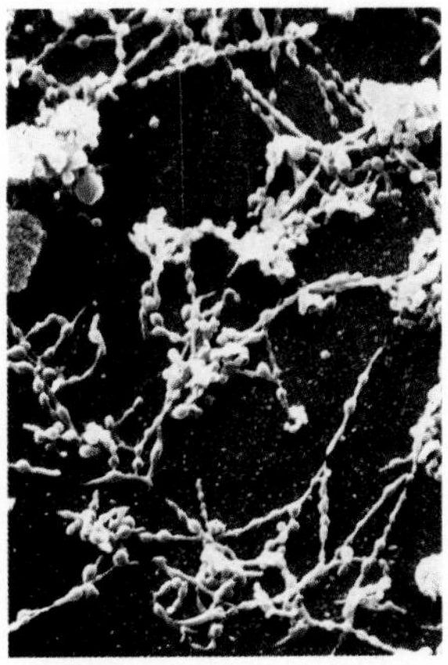

Rickettsiae in a cell.

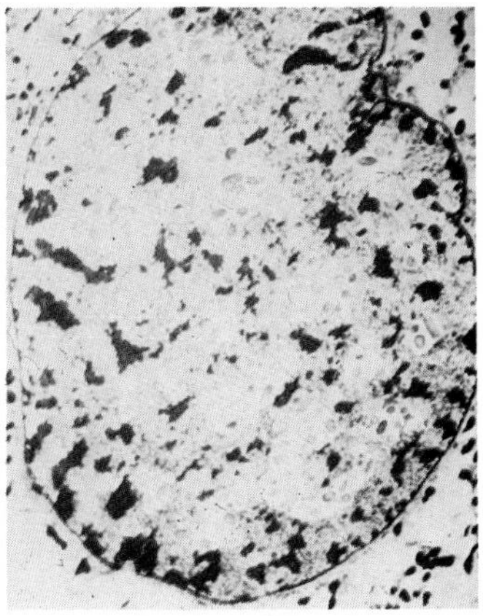

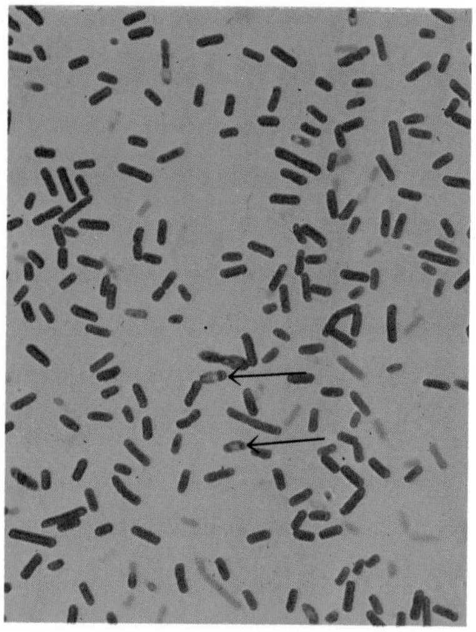

**FIG. 28-12** Endospores of *Bacillus cereus* (arrows).

## DORMANT BACTERIA

Resistant spores are formed in certain bacteria—for example, in *Bacillus* and *Clostridium*—that enable them to survive under adverse environmental conditions. Such **endospores** are formed within the cell by invagination of the plasma membrane and the partitioning off of part of the cell within a thickened coat (Fig. 28-12). Bacterial spores may lie dormant for years before germinating into vegetative, growing cells. Spores that are resistant to heat are the source of the bacterial populations that cause food poisoning from improperly preserved foods. Botulism, caused by *Clostridium botulinum,* was more of a problem in home canning before freeze processing became common. Botulinum toxin is a potent neurotoxin that acts specifically on nerve cells. It is lethal in very small amounts.

## BIOCHEMICAL DIFFERENCES AMONG BACTERIA

While morphology provides a primary basis for classification, biochemical characteristics are also important in distinguishing types of bacteria. Such characteristics as the chemistry of cell walls and storage products, pigments, nutritional requirements, energy sources in autotrophic forms, products of fermentation, antibiotic sensitivity, immunological responses, and pathogenicity are all used in bacterial taxonomy. In addition, the diversity of bacterial metabolism allows some bacteria to be classed as obligate aerobes, others as obligate anaerobes, and some as facultative anaerobes, which are able to carry on metabolism without oxygen but also able to use oxidative pathways in the presence of oxygen. These respiratory pathway options for various bacteria have enabled some to survive in habitats not open to most other organisms. In addition, their versatility in regard to exploiting various sources of energy and nutrition have made bacteria the most widespread of all organisms.

***Heterotrophic Bacteria.*** Most bacteria are **heterotrophs,** obtaining food from organic compounds of other organisms. **Saprobes** are heterotrophs that decompose dead organisms and absorb the products. They secrete enzymes that hydrolyze large molecules to those of intermediate size, which can be absorbed by the bacterial cell and metabolized. **Parasites** live in or on living hosts, utilizing various organic molecules available and bringing harm to their hosts. Parasitic bacteria are responsible for numerous diseases of animals and plants.

***Phototrophic and Chemotrophic Bacteria.*** Autotrophic organisms, which can synthesize carbohydrate from carbon dioxide and water, comprise most of the producers in nature's economy. They probably became essential early in evolution; otherwise, all the organic matter of the primeval seas, produced by chemical evolution, would have been consumed. It is not surprising, then, to find primitive forms of autotrophy among bacteria, since they include the earliest organisms.

Certain pigmented bacteria are able to utilize light energy in the reduction, or fixation, of carbon dioxide to carbohydrate ($CH_2O$); that is, they carry out photosynthesis. Except in cyanobacteria, the chlorophylls of these pigmented bacteria differ slightly from the chlorophyll *a* of algae and higher plants (Fig. 28-13). The bacterial chlorophylls and chlorophyll *a* have different absorption spectra (see Chapter 7). The

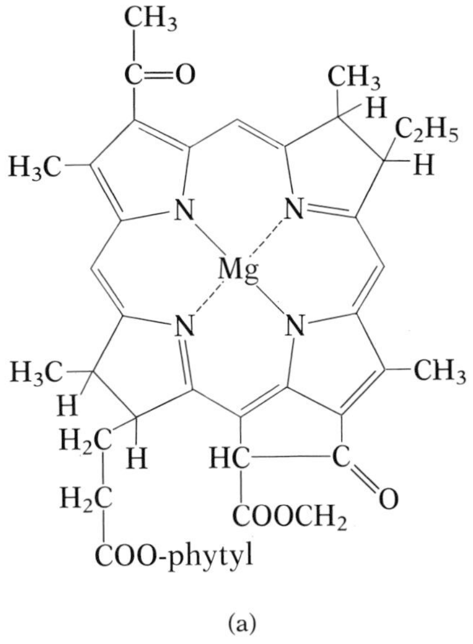


**FIG. 28-13** Molecular formulae of (a) bacteriochlorophyll and (b) chlorophyll *a*.

photosynthetic bacteria use longer wavelengths of light than do plants. Furthermore, photosynthetic bacteria use substances other than water as reducing agents. Therefore, bacterial photosynthesis, unlike plant photosynthesis, does not release molecular oxygen. The green and purple photosynthetic sulfur bacteria, for example, obtain electrons from hydrogen sulfide, producing sulfur, which accumulates in the cell:

$$CO_2 + 2H_2S \longrightarrow (CH_2O) + H_2O + 2S$$

Another group of purple bacteria use organic compounds such as alcohols and keto acids as electron donors in photosynthesis.

A few species of bacteria are able to fix $CO_2$ (produce carbohydrate) by using the energy released by chemical reactions instead of light. They derive energy from the oxidation of inorganic substances, including hydrogen sulfide, nitrogen, sulfur, iron, and gaseous hydrogen. Because these bacteria obtain their energy from inorganic chemicals instead of organic matter or light, they are said to be **chemolithotrophs** (see Chapter 9).

## ARCHAEBACTERIA

Recently, microbiologists who work with a group of anaerobic bacteria that produce methane as a result of the reduction of carbon dioxide by gaseous hydrogen have suggested that these organisms, long known to be primitive autotrophs, along with certain other bacteria of unique environments, comprise a third line of bacteria, distinct from other prokaryotes. Called archaebacteria (''ancient bacteria''), they differ in several ways from true bacteria. Their cell walls are of different composition lacking the usual peptidoglycan, their metabolism is distinctive, fixing carbon dioxide in a cycle involving organic acids instead of using the Calvin cycle, in some cases their ribosomes are different from those found in other organisms, and the base sequence of their ribosomal RNA is different. Their occurrence in environments regarded as representing primitive earth conditions, along with their distinct RNA sequences and primitive autotrophy, support their designation as ancient bacteria. It has been suggested that Archaebacteria be considered a new kingdom, but some microbiologists have argued that the archaebacteria are simply an offshoot of bacteria without special evolutionary significance. Because of their role in the production of methane from organic wastes, they are of interest as a potential source of fuel for human use.

## CYANOBACTERIA

The possession of chlorophyll *a* and the production of oxygen in photosynthesis distinguish the cyanobacteria from other photoautotrophic bacteria that use light to synthesize food. An emphasis on these distinctions leads plant biologists to regard them as algae in the plant division Cyanophyta. Present-day knowledge of their cellular structure, however, tends to place them with the bacteria.*

### THE FORM OF CYANOBACTERIA

Like the bacteria, the cyanobacteria divide by simple fission and come in a variety of forms (Fig. 28-14). In some genera, the cells separate into single cells after cell division and may stay together in a gelatinous mass, while in others the cells form a filament, consisting of a chain of cells, called a **trichome,** enveloped in a mucilaginous sheath. The trichome may be a chain of spheres or of cylinders or a stack of discs. In some genera, it tapers from the attachment end and forms narrow cells at the free end. At specific positions in the trichome, there may be specialized cells, called **akinetes** and **heterocysts** (Fig. 28-14b). These specialized resistant cells can reproduce a new filament after extended dormancy. Heterocysts also fix nitrogen, the important process by which nitrogen of the air is incorporated into organic nitrogen. Trichomes of *Oscillatoria* break into short, growing segments when intervening cells die, thereby increasing the population. Cell division in diverse planes in some cyanobacteria gives rise to clusters of cells held together in colonies. More regular planes of division produce two-dimensional sheets and three-dimensional cubes.

### UNIQUE MOTILITY

None of the cyanobacteria has flagella, but some of the filamentous forms display a forward and backward gliding motion or a peculiar waving or oscillating motion (the source of the name of *Oscillatoria,* Fig. 28-14d). Biologists have long wondered how these motions are achieved. The gliding motion in *Oscillatoria* is believed to involve a layer of fibrils that wrap the peptidoglycan wall layer in a helical pattern. As the trichome glides, it rotates; the rotation pattern matches the angle of fibril orientation. It is thought that waves progressing along these fibrils cause the motion.

* A contemporary classification (Bold, H.C. et al 1980) places the cyanobacteria in the prokaryote division Cyanochloronta (Greek, for the "blue-green ones") to distinguish them from the eubacteria.

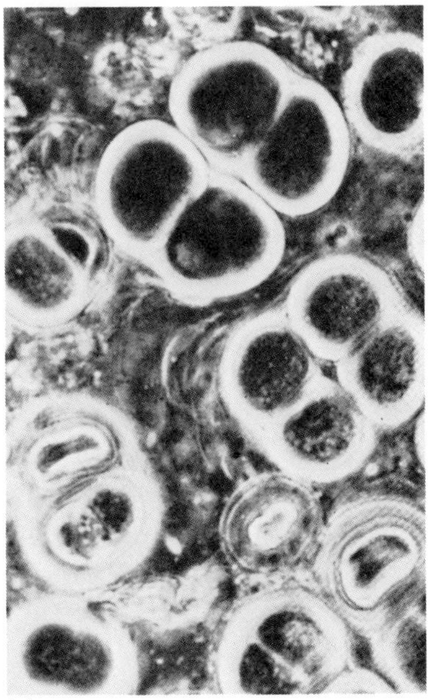

(a)

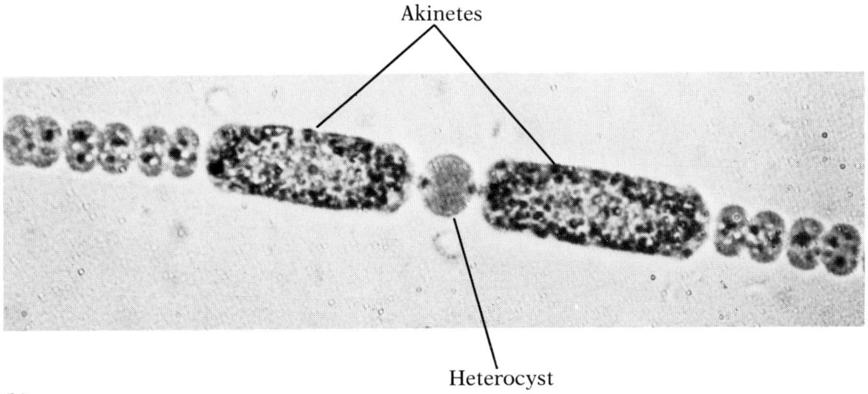


(b)

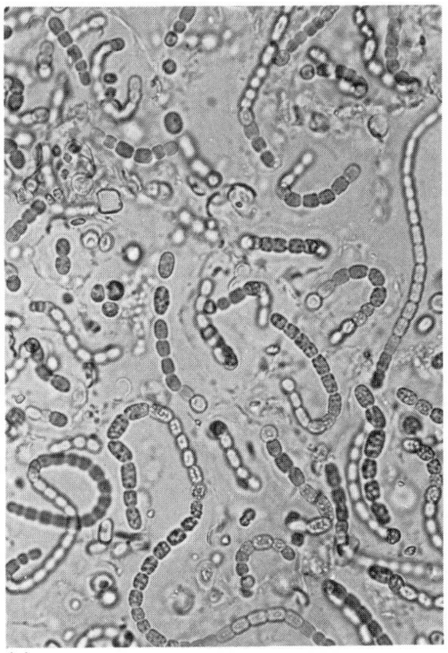

(c)

(d)

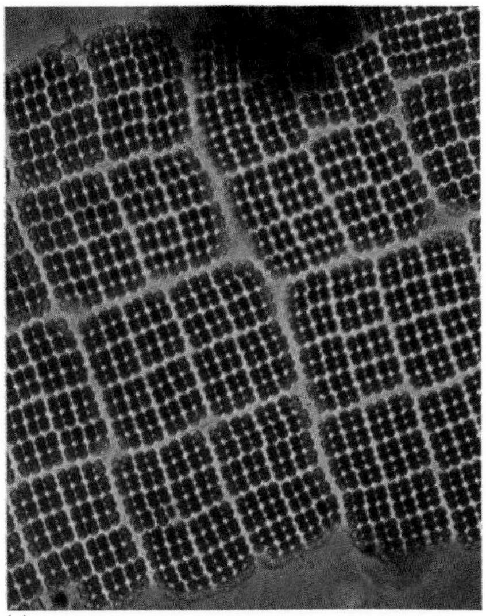

(e)

**FIG. 28-14** Form in cyanobacteria. (a) Clusters of *Gloeocapsa*, a spherical cyanobacterium. (b) Filament of *Anabaena* showing heterocysts and akinetes. (c) Nostoc, gelatinous sphere of filaments. (d) *Oscillatoria*, the dark-green cluster at left, and *Lyngbya*, the finer filaments. (e) *Merismopedia*.

## PIGMENTS AND PHOTOSYNTHESIS

As we have noted, the photosynthetic process in cyanobacteria uses water as a source of electrons and thus generates oxygen, unlike bacterial photosynthesis. The cyanobacteria are the simplest organisms to release oxygen into the atmosphere by photosynthesis. Their presence among the oldest known fossils suggests that such forms may have been the first to provide oxygen for aerobic life on earth.

Like the higher plants, the cyanobacteria contain chlorophyll *a* and carotenoid pigments. In addition, they have a group of unique pigments, the **phycobilins** (Fig. 28-15a). The phycobilins include two blue pigments (c-phycocyanin and c-allophycocyanin) and a red one (c-phycoerythrin). Varying proportions of the pigments in the cells of different species cause them to range in color from blue-green to red to dark brown or black. The varied absorption spectra of the auxilliary pigments (Fig. 28-15b) enable cyanobacteria to harvest light over much of the visible spectrum, contributing to their efficiency in the filtered light in water. The light energy absorbed is transferred to chlorophyll *a*, located on the photosynthetic membranes. Energy transfer is enhanced by phycobilin-containing bodies called **phycobilisomes,** associated with the photosynthetic membranes (see Fig. 28-16).

While the cyanobacteria have chlorophyll *a* in common with green algae and all the higher plants, they lack chlorophyll *b*. Recently, a

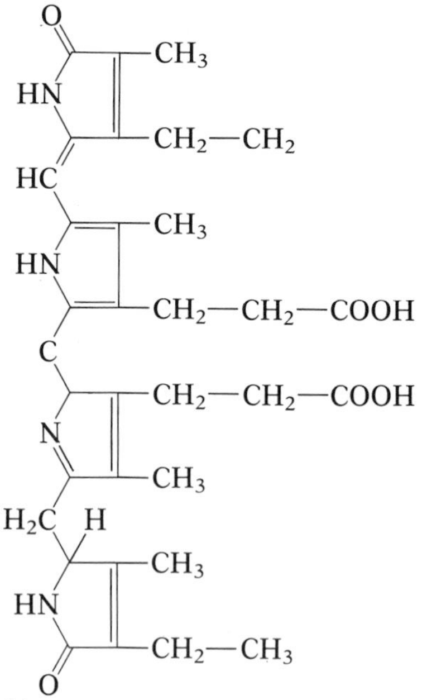

(a)

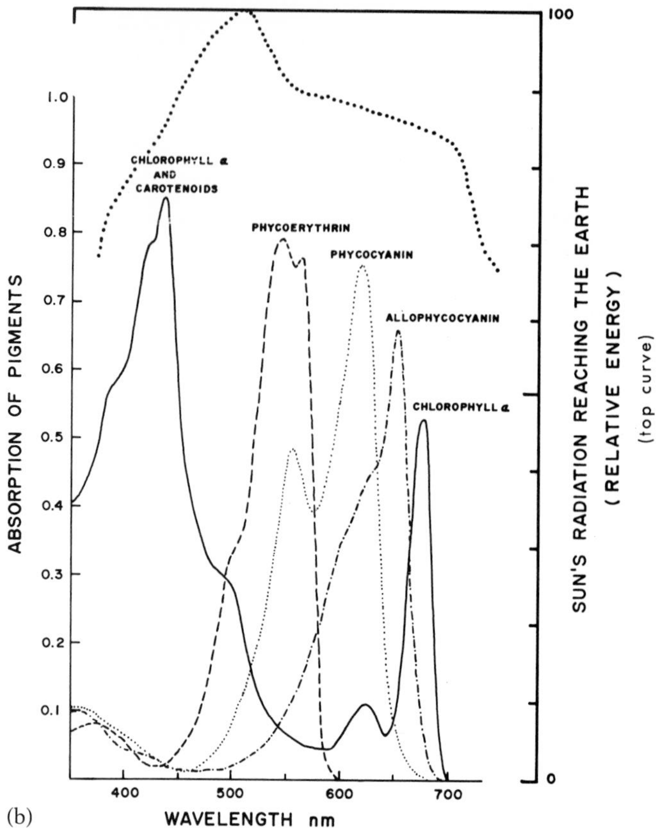


(b)

**FIG. 28-15** (a) Molecular structure of a phycobilin. (b) Absorption spectra of phycobilins and chlorophylls.

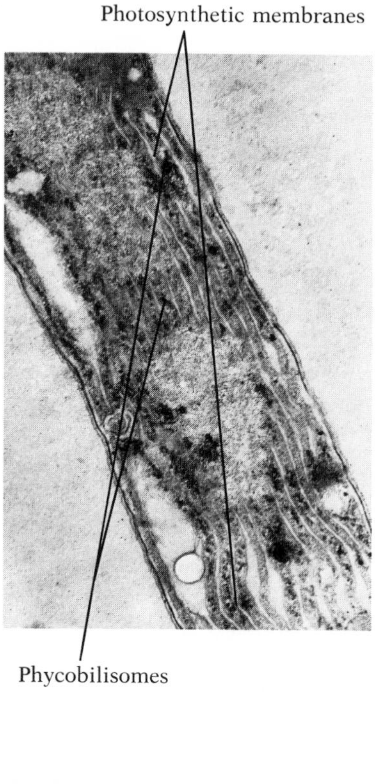


**FIG. 28-16** Electron micrograph of the cyanobacterium *Synechococcus* showing photosynthetic lamellae (thylakoids) with phycobilisomes.

unique alga (Fig. 28-17) with the cytological features of cyanobacteria but containing chlorophyll *b* has been found growing symbiotically in sea squirts (ascidians) in several places in the Pacific Ocean. It has been suggested that this organism, named *Prochloron*, may represent an evolutionary link between the cyanobacteria and the green algae.

# THE FIRST ORGANISMS

As organisms with the simplest cellular organization, the prokaryotes might be expected to be the most primitive, and therefore to be found among the oldest fossils on earth. In 1954, Stanley A. Tyler, a geologist at the University of Wisconsin, and Elso S. Barghorn, a paleobotanist at Harvard University, reported the discovery of microfossils in 1.9-billion-year old rocks of the Gunflint Formation of Ontario, Canada, along the western shore of Lake Superior. Microscopy of thin sections of rock revealed spheroids, filaments, and more complex structures, some of them resembling present day bacteria and cyanobacteria (Fig. 28-18). The Gunflint Formation consists of alternating layers of rock containing organic matter and iron oxide. It has been suggested that the sediments that formed the Gunflint were subjected to changing atmospheric conditions, as pockets of the earth's atmosphere shifted from reducing conditions, when the organic matter was preserved, to oxidizing conditions, when the iron oxide was formed. The presence of the cyanobacteria responsible for the oxygenation of the atmosphere is thus correlated with localized pockets of oxygen and the layered appearance of the Gunflint Formation.

Since the discovery of microfossils in the Gunflint Formation, many

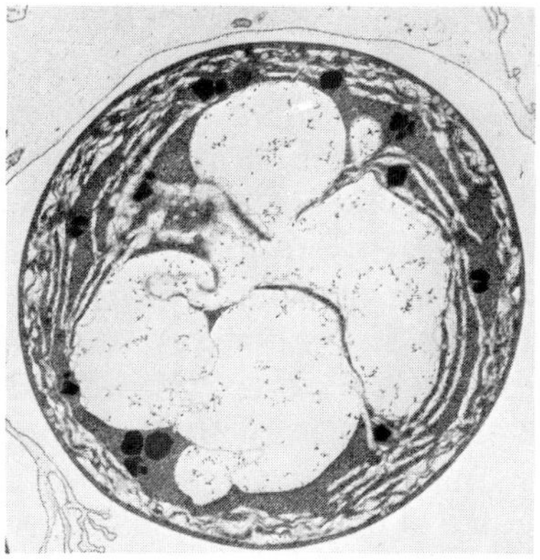

**FIG. 28-17** Electron micrograph of *Prochloron,* a possible missing link between prokaryotes and the algae.

other microfossil-bearing rocks have been found around the world. The oldest fossils now known are in 3.1-billion-year old rocks of the Fig Tree Formation of southeast Africa. Although the fossils resemble bacteria and cyanobacteria, they may actually represent prebiotic chemical fossils, according to J. William Schopf of the University of California Los Angeles, who has studied many of the earliest fossil-bearing formations of Africa, Australia, and Canada. Many of the fossil cyanobacteria are found in unusual layered, pillow-shaped or draping formations called stromatolites. Stromatolites are thought to have been formed from great mats of cyanobacteria in primordial seas. Indeed, they are being formed today at Shark Bay in the Indian Ocean along the coast of Australia, where cyanobacteria deposit calcium carbonate in layers, as the colonies grow at the surface of the stromatolite cushions (Fig. 28-19). Recently, great mats of stromatolite-forming cyanobacteria have also been found growing beneath the ice of permanently ice-covered lakes in Antarctica; this may represent the type of situation in which the extensive Precambrian stromatolites were formed.

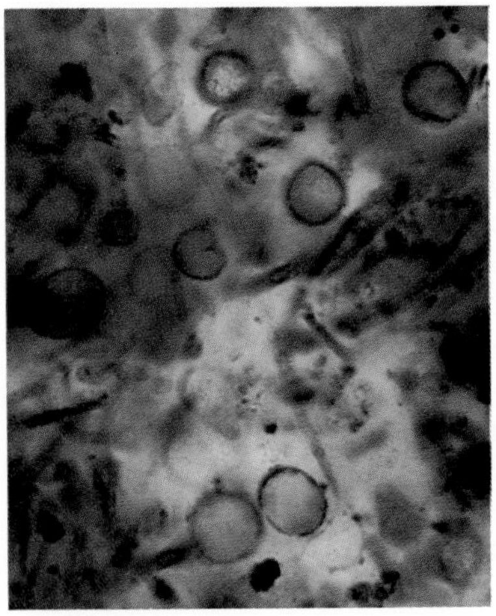

**FIG. 28-18** Microfossils of the Gunflint Formation.

## HOW PROKARYOTES AFFECT OUR LIVES

Prokaryotes are important to humans not only because bacteria are a cause of disease in animals and plants, but also because they play a positive role in the cycling of mineral and organic nutrients. Of special importance to agriculture is the role of both bacteria and cyanobacteria in making nitrogen available to plants through nitrogen fixation. Furthermore, as autotrophs, the cyanobacteria are an important part of the food chain. A recent discovery has shown that the picoplankton, minute particles that pass through the 1 micrometer pores of very fine mesh nets, could account for 60% of the primary production of the open ocean along the Mid-Atlantic Ridge. When viewed in the scanning electron microscope, the picoplankton are found to consist of spherical unicellular cyanobacteria.

### NITROGEN FOR PLANT GROWTH

The fertility of soil depends largely on the activity of prokaryotes, especially bacteria. One such role is the recycling of essential elements contained in organic matter. As they decompose organic matter, an aspect

**FIG. 28-19** Living stromatolites at Shark Bay, Australia.

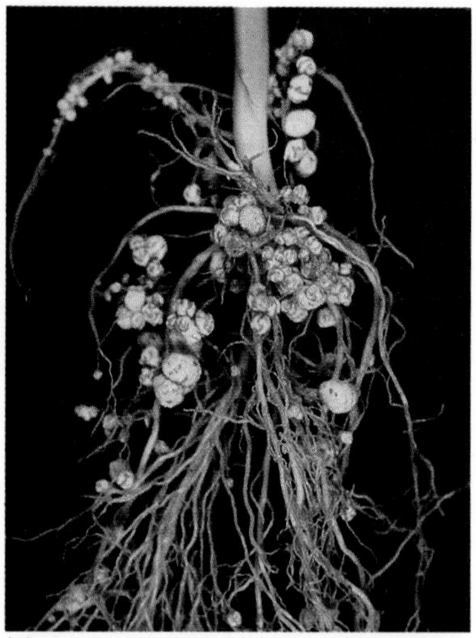
(a)

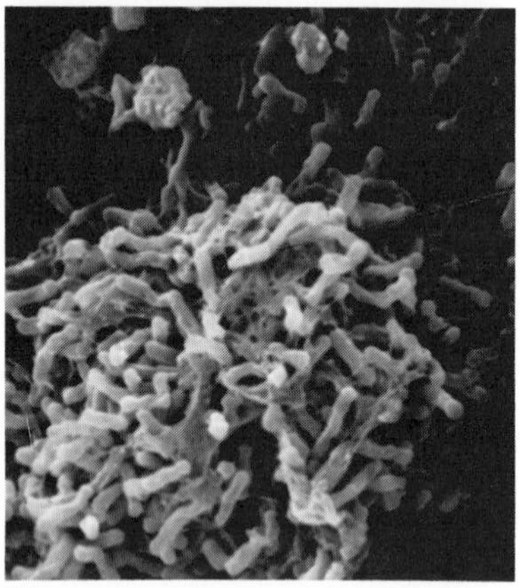
(b)

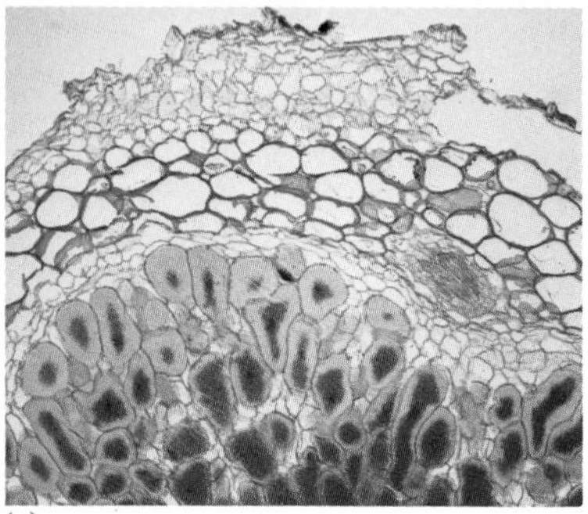
(c)

**FIG. 28-20** Symbiotic nitrogen fixation. (a) Bean roots bearing nodules. (b) Scanning electron micrograph of *Rhizobium* bacteria in an inner cell of a nodule. (c) Section of a nodule filled with *Rhizobium*.

of their nutrition, the bacteria release inorganic compounds back to the soil. This process is especially significant in the case of compounds of nitrogen, an element required in large amounts for plant growth and thus for agriculture. Replenishment of nitrogen in the form of nitrate, the most common form used by plants, usually requires several steps. First, various bacteria free the ammonium ion ($NH_4^+$) from amino acids of proteins, a process known as **ammonification** or **deamination.** The ammonium ion is then converted to nitrite ($NO_2$) by chemoautotrophic soil bacteria of the genus *Nitrosomonas.* Nitrite may then be oxidized to nitrate ($NO_3$) by *Nitrobacter* as part of this process of **nitrification.**

***Nitrogen Fixation.*** Despite the regular return of nitrogen to the soil by the action of bacteria, protozoans, and fungi, the availability of nitrogen is commonly a major limiting factor in plant growth. Agricultural demands challenge our resources; adequate supplies of nitrogen for fertilizer require mining the earth's deposits of nitrates such as Chilean saltpeter and guano (bird dung) and expending energy to prepare ammonia fertilizers. This is true even though nitrogen ($N_2$) is abundant in the atmosphere, which is 78 percent $N_2$ by weight. Plants, of course, cannot use $N_2$ directly. Fortunately, certain prokaryotes are able to convert the $N_2$ of the atmosphere to ammonia and organic nitrogen, a process known as nitrogen fixation. Nitrogen fixation is carried out by a few species of free-living soil bacteria, certain cyanobacteria, and another group of bacteria living symbiotically in the roots of certain plants.

The bacterium *Rhizobium* (Fig. 28-20) induces the formation of swellings called **nodules** in the roots of legumes such as beans and peas. The bacteria occupy cells within the nodule (Fig. 28-20c, d), where they convert atmospheric nitrogen into ammonia, made possible through activity of the enzyme nitrogenase, produced and located in the bacterial cells of the nodule. In the host cells the ammonia is incorporated into glutamate, thereupon available to the host plant as amino acids. When the legume crop is harvested, the proteins of the legume roots and nodules left in the soil also become available to other plants through the action of the previously described ammonification and nitrification bacteria.

The symbiotic association of the nodule is of interest from the standpoint of what the partners contribute. While nitrogenase is produced by the bacterial cells, the substrate and the energy, along with some cofactors, are provided by the host cell cytoplasm. A red pigmented protein, leghemoglobin, similar to the oxygen-carrying hemoglobin of blood is found in the nodule. Leghemoglobin is involved in protecting nitrogenase from oxygen, which oxidizes it. Leghemoglobin in the host cytoplasm may act as a screen, and only a low level of oxygen reaches the bacteria. It is notable that the protein part of the molecule is made by the host cell ribosomes, but the heme moiety is made by the *Rhizobium.* Thus, leghemoglobin occurs only in the nodules and not in either the host or the bacteria alone.

Some of the cyanobacteria also are able to fix nitrogen through the action of their heterocysts: their thickened walls block oxygen uptake, and they lack photosystem II. They are thereby unable to produce oxygen in photosynthesis. Although some cyanobacteria without heterocysts also fix nitrogen, they do so only under anaerobic conditions. It is thought that the heterocysts provide the anaerobic environment necessary for nitrogenase to function.

Although the amount of nitrogen fixed by the cyanobacteria on earth is less than that fixed by nitrogen-fixing bacteria, the cyanobacteria are nevertheless beneficial to agriculture, particularly to rice cultivation. Rice is grown in southeast Asian rice paddies without fertilizer. Presumably, this is possible because the nitrogen is restored by cyanobacteria living in the rice paddies. Cyanobacteria that form symbiotic associations with higher plants may also provide fixed nitrogen to their hosts. One example is the cyanobacterium *Anabaena,* which lives in leaves of the aquatic fern *Azolla.* The *Azolla-Anabaena* symbiote is cultivated in Vietnam on fallow ground to replenish nitrogen. The practice is now being promoted for use in rice fields in the United States.

## THE CYCLING OF SULFUR BY BACTERIA

Even though higher plants are unable to use elemental sulfur, the sulfates they utilize are seldom in limiting supply in the soil. This is because bacteria in the soil convert sulfur containing compounds to the sulfate form useful to plants. The sulfur in proteins is first converted to hydrogen sulfide ($H_2S$) by various decomposer bacteria. Under aerobic conditions in the soil, chemolithotrophic bacteria oxidize the $H_2S$ to sulfate, a reaction that also occurs spontaneously but more slowly in nature.

In waterlogged soils and in stagnant water or bottom mud, oxygen is lacking; under such conditions, hydrogen sulfide accumulates through the action of other bacteria. Hydrogen sulfide is toxic to higher plants and may become a problem during flooding of farmlands. The $H_2S$ may also be oxidized to elemental sulfur spontaneously or by certain aerobic or photosynthetic bacteria. In this way economically useful deposits of sulfur have been formed.

## PROKARYOTES AND PLANT DISEASE

Bacteria cause a number of plant diseases, including several that are destructive of crop plants, fruit trees, and ornamental plants. The diseases, named for the principal symptoms they induce in the host, include leaf spot diseases, rots of roots and fruits, wilts, destructive blights, and proliferative diseases such as cankers or swellings. Crown gall, a proliferative disease that generates swellings on stems, for example, is caused by the bacterium *Agrobacterium tumefaciens.* The disease is actually induced by a plasmid carried by the bacteria. Because of the plasmid's ability to become incorporated into the plant host cell genome, it is of interest in the genetic engineering of plants as a potential vehicle for introducing new genes into plants such as the genes involved in nitrogen fixation.

Plant pathologists, in attempting to understand the mechanisms of plant disease response, have found that pathogenic bacteria induce symptoms in plants by releasing toxins. One way of combatting plant disease has been the breeding of plant strains resistant to pathogens. Resistant plants have been found to produce substances called **phytoalexins** ("warding off substances") that enable the host plant to wall off the infection site by changes in cells near the infection. (See Panel 29-4 on the fungi in the next chapter, page 822.) In recent years it has been shown that certain pathogens produce substances, called **elicitors,** that induce the host cells of the plant to make phytoalexins.

## PROKARYOTES AND HUMAN ILLNESS

Each of us has no doubt been the host to a number of bacterial pathogens. Numerous human diseases are caused by bacteria or their relatives. Once pathogenic bacteria enter the body, they may initiate immunologic reactions or poison the body with toxins (see Chapter 12). Some bacteria release **exotoxins,** which may be carried to other parts of the body from the site of infection, while others produce **endotoxins,** which are within the cell or bound to the wall and are released when the bacterial cells lyse. Bacterial toxins have diverse effects on the host. For example, the toxin of the diphtheria pathogen *(Corynebacterium diphtheriae)* kills cells by blocking an enzyme involved in protein synthesis. Tetanus toxin, on the other hand, binds to certain nerve synapses, blocking nerve impulse transmission and causing spastic muscular paralysis. Other pathogens may act on plasma membranes, causing cell lysis. Endotoxins are lipopolysaccharide components of the bacterial cell wall that are released when the bacterial cells lyse. They are produced, for example, by intestinal bacteria, such as *Salmonella,* that are common in cases of food poisoning. Such endotoxins affect cell permeability in capillaries, causing inflammation. Fever results when host cells release substances that affect the temperature control center of the brain.

***Suppression of Bacterial Growth by Antibiotics.*** Fortunately, many bacterial diseases can be cured by antibiotics, natural cellular products that kill other organisms or limit their growth. Antibiotics were first introduced into the practice of medicine during World War II following the discovery of penicillin. Although penicillin is derived from the fungus *Penicillium,* most antibiotics are obtained from bacteria and actinomycetes. The antibiotics bacitracin, polymixin, tyrocidin, and gramicidin are derived from bacilli; the actinomycete *Streptomyces* provides us with the antibiotics streptomycin, erythromycin, and tetracycline.

Antibiotics suppress bacteria in various ways. **Bacteriostatic agents** merely inhibit the growth of bacteria without killing them. Examples include chloramphenicol, erythromycin, and tetracycline, which block protein synthesis by binding to ribosomes and preventing translation. **Bactericidal agents,** such as streptomycin and neomycin, actually kill bacteria by binding irreversibly to ribosomes, thereby preventing bacterial protein synthesis essential to growth. Other bactericidal agents, such as actinomycin and mitomycin, react with the DNA of the bacteria to prevent transcription and replication. **Bacteriolytic agents,** such as penicillin, lyse the bacteria by blocking cell wall synthesis or affecting the plasma membrane.

***Toxic Cyanobacteria.*** Occasional epidemics of gastrointestinal illness have been correlated with blooms of cyanobacteria in the water supply. Also, large-scale kills of fish and waterfowl are often associated with certain cyanobacterial blooms. Even livestock have been killed by drinking water from troughs contaminated with certain species of cyanobacteria.

Several highly toxic proteins have been isolated from various species of cyanobacteria. When injected into laboratory mice, ducks, or calves, the toxins are lethal, even in small doses. Among such toxins is one from *Anabaena* and another isolated from shellfish occasionally contaminated by blue-green bacteria.

## BACTERIAL BENEFITS

Besides benefiting us by their production of antibiotics, bacteria provide us with other useful metabolic products. For example, the fermentation accomplished by lactic acid bacteria is essential to the manufacture of buttermilk, cheese, and yogurt. Bacterial fermentation is also used industrially in the synthesis of such commercially useful products as acetic, lactic, and citric acids and the solvents acetone and butanol. Vitamins such as vitamin $B_{12}$ and riboflavin and some amino acids are also produced commercially by bacterial processes.

Bacteria that utilize hydrocarbons affect our supplies of petroleum. They may even have participated in its formation. Certain bacteria are able to grow in petroleum and in kerosene; some can create problems by contaminating jet aircraft fuel. On the other hand, in the presence of oxygen, bacteria that eventually decompose petroleum accidentally spilled into our environment, benefit us. Unfortunately, this usually occurs only after great ecological damage has been done. Recombinant DNA technology has been used to develop bacteria that are efficient in metabolizing petroleum products. Because oxygen is required for hydrocarbon decomposition, our petroleum reserves, stored under anaerobic conditions in the earth, are safe from decay by bacteria.

When we add to these benefits the role of bacteria in food production, in recycling nutrients in the soil, and in antibiotic production, their balance between burden and benefit for humans is surely weighted toward the latter.

# VIRUSES

Although viruses cannot be classified as prokaryotes—in fact, few would regard them as organisms—this chapter seems the most appropriate place to discuss them. Viruses are usually not regarded as organisms because they reproduce only within living host cells and because they lack cytoplasm and therefore the machinery to synthesize proteins or release energy. The host cell provides the building materials and energy as well as the enzymes needed for the synthesis of new virus particles. However, these enzymes are made according to specifications provided by the virus genes. The fact that a virus can code for new viral nucleic acid and protein, and thus undergo reproduction, has suggested that they might be degenerate living organisms. Because of their nucleic acid core and ability to code for protein synthesis, viruses have sometimes been called escaped genes. Their actual origin is unknown.

Since the size of viruses is below the resolution of the light microscope, they were not seen until development of the electron microscope revealed them, even though their occurrence as infective agents was known. The first virus to be purified was the tobacco mosaic virus (TMV) in 1935. An extract of infected tobacco plants yielded needlelike nucleoprotein crystals that could be stored indefinitely without loss of infectious properties, hardly the nature of a living organism. When the crystals were introduced into abrasions of the leaves of the tobacco plants the mottling symptomatic of the tobacco mosaic disease soon appeared.

## THE STRUCTURE OF VIRUSES

A virus consists of a nucleic acid core and a protein coat, or **capsid** (Fig. 28-21a). The nucleic acid may be RNA, as in most plant viruses, or DNA, as in bacterial and animal viruses. Despite their simplicity, viruses vary considerably in form. A simple virus may consist of **capsomeres,** or subunits of a single kind of protein wrapped around the nucleic acid core (Fig. 28-21a, b). The morphology of some of the phages is more complex. For example, the $T_4$ virus has a head, containing DNA, a sheath, and a tail piece with attached tail fibers (Fig. 28-22a). Several kinds of proteins are represented in these parts.

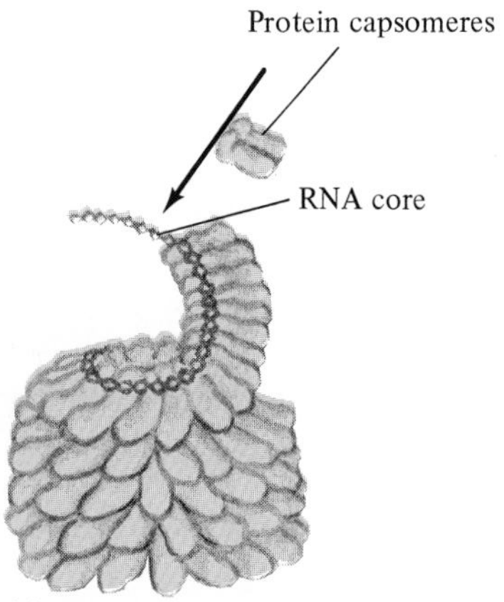


(a)

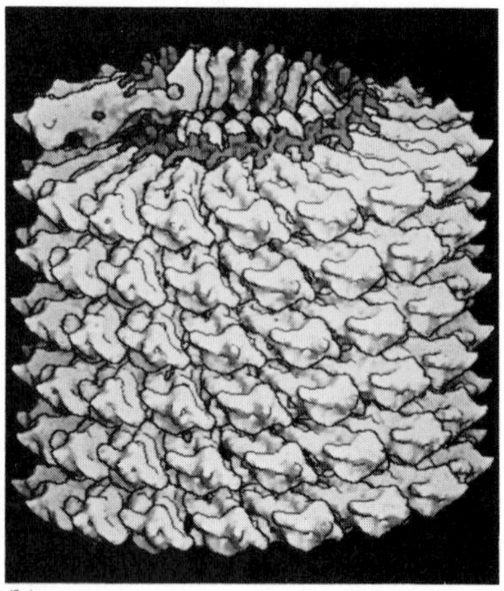

(b)

**FIG. 28-21** (a) Assembly of the tobacco mosaic virus, with RNA core and protein capsomeres being added. (b) A computer model of portion of the virus.

## THE KINDS OF VIRUSES

Viruses are named or classified on a number of bases. In some cases, the name is based on the host organ or tissue they infect. For example, adenoviruses infect the adenoid lymph glands, tobacco mosaic virus infects tobacco plants. In other cases, the name is based on the disease the virus induces, for example, poliovirus. Classes of viruses are named for the kind of organism they infect, for example, insect viruses or plant viruses. Viruses that infect bacteria, called bacteriophages or, simply, phages, are designated by a number. Other viral classifications are based on a variety of other characteristics, including the nature of their nucleic acid (RNA or DNA), their structure, size, host specificity, and the ability of specific immunoglobulins to react with their protein coats.

## VIRAL INFECTION OF HOST CELLS

To invade a host cell, a DNA virus must attach to a specific receptor site on the cell surface. This requirement provides a basis for specific recognition of a host by a virus and for host resistance to viruses.

When the $T_4$ phage infects a bacterial cell, the tail fibers attach to the cell's surface first (Fig. 28-22b). They then contract, pulling the tail into contact with the cell surface. Next a hole is made in the bacterial cell wall, whereupon the tail sheath contracts, injecting only the DNA into the bacterial cell. Animal viruses, on the other hand, enter their host cells intact; the nucleic acid and protein coat separate after entering the cell. Plant viruses lack special attachment mechanisms; they pass intact through the cell wall and into the cell. However, in some cases, they are

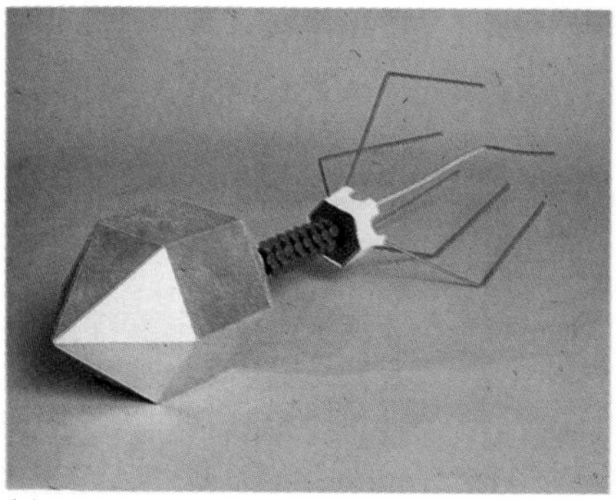

(a)

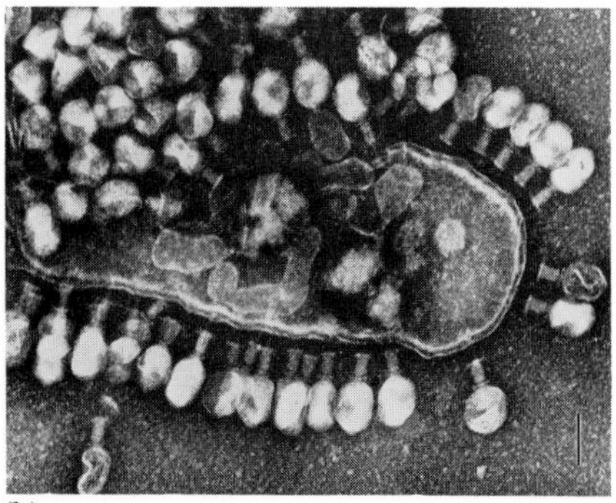

(b)

**FIG. 28-22** (a) Model of the $T_4$ bacteriophage. (b) Bacteriophages attacking *E. coli*. The viruses attach to the bacterium by their tails, then inject their DNA into the cell.

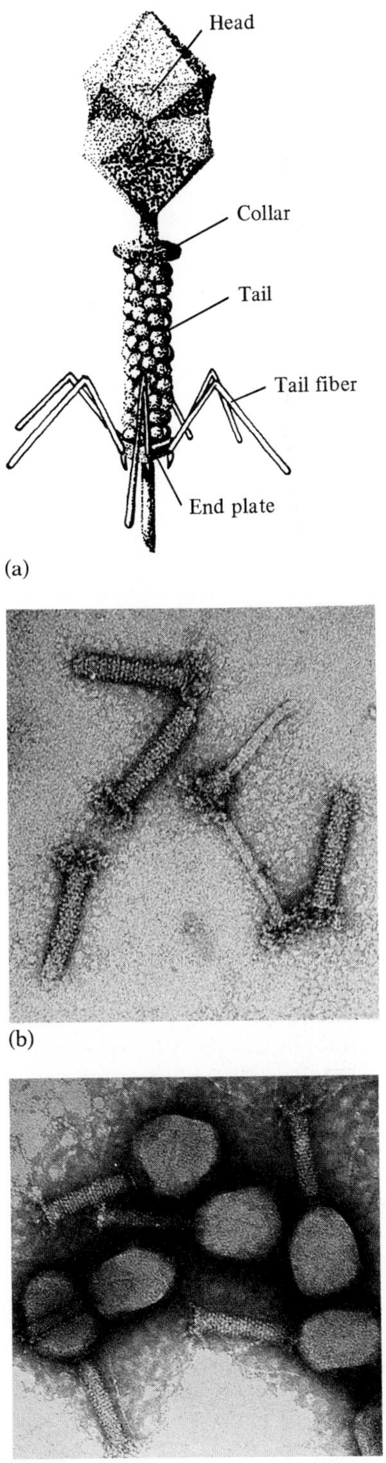


FIG. 28-23 Self assembly of the $T_4$ virus. (a) Structure of a completed virus. Electron micrographs of (b) assembled tail pieces and (c) completed $T_4$ viruses.

injected by insects. Once inside the plant host, viruses are transported from one part of the plant to another, mostly via the phloem.

## VIRAL REPRODUCTION

When a virus enters a host cell and reproduces, it requires enzymes for the synthesis of viral capsid, or coat, proteins. It must therefore make messenger RNA that codes for both the enzymes and the coat proteins. It also needs to replicate its nucleic acid for new viral units. All the raw materials and energy for synthesis, as well as the machinery for protein synthesis, are furnished by the host cell.

Different viruses not only have different nucleic acid cores, some with single stranded and some with double stranded RNA, and some with single stranded and some with double stranded DNA, but the method of replication and gene expression also varies with the different types. Some, such as the influenza virus, have a core of single-stranded RNA that does not directly code for proteins, but with the help of the enzyme **replicase,** complementary RNA strands are made that code for viral proteins. These strands thus act as messenger RNA, coding both for more replicase and for virus coat proteins. Because the infecting form of the RNA has no coding properties for either type of protein, these viruses are termed **negative strand viruses.** Another group of RNA viruses, the **retroviruses,** makes DNA that is complementary to its RNA genome by directing its host cell to synthesize the enzyme **reverse transcriptase.** A second DNA strand is made, and the resulting double-stranded DNA is integrated into the host genome. There it later serves as a template for RNA copies for making new viruses during reproduction of the virus.

Viruses with single-stranded DNA first replicate their circular DNA molecule within the host by making a complementary DNA strand. The new strand then becomes the template for the DNA strands of the new viruses, as well as for transcribing the mRNA that codes for the viral enzymes made by host ribosomes. Double-stranded DNA viruses make the mRNA necessary for a number of viral enzymes directly from both strands, including those needed for replication of the viral DNA and for the various structural proteins of the coat.

The complex $T_4$ phage, with its several parts, is assembled in a process called **self-assembly,** the stepwise fitting together of the parts once they are made. Many of the steps in assembly do not require an enzyme or a template, hence the designation self-assembly. The viral genome codes for the formation of the various parts, which can be seen in electron micrographs of the host cell prior to their assembly (Fig. 28-23a–c). When the parts are completed, the virus is assembled. In this case, the completed viruses are released when the cell is lysed by an enzyme coded for by the virus.

## PROVIRUSES

Although a virus infecting a bacterium usually replicates to produce millions of viruses that soon destroy the host (during what is called the lytic phase), under some conditions the viral genome is integrated into the chromosome of the host. Such a virus is then designated a **provirus** and is said to have entered the **lysogenic pathway.** Once integrated into the host chromosome, the viral genome is replicated only as often as the host DNA is replicated. Thus, each daughter cell receives one copy of the viral genome as well.

Under some conditions, such as irradiation, the provirus may become excised from the host genome and begin to proliferate. It then enters the lytic phase, causing the host cell to burst, or lyse. The advantage to the virus in remaining "dormant" as a provirus is that it can reproduce indefinitely without creating any symptoms in the host that would elicit defensive reactions against it.

An interesting idea that has emerged from studies of viral integration into the host genome and their subsequent release under certain conditions is the concept of viruses as mobile genes. As a viral chromosome breaks away from the host genome, some of the host genes may accompany it. Some of the genes of cancer-causing viruses have been found incorporated into human and other vertebrate chromosomes. These genes are regarded as vertebrate genes that became incorporated into a viral chromosome as the virus left a vertebrate genome at some time in the past.

## VIRAL DISEASES

Those viruses that readily kill cells are said to be virulent. They cause such human diseases as colds, influenza, polio, measles, mumps, chicken pox, and some forms of cancer. In some bacterial diseases, such as diphtheria, caused by *Corynebacterium diphtheriae,* the bacteria are virulent only if associated with a bacteriophage. A number of viruses cause diseases in plants, often at considerable economic loss to farmers. These include a number of mosaic diseases, such as tobacco mosaic, in which the infected leaves develop mottled patterns. Peach yellows, curly top in beets, phloem necrosis, and the wound tumor are other viral diseases of plants named for the symptoms they induce. Many viral diseases of plants are transmitted by insects within which the virus may reproduce without apparent harm to the insect.

***Control of Human Viral Disease.*** Although the fact that host cells affect virus entry and replication suggests that there are potential means of controlling human viral diseases, only limited methods of medical treatment are available. Antibiotics are, in general, ineffective; they act on bacterial cell structures, such as cell walls or bacterial ribosomes, but do not affect viruses. The substance interferon has potential for combating viral diseases by rendering cells less susceptible to viral infection, but it has been used with only limited effectiveness thus far (see Chapter 12). The only effective means yet found for controlling some viral diseases is immunization (see Chapter 12). However, viral mutation, with the generation of new kinds of viruses, presents a continuing medical threat. Immunity to one strain does not necessarily confer immunity to all strains. This is evident in people's repeated susceptibility to the common cold.

## PRIONS

A recently described infectious agent, the **prion,** should be mentioned here, even though it is not a virus. The prion was first isolated from sheep and goats suffering from a neurological disease called scrapie.

Prions have also been found in the brains of human patients with the disorders known as kuru and Creutzfeldt-Jakob disease, both degenerative diseases of the brain. A remarkable aspect of the prion is that it is a protein rod with no associated nucleic acid, which raises the question of how prions replicate. Resemblance of prion rods to amyloids, brain proteins that accumulate in other degenerative diseases of the brain, have suggested possible relationship to Alzheimer's disease, which threatens senior citizens in our society.

## Summary

The prokaryotes include bacteria, some related forms, and cyanobacteria, or blue-green bacteria, together the most widespread and abundant organisms on earth and comprising the oldest fossils. The simplest organisms, they lack nuclei and membrane-bound organelles. The chromosome lacks associated proteins; it is double-stranded DNA. In photosynthetic species, the photosynthetic membranes or lamellae are free in the cell instead of in chloroplasts. The cell wall includes a peptidoglycan framework; the number of cell wall layers varies among different groups of bacteria.

Bacteria vary in form from spheres (individuals or in chains) to rods, to cylinders, to aggregates in the form of filaments, plates, or irregular clusters. Actinomycetes, mycoplasmas, and rickettsiae are regarded as related to bacteria. Some bacteria have flagella, which provide motility; some move by other means; some are immobile. Some bacteria secrete polysaccharide sheaths or other covers that have a protective function. The nutrition of bacteria varies from heterotrophic to chemotrophic to photoautotrophic. Some species exhibit more than one type.

The cyanobacteria are much like other bacteria except that they release oxygen during photosynthesis as a result of having chlorophyll *a*. They also contain additional light-absorbing pigments: carotene and phycobilins. Cyanobacteria divide by fission, as do other bacteria; like some bacteria, some cyanobacteria have special resistant reproductive cells; like bacteria, they exist singly, in filaments, or in colonies of diverse forms. Cyanobacteria lack flagella, but some move by other means.

Prokaryotes affect our lives in numerous ways. Bacteria convert organic nitrogen in the soil to ammonia and nitrate, forms usable by plants. Certain free-living bacteria in the soil and certain cyanobacteria are able to fix nitrogen, that is, incorporate atmospheric $N_2$ into useful compounds. One genus of bacteria, *Rhizobium,* induces swellings or nodules on the roots of leguminous plants within which they live symbiotically. There they convert $N_2$ to amino compounds, which are used by the host plant. In removing $H_2S$ from oxygen-poor soils, some species of bacteria release sulfate, useful for plants, or elemental sulfur, the source of sulfur deposits.

Pathogenic bacteria affect our lives adversely, causing various diseases in plants and animals. Pathogenic bacteria produce toxins. The toxins of some species circulate in the host's body; those of others act locally. Certain cyanobacteria also cause human and animal illness through the production of toxins. In contrast, some bacteria enable us to combat bacterial disease by producing antibiotics, substances that inhibit growth of other bacteria.

Bacteria are useful in the production of certain dairy products and in the production of organic solvents and acids. Bacteria are sometimes used to clean up petroleum spills; some may have participated in petroleum formation.

Viruses are distinct entities from bacteria, unique in that they consist of a nucleic acid genome with a protein coat but no cytoplasm or organelles. They can reproduce only in host cells, which make viral proteins under the direction of viral RNA. Some viruses have an RNA core, others DNA, and they vary in complexity. Hosts infected by viruses include certain bacteria, plants, and animal cells in which the viruses multiply, causing a wide variety of diseases, including several types of cancer.

## Recommended Readings


BARKSDALE, L., and K. S. KIM. 1977. "Mycobacterium." *Bacteriological Reviews*, 41:217–372. A comprehensive review of the biology and pathogenesis of mycobacteria.

BOLD, H. C., C. J. ALEXOPOULOS, and T. DELEVORYAS. 1980. *Morphology of Plants and Fungi;* 4th ed. Harper and Row, New York.

BRILL, W. J. 1977. "Biological nitrogen fixation." *Scientific American*, 236 (March). pp. 68–81. A review of nitrogen fixation by free-living and symbiotic processes by an authority in the field.

BRILL, W. J. 1981. "Agricultural microbiology." *Scientific American*, 245 (September). pp. 199–215. The importance of nitrogen fixation and an account of the research into the genetic engineering of symbiotic nitrogen fixation by an eminent participant.

BROCK, T. D., D. W. SMITH, and M. T. MADIGAN. 1984. *Biology of Microorganisms;* 4th ed. Prentice-Hall, Englewood Cliffs, N.J. A clear and interesting account of the biology of microorganisms, their ecology and evolution, and their place in human society.

FOGG, G. E., W. D. STEWART, P. FAY, and A. E. WALSBY. 1973. *The Blue-green Algae.* Academic Press, New York. A synthesis of the information on blue-green algae, with many items of general interest.

HOLT, J. G., ed. 1977. *Bergey's Manual of Determinative Bacteriology*, 8th ed. Williams & Wilkins, Baltimore. A shorter version of the authoritative guide to species of bacteria. A survey, including keys, of families and genera of bacteria, with abundant figures.

LURIA, S. E., and J. E. DARNELL. 1978. *General Virology.* 3d ed. Wiley, New York. A basic treatment of viruses, with emphasis on classic studies and methodology in viral research.

STANIER, R. Y., E. A. ADELBERG, and J. INGRAHAM. 1976. *The Microbial World*, 4th ed. Prentice-Hall, Englewood Cliffs, N.J. An account of the prokaryotic organisms, with special attention to molecular aspects of their biology and interrelationships.

STARR, M. P., et al., eds. 1981. *The Prokaryotes.* Two volumes. Springer-Verlag, New York. An authoritative treatment of the various groups of prokaryotes by scholars.

WOESE, C. R. 1981. "Archaebacteria." *Scientific American*, 244 (June). pp. 98–125. The author presents evidence for a separate kingdom distinct from the eubacteria and for its possible role in evolution.

# 29 Fungi, Slime Molds, and Lichens

OUTLINE



The Fungi are a kingdom of heterotrophic organisms once considered plants because they have cell walls and produce spores, but they are now separated from plants on the basis of their type of nutrition. As decomposers, they participate with bacteria in recycling inorganic and organic molecules in the environment. Thus, they are important from an ecological point of view. As parasites, they cause several diseases of animals, including humans, and they are responsible for numerous plant diseases, destroying billions of dollars' worth of crops each year. However, some fungi are useful to us as food sources. In addition to discussing fungi, this chapter considers two other groups: slime molds, a group of heterotrophs distinct in form from the fungi; and lichens, symbiotic associations between fungi and algae, unique because each particular association has a distinctive form.

## FUNGI

***The Fungal Body.*** The body of a fungus consists of a **mycelium:** a mass of interconnected filaments called **hyphae** (hypha, singular). Each hypha is bounded by a cell wall of **chitin,** a polysaccharide also found in the exoskeleton of arthropods and certain other animals. In some lower fungi, the wall is composed of cellulose, suggesting a relationship of the fungi to plants. Chitin differs in composition from cellulose, a polymer

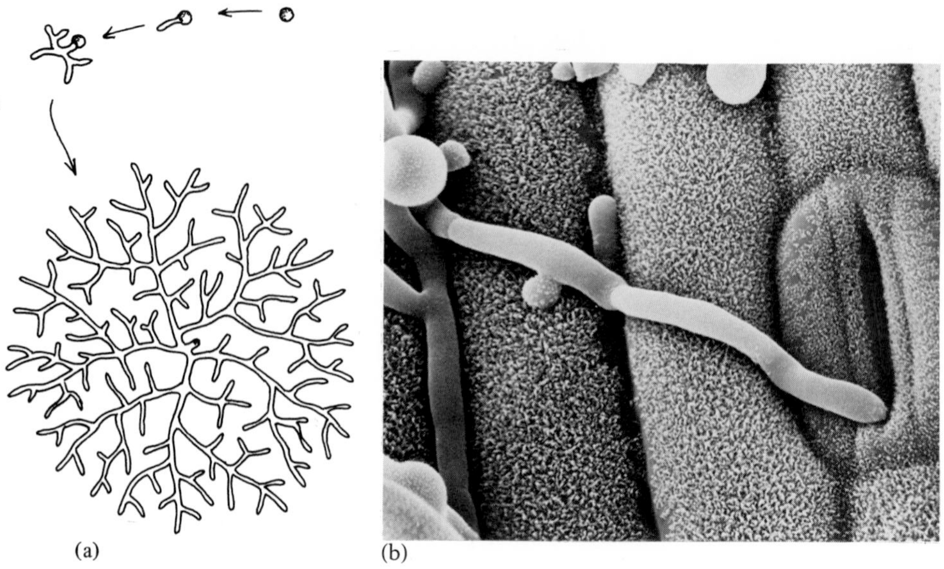

FIG. 29-1 (a) A germinating spore produces a hypha that branches into a mass of mycelium. (b) A hypha of the fungus *Erysiphe* approaches a stomate on the surface of a leaf.

of glucose, in that it is made up of a nitrogen-containing sugar, acetylglucosamine. Acetylglucosamine is also a component of the prokaryote cell wall, suggesting a relationship of the fungi to bacteria. In some fungi, protoplasm streams through hyphae, uninterrupted by cross walls. Such multinucleate organisms are called **coenocytes** (from the Greek *koinos,* "shared in common," and *kytos,* "hollow vessel"). Although the hyphae of higher fungi have cross walls, their complex pores are large enough to permit cytoplasmic flow; even nuclei may squeeze through the pores and pass from cell to cell. Fungi with cross walls are termed **septate.**

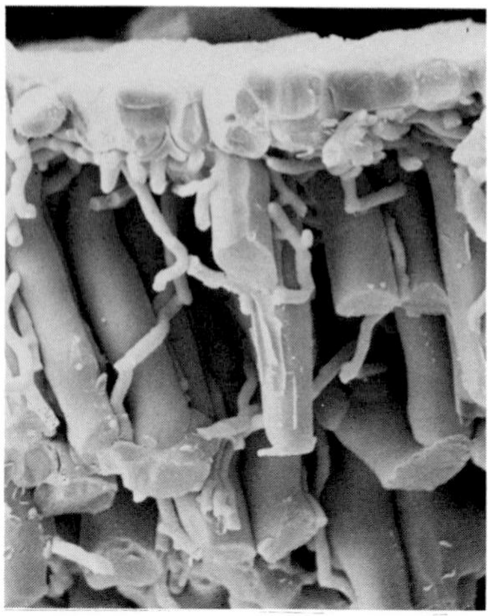

FIG. 29-2 Scanning electron micrograph of bean rust mycelium in a leaf.

A hypha originates from a germinating spore and grows from the tip with the addition of new wall materials (Fig. 29-1). Time-lapse films of growing fungal cultures show the hyphae at the advancing front of the culture extending, branching, and sometimes rejoining, to form a mass of mycelium. Following an initial phase of unorganized vegetative growth in which the hyphae spread through a nutrient-supplying substratum (Fig. 29-2), a reproductive phase occurs. During this phase, the hyphae in the higher fungi produce an organized fruiting body of characteristic form, such as the mushroom. Fungi in which complex fruiting bodies do not form are commonly called **molds.** They are sometimes seen on stale bread or rotting fruits. Yeasts and certain aquatic fungi ordinarily do not have hyphae, but because they either have filamentous relatives or can be induced to form filaments under certain environmental conditions, they are classed as fungi.

***Fungal Nutrition.*** The heterotrophic and absorptive, or **saprobic,** nutrition of the fungi places them, along with bacteria and various protists, in the role of decomposers. Saprobes secrete digestive enzymes into dead plant and animal tissues or their organic products and then absorb and use the digestion products as nutrients. At the same time, much of the matter of the dead tissue is released into the environment in mineral form, becoming available once more to producers and other organisms. Parasitic fungi inhabit living organisms, feeding on their tissues or cellular contents. Often they cause diseases of plants and, sometimes, of animals.

While some fungi spoil our food, others provide us with such epicurean foods as mushrooms and truffles, and flavor our cheeses. With

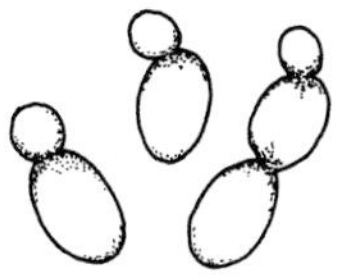

**FIG. 29-3** Asexual reproduction in fungi—budding in yeast.

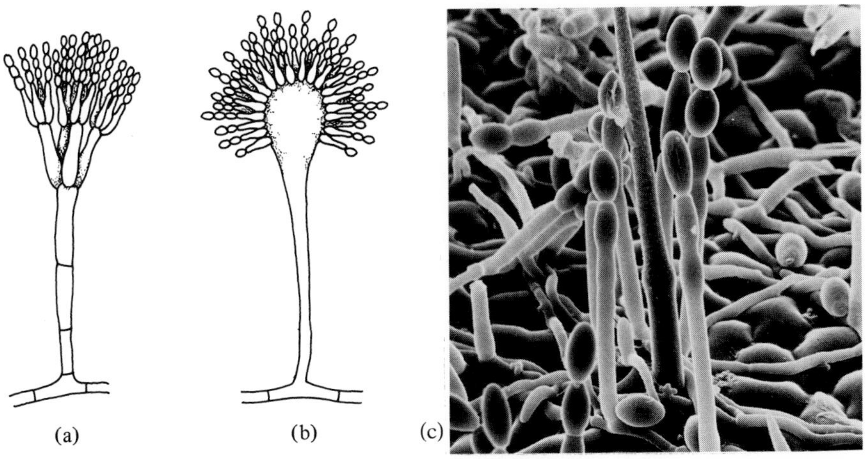


**FIG. 29-4** Conidiospores being produced on conidiophores: (a) in *Penicillium;* (b) in *Aspergillus;* (c) in downy mildew as seen in SEM of frozen leaf.

their metabolic products, types of fungi known as yeasts serve us in the processes of bread making and in the production of alcoholic beverages. (See the discussion of alcoholic fermentation, Chapter 6.)

**FIG. 29-5** The water mold *Achlya* growing on a dead frog.

***Reproduction Through Spores.*** Most fungi are very successful at asexual reproduction, producing new individuals in great numbers. Yeast cells in warm bread dough, for example, double their numbers every few minutes by budding, a process in which a new cell is pinched off from a parent cell (Fig. 29-3). Aquatic fungi arise from motile zoospores produced asexually in great numbers; terrestrial fungi produce innumerable airborne spores that propagate the species asexually. Spores, called **conidiospores** or **conidia** (Fig. 29-4), may be pinched off singly at the tips of special hyphae, or in numbers within a cell called a **sporangium** (sporangia, plural). The higher fungi produce spores mainly as the result of sexual reproduction and following meiosis. The spores of such an individual thus differ in genotype, and their dissemination promotes diversity. Specific forms of fungal reproduction are described later in the chapter.

## TYPES OF FUNGI

The fungi are divided into five classes (or, in some classification schemes, divisions) based largely on modes of sexual reproduction, but utilizing other features as well. The three simplest groups are coenocytes, multinucleate cells without cross walls; these groups include two types of water molds and the bread molds. The higher fungi are all septate, that is they have cross walls. Most also have fruiting bodies of specific form, such as the familiar mushrooms and toadstools, that develop from vegetative hyphae. Higher fungi include the sac fungi and the club fungi. A sixth category, the Fungi Imperfecti, includes fungi that are not known to reproduce sexually and thus cannot be placed within any of the five classes.

**FIG. 29-6** The chytrid *Rhizophydium globosum* growing in a leaf cell of an aquatic plant. By releasing zoospores, it can infect other plants.

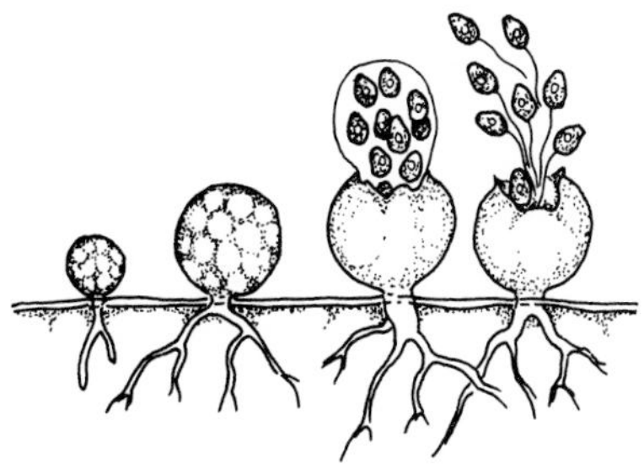

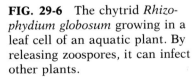

***Water Molds.*** Two groups of fungi, the **Chytridiomycetes** and the **Oomycetes,** are found predominantly in water and are often called water molds. They parasitize fish and other aquatic animals or grow on them after death (Fig. 29-5). Some live within other fungi; others live within the cells of protozoans, algae, and submerged plants. The spores

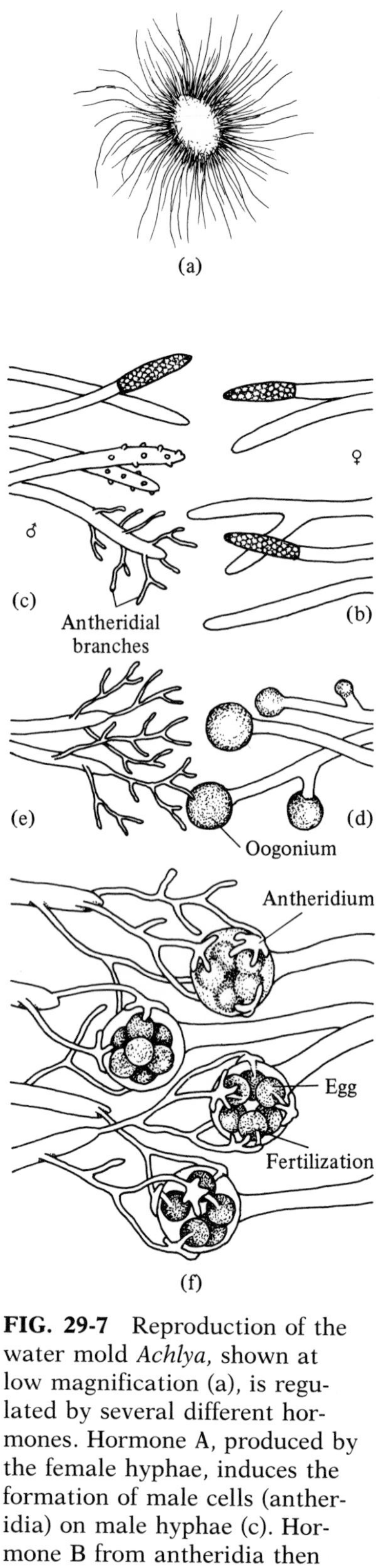


**FIG. 29-7** Reproduction of the water mold *Achlya*, shown at low magnification (a), is regulated by several different hormones. Hormone A, produced by the female hyphae, induces the formation of male cells (antheridia) on male hyphae (c). Hormone B from antheridia then induces initiation of female cells (oogonia) (d). Hormone C from oogonia attracts male branches (e). Male branches produce hormone D, which initiates the formation of eggs in oogonia (f).

of these water molds are motile zoospores with a single posterior flagellum. Like the walls of plant cells and unlike those of other fungi, the cell walls of water molds are composed of cellulose. Some water molds are single-celled organisms, known as **chytrids** (Fig. 29-6), but most form mycelia, as in the case of *Allomyces*. Several species of water molds are parasites on terrestrial plants and cause an assortment of serious plant diseases (Panel 29-1). They are thus not strictly aquatic and, in fact, are dispersed to other plants by airborne sporangial cells. Upon landing on a leaf of the new host, the sporangia release zoospores into drops of water or, if dry, may produce hyphae that penetrate the leaf. Zoospores encyst, retracting their flagella and forming a cell wall, and then produce infective hyphae.

SEXUAL REPRODUCTION AND SPECIAL HORMONES. In all fungi gametes are produced within a single cell, the **gametangium** (see Chapter 21). Sexual reproduction in the Chytridiomycetes involves the union of motile gametes of unequal size, a form of reproduction known as **anisogamy.** The female gametes of *Allomyces* release a pheromone called sirenin (named for the Sirens of Greek mythology), which attracts the male gametes.

In the Oomycetes, gametes are differentiated as egg and sperm, a condition known as **oogamy.** The male individual produces gametes in gametangia called **antheridia,** at the tips of special hyphae. The male hypha grows toward a female hypha and contacts the swollen female gametangium or **oogonium,** which contains the eggs. Male nuclei and cytoplasm then move through a fertilization tube that grows to each egg in the oogonium.

A series of hormones regulates successive reproductive events in the water mold *Achlya* (Fig. 29-7). Some substances direct the growth of hyphae, while others direct the formation of sexual structures. Two of these hormones have been isolated, chemically characterized, and synthesized in the laboratory. One, known as hormone A, is produced by the female mycelium and induces formation of antheridia on male mycelia. The other, hormone B, is produced by the male mycelium and induces formation of oogonia. Still other hormones, designated C and D, regulate other events in the reproduction of this fungus, but have not been isolated.

***Bread Molds.*** Another group of fungi, the **Zygomycetes,** includes the black bread mold (*Rhizopus stolonifer*) common on stale bread before bakers began using fungal retardants. Its hyphae rapidly grow over the surface of a slice of bread and extend special hyphae called **rhizoids** into the bread (Fig. 29-8). Enzymes secreted by the rhizoids and hyphae digest the food to a form that can diffuse into the hyphae. Erect hyphal branches produce asexual sporangia at their tips, which are filled with black spores that give the mold its common name. Each spore is capable of starting a new mycelium.

SEXUAL REPRODUCTION IN THE BREAD MOLDS. Sexual reproduction in this group occurs when two compatible strains of opposite mating type are growing in the same vicinity. There are no motile gametes. Instead, hyphae emerge laterally along the mycelium of each strain and meet (Fig. 29-8). At the tip of each pairing hypha, a gametangium is produced. Compatible gametangia fuse in plasmogamy (Chapter 21) and the nuclei of the two gametes unite in karyogamy, forming a zygote. A rough, thick wall forms around the zygote. This resistant **zygospore,**

# Panel 29-1

## Plant Diseases That Have Influenced History

In the 1840s, the potato crops of Ireland and Europe were devastated by a fungus that spread through the potato fields so rapidly that the entire crop was destroyed before tubers, the potatoes, could be formed. The fungus was the oomycete *Phytophthora infestans,* the cause of the disease known as late blight of potato. A million people in Ireland starved to death as a result of the destruction of this staple of their diet. Many others emigrated to America rather than face starvation.

Under moist conditions, this fungal pathogen spreads rapidly through fields of nonresistant varieties of potatoes. Great numbers of sporangia are produced by the mycelium on the leaves and are blown or splashed by rain to other plants. Each sporangium can produce a number of zoospores, which encyst and in a few hours extend fine hyphae into the leaf. In parasitic fungi, special hyphae called **haustoria** enter the leaf cells and take up nutrients. The mycelium may grow through the potato plant and even into tubers already formed, producing more and more sporangia. Even stored potatoes were destroyed in the Ireland infestation.

Another disease caused by an oomycete is downy mildew of grapes. Grape stocks exported to France from America in the 1860s to combat a root disease of the native grapes carried the mycelium of *Plasmopara viticola.* The nonresistant French grape varieties grafted to the American stocks were devastated by the fungus, nearly destroying the French wine industry. An observant professor at the University of Bordeaux noticed that grape vines sprayed with a mixture of copper sulfate and lime to deter fruit thefts were spared from the fungus. His experiments led to the first fungicide, called Bordeaux mixture.

Late blight of potato. (a) Leaf and (b) tuber of the potato showing the effects of the fungus. (c) The mycelium of *Phytophthora infestans* in the potato leaf, and sporangia-bearing hyphae extending from stomata of the leaf. Sporangia are disseminated and produce zoospores, which reinfect other leaves.

(a)

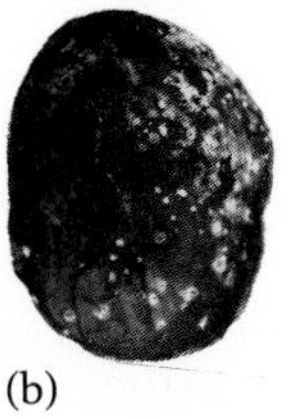
(b)

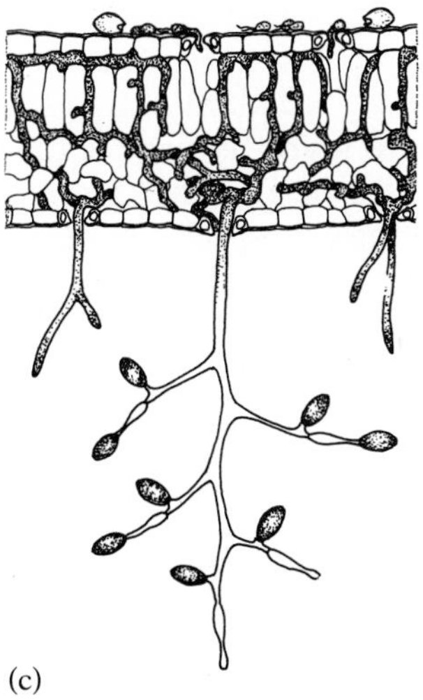
(c)

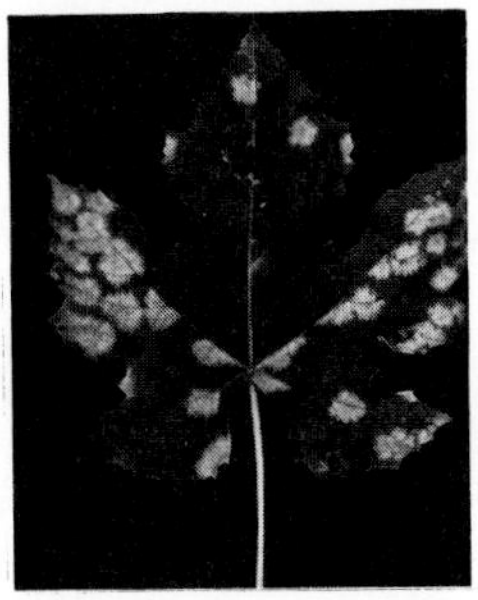
(d)

(e)

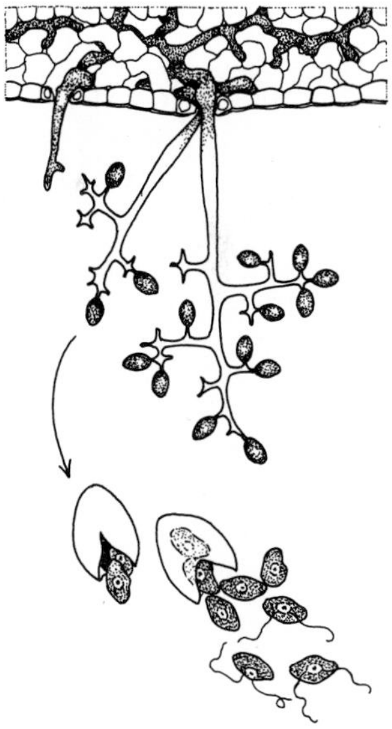
(f)

Downy mildew of grapes. The fungus *(Plasmopara viticola)* on (d) leaves and (e) fruits of the grape. (f) Zoosporangia on hyphae extending from the leaf spread to other leaves and fruits. If dry, they grow infection tubes that penetrate the leaf; if water is present, they germinate to form zoospores that reinfect new leaves.

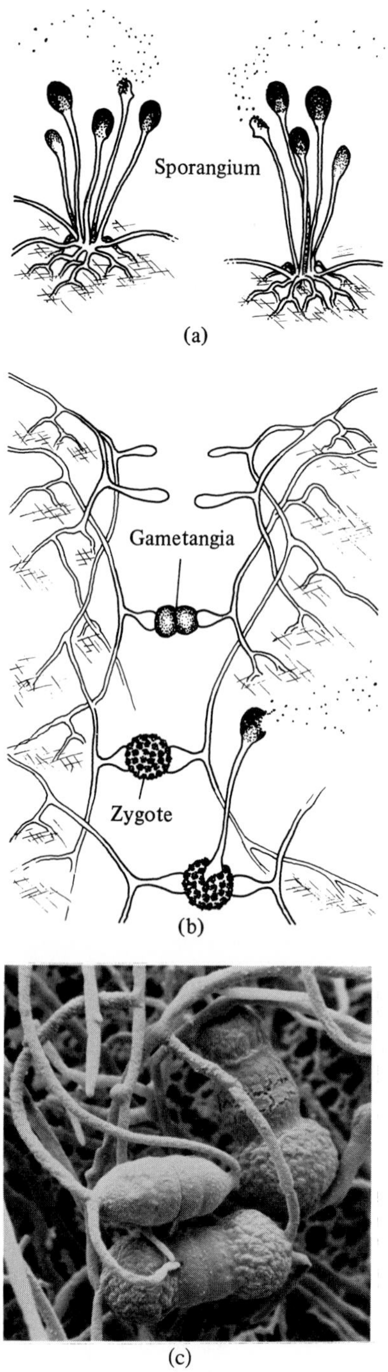


**FIG. 29-8** *Rhizopus stolonifer,* the black bread mold, showing asexual (a) and sexual (b) reproduction. (c) SEM of zygote formation.

suspended between the two hyphal branches, is characteristic of this group of fungi. It may lie dormant for long periods before undergoing meiosis and producing a new hypha, which in turn produces a sporangium and haploid spores.

Zygomycete species live on a variety of foods besides bread. They are also commonly found on organic matter in the soil and on feces. Some species are parasites of plants, causing diseases; other parasitize small soil animals.

***Higher Fungi.*** The features that distinguish the higher fungi, the Ascomycetes and the Basidiomycetes, from the fungi already discussed are the formation of fruiting bodies and the manner in which the spores resulting from sexual reproduction are borne. Following a period of vegetative growth in which the mycelium grows randomly through a nutrient substratum, the mycelium of the higher fungi develops into a characteristic fruiting body of specific form generated through the compact, organized growth of hyphae to form a tissue. What we know as a "mushroom," for example, is really the fruiting body of a fungus whose vegetative mycelium inhabits the soil. Ultimately, as the result of sexual reproduction, spores are produced in the fruiting body along exposed surfaces. In the Ascomycetes, the spores, called **ascospores,** are produced within a sac-like cell or **ascus** (asci, plural). In the Basidiomycetes, the spores, called **basidiospores,** are produced as protuberances of a club-shaped cell or **basidium** (basidia, plural). The Ascomycetes are thus known as sac fungi; the Basidiomycetes as club fungi. The fruiting bodies are designated according to the type of spores they produce, those of the Ascomycetes being called **ascocarps,** those of the Basidiomycetes, **basidiocarps.** Mushrooms are sometimes found in a circle, or "fairy ring," in a lawn as a result of the formation of basidiocarps at the periphery of an outward spreading circle of mycelia.

SAC FUNGI. **Ascomycetes,** or sac fungi, are found in soil and feces, on litter of the forest floor, on the twigs and leaves of plants, and on certain foods. Their ascocarps range in form from the cups of *Peziza* and the spongelike, tasty morels (*Morchella*), through the small, beaked, flask-shaped structures of the pink bread mold *Neurospora,* to the ornamented microscopic structures of *Microsphaera,* the pathogen of powdery mildew disease. Representative Ascomycetes are shown in Figure 29-9.

A number of ascomycetes are plant pathogens, causing such diseases as powdery mildew; apple scab; brown rot of stone fruits; Dutch elm disease, which is now destroying the American elm; and chestnut blight, which years ago eliminated the American chestnut. Ergot, a disease of cereal grains, killed thousands of people during the Middle Ages. Consumption of flour from ergotized rye led to epidemics of "St. Anthony's Fire," a disease that swept through European communities, causing hallucinations, gangrene, convulsions, and death. Ergot can also induce abortion in cattle that consume infected grain. The fungus that causes ergot is *Claviceps purpurea;* it infects rye plants, and ultimately its spores fill the grains. Although the level of contaminated rye marketed today is controlled and there are few reports of human ergotism, the disease still occasionally shows up in cattle. *Claviceps* was also the initial source of the psychedelic drug lysergic acid diethylamide (LSD). In controlled dosages, ergot is used in medicine to induce labor, lower blood pressure, and ease migraine headaches.

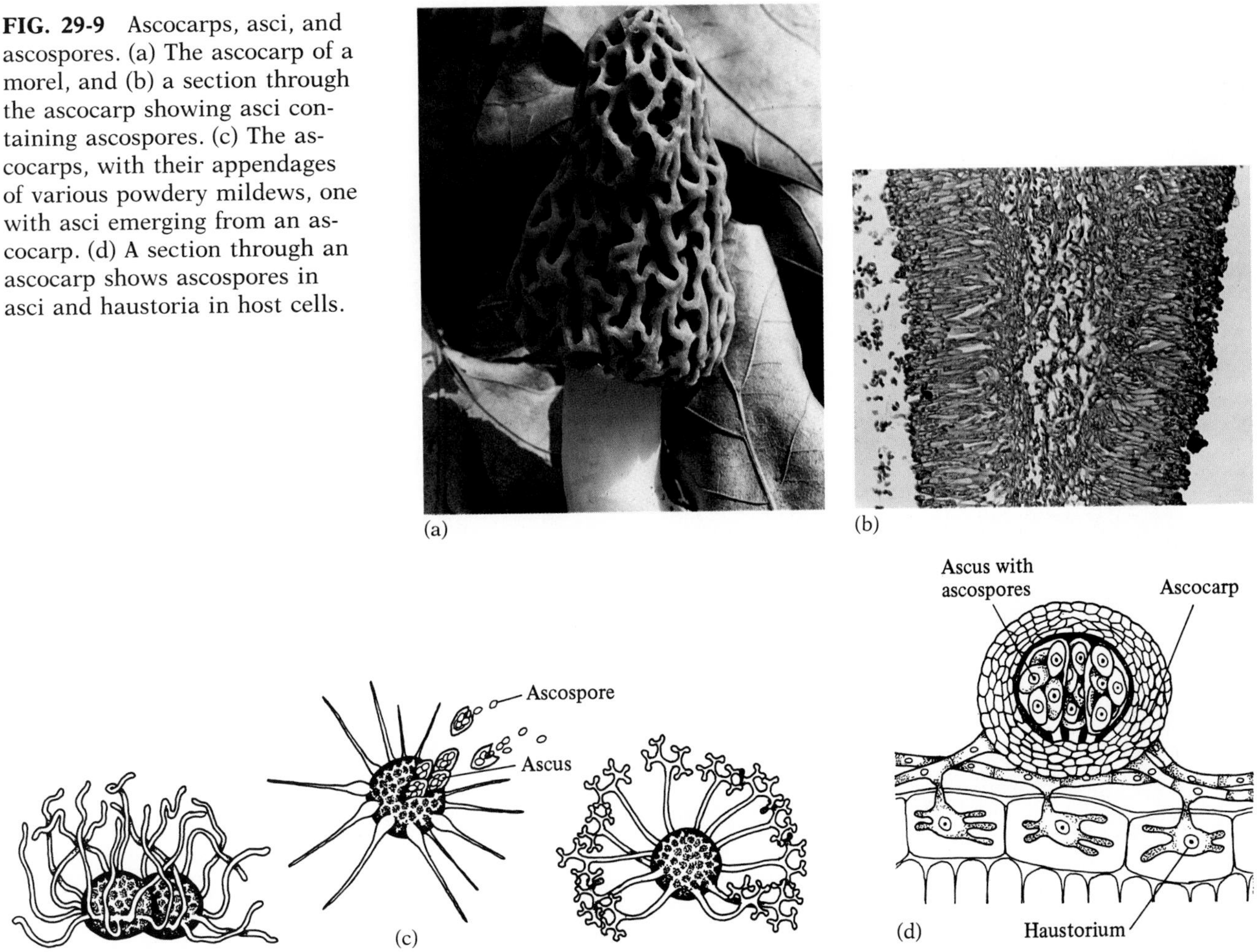


**FIG. 29-9** Ascocarps, asci, and ascospores. (a) The ascocarp of a morel, and (b) a section through the ascocarp showing asci containing ascospores. (c) The ascocarps, with their appendages of various powdery mildews, one with asci emerging from an ascocarp. (d) A section through an ascocarp shows ascospores in asci and haustoria in host cells.

YEASTS. While the term yeast is applied in a general sense to all unicellular fungi that produce by budding or fission, a number of yeast genera are classified as Ascomycetes because they form asci containing ascospores in their sexual reproduction (Fig. 29-10). These yeasts are beneficial to humans (Panel 29-2) as a result of their fermentative metabolism: they produce alcohol and other valuable solvents and emit carbon dioxide, which when released into bread dough causes it to rise. A number of other yeasts do not reproduce sexually, do not produce ascospores, and are classified in the Fungi Imperfecti. Many of these are human pathogens.

Asexual reproduction by budding predominates in rapidly growing cultures, generating tremendous cell numbers in dough or fermentation cultures. In budding, a small bulge appears on the parent cell and expands into a new yeast cell. The nucleus divides without dissolution of the nuclear membrane, and no chromosomes are evident as the DNA is partitioned to the daughter cells. Ultimately, the bud pinches off, leaving a scar on the parent cell, which may bud again and again.

Sexual reproduction in baker's yeast, *Saccharomyces cerevisiae*, requires the presence in the yeast culture of two different mating types, designated "a" and "alpha" (Fig. 29-10). In a yeast culture, large, ovoid cells are diploid. When conditions become unfavorable for asexual budding, as when nutrients become limited, diploid cells undergo meiosis and become asci: they form four haploid ascospores, two of each mating type. The ascospores are released from the ascus and begin asexual bud-

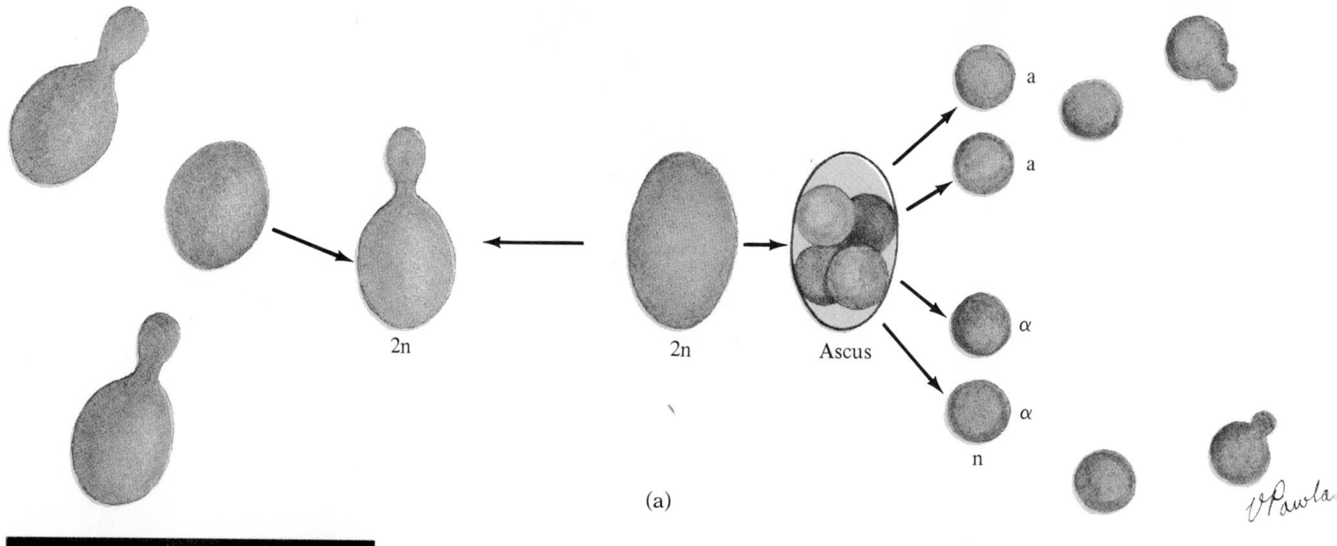


(a)

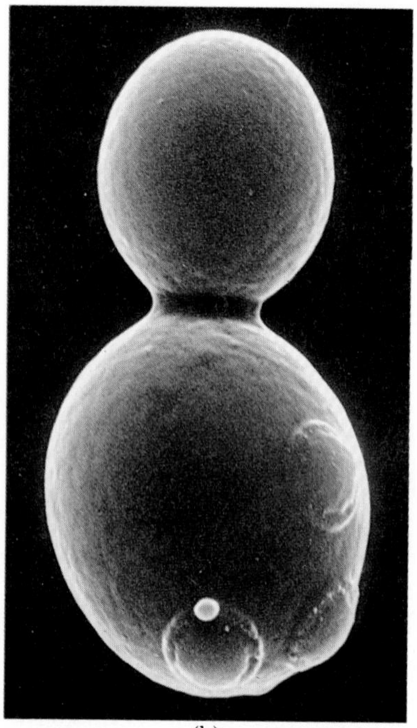

(b)

**FIG. 29-10** Reproduction in the yeast *Saccharomyces cerevesiae.* (a) Diploid ovoid cells and haploid spherical cells reproduce asexually by budding. (b) Budding yeast cell. Scars of previous buds remain on the cell surface.

ding, building a population of small, spherical haploid cells of the two mating types. Mating of "a" and "alpha" cells forms the larger diploid cell, which then may reproduce by budding.

Whether a yeast cell is "a" or "alpha" is determined by which of two genes, "a" or "alpha," occupies the mating type locus on the yeast chromosome. Appearance of haploid cells of one mating type in a culture of the opposite mating type led geneticists to the discovery that the mating type genes are exchanged, the "alpha" gene being replaced by an "a" gene, or vice versa. The replaced gene is replicated from a nonexpressed "silent master" gene at another locus in the chromosome. Because replacing one gene with another is like exchanging one tape cassette for another, this switching of mating type genes in yeast is called a cassette mechanism. The cassette type mechanism has been of interest to molecular biologists because it could explain selective activation of genes in eukaryotes.

*Saccharomyces* has been especially favorable for the study of the role of mitochondrial DNA and nuclear DNA in mitochondrial function. Haploids and diploids may be selected and clones propagated by budding. Since *Saccharomyces* is able to grow by glycolysis without using its mitochondria, reduced numbers of mitochondria, or mutations causing defective mitochondria, need not kill the yeast cells, allowing them to be studied. Mutants with defects in mitochondrial proteins, and therefore strictly fermenting cells, are small and are called petites. Petite mutants have mitochondria, but the mitochondria have poorly developed cristae. It is thus possible to isolate mutants with defective mitochondria and determine whether the mutant gene affecting mitochondrial proteins occurs in the mitochondrial DNA or the nuclear DNA. One finding of such studies is that many mitochondrial enzymes are made on cytoplasmic ribosomes from nuclear genes. The petite mutants of *Saccharomyces* continue to be an important tool in studying the molecular biology of mitochondria.

CLUB FUNGI. The **Basidiomycetes,** or club fungi, are best known by their fruiting bodies, or basidiocarps. They include the assorted mushrooms, toadstools, shelf fungi, coral fungi, and puffballs found on dung, soils, leaf litter, and rotting trees of moist woods (Fig. 29-11). The group also includes some of our most serious plant pathogens, the smuts and rusts.

## PANEL 29-2

# Yeasts and People

Yeasts have probably been beneficial to humans since the beginnings of agriculture. It has been speculated that making fermented beverages derived from grain in the presence of yeasts may have preceded the use of grain and yeast in bread making. Although a number of species of yeast occur in nature—39 genera and 350 species—primarily strains of one species, *Saccharomyces cerevisiae,* are used today in bread making and fermentation. Certain genetic strains, known as brewer's yeast, have been selected for brewing beer and as starters in making wine and distilled liquor. Other varieties, called baker's yeast, have been selected for bread making.

In the brewing of beer, barley is first malted. In this process, the barley grains are allowed to germinate for 5 to 7 days, during which time amylases and proteases are produced in the grains and starch is digested to sugars that can be metabolized by yeast cells. Germination is then stopped by drying. When the malted barley is crushed and remoistened, the soluble extract, called wort, is filtered off to brew kettles, where special strains of yeast are added. For 5 to 7 days, the yeast cells ferment the sugars, producing alcohol. After several weeks of lagering, during which time solids settle out, the clear fluid is drawn off, carbonated, pasteurized, and packaged.

Some yeasts of the Fungi Imperfecti are pathogenic to humans and other animals; among these is *Candida,* which causes serious diseases (candidiasis and moniliasis) of the skin, lungs, and other organs. In cryptococcosis, caused by *Cryptococcus,* an initial skin infection becomes systemic, often resulting in death.

**FIG. 29-11** Assorted basidiocarps. (a) The poisonous gill mushroom *Amanita muscaria.* (b) Gills of the mushroom are lined with basidia in various stages of development. (c) *Boletus,* a polypore. (d) The pores are lined with basidia. (e) A puffball. (f) A coral fungus.

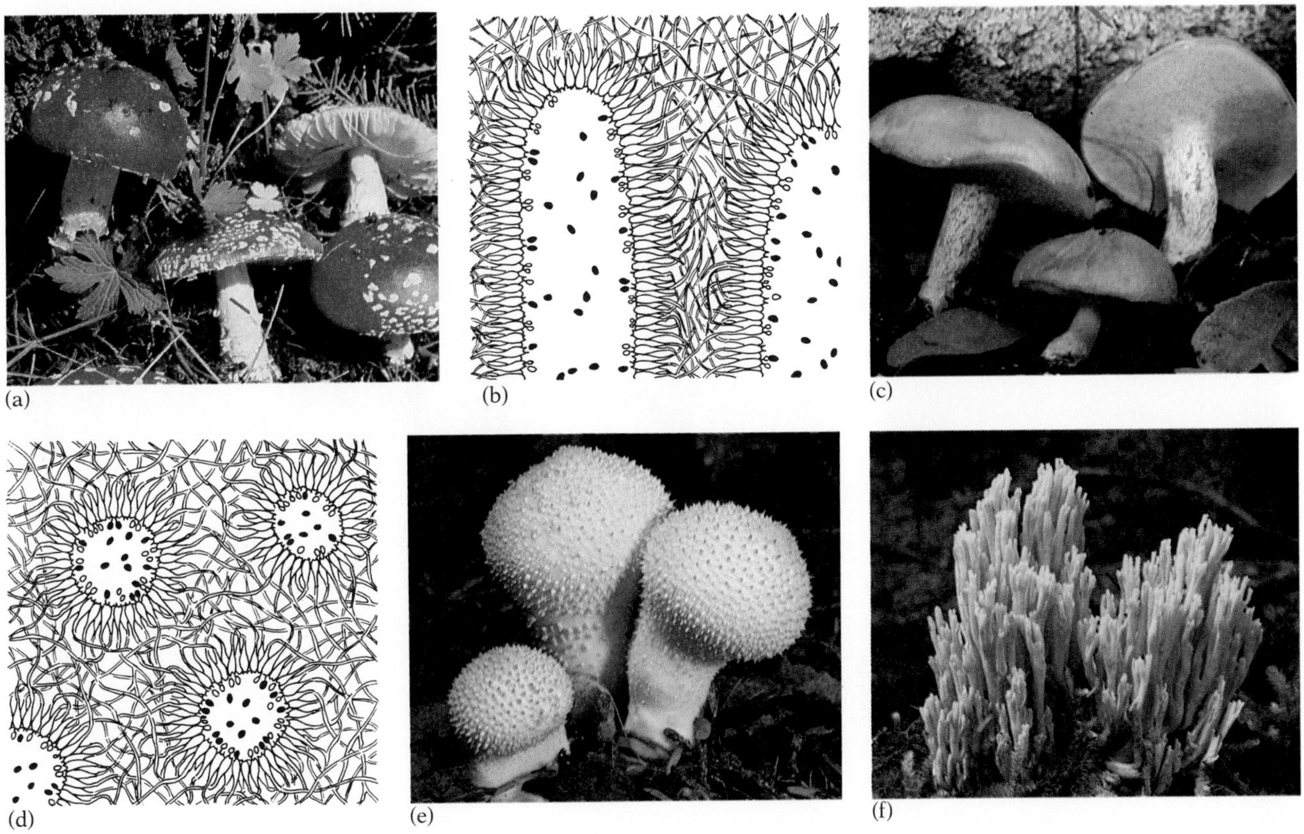

**FIG. 29-12** Mycorrhizal roots of pine. Note absence of root hairs, short thick root tips.

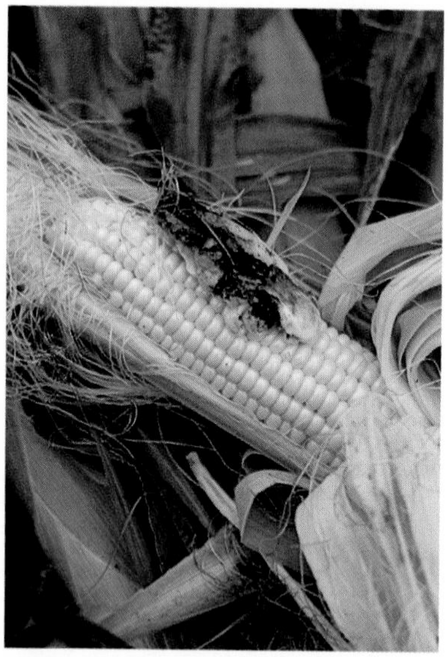

**FIG. 29-13** Corn smut fills corn grains with black spores.

Some basidiocarps have numerous gill-like projections, actually called gills (Fig. 29-11a). Other basidiocarps, called **polypores,** have pores (Fig. 29-11c). Gills and pores have extensive surface areas from which the characteristic sporangia, the basidia, discharge their basidiospores.

When basidiocarp formation commences, the randomly growing mycelium becomes organized and develops into a basidiocarp. Within the basidiocarp, the basidia develop over the surface of the gills of the gill fungi or around the periphery of tubules that open to pores on the underside of fungi of the polypore type. The basidiospores that are produced on the basidia fall to the ground from the gills or pores. They will form a spore print if a piece of paper is placed beneath a basidiocarp at this time. Such spore prints are of a characteristic color and are used in the identification of fungi.

Some basidiomycetes of the forest form associations with tree roots, in which the fungal mycelium penetrates the outer cells of the root and forms a sheath around the root tip. These associations are called **mycorrhizae.** Mycorrhizal roots are short and stumpy and lack root hairs (Fig. 29-12). The hyphae function as root hairs, taking up water and minerals and supplying the tree with them. Many of our forest tree species have mycorrhizae, apparently thriving on the symbiotic association. In fact, mycorrhizae have been artificially induced on trees by introducing fungi into the soil and have been shown to greatly benefit tree growth. Such treated trees planted on the tailings, or residues, of certain mine operations grow more rapidly than untreated trees, thereby accelerating reclamation of such waste sites.

Some of our most serious plant diseases are the rusts and the smuts of cereal crops. For example, one of the smuts commonly forms unsightly black swellings on ears of corn where it fills the grains with sooty spores (Fig. 29-13). Stem rust of cereal crops (Fig. 29-14) results in millions of dollars' worth of crop losses annually (Panel 29-3). Other rusts such as white pine blister rust are scourges of some timber trees.

MUSHROOM TOXINS. Many mushrooms produce alkaloids, some of which are hallucinogens, others deadly poisons. Probably the most dangerous mushroom is the deadly *Amanita phalloides,* the "destroying

**FIG. 29-14** (a) The rust *Puccinia* growing on a leaf of barley. (b) Freeze-fracture of barley leaf showing mycelium of *Uromyces* rust.

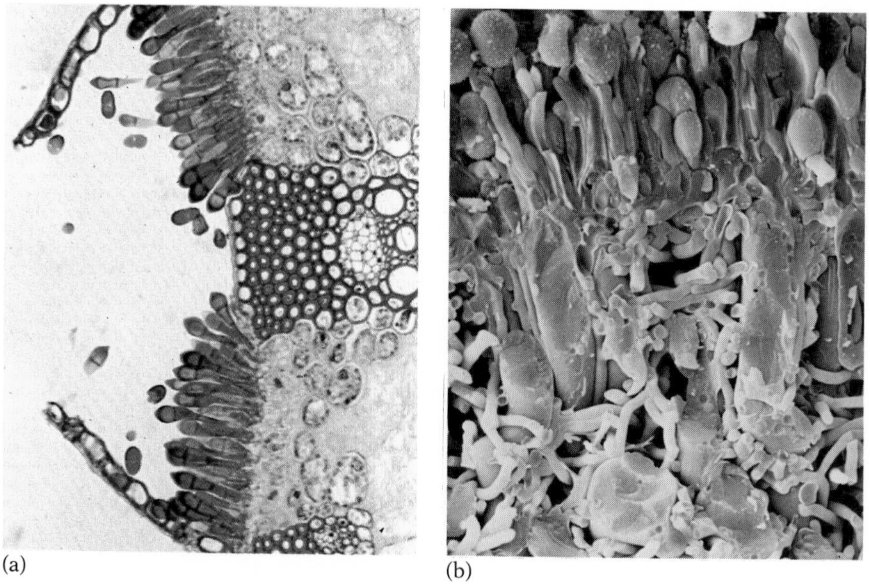

(a) (b)

## PANEL 29-3

# Rust Diseases of Plants

Rusts are of biological interest because of the requirement for specific alternate hosts in order to complete their life cycle. Rusts live one phase of their life cycle on one host plant and a second phase on another species. Cedar-apple rust, for example, occurs on juniper, and on apple or hawthorn trees. White pine blister rust infects the gooseberry and the white pine. Many pine trees are destroyed annually in the eastern United States by this fungus. *Puccinia graminis,* which causes stem rust of wheat, requires the European barberry as its alternate host. Besides wheat, a number of other grasses are attacked by varieties of *Puccinia graminis.* Each rust species produces two or more types of spores, each of which is specific for one of the hosts. In the wheat rust, basidiospores produced on wheat stubble in the spring grow only on barberry. The spores produced on barberry, called **aeciospores,** grow only on wheat. However, one type of spore produced on wheat, the **uredospore,** grows only on wheat.

Such host specificity can provide a means of controlling rust disease: eliminate one of the alternate hosts. Since 1917, the United States Department of Agriculture has been attempting to eradicate the European barberry from the country. Unfortunately, reducing the barberry population, the source of aeciospores that infect wheat in the spring and summer, does not eliminate another source of infection of wheat fields in the spring. Uredospores, which can start infections on wheat, are ordinarily killed in the winter, but uredospores blown in from warmer climates in the American southwest infect the new wheat plants in the north.

The most effective means of controlling stem rust of wheat has been the breeding of resistant strains of wheat. Usually, a given wheat variety resists infection for years, until a mutation in the rust fungus gives rise to a strain capable of growing on the resistant wheats. Genetic engineering also has potential for introducing genes for disease resistance from various wild species of grasses into wheat. The reservoir of the many naturally occurring disease-resistant species in wild populations is one of many reasons for preserving our remaining wilderness.

angel." When ingested even in small amounts, it is nearly always fatal. The symptoms do not appear immediately, but by the time they develop, 24 hours later, it is usually too late to treat the illness.

Two groups of toxins, named **phalloidin** and **α-amanitin,** have been isolated from *Amanita phalloides.* Both attack the liver, but by different mechanisms. Phalloidin binds to microfilaments and to the plasma membranes of cells and to organelle membranes, causing leakage of their contents. It is less toxic than amanitin, which acts in the cell nucleus, blocking nucleolar function. Amanitin has been shown to bind to an RNA polymerase and thus blocks messenger RNA synthesis. In addition, α-amanitin causes kidney damage and attacks the stomach and intestines. It is the toxin responsible for the severe gastrointestinal symptoms that appear several hours after eating *Amanita.* In recent years, some success in treating mushroom poisoning has been reported by using thioctic acid, a vitamin in some animals, as an antidote. The fatal consequences of consuming even one *Amanita phalloides* make it imperative that wild mushroom gatherers know their mushrooms.

Sexual Reproduction in Higher Fungi. Fertilization or syngamy in the sexual reproduction of the higher fungi is an unusual process in which the parental nuclei, following fusion of compatible hyphae, or **plasmogamy,** divide repeatedly. Nuclear fusion or **karyogamy,** which occurs later, is thus separated in time from the initial cytoplasmic union of gametes. In these higher fungi the uniting cells are the tips of hyphae. In the Ascomycetes the hyphae are specialized as male and female structures (antheridium and ascogonium), but in the Basidiomycetes undifferentiated somatic hyphae of opposite mating type unite. When the nuclei from each parent have multiplied they become distributed in pairs, one nucleus from each parent to each cell of the mycelium. The mycelium is said to be **dikaryotic,** meaning that there are two nuclei in each cell. Dikaryotization occurs after the ascocarp is formed in the Ascomycetes, but at the start of basidiocarp formation in Basidiomycetes; the basidiocarp is thus entirely dikaryotic.

When the fruiting body is complete, karyogamy occurs in the hyphal tips that will become asci or basidia. Following karyogamy in the potential sporangium, meiosis occurs. Four nuclei result in the basidium and migrate to its tip where they are pinched off into basidiospores. In the ascus the four nuclei that result divide mitotically, and the ascus is cleaved into eight ascospores, usually arranged in a row. Packed together, the asci line the cup-like ascocarp; in the basidiocarp, the basidia cover the surface of the gills or pores. Reproduction of Ascomycetes and Basidiomycetes is presented in Figures 29-15 and 29-16.

Basidiocarps begin to form only after compatible strains of hyphae of opposite mating type have undergone plasmogamy. The uniting hyphae are indistinguishable in appearance from the rest of the mycelium. The dikaryotization that follows hyphal union and mitosis is an unusual phenomenon: the nuclei from each parent migrate through the cells of the other. Eventually, all cells of each parent become dikaryotic, with each cell bearing its own nucleus and one migrated from the other strain. As the mycelium continues to grow, another unusual phenomenon maintains this dikaryotic condition: at each mitosis, both nuclei of a cell and the cell itself divide, and three of the nuclei become separated by a new cell wall from the nucleus remaining in the precursor cell. Then one daughter nucleus migrates back through a tube, called a **clamp connection,** to the precursor cell, where it joins the remaining nucleus. The process leaves a pair of nuclei, one from each parental strain, in each cell, the terminal cell now being ready to divide again (Fig. 29-16c).

***The Fungi Imperfecti.*** In a large number of fungi a sexual or "perfect" stage has never been observed. Some may have lost this capacity. Be-

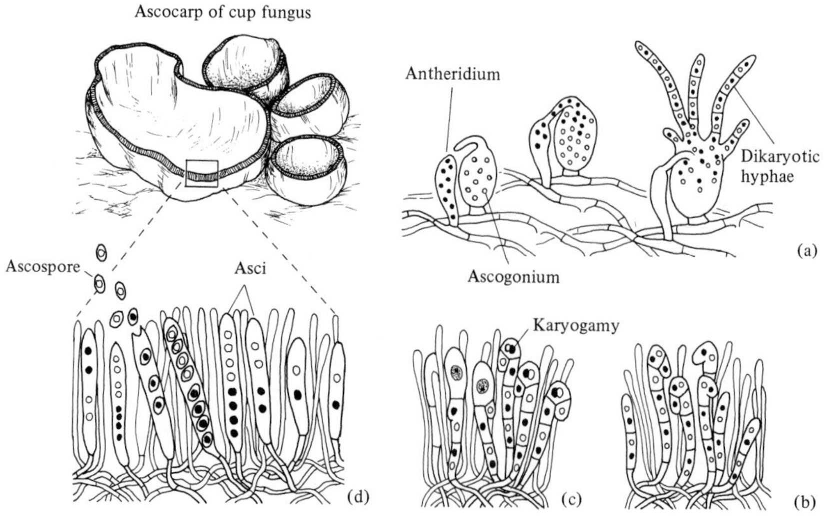


**FIG. 29-15** Reproduction in Ascomycetes. Ascospore formation results when a special hypha (of an ascogonium) unites with an antheridium (a). The resulting dikaryotic hyphae produce asci over the surface of the ascocarp (b–d). Karyogamy leads to zygote formation in the terminal cell (c) followed by meiosis and ascospore formation (d).

**FIG. 29-16** Reproduction in Basidiomycetes. (a) Plasmogamy of compatible strains leads to the dikaryotic condition. (b) Dikaryotic mycelium initiates basidiocarp. (c) Dikaryotization is maintained by clamp connections. (d) Basidium formation, accompanied by karyogamy, meiosis, and basidiospore formation.

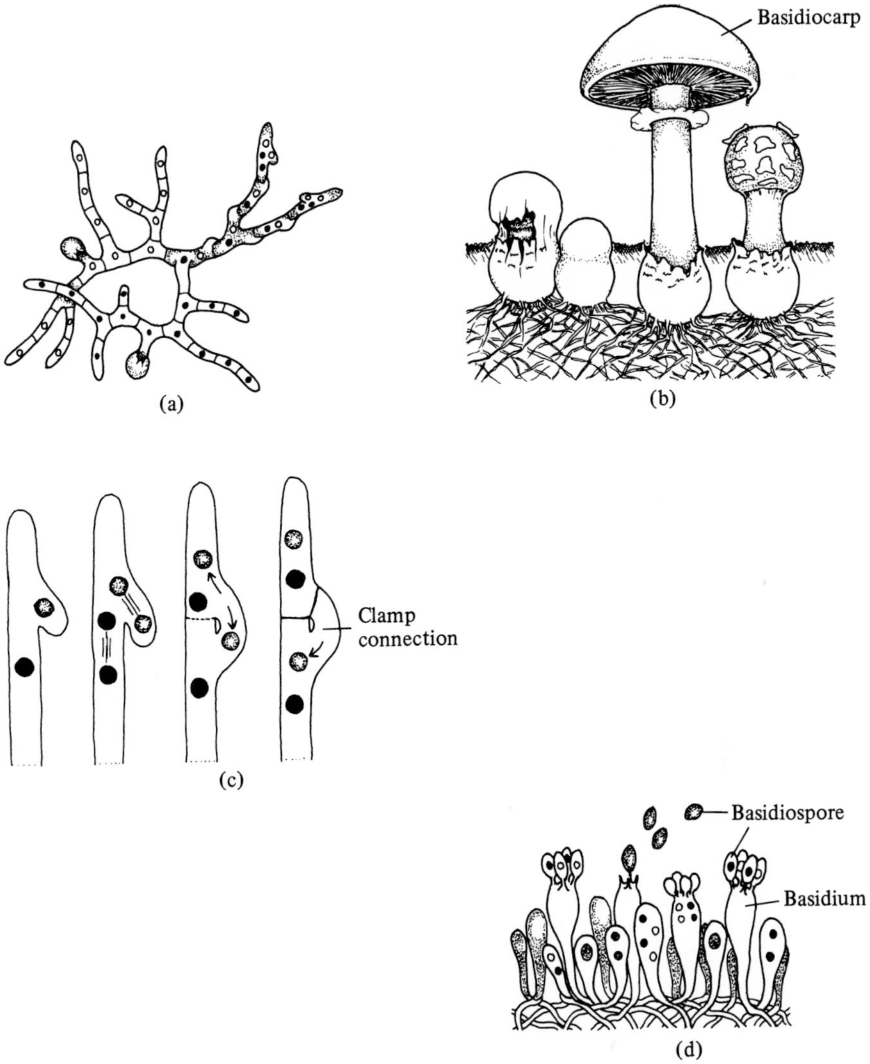


**FIG. 29-17** A nematode-trapping fungus, *Dactylella bembicodes.*

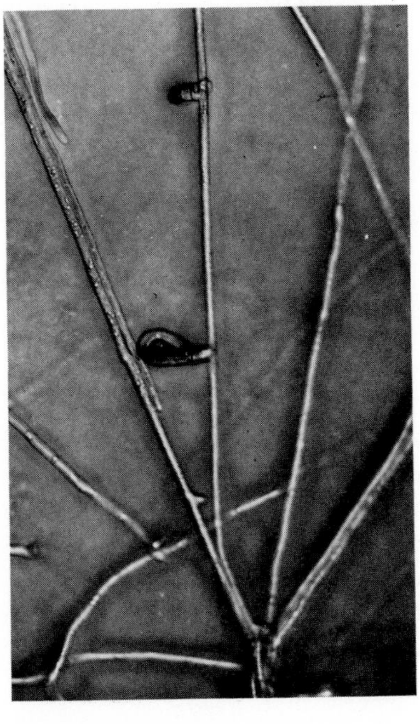

cause fungal classifications are based on modes of sexual reproduction, it is impossible to classify them. All fungi never observed to reproduce sexually are accordingly relegated to a catch-all category, the Fungi Imperfecti. Occasionally, as in the case of *Ajellomyces,* the pathogen of blastomycosis, the sexual stage is discovered and the fungus is then reclassified.

Some of these Fungi Imperfecti are human pathogens, causing such human skin diseases as athlete's foot and scalp "ringworm." Others cause serious systemic diseases, such as histoplasmosis. A number of plant diseases are also caused by the Fungi Imperfecti. Study of some of these pathogens is now revealing how fungal toxins produce disease symptoms and, more important, how disease-resistant host plants protect themselves (Panel 29-4).

Some species of Fungi Imperfecti that reside in the soil are truly remarkable in that they develop prehensile structures that trap nematodes when these worms are present (Fig. 29-17). The worms then become a source of nutrients for the fungus. Polypeptides extracted from nematodes have been shown to induce such fungi to form traps when added to the culture medium in which the fungus is growing. The amino acid valine appears to be the active factor, for when added alone to the culture medium, it induces trap formation in predatory fungi.

## PANEL 29-4

# Plants' Defense Against Disease

Unlike animals, plants lack a specific immune system. Nevertheless, they have evolved protective mechanisms against fungal pathogens. The simplest defenses are mechanical: the waxy cuticle, composed of cutin, that covers the surface of leaves and young stems; and the cork, with its suberin walls, which protects older stems. Fungal hyphae can enter the leaves through the stomata, however, whereupon they penetrate and parasitize the plant cells within the leaf. Some fungal pathogens secrete an enzyme that digests the cuticle and enables direct penetration of the epidermal cells of leaves at any point.

Another line of defense against fungi is the walling off of the invading fungus by the host plant. In such cases, the plant cells surrounding the lesion produced by the fungus develop thick cell walls and produce inhibitory substances that block further fungal growth.

The death of infected plant cells often results from toxins produced by the fungus. Some toxins bind to certain proteins in the plasma membrane of host cells; the complex is thought to kill by altering host cell permeability. The cell membranes of resistant hosts apparently lack these specific binding proteins, accounting for some of the specificity of plant pathogens for certain plant species.

Another form of resistance to fungal pathogens involves **phytoalexins** (from the Greek *alexin*, "warding-off substance"), substances produced by the host plant in response to the presence of a pathogen. They were discovered in 1940 in potato varieties resistant to the potato blight pathogen. In such cases, it was noticed that the fungus growth was limited to the infection site, an area of hypersensitivity in which there is rapid host cell death. This observation led to the discovery that the wounded host cells produced a chemical in the hypersensitivity reaction that killed or inhibited the fungus. The substance was called phytoalexin. Over 100 different phytoalexins have since been identified from various plants.

Recently it has been shown that phytoalexins are synthesized in the host in response to the presence of a product of the fungal pathogen. A factor isolated from culture filtrates of pathogenic fungi induces phytoalexin production when introduced into the plant host. Such a fungal substance is called an **elicitor.** The elicitor from a fungus (*Colletotrichum*) that causes smudge disease of peas is a glycoside. It induces pea cells to produce the phytoalexin phaseolin (from the Latin *phaseolus*, "pea").

The possibility of introducing genes for production of phytoalexins as a protection against plant pathogens has generated great interest among plant pathologists and molecular geneticists. If the genes for phytoalexins could be introduced into crop plants it would be of considerable value in agriculture.

Other substances with potential for providing disease resistance in plants are the lectins. One of these naturally occurring glycoproteins, concanavalin (from the bean, *Canavallia*), was shown in 1889 to bind to human red blood cells. These substances were first named phytohaemagglutinins for this reason. Lectins have been used to identify some human blood groups because of their specificity for glycoproteins. Although their role in plants remains a puzzle, it now seems possible that they may function in recognition and defense against certain plant pathogens.

## SLIME MOLDS

The slime molds include two groups of unique organisms—the plasmodial slime molds (Myxomycetes) and the cellular slime molds (Acrasiomycetes)—that were once classified as fungi of the class Myxomycetes. In the vegetative or feeding phase, they lack cell walls and absorb food or engulf food particles in amoeboid fashion; hence, they are similar to protozoa in their nutrition. However, they form cellulose during the reproductive phase, and produce walled spores within rather intricate sporangia, and thus resemble fungi.

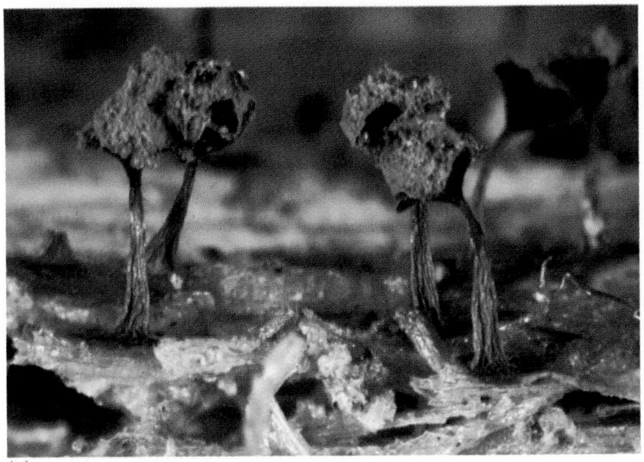
(a)

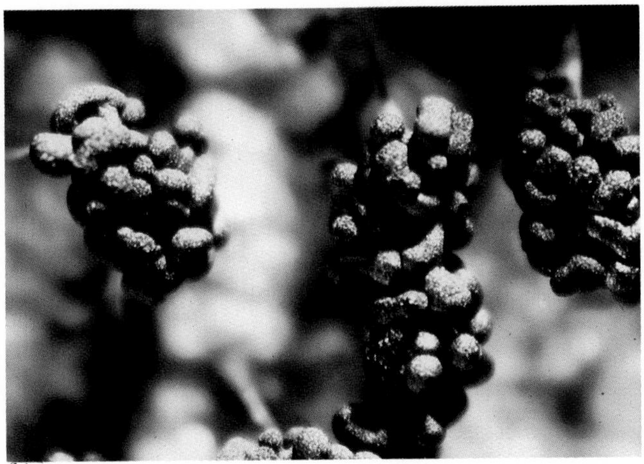
(b)

**FIG. 29-18** Fruiting bodies of 2 species of slime molds: (a) *Trichia;* (b) *Physarum.*

## PLASMODIAL SLIME MOLDS

In the vegetative phase, a plasmodial slime mold, lacking cell walls, consists of a mass of naked protoplasm, the multinucleate **plasmodium.** The plasmodium flows over moist, rotting logs, leaves, or other organic matter, taking up food. Under certain conditions, delicate fruiting bodies small enough to require a lens to be seen are formed (Fig. 29-18).

One species, *Physarum polycephalum,* is easily grown in the laboratory on oatmeal. In the vegetative state, the plasmodium (Fig. 29-19) provides a striking demonstration of protoplasmic streaming, and in the past it has been used to study the properties of living protoplasm. Also striking is the fact that its many hundreds of nuclei undergo mitosis almost simultaneously, and *Physarum* is now used in the study of mitosis. Ultimately, small black fruiting bodies develop in laboratory cultures. When the spores produced in these sporangia are released, they germinate into haploid amoebae, which may develop a flagellum. These amoebae may act as gametes, unite, and start new plasmodia.

**FIG. 29-19** (a) The plasmodium of the slime mold *Physarum* growing in the laboratory. (b) Nuclei within the plasmodium.

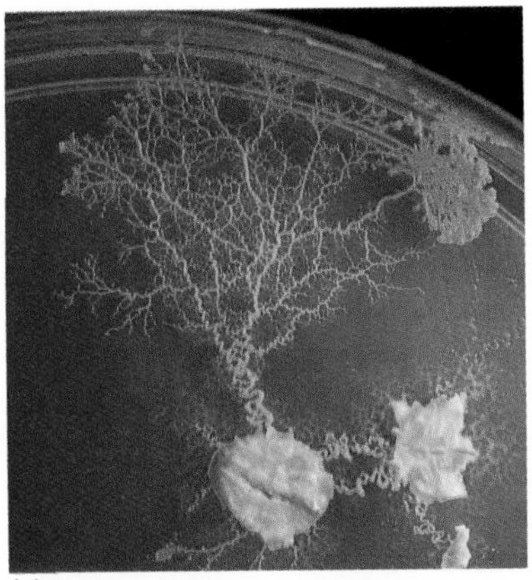
(a)

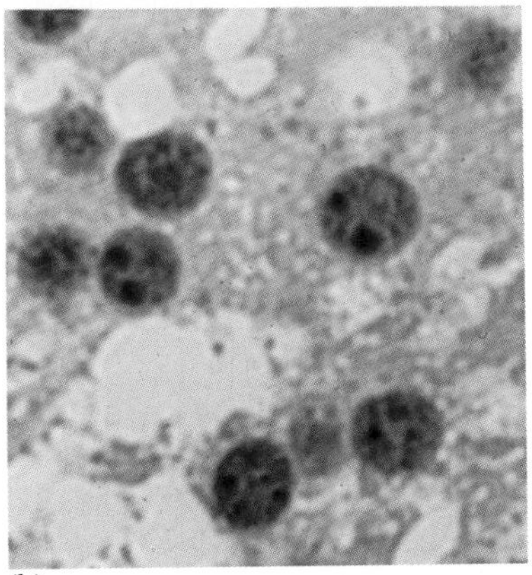
(b)

## CELLULAR SLIME MOLDS

The cellular slime molds differ from the plasmodial slime molds in that the vegetative stage consists of individual amoeboid cells, called **myxamoebae.** In laboratory cultures, myxamoebae feed on bacteria by engulfing them. In nature, the cellular slime molds are found on decaying organic matter and often on dung. Many species are limited in distribution to the dung of a particular animal species. One species, for example, has been found only on the dung of the howler monkey, while another occurs only on the dung of a species of eagle. As the myxamoebae feed, they continue to divide, increasing the population of cells before they enter the next reproductive phase.

A remarkable thing happens when the food supply becomes exhausted. The amoebae aggregate in great numbers, forming a sluglike **pseudoplasmodium** that, though composed of discrete cells, behaves as an individual organism, responding to environmental stimuli. For example, it may move toward warmth and away from light, with its component amoebae, initially independent cells, now behaving like cells of an organism. Ultimately, the pseudoplasmodium reorients and forms an erect fruiting body. At this stage, the myxamoebae have become dif-

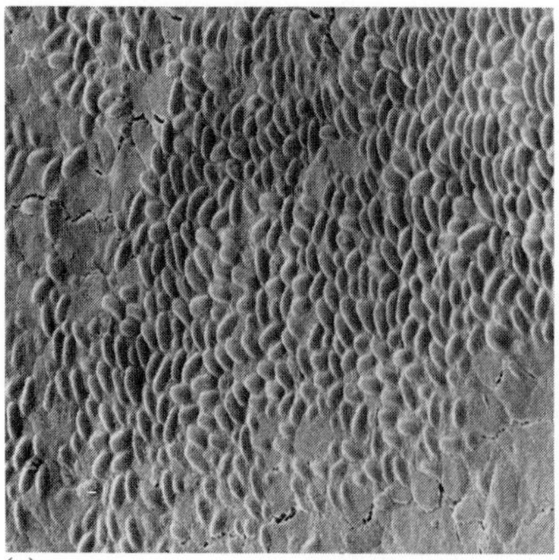

(a)

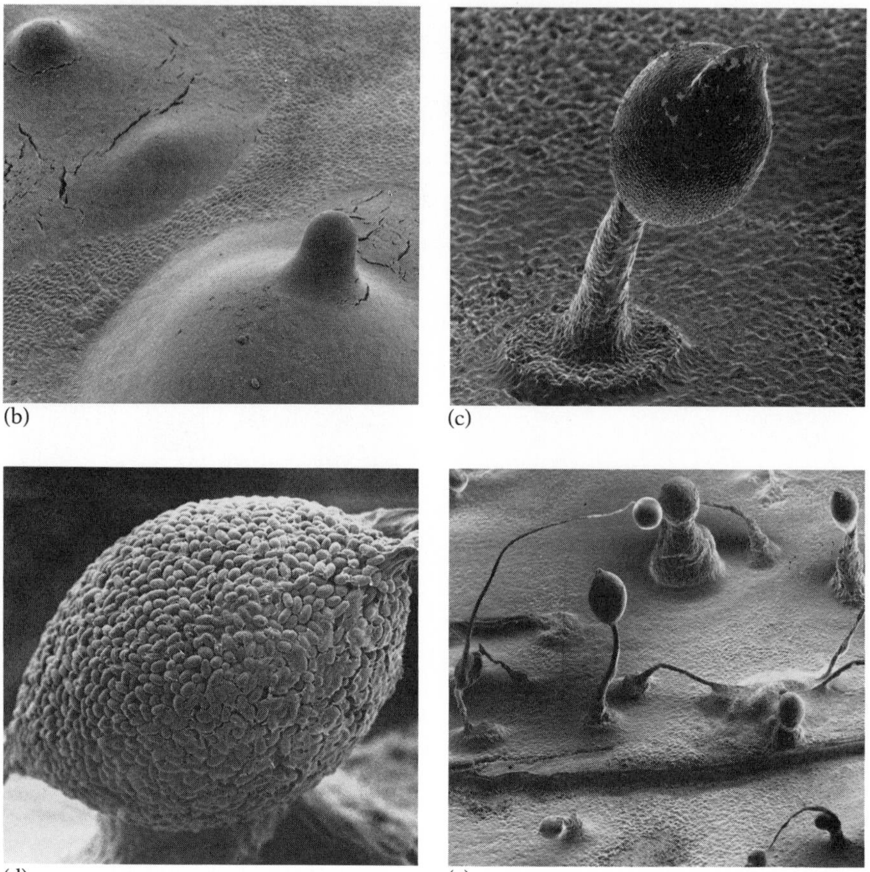

(b) (c) (d) (e)

**FIG. 29-20** Aggregation and fruiting body formation in the cellular slime mold *Dictyostelium.* (a) Aggregating amoebae. (b) Start of sorocarp formation. (c) Sorocarp, or fruiting body. (d) Sporangium showing spores. (e) Several sorocarps.

ferentiated into two kinds of cells: stalk cells and spores. The life cycle of a cellular slime mold is presented in Figure 29-20.

A number of interesting experiments have been performed on the development of the cellular slime molds. The myxamoebae have been shown to be attracted to a substance released by some of the amoebae and now identified as cyclic AMP, the "second messenger" of animal cells (see Chapter 14). This is the factor that induces them to aggregate. The amoebae of a given species are able to recognize one another and exclude foreign amoebae from the aggregation. This recognition of assembling cells has been shown to involve glycoproteins present in the plasma membrane only at certain stages of development. The cells of the pseudoplasmodium, which are initially identical, begin to differentiate into prestalk and prespore cells as they aggregate into the pseudoplasmodium. Their differentiation has been shown to require the synthesis first of specific messenger RNAs and then of specific proteins as the genetic program of development is transcribed and translated.

## LICHENS

Lichens are unusual "organisms" in that they represent a symbiotic association of two species: a fungus and an alga. The remarkable aspect of the association is that its form is so constant that lichens are assigned generic and specific names. The characteristic morphology of a given lichen is a property of the association and is not exhibited by either

**FIG. 29-21** Lichens. (a) Crustose lichen. (b) A foliose lichen. (c, d) Fruticose lichens.

symbiont individually. Lichen forms are classified as **crustose,** meaning that they are compact and appressed to a substratum (Fig. 29-21a); **foliose,** or leaflike (Fig. 29-21b); or **fruticose,** with a shrubby shape (Fig. 29-21c, d).

In a lichen, the fungal partner, or **mycobiont,** is commonly one of the ascomycetes, although, in a few genera, basiodiomycetes participate. The algal partner, or **phycobiont,** is commonly a green alga, although the cyanobacteria are phycobionts in some lichens.

Algal cells are dispersed in a regular pattern among the hyphae of the fungal mycelium. Certain fungal hyphae form clasping associations with the algal cells (Fig. 29-22), and obtain carbohydrate and other or-

**FIG. 29-22** A section of a lichen reveals the algae among the fungal mycelium. The fungus produces its characteristic spores. Fungal hyphae are closely appressed to the algal cells.

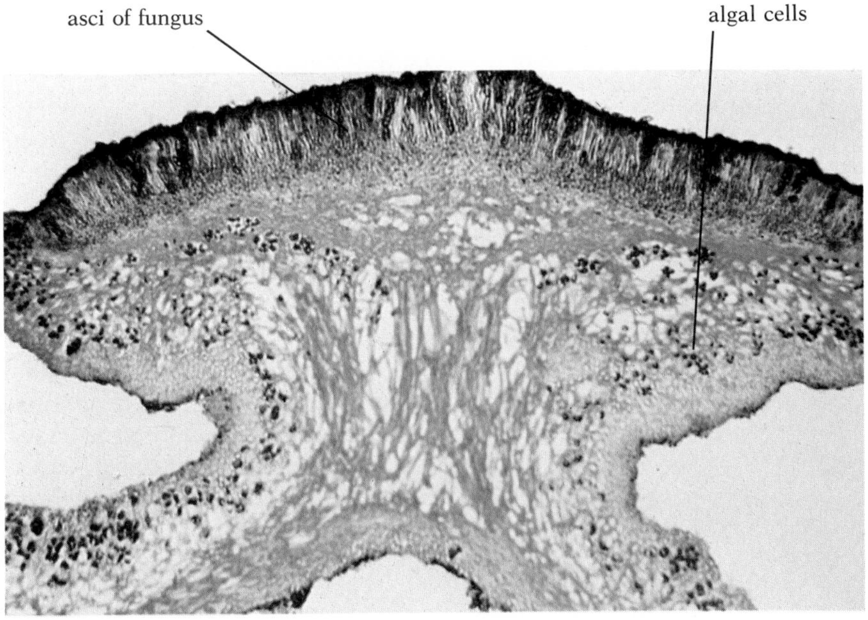

ganic substances from the alga, while the fungus provides water and minerals to the alga. The lichen association is extremely resistant to desiccation, allowing lichens to survive long, dry periods.

Lichens occur on tree trunks and branches, soil, and rocks. On rocks, they may act as pioneers, the first colonizers of new habitats as they gradually break down the rock through the action of the acids they secrete, thereby generating new soil particles. The organic material from decaying lichens adds a valuable component to the soil thus produced. The kinds of acids secreted vary with the species and are used in lichen identification.

Lichens grow at extremely slow rates: crustose lichens as little as 0.1 to 10 mm a year, and fruticose lichens as little as 2 to 4 cm annually. A large patch of a crustose lichen on a rock may represent more than 4,000 years' growth. Lichens are especially vulnerable to air pollutants and thus are sensitive indicators of air quality. We must usually venture several miles from a large city to find lichens growing.

Lichens reproduce only asexually, although the component ascomycetes may sexually produce asci. Fragments of a lichen may break off and start new individuals. Alternatively, some species produce special reproductive bodies called **soredia,** small clusters of algal cells surrounded by hyphae (Fig. 29-22). Soredia are disseminated by wind and rain.

In order to understand the nature of lichen associations and explain their origin, botanists have for years experimentally attempted to synthesize lichens from fungal and algal components. Although both components have been cultured singly, reconstitution of a lichen has been difficult. In recent years, some associations have been established by repressing the growth rates of both the alga and the fungus. Although some of these associations somewhat resemble lichen bodies, characteristic lichen structures have not resulted. The question of whether new lichens are being established in nature today from the association of algae and fungi is also unknown.

Lichens, in particular those known as reindeer moss, are a major component of the arctic tundra and are the predominant food of caribou. Thus, lichens are of prime importance to the Eskimos and to the Lapps, the people of northern Sweden, Norway, and Finland, who depend on the caribou for food and other products.

In addition to their role in soil formation and their importance as food for caribou, the lichens have been used by humans in the production of perfumes and as a source of dyes.

## Summary

The Fungi are important decomposers, breaking down nonliving organic matter into mineral form or living as parasites and causing disease. Most fungal bodies are hyphae, branching filaments that make up a tangled mycelium. The hyphae of simpler groups of fungi—the Chytridiomycetes, Oomycetes, and Zygomycetes—lack cross walls and are thus multinucleate. Hormones have been shown to regulate sexual reproduction in these groups. They also reproduce asexually by forming motile zoospores, as in Chytridiomycetes and Oomycetes, or airborne spores, as in Zygomycetes. The higher fungi—Ascomycetes and Basidiomycetes—are distinguished by hyphae with cross walls and by the formation of characteristic fruiting bodies in which sexual reproduction leads to formation of special spores. Union of hyphae, a process known as plasmogamy, leads to the formation of mycelia with two unfused nuclei per cell, called dikaryotic mycelia, which participate in formation of the fruiting bodies. In Ascomycetes, nuclear fusion, called karyogamy, leads to formation of a cigar-shaped sac, or ascus, which becomes filled with eight haploid ascospores following meiosis. In the Basidiomycetes, karyogamy occurs in a club-shaped cell, or basidium, which produces four basidiospores as a result of meiosis. The behavior of the nuclei in forming and maintaining a dikaryotic mycelium is unusual, including, as it does, nuclear migration and clamp connection formation.

The Basidiomycetes influence our lives more than the other fungi. They are the source of edible, hallucinogenic, and poisonous mushrooms. Other species form symbiotic mycorrhizae with tree roots. Some cause smuts and the rusts, destructive diseases of cereal grains.

The Fungi Imperfecti include those fungi never observed to undergo sexual reproduction. They include molds that cause diseases in animals and plants, and unusual species that trap and consume nematodes.

Not true fungi, the unique slime molds are of interest for their unusual morphology. They are animal-like in nutrition, but form fungus-like sporangia and spores. The plasmodial slime molds provide a view of naked protoplasm and its properties. The cellular slime molds exist as one-celled organisms called myxamoebae, which feed and then aggregate to form a multicellular organism. Problems of cellular recognition and differentiation can be studied at a simple level in these cellular slime molds.

A lichen is a symbiotic organismal association between a fungus and an alga. Despite being composed of two organisms, the lichens exhibit a consistent form that is identifiable as to genus and species. They reproduce from fragments containing the algal and fungal components. Presumably, lichens arise when a fungus and alga grow together in nature. Attempts to reconstruct lichens in the laboratory have not met with complete success. Lichens play a significant role in the food chain of the arctic tundra, a chain that leads quickly to humans.

## Recommended Readings

AINSWORTH, G. C., and A. S. SUSSMAN. 1965–1973. *The Fungi: An Advanced Treatise,* 4 vols. Academic, New York. A complete treatment of the groups of fungi by experts in the field and a valuable reference source.

ALEXOPOULOS, C. J. and C. W. MIMS. 1979. *Introductory Mycology,* 3d ed. Wiley & Sons, New York. A revised version of well-known text on fungi, with emphasis on the classification and structure of fungi and their role in human life.

BAILEY, J. A. and J. W. MANSFIELD. 1982. *Phytoalexins.* Wiley, New York. A collection of scholar's articles on these molecules of current interest in the control of plant diseases.

BOLD, H. C., C. J. ALEXOPOULOS, and T. DELEVORYAS. 1980. *Morphology of Plants and Fungi.* Harper & Row, New York. Includes a general treatment of the fungi, with emphasis on structure and reproduction.

BONNER, J. T. 1968. *The Cellular Slime Molds,* 2d ed. Princeton University Press, Princeton, N.J. A readable account of the nature of these organisms and some of the early research dealing with their development.

DEVERALL, B. J. 1977. *Defense Mechanisms of Plants,* Cambridge Monographs in Experimental Biology. Cambridge University Press, Cambridge, England. A monograph dealing with host-parasite interactions, fungal infection, and host resistance mechanisms of plants.

HALE, M. E., JR. 1975. *The Biology of Lichens,* 2d ed. University Park Press, Baltimore, Md. A general consideration of the lichens, covering structure and physiology, ecology, and human uses of lichens.

KLEIN, R. M. 1979. *The Green World: An Introduction to Plants and People.* Harper & Row, New York. An interesting introduction to plants, with emphasis on their importance to people.

PATRUSKY, B. 1983. "Plants in their own behalf." *Mosaic,* 14:(2)32–39. A current account of research into the defenses of plants against disease.

SCAGEL, R. F., R. J. BANDONI, J. R. MAZE, G. E. ROUSE, W. B. SCHOFIELD, and J. R. STEIN. 1982. *Nonvascular Plants: An Evolutionary Survey.* Wadsworth, Belmont, Calif. A detailed consideration of fungi and slime molds, as well as of prokaryotes and algae, by experts in each area.

# 30 Plant Diversity and Evolution

OUTLINE



The plant kingdom includes all the multicellular photosynthetic organisms. All plants have in common the presence of chlorophyll *a*, essential for oxygen-yielding photosynthesis, as well as other chlorophylls and additional pigments that absorb light and contribute to the photosynthetic process. Plants range in complexity from filaments, colonies, and sheets of the simple green algae to the most advanced, the flowering plants, which, with their roots, stems, leaves, and characteristic flowers, are the most recent group of plants to have evolved on earth.

The simpler plants include several groups of algae, plants found predominantly in water and thus not subjected to the rigors of dry-land existence. The algae are distinguished from other plants in that their gametes are produced within single cells instead of in multicellular organs. Phycologists, scientists who study algae, separate the algae into several divisions* based largely on the pigments they contain and hence on their color.

The higher plants, with more complex tissues and greater specialization of cells, are found predominantly on the land. They possess characteristics that enable them to survive out of water. One of these is the retention of progeny for a time as an embryo within the maternal parent. Because of this embryonic phase, the higher plants are sometimes called **embryophytes** (from the Greek *phyte,* "plant"). The embryophytes most successfully adapted to land have vascular tissues, specialized for the conduction of water and the transport of food; these are the vascular plants (division Tracheophyta). The embryophytes that lack vascular tissues, the mosses and liverworts (division Bryophyta), are small in comparison to vascular plants, restricted in size by their limited ability to transport water.

Evolution among vascular plants led to the appearance of the seed, a structure that enabled the embryo to survive under adverse conditions. This protective structure was a decided advantage to plants on land, where seasonal changes in environment—such as freezing or desiccation when soil dries or soil water turns to ice—may challenge an organism. Two groups of seed plants emerged as the vascular plants evolved. The most advanced seed plants, the angiosperms (from the Greek *angios,* "vessel or container," and *sperm,* "seed"), or flowering plants, have seeds that are enclosed in a container, the fruit, which aids in their dissemination. Several groups of seed plants have unenclosed, or naked, seeds; these are the gymnosperms (from the Greek *gymnos,* "naked," and *sperm,* "seed").

## SOME DIVISIONS OF ALGAE

While multicellular algae are classified as plants in this text, the single-celled algae are placed in the kingdom Protista in accordance with current practice. For convenience, however, and because of obvious evolutionary relationships between protistan and multicellular algae, we consider both types in this chapter.

The divisions of algae differ from one another not only in the pigments they contain, but in chloroplast organization, cell wall composition, the form in which food is stored, and various structural features (Table 30-1). While all divisions of algae have chlorophyll *a*, they differ in other chlorophyll molecules as well as in accessory pigments that contribute to photosynthesis (Chapter 7). The assembly of pigments imparts a characteristic range of colors. In algae that are grass green in color (divisions Chlorophyta, Charophyta, Euglenophyta), chlorophylls *a* and *b* predominate. Algae that are yellow to dark brown or reddish (divisions Phaeophyta, Chrysophyta, Pyrrophyta) have chlorophylls *a* and *c*, masked by various xanthophyll pigments. The red algae (division Rhodophyta) have chlorophyll *a*, hidden by phycobilin pigments similar to those found in the prokaryotic Cyanobacteria.

In addition to the biochemical distinctions among the divisions of algae, the groups exhibit differences in morphology, mode of reproduction, and form of gametes. Each division also exhibits a range in form, often from single cells to complex bodies. Life cycles vary within each division, as do the types of gametes, which may differ in the number, position, and motion of their flagella.

* Botanists subdivide the plant kingdom into divisions; zoologists subdivide the animal kingdom into phyla.

**TABLE 30-1 Comparative Summary of Characteristics in Divisions of Eukaryotic Algae**

| *Division* | *Pigments in addition to chlorophyll* a | *Number of thylakoids per group in chloroplast* | *Storage products* | *Cell wall* | *Flagella* |
|---|---|---|---|---|---|
| Green algae (Chlorophyta) | Chlorophyll *b*, lutein* | Two to many | Starch | Cellulose, mannan, or none; protein, $CaCO_3$ in some | One to many apical |
| Euglenophyta | Chlorophyll *b* | Usually three | Paramylon, lipids | None (protein pellicle) | One to two apical |
| Stoneworts (Charophyta) | Chlorophyll *b* | Many | Starch | Cellulose; $CaCO_3$ in some | Two subapical |
| Yellow-green algae (Xanthophyta) | Chlorophyll *c* | Three | Chrysolaminarin (possibly), lipids | Cellulose or none | One to many apical |
| Golden-brown algae (Chrysophyta) | Chlorophyll *c*, fucoxanthin* | Three | Chrysolaminarin, lipids | Cellulose or none; silica in diatoms; $CaCO_3$ in some | One to three apical or absent |
| Dinoflagellates (Pyrrophyta) | Chlorophyll *c*, peridinin* | Three | Starch, lipids | Cellulose plates or none | Two lateral |
| Brown algae (Phaeophyta) | Chlorophyll *c*, fucoxanthin* | Three | Laminarin, mannitol, lipids | Cellulose, alginates | Two lateral |
| Red algae (Rhodophyta) | Chlorophyll *d*, phycobilins | One | Glycogenlike starch | Cellulose or xylan; agar; carrageenan; $CaCO_3$ in some | Absent |

* Xanthophyll pigments in abundance in addition to other carotenoids.
NOTE: Modified from table in Ray, Steeves, and Fultz. *Botany*. Saunders.

The various characteristics of present-day algae are used in making deductions about algal interrelationships and evolution. This deductive approach is necessary because the fossil record for eukaryotic algae is quite sparse, even though their blue-green prokaryotic ancestors are recorded in rocks 2 billion years old.

## ALGAE WITH GREEN COLORATION

The three divisions of green-colored algae included here differ enough to have led phycologists to place them in separate divisions despite their common pigments. The Chlorophyta include a diversity of forms and cytological features important in determining their evolutionary relationships. The other two divisions are each homogeneous and therefore fit quite easily into separate groups.

### DIVISION CHLOROPHYTA: THE GREEN ALGAE, ANCESTORS OF THE HIGHER PLANTS

Of all the algal divisions, the green algae (division Chlorophyta) appear to be the most likely ancestors of the higher plants. Like the higher plants, they have chlorophylls *a* and *b*, and they store carbohydrate as starch. Many have cell walls of cellulose, as do the higher plants.

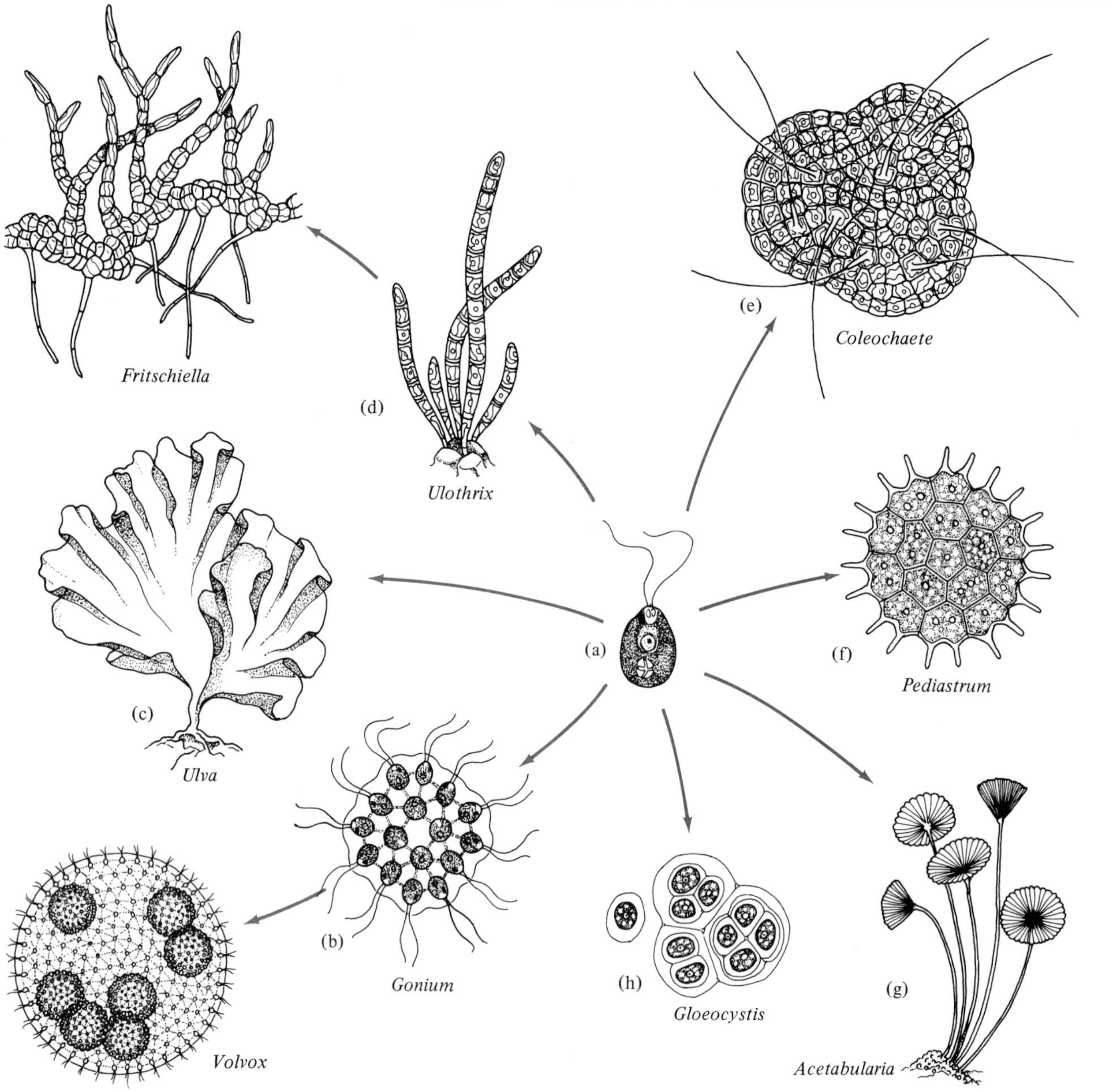


**FIG. 30-1** Diverse forms in the green algae. It is thought that from solitary flagellated unicells (a) have evolved (b) motile colonies of unicells; (c) multicellular sheets of one or two cell layers; (d) unbranched and branched filamentous forms; (e) forms with vegetative cell division forming tissue (parenchyma); (f) colonies formed by aggregating zoospores; (g) multinucleate tubular algae; (h) nonmotile unicells and colonies.

***Diversity of Form.*** The green algae include a wide diversity of forms, ranging from unicellular individuals, or **unicells,** to large multicellular types (Fig. 30-1). From the many solitary, motile unicellular algae arose several colonial species, loosely organized associations of similar or even identical cells that remain together after cell division. Among multicellular species are sheets of cells, as well as unbranched and branched filamentous types. Non-motile colonies of cells include some that have lost their flagella and others formed by aggregation of zoospores. Other algal species of interest include tubular **coenocytes,** in which cytokinesis fails to occur or is infrequent, resulting in a plant that is essentially a large, branched, multinucleate cell or a body of multinucleate cells.

***Diversity of Habitat.*** Green algae are found in fresh and salt water, in the soil, on other organisms, and within other organisms as symbionts. Unattached microscopic forms constitute a significant part of **plankton** (from the Greek *planktos,* "wandering"), the free-floating aquatic organisms so important at the base of the aquatic food chain (see Chapter 33).

***Diversity in Reproduction.*** Sexual reproduction in the green algae spans the scope of gamete types and life cycle (life history) variations discussed in Chapter 21. Examples of species with isogamy, heterogamy, and oogamy can be found even within the same genus. Life cycles may be haplontic or diplontic with single, free-living individuals or may entail alternation of haploid and diploid generations (alternation of generations) in the completion of sexual reproduction.

***Cell Structure in Green Algae.*** Single-celled green algae with a pair of flagella, similar to *Chlamydomonas,* have been thought by some to be ancestral to other green algae. Because of its ancestral nature, the cell of *Chlamydomonas* is of special interest.

A *Chlamydomonas* cell has a pair of anterior flagella protruding through the thin cell wall of glycoprotein (Fig. 30-2). A single, cup-shaped chloroplast occupies the peripheral cytoplasm. The nucleus is located in the cytoplasm of the "cup." A contractile vacuole expels water at the anterior end. A light-sensitive eyespot lies within the chloroplast at the base of the flagella; with this eyespot and its flagella, the cell is able to orient itself and swim in relation to light. Also within the chloroplast is a **pyrenoid,** a body associated with starch storage and unique to certain algal cells.

Biflagellate cells of the *Chlamydomonas* type are found in several different colonial algae that form a line leading to *Volvox,* a rolling ball of cells (Fig. 30-1b). Furthermore, such biflagellate cells occur in the life cycles of many other green algae in the form of reproductive zoospores and gametes.

The nonreproductive vegetative cells of nonmotile and multicellular green algae differ from the *Chlamydomonas* cell in their lack of flagella, in cell shape, and in the presence of a thicker cell wall and the

**FIG. 30-2** Electron micrograph of *Chlamydomonas* in division.

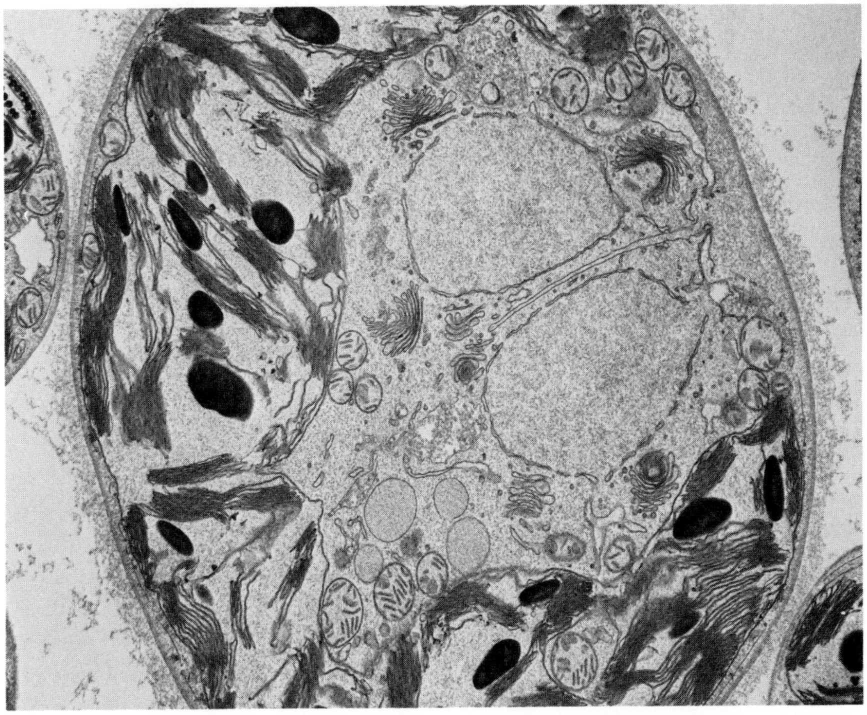

large central vacuole typical of many higher plants. An example is *Ulothrix*, a filamentous alga (Fig. 30-1d).

Chloroplasts in the green algae vary in number from one to many per cell and in form from one species to another. In some, chloroplast shape conforms to the outline of the cell, as in *Chlamydomonas*. In other species, chloroplasts may be ribbonlike, cylinders, bands, lobed bodies, or small ovoid plastids, as in higher plants. Thylakoids of the chloroplasts are grouped in layers of two or more, and the entire chloroplast is bounded by a double membrane.

***Cell Division.*** In the green algae, zoospore or gamete formation occurs when numerous mitoses within a parental cell are followed by multiple cleavages of the cytoplasm as a plasma membrane is formed about each cell. In contrast, cytokinesis (cell division) associated with vegetative growth in multicellular algae involves formation of a new cell wall between the two daughter cells immediately following mitosis. Cytokinesis may occur either by a furrowing inward of the plasma membrane from the cell's periphery or by cell plate formation, as described in Chapter 4.

Electron microscope studies of green algae have shown that they can be divided into two distinct groups on the basis of the distribution of microtubules during cytokinesis. In one group, the cell plate forms when the Golgi vesicles assemble on microtubules of the mitotic spindle, forming a structure called a **phragmoplast** (Fig. 30-3b). A phragmoplast is typical of cytokinesis in all higher plants. In the other group, after the spindle microtubules have disappeared, a new set of microtubules appears perpendicular to them, that is, in the plane of the cell plate (Fig. 30-3c–e). This arrangement is called a **phycoplast.** In the phycoplast type of division, the daughter nuclei come close together, while in the phragmoplast type, the persistent spindle microtubules are thought to hold the nuclei apart. The two types of microtubule distribution are also found in those green algae where furrowing occurs during cell division. Certain other features accompany these patterns of cell division, such as the manner in which flagellar microtubules are anchored, and further distinguish the two groups of algae. Since only the phragmoplast has persisted in higher plants, they are thought to have evolved from the line of green algae with phragmoplasts.

Centrioles are associated with mitosis in some of the green algae, but they do not persist through interphase, as they do in animal cells. The centrioles do not seem to be involved in formation of the mitotic spindle but serve instead as microtubule-organizing centers during flagella formation in zoospores and gametes.

***The Origin of Organelles in Eukaryotes.*** The presence of membrane-bounded organelles, such as the nucleus, mitochondria, and chloroplasts, distinguishes green algal cells and those of all eukaryotes from cells of the prokaryotic bacteria and Cyanobacteria. The evolution of organelles enabled metabolic processes to become localized within the cell and represented a major evolutionary advance. The nature of their origin is thus of interest. The structural resemblance of mitochondria to bacteria and of chloroplasts to cyanobacteria led to the formulation of the **endosymbiont theory.** According to this hypothesis, mitochondria and chloroplasts originated by the invasion of early nucleated cells by prokaryotes that continued to reproduce in synchrony with the host and persisted as **endosymbionts.** Several lines of evidence support this the-

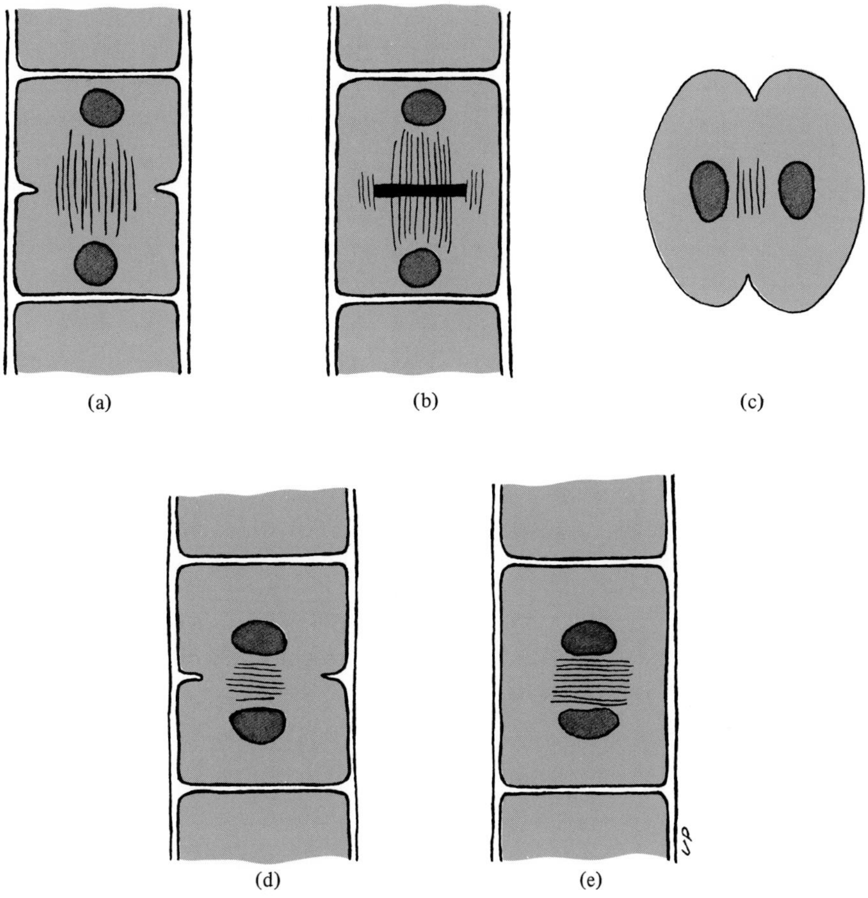


**FIG. 30-3** Two types of cell division in the green algae. In the phragmoplast type, found in higher plants, furrowing (a) and cell plate formation (b) during cytokinesis are associated with the spindle fiber microtubules. In the phycoplast type found in chlamydomonad protistans (c) and in filamentous forms (d, e), a new set of microtubules forms in the plane perpendicular to where the spindle microtubules had been, both in furrowing (c, d) and in cell plate formation (e).

ory. For example, the DNA in mitochondria and chloroplasts is circular, as in prokaryotes. Furthermore, ribosomes in organelles and prokaryotes are of similar size, smaller than that of eukaryote cells. Whether or not evolution of the green algae from prokaryotes involved endosymbionts, the evolutionary step from the one cell type to the other was a major one.

***Evolutionary Trends in the Green Algae.*** The great diversity among the green algae has encouraged phycologists to attempt to classify them so as to indicate evolutionary relationships. In that scheme, one interesting trend is evident in the line leading to *Volvox*. The cells in each of the colonial forms in that line is identical in appearance to a *Chlamydomonas* cell. The trend is one of increasing complexity, from few-celled colonies such as *Gonium* to the many-celled colony of *Volvox*, in which there has evolved a division of labor among cells.

Of course, not all phycologists agree on such evolutionary sequences as the *Chlamydomonas-Volvox* series. Indeed, whether the chlamydomonad type of green flagellate gave rise to the higher plants is questioned. It is probable that another group of green flagellates with phragmoplast division and in which the cell and flagella are covered with minute scales are the likely progenitors of higher plants. However, no matter what the evolutionary pathway, flagellated green algae seem to have been the starting point of plant evolution.

## DIVISION CHAROPHYTA: THE STONEWORTS

**FIG. 30-4** The charophyte *Chara.*

At one time, a group of complex multicellular algae with biochemical properties matching those of the higher plants was classified with the green algae. They have chlorophylls *a* and *b* and store starch. However, the four genera, including the more common *Chara* and *Nitella,* are now grouped in a separate division, the Charophyta (Fig. 30-4). With their upright "stems" and whorls of filamentous branches, charophytes form a dense covering on the bottom of shallow ponds and aging alkaline lakes. The fact that some species precipitate calcium carbonate from the water to form a limestone covering suggested their common names: stoneworts or brittleworts. Although biochemically like green algae, the complex multicellular organization, including an apical meristem and unique reproductive structures, have caused them to be separated from the green algae. Since their egg is enclosed in a jacket of cells, more characteristic of higher plants than of algae, some biologists even question whether *Chara* is an alga. However, their multinucleate cells and the unicellular nature of the gametangia reveal their algal level of organization. The occurrence of phragmoplasts in the Charophyta in addition to their structural complexity provides evidence of an evolutionary relationship with the land plants.

## DIVISION EUGLENOPHYTA

The euglenoids are protists and distinct enough from other algae to be classified by phycologists in a separate division that includes some 40 genera of unicellular and colonial forms, all lacking cellulose cell walls (Fig. 30-5). As planktonic algae, they often form a green scum on stagnant ponds. Despite having chloroplasts, they possess some traits associated with animals, such as the ability to ingest, digest, and metabolize food. Multiple chloroplast envelopes suggest that the chloroplasts may have originated as endosymbionts. If their chloroplasts are eliminated, which can be accomplished in the laboratory by ultraviolet irradiation, high temperature, or culture in the dark in the presence of streptomycin, the euglenoids can survive on other food sources. Such cells can reproduce and can be maintained in culture on complex organic media. No longer green, they are indistinguishable from certain naturally occurring colorless protozoa that absorb organic nutrients or ingest food particles.

**FIG. 30-5** *Euglena.*

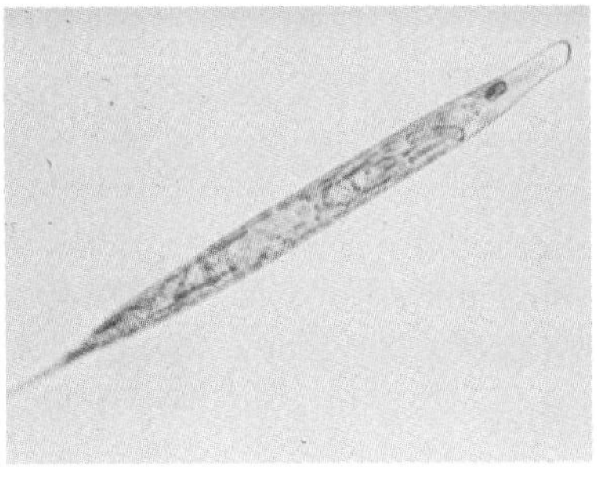

Lacking a cell wall, the euglenoid cell is bounded by the plasma membrane. Inside the plasma membrane is a grooved proteinaceous membrane called a **pellicle,** elastic enough to enable turning and flexing of the flagellated cell. Each cell has several chloroplasts, which contain chlorophylls *a* and *b*, as in the green algae, but the photosynthate produced is stored, not as starch within the chloroplast, as in green algae, but in the cytoplasm as **paramylon,** a polysaccharide unique to euglenoids.

FIG. 30-6 *Laminaria.*

# ALGAE GOLDEN-BROWN TO BROWN IN COLOR

## DIVISION PHAEOPHYTA: THE BROWN ALGAE

Algae that are colored shades of golden brown to brown and even reddish have chlorophylls supplemented by various different xanthophyll molecules, usually named for the particular genus in which the pigment is found or in which it was initially identified. Xanthophylls are carotenoids that differ from carotene in having oxygen in the molecule (see Chapter 7).

***Distinguishing Features.*** Among the largest organisms in the world are the giant kelps, marine representatives of the brown algae. The Pacific Ocean kelp *Macrocystis* may be over 100 meters long. Almost all brown algae are saltwater dwellers, and many, such as *Fucus* and *Laminaria* (Fig. 30-6) float with the tides while attached by specialized holdfasts to rocks or other substrata. The Sargasso Sea in the Atlantic Ocean is named for floating mats of the brown alga *Sargassum,* formed from plants that have broken away from their holdfasts.

Associated with the large size and complexity of brown algae is the presence of meristems that add cells to the growing blades. Also related to their large size is a specialization of cells. In *Laminaria,* for example, strands of conducting cells transport photosynthetic products, much as do sieve tubes of vascular plants. Not all brown algae are large organisms, however. *Ectocarpus* is a small, branched filament with erect branches arising from a prostrate mass of filaments.

Diverse life cycles are found among the brown algae (see Chapter 21). In *Laminaria,* the life cycle is a heteromorphic alternation of generations, with its microscopic gametophyte distinctly different in form from its large sporophyte. The gametophyte consists simply of a filament of cells bearing sperm-producing antheridia and egg-bearing oogonia. But in *Fucus,* with a diplontic life cycle, a free-living diploid generation produces sperm and eggs following meiosis. The diploid zygote develops directly into the *Fucus* thallus.

In addition to the distinct pigments and reserve foods found in brown algae (Table 30-1), the cell walls differ from those of other algae in that they contain alginic acid. This gelatinous hydrophilic colloidal substance can absorb 20 times its weight in water, a factor in preventing the drying out of algae attached to rocks in intertidal zones. Extracted from harvested algae, alginic acid has many economic uses.

***Commercial Uses.*** The brown algae are useful to humans not only as food but also for products derived from their cell walls. They are used as food largely in Asia. Kombu, for example, prepared from pressed and shredded *Laminaria,* is a food staple in Japan.

Among the products made from the alginic acid extracts of the kelp *Macrocystis* are colloidal liquids used in fireproofing fabrics and as emulsifiers and thickeners in ice cream, paint, toothpaste, and lipstick. Those that solidify are used in dental materials, in plastics, to coat paper, and even to make buttons.

The commercial value of several brown algae has led to their cultivation in the sea, an example of marine agriculture, or **mariculture.** *Laminaria* is cultivated on ropes strung on bamboo poles driven into the sea bottom. The giant kelp *Macrocystis* is being machine-harvested off the California coast for its products and cultivated experimentally as a potential source of methane fuel. Growing to 100-foot size at the rate of 2 feet a day, these algae may make mariculture profitable in the United States.

## DIVISION CHRYSOPHYTA: GOLDEN-BROWN ALGAE

The Chrysophyta include filamentous and protistan algae that are golden or yellow-brown in color. They contain pigments similar to those of the brown algae, and they store food in a form similar to the reserve food of the brown algae. They also store oil, visible as large droplets in diatoms, the ubiquitous, glass-walled unicells. The oil from diatoms may have contributed to petroleum deposits. Today, where these algae abound, the oil imparts an unpleasant flavor to water supplies. Chrysophytes are found in fresh and sea water, where many are important in the food chain as major components of plankton. Some grow attached to the substratum.

***Some Protistan Chrysophytes.*** Many chrysophytes are flagellated cells that lack cell walls, but some have cell walls composed of pectic compounds. Flagellated chrysophytes have been of special interest cytologically ever since the electron microscope revealed intricately patterned scales or plates on the surfaces of the cell bodies and the flagella of unicellular forms (Fig. 30-7). The scales, which are carbohydrate in nature, are embedded in the pectin wall but synthesized within the Golgi apparatus and transported from there to the cell surface in vesicles. Several chrysophytes have "tinsel" flagella, which are covered with hairs called **mastigonemes.** Like the scales, the mastigonemes are formed within the Golgi apparatus and incorporated into the surface of the growing flagellum at its base as the Golgi vesicle unites with the plasma membrane (see Chapter 4).

***Diatoms.*** Diatoms are circular or oblong unicells (Fig. 30-8), sometimes attached together in filaments and branched colonies. The unique wall of all diatoms consists of two valves or plates that overlap (hence

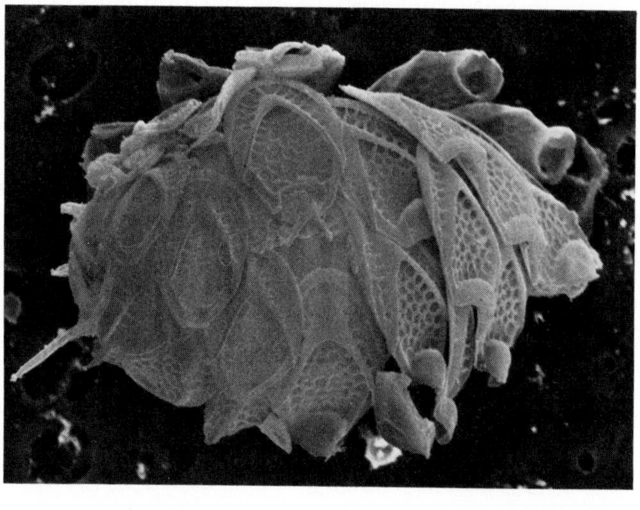

**FIG. 30-7** Scanning electron micrograph of chrysophyte, *Mallomonas,* showing its silica scales.

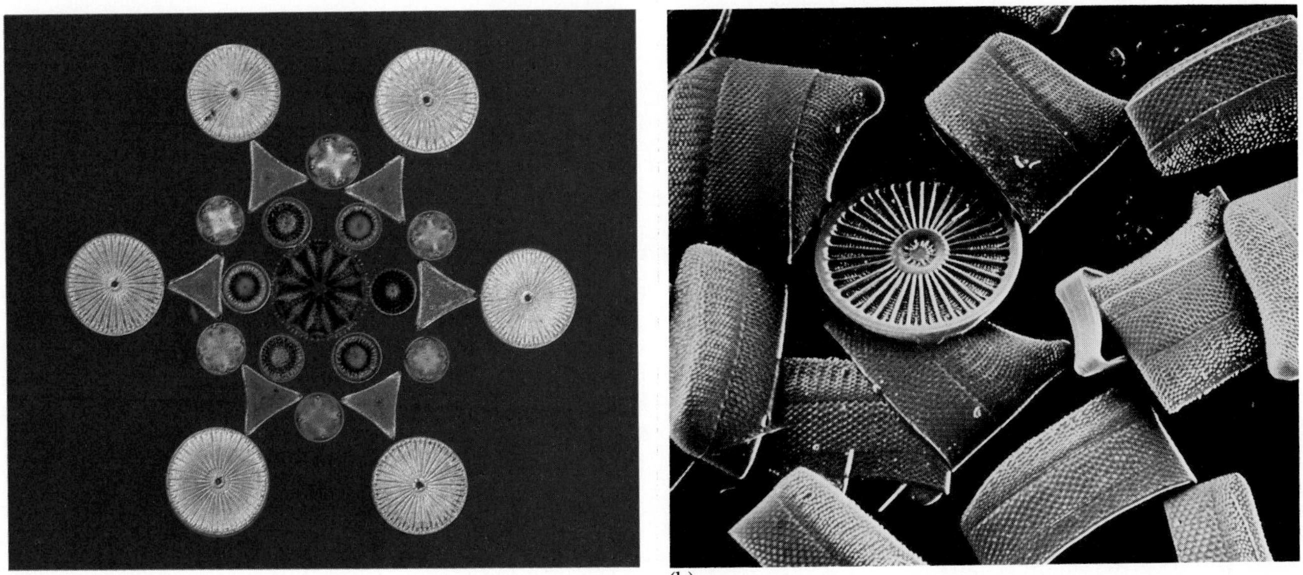

(a) (b)

**FIG. 30-8** (a) Assorted diatoms as arranged by a microscopist. (b) Scanning electron micrograph of diatoms.

their name, from the Greek *diatomos,* "cut-in-two"). When the cell divides, each half builds a new half wall that fits within the margin of the parental half wall. This makes smaller diatoms at each asexual generation. Sexual reproduction ultimately restores the larger size. The walls are of silica in crystalline form, with very fine markings. Each silica plate is formed in a large vesicle at the cell surface, and the process apparently imparts to the wall the distinct pattern of each species. The patterns in the wall are not only specific enough to be utilized, along with cell shape, in diatom identification but are fine enough to be used by microscopists to evaluate the resolution of microscopes.

Diatoms are found in plankton in astronomical numbers. That this was also true in the past is evidenced by extensive deposits of **diatomaceous earth,** sea-floor accumulations of the decay-resistant diatom cell walls that date back to the Jurassic Period. An indication of the abundance of diatoms in the seas of the past is the extent of diatomaceous deposits in the oil fields of California. They extend for miles and may be as much as 1,000 meters thick. It has been estimated that there are 6 million diatom cell walls in 1 $mm^3$. The silica of the cell walls in diatomaceous earth makes it commercially valuable, and it is mined and processed for firebrick, for use in the filtration of liquids (as in refining sucrose), and as an abrasive in silver polish.

## DIVISION PYRROPHYTA: THE DINOFLAGELLATES

Like the chrysophytes and diatoms, the dinoflagellates (Fig. 30-9) are a group of protistan algae that are important components of fresh- and salt-water plankton and are thus at the base of the food chain. Some of the marine forms produce toxins, and a bloom, or massive expansion in population, of such species may kill great numbers of fish and shellfish and poison people who eat them. Because of the reddish pigmentation in some genera, some blooms are known as red tides. Some dinoflagellates are luminescent when disturbed, and during a bloom, the sea appears to glow in response to wave action or in the wake of a ship.

The dinoflagellates comprise the class Dinophyceae (from the Greek *dinein,* "to whirl"), so named because of the spinning of the cell as they

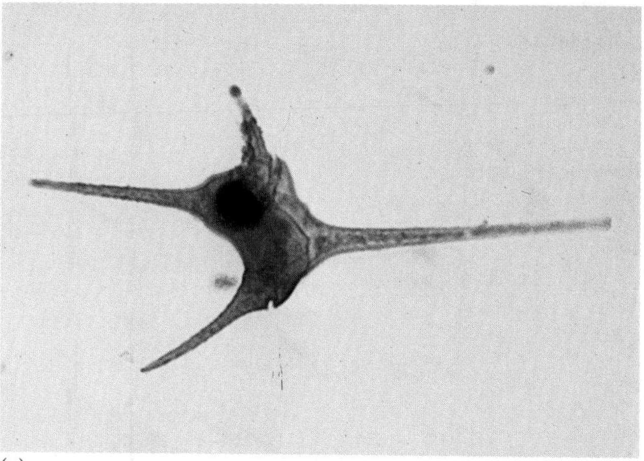
(a)

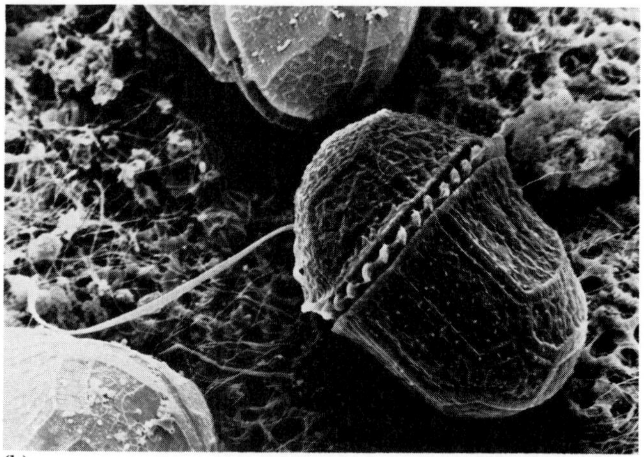
(b)

**FIG. 30-9** (a) The dinoflagellate *Ceratium*. (b) Scanning electron micrograph of the dinoflagellate *Gymnodinium*. The cell is armored with plates of cellulose. Two flagella lie in grooves that run between the plates.

swim. Many are cloaked with patterned cellulose plates (Fig. 30-9b). In some genera, two grooves girdle the cell, one transverse, the other longitudinal. In each groove lies a flagellum, the longitudinal one extending posteriorly as a rudder, the transverse one coiled in a helix so that it propels the cell forward while causing it to spin.

The dinoflagellates have attracted the attention of phycologists in recent years because of the unusual nature of their chromosomes. The chromosomes remain condensed throughout the cell cycle and are thus always visible within the nucleus. They resemble prokaryotic chromosomes in that they lack histones. Because of this, the dinoflagellates have been called **mesokaryotes** to indicate that they are midway between prokaryotes and eukaryotes.

In some species, electron microscopy has shown an additional nucleus that is eukaryotic in nature, apparently that of an endosymbiont alga. The eukaryotic nucleus and the chloroplasts have been shown in electron micrographs to be separated from the dinoflagellate cytoplasm by a membrane system, apparently the plasma membrane of the presumed endosymbiont. This invasion of a mesokaryote by a eukaryotic cell is further support of the endosymbiont theory.

# ALGAE WITH PHYCOBILIN PIGMENTS

## DIVISION RHODOPHYTA: THE RED ALGAE

**FIG. 30-10** The red alga *Ceramium rubrum*.

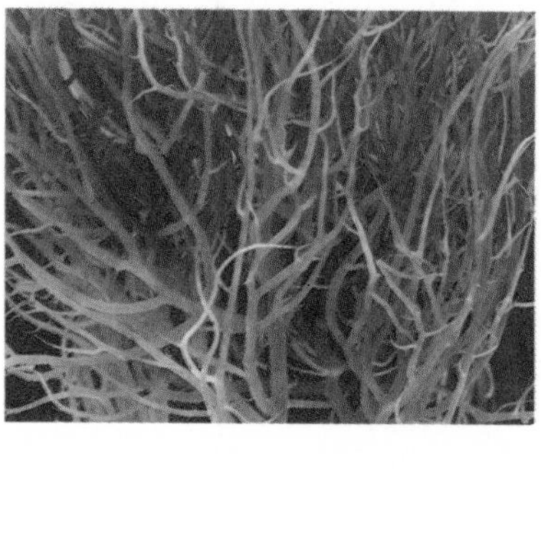

The red algae (Fig. 30-10) contain chlorophyll *a* and phycobilin pigments similar to those of the Cyanobacteria, suggesting an origin from these prokaryotes (Table 30-1). The algae may be decidedly red when the phycobilin phycoerythrin predominates over the other pigments, but shades of brown and dark green also occur due to photodestruction of phycoerythrin in bright light. They tend to be redder in shaded situations and at greater depths. (Phycoerythrin, though reflecting red light, absorbs the shorter-wavelength light in the blue portion of the spectrum and enhances photosynthesis below the surface).

Although red algae vary in form from single-celled types to filamentous forms, sheets, and more complex forms differentiated into tissues, they have a common body plan of connected filaments associated in diverse patterns. Even when the alga has the form of a sheet or a cylinder, the component cells are seen under the microscope to be arranged in filaments. The entire organism is encased in a cover of mucillaginous cell wall material, a sulfated polymer of galactose of commercial value. A number of red algae secrete calcium carbonate into their cell walls and become encrusted with limestone. These so-called coralline algae participate in the formation of coral reefs in tropical seas.

***Useful Products.*** Like the brown algae, the red algae are used as food and are raised in Asian mariculture. Dried *Porphyra*, for example, is a popular food item in Japan. The gelatinous cell wall material of some species can be converted into agar, used extensively in the laboratory as a bacterial and tissue culture medium, and in slow-release pills and medication for intestinal disorders. Carrageenans and gelans are colloids derived from the mucillaginous walls of red algae and used as emulsifiers in dairy products and paint.

***Reproduction.*** The mode of reproduction of the red algae is unique among the algae (Panel 30-1). Not only are the gametes nonmotile, but, most remarkably, in some cases the zygote nucleus migrates from one cell to an adjacent auxiliary cell, where it proceeds to divide within the enlarging auxiliary cell. Spores formed from this multinucleate cell start a new diploid generation.

## EVOLUTION ONTO THE LAND

Compared to aquatic plants, land plants have the advantages of being provided with sunlight unfiltered by water and a less limited supply of carbon dioxide and oxygen. However, they face desiccation, a problem that aquatic plants seldom encounter. Necessarily then, the first successful land plants must have had to evolve features that enabled them to obtain water from the substratum, transport it to all their living cells, and conserve it against the drying effects of the air. Several structural features have made all this possible in land plants. One adaptation was the development of water-absorbing cells, either rhizoids, or root hairs; another was the evolution of water-conducting cells or tissues, the xylem of vascular plants; and another was the development of a specialized covering for aerial surfaces, a waxy cuticle essentially impervious to water. But in retarding water loss, such covered surfaces also had to permit exchange of carbon dioxide and oxygen in photosynthesis. This function was fulfilled by openings in leaf surfaces, the stomata, and by pores in older stems, the lenticels. Dissemination of propagating structures in the air instead of motile cells in water would also be important in a successful land plant.

Two groups of land plants, the bryophytes and the vascular plants, evolved from the green algae. The major difference between the two is the occurrence of water- and food-conducting tissues in the vascular plants. This feature enabled vascular plants to grow to greater size and, with the evolution of roots and seeds, to escape direct contact with open water and to colonize dry land.

Alternation of generations is the life cycle pattern in all land plants. Two generations alternate in the life cycle, the haploid gametophyte and the diploid sporophyte (see Chapter 21). This pattern had already been established in the algae and was retained in the plants that

PANEL 30-1

# Reproduction in *Polysiphonia*

*Polysiphonia* illustrates the uniqueness of reproduction in the red algae. Many of the reproductive cycles in red algae are so complex that they are just now being worked out. In *Polysiphonia,* a diploid stage, called the **tetrasporophyte,** produces meiospores in groups of four within sporangia. The spores germinate into morphologically similar haploid male or female gametophyte plants. Male plants produce colorless, nonmotile gametes called **spermatia** that are released into the seawater and carried to the female plants. At the ends of filaments female plants develop special beaklike cells called **trichogynes** to which the spermatia may attach.

The life cycle of *Polysiphonia.*

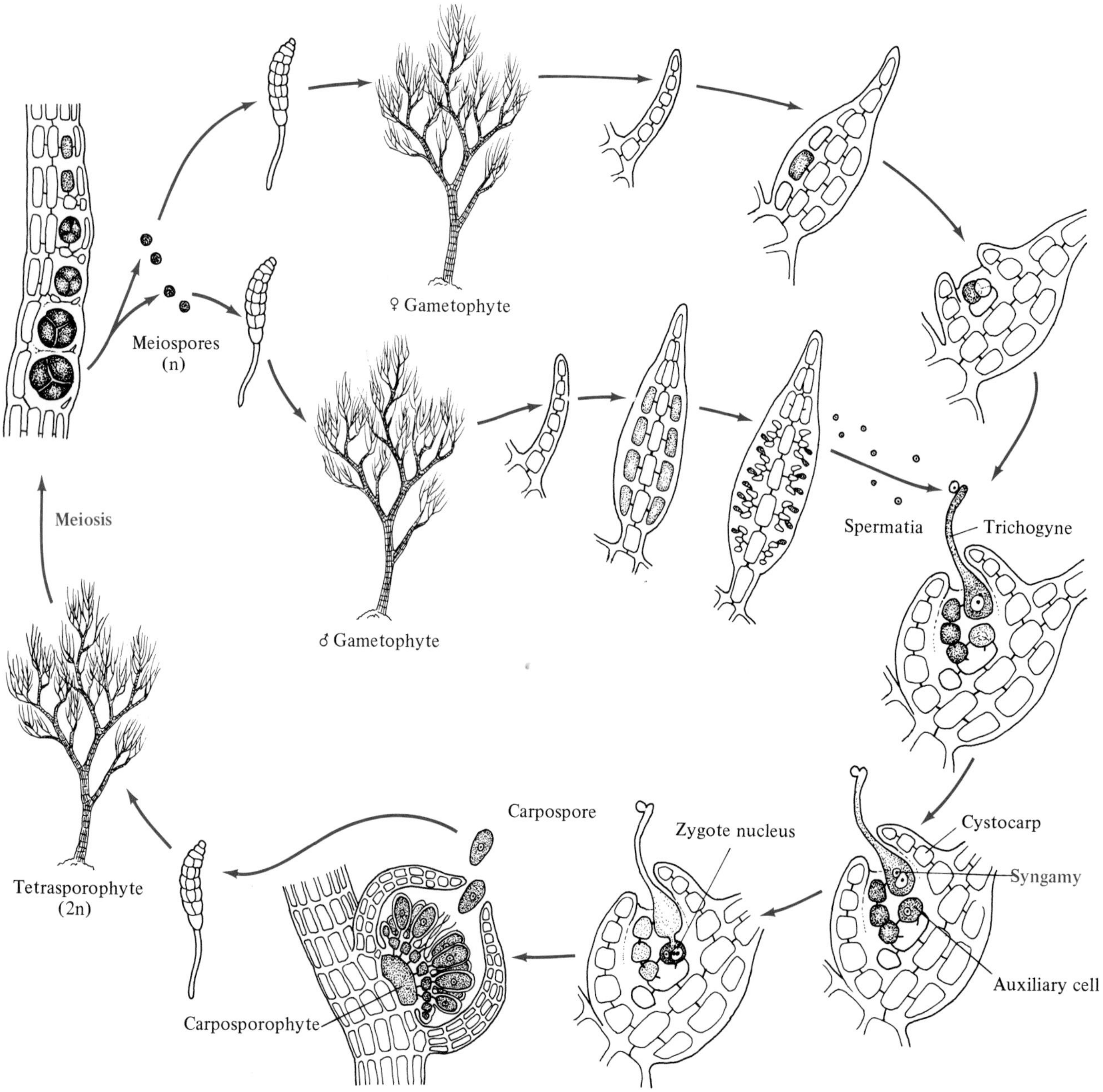

When a spermatium comes in contact with a trichogyne, it fuses with the female cell. This activates division of a supporting cell, which forms an auxiliary cell by division. Meanwhile, the male nucleus migrates downward through the trichogyne to the bottom of the cell. There, the spermatial nucleus and the female nucleus unite to form the zygote nucleus. The zygote nucleus migrates into the adjacent auxiliary cell, which has fused with the zygote cell. Cells surrounding the auxiliary cell grow up about it in the form of a container, the **cystocarp.** Then the zygote nucleus in the auxiliary cell begins to divide, forming numerous diploid nuclei. Another unusual phenomenon is the fusion of adjacent cells with the auxiliary cell to form a large cell, the **carposporophyte,** containing the diploid nuclei resulting from divisions of the zygote nucleus. Diploid spores, or **carpospores,** bud off from the carposporophyte and are liberated into the water from the cystocarp. Upon attachment and germination, a carpospore produces the diploid tetrasporophyte plant, and the life cycle is completed.

evolved from them, the first plants to succeed on land. In the bryophytes, the phototrophic gametophyte is the predominant generation, with the sporophyte small and supported upon it. In contrast, the sporophyte predominates in the vascular plant life cycle. It has vascular tissues, roots, stems, and leaves. The gametophyte is small and autotrophic in some and heterotrophic in a few, but in the seed plants, it is a microscopic, parasitic generation.

## DIVISION BRYOPHYTA

The bryophytes, including the liverworts, mosses, and hornworts, are limited in size because of their lack of vascular tissues, and in distribution because of their dependency on water for fertilization. Most live in aquatic or moist environments, but some mosses have become adapted to dry environments with only seasonal water supplies and survive in a desiccated state until a rainfall permits fertilization and another period of growth. Moss plants grow close together in clumps, which enhances water retention. To reach parts that extend into the air, water must be transported from cell to cell or over the exterior of the plant by capillary movement along the fine, overlapping leaves. This greatly limits the size of the plant body. The largest known moss is only 30 cm high. Water is also essential in sexual reproduction of bryophytes, since the male gamete must swim to the female sex organ.

**FIG. 30-11** The leafy gametophytes of mosses bear stalked sporophytes.

***Mosses.*** The familiar leafy structure we recognize as the moss plant is the mature stage of the gametophyte generation, the stage that bears the sex organs. Following fertilization, the sporophyte grows out of the female sex organ, or **archegonium,** carrying its top upward as a papery cover (Fig. 30-11). In some mosses, the leafy gametophyte is perennial, adding increments to the tips from year to year. The gametophyte generation undergoes a remarkable change in form following its origin in the spore. The spore produces an algalike filamentous phase, or **protonema** (Panel 30-2). In the distinctive bog moss *Sphagnum,* spore germination produces a flattened, irregular thallus only one cell thick. Buds that appear on the protonema give rise to the leafy gametophytes. The meristem of the moss gametophyte has an apical cell in the form of an inverted pyramid at its tip. Regular divisions at the three sides of the pyramid form the leaves in three vertical rows, producing the typical radial symmetry.

PANEL 30-2

# The Life Cycle of Bryophytes: Alternation of Generations

In bryophytes, the predominant gametophyte generation bears the reproductive organs, the **antheridia** and **archegonia.** In contrast to the unicellular antheridia and oogonia of algae, the bryophyte organs are multicellular. They develop in various localized areas in the diverse gametophytes of the liverworts and mosses. In mosses, these organs develop at the tips of the leafy gametophytes. Antheridia and archegonia are remarkably uniform throughout the bryophytes. Antheridia are spherical or oblong structures in which a single layer of cells encloses numerous spermatogenous cells, each of which differentiates into a sperm. Archegonia are flask-shaped, with a wide base, or **venter,** containing the egg, and an extended neck.

Life cycle of a moss.

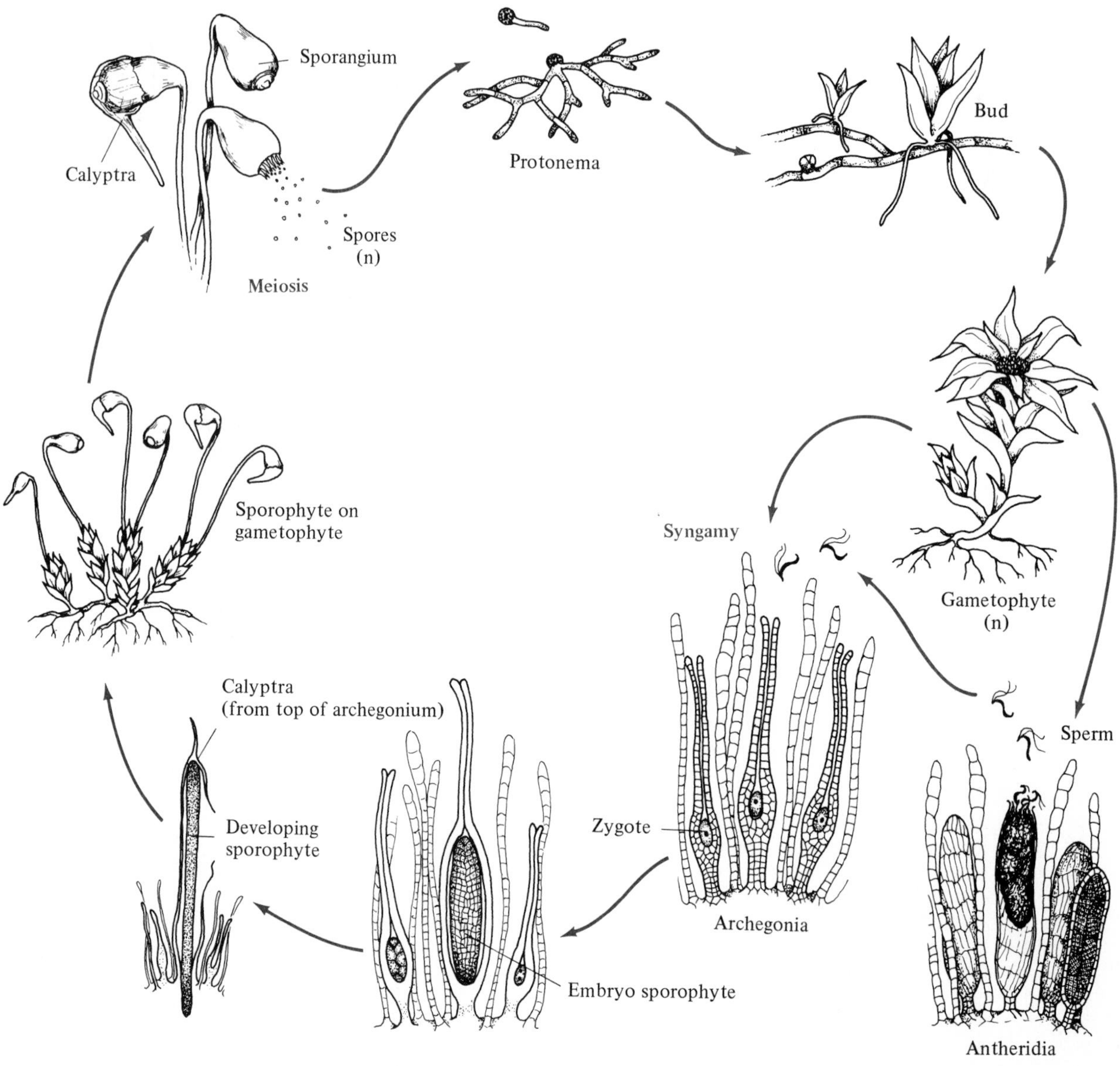

The flagellated male gametes are released into the water around the plants and swim to the neck of the archegonia, to which they are attracted by organic acids that are released when neck canal cells disintegrate. When mature, the archegonia have open canals extending down the neck to the egg. Sperm enter the neck canal, and fertilization occurs in the base, where the zygote develops into the diploid sporophyte generation. As the sporophyte develops, it carries upward the tip of the archegonium which covers the capsule, or **sporangium,** of the sporophyte. Within the sporangium, meiosis occurs, and the haploid spores are disseminated from the specially adapted capsule. Germinating spores produce a filamentous protonema. Buds form on the protonema and develop into the leafy gametophyte shoots, which produce the sex organs.

The gametophyte of *Sphagnum* is highly adapted to a bog environment. It continues to grow from year to year and becomes quite long. The lower, older portions die, and because they undergo little decomposition in the anaerobic, acid bog water, they form the peat typical of the *Sphagnum* bog. The mat of erect plants obtains water from below by capillary action. The cells of the stem include elongate cells at the center, which transport water, and unusual water storage cells in the stem surface (these are called **retort cells** because they are shaped like the flasks called retorts in a chemistry laboratory). The leaf, only one cell layer thick, has a unique, regular pattern of chloroplast-containing photosynthetic cells interspersed with empty water-storing cells with pores in the walls.

The sporophyte generation of mosses grows atop the gametophyte as a stalked capsule with its foot embedded in the gametophyte. At first, the sporophyte is green, but on maturing, it becomes brown and nonphotosynthetic. Its capsule is structurally adapted to spore dispersal under dry conditions, which are favorable for wide dissemination. A circle of teeth covers the open end of the capsule, or clasps a plate or diaphragm at its top, which is covered until mature by a lid. When the capsule is mature, the lid pops off; drying contracts the teeth, leaving spaces between them through which the spores are released. In *Sphagnum,* when the lid pops off under pressure from within, the spores are shot into the air.

**FIG. 30-12** (a) Thallus of the liverwort *Marchantia.* (b) Special branches bearing the antheridia and archegonia of *Marchantia* are modified tips of the ribbon shaped thallus.

(a)

(b)

***Liverworts and Hornworts.*** The liverworts (Division Hepatophyta, in some classifications) are ribbon-shaped or leafy, dorsiventral (flattened) plants that float on water or grow appressed to the soil. They branch in dichotomies, equal divisions of the thallus tips that fork as does the letter Y (Fig. 30-12a). A gametophyte shows a very simple tissue organization. Photosynthetic cells occur in various arrangements that enhance their exposure to light. Intercellular spaces allow diffusion of carbon dioxide and oxygen; these spaces communicate directly with the outside in some species and through permanently open epidermal air pores in others. Water enters the plant all along the under surface through elongated cells, or rhizoids, sometimes interspersed with thin scales. The tissues of the ribbon-shaped gametophyte have their origin in a meristem at the notch between the tip lobes. Some liverworts have special cuplike structures in which new plantlets, or **gemmae,** which asexually propagate the plants, are continually being formed.

The sex organs are borne on unique modified lobes of the gametophyte (Fig. 30-12b). At maturity, the sporophyte generation is little more than a sac of spores, the sporangium, connected in some species by a stalk to a foot embedded in the gametophyte. When the spores are ripe, special cells (elaters) among them twist and turn, expelling the spores.

Hornworts resemble liverworts in that the gametophyte is dorsiventral. The sporophyte, however, is hornlike, with its foot embedded in the gametophyte and a collar ensheathing its base. At its tip the sporophyte splits as the capsule ripens, and the halves twist as the spores are disseminated.

***Uses of the Bryophytes.*** *Sphagnum* is the only commercially useful bryophyte. Because the empty cells of stem and leaves and the external capillary channels hold great quantities of water, *Sphagnum* is used in the packaging of plants as a source of moisture during shipment. Partially decomposed sphagnum, or peat, also retains moisture and, when added to soil, improves its structure and water-holding capacity. Great amounts of peat are harvested and baled in North America for horticultural use.

The peat of ancient bogs has been an important fuel in once-glaciated parts of the world, including Ireland, Wales, Scotland, Denmark, and Finland.

## THE VASCULAR PLANTS

Although the bryophytes have succeeded on land, their success as measured by proportion of the total flora, numbers of species, or individual size in no way matches that of the vascular plants. This suggests that the presence of specialized conducting tissues, the common characteristic of the vascular plants, contributed to their success. Another major difference between the bryophytes and the vascular plants is the predominance of the diploid sporophyte generation in the alternation of generations in the latter. The gametophyte generation is reduced in size, in some cases to only a few cells. The complex diploid sporophyte generation, with its great assemblage of genetic traits and diversity, gave vascular plants great potential for evolutionary adaptation. Later, we shall consider why one group, the flowering plants, far exceeds all others in diversity, or number of species.

The vascular plants not only have roots that tap the water supply in the soil but also have the means of transporting water and minerals and the products of photosynthesis throughout the plant. Mechanical tissues and the turgor of its cells keep the plant erect. A floating alga or a prostrate liverwort survives without erect parts, but more efficient utilization of space and light accompanies the erect growth habit. Large, flattened leaves expose a maximum surface to sunlight, favoring more efficient energy capture, but they also increase susceptibility to the drying effects of the air. Water loss is countered by the waxy cuticle layer covering the surface, as in bryophytes, but the stomata are superior to the pores of bryophytes. Guard cells regulate the opening, permitting maximal entry of carbon dioxide into the leaves while the inevitable water loss is minimized (see Chapter 7).

The vascular tissues include the water-conducting xylem, and the phloem in which carbohydrates are translocated (see Chapter 10). Study of the anatomy of plant fossils and of the comparative morphology of present-day vascular plants has led to the widely held conclusion that the simplest and most primitive water-conducting cell is the tracheid. According to this view, vessel elements, the more efficient water-

**TABLE 30-2 A Key to the Vascular Plants**

| *Type* | *Sporophyte characteristics* | *Origin, duration* |
|---|---|---|
| Plants without seeds | Lower tracheophytes | |
| Division Psilophyta—psilophytes | No roots; leaves; small herbs<br>Terminal sporangia | Silurian to Devonian (extinct) |
| Division Microphyllophyta—club mosses | Microphylls; herbs; trees<br>Strobili | Devonian to present |
| Division Arthrophyta—horsetails | Jointed stems; whorled leaves; herbs; trees<br>Strobili | Carboniferous to present |
| Division Pteridophyta—ferns | Megaphylls; herbs; trees; sporangia on sporphylls | Devonian to present |
| Plants with seeds | Seed plants | |
| Gymnosperms | Plants with naked seeds; usually strobili or cones | |
| Division Progymnospermophyta | Trees | Devonian to Carboniferous |
| Division Pteridospermophyta—seed ferns | Trees; vines | Carboniferous to Permian (extinct) |
| Division Cycadeophyta—cycadeoids | Squat trunks; large compound leaves | Mesozoic (extinct) |
| Division Cycadophyta—cycads | Squat trunks; large compound leaves | Mesozoic to present |
| Division Ginkgophyta—ginkgos | Trees; simple leaves | Mesozoic to present |
| Division Coniferophyta—conifers | Trees; needle or scale leaves | Carboniferous to present |
| Division Gnetophyta | Trees; shrubs; simple, minute or strap-shaped leaves | Unknown to present |
| Angiosperms | Plants with enclosed seeds; flowers; megaphylls | |
| Division Anthophyta | | |
| Monocotyledonae | Herbs; trees | Cretaceous to present |
| Dicotyledonae | Herbs; shrubs; trees | Cretaceous to present |

conducting cells of flowering plants, evolved from the tracheid. Fibers, which are elongated supporting cells, also are thought to have evolved from the tracheid. Similarly, the primitive phloem element is thought to be the sieve cell, which evolved into the sieve tube elements that make up the sieve tubes of flowering plants. The common occurrence of the tracheid in all vascular plants has led to the designation of vascular plants as **tracheophytes.** In some classification schemes, vascular plants are grouped in a single division: Tracheophyta. Present thought, however, generally subdivides the tracheophytes into several divisions (Table 30-2).

The simpler vascular plants, the lower tracheophytes, lack seeds. Evolution of the seed provided selective advantages to the plants that bore them: it afforded a mechanism for the embryonic sporophyte to subsist during adverse seasons and provided nutrients for the start of its growth. Among the seed plants, the enclosure of the seed in a vessel or container promoted seed distribution. The **gymnosperms,** plants with unenclosed seeds, as in the pine cone, are represented by several groups, many of them extinct (Table 30-2). The **angiosperms,** plants with seeds produced inside a closed vessel, as in the fruit of a tomato, and bearing flowers, have evolved most recently and are the most successful plants. They are placed in one division, the Anthophyta.

PANEL 30-3

# The Life Cycle of a Fern

The diploid sporophyte has roots, stems or rhizomes, and leaves. The sporangia are arranged in clusters called **sori** (sorus, singular) on **sporophylls.** Distribution patterns of the sporangia on the sporophyll are used in fern classification. Within the sporangium, spore mother cells called **sporocytes** undergo meiosis to form haploid spores. Special splitting, or dehiscence, mechanisms open the sporangium and disperse the spores under dry conditions. The spores germinate and, in a favorable environment, develop into haploid gametophytes called **prothalli.** On the underside of the gametophytes are rhizoids, which take up water, and the sex organs, the antheridia and archegonia. Male gametes released from the antheridia swim to the archegonia, to which they are chemically attracted. After syngamy, the zygote develops into an embryo sporophyte that remains attached to the parent gametophyte until it has developed root, stem, and leaves.

Life cycle of a fern.

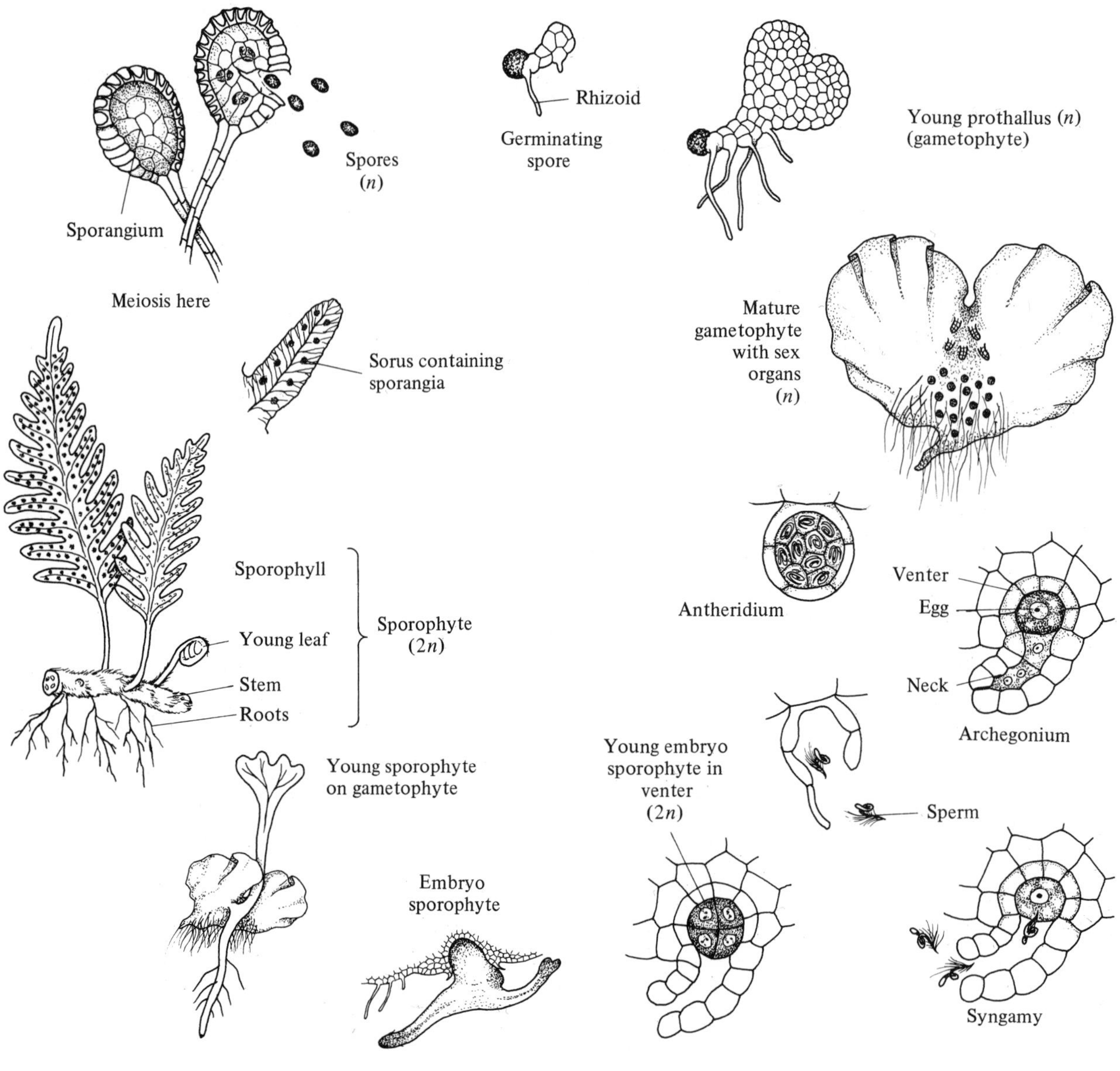

(a) (b)

**FIG. 30-13** The sporophytes of ferns. (a) A tree fern. (b) The aquatic fern *Azolla*.

**FIG. 30-14** The club moss *Lycopodium* branches dichotomously and bears "clubs," or strobili, composed of sporophylls.

**FIG. 30-15** *Equisetum* is the only surviving genus of horsetails.

***The Lower Tracheophytes.*** The lower tracheophytes are represented in present-day flora by three distinct groups: the **ferns** (Fig. 30-13), the **club mosses** (Fig. 30-14), and the **horsetails** (Fig. 30-15). One other group, the **psilophytes,** is known only as fossils, an assortment of various forms from which other vascular plants probably evolved.* The four groups have in common the vascular system that makes them tracheophytes, basically similar life cycles, and similar reproductive organs. A representative life cycle, that of a fern, is described in Panel 30-3.

SPOROPHYTES OF LOWER TRACHEOPHYTES. The sporophyte plant in most lower tracheophytes is represented by a small, erect leafy stem, or a horizontal **rhizome** bearing leaves or **sporophylls,** leaves that bear sporangia (Panel 30-3). Adventitious roots, those that arise along the stem, supply water and minerals. If the stems branch, it is by dichotomous branching, which results when the apical meristem is subdivided. While most modern lower tracheophytes are low, trailing plants, a few tropical tree ferns attain a height of 75 feet. Some of the extinct club mosses and horsetails also attained the form of trees several meters tall with vascular cambium and woody trunks. They were major components of the ancient Carboniferous forests. Their stems and roots are often found as fossils. Since characteristic patterns occur in the vascular tissues of stems and roots, cross sections of these organs are sometimes used in identification.

The leaves of the club mosses and horsetails are small **microphylls,** in contrast to the large, often compound (subdivided into leaflets) **megaphylls** of ferns. Microphylls are thought to have originated as outgrowths of the stem surface, as in some fossil psilophytes, and to have become vascularized during their evolution. According to the Telome theory, described in Fig. 30-16, megaphylls evolved from the dichotomous branches of the primitive psilophytes. The three-dimensional branches became flattened in one plane, (a process called planation), and filled in between distal branches with photosynthetic tissue (a process called webbing).

SPORANGIA: THE BEARING AND DISPERSAL OF SPORES. Each of the lower tracheophyte groups has a characteristic manner of bearing sporangia. In ferns they occur in clusters on the sporophylls. Sporangia are espe-

* Two living genera, *Psilotum* and *Tmesipteris,* are included in the psilophytes by some botanists, but others regard them as ferns.

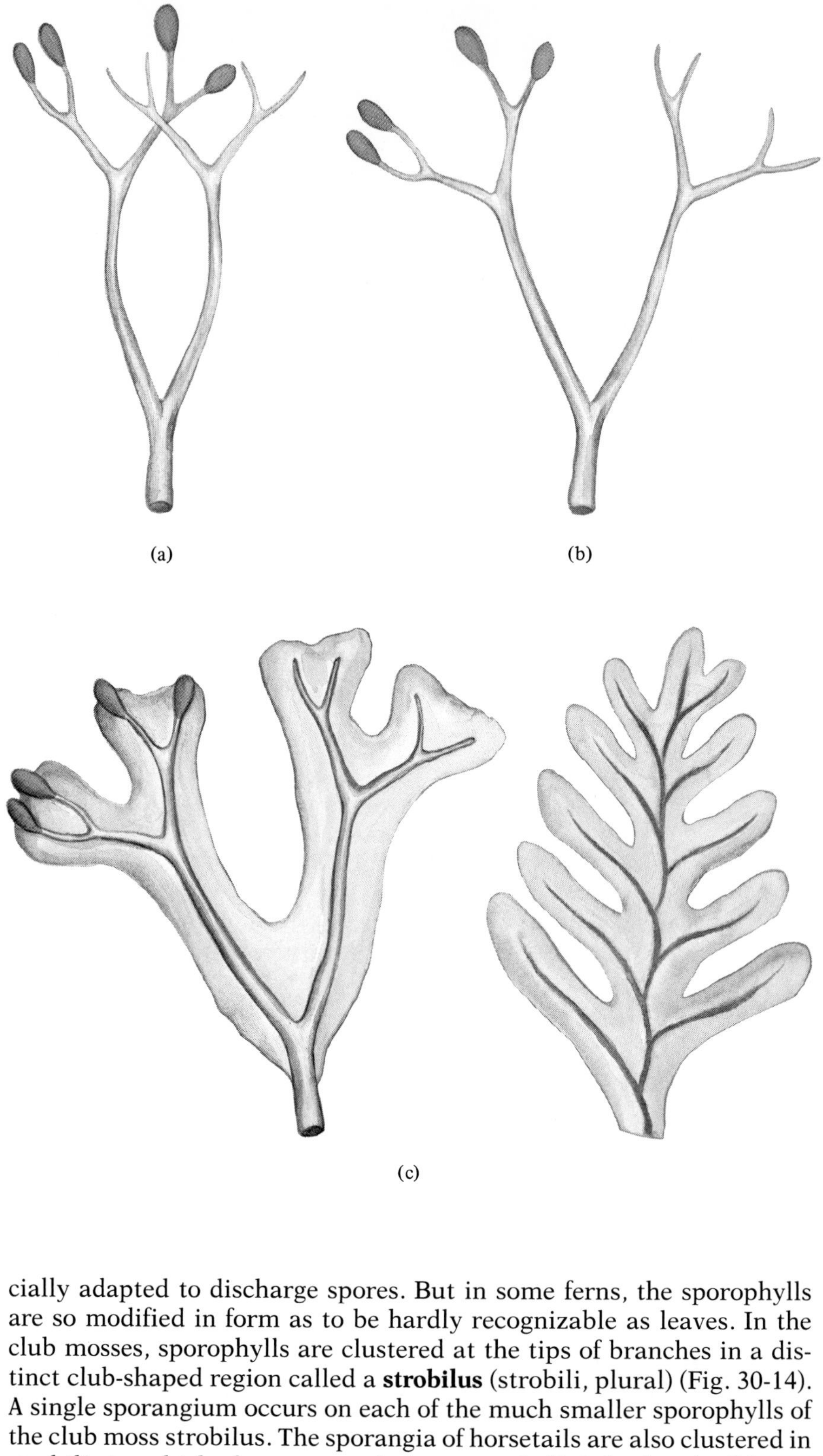


**FIG. 30-16** The Telome Theory attempts to explain the evolution of the various vascular plant groups from (a) primitive psilophytes by certain "elemental processes," including (b) flattening of the dichotomous branching system (planation), and (c) filling in of tissue between the branches (webbing) to produce a fern leaf with marginal sporangia.

cially adapted to discharge spores. But in some ferns, the sporophylls are so modified in form as to be hardly recognizable as leaves. In the club mosses, sporophylls are clustered at the tips of branches in a distinct club-shaped region called a **strobilus** (strobili, plural) (Fig. 30-14). A single sporangium occurs on each of the much smaller sporophylls of the club moss strobilus. The sporangia of horsetails are also clustered in strobili, in which the nature of the sporophyll (whether leaflike or branchlike in origin) is scarcely recognizable (Fig. 30-15).

PSILOPHYTES: THE EARLIEST VASCULAR PLANTS. Attempts to trace back the lower tracheophytes in the fossil record for clues to the ancestry of vascular plants have led to a group of small, simple plants with a cen-

**FIG. 30-17** Psilophytes, forerunners of the Lower Tracheophytes, in a reconstruction of a Devonian scene.

tral strand of vascular tissues and dichotomous branching, and lacking roots or leaves. These are the psilophytes (Fig. 30-17). Their sporangia occur at or along the tips of the branches. An assortment of psilophytes early in the Devonian period is thought to have given rise to the ferns, club mosses, and horsetails. Tracing back still further in the fossil record, paleobotanist Harlan Banks, of Cornell University, suggested late in the 1960s that the psilophytes may have come from such plants as *Cooksonia,* a fossil from the Upper Silurian period, 5 million years earlier.

By the end of the Devonian period, all major lines of vascular plants, except the flowering plants, had been established, including the earliest seed plants, the gymnosperms.

GAMETOPHYTES. The haploid spores, produced in the sporangia as a result of meiosis, are disseminated and initiate the gametophyte generation. In the ferns and horsetails, the gametophyte is a small, green, flattened photosynthetic plant less than a centimeter across (see Panel 30-3). In the club mosses, the gametophyte is also small, but cylindrical with aerial green branches or carrot-shaped and buried in the soil. The nongreen types are heterotrophs and require the presence of a mycorrhizal fungus for growth.

An unusual phenomenon occurs in ferns under certain conditions, as when gametophytes are grown in culture. Sporophytes may appear directly from gametophyte cells without fertilization, in the process of **apogamy,** meaning "without gametes." Though haploid in chromosome number, they have the form of sporophytes. Their existence demonstrates that the forms of the gametophyte and the sporophyte are not controlled by the chromosome complement.

The gametophyte plant of the life cycle produces the gametes in multicellular sex organs, the antheridia and archegonia (see Panel 30-3). These organs are remarkably similar in structure in all the lower tracheophytes. Following syngamy, the zygote develops into a diploid embryo sporophyte within the archegonium (Panel 30-3).

***The Seed Plants: Gymnosperms and Angiosperms.*** While vascular tissues were an evolutionary success, the appearance of the seed led to further advantages which are evident in the success of the seed plants, the gymnosperms and angiosperms.

The gymnosperms (from the Greek *gymnos,* "naked," and *sperm,* "seed") were once considered a homogeneous taxonomic group in which the seed is naked, that is, not enclosed in a container and free to fall from the plant when ripe. But today the consensus is that different groups of gymnosperms are distinct enough to be separate divisions. Some are known only as fossils, and among the survivors some are almost extinct. Gymnosperms are thought to have evolved from the Devonian **progymnosperms,** some with coniferlike wood and fernlike leaves, that became extinct in the Carboniferous period. The seed ferns, so-named because they had seeds borne on fernlike sporophylls, appeared in the Carboniferous period and also became extinct early, but not before they gave rise to the **cycad line,** with its large, fernlike leaves. The **conifer line,** trees usually with fine, needlelike or scalelike leaves, appears to have come from the progymnosperms by another pathway.

The angiosperms (Division Anthophyta), or flowering plants, are relative newcomers to the flora. Fossil angiosperms have been identified with certainty only back to the Lower Cretaceous period. Yet, the angiosperms represent the dominant component of our flora today, with 235,000 species in 12,000 genera, distributed over all the continents, and variously adapted to most of the ecological niches on earth. The postulated evolutionary relationships among these groups are illustrated in Figure 30-18.

THE NATURE OF THE SEED. To appreciate the distinction between the seed plants and the lower tracheophytes, we must be clear on the nature

**FIG. 30-18** Postulated evolution of the vascular plants from the psilophytes.

Suggested evolution of vascular plants, and their distribution in geological time.

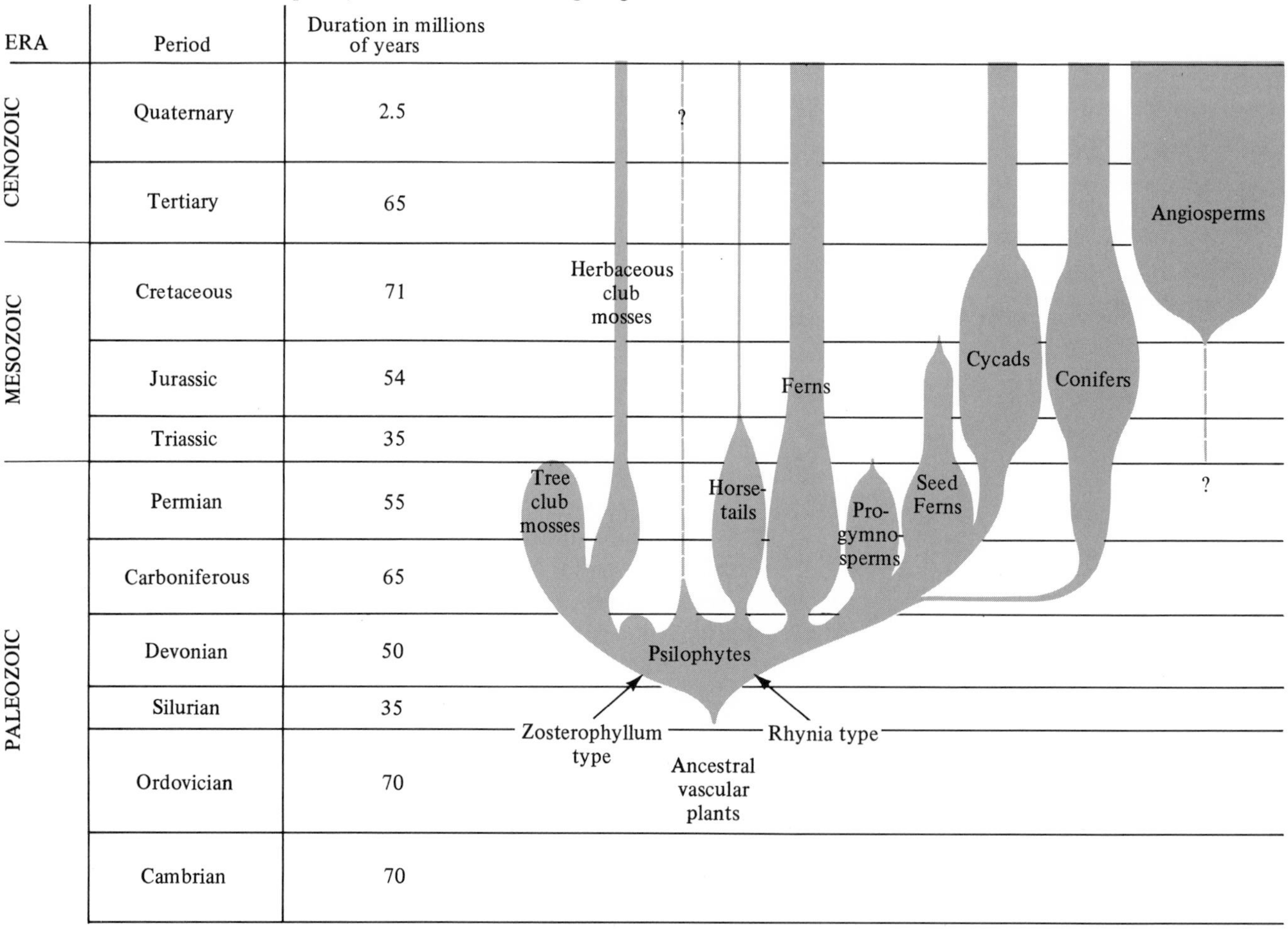

**FIG. 30-19** Early seed plants. (a) The progymnosperm *Archeopteris.* (b) The seed fern *Medullosa,* a viney tree about 15 feet tall. (c) Seed on leaf of *Alethopteris,* another seed fern.

(a)

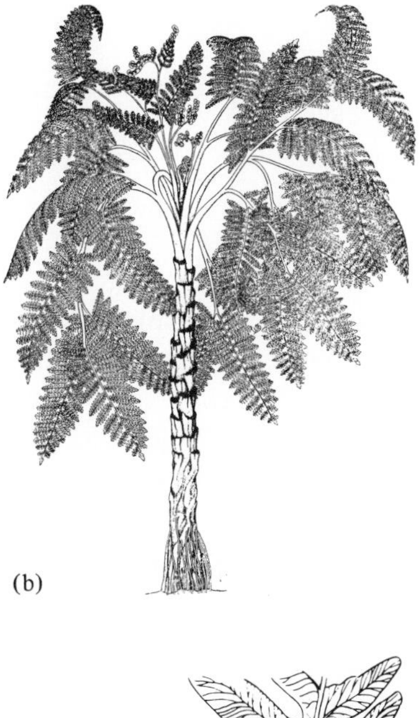

(b)

(c)

of the seed, which first appeared in certain woody plants of the Devonian period. The nature of a seed and how one is formed is illustrated in the life cycle of pine (Panel 30-4).

Evolution of the seed had to be preceded by evolution of **heterospory,** the occurrence of two different types of spores in the life cycle. (This condition is in contrast to **homospory** in which all the spores are alike, as in most lower tracheophytes). In heterospory, large megaspores give rise to megagametophytes as the result of cell division, which takes place within the spore. The resulting haploid tissue produces the egg-bearing archegonia. The smaller microspores divide to form microgametophytes, which ultimately produce male gametes. In the seed plants, microgametophytes became the pollen grains. Heterospory still occurs in a few lower tracheophytes, such as the club moss *Selaginella,* and in some aquatic ferns.

The seed is a structure which protects the embryo sporophyte, enclosed within nutritive tissue and a covering, the seed coat. It develops from an **ovule,** an unopened sporangium in which a megaspore has developed into a megagametophyte. The megagametophyte consists of a number of cells among which an egg is produced. Fertilization can occur in seed plants only when the male gametes are delivered to the egg since the megagametophyte remains within the sporangium. Gamete transfer is the function of the pollen tube, an outgrowth of the pollen grain (microgametophyte).

It is of evolutionary interest that a few gymnosperms (cycads and *Ginkgo*), though having pollen tubes that deliver the gametes, nevertheless still have male gametes with flagella. Ovules, like their forerunners, sporangia, are borne on sporophylls: on the strobilus in cycads and on modified sporophyll-bearing branches of the cone in conifers. In angiosperms, they are contained within a structure called a carpel, postulated to be a rolled up sporophyll. The leafy nature of the sporophyll is particularly evident in some of the cycads, where the ovules occur along the sides of large leaflike sporophylls of a strobilus. The relationship is less apparent in conifers and angiosperms, where evolutionary changes have been greater.

## THE GYMNOSPERMS

The earliest known fossil seed is from the Upper Devonian. The plant it came from is unknown—the seed was unattached. But a number of plants in the Carboniferous period had seeds, and because of their fern-like foliage are called seed ferns (Fig. 30-19). They are the earliest undisputed gymnosperms. Still earlier fossils with wood anatomically similar to that of conifers are classified as progymnosperms, but their position is uncertain because the fossil stems and leaves had no seeds attached.

### THE CYCAD LINE AND THE CONIFER LINE

Out of the progymnosperms apparently evolved several groups of seed plants that formed a significant part of the flora of the Mesozoic era, a time that has become known as "the age of the gymnosperms." Not only the distinctly different cones of cycads and conifers, but the general form of these plants suggests that two lines of gymnosperms evolved.

# PANEL 30-4

## The Life Cycle of Pine

Female pine cones bear numerous cone scales, each with two ovules, only one of which is shown in the figures on p. 855. Each ovule contains one megasporocyte, or megaspore mother cell. Meiosis in the megasporocyte results in four haploid megaspores, three of which degenerate.

Male cones bear many microsporophylls, each with two elongate microsporangia, which initially contain numerous microsporocytes, or microspore mother cells. Microsporocytes undergo meiosis, each producing four haploid microspores. Each microspore develops into a pollen grain, or microgametophyte, a four-celled structure with an inflated ("winged") cell wall. When microsporangia open, they release clouds of pollen grains. The pollen grains are transferred by wind to the ovules, the process known as **pollination.** When the pollen grains land on female cones, they sift into fissures between the scales and settle on the surface of pollination droplets, fluid secreted by each ovule. As the droplets dry, the pollen grains are drawn into contact with the megasporangium, or **nucellus,** of the ovule.

Within each ovule the surviving megaspore begins a series of mitotic divisions, producing the megagametophyte, which contains two or three archegonia. Each archegonium contains an egg. The pollen grains develop pollen tubes which slowly digest their way through the nucellus. In each tube, one of the four nuclei divides to form two male gametes. In twelve months, when a growing pollen tube reaches the megagametophyte, its tip ruptures, releasing the two male gametes. One enters an archegonium and fertilizes its egg; the other degenerates. The result of syngamy is the formation of a diploid zygote, the nucleus of which divides repeatedly, producing up to 2000 free nuclei. With the occurrence of cytokinesis an embryo is organized. As the embryo grows, it is in contact with the megagametophyte, which supplies it with nutrients. After a brief period of growth, the embryo becomes dormant. Meanwhile, the integument has thickened and hardened to form the seed coat. The ovule has thus become a seed.

Mature pine seeds are shed 18 to 20 months after pollination. Attached to each seed is a wing, a strip of tissue derived from the cone scale, that facilitates its dissemination by wind. Upon germination the seedling is capable of photosynthesis.

Cycads have short, broad, usually unbranched trunks covered with the bases of the large compound leaves that have fallen off (Fig. 30-20). Conifers are trees with characteristic woody trunks and branches and small, usually needlelike or scalelike leaves (Fig. 30-21).

The cycads flourished in the Mesozoic era. Today, the nine surviving cycad genera are distributed around the equator in the subtropics. They

(a)

(b)

**FIG. 30-20** (a) The short stumpy trunk of the cycad *Zamia*. (b) Leaflike sporophylls from the strobilus of *Cycas*.

Life cycle of a pine.

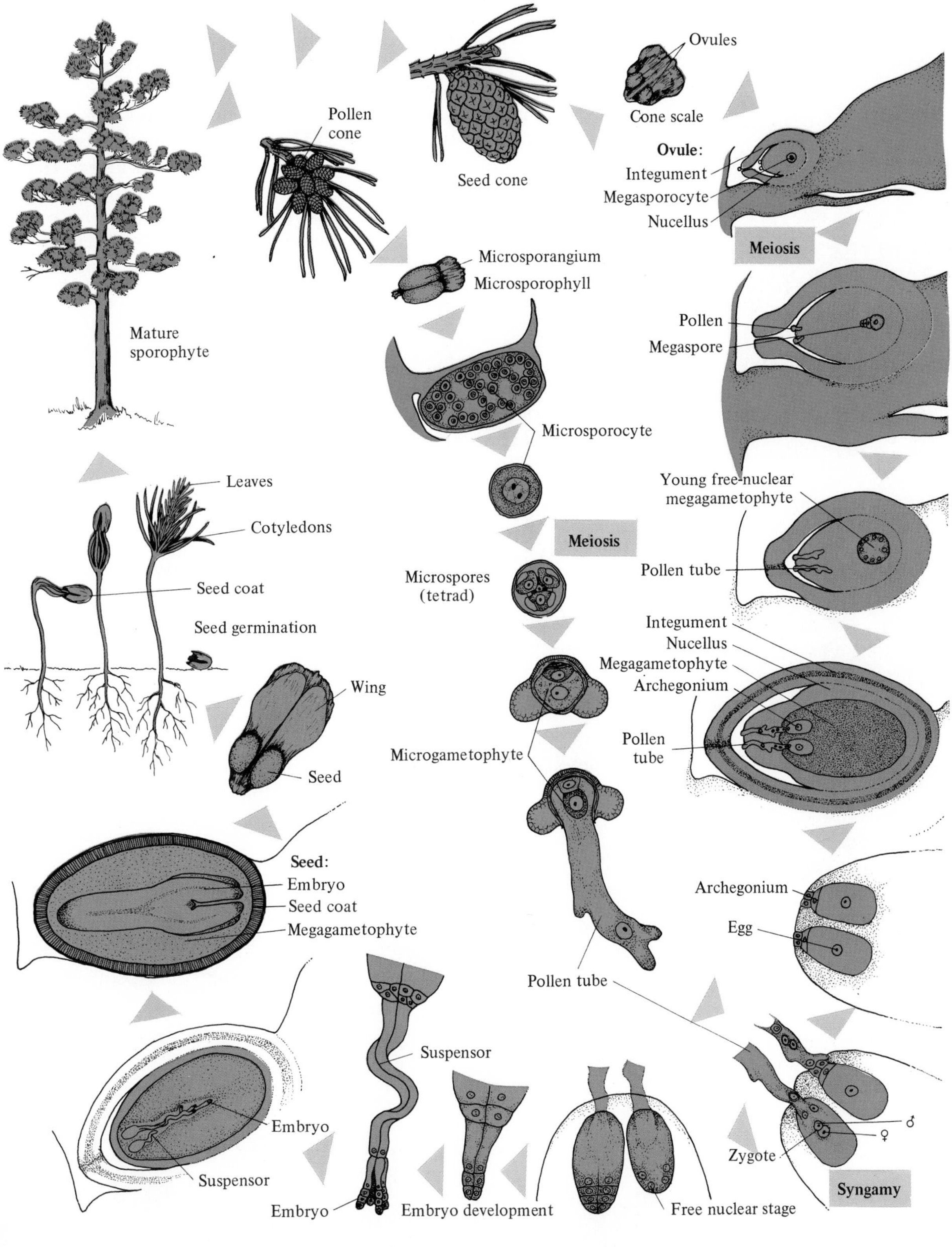
Mature sporophyte
Pollen cone
Seed cone
Ovules
Cone scale
Ovule:
Integument
Megasporocyte
Nucellus
Meiosis
Microsporangium
Microsporophyll
Pollen
Megaspore
Microsporocyte
Leaves
Cotyledons
Seed coat
Seed germination
Young free-nuclear megagametophyte
Meiosis
Microspores (tetrad)
Pollen tube
Integument
Nucellus
Megagametophyte
Archegonium
Wing
Seed
Pollen tube
Microgametophyte
Seed:
Embryo
Seed coat
Megagametophyte
Archegonium
Egg
Pollen tube
Suspensor
Embryo
Suspensor
♂
♀
Zygote
Syngamy
Embryo
Embryo development
Free nuclear stage

are found in Australia, Asia, South Africa, North America, and Cuba, land masses now separated by continental drift. One group of cycadlike plants, the cycadeoids, coevolved with cycads but, unlike them, became extinct.

The conifers are the most successful gymnosperms, having survived in greater numbers and with the widest diversity. They include the giant redwoods of western North America, the softwood trees of commerce (pine, hemlock, Douglas fir, spruce, and so on), the cypress of southern swamps, the cedars, the southern hemisphere genera *Araucaria* and *Podocarpus*, and such ornamental shrubs and trees as the yew, juniper, and arbor vitae. Study of extinct fossil conifer ancestors has indicated the possible evolution of the cone in modern conifers. Branches that bore sporophylls in ancestral forms are thought to have become reduced to the woody scales that comprise the pine cone.

(a)

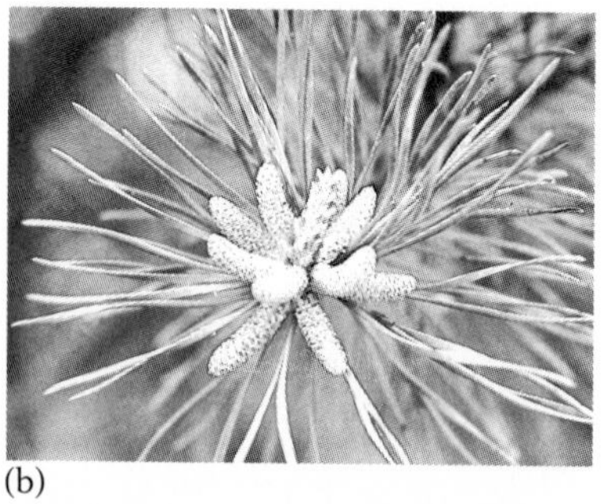

(b)

(c)

**FIG. 30-21** (a) The bristle cone pine, the oldest living organism. (b) Male (pollen) cones of a pine. (c) Female (seed) cone of a pine.

## GINKGO

*Ginkgo biloba* is the only surviving species of a unique group of gymnosperms. Although its fanlike leaves and plumlike seeds are very unlike the leaves and seeds of most gymnosperms, its coniferlike wood and its reproduction indicate its gymnosperm affinities (Fig. 30-22). *Ginkgo* trees have been planted in cities in North America and Europe from seeds brought originally from its native China, where it has survived in cultivation. Although *Ginkgo* grows well as an attractive and exotic shade tree, the female trees are objectionable because decaying seeds on the ground contain butyric acid and emit a putrid odor. However, male trees can be propagated from cuttings.

## THE UNIQUE GNETOPHYTA

One of the most unusual seed plants known is *Welwitschia*, of the division Gnetophyta, a group of plants in which the genera differ greatly in appearance but are similar in reproduction. *Welwitschia*, which grows

**FIG. 30-22** Leaves and seeds of *Ginkgo*.

**FIG. 30-23** *Welwitschia,* of the Gnetophyta, with its pair of strap-shaped leaves.

only in an arid coastal region of southwestern Africa, has a woody underground stem that produces only two leaves in its lifetime (Fig. 30-23). The leaves continue to grow out as straps, ultimately reaching several feet and becoming frayed with age. Only its reproductive cones reveal a relationship to *Gnetum* and *Ephedra,* the other two genera of the Gnetophyta. *Gnetum* is a vine or tree with leaves like those of an angiosperm, while *Ephedra,* the natural source of the drug ephedrine, is an arid-region shrub with small, barely distinguishable leaves. These gymnosperms of the Gnetophyta are unique in that their wood has evolved vessels, the kind of water-conducting cells found in angiosperms. Some botanists have suggested that gnetophytes may have given rise to the angiosperms, but compelling evidence for this is lacking.

## THE ANGIOSPERMS

The flowering plants are the most successful plants on earth. Angiosperms occur as aquatic plants, herbs, shrubs, vines, and trees. An indication of angiosperm accommodation to environment is their adaptation to the seasonal aspects of the temperate zone climate. The **annuals,** or **monocarpic plants,** complete their life cycle and die in a year; the **biennials** do so in two years; and the **perennials** continue to live from year to year, reproducing annually. Perennials may be woody, with above-ground parts and buds that survive adverse seasons; or their aerial parts may die back (in which case they are said to be herbaceous) to underground parts, such as rhizomes, bulbs, corms, and tubers, that remain alive, enabling these herbaceous perennials to survive the winter.

Other adaptations have permitted angiosperms to occupy such various environments as the tropics, the desert, and the tundra. Some have ventured back into the water. Some, such as the dodder, have even lost the ability to make chlorophyll and have become parasites on other plants (Fig. 30-24). Others such as the Indian pipe, are saprophytes that

**FIG. 30-24** Dodder, a parasitic angiosperm. The orange dodder obtains nutrients from the green host.

**FIG. 30-25** Indian pipe, a saprobe.

grow on organic matter of the forest floor (Fig. 30-25). One unusual group, the insectivorous plants, is adapted to the bog environment (see Chapter 11). The anaerobic environment and low pH of bogs inhibits decomposition of organic matter, and sources of nitrogen are limited for bog plants. In several families the leaves have become modified to trap insects and digest their proteins, freeing amino acids for nutrition of the plant.

The diversity of angiosperms is especially remarkable since they are the most recently evolved of the plant groups. There is no certain record of fossil angiosperms prior to the Cretaceous period. The origin of the angiosperms is as much a problem today as it was when Darwin called their sudden emergence and rapid evolution an "abominable mystery."

## THE DISTINCTIVE FEATURES OF ANGIOSPERM REPRODUCTION

The features that distinguish angiosperms from other seed plants have largely to do with reproduction, although the occurrence of vessels and sieve tubes in the vascular tissues is, with a few exceptions, also characteristic. Angiosperms developed the flower, with its ovules contained within a carpel; an added pathway for the pollen tube; and the phenomenon of **double fertilization,** with the formation of a polyploid nutritive tissue for the embryo, the **endosperm.** The major reproductive structures and events of the angiosperm life cycle now to be described are illustrated in Panel 30-5.

***The Flower.*** The flower, with its special reproductive parts, and the fruit that develops from it are distinguishing features of angiosperms. Flower morphology is the basis of angiosperm classification. While flowers vary greatly in different taxa, within species the structure is remarkably stable and is thus regarded as a conservative and reliable indicator of taxonomic relationships.

The typical flower consists of clusters, or whorls, of sepals, petals, stamens, and pistils, arranged in characteristic order and attached to the receptacle, the end of the stem. A flower with all of these parts is said to be **complete.** In a given species, however, any of the whorls of parts may be missing, making the flower **incomplete.** The stamens and pistils are essential organs, since they are the source of the gametes, but one of the other of these whorls may be missing from a particular flower type, which is said to be **imperfect.** Some species may have separate staminate and pistillate flowers. If both types of flower (male and female) exist on the same plant, as in corn, the plant is said to be **monoecious** (from the Greek *monos*, "single"; *oikos*, "house," or "one household"). If staminate and pistillate flowers are on separate plants, as in the date palm and the holly, the plant is termed **dioecious** ("two households"). Beyond these differences, flowers in different species may vary in symmetry, in the number, arrangement, and form of flower parts, and in the extent in which various flower parts are fused together.

***Enclosed Ovules.*** In addition to the presence of the flower, other features of reproduction distinguish the angiosperms from the gymnosperms. One of these features is the enclosure of the ovules within a carpel. The carpel is thought to represent a leaf or sporophyll that in its evolution became rolled into a container. Support for this idea comes from the carpels of angiosperms regarded as primitive. They are leaflike, with ovules attached along the margins of the folded leaf, just as sporangia occur in some ferns and ovules in some gymnosperms. The carpel is a subunit of the pistil. If there is a single carpel, the pistil is a simple pistil; if two or more carpels are joined, they form a compound pistil.

**Stamens** are also leaflike structures in primitive angiosperms, producing pollen in four embedded microsporangia. More commonly, the stamen is differentiated into stalk, or filament, and anther; the anther contains the microsporangia. Pollen released from the anther is transported to the pistil. In primitive flowers, it is deposited along the fused margins of the carpel and grows between the joined carpel edges to the ovules. In more advanced flowers, the surface where pollen adheres and germinates, the **stigma,** is limited to an area of special cells at the tip of the pistil instead of all along the margin.

In the angiosperms, enclosure of the ovules within the carpel imposes an additional route for the pollen tube to traverse as it grows from the pollen grain.

***Double Fertilization.*** A further distinguishing feature of the angiosperms is the functioning of the two male gametes in double fertilization. Both unite with nuclei of the female gametophyte. One gamete fertilizes the egg to form the zygote. The other unites with the two polar nuclei, producing a polyploid nutritive tissue, the endosperm, which nourishes the developing embryo. The ovule develops into the seed within the carpel, which alone or with enclosing flower parts becomes the fruit.

The gametophyte generation is reduced in the angiosperms to a few cells. The male gametophyte is the pollen grain and tube, with its two male gametes and an additional nucleus in the pollen tube. The female gametophyte, or embryo sac, is reduced to seven cells within the ovule.

***Fruit and Seed.*** Because angiosperm seeds are formed within the tissues of the ovary, they remain enclosed within the ovarian tissues as the

PANEL 30-5

# Angiosperm Reproduction

The parts of the flower involved in reproduction are the stamens and the pistil. The pistil may be made up of one carpel, as shown, or more than one joined together in a compound pistil.

## OVULE DEVELOPMENT: MEGASPOROGENESIS AND MEGAGAMETOGENESIS

Within the ovary at the base of the pistil are one or more ovules. The ovule consists of a megasporangium, or nucellus, containing a single **megasporocyte.** A collar consisting of a pair of integuments grows up around the nucellus as the ovule develops. Within, the megasporocyte undergoes meiosis, and three of the resulting megaspores abort in one common pattern of ovule development, leaving one megaspore to develop into the **megagametophyte.** (In some species, all four haploid spores participate in formation of the megagametophyte.) The persistent megaspore begins mitosis, and eight nuclei result. These are arranged in specific positions in the megagametophyte, also called the embryo sac, and cell walls develop about most of them. At one end of the megagametophyte (toward the micropyle, the small opening where the pollen tube enters the ovule) are two **synergid cells,** cells associated with fertilization, and the egg. At the opposite end are three **antipodal cells** that ordinarily do not participate in reproduction. Two **polar nuclei,** which migrated from each pole of the megagametophyte, are associated in the **central cell.**

## POLLEN DEVELOPMENT: MICROSPOROGENESIS AND MICROGAMETOGENESIS

The **anther** at the tip of the stamen bears four lobes, within each of which is a microsporangium containing several **microsporocytes** surrounded by a nutritional layer, the **tapetum.** The microsporocytes undergo meiosis, each forming a tetrad of four microspores. Each microspore undergoes mitosis to form a two-celled **microgametophyte,** or **pollen grain.** The pollen grains are disseminated and land on the stigma of a pistil in the process of **pollination.** A pollen tube emerges and grows down the style and into the ovule, through the micropyle and nucellus, and into a synergid cell. Experiments with cultured, growing pollen grains show that the pollen tube curves toward a chemotropic factor from the ovule, explaining its pathway into the ovule. One cell of the pollen grain, the generative cell, has divided to form two male gametes. They follow the other nucleus of the pollen grain down the pollen tube and are discharged into a synergid cell of the megagametophyte. The synergid cells contain a special deposition of wall material, the **filiform apparatus,** which may have a role in fertilization.

## DOUBLE FERTILIZATION AND EMBRYO DEVELOPMENT

Each male gamete functions in fertilization, one uniting with the egg to form the zygote, the other uniting with the two polar nuclei in the central cell to form the **primary endosperm nucleus.** This process is known as **double fertilization.** The primary endosperm nucleus divides to build a new tissue, the **endosperm,** which provides nutrients to the developing embryo. The zygote develops into an embryo, consisting initially of a **proembryo,** an undifferentiated globe of cells and a structure at its base, the **suspensor.** As the embryo grows, the enlarging suspensor keeps it in the reservoir of nutrients provided by the endosperm. When mature, the embryo has a lower region, the **hypocotyl,** with a root end, an **epicotyl** with a shoot apex between the pair of cotyledons, and, in many cases, a young leaf or two. The integuments have differentiated into the seed coat enclosing the embryo. The ovule has become a **seed.** The ovary containing the ovule or ovules in the meantime has developed into the fruit.

ovary develops into a fruit. In many genera, other parts of the flower, fused to the ovary, participate in formation of the fruit. In the evolution of angiosperms, an enormous variety of fruits have developed. In the apple, for example, the bases of the stamens, petals, and sepals, along with the receptacle, contribute to the edible parts of the fruit. In the pineapple, all the numerous flowers become fleshy and form the "multiple" fruit. Many fruits are attractive food for animals; the seeds are

The angiosperm life cycle.

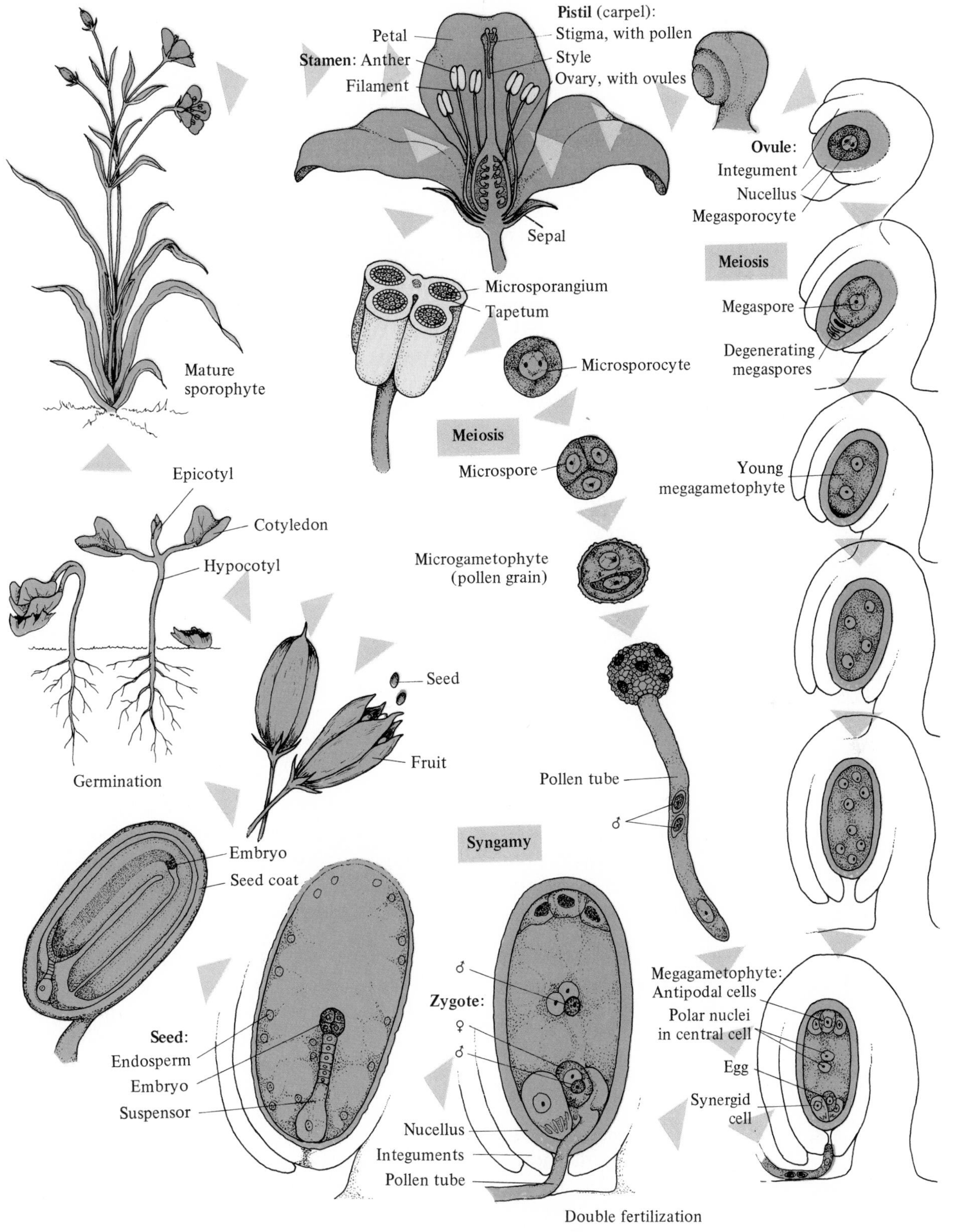

Pistil (carpel):
Stigma, with pollen
Style
Ovary, with ovules
Petal
Stamen: Anther
Filament
Sepal
Ovule:
Integument
Nucellus
Megasporocyte
Meiosis
Megaspore
Degenerating megaspores
Young megagametophyte
Microsporangium
Tapetum
Microsporocyte
Meiosis
Microspore
Microgametophyte (pollen grain)
Mature sporophyte
Epicotyl
Cotyledon
Hypocotyl
Germination
Seed
Fruit
Pollen tube
♂
Syngamy
Embryo
Seed coat
Seed:
Endosperm
Embryo
Suspensor
Zygote:
♂
♀
♂
Nucellus
Integuments
Pollen tube
Double fertilization
Megagametophyte:
Antipodal cells
Polar nuclei in central cell
Egg
Synergid cell

usually unharmed by passage through the digestive tract of a mammal or bird and thus become widely disseminated in animal feces. Some seeds fail to germinate unless they have undergone such passage. Some fruits, such as those of milkweed, form plumes, or winglike structures, as in maple, capable of wafting seeds on the wind. Other fruits possess burrs that cling to the fur of mammals. Some fruits, such as coconuts, when dry, can float for hundreds of kilometers on the sea's surface, later to germinate upon reaching land. Such varied means of seed dispersal have contributed to making the angiosperms the most successful of all vascular plants.

## ORIGIN AND RAPID EVOLUTION OF THE ANGIOSPERMS

Evolution of the angiosperms has been of great interest to botanists, not only because of their relatively recent origin and rapid diversification, but because so little is known of the ancestors of angiosperms. It is assumed that angiosperms originated late in the Mesozoic era and had diversified considerably by the Cretaceous period, from which the first authenticated angiosperm fossils date. These fossils include pollen, leaves, and wood with vessels from the Atlantic coastal plain of New Jersey and the vicinity of Washington, D.C., and from England, Brazil, Africa, Israel, and continental Europe. Occasionally, finds of earlier Triassic leaf fossils resembling palms create problems in interpretation: are they angiosperm leaves? Geographic distribution of early fossils suggests that the angiosperms originated in tropical and semitropical regions and spread toward the poles.

The time of early angiosperm evolution coincides with the breakup of the original land masses into continents, as postulated in the theory of continental drift (see Chapter 27). Strong support for the continental drift theory comes from the distribution of plant families on the various continents now separated by oceans. Similarities between the flora of eastern Brazil and western Africa, for example, suggest that these continents were once contiguous. (The plants in question are never disseminated over such wide expanses of ocean.) As the continents drifted apart, isolated floras of common origin continued to evolve independently. This led to the evolution of distinct floras on different continents.

By the end of the Cretaceous period, most modern angiosperm families are represented in the fossil record. It has been suggested that the present flora had evolved by the Pliocene period, 7 million years ago.

***Evolution Within the Angiosperms.*** The rapid evolution of the angiosperms means that the fossil record, condensed into limited geologic strata, can disclose little of angiosperm phylogeny and may never reveal finely detailed evolutionary relationships among the major modern families. As a result, comparative morphology and anatomy of living species has been an important tool in estimating evolutionary relationships. Studies of wood anatomy, for example, have provided a basis for evaluating relative primitiveness and evolutionary advances. Plants known to be primitive have longer tracheids and vessel elements. Other anatomical features have been correlated with length of the xylem elements, which thus has also become an indicator of phylogeny. On this basis, one could study the xylem of a plant and determine whether the species, living or fossil, represented a primitive or advanced family.

**FIG. 30-26** *Magnolia,* a primitive type flower.

***The Primitive Flower.*** In a similar way, flower morphology has been correlated with evolution. Several lines of evidence, including the fossil record, wood anatomy, and floral and pollen morphology, have led to the widely held theory that the most primitive flower is represented in a group of woody plants, the *Ranales,* many of them natives of the tropical Pacific islands but including the more familiar *Magnolia* family (Fig. 30-26). A few of the species are vesselless angiosperms: they have wood without vessels, as one might expect in an early representative of a line evolved from the lower tracheophytes, with only tracheids as water-conducting cells. The primitive flower, according to this theory, is a shortened shoot consisting of appendages that evolved from leaves.

***Angiosperm Families: Monocotyledons and Dicotyledons.*** From the primitive type of flower, with its many unfused parts in separate whorls, evolved the various flower forms represented in the numerous families of angiosperms.

Two major groups, the **monocotyledons,** or monocots, and the **dicotyledons,** or dicots, are distinguished in the angiosperms (Fig. 30-27). While these names are based on the number of cotyledons in the embryo, several other features distinguish monocots from dicots. Monocots, notably the grasses, commonly have parallel veins in their leaves, in contrast to the network pattern of veins in dicots. Monocots have flower parts in whorls of three—for example, three sepals, three petals, stamens in two whorls of three each, and a tri-lobed pistil. Anatomical evidence indicates that the monocots emerged from dicot ancestors early in angiosperm evolution.

The great variety and adaptability of the angiosperms, which enabled them to colonize new environments and become the predominant flora of the world, can be attributed to their genetic diversity. Genetic diversity in a population is enhanced by factors that promote cross-fertilization. In the angiosperms, a number of devices have evolved that promote cross-pollination and prevent self-fertilization. Among these are (1) separate staminate and pistillate flowers, (2) the maturing of stamens and pistils of the flower at different times, and (3) a genetic self-incompatibility mechanism that prevents pollen of one plant from growing on its own pistil. Important to the evolution of cross-pollination mechanisms was the coevolution of certain animals, especially the

**FIG. 30-27** Two groups, the monocotyledons and the dicotyledons, are thought to have evolved from the primitive angiosperm. Characteristic monocot flower, (a) *Trillium,* and dicot flower, (b) *Dahlia,* are shown.

(a)

(b)

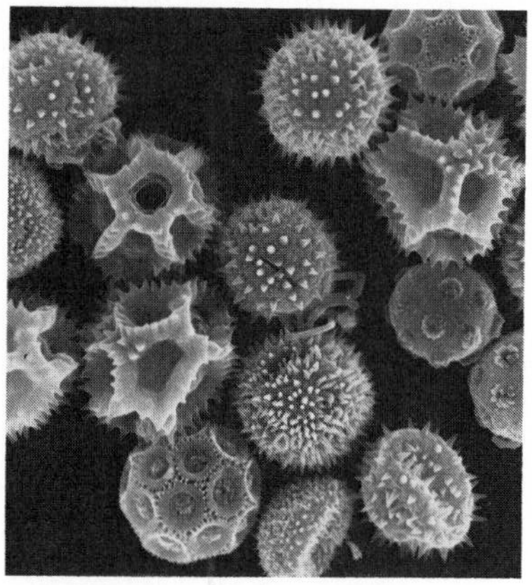

**FIG. 30-28** An assortment of angiosperm pollen.

insects. Insects, as well as birds and bats, are attracted to flowers for pollen and nectar by such devices as petal color and pattern and aromatic and nutritive secretions of nectar producing nectaries and glands. In visiting flowers, animals transfer pollen from one flower to another. Although botanists once thought that the first angiosperms were wind pollinated, as are gymnosperms, the consensus now is that the earliest angiosperms were insect pollinated, probably initially by beetles. Some remarkable adaptations associated with pollination have evolved in insects and in flower structure. Wind-pollinated plants produce enormous amounts of pollen (well-known to allergists) and have exposed stigmas or modified stigmas with hairs to which pollen adheres.

Pollen grains are finely sculptured in patterns often distinct enough to enable identification of the genus or even species of the plant from which they came. Palynologists, scientists who study pollen, by studying fossil pollen in ancient bogs and sediments, have been able to determine many of the plants that existed in certain floras of the past.

***Chemical Evidence for Phylogeny.*** In recent years, additional criteria have been utilized to gain information on phylogeny, the domain of plant taxonomists. Perhaps the most convincing support comes from **chemotaxonomy,** the study of the distribution among families of natural plant products such as alkaloids, terpenes, and phenolics. In addition to the tool of organic chemical analysis, methods from immunological research are used in which antigens extracted from seeds, pollen, and leaves are reacted with antibodies to determine similarities, if any, of their molecules. Amino acid sequences in common protein molecules, such as cytochrome *c*, and hybridization of nucleic acids (see Chapter 24) are also being used to provide information about relationships among taxa.

Still another technique to assess the closeness of species utilizes slight differences in the DNA of mitochondria or of chloroplasts. Restriction endonucleases (Chapter 24) are used to cut apart the mitochondrial and chloroplast DNA of each species. When the fragments from one species are compared with those of another after separation by gel electrophoresis, their closeness of kinship can be detected: the more identical the bands on the gel, the closer the two species are genetically.

Using these diverse methods, taxonomists have attempted to construct phylogenetic trees for the evolution of the families of angiosperms. The evolutionary tree is a much-branched shrub, and without an adequate fossil record, the branches cannot be connected. While relationships between major modern angiosperm groups can be indicated, the connecting pathways are mostly speculative. As new studies are made, of course, occasional changes may be indicated in the phylogenetic shrub, although it has remained quite stable. Nevertheless, in attempting to reconstruct the phylogeny of angiosperms, it must be remembered that our evidence comes largely from living families, which evolved from common ancestors and not from each other.

## ECONOMIC IMPORTANCE OF ANGIOSPERMS

Angiosperms provide so many of our needs that it would require volumes to enumerate their uses. The field of **economic botany** is devoted to this study. We need only think of our sources of food, direct and indirect, to realize our reliance on angiosperms. They also provide us with numerous secondary plant products, such as timber, paper, textiles, waxes, oils, tannins, pesticides, alkaloids, drugs, beverages, spices, and tobacco.

The study of the utilization and domestication of plants by early human cultures is the domain of **ethnobotany.** Such fascinating stories as that of the origin of wheat from weeds in the "fertile crescent" of the Middle East (the valleys of the Tigris and Euphrates rivers), and the evolution of maize in Mexico and Central America are documented in several ethnobotany texts.

## Summary

Several groups of algae, distinct from one another on the basis of pigments and other biochemical characteristics, morphology, and reproduction arose from the Cyanobacteria, or blue-green algae. Many brown algae are large marine organisms; the red algae, also mostly marine, exhibit unique cellular organization and reproduction. Both groups are sources of food and useful products. Other groups of algae possess distinctive features that make each unique. The diatoms, for example, are protists with walls of silica. The dinoflagellates have flagella that cause them to whirl; these protists cause toxic blooms that are sometimes poisonous to swimmers and to consumers of seafood. The green algae display an assortment of forms and a variety of gamete types and life cycles. Alternation of generations, the type of life cycle exhibited by some green algae, became the pattern in the land plants, which are thought to have evolved from green algae.

Two groups of land plants evolved: the bryophytes and the more successful vascular plants. The bryophytes—consisting of mosses and liverworts—have been limited in size, reproduction, and distribution by their water requirements; the gametophyte is the predominant generation in their life cycle. Their photosynthetic gametophytes lack specialized vascular tissues, and they grow clustered together or prostrate in contact with a water supply. The male gametes are motile and swim to the female reproductive organ in which the egg is contained. The sporophyte is smaller and supported by the gametophyte; its primary function is the production and dissemination of spores. When partially decomposed to peat, the moss *Sphagnum* has proven useful to humans as fuel and in horticulture.

Vascular plants have been able to achieve a much larger size than the bryophytes, and several groups have members with the stature of trees, even in the ancient past. The larger size was possible because of their water-conducting xylem tissue and the food-transporting phloem. The predominant generation in the life cycle of the vascular plant is the sporophyte, a diploid generation, which, because of its genetic diversity, functions to accelerate evolution. Several groups of lower tracheophytes evolved from a group of leafless, rootless psilophytes that appeared in the Silurian period of geologic history. Although most abundant in the Carboniferous period, a few relic genera of lower tracheophytes—the horsetails, club mosses and ferns—survive today.

The seed plants evolved from the lower tracheophytes. One group, the gymnosperms, appeared early, in the Devonian period, and achieved maximal development in the Mesozoic era. Some of these gymnosperm groups are now extinct; others have survived with limited numbers of species. The cycads, plants with large, fernlike leaves and short trunks producing strobili or cones, are now reduced to nine genera. *Ginkgo biloba,* an ornamental shade tree, is the sole surviving species of an otherwise extinct order. Of all the gymnosperms, the conifers are the most abundant and familiar. They are the timber trees and ornamental shrubs of the northern and southern hemispheres. A group of gymnosperms with three unusual genera, the Gnetophyta, is of uncertain evolutionary position.

By far the most successful of the seed plants, the angiosperms are not identified with certainty in the fossil record until the Cretaceous period. Their rapid evolution as the continents drifted apart led to a great diversity of families all over the earth. Their unique flowers and their reproductive process provided an immense diversity on which the selective forces of nature could act. The angiosperms have filled innumerable terrestrial and some aquatic ecological niches of the earth. They are primary sources of food for all animals, including *Homo sapiens,* who learned early to cultivate them and to breed and select them for their many uses.

## Recommended Readings

BANKS, H. P. 1970. *Evolution and Plants of the Past.* Wadsworth, Belmont, Calif. An analysis of the fossil record of plants, with particular emphasis on the lower tracheophytes and the origin of vascular plants.

BOLD, H. C., C. J. ALEXOPOULOS, and T. DELEVORYAS. 1980. *Morphology of Plants and Fungi,* 4th ed. Harper & Row, New York. A revised edition of a widely used textbook covering fungi, algae, and plants in depth.

BOLD, H. C., and M. J. WYNNE. 1985. *Introduction to the Algae,* 2nd ed. Prentice-Hall, Englewood Cliffs, N.J. An authoritative and detailed treatment of the biology of the groups of algae.

FOSTER, A. S., and E. M. GIFFORD. 1974. *Comparative Morphology of Vascular Plants.* Freeman, San Francisco. A standard text for courses in vascular plant structure and reproduction.

HEISER, C. B., Jr. 1973. *Seed to Civilization: The Story of Man's Food.* Freeman, San Francisco. The origin of agriculture and domestication of plants and animals, with emphasis on food plants.

JENSEN, W. A., and F. B. SALISBURY. 1984. *Botany,* 2nd edition. Wadsworth, Belmont, Calif. A new revision of a widely used text in botany.

KIERMEYER, O., ed. 1981. *Cytomorphogenesis in Plants.* Springer-Verlag, New York. A monograph dealing with structure and development of unicellular algae and cells of higher plants.

LEE, R. E. 1980. *Phycology.* Cambridge University Press, New York. A concise account of the biology of the algae as revealed by the various families.

PICKETT-HEAPS, J. D. 1975. *Green Algae: Structure, Reproduction, and Evolution in Selected Genera.* Sinauer Associates, Sunderland, Mass. An exceptional collection of light and electron micrographs of the green algae with descriptions and discussion.

RAVEN, P. H., R. G. EVERT, and H. CURTIS. 1981. *Biology of Plants.* Worth, New York. A popular introductory text in botany.

RAY, P. M., T. A. STEEVES, and S. A. FULTZ. 1983. *Botany.* Saunders, Philadelphia. A revised and current version of an older, widely used text.

RICHARDSON, D. H. S. 1981. *The Biology of Mosses.* Wiley, New York. A recent look at the mosses, including research on their physiology and ecology.

SCAGEL, R. F., R. J. BANDONI, J. R. MAZE, G. E. ROUSE, W. B. SCHOFIELD, and J. R. STEIN. 1982. *Nonvascular Plants: An Evolutionary Survey.* Wadsworth, Belmont, Calif. An updated survey of nonvascular plants with excellent, mostly original illustrations by a group of authors thoroughly familiar with the plant groups on which they write.

STEBBINS, G. L. 1974. *Flowering Plants: Evolution Above the Species Level.* Harvard University Press, Cambridge, Mass. A consideration of adaptations in the angiosperms and the basis of angiosperm evolution by an eminent geneticist and botanist.

TAYLOR, T. N. 1981. *Paleobotany: An Introduction to Fossil Plant Biology.* McGraw-Hill, New York. A recent account of the known plant fossils and their relationships.

WATSON, E. V. 1971. *The Structure and Life of Bryophytes,* 3d ed. Hutchinson, London. An interesting description of the mosses and liverworts.

31

# Animal Diversity: The Lower Animals

OUTLINE



Only relatively recently have biologists begun to realize the true extent of animal diversity. The more than 1.2 million species of animals that have been described to date have been largely those that have happened to come to our attention. Most of the large species, particularly the vertebrates—fish, amphibia, reptiles, birds, and mammals—have probably already been discovered. The number of species of animals and the animallike protists, the Protozoa, that have never been officially named and described must number in the millions and may even be as high as 10 million. There are several reasons for believing this. First, remote islands that have not previously been studied have always been found to harbor large numbers of previously unknown species. In addition, any species of vertebrate that has never been checked for parasites has always been found to host species of worms unique to it. For example, over 100 species of worms are known to parasitize humans; many of them live only in us. *Homo sapiens* also provides a home for dozens of species of protozoan parasites, many of them also incapable of surviving in any other host species.

Each environmental niche on the face of the earth, whether an island, an animal body, a mountaintop, or a forest, has been colonized by a great number of animal and protozoan species. To the extent that a particular environment has been isolated from others, even if only by distance, it is found to harbor unique species. We are thus quite confident in our prediction that there are many millions of animal species yet to be named and described. In 1983 even a new phylum, Loricifera, was discovered. This phylum includes a group of small invertebrates living in marine sand up to hundreds of meters below the sea surface. Their features are so different from those of all other animals thus far described that a new phylum was created to accommodate a number of these new species.

As noted in Chapter 1, this text uses a classification system that places all organisms in one of five kingdoms: Monera, Protista, Fungi, Plantae, or Animalia. The classification of all multicellular organisms, such as plants, fungi, and animals, is based on their multicellularity and type of nutrition. The kingdom Monera—consisting of bacteria and blue-green bacteria, or cyanophytes—includes all the prokaryotes, that is, one-celled organisms that lack an organized nucleus and most membranous organelles. The remaining kingdom, Protista, includes all of the unicellular eukaryotes; like plants, fungi, and animals, protists have organized nuclei and membrane-bounded organelles.

Protists, however, are quite diverse. As noted in Chapter 30, algal protists are clearly related to multicellular algae of the kingdom Plantae and, indeed, are included with the plants in the classification of some authors. Similarly, many of the Protozoa, the animallike protists, appear to be closely related to some members of the kingdom Animalia. Like animals generally, most protozoans exhibit **holozoic nutrition;** that is, they ingest particulate food. In some other taxonomic systems, the Protozoa are classified as a subkingdom of the kingdom Animalia, with most multicellular animals placed in the subkingdom Metazoa.

Just as the algal protists, unicellular algae, while not classified as plants, are discussed in the opening pages of the chapter on plant diversity, so the animallike protists, the Protozoa, are the first group considered in this chapter. Later on, in addition to protozoans, lower animals are also described.

## SUBKINGDOM PROTOZOA: THE ANIMALLIKE PROTISTS

Protozoa may be defined as motile eukaryotic unicellular organisms. Classical biologists considered motility to be a major characteristic distinguishing animals from plants, whereas the ability to photosynthesize was regarded as a characteristic of plants. Hence the Protista, which includes all unicellular eukaryotes, was divided into protozoa and protophyta. Although these criteria hold well in most cases, a few protozoans exhibit both locomotory and photosynthetic abilities. According to the modern five kingdom classification scheme that is followed in this text, each major branch of organisms represents a grade of organization rather than a group of organisms with a common ancestry.

At the present state of our knowledge it is not possible to be certain about the phylogeny of protozoans, let alone all protists. Thus the kingdom Protista and the subkingdom Protozoa include organisms all of which are unicellular. Yet they do not necessarily represent a common evolutionary origin.

## PROTOZOAN TAXONOMY

In the past, the Protozoa were usually classified as a single phylum of one-celled animals, divided into four classes. Since 1980, however, experts have regarded the Protozoa as a subkingdom consisting of seven separate phyla within the kingdom Protista. These are the phylum Sarcomastigophora, consisting of flagellates and amoebae, with a single type of nucleus, formerly included in the classes Sarcodina and Mastigophora; the phyla Labyrinthomorpha, Apicomplexa, Microspora, Ascetospora, and Myxozoa, spore-forming parasites previously combined as the class Sporozoa; and the phylum Ciliophora, cilia bearers with two types of nuclei, formerly the class Ciliata. Unlike earlier arrangements into classes on the basis of locomotor organelles, this division into phyla is based primarily on types of nuclei and mode of reproduction.

Many Protozoa appear to be simple forms of life, and in some ways they are, since the single protozoan cell must perform all vital functions: food acquisition, digestion, synthesis, respiration, excretion, ionic regulation and osmoregulation, locomotion, and reproduction. Because multicellular organisms have various specialized cells that perform these tasks, most protozoan cells are more complex than any of ours. The simplicity of protozoa is thus more apparent than real. Protozoan cells usually contain most of the organelles found in metazoan cells, such as nucleus, mitochondria, endoplasmic reticulum and Golgi complex, as well as others that are unusual, such as contractile vacuoles and food vacuoles (Fig. 31-1). Some protozoans have a permanent gullet, or cytopharynx, that leads into the buccal cavity. The cilia or flagella, when present, are anchored on basal bodies, which are similar to centrioles.

## ORGANELLES

As with eukaryotic cells generally, the protozoan cell body is bounded by a plasma membrane. The cytoplasm immediately underlying the cell surface, however, is often semisolid or gelatinous in texture, giving some rigidity to the cell body. It is also common for the protozoan cell

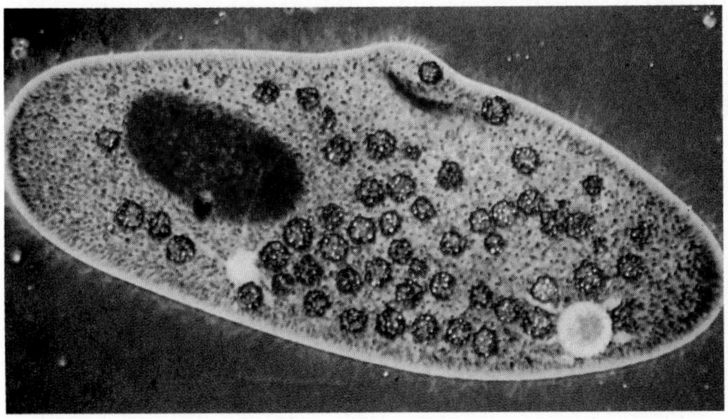
(a)

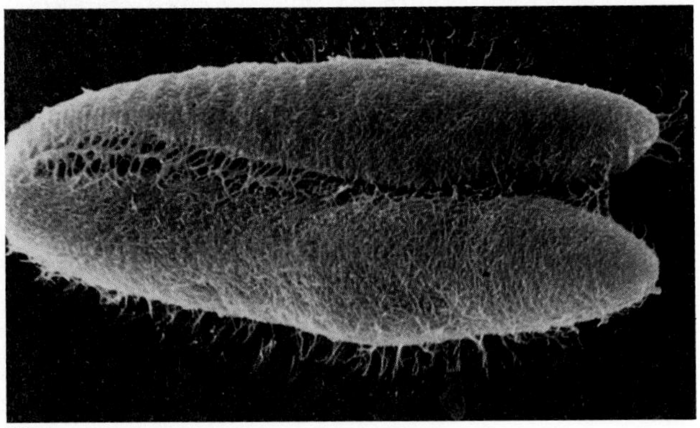
(b)

**FIG. 31-1** (a) *Paramecium caudatum*, a ciliate. Note the organelles found within the cell. (b) *Paramecium* in conjugation.

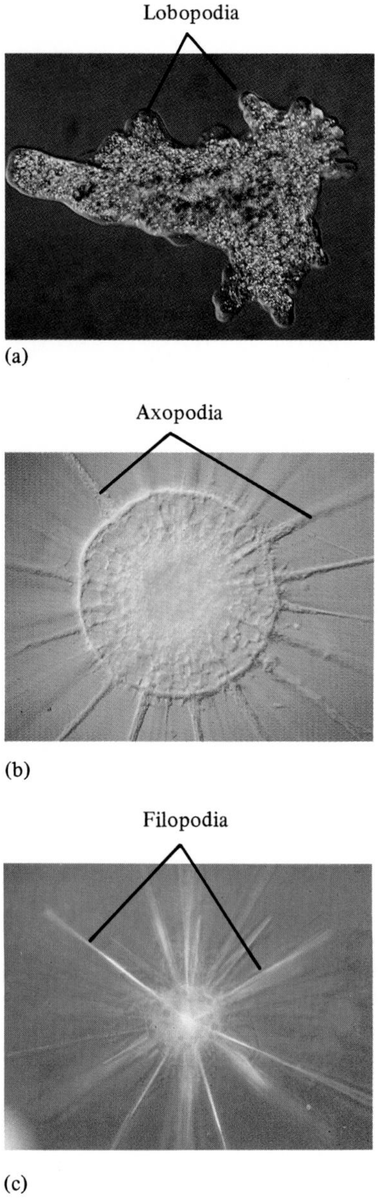


**FIG. 31-2** Representative Sarcodinan protozoans. (a) *Amoeba proteus* with blunt pseudopodia. (b) *Actinophrys* with axopodia. (c) A radiolarian with filopodia.

to be enclosed in a nonliving protective secretion of some type. Many flagellates and all ciliates are enclosed in a pellicle that gives them a permanent shape. Even the common amoeba, with its fluid locomotion and ever-changing shape, bears a fringe of threadlike extensions on the surface of its plasma membrane, similar to extensions found in some metazoan cells.

***Locomotor Organelles.*** Since protozoans are generally distinguished from the Protophyta by their locomotory ability, early systematists used the type of locomotory organelle as the major criterion in the classification of this group. There are three major types of locomotory organelles, pseudopodia, flagella, and cilia. Although taxonomically speaking, locomotory organelles are not as important as they once were, nonetheless they are important morphological features that throw light on the mode of life of the protozoans.

Sarcodinans, which comprise one of the subphyla of the phylum Sarcomastigophora, move and capture food by means of various kinds of **pseudopodia** (from the Latin, "false feet"). These are essentially extensions of the cytoplasm. Pseudopodia of various types—lobopodia (blunt lobelike), filopodia (threadlike), rhizopodia (branched), and axopodia (supported by a rigid structure)—are shown in Figure 31-2.

Flagellates and ciliates move by moving their flagella, which usually "scull," in the fashion of a Venetian gondolier, and cilia, which "row," in the manner of a varsity crew member. Flagella and cilia are also used to generate feeding currents. The structure of flagella and cilia in protozoans is identical to that of other eukaryotes. Each cilium or flagellum has the characteristic arrangement of 9 doublets of microtubules at the periphery and 2 central microtubules (see Chapter 4).

Most sporozoans are nonmotile, although some exhibit amoeboid movement, that is, they move by putting forth pseudopodia, during a brief phase of their life cycle.

***Food Vacuoles.*** Most protozoans ingest particulate food, which, in some, is taken in by a **cytostome** (from the Latin, "cell mouth"), permanently located in one part of the cell. The ingested food, usually other microorganisms or detritus (partially decayed organic matter), is immediately enclosed in a food vacuole. Acid and digestive enzymes are secreted into the vacuole, and the products of digestion are absorbed into the cytoplasm by pinocytosis. Undigested residues are ejected at the cell surface.

## GAS EXCHANGE, EXCRETION, AND OSMOREGULATION

The exchange of oxygen and carbon dioxide with the environment in aerobic forms occurs by simple diffusion. Excretion of ammonia is also accomplished by diffusion. Protozoa living in hypotonic media, such as lakes and streams, maintain osmotic balance by water expulsion through contractile vacuoles or water expulsion vesicles, which collect the excess water that osmotically enters the cell and discharge it at the cell surface. Marine protozoa are isotonic to their environment and lack such vacuoles.

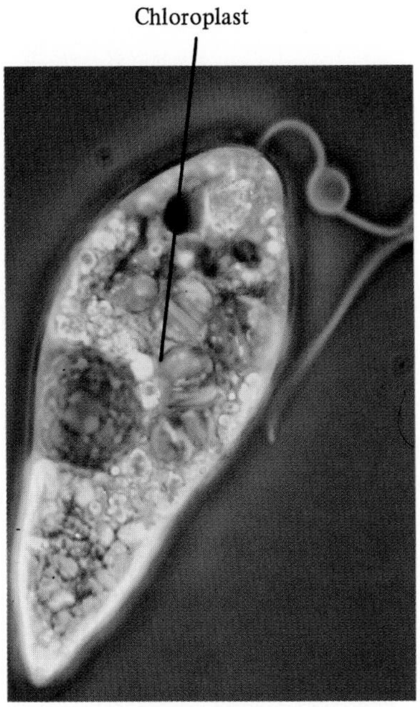


**FIG. 31-3** *Euglena* sp., a protistan flagellate with chloroplasts.

## REPRESENTATIVE PROTOZOA

Protozoology, the study of protozoa, has been long recognized as an independent branch of biology. Largely because of the many parasitic protozoans that cause havoc to human health the study of protozoa has been vigorously pursued since the invention of the light microscope.

***Mastigophora.*** According to the 1980 classification, Mastigophora is one subphylum of the phylum Sarcomastigophora. Mastigophora, as the term implies, all bear flagella. Their life styles are, however, highly varied. Some are holozoic, ingesting solid food usually through a fixed mouth. Some are parasitic or saprobiotic, whereas others are photosynthetic autotrophs and are also included in the Protophyta. This heterogeneity in nutrition has caused, and continues to cause, problems to systematists. There is reason to believe that some of the flagellates are the most primitive of all eukaryotic organisms and that all multicellular organisms—plants, fungi, and animals—evolved from protistan flagellate ancestors. Green (photosynthetic) flagellates, also called algal flagellates or algal protists, many of which form colonies, are obviously closely related to lower plants, whereas several species of protozoan flagellates closely resemble cells of lower animals (Fig. 31-3). It is also logical to suppose that most other protozoans arose from flagellated ancestors. Some sarcodinans, for example, pass through flagellated stages during their life cycles, and ciliates can easily be regarded as greatly modified flagellates, since cilia and flagella have basically the same ultrastructure (see Chapter 4).

It is also significant that the flagellated protists lie directly on the border between plant and animal life. A dramatic demonstration of this is provided by the descriptions of the flagellate genera *Euglena* and *Astasia* given in Chapter 30. Except that it possesses chloroplasts, one species of the green flagellate *Euglena* (Fig. 31-3) is indistinguishable from a species of the protozoan *Astasia*. When treated with streptomycin at high temperatures, *Euglena* loses all of its chloroplasts, whereupon it is entirely indistinguishable from *Astasia*.

All flagellates possess at least one flagellum; some have dozens. They are all capable of asexual reproduction, accomplished by dividing longitudinally while undergoing mitosis. Some species reproduce sexually as well, with the entire cell acting as a gamete; after undergoing meiosis, the cell fuses with a similar cell of opposite mating type ("sex"). Although most flagellates are free-living, many are parasitic, including *Trypanosoma*, which causes African sleeping sickness in humans.

***Sarcodina.*** A second subphylum of Sarcomastigophora, Sarcodina, includes both free-living and parasitic forms that bear pseudopodia as locomotory organs. *Amoeba*, found in freshwater ponds the world over, is a fairly typical sarcodinan. Its locomotion, by means of broad pseudopods, has long fascinated those who have attempted to understand the various factors involved in the locomotory process (Fig. 31-2a). Many sarcodinans have shells or other protective coverings. Some, the Foraminifera and Radiolaria, have, over millions of years, left enormous deposits of their shells (called foraminiferan and radiolarian ooze) on the ocean floor. The parasitic sarcodinans include *Entamoeba histolytica*, which produces the often fatal amoebic dysentery. Although sexual reproduction occurs in many groups, a large number of sarcodinans reproduce only asexually.

***Opalinatea.*** A third subphylum of Sarcomastigophora, Opalinatea, are a ciliated group of protists such as *Opalina*. This genus was formerly included in the Ciliophora, but because, unlike ciliates, its members have only one kind of nucleus and divide longitudinally, they are now rightfully classified as a type related to flagellates and sarcodinans.

***The Sporozoans.*** The former class or phylum Sporozoa, whose representatives all are parasites, are now divided into five separate phyla: Labyrinthomorpha, Apicomplexa, Microspora, Ascetospora, and Myxozoa. All produce spores at some stage of their life cycle. The type of spore and the manner of its production are highly variable. A fairly typical sporozoan life cycle is that of *Plasmodium*, the causative agent of malaria, which is transmitted by mosquitoes (Fig. 31-4). It is doubtful that any disease has caused as much suffering and death as malaria. Even today, with all that is known about its control, a million children die of it each year in Africa alone. Some strains of malarial parasites

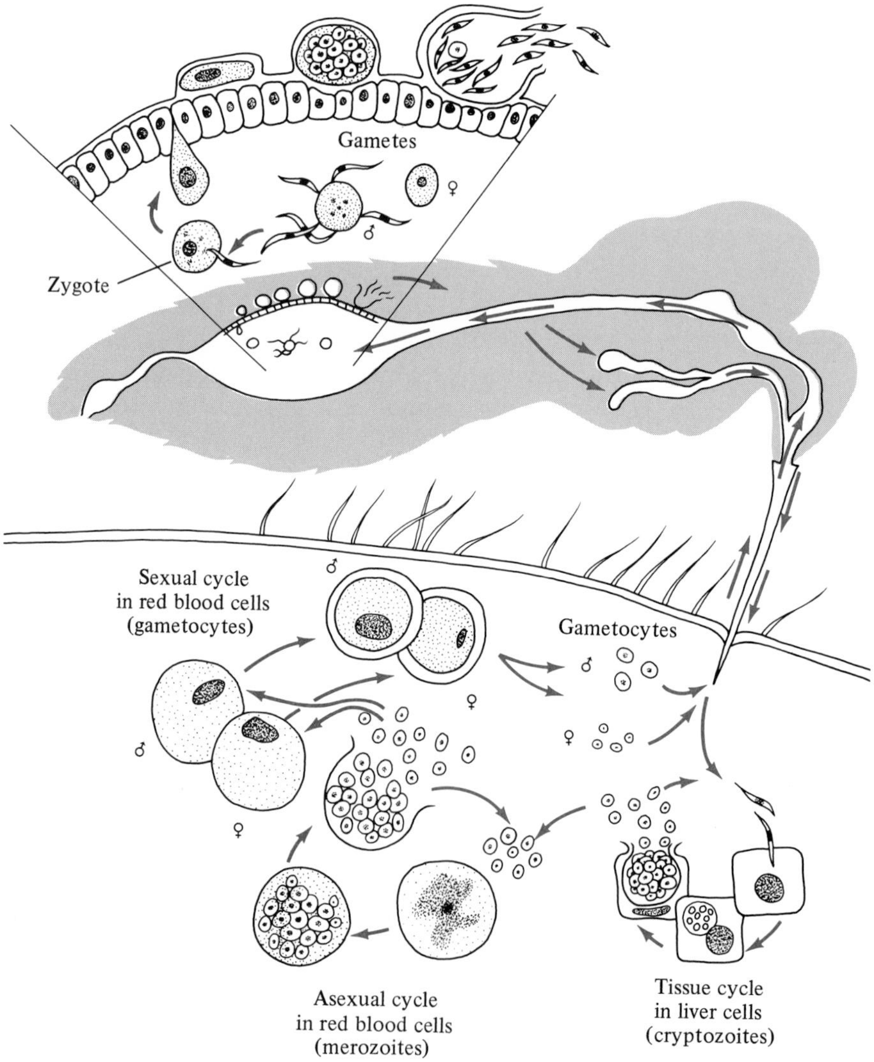


**FIG. 31-4** Life cycle of *Plasmodium* sp., the malarial parasite.

have become resistant to quinine derivatives, the only drugs effective on the malarial parasite. Similarly, mosquitoes, the vectors of the malarial parasite, have become resistant to insecticides. Unless new methods of control are developed, malaria may once again become a major health problem to mankind.

The life cycle of a malarial parasite is shown in Figure 31-4. The parasites enter the human blood stream through the bite of an infected mosquito. After that, the sporozoites enter the liver and are not found in circulating blood. At this stage, when the parasites are called cryptozoites, they are very sensitive to drug treatment. After about 10 days, they reenter the circulation and reproduce asexually within red blood cells, as intracellular parasites. As merozoites, the products of asexual reproduction are released by the breakdown of the infected red blood cells. Some toxins are also released into the circulation. These toxins are the direct cause of malarial fever and chills. After several rounds of asexual reproduction, some of the parasites are passively taken in a blood meal by a mosquito. Now they enter their sexual reproductive phase. They are now called gametocytes. The gametes mature in the mosquito's stomach where they fuse to form a zygote. The zygote then traverses the stomach wall. By asexual reproduction the zygote produces sporozoites which find their way to the salivary glands and thus to another human host.

***The Ciliates.*** Clearly the most highly evolved of the protozoa, the ciliates exhibit a wide range of forms and occupy a great variety of environments. Many parasitic species are known, including some that live in humans. Both sexual and asexual reproduction occur. Mating of opposite types, termed + and −, occurs under certain conditions. Typically, two ciliates fuse, a process called conjugation, at their anterior ends. In each cell, meiosis occurs, producing two haploid gametic nuclei, one of which, known as the migratory nucleus, moves across a cytoplasmic bridge connecting the two conjugants. Upon crossing the bridge, each migratory nucleus fuses with the stationary nucleus of the other cell. This produces a diploid zygotic nucleus in each conjugant. Having exchanged genetic material, the two conjugants separate and resume their usual asexual reproduction by repeated transverse cell divisions.

## SUBKINGDOM PARAZOA

Parazoa are the simplest multicellular animals. This group is represented today by a single phylum, Porifera (from the Latin for "porebearers,") which includes all sponges. Sponges are multicellular organisms in which cells are not organized into tissues and organs. That is to say that although the individuals are composed of many cells, such cells are not integrated into specific structures. These organisms lack a means of internal communication, such as a nervous system and a circulatory system. Thus the individuals are more like colonies of cells.

Multicellularity probably evolved several times among different organisms. It arose in plants from some protophytes, which are photosynthetic autotrophs, and this line of evolution gave rise to the higher plants. It also arose from protozoans, and this line gave rise to Parazoa. But Parazoa, because of their organization, are not regarded as ancestral to other multicellular animals. The reasons for believing that parazoans are an independent offshoot unrelated to the evolution of Metazoa becomes obvious as we study their organization and physiology.

## PHYLUM PORIFERA

In sponges various physiological functions, such as digestion and reproduction, are performed by a large variety of specialized cells that function somewhat independently of each other. The way in which these cells are organized to form the body of a sponge is exemplified by the **asconoid** sponge, which exhibits the simplest body type (Fig. 31-5, 7a). Most other sponges are modifications of this basic form.

The body wall of the asconoid sponge is composed of two prominent layers, an outer epidermis of cells called **pinacocytes** and an inner layer of **choanocytes,** or "collar cells." Each choanocyte bears a flagellum surrounded by a collar of numerous fine cytoplasmic extensions of its cell body (Fig. 31-6a). By their constant beating, the flagella maintain a water current through the sponge body, bringing in food and oxygen and removing wastes. Choanocytes, which resemble a flagellate protozoan known as a choanoflagellate, constitute one of the links indicating that sponges arose from Protozoa (Fig. 31-6b).

Between the pinacocyte and choanocyte layers is an intermediate layer of irregularly shaped mesenchyme cells embedded in a gelatinous matrix called **mesoglea.** Various cells capable of amoeboid movement, called amoebocytes, are also found in this layer, as are many **spicules,** needlelike secretions, usually of calcium carbonate or silicates.

Perforating the surface of the asconoid sponge body are numerous **ostia,** which are openings through doughnut-shaped cells called **porocytes.** Water enters the sponge through the ostia, the incurrent openings, then passes into a central cavity, the **spongocoel,** exiting by way of an excurrent opening, the **osculum.**

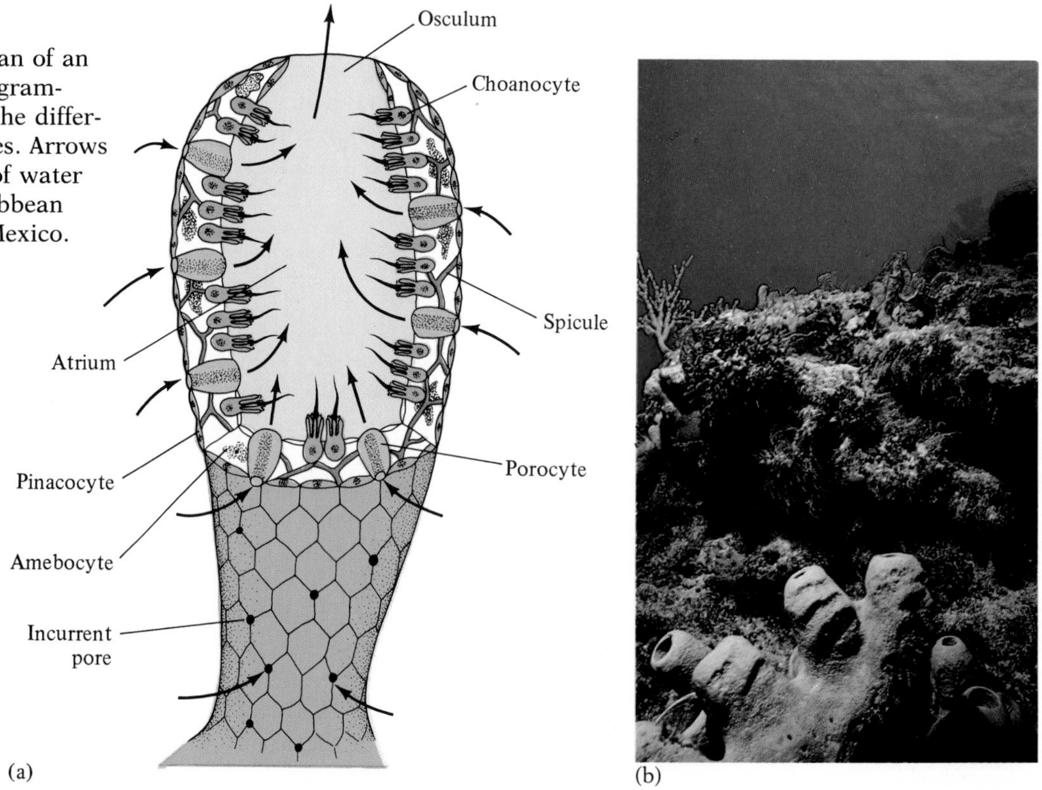


**FIG. 31-5** (a) Body plan of an asconoid sponge. A diagrammatic sketch showing the different cell types of sponges. Arrows indicate the direction of water flow. (b) Group of Caribbean sponges off Cozumel, Mexico.

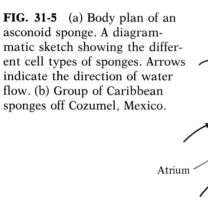

**FIG. 31-6** Comparison of choanocytes of sponges (a) and (b) a choanoflagellate protozoan, *Proterospongia.*

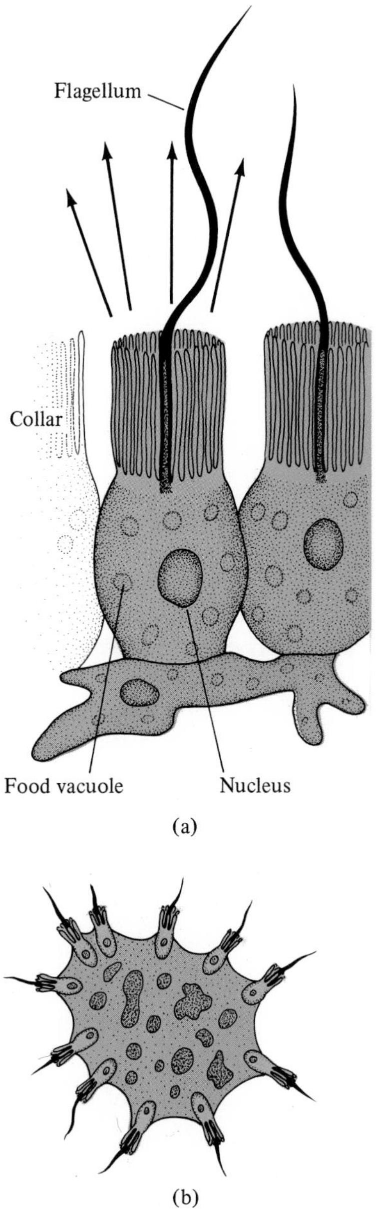


Sponges have no mouths, and no digestive cavities. Food particles, such as bacteria, small protists, and bits of detritus, are engulfed by the choanocytes, which enclose them in food vacuoles. After partial digestion, the particles are passed on to amoebocytes, which complete their digestion, also within food vacuoles. Large food particles may also be captured and engulfed directly by amoebocytes. Intracellular digestion (within food vacuoles) by sponges proceeds in the same way as it does in Protozoa. The products of digestion diffuse from one cell to another throughout the body, since sponges have no circulatory system.

Besides bringing in a food supply, the current of water passing through a sponge provides oxygen, minerals, and, at one time of the year, spermatozoa that are needed to fertilize ova. The exhalant water carries out carbon dioxide, ammonia, and other wastes. About 4,000 liters (4 tons) of water must filter through a sponge's body per day to provide enough food and minerals to produce only 100 g (about 4 ounces) of new skeleton in a year. On the average, sponges process nearly 1,000 l (over 250 gallons) of water a day, contributing significantly to the recycling of materials essential to the life of seashore communities.

A form of sponge more advanced than the asconoid is the **syconoid** type, which apparently evolved by a folding of the asconoid body wall that created numerous choanocyte-lined chambers, each equivalent to the body of an asconoid sponge (Fig. 31-7b). The resulting increase in the ratio of choanocytes to body volume (and thereby the volume of

**FIG. 31-7** Organization of the sponge body. (a) Asconoid. (b) Syconoid. (c) Leuconoid. (d) Rhagon. The layer of choanocytes is shown in a darker shade. Arrows indicate the direction of flow of nutrient and oxygen-bearing water.

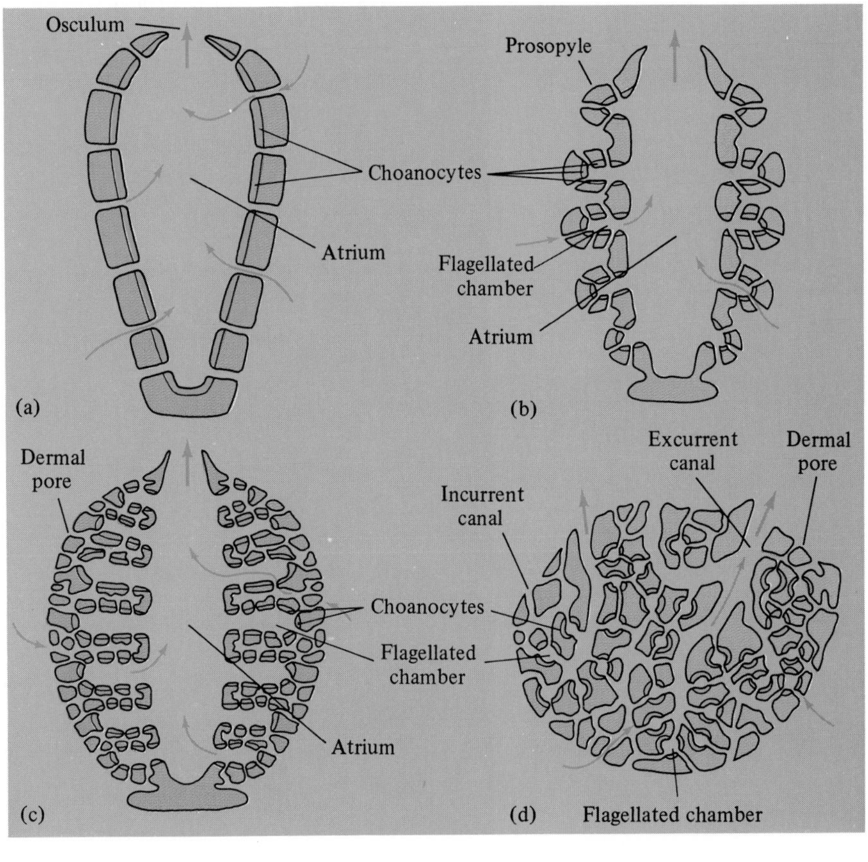

water that could be driven through a sponge of a given size) provided greater efficiency (greater flow of water). The space between the neighboring flagellated chambers, called the incurrent canal, is in open communication with the surrounding water. The water pore leading to the flagellated chamber is called the **prosopyle.**

An even more advanced type of sponge body is the **leuconoid** type (Fig. 31-7c), essentially composed of numerous syconoid units, each bearing many flagellated chambers, and providing an even greater efficiency. The incurrent canals are enclosed in the body, and the opening to the exterior is by the **dermal pore.** In the primitive leuconoid type, the flagellated chamber leads to an excurrent channel, which in turn leads to the osculum via a spongocoel. In the more advanced leuconoid, the **rhagon** type, there is no central spongocoel; the flagellated chambers are connected to the incurrent and excurrent canals (Fig. 31-7d). The greater efficiency of the leuconoid sponge probably accounts for the fact that species of leuconoid sponges are the most adaptable and the most successful—the most numerous—of all.

## SPONGE TAXONOMY

The three sponge body types described above are not taxonomic divisions; they are stages in the evolution of the complex sponge body. Indeed, some taxonomic groups include more than one body type. Sponge taxonomy is based on the form of the skeletal spicules, a unique characteristic of the phylum (Fig. 31-8). The entire sponge body is strengthened and supported either by various types of spicules or by a network of a fibrous protein called **spongin.** (It is the spongin skeleton that is the marketable portion of commercial sponges.)

The phylum is divided into three taxonomic classes, as follows:

1. *Class Calcarea* consists of calcareous, that is, chalky or limy, sponges. The class includes all sponges with calcium carbonate spicules.
2. *Class Hexactinellida* contains the glass sponges, characterized by six-rayed siliceous spicules that are made principally of glassy silicon dioxide (Fig. 31-9).
3. *Class Demospongiae* (from the Greek *demos,* "people," and *spongus* "sponges") is a catchall group. To it belong the horny sponges, which have spongin skeletons; some sponges with siliceous spicules; some with both spongin skeletons and siliceous

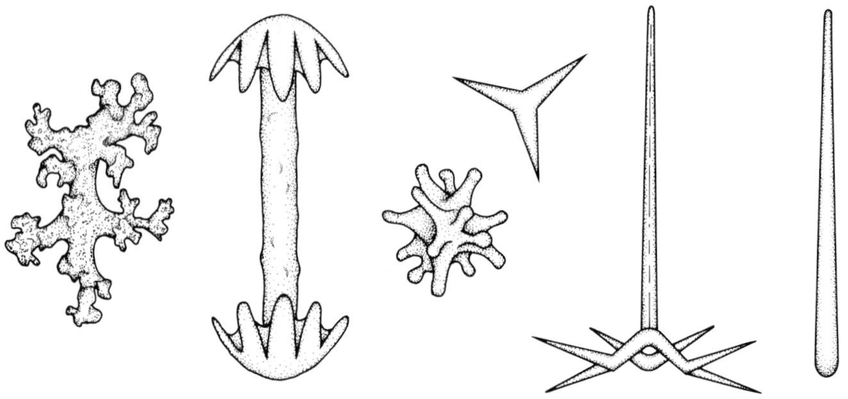

**FIG. 31-8** Representative types of sponge spicules.

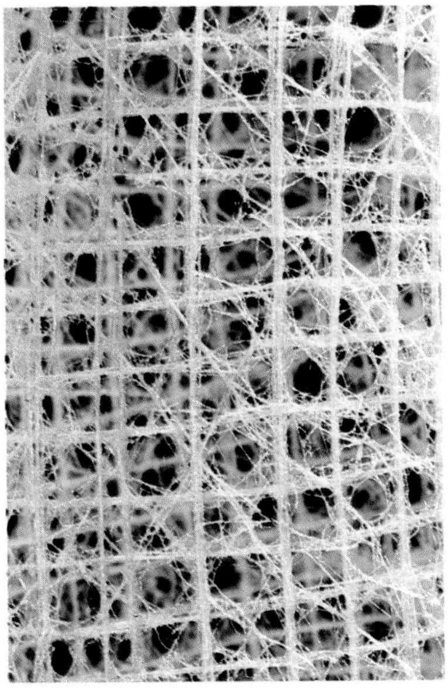

**FIG. 31-9** Skeleton of the glass sponge, *Euplectella.* Note the intricate arrangement of spicules.

spicules; and a few with no skeletons. To this large class belong 80 percent of all sponges, including commercial sponges and a small group of 150 species, found in fresh water.

## INDEPENDENCE AND COOPERATIVITY OF SPONGE CELLS

From one perspective, sponges are very little advanced beyond colonial protozoa. In an easily performed experiment that tends to support this view, a sponge is broken up into individual cells by squeezing it through the fine meshwork of a piece of silk or nylon cloth. The dissociated cells that emerge on the other side of the cloth are placed in water and observed under a microscope. Like amoeboid protozoans, the different kinds of cells begin to wander about the dish until they encounter another cell or a clump of cells. Upon coming in contact, the cells stay together, forming larger and larger clumps as more and more cells and clumps come in contact with them. Within the clumps, the various types of cells sort and arrange themselves until, after several hours, the clumps take on the appearance of young sponges. The "young" organisms then begin to feed, grow, and develop into normal sponges. Thus, sponge cells can survive, at least in a laboratory, either as isolated cells or as components of an organism.

## REPRODUCTION AND DEVELOPMENT

Although sponges lack cells organized into tissues or organs, they exhibit modes of reproduction similar to those of higher organisms. The fact that even the most primitive multicellular animals possess gametes that are similar to those of higher animals suggests that sexual reproduction involving specialized gametes was an early evolutionary innovation and is characteristic of all multicellular animals.

***Asexual Reproduction.*** Like many other lower organisms, sponges reproduce asexually; that is, without the genetic benefits entailed in the union of eggs and sperm from two different parents. Although not truly an asexual method of reproduction, the phenomenon of regeneration in sponges is often used as a means of asexual reproduction. For animals like sponges, which may live in strong currents or on wave-battered shores, the ability to regenerate is not only important for survival but becomes a way of multiplying.

Regeneration is a fairly simple matter in sponges. Archaeocytes, cells that are able to develop into any cell type in the sponge body, are involved in regeneration. The ability of all sponge cells to migrate and to rearrange themselves if disturbed is also useful in repairing damage and in regeneration. The ability of sponges to reconstitute their bodies after being squeezed through cloth emphasizes how good they are at recovering from severe damage.

BUDDING. A common form of asexual reproduction in sponges is **budding.** In asconoid and syconoid sponges especially, a protuberance appears at almost any place on the body surface, but usually at its base. The bud enlarges and develops for a time before detaching. The new individual may be carried by water currents for some distance before it attaches to a rock or other object. In many cases, the newly budded sponges remain attached to the parent sponge, giving rise to a colony with connecting canals or spongocoels. As this process continues in syconoid sponges, the result looks much like an upside-down bunch of tiny bananas.

GEMMULE FORMATION. Freshwater sponges have a special means of asexual reproduction: **gemmule** formation. In early fall, in temperate climates and before the dry season in tropical climates, amoebocytes enriched with food stores aggregate and become covered with a spicule-reinforced wall. This dormant stage, the gemmule, is resistant to both freezing and desiccation. After its gemmules are formed, the sponge body disintegrates. In the spring, the amoebocytes emerge from each gemmule and develop into adult sponges.

***Sexual Reproduction.*** Most sponges are hermaphroditic, or monoecious. Ova and spermatozoa produced by one individual usually mature at different times, however, thus promoting cross-fertilization rather than self-fertilization. Apparently, certain archaeocytes develop into spermatozoa and ova, although choanocytes have also been reported to do so. The maturation of the egg and sperm and the meiotic divisions of oocytes and spermatocytes appear to follow courses similar to those in higher animals and in some of the Protozoa. Ova develop among mesenchyme cells just beneath the choanocyte layers. Spermatozoa develop in similar locations but are released and make their way into open water by way of the osculum.

Spermatozoa released into the ocean have little chance of finding their way into the incurrent ostia of another sponge of the same species. But enough spermatozoa are produced to ensure that many do get drawn into conspecific sponges (belonging to the same species) with mature ova. When the spermatozoa enter the sponge, they are soon ingested by choanocytes. But rather than digesting it, a choanocyte that has engulfed a spermatozoon of its own species resorbs its own collar and flagellum, becomes amoeboid, and delivers the spermatozoon to the nearest ovum! The spermatozoon is actually expelled by the choanocyte and is promptly engulfed by the ovum. The sperm and egg nuclei fuse, and fertilization is complete.

**FIG. 31-10** Amphiblastula larva of sponges bearing flagellated and unflagellated cells.

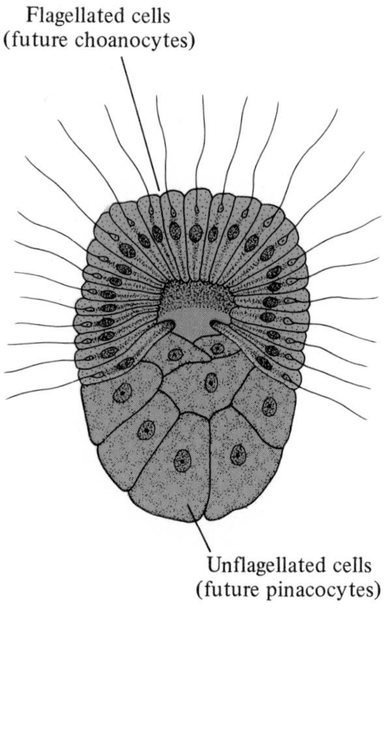


DEVELOPMENT. Following the first cleavages, the embryo in most species becomes flagellated and breaks free of its parent, emerging from the osculum as a larva, called the **amphiblastula,** a free-living stage of development that does not resemble the adult (Fig. 31-10). After a brief existence as a larva, the developing individual, now resembling an asconoid sponge, attaches to a shell, rock, or other object, where it completes its development. Larval leuconoid calcareous sponges and some Demospongiae pass through an asconoid and then a syconoid stage of development before reaching their adult form.

## SUBKINGDOM METAZOA

In contrast to Parazoa, the Metazoa includes all the other advanced groups of multicellular animals. In metazoans, the cells are organized into tissues and organs. In the simplest metazoans only, a distinct outer covering layer, the epidermis, and an inner layer, the gastrodermis, which lines internal cavities, can be distinguished. Probably associated with the efficiency in acquiring nutrients that results from organization of cells into tissues, metazoans evolved into a greater variety of forms.

Theoretically, multicellularity could have evolved in animals in one of two ways. According to one view, multicellularity has arisen from colonial protozoans in which individual cells of the colony serve special functions. Gradually these individuals lose their ability for independent life. They can survive only as a part of the organism. A second view is that metazoans evolved from multinucleate protozoans. In the course of evolution, cell membranes appeared between the nuclei, rendering the organism multicellular. Other possibilities include combinations of the two types. These hypotheses cannot be tested, since they attempt to explain historical incidents. Hence experimental proof is not possible. But they make sense.

## PHYLUM CNIDARIA

Cnidaria (Greek for "nettle") are tentacle-bearing, radially symmetrical animals in which cells are organized into tissues and possess a mouth, a digestive cavity, and stinging capsules called **nematocysts.*** Cnidarians include polyps, jellyfishes, corals, and sea anemones. All but a few are marine. Their general body plan, exemplified by that of the common freshwater polyp *Hydra,* consists of a body wall enclosing a **gastrovascular cavity** (Fig. 31-11). A ring of tentacles surrounds the mouth, located on a hypostome, the only opening into the gastrovascular cavity. The body wall consists of an outer epidermis and inner **gastrodermis,** with an intermediate layer of jellylike material, mesoglea, which is of variable thickness in the different groups. Although in most cnidarians the mesoglea contains various types of cells and fibers, in some the layer is essentially a cell-free jelly. In jellyfishes, this layer makes up most of the body mass.

The gastrovascular cavity, so-named because it serves both as a digestive and a circulatory system, may or may not extend into the tentacles, depending on the group. Its lining, the gastrodermis, consists of a variety of columnar cells, some secretory, some contractile (muscle cells), and most capable of engulfing food particles and digesting them in food vacuoles, much as happens in sponges and protozoans. Initial digestion of large particles and entire prey organisms occurs within the gastrovascular cavity under the action of digestive secretions produced by gastrodermal cells.

The cells of the epidermis vary in shape, from cuboidal, as in *Hydra,* to columnar, in anemones, to thin and flat, in jellyfishes. The epidermal cells may be ciliated. As with some gastrodermal cells, epidermal cells may have contractile bases and function as muscle cells. Other cells are glandular and still others, the **cnidoblasts,** give rise to nematocysts. Cnidoblasts are of many types and are especially concentrated around the mouth and on the tentacles. Nematocysts function both offensively

* At one time, cnidarians were included, along with sponges and comb jellies (ctenophores) in a phylum called Coelenterata. The phylum was later divided into three phyla: Porifera, Cnidaria, and Ctenophora. Although some authors continue to refer to cnidarians as coelenterates, most experts prefer the designations used in this book.

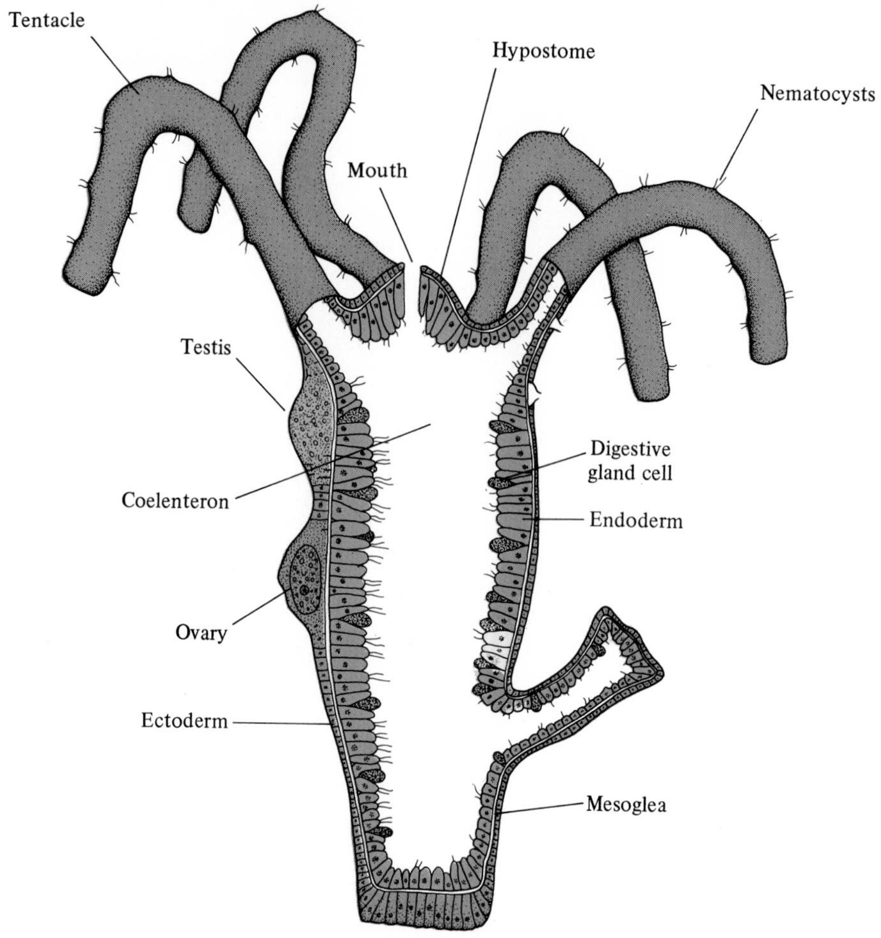


**FIG. 31-11** Diagrammatic sketch of *Hydra,* a coelenterate polyp.

and defensively (Fig. 31-12a–b); some are so poisonous as to cause the death of swimmers who encounter them. Upon contact with prey or predator, the nematocysts discharge, some of them piercing the victim's skin with a volley of poison.

It is easy to imagine the kinds of evolutionary modifications that led to the great diversity of forms among the cnidarians. Not only could all adult cnidarian forms, whether of polyp or of medusa (jellyfish) form, be derived from each other (Fig. 31-13), but in many cases both forms occur at different stages in the life cycle of the same species. When both polyp and medusa stages occur within the same life cycle, the medusa represents the sexual stage.

Most species of cnidarians are dioecious, with the medusae producing either eggs or sperm but not both. In most cases, a small, ciliated larva, the **planula,** develops from the fertilized egg. After a brief free-living existence, during which a gastrovascular cavity appears within the planula, it attaches to the substratum and develops directly into the polyp stage. Because of the planula's close resemblance to a colonial flagellated protozoan and because a planula stage occurs in the life cycles of most cnidarians, including the most primitive, it is believed that cnidarians arose from a planulalike ancestor that was probably derived from colonial flagellated protists. Indeed, with minor modifications, some modern colonial protists, such as *Volvox,* could pass for a planula (Fig. 31-14).

Cnidaria possess neurons, a type of cell not found in sponges. Be-

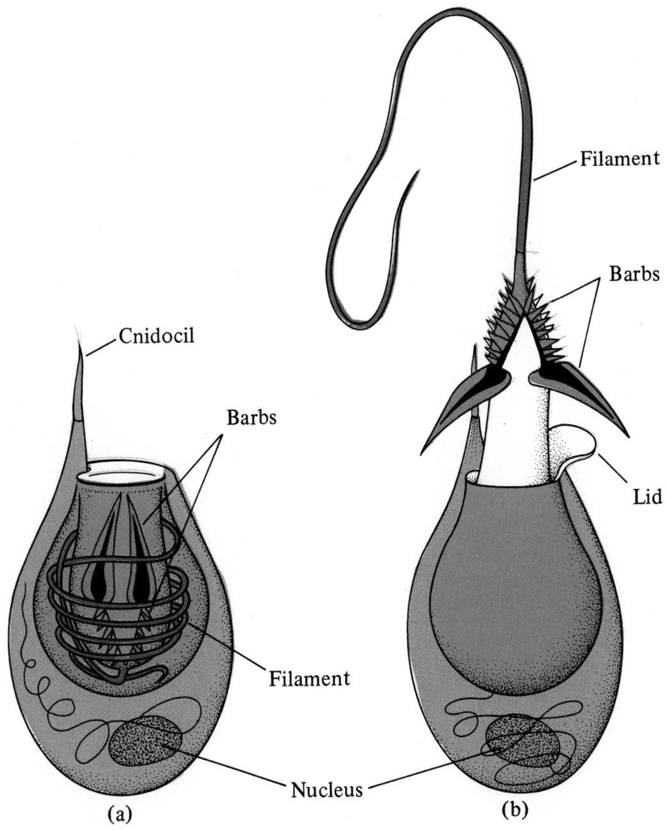


**FIG. 31-12** Nematocyst. (a) Cnidoblast with the enclosed nematocyst. (b) Discharged nematocyst.

cause cnidarians are the lowest form of life to have a nervous system, their neurophysiology and behavior have been intensively studied, providing us with insights into some basic principles of nervous integration and the evolution of nervous systems.

***Taxonomy.*** Cnidaria are divided into three classes: Hydrozoa, Scyphozoa, and Anthozoa (Fig. 31-15). Most hydrozoans are **polymorphic,** that is, they include both a medusa and a polyp stage in their life cycle, the latter often being a colonial form (See Figs. 21-12 and 31-16). In *Obelia,* the species illustrated in Figure 31-16, two types of polyps occur: **hydranths,** or tentacled polyps; and **gonangia,** or reproductive polyps. Some hydrozoans have specialized in being polymorphic, giving rise to complex colonies with six or more different types of polyps, each performing specialized functions. The beautiful and highly poisonous Por-

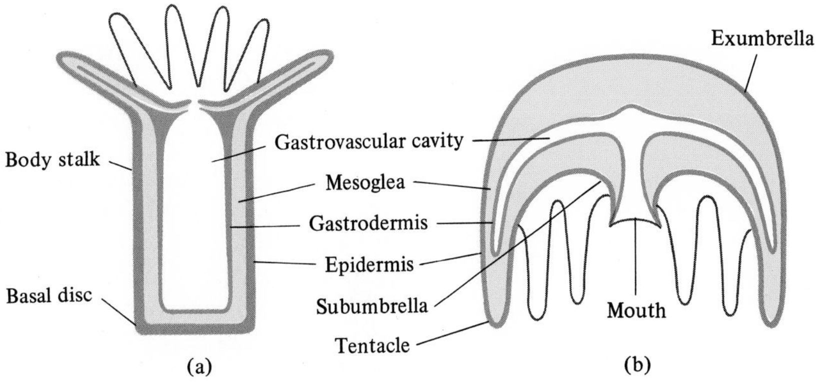


**FIG. 31-13** Diagrammatic sketches of the body plan in (a) polypoid and (b) medusoid cnidarians.

tuguese man-of-war is actually a hydrozoan colony (Fig. 31-17). The gastrovascular cavity is continuous throughout a colony of polyps, providing nutrition to all individuals, whether they are feeding polyps or not.

By asexual reproduction, gonangia produce medusae. Newly released *Obelia* medusae lead a free-living existence for a time, producing eggs or sperm at maturity. (Cnidarian gonads are not permanent organs, but regions of tissue capable of producing gametes.) Following fertilization, the zygote develops into a planula, which soon attaches and develops into a colony of polyps. Thus, there is an alternation of asexual and sexual reproductive phases; this type of alternation of generations is called **metagenesis.**

In contrast to hydrozoan medusae, which are usually minute, scyphozoan medusa is a prominent stage of the life cycle, some measuring 2 m in diameter. The medusoid form reproduces sexually. The fertilized ova develop into planula larvae. The planula stage, upon attaching to a substratum, develops into an inconspicuous polyp called a **scyphistoma**. During the course of its life, the scyphistoma repeatedly reproduces asexually by budding off small medusoidlike larvae called ephyra larvae that mature into adult medusae (Fig. 31-18). Although there are only about 200 species of scyphozoans, they occur worldwide, especially in warm seas, at times in great abundance.

As their name might suggest, anthozoans (from the Greek *anthos*, "flower," and *zoion* "animal") are beautiful creatures (Fig. 31-19). Most sea anemones are sessile, more or less permanently attached to a substratum. These solitary polyps are of rugged construction, with heavy connective tissue fibers and muscle layers giving strength to their bodies. The corals, in contrast, are delicate-bodied polyps that are protected by skeletons of calcium carbonate into which they quickly withdraw in time of danger. Anthozoans lack a medusa stage; eggs and

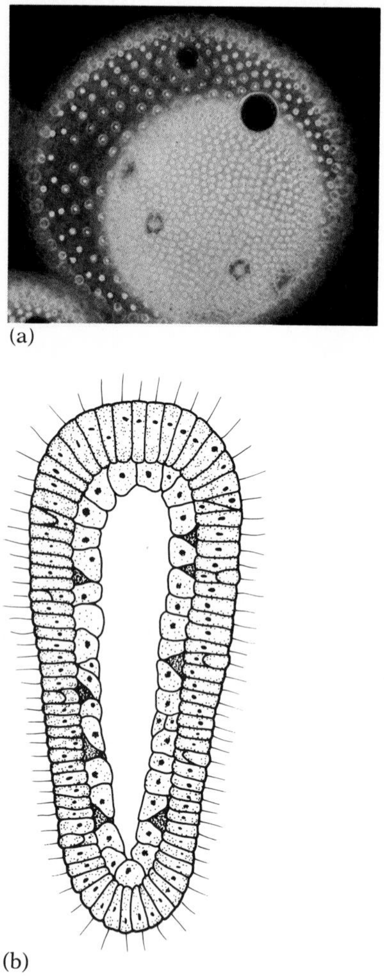

(a) (b)

**FIG. 31-14** (a) *Volvox*, a colonial flagellate. (b) Planula larva of cnidarians.

**FIG. 31-15** Representative cnidarians. (a) *Tubularia*, a hydrozoan. (b) *Chrysaora*, a scyphozoan. (c) *Cerianthus*, an anthozoan.

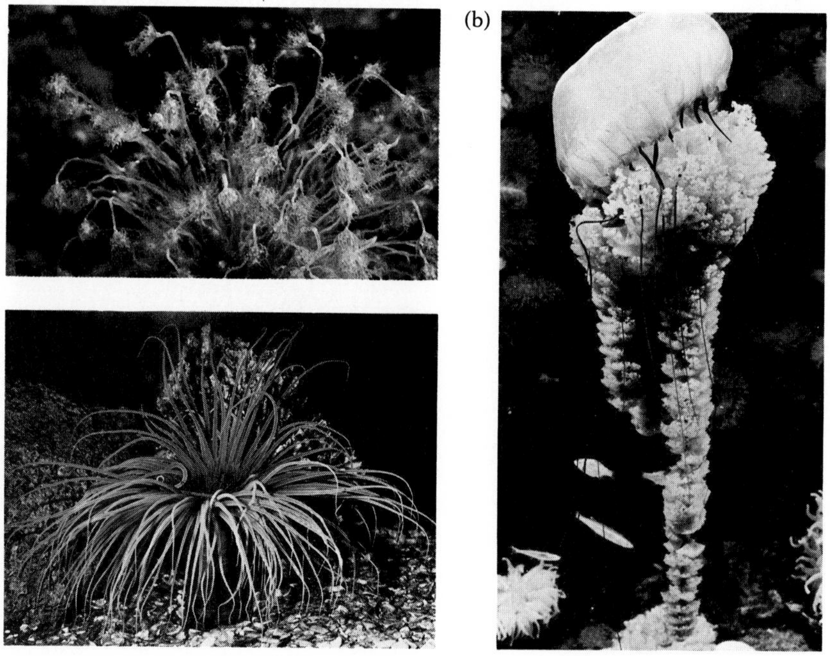

(a) (b) (c)

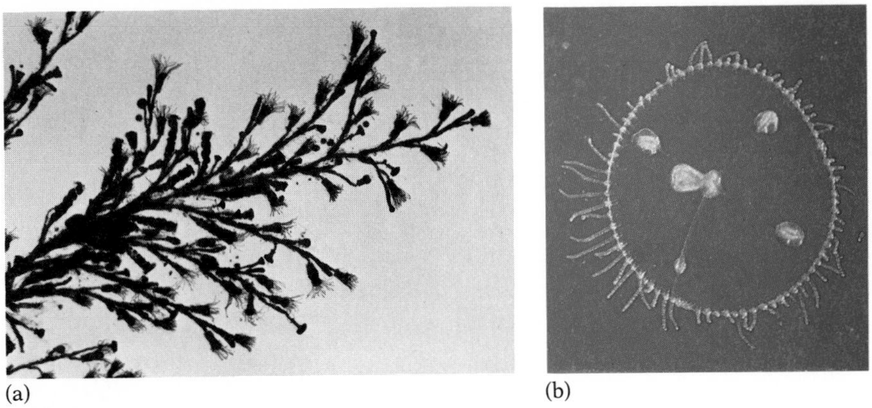

**FIG. 31-16** Hydrozoan life cycle. (a) Asexual stage of the colonial *Obelia*. (b) The medusa stage.

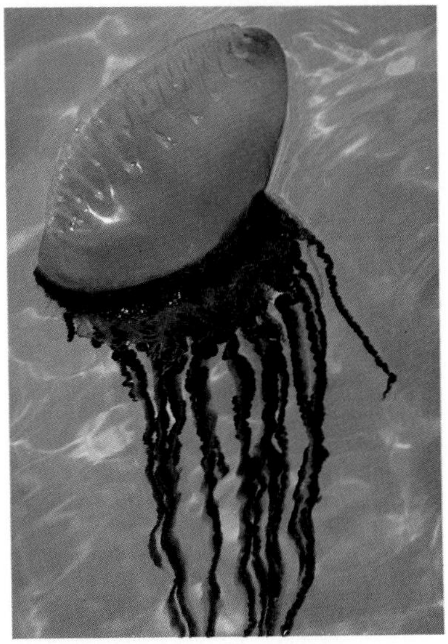

**FIG. 31-17** *Physalia*, the Portuguese man-of-war, a colonial planktonic hydrozoan. This cnidarian produces a highly toxic venom in the nematocysts that can cause death in humans.

sperm are produced directly by the polyp. The planula larva attaches and develops into a polyp in typical cnidarian fashion.

The beauty of many corals is enhanced by their many delicate shades of color. Many empty coral skeletons are beautiful as well (Fig. 31-20). Corals contribute significantly to marine ecology. They form the raw material for the soil of most oceanic islands. Coral reefs, although a

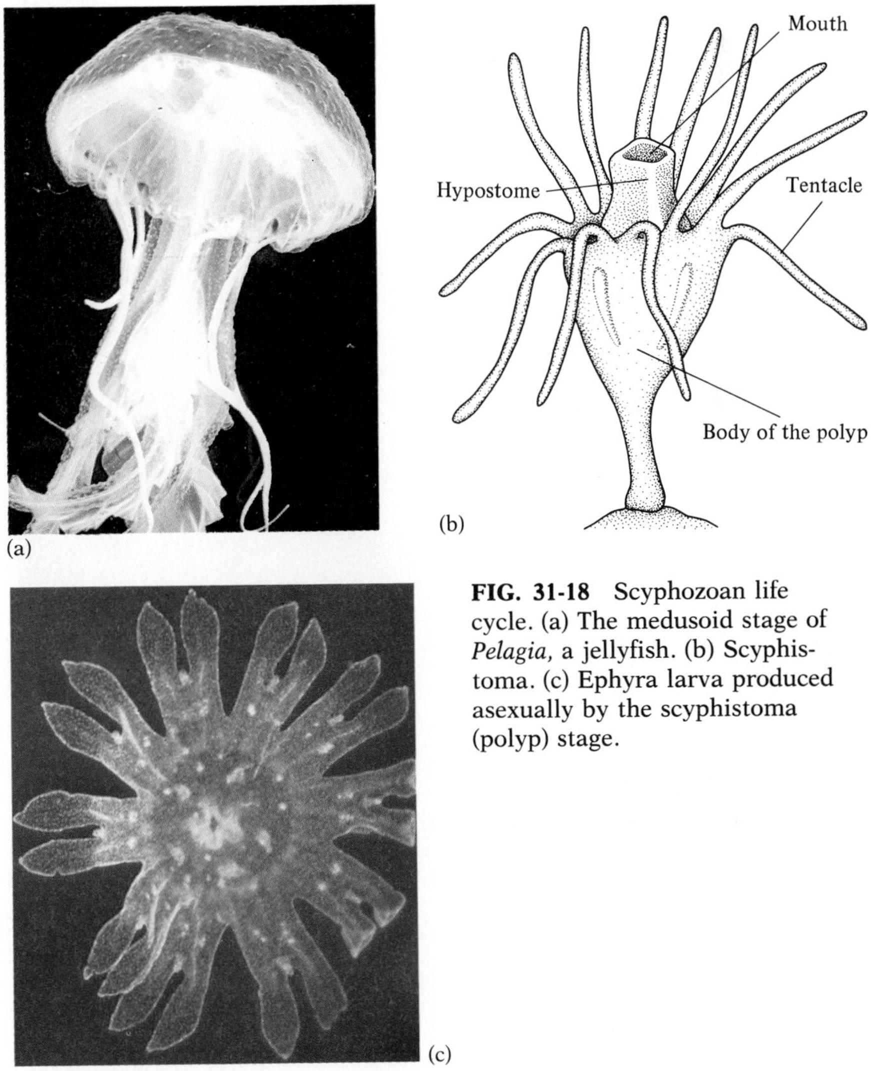


**FIG. 31-18** Scyphozoan life cycle. (a) The medusoid stage of *Pelagia*, a jellyfish. (b) Scyphistoma. (c) Ephyra larva produced asexually by the scyphistoma (polyp) stage.

**FIG. 31-19** A tube anemone, an anthozoan.

serious hazard to shipping, provide an important shelter against wave action for islands while providing a home for trillions of organisms of untold numbers of species. Their importance is dramatically exemplified by the largest of them, the Great Barrier Reef of Australia, which is over 2,000 km (1,200 miles) long and provides shelter for millions of organisms. Reef corals grow best in shallow waters (30 m or less) of warm seas.

## PHYLUM CTENOPHORA

The ctenophores (from the Greek *ktenos,* "comb," and *phoros,* "bearing"), or comb jellies, are closely related to the cnidarians, resembling the medusa form to a degree. There is no polyp stage and few species possess nematocysts. Their name derives from their eight rows of combs, bands of fused cilia used in locomotion. The combs luminesce

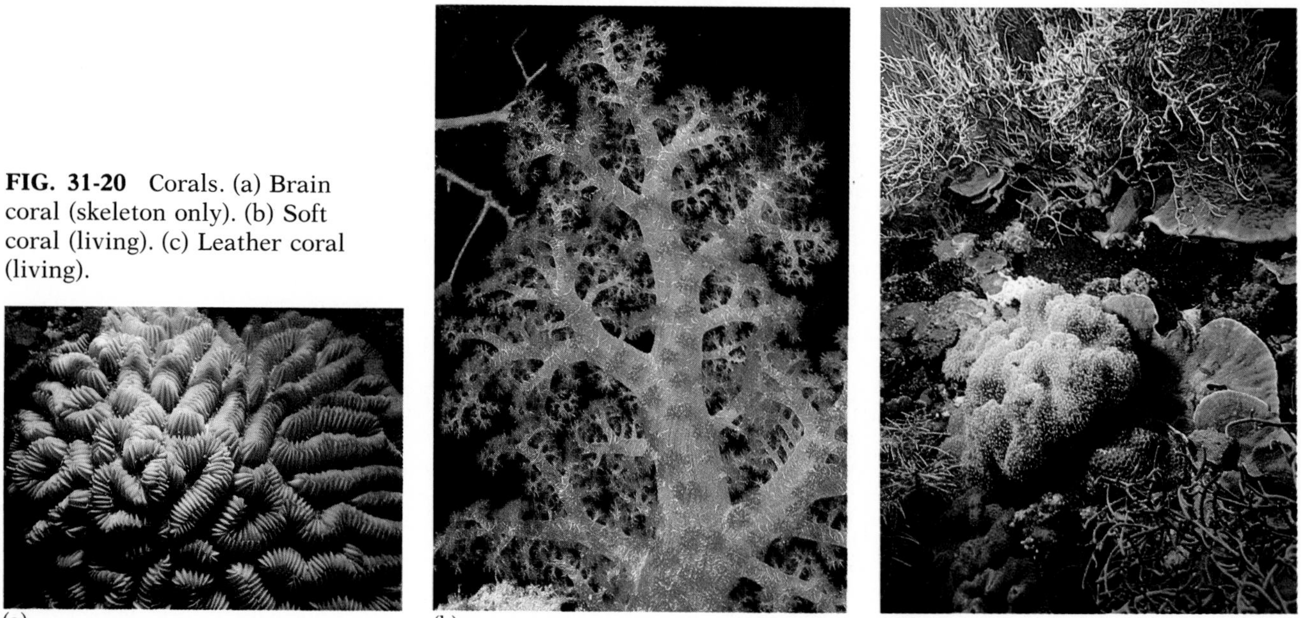

(a) (b) (c)

**FIG. 31-20** Corals. (a) Brain coral (skeleton only). (b) Soft coral (living). (c) Leather coral (living).

**FIG. 31-21** *Pleurobrachia,* a ctenophore.

when the animals are disturbed, adding to their delicate beauty (Fig. 31-21). Although there are less than 100 known species, comb jellies often occur in great abundance in warm seas.

## FLATWORMS AND NEMERTEANS

Flatworms represent a level of organization that is more advanced than that of cnidarians and ctenophores. A characteristic of flatworms is that their embryos bear a well developed middle layer of cells sandwiched between the outer ectoderm and the inner endoderm, which may have evolved from the less conspicuous mesoglea of cnidarians. According to this view, flatworms evolved from a probable planulalike ancestor.

Another view of the origin of flatworms and related animals is that in the course of evolution the metazoan condition evolved more than once. The situation is analogous to the apparent evolution of multicellularity in two unrelated groups of animals: namely, parazoa and other multicellular animals. According to this view, flatworms represent a line of evolution independent of cnidarians. As in most controversial matters of this nature the only reliable evidence in favor of one or the other theory is fossil evidence. But, unfortunately, these early invertebrates had no skeleton; no fossils have been discovered to date, and there is very little hope of finding them.

### PHYLUM PLATYHELMINTHES

The flatworms, of the phylum Platyhelminthes (from the Greek *platy,* "flat," and *helminthes,* "worm"), represent several distinct advances over the cnidarians. They are the first phylum in which all members exhibit bilateral symmetry, an arrangement that makes it possible to divide the body into right and left halves that are mirror images of each other. This development was of major importance because the body had right, left, dorsal, and ventral sides and anterior and posterior ends, making various types of specialization—a key to evolutionary progress—possible for the first time. The dorsal surface could specialize in such functions as protection and concealment, while the ventral surface specialized in locomotion. The anterior end, being directed forward in locomotion, became a region in which special sense organs were concentrated. This led, in turn, to anterior concentrations of sensory and motor neurons, ganglia, and, eventually, the formation of a brain. This trend, characterized by formation of a head, is termed **cephalization** (from the Greek *kephalos,* "head"). Once a brain had evolved, the body's actions became ever more efficiently controlled by this command center of growing complexity. More complex movements were now possible, leading to further evolutionary advancement.

A development concurrent with cephalization was the origin of the third, well-defined, middle germ layer, the **mesoderm,** in addition to the ectoderm and the endoderm. Although sponges, cnidarians, and ctenophores have an intermediate layer that contains cells of ectodermal origin, this layer is not regarded as homologous to the mesoderm of the animals with bilateral symmetry, which originates from or in association with the endoderm.

Few occurrences in animal evolution are of such importance as the development of a mesoderm. From it eventually arose muscles and blood vessels and, in vertebrates, kidneys, skeletons, and the lower layers of the skin. In time, the evolution of mesoderm was accompanied by another advance, the evolution of a major body cavity, the **coelom,** entirely lined by mesoderm or tissue derived from this layer.

Although flatworms possessed both bilateral symmetry and mesoderm, little of the rich potential of these acquisitions was realized in them. A coelom did not evolve. As an acoelomate group, platyhelminthes do not exhibit great variation in their morphology. For example, of the three classes, Turbellaria, Trematoda, and Cestoda, only the turbellarians, made up of free-living forms, possess anything remotely resembling a head. Moreover, no flatworm has a coelom; instead, the space between organs is filled by mesodermally derived mesenchyme and muscles. The flatworm digestive process is as primitive as that of cnidarians, with initial stages of digestion occurring in the gut cavity, which lacks an anus, and the final stages taking place in food vacuoles of cells lining the gut (Fig. 31-22).

Nor is there a circulatory system. Although this might seem to cause a problem for animals as large and active as many flatworms, the problem is avoided by the fact that their bodies are flat and usually no more than 1 mm thick. Half a millimeter is the maximum distance through which diffusion can provide oxygen and remove wastes such as ammonia with much efficiency (Fig. 31-23). Some wastes are also expelled by osmoregulatory-excretory structures known as protonephridia, which end in a special type of cells called flame cells, which are found throughout the body (see Chapter 13).

The platyhelminth nervous system does represent an advance over that of cnidarians.

***Variations Within the Phylum.*** The flatworms hold great interest for biologists for several reasons. One is that two of their three classes are composed entirely of parasitic forms, many with humans or domestic animals as their hosts. Another is that various of these parasites exhibit varying degrees of specialization, and the many transitional forms indicate how some species probably evolved into others.

Locomotion is achieved either by ciliary action of the ventral body surface or by rhythmic contractions of body muscles. Besides well-developed layers of circular and longitudinal muscles in their body walls, flatworms possess diagonally arranged sets of muscles and other muscles as well.

Because the classes differ from each other as much as they do, additional generalizations for the entire phylum are difficult to make.

***Class Turbellaria.*** The free-living flatworms of the class Turbellaria are found in marine and freshwater habitats and in the soil and ground litter of moist forests. The typical animal has a mouth, which may be

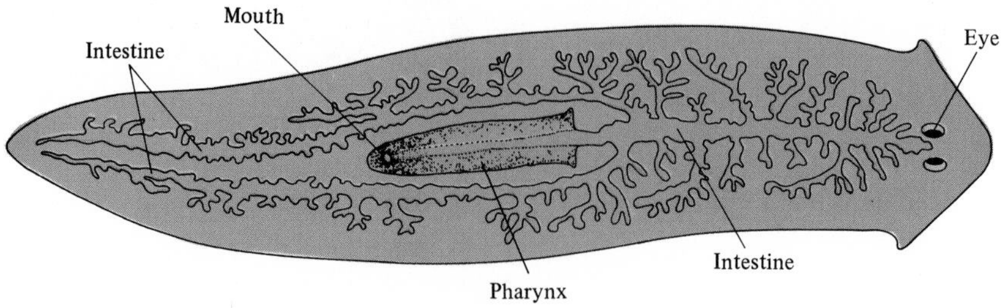


**FIG. 31-22** The digestive organs of a flatworm.

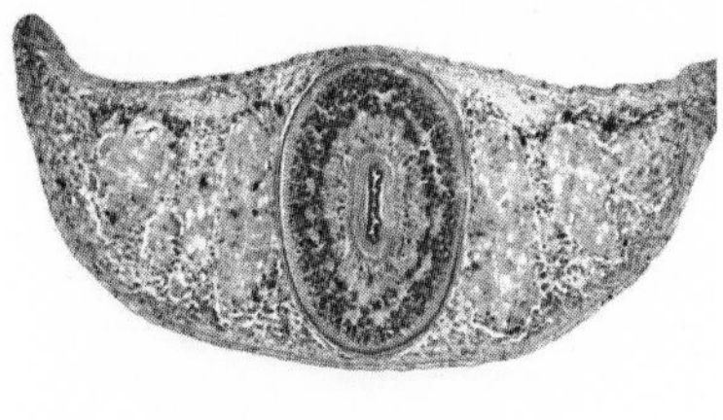

**FIG. 31-23** A transverse section of a planarian. Note the absence of any space (body cavity) other than the pharyngeal lumen.

located at any point on the ventral surface, and a branched gut. Most are carnivorous and feed on smaller animals. All have at least one pair of primitive eyes, pigmented cups containing light-sensitive neurons that can distinguish lights and shadows but do not form images.

Turbellarians reproduce asexually by fission or fragmentation, with each piece regenerating all the missing parts. The animals can also survive long periods of starvation by resorbing nearly all their internal organs. When the starved animal is fed, all the organs regenerate. Their regenerative abilities have made one group of turbellarians, the planarians, a favorite subject for the study of this process (Fig. 31-24).

In contrast to cnidarians, flatworms have evolved a remarkable array of sexual organs. Nearly all species of all three classes are monoecious. Typically, each worm contains two to many testes and ovaries and a variety of accessory organs. Cross-fertilization is the rule in planarians because self-fertilization is physically impossible. Primitive turbellarian species have yolky eggs that exhibit spiral, determinate cleavage. Although some have a larval stage, in most turbellarians several fertilized yolkless eggs are enclosed in a cocoon along with hundreds of yolk cells. The cocoon is attached to rocks or vegetation. Development proceeds in the cocoon, the embryos feeding on the yolk cells until the young worms emerge, thereby obviating the need for a free-living larval stage.

***Trematoda.*** All trematodes, or flukes, lead parasitic lives, usually completing their life cycle within two or more alternate hosts. *Opisthorchis (Clonorchis) sinensis*, the oriental liver fluke that parasitizes humans, is a well-known example (Fig. 31-25a). Like other species of fluke, the adult has a mouth and a muscular pharynx that leads into a simple intestine with two long branches. Some digestion is extracellular, some intracellular. The pumping action of the pharynx draws in bits of the tissue, blood, and so on, on which the parasite feeds. All species have one or more ventral adhesive organs, or suckers.

The 15 mm-long adult liver fluke parasitizes the bile passages of its human host, where it mates and lays shelled eggs that pass into the intestinal lumen and are passed out of the host with the feces. (Fluke eggs are also yolkless; yolk cells are enclosed in the shell with each egg.) The next stage requires that the eggs reach water, where a larval form, the **miracidium,** hatches from the egg (Fig. 31-25b). If it finds a snail of

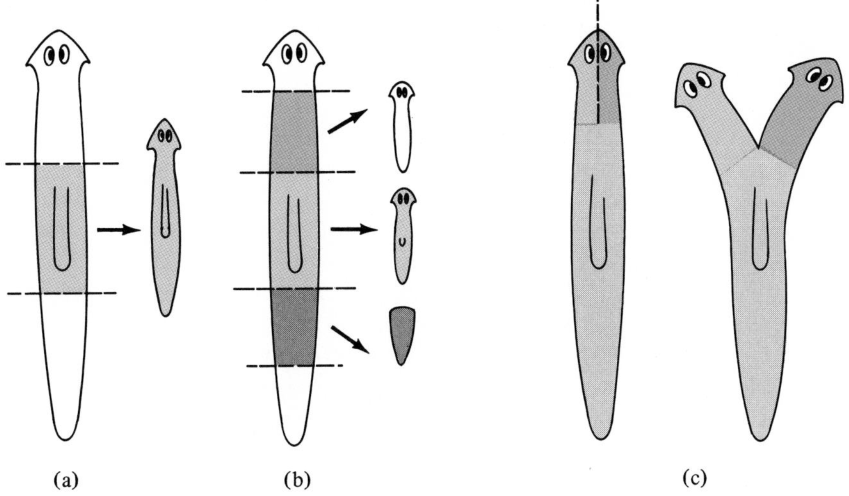


**FIG. 31-24** Experimental design to demonstrate the regenerative capacity of a planarian. (a) A piece without head or tail regenerates a complete planarian. (b) A smaller piece than in (a) can regenerate. (c) If the anterior end is cut longitudinally, two separate heads regenerate.

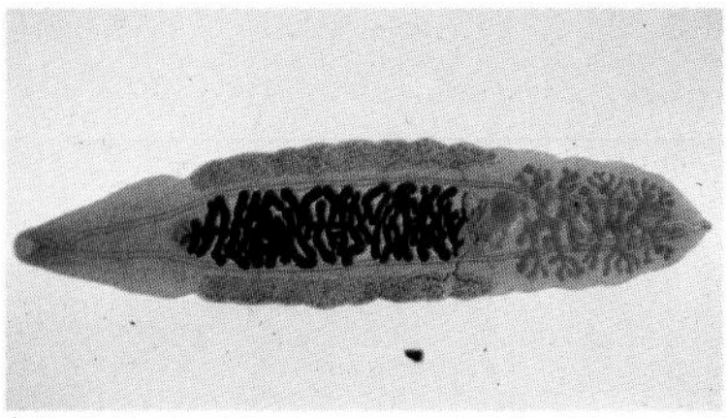

(a)

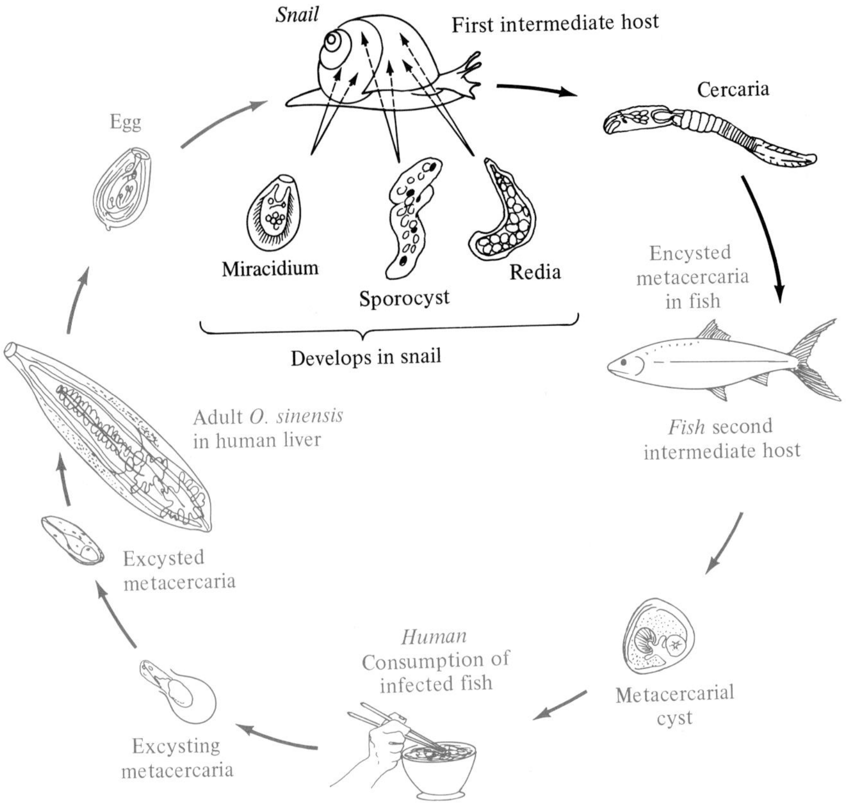


(b)

**FIG. 31-25** The oriental liver fluke. (a) Adult. (b) Life cycle.

the right species within the next few hours, it enters (by penetrating the snail's tissues) and develops into a **sporocyst.** Within the sporocyst develop numerous larvae called **rediae,** which comprise the next developmental stage. The rediae emerge from the sporocyst, still within the snail. Within the rediae develop yet another generation of larvae, the **cercariae.** When the cercariae emerge from the rediae, they leave the snail and enter the water, where, if one of several suitable fish species is encountered, they penetrate its tissues and **encyst** (that is, secrete a protective covering and then become quiescent). The encysted form is called a **metacercaria.** The life cycle is completed if a human eats the fish without thoroughly cooking it. In the human's digestive tract, the metacercariae excyst (digest away the protective covering and resume normal physiology) and migrate to the bile ducts, where they mature to adulthood. Some less highly evolved trematodes have simpler life cycles and parasitize but a single host species.

***Class Cestoda.*** Tapeworms are the most highly specialized of all the flatworm parasites. The adult has no digestive system and relies entirely on the absorption of the products of digestion provided by its vertebrate host. Whereas one or another species of adult trematoda parasitizes almost every type of host tissue, adult tapeworms, because of their need to be bathed in nutrients, are found only in the intestines of their hosts. All have at least one arthropod or vertebrate intermediate host; some have several. The body of a mature tapeworm consists of an attachment organ called a **scolex,** which attaches with the help of suckers to the intestinal wall of the host, and a series of segments called **proglottids.** In some species, a part of the scolex called the **rostellum,** is adorned with a circle of hooks which help in attaching to the host.

(a)

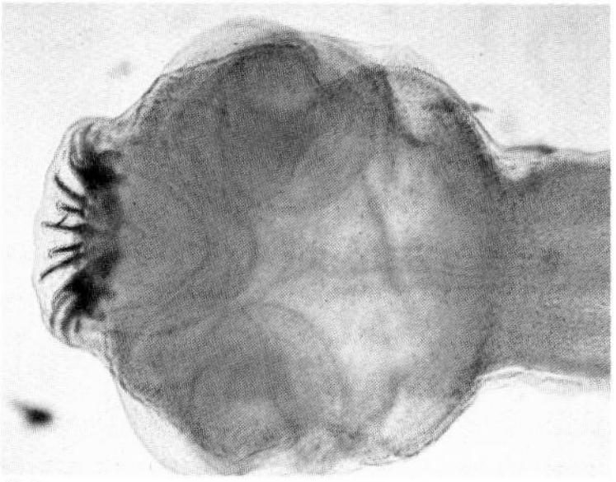

(b)

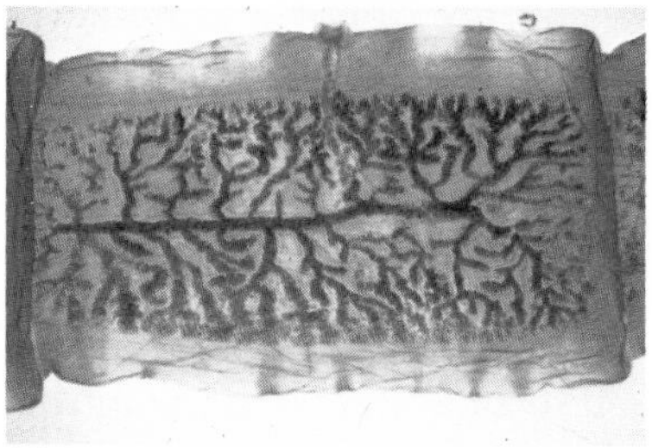

(c)

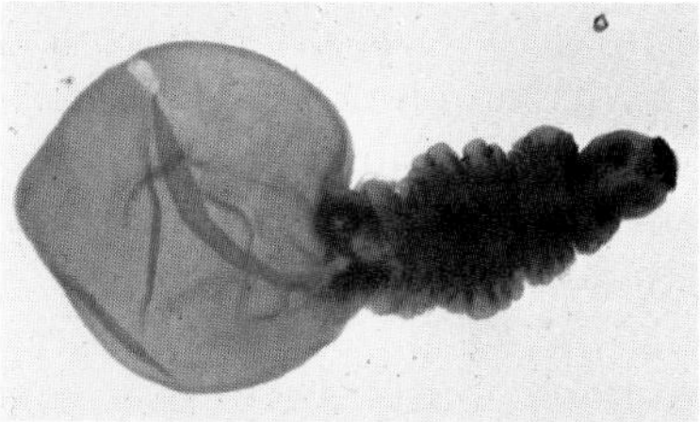

(d)

**FIG. 31-26.** The pork tapeworm, *Taenia solium.* (a) Adult individual. (b) Scolex enlarged. (c) Proglottid of a related species. (d) Evaginated cysticercus larva.

The shelled, fertilized tapeworm eggs undergo partial development before being shed in the feces. The life cycle of the pork tapeworm of humans, *Taenia solium,* is fairly typical (Fig. 31-26). If the eggs released in the feces of a human host are eaten by a pig or other animal (including the same human), a minute **oncosphere larva** bearing hooks, hatches in the digestive tract, bores into the intestinal wall, and makes its way via the circulatory system to a muscle. There it encysts and metamorphoses within the cyst into a bladderworm, or **cysticercus,** larva, the usual infective stage for humans.

Eating insufficiently cooked pork (or beef, in the case of the beef tapeworm) that contains bladderworms is essential for the next stage to occur. The cysticercus is essentially a bladderlike structure containing an inverted scolex. In the intestine of the definitive host, the cyst wall of the cysticercus is digested away and the scolex of the cysticercus everts. The young, unsegmented worm then attaches to the intestinal wall. As it absorbs nutrients, it grows, continually budding off proglottids, which usually remain attached to the scolex.

Each proglottid contains a complete set of reproductive organs and can be considered autonomous except for (1) a pair of longitudinal nerve cords that unite all the segments and (2) a pair of longitudinal excretory canals into which the many protonephridia of each segment empty the fluid they collect. Some biologists regard the tapeworm as a colony of individuals. Although the worm can perform coordinated writhing movements and, in fact, accomplishes cross-fertilization between proglottids by folding upon itself, there is little else to suggest that it is a single individual. Although it contains a pair of ganglia, the scolex is regarded, not as a head, but merely as an attachment organ. There are reasons to believe that the scolex evolved from an attachment organ at the posterior end of some ancestral flukes. Most of the interior of each proglottid is given over to reproductive organs. In mature proglottids, the uterus becomes packed with shelled eggs and expands to occupy almost the entire proglottid.

### NEMERTEANS

Proboscis worms, or nemerteans, are a rather small phylum (650 species), mostly inhabiting the sand or mud of the seashore. Their many similarities to flatworms indicate a close kinship with them. One difference, however, is of special interest. The nemerteans are the most primitive animals to possess a one-way digestive tract, with both a mouth and an anus. This important evolutionary advance, found in all higher animals, greatly increased the efficiency of processing food. Various regions of the digestive tract could now specialize in food capture, ingestion, mastication (chewing), storage, various stages of digestion, absorption of the products of digestion, reclamation of water, formation and storage of feces, and defecation. Only some of this potential was realized in nemerteans, however.

## THE PSEUDOCOELOMATES

Several small phyla of minute worms and wormlike animals possess a minimal degree of cephalization, that is, formation of a distinct head bearing sensory organs. Most are dioecious, and all have determinate

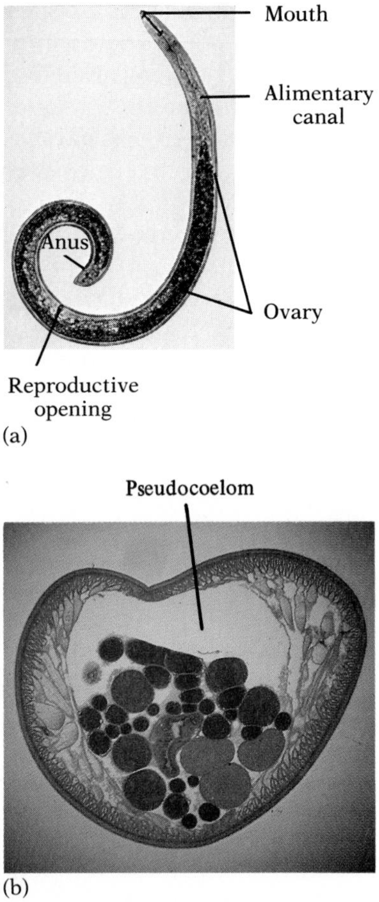


**FIG. 31-27** (a) External and internal features of a female nematode. (b) Cross section of a nematode. Note the presence of a body cavity, the pseudocoelom. Compare to Figure 31-23, a cross section of a planarian.

cleavage. They are grouped together because all possess a major body cavity in which only the outer portion is lined with mesodermally derived tissue. Such a cavity is called a **pseudocoelom** to distinguish it from the true coelom, which is entirely lined by mesodermally derived tissue.

## NEMATODA

The only pseudocoelomates that will be considered here are those in the phylum Nematoda, the unsegmented roundworms. Although 10,000 nematode species have been named and described, many thousands more must exist. Free-living forms occur in every environment, and several parasitic species have been found in every species of plant and animal ever examined for them. The abundance of individual worms is often enormous. A square meter of bottom mud in the sea may contain millions, and an acre of farm soil many billions. One rotting apple lying under an apple tree has yielded 90,000 nematode worms. Nematodes are therefore important as decomposers, an essential component of all ecosystems. As parasites, nematodes have also been responsible for much human misery and death. Ascarid worms are important parasites of humans, and of domestic animals, a fact to which people who have owned dogs can attest. In an unusual instance, autopsy of a fatal human case of an intestinal ascarid infection disclosed over 1,400 worms in the digestive tract. Trichinosis, hookworm, filariasis, and river blindness are among the many serious nematode diseases that afflict humans.

Most nematodes are less than 1 mm long, but some range up to 2 m in length. The body is elongate and enclosed in a complex, tough, multilayered cuticle produced by the epidermis (Fig. 31-27). Muscles comprise the innermost layer of the body wall. Nematode muscle cells are unique in that their inner portion is drawn out to form a long, slender, noncontractile fiber that leads to the nerve cord and functions as a neuron. The muscle cells are arranged longitudinally only, limiting the worms to undulating or thrashing movements. Except for a muscular pharynx, the digestive tract lacks muscle and other mesodermal derivatives. Nematodes lack circulatory systems; the fluid-filled pseudocoelom functions in this capacity. When the body muscles contract, the pressure they exert on the pseudocoelomic fluid, coupled with the resistant nature of the cuticle, forms what is termed a **hydrostatic skeleton,** which stiffens the body. The strong muscles; the tough, smooth cuticle; and the streamlined body, with its pointed anterior end, allow nematodes to move through most soils and tissues with ease.

The nematode cuticle is highly resistant to many agents. Eggs of most nematode species are also provided with impervious shells that allow for widespread dissemination of the species and survival under severely adverse conditions. This combination of impervious cuticle and eggshell has probably been very important in the great success of the nematodes.

Nematodes apparently abandoned the protonephridium early in their evolution. Unique **renette cells** alone act as excretory structures, even in primitive species. More advanced species possess a pair of long excretory tubules that run the length of the body and empty to the exterior by a common duct and pore. Most nitrogen excretion is in the form of ammonia, which diffuses into the intestine and is eliminated with the feces. Although many aquatic and soil nematodes normally conduct osmoregulation through their cuticle and body wall, many of these species are able to adjust to and survive years of desiccation.

Nematodes have a well-defined brain in the form of an anterior nerve ring with distinct dorsal ganglia. A prominent double, but fused, ventral nerve cord, essentially a chain of ganglia, runs the length of the body. A smaller dorsal cord composed entirely of motor fibers parallels the ventral cord. Smaller lateral cords parallel the other two. Several types of sense receptors occur on the body surface and are especially concentrated at the anterior end. Some nematodes have simple, light-sensitive eyespots.

Most nematodes are dioecious and sexually dimorphic. Males, usually the smaller of the two sexes, have one or two tubular testes, each of which leads to a vas deferens that opens on a genital papilla. The females usually possess two tubular ovaries (see Fig. 21-14b) and prominent uteri, the upper ends of which store sperm after copulation and are the site of fertilization prior to the formation of the eggshell. Cleavage is determinate and asymmetrical, although not spiral. The fertilized eggs may be stored in the uteri for a while before being shed, by which time they may be far along in development. In the few ovoviviparous species, the eggs hatch within the uterus, and the young individuals emerge through the female genital pore.

## Summary

Of the perhaps 10 million species of animals and the protozoans, only a little more than a million have been officially named and described, most of them being prominent or abundant forms not likely to have escaped our notice. Although some of the Protozoa are close relatives of certain unicellular algae, most are clearly animallike, exhibiting holozoic nutrition and locomotory ability. Some protozoan genera are identical in construction to the choanocytes of sponges.

Most students of sponges regard them as an unusual side branch of the main line of animal evolution. They are organized primarily at the level of specialized cells rather than of tissues or organs. They are unique in so many fundamental ways as to be placed in their own subkingdom, further supporting the view that multicellularity must have evolved at least twice among animals.

Although possessing few structures that could be called organs, cnidarians and their close relatives, the ctenophores, have reached what is regarded as the tissue grade of organization. They are the simplest forms to possess either a digestive cavity or a nervous system. Not only is their digestion largely intracellular, however, but it suffers from the inefficiency of having the mouth as the only opening into the gastrovascular cavity. The nervous system is primitive, lacking any type of concentration of neurons into a brain. Three major specializations of cnidarian morphology are seen in the three taxonomic classes: Hydrozoa, Scyphozoa, and Anthozoa. In hydrozoans, the prominent morphological type is a tubular polyp, which reproduces asexually by budding, while in scyphozoans, it is the inverted saucer-shaped medusa which is the sexual reproductive stage. In both these classes, both structural types are present at comparable stages of the life cycle. Anthozoans, on the other hand, exhibit only the polyp type of organization. All three classes have planula larvae.

Although flatworms exhibit three important evolutionary advances—bilateral symmetry; cephalization; and possession of a mesoderm, a cellular layer between the ectoderm and the endoderm—the rich potential of these developments was little realized in members of the phylum. The flatworm nervous system, however, which includes a primitive brain in the turbellarians and trematoda, reflects advancement beyond the cnidarians. Primitive excretory-osmoregulatory organs, the protonephridia, also appear in flatworms for the first time. Probably as much as any other factor, the lack of a circulatory system has limited flatworms to remaining flat and very thin, thereby especially restricting the evolution of the free-living species. Variations in the morphology of turbellarians occur mainly in the shape of the body, the position of the gut, the shape of the gut, and, perhaps more important, in the degree of development of the nervous system. Adaptation to a parasitic existence, however, was a successful evolutionary venture for this phylum. Many degrees of adaptation and specialization, from a completely free-living to an obligate, anaerobic, parasitic life are exhibited by the various flatworm species. Nemerteans, or the proboscis worms, are the simplest metazoans with a one-way digestive tract.

Pseudocoelomate animals can be regarded as an evolutionary stage between the acoelomate flatworm, in which spaces between internal organs are filled with mesenchyme and other mesodermal derivatives, and the coelomate phyla, in which the major body cavity, the coelom, is entirely lined by mesodermally derived tissue. Pseudocoelomates possess a body cavity, the pseudocoelom, only partially lined by mesodermal derivatives. Nematodes have a body musculature that permits better-coordinated body movements than that of flatworms. Their gut, however, lacks muscle tissue, which no doubt limits its efficiency. Despite this, nematodes are a very successful phylum, with species inhabiting every environment known to support life.

## Recommended Readings

(For chapters 31 and 32)

ATTENBOROUGH, D. 1979. *Life on Earth: A Natural History.* Little, Brown, Boston. A beautifully illustrated account of the evolution of life on our planet. Written in an engaging and often exciting style, the 13 chapters parallel the 13 programs in the PBS television series of the same name.

BARNES, R. D. 1980. *Invertebrate Zoology,* 4th ed. Saunders, Philadelphia. An excellent introduction to the structure, biology, and phylogenetic relationships of invertebrate animals. Very well illustrated and eminently readable. The book is not only the most suitable full-sized text for any beginning student, but is an excellent reference work.

BORRADAILE, L. A., F. A. POTTS, L. E. S. EASTHAM, and J. T. SAUNDERS. Revised by G. A. Kerkut. 1961. *The Invertebrates.* 2nd ed. Cambridge University Press, London. A classic work on the biology of invertebrates. The book emphasizes the phylogenetic affinities of the various invertebrate phyla.

FINGERMAN, M. 1981. *Animal Diversity,* 3d ed. Saunders, Philadelphia. A succinct, clear introduction to the major phyla and classes of animals. Available in paperback.

ORR, R. T. 1982. *Vertebrate Biology,* 5th ed. Saunders, Philadelphia. An excellent account of the anatomy and biology of vertebrates.

ROMER, A. S., and P. T. S. PARSONS. 1977. *The Vertebrate Body,* 4th ed. Saunders, Philadelphia. A very concise, clear, and interesting account of the comparative anatomy of the vertebrates, with excellent discussions of the evolutionary trends in vertebrates.

YOUNG, J. Z. 1981. *The Life of Vertebrates.* 3d ed. Oxford University Press, London. An outstanding account of vertebrate morphology combined with information on the mode of life of vertebrates.

# 32 Animal Diversity: The Coelomata

OUTLINE



The largest categories of a taxonomic system are always based on the most fundamental differences in the organisms being classified. Thus, the kingdom Animalia is divided into Parazoa, which include the radially symmetrical sponges, and the Metazoa, which include the radially symmetrical cnidarians and ctenophores, and the Bilateria, which include the rest of the animal kingdom, those with a fundamental bilateral symmetry. Bilateria also differ from radiate metazoan animals by the possession of a well-defined, endodermally derived, middle embryonic germ layer: the mesoderm.

Bilateria can be further divided into two large branches, Protostomia and Deuterostomia, which differ from each other in fundamental

**FIG. 32-1** Phylogenetic tree of the animal kingdom showing probable kinships between the major phyla.

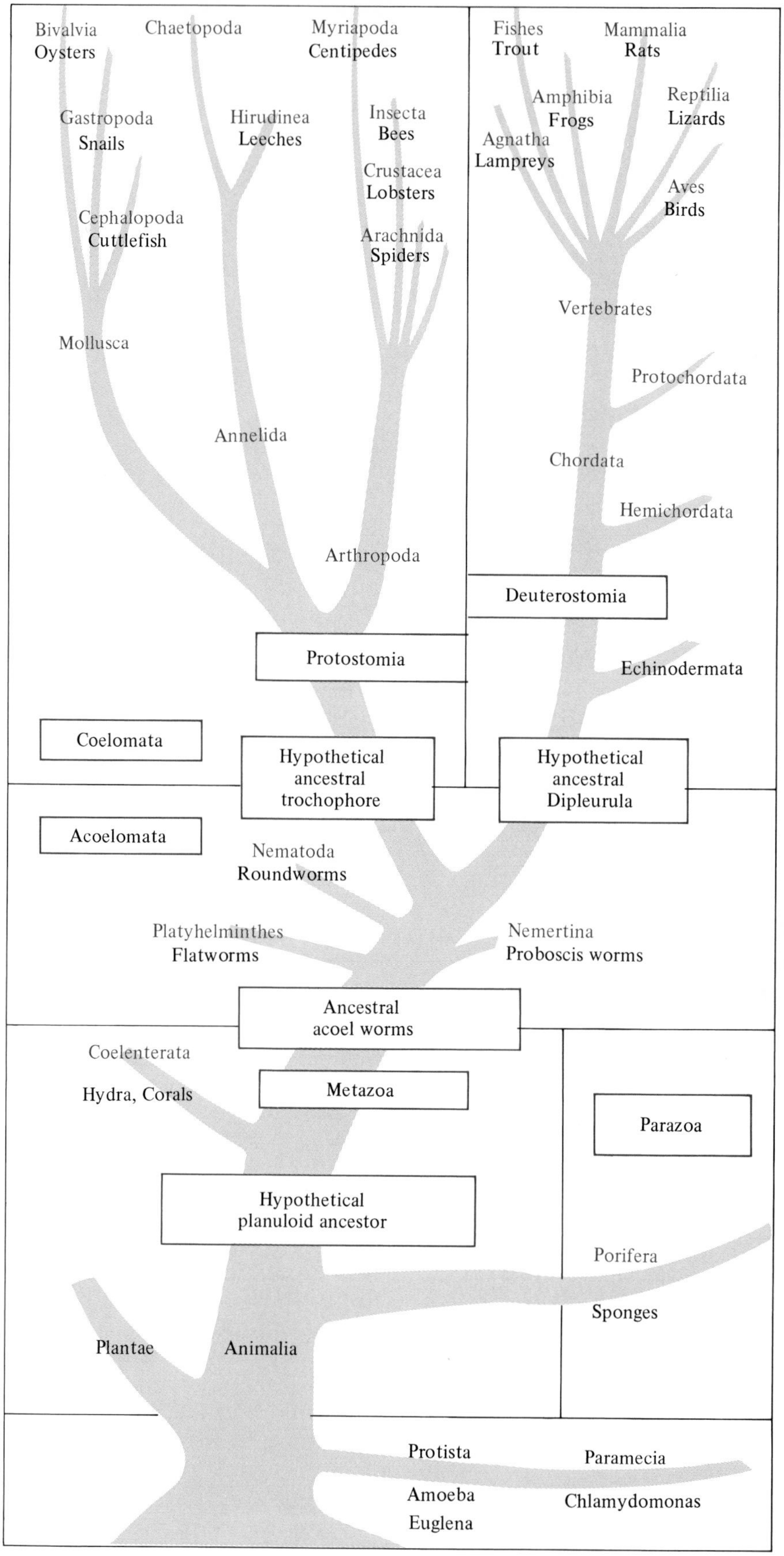

ways (Fig. 32-1). For example, most early embryos of protostomes exhibit spiral cleavage, whereas typical deuterostome embryos exhibit radial cleavage. In spiral cleavage, the cells, or blastomeres, of one tier are staggered in relation to those of adjacent tiers, as are bricks in a wall. In radial cleavage, however, each blastomere is immediately above or below one in an adjacent tier. In addition, spiral cleavages are determinate: they divide an egg into blastomeres whose developmental fate is already determined; thus, for example, if one cell of the four-cell stage is removed, a corresponding defect appears in the embryo. Radial cleavages, on the other hand, are indeterminate; thus, a four-cell stage embryo can be separated into four cells, and each cell will develop into a complete, albeit small, individual.

In the Protostomia, the blastopore, the opening into the primitive gut in the gastrula stage, becomes the mouth of the larva. In Deuterostomia, however, the blastopore usually becomes the anus (see Fig. 25-14). Furthermore, the mesoderm in deuterostomes arises either as outpocketings of the endoderm or in association with the endoderm.

Protostomia includes the acoelomate phyla, such as flatworms and nemerteans, which have no body cavity other than the gut; the pseudocoelomates, such as nematodes, which have a pseudocoelom; and the schizocoelomates, such as annelids, arthropods, and molluscs, in which the coelom arises as splits inside mesodermal bands. Deuterostomes include the following phyla: Chaetognatha, such as arrow worms; Echinodermata, such as starfishes, sea urchins, and brittle stars; Hemichordata, such as acorn worms; and Chordata, such as protochordates, fishes, amphibians, reptiles, birds, and mammals. The actual ancestors of deuterostomes are unknown. But a hypothetical ancestral form, designated as a dipleurula, incorporates the major anatomical features of such an ancestral form during the larval stage.

## *I. COELOMATE PROTOSTOMIA*

Protostome coelomates evolved before the deuterostomes. Both in fossil records and in their morphological and embryological features, primitive coelomate protostomes are closely related to the platyhelminthes.

### PHYLUM ANNELIDA

Of the many kinds of worms on earth, the most familiar are the segmented worms, the annelids (Fig. 32-2). Of the over 8,500 species in this phylum, the best known are the earthworms that are turned up in almost any spadeful of moist garden soil. Earthworms belong to the class Oligochaeta (from the Greek *oligo,* "few," and *chaeta,* "bristles"), which inhabit soil and fresh water. Oligochaete species are especially abundant in lakes and streams that are rich in organic matter. They contribute significantly to the decay of animal and vegetable material, and some species actually constitute an integral part of sewage treatment plants. Charles Darwin was so impressed with the contribution of earthworms to the improvement of the soil that he asked whether "there be any other animals which have played so important a part in the history of the world." A second class of familiar annelids, the Hirudinea (from the Latin *hirudo,* for "leech"), include the blood-sucking leeches some-

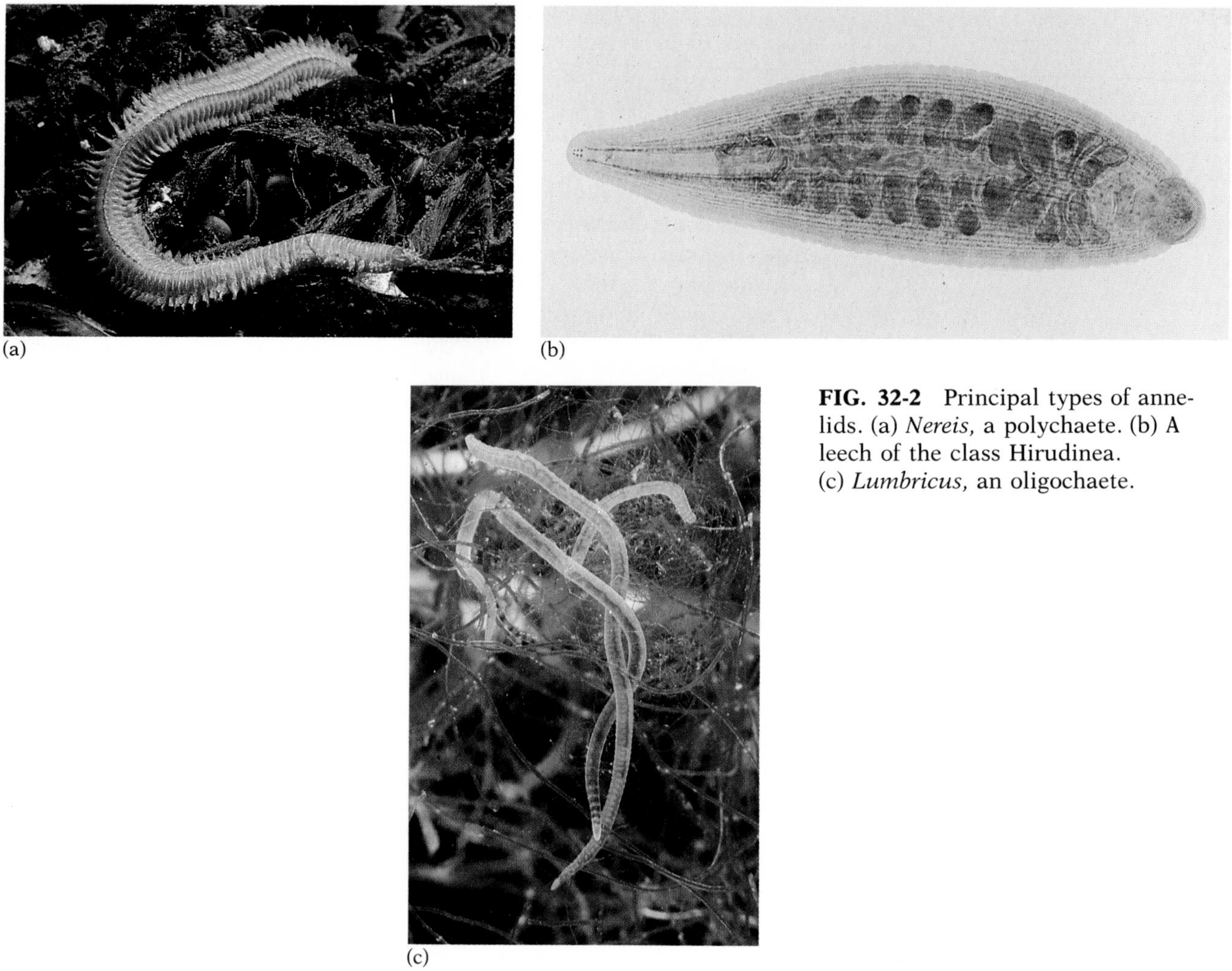

FIG. 32-2 Principal types of annelids. (a) *Nereis,* a polychaete. (b) A leech of the class Hirudinea. (c) *Lumbricus,* an oligochaete.

times found clinging to our skin after wading or swimming in some lakes or sluggish creeks or rivers.

Although the least familiar to most people, the most abundant annelids are members of the class Polychaeta (from the Greek *poly,* "many," and *chaeta,* "bristles"), almost all of which are marine. Despite the great variety and abundance of polychaetes, the casual observer of the ocean shore might never notice these worms. But even a few minutes of digging in the sand or mud in an intertidal zone usually reveals several species.

Polychaetes are composed of two large subclasses, Errantia, which includes various swimming and burrowing species, and Sedentaria, consisting primarily of tube-dwelling or otherwise sedentary species that live on the sea floor in relatively shallow water.

Like earthworms, the marine annelids that inhabit the sand and mud of the seashore make important contributions to their environment by recycling organic matter. Where populations of the lugworm, *Arenicola,* are large, the amount of soil they work over may be close to 800 tons per hectare (2.5 acres) per year (Fig. 32-3). The lugworm's effect on the ecological economy of tidal mud flats, and therefore on the welfare of commercially valuable seafood animals in their vicinity, can be great indeed.

Of all annelids, the polychaetes bear the closest resemblance to the

**FIG. 32-3** *Arenicola,* the lugworm. Note the external gills.

primitive fossil forms. However, the polychaetes exhibit many advanced features over the more primitive annelids, such as specialized structures in the head region. Annelid evolution appears to have followed three major trends, as represented by the following: (1) the polychaetes, which are free-swimming or form tubes or burrows of various types; (2) the oligochaetes, which adapted to fresh water and land; and (3) the leeches, which adapted to a carnivorous or ectoparasitic life. In addition, there is a primitive marine group called Archiannelida.

## THE GENERAL BODY PLAN OF ANNELIDS

The annelid body provides a typical example of **metamerism,** an arrangement in which the body is composed of a series of short cylindrical units called **metameres,** or **segments,** that are repeated along the longitudinal axis of the animal. In the most completely metameric animals, each body segment contains basically the same structures found in nearly all the other segments (Fig. 32-4).

The structures that tend to be repeated in metameric animals are nerve ganglia, muscles, and pairs of the following: locomotory appendages, excretory organs or other ducts, gonads, lateral blood vessels, and nerves. The digestive tract, on the other hand, runs the length of the body without segmentation. In addition, the sense organs such as antennae, palps, eyes, and cirri, are concentrated in some regions, especially the head.

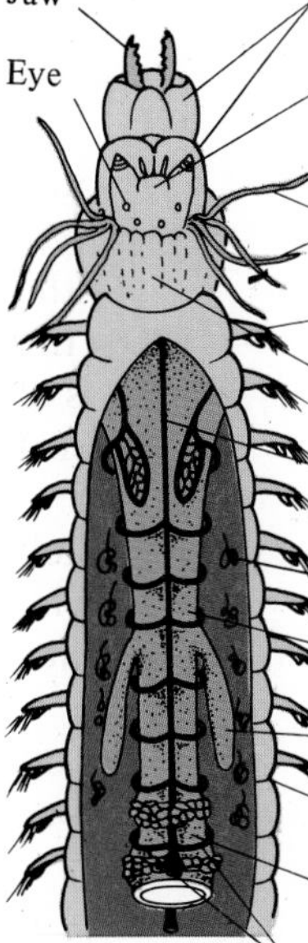


**FIG. 32-4** Diagrammatic sketch of the anatomy of *Nereis,* a polychaete.

Internally, annelid segments are usually marked by **septa,** thin walls that separate the body cavity into a series of compartments connected to one another by pores. The body cavity is a true coelom, being completely lined by tissue derived from mesoderm. Within the coelom, almost every segment contains a pair of excretory organs, the **nephridia.** The digestive tract extends from the mouth, located in the anterior segment called the peristomium, to the anus, in the terminal segment. In front of the mouth is an incomplete segment, the prostomium. The digestive tract is divided into a number of regions or organs specialized in the ingestion, storage, trituration (grinding), digestion, and absorption of food.

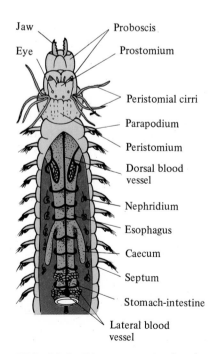

The circulatory system is usually a completely closed system; that is, all the blood is contained in blood vessels, not in cavities or sinuses. The principal pumping organ is the dorsal blood vessel, which moves the blood forward by waves of peristaltic contraction. The dorsal blood vessel may be assisted in pumping by several pairs of contractile vessels, called hearts, that connect the dorsal and the ventral blood vessels. The details of both polychaete and oligochaete circulatory systems exhibit considerable variation. Some species have a partially open system in which the intestine is surrounded by a blood sinus rather than being vascularized by a network of capillaries. In leeches, the circulatory system is reduced or absent.

On almost every segment in the typical polychaete annelid is a pair of locomotory appendages. In their most elaborate form, they occur as large, laterally flattened flaps of tissue called **parapodia** (Fig. 32-5). Being richly supplied with blood vessels, they also function in respiration. Typically, each parapodium branches into a dorsal notopodium and a ventral neuropodium, both supported internally by chitinous* rods. At the tips of both parts are tufts of chitinous bristles, called **setae** or chaetae. The parapodia also bear a sensory organ called a cirrus. As an adaptation to a land existence, the parapodia have been lost in the oligochaetes, but the setae are retained in all but the leeches and continue to function in locomotion.

The annelid body wall consists of an outer nonchitinous cuticle of collagenous protein and polysaccharide; the epidermis, which secretes the cuticle; and two or more layers of muscle. The surface of the cuticle is kept moist in terrestrial worms by liquid secretions of the epidermis. Aquatic annelids also secrete mucus, which is not only protective but can be used in tube building or in feeding. The outer layer of muscles is circular, the inner longitudinal. These muscle layers function in locomotion by alternately extending and contracting the body, which helps in burrowing, and by causing undulating body movements, which can be used in swimming and in creating water currents important for filter feeding and respiration. The muscle layers are supported by layers of connective tissue. Although annelids lack a skeleton, the body wall musculature, by exerting pressure on the coelomic fluid, serves as a hydrostatic skeleton, providing turgidity and rigidity when and where required. Even in leeches, in which the coelom is greatly reduced, the

* Chitin is a mucopolysaccharide, a polymer of acetylglucosamine, that hardens after it is secreted.

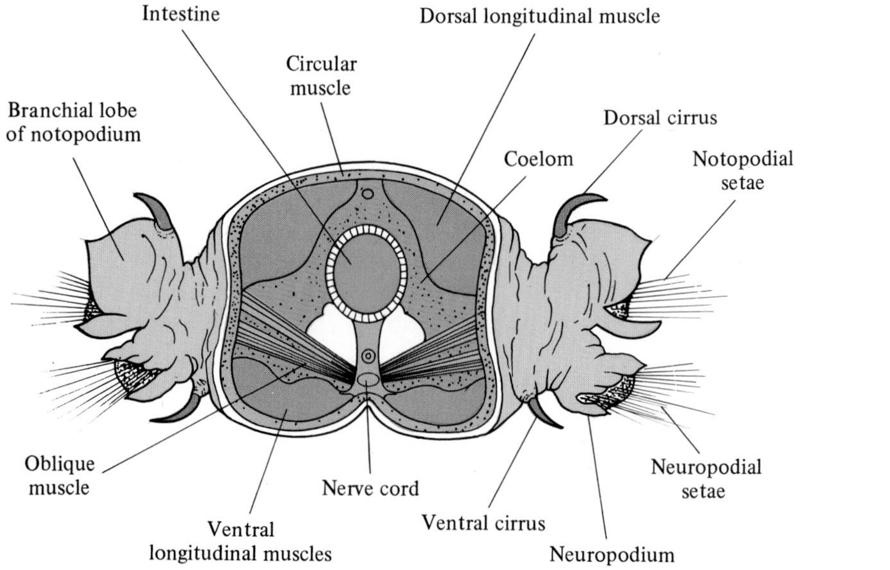


**FIG. 32-5** Cross section of *Nereis* showing the arrangement of the internal organs and a detailed structure of the parapodia.

mesenchyme, which occupies the same space, acts as a liquid under pressure, thereby functioning as coelomic fluid.

## THE ANNELIDS AS AN EVOLUTIONARY WAY STATION

The annelids occupy an important position in the animal kingdom because they continued the exploitation of the one-way digestive tract—one with both a mouth and an anus—and evolved several specialized digestive regions and organs. Mesoderm, which invests the annelid digestive tube, provides it with musculature, improving the efficiency of the digestive system.

Annelids were also the first to exhibit metamerism. Since the relatively simple body was divided into a series of segments each of which was capable of carrying on most bodily functions, the animal was able to survive a variety of evolutionary situations. It is conceivable that the early evolutionary changes affected only some of the segments. Since most mutations are deleterious, changes in a few segments were not lethal because other segments could carry on the body functions. When the mutated segments were beneficial, the individuals possessing them were more successful in reproduction and the trait became more common in the population. Thus, even new species, genera, and families could have evolved. Metamerism may therefore have been the most important of the evolutionary advances made by annelids. For it was from the ancient Annelida, well over 500 million years ago, that there arose other animals that exploited metamerism about as fully as possible. These were the arthropods. They turned metamerism to full advantage by developing a series of highly specialized head and body parts, all invested with a more resistant cuticle.

Another evolutionary advance of importance in the annelids was the development of the coelom, the secondary body cavity that is lined by mesoderm and is filled with fluid. Although acquisition of a coelom was probably important in their early evolution, its benefits were never fully realized by many of the protostome phyla (see Panel 32-1).

By pioneering in the development of the coelom, in the exploitation of the one-way digestive tract, in the evolution of specialized organs of excretion and respiration, and in the development of metamerism, annelids represent a crucial evolutionary step that formed the basis of many subsequent advances. Thus, the annelids can truly be said to occupy a central position among the higher Protostomia.

***Locomotion.*** Annelid locomotion represents a distinct advance over that of lower life forms. The three major groups of annelid types evolved different locomotor methods. Polychaetes have well-developed parapodia that assist in creeping or swimming, achieved by undulations of the body. Although tubiculous annelids may remain in their burrows or tubes during much or all of their lives, many leave them and swim up to swarm at the sea's surface during the reproductive phase.

Earthworms move through the soil or over its surface by first extending their bodies by a contraction of their circular muscles, next anchoring the setae of the anterior segments in the substratum, and then contracting their longitudinal muscles, pulling the rear part of their body forward. Leeches can move by alternately attaching and detaching the anterior and posterior suckers, as does an inchworm. They also swim by body undulations.

## PANEL 32-1

# The Advantages of Having a Coelom

The coelom has proved to be a most useful possession, and its several advantages are exploited in interesting ways by various groups of animals. Some of its principal benefits follow.

1. *Digestive system:* by freeing the internal organs from attachment to one another and the body wall, the development of the coelom permitted their further evolution. For example, it permitted the coiling of the digestive tract, which provided increased area for absorption and secretion.
2. *Movement:* a coelom permits freer bodily movement than when the organs are enmeshed in the mesenchyme.
3. *Respiration and circulation:* the coelom constitutes a fluid-filled system that provides sustenance to the cells lining the cavity. The exchange of oxygen and carbon dioxide between tissues and the environment is facilitated by the ability of the coelomic fluid to dissolve these gases and to aid in their circulation. In the primitive state, the coelomic cavity is connected to the exterior by small pores. Many small oligochaetes and a number of leeches lack a blood circulatory system and rely entirely on the coelomic fluid for circulation of nutrients, respiratory gases, and other metabolites. Coelomic fluid, like blood, may contain clotting factors.
4. *Excretion:* the excretory products resulting from metabolic activity in cells lining the coelom are released into it. Most annelid segments contain pairs of ducts that convey fluid from the coelom to the exterior of the worm. In most species, these ducts take the form of nephridia of diverse kinds—organs that, by processing coelomic fluid, perform many of the functions of the vertebrate kidney.
5. *Reproduction:* in the more primitive annelids, gonads are found in most segments, usually in close association with the metanephridia. In polychaetes, eggs or spermatozoa are shed into the coelom, where they continue their maturation. The coelom may become completely filled with eggs or spermatozoa. At the time of mating, the eggs or sperms are shed into the ocean. Release of ova from the ovary into the coelom is a situation that persists in many higher groups, including mammals, where the ovum is, for a moment before it enters the oviduct, free in the coelomic cavity.
6. *Skeletal support:* as we have noted, the coelomic fluid, confined in a space and placed under pressure by contraction of the body wall muscles, comprises a hydrostatic skeleton. In the annelids, this hydrostatic skeleton is of great importance in burrowing.

***Excretion.*** A further evolutionary advance in annelids is represented by their excretory organs. Besides the simpler protonephridia found in some polychaetes, many annelids possess elaborate metanephridia. In more primitive species, the metanephridia occur as a pair in nearly every segment; in some advanced species, there is a single pair for the entire animal.

Metanephridia and protonephridia differ in several ways. Metanephridia (see Chapter 13), which are open at both ends, process coelomic fluid much as a nephron processes the filtrate of blood formed in the glomeruli of the vertebrate kidney. Protonephridia end blindly in cells that are specialized for the collection of metabolic wastes.

***Nervous System and Sense Organs.*** The nervous system of annelids represents another major advance over the level of development exhibited by flatworms. The largest and most complex brains among annelids are found in free-swimming polychaetes, in which the sense organs are also the most highly developed. In these species, the brain may nearly fill the first few segments, giving off up to 16 pairs of nerves to sense receptors.

In all annelids, the brain connects with a prominent paired ventral nerve cord that runs the length of the body. A large ganglion and lateral nerves are located in each segment. The brain serves as an inhibitory center in that it prevents autonomous actions of diverse organs from occurring. Most annelids are able to regenerate a brain if the anterior end of the body is accidentally cut off.

Some Errantia have highly developed eyes composed of a retinal cup, a vitreous body, a lens, and a highly complex accommodation mechanism, all indicating that the eye forms an image, somewhat as ours does. But most annelid eyes are much simpler than this, and many species lack eyes altogether. All annelids possess photosensitive cells on their body surface, however, as anyone hunting night crawlers can surmise from the response when a flashlight beam is focused directly on a worm that has partially emerged from its burrow.

***Reproduction.*** Reproduction in annelids assumes several forms. Many reproduce asexually by budding off new individuals. In some species, the new worms do not immediately separate from the parent, and up to a dozen worms can remain attached temporarily to each other. Other species undergo fission, as planarians do, pinching apart into two or more pieces, with each piece regenerating all missing parts.

Depending principally on the class to which they belong, annelids are either monoecious or dioecious. The reproductive organs of some species represent the most important departures from total metamerism. Although in polychaetes gametes may be produced in most of the body segments, the gonads of tubiculous species are few in number and restricted to a few posterior segments; a single gonad is sometimes so large as to occupy more than one segment.

Sexual reproduction is seasonal in dioecious polychaetes, usually synchronized with a phase of the moon or tide, adaptations that maximize the chances that male and female worms of the same species and in breeding condition will be brought together. Such worms inherit a drive to release the gametes at the surface of the sea on a specific date, such as at the first quarter of the June moon.

## PHYLUM ARTHROPODA

Members of the phylum Arthropoda (from the Greek *arthron,* "joint," and *podos,* "foot") have jointed legs and a chitinous exoskeleton. The group includes a heterogeneous assemblage of animals that live in every conceivable type of environment, including aquatic, terrestrial, and parasitic. All have a segmented body, with some or all of the segments bearing jointed appendages. The coelom is much reduced, and the blastocoel persists as a prominent **hemocoel,** filled with hemolymph, which functions as blood. Cilia are entirely lacking.

### THE ADVANTAGES OF HAVING AN EXOSKELETON

The exoskeleton has allowed the arthropods to adapt to many ecological niches not otherwise open to them, and, as a result, they have become the most diversified and abundant of all marine and terrestrial animals. The exoskeleton enabled the arthropods to attain a larger size than the annelids; some of the Paleozoic eurypterids, an extinct group of primitive arthropods, reached a length of nearly 3 m (9 feet). With increased size came greater variety and specialization. The exoskeleton and its remarkable joints between segments provided opportunities for the evolution of muscle attachments that permitted movements impossible for a soft-bodied worm (Fig. 32-6a). A whole new variety of locomotory movements developed, as useful in catching prey as in escaping predators (Fig. 32-6b).

Some of the very reasons that enabled arthropods to become the most successful of marine animals also made it possible for them to be the first truly terrestrial animals. The strong, rigid cuticle encasing their limbs enabled them to support the weight of their bodies on land and in the air. The impermeability of the cuticle, which so helpfully limited the gain and loss of water in freshwater and brine pools, could be modified to limit or prevent water loss in air, an essential requirement for a terrestrial animal. Subsequently, the modification of cuticular extensions of the thorax into strong but lightweight wings opened for them yet another world, one that at first was free of predators and competitors.

The opportunity to be the first to invade the land (390 million years ago) and the first to fly (250 million years ago), were golden chances for

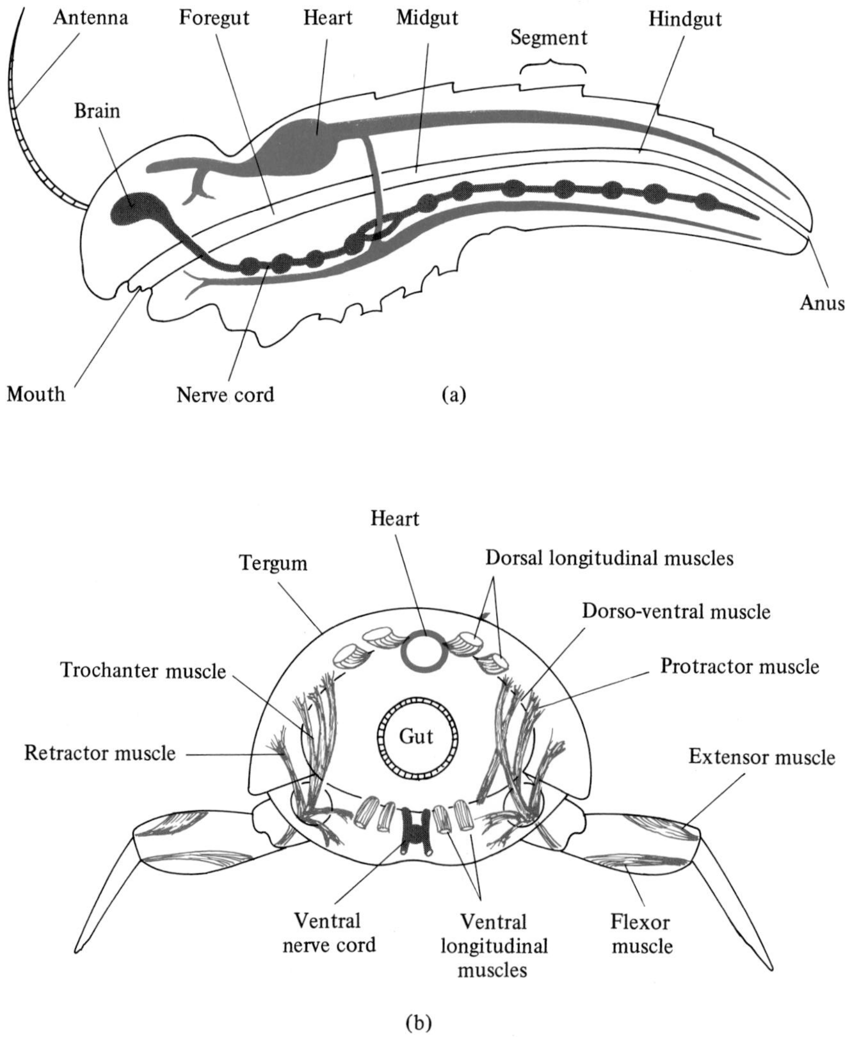


**FIG. 32-6** Arthropod body plan. Diagrammatic sketches of (a) longitudinal and (b) cross sections through the body. Note the segmented body, the chitinous exoskeleton, and the jointed appendages.

any group that could exploit them. The insects were the result. Perhaps the long span of time in which they were the unchallenged animal inhabitants, first of the land and then of the air, allowed the insects to evolve in as many different directions as they have. Among the more remarkable of those early experiments was a giant dragonfly (*Meganeura*), with a wing span of nearly a meter; it is estimated to have whistled through the Paleozoic sky at speeds of 65 km per hour. Insects with wings swarmed through the air for almost 100 million years before the first reptile or bird became airborne.

## EVOLUTION OF THE ARTHROPOD BODY AND APPENDAGES

The hypothetical first arthropods were less specialized than many modern polychaetes but were like them in most respects. Their body was probably segmented, with a pair of appendages on all but the last segment. The head bore a mouth, probably two pairs of eyes, and a pair of fingerlike head appendages inherited from annelids, the **prostomial palps,** which could be easily modified into little antennae, or **antennules.** Exactly what the early arthropods looked like is not known. They arose during a period for which our fossil record is poor. At a very early stage in arthropod history, however, at least two evolutionary directions were taken by the jointed-legged animals. Or perhaps it was that the arthropods arose at least twice from the annelids, each time from a somewhat different group; this point may never be settled. In any case, arthropods today can be grouped into two large subphyla that can be traced back for more than 500 million years.

One of these subphyla, **Mandibulata,** includes Crustacea, Myriapoda (millipedes, centipedes), and Insecta (Fig. 32-7), all of which typically possess antennae, mandibles (chewing mouth parts), and maxillae as head appendages. All of the Mandibulata, except the crustaceans, have apparently retained the ancient palplike antennules as their antennae. In the crustaceans, these appendages are still called antennules. The many additional head appendages seen in mandibulate arthropods were apparently acquired by fusion to the head of four or more body segments, together with their variously modified jointed appendages. Thus, in Crustacea, the first body segment provided the **antennae,** the second the mandibles, and the third and fourth the maxillae. Identical or similar adaptations have occurred in the other Mandibulata. However, the nerve connections to such altered and displaced appendages remain as they were before the fusions, testifying to the evolutionary history of these head parts.

The second arthropod subphylum is **Chelicerata,** to which belong the horseshoe crab; the extinct eurypterids; and the class Arachnida, including scorpions, spiders, ticks, and mites (Fig. 32-8). All exhibit a more extensive fusion and reduction of segments than in Mandibulata. The head, or **cephalothorax,** is regarded by some experts as being composed of six fused segments, and the body, or **abdomen,** a fusion of 12 or 13 segments. The appendages of the cephalothorax consist of a pair of mouthparts called **chelicerae** and another pair called **pedipalps** (both structures are used in seizing and killing prey), in addition to four pairs of legs. The chelicerae are believed to be homologous to the crustacean antennae, which are appendages of the first body segment of ancestral

(a) (b) (c) (d)

**FIG. 32-7** The principal types of mandibulate arthropods. (a) Prawn, a crustacean. (b) Giant centipede (note the shed cuticle), a myriapod. (c) Giant millipede, another myriapod. (d) Dragonfly, one of the primitive flying insects.

arthropods; the other appendages are regarded as being derived from trunk appendages.

Two other groups of arthropods should be mentioned. The once flourishing trilobites, now extinct, might be, on the basis of the age of the fossils, the group from which the Chelicerata and Mandibulata arose. But there seem to be no compelling morphological reasons to believe this. The other group, the very primitive (or perhaps very degenerate) sea spiders, or pycnogonids, is not clearly related to any modern group and may have been yet another product of early arthropod evolution. This group is regarded as an aberrant arthropod not in the main line of arthropod evolution.

## CEPHALIZATION AND SPECIALIZATION OF APPENDAGES

In both the Mandibulata and the Chelicerata we note an increase in the functional ability of the head, by the addition not only of mouthparts but also of sensory and nervous elements. This trend toward forming a more definite head, together with the accompanying concentration of

(a)

(b)

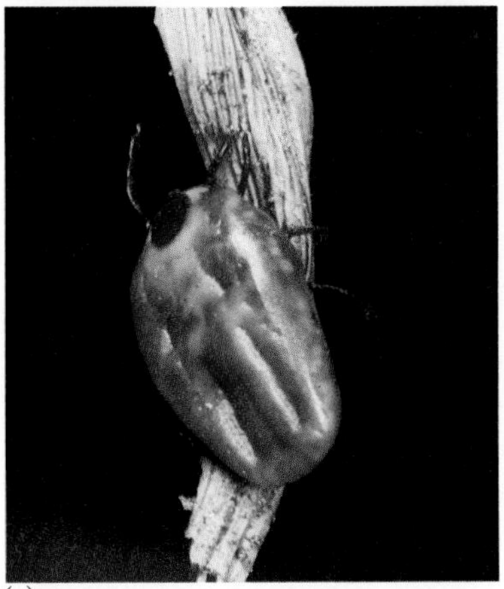
(c)

**FIG. 32-8** The principal types of chelicerate arthropods. (a) Scorpion. (b) Tarantula. (c) Dog tick.

neural elements, the beginnings of which were first noted in the platyhelminthes, is evidence of increasing cephalization. It is dramatically demonstrated by the arthropods. Their ability to convert neighboring segments and their appendages and nervous elements into head components has probably been one more of the important factors in the success of the arthropods. In crustaceans, the head and thorax are together called the cephalothorax, while in insects, the two regions are very distinct. In both groups, the abdomen is a distinct body region.

The other major trend seen in arthropod evolution is no less important: the modification of trunk appendages to perform many tasks. In both crustaceans and insects the original series of identical segments and locomotory appendages are modified into an astoundingly large variety of organs. Once again, annelid metamerism seems to have provided a most useful theme upon which innumerable variations were made. The many similar appendages are rather like basic construction kits that can be made into a number of things. In a primitive metameric animal, the segments are so numerous that almost any chance variation might be turned to some use in time or perhaps in a new environment, leaving more than enough unchanged legs to attend to the important matter of locomotion. For example, not all of the many legs of millipedes and centipedes are necessary; the three pairs of legs of an insect actually function much more efficiently.

Arthropod appendages have evolved into many forms and serve various uses. For example, the lobster or crayfish has a number of different appendages, each with a specific function (Fig. 32-9):

- Antennule and antennae: sense reception
- Mandibles: tearing and chewing
- Maxillae: seizing
- Maxillipeds: respiratory movements
- Pereiopods: walking and respiration
- Pleopods: swimming, aeration of eggs
- Uropods: swimming, assisted by the telson, a postsegmental region at the end of the body.

The appendages listed on page 908 are a series of basically similar structures that have been modified in various ways in a single organism. Such structural modification is an example of **serial homology.**

Not only have appendages become remarkably modified within the same arthropod, but far greater variations are found from one kind of arthropod to another. The modification and distribution of appendages account for many of the morphological differences among arthropods. For example, all insect variations are upon a basic theme of a tripartite body consisting of a head, thorax, and abdomen. The thorax is made up of three segments bearing three pairs of legs and usually one or two pairs of wings. It is the ability to fly that most obviously distinguishes insects from all other arthropods. Perhaps this trait, more than any other, is responsible for their great evolutionary success. Over three

PANEL 32-2

# Insect Flight

An airplane has one mechanism for propulsion (the propeller or turbojet), another to provide lift (the wings), and yet others to control the attitude and direction of flight (the rudder, elevators, and ailerons). In the insect, all three functions are performed by the wings. For propulsion, the wings actually perform very much like a propeller. Rather than being rotated, however, they vibrate up and down. Both strokes are effective because the leading edges of the wings are held rigid while the rest of the wing is bent downward on the upstroke and upward on the downstroke. Because of the way in which the wings are attached to the body, this change in their tilt, or pitch, occurs automatically during each stroke, propelling the insect forward. Controlling muscles do determine the plane in which the wings vibrate, which, in turn, determines the direction of lift and thus the direction of flight. The angle can be varied so as to allow flight in any direction or, in some species, to permit hovering.

Some of the smaller insects are able to beat their wings at very high frequencies: flies at 200 beats per second, mosquitoes at 600 beats per second, and some midges at 2,200 beats per second. No bird can beat its wings at anything approaching these frequencies, nor, for that matter, can the less efficient fliers among insects, such as dragonflies and butterflies, whose top frequencies are little more than about 40 beats per second. The reason some insects can beat their wings so much faster than others is twofold, involving the nature

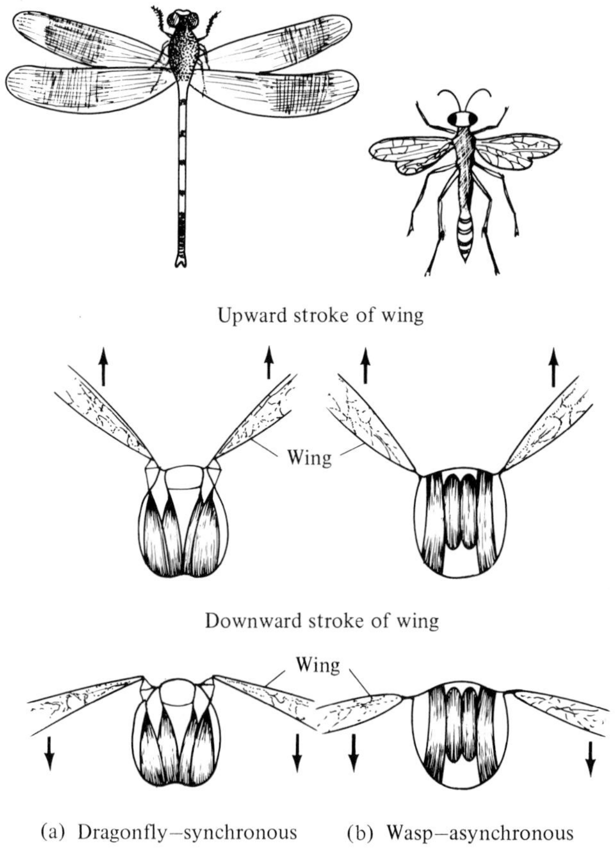


Arrangement of flight muscles in insects with synchronous (a) and asynchronous (b) types of wing beat.

quarters of all animal species known today are insects. A brief account of the physiological and anatomical basis of insect flight is given in Panel 32-2. A variety of head parts, some of them highly specialized, and a pair of compound eyes are present on the head. In some insect groups, only one pair of wings is present. In flies, the stubs of reduced wings are used as balancers called **halteres.** In beetles, the forewings have been converted to thick cases that shield the rear pair of wings, which alone are used for flight. Fleas are adapted to a parasitic existence among the hairs or feathers of their hosts by developing bloodsucking mouthparts, great jumping ability, a laterally flattened body that can slip between hairs, and by a loss of wings. The amount of variation in mouthparts is probably greater in the insects than in any other arthropod group (Fig. 32-10).

of the flight muscles and the mechanism by which their contractions cause the wings to vibrate as shown in the accompanying figure.

Two types of mechanisms are involved in the generation of wing power in insects. In the primitive insects, each wing beat is initiated by a synchronous volley of nerve impulses to the flight muscles. In others with high frequency wing beat the motor nerve impulses supplying the main flight muscles are asynchronous with the beat. In the blow fly, the motor nerve impulse frequency of 3 per second is accompanied by a wing beat frequency of 120 per second. Thus the flight muscles of rapidly flying insects actually contract with a much greater frequency than the frequency of motor impulses. This asynchrony is possible because the wing beat is generated by the muscles stimulating each other by their alternating contraction and relaxation. This type of rhythmic mechanism has been termed myogenic, or asynchronous, motor rhythm.

Rather than being attached to the wings directly, as in birds and the lower insect orders, the flight muscles of Diptera (flies) and Hymenoptera (wasps and bees) are attached to the cuticular elements of the thorax. Contraction of the opposing sets of flight muscles changes the shape of the elastic thorax exoskeleton. The wings are attached to the thorax in such a way that when one set of muscles contracts, the wings snap upward, and when the other set contracts, they snap downward. As the flight muscles exert force on the elastic and springlike exoskeleton, it in turn energizes what amounts to a set of levers, parts of which are the wings themselves. The situation is remotely like that in which a stiff card is held at its edges between the thumb and fingers. Compressing the card causes it to bow outward. If some force is then applied to the bowed-out portion, the card resists it for a time and then suddenly snaps in the direction of the force, suddenly becoming bowed in. Such a device, called a **click mechanism,** operates the wings of flies and bees. By making use of the springlike qualities of the cuticle, the muscles that cause the downstroke need contract only long enough to reach the click point. The snap of the thorax and wings then occurs, which stretches the opposing muscles, thereby stimulating them to contract. They too contract just long enough to reach the click point of the upstroke. This gives both sets of muscles sufficient rest between contractions to permit them to perform for long periods of time without fatigue. Because the stimulus to contract is provided by stretching, the flight muscles are able to contract at their own top frequency—a frequency far greater than that at which nerve impulses are delivered. The nerve impulses serve only to keep them contracting and do not determine the frequency of the contraction cycles.

It has been suggested that insect wings arose first as an adaptation to gliding on air currents or leaping. The wings in insects are an outgrowth of the thoracic cuticle, not a modified limb, as in reptiles, birds, and bats. The first improvement on the gliding insects was probably the addition of muscles to control the attitude of the wings, thereby giving direction to the glide. As these muscles evolved, some were able to move the wings up and down, adding a propulsive force; this primitive situation persists in dragonflies. The next evolutionary step was the addition of muscles that folded the wings back and out of the way when not in use. Last came the remarkable rapid-flight muscles, which, by distorting the thorax, caused the wings to snap up and down at high frequencies.

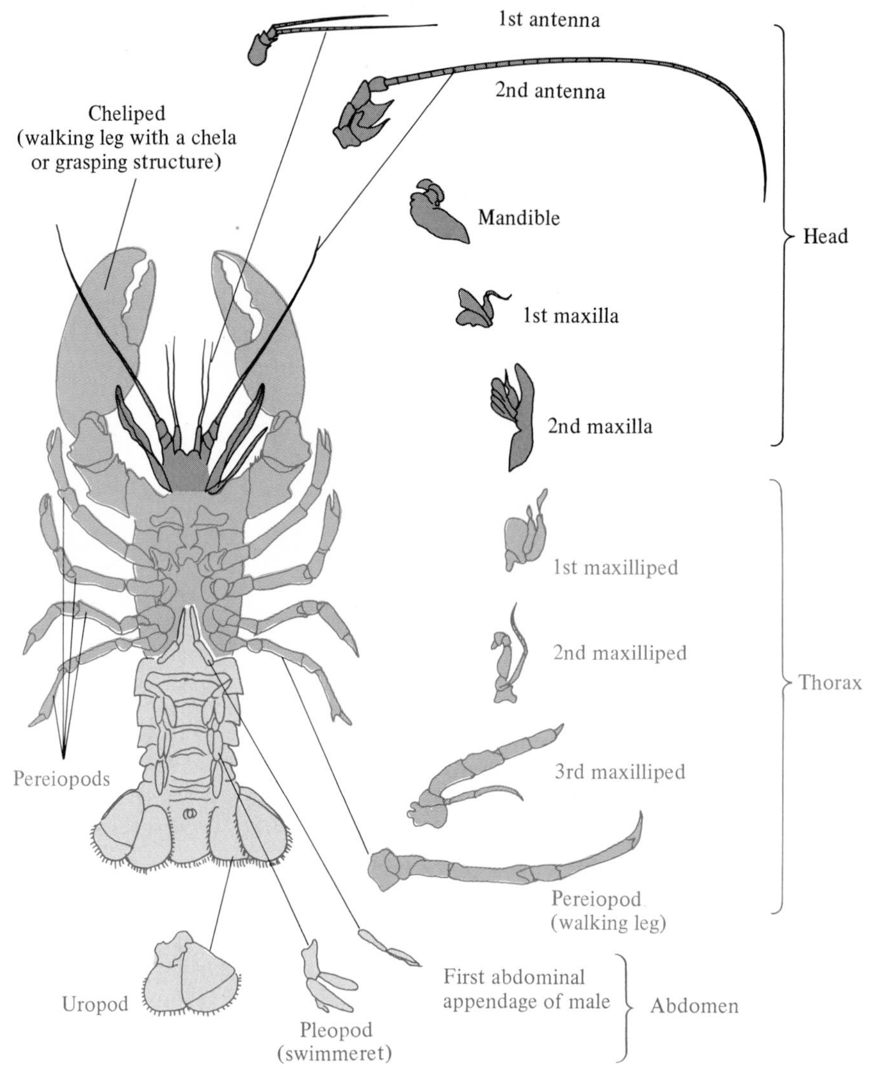


**FIG. 32-9** Appendages of a lobster are shown in situ in the body and also individually.

## PROBLEMS ASSOCIATED WITH AN EXOSKELETON: RESPIRATION

Having a rigid, impermeable exoskeleton causes several problems. Respiration, for example, cannot be conducted merely by diffusion through the body surface, as in annelids and flatworms, especially since many arthropods are terrestrial in habitat. In arthropods, this problem was solved by the evolution of diverse aerial respiratory organs. Insects, myriapods, and some arachnids, mainly terrestrial arthropod groups, evolved unique structures known as **tracheae,** tubular ingrowths that branch repeatedly into fine tracheoles supplying air directly to each cell (see Fig. 8-14). Although air enters and leaves the system as a result of alternate relaxation and contraction of abdominal muscles, oxygen and carbon dioxide move within the system by diffusion. This effectively limits the body size of any arthropod with a tracheal system.

In aquatic crustaceans, respiration is carried out by gills, thin vascularized outgrowths of the appendages. The gills are covered by a very thin cuticle that is permeable to gases, allowing gaseous exchange. Although most insects lack respiratory pigments, the hemolymph of crustaceans and arachnids usually contains **hemocyanin,** a blue blood pigment containing copper that can be used in transport of oxygen.

## THE EFFICIENT INTERNAL ORGANS OF ARTHROPODS

The arthropod heart is a pulsating tubular structure located along the mid dorsal line (Fig. 32-11). Blood leaving the heart flows anteriorly in the dorsal vessel, as in annelids; but here the similarity between the arthropodan and annelidan vascular systems ends. The arthropod body cavity is a hemocoel without blood vessels, and the blood merely flows through the spaces between organs—a classic example of an open circulatory system.

The excretory organs of crustacea are a single pair of modified nephridia associated with the appendages of the head (see Fig. 13-6a), demonstrating the affinity between arthropods and annelids. In terrestrial arthropods, excretion is accomplished by tubular outgrowths of the gut, the Malpighian tubules (see Fig. 13-7), an innovation in arthropods, in that no homologs of these organs occur in any other animals.

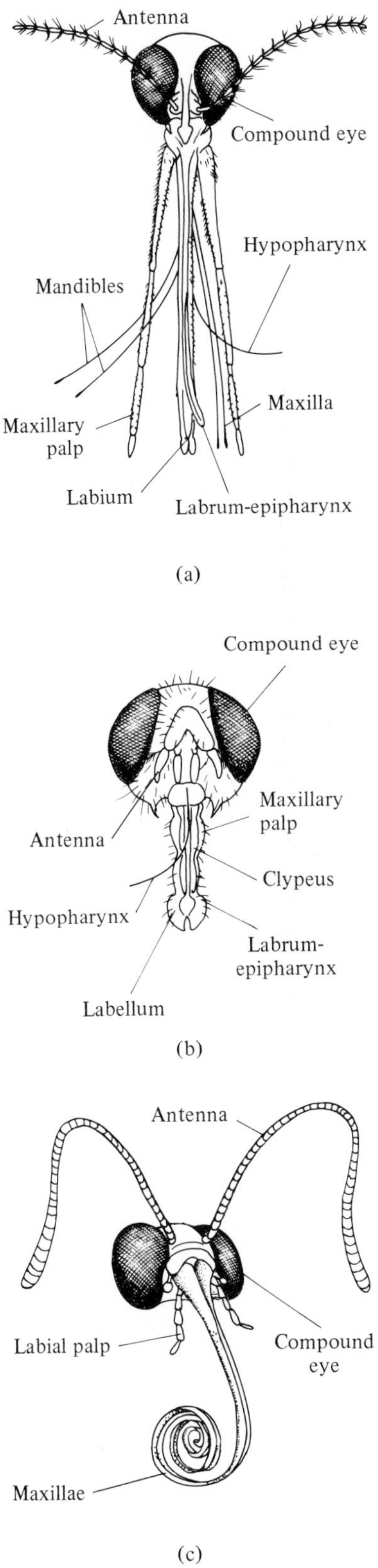


**FIG. 32-10** Mouth parts of three insects adapted for a liquid diet. (a) Mosquito. (b) Housefly. (c) Butterfly.

**FIG. 32-11** Arrangement of central nervous system and digestive and circulatory organs in (a) crayfish and (b) insects.

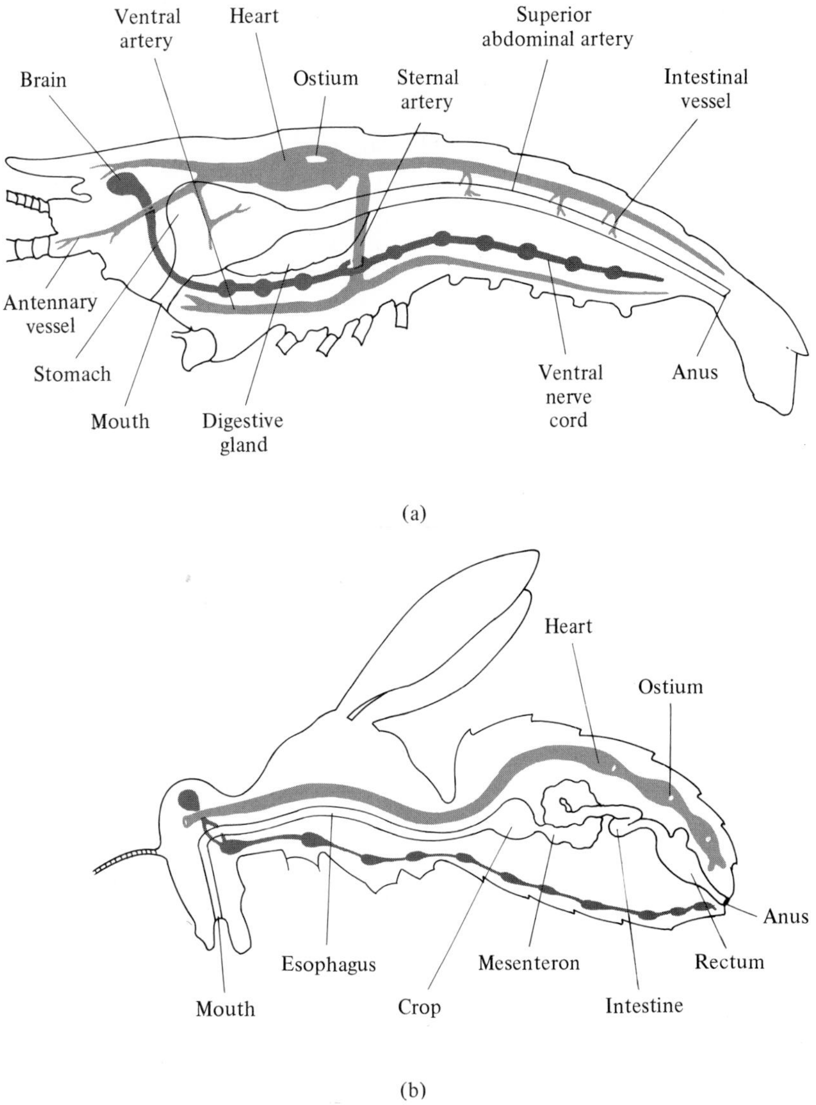

**FIG. 32-12** Nervous system of the crayfish.

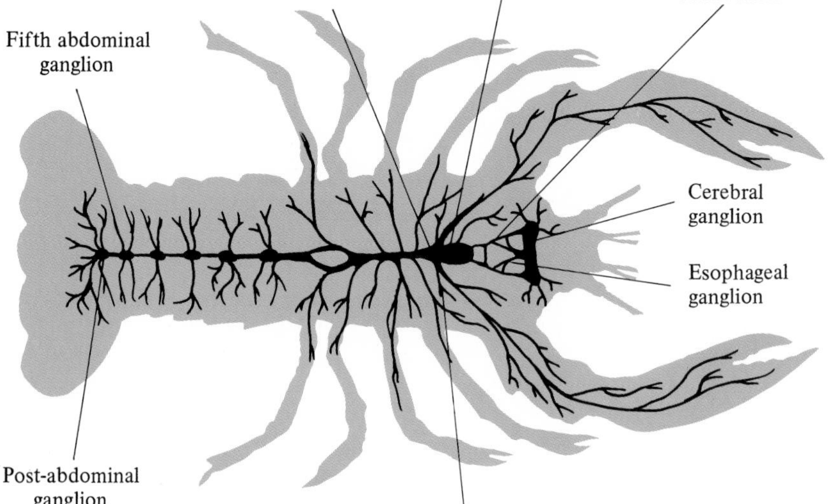


The arthropod nervous system includes a complex dorsal brain and paired or fused ventral ganglionated nerve chains. In primitive forms, there is a ganglion in each segment. But in most arthropods, the ventral ganglia of several groups of neighboring segments are fused, forming composite ganglia. Each set of segmental muscles receives nerve fibers from the ventral ganglion of that segment (Fig. 32-12). The brain is a composite structure formed by the fusion of ganglia from the three most anterior segments. These three ganglia supply nerves to the upper lip (labrum), antennae, and eyes in insects, suggesting that the insect brain evolved by fusion of the segmental ganglia of several anterior segments.

As in the annelids, the brain in arthropods is largely an inhibitory center that either inhibits actions or, by ceasing its inhibition, "permits" an action to occur. Locomotory responses are directly controlled by the segmental ganglia. Some insects can even walk, fly, or mate after decapitation. If a leg is removed from a decapitated insect, the body may also exhibit compensatory behavior for the lack of that leg.

In addition to the brain and the ventral ganglia of the central nervous system, most insects and some other arthropods have a sympathetic nervous system composed of both motor and sensory neurons. These innervate the gut, heart, and reproductive organs.

## DEVELOPMENT AND GROWTH

The presence of a rigid, inflexible cuticle has a profound influence on the developmental pattern of arthropods. Because the exoskeleton prevents growth from the inside, an arthropod must shed, or **molt,** its cuticle periodically to permit growth. The process is under endocrine control and has been intensively studied in insects.

One or more larval stages intervene between the egg and adult in-

(a)

(b)

(c)

(d)

(e)

**FIG. 32-13** Life history of a hemimetabolous insect, the desert locust, showing (a) egg hatching; (b) an early instar nymph; (c) last instar nymph; (d) metamorphic molt; and (e) an adult.

sect. A larva hatches from an egg, feeds, grows, and molts into a larger larva. After several larval molts, during which the old, smaller exoskeleton is replaced by a new, larger one, the insect larva sheds its cuticle and emerges as an adult. These changes in form during development are called **metamorphosis** (see Chapter 25). Insects are classified as (1) **ametabolous,** meaning they do not undergo metamorphosis; (2) **hemimetabolous,** meaning they undergo partial metamorphosis; or (3) **holometabolous,** meaning they exhibit complete metamorphosis.

Ametabolous insects are primitive, wingless species such as silverfish. The young, called **juveniles,** are miniature versions of the adults, differing only in size and by being sexually immature. In hemimetabolous species—such as cockroaches, grasshoppers, and crickets—the larvae, called **nymphs,** bear a close resemblance to the adults, except for the absence of wings and genitalia (Fig. 32-13b). The larvae of holometabolous insects are strikingly different from the adults, or **imagoes,** in their morphology (Fig. 32-14). The larvae may be given special names, such as caterpillars, the larvae of moths and butterflies; maggots, the larvae of flies; and grubs, the larvae of beetles. Larvae increase in size at each molt, with few clues to what the adult form will be like. The primordia of adult organs such as legs and antennae are represented in the larvae as groups of undifferentiated cells called **imaginal disks.** Once a larva has grown to its maximum size, it metamorphoses into a **pupa.** At this molt, the imaginal disks evert from inside the body. As a result, legs, wings, and antennae can be seen on the surface of the pupa in gross outline (Fig. 32-14d). However, the incompletely developed appendages are nonfunctional.

After the pupa is formed, the imaginal structures undergo accelerated development. At the same time, the larval tissues disintegrate, providing a broth of nutrients from which the developing adult structures form. Within the pupal case, a most remarkable transformation occurs: the set of genes that determine the structure and behavior of the larva are suppressed, while the set that specify the structure and function of the adult are activated. The entire process is under hormonal control.

What could be more different than a caterpillar and a butterfly? The two not only differ in morphology but also occupy different habitats, eat different foods, exploit different environments, and in almost every way exhibit different modes of life. Yet they must cooperate by sharing certain vital tasks. In most moths, for example, the larva does the eating and growing, while the adult does the mating and reproducing. It makes sense: eating and growing are best done by a relatively inconspicuous, wormlike organism in out-of-the-way locations that are perhaps safer from predators than is a flying insect.

Besides metamorphosis, some insect species exhibit another kind of polymorphism. For example, in the colonial insects, such as honeybees, ants, and termites, different types of individuals are found in the same colony. There are the reproductive queen and drones (kings), workers, and soldiers, each with a different morphology. During the course of development, the appropriate sets of genes are activated in each case to produce the specific morphology of the individual polymorphic type. All types of individuals in a colony except males bear the same genetic information. The male bees are haploid, while the others are diploid organisms. But in each type, only certain specific sets of genes are functional.

## PHYLUM MOLLUSCA

Another group to evolve from the annelidan stock was Mollusca (from the Latin *mollis,* "soft"), a phylum that includes clams, snails, squids, and octopuses (Fig. 32-15). Over 100,000 living species and over 35,000 fossil forms of molluscs have been described. At first glance, they appear to be a heterogeneous group: clams, for example, bear no structural similarity to squids or snails. But a closer look reveals several common features. For example, most molluscs have a calcareous shell (made of lime), and their bodies are all divisible into three parts: head, foot, and visceral mass. Covering the visceral mass is a soft membrane, the mantle, which usually secretes the shell and extends as flaps beyond the visceral mass, forming a mantle cavity. In most molluscs, the nervous system lacks an organized brain and consists of three pairs of ganglia: cerebral, pedal, and visceral, providing nerves to the head, foot, and visceral mass, respectively. The brain of the octopus, however, is very highly evolved, a fact associated with their muscular movement, complex behavior, and learning ability.

Except for a few species of land snails and slugs, all molluscs are aquatic. Most are bottom dwellers and possess poor locomotory powers. Some species of clams and oysters remain permanently anchored to rocks in shallow seas. Snails, which exhibit greater locomotory ability, are proverbially slow. Evolution of the ability to secrete a protective shell that enclosed the visceral mass and into which the head could be withdrawn probably enabled the ancient molluscs to be highly successful. But the heavy shell limited their locomotory powers and usually restricted them to shallow waters.

**FIG. 32-14** Life history of a holometabolous insect, the monarch butterfly, showing (a) egg hatching; (b) an early instar larva; (c) last instar larva; (d) pupa; (e) an adult emerging from the pupa; and (f) the adult butterfly.

Despite obvious differences from adult annelids, we have several reasons to believe that molluscs are related to them. One is that eggs of both molluscs and annelids undergo spiral, determinate cleavage. In both phyla, the mesoderm arises in the same way, and marine species of both groups have a **trochophore** larva (see Fig. 27-5). The blastopore becomes the mouth in the larvae of both groups. Also, the excretory organs of molluscs and annelids are of similar construction. However, most mollusc and annelid adults bear no resemblance to one another. While all annelids are metamerically segmented, remnants of metamerism are seen in only the most primitive of living molluscs. While the coelom is prominent in annelids, in most molluscs it has been reduced to a small space around the heart.

Molluscs have great commercial importance both as pests and as marketable food. Clams, oysters, and squids form the basis of a multimillion dollar seafood industry. But damage caused by wood-boring molluscs to wharves or by other molluscs to seafood species reaches many millions of dollars annually in the United States alone. Wood-boring molluscs were the bane of sailors when ships were made of wood.

## CLASSIFICATION

Molluscs are now placed in seven classes, as follows:

- *Aplacophora*, or solenogasters, are small wormlike marine forms often found at great depths.
- *Polyplacophora* (formerly Amphineura), or chitons, are oval-shaped, small-headed, marine forms of simple construction, with a composite shell of eight plates.

- *Monoplacophora* include one living genus, *Neopilina,* discovered in 1952. Long thought to be extinct, it is the only mollusc with traces of metamerism.
- *Gastropoda* include snails, slugs, and limpets. These most highly diversified of all molluscs are found in marine, freshwater, and terrestrial environments.
- *Scaphopoda,* or tooth shells have small shells that resemble tusks. They were used as wampum by Indians of the Pacific coast of North America.
- *Bivalvia* (formerly Pelecypoda) include clams and oysters. The shell has two valves, a left and a right, that enclose the prominent foot, the gills, and the visceral mass. Most bivalves are either permanently sedentary, slow in locomotion, or burrowers. Some scallops are active swimmers.
- *Cephalopoda* possess several arms associated with the head; these evolved from the foot. The eyes and brain mark the cephalopods as among the most highly evolved of any invertebrate animals.

### CEPHALOPOD EVOLUTION

The evolution of cephalopods, such as squids and octopuses, spans 550 million years, originating in animals with heavy shells, somewhat like today's giant sea snails. The first step in this evolution was, not to get rid of the shell, but to so increase buoyancy that mobility was easily achieved. This was done by secreting gas into a rear chamber of the shell. Additional growth and increase in shell mass were compensated for by merely adding more gas-filled chambers to the shell. By vigorously forcing water through its siphon, the animal was able to move by a form of jet propulsion. A similar animal survives today: the pearly nautilus (Fig. 32-16a). From those first nautiluslike forms evolved a great variety of shelled cephalopods, including enormous numbers of a now extinct group, the ammonites, which seem to have ruled the seas until fishes appeared. Some ammonites evolved into organisms with reduced shells and then lost them, relying on the speed made possible by jet propulsion and a streamlined body to capture food and escape predators. From these ammonites evolved the modern squids, cuttlefish, and octopuses (Fig. 32-16b). Squids and cuttlefishes still retain an internal remnant of a shell—called a **pen** and **cuttlebone,** respectively—but in octopuses, all vestiges of the shell are gone. Except for nautiloids, all shelled cephalopods are extinct.

The mobility of squids is remarkable. By directing their funnel-shaped siphon forward or backward, they can swim rapidly in either direction, reaching high speed in an instant and stopping just as quickly. Like flying fish, squids frequently become airborne and have been clocked at air speeds of 26 km per hour (16 miles per hour) and at heights of 3.5 m (12 feet) above the sea's surface.

## *II. DEUTEROSTOMIA*

Besides a few minor phyla, deuterostomes include two major phyla, the Echinodermata and the Chordata, the phylum to which we belong. One minor phylum, the Hemichordata, represents an evolutionary link between the chordates and echinoderms.

(a) (b) (c) (d) (e)

**FIG. 32-15** Representatives of the principal classes of molluscs. (a) Dentalium shells (Scaphopoda). (b) Chiton (Polyplacophora). (c) Triton (Gastropoda). (d) Scallop (Bivalvia). (e) Squid (Cephalopoda).

# PHYLUM ECHINODERMATA

Echinoderms (from the Greek *echinos,* "hedgehog," and *dermis,* "skin") are characterized by several unique features: **pentamerous,** or five-sided, radial symmetry as adults; a calcareous skeletal system; an outer surface covered with spines; and a **water vascular system** including tube feet, or **podia,** which function in locomotion, capture of food, and attachment. All species are marine (Fig. 32-17). Of all phyla, Echinodermata exhibit the least diversity in morphology and mode of development. Although pentaradial in symmetry, echinoderms appear to have evolved from bilaterally symmetrical ancestors. One indication of such an evolution is that echinoderm larvae are bilaterally symmetrical. Although the echinoderm ancestry is not clear, their relationship to the chordates is based on their mode of coelom formation.

Living species of echinoderms are placed in five classes, as follows:

- *Crinoidea* include stalked, sessile forms, such as the sea lilies, which are permanently attached to a substratum, and the motile feather stars. Most crinoids are known only as fossils.

FIG. 32-16 Two cephalopods: (a) pearly nautilus and (b) octopus.

- *Holothuroidea,* or sea cucumbers, have bodies that resemble a cucumber and a mouth surrounded by tentacles, which are modified podia. Some holothuroids are able to swim and others are adapted for a planktonic (floating) life.
- *Echinoidea* includes sea urchins and sand dollars, spherically or radially symmetrical animals with movable spines.
- *Asteroidea* consists of the sea stars, or starfishes.
- *Ophiuroidea* contains the brittle stars and serpent stars.

In most echinoderms, the oral surface is directed downward when the animal is in the normal orientation. The exceptions are the crinoids, in which it faces upward, or sideways, when there is a current and the holothuroids, in which it faces to the side. In echinoderms with arms, the number of arms is always five or a multiple of five up to 50. The bodies of most echinoderms are covered with cilia, which, in many species, aid in feeding, often with the help of mucous secretions. Echinoids have a complex masticatory apparatus, called **Aristotle's lantern,** which grinds up the food before it is ingested. They eat anything organic.

Asteroids are usually carnivorous. Clams and oysters are their major food source and thus they are a bane to oyster beds. Their stomach communicates with side branches, the **pyloric cecae,** in which the final stages of digestion and most absorption occur. Although some sea stars swallow prey whole, those with flexible arms grasp the prey and evert their own stomach, which either engulfs the prey or digests it while the stomach is still outside the sea star. In the latter case, digestion products are drawn into the pyloric cecae by ciliary action. Ophiuroids, which lack pyloric cecae, feed by a variety of methods, sometimes involving ciliary and podial action.

## REPRODUCTION AND DEVELOPMENT

Nearly all echinoderms are dioecious, and eggs and sperm are usually shed directly into the water, where fertilization occurs. The embryo hatches as a larva which is bilaterally symmetrical. Two features of embryonic development need special mention. The blastopore in the echinoderm embryos marks the site of the anus in the larva and in the adult, one reason that echinoderms are included in deuterostomes. Furthermore, the coelom develops as an outpocketing of the primitive gut, hence the echinoderms are also called enterocoels.

Although larval types of echinoderms vary widely, the **bipinnaria** larva of sea stars closely resembles the **tornaria** larva of the acorn worm, a hemichordate. A tornaria has an additional ciliated band not present in a bipinnaria. In addition, the coelom arises in tornaria and bipinnaria larvae in a similar manner. In the bipinnaria, it arises as a single pouch which later subdivides, while in the tornaria it arises as two paired and one unpaired pouch. These similarities strongly suggest that the two groups had a relatively recent common ancestor.

**FIG. 32-17** Representatives of the five classes of Echinoderms. (a) Crinoidea, feather star. (b) Holothuroidea, sea cucumber. (c) Echinoidea, sea urchin. (d) Asteroidea, starfish. (e) Ophiuroidea, brittle star.

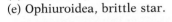

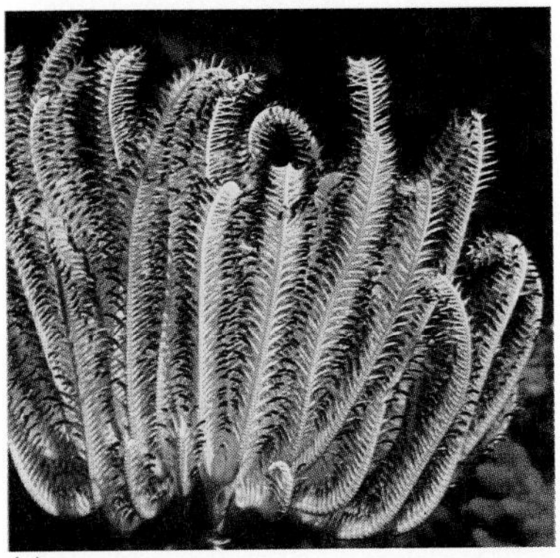

(a)

## PHYLUM HEMICHORDATA

At one time, the hemichordates (from the Greek *hemi,* "half," and *chorda,* "gut" or "cord") were included in the phylum Chordata, a situation that continually provoked arguments among biologists. Of the two classes in this small phylum, the most advanced are members of the class Enteropneusta, the acorn worms, a name suggested by the shape of their prominent proboscis. Two other parts, the collar and trunk, comprise the rest of the body (Fig. 32-18). The pharyngeal region of the trunk has paired slits similar to the pharyngeal gills of chordates with

(b)

(c)

(d)

(e)

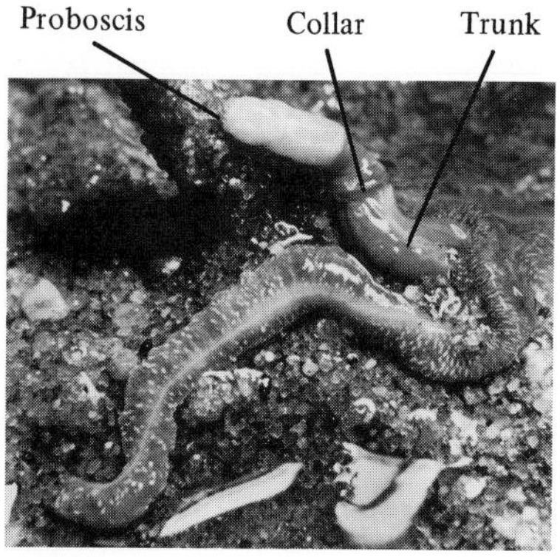


**FIG. 32-18** Acorn worm, an enteropneust.

gills. The nervous system includes a dorsal, longitudinal nerve cord, which, in some acorn worms, is tubular in the region of the collar. An outpocketing of the mouth cavity was once described as a short notochord (hence the name *hemichordate*) but is no longer regarded as homologous to the notochord.

The close similarity of acorn worms to lower chordates strongly suggests that they share a common ancestry. The close similarity of the acorn worm's larva to that of the sea star, on the other hand, argues for a common ancestry with that group as well. Our interest in this small group of worms rests almost entirely on these probable kinships.

## PHYLUM CHORDATA

All members of the phylum Chordata possess a dorsal, tubular nerve cord; segmental muscles along the trunk; and, at least sometime in their lives, two other structural features: pharyngeal gill pouches and furrows; and a stiff, dorsal rod, the **notochord.** In fishes and lower chordates, the gill pouches and furrows of the embryo develop into the gills of the adult. In all vertebrates, the notochord is replaced, or partially replaced, during development by a vertebral column of cartilage or bone. In some protochordates, which include ascidians, or sea squirts, and Amphioxus, the notochord remains throughout life.

The chordates are divided into three subphyla: Urochordata and Cephalochordata, which are together grouped under Protochordata; and Vertebrata.

### SUBPHYLUM UROCHORDATA

The tunicates, or Urochordata, include 2,000 species, most of which possess all three chordate characteristics—pharyngeal gills; notochord; and dorsal, tubular nerve cord—as larvae but many of them lose the notochord and dorsal nerve cord during metamorphosis into adulthood (Fig. 32-19). All tunicates are marine; some are colonial. Many of them exhibit asexual reproduction, which is extremely uncommon among chordates.

### SUBPHYLUM CEPHALOCHORDATA

A small group (30 species) of marine protochordates, the members of the subphylum Cephalochordata are usually known as Amphioxus. They have a fishlike shape and segmental muscles similar to those of fish and retain the three main chordate characteristics throughout life (Fig. 32-19). They live partially buried in the sea floor, filter feeding with the help of a unique mucus-secreting organ in the pharynx, the **endostyle.** Water, bearing nutrients and oxygen, enters through the mouth and escapes through the gills of the pharynx. The food particles are first trapped in a mucus rope secreted by the endostyle and then swallowed. The excretory organs of amphioxus were long believed to be protonephridia with solenocytes. But it is now known that they bear **podocytes,** a type of cell characteristic of vertebrate kidneys, providing reason to

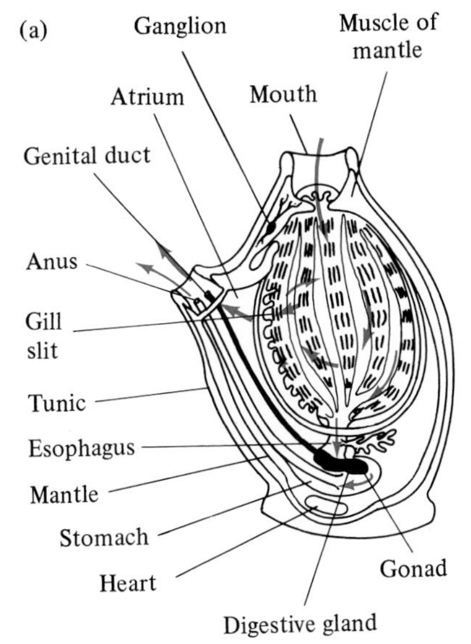


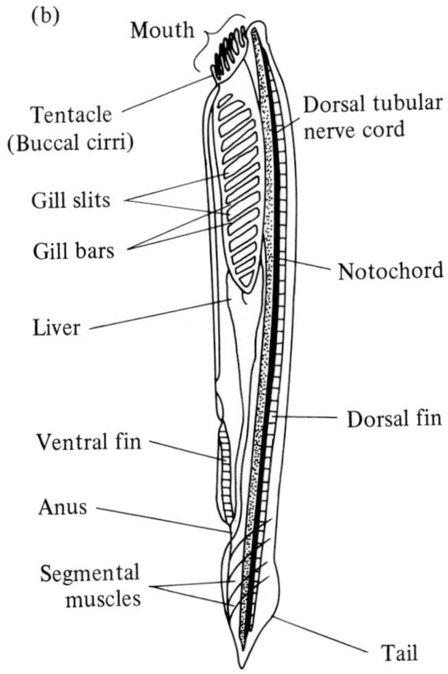


**FIG. 32-19** Protochordates. (a) Sea squirt, a urochordate. (b) Amphioxus, a cephalochordate.

believe that Amphioxus and its relatives are not far from the ancestral line of the vertebrates. In fact, early systematists quipped that if amphioxus had never existed, it would have been hypothesized as the ancestral vertebrate.

## SUBPHYLUM VERTEBRATA

The subphylum Vertebrata includes all chordates with **vertebrae,** or backbones, of cartilage or bone. Vertebrae provide a skeletal support for the attachment of the segmental muscles of the trunk, thereby making possible highly efficient locomotion. Superior locomotory abilities, more than any other characteristic, aided the vertebrates in escaping predators and capturing prey. Associated with rapid and efficient locomotion, the primitive vertebrates developed an efficient circulatory system including a muscular, ventrally placed heart. Their specialized, hemoglobin-bearing red blood cells made their circulatory system the most efficient system of all animals. Given all these advantages, it is not surprising that the early vertebrates, with their cartilaginous skeletons, rapidly diversified. Vertebrates probably arose in fresh waters during the Ordovician Period (about 500 million years ago). These were the jawless, or agnathan fishes, the evolutionary predecessors of the bony fishes, osteichthyes, a more advanced group that came to dominate not only fresh water, but the seas as well. From osteichthyes evolved all the higher vertebrates.

The vertebrates are divided into eight classes: Agnatha, Placodermi, Chondrichthyes, Osteichthyes, Amphibia, Reptilia, Aves, and Mammalia (Fig. 32-20). Seven of the eight have modern representatives, while the placoderms all became extinct very early in the history of vertebrate evolution. All except Agnatha have jaws with skeletal elements and are included in the superclass Gnathostomata.

***Agnatha.*** The most primitive of living vertebrates, members of the class Agnatha, are jawless fishes. Although agnathans, also known as cyclostomes, were very abundant in ancient seas, today they are represented by only two groups: the hagfishes and the lampreys (Fig. 32-21). The lamprey *Petromyzon* is a parasite on other fishes and has played havoc with some commercial fisheries. Like all chordates, lampreys and hagfishes have a dorsal tubular nerve cord, segmental muscles, pharyngeal gill pouches, and a notochord, the latter being partly replaced by cartilaginous vertebrae in the adult. The agnathan circulatory system, consisting of a ventral, two-chambered, muscular heart, and an excretory system made up of kidney tubules, resembles the vertebrate plan. But unlike all other vertebrate classes, agnathans lack jaws; their mouth is a circular opening that functions as a sucker. Lampreys attach themselves to their hosts by their suctional mouth, which contains a rasping organ that scrapes the host tissues.

Although the adult lamprey resembles a fish in its external appearance, the lamprey larva, known as an **ammocetes larva,** (Fig. 32-22) closely resembles the adult amphioxus in all major features. This suggests a phylogenetic relationship between the cephalochordates and these most primitive living vertebrates. However, the fossil record shows that the modern lampreys were preceded by a now extinct group of jawless fishes, the ostracoderms, covered by bony plates. The exact relationships among lampreys, ostracoderms, and cephalochordates are uncertain.

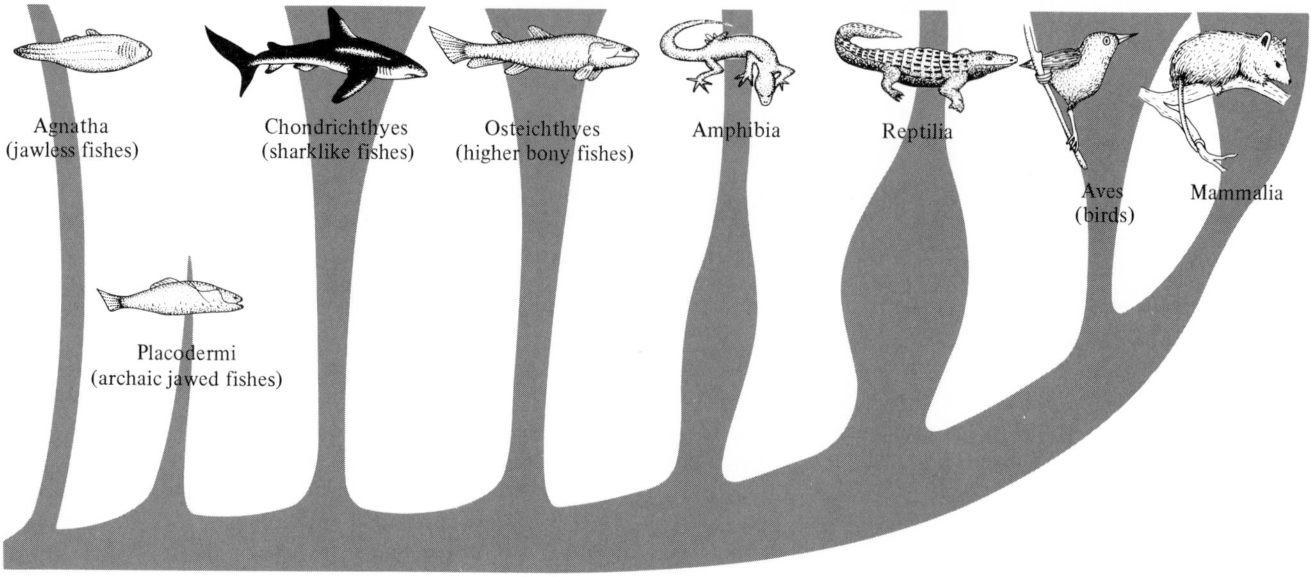


FIG. 32-20 General classification of the vertebrata and their probable phylogenetic relationships.

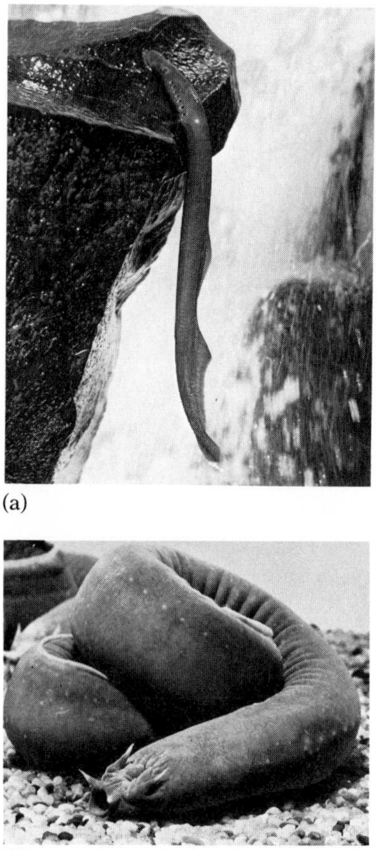

(a)

(b)

FIG. 32-21 Agnatha (Cyclostomes). (a) Petromyzon (lamprey). (b) Myxine.

***Placodermi.*** The placoderms, the first fishes to have paired fins and jaws, were covered with bony plates. Although their jaws and fins were primitive, these organs enabled the placoderms to be more successful than their jawless ancestors. After achieving great diversity, however, all placoderms became extinct, but apparently not before some of them lost their bony armor and gave rise to two other classes, the Chondrichthyes (from the Greek *chondros,* "cartilage," and *ichthyes,* "fish"), sharklike fishes; and the Osteichthyes (from the Greek *osteon,* "bone," and *ichthyes,* "fish"), fishes with bony skeletons.

***Chondrichthyes.*** The class Chondrichthyes consists of fishes with cartilaginous skeletons. It is represented today by sharks, skates, and rays, nearly all of which are denizens of the seas (Fig. 32-23). The body organization of sharks and other Chondrichthyes exhibits an improvement over that of agnathans by the presence of jaws and over that of the placoderms in that the jaws are attached to the skull by other skeletal elements. This configuration has resulted in the powerful jaws of sharks.

An interesting evolutionary progression is associated with the origin of jaws. In agnathan fishes, seven pairs of pharyngeal pouches arise in the embryo, and the furrows between them develop into the gill slits, each supported by a cartilaginous gill arch, together known as the branchial skeleton (Fig. 32-24). Placoderms have only six gill arches; what in agnathans was the first pair has been converted to the skeletal framework of the jaw. Sharks and other chondrichthyes have five pairs of gill arches and a small opening, the **spiracle,** which is a rudimentary, nonfunctional gill. In sharks, what was the second pair of gill arches in agnathans has become jaw supports.

Because sharks have retained their primitive cartilaginous skeletons, they are regarded as a side branch of vertebrate evolution. Although, unlike osteichthyes, Chondrichthyes did not evolve into a more advanced group of vertebrates, they embarked on an adaptive radiation that led to the evolution of active swimmers (sharks), bottom feeders (skates), and intermediate types (rays).

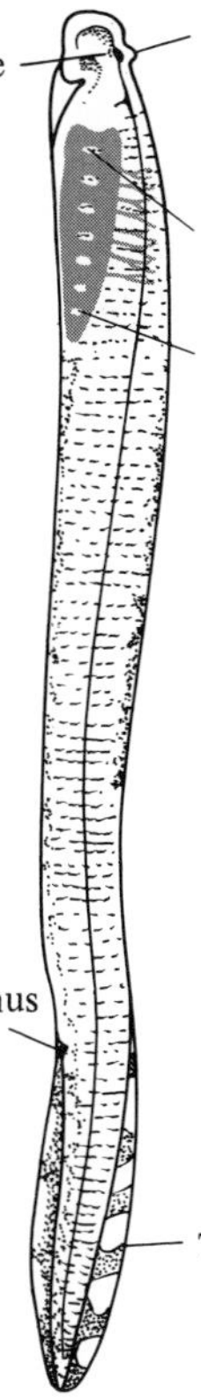

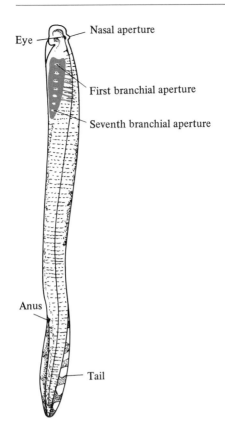

**FIG. 32-22** Ammocetes larva of lamprey. Note similarity in appearance to *Amphioxus*.

***Osteichthyes.*** The vast majority of the present-day fishes are included in the class Osteichthyes (Fig. 32-25). As the name suggests, these fishes have a bony skeleton. They also have an operculum, a protective covering of the gill openings. With a few notable exceptions, all bony fishes are gill breathing, aquatic vertebrates. They are found both in fresh water and the sea. As a class, Osteichthyes is the most successful of all vertebrates. One subclass, Teleostei, to which the salmon, perch, herring, and most other game and food fish belong, includes over 30,000 species. However, because of the selection pressure exerted by their watery medium, the general organization in Osteichthyes, and in fact in most fishes, varies little. Being adapted for swimming, they all have a streamlined body that ends in a powerful **caudal fin,** or tail fin. All possess two sets of paired fins, pectoral and pelvic. They also have unpaired dorsal and anal fins, on the middorsal and midventral lines of the body, respectively.

Despite this common general body plan, some modifications in morphology do occur. For example, in the flounder, which is adapted for living on the sea floor, both eyes are on the same side of the body. Eels have long slender snakelike bodies. Sea horses derive their name from the remarkable resemblance of their body to that of a horse. The swordfish bears an elongate upper jaw that serves as an organ of offense and defense. The flying fish has highly enlarged pectoral fins that function as gliding devices when it leaps out of the water.

Osteichthyes includes one interesting group, the Dipnoi, which is represented today by three genera, one each in Australia, South Africa, and South America. These are fresh water fishes with well developed lungs. Although they have functional gills, dipnoans literally drown if submerged under water for long periods. They come to the surface of the water for a gulp of fresh air every so often to keep their hemoglobin in the oxygenated condition.

Osteichthyes also includes another very interesting group: the coelacanth crossopterygians. Several specimens of the genus *Latimeria,* once thought to be extinct for 70 million years, were caught off the coast of South Africa in 1939. These fishes are important for an understanding of the evolution of the land vertebrates (Fig. 32-25d). Crossopterygians bear a stalklike structure at the base of each of the paired fins. The arrangement of the bones in this lobed fin is very similar to that of primitive tetrapod limbs, suggesting that fishes with limbs similar to those of crossopterygians were ancestral to amphibians and other tetrapods. However, because coelacanth crossopterygian skull bones

**FIG. 32-23** Representative cartilaginous fishes. (a) Lemon shark. (b) Eagle ray. (c) Texas skate.

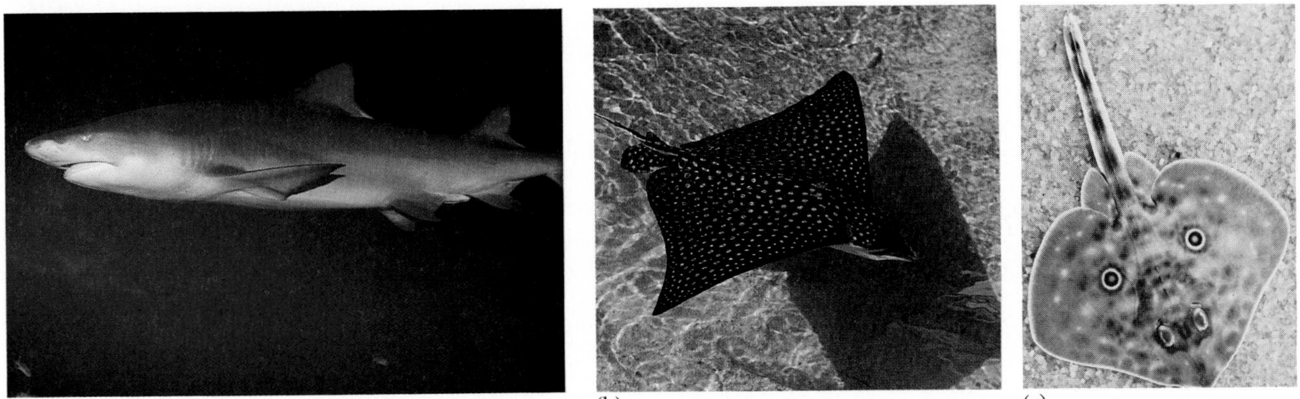

(a) (b) (c)

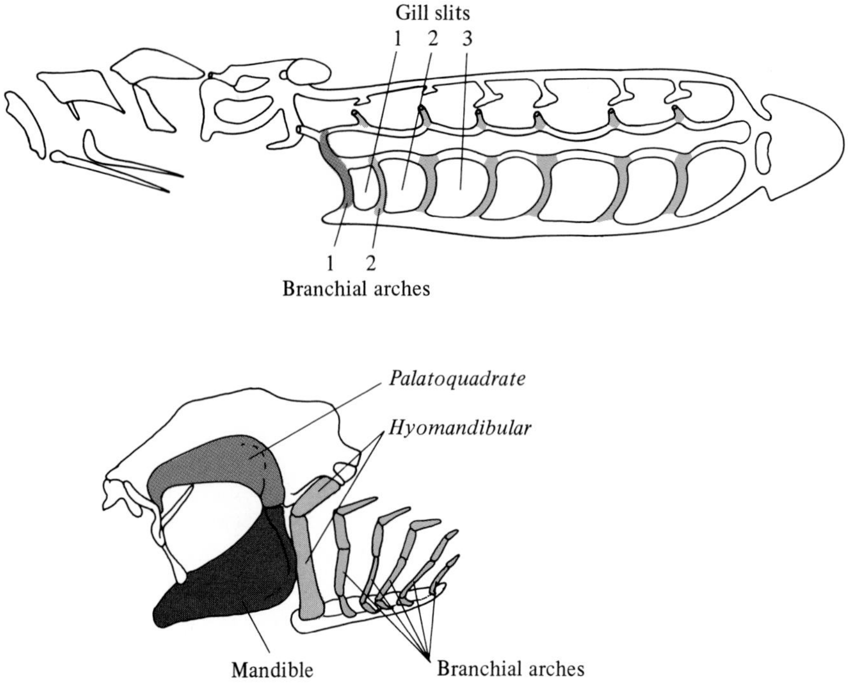


**FIG. 32-24** Branchial skeleton in (a) agnathan fishes (lamprey) and (b) cartilaginous fishes (shark). Note the arrangement of gill arches, gill slits, and jaws shown in the diagrammatic sketches.

are so different from those of the early amphibians, we know that these fishes could not have been their direct ancestors.

***Amphibia.*** The class Amphibia includes the Anura (from the Latin *an*, "without," and *ura*, "tail"), consisting of frogs and toads; the Urodela, made up of salamanders and newts; and the Apoda, or coecilians, a group of blind, limbless burrowing relatives of the urodeles (Fig. 32-26). Amphibians represent a transitional state between the aquatic fishes and the truly terrestrial land vertebrates. Like fishes, most amphibians lay their eggs in water, where the embryo develops into a fishlike larva, the **tadpole.** The larva has functional gills, fins, and various physiological adaptations for aquatic life. The larva metamorphoses into a tetrapod, air-breathing, terrestrial animal. The amphibian skin, however, is not protected against drying and needs to be kept moist. More important, amphibian eggs, which are not protected by shells, cannot survive on land. Their development is possible only in water or in very moist environments. Although amphibians forage on land for brief periods, they must return to water or moist shelters to prevent desiccation and for breeding. As representatives of an evolutionary transition, amphibians are not adapted perfectly for either aquatic or terrestrial life. Consequently, any diversity they exhibit reflects this fact.

Although only three separate groups of amphibians have survived to modern times, many variations of the general amphibian theme were evolved in the past. One group of ancient amphibians evolved a shelled egg that could survive on land. This group later evolved a dry skin that was resistant to desiccation, and it eventually gave rise to the first truly terrestrial vertebrates: the reptiles, from which birds and mammals arose.

***Reptilia.*** One important reptilian innovation is the amniote egg, with its extraembryonic membranes: amnion, yolk sac, chorion, and allantois. The amnion provides a private artificial pond in which the embryo

(a)

(b)

(c)

(d)

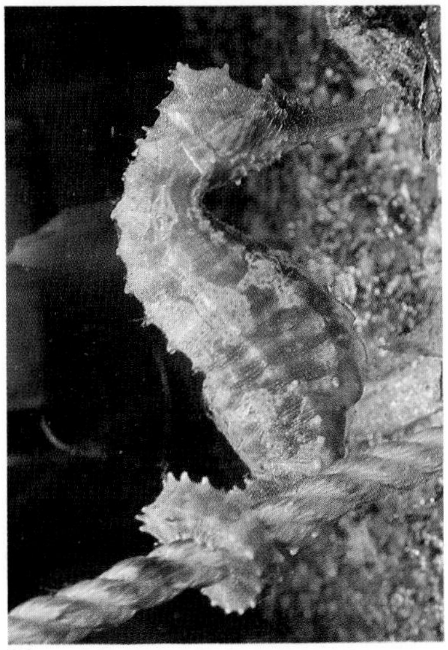
(e)

**FIG. 32-25** A variety of bony fishes. (a) Salmon. (b) Moray eel. (c) Lung fish *(Neoceratodus).* (d) Coelacanth. (e) Sea horse.

develops. This form of development, together with a scaly skin that resists desiccation, enabled the reptiles to be truly terrestrial. Other improvements evolved in the circulatory, respiratory, reproductive, and excretory systems as well.

When the reptiles first appeared, in the late Paleozoic era, they had no important rivals on land. In the absence of competition and with abundant terrestrial plant and animal life as sources of food, the reptiles diversified rapidly. Reptiles of the Mesozoic era ruled the land and air and even reentered the seas. The plesiosaurs and ichthyosaurs were aquatic reptiles with streamlined bodies, somewhat reminiscent of modern whales. Some pterosaurs had enormous wingspans and were probably the only vertebrate fliers of those times. During the entire Mesozoic era, which lasted about 90 million years, reptiles were the masters of the land. That era has been rightfully called the golden age of reptiles. Several large reptiles, commonly known as **dinosaurs** lived during this period (Fig. 32-27).

Today, reptiles are a relatively impoverished group, including only the lizards, crocodiles, snakes, turtles, and an aberrant form, *Sphenodon* (Fig. 32-28). Commonly known as the tuatara, the lizardlike *Sphenodon* is found only in New Zealand. Besides exhibiting several amphibianlike characteristics, this reptile has a "third eye", called a **pineal eye,** on top of its head.

***Aves.*** Birds, class Aves, may be defined from a taxonomic point of view as homeothermic, feathered, bipedal vertebrates in which the forelimbs are usually modified into wings (Fig. 32-29). In addition, all modern

(a)

(b)

**FIG. 32-26** Representative amphibians. (a) Spotted salamander *(Ambystoma maculatum)*. (b) Coecilian or blind worm *(Ichthyophis glutinosa)*.

birds bear toothless jaws. There are two major subclasses of birds: the runners, or nonflying birds (Ratitae); and the flying birds (Carinatae). Systematists, scientists who study diversity and kinships among organisms, have referred to birds as glorified reptiles because their skeleton is quite similar to that of lizardlike reptiles. Birds have also retained reptilian scales on their feet and legs and lay reptilian type eggs. Of course, they exhibit many adaptations for aerial life. The bones of flying birds contain air spaces that make them very light. As noted in Chapter 11, their unique heart is completely divided into left and right halves. And, as in reptiles, the right aortic arch conducts blood to the body. (In mammals, the left arch serves this function.) Although all modern birds are toothless, the skeletal elements of the jaws of all birds and the lizardlike reptiles are identical.

The body organization of various species of birds provides numerous examples of adaptations to special modes of life. Although many constraints have been placed on the evolution of body shape due to the fact that birds are flying machines, their beaks and feet exhibit great variation. For example, many birds of prey, such as eagles and owls, bear talons and have curved beaks (Fig. 32-29b,c). Water birds, such as ducks, geese, and cormorants, have webbed feet and flat beaks (Fig. 32-30). Many other modifications are adaptations for catching and eating flying insects, burrowing insects, worms, shelled molluscs, cereal grains, seeds, fruits, and so on. Penguins have even adapted to a true aquatic life, losing the power to fly in the process.

Flying requires efficient organization, and, in their physiology, birds are at least as efficient as the most highly evolved mammals. Yet, in body form, they do not exhibit the same degree of adaptive radiation as do mammals. Fishes are subject to a similar constraint, one imposed by an aquatic life. Reptiles, on the other hand, are primarily land dwellers and exhibit great variation in body shape.

***Mammalia.*** The class of vertebrates to which humans belong first made its appearance on earth in the mid-Mesozoic era, about 190 million years ago, when reptiles ruled the world. At that time, one group of mammallike reptiles, the therapsids, arose, flourished for a time, then

**FIG. 32-27** Reconstructions of different dinosaurs. (a) Tylosaurus and protostega. (b) Dimetrodon and edaphosaurus.

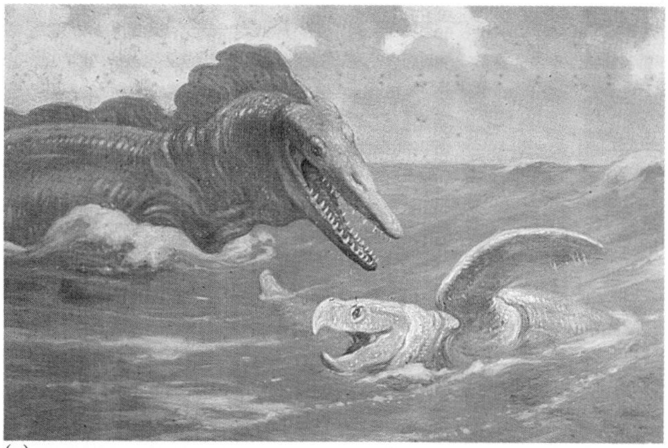
(a)

(b)

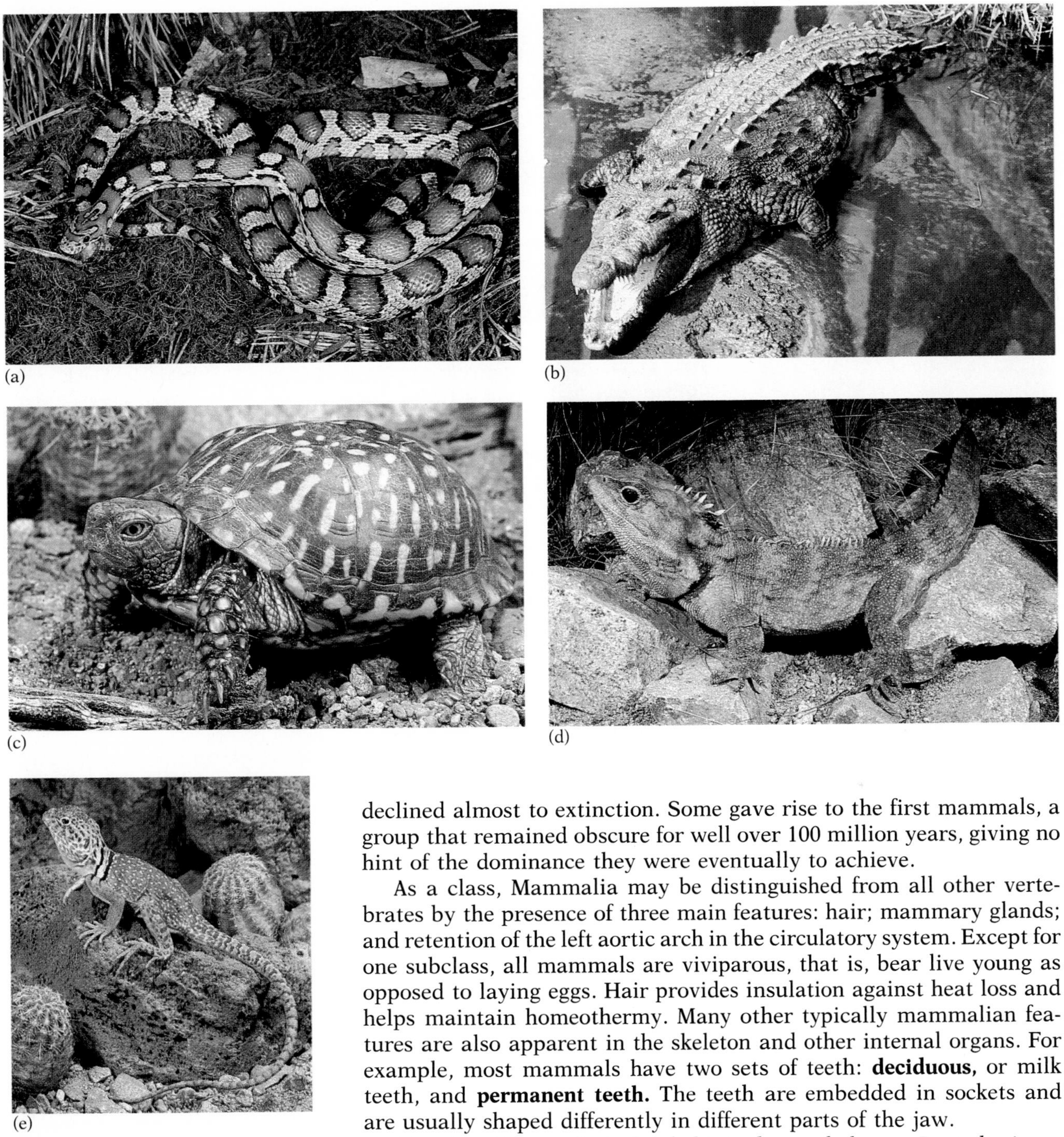

FIG. 32-28 Representative reptiles. (a) Snake. (b) Crocodile. (c) Turtle. (d) *Sphenodon*. (e) Lizard.

declined almost to extinction. Some gave rise to the first mammals, a group that remained obscure for well over 100 million years, giving no hint of the dominance they were eventually to achieve.

As a class, Mammalia may be distinguished from all other vertebrates by the presence of three main features: hair; mammary glands; and retention of the left aortic arch in the circulatory system. Except for one subclass, all mammals are viviparous, that is, bear live young as opposed to laying eggs. Hair provides insulation against heat loss and helps maintain homeothermy. Many other typically mammalian features are also apparent in the skeleton and other internal organs. For example, most mammals have two sets of teeth: **deciduous,** or milk teeth, and **permanent teeth.** The teeth are embedded in sockets and are usually shaped differently in different parts of the jaw.

The class Mammalia is divided into three subclasses: Prototheria, or Monotremata; Metatheria, or Marsupialia; and Eutheria, or placental mammals. The classification is based on reproductive processes. The prototherians are egg layers. The metatherians have only a transient yolk sac placenta. The eutherians have a functional true placenta through which the developing embryo is nourished by the mother.

SUBCLASS PROTOTHERIA: MONOTREMES. Although highly specialized, today's most primitive mammals give us at least some idea of how mammals might have arisen from reptiles. These primitive mammals are the monotremes, subclass Prototheria. Perhaps the most interesting of

FIG. 32-29 Representative members of a few orders of birds. (a) Duck. (b) Barn owl. (c) Hawk. (d) Toucan.

these, as far as evolution is concerned, is *Ornithorhynchus* (Greek for "bird bill"), usually known as the duckbill platypus. It and the echidna, or spiny anteater, *Tachyglossus* (Greek for "swift tongue"), are the only genera in this mammalian subclass (Fig. 32-31). In both these mammals, as in their reptilian ancestors, the terminal gut segment is a urogenital-enteric cavity, the cloaca, and both lay eggs. Yet both have hair and care for their young, nourishing them with milk. Platypus eggs are laid in a nest and tended until hatching, whereupon the very immature young are provided an extended period of maternal care, a practice generally true of mammals. Although echidnas have much in common with platypuses, they differ in what can be regarded a significant way: newly laid echidna eggs, which have moist, sticky shells, are immediately placed in a temporary pouch on the mother's body. Within 10 days, they hatch and are thereupon provided with milk from the nearby mammary glands. They develop within the pouch for another 7 weeks before being ejected. Their mother continues to nurse them for several more weeks, however, before they are weaned. Use of the mother's belly pouch as an accommodation for the immature echidna hatchlings is carried a step further in the marsupials, a much more successful group than the monotremes ever were.

SUBCLASS METATHERIA: MARSUPIALS. The great reptiles faded into extinction 65 million years ago, after over 180 million years of existence and 90 million years of dominance. Thereupon, within only a few hundred thousand years, the small, squirrel-sized mammals that had lived for so long in the shadow of the great reptiles underwent an explosive evolution, filling nearly all the ecological niches left vacant by departed reptiles. Among the most successful was a group that contains nearly all Australian mammals: the marsupials (Fig. 32-32). With only one exception, the opossums of North and South America, today's marsupials (and monotremes as well) are found only in Australia and nearby islands. Fossil records indicate that the marsupials once flourished and diversified in what is now South America, where they seem to have been abundant until a land bridge with North America was formed. The land bridge made possible an invasion of an even more highly evolved group: the placental mammals. Except for the persistent opossums, some of which later invaded North America, the competition with the placental mammals was too much for the marsupials. They became extinct every-

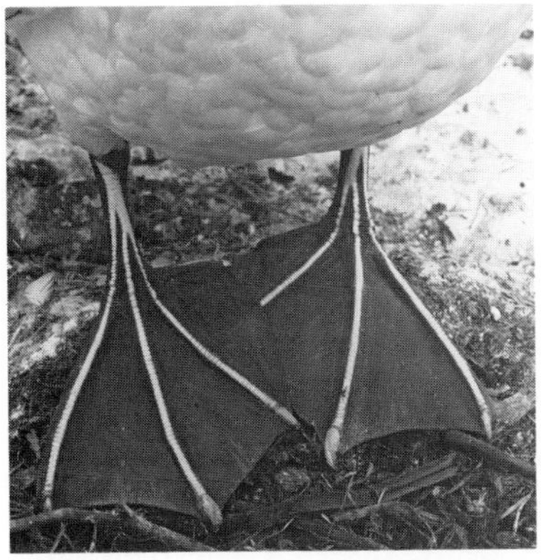

(a)

(b)

(c)

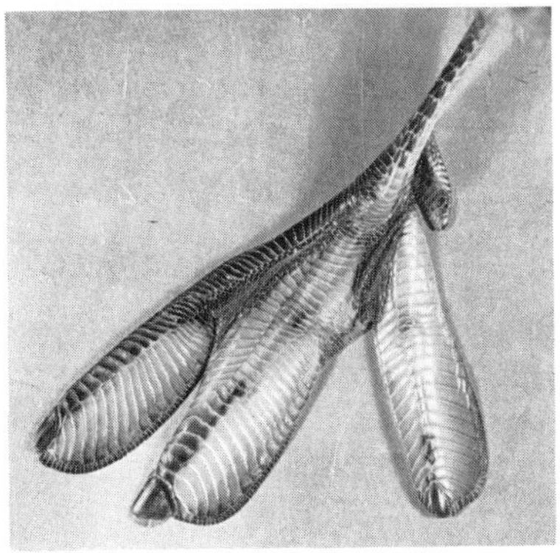

(d)

**FIG. 32-30** Adaptive modifications in birds. (a) Webbed (gannet), for swimming. (b) Two-toed (ostrich), for running. (c) Basic type (grouse), for a terrestrial existence. (d) Lobed toes (grebe), for wading.

where except in a huge land mass that had earlier detached and drifted north and east into the Pacific basin. There its marsupials and monotremes remained isolated from all competition from placental mammals—until humans arrived. That land mass is Australia.

Here marsupials diversified. For example, marsupial carnivores, such as the marsupial wolf, or tiger *(Thylacinus)*, and the Tasmanian devil *(Sarcophilus)*, have powerful jaws, sharp teeth (usually modified molars), and claws much like those of modern wolves and tigers. Similarly, the marsupial anteater *(Myrmecobius)*, like the modern anteater, has a long muzzle. The Australian bandicoot *(Perameles)* is a marsupial rabbit resembling the true rabbit in appearance and habits. It has large ears and digs burrows, in which it lives. There are many such superficial similarities between marsupial and comparable eutherian species. For example, marsupial badgers (wombats), bears (koalas), and moles *(Notorcytes)* resemble placental counterparts in appearance and habits. The environment to which the placentals are adapted is similar, as are their prey or food and their predators. Thus, it is not surprising that similar traits were favored by selection during their evolution.

SUBCLASS EUTHERIA: PLACENTAL MAMMALS. In the eutherians, the most successful and most highly evolved of the mammals, the embryo undergoes an extensive period of gestation within the uterus, nourished and otherwise provided for by the placenta, an indirect connection with the mother's body (Fig. 32-33). As the great reptiles underwent extinction, the mammals, which had for so long remained in the background during the Mesozoic era, underwent a rapid diversification into about two dozen orders. In time, they became adapted to every major type of habitat.

Eutherian mammals constitute a classic example of adaptive radiation. Most of the adaptations involve changes in the limbs and locomotor apparatus. Their types of diet are reflected in various specializations of their teeth. The basic skeletal plan, however much modified, is the same in all placentals.

Small mammals that scurry away from danger, such as rats and mice, lack specializations in their limbs. In ungulates, herbivorous animals with hooves—including cattle, deer, camels, horses, and donkeys—the limbs are adapted for running. In this type of limb, known as the cursorial type, the foot and hand parts of the limbs have lengthened into long, rodlike bones providing an extra joint. The number of digits is decreased from the standard (five digit) pattern, and in the most efficient of runners, the horses, each limb has only one digit, equivalent to the middle finger or toe. Ungulates thus walk or run on tiptoe. The ungulate's long legs and hooves are an adaptation to a grassland environment, where a major defense against predators is speed. However, given an environment with an abundant supply of food, some of the ungulates and their relatives have evolved to a larger size and use size as a defense. Such animals include the elephant, rhinoceros, and hippopotamus, which have large limbs to carry their enormous bodies.

The limbs of moles and anteaters, which bear large claws, are highly adapted to digging. Many of these animals have sicklelike claws. However, the most dramatic adaptation of a mammalian limb to a particular environment is seen in whales, whose forelimbs are modified to a pair of paddles. The probable stages through which whale paddles evolved can be perceived from an examination of the forelimbs of otters, sea lions, dugongs (sirenians), seals, and dolphins. Hind limbs in whales are not of survival value and are represented by a vestigial pelvic girdle at most.

In bats and other chiropterans, the forelimbs are modified for flight. The hind limbs are adapted for hanging upside down. The modifications are so extreme that the limbs are not useful for terrestrial locomotion.

**FIG. 32-31** Egg laying mammals, or Prototheria. (a) Echidna. (b) Duckbill platypus.

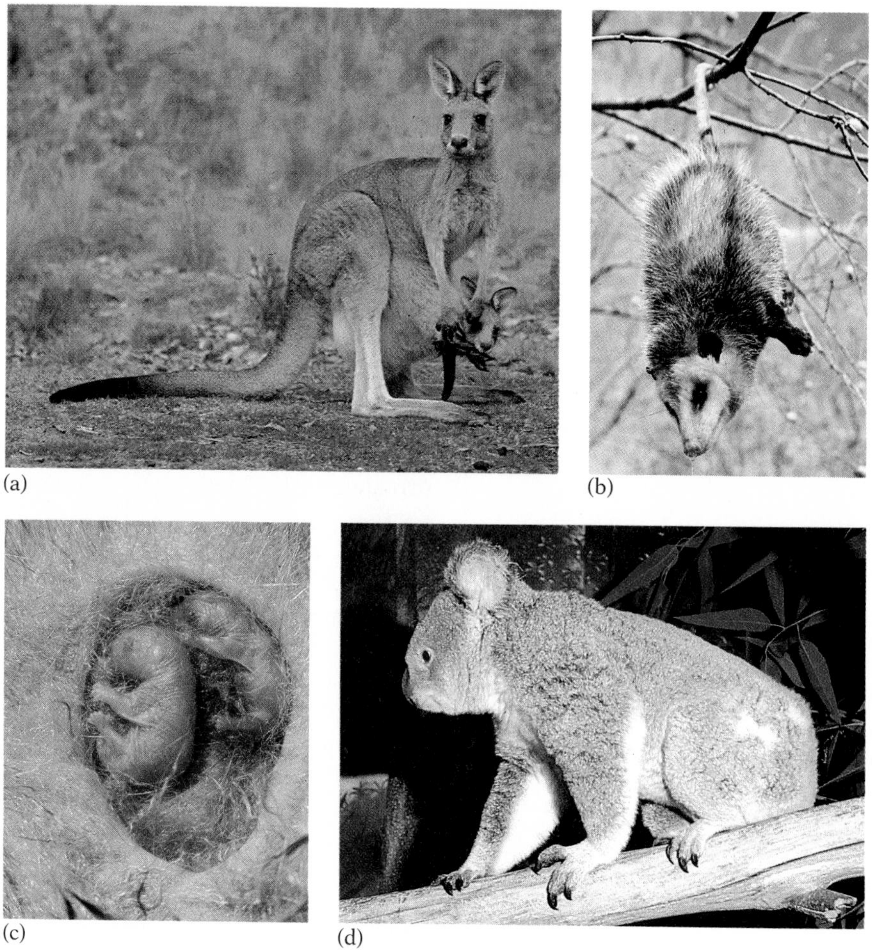

**FIG. 32-32** Marsupials, or Metatheria. (a) Kangaroo. (b) Opposum. (c) Baby opposum in its mother's pouch. (d) Koala bear.

Finally, we turn to a group of arboreal, or tree-climbing mammals, the primates, which include such species as the tree shrews and lemurs. As an adaptation for arboreal life the primates developed a prehensile thumb, a characteristic not found in other mammals. The thumb is opposable to the other digits and thus can be used for grasping. Some primates have developed a prehensile foot with which they can hang from tree branches. These adaptations, particularly in the forelimb, had a great potential for further evolution. It is from similar stock that the acrobatic gibbons and other anthropoid apes probably arose. From these apes came the pen- and sword-wielding humans. A somewhat detailed account of human evolution is found in Chapter 27. The family Hominidae has, at most, been on earth no more than 12 million years. Compared to the 180-million-year tenure of the reptiles, this is but a bare beginning. There is no compelling reason to believe that we will last as long, despite our capacity to take intelligent precautions that were not possible for reptiles. On the contrary, there is reason to believe that we are rapidly making our planet unlivable for vast numbers of species, including our own.

## Summary

Although representing the greatest diversity found in the animal kingdom, the coelomate Protostomia—annelids, molluscs, and arthropods—are a highly unified group of phyla, exhibiting close kinships to each other. Not only do marine annelids and molluscs both exhibit spiral, determinate cleavage, but both have trochophore larvae. In both phyla, the mesoderm arises as bands of tissue within which splits occur to form the coelom. Although arthropod development represents a distinct departure from these ancient patterns, the adult arthropods have obvious close affinities with adult annelids. Both have similar metamerism, nervous systems, and excretory organs. Despite their many adaptations to various modes of life, molluscs also retain clear affinities to annelids, as evidenced by a comparison of the nervous, circulatory, and excretory systems of the primitive molluscs and the annelids.

Kinships among all phyla within Deuterostomia are not nearly as evident as among protostome coelomate phyla. Evidence for a common ancestry of all deuterostomes includes a tendency toward radial, indeterminate cleavage and a mesoderm that forms either as outpocketings from the endoderm or in close association with the endoderm. Unlike the pattern in protostomes, the blastopore does not become the mouth; it either persists to become the anus, or the anus arises in the area where the blastopore had been.

Each level of taxonomic classification among the protostomes and deuterostomes exhibits a diversity of basic body plan. Such variations on the main theme appear to have evolved in the course of adaptation of these organisms to specific ecological niches. The basic body plan of the annelid is metameric segmentation, with each segment bearing all essential organs. The common earthworm comes closest to being an example of this level of organization. In the marine polychaetes that swim, the body is laterally extended to form parapodia, which aid in swimming and crawling. These polychaetes also exhibit a greater degree of concentration of sensory organs at the anterior end of the body. The leeches, which belong to the class Hirudinea, are also segmented. But the segments at the anterior and posterior ends of the body are modified to form suckers.

In the phylum Mollusca, the presence of a protective shell affords its members greater protection against predators than the annelids have, but at the sacrifice of active locomotion. Only in the cephalopods, such as the pearly nautilus, the octopus, and the squid, in which the shell was made buoyant by gas chambers, was reduced in size or eliminated, did active locomotion evolve. Associated with the active locomotion of cephalopods, more efficient sense organs evolved, including eyes very similar to those of vertebrates and an advanced nervous system. In all other molluscs, particularly in bivalves (oysters, mussels, clams, and so on) sensory perception and the nervous system are poorly developed.

Arthropoda is the most successful of all animal groups. In numbers of species, numbers of individuals within a species, as well as in the variety of ecological niches they occupy, arthropods surpass all other animals. Their success is due in part to the presence of a protective rigid exoskeleton, or cuticle. Made up of chitin and protein, the cuticle not only provides rigid structures to which muscles can be attached but also protects the animal from desiccation in arid climates, from osmotic inflow of water in freshwater ponds and lakes, and from osmotic outflow of water in seawater. In addition to the cuticle, arthropods are characterized by the presence of jointed legs that permit very efficient and rapid movement. Crustacea, Insecta, Myriapoda, and Arachnida are the major classes of Arthropods.

The deuterostomes include the phylum Echinodermata and the phylum Chordata as well as a few minor phyla with uncertain kinships. The Echinodermata, as the name indicates, are characterized by a spiny skin. They are all pentaradially symmetrical: the body is divisible into five more or less equal parts. Sea stars, brittle stars, sea urchins, sea cucumbers, and sea lilies are examples of the five classes of echinoderms, found on seashores all over the world.

Hemichordata is an aberrant group of deuterostomes represented by marine, wormlike animals. The acorn worm is a common example of this phylum. The strongest evidence suggesting the common ancestry of hemichordates and echinoderms is the close similarity of early development, including similar larval stages in some species. The hemichordate kinship to chordates is based on the presence of pharyngeal gill slits and a dorsal nerve cord in both groups.

Chordata appeared early in the Paleozoic era and are represented today by a wide variety of organisms. The primitive, marine *Amphioxus,* which belongs to the Cephalochordata, has the typical chordate characteristics: a tubular dorsal nerve cord, a notochord, and pharyngeal gill slits. The lamprey *Petromyzon* and other jawless fishes (class Agnatha) are the most primitive of today's vertebrates. The Gnathostomata include all other vertebrates. Chondrichthyes, the cartilaginous fishes, and Osteichthyes, the bony fishes, are both very successful aquatic groups. Over twenty thousand species of bony fishes are living today and many more are represented in fossil records. Probably as an adaptation to an aquatic habitat, all fishes share a basic body plan. Two pairs of fins, which serve as balancers; a caudal fin, used in propulsion; and a spindle-shaped body are common to all.

The Amphibia are the most primitive of the tetrapod vertebrates. They can survive on land because they have limbs and lungs. But their skin needs to be kept moist to prevent desiccation. More important, their eggs can develop only in water or in very moist environments. Consequently, amphibians have never truly colonized the land.

Evolution of the amniote egg—in which the developing embryo floats in an artificial pond formed by extraembryonic membranes, the chorion and amnion—enabled the amniotes, which include reptiles, birds, and mammals, to become true terrestrial animals. As the earliest land vertebrates, reptiles diversified extensively. They include not only the dinosaurs and other giant reptiles, which are all extinct, but also many groups of smaller reptiles, which gave rise to mammals, birds, and modern reptiles.

Birds are specialized for flying, a mode of locomotion that places great evolutionary constraints on basic body plan. The homogeneous appearance of birds is testimony to this constraint.

Unlike the birds, mammals have basically adapted to terrestrial life, with a few important exceptions, such as whales, seals, and bats. All mammals share three features: mammary glands, hair, and the retention of the left aortic arch for circulation of blood to posterior parts of the body. The great success of mammals may be attributed to their homeothermy and their placental mode of nourishing the embryo, that is, vivipary. The placenta, however, is not universal among mammals. Prototherians—echidna and the duckbill platypus—lay eggs. Marsupials, such as kangaroos and opossums, have a transient placenta called the yolk sac placenta. Only in eutherian mammals is a true allantoic placenta found. This diversity among mammals clearly suggests a lack of direction in evolution of organisms other than by natural selection.

One group of modern mammals, the primates, characterized by the presence of binocular vision and the presence of nails instead of claws on their fingers, separated from the rest of the mammals about 60 million years ago. This group diversified gradually into lemurs, monkeys, and apes and eventually into humans.

As we look to the past, we see that most major groups of vertebrates are represented today by only a few remnants of once very successful groups. Some groups, such as the dinosaurs and placoderms, have become extinct. Looking to the distant future, we have no reason to believe that any of the current dominant species will survive.

## Recommended Readings

A list of recommended readings for this chapter appears at the end of Chapter 31.

# UNIT VII

# Biology of the Environment

CHAPTERS



Integral to any understanding of the living world is a knowledge of the many interactions that occur among organisms and between organisms and their environment. The study known as ecology represents that body of knowledge. Because the adaptations represented by the many types of interactions exhibited by organisms are shaped by natural selection, a knowledge of ecology is essential to a full understanding of evolution. Ecology, moreover, integrates concepts, principles, and understanding, not only from all major areas of biology, but from such physical sciences as chemistry, physics, and geology, and even from the social sciences.

The preface to Unit I noted that a knowledge of general biology provides not only an understanding of ourselves as part of the living world but is also a potential basis for good citizenship. Nowhere are these benefits more applicable than to our knowledge of ecological principles and the impact we humans have upon our environment.

In this, the last unit of the book, the growth of ecology as an area of biological inquiry is briefly traced. This is followed by accounts of basic ecological principles. Population and community ecology are considered next. The closing chapter describes major effects that humans have upon the ecosystem of which they are a part.

# 33 Basic Ecological Concepts

OUTLINE



A widely accepted definition of **ecology** (from the Greek *oikos,* "house" or "living space") is the study of the interrelationships of organisms with their environment and with each other. The environment includes factors such as sunshine, soil, minerals, atmosphere, and moisture (often termed the abiotic factors), and, according to some definitions, other organisms (sometimes called the biotic factors). Ecological studies are carried out at several levels: on individual organisms; on **populations,** consisting of all the organisms of one species that live in a certain area; on certain species of organisms; and on **communities,** consisting of the organisms of all species that live in a certain area. An entire community, together with its physiochemical environment, constitutes an **ecosystem.** It is at the level of the ecosystem that a complete synthesis of ecological understanding is achieved.

The space of an animal's mouth, a small pool of water in a hollow tree stump, and the volume of an ocean can all be regarded as constituting the bounds of ecosystems. A moderately large ecosystem, as represented by a lake or a forest area and its inhabitants, is essentially a self-sufficient unit. However, although we may think of a lake and its organisms as an autonomous ecosystem, it is certainly affected by other ecosystems around it and, indeed, by some factors that operate thousands of miles away. In this sense, all of the life-supporting soil, water, and air on the earth's surface, sometimes referred to as the **biosphere,** in addition to its organisms, constitute one large ecosystem. This point is dramatized whenever a major volcanic eruption alters weather patterns around the world, sometimes for years. Local events can have worldwide effects.

The addition to or removal from an ecosystem of materials or organisms may affect the entire system. The effect of such changes depends on a number of factors, including the number and kinds of organisms present, their physiological state at particular times of the year, and what and how much is added or removed in relation to the size of the ecosystem. One milligram of a poison diluted in an ocean would not even be detectable; the same amount in a small tide pool of that ocean might kill the organisms in it. Some effects on ecosystems are subtle and may go undetected for years; others are rapid and dramatic.

The materials within ecosystems are continually being recycled. For example, under the action of such organisms as fungi and bacteria, all the organic matter of dead organisms is usually decomposed to inorganic form. In fully decomposed form, most of these substances are absorbed by photosynthetic organisms and resynthesized into living matter. An outside source of energy, sunlight, is required for this synthesis. An ecosystem is thus a highly organized processor of energy and materials. Its organisms perform tasks essential for the long-term survival and success of their ecosystem.

## MAJOR BIOTIC COMPONENTS OF ECOSYSTEMS

Few, if any, of the roles an organism plays in its community are as important as those affecting its community's nutrition. In general, these roles can be classified as **producer, consumer,** and **decomposer** (see Fig. 33-1). All life depends on the producers, nearly all of which are photosynthetic organisms; this category includes several prokaryotes, many protists, and all plants. Photosynthesis requires carbon dioxide, water, sunlight, and an adequate supply of certain mineral nutrients. Shortages of any one of these nutrients limits an ecosystem's production of new living material, or **biomass.**

All consumers depend on the productivity of the producers. Some consumers, the herbivores, eat producers directly (Fig. 33-2); others, the carnivores, eat herbivores as well as other carnivores. Omnivores eat both producers and other consumers. The study of such relationships has led to development of the concepts of food chains, food webs, and food pyramids.

### FOOD CHAINS, WEBS, AND PYRAMIDS

The nutritional relationships among producers; the animals that eat them, called **primary consumers;** and the animals that feed on primary consumers, called **secondary consumers,** are often represented by a food

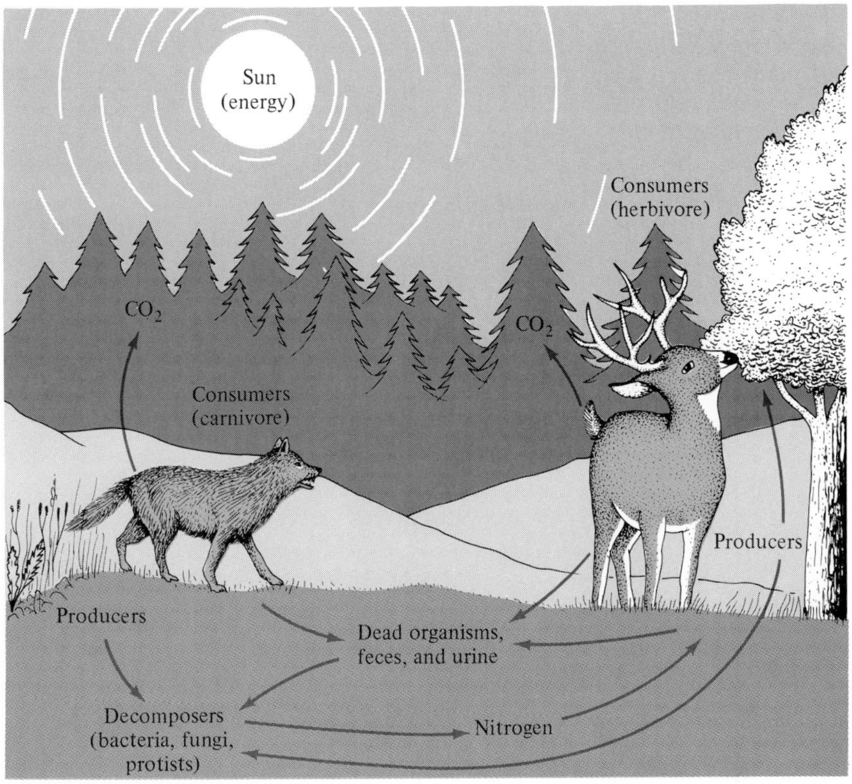


**FIG. 33-1** The major biotic components of an ecosystem. With few exceptions, the producers are photosynthetic organisms, obtaining their energy supply from the sun. Feeding on them are the herbivore and omnivore consumers. The carnivores feed exclusively on animals. All groups eventually become food for the decomposers—bacteria, fungi, and various protists—which return the materials of which organisms are composed to the soil, water, and air, whence it is converted into living matter, or biomass, by the producers.

chain. Figure 33-3 depicts a food chain for the tiny marine crustaceans known as copepods. They feed on photosynthetic protists called **phytoplankton*** and are in turn eaten by herring, which are eaten by tuna, some of which are eaten by humans. Although a chain correctly depicts the principal foods of each member, better representation of the nutritional interdependence of most organisms is provided by a food web.

Simplified webs for a seashore and a shallow-water community are shown in Figure 33-4. Food webs strike us as quite complex compared with food chains. But food webs are also simplifications of the complex relationships existing even in an association as small as a pond community. For example, a single species may harbor dozens of species of bacteria, protozoans, and worms, some of them pathogens and many of them harmless or beneficial to the host. For each plant and animal species on earth, there are many other species that at some time or another may live on or in members of that species and depend on them for shelter or nutrition. A food web not only ignores these relationships, but it seldom portrays the complex roles of the decomposers: those final processors that maintain themselves by ingesting or absorbing various products of organic decomposition. Like animals, decomposers incorporate some material into their bodies and use some for energy needs. In addition, they release various organic and inorganic substances, including ions and salts into the environment, which thereby become avail-

**FIG. 33-2** Strict herbivores, such as the moose, found in northern regions of both the western and eastern hemispheres, depend entirely on producers for their nutrition.

* Plankton (sing., plankter) (from the Greek *planktos*, "wandering") are minute aquatic organisms that are either nonmotile or such weak swimmers that they are always carried by currents. Photosynthetic plankton is termed phytoplankton; the plankton that feed on them, such as protozoans and microcrustaceans, are called zooplankton.

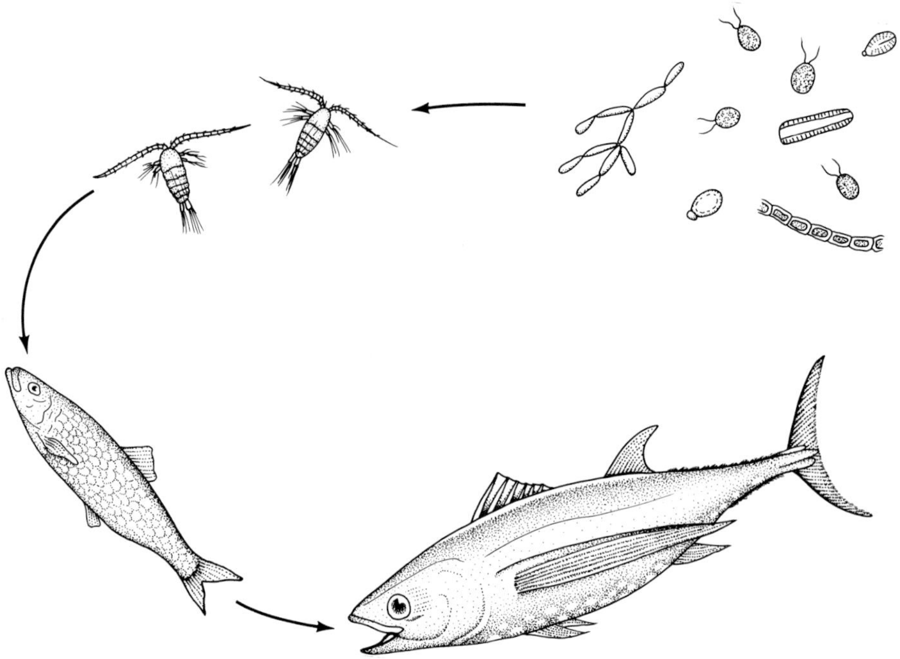

**FIG. 33-3** A marine food chain. The producers in this case are phytoplankton; they are fed upon by small crustaceans called copepods. Copepods form the food of herring, which are fed upon by tuna.

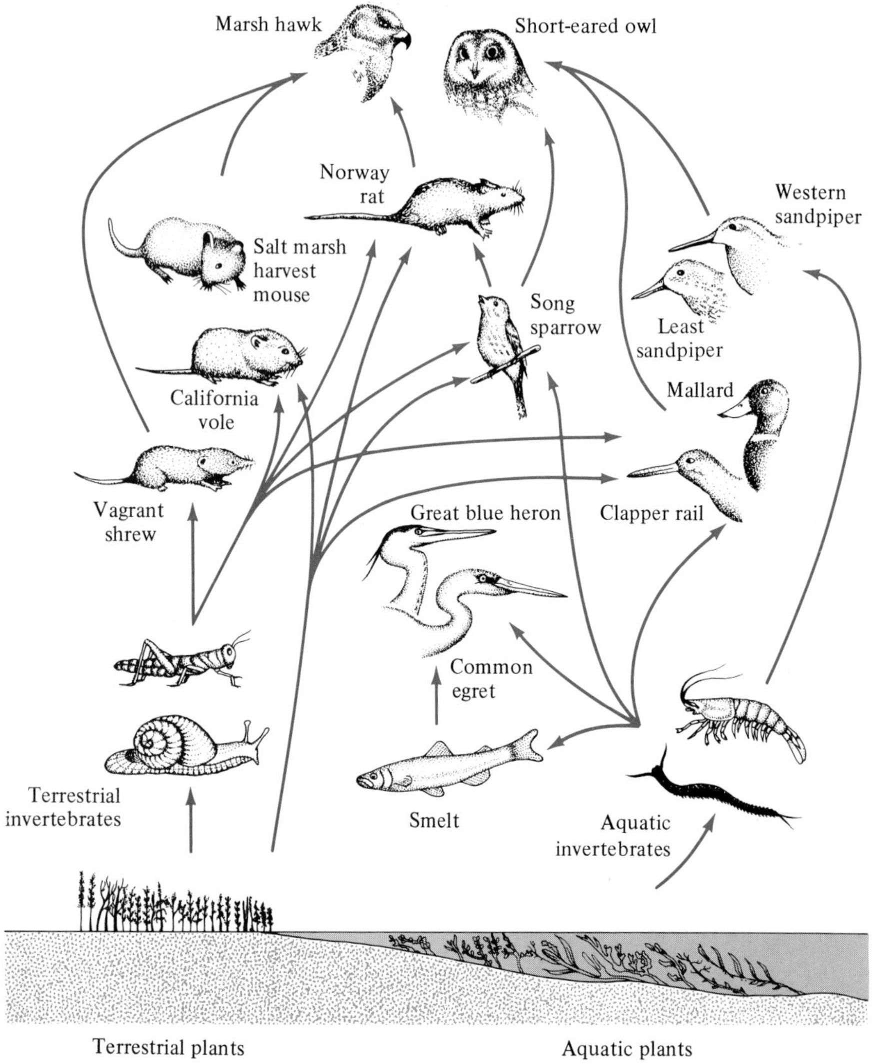


**FIG. 33-4** Two partial, overlapping food webs of a shore and of a shallow-water community. Arrows indicate the flow of energy.

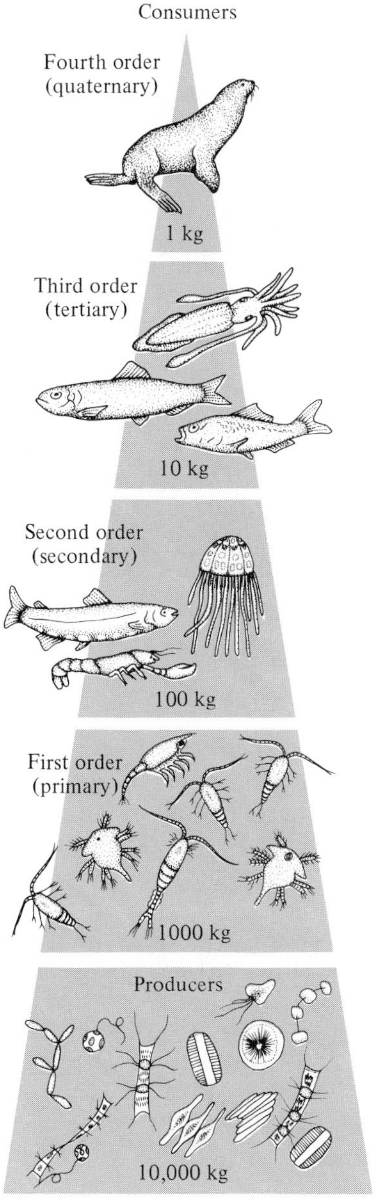


**FIG. 33-5** A marine food pyramid. This particular pyramid has four levels of consumption, termed primary, secondary, tertiary, and quarternary consumers.

able to the producers again, enabling the ecosystem's matter to be recycled indefinitely.

Another useful representation of nutritional relationships is the **food pyramid,** in which the number of individuals, the mass of food, or the food's energy content is usually given for each nutritional level. The producers are shown at the base of the pyramid, primary consumers just above them, followed, in order, by secondary, tertiary, and quaternary consumers. The pyramid of mass in Figure 33-5 represents a marine community in which a seal feeds on fish and squids, which eat smaller fish, jellyfish, and shrimp, which in turn feed on copepods or other harvesters of phytoplankton. The mass (or the energy content) of phytoplankton present in such a community at any one time, called the **standing crop,** is usually less than the mass (or the energy content) of the primary consumers feeding on them. This imbalance is compensated for by the high productivity of phytoplankton. Presenting these relationships in a food pyramid shows that, at each level of consumption, up to 90 percent of the food eaten (or its energy content) does not leave that level; about 50 percent is lost as heat and in such activities as growth and development. The rest is represented by dead bodies, which are acted on by decomposers. As little as 10 percent or less reaches the next nutritional level. In the food pyramid in Figure 33-5, which is a pyramid of mass, about 10,000 kg of phytoplankton are required at the bottom of the pyramid to support 1 kg of growth of a seal at the top of the pyramid. If this were a pyramid of energy, each level would be represented in terms of calories or other units of energy.

Food pyramids correctly represent the dependence of larger animals on smaller ones to harvest the tiny producers. However, natural relationships are much more complex than is indicated by a pyramid. The dominant animal at the top of the pyramid, whether a bird, a bear, or a human being, almost never eats only the consumers just below it in the pyramid. As depicted in the food web represented in Figure 33-3, many animal species feed on over a dozen different kinds of organisms, not all of which are at the next lower level of the pyramid. Moreover, any one animal may be food for many other species of consumers.

## THE ECOLOGICAL NICHE

The fact that each major type of organism makes a specialized contribution to the functioning of its ecosystem led to the development of the concept of the **ecological niche,** defined as the organism's role in its community as well as its tolerances for various kinds of stress. The niche occupied by an organism is determined primarily by nutrition; however, a full description of an organism's niche would include at least the following characteristics:

1. the organisms's habitat requirements: the kind of place in which it prefers to live, for example a hollow tree, under a rock, in a pond, in decaying bark, or in some other natural location;
2. the ways in which it affects other kinds of organisms in the community: what it prefers to eat; what it can eat (Fig. 33-6); what eats it; whether it uses or produces oxygen, alters the pH of the soil or water, and contributes to or depletes various nutrients; and so on;

**FIG. 33-6** A woodpecker finch of the Galapagos Islands. On these isolated islands, one species of finch has come to partially fill the niche ordinarily occupied by woodpeckers on the South American mainland. Using a cactus spine as a woodpecker would use its beak and long tongue, the bird probes for insect larvae in the bark of trees.

3. its physical requirements and tolerances: how much light and moisture it needs and can tolerate, what temperature and pH ranges it can tolerate, and the like;
4. its behavioral patterns: how it becomes distributed; whether and where it migrates; whether it is nocturnal or diurnal (active at night or during the day); whether it hibernates or estivates (becomes dormant in winter or summer); whether it is solitary or social; and so on.

Two closely related species typically occupy a similar niche or niches that substantially overlap each other. However, many unrelated species occupy niches that slightly overlap each other. For example, people and mice living in the same house share parts of the habitat and some of the food. Two species that occupy identical kinds of niches are rarely found in the same geographic region. Where they are, one is usually on the decline, at least at a given time of year.

## ENERGY FLOW IN ECOSYSTEMS

No living organism can function or even live without an energy supply. For photosynthetic organisms, this need is met by sunlight. Photosynthetic organisms convert the radiant energy of the sun to chemical energy by synthesis of energy-containing organic compounds such as carbohydrates, lipids, and proteins. Consumers obtain their energy by eating or absorbing and then oxidizing this organic matter. Decomposers obtain their energy through oxidations associated with the decay of organic matter. At each step in the producer-consumer-decomposer cycle, in obedience to the second law of thermodynamics, some energy is lost as heat, never to be recovered. Because of this unavoidable loss, energy cannot be recycled within ecosystems. It must be fed into every ecosystem, if not from the sun, then in the form of energy-containing molecules from some source. If the outside source of usable energy is cut off, an ecosystem's existence soon comes to an end.

**FIG. 33-7** Although earth receives but a small fraction of the energy radiated by the sun and reradiates most of it back into space, nearly all life on earth depends on the sun as its ultimate energy source.

## PHOTOSYNTHESIS

Of all the energy emitted by the sun, only a small fraction falls on its third planet (Fig. 33-7). However, the amount of solar energy that strikes the United States every 20 minutes is equivalent to its entire power needs for 1 year. Most of this abundant radiant energy is lost by being reflected from the earth or is first absorbed and later lost as heat. (The earth continually radiates heat out into space.)

Although most of the sunlight that strikes plants is reflected, much of it drives transpiration, the evaporation of water by leaves. Usually, less than 1 percent of the light that strikes a land plant is absorbed by its chlorophyll. Marine phytoplankton fares even worse than land plants in capturing the sun's energy because almost all the sunlight striking the surface of a calm sea is reflected. Of the amount that does penetrate the water, less than 0.2 percent is absorbed by the chlorophyll of phytoplankton. Once absorbed and converted into chemically bound energy by chlorophyll, this energy takes one of two possible routes: up to 50 percent or more ultimately drives the photosynthesizer's own metabolism; the rest becomes incorporated into the growing substance of its cells, that is, into its biomass.

## ENERGY PYRAMIDS

Substantial losses of energy are characteristic of the nutrition of animals. This energy loss can be represented in an energy pyramid. As in a food pyramid, the producers occupy the bottom. The energy they provide to the level above is represented as caloric content. Also as in food pyramids, the actual amount of energy present at any one time in the lower levels may be less than that in the levels above. Such a pyramid represents, not the amount of energy present at any given level, but the amount being supplied to it in a given period of time, or the rate of energy flow from level to level. This rate declines at each successive level.

For example, the following data from a study of a lake community in upper New York State could easily be represented by an energy pyramid. In Lake Cayuga, it was found that 1,000 calories derived from algae were needed to support 150 calories of growth of larvae and other zooplankton that fed on the algae. Smelt (a small fish) that lived on these tiny animals grew by only 30 calories for every 150 calories of this food they ate. Trout that ate 30 calories of smelt increased by a mere 6 calories. Finally, trout-eating animals, such as humans, grew (increased their energy content) by only 1.2 calories for every 6 calories of trout eaten.

Much of the inefficiency in an animal's extraction of energy from food is represented by energy lost in its feces.* Herbivores extract only about 50 percent of the calories contained in the plants they eat, a measure of the inefficiency of their digestive systems in breaking down plant tissue. Moreover, only 5 to 8 percent of the calories extracted from plants are usually incorporated into an herbivore's biomass—its meat, bones, and other tissues. Somewhat better efficiency is achieved in carnivore nutrition; animals extract more calories per unit from meat than from comparable units of plant tissue.

* That feces, especially of herbivores, are rich in calories and nutrients is evident from the large number of animals that feed on them. Many insects feed only on feces. A great variety of herbivores and omnivores eat their own feces or those of other animals, a practice that makes an important contribution to the overall efficiency of an ecosystem.

## ESSENTIAL CHEMICAL ELEMENTS OF ECOSYSTEMS

Some 2 dozen chemical elements are essential components of organisms. In most cases, there are no shortages of these elements on our planet. However, a few are used in such large amounts by organisms or are in such short supply in many ecosystems as to sometimes limit growth. The cyclic passage of these elements between the abiotic (nonliving) and biotic phases of the earth's ecosystems thus has special relevance to ecology. As noted in Chapters 2 and 9, the most important of the elements essential for life are carbon, hydrogen, oxygen, nitrogen, phosphorus, and sulfur. The cycles of carbon, nitrogen, and phosphorus are presented here as models to illustrate the principles involved in the cycling of all elements.

### THE CARBON CYCLE

Carbon is not only essential to life; it constitutes the bulk of living matter. Carbon chains are the chemical backbone of carbohydrates, fats, proteins, and nucleic acids, the major molecules of life. The availability of carbon and its fate are thus of considerable interest to ecologists.

***Carbon Sources.*** Producers obtain their carbon as carbon dioxide ($CO_2$), which they remove from the air or water and incorporate into living matter, usually by photosynthesis. As depicted in Figure 33-8, consumers obtain their carbon by eating producers or other consumers. Much of this carbon is eventually returned to the air as $CO_2$ from the respiration or decay of organisms. Not all carbon is returned immediately, however. A sizable amount remains bound in the form of limestone shells. In the exceptional case of **fossil fuels**—coal, oil, and natural gas derived from the remains of ancient plants and other organisms—carbon remains within them for millions of years or indefi-

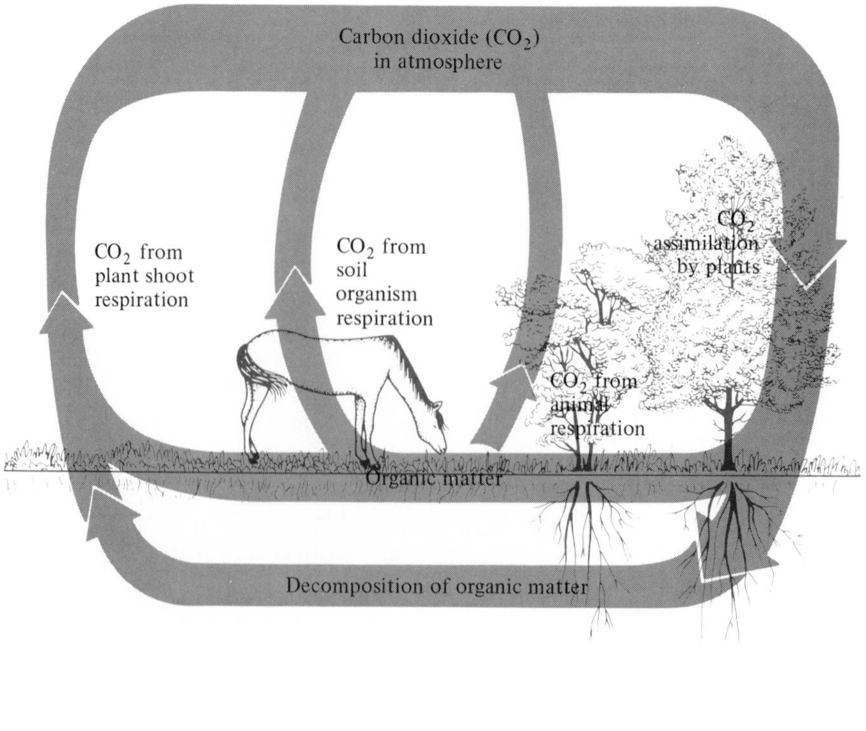


**FIG. 33-8** The carbon cycle. Atmospheric carbon dioxide is fixed by photosynthesis of plants and phytoplankton and enters the myriad organic compounds these organisms synthesize. Some returns to the air as carbon dioxide, a product of plant respiration. The various consumers (animals) acquire the organic compounds by eating plants or other animals; their respiration also releases carbon dioxide into the atmosphere. Dead plants and animals are decomposed by soil and aquatic microorganisms (bacteria, fungi, and protists), returning more carbon dioxide to the air as well as other elements to the soil.

PANEL 33-1

# Effects of $CO_2$ Accumulation in the Atmosphere

Since the mid 1800s, humans have steadily increased the rate at which they burn fossil fuels. Largely for this reason, the concentration of $CO_2$ in the atmosphere rose from an average of 0.029 percent in 1850 to 0.034 percent in the early 1980s. One-fifth of this entire increase occurred during the last 15 years of that 130-year span. Other human activities besides fuel combustion contribute to the $CO_2$ in the atmosphere. According to one estimate, of the 8 billion tons of $CO_2$ added to the atmosphere each year, some 6 billion tons come from the burning of fossil fuels and 2 billion tons come from agriculture. For example, frequent plowing and tilling of large acreages apparently releases more $CO_2$ into the atmosphere from a given amount of soil than is removed from the air by the crops being grown on that soil. The steady destruction of the Amazon forests now being done principally by lumbering and other industries is also expected to add substantially to the atmospheric $CO_2$ content.

## THE GREENHOUSE EFFECT

At its present rate of accumulation, the concentration of $CO_2$ in the atmosphere is expected to reach 0.037 to 0.040 percent by the year 2000. The various effects of this increase, particularly on the earth's climate, are unknown. However, $CO_2$ in the atmosphere acts somewhat as does the glass roof of a greenhouse: less radiation escapes than enters. For this reason, it is feared that the overall temperature of the earth's surface will rise with increased atmospheric $CO_2$. This enhanced "greenhouse effect" might result in the melting of the polar ice caps, an event that could occur within 400 years, with appreciable changes evident in as few as 20 to 30 years. If the polar ice caps were to melt completely, sea level would rise by 65 m, putting such coastal cities as New York, Los Angeles, and Tokyo under water.

## WILL PARTICULATE MATTER IN THE ATMOSPHERE COMPENSATE FOR INCREASED $CO_2$?

Another relevant factor being studied in this context is the increased amounts of particulate matter, such as smoke and dust, that modern societies send into the atmosphere. Because these particles tend to reflect the sun's rays, it is possible that their presence might fully compensate, or even overcompensate, for the increase in $CO_2$. Rather than experiencing a warmer climate, we might be about to enter an era of much colder weather.

While we know that changes in our atmosphere are occurring at an unprecedented rate, we lack enough information on this important subject to predict with any confidence just what their effects will be.

nitely. Probably at least 50 times more carbon is locked in fossil fuels than exists in all living organisms. With the large amounts of fossil fuels burned over the past century, much of this carbon has been returned to the atmosphere as $CO_2$ (see Panel 33-1).

***Turnover Rates.*** It has proved difficult to estimate how much $CO_2$ is removed from the atmosphere by photosynthesis. One technique used in the past to measure its disappearance from the atmosphere was the monitoring of radioactive fallout. Until the early 1960s, nuclear weapons were being tested by exploding them in the air, and one of the fallout products was radioactive carbon ($^{14}C$). The rate at which $^{14}C$-contaminated $CO_2$ disappeared from the atmosphere in the years immediately after the last aerial bomb test, in 1963, was used to measure the rate of turnover of atmospheric $CO_2$. The studies showed that a molecule of $CO_2$ usually remained in the air for about 5 to 10 years, a turnover rate of 10 to 20 percent per year.

The fallout data also showed that about 100 billion ($10^{11}$)* tons of $CO_2$ enter the sea annually; about the same amount is believed to be given off by the sea into the atmosphere each year. Not only does $CO_2$ readily dissolve in water, it also combines with it to form carbonic acid ($H_2CO_3$), which in turn produces carbonates (—$CO_3$) and bicarbonates (—$HCO_3$). As a result, seawater contains a considerable reservoir of $CO_2$, by some estimates about 130,000 billion tons of it. Of the $CO_2$ absorbed by the sea, an estimated 40 to 50 billion tons is used by marine phytoplankton in photosynthesis. Land plants use another 20 to 50 billion tons a year. The entire system within which $CO_2$ is exchanged between organisms and the environment is very complex and not well understood.

***Carbon Cycles in Terrestrial Communities.*** In the billions of years since the evolution of photosynthesis, the production of oxygen has greatly exceeded the amount of $CO_2$ produced by respiration. As a result, the atmosphere has accumulated oxygen to nearly 21 percent of its volume while carrying a relatively low level of $CO_2$—about 0.034 percent, or 340 parts per million (ppm). The total atmospheric content of carbon (as $CO_2$) is estimated to be less than 700 billion tons. A low concentration of $CO_2$ can make it a limiting factor in photosynthesis in some circumstances. An example of the local, temporary impact of photosynthesis on atmospheric $CO_2$ concentrations is given by the following data collected in a forest over several days.

At sunrise, the average concentration of $CO_2$ in the air of the forest was found to be about 320 to 325 ppm. During the morning on a sunlit day, this concentration gradually fell to about 305 ppm at treetop level and to 310 ppm just above the ground (Fig. 33-9). As the temperature rose during the afternoon and respiratory rates of the trees increased, the concentration of $CO_2$ gradually rose again. At sundown, photosynthesis ceased rather abruptly, while respiration continued. Atmospheric $CO_2$ concentration then climbed rapidly, and close to the ground it usually reached 340 ppm by midnight after a warm day, declining slightly toward dawn.

Among land communities, forests are both the largest users of $CO_2$ and the largest organismal reservoirs of carbon. The carbon contained in the world's forests amounts to 400 to 500 billion tons. As the leaves and wood of dead plants decompose, their carbon is returned to the atmosphere as $CO_2$. This process is very rapid in tropical rain forests but very slow in cold or dry regions.

## THE NITROGEN CYCLE

Another key constituent of living systems is nitrogen, which occurs in such vital molecules as proteins, nucleic acids, heme, chlorophyll, and several coenzymes. Nitrogen has many unusual properties. Although it is the most abundant element in the atmosphere (constituting 78 percent of it), in the molecular form ($N_2$) in which it occurs in air, it is unavailable for all but a very few species of specialized bacteria. Another remarkable quality of nitrogen is that it can exist at several oxidation levels, or chemical valences.

Early in the course of life's evolution on this planet, various microor-

* In this book, billion is used in the American sense of a thousand million.

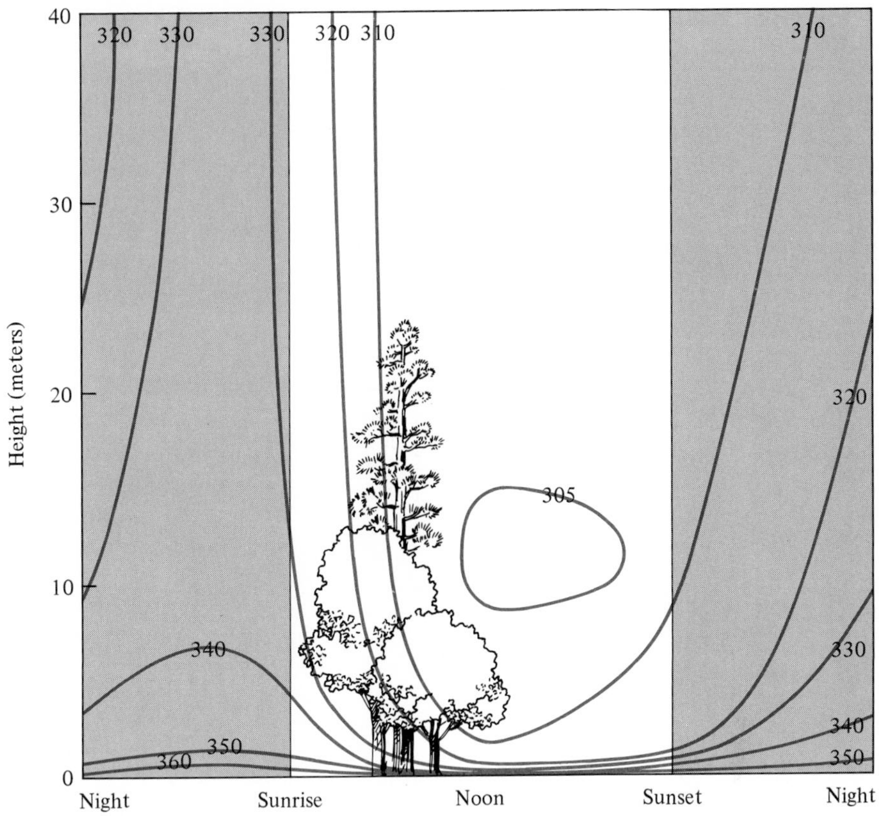


**FIG. 33-9** Variations in the concentration of carbon dioxide at different heights in a forest and at different times of day and night. Concentrations are given in parts per million.

ganisms apparently became specialized in extracting energy from the oxidation of various nitrogenous compounds, while others evolved an ability to use nitrogenous compounds as electron acceptors (oxidizing agents) when oxidizing other energy-rich substances. There has thus evolved a remarkable chain of nitrogen-metabolizing microorganisms. That chain provides an essential component of the nitrogen cycle (Fig. 33-10).

***The Nitrogen Fixers.*** As noted in Chapter 28, atmospheric nitrogen ($N_2$) is fixed by being incorporated into ammonia ($NH_3$) by two general types of nitrogen-fixing microorganisms: free living and symbiotic. The former group includes several species of soil bacteria and the cyanobacteria, or blue-green bacteria. The mutualistic nitrogen fixers inhabit root nodules of several species of plants, including some trees (alder, ginkgo); a shrub (buckthorn); and nearly an entire taxonomic family of plants, the legumes (peas, beans, alfalfa, clover, and peanuts) (Fig. 33-11). Of all nitrogen fixers, those in legume root nodules are by far the most important to the world's ecosystems. After fixing nitrogen by converting it into ammonia ($NH_3$), these bacteria incorporate it into amino acids and other nitrogenous organic compounds, which are of benefit not only to the bacteria but to their host as well. The host contributes to the mutualistic relationship by supplying the bacteria with glucose and other nutrients.

***The Processors of Nitrogenous Compounds.*** Besides the nitrogen fixers, the soil contains nitrifying bacteria that live on the energy that is derived from oxidizing ammonia to nitrite ($NO_2$). Another important source of ammonia for them is the decay of urine, feces, and the dead

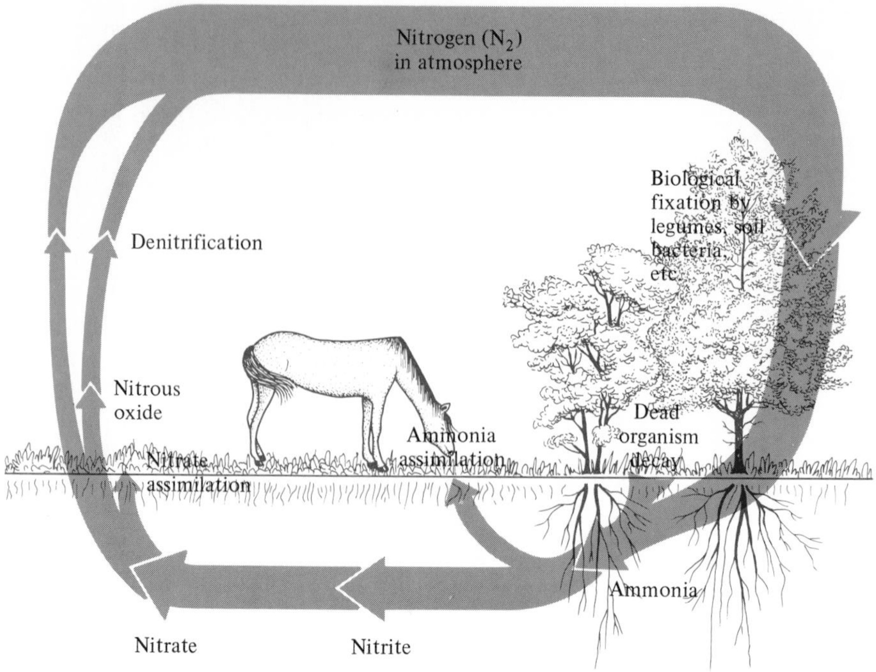


**FIG. 33-10** Major components of the nitrogen cycle. Atmospheric nitrogen ($N_2$) is incapable of being assimilated by most organisms. This task is accomplished by nitrogen-fixing microorganisms, some of which are symbionts in the roots of certain plants. Through the manufacture of fertilizers and the large-scale cultivation of legumes, human intervention has more than doubled the amount of fixed nitrogen available to agriculture. A similar cycle also operates in the oceans.

bodies of organisms. Yet other specialists derive their energy by oxidizing the nitrites produced by the nitrifying bacteria to the more fully oxidized nitrates ($NO_3$), the form in which plants absorb their nitrogen from the soil most readily and most abundantly.

***How Atmospheric Nitrogen Is Replenished.*** Some of the nitrogen in nitrogenous compounds is returned to the atmosphere as $N_2$ by the action of denitrifying bacteria. These organisms convert ammonia, nitrites, and nitrates to $N_2$, releasing it into the soil and the air. It is of historical interest that denitrifying bacteria were discovered in the late 1800s—before the existence of nitrogen fixation was suspected. It was noted that, in the absence of any reverse process, denitrifying bacteria would eventually deplete the soil of nitrogen. The discovery of the nitrogen fixers a few years later eased this concern. However, soil scientists are now studying ways to limit the action of denitrifying bacteria as a way of conserving soil nitrogen.

**FIG. 33-11** Root nodules of the crimson clover, a legume. The nodules contain nitrogen-fixing bacteria.

## THE PHOSPHORUS CYCLE

In the form of phosphate ($—PO_4$), phosphorus is a key substance in energy transfer, forms part of the backbone of the nucleic acids, and is a major component of such structures as shells, skeletons, teeth, and biological membranes. Along with nitrogen, phosphate is often a limiting factor in the growth and reproduction of organisms.

***The Limited Availability of Phosphate.*** A massive reservoir of phosphate exists in sedimentary rock. This phosphate is slowly eroded and dissolved and thus is gradually returned to the oceans, lakes, and soils of the world. At the same time, much phosphate is lost, as far as organisms are concerned, in the dead cells, tissues, and bodies that settle to the bottom of the sea. Although some of this returns to the surface water in areas of upwelling, more settles to the bottom of the sea than is returned in this way. An important exception to this one-way flow of phosphate occurs in areas of massive uplifting of the earth's crust.

**FIG. 33-12** Eutrophication. An algae-choked swamp in New Jersey.

***Phosphate and Agriculture.*** Because of its relative scarcity in available forms, phosphate is often the most limiting factor in plant growth. In regions of high human population density and intensive soil cultivation, as in India, phosphate shortage is an agricultural problem. It has been impractical to apply artificial fertilizers in sufficient quantity to keep up with the rate of agricultural phosphate depletion occurring in many tropical areas.

***Phosphate-Induced Algal Blooms in Freshwater Lakes.*** In most lakes, the growth of phytoplankton and aquatic plants is restricted by the limited availability of phosphate and a usable form of nitrogen. A costly modern problem occurs in populous areas, however, when both of these nutrients enter lakes in the effluent from sewage disposal plants and as farmland runoff of animal wastes and excess fertilizer. Accumulation of massive amounts of these nutrients in a body of water is termed **eutrophication.** It usually results in algae-choked waters (Fig. 33-12). The algae emit, not only noxious odors, but toxins that are especially harmful to fish. Eventually, large amounts of algae die and decompose after which the aerobic microorganisms of decay deplete the lake's dissolved oxygen, causing die-offs of fishes and other aerobic organisms. Such polluted lakes have sometimes become useless for swimming, fishing, and boating within a few years.

## THE LIMITING FACTORS CONCEPT

As noted above, the growth of aquatic plants is often limited by a shortage of a particular nutrient, such as phosphate. The principle involved was expressed as a law by the German biologist Justus Liebig in 1840. Known as Liebig's Law of the Minimum, it states: "Growth of a plant is dependent on the amount of nutrient which is presented to the plant in minimum quantities."

Early in this century, the American ecologist Victor Shelford expanded this concept to include the limiting effect on any organism's growth or distribution by any factor, whether it be in short supply or in excess. This principle, known as Shelford's Law of Tolerance, states: "The abundance or distribution of an organism can be controlled by factors exceeding the maximum or minimum levels of tolerance for that organism." Thus, limiting factors might include too much or not enough light, too high or too low a temperature, too much or not enough moisture, too much of a poison, too many parasites or predators, or too little of an important nutrient. One area to which Shelford's law applies is that of endangered species (Panel 33-2).

### THE CONTEXT OF LIMITING FACTORS

The effects of a limiting factor should never be judged by studying it in isolation. Combined factors can limit populations when none of them alone would have done so. For example, the house mouse survives all summer outdoors in temperate climates. With adequate food, it might also survive the winter in a forest. But during a severe winter and food shortages, usually only those mice that find their way indoors survive.

## PANEL 33-2

# Endangered Species as a Problem of Limiting Factors

As more of the earth's natural ecosystems are transformed into cities, farms, and artificial lakes, and as a steadily growing amount of poisonous by-products of our technology spreads through the soil, air, and water, the extinction of vulnerable species accelerates. The problem is quite complicated. The most common immediate cause of endangerment is probably the loss of adequate areas for special kinds of habitat. Many species have limited tolerances for change or require special conditions. A typical example is provided by the ivory-billed woodpecker, at one time common in southern Louisiana but long thought to be extinct. This largest woodpecker in North America was recently rediscovered in a wooded area of eastern Texas. Investigation has shown that each pair of birds requires 6.5 $km^2$ to 7.8 $km^2$ (2.5 to 3.0 square miles) of undisturbed land near a river containing a minimum number of standing dead trees, which these birds use for nesting. When the minimum area of undisturbed woodland with standing dead trees is no longer available, this species stops reproducing. Several other species of birds have a similar requirement for standing dead trees.

Other examples of species that are disappearing from the earth as the kinds of habitat they need undergo change are the following:

A pair of mounted ivory-billed woodpeckers (from a museum exhibit).

1. The North American mountain goat is now found only in Alaska, western Canada, Washington, Idaho, and Montana.

Most of our understanding of the distribution of organisms has resulted from a combination of studies in ecology, physiology, and biogeography. A knowledge of the following principles derived from all these fields is important to agriculture, wildlife management, and public health:

1. The climatic conditions tolerated by an agricultural pest determine its spread.
2. The conditions tolerated by a new strain of cereal grain determine the parts of the world into which it can be successfully introduced.
3. The oxygen concentration required for the survival of various game fish determines the level of thermal (heat) pollution from industry those fish can tolerate before a river is spoiled for fishing.
4. The spread of communicable diseases is determined in part by limiting ecological factors. For example, bubonic plague does not spread in dry areas of the world. Wherever the relative humidity is below 28 percent, the flea that is the vector (transmitter) of the

(a)

(b)

(c)

Three other bird species that nest only in standing dead trees. (a) Eastern bluebird. (b) Spotted owl (owlet). (c) Wood duck.

2. The Columbian ground squirrel is now found only in a small area of Idaho, and in Montana, southern Alberta, and British Columbia.
3. The golden-cheeked warbler is found only in a very small wooded area in Texas. Because the area is soon to be commercially developed, the species is expected to disappear in 10 to 15 years.
4. The lion-tailed macaque, a type of monkey, is now limited to a few isolated pockets of the monsoon forests south of Bombay, India. It is expected to be extinct in 10 to 15 years.

Some of the endangered species that are better known to the public and are often mentioned in news reports include dolphins, whales, elephants, gorillas, whooping cranes, bald eagles, and the California condor.

Because these animals are relatively large and conspicuous and have been studied extensively, various biologists are aware of many of their requirements and of some of the threats to their continued existence. But we have no knowledge at all of the tolerance limits of most organisms, nor do we know the exact distribution of many of them. In most cases, such as in the tropical rain forests, the endangered organisms have not even been identified. Because estimates of the number of living species range from 3 million to over 10 million, and only about 1.5 million species of organisms have thus far been officially named and described, we are truly in the dark when it comes to knowing what is happening to the endangered species of the world.

disease survives for less than 1 day away from its rat or human host. This explains why epidemics of plague occur only in humid regions of India while being virtually absent in dry areas.

## CLIMATIC FACTORS IN ECOSYSTEMS

Climate determines what kinds of organisms can succeed in a given region and thus profoundly affects the composition of communities. The most important climatic factors are the seasons, winds, humidity, rain, and temperature, all directly or indirectly generated by the sun's energy. Thus, in addition to its role as a primary source of energy for ecosystems, sunlight is crucial to life on earth through its influence on climate. The major ocean currents, which not only affect climate, but determine the geographic distribution of many organisms, are also generated by solar energy. How sunlight produces these important contributions to the physical components of ecosystems is briefly described in the following sections.

## THE SEASONS

As is well known, the earth rotates on its axis once every 24 hours and orbits the sun once every 365.25 days. Because the earth is tilted 23.5° on its north-south axis in reference to the sun, the amount of solar energy received by a given region varies during the year (Fig. 33-13). The effects of the earth's tilt on the seasons are not quite as dramatic elsewhere on earth as they are near the poles. On June 21, a midnight sun shines at the North Pole, and there is darkness at noon at the South Pole. The reverse is true on December 22. Seasonal changes are the least noticeable at the equator, where the temperature varies only about 4°C the year round for a given time of day or night. Temperate zones, on the other hand, have distinct seasons. Thus, organisms living in temperate zones must either be highly adapted physically for surviving both the severe winters and the summers or be able to adapt behaviorally, such as by hibernation, estivation, or annual migration to a more favorable climate.

The sunlight, and thus the energy received by a given area, is related not only to the average number of hours of daylight but, more important, to the angle of incidence of the solar radiation. The light striking the earth's surface in the Arctic has, for example, been filtered through much more heat-absorbing atmosphere and is spread over a much greater land surface than the more direct light striking the tropics (Fig. 33-14).

## WINDS AS ECOLOGICAL FACTORS

Because the tropics receive more sunlight than other parts of the earth, the air there gets hotter than it does elsewhere. This heated air expands and, becoming lighter, rises, forming a "low," or low pressure region. As the air reaches higher altitudes near the equator, it cools, causing its moisture to condense and producing the heavy rains typical of equatorial regions. The rising air is continually replaced by cooler surface air that flows in from both north and south. The result is the generation of two massive high altitude wind currents flowing away from the equa-

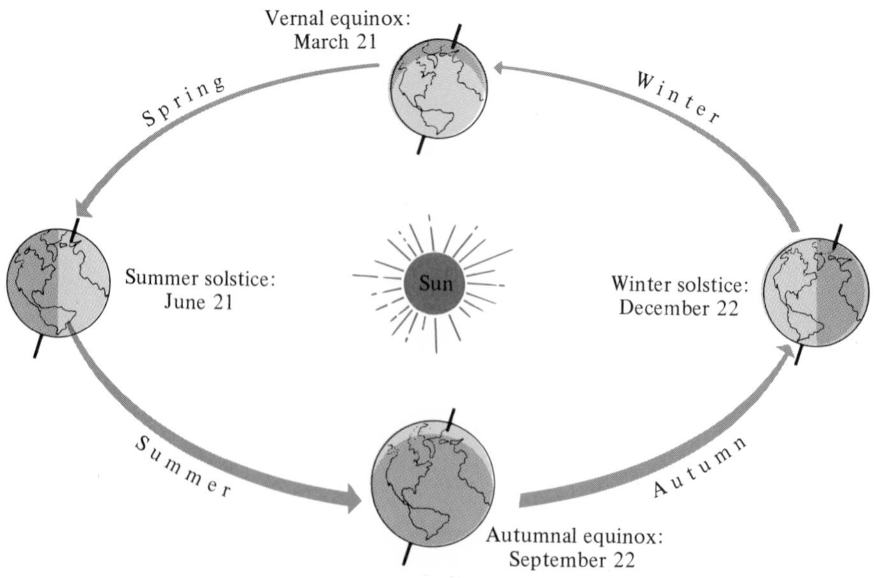


**FIG. 33-13** Variations in the amount of sunlight received at different times of the year by different regions of the earth. The amount of illumination falling on a given latitude also represents the number of hours of daylight it receives at that time of the year.

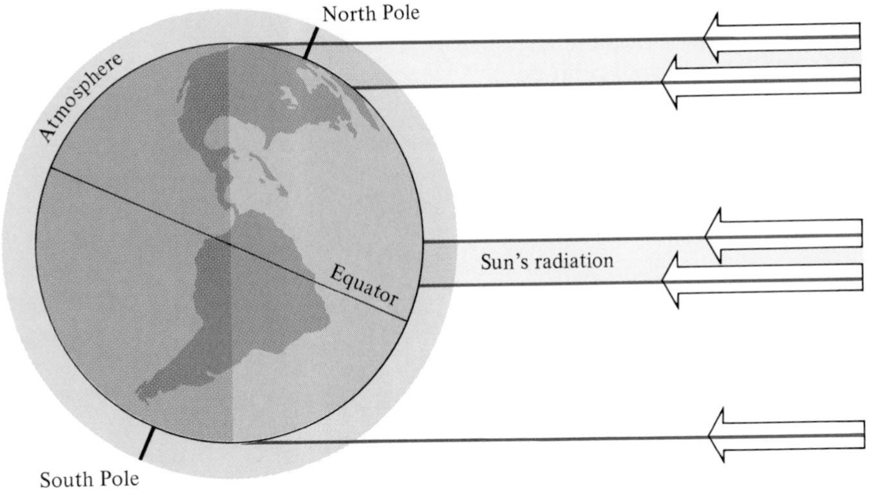


**FIG. 33-14** Variation in the amount of sunlight received and the amount of atmosphere through which it is filtered in different latitudes at the time of the summer solstice.

tor, one flowing north and one south (Fig. 33-15). These masses are replaced by others at the surface of the earth flowing toward the equator, one flowing from the north and one from the south. These air movements form the basis of the earth's major prevailing wind systems.

***The Coriolis Force.*** Winds are generated by the uneven heating of the earth's surface, but the direction of wind movement is dictated by the rotational Coriolis force, the lateral motion imparted to air by the rotation of the earth on its axis. The earth rotates from west to east at over 1,600 km (1,000 miles) per hour at the equator; at latitudes closer to the poles, the rotational velocity of the earth's surface is proportionately less, just as the speed of rotation of a phonograph record is proportionately less closer to its center. As it moves toward the poles at high altitude, the air above the equator also moves eastward. As it moves farther north (or south), the moving air passes over land and sea masses that, are traveling eastward at rates successively slower than its own eastward movement. Those regions of the earth's surface therefore experience high altitude westerly (west-to-east) winds. In actuality, the earth's surface is merely moving eastward more slowly than the air mass above it.

**FIG. 33-15** The probable flow of air from the equator to the poles and back again if there were no Coriolis force.

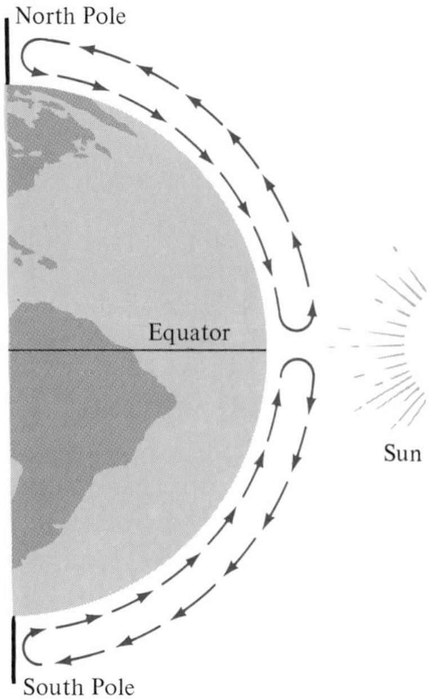


The major prevailing winds, shown in Figure 33-16, are as follows:

1. The north and south trade winds are steady easterly (east-to-west) winds just above and below the equator; except where they pass over islands, they are characterized by little rain. On the equator, between the two belts of trade winds, are the doldrums: calm, hot areas with heavy rains.
2. The north and south westerlies generally flow in a west-to-east direction, but temperate zones are characterized by changeable, stormy weather. Areas of calm, the horse latitudes, lie between the westerlies and the trade winds at 30° to 35° north and south latitude.
3. The easterlies are cold winds that prevail above 60° north and south latitude.

***The Effect of Wind Patterns on Productivity.*** The amount of rainfall received by a given area of the earth, and thus its productivity of biomass, depends primarily on its position in reference to the major prevailing wind patterns. As moisture-laden air rises to high altitude near

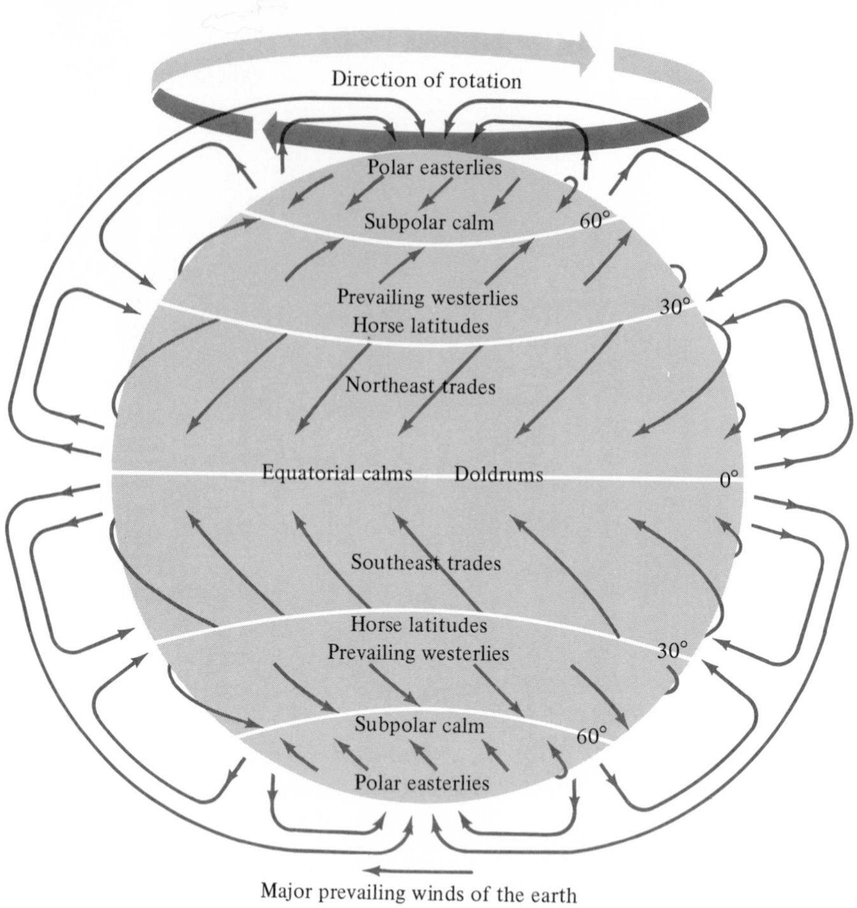


**FIG. 33-16** The directions of the earth's major prevailing winds and the areas of calm between them.

the equator, it cools, causing it to lose great amounts of moisture in torrential tropical rains; it then moves poleward. Figure 33-17, which depicts world (land) patterns of precipitation, shows the heaviest rainfall to occur mostly just south and north of the equator. These are areas of high productivity.

At about 30° to 35° north and south latitude, much of this now drier, cooler (and therefore denser) air begins to sink down toward the earth's surface. As it nears the surface, it warms and rapidly absorbs moisture, with the result that in this belt are found most of the world's great deserts, all areas of low productivity.

***The Effect of Wind Patterns on the Distribution of Biotic Communities.*** The hot deserts in Australia, Africa, and North and South America are located at 30° north or south latitude (Fig. 33-18), regions of high pressure and little rain. Similar patterns of moisture absorption and precipitation by the major prevailing winds account for the fact that the Arctic and Antarctic are, by definition, deserts as well. The existence of scattered deserts in other parts of the earth is determined by a second important factor: the location and altitude of land masses.

Areas on the windward sides of mountains receive above-average rainfall. As moisture-laden air is forced to rise to higher altitudes, as in northeast India and the northwest United States, its moisture condenses and falls as rain or snow. As the now drier, colder, and denser air reaches the leeward side of a mountain range, it drops down, absorbing moisture instead of losing it, producing desert conditions. Such dry land areas are said to be in the rain shadow of the mountains. This

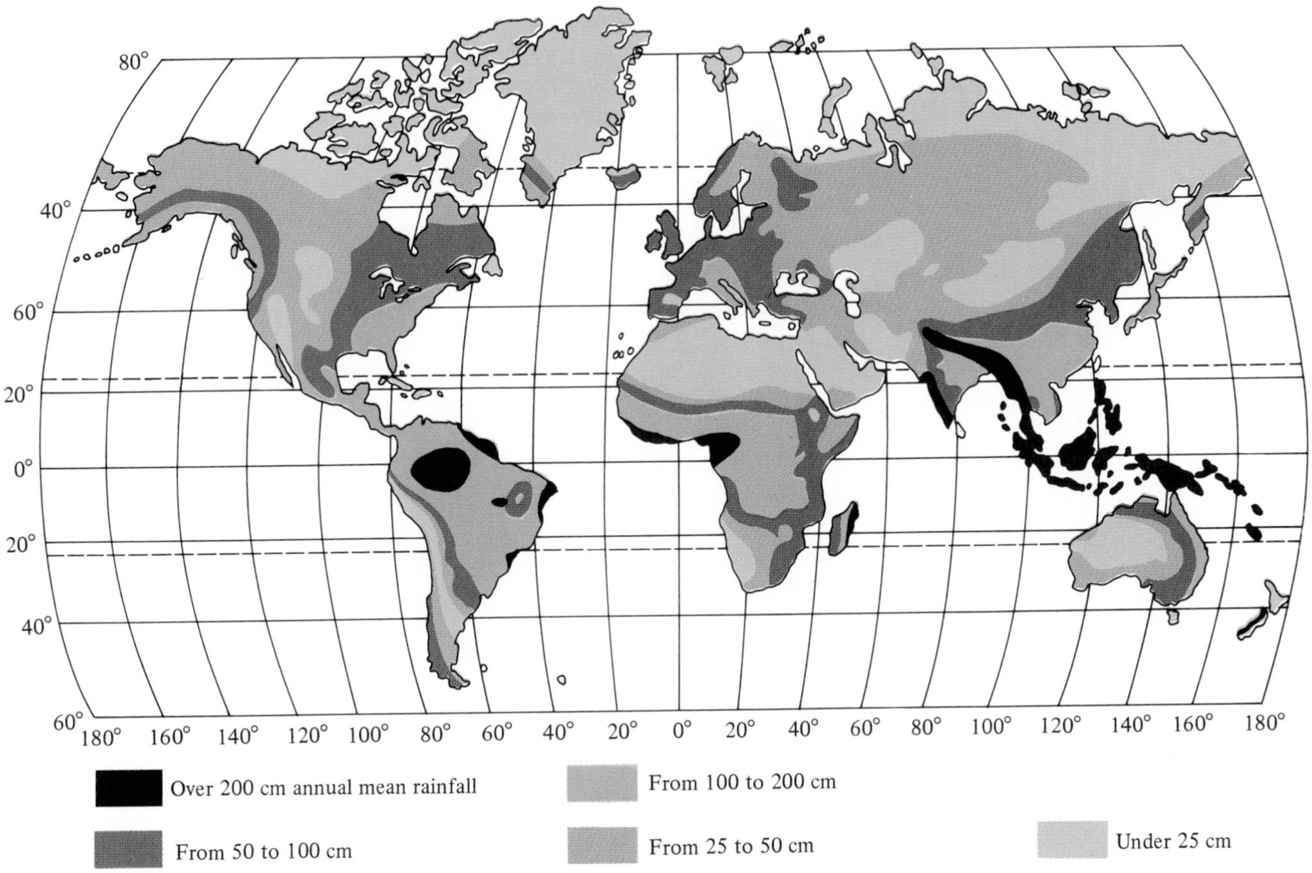


**FIG. 33-17** Worldwide precipitation patterns.

phenomenon accounts for many of the world's temperate and cold deserts.

Chapter 35 describes the world's **biomes,** or major terrestrial plant-animal communities. Because climate is so important in determining what organisms can succeed in a given area, biome distribution at a given latitude is strongly influenced by the area's annual rainfall and, thus, by wind patterns. Along with soil conditions, climate determines what plant communities can grow in an area. Temperate deciduous forests (those that shed their leaves every year), for example, thrive only where the annual rainfall is at least 75 to 100 cm and is somewhat

**FIG. 33-18** A camel caravan crossing the Sahara desert (Libya).

evenly spaced throughout the year. Where annual rainfall is only 25 to 75 cm and where drought years occasionally occur, forests give way to grasslands. Where rainfall is even less, deserts result. In each biome, the predominant plant life determines the kind of animals that can thrive there. The combination of climate and vegetation is also the major factor in determining the type of soil that forms in a region.

### THE DEPENDENCE OF THE HYDROLOGIC CYCLE ON THE SUN AND WINDS

The distribution and success of terrestrial organisms are directly linked to the availability of water by its evaporation into the atmosphere and its return as precipitation, the process known as the hydrologic cycle (Fig. 33-19). This cycle constitutes the most massive physical event on the planet, one in which the sun and winds play a major role. Whether a wind brings rain depends on such factors as whether it has recently passed over a body of water and whether the water was warm or cold, whether the wind has passed over a mountain and lost its moisture on the windward side, and so on. The importance of the hydrologic cycle to human life is especially evident in highly populated regions with low average rainfall or periods of drought (see Chapter 36).

## EFFECTS OF OCEAN CURRENTS ON ECOSYSTEMS

Of great importance to the climate of many land masses, as well as to the productivity of marine communities, are the major ocean currents. These currents are produced by the winds that blow across the surface

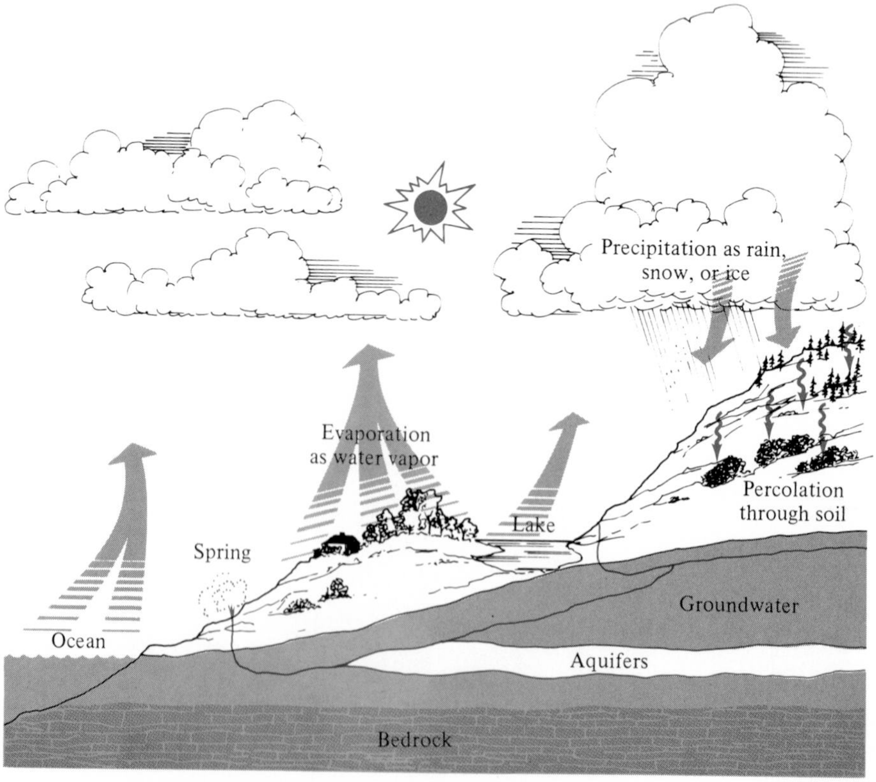


**FIG. 33-19** The hydrologic cycle.

of the seas and thus are another result of the uneven heating of the earth by the sun. A map of the world's ocean currents shows that most of them coincide with the direction of the prevailing winds (Fig. 33-20). Deviations from this rule are due to diversions produced by land masses or to seasonal differences in the heating of the sea and land masses. The directions and strengths of ocean currents have profound influences on our planet's life and thus are of deep interest to ecologists. Two of these influences—the effect on geographic distribution of organisms and the concentrating of productivity in certain seas—are described here.

## GEOGRAPHICAL DISPERSAL OF ORGANISMS

One reason that ocean currents are important to life is that they govern the distribution of so many organisms, a factor that in many cases has profoundly affected the course of their evolution. This is especially true of nonmotile species and weak swimmers. The distribution within an archipelago (island group) or from one archipelago to another of land species that depend on wind and currents for their dispersal is directly related to the direction of the prevailing winds and currents; this is especially true in tropical and subtropical areas in which there is little seasonal variation in wind and ocean current direction. Only rarely do migrations of organisms occur against the prevailing winds and currents.

## UPWELLINGS AND PRODUCTIVITY

Ocean currents also influence the fertility of the sea, that is, the degree to which it supports growth of marine producers and their consumers. In several areas in which ocean currents and winds flow toward the

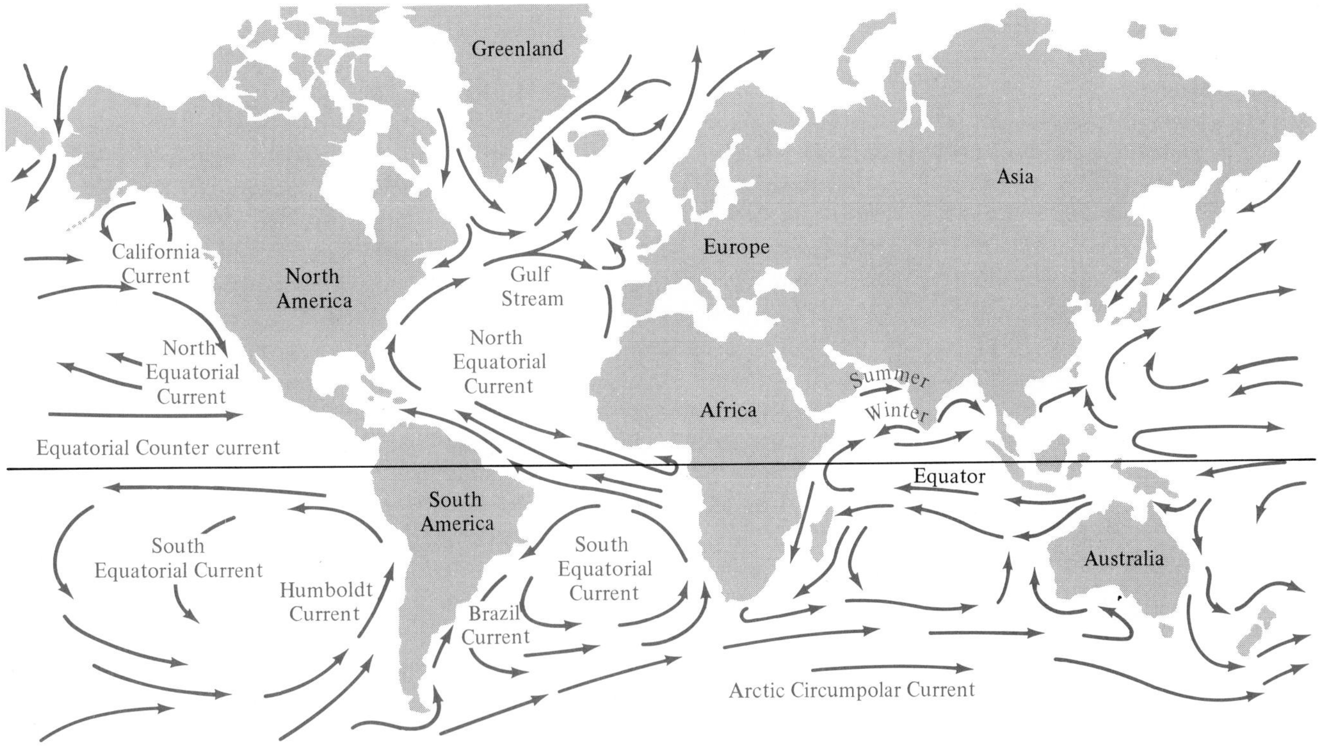


**FIG. 33-20** Major ocean currents of the world. Antarctic Circumpolar Current, Peru (Humboldt) Current, South Equatorial Current, Equatorial Countercurrent, North Equatorial Current, California Current, Brazil Current, Gulf Stream.

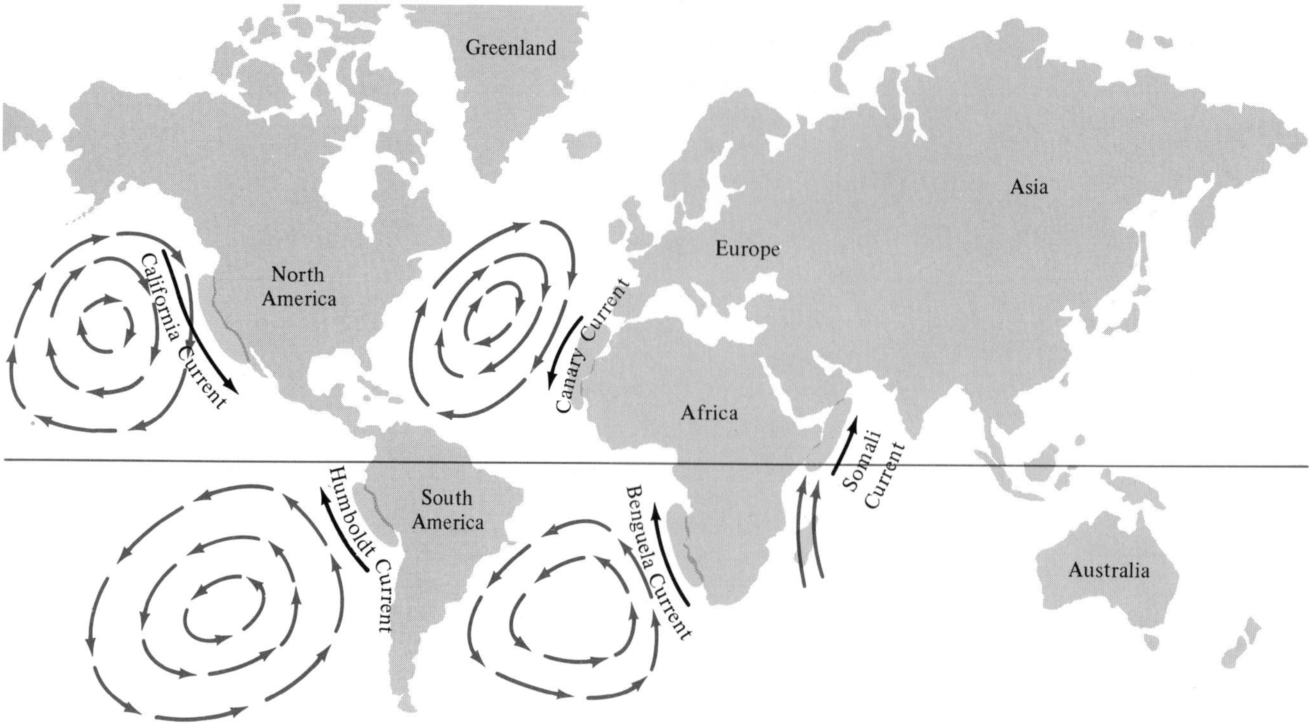


**FIG. 33-21** Major coastal upwelling regions of the world and the ocean currents that contribute to them.

equator along a continental coast, upwellings of cold, mineral-laden water from the sea's depths are produced. Such a situation exists off the coasts of Morocco and the Iberian peninsula and the western coasts of Africa, the United States, and South America (Fig. 33-21). Although these areas represent only 1 percent of the world's oceans, they account for half of the resources of the world's commercial fisheries. One of these ocean currents, the Humboldt Current, sometimes called the Peru Current, has attracted great interest on the part of ecologists.

## UPWELLING IN THE HUMBOLDT CURRENT

**FIG. 33-22** Millions of nesting cormorants on the Peruvian coast.

The Humboldt Current passes northward along the west coast of South America in a broad band several hundred kilometers wide (Figs. 33-20 and 33-21). As with other currents flowing toward the equator along a western coast, the Coriolis force gives this 60-km–wide surface current a westward thrust as well. As the surface water moves north and west, much of the water that replaces it rises from the depths of the Pacific; this water is especially rich in nitrates and phosphates, the nutrients usually in shortest supply in surface waters. The result is a rich bloom of such phytoplankton as diatoms and dinoflagellates. On these feed enormous numbers of crustaceans and a well-known small fish, the anchoveta, or anchovy. Feeding on the crustaceans and anchovies are larger fish, whales, and millions of birds, such as penguins, pelicans, boobies (close relatives of the gannet), and cormorants (known as guanays in Peru) (Fig. 33-22). Deposits of bird droppings, called guano, on seaside cliffs and offshore islands have accumulated in such massive amounts over the millennia that the gathering and marketing of guano for use as a fertilizer and in munitions has been an important Peruvian industry for many years. Most of the ancient accumulations have been harvested, however, and the industry now depends on the annual production of guano. In recent years, a disaster largely of human making has significantly reduced both the fishing and guano harvesting industry of Peru (see Panel 33-3).

PANEL 33-3

## El Niño Plus Overfishing: Recipe for Disaster

Annually, at Christmastime, a small branch of the warm Equatorial Countercurrent (text Fig. 33-20), instead of joining the westward, Equatorial Current, splits off and makes its way southward along the coasts of Ecuador and Peru. Peruvians call the phenomenon El Niño (Spanish for "the child") because its appearance coincides with the celebration of Christ's birth. Every 10 to 15 years, an exceptionally strong current develops, sending a wide band of warm water flowing southward along the Peruvian coast for 1,000 km or more. The colder Humboldt Current is displaced by these warmer waters. Although upwelling continues, most of the water that rises to the surface is from the warmer equatorial regions. It is low in nutrient content and supports little primary productivity (phytoplankton growth). As a result, the phytoplankton and anchovies disappear and the birds starve, their numbers plummeting from 30 million to 5 million or less. Dead fish and birds accumulate, and the waters reek of death. Clouds of hydrogen sulfide from decaying animals mix with fog to blacken ships, cars, and homes. Sharks appear in large numbers. The fishing and guano industries are idled, and personal hardships multiply.

***Overfishing at the Wrong Time.*** The effect of the El Niño event that occurred in late 1972 was a major disaster, largely because it coincided with the close of a decade of unprecedented expansion of the Peruvian fishing industry. Despite repeated warnings, the industry continued to expand until 1970, when an all-time record catch of 12.4 million metric tons (mmt)* was made, representing over one-fifth of the world's harvest of fish protein. However, in 1971, with no reduction in the intensity of fishing, the catch declined to 11 mmt, a likely sign that the waters were being overfished. Although the next year, 1972, produced a major El Niño event, it went unnoticed at first, since many ships were reporting record catches within short periods of time. But these were made close to shore, where anchoveta were gathering in high concentrations in the few remaining regions in which an upwelling of mineral-rich water persisted. By season's end, the total catch amounted to only 5 mmt, less than half the 1971 harvest. The 1973 catch was only 1.8 mmt. A brief recovery followed, to 4 mmt in 1974, followed by further collapse to 1 mmt in 1977. The 1979 catch was also only 1 mmt of anchoveta but included 1 mmt of sardines, which increase in numbers as anchoveta populations decline. If both fish are competing for the same food, it is now possible that the anchoveta population may fail to recover. Annual catches have remained low. Meanwhile, the guanay, a bird that feeds almost exclusively on anchovies, is also in decline.

***What of the Future?*** Surprisingly, it is not yet clear what lessons have been learned from these experiences. Slow progress is being made in attempts to predict the El Niño phenomenon. It is now evident that El Niño events are not limited to the west coast of South America but are part of global disturbances in weather patterns. The matter is very complex, and it may be years before it is possible to provide even a few months warning of a major El Niño event. An important question is whether the fishing industry can be made to refrain from fishing in major El Niño years or from overfishing in any year. Even if a long moratorium were to be imposed on anchovy fishing, it

* A metric ton is 1,000 kg, virtually 2,204 pounds, or a British ton.

*(Continued)*

# PANEL 33-3 *(continued)*

is questionable whether the anchovy population can ever recover.

Although the future of the productive waters off Peru is uncertain, the situation has attracted the attention of international commissions that are attempting, with the use of computerized mathematical models and satellite weather mapping, to predict—and thus be able to prepare for—future changes in winds and currents. There is also much interest in the possibilities of controlling upwelling. In experiments conducted off the coast of the Virgin Islands, in the Caribbean, cold water was artificially pumped up from depths of 900 m to fertilize surface waters; phytoplankton productivity was increased 1,000 percent. It remains to be seen how practicable and how economically feasible this approach might be for the fishing industry.

The meteoric expansion and collapse of the Peruvian fishing industry. The 1979 harvest of anchovies fell back to 1 million metric tons but was supplemented by 1 million metric tons of herring. Catches have continued to remain low ever since, being further depressed by a major El Niño event in 1983.

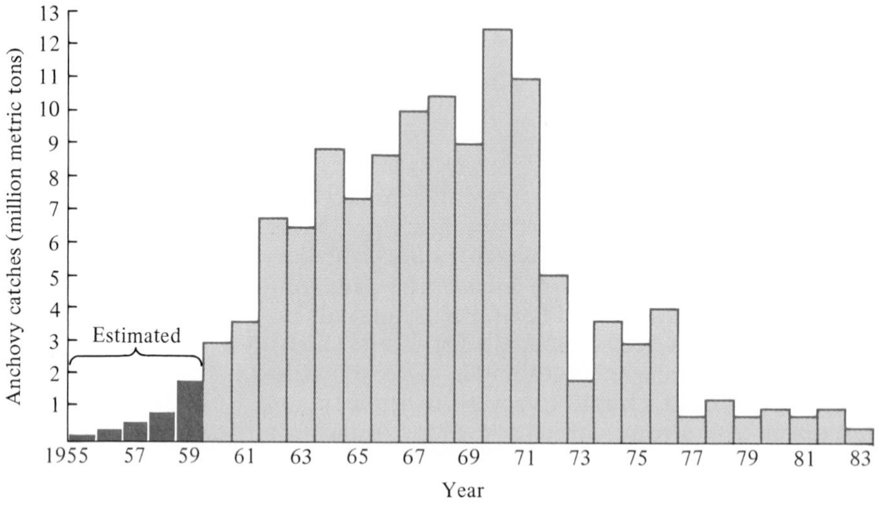


## Summary

Ecology can be defined as the study of the interrelationships of organisms with their physical environment and with each other. A population is all the organisms of the same kind living in a particular area, whereas a community is all the organisms of every kind living in an area. A community and its physical environment comprise an ecosystem, a highly organized processor of energy and materials. The capture and processing of energy and the continued recycling of materials within an ecosystem are performed by its organisms.

All life depends on the community's producers, mostly photosynthetic organisms. Primary consumers eat only producers, whereas secondary, tertiary, and quaternary consumers may eat both producers and consumers. The simplest representation of nutritional relationships is given by a food chain. Two more useful ways of representing such relationships are the food web, showing most of these complex relationships, and the food pyramid, in which the major sources of consumers' food are represented, usually in terms of mass or caloric value.

The ecological niche of an organism is defined by its role in and tolerances for its community. A niche includes an organism's type of habitat, its prey and predators, any of its other effects on the ecosystem, its own requirements and tolerances, and its behavior patterns.

All life needs a source of energy. With rare exception, photosynthetic organisms capture the energy (from sunlight) upon which all the rest of the community depends. Consumers obtain their energy either directly or indirectly from the molecules synthesized by producers. Decomposers reduce the molecules of dead organisms to simple materials that are recycled as producers incorporate them into new living substance. Energy, however, becomes dissipated as heat at each step and cannot be recycled. A supply of energy from outside each organism is essential to its life.

The most important of the elements essential to life are carbon, hydrogen, oxygen, nitrogen, phosphorus, and sulfur. Next to water, carbon constitutes the bulk of living matter. Producers obtain carbon as atmospheric $CO_2$. Consumers obtain it from producers or other consumers. $CO_2$ can be a limiting factor in photosynthesis at times, such as in a forest, late on a bright, cool, humid summer day. Worldwide, however, atmospheric $CO_2$ concentration is rapidly increasing, creating the potential danger of a greenhouse effect that could lead to a warming of the planet and the rise of the oceans as polar ice melts.

Nitrogen recycling involves several kinds of bacteria, some free living and some that are symbionts in root nodules of certain plants, notably legumes. Some "fix" $N_2$ into $NH_3$, some oxidize $NH_3$ to $NO_2$, and others oxidize $NO_2$ to $NO_3$, the form of nitrogen most abundantly used by plants. Decomposing organisms are another important source of $NH_3$. Denitrifying bacteria convert $NH_3$, $NO_2$, and $NO_3$ back to $N_2$.

Although phosphate ($—PO_4$) compounds exist in massive amounts in sedimentary rock, this essential nutrient is in short supply in available forms. Phosphate depletion of arable soils is an agricultural problem in many parts of the world. Because phosphate is a limiting factor in most lakes, the regular entrance of effluent containing phosphates into a lake usually results in its becoming weed choked in a few years.

Limiting factors must be considered in relation to an organism's tolerance; an organism can get too little or too much of a given factor, either of which could limit its growth or survival. Limiting factors must also be considered in context, not in isolation. Combined factors may achieve effects not possible for any one factor alone. A broad view of the factors that determine the distribution of organisms is needed for a full understanding of biologic distribution. Such understanding is important to agriculture, wildlife management, and public health. The endangered species problem is best approached as a problem of limiting factors.

Climate has profound effects on life and determines what organisms can succeed in various regions of the world. The most important climatic factors are seasonal changes, wind, humidity, rain, and temperature. All are at least indirectly generated by the energy of sunlight. The tilt of the earth on its axis in reference to the sun is responsible for seasonal changes, the most dramatic of which occur at the poles. The differences in the angles of incidence of sunlight striking the earth at the equator and at various regions between the equator and the poles result in a substantial warming of the air at the equator and less warming at various distances above and below it. This uneven heating generates the major prevailing winds of the earth. Wind direction is largely determined by the Coriolis force, due to the combined effects of gravity and the earth's spin on its axis. Because wind patterns are responsible for ocean currents and for rainfall, they profoundly affect the productivity of land masses and ocean regions. They are responsible for deserts as well as for moist forests. The productivity of a particular area is a major factor in determining what kind of ecological community can thrive there. The hydrologic cycle, which—like the seasons and the winds—

depends on the sun, is also greatly affected by wind patterns in particular areas.

Ocean currents affect life by governing biogeographic distribution. Open ocean is a barrier to distribution, usually insurmountable if currents are adverse. The effects of ocean currents on the sea's fertility are seen when currents flow toward from the equator along a western continental coast. The Coriolis effect brings up deep, nutrient-laden waters in which phytoplankton and all the life that depends on it can thrive in enormous numbers.

## Recommended Readings

(For Chapters 33, 34, 35.)


CANBY, T. Y. 1984. "El Niño's ill wind." *National Geographic,* 165(2):144–183. An updated account of worldwide aspects of the devastation wrought by recent El Niño events.

COLINVAUX, P. 1973. *Introduction to Ecology.* Wiley, New York. Very well written, interesting to read, with primary emphasis on major concepts.

COLINVAUX, P. 1978. *Why Big Fierce Animals Are Rare.* Princeton University Press, Princeton, N.J. Important essays, written in an engaging style, dealing with basic ecological principles. The author makes an excellent case for reexamination of several long-held ecological views.

COLLIER, B. D., G. W. COX, A. W. JOHNSON, and P. C. MILLER. 1973. *Dynamic Ecology.* Prentice-Hall, Englewood Cliffs, N.J. A serious, professional introduction to the science of ecology. An excellent reference work.

EMMEL, T. C. 1973. *An Introduction to Ecology and Population Biology.* Norton, New York. A concise elementary presentation of ecology, including some attention to human applications. Available in paperback.

GRASSLE, J. F., et al. 1979. "Galapagos '79: initial findings of a deep-sea biological quest." *Oceanus,* 22(2):2–10.

KORMONDY, E. J. 1984. *Concepts of Ecology,* 3d ed. Prentice-Hall, Englewood Cliffs, N.J. A brief but excellent introduction to ecology. Available in paperback.

LABASTILLE, A. 1981. "Acid rain: how great a menace?" *National Geographic,* 160(5):652–681.

LEVINTON, J. S. 1982. *Marine Ecology.* Prentice-Hall, Englewood Cliffs, N.J. An advanced textbook relating basic ecological principles to ocean ecosystems for students wishing to develop a professional interest in marine ecology.

NEBEL, B. J. 1981. *Environmental Science: The Way the World Works.* Prentice-Hall, Englewood Cliffs, N.J. An excellent introductory text with emphasis on the role of natural selection in ecosystems and on human influences on the environment.

ODUM, E. P. 1971. *Fundamentals of Ecology,* 3d ed. Saunders, Philadelphia. A classic introductory text of highest quality. An excellent introduction to ecology written in clear, interesting style.

ODUM, E. P. 1983. *Basic Ecology.* Saunders, Philadelphia. This excellent introductory text is an extensively updated and rewritten version of Part I of the author's *Fundamentals of Ecology.* Part II ("Major Ecosystem Types") of that book has been condensed and included as an appendix.

RICKLEFS, R. E. 1979. *Ecology,* 2d ed. Chiron Press, Newton, Mass. This comprehensive and well-organized introductory text has been acclaimed by many as the best of all. Good emphasis is given to evolutionary aspects of ecology.

*Scientific American,* 221(3). 1969. "The ocean." The entire issue is devoted to the ecology of the oceans. Available in paperback from Freeman, New York.

*Scientific American,* 223(3). 1970. "The biosphere." The entire issue is devoted to energy pathways and biogeochemical cycles. Available in paperback from Freeman, New York.

34

# The Ecology of Populations

OUTLINE



A population is an interacting and interbreeding group of individuals of the same species, for example, all the perch in a small lake. Although they may also occupy a particular area—in this example, the lake—certain populations, such as the elk, a migratory species, live in different locales at different times of the year (Fig. 34-1). Because the members of a population interbreed, they have a common gene pool; thus, populations are the units upon which natural selection acts. Ecologists are interested primarily in **population dynamics:** the various factors contributing to the increase, stability, decline, and extinction of populations, including those of our own species. Understanding these factors is not only important in itself but is essential to an understanding of evolution.

The term *population* is sometimes used to refer to larger or smaller groups than the fishes in a small lake. For instance, all the foxes in England or all the beech trees in North America can be regarded as populations. Even such small groups as all the bacteria living in someone's mouth or all the ascarid worms parasitizing a dog can be considered populations.

Although we often refer to the human population of a country, its members not only interact with but even interbreed with individuals of other countries. Moreover, within a single country, people may be divided into various subpopulations, some of which are quite isolated

**FIG. 34-1** A population of elk (wapiti). Every winter, a population of about 8,000 elk descends from the Teton Mountains to the National Elk Refuge at Jackson Hole, Wyoming.

from each other. The term *population* thus has several connotations. The connotation intended is usually clear from the context in which the term is used.

## POPULATION GROWTH AND STRUCTURE

The growth of a population in a given area is the result of two opposing forces: (1) the biotic potential, or reproductive capacity, of the species and (2) the environmental resistance. **Biotic potential** is the maximum rate of population increase under ideal conditions, which are seldom realized in nature for long periods of time. **Environmental resistance** includes all factors that limit population size, such as nutrient and habitat availability, disease, and predation.

### EXPONENTIAL GROWTH VERSUS POPULATION SIZE STABILITY

An important basis for the natural selection theory of Darwin and Wallace was the fact that all organisms are capable of exponential, or logarithmic, growth in numbers. In a population growing exponentially, each successive total is a multiple of the one preceding it. Put another way, populations are capable of doubling at a constant rate, such as once every so many hours, weeks, or years. (An example of logarithmic increase is provided by the series 2, 4, 8, 16, 32, 64, and so on. An arithmetic increase is an increase by a constant number, exemplified by the series 10, 20, 30, 40, 50, and so on.) Many organisms have enormous reproductive potential. Some bacteria can double in number every 15 minutes. A single mushroom produces hundreds of thousands of spores, each of which could, if provided the space, nutrients, and so on, give rise to a huge colony of mushrooms, each of which could produce many more mushrooms. In many insect and fish species, one individual can lay thousands of eggs. In some species of clams, one individual can produce 3 million eggs in a season. A single pair of mice, given the necessary food and space and an absence of disease and predators, could, theoretically, have 3,000 descendants in one year—or 6.75 million in 3 years! To emphasize the universality of the potential for logarithmic increase among organisms, Darwin analyzed the most slowly reproducing animal species, elephants, and found that even they were potentially capable, within a few years, of producing more offspring than the earth could accommodate. However, as we are well aware, this enormous potential is, except for brief periods, never realized.

Darwin further observed that, with occasional fluctuations, populations of a given species tend to remain about the same size from one year to the next. He concluded that "this geometric tendency to increase must be checked by destruction at some period of life." In the more than a century since those words were written, we have learned much—but far from all—about the factors that regulate population growth.

### ANALYSIS OF EXPONENTIAL GROWTH

Some insight into factors that determine population size has been achieved by studying simplified ecosystems. For example, when a culture vessel containing a warm broth composed of a balanced mixture of all the nutrients needed for the growth of one particular species of bacterium is inoculated with a few hundred individuals derived from a thriving culture, the cells continue dividing at a steady rate (Fig. 34-2a, b). If the culture's growth is graphed as the logarithm of the number of cells against time, the result is a straight line (Fig. 34-2c). Such a culture is said to be in exponential growth, in a logarithmic growth phase, or, simply, in log growth.

To predict the growth of a given population over a particular time period, we need to know the initial size of the population (symbolized by $N$) and the potential growth rate of that species, called its **intrinsic rate of increase** (symbolized $r$). Intrinsic rate is the difference between the population's **natality**—the annual rate of births, hatchings, and the like—and its **mortality**—the annual rate of deaths—under what could be called ideal conditions, with adequate food and shelter, no predation, and so on. Both natality and mortality rates are largely genetically determined and can differ greatly from one species to another.

Expressed mathematically, the increase in size of a population undergoing exponential growth is stated as $\Delta N/\Delta t = rN$, in which $\Delta N$ represents the difference ($\Delta$) in population size ($N$), $\Delta t$ is the length of the time interval (difference in time) during which exponential growth occurred, and $rN$ is the intrinsic rate of increase ($r$) for a starting population of size $N$. Stated in words, the equation reads as follows: the actual increase in numbers ($\Delta N$) during a time interval of $\Delta t$ (some fraction of a year) equals the potential maximum rate of increase per year ($r$) times the number of individuals ($N$) at the beginning of the time interval $\Delta t$. So it can be seen that, under ideal conditions, the actual amount of increase may equal the potential growth rate and, for brief periods, this may be the case.

**FIG. 34-2** (a) Scanning electron micrograph of the common intestinal bacterium *Escherichia coli* from a broth culture. (b) Growth of a bacterial culture plotted arithmetically and (c) logarithmically. Note that the log plot of a population growing at a steadily increasing rate is a straight line.

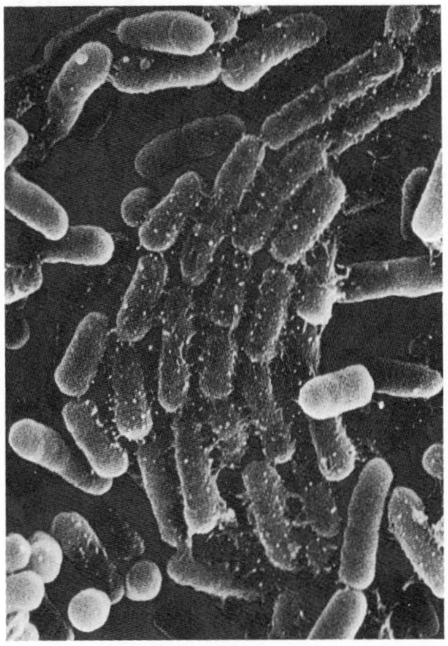
(a)

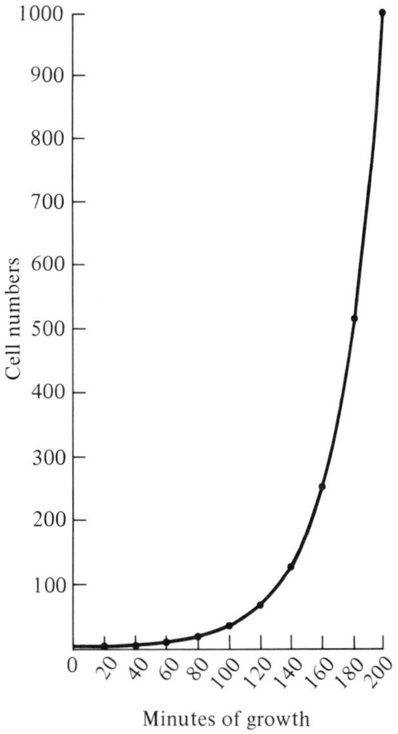

(b)

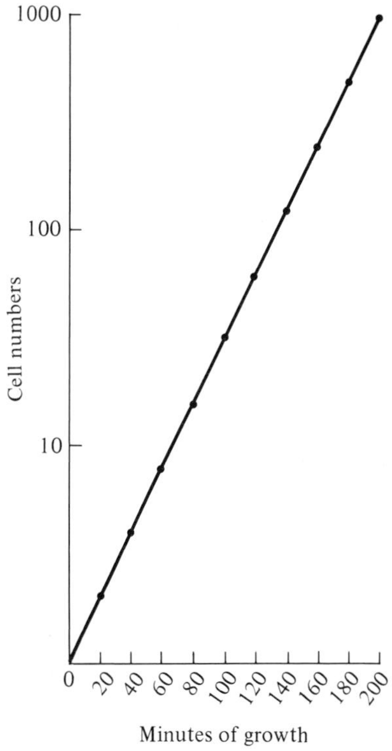

(c)

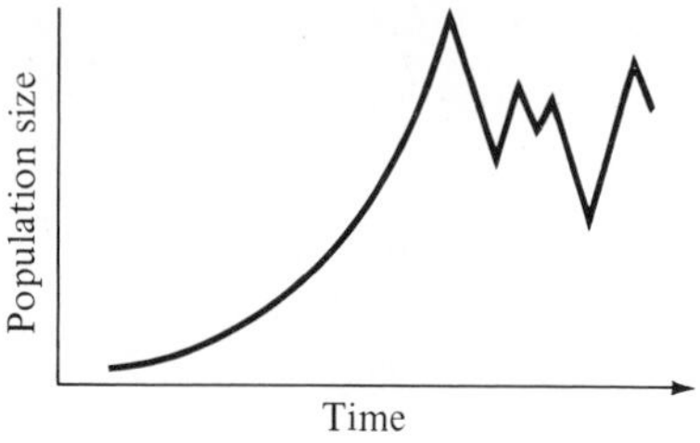


**FIG. 34-3** Malthusian, or irruptive, growth curve. The upslope of the curve is steep and is followed by a series of partial population collapses.

## MALTHUSIAN VERSUS LOGISTIC GROWTH

An early attempt to analyze population growth patterns and thus to derive from them some principles applicable to humans was made by the English economist Thomas R. Malthus in 1798. He noted that populations tended to outgrow their food supply and predicted that human populations would eventually become drastically reduced by famine, disease, war, and similar catastrophes. Later, a differing concept, termed **logistic growth,** emerged. According to it populations eventually tend to level off at a relatively high level.

***Malthusian Growth.*** The Malthusian doctrine can be illustrated by the hypothetical growth curve shown in Figure 34-3, in which a period of exponential growth is climaxed by a series of catastrophic reductions in population size due to massive famine, disease, and other factors. This pattern is referred to as **irruptive growth** or **Malthusian growth.** Malthus' idea about humanity's future was poorly received, not only at the time it was published but also in subsequent years. The concept was decried by religious leaders as an irreverent attempt to curb human procreation and by scientists as a lack of faith in the capabilities of science. Malthus was viewed as an alarmist, and his idea was generally ignored. Only in the latter half of this century, as his predictions have begun to come true, has his theory been accorded much respect.

The strongest reason for ignoring Malthus' idea during the nineteenth century was cogent: the experience of so many observers to the contrary. By the mid-1800s, considerable data had accumulated on the growth of natural and laboratory populations of many species of organisms. In natural populations, in which predation and other limiting factors were operative, growth of the population often leveled off for an indefinite period, exhibiting a sigmoid (S-shaped) growth pattern (see Fig. 34-4a). In laboratory cultures of microorganisms, a similar sigmoid growth curve was often obtained (Fig. 34-4b). Unfortunately, these bacterial experiments were terminated before what is called the death phase began; otherwise, the results would have been more typical of Malthusian predictions.

***Logistic Growth.*** The many examples of sigmoid growth curves in laboratory and natural populations led to the concept of logistic growth. This is a pattern typified by a slow beginning, a log growth phase, and a gradual slowing to a smooth, orderly, predictable pattern of population limitation at a level indefinitely supported by the ecosystem, said to be its **carrying capacity.** Although at first little attention was paid to this concept, it eventually received wide support, especially in the early 1900s. Some even hailed it as a universal law and applied it to the growth of all populations, including humans. One enthusiast, invoking the principle of logistic growth, predicted that the world human population would naturally stabilize at 2.6 billion ($2.6 \times 10^9$) people in AD 2100. As we know, due to changes in several factors since that time, this population size was exceeded before 1960, passed the 4 billion mark in 1975, and is still increasing rapidly.

In a bacterial culture, after a period of log growth, the length of which depends on the size of the initial population and the total volume and nutrient concentration of the broth, growth of the culture slows. It then gradually enters a period of variable length in which the number of cells remains relatively constant. In a bacterial culture, this is termed

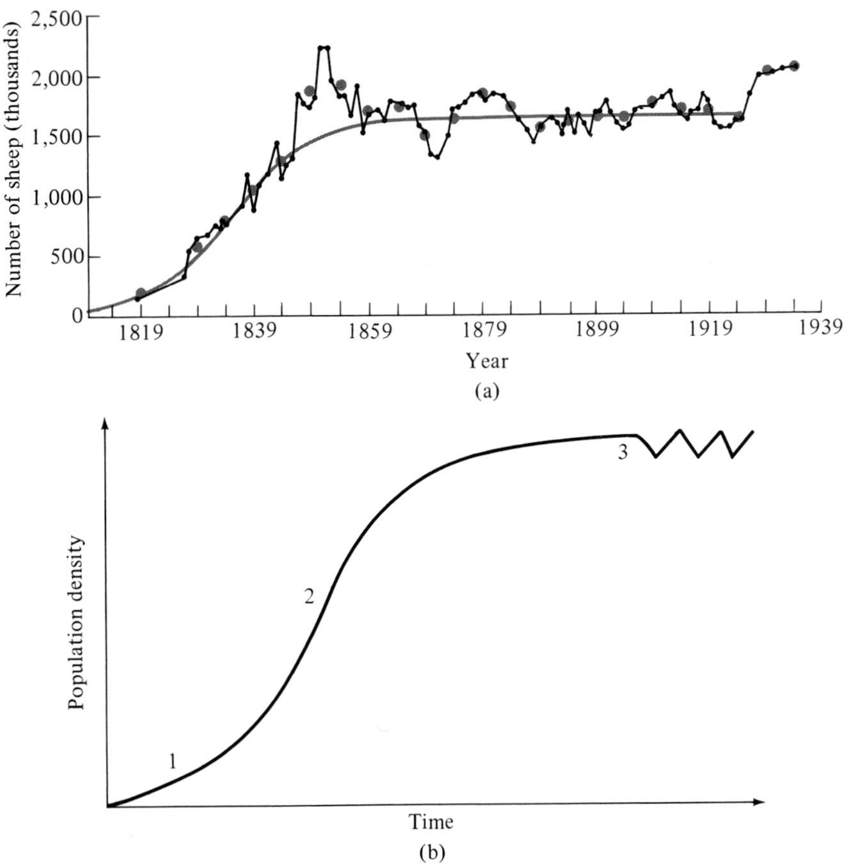


**FIG. 34-4** Examples of the achievement of carrying capacity in two different environments. (a) In the early 1800s, sheep were introduced to the island of Tasmania. The population exhibited a roughly sigmoid growth curve, stabilizing at about 1.7 million in the 1950s, after exceeding the carrying capacity for a few years. The fluctuations exhibited thereafter are not as great as in most natural populations. (b) Sigmoid growth curve of a bacterial culture showing (1) the lag phase; (2) the logarithmic growth phase; and (3) the maximum stationary phase.

the **maximum stationary phase.** It represents a balance between the production of new cells and the death of older ones. A similar pattern is exhibited by populations of most species in most environments. The factors that control population size in such situations are examined later in this chapter.

The final population size achievable by a species in a particular environment, that is, the carrying capacity of that environment, or the environmental resistance, is symbolized as $K$. Although the carrying capacity of a given environment is a constant, it is subject to change as the long-term climatic or certain other factors in an environment may change over the years. Moreover, by its evolutionary adaptation to a particular environment, a species can, over many years, increase the value of $K$.

***The Logistic Growth Equation.*** Stated mathematically, the difference between an environment's carrying capacity and the number of individuals already present in it (or the environment's unused capacity) can be stated as $K - N$. As the population increases, $N$ comes increasingly closer to being equal to $K$, and population growth gradually slows. This relationship can be expressed by the ratio

$$\frac{K - N}{K}.$$

When the population ($N$) is small, the value of this ratio is high, close to 1.0. As population size increases, the value of the ratio correspondingly lessens.

Stated in this form, the adverse effect of environmental resistance can be incorporated into the growth equation, thereby indicating the actual growth rate of a population for various population sizes and environments of various carrying capacities:

$$\Delta N/\Delta t = rN\frac{(K - N)}{K}$$

In this form, it is known as the logistic, or differential, growth equation. It expresses the negative feedback relationship between a population and the environmental resistance.

## THE STRUCTURE OF POPULATIONS

Additional important information about a population's prospects for increase, decrease, or failure can be revealed by a survivorship curve. For natural populations, this is usually obtained by plotting data derived from repeated samplings of marked individuals. Figure 34-5 depicts three types of curves, one approximating that for humans, another for the cnidarian *Hydra,* and a third for oysters. Because they are not cared for, the mortality rate for oyster eggs and larvae is extremely high, although any that survive early development have a good chance of living to maturity. Humans, on the other hand, with extended parental care, have a relatively high life expectancy at birth, with a sharp increase in mortality in the last 15 of their expected 70 years. Hydras exhibit the same death rate at every age.

Although survivorship curves provide some insights into the prospects of a population of a given species for survival, much more useful in this regard are the age structures of particular populations. In one way of representing age structure, the population is divided into three age groups, each a percentage of the entire population: (1) prereproductives, or immature individuals; (2) reproductives, or fecund adults; and (3) postreproductives, or nonfecund adults. The proportions of a population in each class are key factors in determining whether that population can be expected to increase, decrease, or stay the same.*

Figure 34-6 diagrams the three different distributions. The diagram for an expanding population, such as the present worldwide human population, has a pyramidal outline, its base being composed of prereproductives (Fig. 34-6a). The diagram for a stable population has a bell-shaped outline (Fig. 34-6b); that for a declining population has an urn shape (Fig. 34-6c). An urn-shaped distribution, with its small proportion of prereproductives, usually indicates that a population is dying off. While greatly affected by environmental factors, age structures are also genetically influenced and can differ greatly from species to species. Some insects, for example, spend months or even years as larvae and then, as adults, as little as one day in reproduction. In contrast, the reproductive years of humans in a favorable environment comprise about 42 percent of a mean life expectancy of 70 years. Some species reproduce throughout adult life, and many animals, including various worms, insects, and fishes, die almost immediately after reproducing.

**FIG. 34-5** Survivorship curves for humans, hydras, and oysters.

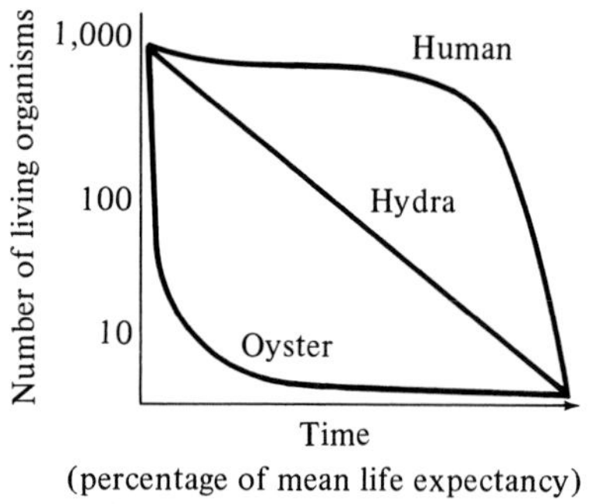


* In the case of the worldwide human population, because of the high percentage of young individuals, even if all families of reproductive-age individuals had only two children from now on (the so-called zero population growth, or ZPG), the world's human population would continue to increase well into the next century.

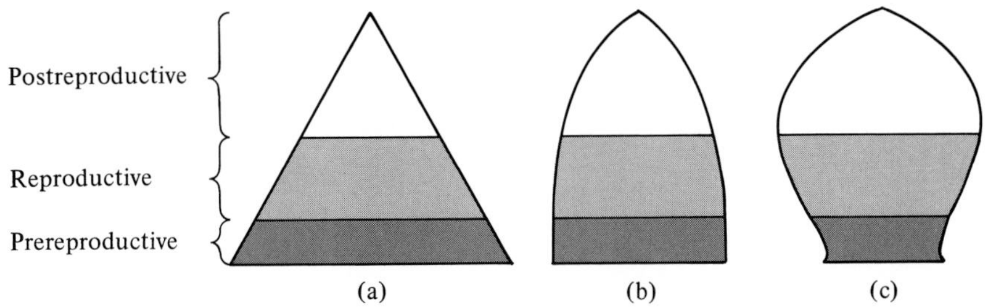


**FIG. 34-6** Age structures for (a) an expanding population, (b) a stable population, and (c) a declining population.

Populations of such species have no postreproductive component in their age structure.

The relative proportions of the component stages in the life history represented by each of the three classes can be greatly affected by both biotic and environmental factors and may therefore differ geographically for a given species. With many human populations, improvements in medicine, sanitation, and health care have increased the proportion of children living to reproductive age and the number of adults reaching old age. (Although life expectancy of North Americans in 1900 was less than 50 years, it is now about 70 years. In the heyday of the Roman Empire, the mean life expectancy of its citizens was only 25 years.) In some economically depressed human populations today, only 50 percent of all children survive their fifth birthday, with very few living beyond their reproductive years. These examples emphasize that the growth rate of a particular population is the result of an interaction between its intrinsic growth rate and the resistance afforded by the local environment.

## LIFE HISTORY PATTERNS

As noted in Chapter 21, organisms exhibit two basic approaches to reproduction. One involves an investment by the parents in large numbers of spores, seeds, eggs, or offspring that receive little or no care (Fig. 34-7). In such species, sheer numbers tend to ensure that some individuals survive to reproductive age. The other approach involves the production of a few offspring that receive substantial parental assistance. In the former case, much of the parental body is given over to reproductive structures; in the latter, most of the body is vegetative, that is, nonreproductive, biomass. This principle applies to all forms of life. Examples of plants that produce few individuals but invest much in each of them are those that send out either stolons (runners) or rhizomes (underground stems) from which new individuals bud off and reach maturity while still attached to the parent plant. An example of animals that produce few individuals are birds that lay one to three eggs and provide the hatchlings with food and other parental care until they are almost fully grown.

**FIG. 34-7** The gills of the mushroom *Laccaria ocheopurpurea*. In some species, the gills produce billions of spores per mushroom. Sheer numbers tend to ensure that some will survive to reproduce.

Between the two extremes of one offspring that receives much care and many offspring that receive no care is a full spectrum of intermediate situations. For example, some organisms produce a moderately large number of offspring that receive some care and protection in the early stages of development. All of these behaviors, of course, are genetically determined, although learned components are often incorporated into them. Like many other hereditary traits, they represent adaptations to particular environmental situations.

What determines, for a particular species, which of the two opposite approaches is selected or whether the species will exhibit an intermediate pattern that incorporates elements of both? The answer is not always clear, and in some situations, it may not matter which approach is followed; in many cases, probably either would serve. However, various environmental situations are most successfully taken advantage of by one or the other of the two approaches.*

***r* Adaptation.** Let us first consider the case of organisms most of whose developmental stages, such as spores, seeds, eggs, or larvae, are eaten by predators or infected by disease organisms. Many such species reproduce only once a year and have adapted to predation and disease by producing more spores, seeds, eggs, larvae, or hatchlings than the existing predators or disease organisms can eat or infect at one time. Such an approach is termed *r* adaptation or *r* selection because it depends on a rapid, exponential increase in numbers at a rate close to the organism's intrinsic rate of increase (*r*). Such high productivity at one time has the double effect of increasing the chances that some offspring will survive to maturity and of discouraging predators from specializing on a type of prey that is briefly available only once a year. However, because parasites and other disease organisms also can evolve, many have adapted their own life cycles to those of their hosts, multiplying and even becoming dormant in concert with the hosts.

An elaboration on the once-a-year reproductive pattern is the **mast crop,** characteristic of many tree species in temperate zones (Fig. 34-8). In these species, 2 to several years may pass with little or no production of seeds or fruits, followed by a single season in which all such trees produce a great abundance of seeds. The populations of seed-eating animals, such as birds, squirrels, and insects, that depend heavily on these particular seeds or fruits may decline in the area during the trees' nonreproductive years. Then, suddenly, the reduced predator population is given far more food than it can eat; the next year's population of preda-

* The term *reproductive strategies* is often applied to these various ways in which organisms adapt to their environments. Because the word *strategy* connotes an intelligent planning by the organism and tends to be misleading, its use is avoided in this book.

(a)

(b)

**FIG. 34-8** (a) A stand of oaks in Comanche County, Oklahoma. Some species of oaks have mast years in which they produce an unusual abundance of acorns, up to 100,000 per tree in some cases, whereas in other years, the number declines greatly. (b) This tends to reduce populations of those animals, such as the gray squirrel, that feed on acorns.

tors, now larger, faces the beginning of another cycle of several lean years.

As might be expected, a selection pressure exists for various species of mast-crop-producing trees that inhabit the same environments to synchronize their mast crop production. Moreover, there would be less selection pressure for a tree to produce a mast crop if similar species in the area produced their mast crops in alternate years to its crop years. For several tree species to synchronize their mast crops and thereby discourage predation of all of them permits a much higher population density of such trees and thus a greater overall efficiency in the capture of energy and the production of biomass. The trees do not need to communicate with each other in any way. Natural selection works against those that fail to synchronize their crops.

Selection pressure for *r* adaptation is also exerted by environments that place a premium on the ability to produce large numbers of offspring within a short time. Two classic examples are the annual plants and the insects that live in deserts and those that live in the Arctic. These organisms develop from eggs, seeds, or other dormant stages whose development is triggered by the arrival of favorable conditions. They begin development, grow to maturity, reproduce, and set seed or lay eggs all within the space of a few weeks each year (Fig. 34-9). Thus, the *r* adaptation tends to be selected for by an environment with a brief growing season followed by conditions in which a high degree of mortality is independent of population density, such as occurs with a sudden, severe onset of desert aridity or an Arctic winter.

Another situation that selects for *r* adaptation is the early stage of colonizing a new island or a burned-over area. Species that are the first to colonize such an area, or pioneer species, exhibit *r* adaptation. Organisms with a high intrinsic rate of population increase are usually small, develop rapidly to reproductive maturity, and have short life spans.

***K* Adaptation.** What kind of a situation might select for organisms that produce only one or a few offspring at a time? Most known examples occur in situations that exert strong negative-feedback controls on pop-

**FIG. 34-9** Poppies *(Eschscholtzia)* carpeting an Arizona desert after a March rain. Nine weeks later, all will have set seed and died. The brief rainy season selects for species that can germinate, mature, flower, and set seed within a short time.

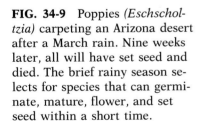

ulation growth. Under relatively constant year-round conditions in which an environment's carrying capacity for a particular species is often reached, competition for limited resources among species with overlapping needs tends to be high; in such circumstances *K* adaptation, or *K* selection, a term derived from the symbol for environmental resistance, tends to be favored. Such are the conditions in tropical rain forests, where birds produce relatively small clutches of eggs. But other factors can also contribute to *K* adaptation. For example, larger, longer-lived organisms, such as the blue whale and the emperor penguin, are usually *K* adapted (Fig. 34-10). A longer life span permits **iteroparity,** or repeated reproduction, which compensates for raising only one offspring at a time.

**FIG. 34-10** An emperor penguin feeding its half-grown (12-week old) chick on Cape Crozier, Antarctica. An emperor penguin pair raises only one offspring a year from an egg laid in the dead of winter on the Antarctic shelf ice, where fierce storms rage and temperatures drop to −70°C (−94°F) or lower. The male incubates the single, large egg for 2 months by holding it atop his feet and completely covering it with a warm, pouchlike flap of belly skin. For 5 months or more after the egg hatches, the parents take turns feeding the young. Although a breeding pair of emperor penguins produces no more than one offspring per year, the slow breeding is compensated for by a long life span (25 years) and iteroparity.

## POPULATION REGULATION

Simple ecosystems represented by cultures of bacteria and protozoa have proven useful in analyzing factors that regulate population size. Among the reasons that cells begin to die in thriving cultures of microorganisms is starvation. This eventually results from depletion of essential nutrients and is an example of a **density-dependent factor** affecting population size. A factor, such as a disease epidemic, that exerts its effect on a population irrespective of its density is a **density-independent factor.**

Density-dependent factors, such as nutrient supply depletion, constitute negative feedback controls over a population's absolute size and account for an increasing percentage of mortality as density increases. The total amount of a key nutrient present in an environment may be quite unimportant when a population is small; it becomes increasingly important and eventually limiting as the population increases in density. In a bacterial culture that has exhausted one or more of its essential nutrients, when the starved cells die, they disintegrate, releasing their nutrient components into the medium. This new supply of nutrients permits some of the other starving cells to resume growth and division. Therefore, at this stage of the culture's existence, the number of new cells tends to equal the number of dying cells, resulting in the upper plateau of the sigmoid growth curve.

If a new culture is begun by introducing into fresh broth medium several cells from an old culture that had reached maximum stationary phase, a lag phase at first occurs—that is, little or no growth takes place for a time—followed by a period of exponential growth and, as the nutrients become used up, culminating once more in a maximum stationary phase (see Fig. 34-4).

The maximum stationary phase of a bacterial culture may last for hours, days, or weeks. But eventually, if nothing is added or removed from the culture, this phase ends as the number of deaths begins to exceed the number of cell divisions. Thus begins the death, or negative growth, phase. What happens then depends on several factors that vary under different conditions and from one species to another. Some species exhibit a log death phase that can be as steep as or steeper than the log growth rate. In other cases, every cell dies in only a few days. In yet other instances, the number of cells drops to a new low level, but the culture continues to survive, its numbers oscillating irregularly at the lower level. Such a culture may persist in this condition for months or even years.

## DENSITY-INDEPENDENT FACTORS

As a general rule, density-independent factors usually reduce a population by a certain constant percentage no matter what the size of the population is.

One type of factor that, despite adequate nutrients, can slow, stop, or reverse the growth of a culture of microorganisms is the accumulation of toxic wastes. This factor can be much more serious for the ultimate survival of the culture than can starvation. In old cultures, enough nutrients may be released from the disintegration of dead cells to support growth and reproduction of other cells indefinitely. But the concentration of inhibitory substances that can neither diffuse out of the culture nor be decomposed or detoxified by the organisms in the culture, although initially having no appreciable effect, may finally reach a level incompatible with the life of any of the cells.

A toxic material that has reached a concentration that kills most of the members of a culture is usually a density-independent factor for population growth. The resulting irruptive growth curve is strikingly different from the sigmoid curve; it is said to be J shaped (Fig. 34-11). In nature, examples of density-independent factors include droughts and cold spells, which by arresting growth, may bring about a nearly total population collapse. Although a sudden drop in population size results in such cases, usually not all individuals die, and, when favorable conditions resume, a second period of log growth usually occurs. Massive death from brief disease epidemics, unfavorable climatic conditions, or temporary periods of starvation also produces J-shaped growth patterns.

Not all growth curves are readily classified as either sigmoid or J shaped. Nor are all factors affecting population growth easily catego-

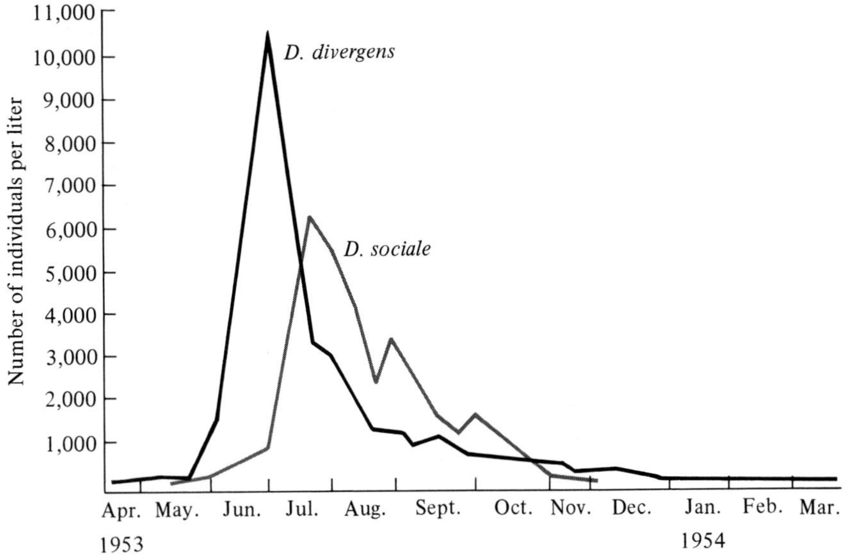


**FIG. 34-11** J-shaped growth curves for two species of algae, *Dinobryon divergens* and *D. sociale*. The J refers to the general shape of the upslope.

rized. Both starvation and predation, which usually function as density-dependent factors, can sometimes cause population collapses resembling the downslope of a J-shaped growth curve.

## INTRASPECIFIC RELATIONSHIPS AND CARRYING CAPACITY

From its first enunciation, some biologists criticized the concept of logistic growth as simplistic. Yet it is certainly true that natural populations maintain relatively stable levels year after year. A controversy also exists over whether density-dependent intraspecific competition or density-independent extrinsic factors are of primary importance in the regulation of population size. Careful analysis of stable populations, both in natural situations and under laboratory conditions, has revealed a wide variety of factors, both density-dependent and density-independent, that affect a population's size once it reaches a certain level. Some of the density-dependent factors have also been revealed by laboratory experiments such as the following on beetles, mice, and rats.

In separate experiments, adults of several species of flour beetles were placed in sifted, humidified flour mixed with yeast (Fig. 34-12). The beetles burrowed into the flour at once and laid eggs in the tunnels they made. The larvae, or meal worms, which hatched 5 to 10 days after egg deposition, also burrowed into the flour. Both the adults and the larvae ate any eggs and pupae they encountered. Male and female adult encounters, however, often resulted in copulation and the laying of more eggs. As the population increased, inhibitory substances released by the beetles accumulated. As the amount of these substances increased, females produced fewer eggs, larvae developed more slowly, and adults grew to a smaller size. A steady population level was finally reached, characteristic of the particular species of flour beetle being tested. The carrying capacity of that environment for that species had

**FIG. 34-12** A common pantry pest, the flour beetle *Tribolium confusum*.

therefore been reached. A typical result was a steady increase of population to a stable level characterized by a fixed ratio of adults to larvae and a 90-percent or higher mortality rate of eggs through cannibalism. But some species never equilibrated; instead, their population levels oscillated repeatedly in a pattern reminiscent of a Malthusian prediction. Such oscillations can result when large cohorts, groups of individuals of the same age, reduce the size of succeeding cohorts by cannibalism. The smaller cohort is incapable of having the same effect on the large cohort that follows it. Such an effect can persist indefinitely. Although other factors are also involved, the beetle experiment is primarily a case of population regulation by density-dependent intraspecific competition.

In another series of experiments, separate mouse populations with unlimited food and water were established in six 2-by-8-meter pens. In three of the pens, the food, water, and nesting boxes were scattered throughout the pens; in the other three pens, all the food and water was placed at one end, and all the nesting boxes were crowded together at the other. The maximum number to which each population rose differed from 25 to 138 among the six pens. Although all were adversely affected by the crowding, the particular level that was reached in each case seemed related more to individual temperamental differences in the animals than the immediate effects of the physical conditions.

As the population grew, so did the fighting among adults (Fig. 34-13). One population showed a decline in birth rate and a high mortality in the litters. In three of the six populations, the primary limiting factor was an increased mortality of newborn. Fighting often was followed by trampling of nests, which led to desertion or cannibalism of the litters by the females. Although food was abundant, food intake declined as the population increased. In some populations, food intake was so reduced as to lower normal female fecundity, the ability to produce offspring. But even in populations in which enough food was eaten to maintain fecundity, birth rates dropped off. This appeared to be due to less frequent copulation under the crowded conditions, increased spontaneous

**FIG. 34-13** Fighting between mice. Normally, only adult males fight. Under crowded conditions, all adults may fight.

abortions (miscarriages), or both. Levels of female sex hormones were shown to be affected by increased odors resulting from crowding, with the result that ovulation ceased and sexual behavior was inhibited.

Similar studies were performed on rats in crowded conditions with unlimited food and water. In these experiments, population size proved to be regulated principally through deaths from fighting (Fig. 34-13). Most deaths were due to wound infections or other aftermaths of fighting. The mouse and rat experiments are cases of density-dependent population size regulation by behavioral and physiological means.

Such studies of intraspecific relationships and carrying capacity have revealed some important principles:

1. Even under controlled conditions, it is difficult and sometimes impossible to predict (a) whether a given population of any species will reach a stable level; (b) whether it will oscillate slightly about a certain level; or (c) whether it will exhibit a Malthusian pattern of catastrophic declines alternating with periods of exponential growth.
2. The natural methods by which populations are limited vary greatly from species to species and sometimes from population to population within a species. Besides such factors as limited food supplies and predation, controls that limit population size in nature appear to include (a) cannibalism of the young; (b) deaths of adults from increased fighting; and (c) reduced fecundity, birth rates, and rates of development as a result of the accumulation of toxic substances or other effects of crowding. Nature's way of regulating population sizes obviously includes methods that are not regarded as desirable or acceptable ways of limiting human populations.

## INTERSPECIFIC RELATIONSHIPS OF POPULATIONS

We have seen how some insight into population dynamics can be gained by studying experimental ecosystems containing single species of organisms. An even better understanding of natural controls on populations can be obtained by studying cultures of more than one species. Two rather different examples are (1) cultures in which two similar species compete for the same resource and (2) cultures made up of a predator and its prey.

***Competition Between Similar Species.*** In ecological terms, competition is the interaction of two organisms seeking the same things, such as food, shelter, space, or mates. The keenest competition most often occurs between members of the same species or two similar species. A 1934 experiment performed on two species of ciliate protozoans by the Russian biologist G. F. Gause nicely demonstrated some of the effects of competition on population dynamics.

Gause found that separate cultures of *Paramecium aurelia* and *Paramecium caudatum* reached a high maximum stationary phase when raised on bacteria (Fig. 34-14). When he grew both species in the same culture vessel, they competed for the available food. As long as food was abundant, both populations grew exponentially; but as food became depleted, the growth of both populations dropped to a slower than normal rate (Fig. 34-14c). Later, the population of *P. caudatum* severely declined, while that of *P. aurelia* continued to increase. Eventually, the

**FIG. 34-14** (a) *Paramecium caudatum* and (b) *Paramecium aurelia.* (c) Growth curves for *P. aurelia* and *P. caudatum* when grown in separate cultures and when grown together.

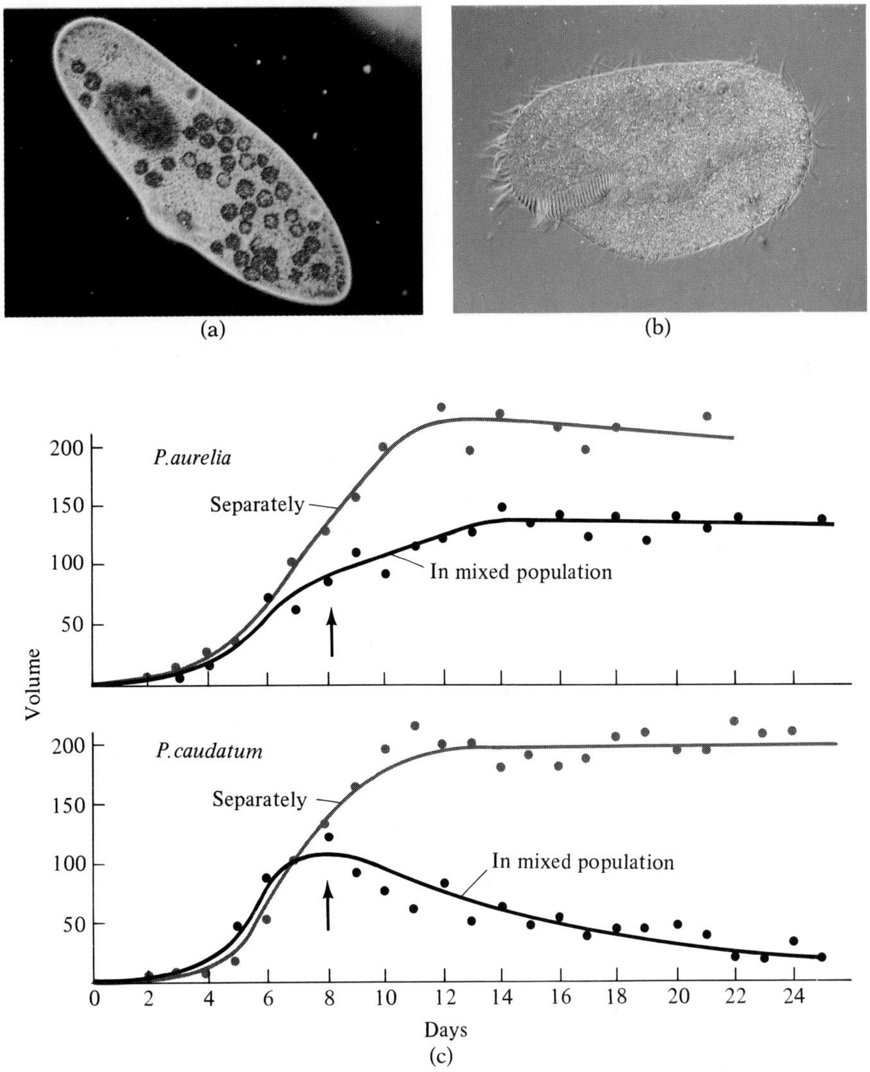


population of *P. aurelia* reached about the same size it had achieved when grown alone.

Gause obtained similar results when he grew other species of competing organisms. These experiments and various field studies led to the development of what came to be known as Gause's rule or the **competitive exclusion principle,** which states that in any one geographical area, an ecological niche cannot be fully and simultaneously occupied by stabilized populations of more than one species. This means that if two very similar species with very similar habits are ever found in the same locality—a rare occurrence—it can be presumed that the situation is unstable and that one of the species may soon disappear from the area. Occasionally, seeming exceptions to this rule have been observed; on closer examination, however, they have usually proved to be cases in which the niches of the two species did not overlap completely. For example, if one of two competing species of birds occasionally leaves its preferred food to its competitors and feeds on another food, the two do not occupy the same niche. In such a case, the two species may be able to coexist indefinitely in the same area. The longer two such species live in the same area, the less overlap they exhibit in appearance and behavior, a phenomenon called **character displacement.** The less similar they become the less they will compete with each other. Such realizations

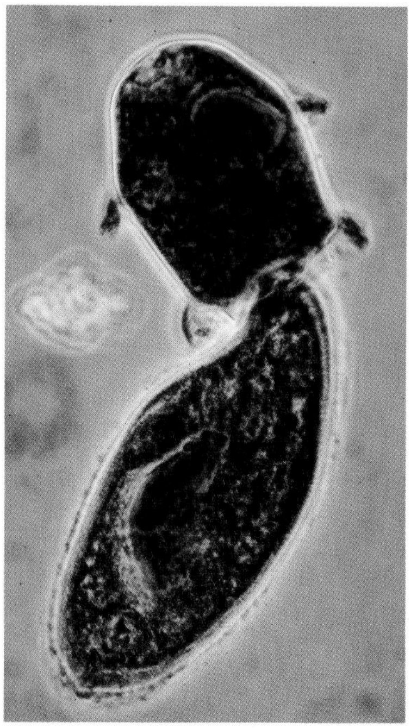

**FIG. 34-15** *Didinium nasutum* attacking a paramecium.

led to the development of a corollary to the exclusion principle, namely, the **coexistence principle,** which states that different species that coexist indefinitely in the same ecosystem usually have different ecological niches.

Exceptions to the competitive exclusion principle occur in at least two types of situations: (1) when a predator regularly feeds on two competing species, thereby preventing the potentially dominant one from excluding the other; and (2) when regular environmental disturbances, including those of human origin, favor the survival of the competing species that would have been excluded in the absence of the disturbance.

***Predator-Prey Relationships.*** Other experiments performed by Gause shed light on predation as one of the checks on population size and density. Gause cultured *Paramecium caudatum* along with one of its most voracious predators, *Didinium nasutum,* another ciliate (Fig. 34-15). First, Gause prepared several cultures and introduced five paramecia into each. (In the absence of predators, paramecia double in number as often as every 6 hours.) Two days later, he added three didinia to each culture. Within a few days, the didinia had eaten all the paramecia, and a few days after that, all the didinia had died of starvation. Gause repeated this experiment but placed sediments at the bottom of the culture vessels to provide hiding places for the paramecia. As before, the population of paramecia increased; when he added the didinia, they fed on the paramecia, drastically reducing their numbers. But in these cultures, some paramecia escaped by hiding. The didinia, which need a steady diet, eventually all starved and, in the total absence of didinia, the surviving paramecia increased exponentially and achieved the same dense concentration as they usually did in the predator's absence.

Gause then repeated the first experiment, with no hiding places provided, but this time, he added one paramecium and one didinium to the culture every third day. This procedure was intended to simulate the occasional immigration of individuals from other areas. The result was that both species survived indefinitely in sizable numbers, although a series of population oscillations of both occurred.

One principle derived from these and similar experiments is that it is not beneficial for a predator species to eat all of its prey individuals. Field studies have shown repeatedly that, in nature, most predators feed on several prey species, but they almost always feed most heavily on whatever prey species is most abundant at any given time. When the numbers of a particular type of prey decline, the predator usually switches to the prey species that is then most abundant. This behavior pattern among some predators apparently enables both prey and predator populations to survive indefinitely. It probably accounts for some of the oscillations in population size so often observed in nature.

Population oscillations are particularly evident when the numbers of all of a predator's prey species are so reduced as to be unable to support a large population of that predator; then the predator usually starves in massive numbers. But it seldom becomes extinct in the area because some individuals usually find enough food to survive.

The population oscillations that commonly occur in many prey and predator species are especially evident in Arctic and sub-Arctic regions, which are subject to extreme annual climatic changes and in which the animal communities are composed of relatively few species. Among the factors that help both predator and prey to survive are hiding places for the prey, a variety of prey species on which the predator can feed, occasional declines in predator populations, and occasional immigration of both prey and predator from other areas. Some oscillation of prey species also occurs in the absence of predation, indicating that predation is only one, and perhaps not even the most important, of the factors that keep natural populations in check. Oscillations in food availability for herbivores, for example, can have dramatic effects. Nevertheless, an important key to the survival of both a predator and its prey in a particular area seems to be the safeguarding of the prey from predation when its populations are at their lowest level.

PREDATION OF CANADIAN LYNXES ON SNOWSHOE HARES. Although careful field studies of fluctuations in prey and predator population sizes have been made only in the last half century, data useful for such a study had been collected since the early nineteenth century across an extensive area of northern Canada. This fortunate circumstance came to light when it occurred to D. A. MacLulich to examine the old sales and purchase records of Canada's largest fur dealer, the Hudson Bay Company. He discovered that two of the animals long trapped for their fur were the Canadian lynx and its most important prey, the snowshoe hare (Fig. 34-16a). The number of pelts of each that were bought and sold in a

**FIG. 34-16** (a) A Canadian lynx and a snowshoe hare. (b) Historical fluctuations in Canadian lynx and snowshoe hare populations as derived from sales records of the Hudson Bay Fur Company.

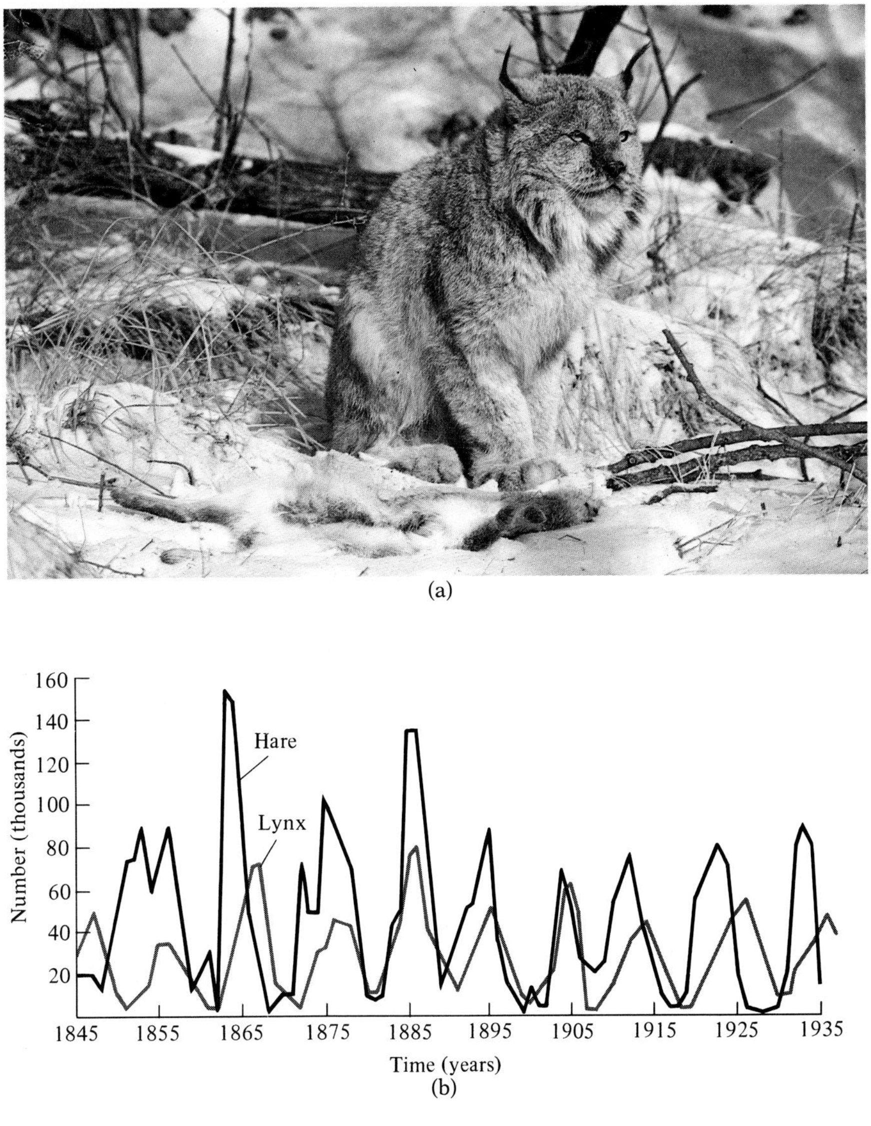

given year represented trapped samples of these two populations in northern Canada. Pelt sales probably reflected the oscillations in the populations of the two species. The fur company records suggested that the size of the lynx population was directly related to the size of the hare population (Fig. 34-16b). When there were many hares, the lynxes also thrived. Then, perhaps because the lynxes ate the hares at a faster rate than the hares reproduced, many lynxes began to starve, and their numbers fell; this would have enabled the hare population to increase. As the graph based on pelt sales indicates, the lynx population peaks always followed the hare population peaks by about a year. Hare population peaks occurred only after a decline in lynx numbers.*

The important principle illustrated by this example is that predation by a carnivore is probably an important immediate check on the population size of its prey species. However, the principle is not as applicable to predation on prey that is larger and more powerful than the

* Predation by lynxes is not the only factor in hare population cycles; hare populations also oscillate in the absence of lynxes.

PANEL 34-1

# The Effects of Predation: Three Examples

In the past half century, there has been a widespread eradication of many predatory animals in various parts of the world. This has been especially true where the predator is believed, rightly or wrongly, to be a threat to the lives or safety of humans or their livestock. In many cases, the belief that a particular "varmint" is a major threat to livestock has been altogether erroneous. As a consequence of such fears, wolves, bears, coyotes, eagles, and many other predators have been eliminated from large areas of the United States, Asia, and Africa.

## TIMBER WOLVES AND WHITE-TAILED DEER

After timber wolves, the natural predator of white-tailed deer were deliberately eliminated from northern Wisconsin by unrestricted hunting, the deer population rose to unprecedented levels. Shortly after World War II, the available vegetation was no longer adequate to support such large numbers, and thousands of deer starved to death despite the fact that emergency feeding stations were provided by the state government. The deer population is now kept down by controlling the length of the hunting season and allowing both bucks and does to be shot in some years. Wolves are also being reintroduced in some areas.

Timber wolf with white-tailed deer.

## THE DINGO

Another example of a campaign to eliminate a predator is that waged against the dingo, the yellow dog of Australia. This wild dog was apparently introduced 8,000 to 10,000 years ago by human immigrants from Asia. It is hated by ranchers, in part for preying on beef cattle, an important export commodity in Australia. One early attempt at dingo control following World War II involved dropping poisoned meat onto cattle ranges from airplanes. Wary adult dogs were seldom poisoned by this technique, although predator. There is a growing amount of evidence to indicate that predation by carnivores on such animals as moose, mountain sheep, and deer is largely limited to the old, the sick, and the young. Any substantial limitation of the prey population by predation in such instances is apparently achieved primarily by reducing the number of young individuals.

BENEFITS OF PREDATION. Paradoxically, despite limiting its size, predation usually tends to be beneficial to a prey species population. When the old, weak, diseased, and crippled are eliminated by predation, the surviving population tends to consist of the healthiest and most vigorous individuals. Eliminating large numbers of young during times of population expansion also reduces competition among the remaining individuals for existing food and other resources. The elimination of a predator from a region has sometimes been followed by a dramatic increase in the prey population, often beyond the carrying capacity of the area (Panel 34-1).

The yellow dingo. One of Australia's few placental mammals, this wild dog is thought to have been introduced by human migrants from the Malay archipelago 8,000 to 10,000 years ago. Although ranchers regard it as a threat to beef cattle, studies indicate it preys mostly on small mammals.

young ones were. A more detailed study of the dingo's eating habits then showed that they subsisted mostly on small animals such as rabbits. It now appears that only during times of severe drought on the range do the dingoes, naturally solitary hunters, form packs to prey on cattle, and then they usually kill only calves. When interviewed, some ranchers noted that the killing of calves during a long drought improves the chances of a herd's survival until the drought is over. Most ranchers, however, continue to war against the dingo.

## STARFISHES VERSUS BARNACLES AND MUSSELS

Among the benefits of predation is that it sometimes prevents local elimination of one or more prey species, a fact already noted as an exception to the exclusion principle. This unanticipated effect on species diversity was revealed by a study of the populations of barnacles and mussels in seashore areas from which all sea stars (starfish) had been removed. The starfish normally feed on both barnacles and mussels. When the starfish were removed, the numbers of barnacles and mussels increased greatly. As they competed with each other for the available spaces on rocks in the intertidal zone, one species of mussel preempted the space, reducing the number of species of marine invertebrates in these areas from fifteen to eight. Because the predator had prevented competitive exclusion from running its course, it thereby enhanced species diversity.

**FIG. 34-17** Small predators, such as the banded garden spider shown here feeding on a silk-wrapped grasshopper, do not prey only on young, old, or ill prey. A spider usually feasts on any insect that is caught in its web.

DIFFERENT RULES FOR SMALL PREDATORS. Although large carnivores that prey on species that can injure them may usually limit their attacks to the old, sick, crippled, and young, this is not the case with small predators, especially when the prey is relatively defenseless. A snake is likely to eat any frog, an owl any mouse, and a spider a wide variety of insects (Fig. 34-17). Moreover, some small predators feed only on a single prey species, a situation that is quite unusual. Being limited to a single prey species means that the predator depends on the continued existence of that prey species for its own existence. If the predator is able to attack members of the prey species of all ages with impunity, both predator and prey could be liable to extinction in areas in which both are found. The same principle also applies to species-specific parasites and to any animal that depends on a single plant species for food.

In most situations, a predator species that enters an area in which its only prey species exists may virtually eliminate that species from that area. Only chance migrants escape to establish colonies elsewhere. The predator then also tends to be eliminated from the area by starvation, with perhaps a few individuals surviving by encountering remaining prey or by migrating to other areas in which the prey is found. The result of such a relationship is a wide scattering of both the prey species and its predator, and both are difficult to find in any one area. Occasionally, local population explosions of the prey species may occur, to be followed, sooner or later, by a corresponding increase in the predator species after some happen into the area. Both may then largely disappear again from the area for a time.

## SYMBIOSIS

Some organisms live much of their lives in a very special ecological relationship: an important part of their environment is a member of another species. In the traditional definition, a relationship in which members of two different species living in intimate relationship to each other, is termed **symbiosis** (from the Greek *sym*, "together," and *bios*,

(a)

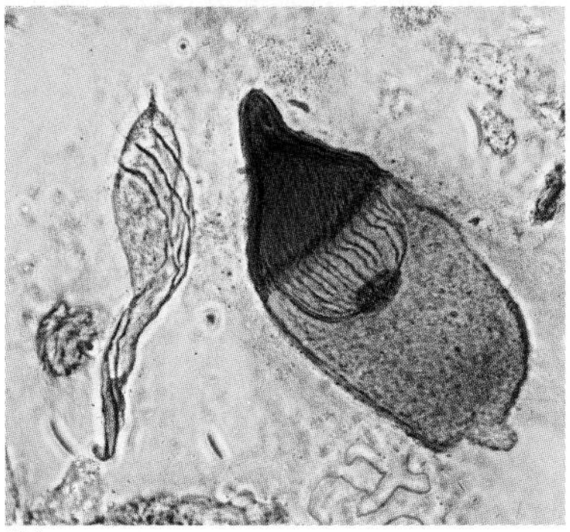

(b)

**FIG. 34-18** Mutualism. The diet of termites includes cellulose, a substance they cannot digest. The wood and grass particles they eat are ingested and digested by flagellate protozoans they harbor in their gut. (a) Workers of the genus *Reticulotermes.* (b) Two of the flagellates inhabiting the gut of *Reticulotermes.* The larger one, which is ingesting a wood particle at its lower end, is probably *Trichonympha.*

"life"). Symbiotic relationships, which are always beneficial to at least one member of the association, may vary from associations of extreme intimacy and interdependency to relatively casual associations and from very temporary relationships to those involving the entire life span of the symbionts. Some relationships are so intimate that one member lives on or in the other. In such relationships, the partner that lives on or in the other organism is termed the **symbiont,** while the one on or in which the symbiont lives is called the **host.** Such highly intimate symbioses can be classified as commensalism, mutualism, or parasitism.*

## COMMENSALISM

**Commensals** (from the Latin *com,* "together," and *mensa,* "table") may share the same food, and the symbiont may gain shelter by living on or in the host. But even if the symbiont shares the host's food, it causes no particular harm. For example, the bacterium *Escherichia coli* lives in the human colon; it benefits from the nutrients, warmth, and shelter found there, but causes no disease or even any discomfort. Similarly, the protozoan *Opalina ranarum* lives harmlessly in the colon of the frog. In both cases the symbiont depends on the host for its supply of food.

## MUTUALISM

In **mutualism,** some reciprocal benefit accrues to both partners. The flagellate protozoans that digest the cellulose of wood particles while living in the gut of termites are good examples (Fig. 34-18). The termite needs the protozoans to digest the wood it eats; the protozoans need the termite for the shelter and the nutrients it provides. When termites are experimentally rid of their protozoans, they soon begin to starve, even when provided with all the wood they can eat.

Many highly evolved mutualistic associations occur in nature, some of them very interesting. For instance, lichens are associations between certain kinds of algae and fungi (see Chapter 29). Other classic examples are provided by several species of tropical acacia trees that are hosts to particular species of ants. The trees provide the ants not only with a safe home (their sharp thorns, which the ants hollow out) but with two types of food as well (special leaf products rich in proteins, fats, and vitamins, and a special nectar rich in sugar). The ants, for their part, destroy injurious insects that attack their tree host. Moreover, ant patrols destroy other plants, including other acacia trees, in the immediate vicinity, thereby ensuring their host tree access to adequate sunlight, soil nutrients, and moisture.

Mutualism combined with parasitism exists between some freshwater clams and fish. Most species of freshwater clams have a larval stage, the glochidium, which is a parasite of freshwater fish. The glochidia are discharged from the female clam and usually settle to the bottom, where, upon encountering a fish of the right species, they attach to its gills or other parts of its body (Fig. 34-19). After 10 to 30 days as a

* Some ecologists have expanded the traditional definition of symbiosis to include all regularly occurring interspecific relationships, whether they are beneficial, neutral, or detrimental (as in parasitism and predation).

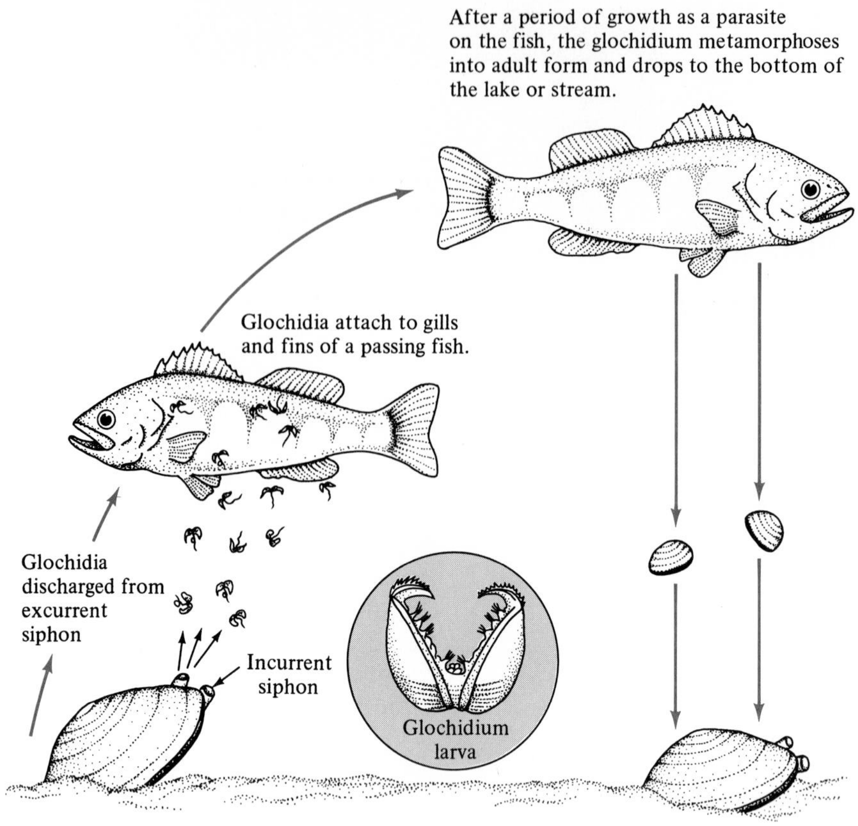
After a period of growth as a parasite on the fish, the glochidium metamorphoses into adult form and drops to the bottom of the lake or stream.
Glochidia attach to gills and fins of a passing fish.
Glochidia discharged from excurrent siphon
Incurrent siphon
Glochidium larva

**FIG. 34-19** The freshwater clam spends its larval stage, the glochidium, as a parasite on the gills and fins of a fish. At metamorphosis, it leaves the fish to take up its adult life.

parasite, they metamorphose and leave the fish as juvenile clams. The parasitic infection does not greatly harm adult fish, and the lesions soon heal after the glochidia drop off. With some species of freshwater clams and fishes, such a parasitic relationship also involves mutualism. The siphons of the female clams become conspicuously colored when the glochidia are ready to be expelled. The coloration attracts females of certain species of fishes when their fertilized eggs are ready for deposition. The fish inserts its ovipositor, a tube-like organ used in egg laying, into the clam's incurrent siphon and releases its eggs, which triggers the expulsion of glochidia by the clam. The fish's eggs are thereby provided a safe, well-aerated place in which to develop until hatching. The glochidia not only derive nutrients from their host but are also given a rapid and wide dispersal not otherwise available to a slow-moving clam, thereby greatly increasing the species' chances of survival.

## PARASITISM

**Parasites** receive benefit from their hosts while harming them. For example, the protozoan *Plasmodium,* the causative agent of malaria, parasitizes both mosquitoes and vertebrates, each at a different stage in its life cycle. Another parasitic protozoan, *Trypanosoma,* produces African sleeping sickness; it is transmitted by the bite of the tsetse fly.

The usual host of a particular species of parasite can be expected to exhibit a certain amount of resistance to the parasite; that is, it limits the degree of infection and is usually not quickly killed by the parasite. Parasites that always kill their hosts soon after invading them tend to become extinct.

## THE EVOLUTION OF SYMBIOTIC RELATIONSHIPS

All three types of symbiosis described above can be characterized by every degree of intimacy, from a relatively casual, temporary association that is not essential to the symbiont's survival to an **obligate symbiotic relationship,** in which it is essential to the symbiont's life that it spend an important part of its life on or in a host. A **facultative symbiotic relationship** is one that is not at all obligatory for the symbiont.

Symbiotic relationships probably evolved from the accidental association of free-living organisms with other species. If shelter, nutrients, or other essentials were provided to the potential symbiont by such an association and if both the potential symbiont and the host were able to survive the experience, a facultative association may eventually have become established. Mutations that improved the symbiont's ability to survive on or in the host and to exploit the resources the host offered might have occurred from time to time. Competition with other symbionts of the same species would have resulted in a selection pressure for an increasingly specialized association, such as the examples of mutualism previously cited. Such an evolution is similar to that of chance immigrants to an oceanic island that are able to survive and eventually give rise to new species adapted to the niches found there. In symbiosis, the role of the island is taken by the host species.

## CLASSIFICATION OF INTERACTIONS BETWEEN TWO SPECIES

Table 34-1 summarizes a way of classifying interactions between populations of two different species devised by the American ecologist Eugene P. Odum. The table lists six types of interactions and rates each species +, −, or 0 according to whether its growth, survival, or other population attributes benefited were, inhibited, or not affected by the interaction. The concept applies to all types of interactions between

**TABLE 34-1 Analysis of Some Interactions Between Populations of Two Different Species**

| *Type of interaction* | *Species 1* | *Species 2* | *General nature of interaction* |
|---|---|---|---|
| Neutralism | 0 | 0 | Neither population affects the other. |
| Competition: direct interference type | − | − | Each species directly inhibits the other. |
| Competition: resource use type | − | − | Each species indirectly inhibits the other when a common resource is in short supply. |
| Amensalism | − | 0 | Population 1 is inhibited; population 2 is not affected. |
| Predation (including parasitism) | + | − | Population 1 is the predator; population 2 is the prey. |
| Commensalism | + | 0 | Population 1, the commensal, benefits; population 2, the host, is not affected. |
| Mutualism | + | + | Interaction is favorable to both. |

FROM: Odum, E. P. 1983. *Basic Ecology*. Saunders. Philadelphia.

NOTE: 0 indicates no significant interaction; + indicates growth, survival, or other population attribute benefited; − indicates growth, survival, or other population attribute inhibited.

species, including (1) **neutralism,** which has no effect on either species; (2) competition for any limited resource, such as habitat or food; (3) **amensalism,** in which one species is inhibited and the other is unaffected; (4) predation, including, in this simplified version, parasite-host relationships, in which one species benefits and the other is inhibited; (5) commensalism, in which one species benefits and the other is unaffected; and (6) mutualism, in which both species benefit. Some categories can be broken down into subdivisions, such as types of competition and whether a symbiotic relationship is facultative or obligatory.

Predator-prey relationships such as that of starfish that feed on barnacles and mussels in intertidal zones make apparent that benefits to and inhibitions of population growth are not always easily determined (see Panel 34-1). In our example, the starfish limited the sizes of the barnacle and mussel populations, but removal of the starfish allowed the mussels to crowd out the barnacles. The starfish, a predator of both species, was thus indirectly beneficial to one of them. In fact, many interactions between two or more species involve multiple beneficial or harmful effects. Tropical communities in particular exhibit many such relationships.

## Summary

Populations are of biological interest for their dynamics—the factors contributing to their increase, stability, decline, and extinction—all of which affect evolution, since populations are the units upon which natural selection acts. Population increase is a result of the interaction between a population's biotic potential for increase and the environmental resistance to achieving that potential by virtue of nutrient and habitat availability, disease, and predation. Most populations remain stable on a long-term basis, usually with temporary fluctuations. Although all species can increase logarithmically, none do, except for brief periods of time. This fact is a key element in Darwin's natural selection theory.

Cultures of single species of bacteria and protozoans with enough food, space, and other necessities soon begin to double in size at a regular rate, termed exponential or logarithmic growth. A population's actual growth over a given time period can be predicted from its initial size and its potential growth rate, or intrinsic rate of increase. In time, exponential growth of a population declines as resources of food, oxygen, space, and the like are exhausted or as toxic wastes accumulate. Eventually, the population usually stops increasing and fluctuates around a certain level. Alternatively, it may undergo a catastrophic crash due to food depletion, waste accumulation, and so on. In 1798, Thomas Malthus predicted such a crash for the earth's human population but was generally ignored until recently. His critics contended that human populations, like all natural populations, would exhibit logistic growth, typified by an initial exponential increase that gradually slows until the population levels off and maintains itself at a high level. However, the leveling off of natural populations is due to such factors as starvation, disease, predation, and fierce competition for available resources.

In addition to the study of population size and growth, population dynamics research involves analysis of survivorship curves, indicating life expectancy at various stages in life, and analysis of the age structure of populations. In one type of age structure analysis, a population is divided into three age groups: immature individuals, fecund adults, and nonfecund adults. The proportions of a population in each class constitute a measure of the population's chances for increase, stability, or decline.

Life history pattern analysis provides other insights into population dynamics showing that all organisms fall somewhere on a spectrum between those that produce numerous offspring to which they give little or no care or protection and those that produce very few offspring in which they invest much of themselves. The former group is said to exhibit *r* adaptation, a term derived from the logistic growth equation. By producing offspring in great abundance, this group compensates for heavy predation, disease, and the like that befall its offspring. Many plants further adapt to such situations by reproducing in abundance only once a year or only in certain years. The once-a-year approach enables organisms to adapt to life in the Arctic, with its short summer season, or in deserts, with one brief period of rainfall.

*K* adaptation, a term also derived from the logistic growth equation, describes the approach of organisms that produce only one or a few offspring a year but invest much in them by way of providing food, protection, and the like. This approach is especially well suited to environments with limited resources that have a low carrying capacity for such organisms. Large, long-lived organisms are also usually *K* adapted.

Population size can be regulated by density-dependent factors, which become limiting only when the population concentration reaches a certain level; or by density-independent factors, which inhibit growth or reproduction of even a few individuals. A slowly reproducing population of food plants or animals is an example of the former; a high concentration of a toxic substance, a bad fire, or a severe drought would qualify as the latter.

Competition among members of the same population is at times a significant but unpredictable factor that varies greatly from species to species and even from one population to another of the same species. In experiments on beetles, mice, and rats, accumulation of wastes, cannibalism, fighting, neglect of offspring, infertility, and reduced food intake played roles in population size regulation. Unpredictably, some of these populations leveled off, while others of the same species oscillated greatly.

The fact that interspecific competition is usually keenest between closely similar species led to the formulation of Gause's rule, or the competitive exclusion principle: in any one area, an ecological niche cannot be fully and simultaneously occupied by more than one species. Situations that appear to violate the rule are transitory, occur in a fluctuating environment that alternately favors one and then the other of the two competitors, or are cases in which the same niche is not fully occupied by both species.

Predator-prey studies have shown that, in nature, most predators feed on more than one prey species, usually switching to the most abundant one as others decline in number. Such relationships enable both the predators and their prey to coexist indefinitely. Population oscillations are greatest in Arctic regions, where the number of species is few and annual climatic changes are extreme. Oscillations of animal species also occur in the absence of any predators, however, showing that predation is only one factor controlling natural populations. Predation by a large carnivore is probably a significant check on the population size of a smaller prey species, but may not significantly check the population of a large prey species. Predation usually benefits a prey population by eliminating its old, weak, and diseased individuals and large numbers of young from expanding populations. Besides enhancing the vigor of the population, predation prevents it from exceeding the carrying capacity of the area.

Traditionally, symbiosis is defined as an intimate association between members of two different species that is of benefit to at least one of them. If the symbiont merely shares the food of its host or is given shelter but does not harm the host, the relationship is known as commensalism. If both symbiont and host derive benefit, the relationship is termed mutualism. If the host is harmed, the relationship is an example of parasitism. Symbiotic relationships range from very temporary to ones lasting for an entire life cycle.

Eugene Odum has classified all types of interaction between two species into six categories. In this system, an interaction between species is rated as positive, negative, or neutral according to how the fitness of each species is affected by the relationship. However, the nature of such relationships is not always easy to determine.

## Recommended Readings

A list of recommended readings appears at the end of Chapter 33.

35

# The Ecology of Communities

OUTLINE



A biotic community is composed of all populations living within a particular area. The structure of a community consists of its species diversity and the patterns of distribution of those species. Species diversity has two components: **richness,** the total number of species present, and **evenness,** their relative abundances. Patterns of distribution include vertical layering, zonation, and a combination of the two.

Richness increases from the poles toward the equator; thus Arctic regions have the fewest species and the tropics the most. The size of the habitat area and the length of time available for speciation both affect richness. Thus, large islands have more species than do smaller islands, and old islands have more species than do younger islands of the same size. Richness can also be affected by interspecific interactions, such as competition, predator-prey relationships, parasitism, and other symbiotic relationships. Any one of these factors can ultimately determine a community's structure. For example, the presence of a predator that feeds upon the more abundant of two competing species tends to ensure that both will remain in the area.

Warmth, moisture, and stability of climate all promote a community's evenness of species. Evenness is also promoted by biological interactions such as interspecific competition for resources and predation. Most communities have a few dominant species that far exceed, in

**FIG. 35-1** Zonation in the intertidal zone of Boston Harbor. The extensive white area of the upper intertidal zone is due to the presence of one genus of barnacle (probably *Chthamalus*) while another kind of barnacle (probably *Balanus*) accounts for the scattered white patches in the lower intertidal zone.

number of individuals, any one of the others, some of which may be only rarely encountered. Northern latitudes and tropics with alternating wet and dry seasons tend to fit such a pattern. The wet tropics have the largest number of species, each in relatively low abundance.

A community's species are not uniformly distributed throughout it but occur in various patterns. Species are often arranged according to differing physical conditions and other factors within the habitat area, a pattern of distribution called **zonation.** Zonation is obvious even to a casual observer on a rocky seashore (Fig. 35-1). It is also evident on the sides of some mountains (Fig. 35-2). Each organism tends to become established in a zone corresponding to its niche and to which it is better adapted than other competitors. Because of such factors as competition from similar species and predation, the zone a species occupies is not necessarily the one that provides it with optimal conditions of food, light, and moisture and other resources.

**Stratification** is a pattern of distribution characterized by vertical layering. Examples include the layers of different species found in soil communities and in mature forests. Each level of soil in a prairie, for example, reveals a different species composition, although some species may occur in more than one layer. Such factors as soil texture, moisture, and nutrient availability determine the soil depth at which a given species is found. Stratification of species in a mature forest is often dramatic (Fig. 35-3). Some species of plants and animals live their entire lives only in the treetops, others only on the forest floor, and others only in intermediate layers.

Among other factors contributing to patterns of species distribution in a community are some of the daily or seasonal activities of animals—such as feeding, migratory, and reproductive behavior—and social behavior, such as formation of herds. A community's age is also a significant factor in community structure. For example, young communities are likely to be undergoing the process of ecological succession, different stages of which exhibit different community structures.

**FIG. 35-2** Zonation on mountains of western North America. The sketch is of a pattern typical of the central Rocky Mountains. Other regions show other patterns.

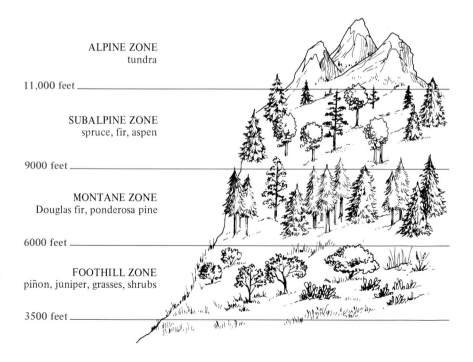

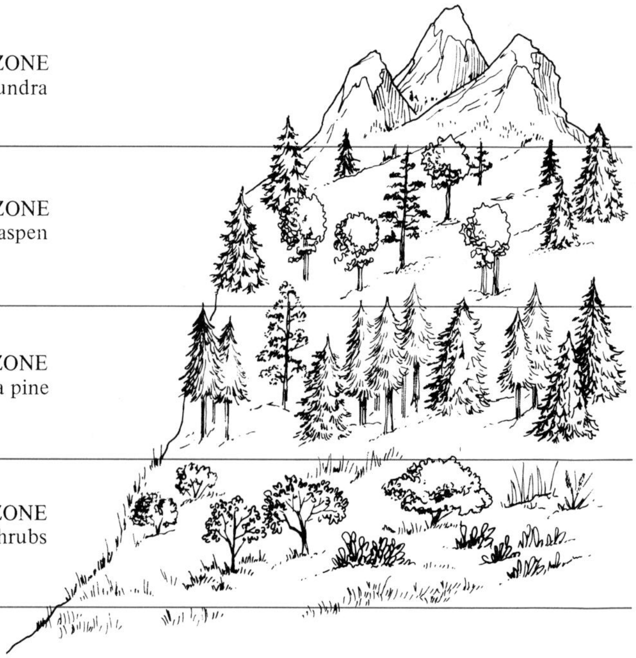

**FIG. 35-3** (a) Profile sketch of a South American rain forest, showing stratification of vegetation. Plants growing in the D and E stories must be able to thrive on the scattered light that filters through the loose canopy formed by the B story and the very dense C understory. In the A story are the scattered emergents, some of which are much taller than those shown. Rain forests in other parts of the world exhibit a similar pattern. (b) The edge of a rain forest on the island of Martinique, West Indies. Three strata are visible in the picture: A story (emergents), B story, and C story.

## ECOLOGICAL SUCCESSION

Aquatic and terrestrial areas from which organisms are removed—either by natural factors, such as volcanism, fire, and flooding, or by human disturbances—are soon repopulated by a variety of organisms. The initial colonization is usually followed by a series of stages, a process known as ecological succession. The entire sequence of stages in a succession is known as a **sere,** and each stage is a **seral stage.** One of the more dramatic demonstrations of succession results when a temperate forest has been cleared by deliberate cutting or by a severe forest fire.

### SUCCESSION IN A WOODLAND COMMUNITY

Most of us have witnessed the earliest stages of succession of a woodland community. These initial stages occur whenever a cleared field or roadside area goes untouched for 2 or 3 years.

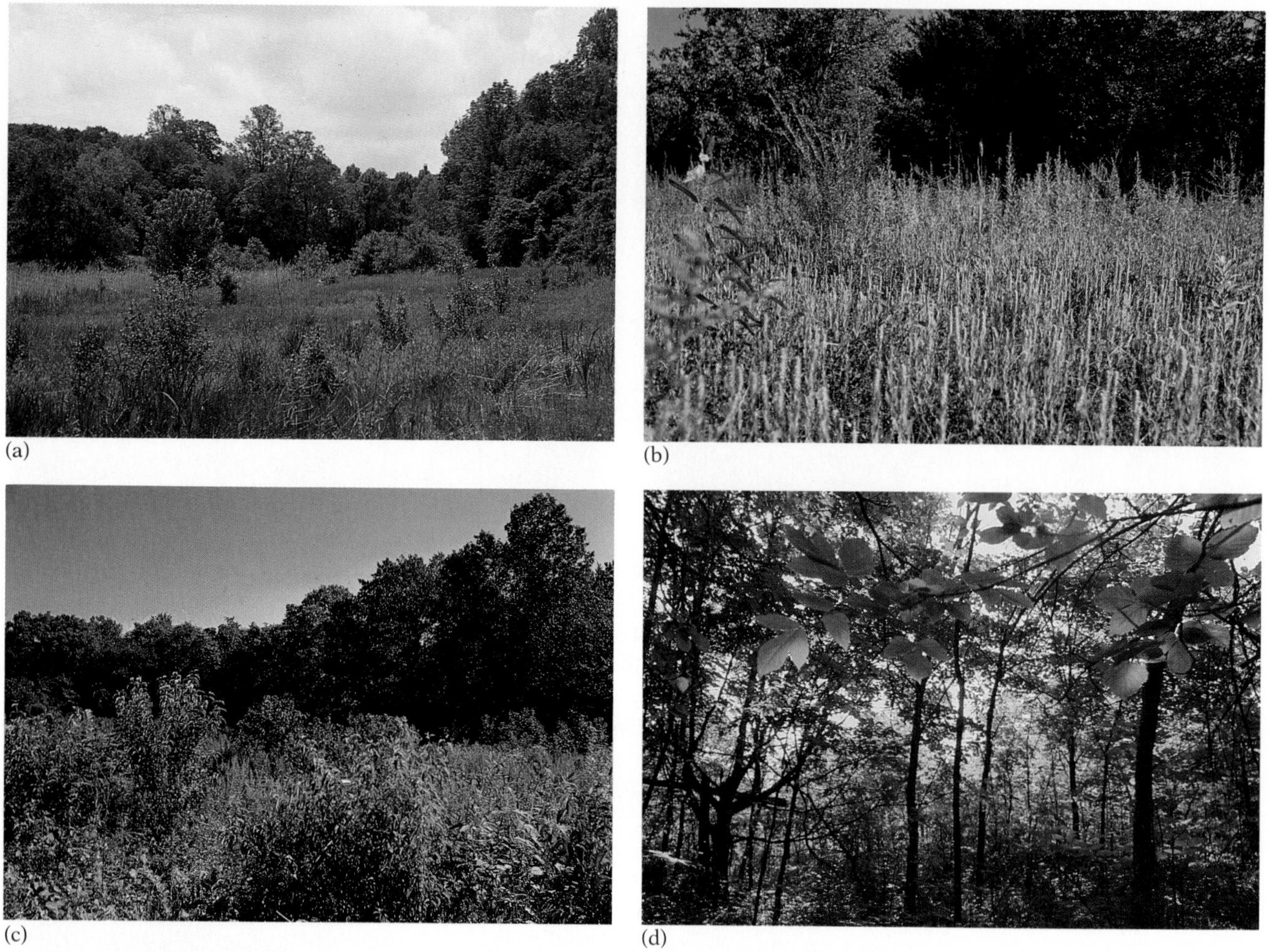


**FIG. 35-4** Succession in a woodland community. (a) In the first few years of succession annual plants predominate, although shrubs may be present, as on this fairway of an abandoned golf course. (b) In the next several years, perennial plants, including some young trees, become more numerous. (c) Shrubs, mixed with young trees, are the next to predominate. (d) In the early tree stage, plants that thrive only in full sun begin to give way to ferns and others that can grow in shade.

***Early Stages of Succession.*** Left alone, a denuded woodland area is soon covered by a variety of plants. Some spring up from seeds or roots that survived the cutting or fire; others are blown or brought in from elsewhere. The type of soil and the climate are important factors in determining which plant species will grow and survive. Some of the first plants to spring up are the same herbaceous (nonwoody) weeds found in gardens and neglected vacant lots in the vicinity. For the most part, these are plants that are readily dispersed and quickly grow to maturity (Fig. 35-4a).

Given sufficient moisture, some herbaceous species grow to much greater heights than others, providing shade and conserving the moisture of the area. Then, species such as ferns that were unable to survive on bare, unshaded soil get a start. Next, various slow-growing woody plants become established, particularly those capable of withstanding the harsh conditions that prevail in a new community (Fig. 35-4b,c). Ants and flying insects arrive daily, becoming increasingly diverse and numerous as time goes on. These first plants and animals to become established in a cleared area are known as **pioneer species.**

Within one or two growing seasons, a complex assemblage of organisms may become established in the once denuded area. Birds appear, nest, and begin to feed on seeds and insects. Small mammals may find enough cover to enter the area. Both plants and animals add organic matter to the soil, further conditioning it for other species.

***Later Stages of Succession.*** As the fastest-growing woody shrubs and seedlings of such trees as sumac and locust spring up, their shade adversely affects some of the pioneer species, often including their own seedlings. Many species that thrived best in the direct sun prove to be poor competitors in shade. Densely shaded areas come to contain rather different plant species than the more open areas (Fig. 35-4d).

As the years go by, many of the pioneer plant species seem to disappear entirely. Although their seeds or spores continue to enter the area and may even germinate, conditions are no longer favorable for their growth. But without the pioneer species, some of the species of the later seral stages might not have become established. This point has been one of the widely held, logical inferences drawn from the succession concept since early in this century. However, the view has lately been challenged.* While it is clear that the appearance of certain heterotrophic species in early stages of succession requires the prior presence of certain other species, it is not clear that a high degree of predictability is associated with the later stages of succession. For example, it has been proposed that slow-growing, long-lived tree species are the last to become established and then come to dominate what is known as the **climax community** because they are slow-growing and long-lived, not because other species had preceded them.

***Development of the Climax Community.*** The shade provided by large herbaceous and woody plants seems to favor slow-growing tree species. As the years pass, the shorter-lived locust and sumac begin to disappear as coniferous (cone-bearing) trees, such as pines and cedars, take over. Depending on conditions, when the forest reaches the age of 20 to 50 years, it begins to form a canopy some 15 to 20 m above the ground. Except where there are rifts in the canopy, only shade-tolerant species grow well on the forest floor. Then, if the winters are not too severe, the conifers may begin to be replaced by such hardwoods as oak, hickory, beech, and maple. Different insect and bird populations are now found in the area. Stratification is now evident: some bird species nest and seek food only along the ground, some only at intermediate levels in the understory, and others mostly in the treetops. Each of the species now

* For a reassessment of the succession theory based on many recent observations, see Connell, J. H., and R. O. Slayter, 1977, "Mechanisms of succession in natural communities and their role in community stability and organization," *American Naturalist,* 111(982):1119–1144.

**FIG. 35-5** White-tailed deer bucks at the edge of a hardwood forest.

**FIG. 35-6** A hardwood forest of Michigan's Upper Peninsula in an area where several trees have fallen. The resulting break in the canopy has allowed light to strike the forest floor and encouraged the growth of plants characteristic of an earlier seral stage.

present is usually more specialized in eating only certain types of foods than were the species in earlier stages. Depending on its size and proximity to centers of distribution, the forest now provides food and shelter for larger herbivores, such as deer and moose (Fig. 35-5), and carnivores, such as wolves and bears.

As the forest nears 100 years of age, it begins to achieve a somewhat steady state, described as full maturity. But as much as 200 years or more may be required to reach this stage. At full maturity, succession ceases for an indefinite period, with biomass remaining stable and no new species replacing older ones. The forest is now a **climax community.**

***Characteristics of a Temperate Forest Climax Community.*** In a mature temperate forest, the tallest trees range up to 25 m in height. The interior of the forest has a cathedrallike appearance; it may be deeply shaded and cool—even on the hottest days. Only in the spring, before the leaves on the trees of a deciduous forest* have grown in, does the forest floor experience a brief period of lush growth. The relatively sparse vegetation of the mature forest floor bears little resemblance to the pioneer community. But here and there, a giant tree may have died or been struck by lightning and come crashing down, perhaps taking some others with it. In the resulting open area, where the sun sends a shaft of light down to the forest floor, a small community of plant species typical of an earlier seral stage appears (Fig. 35-6). Succession begins again in such spots, and in time the gap in the canopy disappears as late-stage species fill it once more.

***The Balance Between Production and Respiration.*** In early seral stages, the production of biomass in a community far exceeds the amount needed for maintenance, as represented by the amount of respiration it conducts. Put another way, the community's biomass gradually increases during succession. Once maturity is reached, production of new biomass tends to equal the amount of material used in maintenance and undergoing decomposition.

***Lake Succession.*** Succession also occurs in lakes and ponds. The virgin, or uncut, forests of temperate zones long ago reached maturity, following the last retreat of the glaciers, while some of the lakes continue a slow progress toward an equilibrium stage. At first a lake's bottom is rocky and without sediments. The nutrient content of the water is low and supports only sparse growth of plants and phytoplankton, a condition termed **oligotrophic** (from the Greek *oligo,* "few" or "little" and *trophos,* "nourishment"). As each year passes, a lake may produce a little more vegetation than is consumed or decomposed. This additional material dies and settles to the bottom along with soil particles carried into the lake by runoffs of rain and melting snow. Each year, such a lake becomes slightly shallower and slightly smaller in area as benthic, or bottom, sediments accumulate (Fig. 35-7). If the process continues and the lake has a drainage outlet, it becomes a marsh; without an outlet it becomes swamplike. In either case, it is finally overgrown by the tree species characteristic of the surrounding forest. The soil may remain boglike, however.†

* Deciduous trees are those that lose their leaves every year.

† A **marsh** is a wetland containing grasses, sedges, and the like but lacking trees, whereas a **swamp** is characterized by the presence of trees. A **bog** is a tract of spongy earth that at one time had been a marsh or swamp.

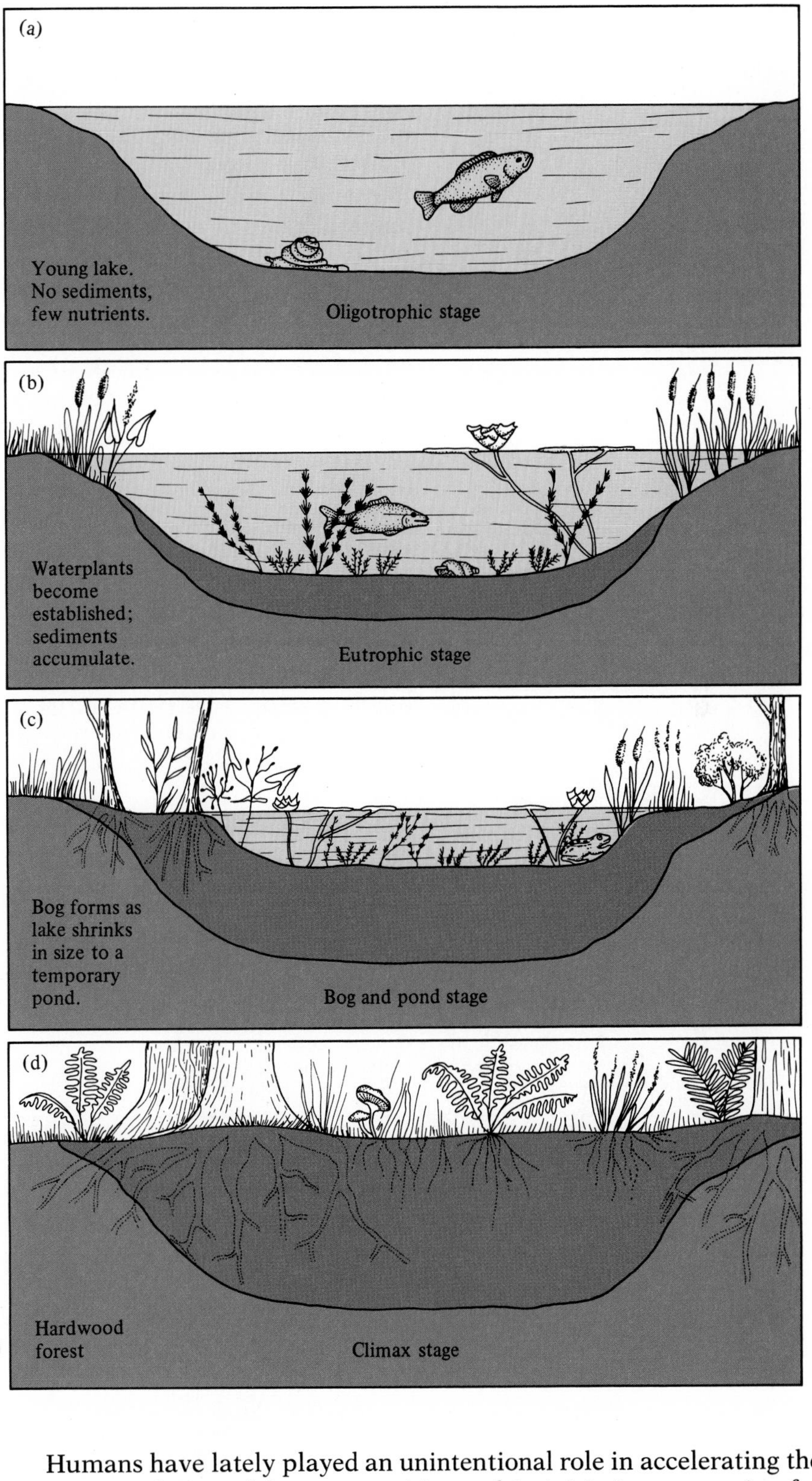


**FIG. 35-7** Lake succession in a forested area. A lake formed by glaciation may gradually fill in as sediments accumulate on its bottom and at its sides. In time, it becomes a bog and then a woodland. (a) In its earliest, oligotrophic, state, the lake is poor in nutrients and the bottom is bare of sediments and vegetation. (b) As sediments accumulate, eutrophic conditions result. Submerged vegetation appears, followed by such emergent plants as water lillies and, in shallows, sedges and cattails. (c) As the lake continues to be filled with sediments it becomes shallower and smaller, the shallows being replaced by a bog on which a prairie community becomes established. The lake-become-pond may even disappear during dry periods. (d) The prairie flora is next replaced by fast-growing coniferous trees and shrubs. In temperate regions, these are eventually replaced by hardwood trees, such as beech and maple.

Humans have lately played an unintentional role in accelerating the succession of many lake communities and in initiating succession for lakes that have shown no signs of change over a very long period of time. When nutrient-laden sewage effluent and runoff from heavily fertilized farms and densely populated feedlots pour into a lake in great quantity, the lake, thus fertilized, becomes an enormous culture vessel for photosynthetic organisms. Whatever the source of the nutrients, this process is known as **eutrophication** (from the Greek *eutrophos*, "well-nourished") and results in a prodigious growth of algae and aquatic plants. When decomposition of this dense growth occurs, aerobic bacteria deplete the water's oxygen, producing massive fish die-offs. When production greatly outstrips consumption, the lake rapidly becomes a marsh, swamp, or bog. Thus, a process that takes many centuries, or in many lakes appears not to occur at all, is sometimes driven to completion within a few decades. This process has been avoided in certain managed lakes by controlling the rate of fertilization and the harvesting of fish.

## MAJOR AQUATIC COMMUNITY TYPES

The following somewhat detailed accounts of nutritional and other aspects of a seashore community and three types of open-ocean communities illustrate the general principles that apply to all community types. The more condensed descriptions of freshwater aquatic communities and of the major terrestrial community, or biome types should be read in light of those principles. One of the principles illustrated is zonation. Another is the dependence of the producers of most communities on the nutrient supply released by decay of dead organisms and on the presence of sunlight, an exception being the chemolithotrophic producers living near volcanic vents on the sea floor. Finally, a dramatic example of the effect of seasonal cycles on productivity is seen in the description of aquatic life of the Antarctic (Fig. 35-8).

### SEASHORE COMMUNITIES

The community zonation often evident on a rocky seashore has already been noted (see Fig. 35-1). Kelp and other seaweeds can be observed in waters where rocks furnish attachment surfaces. In the littoral, or intertidal, zone of a rocky shore—that is, between the high and low tide

**FIG. 35-8** The marine realms diagrammatically represented (not drawn to scale).

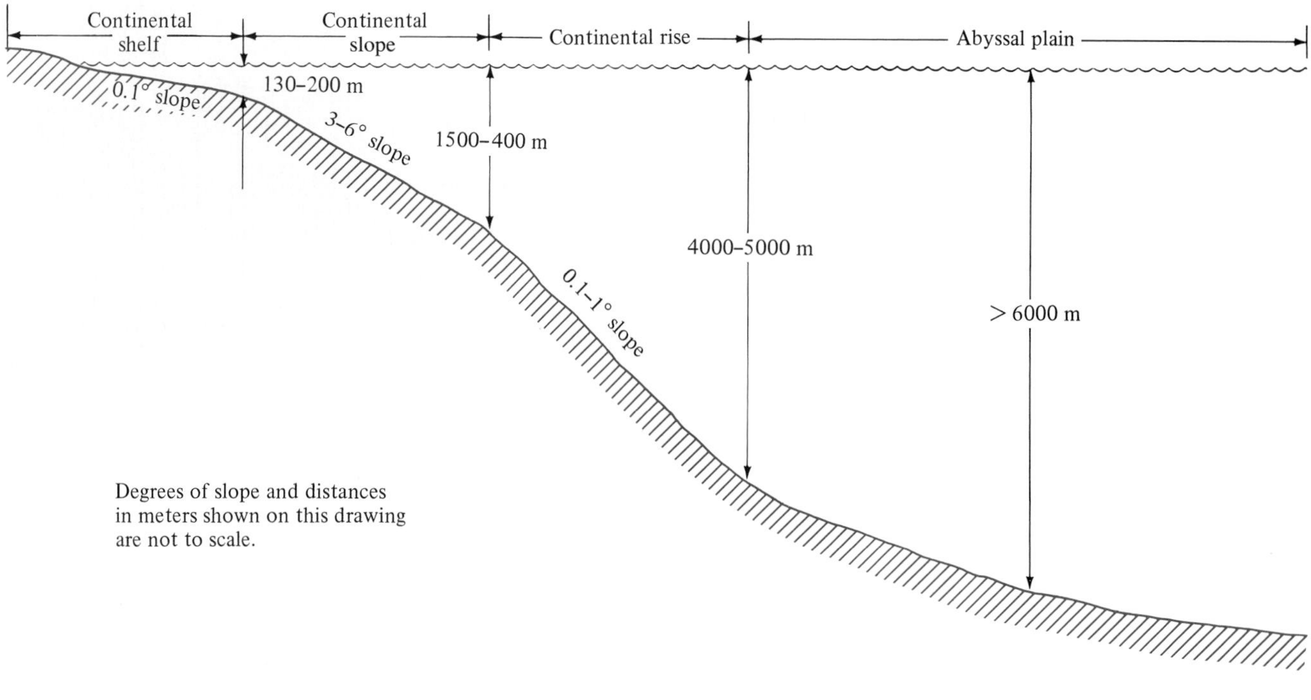

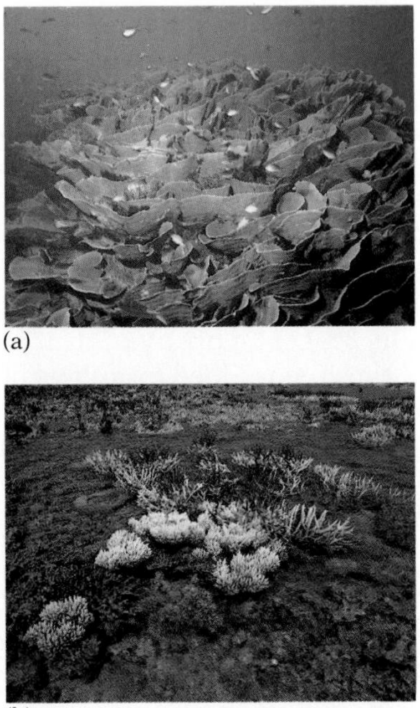

(a)

(b)

**FIG. 35-9** (a) A portion of a shallow coral reef exposed to the air by low tide. (b) Some indication of the rich variety of life to be found on a reef is provided by this underwater view.

marks—as well as just above this line, in the splash zone, a series of distinct subdivisions can be recognized. Different species of organisms, each highly specialized for surviving under particular conditions, are found in each of the smaller zones; some of these species are illustrated in Figure 35-1. Such organisms are adapted to withstand alternating exposure to water and air or the pounding or sweep of waves. Below the low tide mark are plants that can thrive only without ever being exposed to air. Only organisms that can withstand brief, intermittent exposure to air are found just above the low tide mark.

In addition to the profusion of plants that often meets the eye at the sea's edge is a more important group of producers, ones we cannot see without a microscope; these are the photosynthetic protists, or phytoplankton.* Most phytoplankton are **diatoms,** the one-celled algae enclosed in minute two-part silica shells. Another abundant group, the dinoflagellates, may at times outnumber the diatoms. (Diatoms and dinoflagellates are described in Chapter 30.)

The edge of the sea is much more productive than the open seas. The seashore community is the first to receive the rich contribution of minerals and organic matter contained in runoff from the land. Moreover, in relatively shallow water, none of the minerals released by the decomposers are lost to the water below the **euphotic zone,** the upper, sunlit layer of water in which photosynthesis can occur. Instead, these minerals are immediately available once again to the local producers, the phytoplankton and larger algae, ensuring that productivity is as high as light and temperature will allow. For these reasons, the warm waters of tidal flats, and to a lesser extent the shallower regions of the continental shelf, are home to a great abundance of life. Such high productivity also characterizes coral reefs (Fig. 35-9). In the open ocean, however, much nutrient material settles into the depths and is lost to the surface waters, greatly reducing their productivity.

Among marine animals, coelenterates, molluscs, annelids, arthropods, and echinoderms predominate at the seashore. Most of them feed on plankton, algae, or microcrustaceans, those invisible to the unaided eye. Snails rasp algae from hard substrata, while bivalve molluscs consume bacteria, plankton, and other nutrient materials suspended in the water. Many crabs eat algae and other seaweeds, as do many fishes. The tiny crustaceans known as copepods and a variety of animal larvae abound, feeding on bacteria, plankton, and detritus.

As an adaptation to the many types of food available, the animals of seashore communities exhibit a great variety of feeding methods, each of which is an adaptation to a specific ecological niche. One general class of nutrients consists of the fine particles of organic material, living and dead, suspended in the shallow waters at the sea's edge. This constitutes a major source of nutrients and a large variety of animals are adapted to feeding on these particles.

***Filter Feeders.*** Fine materials are most efficiently removed from suspension by some method of **filter feeding.** What is removed from the water and eaten depends on the method employed. In many filter feeders a current of water produced by the beating of cilia directs the food-laden water toward the mouth. As the current bearing the food particles approaches the mouth, a sorting mechanism usually routes particles of the right size and density to the mouth and sends the others away.

* Plankton (from the Greek *planktos,* "wandering") are free-floating aquatic organisms, which cannot make headway against currents. Nekton (Greek for "swimming") are animals, such as fish and squids, that swim independently of currents.

Feeding currents are used by such diverse organisms as ciliate protozoans, sponges, corals, clams, and the minute, ciliated metazoans called rotifers (Fig. 35-10). In one rotifer, sorting is based on whether the particles are small enough to drop into a ciliated groove (Fig. 35-10b). Those that are small enough are passed to the mouth, and the larger ones are rejected. Various filter feeders are specialized for eating particles of different sizes.

In place of cilia, the copepods, shrimps, barnacles, and other small crustaceans that filter feed have fine bristles on their appendages that produce feeding currents and trap food (Fig. 35-11). In bivalves, such as oysters and clams (Fig. 35-12), the work of the cilia is aided by a sheet of mucus produced by the gills. This mucus serves to trap the finest food particles, including bacteria. Both the mucus and its entrapped food particles are digested, much as humans digest the proteins in the saliva we swallow with our food. Heavier sand grains and other such particles are sorted out by weight before reaching the mouth; they drop to the bottom of the mantle cavity, from which they are removed. Many types of clams extend a tubelike siphon up into the food-laden water above them while lying buried in the sand (Fig. 35-13). Some tube-dwelling annelids, such as the parchment worm (Fig. 35-14), trap fine particles by means of a balloon of mucus, which they then ingest.

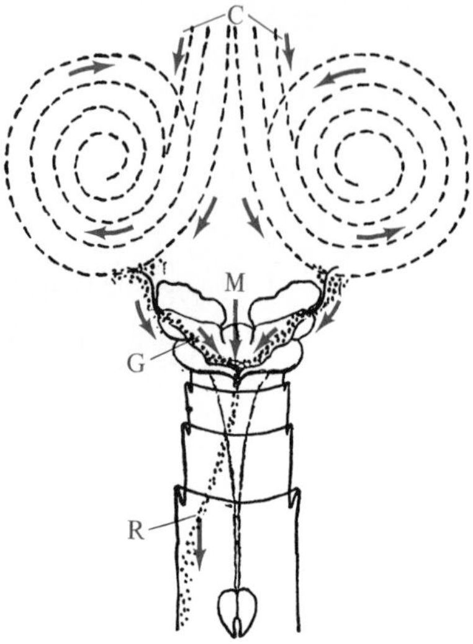


**FIG. 35-10** Feeding currents of a rotifer (C = current of food toward mouth; G = groove; M = mouth; R = rejected particles).

**FIG. 35-11** Barnacles, such as the rock barnacle *Balanus* seen here, remain attached to rocks, docks, ships, whales, crabs, or other submerged objects during their entire adult lives. Some have opened their plates and are combing the water with their bristle-equipped feet, straining out all edible particles.

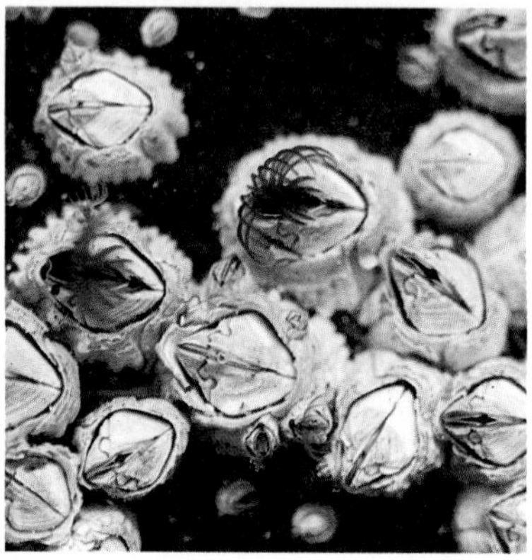

***Deposit Feeders.*** Although filter feeders remove much of the suspended matter from the water before it settles to the bottom, bottom dwellers, such as sand shrimp, feed on the rich deposits of detritus and microorganisms that reach the bottom; the process is termed **deposit feeding.** The material that escapes the bottom dwellers and becomes mixed with the sand and mud meets another community, the burrowing animals. Many of these eat the entire mixture of food particles and mud or sand and digest the organic matter it contains.*

* Not all burrowers are deposit feeders. Many take advantage of the protection afforded by burrowing but rise at high tide to eat as filter feeders at the sand's surface. Others, including many clams, filter feed through siphons extended up through the sand. Many predators burrow through the sand and mud, seizing and eating worms, crustaceans, and other burrowers encountered in their foraging.

**FIG. 35-12** Feeding currents in a clam. Cilia on the gills and palps carry the food particles toward the mouth (indicated by solid arrows); rejected particles are removed (indicated by dashed arrows).

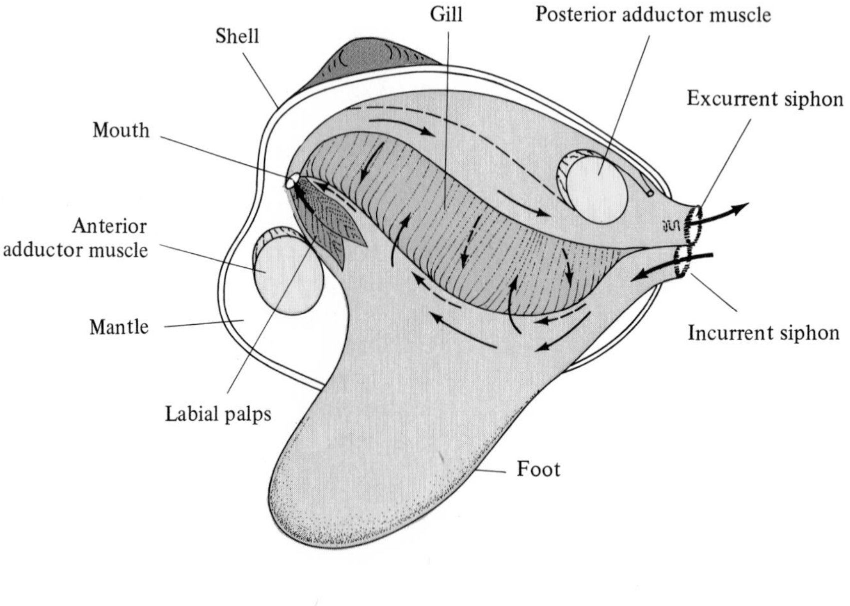

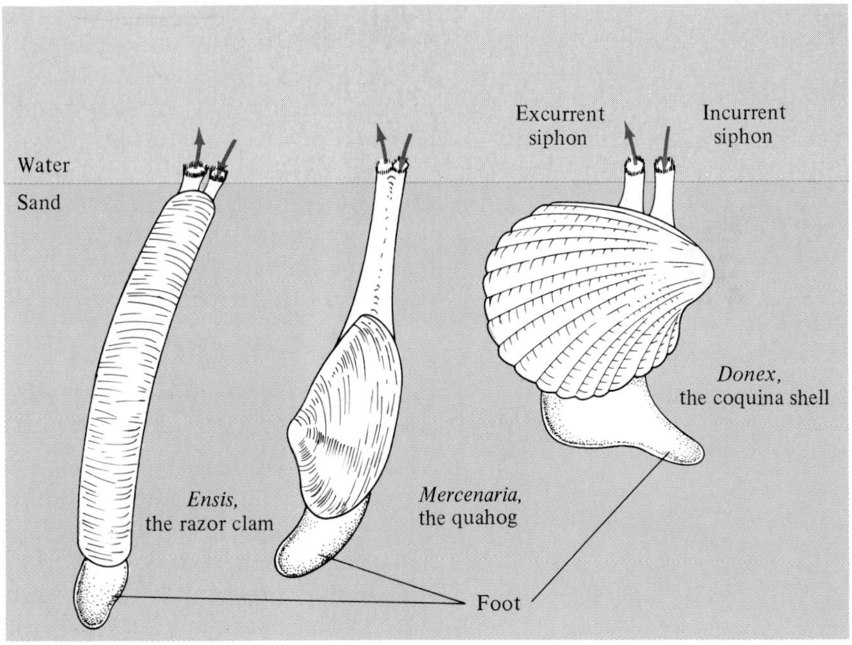


**FIG. 35-13** Some common types of marine clams engaged in filter feeding. Feeding is conducted while tide is in and discontinued when tide is out.

Among the larger burrowers of the seashore are many kinds of worms, such as the deposit-feeding lugworm. A temperate zone population of lugworms can process some 2,000 tons of beach sand and mud per acre per year, thereby altering its character and its suitability as a habitat for other organisms. Like earthworms, lugworms eat the soil, digest the nutrients it contains, and eliminate the residue. In sandy areas, they usually dig a U-shaped burrow with two openings; the sand and detritus that fall into the entrance are eaten, and the undigested residues are discharged via the burrow's exit.

***The Special Roles of the Decomposers.*** Most of the nutritional specialists of the seashore are its microscopic decomposers: hordes of bacteria, protists (including protozoa and phytoflagellates), and fungi. These de-

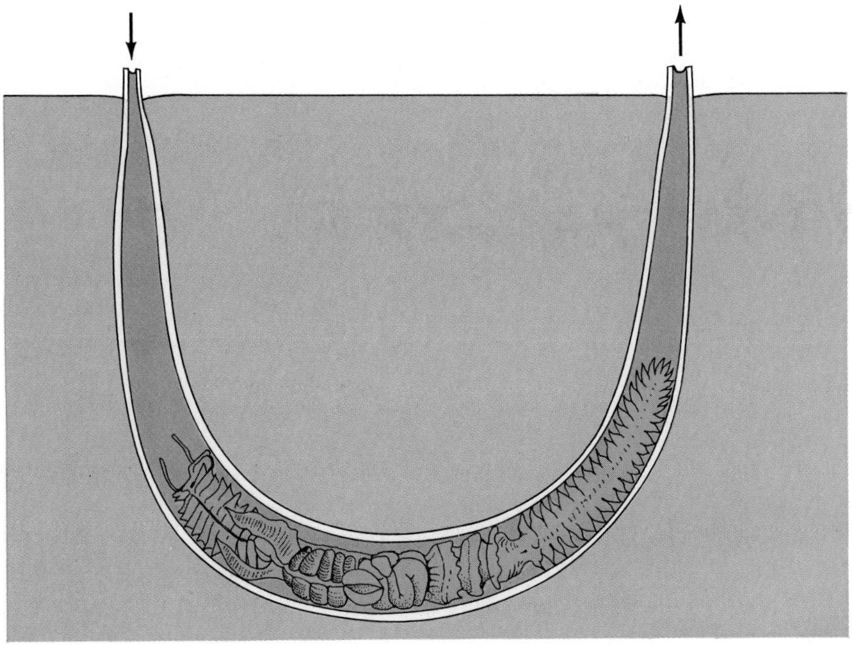

**FIG. 35-14** Feeding current of a parchment worm. The worm attaches to the inside of its tube by means of suckerlike appendages and then creates a current by undulating its fans, which are greatly modified posterior segments.

composers accomplish the decay of dead organisms and most dissolved or suspended organic matter. The mud and sand of the seashore teem with these microorganisms. One or more types of protists, bacteria, or fungi are apparently capable of splitting every kind of organic chemical bond that occurs in nature.* Every type of bond is attacked by one microbe or another until all molecules are reduced to simple materials. The immediate beneficiaries of all of this decomposition—besides the microorganisms themselves, which derive their energy and substance from the materials they degrade—are the photosynthesizers, which depend on this process for their nutrients.

## OPEN-OCEAN COMMUNITIES

Most photosynthesis in the open sea is accomplished by the diatoms and dinoflagellates of phytoplankton (see Fig. 35-8). A measure of the great abundance of marine phytoplankton, each individual of which is no larger than the smallest dust mote that can be seen in a bright ray of sunshine, is provided by the bioluminescent species of dinoflagellates (Fig. 35-15). These cells emit a brilliant white flash of light that lasts for less than a second. Sparkling breakers crashing on a reef or shore on an overcast night in late summer provide a breathtaking sight. Dinoflagellates left on the beach by a wave light up again when new waves strike them. People walking along the wet sand leave glowing footprints behind. The bodies of swimmers glow brightly, often in various hues. Although dinoflagellates are especially numerous in nutrient-rich coastal waters, a demonstration of their abundance in the open ocean can be witnessed from the deck of a ship under way on a summer night. From either side of the bow there tumbles a sparkling, silvery effervescence, and the ship's wake persists as a broad, silvery highway, easily visible at a distance. The beautiful sight may be seen night after night across thousands of miles of open sea.

* Some species of bacteria can even metabolize petroleum. Large cultures of them are used to help clean up oil spills.

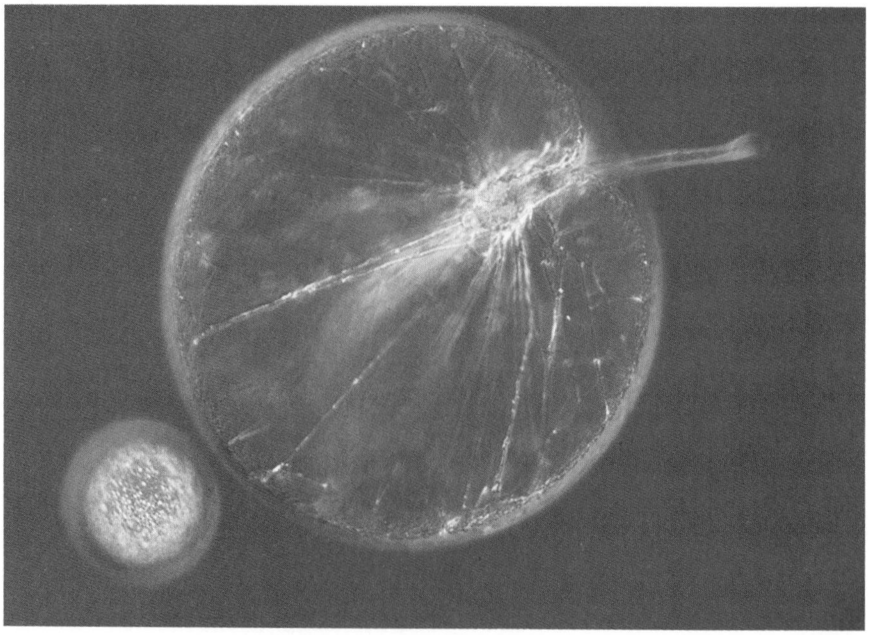

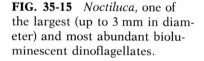

**FIG. 35-15** *Noctiluca,* one of the largest (up to 3 mm in diameter) and most abundant bioluminescent dinoflagellates.

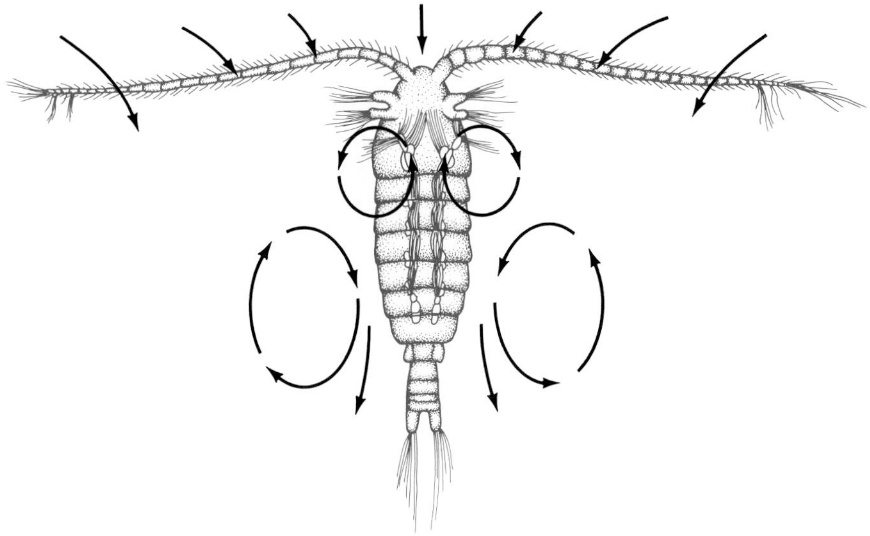

**FIG. 35-16** The feeding currents of the most common of the copepods, *Calanus.* The currents are created as the animal swims.

***Copepods as Harvesters.*** Feeding on the phytoplankton are the zooplankton: colorless species of dinoflagellates, many other protozoans, a host of invertebrate larvae, and large numbers of newly hatched fishes. However, the most abundant of the zooplankton harvesters of phytoplankton are the tiny, filter-feeding copepods, of which *Calanus* is one of the most common (Fig. 35-16).

Copepods typically harvest phytoplankton by setting up a current with three pairs of head appendages. These vibrate very rapidly—up to 40 times per second—in a rowing motion that propels the copepod (Greek for "oar foot") smoothly through the water. The inner portion of the current created on either side of the body is directed across a mesh of bristles that trap plankton. From time to time, the collected microorganisms are scraped off and then passed to the mouth by other appendages.

Filter feeding by zooplankton in a sense performs the task of harvesting and processing the phytoplankton for the larger animals, such as fishes, that harvest copepods. For large animals to live by straining phytoplankton from seawater would usually involve the expenditure of more energy than they could obtain from their food. In nutrient-rich, highly productive regions of the sea, the role of the copepod is assumed by larger primary consumers, such as the shrimplike crustacean euphausid (Fig. 35-17) of Antarctic waters and the small fish anchoveta in areas of upwelling (see Chapter 33).

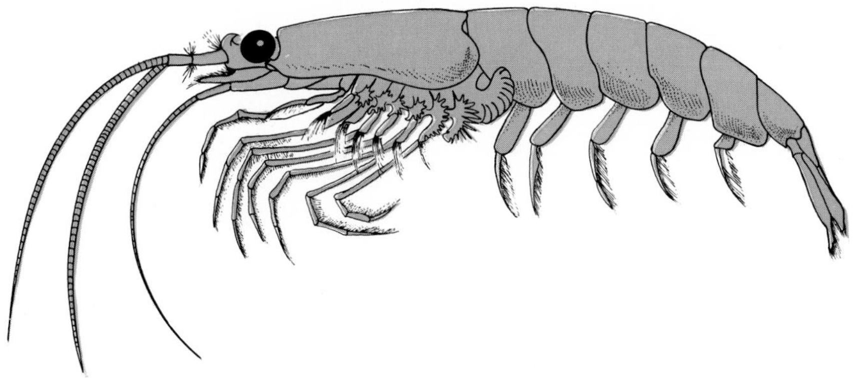

**FIG. 35-17** Euphausid (krill).

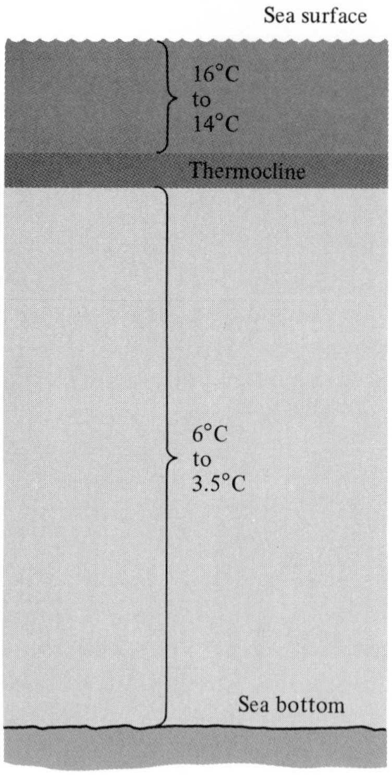


**FIG. 35-18** A thermocline in the sea. Thermoclines form when a layer of warm (for example, 16°C) and less dense water overlies colder (6 to 3.5°C) and denser water below. As long as the heating of the surface layer continues, the thermocline persists; little or no exchange of water or solutes occurs between the two layers.

***Tropical Seas.*** Despite an abundance of sunlight, the open tropical seas are barren compared to temperate seas because of a shortage of nutrients. As in other parts of the ocean, nutrients settle to the bottom. But, except along shores and over reefs, only a limited return of these nutrients occurs. Seasonal gales that might bring up mineral-laden water from below are generally absent in the tropics. As in many other regions, a discontinuity layer, or **thermocline,** usually prevails—a sharp drop in temperature occurring in a zone between a layer of warm water that overlies colder, denser water below (Fig. 35-18). Because layers of two different densities seldom mix or exchange solutes, the upper layer is deprived of minerals. The growth of phytoplankton is therefore limited by whatever nutrients are present in the surface layers. The nutrient-rich deeper waters lie below the euphotic zone, in which sunlight effectively promotes photosynthesis. The result is a relatively scanty crop for the deep blue open seas of the tropics. Nutrient-poor waters are typically clear and, like the sky, a deep blue in color; shallower, productive waters near shore, over a reef, and in an area of upwelling are cloudy and, due to the presence of phytoplankton, a distinct green.

## LIFE AT THE BOTTOM OF THE SEA

Because living matter and its products are generally denser than water, planktonic organisms tend to sink. Both phytoplankton and zooplankton have evolved a variety of hairlike spines and bristles or other buoyant devices. These add much to the organism's surface area but little to its mass, thereby decreasing its overall density and tendency to sink. Even the gentlest currents thus suffice to keep these tiny organisms near the surface of the ocean, in obedience to the same principle by which a dandelion plume is easily wafted about by a gentle breeze. But just as the dandelion plume falls eventually to the ground, all plankton that is not eaten finally falls to the ocean bottom, where it slowly decays. The detritus derived from the death and decay of nonplanktonic organisms also falls to the bottom.

The bacteria that accomplish most of the decay of this material in the dark, icy depths release a wealth of nutrient minerals into the deeper layers (Fig. 35-19). All across the open oceans, covering 70 percent of the earth, the surface layers continually lose nutrient materials to the lower levels of the sea. Only in areas of upwelling (see Chapter 33) and in regions subject to violent seasonal storms are some of these valuable minerals regularly returned to the surface, where most of them are once more incorporated into phytoplankton.

***Animal Life in the Depths.*** Despite tremendous pressure, near-freezing temperatures (2 to 4°C), and a complete lack of sunlight, the depths of the sea support animal life. The ultimate source of food for nearly all deep-sea animals is the "rain" of dying, dead, and decayed plankton and detritus that drops down to them. As in other communities, some of the animals filter feed, some eat deposits, and others prey on the filter and deposit feeders. But so little particulate matter reaches the bottom that only sparse populations can be supported by it. From concentrations as high as 10,000 kg/km$^2$ of animal life in shallow water, the density of animal life drops to an average of 520 gm/km$^2$ far out on the continental shelf. Beyond the shelf (see Fig. 35-13), the concentration of animal life averages only 2.5 mg/km$^2$, although it can be remarkably abundant on the continental slope in regions where the surface waters are highly productive. However, the diversity of bottom-dwelling, or

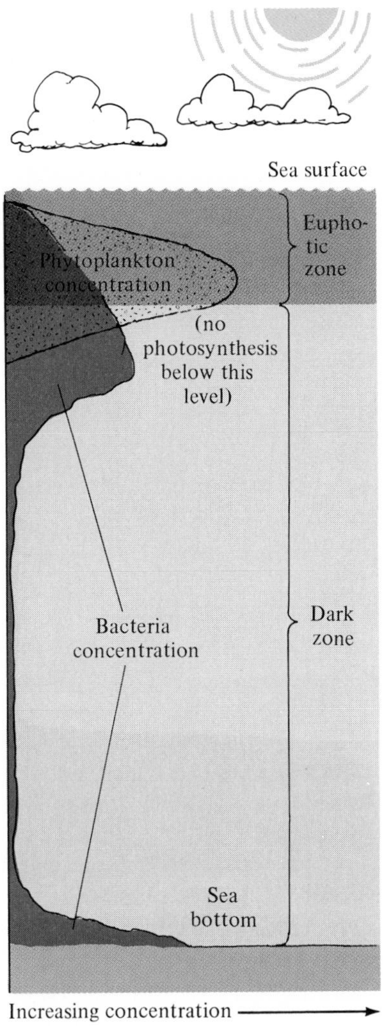


**FIG. 35-19** Distribution of relative concentrations of decomposer bacteria in the open sea. Note the high concentrations found just below the level of maximum phytoplankton production and at the sea's bottom. At the bottom, the greatest concentration is on the surface of the bottom deposits.

benthic, animals increases with depth (Fig. 35-20). This phenomenon is related to the far greater stability of conditions in deeper waters and the great area of the abyssal plain (see Fig. 35-18).

Predominant benthic animals include echinoderms such as starfish, sea urchins, and sea cucumbers. One entire class of sponges lives only in deep water. A variety of deep-sea nekton, such as fish and squids, is also to be found. In 1960, the Swiss oceanographer Jacques Picard was surprised to see a fish 30 cm (1 foot) long swim by the window of his diving bell *Trieste* as it settled in the deepest part of the sea, the bottom of the Mariana Trench, 11 km (7 miles) below the surface.

***Hydrothermal Vent Communities.*** What some have called the oceanographic discovery of the century was made in 1977 by a Woods Hole Oceanographic Institute (WHOI) expedition to the Galapagos Rift, in the Pacific Ocean. The rift is 2,500 to 2,700 m (over 1.5 miles) below the surface, 380 km (236 miles) northwest of the Galapagos Islands, and 1,000 km (620 miles) west of Ecuador (Fig. 35-21). The rift is part of an extensive global midoceanic ridge system that has developed where the tectonic plates of the earth's crust are moving apart. In such places, a flow of lava occasionally emerges and **hydrothermal vents** spew out hot water rich in $H_2S$ and other minerals.

The WHOI expedition of 1977 and a 1979 expedition using very sophisticated equipment studied the Galapagos Rift hydrothermal vent ecosystem in detail, measuring the temperature and the mineral content of the water and photographing and collecting specimens of more than a dozen species of organisms found around the vents. Whereas water temperature in the sea's depths is usually 2.0°C, it was typically 10 to 15°C at hydrothermal vents along the rift. Some chimneylike volcanic vents where the water reached 300°C were also discovered. One unexpected finding was that the life of a vent community is based, not on the "rain" of material generated by the producers in the euphotic zone above, but on a rich community of chemolithotrophic bacteria (see Chapter 9) that derive all the energy they need from oxidation of inorganic compounds such as the hydrogen sulfide that emerges from the earth's interior. This discovery was truly revolutionary. The bottom of the sea has long been classified as an incomplete ecosystem because it lacks photosynthetic producers. This view must now be modified for the regions around hydrothermal vents; although no photosynthesis occurs there, a thriving population of chemolithotrophic producers does.

Bacterial counts for water samples taken 1 m above vents were $10^5$ to $10^6$ cells/ml. Around each active vent, the bottom was covered with hundreds of clams, mussels, enteropneusts (a type of hemichordate), and many large, tube-dwelling worms (Fig. 35-22). These worms, some 1.5 m long, live only in the warm water, but most of the animal species live in the cold (2.0°C) water nearby. Animals thus far encountered are unusual and include a new phylum of worms. One species of crab can survive only under pressure at least 125 times atmospheric pressure.

## ANTARCTIC SEAS

The word *Antarctic* usually conjures up thoughts of sunless winters in which 40° below zero—on either the Celsius or the Fahrenheit scale—would be a warm day. Yet Antarctic seas support much life.* Great

* Unlike the Arctic, the Antarctic lacks a terrestrial flora and fauna in any usual sense. Its land mass is nearly covered by an ice sheet up to three miles thick. Birds, such as penguins, and mammals, such as seals, depend on the sea for their food.

**FIG. 35-20** A long-stemmed marine polyp photographed on the ocean floor at 5000 m about 600 km west of Africa.

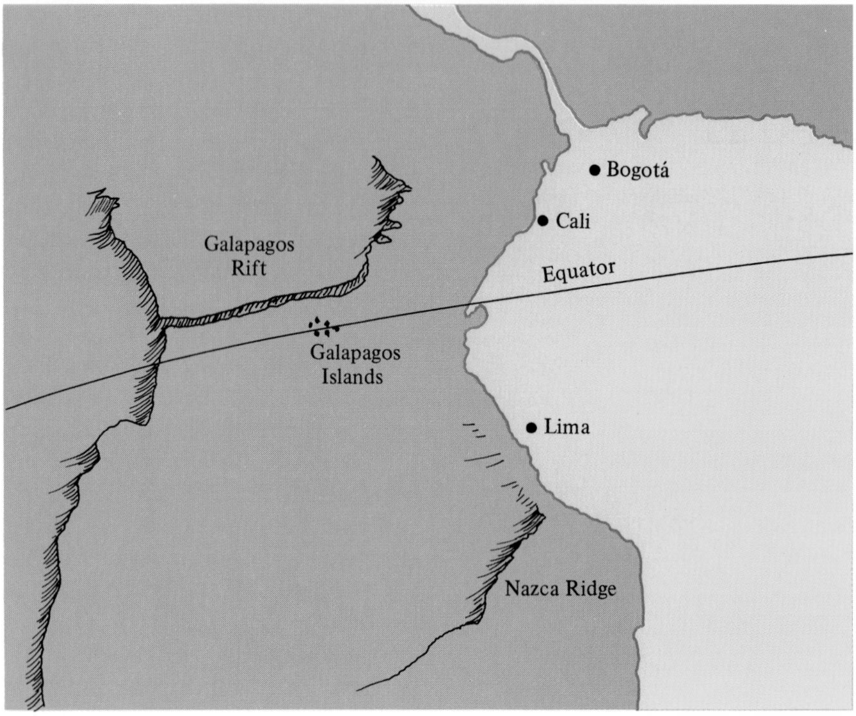


**FIG. 35-21** The Galapagos Rift, located 380 km northwest of the Galapagos Islands, is the site of an extensive ocean bottom community living on inorganic nutrients issuing from hydrothermal vents.

mats of cyanophytes, or blue-green bacteria, are sometimes seen beneath the ice. Fish are plentiful, and several kinds of birds, seals, and whales thrive on them. The waters are usually quite rich in nutrients, the key to their fertility. Unlike Peruvian waters (see Chapter 33), however, there is a sharp seasonal fluctuation in productivity due principally to a lack of sunlight during much of the year.

***The Annual Cycle.*** In the Antarctic spring, melting ice and snow create a slow surface current of dilute seawater moving toward the equator.

**FIG. 35-22** Life around hydrothermal vents in the Galapagos Rift. Tube worms and crabs can be found close to the vents.

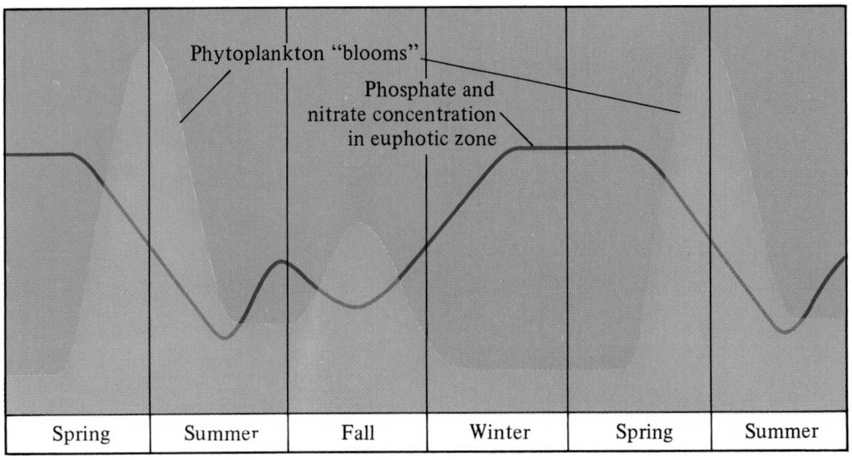


**FIG. 35-23** The annual Antarctic sea cycle. Violent winter storms and the movement of melting ice and snow away from the poles in the spring bring up mineral-laden water from the depths. The fertile seas produce an enormous diatom harvest by late spring, ending when nutrients are depleted. When nutrients increase again in late summer, a second but less extensive diatom increase occurs. This increase stops as the early Antarctic winter begins.

This creates an equally slow countercurrent of denser (because it is saltier) mineral-laden water just below it moving toward the South Pole and rising to the surface to replace the water that moves away. The bright spring sun shining on the mineral-rich water produces a bountiful harvest of diatoms, whose numbers double daily for several weeks (Fig. 35-23). By early summer, the shrimplike euphausids, which feed on diatoms, increase to the point at which the sea is a thick soup of krill, as the Norwegian whalers call these small crustaceans, which they encounter in northern waters. But then the phytoplankton begin to decline, both because they have been eaten in large numbers by the euphausids and because their nutrients—nitrates and phosphates especially—have begun to be exhausted. A succession of different types of zooplankton and fishes follows, with larger animals feeding on smaller ones and the young fishes increasing in size daily.

As fall makes its early appearance and the slanting rays of the southern sun no longer warm the waters, the surface layer cools and mixes with the lower layers. Fall gales further mix the deeper mineral-laden water with the surface layer, and diatom growth resumes, but only briefly and never reaching its springtime peak (Fig. 35-23). Soon, falling temperatures and the darkness of the early winter halt photosynthesis. The krill no longer congregates at the surface. But the surface layer is already fertilized and ready for the spring bloom when the sun rises high in the sky again the following year.

***Antarctic Food Pyramids.*** In Antarctic seas, as elsewhere, birds are often the dominant animal atop food pyramids. But two other, strikingly different food pyramids exist in the Antarctic. In one, the killer whale, which feeds on porpoises and seals, occupies the top position (Fig. 35-24). Killer whales are voracious eaters and have been known to swallow 32 full-grown seals in one meal. This is expensive food in that, for every 10 kg a killer whale grows eating such food, up to 100,000 kg of phytoplankton are needed at the bottom of the pyramid. This, of course, puts a low practical limit on the number of killer whales that can be supported by eating seals alone. The baleen, or whalebone, whales, which include the humpback and the blue whale, the largest animal that has ever lived on earth, have evolved a different approach to nutrition. Full-grown blue whales exceed 30 m (98 feet) in length and 150 tons in weight. What food could support the life of such a creature? The answer is something of a surprise.

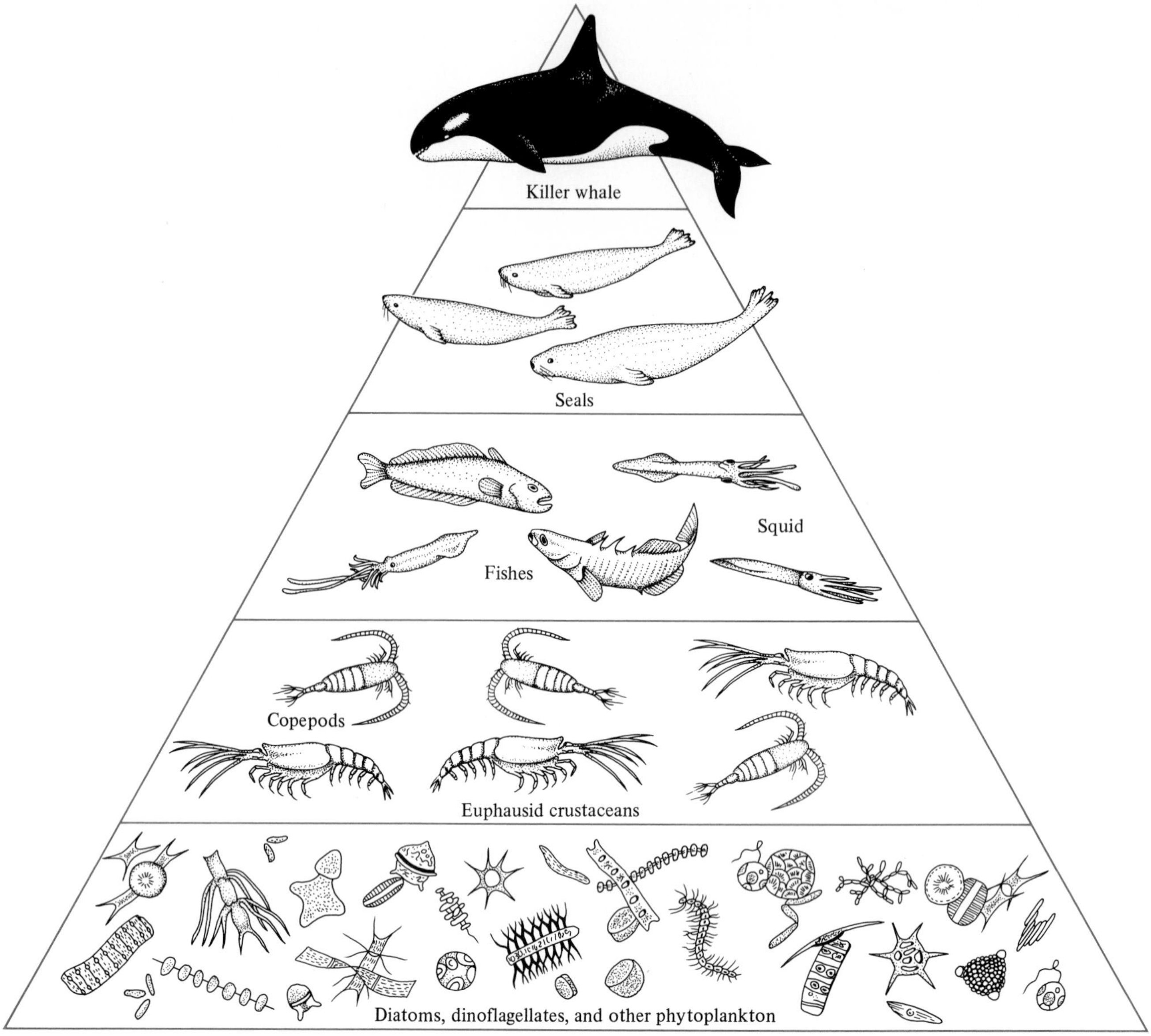


**FIG. 35-24** The killer whale food pyramid. For a killer whale to gain 10 kg, up to 100,000 kg of phytoplankton must be grown and harvested. At each level of consumption, up to 90 percent of the mass and energy of the food organism is used to maintain the consumer.

Baleen whales are named for the fringes of baleen, or whalebone (not true bone), that hang down in mustachelike sheets inside their mouths (Fig. 35-25). Some species of these whales, including the blue whale, swim through dense schools of krill, straining huge amounts of water through the baleen meshwork. A large whale's expandable mouth holds 60 $m^3$, or 60 tons, of water, which it filters of krill in a few seconds. The filtering process allows the phytoplankton to escape with the seawater while the krill is retained to be swallowed at leisure. Tons of krill can be eaten by a whale in a short time. By eating primary consumers directly, the blue whale and several other baleen whales have cut out the intermediate levels of the food pyramid, a most efficient approach to nutrition. Thus, 100 million kg of phytoplankton production supports up to 1,000 kg of blue whale growth, compared to only 10 kg of killer whale growth (Fig. 35-26). This efficiency has provided enough food to support the growth of individual blue whales to enormous size and the growth of their population to hundreds of thousands. However, in recent years, due to severe overhunting, populations of the blue whale and several

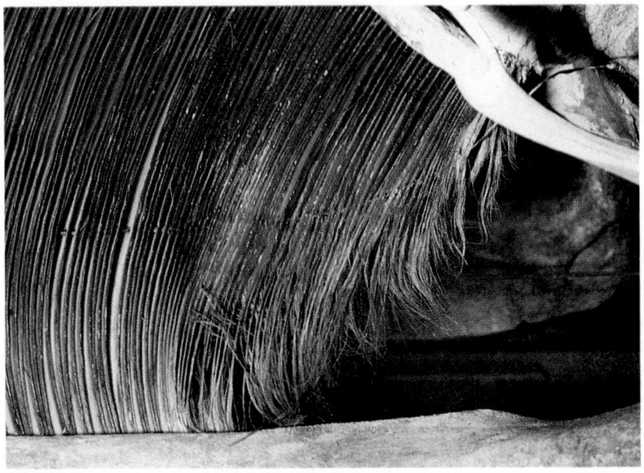
(a)

(b)

**FIG. 35-25** (a) Baleen from a pygmy right whale. (b) A hump-back whale feeding.

other whale species have dropped so low that reproduction has become affected and some species are not expected to survive.

## FRESHWATER COMMUNITIES

Compared to marine and terrestrial communities, freshwater ecosystems occupy a rather small area of the earth's surface. Their importance is quite out of proportion to their size, however. Because water is so resistant to temperature change, the presence of large bodies of water prevents drastic daily fluctuations in temperature on nearby land. Mod-

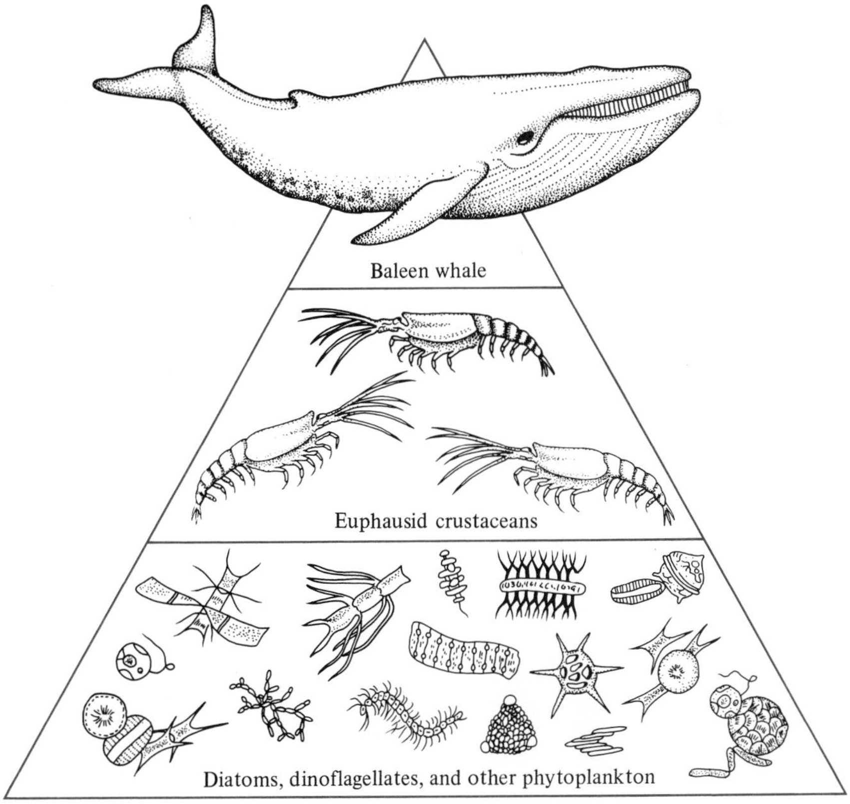


**FIG. 35-26** The baleen whale food pyramid. The hump-backed whale, a type of baleen whale, eats krill directly. Two higher orders of consumers found in the killer whale pyramid are thereby eliminated, resulting in great nutritional efficiency.

eration of seasonal temperatures occurs for the same reason. Because rain and melting snow that soak into the ground reappear at lower altitudes in spring-fed streams, marshes, swamps, ponds, and lakes, such bodies are an integral component of the hydrologic cycle (see Fig. 33-19). Freshwater ecosystems are classified as **lentic,** including ponds and lakes; **lotic,** including streams and rivers; and **wetlands,** including marshes, bogs, and swamps.

***Lentic Ecosystems.*** Compared to mountains and the oceans, ponds and lakes are very young. Some ponds appear and disappear annually, whereas others persist for hundreds of years. Most of the oldest lakes arose as the last great glaciation retreated some 10,000 years ago—only yesterday, geologically speaking. The recent connections of these lakes with each other account for the continent-wide occurrence of many species of freshwater organisms, such as fishes, that probably could not have been so widely distributed in any other way.

Zonation and stratification are characteristic of all lakes, with the highest productivity usually occurring in shallow waters along the littoral zone, or shore. The productivity of open water depends on seasonal or regional conditions that affect the mixing of warmer, nutrient-poor upper layers with colder, nutrient-rich deeper layers. In temperate zones, differential heating and cooling in summer and winter produces a thermocline with an upper layer of different density than the lower layers, thereby preventing mixing. In fall and spring as the temperature of the two zones equalizes, mixing occurs, usually followed by a bloom of phytoplankton.

***Lotic Ecosystems.*** Running water, such as that found in streams and rivers, has several special characteristics. First, the oxygen content of streams and rivers is usually high, and the proportion of their nutrients obtained from adjacent lands is great. Depending on its velocity of flow, moving water may severely limit the kinds of organisms that can survive in it. Small, rapidly flowing headwaters, or upper tributaries of a river, are often shaded from sunlight and therefore poor in productivity and species diversity. The consumers of headwater communities depend largely on detritus, leaves, insects, and other organic materials that fall into them or enter them with runoff (Fig. 35-27). In contrast, the wider midsections of large rivers are characterized by a high productivity of algae and a predominance of filter feeders and other primary consumers. Productivity and species diversity both reach their peak in these midsections. Long rivers become increasingly muddy, which reduces the penetration of sunlight, and hence the productivity and species diversity that depend on it. The rate of productivity in muddy waters is lower than the rate of consumption. Rivers slow as they approach their mouths, resulting in the settling out of many of the heavier particles of silt and sand. The great volume of sediment that accumulates near the mouth of a large river forms a delta. The "toe" of the boot-shaped state of Louisiana was formed in this way (Fig. 35-28).

**FIG. 35-27** A creek in Ricketts Glen, Pennsylvania.

A great variety of conditions can be found along the length of a river. The shallows of a rapid river can be as quiet as any pond. Where current velocity slows in deep pools, sufficient sand may settle out to provide a habitat for various burrowing animals. During flooding, a river may directly connect with marshes or other wetlands, with which it exchanges nutrients and various species of organisms.

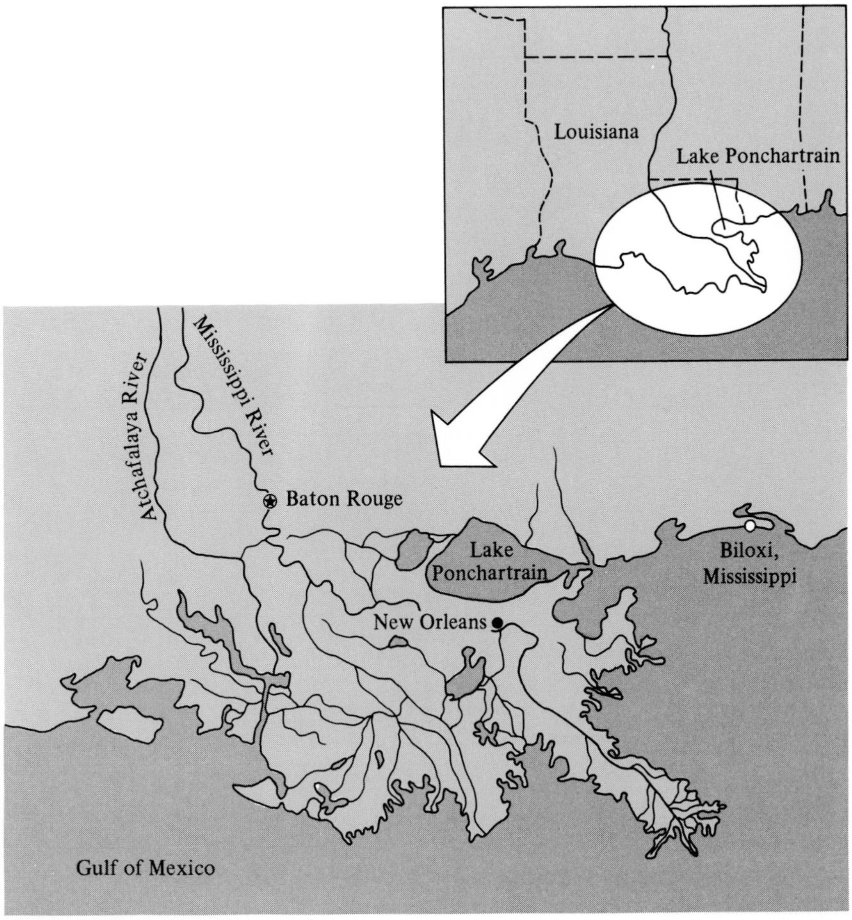


**FIG. 35-28** The delta of the Mississippi River. Much of southern Louisiana south of Baton Rouge was formed from Mississippi River sediments.

***Freshwater Wetlands.*** The two major types of freshwater wetlands are marshes and swamps. Swamps are populated by trees, whereas marshes are not. Wetland sediments of marshes and swamps form two principal layers, the uppermost of which is a relatively shallow zone in which production occurs and oxygen is sufficiently abundant to support aerobic decomposers. A much deeper anaerobic zone of muck lies below the upper layer, accommodating several types of anaerobic bacteria that reduce organic matter to such gaseous forms as carbon dioxide ($CO_2$), molecular nitrogen ($N_2$), hydrogen sulfide ($H_2S$), and methane ($CH_4$). These gases rise to the surface and are released into the atmosphere, thereby completing the cycle begun when they were incorporated into organic matter. Hydrogen sulfide is responsible for the "rotten egg" odor of some marshes, and methane, when ignited by spontaneous combustion, accounts for the pale blue flames, called will-o'-the-wisp, sometimes seen in marshes and swamps at night.

Although coastal and freshwater wetlands occupy only about 2 percent of the earth's surface, their importance in recycling the world's sulfur, nitrogen, phosphorus, and carbon is quite out of proportion to the area they occupy. Wetlands are also breeding grounds for many animal species and feeding grounds for billions of migratory waterfowl (Fig. 35-29). Only recently have public attitudes toward wetlands begun to change from regarding them as wasted areas that should be drained and "improved" for commercial purposes to the realization that they are valuable resources worth more to us in their natural state than as agricultural or commercial developments.

**FIG. 35-29** Wetlands provide stopping and feeding grounds for billions of migratory waterfowl, such as these snow geese and Canada geese on Goose Lake, Iowa.

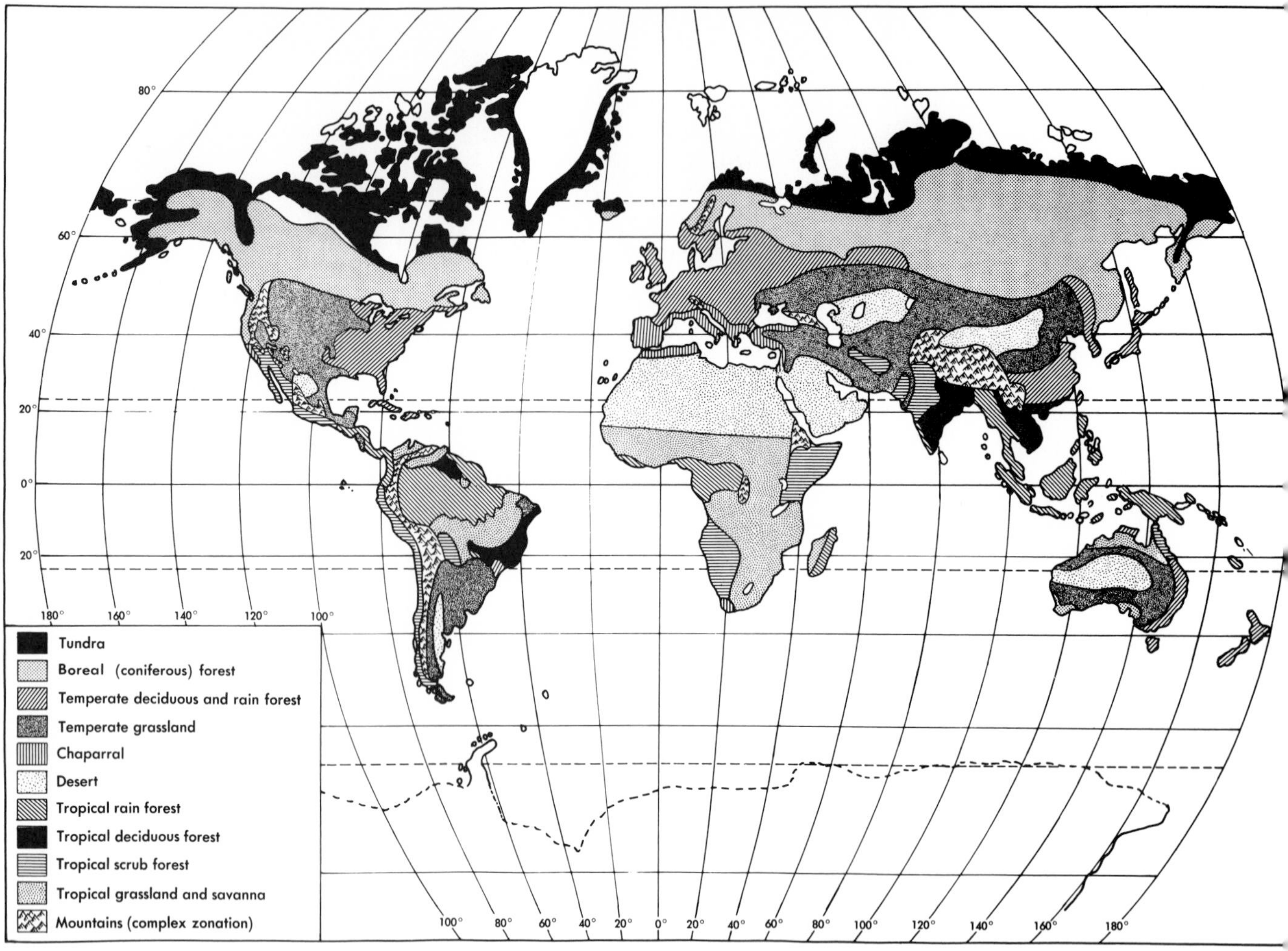


**FIG. 35-30** The major biomes of the world.

## MAJOR TERRESTRIAL BIOME TYPES

Primarily because of differences in climate, major climax communities found in different regions often differ markedly in composition. The largest such terrestrial climax communities are termed **biomes** (Fig. 35-30). The global pattern of their distribution, usually in extensive belts, is principally a result of the wind patterns described in Chapter 33. One example of a terrestrial biome type in the United States is the deciduous forest, made up of trees that shed their leaves annually. Deciduous forest biomes cover much of the eastern states. Another example is the grassland biome of the western plains. Deciduous forest and grassland biomes also occur in other parts of the world wherever soil and climate are favorable for their development. For a given biome type, the particular species of trees, grasses, and other plants, as well as the animal species associated with them, differ from one part of the world to another. Such differences are especially evident where biomes of the same type have been isolated from each other over geologic units of time. In addition to deciduous forest and grasslands, the most important major terrestrial biome types are tundra, taiga, desert, and tropical rain forest.

**FIG. 35-31** A view of Arctic tundra in summer near Fairbanks, Alaska. The Alaska Range is visible in the distance.

## TUNDRA

Above 60° north latitude stretches an extensive, treeless plain, the Arctic **tundra** (Russian for "marshy plain") (Fig. 35-31). The soil of the tundra is frozen for all but 6 weeks or so of every year. Even during this short summer season, the tundra soil, a marshy substratum of partially decomposed plants and lichens, remains permanently frozen a short distance below the surface, a layer known as **permafrost.** The brief summer brings a massive thaw, and many patches of water appear in the low-lying areas (Fig. 35-32). Although, because of wind patterns, precipitation levels are very low in polar regions, evaporation is also low, and drainage is poor. The warm weather is accompanied by the appearance of billions of insects, a colorful display of flowers, and the arrival of millions of migratory birds that feed on the insects and vegetation.

The tundra's most common photosynthetic organisms are lichens, the mutualistic associations of algae and fungi often referred to as rein-

**FIG. 35-32** A view of the tundra of Alaska's North Slope in spring after melting snow has produced extensive patches of water. Masses of sea ice can be seen in the distance.

deer moss (see Chapter 29). Grasses, sedges, and some dwarf shrubs, such as cranberry, are adequate to support a variety of animals. However, the number of species of tundra animals is quite small. The dominant mammals include the caribou (domestic strains are called reindeer), arctic hare, lemming, and arctic fox. Only a few species of predatory birds, such as the snowy owl, are found in the tundra year-round. But in summer, a large variety of migratory birds arrives to nest and breed in the relative absence of predators. In some areas, the nests are numerous, spaced only a meter or so apart. The large animals, such as caribou, are also largely migratory. Populations of small animals tend to fluctuate in cycles. The vast herds of caribou supported by the tundra's brief but productive growing season have long provided sufficient food and materials to allow humans to make the Arctic their permanent home.

***Alpine Tundra.*** Climatic conditions very similar to those of the Arctic prevail at the tops of high mountains, in the alpine tundra biome type. The kinds of organisms found in the alpine tundra are also similar, but not identical, to those of the Arctic. The climate of alpine tundra differs from that of Arctic tundra in that it lacks moisture because of the rapid runoff of rain and melting snow. In contrast, the Arctic tundra, despite its low annual precipitation, has little runoff, which accounts for the marshy conditions that prevail during the summer thaw.

## TAIGA

The **taiga,** also termed the northern coniferous forest or boreal forest, stretches in broad belts across North America and Asia, regions characterized by low atmospheric pressure and a resultant moist, cool climate. The underbrush of the taiga includes deciduous shrubs and other perennial plants that are poorly developed because they are shaded year-round by a dense canopy of evergreens. The most prominent trees of the taiga are spruce, firs, and pines (see Fig. 35-33). The needles of these conifers represent an adaptation that conserves water, withstands freezing, and reaps maximum advantage of sunshine for photosynthesis during the short summer and the warm periods of spring and fall. When

**FIG. 35-33** Winter in a taiga community of central Alaska.

weighted with snow, the branches of conifers bend down, allowing the snow to tumble off. The general shape of coniferous trees also discourages the accumulation of snow.

The taiga is often referred to as the spruce-moose biome, after its two most prominent organisms. Besides moose, prominent taiga animals include snowshoe hare, pine marten, and grouse. These animals are able to subsist during the winter by eating pine cones and the young needles at the tips of conifer branches. But they depend heavily for food on broad-leaved plants at other times of the year. In addition to such predators as timber wolves, taiga provides habitats for a great variety of insects and birds. As with tundra, the relative simplicity of the taiga community is related to certain instabilities, for example, fluctuations in the lemming populations and occasional insect infestations of trees.

Because conifer needles resist decay and because the climate is cold, the floor of a taiga forest is a thick, spongy mat of needles in all stages of decomposition. Decaying needles are quite acidic. This condition, coupled with the substantial annual precipitation and limited evaporation, contributes greatly to the leaching out of nutrients from the top layer of taiga soil. The acidic soil of a coniferous forest is thus quite poor for almost anything but growing conifers; cleared taiga makes poor farmland. Its topsoil is likely to be thin and underlain with glacial deposits of sand and gravel. Like tundra biomes, taiga communities are also found on mountains, just below alpine tundra biomes (see Fig. 35-2). The timberline is the point at which taiga ends and alpine tundra begins.

## TEMPERATE DECIDUOUS FORESTS

As one travels south from the Arctic—or down the slopes of a mountain—there is a perceptible increase in diversity, not only of trees, but of shrubs and herbaceous, or ground-cover, plants as well. Such diversity is characteristic of the **temperate deciduous forest.** The increased richness of plant diversity creates a greater diversity of habitats, which further increases the variety of species. Correlated with the increased diversity are greater efficiency of nutrient cycling and community stability. One requirement for the success of a temperate deciduous forest is abundant rainfall (75 cm to 150 cm, or 30 to 60 inches, annually) fairly evenly distributed throughout the year. The organisms of this biome type are adapted to a relatively uniform, temperate climate (−30°C to 37°C) with distinct seasonal changes.

A much greater variety of plants and animals is found in temperate deciduous forests than in colder biomes. The soil is better than that of the taiga; in the warmer climate, deciduous leaves decay more readily and are less acidic than needles. The soil thus supports a much larger variety of soil organisms. Because of all these factors, a cleared deciduous forest makes better farmland than does cleared taiga.

In various regions, even of the same country, different species of deciduous trees have become dominant in climax forest communities. Beech and maple typify the climax community of the north central United States, for example, while oak and hickory predominate in the southern and western communities of this biome. These communities contain different species of other organisms as well. Such variation in species among biomes of the same type occurs around the world. Serving as a reservoir of valuable species is but one of the many benefits provided by the world's forests (see Panel 35-1).

PANEL 35-1

# The Value of a Forest

Not only are forests valuable for wood and wood products, but they provide many other important benefits, some of them of immense economic worth. Few of these are known or appreciated by the general public. A brief survey of some of them is appropriate here.

A creek in the Podocare Forest of New Zealand.

## RESERVOIR OF VALUABLE SPECIES

The complex mature forest is a reservoir of biologic diversity. One benefit of this fact is that forests form a base of supply of species for communities undergoing succession following a fire or other devastation.

## DETOXIFIER OF POLLUTANTS

With their many decomposers, mature forest communities purify the air and water of toxic pollutants. Modern industries add an enormous burden of pollutants to the global ecosystem. A large number of these substances are detoxified by forests, particularly by the more complex wilderness communities. One or another of the wide variety of species of plants, fungi, protists, bacteria, worms, and insects that is found in a mature forest community can usually detoxify almost any—but unfortunately not all—of the poisonous wastes produced by humans.

Extensive erosion of a hillside in northern Idaho following a forest fire. Immediate replanting would have prevented much of the damage.

## STABILIZER OF THE LANDSCAPE

Forests stabilize the landscape by preventing wind and water erosion. Unless a hillside that has suffered a devastating forest fire is replanted at once, rains soon wash away its topsoil or produce destructive mudslides. Denuded mountainsides allowed to lose their topsoil after having been clear-cut or after a fire seldom recover.

## RECHARGER OF GROUNDWATER

Heavy rains tend to run off most soil surfaces into lakes and rivers. However, trees slow this runoff, giving the water a chance to percolate slowly through the soil. The root systems of trees keep the soil from compacting, which enables water to seep into the ground. Thus, forests function to raise the water table, keep wells from going dry, and promote the flow of springs. Floods are always more frequent and severe downstream from a cut- or burned-over forest than they were before the forest was cut or burned.

## FILTERER OF SOIL WATER

Rivers downstream from extensive mature forests receive clear water that has been filtered by its passage through the forest soil. Such water supports a great variety of fish and other animals. But downstream from extensively cut or burned areas, the water entering rivers has run off the soil surface, taking soil particles with it. Such rivers are murky with sediments that limit light penetration and reduce photosynthesis; they can support only the few species of fish that tolerate such conditions. Dam reservoirs on such rivers become filled with silt, sometimes within less than 10 years of being built. A sediment-filled river needs dikes and levees to prevent flooding of river communities. Billions of dollars have been spent on dams, dikes, and repair of flood damage after deforestation of upstream watersheds.

## RECYCLER OF ATMOSPHERIC WATER

A forest plays an extremely important role in the water cycle of its region. To appreciate the magnitude of this contribution, it should be noted that only about 0.6 to 0.7 percent of the annual amount of sunlight falling on a forest or other plant community is used to drive the reactions of photosynthesis. Only about a half of this energy goes into production of new plant biomass—wood, if we think in terms of the forest's main cash crop. The other half of the 0.6 to 0.7 percent of the incident sunlight's energy absorbed by chlorophyll ultimately goes to support the rest of the life of the community. Much of the other 99 percent of the sunlight's energy falling on the forest is reflected and lost immediately as heat radiated into space. A substantial amount of heat, however, is temporarily absorbed and warms the trees. This energy increases the temperature of the water in the trees' xylem vessels and contributes directly to the process of transpiration.

Transpiration is essentially the evaporation of water from leaves (see Chapter 10). Recall that soil water is absorbed by tree roots, rises in the xylem, and is lost as vapor from the leaves. To convert water from liquid to vapor requires much energy. Whereas it takes only 1 calorie of heat energy to raise the temperature of 1 g of water 1°C, or 100 calories to raise it from freezing to boiling (0° to 100°C), it takes 540 calories to convert 1 g of liquid water to 1 g of water vapor without raising its temperature. (Because water has such an unusually high heat of vaporization, our watery planet's surface is very resistant to rapid changes in temperature, especially in humid regions.)

A tree thus uses an enormous amount of solar energy to absorb water from the soil and convert it to vapor. Only lakes and oceans are more effective in this regard. In 1 year, a hectare (2.5 acres) of mature deciduous forest brings up an average of 3 million l (790,000 gallons) of water from the soil and releases it into the air by transpiration. Calculated on an annual basis, over 12 percent of the radiant energy striking the forest is used in this way. If the amount of solar energy absorbed and used to vaporize water is calculated only for that part of the year when leaves are actually present and functioning, we find a forest's efficiency to be an astounding 38 percent! Forests thus make a most valuable contribution to the functioning of the water cycle by returning precipitation slowly to the atmosphere so that it can fall again as rain or snow. To determine the economic value of this benefit it should be possible to estimate what it would cost to perform these tasks with mechanical equipment. The figure would be enormous.

## MODERATOR OF TEMPERATURES

Besides benefiting the water cycle, trees moderate temperatures by using the sun's energy for transpiration while providing shade, which is of great value in preserving the lives of many organisms. Some measure of the extent of this contribution is dramatically evident to anyone who has ever entered the coolness of a mature forest on a blistering hot day. Moreover, a few large trees sheltering a house usually make expensive air conditioning unnecessary.

## A PLACE FOR RECREATION

Of course, forests provide other benefits not as easily priced as some of those mentioned above. For example, they are favorite locales for a wide variety of recreational activities. Although some monetary figure might be assigned to recreational value, especially when commercialized, who can put a price on the joy provided by the beauty of the world's forests?

(a)

(b)

**FIG. 35-34** (a) A herd of bison (buffalo) *(Bison bison)* in Custer State Park, South Dakota.
(b) Pronghorn antelope *(Antilocarpa)* in Yellowstone National Park, Wyoming.

## GRASSLANDS

Temperate zone regions in which rainfall is not only sparse (about 25 cm to 75 cm, or 10 to 30 inches, per year) but unevenly distributed throughout the year, and in which occasional drought years occur, are characterized by grasslands rather than forests. Droughts render grasslands particularly subject to fires, to which grasses and other monocots are well adapted. Whereas trees and shrubs grow from their apical meristems, grasses grow at the bottom of their shoots, enabling them to withstand not only fires but extensive grazing as well. Fires, grazing, and droughts all discourage tree growth.

***Characteristic Species of Grasslands.*** Not surprisingly, various grass species are the most prominent plants found in grasslands, but many other plant species, such as the composites,* sunflowers and clovers, are found as well. However, the latter constitute only a small part of the producer biomass. Grasslands also support legumes, which are important in soil-building. The most prominent grazing animals of the United States grasslands have been the bison and the pronghorn antelope (Fig. 35-34). The bison was once nearly extinct; small herds are now preserved in protected areas and in zoos. Several herbivorous burrowing rodents, such as prairie dogs and gophers, and carnivores, such as coyotes and badgers, are also part of the grassland community. Grassland birds include the prairie chicken, the meadowlark, and a variety of hawks. The nature of a grassland dictates the kind of larger animals that can survive in it; the herbivores must be fleet-footed or burrowers, and the carnivores must catch the herbivores or other carnivores.

In regions where rainfall is least and grasslands begin to intergrade with desert, the natural community is complex and highly adapted to the conditions. Some hardy grass species have roots that reach down 2 m into the soil (Fig. 35-35) and outweigh the rest of the plant several times over. Many species send out rhizomes, or underground stems, which hold down the soil and thus retard wind and water erosion. Many species become dormant during the heat of summer but resume growth in the fall, remaining green throughout the early frosts. Other species grow well during hot weather but become dormant in the fall. During

* Composites are members of the Compositae, a plant family characterized by multiple small flowers arranged in a compact head, as in sunflowers and dandelions.

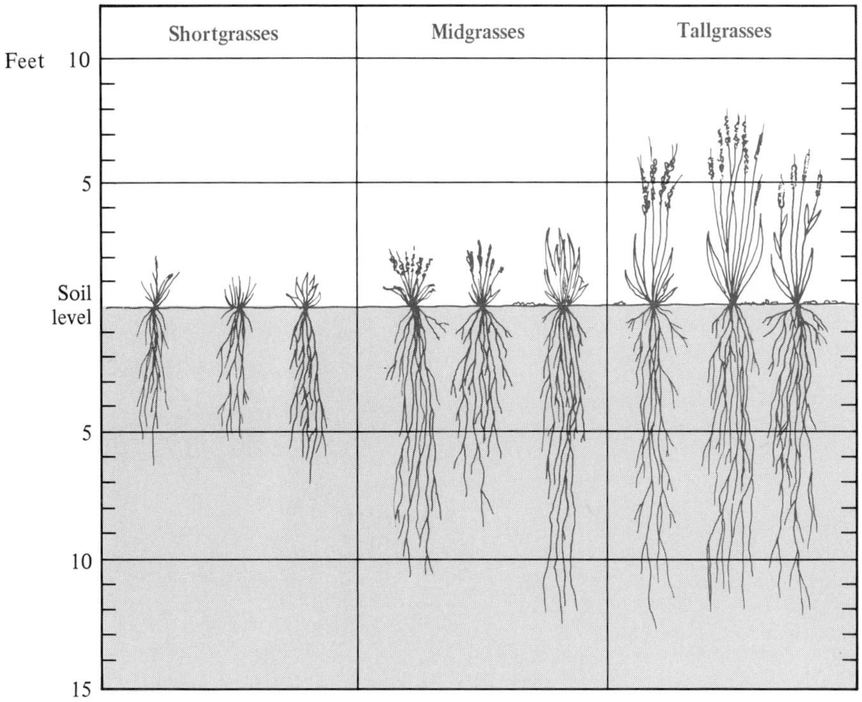


**FIG. 35-35** Representative types of grass species in various grassland communities. Short-bladed, deep-rooted species predominate where grassland merges with desert; note the proportion of the root system to the leaf system.

long dry spells, the tall grass blades die and form a protective mat over the soil surface and root systems; this conserves moisture, prevents wind erosion, and keeps the dormant plants alive.

***The Value of Grasslands.*** As a source of food, few biome types are more important to modern humans than grassland (Fig. 35-36). Besides furnishing grazing land for sheep and for beef and dairy cattle, grasslands have become the world's richest croplands. Moreover, the world's most important foods include crops, such as maize (American corn) and wheat, that were derived from grass species by domestication. Some grasslands receive little rainfall and are subject to recurring drought years; when this fact is ignored serious problems often occur with the human use of grassland for cattle grazing and farming. Overgrazing, by definition, removes more forage than regrows, resulting in severe wind erosion and invasions of a range by undesirable plant species. Annual planting of short-rooted crop plants such as wheat over several years may deplete the soil of moisture and lead to dust-bowl disasters (see Panel 36-1).

## DESERTS

Like other biome types, deserts occur in worldwide belts, receiving little rain because of the patterns of prevailing winds (see Chapter 33). There are two major types of desert; hot and cold. The climatic feature common to all is an annual rainfall of 25 cm (10 inches) or less. Only the central Sahara and northern Chilean deserts receive no rain at all. The plants found in deserts are able to stand long periods without any rain. As in grasslands, the amount of biomass produced by the community is directly proportional to the amount of rainfall.

***How Desert Plants Survive.*** Plants survive in the desert by one of three major adaptations to lack of water (Fig. 35-37):

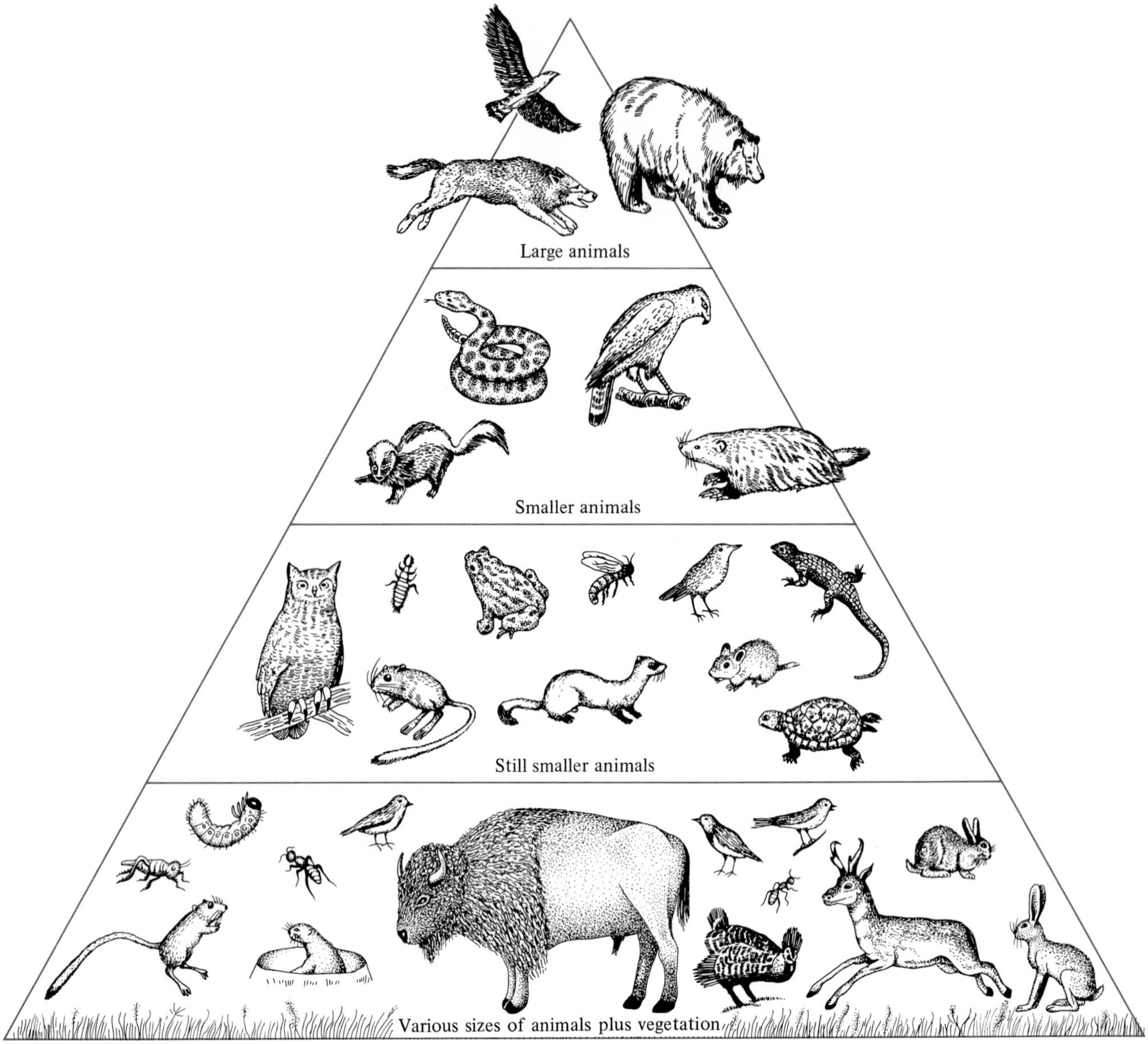


**FIG. 35-36** The complex pyramid of a United States grassland community. The species are arranged roughly in the form of a food pyramid, with grasses and other plants at the bottom, primary consumers next, then secondary consumers, and tertiary consumers on top. Some of the original members, such as grizzly bears and bison, are now extinct in the region, or nearly so.

**FIG. 35-37** Death Valley in bloom following a heavy rain in April.

**FIG. 35-38** Irrigation yields high productivity in the deserts of the southwestern United States.

1. Some annuals germinate, grow, bloom, and set seed within a very short period of time, but only after sufficient rain has fallen to support their growth.
2. Cacti are **succulents;** that is, they store enough water to last through dry periods. The green, leafless stems conduct their photosynthesis.
3. Deciduous shrubs shed their thick leaves in dry periods.

Desert plants are typically widely spaced, permitting individual plants to get the full benefit of any moisture present in their immediate area. Spacing is sometimes accomplished by the production of antibiotics that inhibit the growth of many types of plants in the immediate vicinity of the plant producing the antibiotic. However, much of what appears to be bare ground proves, upon close examination, to be covered with a crust of mosses, lichens, and algae. Some desert soils are rich in nutrients and when irrigated can be highly productive. (Fig. 35-38).

***Adaptations of Desert Animals.*** Desert animals are highly adapted to arid conditions by behavior, physiology, or both. Most hot desert rodents, for example, are nocturnal, remaining in their humid burrows during the day. One of their striking physiological adaptations is the ability of the kangaroo rat to survive indefinitely on such food as dry seeds without ever drinking a drop of water. All needed water is obtained from the water of metabolism: the water produced chemically as a result of cellular respiration (see Chapter 6). Desert animals that remain above ground during the day, such as camels, must drink water at least occasionally, but they too can survive long periods without drinking.

## TROPICAL RAIN FORESTS

No biome type is more complex than the tropical rain forest. Far more species of plants and animals are found in these communities than in any other. Tropical temperatures are high and remarkably constant, in

PANEL 35-2

# Comparative Productivity of Biomes

All the living material in a biome makes up its biomass. The dead material is classified as either humus or litter. Litter is recently dead material the origin of which can be recognized; leaves, twigs, logs, and animal bodies are examples. Humus is organic matter that has reached a stage of decay at which its origin is no longer apparent. Insight into some of the ecological principles of biome types can be gained by comparing some representative biomes as to annual productivity of different types of biomass and annual accumulation of litter and humus.

## PRODUCTIVITY OF BIOMASS

Biomass is classified in four categories: green parts of plants, perennial aerial parts of plants (always the predominant component in a forest), litter, and roots. The accompanying graph depicts the percentage of each biome made up of each of these components and gives the absolute amounts of annual productivity and total accumulation in kilograms per hectare.

In all three types of forest, 70 to 74 percent of the community's biomass is represented by the trunks or stems and the branches of trees and shrubs. In the tundra and the desert, the greatest amount of biomass is represented by roots. In tundra plants, this distribution of biomass constitutes an adaptation to the severity of the winters; in desert plants, an adaptation to aridity.

## THE RATIO OF HUMUS TO LITTER

Table 35-1 compares the productivity and accumulation of litter and humus in representative biomes. The total accumulation of litter increases with temperature from tundra to tropical forest (2,400 kg/hectare for tundra and 25,000 kg/hectare for tropical rain forest) except where a lack of moisture prevents it in deserts. However, despite the enormous productivity of the rain forest, the humus content of its soil, even in absolute figures, is but a small fraction of that found in the soil of colder climates. Humus is generally absent from desert biomes. When the proportion of humus to litter is compared, the difference among biomes is

A rotting tree stump and typical litter on the floor of a temperate deciduous forest. The presence of saprophytes (mushrooms, Indian pipe) indicates that decay is occurring.

some regions varying by only 4°C to 5°C for a given time of day or night the year round. Because the tropics lie on or near the equator, wind flow patterns and low pressure combine to produce heavy rainfall (see Chapter 33). It may rain steadily during much of the day for many weeks on end. Drier seasons, with a monthly rainfall of 12 cm or less, occur; but annual rainfall always exceeds 200 cm (80 inches).

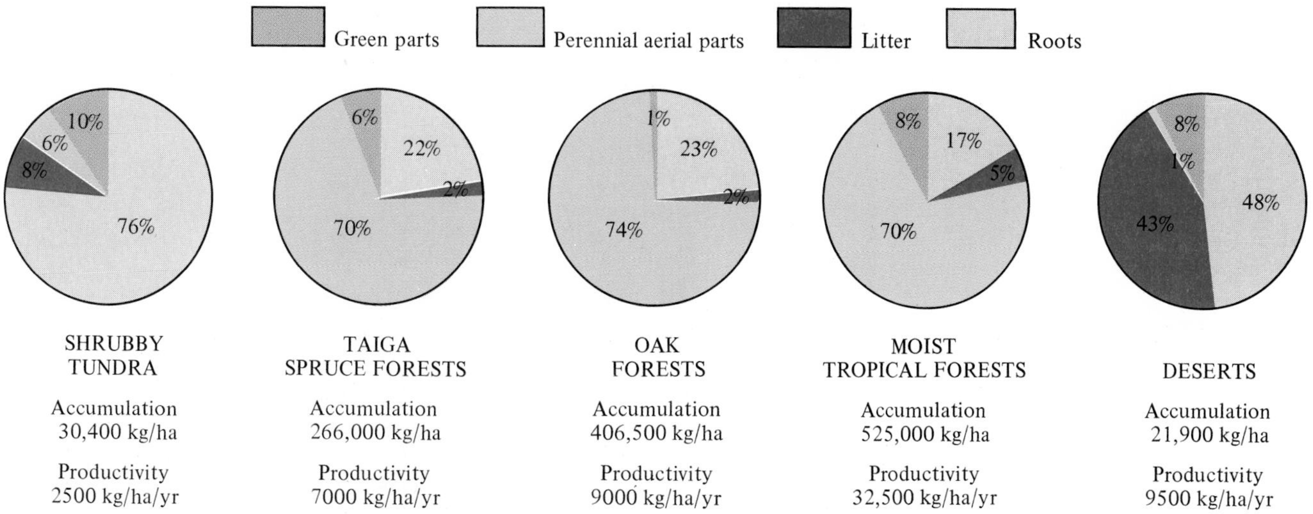


A comparison of the proportion of various categories of biomass within and among selected major biomes. Total biomass productivity and accumulation are also shown.

even more striking: the mass of humus in the tundra is 92 times greater than the mass of litter, while the mass of humus in the tropical forest is one-tenth the amount of green (newly fallen) litter. Thus, although the percentage of a biome's production represented by litter is nearly the same in all climates, the rate of decomposition of humus is high in rain forests and deserts, where high temperatures speed the recycling of minerals.

**TABLE 35-1 Distribution of Organic Matter Between Humus and Litter in Selected Biomes (kg/hectare)**

| | *Shrubby tundra* | *Boreal spruce forests* | *Oak forests* | *Moist tropical forests* | *Deserts* |
|---|---|---|---|---|---|
| Litter | 2,400 | 5,000 | 6,500 | 25,000 | 9,400 |
| Annual net increment in litter | 100 | 2,000 | 2,500 | 7,500 | 100 |
| Percent of productivity | 4 | 29 | 28 | 23 | 1 |
| Dead organic residues (humus) | 83,500 | 45,000 | 15,000 | 2,000 | — |
| Ratio of organic residues to green litter | 92:1 | 15:1 | 4:1 | 0.1:1 | — |

SOURCE: Rodin, L. E., and N. I. Bazilevic. 1964. *Doklady Akademii Nauk SSSR*, 157:215–218.

***Flora.*** The plants of moist tropical forests are deciduous, but almost all shed their leaves continuously, a few at a time, rather than all at once; the forest thus remains green all year round. The trees that form the forest canopy are at least 25 m to 35 m (80 to 100 feet) tall (see Fig. 35-3). Very tall **emergent trees** extend above the canopy here and there, some to a height of 70 m. Some of the emergents are exceptional in that

**FIG. 35-39** A break in the canopy of a virgin tropical rain forest on the island of Borneo allows a shaft of sunlight to reach the forest floor. As in a temperate forest, the full sun allows species characteristic of an earlier seral stage to spring up; eventually the space will be filled by species typical of later stages, and the opening will close.

they shed their leaves during dry periods. The understory of trees, ones smaller than those forming the canopy, become prominent only where a break in the canopy allows the sun to penetrate (Fig. 35-39). Shrubs and herbs that tolerate dense shade form the bottom layer. The number of vines and other climbing plants in the rain forest is enormous; sometimes they completely cover the trunks of trees. Although many of these climbing plants have roots in the soil, the epiphytes, which include the orchids, are rooted in the material that collects in the crotches of the trees on which they grow. Parasitic plants, ones that feed on the host tree, are also to be found.

As in the mature temperate forest, the dark interior of the rain forest has open, parklike areas. The "dense jungle," penetrated only by wielding a machete, is not a climax forest; it is an early seral stage found only at the edges of mature rain forests.

***Tropical Efficiency.*** The abundant moisture and warmth of the mature rain forest, coupled with its extremely diverse biota, make it the most efficient processor of energy and materials among the biome types. Whereas the annual productivity of a tropical rain forest is the highest of any ecosystem, the relative amount of partially decayed matter, called **humus,** is the lowest of all biome types (see Panel 35-2). A lost feather, a fallen leaf or branch, a dead insect, bird, or mammal—anything that has died—is set upon by a wide variety of insects, worms, fungi, protists, and bacteria; in the warm, moist environment of the tropics, such matter is rapidly reduced to its elements. No sooner do any of these released nutrients enter the soil than they are absorbed by roots and incorporated into biomass again. Because plant roots remove almost all salts as the water seeps through the soil, the streams in tropical forests are almost pure rainwater. It is thus typical of the tropical rain forest that almost all of its nutrients are tied up in its biomass; all growth is entirely dependent on the continual recycling of materials. When, as is done from time to time, a rain forest is cleared and traditional farming attempted on the site, crop failure is the usual result, if not the first year, then soon thereafter. Not only do the heavy rains promote a rapid leaching of nutrients from the soil, but the soil tends to harden when exposed to air. The nutrient-poor soil of the rain forest is badly suited to most types of farming.

## THE STABILITY OF MATURE TERRESTRIAL COMMUNITIES

Ecologists have long believed that the very complexity of a mature community contributes to its stability. In a climax community, each stage in the processing of energy and materials is divided among many different species of organisms, sometimes representing the maximum possible diversity for the given climatic conditions. Most factors detrimental to one or a few species in such a complex community can logically be expected to have a relatively small effect on the community as a whole. Based on this assumption, only the destruction of large numbers of key members of a mature community—such as the clear-cutting or defoliation of all the large trees in a forest—would be expected to ruin the community's structure. Moreover, mature communities have a tendency toward metabolic equilibrium: the point at which total gross production and respiration are about equal. Because tropical forests are the most complex, they have been considered the most resistant to change. In contrast, because of their simplicity, the relatively simple terrestrial communities of the Arctic have long been regarded as fragile, that is, far less able to withstand changes.

The validity of this logical stability-from-complexity hypothesis has lately been challenged, largely because of a general lack of evidence to support it. Laboratory studies of experimental communities, field studies, and studies of computer models have yet to show any positive relationship between community diversity and community stability. Moreover, in some instances, it has been shown that some environmental stresses can be amplified in proportion to a community's complexity. These various observations have caused the entire question to be examined more closely.

One fact that indirectly suggests that the older idea might be wrong is that Arctic and temperate zone organisms have undergone natural selection for their ability to withstand the severe environmental stresses associated with great fluctuations in climate. Tropical species, on the other hand, are adapted to a climate that remains uniform all year long, year after year. The reason that complex tropical communities appear to be more stable may be the year-round stability of the climate that favors the stability of whatever organisms live there, while at the same time encouraging the evolution of great variety. Disturbances such as the construction of the Trans-Amazonian Highway (Fig. 35-40) may yet prove to be far more destructive of those tropical communities than such projects as the building of pipelines and highways through the breeding grounds of major caribou herds on the Alaskan North Slope. Another reason for a cautious reexamination of the older position is that there has been some inadvertent bias in the gathering of data. The amount of data obtained from the northern temperate zone, where most ecologists live and work, is far greater than that gathered from either tropical or Arctic zones.

**FIG. 35-40** Construction site on the Trans-Amazonian Highway in Brazil.

## Summary

The structure of a biotic community includes its species diversity and the patterns of their distribution. Diversity is composed of richness, or the total number of species, and evenness, their relative abundance. Richness declines from the equator to the poles, with habitat area, and with the length of time available for speciation. Evenness is largely a function of warmth, moisture, and stability of the climate. Fluctuating seasons reduce numbers and evenness. Patterns of distribution include zonation, which is affected by physical factors, competition, and predation; and vertical stratification, which is especially evident in soil communities and mature forests.

Ecological succession is typified by the series of stages that occur in the years following a severe fire or clear-cutting of a forest. The early stages are characterized by a heavy growth of herbaceous plants that grow to maturity quickly and tolerate bare, unshaded soil. In time, taller species and slow-growing woody species appear, providing shade for still other species of plants and animals that could not tolerate its lack. Some birds and small mammals next appear. Fast-growing trees then emerge, making the area less suitable for pioneer plant species that cannot tolerate shade. In time, short-lived trees disappear and are replaced by slow-growing, long-lived ones, which provide habitats for a more varied insect, bird, and mammal population. Succession eventually ceases as the forest reaches full maturity and the production of new biomass equals the amount that disappears. In the process of lake succession, a lake is filled in and becomes a marsh, swamp, or bog, which eventually becomes overgrown by the surrounding forest or other community. Lake succession is accelerated by the entrance of sewage effluent and farmland runoff.

Animals of seashore communities are adapted to the available food materials by a great variety of feeding methods, including filter feeding, deposit feeding, and raptorial (seizing) methods. Great variety is also evident among the decomposers: bacteria, protists, and fungi. The ammonia, nitrates, and other simple materials that they release are immediately available to the producers of the shallow waters. Because their nutrients are confined to the euphotic zone, the edge of the sea and coral reefs are highly productive.

In the open sea, production is almost exclusively accomplished by phytoplankton. Zooplankton, such as protozoans, invertebrate larvae, and copepods, harvest phytoplankton. Small fishes, which feed on zooplankton, are eaten by larger fishes and other animals.

Because much nutrient material settles to the bottom, the surface layers of tropical open seas are low in nutrients and in productivity. On the bottom, the settled material slowly decays, primarily by bacterial action, releasing its mineral nutrients. The bottom supports a varied animal life that depends entirely on the material descending from the surface layers. Exceptions to this rule are the hydrothermal vent communities, which are based on the production of chemolithotrophic bacteria that derive energy from oxidation of the inorganic compounds issuing from the vents.

Antarctic communities, which are based on phytoplankton growth, experience sharp fluctuations in productivity resulting from seasonal variations in sunlight. Phytoplankton blooms feed euphausid crustaceans called krill, which are eaten by larger animals, including baleen whales. Antarctic food pyramids include one in which birds are the dominant group; one in which the killer whale, which feeds on seals and penguins, has the top position; and a third in which the baleen whales depend on krill as their principal food.

Freshwater ecosystems are classified as lentic, including ponds and lakes; lotic, including streams and rivers; and wetlands, including marshes, swamps, and bogs. Each presents its own special conditions and provides habitats and niches for specialized kinds of flora and fauna.

Different types of major terrestrial communities, or biomes, owe their unique characteristics mostly to climatic differences. Arctic tundras are treeless plains whose surface is frozen for all but a few weeks of the year, during which time insects and flowers appear in profusion, along with millions of migratory birds. The main producers are lichens, grasses, sedges, and small shrubs. Only caribou, hares, lemmings, foxes, and a few species of predatory birds stay all year.

A taiga is marked by conifers, which thrive in cool, moist climates with long, cold winters and much snow, conditions that discourage the growth of deciduous trees. Taiga fauna include moose, snowshoe hares, pine marten, and grouse. Predators include timber wolves. Because slowly decomposing conifer needles produce acid, which leaches valuable nutrients from the soil, taiga makes poor farmland.

Temperate regions with a well-distributed annual rainfall of 75 cm to 150 cm support deciduous forest biomes. Besides their commercial value, forests provide reservoirs of valuable species, detoxify pollutants, recharge groundwater, stabilize the landscape, filter soil water, recycle atmospheric water, moderate air temperature, and provide recreation.

Grassland biomes thrive in temperate regions with unevenly spaced, sparse (25 cm to 75 cm) annual rainfall with occasional droughts. Grasses are well adapted to the fires that occur in drought years and to grazing of bison and pronghorn antelopes. Burrowing rodents such as gophers, carnivores such as coyotes, and birds such as prairie chickens and hawks, are also prominent. Grassland makes the best cropland.

Deserts exist where annual rainfall is 25 cm or less. Their floras are highly adapted to withstand a year or more without rain. Their annuals must germinate, grow, mature, bloom, and set seed within a very short time. Succulents survive by storing water, deciduous shrubs by shedding leaves in dry periods. Desert animals have adapted to the environment through behavior, physiology, or both.

Tropical rain forests thrive in climates with a constant high temperature and an annual rainfall of over 200 cm. All perennial plants are deciduous but usually shed leaves continually. Tall trees, vines, and epiphytes are common. All dead organisms undergo rapid decomposition, and the released nutrients are immediately reincorporated into biomass. For this reason, cleared jungle makes the worst farmland.

Ecologists have long believed that the most complex communities, those of the tropical forests, are the most stable and therefore the most resistant to change. Simpler communities, such as those in the Arctic Circle, have been considered more fragile, that is, less able to withstand changes and survive. However, doubt has been cast on at least the universality of the principle because no positive evidence supporting it has been produced by several studies of the question. It may be that the stability of tropical communities is due to their stable climate, not their complexity.

## Recommended Readings

A list of recommended readings appears at the end of Chapter 33.

36

# Human Impact on World Ecosystems

OUTLINE



Determining the full effect of human activities on the environment is a difficult task. One reason for this is its magnitude: the amounts of food, fuel, and other raw materials that humans consume; the physical destruction of landscapes and natural ecosystems that they accomplish; and the volume of wastes, many highly toxic, that they discharge into the environment are so enormous as to tax our capacity to measure them. The task is further complicated by the fact that modern technology is producing environmental changes with unprecedented and unanticipated speed. Moreover, the toxic effects of many harmful wastes do not become evident until the environment has been exposed to them for 10 to 20 years.

Although some human activities benefit the environment, their net effect has been its increasingly rapid deterioration, a serious matter in that, as aerobic heterotrophs, humans are an integral part of the ecosystem in which they live. We will forever depend for our good health and existence upon an environment that is relatively free of lethal and disease-producing poisons and that will provide us with fresh, oxygen-rich air; pure drinking water; a large variety of pure foods; and raw biological materials required for clothing, shelter, and other necessities.

This chapter describes some of the kinds of damage that humans have done and are doing to their environment, some of the ways in which the lessening of environmental quality threatens the existence of earth's organisms, and some of the means that are or can be employed to arrest environmental deterioration.

## EXHAUSTION OF NATURAL RESOURCES

Utilization of natural resources has enormous impact on environmental quality in several ways. Overfishing, overgrazing, and overlumbering, for example, have often so severely depleted natural populations that they can no longer serve as sources of fish, forage, or wood. Moreover, depletion of their populations lessens the ability of natural ecosystems to recycle air and water and thus purify them of toxic wastes. The use of mineral resources can also adversely affect environmental quality, not only during mining and processing, but also in the use of the products derived from them. For example, discarded mine residues, or tailings, often produce acid runoffs that poison lakes and streams; offshore oil well blowouts and tanker wrecks pollute the sea; and processing plants produce toxic wastes that contaminate water supplies. As deposits of a particular mineral become depleted, new technologies for extracting the mineral from lower-grade ores have often proved to be more polluting than traditional methods. The use of manufactured products, such as insecticides and fuels, also often pollutes air, soil, and water. Unfortunately, examples of the adverse effects of resource utilization on ecosystems are numerous.

### FOSSIL FUELS

Coal, natural gas and crude oil, or petroleum, the major fossil fuels, are all derived from organic matter. Most coal was formed during the Pennsylvanian period, 320 to 280 million years ago, from buried vegetation of the great club moss and tree fern swamps of that time. Petroleum is apparently composed of the decomposition products of marine phyto- and zooplankton trapped in the bottom muds of inland seas. Because of their porosity, sandstone and limestone **sedimentary rocks*** are the principal reservoirs of petroleum.

***Coal.*** Following World War II, consumption of coal, once the most widely used fossil fuel, dropped to second and then third place among fossil fuels in many industrialized countries as oil and then natural gas became more readily available. As oil and gas reserves decline, many expect coal to again become the predominant fuel of industrialized nations. Mining and use of coal, however, have usually exacted a high cost in air and water pollution—including that by acid rain—and landscape destruction by strip mining (open pit mining) and similar practices, in addition to the diseases and accidents associated with mining.

***Crude Oil.*** While the United States has enough coal for several hundred years of expanded energy use, like most oil-producing countries, it

* Layers of such hard materials as seashells and sand that accumulate at a seashore or river mouth become compressed in time by the weight of other layers above them. The type of sedimentary rock formed by this pressure, such as limestone or sandstone, depends on the nature of the sediments.

is rapidly depleting its crude oil reserves. Although it is the world's third largest producer of oil, the United States is estimated to have only about a 20-year supply of readily extractable oil. The world's known reserves of crude oil are expected to be depleted by the year 2010. Because the world has come to depend so heavily on crude oil, not only for fuel but for such products as lubricants, fertilizers, and plastics, we have become especially conscious of these dwindling petroleum reserves. New interest has also developed in synthetic fuels, or **synfuels,** such as gasified coal, coal which has been converted to methane or other fuel.

***Oil Sands and Oil Shale.*** Very large deposits of oil sands exist in Canada and some western areas of the United States. Because crude oil is less expensive to obtain and process than oil extracted from sands, this resource has not yet been exploited. Oil shale is a solid that, theoretically, can be profitably processed to separate its gas, oil, and various other fractions, such as sulfur. Oil shale deposits in the United States are 20 times larger than the country's reserves of oil sands. Despite continuing efforts, however, no satisfactory process has yet been developed to extract shale oil economically. Environmental problems are also involved, the most serious of which is the possibility that the well water and rivers of several western states would be poisoned by byproducts of the extraction process. The use of oil and its products as fuel, especially types containing a substantial amount of sulfur, contributes greatly to air and water pollution.

## URANIUM

Supplies of uranium ore, upon which the operation of conventional nuclear power plants depends, are severely limited. Conventional reactors, as opposed to breeder or fusion reactors,* use uranium-235 ($^{235}U$) as fuel. However, the most abundant form of uranium in most uranium ores is uranium-238 ($^{238}U$). Uranium-235 makes up only 0.7 percent of the ore's uranium fraction. This means that uranium ore must be enriched, or processed to increase its proportion of $^{235}U$ to a level at which it can sustain a chain reaction and thus be used as reactor fuel.

At the originally intended rate of expansion of the United States nuclear power industry, severe shortages of domestic uranium ore would have occurred by 1990 or 1995 if reliance on conventional reactors continued; virtual exhaustion of $^{235}U$ would have occurred by 2020. The $^{235}U$ ore reserves in the United States amount to only 1.5 million tons; with conventional reactors, by the year 2000, some 2.4 million tons would have been used. Without an eventual switch to breeder-reactor technology or an unlikely breakthrough in the perfection of fusion reactors, nuclear power in the United States was to have had a rather brief history. However, the discovery of several additional world deposits of uranium ore in the early 1980s has extended its life somewhat.

Uranium mining, processing, transport, and use all pose some risks of radioactive pollution of the environment. More serious, however, are the environmental hazards associated with the storage and reprocessing of high level radioactive wastes and the storage of the radioactive components of decommissioned nuclear power plants.

* A breeder reactor produces more radioactive fuel, in the form of plutonium, than it consumes. A fusion reactor is essentially a harnessed H-bomb in which deuterium nuclei are fused with tritium nuclei.

**FIG. 36-1** Manganese nodules from the ocean floor. The nodules are as much as 25% manganese and 1.2% copper.

## OTHER MINERALS

The question of the adequacy of the world's supply of mineral resources has long been controversial. Many of the minerals used in industry are nonrenewable or can be recycled only at great expense. The total reserves of many valuable minerals are largely unknown. The difficulty of estimating mineral reserves is demonstrated by the fact that known reserves of several elements once thought to be inexhaustible were judged to be dwindling dangerously by the late 1960s and early 1970s. By the late 1970s, however, the discovery of several new sources and the development of new methods of ore extraction had radically improved the outlook for several of these elements. For example, a new process for extracting iron from the ore taconite removed the looming threat of iron shortages in the United States. Also, rich sources of iron and nickel were discovered in the 1960s in Australia and more recently in Brazil. In addition, it was discovered that enormous quantities of manganese, copper, and other minerals are deposited as nodules on the sea floor (Fig. 36-1). Given the present rates of consumption, these nodules contain enough copper to last 6,000 years, enough nickel to last 150,000 years, and enough manganese to last 400,000 years.

For the most part, we have mined the richest ores and are now extracting elements from lower-grade ores through the use of advanced technology. However, such processes usually raise the cost of the elements, use more water and energy than do traditional methods, and increase rates of pollution. Furthermore, the discovery of new sources and the development of new extraction technologies cannot change the fact that the mineral reserves of our planet are finite. One of the best ways to meet increased demands for many minerals, while taking into account their limited supply, is by recycling. This approach reduces pollution by removing discarded items from the environment and by lessening the need for, and the pollution resulting from, mining and processing ore. Although extracting metals from scrap generates its own pollution and requires considerable energy, the energy efficiency of recycling can be dramatic. For example, the production of aluminum from discarded cans saves up to 95 percent of the energy necessary to extract the same amount of aluminum from ore. Although iron extraction requires large amounts of energy, the production of steel from some scrap iron can save as much as 75 percent of the energy cost of extraction from iron ore. Another way to conserve resources, reduce energy use, and abate pollution is to produce goods that are durable or can be repaired.

## TOPSOIL

It is easy to see why we should conserve short supplies of nonrenewable resources. Yet slowly renewable resources need to be guarded just as closely. Topsoil, for example, that upper portion of the soil from which the roots of food crops absorb their nutrients, is not only essential for growing such crops but for maintaining natural communities. It accumulates very slowly—in grasslands, at a rate of only a few inches per century. Without proper protection, however, a single storm can wash or blow away a layer of topsoil several inches deep. Most of this soil is then lost to lakes or oceans.

***Topsoil Destruction.*** Although the loss of the earth's topsoil is reaching crisis proportions in many parts of the world, it is by no means a modern problem. From the first gathering of wild crops by primitive nomads, from the first tilling of the soil by ancient farmers, and from the first felling of trees for firewood and building materials, the soil has been abused. Whenever in human history slopes were overgrazed or stands of trees on hillsides were cut or burned away, rains soon began to wash the topsoil into the valleys below.

In moderation, woodcutting and grazing need not destroy a forest or the watershed (drainage area) it covers. However, extensive clear-cutting, burning, and overgrazing and any cutting of trees on steep slopes are invariably destructive unless the slopes are replanted at once (Fig. 36-2). Denuded of trees, a hillside no longer retains rainwater, and lowland springs once fed by the rainwater that percolated through the soil of the forest cease to flow. Moreover, rivers dwindle to a trickle in dry seasons and become destructive, raging torrents that overrun their banks when it rains.

Modern-day deforestation and overgrazing are producing a worldwide pattern of topsoil destruction similar to that which centuries ago occurred in most of the Mediterranean countries. Previously wooded,

**FIG. 36-2** Clearcutting on a slope can lead to soil erosion if conservation measures are not taken. This example is from Siuslaw National Forest, Oregon.

**FIG. 36-3** Rice terraces near Katmandu, Nepal.

steep mountain slopes have been reduced to bare bedrock in large areas of Greece, Turkey, North Africa, Spain, Italy, and Yugoslavia. The burgeoning population of Nepal, another example, has been forcing hungry peasants farther and farther up the slopes of the Himalayas in search of cropland (Fig. 36-3). As firewood has become increasingly hard to find in Nepal, even newly planted trees are being used for fuel, as are grasses and other hillside plants. Destructive landslides, as well as the silting and flooding of lowlands, have been increasing as a result. Thirty-eight percent of the most densely populated areas of Nepal now consists of abandoned farmland that can no longer grow any crops—an astounding statistic. As wood supplies have declined, people have also begun a more extensive use of cattle dung for fuel. Although always used as fuel to some extent, dung formerly was used mostly to fertilize farm soil and to improve its water- and nutrient-holding capability.

Similar patterns of soil destruction exist on mountainous slopes elsewhere in the world. In South America, the densest populations have always been in the highlands and plateaus of the Andes. It has always been possible to farm the steep Andean slopes as long as strict rules are observed. Besides being terraced to control the runoff of rainwater, the land must be allowed to remain fallow—that is, return to its natural condition—for 8 to 12 years or more between crops. Only this extraordinary precaution enables the Andean soil to absorb enough water and its organic content to increase sufficiently so that food crops can be grown on it for a year or two before it is again allowed to lie fallow. When this practice is not observed, the soil's structure deteriorates, and then water erosion takes a heavy toll. In the hills of Peru and Colombia, for example, major landslides have become more frequent, floods have worsened, and massive amounts of silt have been carried down into the valleys. As silt fills the rivers, they overflow their banks or change their courses, usually with destructive results. Several multimillion-dollar dams built to produce electric power have become useless within only a few years because their reservoirs are filled with silt.

***Topsoil Depletion in the United States.*** It is widely believed that North Americans learned long ago to control soil erosion and that the problem no longer exists in Canada and the United States. Although the causes and means of controlling soil erosion are indeed known to scientists and government agencies, their well-established theories have proved difficult to put into practice.

OVERGRAZING. The allotment of more animals to a pasture than it can support results in overgrazing. Grass that is continually cropped very closely cannot survive dry periods, and eventually the soil becomes exposed to wind erosion. As a very general rule, it has proved best to remove no more than one-third of the edible productivity of a range in any one year. Such an approach actually improves the range and its cattle during years of adequate rainfall. It also ensures a reserve of soil, root systems, and moisture that can save the range during drought years. Once extensive overgrazing has occurred in regions with low average rainfall, the transformation of fertile, arable land into desert results. Over 10 percent of the world's deserts have been created by human abuse of crop- and rangeland.

WIND AND WATER EROSION OF FARM SOIL. Loss of soil moisture can also be a serious problem for farms in grassland regions; moisture evaporates rapidly from bare soil, especially after it has been tilled. Harvesting of crops always removes soil moisture. Only if native prairie grasses or other soil-building plants are allowed to grow during occasional fallow years will soil moisture be restored.

Even the best-managed farms, those utilizing windbreaks of trees, **contour cultivation,*** terracing, and winter crops to hold snow and retard wind erosion lose at least 1 percent of their topsoil each year. Thus, even a well-managed farm that is kept in continuous cultivation can be ruined eventually; how soon this happens depends largely on the original depth of the natural topsoil.

Many United States farms are badly treated, despite such terrible experiences as the dust bowl of the 1930s (Panel 36-1). At the close of the 1970s, many agricultural and economic experts in the United States, responding to soaring wheat prices and a dwindling world food reserve, actually called for fence-to-fence planting of wheat. While this practice may indeed feed the world's hungry for a few years and even help control economic inflation, it will eventually produce even greater disasters. Resulting dust bowls could turn the Great Plains into a barren desert unsuitable even for light grazing.

The greatest threat to topsoil in the United States at present is water erosion in the corn belt (Fig. 36-4). Most severely affected are southern Iowa, northern Missouri, western Tennessee, western Texas, and the Mississippi basin. Most of the lost topsoil is entering the Mississippi River and being carried into the Gulf of Mexico at an average rate of 15 million tons a minute!

By 1977, about one-third of all land in the United States was being eroded at a rate that was noticeably reducing its productivity. In 1980 and 1981, an estimated 19.2 million hectares (48 million acres)—10 percent of the total land area—were losing almost 6 tons of topsoil per hectare per year, each year's loss representing 9 years' accumulation under the best possible conditions. Much of this erosion is controllable and need not occur.

* In contour cultivation, tillage and planting are done across slopes along contour lines rather than up and down slopes and hills.

**FIG. 36-4** What began as sheet erosion (loss of surface soil) has formed a small gully on this northeastern Wisconsin farm, exposing the subsoil. Remedial action could prevent further damage.

PANEL 36-1

# The Dust-bowl Disaster of the 1930s

Large numbers of settlers began to arrive in the Great Plains in the 1880s. Each homesteader was deeded 64 hectares (160 acres) by the United States government, an amount judged to be more than enough for a family farm; 64 hectares was fully adequate in the eastern United States, but that area had a much higher annual rainfall than the Great Plains. The settlers plowed up the thick turf, with its long grasses and extensive root systems, and replaced it with corn and short-rooted cereal grains, reserving some of the land for pasture. Harvests were bountiful, and the future seemed bright. But the 1890s brought the first drought, and many settlers were forced to sell out and move away. In time, however, the rains returned and hopes rose once more. Then, the year 1910 brought another drought and with it the first huge dark clouds of dust. The plains states have the highest prevailing wind velocity of any part of the country. The winds tear at exposed, dry soil, lifting the fine silt particles high into the air, tumbling the larger particles into smothering drifts. More farmers left in 1910 and 1911, but many were encouraged to stay and even expand their holdings by two important developments: the advent of gasoline-powered tractors and harvest combines, and the soaring wheat prices caused by the disruption of European farming during World War I.

The United States dust-bowl region of the 1930s, showing areas of severest wind erosion.

## DROUGHT, DUST, AND DISASTER

The 1920s were years of increasingly heavy investment in raising wheat and cattle. Ranches and farm ranges were extended farther and farther into arid lands. The next drought began in 1931; it was to last through 1934. By the summer of 1933, millions of hectares of brown, shriveled plants stood unharvested. Water holes dried up, cattle died, and thousands of people began a sad exodus to California, a tragedy made notorious by John Steinbeck's classic novel *The Grapes of Wrath.*

The spring of 1934 brought yet another crop failure and the first major dust storm. On April 14, large regions of Kansas and Colorado acquired an apocalyptic atmosphere as total darkness descended in the middle of the day. Dead birds and rabbits littered the fields. A month later, on the single day of May 11, a violent windstorm swept an estimated 350 million tons of rich organic silt high into the upper atmosphere,

Colorado dust storm of the 1930s.

where it was caught by the high winds of the jet stream. New York City's skies darkened, and dust settled on ships hundreds of miles at sea. On that day Chicago received a fallout of 12 million tons of dust, some 4 tons per person!

## ATTEMPTS AT CONSERVATION

Out of the dust-bowl disaster was born, in 1935, the United States Soil Conservation Service. Without question, this agency has done much to retard soil erosion and, in some areas, has even reversed the trend. Millions of hectares of land most vulnerable to erosion have been returned to pasture. Practices such as fallowing; **strip-cropping,** the practice of alternating rows of cash crops with rows of sod-forming crops, such as hay; and leaving stubble and other crop residues behind at harvest to hold moisture and retard wind erosion during winter have been promoted. Terracing and contour cultivation of slopes and the planting of windbreaks of trees between fields have reduced wind erosion. Ranchers have been encouraged to limit herd densities to reduce overgrazing.

The aftermath of a dust storm. While fine silt particles are blown hundreds of miles away, heavier sand particles accumulate in drifts that smother surviving crops and ruin the land for cropland or pasture.

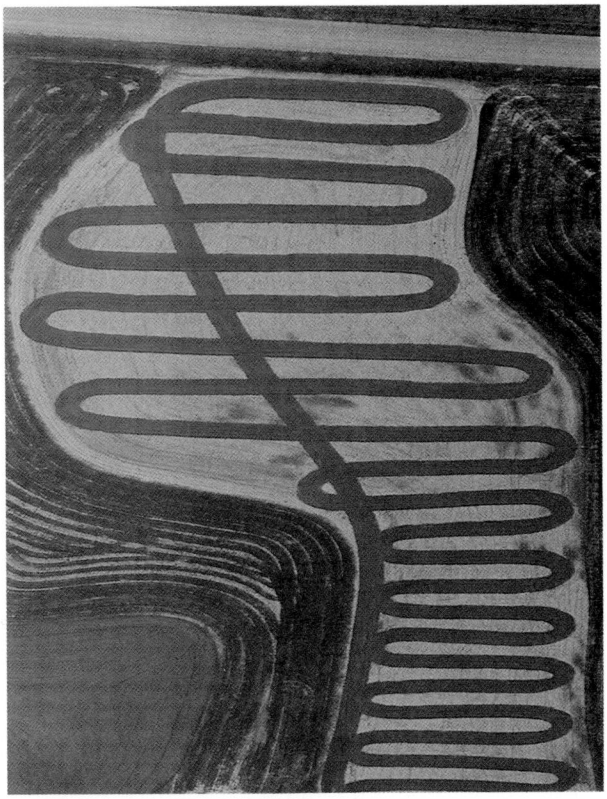

Contour plowing, strip-cropping, and tree windbreaks on Wisconsin farms.

## MORE TROUBLE IN THE FUTURE?

Drought hit the plains again in the 1950s, and, although dust-bowl conditions did not develop, damage was severe. However, rains soon returned, and reasonably good yields were obtained again. But a new dust bowl may be in the making. As reserves of wheat to feed a hungry world have dwindled and prices have risen, the pressure to plant wheat has increased. In 1973 and 1974, 4 million hectares of pastures, woodlands, and idle fields were plowed and planted for the first time. Over half of these lands have been badly treated, and their annual loss of topsoil has averaged 27 tons per hectare. No more than 12 tons per hectare per year is regarded by government soil experts as a tolerable loss for cultivated land. In the southern Great Plains, the 1974 losses on 27,000 hectares of newly planted land ranged from 34 to 314 tons per hectare! In the early 1980s, as grain prices rose and beef prices fell, strip-cropping and fallowing were being eliminated, and dry rangeland was being planted to wheat. The cutbacks that began in the mid-1980s were related to economic conditions rather than being based on environmental concerns.

## WATER SUPPLIES

The supply of fresh water upon which most terrestrial life depends is provided by the hydrologic cycle, described in Chapter 33. Evaporated water soon condenses and falls to earth as rain, sleet, or snow. One-fourth of it falls on land. However, precipitated water is but 0.0007 percent of all the water on earth. When precipitated water seeps into the ground, it enters vast **aquifers** (from the Latin *aqua*, "water," and *ferre*, "to carry"), underground systems of porous, water-bearing rock. The flow of water in an aquifer can be as slow as a few inches per year. A continent's aquifers contain about 30 times as much water as all of its lakes, rivers, and streams. Eventually, precipitated water makes its way into the lower levels of aquifers and from there into the springs, rivers, and lakes. The existence of these sources of fresh water also depend on the hydrologic cycle.

***Meeting Water Demand.*** The amount of the world's water supply is fixed; we cannot control the amount of precipitation, although to some extent we can channel the water once it falls to earth. The total amount of precipitation and its distribution throughout the year differ greatly around the world. These two factors are the key to the adequacy of the water supply for any region. The world's water supply has been calculated as $13.63 \times 10^{20}$ l ($3.59 \times 10^{20}$ gallons) a day. Of this amount, the United States receives a sizable share, some $16.24 \times 10^{12}$ l ($4.3 \times 10^{12}$ gallons) a day in rainfall and another $24.59 \times 10^{11}$ l ($6.5 \times 10^{11}$ gallons) a day available from lakes and rivers. This is about 50 times the country's average yearly demand and thus, in total, would seem to be far more than adequate. Yet water shortages occasionally occur in several parts of the country. One reason is that some high population densities have developed in regions with very low rainfall, necessitating the piping in of water from great distances. Another is that irrigation, which typically either uses more well water than is replaced by precipitation or uses water brought from distant sources, forms the basis of agriculture in several regions.

***Wastefulness Versus Conservation.*** A major reason for shortages in densely populated areas is that people use highly purified drinking water inappropriately—for such needs as washing cars, floors, sidewalks, and dogs; watering lawns; and the like. The average United States household uses about 680 l (180 gallons) a day! Thus, most threatened water shortages in the United States are more accurately threats to a tradition of wastefulness. In an experiment, waste water from bathing and laundry was stored and used for flushing toilets; total consumption was reduced by 39 percent. Traditionally, industry has also used far more water than necessary. This is demonstrated by the fact that when the Kaiser Steel Company of California decided to recycle water, it cut its use from 246,000 l (65,000 gallons) per ton of steel to some 5,300 l (1,400 gallons) per ton.

Worldwide, total water use is actually less than a third of the amount annually available. But demand is increasing and is expected to exceed 50 percent of availability by the year 2000. It is thus obvious that more stringent conservation efforts are necessary. In addition, the cost of piping massive amounts of water from distant sources will eventually exceed what the public is able or willing to pay. This may impose local limits on population expansion in some areas. Los Angeles, California, for example, has an annual rainfall of 38 cm (15 inches), enough to support only a small city. In 1920, an aqueduct 483 km (300 miles) long was built, bringing water to the city from Owens Canyon, high in the Sierra Nevada range. In the 1930s, increased demand for water led to construction of a second aqueduct, 322 km (200 miles) long, from the Colorado River to Los Angeles. In the 1960s, the Feather River, a tributary of the Sacramento River, which was already supplying water to San Francisco, was tapped to meet Los Angeles' ever-growing demand for water. Diverting water from its normal course has been destructive of natural ecosystems and has severely limited crop productivity in regions such as Owens Canyon. Each of these developments was marked by bitter legal battles fought by the angry residents of the areas deprived of water by the projects.

City officials in Los Angeles are now seeking additional sources of water from the low-population wilderness areas in the northernmost part of the state. They hope to divert the Eel and Klamath rivers, the last free-flowing rivers in the state, into the Sacramento River system to provide additional water to both the Los Angeles and the San Francisco areas. Rivers are usually diverted by damming them. Dams have their own environmental impact (see Panel 36-2).

## PROBLEMS ACCOMPANYING IRRIGATION

Irrigation is an ancient practice, probably dating from the time that humans first switched from hunter-gatherer cultures to ones based on agriculture. In lands where rainfall is sparse or irregular, crops need watering if they are to thrive. Two sources are available; surface waters, consisting of lakes and rivers; and groundwater, or well water. With either source, problems have usually accompanied the prolonged practice of irrigation.

***Soil Salinization.*** Eventually, salinization of the soil occurs in any irrigated area that is warm and dry and is poorly drained. Irrigated farms downstream from the Aswân High Dam are being ruined by salt residues, and many are no longer cultivated. Many of the fertile farms of California's hot and arid Imperial Valley have also been abandoned because of salinization. Irrigation water from the Colorado River, which is not unusually salty, deposits 3 million tons of salt in the valley's soil each year. The problem has been attacked in many parts of the valley by the construction of miles of costly drainage ditches into which the accumulated salts are flushed periodically. This salt-laden water is returned to the Colorado River, increasing its salinity by 30 percent in the last 20 years. In response to complaints of Mexican farmers, much of this water is now being desalinized at considerable expense before being returned to the river. The salt is being discharged into the Bay of California.

***Depletion of Groundwater.*** Many areas of the Great Plains in the United States obtain up to 75 percent of their water from wells. For many years, the amount of water drawn from wells had no appreciable impact on the level of the water table, the depth below which the ground is saturated with water. However, with the expansion of irrigation agriculture since the 1950s, this has changed. In southern Texas, which by the 1980s was growing 25 percent of the cotton produced in the United States, the number of water wells multiplied 15-fold, from 2,000 in 1946 to 30,000 in 1966. During that period, the water table dropped 120 m (400 feet) in regions of Arizona that used well water for

PANEL 36-2

# The Mixed Blessings of Dams

Damming rivers is an ancient practice that serves several useful purposes, such as flood control, irrigation of arid lands, and providing a nonpolluting source of electric power. Dam reservoirs may also provide navigable waterways suitable for recreation and fishing. However, dams can create a number of ecological problems.

## ECOSYSTEM DESTRUCTION

Flooding the area upstream from a large dam greatly alters the character of its ecosystems. Some alterations may be desirable, but many are not. For example, thousands of acres of valuable forest or cropland may be flooded, and some species may become extinct.

## DAMAGING DRAWDOWNS

If a dam is to be effective as a flood control facility, it must be drawn down before heavy rains occur, that is, the reservoir level must be lowered by opening the flood gates. This procedure destroys much shallow-water aquatic life along the shores of the reservoir and, in many cases, for miles upstream of the dam.

## DOWNSTREAM EFFECTS

After a dam is built, the reduced flow of water downstream carries far less suspended material, such as the nutrient particles that wash down from the mountains. As a result, the character of the downstream biota usually changes drastically. If the river empties into the sea, the fertility of the estuary and of the sea at the river's mouth is reduced; this is usually reflected in a reduced catch of fish and shellfish. The estuary often becomes more saline, which eliminates some of its species. Instead of continuing to grow, the delta may erode, with seawater intruding into the freshwater marshes and aquifers, rendering coastal areas unsuitable for growing crops and their well water unsuitable for drinking.

All of these problems and more have resulted from Egypt's damming of the Nile River at Aswân, thereby creating Lake Nasser, 150 km long and 10 km wide. Late summer rains in the Nile's headwaters on the nutrient-rich volcanic slopes of Uganda and Ethiopia cause flooding all along the Nile's great length. Before the dam was built, as the Nile overflowed its banks onto the broad flood plains of Egypt each year, it deposited a layer of nutrient-laden silt that formed the basis for Egypt's productivity and survival. The Nile's huge delta has also served as a rich cropland, and the fertilized waters of the Mediterranean at the Nile's mouth provided a thriving fishing industry. Lake Nasser, which was to have been filled by 1970, loses much water by seepage and evaporation and was only about half-filled by the early 1980s. Irrigation has therefore been extended to only about half the planned areas. It is also now necessary to fertilize the farmlands in the Nile flood plain below the dam, which no longer receive their annual deposit of nutrient-rich silt.

Deprived of the fertilizing effect of Nile-borne nutrients, the Egyptian Mediterranean sardine industry has collapsed. Sardine catches of 18,000 tons per year dropped by 97 percent in the first 3 years after the dam was built. Lobster, shrimp, and mackerel catches also fell. The Mediterranean's salty waters are now intruding into delta soil, rendering more and more of it unfit for growing crops.

## SILTATION

Depending primarily on the type of soil in the watershed drained by the river but also on the watershed's topography and amount of vegetation, siltation of the dam's reservoir is sometimes a problem. The Grand Coulee Dam reservoir, on the Columbia River, is not expected to fill with silt for 1,000 years. However, Lake Mead, behind the Hoover Dam, on the Colorado River, is expected to be silt-filled in about 250 years. Lake Nasser is expected to fill in the same length of time. The Mono Reservoir, above Santa Barbara, California, filled with silt in less than 20 years.

## EVAPORATIVE LOSS

Hot, dry, windy regions sustain high evaporative losses of dammed water. For example, Lake Mead loses up to $3.4 \times 10^9$ liters ($8.9 \times 10^8$ gallons) a day in this way, and other reservoirs sustain even greater losses.

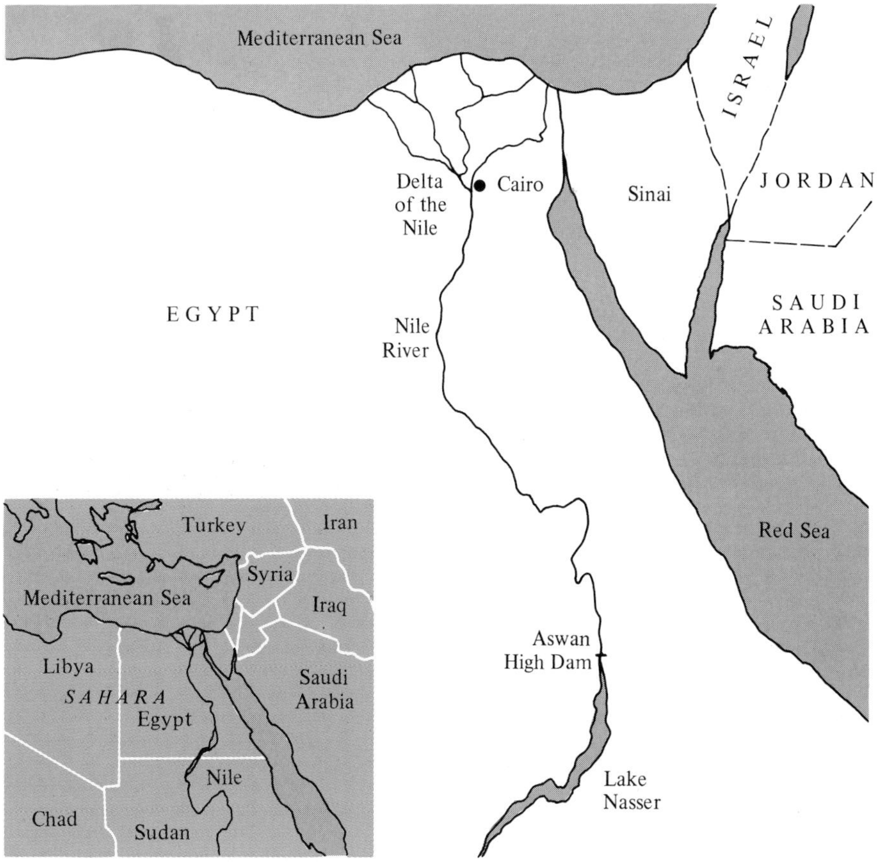


The Nile River. The tributaries of the Nile arise in the nutrient-rich volcanic soil of Uganda and Ethiopia.

Irrigation ditches provided with water from the Aswan High Dam. Although crop production in upper Egypt has increased by 40 percent as a result of irrigation, the practice has also produced several adverse effects on the environment and public health.

Hoover Dam and Lake Mead, on the Colorado River. The dam is 380 m (1,244 feet) long and 220 m (726 feet) high. Lake Mead is over 160 km (100 miles) long and up to 13 km (8 miles) wide.

**FIG. 36-5** Land subsidence in California.

extensive irrigation. As a result, some 128,000 hectares (540 square miles) of cropland had to be abandoned.

The high plains of western Texas and Arizona draw upon the Ogallala Aquifer for their water. This mammoth underground supply represents water that has underlain the southwestern United States since the melting of the last of the great glaciers to cover much of the country, thousands of years ago. Because it is not replenished by precipitation, this water supply, like deposits of fossil fuel, must be regarded as a nonrenewable resource.

California's San Joaquin Valley also makes extensive use of well water to maintain its high productivity; by the early 1980s, it was producing an annual overdraft* of 246 million l (65 million gallons). Many years of overdrafts have resulted in serious soil subsidence, or settling, in the San Joaquin Valley; some areas have sunk by as much as 10 m (33 feet) but other examples can be found elsewhere in California as well (Fig. 36-5). Such extensive subsidence compacts the soil, rendering it incapable of absorbing the amount of water it originally contained, even if irrigation were halted.

On a nationwide basis, the United States withdraws twice as much water from its aquifers each year as is restored by the hydrologic cycle. Land subsidence has been particularly destructive to some residential and industrial areas. Some homes near Houston, Texas, for example, have sunk to a point at which they are often flooded by ocean tides. Moreover, coastal areas are experiencing extensive saltwater intrusion into the land, which ruins its wells and destroys its capability of supporting almost all types of life long adapted to it (Fig. 36-6).

The wisdom of irrigation is now being seriously questioned. Farms are using water at an unprecedented rate, four times the amount used by cities. Although the water used in irrigation is only 60 percent that used in industry, most industrial water can be recycled; irrigation water, however, is lost. This makes growing food by irrigation methods very expensive. According to one estimate, it requires 10,000 to 50,000 tons of water to raise each ton of food produced by an irrigated farm in an arid region.

* An overdraft is the amount of water removed in excess of that returned by precipitation.

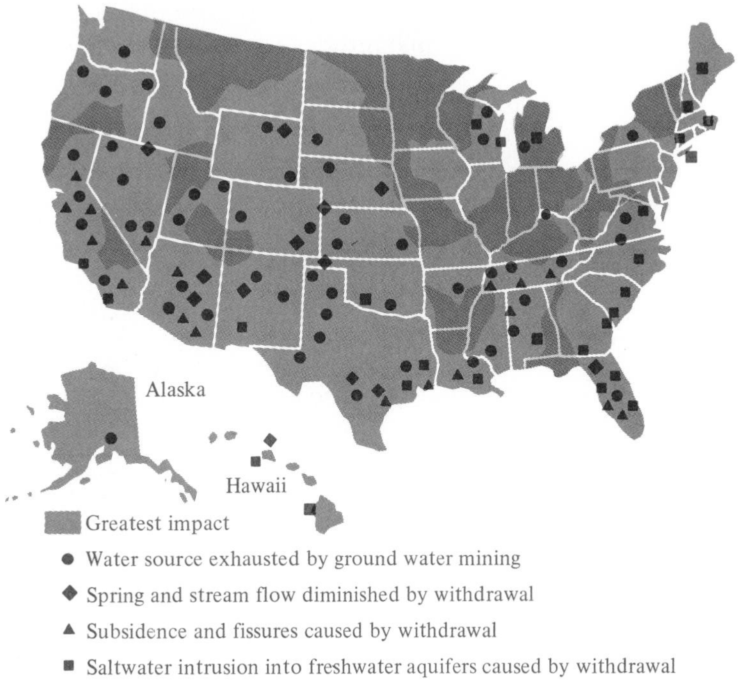


**FIG. 36-6** Areas of major groundwater depletion and sea-water intrusion in the United States.

## DESTRUCTION OF FORESTS

Few human depredations can match what has been done and is still being done to the world's forests. In the nineteenth century, when there were no federal or state agencies overseeing the use of public lands, it was common for small American lumber companies to obtain lumber rights over a region, clear-cut all its marketable trees, and leave behind a denuded area whose topsoil was soon eroded by rains. The slash—consisting of cut-away branches and foliage—left behind after the logs were dragged out soon became tinder dry and was easily ignited into roaring fires by lightning (Fig. 36-7). The fires that resulted from igni-

**FIG. 36-7** A fire in a conifer forest on Bull Mountain, Colorado.

tion of dried slash between 1850 and 1900 were of an unprecedented magnitude. The great Peshtigo, Wisconsin, fire in October 1871, which was fed by dried slash, burned over 400,000 hectares (1 million acres) of forest and claimed between 1,200 and 1,500 lives. In Wisconsin alone, there were 2,500 fires from 1880 to 1890, burning an average of 200,000 hectares (half a million acres) each year.

***Modern Conservation in the United States.*** Government agencies, such as the United States Forest Service, and private citizens' groups, such as the Sierra Club, now exert moderately effective control over American forests. One practice aimed at preventing soil erosion is the selective cutting of only some of the marketable trees from any one stand. Those allowed to remain provide the seeds for new growth. Such a system ensures a sustained yield, the forests being cut selectively about every 10 years. The technique is expensive, however, and works well only with hardwoods, such as beech and maple, whose seedlings can grow in shade.

Another method of forest conservation, called **monoculture,** involves clear-cutting (Fig. 36-8) stands of trees of about the same age and then reseeding or replanting the area with one or a few species. This tree-farming technique permits harvesting such giants as the Douglas fir, whose selective cutting is impractical because of the damage caused to other trees when the firs are felled. Moreover, their seedlings do not grow in shade. Clear-cutting is controversial, however, in part because of the erosion and loss of soil nutrients that invariably follow when trees are cut from slopes. Monoculture also makes trees more susceptible to disease and insect pest epidemics than are natural populations, in which individuals of the same species and age are widely dispersed.

Despite all regulations, the United States still cuts down more trees than it regrows. Current projections are that the country will experience a 265-million-$m^3$ (9.3-billion-cubic-foot) annual deficit by 2010, with demand exceeding supply by 50 percent. Forest conservation needs to be intensified, not only to meet demands for wood and wood products, but because forests constitute an important part of the world's ecosystems.

**FIG. 36-8** The beginning of a clear-cutting operation on a stand of young pine.

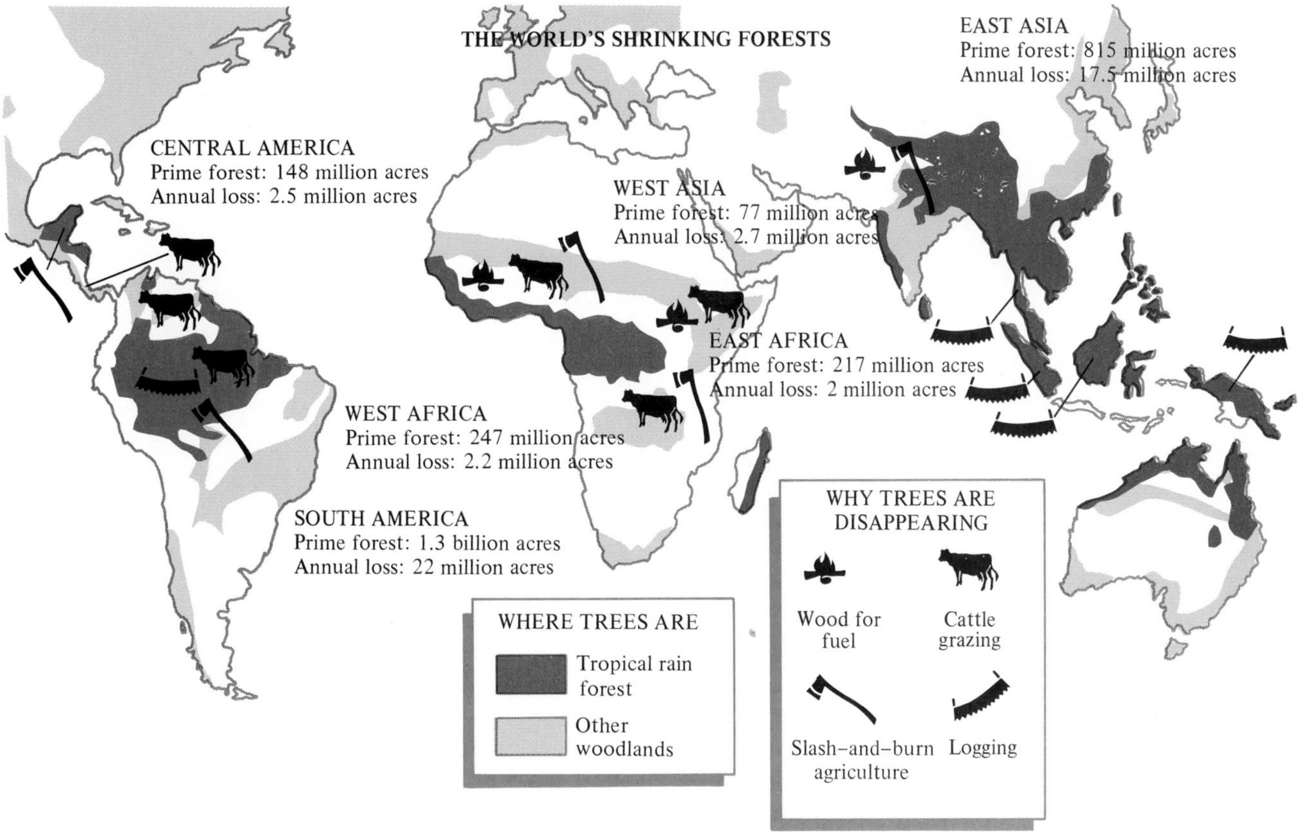


**FIG. 36-9** World regions undergoing the most severe depletion of forests in the early 1980s.

***Worldwide Destruction of Forests.*** Close to half of all the world's trees have been cut down since 1950. Two-thirds of Latin America's forests and half of Africa's have been depleted. During the 1970s alone, Thailand lost one-fourth of its trees, and the Philippines one-seventh. One reason for the destruction is **slash-and-burn agriculture,** an ancient practice which has been increasing as world populations have grown. It involves cutting down all the trees in a chosen area, burning them, and raising crops for a year or two in the ash-enriched soil. Overall, there is a net loss of between 1 and 2 percent of the world's forests each year. This rate is greater than could be compensated for by even the best efforts at reforestation (Fig. 36-9).

A growing need for firewood, which for 75 percent of the world is the only fuel used for heating and cooking, is one reason for so much destruction. For example, over 90 percent of all wood cut in Africa is burned for fuel. The worldwide shortage of firewood is a far greater energy crisis than the oil shortage. By the year 2000, firewood needs are expected to exceed supply by 25 percent. Where firewood shortages have developed, the substitution of dried animal manure as fuel has reduced African food production by 20 million tons of grain a year. In Nepal, the floods caused by denuding slopes of their trees for use as firewood cost millions of dollars in damage and kill thousands of Nepalese and Indians each year.

## WILDERNESSES AS NATURAL RESOURCES

A wilderness is best defined as an extensive, compact tract of land into which human intrusion by roads, motorized vehicles, prospecting, mining, lumbering, management, or control does not occur (Fig. 36-10).

**FIG. 36-10** Although the number of areas officially designated as wildernesses has increased since 1960, the amount of actual wilderness has decreased since then. Half Dome, Yosemite National Park, California.

A true wilderness forest must be free to burn and regrow, even to die from diseases. Fallen dead trees must be left to rot, rivers to flood or change their courses, and moose and deer to starve to death when caught in deep snow. These are all natural processes to which wilderness communities have become adapted without human assistance in the course of the evolution of their various members.

Why preserve such extensive tracts of land, especially ones that could serve as sources of greatly needed wood, oil, or valuable minerals? Scenic beauty is one reason that is often given as the answer. Yet, as important as is the natural beauty of a wilderness for its recreational value and its capacity to gratify our aesthetic sense, these functions could probably be served by far less than the many millions of hectares now officially preserved as wildernesses. The preservation of extensive wilderness areas is much more crucial because of their role in ensuring the survival and variety of species, a point briefly mentioned in Chapter 35.

***The Need for Extensive Tracts as Species Reservoirs.*** One of the important functions served by wildernesses is as reservoirs of many species that would otherwise become extinct. Disturbing a wilderness or reducing its size below a certain level is bound to reduce the chances of many of its species for survival. For example, to support a family of four grizzly bears in Yellowstone National Park requires an average of 300 $km^2$ (115 $mi^2$) of woodland. Yet, of the 89 officially protected wilderness areas in the western United States in the late 1970s, only 10 were larger than 1,000 $km^2$ (386 $mi^2$).

The territory sizes required by birds differ greatly from one species to another. In one study, eastern robins were found to need only 1,200 $m^2$, and red-winged blackbirds only 3,000 $m^2$. The great horned owl was found to require 50 hectares (500,000 $m^2$), and a family of golden eagles a territory of up to 9,300 hectares. As the quality of an environment deteriorates, the size of the territory required by a given species usually increases; robins in a city need more area than do robins in the countryside.

The examples given here are of large and conspicuous species whose requirements have been discovered by painstaking ecological investigations. Wildernesses contain many species whose essential requirements have never been determined and, indeed, many species that have not yet been discovered, named, and described.* Tropical forests in particular are valuable species reservoirs. Although they cover only 10 percent of the earth, they harbor nearly half of all species. If present practices of cutting down rain forests continue, some 20 percent of these species are expected to become extinct within the next 20 years. Thus, the problem of survival affects more than the few dozen well-known endangered species, such as the redwoods, whooping cranes, elephants, and whales; it concerns hundreds of thousands of unique and irreplaceable forms of life that will have vanished from the earth forever.

***The Need to Preserve a Great Variety of Species.*** Ecosystems require the presence of many species for several important reasons. In a complex natural ecosystem, each type of organism has its own approach to processing the forms of energy and materials upon which it subsists. Various species of insects, worms, fungi, protists, bacteria, and even the symbionts inhabiting the intestines of cockroaches and termites each play a role in the constant recycling of the materials present in the environment.

PURIFIERS OF POLLUTED AIR AND WATER. Many of the materials found in nature, including many produced by organisms, are toxic for some kinds of organisms. Yet these same materials, many of them toxic for humans and domestic animals, are metabolized and degraded to simpler, nontoxic forms by various other organisms. Only complex communities contain sufficient diversity of species to accomplish these detoxification processes efficiently.

Wildernesses are highly effective in metabolizing the many poisonous substances with which we pollute our air and water, returning the air and water to us in purified condition. Despite its large size, New York's Central Park does not qualify as a wilderness. Yet it serves a similar purpose to a limited degree. For example, the air's sulfur dioxide ($SO_2$) concentration over the park is measurably below that of adjacent areas (Fig. 36-11). When a breeze blows away any city's accumulation of smog or when a rain washes its air of noxious fumes, these materials are carried to some other part of the planet, where they are usually degraded or otherwise detoxified. Wildernesses play a large role in this valuable process.

VITAL NEEDS IN BIOLOGICAL RESEARCH AND MEDICINE. A second important reason for preserving as many species as possible is that so many of them are useful in biological research or in medicine. Some species of little or no commercial value have often proved to be the material of choice for the investigation and discovery of basic biological mechanisms. Fruit flies, bread mold, and sewage bacteria are the materials on which several Nobel-prize–winning investigations of genetic mechanisms were conducted. Important breakthroughs in our understanding of nervous function have resulted from studies of certain squids, sea slugs, and jellyfish. Our knowledge of embryonic development has been

* Best estimates place the total number of species on earth between 3 and 10 million, with only 15 to 20 percent having thus far been named and described by scientists.

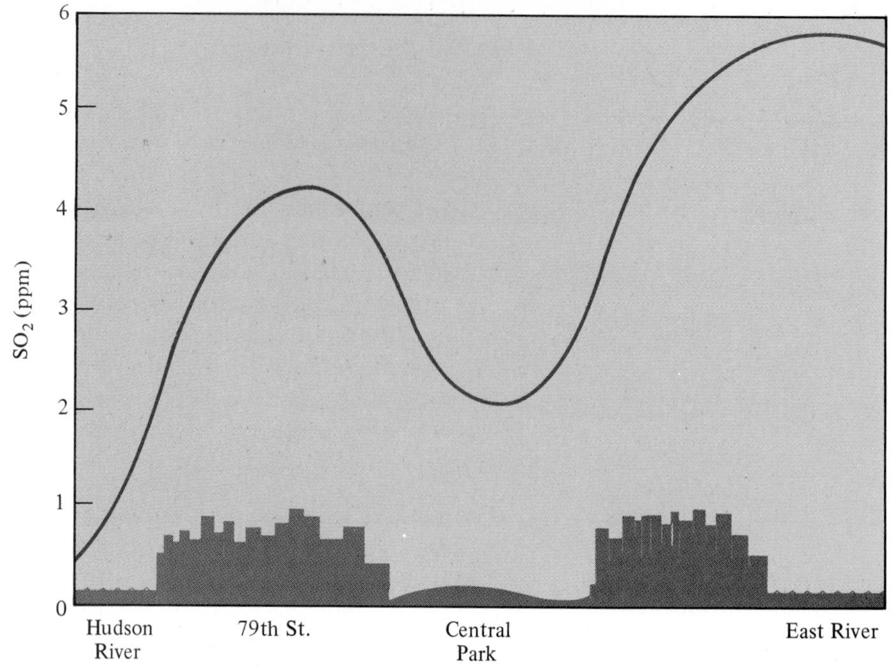


**FIG. 36-11** Atmospheric $SO_2$ concentrations over New York City's Central Park and adjacent areas, from a 1971 report.

advanced through investigations of African frogs, salamanders, and sea urchins. A species of coral was recently found to be an excellent source of precursor molecules used to synthesize prostaglandins, a valuable drug.

How Wilderness Species Benefit Other Ecosystems. The many species in our wildernesses constitute a reservoir of adaptable organisms capable of recolonizing damaged and devastated areas. For example, supplies of readily available species capable of adapting to a particular situation have sometimes restored areas to more natural conditions; again, New York City's Central Park is a case in point (Fig. 36-12). The species of plants, birds, and small mammals that recolonized this area, which in 1850 had been virtually stripped of its trees and shrubs, came mostly from wilderness areas. Even if all of the world's species were known to us, only a small fraction of them could ever be conserved in botanical gardens and zoos. Moreover, only a few animal species are able to reproduce in captivity, and, of those that do adapt to life in a cage, few would survive if returned to the wild. As our wildernesses deteriorate, the ability of other ecosystems to recover from damage or even to maintain their own viability is being seriously jeopardized.

**FIG. 36-12** Central Park, New York City. A portion of the park was planned as a natural area by the park's designers, Calvert Vaux and Frederick L. Olmsted, who, in 1866, planted it with mostly native species of trees and shrubs.

The Need for Concern. Some of those who believe that efforts to halt the extinction of species are misplaced have pointed out that, over millions of years, far more species have become extinct through natural causes than now exist on earth. What harm, it is said, is there in the extinction of a few more? The answer lies in the considerable difference between these two modes of extinction. Most species that become extinct naturally do so in a slow, gradual process over centuries and millenia. This permits evolutionary adaptations of other species to the altered situation. Indeed, the adaptation of competing species is a major cause of such extinctions. However, the changes now being wrought are unprecedented. Between 1600 and 1900, *known* mammals and bird species were eradicated at a rate of one species every 4 years. During the last 80 years, the rate increased to one species every year. However, these were only the known species; the actual number of fish, invertebrate, and plant species now undergoing extinction cannot be determined.

It is a well-established principle that perturbations of one parameter in a natural ecosystem not only can be felt throughout the ecosystem but often become amplified. Many of the species of a large ecosystem have undergone a long coevolution, culminating in a beautifully harmonious adaptation to each other. Extensive destruction of a large ecosystem or extermination of many of its species upsets this balance, leading to overproduction of some species and underproduction of others. The ultimate effects upon surviving species are unknown.

Another reaction to the movement to preserve wildernesses is the suggestion that some wildernesses are of little or no value and ought not to be protected. The ones usually mentioned are deserts, marshes, swamps, and tundras. Again, we do not know what effects—desirable or undesirable—would result from extensive losses of these ecosystems.

## DIRECT DESTRUCTION OF NATURAL ECOSYSTEMS

Many commercial and public works projects have substantially altered and even destroyed unique natural ecosystems. Many of these systems are—or were—the homes or breeding grounds of unique species. Mining, the elimination of wetlands, and flood control measures have all resulted in the direct destruction of natural ecosystems.

***Mining.*** Besides contributing to a runoff of toxic substances, the tailings left by mining operations represent a destruction of both the natural role and the beauty of an area (Fig. 36-13). Strip mining is destructive because it removes whatever type of habitat is present. However, some parts of the United States damaged by strip mining are being reforested, and ponds created by the mining are being stocked with game fish.

**FIG. 36-13** Strip mining. Many states now require that any newly mined areas be regraded and replanted.

**FIG. 36-14** This bayside complex at Dinner Key, Florida was built on landfill largely derived from dredging portions of the shallow bay.

***Draining and Filling in Wetlands.*** Among the most harmful projects from the standpoint of adverse environmental impact are the draining and filling in of marshes, shallow bays, and other wetland areas. Besides destroying wildlife that uses a coastal bay the year-round, dredging and filling ruin its usefulness as a seasonal spawning ground for many marine animals and as a stopping place for migratory waterfowl. Three rather well-known examples of such landfill operations are those of California's San Francisco Bay and Florida's Tampa and Biscayne bays (Fig. 36-14). Tampa Bay is more than 20 percent filled. By the middle 1960s, 60 percent of San Francisco Bay had been filled.

The fill used in such projects is usually obtained by dredging other areas of the same bay, a process in itself destructive of the valuable wildlife habitats the bay provides. All along both United States seaboards, the draining, dredging, and filling of wetlands has been used as a way of creating commercially valuable land. A 1980 study reported that 40 percent of the coastal wetlands of the 48 contiguous states has now been destroyed. The effect of such massive removal of ecosystems upon the rich fauna and flora of the Atlantic and Pacific coastal waters is thought to be enormous.

***The Effects of Flood Control on One Ecosystem.*** Flood control and swamp drainage projects in central and southern Florida have had adverse ecological effects. The southernmost 160 km (100 miles) of the Florida peninsula, south of Lake Okeechobee, constitutes a unique subtropical ecosystem of great complexity: the Everglades. This region enjoys a moderately warm climate and a heavy annual summer rainfall that until early in this century produced a substantial overflow of Lake Okeechobee. This runoff moved slowly southward in a broad shallow sheet that covered much of the land immediately south of the lake (Fig. 36-15). The marsh ecosystem supported by this abundant water was marked by numerous small islands, occasional sloughs (marshy inlets) and "gator holes" (deep depressions in which alligators live), and extensive dense growths of tall saw grass rising from the muck soil of the marshes. At the peninsula's tip, where the flow of fresh water meets and blends with seawater, the saw grass marsh gives way to heavy stands of

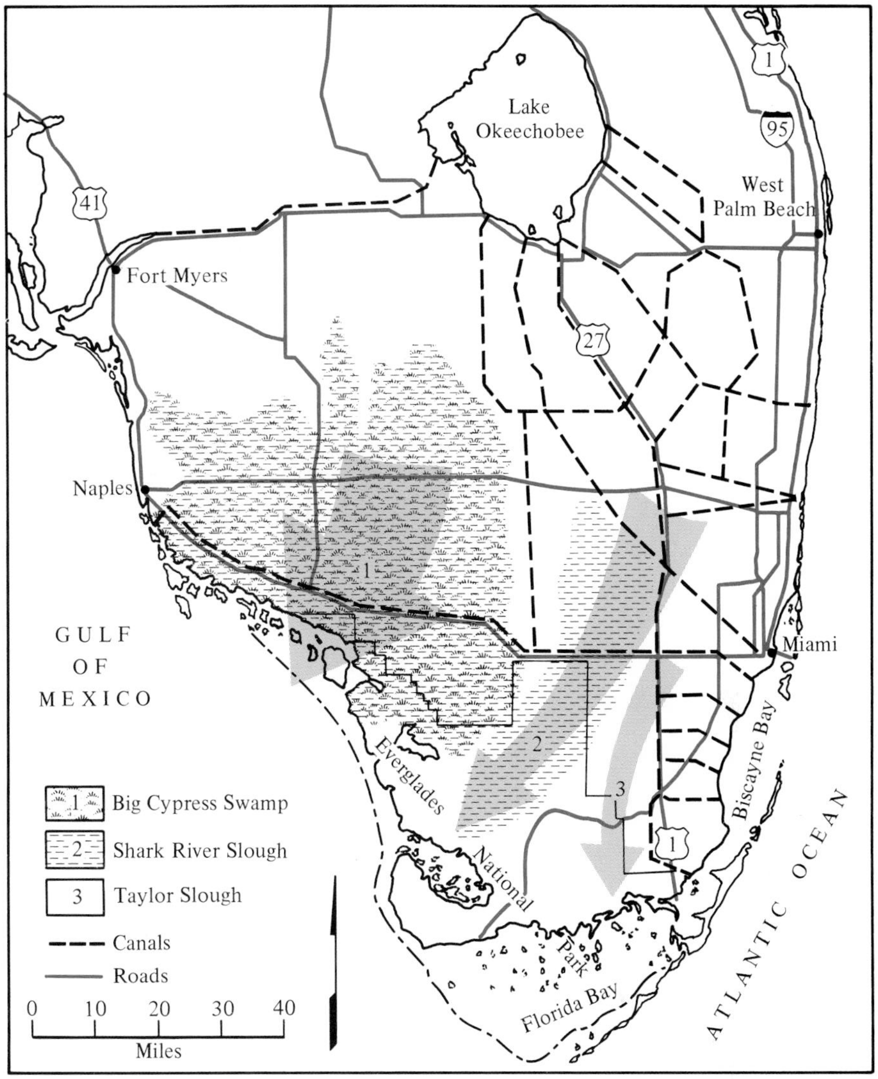


**FIG. 36-15** The three major natural drainage systems of Florida's Everglades–Big Cypress region. Artificial canals and channels are indicated by dashed lines, roads by solid lines.

mangrove trees. The brackish waters of the mangrove swamps provide spawning grounds for a great variety of fish and invertebrates. To the southwest of Lake Okeechobee is the extensive Big Cypress Swamp, its dark waters supporting another complex community of organisms (Fig. 36-16).

The dominant animals of this ancient ecosystem were the alligator and the panther. Although poaching is restricted, alligator populations continue declining and the panther is now seldom seen. Eighty years of "development" of Florida have rendered this highly mature community very unstable; its disintegration is well under way.

Records of engineers examining the commercial potential of the Everglades about 1900 described them as beautiful and awesome but also as "entirely worthless to civilized man in its present condition." In 1906, construction was begun on a canal to drain Lake Okeechobee into the Atlantic. This project was the start of what was to become a vast network of canals, levees, dikes, and spillways (shown as dashed lines in Fig. 36-15). The purpose of this project was to drain the marshy soil for agricultural use and residential development. The system subsequently has proved inadequate to handle the runoff during wet years, yet it overdrains the land during drier years. During droughts that hitherto

**FIG. 36-16** A portion of a Florida swamp.

had little effect on the glades because of the flow of water from the north, the soil dries out. Normally, the peat muck soil* of the glades acted as a giant sponge that soaked up water during rainy periods and slowly released it during dry periods. However, diverting Lake Okeechobee water to the ocean during dry periods has resulted in dried, cracked soil; the death of natural communities; and crop failures. Attempts to regulate the flow to downstate regions have often proved unsatisfactory. The adverse effects include the following:

- *Peat muck fires:* before the drainage canals were built, saw grass fires† were common during the dry season; on occasion, the exposed peat muck also caught fire and smoldered for a brief time. Since the land was drained, however, peat fires often burn uncontrollably for months at a time. Fort Lauderdale and Miami have many dark days because of muck fire smoke. The fires destroy much natural wildlife and valuable agricultural soil.
- *Soil subsidence:* once muck soil is exposed to air, it undergoes both oxidation and drying, shrinking and compacting as a result. Although muck soil requires about 400 years to form to a depth of 33 cm (1 foot), it has been disappearing at the rate of 33 cm every 10 years.
- *Depletion of "gator holes":* at one time, there were an estimated 1 million alligators in southern Florida. Many lived in "gator holes," which also provide a habitat for many unique species of fishes, birds, and invertebrates. Few of these communities remain.
- *Saltwater intrusion:* because the Everglades are barely higher than sea level, the fresh water that normally flowed down from Okeechobee prevented the intrusion of seawater into the soil of southern Florida. After initiation of drainage, salt water began to seep into the freshwater aquifers nearest the sea coast, resulting in contamination of wells with salt and salinization of the soil, which is destructive of natural terrestrial communities and harmful to crops. Natural aquatic communities are also being destroyed as seawater intrudes into the fresh water of brackish swamps and marshes.

* Peat muck is a spongy organic soil so rich in partially decayed vegetation as to be combustible when dry.

† Occasional fires are one of the factors that maintain a grassland community, discouraging growth of trees.

# POLLUTION

Environmental pollution can be defined as the direct or indirect impairment of the environment's suitability for supporting life by harmful concentrations of materials, whether or not the materials are toxic. Thus, a very high concentration of a nutrient might be polluting, whereas a deadly poison in great dilution is not.

Earlier in this century, wastes could be discharged into the environment with relative impunity. Two factors have changed this. One is the population explosion; more people mean more pollution. The other is the rapid advance of technology and affluence that has occurred in the industrialized nations and which has increased the production of toxic pollutants.

In the course of evolution, the wastes produced by one kind of organism eventually became the life-sustaining nutrients of various other organisms. Modern human activity, however, has altered this relationship. As world populations have increasingly become concentrated in cities*, the natural wastes they produce have contributed significantly to environmental pollution. Yet natural wastes are not the most harmful pollutants being generated.

Unlike biological wastes, the wastes that result from mining, processing of raw materials, and manufacturing are often highly toxic. Even when they are biodegradable, that is, can be degraded by natural decomposers in the soil, rivers, lakes, and oceans, some wastes are released into the environment in concentrations and volumes that overwhelm the capacity of the decomposers to degrade them, thereby threatening the existence of many aquatic organisms.

Liquid wastes that are both nonbiodegradable and toxic to one or more forms of life pose more persistent problems. If released into waterways, they kill or prevent the reproduction of much aquatic life and render the water unfit as a source of food organisms and for drinking and recreation. If stored on land, they may contaminate groundwater and thus poison springs, wells, and spring-fed lakes.

Some specific examples of environmental pollution, some of their known ecological effects, and a few approaches to solving the problems they create are briefly examined in the following sections.

## AIR POLLUTION

Because of air's mobility, it has a high natural capability for reducing the impact of air pollution on ecosystems. However, when winds are calm in areas of high population density, air pollution can be extremely serious and even life-threatening. For example, during one brief period in 1952, smoke density and sulfur dioxide ($SO_2$) concentrations in London rose above normal, and death rates soared from about 250 per day to a peak of nearly 1,000. In 6 days, 4,000 more people had died than was normal for the period, and during the next 2 months, 8,000 more people than usual died.

Most people can tolerate some air pollution for extended periods of time. However, long-term exposure to air pollutants has many adverse effects on health. The largest volume and the most serious air pollu-

* Early in this century, London became the first city to reach a population of 5 million. By the year 2000, there will be 60 cities with populations of over 5 million, some of them 20 or even 30 million in size. Over half of the world's population will then be living in cities.

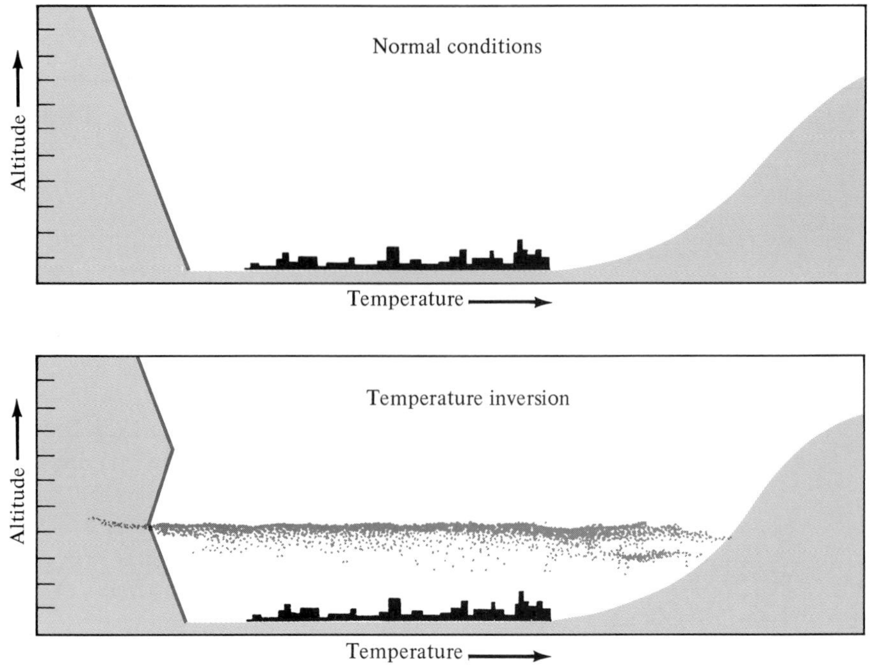


**FIG. 36-17** On clear days in Los Angeles, temperature decreases with altitude and pollutants dissipate. When a layer of warm air stalls over the city and a layer of cool, moist air flows in under it, pollutants become trapped below the inverted layers, which do not mix. This prevents dissipation of pollutants generated at ground level.

tants, many of which are carcinogenic, are emissions from the burning of fossil fuels. Moreover, when various hydrocarbons and nitrogen oxides from automobile emissions combine under the action of sunlight, they form ozone ($O_3$) and the other health-threatening constituents of the atmospheric haze known as **photochemical smog.**

One of the areas most notorious for smog episodes is Los Angeles, California, whose many automobiles and abundant sunshine produce substantial amounts of smog. The problem is serious because of the city's frequent **temperature inversions,** which result when a cool, humid breeze flows in under a stationary high-pressure layer of warm, dry air (ordinarily, air temperature decreases with altitude). Pollutants generated at ground level become trapped under the impenetrable layer of warm air (Fig. 36-17). The mountains that form a broad semicircle to the north of the Los Angeles area enhance the polluting effect by occasionally trapping the inversion in the basin for many days at a time, with air quality worsening daily (Fig. 36-18).

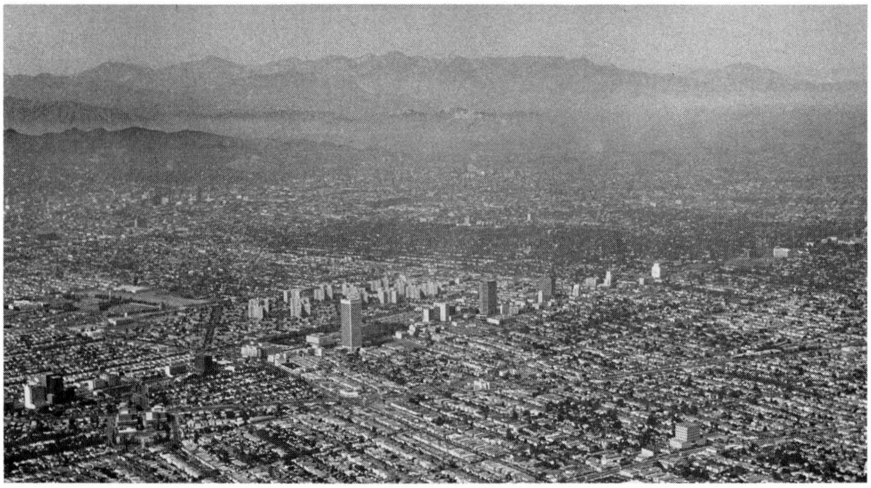

**FIG. 36-18** Los Angeles, with the San Bernardino Mountains in the background. The mountains form a semicircle that tends to trap temperature inversions for days at a time.

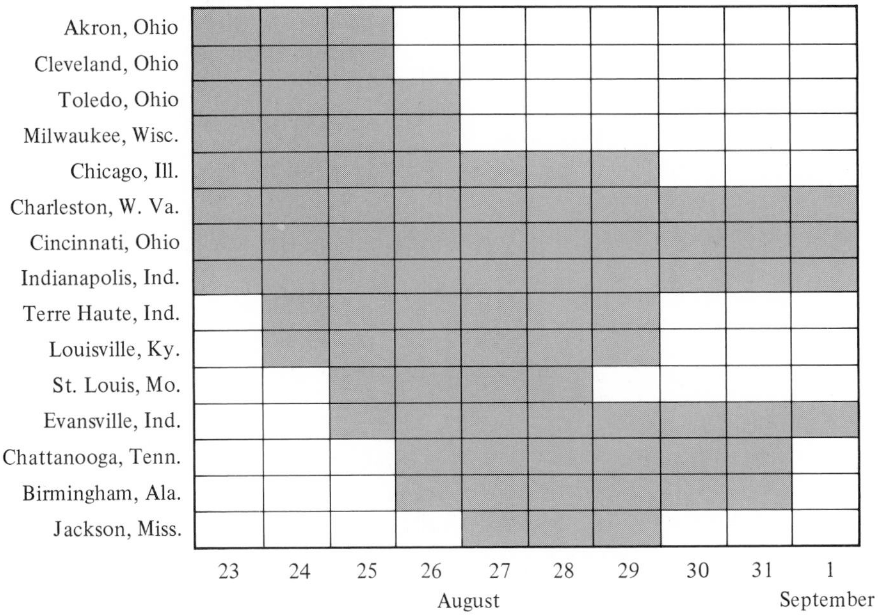


**FIG. 36-19** Cities for which data was collected during HAPP episode 104, in August 1969. Darkened blocks indicate the days a city was placed under a HAPP advisory by the Environmental Science Services Administration.

Temperature inversions and health-threatening air quality crises also occur in the midwestern United States whenever highs (high-pressure weather systems) stall for several days during warm weather. A major episode occurred in August 1969 that affected several large cities for up to 10 days. The crisis is listed as episode 104 in United States Weather Bureau records because it was the one hundred and fourth high-air-pollution-potential (HAPP) episode that covered more than 194,000 km$^2$ (75,000 square miles) and lasted more than 36 hours since monitoring began in the early 1960s (Fig. 36-19). HAPP episodes covering less area are not numbered; they occur more than 25 percent of the time over almost all of the United States (Fig. 36-20).

***Long-Term Effects of Moderate Air Pollution on Health.*** Long-term, moderate air pollution actually causes more disease and death than do acute crises. Long-term exposure to air pollutants increases the risks of contracting cancer and of developing respiratory diseases such as em-

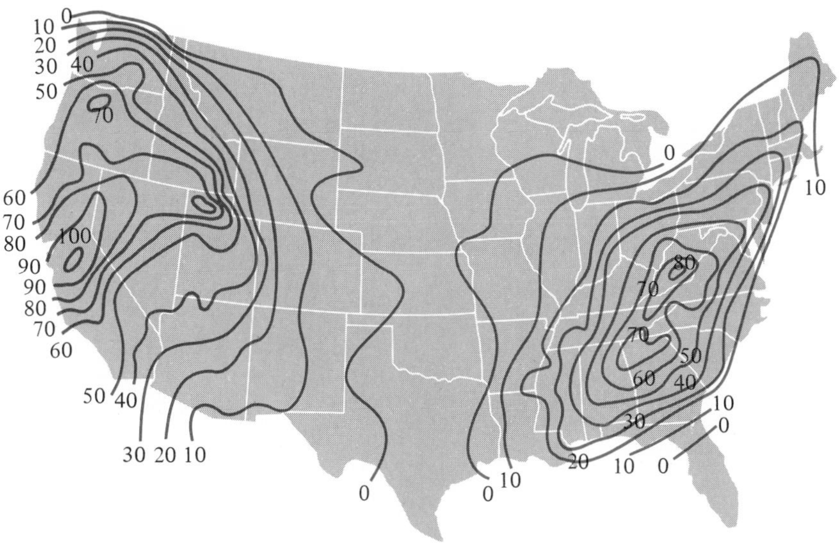

**FIG. 36-20** An Environmental Sciences Services Administration map showing the number of days during which HAPP was forecast for various regions of the United States between August 1, 1960, and April 3, 1970, in the east and between October 1, 1963, and April 3, 1970, in the west. Not indicated are days of unhealthful air conditions affecting areas of less than 75,000 square miles or less than 36 hours in duration.

physema. Although precise data about the long-term effects on health of each of the various pollutants are difficult to obtain, it is well agreed that air pollution is harmful to health. One somewhat useful approach to studying its effects has been to compare mortality (deaths) and morbidity (serious illness) data for a given city with the concentrations of various pollutants found in its air. For example, there is a high correlation between the concentration of heavy industries and the number of deaths due to cancer. World Health Organization data indicate that 80 to 90 percent of all human cancers are environmentally related or induced. Certain types of cancers occur in greatest frequency near factories producing certain products, for example, liver cancer incidence is high near synthetic rubber factories. Air pollution is the suspected cause.

Our understanding of the injurious effects of long-term exposure to some of the airborne pollutants in cities has been enhanced by studies of cigarette smokers. Cigarette smoke contains many of the same pollutants found in the air of large cities, for example, carbon dioxide ($CO_2$), carbon monoxide (CO), and at least seven polycyclic (multiple ring) hydrocarbons that can produce cancer in animals under certain conditions. Numerous studies have confirmed that the incidence of lung cancer, emphysema, and coronary artery disease is far greater in smokers, whether they live in cities or in rural environments, and is in direct proportion to the number of cigarettes they smoke per day and the number of years they have smoked.

**FIG. 36-21** Solar collectors of an apartment complex in El Toro, California. The system furnishes 70% of the annual energy needed for hot water for 32 apartments.

***Acid Rain.*** An increasingly serious problem has lately resulted from fossil fuel combustion products, the most troublesome of which are nitrogen oxides (NO and $NO_2$), hydrogen sulfide ($H_2S$), and sulfur dioxide ($SO_2$). Dissolved in atmospheric moisture, these compounds produce nitric acid ($HNO_3$) and sulfuric acid ($H_2SO_4$), both strong mineral acids. The resulting acid rain has destructive effects on freshwater and soil ecosystems, including ones thousands of miles from the pollution source (Panel 36-3).

***The Control of Air Pollution.*** There are two basic, practicable approaches to controlling the quality of the air we breathe. One is to limit the pollutants emitted into the air, for example, by the burning of sulfur-free oil and coal, or by using sophisticated, expensive technology that is still being developed. The other approach is to choose alternative technologies or sources of energy. Wind, water, and solar power, for example, are far less polluting than is the burning of any fossil fuels (Fig. 36-21). Likewise, the use of bicycles and battery-powered cars is far less polluting than the use of vehicles powered by internal combustion engines.

## WATER POLLUTION

Water pollution is usually defined as the addition of materials to water in such quantity as to lessen its suitability for the life of aquatic organisms, for irrigation, for recreation, or for drinking. Some water pollutants are inherently toxic to one or more forms of life. While seldom intrinsically toxic, nutrients may be toxic in high concentrations or may produce so much growth of bacteria or other aquatic life as to make life impossible for other aquatic organisms (Fig. 36-22).

Sources of water pollution include human wastes; runoffs of indus-

**FIG. 36-22** Farmland runoff may introduce excessive nutrients into the water supply and hinder the growth of certain forms of aquatic life.

trial processes, farmlands, feedlots, and mines; air pollutants that find their way into lakes and rivers; accidental and deliberate discharges of petroleum; and radioactive wastes. Farm runoffs include fertilizer, manure, insecticides, and herbicides.

The volume of water pollutants in the United States is huge. Runoffs into the Ohio River from mines in Pennsylvania and West Virginia, for example, account for a daily outpouring of 200,000 tons of sulfuric acid. The Detroit River alone dumps 20 million tons of miscellaneous waste into Lake Erie every day. Cleveland, Buffalo, and Toledo all add their sewage effluent and industrial discharges to the same lake. By 1980, 11 regions along the shores of the Great Lakes alone had been designated areas of major pollutions (Fig. 36-23).

Several major water pollutants are considered in the sections that follow.

***Biodegradable, Nontoxic Organic Wastes.*** The ways that sewage is usually treated in American cities tells something of the nature of the problem of pollution by biodegradable, nontoxic organic wastes. Raw

**FIG. 36-23** The dots indicate areas of the Great Lakes affected by major water pollution.

PANEL 36-3

# Acid Rain

One problem associated with the burning of fossil fuels has greatly intensified within the last two to three decades. Because atmospheric $CO_2$ dissolves in rain, forming carbonic acid ($H_2CO_3$), even the purest rain is acidic, with a pH of 5.6. The pH of rain may be close to neutrality (pH 7.0) in desert regions when calcium carbonate dust is airborne. In areas where coal- and oil-burning industries and motor vehicles abound, the pH of rain drops to values well below 5.6, in some cases to as low as 2.0, the concentration in automobile batteries. Together, these sources account for over two-thirds of the acid found in rain and snow. They are the sources of nitrogen oxides, hydrogen sulfide, and sulfur dioxide, which, when dissolved in rain, are converted to nitric acid ($HNO_3$) and sulfuric acid ($H_2SO_4$). Unlike carbonic acid, these strong mineral acids ionize completely in dilute solutions (see Panel 2-1).

## A WORLDWIDE PROBLEM

Environmental contamination from acid rain due to smelting and other heavy industries began with the industrial revolution of the nineteenth century. Until recently, it was largely confined to areas near the sources of pollution, often with such devastating effects as lethal smog crises and the destruction of forests and crops. To avoid these local effects, there developed a trend to build higher smokestacks, many over 300 m (1,000 feet) tall. The result was that the pollutants were caught in the earth's prevailing winds (see Chapter 33) and transported hundreds of miles, often to other countries. Acid-forming substances in industrial emissions from the United States contribute to acid rain in Canada and vice-versa. Britain and northern Europe are responsible for acid snow and rain falling on Norway and Sweden.

The magnitude of acid precipitation is not fully known. Besides the fact that monitoring acid rain is expensive and has only recently begun in a relatively few areas, the situation is very complex, involving, as it does, variable mixtures of air pollutants and the changeable weather of the northern temperate zone. But enough data have been gathered to make several generalizations. Taking a pH of 5.6 as the lowest value expected for unpolluted rain, large regions of the world now average from 5 to 30 times more acid in their rain than this standard. Individual rainfalls and snows can be several hundred to several thousand times more acidic than the expected value. Most severely affected are mountain foothills downwind of the source of pollution. There, as moisture-laden air rises to higher altitudes, it condenses to fall as rain or snow, dropping much of its load of pollutants. For this reason, the mountainous Norwegian coast, southern Germany, and Switzerland receive a disproportionately large share of the British Isles' pollution. Another severe, somewhat localized effect occurs when acid snow accumulations melt in the spring. The melted snow can deliver an acid shock to a lake by infusing it with up to 1,000 times the amount of acid it normally receives during a single rainfall. Even in lakes with good buffering capacity, the pH may remain low for several days before being neutral-

Tall industrial smokestacks may solve air pollution problems for the immediate neighborhood but contribute to acid rain thousands of miles away.

ized. Despite many efforts to clean up emissions at their sources, both the acidity of rain and the number of regions suffering this type of pollution have been increasing.

## BIOLOGICAL EFFECTS

The many biological effects of acid rain are far from being known. Acid precipitation may be the primary cause of the worsening damage to the forests of Germany. It is thought to be responsible for the widespread death of trees now occurring in the eastern United States. Although many adverse effects of acid rain on plants and animals can be demonstrated under laboratory conditions, the complexities of the natural environment make it difficult to be certain what is happening in individual instances. Nevertheless, it seems clear that freshwater ecosystems that overlie granite and other siliceous types of bedrock, which are nonbuffering, are particularly vulnerable to additions of acid. In sharp contrast to the sea, whose waters are rich in minerals and have great buffering capacity, the waters of such lakes and streams contain few ions from weathering of the regional bedrock and are therefore weakly buffered. Acid rain falling on lakes in such areas or entering them as runoff from adjacent watersheds are therefore not fully neutralized.* The pH of the soil water and lake drops accordingly. Although mature fishes seem to survive in acidified lakes they fail to reproduce. Lakes being damaged by acid rain may have mature adults but they commonly contain no young fish.

In large areas of northern North America and Europe known to be vulnerable to acid rain, the most obvious effect is that freshwater fish populations are virtually wiped out or severely reduced in the acidified waters. Thousands of lakes and streams in southern Norway and the Adirondack Mountains of New York State have already been so affected. Studies have indicated that in lightly buffered lakes, which number in the hundreds of thousands in northern Europe and North America, fishes will be eliminated if the precipitation the lakes receive over several years has a pH of 4.3 or lower. This level of pollution is probably injurious to vegetation as well.

* Merely because the underlying bedrock is siliceous does not ensure that the acids will never be neutralized. Such factors as rocks and gravel brought into an area by glaciation, the dominant plants of the region, and the soil type also affect a lake's buffering capacity.

Acid rain is the suspected cause of the deaths of the spruce trees seen here below the summit on the windward side of Camel's Hump Mountain in the Green Mountains of Vermont.

Areas of North America especially vulnerable to acid rain damage to freshwater communities.

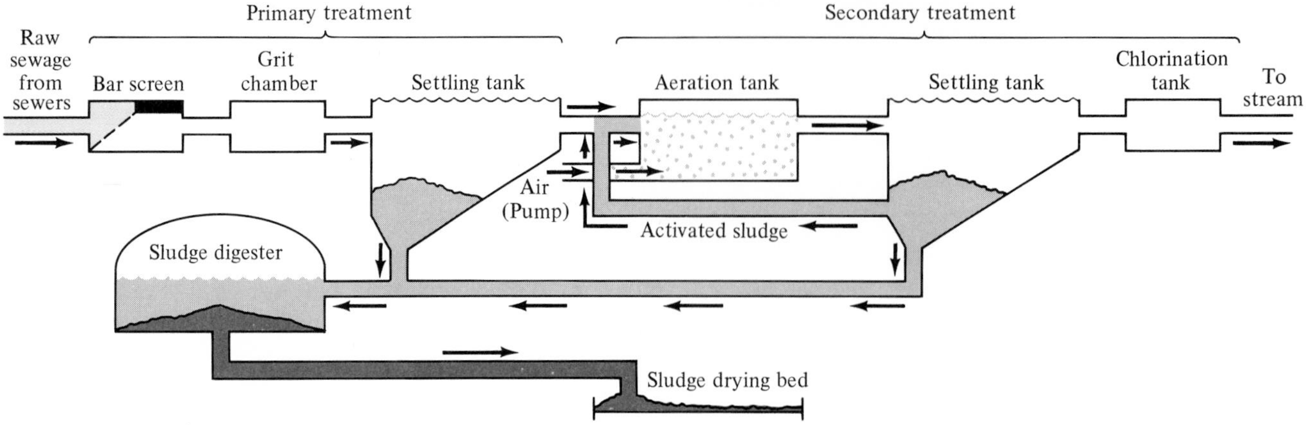


**FIG. 36-24** Sketch of a conventional city sewage treatment plant employing primary and secondary treatment. Secondary treatment involves aeration of the effluent, which promotes the growth of aerobic decomposers. The effluent is then treated with chlorine to kill pathogenic bacteria before being discharged into a stream, lake, or ocean.

sewage from sanitary sewers is first allowed to stand in large tanks until a sludge of solids has settled out; this process is termed **primary treatment** (Fig. 36-24). The sewage sludge is dumped at sea, buried in landfills, burned, or even made into fertilizer if it contains few toxic components. The liquid sewage is drained into incubation tanks for **secondary treatment,** which involves allowing bacteria, sludge worms, and other decomposers to reduce its content of organic matter.

After secondary treatment has greatly reduced biological oxygen demand (BOD), this sewage effluent is allowed to enter a river, lake, or ocean. The BOD of water is defined as the amount of oxygen that is removed from the water by the respiration of sewage bacteria in the course of 5 days at 20°C as they decompose its organic matter; as such, it is an index of the water's content of biodegradable organic matter. Thus, the greater the bacterial growth in a water sample, the more oxygen is used, and the greater its BOD. A high BOD and consequent high depletion of oxygen will slow the process of sewage degradation. Secondary treatment usually reduces the BOD by 60 to 90 percent. The remaining organic matter in the effluent then decomposes in the lake or river into which the effluent is discharged, usually without promoting enough bacterial growth or using enough of the water's oxygen to jeopardize its aquatic life.

BOD, however, is only one index of water quality. The bacteria that decompose the organic matter release inorganic nutrients, such as phosphates and nitrates, into the effluent. This promotes eutrophication of lakes and streams in which such nutrients have been the limiting factors in algal growth. Lakes, especially those with weak currents or still waters, may experience growths of algae massive enough to form a heavy surface scum that reduces the penetration of light (Fig. 36-25). The excess algal growth dies and decomposes, a process that depletes the water's oxygen, thereby killing aerobic aquatic life. The scum and its odor limit the lake's usefulness for recreational purposes and as a community water supply. As noted in Chapter 35, this process may initiate or accelerate ecological succession. Many communities now follow secondary treatments of sewage with **tertiary treatment,** a process that removes toxic metals and 90 percent or more of the phosphates, nitrates, suspended solids, and bacteria.

**FIG. 36-25** A lake with a heavy scum of algal growth resulting from the discharge of sewage effluent rich in nitrates and phosphates.

The burden of biodegradable organic pollution of waterways in the United States has risen steadily in the last 25 years. Only about 15 percent of this, however, can be attributed to population increase. The rest is due partly to increased use of disposable products that end up in

FIG. 36-26 Industrial discharge into a waterway, a primary source of water pollution.

sewage but mostly to the increased industrial pollution produced in the manufacture of these and other modern products. With few exceptions, new technologies tend to use and thus pollute more water than older methods of manufacture. For example, modern commercial canning and freezing of foods involves several methods that increase pollution. One is the "peeling" of tomatoes, peaches, and other soft fruits by immersion in hot lye solutions. Much of the lye and all of the dissolved peelings either directly pollute waterways or further burden sanitary sewer systems.

***Toxic Wastes.*** Thousands of kinds of toxic materials find their way into the waterways of the United States (Figs. 36-26 and 36-27). The volume of any one waste can vary from traces to tons per day. The sources of the hundreds of different types of toxic wastes found in the lakes and rivers of the United States, in decreasing order of their importance as polluters, are chemical plants, petroleum plants, plastics factories, tar and gas plants, rubber factories, dye and tanning plants, ore processing plants, textile factories, paper pulp mills, pharmaceutical plants, and farms. Where toxic pollutants are concerned, sophisticated tests are needed to determine the presence and concentrations of specific materials, some of which may be nonbiodegradable and hazardous to life in very low concentrations.

Globally, over 100 million tons of organic chemicals were being produced annually at the beginning of the 1980s; it is feared that about a third of these products are entering the environment. Much of this material is toxic, and much of it poorly biodegradable. The worldwide output is expected to quadruple by 1995.

One problem with nonbiodegradable toxic pollutants is that they undergo **biological magnification.** Plants that absorb such substances are eaten in large quantity by their primary consumers, the material thereby becoming concentrated in the consumer. Secondary consumers, in turn, eat large quantities of the primary consumers, thereby concentrating the substance further, and so on up the food chain. A

FIG. 36-27 The spraying of pesticides, which become incorporated into groundwater or cropland runoff, contributes to the burden of pollution.

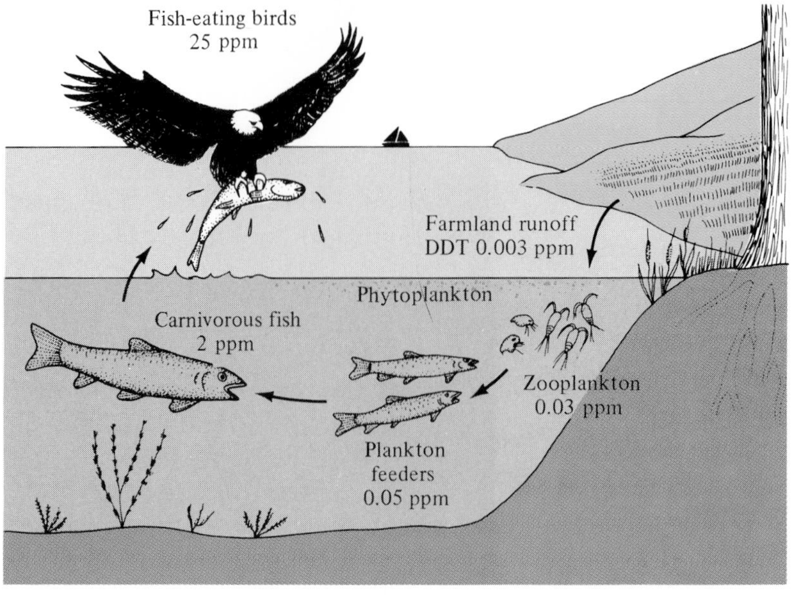


**FIG. 36-28** Pesticides such as DDT that are also toxic to forms of life other than pests, and that are poorly biodegradable, not only persist in the ecosystem but become more concentrated as they move up a food chain.

substance present in great dilution in the environment may thus reach lethal concentration in the top consumers (Fig. 36-28). For this reason the public is often warned against eating large freshwater fish caught in polluted waters.

Much of the toxic waste that pollutes water supplies does not enter lakes, rivers, or the sea. United States industries alone dispose of or store some $190 \times 10^9$ l ($50 \times 10^9$ gallons) of liquid waste each year in dump sites or surface impoundments (pits, ponds, and so on). The United States Environmental Protection Agency (EPA) estimates that until the late 1970s, up to 90 percent of these wastes had been disposed of improperly (Fig. 36-29). By 1981, the EPA had registered over 52,000 hazardous waste dump sites in the United States. Over 12,000 of these were said to pose a "substantial and imminent threat to human health through contaminated groundwater, excessive radiation, or fire and explosion." Of 27,000 such impoundments inspected in 1980, 8,000 were found to be located on permeable soils above usable groundwater sup-

**FIG. 36-29** The Love Canal, in Niagara Falls, N.Y., has become a symbol of an environmental disaster resulting from improperly disposed toxic wastes.

plies. Over half of the groundwater thus far tested in these areas was already contaminated in excess of acceptable health standards. The seriousness of this problem stems from the fact that half of the country's drinking water is obtained from wells and springs supplied by groundwater. The United States Council on Environmental Quality reported in 1981 that wells in 40 states had been shut down in the preceding 3 years because of contamination with toxic organic chemicals. By 1981, all three major aquifers on Long Island, New York, which serve nearly 3 million people, were contaminated by organic chemicals, which resulted in the closing of 36 of the area's public water supplies.

Although sunlight and aquatic organisms may eventually purify surface waters of toxic wastes, aquifers lack these benefits. Once contaminated, an aquifer is likely to remain so for decades. The toxic waste problem, it is widely agreed, is the most serious of all environmental problems now faced by the United States. One of the last acts of the Ninety-sixth Congress in 1980 was to establish a superfund to clean up the worst of the dump sites. However, a 1976 act intended to regulate the disposal of hazardous wastes had not yet been fully implemented by 1985.

***Thermal Pollution.*** Because warm water holds less oxygen and other soluble gases than does cold water, the heated water discharged into rivers and lakes by some industries makes life impossible for many kinds of fishes and other aerobic aquatic life. Such thermal pollution also prevents or delays a lake's springtime turnover—the mixing of surface and lower layers; it thus retards normal planktonic growth. Although warm water itself does not kill certain species of fishes, some lethal fish diseases become more prevalent in warm water. Moreover, warm temperature is often the stimulus for breeding, thereby promoting reproduction at the wrong time of year. The eggs or fry of many fish species fail to develop in warm water. Warm water is a barrier to the spawning migrations of some fishes and can fatally mislead some fishes that normally migrate toward warmer waters at certain times of the year.

How much thermal pollution a particular lake or stream community can withstand depends on such factors as its volume, normal temperature, and importance as a species reservoir. Most of the harm from thermal pollution is a result of the fluctuating nature of such pollution. Whenever an industrial or power plant reduces or shuts down operations, water temperature suddenly drops, only to rise again when operations resume. Few members of warm or cold water communities can survive such shocks.

***Oil Pollution.*** The phrase *oil pollution* usually evokes an image of a shipwreck, perhaps that of a supertanker, or the blowout of an offshore drilling operation, with hundreds of thousands or even millions of liters of crude oil pouring into the ocean. Such accidental spills have indeed contributed significantly to the pollution of the world's oceans. However, despite the drama attending such disasters, they account for only 10 to 14 percent of all the oil spilled into the ocean. Most of the rest comes from such practices as the deliberate dumping of waste oil, the discharging of oil from refineries, the flushing of ships' bilges, and much secret, illegal dumping. About 600,000 tons of oil a year also enter the sea as natural seepage from underground deposits. Additional sources of the hydrocarbons polluting the world's oceans are automobile and other internal combustion engines. Some 90 million tons of hydrocar-

**FIG. 36-30** The environmental hazards of an oil spill are evident in this photo of an oil-covered gull.

bons enter the air each year from these two sources, and up to 50 percent of that is believed to end up in the sea.

The effects of oil pollution are only beginning to be known. The damage easiest to determine is that done to fishes, birds, and seashore communities devastated by a crude oil spill (Fig. 36-30). Besides immediate deaths of fishes and birds, the spawning grounds of oysters, fishes, and other commercially valuable species may be ruined for many years after a major oil spill in their vicinity.* Studies of the effects of oil pollution on the hundreds of thousands of other species that comprise marine communities have only begun. Many of the studies are directed toward establishing a fund of information about the kinds and densities of species in various types of marine ecosystems to provide a basis of comparison for studying the effects of future oil spills. If industries and governments are to be made to bear the cost of oil pollution abatement, they must first be presented with the hard facts concerning what damage the pollution has done.

Except for our knowledge of the severe damage to coastal communities caused by major oil spills, we are largely ignorant of the toxic effects of hydrocarbons on the world's marine ecosystems. The greatest fear is that by the time the adverse effects of marine hydrocarbon pollution are finally understood, the pollution burden may have already passed a critical stage. Because hydrocarbons degrade very slowly, their effects would probably continue to be felt for many years to come. The productivity of the sea is dwindling. Although this is no doubt partially due to overfishing, it is important that we know how much is also due to various types of pollution.

## RADIOACTIVE POLLUTION

Environmental contamination by radioactivity is feared because of its mutagenic and lethal effects. The higher the dosage sustained or the longer it is sustained, the greater the biological damage (Fig. 36-31). Radioactive damage to an ecosystem is not temporary, since radioac-

* Incidentally, the use of detergents to rid beaches of unsightly and noxious crude oil after a spill has proven much more harmful to marine organisms than the oil itself.

**FIG. 36-31** Effects of a long-term exposure of a forest on Long Island, New York, to the ionizing radiation of gamma rays. The trees surrounding the radioactive material placed on the post in the center of the picture were killed after being exposed to the radioactivity for 6 months. The experiment is being continued in order to study the mutagenic effects on the surviving plants.

tive pollutants may persist for thousands of years. Some forms of radioactivity become more concentrated as they are transferred up a food chain. Moreover, the individual atoms are recycled repeatedly, with the potential for doing damage at every step along the way.

***Operational Contamination of the Environment.*** One type of radioactive pollution is the contamination that arises from normal operations in the production of nuclear weapons and the operation of nuclear power plants. These operations include the mining, processing, transporting, and storing of radioactive ores; the normal operating of power plants and weapons arsenals; the storage and reprocessing of spent fuel; and the storage of the radioactive components of decommissioned overage facilities. The possibilities for environmental contamination during the process of producing electricity from nuclear fuel can occur (1) during the mining and processing of ore, (2) during the operation of nuclear power plants, (3) during the reprocessing of spent fuel, and (4) in the storage of radioactive wastes.

Because the mine tailings of uranium ore extraction are themselves radioactive, they pose a threat to the health of miners and anyone else who lives or works in their vicinity for a time. The processing of uranium ore involves its enrichment to increase the proportion of uranium-235 ($^{235}U$) by three to four times the natural level. Radioactive contamination of the environment generated by the enrichment process is minimal and not regarded as hazardous.

In the operation of nuclear power plants, minor defects in boiling-water nuclear reactors result in small leakages of radioactivity into the cooling water circulating through the reactor core and thus into the environment. Some krypton-85 also escapes as a gas. Both of these "routine emissions," however, are at such a low level as to be comparable to the normal background of radioactivity from rocks and the cosmic rays from outer space. Barring accidents, nuclear power plant operation is not regarded by the industry or the United States government as a significant threat to health (Fig. 36-32).

In a conventional reactor, after a year or more of operation, fission products accumulate in the fuel rods in amounts sufficient to slow the reaction. The fuel is then regarded as spent and must be replaced. The

**FIG. 36-32** The Three Mile Island power station on the Susquehanna River near Harrisburg, Pennsylvania, scene of the March 1979 accident that produced a partial meltdown of the fuel core.

**FIG. 36-33** Storage of low-level nuclear waste, Hanford, Washington.

spent rods contain high concentrations of highly radioactive fission products of $^{238}U$ and $^{235}U$. From a health and environmental standpoint, the question of what to do with these and the even more abundant radioactive wastes from military sources has proved to be by far the most difficult problem of the nuclear age (Fig. 36-33). The original goal was to reprocess the waste, remove its uranium and plutonium for reuse, and then safely store the many remaining by-products of radioactive fission for an indefinite period. In the case of the power plants alone, it was anticipated that for each reactor in full operation, some 10 to 60 shipments per year would be made to reprocessing plants; until 1972, specially designed casks were used in shipping, and most shipments were made by truck or rail.

Reprocessing of fuel rods involves dissolving them in concentrated acid and recovering the uranium and plutonium. This procedure leaves a waste solution that remains highly radioactive for as long as a million years. The liquid, which is said to boil "like a teakettle," was to be temporarily stored in huge tanks until it cools. Reprocessing regularly results in the emission of radioactive gases (for example krypton-85) more long-lived than those emitted during the operation of a power plant. A slight increase in cancer deaths—well under 1 percent—was expected to result in those exposed to this pollution source.

Regulations provided that the liquid be converted to a solid form within 5 years and shipped to an approved depository within 10 years. What some regard as the main problem with nuclear power occurs at this point: no generally satisfactory way has been found of permanently storing these wastes for as long as thousands of years, let alone hundreds of thousands. Storage of the liquid in tanks has not proved practicable. Not only is it impossible to design a tank that will last for even 100 years, but between 1969 and 1980, 16 cases of leakage from tanks occurred in the United States, permitting the escape of over 1.3 million l (350,000 gallons) of radioactive liquid into the environment.

The impasse over the storage problem and problems associated with reprocessing nuclear fuel in the United States has halted all reprocessing since 1972. One approach to storage is to incorporate the liquid into solid blocks of glass for storage in abandoned salt mines equipped with remote-control television cameras for continual monitoring. Not only does salt conduct heat well—the glass blocks will remain hot for some time—but most salt mines are dry and sustain little damage from earthquakes. Among other proposals is one to drop the blocks of glass into the ocean at some of the deeper and quieter locations or sink them into deep holes bored in the ocean floor. Partly because of the lack of agreement on how and where to store high-level radioactive wastes, by 1985 over 400,000 tons of spent fuel rods had accumulated at power plants in the United States.

***Nuclear Accidents.*** In addition to the operational contamination of the environment, another area of concern is the possibility of a major accident at a nuclear facility, with the potential for deadly contamination of air, soil, and water in a wide area surrounding the accident site. The only accident that has thus far resulted in massive contamination of a large area apparently took place in 1957 in the Ural Mountains, in the Soviet Union. According to a 1980 report of the Oak Ridge National Laboratory, in Oak Ridge, Tennessee, an accidental explosion occurred during nuclear waste processing at a remote Soviet weapons plant, dispersing millions of curies* of strontium-90 and other nuclides, contaminating over 1,000 $km^2$, including 14 lakes. Like most details of the accident, the number of casualties has been kept secret. Thirty towns were permanently evacuated and subsequently eliminated from Soviet maps, and 60,000 survivors were relocated.

## THE HUMAN POPULATION EXPLOSION

If the world population were to stop increasing at this moment or were even to decrease to half its present size, the problems of pollution and ecosystem destruction would still be with us. Thus, solving the serious problem of the world population explosion could not in itself solve the problem of environmental deterioration. However, the rapid expansion of populations in countries with widespread poverty makes it extremely difficult to arrest some types of environmental deterioration in those countries.

The world's human population is indeed exploding. Although *Homo sapiens* has lived on earth for at least 300,000 years, it was not until 1800 that we attained a population of 1 billion. Then, within only 130 years, the population increased to 2 billion. In only 30 more years (1930 to 1960), a third billion was added, and in another 15 years (1975), a fourth! By 1985, the rate of increase had begun to slow to about 1.8 percent per year.† Nevertheless, the world's human population is expected to reach 6.35 billion by the year 2000, a further increase of over 50 percent (Fig. 36-34). The human population explosion exemplifies the exponential growth capability possessed by all species. A companion principle dictating that when population size exceeds carrying capacity massive mortality results is also being demonstrated by the explosion, but at regional rather than global levels.

The rapid population expansion since 1800 has not been due to an increase in birth rate, that is, to an increase in the number of births per 1,000 population. Rather, it has been due mostly to a sudden decrease in death rate, due to such factors as the development of relatively inexpen-

* The curie, a unit of radioactivity, is defined as $3.7 \times 10^{10}$ nuclear disintegrations per second.

† The global decline in the rate of increase means that world population is increasing at a slower rate, not that the world population itself is decreasing.

**FIG. 36-34** Growth curve of the world population of *Homo sapiens* from the year 1000 to 1990.

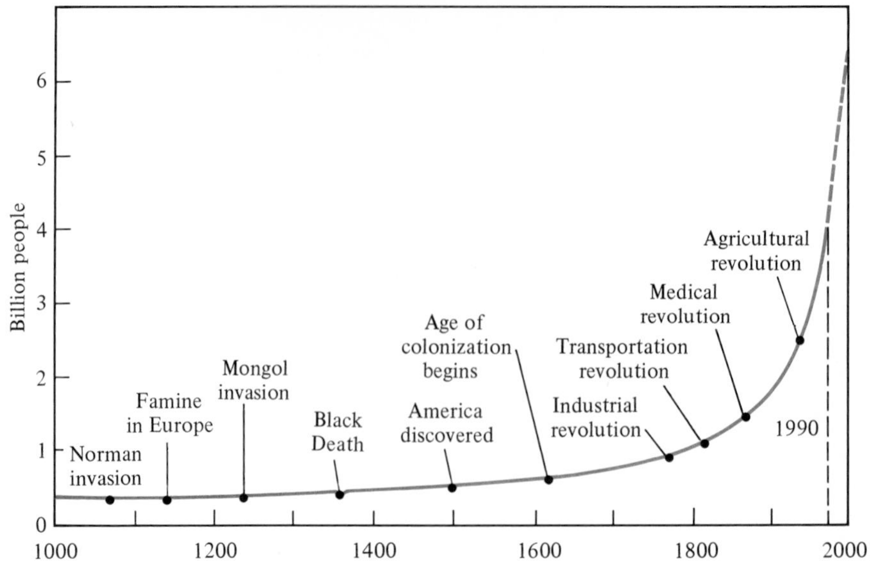


sive medicines, methods of sanitation, and other means of disease control. Also contributing to the problem is the persistent cultural practice of raising large families, a custom influenced in the past by very high infant and childhood morality rates. Now that in most countries children usually live to have children of their own, populations have soared. Cultural adjustment to sudden change is typically slow, and adjustment to longer life expectancy has been no exception.

## DIMENSIONS OF THE PROBLEM

Although the worldwide effect of the decreased death rate has been a population explosion, population growth in numbers is very uneven from one country to the next, with 90 percent of it occurring in the "less developed" countries. For example, while populations in Japan and the United States are growing at an annual rate of 0.1 percent, until recently Mexico's has been increasing at a rate of nearly 3 percent per year, as have the populations of Iran and Iraq. Many of the world's cities, to which people move in attempts to improve their lot in life, are turning into nightmarish jungles with growing shantytowns housing millions upon millions of inadequately fed and clothed people who lack medical care. There is a higher incidence of malnutrition in such cities than in the rural areas from which the migrants came.

## FERTILITY CONTROL

As a general rule, the countries in which fertility control has been most effective are those in which the central government has backed a program of public information about and encouragement in family planning, usually by contraception. For example, the government of Mexico has claimed that, through such programs, it reduced the country's rate of growth from 3.4 percent to 2.7 percent per year between 1977 and the end of 1980. The problem of limiting population growth is complex and includes cultural, religious, political, and economic factors, any one of which can block solutions. Moreover, birth control programs alone,

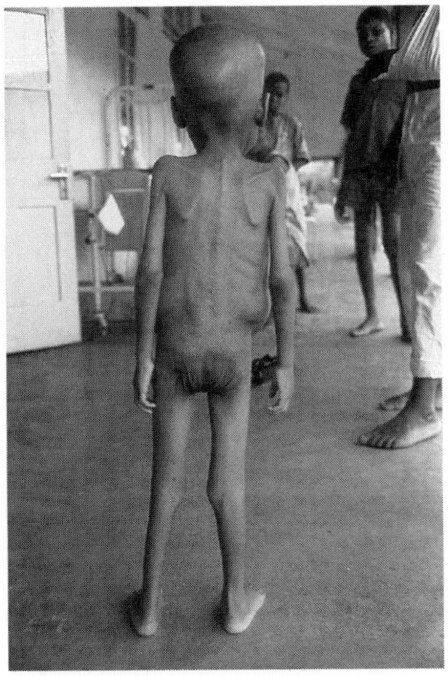

**FIG. 36-35** A starved African child, one of the world's 16 million refugees.

without education, health care, and economic opportunity for the majority of a country's population, have rarely been effective. Since the beginning of the 1970s, however, there has been a remarkably sharp drop in fertility in some areas of the world, largely in response to the changes in public attitudes that began in the early 1960s. This decrease has coincided with the rapidly growing use of oral contraceptives and an increase in induced abortions.

## THE GLOBAL FOOD PROBLEM

Despite the declining fertility of the remaining arable topsoil, world food output is expected to keep pace with population growth into the next century. Stating the matter this simply, however, is deceptive. In many cases, the increased food production, while it has prevented massive starvation, has depended heavily on expensive fertilizers, pesticides, and irrigation. It is feared that this may only set the stage for even greater population crashes in the future. And although overall food production has kept pace with overall population growth, little food usually reaches those who need it the most. The famines that the world experiences are not general, but regional. Millions have starved: in Biafra in the 1960s, in Bangladesh in 1971, in the sub-Sahara Sahel in the mid-1970s, in Cambodia in 1980, and in Uganda, Ethiopia, and Somalia in the 1980s (Fig. 36-35). While surplus food can almost always be obtained by those who can afford to buy it, little of it reaches the starving poor who cannot.

Thus, the world's ability to raise enough food to feed everyone could not feed the over 16 million refugees from war and famine in the early 1980s and cannot feed the 500 million people who are hungry every day of their lives (Fig. 36-36). Although global relief programs may often compensate somewhat for these situations on an emergency basis, populations continue to grow and poor people are increasingly cultivating the remaining marginal lands. Both trends forbode unprecedented ecological disasters for the future. There appears to be little question that world population will exceed 6 billion by the year 2000, with the increase occurring largely in those areas already suffering most from hunger, malnutrition, and starvation, namely, parts of Asia, Latin America, and Africa.

**FIG. 36-36** Total world food production and per capita food production in developed and developing countries from 1961 to 1978.

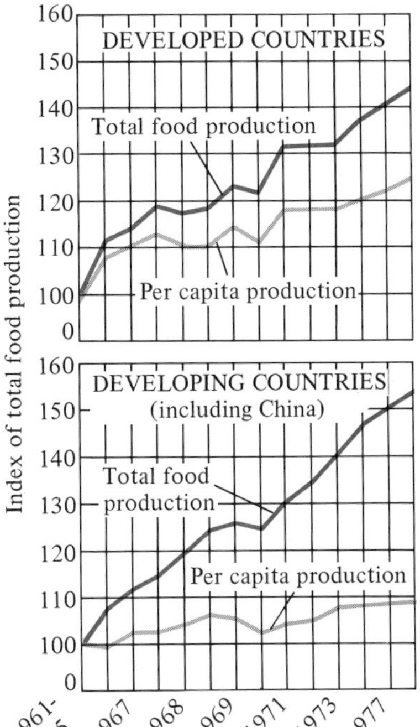


# THE FUTURE

Is there hope for the future? It is easy to feel overwhelmed by the magnitude of the many environmental problems that confront us—to feel that one person is helpless against all the forces that threaten our planet. Although one person might indeed be helpless, literally hundreds of thousands of people are attempting to solve these problems. Moreover, they are achieving success on many fronts; to recount their achievements would fill at least another chapter. The environmental crisis has many facets, and many specialized groups have arisen to meet them. Some organizations are international in scope; some are local. Some appeal to the public, some lobby legislatures, and some act directly by buying up tracts of land to preserve them from reckless development. Some have broad politico-economic perspectives in which respect for the environment is an integral part.

In a number of cases, the action of citizens' groups, of industries, or of governments at the urging of citizens has halted needless destruction of ecosystems or restored severely damaged areas. Although such cases are exceptional and there is much work yet to be done, the success that has been achieved is proof that it is possible. Alternatively, to yield to despair is to guarantee disaster. For those interested in a never-ending battle, we suggest joining one or more of the many agencies and citizen groups that seek to divert us from the dangerous path down which our civilization is heading.

## Summary

Human activities have adversely affected world ecosystems to the extent that our supplies of vitally important materials, including food, air, and water, are in jeopardy. Our use of natural resources, especially nonrenewable resources, contributes to this problem in several ways. For example, the use of mineral resources pollutes the environment with discarded mine tailings, acid runoffs, oil spills, and toxic wastes. In the case of the nuclear power and weapons industries, hazards to the environment have arisen principally from the mining and processing of the ore and the transport and use of the enriched ore. The most serious problems lie ahead: in the storage of highly radioactive wastes, waste reprocessing, and decommissioning of over-age plants.

Many minerals used in industry are nonrenewable or can be recycled only at great expense. Most of the richest ores appear to have been mined. New technologies have enabled industries to use some low-grade ores, usually at a high cost of energy and pollution. Although also expensive and polluting, recycling needs to be encouraged.

Topsoil, which is essential for agriculture and maintaining natural communities, accumulates at the rate of a few inches per century. All around the world, topsoil is being washed or blown away at an unprecedented rate. Overgrazing, imprudent farming, and deforestation of slopes are causing erosion that is destroying rangeland, cropland, and watersheds. In the United States, most topsoil conservation is done on a voluntary basis.

The water on which most terrestrial life depends is provided by the hydrologic cycle. This supply includes the water of lakes and rivers as well as that of aquifers, from which well water is obtained. Because many large cities are located in regions of low supply, water must be brought in to them from elsewhere, often from rivers on which other communities depend. Wastefulness is a significant factor contributing to water shortages. In many regions, aquifers are tapped for irrigation, and water is withdrawn at a much higher rate than it is replenished. Damming of rivers often is destructive of ecosystems above and below the dam. If downstream from deforested slopes, a dam may fill with silt in 50 years or less. Besides depleting aquifers of their water, irrigation on a massive scale has produced subsidence and salinization of the soil. When groundwater is depleted in a coastal region, seawater intrusion has often spoiled the well water and the soil for growing crops.

The forests of the world are being cut down faster than they regrow. This is true in the United States and Canada as well as in many other countries. The worldwide net loss of forests is 1 to 2 percent a year and increasing. About half of all of the world's trees have been cut down since 1950. Besides lumber, much wood is used as fuel; for over 75 percent of the world, wood is the only fuel. A cascade of adverse effects is resulting from the destruction of the world's forests.

A wilderness is an extensive, compact area preserved from all human intrusion and control. Wildernesses are important as reservoirs of valuable species, as purifiers of polluted air and water, as resources for biological research and medicine, and as support systems for other ecosystems.

Human activities often cause direct destruction of valuable, natural ecosystems. Mines of all kinds add to air and water pollution, and strip mining destroys habitats as well. Draining and "developing" wetlands has been extensive, depriving waterfowl and much aquatic life of breeding and feeding grounds. Flood control can be very destructive of habitats, often unique ones, as in the Everglades.

Pollution is the impairment of the environment's suitability for supporting life by harmful concentrations of materials. The population explosion and advances in technology and in affluence in industrialized nations have created toxic pollution of crisis proportions. The largest volume of air pollutants is from the burning of fossil fuels by industries and automobiles. In cities where temperature inversions occur, air pollution can be life-threatening. World Health Organization studies indicate that 80 to 90 percent of all cancers are environmentally related. One of the results of air pollution is acid rain, which destroys forests and lakes thousands of miles from the pollutant source.

Water pollution is caused by a great variety of substances, including industrial wastes, sewage effluent, runoff from farmland, spillage of petroleum, and radioactive wastes. In high concentration, sewage and farmland runoff, composed of biodegradable waste, may deplete aquatic systems of oxygen as bacteria decompose them or may produce eutrophication of lakes, thereby accelerating community succession and ruining the lakes' general usefulness. Toxic wastes are a more serious matter. Almost all are industrial by-products. Of the 100 million tons of organic chemicals produced annually, many of them toxic, about one-third may enter world ecosystems. The output is expected to quadruple by 1995. The worst problem arises from toxic wastes stored in leaky impoundments, which seep into the groundwater, poisoning it for decades. In the United States, about 90 percent of all toxic wastes are stored improperly. By 1981, some wells in 40 states had been shut down because of toxic contamination. Half of the drinking water in the United States is obtained from wells.

Thermal pollution of water occurs near certain industrial and power plants. It has many adverse effects on local natural populations. Oil pollution is mostly deliberate, with only 10 to 14 percent due to shipwrecks, oil rig blowouts, and other accidents. About 600,000 tons of oil also enter the sea from natural seepage. In addition, about 50 percent of the 90 million tons of unburned hydrocarbons from combustion engine exhausts enter the sea each year. Little is known of the long-term effects of hydrocarbons on marine ecosystems.

Radioactive pollution of the environment is primarily a potential problem rather than a current one. The problem of storage of highly radioactive waste, most of it from military sources, is becoming more serious as more of it accumulates.

The world's human population is exploding. A population of 1 billion was reached in 1800, of 2 billion within the next 130 years, of 3 billion within the next 30 years, and of 4 billion within the next 15 years. World population is expected to reach 6.35 billion by the year 2000. The main reasons for this rapid growth have been improvements in medicine, sanitation, and disease control. Fertility control programs have not been generally effective, especially in regions where poverty and illiteracy are widespread. Some worldwide change in attitude and a slowing of the rate of increase have occurred in recent years, however. Although world food supplies are currently adequate overall, regional famines resulting in millions of deaths occur in Asia and Africa. Surplus food seldom reaches those who need it most.

Rather than being overwhelmed by the magnitude of the problems besetting our planet and its inhabitants, we can be encouraged by the limited solutions thus far achieved, and we can join with those who are working to stave off disaster.

## Recommended Readings

EHRENFELD, D. W. 1970. *Biological Conservation.* Holt, Rinehart and Winston, New York. A brief but excellent introduction to conservation principles. Illustrated with dramatic examples, in paperback.

GIBBONS, B., and S. C. WILSON. 1984. "Do we treat our soils like dirt?" *Natural Geography,* 166(3):350–389. An examination of current and future problems of soil erosion as seen through many eyes.

HODGES, L. 1977. *Environmental Pollution,* 2d ed. Holt, Rinehart and Winston, New York. A brief but comprehensive and interesting account of various types of pollution, their causes, biological effects, and means of abatement. Rich in specific data of significance and in references to sources of additional information.

OWEN, O. S., 1980. *Natural Resource Conservation: An Ecological Approach,* 3d ed. Macmillan, New York. A deservedly successful textbook that integrates ecology and conservation with primary emphasis on human environmental impact. Well written and illustrated, with an extensive bibliography.

REVELLE, P., and C. REVELLE. 1984. *The Environment: Issues and Choices for Society,* 2d ed. Willard Grant Press, Boston. A well-documented introduction to the principal problems that humans have created for their environment.

SOUTHWICK, C. H. 1976. *Ecology and the Quality of Our Environment,* 2d ed. Van Nostrand, New York. A textbook of ecology with emphasis on problems affecting the environment. Available in paperback.

TAYLOR, J. WOLFRED, ed. 1984. *The Acid Test.* The November–December, 1984 issue of *Wisconsin Natural Resources* includes a 40-page supplement on acid rain and the $2 million project to study it in Wisconsin. *Wisconsin Natural Resources,* 8(6):suppl.

TROEH, F. R., J. A. HOBBS, and R. L. DONAHUE. 1980. *Soil and Water Conservation for Productivity and Environmental Protection.* Prentice-Hall, Englewood Cliffs, N.J. An authoritative, comprehensive, and up-to-date account of basic principles of natural resource conservation. Well illustrated.

TURK, J., A. TURK, and K. ARMS. 1984. *Environmental Science,* 3d ed. Saunders, Philadelphia. A copiously illustrated introductory textbook of ecology written around the theme of the environmental impact of human activity.

WAGNER, R. H. 1978. *Environment and Man,* 3d ed. Norton, New York. With a brief introductory section on ecological principles, this book is devoted to the effects of human activities on world ecosystems.

# Appendix A

## Classification of Organisms

The following classification is somewhat more detailed than that appearing elsewhere in the book and is intended to provide only the level of detail that is likely to prove useful to the beginning student. The genera mentioned are those evolutionarily significant, used in research, or ones mentioned elsewhere in the book. With a few exceptions, only groups with living representatives are listed. Anyone interested in more detailed classifications should consult advanced textbooks or monographs in their areas of interest.

A concerted effort has been made to respect expert opinion on how various groups should be classified and named while striving for consistency in the overall classification. The widely accepted five-kingdom system of Robert Whittaker has been followed. However, any classification that includes a Kingdom Protista is artificial to some degree. Several of its groups are admittedly more closely related to members of another kingdom than they are to other protists.

### KINGDOM MONERA

DIVISION SCHIZONTA*
- Class Mycoplasmata. *Mycoplasma*
- Class Rickettsiae. *Rickettsia, Coxiella*
- Class Chlamydiae. *Chlamydia*
- Class Actinomycetes. *Actinomyces, Streptomyces*
- Class Eubacteria. *Escherichia, Rhizobium, Spirillum, Salmonella, Nitrosomonas, Streptococcus, Staphylococcus*
- Class Myxobacteria. *Myxococcus*
- Class Spirochetes. *Leptospira, Spirocheta, Treponema*

DIVISION CYANOCHLORONTA*
- Class Cyanobacteria. *Oscillatoria, Nostoc*

DIVISION ARCHAEBACTERIA

### KINGDOM PROTISTA

SUBKINGDOM PROTOPHYTA: Algal protists

DIVISION EUGLENOPHYTA. Euglenoids. *Euglena, Phacus.*

DIVISION CHRYSOPHYTA. Algae golden to yellow in color.
- Golden-brown algae; haptophytes; coccolithophores, yellow-green algae, chloromonads, diatoms.

DIVISION PYRROPHYTA. Dinoflagellates. *Gonyaulax, Gymnodinium, Ceratium.*

DIVISION CRYPTOPHYTA. Cryptomonads. *Cryptomonas, Chilomonas.*

* To avoid using the word ending -phyta ("plants") for organisms such as bacteria that are not in the plant kingdom, some botanists replace it with -onta (Greek for "things that exist," or, roughly, "the ones"). Schizophyta thus become Schizonta and Cyanophyta is Cyanochloronta, literally "the blue-green ones."

SUBKINGDOM PROTOMYCOTA: Fungal protists.* Hyphochytrids, Chytrids.

SUBKINGDOM GYMNOMYCOTA: Slime molds*

DIVISION PLASMODIOPHORAMYCOTA. Endoparasitic slime molds. *Plasmodiophora.*
DIVISION LABYRINTHULOMYCOTA. Net slime molds. *Labyrinthula.*
DIVISION ACRASIOMYCOTA. Cellular slime molds. *Dictyostelium.*
DIVISION MYXOMYCOTA. True (plasmodial) slime molds. *Physarum.*

SUBKINGDOM PROTOZOA. Animallike Protists

PHYLUM SARCOMASTIGOPHORA
- Subphylum Mastigophora. Flagellates.
  - Class Phytomastigophoreae.** Typically, but not necessarily with chloroplasts. Cryptomonads, dinoflagellates, euglenoids, chrysomonads, volvocids, etc.
  - Class Zoomastigophoreae. Colorless flagellates.
  - Choanoflagellates, trypanosomes, monads.
- Subphylum Opalinata
  - Class Opalinatea. *Opalina.*
- Subphylum Sarcodina. Amoeboid protozoans.
  - Superclass Rhizopoda. *Amoeba, Entamoeba, Arcella,* foraminiferans.
  - Superclass Actinopoda. Radiolarians, heliozoans.

PHYLUM APICOMPLEXI.† Gregarines, coccidians *(Plasmodium),* piroplasms *(Babesia).*
PHYLUM MICROSPORA.† *Thelohania, Nosema.*
PHYLUM ASCETOSPORA.†
PHYLUM MYXOZOA.†
PHYLUM CILIOPHORA. Ciliates: *Coleps, Didinium,* suctorians, *Tetrahymena, Paramecium, Vorticella, Stentor, Halteria, Kerona.*

## KINGDOM PLANTAE

DIVISION CHLOROPHYTA. Green algae. *Ulothrix, Spirogyra, Ulva.*
DIVISION CHAROPHYTA. Stoneworts. *Chara, Nitella.*
DIVISION PHAEOPHYTA. Brown algae. *Fucus, Sargassum, Laminaria.*
DIVISION RHODOPHYTA. Red algae. *Polysiphonia, Nemalion.*
DIVISION BRYOPHYTA.
- Class Hepaticae. Liverworts. *Marchantia.*
- Class Anthocerotae. Hornworts. *Anthoceros.*
- Class Musci. Mosses. *Polytrichum, Sphagnum.*

DIVISION TRACHEOPHYTA.
- Subdivision Psilopsida. Psilophytes (extinct).
- Subdivision Pteropsida. Ferns. *Polypodium, Botrychium, Pteridium.*
- Subdivision Lycopsida. Club mosses. *Lycopodium, Selaginella.*
- Subdivision Sphenopsida. Horsetails. *Equisetum.*
- Subdivision Spermopsida. Seed plants.
  - Class Progymnospermae. Pregymnosperms (extinct).
  - Class Pteridospermae. Seed ferns (extinct).
  - Class Cycadae. Cycads and extinct cycadlike members. *Zamia.*
  - Class Ginkgoae. *Ginkgo.*
  - Class Coniferae. Conifers and extinct coniferlike members. *Pinus, Taxus, Sequoia.*

* Protomycota and Gymnomycota are included in the Kingdom Fungi in many classifications. Gymnomycota (slime molds) are regarded as amoeboid protozoans in others.

** Notice that this group, logically included as belonging to Protozoa, is equivalent to the Subkingdom Protophyta, the algal protists. As noted earlier, a similar case can be made for listing slime molds either as protophytans, protozoans, or fungi.

† Formerly included in Class Sporozoa, spore-forming parasitic protozoans.

Class Gneteae. *Gnetum, Ephedra, Welwitschia.*
Class Angiospermae. Flowering plants.
Subclass Dicotyledoneae. Dicots. *Acer, Quercus, Magnolia, Rosa, Aster, Pisum, Ranunculus.*
Subclass Monocotyledoneae. Monocots. *Lilium, Tulipa, Zea, Triticum, Yucca.*

## KINGDOM FUNGI

DIVISION OOMYCOTA. Water molds, white rusts, downy mildews. *Saprolegnia, Phytophthora.*
DIVISION ZYGOMYCOTA. Conjugation fungi. *Rhizopus, Phycomyces.*
DIVISION ASCOMYCOTA. Sac fungi: Cup fungi, yeasts, powdery mildews, fruit molds *(Penicillium)*, bread molds *(Neurospora).*
DIVISION BASIDIOMYCOTA. Club fungi: Rusts and smuts, toadstools, mushrooms, shelf fungi, puff balls.
DIVISION DEUTEROMYCOTA. Fungi imperfecti.

## KINGDOM ANIMALIA

### SUBKINGDOM PARAZOA

PHYLUM PORIFERA. Sponges.
Class Calcarea. Calcareous (limey) sponges. *Leucosolenia, Grantia.*
Class Hexactinellida. Glass sponges. *Euplectella.*
Class Demospongiae. *Euspongia, Spongilla.*
Class Sclerospongiae. Coralline sponges.

### SUBKINGDOM METAZOA

#### SECTION RADIATA

PHYLUM CNIDARIA (or COELENTERATA)
Class Hydrozoa. Hydrozoans. *Hydra, Obelia, Physalia.*
Class Scyphozoa. Jellyfishes. *Aurelia, Cyanea.*
Class Anthozoa. Sea anemones and corals. *Metridium, Gorgonia, Astrangia.*
PHYLUM CTENOPHORA. Comb jellies. *Pleurobrachia, Mnemiopsis, Beroe.*

#### SECTION PROTOSTOMIA

PHYLUM PLATYHELMINTHES. Flatworms.
Class Turbellaria. Free-living flatworms. *Planaria, Dugesia.*
Class Trematoda. Flukes. *Clonorchis, Schistosoma.*
Class Cestoda. Tapeworms. *Taenia, Dipylidium.*
PHYLUM GNATHOSTOMULIDA. *Gnathostomula.*
PHYLUM NEMERTINA (or Rhynchocoela). Proboscis worms. *Cerebratulus.*
PHYLUM ACANTHOCEPHALA. Spiny-headed worms. *Echinorhynchus.*
PHYLUM ASCHELMINTHES
Class Rotifera. Rotifers. *Hydatina.*
Class Gastrotricha. *Chaetonotus.*
Class Kinorhyncha. *Echinoderes.*
Class Nematoda. Round worms. *Ascaris, Trichinella, Necator, Enterobius.*
Class Nematomorpha. Horsehair worms. *Gordius.*
PHYLUM ENTOPROCTA. *Loxosoma.*
PHYLUM PRIAPULIDA. *Priapulus.*
PHYLUM ECTOPROCTA (or Bryozoa). Bryozoans, moss animals. *Bugula, Plumatella.*
PHYLUM PHORONIDA. *Phoronis.*
PHYLUM BRACHIOPODA. Lamp shells. *Lingula, Terebratula.*

PHYLUM MOLLUSCA. Molluscs.
- Class Aplacophora. Solenogasters. *Neomenia.*
- Class Polyplacophora. Chitons. *Chaetopleura.*
- Class Monoplacophora. *Neopilina.*
- Class Gastropoda. Snails and their allies (univalve molluscs). *Helix, Busycon, Crepidula, Haliotis, Littorina, Doris, Limax.*
- Class Scaphopoda. Tusk shells. *Dentalium.*
- Class Bivalvia. Bivalve molluscs. *Mytilus, Ostrea, Pecten, Mercenaria, Teredo, Anodonta.*
- Class Cephalopoda. Squids, octopuses, etc. *Loligo, Octopus, Nautilus.*

PHYLUM POGONOPHORA. Beard worms. *Polybrachia.*

PHYLUM SIPUNCULIDA. *Sipunculus, Golfingia (Phascolosoma).*

PHYLUM ECHIUROIDA. *Echiurus, Urechis.*

PHYLUM ANNELIDA. Segmented worms.
- Class Polychaeta (including Archiannelida). Sandworms, tubeworms, etc. *Nereis, Chaetopterus, Aphrodite, Arenicola.*
- Class Oligochaeta. Earthworms and many freshwater annelids. *Tubifex, Lumbricus.*
- Class Hirudinea. Leeches. *Hirudo, Macrobdella.*

PHYLUM ONYCHOPHORA. *Peripatus.*

PHYLUM TARDIGRADA. Water bears. *Macrobiotus.*

PHYLUM PENTASTOMIDA. Tongue worms. *Linguatula.*

PHYLUM ARTHROPODA
- Subphylum Trilobita. Trilobites (extinct).
- Subphylum Chelicerata
  - Class Eurypterida (extinct).
  - Class Xiphosura. Horseshoe crabs. *Limulus.*
  - Class Arachnida. Spiders, ticks, mites, scorpions. *Aranea, Ixodes, Sarcoptes, Scorpio.*
  - Class Pycnogonida. Sea spiders. *Nymphon.*
- Subphylum Mandibulata
  - Class Crustacea. Lobsters, crayfishes, crabs, shrimps, barnacles, water fleas, etc. *Homarus, Astacus, Uca, Penaeus, Balanus.*
  - Class Chilopoda. Centipedes.* *Scutigera, Lithobius.*
  - Class Diplopoda. Millipedes.* *Iulus.*
  - Class Pauropoda. *Pauropus.**
  - Class Symphyla. *Scutigerella.**
  - Class Insecta. Insects.
    - Subclass Apterygota (Ametabola). Wingless insects without metamorphosis.
      - Order Collembola. Springtails. *Podura.*
      - Order Protura. *Acerentomum.*
      - Order Diplura. *Campodea.*
      - Order Thysanura. Bristletails, silverfish. *Lepisma.*
    - Subclass Pterygota. Winged insects.
      - Section I. Exopterygota (Hemimetabola). Insects with incomplete metamorphosis.
        - Order Ephemeroptera. Mayflies. *Chloeon.*
        - Order Odonata. Dragonflies, damselflies. *Aeschna.*
        - Order Orthoptera. Grasshoppers, crickets, walking sticks, mantids, cockroaches. *Gryllus, Locusta, Melanoplus, Blatta.*
        - Order Isoptera. Termites. *Termes.*
        - Order Dermaptera. Earwigs. *Forficula.*
        - Order Embioptera. Webspinners. *Embia.*
        - Order Plecoptera. Stoneflies. *Perla.*
        - Order Zoraptera. Zorapterans. *Zorotypus.*

* Formerly under a single class, Myriapoda.

Order Psocoptera. Book lice. *Polypsoca.*
Order Mallophaga. Chewing lice. *Menopon.*
Order Anoplura. Sucking lice. *Pediculus.*
Order Thysanoptera. Thrips. *Kakothrips.*
Order Hemiptera. True bugs. Cicadas, aphids, leafhoppers, scale insects, etc. *Aphis, Rhodnius.*
Section II. Endopterygota (Holometabola). Insects with complete metamorphosis.
Order Neuroptera. Dobsonflies, lacewings, etc. *Sialis, Myrmeleon.*
Order Mecoptera. Scorpionflies. *Panorpa.*
Order Trichoptera. Caddisflies. *Phryganea.*
Order Siphonaptera. Fleas. *Pulex.*
Order Coleoptera. Beetles, weevils. *Carabus.*
Order Hymenoptera. Wasps, bees, ants, sawflies. *Apis.*
Order Diptera. True flies, mosquitoes. *Musca, Drosophila, Aedes, Anopheles.*
Order Lepidoptera. Moths, butterflies. *Vanessa, Manduca.*

SECTION DEUTEROSTOMIA

PHYLUM CHAETOGNATHA. Arrow worms. *Sagitta.*
PHYLUM ECHINODERMATA
Class Crinoidea. Crinoids, sea lilies. *Antedon, Metacrinus.*
Class Asteroidea. Sea stars. *Asterias.*
Class Ophiuroidea. Brittle stars, serpent stars, basket stars, etc. *Ophioderma.*
Class Echinoidea. Sea urchins, sand dollars. *Arbacia, Strongylocentrotus, Echinarachnius.*
Class Holothuroidea. Sea cucumbers. *Cucumaria, Thyone.*
PHYLUM HEMICHORDATA. Acorn worms. *Saccoglossus, Balanoglossus.*
PHYLUM CHORDATA. Chordates
Subphylum Urochordata (or Tunicata). Tunicates.
Class Ascidiacea. Ascidians or sea squirts. *Molgula, Ciona.*
Class Thaliacea. Salps. *Salpa, Pyrosoma.*
Class Larvacea. *Oikopleura.*
Subphylum Cephalochordata. Amphioxus. *Branchiostoma.*
Subphylum Vertebrata. Vertebrates.
Class Agnatha. Jawless fishes. *Petromyzon, Myxine.*
Class Placodermi (extinct).
Class Chondrichthyes. Cartilaginous fishes. *Squalus, Raja, Scoliodon.*
Class Osteichthyes. Bony fishes.
Subclass Sarcopterygii
Order Crossopterygii. Coelacanths or lobe-fins. *Latimeria.*
Order Dipnoi. Lungfishes. *Neoceratodus, Protopterus, Lepidosiren.*
Subclass Brachiopterygii. Bichirs. *Polypterus.*
Subclass Actinopterygii. Higher bony fishes. *Amia, Perca, Salmo.*
Class Amphibia
Order Anura. Frogs and toads. *Rana, Hyla, Bufo.*
Order Urodela. Salamanders, newts. *Necturus, Triturus, Ambystoma.*
Order Apoda. Blind worms. *Ichthyophis.*
Class Reptilia
Order Chelonia. Turtles and tortoises. *Chelydra, Terrapene, Chrysemys.*
Order Rhynchocephalia. Tuatara. *Sphenodon.*
Order Crocodilia. Crocodiles and alligators. *Crocodylus, Alligator.*
Order Squamata. Snakes and lizards. *Crotalus, Natrix, Coluber, Anolis, Phrynosoma.*

Class Aves. Birds
- Subclass Archaeornithes. Primitive birds. *Archaeopteryx* (extinct).
- Subclass Neornithes. Modern birds.
  - Superorder Odontognathae. Toothed birds (extinct).
  - Superorder Paleognathae (Ratitae). Flightless, running birds. Ostrich, rhea, emu, kiwi.
  - Superorder Neognathae. Flying birds. *Columba, Gallus, Passer, Corvus.*

Class Mammalia. Mammals.
- Subclass Prototheria
  - Order Monotremata. Egg-laying mammals. *Ornithorhynchus, Tachyglossus.*
- Subclass Theria. Marsupial and placental mammals.
  - Order Marsupialia. Marsupials. *Didelphys, Macropus, Phascolarctus.*
  - Order Insectivora. Insectivores (moles, tree shrews, etc.). *Chrysochloris, Tupaia.*
  - Order Dermoptera. Flying lemurs. *Galeopithecus.*
  - Order Chiroptera. Bats. *Synotus, Myotis.*
  - Order Primates. Lemurs, monkeys, apes, humans. *Pongo, Pan, Homo.*
  - Order Edentata. Sloths, anteaters, and armadillos. *Bradypus, Dasypus.*
  - Order Pholidota. Scaly anteaters. *Manis.*
  - Order Lagomorpha. Rabbits, hares, pikas. *Lepus.*
  - Order Rodentia. Rodents. *Sciurus, Castor, Mus.*
  - Order Cetacea. Whales, dolphins, porpoises. *Balaena, Phocaena.*
  - Order Carnivora. Carnivores. *Felis.*
  - Order Tubulidentata. Aardvark. *Orycteropus.*
  - Order Proboscidea. Elephants. *Loxodonta, Elephas.*
  - Order Hyracoidea. Coneys. *Dendrohyrax.*
  - Order Sirenia. Manatees (sea cows).
  - Order Perissodactyla. Odd-toed ungulates. *Equus, Rhinoceros.*
  - Order Artiodactyla. Even-toed ungulates. *Bison, Ovis, Bos, Sus.*

# Appendix B

## Milestones in Biology

Listed here are some of the more important highlights in the history of biology. Although the list is lengthy, it is far from complete. In the interest of conserving space, selections had to be made, not all of them to everyone's liking. Moreover, in such limited space we could not begin to represent the explosive increase in significant developments in biology that have occurred since World War II. Most of the scientists who have ever lived on earth are now alive. That fact alone suggests the volume of important research being published today. The Nobel prize referred to in the following accounts is the prize for Physiology or Medicine.

**340 B.C.** *Aristotle* (384–322 B.C.) One of the world's greatest thinkers, Aristotle organized all the knowledge of his time and systematized the study of natural history. He theorized regarding embryonic development and studied development of squids. His books were also to be the only comparative anatomy for many centuries.

**200 A.D.** *Galen of Pergamum* (130–201), Greek physician and anatomist, published 400 works, including a book on human anatomy based largely on a study of the Barbary ape. He was to be the authority on human anatomy and physiology for the next 1000 years.

**400–800 A.D.** During the so-called Dark Ages, the Western world had little interest in material matters; scientific curiosity and progress were generally lacking.

**1543** *Andreas Vesalius* (1514–1564), Italian anatomist, breaks with tradition and makes many of his own observations, many of which he publishes in an excellent human anatomy, *De Humani Corporis Fabrica.*

**1628** *William Harvey* (1578–1657), wealthy English court physician, in a classic monograph, demonstrates that the blood circulates and that the heart is a pump. He employed the first quantitative methods to be used in biological investigations. He concluded that there existed minute vessels that connected arteries with the veins. Capillaries were not seen until four years after his death.

**1648** *Jean Baptiste Van Helmont* (1577–1644), wealthy Belgian philosopher-physician, by quantitative studies on plant growth, shows that plants derive little substance from the soil, thereby demolishing the widely held humus theory. He erroneously concluded that water was the primary source of plant matter.

**1661** *Marcello Malpighi* (1628–1694), Italian microscopic anatomist, who did pioneer descriptions of the tissues of lung, spleen, kidney, liver, and skin, describes capillaries for the first time.

**1662** *René Descartes* (1596–1650), French philosopher and scientist, proposes that animal bodies are mechanisms that can ultimately be explained in terms of physical principles, the only exception being the pineal gland, in which the soul is located.

**1665–1674** *Robert Hooke* (1635–1703) and *Antoni van Leeuwenhoek* (1632–1723), introduce the world to life at the microscopic level. Hooke, English mathematician and physicist, uses the word "cell" (in 1665) to describe the tiny compartments he saw in cork under his microscope. Leeuwenhoek, a Dutch dry goods merchant and lens grinder, "the father of microbiology," was the first (in 1674) to describe red blood cells, spermatozoa, and several species of bacteria and protozoans.

**1662** *Robert Boyle* (1627–1691), British pioneer chemist, shows that animals require a component of air, not just air, to maintain their lives. Oxygen had not yet been discovered.

**1735** *Carolus Linnaeus* (1707–1778), Swedish botanist, popularizes the binomial system of nomenclature, giving substance to the concept of species. His *Systema Naturae* of 1735 contained names and descriptions of all known species of plants and animals.

**1752; 1783** Biologists *René Antoine de Réaumur* (1683–1757), France, and *Lazzaro Spallanzani* (1729–1799), Italy, prove that digestion is neither a purely mechanical process nor merely a process of putrefaction.

**1760** *Joseph Gottlieb Kölreuter* (1733–1806) demonstrates that parental traits are transmitted through both the pollen and the ovules of plants, thereby initiating the experimental study of hybridization that would lead to an understanding of heredity.

**1764** *Albrecht von Haller* (1708–1777), Swiss anatomist and pioneer experimental physiologist, compiles all known physiology in a single 8-volume work.

**1774** *Joseph Priestley* (1733–1804), minister, shows that plants "restore common air" that had been "consumed" by a burning candle or a respiring mouse.

**1777** *Antoine Laurent Lavoisier* (1743–1794), French scientist, proves that respiration involves an intake of oxygen and elimination of carbon dioxide. Later (1780) he would show that respiration and combustion are similar processes.

**1791** *Luigi Galvani* (1737–1798) and later, *Allesandro Volta* (1745–1827), find a connection between nerve and muscle function and electricity, the beginning of a mechanistic approach to nerve and muscle physiology.

**1798** *Edward Jenner* (1749–1823), British physician, demonstrates that contraction of the mild disease cowpox provides immunity to smallpox and that deliberate innoculation with cowpox (vaccination) can confer immunity to smallpox.

**1809** *Jean Baptiste de Lamarck* (1744–1829), pioneer French evolutionist, in his *Philosophie Zoologique*, emphasizes the fundamental unity of all life, the capacity of species to vary (evolve), and the effect of environmental influences upon the evolutionary process (the inheritance of acquired characteristics).

**1809–1838** The Cell Theory is formulated. The findings of numerous biologists, *Lamarck* (1809), *R. J. H. Dutrochet* (1824), *Meyen* (1830), *Robert Brown* (1831), *Dumotier* (1832), *Hugo von Mohl* (1835–39), *Matthias Schleiden* (1838), and *Theodor Schwann* (1839), among others, establish the cell as the fundamental unit of structure and function of organisms. Largely because of the attention attracted to their erroneous theory of cell reproduction, the Germans, Schleiden, a botanist, and Schwann, a zoologist, were mistakenly given full credit for the theory.

**1817** *Georges Léopold Cuvier* (1769–1832), French zoologist and geologist, founds the sciences of comparative anatomy and paleontology. Famous for his theory of catastrophism, Cuvier remained staunchly anti-evolutionist all his life.

**1827** *Karl Ernst von Baer* (1792–1876) describes the mammalian ovum for the first time, nearly 200 years after spermatozoa were described.

**1828** *Wöhler* synthesizes urea, the first artificial synthesis of an organic compound, thereby casting doubt on the theory of vitalism.

**1833** *William Beaumont* (1785–1853), American Army surgeon, publishes the results of years of study on gastric digestion as observed in a patient with a gastric fistula. He establishes beyond all doubt that digestion is principally a chemical process.

**1848** *Emil du Bois Reymond* (1818–1896), Prussian physiologist, shows that nerve impulses have an electrical nature, findings that argue against the theory of vitalism, widely held at the time.

**1850** *Hermann Ludwig Ferdinand von Helmholtz* (1821–1894), German physicist and physiologist, with instruments of his own devising, measures the rate of nerve impulse conduction, showing that it was not merely a passage of current. He also established the nature of hearing and the function of parts of the eye.

**1855** *Claude Bernard* (1813–1878). French physiologist, lays the basis for much of modern mammalian physiology. His investigations, such as those in 1855 that demonstrated the role of the liver in glycogen formation and storage, were important to modern understanding of digestion, circulation, and metabolism. Earlier (1851) he had propounded the concept of the constancy of the animal body's "milieu interieur."

**1858** *Rudolph Virchow* (1821–1902) demonstrates that, as far as can be determined, every cell is a product of another cell ("Omnis cellula e cellula").

**1861** *Paul Broca* (1824–1880), French physiologist, identifies a speech center in the cerebral cortex, prompting a large number of such localization studies by others.

**1859** *Charles Darwin* (1809–1882), English biologist, shows how natural selection is the primary shaper of organic evolution. His *On The Origin Of Species By Means Of Natural Selection* of 1859 was to revolutionize biology in all its aspects.

**1862** *Louis Pasteur* (1822–1895), French chemist and microbiologist, establishes the bacterial origin of many infectious diseases, disproving the spontaneous generation theory for microbes as the Italian, Francesco Redi had done for flies and maggots in 1668. He also developed vaccines against rabies and anthrax and discovered the staphylococcus and pneumococcus bacteria.

**1866** *Gregor Johann Mendel* (1822–1884), Austrian scientist and Augustinian monk, discovers basic laws of genetics through experiments on garden peas, results of which he analyzed statistically.

**1867** *Joseph Lister* (1827–1912), British surgeon, introduces antisepsis into surgery and drastically reduces mortality due to infections.

**1875** *Oscar Hertwig* (1849–1922), German embryologist, shows that the sperm head becomes the male pronucleus and unites with the female pronucleus to form the zygotic nucleus.

**1879; 1883** *Oscar Hertwig* and brother *Richard*, establish the germ layer theory on a solid basis, building on a suggestion made by von Baer (see 1827) in 1829.

**1878** *Robert Koch* (1843–1910), German microbiologist, invents the substage condenser, staining techniques, and methods of pure culture of bacteria that would be essential for modern bacteriology. By developing rules ("Koch's postulates") for proving the causal relationship between specific bacteria and the diseases they produce, he shaped bacteriology into an exact science. He also found the cause of anthrax, tuberculosis, cholera, and other diseases.

**1880** *Alphonse Laveran* (1845–1922), French physician, describes a parasite in the blood of malaria patients. No one believed him.

**1880–1890** Employing improved microscopy and staining techniques, *Walther Flemming* (1843–1905), *Eduard Strasburger* (1844–1912), *van Beneden* (see 1883), and others elucidate the details of cell and nuclear division and emphasize the equal distribution of chromosomes to daughter cells in mitosis.

**1882** *Th. W. Engelmann*, German biologist, determines the relative effectiveness of different wavelengths of light in photosynthesis and shows that oxygen is produced in the region of the chloroplasts.

**1883** *Edouard van Beneden* (1846–1910) demonstrates that the sperm and the egg are haploid, that the zygote is diploid, and that the chromosome number is reduced to the haploid number during gametogenesis in animals.

**1884** *George John Romanes* (1848–1894), pioneer British ethologist, places the study of animal behavior on an evolutionary and comparative basis, drawing on contributions made by Darwin in 1859, 1868, 1871, and 1873.

**1884** *Elie Metchnikoff* (1845–1916) Russian biologist, shows that phagocytic blood cells are involved in the defense of the body in several animal species.

**1888** *Wilhelm Roux* (1850–1924), German embryologist, performs the first surgical experiment on an embryo, the beginning of what he would later (1905) call the science of experimental embryology. He later founded an important journal ("Roux's Archives").

**1890** *Emil von Behring* (1854–1917), Prussian army physician, makes the first injection of immune serum as a way of conferring passive immunity (to diphtheria toxin), the first example of specific therapy for a disease in modern times.

**1893** *Theobald Smith* (1859–1934), American parasitologist and immunologist, becomes the first to demonstrate the transmission of a disease-producing protozoan by an arthropod (Texas cattle fever by ticks). Earlier, *Patrick Manson* and others had implicated arthropods as vectors of filarial worm and tapeworm infections.

**1897** *Eduard Buchner* (1860–1917), German biochemist, excites the scientific world by obtaining active enzymes from yeast in a cell-free preparation, thereby proving that fermentation is a chemical process, not an effect of some mysterious life force.

**1897** *Christiaan Eijkman* (1858–1930) and *Frederick Gowland Hopkins* (1861–1947), establish dietary requirements for vitamins. Eijkman, a Dutch bacteriologist, proved (in 1897) that polyneuritis could be produced in chickens by feeding them incomplete diets similar to those being fed to convicts suffering from a similar disease, beri-beri. Hopkins, in further animal experiments, identified several other essential dietary factors later to be named vitamins.

**1898** *Ronald Ross* (1857–1932), British physician proves the role of *Culex* mosquitoes in transmission of bird malaria and describes all stages in the mosquito.

**1899** *Battista Grassi* (1854–1925) implicates the *Anopheles* mosquito in the transmission of two species of human malaria and describes all the stages in the mosquito.

**1900** *Hugo Marie de Vries* (1848–1935), Holland, *Karl Correns*, Germany, and *E. von Tschermak* (1871–1962), Austria, independently discover Mendel's laws and search the literature for any similar work. All come upon Mendel's 1866 paper. Therefore 1900 is often regarded as the birth year of the science of genetics.

**1900–1951** The rise of modern ethology (see 1884). Its contributions are summarized (1951) in *The Study of Instinct* by Dutch biologist, *Nikolaas Tinbergen* (1907– ). Earlier, the Austrian, *Konrad Zacharias Lorenz* (1903– ), gave the study its name of ethology and contributed many concepts. He emphasized (1937) the phenomenon of imprinting, first described by Heinroth in 1910. In the 1940s and later, the Austrian, *Karl von Frisch* (1886– ) described the sense and language of bees.

**1901** *Karl Landsteiner* (1868–1943), Austrian-American biologist, pioneers in human blood typing.

**1901** *Walter Reed* (1851–1902), American Army physician, identifies the *Aedes aegypti* mosquito as the vector of yellow fever. Years of search for the causative agent of the disease led to the conclusion that it was due to a "filterable virus."

**1902** *Ivan Petrovich Pavlov* (1849–1936), Russian physiologist, develops the concept of the conditioned reflex in experiments on dogs.

**1902** *William Maddock Bayliss* (1860–1924) and *Ernest Henry Starling* (1866–1927), British physiologists, discover "secretin," which is released into the circulation by the intestinal mucosa in response to stimulation by hydrochloric acid from the stomach. It promotes pancreatic secretion, previously believed to be under nervous control. In 1905 Starling introduced the word *hormone* ("exciter") for such substances.

**1906** *Charles Scott Sherrington* (1861–1952), British pioneer neurobiologist, publishes his landmark monograph, *The Integrative Action of the Nervous System*, a systematic analysis of the nervous system's reflexes.

**1907** *Ross Granville Harrison* (1870–1959) and later, *Alexis Carrel* (1873–1944), American biologists, develop reliable, well-defined techniques for culturing tissues outside the animal body.

**1909** *Archibald E. Garrod* (1857–1936), British physician, describes several "inborn errors of (human) metabolism," the first accounts of biochemical mutants.

**1909** *Santiago Ramon y Cajal* (1852–1934), Spanish neuroanatomist, formulates the neuron theory which holds that nerves are made of discontinuous cells, or neurons, that are separated from each other by minute spaces, the synapses.

**1910** *Paul Ehrlich* (1854–1915), after testing 605 drugs, develops a combination, "salvarsan," which specifically kills the syphilis spirochete, with little harm to the patient, the first such "magic bullet."

**1911** *Thomas Hunt Morgan* (1866–1945), American embryologist and geneticist, shows that genes occur in linear array along the lengths of chromosomes and that recombinations of linked genes depend on chromosome breaks occurring between them.

**1921** *Otto Loewi* (1873–1961), German-American pharmacologist, discovers that release of acetylcholine ("Vagusstoff") by the vagus nerve slows the heart, thereby elucidating its role as a neurotransmitter.

**1922** *Frederick Grant Banting* (1891–1941) and *Charles Herbert Best* (1899–1978), Canadian physiologists, discover insulin and, in collaboration with the Scottish physiologist, *John J. R. MacLeod,* isolate it from the pancreas.

**1924** *Hans Spemann* (1869–1941), German experimental embryologist, with Hilde Mangold, publish results confirming his earlier (1921) description of the inductive capability of the dorsal lip of the blastopore in the newt, site of what he termed the "organizer."

**1924** *Joseph Erlanger* (1874–1965) and *Herbert Spencer Gasser* (1888–1963), American neurophysiologists, adapt the cathode ray oscilloscope for measuring nerve potentials, thereby elucidating the nature of the action potential.

**1926** *Friedrich August Went* (1863–1934), Dutch plant physiologist, extracts "auxin," a diffusible substance, from plant tissue. It produces bending in growing oat seedlings.

**1927** *Hermann Joseph Muller* (1890–1967), American geneticist, proves that mutation rates can be increased by ionizing radiations.

**1929** *Alexander Fleming* (1881–1955). Scottish bacteriologist, discovers penicillin, the first known of the antibiotics.

**1930** *Otto Heinrich Warburg* (1883–1970), German biochemist, discovers cytochrome oxidase, the terminal enzyme in the respiratory chain, and demonstrates its role in cellular respiration.

**1937** Final elucidation of the tricarboxylic acid cycle is made by *Hans Adolph Krebs* (1900–1981), German-British biochemist, following pioneer work by *Albert Szent-Györgyi* (1893– ), Hungarian biochemist, in 1935.

**1939** *Robert Hill,* British biochemist, shows that oxygen production is a separate reaction from carbon dioxide fixation in photosynthesis and that it depends upon the presence of an electron acceptor (the "Hill reaction").

**1941** *George Wells Beadle* (1903– ) and *Edward Lawrie Tatum* (1909–1975), American geneticists, show that many genes achieve their effects by regulating the synthesis of specific enzymes. They shared the 1958 Nobel prize with American *Joshua Lederberg* (1925– ), who found that bacteria conjugate and exchange DNA.

**1941** Using a heavy isotope of oxygen, *Samuel Ruben* and coworkers, American biochemists, prove that the oxygen produced in photosynthesis is derived from water, not carbon dioxide.

**1944** *Oswald T. C. Avery* (1877–1955), *Colin Munro MacLeod* (1909– ) and *Maclyn McCarty,* American biochemists, prove that DNA is the hereditary material.

**1945** *Melvin Calvin* (1911– ) and *Andrew A. Benson,* American biochemists, establish the nature of the first product formed when carbon is fixed in photosynthesis by joining carbon dioxide to ribulose biphosphate, thereby elucidating what came to be called the Calvin cycle.

**1948** *Frank Macfarlane Burnet* (1899–1985), Australian physician, who earlier had developed means of immunization to influenza, proposes the concept of acquired immunological tolerance, later to be confirmed by the work of British physician *Peter Brian Medawar* (1915– ).

**1950** *Barbara McClintock* (1902– ), American geneticist, describes examples of transposable genes in maize and other genes that control their transposition. Her work was little appreciated until long after similar transposons had also been established in prokaryotes.

**1950** *Bernhard Katz* (1911– ), British biophysicist, discovers the mechanism for the release of acetylcholine at the myoneural junction.

**1952–1957** Physiologists *John Carew Eccles* (1903– ) (Australian), *Alan Lloyd Hodgkin* (1914– ) and *Andrew Fielding Huxley* (1917– ) (British), discover basic principles of nerve cell excitation and impulse conduction.

**1953** *James Dewey Watson* (1928– ), *Francis Harry Compton Crick* (1916– ), and *Maurice Hugh Frederick Wilkins* (1916– ), American biochemist and British physicists, respectively, establish the molecular structure of DNA.

**1957–1960** British biologists, *Andrew F. Huxley* (1917– ), *R. Niedergerke, Hugh E. Huxley,* and *Jean Hanson,* develop the sliding filament model of muscle contraction.

**1959** *Frank M. Burnet* (see 1948) develops the clonal selection theory of immune response.

**1963** *Robert William Holley* (1922– ), *Har Gobind Khorana* (1922– ), and *Marshall Warren Nirenberg* (1927– ), American biochemists, "crack" the genetic code.

**1976** *Georges Köhler* and *César Milstein,* British immunologists, by fusing mouse leukemic cells with lymphocytes, develop cell clones, each of which produces a single type of antibody (monoclonal antibody), thereby providing a powerful research tool to immunology.

# Appendix C

## Review Questions

### CHAPTER 1

1. Why is biology so interested in the history of life? Why should a modern science look backwards when all interest is focused on what can be learned now and in the future?

2. Is it at all possible to devise experiments today examining the origin of life or must such questions as the origin of reproduction be left to philosophical speculation?

3. Taxonomy is not a musty, antiquarian science at all, but a lively, practical concern of many contemporary biologists. For example, discuss how medical diagnosis of infectious agents (virus, bacteria, etc.) is a study of taxonomy. Use examples from current medical news stories.

4. When you learn the scientific name of an organism (e.g., a toad is *Bufo fowleri* and we are *Homo sapiens*), how much do you know about the organism? How much information is contained within this name? Do you need to understand Latin in order to fully understand the binomial names of species?

5. In the broadest view, evolution itself has evolved over the 4.7 billion year history of this planet. Starting with chemical evolution and continuing the story up to present day examples of evolution, what kinds of evolution have evolved and where in general have these processes led?

6. "Emergent properties" or "the whole is greater than the sum of the parts" may sound mysterious at first, but in chemistry and biology it is a very down-to-earth, everyday way of thinking. Propose and discuss several ordinary examples of emergent processes.

7. Darwin's theory came rather late in the history of science. Far more difficult and counter-intuitive theories were proposed and accepted in preceding centuries (e.g., Newton's laws and his calculus). Why was science "ready" for a theory of evolution by natural selection only in the mid-nineteenth century?

8. The young Darwin was an adventurous explorer, but after his voyage on the *Beagle*, he never traveled again. His life thereafter was physically unadventurous. This sharp change in life style coincided with his realization that species change and that he might be able to discover the mechanism for such change. Do you think that his youthful travels were necessary for his breakthrough solution of the question?

9. Most nonbiologists do not view the world through Darwin's theory and most people can regard the natural world without ever noticing natural selection. Biologists, however, organize virtually every observation of nature into an evolutionary framework with natural selection as the primary operating principle. Why isn't natural selection obvious to every observer of nature? Is there some aspect of natural selection that is difficult to observe?

10. In reading over the basic components of Darwin's theory of natural selection (pg. 1–13), put life into this argument by giving examples of points 1 and 3 which can be observed in a woodlot or an abandoned field.

11. Biological museums maintain collections of "type specimens" for species of all kinds. In most cases, these type specimens were chosen by the taxonomist who originally named and described the species and these specimens are supposed to exhibit the species' characteristics in ideal form. If you are to be absolutely certain that you are identifying a species correctly, you need to journey to the museum containing the type specimen for your alleged species and compare your specimens with the standard. This is a very important function of museums but, paradoxically, it is based on a mistaken view of evolution. Why is the designation of type specimens naive in light of our understanding of evolution?

12. The history and present diversity of life is usually diagrammed as a "phylogenetic tree." This is far from an adequate representation of the entire history of life. However, given a tree, how do you interpret such parameters of this kind of figure such as tree height, foliage width, branchpoints, trunk, and roots?

13. Since we term large subdivisions of biological diversity ***kingdoms***, do we imply some sort of political hierarchy or form of governance within the vast array of living organisms? In these kingdoms are there kings, queens, bishops, knights . . . or representatives of various levels of organization?

14. Teleological thinking, a virtual taboo within biology, is difficult to avoid because human language, in its very grammar, is shot through with teleological assumptions (e.g., "the sun has come to warm our lands" attributes intention to the sun). Usually, the confusion arises over what is cause and what is effect. Do trees lose their leaves in preparation for winter or do prewinter climatic conditions cause trees to lose their leaves? Clearly, the latter has it right; cause precedes effect. Discuss the following three natural processes, expressing them first teleologically and then correctly describe them in mechanistic, cause-and-effect terms.

a) Robins migrate south in autumn, north in spring; this has something to do with temperature.

b) When jogging, the heart beats faster and muscles receive more blood; the two events are clearly related.

c) Everywhere you look, natural species are exquisitly adapted to their environments. Evolutionary change seems to be for the best. What is the relationship between evolutionary change and near perfect quality adaptation to the environment?

15. Whenever a change is observed in a species, one of the first hypotheses to be tested is that of natural selection. Indeed, evolution by natural selection is a predominant concept in biology because it provides a testable hypothesis by which we can understand biological change. For example, when it is found that a strain of pathogenic bacteria has suddenly become resistant to antibiotics, e.g., a *Salmonella* strain which is unaffected by penicillin, how would you design an experiment to find out whether this change is the result of natural selection?

## CHAPTER 2

1. Did life originate on this planet? One source of evidence in examining this question is the relative abundance of elements at the earth's surface. Considering the major elements of which all living systems are composed, is it likely that the earth's surface could have been the chemical source for life's origin?

2. The astronomer Robert Jastrow proposes that our species is in the midst of performing a major evolutionary change in which carbon-based life is transforming itself into silicon-based intelligence which at some future point will become the dominant "life" form in the universe. Develop this argument further by outlining what is meant by carbon-based life and silicon-based intelligence and comment on whether it is conceivable that the latter could ever supplant the former.

3. A long-standing classic of biological chemistry is L. J. Henderson's *The Fitness of the Environment* (1913) which carefully examines the many physical and chemical factors that make this plant hospitable to life. For example, consider the Earth's temperature regime—examining the nature of chemical bonding in living systems and the physical properties of water, how is temperature on this planet optimal for life's chemical processes?

4. Although physicists are rapidly completing an inventory of the subatomic particles such as quarks and confidently looking forward to a complete description of matter and energy, the study of biological mechanisms seems to be largely unaffected by the new particle physics. From your present understanding of basic chemistry as it applies to biology, why is the physics of atomic nuclei of little importance in understanding biological mechanisms?

5. One of the most important scientific works of the twentieth century is *The Nature of the Chemical Bond* by Linus Pauling. The measurement of bond distance, angle, and energy has had an enormous influence on biology. Why should the study of life be at all interested in the properties of chemical bonds?

6. Imagine a small organic molecule with 4 corners to it and then imagine that 100 of these molecules are linked together into a chain so that each member is covalently bonded to a molecule on each side of it using 2 of its corners in attaching to its neighbors. Only the molecule on either end of the chain has 3 free corners. Discuss the shapes this chain can assume if:

a) all of the subunit molecules are identical

b) a few of the subunit molecules carry a positive ionic charge, a few are negatively charged, and the rest are uncharged

c) subunits 1–50 are charged (either + or −) and subunits 51–100 are hydrophobic and the molecule is immersed in water

7. Water is both one of the simplest substances known and at the same time quite complex. It is still beyond the powers of science to predict all the crystal configurations which water will assume under various freezing conditions. Water is a perfect example of "emergent properties" (see Panel 1-2). Knowing something of the properties of hydrogen and oxygen, can you point out the properties which "emerge" when these atoms are bonded to form $H_20$?

8. Frozen ammonia ($NH_3$) is denser than liquid ammonia. On an oxygen-free planet where ammonia is the principal liquid, why is life unlikely to evolve (even though we have on this planet a number of $NH_3$-loving bacteria)?

9. Chemical reactions can be classified in respect to energetics as either exergonic or endergonic. Drawing examples from ordinary experience, propose two analogies for each type of reaction. In addition, as you write your essay are you providing examples of each type of reaction?

10. Protein molecules can function only within a narrow pH range. Realizing that the function of a protein (e.g., the oxygen transporting function of hemoglobin) depends on the shape of the protein molecule and that the shape of a protein depends on hydrogen bonding between regions of the molecule, can you explain why changes in pH in either direction from the optimal pH will interfere with the function of any protein?

11. Many desert animals can live an entire lifespan without ever drinking water. These animals need water just like any other animal but are able to obtain all they need from biochemical reactions. From your current knowledge of biochemical reactions, how is this accomplished?

12. By far the commonest form of biochemical reduction is the addition of a hydrogen atom. For example, oxygen is fully reduced in water. Given the definition of oxidation as a loss of electrons, why is the addition of a hydrogen atom a reduction? Is removal of a hydrogen atom an oxidation? What of removing a proton?

13. Discuss the ways in which the dipolar nature of water results in the biologically critical properties of capillarity, high heat of vaporization, and high heat capacity.

14. Alcohols are any organic molecules bearing an -OH group. The hydrogen atom of an alcohol group readily forms hydrogen bonds (just as does H in $H_2O$). The simplest alcohol, methyl alcohol ($CH_3OH$), is lethal to cells; the next simplest alcohol, ethyl alcohol ($CH_3CH_2OH$), is the alcohol of human consumption, but it too is toxic unless highly diluted. Both of these simple alcohols are used to "fix" tissues for microscopic work; that is, these alcohols terminate all biological activities while preserving much of the cellular structure of the tissues. From your present knowledge of biochemistry can you explain why simple alcohols are such good tissue fixatives?

15. The periodic table of elements can be read as a record of the *evolution* of atoms. Hydrogen and helium were formed in the initial Big Bang, while the other atoms evolved from these two lightest elements within stars ("We are made of star stuff"). What is the principle of evolution of atoms and what is the selective force? Are atoms still evolving?

## CHAPTER 3

1. The major biological macromolecules are polymers, long chains of subunits. What are some of the reasons that linear polymers are so common in living systems?

2. Surveying biomolecules gives us some glimpses into the earliest evolution of life. For example, the primary structure of proteins and nucleic acids, suggests how macromolecules may have evolved once the small, organic subunits had accumulated by chemical evolution.

3. Polysaccharides are used in two different capacities—storage and structure. Is this true of lipids? nucleic acids?

4. Which sugars are sweet and what might be the evolutionary reason for animals possessing a separate taste sense for sweetness?

5. In early American history, the principal energy source for lighting was lipid—tallow candles and whale oil lamps. Considering the chemistry of lipid, why were these animal lipids used for lighting rather than polysaccharide fuel such as cellulose or chitin?

6. Understanding the forces which hold lipid micelles together in aqueous environments, can you suggest how the lipid bilayer structure of cell membranes may have evolved in the earliest phases of life's history?

7. Referring back to the discussion of weak, noncovalent bonds in chapter 2, discuss the crucial importance of such bonds in producing the structure of biological macromolecules.

8. Why is the three-dimensional shape of macromolecules so essential to the biological function of the molecules?

9. How do biological systems recognize isomers such as glucose and fructose as different molecules and process them differently? Similarly, why are cellulose and glycogen, molecules of identical atomic composition, treated so differently by organisms?

10. Proteins are the liveliest molecules—most of them *move* in carrying out their functions. Discuss how a protein molecule can move. Do you see how the movements of cells and the contraction of muscles can evolve from such protein activities?

11. With but minor exceptions, all enzymes are proteins. Why has this critically important function evolved only within polypeptide chain molecules and not within polysaccharides, lipids, or DNA?

12. The many functions of proteins beautifully illustrate the concept of emergent properties (see Panel 1-2). How do the functions of proteins "emerge" as greater than the functions of their separate amino acids?

13. Although we can separately designate four levels of protein structure, all are decreed by the primary structure. How can mere sequence of amino acids be responsible for structuring a complex, three-dimensional molecule such as hemoglobin?

14. If enzymes had never evolved, what would be the condition of life today?

15. Throughout the history of biology, structure and function have been the twin goals of investigation. At the level of macromolecules, structure and function are virtually identical. Choosing examples from proteins and nucleic acids, examine what is meant by a molecule's structure and function. Are they synonymous?

## CHAPTER 4

1. The cell theory proposes that the cell is the unit of life. Discuss what this means in terms of how organisms function.

2. Biology is a most visual science. Modern cell biology is attempting to map biochemical functions within the structure of cell organelles. Explain with examples how it is possible to visualize where several biochemical activities occur within the subcellular structure of a cell.

3. Transmission electron microscopy has revealed that cells are complexly structured internally. In general, what are the basic principles of this structuring? In what way is the cell far different from an aqueous solution?

4. An enormous research effort has been invested in developing separation techniques for purifying cellular organelles and macromolecules. Why is separation science a major component of modern biochemistry?

5. How have the techniques of electron microscopy and of biochemistry been combined to work out the structure and functions of membranes?

6. Many animal cell types are predatory, having the ability to capture food particles and digest them through an intracellular digestive system. Describe this process and note the intracellular equivalents of a mouth, a gut, digestive enzymes, absorption sites, and an anus.

7. One of the larger facets of modern biology can be described as "membrane biology", the study of the many functions of membranes. It is worthwhile to set down a list of the many essential functions performed by membranes and to add to the list as your knowledge expands. Are there any common principles which unite the diverse functions of membranes?

8. Virus particles are quite inert; they have no energetics and yet once they encounter a vulnerable cell surface they readily get inside the cell. How can a mere "lump" of protein and nucleic acid get itself inside cells?

9. Membranes are in continuous interaction with vesicles. Membranous sheets proliferate spherical bubbles and vesicles merge and coalesce with sheets. How does this dynamic interaction operate and what are some of the major functions which it accomplishes?

10. Why is the current model of biological membranes called the "fluid mosaic" model? What is fluid and in what sense is it fluid? What is mosaic? How dynamic is the relationship between mosaic and fluid?

11. Of the major cell functions, which ones are coupled to the cell cycle and occur only periodically? Which major functions are not time-linked to the cell cycle? Is there a general rule as to which types of cell functions are timed by the cell cycle?

12. Construct a table showing the combinations of cell structures which are unique to organisms in each of the five kingdoms. Are these characteristics diagnostic, i.e., could these structural differences serve to classify an organism as to kingdom?

13. Why is mitosis so intensely studied by cell biologists? Is it all that important to understanding life to figure out how cells divide? Wouldn't it be more efficient to focus all our research on the undividing cell and learn how cells operate before worrying about how one cell makes two cells?

14. Cytokinesis appears to be far less quantitative in dividing cytoplasmic components between daughter cells compared to mitosis which precisely apportions chromosomes between the two resulting cells. Why is there such a close accounting of chromosome distribution while one daughter cell might inherit more cytoplasmic equipment than the other?

15. Given the extensive evidence that mitochondria and chloroplasts might have evolved from ancient prokaryote invaders of eukaryotic cells, what evidence argues against other organelles such as microtubules, Golgi complex, and lysosomes being descendants of invading organisms?

## CHAPTER 5

1. Each human red blood cell contains some 300 million molecules of hemoglobin. Are there osmotic reasons why evolution favored the containment of hemoglobin within cells rather than allowing hemoglobin to be directly dissolved in the blood?

2. If you could visualize osmosis, seeing the solute and solvent molecules as individual particles, what would an osmotic gradient look like? If you could see only a short distance, maybe 10 molecules in any direction, could you be certain you were witnessing an osmotic gradient?

3. The force of osmotic pressure is dramatically illustrated by plant root growth. Explain how osmotic pressure can "plow" the soils and "build" upward against the force of gravity.

4. Osmotic pressure for all its force is just a statistical result of random, kinetic activity. How can random, nondirectional molecular movement result in a strong directional force?

5. Sodium is the principal cation of the ocean, but potassium is the major cation of soil water. How does this ionic difference compare with the extracellular and intracellular cation environments of animal cells? What does this suggest about the origin of life?

6. Habitually drinking hypertonic sea water or, at the other extreme, drinking doubly-distilled water would not be a good idea. Why must drinking water avoid such osmotic extremes?

7. A large fraction of an organism's energy expenditure is utilized in transporting materials across membranes. How many different energy-requiring transport mechanisms can be tabulated? Can you organize this list into functional categories?

8. Why do membranes "go to so much trouble" to separate ions? Wouldn't it make better sense for intracellular mechanisms to adapt to extracellular ionic conditions and save all the effort of moving ions around?

9. Biological membranes exhibit many cases of membrane fusion in which the membranes of different structures are able to coalesce. Describe several cases of this and speculate on how this operates at the molecular level.

10. What is the significance of transporting materials against a concentration gradient? Could cells survive without this capability?

11. From your understanding of passive and active transport across the plasma membrane, do you think that similar processes occur in intracellular membranes?

12. There are some large animals lacking a circulatory system. For example, coral colonies. Close inspection shows that such animals are composed of very thin living tissues stretched over noncellular materials (jellyfish are another example). The absence of an internal circulatory system restricts the movement of water, gases, ions, and important molecules to diffusion and membrane transport. Explain why this condition limits such animals to an anatomy based on cellular sheets.

13. Bulk transport is relatively unselective and, in conveying a sample of the external medium into the cell, brings in all sorts of materials into the carefully controlled intracellular environment. How could a cell protect itself from being "swamped" by all the unneeded, miscellaneous material that is internalized by bulk transport?

14. The plasma membrane is life's indispensable frontier. Any puncture or breach of the plasma membrane is fatal. From your study of membrane transport phenomena can you speculate why the integrity of the membrane is so essential to a cell's survival?

15. In addition to its many transport functions, the plasma membrane is also a "sensory" surface monitoring the chemical environment surrounding the cell. Some of this capability is directly associated with transport functions. Speculate on ways that membrane transport mechanisms also serve to measure and respond to the immediate chemical environment of the cell.

## CHAPTER 6

1. Why are the two laws of thermodynamics *laws* while the central organizing concept of biology, evolution, is labeled a *theory*? Does this mean that evolution is of lower credibility than thermodynamics?

2. Biological systems are continuously *transducing* energy via many different mechanisms. From your understanding of energy metabolism, describe several cases in which one form of energy is transduced into another form of energy.

3. As you read this you are losing energy and, at the risk of losing even more, describe how you are losing energy; biochemically speaking, is it worth it?

4. Energy flows through food chains. In fact, the organisms we find in any natural environment are linked together by various patterns of such energy flow. Choosing any natural or man-made environment where organisms live, describe from start to finish how energy flows through the various food chain.

5. No energy transfer is 100 percent efficient. In the biological energy transfers that occur in food chains, what sort of efficiency would you predict? Do you think that human agriculture and the American diet are efficient energy transfers?

6. The second law of thermodynamics has been described as "time's arrow" because it infallibly indicates the direction along which change can occur. Can you give examples of this insight in terms of everyday experience?

7. Astrophysicists visualize the origin of the universe as a "big bang" exploding from a very small point. As to whether the universe will continue to expand or will reach a point from which it begins to contract back to a central point is a current uncertainty. How do the laws of thermodynamics fit into this broadest of views of the universe?

8. What is the relationship between entropy and biological information as stored in nucleic acids and proteins? Does it require more energy to construct information-bearing molecules than other molecules?

9. Why is ATP the principal energy-carrying molecule in all living organisms? Considering the stupendous diversity of organism from bacteria to butterflies, etc., why have not all sorts of different energy-carriers evolved?

10. Understanding energy metabolism immediately reveals that many enzymes require cofactors or coenzymes to perform their function. Give two examples of essential cofactors and speculate on why these cofactors are not permanently built into the enzymes.

11. The aerobic metabolism of glucose generates exactly the same chemical end-products and the same amount of energy as when glucose is burned. The living "combustion" of glucose, however, is characterized by many small steps. Why does life divide glucose oxidation into so many fine substeps?

12. If you could visualize the Krebs cycle at a magnification where a glucose molecule was the size of a baseball, what would the Krebs cycle look like? Would it look something like a merry-go-round?

13. Each step of glycolysis and of the Krebs cycle is catalyzed by a specific enzyme. How do so many different enzymes work together? What is the principle underlying the cooperation of enzymes?

14. Considering the chemiosmotic theory, how does the presumed energy coupling function of mitochondrial inner membrane compare with the transport function of plasma membrane? Do you suspect an evolutionary relationship here?

15. One of the largest assumptions of modern biology is that life in all its properties can be explained in terms of chemical reactions. We may be far from translating the workings of our minds to chemical equations, but the progress made so far on this biochemcial explanation of life encourages boldness. As a

measure of this view of life, how do modern biologists explain the "vital force" that turns unliving matter into living activity?

## CHAPTER 7

1. How would you explain photosynthesis to a child? Start with the notion that plants weave light and air and water into sugar and try to include as much real fact as possible in this magical tale.

2. The research history of photosynthesis strongly illustrates the necessity of choosing from the vast diversity of organisms the ideal organism for study. Even though our practical goal may be improving agriculture, crop plants have not often been the organisms used in the major studies of photosynthesis. Give several examples of breakthroughs in this area which depended on using diverse photosynthetic organisms.

3. Plants produce oxygen and carbohydrates not for the benefit of animals, but for their own selfish reasons. What are these reasons?

4. From our present biochemical knowledge concerning photosynthesis, glycolysis, and aerobic metabolism, how might these metabolic pathways have arisen in evolution? Is it conceivable that parts of modern photosynthesis may have evolved independently?

5. Chloroplasts, just like mitochondria, have many characteristics suggesting that they may have evolved from prokaryotes which at some ancient date invaded eukaryote cells. From the point of view of their metabolism, could chloroplasts have once existed as independent organisms?

6. Starting with Priestley, photosynthesis has been measured by changes in gas volume within an enclosed chamber. However, an increase in gas volume produced by an illuminated plant in such an enclosure is not a direct measure of oxygen production, but must be corrected for a number of factors. How do you determine oxygen production from gas volume data?

7. The equation usually presented to illustrate photosynthesis (see page 188) is an extreme simplification, omitting all of the actual reactions. Draw up a better representation of photosynthesis; include the major reactions, but do not overwhelm the reader with all the details. Try to do this on a single sheet of paper.

8. Before the advent of life on this planet, what were the planet-wide patterns of light absorption and emission? How has life modified the energy budget of earth?

9. Both aerobic respiration and photosynthesis are based on enzymatic pathways that are cycles. What do the cycles accomplish in these two major domains of metabolism? Are there features common to them both?

10. Why is the theory of chemiosmosis as important in understanding photosynthesis as it is in explaining respiration?

11. The further we explore cell biology, the more significance we discover in the structure and function of membranes. How does the fluid mosaic model of membrane structure help explain how chloroplast membranes capture light and generate energy?

12. Wouldn't just one photosystem be sufficient? Is there any reason why two photosystems have evolved?

13. Considering the variety of two-dimensional shapes exhibited by leaves of various species, why is there no diversity in the third dimension? Leaves are always flat (or in the case of conifers, needle-shaped). Is there some functional limitation on the depth of leaf structure?

14. How should we regard the evolution of photorespiration? It is a waste of energy without any benefit. Does this mean that inferior, inefficient organisms can result from natural selection or could there be some physical reason that oxygen must interfere with $CO_2$ fixation, leaving photorespiration as the best possible compromise?

15. How has photosynthesis affected the evolution of life on this plantet? If photosynthesis had never evolved, would life have persisted? What if photosynthesis had been confined to small populations of organisms, e.g. the bacteria of hot springs—would the present diversity of organisms have evolved? Estimate the dimensions of life's debt to this energy—recruiting mechanism.

## CHAPTER 8

1. In a very few instances, we find large animals which lack respiratory and circulatory organ systems. The Portugese man-of-war and some other coelenterates have tentacles reaching over 50 feet and though these are big animals they have no system for transporting oxygen to their tissues. Referring to pages 222–23, explain how these animals obtain oxygen.

2. The area of an animal's respiratory surface is usually directly proportional to the body weight of the animal. What does this suggest about the mechanism of gas exchange across a respiratory surface?

3. Historically, pneumonia was the leading cause of death until antibiotics were developed. Pneumonia, if unchecked, is still lethal due to filling of the alveoli with fluid in reaction to a bacterial infection. Why is fluid so dangerous in the alveoli and why are there no other treatments for pneumonia other than antibiotics?

4. The oxygen dissociation curve for hemoglobin (Fig. 8-29) is sigmoid (S-shaped)—somewhat like a ski slope. Discuss the meaning of this type of curve by imagining what it is like to be an oxyhemoglobin molecule "riding down" the dissociation curve as your red blood cell is carried deeper into an oxygen-hungry, active tissue such as the brain.

5. Some swimming coaches advocate hyperventilation before a race; others stress the hazards of forced deep breathing—you can easily faint from hyperventilation. Considering the several monitoring systems which control ventilation, why is hyperventilation risky?

6. When we read that a conformational change increases hemoglobin's affinity of the remaining vacant sites for oxygen (page 242) what would this look like if you could stand before a hemoglobin molecule and witness its allosteric changes as the $pO_2$ varies? Is it too much to say that the molecule appears to be breathing?

7. Membranes—even in the red blood cell, that most simplified cell in the human body, we find that the membrane has a specialized function in maintaining the pH of the blood. Recalling the structure of membranes, how does the erythrocyte membrane accomplish this crucial control activity?

8. Hemoglobin is found in a remarkable diversity of animals—all vertebrates, some arthropods (*Daphnia*, brine shrimp), a few molluscs (red ramshorn aquarium snails) are examples. The conclusion drawn from this comparative data is that hemoglobin must have evolved independently in each of these different groups. Why has hemoglobin been an "easy" molecule to evolve? Are there any possible prototypes of this kind of molecule available in animal cells from which hemoglobin might evolve as needed?

9. The dark and white meat of a turkey results from the presence or absence of myoglobin in the muscle cells (cooked myo-

globin is dark colored). Wild turkeys have no white meat; neither does wild duck. Explain all this in terms of gas exchange physiology.

10. Most people think of respiration and ventilation as the same process. Biologists, however, draw a sharp distinction between the two. Why is such a distinction useful?

11. If you were engineering the oxygen affinity capabilities of hemoglobin molecules for different species, how differently would you design hemoglobin for an aquatic species versus a terrestrial species. Would you choose to make different molecules for a tadpole compared to the frog, for a mammalian fetus compared with a newborn baby?

12. How does the control of breathing illustrate the concept of homeostasis? Is negative feedback an important component here?

## CHAPTER 9

1. "You are what you eat" is a common saying which may be utter nonsense. On the basis of our understanding of digestion and nutrition, discuss the validity of this saying.

2. Although bacteria and fungi seem to covet every scrap of available food, producing rot and mildew at the first opportunity, some of these organisms have entered into cooperative feeding arrangements with plants and animals. Describe two such cases and speculate how such a symbiotic association might have evolved.

3. Plants require a large number of inorganic nutrients but the concentrations of these nutrients within the plant are usually far different from the concentration of these nutrients in the soil. Since plants are stationary, how can they pick and choose among soil nutrients?

4. How different are the mineral needs of plants from those of animals? Do plants use minerals differently than do animals?

5. Given that plants manufacture sugars in photosynthesis, how do they obtain other biomolecules such as lipids, proteins, and nucleic acids?

6. What advantages are there to digestion? Would it not be simpler for animals to absorb proteins, sugars, and nucleic acids from their food and use these molecules without breaking them down?

7. Once organic molecules such as sugars are obtained by a plant or an animal, is there any major difference in the way that plants and animals utilize these molecules?

8. What might have been the evolutionary pressures acting on animals which led to the internalization of digestive systems? Why would not external digestion as found in fungi be quicker and less vulnerable to internal problems?

9. Most people overrate the importance of the stomach. It is commonly thought that the stomach occupies most of the lower abdomen ("he's got a big stomach on him") and that digestion occurs primarily in the stomach. Put the stomach into perspective as but one part of the digestive system.

10. With our present knowledge of human nutrition it is feasible to devise an entirely artificial human consisting of "foods" which have all been synthesized from laboratory chemicals. Outline the categories of substances that would be needed in this purely chemical diet.

11. What do you see as the rationale for taking megadoses of vitamins? From our understanding of the biochemical functions of vitamins why is it controversial whether it is useful to take large doses of vitamins, far exceeding the daily dietary requirements?

12. The seeds of many plants are dispersed by birds and mammals and commonly, these seeds do not germinate unless they have passed through the digestive system of their animal "carrier". What kinds of signals occurring in the digestive tract might seeds react to in order to "know" that it is time to germinate?

## CHAPTER 10

1. Granted that plants obtain their energy from sunlight, what are the temperature problems encountered by plants "standing" in the sun all day? How do plants dissipate heat?

2. The transport tissues of plants have peculiar specialization: many of the xylem cells develop their functional ability by dying (!) and phloem cells lose their nuclei and other organelles. Can you account for these oddities?

3. A good number of tropical plants grow downward—some of these hanging plants are popular houseplants—how can transpiration and translocation operate in a shoot that is pointng downward?

4. Is there any reason why phloem should be partitioned into separate cell compartments when it is obvious that the cytoplasm of phloem tubes is continuous? Why is not phloem composed of continuous channels similar to the condition of xylem?

5. Why does the root system of most plants branch as extensively as the shoot?

6. The hydrogen ions of acid bind readily to the cation binding sites of soil and displace other cations. What effects does this have on soil fertility and the chemistry of water runoff in areas affected by acid rain?

7. Why is it necessary for plant roots to actively transport ions from soil when water can be obtained simply by osmosis?

8. Comparing plants with animals, which is more similar in function to the vertebrate circulatory system—the xylem or the phloem? Why?

9. Flooding agricultural land with sea water kills all terrestrial vegetation and destroys the use of the soil for generations. Why are terrestrial plants so sensitive to salt-laden soil?

10. Explain how transpiration of water by plants is based on the unique physical properties of water such as hydrogen bonding.

11. Vegetation, either natural or agricultural, plays a prominent role in the water cycle (the movement of water between the land, various bodies of water, and the atmosphere). How is this water cycle affected when large areas of vegetation are paved over for man's "benefit." In other words, what is the difference in the local water cycle between a woodlot and a parking lot?

12. Life is never passive; even plants respond to their environments and their experiences. For example, how does a tree respond to the sawing off of one of its living branches?

## CHAPTER 11

1. Nonbiologists often doubt the importance of evolution by asking what they think are unanswerable questions such as "how could anything so intricate and perfectly designed as the human heart have evolved? Wouldn't it have to be perfect from its very beginning?" Biologists can provide a very good answer to this question—what is it?

2. In many invertebrates the body cavity is the circulatory system. In humans, the body cavity plays no role whatsoever in circulation. Why?

3. How might a cost/benefit analysis look at the human circulatory system? Make a list of the benefits, most of which should be obvious, but emphasize the costs such as the energy required to operate the system, the defects that can occur, etc.

4. The vertebrate circulatory system is not symmetrical but shows considerable right-left asymmetry. Describe this asymmetry and comment on whether it leads to any functional difficulties.

5. Artificial hearts are made of nonbiological materials to avoid immune rejection. Are any of the normal control systems that regulate a real heart effective in controlling artificial hearts? What kinds of artificial control systems would need to be added to an artificial heart to fully imitate the natural condition?

6. What are the dangers if blood pressure rises above or fails below the normal range?

7. Since the beginning of human culture, the heart has been regarded with awe as the seat of vitalistic powers. Love and courage are but two affairs of the heart. From a biological viewpoint, why was the heart singled out for such attention?

8. The astounding dynamics of red blood cell turnover seems, at first sight, inefficient. Why make so many new cells only to destroy them? Can you speculate on the body's "wisdom" in maintaining this massive turnover of red blood cells?

9. Blood clotting is absolutely essential for maintaining a vertebrate circulatory system, but inappropriate clotting is a constant threat to the circulatory system. Explain how clotting is both indispensable and dangerous. What are some of the mechanisms that control it?

10. Blood pressure is measured as two values: systolic and diastolic pressure. Why is it important to take two readings of blood pressure?

11. The capillary wall is the exchange surface for *all* materials carried by the blood to the cells. How is this last step in transporting needed materials accomplished? What kinds of materials are transferred across capillary walls?

12. Describe the homeostatic functions of the vertebrate circulatory system. What factors are maintained at relative stability? What kinds of control loops and set points are apparent?

## CHAPTER 12

1. One of the most important goals of immunological research is developing techniques for transplanting skin between individuals. Why is this so important and why is it such a difficult problem?

2. Interferon is produced only by virus-infected cells and thus is present only occasionally. Considering its function, is there some reason why interferon should not be produced constantly and does this recommend against long-term treatment with interferon as a medical procedure?

3. How is the vertebrate immune system related to the circulatory system? Are they the same thing?

4. Passive immunity, such as injecting snakebite victims with antivenins (horse antibodies against the venom), is an emergency procedure that cannot be repeated or boosted. Why not?

5. The immune system in any normal person is capable of generating a million or more different antibody types. How can such incredible diversity be programmed into a population of cells?

6. The clonal selection theory in immunology was inspired partly by the central paradigm of biology, natural selection. Compare these two mechanisms; how are they similar?

7. It is disturbing to realize that circulating among our blood proteins is a group of cellular assassins, the proteins of the complement system. These proteins have as their function the destruction of cells. How are these dangerous proteins targeted and guarded by fail-safe procedures to prevent them from attacking our own cells?

8. All antibodies are proteins. Is there some property of proteins that has favored the evolution of antibodies within this group of biomolecules rather than in others?

9. Although cell-mediated immunity was discovered because of its role in transplantation immunity, it is obvious that this form of immunity did not evolve as a mechansim for frustrating tissue transplantation. What evolutionary reasons might have led to the evolution of cell-mediated immunity?

10. Immunology has contributed a new depth of understanding to our knowledge of membrane function. How has the study of lymphocytes deepened our understanding of membrane biology?

11. Why are swollen lymph nodes a prominent symptom of infectious disease?

12. How did the clonal selection theory inspire the development of monoclonal antibody techniques?

## CHAPTER 13

1. Explain why the excretory products of animals, both carbon- and nitrogen-based compounds, are nutrients for plants.

2. Why is there a continuous supply of waste nitrogen from amino acid and nucleic acid metabolism? Is there some reason why such valuable compounds are broken down in large quantities?

3. In many animal groups there is an association between the excretory and the reproductive systems. Vertebrates have a urogenital system. Can you think of any reason to account for the evolutionary association between these two functions?

4. The highest blood pressure in the human circulatory system is in the renal artery. Moreover, there is a hormonal feedback loop to maintain a constant, high pressure in this artery. Why is this necessary?

5. The diagnostic symptom of diabetes mellitus is sugar in the urine. Why is this a sign of abnormal, dangerously elevated blood sugar?

6. Artificial kidneys operate by dialyzing the patient's blood against isotonic saline (see Panel 13-1). Does this artificial system imitate the operating procedure of the kidney or is it more comparable to the functioning of a capillary?

7. If marooned on a desert isle, do not drink sea water; it is better to go thirsty. Why?

8. The function of the nephron tubule is based on a series of membrane specializations. Describe the unique features of the membrane in each region of the nephron.

9. It is naive to think that each organ has but one function. As examples, the mamallian kidney might be thought of as an excretory organ and the fish gill thought of as a respiratory organ, but in both cases, these organs have more than the one commonly known function. Describe the multifunctional nature of gills and kidneys.

10. With increasing public concern over drug abuse, many organizations are using urine testing to identify drug abusers. Why is urine used to monitor drug usage and will this technique work for all drugs?

11. Does the kidney, as most people believe, literally "filter the blood"?

12. If you were a sanitary engineer designing a small excretory system for terrestrial vertebrates it is unlikely that you would assemble the system in the same way that evolution has proceeded. For instance, why bother with fish kidneys built for an entirely different purpose? Using the vertebrate excretory system as your example, discuss how evolution follows an opportunistic pathway, working with what is available, rather than starting over each time and "designing" a perfectly logical system.

## CHAPTER 14

1. How is information encoded by hormones? How do cells "know what to do" in response to hormonal information?

2. Physiological processes which respond to a number of environmental signals are usually regulated by hormonal systems with some type of neurosecretory component. Molting in insects and reproductive cycles in vertebrates are examples. Why is neurosecretion so important in adapting internal physiological processes to external stimuli?

3. Circulating hormones have a brief half-life. Once secreted, hormones are either degraded or excreted rather rapidly. There is a good functional reason for this which can be imagined if you consider the consequences of a continuous buildup of hormone. Discuss how inactivation of hormone is as essential as hormone secretion in conveying information to tissues.

4. Mental states strongly affect the function of many endocrine glands. For example, any type of stress elevates the secretion of adrenal glucocorticoids. This link between the mind and the body occurs in the hypothalamus. Can you describe how thoughts are transformed into physiological responses in the hypothalamus?

5. Compared to enzymes and genes, hormones are remarkably small molecules. The steroids and the amino acid derivatives are closer in size to the subunits of macromolecules and even the protein hormones are small peptides. Is there some type of size limit imposed on hormones? Wouldn't larger molecules be able to carry more information?

6. The binding of hormones to membrane receptors is transient. That is, a hormone molecule weakly binds for a very short time period to the membrane receptor molecule. This "on-for-a-moment" then "off" sort of binding is absolutely essential for hormonal control of cell function; if binding were strong and permanent, hormonal regulation via the membrane would fail. Explain.

7. Hormone regulation has often suggested musical analogies. The anterior pituitary has been dubbed the "conductor of the endocrine symphony" and a healthy endocrine system regulates the "tone" of tissues. Imagining yourself to be a target organ regulated by various hormones, do you "hear music" when you set about describing the amplitude and tempo of hormonal regulation?

8. Realizing that most peptide hormones (those derived from amino acids) act upon specific membrane receptors, highlights the importance of the fluid mosaic model of the plasma membrane. How does the "mosaic" of the membrane function in cellular responses to peptide hormones?

9. Hormones are exceedingly potent. That is, a very small shift in the extracellular concentration of a specific hormone provokes an enormous change in the largest function of the responding cell. The hormone molecule does not supply the energy for this large response. How then can a slight shift in hormone concentration produce a major change in the responding cell?

10. Homeostasis results from negative feedback mechanisms. Many examples occur among endocrine systems. Describe one.

11. All cells secrete (or excrete) molecules and all cells respond to certain biochemical factors in their external environments. Could the origin of endocrine control systems lie within such ordinary cellular events? How might the earliest multicellular organisms have evolved some sort of endocrine coordination?

12. Summarize your knowledge of how endocrine systems work by describing the "life" of a hormone molecule from its moment of secretion to its final degradation or loss. Does it unerringly hone in on its target organ? How many different biological compartments might it enter?

## CHAPTER 15

1. Knowing that a protozoan such as *Paramecium* readily responds to its environment and carries on a lively behavior without a nervous system, what then is the general function of nervous sytems, the function that accounts for the evolution of nervous systems throughout the animal kingdom.

2. As a corollary to the previous question, imagine a multicellular animal without a nervous system. How would it live? (Hint: there is one animal phylum which lacks a nervous system).

3. Would it not be more efficient to construct a reflex from *one* all-purpose neuron rather than interrupting the circuit with one or more synapses?

4. It is possible to modify reflexes by learning. For example, boxers are trained to suppress blinking when punched in the face. What does this flexibility of reflex responses tell us about the neurons involved in a reflex?

5. Why does every reflex arc begin with a sensory neuron?

6. Many nerve cells of the central nervous system are spontaneously active; in the absence of any stimulation these neurons fire impulses at random intervals. How is this property of spontaneous activity used in neuronal control systems?

7. How does a visceral organ "know" whether it is to respond to a sympathetic or parasympathetic signal and how can a visceral organ "decide" how to balance its response between the opposing demands of the two branches of the autonomic nervous system?

8. Guilt can be determined by drugs as seen in the witch trial proceedings of traditional west African folk medicine. An accused witch is made to eat a drug which inactivates the acetylcholine-secreting synapses of the parasympathetic nervous system. If the accused is innocent, they survive; if they are guilty, they die from the drug. How does this work?

9. Surveying the functions of the evolutionarily oldest parts of the vertebrate brain gives us some idea of original functions of the brain. Examine this idea.

10. Acetylcholine was the first transmitter agent discovered and norepinephrine was the second. Since then dozens of transmitters have been identified, but, from examining the functions of the first two transmitters known, we can gain a partial answer to the question of why there is more than one transmitter agent. Why does the autonomic nervous system require at least two transmitter agents?

11. How do we determine the functions of the various brain regions?

12. The human cerebral cortex is asymmetrical in terms of function and cellular anatomy. Discuss this asymmetry and note how far asymmetry extends into the rest of the human nervous system.

## CHAPTER 16

1. All cells maintain an internal chemical environment differing in many respects from the external environment in which the cell resides. In the case of excitable cells, how does this basic property of cells allow for the communication of information?

2. The plasma membrane of excitable cells displays a marked asymmetry. What are the major structural and functional differences between the outer and inner surfaces of a neuron plasma membrane? Consider both an axon membrane and a postsynaptic membrane of a synapse.

3. In designing machines to encode and manipulate informa-

tion, engineers commonly choose between analog or digital devices. Analog mechanisms record continuous variation, a mercury thermometer, for example, while digital devices record a discrete on-or-off condition—a light switch is a simple example. The neuronal mechanisms responsible for processing information can also be classified as either analog or digital. Considering how a single neuron processes information, explain how each event from postsynaptic potentials through synaptic transmission is either analog or digital. Realizing that in modern machines analog devices are being replaced by digital systems, can you speculate why certain neuronal mechanisms are analog and others are digital?

4. The function of the nervous system may be defined as the integration of information. This means that the nervous system gathers information and "decides" on the appropriate response. Attempting to understand this general function in terms of neuron action has revealed that each individual neuron performs integration. Explain how a typical neuron can integrate information, "deciding" on the proper output for a given pattern of inputs.

5. Genetic engineering techniques have allowed the isolation of the genes coding for the acetylcholine receptor. To fully prove that the genes are correctly identified, the genes have been inserted into a cell type which lacks ACh receptors, the oocyte, and these genetically engineered cells are tested to see if the inserted genes produce membrane proteins which respond as does the ACh receptor.

a) How would you test such an altered oocyte for acetylcholine receptors?

b) What other types of neuronal membrane proteins could be similarly tested in frog oocytes?

6. Compare bioelectricity as generated and used by a neuron with the electricity that operates electronic devices (e.g., a walkman powered by two 1.5 volt batteries).

7. All cells are continuously transducing signals from one type of energy into another. Neurons provide a particularly striking example of such transductions when we realize that various types of chemical and electrical energy are transduced as a neuron performs its normal function. Tracing the passage of a single neuronal message from start to finish, describe each transduction that occurs.

8. One theory of how the central nervous system works assumes that the CNS is an extremely complicated set of electrical circuits hardwired to perform all functions. In this view the neurons are connected together in early development and the connections never change; each neuron behaves in an entirely predictable manner—for a given input, each neuron will produce a predictable output. Such a rigidly determined "machine" might account for simple, mechanical behaviors such as reflexes, but could a hardwired brain produce flexible behavior such as learning and mood change?

Since this hardwired model imagines the CNS to be a massive electronic circuit, how might this model be modified to include the importance of chemical factors in brain function? Might transmitters, hormones, and their receptor molecules account for some of the flexibility of a hardwired brain?

9. The specialized function of neurons is information processing. Describe how information is coded at each step in a typical neuron's function and explain how the information is quantified.

10. Neurons have a relatively high metabolic rate; they require a constant source of glucose to fuel their activities. Examining the function of a typical neuron, how is energy used? Which processes require a continuous supply of energy and which mechanisms are initially constructed using energy but function without a continuous energy input?

11. Nerve cells are exceptionally vulnerable to a number of poisons and examining the mode of action of these poisons is an important research area. Discuss how the following poisons affect nerve cell function:

a) potassium injections are lethal above a certain concentration

b) pufferfish toxin (tetrodotoxin) blocks sodium gates

c) certain nerve gases block acetylcholinesterase

12. A large fraction of the synapses in the vertebrate brain are inhibitory rather than excitatory. At the level of a single neuron, why is inhibition so important?

13. Many central neurons are incapable of generating action potentials. These spikeless neurons such as the horizontal cells of the retina have short axons which make synaptic contact with adjacent cells. These neurons perform integration and transmit information to their postsynaptic cells as do all neurons. Discuss how such spikeless neurons can perform these functions without generating action potentials and comment on the functional importance of action potentials in general.

14. Anatomically, neurons are peculiar cells in that long fiber-like processes extend outward for considerable distances. Protein synthesis occurs only within the cell body and thus all the proteins of the axon and the synaptic terminals are transported through the axoplasm. Discuss what types of protein are carried by this axoplasmic transport system.

15. If you were to design a nervous system would you think of making the cell-to-cell communications *chemical* as we see in real synaptic transmission? Wouldn't electrical synapses, allowing action potentials to freely cross from one cell to the other, be superior? Have we found a design flaw here in nervous systems or is there something uniquely important about the function of chemical synapses?

## CHAPTER 17

1. Sensory transduction exhibits a large amplification factor in that a small stimulus produces a strong biological response. What supplies the energy for this amplification?

2. How is a specific sensory modality encoded in the central nervous system? How does the central nervous system "know" whether it is hearing from a visual nerve cell or an audio nerve cell?

3. Sensory neurons are often spontaneously active, firing occasional impulses at random intervals. How is this "noise" treated by the central nervous system?

4. The traditional five senses—sight, hearing, taste, smell, and touch—omit several indispensable senses. Describe the importance of these unsung senses.

5. Compare the neuromasts of fish lateral line systems with the organs of Corti in the human ear. Do you sense an evolutionary relationship here?

6. The comparison of the vertebrate eye to a camera is not very accurate. In what ways do our eyes differ from film and video cameras?

7. Most mammals lack cones—we primates are exceptions. How then does a dog, for example, view the visual world?

8. Discuss how cilia have sensory functions in addition to their well known effector function.

9. Choosing any sensory system, describe how the membrane of the sensory cells is specialized to respond to the particular stimulus appropriate to the cell.

10. Compare generator potentials with postsynaptic potentials. Are they similar in their electrical properties? Are they similar in their effects within the neuron?

11. Why is the detection of gravity considered a sense of equi-

librium? What happens to an invertebrate such as a crayfish when its statocysts are removed?

12. Discuss the distinction between contact chemoreception and distance chemoreception in humans and in an invertebrate of your choice.

## CHAPTER 18

1. Surveying the wide distribution of microfilaments throughout many types of eukaryotic cells suggests that muscle cells may have evolved as a specialization of microfilament function. What evidence supports this view?

2. Exercise increases the size and strength of skeletal muscle but not by increasing the number of muscle cells. Striated muscle cells never undergo mitosis. How then does long-term exercise strengthen and enlarge muscles?

3. Muscle cells transduce chemical energy into mechanical energy. Present a step-by-step account of how this energy transformation is accomplished.

4. Nerve and muscle cells are both *excitable* cells, a property based on the specialization of certain plasma membrane mechanisms. Compare the membrane functions of a motor neuron and a striated muscle cell. Are there more similarities or more differences?

5. Why is smooth muscle called "involuntary" and striated muscle termed "voluntary"? How does cardiac muscle fit into this distinction? Is "voluntariness" a property of the muscle cell?

6. Summation is of central importance in both the control of muscle contraction and in the integration of information by nerve cells. Are the concepts of spatial and temporal summation (as described in chapter 16) useful in explaining the coordination of muscle contraction?

7. When chilled, we shiver, that is, we rapidly contract and relax the skeletal muscles of our limbs. Do we derive any benefit from this muscular activity or does it literally get us nowhere?

8. After we understand how sliding filaments shorten sarcomere length, it still may be uncertain how this results in mechanical force. How does shortening the dimensions of myofilament bundles generate force?

9. An insect striated muscle cell is innervated by 3 nerve cells—2 excitatory neurons and an inhibitory neuron. The excitatory inputs differ in the speed of the contraction which they stimulate; one is "fast," stimulating a twitch and the other is "slow," producing a tonic contraction. Compare this with the innervation of a vertebrate striated muscle cell. Have we vertebrates been short-changed by evolution? Should we engineer an insect-style musculature?

10. Curare, the poison used by Amazonian natives to coat their blowgun darts, specifically blocks acetylcholine receptors on striated muscle so that the motor nerve has no effect on the muscle. Describe the symptoms of a blowgun dart injection of curare. Must it be fatal? Is there any use for curare in modern medicine?

11. Do effectors always use ATP as their energy source? List as many biological effectors as you can and examine whether they all depend on ATP.

12. Examining the role of calcium ions in muscle contraction and recalling the function of $Ca^{2+}$ in neuron synaptic transmission, is there a common function here that points to a general principle of $Ca^{+2}$ activity in eukaryotic cells?

13. Rigor mortis is a rigid, unbendable fixation of the muscles that develops some hours after death and then gradually dissipates several days later. Realizing that enzyme activity persists well after an organism's death, can you provide a biochemical explanation for the onset and eventual relaxation of rigor mortis?

14. Is there much basis for comparing cilia (and flagella) with muscle cells? Are there common features in structure, function, and phylogenetic distribution?

15. How do all forms of biological locomotion "emerge" from the basic property of proteins—conformational shape change?

## CHAPTER 19

1. The distinction between a taxis and a kinesis is often overlooked. Using examples, provide a clear definition for both behavioral patterns.

2. The science of ethology explains purposive behavior in a scrupulously non-teleological manner. Rather than claiming that a cricket sings *in order to* attract mates, ethology presents a mechanistic explanation for the occurrence of cricket song. Discuss how ethology has explained behavior without resorting to teleology.

3. Ethology asks very specific questions of neurophysiology, e.g., how are releasers and fixed action patterns built into the nervous system? Speculate on the kinds of neural mechanisms that could be responsible for generating these ethological component.

4. Under what circumstances would you think that stereotyped, genetically closed behaviors would be more adaptive than learned, genetically open behaviors?

5. Why is habituation classified as a type of learning rather than as fatigue or inability to respond?

6. Examine imprinting and point out which aspects are closed and which are open types of behavior.

7. If it is possible to "dissect" a particular animal behavior into its components parts, will identification of the releaser and the fixed action pattern provide a complete explanation? Is motivation an essential part of explaining animals' behavior?

8. Chronobiology, a new field of study, seeks to understand the many functions of biological clocks. How important do you think chronobiology will be for human medicine?

10. To what extent do biological clocks allow an animal to anticipate the future? Is this anticipation comparable to your preparing for a trip? Explain.

11. Does appetitive behavior occur in humans? If you think so, give an example of appetitive behavior from your own experience.

12. Ethology presents very compelling arguments for a genetic basis of animal behavior, but how could DNA molecules directing the synthesis of proteins actually create behavior? Speculate on how genes could direct behavior.

## CHAPTER 20

1. The cooperative association of cells to form a multicellular organism can be considered social behavior on the part of the associated cells. Discuss whether the cells of an individual are altruistic or aggressive towards one another?

2. In mammals, what role does sound communication play in maternal-young behavioral interactions?

3. Can you imagine how the honeybee forager's dance could have evolved? Is it likely that scent could have been the original communciation channel?

4. Donald Griffen in his two books examining whether animals other than humans have conscious minds recommends that honeybee dances be examined as possible symbolic languages. Do you think that honeybees convey information through these dances in a manner comparable to human languages?

5. Using Mayr's analysis of behavior as either closed or open, how would you view dominance hierarchy behaviors?

6. Considering human use of animals, has it been easier to domesticate animal species with dominance hierarchy behavior or animals that exhibit individual territorial behavior?

7. What adaptive significance can there be for courtship behavior? Would it not be more efficient for mating to occur immediately upon first encounter between a male and a female without wasting time and energy on courtship displays? (Answer for a species other than our own.)

8. The territorial behavior of an individual is a complicated behavioral response. What types of releasers and drives would be involved in territorial behavior?

9. Pair bonding between male and female parents is very common among songbird species, but surprisingly rare among nonprimate mammals. Is there a correlation here between the difficulty of feeding the young and male "altruism"?

10. The abundance of fishes in tropical reefs is often dependent on the population of cleaner fishes and shrimp. If the cleaners are eliminated, fish populations decline along the reef. Why are the cleaners so important to the ecology of a reef?

11. If sociobiology proves to be a valid interpretation of social behavior, what kinds of genetic discoveries might we anticipate as behavior is analyzed with the techniques of molecular genetics?

12. Why should there be any controversy concerning a genetic component of human social behavior? Genetic control of anatomical and physiological aspects of our species excite little controversy, why is social behavior different?

## CHAPTER 21

1. What are the advantages and disadvantages of the two forms of reproduction, asexual and sexual? Draw up a list of the "cost/benefit" relationships for such factors as energy requirements, genetic diversity, and rate of population growth.

2. Is reproduction life's death-defying escape from inevitable destruction? Compare the life history of a population of living cells with a population of nonliving and non-aging objects such as teacups. Which shall survive the hazards of existence?

3. If the function of meiosis is only to make haploid cells from a diploid cell, this could be accomplished in a single cell division—cancelling chromosome replication and separating homologous chromosomes would produce two haploid gametes from each diploid cell. What important aspects of meiosis would be omitted by this simpler procedure and could these processes be responsible for the evolution of meiosis as it now exists?

4. How does meiosis fit into the cell cycle (see pages 136–37)?

5. The fundamental distinction between the two genders, male and female, began with the evolution of anisogamy. What is the distinctive difference between the sexes? How do you tell which is which in algae, fungi, plants, and animals? How does a zygote "know" what it owes to its mother and to its father?

6. Being diploid organisms ourselves, we are biased in favor of diploidy over haploidy. However, what are the biological advantages of having diploid cells? Are there advantages for some species in being haploid?

7. For our domestic plants and animals, we strive for purebred, low variability strains and yet nature proceeds in the opposite direction—variation abounds in natural species. Why is biological variation within a species favored in nature?

8. Germ cells do not differentiate into specialized cells and in fact contribute nothing to the maintenance of the individual. Can you account for this seemingly inefficient situation?

9. In every species with male and female reproductive systems (including monoecious species) the male system produces far more gametes than does the female system. Since in most cases only one male gamete can fertilize one female gamete, it seems as if males are wasting a good deal of energy in the overproduction of useless gametes. Is there any other interpretation of this male hyperactivity?

10. Compare the importance of spores with that of gametes in the life cycles of fungi and lower plants.

11. Animals which reproduce immense numbers of offspring are oviparous, whereas viviparous animals produce only a small number of offspring per pair. What is the basis for this?

12. Looking at a variety of animals and plants, what are the advantages to restricting reproduction to a limited time period? Why do so many organisms have a sharply defined reproductive season during the year?

## CHAPTER 22

1. Compare the cellular dynamics of the human testes and ovaries. How does the rate of cell division and the timing of the cell cycle differ between male and female germ cell lines?

2. What is meant when it is said that the clitoris is *homologous* to the glans penis?

3. Is the fertility of a woman affected by the length of a given menstrual cycle or whether menstrual cycles are regular or irregular? Explain.

4. Women sharing the same living quarters such as a college dormitory often find that their menstrual cycles become synchronized. Can you suggest an explanation for this well-documented phenomenon?

5. The menstrual cycle, like many complex physiological processes, is quite sensitive to mental states. Emotions, stress, hopes, and attitudes can affect the timing and process of menstruation. Where in the neuroendocrine control of the menstrual cycle does the human mind intervene?

6. As odd as it may sound, there is uncertainty as to the function of sexual excitement particularly in the female. Fertilization of an ovum is not dependent on female sexual excitement. How has the sensory aspect of copulation in humans evolved in relation to the mechanics of ovum fertilization?

7. How does the ovarian follicle integrate hormonal information? Describe how the follicle "decides" on the proper response amidst the fluctuations of gonadotropic hormones.

8. Understanding the anatomy and physiology of human copulation, is there any factual basis for the folklore assumption that the size of the penis is related to the fertility of a male?

9. Occasionally, a fertilized egg escapes from the oviduct and implants in the body cavity. How do such ectopic pregnancies receive proper hormonal maintenance to survive?

10. Artificial insemination is a common practice in agricultural animals and is applicable to humans as well. Given a supply of viable sperm, how complicated is this procedure in humans?

11. Vasectomy should not be confused with castration (removal of the testes). Clarify the anatomical and physiological differences between these two operations.

12. Do you think that textbook discussions of human sexual reproduction and contraception teach new or useful knowledge?

## CHAPTER 23

1. What led Mendel to abandon the assumption that inheritance was a liquid blending of "blood lines." What in his experiments was irrefutably particulate?

2. Is the full range of Mendel's discoveries summarized in his two laws or did he contribute more to our understanding of genetics than stated in his laws?

3. Mendel's two laws are the only *laws* in the science of Biology. Even natural selection is dubbed a theory. What is so eminent about these two statements that qualifies them as laws?

4. Why did Mendel conclude that his factors (genes) were paired?

5. Did the discovery of linkage repeal Mendel's second law?

6. Why, in cases of chromosomal sex determination, is the gender with hemizygous genes always vulnerable to more genetic diseases than the other gender?

7. Linkage maps are constructed from breeding experiment data. The frequency of appearance of phenotypes is somehow converted into a measure of linear distance along the chromosome. How are statistical probabilities transformed into distance measures?

8. Since humans have never been subjected to dihybrid crosses as were Mendel's peas, how can we know that humans have genes transmitted by Mendelian principles?

9. Rarely do students of introductory biology develop a clear definition of the terms "allele" and "gene." Can you dispell the confusion between these two terms?

10. Mutations are random. What do we mean by "randomness" in mutagenesis?

11. What is the mechanism of gene dominance? Is it a direct relationship between alleles?

12. Why is it important to keep in mind the distinction between genotype and phenotype? Does knowing the genotype unerringly predict the phenotype of an individual?

## CHAPTER 24

1. One of the distinctions between chemistry and biochemistry is the importance of *coding*. Discuss how information is encoded in nucleic acids and in proteins and why the words "transcription" and "translation" accurately specify coding procedures.

2. When the tobacco mosaic virus was first artificially reconstituted from protein and RNA extracts, newspaper headlines proclaimed "Life Created in Test Tube." Is that a valid conclusion?

3. Is it a paradox that proteins are produced by DNA, but DNA requires proteins for its own synthesis? Where does the DNA within a cell get the proteins (and RNA) necessary to synthesize more DNA?

4. Explain the importance of complementary base pairing in both transcription and translation.

5. Following the elucidation of DNA structure, Francis Crick predicted on theoretical grounds that a messenger molecule must mediate between DNA and protein synthesis. How did he deduce the existence of what turned out to be *m*RNA?

6. Where is energy invested in protein synthesis?

7. Point mutations affecting the accuracy of aminoacyl synthetase enzymes result in myriads of mistakes in protein amino acid sequences. How can one such point mutation produce a population explosion of many mutant proteins?

8. The double helix for any gene consists of one "sense" strand and one "nonsense" strand. Explain how the gene resides on only one strand of its double helical structure.

9. When recombinant DNA techniques were first devised, there was intense apprehension that inserting human genes into bacteria might create new disease organisms. One of the evolutionary arguments against this possibility was that bacteria associated with humans have been picking up our genes throughout our mutual evolutionary history and that the appearance of any new recombinant would be unlikely. How do you evaluate that argument?

10. Discuss some of the major advances made possible by the technique of gene cloning.

11. The DNA nucleotide sequences within introns and in between genes appears to carry no information. Thus, a good portion of the cell's DNA is suspected of being "genetic junk". How could natural selection "tolerate" useless DNA?

12. Diagram eukaryotic gene expression in terms of molecular events, noting the steps at which gene expression can be regulated.

## CHAPTER 25

1. In early development, gene transcription does not begin until the late gastrula stage. Protein synthesis during cleavage is exclusively directed by *m*RNA templates already present in the zygote cytoplasm. How is the information directing early cleavage inherited?

2. The response of the egg surface to fertilization is a most impressive example of membrane activity. How can this response be interpreted in terms of membrane receptors, channel proteins, and cytoplasmic motility?

3. The shaping of the embryo through the various stages—blastula, gastrula, and so on—looks if some external agent were molding the embryo. Some people see the embryo as clay responding to a potter's hand. But embryonic cells are alive and entirely responsible for the forces shaping the embryo's form. How do "simple" cells accomplish this?

4. That "ontogeny recapitulates phylogeny" is an intriguing notion. In watching human development are there stages reminiscent of fish, amphibian, reptile, and mammalian development, suggesting the course by which we have evolved?

5. Discuss embryonic induction as a form of intercellular communication and compare it with other such events, e.g., fertilization, endocrine control, and synaptic transmission.

6. Distinguish between determination and differentiation during vertebrate development.

7. Even though we do not understand the molecular mechanism of embryonic induction, it is clear that induction is a *trigger phenomenon* in which the energy and genetic information for the response is built into the inducible tissue. The "inducer" merely provides the trigger which releases the induced response. What evidence supports this view?

8. Is the development of gametes typical of the differentiation of adult cell types?

9. It is surprising that less than 10 percent of a cell's genes are active at any given time. Some of these genes must be genes that are active in all cells—these are sometimes called "housekeeping" genes—and other genes which are active only in a specific cell type. Choose a cell type and discuss these two classes of active genes and list some of its inactive genes.

10. The mysteries of development have been fertile ground for vitalistic and teleological speculations. For example, the embryo seems to be "striving" to become an adult. Discuss how modern biology replaces teleological speculation with causal, mechanistic explanations.

11. How does the human fetus abruptly change from an aquatic to a terrestrial organism?

12. Can you speculate on how the linear, essentially one-dimensional information of the gene is transformed into the three-dimensional shape of an animal organism?

## CHAPTER 26

1. During animal embryology, the shape of organs is developed largely by the movement of cells. How do plants develop shape from stationary cells?

2. The growth of plants is confined to specialized growth zones. Is this entirely different from growth processes in animals or do certain animal tissues grow in a similar fashion?

3. How does plant root growth develop sufficient force to penetrate compacted soil?

4. Is bark the skin of a tree? Considering the many functions of vertebrate skin, how similar are the functions of tree bark?

5. Is geotropism as important to roots as phototropism is to shoots?

6. In spring, tree leaves burst from the bud and trees "leaf out" far faster than mitotic growth could accomplish. How do spring leaves appear so rapidly?

7. Is the autumn abcission of leaves a model for aging and death in animals?

8. Do plants measure day length or night length? How is this question answered experimentally?

9. Germination is the target of several plant hormones. What is the significance of germination in a plant's life cycle and why must it be controlled in such a complex fashion?

10. A quick look at plant hormones suggests that auxin affects practically everything. How does a plant grow in orderly fashion amidst these myriad effects of auxin?

11. Is there a homeostatic feedback system controlling phototropism in plants?

12. Do plants go through embryological stages equivalent to the animal blastula and gastrula?

## CHAPTER 27

1. Lamarck's theory of the inheritance of acquired characteristics was taken seriously by biologists throughout the nineteenth century until embryologists discovered that germ cells develop separately from somatic cells. How does this observation refute Lamarck's theory?

2. In reviewing the abundant evidence in favor of organic evolution some people may discount the importance of all the biochemical data showing that life has a common chemistry. It could be argued that there is only one way to construct a living organism out of molecules and that living specie have been separately created millions and millions of times using the same chemical principles. From your biochemical knowledge evaluate this anti-evolutionary argument.

3. Examining the embryological evidence for the evolution of vertebrate animals suggests compelling arguments against the anti-evolutionary position described in the question above. Does the embryological development of vertebrates appear to be the only way that vertebrate organisms can be constructed from cells?

4. The biogeography of the Galapagos deeply fascinated Darwin. He noted that all the bird species there were related to South American birds but the species unique to the Galapagos were land birds; the oceanic birds there were found elsewhere. He was most astounded by the absence of frogs and toads on the islands; although there were ideal habitats for these animals, none were present. How are these biogeographical observations explained?

5. What is the fossil record? Where is it, what is its size and how does it change with depth?

6. Compressing natural selection into the statement "the survival of the fittest" has excited the philosophic criticism that the statement is circular reasoning. "Survival" really means surviving to reproduce and "fittest" means those who survive to reproduce most, so "survival of the fittest" boils down to "reproduction by those who reproduce most". Is this just playing with words or does it tell us something about natural selection?

7. Does the process of natural selection guarantee that organisms are perfect—that their physiology and anatomy are the most efficient that could be rationally engineered? (If you choose our own species as an example, you might want to check with your dentist, optometrist, or physician.)

8. Given the biological definition of species, what is a subspecies or race? Compare the biological significance of races in any nonhuman species with the peculiar social importance that many humans attach to racial distinctions.

9. The evidence for punctuated equilibrium comes from fossil material which of course is anatomical evidence. Do you think that the *development* of anatomical structures, the genetics of which is largely unknown, allows the possibility of abrupt, major change within a species or does our knowledge of embryology argue in favor of gradual, small-scale modifications?

10. Darwin's theory of natural selection has been misapplied to human societies with tragic consequences. A movement calling itself "social Darwinism" (Darwin took no part in this) argued that the wealthy and powerful were the fittest individuals of our species and that economic selection was naturally and rightfully exterminating the poor so as to improve the human gene pool. Government should not interfere with this natural process. Explain how this cruel political philosophy entirely misrepresented Darwin's theory.

11. What would modern biology be like if Darwin and Wallace had gone unread and natural selection had disappeared from the thought process of science? Would we have had the same advances in twentieth-century biology, medicine, and agriculture?

12. Speculate on the future evolution of our species, but, in contrast to most futurologists, do so with an accurate understanding of the mechanism of natural selection. (No giant-brained weaklings evolved from excessive computer use.)

## CHAPTER 28

1. Why are there so many different kinds of bacteria? What characteristics of this group have allowed them to radiate into every conceivable habitat including all the recent man-made environments?

2. A small number of biologists hold the opinion that bacteria could have originated somewhere else in the universe and arrived here from space. Could any of our present bacteria have survived such a journey?

3. Antibiotic's are useful because they selectively kill bacteria within human tissues. What unique features of bacteria are exploited by antibiotics?

4. Are bacteria primitive? Given the evidence that bacteria have existed on this planet for over 3 billion years, have they not had the longest opportunity of any organisms to change and improve any primitive features?

5. What advantages are there for bacteria which adhere to substrates by forming a sticky glycocalyx?

6. What is meant by a *colony* of bacteria? Why is most identification and experimentation performed on colonies?

7. The discovery of the rich assemblage of hydrothermal vent

animals living at great depths in the ocean was a surprise, because no photosynthesis occurs at that depth. It is now realized that hydrogen sulfide bacteria are the producers upon which the hydrothermal food chain is based. How can bacteria take the place of green plants in this deep ocean ecosystem?

8. The foul-smelling scums resulting from sewage pollution of lakes and streams are in most cases massive blooms of cyanobacteria. These organisms thrive in the worst aquatic conditions such as anaerobic, nitrogen-depleted waters. Why are these blue-green bacteria such "tough," indestructible aquatic organisms?

9. Should we consider the nitrogen bacteria in soil as altruistic? Are the several types of nitrogen bacteria carrying out their metabolisms for the benefit of plants?

10. It may be feasible to produce protein at very low cost by mass culturing microorganisms and extracting their edible proteins. Surveying the monera, which groups would be the least expensive to raise in large quantities? Why?

11. Compare the lytic and lysogenic cycles of viruses. Which strategy is best suited for evading a vertebrate animal's immune system?

12. Evaluate the "runaway gene" theory of virus origins. Besides "breaking away," what other events would need to fall in place if an escaped gene were to become a virus?

## CHAPTER 29

1. "Mushrooming" is a proverbial description for explosive growth. Is this an accurate metaphor?

2. Decomposers face severe competition from the many other decomposer organisms—bacteria, fungi, etc. How does a zygomycete compete to rapidly exploit any bit of organic matter that becomes available?

3. Considering how rapidly fungi can reproduce by means of spores, what adaptive "use" might there be for adding on a sexual phase?

4. Parasitism has evolved many times among the fungi. What are some of their characteristics that have facilitated the evolutionary step into parasitism? Are free-living fungi "almost parasitic"?

5. We have no antibiotics—nothing equivalent to penicillin or streptomycin—to control fungi in medicine and agriculture. Recalling the mechanisms by which antibiotics affect bacteria, why has there been no similar success in the chemical control of fungi which infect plant and animal tissues?

6. How do plant phytoalexins compare with vertebrate antibodies?

7. Various fungi exhibit peculiar cases of nucleus-cytoplasmic relationships. Describe two such cases and suggest how such peculiarities can be used to study the functions of nuclei throughout the living kingdoms.

8. Why have the cellular slime molds been classified as Fungi rather than as animals or protists?

9. Are lichens a mutualistic symbiosis? If so, what does the mycobiont contribute to the partnership?

10. How is nutrition in fungi a property of membrane function?

11. Survey the biogeography of fungi. Why are most fungal species not confined to a particular geographical region?

12. Why is the evolutionary history of the fungi obscure? Even though there is little fossil record of this group, do you think the fungi are an ancient or a relatively modern group?

## CHAPTER 30

1. Although multicellularity must have originated many times in the different multicellular groups, discuss how present-day green algae present an especially clear example of the probable origin of multicellularity.

2. The phytoplankton of the open ocean are predominately algae and these cells form the base of oceanic food chains. Which characteristics of algae are adaptations to a planktonic (floating) habitat?

3. Freshwater algae species are distributed worldwide and the common species rapidly colonize man-made lakes and water impoundments. How do algae achieve such widespread dispersal?

4. Aquatic algae require the same nutrients as do terrestrial plants. How do freshwater algae obtain these nutrients?

5. Why are algae difficult to classify within a single kingdom?

6. The alternation of generations, when first encountered, seems foolish—why have a haploid and then a diploid phase when, as we see, the diploid sporophyte proved to be the winner in the long run? There must be some advantage to the gametophyte stage since it has been maintained by lower plants, but what could it be? (Hint: could the haploid phase be the sporophyte's method of purging dangerous alleles?)

7. How have bryophyte species survived the competition of the numerous tracheophyte species? In what kinds of habitats are bryophytes found and have they been "pushed" into these habitats by tracheophytes?

8. Evaluate the adaptive significance of seeds by explaining why all our plant competitors, i.e., weeds, are angiosperms and how their seeds allow them to colonize and occupy our cultivated habitats.

9. To describe pollen grains as plant sperm cells is a mistake. Why?

10. What do the odd, relict plant groups (e.g., horsetails or gnetophyta) tell us about the large-scale pattern of plant evolution? Is there a linear progression from simpler to higher plants?

11. Given the mechanism of pollination it would seem that cross-species fertilization would be quite frequent. But most angiosperm species reproduce only within the species. How does the flower restrict fertilization to its own species?

12. Discuss how the evolution of angiosperms has been strongly affected by animals. Would flowering plants have evolved a large variety of flowers if it were not for insects?

## CHAPTER 31

1. There is good reason to suspect that viruses are "runaway genes." Is there any possibility that Protozoa evolved as "runaway metazoan cells"?

2. Why are there so many different kinds of protozoans? Does the diversity of Protozoa indicate that there are many different ecological opportunities for single-celled eukaryotes and that various protozoans have evolved to fill these possibilities? If so, describe some of the different "lifestyles" of Protozoa.

3. Sponges are exclusively aquatic organisms; most species are marine, but there is one family of freshwater sponges. Why has no terrestrial sponge evolved and why is it probable that no land-dwelling sponge ever will evolve?

4. Sponges and cnidarian polyps are sessile. That is, they are attached permanently to one site or, if they move, they do so slowly. How do unmoving animals feed, protect themselves, reproduce, and disperse their offspring?

5. It is interesting that the cells of sponges cooperate during development, but do not integrate their physiological activities to form functional tissues. Comparing the sponges with all higher animals, what kinds of physiological and behavioral activities are unavailable to these animals without tissues?

6. The maintenance of such a profound polymorphism as the

polyp and medusa phases of Cnidaria must be under a very strong selection pressure. What could be the adaptive reasons for this polymorphism in Hydrozoa and why has there been a differing emphasis on each of these forms in the other two classes of Cnidaria?

7. Compare the behavior of a) a ciliate protozoan, b) Hydra, and c) a planarian worm. How different is the behavior of each of these animals; and what does this comparison tell us about the function of nervous systems and of brains?

8. What does it take to be a parasite? Surveying the parasitic groups among the lower animals, what are their common features which are essential adaptations to a parasitic mode of life?

9. Of the lower animals studied in this chapter, which ones are capable of asexual reproduction? What might be the evolutionary advantages for asexual reproduction?

10. Compare the feeding methods between sponges, anthozoan cnidarians, and turbellarian flatworms. Is there a correlation between the symmetry of these animals and their general method of food capture?

11. Of all animal groups, Nematoda is without doubt the most peculiar. Their many unique features give us no clue as to their evolutionary origin—for all we can tell they could just as well have arrived from outer space as evolved here on earth. In what ways are the nematodes unrelatable to other lower animal groups?

12. State the economic impact, or the human relevance, of these invertebrate groups: Sarcodina, Sporozoa, Anthozoa, Trematoda, Cestoda, and Nematoda.

## CHAPTER 32

1. Explain how comparing the embryological development of different animal phyla reveals evolutionary relationships that are not easily seen when comparing adult forms. Why are embryological comparisons between animal groups usually better indicators of evolutionary relationships than anatomical comparisons?

2. Do you think that the evolutionary origin of body cavities and metamerism was associated with the evolution of larger animals? In other words do the largest acoelomate animals attain the size of typical coelomate animals?

3. How does the hydrostatic skeleton of invertebrate animals serve as a skeleton in converting muscular contraction into locomotion? Describe the operation of the hydrostatic skeleton in an annelid and in a mollusc.

4. Which are the metameric phyla? Are they all directly related in their evolutionary history?

5. The two largest classes of modern arthropods, the Crustacea and the Insecta, each dominate certain ecological communities. Surveying major ecological habitats such as marine intertidal communities, oceanic open water, freshwater lakes, streams, and forests, explain which habitats are dominated by each of these very successful animal groups. Can you account for the ecological pattern of crustacean and insect evolution?

6. In the evolution of chordates, did newly evolved classes always replace existing classes, driving them to extinction? If not, explain why the newer, "improved" animals did not always outcompete the older groups.

7. We live upon the blue planet, the water planet, and, perhaps this explains why most animal groups are aquatic. But is there a deeper reason, other than the proportion of water and land, which accounts for the preponderance of aquatic animal groups?

8. The immensity of evolutionary time is not comprehensible by merely stating millions or billions of years. A fuller conception of the span of evolutionary history can be attained by constructing a "Year of Life," equating the approximately 3.7 billion years of life's history to the 365 days of a year. On this scale, each day is 10 million years and each hour is 420,000 years. On this basis, date some of the more important events in life's history and compare the fabled reign of the dinosaurs with the current supremacy of humans.

9. Which came first, the chicken or the amniote egg? Why do you say so?

10. Is there any reason why all echinoderms are marine? It is rare for an entire phylum to be restricted to one ecological domain, but the echinoderms have never evolved freshwater or terrestrial species. Why not?

11. Can you account for the evolution of a large brain and complex behavior, including learned behavior, in cephalopods? If you cannot, do you think an octopus could?

12. Is evolutionary history logical and predictable? An interesting way to evaluate this question is to speculate whether the origin of life on some other, similar planet would produce the same sequence of evolutionary events. Would there be insects, would there be fishes, amphibians, reptiles, birds and mammals . . . and humans?

## CHAPTER 33

1. In engineering, a system is defined as any set of interactions in which inputs are systematically connected to outputs. The circuitry of a computer is such a system. In ecology, is an ecosystem a *system* in this engineering sense?

2. How do ecologists quantify energy flow through natural ecosystems? How can food webs be traced and the trophic interactions measured?

3. Why are food chains relatively rare in nature, food webs being far more common?

4. What roles do microorganisms play in the carbon cycle?

5. Choosing a particular species, construct a realistic description of an ecological niche in both a terrestrial and an aquatic ecosystem. What does the niche "look like" from the species' point of view?

6. Considering the typical American diet, discuss the energy pyramid supporting this diet. Note the major producers and the prominent consumer levels. Is this an efficient use of biological energy?

7. How is it that the evolution of a species results in the species being vulnerable to limiting factors? Why does natural selection fail to remove such barriers to a species' distribution?

8. All the nutrients necessary for life, C, H, N, etc., cycle through ecosystems but energy does not cycle. Why not?

9. Attempting to predict the consequences of a nuclear missile attack has emphasized the global catastrophe which would result from a large scale injection of smoke and dust into the atmosphere. How might such a massive particulate cloud produce a "nuclear winter"?

10. If you were designing life-support systems for a long space voyage, what would be the simplest, self-maintaining ecosystem that could be built into a spaceship to provide total nutrition for the crew? (Note: this system must maximize the recycling of nutrients.)

11. Why are deep-sea waters rich in nutrients? Where do these nutrients come from?

12. Does our species have an ecological niche? If so, has it changed through the history of our species?

## CHAPTER 34

1. At every level, biology encounters "emergent properties", unique characteristics which arise from the complexity of a system rather than from the individual parts of the system. Is

evolution an emergent property of populations or does it occur in individuals?

2. Graphing biological processes rarely yields straight-line relationships. Curves are far more common, especially exponential curves. Discuss the prominence of exponential curves in the study of population growth.

3. Explain why the economic "doomsday" argument of Thomas Malthus is relevant to the everyday existence of organisms in nature. Is every day a doomsday for natural species?

4. The survivorship curve for hydra (Fig. 34-5) is a straight line. In contrast with the human survivorship curve, hydra populations do not exhibit any trace of aging. Discuss how population survivorship curves illustrate and quantify aging.

5. Why are our most persistent biological competitors, weeds and pests, r-strategists?

6. What do parents invest in their young?

7. Explain how interspecific competition is a density-dependent population regulating factor.

8. Are disease-causing bacteria predators? Ecologically, do pathogenic bacteria populations behave like predatory animals?

9. As a general rule, infectious diseases that are commonly fatal are newly evolved relationships between the parasite and the host. Over the course of many generations, a more benign interaction evolves in which the disease is less dangerous. Discuss why pathogens usually evolve toward less lethal forms while hosts evolve toward more resistant forms?

10. What are the genetic consequences of population oscillations? Does the frequency of unfavorable alleles increase during rapid population growth? What happens to the gene pool during sharp population decreases?

11. Demonstrate that the factors regulating population levels are primary forces in natural selection.

12. How have humans uniquely blunted the effects of density-dependent and density-independent population regulation factors on our species?

## CHAPTER 35

1. One of the first things an ecologist does in studying any habitat is map the distribution of important organisms. Why is mapping a primary tool in ecology?

2. Viewing the course of an ecological succession, do species with r-adaptations play different roles in succession than do species with K-adaptations?

3. Geologically (and biologically), rivers are far older than lakes. Why?

4. Many intertidal marine invertebrates are sessile, permanently attached to the substrate. How is such plant-like behavior adaptive for these marine animals?

5. The generalization that "nature heals her wounds" through ecological succession may not always apply. What types of environmental damage would not be restored by ecological succession?

6. How are forest and stream ecosystems interconnected?

7. Why are oil spills so devastating to shallow water seashore communities?

8. Regarding the ecological determinants which explain how large size could evolve in the baleen whales, are there comparable examples of this phenomenon among terrestrial animals?

9. To preserve patches of grassland biome, it is necessary to set fire to the grassland every year or so. Why is such destructive action necessary to maintain the natural species of a grassland biome?

10. Describe the ecological characteristics of your region 10,000 years ago during the last glaciation. Has your region changed biomes since then?

11. The forests of tropical regions are the last extensive forests on earth and are being destroyed systematically as were the temperate-zone forests. What are the global consequences of this deforestation?

12. Eugene Odum has argued that human exploitation of natural communities invariably reverses the direction of ecological succession. Explore this idea by comparing agricultural fields with natural climax communities.

## CHAPTER 36

1. How many of the ecological problems facing us result from human distortions of biogeochemical cycles?

2. How can the study of natural ecosystems contribute to discovering solutions to our present, human-made ecological problems? Concentrate your answer on one particular issue.

3. Why is topsoil depletion not a problem in natural (uncultivated) grasslands and forests?

4. Discuss how ecological problems extend beyond human governmental boundaries. Are ecological problems ever local?

5. What is the relationship between economic value and ecological value?

6. Choose any issue of pollution and discuss how you as an individual could make a significant contribution toward correcting the problem.

7. Outline the methods for treating the four types of water pollution. Are they equally amenable to treatment?

8. Considering the large number of ecological tragedies confronting the human species, why has the human population continued to grow at an exponential rate?

9. What are the incentives to individuals for controlling population growth and why are they so little successful?

10. Why do some substances such as DDT and strontium-90 accumulate in ecosystems and become most concentrated in the highest trophic levels?

11. Does the human population explosion result from our ability to evade and thwart the pressures of natural selection? Are we the first species to remove itself from natural selection?

12. To what extent has the human species become responsible for the evolution and survival of all species on this planet?

# Glossary

**abscisic acid (ABA):** A plant hormone that acts as a growth inhibitor, causing dormancy in buds and seeds. It induces abscission of buds and participates in root geotropism and stomatal closing during water stress.

**abscission** [L. *abscissus,* to cut off]: The separation and dropping of leaves, flowers, fruits, or other parts from plants.

**absorption spectrum:** The portion of the spectrum absorbed by a given substance.

**accommodation** (of the eye) [L. *accomodare,* to adapt]: Adjustment of the ciliary muscle and the lens to bring objects into focus.

**acetylcholine (Ach):** The acetic acid ester of choline that serves as a neurotransmitter at the neuromuscular junction and at many synapses.

**acetylcholinesterase:** An enzyme that catalyzes hydrolysis of acetylcholine.

**Ach** (see Acetylcholine).

**acid** [L. *acidus,* sour]: A substance that undergoes ionization in aqueous solution with a release of hydrogen ions ($H^+$) and has a pH less than 7.

**acid rain:** Rain or other precipitation containing nitric and sulfuric acids produced from industrial and automobile emissions.

**acidosis:** Decrease in blood pH.

**ACTH:** Abbreviation for adrenocorticotropic hormone.

**actin** [L. *actus,* motion]: The protein that forms the thin filaments of muscle fibrils.

**action potential** (spike potential): The changes in membrane potential that characterize the nerve impulse, essentially the depolarization and repolarization of a neuron or muscle fiber.

**action spectrum:** The portion of the spectrum that elicits a particular reaction.

**activation energy:** The amount of energy required to bring interacting molecules, or reactants, to an energy state conducive to a chemical reaction.

**active transport:** The transport of substances across a cellular membrane, usually from a region of lower concentration to a region of higher concentration. This process requires an expenditure of energy.

**adaptive radiation:** The evolution of a relatively unspecialized stock of organisms into several groups, each specialized for a different ecological niche.

**adenylate cyclase:** The enzyme that catalyzes the conversion of ATP to cAMP (cyclic AMP).

**adenosine triphosphate (ATP):** A mononucleotide consisting of the purine base adenine, the sugar ribose, and three phosphate groups. The energy released during the hydrolysis of the high energy bonds associated with two terminal phosphate groups is utilized for cell metabolism.

**adenylate** (adenylic acid, adenosine monophosphate, or AMP): A mononucleotide consisting of adenine, ribose, and one phosphate group; formed by the removal of two phosphate groups from an ATP molecule.

**ADH:** Abbreviation for antidiuretic hormone.

**adhesion** [L. *adhaerer,* to stick to]: The tendency of unlike molecules to stick together.

**adrenal cortex:** The outer layer of the adrenal gland, itself an endocrine gland that produces several hormones.

**adrenal medulla:** The inner portion of the adrenal gland, which produces epinephrine (adrenaline).

**adrenaline** (see Epinephrine).

**adrenergic:** An adjective applied to axons that produce norepinephrine (noradrenaline) at their tips or to receptors that are responsive to epinephrine or norepinephrine.

**adrenocorticotropic hormone (ACTH):** A hormone of the anterior lobe of the pituitary gland that stimulates the adrenal cortex to secrete its hormones.

**adventitious root:** A root that arises from a plant part other than a root.

**aerobe** [Gk. *aer,* air + *bios,* life]: An organism, usually a microorganism, that requires oxygen (or air) to live.

**age structure** (of a population) (see Population structure).

**aggression** [L. *aggressio,* to approach; to attack]: An act or attitude of hostility not necessarily followed by an actual attack.

**Agnatha:** Jawless fishes; the earliest vertebrates, represented today by lampreys and Myxine.

**agonist** [Gk. *agonistes,* combatant, competitor]: 1. A muscle whose action is opposed by another (antagonistic) muscle. 2. A drug that can initiate an effect.

**air capillaries:** The finest of the air passages in bird lungs.

**alecithal:** An adjective applied to an egg having no yolk.

**algae** (sing. alga): A somewhat imprecise term applying to divisions of simpler plants and to all photosynthetic protists. The term was also formerly applied to the cyanophytes or blue-green bacteria ("blue-green algae").

**algal bloom:** A population explosion of algae, usually algal protists.

**allantois:** One of the extraembryonic membranes that arises as an outgrowth of the embryonic gut in amniotes. It contributes to the placenta in mammals.

**all-or-none phenomenon:** The response of a single muscle or nerve cell to an adequate stimulus, which occurs either completely or not at all.

**alleles** [Gk. *allelon,* of one another]: Alternative forms of a gene.

**allopatric speciation:** Origin of new species from populations of a single species which were physically isolated from each other.

**allosomes** [Gk. *allos,* other + *soma,* body]: The sex chromosomes. The chromosome pair involved in sex determination; usually the members of the pair differ in their morphology. The sex chromosomes are designated X and Y in mammals and other animals in which the male is heterogametic, and W and Z in birds and amphibians and others in which females are heterogametic.

**aleurone** [Gk. *aleuron,* flour]: The outermost cell layer of the endosperm of cereal grains.

**alpha rays:** Streams of helium nuclei spontaneously emitted by radioactive isotopes. The rays travel at about 1/20 speed of light and have relatively little penetrating ability.

**alpine tundra:** A mountain biological community resembling the tundra in its composition and climate (see Tundra).

**alternation of generations:** A life cycle in which alternation of haploid and diploid generations is required for completion of sexual reproduction.

**altruism:** A self-sacrificing behavior for another's benefit. To be altruistic, the behavior must increase the evolutionary fitness of the recipient while decreasing the fitness of the altruist.

**alveolus** [L. diminutive of *alveus,* a hollow or trough]: 1. The smallest of the air pockets in the mammalian lung. 2. Any cavity, pit, or depression.

**ametabola:** Those insects which do not undergo metamorphic changes in their life cycle.

**amoebocyte, amebocyte** [Gk. *amoibe,* change]: 1. Any cell that is amoebalike in that it can change shape. 2. A type of leukocyte.

**amino acids** [Gk. *Ammon,* Egyptian sun god, near whose temple ammonium salts were first extracted from camel dung]: Organic acids possessing both an amino ($NH_2$) and a carboxyl group (COOH); they form the building blocks of proteins.

**ammocetus larva:** The larval stage of a lamprey that bears close resemblance to Amphioxus, a cephalochordate.

**ammonification:** The conversion of proteins to ammonia by bacteria in the soil.

**ammonotely or ammonotelic excretion** [Gk. *tele,* end; ammono, abbreviation for ammonia]: Excretion of ammonia as the endproduct of nitrogen metabolism.

**amniocentesis** [Gk. *amnion,* lamb; *kentesis,* perforation]: Surgical perforation of the uterus through the abdomen to obtain amniotic fluid.

**amnion:** One of the embryonic membranes that, together with the chorion, provides a fluid-filled chamber surrounding the embryo in reptiles, birds, and mammals (amniotes).

**Amniota:** The group of vertebrates (reptiles, birds, and mammals) whose embryos bear the extraembryonic membranes, amnion, chorion, allantois, and yolk sac.

**anabolism** [Gk. *anabole,* a throwing up]: The sum of all of the biosynthetic reactions in the cells of an organism, in which complex compounds are synthesized from simple substances.

**anaerobe** [Gk. *an,* without + *aer,* air]: A cell or organism that can live and grow in the complete, or almost complete absence of molecular oxygen. Obligate anaerobes can live only in the complete absence of molecular oxygen, whereas facultative anaerobes are able to live with or without oxygen.

**anaerobic** [Gk. *an,* without + *aer,* air + *bios,* life]: That which can occur in the absence of oxygen.

**analogy** [Gk. *ana,* according to + *logos,* study]: Similarity of function of two or more organs which is not associated with their developmental origin.

**anamnestic response** [L. *anamnesis,* to recall]: The heightened immune response to a secondary or subsequent contact with the same foreign substance (antigen). Also called immunologic memory.

**anaphase** [Gk. *ana,* up + *phasis,* form]: That stage in mitosis or meiosis during which the chromosomes arranged at the equator separate and move to opposite poles.

**anaphylaxis** [Gk. *ana,* against + *phylaxis,* protection]: A reaction of immediate hypersensitivity that occurs following exposure to an antigen in those individuals who have become sensitized to it, such as in the hypersensitivity to penicillin and bee or wasp stings.

**androgen** [Gk. *andros,* male or man + *genesis,* creation]: Steroid hormone(s) that promote development of male sexual characteristics.

**anemia** [Gk. *anaimia, an,* not + *haima,* blood]: A below normal concentration of erythrocytes or hemoglobin of the blood.

**aneuploidy:** Presence of more or less than the normal diploid number of chromosomes.

**angiography** [Gk. *angeion,* vessel + -graphy]: An invasive determination of the arrangement and condition of blood vessels, usually by injection of a radio-opaque substance followed by photography.

**angiosperm** [Gk. *angeion,* vessel or container + *sperma,* seed]: One of the flowering vascular plants of the division Anthophyta, in which the seeds are enclosed in the carpels that develop into fruits.

**angstrom (Å or A):** A unit of measurement equal to $1 \times 10^{-10}$ meter in length.

**Annelida:** Metamerically segmented coelomate animals in which nephridia serve as excretory organs.

**anion** [Gk. *ana,* up + *ion,* going]: A negatively charged ion that migrates to the positive electrode (anode) when exposed to an electrical field.

**anisogamy** [Gk. *aniso,* unequal + *gamos,* marriage]: The occurrence of gametes that differ in size, both being motile, in sexual reproduction.

**anisotropic** (see Birefringent).

**annual plant** [L. *annus,* year]: A plant that lives one year; it completes its life cycle and produces seed in one growing season and then dies. A monocarpic plant.

**annual ring or growth ring:** The ring in the wood of a plant stem or root representing a year's growth.

**antagonist** [Gk. *antagonistes*, opponent]: A muscle or a drug that opposes the effects of another muscle or drug (see also Agonist).

**antenna:** A head appendage found in Crustacea, Insecta, and Myriapoda, and which bears various sense receptors.

**antennule:** The first head appendage of crustaceans.

**antheridium** (pl. antheridia) [Gk. *anthos*, flower]: A structure producing male gametes; antheridia are unicellular in algae and fungi, but multicellular in plants.

**antibiotic** [Gk. *anti*, against + *bios*, life]: A substance produced by microorganisms that inhibits the growth of other organisms.

**antibody** [Gk. *anti*, against]: A protein produced in response to the introduction of an antigen; it has the ability to combine with the antigen that stimulated its production.

**anticodon:** The trinucleotide sequence, present in a transfer RNA molecule, which is complementary to a codon.

**antidiuretic hormone (ADH):** A peptide hormone that is produced by the hypothalamus and is stored in and released from the posterior pituitary. It raises blood pressure and inhibits urine production. Also called vasopressin.

**antigen** [L. *anti*, against + *gennan*, to produce]: Foreign substances that induce an immune response and interact with the specific antibodies that are produced.

**antimetabolite:** A substance with a close structural resemblance to one required for normal physiological functioning. It exerts its effect by interfering with the utilization of the essential metabolite.

**aorta** [Gk. *aorte*, something suspended]: The large artery leading from the left ventricle of the mammalian heart or the corresponding vessel in lower vertebrates.

**aortic bodies:** Clusters of chemoreceptive cells in the walls of the subclavian arteries where they branch from the aorta. In response to increased blood pressure, these cells generate nerve impulses that inhibit centers in the medulla oblongata controlling blood pressure and cardiac output.

**apical dominance:** The phenomenon in which the shoot tip suppresses the growth of lateral buds in plants.

**apical meristem:** The region of dividing cells at the tip of roots and shoots of plants.

**apocrine glands** [Gk. *apo*, from + *krinein*, to separate]: Glands that secrete their products by shedding portions of their cells.

**apogamy** [Gk. *apo*, without + *gamos*, marriage]: A process occurring in ferns in which haploid sporophytes arise from gametophytes without fertilization.

**apoplast** [Gk. *apo*, away from + *plastos*, molded]: The most common pathway of water absorption by root tissues of plants.

**Arachnida:** A chelicerate class of arthropods including scorpions, spiders, ticks, and mites.

**archegonium** [Gk. *archegonos*, first of a race]: The multicellular female egg-bearing structure in bryophytes and lower vascular plants.

**arterioles:** The small, unnamed arteries of the body.

**artery:** A blood vessel that carries blood away from the heart.

**Arthropoda:** Segmented animals with jointed appendages.

**asconoid:** The type of sponge body in which the flagellated choanocytes line the inner cavity, called the atrium or spongocoel. See Syconoid and Leuconoid.

**ascus** (pl. asci) [Gk. *askos*, bladder]: The sac-shaped, spore-producing cells of the Ascomycota associated with sexual reproduction. The resulting spores are ascospores.

**asexual reproduction:** Reproduction that does not involve the union of gametes.

**association cortex:** The area of the cerebral cortex that is concerned with higher mental activities, such as integration and interpretation, and with the ability to carry out complex tasks, such as speaking or writing.

**aster** [Gk. *aster*, star]: The system of microtubules that radiates in all directions from each pair of centrioles to form a starlike structure during mitosis.

**atherosclerosis** [Gk. *athero*, porridge + *sklerosis*, hardening]: A thickening of the walls of arteries and the consequent narrowing of their lumen due to a deposition of lipids, fibrous tissue, and other matter.

**atomic number:** The number of protons in the nucleus of an atom. It is represented by the subscript to the left of the symbol for an element.

**ATP:** Abbreviation for adenosine triphosphate.

**atrioventricular node (AV node):** A small mass of neuromuscular conducting tissue in the inner wall of the right atrium, near the junction of the atria with the ventricles. It serves as the secondary pacemaker of the heart (see also Sinoatrial node).

**atrioventricular valve:** One of the valves (mitral and tricuspid valves) opening into the ventricles from the atria of the heart.

**atrium** [L. *atrium*, hall]: 1. One of the two chambers of the heart that receive blood from the veins. 2. Any small cavity in the body.

**auricle** (pinna) [L. *auricula*, external ear]: The pinna, or external part, of the ear. Also, the earlike appendage to the atrium of the mammalian heart.

**autoimmunity:** An immune response to the body's own cells and tissues.

**autonomic nervous system** [Gk. *autonomos*, that which happens by itself]: The portion of the nervous system that serves the smooth muscles of the body's internal organs. It is divided into the sympathetic and parasympathetic divisions.

**autoradiography:** A process in which a picture of an object or a tissue is produced by the exposure of a photographic film by the rays emitted from a radioactive substance localized in that object or tissue.

**autosome:** A chromosome other than a sex chromosome. See Allosomes.

**autotroph** [Gk. *auto,* self + *trophos,* feeder]: An organism that is able to grow on simple substances such as ammonia and carbon dioxide without needing to be supplied with organic matter.

**auxin** [Gk. *auxein,* to increase]: A plant hormone involved in growth and differentiation.

**auxotroph** [Gk. *auxein,* to increase + *trophe,* food]: A mutant organism that requires one or more nutrients not required by the wild type.

**Aves:** The feathered, warm-blooded, bipedal vertebrates in which the forelimbs are modified into wings.

**AV node** (see Atrioventricular node).

**axillary bud** [Gk. *axilla,* armpit]: A bud situated in the axil of the leaf, the angle where it joins the stem.

**axon** [Gk. *axon,* axis]: The fiber of a neuron that conducts impulses away from the soma.

**axon hillock:** The conical portion of a nerve cell body from which the axon originates.

**back cross or test cross:** A cross between the recessive parental type and an individual with dominant phenotype, which is used to determine the genotype of the latter.

**bacteria** [Gk. *bacterion,* little rod]: Prokaryotic microorganisms that are chiefly parasitic or saprophytic.

**bacterial antagonism:** The inhibition of the growth and proliferation of other microorganisms by bacteria.

**bacteriophage:** A bacterial virus. Also called phage.

**baleen** (whalebone) [L. *balena,* whale]: The elastic, horny substance hanging in thin, parallel plates from the inside of the upper jaw of certain whales, which the whales use to strain food organisms from the water.

**baroreceptor** [Gk. *baro,* weight, pressure + receptor]: A pressure receptor.

**basal metabolic rate (BMR):** The rate at which energy is expended within a certain time while at rest, such as kilocalories per square meter of body surface per hour.

**base:** Substance that releases hydroxide, or hydroxyl ions ($OH^-$), in aqueous solution and has a pH greater than 7.

**basidium** (pl. basidia) [L. *basidium,* a little pedestal]: The club-shaped, spore-forming cell of the Basidiomycota, which, as a result of sexual reproduction and meiosis, produces basidiospores at its tip.

**B cell:** A type of lymphocyte derived from bone marrow stem cells that matures into an immunologically competent cell under the influence of the bursa of Fabricius in the chicken, and bone marrow in nonavian species. Following interaction with antigen, they become plasma cells, which synthesize and secrete antibody molecules involved in humoral immunity (see T cell).

**bends** (see Decompression sickness).

**benthic** [Gk. *benthos,* depth (of the sea)]: Pertaining to the bottom of the sea.

**beta rays:** Streams of electrons spontaneously emitted by radioactive isotopes, that travel at about 99 percent the speed of light and have great penetrating ability.

**biennial plant:** A plant that requires two growing seasons to complete its life cycle, its flowering and fruiting occurring in the second year.

**bile salts:** The sodium salts of the bile acids, constituents of the bile, formed by conjugation of glycine and taurine with cholic acid. Because of their detergent effect they act as emulsifying agents.

**bilateral symmetry:** Organization of parts in a body so that it is able to be divided into two equal halves.

**Bilateria:** The animals that exhibit bilateral symmetry.

**binary fission** [L. *binarius,* of two parts + *fissus,* split]: Asexual reproduction in which a cell or an organism separates into two cells.

**biodegradable:** Capable of being degraded or decomposed by organisms.

**biological clock:** An internal mechanism that regulates rhythmic biological activities in the absence of external stimuli.

**biological determinism:** A school of thought that in its extreme form contends that human behavior is determined entirely by natural laws over which individuals have no control.

**biological magnification:** The increasing concentration of a nondegradable substance at each stage in a food chain.

**biomass:** The total mass or weight of a group of organisms living in a particular location.

**biome:** One of the earth's major terrestrial communities, typified by the dominant plants growing in the area, for example, taiga (coniferous forest) and grassland.

**biosphere:** The entire area of the surface of the earth, including its soil, water, and air, in which life is maintained.

**biotic potential** (reproductive capacity): The rate at which a species can reproduce under ideal conditions.

**birefringent** (anisotropic): Doubly refractive. Said of a substance that separates transmitted light into two unequally refracted (bent) rays of plane-polarized light.

**Bivalvia:** Molluscs in which the shell has two valves, such as the oysters and clams.

**blastocyst:** An early stage in the embryonic development of mammals. A hollow sphere composed of the trophoblast and the inner cell mass.

**blastomere** [Gk. *blasto,* a sprout or bud + *meros,* part]: The name given to a cell that forms by division in the early stages of cleavage of a zygote.

**blastula:** The embryonic stage in animals that is usually a hollow sphere with a wall one cell thick.

**blind spot:** The light-insensitive spot on the retina of the eye that lacks rods and cones because of the presence there of the optic nerve.

**blood-brain barrier:** The functional barrier between the capillaries supplying the brain and the brain tissue, created by the membranes found there. The effect is to permit the passage of some substances but to restrict or prevent the passage of others.

**blood sinus:** A large channel or cavity containing blood. In some invertebrates, blood sinuses serve the functions of capillaries.

**blood vascular system:** A system of vessels, usually provided with one or more hearts or other pumping organ, that enables blood to circulate.

**BMR** (see Basal metabolic rate).

**bog:** An area of spongy soil that previously had been the site of a marsh or swamp.

**Bohr effect** [named for Christian Bohr, Danish physiologist and father of Niels Bohr]: The increased dissociation of oxygen from hemoglobin in the presence of carbon dioxide.

**book gills:** The booklike gills of the horseshoe crab, *Limulus*.

**book lungs:** The booklike lungs of spiders, thought to have evolved from book gills.

**Bowman's capsule:** The enlarged portion of the blind end of a nephric tubule in the vertebrate kidney.

**brachiocephalic artery:** The largest branch of the aorta. It divides into the right common carotid and right subclavian arteries. Also called the innominate artery.

**brackish:** Said of water which is a mixture of sea water and fresh water.

**brain nuclei:** Areas in the brain marked by concentrations of nerve cell bodies.

**brainstem:** The posterior region of the vertebrate brain, including the midbrain, pons, and medulla oblongata.

**Broca's area** [after Paul P. Broca, French surgeon and anthropologist, 1824–1880]: A posterior region of the cerebrum; the area of articulate speech.

**bryophyte:** A member of the division Bryophyta, which contains the mosses, liverworts, and hornworts.

**budding:** Asexual reproduction in which a new cell is pinched off a parent cell, as in yeast, or in which a new individual is produced from part of another, as in *Hydra*.

**bulk transport:** The transport of large molecules and liquids into and out of cells.

**cDNA:** Copy DNA synthesized on an RNA template.

**$C_3$ pathway** (see Calvin cycle).

**$C_4$ pathway:** The metabolic pathway in certain plants ($C_4$ plants) by which carbon dioxide is fixed in a four-carbon organic acid before being reduced in the Calvin cycle.

**CAD** (see Coronary artery disease).

**caisson disease** (see Decompression sickness).

**callus** [L. *callos,* hard skin]: The mass of cells produced by wounding plant parts or generated from explants in tissue culture.

**calorie** [L. *calor,* heat]: A calorie (c) is the amount of energy in the form of heat required to raise the temperature of one gram of water by 1°C from 14.5°C to 15.5°C. The Calorie (C) used in metabolic studies is the kilocalorie (1,000 c), the amount of heat required to raise the temperature of 1,000 grams of water from 14.5°C to 15.5°C.

**Calvin cycle** [after Melvin Calvin, American biochemist (1911–      )]: The series of reactions in which carbon dioxide is reduced to carbohydrate in the light-independent part of photosynthesis.

**cambial initials:** The dividing cells of the vascular cambium that give rise to cells of the secondary xylem and phloem.

**cambium** [L. *cambiare,* to exchange]: The lateral meristems of the plant, the vascular cambium and cork cambium, that produce the secondary tissues.

**cAMP:** Abbreviation for cyclic adenosine monophosphate.

**CAM plants:** Plants of the Crassulaceae family; they are able to fix carbon dioxide at night in organic acids and use it during the day while the stomata are closed.

**capillary** [L. from *capillus,* hair]: The finest of the blood vessels.

**capillary action:** The ability of water or any liquid to rise against gravity along a surface because of a combination of cohesiveness and adhesiveness (attraction between molecules of different substances).

**capsule** [L. *capsula,* a small chest]: 1. The layer around the walls of certain bacteria. 2. The sporangium of bryophytes. 3. Seed pod.

**carbohydrate** [L. *carbo,* carbon + *hydro,* water]: An organic compound with a chain or ring of carbon atoms to which are attached hydrogen and oxygen in a ratio of 2:1, the same as that of water $(CH_2O)_n$. The most important carbohydrates are the sugars, starches, cellulose, and glycogen.

**carbonic anhydrase:** The enzyme that catalyzes the combining of carbon dioxide with water to form carbonic acid, or the reverse of that reaction.

**carbon monoxide poisoning:** The combining of carbon monoxide (CO) with hemoglobin, thereby limiting its function as a carrier of oxygen.

**carcinogenic:** Capable of inducing malignant tumors, or cancer.

**cardiac muscle** (myocardium): The muscle of the heart. It is characterized by having discs in the myofibrils, centrally placed nuclei, and anastamosing fibers.

**cardiac output:** The volume of blood in liters expelled per minute by the left ventricle.

**caribou:** Any of several large North American deer of the genus *Rangifer*. They are related to the reindeer of Europe and Asia.

**carnivore:** An animal whose diet consists mostly of meats.

**carotenoids** [L. *carota,* carrot]: A class of yellow to red pigments found in plastids and including carotenes and xanthophylls.

**carotid bodies:** Patches of receptor cells located at the points where the common carotid arteries branch; they monitor the concentration of oxygen and, to a lesser extent, of carbon dioxide and hydrogen ions in the blood.

**carpel:** The structure in the flower enclosing the ovule(s). One or more comprise the pistil of the flower.

**carrying capacity:** The largest number of organisms of a particular species that an area can support for an indefinite period of time.

**Casparian band** [after Robert Caspary, German botanist]: A water-tight strip in the cell walls of endodermal cells that encircles the cell in the radial and transverse walls and contains suberin.

**catabolism** [Gk. *katabole,* a throwing down]: The sum of all of the chemical reactions in the cells of an organism, in which complex compounds are broken down into simpler substances.

**catalyst** [Gk. *katalysis,* dissolution]: A substance that increases the rate of a chemical reaction and remains unchanged in the reaction. Enzymes are catalysts.

**catecholamines:** A group of amines that includes epinephrine, norepinephrine, and dopamine.

**cathode ray oscilloscope (CRO):** An instrument that traces on a fluorescent screen, by means of a beam of electrons (cathode ray), variations in either of two variables. By its lack of friction it can accurately graph small, rapid changes in electric potentials.

**cation** [Gk. *kata,* down + *ion,* going]: A positively charged ion; it migrates to the negative electrode (cathode) when exposed to an electrical field.

**cauda equina:** The spinal roots of the sacral and coccygeal nerves, so-called because of their resemblance to a horse's tail.

**caudal** [L. *cauda,* tail]: Pertaining to the tail or posterior end of an animal (see also Cephalic).

**cell** [L. *cella,* a chamber]: The basic structural unit of organisms. In eukaryotic organisms, a cell consists of one or more nuclei surrounded by a cytoplasm bounded by a cell membrane. In prokaryotic organisms, there is no membrane-bound nucleus.

**cellular oncogene:** See Protooncogene.

**cellulose** [L. *cellula,* cell]: A polysaccharide consisting of unbranched chains of glucose units; a major structural component of plant cell walls.

**cell wall:** The distinguishing characteristic of all plant cells, the cells of prokaryotes, algal, and fungal protists, and of multicellular algae. Cell walls consist largely of cellulose, which gives plants a kind of structural rigidity.

**centimorgan:** The unit of distance between loci on a chromosome as measured by percent recombinants formed by crossing over. Named after the American Nobel laureate Thomas Hunt Morgan.

**central nervous system (CNS):** The brain and spinal cord of a vertebrate, or their equivalent in invertebrate animals.

**centriole** [Gk. *kentron,* center]: The two or more minute organelles located near the nucleus of most animal cells and the cells of algae and other organisms with motile gametes. During mitosis the centrioles divide and become associated with the spindle.

**centrolecithal:** Having yolk deposited in the center of the egg.

**centromere** [Gk. *kentron,* center + *meros,* apart]: The region of indentation in a chromosome that holds the two chromatids together.

**cephalic** [Gk. *kephale,* head]: Pertaining to the head or anterior end of an animal.

**cephalization** [Gk. *kephale,* head]: The evolutionary trend that concentrates important organs at the anterior end of the animal body, leading to the formation of a head.

**Cephalochordata:** The primitive chordate subphylum characterized by the extension of the notochord anterior to the brain. Relatives of these animals are believed to be ancestral to the vertebrates.

**Cephalopoda:** The class of molluscs in which the foot is modified into several arms that surround the head.

**cercaria:** The fluke larva that develops from a redia.

**cerebellum:** The 3-lobed portion of the brain that lies below the cerebrum and above the pons and medulla oblongata.

**cerebral cortex:** The outer, gray layer of the cerebrum, the largest part of the mammalian brain.

**cerebrospinal fluid (CSF):** The tissue fluid of the central nervous system, occupying the ventricles of the brain, the cerebrospinal canal, and the space between the meninges and the brain.

**chaetae:** The bristles of annelids.

**character displacement:** The evolutionary tendency of closely related species living in the same area to diverge from each other in regard to those traits in which they are similar or in which their niches overlap.

**Chelicerata:** The group of arthropods in which chelicera is the main masticatory organ; includes horseshoe crabs, arachnids, and the extinct eurypterids. See Mandibulata.

**chemiosmosis:** The process in which energy from an electron transport chain is used to establish a proton gradient across the membranes of mitochondria and chloroplasts. Energy generated by the proton gradient is used to synthesize ATP.

**chemolithotroph** [Gk. *cheimon,* storm + *lethos,* stone + *trophos,* feeder]: Autotrophic organisms, such as certain bacteria, that derive their energy from inorganic compounds.

**chemoorganotroph:** An organism that obtains its energy from organic matter. A heterotroph.

**chemotaxis** [Gk. *cheimon,* storm + *taxis,* arrangement or order]: The movement of cells or organisms toward certain chemical substances (chemotaxins) to which they are attracted.

**chemotroph:** An organism that obtains its energy from oxidation of inorganic or organic matter.

**chief cells:** Cells of the stomach that produce pepsinogen, the inactive form of pepsin.

**chitin** [Gk. *chiton,* tunic]: A linear, unbranched polysaccharide that forms the tough outer skeleton of arthropods and the cell walls of molds.

**chloride shift:** The reversible exchange of chloride ions and bicarbonate ions between erythrocytes and the plasma in response to the ionic imbalances resulting from carbon dioxide transport during respiration.

**chlorophyll** [Gk. *chloros,* green + *phyllon,* leaf]: The green photosynthetic pigments of algae and plants.

**chloroplast** [Gk. *chloros,* green + *plastos,* formed or molded]: A membrane-bound organelle that contains the photosynthetic pigment chlorophyll; it is found in algae and plants.

**choanocyte:** A flagellated sponge cell type that is specialized for ingestion of food.

**cholinergic:** An adjective applied to axons that produce acetylcholine at their tips or to receptors that are responsive to acetylcholine.

**chordae tendineae** [L., tendinous cord]: The cordlike attachments holding the atrioventricular valves to the walls of the ventricles.

**Chordata:** Coelomates which possess a dorsal tubular nerve cord, a ventral pulsatory heart and, at some stage in their lives, pharyngeal gill pouches and a dorsal notochord.

**chorion:** [Gk. *chorion,* protective tissue of eye]: 1. The name given to the outer protective covering of eggs in insects and other invertebrates. 2. An extraembryonic membrane present in embryos of reptiles, birds, and mammals. In modern mammals, the chorion serves as an endocrine organ and forms a part of the placenta.

**choroid layer** (choroid membrane): The vascular (blood vessel) layer of the eye, continuous with the ciliary body in the front of the eye.

**chromatin** [Gk. *chroma,* color]: The deeply staining fibrils found in the chromosomes of eukaryotic cells. Chromatin consists of double-stranded DNA to which large amounts of protein and small amounts of RNA are bound.

**chromonema:** A threadlike, partially condensed, chromosome strand visible during nuclear division.

**chromosomal aberration:** Any deviation from the normal arrangement of genes on chromosomes, involving deletion, duplication, inversion, and/or translocation of parts of or entire chromosomes.

**chromosomal locus:** The point on a chromosome at which a particular gene or one of its alleles is located.

**chromosome** [Gk. *chroma,* color + *soma,* body]: Structures in the nucleus that contain DNA; chromosomes carry the hereditary information of the cell and are most readily visible during nuclear division.

**chyme** [Gk. *chymos,* juice]: The viscous fluid products of gastric digestion as they enter the duodenum.

**cilium** (pl. cilia) [L. *cilium,* eyelash]: The fine, short, hairlike processes extending from the surface of many types of eukaryotic cells; they are involved in locomotion or movement of materials across cell surfaces. Cilia contain two central microtubules surrounded by nine pairs of microtubules.

**circadian rhythm** [L. *circa,* about + *dies,* day]: A biological rhythm that displays cycles of about 24 hours in length.

**circulatory system:** A blood vascular or similar system in which the fluid follows a roughly circular pattern through the body.

**cistron:** The section of the DNA molecule that specifies a particular polypeptide; a gene.

**citric acid cycle** (see Krebs cycle).

**cladistics:** A taxonomic method that classifies organisms primarily according to what are perceived as the historical points at which evolutionary divergences have occurred.

**classical** (Pavlovian) **conditioning:** Conditioning in which a response to a stimulus is paired with another stimulus that normally elicits no response. In time (with conditioning), the second stimulus will elicit the response normally expected from the first stimulus.

**cleavage:** Cell division of early stage animal embryos in which the cytoplasm is divided by pinching in of the plasma membrane.

**climacteric** [Gk. *klimakter,* klimaxladder]: The end of the fertile period in the human female, characterized by irregular menstrual cycles.

**climax community:** In ecological succession, the relatively stable community that results when succession has run its full course.

**clone** [Gk. *klon,* twig]: A line of genetically identical cells derived from one cell by mitosis, or a population of individuals derived asexually from one individual.

**cloning** [Gk. *klon,* shoot]: Production of many identical copies of a gene or of an individual by asexual methods.

**closed program:** In animal behavior, a pattern of behavior that cannot be modified by learning.

**CNS** (see Central nervous system).

**cochlea** [Gk. *kochlos,* shellfish]: The coiled housing of the osseous and membranous labyrinths of the ear in which the organ of hearing (organ of Corti) resides.

**codon:** The triplet code word (that is, the trinucleotide sequence) in a DNA or RNA molecule that specifies a certain amino acid or stop signal in the sequence of a polypeptide.

**coelom** [Gk. *koiloma,* a cavity]: Body cavity or space between the body wall and intestine lined by cells derived from mesoderm.

**coelomoduct:** A duct connecting the coelomic body cavity to the exterior of the body.

**coenocyte** [Gk. *koinos,* shared in common + *kytos,* hollow vessel]: A multinucleate organism or one made up of multinucleate cells.

**coenzyme** [L. *co,* together + Gk. *en,* in + *zyme,* leaven]: A nonprotein organic molecule that plays an accessory role in enzyme-catalyzed reactions. Some coenzymes are vitamins, some are synthesized from vitamins, and others contain vitamins.

**cofactor:** An organic molecule (coenzyme) or metal ion that plays an accessory role in enzyme-catalyzed reactions.

**cohesion** [L. *cohaerere,* to stick together]: The tendency of like molecules to hold together.

**collenchyma** [Gk. *kolla,* glue]: A supporting tissue of living cells with reinforced walls found in elongating stems and petioles of plants.

**colon** [Gk. *kolon,* limb]: The large intestine exclusive of the rectum.

**colostrum:** The milk produced by the mammary glands immediately after birth. It is rich in antibodies that protect newborns against infection in their first few months of life, when their own immune system is not completely developed.

**commensalism** [L. *com,* together + *mensa,* table]: A symbiotic association between two or more organisms without harm to either and with some benefit to the symbiont.

**commissure** [L. *commisura,* a joining together]: 1. A group of nerve fibers making cross connections between two portions of the nervous system. 2. The region of union of any two structures, such as the lips.

**common carotid arteries:** The two large arteries of the neck, which branch into the right and left internal and external carotid arteries.

**community:** A varied assemblage of interacting organisms living at a given location, such as in a particular pond, woods, or the like.

**community structure:** The diversity of a community's species and the pattern of their distribution. Diversity includes richness, the total number of species, and evenness, their relative abundances.

**competitive exclusion principle** (Gause's rule): The hypothesis that two or more species occupying identical ecological niches cannot exist for long in the same area before one of them succeeds to the exclusion of the other or others.

**conception:** A term applied to fertilization of the ovum by the sperm in humans.

**conditioned stimulus:** A stimulus to which a conditioned (Pavlovian) reflex has developed.

**cones:** 1. Reproductive structures of conifers (strobili). 2. The light-sensitive tips of receptor neurons of the vertebrate retina that are concerned with color vision.

**conjugation** [L. *conjugate,* together]: Temporary fusion of cells during which genetic material is exchanged between the conjugants. Common in ciliates, bacteria, and algae.

**coniferous tree:** A cone-bearing tree, such as a pine.

**connecting strands:** Strands of cytoplasm extending through pores in the sieve areas of phloem sieve tubes.

**consumer:** In ecological systems, an organism that derives its energy by consuming others. Primary consumers are the plant-eating animals, or herbivores. Secondary consumers, in turn, feed upon the primary consumers.

**contour cultivation:** Agricultural tilling and planting along the contours of slopes, in contrast to planting in rows that run up and down slopes.

**contraception:** A physical or chemical method of preventing conception during intercourse.

**conus arteriosus:** The largest artery in the amphibian body. It receives blood from the single ventricle and forms the trunk from which the two aortic arches arise.

**convergent evolution:** The evolutionary process resulting in similarity in structural features of organisms that occupy similar habitats.

**copepod** [Gk. *kope,* oar + *pod,* foot]: Various small freshwater or marine crustaceans, of chief importance in the food chain as food for small fish.

**copy DNA:** See cDNA.

**cork (phellem):** The outer cell layers of the periderm, or covering, of stems and some roots.

**cork cambium:** The lateral meristem that produces the cork of the plant stem.

**Coriolis force** (Coriolis effect): The apparent deflection of a mass of air or water that is moving toward or away from the equator, due to the fact that the earth is rotating more rapidly or slowly than the mass is.

**cornea** [L. *corneus,* horny]: The transparent anterior face of the eyeball, continuous with the sclera.

**coronary** [L. from *corona,* garland]: Pertaining to vessels or nerves that encircle an organ, especially the heart.

**coronary artery disease (CAD):** The thickening and hardening of the walls of the coronary arteries, and the consequent narrowing of their lumens, that results from depositions of lipids, fibrous tissue, and other materials.

**corpora allata:** The endocrine glands of insects that secrete juvenile hormone.

**corpora cardica:** The two groups of nerve cells in the insect that release the brain hormones.

**corpus luteum** [L. *corpus,* body + *luteus,* yellow]: The yellow endocrine tissue formed within the vacated graafian follicle following ovulation. Its product is progesterone.

**cortex** [L. bark]: 1. In plants, ground tissue between the epidermis and vascular cylinder. 2. The outer portion of any structure such as a cell or organ.

**cortical granules:** The vesicles at the periphery of animal oocytes that break open at fertilization and help form fertilization membrane.

**corticosteroids:** Any of the several steroid hormones produced by the cortex of the adrenal gland.

**cotyledons** [Gk. *kotyledon,* cup-shaped place]: The first leaf (monocots) or leaf pair (dicots) of the plant embryo in the seed.

**countercurrent exchange:** A system consisting of paired, permeable, walled vessels that carry fluid or gases in opposite directions, thereby permitting maximum efficiency in exchange of their heat, oxygen, solutes, etc.

**courtship:** The various activities preceding mating in animals that are essential if mating is to occur. The activities serve various functions, such as quieting aggressive tendencies and promoting sexual arousal.

**covalent bond** [L. *con,* together + *valere,* to be strong]: The stable chemical bond formed between two atoms as a result of sharing one or more pairs of electrons.

**cranial nerves:** The twelve pairs of nerves that issue from the mammalian brain.

**cristae:** The folds formed by the inward folding of the inner membrane of a mitochondrion.

**CRO** (see Cathode ray oscilloscope).

**cross-fertilization:** Union of gametes from two different individuals.

**crossing over:** A term to describe the meiotic (rarely mitotic) events in which homologous chromosomes exchange genetic material.

**cross-pollination:** Pollination of the flowers of one individual by pollen from another.

**CSF** (see Cerebrospinal fluid).

**Crustacea:** Aquatic arthropods in which the body is divisible into two regions, the cephalothroax and abdomen. Includes shrimps, lobsters, and crabs.

**Ctenophora** [Gk. *ktenos,* comb + *phoros,* bearing]: Comb jellies, a group of marine animals closely allied to cnidarians.

**cuticle:** 1. The lipid covering layer of cutin on the epidermis of plants. 2. The chitinous exoskeleton of arthropods. Any body covering such as that found in nematodes.

**cutin** [L. *cutin*, skin]: The lipid material deposited on the outer walls of epidermal cells of plants and forming the cuticle.

**cyclic AMP** (cyclic adenosine monophosphate or cAMP): An energy-rich mononucleotide that is produced from ATP by the removal of its two terminal phosphate groups by the enzyme adenylate cyclase; results in the formation of a six-member ring. This molecule is important in the regulation of many cellular functions.

**cysticercus:** An encysted developmental stage in the life cycle of a tapeworm.

**cytochromes** [Gk. *kytos*, vessel + *chroma*, color]: A class of heme-containing proteins whose principal biologic function is electron transport; they are therefore important components of cellular respiration and photosynthesis.

**cytogenetics:** The study of the physical basis of genetic phenomena by the combined use of the methods of cytology and genetics.

**cytokinesis** [Gk. *kytos*, vessel + *kinesis*, movement]: The division of the cytoplasm following the division of the nucleus in both prokaryotic and eukaryotic cells.

**cytokinins** [Gk. *kytos*, vessel + *kinesis*, motion]: A group of plant hormones, including kinetin, associated with cell division and other effects.

**cytoplasm** [Gk. *kytos*, vessel + *plasma*, anything formed or molded]: All of the protoplasm of a cell except for the nucleus.

**cytostome:** A permanent site in the ciliate body through which food is ingested.

**dalton** [named after John Dalton, the English chemist who developed the atomic theory of matter, 1766–1844]: A unit of mass equal to the mass of one hydrogen atom, or $1.65 \times 10^{-24}$ gram.

**deciduous tree:** A tree that annually sheds its leaves.

**decomposer:** Any of the many small organisms that promote the decay of organic matter. Included are bacteria, fungi, protists, insects, and various worms and larvae.

**decompression sickness** ("bends" or caisson disease): The cramps and other symptoms resulting from the formation of nitrogen bubbles in the bloodstream when atmospheric pressure is rapidly reduced.

**denaturation:** The alteration of the secondary and tertiary structure of large molecules when subjected to unusually acidic or basic conditions, heat treatment, or certain denaturing chemicals. Denatured molecules usually lose their unique biological properties.

**dendrite** [Gk. *dendron*, tree]: A filamentous process, usually branched, that carries nerve impulses toward the nerve cell body of a neuron.

**density-dependent factor:** A factor restricting population size that is effective only as the population increases in size.

**density-independent factor:** A factor restricting population size that operates independently of population size, such as a severe winter, drought, or the like.

**deoxyribonucleic acid (DNA):** A macromolecule consisting of two complementary chains of polynucleotides wound in a double helix and held together by hydrogen bonds. DNA is capable of self-replication, serves as the template for the synthesis of RNA, and is the carrier of genetic information in cells.

**dephosphorylation:** The chemical removal of a phosphate group from a molecule.

**depolarization:** The partial or complete neutralization of a polarized condition, as in the reduction of the resting potential in a nerve or muscle cell.

**deposit feeder:** An animal that feeds on the bacteria, detritus, or the like that settle to the bottom of a body of water.

**dermis** [Gk. *derma*, skin]: The layer just beneath the epidermis of the skin.

**desertification:** The process by which an area is changed from what is usually marginally suitable land for agriculture into a desert, by overgrazing or other abuses.

**Deuterostomia** [Gk. *deuteros*, second + *stoma*, mouth]: The bilaterally symmetrical animals in which the anus arises from the blastopore of the gastrula or near where it had been.

**dialysis** [Gk. *dialysis*, separation or dissolution]: A procedure to remove small molecules, such as salts and urea, from a solution containing macromolecules, such as proteins and nucleic acids, by exploiting their differing diffusibility through a porous membrane.

**diaphragm** [Gk. *diaphragma*, partition]: The complex of muscles and tendons that forms the partition between the thorax and abdominal cavity in mammals and is the chief muscle of respiration.

**diastole** [Gk. *dia*, through + *stellein*, to set]: The relaxation phase of the functioning of one of the chambers of the heart during which it fills with blood. Also, the phase of the cardiac cycle during which the ventricles are undergoing dilatation and filling.

**diatoms** [Gk. *diatomos*, cut in two]: Algal protists with siliceous shells. They constitute a substantial fraction of the phytoplankton.

**dichotomous branching:** Primitive type of branching in which the axis forks into two branches.

**dicotyledon** (dicot): A plant of the class Dicotyledonae of the flowering plants in which the embryos have two cotyledons.

**differentiation:** The progressive diversification of relatively unspecified cells and tissues into more specialized cells and tissues.

**diffusion** [L. *diffundere*, to pour out]: The spontaneous movement of molecules or other small particles from a region in which they are in high concentration to one in which their concentration is lower.

**diffusion lung:** In land snails, a respiratory chamber that depends to a great extent upon diffusion, rather than upon ventilation, for its functioning.

**dihybrid ratio:** The ratio of various phenotypes (rarely, genotypes) to each other in an $F_2$ generation resulting from a cross between two parental-generation individuals differing in respect to two pairs of alleles.

**dikaryon** [Gk. *di*, two + *karyon*, kernel]: A cell in the fungi containing two infused nuclei derived from different parents.

**dinoflagellates** [Gk. *dinos,* dizziness + flagellate]: A group of algal protists characterized by two flagella used in swimming in a spinning motion. They are important as part of the marine phytoplankton. Many are bioluminescent.

**dioecious** [Gk. *di,* two + *oikos,* house]: Said of organisms in which the male and female are separate individuals.

**diploblastic:** The embryonic animal organization in which only two layers, ectoderm and endoderm, are found.

**diploid** [Gk. *diploos,* twofold]: Having a double set of chromosomes (called the $2n$ number), as normally found in the somatic cells of higher organisms.

**disynaptic arc:** A reflex arc composed of an afferent or sensory neuron, an interneuron, and an efferent or motor neuron, and the connecting synapses that link the three cells. (See also monosynaptic arc.)

**DNA** (see Deoxyribonucleic acid).

**DNA ligase:** The enzyme that can ligate two DNA molecules end-to-end.

**DNA polymerases:** A group of enzymes that catalyze the polymerization of deoxynucleotides on a DNA template.

**doldrums:** A belt of calm and light winds north and south of the equator.

**dominance hierarchy:** A social pattern in which one, or sometimes a very few, animals dominate the rest of the group. Levels of dominance often exist within the group, with older, stronger males usually holding the highest positions in the hierarchy.

**dominant allele:** The allele that is expressed when present in heterozygous condition; opposite of recessive.

**dorsal root ganglion:** A ganglion representing an aggregation of nerve cell bodies situated in the dorsal root of a spinal nerve.

**drive** (motivation): A basic, inner urge that incites or represses action, as in the sex drive or hunger drive.

**duodenum** [L. *duodin,* twelve each, so-called because of its length in humans, about twelve finger breadths]: The first 20 to 25 cm of the small intestine. It contains the openings of the pancreatic and bile ducts.

**dyad:** A chromosome of Prophase II of meiosis, in which its chromatids are visible.

**ear ossicles** [L. *ossiculum,* small bone]: The three tiny bones of the tympanic cavity or middle ear: the malleus, incus, and stapes (mallet, anvil, and stirrup).

**ECG** (see Electrocardiogram).

**Echinodermata:** Pentaradially symmetrical deuterostome coelomates with a calcareous skeleton and a unique water vascular system.

**ecological niche:** The position an organism occupies in its community: for example, what it eats, what eats it, what effects it has on its environment, and its kind of habitat.

**ecological succession:** The process by which the composition and proportion of species within a community changes with time. It may take hundreds of years to reach completion.

**ecology** [Gk. *oikos,* house + *logos,* study]: The study of the relationship of organisms to their environment.

**ecosystem:** The entire complex of a community of organisms and the physical environment with which they interact.

**edema** [Gk. *oidema,* swelling]: Excessive accumulation of tissue fluid in intercellular spaces by excessive loss of fluid from the bloodstream or by obstruction of its return via the lymphatics.

**egg:** A common term to describe the female reproductive cell or gamete.

**EKG** (see Electrocardiogram).

**electric dipole:** The weak positive and negative charged ends of polar molecules.

**electric organ:** In certain fishes, the group of modified muscle or nerve cells that enable the animal to produce an electric field in its vicinity. It can be used in shocking prey or predators or, in some, as an organ of electrolocation.

**electrocardiogram (ECG or EKG):** A graphic display of the electrical activity of the heart as detected by the application of recording electrodes to three or more areas on the surface of the body. The record reflects the electrical potentials associated with each phase of the cardiac cycle.

**electrocyte:** One of the modified muscle or nerve cells that constitute the units of an electric organ.

**electrolocation:** The capacity of certain fishes to locate objects in the vicinity by detecting the variations they produce in an emitted electric field.

**electrolyte** [Gk. *elektron,* amber + *lytos,* that may be dissolved]: A substance that dissociates into ions when in solution, and thus becomes capable of conducting an electrical current.

**electromagnetic spectrum:** The array of forms of energy reaching earth from the sun, usually classified according to their wavelengths or frequencies.

**electrotonic potential:** A local, graded, passively conducted depolarization or hyperpolarization of the resting potential of a neuron or other cell that decays with time and distance. If sufficiently depolarizing, it may develop into an action potential.

**elicitors:** Substances produced by plant pathogens that induce the host plant to produce resistant phytoalexins.

**El Niño** [Sp. the child]: A worldwide weather pattern characterized by changed ocean currents along the Peruvian coast. The changes, which occur every few years, are disruptive of normal marine productivity.

**embryogeny:** The development of an embryo.

**embryoid:** An embryolike structure that appears in certain plant tissue cultures and may give rise to a plant.

**embryonic induction:** Interaction of two embryonic tissues resulting in mutual or unidirectional alteration in the developmental potential of the tissues as exemplified by induction of a neural plate by the notochord, and the lens by the optic cup.

**emergent properties:** A theory that holds that new properties arise out of (emerge from) matter when it is joined to other matter.

**emergent tree:** In a forest, a tree that stands taller than the trees that form the canopy. Usually referred to merely as "an emergent."

**emphysema** [Gk. *emphysema,* inflation]: A lung disease characterized by an abnormal enlargement of the air spaces, a general loss of elasticity of the lung tissue, and destruction of alveoli.

**endergonic reaction** [Gk. *endon,* within + *ergon,* work]: A chemical reaction that requires an input of energy in order to go to completion.

**endocrine gland** [Gk. *endo,* within + *krinein,* to separate]: A gland that secretes hormones directly into the blood or other body fluids.

**endocytosis** [Gk. *endon,* within + *kytos,* vessel]: The cellular engulfment of materials by the formation of an invagination or inward depression of the plasma membrane. The invagination is pinched off as a vesicle that is released into the cytoplasm. The process includes both pinocytosis and phagocytosis.

**endodermis** [Gk. *endon,* within + *derma,* skin]: The innermost layer of cells of the plant root cortex; it surrounds the vascular cylinder and its cells contain the Casparian bands.

**endometrium:** The vascularized epithelial layer lining the uterus.

**endoplasmic reticulum (ER)** [Gk. *endo,* within + *plasma,* anything formed or molded; L. *reticulum,* network]: A system of interconnecting, flattened membranous sacs found in the cytoplasm of most eukaryotic cells. In regions of a cell that are actively synthesizing protein, the outer surfaces of the ER are covered with ribosomes in which case it is called granular or rough ER. The ER that lacks ribosomes is called agranular or smooth ER.

**endosperm** [Gk. *endon,* within + *sperma,* seed]: The triploid nutritive tissue of angiosperm seeds.

**endorphin:** A class of peptides found in the brain and pituitary gland and which have opiatelike activity.

**endostyle:** A glandular, ciliated organ on the midventral side of the pharynx; present in Amphioxus and ammocetes larvae.

**endosymbiont** [Gk. *endon,* within + *sym,* with or together + *bios,* life]: An organism that lives within another organism in a symbiotic association.

**energy:** The capacity for activity or work.

**energy pyramid:** A food pyramid in which the various nutritional relationships are expressed as the amount of energy needed to support each level of consumption.

**engram:** The alteration produced in nervous tissue that accounts for memory.

**enterocoelic coelom:** The type of coelom that arises as an outpouching of the embryonic gut. See also Schizocoelic coelom.

**enkephalins:** One of the classes of endorphins.

**entropy** [Gk. *en,* in + *trope,* turning]: The measure of the heat or energy of a system which is unavailable to perform work.

**environmental resistance:** The carrying capacity of an environment for a species, expressed as the environment's resistance to population increase.

**enzyme** [Gk. *en,* in + *zyme,* leaven]: The protein molecules that catalyze most of the chemical reactions in living systems.

**epidermis** [Gk. *epi,* on or over + *derma,* skin]: The outermost layers of cells in both plants and animals.

**epididymis** [Gk. *epi,* above + *didymis,* twin]: Coiled part of the vas deferens forming a crescent-shaped body next to the testis.

**epigenesis** [Gk. *epi,* upon or after + *genesis,* formation]: A theory that the development of an embryo consists of the gradual production and organization of parts by an elaborative process; the antithesis of preformation.

**epinephrine** (adrenaline) [Gk. *epi,* on or upon + nephr, from *nephros,* kidney + ine]: The hormone of the adrenal medulla. One of the catecholamines. (See also Norepinephrine.)

**epiphyte** [Gk. *epi,* upon + *phyte,* plant]: A plant that grows upon another plant, for example, orchids and bromeliads.

**epistatic gene:** A gene whose effect masks the effect of another gene, the hypostatic gene.

**EPSP** (see Excitatory postsynaptic potential).

**equilibrium** [L. *aequus,* equal + *libra,* balance]: In a system, the state of balance between opposing forces.

**erythrocyte** [Gk. *erythros,* red + *kytos,* vessel]: A red blood cell.

***Escherichia coli (E. coli)*:** The common species of enteric coli used extensively in biological research.

**estradiol:** One of the estrogens, or female sex hormones, responsible for female secondary sexual characteristics and regulation of the menstrual cycle.

**estrus (oestrus)** [Gk. *oistros,* mad desire]: Periodic growth of endometrial cells and associated sexual arousal in female mammals.

**ethnobotany:** The study of plants in relation to human culture.

**ethology** [Gk. *ethos,* custom + *logos,* study]: The study of animal behavior, stressing evolutionary significance of particular patterns, especially of stereotyped, innate behavior.

**ethylene:** A simple hydrocarbon ($H_2C{=}CH_2$) that functions as a plant growth substance and in fruit ripening and abscission.

**euchromatin** [Gk. *eu,* good + *chroma,* color]: The loosely coiled and therefore less densely staining portions of a chromosome. These portions are involved in RNA synthesis.

**eukaryote** [Gk. *eu,* true + *karyon,* nut or kernel]: An organism whose cells have a membrane-bound nucleus and membrane-bound organelles.

**euphausid** (see Krill).

**euphotic zone:** The surface layers of a body of water into which the sunlight penetrates far enough to permit photosynthesis to occur.

**euploidy** (see Polyploidy).

**Eustachian tube** [after Bartolommeo Eustachio, sixteenth-century Italian anatomist]: The tube connecting the pharynx with the middle ear.

***Eutheria*** [Gk. *eu,* modern + *theria,* wild beasts]: Modern mammals which are characterized by a well-developed placenta.

**eutrophication** [Gk. *eus,* good + *trophe,* nutritive]: The condition of an aquatic ecosystem characterized by an abundance or overabundance of nutrients.

**evocation:** The induction of flowering.

**evolution** [L. *evolvere,* to unroll]: The origin of present day organisms by descent with modification from preexisting species.

**evolutionary fitness** (fitness): The extent to which the genes of an individual tend to survive in its descendants.

**excitatory postsynaptic potential (EPSP):** A depolarizing electrotonic potential generated in a soma (nerve cell body) or dendrite with which an axon tip has synaptic contact. If of sufficient strength and duration, it will generate an action potential in the postsynaptic cell.

**excretion:** Elimination of metabolic wastes.

**exergonic reaction:** A chemical reaction that is accompanied by the release of energy.

**exocrine glands** [Gk. *ex,* out + *krinein,* to separate]: Glands, such as the sweat glands, that secrete their products into ducts.

**exocytosis** [Gk. *ex,* out + *kytos,* vessel]: The transport of substances in bulk from the cytoplasm out of the cell.

**exons:** The expressed segments of a gene separated from one another by unexpressed sequences, the introns.

**experiment:** The alteration of a state or process, followed by a study of the effects of the alteration.

**experimental control:** A duplicate of an experiment that repeats, as far as is practicable, everything that had been done in the experiment proper except for the one factor, or variable, being studied in the experiment.

**exponential (logarithmic) growth:** A phase of population growth in which the size of the population regularly doubles within a certain time period, such as daily, monthly, or annually.

**exteroceptor** (exteroreceptor) [L. *exter,* outward + *recipere,* to receive]: A sense organ or other receptor that receives stimuli from outside the body.

**extinction** (of conditioning): The process by which conditioned reflexes fail to be exhibited following an extended period of time during which an action was not rewarded when performed or an unconditioned stimulus was not paired with the conditioning stimulus.

**$F_1$ spheres:** The lollipoplike structures on the mitochondrial cristae which have ATP synthetase and ATPase activity.

**facilitated diffusion:** The transport of substances across the plasma membrane of cells from a region of higher concentration to a region of lower concentration by a mechanism involving membrane-bound carrier molecules.

**facilitation:** The process by which responsiveness to a particular stimulus is enhanced by a similar previous experience.

**facultative symbiont** [Fr. *facultatif,* capability, opportunity + symbiont]: A symbiont that is capable of either a symbiotic or an independent existence.

**fallopian tube:** The name given to the oviduct in the human female.

**FAD** (see Flavine adenine dinucleotide).

**fallow:** Said of agricultural land that has been left uncultivated for one or more growing seasons.

**fecundity** [L. *fecundus,* fruitful]: The ability to produce offspring.

**feeding center:** A region of the medulla oblongata which controls hunger and satiety.

**fermentation** [L. *fermentatio,* leaven]: The enzymatic decomposition, especially of carbohydrates, as used in the production of alcohol, bread, vinegar, and other food or industrial materials. It is an energy-yielding metabolic process and can occur anaerobically.

**ferredoxin:** An iron-sulfur protein that serves as an electron acceptor in the electron transport pathway of photosynthesis.

**fertilization** [L. *fertilis,* to make fertile]: The process of fusion of the male and female gametes; syngamy.

**fertilization membrane:** The vitelline membrane after fusing with the cortical granule exudate following fertilization.

**fetus:** The young placental mammal, after having developed to the point at which its species is recognizable, in humans at the start of the ninth week after fertilization.

**fiber:** A needle-shaped, thick-walled supporting cell of vascular plants, hollow at maturity.

**fibrinogen:** The soluble form of fibrin, the insoluble protein that is the basis of blood clots.

**filter feeder:** An animal that by any of several devices removes finely suspended particulate matter from water as the source of its food.

**filum terminale:** The threadlike, nonnervous caudal end of the spinal cord of vertebrates.

**fingerprinting:** The procedure employed to identify the composition of a polymeric compound (for example, proteins or nucleic acids) by determining the electrophoretic mobility of its component units.

**fitness** (see Evolutionary fitness).

**fixed action pattern:** A pattern of animal behavior that is stereotyped, always being performed in a similar way by all members of a species, and is typically elicited by a specific stimulus, or releaser.

**flagellum** (pl. flagella) [L. *flagellum,* whip]: A long, threadlike organelle found in eukaryotic cells, involved in locomotion or feeding. Like cilia, it consists of two central microtubules surrounded by nine pairs of microtubules.

**flavine adenine dinucleotide (FAD):** A coenzyme that can accept electrons (i.e., it can be reduced) from various substrates during cellular metabolism.

**florigen** [L. *flor,* flower + Gk. *genes,* producer]: The postulated flowering hormone.

**fluorescence:** The emission of light as a result of exposure to energy from another source.

**fluoroscope:** An instrument similar to an X-ray machine but in which a fluorescent screen is substituted for a photographic plate or film, thereby allowing immediate visualization of the object being examined.

**follicle-stimulating hormone** (see FSH).

**food chain:** The flow of energy and matter in living organisms through a producer-consumer sequence.

**food pyramid:** A representation of the dependence of various animals upon other organisms as a source of food, in which the mass, caloric content, or other considerations of each level of consumption are noted.

**forebrain:** The anteriormost region of the embryonic vertebrate brain.

**fossil fuels:** Combustible matter such as coal, petroleum, and natural gas that derive from remains of ancient organisms.

**founder effect (founder principle):** The evolutionary divergence that results when a small, nonrepresentative subpopulation forms the basis for a new, isolated population that then undergoes speciation. An extreme case of genetic drift.

**fovea** [L. *fovea,* a small pit]: 1. An anatomical term referring to any small pit or depression. 2. The fovea centralis, the small depression near the center of the retina of the eye and the region of greatest visual acuity.

**frameshift mutation:** A type of mutation caused by insertion or deletion of usually a single nucleotide from the coding sequence of a gene resulting in the garbling of all the succeeding codons.

**free energy:** The energy in a system that is available to do work.

**fruit** [L. *fructus,* fruit]: The ripened ovary of the flower and associated parts containing the seeds.

**FSH** (follicle-stimulating hormone): The hormone of the anterior lobe of the pituitary gland, which promotes growth and functioning of ovarian follicles in vertebrate females and spermatogenesis in males.

**Fungi:** The phylogenetic kingdom that includes molds, mushrooms, and other plantlike organisms without chlorophyll that derive their nutrition from products of decay.

**GABA** (see Gamma-aminobutyric acid).

**gametangium** (pl. gametangia) [Gk. *gamein,* to marry + L. *tangere,* to touch]: A gamete-producing cell or structure in plants.

**gamete** [Gk. *gamete,* wife]: A haploid reproductive cell that unites with another cell in sexual reproduction.

**gametogenesis:** The process of gamete formation.

**gametophyte:** The haploid, gamete-producing generation in alternation of generations.

**gamma-aminobutyric acid (GABA):** One of several neurotransmitters of the CNS.

**gamma rays:** High-energy electromagnetic radiations that are similar to X-rays but with shorter wavelengths; they travel at the speed of light and have great penetrating power.

**ganglion** [Gk. *ganglion,* mass]: The thickened region of a nerve that contains a cluster of nerve cell bodies.

**gap junction** (nexus): One of the specialized connecting structures between adjacent animal cells. Gap junctions provide channels for passage of moderately large molecules. They constitute electrical synapses between individual cardiac muscle fibers.

**gastrin:** A polypeptide hormone secreted by the gastric mucosa in response to distension of the stomach by food. It elicits secretion of HCl and pepsin by the stomach.

**Gastropoda:** Molluscs that have a spirally coiled shell and a ventral, flat, usually retractable, muscular foot.

**gastrovascular cavity:** The only cavity of cnidarians (coelenterates), serving the functions of both a digestive and vascular system. It has but one opening, the mouth.

**gastrula:** An embryo at the stage in which it is a two- to three-layered sphere, usually formed by the invagination of the wall of the blastula.

**gastrulation:** The process of forming a gastrula from a blastula.

**gemma** (pl. gemmae) [L. *gemma,* bud]: An asexually produced miniature of an organism, a common form of reproduction in bryophytes.

**gene** [Gk. *genos,* birth]: The hereditary unit, composed, except for some viruses, of the polynucleotide DNA.

**gene amplification:** Selective multiple replication of certain genes.

**gene pool:** The total genetic information present in all members of a species.

**generator potential:** The electrotonic potential produced at the receptor region of a sensory neuron in response to a stimulus. If the stimulus is of adequate (threshold) intensity, the generator potential will induce an action potential in the neuron.

**genetic drift:** Fluctuations in the distribution of gene frequencies caused by isolation of nonrepresentative samples of the founding population.

**genetics:** The study of heredity and variation.

**genome:** All of the genes possessed by an organism.

**genotype:** The genetic constitution of an organism with respect to certain traits.

**geotropism** (gravitropism) [Gk. *ge,* earth + *tropes,* turning]: Curvature of a structure such as the root or shoot, in response to gravity.

**germination** [L. *germinare,* to sprout]: The beginning of growth, as in a spore or seed.

**germ layers:** The three layers, ectoderm, mesoderm, and endoderm, into which embryonic cells are organized at gastrulation.

**germ plasm:** 1. The cytoplasmic determinants responsible for development of germ cells. See also polar granules. 2. The germ cells and the cells that give rise to them.

**gibberellin:** One of a group of plant hormones primarily associated with growth of shoots.

**gill:** 1. A respiratory organ of an aquatic animal. 2. The spore-bearing filaments on the underside of a mushroom.

**glomerulonephritis:** A kidney disease affecting glomerular function.

**glomerulus** [L. *glomeris,* ball or round knot]: The capillary tuft present at the head of a nephron in vertebrate kidneys.

**glucagon** [Gk. *glukus,* sweet + *ago,* to lead toward]: The pancreatic hormone that acts to increase the concentration of blood glucose.

**glucocorticoids:** The class of steroid hormones produced by the adrenal cortex that affect carbohydrate, lipid, and protein metabolism, for example, by promoting glucose production from protein and fat.

**glucose** [Gk. *glykys,* sweet]: A six-carbon sugar ($C_6H_{12}O_6$), which is the most abundant monosaccharide in animals.

**glycogen** [Gk. *glykys,* sweet + *gennan,* to produce]: A polysaccharide that yields glucose on hydrolysis; one of the main storage carbohydrates of animal cells.

**glycolysis** [Gk. *glykys,* sweet + *lysis,* dissolution]: The series of enzyme-catalyzed reactions by which cells break down glucose molecules into simpler molecules, chiefly pyruvate or lactate. This reaction sequence occurs in the presence or absence of oxygen and results in the liberation of small amounts of useful energy.

**glyoxisome:** A membrane-bound cellular organelle containing enzymes that convert fat into carbohydrate. They are found in microorganisms and in plant cells.

**Gnathostomata:** The second major subgroup in the Vertebrata along with Agnatha. All vertebrates bearing jaws.

**goiter:** An enlarged thyroid gland. The enlargement may be related to an underproduction of thyroxine (hypothyroid goiter) or an overproduction (hyperthyroid goiter).

**Golgi complex** [Camillo Golgi, Italian histologist, 1844–1926]: An organelle consisting of a system of flattened disc-shaped membranes and vesicles that is located in many eukaryotic cells. Golgi complexes function as a site for the packaging of newly synthesized proteins in membrane-bound vesicles for transport to other parts of the cell or outside the cell.

**gonad:** A primary reproductive organ in which gametes are produced.

**gonadotropin:** A pituitary hormone that promotes growth of gonads.

**Gondwanaland:** The southern land mass formed by the separation of Pangaea into two land masses.

**graafian follicle:** A mature ovarian follicle in the mammals that contains a fully developed oocyte.

**grafting:** The fusing together of the parts of two organisms, in the case of plants, usually a shoot or bud, the scion, on a stem or root, the stock.

**grana** (singular, granum) [L. *granum,* grain]: Structures within the chloroplast, consisting of stacks of appressed thylakoids.

**gray matter:** Regions of the CNS in which most of the neurons are unmyelinated, in contrast to white matter, the color of which is due to myelin sheaths on the neurons.

**greenhouse effect:** The phenomenon in which shortwave solar radiation reaches the earth's surface by passing through the carbon dioxide of the atmosphere while the long-wave radiation (heat) into which it is converted is largely prevented from leaving by the same gas, thereby raising the average temperature of the earth's surface.

**ground tissues:** One of the fundamental tissue systems of plants, including cortex and pith.

**groundwater:** Water of the soil and the aquifers beneath the topsoil.

**group selection:** A theory that implies that individual animals perform altruistic acts that benefit the group or the entire species and thus promote the group's survival. Compare with kin selection.

**guano** [Sp. *guano,* dung]: The droppings of birds, especially sea birds of the Pacific. The material, which includes urine and feces, is valuable as fertilizer.

**guard cells:** Paired cells of the leaf and stem epidermis of plants that regulate the opening and closing of the stomata.

**guttation** [L. *gutta,* a drop]: The exudation of water from leaves, seen as droplets at the tip or along the leaf.

**gymnosperm** [Gk. *gymnos,* naked + *sperma,* seed]: A plant of the class Gymnospermae, a category of vascular plants with naked or unenclosed seeds.

**gyrus** [Gk. *gyros,* circle, ring]: A convolution on the surface of the cerebrum. See also Sulcus.

**habitat** [L. *habitare,* to live in]: The particular place in which an animal lives or can usually be found, such as in a certain cave, under a certain rock, and the like.

**habituation** [L. *habitus,* condition]: The coming to ignore a particular stimulus to which an animal had previously responded, after it has been repeatedly administered.

**half-life:** The average time required for the decay or disappearance of one-half of any amount of a given substance or some property of it.

**haltere:** The reduced hind pair of wings found in dipteran flies.

**haploid** [Gk. *haploides,* single]: Having a single set of chromosomes (called the *n* number), the number normally carried by a gamete.

**HAPP** (high-air-pollution potential): A condition in which the atmosphere has become unhealthful by containing excessive amounts of one or more of seven monitored pollutants, and which condition has lasted more than 36 hours and extends across at least 194,000 square kilometers.

**Hardy-Weinberg equilibrium:** The condition in which gene frequencies and genotype frequencies remain unaltered from generation to generation in a panmictic population.

**haustorium** (pl. haustoria) [L. *haustus,* to drink or draw]: An absorptive fungal hypha or plant part that penetrates the substratum or host and takes up nutrients.

**heat of vaporization:** The amount of heat required to raise the temperature of a given amount of liquid from its boiling point to the point at which it becomes a gas.

**Hemichordata:** An aberrant group of coelomate worms with probable kinship to echinoderms and true chordates.

**hemimetabola:** Insects which undergo partial metamorphosis, e.g., cockroaches and bedbugs.

**hemizygote:** An individual of the heterogametic sex (the male in XY sex inheritance) that therefore carries only one of a pair of sex-linked alleles.

**hemodialysis:** Dialysis of blood to remove diffusible materials, such as urea.

**hemoglobin** [Gk. *haima,* blood + globin]: Any of several respiratory pigments in which the protoporphyrin heme is conjugated to a globin protein. In most cases the pigment is contained within erythrocytes.

**hemolymph** [hemo + lymph]: The counterpart of blood in arthropods.

**hemolysis** [Gk. *haima,* blood + *lysis,* dissolution]: The disruption of the plasma membrane of red blood cells accompanied by the release of hemoglobin.

**herbaceous** [L. *herba,* grass]: Denotes a plant in which at least the above-ground parts die back each year.

**herbivore:** An animal whose diet is chiefly vegetation. A primary consumer.

**heredity:** Transmission of physical, physiological and behavioral traits from one generation to the next.

**heterochromatin** [Gk. *heteros,* other + *chroma,* color]: The tightly coiled and therefore densely staining portions of a chromosome. These portions contribute to the structural framework of chromosomes and are believed to have a role in nuclear division.

**heterocyst** [Gk. *heteros,* different + *cystis,* a bag]: A unique cell formed by certain cyanobacteria that functions in nitrogen fixation.

**heterogametic sex** [Gk. *hetero,* one or other + *gamete,* wife]: The sex that produces two types of gametes with respect to their sex chromosomes; X and Y in male mammals.

**heterogeneous RNA (HnRNA):** High-molecular weight, heterogeneous size RNA; present only in the nucleus.

**heterospory** [Gk. *heteros,* other + *spora,* seed]: The occurrence of spores of two sizes, microspores and megaspores, in the alternation of generations of certain vascular plants.

**heterotroph** [Gk. *hetero,* other, different + *trophos,* feeder]: An organism that feeds on other organisms, living or dead.

**heterozygote** [Gk. *hetero,* of one another + *zygos,* yoked]: A diploid organism with two different alleles occupying a specific gene locus.

**hindbrain:** The posteriormost region of the embryonic vertebrate brain.

**histocompatibility antigens:** Antigens on the surface of cells of tissues and organs that are recognized by the immune response and therefore are important in graft rejection.

**holoblastic cleavage** [Gk. *holo,* complete + *blastos,* bud or shoot]: Undergoing complete cleavage of the zygote.

**holometabola:** Insects which undergo total metamorphosis; for example, butterflies, houseflies, beetles, and honeybees.

**homeostasis** [Gk. *homoios,* like, similar + *stasis,* stoppage]: The maintenance of steady states of pH, glucose concentration, and the like within the organism by coordinated physiological and biochemical processes.

**homeotherm** [Gk. *homoios* + *therm,* heat]: An animal that maintains a relatively steady body temperature.

**homogametic sex** [Gk. *homo,* same + *gametes,* husband; *gamete,* wife]: The sex that produces only a single type of gamete with respect to sex chromosomes; XX in female mammals.

**homology:** [Gk. *homoios,* like or similar]: Similarity in structure, position and developmental origin of parts of different organisms.

**homozygote** [Gk. *homoios,* same + *zygos,* yoked]: A diploid organism bearing two identical alleles in the gene locus under study.

**hormone** [Gk. *hormon,* to excite]: A substance produced by certain cells or an organ of the body and which has a regulatory effect upon cells at a more remote site.

**horse latitudes:** The latitudes of about 30 degrees north and south at the edges of the trade wind belt characterized by high pressure, calms, and light, variable winds.

**humus** [L. *humus,* earth, soil]: Organic components of the soil which have barely reached a level of decay at which it can no longer be seen what they had been, such as a leaf, skin, or the like.

**hunger center:** A portion of the feeding center of the hypothalamus adjacent to the satiety center; if stimulated, it prompts the animal to endure pain, if necessary, to eat food.

**hybrid** [L. *hibrida,* offspring of a tame sow and wild boar]: The offspring of two parents that differ in one or more heritable features.

**hydration** [Gk. *hydor,* water]: The act of combining or causing to combine with water.

**hydrogen bond:** A weak molecular bond between a hydrogen atom covalently bound to an oxygen or nitrogen atom on one molecule and an oxygen or nitrogen atom on the same or another molecule.

**hydrologic cycle** (water cycle): The cycle followed on earth by water: it evaporates, falls to earth as rain or snow, percolates through the soil, enters rivers, lakes, and the sea and again evaporates.

**hydrolysis** [Gk. *hydor,* water + *lysis,* dissolution]: The splitting of a molecule by the addition of water, the hydroxyl group being incorporated into one of the split products and the hydrogen atom into the other.

**hydrophobic** [Gk. *hydor,* water + *phobos,* fear]: Having little or no affinity for water as applied to nonpolar molecules or nonpolar regions of molecules.

**hydrothermal vent:** A volcanic opening on the sea floor occasioned by tectonic plates pulling apart and that issues hot water rich in minerals.

**hypermetropia** [Gk. *hyper,* beyond, above, over + *metr,* measure, distance + *ops,* eye]: A defect of the eye in which the image comes into focus behind the retina, usually because of abnormally short eyeballs. Farsightedness.

**hyperosmotic** [Gk. *hyper,* above + *osmos,* impulsion]: Denotes a solution having a higher concentration of solute particles as compared to another solution.

**hypertonic** [Gk. *hyper* + *tonos,* tension]: Denotes a solution that osmotically gains water from another (hypotonic) solution from which it is separated by a particular differentially permeable membrane. Depending on the membrane and the solutes involved, the hypertonic solution might or might not be hyperosmotic to the hypotonic solution. See Hyperosmotic.

**hyperpolarization:** A state usually due to an increased positive charge on the exterior of a cell membrane, especially of an excitable cell, which makes the cell less likely to undergo depolarization.

**hypha** [Gk. *hyphe*, web]: A filament of the fungus body. The hyphae collectively comprise the mycelium, or interwoven strands, of the fungus body.

**hypophysectomy** [Gk. *hypophysis*, outgrowth + *ect*, out, outside + *tomos*, cutting]: Surgical removal of the pituitary gland or embryonic hypophysis that gives rise to the gland.

**hypophysiotropic factors:** The short-chain polypeptide hormones produced by the hypothalamus, which are stored in and released by the posterior lobe of the pituitary gland; they stimulate the release of the various hormones of the anterior lobe of the pituitary. Most are named for the hormone whose release they trigger: thyrotropin releasing hormone (TRH); gonadotropin releasing hormone (GnRH); and so on.

**hypophysis:** The embryonic rudiment of the pituitary gland.

**hypothalamus** [Gk. *hypo*, below + thalamus]: The small region of the brain lying just below the thalamus.

**hypothesis:** A tentative, unproven explanation for a phenomenon, or natural event, which is usually phrased so as to make it testable by experiment.

**hyposmotic** (also hypoosmotic) [Gk. *hypo*, below + *osmos*]: Denotes a solution having a lower concentration of solute particles as compared to another solution.

**hypostatic gene:** A gene whose effect is masked by another gene, the latter being called an epistatic gene.

**hypotonic** [Gk. *hypo* + *tonos*, tension]: Denotes a solution that osmotically loses water to another (hypertonic) solution from which it is separated by a particular differentially permeable membrane. Depending on the membrane and the solutes involved, the hypotonic solution might or might not be hyposmotic to the hypertonic solution. See Hyposmotic.

**ileum** [Gk. *eileon*, twisted]: The last portion of the small intestine, extending from the jejunum to the colon.

**imbibition** [L. *imbibere*, to drink]: The taking in of liquid.

**immunity** [L. *immunis*, free of taxes or burden]: Resistance of an organism to invasion by foreign organisms or to the effects of their products.

**implantation:** A term used to describe the mammalian embryo's attachment to, and burrowing into, the uterine endometrium.

**imprinting:** A special type of learned behavior in which the learning occurs only during a brief, sensitive period early in the animal's life; it usually cannot be unlearned. It may involve an attachment for another individual regarded as the animal's mother and may influence its choice of a mate later in life.

**inclusive fitness:** The evolutionary fitness of an individual plus the fitness of its close relatives (see also Evolutionary fitness).

**individual fitness** (see Evolutionary fitness).

**individual space:** The space immediately surrounding an animal into which it usually does not permit others to enter.

**inducer:** The regulatory molecule that promotes transcription of a gene, usually by decreasing affinity of the repressor to the operon.

**inductive reasoning:** The logical process of judging the truth of a conclusion based on a sample of cases as applicable to all similar cases.

**infarction** [L. *infarciri*, to stuff into]: The death of a localized area of a tissue or organ due to lack of blood supply.

**inflammation** [L. *inflammare*, to set on fire]: A localized nonspecific defense reaction of the body to invasion by organisms or substances or to injury or destruction of tissues.

**inhibitory postsynaptic potential (IPSP):** A potential generated in a neuron receiving synaptic input, which makes that neuron less likely to generate action potentials.

**innate behavior** [L. *innatus*, inborn]: A genetically determined, unlearned behavior.

**inner cell mass:** The group of cells in a blastocyst that gives rise to the mammalian embryo.

**inner ear** (internal ear): The labyrinth, containing the organs of equilibrium and of hearing.

**Insecta:** Arthropods with three pairs of legs and in which the body is divisible into distinct head, thorax, and abdomen.

**insectivorous plant:** A plant, usually one living in an environment low in nitrogen, that supplements its nutrition with insects captured by one or another of various devices.

**insertional mutation:** A point mutation that results when one or two additional nucleotides are inserted into a DNA or mRNA molecule, the consequence being a frameshift mutation.

**insulin** [L. *insula*, island]: The hormone of the beta cells of the islands of Langerhans in the pancreas. A small protein, it participates in regulating fat and carbohydrate metabolism.

**integration** [L. *integer*, whole]: 1. The coordination of the activities of an organism. 2. The receiving and ordering of various inputs of information and the response to them, as in nervous or hormonal integration.

**interferon:** A protein synthesized by virus-infected cells that inhibits the multiplication of viruses.

**interneuron:** A neuron that is located between a sensory neuron and a motor neuron in a disynaptic arc or other neuronal pathway.

**internode:** The region of the plant stem between two successive nodes.

**interoceptor:** A receptor that is sensitive to changes and stimuli originating from within the body.

**interphase:** The period of a cell cycle between successive mitoses. It is a period of high metabolic activity during which DNA replication occurs.

**interstitial fluid** (see Tissue fluid).

**intertidal zone** (see Littoral zone).

**intervening sequence:** See Intron.

**intrinsic rate of increase:** The rate at which a population of a particular species is capable of growing under ideal conditions.

**intron:** The noncoding segment of DNA embedded in a gene. It does not form part of the genetic information represented in the mRNA. Also known as intervening sequence.

**in vitro** [L. in glass]: Denotes the culturing of cells or organisms or their parts within flasks.

**ion** [Gk. *ion,* going]: An atom or molecule that bears net positive or negative charges as a result of having lost or gained one or more electrons.

**ionic bond:** Electrostatic forces between atoms or molecules of opposite charge that hold them together.

**ionic regulation:** Regulation of salt concentration of cell cytoplasm and body fluids.

**ion pump** (see Sodium-potassium pump).

**IPSP** (see Inhibitory postsynaptic potential).

**iris** (pl. irides) [Gk. *iris*]: The colored, perforated disc suspended in the aqueous humor of the eye (the perforation being the pupil). The size of the pupil is controlled by two sets of smooth muscles under autonomic control.

**irruptive growth** (Malthusian growth) [L. *irruptio,* to intrude suddenly]: A pattern of population growth that is marked by a period of exponential growth followed by a series of catastrophic reductions in population size due to famine, disease, and other factors.

**ischemia** [Gk. *ischein,* to check + *haima,* blood]: Local insufficiency of blood supply.

**islets of Langerhans** [after P. Langerhans, nineteenth century German anatomist]: The patches of endocrine tissue in the pancreas that are the site of insulin and glucagon production.

**isogamy** [Gk. *isos,* equal + *gamos,* marriage]: The occurrence of morphologically identical gametes in sexual reproduction.

**isolation:** Separation of populations of a species either by geographical, ecological, behavioral, physiological, or anatomical factors to prevent interbreeding.

**isomer** [Gk. *isos,* equal + *meros,* part]: A molecule that contains the same molecular formula and identical number of atoms as another molecule but in a different arrangement.

**isometric muscle contraction:** The contraction of a muscle that is not allowed to shorten during the contraction.

**isosmotic** [Gk. *isos,* equal + *osmos*]: A solution having the same concentration of solutes as some other solution with which it is compared, such as a physiologic salt solution and blood serum.

**isotonic** [Gk. *iso* + *tonos*]: Denotes a solution that undergoes neither a net gain nor loss of water by osmosis when separated by a particular differentially permeable membrane from another solution, the two solutions being said to be isotonic to each other. Depending on the membrane and the solutes involved, the two solutions might or might not be isosmotic to each other. See Isosmotic.

**isotonic muscle contraction:** The contraction of a muscle during which the tension remains the same.

**isotope** [Gk. *isos,* equal + *topos,* place]: Atom of an element that differs from other atoms of the same element in the number of neutrons in the nucleus of the atom. Such isotopes have the same atomic number but different atomic weights.

**jejunum** [L. *jejunus,* fasting]: The approximately two-fifths of the small intestine that lies between the duodenum and the ileum.

**J-shaped growth curve:** An irruptive population growth curve characterized by a rapid, exponential increase followed by a population collapse. Malthusian growth is essentially a series of J-shaped growth curves.

**K-adaptation** (K-selection): The life pattern typical of large, long-lived organisms, especially of those that live in stable climates, so-called because such species tend to maintain a population size close to their environment's carrying capacity (K).

**karyogamy** [Gk. *karyo,* nucleus or nut + *gamos,* marriage]: Cell conjugation with union of nuclei.

**karyotyping** [Gk. *karyos,* nucleus or nut + *typos,* mark]: The identification and numbering of the chromosomes of a species according to size.

**kinesis** [Gk. *kinema,* movement]: An alteration in the rate of an animal's locomotion in response to a stimulus.

**kinetic energy:** The energy of motion.

**kin selection:** A behavior implying solicitude for one's close relatives, which thereby tends to ensure the survival of the individual's own genes in others.

**Krebs (tricarboxylic or citric acid) cycle:** Named after Hans Krebs, its discoverer, it is the energy-yielding cyclic set of chemical reactions in all aerobic cells by which carbon chains derived from sugars, fatty acids, and amino acids are broken down to yield carbon dioxide and hydrogen atoms. The electrons of the hydrogen atoms, which are at a high energy level, are then transferred to the respiratory chain, which uses that energy to synthesize high-energy phosphate bonds.

**krill** (euphausids) [Norw. *kril,* fry (young fish)]: Various shrimplike crustaceans of the family Euphausiidae that form the staple food of baleen whales and many fishes.

**lactation:** Milk production by female mammals.

**ladder type** (of nervous system): In Platyhelminthes (flatworms), a type of nervous system characterized by two longitudinal nerve cords with a series of prominent commissures connecting them with each other.

**larynx** [Gk. *larynx,* the voice box]: The organ of voice; Adam's apple.

**lagging strand:** The DNA strand on which synthesis of the new strand is discontinuous during replication of DNA.

**lampbrush chromosome:** Meiotic chromosome, found during oogenesis of many animals, in which the chromonema is decondensed to form loops projecting from a central axis.

**larva:** A motile, developmental stage of an organism that differs morphologically from the adult. It also differs from the embryo in that it can secure its own nutrients.

**latent learning:** Learning that has no apparent immediate function at the time it is acquired, but which may be useful at a future time.

**lateral line system:** In certain fishes, a system of grooves in the body surface that contain receptors that allow the fish to detect the presence of objects in the water about it.

**lateral meristem:** A dividing region along the axis of the plant that produces the secondary tissues; the vascular cambium and cork cambium.

**latex:** A milky secretion of certain plants; the source of natural rubber.

**laticifer:** Elongated branching plant cells that produce latex.

**Laurasia:** The northern land mass formed by the separation of Pangaea into two land masses.

**law of segregation:** The principle by which the alleles present in an organism maintain their identity and segregate into different gametes.

**leading strand:** The DNA strand on which synthesis of the new strand during DNA replication is continuous.

**leaf axil:** The juncture of leaf and stem.

**leaf primordium:** An embryonic leaf.

**learned behavior:** Behavioral patterns that develop as the result of experience. (See also Innate behavior).

**lectins:** Glycoproteins produced by plants and that are thought to play a protective role against pathogens. Lectins bind to red blood cells and are used in blood typing.

**lentic ecosystem** [L. *lentus,* slow, motionless]: A pond or lake ecosystem.

**lenticels** [L., little lentil, a lens-shaped seed]: The lens-shaped pores on the surface of plant stems.

**leuconoid:** The type of sponge body in which the flagellated chambers open to the exterior via excurrent canals. See also Asconoid and Syconoid.

**leucoplast** [Gk. *leukos,* white + *plastos,* formed or molded]: Colorless plant organelles that contain starch: They are particularly common in storage roots and stems of such plants as turnips and potatoes and in seeds.

**leukemia:** Any disease characterized by an uncontrolled production of leukocytes.

**leukocyte** [Gk. *leukos,* light, white + *kytos,* vessel, container]: A white blood cell.

**Leydig cells:** Endocrine cells that are interspersed between the seminiferous tubules in testes and that secrete testosterone.

**LH** (= ICSH) (luteinizing hormone): A hormone of the anterior lobe of the pituitary that stimulates maturation of ovarian follicles and the development and functioning of the corpus luteum following ovulation. In males, in which it is known as interstitial cell stimulating hormone (or ICSH), it stimulates the interstitial cells of the testis to produce testosterone.

**lichen:** An organism composed of a fungus and an alga in a symbiotic association.

**light meromyosin (LMM):** Most of the length of the rod-like portion of the myosin molecule.

**lignin:** A component of the cell wall in sclerenchyma cells and cells of the wood of plants.

**limbic system** [L. *limbus,* border]: A ring of cerebral cortical tissue, phylogenetically the oldest portion of the cortex, precursors of which are found in fishes. It controls emotional patterns and drives.

**linkage** (of genes): The tendency of genes to be inherited together by virtue of their location on the same chromosome.

**lipase:** An enzyme which hydrolyzes or synthesizes lipids.

**lipid** [Gk. *lipos,* fat]: A diverse group of biochemical substances that are only slightly soluble or are insoluble in water and are soluble in organic solvents such as acetone, ether, and chloroform; examples are the fats, oils, waxes, phospholipids, and steroids.

**lipolysis:** Hydrolysis of a lipid.

**litter:** An ecological term referring to dead animal and vegetable matter that has not yet decomposed to a point at which its origin cannot be determined. (See also Humus.)

**littoral (intertidal) zone:** The seashore zone between high and low tide marks.

**LMM** (see Light meromyosin).

**logistic growth:** A kind of population growth that is characterized by a slow increase in rate, followed by a period of logarithmic growth in which the population doubles at a regular rate, and culminating in a period of no growth during which the population fluctuates about a particular size.

**logistic growth equation:** $\Delta N/\Delta t = rN\frac{(K - N)}{K}$, in which K is the environmental carrying capacity; N is the number of organisms in the population at the beginning of growth; r is the intrinsic rate of growth; and $\Delta N/\Delta t$ is the change in the population size (N) during time t.

**loop of Henle, or Henle's loop:** A narrow U-shaped connecting tube between the proximal and distal tubules in a mammalian nephron.

**lotic ecosystem** [L. *lotus,* a bathing]: A running water ecosystem, such as a creek or river.

**lumen** (pl. lumina) [L. *lumen,* light, opening]: The space inside a tubular structure such as a tube, vessel, or duct.

**lungfish:** Any of three major types of air-breathing fishes in which the swim bladder is modified to form a lung.

**lymph** [L. *lympha,* clear water]: The fluid carried in the lymphatics or lymph vessels. It is the name given to the interstitial fluid that has entered the lymphatics.

**lymphatics:** The extensive system of vessels paralleling the blood vascular system of vertebrates and that collects and returns lymph to the blood stream.

**lymphocyte:** A type of small leukocyte that is intimately involved in the immune response. Of the two major types, B cells produce antibodies and T cells attack diseased or foreign cells. See B cell and T cell.

**lysosome** [Gk. *lysis,* loosening + *soma,* body]: A membrane-bound cellular organelle containing various hydrolytic enzymes.

**lysozyme** [Gk. *lysis,* dissolution + *zyme,* leaven]: An enzyme produced by certain tissues that is highly effective in digesting the cell walls of many bacteria.

**macroevolution:** Large-scale evolutionary change, occurring over a long period of time and involving the origin of major taxa.

**macrophage** [Gk. *macros,* large + *phagein,* to eat]: A large white blood cell that possesses marked phagocytic and digestive powers. These cells are widely distributed in tissues and body fluids where they serve to remove foreign materials and dead and dying cells and act as accessory cells in the immune response.

**macromolecule:** Any of several types of large molecules, such as proteins, polysaccharides, and nucleic acids.

**macronutrients:** The nutrients required in relatively large amounts by plants. They include N, P, K, Ca, Mg, and S.

**magnetite:** An iron ore with magnetic properties; the mineral in lodestones. Granules of magnetite stored by some organisms (birds, bacteria) enable them to orient their movements in reference to the earth's magnetic field.

**malignant** [L. *malignans,* acting maliciously]: Tending to become progressively worse and to result in death, as in the case with many tumors.

**Malpighian corpuscle** (named after its discoverer Marcello Malpighi, seventeenth century Italian anatomist and physiologist): A component of the vertebrate nephric tubule consisting of a Bowman's capsule and the glomerulus.

**Malpighian tubule:** The excretory organs of insects and a few other arthropods.

**Malthusian growth** (see Irruptive growth).

**Mammalia:** The homeothermal vertebrates characterized by the presence of hair and mammary glands.

**mammary gland:** Milk-producing gland of mammals.

**Mandibulata:** The main subdivision of arthropods, characterized by the presence of mandibles. Includes crustaceans, insects, and myriapods.

**mantle:** The membranous organ that secretes the shell in molluscs. It is modified in cephalopods (squids, octopods) to provide a form of jet propulsion.

**marginal meristems:** The paired rows of dividing cells along either side of a leaf primordium that produce the leaf blade.

**Marsupialia:** See Metatheria.

**mass number:** The total number of protons and neutrons in an atomic nucleus.

**mast crop:** A crop of fruit produced by certain species of trees on other than an annual basis, such as every two, three, or four years.

**Mastigophora:** A group of protozoans in which a flagellum is the locomotory organelle.

**mechanism:** A philosophical school of thought that contends that science can analyze living systems only in regard to their physical features, ultimately explaining their origin and functioning in physical and chemical terms. (See also Vitalism.)

**mechanoreceptor:** A receptor that is sensitive to mechanical deformation, such as pressure, stretching, and touch.

**medulla oblongata:** The lowest region of the vertebrate brain, continuous with the spinal cord, where centers controlling such vital functions as cardiac and respiratory cycles are located.

**meiosis** [Gk. *meioun,* to make smaller]: Reduction in the chromosome number of a nucleus, usually from diploid to haploid, in two successive divisions, forming four haploid nuclei.

**melanocyte** [Gk. *melas,* black + *kytos,* hollow vessel]: Cells located deep in the epidermis of the skin that produce the pigment melanin.

**menarche:** The time of the first menstruation, signifying sexual maturity.

**meninges** (sing., meninx) [Gk. *meninx,* membrane]: The three membranes enclosing the brain and spinal cord of vertebrates: the dura mater, pia mater, and arachnoid.

**menopause** [Gk., *men,* month + *pausis,* cessation]: Cessation of menstrual cycles, marking the end of reproductive years.

**menstrual cycle:** The periodic cycle of growth, vascularization, and sloughing of the endometrium.

**menstruation** [L. *menstruus,* monthly]: The periodic flow of blood, fluid, and cellular debris from the uterus that occurs during a woman's reproductive years.

**meroblastic cleavage:** [Gk. *meros,* part + *blastos,* bud]: Cleavage restricted to the cytoplasmic region of a yolk-laden egg.

**mesophyll** [Gk. *mesos,* middle + *phyllon,* leaf]: The photosynthetic tissue of the leaf, between the upper and lower epidermis.

**messenger RNA (mRNA):** The RNA that is synthesized in the nucleus and transported to the cytoplasm, where it participates in protein synthesis together with ribosomes.

**metabolism** [Gk. *metabole,* change]: The sum total of all chemical reactions in an organism's cells and tissues. It consists of the synthetic processes of anabolism and the degradative reactions of catabolism.

**metamerism** [Gk. *meta,* after + *meros,* a part]: Segmentation of the body along the primary axis to produce a series of homologous parts.

**metamorphosis:** Transformation of form, particularly from the larval or the nymph form to the adult form.

**metanephridium:** An advanced type of nephridium with an internal opening, the nephrostome, and an external opening, the nephridiopore.

**metaphase** [Gk. *meta,* middle + *phasis,* form]: The stage in mitosis or meiosis in eukaryotic cells in which the chromosomes become arranged in an equatorial plane that lies in the middle of the spindle.

**micelle** [L. *micella,* small crumb]: An orderly aggregate of molecules, such as ones that may have formed the basis for the origin of living systems.

**Metatheria** [Gk. *meta,* after + *theria,* wild beast]: The mammals in which the immature fetus is accommodated in a belly pouch and nourished until fully grown. Also called Marsupialia because of the presence of the marsupium or belly pouch.

**Metazoa** [Gk. *meta,* after + *zoion,* animal]: Multicellular animals in which cells are organized into tissues and organs. Includes all multicellular animals other than sponges. See also Parazoa.

**microfilaments** [Gk. *mikros,* small + *filare,* to spin]: Threadlike fibrils found in the cytoplasm of cells, they are involved in cellular movements and form the cytoskeleton.

**micrometer** ($\mu$m, formerly called micron): A unit of measurement equal to one millionth of a meter ($10^{-6}$) in length.

**micron** ($\mu$): See Micrometer.

**micronutrients:** Those nutrients required in minute amounts by plants.

**microtubules** [Gk. *mikros*, small + L. *tubus*, tube]: The long, cylindrical submicroscopic structures found in the cytoplasm of cells. They are involved in maintaining the shape and rigidity of cells, in cytoplasmic movements, and in the formation of secondary wall thickenings in certain plant cells.

**microvilli** [Gk. *mikros*, small + L. *villus*, shaggy hair]: The numerous submicroscopic cytoplasmic extensions of the border of absorptive and secretory cells.

**midbrain:** The middle portion of the embryonic vertebrate brain.

**middle ear:** The tympanic cavity, in which are located the ear ossicles: the malleus, incus, and stapes.

**middle lamella:** The distinct layer between adjacent plant cell walls.

**mineralcorticoids:** The group of steroid hormones produced by the cortex of the adrenal gland and that primarily regulate mineral metabolism and fluid balance. (See also Glucocorticoids.)

**miracidium:** The larval stage into which eggs of flukes hatch.

**missense mutation:** A point mutation that results in the replacement of one amino acid by another.

**mitochondrion** [Gk. *mitos*, thread + *chondrion*, granule]: A membrane-bound spherical or rod-shaped cellular organelle. This organelle is the principal site where energy is transduced in the form of ion gradients and adenosine triphosphate (ATP). It contains the enzymes of the Krebs tricarboxylic acid cycle and of the respiratory pathway.

**mitosis** [Gk. *mitos*, thread]: In eukaryotic cells, the division of the nucleus resulting in the formation of two daughter nuclei, each with the same number and kinds of chromosomes as the parent nucleus.

**mitral (bicuspid) valve:** The atrioventricular valve between the left atrium and left ventricle of the vertebrate heart.

**mobile genetic elements:** DNA sequences that are capable of being transposed from place to place within the genome. Also known as transposons.

**molar** [L. *moles*, mass]: Denotes a solution containing one mole or gram molecular weight of dissolved solute per liter of that solution.

**mole** [L. *moles*, mass]: The amount of a chemical compound whose mass is equivalent to its molecular weight, expressed in grams.

**molecular clock:** A hypothetical principle that measures the apparent rate of nucleotide substitutions in some genes during the course of evolution.

**molecular drive:** A theory to explain speciation associated with large-scale change in the nucleotide sequence of repetitive DNA.

**molecular hybridization:** A method to determine sequence similarity between two nucleic acids; it is based on complementarity of nucleotides.

**Mollusca:** Soft-bodied coelomates in which the outer body wall, the mantle, usually secretes a protective, calcareous shell.

**Monera:** The taxonomic kingdom that includes the bacteria and cyanobacteria, one-celled organisms lacking an organized nucleus.

**monocotyledon** (monocot): A plant of the class Monocotyledonae, of the flowering plants, in which the embryo has a single absorbing cotyledon.

**monoculture:** The raising of a single species of organism, usually an agricultural crop of commercial value, such as wheat, corn, or trees.

**monoecious** [Gk. *monos*, single + *oikos*, house]: Denotes an organism in which male and female gonads are located in the same individual.

**monohybrid ratio:** The statistical ratio of the distribution of dominant and recessive alleles when two individuals heterozygous in a locus interbreed.

**monosaccharide** [Gk. *monos*, single + *sakcharon*, sugar]: A simple sugar, such as glucose or fructose.

**monosynaptic arc:** A reflex arc composed of an afferent or sensory neuron and an efferent or motor neuron and the single synapse between them. (See also Disynaptic arc.)

**monotremes:** See Prototheria.

**morphogenesis** [Gk. *morphe*, form + *genesis*, production]: The origin and development of a part, organ, or organism.

**mortality:** In population ecology, the rate at which members of a population die, usually given in deaths per ten thousand individuals per annum in the case of human populations.

**mosaic egg:** The type of egg in which the developmental potential of the different regions are unalterably fixed before cleavage.

**motor end plate** (myoneural junction): The specialized synapse between the branch of a motor neuron and a muscle fiber.

**motor homunculus:** A sketch of the anterior edge of the human brain's parietal lobe representing the parts that respond when those particular parts of the lobe are stimulated.

**motor neuron:** A neuron carrying impulses from the CNS to an effector, usually a muscle.

**motor unit:** The group of muscle fibers served by all the branches of a single motor neuron, together with that neuron.

**multiple alleles:** The condition in which more than two alleles may be present at a chromosomal locus in a population.

**mutagen:** A chemical or physical agent that causes mutations.

**mutation:** A change in the genetic material that, in the homozygous condition, brings about a change in an individual's phenotype. The change can occur in the sequence of nucleotides in the DNA molecule (gene or point mutation) or can be a gross rearrangement of the chromosomes.

**mutualism:** A type of symbiotic relationship in which both the host and the symbiont benefit.

**mycorrhizae** [Gk. *mykos*, fungus + *rhiza*, root]: A symbiotic association between a fungus and the root of a vascular plant; this association enhances the plant root's uptake of water and nutrients.

**myelin** [Gk. *mylos,* marrow]: The white, fatty substance of the sheath of some neurons; the so-called white matter of the CNS.

**myelinated neuron:** A neuron covered with a myelin sheath (see Myelin).

**myocardium** (see Cardiac muscle).

**myofibrils** [Gk. *myos,* mouse, muscle + fibril]: The microscopically fine longitudinal fibrils seen in muscle fibers. In striated muscle, they bear the cross striations.

**myofilaments:** The submicroscopic filaments of myosin and actin that constitute the contractile component of the muscle fiber.

**myoglobin:** A kind of hemoglobin found in muscle. Like blood hemoglobin, it also binds oxygen.

**myopia** [Gk. *myein,* to close + *ops,* eye (squint)]: Nearsightedness; the optical defect is usually due to abnormally elongate eyeballs, in which the image is focused in front of, rather than on, the retina.

**myosin:** The largest protein component of the muscle fiber and the one constituting the larger of the two myofilament types.

**Myriapoda:** Arthropods in which the segments are not specialized for specific functions.

**NAD** (see Nicotinamide adenine dinucleotide).

**NADP** (see Nicotinamide adenine dinucleotide phosphate).

**nanometer** (nm, formerly called a millimicron): A unit of measurement equal to $1 \times 10^{-9}$ meter in length.

**natality** [L. *natalis,* of birth]: In population ecology, the rate at which new individuals are produced in a population.

**natural selection:** The evolutionary process by which each environment determines ("selects") which of the heritable variations arising in populations will, because they best adapt their possessors to that environment, be the ones to be passed on to future generations.

**natural system of classification:** A taxonomic system that classifies organisms on the basis of their most fundamental similarities and differences, rather than on superficial ones of color, size, and the like. In modern taxonomy, a system that is based on kinships.

**negative feedback:** A regulatory process in which the response that follows a particular stimulus or triggering action inhibits the stimulus or trigger.

**negative reinforcement:** In animal behavior, the encouragement (rewarding) of an animal in the habitual performance of an act by removing an aversive (uncomfortable or painful) condition following each performance of the action.

**nephridiopore:** The external opening of a nephridium.

**nephridium** [Gk. *nephridion,* diminutive form of nephros, kidney]: An individual excretory organ composed of a single tubular structure, usually a metanephridium.

**nephritis** [Gk. *nephros,* kidney + *itis,* suffix meaning inflammation]: Disease of the kidney.

**nephron** [Gk. *nephros,* kidney]: The functional unit of a vertebrate kidney; it consists of Bowman's capsule and a renal tubule.

**nephrostome:** The internal opening of a nephridium.

**nerve:** A group or bundle of neurons outside the CNS (see Nerve tract).

**nerve cell body** (see Soma).

**nerve impulse** (see Action potential).

**nerve net:** A nervous system characteristic of cnidarians (coelenterates) in which the neurons are arranged in a netlike form. An individual may possess more than one such net.

**nerve tract:** A bundle of neurons inside the CNS (see Nerve).

**neuromast:** A type of mechanoreceptor common in aquatic vertebrates. It consists of a cupula, a column of gelatinous material containing sensory hairs.

**neuromuscular spindle:** A group of specialized muscle fibers, and their innervation, that are sensitive to the rate and extent to which stretching occurs in a muscle.

**neuron** [Gk. *neuron,* sinew, nerve]: A nerve cell, usually consisting of a nerve cell body, or soma, an axon, and one or more dendrites.

**neurosecretory cells:** Neurons that produce and release a hormone or hormones into the blood or hemolymph.

**neurotransmitter** (transmitter substance) [Gk. *neuron,* nerve + L. *trans,* across + *mittere,* to send]: A chemical substance released by a neuron at a synapse, which stimulates another cell, usually another neuron or a muscle cell.

**nexus** (see Gap junction).

**nicotinamide adenine dinucleotide (NAD):** A major electron acceptor in the aerobic or anaerobic oxidation of fuel molecules. It is reduced by accepting a hydrogen ion (proton) and two electrons, the reduced form being designated as NADH or $NAD_{red}$.

**nicotinamide adenine dinucleotide phosphate (NADP):** In its reduced form, NADPH, or $NADP_{red}$, a major electron donor in reduction-dependent biosyntheses, such as carbon dioxide fixation in the Calvin cycle of photosynthesis.

**nitrification:** The oxidation of ammonia to nitrite and nitrate by soil bacteria.

**nitrogen fixation:** The process in which atmospheric nitrogen ($N_2$) is reduced to ammonia by certain free-living bacteria and cyanobacteria and by symbiotic bacteria found in root nodules of certain plants.

**node of Ranvier:** Any of the constrictions in the myelin sheaths of myelinated neurons at which points the plasma membrane is exposed to the tissue fluid, thereby making saltatory conduction of action potentials possible.

**nondisjunction:** Failure of separation of homologous chromosomes during meiosis or mitosis.

**nonsense mutation:** A point mutation that generates a stop codon and thus results in synthesis of a shortened polypeptide.

**noradrenaline** (see Norepinephrine).

**norepinephrine** (noradrenaline). Demethylated epinephrine (adrenaline): A neurotransmitter elaborated by nerve endings of postganglionic sympathetic neurons. Its action is similar to that of epinephrine. It is also produced by the adrenal medulla.

**notochord** [Gk. *notos,* back + *chorde,* string]: A rod of vacuolated cells on the middorsal line beneath the nerve cord that forms a rigid structure in vertebrate embryos and adult protochordates.

**nucleic acid** [L. *nucleus,* kernel]: The inherited macromolecules that direct the synthesis of all substances required for an organism's growth and vital processes.

**nucleoid:** Region of a prokaryotic cell that contains the DNA molecule.

**nucleolus** [L. *nucleolus,* a small kernel]: A small, dark body located in the nucleus of eukaryotic cells. It contains DNA, RNA, and protein and is the site of ribosomal RNA synthesis.

**nucleosome:** The chromatin subunit composed of 2 copies each of $H_2A$, $H_2B$, $H_3$, and $H_4$ histones, an up-to 200-base-pair-long DNA, and one molecule of $H_1$ histone.

**nucleus** [L. *nucleus,* kernel]: 1. The organelle that contains most of a eukaryotic cell's DNA and the enzymes for replication and transcription. It is separated from the cytoplasm by a double-membraned nuclear envelope. 2. Any of several concentrations of somas (nerve cell bodies) of neurons in the central nervous system. 3. The core of an atom, consisting of protons and neutrons.

**obligate symbiont** [L. *obligatus,* bound, obliged + symbiont]: A symbiotic organism that can survive only in association with its host.

**Okazaki fragments:** The short fragments of single-stranded DNA synthesized on the lagging strand.

**olfaction** [L. *olfacere,* to smell]: The sense or action of smelling.

**oligolecithal:** Denotes an egg with a small amount of yolk.

**oligotrophic conditions** [Gk. *oligos,* few, little + *trophos,* feeder]: A situation in which little nutrient is available to an ecosystem's organisms.

**omnivore** [L. *omnis,* all + *vorare,* to devour]: An animal that subsists on a variety of animal and vegetable matter.

**oncogene:** Genes identified with tumor induction.

**oncosphere:** The larva of certain tapeworm species.

**oocyte:** A female reproductive cell that gives rise to the egg.

**oogamy** [Gk. *oion,* egg + *gamos,* marriage]: The occurrence of gametes differentiated as egg and sperm.

**oogenesis:** The formation of an egg from a primary oocyte.

**oogonia** [Gk. *oion,* an egg + *gonos,* procreation, seed, or semen]: Mitotically dividing female germ line cells that produce primary oocytes (see also Spermatogonia).

**ootid:** The mature female reproductive cell; synonymous with ovum or egg.

**oogonium:** Singular of oogonia.

**open program:** In animal behavior, a behavioral pattern that is subject to modification through experience.

**operant (instrumental) conditioning:** A learning situation in which an animal is encouraged to repeat a behavior by being rewarded in some way for performing it.

**operator:** The region of the bacterial chromosome that specifically interacts with the regulatory molecule(s) and thus regulates the function of the adjacent cistrons.

**operon:** A group of bacterial genes that function coordinately and are regulated together by a regulator gene.

**opiate** [L. *opiatus,* made from opium]: Any of several drugs, such as heroin, morphine, and codeine, that are derived from the opium poppy.

**optic chiasm** [Gk. *optos,* seen, visible + *chiasma,* crosspiece of wood]: The X-shaped commissure anterior to the pituitary body in which there is a partial crossing over of the optic nerves.

**orbital** [L. *orbis,* circle]: The space in which an electron of an atom will be found 90 percent of the time.

**organ** [Gk. *organon,* instrument]: A differentiated part of a multicellular organism composed of tissues.

**organelle** [Gk. *organon,* tool or instrument]: The organized structures found in cells. These include the Golgi complex, mitochondria, chloroplasts, lysosomes, ribosomes, and endoplasmic reticulum.

**organism:** A living thing. Any living entity with differentiated, interdependent parts, organized into a unified whole that can independently carry on life's functions.

**organ of Corti:** The organ of hearing in the mammalian ear.

**organogenesis:** The embryonic processes involved in shaping of the organs.

**orientation:** In animal behavior, the adjusting by an animal of its position in reference to its surroundings.

**osculum:** The excurrent opening of the spongocoel.

**osmolar:** One liter (1000 ml) of solution containing $6.02 \times 10^{23}$ molecules of a solute or mixture of solutes of any kind.

**osmoregulation** [Gk. *osmos,* thrust]: Control of water content in cells and organisms.

**osmosis** [Gk. *osmos,* thrust]: The passage of water or other solvent from a solution of lesser to one of greater solute concentration when the two solutions are separated by a membrane that selectively prevents the passage of solute molecules but is permeable to the solvent.

**osmotic pressure:** The hydraulic force developed when two solutions of different concentrations are separated by a membrane that is permeable only to the solvent.

**ovary** [L. *ovum,* egg]: 1. In the flower, the basal part of the carpel or carpels (or of the pistil) containing the ovules. 2. In animals, the female gonad, the organ that produces eggs.

**oviparous:** Denotes an animal that produces eggs that develop into young individuals outside the mother's body.

**ovoviviparous:** Denotes an animal that produces eggs that are retained in the parent's body until hatched.

**ovule** [L. *ovum,* egg]: In seed plants, the megasporangium containing the megagametophyte, enclosed in the integument. The ovule becomes the seed.

**ovum:** See Ootid.

**oxidation:** The loss of one or more electrons by an atom, ion, or molecule. A substance can lose or donate electrons only if another substance is at hand to accept them. Therefore, all oxidations are accompanied by reductions and vice-versa.

**oxidative phosphorylation:** The process by which the electrons released during cellular respiration pass down the electron transport chain of mitochondria, where they are used to add a phosphate group to (phosphorylate) ADP to form ATP.

**oxygen tension:** The partial pressure, usually expressed in millimeters of mercury, to which an amount of oxygen dissolved in a liquid corresponds.

**oxytocin** [Gk. *oxy,* quick + *tokos,* childbirth]: A short-chain polypeptide, released from the posterior lobe of the pituitary gland, which stimulates uterine contractions (hastening childbirth) and milk secretion.

**Pacinian corpuscle:** A mechanoreceptor of the mammalian skin. Also found associated with muscles and tendons.

**pair bonding:** Behavior of a mating pair of animals that strengthens their association. It is particularly evident in species in which much parental care is provided.

**palisade parenchyma:** The columnar photosynthetic cells of leaves.

**pallium** [L. mantle, cloak]: The cerebral cortex and superficial white matter of a cerebral hemisphere.

**pancreas** [Gk. *pan,* all + *kreas,* meat, flesh]: The mixed gland of vertebrates that provides several digestive enzymes and two hormones, insulin and glucagon.

**Pangaea:** The name given to the theoretical single land mass of the earth in the geological past.

**parapodium** [Gk. *para,* near, beside + *podium,* foot]: Primitive appendages that serve as organs of locomotion and respiration in annelid worms.

**parasite** [Gk. *para,* beside + *sitos,* food]: A heterotrophic organism that obtains food from another living organism while living on it or within it.

**parasitism** [Gk. *para,* beside + *sitos,* food]: A symbiotic relationship in which the symbiont lives at the expense of and with some damage to its host.

**parasympathetic division of the autonomic nervous system:** The system of neurons emerging from cranial and sacral nerves that generally serves the same organs served by the sympathetic division but which has opposite effects to it. The parasympathetic postganglionic cells release acetylcholine, whereas norepinephrine is released by the postganglionic fibers of the sympathetic division.

**Parazoa:** The subkingdom that includes the simplest multicellular animals, represented by sponges, in which the cells are not organized into tissues and organs.

**parenchyma** [Gk. *para,* beside + *en,* in + *chein,* to pour]: The thin-walled cells of plants, or the tissue made up of them.

**parietal cells:** The cells of the stomach wall that release hydrochloric acid into the stomach lumen.

**parthenogenesis** [Gk. *parthenos,* a virgin + *genesis,* production]: Development of an embryo from an unfertilized ovum.

**partial pressure** (of a gas): In a mixture of gases, the amount of pressure that is due to one of them.

**parturition** [L. *parturire,* to bring forth]: Separation of the fetus from maternal tissue; delivery of the fetus.

**peptide bond** [Gk. *peptos,* cooked or digested]: The covalent bond formed between the carboxyl group of one amino acid and the amino group of a neighboring one.

**perennial** [L. *per,* through + *annus,* year]: Denotes a plant that persists from year to year, usually producing flowers and fruits each year.

**pericycle** [Gk. *peri,* around + *kykos,* circle]: The outermost tissue of the vascular cylinder, in which branch roots arise; in roots, the source of the cork cambium.

**periderm** [Gk. *peri,* around + *derma,* skin]: Outer protective tissue that replaces the epidermis in widening stems; the cork, cork cambium, and cork parenchyma.

**peripheral chemoreceptors** [Gk. *periphereia,* circumference + chemoreceptor]: Chemoreceptors located on the body's exterior, such as in the skin.

**peripheral nervous system (PNS):** The cranial and spinal nerves plus, in some classifications, the autonomic nervous system.

**peristalsis** [Gk. *peri,* around + *stellein,* to gather, constrict]: A progressive wave of relaxation followed immediately by a progressive wave of constriction of a tubular organ. The action, which can proceed in either direction once begun, serves to move the contents along in the direction in which the wave moves.

**peritoneal dialysis:** Removal of urea and other metabolic wastes by dialyzing body fluids through the peritoneum.

**permafrost:** The perennially frozen subsoil of Arctic or subarctic regions.

**peroxisome:** A membrane-bound cellular organelle containing one or more enzymes that digest amino acids, lactic acid, and other substances, to smaller products. They are active in yeast, protozoans, kidney cells, mammalian liver cells, and plant cells.

**pH:** The symbol used to indicate the concentration of hydrogen ions in a solution. It is defined as the negative logarithm of the hydrogen ion concentration.

**phagocyte** [Gk. *phagein,* to eat + *kytos,* vessel]: A type of white blood cell which is able to engulf solid particles and break them down to molecules.

**phagocytosis** [Gk. *phagein,* to eat + *kytos,* vessel]: The engulfing of microorganisms, other cells, and foreign particles by phagocytes.

**phasic receptors:** Receptors that only respond initially to a stimulus by generating a short burst of action potentials and do not respond to continued stimulation of the same type and intensity (see also Tonic receptors).

**phellem** (see Cork).

**phellogen:** See Cork Cambium.

**phenetics:** A taxonomic method that classifies organisms primarily according to the number of characters they have in common, equal weight being assigned to every character irrespective of how fundamental or superficial it may appear to be.

**phenotype:** Observable characters of an organism.

**pheromone** [Gk. *pherein,* to carry + *homon,* to unite]: A substance produced and released by one individual, which serves to communicate with or elicit a specific behavior from another member of the same species.

**phloem** [Gk. *phloos,* bark]: The carbohydrate-transporting vascular tissue of plants.

**phosphoglyceride** (phospholipid: Similar to triacylglycerols (fats) in structure, except that a phosphoric acid is substituted for one of the three fatty acids. Phosphoglycerides form a major part of cellular membranes.

**phosphorylation:** The addition of one or more phosphate groups to a molecule.

**photoauxotroph:** A phototroph, requiring a supply of some organic nutrients for growth.

**photolithotroph:** A phototroph, requiring only inorganic matter and carbon dioxide as nutrients for growth.

**photon** [Gk. *photos,* light]: The energy particle of light.

**photoperiod:** The interval in a 24-hour day during which an organism is exposed to light and which serves to regulate some function, such as flowering or migration. The actual stimulus, however, in some cases, is an uninterrupted period of darkness.

**photophosphorylation** [Gk. *photos,* light + *phosphoros,* generating light]: The process in which ATP is generated by light in photosynthesis.

**photorespiration:** Light-stimulated release of $CO_2$ by the oxidation of glycolic acid; the process is generally wasteful of energy.

**photosynthesis** [Gk. *photos,* light + *syn,* together + *tithenai,* to place]: The process in which light energy is converted to chemical energy in the presence of chlorophyll. Carbon dioxide is fixed in carbohydrates.

**photosystem:** A photosynthetic unit of the chloroplast thylakoids involved in the capture and transfer of energy.

**phototroph:** An organism for which light is the principal source of energy.

**phototropism** [Gk. *photos,* light + *trope,* turning]: The curvature of a plant during growth in response to light.

**phycobilins:** A class of water soluble pigments found in cyanobacteria and red algae and related to the biliproteins of animals.

**phycobilisomes:** Structures on the photosynthetic membranes of cyanobacteria and red algae that contain the phycobilin pigments.

**phycology** [Gk. *phykos,* seaweed + *logos,* word]: The study of algae.

**phyllotaxy** [Gk. *phyllon,* leaf + *taxis,* arrangement]: The arrangement of the leaves on a shoot.

**phylogenetic tree** [Gk. *phylon,* race, tribe + *genes,* born, generated]: A diagrammatic representation of the evolutionary history of major groups of organisms.

**phylogeny** [Gk. *phylon,* tribe + *genesis,* generation]: Evolutionary history of a taxonomic group.

**phytoalexins** [Gk. *phyton,* plant + *alexein,* to ward off]: Substances produced by plants in response to pathogenic agents that block the infection by those agents.

**phytochrome:** The phycobilinlike pigment in plants that senses red and far-red light and participates in photoperiod responses.

**pinocytosis** [Gk. *pinein,* to drink + *kytos,* vessel]: The intake of liquids by a cell by the formation of invaginations or channels in the plasma membrane; these invaginations close to form fluid-filled sacs called pinosomes.

**pioneer species:** The first species to populate an area, such as one devastated by fire or previously unpopulated, as in a newly-formed volcanic island.

**pith:** The ground tissue at the center of the stem or root of the plant.

**pituitary gland** (hypophysis) [L. *pituita,* phlegm]: In vertebrates, the endocrine gland lying at the base of the brain and composed of two prominent lobes, the anterior and the posterior lobe, each of which secretes several hormones. A third lobe, the intermediate, is present in certain animals.

**placenta** [L., cake]: The organ by which the embryo of mammals above the marsupials attaches to the wall of the uterus. It also acts as an endocrine organ. It is formed in part from embryonic tissue and in part of uterine tissue.

**Placentalia:** See Eutheria.

**plankton** [Gk. *planktos,* wandering]: Collective term (sing. plankter) for the small, free-floating aquatic organisms that are carried by lake, river, or ocean currents.

**planula:** The larva of cnidarians.

**plasma:** The fluid portion of the blood: Blood minus its cells and platelets.

**plasma cell:** A fully differentiated cell derived from B lymphocytes following their interaction with an antigen. These cells synthesize and secrete large amounts of antibody that is specific for that antigen.

**plasma (cell) membrane:** The membrane surrounding the cytoplasm of a cell. It is composed in part of a bimolecular layer of lipid molecules.

**plasmid:** An extrachromosomal, circular piece of DNA that replicates independently within its host cell.

**plasmodesmata** (sing., plasmodesma) [Gk. *plassein,* to mold + *desmos,* a band or bond]: In plants, fine threads of cytoplasm extending through the cell walls from cell to cell and connecting their protoplasts.

**plasmogamy** [Gk. *plasma,* anything formed + *gamos,* marriage]: Cytoplasmic fusion of cells.

**plasmolysis** [Gk. *plasma,* anything molded or formed + *lysis,* dissolution]: In hypertonic solutions, the process by which the plant cell protoplast shrinks away from the cell wall as a result of water loss. If plasmolysis of its cells is prolonged, the plant will wilt and die.

**plastid** [Gk. *plastos,* formed or molded]: A cytoplasmic membrane-bound organelle in plant cells. Chloroplasts, leucoplasts, and chromoplasts are examples of plastids.

**plastocyanin:** A copper protein capable of accepting electrons in the electron transport pathway in photosynthesis.

**platelet:** In mammals, the small cell fragments that participate in blood clot formation; in lower vertebrates, the cells called thrombocytes that perform that function.

**Platyhelminthes** [Gk. *platys,* flat + *helminthos,* worm]: The acoelomate triploblastic worms commonly called the tapeworms, flukes, and flatworms.

**pleiotrophy** [Gk. *pleion,* many + *tropos,* turning]: The influence of a single gene on more than one phenotypic trait.

**plexus** [L. *plexus,* network]: A network of interlacing or anastamosing nerves, blood vessels, or lymphatics.

**pneumatophore:** In plants, the aerial extensions of the root system, such as those of the mangrove, that allow oxygen to reach submerged roots. In colonial hydrozoans (Coelenterata; Cnidaria), the individual that is modified to form a gas-filled float that keeps the colony at the surface of the water.

**PNS** (see Peripheral nervous system).

**podocyte** [Gk. *podos,* foot + *kytos,* cell]: Highly branched cells that line the Bowman's capsule. They form a lattice through which the glomerular filtrate passes.

**polar body:** One of the nonfunctional cells produced by the unequal meiotic divisions of oogenesis in animals.

**polar covalent bond:** A covalent bond in which the electrons are shared unequally between two atoms so that the molecule that is formed has regions of slight positive and negative charge.

**polar granules:** Minute cytoplasmic inclusions of the eggs of many animals (mainly insects) that determine the germ or reproductive cells.

**pollination** [L. *pollen,* fine dust]: The transfer of pollen to the cone of a gymnosperm or the stigma of a flower.

**pollution** [L. *polluere,* to soil or defile]: The condition caused by the addition to the soil, air, or waterways of sufficient materials as to reduce or destroy their normal usefulness.

**polygenes:** Genes occurring at two or more loci and that affect a single heritable trait.

**polymorphism:** The existence of two or more morphologically or physiologically different types of individuals in a species.

**polypeptide** [Gk. *polus,* many + *peptos,* cooked or digested]: A molecule consisting of a long chain of amino acids joined together by peptide bonds.

**Polyplacophora:** The group of primitive molluscs in which the shell is made up of several plates.

**polyploidy** [Gk. *poly,* many + *eidos,* form]: Presence of more than two sets of chromosomes.

**polyribosome** (polysome): The cellular organelle consisting of many ribosomes attached to a single strand of messenger RNA.

**polysaccharides** [Gk. *polus,* many + *sakcharon,* sugar]: A long-chain carbohydrate that yields many monosaccharides on hydrolysis. The most important polysaccharides in living systems are starch, glycogen, and cellulose.

**polytene chromosome:** Giant interphase chromosomes in which up to one thousand copies of chromonemata are arranged in parallel as a result of endoreduplication of the chromosome, as in the salivary glands of fruit fly larvae.

**polyunsaturated fatty acid** (see Unsaturated fatty acid).

**pons** [L. *pons,* bridge]: The convex, white, enlarged region at the base of the brain immediately anterior to the medulla oblongata.

**population:** A group of organisms of the same kind, usually of the same species, living in a given area.

**population dynamics:** The factors contributing to the increase, stability, decline, and extinction of populations.

**population structure:** The relative proportions of a population's immature individuals, reproductive adults, and nonreproductive adults.

**positive feedback:** A process in which an organ or other agency initiating an action is stimulated by the action it initiates. Unless ultimately controlled, positive feedback tends to be destructive.

**positive reinforcement:** In animal behavior, the rewarding of an animal for performing an action, and thus encouraging its repetition.

**postganglionic neuron:** In the autonomic nervous system, the neuron that actually innervates the regulated organ.

**postsynaptic cell:** The neuron that receives input at a synapse from the axon tip of another neuron, termed the presynaptic cell.

**potential energy:** Energy stored in a body by virtue of its composition or its position in space.

**preformation:** A theoretical concept that a plant or animal is preformed in the ovum (ovists) or sperm (animalcultists); the antithesis of epigenesis.

**preganglionic neuron:** In the autonomic nervous system, the neuron that delivers input received from the CNS, where it has its cell body; such a neuron terminates in an autonomic ganglion.

**presynaptic cell:** A neuron that terminates in a synapse where it delivers input to another neuron, termed the postsynaptic cell.

**primary motor cortex:** The area immediately anterior to the central sulcus in both hemispheres of the cerebrum, that is, the rear edge of the frontal lobe. The area controls movement of various parts of the body.

**primary somatosensory cortex:** The area in both hemispheres immediately behind the central sulcus of the cerebrum in which sensory information from various parts of the body is recorded. (See also Primary motor cortex.)

**primary treatment** (of sewage): In sewage disposal, the phase of treatment in which the solids are allowed to settle out, the sludge then being removed from the liquid fraction (see also Secondary treatment).

**primitive streak:** An area of the early embryo, in chick and other telolecithal eggs, equivalent to the blastopore. The cells from the epiblast in this region migrate inwards in a concerted manner.

**procambium** [L. *pro,* before + *cambiare,* to exchange]: In plants, the meristematic tissue originating in the apical meristems that gives rise to the vascular tissues and the vascular cambium. Also called provascular tissue.

**producer:** In ecological systems, an organism that captures the energy of sunlight by means of photosynthesis and converts this energy to chemical energy in the form of carbohydrates and other molecules.

**progesterone** [L. *pro,* before + *gestare,* to bear]: The steroid hormone secreted by the corpus luteum of the ovary. It is essential for implantation of the fertilized ovum and for maintenance of pregnancy.

**proglottid:** A segment of a tapeworm.

**prokaryote** [L. *pro,* before + Gk. *karyon,* nut or kernel]: A one-celled organism whose cells lack a membrane-bound nucleus. The nuclear material is either scattered in the cytoplasm or gathered in a nucleoid region of the cell. Includes the bacteria and cyanobacteria (formerly called the blue-green algae).

**prolactin:** A hormone of the posterior lobe of the pituitary that promotes milk production, uterine contractions, and functioning of the corpus luteum.

**promoter:** The region of the gene to which RNA polymerase binds as a prelude to initiation of transcription.

**pronucleus:** The decondensed haploid nucleus of a gamete.

**prophase** [Gk. *pro,* before + *phasis,* form]: The first stage in the nuclear division of eukaryotic cells, which is characterized by the coiling and folding of the chromatin into discernible chromosomes.

**proprioceptor** [L. *proprius,* one's own + receptor]: A receptor, located in a muscle, that provides information about the location of the body or its appendages in space.

**prostaglandins** [Gk. *prostas,* porch + L. *glaus,* acorn]: A group of long-chain unsaturated fatty acids produced by most mammalian tissues and that function as chemical messengers.

**prostate gland:** An accessory male reproductive organ that secretes a viscous fluid which forms the bulk of the seminal fluid.

**protein** [Gk. *proteios,* primary]: A complex organic compound consisting of one or more chains of amino acids joined together by peptide bonds.

**proteinoids** [Gk. *proteios,* primary + *oid,* like]: Polymers of amino acids that result from heating amino acids at moderate temperatures. When placed in water, they form microspheres, regarded as one of the possible forerunners of living systems.

**prothoracic glands:** In insects, the source of the molting hormone, ecdysone.

**Protista:** The taxonomic kingdom that includes all one-celled, eukaryotic organisms, including many colonial forms.

**Protochordate:** A primitive chordate such as a sea squirt and Amphioxus.

**protoderm** [Gk. *protos,* first + *derma,* skin]: The meristematic tissue of the apical meristems that gives rise to the epidermis.

**proton** [Gk. *protos,* first]: A subatomic particle, located in the nucleus of an atom; its positive charge equals in magnitude the negative charge of an electron.

**protonema** [Gk. *protos,* first + *nema,* a thread]: The filamentous early form of the gametophyte of mosses.

**protonephridium:** A primitive nephridium with no internal opening.

**protooncogene:** A gene that putatively regulated specific developmental event(s) but causes cancer if activated at an inappropriate developmental stage or tissue.

**protoplasm** [Gk. *protos,* first + *plasma,* anything formed or molded]: Living matter. The viscous ground material of all cells.

**protostele** [Gk. *protos,* first + *stele,* pillar]: A type of vascular cylinder consisting of a solid column of xylem and phloem; it is thought to be the most primitive type of vascular tissue.

**Protostomia** [Gk. *protos,* first + *stoma,* mouth]: The group of animals in which the blastopore of the gastrula develops into the mouth. See Deuterostomia.

**Prototheria:** The primitive egg-laying mammals, also called monotremes because of the presence of a common urogenital opening.

**Pseudocoelomata:** Animals with a body cavity, between the intestine and body wall, that is not entirely lined by mesodermally derived tissue.

**psilophytes** [Gk. *psilo,* bare + *phyton,* plant]: A group of primitive vascular plants known only from fossils.

**ptyalin** (see Salivary amylase).

**pulmonary arteries:** In higher vertebrates, the arteries that carry blood to the lungs.

**pulmonary circulation:** The arteries, veins, and capillaries that serve the lungs.

**punctuationism:** The theory that evolution occurred at variable rates at different times in the geological history of the earth.

**punishment:** In psychology, the inflicting of pain or other aversive stimulus, following the performance of a behavior. It tends to discourage the behavior. It is to be distinguished from negative reinforcement in which an aversive stimulus or condition is removed following the performance of a behavior.

**Punnett's square:** A nomogram to illustrate the distribution of alleles in the offspring of heterozygotes. Both genotypes and phenotypes can be predicted by this method.

**purine** [L. *purum* + urine]: The nitrogenous bases adenine and guanine, which form part of the backbone of nucleic acid molecules.

**Purkinje fibers:** The neuromuscular fibers that branch from the bundle of His; they deliver pacemaking impulses to the muscle cells of the vertebrate heart.

**pus:** The yellowish fluid exuding from a wound, consisting of dead cells and large numbers of dead and dying leukocytes.

**pylorus** [Gk. *pyloros,* gatekeeper]: The narrow region of the stomach that opens into the duodenum.

**pyrenoid** [Gk. *pyren,* the stone of a fruit + L. *oides,* like]: An organelle in the chloroplast of certain algae; associated with starch storage.

**r-adaptation** (r-selection): A reproductive pattern marked by a rapid, exponential increase in numbers of individuals at a rate close to the organism's intrinsic rate of increase (r), or maximum possible rate.

**radial symmetry:** Animal organization that allows division of the body into two equal longitudinal halves through any axis.

**radioactive isotope** (radioisotope): An isotope whose nucleus is unstable and as a result emits particles or rays spontaneously as it disintegrates.

**radiometric method:** The procedure used to determine the age of rocks on the basis of the proportion of radioisotope and its degradation product present in the particular rock formation.

**rain forest** (moist forest): A woodland ecosystem that receives more than 200 cm of rain per year.

**rain shadow:** Regions on the leeward side of a mountain range which receive little rain as a consequence of the air having lost most of its moisture as rain or snow as it flowed up the windward side of the range, are said to be in the mountain's rain shadow. The higher the mountain, the greater the effect.

**reaction center:** In photosynthesis, the small group of chlorophyll molecules to which the (light) excitation energy absorbed by most of the chloroplast's chlorophyll molecules is transferred and within which the energy is actually transformed into chemical energy.

**recessive allele:** A gene whose expression is masked when present in heterozygous condition.

**reciprocal altruism:** A symbiotic relationship in which an animal of one species performs an action of benefit to a member of another species and, in turn, benefits from the behavior.

**redia:** The larval stage in the life cycle of flukes that develops within sporocysts and that produces cercariae.

**reduction** [L. *reducere,* to lead back]: The gain of one or more electrons by an atom, ion, or molecule (see Oxidation).

**reflex arc:** A system of two or more synapsing neurons that deliver sensory input into the CNS and route reflex response to the input directly to an effector, such as a muscle.

**refractory period:** The period following reception of a stimulus during which a neuron, receptor, or muscle is unable to respond to a second stimulus.

**regulative egg:** The type of egg in which the developmental potential of its different regions is plastic.

**reindeer moss:** Any of various lichens that grow in arctic and subarctic regions and serve as food for caribou and moose during winter.

**relative refractory period:** The portion of the refractory period that can be shortened by increasing the intensity of a stimulus.

**releaser** (sign stimulus): The triggering stimulus for a fixed action pattern in animal behavior.

**renal cortex** [L. *renes,* kidney]: The outer portion of a vertebrate kidney in which Malpighian corpuscles are located.

**renal medulla:** The inner part of a vertebrate kidney composed of renal pyramids.

**renal papilla:** The common opening of several collecting ducts into the renal pelvis.

**renal tubule:** The tubular part of a nephron; it consists of a proximal and a distal tubule.

**repetitive DNA:** The DNA sequences which are represented in multiple copies in a haploid genome.

**repressor:** The regulatory molecule that prevents transcription when bound to the operon.

**reprocessing** (of spent radioactive fuel): The removal of plutonium to allow its being used as reactor fuel and the preparation of other radioactive fractions for storage.

**resolving power:** The ability of a magnifying lens to separate clearly the individual parts of an image.

**respiration** [L. *respirare,* to breathe]: The process by which organisms take in oxygen and release carbon dioxide. At the cellular level, it is the series of oxygen-requiring chemical reactions involved in the breakdown and release of energy from fuel molecules.

**respiratory center:** The region of the medulla oblongata that controls external respiration (gas exchange with the environment).

**respiratory pigment:** Any of several pigments, usually conjugated to a protein, which serve in some way to increase the oxygen-carrying capacity of blood or other tissue.

**resting potential:** The difference in electrical potential between the inside and the outside of a neuron, muscle fiber, or other cell, when it is not carrying an electrotonic or action potential.

**restriction enzyme:** An endonuclease that cuts double-stranded DNA at sites with specific nucleotide sequences.

**reticular formation:** Most of the central portion of the vertebrate brainstem, a region contributing to alertness.

**retina:** The multilayered, light-sensitive rear wall of the vertebrate eye.

**reverse transcriptase:** The enzyme that catalyzes the synthesis of DNA from an RNA template.

**rhizoid** [Gk. *rhizo,* root]: A water-absorbing cell in fungi and gametophytes of bryophytes and lower vascular plants.

**rhizome** [Gk. *rhizoma,* mass of roots]: A horizontal stem in vascular plants at or beneath the surface of the soil.

**ribonucleic acid (RNA):** A single-stranded macromolecule consisting of nucleotides, each composed of the pentose sugar ribose and a phosphoric group attached to one of the nitrogenous bases adenine, guanine, cytosine, and uracil.

**ribosomal RNA (rRNA):** The RNA which is a major component of ribosomes, the cellular organelles on which proteins are synthesized.

**ribosomes:** Particles located in the cytoplasm of cells; composed of ribonucleic acid (RNA) and protein. In eukaryotic cells, they are often attached to the endoplasmic reticulum, where they serve as a site of translation in protein synthesis.

**ribulose biphosphate carboxylase** (RuBP carboxylase): In photosynthesis, the enzyme that catalyzes the addition of $CO_2$ to the 5-carbon acceptor, ribulose biphosphate.

**RNA:** Abbreviation for ribonucleic acid.

**rods:** The light-sensitive cells at the periphery of the retina that are usually inactive in strong light but are able to function in dim light (night vision).

**R-loop:** The structure formed on displacement of one of the two strands of a DNA double helix upon hybridization to its complementary RNA.

**r-selection** (see r-adaptation).

**rumen** [L. *rumen,* throat]: The first compartment of the ruminant's stomach where food that has been eaten undergoes fermentation before it is regurgitated to be chewed again.

**ruminant animal:** A cud-chewing animal (all even-toed hooved animals except swine and hippopotamuses) with a modified esophagus and stomach comprising four chambers (see Rumen).

**salivary amylase** (ptyalin): A salivary enzyme that hydrolyzes starch to maltose and glucose.

**saltatory conduction:** The rapid conduction of action potentials by myelinated neurons in which action potentials are generated only at the nodes of Ranvier. Conduction between nodes is by a flow of ionic current.

**saltwater intrusion:** In seacoast regions, the seepage of seawater into freshwater aquifers.

**SA node** (see Sinoatrial node).

**saprobe** [Gk. *sapros,* rotten + *bios,* life]: An organism that absorbs decaying organic matter as a source of food.

**saprotroph** [Gk. *sapros,* rotten + *trophe,* nutrition]: An organism that obtains its nourishment by absorption of partially decayed matter. A saprobe.

**sarcolemma** [Gk. *sarkos,* flesh + *lemma,* peel]: The thin membrane covering a muscle cell, consisting of the plasma membrane and an overlying lamina.

**sarcomere** [Gk. *sarkos,* flesh + *mere,* segment]: The segmental units of a striated muscle fiber, separated by Z discs.

**sarcoplasmic reticulum:** The endoplasmic reticulum in striated muscle cells.

**satiety center:** The region in the brain's hypothalamus that controls appetite; if destroyed, an animal will overeat to the point of obesity if food is available.

**schizocoelic coelom:** The type of coelom that arises as a slit in the mesoderm. See also Enterocoelic coelom.

**scion:** Part of a plant grafted to another part, known as the stock, as in the case of a bud grafted to a stem.

**sclera** [Gk. *skleros,* hard]: The tough, fibrous, outer layer of the eyeball, continuous with the cornea at the front of the eye.

**sclereid** [Gk. *skleros,* hard]: A plant cell with a thick secondary wall and hollow center.

**sclerenchyma** [Gk. *skleros,* hard + L. *enchyma,* infusion]: In plants, a supporting tissue of thick-walled cells containing fibers or sclereids.

**scolex:** The sucker-bearing attachment organ of a tapeworm.

**secondary response** (see Anamnestic response).

**secondary sexual characteristics:** The characteristics of the particular sex of a given species exclusive of the gonads.

**secondary tissues:** Plant tissues derived from the lateral meristems, that is, the vascular cambium and cork cambium.

**secondary treatment** (of sewage): The incubation of the liquid fraction of raw sewage to enable bacteria and other decomposers to break down the organic matter in it.

**secretin:** A hormone produced by the wall of the duodenum in response to the entrance of acid from the stomach and which stimulates the pancreas to release its digestive enzymes.

**seed:** In plants, the structure derived from an ovule; it contains the embryo and a supply of food.

**selection pressure:** The advantage of possessing a particular genotype relative to alternative possibilities in a particular environment. If an environment favors a particular genotype, it is said to exert a selection pressure in favor of it.

**self-fertilization:** Union of gametes from the same individual.

**self-pollination:** Pollination of a flower by pollen from anthers of the same flower or others on the same plant.

**semicircular canals:** The organs of equilibrium of the vertebrate ear or the ducts within which they are contained.

**semilunar valves:** The valves between the ventricles and the aorta (aortic) and the pulmonary arteries (pulmonic), so-called because of their half-moon shape.

**seminiferous tubule:** Tubular reproductive tissue in the vertebrate testis. It contains developing spermatozoa and Sertoli cells, or "nurse cells."

**sensory adaptation:** The failure of a receptor to respond to its specific stimulus after the stimulus has been repeatedly administered for a time. Following a period in which it is not stimulated, it will once again become sensitive to the stimulus.

**sensory homunculus:** A sketch of various parts of the human body with the parts arranged along the posterior edge of the frontal lobe of the cerebrum and which represents the specific regions of the lobe in which electrical activity is evoked when a particular body part is stimulated.

**sensory neuron:** A neuron in which action potentials are generated by specific stimuli. Most receptors are modified endings of sensory neurons.

**sepal** [L. *sepalum,* a covering]: One of the outermost parts of the flower; part of the calyx.

**seral stage:** Any of the stages in ecological succession of a particular ecosystem.

**sere:** All of the stages in the ecological succession of an ecosystem.

**Sertoli cells:** Somatic cells in the seminiferous tubules that nourish spermatogonial cells.

**sexual dimorphism:** The occurrence of two distinctly different forms in male and female individuals of the same species.

**sexual reproduction:** Reproduction involving the union of gametes in fertilization or syngamy and the process of meiosis.

**sickle-cell anemia:** A disease marked by the collapse ("sickling") of erythrocytes under oxygen stress (shortage of oxygen). It is due to a mutant form of hemoglobin.

**simple reflex:** A reflex involving a minimum number of neurons, often only two or three.

**sinoatrial node** (SA node): The dense nodule of Purkinje fibers located at the junction of the anterior vena cava and right atrium and constituting the primary pacemaker of the mammalian heart.

**sinus venosus:** In the amphibian heart, the chamber that receives blood from the anterior and posterior venae cavae.

**siphonostele** [Gk. *siphon*, pipe + *stele*, pillar]: In plants, a type of vascular cylinder in which the vascular bundles encircle the pith.

**slash-and-burn agriculture:** The practice of cutting down a portion of a forest, burning the wood and brush, and raising crops on the ash-enriched soil until its nutrients are depleted, at which point the area is abandoned in favor of another such region.

**smooth (involuntary) muscle:** The spindle-shaped nonstriated muscle fibers of the blood vessels and internal organs of the body.

**sociality:** The gathering of members of the same species into cooperative groups.

**sociobiology:** The study of the biological basis of behavior, especially the population structure and communication within animal societies and the physiology underlying social adaptations.

**sodium current:** A flow of sodium ions, usually through a membrane, in response to an electrical imbalance.

**sodium-potassium pump:** The active transport mechanism in membranes of many cells by which sodium ions are secreted and potassium ions are absorbed, usually in a ratio of 3 sodium ions to 2 potassium ions.

**soil structure:** A term applied to the size of granular aggregates of a soil. Structure is important to a soil's water-holding capacity and the ease with which it is penetrated by roots.

**soil texture:** The relative proportions of sand, clay, and silt in a soil, a rough estimate of which is obtained by feeling the soil by rubbing some between the thumb and forefinger.

**solenocyte** [Gk. *solenos*, channel + *kytos*, cell]: A component of a protonephridium consisting of an elongated bulblike cell in which one or more flagella beat in the lumen.

**solute:** A substance which is dissolved in a solvent.

**solvent:** A liquid that dissolves or is capable of dissolving a substance.

**soma** (nerve cell body): The cell body of a neuron, the portion exclusive of its axon and dendrites and in which the nucleus is located.

**speciation:** The process by which new species evolve. See also Allopatric and Sympatric speciation.

**species:** One or more populations of a type of organism, members of which can interbreed and produce fertile offspring.

**spermatid:** Haploid male reproductive cell before it is transformed into a spermatozoon.

**spermatocyte:** A cell that is produced by divisions of a spermatogonium. A *primary* spermatocyte is diploid. At the first meiotic division it produces two haploid *secondary* spermatocytes.

**spermatogenesis:** The process of formation of spermatids from primary spermatocytes.

**spermatogonia** [Gk. *sperma*, seed or germ + *gonos*, procreation]: Mitotically dividing male germline cells that produce spermatocytes.

**spermiogenesis:** The process of formation of spermatozoa from spermatids.

**spike potential** (see Action potential).

**spinal nerves:** Nerves that arise from the spinal cord. In humans, there are 31 pairs.

**spiracles:** In arthropods, the openings of the tracheae to the exterior.

**spiral valve:** In the amphibian heart, the large flap of tissue in the conus arteriosus that routes blood exiting the ventricle, first into the pulmocutaneous arch and then into the systemic arch.

**splash zone:** The seashore zone just above the high tide mark.

**split-brain individual:** An individual in which communication between cerebral hemispheres has been reduced by severing the corpus callosum and other commissures.

**spongy parenchyma:** In plants, the photosynthetic tissue of leaves in which the cells are loosely arranged.

**sporangium** (pl. sporangia) [Gk. *spora*, seed]: The structure in which spores are produced. They are unicellular in algae and fungi and multicellular in plants.

**spore** [Gk. *spora*, seed]: A reproductive cell other than a gamete capable of producing a new individual without union with another cell.

**sporic meiosis:** In sexually reproducing organisms, the occurrence of meiosis during spore formation.

**sporocyst:** The larval stage of a fluke into which the miracidium develops following penetration of the first intermediate host.

**sporogenesis:** The process by which spores are formed. In most plants, it involves meiosis.

**sporophyll** [Gk. *spora*, seed + *phyllon*, leaves]: A sporangium-bearing leaf.

**sporophyte** [Gk. *spora*, seed + *phytos*, growing]: In plants, the diploid, spore-producing generation in alternation of generations.

**staircase effect** (treppe) [Germ., *treppe*, steps]: The series of increases in the extent of contraction of an entire muscle in response to a series of stimuli delivered at brief intervals.

**stamen** [L. *stamen*, thread]: The flower part producing the pollen, consisting of an anther and filament.

**statocyst** [Gk. *statos*, standing + *kystis*, sac]: An organ of equilibrium in invertebrate animals, in which various concretions or other structures stimulate sensitive hairs during body movements.

**stele** [Gk. *stele*, a pillar]: The cylinder of vascular tissues in the roots and stems of plants.

**stenosis** [Gk. *stenosis,* narrowing]: The narrowing of an opening or of a tubular organ.

**stereotyped behavior:** Innate behavior that is exhibited in the same form by all members of a species and which is usually elicited by the same kind of stimulus.

**steroids:** A group of fatlike compounds consisting of four interconnected carbon rings; they are insoluble in water and soluble in organic solvents. Common steroids are cholesterol and the sex hormones.

**stipules:** In plants, paired outgrowths of the stem at the base of a leaf, often leaflike.

**stolon** [L. *stolonis,* stem or branch]: A specialized tubular organ from which buds arise during asexual reproduction in cnidarian animals. In plants a thin stem or "runner".

**stomata** [Gk. *stoma,* mouth]: Openings or pores of leaves and stems through which gasses are exchanged.

**stratification:** In a natural community, the layering of various subcommunities above one another, as in soil layers or at different heights in a forest.

**striated (skeletal) muscle:** The principal musculature of the body, characterized by alternating light and dark cross bands. Also called voluntary muscle.

**stridulation:** The audible vibrations produced by insects, usually by rubbing legs on wing cases. It is usually part of courtship.

**strip-cropping:** The practice of planting two kinds of crops in alternating strips in which one is a tall crop and thus gives protection against wind erosion of the soil in which the shorter plants are growing. Crop rotation is usually practiced between the rows.

**stroma** [Gk. *stroma,* a bed]: The matrix which surrounds the membranous thylakoids in a chloroplast.

**subclavian arteries:** The arteries that supply the forelimbs, so-called because they underlie the clavicle, where they arise from the aorta.

**suberin** [L. *suber,* the cork tree]: A fatty substance in the walls of cork cells and in the Casparian band of endodermal cells.

**substrate** [L. *sub,* under + *stratum,* layer]: A substance upon which an enzyme acts.

**succulent plant** [L. *succus,* juice]: A plant specialized to live in an arid climate. It is usually characterized by having few or no leaves and fleshy stems and branches that function to store water.

**sulcus** [L. *sulcus,* furrow]: Any anatomical groove or furrow. In the mammalian brain, any of the shallow grooves separating the gyri.

**summation:** The accumulation, in space or time, of the effects of action potentials or sensory stimuli.

**surface tension:** The tension due to hydrogen bonds between molecules of a liquid that makes the surface of the liquid behave as if it were an elastic membrane. Water has a very high surface tension.

**swamp:** A wetland characterized by the presence of trees.

**swarming:** The gathering together of large numbers of the same animal species, the term usually being applied to invertebrates, such as insects or worms.

**swim bladder:** A dorsally located sac into which gas is secreted or from which it is absorbed, allowing a fish to regulate its specific gravity and thus its buoyancy.

**syconoid:** The type of sponge body in which the choanocyte-lined flagellated chambers open directly into the spongocoel. See Asconoid and Leuconoid.

**symbiosis** [Gk. *symbiosis,* living together]: A continuing, intimate association between two organisms of different species.

**sympathetic division of the autonomic nervous system:** The sympathetic trunk, sympathetic plexuses, and associated sympathetic pre- and postganglionic neurons. See also Parasympathetic division.

**sympathetic trunk:** The interconnected chain of sympathetic ganglia along each side of the vertebral column.

**sympatric speciation:** Origin of new species from populations of a single species which have been genetically isolated because of physiological or behavioral incompatibility.

**symplast** [Gk. *sym,* together with + *plastos,* molder]: In plant roots, the water pathway through the living cells and their cytoplasmic connections.

**synapse** [Gk. *synapsis,* contact]: The region of communication between neurons or between a neuron and a muscle fiber or other effector.

**synapsis:** Alignment and close association of homologous pairs of chromosomes.

**synaptic delay:** The brief delay in nerve impulse transmission that occurs at chemical synapses.

**synaptic knobs** *(bouton terminaux):* The small terminal enlargements of the branches of axons where synaptic contact is made with other neurons.

**synaptic vesicles:** The vacuoles, 300 to 600 Å in diameter, occurring in synaptic knobs and containing transmitter substances.

**synaptonemal complex:** A ladderlike tripartite structure that binds the paired homologous chromosomes during meiotic prophase.

**syngamy** [Gk. *syn,* together + *gamos,* marriage]: The fusion of gametes, usually an ovum and a spermatazoon, in fertilization.

**systematics:** Taxonomy. The term systematics usually connotes the application of many disciplines to the problems of classifying organisms.

**systemic circulation:** The arteries, veins, and capillaries supplying the vertebrate body, exclusive of the pulmonary or gill circulation.

**systole** [Gk. *systole,* contradiction]: The contraction phase of the cardiac cycle, that is, the contraction of the ventricles.

**taiga:** A northern forest in which the dominant species of trees are conifers.

**taxis** (pl. taxes) [Gk. *taxis,* arrangement]: The movement of a protist or an animal toward (positive taxis) or away from (negative taxis) a stimulus.

**taxonomy:** The science of classifying organisms (see also Systematics).

**T cell:** A type of lymphocyte derived from bone marrow stem cells that matures into an immunologically competent cell under the influence of the thymus. T cells are involved in a variety of cell-mediated immune reactions. See B cell.

**teleology:** A philosophical view of nature that explains the existence and functioning of organisms and their organs as fulfilling various purposes of a creator.

**telolecithal:** Denotes the presence of an abundant supply of yolk in an egg, especially at one pole. See also Alecithal, Oligolecithal and Centrolecithal.

**telophase** [Gk. *telos,* end + *phasis,* form]: The last of the four stages in mitosis and of each of the divisions of meiosis. It begins when the chromosomes arrive at the cell poles.

**temperature inversion:** An atmospheric phenomenon characterized by a layer of cold air coming to lie below a layer of warm air. Because cold air is heavier than warm air, it does not rise; as a result pollutants generated during the inversion become trapped in the layer of cold air.

**temporal lobe:** A portion of a cerebral hemisphere below the lateral sulcus and continuous with the occipital lobe.

**tendon** [L. *tendere,* to stretch]: A band of dense, fibrous connective tissue that attaches a muscle to a bone.

**territoriality:** The establishing and defending of a limited area, usually that surrounding a nesting, breeding, or feeding site. Other members of the same species and sex are driven away.

**tertiary treatment** (of sewage): A process designed to remove toxic metals and 90% or more of the phosphates, nitrates, solids, and bacteria from sewage effluent.

**testis:** The male gonad.

**testosterone** [L. *testis,* witness]: The male sex hormone, secreted by the testes.

**tetanus** [Gk. *tetanos,* stretched, rigid]: A tense, contracted state of muscle. Also, the disease characterized by muscular spasms and extreme rigidity of the body.

**Tethys Sea:** The sea separating Laurasia and Gondwanaland. The Mediterranean Sea is believed to be a remnant of this sea.

**tetrad:** A pair of homologous chromosomes in synapsis in Prophase I or Metaphase I of meiosis, the four chromatids of which are visible. Also called a bivalent.

**tetrapod** [Gk. *tetra,* four + *podos,* foot]: A four-legged, or four-limbed, vertebrate animal.

**thalamus** [Gk. *thalamos,* inner chamber]: Either of the two masses of gray matter located at the sides of the third ventricle of the vertebrate brain.

**thermal pollution:** An artificially caused excessive increase in the temperature of an ecosystem such as is induced by waste industrial heat.

**thermocline:** A sharp temperature discontinuity that develops in bodies of water during the summer between the surface layer of warm water and the colder water below it. Because they are of different densities the two tend not to mix, thereby depriving the surface water of nutrients and the colder water of oxygen from the atmosphere.

**thermodynamics** [Gk. *therme,* heat + *dynamis,* power]: The branch of physics that concerns transfers and conversions of energy. It owes its name to the fact that its laws once dealt only with exchanges of heat within and between systems.

**threshold level intensity:** The minimum level of intensity of a stimulus required to elicit a response.

**thylakoid** [Gk. *thylakos,* small bag]: The parallel flattened sacs or vesicles that form part of the internal membrane structure of chloroplasts. In certain regions of a chloroplast, the thylakoids are arranged in stacks called grana.

**thymus** [Gk. *thymos*]: A lymphoid organ located in the thorax that controls the maturation of bone marrow stem cells into immunologically competent T lymphocytes.

**thyroid gland** [Gk. *thyreos,* shield]: The two-lobed endocrine gland lying across the ventral surface of the mammalian trachea just below the larynx. It is the source of the hormone thyroxine.

**thyroxine:** The hormone of the thyroid gland. It consists of two iodated molecules of the amino acid tyrosine; its action consists of increasing the body's basic metabolic rate.

**tide pool** (tidal pool): A pool in the intertidal zone of the seashore that regularly receives seawater at high tide but is subject to evaporative loss during low tide.

**tissue fluid** (interstitial fluid): The fluid surrounding the tissue cells of the vertebrate body. It is essentially a filtrate of the blood, being the fluid portion of the blood minus its cells and large proteins. As it returns to the blood stream by way of the lymphatic system, it is called lymph.

**tonic muscle contraction:** The slight, constant contraction exhibited by most of the body's muscles.

**tonic receptors:** Those receptors that continue to generate action potentials during a prolonged period of stimulation (see also Phasic receptors).

**trachea** (pl. tracheae) [Gk. *tracheia,* rough]: In vertebrates, the windpipe, the cartilage-supported tube extending from the larynx to its branches, the bronchi. In insects and some other arthropods, one of the air-conducting tubes that constitute their respiratory system.

**tracheid** [Gk. *tracheia,* rough]: In plants, an elongated, thick-walled and hollow water-conducting cell of the xylem, having pits in its walls.

**tracheoles:** The fine branches of arthropod tracheae.

**tracheophyte** [Gk. *tracheia,* rough + *phyton,* plant]: A plant of the division Tracheophyta, the plants with vascular tissues, xylem, and phloem.

**trade winds:** The steady easterly (east to west) prevailing winds occurring just above and just below the equator.

**transcription:** Transfer of genetic information encoded in the nucleotide sequence of the DNA molecule into the nucleotide sequence of a complementary RNA molecule by synthesis of the latter.

**transduction:** Modification of the genetic constitution of an organism by introduction of foreign DNA via an infectious agent.

**transfer cell:** In plants, a parenchyma cell with wall ingrowths that increase the surface of the adjacent plasma membrane. Transfer cells function in secretion and solute transfer.

**transfer RNA (tRNA):** A kind of RNA involved in the transport of amino acids. This RNA exists in several forms, each of which is able to bind to a specific amino acid, which is then delivered to the ribosomes for the synthesis of proteins (see also RNA).

**transformation:** Modification of the genetic constitution of an organism by uptake of foreign DNA.

**translation:** Transfer of the genetic information encoded in the nucleotide sequence of RNA molecules into the amino acid sequence of a corresponding polypeptide chain by synthesis of the latter.

**translocation:** 1. A step in protein synthesis at which the peptidyl tRNA present in the A site is moved to the P site. 2. A chromosomal aberration in which a piece of one chromosome becomes attached to another.

**transmitter substance** (see Neurotransmitter).

**transpiration** [Fr. *transpirer,* to perspire]: The loss of water vapor from plants.

**treppe** (see Staircase effect).

**triacylglycerol** (triglyceride): A kind of lipid that yields glycerol and three fatty acids on hydrolysis; it is stored in mammalian fat cells.

**tricarboxylic acid cycle** (see Krebs cycle).

**trichome** [Gk. *trichos,* hair]: A filament of cells in algae, or an epidermal hair or outgrowth in vascular plants.

**tricuspid valve** [L. *tri,* three + *cuspis,* point]: The three-cusped valve between the right atrium and right ventricle in the mammalian heart.

**triglyceride** (see Triacylglycerol).

**triploblastic:** The animal organization in which the three layers—ectoderm, mesoderm and endoderm—occur.

**trisomy** [Gk. *tri,* three + *soma,* body]: An aneuploid condition in which a chromosome is represented three times instead of the usual two. It is caused by nondisjunction during meiosis.

**trochophore:** The larval stage of most annelids and molluscs.

**trophoblast** [Gk. *trophos,* a feeder + *blast,* sprout]: The outer layer of the mammalian embryonic stage called the blastocyst.

**tropomyosin:** A long fibrous protein of muscle associated with actin filaments; it inhibits formation of cross-bridges between actin and myosin in the absence of calcium ions.

**troponin:** A globular protein associated with the tropomyosin-actin complex.

**TSH:** Abbreviation for thyrotropin or thyroid stimulating hormone.

**tubulonephritis:** A kidney disease affecting the nephric tubules.

**tundra:** An arctic community characterized by treeless plains, lichens, caribou, and permafrost.

**turgor pressure:** The pressure created in plant cells by the presence of a cell wall which restricts the enlargement of the cell following the osmotic uptake of water. Turgor pressure gives rigidity to the plant body.

**twitch:** A brief, all-or-none contraction of a single muscle fiber or the brief, sudden contraction of a small group of muscle fibers in response to a single maximal stimulus.

**tympanum** (ear drum) [Gk. *tympanum,* drum]: 1. The ear drum. 2. The middle ear.

**unconditioned stimulus:** A stimulus that normally elicits a precise, unlearned response.

**unsaturated fatty acid:** A fatty acid possessing one or more sets of double bonds between its carbon atoms.

**upwelling:** The rising of deep, nutrient-laden waters from the sea's depths as a result of wind patterns and other forces.

**uremia:** Abnormal retention of urea in the blood.

**ureotely:** Excretion of urea as the major endproduct of nitrogen metabolism.

**ureter:** One of the two ducts leading from the mammalian kidney pelvis to the urinary bladder.

**urethra** [Gk. *ourethra,* derived from ouron, urine]: The duct leading from the urinary bladder to the exterior.

**uricotely:** Excretion of uric acid as the major endproduct of nitrogen metabolism.

**urinary bladder:** The organ in which urine is stored; it is found in most vertebrates, birds being an exception.

**Urochordata:** The primitive chordate group in which the notochord is limited to the tail region.

**uterus:** An accessory female reproductive organ in which embryos develop.

**vaccination** [L. *vacca* cow]: The injection of a suspension of attenuated or killed microorganisms (vaccine) for the purpose of inducing a specific immunity.

**vacuole** [L. *vacuus,* empty]: The cytoplasmic sacs of most plant cells and some animal cells. The membrane-bound vacuoles of plant cells contain salts, organic molecules, and waste products of cell metabolism.

**vagus nerve:** The tenth (X) cranial nerve, a mixed nerve serving the larynx, pharynx, meninges, external ear, and various viscera.

**variation:** Divergence in morphology, physiology or behavior among individuals of a species.

**vasa vasorum** [L. *vas,* vessel]: The blood vessels that supply and drain the walls of the larger arteries and veins.

**vascular tissues:** The water- and food-conducting tissues of the plant: the xylem and phloem.

**vasopressin:** ADH (see Antidiuretic hormone).

**vein:** A blood vessel that carries blood toward the heart.

**vena cava** (anterior and posterior venae cavae): One of the large veins that empties blood into the right atrium of the mammalian heart or into the sinus venosus of the amphibian heart.

**ventilation** [L. *ventilare,* to wave in the air]: The process by which fresh air is supplied to an organism or its respiratory organs.

**ventricle** [L. *ventriculus,* small belly]: 1. The largest of the chambers of the heart. 2. Any of the four cavities of the brain. 3. Any small cavity.

**venules:** The small, unnamed veins.

**vessel** [L. *vasculum,* a small vessel]: In plants, a water-conducting tube of the xylem composed of thick-walled hollow cells (vessel elements) lacking end walls and arranged end to end.

**vessel element:** In plants, one of the cells comprising a vessel in the xylem.

**villus** (pl. villi) [L. *villus,* shaggy hair]: Any of the minute, hairlike projections from the surface of a membrane, especially those of the small intestinal lining.

**vitalism:** The theory that the activities of organisms and their organs are partially under the control of an agency other than physical forces.

**vitellin:** See Yolk.

**vitelline membrane:** A protective covering membrane of the egg.

**vitellogenesis** [L. *vitellus,* yolk]: Production and accumulation of yolk by the oocyte.

**viviparous** [L. *vivus,* alive + *parere,* to produce]: Producing living young, with embryogenesis occurring in the mother's body.

**vocal cords** (see Vocal folds).

**vocal folds** (vocal cords): The two laryngeal, mucous membrane-covered folds of tissue enclosing the glottis that give rise to vocal sounds when air is forced past them.

**water table:** The level below which the soil's aquifers are saturated with water. Water table levels drop if there is excessive use of well water.

**wetland:** A low-lying area characterized by standing water for much or all of the year. A swamp or marsh.

**white matter:** The portions of the CNS composed predominantly of myelinated neurons.

**whorl:** In plants, a circle of leaves or flower parts.

**xylem** [Gk. *xylon,* wood]: The water-conducting tissue of the plant.

**yolk:** The phosphoprotein stored nutrients of oocytes; also called vitellin.

**zeitgeber** [Germ. *Zeit,* time + *Geber,* giver]: An environmental stimulus, such as day length, that entrains an organism's biological clock.

**zonation:** The division of a natural community into several subcommunities, typically related to variations in physical conditions, as in the different subzones within the littoral zone of the seashore or at different elevations on the sides of a mountain.

**zoospore** [Gk. *zoe,* life + *spora,* seed]: In algae and fungi, a flagellated, motile reproductive cell involved in asexual reproduction.

**zygote** [Gk. *zygotos,* yoked]: The unicellular embryo resulting from the fusion of gametes.

**zygotic meiosis:** In sexually reproducing organisms, the occurrence of meiosis during division of the zygote.

# Credits

## LINE ART

CHAPTER 1 **1-3** Adapted from F. C. Olson and J. A. Robinson, *Concepts of Evolution* (Columbus: Merrill, 1975), p. 43, and S. Salthe, *Evolutionary Biology* (New York: Holt, 1972), pp. 32, 35. **1-12** Jack W. Hudson and Irena M. Scott, "Daily torpor in the laboratory mouse, *Mus musculus*, var. albino," *Physiological Zoology* 52 (1979) :205-218, U. of Chicago Press.

CHAPTER 3 **3-20** From *Tissues and Organs: A Text-Atlas of Scanning Electron Microscopy*, by Richard G. Kessel and Randy H. Kardon. (New York: W. H. Freeman, and Company. Copyright © 1979).

CHAPTER 4 **4-39** Adapted from *National Geographic* 150, no. 3 (September 1976).

CHAPTER 7 **7-3** Adapted from William K. Purvis and Gordon H. Orians, *Life: the Science of biology* (Sunderland, MA: Sinauer, 1983), p. 191.

CHAPTER 8 **8-14** (a) and (c) After R. D. Barnes, *Invertebrate Zoology* (Philadelphia: Saunders, 1980) and H.H. Ross *et al., A Textbook of Entomology*, 4th ed. (New York: Wiley, 1982). (b) (1936) After K. Dreher, *Z. Morph. Dekol. Tiere* 31 (1936): 608-672. **8-17** Redrawn and adapted from A. S. Romer and T. S. Parsons, *The Vertebrate Body*, 5th ed. (Philadelphia: Saunders, 1977). **8-24** After J. E. Crouch, *Functional Human Anatomy*, 3rd ed. (Philadelphia: Lea and Febiger, 1978).

CHAPTER 9 **9-12** From M. S. Gardiner, *The Biology of Invertebrates* (New York: McGraw-Hill, 1972). **9-13** (b), (c), and (d) After A. S. Romer and T. S. Parsons, *The Vertebrate Body*, 5th ed. (Philadelphia: Saunders, 1977). **9-19** (a) After J. E. Crouch, *Functional Human Anatomy*, 3rd ed. (Philadelphia: Lea and Febiger, 1978).

CHAPTER 11 **11-4** After A. S. Romer and T. S. Parsons, *The Vertebrate Body*, 5th ed. (Philadelphia: Saunders, 1977). **11-6** After T. C. Kramer in *American Journal of Anatomy* 71 (1942): 343-370. **11-8** After S. Grollman, *The Human Body, Its Structure and Physiology*, 2nd ed. (New York: Macmillan, 1969). **11-15** After D. Shepro, F. Belamarich, and C. Levy, *Human Anatomy and Physiology, A Cellular Approach* (New York: Holt, 1974). **11-19** (b) After C. P. Anthony and N. J. Kolthoff, *Textbook of Anatomy and Physiology*, 11th ed. St. Louis: Mosby, 1983).

CHAPTER 13 **13-8** After H. W. Smith, *From Fish to Philosopher* (Boston: Little, Brown, 1953). **13-10** After C. L. Prosser, *Comparative Animal Physiology*, 3rd ed. (Philadelphia: Saunders, 1973). **13-12, 13-20** After L. Goldstein, *Introduction to Comparative Physiology* (New York: Holt, 1977). **13-21** After R. W. Hill, *Comparative Physiology of Animals. An Environmental Approach* (New York: Harper & Row, 1976). **Panel 13-1** After "Kidney and urinary tract infections" (Indianapolis: Lilly Research Laboratory, 1971).

CHAPTER 14 **14-1** (a) After P. A. Meglitch, *Invertebrate Zoology* (New York: Oxford University Press, 1967). (b) After M. S. Gardiner, *Biology of Invertebrates* (New York: McGraw-Hill, 1972). **14-13** After L. Stryer, *Biochemistry*, 2nd ed. (New York: Freeman, 1981).

CHAPTER 15 **15-3** After T. H. Bullock, *Introduction To Nervous Systems* (New York: Freeman, 1977). (a) From G. A. Horridge, *Quart. J. Micr. Sci.* 97 (1956): 59-74. (b) From E. J. Batham et al., *Quart. J. Micr. Sci.* 101 (1960): 487-510). **15-4** From T. H. Bullock and G. A. Horridge. *Structure and Function in The Nervous Systems of Invertebrates* (New York: Freeman, 1965). **15-5** From T. H. Bullock, *Introduction To Nervous Systems* (New York: Freeman, 1977); after D. Hadenfeldt Z. *Wiss. Zool.* 133 (1929): 586-638. **15-8** After M. S. Gardiner, *The Biology of Invertebrates* (New York: McGraw-Hill, 1972; from B. Hanström, 1928. **15-16** After A. S. Romer and T. S. Parsons, *The Vertebrate Body*, 5th ed. (Philadelphia: Saunders, 1977) **15-18** After A. S. Romer and T. S. Parsons, *The Vertebrate Body*, 5th ed. (Philadelphia: Saundeers, 1977). **15-24, 15-25** After R. W. Sperry, "The great cerebral commissure." *Scientific American* 210 (1964): 42-52.

CHAPTER 16 **16-1** From T. H. Bullock, *Introduction to Nervous Systems* (New York: Freeman, 1977); after Ramon Y. Cajal, 1909. **16-7** After W. S. Beck, *Human Design: Molecular, Cellular, and Systematic Physiology* (New York: Harcourt, 1971). **16-12** After C. R. Noback and R. J. Demarest, *The Human Nervous System: Basic Principles of Neurobiology*, 2nd ed. (New York: McGraw-Hill, 1975).

CHAPTER 17 **17-5** After W. S. Hoar, *General and Comparative Physiology*, 3rd ed. (Englewod Cliffs, NJ: Prentice-Hall, 1983) (From Iwai, IN P.H. Cahn, ed., *Lateral Line Detectors*. Bloomington, IN: Indiana University Press, 1967.) **17-7** After W. S. Hoar, *General and Comparative Physiology*, 3rd ed. (Englewood Cliffs, NJ: Prentice-Hall, 1983); 1957 from Heidermann's Grundzüge der *Tierphysiologie* (Stuttgard: G. Fischer 1957). **17-11** After F. Netter, *The Ciba Collection of Medical Illustrations*, Vol. 1. (Montreal: Ciba, 1953) **17-13** After W. S. Hoar, *General and Comparative Physiology*, 3rd ed. (Englewood Cliffs, NJ: Prentice-Hall, 1983). (Based on Lorenzo, in Zotterman, ed., *Olfaction and Taste* (New York: Pergamon, 1963). **17-14** After M. M. Novikoff, "Regularity of form in organisms." *Systematic Zoology* 2 (1953):57-62. **17-18** From R. F. Schmidt, ed., *Fundamentals of Sensory Physiology* (New York: Springer Verlag, 1978), after Boycott and Dowling, (Proc. R. Soc. London (Biol.) 166 (1966):80-111)

CHAPTER 18 **18-2** After E. J. DuPraw, *The Biosciences: Cell and Molecular Biology* (Stanford, CA: Cell and Molecular Biology Council, 1972). **18-7** After William S. Beck, *Human Design* (New York: Harcourt, 1971). **18-10, 18-17** After William Bloom and Don Fawcett, eds., *A Textbook of Histology*, 10th ed. (Philadelphia: Saunders, 1975). **18-14, 18-20** After John M. Murray and Annemarie Weber, "The cooperative action of muscle proteins." *Scientific American* 230 (1974): 58-71. **18-15** From H. E. Huxley, "The mechanisms of muscular contractions." *Scientific American* (December 1965). **18-18** After Carolyn Cohn, "The protein switch of muscle contraction." *Scientific American* 233 (1975): 36-45. **18-19** After C. R. Noback and R. J. Demarest, *The Human Nervous System: Basic Principles of Neurobiology*, 2nd ed. (New York: McGraw-Hill, 1975. **18-21, 18-23** After P. Satir, "How Cilia Move." *Scientific American* 231 (1974): 44-63. **18-25** After William S. Hoar, *General and Comparative Physiology*, 3rd ed. (Englewood Cliffs, NJ: Prentice-Hall, 1983), p. 367.

CHAPTER 19 **19-3** After E. J. W. Barrington, *Invertebrate Structure and Function* (Boston: Houghton, Mifflin, 1967), and Evans, *Journal of Animal Ecology* 20 (1951):1-110. **19-10(c)** From N. Tinbergen, *Animal Behavior* (New York: Time-Life Books, 1965). **19-11** After K. Lorenz, "The evolution of behavior." *Scientific American* (December 1958), pp. 67-78. **19-15(b)** After Peter Marler and William J. Hamilton III, *Mechanisms of Animal Behavior* (New York: Wiley, 1966). **19-18** After Lilli Koenig, *Studies in Animal Behavior*, trans. Marjorie Latzke (New York: Crowell, 1958).

CHAPTER 20 **20-2** After N. Tinbergen, *The Study of Instinct*, rev. ed. (New York: Oxford University Press, 1969). **20-12** After J. P. Scott, *Animal Behavior* (Chicago: University of Chicago Press). **20-18** From N. Tinbergen, *Social Behavior in Animals* (London: Methuen, 1953). **20-20(b)** From S. A. Barnett, *Instinct and Intelligence—Behavior of Animals and Man* (Englewood Cliffs, NJ: Prentice-Hall, 1967). **20-21** After A. Jolly, *The Evolution of Primate Behavior*, 2nd ed. (New York: Macmillan, 1985). **Panel 20-1** After J. L. Brown, *The Evolution of Animal Behavior (New York: Norton, 1975), and Karl v. Frisch (1967).*

CHAPTER 21 **21-1 (b-d), 21-2, 21-8, 21-9, 21-10** Barbara L. Schmidt. **21-11 (a), (b)** Redrawn and modified from William A. Jensen and Frank B. Salisbury, *Botany: An Ecological Approach* (Belmont, CA: Wadsworth, 1972).

CHAPTER 22 **22-1, 22-4, 22-6** After C. D. Turner and J. T. Bagnara, *General Endocrinology*, (Philadelphia: Saunders, 1976). **22-5, 22-9** After M. E. Hadley, *Endocrinology* (Englewood Cliffs, NJ: Prentice-Hall, 1984).

CHAPTER 23 **23-12** After C. B. Bridges, 1936, from E. J. Gordner and D. P. Snustand, *Principles of Genetics*, 6th ed. (New York: Wiley, 1981).

CHAPTER 24 **Panel 24-1, 24-7** After B. Alberts *et al.*, *Molecular Biology of the Cell* (New York: Garland, 1983). **24-14** After B. Lewin, *Gene Expression*, 2nd ed. (New York: Wiley, 1980). **24-15** After J. D. Watson, *Molecular Biology of the Gene* (Menlo Park, CA: Benjamin, 1976).

CHAPTER 25 **25-1** After M. H. Burgos and D. W. Fawcett, in *J. Biophys. Biochem. Cytol.* 1 (1955): 287. **25-6** After R. Rugh, *Vertebrate Embryology: The Dynamics of Development* (New York: Harcourt, 1964). **25-24, 25-26** After H. Spemann, *Embryonic Development and Induction* (New Haven: Yale, 1938). **25-25** After N. K. Wessells, *Tissue Interactions and Development* (Menlo Park, CA: Benjamin, 1977). **25-27** After J. B. Gurdon, "Egg cytoplasm and gene control in development." Proc. Roy. Soc. L. B. (1977): 198,211 **Panel 25-1** After A. L. Colwin and L. H. Colwin, *J. Cell Biol.* 19 (1963):477.

CHAPTER 26 **26-1 (a), (b)** Courtesy of Dr. E. Gifford, University of California, Davis. From Foster And Gifford, *Morphology of Vascular Plants* (New York: Freeman, and W. Troll, *Vergleichend Morphologie der hoheren Pflanzen* (Berlin: Gebruder Borntraeger. (c) Dr. J. A. Sargent, Hexland Ltd., London. **26-2, 26-3, 26-4, 26-8** Drawings by Valerie Pawlak after Esau, *Plant Anatomy*, 2nd ed. (New York: Wiley,). **26-5** Dr. A. J. Foster and the *American Journal of Botany*. **26-10, 26-12, 26-21, 26-23, 26-24, 26-25** Drawings by Barbara L. Schmidt. **26-29** Drawing by Barbara L. Schmidt after Ray, Steeves, and Fultz, *Botany* (Philadelphia: Saunders). **26-37** Courtesy of Prof. U. Zimmerman, University of Wurzburg, West Germany, and GCA Corporation.

CHAPTER 27 **27-3** After R. E. Dickerson, "X-ray studies of protein mechanisms." *Ann. Rev. Biochem.* 41 (1972): 815-842. **Panel 27-2** After H. G. Owen, *Atlas of Continental Displacement 200 Million Years to the Present* (Cambridge, England: Cambridge University Press, 1983). **28-15(b)** Courtesy of Dr. Elizabeth Gantt, Smithsonian Environmental Research Center from *BioScience* 25:781. **28-23 (a)** Courtesy of Paul Driftmier.

CHAPTER 29 **29-1 (a), 29-4, 29-6, 29-7, 29-8(a), 29-9 (c, d), 29-11** (b, d), **29-15, 29-16, Panel 29-1 (c, f)**, Drawings by Barbara L. Schmidt. **29-10 (a)** Drawing by Valerie Pawlak.

CHAPTER 30 **30-1, Panel 30-2, Panel 30-3, Panel 30-4, Panel 30-5** Drawings by Barbara L. Schmidt. **30-3** Drawing by Valerie Pawlak. **30-16** Drawing by Valerie Pawlak after Dr. Wilson N. Stewart, University of Saskatchewan. **30-17** Courtesy of Milwaukee Public Museum. **30-18** Adapted and redrawn from H. Banks, *Evolution and Plants of the Past* (Belmont, CA: Wadsworth, 1970), and Jensen and Salisbury, *Botany* (Belmont, CA: Wadsworth, 1972). **30-19 (a)** After Beck with permission of *American Journal of Botany* 49:376. (b) After W. N. Stewart and T. Delevoryas, *Morphology and Evolution of Fossil Plants* (New York: Holt, 1962), and Bold, *The Plant Kingdom* (Englewood Cliffs, NJ: Prentice-Hall, 1977), p. 229. (c) After Andrews, *Studies in Paleobotany* (New York: Wiley, 1961), p. 153. **Panel 30-1** Drawing by Barbara L. Schmidt after Scagel *et al.*, *An Evolutionary Survey of the Plant Kingdom* (Belmont, CA: Wadsworth, 1965).

CHAPTER 31 **31-6** After L. H. Hyman, *The Invertebrates: Protozoa through Ctenophora* (New York: McGraw-Hill, 1940). **31-24** After R. J. Goss, *Principles of Regeneration* (New York: Academic Press, 1969).

CHAPTER 32 **32-9** After R. Buchsbaum, *Animals Without Backbones: An Introduction to the Invertebrates* (Chicago: University of Chicago Press, 1958). **32-11** After M. S. Gardner, *The Biology of Invertebrates* (New York: McGraw-Hill, 1972). **32-24** After A. Sedgloick, *A Student's Textbook of Geology* (New York: Macmillan, 1932). **Panel 32-2** After D. S. Smith, "The flight muscles of insects." *Scientific American* 212 (1965): 77-88.

CHAPTER 33 **33-9** Redrawn from B. Bolin, "The carbon cycle." *Scientific American* 223 (3):124-135. **33-17** After W. G. Kendrew, *Climate* (New York: Oxford University Press, 1930).

CHAPTER 34 **34-4** After T. C. Emmel, *An Introduction to Ecology and Population Biology* (New York: Norton, 1973). **34-11** After E. J. Kormondy, *Concepts of Ecology*, 3rd ed. (Englewood Cliffs, NJ: Prentice-Hall, 1984), and S. Stankovic, *Monographiae Biologicae* (1960), 1-357. **34-14** After E. J. Kormondy, *Concepts of Ecology*, 3rd ed. (Englewood Cliffs, NJ: Prentice-Hall, 1984); after G. F. Gause, *The Struggle for Existence* (Baltimore: Williams and Wilkins, 1934). **34-16(b)** After E. J. Kormondy, *Concepts of Ecology*, 3rd ed. (Englewood Cliffs, NJ: Prentice-Hall, 1984); from D. A. MacLulich, *University of Toronto Studies, Biological Series No. 43*, 1937.

CHAPTER 35 **35-8** After P. J. Herring and M. R. Clarke, eds., *Deep Oceans* (New York: Praeger, 1971). **35-10** After G. S. Carter, *A General Zoology of the Invertebrates* (London: Sidgewick and Jackson, 1946) **35-16** After W. D. Russell-Hunter, *A Life of Invertebrates* (New York: Macmillan, 1979). **35-19** After W. D. Russell-Hunter, *Aquatic Productivity* (New York: Macmillan, 1970). **35-30** After E. J. Kormondy, *Concepts of Ecology*, 3rd ed. (Englewood Cliffs, NJ: Prentice-Hall, 1984).

CHAPTER 36 **36-6, 36-23** After *Environmental Quality*, Eighth Annual Report of the Council on Environmental Quality (Washington: Government Printing Office, 1978). **36-9** Redrawn and adapted from *Newsweek*, November 11, 1980. **36-11** After R. H. Wagner, *Environment and Man*, 3rd ed. (New York: Norton, 1978), and Joseph J. Shomon, *Open Land for Urban America* (Baltimore: Johns Hopkins University Press, 1971). **36-15** After J. Carter, *The Florida Experience, Land and Water Policy in a Growth State* (Baltimore: Johns Hopkins University Press, 1974). **36-19, 36-20** From V. Brodine, ed., *Air Pollution*, Environmental Issues Series (New York: Harcourt Brace Jovanovich, 1972).

## PHOTOGRAPHS

**Unit I Hank Morgan, Photo Researchers, Inc.**
CHAPTER 1 **1-1a** Russ Kinne, Photo Researchers, Inc.; **1-1b** Steven J. Krasemann, Photo Researchers, Inc.; **1-1c** Tom McHugh, Photo Researchers, Inc.; **1-2a** Stern, Black Star; **1-2b** Clyde H. Smith, Peter Arnold, Inc.; **Panel 1-2** Robert Caputo, Photo Researchers, Inc.; **1-4** S.J. Krasemann, National Audubon Society, Photo Researchers, Inc.; **1-5** Tom McHugh, Photo Researchers, Inc.; **1-6** Culver Pictures, Inc.; **1-7a** The Bettmann Archive, Inc.; **1-9** Peter Arnold; **1-10** Robert Ashworth, Photo Researchers, Inc.; **1-11a** Douglas Kirkland, Woodfin Camp & Associates; **1-11b** Douglas Kirkland, Woodfin Camp & Associates; **1-11c** Jacques Jangoux, Peter Arnold, Inc.; **1-11d** Michael Heron, Woodfin Camp & Associates; **1-11e** Hank Morgan, Photo Researchers, Inc.; **1-13** Culver Pictures, Inc.

**Unit II P. Dayanandan, Photo Researchers, Inc.**
CHAPTER 2 **2-1** Photo courtesy U.S. Navy; **2-2** Photo courtesy NASA; **2-6** Larry Mulvehill, Photo Researchers, Inc.; **2-9** Rip Griffith, Photo Researchers, Inc.; **2-19** S. Green—Armytage, Photo Researchers, Inc.; **2-21** USDA-SCS Photo by Robert Bransteed; **2-24** N. E. Beck Jr., National Audubon Society, Photo Researchers, Inc.; **2-25** Hans Pfletschinger, Peter Arnold, Inc.; **2-26** Ken Kay, Time-Life Books; **2-27** Lynwood M. Chace, National Audubon Society, Photo Researchers, Inc.; **2-28** Chester Higgins, Jr., Photo Researchers, Inc.; **2-29** J. Goodenough.

CHAPTER 3 **3-9** Jerome Wexler, Photo Researchers, Inc.; **3-10** Courtesy USDA; **3-13** Dr. Don Fawcett, Photo Researchers, Inc.; **3-14** Biophoto Assoc., Photo Researchers, Inc.; **3-16a** B. Berg, B.V. Hofstein, and G. Petterson, 1972. *J. Applied Bacteriology*, 35: 215–219; **3-16b** Betty Faber; **3-17** Grant Heilman, Runk Schoenberger; **3-18** Dr. Don Fawcett, Photo Researchers, Inc.; **3-20** Courtesy Richard G. Kessel, Univ. of Iowa; **3-24** W. H. Hodge, Peter Arnold, Inc.; **3-28** Biophoto Associates, Photo Researchers, Inc.; **Panel 3-1** Bill Longcore, Photo Researchers, Inc.; **Panel 3-2** Dr. M.F. Perutz, Medical Research Council Centre, England; **3-40** Courtesy New York Public Library Picture Collection; **3-45** A. C. Barrington Brown. From J.D. Watson, *The Double Helix*, Atheneum, New York, p. 215, © 1968 by J.D. Watson; **3-49** D.L. Miller, Jr., Barbara A. Hamkalo, C.A. Thomas, Jr., *Science*, 169, 1970.

CHAPTER 4 top, **4-1b** The Bettmann Archive, Inc.; bottom, **4-2a** The Bettmann Archive, Inc. **4-2b** The Bettmann Archive, Inc.; **4-3b** Walter Dawn, National Audubon Society, Photo Researchers, Inc.; **4-6b** Keith Porter, Photo Researchers, Inc. **4-6d** Science Source, Photo Researchers, Inc. **4-7** E. Kellenberger, Omikron, Photo Researchers, Inc.; **4-8** Bill Longcore, Photo Researchers, Inc.; **4-10a** Copyright © David Scharf, 1978, All rights reserved; **4-10b** Omikron, Photo Researchers, Inc.; **4-11** Keith Porter, Photo Researchers, Inc.; **4-12** Dr. Eugene Vigil; **4-14** Dr. Don Fawcett, Photo Researchers, Inc.; **4-17** Biophoto Associates, Photo Researchers, Inc.; **4-19a** Biophoto Associates, Photo Researchers, Inc.; **4-20b** M.M. Rhoades and D.T. Morgan, Jr.; **4-21** Dr. Don Fawcett, Photo Researchers, Inc.; **4-22a** Dr. Don Fawcett/R. Bolender, Photo Researchers, Inc.; **4-23** Omikron, Photo Researchers, Inc.; **4-24a** Biophoto Associates, Photo Researchers, Inc.; **4-26** Manfred Kage, Peter Arnold, Inc.; **4-29** Dr. Eugene Vigil, Biology Media 1978, Photo Researchers, Inc. **4-30** Keith Porter, Photo Researchers, Inc.; **4-31** Dr. Don Fawcett, Photo Researchers, Inc.; **4-32a** Omikron, Photo Researchers, Inc.; **4-32b** Omikron, Photo Researchers, Inc.; **4-34** Dr. Don Fawcett/Gaddum-Rosse, Photo Researchers, Inc.; **4-35a** Dr. Ann Smith, Science Photo Library, Photo Researchers, Inc.; **4-35b** Omikron, Photo Researchers, Inc.; **4-35c** Dr. Don Fawcett, Photo Researchers, Inc.; **4-37a** Dr. Don Fawcett, Photo Researchers, Inc.; **4-37b** Dr. Don Fawcett/S. Ito, J. André, Photo Researchers, Inc.; **4-38** Omikron, Photo Researchers, Inc.; **4-40a** Lee D. Simon, Photo Researchers, Inc.; **4-40b** Lee D. Simon, Photo Researchers, Inc.; **4-42a–h** Photo Researchers, Inc.; **4-43a** Marjorie Shaw, M.D., Anderson Hospital, Houston, Texas; **4-44a** Dr. G. Schatten, Science Photo Library, Photo Researchers, Inc.; **4-46** Jeremy Pickett-Heaps, Photo Researchers, Inc.

CHAPTER 5 **5-4** George Whiteley, Photo Researchers, Inc.; **5-9** Dr. Don Fawcett/S. Ito, Photo Researchers, Inc.; **5-11** Dr. Don Fawcett, Photo Researchers, Inc.; **5-13** Dr. Don Fawcett/M. Neutra & R. Specian; Photo Researchers, Inc.

CHAPTER 6 **6-1** Bruce Roberts, Rapho Photo Researchers, Inc.; **6-4** M. Abbey, Photo Researchers, Inc.; **6-5** Omikron, Photo Researchers, Inc.; **6-9** Bettye Lane, Photo Researchers, Inc.; **6-10** Joe Munroe, Wine Institute; **6-13** Dr. Don Fawcett, Photo Researchers, Inc.; **6-15** Omikron, Photo Researchers, Inc.

CHAPTER 7 **7-10a, 7-10b** Courtesy of Dr. James A. Bassham, Lawrence Berkeley Laboratory; **7-12b** Biophoto Associates, Photo Researchers, Inc.; **7-12c** Photo Researchers, Inc.; **7-13b** Biophoto Associates, Photo Researchers, Inc.; **7-14a** Prof. G.F. Leedale, Biophoto Associates, Photo Researchers, Inc.; **7-14b** Biophoto Associates, Photo Researchers, Inc.; **7-15a** Biophoto Associates, Photo Researchers, Inc.; **7-17** Russ Kinne, Photo Researchers, Inc.

**Unit III George Holton, Photo Researchers, Inc.; Unit III-1,2 The Bettmann Archive, Inc.**
CHAPTER 8 **8-1** Thomas Hopker, Woodfin Camp & Associates; **8-2b** Peru National Tour Office; **8-4** Russ Kinne, Photo Researchers, Inc.; **8-6** Tom McHugh, Photo Researchers, Inc.; **8-7** Grant Heilman, Agricultural Photograph, Runk/Schoenberger; **8-8** R.F. Head, National Audubon Society, Photo Researchers, Inc.; **8-25** Science Source, Photo Researchers, Inc.; **8-26** Science Photo Library, Photo Researchers, Inc.; **8-27** American Cancer Society.

CHAPTER 9 **9-1** Tom Smylie, Fish & Wildlife Service; **9-2** Hank Morgan, Photo Researchers, Inc.; **9-4** Eric Kroll, Taurus Photos; **9-5a** Dr. Wm. M. Harlow, Photo Researchers, Inc.; **9-5b** J.H. Robinson, Photo Researchers, Inc.; **9-6a** S.J. Krasemann, National Audubon Society, Photo Researchers, Inc.; **9-6b** Tom Branch, Photo Researchers, Inc.; **9-7** J.L. Lepore, Photo Researchers, Inc.; **9-11a** M.I. Walker, Photo Researchers, Inc.; **9-11b** Russ Kinne, Photo Researchers, Inc.; **9-11c** Courtesy of R.D. Lumsden, Tulane University; **9-17** Biophoto Associates, Photo Researchers, Inc.; **9-18** Martin M. Rotker, Photo Researchers, Inc.; **9-19b** J.F. Gennaro, Jr., Photo Researchers, Inc.; **Panel 9-3** The Bettmann Archive, Inc.

CHAPTER 10 **10-4a–d** P. Dayanandan, Photo Researchers, Inc.; **10-5** Oxford Scientific Films; **10-9** Hans Pfletschinger, Peter Arnold, Inc.; **10-10a** Dr. Martin H. Zimmermann, Harvard University; **10-10b** George A. Schaeffers, N.Y. State Agriculture Experiment Station; **10-12** James Cronshaw.

CHAPTER 11 **11-3** Animals Animals; **11-14** Courtesy American Heart Association; **11-16** S. Peiper/Marshall Sklar, Photo Researchers, Inc.; **11-17** Biophoto Associates, Photo Researchers, Inc.; **11-18** Biophoto Associates, Photo Researchers, Inc.; **Panel 11-b** Manfred Kage, Peter Arnold, Inc.

CHAPTER 12 **12-3** Eric Gravé, Photo Researchers, Inc.; **12-4a,b** Dr. Tony Brain; **12-5** United States Department of the Interior, Fish & Wildlife Service; **12-6a,b** Biophoto Associates, Photo Researchers, Inc.; **12-7** WHO Photo by P. Almsy; **12-8** The Bettmann Archive, Inc.; **12-9** Slim Aarons, Photo Researchers, Inc.; **12-11** From *The Cell*, 5 ed., Carl P. Swanson, Peter L. Webster; **12-17** Dr. Don Fawcett, Photo Researchers, Inc.; **12-10b** Runk/Schoenberger, Grant Heilman; **12-22** Courtesy of E. Munn; **12-25** Craig Hammell, The Stock Market; **12-27** Courtesy of NASA; **Panel 12-3** Jay A. Levy, Cancer Research Institute, Univ. of California School of Medicine; **12-28** Dr. Rosalind King, Science Photo Library, Photo Researchers, Inc.

CHAPTER 13 **13-16a** Walker, Photo Researchers, Inc.; **13-16b** Biophoto Associates, Photo Researchers, Inc.; **13-17** Courtesy Richard G. Kessel, Univ. of Iowa; **13-18** Dr. G. F. Leedale, Biophoto Associates, Photo Researchers, Inc.; **13-19** Courtesy Richard G. Kessel, Univ. of Iowa.

CHAPTER 14 **14-3** Biophoto Associates, Photo Researchers, Inc.; **14-4** Dr. Ernst Knobil; **14-7** Scala, Art Resource; **14-8** Biophoto Associates, Photo Researchers, Inc.; **14-10** Biophoto Associates, Photo Researchers, Inc.

CHAPTER 16 **16-1** Omikron, Photo Researchers, Inc.; **16-11a** Omikron, Photo Researchers, Inc; **16-11d** Dr. Don Fawcett, J. Heuser, T. Reese, Photo Researchers, Inc.

CHAPTER 17 **17-1** Courtesy of Ted Levin; **17-2a** Biophoto Associates, Photo Researchers, Inc; **17-4b** L.L.T. Rhodes, Taurus Photo; **17-6b** John Lidington, Photo Researchers, Inc; **17-10b** Dr. G. Bredberg, Photo Researchers, Inc; **17-17c** Ralph C. Eagle, Jr., M.D., Photo Researchers, Inc.; **17-19** Courtesy of Ron Willocks.

CHAPTER 18 **18-1a** Manfred Kage, Peter Arnold, Inc.; **18-1b** Biophoto Associates, Photo Researchers, Inc.; **18-1c** M. Abbey, Photo Researchers, Inc.; **18-3** Tom McHugh, Photo Researchers, Inc.; **18-4a** Biophoto Associates, Photo Researchers, Inc.; **18-6** Biophoto Associates, Photo Researchers, Inc.; **18-9** Dr. Don Fawcett, Photo Researchers, Inc.; **18-11** Biophoto Associates, Photo Researchers, Inc.; **18-13** Dr. Don Fawcett, Photo Researchers, Inc.; **18-15** Courtesy of H.E. Huxley, *J. Mol. Biol.*, 7:281 (1963), Academic Press Inc., London Ltd.; **18-16** From R.H. Depue and R.V. Rice, *J. Mol. Biol.*, 12:32 (1965) Academic Press Inc., London; **18-21** Gary W. Grimes, Taurus Photos, Ltd.; **18-22** Dr. Don Fawcett, Photo Researchers, Inc.

CHAPTER 19 **19-1** Courtesy of American Museum of Natural History; **19-3** Oxford Scientific Films, Animals Animals; **19-4** George Holton, Photo Researchers, Inc.; **19-5** E. Hanumantha Rao, Photo Researchers, Inc.; **19-6** Carson Baldwin Jr., Animals Animals; **19-7** Carl H. Maslowski, Photo Researchers, Inc.; **19-8a,b** William D. Griffin, Animals Animals; **19-9** Rohn Engh, Photo Researchers, Inc.; **19-10a,b** William Vandivert; **19-10c** Martin Rogers, Woodfin Camp & Associates; **19-11a,b** Lorenz, Konrad, The Evolution of Behavior, *Scientific American*, December, 1958, pp. 67–78; **19-12a** George Holton, Photo Researchers, Inc.; **19-12b** James Hancock, Photo Researchers, Inc.; **19-12c** F. Gohier, Photo Researchers, Inc.; **19-12d** Bruce Piatcher, Photo Researchers, Inc.; **19-12e** Russ Kinne, Photo Researchers, Inc.; **19-12f** Dan Guravich, Photo Researchers, Inc.; **19-14** Vema R. Johnston, Photo Researchers, Inc.; **19-15** G. Ronald Austing, National Audubon Society, Photo Researchers, Inc.; **19-16** R. Blakemore, University of New Hampshire; **19-17** H.W. Silvester, Photo Researchers, Inc.; **19-18** Thomas McAvon, Life Magazine © 1955 Time, Inc.; **19-19** The Bettmann Archive; **19-20a** Christopher S. Johns; **19-20b** B.F. Skinner.

CHAPTER 20 **20-3** Leonard Lee Rue III, Photo Researchers, Inc.; **20-4** Frank Roberts, Animals Animals; **20-5a** Z. Leszczynski, Animals Animals; **20-6** John Bovca, Photo Researchers, Inc.; **20-7** Mitchell Campbell, From National Audubon Society, Photo Researchers, Inc.; **Panel 20-1a** Oxford Scientific Films, Animals Animals; **20-8** Z. Leszczynski, Animals Animals; **20-9a** F. Allan, Animals Animals; **20-9b** S.A. Grimes, From National Audubon Society, Photo Researchers, Inc.; **20-10** Frank Roche, Photo Researchers, Inc.; **20-11** The Humane Society of the United States; **20-13a** © Grapes, Michaud, Photo Researchers, Inc.; **20-13b** Arthur C. Twomey, Photo Researchers, Inc.; **20-14** Robert W. Hernandez, Photo Researchers, Inc.; **20-15** Tom McHugh, Photo Researchers, Inc.; **20-16a** Tom McHugh, Photo Researchers, Inc.; **20-16b** Brian J. Coates, Bruce Coleman, Inc.; **20-16c** J.A. Robinson, Photo Researchers, Inc.; **20-16d** Gilbert S. Grant, Photo Researchers, Inc.; **20-17** Tom McHugh, Photo Researchers, Inc.; **20-19a–d** Oxford Scientific Films, Animals Animals; **20-20** Jack Novak, Photo Researchers, Inc.; **20-21** George Holton, Photo Researchers, Inc.; **20-22a,b** Toni Angermayer, Photo Researchers, Inc.; **20-23** University of Wisconsin/Primate Laboratory, Madison, Wisconsin; **20-24** University of Wisconsin/Primate Laboratory, Madison, Wisconsin; **20-25** Carl Roessler, Animals Animals; **20-26** Marc and Evelyne Bernheim.

**Unit V NIH Science Source, Photo Researchers, Inc.**
CHAPTER 21 **21-3a,b** Courtesy of Carolina Biological Supply Company; **21-4a** Jerome Wexler, National Audubon Society, Photo Researchers, Inc.; **21-4b** Plantek, Photo Researchers, Inc.; **21-5b** Walker England, Photo Researchers, Inc.; **21-5c** Courtesy Nova Scientific Corp., Oxford Scientific Films.

CHAPTER 22 **22-2** Biophoto Associates, Photo Researchers, Inc.; **22-3a** Biophoto Associates, Photo Researchers, Inc.; **22-3b** C.W. Brown, Photo Researchers, Inc.; **22-6a** Biophoto Associates, Photo Researchers, Inc.; **22-6b** Ed Rechur, Peter Arnold, Inc.; **22-6c** Manfred Kage, Peter Arnold, Inc.; **22-6d** Biophoto Associates, Photo Researchers, Inc.; **22-7** P. Bagavandoss, Photo Researchers, Inc.; **22-8a,b** Biophoto Associates, Photo Researchers, Inc.; **22-11** Teri Leigh Stratford, materials courtesy of Planned Parenthood, New York.

CHAPTER 23 **23-1** The Bettmann Archive, Inc.; **23-2** Russ Kinne, Photo Researchers, Inc.; **23-3** Runk/Schoenberger, Grant Heilman; **Panel 23-1a** M. Abbey, Photo Researchers, Inc.; **Panel 23-1b** Russ Kinne, Photo Researchers, Inc.; **23-11** Omikron, Photo Researchers, Inc.; **23-15b** Omikron, Photo Researchers, Inc.; **23-16** Dr. M.M. Rhoades, Indiana University; **23-18b** Bruce Roberts, Photo Researchers, Inc.

CHAPTER 24 **24-2a** S.P.L. Brochure, Photo Researchers, Inc.; **24-2b** Dr. Gopal Murti, Photo Researchers, Inc.; **24-5d** NHI, Dr. Kakefuda, Science Source, Photo Researchers, Inc.; **Panel 24-1d** Science Source, Photo Researchers, Inc.; **Panel 24-2b** NHI, Dr. Kakefuda, Science Source, Photo Researchers, Inc.; **24-9** O.L. Miller, Jr. and B.R. Beatty, *J. Cell Physiol.* Vol. 74 (1969); **24-12** Alexander Rich, Omikron, Photo Researchers, Inc.

CHAPTER 25 **25-1a** Biophoto Associates, Photo Researchers, Inc.; **25-1b** Dr. G. Schatten, Science Photo Library, Photo Researchers, Inc.; **25-3** Dr. Don Fawcett/O. Miller, Photo Researchers, Inc.; **25-4** Joesph G. Gall, Omikron, Photo Researchers, Inc.; **Panel 25-1a** Landrum B. Shettles, M.D., Las Vegas, Nev.; **25-5** Carolina Biological Supply Co.; **25-8a–g** Carolina Biological Supply Co.; **25-9a** Dr. L.B. Shettlos, Las Vegas, Nev.; **25-9b–f** Carolina Biological Supply Co.; **25-10a,b** Carolina Biological Supply Co.; **25-13a,b** Carolina Biological Supply Co.; **25-15a,b** Carolina Biological Supply Co.; **25-16a–c** Carolina Biological Supply Co.; **25-19a–c** Carolina Biological Supply Co.; **25-20a–d** Carolina Biological Supply Co.; **25-21d** Carolina Biological Supply Co.; **Panel 25-2-a** Omikron, Photo Researchers, Inc.; **Panel 25-2-b** Biophoto Associates, Photo Researchers, Inc.; **Panel 25-3-a** © C.G. Maxwell, From National Audubon Society, Photo Researchers, Inc.; **Panel 25-3-b** © A.W. Ambler, From National Audubon Society, Photo Researchers, Inc.; **Panel 25-3-c** © George Porter, National Aububon Society, Photo Researchers, Inc.; **Panel 25-3-d** © N.E. Beck Jr., National Audubon Society, Photo Researchers, Inc.; **25-29c** Photo Researchers, Inc.; **25-29f** Omikron, Photo Researchers, Inc.; **25-31a** Photo Researchers, Inc.; **25-31b–h** Omikron, Photo Researchers, Inc.; **25-32a–d** From Omikron, Photo Researchers, Inc, **Panel 25-4a–c** R. Knauft, © 1979, Biology Media, Photo Researchers, Inc.; **Panel 25-4d** © 1983, Chuck Brown, Photo Researchers, Inc.; **Panel 25-4e** © 1978, Biology Media, Photo Researchers, Inc.; **Panel 25-4f** © D. Fawcett, J. Heuser, Photo Researchers, Inc.

CHAPTER 26 **26-1a,b** Courtesy of Dr. E. Gifford. From Foster and Gifford, *Comparative Morphology of Vascular Plants*, Freeman; **26-1c** Courtesy of Dr. J.A. Sargent, EM technology, Hexland, Ltd. England; **26-6** P. Mahlberg, *American Journal of Botany*. **26-7** Courtesy of Drs. B.E.S. Gunning and M.W. Steer; **26-8d** Biophoto Associates, Photo Researchers, Inc.; **26-9a,b** Courtesy Dr. R.F. Evert, University of Wisconsin; **26-11** Photo Researchers, Inc.; **26-13a** W.F. Millington, Photo Researchers, Inc.; **26-13b** Courtesy of Drs. A. Wilson and A.W. Robards, University of York, England; **26-14a–c** Photo Researchers, Inc.; **26-15a,b** W.F. Millington; **26-16a** W.F. Millington, Photo Researchers, Inc.; **26-17a–d** W.F. Millington, Photo Researchers, Inc.; **26-18a,b** W.F. Millington; **26-19a** W.F. Millington; **26-20a** W.F. Millington; **26-20b** Biophoto Associates, Photo Researchers, Inc.; **26-22a** Dr. Shirley Tucker, *American Journal of Botany*; **26-22b** W.F. Millington, Photo Researchers, Inc.; **26-27c–e** From J.B. Nitsch, *American Journal of Botany*; **26-28** W.F. Millington; **26-32** Courtesy of Drs. Folke Skoog, University of Wisconsin, and Carlos Miller, Indiana University, from Symposium of Society for Experimental Biology.

CHAPTER 27 **27-1** The Bettmann Archive; **27-2** The Bettmann Archive; **27-9a** Michael Lustbader, Photo Researchers, Inc.; **27-9b** Joseph T. Collins, Photo Researchers, Inc.; **27-9c** Tom McHugh, Photo Researchers, Inc.; **27-10a** Tom McHugh, Steinhart Aquarium, Photo Researchers, Inc.; **27-10b** Tom McHugh, Photo Researchers, Inc.; **27-10c** Tom McHugh, Marineland of the Pacific, Photo

Researchers, Inc. **27-12a,b** M.W.F. Tweedie, Photo Researchers, Inc.; **27-13a,b** Robert Maier, Animals Animals; **27-14a** R.T. Peterson, Photo Researchers, Inc.; **27-16a,b** Tom McHugh, Photo Researchers, Inc.; **Panel 27-1,2** The Bettmann Archive; **27-17a–d** By permission of the Trustees of the British Museum, Natural History; **27-17e** A.T.&T. Co. Photo Center.

**Unit VI Scott Johnson, Animals Animals**

**CHAPTER 28 28-1** Lee D. Simon, Photo Researchers, Inc.; **28-2** Biophoto Associates, Photo Researchers, Inc.; **28-3** Omikron, Photo Researchers, Inc.; **28-4b** From T.D. Brock and S.F. Conti, *Archiv. Mikrobiol*, 1969; **28-4e** Biophoto Associates, Photo Researchers, Inc.; **28-5** Courtesy of Dr. G.J. Schumacher, SUNY at Binghamton; **28-6** M. Abbey, Photo Researchers, Inc.; **28-7** Biophoto Associates, Photo Researchers, Inc.; **28-9a** Eric Gravé, Photo Researchers, Inc.; **28-9b** Michael Abbey, Photo Researchers, Inc.; **28-10** John Walsh, Science Photo Library, Photo Researchers, Inc.; **Panel 28-1a** Eric V. Gravé, Photo Researchers, Inc.; **Panel 28-1b** Dr. J. Burgess, Science Photo Library, Photo Researchers, Inc.; **Panel 28-1c** Science Source, Photo Researchers, Inc.; **Panel 28-1d** American Society for Microbiology, Photo Researchers, Inc.; **28-11a** Dr. Tony Brain, Science Photo Library, Photo Researchers, Inc.; **28-11b** Prof. L. Caro, Photo Researchers, Inc.; **28-12a** Science Photo Library, Inc., Photo Researchers, Inc.; **28-12b** Biophoto Associates; **28-14a** Biophoto Associates; **28-14b** Omikron, Photo Researchers, Inc.; **28-14c** Hugh Spencer, National Audubon Society, Photo Researchers, Inc,; **28-14d** Kim Taylor, Bruce Coleman, Inc.; **28-24e** Dr. J. Metzner, Peter Arnold, Inc.; **28-16** Omikron, Photo Researchers, Inc,; **28-17** Courtesy of Drs. T.D. Pugh and E.H. Newcomb, Univ. of Wisconsin/Madison; **28-18** M. Abbey, Photo Researchers, Inc.; **28-19** Australian Information Service Photograph; **28-20a** W.F. Millington; **28-20b** Courtesy of Dr. Stewart Smith, The Nitragin Co., Milwaukee; **28-20c** Runk/Schoenberger from Grant Heilman. **28-21b** Courtesy of Drs. K. Namba, D.L.D. Caspar, and G.J. Stubbs from *Science* 227:775; **28-22a** Biophoto Associates, Photo Researchers, Inc.; **28-22b** Lee D. Simon, Photo Researchers, Inc.; **28-23b,c** Courtesy of Drs. Jonathan King and Erika Hartwieg, Electron Microscopy Facility of the Biology Department, Massachusetts Institute of Technology.

CHAPTER 29 **29-1** Courtesy of Dr. J.A. Sargent, Hexland Ltd., England; **29-2** Biophoto Associates/Photo Researchers, Inc.; **29-5** Dr. M. Fuller, Univ. of Georgia; **Panel 29-1a,c,d** U.S. Dept of Agriculture; **29-8b** Biophoto Associates, Photo Researchers, Inc.; **29-9a,b** Photo Researchers, Inc.; **29-10b** J. Forsdyke, Photo Researchers, Inc.; **29-11a,c** W.F. Millington; **29-11e** J.L. Lepore, Photo Researchers, Inc.; **29-11g** Robert Lee, Photo Researchers, Inc.; **29-12** J.A. Menge; **29-13** Patrick Grace, Photo Researchers, Inc.; **29-14a** W.F. Millington; **29-14b** Biophoto Associates, Photo Researchers, Inc.; **29-17** Eric Gravé, Photo Researchers, Inc.; **29-18a** Biophoto Associates, Photo Researchers, Inc.; **29-18b** Carolina Biological Supply Co.; **29-19** Ray Simons, Photo Researchers, Inc.; **29-20a–e** David Scharf, Peter Arnold, Inc.; **29-21a** Robert Lee, Photo Researchers, Inc.; **29-21b** S.J. Krasemann, National Audubon Society, Photo Researchers, Inc.; **29-21c** Winton Patnode, Photo Researchers, Inc.; **29-21d** Michael Lustbader, Photo Researchers, Inc.; **29-22a** Carolina Biological Supply Co.

CHAPTER 30 **30-2** Biophoto Associates, Photo Researchers, Inc.; **30-4** W.F. Millington; **30-5** Carolina Biological Supply Co.; **30-6** Carolina Biological Supply Co.; **30-7a** Dr. Ann Smith, Science Photo Library, Photo Researchers, Inc.; **30-8a** M.I. Walker, Photo Researchers, Inc.; **30-8b** Manfred Kage, Peter Arnold, Inc.; **30-9a** Carolina Biological Supply Co.; **30-9b** J. Pickett-Heaps, Photo Researchers, Inc.; **30-10** Patrick Lynch, Photo Researchers, Inc.; **30-11** Carolina Biological Supply Co.; **30-12a** Richard Carlton, Photo Researchers, Inc.; **30-12b** Wm. Harlow, Photo Researchers, Inc.; **30-13a** Photo by Jacques Jangoux, Peter Arnold Inc.; **30-13b** © K.G. Preston-Mafham/Earth Scenes; **30-14** P.W. Grace, Photo Researchers, Inc.; **30-15** Ken Brate, Photo Researchers, Inc.; **30-17a,b** W.F. Millington; **30-18** © M.I. Walker, Photo Researchers, Inc.; **30-20a** © George K. Bryce, Earth Scenes; **30-20b** Dr. J. Metzner, © Peter Arnold Inc.; **30-21a** © Charlie Ott, Photo Researchers, Inc.; **30-21b,c** Carolina Biological Supply Co.; **30-22** © A-Z Collection, Photo Researchers, Inc.; **30-23** © M.P. Kahl, Photo Researchers, Inc.; **30-24** © Noble Proctor, Photo Researchers, Inc.; **30-25** Louis Quitt, Photo Researchers, Inc.; **30-26** © Noble Proctor, Photo Researchers, Inc.; **30-27a** © M. dos Passos, Photo Researchers, Inc.; **30-27b** © Jerome Wexler, Photo Researchers, Inc.

CHAPTER 31 **31-1a** Eric V. Gravé, Photo Researchers, Inc.; **31-1b** Carolina Biological Supply Co.; **31-2a** Eric V. Gravé, Photo Researchers, Inc.; **31-2b** M.I. Walker, Photo Researchers, Inc.; **31-2c** © 1983 Rex Educational Resources Co., Bruce Coleman Ltd.; **31-3** Biophoto Associates, Photo Researchers, Inc.; **31-5b** Tim Rock, Animals Animals; **31-9** G.I. Bernard, Oxford Scientific Films, Animals Animals; **31-14a** Science Source, Photo Researchers, Inc.; **31-15a** Z. Leszczynski, Animals Animals; **31-15b** Tom McHugh, National Audubon Society, Photo Researchers, Inc.; **31-15c** A.W. Ambler, National Audubon Society, Photo Researchers, Inc.; **31-16a** Russ Kinne, Photo Researchers, Inc.; **31-16b** Oxford Scientific Films, Animals Animals; **31-17** Oxford Scientific Films, Animals Animals; **31-18a,b** Oxford Scientific Films, Animals Animals; **31-19** Steve Earley, Animals Animals; **31-20a** Tim Rock, Animals Animals; **31-20b** Carl Roessler, Animals Animals; **31-20c** Tim Rock, Animals Animals; **31-21** © Jack Dermid, from National Audubon Society, Photo Researchers, Inc.; **31-22** Michael Abbey, Photo Researchers, Inc.; **31-23** Photo Researchers, Inc.; **31-25a** Eric V. Gravé, Photo Researchers, Inc.; **31-26a** Omikron, Photo Researchers, Inc.; **31-26b** Noble Proctor, Photo Researchers, Inc.; **31-26c** Eric V. Gravé, Photo Researchers, Inc.; **31-26d** Biophoto Associates, Photo Researchers, Inc.; **31-27a** W.F. Mai, Photo Researchers, Inc.; **31-27b** Russ Kinne, National Audubon Society, Photo Researchers, Inc.

CHAPTER 32 **32-2a** Breck P. Kent, Animals Animals; **32-2b** Russ Kinne, Photo Researchers, Inc.; **32-2c** G.I. Bernard, Animals Animals; **32-3a** Dr. J.A.L. Cooke, Animals Animals; **32-7a** Oxford Scientific Films, Animals Animals; **32-7b** Z. Leszczynski, Animals Animals; **32-7c** Michael Fogden, Oxford Scientific Films, Animals Animals; **32-7d** Ralph A. Reinhold, Animals Animals; **32-8a,b** Z. Leszczynski, Animals Animals; **32-8c** Dr. J.A.L. Cooke, Animals Animals; **32-8d** M.A. Chappell, Animals Animals; **32-13a–e** Carolina Supply Company; **32-14a–f** Carolina Biological Supply Co.; **32-15a** Patti Murray, Animals Animals; **32-15b** Anne Wertheim, Animals Animals; **32-15c** Carl Roessler, Animals Animals; **32-15d** Z. Leszczynski, Animals Animals; **32-15e** Oxford Scientific Films, Animals Animals; **32-16a** Jack Wilburn, Animals Animals; **32-16b** U.K. Devon, Animals Animals; **32-17a** C. Roessler, Animals Animals; **32-17b** E.R. Debbinger, Animals Animals; **32-17c** Steve Earley, Animals Animals; **32-17d,e** Z. Leszczynski, Animals Animals; **32-21** Dr. Tom McHugh, Photo Researchers, Inc.; **32-23a** M. Austerman, Animals Animals; **32-23b** Z. Leszczynski, Animals Animals; **32-23c** Jen and Des Bartlett, Photo Researchers, Inc.; **32-25a,b** Z. Leszczynski, Animals Animals; **32-25c** Tom McHugh, Steinhart Aquarium, Photo Researchers, Inc.; **32-25d** Tom McHugh, Photo Researchers, Inc.; **32-25e** R.F. Head, Animals Animals; **32-26a** Z. Leszczynski, Animals Animals; **32-26b** Tom McHugh, Photo Researchers, Inc.; **32-27a,b** Field Museum, Photo Researchers, Inc.; **32-28a** Z. Leszczynski, Animals Animals; **32-28b** Lynn M. Stone, Animals Animals; **32-28c** Breck P. Kent, Animals Animals; **32-28d** Z. Leszczynski, Animals Animals; **32-28e** Breck P. Kent, Animals Animals; **32-29a** James R. Fisher, Photo Researchers, Inc.; **32-29b** Leonard Lee Rue III, Photo Researchers, Inc.; **32-29c** Tom McHugh, Photo Researchers, Inc.; **32-29d** Alan G. Nelson, Photo Researchers, Inc.; **32-29e** Townsend P. Dickinson, Photo Researchers, Inc.; **32-30a** Leonard Lee Rue III, National Audubon Society, Photo Researchers, Inc.; **32-30b–d** Leonard Lee Rue III, Photo Researchers, Inc.; **32-31a,b** Hans & Judy Beste, Animals Animals; **32-32a** Hans & Judy Beste, Animals Animals; **32-32b** Stouffer Productions, Ltd., Animals Animals; **32-32c** Leonard Lee Rue III, Animals Animals; **32-32d** Miriam Austerman, Animals Animals.

**UNIT VII NASA**

CHAPTER 33 **33-2** Stephen J. Krasemann, Photo Researchers, Inc.; **33-6** Miguel Castro, Photo Researchers, Inc.; **33-12** Anne Hubbard, Photo Researchers, Inc.; **33-18** Tom Hollyman, Photo Researchers, Inc.; **Panel 33-1** Tom McHugh, Photo Researchers, Inc.; **Panel 33-2a** Gregory K. Scott, Photo Researchers, Inc.; **Panel 33-2b** E.H. Rao, Photo Researchers, Inc.; **Panel 33-2c** Leonard Lee Rue III, Photo Researchers, Inc.

CHAPTER 34 **34-2a** Dr. Tony Brain, Photo Researchers, Inc.; **34-7** Mr. Peter Katsaros, Photo Researchers, Inc.; **34-8a** Tom McHugh, Photo Researchers, Inc.; **34-8b** Steve Maslowski, Photo Researchers, Inc.; **34-9** Charlie Ott, Photo Researchers, Inc.; **34-10** Michael C.T. Smith, National Audubon Society, Photo Researchers, Inc.; **34-12** Grant Heilman; **34-13** Tom McHugh, Photo Researchers, Inc.; **34-14a** Eric Gravé, Photo Researchers, Inc.; **34-14b** M. Abbey Photo, Photo Researchers, Inc.; **34-15** Eric Gravé, Photo Researchers, Inc.; **34-16a** E.R. Degginger, Animals Animals; **34-17** Robert Noonan, Photo Researchers, Inc.; **34-18a** Tom Myers, Photo Researchers, Inc.; **34-18b** M. Abbey, Photo Researchers, Inc.; **Panel 34-1a** © C.C. Lockwood, Animals Animals; **Panel 34-1b** Russ Kinne, Photo Researchers, Inc.

CHAPTER 35 **35-1** Townsend P. Dickinson, Photo Researchers, Inc.; **35-3b** Jon Wrice, Photo Researchers, Inc.; **35-4a** Renee Purse, Photo Researchers, Inc.; **35-4b–d** James P. Jackson, Photo Researchers, Inc.; **35-5** © Leonard Lee Rue III, Photo Researchers, Inc.; **35-6** M.J. Manuel, National Audubon Society, Photo Researchers, Inc.; **35-9a** © Douglas Faulkner, Photo Researchers, Inc.; **35-9b** © David Hall, Photo Researchers, Inc.; **35-11** Robert Dunne, Photo Researchers, Inc.; **35-15** © Manfred Kage, Peter Arnold, Inc.; **35-20** Naval Photographic Center, Naval District Washington; **35-22** Woods Hole Oceanographic Institution; **35-25a** © Dr. Charles R. Belinsky, Photo Researchers, Inc.; **35-25b** © Townsend P. Dickinson, Photo Researchers, Inc.; **35-27** Susan McCartney, Photo Researchers, Inc.; **35-29** James P. Rod, Photo Researchers, Inc.; **35-31** Charlie Ott, Photo Researchers, Inc.; **35-32** Philippa Scott, Photo Researchers, Inc.; **35-33** Charlie Ott, Photo Researchers, Inc.; **35-34a** John M. Burnley, National Audubon Society, Photo Researchers Inc.; **35-34b** G.C. Kelley, Photo Researchers, Inc.; **35-37** Allen Rokach, National Audubon Society, Photo Researchers, Inc., **35-38** Joe Munroe, Photo Researchers, Inc.; **35-40** S.J. Krasemann, National Audubon Society, Photo Researchers, Inc.; **Panel 35-1a** Omikron, Photo Researchers, Inc.; **Panel 35-1b** Brian Enting; Photo Researchers, Inc.; **Panel 35-2** National Audubon Society, Photo Researchers, Inc.

CHAPTER 36 **36-1** Tom McHugh, Photo Researchers, Inc.; **36-2** Pat and Tom Leeson, Photo Researchers, Inc.; **36-3** Jack Fields, Photo Researchers, Inc.; **36-4** D.P. Burnside, Photo Researchers, Inc.; **Panel 36-1b,c** Omikron, Photo Researchers, Inc.; **Panel 36-1d** Georg Geister, Photo Researchers, Inc.; **Panel 36-2b** Jack Fields, Photo Researchers, Inc.; **Panel 36-2c** Lowell Georgia, Photo Researchers, Inc.; **36-5** State of California, Department of Water Resources; **36-7** Dr. Charles R. Belinky, Photo Researchers, Inc.; **36-8** Kent and Donna Dannen, Photo Researchers, Inc.; **36-10** National Parks Service; **36-12** Peter B. Kaplan, Photo Researchers, Inc.; **36-13** Milton Rogovin, Photo Researchers, Inc.; **36-14** Russ Kinne, Photo Researchers Inc.; **36-16** Carleton Ray, Photo Researchers, Inc.; **36-18** Georg Geister, Photo Researchers, Inc.; **36-21** Tom McHugh, Photo Researchers, Inc.; **Panel 36-3a** Gordon Smith, National Audubon Society, Photo Researchers, Inc.; **36-3b** Townsend P. Dickinson, Photo Researchers, Inc.; **36-22** John Colwell from Grant Heilman; **36-25** Lynwood M. Chace, National Audubon Society, Photo Researchers, Inc.; **36-26** Michael Philip Manheim, Photo Researchers, Inc.; **36-27** Georg Geister, Photo Researchers, Inc.; **36-29** Barbara Burnes, Photo Researchers, Inc.; **36-30** © Tom Myers, Photo Researchers, Inc.; **36-32** Georg Geister, Photo Researchers, Inc.; **36-33** James Mason, Black Star; **36-35** Dordin, Photo Researchers, Inc.

# Index

*Italic* page numbers indicate references to figures and tables; **boldface** page numbers indicate major discussions.

## A

# C

# E

# F

## G

# H

# I

# J



# K



# L

# M

# N

## O

# P

# Q



# R

# S

# T

# GEOLOGICAL TIME CHART

| *ERA (duration in millions of years)* | *PERIOD* | *EPOCH* | *Millions of years ago (from start of period to now)* | *Duration (in millions of years)* | *Major events of biological importance* |
|---|---|---|---|---|---|
| CENOZOIC (70) "Age of Mammals" | Quaternary | Recent | 0.01 | 0.01 | Human dominance. |
| | | Pleistocene | 2.5 | 2.5 | Four great glaciations ("Ice Age"); Sierra Nevada, Cascade, Alps, Himalaya ranges rise; first modern humans. |
| | Tertiary | Pliocene | 7 | 4.5 | Large carnivores; first hominids; extinction of early mammals. |
| | | Miocene | 26 | 19 | Forests give way to grasslands; further rise of Rockies; Ramapithecus. |
| | | Oligocene | 38 | 12 | First anthropoid apes; whales and ungulates thrive. |
| | | Eocene | 54 | 16 | First (tiny) horses and camels; first cetaceans; angiosperms and gymnosperms dominant; all mammalian orders present; giant birds. |
| | | Paleocene | 65 | 11 | First primates and carnivores; first modern invertebrates; most modern angiosperms present. |
| *Massive extinctions: nearly all great reptiles, 70% of all animal species disappear.* | | | | | |
| MESOZOIC (155) "Age of Reptiles" | Cretaceous | | 136 | 71 | Last of the great reptiles; angiosperms arise and expand; decline of gymnosperms; second expansion of insects; Andes and Rockies begin to form. |
| | Jurassic | | 190 | 54 | First birds; reptiles dominant; last of seed ferns; cycads, ferns present; Pangaea breaks up. |
| | Triassic | | 225 | 35 | Continental drift begins; first dinosaurs; gymnosperm forests; Appalachians formed. |
| PALEOZOIC (375) | Permian | | 280 | 55 | Reptiles expand; seed ferns, lycopsids, and sphenopsids declining; Urals, Appalachians forming, climate cools; mass extinctions. |
| | Carboniferous* | | 345 | 65 | Age of amphibians; first reptiles; expansion of insects; large, coal-forming lycopsids, sphenopsids, ferns, and gymnosperms. |
| | Devonian | | 395 | 50 | Age of fishes; first insects; first amphibians; most of North America covered by sea; early vascular plants. |
| | Silurian | | 430 | 35 | First terrestrial arthropods; first vascular plants; modern algae and fungi; mountains forming in Europe. |
| | Ordovician | | 500 | 70 | First vertebrates (Agnatha); invertebrates and algae abundant. |
| | Cambrian | | 570 | 70 | Most invertebrate phyla represented; cyanophytes and primitive algae abundant. |
| PRECAMBRIAN (4030) | Edicarian | | 700 | 130 | First large, soft-bodied invertebrates; Laurentian Mountains rise. |
| | Proterozooic | | 3400 | 2700 | Prokaryotes, protists. |
| | Archean | | 4600 | 1200 | Formation of the earth; pre-biotic biochemical evolution; origin of life; first organisms. |

* The Carboniferous is sometimes divided into early (Pennsylvanian) and late (Mississippian) periods.